T0344656

MOLECULAR GENETICS OF
MYCOBACTERIA
2ND EDITION

Molecular Genetics of
MYCOBACTERIA
2ND EDITION

EDITED BY

●　　　●　　　●　　　●　　　●　　　●　　　●

GRAHAM F. HATFULL
Department of Biological Sciences,
University of Pittsburgh, Pittsburgh, PA 15260

AND

WILLIAM R. JACOBS, JR.
Howard Hughes Medical Institute,
Albert Einstein College of Medicine, Bronx, NY 10461

ASM Press
Washington, DC

Library of Congress Cataloging-in-Publication Data

Molecular genetics of mycobacteria / edited by Graham F. Hatfull, Department of Biological Sciences, University of Pittsburgh, Pittsburgh, PA, and William R. Jacobs, Jr., Howard Hughes Medical Institute, Albert Einstein College of Medicine, Bronx, NY. – Second edition.
   pages cm
 Includes index.
 ISBN 978-1-55581-883-8 (print) – ISBN 978-1-55581-884-5 (electronic)
 1. Mycobacteria. 2. Bacterial genetics. 3. Tuberculosis–Genetic aspects.
 I. Hatfull, Graham F., editor. II. Jacobs, William R., Jr, editor.
  QR82.M8M64 2014
  579.3'74–dc23
                                    2014016666
 doi:10.1128/9781555818845

Printed in the United States of America

10 9 8 7 6 5 4 3 2 1

Address editorial correspondence to: ASM Press, 1752 N St., N.W., Washington, DC 20036-2904, USA.

Send orders to: ASM Press, P.O. Box 605, Herndon, VA 20172, USA.
Phone: 800-546-2416; 703-661-1593. Fax: 703-661-1501.
E-mail: books@asmusa.org
Online: http://www.asmscience.org

Contents

v

VII.   GENETICS OF MACROMOLECULAR BIOSYNTHESIS

VIII.   THE MYCOBACTERIAL LIFESTYLE, PERSISTENCE, AND MACROPHAGE SURVIVAL

Contributors

RUEDI AEBERSOLD
Department of Biology, Institute of Molecular Systems Biology, ETH Zurich,
Wolfgang-Pauli Strasse 16, 8093 Zurich, and Faculty of Science, University of
Zurich, 8057 Zurich, Switzerland

BREE B. ALDRIDGE
Department of Molecular Biology & Microbiology and Department of Biomedical
Engineering, Tufts University, Boston, MA 02111, and Medford, MA 02155

REEM AL-MUBARAK
Department of Microbiology, Immunology, and Pathology, Colorado State
University, Fort Collins, CO 80523

PRAVEEN ANAND
Department of Biochemistry, Indian Institute of Science, Bangalore, Karnataka
560012, India

KRISTINE B. ARNVIG
Mycobacterial Research Division, National Institute for Medical Research,
London NW7 1AA, United Kingdom

ANTHONY D. BAUGHN
Department of Microbiology, University of Minnesota, 420 Delaware St. SE,
MMC196, Mayo Building Room 1020, Minneapolis, MN 55455

MARCEL A. BEHR
McGill International TB Centre, Montreal, Quebec, Canada, H3G 1A4

JOHN T. BELISLE
Department of Microbiology, Immunology, and Pathology, Colorado State
University, Fort Collins, CO 80523

MICHAEL BERNEY
Department of Microbiology and Immunology, Albert Einstein College of
Medicine, Bronx, NY 10461

GURDYAL S. BESRA
School of Biosciences, University of Birmingham, Edgbaston, Birmingham, B15
2TT, United Kingdom

HELEN L. BIRCH
School of Biosciences, University of Birmingham, Edgbaston, Birmingham, B15
2TT, United Kingdom

WILLIAM R. BISHAI
Johns Hopkins School of Medicine, The Center for TB Research, 1550 Orleans
St., CRBII, Room 103, Baltimore, MD 21287

NADINE J. BODE
Department of Microbiology, New York University School of Medicine, 550
First Avenue, MSB 236, New York, NY 10016

DARIA BOTTAI
Dipartimento di Ricerca Traslazionale e delle Nuove Tecnologie in Medicina e
Chirurgia Università di Pisa, Pisa, Italy

ROLAND BROSCH
Institut Pasteur, Unit for Integrated Mycobacterial Pathogenomics, Paris, France

NAGASUMA CHANDRA
Department of Biochemistry, Indian Institute of Science, Bangalore, Karnataka
560012, India

KEIRA A. COHEN
KwaZulu-Natal Research Institute for TB and HIV (K-RITH), Nelson R.
Mandela School of Medicine, 719 Umbilo Road, Durban, South Africa, and
Brigham and Women's Hospital, Harvard Medical School, 75 Francis Street,
Boston, MA 02115

GREGORY M. COOK
University of Otago, Department of Microbiology and Immunology, Dunedin,
New Zealand

TERESA CORTES
Mycobacterial Research Division, National Institute for Medical Research,
London NW7 1AA, United Kingdom

DEAN C. CRICK
Mycobacteria Research Laboratories, Department of Microbiology,
Immunology and Pathology, Colorado State University, Fort Collins, CO
80523-1682

BRIDGETTE M. CUMMING
KwaZulu-Natal Research Institute for Tuberculosis and HIV (K-RITH),
Durban, South Africa

MAMADOU DAFFÉ
CNRS, Institut de Pharmacologie et de Biologie Structurale, Département
Mécanismes Moléculaires des Infections Mycobactériennes, and the Université
de Toulouse Paul Sabatier, F-31077 Toulouse, France

K. HERAN DARWIN
Department of Microbiology, New York University School of Medicine, 550
First Avenue, MSB 236, New York, NY 10016

KEITH M. DERBYSHIRE
Division of Genetics, Wadsworth Center, New York State Department of
Health, and Department of Biomedical Sciences, University at Albany,
Albany, NY 12201

SABINE EHRT
Department of Microbiology and Immunology, Weill Medical College, and
Program in Immunology and Microbial Pathogenesis, Weill Graduate School
of Medical Sciences of Cornell University, New York, NY 10065

JOANNA C. EVANS
Molecular Mycobacteriology Research Unit, Institute of Infectious Disease and
Molecular Medicine, University of Cape Town, P/Bag X3, Rondebosch 7700,
South Africa

RUTH J. FAHEY
Microbiology and Immunology, College of Veterinary Medicine, Cornell
University, Ithaca, NY 14853

SARAH M. FORTUNE
Department of Immunology and Infectious Diseases, Harvard School of Public
Health, Boston, MA 02115

JAMES GALAGAN
Department of Biomedical Engineering, Bioinformatics Program, and National
Emerging Infectious Diseases Laboratory, Boston University, Boston, MA
02215, and Broad Institute of MIT and Harvard, Cambridge, MA 02142

MARTIN GENGENBACHER
Max Planck Institute for Infection Biology, Department of Immunology,
Charitéplatz 1, 10117 Berlin, Germany

MICHAEL S. GLICKMAN
Immunology Program, Memorial Sloan Kettering Cancer Center, 1275 York
Ave., New York, NY 10803

MICHAEL F. GOLDBERG
Department of Microbiology and Immunology, Albert Einstein College of
Medicine, Bronx, NY 10461

ANTONIO GOMES
Bioinformatics Program, Boston University, Boston, MA 02215

TODD A. GRAY
Division of Genetics, Wadsworth Center, New York State Department of
Health, and Department of Biomedical Sciences, University at Albany,
Albany, NY 12201

SHIPRA GROVER
School of Biosciences, University of Birmingham, Edgbaston, Birmingham, B15 2TT, United Kingdom

KIEL HARDS
University of Otago, Department of Microbiology and Immunology, Dunedin, New Zealand

TRAVIS HARTMAN
Department of Microbiology and Immunology, Albert Einstein College of Medicine, Bronx, NY 10461

GRAHAM F. HATFULL
Department of Biological Sciences, University of Pittsburgh, Pittsburgh, PA 15260

ROBERT N. HUSSON
Division of Infectious Diseases, Boston Children's Hospital, Harvard Medical School, 300 Longwood Ave., Boston, MA 02115

MARY JACKSON
Mycobacteria Research Laboratories, Department of Microbiology, Immunology and Pathology, Colorado State University, Fort Collins, CO 80523-1682

WILLIAM R. JACOBS, JR.
Howard Hughes Medical Institute, Department of Microbiology and Immunology, Albert Einstein College of Medicine, Bronx, NY 10461

SUMA JAINI
Department of Biomedical Engineering and National Emerging Infectious Diseases Laboratory, Boston University, Boston, MA 02115

MONIKA JANKUTE
School of Biosciences, University of Birmingham, Edgbaston, Birmingham, B15 2TT, United Kingdom

RAINER KALSCHEUER
Institute for Medical Microbiology and Hospital Hygiene, Heinrich-Heine-University Duesseldorf, Universitaetsstr. 1, 40225 Duesseldorf, Germany

STEFAN H. E. KAUFMANN
Max Planck Institute for Infection Biology, Department of Immunology, Charitéplatz 1, 10117 Berlin, Germany

IRIS KEREN
Antimicrobial Discovery Center and Department of Biology, Northeastern University, Boston, MA 02115

GWENDOWLYN S. KNAPP
Wadsworth Center, New York State Department of Health, 120 New Scotland Avenue, PO Box 22002, Albany, NY 12201-2002

HENDRIK KOLIWER-BRANDL
Institute for Medical Microbiology and Hospital Hygiene, Heinrich-Heine-University Duesseldorf, Universitaetsstr. 1, 40225 Duesseldorf, Germany

LAURENT KREMER
Laboratoire de Dynamique des Interactions Membranaires Normales et
Pathologiques, Université de Montpellier 2 et 1, CNRS; UMR 5235,
case 107; and INSERM, DIMNP, Place Eugène Bataillon, 34095 Montpellier
Cedex 05, France

DIRK LAMPRECHT
KwaZulu-Natal Research Institute for Tuberculosis and HIV (K-RITH),
Durban, South Africa

EMILIE LAYRE
Division of Rheumatology, Immunology and Allergy, Brigham and Women's
Hospital, Harvard Medical School, Boston, MA 02115

WONSIK LEE
Microbiology and Immunology, College of Veterinary Medicine, Cornell
University, Ithaca, NY 14853

JUN LIU
Department of Molecular Genetics, University of Toronto, Toronto,
Ontario, Canada M5S 1A8

ANNA LYUBETSKAYA
Bioinformatics Program, Boston University, Boston, MA 02215

SEBABRATA MAHAPATRA
Mycobacteria Research Laboratories, Department of Microbiology,
Immunology and Pathology, Colorado State University, Fort Collins, CO 80523

RICCARDO MANGANELLI
Department of Molecular Medicine, University of Padova, Italy

CLAUDINE MAYER
Unité de Microbiologie Structurale, Institut Pasteur; UMR 3528 du CNRS; and
Université Paris Diderot, Sorbonne Paris Cité, Cellule Pasteur, 75015,
Paris, France

JAMES H. MAZORODZE
KwaZulu-Natal Research Institute for Tuberculosis and HIV (K-RITH),
Durban, South Africa

KATHLEEN A. MCDONOUGH
Wadsworth Center, New York State Department of Health, 120 New Scotland
Avenue, PO Box 22002, and Department of Biomedical Sciences, University at
Albany, Albany, NY 12222

DENIS MITCHISON
Centre for Infection, St. George's, University of London, Cranmer Terrace,
London SW17 0RE, United Kingdom

VALERIE MIZRAHI
Molecular Mycobacteriology Research Unit, Institute of Infectious Disease and
Molecular Medicine, University of Cape Town, P/Bag X3, Rondebosch 7700,
South Africa

D. BRANCH MOODY
Division of Rheumatology, Immunology and Allergy, Brigham and Women's Hospital, Harvard Medical School, Boston, MA 02115

JEPPE MOURITSEN
Department of Biology, Institute of Molecular Systems Biology, ETH Zurich, Wolfgang-Pauli Strasse 16, 8093 Zurich, Switzerland

OLIVIER NEYROLLES
Centre National de la Recherche Scientifique & Université de Toulouse, Université Paul Sabatier, Institut de Pharmacologie et de Biologie Structurale, Toulouse, France

ANIL K. OJHA
Department of Infectious Diseases and Microbiology, Graduate School of Public Health, University of Pittsburgh, Pittsburgh, PA 15261

TANYA PARISH
Infectious Disease Research Institute, Seattle, WA 98102, and Queen Mary University of London, London, United Kingdom

SANG TAE PARK
Macrogen Clinical Laboratory, Macrogen Corp, Rockville, MD 20850

MARTIN S. PAVELKA, JR.
Department of Microbiology and Immunology, University of Rochester Medical Center, Rochester, NY 14642

JAKUB PAWEŁCZYK
Institute for Medical Biology, Polish Academy of Sciences, Lodz, Poland

MATTHEW PETERSON
Department of Biomedical Engineering, Boston University, Boston, MA 02215

MADALENA PIMENTEL
Centro de Patogénese Molecular, Unidade dos Retrovirus e Infecções Associadas, Faculty of Pharmacy, University of Lisbon, Lisbon, Portugal

MARIA PODINOVSKAIA
Microbiology and Immunology, College of Veterinary Medicine, Cornell University, Ithaca, NY 14853

STEVEN A. PORCELLI
Department of Microbiology and Immunology and Department of Medicine, Albert Einstein College of Medicine, Bronx, NY 10461

SLADJANA PRISIC
Division of Infectious Diseases, Boston Children's Hospital, Harvard Medical School, 300 Longwood Ave., Boston, MA 02115

ALEXANDER S. PYM
KwaZulu-Natal Research Institute for TB and HIV (K-RITH), Nelson R. Mandela School of Medicine, 719 Umbilo Road, Durban, South Africa

SAHADEVAN RAMAN
National Emerging Infectious Diseases Laboratory, Boston University, Boston, MA 02215

KYU Y. RHEE
Department of Medicine, Weill Cornell Medical College, 1300 York Avenue
A-431A, New York, NY 10065

JACOBS P. RICHARDS
Department of Infectious Diseases and Microbiology, Graduate School of Public
Health, University of Pittsburgh, Pittsburgh, PA 15261

G. MARCELA RODRIGUEZ
Public Health Research Institute Center & Department of Medicine, University
of Medicine and Dentistry of New Jersey, Newark, NJ 07103

ERIC J. RUBIN
Harvard School of Public Health, 677 Huntington Avenue, Boston, MA 02115

DAVID G. RUSSELL
Microbiology and Immunology, College of Veterinary Medicine, Cornell
University, Ithaca, NY 14853

VIKRAM SAINI
Department of Microbiology, University of Alabama at Birmingham,
Birmingham, AL 35294

NEERAJ K. SAINI
Department of Microbiology and Immunology, Albert Einstein College of
Medicine, Bronx, NY 10461

SANKARAN SANDHYA
Department of Biochemistry, Indian Institute of Science, Bangalore, Karnataka
560012, India

CHRISTOPHER M. SASSETTI
University of Massachusetts Medical School, 55 Lake Avenue North, Worcester,
MA 01655

DIRK SCHNAPPINGER
Department of Microbiology and Immunology, Weill Medical College, and
Program in Molecular Biology, Weill Graduate School of Medical Sciences of
Cornell University, New York, NY 10065

GARY SCHOOLNIK
Department of Medicine and Department of Microbiology and Immunology,
Stanford Medical School, Stanford, CA 94305

OLGA T. SCHUBERT
Department of Biology, Institute of Molecular Systems Biology, ETH Zurich,
Wolfgang-Pauli Strasse 16, 8093 Zurich, Switzerland

WANLIANG SHI
Department of Molecular Microbiology and Immunology, Bloomberg School of
Public Health, Johns Hopkins University, Baltimore, MD 21205

ADRIE J. C. STEYN
KwaZulu-Natal Research Institute for Tuberculosis and HIV (K-RITH),
Durban, South Africa, and Department of Microbiology, University of Alabama
at Birmingham, Birmingham, AL 35294

TIMOTHY P. STINEAR
Department of Microbiology and Immunology, University of Melbourne, Parkville, Australia

NEELIMA SUKUMAR
Microbiology and Immunology, College of Veterinary Medicine, Cornell University, Ithaca, NY 14853

PHILIP SUPPLY
CNRS UMR 8204; INSERM, U1019; Center for Infection and Immunity of Lille, Institut Pasteur de Lille; and Université Lille Nord de France, Lille, France

HOWARD TAKIFF
Laboratorio de Genética Molecular, CMBC, IVIC, Caracas, Venezuela

SHUMIN TAN
Microbiology and Immunology, College of Veterinary Medicine, Cornell University, Ithaca, NY 14853

VANESSA TRAN
Department of Molecular Genetics, University of Toronto, Toronto, Ontario, Canada M5S 1A8

ANDREJ TRAUNER
Harvard School of Public Health, 677 Huntington Avenue, Boston, MA 02115

BRIAN C. VANDERVEN
Microbiology and Immunology, College of Veterinary Medicine, Cornell University, Ithaca, NY 14853

CATHERINE VILCHÈZE
Howard Hughes Medical Institute, Department of Microbiology and Immunology, Albert Einstein College of Medicine, Bronx, NY 10461

DIGBY F. WARNER
Molecular Mycobacteriology Research Unit, Institute of Infectious Disease and Molecular Medicine, University of Cape Town, P/Bag X3, Rondebosch 7700, South Africa

RYAN M. WELLS
KwaZulu-Natal Research Institute for Tuberculosis and HIV (K-RITH), Durban, South Africa, and Department of Microbiology, University of Alabama at Birmingham, Birmingham, AL 35294

DOUGLAS B. YOUNG
Mycobacterial Research Division, National Institute for Medical Research, London NW7 1AA, United Kingdom

YING ZHANG
Department of Molecular Microbiology and Immunology, Bloomberg School of Public Health, Johns Hopkins University, Baltimore, MD 21205, and Department of Infectious Diseases, Huashan Hospital, Fudan University, Shanghai, China

WENHONG ZHANG
Department of Infectious Diseases, Huashan Hospital, Fudan University, Shanghai, China

Preface

Fourteen years have passed since the first edition of *Molecular Genetics of Mycobacteria* was published in 2000, and the mycobacterial field has exploded in the intervening time. In 2000 the *Mycobacterium tuberculosis* genome sequence had recently been reported, and there was considerable optimism for the advances in tuberculosis genetics that this would stimulate. This Second Edition of *Molecular Genetics of Mycobacteria* offers insights into how these promises have been realized, as well as the substantial impact of the numerous new molecular tools developed over the past dozen years. The field of mycobacterial genetics has thus expanded dramatically, with investigations into new areas of mycobacterial growth, replication, metabolism, physiology, drug susceptibility, and virulence.

The size and scope of *Molecular Genetics of Mycobacteria*, Second Edition, reflect this rapidly expanding field. This new edition contains double the number of chapters in the first edition and includes many topics not discussed there. The book is divided into eight main sections that focus on genomics and genetic exchange, gene expression, the proteome, metabolism, drug resistance, cell wall biosynthesis, macromolecular biosynthesis, and growth and persistence. Each contains several chapters written by leading experts in the field and includes a genetic perspective on the various topics discussed. The field is growing so rapidly that there are undoubtedly some specific topics and areas—especially those developed over the past year—that we have not been able to include and will have to await another edition.

Although *M. tuberculosis* is now fully tractable to genetic manipulation, tuberculosis the disease advances with little abatement of its impact on human health. Better clinical management across the world has led to steadying of the numbers of new cases reported each year, tuberculosis mortality, and the total number of infected people. Nonetheless, most of the problems in tuberculosis control that existed in 2000 are still with us today. The only available vaccine is BCG, with its dubious efficacy against adult pulmonary tuberculosis; drug resistance continues to grow; antituberculosis drug regimens have barely changed; and diagnosis is either slow or costly. The good news is that the advances in mycobacterial genetics are beginning to be reflected in exciting recent developments. New diagnostic approaches can determine rifampin resistance within a few hours, promising new drugs are progressing through the pipeline and into the clinic, and a range of newly developed vaccines are being evaluated. The fruits of 30 years of intensive genetic investigations are finally beginning to emerge. But there remains much to learn about the mycobacteria and their curious but deadly habits and habitats. We anticipate that molecular genetic approaches will blunt the defenses of humanity's deadliest microbial enemies over the next dozen years. It is our hope that this book inspires both newcomers to the field and veterans in tuberculosis research alike to think about tuberculosis problems with fresh perspectives and understanding.

We would like to thank Ellie Tupper of ASM Press for her tireless efforts; Greg Payne, ASM Press, for his continual encouragement and advice; and our exceptionally gifted and dedicated authors who contributed so splendidly to this book.

GRAHAM F. HATFULL
WILLIAM R. JACOBS, JR.

Genomes, Genomics, and Genetic Exchange

I

Molecular Genetics of Mycobacteria, 2nd Edition
Edited by Graham F. Hatfull and William R. Jacobs, Jr.
© 2014 American Society for Microbiology, Washington, DC
doi:10.1128/microbiolspec.MGM2-0037-2013

William R. Jacobs, Jr.[1]

Gene Transfer in *Mycobacterium tuberculosis*: Shuttle Phasmids to Enlightenment

1

> "I had to know my enemy in order to prevail against him."
>
> Nelson Mandela

Infectious diseases have plagued humankind throughout history and have posed serious public health problems. Yet vaccines have eradicated smallpox and antibiotics have drastically decreased the mortality rate of many infectious agents (1). Although the precise viral agents had not yet been characterized, the smallpox vaccine work of Edward Jenner was critical in demonstrating that inoculation with pus from cowpox lesions could protect from a subsequent challenge with smallpox. These pioneering transfer experiments laid the groundwork for the eventual eradication of smallpox, as announced by the World Health Organization in 1979. The discovery of DNA as genetic material and the understanding of how this information translates into specific phenotypes have changed the paradigm for developing new vaccines, drugs, and diagnostic tests. Knowledge of the mechanisms of immunity and mechanisms of action of drugs has led to new

vaccines and new antimicrobial agents. For example, the discovery of the Australia antigen (HBsAg) led to the subsequent engineering of the first recombinant vaccine, whose remarkable efficacy offers hope that eradication of hepatitis B in humans is not an unreasonable expectation (1, 2). Similarly, HIV infections, which not so long ago were uniformly fatal, are now controlled with drugs that were developed by understanding the HIV genome and gene products required for the HIV life cycle. The key to the acquisition of the knowledge of these mechanisms has been identifying the elemental causes (i.e., genes and their products) that mediate immunity and drug resistance. The identification of these genes is made possible by being able to transfer the genes or mutated forms of the genes into causative agents or surrogate hosts. Such an approach was limited in *Mycobacterium tuberculosis* by the difficulty of transferring genes or alleles into *M. tuberculosis* or a suitable surrogate mycobacterial host. The construction of shuttle phasmids, chimeric molecules that replicate in *Escherichia coli* as plasmids and in mycobacteria as mycobacteriophages,

[1]Howard Hughes Medical Institute, Albert Einstein College of Medicine, 1301 Morris Park Avenue, Bronx, NY 10461.

was instrumental in developing gene transfer systems for *M. tuberculosis*. This review will discuss *M. tuberculosis* genetic systems and their impact on tuberculosis (TB) research.

OVERVIEW OF KEY MUTATIONS THAT FACILITATED GENE TRANSFER IN *M. TUBERCULOSIS*

I distinctly remember sitting in my first bacterial genetics course, fascinated with the isolation of mutant bacteria. We had just read François Jacob and Jacques Monod's (3) model of the lactose operon—an elegant hypothesis in which they postulated that *Escherichia coli* could selectively regulate the transcription of a set of genes required to degrade the disaccharide, lactose. They imagined the existence of a repressor protein that prevented the expression of the genes encoding lactose degradation when lactose was not available to the cell. The model was conceptualized and validated with the isolation and characterization of mutants that either had lost their ability to utilize lactose or lost their ability to regulate the degradation of lactose. Certainty of the causality of a phenotype by a mutation in a single gene was proven with gene transfer experiments (3). Key to this conceptualization was the isolation of the mutants with mutations in the specific genes. The lactose operon was a novel paradigm for gene regulation that arose from the study of bacterial mutants.

If we were to make similar advances in understanding how *M. tuberculosis* so successfully infects humans, we would need new tools to manipulate this once genetically intractable pathogen. I consider the following the key developments.

1. TM4::pHC79—introducing foreign DNA into mycobacteria. The shuttle phasmid phAE1 contained the cosmid pHC79 inserted in the nonessential region of the phage TM4 (4). I had hypothesized that I should be able to stably clone the entire TM4 genome into an *E. coli* cosmid and it would replicate in *E. coli* as a plasmid but replicate in mycobacteria as a phage. Furthermore, this chimeric molecule would be packageable in either bacteriophage lambda particles or TM4 mycobacteriophage particles. Although this first-generation shuttle phasmid was not useful for additional cloning experiments, it was a proof of principle that creation of shuttle phasmids was possible.

2. *attL^{L5}*::L5::*kan*::*attR^{L5}*—stable integration into mycobacterial chromosomes. This genetic alteration of mc²6, in which an L5 shuttle phasmid had integrated a kanamycin resistance gene stably into the chromosome

of *Mycobacterium smegmatis*, provided the first proof of principle that it was possible to use kanamycin and the *aph* gene as a selection system for mycobacteria (5). This insertion results when the *attP* site of the mycobacteriophage L5 integrates into the *attB* site of *M. smegmatis* or *M. tuberculosis*. The combination of DNA sequence analyses and genetic engineering of this region by Graham Hatfull's lab led to the development of integration proficient vectors with which small or large fragments of foreign DNA could be stably integrated into the genomes of many different mycobacteria (6). These integration proficient vectors provided a means to stably integrate foreign antigens into bacille Calmette-Guérin (BCG) (7) or a means to screen cosmid-size fragments of the *M. tuberculosis* chromosome for virulence functions in mice by virtue of its stability and ease of use (8–10).

3. *ept-1*—plasmid transformation of mycobacteria. An efficient plasmid transformation (*ept*) phenotype allowed for the first plasmid transformation system of mycobacteria using pAL5000 plasmids and *ept* mutants of *M. smegmatis* (5, 11). As mentioned earlier, 25 years later we've now discovered that the *ept* phenotype results from a single point mutation causing a loss of function that normally prevents replication of pAL5000 plasmids. Despite not knowing the mechanism, the availability of a plasmid transformation system for *M. smegmatis* mc²155 provided a gene transfer system to facilitate the analysis of mycobacterial genes. It allowed for the development of plasmid expression vectors (12) and a system to analyze the functions required for pAL5000 replication (13–17). Importantly, it provided a surrogate host for the study of genes from the slow-growing pathogenic mycobacteria in a fast-growing, nonpathogenic mycobacterium which, unlike *E. coli*, had similar metabolic pathways allowing for the analyses of genes encoding complex carbohydrates (18) and previously undiscovered genes associated with resistance to isoniazid (INH) (19, 20), ethionamide (ETH) (19, 21, 22), and ethambutol (23, 24). Plasmid transformation was indispensable for identifying the activator of the prodrug pyrazinamide (25) and the common target for isoxyl and thiacetazone (26, 27). There are hundreds of papers that have used mc²155 as a system to study the biology of mycobacteria and their phages. In addition, it is noteworthy that bedaquiline, the first new TB drug to be FDA approved in 40 years, was identified using *M. smegmatis* mc²155, which was subsequently used to identify and validate the target (28).

4. *leuD*::IS*1096*—auxotrophic mutants of *M. tuberculosis*. Airborne pathogens present particular challenges in that they require biosafety level III (BSL3) containment. We reasoned that it might be possible to generate *M. tuberculosis* mutants that we could work with in a BSL2 laboratory if we introduced mutations that prevented the mutant from growing in mammals. This was the strategy that was used to biologically contain *E. coli*. We set out to delete the genes required to make diaminopimelic acid (Dap), an essential metabolite that is not synthesized by mammals. Moreover, Dap auxotrophs of *E. coli* undergo Dap-less death when exponentially growing cells are transferred from media with Dap to media without Dap. For *E. coli*, or *Salmonella*, the deletion of either of the genes encoding aspartokinase or aspartate semialdehyde dehydrogenase can be obtained in the presence of methionine, threonine, lysine, and Dap, but we were unable to obtain these mutants in *M. smegmatis* (29). In the process of testing whether there was a problem with recombination in this region of the *M. smegmatis* chromosome, we set up a screen to measure rare recombination events using the loss of the *lacZ* gene, encoding beta-galactosidase, and we serendipitously discovered a novel insertion element, IS*1096*, that was present in the *M. smegmatis* chromosome (30). This insertion element transposed with a simple cut-and-paste mechanism and was used to make an efficient tool for transposon mutagenesis that we used to generate mutants of BCG. We screened the library of insertions for auxotrophic mutants and found a leucine auxotrophic mutant in which the transposon had jumped into the *leuD* gene (31). What makes this auxotrophic mutant particularly noteworthy is that unlike the parental BCG strain, this mutant had lost its ability to replicate in mice. This was surprising, as *Legionella*, another intracellular pathogen, is naturally auxotrophic for several amino acids including leucine (32, 33), suggesting that leucine and other amino acids are not limited in the mouse. We tested a number of additional auxotrophic mutants of BCG and found that they lost their ability to replicate not only in immunocompetent mice, but also in immunocompromised mice lacking T and B cells, and yet still provided protection comparable to BCG against virulent *M. tuberculosis* challenge (34). The inability of these auxotrophic mutants of BCG to grow in mice suggested that they lived in a niche that differs from *Legionella pneumophila*, even though both are known to live in phagocytic vacuoles. Since it had been hypothesized that *M. tuberculosis* might escape into the macrophage cytoplasm, and BCG could not (35), it was unclear whether auxotrophic mutants of *M. tuberculosis* would be able to grow in mice. Leucine auxotrophs of *M. tuberculosis* failed to grow in mice as well (36). In fact,

we discovered that *M. tuberculosis* mutants defective in making the amino acid lysine (37) or the vitamin pantothenate also were highly attenuated (38). After constructing double auxotroph of mutants of *M. tuberculosis*, we demonstrated that they are very safe and have the ability to generate protection comparable to BCG (39–42).

5. PH101—a conditionally replicating mutant of TM4. While shuttle phasmids provided a way to deliver foreign DNA into *M. tuberculosis* and other mycobacteria, since TM4 is a lytic phage, the recipient cell will be lysed, so we needed a way to prevent this cell death. We reasoned that we could isolate mutants of TM4 that could replicate at 30°C, but not at 37°C. To achieve this goal, we mutagenized TM4 phage and then screened plaques for those that replicated at 30°C, but not at 37°C. While nearly 100 plaques were identified that had this phenotype, all but one likely had a single mutation causing temperature-sensitive growth and thus reverted at frequencies of approximately 10^{-4} to 10^{-5}. However, we identified one mutant that we named PH101 (106, 107), which reverted at a frequency of less than 10^{-8}. We reasoned that this mutant phage must have contained at least two different point mutations that caused this phenotype. It is gratifying that reversion analyses and sequencing of the PH101 parent and revertants have demonstrated that point mutations in gp49 and gp66 are necessary and sufficient for this temperature-sensitive phenotype with the low reversion rate. While the function of the two proteins encoded by these two genes is yet unknown, this mutant phage has been a necessary component of efficient transposon mutagenesis, enhanced reporter phages, and specialized transduction.

TB TRANSMISSION ESTABLISHED TB CAUSALITY

In 1868, the cause of TB was unknown. Controversy followed the conclusion of Jean Antoine Villemin, first stated in his book *Etudes sur la Tuberculose* (43), that TB was caused by a virus. Most scientists at that time believed that TB was a hereditary disease that resulted in cancerous lesions. As a young physician in the French Army, Dr. Villemin had observed that many new recruits developed consumption (as TB used to be called) after living in confined quarters. He hypothesized that consumption was an infectious disease, a disease that was transmissible from one human to another by a virulent biological agent. He further postulated that these lesions were not cancerous, based on the observation that many of the lesions were "tubercules that looked like small, pearl-shaped globules or 'matière caséeuse'"—a cheese-like matter (43). To test if these lesions contained a putative virus that caused consumption, Villemin isolated the tuberculous lesions (now known as granulomas) from patients that had died of consumption, cut the granulomas into pieces, and then transferred them to rabbits via subcutaneous injection. Invariably, these inoculated rabbits developed consumption-like disease, forming the tubercles he had seen in the consumptive human patients. Moreover, the transfer of tuberculous granulomas among rabbits resulted in the transmission of consumptive disease to the uninfected recipients. In contrast, the transfer of cancerous cells from deceased soldiers that had died of various neoplasms did not yield tuberculous lesions or disease.

Thus Dr. Villemin provided compelling evidence that TB was a transmissible infection, the first such published report of a communicable disease in humans. He concluded that TB was caused by a virus—a poorly defined entity at that time. The innovation of this work was the successful and reproducible transfer of infectious material from humans to animals. (Since this magnificent 600-page book is only available in French, we have relied on a translation from Dr. Catherine Vilchèze, who was impressed by the genuine passionate enthusiasm and excitement conveyed by Dr. Villemin for the scores of experiments he had performed to repeat the successful transfer.) Despite his conviction in his prescient conclusion, this work was not readily accepted by his peers.

Dr. Villemin's transfer studies were noticed, reproduced, and extended by Dr. Robert Koch as described in his landmark presentation of "The Etiology of Tuberculosis" on March 24, 1882. Not only did Robert Koch reproduce the granuloma transfer experiments, but he was also able to develop (i) a staining procedure to visualize the tubercle bacilli in infected granulomas and (ii) a solid growth medium (bovine serum that was repeatedly heated and cooled) that allowed him to culture colonies of the tubercle bacilli from human tissue. By transferring the organisms that had been grown in pure culture to numerous types of animals, Dr. Koch established that TB was caused by tubercle bacilli and therefore unequivocally established that TB was an infectious disease (44).

Koch elegantly wrote, "To prove that tuberculosis is a parasitic disease, that it is caused by the invasion of bacilli and that it is conditioned primarily by the growth and multiplication of the bacilli, it was necessary: 1) to isolate the bacilli from the body; 2) to grow them in pure culture until they were freed from any disease product of the animal organism which might adhere to them; and, 3) to administer the isolated bacilli to animals to reproduce the same morbid condition which, as known, is obtained by inoculation with spontaneously developed tuberculous material." We call this philosophical argument Koch's postulate, whose fulfillment can directly link a pathogen as the causative agent to an infectious disease. (Many textbooks call this Koch's postulates, but it should be the singular form postulate. The postulate or "deduced truth" is that the bacilli cause the specific disease if three conditions are met.) Since its publication, this has been the paradigm that establishes the causative agents of infectious diseases. The innovations of Koch's work were in his development of bacteriological methods—staining and bacterial culture—that allowed him to clone the tubercle bacilli. Moreover, he extended the transfer experiments of Villemin by transferring isolated bacilli to allow for the unequivocal conclusion that TB is caused by tubercle bacilli. This knowledge meant that controlling TB involved not cancer therapy, but rather killing tubercle bacilli.

GENE TRANSMISSION ESTABLISHED PHENOTYPE CAUSALITY

In 1941, the molecular basis of genes was unknown. In fact, we did not know that a specific gene encoded a single polypeptide. The discovery of this important fact came from a gene transfer experiment. George Beadle and Edward Tatum established that a single enzyme was likely encoded by a single gene. They achieved this success by isolating mutants of *Neurospora crassa* that required specific vitamins (45). This accomplishment began with the hypothesis that it might be possible to isolate mutants of this fungus generated by X rays that would be defective for making a specific nutrient. By treating spores with X-ray radiation and plating for isolated clones, they were able to discover three mutants that required specific vitamins to grow. By performing a genetic cross with a wild-type strain, they were able to establish that the enzymatic defect was due to a mutation in a single gene, allowing them to conclude that a single gene encoded a single polypeptide. Joshua Lederberg collaborated with Edward Tatum to isolate amino acid– and vitamin-requiring mutants of *E. coli*. By isolating strains of *E. coli* that contained at least two different auxotrophic mutations, it was possible to demonstrate that these bacteria could mediate genetic recombination (46). Gene transfer confirmed that these specific enzymatic defects were due to precise mutations in particular genes.

In 1928, Frederick Griffith, an epidemiologist and bacteriologist, made the serendipitous discovery of transformation, the first identified process of gene transfer in bacteria (47). He had been studying the virulence of the pneumococcal bacterium, now known as *Streptococcus pneumoniae*. He was investigating the basis of virulence with two colonial morphotypes of *S. pneumoniae*, a smooth (S) virulent strain and a rough (R) avirulent strain. He had made the observations that the injection of either the R strain or the heat-killed S strain into mice was not lethal. Surprisingly, if he mixed the heat-killed virulent strain with the rough avirulent strain and injected this mixture into a mouse, mortality was caused, but not by either individual component alone. An even greater surprise was that when he plated out the bacteria from the dead mice, the resulting strain no longer looked like the viable rough strain, but rather had been transformed to resemble the smooth strain. Griffith also showed that bacilli could revert from one phenotype back to the other. He concluded that some sort of "transforming" event had occurred that changed the avirulent strain into a virulent strain.

Dr. Oswald Avery, a leading pneumococcal researcher at the Rockefeller Institute for Medical Research, now Rockefeller University, was very skeptical of the studies and was convinced there must be some sort of contamination. However, researchers in his lab had confirmed the transformation process. This led to Dr. Avery's extending the work to demonstrate that the transformation event can be selected for by treatment with an antibody *in vitro*. In their 1944 paper, Avery, MacLeod, and McCarty concluded from their data that the transforming principle was DNA, stating, "the active fraction contains no demonstrable protein, unbound lipid, or serologically reactive polysaccharide and consists principally, if not solely, of a highly polymerized, viscous form of deoxyribonucleic acid" (48). Acceptance that DNA, not protein, was genetic material would have to wait for the Hershey and Chase experiment using radiolabeled DNA, which revealed that the transfer of nucleic acid correlated with the transfer of genetic material (49). Again, the theme common to these seminal experiments was the transfer of the causative entity (much like in the experiments conducted by Koch that led to the establishment of the Koch's postulate), which in this case was DNA.

The transforming principle transfer experiment of Griffith for *S. pneumoniae* parallels the Villemin transfer experiment in that both researchers discovered that specific phenotypes were the cause of a particular disease. The Avery, McLeod, and McCarty transfer experiment parallels the Koch transfer experiment because in both cases the fundamental cause of the disease or a virulent phenotype had been purified and transferred. This knowledge of causality has served to focus modern biology to build on this basic principle and elucidate the mechanisms by which DNA functions as genetic material that replicates and encodes RNA and protein products. By focusing on elucidating the structure of DNA, James Watson and Francis Crick (50, 51) provided an explanation for how genetic material was faithfully replicated. Moreover, this knowledge led to the central dogma of biology in which DNA is transcribed into RNA, which is translated into proteins (reviewed in reference 52).

As a graduate student, I was taught by Roy Curtiss III that the Avery, MacLeod, and McCarty experiment was a fulfillment of the molecular Koch's postulate, which was brilliantly described by Stanley Falkow (53). In other words, it is the way in which we know if a phenotype is caused by a genotype. To paraphrase Robert Koch, to prove that a phenotype is caused by a genotype, it is necessary to (i) identify a mutant with an altered phenotype, (ii) clone the genotype, and (iii) transfer the mutant genotype into the wild-type strain and demonstrate that the recipient strain acquires the mutant phenotype. The gene transfer was observed in *S. pneumoniae* because the bacteria had a natural transformation system. Although this transformation study established that DNA is genetic material, linkage analyses or DNA sequencing would be required to prove that a specific gene is required to make an enzyme to make a complex polysaccharide. Linkage analyses—conducted by measuring frequencies of recombination mediated by either conjugation of cells to cells (46) or DNA transfer using bacterial viruses (i.e., transduction)—first described in *Salmonella* by Norton Zinder (54) provided a means to identify specific genes and to transfer them. Thus, the three transfer systems in bacteria, of transformation, conjugation, or transduction, allowed for a rapid accumulation of basic biological knowledge by fulfilling the molecular Koch's postulate.

ELUCIDATING THE MECHANISMS OF DRUG KILLING: THE STREPTOMYCIN EXAMPLE

Gene transfer, in combination with the ability of a bactericidal drug to select for independent mutants, constitutes an essential system for identifying drug targets, a critical first step in characterizing the mechanism of drug action. Knowledge of the mechanism of action of a drug provides the basis for optimizing the efficacy of the agent being studied. Equally as important is discovering the mechanism of resistance of the organism. Both mechanisms lead to the identification of new and improved chemotherapeutic agents and the development of rapid susceptibility testing. This knowledge can be acquired by the isolation and characterization of mutants that are resistant to the death-inducing action of drugs. Mutations causing drug resistance can be used to identify at least four classes of genes including those encoding (i) drug activators, (ii) drug targets, (iii) modulators of drug action, and (iv) drug-degrading enzymes. Isolates whose resistance to a specific agent is due to a single mutation provide isogenic sets of strains that can be rigorously compared in various biochemical, physiological, and cellular assays.

One of the questions that Robert Koch addressed was whether different bacteria cause different diseases or whether one or a limited number of bacteria cause many different diseases. Implicit within Koch's postulate is the premise that if the three conditions are met, the conclusion is that the particular infectious disease under investigation is caused by the pathogen being examined. In parallel reasoning, the cloning of a particular drug resistance-conferring genotype or an allele by gene transfer experiments allows for the conclusion that the cloned gene is the basis for the specific resistance. Once a bacterium or virus has been identified and proven to be the causative agent for an infectious disease by fulfilling the Koch's postulate, experiments aimed at comparing virulence, the pathological reaction, and immune responses elicited in the host among various specific pathogens are possible. Analogously, once an allele of a gene has been demonstrated to be the cause of a drug resistance phenotype by fulfilling the molecular Koch's postulate, studies can be conducted comparing the biochemistry, molecular biology, and physiology of cells treated with the agent with those of others.

Streptomycin is a superb example of a drug in which genetic analysis led to an understanding of its mechanism of action of killing a bacterial cell and subsequently the cell's mechanisms of resistance. This drug, discovered by David Schatz and Salman Waksman (55), was the first antimicrobial agent found to be highly active against the tubercle bacilli. It is also effective against Gram-negative and Gram-positive organisms (56, 57). In contrast to the modern approach to drug development, which is largely based on targeting essential gene products, streptomycin was identified by empirically screening bacteria from soil for antimicrobial activities. As a result of this method of

discovery, neither its mechanism of action nor its target was known. Streptomycin was first used as a selective tool to probe for the development of specific mutations and hence to measure mutation rates under defined experimental conditions, similar to studies that used lytic phages (58). Elucidation of the target of streptomycin was not mentioned in these early papers, but the unambiguous nature of the high-level streptomycin resistance phenotype provided an attractive lead for further genetic studies. Interestingly, two types of mutations were identified when streptomycin-resistant mutants of *E. coli* were isolated (58). One class conferred high-level streptomycin resistance and the other, streptomycin dependence. Gene transfer experiments using either conjugation or transduction in *E. coli* were able to demonstrate that the three phenotypes (streptomycin susceptibility, streptomycin resistance, and streptomycin dependence) all mapped to a single gene, designated *strA* (58–61). Results of the gene transfer studies fulfilled the molecular Koch's postulate, because linkage analysis provided evidence that the specific allele transferred via conjugation or transduction was responsible for the observed phenotype.

Spotts and Stanier grasped the significance of these gene transfer experiments because the streptomycin-resistant, -susceptible, or -dependent phenotypes of *E. coli* that were generated resulted from different alleles of a single gene (62). Based on biochemical analysis of these isogenic strains, they hypothesized that this gene likely produced a polypeptide or regulated the expression of a protein that was part of the *E. coli* ribosome and that mutations in specific domains of this target resulted in distinct structures that either bound or failed to bind streptomycin. This idea of a specific target was a revolutionary and unifying hypothesis. In support of this hypothesis, they cited the data of Erdos and Ullman (63, 64) that suggested that streptomycin inhibited protein synthesis in a sensitive strain but failed to do so in a resistant isolate. Erdos and Ullmann had used *Mycobacterium friburgenesis* for their study, a species for which there were no genetic tools, and thus they could not conclude that their resistance mutation mapped to a single specific gene and hence a specific target.

When Julian Davies isolated ribosomes from isogenic streptomycin-resistant or -susceptible strains that differed by a single allele, he was able to demonstrate that protein synthesis was inhibited by streptomycin treatment of the ribosomes from the susceptible but not from the resistant strain (65). Moreover, he was able to map the resistance or susceptibility phenotype to the 30S subunit of the ribosome. Five years later,

Ozaki, Mizushima, and Nomura were able to isolate the specific protein (which had been encoded by *strA*) of the 30S ribosome subunit that mediated resistance to streptomycin (66). It would be another 21 years before the elucidation of the structure of the ribosome allowed for direct visualization of the streptomycin bound to the ribosome (67). In *M. tuberculosis*, streptomycin resistance also maps to the gene encoding the 16S rRNA (68), a result that not unexpectedly was not observed in *E. coli*, which has seven copies of the rRNA genes. In addition, the crystal structure of streptomycin bound to the ribosome revealed how specific mutations in the 16S rRNA would mediate resistance (68).

The mechanism of action of streptomycin is one of the most definitive in the history of the study of drugs aimed at defining the basis for the observed antimicrobial effect, because it was verified by genetic, biochemical, and X-ray crystallographic analysis. Clearly, the use of mutant isolation and the fulfillment of the molecular Koch's postulate using the transfer systems of *E. coli* were seminal in these efforts.

TRANSFER OF GENES FROM PATHOGENIC MYCOBACTERIA INTO THE SURROGATE HOST *E. COLI*: GREAT PROMISE WITH LIMITATIONS

Transfer of any gene became relatively easy with the advent of recombinant DNA technologies (69–71) and the development of plasmid (72–76) and phage (77–80) cloning vectors for *E. coli*. The combination of recombinant DNA technologies with DNA sequencing technologies (81, 82) opened up entirely new approaches to the study of any organism, facilitating an explosion in new biological knowledge. While cloning technology offered great promise, it also represented potential new hazards. Would the addition of virulence genes from other pathogens into *E. coli* render this bacterium hazardous? This issue was discussed at a conference at Asilomar in 1973, which generated a set of guidelines and proposals to move forward in a safe and thoughtful way with experiments involving recombinant DNA technologies (83). The guidelines concerned safe practices for experimentation both with recombinant DNA and with physical and biological containment issues. The latter included engineering a bacterium that would be unable to replicate in mammalian hosts. Thus, the prototype strain χ1776 was genetically engineered to be unable to replicate outside the laboratory due to auxotrophic requirements for diaminopimelic acid, which is absent from mammalian

hosts, as well as a mutation in thymidylate synthase that induced bacterial death when the organism was starved for thymidine (84). Moreover, the strain was found to be sensitive to killing by complement in human serum (85). Based on the safety experiences with *E. coli* of researchers throughout the world, guidelines were relaxed over the years. The concepts of generating strains of bacteria that would be unable to replicate in mammalian hosts were influential in developing strains of *M. tuberculosis* that could safely be used in a BSL2 laboratory (see below).

Despite the availability of recombinant DNA technologies, research on leprosy bacillus and the tubercle bacillus were limited until the second half of the 1980s. First, both of these mycobacteria are BSL3 pathogens, whose manipulation mandates a tedious biocontainment environment that significantly curtails efficiency. More limiting was the fact that the leprosy bacillus cannot be grown in any artificial media. It was not until 1960 that Charles Shepard demonstrated that the leprosy bacillus procured from humans could be grown in mouse footpads (86). In this seminal work, Shepard showed an increase from 1,000 acid-fast bacilli to 1 million in 6 months' time. Despite this advance, this system, which required growth monitoring by microscopic inspection of tissue samples, is limited by its inefficiency in terms of time and bacterial yield, the latter significantly hampering efforts to generate genomic libraries. Subsequently, it is debated whether the Shepard experiments fulfilled Koch's postulate for the leprosy bacillus since the growth in the mouse does not reproduce any neurological symptoms typically associated with this pathogen. Nevertheless, the Shepard model provided a means to culture various *M. leprae* strains and to test drugs and vaccines (87).

The next significant advance for leprosy research was in 1971, when Wilemar Kirchheimer and Eleanor Storrs demonstrated that the leprosy bacillus replicated in nine-banded armadillos (88). In contrast to the mouse footpad model, infection of the nine-banded armadillo reproduces much of the clinical pathology associated with leprosy. Charles Shepard not only reproduced this work, but also showed it was possible to isolate 10^{10} to 10^{11} leprosy bacilli from the armadillo liver following a two-year infection. Under the supervision of my comentors Josephine E. Clark-Curtiss and Roy Curtiss III, I had generated cosmid genomic libraries of *M. leprae*. These reagents enabled attempts to characterize specific biosynthetic pathways of *M. leprae* by complementation studies using auxotrophic strains of *E. coli* (89, 90). I failed to obtain complementation with these cosmids and was only later successful when I used a promoter that was highly active in *E. coli*, allowing me to complement a mutation in the citrate synthase gene (91). The conclusions from these studies posited that promoters of mycobacterial genes did not function well in *E. coli*. Numerous other groups had found that genes from organisms with a high guanine plus cytosine content failed to express well in *E. coli*. To overcome this limitation, Richard Young and Ron Davis developed an elegant strategy to express mycobacterial genes in *E. coli* using a highly active promoter from λgt11, a bacteriophage lambda vector (92), to efficiently express proteins in *E. coli*. This bacteriophage-based system enables identification of major protein antigens of *M. leprae* (93) and *M. tuberculosis* (94) that elicit human humoral immune response by screening of the expression library using sera from infected individuals. The identification of these antigens provided novel opportunities to study the immune responses in leprosy and TB. However, identification of the genes encoding the targets of specific antibiotics used for the treatment of TB and leprosy—or elucidation of mycobacterial virulence factors—would still need the development of a gene transfer system in mycobacteria.

INTRODUCTION OF FOREIGN DNA INTO MYCOBACTERIA: SHUTTLE PHASMIDS AND THE DEVELOPMENT OF A PLASMID TRANSFORMATION SYSTEM FOR *M. SMEGMATIS*

Mycobacteria are substantially different from *E. coli*, and there were numerous questions that were unanswerable using *E. coli* as a surrogate host. For example, in 1987, we did not know (i) the targets of the TB-specific drugs INH, ETH, ethambutol, thioacetazones, and pyrazinamide; (ii) the genetic basis of acid-fast staining; (iii) if it was possible to isolate auxotrophic mutants of *M. tuberculosis;* or (iv) the basis for attenuation of BCG. This lack of knowledge was due to the inability to fulfill the third condition of the molecular Koch's postulate, because at the time, an effective DNA transfer system for *M. tuberculosis* did not exist. For example, while it was feasible to isolate INH-resistant mutants of *M. tuberculosis* by plating a large number of wild-type bacilli on INH-containing plates, the lack of a DNA transfer system for the tubercle bacillus at the time precluded the mapping of the resistance phenotype to a specific gene. Although generalized transduction (95) and conjugation (96, 97) for fast-growing *M. smegmatis* had been described, the inability to transfer genes from *M. tuberculosis* or

M. leprae into *M. smegmatis* limited the use of these tools for analyzing these pathogens. Even if conjugation or generalized transduction had been discovered for *M. tuberculosis* or BCG, the slow replication time of these strains of 16 to 24 h would have made genetic analysis difficult. An effective genetic system that enables transfer of genes into the slow-growing *M. tuberculosis* was needed for study of this global pathogen.

The cloning of genes into mycobacterial plasmids offered the simplest approach to identify drug targets or virulence factors. Although mycobacterial plasmids were discovered in 1979 (98) and 1985 (99), no successful transformation of any mycobacterial cell had been achieved when I started in the laboratory of Barry Bloom in 1985. Barry had wanted to develop the TB vaccine strain, BCG, into a recombinant vaccine vector, and I wanted to develop a system by which the genotype-phenotype causality can be proven according to the molecular Koch's postulate. These shared goals both required the development of gene transfer for *M. tuberculosis*. Plasmid transformation was the obvious choice, but numerous individuals had told me they had been unsuccessful. While transformation in *M. tuberculosis* was the ultimate goal, the process was not straightforward. First, studies of *M. tuberculosis* require BSL3 containment. Second, *M. tuberculosis* has a generation time of 16 to 24 h, requiring 3 to 4 weeks to form colonies on a plate. This slow growth was enough of a deterrent to spawn the consideration of the use of a more rapidly growing non-BSL3 mycobacterium. For these reasons, *M. smegmatis* was an attractive surrogate host for the study of genes from *M. tuberculosis* and *M. leprae*. It is fast-growing with a generation time of 2.5 to 4 h and thus yields colonies from single cells in 3 to 4 days; it is nonpathogenic, and the ATCC607 strain is an excellent host for mycobacteriophage isolation. Dr. Wilbur Jones from the CDC shared this strain with me as well as the mycobacteriophages D29 and 33D.

The simplest vector to imagine would have been to construct an *E. coli*–mycobacterial shuttle plasmid that could transform *E. coli* and mycobacteria including *M. smegmatis*. While the specific DNA fragment required for replication of a plasmid in these host cells was not known, the construction of a random insertion of an *E. coli* plasmid into a mycobacterial plasmid was easily done, but this did not yield transformants into *M. smegmatis*. We reasoned that the failures of others could have been due to four possible limitations: (i) the inability of naked DNA to enter the mycobacterial cell, (ii) degradation by a restriction system of the *E. coli* DNA seen as foreign, (iii) failure to have a functional selection system (appropriate concentrations of antibiotic and sufficiently high levels of expression of the selectable marker gene), and/or (iv) the inability of the mycobacterial plasmid to replicate and segregate into *M. smegmatis* cells. By making my own bacteriophage lambda *in vitro* packaging mixes in the Curtiss laboratory, I had come to appreciate the utility and power of phage vectors because phages had developed highly efficient means to deliver DNA into a cell. I therefore decided to focus on developing recombinant DNA phage vectors for the genetic manipulation of mycobacteria. Using the protoplast methodologies developed by Mervyn Bibb and David Hopwood (100), I focused on optimizing DNA transfer into *M. smegmatis* by assaying infectious centers formed after transfecting naked D29 phage DNA (4). The advantage of this choice was that phage plaques could be observed in 24 h, allowing me a rapid readout for optimizing protoplast generation protocols and to conduct polyethylene glycol (PEG) lot evaluations. I routinely obtained greater than 1,000 plaque forming units (PFU) per microgram of DNA (4). As I was optimizing my protoplast regeneration methods from the culture of *M. smegmatis* ATCC607 (which I had stocked as mc^{2}1; I had decided to name my mycobacterial culture collection the mc^2 collection in honor of the Einstein equation), I observed three colonial morphotypes that I cloned and stocked, the predominant one being orange rough (mc^{2}6), with the other two being white rough (mc^{2}21) and orange smooth (mc^{2}22). I chose to use the predominant mc^{2}6 for subsequent experiments.

Having established an efficient transfection assay, I decided to focus on the TM4 mycobacteriophage (101) that had been isolated from *Mycobacterium avium* in the laboratory of Patrick Brennan and was likely a temperate phage. I reasoned that a temperate phage vector could be used to introduce a stable selectable marker gene into the mycobacterial genome. By analyzing the restriction digests of TM4 phage DNA, I had discovered the phage was very similar to 33D, had a genome size of approximately 54 kb, and possessed cohesive ends. To construct a mycobacteriophage cloning vector patterned after what had been done with bacteriophage lambda, a unique restriction enzyme site would have to be introduced in a deleted nonessential region of the TM4 genome. Since I had discovered in my doctoral work that mycobacterial DNA did not express well in *E. coli*, I reasoned I could clone an entire mycobacteriophage genome into an *E. coli* plasmid and it would not kill the *E. coli* cell. Moreover, I reasoned that if I included a bacteriophage lambda *cos*

site, I would have a shuttle phasmid—a chimeric molecule that had four important attributes that would allow it to (i) be packaged into bacteriophage lambda particles, (ii) be packaged into mycobacteriophage particles, (iii) replicate in *E. coli* as a plasmid, and (iv) replicate in mycobacteria as a phage (Fig. 1).

To achieve this goal, TM4 phage DNA was ligated to form long concatemers, partially digested with a frequent-cutting DNA restriction enzyme, ligated to a bacteriophage lambda–based cosmid, *in vitro* packaged, and transduced into *E. coli*. Restriction analysis of individual *E. coli* cosmid clones showed I had generated a library of cosmid constructs containing 45-kb fragments of the TM4 genome. From the TM4 viewpoint, I had inserted cosmids at random sites around its genome with concomitant small deletions. I hypothesized that shuttle phasmids could be identified by transfecting the cosmid library into *M. smegmatis* protoplasts and assaying for plaques because I reasoned that the only molecules that could yield a plaque would be those in which the cosmid had inserted in the nonessential region of TM4. To my astonishment, the restriction fragment analysis of the DNA isolated from these plaques revealed the presence of the parental TM4 DNA with no cosmid. After reflecting on this result, I deduced that the wild-type TM4 phage had been generated by two or more TM4-cosmid hybrid molecules that underwent recombination to generate wild-type TM4 DNA. (It is worth noting that this high frequency of recombination yielding wild-type TM4 plaques is likely a result of a recombination function within TM4, as we have not observed this phenomenon with D29, L5, or DS6A. The possibility of this recombination function prompted us to develop specialized transduction, described in a later section, with the hope that the TM4 recombination functions could promote homologous recombination events.) Plaque hybridizations revealed that 4% of the plaques hybridized with the cosmid DNA, and these recombinant molecules were *bona fide* shuttle phasmids that could replicate in *E. coli* as a plasmid and in a mycobacterium as a phage (4). The mycobacteriophage-packaged cosmids readily infected BCG and *M. tuberculosis* (4).

These TM4 shuttle phasmid studies yielded important new knowledge. First, they provided the means to introduce foreign DNA into *M. smegmatis*, BCG, and *M. tuberculosis* for the first time. Second, they demonstrated not only that transfection of *M. smegmatis* protoplasts was highly reproducible, but also that DNA molecules propagated in *E. coli* were not being degraded by a restriction modification system of *M. smegmatis*.

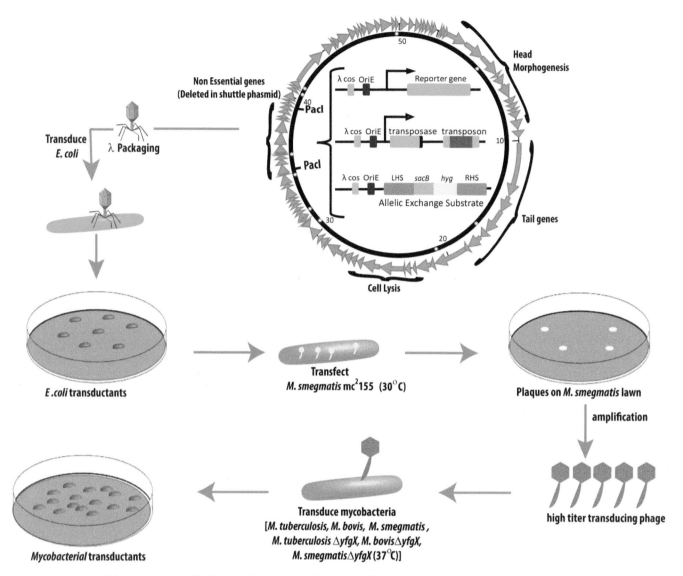

Figure 1 Specialized transduction is outlined as follows: the center plasmid represents the shuttle phasmid phA159, which contains 90% TM4 phage DNA and 10% plasmid DNA. The stars mark the sites of the mutations in the TM4 genome. The nonessential genes that are deleted to create the shuttle phasmid are noted in the picture, flanked by PacI sites. This site can be replaced with one of three things: (i) a reporter gene such as green fluorescent protein (GFP), (ii) an allelic exchange substrate (AES) that contains an antibiotic resistance marker, or (iii) a transposase gene to facilitate transposon mutagenesis. Going counterclockwise in this schematic, the recombinant cosmid can be packaged into phage heads using an *in vitro* packaging mix, and the subsequent phages can be used to transduce *E. coli* to create *E. coli* transductant colonies. Going clockwise from the shuttle phasmid, one can transfect *M. smegmatis* mc²155 at 30°C to yield plaques on an *M. smegmatis* lawn, resulting from lysis of the cells. The plaques can then be purified and amplified to obtain a high-titer phage lysate that can subsequently be used to transduce any mycobacterial species. doi:10.1128/microbiolspec.MGM2-0037-2013.f1

Using the first-generation TM4 shuttle phasmids, a gene encoding kanamycin resistance was cloned into the shuttle phasmid in *E. coli*. This construct, however, failed to yield stable kanamycin-resistant transductants in *M. smegmatis*. We concluded that the TM4 shuttle phasmid was not able to lysogenize, and I started screening for mycobacteriophages that could site-specifically integrate into mycobacterial chromosomes.

Numerous groups had identified mycobacteriophages that formed turbid plaques on *M. smegmatis*, so I obtained many such phages, characterized their genomes and screened them for the ability to site-specifically integrate into mc²6. The mycobacteriophages L1 and L5 had been well characterized by Margaret Sellers (7). I demonstrated by performing a Southern blot on an L5 lysogen of mc²6 that this phage clearly integrated site-specifically into *M. smegmatis* (5). It was because of this result that Graham Hatfull decided to determine the DNA sequence of this particular phage (102). Scott Snapper and I generated shuttle phasmids from L5 and demonstrated that these could transfer a kanamycin resistance gene stably into the chromosome of mc²6 (5). This result allowed us to establish the concentration of kanamycin to be used to select for the presence of a newly introduced kanamycin resistance gene. We had then overcome three of the four limitations we proposed to be possible reasons for the difficulty in obtaining *M. smegmatis* transformants.

Tobias Kieser had made a library of insertions at random sites in the 5-kb plasmid pAL5000, which had been isolated from *Mycobacterium fortuitum* (99) with an *E. coli* plasmid bearing a kanamycin resistance gene. The use of a library was important because the DNA sequence that mediated plasmid replication was not known. Having the knowledge that we could introduce DNA into mycobacteria, not worry about restriction, and use kanamycin as a selection, Scott Snapper set up to transform *M. smegmatis* mc²6 using this library. The one concern of using protoplasts was that regeneration could be an inefficient process. At the time, the technique of electroporation was being used to transform bacteria (103), thereby circumventing the regeneration step. The electroporation of mc²6 cells with recombinant DNAs of the *E. coli* plasmid pAL5000 chimeric library yielded three kanamycin-resistant transformants. From these *M. smegmatis* transformants, plasmid DNA was isolated, revealing extrachromsomal chimeric plasmids. These plasmids yielded transformants in both *E. coli* and BCG (5), thus validating the first plasmid transformation system for mycobacteria. Scott Snapper and I hypothesized that the transformants contained a mutation that allowed for replication of the pAL5000 replicon. To test this hypothesis, we cured one of the original transformants (namely, mc²154) of its plasmid by propagating the strain in the absence of kanamycin. A kanamycin-sensitive clone was stocked as mc²155. Astonishingly, transformation of mc²155 yielded 10^4 to 10^6 transformants compared to mc²6. A series of experiments allowed us to demonstrate that this efficient plasmid transformation phenotype was not due to DNA uptake, restriction, or selective expression of kanamycin resistance genes. We concluded that it was due to a mutation that specifically affected the replication of pAL5000 plasmids (11).

Despite not knowing the mechanism underlying the significantly enhanced transformation efficiency observed, the availability of mc²155 revolutionized the field of mycobacteria because it provided the surrogate mycobacterial host to study the genes of the pathogenic mycobacteria as well as a genetically tractable organism for studying mycobacterial biology. It is gratifying that 25 years after the discovery of mc²155, we have just recently found a single-point mutation that confers the efficient plasmid transformation phenotype. The mutation causes a loss of function of the DNA binding protein that prevents the replication of pAL5000 (M. Panas, P. Jain, and W. R. Jacobs, Jr., submitted). Shuttle phasmids showed us that *M. smegmatis* does not restrict *E. coli* DNA and enabled the first successful transfer of a selectable marker gene to *M. smegmatis*, leading to the development of a plasmid transformation system for mycobacteria.

SECOND GENERATION SHUTTLE PHASMIDS: REPORTER MYCOBACTERIOPHAGES, TRANSPOSON DELIVERY, AND SPECIALIZED TRANSDUCTION

Conditional Replication and PacI Excisable Cosmids

Shuttle phasmids packaged in TM4 mycobacteriophage particles facilitate the transfer of DNA into *M. tuberculosis*, *M. smegmatis*, and many other mycobacteria, because it simply requires a phage infection. One of the limitations of this system was that TM4 is a lytic phage that lyses and kills the infected cells. In retrospect, it was fortuitous the TM4 had lost its ability to be a temperate phage because it allowed us to focus on delivering DNA transiently. Sequence analysis of the Cluster K family phages has revealed that the genes encoding the integrase and repressor genes (104) are deleted from TM4, thereby making it unable to stably integrate into the mycobacterial chromosome. Conditionally replicating lytic phages, such as repressor and integrase-deleted mutants of bacteriophage lambda, have proven to be very useful for delivery of transposons to *E. coli* (105). We reasoned that we could isolate mutations that allowed the TM4 mycobacteriophage to replicate at 30°C but not at 37°C. Stoyan Bardarov, Jordan Kriakov, and Christian Carriere had screened a large number of TM4 mycobacteriophage mutants for those that were not able to replicate at 37°C but were able to replicate at 30°C and reverted at frequencies of less than 1 in 10 million (106, 107). One such phage mutant, named PH101, was identified out of hundreds that were temperature-sensitive for growth and reverted at frequencies of 1 in 10,000. PH101 has recently been sequenced and found to have nine independent point mutations, of which two were found by Graham Hatfull and colleagues to be necessary and sufficient for the low reversion frequency temperature-sensitive phenotype (see Fig. 1).

Since it was desirable to be able to test a variety of different transposons for mutagenesis, we developed a very simple system for replacing the cosmid that had been used to create new TM4 shuttle phasmids by flanking it with unique restriction enzyme sites. We took advantage of the fact that the restriction enzyme PacI recognizes the sequence TTAATTAA, which is rarely found in mycobacteriophages that generally (but with some notable exceptions; see reference 144) have a GC content of greater than 60%. Shuttle phasmids are constructed by partial digestion of concatemerized phage genomes with the frequent-cutting restriction enzyme Sau3A. These fragments are then ligated to a bacteriophage lambda cosmid, pYUB328, which has a unique restriction enzyme site that is compatible with Sau3A like BamHI or BglII. We engineered this cosmid to have PacI sites flanking the unique BamHI or BglII site, and as such, shuttle phasmids generated with pYUB328 allowed easy excision and replacement with other cosmids. The ability to package the shuttle phasmids into bacteriophage lambda heads makes the cloning of any cosmid with a unique PacI site a relatively easy construction. The transfection of the resulting shuttle phasmid into M. smegmatis allows for the generation of mycobacteriophage particles that can infect most other mycobacterial strains. The conditionally replicating TM4 phage provides a means to deliver virtually any DNA fragment in a way that does not kill the recipient cells. We have utilized these shuttle phasmids to deliver reporter genes, transposons, and allelic exchange substrates to develop novel diagnostic tests, transposon libraries, and specialized transductants (Fig. 1).

Reporter Mycobacteriophages for Rapid Drug Susceptibility Testing of M. tuberculosis
Increasing numbers of drug-resistant M. tuberculosis strains are compromising first-line therapies. Unfortunately, determination of drug resistances can require a minimum of 6 to 12 weeks if done by culture methodologies. The recent development of a molecular beacon-based test has made diagnosis of M. tuberculosis infection and rifampin resistance available in 90 min to 48 h, but knowledge of viability and resistance to other first-line drugs is also important for proper treatment. Shimon Ulitzer and Jonathan Kuhn had proposed the idea of introducing a bioluminescence gene to identify bacteria in any sample (108). Since the TM4 shuttle phasmids infected M. tuberculosis strains and had a cloning capacity of over 5 kb of DNA, it was possible to clone the firefly luciferase gene into a TM4 shuttle phasmid to develop luciferase reporter mycobacteriophages (109). In the presence of exogenous luciferin, we were able to detect light production from M. smegmatis or M. tuberculosis cells within 30 min following the addition of the phage. We reasoned that since the luciferase reaction required ATP for light production, luciferase reporter mycobacteriophages would be great tools for assessing drug susceptibilities. Rather than waiting a month for M. tuberculosis cells to grow, then another month to do drug susceptibility testing, we hypothesized that it would be possible to assess drug action by incubating samples of M. tuberculosis cells with or without drugs and then adding the luciferase reporter mycobacteriophage. If the drug was effective, we reasoned that no light would be emitted. If the strain was drug resistant, it would become luminescent.

This technique works very well in ideal laboratory conditions but has proven challenging in a clinical laboratory in the developing world (110–114). To circumvent some of these issues, we made a number of innovative improvements including the development of the "Bronx Box," which used a Polaroid film–based assay (115) and then the isolation of a temperature-sensitive mutant that prevented the killing of the infected cells (107). The incorporation of the gene encoding the green fluorescent protein (116, 117) with a highly expressed phage promoter (118) has provided the means to visualize individual cells and the potential to measure fractional drug resistance. We are collaborating with Dr. Alex Pym at the K-RITH laboratories in Durban, South Africa, to test the utility of these next-generation fluorophages for rapidly assessing drug susceptibilities of patient samples.

Transposon Mutagenesis
Transposons, genetic elements that can jump from one piece of DNA to another, have proven to be indispensable tools for the genetic analysis of organisms, because they generate random insertions into genes,

thereby creating a mutation that often inactivates the target gene (105). Transposition was first observed in mycobacteria in the laboratory of Brigitte Gicquel using a new transposon that they had identified in *M. fortuitum* (119). The Gicquel laboratory went on to develop a temperature-sensitive plasmid that replicated at 30°C but not at 37°C as a system for efficient transposon mutagenesis (120). Unfortunately, these initially isolated transposons did not transpose by a simple cut-and-paste mechanism or in a random manner, limiting transposon mutagenesis as a genetic tool. Serendipitously, while trying to understand the inability to disrupt the gene encoding aspartate semi-aldehyde dehydrogenase, Jeffery Cirillo discovered a novel insertion element in *M. smegmatis*, which we named IS*1096* (30). IS*1096* is 2.3 kb and possesses two open reading frames (ORFs). By cloning a gene encoding kanamycin resistance into various sites within IS*1096*, we were able to identify which gene was responsible for transposition and also to generate a transposon library of BCG (31). Following transformation of the IS*1096* transposon, the cells were plated on media with Casamino Acids, and then the resulting colonies were screened for mutants that failed to grow on minimal media. Three auxotrophic mutants were found: two were leucine auxotrophs and one was a methionine auxotroph. These transposon mutants and the illegitimate recombination mutants discussed below were the first auxotrophic mutants isolated for BCG or *M. tuberculosis*. The studies also established the IS*1096*-derived transposons as an excellent system for transposon mutagenesis.

Since phages can deliver their DNA genomes to every cell in a population of bacteria, the goal of developing a conditionally replicating phage that could deliver a transposon transiently to every cell, but not kill the infected cells, was highly desirable. As described above, we isolated PH101, a temperature-sensitive mutant of TM4 that failed to form plaques at 37°C and had a low reversion frequency. It is worth noting that, in addition to PH101, we isolated numerous independent temperature-sensitive mutants of TM4 and D29 shuttle phasmids, all with reversion frequencies of less than 1 in 100,000, that all proved unsuitable for transposon mutagenesis experiments. Two shuttle phasmids, phAE87 and phAE159, were constructed using the PacI excisable cosmid for efficient transposon mutagenesis. The shuttle phasmid phAE87 was used to make our first library of thousands of transposon mutants in *M. tuberculosis* with Tn*5367* (an IS*1096*-derived transposon [106]). The shuttle phasmid was useful for generating transposon libraries

in *M. avium* (121), *Mycobacterium ulcerans* (122), and *Mycobacterium paratuberculosis* (123). Michael Glickman screened a library of transposon mutants of *M. tuberculosis* and identified one that failed to cord, a property that has long been associated with virulence (124). Using the clever strategy of signature-tagged mutagenesis developed by the laboratory of David Holden (125), in which bar codes were incorporated into transposons to screen for underrepresented mutants, Jeffery Cox used tagged Tn*5367* to identify a novel lipid secretion system for *M. tuberculosis* (126). Sequence analysis of a large number of transposon insertions revealed a bias in IS*1096* transposition, but it was still useful in revealing a number of new genes required for virulence (127). John McKinney's laboratory used this system to screen for *M. tuberculosis* counter-immune mutants (128). The selection for underrepresented clones was improved by Christopher Sassetti and Eric Rubin by cloning the Himar transposon into the conditionally replicating shuttle phasmid phAE87 and performing TraSH (transposon site hybridization) (129). TraSH has been used extensively to identify the essential genes under various growth conditions *in vitro* (130) and *in vivo* (131). Common to all the transposon studies in slow-growing mycobacteria is the use of the conditionally replicating shuttle phasmid to generate transposon libraries.

Specialized Transduction and Linkage Cotransduction

Phages were discovered to be capable of transferring DNA from one strain to another to repair auxotrophic mutations. This process of transduction was characterized as being generalized if a single phage could transfer genes to a diverse set of genes around the chromosome as for P22 of *Salmonella* (54) and P1 of *E. coli* (60). In contrast, lysates induced from bacteriophage lambda lysogens were found to be able to specifically transduce genes involved the galactose utilization (132). This was called specialized transduction. Generalized transducing phages package by headful mechanisms and occasionally package genome fragments of the bacterial chromosome. Thus, in lysates of generalized transducing phages, a small percentage of phage particles exist that contain headful packaged lengths of the bacterial chromosome. In contrast, because specialized transducing particles from lambda phage in *E. coli* are made from aberrant excisions of the integrated lysogenized phage DNA, these rare particles only contain a small fraction of the chromosome adjacent to the integrated phage.

For generalized or specialized transduction, the DNA from the bacterial chromosome has the ability to recombine by homologous recombination into the recipient cell. We hypothesized that it should be possible to deliver homologous recombination substrates efficiently to the *M. tuberculosis* chromosome and select for the introduction of precise null deletions of any gene of any mycobacteria (133). We demonstrated the proof of principle by disrupting numerous genes encoding amino acid or vitamin biosynthetic functions in *M. tuberculosis*, BCG, and *M. smegmatis* (133) and the gene encoding a mycolic acid biosynthesis enzyme that is required for the cording phenotype of *M. tuberculosis* (124).

The robustness of specialized transduction was demonstrated with the deletion of not only a single gene, but also all 11 genes of the *esx1* locus (the primary attenuating mutation of BCG) in three different *M. tuberculosis* strains and *M. bovis* (134). Ken Stover's group had performed subtractive hybridizations and found many deletions that were absent from BCG yet present in the *M. tuberculosis* genome. The one deletion that was common to all the BCG strains was *RD1*, later named *esx1* (135). Tsungda Hsu had been successful in obtaining one deletion mutant of this region in *M. tuberculosis* that was unmarked by a plasmid transformation. He was trying to re-create the primary attenuating mutation of BCG that had been hypothesized to be the deletion of *esx1* since it is common to all BCG strains. But Drs. Calmette and Guérin had isolated this mutant from *Mycobacterium bovis*, so Dr. Hsu wanted to make the strain in *M. bovis*, not just *M. tuberculosis*. Dr. Hsu performed over 100 independent transformations with *M. bovis* over a 5-month span, and although he obtained hundreds of hygromycin-resistant colonies, he obtained no knockouts of *esx1*. We were unsure if specialized transduction would be successful in making such a large deletion. To our delight, the *esx1* deletion was successfully constructed in three *M. tuberculosis* strains (H37Rv, Erdman, and CDC1551) as well as in *M. bovis* (Fig. 2). All four of these strains had demonstrated a marked attenuation in both immunocompetent C57BL/6 mice and immunocompromised SCID mice and could be fully complemented back to virulence with a cosmid that spanned this region (134). Interestingly, we had shared our SCID data and the complementing cosmid with Stewart Cole, and he transformed BCG and showed that it was able to restore virulence in SCID mice but not in immunocompetent mice (136). Specialized transduction allowed us to fulfill the

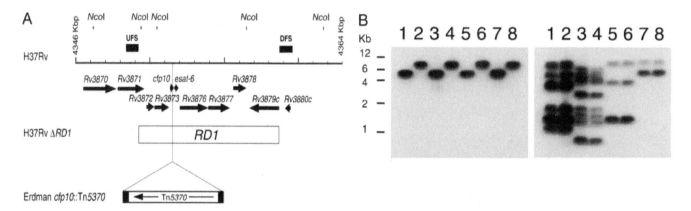

Figure 2 Generation of mutants in the *RD1* region of *M. tuberculosis* and *M. bovis*. (**A**) Schematic of *M. tuberculosis* H37Rv *RD1* region showing predicted NcoI sites. Arrows at the top represent the genes in this region. Upstream flanking sequences (UFS) and downstream flanking sequences (DFS) used to generate the knockout are indicated as filled bars above the grid line. Each increment in the grid line represents 1 kbp. The *RD1* sequence deleted from *M. bovis* BCG is represented by an open bar spanning from *Rv3871* to *Rv3879c*. The site of the insertion of transposon Tn*5370* is also indicated. (**B**) Southern analysis of the NcoI-digested genomic DNA isolated from the wild type and the ΔRD1 mutants generated by using specialized transduction in *M. tuberculosis* and *M. bovis*. Lane 1, *M. tuberculosis* H37Rv; lane 2, *M. tuberculosis* H37Rv ΔRD1; lane 3, *M. tuberculosis* Erdman; lane 4, *M. tuberculosis* Erdman ΔRD1; lane 5, *M. tuberculosis* CDC1551; lane 6, *M. tuberculosis* CDC1551 ΔRD1; lane 7, *M. bovis* Ravenel; lane 8, *M. bovis* Ravenel ΔRD1. The probe used in the Southern analysis was either DFS (left), demonstrating the deletion of *RD1*, or IS*6110*-specific (right). The IS*6110* probe is used to characterize the four strains. Reprinted with permission. doi:10.1128/microbiolspec.MGM2-0037-2013.f2

molecular Koch's postulate for the primary attenuating mutation of BCG.

Specialized transduction also enabled us to transfer a specific point mutation within an essential gene by linkage cotransduction. As shown in the streptomycin section above, conjugation or generalized transduction had allowed investigators to transfer specific mutant alleles within a single defined gene by linking this allele to some other mutation. The end result was the generation of isogenic strains that differed by a single point mutation. We had identified a single point mutation in the *inhA* gene in *M. smegmatis* and had linked that mutation to a kanamycin resistance gene on a cosmid, transformed it into INH-susceptible *M. smegmatis*, and

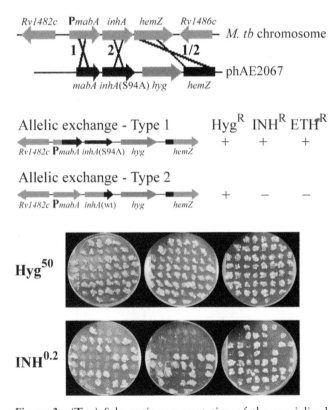

Figure 3 (**Top**) Schematic representation of the specialized transducing phage. A replicating shuttle phasmid phAE2067 containing *mabA*, *inhA* carrying the S94A mutation, a *hyg* resistance cassette, and *hemZ* was used to transduce *M. tuberculosis* (*M. tb*). The two possible sites of recombination are marked 1 and 2. (**Middle**) The recombination can occur either before the point mutation (crossover type 1), resulting in an INH-resistant and ETH-resistant recombinant carrying the S94A mutation, or after the point mutation (crossover type 2; the strain contains a wild-type *inhA* gene). (**Bottom**) Individual *M. tuberculosis* H37Rv *inhA*(S94A) transductants (*n* = 150) were screened by picking and patching onto plates containing either hygromycin (50 µg/ml) or INH (0.2 µg/ml). Reprinted with permission. doi:10.1128/microbiolspec.MGM2-0037-2013.f3

selected for kanamycin resistance transformants. By this method, we had linked a kanamycin resistance gene to *inhA* and then generated INH-susceptible or INH-resistant *M. smegmatis* strains. Due to inefficient allelic exchange in *M. tuberculosis* and the high degree of illegitimate recombination, this approach was not possible in *M. tuberculosis*. However, by generating a specialized transducing phage in which we had genetically engineered a hygromycin resistance gene closely linked to the *inhA* Ser 94 Ala allele, we were able to obtain hygromycin-resistant transductants that yielded INH-resistant or INH-susceptible phenotypes (Fig. 3). Thus, we were able to generate isogenic strains of *M. tuberculosis* that differed by a single point mutation in the *inhA* allele. Specialized transduction allowed us to prove that a single point mutation within an essential gene caused an INH-resistant phenotype, consistent with the hypothesis that InhA was the target of INH and ETH (137).

HIGH-THROUGHPUT SPECIALIZED TRANSDUCTION

The generation of the complete genome sequences of a bacterium, the first being *Haemophilus influenzae* (138), altered the way we investigated organisms. The first eukaryotic organism to have its genome fully sequenced and annotated was *Saccharomyces cerevisiae* (139). From such a sequence, it was possible to predict the number of ORFs and the numbers of functional RNA molecules. These predictions came from our understanding of genes, gene structures, ORFs, and operons in bacteria, primarily in *E. coli*. Although it was possible to assign putative functions to a subset of ORFs based on homology to related genes with proven functions, 40% of the ORFs in the *H. influenzae* genome had no known homology. It is important to note that all functional phenotypes are concluded to be assigned to a specific gene by the fulfillment of the molecular Koch's postulate. Functions had been identified by isolating mutants with defined phenotypes in the organism of interest and then proving that the specific phenotype was caused by a specific genotype by performing a gene transfer.

The most direct way to determine the function of a gene from an organism is to generate a precise null deletion of an ORF from the organism with the concomitant introduction of a selectable marker gene. Such mutated alleles can be readily made *in vitro* using PCR (140, 141). Again, the operative theme is gene transfer, where a mutated allele is introduced into a chromosome to generate a mutant strain. *S. cerevisiae* was the first organism for which a complete set of precise null deletion mutants were made, primarily because it has a highly efficient allelic exchange system that enabled *in vitro* null deletion alleles containing a selectable marker gene to be readily introduced into the genome (142). Another property that made *S. cerevisiae* attractive for this analysis was that the strain was a donor diploid strain. By allowing it to sporulate and form haploid spores, tetrad analyses made it possible to test whether genes were essential, depending on if it produced four viable spores. Moreover, the ability to mate the knockout mutated alleles with any knockout recipient query strain allowed for the screening of synthetic lethal alleles and provided a new way to observe previously unknown genetic interactions (147). In addition, by introducing a unique bar code into every knockout allele, screens for mutants that are underrepresented in specific growth conditions can be readily identified. Such a strategy had been first employed by David Holden's group using bar-coded transposons for *Salmonella* (125).

The completion of the genomic sequence of *M. tuberculosis* was first achieved by Stewart Cole and colleagues on the strain H37Rv (148) and later by others on CDC1551 (143). These sequences reveal the presence of over 4,000 ORFs, with over 40% of unknown functions. In a collaboration with the Genomic Foundation of Novartis, Graham Hatfull, and William Bishai, we have initiated the generation of the entire set of specialized transducing phages to make precise null deletions of every gene of *M. tuberculosis*. To achieve this knockout set:

1. We have developed a high-throughput method to construct each allelic exchange substrate (AES) for each ORF deletion mutation with the concomitant *hyg* selectable marker gene that possesses 500- to 1,000-bp homologous DNA fragments flanking the deletion. Included in each construct is a unique bar code that associates with each deletion mutation as well as a system to excise the *hyg* gene from the resulting *M. tuberculosis* mutant to make unmarked mutations that retain their bar codes. The AESs are generated *in vitro*, cloned into a PacI cosmid, sequence verified, and then stored in *E. coli* as plasmid transformants.

2. We have constructed AESs that can be readily introduced into any PacI-flanked conditionally replicating shuttle phasmid, such as phAE159, and then introduced into *E. coli* using the *in vitro* lambda packaging system.

3. We then electroporate the shuttle phasmid form of the specialized transducing phages into *M. smegmatis*, yielding plaques that then can be amplified at 30°C to generate high-titered lysates of TM4 packaged specialized transducing phages.

4. We can infect almost any *M. tuberculosis* mutant or strain with conditionally replicating specialized transducing phages. (We have screened hundreds of clinical isolates of *M. tuberculosis* to date with the reporter phage assay and have yet to find any strain that is not infected.)

5. We can store at −80°C or lyophilize the TM4 phages to retain stability.

6. We have shown that the yields of specialized transductants can be enhanced 50- to 400-fold with recombineering containing hosts (unpublished data), making them amenable for a high-throughput specialized transduction approach in microtiter plates.

Specialized transduction is particularly well suited to this purpose because of the high efficiency of gene transfer (Table 1). Our recent work showing signi-

Table 1   Improvements for high-throughput specialized transduction

Steps for specialized transduction	Optimization for high-throughput knockouts
Construction of plasmid with AES (500 to 1,000 bp of upstream and downstream sequence of gene targeted for deletion). AES flanks a hygromycin (*hyg*) and *sacB* cassette	Improve efficiency of four-component ligation by use of type IIP restriction enzymes that recognize symmetric palindromic sequence interrupted by a few degenerate interior base pairs
Cloning of AES plasmid into temperature-sensitive phage backbone to create AES phasmid	Increased cloning capacity of phasmid
Transfection of *M. smegmatis* with plasmid DNA to recover plaques	Adapted for 96-well electroporation format
Production of high-titer phage lysate	Adapted protocol for high-volume high-titer lysate production
Transduction of *M. tuberculosis*	Optimized transduction conditions for smaller volumes and 96-well format
Screening of potential *M. tuberculosis* knockouts	Unique barcodes in each AES allow for efficient screening
Removal of hygromycin cassette for unmarked knockout strain	Phage-based delivery of resolvase to efficiently unmark knockout strains

ficantly enhanced recombination by combining specialized transduction with recombineering plasmids convinced us that this is amenable to high throughput. We believe that this set of phages will be useful for the TB research community as the donor transfer has been for the yeast community.

The use of defined bar-coded deleted sets should allow for screening of genes required for growth and virulence in diverse animal models. By using such a set, we can expedite screens in animals for loss of virulence and specific immune evasion function(s). In addition, comprehensive screens looking for mutants that are defective in persisting in the presence of drugs *in vitro* and *in vivo* should be feasible.

Lastly, the generation of this complete set of specialized transducing phages will provide many new avenues for exploring the biology of *M. tuberculosis* because specific phenotypes can be tested with synthetic lethal screens. Synthetic lethality is the observation that two independent nonessential genes cannot be disrupted. In the process of defining interactive genetic pathways, the combination of making the whole set of mutants in a specific set of selected unmarked mutants will allow for the identification of analogous synthetic phenotypes in virulence, persistence, and survival.

IMAGINING A WORLD WITHOUT TB

Albert Einstein wrote, "Imagination is more important than knowledge because knowledge tells you what is and imagination tells you what can be." Before we had gene transfer for *M. tuberculosis*, we did not know the targets of TB-specific drugs, why *M. tuberculosis* stains acid-fast, if it were even possible to isolate auxotrophic mutants, and why the BCG vaccine strain that had been given to half the world's population was attenuated. By developing gene transfer for *M. tuberculosis*, we now know exactly how INH kills *M. tuberculosis*; we know what genes are involved in INH resistance and which enzymes when inactivated lead to the death of the tubercle bacillus. We now know that if the KasB enzyme is inactivated or deleted, it causes *M. tuberculosis* to lose its ability to be stained acid-fast and that such mutants can cause a truly latent infection in the mouse (145). We now know that auxotrophic mutants of *M. tuberculosis* are unable to grow in mammals, a somewhat surprising finding since other intracellular pathogens are naturally auxotrophic. Nevertheless, this knowledge has allowed us to make new vaccine candidates that are safer than BCG and precisely defined deletion mutants of *M. tuberculosis* that because of their safety are allowed to be used in a BSL2 lab,

thereby making work on these *M. tuberculosis* strains much easier. We now know that the primary attenuating mutation of BCG is the loss of a specialized secretion system that is required for *M. tuberculosis* to exit phagosomes and in the right oxygen conditions causes *M. tuberculosis* to form a DNA net from an infected macrophage (146). These nets may be ideal substrates for *M. tuberculosis* to form a biofilm *in vivo*.

During the time that gene transfer systems were being developed for *M. tuberculosis*, we saw a rise in the HIV epidemic—an epidemic that collided with the existing TB epidemic. The result has been a syndemic that has caused the worsening of each disease when a patient is co-infected. Unfortunately, this has led to an increased spread of TB and the emergence of strains of *M. tuberculosis* that are resistant to multiple or even all of the existing anti-TB drugs. Clearly, there is an unquestionable need for more research to develop new vaccines, new drugs, and new diagnostic tests for TB. While the emergence of multiple, extensively, or totally drug-resistant strains may seem ominous for health care in impoverished areas of the world, the ability to discover new ways to battle TB has never been greater. Having studied TB for the last 25 years, I believe that there are a few questions that are paramount to the control and eradication of the disease. I would like to end with these questions: (i) What is the molecular basis for the slow growth phenotype of *M. tuberculosis*? I believe this is a strategy that *M. tuberculosis* has employed to resist death. (ii) What are the genes required for *M. tuberculosis* to enter into a persistent state? We know that if you add a bactericidal drug to an actively growing culture of *M. tuberculosis* you can only kill 99.9% of the cells in 4 to 5 days. The remaining 0.1% of cells are persisters that resist killing. Is it possible that resisting killing by antibiotics overlaps with resisting killing by an adaptive immune response? We will have to understand this process of persistence in *M. tuberculosis*, the process by which *M. tuberculosis* cells enter into a slowly replicating state. This might be induced by hypoxia, nutrient limitation, or growth in a biofilm. The fact is, we just don't know, but now we have the power to find out. (iii) Is it possible to elicit a bactericidal immune response against *M. tuberculosis*? A number of new vaccine candidates have been developed that elicit better T-cell responses to *M. tuberculosis* than the BCG vaccine, yet we do not know how to engender an immune response that is able to sterilize an *M. tuberculosis* infection. Is this because persister cells are totally refractory to any T-cell–mediated or macrophage-mediated immune response? Is this because *M. tuberculosis* can persist in an extracellular biofilm in infected animals and thereby be hidden from an adaptive immune response? Do we need to understand how to make an antibody response that prevents the formation of these extracellular biofilms?

I believe that the answer to the control of TB will come from continued rational studies and serendipitous observations. The word *serendipity* comes from a Persian tale of the Three Princes of Serendip, who would discover new things while pursuing other quests. Louis Pasteur described this process in a different way when he said, "Chance favors the prepared mind." The rational project to delete every gene from *M. tuberculosis* will undoubtedly result in a number of serendipitous observations. This endeavor along with knowledge and wisdom might lead to the eradication of TB.

I thank David Hopwood, Ruth McAdams, Michelle Larsen, Catherine Vilchèze, Derek Essegian, and John Chan for critical reading of the manuscript and Paras Jain and Hui Yang for help with the figures.

Citation. Jacobs WR Jr. 2014. Gene transfer in *Mycobacterium tuberculosis*: shuttle phasmids to enlightenment. Microbiol Spectrum 2(2):MGM2-0037-2013.

References

1. Kaji M. 1972. Prevention of viral hepatitis—with special reference to the possibility of development of a vaccine in relation to Australia antigen. *Nihon Rinsho* 30:1159–1163. (In Japanese.)

2. Yap SF. 2004. Hepatitis B: review of development from the discovery of the "Australia Antigen" to end of the twentieth century. *Malaysian J Pathol* 26:1–12.

3. Jacob F, Monod J. 1961. Genetic regulatory mechanisms in the synthesis of proteins. *J Mol Biol* 3:318–356.

4. Jacobs WR Jr, Tuckman M, Bloom BR. 1987. Introduction of foreign DNA into mycobacteria using a shuttle phasmid. *Nature* 327:532–535.

5. Snapper SB, Lugosi L, Jekkel A, Melton RE, Kieser T, Bloom BR, Jacobs WR Jr. 1988. Lysogeny and transformation in mycobacteria: stable expression of foreign genes. *Proc Natl Acad Sci USA* 85:6987–6991.

6. Lee MH, Pascopella L, Jacobs WR Jr, Hatfull GF. 1991. Site-specific integration of mycobacteriophage L5: integration-proficient vectors for *Mycobacterium smegmatis*, *Mycobacterium tuberculosis*, and bacille Calmette-Guerin. *Proc Natl Acad Sci USA* 88:3111–3115.

7. Stover CK, Delacruz VF, Fuerst TR, Burlein JE, Benson LA, Bennett LT, Bansal GP, Young JF, Lee MH, Hatfull GF, Snapper SB, Barletta RG, Jacobs WR, Bloom BR. 1991. New use of BCG for recombinant vaccines. *Nature* 351:456–460.

8. Collins DM, Kawakami RP, de Lisle GW, Pascopella L, Bloom BR, Jacobs WR Jr. 1995. Mutation of the principal sigma factor causes loss of virulence in a strain of the *Mycobacterium tuberculosis* complex. *Proc Natl Acad Sci USA* 92:8036–8040.

9. Pascopella L, Collins FM, Martin JM, Jacobs WR Jr, Bloom BR. 1993. Identification of a genomic fragment of *Mycobacterium tuberculosis* responsible for in vivo growth advantage. *Infect. Agents Dis* 2:282–284.

10. Pascopella L, Collins FM, Martin JM, Lee MH, Hatfull GF, Stover CK, Bloom BR, Jacobs WR Jr. 1994. Use of in vivo complementation in *Mycobacterium tuberculosis* to identify a genomic fragment associated with virulence. *Infect Immun* 62:1313–1319.

11. Snapper SB, Melton RE, Mustafa S, Kieser T, Jacobs WR Jr. 1990. Isolation and characterization of efficient plasmid transformation mutants of *Mycobacterium smegmatis*. *Mol Microbiol* 4:1911–1919.

12. Stover CK, de la Cruz VF, Fuerst TR, Burlein JE, Benson LA, Bennett LT, Bansal GP, Young JF, Lee MH, Hatfull GF, Snapper SB, Barletta RG, Jacobs WR Jr, Bloom BR. 1991. New use of BCG for recombinant vaccines. *Nature* 351:456–460.

13. Labidi A, Mardis E, Roe BA, Wallace RJ Jr. 1992. Cloning and DNA sequence of the *Mycobacterium fortuitum* var *fortuitum* plasmid pAL5000. *Plasmid* 27:130–140.

14. Ranes MG, Rauzier J, Lagranderie M, Gheorghiu M, Gicquel B. 1990. Functional analysis of pAL5000, a plasmid from *Mycobacterium fortuitum*: construction of a "mini" mycobacterium-*Escherichia coli* shuttle vector. *J Bacteriol* 172:2793–2797.

15. Stolt P, Stoker NG. 1996. Protein-DNA interactions in the *ori* region of the *Mycobacterium fortuitum* plasmid pAL5000. *J Bacteriol* 178:6693–6700.

16. Stolt P, Stoker NG. 1997. Mutational analysis of the regulatory region of the *Mycobacterium* plasmid pAL5000. *Nucleic Acids Res* 25:3840–3846.

17. Villar CA, Benitez J. 1992. Functional analysis of pAL5000 plasmid in *Mycobacterium fortuitum*. *Plasmid* 28:166–169.

18. Belisle JT, Pascopella L, Inamine JM, Brennan PJ, Jacobs WR Jr. 1991. Isolation and expression of a gene cluster responsible for biosynthesis of the glycopeptidolipid antigens of *Mycobacterium avium*. *J Bacteriol* 173:6991–6997.

19. Banerjee A, Dubnau E, Quemard A, Balasubramanian V, Um KS, Wilson T, Collins D, de Lisle G, Jacobs WR Jr. 1994. *inhA*, a gene encoding a target for isoniazid and ethionamide in *Mycobacterium tuberculosis*. *Science* 263:227–230.

20. Zhang Y, Heym B, Allen B, Young D, Cole S. 1992. The catalase-peroxidase gene and isoniazid resistance of *Mycobacterium tuberculosis*. *Nature* 358:591–593.

21. Baulard AR, Betts JC, Engohang-Ndong J, Quan S, McAdam RA, Brennan PJ, Locht C, Besra GS. 2000. Activation of the pro-drug ethionamide is regulated in mycobacteria. *J Biol Chem* 275:28326–28331.

22. DeBarber AE, Mdluli K, Bosman M, Bekker LG, Barry CE 3rd. 2000. Ethionamide activation and sensitivity in

multidrug-resistant *Mycobacterium tuberculosis*. *Proc Natl Acad Sci USA* 97:9677–9682.

23. Belanger AE, Besra GS, Ford ME, Mikusova K, Belisle JT, Brennan PJ, Inamine JM. 1996. The *embAB* genes of *Mycobacterium avium* encode an arabinosyl transferase involved in cell wall arabinan biosynthesis that is the target for the antimycobacterial drug ethambutol. *Proc Natl Acad Sci USA* 93:11919–11924.

24. Telenti A, Philipp WJ, Sreevatsan S, Bernasconi C, Stockbauer KE, Wieles B, Musser JM, Jacobs WR Jr. 1997. The *emb* operon, a gene cluster of *Mycobacterium tuberculosis* involved in resistance to ethambutol. *Nature Med* 3:567–570.

25. Scorpio A, Zhang Y. 1996. Mutations in *pncA*, a gene encoding pyrazinamidase/nicotinamidase, cause resistance to the antituberculous drug pyrazinamide in tubercle bacillus. *Nature Med* 2:662–667.

26. Gannoun-Zaki L, Alibaud L, Kremer L. 2013. Point mutations within the fatty acid synthase type II dehydratase components HadA or HadC contribute to isoxyl resistance in *Mycobacterium tuberculosis*. *Antimicrobial Agents Chemother* 57:629–632.

27. Grzegorzewicz AE, Kordulakova J, Jones V, Born SE, Belardinelli JM, Vaquie A, Gundi VA, Madacki J, Slama N, Laval F, Vaubourgeix J, Crew RM, Gicquel B, Daffe M, Morbidoni HR, Brennan PJ, Quemard A, McNeil MR, Jackson M. 2012. A common mechanism of inhibition of the *Mycobacterium tuberculosis* mycolic acid biosynthetic pathway by isoxyl and thiacetazone. *J Biol Chem* 287:38434–38441.

28. Andries K, Verhasselt P, Guillemont J, Gohlmann HW, Neefs JM, Winkler H, Van Gestel J, Timmerman P, Zhu M, Lee E, Williams P, de Chaffoy D, Huitric E, Hoffner S, Cambau E, Truffot-Pernot C, Lounis N, Jarlier V. 2005. A diarylquinoline drug active on the ATP synthase of *Mycobacterium tuberculosis*. *Science* 307:223–227.

29. Pavelka MS Jr, Jacobs WR Jr. 1996. Biosynthesis of diaminopimelate, the precursor of lysine and a component of peptidoglycan, is an essential function of *Mycobacterium smegmatis*. *J Bacteriol* 178:6496–6507.

30. van Kessel JC, Hatfull GF. 2007. Recombineering in *Mycobacterium tuberculosis*. *Nat Methods* 4:147–152.

31. McAdam RA, Weisbrod TR, Martin J, Scuderi JD, Brown AM, Cirillo JD, Bloom BR, Jacobs WR Jr. 1995. In vivo growth characteristics of leucine and methionine auxotrophic mutants of *Mycobacterium bovis* BCG generated by transposon mutagenesis. *Infect Immun* 63:1004–1012.

32. George JR, Pine L, Reeves MW, Harrell WK. 1980. Amino acid requirements of *Legionella pneumophila*. *J Clin Microbiol* 11:286–291.

33. Tesh MJ, Miller RD. 1981. Amino acid requirements for *Legionella pneumophila* growth. *J Clin Microbiol* 13:865–869.

34. Guleria I, Teitelbaum R, McAdam RA, Kalpana G, Jacobs WR Jr, Bloom BR. 1996. Auxotrophic vaccines for tuberculosis. *Nat Med* 2:334–337.

35. McDonough KA, Kress Y, Bloom BR. 1993. Pathogenesis of tuberculosis: interaction of *Mycobacterium tuberculosis* with macrophages. *Infect Immun* 61:2763–2773.

36. Hondalus MK, Bardarov S, Russell R, Chan J, Jacobs WR Jr, Bloom BR. 2000. Attenuation of and protection induced by a leucine auxotroph of *Mycobacterium tuberculosis*. *Infect Immun* 68:2888–2898.

37. Pavelka MS Jr, Chen B, Kelley CL, Collins FM, Jacobs WR Jr. 2003. Vaccine efficacy of a lysine auxotroph of *Mycobacterium tuberculosis*. *Infect Immun* 71:4190–4192.

38. Sambandamurthy VK, Wang X, Chen B, Russell RG, Derrick S, Collins FM, Morris SL, Jacobs WR Jr. 2002. A pantothenate auxotroph of *Mycobacterium tuberculosis* is highly attenuated and protects mice against tuberculosis. *Nat Med* 8:1171–1174.

39. Jensen K, Ranganathan UD, Van Rompay KK, Canfield DR, Khan I, Ravindran R, Luciw PA, Jacobs WR Jr, Fennelly G, Larsen MH, Abel K. 2012. A recombinant attenuated *Mycobacterium tuberculosis* vaccine strain is safe in immunosuppressed simian immunodeficiency virus-infected infant macaques. *Clin Vaccine Immunol* 19:1170–1181.

40. Sampson SL, Dascher CC, Sambandamurthy VK, Russell RG, Jacobs WR Jr, Bloom BR, Hondalus MK. 2004. Protection elicited by a double leucine and pantothenate auxotroph of *Mycobacterium tuberculosis* in guinea pigs. *Infect Immun* 72:3031–3037.

41. Sampson SL, Mansfield KG, Carville A, Magee DM, Quitugua T, Howerth EW, Bloom BR, Hondalus MK. 2011. Extended safety and efficacy studies of a live attenuated double leucine and pantothenate auxotroph of *Mycobacterium tuberculosis* as a vaccine candidate. *Vaccine* 29:4839–4847.

42. Zimmerman DM, Waters WR, Lyashchenko KP, Nonnecke BJ, Armstrong DL, Jacobs WR Jr, Larsen MH, Egan E, Dean GA. 2009. Safety and immunogenicity of the *Mycobacterium tuberculosis* DeltalysA DeltapanCD vaccine in domestic cats infected with feline immunodeficiency virus. *Clin Vaccine Immunol* 16:427–429.

43. Villemin JA. 1868. *Etudes sur la Tuberculose*. J.-B. Baillière et Fils, Paris.

44. Koch R. 1882. Die Aetiologie der Tuberkulose. *Berl Klin Wochenschr* 19:221–230.

45. Beadle GW, Tatum EL. 1941. Genetic control of biochemical reactions in *Neurospora*. *Proc Natl Acad Sci USA* 27:499–506.

46. Lederberg J, Tatum EL. 1946. Gene recombination in *Escherichia coli*. *Nature* 158:558.

47. Griffith F. 1928. The significance of pneumococcal types. *J Hyg* 27:113–159.

48. Avery OT, Macleod CM, McCarty M. 1944. Studies on the chemical nature of the substance inducing transformation of pneumococcal types: induction of transformation by a desoxyribonucleic acid fraction isolated from pneumococcus type III. *J Exp Med* 79:137–158.

49. Hershey AD, Chase M. 1952. Independent functions of viral protein and nucleic acid in growth of bacteriophage. *J Gen Physiol* **36**:39–56.

50. Watson JD, Crick FH. 1953. Genetical implications of the structure of deoxyribonucleic acid. *Nature* **171**:964–967.

51. Watson JD, Crick FH. 1953. Molecular structure of nucleic acids; a structure for deoxyribose nucleic acid. *Nature* **171**:737–738.

52. Crick F. 1970. Central dogma of molecular biology. *Nature* **227**:561–563.

53. Falkow S. 1988. Molecular Koch's postulates applied to microbial pathogenicity. *Rev Infect Dis* **10**(Suppl 2):S274–S276.

54. Zinder ND. 1992. Forty years ago: the discovery of bacterial transduction. *Genetics* **132**:291–294.

55. Waksman SA, Schatz A. 1943. Strain specificity and production of antibiotic substances. *Proc Natl Acad Sci USA* **29**:74–79.

56. Schatz A, Waksman SA. 1945. Strain specificity and production of antibiotic substances. IV. Variations among actionomycetes, with special reference to *Actinomyces griseus*. *Proc Natl Acad Sci USA* **31**:129–137.

57. Waksman SA, Reilly HC, Schatz A. 1945. Strain specificity and production of antibiotic substances. V. Strain resistance of bacteria to antibiotic substances, especially to streptomycin. *Proc Natl Acad Sci USA* **31**:157–164.

58. Demerec M. 1951. Studies of the streptomycin-resistance system of mutations in E. coli. *Genetics* **36**:585–597.

59. Hashimoto K. 1960. Streptomycin resistance in *Escherichia coli* analyzed by transduction. *Genetics* **45**:49–62.

60. Lennox ES. 1955. Transduction of linked genetic characters of the host by bacteriophage P1. *Virology* **1**:190–206.

61. Newcombe HB, Nyholm MH. 1950. The inheritance of streptomycin resistance and dependence in crosses of *Escherichia coli*. *Genetics* **35**:603–611.

62. Spotts CR, Stanier RY. 1961. Mechanism of streptomycin action on bacteria: a unitary hypothesis. *Nature* **192**:633–637.

63. Erdos T, Ullmann A. 1959. Effect of streptomycin on the incorporation of amino-acids labelled with carbon-14 into ribonucleic acid and protein in a cell-free system of a mycobacterium. *Nature* **183**:618–619.

64. Erdos T, Ullmann A. 1960. Effect of streptomycin on the incorporation of tyrosine labelled with carbon-14 into protein of *Mycobacterium* cell fractions in vivo. *Nature* **185**:100–101.

65. Davies JE. 1964. Studies on the ribosomes of streptomycin-sensitive and resistant strains of *Escherichia coli*. *Proc Natl Acad Sci USA* **51**:659–664.

66. Ozaki M, Mizushima S, Nomura M. 1969. Identification and functional characterization of the protein controlled by the streptomycin-resistant locus in E. coli. *Nature* **222**:333–339.

67. Carter AP, Clemons WM, Brodersen DE, Morgan-Warren RJ, Wimberly BT, Ramakrishnan V. 2000. Functional insights from the structure of the 30S ribosomal subunit and its interactions with antibiotics. *Nature* **407**:340–348.

68. Finken M, Kirschner P, Meier A, Wrede A, Bottger EC. 1993. Molecular basis of streptomycin resistance in *Mycobacterium tuberculosis*: alterations of the ribosomal protein S12 gene and point mutations within a functional 16S ribosomal RNA pseudoknot. *Mol Microbiol* **9**:1239–1246.

69. Jackson DA, Symons RH, Berg P. 1972. Biochemical method for inserting new genetic information into DNA of Simian Virus 40: circular SV40 DNA molecules containing lambda phage genes and the galactose operon of *Escherichia coli*. *Proc Natl Acad Sci USA* **69**:2904–2909.

70. Lark C, Arber W. 1970. Host specificity of DNA produced by *Escherichia coli*. 13. Breakdown of cellular DNA upon growth in ethionine of strains with r plus-15, r plus-P1 or r plus-N3 restriction phenotypes. *J Mol Biol* **52**:337–348.

71. Smith HO, Wilcox KW. 1970. A restriction enzyme from *Haemophilus influenzae*. I. Purification and general properties. *J Mol Biol* **51**:379–391.

72. Chang AC, Lansman RA, Clayton DA, Cohen SN. 1975. Studies of mouse mitochondrial DNA in *Escherichia coli*: structure and function of the eucaryotic-procaryotic chimeric plasmids. *Cell* **6**:231–244.

73. Cohen SN, Chang AC, Boyer HW, Helling RB. 1973. Construction of biologically functional bacterial plasmids *in vitro*. *Proc Natl Acad Sci USA* **70**:3240–3244.

74. Kedes LH, Chang AC, Houseman D, Cohen SN. 1975. Isolation of histone genes from unfractionated sea urchin DNA by subculture cloning in E. coli. *Nature* **255**:533–538.

75. Morrow JF, Cohen SN, Chang AC, Boyer HW, Goodman HM, Helling RB. 1974. Replication and transcription of eukaryotic DNA in *Escherichia coli*. *Proc Natl Acad Sci USA* **71**:1743–1747.

76. Ratzkin B, Carbon J. 1977. Functional expression of cloned yeast DNA in *Escherichia coli*. *Proc Natl Acad Sci USA* **74**:487–491.

77. Borck K, Beggs JD, Brammar WJ, Hopkins AS, Murray NE. 1976. The construction in vitro of transducing derivatives of phage lambda. *Mol Gen Genet* **146**:199–207.

78. Hohn B, Murray K. 1977. Packaging recombinant DNA molecules into bacteriophage particles in vitro. *Proc Natl Acad Sci USA* **74**:3259–3263.

79. Murray NE, Murray K. 1974. Manipulation of restriction targets in phage lambda to form receptor chromosomes for DNA fragments. *Nature* **251**:476–481.

80. Struhl K, Cameron JR, Davis RW. 1976. Functional genetic expression of eukaryotic DNA in *Escherichia coli*. *Proc Natl Acad Sci USA* **73**:1471–1475.

81. Maxam AM, Gilbert W. 1980. Sequencing end-labeled DNA with base-specific chemical cleavages. *Methods Enzymol* **65**:499–560.

82. Sanger F, Donelson JE, Coulson AR, Kossel H, Fischer D. 1973. Use of DNA polymerase I primed by a synthetic oligonucleotide to determine a nucleotide sequence in phage fl DNA. *Proc Natl Acad Sci USA* **70**:1209–1213.

83. Berg P, Baltimore D, Brenner S, Roblin RO, Singer MF. 1975. Summary statement of the Asilomar conference on recombinant DNA molecules. *Proc Natl Acad Sci USA* **72**:1981–1984.

84. Curtiss R 3rd. 1978. Biological containment and cloning vector transmissibility. *J Infect Dis* **137**:668–675.

85. Alexander WJ, Alexander LS, Curtiss R 3rd. 1980. Bactericidal activity of human serum against *Escherichia coli chi*1776. *Infect Immun* **28**:837–841.

86. Shepard CC. 1960. The experimental disease that follows the injection of human leprosy bacilli into foot-pads of mice. *J Exp Med* **112**:445–454.

87. Shepard CC. 1971. The first decade in experimental leprosy. *Bull World Health Organ* **44**:821–827.

88. Kirchheimer WF, Storrs EE. 1971. Attempts to establish the armadillo (*Dasypus novemcinctus* Linn.) as a model for the study of leprosy. I. Report of lepromatoid leprosy in an experimentally infected armadillo. *Int J Lepr Other Mycobact Dis* **39**:693–702.

89. Clark-Curtiss JE, Jacobs WR, Docherty MA, Ritchie LR, Curtiss R 3rd. 1985. Molecular analysis of DNA and construction of genomic libraries of *Mycobacterium leprae*. *J Bacteriol* **161**:1093–1102.

90. Jacobs WR, Barrett JF, Clark-Curtiss JE, Curtiss R 3rd. 1986. In vivo repackaging of recombinant cosmid molecules for analyses of *Salmonella typhimurium*, *Streptococcus mutans*, and mycobacterial genomic libraries. *Infect Immun* **52**:101–109.

91. Jacobs WR, Docherty MA, Curtiss R 3rd, Clark-Curtiss JE. 1986. Expression of *Mycobacterium leprae* genes from a *Streptococcus mutans* promoter in *Escherichia coli* K-12. *Proc Natl Acad Sci USA* **83**:1926–1930.

92. Young RA, Davis RW. 1983. Efficient isolation of genes by using antibody probes. *Proc Natl Acad Sci USA* **80**:1194–1198.

93. Young RA, Mehra V, Sweetser D, Buchanan T, Clark-Curtiss J, Davis RW, Bloom BR. 1985. Genes for the major protein antigens of the leprosy parasite *Mycobacterium leprae*. *Nature* **316**:450–452.

94. Young RA, Bloom BR, Grosskinsky CM, Ivanyi J, Thomas D, Davis RW. 1985. Dissection of *Mycobacterium tuberculosis* antigens using recombinant DNA. *Proc Natl Acad Sci USA* **82**:2583–2587.

95. Raj CV, Ramakrishnan T. 1970. Transduction in *Mycobacterium smegmatis*. *Nature* **228**:280–281.

96. Mizuguchi Y, Tokunaga T. 1970. Genetic recombination between *Mycobacterium smegmatis* (strains jucho and lactocola). *Igaku to Seibutsugaku.* **81**:163–167. (In Japanese.)

97. Takahashi M, Tanaka M, Okudera T, Mihara K, Tokunaga M. 1973. Angiography with a new contrast medium, Isopaque. *Rinsho Hoshasen* **18**:1001–1006.

98. Crawford JT, Bates JH. 1979. Isolation of plasmids from mycobacteria. *Infect Immun* **24**:979–981.

99. Labidi A, David HL, Roulland-Dussoix D. 1985. Restriction endonuclease mapping and cloning of *Mycobacterium fortuitum* var. *fortuitum* plasmid pAL5000. *Ann Inst Pasteur Microbiol* **136B**:209–215.

100. Bibb MJ, Ward JM, Hopwood DA. 1978. Transformation of plasmid DNA into *Streptomyces* at high frequency. *Nature* **274**:398–400.

101. Timme TL, Brennan PJ. 1984. Induction of bacteriophage from members of the *Mycobacterium avium*, *Mycobacterium intracellulare*, *Mycobacterium scrofulaceum* serocomplex. *J Gen Microbiol* **130**:2059–2066.

102. Hatfull GF, Sarkis GJ. 1993. DNA sequence, structure and gene expression of mycobacteriophage L5: a phage system for mycobacterial genetics. *Mol Microbiol* **7**:395–405.

103. Miller JF, Dower WJ, Tompkins LS. 1988. High-voltage electroporation of bacteria: genetic transformation of *Campylobacter jejuni* with plasmid DNA. *Proc Natl Acad Sci USA* **85**:856–860.

104. Pope WH, Ferreira CM, Jacobs-Sera D, Benjamin RC, Davis AJ, DeJong RJ, Elgin SC, Guilfoile FR, Forsyth MH, Harris AD, Harvey SE, Hughes LE, Hynes PM, Jackson AS, Jalal MD, MacMurray EA, Manley CM, McDonough MJ, Mosier JL, Osterbann LJ, Rabinowitz HS, Rhyan CN, Russell DA, Saha MS, Shaffer CD, Simon SE, Sims EF, Tovar IG, Weisser EG, Wertz JT, Weston-Hafer KA, Williamson KE, Zhang B, Cresawn SG, Jain P, Piuri M, Jacobs WR Jr, Hendrix RW, Hatfull GF. 2011. Cluster K mycobacteriophages: insights into the evolutionary origins of mycobacteriophage TM4. *PLoS One* **6**:e26750.

105. Kleckner N, Roth J, Botstein D. 1977. Genetic engineering in vivo using translocatable drug-resistance elements. New methods in bacterial genetics. *J Mol Biol* **116**:125–159.

106. Bardarov S, Kriakov J, Carriere C, Yu S, Vaamonde C, McAdam RA, Bloom BR, Hatfull GF, Jacobs WR Jr. 1997. Conditionally replicating mycobacteriophages: a system for transposon delivery to *Mycobacterium tuberculosis*. *Proc Natl Acad Sci USA* **94**:10961–10966.

107. Carriere C, Riska PF, Zimhony O, Kriakov J, Bardarov S, Burns J, Chan J, Jacobs WR Jr. 1997. Conditionally replicating luciferase reporter phages: improved sensitivity for rapid detection and assessment of drug susceptibility of *Mycobacterium tuberculosis*. *J Clin Microbiol* **35**:3232–3239.

108. Ulitzur S, Kuhn J. 1989. Detection and/or identification of microorganisms in a test sample using bioluminescence or other genetically introduced marker. US Patent No. 4,861,709.

109. Jacobs WR Jr, Barletta RG, Udani R, Chan J, Kalkut G, Sosne G, Kieser T, Sarkis GJ, Hatfull GF, Bloom BR. 1993. Rapid assessment of drug susceptibilities of

Mycobacterium tuberculosis by means of luciferase reporter phages. *Science* 260:819–822.

110. Banaiee N, Bobadilla-Del-Valle M, Bardarov S Jr, Riska PF, Small PM, Ponce-De-Leon A, Jacobs WR Jr, Hatfull GF, Sifuentes-Osornio J. 2001. Luciferase reporter mycobacteriophages for detection, identification, and antibiotic susceptibility testing of *Mycobacterium tuberculosis* in Mexico. *J Clin Microbiol* 39:3883–3888.

111. Banaiee N, Bobadilla-del-Valle M, Riska PF, Bardarov S Jr, Small PM, Ponce-de-Leon A, Jacobs WR Jr, Hatfull GF, Sifuentes-Osornio J. 2003. Rapid identification and susceptibility testing of *Mycobacterium tuberculosis* from MGIT cultures with luciferase reporter mycobacteriophages. *J Med Microbiol* 52:557–561.

112. Bardarov S Jr, Dou H, Eisenach K, Banaiee N, Ya S, Chan J, Jacobs WR Jr, Riska PF. 2003. Detection and drug-susceptibility testing of *M. tuberculosis* from sputum samples using luciferase reporter phage: comparison with the Mycobacteria Growth Indicator Tube (MGIT) system. *Diagn Microbiol Infect Dis* 45:53–61.

113. Riska PF, Jacobs WR Jr. 1998. The use of luciferase-reporter phage for antibiotic-susceptibility testing of mycobacteria. *Methods Mol Biol* 101:431–455.

114. Riska PF, Jacobs WR Jr, Bloom BR, McKitrick J, Chan J. 1997. Specific identification of *Mycobacterium tuberculosis* with the luciferase reporter mycobacteriophage: use of *p*-nitro-alpha-acetylamino-beta-hydroxy propiophenone. *J Clin Microbiol* 35:3225–3231.

115. Riska PF, Su Y, Bardarov S, Freundlich L, Sarkis G, Hatfull G, Carriere C, Kumar V, Chan J, Jacobs WR Jr. 1999. Rapid film-based determination of antibiotic susceptibilities of *Mycobacterium tuberculosis* strains by using a luciferase reporter phage and the Bronx Box. *J Clin Microbiol* 37:1144–1149.

116. Piuri M, Jacobs WR Jr, Hatfull GF. 2009. Fluoromycobacteriophages for rapid, specific, and sensitive antibiotic susceptibility testing of *Mycobacterium tuberculosis*. *PLoS One* 4:e4870.

117. Rondon L, Piuri M, Jacobs WR Jr, de Waard J, Hatfull GF, Takiff HE. 2011. Evaluation of fluoromycobacteriophages for detecting drug resistance in *Mycobacterium tuberculosis*. *J Clin Microbiol* 49:1838–1842.

118. Jain P, Hartman TE, Eisenberg N, O'Donnell MR, Kriakov J, Govender K, Makume M, Thaler DS, Hatfull GF, Sturm AW, Larsen MH, Moodley P, Jacobs WR Jr. 2012. phi(2)GFP10, a high-intensity fluorophage, enables detection and rapid drug susceptibility testing of *Mycobacterium tuberculosis* directly from sputum samples. *J Clin Microbiol* 50:1362–1369.

119. Martin C, Timm J, Rauzier J, Gomez-Lus R, Davies J, Gicquel B. 1990. Transposition of an antibiotic resistance element in mycobacteria. *Nature* 345:739–743.

120. Guilhot C, Otal I, Van Rompaey I, Martin C, Gicquel B. 1994. Efficient transposition in mycobacteria: construction of *Mycobacterium smegmatis* insertional mutant libraries. *J Bacteriol* 176:535–539.

121. Otero J, Jacobs WR Jr, Glickman MS. 2003. Efficient allelic exchange and transposon mutagenesis in *Mycobacterium avium* by specialized transduction. *Appl Environ Microbiol* 69:5039–5044.

122. Stinear TP, Mve-Obiang A, Small PL, Frigui W, Pryor MJ, Brosch R, Jenkin GA, Johnson PD, Davies JK, Lee RE, Adusumilli S, Garnier T, Haydock SF, Leadlay PF, Cole ST. 2004. Giant plasmid-encoded polyketide synthases produce the macrolide toxin of *Mycobacterium ulcerans*. *Proc Natl Acad Sci USA* 101:1345–1349.

123. Harris NB, Feng Z, Liu X, Cirillo SL, Cirillo JD, Barletta RG. 1999. Development of a transposon mutagenesis system for *Mycobacterium avium* subsp. *paratuberculosis*. *FEMS Microbiol Lett* 175:21–26.

124. Glickman MS, Cox JS, Jacobs WR Jr. 2000. A novel mycolic acid cyclopropane synthetase is required for cording, persistence, and virulence of *Mycobacterium tuberculosis*. *Mol Cell* 5:717–727.

125. Hensel M, Shea JE, Gleeson C, Jones MD, Dalton E, Holden DW. 1995. Simultaneous identification of bacterial virulence genes by negative selection. *Science* 269:400–403.

126. Cox JS, Chen B, McNeil M, Jacobs WR Jr. 1999. Complex lipid determines tissue-specific replication of *Mycobacterium tuberculosis* in mice. *Nature* 402:79–83.

127. McAdam RA, Quan S, Smith DA, Bardarov S, Betts JC, Cook FC, Hooker EU, Lewis AP, Woollard P, Everett MJ, Lukey PT, Bancroft GJ, Jacobs WR Jr, Duncan K. 2002. Characterization of a *Mycobacterium tuberculosis* H37Rv transposon library reveals insertions in 351 ORFs and mutants with altered virulence. *Microbiology* 148:2975–2986.

128. Hisert KB, Kirksey MA, Gomez JE, Sousa AO, Cox JS, Jacobs WR Jr, Nathan CF, McKinney JD. 2004. Identification of *Mycobacterium tuberculosis* counterimmune (*cim*) mutants in immunodeficient mice by differential screening. *Infect Immun* 72:5315–5321.

129. Sassetti CM, Boyd DH, Rubin EJ. 2001. Comprehensive identification of conditionally essential genes in mycobacteria. *Proc Natl Acad Sci USA* 98:12712–12717.

130. Griffin JE, Gawronski JD, Dejesus MA, Ioerger TR, Akerley BJ, Sassetti CM. 2011. High-resolution phenotypic profiling defines genes essential for mycobacterial growth and cholesterol catabolism. *PLoS Pathog* 7:e1002251.

131. Sassetti CM, Rubin EJ. 2003. Genetic requirements for mycobacterial survival during infection. *Proc Natl Acad Sci USA* 100:12989–12994.

132. Morse ML, Lederberg EM, Lederberg J. 1956. Transduction in *Escherichia coli* K-12. *Genetics* 41:142–156.

133. Bardarov S, Bardarov S Jr, Pavelka MS Jr, Sambandamurthy V, Larsen M, Tufariello J, Chan J, Hatfull G, Jacobs WR Jr. 2002. Specialized transduction: an efficient method for generating marked and unmarked targeted gene disruptions in *Mycobacterium tuberculosis*, *M. bovis* BCG and *M. smegmatis*. *Microbiology* 148:3007–3017.

134. Hsu T, Hingley-Wilson SM, Chen B, Chen M, Dai AZ, Morin PM, Marks CB, Padiyar J, Goulding C, Gingery M, Eisenberg D, Russell RG, Derrick SC, Collins FM, Morris SL, King CH, Jacobs WR Jr. 2003. The primary mechanism of attenuation of bacillus Calmette-Guerin is a loss of secreted lytic function required for invasion of lung interstitial tissue. *Proc Natl Acad Sci USA* **100:** 12420–12425.

135. Mahairas GG, Sabo PJ, Hickey MJ, Singh DC, Stover CK. 1996. Molecular analysis of genetic differences between *Mycobacterium bovis* BCG and virulent *M. bovis*. *J Bacteriol* **178:**1274–1282.

136. Pym AS, Brodin P, Brosch R, Huerre M, Cole ST. 2002. Loss of RD1 contributed to the attenuation of the live tuberculosis vaccines *Mycobacterium bovis* BCG and *Mycobacterium microti*. *Mol Microbiol* **46:**709–717.

137. Vilcheze C, Wang F, Arai M, Hazbon MH, Colangeli R, Kremer L, Weisbrod TR, Alland D, Sacchettini JC, Jacobs WR Jr. 2006. Transfer of a point mutation in *Mycobacterium tuberculosis inhA* resolves the target of isoniazid. *Nat Med* **12:**1027–1029.

138. Fleischmann RD, Adams MD, White O, Clayton RA, Kirkness EF, Kerlavage AR, Bult CJ, Tomb JF, Dougherty BA, Merrick JM, et al. 1995. Whole-genome random sequencing and assembly of *Haemophilus influenzae* Rd. *Science* **269:**496–512.

139. Anonymous. 1997. The yeast genome directory. *Nature* **387:**5.

140. Mullis K, Faloona F, Scharf S, Saiki R, Horn G, Erlich H. 1986. Specific enzymatic amplification of DNA in vitro: the polymerase chain reaction. *Cold Spring Harbor Symp Quant Biol* **51**(Pt 1):263–273.

141. Saiki RK, Scharf S, Faloona F, Mullis KB, Horn GT, Erlich HA, Arnheim N. 1985. Enzymatic amplification of beta-globin genomic sequences and restriction site analysis for diagnosis of sickle cell anemia. *Science* **230:**1350–1354.

142. Giaever G, Chu AM, Ni L, Connelly C, Riles L, Veronneau S, Dow S, Lucau-Danila A, Anderson K, Andre B, Arkin AP, Astromoff A, El-Bakkoury M, Bangham R, Benito R, Brachat S, Campanaro S, Curtiss M, Davis K, Deutschbauer A, Entian KD, Flaherty P, Foury F, Garfinkel DJ, Gerstein M, Gotte D, Guldener U, Hegemann JH, Hempel S, Herman Z, Jaramillo DF, Kelly DE, Kelly SL, Kotter P, LaBonte D, Lamb DC, Lan N, Liang H, Liao H, Liu L, Luo C, Lussier M, Mao R, Menard P, Ooi SL, Revuelta JL, Roberts CJ, Rose M, Ross-Macdonald P, Scherens B, Schimmack G, Shafer B, Shoemaker DD, Sookhai-Mahadeo S, Storms RK, Strathern JN, Valle G, Voet M, Volckaert G, Wang CY, Ward TR, Wilhelmy J, Winzeler EA, Yang Y, Yen G, Youngman E, Yu K, Bussey H, Boeke JD, Snyder M, Philippsen P, Davis RW, Johnston M. 2002. Functional profiling of the *Saccharomyces cerevisiae* genome. *Nature* **418:**387–391.

143. Fleischmann RD, Alland D, Eisen JA, Carpenter L, White O, Peterson J, DeBoy R, Dodson R, Gwinn M, Haft D, Hickey E, Kolonay JF, Nelson WC, Umayam LA, Ermolaeva M, Salzberg SL, Delcher A, Utterback T, Weidman J, Khouri H, Gill J, Mikula A, Bishai W, Jacobs WR Jr, Venter JC, Fraser CM. 2002. Whole-genome comparison of *Mycobacterium tuberculosis* clinical and laboratory strains. *J Bacteriol* **184:**5479–5490.

144. Hatfull GF. 2014. Molecular genetics of mycobacteriophages. *Microbiol Spectrum* **2**(2):MGM2-0032-2013.

145. Bhatt A, Fujiwara N, Bhatt K, Gurcha SS, Kremer L, Chen B, Chan J, Porcelli SA, Kobayashi K, Besra GS, Jacobs WR Jr. 2007. Deletion of *kasB* in *Mycobacterium tuberculosis* causes loss of acid-fastness and subclinical latent tuberculosis in immunocompetent mice. *Proc Natl Acad Sci USA* **104:**5157–5162.

146. Wong KW, Jacobs WR Jr. 2013. *Mycobacterium tuberculosis* exploits human interferon gamma to stimulate macrophage extracellular trap formation and necrosis. *J Infect Dis* **208:**109–119.

147. Giaever G, Chu AM, Ni L, Connelly C, Riles L, Véronneau S, Dow S, Lucau-Danila A, Anderson K, André B, Arkin AP, Astromoff A, El-Bakkoury M, Bangham R, Benito R, Brachat S, Campanaro S, Curtiss M, Davis K, Deutschbauer A, Entian KD, Flaherty P, Foury F, Garfinkel DJ, Gerstein M, Gotte D, Güldener U, Hegemann JH, Hempel S, Herman Z, Jaramillo DF, Kelly DE, Kelly SL, Kötter P, LaBonte D, Lamb DC, Lan N, Liang H, Liao H, Liu L, Luo C, Lussier M, Mao R, Menard P, Ooi SL, Revuelta JL, Roberts CJ, Rose M, Ross-Macdonald P, Scherens B, Schimmack G, Shafer B, Shoemaker DD, Sookhai-Mahadeo S, Storms RK, Strathern JN, Valle G, Voet M, Volckaert G, Wang CY, Ward TR, Wilhelmy J, Winzeler EA, Yang Y, Yen G, Youngman E, Yu K, Bussey H, Boeke JD, Snyder M, Philippsen P, Davis RW, Johnston M. 2002. Functional profiling of the *Saccharomyces cerevisiae* genome. *Nature* **418:** 387–391.

148. Cole ST, Brosch R, Parkhill J, Garnier T, Churcher C, Harris D, Gordon SV, Eiglmeier K, Gas S, Barry CE 3rd, Tekaia F, Badcock K, Basham D, Brown D, Chillingworth T, Connor R, Davies R, Devlin K, Feltwell T, Gentles S, Hamlin N, Holroyd S, Hornsby T, Jagels K, Krogh A, McLean J, Moule S, Murphy L, Oliver K, Osborne J, Quail MA, Rajandream MA, Rogers J, Rutter S, Seeger K, Skelton J, Squares R, Squares S, Sulston JE, Taylor K, Whitehead S, Barrell BG. 1998. Deciphering the biology of *Mycobacterium tuberculosis* from the complete genome sequence. *Nature* **393:**537–544.

Molecular Genetics of Mycobacteria, 2nd Edition
Edited by Graham F. Hatfull and William R. Jacobs, Jr.
© 2014 American Society for Microbiology, Washington, DC
doi:10.1128/microbiolspec.MGM2-0025-2013

Daria Bottai,[1] Timothy P. Stinear,[2]
Philip Supply,[3] and Roland Brosch[4]

Mycobacterial Pathogenomics and Evolution

2

INTRODUCTION

Mycobacteria are widespread microorganisms characterized by the high G+C content of their genomes and a lipid-rich cell envelope. The genus *Mycobacterium* represents the only entity within the family *Mycobacteriaceae*, which belongs to the order *Actinomycetales* and the phylum *Actinobacteria* (1). Whereas the great majority of the ~130 described species in the genus are harmless environmental saprophytes, some mycobacteria have evolved to be major pathogens. With the exception of *Mycobacterium abscessus*, which is recognized as an emerging human pathogen in cystic fibrosis patients (2–4), the pathogenic species mainly belong to the slowly growing mycobacteria and comprise well-known human pathogens such as *Mycobacterium tuberculosis*, *Mycobacterium leprae*, and *Mycobacterium ulcerans* as well as confirmed animal pathogens such as *Mycobacterium bovis*, *Mycobacterium avium paratuberculosis*, and *Mycobacterium marinum* (5).

Their slow axenic growth, their pathogenicity, and their particular physiology make these bacteria quite difficult to work with. However, in order to prevent the diseases caused by these pathogens, a detailed understanding is required of their genetic and physiological resources and the mechanisms that have contributed to their evolutionary success. The development of mycobacterial genomics and related research disciplines that are building upon "omic" data now provide the scientific basis for the prediction and identification of factors that determine pathogenicity and differentiate mycobacterial pathogens from nonpathogenic strains. In-depth knowledge of these factors is a key outcome of the research that tries to understand the underlying biological mechanisms employed by the bacteria to circumvent host defense strategies and propagate in hostile environments. Here, we focus on insights gained from recent mycobacterial genome and functional analyses and provide an overview of the evolution and the infection strategies employed by selected mycobacterial pathogens, with the main emphasis on *M. tuberculosis*.

MYCOBACTERIAL GENOMICS

Mycobacterial genome research started well before the availability of complete genome sequences. Integrated genome maps that combined restriction digest–based

[1]Dipartimento di Ricerca Traslazionale e delle Nuove Tecnologie in Medicina e Chirurgia, Università di Pisa, Pisa, Italy; [2]Department of Microbiology and Immunology, University of Melbourne, Parkville, Australia; [3]CNRS UMR 8204; INSERM, U1019; Center for Infection and Immunity of Lille, Institut Pasteur de Lille; and Université Lille Nord de France, Lille, France; [4]Institut Pasteur, Unit for Integrated Mycobacterial Pathogenomics, Paris, France.

physical maps (6) of the chromosome with sets of ordered cosmids or bacterial artificial chromosomes were of great use for building the scaffold of whole-genome sequencing projects (7) and allowed an independent estimate of genome organization and genome size of the studied strain. Furthermore, the ordered clone libraries used for constructing the integrated genome maps represent an archived source of mycobacterial DNA and serve as a valuable resource for genetic manipulation and gene complementation studies (8).

The first mycobacterial genome that was completely sequenced and made available to the scientific community was that of *M. tuberculosis* H37Rv. This strain, which was originally isolated from a pulmonary tuberculosis patient in 1905, is a widely studied *M. tuberculosis* reference strain that has retained its full virulence over the years (9). The *M. tuberculosis* H37Rv genome sequencing project was the first in a series of mycobacterial genome projects that were based on a fruitful collaboration involving the Institut Pasteur in Paris, France, and the Sanger Institute in Hinxton, United Kingdom, as core partners, allowing the genome sequences of several important mycobacterial species and strains to be obtained (10–15). The analysis of the genome of *M. tuberculosis* H37Rv showed that this bacterium encodes in its 4.4 megabase (Mb) genome around 4,000 predicted proteins and 50 RNA molecules. The G+C content of almost 66% confirmed the phylogenetic position as part of the G+C rich, Gram-positive bacteria. However, it should be noted that in contrast to true Gram-positive bacteria regrouped within the phylum *Firmicutes*, mycobacteria have a different cell envelope architecture that has been shown to contain an inner and an outer membrane (mycomembrane) (16–18), thereby organizationally resembling more closely the envelope of Gram-negative bacteria than Gram-positive bacteria. In this respect, it is noteworthy that about 8% of the *M. tuberculosis* H37Rv genome encodes proteins involved in lipid metabolism. This finding highlights the importance of this class of molecules for the lifestyle of *M. tuberculosis* as an intracellular pathogen that protects itself with a particular cell envelope that is rich in lipids, glycolipids, lipoglycans, and polyketides (18, 19). The presence of enzymes that are predicted to show lipolytic functions further suggests that *M. tuberculosis* might use host-cell lipids and sterols as energy sources via the β-oxidation cycle, a pathway that is required for lipid catabolism. The *M. tuberculosis* genome contains genes for more than 100 enzymes that might be involved in various lipid oxidation pathways used for metabolizing putative degradation products of host cell membranes (10, 20, 21).

While the analysis of the *M. tuberculosis* H37Rv genome sequence showed that around 3.4% of the genome is composed of mobile elements such as prophages (phiRv1, phiRv2) and insertion sequences (IS) belonging to various families (e.g., IS*3*, IS*5*, IS*21*, IS*30*, IS*110*, IS*256*, and IS*L3*) (22), another key finding was the identification of novel gene and protein families, some of which also contained highly repetitive motifs. Probably, the most surprising of these were the PE and PPE families, which were named according to their characteristic N-terminal motifs ProGlu (PE) or ProProGlu (PPE). There are 169 PE and PPE proteins, representing 7.1% of the genome's coding capacity, in *M. tuberculosis* H37Rv. Whereas PE proteins have a highly conserved N-terminal domain of ~110-amino-acid residues, followed by a C-terminal segment that varies in size and may contain a polymorphic G+C-rich sequence (subfamily PGRS), PPE proteins are characterized by a conserved N-terminal segment of ~180 residues and a variable C-terminal domain that may contain major polymorphic tandem repeats (subfamily MPTR) (10, 23). It seems clear from recent studies that many of the PE and PPE proteins are surface exposed (24–28). Their transport depends on the functionality of a dedicated secretion system named ESX-5, which belongs to the type VII secretion systems of mycobacteria (29–31). The exact biological role of these proteins remains obscure, but it appears that their encoding genes have undergone a dramatic expansion during mycobacterial evolution toward pathogenicity. Indeed, the subgroup of slowly growing, pathogenic mycobacteria harbor a much larger number of PE- and PPE-encoding genes than the fast-growing saprophytic species. Similarly, the slow growers also have genes that encode more complex PE and PPE proteins, which contain the PGRS and MPTR regions, respectively (32). Given the large number of genes that encode PE and PPE proteins in the genomes of *M. tuberculosis* or *M. marinum*, it seems likely that they fulfill important functions, which are, however, difficult to study due to the apparent redundancy of sequence motifs within these proteins.

One of the key questions in this respect is whether some PE/PPE members play a unique role in pathogenesis. The protein PE_PGRS33 (Rv1818c), for example, was described as exerting a cytotoxic effect on host cells (33, 34). Interestingly, the gene encoding PE_PGRS33 in *M. tuberculosis* strains is not present in the genomes of *Mycobacterium canettii* strains, representing early branching tubercle bacilli with smooth colony morphology and showing lower virulence and persistence in animal infection models relative to

M. tuberculosis (35). This gene is also absent from the more distantly related nontuberculous mycobacteria, such as *M. marinum* (15), suggesting that the gene *rv1818c* was specifically acquired by a recent ancestor of the *M. tuberculosis* complex from an unknown source. It remains to be determined whether the acquisition of this gene may have supplied some selective advantage to *M. tuberculosis* for survival inside the host.

Another large gene family identified in the *M. tuberculosis* H37Rv genome concerns, for example, the genes encoding the mammalian cell entry proteins Mce, originally named after the observation that one of the *mce* genes cloned into *Escherichia coli* conferred to the recombinant strain the ability to enter HeLa cells (36). Four *mce* clusters with a total of 24 genes were identified at four positions in the genome of *M. tuberculosis* H37Rv (10, 37), although the number of *mce* clusters is even higher in environmental mycobacteria (38). One Mce operon (Mce3) is deleted in the closely related *M. bovis* lineage due to the deletion of the region of difference (RD) 7 (39, 40). It is noteworthy that most recently, a fifth Mce operon was identified in *M. canettii* strain STB-J of the early branching tubercle bacilli (35). Certain *mce* operons of *M. tuberculosis* have been associated with a role in pathogenicity (41, 42); in particular, the Mce4 operon seems to be required for cholesterol uptake and enhanced persistence of *M. tuberculosis* (43, 44).

Another highlight of the *M. tuberculosis* H37Rv genome project was the identification of numerous ESAT-6 loci, which were named after the paradigm member of this class of proteins defined as Early Secreted Antigenic Target due to its presence in culture filtrate of *M. tuberculosis* and its relatively small size of 6 kDa (45). ESAT-6 is encoded together with its protein partner CFP-10, the 10-kDa culture filtrate protein of *M. tuberculosis* (46) by a so-called *esx* operon in the proximity of the origin of replication. The genome project revealed that apart from these two genes, there were 21 other *esx* genes in the genome, organized mainly in gene couples, including five loci, where the *esx*-gene couples were found within large operon structures (10, 37, 47) together with genes encoding components of dedicated secretion systems involved in their export. As a common sequence motif in this ESX protein family, a specific Trp-Xaa-Gly (W-X-G) motif forming a characteristic hairpin bend between the two helical parts of the proteins was determined (48). Insights into the effects of Esx proteins and their ESX systems on the pathogenicity of mycobacteria are described below in the section on mycobacterial systems discovered by genomics that are involved in pathogenicity.

In the 15 years since the first mycobacterial genome description of *M. tuberculosis* H37Rv, many other mycobacterial genome sequences have been determined, originating from more distantly related mycobacterial species as well as from different tubercle bacilli and *M. tuberculosis* clinical isolates from various parts of the world. The data are thus available for global analyses via systems-biology-based methods. The in-depth analysis of this enormous dataset will allow specific traits linked to pathogenicity, transmissibility, and/or other strain- or strain-lineage-specific particularities to be identified that have likely been selected during mycobacterial evolution.

INSIGHTS INTO THE MACROEVOLUTION OF *M. TUBERCULOSIS* BASED ON COMPARISON WITH *M. MARINUM* AND *MYCOBACTERIUM KANSASII*

Genome comparisons with more distantly related mycobacterial species have been powerful ways to gain deeper insights into the broader genetic changes that led to the emergence of tuberculosis-causing mycobacteria (49). Whole-genome comparisons of *M. tuberculosis* with five selected mycobacterial species such as *M. avium paratuberculosis* (50), *M. leprae* (11), *M. marinum* (15), *Mycobacterium smegmatis* (GenBank accession-no. NC_008596), and *M. ulcerans* (13), for example, revealed 1,072 orthologous genes that are conserved across the six species tested, thus potentially defining the minimum mycobacterial genome (15, 49). Sequence comparisons of the core genes among these strains showed that *M. tuberculosis* is most closely related to *M. marinum* and least closely related to the environmental saprophyte *M. smegmatis* (15), which is in agreement with previous results from partial sequence analysis of the 16S rDNA (5). Phylogenetic reconstructions using these genome sequences strongly suggest a scenario in which *M. tuberculosis*, *M. marinum*, and *M. ulcerans* evolved from a common environmental ancestor, with the tuberculosis-causing mycobacteria undergoing extensive gene loss in parallel to gain of at least 600 new genes via horizontal gene transfer to become a specialized pathogen of humans and certain other mammals (15) (Fig. 1).

In contrast, *M. marinum* has maintained a genome that is 2.2 Mb larger than that of *M. tuberculosis*, which apparently is well adapted to the probably more fluctuating environmental challenges encountered by an aquatic microorganism and facultative pathogen. As an example, *M. marinum* has retained the faculty to produce photochromogenic pigment to protect itself

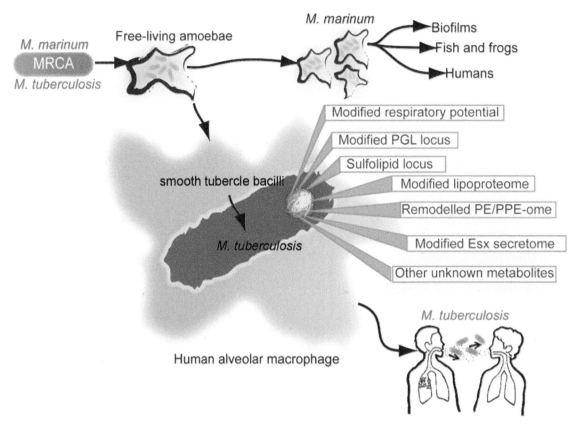

Figure 1 Proposed scenario of pathoadaptation of tubercle bacilli from a hypothetical mycobacterial common ancestor (most recent common ancestor, MRCA), adapted from reference 49. In the proposed evolutionary pathway *M. tuberculosis* strains gain specific functions (via horizontal gene transfer, recombination, mutation and/or gene loss, etc.), which allow them to better replicate and persist under the environment and temperature conditions of a niche such as human macrophages. It is plausible that the pathogenomic adaptation at some stage involved smooth tubercle bacilli/*M. canettii*, which show a broader environmental adaptability and a genetically much larger diversity than the *M. tuberculosis senso stricto* strains (35). doi:10.1128/microbiolspec.MGM2-0025-2013.f1

from light, which is linked to the *crtE-crtY* carotenoid biosynthesis gene cluster that does not have a counterpart in *M. tuberculosis* or other tubercle bacilli (15, 51). However, transposon mutagenesis of *M. marinum* also identified an associated gene named *crtP* to be involved in the regulation of pigmentation, which has homologs in the *M. tuberculosis* genomic region *rv2606c-rv2603c*, which is known to be involved in stress responses (52).

In parallel, a comparison between the 4.4-Mb genome of *M. tuberculosis* and the 6.4-Mb genome of *M. kansasii*, another closely related nontuberculous mycobacterium that can cause pulmonary infections in humans, was also instructive for insights into the macroevolution of tuberculosis-causing mycobacteria. The genomic comparison between the two species predicted that at least 137 genes had been acquired by

M. tuberculosis mainly via horizontal gene transfer since the phylogenetic divergence of the two species, including mainly genes coding for metabolic functions and modification of mycobacterial lipids (53, 54). It is noteworthy that based on the analysis of shared mycobacterial genes, *M. kansasii* appears more closely related to *M. tuberculosis* than *M. marinum* and *M. ulcerans*, which were considered the closest relatives of the tubercle bacilli according to comparisons of the 16S rRNA encoding genes (5, 55). This indicates that depending on the genes considered for the analysis, the relative phylogenetic position of an organism may vary due to the different evolutionary histories of the genes. In any case, it is clear from these analyses that the *M. marinum* (including *M. ulcerans*) and *M. kansasii* phylogenetic lineages represent the most closely related nontuberculous mycobacterial species of *M. tuberculosis* for which genome information is presently available. Although both species can cause opportunistic infections in humans (15, 54), they are far from having acquired the outstanding capacity of *M. tuberculosis* to act as a professional pathogen that causes lung tissue necrosis and cavity formation in order to ensure efficient transmission via aerosol to new hosts.

MICROEVOLUTIONARY GENOMICS OF THE TUBERCLE BACILLI

A better understanding of the more recent evolution of *M. tuberculosis* has come from genome comparisons within the *M. tuberculosis* complex, which comprises a variety of ecotypes that seem to be specifically found in a variety of mammalian hosts (56, 57). Despite this host range, these different tubercle bacilli show an extraordinary genetic homogeneity, with only 0.01 to 0.03% synonymous nucleotide variations between their genomes. Nevertheless, they can be differentiated into some main phylogenetic lineages by the presence or absence of selected genomic regions (39, 57, 58) or, more recently, by comparison of single nucleotide polymorphism (SNP) datasets (59, 60). Based on the distribution of the region of difference 9 (RD9) and the *M. tuberculosis* deleted region TbD1, three main lineages were defined (57). The lineage deleted for RD9 comprises *M. africanum*, prevalent in humans in West Africa, as well as a range of subspecies with animal reservoirs, such as *Mycobacterium microti*, a pathogen originally isolated from voles that also infects cats (61) and sporadically humans (62), *Mycobacterium pinnipedii*, which was isolated from fur seals and sea lions in different continents (63), *Mycobacterium caprae*, isolated from goats and deer (64, 65), and *M. bovis*, the bovine tubercle bacillus, which can infect a wide range of wild and domestic animals (66). The spectrum of RD9-deleted tubercle bacilli was further enriched by the identification of *Mycobacterium mungi* (67), the dassie bacillus (68, 69), a chimpanzee isolate (70), and *Mycobacterium orygis*, which is found in antelopes but was also repeatedly isolated from human tuberculosis cases (40). In contrast to previous hypotheses, the absence of RD9 from these strains suggests that the agent of bovine tuberculosis, *M. bovis*, is not the ancestor of the human tuberculosis agent *M. tuberculosis*, as was thought for a long time (71). This perspective is further supported by results from paleomicrobiological investigations that indicate the presence of *M. tuberculosis* rather than *M. bovis* in ancient human remains (72, 73).

The second major phylogenetic marker identified is the TbD1 region, which is absent from a large cluster of *M. tuberculosis* lineages named "modern" *M. tuberculosis* strains but present in all other members of the *M. tuberculosis* complex (57). This deletion truncates a representative member of the *mmpL* gene family encoding mycobacterial membrane proteins dedicated to the transport of cell wall components (74), whose function is thus likely missing from all modern *M. tuberculosis* strains. When compared to the SNP-based phylogenetic scheme proposed by the team of Gagneux and colleagues, the TbD1-deleted (and RD9-intact) strains correspond to lineages 2, 3, and 4, while the above-mentioned RD9-deleted strains are regrouped in lineages 5 and 6 and the animal strain lineages (56, 59, 60). Finally, lineage 1 in the SNP-based phylogeny comprises *M. tuberculosis* strains that have both regions, RD9 and TbD1, intact and thus resemble the closest last common ancestor of the *M. tuberculosis* complex members with respect to the structure of these genomic regions (Fig. 2). These strains have thus been named "ancestral" or TbD1-intact *M. tuberculosis* strains (56, 57, 59). They are most prevalent in countries bordering the Indian Ocean (75, 76) and the Philippines and have been reported to differ from strains of other lineages by the type of inflammatory response they induce (77). Strain lineages of *M. tuberculosis* isolates were reported to show close association with their host populations over time, so that a patient's region of birth can even be used as a predictor of the strain type a patient might carry (78, 79). As such, TbD1-intact strains are predominantly found in patients of South Asian origin. As an exception, *M. tuberculosis* strains with a preserved TbD1 region belonging to a previously unknown, deep branching lineage, named lineage 7, were recently isolated from patients from around the Horn of Africa (80, 81). According to the wide distribution of the lineage 1 strains in South Asian countries with high tuberculosis prevalence, the total number of infections caused by TbD1-intact strains might be larger than that of cases caused by TbD1-deleted strains (82). Overall, RD and SNP analyses confirmed that *M. tuberculosis* strain lineages and the human host populations are geographically linked. Paleomicrobiological investigations using amplification of ancient *M. tuberculosis* DNA from human remains have further suggested that *M. tuberculosis* strains present in the UK and Hungary in the past centuries were TbD1-deleted strains, which are still the most prevalent genotypes in Europe today (72, 73, 83). This marked phylogeographic structure is consistent with the highly clonal population structure

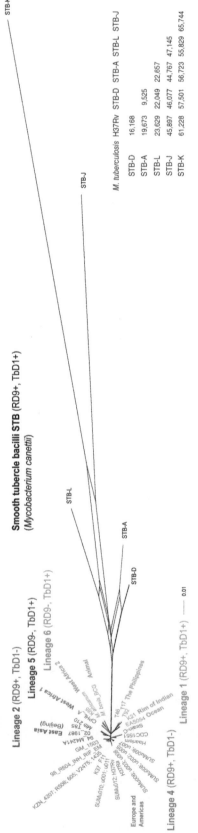

M. tuberculosis	H37Rv	STB-D	STB-A	STB-L	STB-J
STB-D	16,168				
STB-A	19,673	9,525			
STB-L	23,629	22,049	22,657		
STB-J	45,897	46,077	44,767	47,145	
STB-K	61,228	57,501	56,723	55,829	65,744

Figure 2 Network phylogeny inferred among the five *M. canettii* strains subjected to complete genome sequence analysis and 39 selected genome sequences from members of the classical *M. tuberculosis* complex by NeighborNet analysis, based on pairwise alignments of whole-genome SNP data, which in part are also listed in the lower right portion of the figure. The color code and the naming of different phylogenetic lineages within the *M. tuberculosis* complex refer to the nomenclature used in reference 59. doi:10.1128/microbiolspec.MGM2-0025-2013.f2

of *M. tuberculosis* strains, where genotypes are expected to be stable in space and time (84). Initial genome analyses indicated the lack of any significant recent horizontal gene transfer in classical members of the *M. tuberculosis* complex (12, 60, 85), a feature that was also pointed out by Supply and colleagues (84) based on analyses of genetic linkage between loci containing mycobacterial interspersed repetitive unit-variable number tandem repeat (MIRU-VNTR). In addition, recent genome sequencing studies of *M. tuberculosis* strains involved in tuberculosis epidemics made it possible to follow the genetic events during the outbreaks and to confirm that *M. tuberculosis* strains evolve by clonal descent (86, 87).

In contrast, a recent study has reported homoplastic SNP regions with at least two nucleotides concerned, present in different strains of the *M. tuberculosis* complex, suggesting that potential recombination tracts of small sizes might exist within their genomes (88). However, three of the four examples of suggested recombination tracts presented (in Fig. S4 of reference 88), correspond to adjacent base pairs changes, i.e., putative tandem-base mutations, shared by strains within particular subclusters of *M. tuberculosis* lineages (88). While these tandem-base mutations seem perfectly compatible with a strictly clonal evolution scenario, the remaining example of a SNP signature shared among *M. africanum* CPHL_A and the *M. tuberculosis* CDC1551- and C-strains might more likely be the result of a recombination event (88). Because the question of potential recombination episodes among different *M. tuberculosis* strains is also important for evaluating the risk of potential transfer of drug resistance mutations between *M. tuberculosis* strains, further functional studies are needed to unambiguously determine if and how *M. tuberculosis* strains might transfer DNA among each other. However, the previously reported findings did not concern larger genomic segments, such as the RD regions, which thus remain stable markers for differentiating phylogenetic lineages of the *M. tuberculosis* complex (57, 79, 89).

In contrast to *M. tuberculosis* and the classical members of the *M. tuberculosis* complex, the situation with horizontal gene transfer and interstrain-recombination is clearly different for *M. canettii* strains, whose main phenotypic characteristics are their—for tubercle bacilli—unusual, smooth colony morphology and their somewhat faster growth (35, 90, 91). Such smooth tubercle bacilli were first observed in the 1970s by Georges Canetti from a patient with pulmonary tuberculosis and were later named *M. canettii* (49, 92, 93). However, these strains are very rare and geographically restricted. Patients infected with *M. canettii* mostly originate from or have contact with the region of the Horn of Africa. The majority of the less than 100 worldwide available *M. canettii* isolates come from patients from Djibouti, East Africa (35, 55, 91, 94), which corresponds to the same geographical region where the above-mentioned, early branching *M. tuberculosis* strains of lineage 7 were isolated (80). From the 16S rDNA sequence and genome sequencing data, it is clear that the *M. canettii* strains share more than 97 to 99% DNA similarity with *M. tuberculosis* strains and thus can be considered as belonging to the same bacterial species (35, 95). However, when considering the SNP levels reported for *M. tuberculosis*, *M. africanum*, and *M. bovis* genomes in the maximal range of around 2,200 to 2,350 SNPs (12, 60, 85), the observed 16,000 to 65,000 SNPs among *M. tuberculosis* and *M. canettii* strains clearly demonstrate the much greater genetic diversity and phylogenetic distance of the smooth tubercle bacilli compared to the classical members of the *M. tuberculosis* complex (35) (Fig. 2).

The smooth tubercle bacilli share almost the same core genome as *M. tuberculosis* complex members but individually contain many additional genes contributing to an ~20% larger pan-genome of tubercle bacilli than was previously known (35). These findings suggest that the genomes of *M. canettii* strains might mirror the ancestral gene pool of those tubercle bacilli from which the members of the classical *M. tuberculosis* complex evolved by clonal expansion. This model is in agreement with the finding that several previously identified interrupted coding sequences of *M. tuberculosis* (56, 96) were found intact in the *M. canettii* strains as well as in the *M. marinum* and/or *M. kansasii* outgroups, suggesting that the *M. tuberculosis* lineages diverged from the smooth strain lineages before these frameshift mutations had occurred in the most recent common ancestor of the classical *M. tuberculosis* complex (35). Similarly, the RD9 and TbD1 regions are also preserved in smooth strains. Another striking feature of the *M. canettii* genome sequences was the identification of prophages and clustered, regularly interspaced short palindromic repeat (CRISPR) systems that were completely different from those present in the classical members of the *M. tuberculosis* complex strains (35, 97), suggesting the continued genetic changes after the branching of the lineage leading to the classical *M. tuberculosis* complex from the smooth tubercle bacilli.

The large genetic diversity found in smooth strains, all originating from a restricted geographical location, i.e., Djibouti, and the highly reduced diversity within the globally spread *M. tuberculosis* strains and

M. tuberculosis complex members suggest that the close ancestors of tubercle bacilli emerged in East Africa and that a dominant clone, later diversifying into the classical members of the *M. tuberculosis* complex (*M. tuberculosis*, *M. africanum*, *M. microti*, *M. bovis*) then spread throughout the world, possibly carried by waves of human migration. Moreover, from inspection of the genome sequences of *M. canettii* strains it is apparent that a substantial part of the genetic diversity in *M. canettii* strains is due to interstrain recombination events and horizontal gene transfer (35, 55). For the moment, it remains unclear where, when, and by which mechanisms these apparent transfers might have been achieved. The presence of homoplasic regions in different *M. canettii* genomes, in some cases also involving sequences characteristic for *M. tuberculosis*, observed at multiple loci, suggests that smooth tubercle bacilli might have (had) frequent occasion to exchange or uptake DNA with or from other strains (35, 55). It is tempting to speculate that distributive conjugation, recently described for the fast-growing *M. smegmatis* (98), could be involved. However, in *M. smegmatis*, conjugation is enabled by the ESX-1 secretion system of the bacterium, which is present and highly conserved in both *M. canettii* and *M. tuberculosis* and is responsible for secreting proteins involved in virulence, such as the early secreted antigenic target ESAT-6 (99). Alternatively, the specific presence of a large putative complete phage-encoding region in at least one *M. canettii* strain (35) suggests a possible involvement of phages in the horizontal gene exchanges. Thus, more experimental work is needed to get deeper insight into this matter for tubercle bacilli.

Apart from genetically diverse regions, the genome project of smooth tubercle bacilli also identified genomic regions that are largely conserved in *M. canettii* and *M. tuberculosis* strains. Among these, genes encoding proteins harboring recognized human T-cell epitopes have dN/dS ratios that are on average lower than those of genes classified as nonessential and similar or slightly lower than those of the essential genes (Table 1) (35). Overall, these observations are in agreement with results from Comas and colleagues (60), who first reported that human T-cell antigens were more conserved in *M. tuberculosis* strains relative to the rest of the proteome, suggestive of purifying selection acting on these epitopes. While these results could be interpreted to suggest that *M. canettii* might benefit from recognition by human T cells, as was proposed for *M. tuberculosis* (60), an alternative interpretation might be more plausible. It could well be that the conservation of the concerned proteins is caused by

Table 1 Ratios of nonsynonymous versus synonymous SNPs in gene categories[a]

Strain	dN/dS in gene category[b]				
	All	Essential	Nonessential	T-cell antigens	T-cell epitopes
STB-A	0.19/0.15	0.14/0.11	0.21/0.17	0.14/0.12	0.18/0.14
STB-J	0.18/0.13	0.14/0.11	0.19/0.15	0.14/0.11	0.13/0.09
STB-D	0.20/0.16	0.16/0.12	0.22/0.17	0.15/0.12	0.10/0.08
STB-L	0.19/0.15	0.16/0.12	0.21/0.17	0.14/0.12	0.15/0.11
STB-K	0.17/0.13	0.14/0.10	0.19/0.15	0.15/0.12	0.13/0.09
MTBC[c]	ND	0.53	0.66	0.50	0.53–0.25[d]

[a]STB, smooth tubercle bacilli; MTBC, *M. tuberculosis* complex; ND, not done.
[b]dN/dS ratios were calculated on orthologs conserved in all smooth tubercle bacilli strains and *M. tuberculosis* H37Rv, based on pairwise, concatenated codon alignments and using SNAP (value on the left) and PAML maximum likelihood methods (value on the right) as reported in reference 35. *M. tuberculosis* H37Rv T-cell antigen, essential, and nonessential gene categories, as well as T-cell epitope codon concatenates, were constructed as in Comas et al. (60).
[c]dN/dS ratios calculated by Comas et al. (60) from SNPs identified across 21 MTBC strains.
[d]Lower value obtained after exclusion of epitopes of three antigens considered as outliers.

purifying selection due to specific, important functions for the mycobacterial cell (e.g., cell wall functions, secretion systems involved in environmental competition) despite nonessentiality for *in vitro* growth. Indeed, many of the T-cell-epitope containing proteins of *M. tuberculosis* are also conserved in mycobacteria that do not have a long-lasting coevolution/host-pathogen interaction with humans. As an example, the antigen 85 complex, well known for its strong recognition by human T cells, is also highly conserved in *M. marinum* and other mycobacterial species (100). The same is true for ESAT-6 and CFP-10, which are more than 90% conserved in *M. marinum* and some other mycobacteria not necessarily in contact with humans (15). Why the *M. tuberculosis* orthologs were targeted by the human immune system remains an open question; however, their importance for the mycobacterial cell and the implied conservation and reduced possibility of escape variants might have played a role.

Finally, *M. canettii* strains were found to be less virulent in two different mouse models and also showed reduced persistence compared to *M. tuberculosis* H37Rv (35). It should be mentioned that *M. canettii* strains can cause very serious disease in humans (91–93). However, it is intriguing that the lower virulence in mice seems to correlate with the epidemiological situation in humans; i.e., the potentially less virulent *M. canettii* strains cause very few human infections compared to the millions of tuberculosis cases caused by the more virulent *M. tuberculosis* strains. If we consider *M. canettii* strains as the closest relatives of *M. prototuberculosis*, the proposed ancestor of the *M. tuberculosis* complex, it seems likely that the highly prevalent *M. tuberculosis* strains have emerged from low-virulence, potentially environmental, smooth strains by gaining additional virulence and persistence factors (35, 55, 91, 95).

This scenario also fits well with the above-described comparison of *M. tuberculosis* with *M. marinum* (Fig. 1), where the closest related, nontuberculous mycobacteria reside in the aquatic environment, possibly in contact with aquatic protozoa as potential hosts. The interaction with these organisms could have favored the development or adaptation of traits that have later also enabled coping with the intracellular environment in macrophages. These thoughts are, of course, still highly speculative, although the availability of the various genome and physiopathology data allows new insights into the genetic factors that might have contributed to the development of *M. tuberculosis* becoming an obligate, highly efficient pathogen. The availability of these data provide an excellent opportunity to design further experimental work in order to confirm or dismiss such hypotheses.

MYCOBACTERIAL SYSTEMS DISCOVERED BY GENOMICS THAT ARE INVOLVED IN PATHOGENICITY

Progress in mycobacterial genome research has enabled the development of genome-wide methods for identifying genes implicated in *in vitro* and *in vivo* growth. One of the first studies in this respect was carried out by Sassetti and colleagues, who adapted the mariner transposon for high-density mutagenesis in *M. tuberculosis* and developed a method named transposon site hybridization (TraSH). By this method these authors identified around 700 genes that were essential for optimal growth of *M. tuberculosis* H37Rv under *in vitro* conditions (101). In a similar approach, Lamichhane and coworkers identified numerous essential genes in the *M. tuberculosis* CDC1551 strain (102). In addition, almost 200 genes essential for survival under *in vivo* conditions in the mouse infection model were found by adaptation of the TraSH screen to select for *in vivo* growth mutants (103). These findings extended the previously established list of genes identified by gene knockout and signature tagged mutagenesis (104–107). The TraSH approach was recently refined by using next-generation sequencing (NGS): 453 genes were identified as essential for *in vitro* growth both by microarrays and NGS. However, the NGS-based study suggested 321 additional genes that seem essential for *in vitro* growth, depending on the culture media used (44).

In parallel to gene knockout and transposon-insertion approaches, comparative genomics of virulent and attenuated strains from the *M. tuberculosis* complex also provided highly relevant information for the identification of virulence gene clusters. As mentioned above, initial studies identified genomic polymorphisms across different members of the complex. Certain RDs were present in the genome of *M. tuberculosis* H37Rv but missing from attenuated strains (39, 108, 109). As the most prominent example, the 9.5-kb ESAT-6-encoding RD1 region is deleted from the genome of *M. bovis* BCG (RD1^{bcg}), the only currently used attenuated antituberculosis vaccine (108). However, an RD1^{BCG} overlapping portion (RD1^{mic}) is also deleted from *M. microti* strains (110), which were also used as live attenuated vaccines in the 1960s (111, 112). Complementation of BCG and *M. microti* with the extended RD1 locus from *M. tuberculosis* restored ESAT-6 secretion and partially increased the virulence of recombinant BCG and *M. microti* strains (8, 113, 114). Together with results from *M. tuberculosis* RD1 deletion/knockout mutants (115–117), it became clear that RD1 encoded proteins have an important impact on *M. tuberculosis* virulence. Their loss thus seems to have played a key role in the attenuation of BCG- and *M. microti*–based vaccines. However, it should be mentioned that some other rare members of the *M. tuberculosis* complex, similar to *M. microti*, seem to have naturally lost overlapping portions of the RD1 region, as reported for the dassie bacillus (69) and for *M. mungi* (67).

Soon after the genomic identification of the RD1 encoded gene cluster, experimental evidence for a novel specialized secretion system was obtained. The system was named the ESAT-6 secretion system (ESX-1) (113, 118, 119), the Snm system (117), and more recently, the type VII secretion system (120, 121). The four ESX-1 paralogs found in *M. tuberculosis* were named ESX-2 to ESX-5 (10, 37, 47). Because it was first discovered and shown to be a key virulence factor, ESX-1 can be considered the prototype of ESX systems (Fig. 3).

Most of the current knowledge on ESX secretion comes from the characterization of the ESX-1 secretion machinery, although more mechanistic data became recently available for the ESX-5 system (30, 122). Each ESX secretion apparatus is a multiprotein complex, consisting of so-called ESX conserved components (Ecc) and ESX-secretion-associated proteins (Esp) as well as Esx and PE/PPE proteins (121). It is constituted by cytosolic and membrane-anchored ATP-binding proteins (EccA and EccC, respectively) and other proteins containing a number of transmembrane domains (EccB, EccD, EccE), which are thought to mediate the ATP-dependent export of ESX substrates across the cytoplasmic membrane (114, 121) (Fig. 3). To date, the protein pores responsible for translocation of ESX substrates across the mycobacterial outer membrane have not been experimentally identified. However, it has been recently proposed that one of the components of the membrane-anchored complex, namely EccE or EccC, might span both the inner and outer membrane, thus forming a channel in the mycomembrane (122). All ESX systems also include membrane-bound mycobacteria-specific subtilisin-like serine proteases called mycosins (MycP$_1$-MycP$_5$) (123, 124), which might be involved in proteolytic digestion of ESX substrates. The ESX-1-associated MycP1, the only mycosin characterized so far, plays a key role in modulation of ESX-1 secretion activity, via the proteolytic digestion of the ESX-1 substrate EspB (125).

A typical hallmark of Esx proteins and other ESX substrates is the absence of a classical N-terminal signal sequence for secretion (99, 126, 127). A conserved secretion signal has been identified in CFP-10 (EsxB), which is required for CFP-10 being recognized by the FtsK-SpoIIIE ATPase EccC and thus for targeting the protein together with its protein partner ESAT-6 (EsxA) to the corresponding ESX secretion machinery (128). This observation was recently extended by defining a C-terminal domain (YxxxD/E) as a general sequence required for targeting proteins to the ESX/type VII secretion machineries (129). To date, the signal sequences that specifically target each ESX substrate to the corresponding ESX secretion apparatus are still unknown, but it has been hypothesized that selected EspG proteins (e.g., the *M. marinum* ESX-1- and ESX-5-encoded EspG) might act as chaperons in directing type VII substrates to the corresponding ESX machineries (130). In this respect it is intriguing that gene inactivation of EspG1 from the *M. tuberculosis* ESX-1 system results in attenuation of the mutant strain, although the secretion of two of the main substrates, ESAT-6 and CFP-10, was not impaired

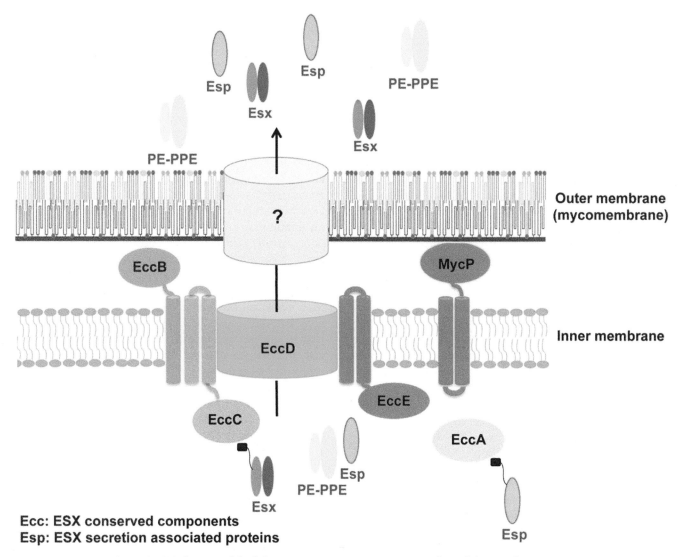

Figure 3 Working model of the type VII secretion apparatus adapted from reference 121. Schematic representation of the core components and their interactions. Various ESX components belonging to different protein families are represented by different colors: orange, amino-terminal transmembrane protein; violet, amino-terminal transmembrane ATPase; green, integral membrane protein; red, mycosin (subtilisin-like serine protease); pink, AAA+ ATPase. Esx secreted substrates, PE and PPE proteins, as well as ESX-1-associated Esp proteins are also represented. Note that the channel drawn in the mycomembrane refers to a hypothetical protein, whose existence has not been experimentally demonstrated, and that the drawing of the mycomembrane follows a schematic representation of reference 17. doi:10.1128/microbiolspec.MGM2-0025-2013.f3

by the deletion process (131). The involvement of an ATPase in the contact between Esx substrates and the secretion apparatus shows similarity with type IV secretion systems in Gram-negative bacteria, where a membrane-bound SpoIIIE/FtsK-like ATPase recognizes an unstructured C-terminal sequence and directs the secreted substrate to the cytoplasmic membrane (132).

Phylogenetic analyses and comparative genomics suggest that the ESX-4 cluster is the most ancestral ESX locus in the genus *Mycobacterium* (47). The other ESX loci are supposed to have evolved from ESX-4-like systems by gene duplication events and, in some cases, insertion of additional genes, with the order ESX-1, ESX-3, ESX-2, and ESX-5 (47). One of the most important gene insertion events is related to *pe* and *ppe* genes, which are absent at the ESX-4 locus and have been inserted during and after the ESX-1 diversification. It is thought that the ESX system contained *pe* and *ppe* genes that are the ancestral representatives of these gene families, from which the other genes have evolved (32).

Although different segments of the ESX-1 locus are deleted in some strains and/or lineages of the *M. tuberculosis* complex (e.g., BCG, *M. microti*, the dassie bacillus, *M. mungi*) (67, 69, 108, 110), ESX-1 is highly conserved in *M. tuberculosis* as well as in the early branching, *M. canettii* strains (35), suggesting that the full ESX-1 encoding region is part of the ancestral genome organization of the tuberculosis-causing mycobacteria. In addition to the ESX-1 encoding genes at the RD1 region, the *espACD* locus, which is situated elsewhere in the genome, is also important for ESX-1 functions in *M. tuberculosis* (133, 134). This locus plays a fundamental role in modulation of ESX-1 (121, 133, 134), because the *espACD* gene products (ESX-1-associated EspA, EspC, and EspD proteins) are secreted in a codependent manner with ESAT-6 and CFP-10 (e.g., EspA) (134) or required for regulating the intracellular levels of other Esp proteins (e.g., EspD) (135). In contrast, the *espACD* operon is not necessary for the secretion of EspB, another ESX-1 substrate (136). The expression of genes of the *espACD* cluster is regulated by a mechanism that involves the activators PhoP, EspR, and/or MprAB (137–139). The association between the PhoP/PhoR two-component regulatory system and ESAT-6 secretion was demonstrated in the *M. tuberculosis* strain H37Ra (137), one of the most widely used attenuated variants of the H37 strain, which apparently lost its virulence during *in vitro* serial passages. Comparative sequence analysis of *M. tuberculosis* H37Rv and H37Ra genomes by microarray-based DNA resequencing techniques identified a point mutation in the predicted DNA binding region of the *phoP* gene in H37Ra (137), which interferes with the DNA binding capacities of the corresponding PhoP protein (140). The complementation of H37Ra with a wild-type copy of the *phoP* gene restored ESAT-6 secretion and partially increased the virulence of the recombinant H37Ra::*phoP* strain (137), thus demonstrating the role of PhoP as a two-component regulator acting indirectly on the modulation of ESX-1 secretion activity. More recently, the PhoP-mediated regulation cascade of ESAT-6 secretion was found to be linked with the nucleoid-associated regulator factor EspR (138), which seems to directly bind and activate the *espACD* operon (138, 141).

It is noteworthy that other ESX-1-associated Esp proteins have a species-specific impact on ESX secretion. This is the case for EspF1 or the above-mentioned EspG1: Unlike observations with the model organisms *M. smegmatis* or *M. marinum*, where orthologous genes are involved in EsxA/B secretion (142, 143), disruption of *espF* and *espG1* in *M. tuberculosis* did not impact secretion, the post-translational modification, or T-cell recognition of ESAT-6/CFP-10, but still caused strong attenuation (131). These findings suggest that, despite the high homology of ESX systems in various mycobacterial species, substantial differences in the secretion machinery exist that might have evolved during the adaptation to the respective hosts (114, 131).

It has been recently reported that the presence of a functional ESX-1 system strongly influences the ability of *M. tuberculosis* to establish infection and its trafficking in host cells. Analysis by electron microscopy techniques of the subcellular localization of different mycobacterial species, carrying an intact ESX-1 system or not, demonstrated a link between ESX-1 and the ability of tubercle bacilli to escape from the phagovacuole of infected cells. While *M. tuberculosis* and a range of pathogenic mycobacterial species (*M. leprae, M. bovis,* and *M. marinum*)—as well as recombinant BCG variants expressing an intact ESX-1 system—were reported to translocate from the phagosome into the cytosol of infected macrophages and dendritic cells, strains with an interrupted ESX-1 system remained enclosed in the phagovacuole of the cells (144–146).

An important factor for the ESX-1-mediated phagosomal rupture appears to be the ability of ESAT-6 to interact with biomembranes, causing their destabilization and lysis (115, 147). It was hypothesized that this process might be pH dependent (147), but it remains unclear if other factors are also involved, because the data obtained from spectroscopic analyses of recombinant ESAT-6 and CFP-10 preparations predicted stability of the complex even under low pH conditions (148). However, a recent study suggested that there were differences among ESAT-6 orthologs from pathogenic and nonpathogenic mycobacterial species in their ability to undergo a conformational change under acidic pH conditions (149). Furthermore, it is not known whether other proteins co-secreted by ESX-1 might also contribute to membrane perforation. For example, ESX-1 systems are widely distributed among pathogenic and nonpathogenic species, although the environmental, nonpathogenic mycobacterial species lack the EspACD region in contrast to pathogenic species (145). Thus, for the moment it remains unclear which effectors of the ESX-1 secretion systems contribute to the rupture of vacuoles containing pathogenic mycobacteria. It seems clear, though, that ESX-1-dependent pore formation and lysis of the vacuolar membrane allows pathogenic mycobacteria and/or bacterial components to gain access to the cytosolic compartment of the host cell at different stages of infection, which accounts for major differences observed between virulent *M. tuberculosis* and attenuated BCG, the latter lacking ESX-1 functions due to the RD1 deletion. Indeed, *M. tuberculosis* and BCG differ in a wide variety of features such as cell-to-cell spread (115, 150), apoptosis (151), autophagy induction, and/or impairment (152, 153), and access of mycobacterial proteins to the class I–processing machinery contained in the proteasome, with impact on NLRP3 inflammasome activation (154–156), type I interferon responses (157), and induction of CD8 T-cell responses (158).

Recent research demonstrated that in addition to ESX-1, ESX-5 is also a key virulence determinant of pathogenic mycobacteria. Insights into the functional role of the ESX-5 system have been obtained for *M. tuberculosis* and *M. marinum* (29–31, 159, 160), which harbors an orthologous ESX-5 system (47). Secretome analysis of a panel of *M. tuberculosis* and *M. marinum* strains, mutated in selected components of the ESX-5 system, demonstrated that ESX-5 mediates the secretion of EsxN/EsxM, the ESX-5-encoded EsxA/EsxB paralogs. ESX-5 also transports a number of PPE and PE proteins, including the representative PE25-PPE41 proteins (29, 30, 122, 129, 160) that are co-secreted as a heterodimer, whose crystal structure is similar to that reported for the ESAT-6•CFP-10 complex. Further substrates of ESX-5 are proteins belonging to the PE-PGRS and PPE_MPTR subfamilies (3, 97), including the mycobacterial lipase LipY (51) that is involved in degradation of long chain triacylglycerols during late phases of infection (161). Species-specific or strain-specific differences among LipY orthologs of *M. tuberculosis* ($LipY_{Mt}$) and *M. marinum* ($LipY_{Mm}$) might exist due to the species-associated differences in glycolipid composition of the cell wall and differences in PE/PPE expression profiles (30, 122, 159). Interestingly, PE_PGRS and PPE_MPTR are the phylogenetically most recent subclasses of PE and PPE proteins, and their emergence and expansion are linked to the ancestral PE and PPE proteins encoded at the ESX-5 locus (32), the last ESX locus supposed to have appeared in the *M. tuberculosis* genome (see above). Consistent with its role as a secretion system specialized for the transport of PE and PPE proteins, ESX-5 is absent in fast-growing saprophytic mycobacterial species (47, 119), which possess only a very limited set of PE/PPE proteins. In contrast,

it is present in various slow-growing human pathogenic species, such as *M. leprae* and *M. ulcerans*, which harbor a wide spectrum of PE/PPE proteins like *M. tuberculosis*.

An intact ESX-5 system is required for optimal *in vitro* growth of *M. tuberculosis* and is also essential for full virulence of tubercle bacilli. Initial attempts to delete a large portion of the ESX-5 locus that encodes key building blocks of the ESX-5 membrane-bound protein complex (e.g., the $eccB_5$-$eccC_5$ operon) were not successful. Deletion of the $eccB_5$-$eccC_5$ genomic segment was only obtained in a merodiploid strain, which carried an additional copy of $eccB_5$-$eccC_5$ genes inserted into the chromosome via an *attB* site (160). In contrast, disruption of another core component of the ESX-5 secretion apparatus, e.g., the predicted transmembrane channel $EccD_5$, did not affect the growth properties of the mutant strain in liquid medium but caused strong attenuation of the mutant, which was not able to replicate in murine macrophages or in severely combined immunodeficiency (SCID) mice (30). The impact of ESX-5 on *M. tuberculosis* viability and virulence might be explained by the involvement of this system in maintaining cell envelope stability and transport of PE and PPE proteins. ESX-5 inactivation causes extensive damage to the mycobacterial cell envelope, as revealed by the increased sensitivity of ESX-5 mutants to detergents and hydrophilic antibiotics to which mycobacteria normally are naturally resistant (30).

Interestingly, the putative nucleoid-associated regulator EspR binds to multiple sites in the ESX-1, ESX-2, and ESX-5 loci as well as upstream of a number of genes encoding enzymes involved in cell-wall biosynthesis (138). This finding provides evidence for a functional link between ESX-mediated secretion and cell-wall biogenesis. In addition, it is also possible that the perturbation of the transport of PE and PPE proteins caused by the inactivation of ESX-5 functions might contribute to the decreased viability of ESX-5 mutants: Accumulation of nonsecreted or falsely localized ESX-5 substrates might have a toxic effect on the mycobacterial cell and diminish fitness. To date, the role of many ESX-5 substrates on mycobacterial virulence has not been investigated. However, it seems clear that upon infection with *M. tuberculosis* the host mounts specific T-cell responses against ESX-5-associated PE/PPE proteins (PPE25, PE18, PPE26, PPE27, PE19) and a number of their non-ESX-5-encoded homologs (31), which are completely lost if the ESX-5 secretion system is nonfunctional, confirming a role for ESX-5 in the transport of these proteins during infection. ESX-5-encoded PE and PPE proteins have been demonstrated to be involved in virulence functions of *M. tuberculosis*. An *M. tuberculosis* Δppe25-pe19 mutant, deleted for the genomic segment encoding all ESX-5-associated PE/PPE proteins, is strongly attenuated in SCID mice and is unable to replicate and disseminate to the spleen in aerosol-infected immunocompetent C57BL mice (30, 31). The observed attenuated phenotype together with the retained ability of secreting important protective ESX-1 antigens makes this strain a potential candidate for further testing as a promising attenuated live vaccine (31). Although more and more aspects of ESX-related features have become known, further research is needed to better understand the many contributions of type VII systems to tuberculosis-related infection and immunity.

CONCLUSIONS AND PERSPECTIVES

The above-described proteins and systems that influence the outcome of *M. tuberculosis* infections represent only a small selection of the large spectrum of potential factors that intervene in host-pathogen interactions during the course of tuberculosis infection. There is a range of numerous lipids from *M. tuberculosis* that are important for the pathogenicity of *M. tuberculosis* (18, 162, 163), but this complex subject is beyond the scope of this chapter. It is also clear that many proteins other than those mentioned here affect the survival and *in vivo* growth of *M. tuberculosis*. Many of these additional virulence candidates were identified by the above-mentioned genetic knockout strategies such as gene deletion, transposon insertion, signature tagged mutagenesis, and TraSH screens (103, 105, 106, 164). It is likely that even more genes will be identified that might be specific for particular physiological states or growth phases, such as dormancy, resuscitation, etc., which are, however, more difficult to model or investigate (165–167). In addition, it is well known that besides the bacterial factors that contribute to virulence, the outcome of infection with *M. tuberculosis* also depends largely on the response of host immunity and underlying host genetics. These two substantial research domains are dealt with in more detail elsewhere and are not discussed further here.

We have described the evolution of *M. tuberculosis* and the genomic factors that seem to have played crucial roles as it has evolved to become such a devastating pathogen. The insights we are continuing to gain from genome data are substantial, particularly nowadays, because the cost to sequence a bacterial genome has become trivial. On this basis, we are living in a scientifically exciting era, progressively revealing long-hidden secrets and solving mysteries surrounding ancient mycobacterial diseases. However, the bottleneck now lies in the time and effort required for the analysis and experimental confirmation of hypotheses raised from genomic data. Our challenge is thus now to translate the new insights from genomics into practical approaches that have the power to prevent tuberculosis and other mycobacterial diseases. It remains to develop a novel vaccination concept that is more protective than the widely used BCG vaccine (14, 113, 168). While some promising candidates are entering clinical trials (169, 170), their protective efficacy in humans still needs to be confirmed. The example of the functional characterization of the ESX-5 locus shows the interest of rational pathogenomic approaches to developing potential additional vaccine candidates (see above).

Another real possibility by which progress in tuberculosis control can be made is the development of new antituberculosis drugs, which are urgently required for the treatment of the emerging drug-resistant variants of the tubercle bacillus. Several new molecules have been presented that are active against *M. tuberculosis* under *in vitro* and/or *in vivo* conditions (171–174). Information on the genome content and "omic" approaches in general are helping to identify the molecular targets of these new molecules (172, 174–177). Preclinical and clinical investigations will then have to show whether some of the drug candidates can be confirmed and developed into new drugs (173, 178, 179). The road ahead appears long, but the darkness has begun to lift.

Special thanks to our many colleagues who contributed the various original data summarized and discussed here. This work was supported in part by the Fondation de Recherche Medicale FRM DEQ20130326471, the European Community's Framework Programme 7 grants 241745 and 260872, the Institut Pasteur (PTR372 and 383), grant 09 BLAN 0400 01 from the Agence Nationale pour la Recherche (France), and the National Health and Medical Research Council of Australia.

Citation. Bottai D, Stinear TP, Supply P, Brosch R. 2014. Mycobacterial pathogenomics and evolution. Microbiol Spectrum 2(1):MGM2-0025-2013.

References

1. **Stackebrandt E, Rainey FA, WardRainey NL.** 1997. Proposal for a new hierarchic classification system, *Actinobacteria classis nov. Int J Syst Bacteriol* **47:** 479–491.

2. **Ripoll F, Pasek S, Schenowitz C, Dossat C, Barbe V, Rottman M, Macheras E, Heym B, Herrmann JL, Daffe M, Brosch R, Risler JL, Gaillard JL.** 2009. Non mycobacterial virulence genes in the genome of the emerging pathogen *Mycobacterium abscessus. PLoS One* **4:**e5660.

3. Bryant JM, Grogono DM, Greaves D, Foweraker J, Roddick I, Inns T, Reacher M, Haworth CS, Curran MD, Harris SR, Peacock SJ, Parkhill J, Floto RA. 2013. Whole-genome sequencing to identify transmission of *Mycobacterium abscessus* between patients with cystic fibrosis: a retrospective cohort study. *Lancet* 381: 1551–1560.

4. Pawlik A, Garnier G, Orgeur M, Tong P, Lohan AJ, Le Chevalier F, Sapriel G, Roux AL, Conlon K, Honoré N, Dillies MA, Ma L, Bouchier C, Coppée JY, Gaillard JL, Gordon SV, Loftus B, Brosch R, Herrmann JL. 2013. Identification and characterization of the genetic changes responsible for the characteristic smooth-to-rough morphotype alterations of clinically persistent *Mycobacterium abscessus*. *Mol Microbiol* 90:612–629.

5. Springer B, Stockman L, Teschner K, Roberts GD, Bottger EC. 1996. Two-laboratory collaborative study on identification of mycobacteria: molecular versus phenotypic methods. *J Clin Microbiol* 34:296–303.

6. Philipp WJ, Poulet S, Eiglmeier K, Pascopella L, Balasubramanian V, Heym B, Bergh S, Bloom BR, Jacobs WR Jr, Cole ST. 1996. An integrated map of the genome of the tubercle bacillus, *Mycobacterium tuberculosis* H37Rv, and comparison with *Mycobacterium leprae*. *Proc Natl Acad Sci USA* 93:3132–3137.

7. Brosch R, Gordon SV, Billault A, Garnier T, Eiglmeier K, Soravito C, Barrell BG, Cole ST. 1998. Use of a *Mycobacterium tuberculosis* H37Rv bacterial artificial chromosome library for genome mapping, sequencing, and comparative genomics. *Infect Immun* 66:2221–2229.

8. Pym AS, Brodin P, Brosch R, Huerre M, Cole ST. 2002. Loss of RD1 contributed to the attenuation of the live tuberculosis vaccines *Mycobacterium bovis* BCG and *Mycobacterium microti*. *Mol Microbiol* 46:709–717.

9. Manca C, Tsenova L, Barry CE 3rd, Bergtold A, Freeman S, Haslett PA, Musser JM, Freedman VH, Kaplan G. 1999. *Mycobacterium tuberculosis* CDC1551 induces a more vigorous host response in vivo and in vitro, but is not more virulent than other clinical isolates. *J Immunol* 162:6740–6746.

10. Cole ST, Brosch R, Parkhill J, Garnier T, Churcher C, Harris D, Gordon SV, Eiglmeier K, Gas S, Barry CE 3rd, Tekaia F, Badcock K, Basham D, Brown D, Chillingworth T, Connor R, Davies R, Devlin K, Feltwell T, Gentles S, Hamlin N, Holroyd S, Hornsby T, Jagels K, Krogh A, McLean J, Moule S, Murphy L, Oliver K, Osborne J, Quail MA, Rajandream MA, Rogers J, Rutter S, Seeger K, Skelton J, Squares R, Squares S, Sulston JE, Taylor K, Whitehead S, Barrell BG. 1998. Deciphering the biology of *Mycobacterium tuberculosis* from the complete genome sequence. *Nature* 393:537–544.

11. Cole ST, Eiglmeier K, Parkhill J, James KD, Thomson NR, Wheeler PR, Honore N, Garnier T, Churcher C, Harris D, Mungall K, Basham D, Brown D, Chillingworth T, Connor R, Davies RM, Devlin K, Duthoy S, Feltwell T, Fraser A, Hamlin N, Holroyd S, Hornsby T, Jagels K, Lacroix C, Maclean J, Moule S, Murphy L, Oliver K, Quail MA, Rajandream MA, Rutherford KM, Rutter S, Seeger K, Simon S, Simmonds

M, Skelton J, Squares R, Squares S, Stevens K, Taylor K, Whitehead S, Woodward JR, Barrell BG. 2001. Massive gene decay in the leprosy bacillus. *Nature* 409: 1007–1011.

12. Garnier T, Eiglmeier K, Camus JC, Medina N, Mansoor H, Pryor M, Duthoy S, Grondin S, Lacroix C, Monsempe C, Simon S, Harris B, Atkin R, Doggett J, Mayes R, Keating L, Wheeler PR, Parkhill J, Barrell BG, Cole ST, Gordon SV, Hewinson RG. 2003. The complete genome sequence of *Mycobacterium bovis*. *Proc Natl Acad Sci USA* 100:7877–7882.

13. Stinear TP, Seemann T, Pidot S, Frigui W, Reysset G, Garnier T, Meurice G, Simon D, Bouchier C, Ma L, Tichit M, Porter JL, Ryan J, Johnson PD, Davies JK, Jenkin GA, Small PL, Jones LM, Tekaia F, Laval F, Daffe M, Parkhill J, Cole ST. 2007. Reductive evolution and niche adaptation inferred from the genome of *Mycobacterium ulcerans*, the causative agent of Buruli ulcer. *Genome Res* 17:192–200.

14. Brosch R, Gordon SV, Garnier T, Eiglmeier K, Frigui W, Valenti P, Dos Santos S, Duthoy S, Lacroix C, Garcia-Pelayo C, Inwald JK, Golby P, Garcia JN, Hewinson GR, Behr MA, Quail MA, Churcher C, Barrell BG, Parhill J, Cole ST. 2007. Genome plasticity of BCG and impact on vaccine efficacy. *Proc Natl Acad Sci USA* 104:5596–5601.

15. Stinear TP, Seemann T, Harrison PF, Jenkin GA, Davies JK, Johnson PD, Abdellah Z, Arrowsmith C, Chillingworth T, Churcher C, Clarke K, Cronin A, Davis P, Goodhead I, Holroyd N, Jagels K, Lord A, Moule S, Mungall K, Norbertczak H, Quail MA, Rabbinowitsch E, Walker D, White B, Whitehead S, Small PL, Brosch R, Ramakrishnan L, Fischbach MA, Parkhill J, Cole ST. 2008. Insights from the complete genome sequence of *Mycobacterium marinum* on the evolution of *Mycobacterium tuberculosis*. *Genome Res* 18:729–741.

16. Hoffmann C, Leis A, Niederweis M, Plitzko JM, Engelhardt H. 2008. Disclosure of the mycobacterial outer membrane: cryo-electron tomography and vitreous sections reveal the lipid bilayer structure. *Proc Natl Acad Sci USA* 105:3963–3967.

17. Zuber B, Chami M, Houssin C, Dubochet J, Griffiths G, Daffe M. 2008. Direct visualization of the outer membrane of mycobacteria and corynebacteria in their native state. *J Bacteriol* 190:5672–5680.

18. Kaur D, Guerin ME, Skovierova H, Brennan PJ, Jackson M. 2009. Chapter 2: biogenesis of the cell wall and other glycoconjugates of *Mycobacterium tuberculosis*. *Adv Appl Microbiol* 69:23–78.

19. Chopra T, Gokhale RS. 2009. Polyketide versatility in the biosynthesis of complex mycobacterial cell wall lipids. *Methods Enzymol* 459:259–294.

20. Cotes K, Bakala N'goma JC, Dhouib R, Douchet I, Maurin D, Carriere F, Canaan S. 2008. Lipolytic enzymes in *Mycobacterium tuberculosis*. *Appl Microbiol Biotechnol* 78:741–749.

21. Layre E, Moody DB. 2013. Lipidomic profiling of model organisms and the world's major pathogens. *Biochimie* 95:109–115.

22. Gordon SV, Heym B, Parkhill J, Barrell B, Cole ST. 1999. New insertion sequences and a novel repeated sequence in the genome of *Mycobacterium tuberculosis* H37Rv. *Microbiology* 145:881–892.

23. Bottai D, Brosch R. 2009. Mycobacterial PE, PPE and ESX clusters: novel insights into the secretion of these most unusual protein families. *Mol Microbiol* 73: 325–328.

24. Sampson SL, Lukey P, Warren RM, van Helden PD, Richardson M, Everett MJ. 2001. Expression, characterization and subcellular localization of the *Mycobacterium tuberculosis* PPE gene Rv1917c. *Tuberculosis* (Edinb) 81:305–317.

25. Delogu G, Brennan MJ. 2001. Comparative immune response to PE and PE_PGRS antigens of *Mycobacterium tuberculosis*. *Infect Immun* 69:5606–5611.

26. Banu S, Honore N, Saint-Joanis B, Philpott D, Prevost MC, Cole ST. 2002. Are the PE-PGRS proteins of *Mycobacterium tuberculosis* variable surface antigens? *Mol Microbiol* 44:9–19.

27. Cascioferro A, Delogu G, Colone M, Sali M, Stringaro A, Arancia G, Fadda G, Palu G, Manganelli R. 2007. PE is a functional domain responsible for protein translocation and localization on mycobacterial cell wall. *Mol Microbiol* 66:1536–1547.

28. Daleke MH, Cascioferro A, de Punder K, Ummels R, Abdallah AM, van der Wel N, Peters PJ, Luirink J, Manganelli R, Bitter W. 2011. Conserved Pro-Glu (PE) and Pro-Pro-Glu (PPE) protein domains target LipY lipases of pathogenic mycobacteria to the cell surface via the ESX-5 pathway. *J Biol Chem* 286:19024–19034.

29. Abdallah AM, Verboom T, Hannes F, Safi M, Strong M, Eisenberg D, Musters RJ, Vandenbroucke-Grauls CM, Appelmelk BJ, Luirink J, Bitter W. 2006. A specific secretion system mediates PPE41 transport in pathogenic mycobacteria. *Mol Microbiol* 62:667–679.

30. Bottai D, Di Luca M, Majlessi L, Frigui W, Simeone R, Sayes F, Bitter W, Brennan MJ, Leclerc C, Batoni G, Campa M, Brosch R, Esin S. 2012. Disruption of the ESX-5 system of *Mycobacterium tuberculosis* causes loss of PPE protein secretion, reduction of cell wall integrity and strong attenuation. *Mol Microbiol* 83: 1195–1209.

31. Sayes F, Sun L, Di Luca M, Simeone R, Degaiffier N, Fiette L, Esin S, Brosch R, Bottai D, Leclerc C, Majlessi L. 2012. Strong immunogenicity and cross-reactivity of *Mycobacterium tuberculosis* ESX-5 type VII secretion-encoded PE-PPE proteins predicts vaccine potential. *Cell Host Microbe* 11:352–363.

32. Gey van Pittius NC, Sampson SL, Lee H, Kim Y, van Helden PD, Warren RM. 2006. Evolution and expansion of the *Mycobacterium tuberculosis* PE and PPE multigene families and their association with the duplication of the ESAT-6 (*esx*) gene cluster regions. *BMC Evol Biol* 6:95.

33. Dheenadhayalan V, Delogu G, Brennan MJ. 2006. Expression of the PE_PGRS 33 protein in *Mycobacterium smegmatis* triggers necrosis in macrophages and enhanced mycobacterial survival. *Microbes Infect* 8:262–272 [Epub Sep 12, 2005].

34. Cadieux N, Parra M, Cohen H, Maric D, Morris SL, Brennan MJ. 2011. Induction of cell death after localization to the host cell mitochondria by the *Mycobacterium tuberculosis* PE_PGRS33 protein. *Microbiology* 157:793–804.

35. Supply P, Marceau M, Mangenot S, Roche D, Rouanet C, Khanna V, Majlessi L, Criscuolo A, Tap J, Pawlik A, Fiette L, Orgeur M, Fabre M, Parmentier C, Frigui W, Simeone R, Boritsch EC, Debrie AS, Willery E, Walker D, Quail MA, Ma L, Bouchier C, Salvignol G, Sayes F, Cascioferro A, Seemann T, Barbe V, Locht C, Gutierrez MC, Leclerc C, Bentley SD, Stinear TP, Brisse S, Médigue C, Parkhill J, Cruveiller S, Brosch R. 2013. Genomic analysis of smooth tubercle bacilli provides insights into ancestry and pathoadaptation of *Mycobacterium tuberculosis*. *Nat Genet* 45:172–179.

36. Arruda S, Bomfim G, Knights R, Huima-Byron T, Riley LW. 1993. Cloning of an *M. tuberculosis* DNA fragment associated with entry and survival inside cells. *Science* 261:1454–1457.

37. Tekaia F, Gordon SV, Garnier T, Brosch R, Barrell BG, Cole ST. 1999. Analysis of the proteome of *Mycobacterium tuberculosis* in silico. *Tuber Lung Dis* 79: 329–342.

38. Casali N, Riley LW. 2007. A phylogenomic analysis of the Actinomycetales *mce* operons. *BMC Genomics* 8:60.

39. Gordon SV, Brosch R, Billault A, Garnier T, Eiglmeier K, Cole ST. 1999. Identification of variable regions in the genomes of tubercle bacilli using bacterial artificial chromosome arrays. *Mol Microbiol* 32:643–656.

40. van Ingen J, Rahim Z, Mulder A, Boeree MJ, Simeone R, Brosch R, van Soolingen D. 2012. Characterization of *Mycobacterium orygis* as *M. tuberculosis* complex subspecies. *Emerg Infect Dis* 18:653–655.

41. Shimono N, Morici L, Casali N, Cantrell S, Sidders B, Ehrt S, Riley LW. 2003. Hypervirulent mutant of *Mycobacterium tuberculosis* resulting from disruption of the *mce1* operon. *Proc Natl Acad Sci USA* 100: 15918–15923.

42. Senaratne RH, Sidders B, Sequeira P, Saunders G, Dunphy K, Marjanovic O, Reader JR, Lima P, Chan S, Kendall S, McFadden J, Riley LW. 2008. *Mycobacterium tuberculosis* strains disrupted in *mce3* and *mce4* operons are attenuated in mice. *J Med Microbiol* 57: 164–170.

43. Pandey AK, Sassetti CM. 2008. Mycobacterial persistence requires the utilization of host cholesterol. *Proc Natl Acad Sci USA* 105:4376–4380.

44. Griffin JE, Gawronski JD, Dejesus MA, Ioerger TR, Akerley BJ, Sassetti CM. 2011. High-resolution phenotypic profiling defines genes essential for mycobacterial growth and cholesterol catabolism. *PLoS Pathog* 7: e1002251.

45. Sorensen AL, Nagai S, Houen G, Andersen P, Andersen AB. 1995. Purification and characterization of a low-molecular-mass T-cell antigen secreted by *Mycobacterium tuberculosis*. *Infect Immun* 63:1710–1717.

46. Berthet FX, Rasmussen PB, Rosenkrandt I, Andersen P, Gicquel B. 1998. A *Mycobacterium tuberculosis* operon

encoding ESAT-6 and a novel low-molecular-mass culture filtrate protein (CFP-10). *Microbiology* **144**: 3195–3203.

47. Gey Van Pittius NC, Gamieldien J, Hide W, Brown GD, Siezen RJ, Beyers AD. 2001. The ESAT-6 gene cluster of *Mycobacterium tuberculosis* and other high G+C Gram-positive bacteria. *Genome Biol* **2**:RE-SEARCH0044.

48. Pallen MJ. 2002. The ESAT-6/WXG100 superfamily—and a new Gram-positive secretion system? *Trends Microbiol* **10**:209–212.

49. Gordon SV, Bottai D, Simeone R, Stinear TP, Brosch R. 2009. Pathogenicity in the tubercle bacillus: molecular and evolutionary determinants. *Bioessays* **31**:378–388.

50. Li L, Bannantine JP, Zhang Q, Amonsin A, May BJ, Alt D, Banerji N, Kanjilal S, Kapur V. 2005. The complete genome sequence of *Mycobacterium avium* subspecies *paratuberculosis*. *Proc Natl Acad Sci USA* **102**: 12344–12349.

51. Ramakrishnan L, Tran HT, Federspiel NA, Falkow S. 1997. A *crtB* homolog essential for photochromogenicity in *Mycobacterium marinum*: isolation, characterization, and gene disruption via homologous recombination. *J Bacteriol* **179**:5862–5868.

52. Gao LY, Groger R, Cox JS, Beverley SM, Lawson EH, Brown EJ. 2003. Transposon mutagenesis of *Mycobacterium marinum* identifies a locus linking pigmentation and intracellular survival. *Infect Immun* **71**:922–929.

53. Veyrier F, Pletzer D, Turenne C, Behr MA. 2009. Phylogenetic detection of horizontal gene transfer during the step-wise genesis of *Mycobacterium tuberculosis*. *BMC Evol Biol* **9**:196.

54. Veyrier FJ, Dufort A, Behr MA. 2011. The rise and fall of the *Mycobacterium tuberculosis* genome. *Trends Microbiol* **19**:156–161.

55. Gutierrez MC, Brisse S, Brosch R, Fabre M, Omais B, Marmiesse M, Supply P, Vincent V. 2005. Ancient origin and gene mosaicism of the progenitor of *Mycobacterium tuberculosis*. *PLoS Pathog* **1**:e5.

56. Smith NH, Hewinson RG, Kremer K, Brosch R, Gordon SV. 2009. Myths and misconceptions: the origin and evolution of *Mycobacterium tuberculosis*. *Nat Rev Microbiol* **7**:537–544.

57. Brosch R, Gordon SV, Marmiesse M, Brodin P, Buchrieser C, Eiglmeier K, Garnier T, Gutierrez C, Hewinson G, Kremer K, Parsons LM, Pym AS, Samper S, van Soolingen D, Cole ST. 2002. A new evolutionary scenario for the *Mycobacterium tuberculosis* complex. *Proc Natl Acad Sci USA* **99**:3684–3689.

58. Mostowy S, Cousins D, Brinkman J, Aranaz A, Behr MA. 2002. Genomic deletions suggest a phylogeny for the *Mycobacterium tuberculosis* complex. *J Infect Dis* **186**:74–80.

59. Hershberg R, Lipatov M, Small PM, Sheffer H, Niemann S, Homolka S, Roach JC, Kremer K, Petrov DA, Feldman MW, Gagneux S. 2008. High functional diversity in *M. tuberculosis* driven by genetic drift and human demography. *PLoS Biol* **6**:e311.

60. Comas I, Chakravartti J, Small PM, Galagan J, Niemann S, Kremer K, Ernst JD, Gagneux S. 2010. Human T cell epitopes of *Mycobacterium tuberculosis* are evolutionarily hyperconserved. *Nat Genet* **42**:498–503.

61. Smith NH, Crawshaw T, Parry J, Birtles RJ. 2009. *Mycobacterium microti*: more diverse than previously thought. *J Clin Microbiol* **47**:2551–2559.

62. van Soolingen D, van der Zanden AG, de Haas PE, Noordhoek GT, Kiers A, Foudraine NA, Portaels F, Kolk AH, Kremer K, van Embden JD. 1998. Diagnosis of *Mycobacterium microti* infections among humans by using novel genetic markers. *J Clin Microbiol* **36**: 1840–1845.

63. Cousins DV, Bastida R, Cataldi A, Quse V, Redrobe S, Dow S, Duignan P, Murray A, Dupont C, Ahmed N, Collins DM, Butler WR, Dawson D, Rodriguez D, Loureiro J, Romano MI, Alito A, Zumarraga M, Bernardelli A. 2003. Tuberculosis in seals caused by a novel member of the *Mycobacterium tuberculosis* complex: *Mycobacterium pinnipedii* sp. nov. *Int J Syst Evol Microbiol* **53**:1305–1314.

64. Aranaz A, Cousins D, Mateos A, Dominguez L. 2003. Elevation of *Mycobacterium tuberculosis subsp. caprae* Aranaz et al. 1999 to species rank as *Mycobacterium caprae comb. nov., sp. nov. Int J Syst Evol Microbiol* **53**:1785–1789.

65. Domogalla J, Prodinger WM, Blum H, Krebs S, Gellert S, Muller M, Neuendorf E, Sedlmaier F, Buttner M. 2013. Region of difference 4 in alpine *Mycobacterium caprae* isolates indicates three variants. *J Clin Microbiol* **51**:1381–1388.

66. Berg S, Garcia-Pelayo MC, Muller B, Hailu E, Asiimwe B, Kremer K, Dale J, Boniotti MB, Rodriguez S, Hilty M, Rigouts L, Firdessa R, Machado A, Mucavele C, Ngandolo BN, Bruchfeld J, Boschiroli L, Muller A, Sahraoui N, Pacciarini M, Cadmus S, Joloba M, van Soolingen D, Michel AL, Djonne B, Aranaz A, Zinsstag J, van Helden P, Portaels F, Kazwala R, Kallenius G, Hewinson RG, Aseffa A, Gordon SV, Smith NH. 2011. African 2, a clonal complex of *Mycobacterium bovis* epidemiologically important in East Africa. *J Bacteriol* **193**:670–678.

67. Alexander KA, Laver PN, Michel AL, Williams M, van Helden PD, Warren RM, Gey van Pittius NC. 2010. Novel *Mycobacterium tuberculosis* complex pathogen, M. *mungi*. *Emerg Infect Dis* **16**:1296–1299.

68. Cousins DV, Peet RL, Gaynor WT, Williams SN, Gow BL. 1994. Tuberculosis in imported hyrax (*Procavia capensis*) caused by an unusual variant belonging to the *Mycobacterium tuberculosis* complex. *Vet Microbiol* **42**:135–145.

69. Mostowy S, Cousins D, Behr MA. 2004. Genomic interrogation of the dassie bacillus reveals it as a unique RD1 mutant within the *Mycobacterium tuberculosis* complex. *J Bacteriol* **186**:104–109.

70. Coscolla M, Lewin A, Metzger S, Maetz-Rennsing K, Calvignac-Spencer S, Nitsche A, Dabrowski PW, Radonic A, Niemann S, Parkhill J, Couacy-Hymann E, Feldman J, Comas I, Boesch C, Gagneux S, Leendertz

FH. 2013. Novel *Mycobacterium tuberculosis* complex isolate from a wild chimpanzee. *Emerg Infect Dis* **19:** 969–976.

71. Stead WW, Eisenach KD, Cave MD, Beggs ML, Templeton GL, Thoen CO, Bates JH. 1995. When did *Mycobacterium tuberculosis* infection first occur in the New World? An important question with public health implications. *Am J Respir Crit Care Med* **151:**1267–1268.

72. Donoghue HD, Spigelman M, Greenblatt CL, Lev-Maor G, Bar-Gal GK, Matheson C, Vernon K, Nerlich AG, Zink AR. 2004. Tuberculosis: from prehistory to Robert Koch, as revealed by ancient DNA. *Lancet Infect Dis* **4:**584–592.

73. Taylor GM, Young DB, Mays SA. 2005. Genotypic analysis of the earliest known prehistoric case of tuberculosis in Britain. *J Clin Microbiol* **43:**2236–2240.

74. Domenech P, Reed MB, Barry CE 3rd. 2005. Contribution of the *Mycobacterium tuberculosis* MmpL protein family to virulence and drug resistance. *Infect Immun* **73:**3492–3501.

75. Gutierrez MC, Ahmed N, Willery E, Narayanan S, Hasnain SE, Chauhan DS, Katoch VM, Vincent V, Locht C, Supply P. 2006. Predominance of ancestral lineages of *Mycobacterium tuberculosis* in India. *Emerg Infect Dis* **12:**1367–1374.

76. Banu S, Gordon SV, Palmer S, Islam R, Ahmed S, Alam KM, Cole ST, Brosch R. 2004. Genotypic analysis of *Mycobacterium tuberculosis* in Bangladesh and prevalence of the Beijing Strain. *J Clin Microbiol* **42:** 674–682.

77. Portevin D, Gagneux S, Comas I, Young D. 2011. Human macrophage responses to clinical isolates from the *Mycobacterium tuberculosis* complex discriminate between ancient and modern lineages. *PLoS Pathog* **7:** e1001307.

78. Hirsh AE, Tsolaki AG, DeRiemer K, Feldman MW, Small PM. 2004. Stable association between strains of *Mycobacterium tuberculosis* and their human host populations. *Proc Natl Acad Sci USA* **101:**4871–4876.

79. Gagneux S, Deriemer K, Van T, Kato-Maeda M, de Jong BC, Narayanan S, Nicol M, Niemann S, Kremer K, Gutierrez MC, Hilty M, Hopewell PC, Small PM. 2006. Variable host-pathogen compatibility in *Mycobacterium tuberculosis*. *Proc Natl Acad Sci USA* **103:** 2869–2873.

80. Blouin Y, Hauck Y, Soler C, Fabre M, Vong R, Dehan C, Cazajous G, Massoure PL, Kraemer P, Jenkins A, Garnotel E, Pourcel C, Vergnaud G. 2012. Significance of the identification in the Horn of Africa of an exceptionally deep branching *Mycobacterium tuberculosis* clade. *PLoS One* **7:**e52841.

81. Firdessa R, Berg S, Hailu E, Schelling E, Gumi B, Erenso G, Gadisa E, Kiros T, Habtamu M, Hussein J, Zinsstag J, Robertson BD, Ameni G, Lohan AJ, Loftus B, Comas I, Gagneux S, Tschopp R, Yamuah L, Hewinson G, Gordon SV, Young DB, Aseffa A. 2013. Mycobacterial lineages causing pulmonary and extrapulmonary tuberculosis, Ethiopia. *Emerg Infect Dis* **19:**460–463.

82. Gagneux S, Small PM. 2007. Global phylogeography of *Mycobacterium tuberculosis* and implications for tuberculosis product development. *Lancet Infect Dis* **7:** 328–337.

83. Chan JZ, Sergeant MJ, Lee OY, Minnikin DE, Besra GS, Pap I, Spigelman M, Donoghue HD, Pallen MJ. 2013. Metagenomic analysis of tuberculosis in a mummy. *N Engl J Med* **369:**289–290.

84. Supply P, Mazars E, Lesjean S, Vincent V, Gicquel B, Locht C. 2000. Variable human minisatellite-like regions in the *Mycobacterium tuberculosis* genome. *Mol Microbiol* **36:**762–771.

85. Bentley SD, Comas I, Bryant JM, Walker D, Smith NH, Harris SR, Thurston S, Gagneux S, Wood J, Antonio M, Quail MA, Gehre F, Adegbola RA, Parkhill J, de Jong BC. 2012. The genome of *Mycobacterium africanum* West African 2 reveals a lineage-specific locus and genome erosion common to the *M. tuberculosis* complex. *PLoS Negl Trop Dis* **6:**e1552.

86. Walker TM, Ip CL, Harrell RH, Evans JT, Kapatai G, Dedicoat MJ, Eyre DW, Wilson DJ, Hawkey PM, Crook DW, Parkhill J, Harris D, Walker AS, Bowden R, Monk P, Smith EG, Peto TE. 2013. Whole-genome sequencing to delineate *Mycobacterium tuberculosis* outbreaks: a retrospective observational study. *Lancet Infect Dis* **13:**137–146.

87. Roetzer A, Diel R, Kohl TA, Ruckert C, Nubel U, Blom J, Wirth T, Jaenicke S, Schuback S, Rusch-Gerdes S, Supply P, Kalinowski J, Niemann S. 2013. Whole genome sequencing versus traditional genotyping for investigation of a *Mycobacterium tuberculosis* outbreak: a longitudinal molecular epidemiological study. *PLoS Med* **10:**e1001387.

88. Namouchi A, Didelot X, Schock U, Gicquel B, Rocha EP. 2012. After the bottleneck: genome-wide diversification of the *Mycobacterium tuberculosis* complex by mutation, recombination, and natural selection. *Genome Res* **22:**721–734.

89. Marmiesse M, Brodin P, Buchrieser C, Gutierrez C, Simoes N, Vincent V, Glaser P, Cole ST, Brosch R. 2004. Macro-array and bioinformatic analyses reveal mycobacterial 'core' genes, variation in the ESAT-6 gene family and new phylogenetic markers for the *Mycobacterium tuberculosis* complex. *Microbiology* **150:**483–496.

90. Fabre M, Hauck Y, Soler C, Koeck JL, van Ingen J, van Soolingen D, Vergnaud G, Pourcel C. 2010. Molecular characteristics of "*Mycobacterium canettii*" the smooth *Mycobacterium tuberculosis* bacilli. *Infect Genet Evol* **10:**1165–1173.

91. Koeck JL, Fabre M, Simon F, Daffe M, Garnotel E, Matan AB, Gerome P, Bernatas JJ, Buisson Y, Pourcel C. 2011. Clinical characteristics of the smooth tubercle bacilli '*Mycobacterium canettii*' infection suggest the existence of an environmental reservoir. *Clin Microbiol Infect* **17:**1013–1019.

92. van Soolingen D, Hoogenboezem T, de Haas PE, Hermans PW, Koedam MA, Teppema KS, Brennan PJ, Besra GS, Portaels F, Top J, Schouls LM, van Embden JD. 1997. A novel pathogenic taxon of the *Mycobacterium*

tuberculosis complex, Canetti: characterization of an exceptional isolate from Africa. *Int J Syst Bacteriol* **47:** 1236–1245.

93. Pfyffer GE, Auckenthaler R, van Embden JD, van Soolingen D. 1998. *Mycobacterium canettii*, the smooth variant of *M. tuberculosis*, isolated from a Swiss patient exposed in Africa. *Emerg Infect Dis* **4:**631–634.

94. Fabre M, Koeck JL, Le Fleche P, Simon F, Herve V, Vergnaud G, Pourcel C. 2004. High genetic diversity revealed by variable-number tandem repeat genotyping and analysis of *hsp65* gene polymorphism in a large collection of "*Mycobacterium canettii*" strains indicates that the *M. tuberculosis* complex is a recently emerged clone of "*M. canettii*." *J Clin Microbiol* **42:**3248–3255.

95. Brisse S, Supply P, Brosch R, Vincent V, Gutierrez MC. 2006. "A re-evaluation of *M. prototuberculosis*": continuing the debate. *PLoS Pathog* **2:**e95.

96. Deshayes C, Perrodou E, Euphrasie D, Frapy E, Poch O, Bifani P, Lecompte O, Reyrat JM. 2008. Detecting the molecular scars of evolution in the *Mycobacterium tuberculosis* complex by analyzing interrupted coding sequences. *BMC Evol Biol* **8:**78.

97. Cain AK, Boinett CJ. 2013. A CRISPR view of genome sequences. *Nat Rev Microbiol* **11:**226.

98. Gray T, Krywy J, Harold J, Palumbo M, Derbyshire K. 2013. Distributive conjugal transfer in mycobacteria generates progeny with meiotic-like genome-wide mosaicism, allowing mapping of a mating identity locus. *PLoS Biol* **11:**e1001602.

99. Simeone R, Bottai D, Brosch R. 2009. ESX/type VII secretion systems and their role in host-pathogen interaction. *Curr Opin Microbiol* **12:**4–10.

100. Wiker HG, Harboe M. 1992. The antigen 85 complex: a major secretion product of *Mycobacterium tuberculosis*. *Microbiol Rev* **56:**648–661.

101. Sassetti CM, Boyd DH, Rubin EJ. 2003. Genes required for mycobacterial growth defined by high density mutagenesis. *Mol Microbiol* **48:**77–84.

102. Lamichhane G, Zignol M, Blades NJ, Geiman DE, Dougherty A, Grosset J, Broman KW, Bishai WR. 2003. A postgenomic method for predicting essential genes at subsaturation levels of mutagenesis: application to *Mycobacterium tuberculosis*. *Proc Natl Acad Sci USA* **100:**7213–7218.

103. Sassetti CM, Rubin EJ. 2003. Genetic requirements for mycobacterial survival during infection. *Proc Natl Acad Sci USA* **100:**12989–12994.

104. Berthet FX, Lagranderie M, Gounon P, Laurent-Winter C, Ensergueix D, Chavarot P, Thouron F, Maranghi E, Pelicic V, Portnoi D, Marchal G, Gicquel B. 1998. Attenuation of virulence by disruption of the *Mycobacterium tuberculosis erp* gene. *Science* **282:**759–762.

105. Camacho LR, Ensergueix D, Perez E, Gicquel B, Guilhot C. 1999. Identification of a virulence gene cluster of *Mycobacterium tuberculosis* by signature-tagged transposon mutagenesis. *Mol Microbiol* **34:**257–267.

106. Cox JS, Chen B, McNeil M, Jacobs WR Jr. 1999. Complex lipid determines tissue-specific replication of *Mycobacterium tuberculosis* in mice. *Nature* **402:**79–83.

107. Hingley-Wilson SM, Sambandamurthy VK, Jacobs WR Jr. 2003. Survival perspectives from the world's most successful pathogen, *Mycobacterium tuberculosis*. *Nat Immunol* **4:**949–955.

108. Mahairas GG, Sabo PJ, Hickey MJ, Singh DC, Stover CK. 1996. Molecular analysis of genetic differences between *Mycobacterium bovis* BCG and virulent *M. bovis*. *J Bacteriol* **178:**1274–1282.

109. Behr MA, Wilson MA, Gill WP, Salamon H, Schoolnik GK, Rane S, Small PM. 1999. Comparative genomics of BCG vaccines by whole-genome DNA microarrays. *Science* **284:**1520–1523.

110. Brodin P, Eiglmeier K, Marmiesse M, Billault A, Garnier T, Niemann S, Cole ST, Brosch R. 2002. Bacterial artificial chromosome-based comparative genomic analysis identifies *Mycobacterium microti* as a natural ESAT-6 deletion mutant. *Infect Immun* **70:**5568–5578.

111. Sula L, Radkovsky I. 1976. Protective effects of *M. microti* vaccine against tuberculosis. *J Hyg Epidemiol Microbiol Immunol* **20:**1–6.

112. Hart PD, Sutherland I. 1977. BCG and vole bacillus vaccines in the prevention of tuberculosis in adolescence and early adult life. *BMJ* **2:**293–295.

113. Pym AS, Brodin P, Majlessi L, Brosch R, Demangel C, Williams A, Griffiths KE, Marchal G, Leclerc C, Cole ST. 2003. Recombinant BCG exporting ESAT-6 confers enhanced protection against tuberculosis. *Nat Med* **9:**533–539.

114. Brodin P, Majlessi L, Marsollier L, de Jonge MI, Bottai D, Demangel C, Hinds J, Neyrolles O, Butcher PD, Leclerc C, Cole ST, Brosch R. 2006. Dissection of ESAT-6 system 1 of *Mycobacterium tuberculosis* and impact on immunogenicity and virulence. *Infect Immun* **74:**88–98.

115. Hsu T, Hingley-Wilson SM, Chen B, Chen M, Dai AZ, Morin PM, Marks CB, Padiyar J, Goulding C, Gingery M, Eisenberg D, Russell RG, Derrick SC, Collins FM, Morris SL, King CH, Jacobs WR Jr. 2003. The primary mechanism of attenuation of bacillus Calmette-Guerin is a loss of secreted lytic function required for invasion of lung interstitial tissue. *Proc Natl Acad Sci USA* **100:**12420–12425.

116. Lewis KN, Liao R, Guinn KM, Hickey MJ, Smith S, Behr MA, Sherman DR. 2003. Deletion of RD1 from *Mycobacterium tuberculosis* mimics bacille Calmette-Guerin attenuation. *J Infect Dis* **187:**117–123.

117. Stanley SA, Raghavan S, Hwang WW, Cox JS. 2003. Acute infection and macrophage subversion by *Mycobacterium tuberculosis* require a specialized secretion system. *Proc Natl Acad Sci USA* **100:**13001–13006.

118. Young DB. 2003. Building a better tuberculosis vaccine. *Nat Med* **9:**503–504 [Epub Apr 14, 2003].

119. Brodin P, Rosenkrands I, Andersen P, Cole ST, Brosch R. 2004. ESAT-6 proteins: protective antigens and virulence factors? *Trends Microbiol* **12:**500–508.

120. Abdallah A, Gey van Pittius N, Champion P, Cox J, Luirink J, Vandenbroucke-Grauls C, Appelmelk B, Bitter W. 2007. Type VII secretion—mycobacteria show the way. *Nat Rev Microbiol* **5:**883–891.

121. Bitter W, Houben EN, Bottai D, Brodin P, Brown EJ, Cox JS, Derbyshire K, Fortune SM, Gao LY, Liu J, Gey van Pittius NC, Pym AS, Rubin EJ, Sherman DR, Cole ST, Brosch R. 2009. Systematic genetic nomenclature for type VII secretion systems. *PLoS Pathog* 5: e1000507.

122. Houben EN, Bestebroer J, Ummels R, Wilson L, Piersma SR, Jimenez CR, Ottenhoff TH, Luirink J, Bitter W. 2012. Composition of the type VII secretion system membrane complex. *Mol Microbiol* 86:472–484.

123. Brown GD, Dave JA, Gey van Pittius NC, Stevens L, Ehlers MR, Beyers AD. 2000. The mycosins of *Mycobacterium tuberculosis* H37Rv: a family of subtilisin-like serine proteases. *Gene* 254:147–155.

124. Dave JA, Gey van Pittius NC, Beyers AD, Ehlers MR, Brown GD. 2002. Mycosin-1, a subtilisin-like serine protease of *Mycobacterium tuberculosis*, is cell wall-associated and expressed during infection of macrophages. *BMC Microbiol* 2:30.

125. Ohol YM, Goetz DH, Chan K, Shiloh MU, Craik CS, Cox JS. 2010. *Mycobacterium tuberculosis* MycP1 protease plays a dual role in regulation of ESX-1 secretion and virulence. *Cell Host Microbe* 7:210–220.

126. Brodin P, Majlessi L, Brosch R, Smith D, Bancroft G, Clark S, Williams A, Leclerc C, Cole ST. 2004. Enhanced protection against tuberculosis by vaccination with recombinant *Mycobacterium microti* vaccine that induces T cell immunity against region of difference 1 antigens. *J Infect Dis* 190:115–122.

127. Champion PA, Cox JS. 2007. Protein secretion systems in mycobacteria. *Cell Microbiol* 9:1376–1384.

128. Champion PA, Stanley SA, Champion MM, Brown EJ, Cox JS. 2006. C-terminal signal sequence promotes virulence factor secretion in *Mycobacterium tuberculosis*. *Science* 313:1632–1636.

129. Daleke MH, Ummels R, Bawono P, Heringa J, Vandenbroucke-Grauls CM, Luirink J, Bitter W. 2012. General secretion signal for the mycobacterial type VII secretion pathway. *Proc Natl Acad Sci USA* 109: 11342–11347.

130. Daleke MH, van der Woude AD, Parret AH, Ummels R, de Groot AM, Watson D, Piersma SR, Jimenez CR, Luirink J, Bitter W, Houben EN. 2012. Specific chaperones for the type VII protein secretion pathway. *J Biol Chem* 287:31939–31947.

131. Bottai D, Majlessi L, Simeone R, Frigui W, Laurent C, Lenormand P, Chen J, Rosenkrands I, Huerre M, Leclerc C, Cole ST, Brosch R. 2011. ESAT-6 secretion-independent impact of ESX-1 genes *espF* and *espG1* on virulence of *Mycobacterium tuberculosis*. *J Infect Dis* 203:1155–1164.

132. Nagai H, Cambronne ED, Kagan JC, Amor JC, Kahn RA, Roy CR. 2005. A C-terminal translocation signal required for Dot/Icm-dependent delivery of the Legionella RalF protein to host cells. *Proc Natl Acad Sci USA* 102:826–831.

133. MacGurn JA, Raghavan S, Stanley SA, Cox JS. 2005. A non-RD1 gene cluster is required for Snm secretion in *Mycobacterium tuberculosis*. *Mol Microbiol* 57:1653–1663.

134. Fortune SM, Jaeger A, Sarracino DA, Chase MR, Sassetti CM, Sherman DR, Bloom BR, Rubin EJ. 2005. Mutually dependent secretion of proteins required for mycobacterial virulence. *Proc Natl Acad Sci USA* 102: 10676–10681.

135. Chen JM, Boy-Rottger S, Dhar N, Sweeney N, Buxton RS, Pojer F, Rosenkrands I, Cole ST. 2012. EspD is critical for the virulence-mediating ESX-1 secretion system in *Mycobacterium tuberculosis*. *J Bacteriol* 194:884–893.

136. Chen JM, Zhang M, Rybniker J, Boy-Rottger S, Dhar N, Pojer F, Cole ST. 2013. *Mycobacterium tuberculosis* EspB binds phospholipids and mediates EsxA-independent virulence. *Mol Microbiol* 89: 1154–1166.

137. Frigui W, Bottai D, Majlessi L, Monot M, Josselin E, Brodin P, Garnier T, Gicquel B, Martin C, Leclerc C, Cole S, Brosch R. 2008. Control of *M. tuberculosis* ESAT-6 secretion and specific T cell recognition by PhoP. *PLoS Pathog* 4:e33.

138. Blasco B, Chen JM, Hartkoorn R, Sala C, Uplekar S, Rougemont J, Pojer F, Cole ST. 2012. Virulence regulator EspR of *Mycobacterium tuberculosis* is a nucleoid-associated protein. *PLoS Pathog* 8:e1002621.

139. Pang X, Samten B, Cao G, Wang X, Tvinnereim AR, Chen XL, Howard ST. 2013. MprAB regulates the *espA* operon in *Mycobacterium tuberculosis* and modulates ESX-1 function and host cytokine response. *J Bacteriol* 195:66–75.

140. Wang S, Engohang-Ndong J, Smith I. 2007. Structure of the DNA-binding domain of the response regulator PhoP from *Mycobacterium tuberculosis*. *Biochemistry* 46:14751–14761.

141. Hunt DM, Sweeney NP, Mori L, Whalan RH, Comas I, Norman L, Cortes T, Arnvig KB, Davis EO, Stapleton MR, Green J, Buxton RS. 2012. Long-range transcriptional control of an operon necessary for virulence-critical ESX-1 secretion in *Mycobacterium tuberculosis*. *J Bacteriol* 194:2307–2320.

142. Converse SE, Cox JS. 2005. A protein secretion pathway critical for *Mycobacterium tuberculosis* virulence is conserved and functional in *Mycobacterium smegmatis*. *J Bacteriol* 187:1238–1245.

143. Gao LY, Guo S, McLaughlin B, Morisaki H, Engel JN, Brown EJ. 2004. A mycobacterial virulence gene cluster extending RD1 is required for cytolysis, bacterial spreading and ESAT-6 secretion. *Mol Microbiol* 53: 1677–1693.

144. van der Wel N, Hava D, Houben D, Fluitsma D, van Zon M, Pierson J, Brenner M, Peters PJ. 2007. *M. tuberculosis* and *M. leprae* translocate from the phagolysosome to the cytosol in myeloid cells. *Cell* 129: 1287–1298.

145. Simeone R, Bobard A, Lippmann J, Bitter W, Majlessi L, Brosch R, Enninga J. 2012. Phagosomal rupture by *Mycobacterium* tuberculosis results in toxicity and host cell death. *PLoS Pathog* 8:e1002507.

146. Houben D, Demangel C, van Ingen J, Perez J, Baldeon L, Abdallah AM, Caleechurn L, Bottai D, van Zon M, de Punder K, van der Laan T, Kant A, Bossers-de Vries R, Willemsen P, Bitter W, van Soolingen D, Brosch R, van der Wel N, Peters PJ. 2012. ESX-1-mediated translocation to the cytosol controls virulence of mycobacteria. Cell Microbiol 14:1287–1298.

147. de Jonge MI, Pehau-Arnaudet G, Fretz MM, Romain F, Bottai D, Brodin P, Honore N, Marchal G, Jiskoot W, England P, Cole ST, Brosch R. 2007. ESAT-6 from Mycobacterium tuberculosis dissociates from its putative chaperone CFP-10 under acidic conditions and exhibits membrane-lysing activity. J Bacteriol 189:6028–6034.

148. Lightbody KL, Ilghari D, Waters LC, Carey G, Bailey MA, Williamson RA, Renshaw PS, Carr MD. 2008. Molecular features governing the stability and specificity of functional complex formation by Mycobacterium tuberculosis CFP-10/ESAT-6 family proteins. J Biol Chem 283:17681–17690.

149. De Leon J, Jiang G, Ma Y, Rubin E, Fortune S, Sun J. 2012. Mycobacterium tuberculosis ESAT-6 exhibits a unique membrane-interacting activity that is not found in its ortholog from non-pathogenic Mycobacterium smegmatis. J Biol Chem 287:44184–44191.

150. Guinn KI, Hickey MJ, Mathur SK, Zakel KL, Grotzke JE, Lewinsohn DM, Smith S, Sherman DR. 2004. Individual RD1-region genes are required for export of ESAT-6/CFP-10 and for virulence of Mycobacterium tuberculosis. Mol Microbiol 51:359–370.

151. Aguilo J, Alonso H, Uranga S, Marinova D, Arbues A, de Martino A, Anel A, Monzon M, Badiola J, Pardo J, Brosch R, Martin C. 2013. ESX-1-induced apoptosis is involved in cell-to-cell spread of Mycobacterium tuberculosis. Cell Microbiol [Epub ahead of print.] doi: 10.1111/cmi.12169.

152. Watson RO, Manzanillo PS, Cox JS. 2012. Extracellular M. tuberculosis DNA targets bacteria for autophagy by activating the host DNA-sensing pathway. Cell 150:803–815.

153. Romagnoli A, Etna MP, Giacomini E, Pardini M, Remoli ME, Corazzari M, Falasca L, Goletti D, Gafa V, Simeone R, Delogu G, Piacentini M, Brosch R, Fimia GM, Coccia EM. 2012. ESX-1 dependent impairment of autophagic flux by Mycobacterium tuberculosis in human dendritic cells. Autophagy 8:1357–1370.

154. Mishra BB, Moura-Alves P, Sonawane A, Hacohen N, Griffiths G, Moita LF, Anes E. 2010. Mycobacterium tuberculosis protein ESAT-6 is a potent activator of the NLRP3/ASC inflammasome. Cell Microbiol 12:1046–1063.

155. Wong KW, Jacobs WR Jr. 2011. Critical role for NLRP3 in necrotic death triggered by Mycobacterium tuberculosis. Cell Microbiol 13:1371–1384.

156. Dorhoi A, Nouailles G, Jorg S, Hagens K, Heinemann E, Pradl L, Oberbeck-Muller D, Duque-Correa MA, Reece ST, Ruland J, Brosch R, Tschopp J, Gross O, Kaufmann SHE. 2012. Activation of the NLRP3 inflammasome by Mycobacterium tuberculosis is uncoupled from susceptibility to active tuberculosis. Eur J of Immunol 42:374–384.

157. Stanley SA, Johndrow JE, Manzanillo P, Cox JS. 2007. The type I IFN response to infection with Mycobacterium tuberculosis requires ESX-1-mediated secretion and contributes to pathogenesis. J Immunol 178:3143–3152.

158. Ryan AA, Nambiar JK, Wozniak TM, Roediger B, Shklovskaya E, Britton WJ, Fazekas de St Groth B, Triccas JA. 2009. Antigen load governs the differential priming of CD8 T cells in response to the bacille Calmette Guerin vaccine or Mycobacterium tuberculosis infection. J Immunol 182:7172–7177.

159. Abdallah AM, Verboom T, Weerdenburg EM, Gey van Pittius NC, Mahasha PW, Jimenez C, Parra M, Cadieux N, Brennan MJ, Appelmelk BJ, Bitter W. 2009. PPE and PE_PGRS proteins of Mycobacterium marinum are transported via the type VII secretion system ESX-5. Mol Microbiol 73:329–340.

160. Di Luca M, Bottai D, Batoni G, Orgeur M, Aulicino A, Counoupas C, Campa M, Brosch R, Esin S. 2012. The ESX-5 associated eccB-eccC locus is essential for Mycobacterium tuberculosis viability. PLoS One 7:e52059.

161. Deb C, Daniel J, Sirakova TD, Abomoelak B, Dubey VS, Kolattukudy PE. 2006. A novel lipase belonging to the hormone-sensitive lipase family induced under starvation to utilize stored triacylglycerol in Mycobacterium tuberculosis. J Biol Chem 281:3866–3875.

162. Russell DG, Mwandumba HC, Rhoades EE. 2002. Mycobacterium and the coat of many lipids. J Cell Biol 158:421–426.

163. Guilhot C, Chalut C, Daffe M. 2008. Biosynthesis and roles of phenolic glycolipids and related molecules in Mycobacterium tuberculosis, p 273–289. In Daffe M, Reyrat JM (ed), The Mycobacterial Cell Envelope. ASM Press, Washington, DC.

164. Joshi SM, Pandey AK, Capite N, Fortune SM, Rubin EJ, Sassetti CM. 2006. Characterization of mycobacterial virulence genes through genetic interaction mapping. Proc Natl Acad Sci USA 103:11760–11765.

165. Schnappinger D, Ehrt S, Voskuil MI, Liu Y, Mangan JA, Monahan IM, Dolganov G, Efron B, Butcher PD, Nathan C, Schoolnik GK. 2003. Transcriptional adaptation of Mycobacterium tuberculosis within macrophages: insights into the phagosomal environment. J Exp Med 198:693–704.

166. Voskuil MI, Schnappinger D, Visconti KC, Harrell MI, Dolganov GM, Sherman DR, Schoolnik GK. 2003. Inhibition of respiration by nitric oxide induces a Mycobacterium tuberculosis dormancy program. J Exp Med 198:705–713.

167. Kana BD, Gordhan BG, Downing KJ, Sung N, Vostroktunova G, Machowski EE, Tsenova L, Young M, Kaprelyants A, Kaplan G, Mizrahi V. 2008. The resuscitation-promoting factors of Mycobacterium tuberculosis are required for virulence and resuscitation from dormancy but are collectively dispensable for growth in vitro. Mol Microbiol 67:672–684.

168. Zwerling A, Behr MA, Verma A, Brewer TF, Menzies D, Pai M. 2011. The BCG World Atlas: a database of global BCG vaccination policies and practices. PLoS Med 8:e1001012.

169. Asensio JA, Arbues A, Perez E, Gicquel B, Martin C. 2008. Live tuberculosis vaccines based on *phoP* mutants: a step towards clinical trials. *Expert Opin Biol Ther* 8:201–211.

170. Grode L, Seiler P, Baumann S, Hess J, Brinkmann V, Nasser Eddine A, Mann P, Goosmann C, Bandermann S, Smith D, Bancroft GJ, Reyrat JM, van Soolingen D, Raupach B, Kaufmann SH. 2005. Increased vaccine efficacy against tuberculosis of recombinant *Mycobacterium bovis* bacille Calmette-Guerin mutants that secrete listeriolysin. *J Clin Invest* 115:2472–2479.

171. Stover CK, Warrener P, VanDevanter DR, Sherman DR, Arain TM, Langhorne MH, Anderson SW, Towell JA, Yuan Y, McMurray DN, Kreiswirth BN, Barry CE, Baker WR. 2000. A small-molecule nitroimidazopyran drug candidate for the treatment of tuberculosis. *Nature* 405:962–966.

172. Andries K, Verhasselt P, Guillemont J, Gohlmann HW, Neefs JM, Winkler H, Van Gestel J, Timmerman P, Zhu M, Lee E, Williams P, de Chaffoy D, Huitric E, Hoffner S, Cambau E, Truffot-Pernot C, Lounis N, Jarlier V. 2005. A diarylquinoline drug active on the ATP synthase of *Mycobacterium tuberculosis*. *Science* 307:223–227.

173. Lee M, Lee J, Carroll MW, Choi H, Min S, Song T, Via LE, Goldfeder LC, Kang E, Jin B, Park H, Kwak H, Kim H, Jeon HS, Jeong I, Joh JS, Chen RY, Olivier KN, Shaw PA, Follmann D, Song SD, Lee JK, Lee D, Kim CT, Dartois V, Park SK, Cho SN, Barry CE 3rd. 2012. Linezolid for treatment of chronic extensively drug-resistant tuberculosis. *N Engl J Med* 367:1508–1518.

174. Makarov V, Manina G, Mikusova K, Mollmann U, Ryabova O, Saint-Joanis B, Dhar N, Pasca MR, Buroni S, Lucarelli AP, Milano A, De Rossi E, Belanova M, Bobovska A, Dianiskova P, Kordulakova J, Sala C, Fullam E, Schneider P, McKinney JD, Brodin P, Christophe T, Waddell S, Butcher P, Albrethsen J, Rosenkrands I, Brosch R, Nandi V, Bharath S, Gaonkar S, Shandil RK, Balasubramanian V, Balganesh T, Tyagi S, Grosset J, Riccardi G, Cole ST. 2009. Benzothiazinones kill *Mycobacterium tuberculosis* by blocking arabinan synthesis. *Science* 324:801–804.

175. Christophe T, Jackson M, Jeon HK, Fenistein D, Contreras-Dominguez M, Kim J, Genovesio A, Carralot JP, Ewann F, Kim EH, Lee SY, Kang S, Seo MJ, Park EJ, Skovierova H, Pham H, Riccardi G, Nam JY, Marsollier L, Kempf M, Joly-Guillou ML, Oh T, Shin WK, No Z, Nehrbass U, Brosch R, Cole ST, Brodin P. 2009. High content screening identifies decaprenyl-phosphoribose 2′ epimerase as a target for intracellular antimycobacterial inhibitors. *PLoS Pathog* 5:e1000645.

176. Hartkoorn RC, Sala C, Neres J, Pojer F, Magnet S, Mukherjee R, Uplekar S, Boy-Rottger S, Altmann KH, Cole ST. 2012. Towards a new tuberculosis drug: pyridomycin: nature's isoniazid. *EMBO Mol Med* 4:1032–1042.

177. Shi W, Zhang X, Jiang X, Yuan H, Lee JS, Barry CE 3rd, Wang H, Zhang W, Zhang Y. 2011. Pyrazinamide inhibits trans-translation in *Mycobacterium tuberculosis*. *Science* 333:1630–1632.

178. Zumla A, Nahid P, Cole ST. 2013. Advances in the development of new tuberculosis drugs and treatment regimens. *Nat Rev Drug Discov* 12:388–404.

179. Diacon AH, Pym A, Grobusch M, Patientia R, Rustomjee R, Page-Shipp L, Pistorius C, Krause R, Bogoshi M, Churchyard G, Venter A, Allen J, Palomino JC, De Marez T, van Heeswijk RP, Lounis N, Meyvisch P, Verbeeck J, Parys W, de Beule K, Andries K, McNeeley DF. 2009. The diarylquinoline TMC207 for multidrug-resistant tuberculosis. *N Engl J Med* 360:2397–2405.

Molecular Genetics of Mycobacteria, 2nd Edition
Edited by Graham F. Hatfull and William R. Jacobs, Jr.
© 2014 American Society for Microbiology, Washington, DC
doi:10.1128/microbiolspec.MGM2-0028-2013

Vanessa Tran[1]
Jun Liu[1]
Marcel A. Behr[2]

BCG Vaccines

3

HISTORY OF BCG SUBSTRAINS

BCG is named for Albert Calmette and Camille Guérin (bacillus of Calmette and Guérin), who derived the original strain between 1908 and 1921 at the Institut Pasteur of Lille. By serially passaging Nocard's strain of *M. bovis* on potato slices soaked in ox bile and glycerol, they observed a gradual loss of its ability to cause disease in various animal models (1, 2), leading finally to the attempt to use it as a vaccine for the prevention of tuberculosis (TB) in humans in 1921 (reviewed in reference 3).

After Calmette and Guérin proved BCG to be attenuated and safe, cultures of BCG strains were sent out from the Institut Pasteur to different laboratories throughout the world starting from 1924; these cultures were further propagated and used for vaccine production. Lyophilization of vaccine lots became available in the latter half of the 20th century, exposing different BCG daughter strains to some 1,000 additional passages of *in vitro* evolution in different laboratories. This means that BCG does not refer to a cloned strain but rather to a family of different BCG seed strains ("substrains") used in BCG manufacture that have acquired particular genomic modifications during passaging in the various laboratories.

Investigation of the records of BCG transfers between laboratories has enabled the reconstruction of the history of the different BCG daughter strains. Although BCG vaccines have been produced in a wide variety of countries (e.g., Australia, Bulgaria, Mexico, Romania, etc.), for the purpose of genetic studies of BCG strains, a representative subset is generally selected, based on strains used in clinical trials or currently in wide use in vaccination programs. Reviews of the history of these strains have been published by different groups and showed agreement of the proposed provenance of these strains (4, 5). The natural historical division that arises from these genealogic trees is the separation of "early strains" obtained in the early 1920s (BCG Russia, BCG Brazil, BCG Tokyo, BCG Sweden, BCG Birkhaug) from "late strains" obtained from the Pasteur Institute in 1927 or later (outlined in Fig. 1). The distinction between early strains and late strains coincides with reports of ongoing attenuation of BCG in the late 1920s (6, 7) and is correlated with severely reduced production of the antigenic proteins MPB70, MPB83, and MPB64 in late strains (8). However, whether this historical distinction is critical for BCG phenotypes remains uncertain, because certain properties, such as production of virulence lipids, have

---

[1]Department of Molecular Genetics, University of Toronto, Toronto, Ontario, Canada M5S 1A8; [2]McGill International TB Center, Montreal, Quebec, Canada, H3G 1A4.

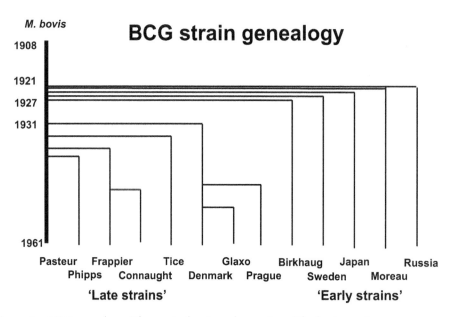

Figure 1 BCG genealogy. The vertical axis scales to time. The horizontal represents movement of vaccines between labs. Strains obtained before 1927 are labeled as "early strains"; strains obtained in 1931 or later are indicated as "late strains."
doi:10.1128/microbiolspec.MGM2-0028-2013.f1

been independently lost in both early and late strains. Moreover, the relevance of an early versus late dichotomy in terms of protective efficacy in humans cannot be ascertained, because randomized clinical trials of BCG vaccination only employed late strains.

Since 1921, an estimated 4 billion doses of BCG have been administered worldwide with remarkably low incidence of major side effects in immune-competent individuals. One survey conducted in France estimated that the rate of BCG-osis is on the order of 1 per million vaccinees, although allowing for under-diagnosis and/or under-reporting, it is possible that the risk is somewhat higher (9). Still, this attack rate remains orders of magnitude lower than that of virulent *M. tuberculosis*, where the widely reported risk of progression from documented infection to disease is about 10% over one's lifetime. Therefore, it is worth reflecting that despite great similarities in genomes between virulent *M. tuberculosis* and BCG, the genomic differences translate into a 4- to 5-log lower risk of disease. These epidemiologic data, compiled over decades in countries worldwide, provide a compelling argument for using BCG as an entrée to study attenuation of virulence during laboratory passage.

In contrast to the indisputable evidence of attenuation, the data on protective efficacy are somewhat more controversial. BCG vaccination is felt to be effective against disseminated forms of TB in children (80%), such as miliary TB and meningitis (10, 11), although the data supporting this protection are derived mostly from observational and case-control studies. In randomized controlled trials, BCG vaccination has provided as much as 80% protection in several studies (12, 13) but fared no better than placebo in others (14, 15), leading to a policy stalemate, where findings from either pole of the efficacy spectrum have been used to justify national immunization guidelines. Adding further confusion, BCG vaccination has also been shown to protect against leprosy in a controlled trial where it did not protect against TB (16). A wide range of BCG schedules have emerged (17), tabulated at the BCG Atlas (http://www.bcgatlas.org/). In countries with low or declining rates of TB, BCG vaccination has either never been used or has been stopped in recent decades. In most other parts of the world, routine BCG vaccination is applied either to all subjects or only in high-risk communities, with schedules that can be divided into three groups: BCG administration only at birth, BCG once in childhood, or repeated/booster BCG. It has been estimated that over 100 million doses of BCG are administered each year (18).

Although vaccines more effective than BCG are needed, BCG vaccination will probably continue for at least the next decade on a worldwide scale until controlled long-term vaccine trials prove that novel anti-TB vaccines offer better protection. Moreover, of novel vaccines in development, candidates that have performed better than BCG in animal models have been modified forms of BCG vaccines (19–21). As such, molecular and immunological characterization of the currently used BCG vaccine strains continues to be an important research topic, not only to elucidate the genetic basis for the attenuation of BCG, but also to learn more about the genetic background of diverse BCG substrains that may serve as the backbone of new vaccines. Recent advances in the field of mycobacterial genetics, comparative genomics, and related techniques have supplied an enormous amount of new information, which will be summarized and discussed in this chapter.

BCG AS AN ATTENUATED MEMBER OF THE *M. TUBERCULOSIS* COMPLEX

M. bovis BCG emerged as an attenuated mutant of *M. bovis*. Thus, comparisons between *M. tuberculosis* and BCG will naturally present two strata of differences: *M. tuberculosis* versus *M. bovis* and *M. bovis* versus BCG. While it is tempting to disregard other naturally occurring variants of the tubercle bacillus, some surprising observations have been made upon genomic study of organisms such as *Mycobacterium microti* (Voles bacillus), *Mycobacterium mungi* (TB in mongooses) and the dassie bacillus (TB in hyraxes). In common with BCG vaccines, these other three subspecies of the *M. tuberculosis* complex are deletion mutants in the region of difference 1 (RD1) locus of the genome (22–24), and consequently each of these organisms lacks a functional ESX-1 secretion system (discussed in greater detail in reference 73). One interpretation of this observation is that the lack of ESX-1 in these organisms explains why they are apparently attenuated for humans. Indeed, we rarely see human TB due to these organisms, and *M. microti* has even been used, with an excellent safety record, in field trials of TB vaccination (13). A further inference is that these organisms are able to cause transmissible disease in their natural hosts without ESX-1. This suggests that either alternative virulence mechanisms have emerged to make up for this defect or that permissive host factors have enabled the propagation of attenuated variants of the tubercle bacillus. In either case, these natural experiments serve to remind us that *M. tuberculosis* virulence depends on more than the presence of ESX-1, such that this locus is perhaps best deemed necessary but not sufficient for full virulence.

Setting aside other members of the *M. tuberculosis* complex, postgenomic study has affirmed that BCG did indeed derive from *M. bovis*, itself comprising at least four different genotypes (25). Since the original strain

of *M. bovis* used to derive BCG was lost during the First World War, genomic studies have compared BCG vaccines to a variety of different *M. bovis* strains until the sequencing of the AF2122/97 strain was completed in 2003 (26). As a consequence, it must be remembered that certain differences uncovered between a particular BCG strain and the chosen *M. bovis* strain could represent *M. bovis*–*M. bovis* variants rather than BCG-specific differences. Fortunately, the existence of a variety of contemporary BCG strains provides opportunities to logically deduce the genetics of the original BCG. For instance, BCG strains obtained before 1926 (BCG Russia, BCG Moreau, and BCG Tokyo) all have an IS*6110* element at base pair 851,592 of the H37Rv genome, unlike other BCG strains, where the sequence at this locus is identical to that of *M. tuberculosis* and *M. bovis* (27). One reason why it is much more likely that the original BCG strain contained this second IS element is that the alternative (and unlikely) event would be that, on three different continents, a transpositional duplication of IS*6110* occurred at precisely the same locus.

Therefore, based on the availability of a number of different BCG strains, and the genomic sequences of *M. tuberculosis* H37Rv and *M. bovis* AF2122/97, it is now possible to describe the genomic evolution of BCG strains in various vaccine laboratories. In the section below we will document what is currently known from such studies about the genome of BCG strains, recognizing that the overriding challenge is not to find genetic differences between strains, but rather to use these data to answer biologic questions, such as (i) why are BCG vaccines attenuated?, (ii) are some strains overattenuated?, and (ii) what are the properties of BCG that offer protection against TB?

MOLECULAR EVOLUTION AND THE DERIVATION OF BCG SUBSTRAINS

As described above, BCG is not a clonal organism, and several substrains exist that may have accumulated individual genetic modifications due to long-term passaging in different laboratories and vaccine production facilities. Using the historical records about when and where BCG strains were distributed, together with information on the genetic particularities of certain BCG substrains, the short-term evolution of BCG strains can be reconstructed. By employing techniques of comparative genome analyses, such as subtractive hybridization, BAC libraries, microarrays, and GeneChips, it was shown that several BCG daughter strains indeed have accumulated additional genomic modifications that apparently were not present in the original culture obtained by Calmette and Guérin (27–31). With the huge amount of genomic data that have been generated in the last few years, including NimbleGen resequencing arrays and whole-genome sequencing, the previously reported genealogy of BCG strains has been retained, with smaller polymorphisms uncovered, specific to each strain. At last count, there are 27 BCG genome sequences deposited at NCBI (search performed July 2013), indicating that all of the most commonly used strains of BCG have been subject to whole-genome sequencing.

Based on the available data, it is now clear which genetic particularities are shared across all BCG strains relative to other *M. bovis* strains and which genetic variations are specific for only certain BCG substrains. Whereas the former may be directly involved in the attenuation of BCG, the latter may account for variation in protective efficacy and/or overattenuation of certain BCG substrains. For the genomic particularities that apply to all BCG substrains and are shown to be implicated in virulence, the example of the RD1 region will be discussed separately below. The variations that were observed for only certain substrains of BCG consist of deletions, duplications, and point mutations. Of note, deletions were discovered about one decade before most point mutations, simply because it is easier to find large chunks of missing DNA than single base-pair substitutions. As a result, there has been more time to pursue functional studies of the deletions than of single nucleotide polymorphisms (SNPs), with an ensuing bias in the kinds of mutations that have been linked to phenotypic consequences.

Among the genomic deletions specific to BCG vaccines, without question the one that has attracted the most attention is the RD1, first identified by subtractive hybridization between BCG and wild-type *M. bovis* (28). When preliminary complementation experiments of BCG with the RD1 from *M. bovis* did not result in increased virulence of the recombinant BCG strain in mice (28), the hypothesis that RD1 could be involved in the attenuation of BCG remained unconfirmed for several years. Yet several years later, RD1 remained the only genomic deletion specific to BCG (29, 30), prompting three independent teams of investigators to revisit the potential role of RD1 loss in the attenuation of BCG. In one study, BCG Pasteur was complemented with an integrative cosmid clone containing the RD1, and the recombinant BCG strain was more virulent in severe combined immune deficient (SCID) mice than the BCG-vector control strain (22). In the other two studies, RD1 was disrupted from *M. tuberculosis*, with both groups reporting decreased virulence of *M. tuberculosis* ΔRD1 compared to the reference strain of *M. tuberculosis* (32, 33). Together, these data implicated RD1 deletion as a key mechanism of BCG attenuation; however, they did not exclude other mutations as also contributing to the loss of virulence described by Calmette and Guérin between 1908 and 1921.

In addition to RD1, several other genomic polymorphisms that are common to all BCG strains have been investigated, including several point mutations that may have contributed to the early evolution of BCG. One example is a SNP in *pykA*, which encodes pyruvate kinase, an enzyme that catalyzes the final step in glycolysis. This point mutation results in an aspartic acid substitution at glutamic acid 220 and renders pyruvate kinase nonfunctional. Interestingly, this SNP is present in *M. bovis* but absent in BCG and was suggested to account for the inability of *M. bovis* to grow on glycerol as a sole carbon source (34). Since all the BCG strains that were tested exhibited functional pyruvate kinase enzyme activity, it was suggested that the original growth conditions (i.e., potato slices soaked in ox bile and glycerol) used in the *in vitro* passaging of *M. bovis* by Calmette and Guérin had selected for a strain of *M. bovis* capable of utilizing glycerol.

Another study implicated a point mutation in the gene that encodes the cyclic AMP (cAMP) receptor protein (CRP) (*Rv3676*) in BCG that could affect the DNA binding activity of this putative global transcriptional regulator and thus contribute to BCG attenuation (35). Follow-up studies using electrophoretic mobility shift assays (EMSA), chromatin immunoprecipitation (ChIP), and DNA microarrays coupled with quantitative reverse transcription PCR (qRT-PCR) showed that the point mutation resulting in a glutamic acid-to-lysine substitution (E178K) present in all BCG strains tested increased DNA binding of CRP to target sites and increased expression of targets (36, 37). However, it was shown that CRP from BCG was able to rescue the growth defect of a CRP mutant of *M. tuberculosis* in macrophages and BALB/c mice, suggesting that this SNP does not necessarily contribute to BCG attenuation. Rather, this SNP may have merely conferred a survival advantage for BCG under the *in vitro* conditions used in the initial derivation by Calmette and Guérin (37).

A recent comprehensive study of SNPs between *M. bovis* and BCG strains uncovered additional point mutations that could also play a role in the early evolution of BCG (38). Two follow-up studies examined these SNPs in BCG. The first examined a SNP found in *BCG3145*, an ortholog of *Rv3124* in *M. tuberculosis*.

BCG3145/Rv3124 are members of the AfsR/DnrI/SARP (*Streptomyces* antibiotic regulatory protein) class of transcriptional regulators and, by microarray and qRT-PCR analysis, were found to positively regulate the *mao1* locus required for molybdopterin biosynthesis (39). Interestingly, the E159G (glutamic acid-to-glycine substitution) mutation in BCG3145 caused a reduction in activity but did not abolish induction of the *moa1* locus. Given that *Rv3124* has not been identified as a virulence factor for *M. tuberculosis*, these authors suggested that this mutation is likely not a key attenuating genetic lesion in BCG, but this has not been formally tested. The second study examined a nonsynonymous SNP in *cycA*, which encodes a D-serine/D,L-alanine/glycine/D-cycloserine transporter. This study showed that the G122S (glycine-to-serine substitution) in *cycA* was responsible for the D-cycloserine resistance seen in BCG strains. Interestingly, expression of a functional copy of *cycA* from *M. tuberculosis* or *M. bovis* increased susceptibility of BCG Pasteur to D-cycloserine, albeit not to the levels of *M. tuberculosis* or *M. bovis*, suggesting that other genetic lesions may contribute to BCG resistance to D-cycloserine (40). Whether this mutation has a functional consequence on BCG virulence remains unknown. Although these SNPs have not been directly implicated in the original attenuation of BCG, it remains possible that the accumulation of multiple SNPs between 1908 and 1921 may have resulted in a compound effect that is difficult to formally prove by piecemeal study.

MOLECULAR EVOLUTION AND THE PROPAGATION OF BCG SUBSTRAINS

Subsequent to the original derivation of BCG, strains were passaged in nonsynthetic media, based on variable growth factors such as the local potatoes obtained at one's farmer's market. Not surprisingly, these laboratory conditions were not uniform, and it was perhaps not surprising that BCG laboratories noticed phenotypic differences between their strains in the 1940s and 1950s. With genomic study, it is now known that different BCG strains differ both from the original BCG of 1921 and from each other, due to deletions, SNPs, and duplications. Thus, there is no "ancestral" BCG in existence. BCG Russia may look more ancestral than BCG Pasteur, because the latter is separated from BCG 1921 by a number of genetic events (e.g., deletion of IS6110, RD2, nRD18, and RD14; SNPs in *mmaA3*, *sigK*, and CRP; duplication of DU1) (Fig. 2). However, BCG Russia is expected to harbor a number of mutations compared to BCG 1921 that are not observed in BCG

Pasteur. Evidence to suggest a closer relationship between BCG Russia and the original BCG progenitor strain has emerged from the discovery that BCG Russia is a natural *recA* mutant (41). This mutation was suggested to have prevented BCG Russia from undergoing further genetic evolution, but whether this occurred soon after the strain arrived in Russia or decades later cannot be determined.

Since the deletion of RD1 coincided with and contributed to the attenuation of virulence, it has been tempting to hypothesize that other deletions have contributed to further attenuation. Of genes contained in BCG deletions, a number stand out as candidates for the ongoing attenuation of BCG strains in the laboratory. The RD2 region, encompassing Rv1978 to Rv1988, has been disrupted from *M. tuberculosis*, and the deletion mutant was shown to be attenuated in murine models of infection (42). Moreover, the RD2 region contains the gene *mpt64*, coding for the antigenic protein MPT64; introduction of *mpt64* into BCG Pasteur improved the immunogenicity of the vaccine strain without providing a protection benefit (43). Rv3405c, annotated as a possible transcriptional regulator, has been disrupted by independent deletions in two different BCG strains (see below). Two other predicted regulators have been disrupted during serial passage of BCG strains. Rv1773, also annotated as a probable transcriptional regulatory protein, has been deleted from the BCG Pasteur strain alone (29); its regulatory role was subsequently confirmed (44). Finally, Rv1189, otherwise known as *sigI* (possible alternative RNA polymerase sigma factor SigI), is missing from all BCG strains obtained after 1933. The loss of regulatory genes from a number of different BCG strains argues that mutation of regulatory genes can be tolerated during the monotonous conditions of laboratory growth.

Three BCG strains (BCG Japan, BCG Glaxo, and BCG Moreau) were found to be naturally deficient in lipid virulence factors, phthiocerol dimycocerosates (PDIMs), and phenolic glycolipids (PGLs). The loss of these lipids correlates with the superior safety records of these strains in clinical studies (45, 46). Furthermore, deletion of PDIMs/PGLs from BCG Pasteur reduces its virulence in SCID mice (V. Tran et al., unpublished data). Genomic analysis revealed that the loss of PDIMs/PGLs in BCG Moreau is due to a 975-bp deletion that affects *fadD26* and *ppsA*, members of the PDIM/PGL biosynthetic operon (47). Similarly, a point mutation in *ppsA* is responsible for the lack of PDIMs/PGLs in BCG Japan (48). The mutation in BCG Glaxo has yet to be discovered. Together, this suggests that these three BCG strains independently acquired

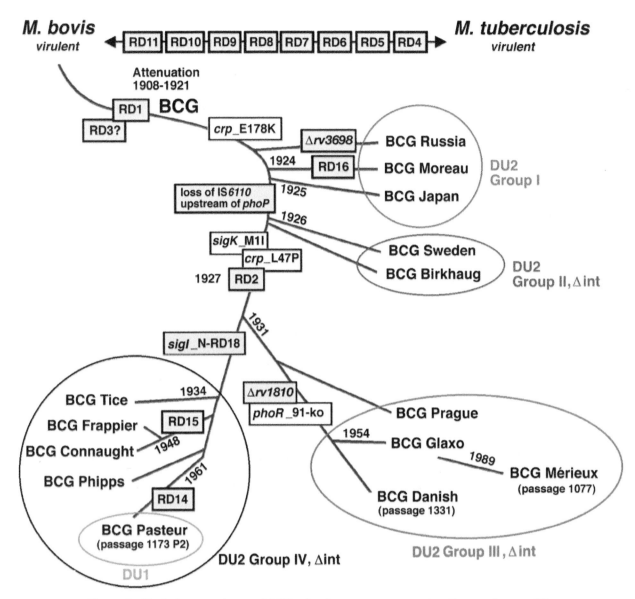

Figure 2 Evolutionary scheme of BCG after its genome sequencing (from reference 58), indicating that *M. bovis* BCG has undergone deletions, duplications, and SNPs since its divergence from *M. bovis*. doi:10.1128/microbiolspec.MGM2-0028-2013.f2

mutations in PDIM/PGL biosynthesis that contributed to their further attenuation. Several BCG strains harbor mutations in the *phoP-phoR* locus, a two-component system known to regulate the expression of multiple genes, including some well-established T-cell antigens (49). Of note, a frameshift mutation present in *phoP* of BCG Prague eliminates the C-terminal binding domain and makes this strain a natural *phoP* mutant (47). It is possible that this could account for this strain's reported low immunogenicity (50, 51). In addition, as mentioned above, three early BCG strains have an IS*6110* insertion in the *phoP* promoter, which could

eliminate the auto-repression regulatory mechanism of this two-component system (52), and five BCG strains contain mutations in the *phoR* gene that may abrogate its function. Unlike other BCG strains, BCG Sweden and BCG Birkhaug also contain mutations in *whiB3*, a transcriptional regulator implicated in virulence of *M. bovis* (53) and *M. tuberculosis* (54), and in *trcR*, the response regulator of the TrcR-TrcS two-component system that controls expression of the *trcRS* operon, including *mmpS5/mmpL5* transporter and *bfrB* bacterioferritin (55). Interestingly, deletion of *trcS* from *M. tuberculosis* produced a hypervirulent phenotype in

SCID mice (56), which suggests a role for this gene in the attenuation of BCG.

Comparative genomics has also uncovered two large tandem duplications of 29 and 36 kb (DU1, DU2) in BCG Pasteur (57, 58). These seem to have arisen independently, as their presence and/or their size varies between the different BCG substrains. While DU1 appears to be restricted to BCG Pasteur, DU2 has been detected in all BCG substrains tested so far (58). Interestingly, DU1 comprises the *oriC* locus, indicating that BCG Pasteur is diploid for *oriC*, and several genes involved in replication (57). For DU2, we know that the tandem duplication resulted in diploidy for 30 genes, including *sigH*, a sigma factor implicated in the heat shock response (59). Gene duplications are a common evolutionary response in bacteria exposed to different selection pressures in the laboratory and presumably in nature, because they provide a means for increasing gene dosage and for generating novel functions from potential gene fusion events at duplication endpoints, and they represent a source of redundant DNA for divergence. As such, the duplication events seen in BCG Pasteur may reflect a common adaptation mechanism of mycobacteria to cope with environmental stress, and they could influence the immunogenicity of a certain vaccine strain. Other duplications specific to certain BCG strains have also been uncovered, such as a 22-kb duplication present only in BCG Tice (DU-Tice) (47). This duplication encompasses *Rv1782-Rv1800* that encodes the ESX-5 secretion system, which is present only in pathogenic mycobacteria and has been shown to facilitate the cell-to-cell spread of *Mycobacterium marinum* in infected macrophages, a function shared by ESX-1 (60).

The search for BCG-associated SNPs has occurred in two waves: phenotype to genotype and, more recently, from whole-genome sequencing. Two examples of phenotype-driven searches involve the variable production of lipids by BCG strains. In the first example, it had been known since the 1980s that production of mycolic acid subsets varied across BCG strains (61). Since the gene *mmaA3* had been shown to be involved in production of methoxymycolic acids (62), it was hypothesized that mutations in this gene might explain this variable phenotype. Indeed, methoxymycolic acid production is restricted to "early strains," and "late strains" all have a guanine-to-adenine replacement at base pair 293 of the *mmaA3* gene (63). By introducing this mutant form of *mmaA3* into *M. smegmatis*, it was possible to verify that ablation of methoxymycolate synthesis occurs with this mutation. Additionally, in a follow-up study, the wild-type sequence of *mmaA3* was complemented into two late strains, BCG Danish and BCG Pasteur, with restoration of methoxymycolate synthesis (64). The second example of such a phenotype-to-genotype search was provided above, in the case of PDIMs and PGLs.

In contrast, the higher throughput means of identifying BCG-specific mutations involves whole-genome sequencing. Prior to the complete closure and annotation of BCG Pasteur in 2007 (58), the shotgun reads available on the Internet helped frame the search for a genetic basis to a transcriptome difference that separated early from late strains. During *in vitro* growth, it was noted that the genes *mpt70* and *mpt83* were differentially expressed in early strains, consistent with the previously published difference in expression of the antigenic proteins that they encode (MPT70, MPT83). The same strains with high expression of these genes had overexpression of a set of genes from *Rv0449c* to *Rv0444c*, a run of genes that includes the gene *sigK*, coding for the extracytosolic sigma factor SigK. By alignment of the *sigK* sequences, it was seen that BCG Pasteur had a start codon polymorphism, prompting targeted sequencing of *sigK* across BCG strains. Because this mutation correlated with the strains obtained in 1931 or later, the wild-type *sigK* was introduced into BCG Pasteur, restoring expression of *mpt70/mpt83* and production of MPT70/MPT83 (65). While the restoration of the wild-type expression profile led to clear differences in immunogenicity of BCG Pasteur, as measured by MPT70-specific interferon-γ production by vaccinated animals, there was no benefit in terms of protective efficacy, in both guinea pigs and calves (M. A. Behr, unpublished data). These findings once again highlight how the documentation of genetic changes among BCG strains has far outstripped our capacity to define the basis of protective efficacy.

Following the sequencing of BCG Pasteur in 2007 and publication of the BCG Pasteur genome, the sequencing of BCG Tokyo was determined in 2009, aiding in the search for further differences between early and late BCG strains (66). The sequencing of four additional BCG strains (BCG Russia, BCG Tice, BCG Denmark 1331, and BCG China) uncovered more than 360 SNPs between these BCG strains compared to *M. bovis* (67). Of these, ~170 SNPs were unique to BCG, and some were exclusive to specific strains, which may have contributed to the evolution of each strain. Whole-genome sequencing of BCG Moreau confirmed the presence of several genetic aberrations including DU2-1, *fadD26-ppsA* 976 bp deletion, RD16 deletion, and in-frame deletion in the *rv3887c* (*eccD2*) gene (68). Interestingly, sequencing of BCG Mexico

1931 repositioned this strain from the DU2-III group into the DU2-IV group that includes BCG Pasteur, BCG Connaught, BCG Tice, BCG Frappier, and BCG Phipps. This analysis revealed that BCG Mexico was not originally derived from BCG Danish, as previously thought, and that this inconsistency was likely attributed to different BCG manufacturing periods in Mexico (69). Additionally, this project uncovered novel deletions in BCG Mexico that shed light on its evolution. Two new RDs unique to BCG Mexico 1931 were discovered and named RDMex02 and RDMex03. RDMex02 encompasses a deletion in *fadD23*, a gene that has been implicated in sulfolipid production and association with macrophages (70). RDMex03 was the largest deletion found in BCG Mexico 1931 and spanned four genes, one of which encodes the transcriptional regulator *whiB6*. It was also shown that BCG Mexico 1931 contains four unique SNPs, one of which occurs in *ponA2*, a gene that has been shown to be involved in sensitivity to heat shock and H_2O_2 and survival of *M. tuberculosis* in mice (71). Finally, whole-genome sequences for BCG Moreau (68) and BCG Korea (72) have also been reported, but comparative genomics studies to uncover novel mutations in these strains have yet to be completed.

With a number of whole-genome sequences of BCG strains now available, it is possible to begin a more in-depth analysis of genetic polymorphisms that affect phenotypic traits of BCG, such as immunogenicity and virulence. A detailed whole-genome sequence comparison of the available BCG strains would shed light on the specific differences among BCG strains and further our understanding of the mechanisms of attenuation specific to each lineage. This would be useful in further delineating BCG strains into phenotypic/genetic categories that could mediate its protective efficacy.

CONCLUSION AND PERSPECTIVES: ATTENUATION OF VIRULENCE AND PROTECTIVE EFFICACY

While it is tempting to focus on the differences across BCG strains, there are certain unifying messages that emerge from the study of BCG. First, in spite of its multibillion-fold application, none of the BCG substrains have ever reverted to the virulent form of *M. bovis*. This finding suggests that during the attenuation process of BCG, multiple irreversible genetic changes have occurred that have permanently disabled BCG's ability to cause disease in the immunocompetent host. This prediction has to a large degree

been confirmed by genomic studies, suggesting that these vaccines can be given safely and can serve as the basis of new, engineered vaccines, whereas the alternative involves the *de novo* generation of new, attenuated strains of *M. tuberculosis* for which there is no track record of safety. The key pending question is not whether BCG vaccines are attenuated, but rather, whether certain strains are overattenuated and whether this would affect their ability to protect against TB.

Another message one can derive from the collective study of BCG strains is that unlike structural molecules (peptidoglycan, mycolic acids, etc.), secreted products (e.g., antigenic proteins, virulence lipids) are apparently dispensable under the conditions encountered during laboratory culture. It can be reasoned that these molecules play a particularly important role when the bacterium interacts with the host, whereas there is no selective pressure to maintain their production in the lab. In the case of antigens, BCG strains obtained after 1931 do not make ESAT-6, CFP-10, MPT64, MPT70, or MPB83; yet, despite the absence of five of the six immunodominant antigens of *M. bovis*, some of these BCG strains have provided protection against TB in multiple trials and case-control studies. A potential message is that the induction of antigen-specific immunity, which does occur with BCG vaccination, is not on the causal pathway of protection. Indeed, BCG has long been used in the treatment of bladder cancer, and a recent study found that BCG protects SCID mice against *Candida albicans* infection (74). These observations suggest (i) that the key biologic effects of BCG occur in inducing innate, rather than adaptive, immune responses, and (ii) that the immunologic effects are nonspecific. In the case of lipids, there is much more to be learned. BCG strains show differential production of triacylglycerols, methoxymycolic acids, and PDIMs/PGLs. Whether reduction or loss of triacylglycerols and methoxymycolates affects protection is unknown. In contrast, emerging data indicate that the presence of PDIMs/PGLs may be important for BCG protection.

A challenge for future research on BCG will be to capitalize on the advances in descriptive and experimental genetics to determine how BCG vaccines differentially engage our immune system. While the historic literature on BCG protection revealed variability, and the contemporary literature on BCG genomics adds further complexity, we should not lose sight of the fact that BCG has provided proof-of-concept that TB is a vaccine-preventable illness. The secret to a TB vaccine is encoded in the genome of BCG.

Citation. Tran V, Liu J, Behr MA. 2014. BCG vaccines. Microbiol Spectrum 2(1):MGM2-0028-2013.

References

1. Calmette A, Guérin C. 1911. Recherches experimentales sur la defense de l'organisme contre l'infection tuberculeuse. *Ann Inst Pasteur* **25**:625–641.

2. Calmette A, Guérin C. 1920. Nouvelles recherches experimentales sur la vaccination des bovides contre la tuberculose. *Ann Inst Pasteur* **34**:553–560.

3. Guérin C, Rosenthal SR. 1957. The history of BCG: early history, p 48–57. *In* Rosenthal SR (ed), *BCG Vaccination Against Tuberculosis.* J&H Churchill, London, United Kingdom.

4. Behr MA, Small PM. 1999. A historical and molecular phylogeny of BCG strains. *Vaccine* **17**:915–922.

5. Oettinger T, Jorgensen M, Ladefoged A, Haslov K, Andersen P. 1999. Development of the *Mycobacterium bovis* BCG vaccine: review of the historical and biochemical evidence for a genealogical tree. *Tuber Lung Dis* **79**:243–250.

6. Dreyer G, Vollum RL. 1931. Mutation and pathogenicity experiments with BCG. *Lancet* **1**:9–14.

7. Streng KO. 1940. Etude des caracteres d'atennuation du bacille BCG suivant le nombre de passages de ce germe sur pomme de terre a la bile de boeuf. *Ann Inst Pasteur* **64**:196–202.

8. Wiker HG, Nagai S, Hewinson RG, Russell WP, Harboe M. 1996. Heterogenous expression of the related MPB70 and MPB83 proteins distinguish various substrains of *Mycobacterium bovis* BCG and *Mycobacterium tuberculosis* H37Rv. *Scand J Immunol* **43**:374–380.

9. Casanova J, Jouanguy E, Lamhamedi S, Blanche S, Fischer A. 1995. Immunological conditions of children with BCG disseminated infection. *Lancet* **346**:581.

10. Colditz GA, Berkley CS, Mosteller F, Brewer TF, Wilson ME, Burdick E, Fineberg HV. 1995. The efficacy of bacillus Calmette-Guerin vaccination of newborns and infants in the prevention of tuberculosis: meta-analysis of the published literature. *Pediatrics* **96**:29–35.

11. Trunz BB, Fine P, Dye C. 2006. Effect of BCG vaccination on childhood tuberculous meningitis and miliary tuberculosis worldwide: a meta-analysis and assessment of cost-effectiveness. *Lancet* **367**:1173–1180.

12. Ferguson RG, Simes AB. 1949. BCG vaccination of Indian infants in Saskatchewan. *Tubercle* **30**:5–11.

13. Hart PD, Sutherland I. 1977. BCG and vole bacillus vaccines in the prevention of tuberculosis in adolescence and early adult life. Final report to the Medical Research Council. *Br Med J* **ii**:293–295.

14. Comstock GW, Palmer CE. 1966. Long-term results of BCG vaccination in the southern United States. *Am Rev Respir Dis* **93**:171–183.

15. Tuberculosis Prevention Trial. 1980. Trial of BCG vaccines in South India for tuberculosis prevention. *Indian J Med Res* **72S**:1–74.

16. Ponnighaus JM, Fine PE, Sterne JA, Wilson RJ, Msosa E, Gruer PJ, Jenkins PA, Lucas SB, Liomba NG, Bliss L. 1992. Efficacy of BCG vaccine against leprosy and tuberculosis in northern Malawi. *Lancet* **339**:636–639.

17. Zwerling A, Behr MA, Verma A, Brewer TF, Menzies D, Pai M. 2011. The BCG World Atlas: a database of global BCG vaccination policies and practices. *PLoS Med* **8**: e1001012.

18. World Health Organization. 2004. BCG vaccine. WHO position paper. *Wkly Epidemiol Rec* **79**:27–38.

19. Horwitz MA, Harth G, Dillon BJ, Maslesa-Galić S. 2000. Recombinant bacillus Calmette-Guerin (BCG) vaccines expressing the *Mycobacterium tuberculosis* 30-kDa major secretory protein induce greater protective immunity against tuberculosis than conventional BCG vaccines in a highly susceptible animal model. *Proc Natl Acad Sci USA* **97**:13853–13858.

20. Pym AS, Brodin P, Majlessi L, Brosch R, Demangel C, Williams A, Griffiths KE, Marchal G, Leclerc C, Cole ST. 2003. Recombinant BCG exporting ESAT-6 confers enhanced protection against tuberculosis. *Nat Med* **9**: 533–539.

21. Grode L, Seiler P, Baumann S, Hess J, Brinkmann V, Eddine AN, Mann P, Goosmann C, Bandermann S, Smith D, Bancroft GJ, Reyrat JM, van Soolingen D, Raupach B, Kaufmann SH. 2005. Increased vaccine efficacy against tuberculosis of recombinant *Mycobacterium bovis* bacille Calmette-Guerin mutants that secrete listeriolysin. *J Clin Invest* **115**:2472–2479.

22. Pym AS, Brodin P, Brosch R, Huerre M, Cole ST. 2002. Loss of RD1 contributed to the attenuation of the live tuberculosis vaccines *Mycobacterium bovis* BCG and *Mycobacterium microti*. *Mol Microbiol* **46**:709–717.

23. Mostowy S, Cousins D, Behr MA. 2004. Genomic interrogation of the dassie bacillus reveals it as a unique RD1 mutant within the *Mycobacterium tuberculosis* complex. *J Bacteriol* **186**:104–109.

24. Alexander KA, Laver PN, Michel AL, Williams M, van Helden PD, Warren RM, Gey van Pittius NC. 2010. Novel *Mycobacterium tuberculosis* complex pathogen, *M. mungi*. *Emerg Infect Dis* **16**:1296–1299.

25. Mostowy S, Inwald J, Gordon S, Martin C, Warren R, Kremer K, Cousins D, Behr MA. 2005. Revisiting the evolution of *Mycobacterium bovis*. *J Bacteriol* **187**: 6386–6395.

26. Garnier T, Eiglmeier K, Camus JC, Medina N, Mansoor H, Pryor M, Duthoy S, Grondin S, Lacroix C, Monsempe C, Simon S, Harris B, Atkin R, Doggett J, Mayes R, Keating L, Wheeler PR, Parkhill J, Barrell BG, Cole ST, Gordon SV, Hewinson RG. 2003. The complete genome sequence of *Mycobacterium bovis*. *Proc Natl Acad Sci USA* **100**:7877–7882.

27. Mostowy S, Tsolaki AG, Small PM, Behr MA. 2003. The *in vitro* evolution of BCG vaccines. *Vaccine* **21**: 4270–4274.

28. Mahairas GG, Sabo PJ, Hickey MJ, Singh DC, Stover CK. 1996. Molecular analysis of genetic differences between *Mycobacterium bovis* BCG and virulent *M. bovis*. *J Bacteriol* **178**:1274–1282.

29. Behr MA, Wilson MA, Gill WP, Salamon H, Schoolnik GK, Rane S, Small PM. 1999. Comparative genomics of

BCG vaccines by whole-genome DNA microarray. *Science* 284:1520–1523.

30. Gordon SV, Brosch R, Billault A, Garnier T, Eiglmeier K, Cole ST. 1999. Identification of variable regions in the genomes of tubercle bacilli using bacterial artificial chromosome arrays. *Mol Microbiol* 32:643–655.

31. Salamon H, Kato-Maeda M, Small PM, Drenkow J, Gingeras TR. 2000. Detection of deleted genomic DNA using a semiautomated computational analysis of GeneChip data. *Genome Res* 10:2044–2054.

32. Lewis KN, Liao RL, Guinn KM, Hickey MJ, Smith S, Behr MA, Sherman DR. 2003. Deletion of RD1 from *Mycobacterium tuberculosis* mimics bacille Calmette-Guerin attenuation. *J Infect Dis* 187:117–123.

33. Hsu T, Hingley-Wilson SM, Chen B, Chen M, Dai AZ, Morin PM, Marks CB, Padiyar J, Goulding C, Gingery M, Eisenberg D, Russell RG, Derrick SC, Collins FM, Morris SL, King CH, Jacobs WR Jr. 2003. The primary mechanism of attenuation of bacillus Calmette-Guerin is a loss of secreted lytic function required for invasion of lung interstitial tissue. *Proc Natl Acad Sci USA* 100:12420–12425.

34. Keating LA, Wheeler PR, Mansoor H, Inwald JK, Dale J, Hewinson RG, Gordon SV. 2005. The pyruvate requirement of some members of the *Mycobacterium tuberculosis* complex is due to an inactive pyruvate kinase: implications for *in vivo* growth. *Mol Microbiol* 56:163–174.

35. Spreadbury CL, Pallen MJ, Overton T, Behr MA, Mostowy S, Spiro S, Busby SJ, Cole JA. 2005. Point mutations in the DNA- and cNMP-binding domains of the homologue of the cAMP receptor protein (CRP) in *Mycobacterium bovis* BCG: implications for the inactivation of a global regulator and strain attenuation. *Microbiology* 151:547–556.

36. Bai G, Gazdik MA, Schaak DD, McDonough KA. 2007. The *Mycobacterium bovis* BCG cyclic AMP receptor-like protein is a functional DNA binding protein *in vitro* and *in vivo*, but its activity differs from that of its *M. tuberculosis* ortholog, Rv3676. *Infect Immun* 75:5509–5517.

37. Hunt DM, Saldanha JW, Brennan JF, Benjamin P, Strom M, Cole JA, Spreadbury CL, Buxton RS. 2008. Single nucleotide polymorphisms that cause structural changes in the cyclic AMP receptor protein transcriptional regulator of the tuberculosis vaccine strain *Mycobacterium bovis* BCG alter global gene expression without attenuating growth. *Infect Immun* 76:2227–2234.

38. Garcia Pelayo MC, Uplekar S, Keniry A, Mendoza LP, Garnier T, Nunez GJ, Boschiroli L, Zhou X, Parkhill J, Smith N, Hewinson RG, Cole ST, Gordon SV. 2009. A comprehensive survey of single nucleotide polymorphisms (SNPs) across *Mycobacterium bovis* strains and *M. bovis* BCG vaccine strains refines the genealogy and defines a minimal set of SNPs that separate virulent *M. bovis* strains and *M. bovis* BCG strains. *Infect Immun* 77:2230–2238.

39. Mendoza LP, Golby P, Wooff E, Nunez GJ, Garcia Pelayo MC, Conlon K, Gema CA, Hewinson RG, Polaina J, Suarez GA, Gordon SV. 2010. Characterization of the transcriptional regulator Rv3124 of *Mycobacterium tuberculosis* identifies it as a positive regulator of

molybdopterin biosynthesis and defines the functional consequences of a non-synonymous SNP in the *Mycobacterium bovis* BCG orthologue. *Microbiology* 156:2112–2123.

40. Chen JM, Uplekar S, Gordon SV, Cole ST. 2012. A point mutation in cycA partially contributes to the D-Cycloserine resistance trait of *Mycobacterium bovis* BCG vaccine strains. *PLoS One* 7:e43467.

41. Keller PM, Bottger EC, Sander P. 2008. Tuberculosis vaccine strain *Mycobacterium bovis* BCG Russia is a natural recA mutant. *BMC Microbiol* 8:120.

42. Kozak RA, Alexander DC, Liao R, Sherman DR, Behr MA. 2011. Region of difference 2 contributes to virulence of *Mycobacterium tuberculosis*. *Infect Immun* 79:59–66.

43. Kozak R, Behr MA. 2011. Divergence of immunologic and protective responses of different BCG strains in a murine model. *Vaccine* 29:1519–1526.

44. Alexander DC, Behr MA. 2007. Rv1773 is a transcriptional repressor deleted from BCG-Pasteur. *Tuberculosis* (Edinb.) 87:421–425.

45. Lotte A, Wasz-Hockert O, Poisson N, Dumitrescu N, Verron M, Couvet E. 1984. BCG complications. Estimates of the risks among vaccinated subjects and statistical analysis of their main characteristics. *Adv Tuberc Res* 21:107–193.

46. Chen JM, Islam ST, Ren H, Liu J. 2007. Differential productions of lipid virulence factors among BCG vaccine strains and implications on BCG safety. *Vaccine* 25:8114–8122.

47. Leung AS, Tran V, Wu Z, Yu X, Alexander DC, Gao GF, Zhu B, Liu J. 2008. Novel genome polymorphisms in BCG vaccine strains and impact on efficacy. *BMC Genomics* 9:413.

48. Naka T, Maeda S, Niki M, Ohara N, Yamamoto S, Yano I, Maeyama J, Ogura H, Kobayashi K, Fujiwara N. 2011. Lipid phenotype of two distinct subpopulations of *Mycobacterium bovis* bacillus Calmette-Guerin Tokyo 172 substrain. *J Biol Chem* 286:44153–44161.

49. Walters SB, Dubnau E, Kolesnikova I, Laval F, Daffe M, Smith I. 2006. The *Mycobacterium tuberculosis* PhoPR two-component system regulates genes essential for virulence and complex lipid biosynthesis. *Mol Microbiol* 60:312–330.

50. Vallishayee RS, Shashidhara AN, Bunch-Christensen K, Guld J. 1974. Tuberculin sensitivity and skin lesions in children after vaccination with 11 different BCG strains. *Bull World Health Organ* 51:489–494.

51. Ladefoged A, Bunch-Christensen K, Guld J. 1976. Tuberculin sensitivity in guinea-pigs after vaccination with varying doses of BCG of 12 different strains. *Bull World Health Organ* 53:435–443.

52. Gupta S, Sinha A, Sarkar D. 2006. Transcriptional autoregulation by *Mycobacterium tuberculosis* PhoP involves recognition of novel direct repeat sequences in the regulatory region of the promoter. *FEBS Lett* 580:5328–5338.

53. Steyn AJ, Collins DM, Hondalus MK, Jacobs WR Jr, Kawakami RP, Bloom BR. 2002. *Mycobacterium tuberculosis* WhiB3 interacts with RpoV to affect host survival

but is dispensable for *in vivo* growth. *Proc Natl Acad Sci USA* **99:**3147–3152.

54. Singh A, Crossman DK, Mai D, Guidry L, Voskuil MI, Renfrow MB, Steyn AJ. 2009. *Mycobacterium tuberculosis* WhiB3 maintains redox homeostasis by regulating virulence lipid anabolism to modulate macrophage response. *PLoS Pathog* **5:**e1000545.

55. Wernisch L, Kendall SL, Soneji S, Wietzorrek A, Parish T, Hinds J, Butcher PD, Stoker NG. 2003. Analysis of whole-genome microarray replicates using mixed models. *Bioinformatics* **19:**53–61.

56. Parish T, Smith DA, Kendall S, Casali N, Bancroft GJ, Stoker NG. 2003. Deletion of two-component regulatory systems increases the virulence of *Mycobacterium tuberculosis*. *Infect Immun* **71:**1134–1140.

57. Brosch R, Gordon SV, Buchrieser C, Pym AS, Garnier T, Cole ST. 2000. Comparative genomics uncovers large tandem chromosomal duplications in *Mycobacterium bovis* BCG Pasteur. *Yeast* **17:**111–123.

58. Brosch R, Gordon SV, Garnier T, Eiglmeier K, Frigui W, Valenti P, Dos SS, Duthoy S, Lacroix C, Garcia-Pelayo C, Inwald JK, Golby P, Garcia JN, Hewinson RG, Behr MA, Quail MA, Churcher C, Barrell BG, Parkhill J, Cole ST. 2007. Genome plasticity of BCG and impact on vaccine efficacy. *Proc Natl Acad Sci USA* **104:**5596–5601.

59. Fernandes ND, Wu QL, Kong D, Puyang X, Garg S, Husson RN. 1999. A mycobacterial extracytoplasmic sigma factor involved in survival following heat shock and oxidative stress. *J Bacteriol* **181:**4266–4274.

60. Abdallah AM, Verboom T, Hannes F, Safi M, Strong M, Eisenberg D, Musters RJ, Vandenbroucke-Grauls CM, Appelmelk BJ, Luirink J, Bitter W. 2006. A specific secretion system mediates PPE41 transport in pathogenic mycobacteria. *Mol Microbiol* **62:**667–679.

61. Minnikin DE, Parlett JH, Magnusson M, Ridell M, Lind A. 1984. Mycolic acid patterns of representatives of *Mycobacterium bovis* BCG. *J Gen Microbiol* **130:**2733–2736.

62. Yuan Y, Zhu Y, Crane DD, Barry CE III. 1998. The effect of oxygenated mycolic acid composition on cell wall function and macrophage growth in *Mycobacterium tuberculosis*. *Mol Microbiol* **29:**1449–1458.

63. Behr MA, Schroeder BG, Brinkman JN, Slayden RA, Barry CE III. 2000. A point mutation in the mma3 gene is responsible for impaired methoxymycolic acid production in *Mycobacterium bovis* BCG strains obtained after 1927. *J Bacteriol* **182:**3394–3399.

64. Belley A, Alexander D, Di Pietrantonio T, Girard M, Jones J, Schurr E, Liu J, Sherman DR, Behr MA. 2004. Impact of methoxymycolic acid production by *Mycobacterium bovis* BCG vaccines. *Infect Immun* **72:**2803–2809.

65. Charlet D, Mostowy S, Alexander D, Sit L, Wiker HG, Behr MA. 2005. Reduced expression of antigenic proteins MPB70 and MPB83 in *Mycobacterium bovis* BCG strains due to a start codon mutation in sigK. *Mol Microbiol* **56:**1302–1313.

66. Seki M, Honda I, Fujita I, Yano I, Yamamoto S, Koyama A. 2009. Whole genome sequence analysis of *Mycobacterium bovis* bacillus Calmette-Guerin (BCG) Tokyo 172: a comparative study of BCG vaccine substrains. *Vaccine* **27:**1710–1716.

67. Pan Y, Yang X, Duan J, Lu N, Leung AS, Tran V, Hu Y, Wu N, Liu D, Wang Z, Yu X, Chen C, Zhang Y, Wan K, Liu J, Zhu B. 2011. Whole-genome sequences of four *Mycobacterium bovis* BCG vaccine strains. *J Bacteriol* **193:**3152–3153.

68. Gomes LH, Otto TD, Vasconcellos V, Ferrao PM, Maia RM, Moreira AS, Ferreira MA, Castello-Branco LR, Degrave WM, Mendonca-Lima L. 2011. Genome sequence of *Mycobacterium bovis* BCG Moreau, the Brazilian vaccine strain against tuberculosis. *J Bacteriol* **193:**5600–5601.

69. Orduna P, Cevallos MA, de Leon SP, Arvizu A, Hernandez-Gonzalez IL, Mendoza-Hernandez G, Lopez-Vidal Y. 2011. Genomic and proteomic analyses of *Mycobacterium bovis* BCG Mexico 1931 reveal a diverse immunogenic repertoire against tuberculosis infection. *BMC Genomics* **12:**493.

70. Lynett J, Stokes RW. 2007. Selection of transposon mutants of *Mycobacterium tuberculosis* with increased macrophage infectivity identifies fadD23 to be involved in sulfolipid production and association with macrophages. *Microbiology* **153:**3133–3140.

71. Vandal OH, Roberts JA, Odaira T, Schnappinger D, Nathan CF, Ehrt S. 2009. Acid-susceptible mutants of *Mycobacterium tuberculosis* share hypersusceptibility to cell wall and oxidative stress and to the host environment. *J. Bacteriol.* **191:**625–631.

72. Joung SM, Jeon SJ, Lim YJ, Lim JS, Choi BS, Choi IY, Yu JH, Na KI, Cho EH, Shin SS, Park YK, Kim CK, Kim HJ, Ryoo SW. 2013. Complete genome sequence of *Mycobacterium bovis* BCG Korea: the Korean vaccine strain for substantial production. *Genome Announc* **1:**e0006913.

73. Bottai D, Stinear TP, Supply P, Brosch R. 2014. Mycobacterial pathogenomics and evolution. *Microbiol Spectrum* **2**(1):MGM2-0025-2013.

74. Kleinnijenhuis J, Quintin J, Preijers F, Joosten LA, Ifrim DC, Saeed S, Jacobs C, van Loenhout J, de Jong D, Stunnenberg HG, Xavier RJ, van der Meer JW, van Crevel R, Netea MG. 2012. Bacille Calmette-Guérin induces NOD2-dependent nonspecific protection from reinfection via epigenetic reprogramming of monocytes. *Proc Natl Acad Sci USA* **109:**17537–17542.

Molecular Genetics of Mycobacteria, 2nd Edition
Edited by Graham F. Hatfull and William R. Jacobs, Jr.
© 2014 American Society for Microbiology, Washington, DC
doi:10.1128/microbiolspec.MGM2-0022-2013

Keith M. Derbyshire[1]
Todd A. Gray[1]

Distributive Conjugal Transfer: New Insights into Horizontal Gene Transfer and Genetic Exchange in Mycobacteria

4

Eukaryotes use meiotic recombination to blend parental genomes and generate genome-wide variation between progeny. By contrast, bacteria divide asexually to generate two clones of the original. The genomic variation present in asexual populations is modest and is limited to those relatively rare events of spontaneous mutation and transposition. Significant variation must await exceedingly rare events wherein DNA from another organism is sporadically introduced into the bacterial genome. The acquisition of foreign DNA by a recipient bacterium provides the genetic diversity needed to facilitate evolution. This process is called horizontal gene transfer (HGT; or sometimes LGT for lateral gene transfer) and is mediated by three distinct mechanisms: conjugation, transformation, and transduction (1, 2). In this chapter we will discuss in detail HGT mediated by conjugation in mycobacteria, its potential evolutionary impact, and its application as a tool for mycobacterial genetics. Transduction is discussed in references 72 and 73 on mycobacteriophages.

Transformation (by electroporation) is a vital laboratory tool, but transformation *per se* is unlikely to play a major role in HGT because mycobacterial species are not known to be naturally competent.

HGT can occur between cells of the same species, different species, and even across kingdoms, thus blurring species boundaries. In the pregenomic era, the impact of HGT was mostly restricted to our knowledge of the movement of mobile elements such as plasmids and phages. However, as the number of bacterial whole-genome sequences increases, it is becoming clear that HGT has significantly influenced the genomes of extant bacteria (3–7). This is also true for the members of the *Mycobacterium tuberculosis* complex (MTBC; *M. africanum*, *M. bovis*, *M. canettii*, *M. caprae*, *M. microti*, *M. pinnipedii*, and *M. tuberculosis*), whose genomes have been extensively studied in an attempt to type and track TB outbreaks. HGT was not originally considered a major factor in the MTBC, because preliminary sequence comparisons

[1]Division of Genetics, Wadsworth Center, New York State Department of Health, and Department of Biomedical Sciences, University at Albany, Albany, NY 12201.

among them indicated extremely low diversity and lacked any clear evidence of HGT (8–11). Consequently, it was postulated that the MTBC is essentially a clonal outgrowth of a particularly successful subtype of a progenitor species (10–12). However, more recent sequence analyses have provided convincing evidence of recombination occurring between isolates of *M. canettii*, the most divergent member and probable progenitor species of the MTBC (13, 14). These sequence analyses, combined with the characterization of an entirely novel mycobacterial conjugation system have forced a rethinking of the clonal paradigm and the role of HGT in shaping mycobacterial genomes (14, 15). In this chapter we will discuss the latest results indicating that HGT, specifically conjugation, is active among mycobacterial species and that it likely has had —and will continue to have—a large impact on mycobacterial evolution.

CLASSICAL *oriT*-MEDIATED CONJUGATION: A BRIEF OVERVIEW

Plasmid Transfer

Bacterial conjugation is a naturally occurring process that involves the unidirectional transfer of DNA from a donor to a recipient and requires cell-cell contact (16, 17). Conjugation generally involves the transfer of plasmids, or integrative and conjugative elements (ICEs). ICEs are found in the chromosome, but excise to form a plasmid-like circle before transferring into a recipient and reintegrating into the chromosome (18). Most of the early characterization of conjugative plasmids and the transfer process was completed in *Escherichia coli*, but the mechanism of transfer is essentially the same for both plasmids and ICEs, regardless of bacterial species (16, 19).

The first step in conjugation is establishment of cell-cell contact between the donor and recipient cells, or mating-pair formation (MPF). In Gram-negative species, MPF is facilitated by the plasmid-encoded pilus that is assembled by a type IV secretion system (T4SS) (20, 21). Gram-positive species also encode many of the T4SS proteins, but they are mainly localized to the cytoplasm or membrane and are therefore proposed to assemble as the conduit of DNA transfer and not to assemble pilus structures (21, 22). Instead, Gram-positive mating pairs are likely formed by other mechanisms including surface-exposed adhesins, which cause aggregation of donor and recipient enterococci cells (23, 24). Once MPF is established, DNA-processing enzymes mediate transfer of a single strand of DNA from the donor to the recipient (Fig. 1A). A key protein is a relaxase, which induces a strand-specific nick within a sequence called the origin of transfer, *oriT* (16). The relaxase, bound to the 5′ end of the nicked DNA, mediates transfer of the DNA through the pore and into the recipient. On completion of plasmid transfer, the relaxase recognizes the 3′ end of *oriT* and recircularizes the single strand by a reversal of the nicking process. Complementary DNA synthesis in the recipient results in a transconjugant containing a copy of the conjugative plasmid and, thus, is now a donor. This replicative process facilitates the rapid dissemination of conjugative plasmids through bacterial populations.

Chromosomal Transfer Mediated by Integrated Plasmids

Despite this plasmid-centric view of conjugation, the first description of conjugation was the transfer of *E. coli* chromosomal DNA by Lederberg and Tatum (25). Unaware that plasmids existed and promoted transfer, Lederberg and Tatum selected for transfer of chromosomal markers. Fortunately, one of the strains they were using had an F plasmid integrated into the chromosome, and thus chromosomal DNA was transferred, because it contained the *cis*-acting *oriT*. The mechanism of transfer by these so-called Hfr strains (high frequency recombination) is identical to that of plasmid transfer, but the outcome is very different (Fig. 1B). Because the chromosome is now conceptually equivalent to a *very* large plasmid, transfer requires a commensurately extended time and depends on nick-free DNA to be pulled along by the relaxase-*oriT* complex. Therefore, transfer is often incomplete, either due to physical dissociation of mating pairs or because of nicks in the strand of DNA being transferred. Consequently, only the segment of the donor chromosome adjacent to *oriT* is reliably transferred. Since partial transfer prevents reconstitution of *oriT* to circularize the chromosome, this linear segment must recombine into the recipient chromosome by homologous recombination for stable inheritance. Thus, chromosomal transfer contrasts with plasmid transfer: (i) host recombination functions are required to integrate transferred DNA into the chromosome; (ii) *oriT* is not regenerated in the recipient, and therefore transconjugants do not become donors, but remain as recipients; and (iii) genes 3′-proximal to *oriT* are transferred most frequently, so the efficiency of transfer of chromosomal genes is location dependent. By taking advantage of these observations, Wollman et al. established the circularity of the *E. coli* genome and mapped genes by minutes around the "100-minute"

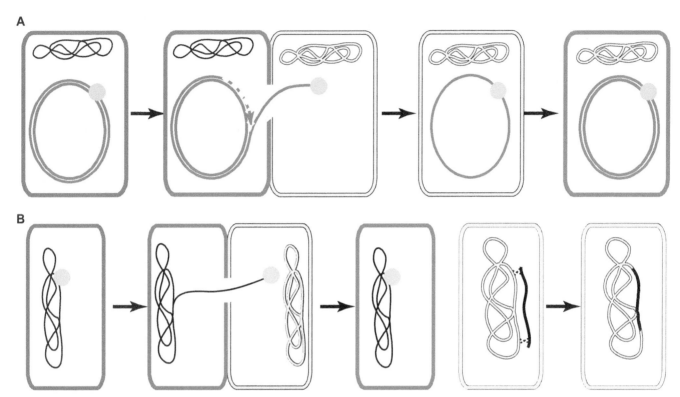

Figure 1 *oriT* directs both plasmid and chromosomal transfer in traditional conjugation systems. (**A**) F episome plasmid transfer. The *oriT* is nicked and a single strand is guided into the engaged recipient by a plasmid-encoded relaxase. Upon transfer, relaxase catalyzes recircularization at *oriT*, and host polymerases synthesize a complementary strand. The recipient chromosome is unaltered, but the cell now exhibits a donor phenotype (blue). (**B**) Hfr strains have a plasmid integrated into the chromosome, shown as a single strand for simplicity. The integrated *oriT* functions as it would in the plasmid scenario above, except that transfer of the chromosome is usually incomplete. The linear chromosomal fragment must be incorporated into the recipient chromosome by homologous recombination for stable inheritance. Homologous recombination excludes the *oriT*, and, while the transconjugant chromosome now has some donor sequences, it retains a recipient phenotype (yellow). For an Hfr transconjugant to become a donor, the entire donor chromosome must be transferred to regenerate *oriT*. doi:10.1128/microbiolspec.MGM2-0022-2013.f1

chromosome (26). As we will describe below, the outcomes and requirements of transfer in *M. smegmatis* substantially differ from *oriT*-initiated transfer, providing the first clues that mycobacterial conjugation occurs by an entirely new mechanism.

MYCOBACTERIAL CONJUGATION

Do Mycobacterial Plasmids Mediate Transfer?

Rather strangely, mycobacteria seem to lack the sheer numbers and classes of plasmids found in almost all other bacteria. While many environmental species contain plasmids (27), many mycobacteria lack plasmids altogether, including members of the MTBC and

M. smegmatis. This paucity of plasmids in mycobacteria explains why all natural antibiotic resistance is chromosomally mediated in *M. tuberculosis*, unlike most other pathogens. We speculate that this is a consequence of the unusual nature of the mycobacterial envelope preventing plasmid transfer and the unique requirements for plasmid replication and maintenance in mycobacteria.

In general, mycobacterial plasmids have only been partially characterized, while the main focus has been their development as vectors (28, 29). Descriptions of plasmids are mostly limited to the environmental species, such as members of the *Mycobacterium avium, intracellulare, scrofulaceum* (MAIS) complex and *Mycobacterium fortuitum* (27, 28). Hybridization studies indicate that some of these plasmids are related despite being found in different species and isolates, which suggests that the plasmids have moved between species in the environment (30). However, the technical problems associated with manipulating members of the MAIS complex and other environmental mycobacteria have made it difficult to experimentally address the mechanism of dissemination. Most of these plasmids can be transformed and maintained in other mycobacterial species, such as *M. smegmatis* and *M. tuberculosis*, indicating that their limited host range is not the result of an inability to replicate, but a consequence of either a physical inability to spread or because their hosts do not occupy the same environmental niches (28, 31, 32).

A few of these characterized mycobacterial plasmids have been sequenced, and these analyses have not identified obvious transfer functions (31, 33, 34). *M. ulcerans* contains a large plasmid, pMUM1001 (>174 Kb), which encodes polyketide synthases that produce the macrolide toxin necessary for its pathogenesis (35). Its large size has prevented functional characterization, but sequence analysis again indicates that pMUM1001 does not encode a classical conjugation system (36). The lack of defined transfer-associated genes on these plasmids suggests that either they are not transferred between cells, plasmid transfer occurs by a novel mechanism, or they are indirectly mobilized between environmental species via chromosomal transfer, as observed in *M. smegmatis* (37).

Sequence analyses of two mycobacterial plasmids indicates that they encode genes related to those found on classical *oriT*-like plasmids, but these plasmids have not been shown to be mobile in mycobacteria. The sequence of the plasmid pVT2 from *M. avium* revealed that it encodes a relaxase, with homology to the relaxases of F, R100, and R388, including conservation of the active site sequence motifs (31). However, pVT2-mediated conjugation was not experimentally demonstrated, and the plasmid lacks other conjugation-like genes. One possible scenario is that pVT2 is a mobilizable plasmid, encoding its own relaxase and *oriT*, but requiring a conjugative plasmid, or equivalent, to establish MPF. More recently, a plasmid was identified in an epidemic strain of *M. abscessus*, which was sequenced and shown to belong to the broad-host-range IncP plasmids found in many Gram-negative bacteria (38). The 56-kb plasmid, pMAB01, contained all the genes needed for MPF and DNA transfer. Importantly, when introduced into *E. coli*, pMAB01 could transfer between strains of *E. coli*, albeit at low frequencies. While this study demonstrated that a large broad-host-range plasmid can get into a mycobacterial cell (proba-

bly by conjugation; also see reference 39), the restriction of pMAB01 to a single epidemic strain suggests that it is unable to mediate transfer efficiently between mycobacteria; i.e., mycobacteria are a dead end for *oriT*-mediated plasmid transfer systems. We speculate that this is because of the novel composition and architecture of the mycobacterial membrane, which prevents assembly of a functional T4SS for pilus synthesis and DNA export. This membrane barrier may also prevent acquisition of broad-host-range plasmids, which are normally capable of transfer into an extraordinarily diverse range of bacterial species.

Chromosomal Transfer in Mycobacteria
Similar to its discovery in *E. coli*, the first definitive proof of conjugation in mycobacteria was that of chromosomal DNA transfer in *M. smegmatis* (40). However, as we will describe, the mechanism of transfer is fundamentally different, requiring a complete rethinking of the conjugation process and its impact on genome dynamics and mycobacterial evolution.

The basics
The first accounts of a conjugation-like process between isolates of *M. smegmatis* were described in the early 1970s (41, 42). However, these initial observations required modern molecular genetic tools to convincingly demonstrate that recombinants were generated by conjugal transfer of chromosomal DNA and did not involve plasmids (40). To monitor DNA transfer, cassettes encoding different antibiotic resistance markers were integrated into the chromosomes of *M. smegmatis* isolates, the differentially marked strains were co-incubated, and transconjugants were identified by selecting for recombinant cells resistant to both antibiotics (Fig. 2). This simple assay confirmed that DNA transfer in *M. smegmatis* satisfies all criteria of conjugation: (i) transfer required co-incubation of both parent cells, and that both parental types were viable; (ii) there are distinct donor and recipient strains; (iii) transformation was excluded because recombinants were isolated in the presence of DNase I; (iv) transduction was also ruled out because transfer did not occur in liquid medium and phage could not be detected in culture filtrates; and (v) cell fusion was ruled out because an episomal plasmid introduced into the donor strain was not transferred in transconjugants even when a plasmid-encoded antibiotic marker was selected (40). One pair of isolates, derivatives of Jucho and P73, while normally acting exclusively as recipients with other donors, could exchange antibiotic markers bidirectionally. This unusual bidirectional transfer could

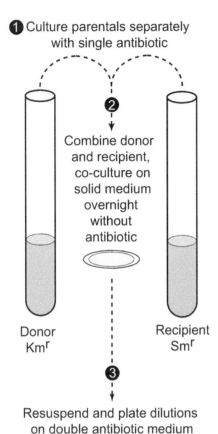

❶ Culture parentals separately with single antibiotic

❷ Combine donor and recipient, co-culture on solid medium overnight without antibiotic

Donor Km^r

Recipient Sm^r

❸ Resuspend and plate dilutions on double antibiotic medium

Km + Sm Plate

Mated Strains	Km^r + Sm^r Colonies	Transcojugants per Donor
Donor only	0	<1 x 10^{-9}
Recipient only	0	<1 x 10^{-9}
Donor x Recip	8 x 10^4	1 x 10^{-4}

Figure 2 An outline of the mycobacterial mating procedure. Selectable markers for donor and recipient are chromosomally encoded. Efficient transfer requires prolonged incubation (overnight) on solid medium. Transfer does not take place in liquid cultures, probably reflecting the need to force cell-cell contact on the solid medium. The transfer frequency is 1 event per 10^4 donors. However, this only reflects successful transfer of *Km^r* and therefore is an underestimate of the total number of events.
doi:10.1128/microbiolspec.MGM2-0022-2013.f2

occur if these two strains are capable of switching mating types; i.e., they contain genetic information for both donor and recipient functions, but only one function is expressed in any one cell. It is possible this switch is relevant to the conversion of transconjugants to donors, which will be discussed later.

Of considerable note to the mycobacterial research community was that mc^{2}155, the widely used laboratory strain, was characterized as a conjugal donor. This had the reciprocal dual benefit of using tools developed in mc^{2}155 to study the conjugation mechanism, while allowing the development of conjugation as a tool to study mc^{2}155.

Initiation of mycobacterial DNA transfer does not occur from a unique site

oriT regions are small (~200 bp) *cis*-acting sequences that include all the necessary sites for the relaxase and accessory proteins to bind, nick, and mediate DNA transfer (16). They are easily isolated because they confer mobility on otherwise nonmobilizable plasmids, when the necessary transfer proteins are supplied in *trans* (43, 44). The pAL5000-based mycobacterial plasmid is not mobilized, even when a plasmid antibiotic marker is selected, indicating that it lacks the *cis*-acting sequences needed to mediate chromosomal transfer into the recipient cell. Wang et al. (37) took advantage of this observation to isolate chromosomal *cis*-acting mobilization sites by constructing a library of chromosomal donor DNA in a pAL5000 vector and selecting for its transfer into the recipient. By analogy with other transfer systems, it was anticipated that a single, *cis*-acting *oriT*-containing DNA segment would be identified in vector derivatives that successfully transferred. Surprisingly, multiple, nonoverlapping segments of donor DNA were identified in plasmid transconjugants, indicating that there are many *cis*-acting sequences in the chromosome. The *cis*-acting sequences were termed *bom* regions, for basis of mobilization, to distinguish them from the smaller, singular, better characterized, *oriT*.

oriT provides two functions: It initiates transfer in the donor and recircularizes plasmid DNA in the recipient to complete transfer (Fig. 1A). In *M. smegmatis*, a study to define functionally comparable *cis*-acting sequences revealed that efficient plasmid transfer required at least 5 kb of chromosomal DNA and often resulted in acquisition of recipient chromosome single nucleotide polymorphisms (SNPs) in the transferred plasmid DNA (37). This simple observation provided key insights for a new model, which proposed that *bom* might initiate transfer, but recircularization of a mycobacterial plasmid required homologous recombination functions in the recipient in the form of gap repair, using the recipient chromosome as a template (Fig. 3A). A gap-repair model encompasses all of the experimental findings: transfer requires a minimum size of *bom* (for homology to promote strand invasion and gap repair), the requirement of RecA in the recipient, and the acquisition of recipient SNPs in the plasmid DNA. In further support of this gap-repair model, plasmids were engineered to contain two neighboring *bom* regions, normally spaced 5 kb apart on the chromo-

some. Transconjugants containing these two-*bom* plasmids were shown to have acquired the entire intervening segment of recipient chromosomal DNA (Fig. 3B) (37, 45). The recovered *bom* sequences were not enriched for elements that might function as a binding site for an enzyme, or have any discernable structural features, similar to *oriT*. In addition, although *bom* initiates plasmid transfer, it is unclear whether initiation involves a nick or a double-stranded break, resulting in single-stranded or double-stranded DNA transfer, respectively.

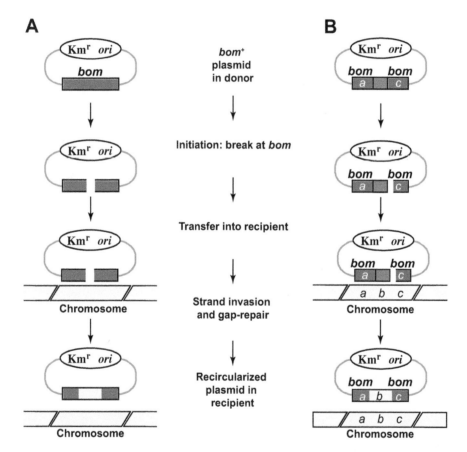

Figure 3 Donor chromosomal *bom* sites mobilize plasmids and mediate recircularization by gap repair in the recipient. Episomal plasmids are not subject to conjugal transfer unless they carry chromosomal DNA segments functionally defined as *bom* (basis of mobility) sites. (**A**) Recovery of the transferred plasmids, and sequencing of the *bom* sites, revealed the presence of embedded recipient SNPs, suggesting a gap-repair mechanism. In this model, transfer would be initiated in the donor via a break in *bom*. Following transfer of the linear plasmid, the homologous region of the recipient chromosome would act as the template for gap repair to seal the break and recircularize the plasmid. As a consequence, recipient SNPs would be incorporated into the plasmid DNA. (**B**) This model was confirmed by the use of two adjacent *bom* sites, separated by a nonhomologous sequence. The region of nonidentity was replaced by the intervening recipient chromosomal sequence upon transfer. In this two-*bom* model a break could occur at both *boms* to create a gap spanning hypothetical gene *b*. Alternatively, a break could occur at just one *bom* site (as shown) and then the nonhomologous region resected by exonucleases to generate ends suitable for gap repair, capture of gene *b*, and plasmid circularization. doi:10.1128/microbiolspec.MGM2-0022-2013.f3

Extrapolating these plasmid findings to a chromosomal context suggests that cleavage at multiple chromosomal *bom* sites could result in transfer of discrete segments of DNA that are then integrated into the recipient chromosome by homologous recombination. In addition, the presence of numerous *bom* sites is consistent with location-independent DNA transfer, in contrast to that observed in Hfr strains (46). Importantly, these mechanistic distinctions are consistent with two very different evolutionary paths to conjugal DNA transfer. Chromosomal transfer by Hfr in *E. coli* is directed by *cis* and *trans* functionality from a plasmid —something that has evolved for self-propagation— simply because incidental integration embedded the plasmid in the chromosome. In contrast, mycobacterial conjugal DNA transfer is apparently plasmid-independent, implying that the necessary constituent conjugation proteins and elements have coevolved for the express purpose of transferring chromosomal genes between cells.

Genetic requirements of mycobacterial conjugation

Bioinformatic searches have failed to identify any orthologs of transfer genes in *M. smegmatis* genomes. In hindsight this is not entirely surprising, given both the unique aspects of mycobacterial transfer and the structure of the mycobacterial cell envelope (47). Further underscoring the unique properties of mycobacterial conjugation is the unanticipated role of the ESX-1 secretion apparatus in DNA transfer. Transposon mutagenesis screens demonstrated very different roles for the ESX-1 secretion apparatus in the donor and recipient. In the donor, ESX-1 functions suppress transfer: *esx1* donor mutants are hyper-conjugative (48). Paradoxically, in the recipient, ESX-1 functions are essential for transfer: *esx1* recipient mutants are nonconjugative (49). This reciprocal *esx1* conjugation phenotype suggests a level of complexity well beyond a simple model of DNA passing through a secretory apparatus.

ESX-1 is the flagship representative of type VII secretion systems. There are five apparently nonredundant paralogous *esx* loci in *M. tuberculosis*, with *esx1* mutants having an attenuated phenotype (discussed in more detail in references 50, 51, and 71). Briefly, ESX-1 is a complex secretion machine with its own specific set of substrates, which include the dominant antigens EsxA and EsxB (previously called Esat6 and Cfp10). The majority of the proteins required for secretion are encoded from a single, multigene locus called *esx1*. The genetic organization and the proteins encoded by *esx1* are highly conserved among mycobacteria, including

M. smegmatis. ESX-1 mutants of *M. smegmatis*, like those of *M. tuberculosis*, fail to secrete EsxA and EsxB and other ESX-1-dependent proteins (49, 52). Thus, a role for ESX-1 in conjugation suggests an overarching function for ESX-1 in regulating extracellular interactions; in *M. smegmatis*, ESX-1 regulates interaction between mating mycobacterial cells, while in *M. tuberculosis*, ESX-1 regulates interactions with the host.

There are many unanswered questions concerning the mechanism and functions of type VII secretion systems, but several are relevant to this chapter. How do the roles of ESX-1 differ between *M. smegmatis* and *M. tuberculosis*? And, within *M. smegmatis*, how do the roles differ between the donor and recipient strains? Given the high level of conservation of the *esx1*-encoded proteins, the most likely model is that the structural apparatus and mechanisms of secretion are the same, but some of the proteins that are secreted differ between species and strains. Thus, *M. tuberculosis* would secrete effector proteins to mediate pathogenesis, while the donor and recipient *M. smegmatis* strains would use the same delivery system to secrete different sets of proteins to regulate or mediate DNA transfer. Whether ESX-1 mediates these disparate functions through secretion of diffusible factors or by decorating or modifying the mycobacterial cell wall is a pressing question.

Could a paradoxical riddle hold the key?

Paradoxes are only paradoxes until the underlying reason is known. The opposing effects of the *esx1* mutation in donor and recipient *M. smegmatis* strains defy most models of orthologous function, in which orthologous genes and their encoded proteins will have similar—not opposing—functions. Solving this paradox will likely go hand-in-hand with understanding the fundamentals of mycobacterial conjugation, but critical pieces of the puzzle are still missing.

In the donor, an active ESX-1 apparatus suppresses DNA transfer, suggesting that the ESX-1 secreted proteins are inhibitors of conjugation (48). One possibility is that ESX-1-secreted proteins coat the surface of the donor and act as a physical barrier to prevent intimate cell contact and the transfer of DNA. An alternative hypothesis is that the secreted proteins are signaling proteins, or quorum sensors, that suppress activation of DNA transfer until suitable recipient cells are present. There is precedent for transfer being activated by secreted peptides in the Gram-positive enterococcus, in which peptides secreted by both donor and recipient regulate the induction and suppression of DNA transfer (24).

In the recipient, ESX-1 secreted proteins may be receptors on the cell surface that promote donor-recipient contact (49). Alternatively, as for the donor, the secreted recipient proteins may actively signal the donor to initiate transfer. A third scenario is that the ESX-1 apparatus itself is responsible for DNA uptake into the recipient. However, this model would imply that the donor ESX-1 apparatus is fundamentally different—donors cannot take up DNA—and that its presence negatively interferes with DNA transfer into the recipient. Regardless of model, DNA transfer provides a simple, sensitive and robust assay for ESX-1 function for the molecular genetic dissection of ESX-1 and its many, apparently disparate, roles in mycobacterial biology.

DISTRIBUTIVE CONJUGAL TRANSFER

Transconjugant Genomes Are Mosaic

Recent revelations about mycobacterial DNA transfer have come from whole-genome sequencing of transconjugant progeny (15). Whole-genome comparison between transconjugant progeny and their parents is feasible because parental strains differ significantly at the nucleotide level (~1 SNP per 56 nucleotides), thus allowing discrimination between the parental origin (donor or recipient) of genomic DNA in transconjugants. The most striking feature of such a comparison is that the progeny genomes are mosaic blends of the parental genomes (Fig. 4). This mosaicism contrasts with the acquisition of a single, *oriT*-linked segment of donor DNA in Hfr transfer (Fig. 1). Thus, while all transconjugants acquire a segment of donor DNA encoding the selected marker (for example, *Km^r*), segments of DNA not selected are also transferred. In fact, on average, 12 additional segments of donor DNA are co-inherited for every selected *Km^r* segment, and these bonus segments are distributed around the genome with no obvious regional biases. The sizes of the donor segments vary dramatically, ranging from 59 bp to 226 kb, with an average size of 44.2 kb and a mean of 13 tracts, totaling 575 kb of transferred DNA per genome (data from 22 transconjugants). To more accurately reflect the products of mycobacterial conjugation, and to distinguish it from the classical *oriT*-mediated transfer, this process was termed distributive conjugal transfer (DCT).

Surprisingly, whole-genome sequencing has not been reported for Hfr transconjugants, preventing a detailed genomic comparison of the two conjugation systems. However, current Hfr models predict that a single segment of donor DNA would be integrated into the recipient chromosome corresponding to donor DNA located 3′ of *oriT* (Fig. 1B). DCT is remarkable compared to other modes of HGT for two main reasons. First, DCT creates unprecedented genome-wide mosaicism within individual transconjugants, and, second, the variation it generates in a single step approaches that seen in sexual reproduction.

Models for DCT

Based on the mosaicism observed in progeny genomes, we speculate that random chromosomal DNA fragments are generated in the donor, some of which are cotransferred into a recipient cell where they replace recipient sequences through homologous recombination (Fig. 5, left). The trigger for chromosome fragmentation is not clear, but the model is consistent with multiple donor *bom* break sites initiating transfer of individual segments (37).

An alternative to the model described above is that a single, large DNA molecule is transferred into the recipient, which is processed into smaller segments before their integration into the recipient chromosome by homologous recombination (Fig. 5, right). This scenario seems less likely because it predicts some progeny would contain exceedingly large chunks of donor DNA (3 to 4 Mb) integrated into the chromosome. These large chunk events would have resulted from recombination close to the ends of the transferred molecule—sites that are known to load recombination machines and promote homologous recombination (53, 54). Based on the expected recombination frequencies, it would seem more parsimonious for these large chunks to be integrated before fragmentation into smaller segments occurred. This large chunk scenario is also less consistent with previous observations, which indicated that the donor chromosome contained multiple initiation sites (37).

The majority of the integration events observed in transconjugants can be explained by homologous recombination promoting double crossover events at either end of the transferred DNA segment (Fig. 6, left). RecA is required for transfer in the recipient and likely works with the RecA-dependent AdnAB recombination system described by Glickman and colleagues (55). Mycobacteria encode a RecBCD complex, but these enzymes perform a different role in mycobacteria (RecA-independent single-strand annealing) than in *E. coli* and, therefore, are unlikely to mediate conjugational recombination as they do in *E. coli* (56, 57). However, regions of microcomplexity that contain extremely short alternating tracts of donor and recipient DNA

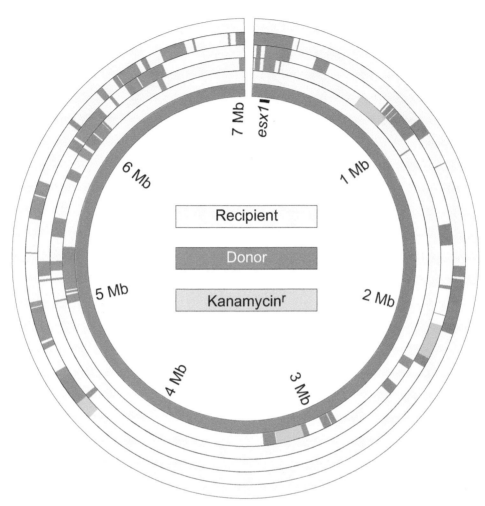

Figure 4 Transconjugant chromosomes are mosaic blends of the parental strains. Whole-genome sequencing of transconjugants and alignment with the parental sequences used the presence or absence of parental SNPs to define tracts of transferred donor DNA. A Circos projection of the 7-Mb chromosome is shown as concentric circles, with the recipient genome on the outside (yellow), the donor on the inside (blue), and four transconjugants in between. The segment containing the Km^r gene used for postmating selection is indicated (green); the recipient antibiotic marker is episomally encoded. The outer three transconjugants have a donor mating identity, and all have inherited the *esx1* locus from the donor strain (indicated at 0.1 Mb on the chromosome). The innermost transconjugant lacks the *esx1* locus and is a recipient strain. These data and additional Circos plots can be seen in reference 15. doi:10.1128/microbiolspec.MGM2-0022-2013.f4

demonstrate that other mechanisms are also at work (Fig. 6, right). These tracts indicate that the recombinant products likely arose from resolution of heteroduplexes between homeologous sequences, rather than simple exchange events. In other bacteria, the mismatch repair proteins recognize such heteroduplex DNA and prevent recombination (58). However, mycobacteria lack mismatch repair genes (59) and thus likely rely on alternative recombination mechanisms to resolve heteroduplexes, offering one explanation for the observed tracts of blended microcomplexity.

Untangling the processes that create these tracts will require a defined genetic approach, identifying the required recombination functions known to be present in mycobacteria (NHEJ, RecBCD, AdnAB, RecO), and equally likely to reveal yet other recombination surprises (57, 60). Regardless of the mechanism, the net effect of microcomplexity is to generate a localized composite blend of nucleotide substitutions. From an evolutionary standpoint, these provide localized subtle diversity, which, for example, could modify the activity or interaction specificity of an enzyme.

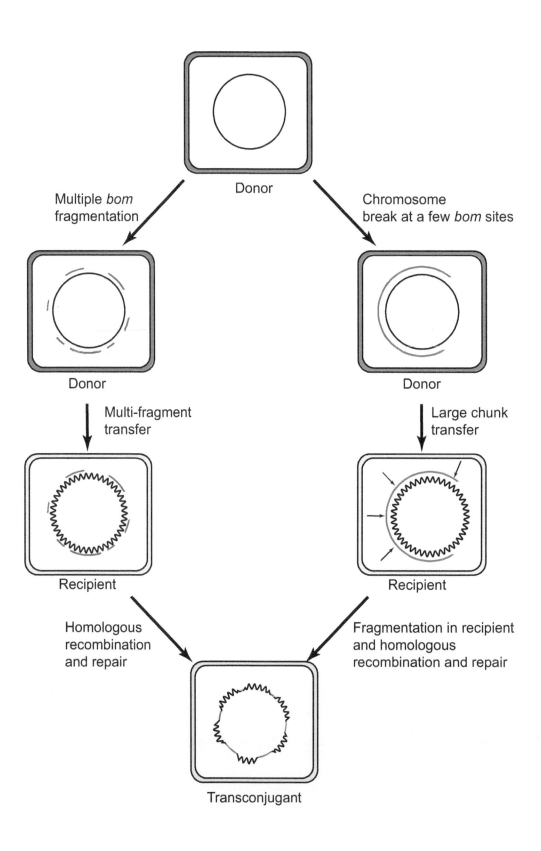

Multiple *bom* fragmentation

Chromosome break at a few *bom* sites

Donor

Donor

Donor

Multi-fragment transfer

Large chunk transfer

Recipient

Recipient

Homologous recombination and repair

Fragmentation in recipient and homologous recombination and repair

Transconjugant

Unanswered Questions

For all that is known about the products of DCT, there is still much to be discovered about the process. Whether single-stranded or double-stranded DNA is transferred is unknown. While *oriT*-mediated transfer occurs via a single-stranded intermediate, conjugation in *Streptomyces* is of double-stranded DNA (61; see below). The absence of known mycobacterial relaxases suggests that DCT might also be double-stranded and obviates the requirement for complementary-strand synthesis in the recipient before recombination.

An intriguing possibility is that DCT is similar to the conjugation system in *Streptomyces*, a fellow actinomycete. *Streptomyces* conjugation is mediated by a single protein, TraB, which is plasmid encoded. No orthologs of the T4SS or pilus protein are encoded from the plasmids or chromosome, most likely because MPF involves hyphal fusion (22). TraB is an ATPase, resembling the DNA translocators FtsK/SpoEIII in its overall structural organization (62). In further functional analogy to FtsK, which binds 8-bp *FtsK* orienting sequences to propel DNA through the septal pore (63), TraB binds a series of 8-bp repeats (TRSs [TraB recognition sequences]). TRSs are found on transferable plasmids associated with TraB and at sites around the chromosome and are thought to provide the specificity and directionality to transfer (62, 64). TraB localizes at the hyphal tips, where it is proposed to form a channel for translocation of DNA (64, 65). It is possible that a chromosomal ortholog of TraB performs a similar role in mycobacteria. If this protein also carried out essential replication functions, it might explain the inability to isolate transfer-defective donors (48). The dual functionality might also allow coordination of transfer and chromosome replication, such that transfer occurs after a round of replication, ensuring that DCT is not a donor suicide process. The lack of a defined mycobacterial *oriT*-like sequence might also be explained by its TraB ortholog recognizing cryptic TRS-like repeats. However, whether DCT is mediated by a TraB ortholog or by an entirely different process will require more definitive experimentation, especially because there are so many differences between mycobacteria and *Streptomyces* (perhaps most significantly, in membrane organization and growth characteristics).

DCT AS AN ENGINE FOR MYCOBACTERIAL EVOLUTION

DCT Dramatically Shortens the Evolutionary Clock

The analysis of transconjugant genomes not only provides insights on the transfer mechanism, but also suggests that mycobacterial DCT can create genome diversity almost instantly. Genomic changes that typically accompany evolution of bacteria are assumed to be a serial accrual of HGT and spontaneous mutation events that occur over many generations and great spans of time (Fig. 7). By contrast, a single-step DCT event between two *M. smegmatis* cells generates a transconjugant that is a mosaic blend of the parental genomes and not merely an incrementally altered derivative (Fig. 4 and 7). Thus, theoretically, DCT could dramatically increase the evolutionary rate estimates for mycobacteria.

Is DCT Still Active Among Native Mycobacterial Species?

Determining whether DCT is still active is extremely difficult to address. Naturally occurring HGT events can only be inferred from anecdotal sequence analyses of extant parental and transconjugant lineages. Because the HGT event may have happened long ago, the participating parental strains may be extinct or may not have been sequenced yet, so the nearest sequenced relative in its lineage must suffice as a stand-in for comparisons. Subsequent changes to either the parental or transconjugant genomes, including possible secondary HGT events, can complicate interpretations. Thus, while genome analysis is the most effective method to detect HGT, definitive proof that DCT is active in other species will require using sequenced, marked, parental strains and selecting for recombination events *in vitro*,

Figure 5 Models for chromosome fragmentation in DCT. DCT could follow one of two pathways, depending on whether fragmentation of the transferred chromosome occurs in the donor cell before transfer (left) or following transfer in the recipient (right and indicated by arrows). The left-hand model posits that multiple chromosome segments are cotransferred into the recipient, where they are recombined into the recipient chromosome to generate a mosaic pattern. Large chunk transfer (right) would be predicted to result in fewer large integrated segments rather than many widely distributed donor segments. Current results, as seen in Fig. 4, are more consistent with the fragmentation-before-transfer model. doi:10.1128/microbiolspec.MGM2-0022-2013.f5

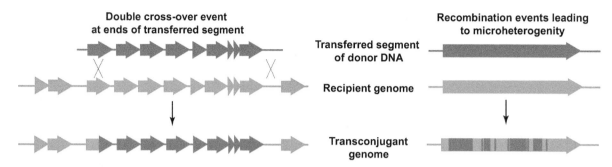

Figure 6 DCT brings large and small changes to transconjugant genomes. Transferred donor segments can be contiguous blocks hundreds of kilobases in length, spanning hundreds of genes. These large blocks can exchange recipient for donor orthologs (left). The large segments may also contain additional genes not present in the original recipient sequence (insertion) or may lack some genes that had been present (deletion). At the opposite end of the size spectrum, donor segments of <100 bp are often found in clusters to generate microcomplexity, with the potential to fine-tune genes or functional elements (right). Comparison of *M. canettii* genomes has identified similar mosaicism, in which sequences of entire genes are identical between some isolates, while other regions contain short regions of exchanged SNPs consistent with the recombinant patterns observed with DCT (14). doi:10.1128/microbiolspec.MGM2-0022-2013.f6

as demonstrated for *M. smegmatis*. However, this also requires that the mating type of the species is known. For example, one would predict that all *M. tuberculosis* isolates are the same mating type, either donor or recipient, because of its clonal nature. Therefore, DCT is unlikely to occur between *M. tuberculosis* isolates. But this does not preclude *M. tuberculosis* exchanging genetic information with other mycobacteria of the opposite mating type. Below, we describe three documented examples of mosaicism in mycobacterial genomes that could have been generated by DCT, although we caution hard conclusions because of the above caveats.

DCT provides a plausible mechanism for the genome mosaicism observed in *M. canettii*

While DCT certainly is active in isolates of *M. smegmatis*, a bigger question is whether it occurs in other mycobacterial species and especially, from a public health perspective, the MTBC. Early sequencing studies of *M. tuberculosis* isolates found extremely low genetic variation, suggesting that *M. tuberculosis* does not undergo HGT, is evolutionarily young, and resulted from a recent clonal expansion (10–12). However, there is convincing evidence for HGT among *M. canettii* and other smooth-colony MTBC strains: genome comparisons between *M. canettii* isolates have identified large numbers of SNPs (see reference 71). Notably, clusters of SNPs are shared between different isolates, indicating that pairs of isolates have undergone recombination events via HGT (13). As a result of these pairwise recombination events, the *M. canettii* isolates display

genome-wide mosaicism similar to that observed following DCT in *M. smegmatis* (13, 14). It was proposed that *M. canettii* strains are extant members of a genetically diverse MTBC progenitor species, *Mycobacterium prototuberculosis*, whose members underwent frequent HGT, and that *M. tuberculosis* emerged from this pool of diverse species by acquisition of enhanced virulence mechanisms (13).

The unspecified HGT process underlying the *M. canettii* genome mosaicism is presumed to result from a series of sequential transfer events. However, DCT involving the ancestral *M. prototuberculosis* offers a plausible and parsimonious explanation for the remarkably similar mosaicism observed among the extant *M. canettii*. We could envision that DCT in *M. prototuberculosis* rapidly incorporated the necessary blend of parental genotypes that drove the emergence of a pathogenic, rough-colony morphology species, like *M. tuberculosis*, allowing their subsequent clonal expansion.

Has DCT occurred between *M. canettii* and *M. tuberculosis*?

While the HGT events proposed for *M. prototuberculosis* are thought to have occurred over 35,000 years ago, there are indicators that gene flux still occurs among the *M. canettii* isolates and *M. tuberculosis*. Sequence comparison of the *M. canettii* isolates with *M. tuberculosis* confirms that their genomes are highly conserved and syntenic, but the two species are significantly more divergent and are easily distinguished by

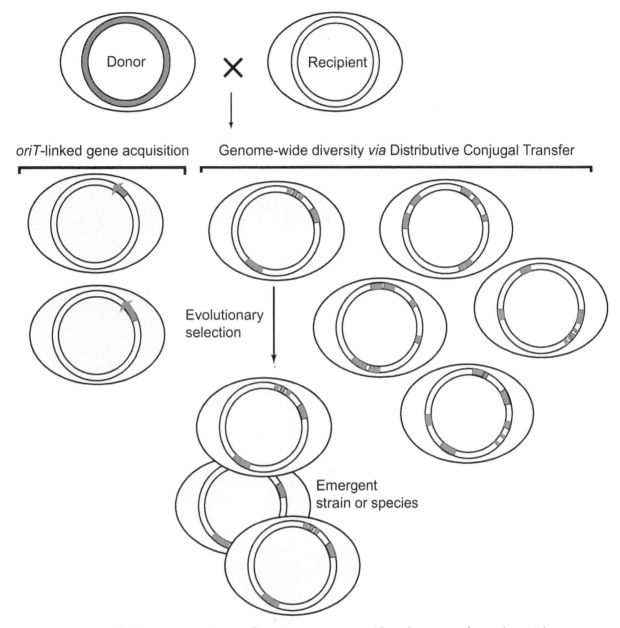

Figure 7 DCT generates instant diversity on a genome-wide scale. Progeny from a bacterial cross are shown for *oriT*-mediated chromosomal transfer (left) and for DCT (right). In a single Hfr cross, transconjugants can acquire segments of DNA proximal to *oriT*. Depending on the length of transfer, all progeny will have overlapping regions of the donor chromosome extending from *oriT*. By contrast, all DCT progeny are different because transfer initiates from sites all around the chromosome. In addition, DCT recombination results in both large segment exchanges and smaller regions of micro-heterogeneity, creating further diversity. The extreme diversity in the transconjugant population allows for rapid expansion under changing selective pressures. doi:10.1128/microbiolspec.MGM2-0022-2013.f7

numerous SNPs. However, there are shared blocks of sequence identity embedded in regions of nucleotide diversity detected between *M. tuberculosis* and individual *M. canettii* isolates (14, 66). These regions are entirely consistent with relatively recent recombination events that have occurred between the two species since their divergence. The strikingly similar pattern of mosaicism observed in our experimental system (Fig. 6), and among extant members of the MTBC, suggests that the latter may have been created by the same mechanism: DCT.

Genome mosaicism observed in *M. avium* could also result from DCT

A study on the glycopeptidolipid (GPL) biosynthetic pathways of four *M. avium* strains belonging to different serotypes led to the sequencing of their *gpl* gene clusters (67). The region was shown to be highly mosaic among the four strains, with multiple SNPs clearly defining shared regions and breakpoints. The lack of insertion sequences and the extended sequence continuity beyond the mosaic region strongly indicated acquisition by some form of HGT. Again, the products are remarkably similar to those generated by DCT.

We suspect that these observations of mosaicism represent the tip of the iceberg that will gradually be exposed as more mycobacterial species are sequenced and their genomes are compared. While the pathogenic MTBC species are frequently sequenced for public health reasons, their extreme similarity undermines their value in identifying HGT events.

DCT Adds Up to Reproductive Success

It is important to keep in mind that, in spite of the many superficial similarities between DCT and the mammalian version of sex, DCT is not reproduction *per se*. Sexual reproduction generally brings together two parents whose haploid genomes combine to create genetically blended progeny. A basic mathematical representation would be $1 + 1 = 3$, where each of the parents and the created single offspring total three individuals: a net population increase of one. Since the viability/fate of the donor bacterium is unknown in mycobacterial mating, DCT might either result in no net increase ($1 + 1 = 2$) or, in the case of donor suicide, a net decrease ($1 + 1 = 1$). Even though the short-term math may not tally, it is likely that in the long term, participants of DCT should eventually be present in greater numbers, as an occasional transconjugant may have a competitive advantage and divide more frequently.

The nearly boundless parameters of DNA transferred by DCT—large and small, isolated tracts or clustered microcomplexity—quickly create so many combinatorial possibilities that it becomes clear that no two transconjugants can be identical. A typical laboratory mating experiment generates >10^4 antibiotic selectable recombinants. If DCT transfers 10% of the donor genome on average (15), the corollary is that 90% of the transconjugants go undetected in this simple assay. Thus, of the >10^5 total transconjugants generated in an experiment, all will be different from each other and from the parental strains.

DCT Complements Spontaneous Mutation

Mycobacteria undoubtedly experience spontaneous mutation, a ubiquitous source of *de novo* genetic diversity. However, only a very small fraction of mutations will be advantageous, and if multiple sequence changes are required for a competitive advantage, the beneficial mutations must occur sequentially in the same lineage to become fixed in the population. This would likely require a very protracted time scale. DCT can expedite this process by actively mixing variants available within a community and letting competition select for the best combinations. Importantly, not only does DCT introduce variation instantly, but it also brings in variant sequences that have already been vetted. They were functional in the context of the donor genome and therefore should have a reduced chance of carrying debilitating mutations (Fig. 6, right). While this consideration might elevate the transconjugant viability rate somewhat, their overall success will ultimately depend on how well the altered genes interact with partner genes or with pathways remaining in the cell and how well the new genome interacts with the environment. This is analogous to trying on new shoes for a specific occasion. The existing pair might suffice, but another pair might be a better match for your clothing (loafers versus sneakers) or the environment (sandals versus boots) or the activity (cleats versus flippers). The rack of shoes from which to choose is limited only by the variation of the neighbors that are willing to donate them (DCT donors) and that the shoes fit (homology to recombine into the recipient). Whereas a random mutation might give rise to a nice fresh set of shoelaces, most mutations will more likely put a hole in the shoe or leave it stuck in the mud. The lack of mismatch repair systems in mycobacteria allows homologous recombination and, thus, further enhances the repertoire of mutations a recipient can acquire from distant relatives.

DCT Can Result in Indels

Significant evolutionary changes require entirely new activities that cannot be created by incremental changes to a finite genome and must be acquired through HGT. Classic horizontal gene transfer imports new genes (having no ortholog in the recipient) that are then integrated into the genome to result in indels (inserts/deletions). Indels initiate potentially major evolutionary leaps, as these might include entire operons that encode complete signaling pathways, biochemical activities, or structural components, which could open up opportunities to populate a new environment. The extremely large sizes of some transferred segments (250 kb) in

DCT allow it to introduce, or remove, nonhomologous segments by bridging to regions of homology (Fig. 6, left). Indeed, insertions of up to ~50 kb were identified in *M. smegmatis* transconjugants, representing new donor DNA sequences that were originally absent in the recipient chromosome (15).

DCT AS A GENETIC MAPPING TOOL

esx1 Encodes a Mating Identity Switch

The mosaic genomes generated in a single step by DCT look remarkably like the products of meiotic recombination in that homologous recombination indiscriminately swaps segments of DNA from both parents to create a new unique chromosome that has tracts from both parents. Single sperm sequencing has shown that spermatogenesis creates an average of 23 crossovers per genome (68), comparable to exchanges by DCT in a mycobacterial genome. This level of genome mixing has been exploited to map genetic traits through association studies in mammals. With the exception of monozygotic twins (and inbred mouse lines), individuals have different combinations of sequence variants in their genomes. In spite of their genome-wide variation, some individuals may share a gene that gives rise to a distinct phenotype. Genome-wide association studies (GWASs) look for variants—usually SNPs—that are overrepresented in the group exhibiting the trait of interest relative to a group that does not have the phenotype. Therefore, SNPs tightly linked with a gene that controls a trait will be highly enriched in the affected group. The overt similarities between meiotic and DCT-generated mosaicism enabled a similar approach to be applied to mycobacteria. A mycobacterial GWAS can theoretically map any genetic trait that differs between the parental strains. This trait could be as simple as colony morphology or color or as complex as a biochemical pathway, as long as there is a measurable phenotype. One practical application might be to map donor genes encoding drug resistance, upon transfer of resistance into the drug-susceptible recipient. Whole-genome sequencing of drug-resistant transconjugants should identify a segment of DNA in common containing the responsible gene(s).

Most relevantly for this review, this GWAS-DCT methodology was used to map a locus that conferred mating identity (*mid*) to *M. smegmatis* transconjugants (15). Unlike *oriT*-mediated Hfr transfer, a subset of mycobacterial transconjugants become donors (46) and therefore likely acquire a donor-conferring locus. Comparison of the genome sequence of just 10 donor-proficient transconjugants identified a single region in common, which encompassed the mc^{2}155 *esx1* locus (Fig. 4), indicating that *esx1* encodes a switch in mating identity from recipient to donor in these transconjugants (15). Despite the previous links between ESX-1 and transfer, this result was not anticipated, since transposon insertions or deletion of *esx1* genes do not result in a switch in mating phenotype. The ability to rapidly associate genes with phenotypes via this GWAS approach provides a simple and effective tool for mycobacterial geneticists. Importantly, DCT can be used to map genes by exchange of function, as opposed to traditional methods that employ loss-of-function mapping. The ease of sequencing bacterial genomes, combined with the dramatic drop in costs, makes this a cost-effective process.

Where Are the *mid* Genes in *esx1*?

In mammalian genetic studies, fine mapping of a genetic determinant can be achieved by performing successive backcrosses to genetically purify a locus in a recipient background. A similar approach was employed using DCT to introgress the *mid* locus in the recipient genome. F1 donors were crossed with the original recipient to generate (N1) progeny that were screened for donor proficiency. N1 donors were then used in a second backcross, and the process was repeated. Six serial backcrosses resulted in a purifying selection of the donor-conferring locus (and the Km^r genes used to select for transfer of donor DNA) such that donor-proficient transconjugants contained as little as 1.5% of donor DNA embedded in a recipient background (Fig. 8). Just as informative are those lines that were reiteratively scored as donors (transferring the donor *mid* region of the *esx1* locus at each generation), up until the last generation when some transconjugants, now scored as recipients, were also sequenced. In this last generation, the recipient-proficient transconjugants invariably retained all or most of the *esx1* genes of the recipient, as donor-specific SNPs throughout this region were not transferred (Fig. 8).

By combining the genome data of both donor-proficient backcross progeny (having donor *mid* genes that were sufficient to confer the donor phenotype) and those that are recipient-proficient (having donor genes that were insufficient to confer the donor phenotype), the *mid* locus was mapped within *esx1* to six genes: *Ms0069-0071* and *Ms0076-0078* (Fig. 8). None of the encoded proteins have been functionally annotated, so their putative role(s) in determining mating identity is currently unknown. The region spanning genes 0069 to 0071 is the most divergent part of the *esx1* locus

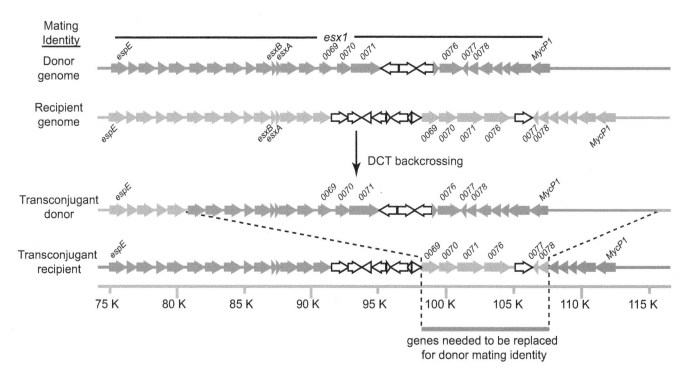

Figure 8 Mating identity is determined by genes within *esx1*. The *M. smegmatis esx1* locus spans 26 homologous genes in the donor (blue) and recipient (orange) parental strains (top). Genes *espE* and *MycP1* define the ends of the locus. *esxA* and *esxB* encode the primary secreted substrates EsxA and EsxB. Whole-genome sequencing and genome-wide association mapping of F1 transconjugants that exhibited a donor phenotype revealed that all donor-proficient transconjugants had an *esx1* locus of donor origin (Fig. 4). Mapping of the mating identity (*mid*) locus—the donor genes associated with the donor conjugal phenotype—to *esx1* was confirmed by successively backcrossing donor-proficient transconjugants with the recipient, while maintaining the donor phenotype (transconjugant donor). Backcrossed transconjugants that showed a recipient conjugal phenotype (transconjugant recipient) all shared a region of their *esx1* locus that was composed of recipient-derived genes, further refining the *mid* locus to six genes (*Ms0069-0071* and *Ms0076-0078*) near the 3′ end of *esx1*. White-filled gene symbols represent repetitive elements.
doi:10.1128/microbiolspec.MGM2-0022-2013.f8

between the donor and recipient. Orthologs of some of these genes are present in sequenced environmental mycobacterial species and to a lesser extent in the MTBC. More accurate predictions of potentially active DCT participants await a more definitive mapping and characterization of the pivotal *mid* activities in the existing *M. smegmatis* DCT model strains or directed empirical testing of new strains for DCT.

How Do the Mid Proteins Confer Mating Identity?

The genetic mapping data and transposon mutagenesis data provide compelling evidence for a role for ESX-1 in DNA transfer. But what is that role and how can it be addressed? Transposon insertions and targeted deletions in recipient genes that map within *esx1*, but

outside *mid*, abolish both DNA transfer and ESX-1 secretion, suggesting that there is a functional requirement for the apparatus *per se*. This requirement for ESX-1 likely reflects the need to secrete some of the Mid proteins to define mating identity. In support of this we, and others, have already shown that Ms0076 (EspB) is secreted (69, 70) and have evidence that at least one other Mid protein, Ms0077, is secreted (J. Krywy, A. Collins, T. A. Gray, and K. M. Derbyshire, unpublished data). Alternatively, Mid proteins could modify other ESX-1 substrates to exert an effect. Either way, any future mechanistic model of DCT is likely to begin with the Mid proteins, as they seem to dictate how a shared apparatus can have opposing activities in conjugation. Conceptually, communication between opposite mating types might be a good checkpoint activity to coordinate MPF and donor genome mobilization to the waiting recipient.

SUMMARY AND FUTURE PROSPECTS FOR DCT

The work described in this chapter demonstrates the unique nature of conjugal DNA transfer in mycobacteria. Our knowledge has expanded dramatically since the chapter on conjugation in the first edition of this book. Perhaps the biggest advance is the realization of DCT's potential impact on all mycobacteria. In 2000, *M. tuberculosis* was considered clonal, evolving only by random genetic drift and selection, untouched by HGT. Mycobacterial conjugation was assumed to be similar to *E. coli* Hfr transfer and restricted to *M. smegmatis* and, perhaps, other environmental mycobacteria.

Now, 13 years later, DCT can be almost envisaged as a fourth mechanism of HGT, quite distinct from *oriT*-mediated conjugation. DCT requires a functional ESX-1 system, previously considered a virulence determinant. The essential role for ESX-1 in DCT has provided an ideal genetic model system to dissect ESX-1 functions. Genome sequence analyses have provided strong, albeit indirect, evidence for extensive HGT among the MTBC, as well as other mycobacteria. The mosaicism observed in some of these genomes is entirely consistent with it occurring by DCT, suggesting that DCT is more prevalent among mycobacteria than previously appreciated. Together, these observations indicate that the MTBC is capable of recombination and that HGT has played a role in its evolution. We anticipate that additional examples from genome sequencing will further reinforce the ability of these species to diversify via HGT, in addition to genetic drift.

What might we expect in the next edition of this book? There are likely to be many advances, especially in defining the transfer genes, their protein structure, and the mechanism of transfer. However, we speculate that the more likely global impact will be on the application of DCT to genetically engineer strains and on understanding gene flow in mycobacteria and related species. Identification of the genes required for mating identity and those that mediate transfer will allow construction of defined mating pairs in all mycobacteria. Thus, DCT could be used to introduce defined mutations into the slow-growing pathogens from the genetically amenable fast-growing species or to create attenuated, hybrid strains specifically designed to enhance live vaccine development. The similarity of DCT to meiotic recombination certainly opens the genetic doors, and in this era of genome-scale science, we predict DCT will continue to be a major contributor to mycobacterial genetics.

The authors gratefully acknowledge input from all members of their laboratory over the years and the generous support of the NIAID.

Citation. Derbyshire KM, Gray TA. 2014. Distributive conjugal transfer: new insights into horizontal gene transfer and genetic exchange in mycobacteria. Microbiol Spectrum 2(1): MGM2-0022-2013.

References

1. **Frost LS, Leplae R, Summers AO, Toussaint A.** 2005. Mobile genetic elements: the agents of open source evolution. *Nat Rev Microbiol* 3:722–732.

2. **Thomas CM, Nielsen KM.** 2005. Mechanisms of, and barriers to, horizontal gene transfer between bacteria. *Nat Rev Microbiol* 3:711–721.

3. **Gogarten JP, Townsend JP.** 2005. Horizontal gene transfer, genome innovation and evolution. *Nat Rev Microbiol* 3:679–687.

4. **McDaniel LD, Young E, Delaney J, Ruhnau F, Ritchie KB, Paul JH.** 2010. High frequency of horizontal gene transfer in the oceans. *Science* 330:50.

5. **Nakamura Y, Itoh T, Matsuda H, Gojobori T.** 2004. Biased biological functions of horizontally transferred genes in prokaryotic genomes. *Nat Genet* 36:760–766.

6. **Ochman H, Lawrence JG, Groisman EA.** 2000. Lateral gene transfer and the nature of bacterial innovation. *Nature* 405:299–304.

7. **Wiedenbeck J, Cohan FM.** 2011. Origins of bacterial diversity through horizontal genetic transfer and adaptation to new ecological niches. *FEMS Microbiol Rev* 35: 957–976.

8. **Smith NH, Dale J, Inwald J, Palmer S, Gordon SV, Hewinson RG, Smith JM.** 2003. The population structure of *Mycobacterium bovis* in Great Britain: clonal expansion. *Proc Natl Acad Sci USA* 100:15271–15275.

9. **Smith NH, Gordon SV, de la Rua-Domenech R, Clifton-Hadley RS, Hewinson RG.** 2006. Bottlenecks and broomsticks: the molecular evolution of *Mycobacterium bovis*. *Nat Rev Microbiol* 4:670–681.

10. **Sreevatsan S, Pan X, Stockbauer KE, Connell ND, Kreiswirth BN, Whittam TS, Musser JM.** 1997. Restricted structural gene polymorphism in the *Mycobacterium tuberculosis* complex indicates evolutionarily recent global dissemination. *Proc Natl Acad Sci USA* 94: 9869–9874.

11. **Supply P, Warren RM, Banuls AL, Lesjean S, Van Der Spuy GD, Lewis LA, Tibayrenc M, Van Helden PD, Locht C.** 2003. Linkage disequilibrium between minisatellite loci supports clonal evolution of *Mycobacterium tuberculosis* in a high tuberculosis incidence area. *Mol Microbiol* 47:529–538.

12. **Brosch R.** 2002. A new evolutionary scenario for the *Mycobacterium tuberculosis* complex. *Proc Natl Acad Sci USA* 99:3684–3689.

13. **Guttierez MC, Brisse S, Brosch R, Fabre M, Omais B, Marmiesse M, Supply P, Vincent V.** 2005. Ancient origin and gene mosaicism of the progenitor of *Mycobacterium tuberculosis*. *PLoS Pathog* 1:1–7.

14. Supply P, Marceau M, Mangenot S, Roche D, Rouanet C, Khanna V, Majlessi L, Criscuolo A, Tap J, Pawlik A, Fiette L, Orgeur M, Fabre M, Parmentier C, Frigui W, Simeone R, Boritsch EC, Debrie AS, Willery E, Walker D, Quail MA, Ma L, Bouchier C, Salvignol G, Sayes F, Cascioferro A, Seemann T, Barbe V, Locht C, Gutierrez MC, Leclerc C, Bentley SD, Stinear TP, Brisse S, Medigue C, Parkhill J, Cruveiller S, Brosch R. 2013. Genomic analysis of smooth tubercle bacilli provides insights into ancestry and pathoadaptation of *Mycobacterium tuberculosis*. *Nat Genet* **45**:172–179.

15. Gray TA, Krywy JA, Harold J, Palumbo MJ, Derbyshire KM. 2013. Distributive conjugal transfer in mycobacteria generates progeny with meiotic-like genome-wide mosaicism, allowing mapping of a mating identity locus. *PLoS Biol* **11**:e1001602.

16. de la Cruz F, Frost LS, Meyer RJ, Zechner EL. 2010. Conjugative DNA metabolism in Gram-negative bacteria. *FEMS Microbiol Rev* **34**:18–40.

17. Firth N, Ippen-Ihler K, Skurray RA. 1996. Structure and function of the F factor and mechanism of conjugation, p 2377–2401. *In* Neidhardt FC (ed), Escherichia coli *and* Salmonella: *Cellular and Molecular Biology*, 2nd ed, vol. 2. ASM Press, Washington, DC.

18. Wozniak RA, Waldor MK. 2010. Integrative and conjugative elements: mosaic mobile genetic elements enabling dynamic lateral gene flow. *Nat Rev Microbiol* **8**: 552–563.

19. Guglielmini J, Quintais L, Garcillan-Barcia MP, de la Cruz F, Rocha EP. 2011. The repertoire of ICE in prokaryotes underscores the unity, diversity, and ubiquity of conjugation. *PLoS Genet* **7**:e1002222.

20. Christie PJ, Atmakuri K, Krishnamoorthy V, Jakubowski S, Cascales E. 2005. Biogenesis, architecture and function of bacterial type IV secretion systems. *Annu Rev Microbiol* **59**:451–485.

21. Bhatty M, Laverde Gomez JA, Christie PJ. 2013. The expanding bacterial type IV secretion lexicon. *Res Microbiol* **164**:620–639.

22. Grohmann E, Muth G, Espinosa M. 2003. Conjugative plasmid transfer in Gram-positive bacteria. *Microbiol Mol Biol Rev* **67**:277–301.

23. Alvarez-Martinez CE, Christie PJ. 2009. Biological diversity of prokaryotic type IV secretion systems. *Microbiol Mol Biol Rev* **73**:775–808.

24. Dunny GM. 2007. The peptide pheromone-inducible conjugation system of *Enterococcus faecalis* plasmid pCF10: cell-cell signaling, gene transfer, complexity and evolution. *Philos Trans R Soc Lond B Biol Sci* **362**: 1185–1193.

25. Lederberg J, Tatum EL. 1946. Gene recombination in *E. coli*. *Nature* **158**:558.

26. Wollman EL, Jacob F, Hayes W. 1956. Conjugation and genetic recombination in *Escherichia coli* K-12. *Cold Spring Harb Symp Quant Biol* **21**:141–162.

27. Crawford JT, Falkinham JO. 1990. Plasmids of the *Mycobacterium avium* complex, p 97–120. *In* McFadden J (ed), *Molecular Biology of the Mycobacteria*. Academic Press, San Diego, CA.

28. Movahedzadeh F, Bitter W. 2009. Ins and outs of mycobacterial plasmids. *Methods Mol Biol* **465**:217–228.

29. Pashley C, Stoker NG. 2000. Plasmids in mycobacteria, p 55–67. *In* Hatfull GF, Jacobs WR Jr (ed), *Molecular Genetics of Mycobacteria*. ASM Press, Washington DC.

30. Jucker MT, Falkingham JO. 1990. Epidemiology of infection by nontuberculous mycobacteria. *Am Rev Respir Dis* **142**:858–862.

31. Kirby C, Waring A, Griffin TJ, Falkinham JO 3rd, Grindley ND, Derbyshire KM. 2002. Cryptic plasmids of *Mycobacterium avium*: Tn552 to the rescue. *Mol Microbiol* **43**:173–186.

32. Picardeau M, Le Dantec C, Vincent V. 2000. Analysis of the internal replication region of a mycobacterial linear plasmid. *Microbiology* **146**:305–313.

33. Le Dantec C, Winter N, Gicquel B, Vincent V, Picardeau M. 2001. Genomic sequence and transcriptional analysis of a 23-kilobase mycobacterial linear plasmid: evidence for horizontal transfer and identification of plasmid maintenance systems. *J Bacteriol* **183**:2151–2164.

34. Rauzier J, Moniz-Pereira J, Gicquel-Sanzey B. 1988. Complete nucleotide sequence of pAl5000, a plasmid from *Mycobacterium fortuitum*. *Gene* **71**:315–321.

35. Stinear TP, Mve-Obiang A, Small PL, Frigui W, Pryor MJ, Brosch R, Jenkin GA, Johnson PD, Davies JK, Lee RE, Adusumilli S, Garnier T, Haydock SF, Leadlay PF, Cole ST. 2004. Giant plasmid-encoded polyketide synthases produce the macrolide toxin of *Mycobacterium ulcerans*. *Proc Natl Acad Sci USA* **101**:1345–1349.

36. Stinear TP, Pryor MJ, Porter JL, Cole ST. 2005. Functional analysis and annotation of the virulence plasmid pMUM001 from *Mycobacterium ulcerans*. *Microbiology* **151**:683–692.

37. Wang J, Parsons LM, Derbyshire KM. 2003. Unconventional conjugal DNA transfer in mycobacteria. *Nat Genet* **34**:80–84.

38. Leao SC, Matsumoto CK, Carneiro A, Ramos RT, Nogueira CL, Lima JD Jr, Lima KV, Lopes ML, Schneider H, Azevedo VA, da Costa da Silva A. 2013. The detection and sequencing of a broad-host-range conjugative IncP-1beta plasmid in an epidemic strain of *Mycobacterium abscessus* subsp. *bolletii*. *PLoS One* **8**:e60746.

39. Gormley EP, Davies J. 1991. Transfer of plasmid RSF1010 by conjugation from *Escherichia coli* to *Streptomyces lividans* and *Mycobacterium smegmatis*. *J Bacteriol* **173**:6705–6708.

40. Parsons LM, Jankowski CS, Derbyshire KM. 1998. Conjugal transfer of chromosomal DNA in *Mycobacterium smegmatis*. *Mol Microbiol* **28**:571–582.

41. Mizuguchi Y, Suga K, Tokunaga T. 1976. Multiple mating types of *Mycobacterium smegmatis*. *Jpn J Microbiol* **20**:435–443.

42. Mizuguchi Y, Tokunaga T. 1971. Recombination between *Mycobacterium smegmatis* strains Jucho and Lacticola. *Jpn J Microbiol* **15**:359–366.

43. Derbyshire KM, Willetts NS. 1987. Mobilization of the non-conjugative plasmid RSF1010: a genetic analysis of its origin of transfer. *Mol Gen Genet* **206**:154–160 (Erratum, **209**:411).

44. Everett R, Willetts NS. 1982. Cloning, mutation and location of the origin of conjugal transfer. *EMBO J* **1**:747–753.

45. Wang J, Derbyshire KM. 2004. Plasmid DNA transfer in *Mycobacterium smegmatis* involves novel DNA rearrangements in the recipient, which can be exploited for molecular genetic studies. *Mol Microbiol* **53**:1233–1241.

46. Wang J, Karnati PK, Takacs CM, Kowalski JC, Derbyshire KM. 2005. Chromosomal DNA transfer in *Mycobacterium smegmatis* is mechanistically different from classical Hfr chromosomal DNA transfer. *Mol Microbiol* **58**:280–288.

47. Daffe M, Reyrat JM. 2008. *The Mycobacterial Cell Envelope*. ASM Press, Washington, DC.

48. Flint JL, Kowalski JC, Karnati PK, Derbyshire KM. 2004. The RD1 virulence locus of *Mycobacterium tuberculosis* regulates DNA transfer in *Mycobacterium smegmatis*. *Proc Natl Acad Sci USA* **101**:12598–12603.

49. Coros A, Callahan B, Battaglioli E, Derbyshire KM. 2008. The specialized secretory apparatus ESX-1 is essential for DNA transfer in *Mycobacterium smegmatis*. *Mol Microbiol* **69**:794–808.

50. Abdallah AM, Gey van Pittius NC, DiGiuseppe Champion PA, Cox J, Luirink J, Vandenbroucke-Grauls CMJE, Appelmelk BJ, Bitter W. 2007. Type VII secretion: mycobacteria show the way. *Nat Rev Microbiol* **5**:883–891.

51. Bitter W, Houben EN, Bottai D, Brodin P, Brown EJ, Cox J, Derbyshire KM, Fortune SM, Gao LY, Liu J, Gey van Pittius NC, Pym AS, Rubin EJ, Sherman DR, Cole ST, Brosch R. 2009. Systematic genetic nomenclature for type VII secretion systems. *PLoS Pathog* **5**:e1000507.

52. Converse SE, Cox JS. 2005. A protein secretion pathway critical for *Mycobacterium tuberculosis* virulence is conserved and functional in *Mycobacterium smegmatis*. *J Bacteriol* **187**:1238–1245.

53. Smith GR. 1991. Conjugational recombination in *E. coli*: myths and mechanisms. *Cell* **64**:19–27.

54. Taylor AF, Smith GR. 1992. RecBCD enzyme is altered upon cutting DNA at a chi recombination hotspot. *Proc Natl Acad Sci USA* **89**:5226–5230.

55. Sinha KM, Unciuleac MC, Glickman MS, Shuman S. 2009. AdnAB: a new DSB-resecting motor-nuclease from mycobacteria. *Genes Dev* **23**:1423–1437.

56. Gupta R, Barkan D, Redelman-Sidi G, Shuman S, Glickman MS. 2011. Mycobacteria exploit three genetically distinct DNA double-strand break repair pathways. *Mol Microbiol* **79**:316–330.

57. Warner DF, Mizrahi V. 2011. Making ends meet in mycobacteria. *Mol Microbiol* **79**:283–287.

58. Rayssiguier C, Thaler DS, Radman M. 1989. The barrier to recombination between *Escherichia coli* and *Salmonella typhimurium* is disrupted in mismatch-repair mutants. *Nature* **342**:396–401.

59. Springer B, Sander P, Sedlacek L, Hardt WD, Mizrahi V, Schar P, Bottger EC. 2004. Lack of mismatch correction facilitates genome evolution in mycobacteria. *Mol Microbiol* **53**:1601–1609.

60. Shuman S, Glickman MS. 2007. Bacterial DNA repair by nonhomologous end joining. *Nat Rev Microbiol* **5**:852–861.

61. Possoz C, Ribard C, Gagnat J, Pernodet JL, Guerineau M. 2001. The integrative element pSAM2 from *Streptomyces*: kinetics and mode of conjugal transfer. *Mol Microbiol* **42**:159–166.

62. Vogelmann J, Ammelburg M, Finger C, Guezguez J, Linke D, Flotenmeyer M, Stierhof YD, Wohlleben W, Muth G. 2011. Conjugal plasmid transfer in *Streptomyces* resembles bacterial chromosome segregation by FtsK/SpoIIIE. *EMBO J* **30**:2246–2254.

63. Lee JY, Finkelstein IJ, Crozat E, Sherratt DJ, Greene EC. 2012. Single-molecule imaging of DNA curtains reveals mechanisms of KOPS sequence targeting by the DNA translocase FtsK. *Proc Natl Acad Sci USA* **109**:6531–6536.

64. Reuther J, Gekeler C, Tiffert Y, Wohlleben W, Muth G. 2006. Unique conjugation mechanism in mycelial streptomycetes: a DNA-binding ATPase translocates unprocessed plasmid DNA at the hyphal tip. *Mol Microbiol* **61**:436–446.

65. Sepulveda E, Vogelmann J, Muth G. 2011. A septal chromosome segregator protein evolved into a conjugative DNA-translocator protein. *Mob Genet Elements* **1**:225–229.

66. Namouchi A, Didelot X, Schock U, Gicquel B, Rocha EP. 2012. After the bottleneck: genome-wide diversification of the *Mycobacterium tuberculosis* complex by mutation, recombination, and natural selection. *Genome Res* **22**:721–734.

67. Krzywinska E, Krzywinski J, Schorey JS. 2004. Naturally occurring horizontal gene transfer and homologous recombination in *Mycobacterium*. *Microbiology* **150**:1707–1712.

68. Wang J, Fan HC, Behr B, Quake SR. 2012. Genome-wide single-cell analysis of recombination activity and de novo mutation rates in human sperm. *Cell* **150**:402–412.

69. McLaughlin B, Chon JS, MacGurn JA, Carlsson F, Cheng TL, Cox JS, Brown EJ. 2007. A mycobacterium ESX-1–secreted virulence factor with unique requirements for export. *PLOS Pathog* **3**:e105.

70. Xu J, Laine O, Masciocchi M, Manoranjan J, Smith J, Du SJ, Edwards N, Zhu X, Fenselau C, Gao L-Y. 2007. A unique mycobacterium ESX-1 protein cosecretes with CFP-10/ESAT-6 and is necessary for inhibiting phagosome maturation. *Mol Microbiol* **66**:787–800.

71. Bottai B, Stinear TP, Supply P, Brosch B. 2014. Mycobacterial pathogenomics and evolution. *Microbiol Spectrum* **2**(1):MGM2-0025-2013. doi:10.1128/microbiolspec.MGM2-0025-2013.

72. Derbyshire KM, Bardarov S. 2000. DNA transfer in mycobacteria: conjugation and transduction, p 93–107. *In* Hatfull GF, Jacobs WR Jr (ed), *Molecular Genetics of Mycobacteria*. ASM Press, Washington, DC.

73. Hatfull GF. 2014. Molecular genetics of mycobacteriophages. *Microbiol Spectrum* **2**(2):MGM2-0032-2013.

Molecular Genetics of Mycobacteria, 2nd Edition
Edited by Graham F. Hatfull and William R. Jacobs, Jr.
© 2014 American Society for Microbiology, Washington, DC
doi:10.1128/microbiolspec.MGM2-0032-2013

Graham F. Hatfull[1]

Molecular Genetics of Mycobacteriophages

5

INTRODUCTION

Mycobacteriophages are viruses that infect mycobacterial hosts including *Mycobacterium smegmatis* and *Mycobacterium tuberculosis*. The first mycobacteriophages were isolated in the late 1940s using *M. smegmatis* as a host (1, 2), followed by isolation of phages that infect *M. tuberculosis* (3). The application of phages with distinct host preferences to typing clinical mycobacterial isolates was recognized, and numerous studies on mycobacteriophage typing were published over the subsequent 30 years (4–12). In the 1950s a variety of further investigations focusing on the biology of these phages and their potential applications were initiated including studies on generalized transduction (13), viral morphology (14, 15), lysogeny (16–20), transfection of phage DNA (21, 22), and other biochemical features (23–40). These early contributions provided a critical foundation for the further characterization and application of mycobacteriophages to tuberculosis research that emerged from them.

Prior to the mid-1990s, the lack of methods for efficient and reproducible introduction of DNA into mycobacteria—coupled with the lack of simple plasmid vectors—represented substantial impediments to the development of facile systems for genetic manipulation of *M. tuberculosis* and other mycobacteria (41).

Bacteriophages played a critical role in overcoming these roadblocks, in part because of the ability to introduce phage DNA into *M. smegmatis* spheroplasts using strategies developed earlier for *Streptomyces* (42, 43) and because of the development of shuttle phasmids, which grow as phages in mycobacteria and as large plasmids in *Escherichia coli* (44, 45). These contributed to the development of methods for more efficient transfection and transformation, demonstration of genetically selectable systems, and general methods for gene transfer into mycobacteria (46–49).

In the early 1990s, the first complete genome sequence of a mycobacteriophage was described (50), followed by another dozen or so over the following decade (51–54). With the advancements in DNA sequence technologies and the development of integrated research and education programs in phage discovery and genomics (55–58), the number of sequenced mycobacteriophage genomes now exceeds 500. These show substantial degrees of both genetic diversity and genetic novelty, providing insights into viral evolution and greatly expanding the potential for developing additional tools for mycobacterial genetics and for gaining insights into the physiology of their hosts (59–61).

In this chapter I review our current understanding of mycobacteriophages including their genetic diversity

[1]Department of Biological Sciences, University of Pittsburgh, Pittsburgh, PA 15260.

and applications for tuberculosis genetics. The focus will primarily be on recent developments, and there are a number of other reviews that the reader may find useful (18, 59, 60, 62–78).

GENERAL ASPECTS OF MYCOBACTERIOPHAGES

Mycobacteriophage Isolation

Bacteriophages can typically be isolated from any environment in which their bacterial host—or close relatives of their host—is present. There is no obvious reservoir of *M. tuberculosis* outside of its human host, but mycobacteriophages have been isolated from various patient samples including stool (79–81). However, because there are numerous saprophytic mycobacterial relatives, mycobacteriophages can be readily isolated from environmental samples such as soil or compost, using *M. smegmatis* or other nonpathogenic mycobacteria as hosts; a subset of these phages also infect *M. tuberculosis* (discussed further below). Phages can also be isolated by release from lysogenic host bacteria (82, 83), and phage genomic sequences can be identified in sequenced mycobacterial genomes (83–88).

M. smegmatis has proven to be a useful surrogate for phage isolation, using either direct plating from environmental samples or by enrichment, in which the sample is first incubated with *M. smegmatis* to promote amplification of phages for that host. Typically, direct plating yields only a small number of plaques from ~10% of the samples that are tested, whereas enrichment generates plaques from a higher proportion of samples, and the phage titers can be greater than 10^9 plaque forming units (pfu)/ml. Although enrichment might be anticipated to reduce the diversity of the phages isolated due to the potential growth advantages that a subset of phages might enjoy under these conditions, there is little evidence to support this with the mycobacteriophages. However, it does appear to alter the prevalence with which different phage types are recovered. For example, phages with one particular genome type (Cluster A; see below for details) represent over 40% of the phages recovered after enrichment, whereas they are only 25% isolated by direct plating. However, phages of a second genome type (Cluster B) are more prevalently recovered by direct plating (27%) than by enrichment (13%). But both of these general phage types themselves encompass considerable diversity and can be divided into several subtypes (i.e., subclusters), and enrichment also influences the abundance of particular subclusters. For example, phages of

Subcluster A4 are relatively abundant in enriched samples (25% of the Cluster A phages, and 10% of the total) relative to direct plating (10% of Cluster A phages and 1% of the total). However, these biases in isolation should be interpreted cautiously, because there are numerous influences on which particular phages are selected for sequencing and further characterization. Additional hosts including *M. tuberculosis* could be used for enrichment, although plating directly on *M. tuberculosis* lawns is challenging because of its slow growth and the propensity for growth of contaminants that increase the difficulty of identifying plaques.

All mycobacteriophages examined to date carry double-stranded DNA (dsDNA). No phages with single-stranded DNA (ssDNA) or RNA genomes—or with virion morphologies other than the caudoviruses (see below)—have been described. It is plausible that they exist in these or other environments but that the isolation methods used to date bias against their growth and recovery. Phages with ssDNA or RNA genomes have been described for many other bacterial hosts, and it is somewhat surprising that none have been recovered for the mycobacteria. However, we note that this is also true for the phages described to date for all other species within the *Actinomycetales*.

Mycobacteriophage Virion Morphologies

The virion morphologies of many mycobacteriophages have been examined, and although three morphotypes of dsDNA tailed phages (*Siphoviridae*, *Myoviridae*, and *Podoviridae*) have been described for other hosts, only phages with siphoviral and myoviral morphologies (with long, flexible, noncontractile tails and contractile tails, respectively) have been identified for the mycobacteria (Fig. 1). No podoviral morphotypes (with short stubby tails) have been described, and with several hundred phages examined microscopically, it seems likely that these are truly excluded from the mycobacteriophage population. Although this could result from evolutionary exclusion, the rampant horizontal genetic exchange common to bacteriophages would suggest that this is unlikely, and it is plausible that a physical barrier, such as the complex mycobacterial cell wall, prevents them from accessing the cell membrane for successful infection. This may also account for why phages other than the tailed dsDNA phages have yet to be discovered.

By far the majority (>90%) of mycobacteriophages have siphoviral morphologies with long flexible tails, and only ~8% have myoviral morphologies, all of which belong to a single genome type (Cluster C; see below). Most of the siphoviral morphotypes have isometric

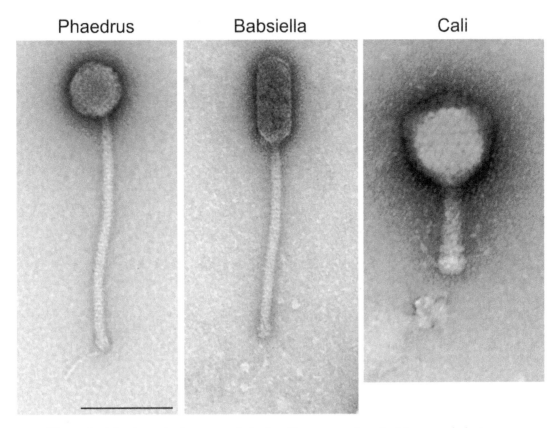

Figure 1 Mycobacteriophage morphologies. Three examples of virion morphologies are illustrated. Phaedrus and Babsiella exhibit siphoviral morphologies with long flexible tails; Phaedrus has an isometric head, whereas the Babsiella head is prolate. Cali is an example of myoviral morphology. Scale bar is 100 nm. doi:10.1128/microbiolspec.MGM2-0032-2013.f1

heads (varying in diameter from 50 to 80 nm), but a few have prolate heads with length:width ratio ranges from 2.5:1 to 4:1 (Fig. 1). There is substantial variation in tail length, ranging from 135 to 350 nm (89). The range in diameters of the isometric capsids—corresponding to differences in genome length—likely reflects different triangulation numbers in the capsid structures, but few have been determined experimentally. An electron micrograph–based reconstruction of mycobacteriophage *Araucaria* shows it has a T = 7 capsid symmetry (83), and a cryo-EM reconstruction of the phage BAKA capsid shows it has a T = 13 symmetry (90).

Phage Discovery and Genomics as an Integrated Research and Education Platform

Because phage genomes are relatively small, sequencing them is much less of a technological feat than it was 25 years ago. The chief limitation thus becomes the availability of phage isolates to sequence and characterize. In developing the Phage Hunters Integrating Research

and Education (PHIRE) program at the University of Pittsburgh, this research mission (i.e., discovering and genomically characterizing new mycobacteriophages) is paired with a mission in science education, specifically to develop programs that introduce young students (high school and undergraduate students) to authentic scientific research (55–58). The success of this program fueled implementation of a nationwide research course for freshman undergraduate students in collaboration with the Science Education Alliance (SEA) program of the Howard Hughes Medical Institute. The SEA Phage Hunters Advancing Genomic and Evolutionary Science (SEA-PHAGES) initiative began in 2008 and has involved over 4,000 undergraduate students at over 70 institutions ranging from R1 research universities to primarily undergraduate teaching colleges. The research accomplishments of the PHIRE and SEA-PHAGES programs are reflected in the massive increase in the number of completely sequenced mycobacteriophage genomes, growing from fewer than 20 to more than 500 in a 10-year span (www.phagesdb.org). The impact on science education is indicated by numerous parameters including student surveys such as the Classroom Undergraduate Research Experience (CURE) instrument (91).

Successful implementation of the PHIRE and SEA-PHAGES programs is predicated on a suite of features that characterize phage discovery and genomics. The enormous diversity of the phage population stacks the deck in favor of individual phage hunters isolating a phage with novel features, in some instances being completely unrelated to previously sequenced phages. This, however, is only apparent from the genomic characterization, because little or nothing can be learned about phage diversity and evolution without the genomic sequence. Students name their phages following a nonsystematic nomenclature that reflects the individualistic aspects of the phage genomes (see below). With over 500 completely sequenced genomes of phages isolated on *M. smegmatis* mc^{2}155, the prospects of isolating a phage with a completely novel DNA sequence is greatly diminished, although it is still extremely rare to find two phages with identical or near-identical genomes. However, the diversity of phages isolated using other—but closely related—hosts is likely to be just as great, and it is easy to envisage that large collections of sequenced phage genomes using numerous hosts within the *Actinomycetales* are likely to substantially advance our understanding of viral evolution.

Seven additional attributes of the PHIRE program have been described that contribute to its productivity (55, 56). These are (i) technical simplicity in beginning the project, (ii) the lack of requirement for prior advanced conceptual understanding, (iii) flexibility in timing and implementation, (iv) multiple milestones for success, (v) a parallel project design with strong peer-mentoring opportunities and the potential to engage large numbers of students, (vi) authentic research leading to peer-reviewed publications, and (vii) project ownership that powerfully engages students in their own individual contributions. These attributes should be transferable to other research and education platforms, and there are several excellent examples (92–95).

MYCOBACTERIOPHAGE GENOMICS

At the time of writing, there are a total of 285 complete mycobacteriophage genomes available in GenBank, and we will focus on these in this chapter (Table 1). The total number of sequenced mycobacteriophages is larger—currently 531 (www.phagesdb.org)—and the outstanding genomes will be available in GenBank pending completion and review of the annotations. Of

the 285, all except one were either isolated on *M. smegmatis* or are known to infect *M. smegmatis*. The single exception is DS6A (61), whose host preference is restricted to the *M. tuberculosis* complex (6, 24).

Grouping of Mycobacteriophages into Clusters and Subclusters

A simple dotplot comparison of the 285 genomes reveals an obvious feature of these genomes: that there is substantial diversity (i.e., many different types) but that the diversity is heterogeneous, and some phages are more closely related to each other than to others (Fig. 2). To recognize this heterogeneity and to simplify discussion, analysis, and presentation, these phages are assorted into groups called clusters (Cluster A, Cluster B, Cluster C, etc.), with the main criterion being that grouping of phages within a cluster requires recognizable nucleotide sequence similarity spanning over 50% of genome lengths with another phage in that cluster (56). Phages without any close relatives are referred to as singletons. For most phages, assignment to a cluster is simple, and extensive DNA sequence similarity is clear and apparent (see Fig. 3). However, assignment of phages to clusters is a taxonomy of convenience and does not reflect well-defined distinctions based on phylogeny or evolutionary relationships. Closer examination of sequence relationships reveals that they are mosaic and are constructed from segments swapped horizontally across the phage population (54, 96). This mosaicism is evident at both the DNA and the gene product levels and imposes challenges to some cluster assignments.

Difficulties in cluster assignment have become more prevalent as the number of sequenced genomes has increased and generally fall into two categories. In the first, there are examples of genomes that are distantly related such that the nucleotide sequence similarity extends over substantial parts of the genomes but is sufficiently weak that it barely rises above the threshold levels of recognizable similarity, whether it is viewed by dotplot analysis or more quantitative methods. One example is phage Wildcat, which is currently classified as a singleton and although it has many similarities to the Cluster M phages, is not so closely related that it warrants inclusion in the cluster. A second conundrum arises from genomes that have segments of strong similarity to other phages, but over a span that either doesn't convincingly meet the 50% threshold or spans more than 50% of one genome but not the other. An example is the inclusion of phage Che9c in Cluster I, where it belongs somewhat tenuously. Notwithstanding these

Table 1 Sequenced mycobacteriophage genomes

Phage	Cluster	Accession #	Length (bp)	GC %
Aeneas	A1	JQ809703	53684	63.6
Bethlehem	A1	AY500153	52250	63.2
BillKnuckles	A1	JN699000	51821	63.4
BPBiebs31	A1	JF957057	53171	63.4
Bruns	A1	JN698998	53003	63.6
Bxb1	A1	AF271693	50550	63.6
DD5	A1	EU744252	51621	63.4
Doom	A1	JN153085	51421	63.8
Dreamboat	A1	JN660814	51083	63.9
Euphoria	A1	JN153086	53597	63.7
Jasper	A1	EU744251	50968	63.7
JC27	A1	JF937099	52169	63.6
KBG	A1	EU744248	53572	63.6
KSSJEB	A1	JF937110	51381	63.6
Kugel	A1	JN699016	52379	63.8
Lesedi	A1	JF937100	50486	63.8
Lockley	A1	EU744249	51478	63.4
Marcell	A1	JX307705	49186	64.0
MrGordo	A1	JN020140	50988	63.8
Museum	A1	JF937103	51426	63.6
PattyP	A1	KC661273	52057	63.6
Perseus	A1	JN572689	53142	63.7
RidgeCB	A1	JN398369	50844	64.0
SarFire	A1	KF024726	53701	63.8
SkiPole	A1	GU247132	53137	63.6
Solon	A1	EU826470	49487	63.8
Switzer	A1	JF937108	52298	63.8
Trouble	A1	KF024724	52102	63.6
U2	A1	AY500152	51277	63.7
Violet	A1	JN687951	52481	63.8
Che12	A2	DQ398043	52047	62.9
D29	A2	AF022214	49136	63.5
L5	A2	Z18946	52297	62.3
Odin	A2	KF017927	52807	62.3
Pukovnik	A2	EU744250	52892	63.3
RedRock	A2	GU339467	53332	64.5
Trixie	A2	JN408461	53526	64.5
Turbido	A2	JN408460	53169	63.3
Bxz2	A3	AY129332	50913	64.2
HelDan	A3	JF957058	50364	64.0
JHC117	A3	JF704098	50877	64.0
Jobu08	A3	KC661281	50679	64.0
Methuselah	A3	KC661272	50891	64.2
Microwolf	A3	JF704101	50864	64.0
Rockstar	A3	JF704111	47780	64.3
Vix	A3	JF704114	50963	64.0
Arturo	A4	JX307702	51500	64.1
Backyardigan	A4	JF704093	51308	63.7
Dhanush	A4	KC661271	51373	63.9
Eagle	A4	HM152766.1	51436	63.9
Flux	A4	JQ809701	51370	63.9
ICleared	A4	JQ896627	51440	63.9
LHTSCC	A4	JN699015	51813	63.9

Table 1 Sequenced mycobacteriophage genomes *(Continued)*

Phage	Cluster	Accession #	Length (bp)	GC %
Medusa	A4	KF024733	51384	63.9
MeeZee	A4	JN243856	51368	63.9
Peaches	A4	GQ303263.1	51376	63.9
Sabertooth	A4	JX307703	51377	63.9
Shaka	A4	JF792674	51369	63.9
TiroTheta9	A4	JN561150	51367	63.9
Wile	A4	JN243857	51308	63.7
Airmid	A5	JN083853	51241	60.0
Benedict	A5	JN083852	51083	59.8
Cuco	A5	JN408459	50965	60.9
ElTiger69	A5	JX042578	51505	59.8
George	A5	JF704107	51578	61.0
LittleCherry	A5	KF017001	50690	60.9
Tiger	A5	JQ684677	50332	60.7
Blue7	A6	JN698999	52288	61.4
DaVinci	A6	JF937092	51547	61.5
EricB	A6	JN049605	51702	61.5
Gladiator	A6	JF704097	52213	61.4
Hammer	A6	JF937094	51889	61.3
Jeffabunny	A6	JN699019	48963	61.6
HINdeR	A7	KC661275	52617	62.8
Timshel	A7	JF957060	53278	63.1
Astro	A8	JX015524.1	52494	61.4
Saintus	A8	JN831654	49228	61.2
Alma	A9	JN699005	53177	62.5
PackMan	A9	JF704110	51339	62.6
Goose	A10	JX307704	50645	65.1
Rebeuca	A10	JX411619	51235	65.1
Severus	A10	KC661279	49894	64.4
Twister	A10	JQ512844	51094	65.0
ABU	B1	JF704091	68850	66.5
Chah	B1	FJ174694	68450	66.5
Colbert	B1	GQ303259.1	67774	66.5
Fang	B1	GU247133	68569	66.5
Harvey	B1	JF937095	68193	66.5
Hertubise	B1	JF937097	68675	66.4
IsaacEli	B1	JN698990	68839	66.5
JacAttac	B1	JN698989	68311	66.5
Kikipoo	B1	JN699017	68839	66.5
KLucky39	B1	JF704099	68138	66.5
Morgushi	B1	JN638753	68307	66.4
Murdoc	B1	JN638752	68600	66.4
Newman	B1	KC691258	68598	66.5
Oline	B1	JN192463	68720	66.4
Oosterbaan	B1	JF704109	68735	66.5
Orion	B1	DQ398046	68427	66.5
OSmaximus	B1	JN006064	69118	66.3
PG1	B1	AF547430	68999	66.5
Phipps	B1	JF704102	68293	66.5
Puhltonio	B1	GQ303264.1	68323	66.4
Scoot17C	B1	GU247134	68432	66.5
SDcharge11	B1	KC661274	67702	66.5
Serendipity	B1	JN006063	68804	66.5

Table 1 *(Continued)*

Phage	Cluster	Accession #	Length (bp)	GC %
ShiVal	B1	KC576784	68355	66.5
TallGrassMM	B1	JN699010	68133	66.5
Thora	B1	JF957056	68839	66.5
ThreeOh3D2	B1	JN699009	68992	66.5
UncleHowie	B1	GQ303266.1	68016	66.5
Vista	B1	JN699008	68494	66.5
Vortex	B1	JF704103	68346	66.5
Yoshand	B1	JF937109	68719	66.5
Arbiter	B2	JN618996	67169	68.9
Ares	B2	JN699004	67436	69.0
Hedgerow	B2	JN698991	67451	69.0
Qyrzula	B2	DQ398048	67188	68.9
Rosebush	B2	AY129334	67480	68.9
Akoma	B3	JN699006	68711	67.5
Athena	B3	JN699003	69409	67.5
Daisy	B3	JF704095	68245	67.6
Gadjet	B3	JN698992	67949	67.5
Kamiyu	B3	JN699018	68633	67.5
Phaedrus	B3	EU816589	68090	67.6
Phlyer	B3	FJ641182.1	69378	67.5
Pipefish	B3	DQ398049	69059	67.3
ChrisnMich	B4	JF704094	70428	69.1
Cooper	B4	DQ398044	70654	69.1
KayaCho	B4	KF024729	70838	70.0
Nigel	B4	EU770221	69904	68.3
Stinger	B4	JN699011	69641	68.6
Zemanar	B4	JF704104	71092	68.9
Acadian	B5	JN699007	69864	68.4
Reprobate	B5	KF024727	70120	68.3
Alice	C1	JF704092	153401	64.7
ArcherS7	C1	KC748970	156558	64.7
Astraea	C1	KC691257	154872	64.7
Ava3	C1	JQ911768	154466	64.8
Breeniome	C1	KF006817	154434	64.8
Bxz1	C1	AY129337	156102	64.8
Cali	C1	EU826471	155372	64.7
Catera	C1	DQ398053	153766	64.7
Dandelion	C1	JN412588	157568	64.7
Drazdys	C1	JF704116	156281	64.7
ET08	C1	GQ303260.1	155445	64.6
Ghost	C1	JF704096	155167	64.6
Gizmo	C1	KC748968	157482	64.6
LinStu	C1	JN412592	153882	64.8
LRRHood	C1	GQ303262.1	154349	64.7
MoMoMixon	C1	JN699626	154573	64.8
Nappy	C1	JN699627	156646	64.7
Pio	C1	JN699013	156758	64.8
Pleione	C1	JN624850	155586	64.7
Rizal	C1	EU826467	153894	64.7
ScottMcG	C1	EU826469	154017	64.8
Sebata	C1	JN204348	155286	64.8
Shrimp	C1	KF024734	155714	64.7
Spud	C1	EU826468	154906	64.8

Table 1 Sequenced mycobacteriophage genomes *(Continued)*

Phage	Cluster	Accession #	Length (bp)	GC %
Wally	C1	JN699625	155299	64.7
Myrna	C2	EU826466	164602	65.4
Adjutor	D	EU676000	64511	59.9
Butterscotch	D	FJ168660	64562	59.7
Gumball	D	FJ168661	64807	59.6
Nova	D	JN699014	65108	59.7
PBI1	D	DQ398047	64494	59.8
PLot	D	DQ398051	64787	59.8
SirHarley	D	JF937107	64791	59.6
Troll4	D	FJ168662	64618	59.6
244	E	DQ398041	74483	63.4
ABCat	E	KF188414	76131	63.0
Bask21	E	JF937091	74997	62.9
Cjw1	E	AY129331	75931	63.7
Contagion	E	KF024732	74533	63.1
Dumbo	E	KC691255	75736	63.0
Elph10	E	JN391441	74675	63.0
Eureka	E	JN412590	76174	62.9
Henry	E	JF937096	76049	63.0
Kostya	E	EU816591	75811	63.5
Lilac	E	JN382248	76260	63.0
Murphy	E	KC748971	76179	62.9
Phaux	E	KC748969	76479	62.9
Phrux	E	KC661277	74711	63.1
Porky	E	EU816588	76312	63.5
Pumpkin	E	GQ303265.1	74491	63.0
Rakim	E	JN006062	75706	62.9
SirDuracell	E	JF937106	75793	62.9
Toto	E	JN006061	75933	63.0
Ardmore	F1	GU060500	52141	61.5
Bobi	F1	KF114874	59179	61.7
Boomer	F1	EU816590	58037	61.1
Che8	F1	AY129330	59471	61.3
Daenerys	F1	KF017005	58043	61.6
DeadP	F1	JN698996	56461	61.6
DLane	F1	JF937093	58899	61.9
Dorothy	F1	JX411620	58866	61.4
DotProduct	F1	JN859129	55363	61.8
Drago	F1	JN542517	54411	61.2
Fruitloop	F1	FJ174690	58471	61.8
GUmbie	F1	JN398368	57387	61.4
Hamulus	F1	KF024723	57155	61.8
Ibhubesi	F1	JF937098	55600	61.2
Job42	F1	KC661280	59626	61.2
Llij	F1	DQ398045	56852	61.5
Mozy	F1	JF937102	57278	61.1
Mutaforma13	F1	JN020142	57701	61.3
Pacc40	F1	FJ174692	58554	61.3
PMC	F1	DQ398050	56692	61.4
Ramsey	F1	FJ174693	58578	61.2
RockyHorror	F1	JF704117	56719	61.1
SG4	F1	JN699012	59016	61.9
Shauna1	F1	JN020141	59315	61.7

Table 1 *(Continued)*

Phage	Cluster	Accession #	Length (bp)	GC %
ShiLan	F1	JN020143	59794	61.4
SiSi	F1	KC661278	56279	61.5
Spartacus	F1	JQ300538	61164	61.7
Taj	F1	JX121091	58550	61.9
Tweety	F1	EF536069	58692	61.7
Velveteen	F1	KF017004	54314	61.5
Wee	F1	HQ728524	59230	61.8
Avani	F2	JQ809702	54470	61.0
Che9d	F2	AY129336	56276	60.9
Jabbawokkie	F2	KF017003	55213	61.1
Yoshi	F2	JF704115	58714	61.0
Angel	G	FJ973624	41441	66.7
Avrafan	G	JN699002	41901	66.6
BPs	G	EU568876	41901	66.6
Halo	G	DQ398042	42289	66.7
Hope	G	GQ303261.1	41901	66.6
Liefie	G	JN412593	41650	66.8
Konstantine	H1	FJ174691	68952	57.4
Predator	H1	EU770222	70110	56.4
Barnyard	H2	AY129339	70797	57.5
Babsiella	I1	JN699001	48420	67.1
Brujita	I1	FJ168659	47057	66.8
Island3	I1	HM152765	47287	66.8
Che9c	I2	AY129333	57050	65.4
BAKA	J	JF937090	111688	60.7
Courthouse	J	JN698997	110569	60.9
LittleE	J	JF937101	109086	61.3
Omega	J	AY129338	110865	61.4
Optimus	J	JF957059	109270	60.8
Redno2	J	KF114875	108297	60.9
Thibault	J	JN201525	106327	60.8
Wanda	J	KF006818	109960	60.8
Adephagia	K1	JF704105	59646	66.6
Anaya	K1	JF704106	60835	66.4
Angelica	K1	HM152764	59598	66.4
BarrelRoll	K1	JN643714	59672	66.6
CrimD	K1	HM152767	59798	66.9
JAWS	K1	JN185608	59749	66.6
TM4	K2	AF068845	52797	68.1
MacnCheese	K3	JX042579	61567	67.3
Pixie	K3	JF937104	61147	67.3
Fionnbharth	K4	JN831653	58076	68.0
Larva	K5	JN243855	62991	65.3
JoeDirt	L1	JF704108	74914	58.8
LeBron	L1	HM152763	73453	58.8
UPIE	L1	JF704113	73784	58.8
Breezona	L2	KC691254	76652	58.9
Crossroads	L2	KF024731	76129	58.9
Faith1	L2	JF744988	75960	58.9
Rumpelstiltskin	L2	JN680858	69279	58.9
Winky	L2	KC661276	76653	58.9
Whirlwind	L3	KF024725	76050	59.3
Bongo	M	JN699628	80228	61.6

Table 1 Sequenced mycobacteriophage genomes *(Continued)*

Phage	Cluster	Accession #	Length (bp)	GC %
PegLeg	M	KC900379	80955	61.5
Rey	M	JF937105	83724	60.9
Butters	N	KC576783	41491	65.8
Charlie	N	JN256079	43036	66.3
Redi	N	JN624851	42594	66.1
Catdawg	O	KF017002	72108	65.4
Corndog	O	AY129335	69777	65.4
Dylan	O	KF024730	69815	65.4
Firecracker	O	JN698993	71341	65.5
BigNuz	P	JN412591	48984	66.7
Fishburne	P	KC691256	47109	67.3
Jebeks	P	JN572061	45580	67.3
Giles	Q	EU203571	53746	67.3
Send513	R	JF704112	71547	56.0
Marvin	S	JF704100	65100	63.4
Dori	Single	JN698995	64613	66.0
DS6A	Single	JN698994	60588	68.4
Muddy	Single	KF024728	48228	58.8
Patience	Single	JN412589	70506	50.3
Wildcat	Single	DQ398052	78296	57.2

caveats, the 285 genomes are grouped into 19 clusters (Clusters A to S) and five singleton genomes (Table 1).

Within some of these groupings, there is further heterogeneity in the degrees and extent of DNA sequence similarities. Thus, some of the clusters can be divided into subclusters, and the groupings are usually apparent by differences in the average nucleotide identity values (56, 89, 97, 98). However, the diversity varies from cluster to cluster, and thus the average nucleotide identity subcluster threshold values are not fixed and are relative to other members of the cluster. The subdivisions are usually apparent by dotplot comparison and, for example, the five subclusters within Cluster B can be easily seen in Fig. 2. Of the current 19 clusters, 9 are divided into subclusters, with the greatest division being the 10 subclusters within Cluster A (Table 1). The total number of different cluster-subcluster-singleton types is 47.

Assortment of Genes into Phamilies using Phamerator

An alternative representative of genome diversity is through gene content comparison (56, 89, 98, 99). This is accomplished with the program Phamerator (100), which sorts genes into phamilies (phams)—groups of genes in which each gene product has amino acid sequence similarity to at least one other phamily member above threshold levels (typically 32.5% amino acid identity or a BLASTP cutoff of less than 10^{-50}) (100). Genomes can

then be compared according to whether they do or do not have a member of each phamily. In a Phamerator database (Mycobacteriophage_285) generated with these 285 genomes there are a total of 3,435 phams, of which 1,322 (38%) are orphams, i.e., they contain only a single gene member. The relationships can be displayed using a network comparison in Splitstree (101) and in general show groupings of related phages that closely mirror the cluster designations from nucleotide sequence comparisons (56). It should be noted, however, that not all of the genes within the genomes necessarily have the same evolutionary histories—as a consequence of the mosaicism generated by horizontal genetic exchange—and thus these relationships only reflect the aggregate similarities and differences among the phages (56).

Comparative Genome Analysis

An especially informative representation of genome comparisons is by alignment of genome maps. These maps can be generated using Phamerator (100) and provide an overview of similarities at both the nucleotide and amino acid sequence levels (Fig. 3). Pairwise nucleotide sequence similarities are calculated using BLASTN, and values above a threshold level are displayed as colored shading between adjacent genomes in the display (Fig. 3); this is especially useful for identifying differences between genomes that have occurred in relatively recent evolutionary time. Because each gene is represented as a box colored according to its phamily

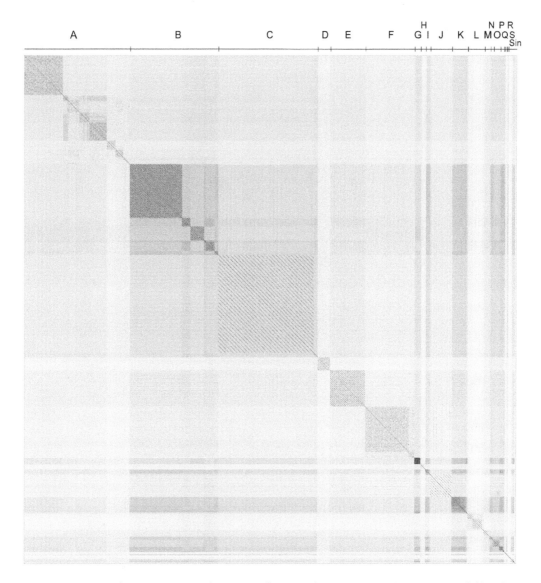

Figure 2 Dotplot comparison of 285 mycobacteriophage genomes. A concatenated file of 285 mycobacteriophage nucleotide sequences was compared against itself using the Gepard program (242) to generate the dotplot. The order of the genomes was arranged such that genomically related phages were adjacent to each other in this file, and the clusters of related phages (Clusters A, B, C, etc.) are shown above the plot. Five of the genomes are singletons with no closely related phages and are denoted collectively as Sin. doi:10.1128/microbiolspec.MGM2-0032-2013.f2

membership, gene content comparisons and synteny also are displayed (Fig. 3).

Comparison of the genome maps of phages Che9d and Jabbawokkie—both members of Subcluster F2 (Table 1)—provides an informative illustration (Fig. 3). A large segment (~22 kbp) in the leftmost parts of the genomes (extending from the left ends through Che9d 25 and Jabbawokkie 27), encoding the virion structure and assembly genes, is extremely similar at the nucleotide sequence level (98% identity), as shown by the

purple shading between the genomes in Fig. 3. The most obvious difference in this region is the insertion of an orpham (a gene that has no close mycobacteriophage relatives; i.e., it is the sole member of that pham) between the left physical end of the genome and a homing endonuclease (HNH) gene (Fig. 3), which is also predicted to encode an HNH endonuclease.

To the right of Che9d 25 and Jabbawokkie 27, there is another region of close similarity that extends to the left of the integrase genes and includes the lysis cassette

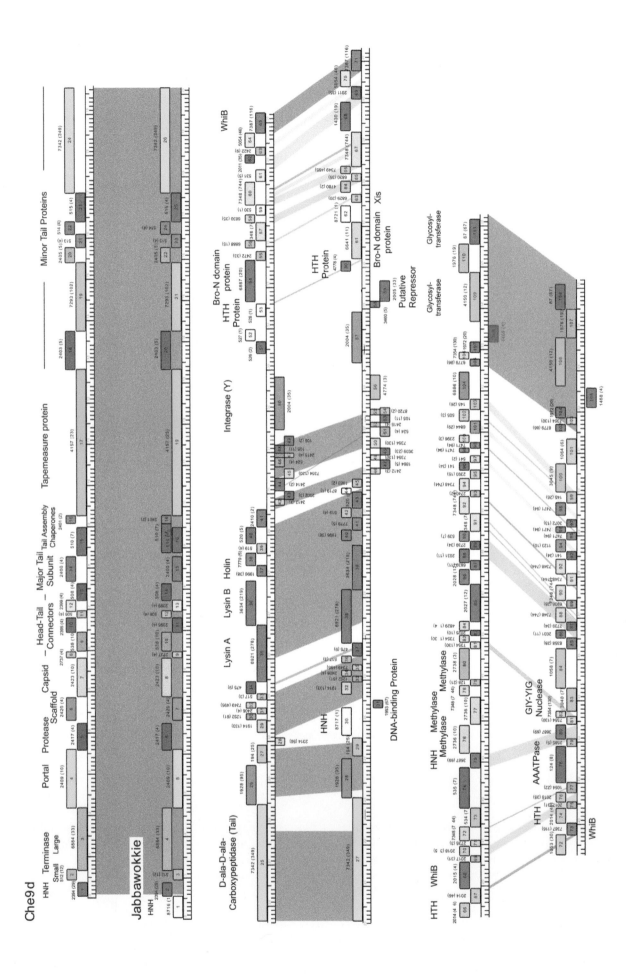

(Fig. 3). Although there are several large interruptions in the alignment, the matching segments are closely related and vary between 90% and 100% nucleotide identity. Apart from the lysis genes and another HNH insertion in Jabbawokkie (gene 30), most of these genes are of unknown function. The entire right parts of the genomes (from ~31 kbp coordinates to the right physical ends) are much less closely related at the nucleotide sequence level, and many regions are sufficiently different that they are below the threshold for displaying any similarity; there are also many small segments of intermediate similarity and a block of closely related sequences (96% identity) at the right end. Another noteworthy feature is that some of the genes encode proteins of the same phamily even though there is little detectable nucleotide similarity. Perhaps the best example is the integrase genes (Che9d 50, Jabbawokkie 57; Fig. 3); the proteins share 45% amino acid identity but do not have recognizable DNA sequence similarity.

This comparison illustrates six general aspects of mycobacteriophage genomes. First, the genes are tightly packed and there is little noncoding space. Second, they are mosaic in their architecture, with different segments having different evolutionary histories (54, 96). Third, among those with siphoviral morphologies, the synteny of the virion structure and assembly genes is well conserved. Fourth, there are large numbers of genes of unknown function (56, 89, 98). Fifth, there is an abundance of small open reading frames, especially in the right parts of the genomes. Sixth, even where putative gene functions can be assigned, it is often unclear what their specific roles are (for example, why are there two *whiB* genes in Che9d, distantly related and sharing only ~25% amino acid sequence identity?).

Virion Structure and Assembly Genes

Phages with siphoviral morphologies (i.e., with a long flexible noncontractile tail)—including all mycobacteriophages except those in Cluster C—have a well-defined operon of virion structure and assembly genes in a syntenically well-conserved order. The genomes are typically represented, following the Lambda precedent, with these genes at the left end of the genome and transcribed rightward (see Fig. 3, 4, and 5). Sometimes this structural operon is situated such that the terminase genes at its left end are close to the physical ends of the genome, but there are many examples (e.g., Cluster A phages) where other genes are situated in this interval. Large subunit terminase genes can be readily identified in most of these phages, and many—but not all—have an identifiable small terminase subunit. The other virion structural genes typically follow in the order: terminase, portal, capsid maturation protease, scaffolding, capsid subunit, 4 to 6 head-tail connector proteins, major tail subunit, tail assembly chaperones, Tapemeasure protein, and 5 to 10 minor tail proteins.

Although the gene order is very well conserved, there is substantial sequence diversity, to the extent that genes conferring some of the specific functions cannot be predicted from their sequence alone, although their position is also informative. However, because it is relatively simple to characterize the virions themselves, by SDS-PAGE, N-terminal sequencing, or mass spectrometry, correlations between the structural genes and proteins can be readily determined (50, 53, 90, 102, 103). In general, these studies show that the major capsid and tail subunits are the most abundant and that in some phages (e.g., L5) the capsid subunit is extensively covalently cross-linked, similar to the well-studied phage HK97 (50, 104). A scaffold protein gene involved in head assembly is absent from some mycobacteriophages, although its function may be provided by a domain of the capsid subunit.

The Tapemeasure protein is simple to identify because it is typically encoded by the longest gene in the genome, and the two reading frames immediately upstream are expressed via a programmed translational frameshift (105). Although the sequences of the Tapemeasure proteins are very diverse, they often contain small conserved domains corresponding to peptidoglycan hydrolysis motifs (54). The precise roles for these is not clear, but in TM4 it has been shown that

Figure 3 Comparison of mycobacteriophage Che9d and Jabbawokkie genome maps. Mycobacteriophages Che9d and Jabbawokkie are grouped into Subcluster F2, and their genome maps are shown as represented by the Phamerator program (100). Each genome is shown with markers, and the shading between the genomes reflects nucleotide sequence similarity determined by BLASTN, spectrum-colored with the greatest similarity in purple and the least in red. Protein-coding genes are shown as colored boxes above or below the genomes, reflecting rightward or leftward transcription, respectively. Each gene is assigned a phamily (Pham) designation based on amino acid sequence similarity (see text), as shown above or below each box, with the number of phamily members shown in parentheses; genes shown as white boxes are orphams and have no other phamily members. Putative gene functions are indicated. doi:10.1128/microbiolspec.MGM2-0032-2013.f3

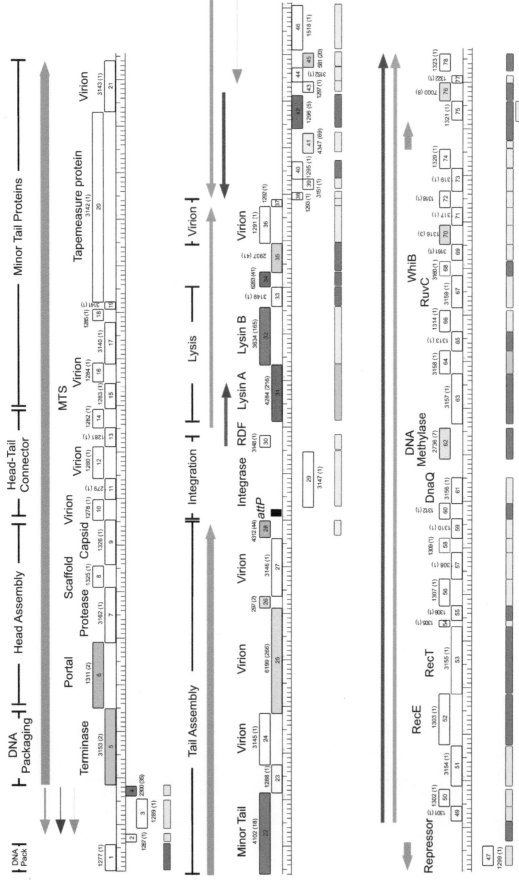

Figure 4 Functional genomics of mycobacteriophage Giles. A map of the mycobacteriophage Giles was generated using Phamerator and annotated as described for Fig. 3. Boxes below the genome indicate whether the gene is nonessential for lytic growth (yellow), likely essential (blue), or essential (green). Arrows indicate genes expressed in lysogeny (red) or early (green) or late (purple) lytic growth, with line thickness reflecting transcription strength. Reproduced with permission from Dedrick et al. (107). doi:10.1128/microbiolspec.MGM2-2013.f4

the motif is not required for phage assembly or growth, and removal of it results in a predictably shorter tail but also a reduction of its efficiency of infection of stationary phase cells (106). Following the *tapemeasure* gene are the minor tail protein genes encoding the proteins that constitute the tail tip structure. These are among some of the most diverse sequences in the mycobacteriophages, with many complex relationships, reflecting recombination and a mutational bias likely associated with host resistance. Many of the phages have a tail protein containing a D-Ala-D-Ala carboxypeptidase motif that presumably promotes infection through enzymatic modification or remodeling of the peptidoglycan. As described below, mutations conferring expanded host range also map to these minor tail protein genes (99).

Phages such as BPs and its Cluster G relatives have a compact virion structure and assembly operons with a total of 26 genes in a ~24-kbp span (103). At the opposite extreme, the Cluster J phages have a similar number of structural genes, but they span >30 kbp as a result of insertions of HNH endonucleases, introns, inteins (see below), and an assortment of other genes including those coding for methyltransferases and glycosyltransferases, whose roles are unknown. In Marvin, there is an unusual genome rearrangement in which a segment encoding several of the minor tail protein genes is displaced and sits among the nonstructural genes in the right part of the genome (102). The organizations of structural genes in the Cluster C phages with myoviral morphologies are much less well characterized.

Nonstructural Genes

The right parts of the genomes encoding nonstructural functions are characterized by an abundance of small open reading frames of unknown functions. They usually include a subset of genes whose functions can be bioinformatically predicted, and these are often associated with nucleotide metabolism or DNA replication. For example, some phages encode a DNA polymerase similar to *E. coli* DNA Polymerase I, whereas others encode alpha subunits of bacterial DNA Polymerase III. While it is simple to reason that these are involved in phage DNA replication, this has not been shown directly, and there are many mycobacteriophages that do not encode their own DNA polymerase at all. So why it would be needed in some phages but not in others is not clear. It is also common to find helicase and primase-like proteins, and sometimes recombination functions, including both *recA* and *recET*-like genes. In phage Giles, proteins of previously unknown

function are implicated in DNA replication from mutagenesis studies (see Fig. 4) (107); D29 gp65—predicted to be part of the RecA/DnaB helicase superfamily—has been demonstrated to be an exonuclease (108). But overall, very little is known about the mechanics of regulation of DNA replication in mycobacteriophages or many of the possible genes that are involved.

Some of the nonstructural genes are cytotoxic and kill the host when expressed or overexpressed. This has been extensively characterized in staphylococcal phages (109), where it has been developed as part of a drug development pipeline, but several mycobacteriophage-encoded cytotoxic proteins have also been identified. Initially, segments of the phage L5 genome were shown to be not tolerated on plasmid vectors and could not be transformed into *M. smegmatis* (110). Further dissection showed that L5 genes 77, 78, and 79 all have cytotoxic properties, with 77 being the most potent (111). Interestingly, expression of L5 gp79 seems to specifically inhibit cell division of *M. smegmatis* and promote filamentation (111). L5 gp77 appears to act by interacting directly with Msmeg_3532, a pyridoxal-5′-phosphate-dependent L-serine dehydratase that converts L-serine to pyruvate (112), although it is unclear what the consequences of the interaction are. The role of such an interaction in the growth of L5—or the many other Cluster A phages encoding homologues of L5 gp77—is unclear.

tRNA and tmRNA Genes

A considerable variety of tRNA gene repertoires are seen among the mycobacteriophages. Many of them do not appear to code for any tRNA genes at all, whereas others have dozens, close to a nearly complete coding set (54). But there are many intermediate variations, with some carrying just a single tRNA gene, and others with five or six. For example, many of the Cluster A phages carry one or more tRNA genes, but their specificities are quite varied. For example, just among the Subcluster A2 phages, L5 has three tRNA genes (tRNA^{Asn}, tRNA^{Trp}, and tRNA^{Gln}) (50), and D29 contains these plus two more (tRNA^{Glu} and tRNA^{Tyr}) (52). But Turbido and Pukovnik have a single tRNA^{Gln} gene, Redrock has a single tRNA^{Trp} gene, and Trixie has both tRNA^{Trp} and tRNA^{Gln} genes. So while there have been several efforts to account for the specific roles of these genes in accommodating phage codon usage requirements (113–116), this variation suggests considerable complexity. It is plausible, for example, that many of them are "legacy genes" that may have been required for growth in a particular unknown host in their recent evolutionary past but play no role in the

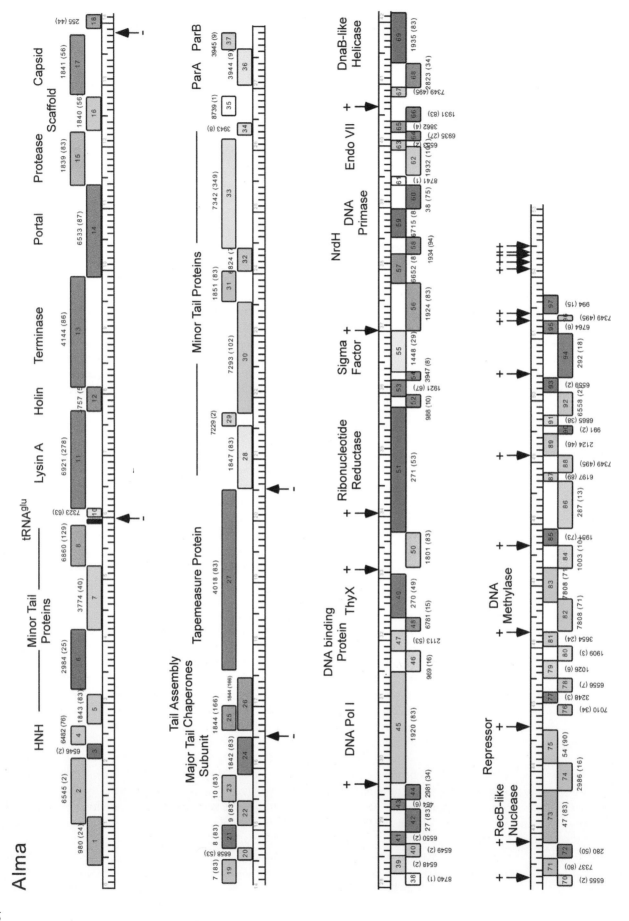

current host (see Fig. 6). Furthermore, there are examples of genes that appear to encode nonsense tRNAs as well as potential frameshifting tRNAs, suggesting regulatory roles in gene expression. The Subcluster K1 phages all encode a single tRNA gene (tRNA^{Trp}) but also carry genes coding for a putative RNA ligase (RtcB), which could a play a role in repair of host-mediated attack of the tRNA (97).

Some phages (e.g., Cluster C) also encode a tmRNA gene, similar to host-encoded tmRNA genes that play a role in the release of ribosomes from broken mRNA. The specific roles of the phage-encoded tmRNAs are unclear, but it is plausible that they enhance the pool of free ribosomes for late gene expression by releasing them from early transcripts when they are no longer required.

Mobile Elements: Transposons, Introns, Inteins, and HNH Endonucleases

The key architectural feature of phage genomes—pervasive mosaicism—is likely generated by illegitimate recombination events that occur at regions of DNA sharing little or no DNA sequence similarity (96). Although most of these may result from replication or repair "accidents," there are several active processes that could contribute, including transposition, intron and intein mobility, and HNH endonuclease activity.

A number of different transposable elements in mycobacterial genomes have been described, although these are generally not present in mycobacteriophage genomes. However, both active transposons and residual segments of transposons have been identified in the phages. Comparative analysis of the Cluster G phages provides compelling evidence for identification of novel ultra-small mycobacteriophage mobile elements (MPMEs), with two closely related subtypes, MPME1 and MPME2 (117). There are at least three instances in which an MPME insertion is present in one genome but absent from others, and because the Cluster G genomes are extremely similar to each other, the preintegration site and the precise insertion can be

readily interpreted. The 439-bp MPME1 elements contain 11-bp imperfect inverted repeats (IRs) near their ends and a small (125-codon) open reading frame encoding a putative transposase. The MPME1 elements in phages BPs and Hope are 100% identical to each other (from IR-L to IR-R), but the insertions differ in a 6-bp segment between IR-L and the preintegration site. The origin of this 6-bp sequence is unclear but does not correspond to a target duplication (117).

MPME1 elements are found in a variety of other mycobacteriophage genomes including those in Clusters F and I and are either identical or have no more than a single base pair difference; there are also truncated versions of the element in the Cluster O phages (117). MPME2 elements are 1 base pair longer and share 79% nucleotide sequence identity to MPME1. They are present in the Cluster G phage Halo and phages within Clusters F, I, and N, being either identical to each other or having no more than a single base pair difference. As yet there is no direct evidence for the mobility of the MPME elements, the frequency of movement, or the mechanism involved, although the comparative genomics suggests that these are active and moving at a respectable rate. No MPME elements have been identified outside of the mycobacteriophages.

Comparative genomics also reveals an IS110-like element in phage Omega, which is absent from phage LittleE, and the genomes are sufficiently similar in these regions to indicate the preintegration site and a 5-bp target duplication (90). There are no closely related copies in other mycobacteriophages, but there are some more distantly related segments in some Subcluster A1 phages (90). It is unclear if this is a remnant or an active transposon.

There are few examples of self-splicing introns in the mycobacteriophages, but two have been recently described, both in Cluster J phages (90) and both within virion structural protein genes. In phage BAKA the intron is small (265 bp) and is within a putative tail protein gene; in phage LittleE the intron is within the capsid subunit gene and is larger (819 bp) due to inclusion of a small open reading frame on the opposite

Figure 5 Genome map of mycobacteriophage Alma. The genome map of mycobacteriophage Alma was generated using Phamerator and is illustrated as described for Fig. 3. Alma is a Subcluster A6 phage and shares the features of other Cluster A phages in having multiple binding sites for its repressor protein (gp75). These stoperator sites are indicated by vertical arrows, and the orientation of the asymmetric sites relative to genome orientation are shown as (-) or (+). Stoperators were identified as sequences corresponding to the consensus sequence 5′-GATGAGTGTCAAG with no more than a single mismatch. Note that the stoperator consensus sequences can differ for different Subcluster A phages (98). doi:10.1128/microbiolspec.MGM2-0032-2013.f5

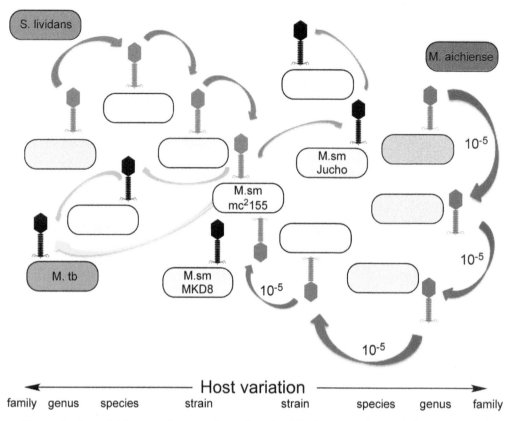

Figure 6 A model for mycobacteriophage diversity. The large number of different types of mycobacteriophages isolated on *M. smegmatis* mc²155 can be explained by a model in which phages can readily infect new bacterial hosts—either by a switch or an expansion of host range—using a highly diverse bacterial population that includes many closely related strains. As such, phages with distinctly different genome sequences and GC% contents infecting distantly related bacterial hosts, such as those to the left (red) or right (blue) extremes of a spectrum of hosts, can migrate across a microbial landscape through multiple steps. Each host switch occurs at a relatively high frequency ($\sim$1 in 10^5 particles, or an average of about one every 10^3 bursts of lytic growth) and much faster than either amelioration of phage GC% to its new host or genetic recombination. Two phages (such as those shown in red and blue) can thus "arrive" at a common host (*M. smegmatis* mc²155) but be of distinctly different types (clusters, subclusters, and singletons). Reproduced with permission from Jacobs-Sera et al. (99). doi:10.1128/microbiolspec.MGM2-0032-2013.f6

strand with similarities to homing endonucleases. Because both these proteins are required for virion assembly, and the capsid must be expressed at very high levels, the splicing events are expected to occur extremely efficiently. They are probably variants of group I introns, but little is known about their mechanisms of splicing (90). Other introns may be present in other genomes but have escaped detection because of insertion in genes of unknown function.

The rarity of self-splicing RNA introns contrasts with the relative abundance of inteins that splice out at the protein level. These are typically identifiable through conserved domains associated with intein splicing and are often apparent by comparative genomics.

Phage ET08 has a total of five inteins—the most in any single mycobacteriophage genome to date (98). These are located within gp3, gp79, gp202, gp239, and gp248, but the functions of these are largely unknown, except for gp3, which is predicted to be a nucleotidyltransferase. For all five, there are examples of intein-less homologues in other phages. Most of the inteins contain an endonuclease domain that is predicted to promote DNA cleavage in intein-less genes followed by repair with the intein-containing DNA copy. The intein in Bethlehem gp51 has been dissected biochemically and shown to represent a new type (type III) of splicing mechanism (118), of which the Omega gp206 intein is an unusual variant (119).

It is common for mycobacteriophage genomes to have one or more putative HNH homing endonucleases encoded as freestanding genes (rather than as part of an intein or coded within an intron). These are usually recognized from the presence of conserved motifs, and none have yet been shown to have nuclease activity. However, a comparison of the freestanding HNH encoded by Courthouse gene *51* and phage Thibault, which lacks it, shows a precise insertion and the presumed location of endonucleolytic cleavage (90). A second example is an HNH insertion in Omega that is somewhat messier and when compared to phage Optimus, lacking the insertion, is associated with a loss of 35 bp, presumably during the process of HNH acquisition (90). While the specific mechanisms generating these events are not clear, it is not difficult to see how these could play important roles in generating genomic mosaicism during phage evolution.

MYCOBACTERIOPHAGE-HOST INTERACTIONS

Mycobacteriophage Host Range and Host Range Expansion

Considerable efforts in using mycobacteriophages for phage typing illustrate that they readily discern between different hosts and that by using panels of phages with defined host ranges, the identities of unknown hosts can be predicted (6). Unfortunately, for the most part the typing phages are no longer available, and the genomic information is lacking. However, the host ranges of a few sequenced phages have been examined and are informative in regard to understanding the genetic diversity of the phages. For example, Rybniker et al. (120) tested the host ranges of 14 mycobacteriophages and showed that L5 and D29 (both Subcluster A2 phages) as well as Bxz2 (Subcluster A3) have broad host ranges and infect *M. tuberculosis*, BCG, *Mycobacterium scrofulaceum*, *Mycobacterium fortuitum*, *Mycobacterium chelonae*, and some strains of both *Mycobacterium ulcerans* and *Mycobacterium avium*, in addition to *M. smegmatis*. Other phages have intermediate host preferences, such as phage Wildcat, which infects *M. scrofulaceum*, *M. fortuitum*, and *M. chelonae*, and some (e.g., Barnyard, Che8, Rosebush) infect only *M. smegmatis* or its substrains, out of the strains tested (120). Jacobs-Sera et al. (99) determined whether a collection of 220 of the sequenced phages are able to infect *M. tuberculosis*. In general, there is a close correlation between the ability to infect *M. tuberculosis* and the cluster or subcluster type. For example, all of the

Cluster K phages—regardless of subcluster—efficiently infect *M. tuberculosis*; among the Cluster A phages, only phages within specific subclusters can infect *M. tuberculosis*, including Subclusters A2 and A3. Interestingly, the Cluster G phages (e.g., BPs) do not efficiently infect *M. tuberculosis*, although mutants that do can be isolated at a frequency of $\sim 10^{-5}$, and the mutants maintain the ability to infect *M. smegmatis*. Several of these mutants have been mapped, and all have single amino acid substitutions in gene *22*, encoding a putative tail fiber protein (107, 117). A simple interpretation is that the mutants overcome the need for a specific interaction with receptors on the *M. tuberculosis* cell surface. However, adsorption assays suggest there is a more complex explanation, because the mutants do not appear to adsorb to *M. tuberculosis* more efficiently than the wild-type parent phage, and surprisingly, they adsorb substantially better to *M. smegmatis*, in some cases, dramatically so (99). This is not easy to explain, because the mutants appear to infect *M. smegmatis* quite normally and were selected for infection of *M. tuberculosis*.

The same set of phages was also tested for their ability to infect two other strains of *M. smegmatis* (99). Many, but not all, of the phages efficiently infect these strains, and although there is a correlation between infection and cluster/subcluster type, it is statically weaker than with *M. tuberculosis* infection. But similarly, there are instances where phages are able to overcome the host barrier at moderate frequencies (10^{-5}), and examination of phage Rosebush mutants capable of infecting *M. smegmatis* Jucho (99) reveals amino acid substitutions in putative tail fibers, reminiscent of the Cluster G mutants that infect *M. tuberculosis*.

Mechanisms of Phage Resistance

Resistance to phage infection can occur by a variety of mechanisms including surface changes, and phages can presumably coevolve to overcome this resistance, reflecting the processes giving rise to the expanded-host-range mutants described above. However, in general, little is known about mycobacterial receptors for phage recognition or the determinants of host specificity. For phage I3 (which is not completely sequenced but is a Cluster C–like phage) resistance is accompanied by loss of cell-wall-associated glycopeptidolipids (121), and an *M. smegmatis* peptidoglycolipid, mycoside C (sm), has been implicated in the binding of phage D4 (which is also genomically uncharacterized) (122). Lyxose-containing molecules have also been proposed as possible receptors for the unsequenced phage Phlei (123, 124). There is clearly much to be learned about mycobacteriophage receptors and how phage tail structures specifically recognize them.

However, there are many other determinants of host range beyond the availability of cell wall receptors, including restriction/modification systems, lysogenic immunity, CRISPR elements (125), and various abortive infection systems including those mediated by toxin-antitoxin systems (126). *M. smegmatis* mc²155 has no known prophages, restriction modification systems, or CRISPR arrays, which may contribute to its suitability as a permissive host for phage isolation. However, *M. tuberculosis* H37Rv contains a type III-A family CRISPR array (127) and more than 80 putative toxin/antitoxin systems (128), perhaps suggesting development within a phage-rich environment in its recent evolutionary history. None of the spacer elements within the *M. tuberculosis* CRISPR array (which are used for spoligotyping *M. tuberculosis* isolates) closely match known phage sequences, and it is unclear whether the CRISPR arrays are currently active or contribute in any way to phage resistance (129). *M. tuberculosis* H37Rv has two small (~10 kbp) prophage-like elements in its genome (φRv1 and φRv2), but it is unlikely that these confer immunity to other phages (96). However, a likely intact 55-kbp prophage has been identified in a *Mycobacterium canettii* strain (88), and other prophages are present in the genomes of *M. avium* 104 (130), *Mycobacterium marinum* (87), *M. ulcerans* (85), and *Mycobacterium abscessus* (83, 84). Resistance of *M. smegmatis* to D29 infection can result from overexpression of the host multicopy phage resistance (*mpr*) gene, perhaps by alteration of the cell surface, although the specific mechanism is not known (131, 132).

Because phages can easily replicate from a single particle to vast numbers (there are 10^6 to 10^8 particles in a typical plaque), and mutants arise at moderate frequencies, host range mutants can readily arise within environmental populations. As such, phages are expected to move from one host to another at frequencies that are vastly greater than the time required to ameliorate their genomic features, such as coding biases and GC%, to a specific host. This leads to a model (Fig. 6) in which two key parameters contribute to the diversity of phages: the ability of phages to rapidly switch hosts, and the availability of a broad spectrum of closely related hosts in the environment from which the phages are isolated (99). This model can therefore account for the nature of mycobacteriophage genomes and predicts that similar phage diversity will be seen using other hosts and similar sampling (99).

Lysis Systems

Host cell lysis is a critical step in phage lytic growth, and in the prototype lambda system is both efficient and precisely timed (133, 134). Understanding the lysis systems of mycobacteriophages is of particular interest in that these can illuminate features of mycobacterial cell walls. Mycobacteriophage lysis systems are described in detail in reference 243 and will be discussed only briefly here.

Most mycobacteriophages encode at least three proteins required for lysis: a peptidoglycan-cleaving endolysin (also called Lysin A), Lysin B, and a holin; a few lack Lysin B, some either lack a holin or it is difficult to identify bioinformatically, but all encode an endolysin (135–138). Phage Ms6 and its relatives encode an additional protein that acts as a chaperone for delivery of the endolysin to its peptidoglycan target (139). The endolysins are diverse and modular in nature, reflecting an intragenic mosaicism that is a microcosm of the generally mosaic nature of the phage genomes (138). Many are composed of three segments: an N-terminal domain with predicted peptidase activity, a central domain that cleaves the peptidoglycan sugar backbone, and a C-terminal cell wall binding domain; there are, however, numerous departures from this general organization (138). In total, there are at least 25 different organizations (Org-A to Org-Y) with unique combinations of the constituent domains. Interestingly, the Ms6 *lysA* (Org-J) gene encodes a second lysis gene that is wholly embedded within *lysA* and is expressed by translation initiation from an internal start codon; phage mutants that express only the longer (Lysin384) or the shorter (Lysin241) endolysin are viable (140). Lysin B encodes an esterase that cleaves the linkage of the mycolyl-arabinogalactan to the peptidoglycan (136, 137) and, unlike lysin A, is dispensable for lytic growth (137). However, it is required for optimal lysis and efficient phage reproduction (137) and likely plays an analogous role to the spanins that facilitate fusion of the inner and outer membranes of Gram-negative bacteria during phage lysis (134, 141). Analysis of the Ms6 gp4 shows that it is a likely holin, with a signal-arrest-release (SAR) domain followed by a transmembrane domain (142). However, Ms6 gp5 also has a transmembrane domain, and gp4 and gp5 interact to facilitate lysis (142).

INTEGRATION SYSTEMS

Phages within more than one-half of the different clusters encode a phage integrase, and there are numerous examples of both tyrosine integrases (Int-Y) and serine integrases (Int-S); these include Clusters A, E, F,

G, I, J, K, L, M, and P and the singletons Dori and DS6A. But the distribution of Int-Y and Int-S types within different clusters is nonrandom. For example, all of the phages within Clusters E, F, G, I, J, K, L, and P (as well as Dori and DS6A) encode tyrosine integrases, and all of the Cluster M phages encode serine integrases. However, the distribution of different integrase types varies among the various subclusters within Cluster A. For example, all of the A1, A5, A7, and A10 phages encode an Int-S, as well as 2 of the 8 A3 phages, 11 of the 14 A4 phages, and 1 of the 2 A8 phages (the other has a deletion in this region of the genome); 7 of the 9 A2 phages, 6 A3 phages, and 3 A4 phages encode Int-Ys. These systems thus may evolve quickly relative to the rest of the genomes, presumably by promoting site-specific or quasi-site-specific recombination events.

Interestingly, two of the A2 phages, the one A9 phage, and all six A6 phages have a parA/B partitioning system at the location where the integrase is in the related phages (67). Presumably, these genomes replicate extrachromosomally during lysogeny, and the parA/B systems ensure their accurate segregation at cell division and conferring prophage maintenance—essentially the same functionality provided by the integration systems (143). How these phages are able to replicate their genomes during lysogeny is unclear, especially as their related genomes that integrate presumably do not.

For most of the genomes encoding a tyrosine integrase, the location of attP can be bioinformatically predicted. Typically, these use a bacterial attachment site (attB) overlapping the 3′ end of a tRNA gene, and to preserve the integrity of the tRNA following integration, attP and attB typically share 30- to 40-bp sequence identity. Thus, a BLASTN search of the phage genome against the M. smegmatis chromosome usually identifies the putative attB site. At least 12 attB sites have been identified this way, and usage of the sites has been shown for at least 8 of these (67, 90, 97, 103, 144–146). For the Cluster E and Cluster L phages, bioinformatic analysis fails to identify putative attB sites, and these await experimental determination.

Phages encoding serine integrases do not integrate at host tRNA genes and often use attB sites located within open reading frames (147, 148). The attP and attB sites typically share only minimal segments of sequence identity (3 to 10 bp) and thus cannot be readily predicted bioinformatically and require experimental analysis. The attB sites for two different mycobacterial Int-S systems have been described: Bxb1, which integrates into the groEL1 gene of M. smegmatis, and Bxz2, which integrates into the gene Msmeg_5156 (145, 147). Bxb1 integration has been especially useful for gaining insights into mycobacterial physiology because Bxb1 lysogens are defective in the formation of mature biofilms (149). This results specifically from inactivation of the groEL1 gene by integrative disruption and led to the demonstration that GroEL1 plays a role as a dedicated chaperone for mycolic acid biosynthesis (149, 150). It is not known if disruption of Msmeg_5156 or any other host gene as a consequence of phage integration via a serine integrase has specific physiological consequences.

Phage integration systems typically are carefully regulated in their recombination directionality, such that integrase catalyzes integrative recombination using attP and attB sites (252 bp and 29 bp, respectively [151, 152]) but in the absence of accessory factors does not mediate excisive recombination utilizing attL and attR sites (which are themselves the products of integration) (153). Directional control is enabled by a recombinational directionality factor (RDF), a phage-encoded accessory actor that determines which pairs of attachment sites can undergo recombination. The L5 RDF has been identified (gp36) and acts by binding to specific DNA sites within attP and attR to bend DNA and alter the nature of higher-order protein-DNA architectures that can be formed by the Int-Y, the RDF, and the host integration factor, mIHF (154–157). Directional control is determined by a different mechanism for the serine integrases, as illustrated for Bxb1. The attachment sites are relatively small (<50 bp), and Int binds as a dimer to each site, and the choice of pairs of sites competent for recombination is determined by which protein-DNA complexes can undergo synapsis, presumably predicated on compatible conformations of Int bound to specific sites (158, 159). A phage-encoded RDF, Bxb1 gp47, associates with Int-DNA complexes not through DNA binding, but by direct protein-protein interactions and presumably alters the conformations such that attL and attR can undergo synapsis, but attP and attB cannot (160, 161). A curious feature of Bxb1 is that the RDF is encoded by gene 47, which is situated among DNA replication genes, and is not closely linked to int (gene 35). Numerous mycobacteriophages have homologues of Bxb1 47, including many (such as L5) that utilize tyrosine integrases and for which there is no obvious role in recombination. Presumably, these genes play alternative roles in phage growth, most likely in DNA replication, and have been coopted by the integration systems for directional regulation (162).

MYCOBACTERIOPHAGE GENE EXPRESSION AND ITS REGULATION

None of the sequenced mycobacteriophage genomes encode single-subunit RNA polymerases, and it is likely that all phage transcription utilizes the host RNA polymerase. In some examples, phage growth has specifically been shown to be sensitive to the transcription inhibitor, rifampin (33, 50). Promoters corresponding to those used by the host major sigma factor (SigA) have also been identified in several mycobacteriophages and can be predicted in many others. For example, the strong P_{left} promoter in phage L5 closely corresponds to −10 and −35 SigA sequences (163), as do both the early lytic promoter (P_R) and the repressor promoter (P_{rep}) of BPs (164), and in all three examples, the transcription start site has been mapped. Interestingly, in both the BPs promoters, these transcripts provide a leaderless mRNA for the first gene in the operon, with the first base of the translation initiation codon corresponding to the 5′ end of the mRNA. The use of leaderless mRNA transcripts has been shown previously in expression of the firefly luciferase reporter gene in an L5 recombinant phage (165). A SigA-like promoter has also been described for expression of the Ms6 lysis cassette, but this does include a leader between the transcription initiation site and the predicted translational start codon (166).

The use of host SigA-like promoters does not, however, appear to be universal for mycobacteriophage transcription. For example, a search for SigA-like promoters in mycobacteriophage Rey reveals no close matches and none that closely correlate with positions where promoters are anticipated to be. In phage Giles, where the transcripts in lysogeny as well as early and late lytic growth have been mapped by RNAseq (Fig. 4), there are no evident SigA-like promoter sequences upstream of where transcription starts. Although late transcripts are likely to utilize phage-encoded activators, it is unclear whether the early promoters in Giles use an alternative host sigma factor or if these also are regulated by phage-encoded functions. There is clearly much to be learned about how transcription initiation is regulated in mycobacteriophage growth. The Cluster A phages all encode a potential sigma factor (Phams 1448, 1922, 2982 in the Phamerator database Mycobacteriophage_285), as indicated by HHPred searches with the closest similarity to SigK of *M. tuberculosis*, which could regulate phage gene expression. Many phages also encode one or more WhiB proteins, but whether these regulate host or phage expression is not clear. In TM4, it has been shown that the phage-encoded WhiB protein is not required for phage growth, although it contains an Fe-S cluster, is a dominant negative regulator of the host *whiB2* gene, and promotes superinfection exclusion (167).

Transcriptomics

RNAseq analysis of transcription in phage Giles is quite informative. In lysogeny, few regions are expressed—as expected—and these include gene *47*, which encodes the phage repressor. Somewhat surprisingly, the repressor is expressed at high levels, equivalent to the 0.5 percentile of highest expressed genes in *M. smegmatis* (107), in noted contrast to the lower expression of other phage repressors (168). Presumably, the Giles repressor binds with relatively low affinity to its binding site (s), although these have yet to be identified. Lysogenic transcription proceeds through the four downstream genes, although at a much lower level (Fig. 4). The leftward-transcribed genes at the left end of the genome (*3–4*) are also expressed at a low level in lysogeny. Perhaps most surprising from this RNAseq analysis is the expression of an apparent noncoding RNA near the right end of the genome (Fig. 4), which is made during both lysogenic and early lytic growth. It is expressed at a high level, but its function is unclear, and the DNA segment encoding it is not required for phage growth or lysogeny.

Immunity Regulation in Cluster A Phages

An unusual system for gene expression is found in the temperate Cluster A phages but has been studied in detail in just a few examples: L1, L5, and Bxb1 (163, 169–177). The repressor is encoded in the right part of the genome (gene *71* in phage L5) and codes for a small protein (183 amino acids) containing a putative helix-turn-helix DNA binding motif near its N-terminus; deletion of the gene interferes with lysogeny and generates a clear plaque phenotype. L5 gene *71* is sufficient to confer immunity to superinfection, and thermo-inducible mutations map in the repressor gene (110, 176, 178). The repressor sequence of the closely related L1 phage is identical to that of L5 (176). The repressor functions by binding to a 13-bp operator site that overlaps the early lytic promoter, P_{left}, situated near the right end of the genome and transcribing leftward (163, 169). It is a two-domain protein with an N-terminal domain (residues 1 to 64) and a C-terminal domain (residues 64 to 183) separated by a hinge region (172), and mutations in either domain influence DNA binding (172, 176). The repressor binds as a monomer and imparts a modest DNA bend (30°) at its binding site (171).

Surprisingly, there are many additional repressor-binding sites situated throughout the L5 genome, and the repressor has been shown to bind to 23 of these in addition to its operator at P_{left} (169). These sites correspond to a tightly conserved asymmetric consensus sequence, 5′-GGTGGc/aTGTCAAG, and all or most of the base positions are critical for repressor binding (169, 171). A clue to the role of these additional binding sites emerges from their genomic locations and orientation. With few exceptions, they are located within short intergenic regions or overlapping translation initiation or termination codons and are oriented in one direction relative to the direction of transcription (169). Thus, in the left arm of L5, in which the virion structure and assembly genes are transcribed rightward, there are four sites in the "-" orientation (and one at the extreme left end in the nontranscribed region in the "+" orientation, where "+" and "-" refer to whether the sequence corresponds to the top or bottom strand of the genome), and in the leftward transcribed right arm there are 17 sites in the "+" orientation (and one at the extreme right end in the nontranscribed region in the "-" orientation). There are no apparent promoters at each of these sites, suggesting a role different from that of the true operator site at P_{left} (169). Insertion of one or more sites between a nonphage promoter (hsp60) and a reporter gene results in a reduction of reporter activity, in a manner that is dependent on both the repressor and the orientation of the site, which is magnified with larger numbers of binding sites (169). This suggests a model in which repressor binding to these sites acts to promote cessation of transcription and facilitate silencing of the L5 genome in the lysogenic state. These sites are thus referred to as "stoperators" (169). An example of stoperator location and orientation in mycobacteriophage Alma is illustrated in Fig. 5.

Mycobacteriophage Bxb1 is heteroimmune to L5 but contains a similar regulatory system (51, 175). The Bxb1 repressor (gp69) shares only 41% amino acid sequence identity to L5 gp71, and there are a total of 34 putative operator/stoperator sites in the Bxb1 genome. As in L5, these sites correspond to a tightly conserved asymmetric consensus sequence, 5′-GTTACGt/ag/aTCAAG, and are located in short intergenic regions in one orientation relative to transcription. The consensus sequences of the L5 and Bxb1 stoperators are closely related but with differences that likely contribute to heteroimmunity of the two phages (175). Genomic analysis of the large number of Cluster A phages shows that they all share this unusual repressor/stoperator system, with variations of the stoperator consensus sequence that correlate with subcluster designation (98). However, the variation typically occurs within positions 2 to 8 of the consensus sequence, and positions 1 (G) and 9 to 13 (5′-TCAAG) are invariant. How

heteroimmunity evolves in this system is unclear, because although amino substitutions within the repressor could readily lead to altered recognition of positions 2 to 8 in the stoperator sites, it is unclear what selection pressure leads to coordinate evolution of over 30 genomically dispersed binding sites to new specificities (see Fig. 6). Interestingly, although the repressor/stoperator system is restricted to the Cluster A phages within the collection of sequenced mycobacteriophages, it appears to be shared with phages of some other Actinomycetales hosts. For example, there are multiple stoperator-like sites in Streptomyces phage R4 and its relatives (244), as well as in Rhodococcus phage RER2 (179); in the latter phage the invariant positions (1, 9 to 13) are the same as in the Cluster A mycobacteriophages, but the R4-like Streptomyces phage stoperators are substantially different. The prevalence of this regulatory system among the broader Actinomycetales phage population remains to be explored.

Although this particular regulatory system is confined to Cluster A phages, within the mycobacteriophages, other phages contain repeated sequences that are likely involved in the regulation of gene expression. One example is the Cluster K phages that contain a conserved sequence (5′-GGGATAGGAGCCC) positioned 2 to 8 bp upstream of putative translation start sites and are thus referred to as start-associated sequences (SASs); there are 10 to 19 copies per genome (97). The sequence is asymmetric and is situated in the expected location of the ribosome binding site, and eight of the conserved positions can pair with the 3′ end of the 16S rRNA. However, there are well-conserved positions at the edges of the site that cannot pair with the 16SrRNA, suggesting that it is not just a common variation of a ribosome binding site. Moreover, the sites are situated exclusively next to nonvirion structural genes, and the consensus site is present neither in other mycobacteriophages nor in the M. smegmatis genome (97). So if these SASs are involved in translation initiation, they likely also require a phage-encoded regulator. An intriguing hypothesis is that the SAS-associated genes are highly expressed in early lytic growth and that the SASs play a role in releasing ribosomes from these transcripts during late lytic growth to optimize ribosome availability for late gene expression. Curiously, a subset of the SAS sites has a second conserved sequence composed of imperfect 17-bp IRs separated by a variable spacer. Because of their tight association with SASs—typically located with a few base pairs upstream—these are referred to as extended start-associated sequences, and their roles are unknown (97).

Integration-Dependent Immunity

A novel system of immunity regulation has been described in several different types of mycobacteriophages, including BPs (Cluster G), Brujita (Cluster I), Charlie (Cluster N) and BigNuz (Cluster P) (164, 180). The characteristic feature—readily recognizable bioinformatically—is that the phage attachment site for integration (*attP*) is located within the coding sequence for the phage repressor (e.g., BPs gp33). In these systems, the repressor is located immediately upstream of the integrase gene (*int*), in contrast to the immunity system of Cluster A phages described above, in which *int* and the repressor are separated by ~20 kbp (see Fig. 6). The immunity functions are thus compactly organized, with all of the required functions situated within ~2 kbp of the genomes. Because of the specific location of *attP* with the repressor, integration of the phage genome plays a central role in the lytic-lysogenic decision, generating two alternative forms of the repressor: a longer viral form (e.g., BPs gp33[136]) and a shorter prophage form (BPs gp33[103]) that lacks the C-terminal 33 residues of the viral form. As expected, the shorter prophage form is active in conferring superinfection immunity, but interestingly, the longer viral form is not. The reason for this is that the extreme C-terminus of the viral form contains an ssrA-like tag that targets it for degradation, presumably by the ClpXP protease (164). This is a functionally important distinction in the two forms of the repressor and directly determines the frequency at which lysogeny is established; a mutant phage encoding a stabilized viral repressor form has a greatly elevated frequency of lysogenization (164). Nonetheless, establishment of immunity is dependent on phage integration, because otherwise the active prophage form of the repressor cannot be expressed (164). The activity of the integration system thus must also play a central role in the lytic-lysogenic decision, because if Integrase (Int) was always expressed and fully active, then lytic growth would not occur.

The resolution to this conundrum is that the integrase also contains a C-terminal ssrA-like tag, such that degradation of Int is anticipated to also determine the lysogenization frequency. This feature is illustrated by the behavior of integration-proficient plasmid vectors (see below) that transform at unexpectedly low frequencies due to Int degradation; stabilization of Int leads to large increases in the transformation frequencies (164). The relative simplicity of these immunity systems with a few genes and DNA sites serving multiple functions suggests that these may have evolved relatively early in the development of phage immunity systems (180).

GENETIC MANIPULATION OF MYCOBACTERIOPHAGES

Four approaches have been described for manipulation of mycobacteriophage genomes. The first is a simple genetic cross in which a phage can acquire a DNA segment by homologous recombination from a plasmid containing homologous phage sequences. Although this should be a generalizable technique, there is only a single example, in which the firefly luciferase gene was crossed onto the L5 genome and recombinants were identified by hybridization (165). A second approach was through direct cloning into the TM4 genome, replacing a small restriction fragment with a PCR-generated substitute containing deletion of gene *49* (167). A third approach is the construction of shuttle phasmids, which is discussed in detail in reference 245. Shuttle phasmids provide information about which phage genome segments are likely to be nonessential for lytic growth and are instrumental for delivery of reporter genes, transposons, and allelic exchange substrates to mycobacteria. However, they are of more limited use for constructing specific mutations in phage genes.

A general method for genome manipulation is the bacteriophage recombineering of electroporated DNA (BRED) technique, which can be used to efficiently construct precise gene deletions (minimizing genetic polarity) and to introduce point mutations (181–183). The technique takes advantage of a mycobacterial recombineering system in which recombinase genes from mycobacteriophage Che9c are inducibly expressed in *M. smegmatis* (184–186). Following coelectroporation of phage genomic DNA and a dsDNA substrate containing the desired mutation (typically ~200 bp), plaques are recovered using an infectious center assay, and individual plaques are screened by PCR for the presence of wild-type and mutant alleles at the site of interest (182). Because the recombination frequencies are high, mutant plaques are typically present in at least 10% of the plaques tested; however, all plaques contain a wild-type signal, likely reflecting the use of replicating genomes as substrates for recombination. If the mutant is viable, it can be recovered from replating of single plaques. If the mutant is not viable, it can usually be identified among the primary plaques due to the presence of wild-type helper phage within each plaque but then is not recoverable from a secondary plating (182).

The BRED approach is applicable to many different types of phage genomes but perhaps not all. For example, it is not possible to recover plaques of Omega by electroporation of phage DNA into *M. smegmatis* (90).

The reason for this is not clear. An intriguing possibility is that a capsid-enclosed protein is required for recircularization of the phage DNA upon injection, which is then absent during electroporation. Phage Omega is unusual in that it has defined genome termini with very short (4-base) single-stranded DNA extension and requires the host nonhomologous end joining (NHEJ) system for efficient infection (187). Omega encodes its own Ku-like protein, which could be required for recircularization, but mass spectrometry of Omega virions failed to identify any such components (90). Some phage genomes are readily engineered with BRED, and Dedrick et al. performed a whole-genome analysis for essential genes in phage Giles (see Fig. 4) (107); the introduction of point mutations into the BPs genome was critical for establishing the mechanisms of integration-dependent immunity (164).

APPLICATIONS FOR MYCOBACTERIAL GENETICS

Mycobacteriophages have proven to be essential tools for the development of mycobacterial genetics and to provide potential clinical tools for tuberculosis control. Given the well-established advantages of viral approaches to a variety of biological problems, mycobacteriophages would seem to offer considerable promise for genetic and clinical applications. Like most viruses, mycobacteriophages show host specificity, efficiently introduce their DNA into the host, replicate efficiently, express genes at high levels, utilize a variety of regulatory strategies, and are simple to grow. Mycobacteriophages offer an additional advantage in that their rapid growth contrasts to the extremely slow growth rate of M. tuberculosis. The tools for mycobacterial genetics are typically of two general classes. First, there are those that take advantage of the phages themselves, such as for efficient delivery of foreign DNA to the hosts. The second is the exploitation of phage components such as plasmid vectors or selectable markers. Applications in clinical microbiology have largely focused on simple and rapid tools for tuberculosis diagnosis.

Generalized Transduction

Generalized transduction of genetic markers in M. smegmatis was one of the earliest applications of mycobacteriophages (13). Although transduction was initially demonstrated using phage I3—which has yet to be fully genomically characterized but likely is grouped within Cluster C—it has also been shown for other Cluster C phages such as Bxz1 (246). Transduc-

tion using these phages typically generates relatively few transductants, depending on the locus, but a number of markers have been moved between M. smegmatis strains. Unfortunately, the Cluster C phages do not infect M. tuberculosis and there currently is no generalized transducing phage for slow-growing mycobacteria. Moreover, all of the phages that infect M. tuberculosis contain genomic cohesive ends and are thus not good candidates for generalized transducers. Presumably phages of M. tuberculosis that have terminally redundant genomes—and are thus candidates for generalized transducers—exist in nature, but have yet to be isolated.

Mycobacteriophages for Efficient Gene Delivery

A key feature of mycobacteriophages is that when added to host cells at a multiplicity of infection (moi) greater than about 3, every cell in the culture is infected. This is vastly more efficient than introducing DNA by electroporation and when using recombinant phages provides a simple means of introducing foreign DNA. The first strategy to exploit this was the construction of shuttle phasmids that replicate as large (~50 kbp) cosmids in E. coli and as phages in mycobacteria (44), and these continue to be indispensible tools of mycobacterial genetics; see reference 245 for a detailed discussion. Shuttle phasmids are typically constructed by ligation of phage DNA to a cosmid vector and packaging into phage lambda particles in vitro. These are used to infect E. coli, the library of colonies is harvested, and DNA is isolated and then used to electroporate M. smegmatis. The plaques recovered are purified and demonstrated to retain the cosmid vector (76). The strategy relies on the use of mycobacteriophages with appropriately sized genomes, as the packaging constraints for phage lambda are ~53 kbp and there must be a region of nonessential genes that can be replaced by the cosmid vector. Shuttle phasmids have been constructed from phages TM4 and D29, both of which satisfy these constraints, as well as having the ability to efficiently infect both M. smegmatis and M. tuberculosis. TM4 shuttle phasmids are the most widely used, and conditionally replicating temperature-sensitive mutants have been developed that facilitate the recovery of survivors following shuttle phasmid infection (49); these mutations have been mapped through analysis of temperature-resistant revertants (97).

Shuttle phasmids can be readily manipulated in E. coli by replacing the cosmid moiety with another cosmid carrying desired genes. The three primary applications have been in the construction of reporter phages for the delivery of luciferase or fluorescent genes (188–190), for the delivery of allelic exchange substrates for constructing *M. tuberculosis* mutants (48), and for transposon delivery (49). Because transposon insertion is typically a low-frequency event, the high efficiency of shuttle phasmid infection is critical and has enabled the construction of high-density transposon libraries and their use for determining gene requirements under specific conditions (191–194).

Recombinant phages have been constructed for the delivery of reporter genes by two other methods. First, a derivative of phage L5 that carries the firefly luciferase gene was constructed by crossing over from a plasmid and identifying the recombinants by hybridization (165); these infect both fast- and slow-growing mycobacteria (195). Second, a reporter phage derivative of D29 has been constructed carrying green fluorescent protein (GFP) using recombineering (196). These approaches are more directed than using shuttle phasmids but contribute toward a suite of strategies for using phages to deliver DNA efficiently to mycobacterial hosts.

Integration-Proficient Plasmid Vectors

Some of the earliest tools developed by exploiting parts of the phage genomes are the integration-proficient plasmids derived from phage L5 (144). The plasmids were generated by inserting a segment of the phage genome carrying the integrase gene and *attP* site (~2 kbp) into a plasmid backbone that has an origin of replication for *E. coli* and a selectable marker for mycobacteria. These plasmids are not capable of extrachromosomal replication in mycobacteria and can only transform mycobacterial cells if the integrase gene is expressed and can mediate site-specific recombination between the plasmid *attP* and host *attB* site (197). In practice, these plasmids transform both *M. smegmatis* mc^{2}155 and *M. tuberculosis* H36Rv efficiently (~10^6 transformants/g of DNA) (144) as a result of a single well-defined predictable plasmid integration event into the host chromosome. These integration-proficient plasmids offer a number of useful attributes for mycobacterial genetic manipulation. Perhaps most importantly, they provide a simple method for inserting genes into single copy on the chromosome, in contrast to the commonly used extrachromosomal plasmids that have copy numbers in excess of 20 (198). This is specifically useful in complementation experiments, where extrachromosomal vectors can present misleading behaviors (199–201). Furthermore, as integration results in reconstruction of the tRNA gene at the site of insertion,

there is little or no disruption to the behavior of the host cell.

An additional attribute is that the integrated vectors are stably maintained in the absence of selection (144), although if the plasmid is expressing genes that disfavor bacterial growth and there is a selection for plasmid loss, then a low level of Xis-independent recombination of *attL* and *attR* gives rise to plasmid loss (202). This issue can be readily addressed by using a two-plasmid system in which the integrase gene is provided on a second, nonreplicating, nonintegrating vector, such that after co-electroporation only the *attP*-containing vector integrates (151). This eliminates the possibility of integrase-mediated events conferring instability, although homologous recombination can still occur between the ~40 bp of homology between *attL* and *attR*.

Because of the large number of sequenced mycobacteriophage genomes and their overall diversity, there are numerous alternative types of integration-proficient vectors that can be constructed and, if they use different *attB* sites, should be fully compatible with each other and could be used in combination to construct complex recombinants. To date, integrating vectors have been constructed from phages Tweety (145), Giles (103), Omega (90), Adephagia (97), Ms6 (146), BPs (164), Charlie (164), Brujita (164), and *Streptomyces* pSAM2 (203) and phage phiC31 (204), all of which use tyrosine integrases and insert into different *attB* sites overlapping tRNA genes. Because there is generally excellent conservation of tRNA genes between *M. smegmatis* and *M. tuberculosis*, these vectors usually work well in both hosts, even though the parent phage may not necessarily infect *M. tuberculosis*. It is important to note that for those systems derived from phages that use integration-dependent immunity systems (e.g., BPs, Charlie, Brujita [164]), transformation frequencies are extremely low unless a stabilized mutant form of the integrase is used (as discussed above). Additional vectors have been constructed from serine-integrase systems, including Bxb1 (147) and Bxz2 (145), although because the *attB* sites are located within open reading frames, the recombinants may have altered physiological features (149).

Further refinements to these vector systems broaden their utilities. For example, Huff et al. (198) constructed a series of related vector systems that enable use of different integration systems and easy subsequent removal of the integrase gene, and Saviola and Bishai (205) described a system for multiple integration events by introducing a new *attB* copy on the integrated vector itself. Taking advantage of the ability to promote excision of the plasmids in the presence of the

L5 RDF (gp36), Parish and colleagues developed systems to test for gene essentiality (206) and for doing integrated plasmid switching (207).

Other Applications for Mycobacteriophage Site-Specific Recombination Systems

The serine-integrase systems have considerable potential for broad use in molecular biology and genetics. Although only a few have been studied in biochemical detail (including phiC31 and Bxb1 [148, 158, 208]), they generally share the features of using small attachments sites (<50 bp), lack requirements for host-encoded accessory factors, and have tightly regulated directionality. As such, they typically function efficiently in other biological contexts, and the mycobacteriophage Bxb1 system has been used in *Plasmodium* (209), *Drosophila* (210), *Arabidopsis* (211), and mammalian cells (212).

The Bxb1 system has also been used to construct microbial data storage and computing systems (213, 214). These take advantage of fine-tuning expression of both the integrase and the RDF (gp47), so as to be able to reproducibly flip chromosomal segments in *E. coli*. Generation of greater complexity to these systems will require exploitation of additional but compatible serine integrases, and phage diversity should more than satisfy this requirement (213).

Selectable Markers

Although they have yet to find widespread use, phage immunity systems provide effective selectable markers that avoid the use of antibiotics and antibiotic resistance genes. These have advantages both for use in combination with other antibiotic resistance genes when constructing complex recombinants and few antibiotic resistance markers are available, as well as for constructing recombinants of virulent strains where introducing additional antibiotic resistance genes can enhance the biosafety concerns. The first example of immunity selection was demonstrated using the L5 repressor, which confers immunity to both L5 and D29 (110), but the strategy is adaptable for any phage in which the repressor has been identified and shown to confer immunity to superinfection (97, 107, 164). Patterns of phage heteroimmunity indicate which of these can be used in combination with each other. In principle, there are dozens of additional immunity systems and thus the potential to create a large suite of compatible selectable systems.

Selection for transformants carrying the immunity marker typically uses a clear-plaque derivative of the parent phage from which the repressor gene was derived (110). Such mutants can usually be isolated easily and efficiently kill nontransformed cells, although phage-resistant mutants can arise at low frequency. Selection can be accomplished by first spreading 10^8 to 10^9 virion particles onto the surface of solid media, followed by plating of the electroporated cells, or by first adding the phage to the recovering electroporated cells and plating the mixture onto solid media (164). When using such selectable schemes, it is helpful to be mindful of the nature of the recombinants recovered. If the phage is competent for integration (the more usual circumstance), then the recombinants recovered will also harbor an integrated complete prophage. For many experimental purposes this may not matter and has the potential advantage of providing a system for complete maintenance of the repressor plasmid, even when replicating extrachromosomally, because should plasmid loss occur, the phage immediately begins lytic growth leading to phage propagation and cell death. If this is undesirable, then phages defective in integration can be used, allowing the recovery of transformants that are completely phage free. The availability of the BRED engineering approach (182) facilitates the construction of such integration-defective mutants (164; G. W. Broussard and G. F. Hatfull, unpublished observations).

Mycobacterial Recombineering

Recombineering—genetic engineering by recombination (215)—is a method developed originally for manipulation of *E. coli*, taking advantage of the ability of phage-encoded recombination systems to greatly stimulate the levels of recombination, such that introduction of DNA substrates containing homology to the chromosome are efficiently incorporated (215, 216). The commonly used recombination systems are the Red function of phage Lambda, which contains three components, an exonuclease (Exo), a DNA pairing enzyme (Bet), and a RecBCD inhibitor (Gam), or the rac prophage-derived RecET system (215, 217). There are anecdotal reports of attempts to use the Lambda Red system in mycobacteria, all without success, although this is perhaps not surprising given the species specificities seen in similar recombineering systems (218).

A mycobacterial recombineering system was developed using recombinases encoded by mycobacteriophages (184–186), and this strategy of using species-cognate recombination systems has been used for other bacteria (219). However, not all phages encode recombination systems that are bioinformatically identifiable, and it was not until a substantial collection of sequenced genomes was available (56) that *recET*-like genes were identified in mycobacterio-phage Che9c (184). Induced expression of Che9c *60* and *61* promotes recombination levels such that recombinants can be recovered following electroporation of a dsDNA substrate with homology to the chromosome, both in *M. smegmatis* and in *M. tuberculosis* (184). The requirements for homology seem less permissive than in *E. coli*, and the frequency of recombination drops substantially once the length falls below about 500 bp. Recombineering can also use ssDNA substrates, and then only Che9c gp61 is required (185). This enables the introduction of point mutations, especially when they can be selected for directly, such as those that confer antibiotic resistance, where the total number of recombinants recovered can be greater than 10^5 (185). A mutant that cannot be selected for directly may also be constructed by co-electroporation with one that does, because a high proportion of the selected transformants also carry the unselected mutation and can be screened for using PCR (185). This approach mirrors that developed for manipulation of phage genomes (182), as described above.

Tools for Diagnosis of Tuberculosis

Several approaches have been described that use phage-based systems for diagnosis of tuberculosis and assessment of drug susceptibility profiles. There is a strong motivation to develop such systems because traditional methodologies are typically very slow and more modern ones are often expensive. Phage-based diagnostics are thus typically aimed toward a low-cost point-of-care application (220). The phage amplified biologically (PhaB) assay (221) takes advantage of the ability of *M. tuberculosis* cells present in a clinical sample to support replication of phage D29, efficient elimination of excess input phage particles, and scoring phage amplification on lawns of *M. smegmatis*. The test requires simple microbiological methods, has reasonable sensitivity, and can be used to distinguish between rifampin-resistant and -sensitive strains (74, 222, 223). More than a dozen evaluations of specificity and sensitivity have been conducted, which have been summarized in meta-analyses reported in 2005 (224) and 2010 (225). The more recent analysis shows that the commercially available FastPlaque assay has rather variable estimates of both sensitivity (81% to 100%) and specificity (73% to 100%) (225).

An alternative approach uses reporter mycobacteriophages to deliver a readily assayable gene such as the firefly luciferase gene or a fluorescent reporter such as GFP (188, 190). Reporter phages have been built using a variety of phage genomes including TM4, D29, and L5, all of which can infect both *M. smegmatis* and

M. tuberculosis (165, 188, 196, 226); the TM4 platform may offer some advantages because of its ability to infect stationary-phase cells, a phenomenon that is dependent on the peptidoglycan hydrolytic motif embedded within the phage Tapemeasure protein (106). The main difference in the choice of the reporter is the detection method used. For luciferase reporter phages, light emission from infected cells can be detected using either a luminometer or photographically (188, 227), whereas fluorescent reporter phage infection can be detected using either microscopy or flow cytometry (189, 190). For optimal signal generation, it is important to prevent phage-mediated lysis, which can be accomplished using conditionally replicating mutants (49, 228, 229). Several evaluations of both luciferase and fluorescent phages have been conducted (230–235), and a meta-analysis of the luciferase studies shows consistent estimates of diagnostic accuracy (with seven of eight studies reporting 100% sensitivity) and specificity (with four of eight reporting 100% specificity) (225). Addition of a tag to the capsid protein provides an approach for the simple recovery of phage and phage-bacterial complexes (236).

Ideally, both PhaB and reporter phage assays could be applied directly to sputum, without the requirement for complex processing methods. Although it has proven tricky to develop methods for efficient sputum processing that provide good recovery of viable mycobacterial cells and inhibition of contaminant growth and that facilitate good phage infection, recent studies show that this is feasible (189), and it now awaits evaluation in larger collections of patient samples.

Phage Therapy?

No discussion of mycobacteriophages is complete without some consideration of their use for direct therapeutic control of mycobacterial infections including tuberculosis. The general concept of phage therapy is an old one and was championed by Felix D'Herelle for several decades following his discovery of bacteriophages (237). Phage therapy has been widely employed in the former Soviet Union but has been slow to gain acceptance elsewhere, although recently, phages have been approved for control of both *Listeria* and *E. coli* contaminations in meat. The use of mycobacteriophages for control of tuberculosis has considerable appeal given the predominance of multidrug-resistant (MDR) strains and the appearance of extensively drug-resistant (XDR) strains (238). Although there has been some laboratory assessment of phage therapy in *M. tuberculosis*–infected guinea pigs (239), and the possible use of surrogate systems for phage delivery to infected macrophages (240, 241), there have been no human trials.

There are two obvious impediments to phage therapy of tuberculosis. The first is the concern that during a pulmonary infection with *M. tuberculosis*, it would prove difficult for phages to efficiently access the bacteria, particularly where they are intracellular or contained with granulomas. This is a serious concern, but one that could be alleviated if there is any significant dynamic exchange of bacteria within an infection and if surrogates were used for phage delivery (240, 241). The second concern is the rapid development of phage resistance, which is expected to occur at frequencies equivalent to that of antibiotic resistance. This could be alleviated by use of multiple phages for which there are different resistance mechanisms, or inclusion of phage mutants isolated in the laboratory that promote escape from host resistance. Although these two problems may be tough, they would not seem insurmountable.

An alternative application might be to use mycobacteriophages prophylactically rather than therapeutically. For example, family members or coworkers of patients recently diagnosed with pulmonary tuberculosis could use aspirated phages to interfere with transmission and acquisition of the disease. This is appealing because the bacteria involved in disease acquisition are more likely to be accessible to phage attack, and typically only relatively small numbers of bacteria are inhaled, so that not only could relatively high multiplicities of infection be achieved, but the incidence of phage resistance would be low due to the relatively small size of the bacterial population under control. Although this would seem like an effective way to disrupt transmission that is both cheap and safe—without interfering with ongoing treatment of infected patients—to our knowledge it has not yet been evaluated.

PERSPECTIVES: PAST, PRESENT, AND FUTURE

In closing a comprehensive review of mycobacteriophages in 1984, Mizugichi concluded, "Studies on mycobacteriophages have contributed little to the knowledge and understanding of general virology especially in the fields of molecular biology and genetics. The future use of mycobacteriophages in studies of molecular biology would appear to be nonprofitable because of the paucity of necessary basic information. It would seem that the thrust of current mycobacteriophage studies should be to clarify the host-virus interaction in mycobacteria. A profound analysis of the relationship between mycobacteria and their phages would certainly enhance our knowledge of genetics, pathogenicity, drug resistance, variation and evolution

of mycobacteria" (18). These words were quite prescient, especially with the emphasis on the need for a stronger basic understanding of mycobacteriophages and the ways in which this would be useful. The discussion of mycobacteriophages in the 2000 first edition of *Molecular Genetics of Mycobacteria* (64) reflected the considerable advances in both basic and applied studies over the subsequent dozen years.

However, the 14 years since the first edition in 2000 have quite dramatically advanced our understanding of mycobacteriophages and ways in which they can be exploited for tuberculosis genetics. In 2000, the complete genome sequences of only four mycobacteriophages (L5, D29, Bxb1, TM4) had been determined, dwarfed now by the ~285 complete genome sequences in GenBank (61, 90, 97, 98) and the total of over 500 (http://www.phagesdb.com). This wealth of genomic data has changed our perspectives of these phages and our expectations for exploiting them for mycobacterial genetics. The diversity is clearly enormous, and although we may now be closer to saturating the isolation of the vast majority of genome types that infect *M. smegmatis* mc²155, the use of other mycobacterial hosts is likely to reveal further expansion of this diversity.

The concluding sentence of the chapter on mycobacteriophages in the first edition stated, "We can therefore expect a rich and exciting future in elucidating mycobacteriophage evolution, novel systems of genetic regulation, tools for genetic and clinical applications, and the role of phages in mycobacterial physiology and virulence" (64). The ensuing 14 years more than adequately met these expectations; phage genomics has elucidated evolutionary mechanisms, integration-dependent immunity is an excellent example of novel regulation, recombineering is an effective genetic application, and Bxb1 integration has revealed aspects of host physiology and biofilm formation.

What will the next period of mycobacteriophage research bring? There are numerous areas in which advances might be anticipated. Currently, we have only limited insights into the host ranges of the currently characterized phages, know very little about what receptors are used and what determines specificity, and have only limited information on the rates and mechanisms by which phages switch host ranges. It is also reasonable to expect substantial advances in determining the functions of mycobacteriophage genes. There are a vast number of genes (>40,000) with unknown functions, and with new mutagenesis approaches we should be poised to find out what these do and why they are carried in phage genomes. Likewise, although we're learning more about phage gene expression and the regulation of gene expression, this is mostly confined to small numbers of phages, and there is a great deal more to learn. The genomic information also suggests numerous ways in which these phages can be exploited for mycobacterial genetics and further insights into mycobacterial physiology.

Finally, the disseminated use of mycobacteriophage research for science education and research training has engaged hundreds of faculty and thousands of undergraduate students. Their involvement has had a huge impact over the past 10 years, and the continued engagement of these students and researchers will secure an active future for mycobacteriophage investigations.

I am grateful to Roger Hendrix, Craig Peebles, and Jeffrey Lawrence for discussions; to all members of my laboratory past and present for their insights and contributions; to Welkin Pope, Bekah Dedrick, Greg Broussard, and Debbie Jacobs-Sera for comments on the manuscript; and to all student phage hunters everywhere. Studies in the Hatfull laboratory were supported by National Institutes of Health grant GM51975 and by the Howard Hughes Medical Institute through its Professorship grant to GFH.

Citation. Hatfull GF. 2014. Molecular genetics of mycobacteriophages. Microbiol Spectrum 2(2):MGM2-0032-2013.

References

1. Gardner GM, Weiser RS. 1947. A bacteriophage for *Mycobacterium smegmatis*. *Proc Soc Exp Biol Med* **66:** 205–206.

2. Whittaker E. 1950. Two bacteriophages for *Mycobacterium smegmatis*. *Can J Public Health* **41:**431–436.

3. Froman S, Will DW, Bogen E. 1954. Bacteriophage active against *Mycobacterium tuberculosis*. I. Isolation and activity. *Am J Pub Health* **44:**1326–1333.

4. Grange JM. 1975. Proceedings: bacteriophage typing of strains of *Mycobacterium tuberculosis* isolated in south-east England. *J Med Microbiol* **8:**ix(2).

5. Grange JM. 1975. The genetics of mycobacteria and mycobacteriophages: a review. *Tubercle* **56:**227–238.

6. Jones WD Jr. 1975. Phage typing report of 125 strains of "*Mycobacterium tuberculosis*". *Ann Sclavo* **17:** 599–604.

7. Jones WD Jr. 1988. Bacteriophage typing of *Mycobacterium tuberculosis* cultures from incidents of suspected laboratory cross-contamination. *Tubercle* **69:**43–46.

8. Snider DE Jr, Jones WD, Good RC. 1984. The usefulness of phage typing *Mycobacterium tuberculosis* isolates. *Am Rev Respir Dis* **130:**1095–1099.

9. Goode D. 1983. Bacteriophage typing of strains of *Mycobacterium tuberculosis* from Nepal. *Tubercle* **64:** 15–21.

10. Murohashi T, Tokunaga T, Mizuguchi Y, Maruyama Y. 1963. Phage typing of slow-growing mycobacteria. *Am Rev Respir Dis* **88:**664–669.

11. Baess I. 1969. Subdivision of *M. tuberculosis* by means of bacteriophages. With special reference to epidemiological studies. *Acta Pathol Microbiol Scand* **76:**464–474.

12. Gangadharam PR, Simmons CS, Stager CE. 1978. Phage typing of mycobacteria using paper discs. *Am Rev Respir Dis* **118:**148–150.

13. Raj CV, Ramakrishnan T. 1970. Transduction in *Mycobacterium smegmatis*. *Nature* **228:**280–281.

14. Sellers MI, Tokuyasu K, Price Z, Froman S. 1957. Electron microscopic studies of mycobacteriophages. *Am Rev Tuberc* **76:**964–969.

15. Buraczewska M, Kwiatkowski B, Manowska W, Rdultowska H. 1972. Ultrastructure of some mycobacteriophages. *Am Rev Respir Dis* **105:**22–29.

16. Jones W, White A. 1968. Lysogeny in mycobacteria. I. Conversion of colony morphology, nitrate reductase activity, and tween 80 hydrolysis of *Mycobacterium* sp. ATCC 607 associated with lysogeny. *Can J Microbiol* **14:**551–555.

17. Mankiewicz E, Liivak M, Dernuet S. 1969. Lysogenic mycobacteria: phage variations and changes in host cells. *J Gen Microbiol* **55:**409–416.

18. Mizuguchi Y. 1984. Mycobacteriophages, p 641–662. *In* Kubica GP, Wayne LG (ed), *The Mycobacteria: A Sourcebook*, part A. Marcel Dekker, New York.

19. Grange JM, Bird RG. 1975. The nature and incidence of lysogeny in *Mycobacterium fortuitum*. *J Med Microbiol* **8:**215–223.

20. Grange JM, Bird RG. 1978. Lysogeny associated with mucoid variation in *Mycobacterium kansasii*. *J Med Microbiol* **11:**1–6.

21. Karnik SS, Gopinathan KP. 1983. Transfection of *Mycobacterium smegmatis* SN2 with mycobacteriophage I3 DNA. *Arch Microbiol* **136:**275–280.

22. Tokunaga T, Nakamura RM. 1968. Infection of competent *Mycobacterium smegmatis* with deoxyribonucleic acid extracted from bacteriophage B1. *J Virol* **2:**110–117.

23. Tokunaga T, Mizuguchi Y, Murohashi T. 1963. Deoxyribonucleic acid base composition of a mycobacteriophage. *J Bacteriol* **86:**608–609.

24. Bowman BU. 1969. Properties of mycobacteriophage DS6A. I. Immunogenicity in rabbits. *Proc Soc Exp Biol Med* **131:**196–200.

25. Buraczewska M, Manowska W, Rdultowska H. 1969. Studies on mycobacteriophages. Adsorption of mycobacteriophages on bacillary cells. *Pol Med J* **8:**1337–1341.

26. Castelnuovo G, Giuliani HI, Luchini de Giuliani E, Arancia G. 1969. Protein components of a mycobacteriophage. *J Gen Virol* **4:**253–257.

27. Juhasz SE, Gelbart S, Harize M. 1969. Phage-induced alteration of enzymic activity in lysogenic *Mycobacterium smegmatis* strains. *J Gen Microbiol* **56:** 251–255.

28. Menezes J, Pavilanis V. 1969. Properties of mycobacteriophage C2. *Experientia* **25:**1112–1113.

29. Bonicke R, Saito H. 1970. Phage conversion of biochemical properties in the genus *Mycobacterium*. *Bull Int Union Tuberc* **43**:217–225.

30. David HL, Jones WD Jr. 1970. Biosynthesis of a lipase by *Mycobacterium smegmatis* ATCC 607 infected by mycobacteriophage D29. *Am Rev Respir Dis* **102**:818–820.

31. Jones WD, David HL, Beam RE. 1970. The occurrence of lipids in mycobacteriophage D29 propagated in *Mycobacterium smegmatis* ATCC 607. *Am Rev Respir Dis* **102**:814–817.

32. Tokunaga T, Kataoka T, Suga K. 1970. Phage inactivation by an ethanol-ether extract of *Mycobacterium smegmatis*. *Am Rev Respir Dis* **101**:309–313.

33. Jones WD Jr, David HL. 1971. Inhibition by fifampin of mycobacteriophage D29 replication in its drug-resistant host, *Mycobacterium smegmatis* ATCC 607. *Am Rev Respir Dis* **103**:618–624.

34. Bowman BU, Newman HA, Moritz JM, Koehler RM. 1973. Properties of mycobacteriophage DS6A. II. Lipid composition. *Am Rev Respir Dis* **107**:42–49.

35. Bowman BU Jr, Fisher LJ, Witiak DT, Newman HA. 1975. Inactivation of phages DS6A and D29 by acetone extracts of *Mycobacterium tuberculosis* and *Mycobacterium bovis*. *Am Rev Respir Dis* **112**:17–22.

36. Wisingerova E, Sulova J, Sassmannova E. 1975. Inactivation of the mycophage activity by lipid solvents. *Ann Sclavo* **17**:634–640.

37. Schafer R, Huber U, Franklin RM. 1977. Chemical and physical properties of mycobacteriophage D29. *Eur J Biochem* **73**:239–246.

38. Soloff BL, Rado TA, Henry BE 2nd, Bates JH. 1978. Biochemical and morphological characterization of mycobacteriophage R1. *J Virol* **25**:253–262.

39. Somogyi PA, Maso Bel M, Foldes I. 1982. Methylated nucleic acid bases in mycobacterium and mycobacteriophage DNA. *Acta Microbiol Acad Sci Hung* **29**:181–185.

40. Reddy AB, Gopinathan KP. 1986. Presence of random single-strand gaps in mycobacteriophage I3 DNA. *Gene* **44**:227–234.

41. Jacobs WR Jr. 2000. *Mycobacterium tuberculosis*: a once genetically intractable organism, p 1–16. *In* Hatfull GF, Jacobs WR, Jr (ed), *Molecular Genetics of the Mycobacteria*. ASM Press, Washington, DC.

42. Hopwood DA, Wright HM. 1978. Bacterial protoplast fusion: recombination in fused protoplasts of *Streptomyces coelicolor*. *Mol Gen Genet* **162**:307–317.

43. Okanishi M, Suzuki K, Umezawa H. 1974. Formation and reversion of streptomycete protoplasts: cultural condition and morphological study. *J Gen Microbiol* **80**:389–400.

44. Jacobs WR Jr, Tuckman M, Bloom BR. 1987. Introduction of foreign DNA into mycobacteria using a shuttle phasmid. *Nature* **327**:532–535.

45. Jacobs WR Jr, Kalpana GV, Cirillo JD, Pascopella L, Snapper SB, Udani RA, Jones W, Barletta RG, Bloom BR. 1991. Genetic systems for mycobacteria. *Methods Enzymol* **204**:537–555.

46. Snapper SB, Lugosi L, Jekkel A, Melton RE, Kieser T, Bloom BR, Jacobs WR Jr. 1988. Lysogeny and transformation in mycobacteria: stable expression of foreign genes. *Proc Natl Acad Sci USA* **85**:6987–6991.

47. Snapper SB, Melton RE, Mustafa S, Kieser T, Jacobs WR Jr. 1990. Isolation and characterization of efficient plasmid transformation mutants of *Mycobacterium smegmatis*. *Mol Microbiol* **4**:1911–1919.

48. Bardarov S, Bardarov S Jr, Pavelka MS Jr, Sambandamurthy V, Larsen M, Tufariello J, Chan J, Hatfull G, Jacobs WR Jr. 2002. Specialized transduction: an efficient method for generating marked and unmarked targeted gene disruptions in *Mycobacterium tuberculosis*, *M. bovis* BCG and *M. smegmatis*. *Microbiology* **148**:3007–3017.

49. Bardarov S, Kriakov J, Carriere C, Yu S, Vaamonde C, McAdam RA, Bloom BR, Hatfull GF, Jacobs WR Jr. 1997. Conditionally replicating mycobacteriophages: a system for transposon delivery to *Mycobacterium tuberculosis*. *Proc Natl Acad Sci USA* **94**:10961–10966.

50. Hatfull GF, Sarkis GJ. 1993. DNA sequence, structure and gene expression of mycobacteriophage L5: a phage system for mycobacterial genetics. *Mol Microbiol* **7**:395–405.

51. Mediavilla J, Jain S, Kriakov J, Ford ME, Duda RL, Jacobs WR Jr, Hendrix RW, Hatfull GF. 2000. Genome organization and characterization of mycobacteriophage Bxb1. *Mol Microbiol* **38**:955–970.

52. Ford ME, Sarkis GJ, Belanger AE, Hendrix RW, Hatfull GF. 1998. Genome structure of mycobacteriophage D29: implications for phage evolution. *J Mol Biol* **279**:143–164.

53. Ford ME, Stenstrom C, Hendrix RW, Hatfull GF. 1998. Mycobacteriophage TM4: genome structure and gene expression. *Tuber Lung Dis* **79**:63–73.

54. Pedulla ML, Ford ME, Houtz JM, Karthikeyan T, Wadsworth C, Lewis JA, Jacobs-Sera D, Falbo J, Gross J, Pannunzio NR, Brucker W, Kumar V, Kandasamy J, Keenan L, Bardarov S, Kriakov J, Lawrence JG, Jacobs WR, Hendrix RW, Hatfull GF. 2003. Origins of highly mosaic mycobacteriophage genomes. *Cell* **113**:171–182.

55. Hatfull GF. 2010. Bacteriophage research: gateway to learning science. *Microbe* **5**:243–250.

56. Hatfull GF, Pedulla ML, Jacobs-Sera D, Cichon PM, Foley A, Ford ME, Gonda RM, Houtz JM, Hryckowian AJ, Kelchner VA, Namburi S, Pajcini KV, Popovich MG, Schleicher DT, Simanek BZ, Smith AL, Zdanowicz GM, Kumar V, Peebles CL, Jacobs WR Jr, Lawrence JG, Hendrix RW. 2006. Exploring the mycobacteriophage metaproteome: phage genomics as an educational platform. *PLoS Genet* **2**:e92.

57. Hanauer D, Hatfull GF, Jacobs-Sera D. 2009. *Active Assessment: Assessing Scientific Inquiry*. Springer, New York, NY.

58. Hanauer DI, Jacobs-Sera D, Pedulla ML, Cresawn SG, Hendrix RW, Hatfull GF. 2006. Inquiry learning. Teaching scientific inquiry. *Science* **314**:1880–1881.

59. Hatfull GF. 2008. Bacteriophage genomics. *Curr Opin Microbiol* **11**:447–453.

60. Hatfull GF. 2010. Mycobacteriophages: genes and genomes. *Annu Rev Microbiol* **64**:331–356.

61. Hatfull GF. 2012. Complete genome sequences of 138 mycobacteriophages. *J Virol* **86**:2382–2384.

62. Hatfull GF. 1994. Mycobacteriophage L5: a toolbox for tuberculosis. *ASM News* **60**:255–260.

63. Hatfull GF. 1999. Mycobacteriophages, p 38–58. *In* Ratledge C, Dale J (ed), *Mycobacteria: Molecular Biology and Virulence*. Chapman and Hall, London.

64. Hatfull GF. 2000. Molecular genetics of mycobacteriophages, p 37–54. *In* Hatfull GF, Jacobs WR, Jr (ed), *Molecular Genetics of Mycobacteria*. ASM Press, Washington, DC.

65. Hatfull GF. 2004. Mycobacteriophages and tuberculosis, p 203–218. *In* Eisenach K, Cole ST, Jacobs WR, Jr, McMurray D (ed), *Tuberculosis*. ASM Press, Washington, DC.

66. Hatfull GF. 2006. Mycobacteriophages, p 602–620. *In* Calendar R (ed), *The Bacteriophages*. Oxford University Press, New York, NY.

67. Hatfull GF. 2012. The secret lives of mycobacteriophages. *Adv Virus Res* **82**:179–288.

68. Hatfull GF, Barsom L, Chang L, Donnelly-Wu M, Lee MH, Levin M, Nesbit C, Sarkis GJ. 1994. Bacteriophages as tools for vaccine development. *Dev Biol Stand* **82**:43–47.

69. Hatfull GF, Cresawn SG, Hendrix RW. 2008. Comparative genomics of the mycobacteriophages: insights into bacteriophage evolution. *Res Microbiol* **159**:332–339.

70. Hatfull GF, Hendrix RW. 2011. Bacteriophages and their genomes. *Curr Opin Virol* **1**:298–303.

71. Hatfull GF, Jacobs WR Jr. 1994. Mycobacteriophages: cornerstones of mycobacterial research, p 165–183. *In* Bloom BR (ed), *Tuberculosis: Pathogenesis, Protection and Control*. ASM Press, Washington, DC.

72. Stella EJ, de la Iglesia AI, Morbidoni HR. 2009. Mycobacteriophages as versatile tools for genetic manipulation of mycobacteria and development of simple methods for diagnosis of mycobacterial diseases. *Rev Argent Microbiol* **41**:45–55.

73. McNerney R. 1999. TB: the return of the phage. A review of fifty years of mycobacteriophage research. *Int J Tuberc Lung Dis* **3**:179–184.

74. McNerney R. 2002. Phage tests for diagnosis and drug susceptibility testing. *Int J Tuberc Lung Dis* **6**:1129–1130; author reply 1130.

75. McNerney R, Traore H. 2005. Mycobacteriophage and their application to disease control. *J Appl Microbiol* **99**:223–233.

76. Jacobs WR Jr, Snapper SB, Tuckman M, Bloom BR. 1989. Mycobacteriophage vector systems. *Rev Infect Dis* **11**(Suppl 2):S404–S410.

77. Redmond WB. 1963. Bacteriophages of mycobacteria: a review. *Adv Tuberc Rev* **12**:191–229.

78. Sarkis GJ, Hatfull GF. 1998. Mycobacteriophages. *Methods Mol Biol* **101**:145–173.

79. Cater JC, Redmond WB. 1961. Isolation of and studies on bacteriophage active against mycobacteria. *Can J Microbiol* **7**:697–703.

80. Cater JC, Redmond WB. 1963. Mycobacterial phages isolated from stool specimens of patients with pulmonary disease. *Am Rev Respir Dis* **87**:726–729.

81. Redmond WB, Carter JC. 1960. A bacteriophage specific to *Mycobacterium tuberculosis* varieties *hominis* and *bovis*. *Am Rev Respir Dis* **82**:781–786.

82. Timme TL, Brennan PJ. 1984. Induction of bacteriophage from members of the *Mycobacterium avium*, *Mycobacterium intracellulare*, *Mycobacterium scrofulaceum* serocomplex. *J Gen Microbiol* **130**:2059–2066.

83. Sassi M, Bebeacua C, Drancourt M, Cambillau C. 2013. The first structure of a mycobacteriophage, the *Mycobacterium abscessus* subsp. *bolletii* phage Araucaria. *J Virol* **87**:8099–8109.

84. Choo SW, Yusoff AM, Wong YL, Wee WY, Ong CS, Ng KP, Ngeow YF. 2012. Genome analysis of *Mycobacterium massiliense* strain M172, which contains a putative mycobacteriophage. *J Bacteriol* **194**:5128.

85. Tobias NJ, Doig KD, Medema MH, Chen H, Haring V, Moore R, Seemann T, Stinear TP. 2013. Complete genome sequence of the frog pathogen *Mycobacterium ulcerans* ecovar Liflandii. *J Bacteriol* **195**:556–564.

86. Cole ST, Brosch R, Parkhill J, Garnier T, Churcher C, Harris D, Gordon SV, Eiglmeier K, Gas S, Barry CE 3rd, Tekaia F, Badcock K, Basham D, Brown D, Chillingworth T, Connor R, Davies R, Devlin K, Feltwell T, Gentles S, Hamlin N, Holroyd S, Hornsby T, Jagels K, Krogh A, McLean J, Moule S, Murphy L, Oliver K, Osborne J, Quail MA, Rajandream MA, Rogers J, Rutter S, Seeger K, Skelton J, Squares R, Squares S, Sulston JE, Taylor K, Whitehead S, Barrell BG. 1998. Deciphering the biology of *Mycobacterium tuberculosis* from the complete genome sequence. *Nature* **393**:537–544.

87. Stinear TP, Seemann T, Harrison PF, Jenkin GA, Davies JK, Johnson PD, Abdellah Z, Arrowsmith C, Chillingworth T, Churcher C, Clarke K, Cronin A, Davis P, Goodhead I, Holroyd N, Jagels K, Lord A, Moule S, Mungall K, Norbertczak H, Quail MA, Rabbinowitsch E, Walker D, White B, Whitehead S, Small PL, Brosch R, Ramakrishnan L, Fischbach MA, Parkhill J, Cole ST. 2008. Insights from the complete genome sequence of *Mycobacterium marinum* on the evolution of *Mycobacterium tuberculosis*. *Genome Res* **18**:729–741.

88. Supply P, Marceau M, Mangenot S, Roche D, Rouanet C, Khanna V, Majlessi L, Criscuolo A, Tap J, Pawlik A, Fiette L, Orgeur M, Fabre M, Parmentier C, Frigui W, Simeone R, Boritsch EC, Debrie AS, Willery E, Walker D, Quail MA, Ma L, Bouchier C, Salvignol G, Sayes F, Cascioferro A, Seemann T, Barbe V, Locht C, Gutierrez MC, Leclerc C, Bentley SD, Stinear TP, Brisse S, Medigue C, Parkhill J, Cruveiller S, Brosch R. 2013. Genomic analysis of smooth tubercle bacilli provides insights into ancestry and pathoadaptation of *Mycobacterium tuberculosis*. *Nat Genet* **45**:172–179.

89. Hatfull GF, Jacobs-Sera D, Lawrence JG, Pope WH, Russell DA, Ko CC, Weber RJ, Patel MC, Germane KL, Edgar RH, Hoyte NN, Bowman CA, Tantoco AT, Paladin EC, Myers MS, Smith AL, Grace MS, Pham TT, O'Brien MB, Vogelsberger AM, Hryckowian AJ, Wynalek JL, Donis-Keller H, Bogel MW, Peebles CL, Cresawn SG, Hendrix RW. 2010. Comparative genomic analysis of 60 mycobacteriophage genomes: genome clustering, gene acquisition, and gene size. *J Mol Biol* 397:119–143.

90. Pope WH, Jacobs-Sera D, Best AA, Broussard GW, Connerly PL, Dedrick RM, Kremer TA, Offner S, Ogiefo AH, Pizzorno MC, Rockenbach K, Russell DA, Stowe EL, Stukey J, Thibault SA, Conway JF, Hendrix RW, Hatfull GF. 2013. Cluster J mycobacteriophages: intron splicing in capsid and tail genes. *PLoS One* 8:e69273.

91. Harrison M, Dunbar D, Ratmansky L, Boyd K, Lopatto D. 2011. Classroom-based science research at the introductory level: changes in career choices and attitude. *CBE Life Sci Educ* 10:279–286.

92. Shaffer CD, Alvarez C, Bailey C, Barnard D, Bhalla S, Chandrasekaran C, Chandrasekaran V, Chung HM, Dorer DR, Du C, Eckdahl TT, Poet JL, Frohlich D, Goodman AL, Gosser Y, Hauser C, Hoopes LL, Johnson D, Jones CJ, Kaehler M, Kokan N, Kopp OR, Kuleck GA, McNeil G, Moss R, Myka JL, Nagengast A, Morris R, Overvoorde PJ, Shoop E, Parrish S, Reed K, Regisford EG, Revie D, Rosenwald AG, Saville K, Schroeder S, Shaw M, Skuse G, Smith C, Smith M, Spana EP, Spratt M, Stamm J, Thompson JS, Wawersik M, Wilson BA, Youngblom J, Leung W, Buhler J, Mardis ER, Lopatto D, Elgin SC. 2010. The genomics education partnership: successful integration of research into laboratory classes at a diverse group of undergraduate institutions. *CBE Life Sci Educ* 9:55–69.

93. Lopatto D, Alvarez C, Barnard D, Chandrasekaran C, Chung HM, Du C, Eckdahl T, Goodman AL, Hauser C, Jones CJ, Kopp OR, Kuleck GA, McNeil G, Morris R, Myka JL, Nagengast A, Overvoorde PJ, Poet JL, Reed K, Regisford G, Revie D, Rosenwald A, Saville K, Shaw M, Skuse GR, Smith C, Smith M, Spratt M, Stamm J, Thompson JS, Wilson BA, Witkowski C, Youngblom J, Leung W, Shaffer CD, Buhler J, Mardis E, Elgin SC. 2008. Undergraduate research. *Genom Educ Partnership Sci* 322:684–685.

94. Smith SA, Tank DC, Boulanger LA, Bascom-Slack CA, Eisenman K, Kingery D, Babbs B, Fenn K, Greene JS, Hann BD, Keehner J, Kelley-Swift EG, Kembaiyan V, Lee SJ, Li P, Light DY, Lin EH, Ma C, Moore E, Schorn MA, Vekhter D, Nunez PV, Strobel GA, Donoghue MJ, Strobel SA. 2008. Bioactive endophytes warrant intensified exploration and conservation. *PLoS One* 3:e3052.

95. Hanauer DI, Frederick J, Fotinakes B, Strobel SA. 2012. Linguistic analysis of project ownership for undergraduate research experiences. *CBE Life Sci Educ* 11:378–385.

96. Hendrix RW, Smith MC, Burns RN, Ford ME, Hatfull GF. 1999. Evolutionary relationships among diverse bacteriophages and prophages: all the world's a phage. *Proc Natl Acad Sci USA* 96:2192–2197.

97. Pope WH, Ferreira CM, Jacobs-Sera D, Benjamin RC, Davis AJ, DeJong RJ, Elgin SCR, Guilfoile FR, Forsyth MH, Harris AD, Harvey SE, Hughes LE, Hynes PM, Jackson AS, Jalal MD, MacMurray EA, Manley CM, McDonough MJ, Mosier JL, Osterbann LJ, Rabinowitz HS, Rhyan CN, Russell DA, Saha MS, Shaffer CD, Simon SE, Sims EF, Tovar IG, Weisser EG, Wertz JT, Weston-Hafer KA, Williamson KE, Zhang B, Cresawn SG, Jain P, Piuri M, Jacobs WR Jr, Hendrix RW, Hatfull GF. 2011. Cluster K mycobacteriophages: insights into the evolutionary origins of mycobacteriophage TM4. *PLoS One* 6:e26750.

98. Pope WH, Jacobs-Sera D, Russell DA, Peebles CL, Al-Atrache Z, Alcoser TA, Alexander LM, Alfano MB, Alford ST, Amy NE, Anderson MD, Anderson AG, Ang AAS, Ares M Jr, Barber AJ, Barker LP, Barrett JM, Barshop WD, Bauerle CM, Bayles IM, Belfield KL, Best AA, Borjon A Jr, Bowman CA, Boyer CA, Bradley KW, Bradley VA, Broadway LN, Budwal K, Busby KN, Campbell IW, Campbell AM, Carey A, Caruso SM, Chew RD, Cockburn CL, Cohen LB, Corajod JM, Cresawn SG, Davis KR, Deng L, Denver DR, Dixon BR, Ekram S, Elgin SCR, Engelsen AE, English BEV, Erb ML, Estrada C, Filliger LZ, Findley AM, Forbes L, Forsyth MH, Fox TM, Fritz MJ, Garcia R, George ZD, Georges AE, Gissendanner CR, Goff S, Goldstein R, Gordon KC, Green RD, Guerra SL, Guiney-Olsen KR, Guiza BG, Haghighat L, Hagopian GV, Harmon CJ, Harmson JS, Hartzog GA, Harvey SE, He S, He KJ, Healy KE, Higinbotham ER, Hildebrandt EN, Ho JH, Hogan GM, Hohenstein VG, Holz NA, Huang VJ, Hufford EL, Hynes PM, Jackson AS, Jansen EC, Jarvik J, Jasinto PG, Jordan TC, Kasza T, Katelyn MA, Kelsey JS, Kerrigan LA, Khaw D, Kim J, Knutter JZ, Ko C-C, Larkin GV, Laroche JR, Latif A, Leuba KD, Leuba SI, Lewis LO, Loesser-Casey KE, Long CA, Lopez AJ, Lowery N, Lu TQ, Mac V, Masters IR, McCloud JJ, McDonough MJ, Medenbach AJ, Menon A, Miller R, Morgan BK, Ng PC, Nguyen E, Nguyen KT, Nguyen ET, Nicholson KM, Parnell LA, Peirce CE, Perz AM, Peterson LJ, Pferdehirt RE, Philip SV, Pogliano K, Pogliano J, Polley T, Puopolo EJ, Rabinowitz HS, Resiss MJ, Rhyan CN, Robinson YM, Rodriguez LL, Rose AC, Rubin JD, Ruby JA, Saha MS, Sandoz JW, Savitskaya J, Schipper DJ, Schnitzler CE, Schott AR, Segal JB, Shaffer CD, Sheldon KE, Shepard EM, Shepardson JW, Shroff MK, Simmons JM, Simms EF, Simpson BM, Sinclair KM, Sjoholm RL, Slette IJ, Spaulding BC, Straub CL, Stukey J, Sughrue T, Tang T-Y, Tatyana LM, Taylor SB, Taylor BJ, Temple LM, Thompson JV, Tokarz MP, Trapani SE, Troum AP, Tsay J, Tubbs AT, Walton JM, Wang DH, Wang H, Warner JR, Weisser EG, Wendler SC, Weston-Hafer KA, Whelan HM, Williamson KE, Willis AN, Wirtshafter HS, Wong TW, Wu P, Yang YJ, Yee BC, Zaidins DA, Zhang B, Zúniga MY, Hendrix RW, Hatfull GF. 2011. Expanding the diversity of mycobacteriophages: insights into genome architecture and evolution. *PLoS One* 6:e16329.

99. Jacobs-Sera D, Marinelli LJ, Bowman C, Broussard GW, Guerrero Bustamante C, Boyle MM, Petrova ZO, Dedrick RM, Pope WH, Science Education Alliance Phage Hunters Advancing Genomics and Evolutionary Science Sea-Phages Program, Modlin RL, Hendrix RW, Hatfull GF. 2012. On the nature of mycobacteriophage diversity and host preference. *Virology* **434**:187–201.

100. Cresawn SG, Bogel M, Day N, Jacobs-Sera D, Hendrix RW, Hatfull GF. 2011. Phamerator: a bioinformatic tool for comparative bacteriophage genomics. *BMC Bioinformatics* **12**:395.

101. Huson DH, Bryant D. 2006. Application of phylogenetic networks in evolutionary studies. *Mol Biol Evol* **23**:254–267.

102. Mageeney C, Pope WH, Harrison M, Moran D, Cross T, Jacobs-Sera D, Hendrix RW, Dunbar D, Hatfull GF. 2012. Mycobacteriophage Marvin: a new singleton phage with an unusual genome organization. *J Virol* **86**:4762–4775.

103. Morris P, Marinelli LJ, Jacobs-Sera D, Hendrix RW, Hatfull GF. 2008. Genomic characterization of mycobacteriophage Giles: evidence for phage acquisition of host DNA by illegitimate recombination. *J Bacteriol* **190**:2172–2182.

104. Duda RL, Hempel J, Michel H, Shabanowitz J, Hunt D, Hendrix RW. 1995. Structural transitions during bacteriophage HK97 head assembly. *J Mol Biol* **247**:618–635.

105. Xu J, Hendrix RW, Duda RL. 2004. Conserved translational frameshift in dsDNA bacteriophage tail assembly genes. *Mol Cell* **16**:11–21.

106. Piuri M, Hatfull GF. 2006. A peptidoglycan hydrolase motif within the mycobacteriophage TM4 tape measure protein promotes efficient infection of stationary phase cells. *Mol Microbiol* **62**:1569–1585.

107. Dedrick RM, Marinelli LJ, Newton GL, Pogliano K, Pogliano J, Hatfull GF. 2013. Functional requirements for bacteriophage growth: gene essentiality and expression in mycobacteriophage Giles. *Mol Microbiol* **88**:577–589.

108. Giri N, Bhowmik P, Bhattacharya B, Mitra M, Das Gupta SK. 2009. The mycobacteriophage D29 gene 65 encodes an early-expressed protein that functions as a structure-specific nuclease. *J Bacteriol* **191**:959–967.

109. Liu J, Dehbi M, Moeck G, Arhin F, Bauda P, Bergeron D, Callejo M, Ferretti V, Ha N, Kwan T, McCarty J, Srikumar R, Williams D, Wu JJ, Gros P, Pelletier J, DuBow M. 2004. Antimicrobial drug discovery through bacteriophage genomics. *Nat Biotechnol* **22**:185–191.

110. Donnelly-Wu MK, Jacobs WR Jr, Hatfull GF. 1993. Superinfection immunity of mycobacteriophage L5: applications for genetic transformation of mycobacteria. *Mol Microbiol* **7**:407–417.

111. Rybniker J, Plum G, Robinson N, Small PL, Hartmann P. 2008. Identification of three cytotoxic early proteins of mycobacteriophage L5 leading to growth inhibition in *Mycobacterium smegmatis*. *Microbiology* **154**:2304–2314.

112. Rybniker J, Krumbach K, van Gumpel E, Plum G, Eggeling L, Hartmann P. 2011. The cytotoxic early protein 77 of mycobacteriophage L5 interacts with MSMEG_3532, an L-serine dehydratase of *Mycobacterium smegmatis*. *J Basic Microbiol* **51**:515–522.

113. Hassan S, Mahalingam V, Kumar V. 2009. Synonymous codon usage analysis of thirty two mycobacteriophage genomes. *Adv Bioinformatics* **2009**:1–11.

114. Kunisawa T. 2000. Functional role of mycobacteriophage transfer RNAs. *J Theor Biol* **205**:167–170.

115. Kunisawa T, Kanaya S, Kutter E. 1998. Comparison of synonymous codon distribution patterns of bacteriophage and host genomes. *DNA Res* **5**:319–326.

116. Sahu K, Gupta SK, Ghosh TC, Sau S. 2004. Synonymous codon usage analysis of the mycobacteriophage Bxz1 and its plating bacteria *M. smegmatis*: identification of highly and lowly expressed genes of Bxz1 and the possible function of its tRNA species. *J Biochem Mol Biol* **37**:487–492.

117. Sampson T, Broussard GW, Marinelli LJ, Jacobs-Sera D, Ray M, Ko CC, Russell D, Hendrix RW, Hatfull GF. 2009. Mycobacteriophages BPs, Angel and Halo: comparative genomics reveals a novel class of ultra-small mobile genetic elements. *Microbiology* **155**:2962–2977.

118. Tori K, Dassa B, Johnson MA, Southworth MW, Brace LE, Ishino Y, Pietrokovski S, Perler FB. 2009. Splicing of the mycobacteriophage Bethlehem DnaB intein: identification of a new mechanistic class of inteins that contain an obligate block F nucleophile. *J Biol Chem* **285**:2515–2526.

119. Tori K, Perler FB. 2011. Expanding the definition of class 3 inteins and their proposed phage origin. *J Bacteriol* **193**:2035–2041.

120. Rybniker J, Kramme S, Small PL. 2006. Host range of 14 mycobacteriophages in *Mycobacterium ulcerans* and seven other mycobacteria including *Mycobacterium tuberculosis*: application for identification and susceptibility testing. *J Med Microbiol* **55**:37–42.

121. Chen J, Kriakov J, Singh A, Jacobs WR Jr, Besra GS, Bhatt A. 2009. Defects in glycopeptidolipid biosynthesis confer phage I3 resistance in *Mycobacterium smegmatis*. *Microbiology* **155**:4050–4057.

122. Furuchi A, Tokunaga T. 1972. Nature of the receptor substance of *Mycobacterium smegmatis* for D4 bacteriophage adsorption. *J Bacteriol* **111**:404–411.

123. Bisso G, Castelnuovo G, Nardelli MG, Orefici G, Arancia G, Laneelle G, Asselineau C, Asselineau J. 1976. A study on the receptor for a mycobacteriophage: phage phlei. *Biochimie* **58**:87–97.

124. Khoo KH, Suzuki R, Dell A, Morris HR, McNeil MR, Brennan PJ, Besra GS. 1996. Chemistry of the lyxose-containing mycobacteriophage receptors of *Mycobacterium phlei*/*Mycobacterium smegmatis*. *Biochemistry* **35**:11812–11819.

125. Barrangou R, Fremaux C, Deveau H, Richards M, Boyaval P, Moineau S, Romero DA, Horvath P. 2007. CRISPR provides acquired resistance against viruses in prokaryotes. *Science* **315**:1709–1712.

126. Fineran PC, Blower TR, Foulds IJ, Humphreys DP, Lilley KS, Salmond GP. 2009. The phage abortive infection system, ToxIN, functions as a protein-RNA toxin-antitoxin pair. *Proc Natl Acad Sci USA* **106**:894–899.

127. Brudey K, Driscoll JR, Rigouts L, Prodinger WM, Gori A, Al-Hajoj SA, Allix C, Aristimuno L, Arora J, Baumanis V, Binder L, Cafrune P, Cataldi A, Cheong S, Diel R, Ellermeier C, Evans JT, Fauville-Dufaux M, Ferdinand S, Garcia de Viedma D, Garzelli C, Gazzola L, Gomes HM, Guttierez MC, Hawkey PM, van Helden PD, Kadival GV, Kreiswirth BN, Kremer K, Kubin M, Kulkarni SP, Liens B, Lillebaek T, Ho ML, Martin C, Martin C, Mokrousov I, Narvskaia O, Ngeow YF, Naumann L, Niemann S, Parwati I, Rahim Z, Rasolofo-Razanamparany V, Rasolonavalona T, Rossetti ML, Rusch-Gerdes S, Sajduda A, Samper S, Shemyakin IG, Singh UB, Somoskovi A, Skuce RA, van Soolingen D, Streicher EM, Suffys PN, Tortoli E, Tracevska T, Vincent V, Victor TC, Warren RM, Yap SF, Zaman K, Portaels F, Rastogi N, Sola C. 2006. *Mycobacterium tuberculosis* complex genetic diversity: mining the fourth international spoligotyping database (SpolDB4) for classification, population genetics and epidemiology. *BMC Microbiol* **6**:23.

128. Ramage HR, Connolly LE, Cox JS. 2009. Comprehensive functional analysis of *Mycobacterium tuberculosis* toxin-antitoxin systems: implications for pathogenesis, stress responses, and evolution. *PLoS Genet* **5**:e1000767.

129. He L, Fan X, Xie J. 2012. Comparative genomic structures of *Mycobacterium* CRISPR-Cas. *J Cell Biochem* **113**:2464–2473.

130. Horan KL, Freeman R, Weigel K, Semret M, Pfaller S, Covert TC, van Soolingen D, Leao SC, Behr MA, Cangelosi GA. 2006. Isolation of the genome sequence strain *Mycobacterium avium* 104 from multiple patients over a 17-year period. *J Clin Microbiol* **44**:783–789.

131. Barsom EK, Hatfull GF. 1996. Characterization of *Mycobacterium smegmatis* gene that confers resistance to phages L5 and D29 when overexpressed. *Mol Microbiol* **21**:159–170.

132. Rubin EJ, Akerley BJ, Novik VN, Lampe DJ, Husson RN, Mekalanos JJ. 1999. In vivo transposition of mariner-based elements in enteric bacteria and mycobacteria. *Proc Natl Acad Sci USA* **96**:1645–1650.

133. Young R. 1992. Bacteriophage lysis: mechanism and regulation. *Microbiol Rev* **56**:430–481.

134. Berry J, Summer EJ, Struck DK, Young R. 2008. The final step in the phage infection cycle: the Rz and Rz1 lysis proteins link the inner and outer membranes. *Mol Microbiol* **70**:341–351.

135. Gil F, Catalao MJ, Moniz-Pereira J, Leandro P, McNeil M, Pimentel M. 2008. The lytic cassette of mycobacteriophage Ms6 encodes an enzyme with lipolytic activity. *Microbiology* **154**:1364–1371.

136. Gil F, Grzegorzewicz AE, Catalao MJ, Vital J, McNeil MR, Pimentel M. 2010. Mycobacteriophage Ms6 LysB specifically targets the outer membrane of *Mycobacterium smegmatis*. *Microbiology* **156**:1497–1504.

137. Payne K, Sun Q, Sacchettini J, Hatfull GF. 2009. Mycobacteriophage Lysin B is a novel mycolylarabinogalactan esterase. *Mol Microbiol* **73**:367–381.

138. Payne KM, Hatfull GF. 2012. Mycobacteriophage endolysins: diverse and modular enzymes with multiple catalytic activities. *PLoS One* **7**:e34052.

139. Catalao MJ, Gil F, Moniz-Pereira J, Pimentel M. 2010. The mycobacteriophage Ms6 encodes a chaperone-like protein involved in the endolysin delivery to the peptidoglycan. *Mol Microbiol* **77**:672–686.

140. Catalao MJ, Gil F, Moniz-Pereira J, Pimentel M. 2011. The endolysin-binding domain encompasses the N-terminal region of the mycobacteriophage Ms6 Gp1 chaperone. *J Bacteriol* **193**:5002–5006.

141. Berry J, Rajaure M, Pang T, Young R. 2012. The spanin complex is essential for lambda lysis. *J Bacteriol* **194**:5667–5674.

142. Catalao MJ, Gil F, Moniz-Pereira J, Pimentel M. 2011. Functional analysis of the holin-like proteins of mycobacteriophage Ms6. *J Bacteriol* **193**:2793–2803.

143. Stella EJ, Franceschelli JJ, Tasselli SE, Morbidoni HR. 2013. Analysis of novel mycobacteriophages indicates the existence of different strategies for phage inheritance in mycobacteria. *PLoS One* **8**:e56384.

144. Lee MH, Pascopella L, Jacobs WR Jr, Hatfull GF. 1991. Site-specific integration of mycobacteriophage L5: integration-proficient vectors for *Mycobacterium smegmatis*, *Mycobacterium tuberculosis*, and bacille Calmette-Guerin. *Proc Natl Acad Sci USA* **88**:3111–3115.

145. Pham TT, Jacobs-Sera D, Pedulla ML, Hendrix RW, Hatfull GF. 2007. Comparative genomic analysis of mycobacteriophage Tweety: evolutionary insights and construction of compatible site-specific integration vectors for mycobacteria. *Microbiology* **153**:2711–2723.

146. Freitas-Vieira A, Anes E, Moniz-Pereira J. 1998. The site-specific recombination locus of mycobacteriophage Ms6 determines DNA integration at the tRNA(Ala) gene of *Mycobacterium* spp. *Microbiology* **144**:3397–3406.

147. Kim AI, Ghosh P, Aaron MA, Bibb LA, Jain S, Hatfull GF. 2003. Mycobacteriophage Bxb1 integrates into the *Mycobacterium smegmatis* groEL1 gene. *Mol Microbiol* **50**:463–473.

148. Smith MC, Brown WR, McEwan AR, Rowley PA. 2010. Site-specific recombination by phiC31 integrase and other large serine recombinases. *Biochem Soc Trans* **38**:388–394.

149. Ojha A, Anand M, Bhatt A, Kremer L, Jacobs WR Jr, Hatfull GF. 2005. GroEL1: a dedicated chaperone involved in mycolic acid biosynthesis during biofilm formation in mycobacteria. *Cell* **123**:861–873.

150. Ojha AK, Trivelli X, Guerardel Y, Kremer L, Hatfull GF. 2010. Enzymatic hydrolysis of trehalose dimycolate releases free mycolic acids during mycobacterial growth in biofilms. *J Biol Chem* **285**:17380–17389.

151. Peña CE, Lee MH, Pedulla ML, Hatfull GF. 1997. Characterization of the mycobacteriophage L5 attachment site, attP. *J Mol Biol* **266**:76–92.

152. Peña CE, Stoner JE, Hatfull GF. 1996. Positions of strand exchange in mycobacteriophage L5 integration and characterization of the attB site. *J Bacteriol* **178:** 5533–5536.

153. Lewis JA, Hatfull GF. 2001. Control of directionality in integrase-mediated recombination: examination of recombination directionality factors (RDFs) including Xis and Cox proteins. *Nucleic Acids Res* **29:**2205–2216.

154. Lewis JA, Hatfull GF. 2000. Identification and characterization of mycobacteriophage L5 excisionase. *Mol Microbiol* **35:**350–360.

155. Lewis JA, Hatfull GF. 2003. Control of directionality in L5 integrase-mediated site-specific recombination. *J Mol Biol* **326:**805–821.

156. Pedulla ML, Lee MH, Lever DC, Hatfull GF. 1996. A novel host factor for integration of mycobacteriophage L5. *Proc Natl Acad Sci USA* **93:**15411–15416.

157. Pena CEA, Kahlenberg JM, Hatfull GF. 1999. Protein-DNA complexes in mycobacteriophage L5 integrative recombination. *J Bacteriol* **181:**454–461.

158. Ghosh P, Kim AI, Hatfull GF. 2003. The orientation of mycobacteriophage Bxb1 integration is solely dependent on the central dinucleotide of attP and attB. *Mol Cell* **12:**1101–1111.

159. Ghosh P, Pannunzio NR, Hatfull GF. 2005. Synapsis in phage Bxb1 integration: selection mechanism for the correct pair of recombination sites. *J Mol Biol* **349:** 331–348.

160. Ghosh P, Bibb LA, Hatfull GF. 2008. Two-step site selection for serine-integrase-mediated excision: DNA-directed integrase conformation and central dinucleotide proofreading. *Proc Natl Acad Sci USA* **105:**3238–3243.

161. Ghosh P, Wasil LR, Hatfull GF. 2006. Control of phage Bxb1 excision by a novel recombination directionality factor. *PLoS Biol* **4:**e186.

162. Savinov A, Pan J, Ghosh P, Hatfull GF. 2012. The Bxb1 gp47 recombination directionality factor is required not only for prophage excision, but also for phage DNA replication. *Gene* **495:**42–48.

163. Nesbit CE, Levin ME, Donnelly-Wu MK, Hatfull GF. 1995. Transcriptional regulation of repressor synthesis in mycobacteriophage L5. *Mol Microbiol* **17:** 1045–1056.

164. Broussard GW, Oldfield LM, Villanueva VM, Lunt BL, Shine EE, Hatfull GF. 2013. Integration-dependent bacteriophage immunity provides insights into the evolution of genetic switches. *Mol Cell* **49:**237–248.

165. Sarkis GJ, Jacobs WR Jr, Hatfull GF. 1995. L5 luciferase reporter mycobacteriophages: a sensitive tool for the detection and assay of live mycobacteria. *Mol Microbiol* **15:**1055–1067.

166. Garcia M, Pimentel M, Moniz-Pereira J. 2002. Expression of mycobacteriophage Ms6 lysis genes is driven by two sigma(70)-like promoters and is dependent on a transcription termination signal present in the leader RNA. *J Bacteriol* **184:**3034–3043.

167. Rybniker J, Nowag A, van Gumpel E, Nissen N, Robinson N, Plum G, Hartmann P. 2010. Insights into the function of the WhiB-like protein of mycobacteriophage TM4: a transcriptional inhibitor of WhiB2. *Mol Microbiol* **77:**642–657.

168. Ptashne M, Jeffrey A, Johnson AD, Maurer R, Meyer BJ, Pabo CO, Roberts TM, Sauer RT. 1980. How the lambda repressor and cro work. *Cell* **19:**1–11.

169. Brown KL, Sarkis GJ, Wadsworth C, Hatfull GF. 1997. Transcriptional silencing by the mycobacteriophage L5 repressor. *EMBO J* **16:**5914–5921.

170. Bandhu A, Ganguly T, Chanda PK, Das M, Jana B, Chakrabarti G, Sau S. 2009. Antagonistic effects Na+ and Mg2+ on the structure, function, and stability of mycobacteriophage L1 repressor. *BMB Rep* **42:** 293–298.

171. Bandhu A, Ganguly T, Jana B, Mondal R, Sau S. 2010. Regions and residues of an asymmetric operator DNA interacting with the monomeric repressor of temperate mycobacteriophage L1. *Biochemistry* **49:**4235–4243.

172. Ganguly T, Bandhu A, Chattoraj P, Chanda PK, Das M, Mandal NC, Sau S. 2007. Repressor of temperate mycobacteriophage L1 harbors a stable C-terminal domain and binds to different asymmetric operator DNAs with variable affinity. *Virol J* **4:**64.

173. Ganguly T, Chanda PK, Bandhu A, Chattoraj P, Das M, Sau S. 2006. Effects of physical, ionic, and structural factors on the binding of repressor of mycobacteriophage L1 to its cognate operator DNA. *Protein Pept Lett* **13:**793–798.

174. Ganguly T, Chattoraj P, Das M, Chanda PK, Mandal NC, Lee CY, Sau S. 2004. A point mutation at the C-terminal half of the repressor of temperate mycobacteriophage L1 affects its binding to the operator DNA. *J Biochem Mol Biol* **37:**709–714.

175. Jain S, Hatfull GF. 2000. Transcriptional regulation and immunity in mycobacteriophage Bxb1. *Mol Microbiol* **38:**971–985.

176. Sau S, Chattoraj P, Ganguly T, Lee CY, Mandal NC. 2004. Cloning and sequencing analysis of the repressor gene of temperate mycobacteriophage L1. *J Biochem Mol Biol* **37:**254–259.

177. Chaudhuri B, Sau S, Datta HJ, Mandal NC. 1993. Isolation, characterization, and mapping of temperature-sensitive mutations in the genes essential for lysogenic and lytic growth of the mycobacteriophage L1. *Virology* **194:** 166–172.

178. Bandhu A, Ganguly T, Jana B, Chakravarty A, Biswas A, Sau S. 2012. Biochemical characterization of L1 repressor mutants with altered operator DNA binding activity. *Bacteriophage* **2:**79–88.

179. Petrovski S, Seviour RJ, Tillett D. 2013. Genome sequence and characterization of a *Rhodococcus equi* phage REQ1. *Virus Genes.* [Epub ahead of print.] doi: 10.1007/s11262-013-0887-1.

180. Broussard GW, Hatfull GF. 2013. Evolution of genetic switch complexity. *Bacteriophage* **3:**e24186.

181. Marinelli LJ, Hatfull GF, Piuri M. 2012. Recombineering: a powerful tool for modification of bacteriophage genomes. *Bacteriophage* **2:**5–14.

182. Marinelli LJ, Piuri M, Swigonova Z, Balachandran A, Oldfield LM, van Kessel JC, Hatfull GF. 2008. BRED: a

simple and powerful tool for constructing mutant and recombinant bacteriophage genomes. *PLoS One* 3:e3957.

183. van Kessel JC, Marinelli LJ, Hatfull GF. 2008. Recombineering mycobacteria and their phages. *Nat Rev Microbiol* 6:851–857.

184. van Kessel JC, Hatfull GF. 2007. Recombineering in *Mycobacterium tuberculosis*. *Nat. Methods* 4:147–152.

185. van Kessel JC, Hatfull GF. 2008. Efficient point mutagenesis in mycobacteria using single-stranded DNA recombineering: characterization of antimycobacterial drug targets. *Mol Microbiol* 67:1094–1107.

186. van Kessel JC, Hatfull GF. 2008. Mycobacterial recombineering. *Methods Mol Biol* 435:203–215.

187. Pitcher RS, Tonkin LM, Daley JM, Palmbos PL, Green AJ, Velting TL, Brzostek A, Korycka-Machala M, Cresawn S, Dziadek J, Hatfull GF, Wilson TE, Doherty AJ. 2006. Mycobacteriophage exploit NHEJ to facilitate genome circularization. *Mol Cell* 23:743–748.

188. Jacobs WR Jr, Barletta RG, Udani R, Chan J, Kalkut G, Sosne G, Kieser T, Sarkis GJ, Hatfull GF, Bloom BR. 1993. Rapid assessment of drug susceptibilities of *Mycobacterium tuberculosis* by means of luciferase reporter phages. *Science* 260:819–822.

189. Jain P, Hartman TE, Eisenberg N, O'Donnell MR, Kriakov J, Govender K, Makume M, Thaler DS, Hatfull GF, Sturm AW, Larsen MH, Moodley P, Jacobs WR Jr. 2012. phi(2)GFP10, a high-intensity fluorophage, enables detection and rapid drug susceptibility testing of *Mycobacterium tuberculosis* directly from sputum samples. *J Clin Microbiol* 50:1362–1369.

190. Piuri M, Jacobs WR Jr, Hatfull GF. 2009. Fluoromycobacteriophages for rapid, specific, and sensitive antibiotic susceptibility testing of *Mycobacterium tuberculosis*. *PLoS One* 4:e4870.

191. Sassetti CM, Boyd DH, Rubin EJ. 2001. Comprehensive identification of conditionally essential genes in mycobacteria. *Proc Natl Acad Sci USA* 98:12712–12717.

192. Sassetti CM, Boyd DH, Rubin EJ. 2003. Genes required for mycobacterial growth defined by high density mutagenesis. *Mol Microbiol* 48:77–84.

193. Lamichhane G, Zignol M, Blades NJ, Geiman DE, Dougherty A, Grosset J, Broman KW, Bishai WR. 2003. A postgenomic method for predicting essential genes at subsaturation levels of mutagenesis: application to *Mycobacterium tuberculosis*. *Proc Natl Acad Sci USA* 100:7213–7218.

194. Murry JP, Sassetti CM, Lane JM, Xie Z, Rubin EJ. 2008. Transposon site hybridization in *Mycobacterium tuberculosis*. *Methods Mol Biol* 416:45–59.

195. Fullner KJ, Hatfull GF. 1997. Mycobacteriophage L5 infection of *Mycobacterium bovis* BCG: implications for phage genetics in the slow-growing mycobacteria. *Mol Microbiol* 26:755–766.

196. da Silva JL, Piuri M, Broussard G, Marinelli LJ, Bastos GM, Hirata RD, Hatfull GF, Hirata MH. 2013. Application of BRED technology to construct recombinant D29 reporter phage expressing EGFP. *FEMS Microbiol Lett* 344:166–172.

197. Lee MH, Hatfull GF. 1993. Mycobacteriophage L5 integrase-mediated site-specific integration in vitro. *J Bacteriol* 175:6836–6841.

198. Huff J, Czyz A, Landick R, Niederweis M. 2010. Taking phage integration to the next level as a genetic tool for mycobacteria. *Gene* 468:8–19.

199. Pascopella L, Collins FM, Martin JM, Lee MH, Hatfull GF, Stover CK, Bloom BR, Jacobs WR Jr. 1994. Use of in vivo complementation in *Mycobacterium tuberculosis* to identify a genomic fragment associated with virulence. *Infect Immun* 62:1313–1319.

200. Banerjee A, Dubnau E, Quemard A, Balasubramanian V, Um KS, Wilson T, Collins D, de Lisle G, Jacobs WR Jr. 1994. inhA, a gene encoding a target for isoniazid and ethionamide in *Mycobacterium tuberculosis*. *Science* 263:227–230.

201. Stover CK, de la Cruz VF, Fuerst TR, Burlein JE, Benson LA, Bennett LT, Bansal GP, Young JF, Lee MH, Hatfull GF, Snapper SB, Barletta RG Jr, Bloom BR. 1991. New use of BCG for recombinant vaccines. *Nature* 351:456–460.

202. Springer B, Sander P, Sedlacek L, Ellrott K, Bottger EC. 2001. Instability and site-specific excision of integration-proficient mycobacteriophage L5 plasmids: development of stably maintained integrative vectors. *Int J Med Microbiol* 290:669–675.

203. Martin C, Mazodier P, Mediola MV, Gicquel B, Smokvina T, Thompson CJ, Davies J. 1991. Site-specific integration of the *Streptomyces* plasmid pSAM2 in *Mycobacterium smegmatis*. *Mol Microbiol* 5:2499–2502.

204. Murry J, Sassetti CM, Moreira J, Lane J, Rubin EJ. 2005. A new site-specific integration system for mycobacteria. *Tuberculosis (Edinb)* 85:317–323.

205. Saviola B, Bishai WR. 2004. Method to integrate multiple plasmids into the mycobacterial chromosome. *Nucleic Acids Res* 32:e11.

206. Parish T, Lewis J, Stoker NG. 2001. Use of the mycobacteriophage L5 excisionase in *Mycobacterium tuberculosis* to demonstrate gene essentiality. *Tuberculosis (Edinb)* 81:359–364.

207. Pashley CA, Parish T. 2003. Efficient switching of mycobacteriophage L5-based integrating plasmids in *Mycobacterium tuberculosis*. *FEMS Microbiol Lett* 229:211–215.

208. Thorpe HM, Smith MC. 1998. In vitro site-specific integration of bacteriophage DNA catalyzed by a recombinase of the resolvase/invertase family. *Proc Natl Acad Sci USA* 95:5505–5510.

209. Nkrumah LJ, Muhle RA, Moura PA, Ghosh P, Hatfull GF, Jacobs WR Jr, Fidock DA. 2006. Efficient site-specific integration in *Plasmodium falciparum* chromosomes mediated by mycobacteriophage Bxb1 integrase. *Nat Methods* 3:615–621.

210. Huang J, Ghosh P, Hatfull GF, Hong Y. 2011. Successive and targeted DNA integrations in the *Drosophila*

genome by Bxb1 and phiC31 integrases. *Genetics* **189:** 391–395.

211. Thomson JG, Chan R, Smith J, Thilmony R, Yau YY, Wang Y, Ow DW. 2012. The Bxb1 recombination system demonstrates heritable transmission of site-specific excision in *Arabidopsis*. *BMC Biotechnology* **12:**9.

212. Russell JP, Chang DW, Tretiakova A, Padidam M. 2006. Phage Bxb1 integrase mediates highly efficient site-specific recombination in mammalian cells. *Biotechniques* **40:**460, 462, 464.

213. Bonnet J, Subsoontorn P, Endy D. 2012. Rewritable digital data storage in live cells via engineered control of recombination directionality. *Proc Natl Acad Sci USA* **109:**8884–8889.

214. Bonnet J, Yin P, Ortiz ME, Subsoontorn P, Endy D. 2013. Amplifying genetic logic gates. *Science* **340:** 599–603.

215. Court DL, Sawitzke JA, Thomason LC. 2002. Genetic engineering using homologous recombination. *Annu Rev Genet* **36:**361–388.

216. Datsenko KA, Wanner BL. 2000. One-step inactivation of chromosomal genes in *Escherichia coli* K-12 using PCR products. *Proc Natl Acad Sci USA* **97:**6640–6645.

217. Murphy KC. 2012. Phage recombinases and their applications. *Adv Virus Res* **83:**367–414.

218. Datta S, Costantino N, Zhou X, Court DL. 2008. Identification and analysis of recombineering functions from Gram-negative and Gram-positive bacteria and their phages. *Proc Natl Acad Sci USA* **105:**1626–1631.

219. Swingle B, Bao Z, Markel E, Chambers A, Cartinhour S. 2010. Recombineering using RecTE from *Pseudomonas syringae*. *Appl Environ Microbiol* **76:**4960–4968.

220. Jain P, Thaler DS, Maiga M, Timmins GS, Bishai WR, Hatfull GF, Larsen MH, Jacobs WR. 2011. Reporter phage and breath tests: emerging phenotypic assays for diagnosing active tuberculosis, antibiotic resistance, and treatment efficacy. *J Infect Dis* **204**(Suppl 4): S1142–S1150.

221. Wilson SM, al-Suwaidi Z, McNerney R, Porter J, Drobniewski F. 1997. Evaluation of a new rapid bacteriophage-based method for the drug susceptibility testing of *Mycobacterium tuberculosis*. *Nat Med* **3:** 465–468.

222. Watterson SA, Wilson SM, Yates MD, Drobniewski FA. 1998. Comparison of three molecular assays for rapid detection of rifampin resistance in *Mycobacterium tuberculosis*. *J Clin Microbiol* **36:**1969–1973.

223. McNerney R, Kiepiela P, Bishop KS, Nye PM, Stoker NG. 2000. Rapid screening of *Mycobacterium tuberculosis* for susceptibility to rifampicin and streptomycin. *Int J Tuberc Lung Dis* **4:**69–75.

224. Pai M, Kalantri S, Pascopella L, Riley LW, Reingold AL. 2005. Bacteriophage-based assays for the rapid detection of rifampicin resistance in *Mycobacterium tuberculosis*: a meta-analysis. *J Infect* **51:**175–187.

225. Minion J, Pai M. 2010. Bacteriophage assays for rifampicin resistance detection in *Mycobacterium tuberculosis*: updated meta-analysis. *Int J Tuberc Lung Dis* **14:** 941–951.

226. Pearson RE, Jurgensen S, Sarkis GJ, Hatfull GF, Jacobs WR Jr. 1996. Construction of D29 shuttle phasmids and luciferase reporter phages for detection of mycobacteria. *Gene* **183:**129–136.

227. Riska PF, Su Y, Bardarov S, Freundlich L, Sarkis G, Hatfull G, Carriere C, Kumar V, Chan J, Jacobs WR Jr. 1999. Rapid film-based determination of antibiotic susceptibilities of *Mycobacterium tuberculosis* strains by using a luciferase reporter phage and the Bronx Box. *J Clin Microbiol* **37:**1144–1149.

228. Bardarov S Jr, Dou H, Eisenach K, Banaiee N, Ya S, Chan J, Jacobs WR Jr, Riska PF. 2003. Detection and drug-susceptibility testing of *M. tuberculosis* from sputum samples using luciferase reporter phage: comparison with the Mycobacteria Growth Indicator Tube (MGIT) system. *Diagn Microbiol Infect Dis* **45:** 53–61.

229. Carriere C, Riska PF, Zimhony O, Kriakov J, Bardarov S, Burns J, Chan J, Jacobs WR Jr. 1997. Conditionally replicating luciferase reporter phages: improved sensitivity for rapid detection and assessment of drug susceptibility of *Mycobacterium tuberculosis*. *J Clin Microbiol* **35:**3232–3239.

230. Banaiee N, Bobadilla-Del-Valle M, Bardarov S Jr, Riska PF, Small PM, Ponce-De-Leon A, Jacobs WR Jr, Hatfull GF, Sifuentes-Osornio J. 2001. Luciferase reporter mycobacteriophages for detection, identification, and antibiotic susceptibility testing of *Mycobacterium tuberculosis* in Mexico. *J Clin Microbiol* **39:** 3883–3888.

231. Banaiee N, Bobadilla-del-Valle M, Riska PF, Bardarov S Jr, Small PM, Ponce-de-Leon A, Jacobs WR Jr, Hatfull GF, Sifuentes-Osornio J. 2003. Rapid identification and susceptibility testing of *Mycobacterium tuberculosis* from MGIT cultures with luciferase reporter mycobacteriophages. *J Med Microbiol* **52:**557–561.

232. Banaiee N, January V, Barthus C, Lambrick M, Roditi D, Behr MA, Jacobs WR Jr, Steyn LM. 2008. Evaluation of a semi-automated reporter phage assay for susceptibility testing of *Mycobacterium tuberculosis* isolates in South Africa. *Tuberculosis (Edinb)* **88:**64–68.

233. Rondon L, Piuri M, Jacobs WR Jr, de Waard J, Hatfull GF, Takiff H. 2011. Evaluation of fluoromycobacteriophages for detecting drug resistance in *Mycobacterium tuberculosis*. *J Clin Microbiol* **49:** 1838–1842.

234. Hazbon MH, Guarin N, Ferro BE, Rodriguez AL, Labrada LA, Tovar R, Riska PF, Jacobs WR Jr. 2003. Photographic and luminometric detection of luciferase reporter phages for drug susceptibility testing of clinical *Mycobacterium tuberculosis* isolates. *J Clin Microbiol* **41:**4865–4869.

235. Lu B, Fu Z, Xu S. 2000. Rapid rifampicin susceptibility test by using recombinant mycobacteriophages. *Zhonghua Jie He He Hu Xi Za Zhi* **23:**480–484 (In Chinese.)

236. Piuri M, Rondon L, Urdaniz E, Hatfull GF. 2013. Generation of affinity-tagged fluoromycobacteriophages by mixed assembly of phage capsids. *Appl Environ*

Microbiol. [Epub ahead of print.] doi:10.1128/AEM .01016-13.

237. **Summers WC.** 1999. *Felix d'Herelle and the Origins of Molecular Biology.* Yale University Press, New Haven, CT.

238. **Chang KC, Yew WW.** 2013. Management of difficult multidrug-resistant tuberculosis and extensively drug-resistant tuberculosis: update 2012. *Respirology* **18:**8–21.

239. **Sula L, Sulova J, Stolcpartova M.** 1981. Therapy of experimental tuberculosis in guinea pigs with mycobacterial phages DS-6A, GR-21 T, My-327. *Czech Med* **4:** 209–214.

240. **Broxmeyer L, Sosnowska D, Miltner E, Chacon O, Wagner D, McGarvey J, Barletta RG, Bermudez LE.** 2002. Killing of *Mycobacterium avium* and *Mycobacterium tuberculosis* by a mycobacteriophage delivered by a nonvirulent mycobacterium: a model for phage therapy of intracellular bacterial pathogens. *J Infect Dis* **186:** 1155–1160.

241. **Danelishvili L, Young LS, Bermudez LE.** 2006. In vivo efficacy of phage therapy for *Mycobacterium avium* infection as delivered by a nonvirulent mycobacterium. *Microb Drug Resist* **12:**1–6.

242. **Krumsiek J, Arnold R, Rattei T.** 2007. Gepard: a rapid and sensitive tool for creating dotplots on genome scale. *Bioinformatics* **23:**1026–1028.

243. **Pimentel M.** 2014. Genetics of phage lysis. *Microbiol Spectrum* **2(1):**MGM2-0017-2013.

244. **Smith MC, Hendrix RW, Dedrick R, Mitchell K, Ko CC, Russell D, Bell E, Gregory M, Bibb MJ, Pethick F, Jacobs-Sera D, Herron P, Buttner MJ, Hatfull GF.** 2013. Evolutionary relationships among actinophages and a putative adaptation for growth in *Streptomyces* spp. *J Bacteriol* **195:**4924–4935.

245. **Jacobs WR Jr.** 2014. Gene transfer in *Mycobacterium tuberculosis*: shuttle phasmids to enlightenment. *Microbiol Spectrum* **2(2):**MGM2-0037-2013.

246. **Lee S, Kriakov J, Vilcheze C, Dai Z, Hatfull GF, Jacobs WR Jr.** 2004. Bxz1, a new generalized transducing phage for mycobacteria. *FEMS Microbiol Lett* **241:** 271–276.

Molecular Genetics of Mycobacteria, 2nd Edition
Edited by Graham F. Hatfull and William R. Jacobs, Jr.
© 2014 American Society for Microbiology, Washington, DC
doi:10.1128/microbiolspec.MGM2-0017-2013

Madalena Pimentel[1]

Genetics of Phage Lysis

6

Bacteriophages, or simply phages—the viruses that infect bacteria—are the most abundant biological entities on Earth, playing a fundamental role in bacterial ecology and evolution (1, 2). To survive they need to infect sensitive bacteria, where they replicate and produce new viral particles. An infectious cycle starts with adsorption of the phage particle to the surface of a specific bacterial host, followed by penetration of the phage genome into the cytoplasm. Once inside the host cell, the genetic information carried in the viral genome is responsible for its own replication and for the synthesis of the components to make new phage particles. These newly assembled progeny virions now need to be released into the environment where host bacteria are potentially available for new infection cycles. Except for filamentous bacteriophages, which are released from their hosts without affecting the cell viability (3), all other phages must lyse the infected bacteria to liberate the virion progeny to the extracellular milieu (4, 5).

Bacterial lysis is thus the final event of a phage lytic cycle. To accomplish this step phages have to overcome the bacterial cell barriers, especially the peptidoglycan (PG) layer, a rigid and stable structure that allows the bacterial envelope to support internal osmotic pressure. For this, phages may use two basic strategies. Phages with single-stranded small genomes, exemplified by phages Qβ (ssRNA) and φX172 (ssDNA), synthesize a single lysis protein termed "amurin," which causes lysis by interfering with PG biosynthesis (4, 6, 7). Double-stranded DNA phages, which represent more than 95% of known bacterial viruses (8), use the holin-endolysin strategy, employing at least two essential proteins whose coordinated action results in well-timed and swift host cell lysis. Endolysins are enzymes that target the integrity of the PG layer, while holins are small membrane proteins that control the activation or its access to the murein at a precisely defined time (4, 5).

The mechanism of lysis performed by bacteriophage λ is by far the most well studied and was, for many years, considered to be universal (4). However, it is now becoming increasingly evident that phages of both Gram-positive and Gram-negative bacteria employ diverse lysis models, reflecting adaptations to their hosts. A remarkable example of such diversity is given by mycobacteriophages, which exhibit different lysis cassettes. These phages infect mycobacterial hosts, bacteria that possess a complex cell envelope with a cell wall core composed of PG covalently linked to arabinogalactan, which is in turn linked to mycolic acids (9). To overcome the disadvantage that this complex envelope represents for a successful infective cycle, these viruses have evolved new lysis systems by acquiring specific lysis genes that confer a substantial selective advantage providing optimal lysis timing in response to environmental factors.

---

[1]Centro de Patogénese Molecular, Unidade dos Retrovirus e Infecções Associadas, Faculty of Pharmacy, University of Lisbon, Portugal.

This chapter will focus on the current knowledge of mycobacteriophage-mediated lysis. An overview of bacteriophage lysis will first be presented to introduce the reader to the subject. Recent reviews discuss phage lysis in other systems (4, 5, 10).

THE LYSIS PLAYERS

Two major aspects are considered for successful bacteriophage-induced lysis: (i) compromise the PG layer and (ii) ensure the proper time of lysis. Determination of the precise time of lysis is achieved by holins. These are small membrane proteins with at least one predicted transmembrane domain (TMD) and a hydrophilic, highly charged C-terminus. According to the number of TMDs (three, two, or one), holins can be classified in classes I, II, or III, respectively (11, 12). Holins accumulate in the bacterial cytoplasmic membrane (CM) during late gene expression without disturbing membrane integrity (4, 13–16). Then at a precise, allele-specific time, the holin triggers to form holes that permeabilize the membrane, resulting in its disruption, and lysis begins (11, 12). According to the size of the holin-mediated hole, holins can be classified in two types: canonical holins (e.g., λ S105), which form large holes of near-micrometer scale (17), and pinholins (e.g., $S^{21}68$ of phage 21), which form heptameric channels of ~2 nm (18). The timing of lysis must be tightly regulated, since early lysis can release few or no phage particles, resulting in an unprofitable infection cycle, while a delay might compromise the availability of new bacterial hosts in the environment. Some phages also encode an antiholin, a negative regulator that controls the functions of all three classes of holins contributing to the precise regulation of lysis (11). In some cases the antiholin is encoded by the holin gene (19), while in others it is encoded by an independent gene (20–22).

Endolysins are PG hydrolases that are produced during the late phase of gene expression and are designed to cleave one or more of the covalent bonds in the PG meshwork, compromising the integrity of the cell wall. They can be classified into five major types (Fig. 1) depending on the specific bonds that are attacked: (i) N-acetylmuramidases (generically termed lysozymes) and (ii) lytic transglycosylases, both cleaving the β-1,4 glycosidic bond between N-acetylmuramic acid (NAM) and N-acetylglucosamine (NAG); (iii) endo-β-N-acetylglucosaminidases, cleaving the other glycosidic bond between NAG and NAM; (iv) N-acetyl-muramoyl-L-alanine amidases, which hydrolyze the amide bond between NAM and L-alanine residues in the oligopeptide cross-linking chains, and (v) endopeptidases, which attack the peptide bonds in the same chains (23). Endopeptidases can be further divided into two groups: enzymes that cut within the stem peptide of the PG and those that attack bonds within the interpeptide bridge (if present) (24). Most endolysins from phages that infect Gram-negative bacteria are relatively small globular proteins composed of only a single catalytic domain, while endolysins from Gram-positive-infecting phages are typically modular, having one or more catalytic domains at the N-terminus and a cell wall–binding domain at the C-terminus (23, 24). These structural differences reflect the variations in cell wall architecture between these major bacterial groups (for an extensive review on bacteriophage endolysins see references 23–26).

The access of endolysins to the PG may occur in two ways: canonical endolysins, such as λR, accumulate fully folded and active in the cytoplasm during late gene expression, until the holin forms holes in the CM large enough to allow the passage of the endolysin through them; secreted endolysins are exported as inactive enzymes in a holin-independent manner, using the host Sec system. Two types of secreted endolysins have been described: (i) endolysins with a Sec-dependent SP sequence, which are processed by the leader peptidase during translocation of the enzyme to the cell wall compartment, e.g., Lys44 of *Oenococcus* phage fOg44 (27), and (ii) endolysins endowed with a signal arrest release (SAR) sequence, which is not removed during translocation, retaining the endolysins in the periplasm as inactive forms tethered to the CM, e.g., R21 of phage 21 and Lyz of phage P1 (28, 29). Although in the former case it is not known how endolysins are activated, in the latter, activation is a consequence of the membrane depolarization by either a canonical holin or a pinholin, which allows the escape of the SAR endolysin from the bilayer and refolding into an active form, ready to attack the PG (30–32).

In addition to the essential lysis proteins, phages infecting Gram-negative hosts encode a third functional class of lysis proteins, the spanins. These proteins span the entire periplasm, allowing the fusion of the inner membrane with the outer membrane (OM) at a third stage of host lysis, where the first stage is disruption of the CM by the holin, followed by degradation of the cell wall by the endolysin. Spanins, which may be synthesized as a sole protein (T1 Gp11) or as two subunits (λ Rz and Rz1 proteins), are responsible for elimination of the third and final barrier to phage release, the OM (33–36).

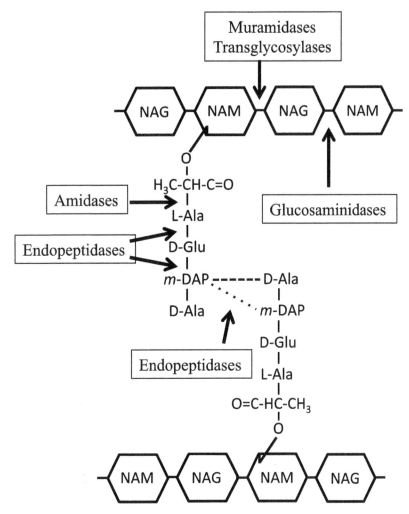

Figure 1 Schematic representation of the endolysins' targets in the bacterial PG. Proposed 4→3 interpeptide bridges between *m*-DAP and D-Ala but also 3→3 *m*-DAP to *m*-DAP bonds in the mycobacterial PG are indicated by dashed lines. NAG, N-acetylglucosamine; NAM, N-acetylmuramic acid. doi:10.1128/microbiolspec.MGM2-0017-2013.f1

MYCOBACTERIOPHAGE-MEDIATED LYSIS

Mycobacteriophages are a remarkably diverse group of viruses that infect mycobacteria. Currently, more than 3,400 mycobacteriophages have been isolated, with more than 270 complete genome sequences available in GenBank (http://www.phagesdb.org). Most of them have *Mycobacterium smegmatis* as their host, and they all share a double-stranded DNA genome (37–39). As dsDNA-tailed phages, mycobacteriophages have to overcome the complex envelope of mycobacteria to achieve lysis of their hosts and release the progeny virions at the end of a lytic cycle. Until recently, little was known about the mechanisms underlying mycobacteriophage-induced lysis of mycobacteria, but studies on mycobacteriophage Ms6 (40–45) provided new insights into the way phages achieve lysis of their

hosts. The first report came from the work of Garcia et al. (46), who described the genetic organization of the lysis module of mycobacteriophage Ms6. A detailed analysis of the Ms6 lysis model is described below.

The Lysis Model of Mycobacteriophage Ms6

Mycobacteriophage Ms6 is a temperate phage, isolated from *M. smegmatis* strain HB5688 (47). The Ms6 lysis cassette is composed of five genes (Fig. 2) clustered downstream of two σ^{70}-like promoters. This promoter region (P_{lys}) is separated from the first lysis gene by a leader sequence of 214 bp, in which was detected a transcription termination signal, suggesting that an anti-termination mechanism is involved in the regulation of Ms6 lysis gene transcription (46). Although the complete genome sequence is not yet available, according

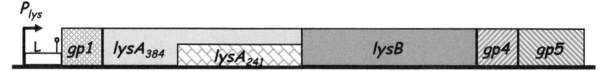

Figure 2 Genetic organization of the Ms6 lysis cassette. Genes are drawn to scale with gene names indicated. Segments of *lysA* generating the full-length Lysin$_{384}$ and the N-terminal truncated version Lysin$_{241}$ are indicated separately. The promoter region P_{lys} is separated from *gp1* by a leader sequence (L). The arrow indicates direction of the transcription from the promoter region P_{lys}. ⌐ indicates the localization of a transcription termination signal. Adapted from reference 46 with permission. doi:10.1128/microbiolspec.MGM2-0017-2013.f2

to the organization of the lysis genes, Ms6 seems to be closely related to phages included in cluster F, subcluster F1 (10, 37). Like all dsDNA phages, Ms6 uses the holin-endolysin strategy to achieve lysis of its host; however, the model of lysis is different from those described for other bacteriophages. Perhaps one of the most striking features of Ms6 is that, in addition to encoding a specialized function related to the nature of the mycobacteria cell envelope, it displays a peculiar mechanism of endolysin export. The access of the Ms6 endolysin (LysA) to the PG was shown to be independent of the holin function; however, LysA does not have a predicted Sec-type SP or a SAR sequence, as do the secreted endolysins described above.

Catalão et al. (42) have shown that LysA export is assisted by a chaperone-like protein (Gp1) encoded by the first gene of the Ms6 lysis cassette. Analysis of the physical and structural properties of Gp1 shows that it shares the properties of molecular chaperones, particularly type III secretion system (TTS) chaperones: *gp1*, positioned immediately upstream of the endolysin gene and overlapped with it, encodes a small protein of 8.3 kDa with an acidic isolectric point of 4.6. In common with molecular chaperones, Gp1 has the ability to oligomerize and was shown to interact with its effector, an interaction that encompasses the N-terminal region of the chaperone and the first 60 amino acids of the Ms6 endolysin (42, 45). Evidence for a role in LysA translocation comes from experiments in *M. smegmatis* that showed an increased alkaline phosphatase activity of a LysA-PhoA' hybrid protein only in the presence of Gp1 (42). Doubts about the *in vivo* role of Gp1 in lysis were resolved when deletion of *gp1* from the Ms6 genome resulted in alterations in the phage lysis phenotype. Although not essential for plaque formation, and thus not essential for lysis, Gp1 is necessary to achieve efficient lysis, since its absence resulted in a decrease of ∼70% in the burst size (42).

Interestingly, and although no secretion signals were predicted in the amino acid sequence of Ms6 LysA,

experiments in *Escherichia coli*, using either a thermosensitive *secA* strain or the SecA inhibitor sodium azide, showed that lysis is blocked in cells expressing Gp1 and LysA, indicating that the *E. coli sec* machinery is involved in Ms6 LysA translocation, as it is with the secretory endolysins described so far (27, 28, 42). However, the involvement of the mycobacteria *sec* system in Ms6 LysA delivery was not yet determined. The Sec-dependent export pathway is highly conserved among different bacteria. Mycobacteria have two homologues of *secA*: *secA1* and *secA2* (48); *secA1* is essential in *M. smegmatis*, while *secA2* is nonessential for viability. Depletion of *secA1* leads to a loss of cell viability. Consequently, the attempts made so far to evaluate the dependence on the *M. smegmatis* SecA1 for Ms6-induced lysis were not successful, so the direct involvement of SecA1 in Ms6 LysA translocation remains to be demonstrated. Although the dependence of SecA2 for export of some proteins has been described (49), infection of an *M. smegmatis* Δ*secA2* mutant strain with Ms6 wild type did not alter the phage plaque ability (42).

Ms6 LysA has a central PG recognition protein conserved domain localized between amino acid residues 168 and 312, containing an amidase-2 domain (pfam01510) (44). Its *N*-acetylmuramoyl-L-alanine amidase activity was recently reported; LysA was shown to cleave the bond between L-Ala of the stem peptide and the lactyl moiety of the muramic acid residues of muramyl pentapeptide and to release up to 70% of the diaminopimelic acid present in the isolated mycobacterial cell wall (50). Interestingly, a recent analysis of the *lysA* gene shows that it generates two proteins designated Lysin$_{384}$ and Lysin$_{241}$ according to the size of the produced polypeptides. Catalão et al. (44) have shown that Lysin$_{241}$ is not a mature form of Lysin$_{384}$, but rather is a result of a second translation event from an initiation codon, in the same reading frame, positioned 143 amino acids away from the first initiation codon. Not surprisingly, both proteins have PG hydrolase activity, since $Lysin_{241}$ retains the PG recognition protein domain.

What is puzzling is that both forms of LysA are necessary for complete and efficient lysis of *M. smegmatis*. Although deletion of the complete *lysA* nucleotide sequence revealed it to be essential for *M. smegmatis* lysis, Ms6 mutants producing only one of the forms of LysA were shown to be viable, albeit defective in the normal timing, progression, and completion of host cell lysis. Lack of $Lysin_{384}$ resulted in a lysis delay of 30 min and in a reduction in the number of phage particles released, while in the absence of $Lysin_{241}$, lysis starts 90 min later with no significant effect on the number of phage particles released (44). Worthy of note is the fact that although $Lysin_{241}$ keeps the enzyme catalytic domain, the N-terminal region that interacts with Gp1 is absent. However, a tight association between Gp1 and $Lysin_{384}$ exists, since Catalão et al. observed that during an Ms6 infection, the synthesis/stability of the larger endolysin is dependent on Gp1 production (44).

Despite the fact that both proteins are necessary for efficient lysis, $Lysin_{241}$ seems to be much more active than $Lysin_{384}$, as revealed by zymogram assays. Contrary to what is observed in an *M. smegmatis* infection with an Ms6 derivative mutant, where the two forms of endolysin are expressed with a hexastidine tag at the C-terminus, in *E. coli*, expression from the full *lysA* gene results in almost undetectable levels of $Lysin_{241}$. However, even a low level of $Lysin_{241}$ showed high hydrolytic activity on *Micrococus luteus* lyophilized cells (44). The same observation was reported for the endolysin produced by the mycobacteriophage Corndog (51). The authors suggested that the N-terminus, present in the larger form of the endolysin, could somehow inhibit catalysis in the zymogram assay. Interestingly, *E. coli* crude extracts containing $Lysin_{384}$ or $Lysin_{241}$ were both shown to inhibit the growth of several bacteria "from without" (44). Although it is clear that Gp1 is essential for translocation of $Lysin_{384}$, it is not known at this time how $Lysin_{241}$ accesses the PG or why Ms6 synthesizes two endolysins.

Even though Ms6 $Lysin_{384}$ is exported in a holin-independent manner, lysis of *M. smegmatis* does not occur until holin triggering. It is now evident that secreted endolysins, once positioned in the extracytoplasmatic environment, must be kept inactive until the proper time of lysis (5, 31, 32). In fact, compromising the murein layer at a time well before the new phage particles are assembled would result in no or too little progeny release. Thus, even for phages encoding secreted endolysins, holins are still crucial for determining

the timing of lysis. Indeed, all phages that encode secreted endolysins also appear to encode a holin-like protein (5). Holin function is thus confined to controlling the access of the endolysin to its target, either by allowing its passage through holes formed in the CM—as happens in the λ model of lysis (5)—or by allowing the activation of endolysin already positioned close to the murein layer at the time of dissipation of the membrane potential—exemplified by the P1 model of lysis (18, 28, 30).

In Ms6, the potential role of the holin function in endolysin activation was supported using nisin, a permeabilizing compound that triggers CM depolarization. In contrast to what happens with *M. smegmatis* cells expressing only Gp1 or LysA, addition of nisin to cells expressing both proteins resulted in complete lysis. This means that the endolysin was already positioned next to its target, the PG, at lysis onset, since the pore diameter produced by nisin (2 nm) should not allow the passage of a protein as large as Ms6 endolysin (42). However, how Ms6 endolysin is kept inactive until the proper time of lysis is a question that remains to be elucidated.

Achieving the Proper Time of Lysis

In Ms6, the regulation of mycobacteria lysis timing also seems to display peculiar features. Achievement of the correct lysis timing was shown to be dependent on the interaction and concerted action of two proteins with holin-like features (43), even though the presence of both is not absolutely required for *M. smegmatis* lysis. Gene *gp4* encodes a small protein of 77 amino acids, sharing structural characteristics with class II holins. A holin function was also supported by Gp4's ability to complement a λ S defective mutant (46). A more recent characterization of Ms6 Gp4 function suggested that Gp4 might function as a pinholin, since the first TMD has characteristics of a SAR domain with a high percentage of weakly hydrophobic or polar residues (43). The presence of a SAR domain followed by a typical TMD is characteristic of the previously described pinholins, such as the holin of phage 21, $S^{21}68$ (18, 27, 52). Coexpression of Ms6 LysA and Gp4 in *E. coli* does not induce bacterial lysis (46); however, changing Ms6 *gp4* by the mycobacteriophage D29 holin gene (*gp11*) results in *E. coli* lysis (43). This is consistent with the idea that, in contrast to the D29 holin, Ms6 Gp4 forms holes too small to allow passage of the 43-kDa Ms6 endolysin. Unexpectedly, deletion of *gp4* from the Ms6 genome results in earlier lysis timing by about 30 min, a lysis phenotype that seems to be more consistent with antiholin function for Gp4 rather than holin (43). The last gene (*gp5*) of the Ms6 lysis module was also shown to encode a holin-like protein. Gp5 is a 124-amino-acid protein with a predicted TMD at the N-terminus and a highly charged C-terminus, fitting the structural characteristics of class III holins; however, it does not complement an S^{λ} defective mutant. Nevertheless, it was shown that Gp5 has a regulatory role in the timing of lysis, since a deletion of *gp5* from the Ms6 genome resulted in a viable phage, but with a delayed time of lysis (43), excluding, in light of current knowledge, the possibility of functioning as an antiholin. It has been described elsewhere that null mutation in antiholin genes causes acceleration of the onset of lysis, while null mutation in holin genes results in a delay of the timing of lysis (4, 11, 53).

The observation that Gp4 interacts with Gp5 supports the notion that, in Ms6, the holin function is a result of the combined action of Gp4 and Gp5, contributing to the precise adjustment of the timing of hole formation and to keep the infected cell productive, allowing the assembly of more virions (43). The regulation of the timing of lysis as the result of activity of a complex is not restricted to a phage that infects a mycobacterial host. Similarly, it was proposed that the holin function of the *Bacillus subtilis* prophage PBSX is a result of two holin-like proteins, XhlA and XhlB, that associate in the membrane to form a holin functional unit (54).

It is now clear that mycobacteriophage Ms6 insults the first cell barriers to phage release through a holin-endolysin strategy, but the access of the endolysin to its target is different from all phage-mediated lysis described so far. Figure 3 represents the proposed Ms6 model to overcome the first cell barriers to phage release: the CM and the cell wall. However, mycobacteriophages still have to face a third barrier: the mycobacterial OM. Although mycobacteria are classified as Gram positive, they have a complex cell envelope, which includes an OM. This is a lipid bilayer, consisting of an inner leaflet rich in mycolic acids that are covalently bound to the PG-arabinogalactan complex via an ester linkage, and an outer leaflet mainly composed of glycolipids, phospholipids, and species-specific lipids (55, 56).

Overcoming the Last Barrier to Phage Release

Similar to what has been described for bacteriophage λ and other phages infecting Gram-negative hosts, which encode spanins to overcome the Gram-negative OM, it is not surprising that mycobacteriophages encode additional lysis proteins targeting the mycobacterial OM. In fact, mycobacteriophage Ms6 synthesizes an additional lysis protein, encoded by *lysB*, a gene positioned

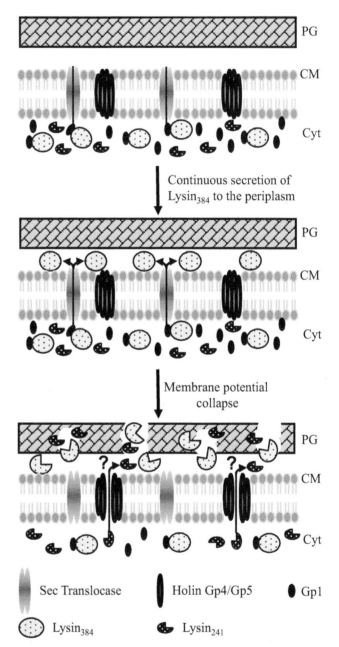

Figure 3 Model for Ms6 endolysin export. Export of the full-length endolysin (Lysin$_{384}$) is assisted by the chaperone Gp1 through the Sec translocase. Once in the cell wall compartment, it is kept in an inactive state until the holin complex Gp4/Gp5 dissipates the membrane potential. The endolysin activation is schematically represented by the change of the enzyme spherical configuration to a "pacman" shape. Lysin$_{241}$ is an N-terminally truncated version of Lysin$_{384}$. (?) indicates that export of this shorter version to the extracytoplasmatic environment is not known. PG, peptidoglycan; CM, cytoplasmic membrane; Cyt, cytoplasm. Adapted from reference 10 with permission.
doi:10.1128/microbiolspec.MGM2-0017-2013.f3

immediately downstream of the *lysA* gene (46), which was shown to have lipolytic activity (40). The Ms6 LysB amino acid sequence contains a pentapeptide, GYSQG, which matches the conserved G-X-S-X-G motif characteristic of lipolytic enzymes. Biochemical characterization of the enzyme showed that Ms6 LysB hydrolyzes both esterase and lipase substrates, showing a higher affinity for long-chain substrates, a characteristic that is in agreement with the highly diverse lipid content of the mycobacteria cell envelope. Importantly, Ms6 LysB was shown to cleave the ester bond between mycolic acids and the arabinogalactan of the mycolyl arabinogalactan-peptidoglycan (mAGP) complex, releasing free mycolic acids (41). However, the activity of Ms6 LysB is not restricted to mAGP. Ms6 LysB was also shown to hydrolyze other mycobacterial lipid components of the cell envelope, particularly the trehalose dimycolate (TDM, a glycolipid involved in the virulence of pathogenic species) of both fast- (*M. smegmatis*) and slow-growing mycobacteria (*Mycobacterium bovis* BCG and *Mycobacterium tuberculosis* H37Ra), indicating that Ms6 LysB activity is not species-specific (41). Nevertheless, due to the importance of the mAGP complex to stability of the mycobacteria cell envelope, it was thus proposed that the mAGP complex is the major target for Ms6 LysB activity. Breaking this linkage allows the separation of the OM from the cell wall, disrupting the last barrier to phage release.

Analogies can be made between the function of LysB against the mycobacterial OM and the λ Rz/Rz1 homologues or spanins of Gram-negative hosts, which mediate the third and final step of host lysis by fusing the inner and outer membranes (33, 35, 57, 58). The Rz and Rz1 λ proteins were long considered to be auxiliary lysis proteins since, in laboratory conditions, *RzRz1*-null mutants are still able to cause lysis unless the OM is artificially stabilized with milimolar concentrations of divalent cations (59). Recently, Young and his collaborators showed that λ lysis actually requires functioning *RzRz1* genes without artificial stabilization of the OM and thus could no longer be considered auxiliary lysis proteins (36).

Under normal infection conditions, LysB is not essential for lysis, since a deletion of *lysB* from the Ms6 (60) genome results in a mutant phage that retains the ability to form plaques. If there are other factors that influence the lysis dependence on LysB, it is not currently known. Despite their different modes of action, both spanins and Ms6 LysB mediate the third and final step in host lysis by eliminating the last barrier to phage release, after holins and endolysins have

disrupted the CM and PG, respectively. At this time it is unknown how Ms6 LysB is localized to its substrate, since no signal sequences allowing its transport across the cell barriers have been identified (40). One can hypothesize that LysB might reach the OM simply by diffusion after PG breakdown by LysA.

Diversity of the Mycobacteriophage Lysis Systems

The availability of more than 270 mycobacteriophage complete genome sequences shows that they are highly diverse; however, some are more closely related than others. Based on nucleotide sequence comparison, these phages have been grouped into clusters and subclusters (for a more detailed explanation see reference 71).

The majority of data regarding the lysis systems of mycobacteriophages, other than Ms6, are restricted to bioinformatic analysis of their genomes. Description and comparison of more than 80 phage genome sequences (37, 38, 61–63) provides insights into lysis cassette diversity (Fig. 4).

All mycobacteriophages described to date encode an endolysin. A bioinformatic analysis of 224 sequenced mycobacteriophage endolysins shows that they are highly diverse and modular in nature, with a large number of domain organizations (38, 51). Interestingly, most of them seem to be composed of three domains: a C-terminal domain that is likely to be associated with binding to the cell wall and two catalytic domains—an N-terminal domain with putative peptidase activity and a central catalytic domain that specifies a glycoside hydrolase (muramidase, transglycosylase) or an amidase (51). This is in contrast to the majority of endolysins from phages infecting Gram-positive or Gram-negative bacteria. In general, endolysins from phages infecting Gram-positive hosts are composed of an N-terminal catalytic domain and a cell wall–binding domain positioned at the C-terminus, while endolysins from phages infecting Gram-negative hosts are mostly small single-domain proteins, usually without a specific cell wall–binding domain module (24).

Some mycobacteriophage endolysins contain both glycosidase and amidase domains, such as Che9d Gp35 and Wildcat GP49 (51). Of note is the fact that for the majority of the mycobacteriophage endolysins, the peptidase motifs were identified through similarities to known peptidases. Except for the M23 domain, peptidase domains were not readily identified by conserved domain searches (51). Although PG hydrolase activity was demonstrated, by zymogram analysis, for the endolysins of TM4, D29, Che8, Bxz1, and Corndog

Cluster/Subcluster Phage — Lysis Genes Organization

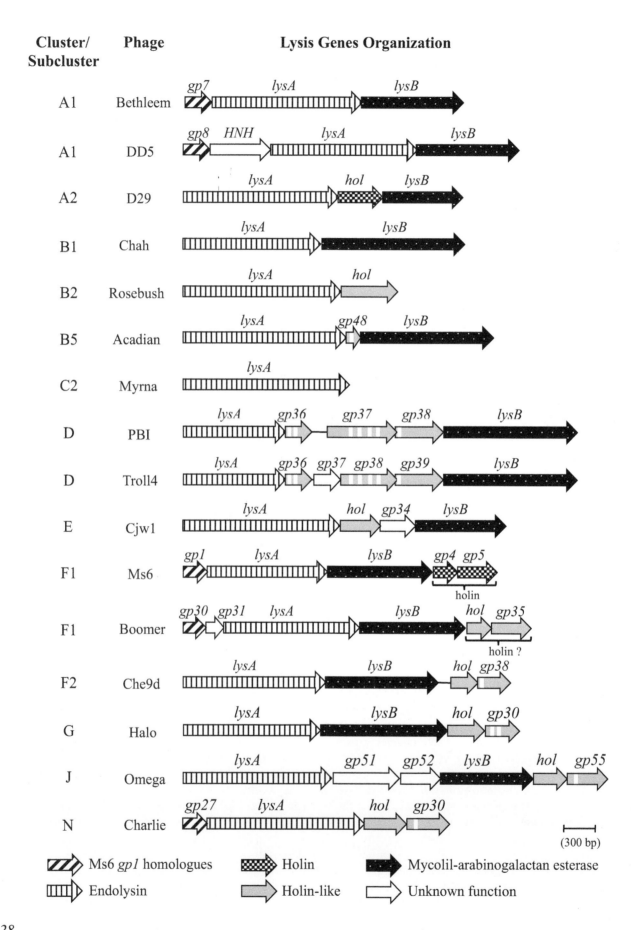

Cluster/Subcluster	Phage
A1	Bethleem
A1	DD5
A2	D29
B1	Chah
B2	Rosebush
B5	Acadian
C2	Myrna
D	PBI
D	Troll4
E	Cjw1
F1	Ms6
F1	Boomer
F2	Che9d
G	Halo
J	Omega
N	Charlie

(300 bp)

Ms6 *gp1* homologues
Endolysin
Holin
Holin-like
Mycolil-arabinogalactan esterase
Unknown function

(58, 64), apart from the Ms6 endolysin (50), none of the predicted proteins were, so far, characterized experimentally.

Apart from Ms6, little is known about how endolysins access the extracytoplasmatic compartment. Ms6 *gp1* homologues have been identified in other mycobacteriophage genomes (Fig. 4), particularly in the lysis cassettes of phages that belong to subclusters A1 and F1 (37, 38). Taking into consideration the high similarity between the lysis genes among phages of the same cluster, one can speculate that phages belonging to subcluster F1 will use the Ms6 strategy to deliver LysAs to their PG substrate. If that is also the case of phages belonging to cluster A1, which although sharing Gp1 homologues, have a different lysis cassette organization, it is a question that deserves further investigation. However, Gp1 homologues are missing from the majority of mycobacteriophage genomes, suggesting that in these phages LysA export occurs in a different way. Similar to Ms6, *lysA* secondary gene products seem also to occur in other mycobacteriophage genomes. Zymogram analysis of *E. coli* extracts containing LysA proteins of Corndog (Gp69), Bxz1 (Gp236), and Che8 (Gp32) show that in addition to the full-length protein, smaller products can also hydrolyze the PG (51). Recently, Payne and Hatfull (51) reported that the sole expression of D29, L5, and Kostya endolysins in *M. smegmatis* mc²155, under the control of the acetamidase promoter, resulted in lysis. This is an intriguing observation since the holin gene was excluded from the expression clone, and none of the proteins is predicted to have secretion signals. As suggested by the authors, it is unlikely that upon a L5 infection, lysis would be holin independent, since a predicted holin gene is localized immediately downstream of the *lysA* gene. For any degree of confidence, the apparent holin-independent lysis must be evaluated in the phage infection context.

In the majority of mycobacteriophages sequenced so far, no holin function was experimentally determined. Homologues of Ms6 Gp4 and Gp5 were found in other mycobacteriophages, with the highest identity within subcluster F1. However, holin proteins are highly diverse and do not share high sequence identity (11),

and thus it is more difficult to identify genes encoding putative holins. Among all mycobacteriophage genomes available in the Genbank database, few have annotated holin genes; however, genes encoding putative proteins with predicted TMDs can be identified in the vicinity of endolysin genes, making them good candidates to encode the lysis-timing regulators. Whenever a holin gene has been assigned, it is closely linked and downstream to *lysA*.

Like mycobacteriophage Ms6, the majority of mycobacteriophages also encode a LysB protein. Structure and amino acid sequence comparisons show that LysB proteins are also diverse, although they all share the characteristic motif, G-X-S-X-G, of lipolytic enzymes (40, 58, 65). A mycolylarabinogalactan esterase activity was also demonstrated for the LysB protein of D29 (58). These authors have also determined the crystal structure of D29 LysB, revealing structural similarities to cutinases (enzymes that also cleave esterase and lipase substrates). At the time of the last mycobacteriophage genome analysis, *lysB* genes were identified in 76 out of 80 phage genomes (38). When present, the *lysB* gene is always positioned downstream of *lysA*—in some cases immediately downstream, while in other cases it is separated from *lysA* by no more than four genes (Fig. 4). With few exceptions (38), all genes positioned between *lysA* and *lysB* also encode lysis proteins, specifically putative holin proteins (Fig. 4). The high incidence of *lysB* genes in mycobacteriophage lysis cassettes suggests that LysB proteins may have an important role in lysis. In fact, Payne et al. (58) showed that a deletion of gene *lysB* from the phage Giles genome delayed the time of lysis by 30 min. It is suggested that this delay results from a defective break of the cell barriers in the absence of LysB. This explanation derives support from the observation that after an infection of *M. smegmatis* with a Δ*lysB* mutant of phage Giles, 45% of the phage particles remained associated with unlysed cells in contrast to the 10% observed in a wild-type infection. It seems, therefore, that in the former, the new phage particles are trapped within cell debris as a result of a deficient elimination of the OM, resulting in a delay of their release into the environment.

Figure 4 Diversity of mycobacteriophage lysis cassettes. The illustration shows representatives of mycobacteriophages with diverse genome organization. Not previously assigned holin-like genes display white bars that represent the number and location of putative TMD coding sequences. Adapted from reference 10 with permission. doi:10.1128/microbiolspec.MGM2-0017-2013.f4

Similar to what happens with phages that infect Gram-negative hosts, lysis induced by mycobacteriophages seems to be a three-step event, with LysB playing a role in the third step: elimination of the OM layer. For the few phages that do not encode a LysB protein (Che12, Rosebush, Quirzula, Myrna, and Charlie) (38, 39, 61), it was suggested that they have evolved a mechanism to utilize a host-encoded cutinase-like protein that would replace the function of LysB (37).

It is interesting to note that *lysB* equivalents were recently described in phages infecting other bacterial groups that have in common with mycobacteria a thick mycolic-acid-containing outer layer covalently linked to the PG—the mycolata group. Examples are the *Rhodococcus equi* phages DocB7, Pepy6, and Poco6 (66). This supports the notion that the presence of LysB-like proteins is a selective advantage for phages infecting members of the mycolata group.

In conclusion, like phages that infect Gram-negative hosts, lysis induced by mycobacteriophages is also a three-step event (Fig. 5). The first step is achieved by holins that compromise the CM, allowing the endolysins to hydrolase, in a second step, the PG layer. Finally, LysB proteins disrupt the mycobacterial OM, eliminating the last barrier to mycobacteriophage release.

APPLICATIONS OF THE LYSIS PROPERTIES

The capacity of bacteriophages to lyse their hosts has been intensively explored, with the main goal of elimination of pathogenic bacteria. These studies are driven either by the use of bacteriophage itself as a therapeutic agent (phage therapy) (67) or for exploring the enzymes (enzybiotics) involved in cell lysis for enzyme therapy (26, 67). Phage-encoded endolysins have shown high potential as antibacterial agents against a number of Gram-positive bacteria. At first glance, mycobacteriophages or their lysis enzymes might be of limited use as therapeutic agents to fight tuberculosis, in part due to the inaccessibility of phages to the pathogenic bacteria located intracellularly. On the other hand, the presence of an OM in the mycobacterial cell envelope constitutes a barrier for endolysins to access the PG from without. However, some studies on the use of mycobacteriophage lysis properties have recently been reported. Broxmeyer et al. (68) have shown the potential of mycobacteriophage TM4 delivered by the nonpathogenic *M. smegmatis* to kill *Mycobacterium avium* and *M. tuberculosis* within macrophages. An *in vivo* assay with mice infected with *M. avium* showed that treatment with *M. smegmatis* transiently infected with TM4 significantly reduced the number of *M. avium* cells in

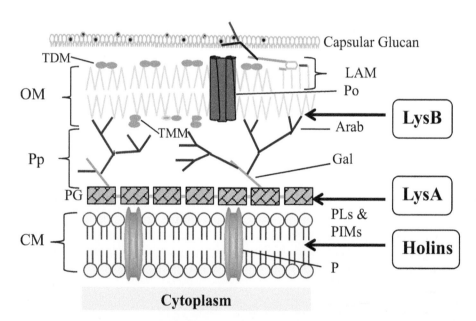

Figure 5 Targets of Ms6 lysis proteins. Schematic representation of the mycobacteria cell envelope, where the target layer of each protein is indicated by an arrow. Arab, arabinan; CM, cytoplasmic membrane; Gal, galactan; LAM, lipoarabinomannan; OM, outer membrane; P, protein; PG, peptidoglycan; PIMs, phosphatidylinositol mannosides; PLs, phospholipids; Po, porin; Pp, periplasm; TDM, trehalose dimycolate; TMM, trehalose monomycolate. doi:10.1128/microbiolspec.MGM2-0017-2013.f5

the spleen (69). More recently Trigo et al. (70) demonstrated for the first time the potential of phage therapy against a *Mycobacterium ulcerans* infection. These authors showed in the mouse footpad model that a single subcutaneous injection of the lytic mycobacteriophage D29 can effectively decrease the proliferation of *M. ulcerans*.

CONCLUDING REMARKS

With the increased knowledge of phage lysis, it has become evident that phages have evolved sophisticated solutions to improve their survival in the biosphere. The studies on the Ms6 lysis model have provided new insights into the mechanisms of bacteriophage-induced lysis. In order to obtain the highest level of fitness, mycobacteriophages have evolved specific lytic functions by acquiring additional and specific lysis genes that confer a selective advantage by allowing efficient disruption of the complex cell barriers. It seems clear that the genetic diversity observed among the mycobacteriophage lysis cassettes reflects the ability of mycobacteriophages to adapt to different environmental conditions. The huge number of isolated mycobacteriophages opens exciting research opportunities to explore the mycobacteriophage-induced lysis pathways, which can give us new clues into mycobacterial secretion systems. Identification of the lysis enzymes that target specific linkages in the cell envelope, and thus compromise the mycobacteria viability, also opens new perspectives to explore these enzymes as new tools to destroy mycobacteria cells.

Citation. Pimentel M. 2014. Genetics of phage lysis. Microbiol Spectrum 2(1):MGM2-0017-2013.

References

1. Rodriguez-Valera F, Martin-Cuadrado AB, Rodriguez-Brito B, Pasić L, Thingstad TF, Rohwer F, Mira A. 2009. Explaining microbial population genomics through phage predation. *Nat Rev Microbiol* 7:828–836.

2. Kutter E, Sulakvelidze A. 2005. Introduction, p 1–4. *In* Kutter E, Sulakvelidze A (ed), *Bacteriophages Biology and Applications.* CRC Press, Boca Raton, Florida.

3. Rakonjac J, Bennett NJ, Spagnuolo J, Gagic D, Russel M. 2011. Filamentous bacteriophage: biology, phage display and nanotechnology applications. *Curr Issues Mol Biol* 13:51–76.

4. Young R, Wang IN. 2006. Phage lysis, p 104–125. *In* R Calendar (ed), *The Bacteriophages*, 2nd ed. Oxford University Press, New York, NY.

5. São-José C, Nascimento J, Parreira R, Santos M. 2007. Release of progeny phages from infected cells, p 309–336. *In* McGrath S, van Sinderen D (ed), *Bacteriophage: Genetics and Molecular Biology.* Caister Academic Press, Wymondham, UK.

6. Bernhardt TG, Struck DK, Young R. 2001. The lysis protein E of phi X174 is a specific inhibitor of the MraY-catalyzed step in peptidoglycan synthesis. *J Biol Chem* 276:6093–6097.

7. Bernhardt TG, Wang IN, Struck DK, Young R. 2001. A protein antibiotic in the phage Qbeta virion: diversity in lysis targets. *Science* 292:2326–2329.

8. Ackermann HW, Prangishvili D. 2012. Prokaryote viruses studied by electron microscopy. *Arch Virol* 157:1843–1849.

9. Brennan PJ. 2003. Structure, function, and biogenesis of the cell wall of *Mycobacterium tuberculosis. Tuberculosis* 83:91–97.

10. Catalão MJ, Gil F, Moniz-Pereira J, Pimentel M. 2013. Diversity in bacterial lysis systems: bacteriophages show the way. *FEMS Microbiol Rev* 37:554–571.

11. Wang IN, Smith DL, Young R. 2000. Holins: the protein clocks of bacteriophage infections. *Annu Rev Microbiol* 54:799–825.

12. Young R. 2002. Bacteriophage holins: deadly diversity. *J Mol Microbiol Biotechnol* 4:21–36.

13. Gründling A, Smith DL, Bläsi U, Young R. 2000. Dimerization between the holin and holin inhibitor of phage lambda. *J Bacteriol* 182:6075–6081.

14. Pang T, Park T, Young R. 2010. Mutational analysis of the S21 pinholin. *Mol Microbiol* 76:68–77.

15. Pang T, Park T, Young R. 2010. Mapping the pinhole formation pathway of S21. *Mol Microbiol* 78:710–719.

16. White R, Chiba S, Pang T, Dewey JS, Savva CG, Holzenburg A, Pogliano K, Young R. 2011. Holin triggering in real time. *Proc Natl Acad Sci USA* 108:798–803.

17. Dewey JS, Savva CG, White RL, Vitha S, Holzenburg A, Young R. 2010. Micron-scale holes terminate the phage infection cycle. *Proc Natl Acad Sci USA* 107:2219–2223.

18. Pang T, Savva CG, Fleming KG, Struck DK, Young R. 2009. Structure of the lethal phage pinhole. *Proc Natl Acad Sci USA* 106:18966–18971.

19. Bläsi U, Young R. 1996. Two beginnings for a single purpose: the dual-start holins in the regulation of phage lysis. *Mol Microbiol* 21:675–682.

20. Walker JT, Walker DH Jr. 1980. Mutations in coliphage P1 affecting host cell lysis. *J Virol* 35:519–530.

21. Ziermann R, Bartlett B, Calendar R, Christie GE. 1994. Functions involved in bacteriophage P2-induced host cell lysis and identification of a new tail gene. *J Bacteriol* 176:4974–4984.

22. Ramanculov E, Young R. 2001. An ancient player unmasked: T4 *rI* encodes a *t*-specific antiholin. *Mol Microbiol* 41:575–583.

23. Loessner MJ. 2005. Bacteriophage endolysins: current state of research and applications. *Curr Opin Microbiol* 8:480–487.

24. Schmelcher M, Donovan DM, Loessner MJ. 2012. Bacteriophage endolysins as novel antimicrobials. *Future Microbiol* 7:1147–1171.

25. Fischetti V. 2010. Bacteriophage endolysins as a novel class of antibacterial agents. *Int J Med Microbiol* 300:357–362.

26. Nelson DC, Schmelcher M, Rodriguez-Rubio L, Klumpp J, Pritchard DG, Dong S, Donovan DM. 2012. Endolysins as antimicrobials. *Adv Virus Res* 83:299–365.

27. São-José C, Parreira R, Vieira G, Santos MA. 2000. The N-terminal region of the *Oenococcus oeni* bacteriophage fOg44 lysin behaves as a bona fide signal peptide in *Escherichia coli* and as a *cis*-inhibitory element, preventing lytic activity on oenococcal cells. *J Bacteriol* 182:5823–5831.

28. Xu M, Struck DK, Deaton J, Wang IN, Young R. 2004. A signal-arrest-release sequence mediates export and control of the phage P1 endolysin. *Proc Natl Acad Sci USA* 101:6415–6420.

29. Park T, Struck DK, Dankenbring CA, Young R. 2007. The pinholin of lambdoid phage 21: control of lysis by membrane depolarization. *J Bacteriol* 189:9135–9139.

30. Xu M, Arulandu A, Struck DK, Swanson S, Sacchettini JC, Young R. 2005. Disulfide isomerization after membrane release of its SAR domain activates P1 lysozyme. *Science* 307:113–117.

31. Sun Q, Kuty GF, Arockiasamy A, Xu M, Young R, Sacchettini JC. 2009. Regulation of a muralytic enzyme by dynamic membrane topology. *Nat Struct Mol Biol* 16:1192–1194.

32. Kuty GF, Xu M, Struck DK, Summer EJ, Young R. 2010. Regulation of a phage endolysin by disulfide caging. *J Bacteriol* 192:5682–5687.

33. Berry J, Summer EJ, Struck DK, Young R. 2008. The final step in the phage infection cycle: the Rz and Rz1 lysis proteins link the inner and outer membranes. *Mol Microbiol* 70:341–351.

34. Krupovič M, Cvirkaite-Krupovič V, Bamford DH. 2008. Identification and functional analysis of the *Rz/Rz1*-like accessory lysis genes in the membrane-containing bacteriophage PRD1. *Mol Microbiol* 68:492–503.

35. Berry J, Savva C, Holzenburg A, Young R. 2010. The lambda spanin components Rz and Rz1 undergo tertiary and quaternary rearrangements upon complex formation. *Protein Sci* 19:1967–1977.

36. Berry J, Rajaure M, Pang T, Young R. 2012. The spanin complex is essential for lambda lysis. *J Bacteriol* 194:5667–5674.

37. Hatfull GF. 2010. Mycobacteriophages: genes and genomes. *Annu Rev Microbiol* 64:331–356.

38. Hatfull GF. 2012. The secret lives of mycobacteriophages. *Adv. Virus Res* 82:179–288.

39. Hatfull GF, Science Education Alliance Phage Hunters Advancing Genomics and Evolutionary Science Program, KwaZulu-Natal Research Institute for Tuberculosis and HIV Mycobacterial Genetics Course Students, Phage Hunters Integrating Research and Education Program. 2012. Complete genome sequences of 138 mycobacteriophages. *J Virol* 86:2382–2384.

40. Gil F, Catalão MJ, Moniz-Pereira J, Leandro P, McNeil M, Pimentel M. 2008. The lytic cassette of mycobacteriophage Ms6 encodes an enzyme with lipolytic activity. *Microbiology* 154:1364–1371.

41. Gil F, Grzegorzewicz AE, Catalão MJ, Vital J, McNeil M, Pimentel M. 2010. The mycobacteriophage Ms6 LysB specifically targets the outer membrane of *Mycobacterium smegmatis*. *Microbiology* 156:1497–1504.

42. Catalão MJ, Gil F, Moniz-Pereira J, Pimentel M. 2010. A chaperone-like protein is involved in the endolysin delivery to the peptidoglycan. *Mol Microbiol* 77:672–686.

43. Catalão MJ, Gil F, Moniz-Pereira J, Pimentel M. 2011. Functional analysis of the holin-like proteins of mycobacteriophage Ms6. *J Bacteriol* 193:2793–2803.

44. Catalão MJ, Milho C, Gil F, Moniz-Pereira J, Pimentel M. 2011. A second endolysin gene is fully embedded in-frame with the *lysA* gene of mycobacteriophage Ms6. *PLoS One* 6:e20515.

45. Catalão MJ, Gil F, Moniz-Pereira J, Pimentel M. 2011. The endolysin-binding domain encompasses the N-terminal region of the mycobacteriophage Ms6 *gp1* chaperone. *J Bacteriol* 193:5002–5006.

46. Garcia M, Pimentel M, Moniz-Pereira J. 2002. Expression of mycobacteriophage Ms6 lysis genes is driven by two sigma (70)-like promoters and is dependent on a transcription termination signal present in the leader RNA. *J Bacteriol* 184:3034–3043.

47. Portugal I, Anes E, Moniz-Pereira J. 1989. Temperate mycobacteriophage from *M. smegmatis*. *Acta Leprol* 7:243–244.

48. Braunstein M, Brown AM, Kurtz S, Jacobs WR Jr. 2001. Two nonredundant SecA homologues function in mycobacteria. *J Bacteriol* 183:6979–6990.

49. Rigel NW, Braunstein M. 2008. A new twist on an old pathway: acessory secretion systems. *Mol Microbiol* 69:291–302.

50. Mahapatra S, Piechota C, Gil F, Ma Y, Huang H, Scherman M, Jones V, Pavelka M Jr, Moniz-Pereira J, Pimentel M, McNeil M, Crick D. 2013. Mycobacteriophage Ms6 LysA is a peptidoglycan amidase and a useful analytical tool. *Appl Environ Microbiol* 79:768–773.

51. Payne KM, Hatfull GF. 2012. Mycobacteriophage endolysins: diverse and modular enzymes with multiple catalytic activities. *PLos One* 7:e34052.

52. Park T, Struck DK, Deaton JF, Young R. 2006. Topological dynamics of holins in programmed bacterial lysis. *Proc Natl Acad Sci USA* 103:19713–19718.

53. Young R. 2005. Phage lysis, p 92–127. *In* Waldor MK, Friedman DI, Adhya SL (ed), *Phages: Their Role in Bacterial Pathogenesis and Biotechnology*. ASM Press, Washington DC.

54. Krogh S, Jørgensen ST, Devine KM. 1998. Lysis genes of the *Bacillus subtilis* defective prophage PBSX. *J Bacteriol* 180:2110–2117.

55. Hoffmann C, Leis A, Niederweis M, Plitzko JM, Engelhardt H. 2008. Disclosure of the mycobacterial outer membrane: cryo-electron tomography and vitreous sections reveal the lipid bilayer structure. *Proc Natl Acad Sci USA* 105:3963–3967.

56. Zuber B, Chami M, Houssin C, Dubochet J, Griffiths G, Daffé M. 2008. Direct visualization of the outer membrane of mycobacteria and corynebacteria in their native state. *J Bacteriol* 190:5672–5680.

57. **Summer EJ, Berry J, Tran TA, Niu L, Struck DK, Young R.** 2007. Rz/Rz1 lysis gene equivalents in phages of Gram-negative hosts. *J Mol Biol* **373**:1098–1112.

58. **Payne K, Sun Q, Sacchettini J, Hatfull GH.** 2009. Mycobacteriophage lysin B is a novel mycolylarabino-galactan esterase. *Mol Microbiol* **73**:367–381.

59. **Zhang N, Young R.** 1999. Complementation and characterization of the nested *Rz* and *Rz1* reading frames in the genome of bacteriophage lambda. *Mol Gen Genet* **262**: 659–667.

60. **Gil F.** 2012. Functional analysis of *lysB* gene from the lysis module of mycobacteriophage. PhD thesis. University of Lisbon, Lisbon.

61. **Hatfull GF, Jacobs-Sera D, Lawrence JG, Pope WH, Russell DA, Ko CC, Weber RJ, Patel MC, Germane KL, Edgar RH, Hoyte NN, Bowman CA, Tantoco AT, Paladin EC, Myers MS, Smith AL, Grace MS, Pham TT, O'Brien MB, Vogelsberger AM, Hryckowian AJ, Wynalek JL, Donis-Keller H, Bogel MW, Peebles CL, Cresawn SG, Hendrix RW.** 2010. Comparative genomic analysis of sixty mycobacteriophage genomes: genome clustering, gene acquisition and gene size. *J Mol Biol* **397**:119–143.

62. **Henry M, O'Sullivan O, Sleator RD, Coffey A, Ross RP, McAuliffe O, O'Mahony JM.** 2010. In *silico* analysis of Ardmore, a novel mycobacteriophage isolated from soil. *Gene* **453**:9–23.

63. **Mageeney C, Pope WH, Harrison M, Moran D, Cross T, Jacobs-Sera D, Hendrix RW, Dunbar D, Hatfull GF.** 2012. Mycobacteriophage Marvin: a new singleton phage with an unusual genome organization. *J Virol* **86**:4762–4775.

64. **Henry M, Neve H, Maher F, Sleator RD, Coffey A, Reynold PR, McAuliffe O, Coffey A, O'Mahony JM.** 2010b. Cloning and expression of a mureinolytic enzyme from the mycobacteriophage TM4. *FEMS Microbiol Lett* **311**:126–132.

65. **Henry M, Coffey A, O'Mahony JM, Sleator RD.** 2011. Comparative modelling of LysB from the mycobacterial bacteriophage Ardmore. *Bioeng. Bugs.* **2**:88–95.

66. **Summer EJ, Liu M, Gill JJ, Grant M, Chan-Cortes TN, Ferguson L, Janes C, Lange K, Bertoli M, Moore C, Orchard RC, Cohen ND, Young R.** 2011. Genomic and functional analyses of *Rhodococcus equi* phages ReqiPepy6, ReqiPoco6, ReqiPine5, and ReqiDocB7. *Appl Environ Microbiol* **77**:669–683.

67. **Hermoso JA, García JL, García P.** 2007. Taking aim on bacterial pathogens: from phage therapy to enzybiotics. *Curr Opin Microbiol* **10**:461–472.

68. **Broxmeyer L, Sosnowska D, Miltner E, Chacón O, Wagner D, McGarvey J, Barletta RG, Bermudez LE.** 2002. Killing of *Mycobacterium avium* and *Mycobacterium tuberculosis* by a mycobacteriophage delivered by a nonvirulent mycobacterium: a model for phage therapy of intracellular bacterial pathogens. *J Infect Dis* **186**: 1155–1160.

69. **Danelishvili L, Young LS, Bermudez LE.** 2006. In vivo efficacy of phage therapy for *Mycobacterium avium* infection as delivered by a nonvirulent mycobacterium. *Microb Drug Resist* **12**:1–6.

70. **Trigo G, Martins TG, Fraga AG, Longatto-Filho A, Castro AG, Azeredo J, Pedrosa J.** 2013. Phage therapy is effective against infection by *Mycobacterium ulcerans* in a murine footpad model. *PLoS Negl Trop Dis* **7**: e2183.

71. **Hatfull G.** 2014. Molecular genetics of mycobacteriophages. *Microbiol Spec* **2**(2):MGM2-0032-2013.

Gene Expression
and Regulation

II

Molecular Genetics of Mycobacteria, 2nd Edition
Edited by Graham F. Hatfull and William R. Jacobs, Jr.
© 2014 American Society for Microbiology, Washington, DC
doi:10.1128/microbiolspec.MGM2-0007-2013

Riccardo Manganelli[1]

Sigma Factors: Key Molecules in *Mycobacterium tuberculosis* Physiology and Virulence

7

Rapid adaptation to changing environments is one of the keys of the success of microorganisms. Such adaptation is achieved by enabling different strategies, the most potent of which is global transcriptional modulation. Bacterial genomes encode several transcriptional regulators that in response to external and internal stimuli rapidly change the transcriptional profile of the cells, allowing maintenance of the homeostasis. Among the several players responsible for global transcriptional regulation, sigma (σ) factors have a prominent role (1). These are small interchangeable subunits of the RNA polymerase (RNAP) holoenzyme that are required for transcriptional initiation and that determine the promoter specificity of the enzyme recognizing specific -35 and -10 consensus promoter sequences (2). All bacterial genomes encode at least one essential σ factor, responsible for the transcription of housekeeping genes, and a variable number of alternative σ factors enabling a rapid transcriptional shift in response to specific stimuli (3, 4).

Two main families of evolutionarily distinct σ factors are encoded in the bacterial pan-genome: σ^{70} and σ^{54}. However, mycobacteria only encode for σ factors of the σ^{70} family (5, 6). These σ factors have four main conserved regions (regions 1, 2, 3, and 4): Region 1 inhibits DNA binding of free σ factors, region 2 is involved in recognition of the -10 consensus sequence and melting of the transcription bubble, region 3 is involved in recognition of the extended -10 consensus sequence, and region 4 is involved in recognition of the -35 consensus sequence (2, 3). σ factors can be divided into four groups (3): Group 1 comprises the primary σ factors. These are essential molecules since they are required for transcription of housekeeping genes. Group 2 includes primary-like σ factors. Its members have a structure very similar to members of group 1, but with some exceptions are involved in stationary phase survival and stress response and are not essential (7). Group 3 contains σ factors involved in cellular differentiation and general stress response of biosynthesis of the flagellum; they miss the conserved region 1. Group 4, also named ECF (extra cellular function) σ factors, miss conserved regions 1 and 3 and are the most numerous and heterogeneous. Although the roles of several of them has not been described, they are often involved in the control of extracellular functions as a response to surface stress or some aspect of the cell surface or transport (7, 8). Several σ factors of this group have been shown to be essential for virulence (9).

[1]Department of Molecular Medicine, University of Padova, Italy.

The alternative σ factor-density (ASFD) of a genome (number of alternative σ factors/Mb) is a measure of the impact of these molecules on the physiology of a microorganism and usually reflects the complexity of its growth cycle and of the environment colonized from the bacterium (10). Consequently, intracellular parasites, endosymbionts, and obligate pathogens, adapted to a stable environment, usually have an ASFD below 2, together with small genomes. With the increased complexity of the environment usually inhabited, both genome size and ASFD increase to reach a value around 8 in genera such as *Bacteroides* and *Streptomyces* (10).

The contemporaneous presence of several alternative σ factors implies that their activity must be fine-tuned to allow coordination of transcription. Moreover, several complex σ factor regulatory networks have been described as the cascade involved in *Bacillus* and *Clostridium* sporulation (10–12).

σ factors can be regulated at different levels (1, 13). (i) Transcriptional regulation: Often σ factor genes are transcribed from multiple promoters, each with a different activity and regulation. Usually, one of these promoters is recognized by the σ factor itself, resulting in a positive feedback loop, while the others can be constitutive (to allow a basal level of expression in the absence of the inducing conditions) or regulated from other transcriptional regulators as two-component systems or other σ factors. (ii) Posttranscriptional regulation: through small noncoding RNA, for which the best example is *Escherichia coli rpoS* regulated by at least three small RNAs (14). (iii) Translational regulation, as in the case of the heat shock responsive σ factor of *E. coli* σ^{32}, whose coding mRNA changes its conformation depending on temperature-regulating translational initiation (15). (iv) Posttranslational regulation: This is usually due to anti-σ factors, proteins that specifically bind to a σ factor, preventing its interaction with the RNA core enzyme (16). Specific environmental signals cause disruption of this interaction, leading to σ factor release. Some anti-σ factors are posttranslationally regulated by anti-anti-σ factors (9, 16). σ, anti-σ, and anti-anti-σ factor activity can be regulated by posttranslational modification, leading to their proteolysis or conformational changes altering their activity (17). Regulation of σ^E activity in the *E. coli* response to stresses involving a damaged cell envelope and preventing outer membrane porins' correct folding is the best-characterized example of posttranslational σ factor regulation (18, 19). In this case, regulated intramembrane proteolysis (RIP) is the signal transduction pathway linking periplasmic stress with σ^E activity.

The structure of this pathway involves the presence of two proteases (S1P and S2P) that act sequentially, processing the transmembrane anti-σ factor RseA, resulting in the release (20) of its cytoplasmic domain from the membrane. However, this truncated form of RseA can still bind σ^E (21) until other ATP-dependent cytoplasmic proteases such as ClpXP totally degrade RseA, releasing the active form of σ^E (22).

Finally, some σ factors can be translated from alternative translational start codons in different environmental conditions with generation of proteins with different vulnerability to proteolytic turnover (23).

THE σ FACTORS OF *MYCOBACTERIUM TUBERCULOSIS*

The genome of *M. tuberculosis* encodes 13 σ factors (5), and its ASFD is 2.8, which makes it the obligate pathogen with the higher density of alternative σ factors (10). Of these 13 σ factors, σ^A belongs to group 1 and represents the primary σ factor. It is the only σ factor to be essential for *M. tuberculosis* viability. σ^B belongs to group 2 (primary-like σ factors), σ^F belongs to group 3, and the other 10 (σ^C, σ^D, σ^E, σ^G, σ^H, σ^I, σ^J, σ^K, σ^L, and σ^M) belong to group 4 (ECF σ factors) (10, 24).

Nearly all the *M. tuberculosis* σ factors (with the exception of σ^A, σ^B, σ^I, and σ^J) have been predicted or shown to be posttranslationally regulated by cognate anti-σ factors (6, 24): σ^E, σ^F, and σ^H anti-σ factors are predicted to be cytoplasmic proteins, while all the other are predicted to be membrane proteins (25) (Table 1). At least three of them (those specific for σ^K, σ^L, and σ^M) were suggested to be regulated by RIP, a common feature of posttranslational regulation in which the anti-σ factor is sequentially cleaved in two sites by two proteases—site 1 protease (S1P) and site 2 protease (S2P)—before being totally degraded (20, 26). The most likely *M. tuberculosis* S1P is the serine-protease HtrA (25), while S2P was recently identified in the membrane-bound metalloprotease Rv2869c (Rip1) (27). RskA, RslA, and RsmA were shown to be subject to proteolytic degradation by Rip1 following exposure to the metal chelator phenanthroline (27). However, it is unknown if they are still able to bind their cognate σ factor until they are totally degraded by other cytoplasmic proteases, as demonstrated for RseA in *E. coli* (Table 1).

RsdA is insensitive to Rip1 cleavage but is sensitive to proteolytic degradation by the cytoplasmic protease ClpX-ClpP1-ClpP2 (28). However, whether its degradation occurs after release from the membrane by RIP is still unknown (Table 1).

Table 1 *M. tuberculosis* σ factors

σ factor	Anti-σ factor	Anti-σ factor putative cellular localization	Anti-anti-σ factor	Inducing conditions[a]
σ^A				Growth in macrophages[b]
σ^B				Heat shock, oxidative stress, surface stress, growth in macrophages
σ^C	Rv0093c[c]	Membrane		Downregulated in stationary phase and in response to several stress conditions
σ^D	RsdA	Membrane		Nutrient depletion; downregulated in response to several stress conditions
σ^E	RseA	Cytoplasm		Surface stress, oxidative stress, heat shock, low phosphate, alkaline pH, vancomycin, growth in macrophages
σ^F	UsfX	Cytoplasm	Rv1635c,Rv3687c, Rv0516c, Rv1364c, Rv1904, Rv2638[d]	Nutrient depletion
σ^G	Rv0181c[c]	Membrane		Growth in macrophages, DNA-damaging agents
σ^H	RshA	Cytoplasm		Heat shock, oxidative stress, growth in macrophages
σ^I				Stationary phase, heat or cold shock
σ^J				Stationary phase, growth in macrophages
σ^K	RskA	Membrane		
σ^L	RslA	Membrane		
σ^M	RsmA	Membrane		Heat shock, late stationary phase

[a]Reviewed in reference 10; underlined anti-σ factors were shown or hypothesized to be regulated by RIP (27). Modified from reference 25.
[b]Only some strains.
[c]Putative anti-σ factors.
[d]The last four genes represent putative anti-anti-σ factors.

Most of the *M. tuberculosis* σ factors have been extensively studied, and for most of them a clear role in pathogenesis was shown, making them an extremely important class of molecules for understanding the mechanism of *M. tuberculosis* adaptation to the host (10, 24, 25).

σ^A

σ^A is the only member of group 1 σ factors encoded in the *M. tuberculosis* genome and is predicted to be an essential gene. While its essentiality was clearly shown in *Mycobacterium smegmatis* (29), no direct demonstration of it has been shown in *M. tuberculosis*. We recently constructed a conditional mutant in which its structural gene (*sigA*) transcription was placed under transcriptional control of a repressible promoter system recently developed in our laboratory (30) and showed that upon its down-modulation the bacterial growth rapidly stops (S. Anoosheh and R. Manganelli, unpublished). Like most of the primary σ factors, its expression is rather constant during exponential growth and decreases under conditions leading to growth rate reduction such as stationary phase or hypoxia (31). Interestingly, the half-life of its mRNA was shown to be unusually long (more than 40 minutes) compared to

that of the primary-like σ factor σ^B (2.4 minutes) (32). It is possible that in strong protein-damaging stress conditions, transcription stops, causing a quick decrease of the mRNA pool. When the intensity of stress decreases, if no functional σ^A is present in the cell, translation of housekeeping genes cannot restart, leading to cell death. If *sigA* mRNA is still present, new σ^A can be translated and both transcription and translation can be restored.

It was recently shown that σ^A has low affinity for the RNAP core enzyme, probably due to the nonoptimal structure of the σ-RNAP-binding interface; however, binding of the small protein RbpA to RNAP was recently shown to modify RNAP structure, leading to increased affinity for σ^A (33). RbpA, first characterized in *Streptomyces*, is an essential protein in *M. tuberculosis* (34), and transcription of its structural gene is induced in several stress conditions such as starvation (35), surface stress (36), and exposure to vancomycin (37). It has been hypothesized that in these conditions RbpA increases σ^A competitiveness for RNAP, guaranteeing the expression of housekeeping genes; even when following exposure to stress, other σ factors with higher affinity for RNAP are present in the cytoplasm (33). Recently, RbpA was shown to bind directly also to both σ^A and σ^B, further supporting this hypothesis (38).

Due to its invariant expression and its long half-life, *sigA* mRNA has been used often as an invariant internal standard in quantitative reverse transcription-PCR (RT-PCR) experiments (31, 39, 40).

Even if σ^A is principally involved in housekeeping gene expression, there are reports of its involvement in virulence modulation. Collins and collaborators (41) showed that an arginine to histidine substitution at residue 515 (R515H) of σ^A in *Mycobacterium bovis* ATCC35723 totally impaired growth in guinea pigs even if not affecting extracellular growth. This mutation, mapping in region 4.2, known to interact with transcriptional activators in other microorganisms (42), was shown to prevent the binding of WhiB3, a transcriptional regulator involved in maintenance of redox homeostasis through regulation of lipid anabolism (43, 44). Moreover, an *M. bovis whiB3* mutant was shown to have the same attenuation phenotype as an R515H *M. bovis* mutant, strongly suggesting that the interaction between σ^A and WhiB3 was the cause of the observed attenuation. Interestingly, when *whiB3* was deleted in *M. tuberculosis* H37Rv, the resulting strain was still able to infect both mice and guinea pigs, reaching the same bacterial burden of the wild type strain in their organs, even if conferring a lower pathology, suggesting a different role of σ^A/WhiB3 interaction in these species (43).

Finally, it was recently shown that in a clinical strain of the Beijing family characterized by enhanced intracellular growth, *sigA* was expressed at a higher level than H37Rv and other clinical strains and was induced upon macrophage infection (45). Moreover, overexpression of *sigA* in H37Rv resulted in better growth of the recombinant strain in both macrophages and mice lungs after aerosol infection as well as increased resistance to superoxide (45). Recently, these phenotypes were correlated to induction of *eis* (46). The product of this gene is an N(ϵ)-acctyltransferase acetylating Lys55 of a JNK-specific phosphatase resulting in down-modulation of autophagy, inflammation, and cell death through inhibition of reactive oxygen species generation (47).

σ^B

σ^B is dispensable for growth in both *M. smegmatis* (48) and *M. tuberculosis* (49) and is very similar to the C-terminal portion of σ^A. Its expression is induced in several stress conditions such as heat shock, surface stress, oxidative stress, starvation, and during macrophage infection (31, 35, 50).

In vitro transcription experiments identified two putative promoters upstream of *sigB*: one recognized by σ^F and one recognized by either σ^E, σ^H, or σ^L (51–53) (Fig. 1). However, a direct role in *sigB* expression *in vivo* has only been confirmed for σ^E, which is responsible for its basal level of transcription and for

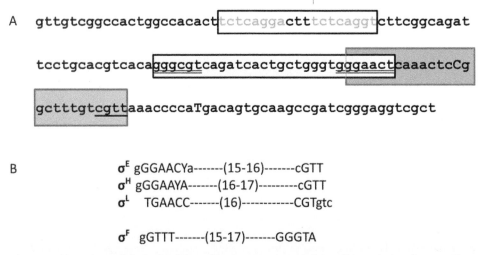

Figure 1 Sequence of the region immediately upstream of the *sigB* translational stat codon. (**A**) White box indicates MprA binding sites. Gray letters indicate residues protected by MprA (55, 56); light gray box indicates the putative σ^F-dependent promoter; dark gray box indicates the σ^E, σ^H, and σ^L putative promoter. −35 and −10 sequences are underlined. Transcriptional start points are shown in capital letters. (**B**) Putative consensus sequences recognized by the four σ factors (53). Modified from reference 51. doi:10.1128/microbiolspec.MGM2-0007-2013.f1

its induction following surface and oxidative stress (36, 54), σ^L (51, 52), and σ^H (R. Chauhan and M. L. Gennaro, personal communication). Another regulator involved in *sigB* regulation is MprAB, a stress-responsive two-component system activated in response to macrophage infection, surface-damaging agents, nutrient limitation, alkaline pH, and other stresses (36, 55–57) that is required for the establishment and maintenance of persistent infection in mice (58). Binding of MprA in the *sigB* upstream region might be required to facilitate access to the promoter of σ^F- or σ^L-containing RNAP due to the poor conservation of their consensus sequences in this region (Fig. 1).

Overexpression of *M. tuberculosis sigB* in *M. smegmatis* resulted in hyperproduction of surface hyperglycosylated polar glycopeptidolipid (GPL), a change in colony morphology, and an extended generation time (59). Since hyperproduction of polar GPL is typical of *M. smegmatis* carbon-starved cultures (60), this phenotype suggests a role of σ^B in the adaptation to starvation and stationary phase (48).

Recently, we reported that an *M. tuberculosis sigB* knockout mutant, although not attenuated in macrophages or during mouse and guinea pig infection, is more sensitive to various stresses, such as surface-damaging agents, heat shock, and oxidative stress (37, 49). Moreover, the mutant was very sensitive to hypoxia, showing a survival about 3-log10 lower than that of the wild type or parental strain, strongly suggesting that it might be involved in survival during latent or persistent infection (49).

Two reports have been published that used DNA microarrays to characterize the σ^B regulon. In the first, the global transcriptional profile of the wild type strain was compared to that of a strain overexpressing σ^B (52). In the second the comparison was performed between the global transcriptional profiles of a *sigB* null mutant and that of its parental strain during exponential growth following exposure to the surface-damaging agent SDS or to the oxidative agent diamide (49). The overlap between the genes identified in these two studies was very low, suggesting that under stress conditions several other regulatory mechanisms may be activated and affect the expression of the regulon. Overexpression of σ^B led to the induction of 72 genes including its own gene, genes encoding the transcriptional regulators WhiB2 and IdeR, and several ESAT-6-like proteins, ribosomal proteins, and PE-PGRS (52). However, during exponential growth only eight genes were downregulated in the *sigB* mutant, while 28 were upregulated, in particular *sigE* and several genes of its regulon, probably representing a compensatory

response to the lack of σ^B. Following surface stress, 72 genes were downregulated in the *sigB* mutant; these included several genes involved in envelope stress response and several encoding transcriptional regulators such as IdeR. Finally, following oxidative stress, 40 genes were downregulated in the *sigB* mutant; these included genes encoding the heat shock proteins Hsp, ClpB, and DnaK, several transcriptional regulators such as FurA, as well as several proteins involved in cysteine and arginine biochemistry (49).

σ^C

The chromosome of *Mycobacterium leprae* was subjected during evolution to massive gene decay and is hypothesized to carry almost only indispensable genes. Interestingly, the only σ factor genes present in its genome are those encoding σ^A, σ^B, and the two ECF σ factors σ^E and σ^C (61), so it is considered very important for mycobacterial physiology and virulence. The *sigC* gene is present in all members of the *M. tuberculosis* complex as well as in the members of the *Mycobacterium avium* complex, *Mycobacterium marinum*, *Mycobacterium ulcerans*, and *Mycobacterium abscessus*, but it is absent in *M. smegmatis* and other nonpathogenic mycobacterial species (6, 62).

The σ^C structural gene is highly expressed in *M. tuberculosis* during the exponential phase, but its level drops about 20-fold during stationary phase and between 5- and 10-fold following exposure of the bacterial culture to starvation or stress conditions such as heat shock, cold shock, and surface stress (31, 35), suggesting that expression of its regulon is needed during active growth and not when stressing conditions decrease or stop bacterial growth. Its transcription starts from two promoters: The dominant promoter has a typical consensus sequence for σ^A, while the second presents a σ^C consensus sequence, suggesting an autoregulatory mechanism. However, transcription from this promoter was not affected in a *sigC* null mutant, which suggests that it is recognized by another, still unrecognized σ factor (63). σ^C was also found to be induced in an *M. tuberculosis* strain overexpressing σ^F, suggesting a role of this σ factor in its expression (52). However, no binding sites for σ^F upstream of *sigC* were found in ChIP-on-chip experiments; this suggests that the effect of σ^F on *sigC* transcription may be indirect (64, 65).

Most ECF σ factors are regulated at the posttranslational level by an anti-σ factor, and usually the genes encoding σ and anti-σ are cotranscribed. The *sigC* gene is not associated with any gene encoding for a putative anti-σ factor, suggesting that it is not posttranslationally regulated. However, a transmembrane protein of unknown function with an anti-σ factor signature (Rv0093c) (Fig. 2) has been predicted *in silico* to bind σ^C, although it was not possible to demonstrate the interaction between these two proteins by copurification (66). Interestingly, we found that in the genomes of bacteria belonging to the *M. avium* complex, *sigC* and *rv0093c* are fused to encode a single protein (unpublished observation) (Fig. 2 and Fig. 3). Further work is needed to understand the physiological role of Rv0093c in *M. tuberculosis* and the role of σ^C-Rv0093c fusion in the *M. avium* complex.

Deletion of the *sigC* gene from *M. tuberculosis* was shown to have little or no effect on growth in axenic culture or in macrophages (67, 68). However, *sigC* mutant–infected mice survived longer than animals infected with wild type or a complemented strain, despite the fact that the three strains were able to reach the same bacterial load in the organs of infected animals (67). Mice infected with the mutant strain showed reduced inflammatory infiltrates and had reduced levels of tumor necrosis factor-α (TNF-α), interleukin-1β (IL-1β), IL-6, and IFN-γ in their lungs, as well as fewer infiltrating neutrophils in bronchoalveolar lavage fluid, suggesting that the lower mortality induced by the *sigC* mutant was due to the failure to elicit the same degree of immunopathology as the wild type (68). Finally, when used to infect guinea pigs, the *sigC* mutant was attenuated and did not induce the formation of necrotic granulomas (68, 69). Taken together, these data demonstrate a role for σ^C in pathology through the modulation of the host immune response.

DNA microarrays were used to define the effect of σ^C on the *M. tuberculosis* global transcriptional profile.

About 200 genes were shown to be downregulated in the *sigC* mutant at different points of the growth curve (67). Some of these genes encode important virulence-associated proteins such as the α-crystalline homolog HspX (70), the response regulator MtrA, involved in cell morphology and division (71), and the sensor histidine kinase SenX, involved in sensing low phosphate levels and in regulation of the stringent response (72). However, whether σ^C directly regulates the expression of these genes is still not clear since their transcriptional start site was not determined and no *in vitro* transcription data are available.

σ^D

σ^D is posttranslationally regulated by the anti-σ factor RsdA, recently shown to be subject to proteolytic degradation by ClpX-ClpP1-ClpP2 (28). The expression of *sigD* decreases after exposure to several stresses such as heat and cold shock, hypoxia, and stationary phase (31), while it was shown to be slightly induced under conditions of nutrient depletion (35). Moreover, its expression is decreased during macrophage infection (50) and in a *relA* mutant (73). RelA is an enzyme that catalyzes the synthesis of hyperphosphorylated guanidine (p)ppGpp, a key molecule for the development of the stringent response, which has been shown to be essential for *M. tuberculosis* pathogenesis (73).

A mutant lacking *sigD* was able to grow in resting and activated macrophages at the same rate as the wild type parental strain but induced a lower level of TNF-α (74). σ^D was shown to be essential for virulence in mice in both H37Rv and CDC1551 backgrounds. Both mutant strains were still able to infect mice and grow in their lungs, reaching the same bacterial burden reached

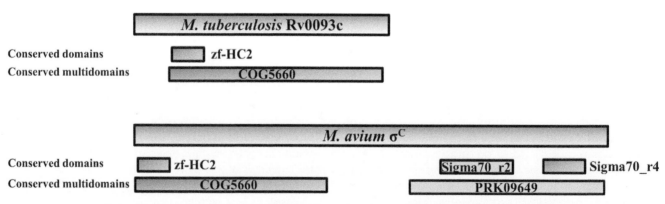

Figure 2 Comparison of conserved domains and multidomains in *M. tuberculosis* Rv0093c and *M. avium* σ^C. zf-HC2, Putative zinc-finger found in some anti-σ factor proteins; COG5660, predicted integral membrane protein; PRK09649, RNA polymerase sigma factor σ^C; Sigma70_r2, σ^{70} region 2; Sigma70_r4, σ^{70} region 4. Alignments were performed at http://blast.ncbi.nlm.nih.gov/Blast.cgi. doi:10.1128/microbiolspec.MGM2-0007-2013.f2

```
Mav   ------------------------------MRND-----------AGMRCDVAREALSARL  20
Mtb   VLAQATTAGSFNHHASTVLQGCRGVPAAMWSEPAGAIRRHCATIDGMDCEVAREALSARL  60
                       *  .:                  **  *:**********


Mav   DGERPQVLAQQVDAHLEACRGCRSWLIGAAVQTRRLASVTPGEGPDLVDKIMASIGEQPT  80
Mtb   DGERAPVPSARVDEHLGECSACRAWFTQVASQAGDLRRLAESR------PVVPPVGRLGI 114
      ****. *  :  :** **   *  .**:*:     .* *:  *   :: ..      ::..:*.


Mav   GRPAWMRWLRSHYRRWGLIGVGLFQVAIAAAQISGIDFGMVAGHMHGAMSGEHLMHESTA 140
Mtb   RRAPRRQHSPMTWRRWALLCVGIAQIALGTVQGFGLDVGLT--HQHPTGAGTHLLNESTS 172
        *..  :      :***.*: **: *:*:.:.*  *:*.*:.  * *  : :* **::***:


Mav   WLLALGLAMIAAGVWPASASGVAAITGVYSVALLGYVIVDAFDGEVTATRIASHMPLLLG 200
Mtb   WSIALGVIMVGAALWPSAAAGLAGVLTAFVAILTGYVIVDALSGAVSTTRILTHLPVVIG 232
      * :***: *:.*.*:**::*:*:*.:   .:  .* *******:.* *:.:*** :*:*::.*


Mav   LAFALLVARERVGSRRPGSSDATADAGFAAWAADAPAGRRRGHLRPINRAAPDPTASSTT 260
Mtb   AVLAIMVWRSASGPRP--RPDAVAAEPDIVLPDNASRGRRRGHLWPT-----DGSAAMTA 285
      .:*::* *.  *.*     .**.*     . .:*. ******* *    * :*: *:


Mav   TRRRIGKPAQLRWLGMIAPSGDEAVTELALSAARGNARALEAFIKATQQDVWRFVAYLCD 320
Mtb   T----------------ASDDEAVTALALSAAKGNGRALEAFIKATQQDVWRFVAYLSD 328
      *                .*.***** *****:**.*********************.*


Mav   AGSADDLPQETFLRAIGAIERFSGRSSARTWLLSIARRVVADHIRHLQSRPRAAVGADPE 380
Mtb   VGSADDLTQETFLRAIGAIPRFSARSSARTWLLAIARHVVADHIRHVRSRPRTTRGARPE 388
      .******.***********  ***.*********:***:********:.:****:: ** **


Mav   HVLRTDRHARGFEDLVEVTTMIASLNPEQREALLLTQLLGLPYADAAAVCGCPVGTIRSR 440
Mtb   HLIDGDRHARGFEDLVEVTTMIADLTTDQREALLLTQLLGLSYADAAAVCGCPVGTIRSR 448
      *:: .   *****************.*.,:*************.******************


Mav   VARARDALLADGERSDLTG 459
Mtb   VARARDALLADAEPDDLTG 467
      ***********.* .****
```

Figure 3   Alignment of *M. avium* σ^C (black) with *M. tuberculosis* Rv0093c (green) and σ^C (red). doi:10.1128/microbiolspec.MGM2-0007-2013.f3

from wild type and complemented strains. However, both histopathology and mortality were reduced in mice infected with the mutant strain, suggesting that σ^D regulates some bacterial component able to modulate immune response (74, 75).

DNA microarrays have been used to determine the σ^D regulon. Calamita and colleagues (74) compared the global transcriptional profile of the *sigD* mutant and of its wild type parental strain in different points of the growth curve (mid-exponential, early and late stationary phase). They reported 61 genes downregulated in the *sigD* mutant, mostly in late stationary phase. Several of them encoded for a set of ribosome-associated proteins usually expressed in stationary phase. Similar experiments were performed by Raman and collabo-

rators (75) in the H37Rv background. These authors analyzed the impact of the lack of σ^D only in cultures growing in mid-exponential phase. They found 51 genes downregulated in the *sigD* mutant. The genes whose transcription was more affected by the lack of σ^D, beyond its own gene, were those encoding the resuscitating factor RpfC; the chaperones GroEL2, GroEL1, and GroES; and the isoniazide (INH)–inducible protein IniB. Rpf proteins, first characterized in *Micrococus luteus*, are peptidoglycan hydrolases and transglycosylases that can degrade the cell wall (76) and are required for resumption of growth from stationary phase (77). The regulation of *rpfC* by σ^D suggests a role for this σ factor in regulating cell wall structure during the entry into or emergence from hypoxia-induced dormancy. However, the complete lack of overlap between the σ^D-regulated genes and the putative σ^D consensus sequences identified in the two studies, even if it may be ascribed, at least in part, to the different genetic backgrounds in which the experiments were performed and to the different experimental conditions used by the two groups, prevents us from having a clear idea of which genes really belong to the σ^D regulon.

σ^E

In *M. tuberculosis sigE* transcription is induced after exposure to heat shock, alkaline pH, detergents, oxidative stress, and vancomycin and during growth in low-phosphate media and in human macrophages (31, 36, 37, 50, 55, 78, 79). *sigE* deletion in *M. tuberculosis* results in several dramatic phenotypes *in vitro* and *in vivo*, underscoring the importance of this σ factor in physiology and virulence: Indeed, a *sigE* null mutant was shown to be more sensitive than the wild type parental strain to heat shock, to the detergent SDS, to oxidative stress (36), and to vancomycin (37), INH, streptomycin, gentamicin, and rifampin (D. Pisu and R. Manganelli, unpublished observation). It was also unable to grow in resting THP-1-derived human macrophages, where it was not able to block phagosome maturation (36) (S. Casonato and R. Manganelli, unpublished observation), and human dendritic cells (80), while its growth was not restricted in the human pneumocyte cell line A549 (Casonato and Manganelli, unpublished observation). Moreover, it was killed more efficiently by activated mouse macrophages (36). Finally, its virulence was strongly attenuated in mice (81, 82) and in guinea pigs (A. Izzo and R. Manganelli, unpublished observation). Recently, a study performed in *M. smegmatis* suggested σ^E as one of the major regulators involved in the development of the stringent response in this organism, demonstrating that *relA* induction in low-phosphate media is σ^E-dependent (78, 79). This is of particular interest since RelA and the stringent response have been recently shown to be required for mycobacterial persistence (73). Preliminary experiments strongly suggest that *sigE* is an essential gene in *M. bovis* BCG, in which we could obtain a *sigE* deletion only in the presence of a second copy of the gene (Casonato and Manganelli, unpublished observation). Whether *sigE* is essential in *M. bovis* or only in *M. bovis* BCG is still unknown.

A clear indication that σ^E is a critical node of the *M. tuberculosis* stress response derives from the intricate regulation of its gene expression and protein activity. Its structural gene is transcribed from three promoters (83). The first one (P1) is constitutively expressed during growth and is probably under the control of the principal σ factor σ^A; the second (P2) is induced in conditions of surface stress mediated by SDS or vancomycin and is autoregulated by σ^E (Chauhan and Gennaro, personal communication), while the third (P3) is induced following oxidative stress and is under the control of σ^H (83, 84). P2, even if autoregulated by σ^E, does not have a canonical σ^E consensus sequence and requires for its activation the two-component system MprAB, whose structural genes are under σ^E control (55, 56); interestingly, one of the two MprA binding sites upstream of P2 overlaps the P1 transcriptional start point, causing the downregulation of this promoter in stress conditions (83) (Fig. 4). It was recently reported that σ^E positively regulates *ppk1* encoding for a polyphosphate kinase (S. Sanyal, S. K. Banerjee, and M. Kundu, presented at Tuberculosis 2012: Biology, Pathogenesis, Intervention Strategies, Paris, France, 2012) and that polyphosphate is the main substrate of MprB (79). So *sigE* is subjected to at least two positive feedback loops: the first involving positive regulation of *mprAB* by σ^E, leading to P2 activation, and the second involving the positive regulation of *ppk1* by σ^E, resulting in higher polyphosphate intracellular levels and consequently a higher rate of MprA phosphorylation by MprB and P2 activation (12).

Another peculiar feature of *sigE* regulation is that P3 is located inside its own open reading frame (at position +63). Using single nucleotide mutagenesis and translational fusion to *lacZ*, we demonstrated that the mRNA transcribed from P3 is translated starting from two noncanonical translational start codons (ATC and TTG) located downstream from P3 (Fig. 4). So σ^E can be present in *M. tuberculosis* in three different isoforms depending on the environmental conditions encountered from the bacteria: one of 257 aa, translated from mRNA transcribed from P1 or P2 in normal physiological conditions and after surface stress, and two nearly identical isoforms of 218 aa and 215 aa, translated from mRNA starting from P3 in conditions of oxidative stress (83). The presence of the σ^H-dependent P3 internal to the *sigE* open reading frame is a general feature of mycobacterial genomes.

However, the presence of two active noncanonical translational start codons in this region might be restricted to the members of the *M. tuberculosis* complex: While the first translational start codon (ATC) is located at the correct distance from the putative ribosome binding site (17 bp), the second one (TTG) is 9 bp downstream due to a 9-bp duplication present only in *M. tuberculosis* complex genomes (Fig. 4). It is possible to hypothesize that before the occurrence of the

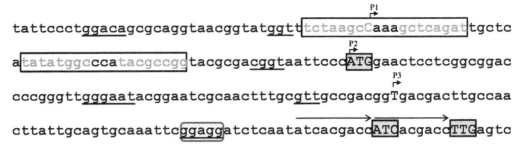

Figure 4 Sequence of the *sigE* upstream region and of the first part of the *sigE* coding region in *M. tuberculosis* containing the three promoters and the translational start sites of its three isoforms. Transcription start points are shown in capital letters, and putative −10 and −35 are underlined. MprA binding sites are boxed. Letters in gray indicate residues protected by MprA binding (55, 56). Putative translation start codons are shown in boxed capital letters, while the putative ribosome binding site (RBS) is shown in boxed plain letters. The two arrows show the 9-bp duplication (ATCACGACC) found in the *M. tuberculosis* complex genomes. Modified from reference 83.
doi:10.1128/microbiolspec.MGM2-0007-2013.f4

duplication, TTG represented the only active translational start codon in this region and translational initiation from the ATC codon began only after it was moved by the duplication (83). Consequently, TTG probably still represents the only translational start codon in mycobacteria not belonging to the *M. tuberculosis* complex in which the 9-bp duplication did not occur.

The translation of different σ^E isoforms in response to specific environmental conditions suggests that these isoforms may have different biological functions. One possibility is that they could recognize different consensus sequences, but this is unlikely since the domains involved in DNA binding are conserved in the three isoforms. Moreover, recent preliminary experiments designed to analyze by DNA microarrays the variation of the *M. tuberculosis* global transcriptional profile following chemical induction of each of the different σ^E isoforms showed that their regulons are superimposable (R. Provvedi, E. Dainese, and R. Manganelli, unpublished data). Another possibility is that the N-terminal region of the larger σ^E isoform (which does not share homology with any other σ factor) could be involved in protein stability, as already shown for σ^R in *Streptomyces coelicolor*, also translated in two different isoforms, one constitutively expressed and the other inducible by stress. In this case the inducible isoform, containing an additional 55 amino acids, was shown to have a half-life 7-fold shorter than that of the constitutive isoform (23).

σ^E activity is also regulated at the posttranscriptional level by the anti-σ factor RseA (83, 85). Its gene is located downstream of *sigE* but belongs to a different transcriptional unit that is constitutively expressed

(86). RseA/σ^E interaction is disrupted in an oxidative environment due to oxidation of cysteines 70 and 73 (85). However, in conditions of surface stress RseA is phosphorylated by the Ser/Thr kinase PknB, which contains PASTA (PBP and serine/threonine kinase-associated) domains that have been hypothesized to bind peptidoglycan to serve as cell wall stress sensors (87), leading to its ClpC1P2-dependent proteolysis. Since transcription of the genes encoding this protease are positively regulated by the global regulator ClgR, a member of the σ^E regulon, this represents a third positive feedback loop involved in σ^E regulation (12). Figure 5 shows the complex regulatory network involved in SigE regulation.

Finally, a small antisense RNA was recently identified overlapping the 5′ of the *sigE* coding region, suggesting the existence of posttranscriptional regulation (88). Experiments are in progress in our laboratory to confirm the role of this small RNA in *sigE* posttranscriptional regulation.

A recent study performed in *M. smegmatis* revealed bimodal distribution of *relA* transcription from its σ^E-dependent promoter, suggesting bistability in the upstream MprAB network (89). Tiwari and collaborators (90) recently tested this hypothesis by constructing a comprehensive mathematical model to analyze the MprAB-σ^E-RseA stress-response network for the presence of bistability. Their results demonstrated that the peculiar network architecture containing multiple positive feedback loops may lead to bistability, making it a good candidate for the persistence switch.

The impact of σ^E in the *M. tuberculosis* transcription profile was first analyzed using DNA microarrays in the absence of specific stimuli inducing *sigE* and

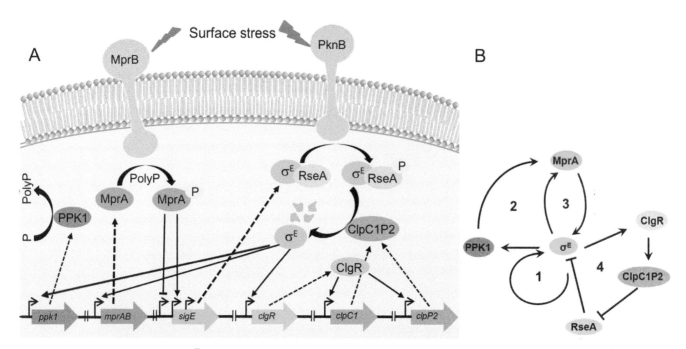

Figure 5 The σ^E network. **(A)** Surface stress promotes PknB-dependent phosphorylation of RseA, leading to cleavage by ClpC1P2, which results in activation of the σ^E regulon. σ^E controls transcription of *clgR*, which in turn induces the *clp* regulon. An increase in ClpC1P2 levels leads to increased RseA degradation and a higher concentration of free σ^E (positive feedback loop) (86). σ^E also controls transcription of *ppk1*: Increased PPK1 levels raise PolyP concentration, which controls *sigE* transcription through MprB-dependent phosphorylation of MprA (positive feedback loop) (85). Also, a σ^E-dependent promoter drives *mprAB* transcription (positive feedback loop) (36). Finally, *sigE* is subject to autoregulation (Chauhan and Gennaro, personal communication). Solid, dashed, and curved arrows represent transcriptional regulation, protein production, and catalytic reactions, respectively. **(B)** Abstraction of feedback loops discussed in panel **(A)**. Modified from reference 12. doi:10.1128/microbiolspec.MGM2-0007-2013.f5

following exposure to SDS-induced surface stress. In the absence of stress 38 genes were shown to be differentially regulated in the mutant. The most affected gene in these conditions was *sigB*, which was strongly repressed in the *sigE* mutant. However, 23 genes belonging to 13 putative transcriptional units failed to be induced in the *sigE* mutant after exposure to stress (36). Among these were *sigB* and the gene encoding the global regulator ClgR. Interestingly, at least five genes involved in lipid catabolism (*aceA*, *gltA1*, *fadB2*, *fadE23*, and *fadE24*) were found to be part of the σ^E regulon. Of particular interest, the first two are involved in the glyoxylate shunt and the methylcitrate cycle, and these metabolic pathways are critical for the metabolic remodeling associated with stress-induced growth arrest of *M. tuberculosis* (91) and are required for persistence and virulence of *M. tuberculosis* (92). The regulation of these two genes was recently further characterized, showing that in both cases it is subject to a parallel feed-forward loop with an AND input

function. The *icl1* loop is directly controlled by σ^E, while the *gltA1* loop is controlled by σ^E through σ^B. Each loop requires a local regulator (LrpI for *iclI* and LprG for *gltA1*) produced from a gene adjacent to the corresponding effector gene and transcribed from σ^E (*lrpI*) or σ^B (*lrpG*) (93). The presence in the σ^E regulon of genes involved in fatty acid degradation also suggests that they not only are involved in the metabolic exploitation of the fatty acids, but also may be responsible for degradation and detoxification of the fatty acids that accumulate following the inhibition of biosynthesis of cell surface lipids or damage to the fatty acid–rich envelope of *M. tuberculosis*. The fact that *fadE23* and *fadE24* were also induced in cells exposed to isoniazid, an inhibitor of mycolic acid biosynthesis (94), supports this hypothesis.

Finally, another operon shown to require σ^E for its induction after surface stress is that including *clgR*, *rv2744c*, and *rv2745c*. While, as already mentioned, the first gene encodes a global transcriptional regulator involved in one of the positive feedback loops regulating *sigE* (12, 85, 95), *rv2744c* encodes a protein highly homologous to PspA (phage shock protein A), a protein involved in homeostasis of the cell membrane and maintenance of proton motive force in Gram-negative bacteria (96), while *rv2745c* encodes a transmembrane protein of unknown function. Recently, Rv2744c was shown to be a substrate of PepD (57), a protease whose structural gene is induced by σ^E, which is involved in processing misfolded proteins, leading to the activation of the MprAB/σ^E signaling pathways (57).

The σ^E regulon was also studied during infection in THP-1-derived human macrophages. Beyond confirming several genes previously described as σ^E dependent, it was possible to identify other genes as *rmlB2*, encoding a putative galactose epimerase that might be involved in the linking of peptidoglycan and mycolic acid and *cyp121*, encoding a cytochrome P450, probably involved in detoxification of fatty acids (97). In the same study the transcriptional response of the macrophage to *M. tuberculosis* infection was analyzed, showing that the σ^E regulon is involved in the modulation of the inflammatory response.

A role of the σ^E regulon in the modulation of the immune response has also been observed in human dendritic cells in which infection with the *sigE* mutant increased the production of IL-10 about 10-fold compared with infection with the wild type parental strain (80). Finally, *sigE* mutant–infected mice, despite the lower bacterial burden in their organs, were shown to produce more protective factors such as IFN-γ, TNF-α, iNOS, and β-defensins than animals infected with the parental or complemented mutant strain (98), suggesting that the *sigE* mutant might induce strong protection against *M. tuberculosis* infection. In support of this hypothesis, when the *sigE* mutant was used to vaccinate mice, it was able to confer better protection against *M. tuberculosis* challenge than *Mycobacterium bovis* BCG. Interestingly, this was more evident when vaccinated mice were challenged with a hypervirulent Beijing strain (98).

σ^F

σ^F was initially studied in *M. bovis* BCG; in this species it was shown to be induced in several experimental conditions such as anaerobiosis, cold shock, oxidative stress, and nutrient depletion and after entry into stationary phase. Moreover, its expression increased after exposure to several antibiotics, suggesting a role for σ^F in the basal level of sensitivity to drug treatment (99, 100). Despite the identity of *sigF* and its upstream region in *M. bovis* BCG and *M. tuberculosis*, in this latter

species it was possible to demonstrate *sigF* induction only in response to nutrient depletion (31, 35).

It was recently shown that in *M. smegmatis* σ^F is required for resistance to hydrogen peroxide, heat shock, and acidic pH (101) and to regulate the biosynthesis of a surface-associated lipidic pigment (102). Moreover, an *M. smegmatis sigF* null mutant had a transformation efficiency four orders of magnitude higher than that of the wild type parental strain, suggesting σ^F involvement in the regulation of cell wall permeability (102).

A *sigF* null mutant of *M. tuberculosis* CDC1551 reached higher density than the wild type strain and did not show any lag phase after dilution into fresh medium, suggesting that σ^F might be involved in negative control of bacterial growth (103), but did not show any sensitivity against several stress conditions such as heat shock, cold shock, hypoxia, and long-term stationary phase. The mutant was found to be more resistant to rifampin than the wild type strain, and this resistance was hypothesized to be the result of reduced permeability to hydrophobic solutes due to a reduced production of cell wall-associated sulpholipids (103). However, characterization of a *sigF* mutant of *M. tuberculosis* H37Rv did not confirm any of these phenotypes, suggesting either that σ^F has different roles in these two strains or that subtle differences in experimental conditions result in different impacts of σ^F in *M. tuberculosis* physiology (104). Experiments run in parallel with the two mutant strains are required to discriminate between these hypotheses.

The CDC1551 *sigF* mutant virulence is clearly attenuated in a mouse infection model, with lower bacterial burden and histopathology in the lungs at the late stage of infection (105). However, no difference from the wild type was shown during infection of human monocytes (103). Finally, a *sigF* mutant of H37Rv was found to be attenuated in guinea pigs (69).

Several transcriptome analyses have been performed using different approaches to reveal the σ^F-dependent genes in strains CDC1551 and H37Rv: (i) Geiman and collaborators (105) compared the transcriptional profiles of a CDC1551 *sigF* mutant with that of the wild type parental strain in different phases of growth using DNA microarrays. They found 38 genes downregulated in the *sigF* mutant in exponential phase and 187 in early and 277 in late stationary phase, suggesting a major role of σ^F in the adaptation to stationary phase. Several σ^F-dependent genes in stationary phase were involved in the biosynthesis and degradation of surface (lipo)-polysaccharides or in the biosynthesis and structure of the cell envelope. In addition, genes expressing transcriptional regulators of the TetR, GntR, and MarR families, usually controlling the expression of efflux pumps in other bacteria were found to be differentially regulated. (ii) Williams and collaborators (106) used a CDC1551 strain in which a copy of *sigF* was placed downstream of the inducible acetamidase promoter to compare its transcriptional profile with that of the control strain at different times after addition of acetamide. The genes differentially regulated following *sigF* overexpression included those encoding several cell-associated proteins such as members of the MmpL, PE, and PPE families. The most upregulated gene was *phoY1*, encoding a transcriptional regulator probably involved in regulation of phosphate uptake. (iii) Homerova and collaborators (107) used an *E. coli* two-plasmid system to identify *M. tuberculosis* promoters recognized by σ^F-RNAP polymerase, identifying five σ^F-dependent promoters containing sequences highly similar to the previously identified σ^F-dependent promoter of *usfX1* (108). (iv) Hartkoorn and collaborators (64) recently published a study in which both chromatin immunoprecipitation (ChIP-on-chip) and microarray analysis were used to identify σ^F-dependent genes using an H37Rv recombinant strain in which *sigF* was inducible by the addition of pristinamyin IA. Integrating the two data sets, it was possible to identify 16 σ^F-binding sites corresponding to RNA transcripts in the sense orientation and 9 σ^F-binding sites associated with antisense transcripts. Beyond its own structural gene, σ^F was found to directly regulate genes encoding proteins involved in lipid and intermediary metabolism and virulence (such as HbHa, which is involved in extrapulmonary dissemination) and transcriptional regulators such as Rv2884 and PhoY1. From these data it was possible to clearly determine the σ^F consensus sequence: GGTTT-N$_{(15-17)}$-GGGTA. For 26 genes found to be overexpressed following *sigF* induction, it was not possible to demonstrate a σ^F binding region by ChIP-on-chip, suggesting an indirect role of σ^F in their regulation.

The overlap of genes reported to be σ^F dependent in this study with those reported by Geiman and collaborators (105) (1/99) and Williams and collaborators (106) (7/70) was quite small, further supporting the hypothesis that CDC1551 and H37Rv might have different physiological traits or that experimental conditions may result in dramatic changes in the impact of σ factors in global gene regulation.

σ^F activity is regulated at the posttranslational level by a complex anti- and anti-anti-σ factor network (Fig. 6). Beaucher and collaborators (108) demonstrated that UsfX is an anti-σ factor able to specifically bind

to prevent its interaction with the RNAP core enzyme. They also demonstrated the presence of two anti-anti-σ factors: RsfA and RsfB are able to interact with UsfX, preventing its binding to σ^F. RsfA was hypothesized to be regulated by redox potential because it was able to bind UsfX only in reducing conditions, while RsfB was hypothesized to be regulated by phosphorylation because a mutation mimicking phosphorylation prevented its binding to UsfX (108). In *Bacillus subtilis* two anti-σ factors belonging to the same family of UsfX (RsbW and SpoIIAB) can phosphorylate their anti-anti-σ factor, leading to their inactivation. However, UsfX was shown to have several unfavorable substitutions in its kinase domain (109) and to be unable to phosphorylate both RsfA and RsfB (110), suggesting the evolution of a different kind of regulation of the UsfX/RsfA and UsfX/RsfB interaction in *M. tuberculosis*. Interestingly, RsfB was shown to be phosphorylated by the Ser/Thr protein kinase E (PknE), suggesting a role of this

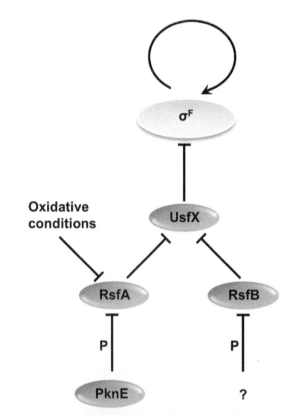

Figure 6 Main regulatory network responsible for SigF regulation. Arrows indicate transcriptional regulation; truncated lines indicate posttranslational regulation. At least four other proteins—Rv0516c, Rv1364c, Rv1904, and RV2638—have been hypothesized to be involved in this network, but their role has not been elucidated yet.
doi:10.1128/microbiolspec.MGM2-0007-2013.f6

protein in the regulation of UsfX/RsfB interactions. At least four other putative anti-anti-σ factors were predicted to interact with either UsfX or σ^F or both in a yeast-two-hybrid system: Rv0516c, Rv1364c, Rv1904, and Rv2638 (111). Among these, Rv0516c was shown to be phosphorylated by PknD, and Rv1904 by PknE (110). Of particular interest, sequence analysis predicted the presence of a phosphatase domain, an anti-σ^F domain, and an anti-anti-σ^F domain in the protein encoded by *rv1364c*. While the anti-σ^F domain showed some unfavorable substitutions inactivating its kinase activity, the other domains retained their activity (109). However, the roles of these four proteins in σ^F regulation have not yet been confirmed.

σ^G

Attention focused on *sigG* after the demonstration of its strong induction after macrophage infection suggested its importance in adaptation to intracellular growth and virulence (112). The effect of *sigG* deletion was studied in two different genetic backgrounds (CDC1551 and H37Rv), generating contrasting results and suggesting that, despite the high conservation of genome sequences, σ factors may have distinct roles in different *M. tuberculosis* strains (113, 114).

The presence in the *sigG* upstream region of a sequence similar to the *recA* promoter (115) suggested the possibility that this σ factor might be involved in the SOS response. Moreover, *sigG* was shown to be induced by mitomycin C, a DNA-damaging agent able to induce the SOS response in H37Rv (116). However, when a CDC1551 *sigG* mutant was exposed to this DNA-damaging agent, it showed marked resistance, suggesting that σ^G might act as a repressor of the SOS response, in apparent contradiction with the induction of its structural gene in the presence of DNA damage (113). Microarray analysis showed a mild downregulation of *lexA* (0.6-fold), encoding the main repressor of the genes belonging to the SOS regulon, and a mild upregulation of *recA* (1.5-fold), which might be the reason for mitomycin C resistance in this strain. Other genes differentially expressed in the *sigG* mutant were *sigH* (repressed) and *sigD* (induced), as well as several genes of their regulons (113). However, which of these effects was due to direct regulation by σ^G or from an indirect effect was not addressed. Finally, the *sigG* mutant was shown to be severely affected for growth in macrophages, again not easily reconciling with a resistance to DNA damage (113).

Smollett and collaborators (114) characterized a *sigG* mutant in the H37Rv genetic background. While they could confirm its induction by several DNA-damaging agents, they failed to demonstrate a different sensibility of the mutant strain to these compounds. Moreover, comparing the global transcriptional profiles of the *sigG* mutant and wild type in the presence and absence of DNA-damaging compounds, they could not find any genes expressed differently between the two strains and concluded that σ^G is not involved in the regulation of RecA-dependent or -independent genes in response to DNA damage in *M. tuberculosis* H37Rv.

Due to the opposite conclusions obtained from the two studies, it is difficult to hypothesize a role for this σ factor. That it could have such different roles in two strains of the same bacterial species is possible but unlikely, given the extreme conservation of the genome sequences in *M. tuberculosis*. Further studies performed in parallel with the mutants in the two genetic backgrounds are needed to shed some light on these intriguing findings.

σ^H

The gene encoding σ^H is induced by heat shock and oxidative stress in *M. smegmatis*, *M. avium* subsp. *paratuberculosis*, and *M. tuberculosis* (54, 84, 117, 118), where it was found to be induced also during macrophage infection (119). Interestingly, it was suggested that the lack of a functional *sigH* gene might be responsible for the temperature sensitivity of *M. leprae* (120). σ^H is posttranslationally regulated by the cytoplasmic anti-σ factor RshA, while its structural gene is subjected to autoregulation. The RshA-σ^H complex is disrupted *in vitro* by elevated temperature and in oxidizing conditions (121). The presence in RshA of a HX$_3$CX$_2$C motif suggested its inclusion in the family of the zinc-associated anti-σ factor (ZAS). However, as was recently shown for *S. coelicolor* RsmA, in RshA this motif coordinates a 2Fe-2S cluster (122, 123), enabling responses to oxidative/reductive stress at a much faster rate than Zn. Finally, both σ^H and RshA are phosphorylated by PknB, a eukaryotic-like serine/threonine protein kinase involved in the regulation of cell wall synthesis and cell morphology. Phosphorylation of RshA, but not of σ^H, was shown to affect the σ^H/RshA interaction, resulting in decreased binding of σ^H by RshA, thus adding a further level of regulation (124). The σ^H ortholog in *S. coelicolor* σ^R was shown to be translated from two translational start sites, resulting in two isoforms: a shorter one, constitutively expressed and with a half-life longer than 70 minutes, and a stress-inducible longer one, whose half-life was about 10 minutes. The presence of two putative translational start sites and of the same regulatory mechanism was hypothesized for several members of the mycobacterial genome including *M. tuberculosis* and demonstrated in *M. smegmatis* (23).

Genes regulated by σ^H were identified in different phases of growth or following diamide-mediated oxidative stress by comparing the global transcriptional profile of σ^H mutants obtained in two different genetic backgrounds (CDC1551 and H37Rv) with that of their parental strains (54, 121, 125). σ^H was shown to modulate the expression of about 180 genes, at least 31 of them in a direct manner. Beyond its own structural gene, σ^H was found to regulate the genes encoding the two σ factors σ^E and σ^B, DNA repair proteins, general stress proteins, enzymes involved in cysteine and molybdenum biosynthesis, and enzymes involved in thiol metabolism such as thioredoxin and thioredoxin reductase. In *S. coelicolor* the ortholog of σ^H was shown to be involved in the synthesis of mycothiol (126), a molecule involved in an alternative pathway to reduce intracellular disulfide bonds typical of actinomycetes, which do not synthesize glutathione (127). Even if no direct evidence of σ^H involvement was found in the mycothiol biosynthesis in *M. tuberculosis*, one of the σ^H-dependent most highly induced genes following diamide exposure (Rv2466c) contains a putative glutaredoxin active site (54), suggesting its involvement in the mycothiol cycle.

The dynamic nature of the σ^H-mediated response to oxidative stress was explored by comparing the variation of the global transcriptional profile of a *sigH* mutant and of its parental strain at different times after a short exposure to diamide. The expression of *sigH*, and *sigE* reached their peak 30 minutes after diamide exposure and then declined to reach the background level after 120 minutes. Several stress-related genes regulated by these σ factors followed a similar dynamic. Interestingly, a large number of genes involved in lipid biosynthesis were downregulated at early time points following oxidative stress, while their expression resumed at later time points (128).

The consensus sequence recognized by σ^H was clearly determined in these studies and found to be very similar to that previously proposed for σ^E (36, 125), suggesting the possibility of an overlap between the regulons of these σ factors. Song and collaborators (53), using mutational and primer extension analyses, confirmed that σ^E and σ^H recognize nearly identical promoters. The sixth position of -35 was shown to have a critical role for discrimination between promoters preferentially recognized by σ^E and σ^H.

The phenotype of different *sigH* null mutants was extensively studied and shown to be more sensitive to oxidative stress and heat shock (54, 84) but still able to replicate in THP-1-derived macrophages and to resist the bactericidal activity of activated murine macrophages (54). The *sigH* mutants were still able to infect mice, reaching the same bacterial burden of that reached from their wild type parental strains (54, 125). However, despite the same bacterial burden as those infected with the wild type strain, mice infected with the *sigH* mutant showed decreased immunopathology, lower recruitment of CD4 and CD8 T cells to the lung in the early stages of infection, and lower mortality, suggesting that the antioxidant activity of the σ^H regulon might have a role in modulating the interaction of the pathogen with the immune system (125). This hypothesis was recently reinforced by the finding that an *M. bovis* BCG mutant defective in oxidative stress response was more immunogenic (129).

A *sigH* mutant was also used to infect primary rhesus macaque bone marrow–derived macrophages. Cells infected with the mutant showed higher levels of expression of β-chemokines and several apoptotic markers relative to cells infected with the wild type parental strain, while the mutant exhibited reduced survival after 72 hours of infection (130). The *sigH* mutant was also shown to be severely attenuated in a rhesus macaque aerosol infection model. In this case, beyond a longer survival and a lower immunopathology, animals infected with the *sigH* mutant also showed a lower bacterial burden in the lungs than those infected with the wild type parental strain (131).

Recently, the role of σ^H in oxidative stress and virulence was also shown in *M. avium* subspecies *paratuberculosis*, suggesting that its role is conserved in different members of the mycobacterial genus (132).

σ^I

sigI is induced in stationary phase and after either heat (133) or cold shock (31). At least in stationary phase its induction probably depends on σ^J (known to be induced in this growth phase), which has been shown to recognize a consensus sequence upstream of *sigI* (134). Recently, an *M. tuberculosis* CDC1551 *sigI* mutant was constructed, and its global transcriptional profile was analyzed in comparison with that of its parental strain. The σ^I regulon included genes encoding an ATP synthase (Rv1304), several heat shock proteins (Rv0440, Rv0350, and Rv3417), and KatG (133). Using *in vitro* transcription, the authors demonstrated that σ^I is directly responsible for initiation of transcription at both *katG* promoters. Downregulation of *katG* in the *sigI* mutant suggested the possibility, confirmed by the authors, that this strain is resistant to isoniazide, which requires KatG for its activation. Finally, the mutant was used to infect mice and, surprisingly, despite the lower *katG* expression, the mutant showed a hyper-virulent phenotype, probably mediated from an increased CREB phosphorylation and TNF-α secretion from infected macrophages (133). Usually, resistance to INH is due to mutations in *katG* and has a high fitness cost that can be alleviated by compensatory mutations in the *ahp* gene (encoding a peroxidase) (135). The possibility of a *sigI* mutant acquiring INH resistance without loss of fitness may represent an alternative pathway toward INH resistance.

σ^J

σ^J is the least-well-characterized *M. tuberculosis* σ factor. Its expression is strongly induced in late stationary phase (136) and in human macrophages (113). However, neither the survival in late stationary phase, nor the pathogenicity in a model of intravenous mouse infection were affected in a *sigJ* null mutant (137), suggesting that despite its induction in stationary phase and in macrophages, σ^J is not involved in pathogenicity or adaptation to prolonged stationary phase cultures. However, the mutant was more sensitive to H_2O_2 than the wild type parental strain, suggesting a role in the defense from reactive oxygen species. Using an *E. coli* two-plasmid system, it was demonstrated that *sigI* can be transcribed by σ^J (134). However, no genome-wide study has been performed to characterize the σ^J regulon.

σ^K

This σ factor is present in all members of the *M. tuberculosis* complex and in *M. marinum, M. ulcerans,* and *Mycobacterium gilvum,* but it is absent in other mycobacteria such as *M. leprae* (in which it is a pseudogene), *M. avium, M. smegmatis,* and *Mycobacterium kansasii* (6, 138).

In 2005 the known decreased expression of the two antigenic proteins MPT70 and MPT83 observed in some strains of *M. bovis* BCG was associated with a polymorphism in the *sigK* gene, resulting in a mutation in its translation start codon (139). DNA microarrays were used to determine the global transcriptional profile of an *M. bovis* BCG strain containing the mutation impairing σ^K translation complemented with a wild type copy of *sigK,* leading to the identification of two chromosomal loci containing σ^K-regulated genes.

The first locus included *sigK* and *rskA,* together with genes encoding a putative amine oxidase, a cyclopropane-fatty-acyl-phospholipid synthase, and four other proteins of unknown function. The second locus included genes encoding the surface-associated lipopro-

tein MPT83; the integral membrane proteins Rv2877c, DipZ, and Rv2876; and finally the secreted proteins MPT70 and MPT53 (139). MPT83 and MPT70 have similar C-terminal domains structurally related to the FAS1 domain, mediating interactions between cells and the extracellular matrix, and might be involved in binding to host cell proteins (140). MPT53 is a secreted DsbE-like protein probably involved in the formation of disulfide bonds in unfolded secreted proteins (141). Rv2877c and DipZ (whose genes are also under σ^L control) have a conserved CcdA domain, usually contained in proteins involved in the transfer of a reducing potential from the cytoplasm to secreted disulfide bond isomerases (142, 143). Their function might be to maintain proper disulfide bond formation in proteins in the periplasmic space of *M. tuberculosis.*

The expression of MPT70 and MPT83 is different in *M. tuberculosis* (low) and *M. bovis* (high). Said-Salim and collaborators (144) identified the anti-σ factor responsible for σ^K posttranslational regulation (Rv0444c, RskA) and analyzed the sequence of its structural gene in both *M. bovis* and *M. tuberculosis.* The results clearly showed that the high level of MPT70 and MPT83 production in *M. bovis* was due to a mutation in *rskA* disrupting the negative regulation of σ^K and resulting in its constitutive activation. It is not clear if the overexpression of MPT70 and MPT83 give a selective advantage to *M. bovis* in its natural hosts or if it is the result of a "founder effect." However, the presence of the same mutation in *Mycobacterium caprae* and of independent mutations in the same gene in the strain infecting the oryx support the idea that high levels of MPT70 and MPT83 might provide a selective advantage when infecting members of the *Bovidae.* RskA was recently shown to be subject to RIP in conditions of oxidative stress induced by metal chelator phenanthroline (27).

Evolutionary analysis of the σ^K regulon showed the presence in several bacteria related to mycobacteria of the minimal set of genes *mpt83-sigK-rskA,* which might represent the nucleus of the σ^K regulon. Before the separation between *Rhodococcus* and *Mycobacterium* this minimal region was subjected to the insertion of six genes that separated *mpt83* from *sigK* and *rskA.* During the evolution of mycobacteria, the locus was separated into two loci, the *sigK-rskA* locus and the *mpt70/83* locus. In slow-growing mycobacteria an additional gene, *dipZ,* was inserted between the two *mpt83* paralogs. Finally, it was shown that σ^K and RskA from the environmental species *M. gilvum* can complement the activity of their orthologs of *M. tuberculosis,* suggesting that although its regulon varies considerably across species, the regulatory system σ^K/RskA is conserved across the *Mycobacterium* genus (138).

σ^L

The gene encoding *sigL* is transcribed from two promoters, the first being constitutive and the second under σ^L control (145). While *sigL* mRNA was shown to be constantly expressed during the growth curve (145), when the *sigL* upstream region was used to drive the expression of *lacZ*, β-galactosidase activity increased with culture density, principally in the transition from lag to exponential phase (51). Like several other σ factors, σ^L is posttranslationally regulated by an anti-σ factor: RslA (51). This is a transmembrane protein that in reducing environments binds Zn^{2+} and sequesters σ^L. In oxidative environments the cysteines present in the CXXC motif of RslA form a disulfide bridge. This results in the release of Zn^{2+} and a conformational change of RslA, leading to a strong decrease in RslA affinity for σ^L and its subsequent release (146). Moreover, RslA was recently shown to be subject to RIP in conditions of oxidative stress induced by metal chelator phenanthroline (27).

Two independently obtained *sigL* mutants of *M. tuberculosis* H37Rv were shown to be attenuated in the intravenous or the aerosol murine infection model (51, 145). As shown for other *M. tuberculosis* mutants, the *sigL* mutants were not impaired for growth in the murine organs but produced a lower pathology and a lower mortality than wild type and complemented strains. When analyzed for in vitro phenotypes, only a very small, but reproducible, sensitivity to the superoxide generator plumbagine and the detergent SDS was detected (51).

Since the environmental conditions able to activate σ^L are still unknown, two similar strategies were implemented to define its regulon by DNA microarrays. Hahn and collaborators (145) constructed an *M. tuberculosis* merodiploid strain by integrating in the chromosome of H37Rv a copy of *sigL* expressed from an acetamide-inducible promoter. Dainese and collaborators (51), however, introduced a copy of *sigL* with its physiologic regulatory region in the chromosome of a strain in which both *sigL* and *rslA* were deleted. In this strain, the presence of σ^L in the absence of its negative regulator RslA led to an overexpression of both *sigL* and the members of its regulon. The first approach identified 18 genes overexpressed in the merodiploid strain upon induction of *sigL* including four two-gene operons (*sigL-rslA*, *mpt53-rv2877c*, *pks10-pks7*, and *rv1138c-rv1139c*) and four genes encoding for PE_PGRS proteins (145). With the alternative approach Dainese and collaborators (51) identified 27 induced genes, included in 13 putative transcriptional units, 6 of which showed a σ^L consensus sequence in their upstream regions. Induction of *pks10* and *pks7* was confirmed, but in this case all the other genes of the operon were identified (*pks8, pks17, pks9,* and *pks11*), strongly suggesting σ^L involvement in the biosynthesis or phthiocerol dimycocerosic acid (PDIM) (a component of the cell envelope that is essential for virulence) (147). However, a *sigL* null mutant of *M. tuberculosis* is not deficient in PDIM synthesis (145), and recent *in vitro* characterization of *fadD26-ppsA-E* suggests that this locus is sufficient for PDIM biosynthesis (148); the nature of the lipids produced by this *pks* gene cluster remain to be determined.

In addition, induction of *rv1138c-rv1139c* (as well as *rv1137c*) and *mpt53* was confirmed. The function of Rv1138c and Rv1139c is still unknown, but Rv1139c has an isoprenyl-cysteine carboxymethyl transferase domain (25, 149), which in eukaryotes is involved in the posttranslational modification of prenylated proteins. Rv1138c, however, is predicted to be an aromatic ring monooxygenase. Other genes showed to be under σ^L control are *mmpL13a* and *mmpL13b*, probably involved in fatty acid transport (150) and *sigB*.

σ^M

Expression of *sigM* was reported to be constant during bacterial growth. Induction was observed after heat shock and in late stationary phase (151). However, since *sigA* mRNA (used as an internal control) is known to decrease in stationary phase (31), the strength of this result is questionable. σ^M is posttranslationally regulated by the anti-σ factor RsmA, recently shown together with RslA and RskA to be subject to RIP in conditions of oxidative stress induced by metal chelator phenanthroline (27). A CDC1551 mutant in which *sigM* was deleted did not show any phenotype when exposed to different stress conditions. Moreover, it was not attenuated for growth in macrophages or in a murine model of infection, suggesting that this σ factor is not involved in stress response or virulence development (151). The regulon of σ^M was assessed in both CDC1551 and H37Rv genetic backgrounds using strains overexpressing *sigM* by an acetamide-inducible promoter (152, 153). Several genes were found to be induced following σ^M overexpression, including those encoding PPE1, a nonribosomal peptide synthetase, and proteins of unknown function. Interestingly, the most evidently upregulated genes in both strains were those encoding the type VII secretion system (T7SS) ESX-4, including EsxT and EsxU, two secreted proteins belonging to the WXG100 family typically associated with these secretion systems. It is also notable that the genes encoding two other proteins belonging to this family, but whose genes are not physically associated with any T7SS gene cluster, EsxE and EsxF, were also shown to be induced.

Recently, *sigM* was shown to be induced together with the genes encoding the ESX-4 T7SS and the genes encoding EsxE and EsxF, in response to the induction of *whiB5*, encoding a transcriptional regulator of the WhiB family (154). A *whiB5* null mutant was more sensitive to *S*-nitrosoglutathione (GSNO) and was less metabolically active following prolonged starvation. Most importantly, it was attenuated in a murine model of progressive infection and was not able to resume growth after reactivation from chronic infection (154). Whether some of these phenotypes were due to the effect of the lack of WhiB5 on *sigM* and its downstream genes is unknown; this will be the subject of future investigations.

Finally, several genes involved in PDIM biosynthesis were shown to be induced in the *sigM* mutant, suggesting that σ^M might have a negative regulatory effect on PDIM biosynthesis. Quantitative analysis of nonpolar surface lipids produced from the *sigM* mutant and its wild type parental strains confirmed this hypothesis (153).

CONCLUDING REMARKS

Since infection is a dynamic process, we predict that *M. tuberculosis* adaptation involves continuous modulation of its global transcriptional profile in response to the changing environment. In the last few years the 13 sigma factors encoded in its genome have been subject to extensive characterization and were shown to be fundamental players in this process. Null mutants for each of them were constructed, with the exception of *sigA* (being an essential gene, only a conditional mutant was obtained) (10, 24). The mutants missing *sigB*, *sigJ*, or *sigM* did not show any virulence attenuation (49, 137, 151, 153), the *sigI* mutant showed a hypervirulent phenotype (133), while no experiments assessing virulence have been reported to date for the *sigK* mutant. All the other mutants were attenuated in mice or in macrophages (10, 24). However, the lack of attenuation in mice does not exclude that σ^B, σ^J, and σ^M can be involved in virulence, given the limitations of this animal model. For example, the *sigB* mutant, even if still able to replicate in macrophages and virulent in mice, was shown to be very sensitive to hypoxia (49), a condition believed to be present in human granulomas and

hypothesized to be important for the development of dormancy and subsequently latent infection. So it is possible to hypothesize that σ^B might be involved in the development of latent, rather than active, infection and that this phenotype was not discovered, due to the animal model used in its characterization. The central position of *sigB* in the most complex regulatory network ever described in mycobacteria involving σ^E, σ^H, σ^L, and probably σ^F (Fig. 7) strongly suggests its importance in *M. tuberculosis* physiology.

Interestingly, mutants missing σ^C, σ^D, σ^H, or σ^L were still able to replicate in mouse lungs at the same rate as the wild type but induced less inflammation and histopathology, suggesting their involvement in the complex interaction between *M. tuberculosis* and the immune system (54, 67, 74, 75, 125, 145). Since tissue damage induced by the immune system is of great importance in tuberculosis progression and transmission, understanding the reason for the lower inflamogenic potential of these strains might be of great help to further understand *M. tuberculosis* pathogenicity and to design a new strategy of intervention against tuberculosis. However, the *sigE* and *sigF* mutants' growth in mouse lungs was impaired, suggesting a different role for their regulon in pathogenesis (81, 105). It is notable that in infected lungs the *sigE* mutant was shown to induce stronger production of protective factors such as IFN-γ, TNF-α, iNOS, and β-defensins than the wild type, despite a lower bacterial load, suggesting that its regulon might be involved in the down-modulation of the host response (98). In this case, a full characterization of this occurrence might help to better understand *M. tuberculosis* pathogenicity and help to design better vaccines against tuberculosis. Accordingly, mouse vaccination with the *sigE* mutant conferred better protection than *M. bovis* BCG from challenge with virulent *M. tuberculosis* (98).

Since 1995, the year in which the first papers on mycobacterial σ factors were published (41, 155, 156), their characterization has been one of the more challenging and productive areas in the mycobacterial field. After 18 years, most of them have been extensively characterized and their role in *M. tuberculosis* physiology and virulence highlighted. Nonetheless, the complexity of the σ factors' role in the global regulation of transcription is still far from fully understood, principally due to their complex regulation, the overlap of some of their regulons, and the existence of regulatory networks including several σ factors. In the last few years the first papers studying mycobacterial σ factor networks using systems biology and mathematical modeling were published (89, 90), laying the tracks for

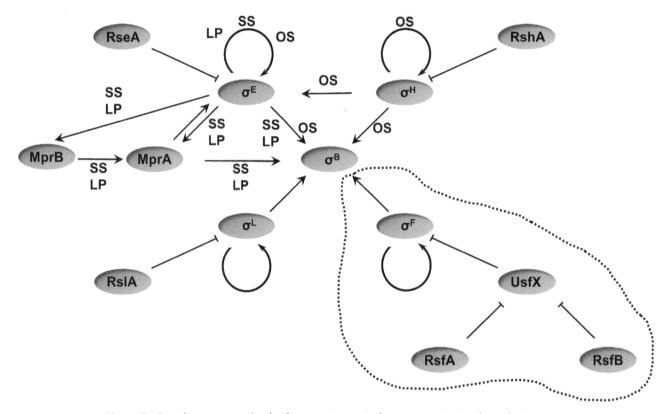

Figure 7 Regulatory network of σ factors. Arrows indicate transcriptional regulation; truncated lines indicate posttranslational regulation. The role of σ^F is still putative. SS, surface stress; LP, low phosphate; OS, oxidative stress.
doi:10.1128/microbiolspec.MGM2-0007-2013.f7

a more comprehensive understanding of their role in the physiology and virulence of one of the most ancient enemies of humankind.

I wish to thank Roberta Provvedi and all the members of my lab for continuous discussion and support, Oleg Igoshin and Marila Gennaro for fruitful discussion and sharing unpublished data, and Issar Smith for invaluable support and for introducing me to the mycobacterial factors world.

Citation. Manganelli R. 2014. Sigma factors: key molecules in *Mycobacterium tuberculosis* physiology and virulence. Microbiol Spectrum 2(1):MGM2-0007-2013.

References

1. Osterberg S, del Peso-Santos T, Shingler V. 2011. Regulation of alternative sigma factor use. *Annu Rev Microbiol* **65:**37–55.

2. Lonetto M, Gribskov M, Gross CA. 1992. The sigma 70 family: sequence conservation and evolutionary relationships. *J Bacteriol* **174:**3843–3849.

3. Gruber TM, Gross CA. 2003. Multiple sigma subunits and the partitioning of bacterial transcription space. *Annu Rev Microbiol* **57:**441–466.

4. Browning DF, Busby SJ. 2004. The regulation of bacterial transcription initiation. *Nat Rev Microbiol* **2:**57–65.

5. Cole ST, Brosch R, Parkhill J, Garnier T, Churcher C, Harris D, Gordon SV, Eiglmeier K, Gas S, Barry CE 3rd, Tekaia F, Badcock K, Basham D, Brown D, Chillingworth T, Connor R, Davies R, Devlin K, Feltwell T, Gentles S, Hamlin N, Holroyd S, Hornsby T, Jagels K, Krogh A, McLean J, Moule S, Murphy L, Oliver K, Osborne J, Quail MA, Rajandream MA, Rogers J, Rutter S, Seeger K, Skelton J, Squares R, Squares S, Sulston JE, Taylor K, Whitehead S, Barrell BG. 1998. Deciphering the biology of *Mycobacterium tuberculosis* from the complete genome sequence. *Nature* **393:**537–544.

6. Manganelli R, Provvedi R, Rodrigue S, Beaucher J, Gaudreau L, Smith I. 2004. Sigma factors and global gene regulation in *Mycobacterium tuberculosis*. *J Bacteriol* **186:**895–902.

7. Helmann JD. 2002. The extracytoplasmic function (ECF) sigma factors. *Adv Microb Physiol* **46:**47–110.

8. Lonetto MA, Brown KL, Rudd KE, Buttner MJ. 1994. Analysis of the *Streptomyces coelicolor sigE* gene reveals the existence of a subfamily of eubacterial RNA polymerase sigma factors involved in the regulation of extracytoplasmic functions. *Proc Natl Acad Sci USA* **91:**7573–7577.

9. Bashyam MD, Hasnain SE. 2004. The extracytoplasmic function sigma factors: role in bacterial pathogenesis. *Infect Genet Evol* **4:**301–308.

10. Rodrigue S, Provvedi R, Jacques PE, Gaudreau L, Manganelli R. 2006. The sigma factors of *Mycobacterium tuberculosis*. *FEMS Microbiol Rev* 30:926–941.

11. Hilbert DW, Piggot PJ. 2004. Compartmentalization of gene expression during *Bacillus subtilis* spore formation. *Microbiol Mol Biol Rev* 68:234–262.

12. Manganelli R, Provvedi R. 2010. An integrated regulatory network including two positive feedback loops to modulate the activity of sigma in mycobacteria. *Mol Microbiol* 75:538–542.

13. Missiakas D, Raina S. 1998. The extracytoplasmic function sigma factors: role and regulation. *Mol Microbiol* 28:1059–1066.

14. Repoila F, Majdalani N, Gottesman S. 2003. Small non-coding RNAs, co-ordinators of adaptation processes in *Escherichia coli*: the RpoS paradigm. *Mol Microbiol* 48:855–861.

15. Morita MT, Tanaka Y, Kodama TS, Kyogoku Y, Yanagi H, Yura T. 1999. Translational induction of heat shock transcription factor sigma32: evidence for a built-in RNA thermosensor. *Genes Dev* 13:655–665.

16. Helmann JD. 1999. Anti-sigma factors. *Curr Opin Microbiol* 2:135–141.

17. Hughes KT, Mathee K. 1998. The anti-sigma factors. *Annu Rev Microbiol* 52:231–286.

18. Ades SE. 2004. Control of the alternative sigma factor sigmaE in *Escherichia coli*. *Curr Opin Microbiol* 7:157–162.

19. Ades SE. 2008. Regulation by destruction: design of the sigmaE envelope stress response. *Curr Opin Microbiol* 11:535–540.

20. Brown MS, Ye J, Rawson RB, Goldstein JL. 2000. Regulated intramembrane proteolysis: a control mechanism conserved from bacteria to humans. *Cell* 100:391–398.

21. Campbell EA, Tupy JL, Gruber TM, Wang S, Sharp MM, Gross CA, Darst SA. 2003. Crystal structure of *Escherichia coli* sigmaE with the cytoplasmic domain of its anti-sigma RseA. *Mol Cell* 11:1067–1078.

22. Chaba R, Grigorova IL, Flynn JM, Baker TA, Gross CA. 2007. Design principles of the proteolytic cascade governing the sigmaE-mediated envelope stress response in *Escherichia coli*: keys to graded, buffered, and rapid signal transduction. *Genes Dev* 21:124–136.

23. Kim MS, Hahn MY, Cho Y, Cho SN, Roe JH. 2009. Positive and negative feedback regulatory loops of thiol-oxidative stress response mediated by an unstable isoform of sigmaR in actinomycetes. *Mol Microbiol* 73:815–825.

24. Sachdeva P, Misra R, Tyagi AK, Singh Y. 2010. The sigma factors of *Mycobacterium tuberculosis*: regulation of the regulators. *FEBS J* 277:605–626.

25. Raman S, Cascioferro A, Husson R, Manganelli R. 2008. Mycobacterial sigma factors and surface biology, p 223–234. *In* Daffe M, Reyrat M (ed), *The Mycobacterial Cell Envelope*. ASM Press, Washington, DC.

26. Makinoshima H, Glickman MS. 2005. Regulation of *Mycobacterium tuberculosis* cell envelope composition and virulence by intramembrane proteolysis. *Nature* 436:406–409.

27. Sklar JG, Makinoshima H, Schneider JS, Glickman MS. 2010. *M. tuberculosis* intramembrane protease Rip1 controls transcription through three anti-sigma factor substrates. *Mol Microbiol* 77:605–617.

28. Jaiswal RK, Prabha TS, Manjeera G, Gopal B. 2013. *Mycobacterium tuberculosis* RsdA provides a conformational rationale for selective regulation of sigma-factor activity by proteolysis. *Nucleic Acids Res* 41:3414–3423.

29. Gomez M, Doukhan L, Nair G, Smith I. 1998. *sigA* is an essential gene in *Mycobacterium smegmatis*. *Mol Microbiol* 29:617–628.

30. Boldrin F, Casonato S, Dainese E, Sala C, Dhar N, Palu G, Riccardi G, Cole ST, Manganelli R. 2010. Development of a repressible mycobacterial promoter system based on two transcriptional repressors. *Nucleic Acids Res* 38:e134.

31. Manganelli R, Dubnau E, Tyagi S, Kramer FR, Smith I. 1999. Differential expression of 10 sigma factor genes in *Mycobacterium tuberculosis*. *Mol Microbiol* 31:715–724.

32. Hu Y, Coates AR. 1999. Transcription of two sigma 70 homologue genes, *sigA* and *sigB*, in stationary-phase *Mycobacterium tuberculosis*. *J Bacteriol* 181:469–476.

33. Hu Y, Morichaud Z, Chen S, Leonetti JP, Brodolin K. 2012. *Mycobacterium tuberculosis* RbpA protein is a new type of transcriptional activator that stabilizes the sigma A-containing RNA polymerase holoenzyme. *Nucleic Acids Res* 40:6547–6557.

34. Forti F, Mauri V, Deho G, Ghisotti D. 2011. Isolation of conditional expression mutants in *Mycobacterium tuberculosis* by transposon mutagenesis. *Tuberculosis* 91:569–578.

35. Betts JC, Lukey PT, Robb LC, McAdam RA, Duncan K. 2002. Evaluation of a nutrient starvation model of *Mycobacterium tuberculosis* persistence by gene and protein expression profiling. *Mol Microbiol* 43:717–731.

36. Manganelli R, Voskuil MI, Schoolnik GK, Smith I. 2001. The *Mycobacterium tuberculosis* ECF sigma factor SigE: role in global gene expression and survival in macrophages. *Mol Microbiol* 41:423–437.

37. Provvedi R, Boldrin F, Falciani F, Palu G, Manganelli R. 2009. Global transcriptional response to vancomycin in *Mycobacterium tuberculosis*. *Microbiology* 155:1093–1102.

38. Bortoluzzi A, Muskett FW, Waters LC, Addis PW, Rieck B, Munder T, Schleier S, Forti F, Ghisotti D, Carr MD, O'Hare HM. 2013. *Mycobacterium tuberculosis* RNA polymerase binding protein A (RbpA) and its interactions with sigma factors. *J Biol Chem* 288:14438–14450.

39. Dubnau E, Fontan P, Manganelli R, Soares-Appel A, Smith I. 2002. *Mycobacterium tuberculosis* genes induced during infection of human macrophages. *Infect Immun* 70:2787–2795.

40. Manganelli R, Tyagi S, Smith I. 2001. Real-time PCR using molecular beacons: a new tool to identify point mutations and to analyze gene expression in *Mycobacterium*

tuberculosis, p 295–310. *In* Parish T, Stoker NG (ed), *Mycobacterium Tuberculosis Protocols*. Humana Press, Totowa, NJ.

41. Collins DM, Kawakami RP, de Lisle GW, Pascopella L, Bloom BR, Jacobs WR Jr. 1995. Mutation of the principal sigma factor causes loss of virulence in a strain of the *Mycobacterium tuberculosis* complex. *Proc Natl Acad Sci USA* **92**:8036–8040.

42. Dove SL, Darst SA, Hochschild A. 2003. Region 4 of sigma as a target for transcription regulation. *Mol Microbiol* **48**:863–874.

43. Steyn AJ, Collins DM, Hondalus MK, Jacobs WR Jr, Kawakami RP, Bloom BR. 2002. *Mycobacterium tuberculosis* WhiB3 interacts with RpoV to affect host survival but is dispensable for in vivo growth. *Proc Natl Acad Sci USA* **99**:3147–3152.

44. Singh A, Crossman DK, Mai D, Guidry L, Voskuil MI, Renfrow MB, Steyn AJ. 2009. *Mycobacterium tuberculosis* WhiB3 maintains redox homeostasis by regulating virulence lipid anabolism to modulate macrophage response. *PLoS Pathog* **5**:e1000545.

45. Wu S, Howard ST, Lakey DL, Kipnis A, Samten B, Safi H, Gruppo V, Wizel B, Shams H, Basaraba RJ, Orme IM, Barnes PF. 2004. The principal sigma factor SigA mediates enhanced growth of *Mycobacterium tuberculosis* in vivo. *Mol Microbiol* **51**:1551–1562.

46. Wu S, Barnes PF, Samten B, Pang S, Rodrigue S, Ghanny S, Soteropoulos P, Gaudreau L, Howard ST. 2009. Activation of the *eis* gene in a W-Beijing strain of *Mycobacterium tuberculosis* correlates with increased SigA levels and enhanced intracellular growth. *Microbiology* **155**:1272–1281.

47. Kim KH, An DR, Song J, Yoon JY, Kim HS, Yoon HJ, Im HN, Kim J, Kim do J, Lee SL, Lee HM, Kim HJ, Jo EK, Lee JY, Suh SW. 2012. *Mycobacterium tuberculosis* Eis protein initiates suppression of host immune responses by acetylation of DUSP16/MKP-7. *Proc Natl Acad Sci USA* **109**:7729–7734.

48. Mukherjee R, Gomez M, Jayaraman N, Smith I, Chatterji D. 2005. Hyperglycosylation of glycopeptidolipid of *Mycobacterium smegmatis* under nutrient starvation: structural studies. *Microbiology* **151**:2385–2392.

49. Fontan PA, Voskuil MI, Gomez M, Tan D, Pardini M, Manganelli R, Fattorini L, Schoolnik GK, Smith I. 2009. The *Mycobacterium tuberculosis* sigma factor SigB is required for full response to cell envelope stress and hypoxia in vitro, but it is dispensable for in vivo growth. *J Bacteriol* **191**:5628–5633.

50. Schnappinger D, Ehrt S, Voskuil MI, Liu Y, Mangan JA, Monahan IM, Dolganov G, Efron B, Butcher PD, Nathan C, Schoolnik GK. 2003. Transcriptional adaptation of *Mycobacterium tuberculosis* within macrophages: insights into the phagosomal environment. *J Exp Med* **198**:693–704.

51. Dainese E, Rodrigue S, Delogu G, Provvedi R, Laflamme L, Brzezinski R, Fadda G, Smith I, Gaudreau L, Palu G, Manganelli R. 2006. Posttranslational regulation of *Mycobacterium tuberculosis* extracytoplasmic-function sigma factor SigL and roles in virulence and in global regulation of gene expression. *Infect Immun* **74**:2457–2461.

52. Lee JH, Karakousis PC, Bishai WR. 2008. Roles of SigB and SigF in the *Mycobacterium tuberculosis* sigma factor network. *J Bacteriol* **190**:699–707.

53. Song T, Song SE, Raman S, Anaya M, Husson RN. 2008. Critical role of a single position in the -35 element for promoter recognition by *Mycobacterium tuberculosis* SigE and SigH. *J Bacteriol* **190**:2227–2230.

54. Manganelli R, Voskuil MI, Schoolnik GK, Dubnau E, Gomez M, Smith I. 2002. Role of the extracytoplasmic-function sigma factor SigH in *Mycobacterium tuberculosis* global gene expression. *Mol Microbiol* **45**:365–374.

55. He H, Hovey R, Kane J, Singh V, Zahrt TC. 2006. MprAB is a stress-responsive two-component system that directly regulates expression of sigma factors SigB and SigE in *Mycobacterium tuberculosis*. *J Bacteriol* **188**:2134–2143.

56. Pang X, Vu P, Byrd TF, Ghanny S, Soteropoulos P, Mukamolova GV, Wu S, Samten B, Howard ST. 2007. Evidence for complex interactions of stress-associated regulons in an *mprAB* deletion mutant of *Mycobacterium tuberculosis*. *Microbiology* **153**:1229–1242.

57. White MJ, He H, Penoske RM, Twining SS, Zahrt TC. 2010. PepD participates in the mycobacterial stress response mediated through MprAB and SigE. *J Bacteriol* **192**:1498–1510.

58. Zahrt TC, Deretic V. 2001. *Mycobacterium tuberculosis* signal transduction system required for persistent infections. *Proc Natl Acad Sci USA* **98**:12706–12711.

59. Mukherjee R, Chatterji D. 2005. Evaluation of the role of sigma B in *Mycobacterium smegmatis*. *Biochem Biophys Res Commun* **338**:964–972.

60. Ojha AK, Varma S, Chatterji D. 2002. Synthesis of an unusual polar glycopeptidolipid in glucose-limited culture of *Mycobacterium smegmatis*. *Microbiology* **148**:3039–3048.

61. Cole ST, Eiglmeier K, Parkhill J, James KD, Thomson NR, Wheeler PR, Honore N, Garnier T, Churcher C, Harris D, Mungall K, Basham D, Brown D, Chillingworth T, Connor R, Davies RM, Devlin K, Duthoy S, Feltwell T, Fraser A, Hamlin N, Holroyd S, Hornsby T, Jagels K, Lacroix C, Maclean J, Moule S, Murphy L, Oliver K, Quail MA, Rajandream MA, Rutherford KM, Rutter S, Seeger K, Simon S, Simmonds M, Skelton J, Squares R, Squares S, Stevens K, Taylor K, Whitehead S, Woodward JR, Barrell BG. 2001. Massive gene decay in the leprosy bacillus. *Nature* **409**:1007–1011.

62. Waagmeester A, Thompson J, Reyrat JM. 2005. Identifying sigma factors in *Mycobacterium smegmatis* by comparative genomic analysis. *Trends Microbiol* **13**:505–509.

63. Chang A, Smollett KL, Gopaul KK, Chan BH, Davis EO. 2012. *Mycobacterium tuberculosis* H37Rv *sigC* is expressed from two promoters but is not auto-regulatory. *Tuberculosis* **92**:48–55.

64. Hartkoorn RC, Sala C, Uplekar S, Busso P, Rougemont J, Cole ST. 2012. Genome-wide definition of the SigF

regulon in *Mycobacterium tuberculosis*. *J Bacteriol* **194:** 2001–2009.

65. Rodrigue S, Brodeur J, Jacques PE, Gervais AL, Brzezinski R, Gaudreau L. 2007. Identification of mycobacterial sigma factor binding sites by chromatin immunoprecipitation assays. *J Bacteriol* **189:**1505–1513.

66. Thakur KG, Joshi AM, Gopal B. 2007. Structural and biophysical studies on two promoter recognition domains of the extra-cytoplasmic function sigma factor sigmaC from *Mycobacterium tuberculosis*. *J Biol Chem* **282:**4711–4718.

67. Sun R, Converse PJ, Ko C, Tyagi S, Morrison NE, Bishai WR. 2004. *Mycobacterium tuberculosis* ECF sigma factor SigC is required for lethality in mice and for the conditional expression of a defined gene set. *Mol Microbiol* **52:**25–38.

68. Abdul-Majid KB, Ly LH, Converse PJ, Geiman DE, McMurray DN, Bishai WR. 2008. Altered cellular infiltration and cytokine levels during early *Mycobacterium tuberculosis sigC* mutant infection are associated with late-stage disease attenuation and milder immunopathology in mice. *BMC Microbiol* **8:**151.

69. Karls RK, Guarner J, McMurray DN, Birkness KA, Quinn FD. 2006. Examination of *Mycobacterium tuberculosis* sigma factor mutants using low-dose aerosol infection of guinea pigs suggests a role for SigC in pathogenesis. *Microbiology* **152:**1591–1600.

70. Hu Y, Movahedzadeh F, Stoker NG, Coates AR. 2006. Deletion of the *Mycobacterium tuberculosis* alpha-crystallin-like *hspX* gene causes increased bacterial growth in vivo. *Infect Immun* **74:**861–868.

71. Plocinska R, Purushotham G, Sarva K, Vadrevu IS, Pandeeti EV, Arora N, Plocinski P, Madiraju MV, Rajagopalan M. 2012. Septal localization of the *Mycobacterium tuberculosis* MtrB sensor kinase promotes MtrA regulon expression. *J Biol Chem* **287:**23887–23899.

72. Rifat D, Bishai WR, Karakousis PC. 2009. Phosphate depletion: a novel trigger for *Mycobacterium tuberculosis* persistence. *J Infect Dis* **200:**1126–1135.

73. Dahl JL, Kraus CN, Boshoff HI, Doan B, Foley K, Avarbock D, Kaplan G, Mizrahi V, Rubin H, Barry CE 3rd. 2003. The role of RelMtb-mediated adaptation to stationary phase in long-term persistence of *Mycobacterium tuberculosis* in mice. *Proc Natl Acad Sci USA* **100:** 10026–10031.

74. Calamita H, Ko C, Tyagi S, Yoshimatsu T, Morrison NE, Bishai WR. 2005. The *Mycobacterium tuberculosis* SigD sigma factor controls the expression of ribosome-associated gene products in stationary phase and is required for full virulence. *Cell Microbiol* **7:**233–244.

75. Raman S, Hazra R, Dascher CC, Husson RN. 2004. Transcription regulation by the *Mycobacterium tuberculosis* alternative sigma factor SigD and its role in virulence. *J Bacteriol* **186:**6605–6616.

76. Mukamolova GV, Murzin AG, Salina EG, Demina GR, Kell DB, Kaprelyants AS, Young M. 2006. Muralytic activity of *Micrococcus luteus* Rpf and its relationship

to physiological activity in promoting bacterial growth and resuscitation. *Mol Microbiol* **59:**84–98.

77. Mukamolova GV, Kaprelyants AS, Young DI, Young M, Kell DB. 1998. A bacterial cytokine. *Proc Natl Acad Sci USA* **95:**8916–8921.

78. Manganelli R. 2007. Polyphosphate and stress response in mycobacteria. *Mol Microbiol* **65:**258–260.

79. Sureka K, Dey S, Datta P, Singh AK, Dasgupta A, Rodrigue S, Basu J, Kundu M. 2007. Polyphosphate kinase is involved in stress-induced *mprAB-sigE-rel* signalling in mycobacteria. *Mol Microbiol* **65:**261–276.

80. Giacomini E, Sotolongo A, Iona E, Severa M, Remoli ME, Gafa V, Lande R, Fattorini L, Smith I, Manganelli R, Coccia EM. 2006. Infection of human dendritic cells with a *Mycobacterium tuberculosis sigE* mutant stimulates production of high levels of interleukin-10 but low levels of CXCL10: impact on the T-cell response. *Infect Immun* **74:**3296–3304.

81. Manganelli R, Fattorini L, Tan D, Iona E, Orefici G, Altavilla G, Cusatelli P, Smith I. 2004. The extra cytoplasmic function sigma factor SigE is essential for *Mycobacterium tuberculosis* virulence in mice. *Infect Immun* **72:**3038–3041.

82. Ando M, Yoshimatsu T, Ko C, Converse PJ, Bishai WR. 2003. Deletion of *Mycobacterium tuberculosis* sigma factor E results in delayed time to death with bacterial persistence in the lungs of aerosol-infected mice. *Infect Immun* **71:**7170–7172.

83. Dona V, Rodrigue S, Dainese E, Palu G, Gaudreau L, Manganelli R, Provvedi R. 2008. Evidence of complex transcriptional, translational, and posttranslational regulation of the extracytoplasmic function sigma factor SigE in *Mycobacterium tuberculosis*. *J Bacteriol* **190:** 5963–5971.

84. Raman S, Song T, Puyang X, Bardarov S, Jacobs WR Jr, Husson RN. 2001. The alternative sigma factor SigH regulates major components of oxidative and heat stress responses in *Mycobacterium tuberculosis*. *J Bacteriol* **183:**6119–6125.

85. Barik S, Sureka K, Mukherjee P, Basu J, Kundu M. 2010. RseA, the SigE specific anti-sigma factor of *Mycobacterium tuberculosis*, is inactivated by phosphorylation-dependent ClpC1P2 proteolysis. *Mol Microbiol* **75:**595–606.

86. White MJ, Savaryn JP, Bretl DJ, He H, Penoske RM, Terhune SS, Zahrt TC. 2011. The HtrA-like serine protease PepD interacts with and modulates the *Mycobacterium tuberculosis* 35-kDa antigen outer envelope protein. *PLoS One* **6:**e18175.

87. Yeats C, Finn RD, Bateman A. 2002. The PASTA domain: a beta-lactam-binding domain. *Trends Biochem Sci* **27:**438.

88. Miotto P, Forti F, Ambrosi A, Pellin D, Veiga DF, Balazsi G, Gennaro ML, Di Serio C, Ghisotti D, Cirillo DM. 2012. Genome-wide discovery of small RNAs in *Mycobacterium tuberculosis*. *PLoS One* **7:**e51950.

89. Sureka K, Ghosh B, Dasgupta A, Basu J, Kundu M, Bose I. 2008. Positive feedback and noise activate the

stringent response regulator rel in mycobacteria. *PLoS One* 3:e1771.

90. Tiwari A, Balazsi G, Gennaro ML, Igoshin OA. 2010. The interplay of multiple feedback loops with post-translational kinetics results in bistability of mycobacterial stress response. *Phys Biol* 7:036005.

91. Shi L, Sohaskey CD, Pfeiffer C, Datta P, Parks M, McFadden J, North RJ, Gennaro ML. 2010. Carbon flux rerouting during *Mycobacterium tuberculosis* growth arrest. *Mol Microbiol* 78:1199–1215.

92. Gould TA, van de Langemheen H, Munoz-Elias EJ, McKinney JD, Sacchettini JC. 2006. Dual role of isocitrate lyase 1 in the glyoxylate and methylcitrate cycles in *Mycobacterium tuberculosis*. *Mol Microbiol* 61:940–947.

93. Datta P, Shi L, Bibi N, Balazsi G, Gennaro ML. 2011. Regulation of central metabolism genes of *Mycobacterium tuberculosis* by parallel feed-forward loops controlled by sigma factor E (sigma(E)). *J Bacteriol* 193:1154–1160.

94. Wilson M, DeRisi J, Kristensen HH, Imboden P, Rane S, Brown PO, Schoolnik GK. 1999. Exploring drug-induced alterations in gene expression in *Mycobacterium tuberculosis* by microarray hybridization. *Proc Natl Acad Sci USA* 96:12833–12838.

95. Mehra S, Dutta NK, Mollenkopf HJ, Kaushal D. 2010. *Mycobacterium tuberculosis* MT2816 encodes a key stress-response regulator. *J Infect Dis* 202:943–953.

96. Darwin AJ. 2005. The phage-shock-protein response. *Mol Microbiol* 57:621–628.

97. Fontan PA, Aris V, Alvarez ME, Ghanny S, Cheng J, Soteropoulos P, Trevani A, Pine R, Smith I. 2008. *Mycobacterium tuberculosis* sigma factor E regulon modulates the host inflammatory response. *J Infect Dis* 198:877–885.

98. Hernandez Pando R, Aguilar LD, Smith I, Manganelli R. 2010. Immunogenicity and protection induced by a *Mycobacterium tuberculosis* sigE mutant in a BALB/c mouse model of progressive pulmonary tuberculosis. *Infect Immun* 78:3168–3176.

99. DeMaio J, Zhang Y, Ko C, Young DB, Bishai WR. 1996. A stationary-phase stress-response sigma factor from *Mycobacterium tuberculosis*. *Proc Natl Acad Sci USA* 93:2790–2794.

100. Michele TM, Ko C, Bishai WR. 1999. Exposure to antibiotics induces expression of the *Mycobacterium tuberculosis* sigF gene: implications for chemotherapy against mycobacterial persistors. *Antimicrob Agents Chemother* 43:218–225.

101. Humpel A, Gebhard S, Cook GM, Berney M. 2010. The SigF regulon in *Mycobacterium smegmatis* reveals roles in adaptation to stationary phase, heat, and oxidative stress. *J Bacteriol* 192:2491–2502.

102. Provvedi R, Kocincova D, Dona V, Euphrasie D, Daffe M, Etienne G, Manganelli R, Reyrat JM. 2008. SigF controls carotenoid pigment production and affects transformation efficiency and hydrogen peroxide sensitivity in *Mycobacterium smegmatis*. *J Bacteriol* 190:7859–7863.

103. Chen P, Ruiz RE, Li Q, Silver RF, Bishai WR. 2000. Construction and characterization of a *Mycobacterium tuberculosis* mutant lacking the alternate sigma factor gene, *sigF*. *Infect Immun* 68:5575–5580.

104. Hartkoorn RC, Sala C, Magnet SJ, Chen JM, Pojer F, Cole ST. 2010. Sigma factor F does not prevent rifampin inhibition of RNA polymerase or cause rifampin tolerance in *Mycobacterium tuberculosis*. *J Bacteriol* 192:5472–5479.

105. Geiman DE, Kaushal D, Ko C, Tyagi S, Manabe YC, Schroeder BG, Fleischmann RD, Morrison NE, Converse PJ, Chen P, Bishai WR. 2004. Attenuation of late-stage disease in mice infected by the *Mycobacterium tuberculosis* mutant lacking the SigF alternate sigma factor and identification of SigF-dependent genes by microarray analysis. *Infect Immun* 72:1733–1745.

106. Williams EP, Lee JH, Bishai WR, Colantuoni C, Karakousis PC. 2007. *Mycobacterium tuberculosis* SigF regulates genes encoding cell wall-associated proteins and directly regulates the transcriptional regulatory gene *phoY1*. *J Bacteriol* 189:4234–4242.

107. Homerova D, Surdova K, Kormanec J. 2004. Optimization of a two-plasmid system for the identification of promoters recognized by RNA polymerase containing *Mycobacterium tuberculosis* stress response sigma factor, sigmaF. *Folia Microbiol (Praha)* 49:685–691.

108. Beaucher J, Rodrigue S, Jacques PE, Smith I, Brzezinski R, Gaudreau L. 2002. Novel *Mycobacterium tuberculosis* anti-sigma factor antagonists control sigmaF activity by distinct mechanisms. *Mol Microbiol* 45:1527–1540.

109. Sachdeva P, Narayan A, Misra R, Brahmachari V, Singh Y. 2008. Loss of kinase activity in *Mycobacterium tuberculosis* multidomain protein Rv1364c. *FEBS J* 275:6295–6308.

110. Greenstein AE, MacGurn JA, Baer CE, Falick AM, Cox JS, Alber R. 2007. *M. tuberculosis* Ser/Thr protein kinase D phosphorylates an anti-anti-sigma factor homolog. *PLoS Pathog* 3:e49.

111. Parida BK, Douglas T, Nino C, Dhandayuthapani S. 2005. Interactions of anti-sigma factor antagonists of *Mycobacterium tuberculosis* in the yeast two-hybrid system. *Tuberculosis* 85:347–355.

112. Cappelli G, Volpe E, Grassi M, Liseo B, Colizzi V, Mariani F. 2006. Profiling of *Mycobacterium tuberculosis* gene expression during human macrophage infection: upregulation of the alternative sigma factor G, a group of transcriptional regulators, and proteins with unknown function. *Res Microbiol* 157:445–455.

113. Lee JH, Geiman DE, Bishai WR. 2008. Role of stress response sigma factor SigG in *Mycobacterium tuberculosis*. *J Bacteriol* 190:1128–1133.

114. Smollett KL, Dawson LF, Davis EO. 2011. SigG does not control gene expression in response to DNA damage in *Mycobacterium tuberculosis* H37Rv. *J Bacteriol* 193:1007–1011.

115. Gamulin V, Cetkovic H, Ahel I. 2004. Identification of a promoter motif regulating the major DNA damage response mechanism of *Mycobacterium tuberculosis*. *FEMS Microbiol Lett* 238:57–63.

116. Rand L, Hinds J, Springer B, Sander P, Buxton RS, Davis EO. 2003. The majority of inducible DNA repair genes in *Mycobacterium tuberculosis* are induced independently of RecA. *Mol Microbiol* **50**:1031–1042.

117. Fernandes ND, Wu QL, Kong D, Puyang X, Garg S, Husson RN. 1999. A mycobacterial extracytoplasmic sigma factor involved in survival following heat shock and oxidative stress. *J Bacteriol* **181**:4266–4274.

118. Talaat AM, Ward SK, Wu CW, Rondon E, Tavano C, Bannantine JP, Lyons R, Johnston SA. 2007. Mycobacterial bacilli are metabolically active during chronic tuberculosis in murine lungs: insights from genome-wide transcriptional profiling. *J Bacteriol* **189**:4265–4274.

119. Graham JE, Clark-Curtiss JE. 1999. Identification of *Mycobacterium tuberculosis* RNAs synthesized in response to phagocytosis by human macrophages by selective capture of transcribed sequences (SCOTS). *Proc Natl Acad Sci USA* **96**:11554–11559.

120. Williams DL, Pittman TL, Deshotel M, Oby-Robinson S, Smith I, Husson R. 2007. Molecular basis of the defective heat stress response in *Mycobacterium leprae*. *J Bacteriol* **189**:8818–8827.

121. Song T, Dove SL, Lee KH, Husson RN. 2003. RshA, an anti-sigma factor that regulates the activity of the mycobacterial stress response sigma factor SigH. *Mol Microbiol* **50**:949–959.

122. Kumar S, Badireddy S, Pal K, Sharma S, Arora C, Garg SK, Alam MS, Agrawal P, Anand GS, Swaminathan K. 2012. Interaction of *Mycobacterium tuberculosis* RshA and SigH is mediated by salt bridges. *PLoS One* **7**:e43676.

123. Gaskell AA, Crack JC, Kelemen GH, Hutchings MI, Le Brun E. 2007. RsmA is an anti-sigma factor that modulates its activity through a [2Fe-2S] cluster cofactor. *J Biol Chem* **282**:31812–31820.

124. Park ST, Kang CM, Husson RN. 2008. Regulation of the SigH stress response regulon by an essential protein kinase in *Mycobacterium tuberculosis*. *Proc Natl Acad Sci USA* **105**:13105–13110.

125. Kaushal D, Schroeder BG, Tyagi S, Yoshimatsu T, Scott C, Ko C, Carpenter L, Mehrotra J, Manabe YC, Fleischmann RD, Bishai WR. 2002. Reduced immunopathology and mortality despite tissue persistence in a *Mycobacterium tuberculosis* mutant lacking alternative sigma factor, SigH. *Proc Natl Acad Sci USA* **99**:8330–8335.

126. Paget MS, Kang JG, Roe JH, Buttner MJ. 1998. sigmaR, an RNA polymerase sigma factor that modulates expression of the thioredoxin system in response to oxidative stress in *Streptomyces coelicolor* A3 (2). *EMBO J* **17**:5776–5782.

127. Fahey RC. 2001. Novel thiols of prokaryotes. *Annu Rev Microbiol* **55**:333–356.

128. Mehra S, Kaushal D. 2009. Functional genomics reveals extended roles of the *Mycobacterium tuberculosis* stress response factor sigmaH. *J Bacteriol* **191**:3965–3980.

129. Sadagopal S, Braunstein M, Hager CC, Wei J, Daniel AK, Bochan MR, Crozier I, Smith NE, Gates HO, Barnett L, Van Kaer L, Price JO, Blackwell TS, Kalams SA, Kernodle DS. 2009. Reducing the activity and secretion of microbial antioxidants enhances the immunogenicity of BCG. *PLoS One* **4**:e5531.

130. Dutta NK, Mehra S, Martinez AN, Alvarez X, Renner NA, Morici LA, Pahar B, Maclean AG, Lackner AA, Kaushal D. 2012. The stress-response factor SigH modulates the interaction between *Mycobacterium tuberculosis* and host phagocytes. *PLoS One* **7**:e28958.

131. Mehra S, Golden NA, Stuckey K, Didier PJ, Doyle LA, Russell-Lodrigue KE, Sugimoto C, Hasegawa A, Sivasubramani SK, Roy CJ, Alvarez X, Kuroda MJ, Blanchard JA, Lackner AA, Kaushal D. 2012. The *Mycobacterium tuberculosis* stress response factor SigH is required for bacterial burden as well as immunopathology in primate lungs. *J Infect Dis* **205**:1203–1213.

132. Wu CW, Schmoller SK, Shin SH, Talaat AM. 2007. Defining the stressome of *Mycobacterium avium* subsp. *paratuberculosis* in vitro and in naturally infected cows. *J Bacteriol* **189**:7877–7886.

133. Lee JH, Ammerman NC, Nolan S, Geiman DE, Lun S, Guo H, Bishai WR. 2012. Isoniazid resistance without a loss of fitness in *Mycobacterium tuberculosis*. *Nat Commun* **3**:753.

134. Homerova D, Halgasova L, Kormanec J. 2008. Cascade of extracytoplasmic function sigma factors in *Mycobacterium tuberculosis*: identification of a sigmaJ-dependent promoter upstream of *sigI*. *FEMS Microbiol Lett* **280**:120–126.

135. Vilcheze C, Jacobs WR Jr. 2007. The mechanism of isoniazid killing: clarity through the scope of genetics. *Annu Rev Microbiol* **61**:35–50.

136. Hu Y, Coates AR. 2001. Increased levels of *sigJ* mRNA in late stationary phase cultures of *Mycobacterium tuberculosis* detected by DNA array hybridisation. *FEMS Microbiol Lett* **202**:59–65.

137. Hu Y, Kendall S, Stoker NG, Coates AR. 2004. The *Mycobacterium tuberculosis sigJ* gene controls sensitivity of the bacterium to hydrogen peroxide. *FEMS Microbiol Lett* **237**:415–423.

138. Veyrier F, Said-Salim B, Behr MA. 2008. Evolution of the mycobacterial SigK regulon. *J Bacteriol* **190**:1891–1899.

139. Charlet D, Mostowy S, Alexander D, Sit L, Wiker HG, Behr MA. 2005. Reduced expression of antigenic proteins MPB70 and MPB83 in *Mycobacterium bovis* BCG strains due to a start codon mutation in *sigK*. *Mol Microbiol* **56**:1302–1313.

140. Carr MD, Bloemink MJ, Dentten E, Whelan AO, Gordon SV, Kelly G, Frenkiel TA, Hewinson RG, Williamson RA. 2003. Solution structure of the *Mycobacterium tuberculosis* complex protein MPB70: from tuberculosis pathogenesis to inherited human corneal desease. *J Biol Chem* **278**:43736–43743.

141. Goulding CW, Apostol MI, Gleiter S, Parseghian A, Bardwell J, Gennaro M, Eisenberg D. 2004. Gram-positive DsbE proteins function differently from Gram-negative DsbE homologs. A structure to function analysis of DsbE from *Mycobacterium tuberculosis*. *J Biol Chem* **279**:3516–3524.

142. Appia-Ayme C, Berks BC. 2002. SoxV, an orthologue of the CcdA disulfide transporter, is involved in thiosulfate oxidation in *Rhodovulum sulfidophilum* and reduces the periplasmic thioredoxin SoxW. *Biochem Biophys Res Commun* 296:737–741.

143. Le Brun NE, Bengtsson J, Hederstedt L. 2000. Genes required for cytochrome c synthesis in *Bacillus subtilis*. *Mol Microbiol* 36:638–650.

144. Said-Salim B, Mostowy S, Kristof AS, Behr MA. 2006. Mutations in *Mycobacterium tuberculosis* Rv0444c, the gene encoding anti-SigK, explain high level expression of MPB70 and MPB83 in *Mycobacterium bovis*. *Mol Microbiol* 62:1251–1263.

145. Hahn MY, Raman S, Anaya M, Husson RN. 2005. The *Mycobacterium tuberculosis* extracytoplasmic-function sigma factor SigL regulates polyketide synthases and secreted or membrane proteins and is required for virulence. *J Bacteriol* 187:7062–7071.

146. Thakur KG, Praveena T, Gopal B. 2010. Structural and biochemical bases for the redox sensitivity of *Mycobacterium tuberculosis* RslA. *J Mol Biol* 397:1199–1208.

147. Sirakova TD, Dubey VS, Cynamon MH, Kolattukudy PE. 2003. Attenuation of *Mycobacterium tuberculosis* by disruption of a mas-like gene or a chalcone synthase-like gene, which causes deficiency in dimycocerosyl phthiocerol synthesis. *J Bacteriol* 185:2999–3008.

148. Trivedi OA, Arora P, Vats A, Ansari MZ, Tickoo R, Sridharan V, Mohanty D, Gokhale RS. 2005. Dissecting the mechanism and assembly of a complex virulence mycobacterial lipid. *Mol Cell* 17:631–643.

149. Mulder NJ, Apweiler R, Attwood TK, Bairoch A, Bateman A, Binns D, Bradley P, Bork P, Bucher P, Cerutti L, Copley R, Courcelle E, Das U, Durbin R, Fleischmann W, Gough J, Haft D, Harte N, Hulo N, Kahn D, Kanapin A, Krestyaninova M, Lonsdale D, Lopez R, Letunic I, Madera M, Maslen J, McDowall J, Mitchell A, Nikolskaya AN, Orchard S, Pagni M, Ponting CP, Quevillon E, Selengut J, Sigrist CJ, Silventoinen V, Studholme DJ, Vaughan R, Wu CH. 2005. InterPro, progress and status in 2005. *Nucleic Acids Res* 33:D201–D205.

150. Sonden B, Kocincova D, Deshayes C, Euphrasie D, Rhayat L, Laval F, Frehel C, Daffe M, Etienne G, Reyrat JM. 2005. Gap, a mycobacterial specific integral membrane protein, is required for glycolipid transport to the cell surface. *Mol Microbiol* 58:426–440.

151. Agarwal N, Woolwine SC, Tyagi S, Bishai WR. 2007. Characterization of the *Mycobacterium tuberculosis* sigma factor SigM by assessment of virulence and identification of SigM-dependent genes. *Infect Immun* 75:452–461.

152. Agarwal N, Raghunand TR, Bishai WR. 2006. Regulation of the expression of *whiB1* in *Mycobacterium tuberculosis*: role of cAMP receptor protein. *Microbiology* 152:2749–2756.

153. Raman S, Puyang X, Cheng TY, Young DC, Moody DB, Husson RN. 2006. *Mycobacterium tuberculosis* SigM positively regulates Esx secreted protein and nonribosomal peptide synthetase genes and down regulates virulence-associated surface lipid synthesis. *J Bacteriol* 188:8460–8468.

154. Casonato S, Cervantes Sanchez A, Haruki H, Rengifo Gonzalez M, Provvedi R, Dainese E, Jaouen T, Gola S, Bini E, Vicente M, Johnsson K, Ghisotti D, Palu G, Hernandez-Pando R, Manganelli R. 2012. WhiB5, a transcriptional regulator that contributes to *Mycobacterium tuberculosis* virulence and reactivation. *Infect Immun* 80:3132–3144.

155. Predich M, Doukhan L, Nair G, Smith I. 1995. Characterization of RNA polymerase and two sigma-factor genes from *Mycobacterium smegmatis*. *Mol Microbiol* 15:355–366.

156. Doukhan L, Predich M, Nair G, Dussurget O, Mandic-Mulec I, Cole ST, Smith DR, Smith I. 1995. Genomic organization of the mycobacterial sigma gene cluster. *Gene* 165:67–70.

Molecular Genetics of Mycobacteria, 2nd Edition
Edited by Graham F. Hatfull and William R. Jacobs, Jr.
© 2014 American Society for Microbiology, Washington, DC
doi:10.1128/microbiolspec.MGM2-0035-2013

Suma Jaini,[1,4] Anna Lyubetskaya,[2] Antonio Gomes,[2]
Matthew Peterson,[1] Sang Tae Park,[3] Sahadevan Raman,[4]
Gary Schoolnik,[5] and James Galagan[1,2,4,6]

Transcription Factor Binding Site Mapping Using ChIP-Seq

8

This chapter describes a chromatin immunoprecipitation followed by sequencing (ChIP-Seq) method tailored for the study of *Mycobacterium tuberculosis* transcription factors (TFs) but amenable for the study of other prokaryotes. Noteworthy features of this method include the following: (i) it is conducted using a standard and readily reproducible growth condition that is the same for each TF studied; (ii) the binding behavior of each TF is studied using a tagged variant of the TF whose production is driven by an inducible promoter; (iii) each TF is studied using the same concentration of an exogenously added chemical inducer where the strength of TF gene induction is varied to systematically measure the effect of TF concentration on binding site strength and location; (iv) the resulting binding site data is spatially well resolved and highly reproducible, and binding strength is correlated with the degree of binding site motif conservation; and (v), because this method does not require knowledge of the physiological conditions that normally cause TF gene expression or of antibodies specific to each TF, it is applicable for the high-throughput study of multiple TFs;

thus far it has generated binding site data for over 119 annotated *M. tuberculosis* TFs. Despite the use of a standard growth condition for all TFs and an inducible promoter system, the binding site data was found to agree well with data from a subset of TFs expressed under their own promoter by physiologically relevant conditions and was immunoprecipitated using specific antibodies to the native TF. The development and use of this method has been combined with the development of a data analysis pipeline; features of this pipeline are described below. Taken together, these binding data provide additional evidence that the long-accepted spatial relationship between TF binding site, promoter motif, and the corresponding regulated gene may be too simple a paradigm, failing to adequately capture the variety of TF binding sites found in prokaryotes.

When this method is combined with genome-wide expression data from the same experiment, ChIP-Seq and expression data for multiple TFs can be used to derive a system-wide regulatory network model. Even though this network model does not include other regulatory effects coming from, for example, the action of

[1]Department of Biomedical Engineering, Boston University, Boston, MA 02215; [2]Bioinformatics Program, Boston University, Boston, MA 02215; [3]Macrogen Clinical Laboratory, Macrogen Corp, Rockville, MD 20850; [4]National Emerging Infectious Diseases Laboratory, Boston University, Boston, MA 02118; [5]Department of Medicine and Department of Microbiology and Immunology, Stanford Medical School, Stanford, CA 94305; [6]Broad Institute of MIT and Harvard, Cambridge, MA 02142.

small regulatory RNAs or epigenetic mechanisms, it was able to predict responses of *M. tuberculosis* to specific conditions of growth (1).

ChIP-Seq METHOD AND PROTOCOL

ChIP-Seq is a well-established method for identifying binding sites for DNA binding proteins (2–4). Chromatin immunoprecipitation (ChIP) is the first step (Fig. 1). Proteins bound to the genomes of these cells are cross-linked to DNA with formaldehyde. The cells are then broken open, and the DNA is sheared through sonication or enzymatic digestion. DNA fragments bound by a protein of interest (a TF, for example) are immunoprecipitated using an antibody to the protein. Cross-linking is then reversed to remove the proteins, and the precipitated DNA fragments are isolated and sequenced

to generate reads from the ends of the fragments. Sequenced reads are aligned to the corresponding genome sequence, and genomic locations from which the DNA fragments are derived are identified as regions that are overrepresented with aligned reads.

Ideally, only genomic regions that are bound by the protein of interest would display read coverage. In practice, nonspecific DNA fragments will also be isolated and sequenced, resulting in a background coverage on genome regions that are not bound. To assess this background coverage, one or more control experiments are typically used. One common example is the use of mock ChIP runs; they include every step of the ChIP process with the exception of the antibody purification. These experiments control for the many nonantibody steps of ChIP that may lead to nonspecific DNA isolation. Antibodies can also be used against epitope tagged proteins. In those cases the control experiment can use

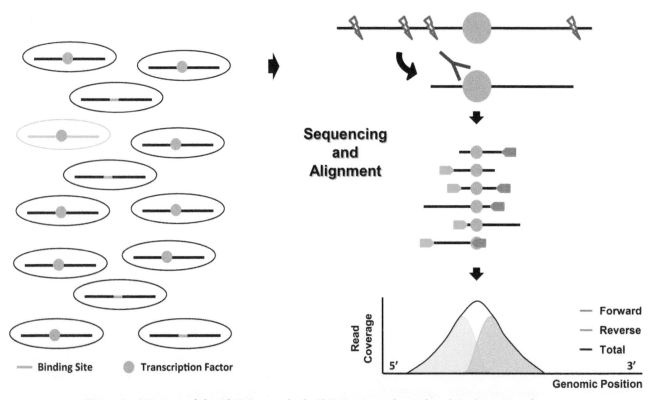

Binding and Crosslinking

Fragmentation and IP

Sequencing and Alignment

— Binding Site   ● Transcription Factor

Read Coverage

5′   3′

Genomic Position

— Forward
— Reverse
— Total

Figure 1 Diagram of the ChIP-Seq method. ChIP-Seq is performed on log-phase *M. tuberculosis* cells. The cross-linked cells are first lysed using BeadBeater. The cells are further lysed, and DNA is sheared using Covaris. Anti-FLAG antibody is used to immunoprecipitate the protein of interest, and the protein-DNA complexes are further captured using protein-G agarose beads. The cross-links are reversed using proteinase K, and the DNA is purified using a PCR purification kit. The standard Illumina protocol is used to prepare the library, which is then sent for next generation sequencing. doi:10.1128/microbiolspec.MGM2-0035-2013.f1

the same antibodies for ChIP in strains that lack the epitope tag. Such experiments assess the degree to which the antibody recognizes nonspecific targets. A genomic DNA preparation can also be generated as a control. Control experiments help to identify the differential efficiency of isolation and sequencing of different locations of the genome. The genome regions that contain binding sites for the protein of interest are characterized by greater coverage in comparison with the background coverage.

ChIP-Seq also produces a strand-specific signature of enrichment that can be used to identify true binding peaks. When DNA fragments from ChIP are sequenced, reads are typically generated from one end of the fragment or the other. Protein binding sites will occur between the ends of this fragment. Reads will thus align to one side or the other of the binding site and thus on different strands. If coverage is visualized for the two strands separately, this process gives rise to a bimodal enrichment profile: the coverage of the forward strand will be shifted upstream with respect to the actual binding site, while the coverage on the reverse strand will be shifted downstream (5). The distance between the forward and reverse coverage profile is determined by the size of the DNA fragment sequence, which can be estimated during the sequencing process. Nonspecific binding, by contrast, often results in enrichment that lacks this bimodal shift.

DNA binding proteins, especially TFs, typically bind to short DNA sequences (on the order of 15 bp or less). Enriched peaks, however, typically span a region of several hundred base pairs as a consequence of the larger fragment size generated during ChIP (typically around 250 bp). Moreover, when multiple closely spaced binding sites for a protein exist in a particular location, the read coverage for these sites can merge into a single broad enriched region.

We have developed a protocol for ChIP-Seq that has been tailored for use with mycobacteria, although we have also successfully utilized the same protocol for other bacteria. The protocol is provided in the appendix. A number of factors contribute to obtaining optimal results with this protocol. The most important factor is the quality of the antibody. Strong binding to the protein of interest and minimal nonspecific binding are critical. During IP, sufficient time should be allowed for the antibodies to interact with their target proteins. Also, incubation with antibodies should be performed at 4°C in order to minimize any protein degradation. During capturing of immuno-complexes, washes should be performed with care to minimize the loss of agarose beads. The nature of protein-G coated

beads to capture the antibody complex may also influence the quality of the preparation. In our experience, protein-G coated agarose beads are found to be superior to protein-G coated magnetic beads. Agarose beads are fragile, however, so care must be taken to maintain the integrity of the beads during centrifugation.

Another critical factor is to have proper lysing and shearing conditions. Cell lysis can be performed separately from DNA shearing or in a single step using Covaris. The time and strength of shearing should be optimized such that most of the sheared DNA is between 200 and 500 bp. Depending on the cell lysis method used, it is important to make sure the cells are lysed efficiently to release the cellular contents. Avoid contamination of the lysate with cellular debris during centrifugation and separation of the lysate.

For library preparation, the last amplification step is important. To prevent amplification of nonspecific DNA, do not exceed 18 cycles of amplification. Also, there should not be a significant amount of primer dimers or adapter dimers. In order to prevent the carryover of primer/adapter dimers, adjust the number of adapters and primers accordingly. The standard Illumina library prep protocol is provided for 10 ng of ChIP DNA. If there is insufficient DNA, ChIP should be performed using more cells.

DATA ANALYSIS

We have developed a computational pipeline to perform data processing and analysis of ChIP-Seq data for mycobacteria and other microbes (Fig. 2). The pipeline consists of two stages: (i) detection of enriched regions from raw ChIP-Seq data and (ii) identification of binding sites and motifs within enriched regions.

Detection of Enriched Regions from Raw ChIP-Seq Sequencing Data

Sequence reads are mapped to the target genome sequence using MAQ (6). The "pileup" command from the SAMtools (7) suite is used to calculate coverage along the forward and reverse strands along the genome. From this coverage, regions of enrichment along the genome are identified using a log normal distribution. The log normal distribution is defined by the probability density function (PDF):

$$f(x) = \frac{1}{x\sigma\sqrt{2\pi}} e^{-\frac{(ln\,(x)-\mu)^2}{2\sigma}}$$

Here, μ and σ are the mean and standard deviation of the natural logarithm of the random variable, respectively. For each experiment, positions along the

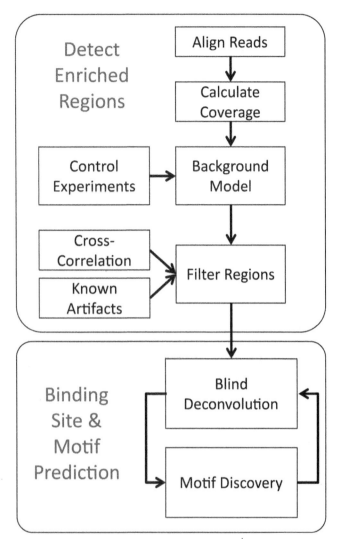

Figure 2 Overview of analysis pipeline. First, sequencing reads are aligned to the genome of the organism, and a profile of coverage is calculated by counting the number of reads that overlap a given position along the chromosome. This profile is used to fit a log normal background model of coverage. This model is used to identify regions of the genome that are statistically enriched relative to background. A cross-correlation filter is applied to identify regions consistent with the bimodal profile associated with ChIP-Seq binding, and comparison to control experiments is used to identify regions specific to the transcription factor of interest. Finally, a blind deconvolution approach combined with motif identification is used to identify binding sites at high resolution. doi:10.1128/microbiolspec.MGM2-0035-2013.f2

genome greater than five times the mean coverage or higher are excluded to avoid fitting outliers, and the parameters of the distribution are estimated using maximum likelihood. Although the log normal distribution has been the best fit for most microbial ChIP-Seq data

in our experience, we have also tested other distributions including the Gaussian, negative binomial, and generalized extreme value distributions.

The resulting distribution is used to score each position of the genome to identify enriched positions along the genome. Positions with a P value of 0.01 or lower are called enriched. Since TF binding is expected to result in contiguous regions of enrichment, only regions of enrichment of a threshold length or longer are included for further analysis. This threshold is set by the user and is typically 100 bp for our analyses.

A cross-correlation filter is then applied to the resulting regions to identify those that have the expected signature of transcription binding in ChIP-Seq experiments, identified by a shift between peaks between the forward and reverse strands. The cross-correlation function is calculated as

$$c[n] = \sum_{m=-L}^{L} f[m]r[n+m]$$

In this function, f represents the coverage on the forward strand, while r is the coverage on the reverse strand, and n is the amount of shift applied to the function. The shift between peaks in the forward and reverse is defined as the value n that maximizes the cross-correlation. Regions with a shift of less than a particular threshold are removed from further analysis.

Regions are then compared against a database of known artifacts that have been identified in the organisms we have worked on. In some cases, these artifacts are general and include annotated repeat regions. In other cases, artifacts can be organism specific. For example, in *M. tuberculosis* we observed that certain regions showed statistical enrichment in nearly all ChIP-Seq experiments regardless of the ChIP'd TF. It was found that most of these regions displayed strong binding by the histone-like protein Lsr2 (8) and may reflect nonspecific interactions. Our pipeline flags these artifacts and removes them for separate consideration. The output at this stage of the analysis pipeline is a list of enriched regions and their locations. However, as noted above, enriched regions can typically span hundreds of nucleotides, while binding sites are more often less than 15 bases long. Moreover, a single enriched region can contain multiple individual binding sites.

Identification of Binding Sites and Motifs within Enriched Regions

To resolve individual binding sites within an enriched region, we treat binding site prediction as a signal detection problem. Conceptually, binding sites can be considered a source of an input signal that, through the process of ChIP-Seq, give rise to the broader output signal observed in the ChIP-Seq coverage. Multiple binding sites give rise, to a first approximation, to an output coverage that is the sum of the outputs of each of the individual binding sites. This process can be modeled as a linear convolution.

In this model, the output signal arising from a point input is called an impulse response (or point spread function). In the case of the ChIP-Seq, the impulse response is a consequence of "transmitting" the input signal through the process of randomly sequencing the ends of large DNA fragments that overlap the point source and "receiving" this transmission in terms of coverage after aligning these reads. The impulse function essentially "blurs" the output signal arising from a point input. Input signals are modeled as the sum of multiple point sources, or impulse functions, that are each scaled to a particular magnitude. In the case of ChIP-Seq, impulse functions correspond to binding sites where the scaling associated with each site corresponds roughly to relative enrichment (and thus relative occupancy). The output signal is then the sum of the correspondingly scaled impulse functions, or a convolution in mathematical terms (specifically, a discrete convolution since DNA coordinates are integer based).

The operation of recovering the binding site locations from regions of enriched coverage is then a deconvolution. In the case where the impulse function is known, this process is straightforward (9). For ChIP-Seq, however, this function is not known and depends in part on the details of each specific experiment. Thus, the impulse function must be estimated at the same time that the coverage signal is being deconvolved.

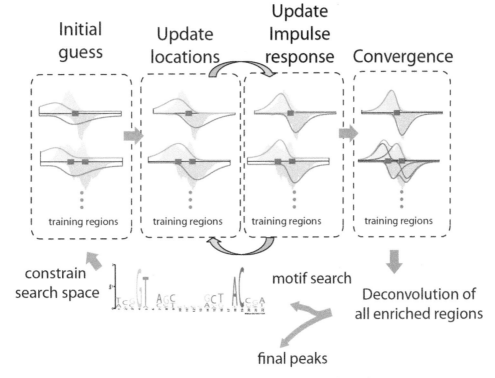

Figure 3 Blind deconvolution analysis of ChIP-Seq data. This schematic representation illustrates the steps of BRACIL (12) in the analysis of ChIP-Seq data. BRACIL uses a blind deconvolution algorithm that takes advantage of ChIP-Seq coverage and genome sequence to refine the resolution of ChIP-Seq binding regions into single-nucleotide resolution. The top panel illustrates the steps of the blind-deconvolution algorithm. The algorithm starts with a guess about the shape of the impulse response as well as the location of binding sites. Both the shape of the impulse response and the binding site locations are updated iteratively until convergence. The predicted binding site locations are used in motif discovery to predict a binding motif. This motif is used to constrain the search space for deconvolution and refine prediction of binding site locations. A set of high-resolution binding site locations is obtained as a final prediction. doi:10.1128/microbiolspec.MGM2-0035-2013.f3

This operation is termed a blind deconvolution (blind because the impulse function is not known *a priori*), and techniques have also been developed to solve this problem (10).

We previously developed a method, CSDeconv, that uses a reestimation method to solve the blind deconvolution problem (11) (Fig. 3). The basic approach begins by generating an initial estimate of the impulse function. This is typically generated by selecting a set of peaks with high coverage that are then used to fit a model of the impulse function based on an initial estimate of the binding site locations in these regions. This estimated impulse function can then be used to perform blind deconvolution on all enriched regions, which results in new binding site locations for all enriched regions. These new binding site locations are then used to fit an updated model of the impulse function, and the process iterates in this fashion until convergence criteria are reached.

We more recently enhanced the basic CSDeconv algorithm in a number of ways to develop a new algorithm called BRACIL (Gomes A, Abeel T, Peterson M, Azizi E, Lyubetskaya A, Carvalho L, Galagan J, submitted for publication). First, we incorporated *de novo* motif discovery into the peak detection process (Fig. 3). By performing motif discovery during the blind deconvolution process, we can improve the prediction of binding site locations. By using binding sites locations predicted by the blind deconvolution algorithm as input, we are able to improve motif discovery. Second, we developed a more principled model for the impulse function; this and other enhancements have improved the accuracy and runtime of the algorithm such that we are able to rapidly analyze binding sites for even large mammalian chromosomes.

We have also incorporated the ability to detect potential cooperative binding. Closely spaced binding sites have been shown to mediate cooperative binding that can substantially alter the apparent affinity of individual sites (13). The deconvolution approach described above implicitly assumes that TFs bind individual sites independently of all other sites. The approach can be generalized, however, to explicitly model dependencies between sites for the same TF. Using this approach, it is possible to predict known cooperative interactions between closely spaced sites for the DosR transcription factor in *M. tuberculosis* (12).

The final output of our blind deconvolution method is a list of binding sites with high spatial resolution and accuracy (11, 12). Based on a test using ChIP-Seq data for the GABP TF in humans and the DosR TF in *M. tuberculosis*, the approach is able to identify binding locations to within an average absolute difference of less than 24 bp (11). Moreover, the method can accurately predict multiply spaced binding sites within the same ChIP-Seq enriched region. In the case of DosR, binding sites located less than 57 bp apart in the same intergenic region could be resolved, while several closely spaced binding sites for GABP were observed, two as close as 20 bp apart.

COMPREHENSIVE MAPPING OF TF BINDING SITES IN *M. TUBERCULOSIS*

As part of a consortium effort (1), we developed a high-throughput system for comprehensively mapping and functionally validating regulatory interactions in *M. tuberculosis*. For this effort, we performed ChIP-Seq (2, 4, 14) using FLAG-tagged TFs episomally expressed under the control of the well-known mycobacterial tetracycline-inducible promoter (15–18). An anhydrotetracycline (Atc)-inducible episomal vector containing a Gateway Recombination™ (Invitrogen) cassette was modified to contain an in-frame N- or C-terminal FLAG epitope tag. TFs from a Gateway entry clone library (supplied by the NIAID-funded PFGRC) were recombined into this vector to create N- or C-terminally epitope-tagged expression vectors. The full complement of these epitope-tagged expression vectors is being made available to all researchers through the NIAID-funded BEI repository.

This system allowed us to map DNA binding of the vast majority of the ~180 *M. tuberculosis* regulators in a consistent and comparable manner independent of regulatory function. The use of tagged TFs obviates the need to develop antibodies to native proteins, an effort that would have been prohibitive for all *M. tuberculosis* TFs. Moreover, we are able to use a ChIP grade antibody to FLAG for IP that enables high sensitivity and specificity of pull down. Control IP experiments in strains lacking FLAG-tagged proteins revealed little nonspecific binding of the antibody used. What nonspecific binding occurred could be systematically removed for all tagged TFs in our analysis pipeline. The antibody used also displayed a high level of enrichment for tagged proteins during IP, leading to high signal to noise in subsequent sequencing and alignment.

The use of an inducible promoter system ensures expression of targeted TFs, which allowed us to study the full complement of *M. tuberculosis* regulators in a standard and highly reproducible baseline state without *a priori* knowledge of the conditions that normally induce their expression. All regulators studied this way were induced with ATc during mid-log-phase growth. We confirmed that protein expression per se did not lead to spurious nonspecific binding. ChIP-Seq of three non–DNA binding proteins using the inducible system resulted in no enriched binding (1). The tetracycline-inducible promoter is a relatively weaker promoter that induces the expression to a moderate level, preventing the overproduction of the protein beyond a certain level. The standard level of induction was 100 ng Atc per ml of culture, although the inducible system provided the ability to induce at multiple levels (see below). The use of 100 ng/ml Atc results in different levels of expression for different TFs relative to their physiological expression levels. For example, in the case of KstR and Rv0081, induction with 100 ng/ml

resulted in roughly four times and three times overexpression, respectively, at the mRNA level relative to the expression observed for known physiological stimuli (1). In contrast, for DosR the same level of Atc resulted in expression induction essentially equivalent to that observed under physiological conditions.

The small size of the *M. tuberculosis* genome (~4.4 Mb) serves as an advantage for ChIP-Seq mapping. ChIP-Seq coverage for binding sites scales with the overall number of reads generated relative to the size of the reference genome. A typical ~36-bp sequencing lane on an Illumina GAIIx is sufficient to provide an average of 500-fold coverage for the *M. tuberculosis*

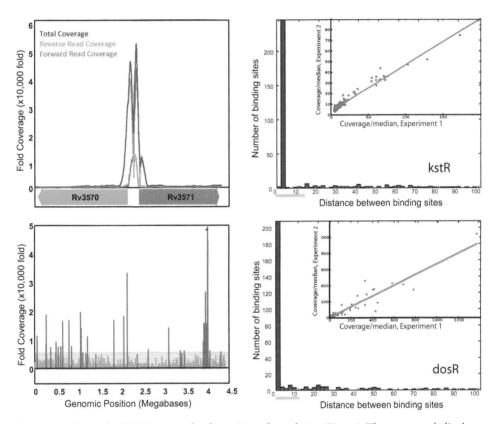

Figure 4 Example ChIP-Seq results from *M. tuberculosis*. (Upper) The top panel displays the fold read coverage for a single binding region with two known binding sites for the TF KstR. ChIP-Seq coverage visually resolves both binding sites and confirms the experimental observation that the site closest to Rv3571 is a weaker affinity. Total coverage is shown in blue, and the forward and reverse coverage is shown in red and green, respectively. The binding event also displays the expected shift in position between the forward and reverse reads. The bottom panel displays the genome-wide fold coverage for the same experiment. Peaks above a coverage threshold are shown in blue. The peak shown in the top panel is marked with a star in the bottom panel. The horizontal gray lines are multiples of the standard deviation of background coverage. (Lower) Binding site identification is highly reproducible. Bar plots show the distance between corresponding sites in two replicates for two TFs. The blue line indicates the length of known motifs. Insets show relationship of peak height between corresponding peaks in two replicates (R2 > 0.83 for all TFs). Figures from reference 1. doi:10.1128/microbiolspec.MGM2-0035-2013.f4

genome. With this degree of coverage, binding sites for individual TFs can be identified with coverage that spans several logs in magnitude (Fig. 4A). We routinely observed enrichments of over 100,000-fold relative to a baseline of 150-fold, while the highest affinity peaks for certain TFs could result in binding regions with over 1 million-fold coverage. The differences in coverage between different binding sites reflect the probability of occupancy of each site in the population of cells on which ChIP-Seq was performed. Occupancy, in turn, reflects a number of factors including the concentration and modification state of the TF, the affinity of the binding site for the TF, the accessibility of the binding site, and the availability of molecular cofactors. The high coverage that can be generated for ChIP-Seq in *M. tuberculosis* provides insights into the variation of these factors on a genome-wide scale with unprecedented resolution (see below).

VALIDATION OF THE ChIP-Seq SYSTEM

The comprehensive mapping of binding sites for all TFs in the *M. tuberculosis* genome provided the opportunity to perform substantial validation of the results of ChIP-Seq as applied to a large number of different TFs in *M. tuberculosis* (1). We highlight two results of this validation here.

Binding Is Sensitive and Reproducible

The inducible and FLAG-tagged ChIP-Seq system displays high sensitivity. For example, we confirmed the sensitivity of the system using high-confidence direct regulation from two well-studied regulators, the activator DosR (Rv3133c) (11, 19–24) and the repressor KstR (Rv3574) (25–28). In both cases we can identify all of the previously reported direct binding sites. The system also correctly identifies the known binding motifs for these factors. Our system also replicates biochemically identified binding sites for Rv2034 reported by Gao and colleagues (29, 30). Using electrophoretic mobility shift assay (EMSA), these authors demonstrated that Rv2034 binds to its own promoter more strongly than to promoters of other genes they tested, binds both upstream of the DosR operon as well as near the DosR gene, binds upstream of the GroEL2 gene, and binds the promoter of PhoP. ChIP-Seq of Rv2034 using the inducible FLAG-tagged system replicates the location and strength of these binding regions (1). ChIP-Seq also recapitulates the core binding motif identified using EMSA. To date, all binding interactions previously published for TFs that we have successfully mapped with our system have been identified in our experiments.

Binding is also highly reproducible. Biological replicates using the inducible FLAG-tagged system show that coverage for enriched sites is highly correlated between experiments ($R^2 > 0.83$ for all TFs with replicates; two examples are shown in Fig. 4B). There is also high reproducibility in binding location, with distances between replicate binding sites less than the length of predicted binding site motifs for the vast majority of sites.

We further confirmed that binding sites were reproducible between different cellular conditions. To comprehensively map all *M. tuberculosis* TFs, a standard normoxic culture condition was selected that was both practical and not specific to any subset of TFs. To test the degree to which the binding sites observed in normoxia could be detected under different physiological conditions, ChIP-Seq was also performed on 11 TFs under hypoxic conditions. Substantial concordance between the binding enrichment in normoxia and hypoxia was seen (Fig. 5). For the large majority of the regions, the evidence of binding in normoxia was also observed in hypoxia. Moreover, all binding sites observed in hypoxia were also detected in normoxia. In addition, the relative coverage of corresponding binding regions was broadly conserved between hypoxia and normoxia conditions.

Finally, we confirmed that the inducible FLAG-tagged system could replicate binding sites found by an independent lab using a different ChIP-Seq system. In a report by Blasco and colleagues, the TF EspR was mapped by ChIP-Seq using antibodies to the wild-type EspR protein and thus with EspR induced from its native promoter (31). To further validate the inducible tagged TF system, we performed ChIP-Seq on EspR using induction with 100 ng/ml Atc and compared the results to those from the Blasco report. As shown in Fig. 6, we observed a high level of agreement between the two methods applied to map the same TF in independent groups. In particular, the inducible FLAG-tagged system recapitulates the distribution of binding locations and heights, the motif, and the specific pattern of binding at individual loci that was found by Blasco.

Binding is Correlated with Bound Sequence

The physiological nature of the binding identified by ChIP-Seq is also supported by an analysis of the sequences underlying each of the bound sites. For nearly all TFs, a consensus motif can be determined based on the ChIP-Seq data, and an instance of this motif can be identified underlying the vast majority of binding sites. Moreover, as a general trend, variations in motif instances can be associated with differences in apparent

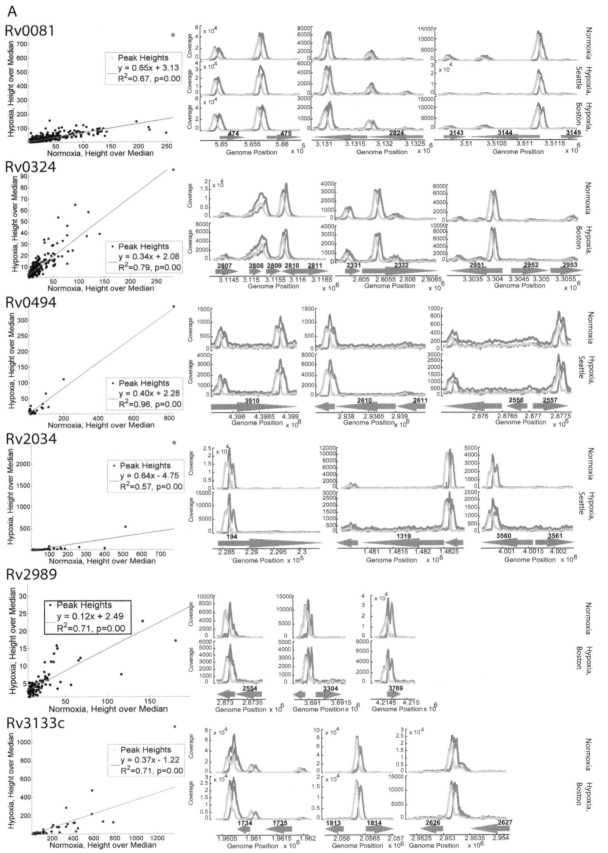

Figure 5 *continues on next page*

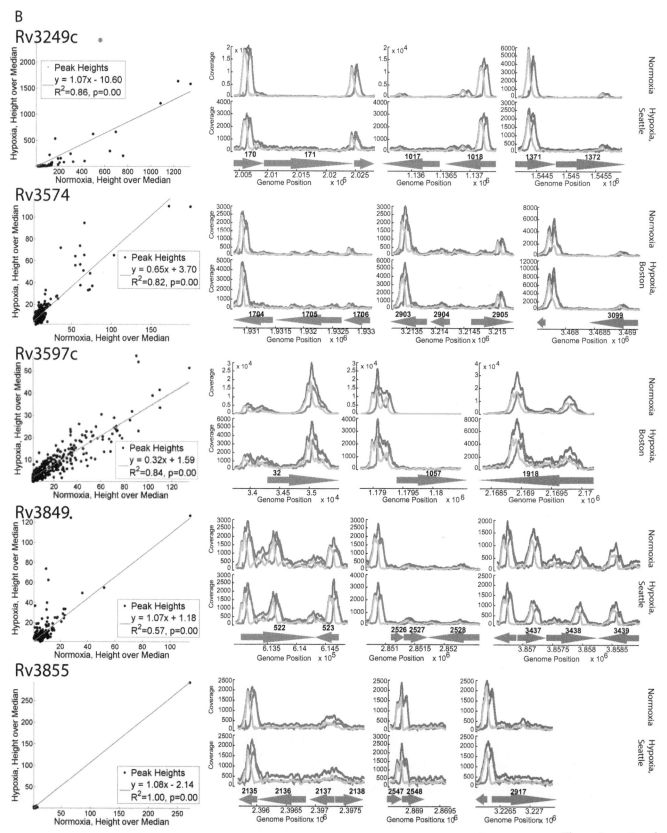

Figure 5 *continued*

binding affinity. An example for KstR is shown in Fig. 7. In the bottom portion of the heat map, each row represents the sequence region for a single binding site, with the nucleotides color coded. The rows are ordered from the binding peaks with the most enrichment at the top down to binding peaks with the lowest enrichment at the bottom. A clear consensus motif can be detected as shown, which corresponds to the published binding motif for KstR (25–28), and the core of this motif is visible in the heat map across all the binding sites. Individual binding sites, however, display differences in their underlying motifs. The strongest binding sites contain sequences that contain not only the core palindrome, but also conserved accessory bases known to play an important role in shaping affinity (32). Less-enriched peaks, by contrast, are often found to lack these accessory bases. Furthermore, in some cases only one-half of the palindrome is present. Binding to such half sites was described in the earliest ChIP-Seq experiments in eukaryotes (4). This pattern suggests that, to a first approximation, differences in enrichment at different binding sites can often be explained by differences in the affinities of the underlying sequence motif.

The difference in affinities of different binding sites is also reflected in experiments in which ChIP-Seq was performed on TFs induced to different levels. ChIP enrichment is a function of the number of cells in which a site is bound (33, 34), which is governed by the affinity of the site and the concentration of the factor. By overexpressing a factor, we expect to increase the occupancy of strong sites up to a saturation limit while also occupying, and thus detecting, weaker affinity sites. This was experimentally confirmed. In the example shown in Fig. 6, when KstR is expressed at a low level, only a small number of low peaks are identified, and a number of gold standard sites are missed. As expression increases, overall peak heights increase while maintaining roughly the same relative heights, and additional lower peaks are revealed. At the highest levels of expression, the increase in enrichment of strong sites begins to level off.

It is important to note that the binding sites detected do not simply reflect the association of induced TFs with all available motifs. For the majority of factors mapped, only a fraction of computational identified instances of motifs in the genome are typically bound based on ChIP-Seq. In the example for KstR in Fig. 7, numerous instances of the strongest KstR motif can be detected throughout the genome that are not bound in any of the ChIP-Seq experiments we have performed. These instances include examples of nearly the identical sequence in two different genomic locations, one of which is bound, while the other is not. Binding of the majority of TFs thus displays genome context specificity: only a fraction of the possible binding sites appear occupied, and this differential occupancy is highly reproducible. This phenomenon is well known in eukaryotic systems (34) but is less well appreciated in bacteria. Examples are known of TFs in prokaryotes that appear to be able to bind to any available motif instance regardless of genomic location (35). However, the data from *M. tuberculosis*, as well as corresponding data from ChIP-Seq in *Escherichia coli* (unpublished results), indicates that for many bacterial TFs, factors other than the underlying sequence determine whether individual motif instances are bound *in vivo*. The determinants of this specificity are not fully understood (34).

THE DIVERSITY OF TF BINDING IN PROKARYOTES

ChIP-Seq has been applied extensively to map TF binding sites in a range of eukaryotic organisms. The data from these studies have provided a wealth of information about the global binding patterns of TFs that has overturned many previously held assumptions about the nature, diversity, and possible functions of TF binding sites (34). The data from *M. tuberculosis* represent the first large-scale application of ChIP-Seq to bacteria (1). The resulting data have confirmed several surprises that have also emerged from mapping eukaryotic cells. These surprises call into question some of the simplifying assumptions held about bacterial transcriptional regulation (36). In particular, the data reveal that binding of TFs in *M. tuberculosis* (i) occurs in many more diverse genomic locations than expected and

Figure 5 Binding sites replicate between normoxia and hypoxia. Each panel compares the results of ChIP-Seq under both normoxic (*x* axis, top traces) and hypoxic (*y* axis, bottom traces). Strong concordance was seen in both peak heights (scatter plots) and coverage profiles (sequencing traces) in experiments performed under both conditions. While no binding sites were identified in normoxia that were not identified in hypoxia, a small number of sites exhibited greatly increased binding under hypoxic conditions. The three sites showing the largest increases are shown in red on the scatter plots.
doi:10.1128/microbiolspec.MGM2-0035-2013.f5

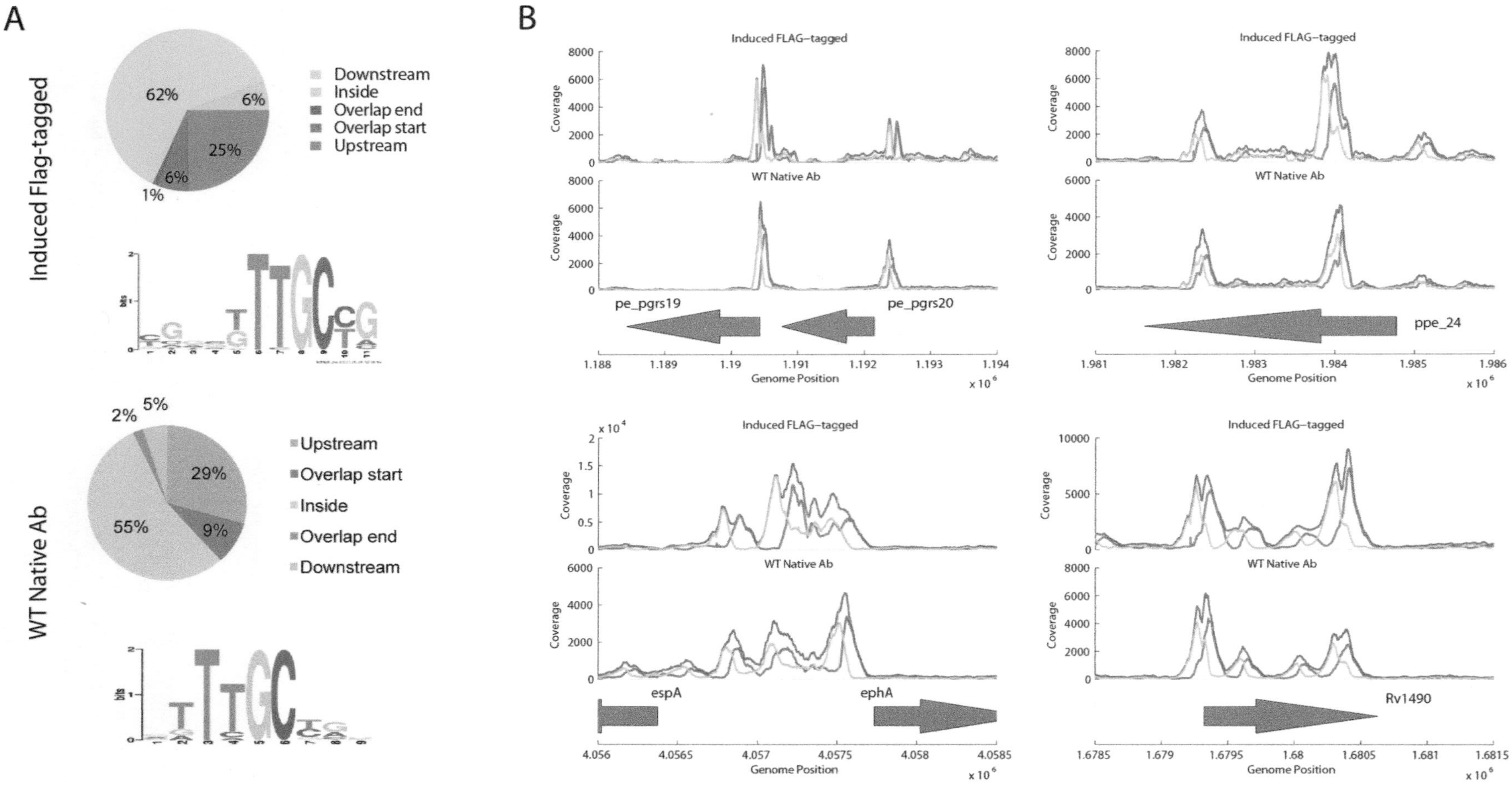

Figure 6 Independent replication of EspR binding sites. The ChIP-Seq experiment performed with the native EspR antibody (31) compares well to the ChIP-Seq with the inducible promoter. (A) Binding sites are categorized by their locations relative to target genes. Motifs and binding site categories detected by independent protocols are very similar. (B) Coverage tracks between two experiments are in concordance. doi:10.1128/microbiolspec.MGM2-0035-2013.f6

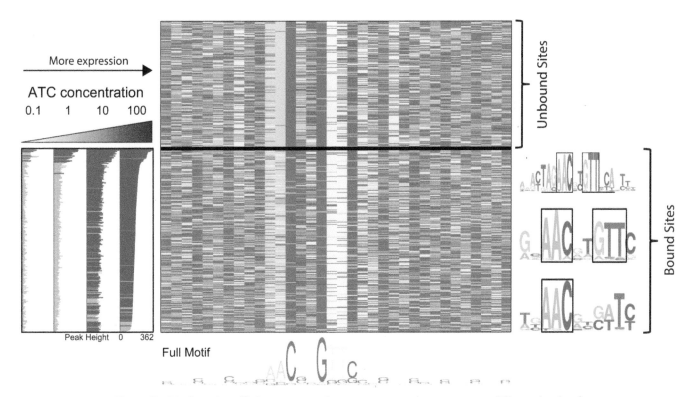

Figure 7 Binding site affinity corresponds to sequence and occupancy at different levels of expression. For transcription factor KstR, the heat map shows experimentally detected binding sites at the bottom (bound sites) and binding sites found by sequence similarity but not experimentally at the top (unbound strong motifs). Each row is a binding site, and each column is a particular position of the binding motif. The KstR binding motif describing all sites in the heat map is shown at the bottom of the figure (full motif). Binding site coverage is shown in four bar plots, corresponding to four induction levels as indicated by Atc concentration. As shown by the arrows, high-coverage binding sites correspond to a wide high-affinity motif, while low-coverage sites correspond to a degraded version of the same motif. doi:10.1128/microbiolspec.MGM2-0035-2013.f7

(ii) involves much more weak binding than was previously appreciated.

NONPROMOTER TF BINDING

The canonical model of bacterial transcriptional initiation focuses on the role of binding in the proximal promoter region. Binding in this region enables TFs to facilitate or directly block the recruitment of the polymerase to the transcription start site. Data from ChIP-Seq mapping of over 119 TFs in *M. tuberculosis* confirms that binding in upstream intergenic regions is enriched over what would be expected by chance. But surprisingly, binding in this region is the exception. As shown in Fig. 8, binding to intergenic regions represents less than 40% of the binding events for any TF. The majority of binding events occur outside of upstream intergenic regions. This has been confirmed by

ChIP-Seq of TFs in both *E. coli* and *Salmonella* (data not shown).

A number of explanations for this observation are conceivable. The most straightforward explanation is the presence of errors in the annotation of coding regions. The majority of genes in all prokaryotic genomes are computational predictions that are known to be error-prone in predicting start codons. In some cases, binding sites that occur at the 5′ ends of annotated coding regions may thus reflect intergenic binding relative to an actual downstream start codon. Another explanation is the well-known fact that promoter regions are not strictly limited to intergenic regions. Although canonical promoter signals are enriched in intergenic regions and may be selected against in genic regions (37, 38), many examples of promoter regions occurring in coding regions have been described in prokaryotes (39–43).

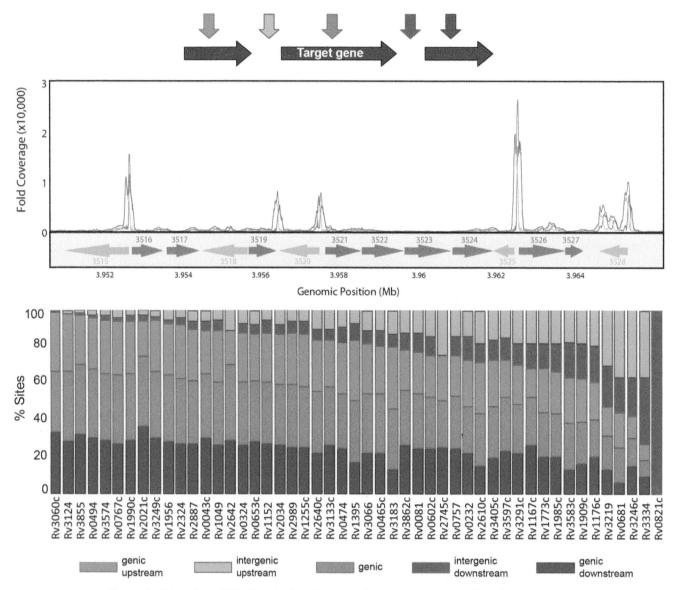

Figure 8 Diversity of binding site locations. Binding sites are assigned to five categories depending on their location relative to the target. The top figure shows a few binding sites located in divergent and covergent areas of the genome. The bottom figure shows the percentage of binding sites in each category for 49 TFs.
doi:10.1128/microbiolspec.MGM2-0035-2013.f8

Binding outside of promoter regions may also reflect long-range interactions between binding sites and the proximal promoter. There is substantial evidence in the literature that functional binding in prokaryotes can and does occur at much larger distances from promoters (40, 44–54). Several examples are known for *M. tuberculosis* in particular. The regulation of GroEL2 by Rv2034 described above occurs from a binding site 746 bp upstream of the start codon for the gene (30). In another example, maximal expression of the espA gene requires binding by EspR to the *espA* activating region located between 1,004 and 884 bp upstream (55). A deletion bringing the *espA* activating region as close as ~200 nucleotides to the promoter abolishes its function.

It is also possible that TF binding sites in prokaryotes may play roles beyond the classical activation or repression of transcriptional regulation. In eukaryotes, TF binding has been shown to modulate higher-order DNA packaging and accessibility through the modulation of chromatin structure (56). Although bacteria lack histones, proteins termed nucleoid-associated proteins (NAPs) have been described in prokaryotes (57–60) that alter the degree of compaction, looping, and DNA supercoiling of bacterial chromosomes through interactions that bend, wrap, or bridge DNA (57). Through such interactions, NAPs can modulate gene expression. Recently, the distinction between proteins that modulate DNA structure and proteins that regulate transcription has blurred (57).

This is well illustrated by the example of EspR (Fig. 6). ChIP-Seq of EspR revealed binding to over 165 loci in the *M. tuberculosis* genome, distributed nearly equally between intergenic and genic regions. Moreover, EspR binding data display substantial overlap with the binding sites for Lsr2, a known nucleoid associate protein in *M. tuberculosis* (8, 61, 62). EspR contains a helix-turn-helix DNA binding domain typical of many TFs and a C-terminal domain that mediates dimerization (63). Structural studies have led to a model in which EspR acts as a dimer of dimers in which each helix-turn-helix domain in a dimer can bind to distantly separated binding sites (64). This model was corroborated by atomic force microscopy that revealed DNA bending and DNA loop formation in conjunction with EspR binding. Together these data led to the hypothesis that EspR does not behave like a traditional TF but instead regulates transcription globally like a NAP through long-range interactions and DNA structure modifications (31). The results of the large-scale mapping of *M. tuberculosis* TFs suggest, however, that EspR may not be altogether unusual. Rather, considered in the context of the diverse binding of the majority of TFs in both prokaryotes and eukaryotes, these results suggest the possibility that other regulators may also act to modulate DNA structure on either a local or global scale.

EXTENSIVE TF BINDING

A nearly universal finding of TF ChIP-Seq studies is the unexpectedly large number of binding sites that are found for many, though not all, TFs. Even well-studied TFs can reveal a large number of novel sites. For example, in the case of DosR and KstR, 47 and 27 sites, respectively, had been associated with these factors. ChIP-Seq mapping identified 361 novel sites for DosR and 310 novel sites for KstR. Across all TFs mapped, the number of binding sites associated with each factor can be roughly fit to a power law distribution ($p[k] \sim k^{-x}$). This pattern reveals that there are some TFs that have numerous binding sites and thus likely interact with many genes. TFs of this sort have been termed "hubs" in the regulatory network literature. In contrast, most other TFs have fewer binding sites, and there is a long tail of TFs that have only a few binding sites. In some examples, we have observed a TF to bind only to its own promoter. Such cases may be explained by the TF modulating the expression dynamics of the operon in which it resides (65) or may reflect regulation of neighboring genes through divergent promoters.

One of the surprises to come out of the first set of 50 TFs mapped using ChIP-Seq in *M. tuberculosis* was the presence of Rv0081 as a hub TF. Rv0081 is the only other TF that is part of the initial hypoxic response (19, 23), but it has not been well studied. Rv0081 binds at 560 regions with 880 predicted binding locations, and its overexpression differentially regulates over 400 genes equally split between activation and repression. Prior to mapping, there were few clues that Rv0081 might have such a large regulatory role. Its identification as a hub, and the more recent identification of other previously unknown hubs, is one benefit of an unbiased and comprehensive approach to mapping *M. tuberculosis* TFs.

For TFs with many binding sites, the distribution of binding affinities often follows a common broad pattern. There is a set of binding peaks with higher apparent affinity and a gradual long tail of binding sites with weaker affinity. It should be cautioned that there are numerous variations within this theme. There are examples of TFs for which a clear set of strong peaks can be differentiated from the set of weak peaks. In other cases, there is simply a gradual decrease in affinities across binding sites. Many other patterns can also be seen, but a common element to most of these patterns is the presence of numerous weaker binding sites.

Extensive weak binding has been observed for TFs in every eukaryotic organism in which ChIP-Seq mapping of TFs has been performed (34, 56, 66–72). The data from *M. tuberculosis* (1) extends this observation to prokaryotes and indicates that extensive weak binding may be a general property of TFs. The physiological significance of this weak binding is extensively debated. As described above, substantial data support the contention that weak binding represents sequence-specific binding sites with true affinity for the corresponding TF. The question remains, however, as to whether such binding sites have physiological significance.

There are specific examples of low-affinity binding sites with confirmed regulatory roles. For both DosR and KstR, although the majority of known regulatory interactions correspond to the strongest binding peaks, many known interactions correspond to weak peaks. For KstR in particular, the majority of novel weaker binding sites have greater affinity than the weakest-affinity binding site for a previously identified regulatory interaction. The work by Gao and colleagues on Rv2034 provides additional examples (29, 30). The confirmed binding of Rv2034 to the DosR promoter is one of the weaker binding sites for Rv2034 in the ChIP-Seq data (1.4% of the maximal binding peak). Rv2034 was also shown to regulate PhoP and bind to its promoter in EMSA experiments, and ChIP-Seq mapping identified a corresponding weakly enriched binding region at the far threshold for calling a binding site.

More broadly, the analysis of potential regulatory interactions for the first set of 50 TFs mapped by the consortium effort revealed a correspondence between binding affinity and the ability to assign a regulatory effect. By integrating expression data derived from the induction of each TF with the ChIP-Seq binding data for the corresponding TF, we sought to identify binding sites that mediated strong regulatory effects (1). This analysis was able to assign putative regulatory roles for 25% of binding sites, a number in line with analyses in previous systems (34). Of particular interest, stronger binding sites were more often associated with regulation than weaker sites. This result suggests a correlation between binding affinity and regulatory impact.

This is consistent with an analysis of binding sites detected in yeast using ChIP-ChIP (68). Consistent with ChIP-Seq data in other organisms, ChIP-ChIP in yeast revealed that weak binding sites likely represented the majority of binding events for most TFs. The substantial noise associated with ChIP-ChIP compared to ChIP-Seq necessitates caution in the analyses of these data. Yet despite this, a clear relationship was observed between the predicted binding energy of promoters for different TFs and the regulatory effect of perturbing that TF. Most notably, this trend was observed even for promoters whose effect fell below standard significance thresholds.

Together with the results shown in Fig. 7, these data suggest a more analog role of TF binding and regulation than typically considered. Rather than a TF modulating the expression of a regulon in a binary manner—in which all regulon genes are bound and induced/repressed in an on/off manner—the analog view suggests a continuum of binding and regulatory impact (68, 69). At one end, very-high-affinity sites may be bound at even low TF concentration and have strong regulatory effects. At the other end, weaker affinity sites may be more selectively bound or only bound in a fraction of cells, because TF concentrations are varied. These sites may have weaker regulatory effects or may serve to fine-tune other regulatory interactions. For example, substantial evidence exists that weak sites may

play a significant role in modulating the effects of other binding sites through cooperative interactions (73). The distribution of binding site affinity may also play a role in sculpting the relative timing of expression of genes within a regulon (74, 75).

These data also suggest why strong sites are more likely to have been previously identified, since most methods for detecting regulation are designed to find substantial changes in gene expression. A common criterion is to select only genes that display a greater than 2-fold change in expression, yet the data above suggest that physiological binding sites may have relevant regulatory effects that are below this threshold. More generally, it is unlikely that any instrumentation-based threshold would happen to match the biological thresholds for physiological relevance for all TFs, even if such biological thresholds existed.

Of course, the null hypothesis is that many of these weak sites may not be functional in any meaningful or detectable sense. They may reflect the limitations of tuning the binding affinity of TFs to exclude spurious sites that are likely to arise in a large genome (although the existence of TFs that only bind a few sites argues this is not impossible). They may also lead to noise in transcription that is simply filtered out by other mechanisms (i.e., through the degradation of unstable transcripts arising from nonstandard transcription start sites). Many hypotheses are possible. The growing data from ChIP-Seq in both eukaryotes and now prokaryotes makes this an important and outstanding question in gene regulation research.

TBDB AND DATA AVAILABILITY

ChIP-Seq data for *M. tuberculosis* TFs generated by the NIAID TB Systems Biology Project have been released through the Tuberculosis Database (TBDB.org) (Fig. 9). TBDB is an online database providing integrated access to genome sequence, expression data, literature curation, and systems biology data for *M. tuberculosis* and related genomes (76). TBDB currently houses genome assemblies for numerous strains of *M. tuberculosis* as well as assemblies for over 20 strains related to *M. tuberculosis* and useful for comparative analysis. It also houses resequencing data for over 31 *M. tuberculosis* strains selected as part of the *M. tuberculosis* Phylogeographic Diversity Sequencing Project. These data provide a global view of the genomic diversity of *M. tuberculosis* at the level of SNPs and indels. TBDB stores pre- and postpublication gene-expression data from *M. tuberculosis* and its close relatives, including over 3,000 *M. tuberculosis* microarrays, 95 RT-PCR

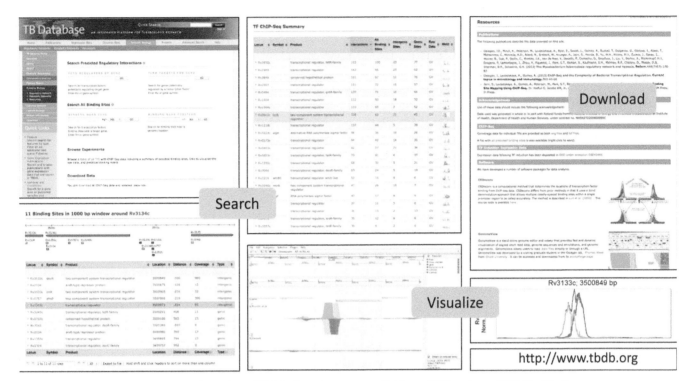

Figure 9 *M. tuberculosis* TF binding data are available at TBDatabase (TBDB). Binding data for 50 TFs generated by the NIAID-funded TB Systems Biology Project have been integrated with the genome sequence and annotation of *M. tuberculosis* and released at TBDB.org. Selected screen shots show online tools available for searching, browsing, and downloading these data. doi:10.1128/microbiolspec.MGM2-0035-2013.f9

datasets, 2,700 microarrays for human and mouse tuberculosis-related experiments, and 260 arrays for *Streptomyces coelicolor*. In addition, metabolic reconstructions have been performed on all organisms in the site, and these models are hosted as Biocyc Pathway/ Genome databases (http://biocyc.org/) in TBDB. To enable wide use of these data, TBDB provides a suite of tools for searching, browsing, analyzing, and downloading the data.

TBDB also provides a growing set of tools for utilizing the *M. tuberculosis* ChIP-Seq data (Fig. 9). Through TBDB, users can search for regulatory binding sites by regulator, by target, or by genomic coordinate. Users can also browse a regulatory network constructed from these data. From the results of any of these searches, users may view the regulatory network for the gene of interest, select and view raw ChIP-Seq data in the dynamic real-time genome browser GenomeView (77), view the summary page for each experiment, or view static images of the ChIP-Seq peak data. Users may browse experiments directly and view the entire genome for each experiment. Users can also download all raw data (Fig. 9).

APPENDIX

Day 1

Protein-DNA cross-linking

- Grow 50 ml of *M. tuberculosis* cells to mid log or OD_{600} = 0.5 to 0.6.
- Add formaldehyde to 1% (final concentration): 1.35 ml of 37% formaldehyde to 50 ml of culture.
- Rock at room temperature (RT) for 30 min.
- Stop cross-linking by adding final 250 mM of glycine (4.17 ml of 3 M glycine to 50 ml); rock at RT for 15 min.
- Spin down the cells at 4°C for 10 min and 3,000 rpm.
- Wash two times with ice-cold 1 × PBS (50 ml) and spin down the cells at 4°C for 10 min and 3,000 rpm.
- Resuspend cells in 0.5 to 0.6 ml of freshly made buffer 1 + PI (protease inhibitor, Complete Mini).
- Lyse the cells using MagNA Lyser four times at 4,000 rpm and 45 sec. Keep the samples on ice between cycles.

- Spin down the cells at 4°C for 10 min and 13,000 rpm.
- Transfer the lysate to Covaris tubes.

Cell lysis and DNA shearing

- Do Covaris for 18 min for lysate, or 25 min if you skip the above three steps (amplitude = 20%, intensity = 5, cycles/burst = 200).
- Spin down the lysate for 10 min at 13,000 rpm and 4°C.
- Transfer supernatant to a new tube.
- Adjust salt concentration to:
- 10 mM Tris HCl pH 8 (Add 10 μl of 1 M Tris/ml of sample.)
- 150 mM NaCl (Add 30 μl of 5 M NaCl/ml of sample.)
- 0.1% NP40 (Add 10 μl of 10% NP40/ml of sample.)
- Invert tubes to mix.

Immunoprecipitation

- Add 5 μl of anti-FLAG antibody to the lysate.
- Incubate overnight at 4°C on the rotating platform.

Day 2

- Rinse 50 μl of protein-G agarose beads (end of the tip [200 μl] should be cut by scissors for taking beads) with 1 ml of IPP150 buffer.
- Spin for 2 min at 2,000 × g to pellet the beads and discard the buffer.
- Transfer the lysate-Ab to the tube containing beads.
- Incubate for 2 hours at RT (0.5 hour at 4°C and 1.5 hours at RT) on a rotating platform.

Washes and elution of protein-DNA complexes

- Wash at least five times with 1 ml of IPP150 buffer. (Invert the tubes for 2 min between the washes and spin for 2 min at 2,000 × g to pellet the beads.)
- Wash at least two times with 1 ml of 1 × TE. (Invert the tubes for 2 min between the washes and spin for 2 min at 2,000 × g to pellet the beads.)
- Add 150 μl of "elution from the beads" buffer.
- Incubate at 65°C for 15 min.
- Spin for 5 min at 2,000 × g to pellet the beads; keep the supernatant (first elution).
- Add 100 μl of 1 × TE + 1% SDS (has to be fresh) to the pellet and incubate at 65°C for 5 min for second elution.
- Spin for 5 min at 2,000 × g to pellet the beads, keep the supernatant, and pool the samples (first and second elution).

Reverse cross-linking

- Add proteinase K (1 mg/ml final concentration) and incubate at 37°C for 1 to 2 hours.
- Incubate at 65°C overnight.

Day 3

DNA purification

- Purify the DNA with the PCR purification kit from Qiagen.
- Elute with 30 μl of EB.

Library preparation

The library is prepared using the standard Illumina protocol (http://www.broadinstitute.org/annotation/tbsysbio /Protocols/ChIPSeq_Protocol_2.pdf).

Specific Buffers

Buffer 1 and buffer 1 + PI[a]

Composition	Stock solutions	Vol to add for 40 ml	Vol to add for 10 ml
20 mM KHEPES pH 7.9	1 M KHEPES	800 μl	200 μl
50 mM KCl	1 M KCl	2,000 μl	500 μl
0.5 mM DTT	0.5 M DTT	40 μl	10 μl
10% glycerol	100% glycerol	4 ml	1 ml
	+ Water	37.2 ml	8.290 ml

[a]To make buffer 1 + PI, take 2 ml of buffer 1 and add a half tablet of Mini Protease Inhibitor.

IPP150 buffer

Composition	Stock solutions	Vol to add for 250 ml	Vol to add for 50 ml
10 mM Tris-HCl pH 8	1 M	2.5 ml	500 μl
150 mM NaCl	5 M	7.5 ml	1.5 ml
0.1% NP40	10%	2.5 ml	500 μl
	+ Water	237.5 ml	47.5 ml

"Elution from beads" buffer

Composition	Stock solutions	Vol to add for 10 ml
mM Tris-HCl pH 8	1 M	500 μl
10 mM EDTA	0.5 M	200 μl
1% SDS	20%	500 μl
	+ Water	8.8 ml

Citation. Jaini S, Lyubetskaya A, Gomes A, Peterson M, Park ST, Raman S, Schoolnik G, Galagan J. 2014. Transcription factor binding site mapping using ChIP-Seq. Microbiol Spectrum 2(2):MGM2-0035-2013.

References

1. Galagan JE, Minch K, Peterson M, Lyubetskaya A, Azizi E, Sweet L, Gomes A, Rustad T, Dolganov G, Glotova I, Abeel T, Mahwinney C, Kennedy AD, Allard R, Brabant W, Krueger A, Jaini S, Honda B, Yu WH, Hickey MJ, Zucker J, Garay C, Weiner B, Sisk P, Stolte C, Winkler JK, Van de Peer Y, Iazzetti P, Camacho D, Dreyfuss J, Liu Y, Dorhoi A, Mollenkopf HJ, Drogaris P, Lamontagne J, Zhou Y, Piquenot J, Park ST, Raman S, Kaufmann SH, Mohney RP, Chelsky D, Moody DB, Sherman DR, Schoolnik GK. 2013. The *Mycobacterium tuberculosis* regulatory network and hypoxia. *Nature* **499:**178–183.

2. Robertson G, Hirst M, Bainbridge M, Bilenky M, Zhao Y, Zeng T, Euskirchen G, Bernier B, Varhol R, Delaney A, Thiessen N, Griffith OL, He A, Marra M, Snyder M, Jones S. 2007. Genome-wide profiles of STAT1 DNA association using chromatin immunoprecipitation and massively parallel sequencing. *Nat Methods* **4:**651–657.

3. Mikkelsen TS, Ku M, Jaffe DB, Issac B, Lieberman E, Giannoukos G, Alvarez P, Brockman W, Kim TK, Koche RP, Lee W, Mendenhall E, O'Donovan A, Presser A, Russ C, Xie X, Meissner A, Wernig M, Jaenisch R, Nusbaum C, Lander ES, Bernstein BE. 2007. Genome-wide maps of chromatin state in pluripotent and lineage-committed cells. *Nature* **448:**553–560.

4. Johnson DS, Mortazavi A, Myers RM, Wold B. 2007. Genome-wide mapping of in vivo protein-DNA interactions. *Science* **316:**1497–1502.

5. Valouev A, Johnson DS, Sundquist A, Medina C, Anton E, Batzoglou S, Myers RM, Sidow A. 2008. Genome-wide analysis of transcription factor binding sites based on ChIP-Seq data. *Nat Methods* **5:**829–834.

6. Li H, Ruan J, Durbin R. 2008. Mapping short DNA sequencing reads and calling variants using mapping quality scores. *Genome Res* **18:**1851–1858.

7. Li H, Handsaker B, Wysoker A, Fennell T, Ruan J, Homer N, Marth G, Abecasis G, Durbin R, 1000 Genome Project Data Processing Subgroup. 2009. The Sequence Alignment/Map format and SAMtools. *Bioinformatics* **25:**2078–2079.

8. Gordon BR, Li Y, Wang L, Sintsova A, van Bakel H, Tian S, Navarre WW, Xia B, Liu J. 2010. Lsr2 is a nucleoid-associated protein that targets AT-rich sequences and virulence genes in *Mycobacterium tuberculosis. Proc Natl Acad Sci USA* **107:**5154–5159.

9. Oppenheim AV, Willsky AS, Nawab SH. 1997. *Signals & Systems,* 2nd ed. Prentice Hall, Upper Saddle River, NJ.

10. Levin A, Weiss Y, Durand F, Freeman WT. 2011. Understanding blind deconvolution algorithms. *IEEE Trans Pattern Anal* **33:**2354–2367.

11. Lun DS, Sherrid A, Weiner B, Sherman DR, Galagan JE. 2009. A blind deconvolution approach to high-resolution mapping of transcription factor binding sites from ChIP-seq data. *Genome Biol* **10:**R142.

12. (Reference deleted.)

13. Chauhan S, Tyagi JS. 2008. Cooperative binding of phosphorylated DevR to upstream sites is necessary and sufficient for activation of the *Rv3134c-devRS* operon in *Mycobacterium tuberculosis:* implication in the induction of DevR target genes. *J Bacteriol* **190:**4301–4312.

14. Mikkelsen TS, Ku M, Jaffe DB, Issac B, Lieberman E, Giannoukos G, Alvarez P, Brockman W, Kim T-K, Koche RP, Lee W, Mendenhall E, O'Donovan A, Presser A, Russ C, Xie X, Meissner A, Wernig M, Jaenisch R, Nusbaum C, Lander ES, Bernstein BE. 2007. Genome-wide maps of chromatin state in pluripotent and lineage-committed cells. *Nature* **448:**553–560.

15. Ehrt S, Schnappinger D. 2006. Controlling gene expression in mycobacteria. *Future Microbiol* **1:**177–184.

16. Ehrt S, Guo XV, Hickey CM, Ryou M, Monteleone M, Riley LW, Schnappinger D. 2005. Controlling gene expression in mycobacteria with anhydrotetracycline and Tet repressor. *Nucleic Acids Res* **33:**e21.

17. Klotzsche M, Ehrt S, Schnappinger D. 2009. Improved tetracycline repressors for gene silencing in mycobacteria. *Nucleic Acids Res* **37:**1778.

18. Guo XV, Monteleone M, Klotzsche M, Kamionka A, Hillen W, Braunstein M, Ehrt S, Schnappinger D. 2007. Silencing *Mycobacterium smegmatis* by using tetracycline repressors. *J Bacteriol* **189:**4614–4623.

19. Park HD, Guinn KM, Harrell MI, Liao R, Voskuil MI, Tompa M, Schoolnik GK, Sherman DR. 2003. Rv3133c/dosR is a transcription factor that mediates the hypoxic response of *Mycobacterium tuberculosis. Mol Microbiol* **48:**833–843.

20. Saini DK, Malhotra V, Dey D, Pant N, Das TK, Tyagi JS. 2004. DevR-DevS is a bona fide two-component system of *Mycobacterium tuberculosis* that is hypoxia-responsive in the absence of the DNA-binding domain of DevR. *Microbiology* **150:**865–875.

21. Flores Valdez MA, Schoolnik GK. 2010. DosR-regulon genes induction in *Mycobacterium bovis* BCG under aerobic conditions. *Tuberculosis (Edinb)* **90:**197–200.

22. Chauhan S, Sharma D, Singh A, Surolia A, Tyagi JS. 2011. Comprehensive insights into *Mycobacterium tuberculosis* DevR (DosR) regulon activation switch. *Nucleic Acids Res* **39:**7400–7414.

23. Sherman DR, Voskuil M, Schnappinger D, Liao R, Harrell MI, Schoolnik GK. 2001. Regulation of the *Mycobacterium tuberculosis* hypoxic response gene encoding alpha -crystallin. *Proc Natl Acad Sci USA* **98:** 7534–7539.

24. Vasudeva-Rao HM, McDonough KA. 2008. Expression of the *Mycobacterium tuberculosis* acr-coregulated genes from the DevR (DosR) regulon is controlled by multiple levels of regulation. *Infect Immun* **76:**2478–2489.

25. Kendall SL, Burgess P, Balhana R, Withers M, Ten Bokum A, Lott JS, Gao C, Uhia-Castro I, Stoker NG. 2010. Cholesterol utilization in mycobacteria is controlled by two TetR-type transcriptional regulators: kstR and kstR2. *Microbiology* **156:**1362–1371.

26. Kendall SL, Withers M, Soffair CN, Moreland NJ, Gurcha S, Sidders B, Frita R, Ten Bokum A, Besra GS, Lott JS, Stoker NG. 2007. A highly conserved transcriptional repressor controls a large regulon involved in lipid degradation in *Mycobacterium smegmatis* and *Mycobacterium tuberculosis*. *Mol Microbiol* **65**:684–699.

27. Nesbitt NM, Yang X, Fontan P, Kolesnikova I, Smith I, Sampson NS, Dubnau E. 2010. A thiolase of *Mycobacterium tuberculosis* is required for virulence and production of androstenedione and androstadienedione from cholesterol. *Infect Immun* **78**:275–282.

28. Uhia I, Galan B, Medrano FJ, Garcia JL. 2011. Characterization of the KstR-dependent promoter of the first step of cholesterol degradative pathway in *Mycobacterium smegmatis*. *Microbiology* **157**:2670–2680.

29. Gao CH, Yang M, He ZG. 2012. Characterization of a novel ArsR-like regulator encoded by Rv2034 in *Mycobacterium tuberculosis*. *PLoS One* **7**:e36255.

30. Gao CH, Yang M, He ZG. 2011. An ArsR-like transcriptional factor recognizes a conserved sequence motif and positively regulates the expression of *phoP* in mycobacteria. *Biochem Biophys Res Commun* **411**:726–731.

31. Blasco B, Chen JM, Hartkoorn R, Sala C, Uplekar S, Rougemont J, Pojer F, Cole ST. 2012. Virulence regulator EspR of *Mycobacterium tuberculosis* is a nucleoid-associated protein. *PLoS Pathog* **8**:e1002621.

32. Fordyce PM, Gerber D, Tran D, Zheng J, Li H, DeRisi JL, Quake SR. 2010. De novo identification and biophysical characterization of transcription-factor binding sites with microfluidic affinity analysis. *Nat Biotechnol* **28**:970–975.

33. Fernandez PC, Frank SR, Wang L, Schroeder M, Liu S, Greene J, Cocito A, Amati B. 2003. Genomic targets of the human c-Myc protein. *Genes Dev* **17**:1115–1129.

34. Farnham PJ. 2009. Insights from genomic profiling of transcription factors. *Nat RevGenet* **10**:605–616.

35. Wade JT, Reppas NB, Church GM, Struhl K. 2005. Genomic analysis of LexA binding reveals the permissive nature of the *Escherichia coli* genome and identifies unconventional target sites. *Genes Dev* **19**:2619–2630.

36. Galagan J, Lyubetskaya A, Gomes A. 2013. ChIP-Seq and the complexity of bacterial transcriptional regulation. *Curr Top Microbiol Immunol* **363**:43–68.

37. Huerta AM, Francino MP, Morett E, Collado-Vides J. 2006. Selection for unequal densities of sigma70 promoter-like signals in different regions of large bacterial genomes. *PLoS Genet* **2**:e185.

38. Froula JL, Francino MP. 2007. Selection against spurious promoter motifs correlates with translational efficiency across bacteria. *PLoS One* **2**:e745.

39. Koide T, Reiss DJ, Bare JC, Pang WL, Facciotti MT, Schmid AK, Pan M, Marzolf B, Van PT, Lo FY, Pratap A, Deutsch EW, Peterson A, Martin D, Baliga NS. 2009. Prevalence of transcription promoters within archaeal operons and coding sequences. *Mol Syst Biol* **5**:285.

40. Czaplewski LG, North AK, Smith MC, Baumberg S, Stockley PG. 1992. Purification and initial characterization of AhrC: the regulator of arginine metabolism genes in *Bacillus subtilis*. *Mol Microbiol* **6**:267–275.

41. Mullin DA, Newton A. 1993. A sigma 54 promoter and downstream sequence elements *ftr2* and *ftr3* are required for regulated expression of divergent transcription units *flaN* and *flbG* in *Caulobacter crescentus*. *J Bacteriol* **175**:2067–2076.

42. Madan Babu M, Teichmann SA. 2003. Functional determinants of transcription factors in *Escherichia coli*: protein families and binding sites. *Trends Genet* **19**:75–79.

43. Collado-Vides J, Magasanik B, Gralla JD. 1991. Control site location and transcriptional regulation in *Escherichia coli*. *Microbiol Rev* **55**:371–394.

44. Dunn TM, Hahn S, Ogden S, Schleif RF. 1984. An operator at −280 base pairs that is required for repression of araBAD operon promoter: addition of DNA helical turns between the operator and promoter cyclically hinders repression. *Proc Natl Acad Sci USA* **81**:5017–5020.

45. Wedel A, Weiss DS, Popham D, Droge P, Kustu S. 1990. A bacterial enhancer functions to tether a transcriptional activator near a promoter. *Science* **248**:486–490.

46. Dandanell G, Valentin-Hansen P, Larsen JE, Hammer K. 1987. Long-range cooperativity between gene regulatory sequences in a prokaryote. *Nature* **325**:823–826.

47. Belitsky BR, Sonenshein AL. 1999. An enhancer element located downstream of the major glutamate dehydrogenase gene of *Bacillus subtilis*. *Proc Natl Acad Sci USA* **96**:10290–10295.

48. Oehler S, Eismann ER, Kramer H, Muller-Hill B. 1990. The three operators of the *lac* operon cooperate in repression. *EMBO J* **9**:973–979.

49. Narang A. 2007. Effect of DNA looping on the induction kinetics of the *lac* operon. *J Theor Biol* **247**:695–712.

50. Flashner Y, Gralla JD. 1988. Dual mechanism of repression at a distance in the *lac* operon. *Proc Natl Acad Sci USA* **85**:8968–8972.

51. Ninfa AJ, Reitzer LJ, Magasanik B. 1987. Initiation of transcription at the bacterial *glnAp2* promoter by purified E. coli components is facilitated by enhancers. *Cell* **50**:1039–1046.

52. Reitzer LJ, Magasanik B. 1986. Transcription of *glnA* in E. coli is stimulated by activator bound to sites far from the promoter. *Cell* **45**:785–792.

53. Ueno-Nishio S, Mango S, Reitzer LJ, Magasanik B. 1984. Identification and regulation of the *glnL* operator-promoter of the complex *glnALG* operon of *Escherichia coli*. *J Bacteriol* **160**:379–384.

54. Ueno-Nishio S, Backman KC, Magasanik B. 1983. Regulation at the *glnL*-operator-promoter of the complex *glnALG* operon of *Escherichia coli*. *J Bacteriol* **153**:1247–1251.

55. Hunt DM, Sweeney NP, Mori L, Whalan RH, Comas I, Norman L, Cortes T, Arnvig KB, Davis EO, Stapleton MR, Green J, Buxton RS. 2012. Long-range transcriptional control of an operon necessary for virulence-critical ESX-1 secretion in *Mycobacterium tuberculosis*. *J Bacteriol* **194**:2307–2320.

56. Cao Y, Yao Z, Sarkar D, Lawrence M, Sanchez GJ, Parker MH, MacQuarrie KL, Davison J, Morgan MT, Ruzzo WL, Gentleman RC, Tapscott SJ. 2010. Genome-wide MyoD binding in skeletal muscle cells: a potential for broad cellular reprogramming. *Dev Cell* **18**:662–674.

57. Dillon SC, Dorman CJ. 2010. Bacterial nucleoid-associated proteins, nucleoid structure and gene expression. *Nat Rev Microbiol* 8:185–195.

58. Browning DF, Grainger DC, Busby SJ. 2010. Effects of nucleoid-associated proteins on bacterial chromosome structure and gene expression. *Curr Opin Microbiol* 13:773–780.

59. Rimsky S, Travers A. 2011. Pervasive regulation of nucleoid structure and function by nucleoid-associated proteins. *Curr Opin Microbiol* 14:136–141.

60. Wang W, Li GW, Chen C, Xie XS, Zhuang X. 2011. Chromosome organization by a nucleoid-associated protein in live bacteria. *Science* 333:1445–1449.

61. Colangeli R, Haq A, Arcus VL, Summers E, Magliozzo RS, McBride A, Mitra AK, Radjainia M, Khajo A, Jacobs WR Jr, Salgame P, Alland D. 2009. The multifunctional histone-like protein Lsr2 protects mycobacteria against reactive oxygen intermediates. *Proc Natl Acad Sci USA* 106:4414–4418.

62. Colangeli R, Helb D, Vilcheze C, Hazbon MH, Lee CG, Safi H, Sayers B, Sardone I, Jones MB, Fleischmann RD, Peterson SN, Jacobs WR Jr, Alland D. 2007. Transcriptional regulation of multi-drug tolerance and antibiotic-induced responses by the histone-like protein Lsr2 in *M. tuberculosis*. *PLoS Pathog* 3:e87.

63. Rosenberg OS, Dovey C, Tempesta M, Robbins RA, Finer-Moore JS, Stroud RM, Cox JS. 2011. EspR, a key regulator of *Mycobacterium tuberculosis* virulence, adopts a unique dimeric structure among helix-turn-helix proteins. *Proc Natl Acad Sci USA* 108:13450–13455.

64. Blasco B, Stenta M, Alonso-Sarduy L, Dietler G, Peraro MD, Cole ST, Pojer F. 2011. Atypical DNA recognition mechanism used by the EspR virulence regulator of *Mycobacterium tuberculosis*. *Mol Microbiol* 82:251–264.

65. Rosenfeld N, Elowitz MB, Alon U. 2002. Negative autoregulation speeds the response times of transcription networks. *J Mol Biol* 323:785–793.

66. Robertson G, Hirst M, Bainbridge M, Bilenky M, Zhao Y, Zeng T, Euskirchen G, Bernier B, Varhol R, Delaney A, Thiessen N, Griffith OL, He A, Marra M, Snyder M, Jones S. 2007. Genome-wide profiles of STAT1 DNA association using chromatin immunoprecipitation and massively parallel sequencing. *Nat Methods* 4:651–657.

67. MacQuarrie KL, Fong AP, Morse RH, Tapscott SJ. 2011. Genome-wide transcription factor binding: beyond direct target regulation. *Trends Genet* 27:141–148.

68. Tanay A. 2006. Extensive low-affinity transcriptional interactions in the yeast genome. *Genome Res* 16:962–972.

69. Rhee HS, Pugh BF. 2011. Comprehensive genome-wide protein-DNA interactions detected at single-nucleotide resolution. *Cell* 147:1408–1419.

70. Zhong M, Niu W, Lu ZJ, Sarov M, Murray JI, Janette J, Raha D, Sheaffer KL, Lam HY, Preston E, Slightham C, Hillier LW, Brock T, Agarwal A, Auerbach R, Hyman AA, Gerstein M, Mango SE, Kim SK, Waterston RH, Reinke V, Snyder M. 2010. Genome-wide identification of binding sites defines distinct functions for *Caenorhabditis elegans* PHA-4/FOXA in development and environmental response. *PLoS Genet* 6:e1000848.

71. Zeitlinger J, Zinzen RP, Stark A, Kellis M, Zhang H, Young RA, Levine M. 2007. Whole-genome ChIP-chip analysis of Dorsal, Twist, and Snail suggests integration of diverse patterning processes in the *Drosophila* embryo. *Genes Dev* 21:385–390.

72. Li XY, MacArthur S, Bourgon R, Nix D, Pollard DA, Iyer VN, Hechmer A, Simirenko L, Stapleton M, Luengo Hendriks CL, Chu HC, Ogawa N, Inwood W, Sementchenko V, Beaton A, Weiszmann R, Celniker SE, Knowles DW, Gingeras T, Speed TP, Eisen MB, Biggin MD. 2008. Transcription factors bind thousands of active and inactive regions in the *Drosophila* blastoderm. *PLoS Biol* 6:e27.

73. Gertz J, Siggia ED, Cohen BA. 2009. Analysis of combinatorial *cis*-regulation in synthetic and genomic promoters. *Nature* 457:215–218.

74. Kalir S, McClure J, Pabbaraju K, Southward C, Ronen M, Leibler S, Surette MG, Alon U. 2001. Ordering genes in a flagella pathway by analysis of expression kinetics from living bacteria. *Science* 292:2080–2083.

75. Zaslaver A, Mayo AE, Rosenberg R, Bashkin P, Sberro H, Tsalyuk M, Surette MG, Alon U. 2004. Just-in-time transcription program in metabolic pathways. *Nat Genet* 36:486–491.

76. Galagan JE, Sisk P, Stolte C, Weiner B, Koehrsen M, Wymore F, Reddy TB, Zucker JD, Engels R, Gellesch M, Hubble J, Jin H, Larson L, Mao M, Nitzberg M, White J, Zachariah ZK, Sherlock G, Ball CA, Schoolnik GK. 2010. TB database 2010: overview and update. *Tuberculosis (Edinb)* 90:225–235.

77. Abeel T, Van Parys T, Saeys Y, Galagan J, Van de Peer Y. 2012. GenomeView: a next-generation genome browser. *Nucleic Acids Res* 40:e12.

Molecular Genetics of Mycobacteria, 2nd Edition
Edited by Graham F. Hatfull and William R. Jacobs, Jr.
© 2014 American Society for Microbiology, Washington, DC
doi:10.1128/microbiolspec.MGM2-0029-2013

Kristine B. Arnvig[1]
Teresa Cortes[1]
Douglas B. Young[1]

Noncoding RNA in Mycobacteria

9

In the 15 years since the publication of the genome sequence of *Mycobacterium tuberculosis*, analysis of transcriptional regulation has provided a dominant framework for rendering genomic information into understanding of the functional biology of the organism. The powerful combination of molecular biology tools for characterization of individual genetic loci alongside the genome-wide perspective provided by microarray analysis has led to the use of transcription profiles as a surrogate for key phenotypic states implicated in pathogenesis, immunogenicity, and response to drug treatment. As technologies move toward definition of phenotypes by more direct measurement of metabolic status and high-resolution ultrastructural imaging, it is important to develop a rigorous understanding of the quantitative relationship between RNA transcript abundance and broader aspects of cellular physiology, both at the level of bulk populations and at the level of individual cells. The growing awareness—driven largely by the application of high-throughput sequencing technologies to the analysis of RNA (RNA-seq)—that bacteria transcribe much more RNA than is required for direct translation into proteins is likely to be important in this context.

Bacteria were largely excluded from debates surrounding the presence of "junk DNA" in eukaryotic genomes; prokaryotic coding sequences are tightly packed with apparently minimal intergenic space. It is now clear that junk DNA of eukaryotes is in fact transcribed into a vast repertoire of noncoding RNAs (ncRNAs), many of which play profound and pervasive roles in the regulation of gene expression. However, bacteria have become active participants as the debate moves from "junk DNA" to "junk RNA." Do these novel ncRNAs in bacteria reflect some form of transcriptional "noise," or do they provide a crucial layer of posttranscriptional regulation with a central—and perhaps manipulable—function in determining the amount of protein that is produced from an individual mRNA transcript? While "both of the above" seems intuitively likely, answering this question will require cataloguing the total noncoding transcriptome together with targeted functional studies. It is clear that computational techniques will be essential for integration of the rapidly accumulating mass of relevant genome-wide datasets.

In describing experimental and computational approaches applied to *M. tuberculosis* and other bacteria,

[1]National Institute for Medical Research, Mycobacterial Research Division, London, NW7 1AA, United Kingdom.

this article summarizes the initial attempts to elucidate the role of ncRNA in mycobacteria.

WHAT IS ncRNA?

ncRNA refers to any transcript that does not code for a protein or peptide. In bacteria this includes a group of well-studied stable RNAs with defined cellular functions, i.e., rRNA (ribosomal RNA), tRNA (transfer RNA), 4.5S RNA, RnpB, and transfer-messenger RNA (tmRNA, although a small part of this actually is coding). Distinct from this group are the regulatory RNAs, which include 5′ and 3′ untranslated regions (UTRs) linked to coding sequences, and a rapidly expanding category of *cis*- or *trans*-encoded transcripts with an ability to base-pair with mRNAs or to interact with proteins (reviewed in, e.g., references 1, 2). Members of this last category are often loosely referred to as small RNAs (sRNAs). Bacterial sRNAs are typically in the range of 50 to 500 nucleotides and hence are comparable to eukaryotic long ncRNAs. There are a few reports of smaller transcripts, i.e., equivalent to eukaryotic micro RNAs, but it remains to be determined if these have any functional significance (3, 4).

It is important to note that the terms "noncoding" and "regulatory" are not interchangeable, although there are many overlaps (Fig. 1). For example, sRNAs can be coding or noncoding and regulatory or not. Some sRNAs have been found to encode small peptides as well as regulate on an RNA level; the best-characterized examples are the SgrS RNA (227 nucleotides) found in enteric bacteria and the unusually large (514 nucleotide) RNAIII found in *Staphylococcus aureus* (reviewed by Vanderpool et al. [5]).

Finally, the definitions of what is regulatory and what is not may be ambiguous. An RNA such as tmRNA is generally considered to be a housekeeping RNA (6) but could be regarded as a stress-induced regulator of translation.

In this article we focus on the potentially regulatory RNA species, i.e., leader regions, UTRs, and antisense and intergenic sRNAs.

ncRNA in *M. tuberculosis*

The experimental identification of noncoding, potentially regulatory RNA in mycobacteria is so far limited to just over half a dozen studies, mostly focused on *M. tuberculosis* (7–14). The annotation of newly identified *M. tuberculosis* ncRNAs was inconsistent until it was addressed recently (15). In this article we use this new annotation, along with the nomenclature used in the primary publications where required for clarification.

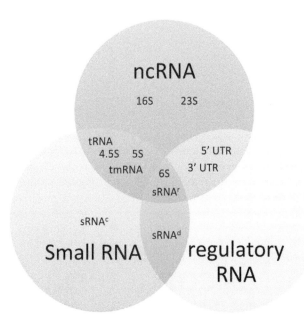

Figure 1 Venn diagram illustrating how the (nomenclature of) different types of ncRNA and regulatory RNAs overlap and, in particular, how sRNAs can be assigned to more than one category. The sRNA subcategories shown are sRNA^r (purely regulatory function), sRNA^d (dual function, i.e., regulatory potential as well as encoding small peptide), sRNA^c (purely coding, i.e., no function as ribo-regulator). Thus, sRNAs can be coding or noncoding, regulatory or not regulatory. Figure modified from reference 15. doi:10.1128/microbiolspec.MGM2-0029-2013.f1

Deep sequencing of *M. tuberculosis* RNA (RNA-seq) demonstrates that during exponential growth about a third of nonribosomal RNA in *M. tuberculosis* is noncoding and hence potentially regulatory. The proportion of ncRNA varies with the growth phase, increasing to more than half in stationary phase and during starvation (7, 16). The quantitative increase in ncRNA during stationary phase is largely due to accumulation of a single, highly abundant sRNA transcript, ncRv13661/MTS2823 (see below). Genome browsers such as Artemis facilitate the visualization of high-throughput RNA-seq data and represent a powerful way of conveying the complexity and flexibility of transcriptomic data (17).

There are significant annotation issues associated with the assignment of coding RNAs versus ncRNAs in *M. tuberculosis*. Short open reading frames (ORFs) are more frequently annotated as coding sequences in *M. tuberculosis* CDC1551 (18, 19) than in *M. tuberculosis* H37Rv (20). For example, a transcript mapping to the intergenic region (IGR) between Rv2395 and Rv2396 in H37Rv has been described as an sRNA (mcr7) (9) but as hypothetical proteins MT2466 and MT2467 in the CDC1551 genome, and was subsequently reported to encode two small acid-inducible proteins (21). Conversely, the region between Rv3661 and Rv3662c has been annotated as encoding the hypothetical protein MT3762 on the minus strand (19) and alternatively as sRNA ncRv13661/MTS2823 (predicted as mpr4) on the plus strand (7, 9). The precise location of translation start sites determines assignment of coding versus noncoding UTR sequences and is subject to ongoing revision in *M. tuberculosis* (22). Application of mass spectrometry technology for proteome analysis (e.g., references 23–25) will play an important role in resolving many ambiguities.

cis-ACTING REGULATORY ELEMENTS

cis-Regulatory RNAs (for definitions, see Glossary) include sequences flanking either end of coding transcripts, referred to as 5′ and 3′ UTRs, or those flanking noncoding transcripts, such as tRNAs and rRNAs. These make an important contribution to the total complement of ncRNA and often play important roles in the expression, processing, and stability of their cognate effector transcript.

The 3′ End

The 3′ UTR often includes the site of transcription termination, thereby providing an opportunity for regulation by modification of transcriptional read-through. Moreover, a number of studies from *Escherichia coli* suggest that canonical or L-shaped intrinsic terminators (i.e., a stem-loop followed by a stretch of U residues [26]) may serve as polyadenylation signals. Unlike the situation in eukaryotes, polyadenylated 3′ UTRs in bacteria tend to facilitate RNA turnover rather than stabilize the transcripts (reviewed by Mohanty and Kushner [27]). So far, the evidence for polyadenylation in mycobacteria is very limited (28), and we do not know how this may affect RNA stability, but it is likely that the same general rules apply as in other bacteria.

A subset of *M. tuberculosis* mRNAs have well-defined 3′ ends associated with canonical intrinsic terminators or with the mycobacterium-specific terminator, TRIT (for tuberculosis rho-independent terminator), which contains a highly conserved sequence but no poly-U stretch (Fig. 2) and which is present in 147 copies in *M. tuberculosis* (29). It remains to be determined if TRIT termination is dependent on additional factors, but the high degree of sequence conservations suggests that this may be the case. The paucity of canonical terminator structures complicates bioinformatic approaches to the prediction of 3′ ends

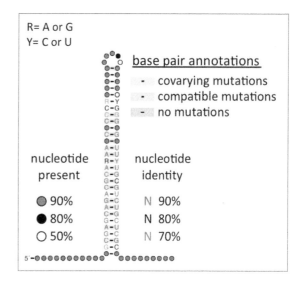

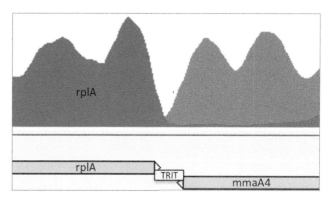

Figure 2 Transcription termination in mycobacteria. The top panel illustrates the consensus sequence and structure of the mycobacterial terminator, TRIT (tuberculosis rho-independent terminator). TRIT is a novel rho-independent terminator with high sequence conservation identified in and specific for mycobacteria (29). The bottom panel illustrates the expression of two converging genes in *M. tuberculosis*, according to RNA-seq and visualized in the Artemis genome browser; blue represents expression from the forward strand (*rplA*), and red represents expression from the reverse strand (*mmaA4*); the height of the trace represents the normalized expression level (reads) over the entire region. The traces demonstrate the termination efficiency exerted by TRIT between the two converging genes.
doi:10.1128/microbiolspec.MGM2-0029-2013.f2

in *M. tuberculosis* and is further reflected in the fact that a significant proportion of *M. tuberculosis* mRNAs have 3′ UTRs extending over more than 50 nucleotides (7). These long 3′ UTRs are particularly intriguing in the context of convergent genes, where they may provide an antisense transcript with perfect complementarity to the adjacent gene. As in the case of the internal antisense transcripts described below, understanding of the functional consequence of overlapping 3′ UTRs will require further experimental analysis, but it is intriguing that the set of genes subject to this form of antisense coverage shows a nonrandom distribution that is skewed toward genes with a shared role in cell wall processes (7).

It has also been demonstrated in both *M. tuberculosis* (ncRv10243/F6/MTS194) and *Salmonella* (DapZ) that the 3′ UTRs of some mRNAs have the potential to function independently as *trans*-acting sRNAs (8, 30). Considering the phylogenetic distance between the two species, these are likely to be examples of a widespread phenomenon.

The 5′ End

In contrast to the 3′ UTR, the role of the 5′ UTR has been extensively characterized in bacteria. A prerequisite for the correct identification of a 5′ leader/UTR is a well-defined transcriptional start site (TSS). A number of methods have been applied to identify primary 5′ ends of transcripts such as RNA-ligase-mediated RACE (RLM-RACE) (31) and, more globally, exploitation of the resistance of 5′ triphosphorylated transcripts to digestion by Terminator 5′-phosphate-dependent exonuclease prior to deep sequencing (32). The 5′ UTR represents a hub for regulation; apart from the ribosome-binding site, which in itself is highly regulated by its intrinsic sequence and hence affinity for ribosomes, the 5′ UTR typically provides binding sites for regulatory proteins, metabolites, or sRNAs that modulate transcription or translation of the downstream sequence. Many 5′ UTRs have an inherent ability to alter their conformation in response to external stimuli such as changes in temperature (33), pH (34), or certain metabolites, in which case the RNA element is defined as a riboswitch (35). The changes in conformation lead in turn to changes in the expression of the downstream gene either by blocking/unblocking of the ribosome-binding site or by transcriptional termination/antitermination (36).

Protein-responsive 5′ UTRs are often associated with operons encoding proteins involved in transcription and translation. This type of element provides negative feedback regulation by free proteins encoded in the operon when these are in excess; this form of regulation is well characterized in the case of ribosomal proteins (reviewed by Lindahl and Zengel [37]). The organization of these operons is highly conserved across bacterial species, and the presence of analogous 5′ UTRs associated with corresponding genes and operons in *M. tuberculosis* suggests that similar circuits play an important role in posttranscriptional regulation (7).

A functionally related example can be seen in the 5′ leaders of rRNA transcripts. The 5′ leader (annotated as mpr7/mcr3 in reference 9) of the *M. tuberculosis* rRNA transcript contains a number of well-characterized and highly conserved regulatory elements (38, 39). In *E. coli* this region has been shown to be essential for efficient and balanced expression of the individual rRNA transcripts due to a modification of the transcription elongation complex to a so-called antitermination (AT) complex that has a higher elongation rate and which can read through rho-dependent terminators (40–42). The modification is induced by the binding of a number of AT factors to a specific region of the rRNA leader, the AT site. The sequences and proteins involved are highly conserved and are believed to function in all bacteria (39). The *M. tuberculosis* AT site has been shown to interact specifically with the NusA transcription factor, leading to an increase in RNA polymerase processivity (38, 43). Moreover, the NusA binding site overlaps with an RNase III processing half-site, which has led to the suggestion that NusA also functions as an RNA chaperone for the correct formation of the RNA duplex recognized by RNase III and hence the correct folding of the rRNA (43, 44).

Riboswitches

Riboswitches can regulate transcription or translation, and they are classified as "ON" or "OFF" switches depending on the outcome of ligand binding; so far, the OFF switch appears to be the most prevalent type (45). A number of highly conserved riboswitches have been identified in *M. tuberculosis* by sequence homology and covariance models (46). Genes involved in the biosynthesis of methionine provide two examples (Fig. 3). The MetC enzyme catalyzes the conversion of O-acetylhomoserine to L-methionine. In *M. tuberculosis* a SAM-IV riboswitch is located in the 5′ UTR of the *metC* mRNA (47, 48). The SAM-IV riboswitch represses expression of the downstream gene upon binding of S-adenosylmethionine (i.e., an OFF switch), which is formed by the addition of ATP to methionine (catalyzed by MetK) and which is required as a cofactor for numerous methyltransferases in *M. tuberculosis* (48). The SAM-IV riboswitch is closely related to a family of seven predicted SAM-I riboswitches in *Listeria monocytogenes*. For two of these, the attenuated riboswitch moiety has been shown to act in *trans* as an sRNA, thereby adding further complexity to the regulation by (and the distinction between) riboswitches and sRNAs (49). Whether this mechanism applies to additional riboswitches or other attenuated

SAM-IV Riboswitch

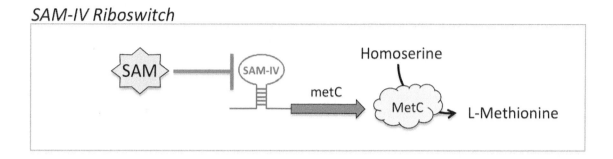

B$_{12}$ Riboswitch

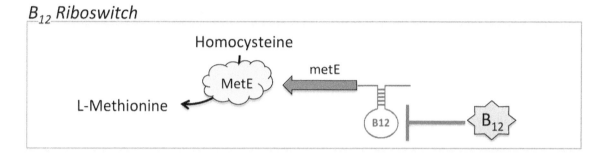

Leaderless regulation

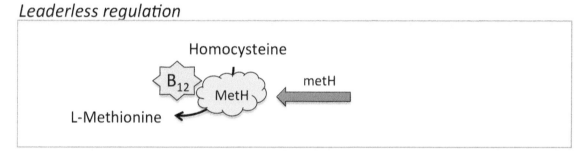

Figure 3 Ribo-regulation associated with methionine biosynthesis in *M. tuberculosis*. At least three enzymes synthesize methionine in *M. tuberculosis*; the expression of two of these is regulated by riboswitches. *metC* expression is regulated by a SAM-IV riboswitch, and the MetC enzyme uses homoserine as substrate. *metE* expression is regulated by a B$_{12}$ riboswitch, and the MetE enzyme uses homocysteine as substrate. The third enzyme, MetH, is a B$_{12}$-dependent isoform of MetE, and MetH also uses homocysteine as substrate; the mRNA of this gene belongs to the category of naturally leaderless mRNAs, which are unusually widespread in *M. tuberculosis* (16). Riboswitch ligands and enzyme cofactors are shown as "stars," genes are shown in blue, and enzymes are shown as green/yellow "clouds." doi:10.1128/microbiolspec.MGM2-0029-2013.f3

transcripts in general, including those in *M. tuberculosis*, remains to be determined.

In addition to the MetC enzyme, *M. tuberculosis* encodes two isoforms of methionine synthase, MetE and MetH. Both catalyze the conversion of homocysteine to L-methionine. However, while the MetH enzyme requires vitamin B$_{12}$ (cobalamin) to function, the MetE enzyme does not. Instead, the *metE* mRNA harbors a B$_{12}$ riboswitch, which is found in two copies in *M. tuberculosis*. This riboswitch represses the expression of the downstream genes in the presence of

B$_{12}$, presumably by occluding the ribosome-binding site, which in turn leads to destabilization of the transcript (50–52). In other words, while the MetE enzyme is only expressed when B$_{12}$ availability is low, it can also function in the absence of this cofactor, whereas the MetH enzyme requires B$_{12}$ to function and is encoded by a leaderless mRNA (16), meaning that *M. tuberculosis* employs at least three distinct adaptations of ribo-regulation for methionine biosynthesis.

In addition to the *metE* riboswitch, a second B$_{12}$ riboswitch is located upstream of an operon comprising

PPE2, *cobQ1,* and *cobU* (50–52). PPE2 belongs to the family of proteins sharing proline-glutamate (PE) or proline-proline-glutamate (PPE) N-terminal motifs that were identified from the *M. tuberculosis* genome sequence and for the most part lack known biological function (reviewed in Akhter et al. [53]). However, in agreement with previous predictions (50, 51), the PPE2 gene was recently shown to encode an ABC-transporter responsible for B_{12} uptake in *M. tuberculosis* (54).

Another riboswitch that is also represented twice in the genome of *M. tuberculosis* is the *ykok* leader or Mbox. This element is often found in the context of magnesium transporters, and like the cobalamin riboswitch, it represses the expression of downstream genes upon binding of its ligand (in this case Mg^{2+}) (55). However, unlike the B_{12} riboswitch, the *B. subtilis* Mbox has been shown to repress gene expression by transcriptional rather than translational attenuation upon Mg^{2+} binding, downregulating Mg^{2+} transport and ensuring homeostasis (56). In *M. tuberculosis* one highly expressed Mbox is found upstream of Rv1535, a conserved hypothetical with some homology to nucleoid-associated proteins, which is induced during Mg^{2+}-starvation (7, 57). Rv1535 is cotranscribed with a downstream Tbox (a tRNA-responsive element [58]) that forms the 5′ UTR of the essential isoleucine tRNA

synthase mRNA (IleS, Rv1536) (7). The second Mbox is located upstream of PE20, followed by a series of genes that are all induced by magnesium starvation (57) (Fig. 4). However, qRT-PCR and TSS mapping indicate that only PE20 and PPE31 are cotranscribed with the Mbox (16; K. B. Arnvig, unpublished). Transcription of PPE32 and PPE33 initiates approximately 26 base pairs upstream of the PPE32 ORF, while expression of Rv1810 and the virulence factor MgtC are directed by individual promoters; the transcription start for Rv1810 maps to an AUG codon 12 codons into the annotated ORF, suggesting that this may be the correct (leaderless) start (16). This suggests that not only is the observed magnesium responsiveness of genes downstream of PPE31 an indirect effect that does not directly involve the Mbox, but also that the two Mboxes in *M. tuberculosis* regulate genes of unknown function (Fig. 4).

A final example is the YdaO riboswitch, which is located upstream of *rpfA* (Rv0867c) in *M. tuberculosis*. The *rpfA* gene product is a member of a family of "resuscitation-promoting factors" that share structural homology with lysozyme and may play a particularly important role in reinitiation of cell division after periods of nonreplication (59–61). The YdaO riboswitch is widespread in Gram-positive bacteria and is

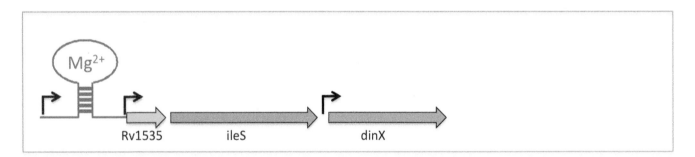

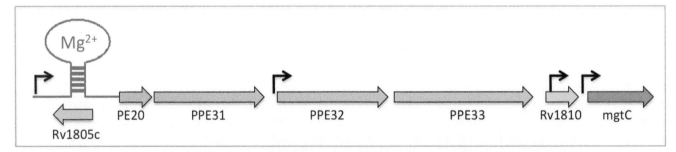

Figure 4 Magnesium-sensing riboswitches in *M. tuberculosis*. The figure illustrates the genomic context of the two identified Mboxes (magnesium-sensing riboswitches) in *M. tuberculosis*. Genes shown in green are conserved hypotheticals, those in orange are information pathways, gray represents PE-PPE genes, and genes in blue are cell wall associated. Black arrows indicate relevant transcription start sites.
doi:10.1128/microbiolspec.MGM2-0029-2013.f4

most often found in the context of genes associated with cell wall metabolism and remodeling (62, 63). However, the ligand for this regulatory element remained elusive until recently, when it was demonstrated that it provides an OFF switch responsive to, and highly selective for, the second messenger cyclic di-AMP (63).

These examples demonstrate an extensive and significant role for riboswitches in regulating *M. tuberculosis* gene expression and growth. It is likely that there are many more as yet undiscovered riboswitches in *M. tuberculosis*; however, due to lack of conservation these will be more difficult to identify (64). Inspection of *M. tuberculosis* RNA-seq profiles identifies numerous truncated/attenuated 5′ transcripts that may be considered candidate riboswitches (7, 16). Care has to be exercised in distinguishing truncated 5′ UTRs from stand-alone intergenic transcripts (sRNAs), a distinction that is not necessarily trivial.

trans-ACTING RNAs

trans-Acting sRNAs (i.e., stand-alone transcripts) include a wealth of different species that either bind to proteins or base-pair with mRNA; the latter category can be further divided into *cis*- and *trans*-encoded (corresponding to antisense or intergenic location). The majority of *trans*-acting RNAs regulate gene expression by base-pairing to one or more target mRNAs, mostly around the translation start site, competing for ribosome binding, and uncoupling transcription from translation, thereby modifying mRNA stability (reviewed in references 6, 65). Protein-binding RNAs in other bacteria include CsrB, CsrC, and 6S RNA. CsrB and CsrC both bind to the carbon storage regulatory protein, CsrA, reviewed by Babitzke and Romeo (66), while 6S RNA binds specifically to the σ^{70}-RNA polymerase holoenzyme, thereby downregulating a large number of housekeeping genes in stationary phase (e.g., 67, 68). To date, there are no known homologues of either 6S RNA or the CsrBC systems in mycobacteria.

cis-ENCODED BASE-PAIRING RNAs

cis-Encoded RNAs are encoded opposite their target RNAs; hence, they have perfect target complementarity forming duplexes between mRNAs and cognate antisense partners representing ideal substrates for RNase III. The interaction between antisense RNAs (asRNAs) and their sense targets does not require the RNA chaperone Hfq (69) (discussed below). asRNAs were initially discovered in the context of plasmid copy number control and foreign DNA elements (69), but the steadily increasing amount of data from studies using tiling arrays, RNA-seq, and TSS mapping has revealed an unexpected extent and heterologous nature of antisense transcriptomes. asRNAs vary in size from ~75 nucleotides to several kilobases, and they play a wide variety of roles in bacterial gene expression (69, 70). In some cases, asRNAs have been found to mediate transcriptional attenuation (e.g., 71, 72, and references therein), but in most cases asRNA is involved in modifying translation and stability of its mRNA targets. Thus, asRNA can increase mRNA stability as seen in the *E. coli* GadY, where base-pairing between asRNA and mRNA leads to an RNase III–mediated cleavage of an unstable polycistronic mRNA into its more stable monocistronic derivatives (73, 74). Another example of asRNA-mediated mRNA stabilization is seen in *Prochlorococcus,* where duplex formation between asRNA and mRNA masks an RNase E cleavage site on the mRNA (75).

More often, however, asRNAs are involved in mRNA destabilization and degradation. Thus, asRNAs can block translation initiation, as seen in the "classical" antisense scenario where the asRNA covers the translation initiation region of the target mRNA, often followed by destabilization of the mRNA in a manner similar to many *trans*-encoded sRNAs (69). More recently, asRNA has been shown to play a significant role in genome-wide RNase III–mediated processing of mRNA in Gram-positive bacteria (76). Consistent with this mechanism is the observation that in *M. tuberculosis* there is an inverse correlation in the abundance of sense and antisense transcripts of individual genes/gene classes (7). The fraction of asRNA remains relatively constant around 10% of total RNA, although individual transcripts show variation between growth phases.

Mycobacterial antisense transcripts have been identified by cDNA cloning followed by sequencing, by computational predictions, by expressing *E. coli* Hfq followed by pull-downs (in *Mycobacterium smegmatis*), by RNA-seq, and by TSS mapping. The sizes of these asRNAs as well as their locations relative to their cognate ORFs vary significantly: some are encoded at the 5′ end of the ORF, some in the center, and some at the 3′ end, and a few cover an entire ORF or more (7–9). asRNAs can be expressed from SNP-generated promoters in clinical isolates of *M. tuberculosis* (77). The SNPs can be synonymous or nonsynonymous with respect to the mRNA but in most cases result in formation of a −10 region with the consensus TANNNT on the antisense strand, which appears to be sufficient for expression of the asRNA in the absence of any obvious −35 motif (16). An interesting example is a mid-ORF asRNA opposite the *ino1* gene (Rv0046c), expressed from a TANNNT promoter generated by an identical SNP acquired independently in sub-branches of *M. tuberculosis* lineages 1 and 4 (including the common lineage 4 laboratory model strain H37Rv). The *asino1* is highly expressed during exponential growth and significantly downregulated in stationary phase (7) (Fig. 5). *ino1* encodes an enzyme that catalyzes the first step in inositol synthesis, and this gene has been shown to be essential for virulence (78). It is currently unknown whether the regulated expression of *asino1* has an effect on the synthesis of inositol. Overall, the degree of conservation of the antisense transcriptome among clinical isolates of *M. tuberculosis* is similar to that observed for the coding transcriptome (77).

The general conception has been that asRNAs target only their cognate mRNA. However, it was recently demonstrated that a *cis*-encoded sRNA from the archaeon *Methanosarcina mazei* had a second *trans*-encoded mRNA target (79). At least two *M. tuberculosis* *cis*-encoded RNAs, both associated with lipid metabolism, have the potential to base-pair with multiple mRNAs generated from duplicated gene sequences (8). One of these, ASpks, is located opposite of *pks12*, which encodes one of multiple polyketide synthases required for biosynthesis of complex lipid molecules

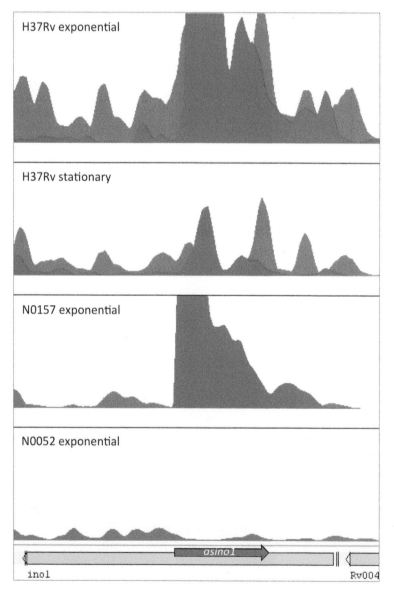

Figure 5 The *asino1* RNA is highly expressed during exponential growth and significantly downregulated in stationary phase in H37Rv. The *asino1* RNA is expressed in the majority of lineage 4 strains (here represented by H37Rv) and in some lineage 1 strains (represented by N0157), but not in other lineages (N0052—lineage 2—has been shown for comparison). RNA-seq data is visualized in Artemis. All reads are normalized to total reads and adjusted to the same scale.
doi:10.1128/microbiolspec.MGM2-0029-2013.f5

involved in pathogenesis (80–82). ASpks is a 75-nucleotide transcript that maps to a duplicated module in the 12.5-kb *pks12* gene (Rv2048). Moreover, it has significant sequence identity with the equivalent modules in *pks7, pks8,* and *pks15* and hence could potentially base-pair with these mRNAs as well (Fig. 6). During oxidative stress, a larger 200-nucleotide ASpks transcript is induced (8), while oxidative stress also results in a downregulation of potential target *pks12, pks8,* and *pks15* mRNAs (83). Further analysis should clarify whether this is due to the increased ASpks levels and whether this has an effect on Pks12-specific host immune recognition (84).

The extent of the antisense transcriptome suggests that it may represent a common component of gene regulation in *M. tuberculosis*, with the potential in some cases for *cis*-regulation of their cognate mRNA partner as well as coordinated *trans*-regulation of related genes.

trans-Encoded Base-Pairing RNAs

trans-Encoded base-pairing RNAs are encoded in IGRs, which in most cases are distinct from the locations of their targets. These sRNAs share the common feature of imperfect complementarity to their mRNA targets, posing a challenge for target prediction (85, 86). Many sRNAs are regulated by stress stimuli and may therefore be associated with host adaptation and virulence in pathogens (reviewed in references 1, 87–89).

Most *trans*-encoded sRNAs have multiple targets and act by modifying the accessibility to the ribosome-binding site of their target mRNAs; hence, they are classified as posttranscriptional regulators of gene expression. However, blocking ribosome entry leads to an uncoupling of transcription from translation, associated with rho-dependent termination of transcription and, in the case of polycistronic mRNAs, polarity. This scenario has been demonstrated for the bicistronic ChiPQ mRNA regulated by the ChiX sRNA but is likely to be of significance in the regulation of other polycistronic transcripts (90).

sRNAs can regulate any gene within any regulon and tend to target genes from different regulons, thus adding complexity to regulation rather than following the same paths as their own transcriptional regulators. An emerging theme is that many sRNAs regulate transcription factors, whereby their regulatory potential is significantly enhanced (65, 91, 92). Examples of this mechanism are the *E. coli/Salmonella* RpoS, targeted by DsrA, ArcZ, and RprA (reviewed by Battesti et al. [93]), and the *S. aureus* Rot, targeted by RNA III (94). Alternatively, an sRNA can repress the expression of

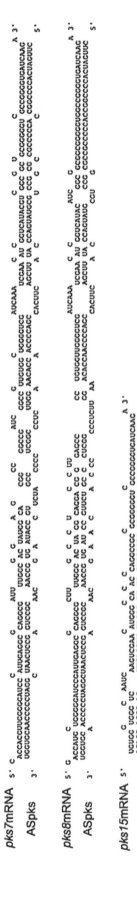

Figure 6 *cis*-encoded RNAs with *trans*-regulating potential. A transcript encoded opposite (antisense) to the *pks12* gene (ASpks) shows high complementarity to three other *pks* mRNAs, namely *pks7, pks8,* and *pks15*. The figure illustrates the predicted base-pairing between ASpks and the three mRNAs. From reference 8. doi:10.1128/microbiolspec.MGM2-0029-2013.f6

one or more regulators that drive the transcription of the same sRNA, thereby ensuring a faster return to homeostasis. Two coexisting examples of this type of mechanism can be found in the *Vibrio cholerae* quorum-sensing circuit in which LuxO and HapR both positively regulate the expression of four sRNAs, Qrr1 to 4, which in turn repress the expression of both LuxO and HapR (95).

Many intergenic sRNA candidates have been identified in mycobacteria, though only a fraction of these transcripts have been verified by Northern blotting, TSS mapping, and/or qRT-PCR (7–14). As noted above, it is important to distinguish intergenic sRNAs from truncated 5′ UTRs and from misannotated short protein ORFs. Table 1 provides a list of partially characterized *M. tuberculosis* intergenic sRNAs. *M. tuberculosis* sRNAs display varying degrees of conservation: some are restricted to closely related members of the *M. tuberculosis* complex; others are present in multiple pathogenic mycobacteria, with *Mycobacterium leprae* being the most distant relative; and finally, some are conserved in all mycobacteria, with a few being identified in other actinomycetes (8, 96).

In Gram-negative bacteria, including many pathogens, the RNA chaperone Hfq plays a central role by facilitating the interaction between *trans*-encoded sRNAs and their targets, with deletion of the *hfq* gene often leading to loss of virulence (reviewed in references 89, 97). However, the role of Hfq in the Gram-positive *S. aureus* is more controversial (98, 99), and the protein is completely absent from several pathogens, ranging from the AT-rich (39% GC) *Helicobacter pylori* to the GC-rich (67% GC) *M. tuberculosis* (97). An obvious question is whether *M. tuberculosis* sRNAs interact with their targets unaided or whether in *M. tuberculosis* there are one or more alternative, hitherto unidentified chaperones. The high GC content in *M. tuberculosis* limits the frequency of AU-rich stretches implicated in conventional Hfq interactions, as does the lack of canonical intrinsic terminators (100, 101). *M. tuberculosis* may have a chaperone that has functional but not sequence homology with Hfq, or regulon-specific chaperones, each of which could interact with a subset of sRNAs and their targets in a manner similar to the FbpABC proteins in *B. subtilis* (102); or it may be that the high GC content facilitates interactions between sRNAs and their mRNA targets independently of chaperone mediation (103). Future investigations, in which tagged sRNAs are used as bait for the isolation of RNA-binding proteins and hence putative chaperones (104), will shed more light on these questions in the near future.

STRESS RESPONSES OF INTERGENIC sRNAs

The abundance of many intergenic sRNAs changes in response to different environmental stresses. Conditions such as starvation in phosphate-buffered saline, stationary phase, or infection have been shown to enhance the expression of numerous sRNAs (7–14). However, all of these conditions represent a mixture of environmental stresses, and the application of specific stresses may tell us more about the regulatory pathways associated with individual sRNAs. The following section provides some examples of changes in *M. tuberculosis* sRNA expression upon the application of separable stresses encountered in the host environment.

Oxidative Stress

Three of the sRNAs listed in Table 1—ncRv10243, ncRv10609A, and ncRv13660c—show increased expression in response to H_2O_2, which mimics the oxidative stress encountered inside the host macrophage; hence, these sRNAs are likely to be induced during the early stages of infection and may play a role in the first stages of host adaptation.

ncRv10243 is encoded between two genes involved in lipid degradation: Rv0243 and Rv0244c. The ncRv10243 promoter has been shown to have the highest occupancy of SigF, suggesting that expression of ncRv10243 has high priority under SigF-inducing conditions (105). Consistent with this prediction, expression of ncRv10243 is virtually abolished in a SigF knockout strain (J. Houghton and K. B. Arnvig, unpublished). Overexpression of ncRv10243 leads to reduced growth of *M. tuberculosis*, but although an ncRv10243 homologue is present in *M. smegmatis*, overexpression of ncRv10243 in *M. smegmatis* has no obvious growth phenotype (8). Overexpression of the *M. smegmatis* homologue has not been tested in either background.

ncRv13660c is ~95 nucleotides in size and is located between Rv3660c, which encodes the septum site determining protein Ssd (106), and Rv3661, which is predicted to be involved in cellular differentiation. Downstream of Rv3661 is another sRNA, ncRv13661 (see below), and the entire locus is highly conserved among mycobacteria, with the notable exception of *M. leprae*. The predicted structure of ncRv13660c contains a so-called 6C motif consisting of two C-rich loops found to be widespread among *Actinobacteria*, and this structure has led to the suggestion that this sRNA may be a structural/protein binding RNA rather than a base-paring regulatory sRNA (47). On the other hand, C-rich loops have been shown to interact specifically with the purine-rich Shine-Dalgarno sequence in

S. aureus, thereby blocking translation initiation (107). Additional *M. tuberculosis* sRNAs are predicted to form similar C-rich loops (8), and due to the high GC content of the bacterium, one can imagine this to be a relatively common feature of *M. tuberculosis* sRNAs. In addition, it is worth noting that while *M. tuberculosis* has no Hfq, the C-rich sRNAs in *S. aureus* are independent of Hfq for function, suggesting the possibility that this feature may be a signature for Hfq-independent sRNAs. *M. smegmatis* harbors an sRNA that is >90% identical to ncRv13660c. Although (strong) overexpression of ncRv13660c in *M. tuberculosis* is lethal, the same level of overexpression of ncRv13660c in *M. smegmatis* is possible but results in a dramatic phenotype: growth is very poor on solid and in liquid media, and the cells display aberrant morphology including irregular and elongated shape, suggesting association with cell wall synthesis and/or cell division (8).

Hypoxia

The master regulator of the hypoxic response in *M. tuberculosis* is DosR (108). The DosR regulon includes genes required for remodeling of protein, lipid, and energy metabolism together with members of the universal stress protein family and multiple proteins of unknown function (83). The DosR genes are highly expressed during infection in mice and in sputum samples from patients, and their expression has been linked to the generation of drug-tolerant persister cells (109, 110). A number of *M. tuberculosis* intergenic sRNAs accumulate to high levels during the transition from exponential to stationary phase, which represents nutrient starvation as well as some degree of hypoxia depending on the method of cultivation (7–9, 12–14). Some of these accumulate to even higher levels during infection, again suggesting a role in host adaptation (7, 14). An example is ncRv11733, which is almost undetectable during exponential growth. Interestingly, the number of ncRv11733 molecules relative to the number of ribosomes, estimated to be around 4,000 per cell during fast growth (111), implies that less than 10% of exponentially growing cells express ncRv11733. It therefore seems likely that the few cells expressing this sRNA are in a different metabolic state than the remaining population, and therefore these could represent putative persister cells. Moreover, the induction of ncRv11733 seen in stationary phase is almost eliminated by deletion of the dormancy regulator, DosR (7). It remains to be determined if ncRv11733 is directly involved in the generation of persister cells or is simply a marker for this subpopulation.

Acid Stress

M. tuberculosis is well known for its ability to inhibit phagosome maturation (reviewed by Pieters [112]). Nevertheless, acid stress and acid resistance still play major roles in *M. tuberculosis* intracellular survival (21, 113). A number of *M. tuberculosis* sRNAs are induced by low pH, including ncRv10243 (8) and ncRv13661 (K. B. Arnvig, unpublished). ncRv13661 is the most highly expressed nonribosomal RNA during all growth conditions investigated, with levels increasing ~10-fold between exponential and stationary phase and abundance in infected mice approaching 1:1 stoichiometry with rRNA (7). Overexpression of ncRv13661 in exponential phase results in a phenotype that resembles stationary phase and models for persistence in terms of a significant downregulation of active growth functions (7, 108, 114, 115). Genes that are most significantly downregulated by ncRv13661 overexpression are associated with the methyl citrate cycle, which may explain the reduced growth on propionate observed in the overexpression strain (7; Houghton and Arnvig, unpublished).

Downregulation of active growth functions during stationary phase transition in bacterial cultures is commonly associated with increased expression of the structurally conserved 6S RNA that binds to RNA polymerase and inhibits transcription of genes controlled only by $\sigma 70$ (67, 68, 116). However, there is to date no evidence of an immediately recognizable 6S homologue, although ncRv13661, or more specifically its *M. smegmatis* homologue, have been shown to have some structural relationship to 6S (96). It remains to be determined if ncRv13661 is a conventional sRNA that acts via base-pairing with specific mRNA or protein targets or whether its mode of action differs entirely from other sRNAs.

So far, the regulatory targets of all *M. tuberculosis* sRNAs remain elusive. Target prediction webservers such as TargetRNA (86), the enhanced version TargetRNA2 (http://cs.wellesley.edu/~btjaden/TargetRNA2/), and RNApredator (http://rna.tbi.univie.ac.at/RNApredator2/target_search.cgi [85]) provide some suggestions; however, these are often not identical and require experimental verification just like other computational predictions.

RNA Expression and Degradation

Increased abundance of sRNAs under different growth conditions may be the result of increased expression and/or decreased degradation. An understanding of the dynamics and mechanisms of RNA degradation in *M. tuberculosis* will be an essential component of the

Table 1 Intergenic *M. tuberculosis* sRNAs[a]

New annotation[b]	Other[c]	Tuberculist[d]	Rfam[e]	Strand	Flanking CDSs	Verified by	5′ (RACE)[f]	5′ (TSS)[g]	3′[b]	Expression[i]	Conservation[i]	Reference
ncRv10243	F6, mcr14, mpr13	MTB000051	RF01791	F	Rv0243; Rv0244c	Northern, RACE, TSS	293604	293602	293705	SigF; H_2O_2, low pH	NPMB	8, 9
ncRv10609A	B55	MTB000052	RF01783	F	Rv0609A; Rv0610	Northern, RACE, TSS	704187	704187	704247	SigA; H_2O_2	MTBC	8
ncRv10685c	—	—	—	R	Rv0685; Rv0686	Northern, TSS	—	786164	786038; 786090	—	MTBC	162 and A. Gaudion, K. B. Arnvig, and D. B. Young, unpublished
ncRv11051	mpr5	MTB000061	—	F	Rv1051c; Rv1052	Northern, TSS	—	1174124; 1175178	—	—	PMB	9
ncRv11144c	MTS0900	—	—	R	Rv1144; Rv1145	Northern, TSS	—	1272168	1271907	SigA; exponential growth	NPM	162 and Gaudion et al., unpublished
ncRv11147c	MTS0930	—	—	R	Rv1147; Rv1148c	Northern, TSS	—	1275957; 1275993	—	SigA; exponential growth	PMB	7
ncRv11264c	mcr11, MTS0997, MT1302	MTB000063	—	R	Rv1264; Rv1265	Northern, RACE, TSS	1413224	1413225	1413107	SigA; CRP, stationary phase, low pH, infection	MTBC	7, 9, 14
ncRv11373	MTS1082*[k]	MTB000076	—	F	Rv1373; Rv1374c	Northern, TSS	—	1547129	1547350**[l]	SigA	MTBC	7, 191
ncRv11435	RNA1	—	—	F	Rv1435c; Rv1436	Northern	—	—	—	—	NPMB	11
ncRv11689c	G2	MTB000054	RF01798	R	Rv1689; Rv1690	Northern, RACE, TSS	1915190; 1915028	1915190; 1915028	1914962	exponential growth	MTBC	8

ncRNA	Name	Identifier	Strand	Flanking ORFs	Method				Regulation	Conservation	Reference
ncRv11733	MTS1338*	MTB000077	F	Rv1733c; Rv1734c	Northern, RACE, TSS	1960601; 1960667	1960667	1960783	DosR; stationary phase, NO, infection	MTBC	7
ncRv11789c	mcr1	—	R	Rv1179; Rv1180	Northern, TSS	—	2028215	—	—	MTBC	9
ncRv12659	MTS2048*	—	F	Rv2659c; Rv2660c	Northern, RACE, TSS	2980911	2980911	2981083	SigA; starvation	MTBC	7, 190
ncRv13059	—	—	F	Rv3059; Rv3060c	Northern, TSS	—		3421223	SigA	MTBC	162 and Gaudion et al., unpublished
ncRv13660c	B11,mpr19	MTB000058 RF01066	R	Rv3660c; Rv3661	Northern, RACE, TSS	4099478	4099480	4099386	SigA; H$_2$O$_2$	NPM	7
ncRv13661	mcr8,mpr4, MTS2823	MTB000078	F	Rv3661; Rv3662c	Northern, RACE, TSS	4100683	4100682	4100987	stationary phase, low pH, infection	NPM	7, 9
ncRv13943	MTS2975	MTB000080	F	Rv3943c; Rv3844	Northern, TSS	—	4317071	—	SigA; exponential growth	PMB	7
ncRv13596	mpr17	MTB000070	F	Rv3596c; Rv3597c	Northern, TSS	—	4040805	—	SigA	NPMB	9

[a]Most of the listed sRNAs have been verified by at least two methods; transcripts listed as sRNAs/sRNA candidates in primary publications that are not listed here include those that are likely to be 5′ UTRs, fragments of mRNAs, those in which the ends are not mapped and cannot be deduced by our TSS mapping, and those that are antisense to annotated ORFs. Exceptions include ncRv11373, ncRv11733, and ncRv12659, for which we have found no evidence for expression from the opposite strand and which have all been further described elsewhere (7, 190, 191).

[b]Refers to annotation rules in Lamichhane et al. (15).

[c]Refers to the names given in previous publications.

[d]Refers to annotation on Tuberculist.

[e]Rfam identifier if present.

[f]Identified by 5′ RACE.

[g]Identified by RNA-seq TSS mapping, all from reference 16.

[h]Approximate 3′ end and identified by RACE or canonical terminator structure (first U) or calculated from 5′ end and size; according to primary publication.

[i]Refers to transcription factor association and inducing conditions; SigA indicates that the SigA consensus TANNNT has been identified <10 base pairs upstream of mapped TSS.

[j]Conservation (BLAST only): MTBC, only identified in M. tuberculosis complex; PMB, also in other pathogenic mycobacteria beyond MTBC; NPMB, also in nonpathogenic mycobacteria.

[k]* Indicates that sRNA is antisense to annotated ORF but that we have seen little (weak TSS, and no total RNA) or (no TSS, no total RNA) evidence of expression from ORF annotation strand.

[l]** As suggested by RNAseq and terminator structure; Northern suggests shorter (processed) transcript.

effort to understand the function of ncRNAs. RNase E is the principal catalytic component of the *E. coli* degradosome. This multicomponent complex is a key player in RNA processing and turnover, including sRNA-mediated mRNA degradation (30). Very little is known about RNase E and RNA metabolism in general in *M. tuberculosis*, although factors such as hypoxia and cold shock have been shown to have a stabilizing effect on mRNA (117). *M. tuberculosis* RNase E appears to interact only with a subset of the degradosome components found in enteric bacteria (118). On the other hand, the Gram-positive *B. subtilis* does not encode an RNase E homologue, and hence the *B. subtilis* degradosome has a significantly different composition (119). This raises a number of interesting questions with respect to mRNA-sRNA interactions,

M. tuberculosis chaperones, and the *M. tuberculosis* degradosome including (i) does *M. tuberculosis* form an RNA degradosome(-like) structure, and (ii) what is its role in RNA metabolism and, in particular, sRNA regulation? The stable secondary structures predicted for many intergenic sRNAs may inhibit their degradation and contribute observed increases in abundance.

OTHER *M. TUBERCULOSIS* ncRNAs

CRISPR

The clustered regularly interspaced short palindromic repeat (CRISPR) locus provides a specialized source for generation of *M. tuberculosis* sRNAs. The CRISPR locus incorporates sequences from phage and other

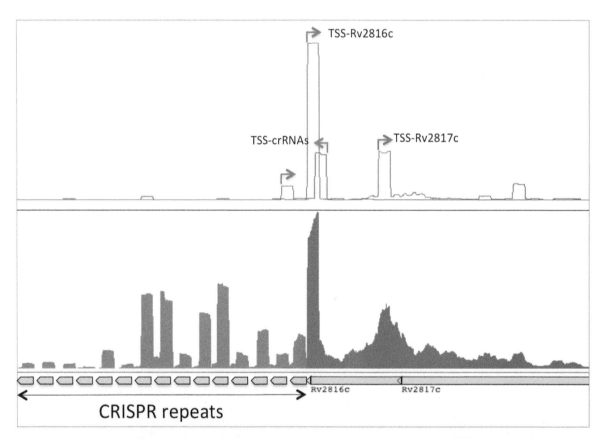

Figure 7 RNA-seq data (visualized in Artemis) of the *M. tuberculosis* CRISPR locus. The upper trace records TSS mapping (i.e., enriched for primary transcripts) from the right-hand side of the CRISPR locus, showing overlapping start sites for the Rv2816c antisense transcript (forward direction in blue) and the single CRISPR-RNA (crRNA) transcript (reverse, in red). The lower trace records sequencing of total RNA, showing the antisense transcript covering Rv2816c and Rv2817c (blue) and illustrating how the single crRNA transcript is processed into a series of mature, smaller crRNAs (red). A similar crRNA profile is seen on the left-hand side of the CRISPR locus, upstream of the Rv2614c/Rv2615c IS6110 insertion sequence (not shown). doi:10.1128/microbiolspec.MGM2-0029-2013.f7

invading genetic elements and repackages them as sRNA defense molecules that confer resistance to reinfection (reviewed by Deveau et al. [120]). The CRISPR loci vary substantially between species of mycobacteria, and the composition of the CRISPR in *Mycobacterium canetti* suggests acquisition by horizontal transfer prior to final branching of the *M. tuberculosis* complex (121). Sporadic deletion of integrated foreign sequence elements from the CRISPR domain has been exploited in a widely used system for differentiation of *M. tuberculosis* strains referred to as "spoligotyping" (122). Phylogenetically, the *M. tuberculosis* CRISPR system belongs to the type III-A family (123 and references therein) and is related to a well-characterized CRISPR system in *Staphylococcus epidermis* (124). Inspection of RNA-seq profiles shows that the CRISPR domain is transcribed and processed into a series of 50 to 60 nucleotide CRISPR-RNAs (crRNAs) that are presumably incorporated into a cascade effector complex. The transcription start site driving crRNA expression overlaps with a divergent start site that generates a long antisense transcript covering Rv2816c and Rv2817c, which code for the Cas2 and Cas1 proteins, respectively, involved in incorporation of novel sequences into the CRISPR locus (Fig. 7) (7).

This profile is consistent with silencing of the genes required to capture novel crRNAs, but functional expression of Cas6 and Cas10 enzymes (Rv2824c and Rv2823c) is required for processing of pre-existing crRNAs. An exception is the sublineage of *M. tuberculosis* known as the Beijing family, which is undergoing a current global expansion in marked association with multiple drug resistance (125). A genomic deletion has removed most of the CRISPR locus and many of the associated Cas genes, eliminating expression of crRNAs in these strains (126). A similar deletion has been found in an unrelated sublineage referred to as "pseudo-Beijing" (127). It is open to speculation whether the presence of a functional CRISPR system in the majority of *M. tuberculosis* strains has current biological relevance or reflects fortuitous retention of a defense system appropriate to an ancestral lifestyle.

HOW TO IDENTIFY ncRNA

Until 2001, only 10 sRNAs had been identified in the model organism *E. coli* through genetic screens or radiolabeling of total RNA and subsequent isolation from gels (128). Over the last decade, a combination of computational prediction and experimental approaches has allowed identification of a wider set of sRNAs in *E. coli* and other closely related bacteria like

Salmonella typhimurium (129). Attempts to predict sRNAs in bacteria involve computational prediction methods that employ RNA sequence homology, comparative genomics, thermodynamically favorable secondary structures, detection of transcription signals, and *ab initio* methods using sRNA-specific features. However, the use of computational methods for the detection of sRNAs in bacterial genomes is not easy because sRNAs are diverse in length (ranging from 50 to 500 nucleotides), they do not share a common secondary structure, they do not exhibit any clear nucleotide biases, and generally they are not well conserved among distantly related bacteria. In combination with the recent development of powerful biocomputing algorithms for the prediction of ncRNA, the advent of high-throughput sequencing technologies, including RNA-sequencing and 5′ transcription start site mapping, has allowed the identification of sRNAs at a genome-wide level followed by *in vivo* validations. This section reviews the past and current methods used for the identification of ncRNAs in bacteria, with a special focus on the identification in mycobacteria.

Computational sRNA Prediction

One of the major challenges in sRNA prediction is that although there is a common acceptance of some shared characteristics such as small size (usually less than 200 nucleotides) and lack of coding capacity, there is a general lack of accepted identifiers for bacterial sRNAs. Consensus structure prediction is a good start for sRNA prediction when a given sRNA secondary structure is known. Tools like RNAMotif (130) can be used to scan defined secondary structures of RNA throughout the genome, but they are limited by the known secondary structure of a specific sRNA family.

Comparative genomics has been extensively applied for *de novo* sRNA prediction in bacterial genomes (131–135). Briefly, shared conserved sequences are first identified in IGRs, followed by clustering and comparison by multiple alignments that are scored based on predicted RNA structural features. Programs like eQRNA (136), ERPIN (137), ISI (138), INFERNAL (139), MSARI (140), and evofold (141) have been used for the detection of bacterial sRNAs by comparative genomics. RNAz is a program that combines comparative genomics with thermodynamic stability values (prediction of conserved stable RNA secondary structures) for predicting structural noncoding RNAs in bacteria (142, 143).

Noncomparative approaches have been based on systematic scanning for transcriptional signals, including conserved promoter sequences, rho-independent/intrinsic terminators, transcription factor binding sites and or-phan transcriptional signals. sRNAPredict (144) is an algorithm for the prediction of putative sRNAs in bacteria, using databases like TRANSTERMHP (145) or TRANSFAC (146). sRNAPredict3 and SIPHT are recent versions of this program that have been used for the prediction of sRNAs in bacteria (147). sRNAscanner (148) has recently been developed to overcome the limitation of the lack of transcription signals predicted in all the available genome sequences and has been proved to be an efficient platform for the prediction of sRNAs in any bacterial genome.

Ab initio sRNA finders are used for the identification of sRNAs based on specific RNA structural elements such as known RNA sequence motifs, di/trinucleotide preferences, uridine loops, or atypical GC content. Programs like RNAGENiE (149), Atypical GC (150), and NAPP (151) can be used.

Low-Throughput Experimental Methods and Computational Predictions

The first 10 ncRNAs discovered were found incidentally following genetic screens in *E. coli*. The first group identified included 4.5S, 6S, tmRNA, RNase P, and Spot 42 detected by metabolic radiolabeling (152–154). OxyS and CsrB sRNAs were detected while mapping the transcription start site of the adjacent regulator gene OxyR (155) and after copurification with overproduced CrsA protein (156), respectively. Finally, some sRNAs (e.g., MicF and DsrA) were identified when genomic fragments harboring these genes were shown to affect gene regulation in *E. coli* (157, 158).

Subsequent discovery of sRNAs has been based mainly on systematic genome-wide screens combining computational and experimental approaches, including microarrays and shotgun cloning (159 and references therein). Identification of ncRNAs through computational approaches has relied mainly on sequence and structural analysis, conservation across species, and position within IGRs. The first computational-based genome-wide screens predicted several hundred new sRNA candidates in *E. coli* (31, 136, 160, 161). Genes encoding novel putative sRNAs were identified by Argaman et al. (31) following a computational strategy centered basically on three main parameters: length (search restricted from 50 to 400 nucleotides), sequence conservation, and presence of transcription signals (search for promoter sequences recognized by the *E. coli* principal RNA polymerase sigma factor σ^{70} and intrinsic terminators). Wassarman et al. (161) employed an algorithm that was also focused on sequence

conservation but further extended the predictions of transcription signals, incorporating sequence elements for potential promoters, terminators, and inverted repeat regions, and a method developed by Chen et al. (160) restricted the detection of σ^{70} conserved promoter sequences within a short distance of a rho-independent terminator. While these methods were mainly based on primary sequence conservation of the identified sRNAs, the QRNA tool, developed by Rivas et al. (136) focused mainly on searching for conserved RNA secondary structures in combination with comparative genomics. Bioinformatic predictions of sRNA candidates identified by these algorithms were verified by Northern blot analyses and mapping of 5′ and 3′ ends, leading to the identification of several new sRNAs in *E. coli*.

The first computational predictions applied for the identification of ncRNAs in the *M. tuberculosis* genome were the result of a comparative genomics analysis carried out using the sRNAPredict2 program (162). The IGRs of 11 diverse pathogens were sought on the basis of sequence conservation and transcription signals (intrinsic terminators and putative promoter sequences) allowing the detection of 56 putative intergenic sRNAs in *M. tuberculosis*. Of the 56 predicted sRNA candidates, 1 (mpr8) was subsequently predicted by a different algorithm, tested by Northern blotting, and annotated as the 5′ UTR of *infC* (7, 9). Of the remaining sRNA candidates, a few have been verified in our laboratory by RNA-seq and/or Northern blotting (Table 1; A. Gaudion, K. B. Arnvig, and D. B. Young, unpublished).

The first experimental evidence for the existence of sRNAs in *M. tuberculosis* came from Arnvig and Young in 2009 and was based on screening of cDNA libraries generated from low-molecular-weight RNA (8). A set of nine ncRNAs was identified, five located within IGRs and four located antisense to annotated coding genes. Northern blot analyses demonstrated variations in expression between exponential and stationary growth phases and in response to environmental stresses. All of the sRNAs displayed predicted stable secondary structures, and some had recognizable intrinsic terminators, but none of them were predicted by the computational method discussed above (162); this may be due to a mixture of the heterologous nature of mycobacterial promoters and the lack of canonical terminator structure.

A mixed approach using a cloning-based screen together with a computational approach was used by DiChiara et al. for the identification of ncRNAs in *M. bovis* BCG (9). They described 37 sRNA candidates in BCG, 34 of which were novel. The computational search was performed using the SIPHT program, in which identification of ncRNAs is based on the presence of intergenic sequence conservation upstream of a putative intrinsic terminator. *In silico* identification of putative sRNAs yielded 144 candidates that were further reduced to 67 candidates partially conserved in mycobacterial species outside of the MTB complex. All 67 candidates were tested by Northern blot analyses and 21 sRNAs were confirmed, of which 5 were intergenic sRNAs. The cloning-based experimental approach identified 60 candidates, 13 located in IGRs and 47 located inside annotated coding genes. Northern blot analysis confirmed 19, of which 4 were intergenic sRNAs. Among the overall set of intergenic sRNAs, only two, ncRv10243 and ncRv13661, were identified by both methodologies. The sRNAs were predicted to be present in a wide range of mycobacterial species, highlighting their potential regulatory role in highly conserved cellular functions. Sequence conservation identified similar sequences for all of them within the *M. tuberculosis* genome, although only 20% were found to be expressed in *M. tuberculosis* under the conditions tested, leading to the identification of 17 novel sRNA candidates.

Recently, Pelly et al. used a cDNA cloning strategy to identify sRNA species in *M. tuberculosis* CDC1551 in the size range of 70 to 200 nucleotides from RNA isolated during late exponential growth (14). After removal of tRNA and rRNA clones, 12 clones representing intergenic sequences were further investigated. The previously described sRNA ncrMT1302 (named mcr11 in reference 9 and MTS0997 in reference 7), located in a locus involved in cAMP metabolism, was further investigated; the study demonstrated that the expression of ncRv11264c responded to changes in pH and cAMP concentration and confirmed that the sRNA is expressed during infection.

High-Throughput Experimental Methods

While low-throughput and computational methods have been mainly useful for the detection of sRNAs in IGRs, high-throughput methods such as tiling arrays and RNA sequencing (RNA-seq) have been very useful for the genome-wide detection of sRNAs and especially for the detection of asRNAs, unraveling a whole new level of antisense regulation in bacterial transcriptomes including *L. monocytogenes* (163), *S. aureus* (164), *B. subtilis* (165), and *Streptococcus pyogenes* (166). A tiling array consisting of overlapping 60-mer probes covering the entire genome was designed for the identification of ncRNAs in *M. leprae* (167).

The application of high-throughput sequencing technologies has revolutionized our understanding of ncRNAs and antisense regulation in bacterial genomes. RNA-seq has allowed the sequencing of whole transcriptomes in a strand-specific manner from total RNA or mRNA-enriched samples. Recent studies generating RNA-seq transcriptomes cover pathogenic bacteria as diverse as *L. monocytogenes* (168), *Bacillus anthracis* (169), *Burkholderia cenocepacia* (170), *Mycoplasma pneumoniae* (171), *Salmonella* spp. (172, 173), *H. pylori* (32), and *Campylobacter jejuni* (174); these are all generating important results that are reshaping our understanding of bacterial gene regulation and adaptation.

The first *M. tuberculosis* transcriptome study using RNA-seq was performed by Arnvig et al. (7), who investigated whole-transcriptome expression during exponential growth and stationary phase. During exponential growth, ~79% of the annotated coding transcriptome had significant levels of expression (defined as reads mapped to that particular genome region), while during stationary phase the transcriptome expression decreased dramatically (only 11% of the genome expressed) and was mainly dominated by expression of DosR regulon genes. Regarding the noncoding transcriptome, this study revealed an extensive presence of ncRNA in the *M. tuberculosis* genome, including long 5′ and 3′ UTRs, antisense transcripts, and intergenic sRNA molecules (7). A substantial portion of antisense transcripts were derived from long 3′ UTRs of converging genes, whereas others were independent of 3′ UTRs and located antisense within annotated coding genes. The dominant sources of intergenic reads corresponded to 5′ UTRs and sRNA molecules. Twenty-one intergenic sRNA candidates were identified, some of which had been identified previously (8, 9); three of these (ncRv11264c, ncRv11733, and ncRv13661) were further characterized and shown to be highly upregulated in a mouse model of infection.

A comparative genomics approach combined with RNA-seq was conducted by Pellin et al. (13) to identify sRNA candidates in the *M. tuberculosis* genome. They constructed the *M. tuberculosis* effective target genome for ncRNAs (excluding regions annotated as coding genes and stable RNAs, hence only keeping IGRs and antisense sequences as possible regions for ncRNA discovery). By combining RNA-seq and IGR conservation analysis they predicted ~2,000 sRNA candidates in *M. tuberculosis*. Although this study has greatly increased the putative number of ncRNAs in the *M. tuberculosis* genome, only 52% of the previously described and verified sRNAs were identified by means of this approach. A subset of these potential sRNAs was further investigated by Miotto et al. (12) through microarray expression analysis. Approximately 19% of the previously predicted transcripts were validated by microarray expression profiling including intergenic sRNA candidates, 5′/3′ UTRs, and antisense sRNAs. These showed enriched higher secondary structure stability, and functional enrichment analyses suggested that antisense sRNAs might target genes involved in two-component systems and membrane activity. Analysis of sequences upstream of the transcription start sites identified a SigA promoter consensus sequence (TANNNT) in the majority of cases and a canonical terminator structure in about one fifth. Twenty transcripts were verified by Northern blot analyses and/or primer extension. Furthermore, some putative riboswitches were described.

Copurification with Proteins

Copurification of sRNAs with proteins has been a successful experimental approach for the detection of sRNAs in bacteria (175). Coimmunoprecipitation (coIP) with Hfq antibodies followed by hybridization to tiling arrays, conventional RNomics, or RNA-seq has been a common technique for sRNA discovery in Gram-negative bacteria including *E. coli* (176), *Salmonella* (177), and *Pseudomonas aeruginosa* (178).

As discussed above, the role of the Hfq protein in Gram-positive bacteria is less clear, and no Hfq orthologs have been described in important bacterial pathogens like *H. pylori* and *M. tuberculosis*. Despite this, the RNA-binding potential of Hfq can still be exploited, and using coIP, Hfq-binding sRNAs have been identified in *L. monocytogenes* (179) and *B. subtilis* (180). Furthermore, by using heterologous Hfq, which presumably has different sequence specificity than the native protein, sRNAs that are not detected under standard assay conditions have been detected in *Salmonella* (181). In addition, Hfq has been expressed in *M. smegmatis*, which does not have Hfq at all (10). The latter example involved the expression of FLAG-tagged Hfq from *E. coli* in *M. smegmatis*, followed by coIP and RNA-seq of Hfq-binding transcripts. This led to the identification of 24 novel sRNAs including both *cis*-acting and *trans*-acting sRNAs targeting genes involved in metabolic pathways, as well as two of the previously described *M. tuberculosis* sRNAs (8). The 24 novel sRNAs have been verified by Northern blot analysis showing differential expression in exponential and stationary phases (10). Prediction of secondary structures using RNAFold (182–184) and centroidFold (185, 186) identified a typical I-shaped terminator or a

stem-loop without a long poly(U) stretch for most of the identified sRNAs. Phylogenetic conservation among mycobacteria revealed that 13 out of 24 were exclusively present in *M. smegmatis,* and only 2 were found in *M. tuberculosis.*

The application of coIP to isolate RNA binding to proteins other than Hfq includes CsrA (66, 187), σ70 RNA polymerase (188), and RNase III (189). However, the approach could in theory be applied to a larger number of RNA-binding proteins, some of which have been described by Pichon and Felden (175). Additional insights into the interactions between regulatory RNAs and proteins as well as regulatory RNAs and other RNAs will take us one step further into the world of mycobacterial regulatory RNA that we have only just entered.

GLOSSARY

3′ UTR   3′ Untranslated region. Part of an mRNA downstream of a stop codon, located between the stop codon and a transcriptional terminator.

5′ UTR   5′ Untranslated region. The 5′ part of an mRNA that is not translated, i.e., located between a transcription start site and translation start site.

Antisense   RNA that can base-pair with sense RNA; the pairing can be imperfect or perfect ("true" antisense).

***cis*-Encoded**   "True" antisense. RNA encoded in the same genomic location but opposite its target. Can cover coding sequence as well as UTRs.

***cis*-Regulatory**   A segment of RNA that regulates expression of another segment of the same transcript, e.g., UTRs, riboswitches, etc.

IGR   Intergenic region. A sequence between open reading frames.

Leader region   Similar to 5′ UTR but part of ncRNAs such as ribosomal RNA (rRNA) or tRNA.

Leaderless mRNA   mRNA that lacks a 5′ UTR and hence the Shine-Dalgarno sequence involved in ribosome binding. In leaderless mRNAs the transcription start site and the translation start site are the same.

ncRNA   Any RNA that does not code for a protein or peptide.

Primary transcript   Unprocessed transcript. Characterized by a 5′ triphosphate, unlike processed transcripts that have a 5′ monophosphate or 5′ OH.

Regulatory RNA   RNA that has regulatory potential in the nucleotide sequence. Usually UTRs and sRNAs that respond to changes in growth conditions, but in essence most RNA harbors regulatory potential in the form of codon bias, Shine-Dalgarno sequences, etc.

Riboswitch *cis*-regulatory 5′ end of mRNA that changes conformation in response to changes in concentration of small molecule ligands; can be on or off switches and can act on either a transcriptional or translational level.

Shine-Dalgarno (SD) sequence AG-rich hexameric sequence located approximately 10 to 20 nucleotides upstream of the start codon. The SD sequence has partial complementarity to 16S rRNA and thus facilitates binding of the small ribosomal subunit to the mRNA.

sRNA Abbreviation of small RNA, mostly refers to small regulatory RNA.

***trans*-Encoded** sRNA encoded in a different genomic location than its target.

Translation initiation region Part of mRNA where translation initiates. Includes Shine-Dalgarno (SD) sequence and initiation/start codon.

TSS Transcription start site. The nucleotide position(s) where RNA polymerase initiates transcription.

Citation. Arnvig KB, Cortes T, Young DB. 2014. Noncoding RNA in mycobacteria. Microbiol Spectrum 2(2):MGM2-0029-2013.

References

1. Gripenland J, Netterling S, Loh E, Tiensuu T, Toledo-Arana A, Johansson J. 2010. RNAs: regulators of bacterial virulence. *Nat Rev Microbiol* 8:857–866.

2. Sorek R, Cossart P. 2010. Prokaryotic transcriptomics: a new view on regulation, physiology and pathogenicity. *Nat Rev Genet* 11:9–16.

3. Yus E, Guell M, Vivancos AP, Chen WH, Lluch-Senar M, Delgado J, Gavin AC, Bork P, Serrano L. 2012. Transcription start site associated RNAs in bacteria. *Mol Syst Biol* 8:585.

4. Kang SM, Choi JW, Lee Y, Hong SH, Lee HJ. 2013. Identification of microRNA-size, small RNAs in *Escherichia coli*. *Curr Microbiol* 67:609–613.

5. Vanderpool CK, Balasubramanian D, Lloyd CR. 2011. Dual-function RNA regulators in bacteria. *Biochimie* 93:1943–1949.

6. Waters LS, Storz G. 2009. Regulatory RNAs in bacteria. *Cell* 136:615–628.

7. Arnvig KB, Comas I, Thomson NR, Houghton J, Boshoff HI, Croucher NJ, Rose G, Perkins TT, Parkhill J, Dougan G, Young DB. 2011. Sequence-based analysis uncovers an abundance of non-coding RNA in the total transcriptome of *Mycobacterium tuberculosis*. *PLoS Pathog* 7:e1002342.

8. Arnvig KB, Young DB. 2009. Identification of small RNAs in *Mycobacterium tuberculosis*. *Mol Microbiol* 73:397–408.

9. DiChiara JM, Contreras-Martinez LM, Livny J, Smith D, McDonough KA, Belfort M. 2010. Multiple small RNAs identified in *Mycobacterium bovis* BCG are also expressed in *Mycobacterium tuberculosis* and *Mycobacterium smegmatis*. *Nucleic Acids Res* 38:4067–4078.

10. Li SK, Ng PK, Qin H, Lau JK, Lau JP, Tsui SK, Chan TF, Lau TC. 2013. Identification of small RNAs in *Mycobacterium smegmatis* using heterologous Hfq. *RNA* 19:74–84.

11. McGuire AM, Weiner B, Park ST, Wapinski I, Raman S, Dolganov G, Peterson M, Riley R, Zucker J, Abeel T, White J, Sisk P, Stolte C, Koehrsen M, Yamamoto RT, Iacobelli-Martinez M, Kidd MJ, Maer AM, Schoolnik GK, Regev A, Galagan J. 2012. Comparative analysis of *Mycobacterium* and related *Actinomycetes* yields insight into the evolution of *Mycobacterium tuberculosis* pathogenesis. *BMC Genomics* 13:120.

12. Miotto P, Forti F, Ambrosi A, Pellin D, Veiga DF, Balazsi G, Gennaro ML, Di Serio C, Ghisotti D, Cirillo DM. 2012. Genome-wide discovery of small RNAs in *Mycobacterium tuberculosis*. *PLoS One* 7:e51950.

13. Pellin D, Miotto P, Ambrosi A, Cirillo DM, Di Serio C. 2012. A genome-wide identification analysis of small regulatory RNAs in *Mycobacterium tuberculosis* by RNA-Seq and conservation analysis. *PLoS One* 7:e32723.

14. Pelly S, Bishai WR, Lamichhane G. 2012. A screen for non-coding RNA in *Mycobacterium tuberculosis* reveals a cAMP-responsive RNA that is expressed during infection. *Gene* 500:85–92.

15. Lamichhane G, Arnvig KB, McDonough KA. 2013. Definition and annotation of (myco)bacterial noncoding RNA. *Tuberculosis (Edinb)* 93:26–29.

16. Cortes T, Schubert OT, Rose G, Arnvig KB, Comas I, Aebersold R, Young DB. 2013. Genome-wide mapping of transcriptional start sites defines an extensive leaderless transcriptome in *Mycobacterium tuberculosis*. *Cell Rep* 5:1121–1131.

17. Carver T, Harris SR, Berriman M, Parkhill J, McQuillan JA. 2012. Artemis: an integrated platform for visualization and analysis of high-throughput sequence-based experimental data. *Bioinformatics* 28:464–469.

18. Chaudhuri RR, Loman NJ, Snyder LA, Bailey CM, Stekel DJ, Pallen MJ. 2008. xBASE2: a comprehensive resource for comparative bacterial genomics. *Nucleic Acids Res* 36:D543–D546.

19. Reddy TB, Riley R, Wymore F, Montgomery P, DeCaprio D, Engels R, Gellesch M, Hubble J, Jen D, Jin H, Koehrsen M, Larson L, Mao M, Nitzberg M, Sisk P, Stolte C, Weiner B, White J, Zachariah ZK, Sherlock G, Galagan JE, Ball CA, Schoolnik GK. 2009. TB database: an integrated platform for tuberculosis research. *Nucleic Acids Res* 37:D499–D508.

20. Cole ST, Brosch R, Parkhill J, Garnier T, Churcher C, Harris D, Gordon SV, Eiglmeier K, Gas S, Barry CE 3rd, Tekaia F, Badcock K, Basham D, Brown D, Chillingworth T, Connor R, Davies R, Devlin K, Feltwell T, Gentles S, Hamlin N, Holroyd S, Hornsby T, Jagels K, Krogh A, McLean J, Moule S, Murphy L, Oliver K, Osborne J, Quail MA, Rajandream MA, Rogers J, Rutter S, Seeger K, Skelton J, Squares R, Squares S, Sulston JE, Taylor K, Whitehead S, Barrell

BG. 1998. Deciphering the biology of *Mycobacterium tuberculosis* from the complete genome sequence. *Nature* 393:537–544.

21. Abramovitch RB, Rohde KH, Hsu FF, Russell DG. 2011. aprABC: a *Mycobacterium tuberculosis* complex-specific locus that modulates pH-driven adaptation to the macrophage phagosome. *Mol Microbiol* 80:678–694.

22. Dejesus MA, Sacchettini JC, Ioerger TR. 2013. Reannotation of translational start sites in the genome of *Mycobacterium tuberculosis*. *Tuberculosis (Edinb)* 93:18–25.

23. Bell C, Smith GT, Sweredoski MJ, Hess S. 2012. Characterization of the *Mycobacterium tuberculosis* proteome by liquid chromatography mass spectrometry-based proteomics techniques: a comprehensive resource for tuberculosis research. *J Proteome Res* 11:119–130.

24. Kruh NA, Troudt J, Izzo A, Prenni J, Dobos KM. 2010. Portrait of a pathogen: the *Mycobacterium tuberculosis* proteome in vivo. *PloS One* 5:e13938.

25. Schubert OT, Mouritsen J, Ludwig C, Rost HL, Rosenberger G, Arthur PK, Claassen M, Campbell DS, Sun Z, Farrah T, Gengenbacher M, Maiolica A, Kaufmann SH, Moritz RL, Aebersold R. 2013. The Mtb Proteome Library: a resource of assays to quantify the complete proteome of *Mycobacterium tuberculosis*. *Cell Host Microbe* 13:602–612.

26. Unniraman S, Prakash R, Nagaraja V. 2001. Alternate paradigm for intrinsic transcription termination in eubacteria. *J Biol Chem* 276:41850–41855.

27. Mohanty BK, Kushner SR. 2011. Bacterial/archaeal /organellar polyadenylation. *Wiley Interdiscip Rev RNA* 2:256–276.

28. Adilakshmi T, Ayling PD, Ratledge C. 2000. Poly-adenylylation in mycobacteria: evidence for oligo(dT)-primed cDNA synthesis. *Microbiology* 146(Pt 3): 633–638.

29. Gardner PP, Barquist L, Bateman A, Nawrocki EP, Weinberg Z. 2011. RNIE: genome-wide prediction of bacterial intrinsic terminators. *Nucleic Acids Res* 39: 5845–5852.

30. Gorna MW, Carpousis AJ, Luisi BF. 2012. From conformational chaos to robust regulation: the structure and function of the multi-enzyme RNA degradosome. *Q Rev Biophys* 45:105–145.

31. Argaman L, Hershberg R, Vogel J, Bejerano G, Wagner EG, Margalit H, Altuvia S. 2001. Novel small RNA-encoding genes in the intergenic regions of *Escherichia coli*. *Curr Biol* 11:941–950.

32. Sharma CM, Hoffmann S, Darfeuille F, Reignier J, Findeiss S, Sittka A, Chabas S, Reiche K, Hackermuller J, Reinhardt R, Stadler PF, Vogel J. 2010. The primary transcriptome of the major human pathogen *Helicobacter pylori*. *Nature* 464:250–255.

33. Narberhaus F, Waldminghaus T, Chowdhury S. 2006. RNA thermometers. *FEMS Microbiol Rev* 30:3–16.

34. Nechooshtan G, Elgrably-Weiss M, Sheaffer A, Westhof E, Altuvia S. 2009. A pH-responsive riboregulator. *Genes Dev* 23:2650–2662.

35. Nahvi A, Sudarsan N, Ebert MS, Zou X, Brown KL, Breaker RR. 2002. Genetic control by a metabolite binding mRNA. *Chem Biol* 9:1043.

36. Breaker RR. 2012. Riboswitches and the RNA world. *Cold Spring Harbor Perspect Biol* 4.

37. Lindahl L, Zengel JM. 1986. Ribosomal genes in *Escherichia coli*. *Annu Rev Genet* 20:297–326.

38. Arnvig KB, Pennell S, Gopal B, Colston MJ. 2004. A high-affinity interaction between NusA and the rrn nut site in *Mycobacterium tuberculosis*. *Proc Natl Acad Sci USA* 101:8325–8330.

39. Arnvig KB, Zeng S, Quan S, Papageorge A, Zhang N, Villapakkam AC, Squires CL. 2008. Evolutionary comparison of ribosomal operon antitermination function. *J Bacteriol* 190:7251–7257.

40. Condon C, Squires C, Squires CL. 1995. Control of rRNA transcription in *Escherichia coli*. *Microbiol Rev* 59:623–645.

41. Quan S, Zhang N, French S, Squires CL. 2005. Transcriptional polarity in rRNA operons of *Escherichia colinus A* and *nusB* mutant strains. *J Bacteriol* 187: 1632–1638.

42. Vogel U, Jensen KF. 1995. Effects of the antiterminator BoxA on transcription elongation kinetics and ppGpp inhibition of transcription elongation in *Escherichia coli*. *J Biol Chem* 270:18335–18340.

43. Beuth B, Pennell S, Arnvig KB, Martin SR, Taylor IA. 2005. Structure of a *Mycobacterium tuberculosis* NusA-RNA complex. *EMBO J* 24:3576–3587.

44. Bubunenko M, Court DL, Al Refaii A, Saxena S, Korepanov A, Friedman DI, Gottesman ME, Alix JH. 2013. Nus transcription elongation factors and RNase III modulate small ribosome subunit biogenesis in *Escherichia coli*. *Mol Microbiol* 87:382–393.

45. Tucker BJ, Breaker RR. 2005. Riboswitches as versatile gene control elements. *Curr Opinion Struct Biol* 15: 342–348.

46. Gardner PP, Daub J, Tate J, Moore BL, Osuch IH, Griffiths-Jones S, Finn RD, Nawrocki EP, Kolbe DL, Eddy SR, Bateman A. 2011. Rfam: Wikipedia, clans and the "decimal" release. *Nucleic Acids Res* 39: D141–D145.

47. Weinberg Z, Barrick JE, Yao Z, Roth A, Kim JN, Gore J, Wang JX, Lee ER, Block KF, Sudarsan N, Neph S, Tompa M, Ruzzo WL, Breaker RR. 2007. Identification of 22 candidate structured RNAs in bacteria using the CMfinder comparative genomics pipeline. *Nucleic Acids Res* 35:4809–4819.

48. Weinberg Z, Regulski EE, Hammond MC, Barrick JE, Yao Z, Ruzzo WL, Breaker RR. 2008. The aptamer core of SAM-IV riboswitches mimics the ligand-binding site of SAM-I riboswitches. *RNA* 14:822–828.

49. Loh E, Dussurget O, Gripenland J, Vaitkevicius K, Tiensuu T, Mandin P, Repoila F, Buchrieser C, Cossart P, Johansson J. 2009. A trans-acting riboswitch controls expression of the virulence regulator PrfA in *Listeria monocytogenes*. *Cell* 139:770–779.

50. Rodionov DA, Vitreschak AG, Mironov AA, Gelfand MS. 2003. Comparative genomics of the vitamin B12

metabolism and regulation in prokaryotes. *J Biol Chem* **278:**41148–41159.

51. **Vitreschak AG, Rodionov DA, Mironov AA, Gelfand MS.** 2003. Regulation of the vitamin B12 metabolism and transport in bacteria by a conserved RNA structural element. *RNA* **9:**1084–1097.

52. **Warner DF, Savvi S, Mizrahi V, Dawes SS.** 2007. A riboswitch regulates expression of the coenzyme B12-independent methionine synthase in *Mycobacterium tuberculosis:* implications for differential methionine synthase function in strains H37Rv and CDC1551. *J Bacteriol* **189:**3655–3659.

53. **Akhter Y, Ehebauer MT, Mukhopadhyay S, Hasnain SE.** 2012. The PE/PPE multigene family codes for virulence factors and is a possible source of mycobacterial antigenic variation: perhaps more? *Biochimie* **94:**110–116.

54. **Gopinath K, Venclovas C, Ioerger TR, Sacchettini JC, McKinney JD, Mizrahi V, Warner DF.** 2013. A vitamin B12 transporter in *Mycobacterium tuberculosis. Open Biol* **3:**120175.

55. **Dann CE 3rd, Wakeman CA, Sieling CL, Baker SC, Irnov I, Winkler WC.** 2007. Structure and mechanism of a metal-sensing regulatory RNA. *Cell* **130:**878–892.

56. **Barrick JE, Corbino KA, Winkler WC, Nahvi A, Mandal M, Collins J, Lee M, Roth A, Sudarsan N, Jona I, Wickiser JK, Breaker RR.** 2004. New RNA motifs suggest an expanded scope for riboswitches in bacterial genetic control. *Proc Natl Acad Sci USA* **101:**6421–6426.

57. **Walters SB, Dubnau E, Kolesnikova I, Laval F, Daffe M, Smith I.** 2006. The *Mycobacterium tuberculosis* PhoPR two-component system regulates genes essential for virulence and complex lipid biosynthesis. *Mol Microbiol* **60:**312–330.

58. **Vitreschak AG, Mironov AA, Lyubetsky VA, Gelfand MS.** 2008. Comparative genomic analysis of T-box regulatory systems in bacteria. *RNA* **14:**717–735.

59. **Biketov S, Potapov V, Ganina E, Downing K, Kana BD, Kaprelyants A.** 2007. The role of resuscitation promoting factors in pathogenesis and reactivation of *Mycobacterium tuberculosis* during intra-peritoneal infection in mice. *BMC Infect Dis* **7:**146.

60. **Keep NH, Ward JM, Cohen-Gonsaud M, Henderson B.** 2006. Wake up! Peptidoglycan lysis and bacterial non-growth states. *Trends Microbiol* **14:**271–276.

61. **Telkov MV, Demina GR, Voloshin SA, Salina EG, Dudik TV, Stekhanova TN, Mukamolova GV, Kazaryan KA, Goncharenko AV, Young M, Kaprelyants AS.** 2006. Proteins of the Rpf (resuscitation promoting factor) family are peptidoglycan hydrolases. *Biochemistry (Mosc)* **71:**414–422.

62. **Block KF, Hammond MC, Breaker RR.** 2010. Evidence for widespread gene control function by the ydaO riboswitch candidate. *J Bacteriol* **192:**3983–3989.

63. **Nelson JW, Sudarsan N, Furukawa K, Weinberg Z, Wang JX, Breaker RR.** 2013. Riboswitches in eubacteria sense the second messenger c-di-AMP. *Nature Chemical Biol* **9:**834–839.

64. **Breaker RR.** 2011. Prospects for riboswitch discovery and analysis. *Mol Cell* **43:**867–879.

65. **Storz G, Vogel J, Wassarman KM.** 2011. Regulation by small RNAs in bacteria: expanding frontiers. *Mol Cell* **43:**880–891.

66. **Babitzke P, Romeo T.** 2007. CsrB sRNA family: sequestration of RNA-binding regulatory proteins. *Curr Opin Microbiol* **10:**156–163.

67. **Wassarman KM.** 2007. 6S RNA: a regulator of transcription. *Mol Microbiol* **65:**1425–1431.

68. **Wassarman KM.** 2007. 6S RNA: a small RNA regulator of transcription. *Curr Opin Microbiol* **10:**164–168.

69. **Thomason MK, Storz G.** 2010. Bacterial antisense RNAs: how many are there, and what are they doing? *Annu Rev Genet* **44:**167–188.

70. **Sesto N, Wurtzel O, Archambaud C, Sorek R, Cossart P.** 2013. The excludon: a new concept in bacterial antisense RNA-mediated gene regulation. *Nat Rev Microbiol* **11:**75–82.

71. **Brantl S, Wagner EG.** 2000. Antisense RNA-mediated transcriptional attenuation: an in vitro study of plasmid pT181. *Mol Microbiol* **35:**1469–1482.

72. **Stork M, Di Lorenzo M, Welch TJ, Crosa JH.** 2007. Transcription termination within the iron transport-biosynthesis operon of *Vibrio anguillarum* requires an antisense RNA. *J Bacteriol* **189:**3479–3488.

73. **Opdyke JA, Fozo EM, Hemm MR, Storz G.** 2011. RNase III participates in GadY-dependent cleavage of the gadX-gadW mRNA. *J Mol Biol* **406:**29–43.

74. **Opdyke JA, Kang JG, Storz G.** 2004. GadY, a small-RNA regulator of acid response genes in *Escherichia coli. J Bacteriol* **186:**6698–6705.

75. **Stazic D, Lindell D, Steglich C.** 2011. Antisense RNA protects mRNA from RNase E degradation by RNA-RNA duplex formation during phage infection. *Nucleic Acids Res* **39:**4890–4899.

76. **Lasa I, Toledo-Arana A, Dobin A, Villanueva M, de Los Mozos IR, Vergara-Irigaray M, Segura V, Fagegaltier D, Penades JR, Valle J, Solano C, Gingeras TR.** 2011. Genome-wide antisense transcription drives mRNA processing in bacteria. *Proc Natl Acad Sci USA* **108:**20172–20177.

77. **Rose G, Cortes T, Comas I, Coscolla M, Gagneux S, Young DB.** 2013. Mapping of genotype-phenotype diversity amongst clinical isolates of *Mycobacterium tuberculosis* by sequence-based transcriptional profiling. *Genome Biol Evol* **5:**1849–1862.

78. **Movahedzadeh F, Smith DA, Norman RA, Dinadayala P, Murray-Rust J, Russell DG, Kendall SL, Rison SC, McAlister MS, Bancroft GJ, McDonald NQ, Daffe M, Av-Gay Y, Stoker NG.** 2004. The *Mycobacterium tuberculosis* ino1 gene is essential for growth and virulence. *Mol Microbiol* **51:**1003–1014.

79. **Jager D, Pernitzsch SR, Richter AS, Backofen R, Sharma CM, Schmitz RA.** 2012. An archaeal sRNA targeting cis- and trans-encoded mRNAs via two distinct domains. *Nucleic Acids Res* **40:**10964–10979.

80. **Hotter GS, Collins DM.** 2011. *Mycobacterium bovis* lipids: virulence and vaccines. *Vet Microbiol* **151:**91–98.

81. Sassetti CM, Boyd DH, Rubin EJ. 2003. Genes required for mycobacterial growth defined by high density mutagenesis. *Mol Microbiol* 48:77–84.

82. Sassetti CM, Rubin EJ. 2003. Genetic requirements for mycobacterial survival during infection. *Proc Natl Acad Sci USA* 100:12989–12994.

83. Schnappinger D, Ehrt S, Voskuil MI, Liu Y, Mangan JA, Monahan IM, Dolganov G, Efron B, Butcher PD, Nathan C, Schoolnik GK. 2003. Transcriptional adaptation of *Mycobacterium tuberculosis* within macrophages: insights into the phagosomal environment. *J Exp Med* 198:693–704.

84. Matsunaga I, Bhatt A, Young DC, Cheng TY, Eyles SJ, Besra GS, Briken V, Porcelli SA, Costello CE, Jacobs WR Jr, Moody DB. 2004. *Mycobacterium tuberculosis* pks12 produces a novel polyketide presented by CD1c to T cells. *J Exp Med* 200:1559–1569.

85. Eggenhofer F, Tafer H, Stadler PF, Hofacker IL. 2011. RNApredator: fast accessibility-based prediction of sRNA targets. *Nucleic Acids Res* 39:W149–W154.

86. Tjaden B. 2008. TargetRNA: a tool for predicting targets of small RNA action in bacteria. *Nucleic Acids Res* 36:W109–W113.

87. Arnvig K, Young D. 2012. Non-coding RNA and its potential role in *Mycobacterium tuberculosis* pathogenesis. *RNA Biol* 9:427–436.

88. Bardill JP, Hammer BK. 2012. Non-coding sRNAs regulate virulence in the bacterial pathogen *Vibrio cholerae*. *RNA Biol* 9:392–401.

89. Papenfort K, Vogel J. 2010. Regulatory RNA in bacterial pathogens. *Cell Host Microbe* 8:116–127.

90. Bossi L, Schwartz A, Guillemardet B, Boudvillain M, Figueroa-Bossi N. 2012. A role for Rho-dependent polarity in gene regulation by a noncoding small RNA. *Genes Dev* 26:1864–1873.

91. Beisel CL, Storz G. 2010. Base pairing small RNAs and their roles in global regulatory networks. *FEMS Microbiol Rev* 34:866–882.

92. Gottesman S, Storz G. 2011. Bacterial small RNA regulators: versatile roles and rapidly evolving variations. *Cold Spring Harbor Perspect Biol* 3.

93. Battesti A, Majdalani N, Gottesman S. 2011. The RpoS-mediated general stress response in *Escherichia coli*. *Annu Rev Microbiol* 65:189–213.

94. Geisinger E, Adhikari RP, Jin R, Ross HF, Novick RP. 2006. Inhibition of rot translation by RNAIII, a key feature of agr function. *Mol Microbiol* 61:1038–1048.

95. Svenningsen SL, Tu KC, Bassler BL. 2009. Gene dosage compensation calibrates four regulatory RNAs to control *Vibrio cholerae* quorum sensing. *EMBO J* 28:429–439.

96. Panek J, Krasny L, Bobek J, Jezkova E, Korelusova J, Vohradsky J. 2011. The suboptimal structures find the optimal RNAs: homology search for bacterial noncoding RNAs using suboptimal RNA structures. *Nucleic Acids Res* 39:3418–3426.

97. Chao Y, Vogel J. 2010. The role of Hfq in bacterial pathogens. *Curr Opin Microbiol* 13:24–33.

98. Bohn C, Rigoulay C, Bouloc P. 2007. No detectable effect of RNA-binding protein Hfq absence in *Staphylococcus aureus*. *BMC Microbiol* 7:10.

99. Liu Y, Wu N, Dong J, Gao Y, Zhang X, Mu C, Shao N, Yang G. 2010. Hfq is a global regulator that controls the pathogenicity of *Staphylococcus aureus*. *PloS One* 5.

100. Otaka H, Ishikawa H, Morita T, Aiba H. 2011. PolyU tail of rho-independent terminator of bacterial small RNAs is essential for Hfq action. *Proc Natl Acad Sci USA* 108:13059–13064.

101. Vogel J, Luisi BF. 2011. Hfq and its constellation of RNA. *Nat Rev Microbiol* 9:578–589.

102. Gaballa A, Antelmann H, Aguilar C, Khakh SK, Song KB, Smaldone GT, Helmann JD. 2008. The *Bacillus subtilis* iron-sparing response is mediated by a Fur-regulated small RNA and three small, basic proteins. *Proc Natl Acad Sci USA* 105:11927–11932.

103. Bandyra KJ, Said N, Pfeiffer V, Gorna MW, Vogel J, Luisi BF. 2012. The seed region of a small RNA drives the controlled destruction of the target mRNA by the endoribonuclease RNase E. *Mol Cell* 47:943–953.

104. Said N, Rieder R, Hurwitz R, Deckert J, Urlaub H, Vogel J. 2009. In vivo expression and purification of aptamer-tagged small RNA regulators. *Nucleic Acids Res* 37:e133.

105. Hartkoorn RC, Sala C, Uplekar S, Busso P, Rougemont J, Cole ST. 2012. Genome-wide definition of the SigF regulon in *Mycobacterium tuberculosis*. *J Bacteriol* 194:2001–2009.

106. England K, Crew R, Slayden RA. 2011. *Mycobacterium tuberculosis* septum site determining protein, Ssd encoded by rv3660c, promotes filamentation and elicits an alternative metabolic and dormancy stress response. *BMC Microbiol* 11:79.

107. Geissmann T, Chevalier C, Cros MJ, Boisset S, Fechter P, Noirot C, Schrenzel J, Francois P, Vandenesch F, Gaspin C, Romby P. 2009. A search for small non-coding RNAs in *Staphylococcus aureus* reveals a conserved sequence motif for regulation. *Nucleic Acids Res* 37:7239–7257.

108. Sherman DR, Voskuil M, Schnappinger D, Liao R, Harrell MI, Schoolnik GK. 2001. Regulation of the *Mycobacterium tuberculosis* hypoxic response gene encoding alpha -crystallin. *Proc Natl Acad Sci USA* 98:7534–7539.

109. Garton NJ, Waddell SJ, Sherratt AL, Lee SM, Smith RJ, Senner C, Hinds J, Rajakumar K, Adegbola RA, Besra GS, Butcher PD, Barer MR. 2008. Cytological and transcript analyses reveal fat and lazy persister-like bacilli in tuberculous sputum. *PLoS Med* 5:e75.

110. Voskuil MI, Schnappinger D, Visconti KC, Harrell MI, Dolganov GM, Sherman DR, Schoolnik GK. 2003. Inhibition of respiration by nitric oxide induces a *Mycobacterium tuberculosis* dormancy program. *J Exp Med* 198:705–713.

111. Beste DJ, Peters J, Hooper T, Avignone-Rossa C, Bushell ME, McFadden J. 2005. Compiling a molecular inventory for *Mycobacterium bovis* BCG at two growth

rates: evidence for growth rate-mediated regulation of ribosome biosynthesis and lipid metabolism. *J Bacteriol* **187:**1677–1684.

112. Pieters J. 2008. *Mycobacterium tuberculosis* and the macrophage: maintaining a balance. *Cell Host Microbe* **3:**399–407.

113. Tan S, Sukumar N, Abramovitch RB, Parish T, Russell DG. 2013. *Mycobacterium tuberculosis* responds to chloride and pH as synergistic cues to the immune status of its host cell. *PLoS Pathog* **9:**e1003282.

114. Betts JC, Lukey PT, Robb LC, McAdam RA, Duncan K. 2002. Evaluation of a nutrient starvation model of *Mycobacterium tuberculosis* persistence by gene and protein expression profiling. *Mol Microbiol* **43:**717–731.

115. Keren I, Minami S, Rubin E, Lewis K. 2011. Characterization and transcriptome analysis of *Mycobacterium tuberculosis* persisters. *mBio* **2:**e00100–e00111.

116. Barrick JE, Sudarsan N, Weinberg Z, Ruzzo WL, Breaker RR. 2005. 6S RNA is a widespread regulator of eubacterial RNA polymerase that resembles an open promoter. *RNA* **11:**774–784.

117. Rustad TR, Minch KJ, Brabant W, Winkler JK, Reiss DJ, Baliga NS, Sherman DR. 2013. Global analysis of mRNA stability in *Mycobacterium tuberculosis*. *Nucleic Acids Res* **41:**509–517.

118. Marcaida MJ, DePristo MA, Chandran V, Carpousis AJ, Luisi BF. 2006. The RNA degradosome: life in the fast lane of adaptive molecular evolution. *Trends Biochem Sci* **31:**359–365.

119. Lehnik-Habrink M, Lewis RJ, Mader U, Stulke J. 2012. RNA degradation in *Bacillus subtilis*: an interplay of essential endo- and exoribonucleases. *Mol Microbiol* **84:**1005–1017.

120. Deveau H, Garneau JE, Moineau S. 2010. CRISPR/Cas system and its role in phage-bacteria interactions. *Annu Rev Microbiol* **64:**475–493.

121. Supply P, Marceau M, Mangenot S, Roche D, Rouanet C, Khanna V, Majlessi L, Criscuolo A, Tap J, Pawlik A, Fiette L, Orgeur M, Fabre M, Parmentier C, Frigui W, Simeone R, Boritsch EC, Debrie AS, Willery E, Walker D, Quail MA, Ma L, Bouchier C, Salvignol G, Sayes F, Cascioferro A, Seemann T, Barbe V, Locht C, Gutierrez MC, Leclerc C, Bentley SD, Stinear TP, Brisse S, Medigue C, Parkhill J, Cruveiller S, Brosch R. 2013. Genomic analysis of smooth tubercle bacilli provides insights into ancestry and pathoadaptation of *Mycobacterium tuberculosis*. *Nat Genet* **45:**172–179.

122. Driscoll JR. 2009. Spoligotyping for molecular epidemiology of the *Mycobacterium tuberculosis* complex. *Methods Mol Biol* **551:**117–128.

123. Makarova KS, Haft DH, Barrangou R, Brouns SJ, Charpentier E, Horvath P, Moineau S, Mojica FJ, Wolf YI, Yakunin AF, van der Oost J, Koonin EV. 2011. Evolution and classification of the CRISPR-Cas systems. *Nat Rev Microbiol* **9:**467–477.

124. Marraffini LA, Sontheimer EJ. 2008. CRISPR interference limits horizontal gene transfer in staphylococci by targeting DNA. *Science* **322:**1843–1845.

125. Hanekom M, Gey van Pittius NC, McEvoy C, Victor TC, Van Helden PD, Warren RM. 2011. *Mycobacterium tuberculosis* Beijing genotype: a template for success. *Tuberculosis (Edinb)* **91:**510–523.

126. Tsolaki AG, Hirsh AE, DeRiemer K, Enciso JA, Wong MZ, Hannan M, Goguet de la Salmoniere YO, Aman K, Kato-Maeda M, Small PM. 2004. Functional and evolutionary genomics of *Mycobacterium tuberculosis*: insights from genomic deletions in 100 strains. *Proc Natl Acad Sci USA* **101:**4865–4870.

127. Fenner L, Malla B, Ninet B, Dubuis O, Stucki D, Borrell S, Huna T, Bodmer T, Egger M, Gagneux S. 2011. "Pseudo-Beijing": evidence for convergent evolution in the direct repeat region of *Mycobacterium tuberculosis*. *PLoS One* **6:**e24737.

128. Wassarman KM, Zhang A, Storz G. 1999. Small RNAs in *Escherichia coli*. *Trends Microbiol* **7:**37–45.

129. Vogel J. 2009. A rough guide to the non-coding RNA world of Salmonella. *Mol Microbiol* **71:**1–11.

130. Macke TJ, Ecker DJ, Gutell RR, Gautheret D, Case DA, Sampath R. 2001. RNAMotif, an RNA secondary structure definition and search algorithm. *Nucleic Acids Res* **29:**4724–4735.

131. Chen Y, Indurthi DC, Jones SW, Papoutsakis ET. 2011. Small RNAs in the genus *Clostridium*. *mBio* **2:**e00340–e00310.

132. Liang H, Zhao YT, Zhang JQ, Wang XJ, Fang RX, Jia YT. 2011. Identification and functional characterization of small non-coding RNAs in *Xanthomonas oryzae* pathovar oryzae. *BMC Genomics* **12:**87.

133. Panek J, Bobek J, Mikulik K, Basler M, Vohradsky J. 2008. Biocomputational prediction of small non-coding RNAs in *Streptomyces*. *BMC Genomics* **9:**217.

134. Schluter JP, Reinkensmeier J, Daschkey S, Evguenieva-Hackenberg E, Janssen S, Janicke S, Becker JD, Giegerich R, Becker A. 2010. A genome-wide survey of sRNAs in the symbiotic nitrogen-fixing alphaproteobacterium *Sinorhizobium meliloti*. *BMC Genomics* **11:**245.

135. Voss B, Georg J, Schon V, Ude S, Hess WR. 2009. Biocomputational prediction of non-coding RNAs in model cyanobacteria. *BMC Genomics* **10:**123.

136. Rivas E, Klein RJ, Jones TA, Eddy SR. 2001. Computational identification of noncoding RNAs in *E. coli* by comparative genomics. *Curr Biol* **11:**1369–1373.

137. Gautheret D, Lambert A. 2001. Direct RNA motif definition and identification from multiple sequence alignments using secondary structure profiles. *J Mol Biol* **313:**1003–1011.

138. Pichon C, Felden B. 2003. Intergenic sequence inspector: searching and identifying bacterial RNAs. *Bioinformatics* **19:**1707–1709.

139. Nawrocki EP, Kolbe DL, Eddy SR. 2009. Infernal 1.0: inference of RNA alignments. *Bioinformatics* **25:**1335–1337.

140. Coventry A, Kleitman DJ, Berger B. 2004. MSARI: multiple sequence alignments for statistical detection of RNA secondary structure. *Proc Natl Acad Sci USA* **101:**12102–12107.

141. Pedersen JS, Bejerano G, Siepel A, Rosenbloom K, Lindblad-Toh K, Lander ES, Kent J, Miller W, Haussler D. 2006. Identification and classification of conserved RNA secondary structures in the human genome. *PLoS Comput Biol* **2**:e33.

142. Gruber AR, Findeiss S, Washietl S, Hofacker IL, Stadler PF. 2010. RNAz 2.0: improved noncoding RNA detection. *Pac Symp Biocomput* **2010**:69–79.

143. Washietl S, Hofacker IL, Stadler PF. 2005. Fast and reliable prediction of noncoding RNAs. *Proc Natl Acad Sci USA* **102**:2454–2459.

144. Livny J, Fogel MA, Davis BM, Waldor MK. 2005. sRNAPredict: an integrative computational approach to identify sRNAs in bacterial genomes. *Nucleic Acids Res* **33**:4096–4105.

145. Kingsford CL, Ayanbule K, Salzberg SL. 2007. Rapid, accurate, computational discovery of Rho-independent transcription terminators illuminates their relationship to DNA uptake. *Genome Biol* **8**:R22.

146. Wingender E, Chen X, Hehl R, Karas H, Liebich I, Matys V, Meinhardt T, Pruss M, Reuter I, Schacherer F. 2000. TRANSFAC: an integrated system for gene expression regulation. *Nucleic Acids Res* **28**:316–319.

147. Livny J, Teonadi H, Livny M, Waldor MK. 2008. High-throughput, kingdom-wide prediction and annotation of bacterial non-coding RNAs. *PloS One* **3**:e3197.

148. Sridhar J, Sambaturu N, Sabarinathan R, Ou HY, Deng Z, Sekar K, Rafi ZA, Rajakumar K. 2010. sRNAscanner: a computational tool for intergenic small RNA detection in bacterial genomes. *PloS One* **5**:e11970.

149. Carter RJ, Dubchak I, Holbrook SR. 2001. A computational approach to identify genes for functional RNAs in genomic sequences. *Nucleic Acids Res* **29**:3928–3938.

150. Klein RJ, Misulovin Z, Eddy SR. 2002. Noncoding RNA genes identified in AT-rich hyperthermophiles. *Proc Natl Acad Sci USA* **99**:7542–7547.

151. Ott A, Idali A, Marchais A, Gautheret D. 2012. NAPP: the Nucleic Acid Phylogenetic Profile Database. *Nucleic Acids Res* **40**:D205–D209.

152. Griffin BE. 1971. Separation of 32P-labelled ribonucleic acid components. The use of polyethylenimine-cellulose (TLC) as a second dimension in separating oligoribonucleotides of "4.5 S" and 5 S from *E. coli*. *FEBS Lett* **15**:165–168.

153. Hindley J. 1967. Fractionation of 32P-labelled ribonucleic acids on polyacrylamide gels and their characterization by fingerprinting. *J Mol Biol* **30**:125–136.

154. Ikemura T, Dahlberg JE. 1973. Small ribonucleic acids of *Escherichia coli*. I. Characterization by polyacrylamide gel electrophoresis and fingerprint analysis. *J Biol Chem* **248**:5024–5032.

155. Altuvia S, Weinstein-Fischer D, Zhang A, Postow L, Storz G. 1997. A small, stable RNA induced by oxidative stress: role as a pleiotropic regulator and antimutator. *Cell* **90**:43–53.

156. Romeo T. 1998. Global regulation by the small RNA-binding protein CsrA and the non-coding RNA molecule CsrB. *Mol Microbiol* **29**:1321–1330.

157. Mizuno T, Chou MY, Inouye M. 1984. A unique mechanism regulating gene expression: translational inhibition by a complementary RNA transcript (micRNA). *Proc Natl Acad Sci USA* **81**:1966–1970.

158. Sledjeski D, Gottesman S. 1995. A small RNA acts as an antisilencer of the H-NS-silenced rcsA gene of *Escherichia coli*. *Proc Natl Acad Sci USA* **92**:2003–2007.

159. Vogel J, Sharma CM. 2005. How to find small noncoding RNAs in bacteria. *Biol Chem* **386**:1219–1238.

160. Chen S, Lesnik EA, Hall TA, Sampath R, Griffey RH, Ecker DJ, Blyn LB. 2002. A bioinformatics based approach to discover small RNA genes in the *Escherichia coli* genome. *Biosystems* **65**:157–177.

161. Wassarman KM, Repoila F, Rosenow C, Storz G, Gottesman S. 2001. Identification of novel small RNAs using comparative genomics and microarrays. *Genes Dev* **15**:1637–1651.

162. Livny J, Brencic A, Lory S, Waldor MK. 2006. Identification of 17 *Pseudomonas aeruginosa* sRNAs and prediction of sRNA-encoding genes in 10 diverse pathogens using the bioinformatic tool sRNA Predict2. *Nucleic Acids Res* **34**:3484–3493.

163. Toledo-Arana A, Dussurget O, Nikitas G, Sesto N, Guet-Revillet H, Balestrino D, Loh E, Gripenland J, Tiensuu T, Vaitkevicius K, Barthelemy M, Vergassola M, Nahori MA, Soubigou G, Regnault B, Coppee JY, Lecuit M, Johansson J, Cossart P. 2009. The *Listeria* transcriptional landscape from saprophytism to virulence. *Nature* **459**:950–956.

164. Pichon C, Felden B. 2005. Small RNA genes expressed from *Staphylococcus aureus* genomic and pathogenicity islands with specific expression among pathogenic strains. *Proc Natl Acad Sci USA* **102**:14249–14254.

165. Silvaggi JM, Perkins JB, Losick R. 2006. Genes for small, noncoding RNAs under sporulation control in *Bacillus subtilis*. *J Bacteriol* **188**:532–541.

166. Patenge N, Billion A, Raasch P, Normann J, Wisniewska-Kucper A, Retey J, Boisguerin V, Hartsch T, Hain T, Kreikemeyer B. 2012. Identification of novel growth phase- and media-dependent small non-coding RNAs in *Streptococcus pyogenes* M49 using intergenic tiling arrays. *BMC Genomics* **13**:550.

167. Akama T, Suzuki K, Tanigawa K, Kawashima A, Wu H, Nakata N, Osana Y, Sakakibara Y, Ishii N. 2009. Whole-genome tiling array analysis of *Mycobacterium leprae* RNA reveals high expression of pseudogenes and noncoding regions. *J Bacteriol* **191**:3321–3327.

168. Oliver HF, Orsi RH, Ponnala L, Keich U, Wang W, Sun Q, Cartinhour SW, Filiatrault MJ, Wiedmann M, Boor KJ. 2009. Deep RNA sequencing of *L. monocytogenes* reveals overlapping and extensive stationary phase and sigma B-dependent transcriptomes, including multiple highly transcribed noncoding RNAs. *BMC Genomics* **10**:641.

169. Passalacqua KD, Varadarajan A, Ondov BD, Okou DT, Zwick ME, Bergman NH. 2009. Structure and complexity of a bacterial transcriptome. *J Bacteriol* **191**:3203–3211.

170. Yoder-Himes DR, Chain PS, Zhu Y, Wurtzel O, Rubin EM, Tiedje JM, Sorek R. 2009. Mapping the *Burkholderia cenocepacia* niche response via high-throughput sequencing. *Proc Natl Acad Sci USA* **106:** 3976–3981.

171. Guell M, van Noort V, Yus E, Chen WH, Leigh-Bell J, Michalodimitrakis K, Yamada T, Arumugam M, Doerks T, Kuhner S, Rode M, Suyama M, Schmidt S, Gavin AC, Bork P, Serrano L. 2009. Transcriptome complexity in a genome-reduced bacterium. *Science* **326:**1268–1271.

172. Perkins TT, Kingsley RA, Fookes MC, Gardner PP, James KD, Yu L, Assefa SA, He M, Croucher NJ, Pickard DJ, Maskell DJ, Parkhill J, Choudhary J, Thomson NR, Dougan G. 2009. A strand-specific RNA-Seq analysis of the transcriptome of the typhoid bacillus *Salmonella typhi*. *PLoS Genet* **5:**e1000569.

173. Sittka A, Lucchini S, Papenfort K, Sharma CM, Rolle K, Binnewies TT, Hinton JC, Vogel J. 2008. Deep sequencing analysis of small noncoding RNA and mRNA targets of the global post-transcriptional regulator, Hfq. *PLoS Genet* **4:**e1000163.

174. Dugar G, Herbig A, Forstner KU, Heidrich N, Reinhardt R, Nieselt K, Sharma CM. 2013. High-resolution transcriptome maps reveal strain-specific regulatory features of multiple *Campylobacter jejuni* isolates. *PLoS Genet* **9:**e1003495.

175. Pichon C, Felden B. 2007. Proteins that interact with bacterial small RNA regulators. *FEMS Microbiol Rev* **31:**614–625.

176. Brennan RG, Link TM. 2007. Hfq structure, function and ligand binding. *Curr Opin Microbiol* **10:**125–133.

177. Uzzau S, Figueroa-Bossi N, Rubino S, Bossi L. 2001. Epitope tagging of chromosomal genes in *Salmonella*. *Proc Natl Acad Sci USA* **98:**15264–15269.

178. Sonnleitner E, Sorger-Domenigg T, Madej MJ, Findeiss S, Hackermuller J, Huttenhofer A, Stadler PF, Blasi U, Moll I. 2008. Detection of small RNAs in *Pseudomonas aeruginosa* by RNomics and structure-based bioinformatic tools. *Microbiology* **154:**3175–3187.

179. Christiansen JK, Nielsen JS, Ebersbach T, Valentin-Hansen P, Sogaard-Andersen L, Kallipolitis BH. 2006. Identification of small Hfq-binding RNAs in *Listeria monocytogenes*. *RNA* **12:**1383–1396.

180. Dambach M, Irnov I, Winkler WC. 2013. Association of RNAs with *Bacillus subtilis* Hfq. *PLoS One* **8:**e55156.

181. Sittka A, Sharma CM, Rolle K, Vogel J. 2009. Deep sequencing of *Salmonella* RNA associated with heterologous Hfq proteins in vivo reveals small RNAs as a major target class and identifies RNA processing phenotypes. *RNA Biol* **6:**266–275.

182. Zuker M, Stiegler P. 1981. Optimal computer folding of large RNA sequences using thermodynamics and auxiliary information. *Nucleic Acids Res* **9:**133–148.

183. McCaskill JS. 1990. The equilibrium partition function and base pair binding probabilities for RNA secondary structure. *Biopolymers* **29:**1105–1119.

184. Schuster P, Fontana W, Stadler PF, Hofacker IL. 1994. From sequences to shapes and back: a case study in RNA secondary structures. *Proc Biol Sci* **255:**279–284.

185. Sato K, Hamada M, Asai K, Mituyama T. 2009. CENTROIDFOLD: a web server for RNA secondary structure prediction. *Nucleic Acids Res* **37:**W277–W280.

186. Hamada M, Sato K, Kiryu H, Mituyama T, Asai K. 2009. Predictions of RNA secondary structure by combining homologous sequence information. *Bioinformatics* **25:**i330–i338.

187. Timmermans J, Van Melderen L. 2010. Post-transcriptional global regulation by CsrA in bacteria. *Cell Mol Life Sci* **67:**2897–2908.

188. Trotochaud AE, Wassarman KM. 2005. A highly conserved 6S RNA structure is required for regulation of transcription. *Nat Struct Mol Biol* **12:**313–319.

189. Lioliou E, Sharma CM, Caldelari I, Helfer AC, Fechter P, Vandenesch F, Vogel J, Romby P. 2012. Global regulatory functions of the *Staphylococcus aureus* endoribonuclease III in gene expression. *PLoS Genetics* **8:** e1002782.

190. Houghton J, Cortes T, Schubert OT, Rose G, Rodgers A, De Ste Croix M, Aebersold R, Young DB, Arnvig KB. 2013. A small RNA encoded in the Rv2660c locus of *Mycobacterium tuberculosis* is induced during starvation and infection. *PLoS One* **8(12):**e80047. doi: 10.1371/journal.pone.0080047.

191. Smollett KL, Smith KM, Kahramanoglou C, Arnvig KB, Buxton RS, Davis EO. 2012. Global analysis of the regulon of the transcriptional repressor LexA, a key component of SOS response in *Mycobacterium tuberculosis*. *J Biol Chem* **287:**22004–22014.

Molecular Genetics of Mycobacteria, 2nd Edition
Edited by Graham F. Hatfull and William R. Jacobs, Jr.
© 2014 American Society for Microbiology, Washington, DC
doi:10.1128/microbiolspec.MGM2-0010-2013

Tanya Parish[1]

Two-Component Regulatory Systems of Mycobacteria

10

INTRODUCTION

Two-component regulatory systems (2CRSs) are key players in bacterial responses to changing environments (1, 2). These systems act to integrate multiple stimuli into coordinated changes in global gene expression. Bacteria normally possess many 2CRSs which respond to specific signals and allow adaptive responses (2–7).

Two-component systems are composed of a sensor histidine kinase (HK) and a response regulator (RR) (1, 2, 4, 7). Changes in the external environment result in activation of the HK, which autophosphorylates on a conserved histidine residue; the HK then mediates phosphotransfer to a conserved aspartate residue on the RR (Fig. 1). Phosphorylation of the RR normally activates this protein, leading to DNA binding and promotion of transcription for a set of genes, termed the regulon.

Sensor HKs

HKs are usually membrane proteins, composed of a receiver domain, which senses the specific signal, and the transmitter domain, with kinase activity (2, 8). The specificity of the system lies in the receiver domain, which senses changes in the environment; HKs can respond to a number of stimuli including phosphate concentration, temperature, light, oxygen availability,

pH, and redox states. The kinase domain is largely conserved and provides both the site of autophosphorylation and the interaction domain with the RR. HKs function as dimers in which one monomer catalyzes phosphorylation of the histidine residue in the other monomers. The C-terminal transfer domain of HKs is generally composed of a dimerization and phosphotransfer (DHp) domain and the catalytic/ATP binding domain. DHp domains are the interaction sites between two monomers of the HK and the interface with the RR; thus, they control the specificity of the HK and RR interaction. A small number of HKs also have phosphatase activity and can mediate both phosphorylation and dephosphorylation of their cognate RRs.

RRs

RRs are the cytoplasmic partners responsible for receiving the signal via phosphorelay from the HK and for effecting a change in gene expression (1, 2, 4, 5). RRs preserve a two-domain structure comprising a receiver domain containing the conserved aspartate residue and an effector domain containing a DNA binding motif. The effector domains are used to group RRs into subfamilies depending on the DNA binding domain; for example, the OmpR family contains a winged helix-turn-helix motif, whereas the NarL family has a

---

[1]Infectious Disease Research Institute, Seattle, WA and Queen Mary University of London, London, United Kingdom.

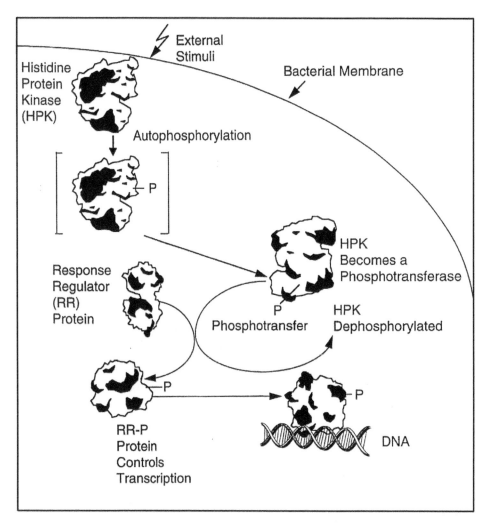

Figure 1 Two-component regulatory systems. From reference 7.
doi:10.1128/microbiolspec.MGM2-0010-2013.f1

four-helix domain and the LuxR family has a tetra-helical helix-turn-helix domain. RRs are normally phosphorylated by their cognate HK partner, but some can also actively catalyze phosphotransfer from acetyl phosphate *in vitro*, although the biological significance of this is far from clear. In addition, most RRs have autophosphatase activity.

Mycobacterial 2CRSs
Mycobacteria have relatively few 2CRSs compared to other bacterial species (3, 6). Most studies have focused on their role in the pathogenic species *Mycobacterium tuberculosis*, although some data are available for the vaccine strain *Mycobacterium bovis* BCG and the saprophyte *Mycobacterium smegmatis*. This chapter will largely focus on the 2CRSs of *M. tuberculosis*, where most studies have been conducted.

M. tuberculosis has 12 complete 2CRSs, comprising an HK and an RR (Table 1). PhoPR, SenX3/RegX3, PrrAB, and MprAB are all involved in virulence. Among their key roles, PhoP regulates cell wall composition and SenX3-RegX3 controls phosphate acquisition. MprAB is part of a complex regulatory cascade involved in the stress response and maintaining cell envelope integrity. MprAB is the only essential system that controls DNA replication and cell division. KdpDE may control potassium uptake. NarL may be responsive to hypoxia and is induced in starvation. The DosRST system incorporates two RRs that interact with a single HK. The function of TrcRS remains unclear. There are also four "orphan" RRs for which no cognate sensory proteins have been identified and about which little is known of their function. The stimuli for the sensors are largely uncharacterized; PhoPR

Table 1 The paired two-component regulatory systems of *M. tuberculosis*[a]

Gene	Rv #	Identity	Role	Stimulus
SenX3	Rv0490	Sensor	Virulence, phosphate uptake, aerobic respiration	Phosphate
RegX3	Rv0491	Regulator		
TcrA	Rv0602c	Regulator	Unknown	Unknown
HK1	Rv0600c	Sensor		
HK2	Rv0601c	Sensor		
PhoP	Rv0757	Regulator	Virulence, cell wall components, secretion of Esx-1 components	pH
PhoR	Rv0758	Sensor		
NarS	Rv0845c	Sensor	Unknown	Unknown
NarL	Rv0844c	Regulator		
PrrA	Rv0903c	Regulator	Adaptation to intracellular infection	Macrophage infection
PrrB	Rv0902c	Sensor		
MprA	Rv0981	Regulator	Sigma factors, persistence *in vivo*, stress response, cell envelope	Detergents
MprB	Rv0982	Sensor		
KdpD	Rv1028c	Sensor	Potassium uptake	Possibly [K]
KdpE	Rv1027c	Regulator		
TrcR	Rv1033c	Regulator	Controls expression of Rv1057	Unknown
TrcS	Rv1032c	Sensor		
MtrA	Rv3246c	Regulator	DNA replication, cell division	Unknown
MtrB	Rv3245c	Sensor		
TcrX	Rv3765c	Regulator	Unknown	Low iron, starvation
TcrY	Rv3764c	Sensor		
PdtaS	Rv3220c	Sensor	Unknown	Unknown
PdtaR	Rv1626	Regulator		

[a]The gene annotations for the laboratory reference strains H37Rv (Rv#) and common names are given. The role describes the pathways or genes controlled by each system. The stimulus is the environmental signal that causes either upregulation or activation of the regulator.

may respond to pH, KdpDE may respond to potassium concentration, and SenX3 is indirectly stimulated by phosphate, although accessory proteins may be involved in actually sensing these parameters.

Although most HKs and RRs function as pairs, crosstalk between systems can occur where an HK phosphorylates a noncognate RR. Several such interactions have been demonstrated between *M. tuberculosis* recombinant proteins (10), although it is not clear what significance these interactions have, nor whether they truly occur in the mycobacterial cell. For example, PhoR and DosT can interact with NarL and PrrA, and PdtaS can interact with TcrX (10). PhoP appears to control the hypoxic response via interaction with DosR (11).

SPECIFIC SYSTEMS

A number of two-component systems have been studied in detail, in particular the DosRST, PhoPR, MprAB, and SenX3-RegX3 2CRSs.

PhoPR

The PhoPR system is one of the most studied of the mycobacterial 2CRSs and plays a key role in virulence. The structure of the PhoP DNA binding domain is

known, comprising the winged helix-turn-helix of the OmpR family (12). PhoR is the cognate sensor and phosphorylates PhoP (13). PhoP and PhoR are cotranscribed; the operon is autoregulated, with PhoP binding to direct repeats in its own promoter (13) and positively autoregulating its expression (14).

As with other regulators, PhoP can bind DNA in a nonphosphorylated state, although phosphorylation induces conformational changes and alters the protein-protein interface (15). PhoP recognizes a 23-bp sequence in its own promoter, composed of three direct repeats. PhoP dimerization can be promoted by DNA binding (as well as phosphorylation), and modeling suggests that the two molecules bind in a head-to-head orientation (15, 16). Dimerization occurs between the receiver domains, which are linked to the effector domain via a short flexible region (17). Phosphorylation may stabilize the dimerization process and increase DNA binding affinity by bringing the two DNA binding domains closer together (17).

The PhoP regulon has been characterized in some detail; the members include genes involved in the synthesis of diacyltrehalose, polyacyltrehalose, and sulfolipids in the *pks2-mmpL8* and *msl3* gene cluster (18). Over 40 genes are upregulated, and 70 are downregulated in the absence of PhoP (18). PhoP controls the hypoxic response through cross-talk with the DosR system (11). PhoP expression is controlled by the ArsF family transcriptional factor in *M. smegmatis* (19), suggesting that it forms part of a regulatory cascade. Rv2034 also regulates DosR expression, suggesting another mechanism for cross-talk between different regulons (20).

The control of transcription by PhoP is complex, since its phosphorylation state determines the specificity and strength of DNA binding; for example, nonphosphorylated PhoP can bind to its own promoter, but only phosphorylated PhoP can bind to the *msl3* promoter (21, 22) or to the *pks* promoter (22). In contrast, PhoP binding to the *mce1* promoter is reduced by phosphorylation (23). A consensus DNA binding motif for PhoP has been constructed that consists of two direct repeats, sometimes associated with a third direct repeat (24); the motif is found upstream of 87 genes (24).

The PhoPR system controls cell wall composition, and deletion of PhoP has pleiotropic effects. Changes in the relative amounts of mannosylated lipoarabinomannan (manLAM) are seen in the PhoP mutant strain, with monoacylated forms predominating over the triacylated forms (25). LAM is a major, essential component of the *M. tuberculosis* cell wall (26, 27) which plays a key role in immunomodulation and virulence by preventing phagolysosome fusion (28). PhoP mutants also have altered colony morphology, smaller cell size, and reduced cording properties resulting from a lack of synthesis of diacyltrehalose, polyacyltrehalose, and sulfolipids, all implicated in virulence (18, 29).

PhoP may also play a role in sensing and responding to pH; at least one pH-responsive gene cluster, *aprABC*, is controlled by PhoP (30). AprABC expression is induced by low pH in macrophages and *in vitro* and is required for survival in macrophages; disruption of AprABC leads to changes in cell wall and storage lipids (30). PhoP also regulates *lipF*, another pH-responsive gene (24).

A single point mutation in PhoP is sufficient to attenuate the virulence of *M. tuberculosis*. The avirulent strain H37Ra has the mutation S219L in the DNA binding domain of PhoP, and this single nucleotide polymorphism (SNP) is partly responsible for loss of virulence (31–33). Mutant PhoP is no longer able to bind to its own promoter (33). Among other changes, the mutation leads to a loss of secretion of the major T cell antigen, ESAT-6. Complementation with a PhoP wild-type allele restores ESAT-6 secretion and induces T cell responses during infection, as well as increasing

virulence; in addition, changes in colony morphology are seen (31). Transcriptome profiles show that many members of the PhoP regulon are underexpressed during intracellular growth in H37Ra (as compared to H37Rv); these include genes involved in complex lipid biosynthesis, such as *papA3* and *fadD21* (34).

Given PhoP's role in regulating key virulence pathways, many studies have characterized PhoP mutant strains in infection models. In particular, much interest has been paid to the potential of PhoP deletion strains for use as live vaccines with superior efficacy compared to the current vaccine, *M. bovis* BCG. PhoP mutants show increased adhesion to macrophages, suggesting increased uptake, as well as reduced intracellular replication that is likely due to the loss of the ability to prevent phagolysosome fusion (35). The latter ability may be affected by changes in the composition of LAM, since it is known to play a major role in phagosome maturation arrest (25). In addition, PhoP mutants fail to induce apoptosis in macrophages *in vitro* or during infection in the lung (36). PhoP mutants are unable to grow in low Mg^{2+} medium, and at least part of the attenuation seen in macrophages is due to this requirement, since growth can be partially restored by adding excess Mg^{2+} (18).

The SO2 strain (PhoP disruption) (37) has been widely tested and is attenuated in murine macrophages and in both immunocompetent and immunocompromised mice (18, 37–39). The strain shows greater attenuation in SCID mice than *M. bovis* BCG (39). Protection from challenge infection in the mouse model is comparable to BCG vaccination, whereas protection in the susceptible guinea pig is greater than that with BCG (39). SO2 showed protective efficacy against *M. tuberculosis* in the rhesus macaque model of infection and is well tolerated, supporting its further development as a live vaccine (40). The SO2 strain is stable *in vitro* and *in vivo*, with no reversion to wild type, even after 6 months, and was demonstrably safe in guinea pigs, with no lesions or histology after 6 months of dosing with 50 times a normal vaccine dose (41). These studies underscore the essential role of the PhoPR 2CRS in virulence.

There is little information about PhoP variation in clinical isolates, although upregulation of PhoP after IS6110 insertion into the promoter has been seen in a particularly virulent clinical isolate (42). However, numerous mutations have been noted in strains of BCG; several mutations would lead to frameshifts in PhoR, with one frameshift in PhoP, and one IS6110 insertion upstream (and divergently transcribed) (43). It is presumed that all mutations lead to inactivation or disruption of the PhoPR system and contribute to the lack of virulence in the vaccine strains (43).

MprAB

The MprAB system appears to be involved in the response to stress conditions, in particular those that affect the cell envelope. MprA is a member of the OmpR family of RR, containing a winged helix-turn-helix DNA binding domain and the conserved aspartate site of phosphorylation (44). MprB is the cognate HK sensor, which autophosphorylates and transduces the signal to MprB (45). MprB activity is dependent on Mg^{2+}, although this HK is one of a small group that can also utilize Mn^{2+} (45).

The *mprAB* operon is autoregulated, and MprA binds to its own promoter; a binding motif (the "Mpr box") consisting of two short direct repeats is located upstream of *mprA* (46). MprA binds to the motif as two monomers and presumably dimerizes (46). Although nonphosphorylated RR can bind to this motif, the affinity is enhanced by phosphorylation, and activation is required for regulation (46).

MprAB is part of a complex regulatory cascade involving several sigma factors that respond to environmental stresses, in particular to envelope stress. MprA controls the expression of the sigma factors *sigE* and *sigB* (47, 48), as well as the serine protease PepD (46), in response to sodium dodecyl sulfate exposure. In view of this, it is surprising that MprAB deletion strains are more resistant to sodium dodecyl sulfate (48), although this may result from upregulation of genes via a compensatory mechanism within the cascade. The interaction between MprAB and SigE/B/H is complex, with many genes in the regulatory cascade being controlled by more than one of the regulators. For example, PepD is controlled by both MprA and SigE (49–51), whereas SigE is controlled by MprA and SigH (47, 48). In *M. smegmatis*, the MprAB-SigE cascade is controlled by the availability of polyphosphate, suggesting this may be the source of the phosphoryl groups utilized by the sensor MprB under stress conditions, rather than ATP (49).

Members of the MprA regulon play key roles in the cell envelope. PepD is a member of the HtrA family and has both chaperone and protease activity, so it can participate in protein refolding and degradation under stress conditions (51, 52). A major substrate of PepD is a homolog of the phage shock protein, PspA (52). Deletion of PepD results in increased sensitivity to cell wall antibiotics and induction of the SigE stress response pathway (51, 52).

Other members of the MprA regulon have been characterized. In particular, two operons originally identified as being in the "Dos" regulon—the formate hydrogenlyase complex (Rv0081 to Rv0088) (53) and the Rv1812c-Rv1813c operon (54)—can be regulated by both MprA and DosR. In both cases there are two motifs upstream, a Dos box, and an Mpr box allowing for binding of either RR independently. Rv0081 negatively autoregulates by binding to its own operon promoter; DosR and MprA have independent binding sites that overlap the Rv0081 binding site and can therefore prevent Rv0081 binding and positively regulate expression (53). Thus, the operon can be switched on by either one of the two systems, thereby responding to alternative environmental signals. The Rv1812c operon is also positively regulated by either DosR or MprA binding (54); although the role of these two proteins is unknown, they are required for virulence in the mouse model of infection.

MprA partly controls the expression of Acr2, a member of the alpha-crystallin family of chaperones (55) involved in stress responses. Rv1057 is controlled by MprA and TrcR, with MprA being responsible for upregulation of Rv1057 under detergent stress and in macrophages (56). MprA may also control the secretion of components of the Esx-1 system, most notably the major T cell antigens ESAT-6 and CFP10. MprA negatively regulates the expression of EspA, binding to several Mpr boxes in the promoter region (57). Deletion of MprAB leads to reduced secretion of a number of proteins, including EspA, ESAT-6, and EspB (57); these changes lead to altered immune responses in macrophages (57).

MprAB has a role during infection with *M. tuberculosis*. Early studies demonstrated that MprA is required to establish a persistent infection in mice, at least in competition experiments with the wild-type strain (44). MprAB expression differs between *M. tuberculosis* and *M. bovis* BCG, and this leads to different phenotypes for deletion strains; for example, the *M. bovis* BCG deletion strain is attenuated in macrophages (45), whereas the *M. tuberculosis* strain is hypervirulent (44).

SenX3-RegX3

The SenX3-RegX3 system was one of the first mycobacterial 2CRSs identified (58). The operon is slightly unusual in that the sensor gene is upstream of the regulator gene, and the two are separated by several copies of a repeat unit (mycobacterial interspersed repeat unit, or MIRU) (59). The number of repeats in MIRUs has been used as a genetic typing method for many studies (60). The biological consequences of the repeat units are unknown, but the number of repeats varies in clinical isolates, and there are no repeats in the *M. smegmatis* operon.

SenX autophosphorylates at His167 in the cytoplasmic domain (61), followed by phosphotransfer to Asp52 in the receiver domain of RegX3 (61). SenX3 is a member of the PhoR/EnZ subfamily of sensors; the protein has at least one transmembrane domain located N-terminal to the sensor domain (62). The stimulus for SenX3 is unknown, although the system responds to low phosphate mediated by a phosphate transporter (63, 64). The sensor has homology with ArcB, the global regulator of aerobic metabolism in *Escherichia coli* (62), and contains an atypical sensor PAS domain, which often acts as sensors of oxygen or redox states.

RegX3 is a member of the OmpR subfamily of regulators with an N-terminal receiver domain and a winged helix-turn-helix DNA binding motif in the C-terminal effector domain (65–67). For most regulators, dimers are the active DNA binding form (68, 69). The structure of full-length RegX3 reveals that the protein dimerizes, as expected for RRs, but that the active form appears to be an unusual dimer in which three-dimensional domain swapping has occurred (70). In this form a domain from each of the two monomers is swapped, a more complex interaction than simple dimerization by interaction at an interface. Achieving domain swapping is likely to be a highly energetic process, since it requires disruption of the interactions within a monomer, as well as partial protein unfolding (in the receiver domain). This leads to the possibility that there may be alternative dimers and these may have different DNA binding motifs and thus regulate alternative regulons (70). The stimuli that could promote domain swapping are unknown.

The regulon controlled by RegX3 is not fully characterized and differs between the pathogenic and nonpathogenic strains *M. tuberculosis* and *M. smegmatis*. Data using deletion strains and gene expression analysis suggest that RegX3 may have a dual role in controlling respiration and phosphate metabolism (63, 64, 71–73). Interestingly, the system is essential in *M. smegmatis*, but not in *M. tuberculosis* (74), and it is one of the five 2CRSs retained in *M. leprae*, suggesting its importance for bacterial survival.

Autoregulation has been demonstrated in *M. smegmatis* (61) with RegX3 binding upstream of the operon promoter, although the fold change is small in comparison to other systems—only about 2-fold upregulation. Autoregulation was not seen in *M. tuberculosis* (71), where deletion of RegX3 resulted in increased promoter activity. The expression of the operon is controlled by growth, with transient upregulation during log phase aerobic growth, but the level of expression is generally low (as compared to *sigA*), and changes in expression levels are small (less than 2-fold) (72).

The regulon controlled by RegX3 is not fully defined; early microarray analysis suggested a group of 50 genes controlled by the regulator under aerobic conditions in *M. tuberculosis* (71). The majority of changes in a SenX3-RegX3 deletion strain were associated with general growth, for example, downregulation of ribosomal proteins (71). Characterization of global gene expression under nonreplicating conditions and complete nutrient starvation demonstrated that genes associated with low-oxygen environments are differentially expressed (71, 72). In contrast to RegX3 binding to its own promoter, the regulator binds to the promoter regions of the regulon members *ald*, *cydB*, and *gltA1* in a phosphorylation-dependent manner (72). Under static conditions, RegX3 acts as a positive regulator of the *cydB* and *gtlA1* operons (72).

In *M. smegmatis*, RegX3 controls the expression of genes involved in phosphate uptake. SenX3 acts as a phosphodonor or as a phosphatase of RegX3, depending on phosphate availability; phosphorylated regX3 binds to an inverted repeat upstream of *phoA* and *pstSCAB* (and its own promoter) genes to activate transcription (63). PhoA encodes an exported alkaline phosphatase, which can release inorganic phosphate from a range of macromolecules (75), whereas the Pst operon encodes a high-affinity phosphate transport system. In addition, RegX3 is implicated in the regulation of an alternative phosphate uptake system, PhnDCE (76).

In *M. tuberculosis* the operon is upregulated by phosphate starvation (2- to 4-fold). In contrast to *M. smegmatis*, there are three Pst high-affinity phosphate uptake systems (PstS1, PstS2, and PstS3) but no Phn system. RegX3 controls the induction of the *pstS3* operon (*pstS3-pstC2-pstA1*) but not *pstS1* or *pstS2* (64). PstA controls expression of the SenX3 operon, exerting a negative effect when phosphate is abundant and preventing the expression of genes required for phosphate acquisition (73). *M. tuberculosis* PstA1 deletion strains are attenuated *in vivo* in an interferon-γ-dependent fashion, but full virulence can be restored by deleting *regX3*, suggesting that dysregulation of phosphate-responsive genes is the cause (73).

The changes induced by phosphate limitation dependent on RegX3 have also been determined, but there is little overlap with those identified in aerobic culture (64, 71–73); the two studies conducted with *M. tuberculosis* in phosphate limitation showed little agreement in differentially expressed genes, except for the Pst transporters (64, 73). These studies demonstrate the difficulty of finding appropriate conditions for expression analyses and the identification of the true regulon. Differences may be explained by the possibility that the alternative dimer forms seen in structural studies control different regulons (70).

The role of the RegX3 system must differ between the nonpathogenic and the pathogenic species, since RegX3 is essential in _M. smegmatis_ (63, 74) but not in _M. tuberculosis_ (62, 71, 77). One major difference is that _M. tuberculosis_ lacks a PhoA homolog, whose expression is RegX3-dependent in _M. smegmatis_ (63).

The model for regulation of phosphate-dependent genes is not complete. The sensor of phosphate is the Pst transporter, but it is not clear how the signal is transduced to SenX3. In addition, mutants with deletion of SenX3 or with transposon disruptions displayed different PhoA expression profiles; deletion led to a complete lack of PhoA expression, whereas transposon insertion lead to upregulation of PhoA under phosphate-replete conditions. Since SenX3 may act as both a kinase and a phosphatase for RegX3, it is possible that the system is very sensitive to the amount of RegX3 that is changed in the transposon mutant (63).

Deletion of the complete SenX3-RegX3 operon was achieved in _M. smegmatis_ at a low frequency, suggesting that suppressor mutations can compensate for loss of the complete operon (74). Mutations in NhaA, among others, are proposed to relieve the essentiality. In this context it is interesting to note that _M. tuberculosis_ does not have a homolog of NhaA and that regulation of intrabacterial pH differs between the species, with _M. tuberculosis_ being much more susceptible to weak acids than _M. smegmatis_ (78). The system is required for normal growth in low phosphate, with SenX3 or operon deletions showing major defects in growth when phosphate is limiting.

The RegX3 system undoubtedly plays a role during infection; several studies have demonstrated that deletion strains have defects in macrophages or in mice (62, 64, 71, 73, 79). Phosphate limitation occurs in macrophages infected with _M. tuberculosis,_ suggesting that these phenotypes relate to the inability to acquire sufficient exogenous inorganic phosphate (64). A partial deletion of SenX3 and RegX3 resulted in attenuation in mouse bone marrow–derived macrophages and human macrophages (THP-1 cell line) (71); attenuation during infection of immune-compromised, SCID mice and immune-competent, DBA mice was seen. In addition, deletion of either RegX3 or SenX3 resulted in reduced replication in BALB/c or nude mice (62). Transposon disruption of RegX3 in the CDC1551 strain also resulted in a reduction in the persistence of the strain in

guinea pig and mouse infections (64). In contrast, in the only study to show no attenuation, disruption of RegX3 in _M. tuberculosis_ MT103 by transposon insertion had no detectable effect on intracellular replication in mouse bone marrow–derived macrophages or in the replication of the organism after aerosol infection of BALB/c mice (77).

PrrAB

The PrrAB system is composed of the regulator PrrA and the sensor PrrB; the structures of both the sensor (80) and the regulator have been determined (81). PrrA is a member of the OmpR family and contains the characteristic winged helix-turn-helix DNA binding domain.

Structural studies reveal that the regulator can exist in either a "closed" or "open" form in which the accessibility of the DNA binding recognition helix changes (81). Activation by phosphorylation results in increased formation of the open form (in which the helix is accessible), presumably resulting in increased DNA binding. As with other members of the OmpR family, PrrA is able to bind to DNA in a nonphosphorylated state, although phosphorylation increases its affinity for DNA, and autoregulation occurs (82), with the regulator being required for transient intracellular induction in macrophages (77, 82).

The PrrAB 2CRS plays a role in early adaptation to intracellular infection (77, 82) and is essential for _in vitro_ survival (83). Disruption of the operon by transposon mutagenesis leads to reduced expression of PrrA and PrrB (83); in this strain, the replication rates in murine bone marrow–derived macrophages are reduced (77). This attenuation is only seen at early time points postinfection (up to 6 days), and strains eventually reach the same plateau numbers as wild type (after 9 days) (77).

PrrAB is expressed as a bicistronic operon at low levels during _in vitro_ growth (83, 84). Expression of the _prrAB_ operon is transiently induced after macrophage infection, with a 2-fold increase in expression seen over the first 3 days of infection, followed by a gradual reduction to baseline (77). _In vitro,_ expression is induced by nitrogen limitation but not by carbon starvation, hypoxia, or acidic pH, suggesting that this may be the stimulus seen in macrophages (83).

MtrAB

MtrA was originally identified as a homolog of _Pseudomonas aeruginosa_ PhoB by DNA hybridization (85) and forms a 2CRS with the sensor MtrB. The genes are located in an apparent operon, with the RR upstream of the HK (86). Phosphorylation of MtrA by a noncognate HK (CheA) was demonstrated, confirming its role as an RR (85), and the RR directly interacts with MtrB and DNA (87). MtrB autophosphorylation and phosphotransfer to MtrA has been demonstrated and requires divalent metal ions, with Mg^{2+} essential for the latter activity (88).

MtrA is a member of the OmpR family and has the characteristic winged helix-turn-helix DNA binding motif. The structure of inactive MtrA has been determined; the two domains of the protein interact extensively (89). Activation of MtrA via phosphorylation and dimerization would require a significant disruption of this interaction, suggesting that activation may be more difficult than for other members of the family and that phosphorylation may be relatively slow (89).

The MtrAB system is essential in *M. tuberculosis* as demonstrated by several studies in which deletion mutants were not viable (86, 90, 91). The MtrAB system is not essential in all mycobacteria. In *Mycobacterium avium*, disruption of MtrB leads to changes in cell morphotype, with cells losing the ability to switch from the red to the white morphotype (92). The red morphotype (defined as staining with Congo red) was exhibited by the MtrB mutant strain, which was also more permeable and more sensitive to antibiotics, presumably due to changes in the cell wall composition controlled by MtrB (92). In addition, the mutant strains were unable to survive in macrophages, showing loss of virulence (92). In *M. smegmatis* MtrB mutants are filamentous, with increased cell clumping and defective septum formation and cell division; in addition, cell wall defects were noted, which gave rise to increased susceptibility to lysozyme (93).

MtrAB expression is controlled by sigma factor C (94), and there is at least one accessory protein, the lipoprotein vLpqB, which interacts with the MtrB extracellular domain. This interaction affects phosphotransfer to MtrA and subsequent downstream effects on the regulon, including changes in *dnaA* expression (95). Disruption of LpqB in *M. smegmatis* leads to pleiotropic effects arising from cell wall changes, including increased cellular aggregation and loss of biofilm formation, as well as changes in motility and cell morphology during growth (95). These changes are dependent on the interaction with MtrB, since they can be reversed by expressing constitutively phosphorylated MtrA (95).

MtrB is localized to the septa and poles in an FtsX-dependent manner (93) consistent with its proposed role in controlling cell division. Unusually, both phosphorylated and nonphosphorylated forms show the same location, although it is proposed that MtrB must locate to the septum for activation and autophosphorylation (93).

Expression of the system differs between the virulent and avirulent species. For example, expression of the operon is induced after infection of macrophages in *M. bovis* BCG (85, 86), but it is constitutively expressed in *M. tuberculosis* (86), consistent with it being essential *in vitro* in the latter. Overexpression of MtrA does not affect growth *in vitro* but has profound effects on virulence, leading to reduced growth in macrophages and reduced virulence in the mouse model of infection (96). Attenuation is associated with loss of the ability to prevent phagolysosome fusion, a hallmark of infection by *M. tuberculosis* (96). MtrA positively regulates the expression of DnaA by binding to the promoter region in a phosphorylation-dependent manner (96). DnaA is required for DNA replication, and overexpression leads to a decreased growth rate (96). Coexpression of MtrB is sufficient to relieve the attenuation in macrophages, suggesting that the ratio of phosphorylated to nonphosphorylated RR is key (96).

The complete regulon of MtrA is not defined, but aside from DnaA, the RR also binds to the promoter of the mycolyl transferase *fbpB* and regulates its expression (97). The MtrA DNA binding motif is comprised of two direct repeats of GTCACAgcg, and these are also found in the origin of replication (*oriC*). MtrA binds to the motifs in the *oriC* region that are required for replication of autonomous plasmids, suggesting a key role for MtrA in chromosomal replication (97). Studies in *M. smegmatis* using unphosphorylated MtrA suggest that the RR may bind to an alternative motif CACGCCG (98); this motif is found upstream of over 400 mycobacterial genes, some of which may represent the true regulon members (98). In *M. smegmatis* the cell wall hydrolase *ripA* is controlled by MtrA (93).

TrcRS

The TrcRS system is another 2CRS that was identified by degenerate PCR before the availability of the complete genome sequence (99). TrcR and TrcS form a classical two-component system; the genes are arranged in a bicistronic operon, with the regulator gene (*trcR*) located upstream of the sensor kinase gene (*trcS*) (100). The sensor kinase, TrcS, has two predicted transmembrane regions that flank the proposed sensor domain in the N-terminal region; the C-terminus carries the kinase transmitter domain with the expected conserved regions and His-287, the site of autophosphorylation (99). The regulator, TrcR, is a member of the OmpR family of regulators. The N-terminal receiver domain contains several conserved residues required for function in this family and the Asp-82 residue, which is the site of phosphotransfer from TrcS (99).

The C-terminal effector domain contains a helix-turn-helix DNA binding motif characteristic of the OmpR family (99).

Autophosphorylation and phosphotransfer have been shown using recombinant proteins *in vitro* with a truncated version of TrcS carrying the cytoplasmic effector domain alone (99). TrcS kinase activity is dependent on the divalent cations Mn^{2+} or Ca^{2+}, but not Mg^{2+} (99). Interestingly, the transmembrane domain of TrcS is toxic for *E. coli* and induces cellular lysis whether expressed as a full-length protein or alone. Thus, the stimulus for TrcS remains unknown, although the system is expressed during aerobic growth in culture and at low levels early after infection of human macrophages (but not at later stages of infection) (100). Transposon mutants with disruption of *trcS* were fully virulent in both murine macrophages and in the acute mouse model of infection, suggesting that this regulon is not involved in response to the intracellular environment, or for expression of pathogenic traits (77), consistent with its lack of expression under those conditions.

The *trcRS* operon is autoregulated, with TrcR binding to an AT-rich region upstream of the promoter leading to >500-fold induction of the operon (100). The *M. tuberculosis* genome is GC-rich (65% GC) (101), such that the AT-rich region of 78% is unusual. Regions of high AT are often associated with promoters or regulatory regions, since the energy required for unwinding and generating the open complex are lower. The minimal binding region for TrcR is a 28-bp region, although footprinting assays reveal that the regulator protects a larger fragment of 58 to 86 bp. Unphosphorylated TrcR is able to bind to the operon promoter, but the phosphorylated regulator binds with a 10-fold higher affinity and protects a larger region of DNA (100). This suggests that, as with other regulators, TrcR dimerizes upon activation.

A single member of the regulon is known; Rv1057 encodes a protein of unknown function, but with an unusual structure of a seven-bladed β-propeller (56, 102). TrcR represses expression of Rv1057, and TrcRS deletion mutants show increased expression (~30-fold) of Rv1057 in culture. Rv1057 is induced during infection of macrophages when expression of TrcRS is expected to be decreasing. As expected, TrcR binds to the AT-rich region upstream of the transcriptional start site (102), although a binding motif is not clearly defined. Regulation of Rv1057 is complex, since it is also dependent on SigE (103) and MprAB (56), both of which are involved in the response to cell envelope stress.

KdpDE

The KdpDE 2CRS is one of the few systems identified with a clear homolog in other bacteria; the sensor KdpD interacts with the regulator KdpE (104, 105). KdpD is unusual, since it has a large cytoplasmic sensor domain. The N-terminal cytoplasmic domain of the sensor KdpD interacts with the lipoproteins LprF and LprJ (105). The role of these accessory proteins may be during signal relay or signal recognition, although this is not yet established (105).

KdpDE controls the expression of the adjacent and divergently transcribed *kdpFABC* operon; expression is controlled in response to K^+ concentration as in other bacterial species (105). The role of KdpDE in other bacteria is to control K^+ uptake by the KdpFABC transport system, and it is assumed that the mycobacterial system performs the same function. The four proteins make up a P-type ATPase K^+ transporter whose role is to control the intracellular concentration of potassium ions to maintain turgor pressure (106).

LprF, LprJ, and KdpE are upregulated in response to starvation, suggesting that they may play a role in persistence (107). Potassium ions may play a role during infection, since they are likely to be at limiting concentrations, although no defect in growth under low K^+ was evident in a deletion strain (90). Deletion of KdpDE resulted in increased virulence in immunocompromised mice (90).

NarL(Rv0844)/Rv0845 2CRS

NarL (Rv0844) is a member of the NarL subfamily of RRs; the structure of the receiver domain has been determined (108). The protein has the typical $(\beta\alpha)_5$ fold of signal receiver domains, with the phosphorylation site located in the loop after the third β strand. As with other members of the family, the active site is structurally similar to haloacid dehalogenases, but with a key substitution of aspartate by arginine.

The stimulus and regulon for NarL remain uncharacterized. The cognate sensor Rv0845 lacks homology with nitrate or nitrite sensors in other bacteria, although it is upregulated in the stationary phase (109). Protein interaction with the HK Rv0845 has been shown, but there also appears to be cross-talk with the DosRST system. NarL can interact with DosT, but not DosS, and *in vitro* phosphotransfer from DosT to NarL occurs, although the rate is relatively slow (10); the specificity of this interaction is controlled by the DHp domain (10). This raises the possibility that the NarL regulon could be responsive to hypoxia as well as to other environmental stimuli sensed by Rv0845.

Other Systems

Several 2CRSs remain poorly characterized in mycobacteria, with little known of the stimulus or regulon. These include TcrXY, the split TcrA system, PdtaRS, and the orphan components.

The HK TcrY and the RR TcrX form a *bona fide* system, with phosphotransfer from the sensor to the kinase demonstrated *in vitro* (110). TcrX binds to two inverted repeats in its own promoter and may autoregulate, although, unusually, binding affinity appears to be the same regardless of phosphorylation state (111). Little is known of the function of the system, although its deletion results in hyper-virulence in immune-compromised mice (90), and the components are expressed *in vivo* (112) and may be upregulated under low iron (113).

Rv0600c (HK1), Rv0601c (HK2), and Rv0602c (TcrA) form an unusual system (114–117); in some strains of *M. tuberculosis*, Rv0600c and Rv0601c are expressed as a single, functional HK, whereas in others two proteins are expressed that can functionally complement each other to generate functional HK activity (118). Rv0601c encodes the membrane sensor domain that is proposed to dimerize and interact with monomers of Rv0600c (114–117). In either case, phosphotransfer to the RR is seen (117).

PdtaR/S form an unusual system in that the RR and HK are not coexpressed. PdtaS (Rv3220) autophosphorylates and then mediates phosphotransfer to PdtaR (Rv1626) (119). There are four orphan RRs found in the genome of *M. tuberculosis*. Rv0195 is a member of the LuxR family; it appears to be induced by sodium azide treatment (120) or nutrient starvation (107). Rv0260c has a transcriptional regulatory motif characteristic of RRs but also has homology with HemD, a uroporphyrinogen-II synthase, and is located upstream of CbiX, which is involved in cobalamin biosynthesis, suggesting it may have a dual function. Rv2884 is a possible homolog of GlnR involved in control of nitrogen metabolism. Rv0818 is an RR of unknown function. In addition, Rv3143 has a receiver domain, but no effector DNA binding domain, suggesting it may play a role as an ancillary protein or in a phosphorelay system.

CONCLUSION

Two-component systems undoubtedly play major roles in the ability of mycobacteria to adapt to external conditions. In particular, several systems are required for pathogenicity and thus pose interesting targets for further study. Although much work has focused on identifying regulon members and understanding the structure and function of the RRs, much less insight has been gained on the role of the sensory proteins and the stimuli to which they respond. Future work to understand the sensor domains and the identifying additional accessory proteins is likely to answer several questions about the role of these important global regulatory systems.

Citation. Parish T. 2014. Two-component regulatory systems of mycobacteria. Microbiol Spectrum 2(1):MGM2-0010-2013.

References

1. Stock AM, Robinson VL, Goudreau PN. 2000. Two-component signal transduction. *Annu Rev Biochem* **69:**183–215.

2. Hoch JA. 2000. Two-component and phosphorelay signal transduction. *Curr Opin Microbiol* **3:**165–170.

3. Wuichet K, Cantwell BJ, Zhulin IB. 2010. Evolution and phyletic distribution of two-component signal transduction systems. *Curr Opin Microbiol* **13:**219–225.

4. West AH, Stock AM. 2001. Histidine kinases and response regulator proteins in two-component signaling systems. *Trends Biochem Sci* **26:**369–376.

5. Jung K, Fried L, Behr S, Heermann R. 2012. Histidine kinases and response regulators in networks. *Curr Opin Microbiol* **15:**118–124.

6. Ashby MK. 2004. Survey of the number of two-component response regulator genes in the complete and annotated genome sequences of prokaryotes. *FEMS Microbiol Lett* **231:**277–281.

7. Barrett JF, Hoch JA. 1998. Two-component signal transduction as a target for microbial anti-infective therapy. *Antimicrob Agents Chemother* **42:**1529–1536.

8. Wolanin PM, Thomason PA, Stock JB. 2002. Histidine protein kinases: key signal transducers outside the animal kingdom. *Genome Biol* **3:**30133011–30133018.

9. Cole ST, Brosch R, Parkhill J, Garnier T, Churcher C, Harris D, Gordon SV, Eiglmeier K, Gas S, Barry CE, Tekaia F, Badcock K, Basham D, Brown D, Chillingworth T, Connor R, Davies R, Devlin K, Feltwell T, Gentles S, Hamlin N, Holroyd S, Hornby T, Jagels K, Krogh A, McLean J, Moule S, Murphy L, Oliver K, Osborne J, Quail MA, Rajandream MA, Rogers J, Rutter S, Seeger K, Skelton J, Squares R, Squares S, Sulston JE, Taylor K, Whitehead S, Barrell BG. 1998. Deciphering the biology of *Mycobacterium tuberculosis* from the complete genome sequence. *Nature* **393:**537–544.

10. Lee HN, Jung KE, Ko IJ, Baik HS, Oh JI. 2012. Protein-protein interactions between histidine kinases and response regulators of *Mycobacterium tuberculosis* H37Rv. *J Microbiol* **50:**270–277.

11. Gonzalo-Asensio J, Mostowy S, Harders-Westerveen J, Huygen K, Hernandez-Pando R, Thole J, Behr M, Gicquel B, Martin C. 2008. PhoP: a missing piece in the

intricate puzzle of *Mycobacterium tuberculosis* virulence. *PLoS One* 3(10):e3496.

12. Wang S, Engohang-Ndong J, Smith I. 2007. Structure of the DNA binding domain of the response regulator PhoP from *Mycobacterium tuberculosis*. *Biochemistry* 46:14751–14761.

13. Gupta S, Sinha A, Sarkar D. 2006. Transcriptional autoregulation by *Mycobacterium tuberculosis* PhoP involves recognition of novel direct repeat sequences in the regulatory region of the promoter. *FEBS Lett* 580:5328–5338.

14. Gonzalo-Asensio J, Soto CY, Arbues A, Sancho J, del Carmen Menendez M, Garcia MJ, Gicquel B, Martin C. 2008. The *Mycobacterium tuberculosis* phoPR operon is positively autoregulated in the virulent strain H37Rv. *J Bacteriol* 190:7068–7078.

15. Sinha A, Gupta S, Bhutani S, Pathak A, Sarkar D. 2008. PhoP-PhoP interaction at adjacent PhoP binding sites is influenced by protein phosphorylation. *J Bacteriol* 190:1317–1328.

16. Gupta S, Pathak A, Sinha A, Sarkar D. 2009. *Mycobacterium tuberculosis* PhoP recognizes two adjacent direct-repeat sequences to form head-to-head dimers. *J Bacteriol* 191:7466–7476.

17. Menon S, Wang S. 2011. Structure of the response regulator PhoP from *Mycobacterium tuberculosis* reveals a dimer through the receiver domain. *Biochemistry* 50:5948–5957.

18. Walters SB, Dubnau E, Kolesnikova I, Laval F, Daffe M, Smith I. 2006. The *Mycobacterium tuberculosis* PhoPR two-component system regulates genes essential for virulence and complex lipid biosynthesis. *Mol Microbiol* 60:312–330.

19. Gao CH, Yang M, He ZG. 2011. An ArsR-like transcriptional factor recognizes a conserved sequence motif and positively regulates the expression of phoP in mycobacteria. *Biochem Biophys Res Commun* 411:726–731.

20. Gao CH, Yang M, He ZG. 2012. Characterization of a novel ArsR-like regulator encoded by Rv2034 in *Mycobacterium tuberculosis*. *PLoS One* 7:e36255.

21. Pathak A, Goyal R, Sinha A, Sarkar D. 2010. Domain structure of virulence-associated response regulator PhoP of *Mycobacterium tuberculosis*: role of the linker region in regulator-promoter interaction(s). *J Biol Chem* 285:34309–34318.

22. Goyal R, Das AK, Singh R, Singh PK, Korpole S, Sarkar D. 2011. Phosphorylation of PhoP protein plays a direct regulatory role in lipid biosynthesis of *Mycobacterium tuberculosis*. *J Biol Chem* 286:45197–45208.

23. Zeng J, Cui T, He ZG. 2012. A genome-wide regulator-DNA interaction network in the human pathogen *Mycobacterium tuberculosis* H37Rv. *J Proteome Res* 11:4682–4692.

24. Cimino M, Thomas C, Namouchi A, Dubrac S, Gicquel B, Gopaul DN. 2012. Identification of DNA binding motifs of the *Mycobacterium tuberculosis* PhoP/PhoR two-component signal transduction system. *PLoS One* 7:e42876.

25. Ludwiczak P, Gilleron M, Bordat Y, Martin C, Gicquel B, Puzo G. 2002. *Mycobacterium tuberculosis* phoP mutant: lipoarabinomannan molecular structure. *Microbiology* 148:3029–3037.

26. Goude R, Amin AG, Chatterjee D, Parish T. 2008. The critical role of embC in *Mycobacterium tuberculosis*. *J Bacteriol* 190:4335–4341.

27. Goude R, Parish T. 2008. The genetics of cell wall biosynthesis in *Mycobacterium tuberculosis*. *Future Microbiol* 3:299–313.

28. Brennan PJ. 2003. Structure, function, and biogenesis of the cell wall of *Mycobacterium tuberculosis*. *Tuberculosis* 83:91–97.

29. Asensio JG, Maia C, Ferrer NL, Barilone N, Laval F, Soto CY, Winter N, Daffe M, Gicquel B, Martin C, Jackson M. 2006. The virulence-associated two component PhoP-PhoR system controls the biosynthesis of polyketide-derived lipids in *Mycobacterium tuberculosis*. *J Biol Chem* 281:1313–1316.

30. Abramovitch RB, Rohde KH, Hsu FF, Russell DG. 2011. aprABC: a *Mycobacterium tuberculosis* complex-specific locus that modulates pH-driven adaptation to the macrophage phagosome. *Mol Microbiol* 80:678–694.

31. Frigui W, Bottai D, Majlessi L, Monot M, Josselin E, Brodin P, Garnier T, Gicquel B, Martin C, Leclerc C, Cole ST, Brosch R. 2008. Control of M. tuberculosis ESAT-6 secretion and specific T cell recognition by PhoP. *PLoS Pathog* 4:e33.

32. Lee JS, Krause R, Schreiber J, Mollenkopf HJ, Kowall J, Stein R, Jeon BY, Kwak JY, Song MK, Patron JP, Jorg S, Roh K, Cho SN, Kaufmann SH. 2008. Mutation in the transcriptional regulator PhoP contributes to avirulence of *Mycobacterium tuberculosis* H37Ra strain. *Cell Host Microbe* 3:97–103.

33. Chesne-Seck ML, Barilone N, Boudou F, Gonzalo Asensio J, Kolattukudy PE, Martin C, Cole ST, Gicquel B, Gopaul DN, Jackson M. 2008. A point mutation in the two-component regulator PhoP-PhoR accounts for the absence of polyketide-derived acyltrehaloses but not that of phthiocerol dimycocerosates in *Mycobacterium tuberculosis* H37Ra. *J Bacteriol* 190:1329–1334.

34. Li AH, Waddell SJ, Hinds J, Malloff CA, Bains M, Hancock RE, Lam WL, Butcher PD, Stokes RW. 2010. Contrasting transcriptional responses of a virulent and an attenuated strain of *Mycobacterium tuberculosis* infecting macrophages. *PLoS One* 5:e11066.

35. Ferrer NL, Gomez AB, Neyrolles O, Gicquel B, Martin C. 2010. Interactions of attenuated *Mycobacterium tuberculosisphoP* mutant with human macrophages. *PLoS One* 5(9):e12978.

36. Aporta A, Arbues A, Aguilo JI, Monzon M, Badiola JJ, de Martino A, Ferrer N, Marinova D, Anel A, Martin C, Pardo J. 2012. Attenuated *Mycobacterium tuberculosis* SO2 vaccine candidate is unable to induce cell death. *PLoS One* 7:e45213.

37. Perez E, Samper S, Bordas Y, Guilhot C, Gicquel B, Martin C. 2001. An essential role for phoP in *Mycobacterium tuberculosis* virulence. *Mol Microbiol* 41:179–187.

38. Asensio JA, Arbues A, Perez E, Gicquel B, Martin C. 2008. Live tuberculosis vaccines based on phoP mutants: a step towards clinical trials. *Expert Opin Biol Ther* 8:201–211.

39. Martin C, Williams A, Hernandez-Pando R, Cardona PJ, Gormley E, Bordat Y, Soto CY, Clark SO, Hatch GJ, Aguilar D, Ausina V, Gicquel B. 2006. The live *Mycobacterium tuberculosisphoP* mutant strain is more attenuated than BCG and confers protective immunity against tuberculosis in mice and guinea pigs. *Vaccine* 24:3408–3419.

40. Verreck FA, Vervenne RA, Kondova I, van Kralingen KW, Remarque EJ, Braskamp G, van der Werff NM, Kersbergen A, Ottenhoff TH, Heidt PJ, Gilbert SC, Gicquel B, Hill AV, Martin C, McShane H, Thomas AW. 2009. MVA.85A boosting of BCG and an attenuated, *phoP* deficient M. *tuberculosis* vaccine both show protective efficacy against tuberculosis in rhesus macaques. *PLoS One* 4:e5264.

41. Cardona PJ, Asensio JG, Arbues A, Otal I, Lafoz C, Gil O, Caceres N, Ausina V, Gicquel B, Martin C. 2009. Extended safety studies of the attenuated live tuberculosis vaccine SO2 based on *phoP* mutant. *Vaccine* 27:2499–2505.

42. Soto CY, Menendez MC, Perez E, Samper S, Gomez AB, Garcia MJ, Martin C. 2004. IS6110 mediates increased transcription of the *phoP* virulence gene in a multidrug-resistant clinical isolate responsible for tuberculosis outbreaks. *J Clin Microbiol* 42:212–219.

43. Leung AS, Tran V, Wu Z, Yu X, Alexander DC, Gao GF, Zhu B, Liu J. 2008. Novel genome polymorphisms in BCG vaccine strains and impact on efficacy. *BMC Genomics* 9:413.

44. Zahrt TC, Deretic V. 2001. *Mycobacterium tuberculosis* signal transduction system required for persistent infections. *Proc Natl Acad Sci USA* 98:12706–12711.

45. Zahrt TC, Wozniak C, Jones D, Trevett A. 2003. Functional analysis of the *Mycobacterium tuberculosis* MprAB two-component signal transduction system. *Infect Immun* 71:6962–6970.

46. He H, Zahrt TC. 2005. Identification and characterization of a regulatory sequence recognized by *Mycobacterium tuberculosis* persistence regulator MprA. *J Bacteriol* 187:202–212.

47. Dona V, Rodrigue S, Dainese E, Palu G, Gaudreau L, Manganelli R, Provvedi R. 2008. Evidence of complex transcriptional, translational, and posttranslational regulation of the extracytoplasmic function sigma factor sigmaE in *Mycobacterium tuberculosis*. *J Bacteriol* 190:5963–5971.

48. Pang X, Vu P, Byrd TF, Ghanny S, Soteropoulos P, Mukamolova GV, Wu S, Samten B, Howard ST. 2007. Evidence for complex interactions of stress-associated regulons in an *mprAB* deletion mutant of *Mycobacterium tuberculosis*. *Microbiology* 153:1229–1242.

49. Sureka K, Dey S, Datta P, Singh AK, Dasgupta A, Rodrigue S, Basu J, Kundu M. 2007. Polyphosphate kinase is involved in stress-induced *mprAB-sigE-rel* signalling in mycobacteria. *Mol Microbiol* 65:261–276.

50. Sureka K, Ghosh B, Dasgupta A, Basu J, Kundu M, Bose I. 2008. Positive feedback and noise activate the stringent response regulator *rel* in mycobacteria. *PLoS One* 3:e1771.

51. White MJ, He H, Penoske RM, Twining SS, Zahrt TC. 2010. PepD participates in the mycobacterial stress response mediated through MprAB and SigE. *J Bacteriol* 192:1498–1510.

52. White MJ, Savaryn JP, Bretl DJ, He H, Penoske RM, Terhune SS, Zahrt TC. 2011. The HtrA-like serine protease PepD interacts with and modulates the *Mycobacterium tuberculosis* 35-kDa antigen outer envelope protein. *PLoS One* 6:e18175.

53. He H, Bretl DJ, Penoske RM, Anderson DM, Zahrt TC. 2011. Components of the Rv0081-Rv0088 locus, which encodes a predicted formate hydrogenlyase complex, are coregulated by Rv0081, MprA, and DosR in *Mycobacterium tuberculosis*. *J Bacteriol* 193:5105–5118.

54. Bretl DJ, He H, Demetriadou C, White MJ, Penoske RM, Salzman NH, Zahrt TC. 2012. MprA and DosR coregulate a *Mycobacterium tuberculosis* virulence operon encoding Rv1813c and Rv1812c. *Infect Immun* 80:3018–3033.

55. Pang X, Howard ST. 2007. Regulation of the alpha-crystallin gene acr2 by the MprAB two-component system of *Mycobacterium tuberculosis*. *J Bacteriol* 189:6213–6221.

56. Pang X, Cao G, Neuenschwander PF, Haydel SE, Hou G, Howard ST. 2011. The beta-propeller gene Rv1057 of *Mycobacterium tuberculosis* has a complex promoter directly regulated by both the MprAB and TrcRS two-component systems. *Tuberculosis* 91(Suppl 1):S142–S149.

57. Pang X, Samten B, Cao G, Wang X, Tvinnereim AR, Chen XL, Howard ST. 2013. MprAB regulates the *espA* operon in *Mycobacterium tuberculosis* and modulates ESX-1 function and host cytokine response. *J Bacteriol* 195:66–75.

58. Wren BW, Colby SM, Cubberley RR, Pallen MJ. 1992. Degenerate PCR primers for the amplification of fragments from genes encoding response regulators from a range of pathogenic bacteria. *FEMS Microbiol Lett* 99:287–292.

59. Supply P, Magdalena J, Himpens S, Locht C. 1997. Identification of novel intergenic repetitive units in a mycobacterial two-component system operon. *Mol Microbiol* 26:991–1003.

60. Allix-Béguec C, Harmsen D, Weniger T, Supply P, Niemann S. 2008. Evaluation and user-strategy of MIRU-VNTRplus, a multifunctional database for online analysis of genotyping data and phylogenetic identification of *Mycobacterium tuberculosis* complex isolates. *J Clin Microbiol* 46:2692–2699.

61. Himpens S, Locht C, Supply P. 2000. Molecular characterization of the mycobacterial SenX3-RegX3 two-component system: evidence for autoregulation. *Microbiology* 146:3091–3098.

62. Rickman L, Saldanha JW, Hunt DM, Hoar DN, Colston MJ, Millar JBA, Buxton RS. 2004. A

two-component signal transduction system with a PAS domain-containing sensor is required for virulence of *Mycobacterium tuberculosis* in mice. *Biochem Biophys Res Commun* 314:259–267.

63. Glover RT, Kriakov J, Garforth SJ, Baughn AD, Jacobs WR Jr. 2007. The two-component regulatory system *senX3-regX3* regulates phosphate-dependent gene expression in *Mycobacterium smegmatis*. *J Bacteriol* 189:5495–5503.

64. Rifat D, Bishai WR, Karakousis PC. 2009. Phosphate depletion: a novel trigger for *Mycobacterium tuberculosis* persistence. *J Inf Dis* 200:1126–1135.

65. Brennan RG. 1993. The winged-helix DNA binding motif: another helix-turn-helix takeoff. *Cell* 74:773–776.

66. Martinez-Hackert E, Stock AM. 1997. Structural relationships in the OmpR family of winged-helix transcription factors. *J Mol Biol* 269:301–312.

67. Kondo H, Nakagawa A, Nishihira J, Nishimura Y, Mizuno T, Tanaka I. 1997. *Escherichia coli* positive regulator OmpR has a large loop structure at the putative RNA polymerase interaction site. *Nat Struct Biol* 4:28–31.

68. Toro-Roman A, Mack TR, Stock AM. 2005. Structural analysis and solution studies of the activated regulatory domain of the response regulator ArcA: a symmetric dimer mediated by the alpha4-beta5-alpha5 face. *J Biol Chem* 349:11–26.

69. Toro-Roman A, Wu T, Stock AM. 2005. A common dimerization interface in bacterial response regulators KdpE and TorR. *Protein Sci* 14:3077–3088.

70. King-Scott J, Nowak E, Mylonas E, Panjikar S, Roessle M, Svergun DI, Tucker PA. 2007. The structure of a full-length response regulator from *Mycobacterium tuberculosis* in a stabilized three-dimensional domain-swapped, activated state. *J Biol Chem* 282:37717–37729.

71. Parish T, Smith DA, Roberts G, Betts J, Stoker NG. 2003. The *senX3-regX3* two-component regulatory system of *Mycobacterium tuberculosis* is required for virulence. *Microbiology* 149:1423–1435.

72. Roberts G, Vadrevu IS, Madiraju MV, Parish T. 2011. Control of CydB and GltA1 expression by the SenX3 RegX3 two component regulatory system of *Mycobacterium tuberculosis*. *PLoS One* 6:e21090.

73. Tischler AD, Leistikow RL, Kirksey MA, Voskuil MI, McKinney JD. 2013. *Mycobacterium tuberculosis* requires phosphate-responsive gene regulation to resist host immunity. *Infect Immun* 81:317–328.

74. James JN, Hasan ZU, Ioerger TR, Brown AC, Personne Y, Carroll P, Ikeh M, Tilston-Lunel NL, Palavecino C, Sacchettini JC, Parish T. 2012. Deletion of SenX3-RegX3, a key two-component regulatory system of *Mycobacterium smegmatis*, results in growth defects under phosphate-limiting conditions. *Microbiology* 158:2724–2731.

75. Kriakov J, Lee SH, Jacobs WR. 2003. Identification of a regulated alkaline phosphatase, a cell surface-associated lipoprotein, in *Mycobacterium smegmatis*. *J Bacteriol* 185:4983–4991.

76. Gebhard S, Cook GM. 2008. Differential regulation of high-affinity phosphate transport systems of *Mycobacterium smegmatis*: identification of PhnF, a repressor of the phnDCE operon. *J Bacteriol* 190:1335–1343.

77. Ewann F, Jackson M, Pethe K, Cooper A, Mielcarek N, Ensergueix D, Gicquel B, Locht C, Supply P. 2002. Transient requirement of the PrrA-PrrB two-component system for early intracellular multiplication of *Mycobacterium tuberculosis*. *Infect Immun* 70:2256–2263.

78. Zhang Y, Zhang H, Zhonghe S. 2003. Susceptibility of *Mycobacterium tuberculosis* to weak acids. *J Antimicrob Chemother* 52:56–60.

79. Carroll P, Faray-Kele M-C, Parish T. 2011. Identifying vulnerable pathways in *Mycobacterium tuberculosis* by using a knockdown approach. *Appl Environ Microbiol* 77:5040–5043.

80. Nowak E, Panjikar S, Morth JP, Jordanova R, Svergun DI, Tucker PA. 2006. Structural and functional aspects of the sensor histidine kinase PrrB from *Mycobacterium tuberculosis*. *Structure* 14:275–285.

81. Nowak E, Panjikar S, Konarev P, Svergun DI, Tucker PA. 2006. The structural basis of signal transduction for the response regulator PrrA from *Mycobacterium tuberculosis*. *J Biol Chem* 281:9659–9666.

82. Ewann F, Locht C, Supply P. 2004. Intracellular autoregulation of the *Mycobacterium tuberculosis* PrrA response regulator. *Microbiology* 150:241–246.

83. Haydel SE, Malhotra V, Cornelison GL, Clark-Curtiss JE. 2012. The *prrAB* two-component system is essential for *Mycobacterium tuberculosis* viability and is induced under nitrogen-limiting conditions. *J Bacteriol* 194:354–361.

84. Graham J, Clark-Curtiss J. 1999. Identification of *Mycobacterium tuberculosis* RNAs synthesized in response to phagocytosis by human macrophages by selective capture of transcribed sequences (SCOTS). *Proc Natl Acad Sci USA* 96:11554–11559.

85. Via LE, Curcic R, Mudd MH, Dhandayuthapani S, Ulmer RJ, Deretic V. 1996. Elements of signal transduction in *Mycobacterium tuberculosis*: in vitro phosphorylation in vivo expression of the response regulator MtrA. *J Bacteriol* 178:3314–3321.

86. Zahrt TC, Deretic V. 2000. An essential two-component signal transduction system in *Mycobacterium tuberculosis*. *J Bacteriol* 182:3832–3838.

87. Li Y, Zeng J, He ZG. 2010. Characterization of a functional C-terminus of the *Mycobacterium tuberculosis* MtrA responsible for both DNA binding and interaction with its two-component partner protein, MtrB. *J Biochem* 148:549–556.

88. Al Zayer M, Stankowska D, Dziedzic R, Sarva K, Madiraju MV, Rajagopalan M. 2011. *Mycobacterium tuberculosis mtrA* merodiploid strains with point mutations in the signal-receiving domain of MtrA exhibit growth defects in nutrient broth. *Plasmid* 65:210–218.

89. Friedland N, Mack TR, Yu M, Hung L-W, Terwilliger TC, Waldo GS, Stock AM. 2007. Domain orientation in the inactive response regulator *Mycobacterium tuberculosis*

MtrA provides a barrier to activation. *Biochemistry* **46**: 6733–6743.

90. Parish T, Smith DA, Kendall S, Casali N, Bancroft GJ, Stoker NG. 2003. Deletion of two-component regulatory systems increases the virulence of *Mycobacterium tuberculosis*. *Infect Immun* **71**:1134–1140.

91. Robertson D, Carroll P, Parish T. 2007. Rapid recombination screening to test gene essentiality demonstrates that *pyrH* is essential in *Mycobacterium tuberculosis*. *Tuberculosis* **87**:450–458.

92. Cangelosi GA, Do JS, Freeman R, Bennett JG, Semret M, Behr MA. 2006. The two-component regulatory system *mtrAB* is required for morphotypic multidrug resistance in *Mycobacterium avium*. *Antimicrob Agents Chemother* **50**:461–468.

93. Plocinska R, Purushotham G, Sarva K, Vadrevu IS, Pandeeti EV, Arora N, Plocinski P, Madiraju MV, Rajagopalan M. 2012. Septal localization of the *Mycobacterium tuberculosis* MtrB sensor kinase promotes MtrA regulon expression. *J Biol Chem* **287**:23887–23899.

94. Sun R, Converse PJ, Ko C, Tyagi S, Morrison NE, Bishai WR. 2004. *Mycobacterium tuberculosis* ECF sigma factor *sigC* is required for lethality in mice and for the conditional expression of a defined gene set. *Mol Microbiol* **52**:25–38.

95. Nguyen HT, Wolff KA, Cartabuke RH, Ogwang S, Nguyen L. 2010. A lipoprotein modulates activity of the MtrAB two-component system to provide intrinsic multidrug resistance, cytokinetic control and cell wall homeostasis in *Mycobacterium*. *Mol Microbiol* **76**:348–364.

96. Fol M, Chauhan A, Nair NK, Maloney E, Moomey M, Jagannath C, Madiraju M, Rajagopalan M. 2006. Modulation of *Mycobacterium tuberculosis* proliferation by MtrA, an essential two-component response regulator. *Mol Microbiol* **60**:643–657.

97. Rajagopalan M, Dziedzic R, Al Zayer M, Stankowska D, Ouimet M-C, Bastedo DP, Marczynski GT, Madiraju MV. 2010. *Mycobacterium tuberculosis* origin of replication and the promoter for immunodominant secreted antigen 85B are the targets of MtrA, the essential response regulator. *J Biol Chem* **285**:15816–15827.

98. Li Y, Zeng J, Zhang H, He ZG. 2010. The characterization of conserved binding motifs and potential target genes for *M. tuberculosis* MtrAB reveals a link between the two-component system and the drug resistance of *M. smegmatis*. *BMC Microbiol* **10**:242.

99. Haydel SE, Dunlap NE, Benjamin WH. 1999. In vitro evidence of two-component system phosphorylation between the *Mycobacterium tuberculosis* TrcR/TrcS proteins. *Microb Pathog* **26**:195–206.

100. Haydel SE, Benjamin WH, Dunlap NE, Clark-Curtiss JE. 2002. Expression, autoregulation, and DNA binding properties of the *Mycobacterium tuberculosis* TrcR response regulator. *J Bacteriol* **184**:2192–2203.

101. Wayne LG, Grosse WM. 1968. Base composition of deoxyribonucleic acid isolated from mycobacteria. *J Bacteriol* **96**:1915–1919.

102. Haydel SE, Clark-Curtiss JE. 2006. The *Mycobacterium tuberculosis* TrcR response regulator represses transcription of the intracellularly expressed Rv1057 gene, encoding a seven-bladed beta-propeller. *J Bacteriol* **188**:150–159.

103. Manganelli R, Voskuil MI, Schoolnik GK, Dubnau E, Gomez M, Smith I. 2002. Role of the extracytoplasmic-function sigma factor sigma H in *Mycobacterium tuberculosis* global gene expression. *Mol Microbiol* **45**:365–374.

104. Singh A, Mai D, Kumar A, Steyn AJ. 2006. Dissecting virulence pathways of *Mycobacterium tuberculosis* through protein-protein association. *Proc Natl Acad Sci USA* **103**:11346–11351.

105. Steyn AJ, Joseph J, Bloom BR. 2003. Interaction of the sensor module of *Mycobacterium tuberculosis* H37Rv KdpD with members of the Lpr family. *Mol Microbiol* **47**:1075–1089.

106. Gassel M, Siebers A, Epstein W, Altendorf K. 1998. Assembly of the Kdp complex, the multi-subunit K+-transport ATPase of *Escherichia coli*. *Biochim Biophys Acta* **1415**:77–84.

107. Betts JC, Lukey PT, Robb LC, McAdam RA, Duncan K. 2002. Evaluation of a nutrient starvation model of *Mycobacterium tuberculosis* persistence by gene and protein expression profiling. *Mol Microbiol* **43**:717–731.

108. Schnell R, Agren D, Schneider G. 2008. 1.9 A structure of the signal receiver domain of the putative response regulator NarL from *Mycobacterium tuberculosis*. *Acta Crystallogr Sect F Struct Biol Cryst Commun* **64**:1096–1100.

109. Hu YM, Movahedzadeh F, Stoker NG, Coates ARM. 2006. Deletion of the *Mycobacterium tuberculosis* alpha-crystallin-like *hspX* gene causes increased bacterial growth in vivo. *Infect Immun* **74**:861–868.

110. Bhattacharya M, Biswas A, Das AK. 2010. Interaction analysis of TcrX/Y two component system from *Mycobacterium tuberculosis*. *Biochimie* **92**:263–272.

111. Bhattacharya M, Das AK. 2011. Inverted repeats in the promoter as an autoregulatory sequence for TcrX in *Mycobacterium tuberculosis*. *Biochem Biophys Res Commun* **415**:17–23.

112. Haydel SE, Clark-Curtiss JE. 2004. Global expression analysis of two-component system regulator genes during *Mycobacterium tuberculosis* growth in human macrophages. *FEMS Microbiol Lett* **236**:341–347.

113. Bacon J, Dover LG, Hatch KA, Zhang Y, Gomes JM, Kendall S, Wernisch L, Stoker NG, Butcher PD, Besra GS, Marsh PD. 2007. Lipid composition and transcriptional response of *Mycobacterium tuberculosis* grown under iron limitation in continuous culture: identification of a novel wax ester. *Microbiology* **153**:1435–1444.

114. Shrivastava R, Das AK. 2007. Temperature and urea induced conformational changes of the histidine kinases from *Mycobacterium tuberculosis*. *Int J Biol Macromol* **41**:154–161.

115. Shrivastava R, Das DR, Wiker HG, Das AK. 2006. Functional insights from the molecular modelling of a

novel two-component system. *Biochem Biophys Res Commun* **344**:1327–1333.

116. **Shrivastava R, Ghosh AK, Das AK.** 2007. Probing the nucleotide binding and phosphorylation by the histidine kinase of a novel three-protein two-component system from *Mycobacterium tuberculosis. FEBS Lett* **581**:1903–1909.

117. **Shrivastava R, Ghosh AK, Das AK.** 2009. Intra- and intermolecular domain interactions among novel two-component system proteins coded by Rv0600c, Rv0601c and Rv0602c of *Mycobacterium tuberculosis. Microbiology* **155**:772–779.

118. **Tyagi JS, Sharma D.** 2004. Signal transduction systems of mycobacteria with special reference to *M. tuberculosis. Curr Sci* **86**:93–102.

119. **Morth JP, Gosmann S, Nowak E, Tucker PA.** 2005. A novel two-component system found in *Mycobacterium tuberculosis. FEBS Lett* **579**:4145–4148.

120. **Boshoff HIM, Myers TG, Copp BR, McNeil MR, Wilson MA, Barry CE.** 2004. The transcriptional responses of *Mycobacterium tuberculosis* to inhibitors of metabolism: novel insights into drug mechanisms of action. *J Biol Chem* **279**:40174–40184.

Molecular Genetics of Mycobacteria, 2nd Edition
Edited by Graham F. Hatfull and William R. Jacobs, Jr.
© 2014 American Society for Microbiology, Washington, DC
doi:10.1128/microbiolspec.MGM2-0018-2013

Dirk Schnappinger[1,2]
Sabine Ehrt[1,3]

Regulated Expression Systems for Mycobacteria and Their Applications

11

GENETIC SWITCHES FOR CONTROLLING GENE EXPRESSION IN MYCOBACTERIA

The Acetamidase System

During growth with short aliphatic amides (e.g., acetamide) as the primary carbon source, *Mycobacterium smegmatis* induces expression of the acetamidase encoded by *amiE* (Fig. 1a) (1–3). The regulatory elements of this gene were utilized to generate the first inducible expression system for mycobacteria (4). The system proved valuable for the production of mycobacterial antigens (4, 5) and enabled the first silencing studies of essential genes (e.g., *whmD* and *dnaA*) in *M. smegmatis* (6, 7). But genetic instability limited the use of this system in *Mycobacterium tuberculosis* (8), and its complexity—regulation of *amiE* involves three regulators (AmiC, AmiD, and AmiA) (9, 10)—prevented its optimization. While the acetamidase system has been largely replaced by other tools, especially in *M. tuberculosis*, a derivative, which incorporated the T7 RNA polymerase (RNAP), remains one of the best tools available to achieve high-level overexpression of a protein in *M. smegmatis* (11).

TetON and TetOFF

Tetracycline resistance of many bacteria is caused by efflux pumps whose expression is—due to the fitness defect the pumps cause in the absence of drug pressure—tightly regulated (Fig. 1b, c). This regulation is mediated by a single repressor protein, the tetracycline repressor (TetR), which specifically binds two operators (*tetO*$_1$ and *tetO*$_2$) in the promoter that drives transcription of the efflux pump (12). In the complex with TetR the tet promoter (P$_{tet}$) is masked from access by RNAP, and initiation of transcription is inhibited. When tetracycline enters the bacterial cell, it binds to TetR and induces transcription of the efflux pump before the drug can inhibit the ribosome. This sensitivity toward low drug concentrations is due to the remarkable affinity of TetR to tetracyclines, which is up to 10^5-fold higher than the ribosome's affinity to tetracyclines (13).

In 2005, three groups independently reported TetR-controlled expression systems for mycobacteria (14–16). The systems shared the same basic design but differed in the origin of their regulatory components:

[1]Department of Microbiology and Immunology, Weill Medical College of Cornell University, New York, NY 10065; Programs in [2]Molecular Biology and [3]Immunology and Microbial Pathogenesis, Weill Graduate School of Medical Sciences of Cornell University, New York, NY 10065.

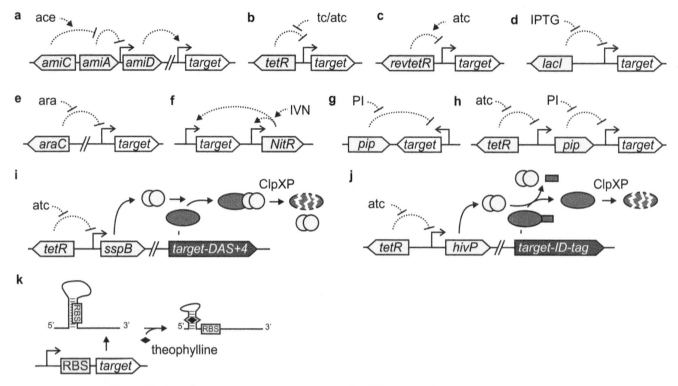

Figure 1 Regulatory systems for mycobacteria. The transcriptional regulatory systems are shown in (a) to (h), the two controlled proteolysis systems in (i) and (j), and the theophylline riboswitch in (k). Dotted lines ending in a perpendicular line indicate negative regulatory interactions; dotted lines ending in an arrow represent positive regulatory interactions. Ace, acetamide; tc/atc, tetracycline/anhydrotetracycline; IPTG, isopropyl β-D 1-thiogalactopyranoside; ara, arabinose; IVN, isovaleronitrile; PI, pristinamycin. doi:10.1128/microbiolspec.MGM2-0018-2013.f1

the TetRs were derived either from the *Corynebacterium glutamicum* resistance determinant TetZ (14, 17) or the *Escherichia coli* transposon Tn*10* (15, 16); the regulated promoters were either also from TetZ (14, 17), derived from the *Bacillus subtilis* xyl promoter (16), or constructed by inserting *tet*Os into a mycobacterial promoter (15). All three systems can be induced with low concentrations of tetracyclines in a dose-dependent manner, with the preferred inducer either being tetracycline (for the TetZ-derived systems) or anhydrotetracycline (for the two systems that utilize the Tn*10* TetR). Because tetracycline/anhydrotetracycline has to be added to induce expression, we refer to these systems as "TetON" systems.

The Tn*10* TetR has been the subject of many mechanistic analyses. In a screening strain that proved particularly useful, TetR controlled expression of β-galactosidase while the lac repressor (LacI) and transcription of *galK*, encoding galactokinase, were repressed by LacI. This allowed the identification of amino acids required for binding of TetR to *tet*O

(mutations in these amino acids led to β-galactosidase-positive and galactokinase-negative colonies without anhydrotetracycline) (18) or for induction of TetR by tetracyclines (19). A mutagenesis originally performed for the latter purpose also identified the first TetR that only bound *tet*O in complex with tetracycline. Such reverse TetRs were later adapted for use in mycobacteria to construct a "TetOFF" switch in which transcription of the target gene is turned off by the addition of anhydrotetracycline (20, 21). Optimization of these repressors for use in mycobacteria included adapting the guanine-cytosine (GC) content of the encoding genes to that of mycobacteria, which increased TetR expression and also led to an improved TetON system (21). TetON and TetOFF have been used by several groups to analyze gene functions in *M. smegmatis* and *M. tuberculosis* (Table 1). They also provide the basis for some of the other regulatory expression systems developed more recently (22–24) and a tunable coexpression system to analyze protein-protein interactions (25).

AraC and LacI

Leakiness, i.e., expression without inducer, is a limitation of many regulated expression systems (Fig. 1d,e). One of the most tightly regulated *E. coli* expression systems is the pBAD system (26). Its promoter, P_{BAD}, is controlled by two regulators: AraC, which represses the promoter without arabinose and activates it in its presence, and the catabolite activator protein, CAP, which acts as a second activating factor (27). Activation of P_{BAD} by CAP increases with the intracellular cAMP concentration. In *E. coli*, activity of P_{BAD} without arabinose can thus be reduced by adding glucose to the growth medium because glucose decreases cAMP levels in this species. Unfortunately, P_{BAD} does not function in *M. smegmatis* as it does in *E. coli* (28), and there is no apparent advantage that pBAD has over the other systems developed for mycobacteria. Tight regulation of P_{BAD} in *E. coli* depends not only on protein-DNA interactions but also on direct protein-protein interactions of AraC and CAP with RNAP, as well as low levels of cAMP. It therefore would be difficult to optimize the pBAD system for use in mycobacteria.

Other frequently used *E. coli* expression systems depend on promoters that are repressed by LacI and induced with IPTG (29). Two studies demonstrated the value of LacI for regulating gene expression in mycobacteria. The first applied LacI to repress a promoter recognized by the T7 RNAP (30); the second inserted a lac operator (*lacO*) downstream of a mycobacterial promoter to impose susceptibility to repression by LacI (31). For both systems little expression was measured without IPTG, but no follow-up studies or applications have been published, and their value for broader studies remains to be determined.

NitR

The saprophytic actinomycete *Rhodococcus rhodochrous* encodes several nitrilases, which detoxify nitriles by hydrolyzing them into their carboxylic acid and ammonia (32) (Fig. 1f). Under optimal conditions *R. rhodochrous* J1 increases nitrilase expression up to ~3,000-fold, which results in the nitrilase encoded by *nitA* accounting for ~35% of total soluble protein (32, 33). This drastic overexpression is achieved via a positive feedback loop controlled by NitR, a member of the AraC family of transcriptional regulators. The molecular mechanism by which NitR acts has not been investigated in detail. But NitR alone is sufficient to mediate induction of P_{nitA} and its own promoter in other bacterial species, most likely functioning as a

direct activator of transcription initiation (34, 35). In *M. smegmatis* NitR strongly activated transcription after addition of either ε-caprolactam or isovaleronitrile, whereas in *M. tuberculosis* only isovaleronitrile was effective (36). The positive feedback loop that is generated by NitR's activation of its own promoter distinguishes this system from all other expression systems available for mycobacteria and has three consequences: (i) induction is strong; (ii) on a single-cell level the switch is either ON or OFF; and (iii) intermediate inducer concentrations create two subpopulations, one that has NitR-controlled gene expression turned fully ON and one that is still in the OFF state. In contrast, intermediate concentrations of anhydrotetracycline partially activate the TetON system so that the average expression level of most cells increases to levels between the OFF and fully induced states (36).

PipON and Tet/PipOFF

Pristinamycin belongs to the streptogramin group of antibiotics, which consist of at least two structurally unrelated but synergistically acting molecules (Fig. 1g, h). In the case of pristinamycin, these two molecules are pristinamycin I and pristinamycin II, both of which inhibit bacterial ribosomes (37). Resistance of *Streptomyces pristinaespiralis* to pristinamycin is due to the pristinamycin resistance gene, *ptr*, which encodes a multidrug efflux pump (38). The *ptr* promoter, P_{ptr}, is repressed by the transcription factor Pip and can be activated with pristinamycin I, pristinamycin II, and several other antibiotics (39, 40). Pip belongs to the TetR family of transcription factors and binds to three sites in P_{ptr}, two of which overlap with the promoters −35 and −10 hexamers (40). P_{ptr} is a strong promoter in *M. smegmatis* and *M. tuberculosis* and can be efficiently repressed by Pip and induced with low concentrations of PI. As a consequence, the PipON system has an excellent regulatory range (41).

The Pip system was also adapted to confer repression upon addition of anhydrotetracycline. In contrast to the TetOFF system, which utilizes a reverse TetR, in Tet/PipOFF Pip is placed under the control of wild-type (wt) TetR so that anhydrotetracycline increases expression of Pip. The target gene is located downstream of P_{ptr} and thus is repressed as a consequence of the increased Pip expression. When desired, PI can be used to overcome the repression caused by anhydrotetracycline (24). A system with a similar regulatory circuit has placed TetR under the control of the acetamidase system (42).

Table 1   Regulated expression systems for mycobacteria and examples of their applications

Expression system	Components	Regulatory range	Applications
Regulation of transcription			
Acetamidase	AmiC; AmiA; AmiD; promoters Pc, P1, P1, and P3 (9,10)	~80-fold induction of *M. leprae* 35 kDa protein in *M. smegmatis* (4); 22-fold induction of FtsZ in *M. smegmatis* (81)	Ectopic expression of mycobacterial antigens (4,5), PknA/B (82,83), and toxin-antitoxin (TA) proteins (63) in *M. smegmatis*; Silencing of *whiB2* (*whmD*) (6), *uvag31* (82,83), *kasA* (84), *inhA* (84), and *dnaA* (7) in *M. smegmatis*
	Acetamidase inducible T7 RNAP	Not reported	Overexpression of putative drug targets in *M. smegmatis* (11)
TetON	TetR and P$_{tet}$ from TetZ (14)	230-fold induction of luciferase activity in *M. smegmatis* (14,17); 13-fold induction of luciferase activity in *M. tuberculosis* (14); 21-fold induction of luciferase activity in *M. bovis* BCG (14)	Ectopic expression of Ms2173 (85) and TA proteins (86–89) in *M. smegmatis*; Silencing of *ftsZ* (14), *uvag31* (82,83), *clpC1*, *pknB*, *msmeg_2694* (90), *parA* (91), and *glmM* (92) in *M. smegmatis*; Silencing of *clpC1* (90), *ppk1* (93), *ppk2* (94), and *dosR* (95) in *M. tuberculosis*
	Tn10 TetR, P$_{myc1}$*tetO* (15)	170-fold induction of GFP activity in *M. smegmatis* (14); 160-fold induction of β-galactosidase activity in *M. tuberculosis* (14)	Ectopic expression of PhoP (96), DosR (97), various sigma factors (64), various transcription factors (65), I-SceI (98,99), PzaA/PncA (100), BirA (101), Pbp1 (102), EspR (103), and TA proteins (86,104–107) in *M. smegmatis, M. bovis* BCG, and/or *M. tuberculosis*; Silencing of *ftsZ* (14), *secA1* (20,108), *pptT* (109), *ripA* (110), *pbp1* (102), *ppm1* (111), *carD* (60), *msmeg_3935* (112), and *clpP* (71) in *M. smegmatis*; Silencing of *icl* (58), *rv3671c* (58), *prcBA* (58,59), *pptT* (56,109), *espA* (113), *bioA* (55), *fba* (114), *esx-3* (115), *carD* (60), *pckA* (57), *clpP* (71), *panC* (79), *lysA* (79), and *dfrA* (116) in *M. tuberculosis*
	Tn10 TetR, P$_{xyl}$ *tetO* (16)	~10-fold induction of GFP activity in *M. smegmatis* and *M. tuberculosis* (16)	Silencing of *trpD* (16), *dprE1* (117), *clpP1* (117), *fadD32, glnA1* (117), *glnE* (117), *pknL* (117), *regX3* (117), and *senX3* (117) in *M. tuberculosis*
TetOFF	Reverse TetR, P$_{myc1}$*tetO* (20,21)	50-fold repression of β-galactosidase activity in *M. smegmatis* (21); 10-fold repression of β-galactosidase activity in *M. bovis* BCG (21)	Silencing of *secA1* (20) in *M. smegmatis*; Silencing of *prcBA* (59), *panC* (79), *lysA* (79), and *icl* (79) in *M. tuberculosis*

System	Components	Activity	Application
pBAD	AraC, P_{BAD} (28)	~3-fold induction of β-galactosidase activity in M. smegmatis (28)	Ectopic expression of Rv1991c (28)
LacI	LacI, P_{T7Lac}, T7 RNAP (30)	40-fold induction of GFP protein in M. tuberculosis (30)	Identification of metabolically active M. tuberculosis in macrophages (30)
LacI	LacI, P_{trc}lacO	30-fold induction of β-galactosidase activity in M. smegmatis (31)	Silencing of ftsZ, gyrA, and gyrB in M. smegmatis (31) Silencing of gyrA, gyrB, inhA, embR, rpoB, rpoC, rplJ, rpsL, and ilvB in M. tuberculosis (31)
NitR	NitR, P_{nitA} (36)	>100-fold induction of XylE activity in M. smegmatis (36) ~100-fold induction of XylE activity in M. tuberculosis (36)	
PipON	Pip, P_{ptr} (41)	52-fold induction of β-galactosidase activity in M. smegmatis (41) 450-fold induction of β-galactosidase in M. tuberculosis (41)	Silencing of fadD32 (41), pknB (41), ftsK (68), glf (68), infB (68), leuA (68), metC (68), rne (68), rv0883c (68), rv1478 (68), rv2050 (68), rv2204c (68), secY (68), and tuf (68) in M. tuberculosis
TetPipOFF	Tn10 TetR, Pip, $P_{myc1}tetO$, P_{ptr} (24)	~60-fold repression of β-galactosidase in M. smegmatis and M. tuberculosis (24)	Silencing of ftsZ (24) in M. smegmatis Silencing of fadD32 (118) in M. abscessus Silencing of fadD32 (24), eccB5 (119), eccC5 (119), and esx-3 (120) in M. tuberculosis
Regulation of protein stability			
DAS+4-tag	Tn10 TetR, $P_{myc1}tetO$, SspB	36- and 250-fold repression of GFP and luciferase activity in M. smegmatis, respectively (23) 7-fold repression of GFP activity in M. tuberculosis (23)	Depletion of RpoB in M. smegmatis (23)
ID-tag	Tn10 TetR, $P_{myc1}tetO$, HIV2 derived protease	~80-fold repression of GFP activity in M. smegmatis (22) >30-fold depletion of Alr protein; ~5-fold depletion of RpoB protein in M. smegmatis (22)	Depletion of Alr (22), DHFR (22), InhA (22), GyrA (22), KasA (22), RpoB (22), and FhaA (72) in M. smegmatis
Repression of translation			
Theophyllin	Riboswitch (53)	65-fold and 89-fold induction of GFP and β-galactosidase activities in M. smegmatis (53) 8-fold induction of GFP activity in M. tuberculosis (53)	Silencing of katG in M. smegmatis (53)

Controlled Proteolysis

Bacterial regulatory circuits often rely on posttranscriptional modifications, which include controlled degradation, to achieve rapid inactivation of a protein (Fig. 1i, j). In fact, posttranscriptional modification is crucial to quickly inactivate proteins with a long half-life because their abundances only change slowly, even after transcription and translation have stopped (43). The recognition sites of bacterial proteases include C-terminal degradation tags (44). One such tag is added to proteins in a process called *trans*-translation and is encoded by the small stable RNA *ssrA* (45). In *E. coli* *ssrA*-tagged proteins are degraded by several proteases including ClpXP, which directly binds to the tag's C-terminal amino acids (46). Affinity of ClpXP to the *ssrA* tag is increased by the adaptor protein SspB, which binds both the tag's N-terminus and ClpX (47, 48). Proteins containing the *ssrA*-derived DAS+4 tag depend on the tethering of ClpXP to the tag by SspB. As a consequence, they are only degraded when SspB is expressed (Fig. 1i). This SspB dependency is due to mutations that change the tag's C-terminal amino acids from Leu-Ala-Ala to Asp-Ala-Ser (hence the "DAS") and weaken the direct interaction with ClpX and an insertion of four amino acids (hence the "+4") that facilitates simultaneous binding of SspB and ClpX (49).

Interestingly, SspB is also capable of delivering DAS+4-tagged proteins to ClpXP in bacteria that do not themselves encode an SspB homolog (50). This provided the mechanistic basis for one type of gene silencing tool that utilizes proteolysis to deplete proteins in mycobacteria (23). A second such tool was developed by placing the *ssrA* tag upstream of a protecting peptide that can be removed by a site-specific protease derived from HIV-2 (labeled *hivP* in Fig. 1j). The resulting tag was named the inducible degradation (ID) tag (22). In both systems degradation of the tagged protein is induced with anhydrotetracycline, which turns on expression of either SspB or the HIV-2-derived protease.

The Theophylline Riboswitch

Riboswitches are regulatory elements in which binding of a small molecule to an RNA aptamer results in a change in gene expression (Fig. 1k) (51). They are entirely RNA-encoded and do not require any trans-factors besides the aptamer-binding ligand, which can simplify transferring functional riboswitches from one species to another (52). The riboswitch adapted for use in mycobacteria is induced by theophylline (53), a methylxanthine drug used to treat pulmonary diseases (54). In the absence of theophylline, the switch forms a secondary structure that masks the ribosome binding site and thus prevents translation. Binding of theophylline stabilizes an alternative secondary structure, which liberates the ribosome binding site and induces translation of the regulated mRNA.

COMMON AND DISTINCTIVE FEATURES OF THE DIFFERENT REGULATORY SYSTEMS AND STRATEGIES

The ideal system for manipulating gene expression would (a) be completely silent under repressing conditions, (b) provide a large (i.e., >1,000-fold) regulatory range that can be adjusted in a dose-responsive manner with a small molecule that has no direct effects other than controlling the targeted gene, (c) not interfere with the target's native regulation under inducing conditions, (d) leave the protein sequence unchanged, and (e) allow rapid gene induction and protein depletion in growing and nonreplicating bacteria *in vitro* and during infections of host cells and animals. Not surprisingly, such a system has yet to be developed. But the available systems approach these features to different degrees.

Regulatory range: The range of regulated expression systems can be easily assessed using reporter gene assays. It is often calculated by dividing the reporter activity under inducing conditions by that measured under maximally repressing conditions. For most systems this has been achieved using either green fluorescent protein, β-galactosidase, or luciferase as the reporter. A regulatory range of >100-fold was measured for several systems (i.e., two of the TetON systems, the NitR system, PipON, and SspB-mediated proteolysis), with the largest range reported for PipON (Table 1).

Leakiness: Identifying expression systems that permit moderate expression without inducer is straightforward and can be achieved using the same reporter gene assays used to measure their regulatory range. However, none of the reporter assays that have been used to characterize mycobacterial expression systems approach single-molecule sensitivity. Lack of detectable reporter activity under repressing conditions, which has been reported for several systems, can therefore not provide proof of complete repression. In fact, all mycobacterial expression systems most likely permit some low level of expression without inducer. Whether or not this leakiness interferes with the goals of an experiment is difficult to predict and depends on the question that is being addressed and the gene under investigation. However, when necessary, the leakiness of an expression system can be reduced by decreasing the efficiency with which the targeted mRNA is translated (55).

Dose-responsiveness: All systems besides the one regulated by NitR have either been demonstrated to be dose-responsive or are likely to be dose-responsive in the sense that intermediate concentrations of the inducer or corepressor result in intermediate expression levels within most of the bacteria. Lack of dose-responsiveness of the NitR system comes with the benefit of achieving very high expression in the induced state.

Invasiveness: Controlling a gene's expression is not possible without changing at least its promoter, the 5′ noncoding end of its mRNA, or the 3′ end of its open reading frame. An alteration of the promoter is required to allow for transcriptional regulation, the incorporation of the riboswitch changes the mRNA's translation initiation sequence and its 5′ end, and the gene's 3′ end and the C-terminus of the encoded protein need to be changed to achieve controlled proteolysis. Fortunately, these modifications have little impact on the function of many genes, but any one of them can prevent complementation of a particular mutant. Strategies that rely on controlling transcription and/or translation have the advantage of leaving the open reading frame of the targeted gene unchanged. Controlled proteolysis, on the other hand, can leave a target's native regulation of transcription and translation intact. The theophylline riboswitch can also be used in combination with a gene's native promoter and does not require changes of the regulated protein. Riboswitches can thus provide the least invasive strategy to artificially control gene expression in bacteria.

Regulation during infections: Providing control over *M. tuberculosis* gene expression during infections is a key ability of expression systems designed for this pathogen. Evidence that this can be achieved during macrophage infections has been obtained for TetON/OFF, PipON, Tet/PipON, and the systems controlled by LacI or NitR. However, only for TetON/OFF have experiments been reported that demonstrated that efficient regulation can be achieved in animal models (55–60).

APPLICATIONS

Ectopic Expression

One motivation for the construction of the acetamidase system was to enable purification of *M. tuberculosis* or *Mycobacterium leprae* proteins from a fast-growing mycobacterial host, which was expected to yield proteins better suited for structural and immunological studies than those expressed in *E. coil* (4). The need for a mycobacterial expression host is supported by the finding that >50% of all *M. tuberculosis* proteins can either not be efficiently produced in *E. coli* or accumulate as insoluble inclusion bodies (61). For these proteins *M. smegmatis* can be a superior expression host because its codon usage is very similar to that of pathogenic mycobacteria, which facilitates high-level expression of proteins encoded by GC-rich mRNAs. Furthermore, proteins that accumulate as insoluble inclusion bodies in *E. coli* can—at least in some cases—be expressed as soluble proteins in *M. smegmatis* (61). Purification of polyhistidine-tagged recombinant proteins from *M. smegmatis* can be complicated by contamination with copurified GroEL1, but this can be avoided by using an *M. smegmatis* strain in which the histidine-rich C-terminus of GroEL1 has been removed (62).

More recently, ectopic expression was also used to analyze gene functions in *M. smegmatis* and *M. tuberculosis*. Many of these studies focused on type I toxin-antitoxin (TA) modules. These modules consist of two proteins that are often encoded by bicistronic operons wherein the 5′ gene encodes the antitoxin and the 3′ gene encodes the toxin. As long as expression of the TA module continues, the toxin is bound and neutralized by its cognate antitoxin. Once expression stops, the inherently unstable antitoxin is degraded, leading to release and activation of the toxin. The *M. tuberculosis* genome encodes 88 putative TA modules, many of which are conserved within the *M. tuberculosis* complex yet absent from other mycobacteria (63). For many of these putative toxins, inducible overexpression was used to confirm that they are indeed functional toxins capable of arresting growth of *M. smegmatis* and/or *M. tuberculosis* (Table 1). This growth arrest generally does not occur upon simultaneous overexpression of the cognate antitoxin, i.e., the antitoxin encoded within the same TA module, but is not relieved by overexpression of other antitoxins (63). Another informative application has been to combine ectopic overexpression of DNA-binding proteins with chromatin immunoprecipitation (ChIP). This was first demonstrated in experiments that defined the *in vivo* binding sites of SigA and several alternative sigma factors (64).

Recently, this approach has been extended to define the binding sites of many DNA-binding proteins in *M. tuberculosis* (65).

Gene Silencing

Controlled gene silencing allows the study of a gene's *in vivo* function under a variety of conditions even if the gene is required for growth. One conceptually attractive strategy to conditionally inactivate a gene is to destabilize and prevent translation of its mRNA with an antisense RNA. This strategy was first applied to reduce expression of AhpC in *Mycobacterium bovis* (66) and has since been used to inactivate several genes in *M. smegmatis* and *M. tuberculosis* (Table 1). One study in particular reported striking phenotypes for antisense-mediated gene silencing in several *M. tuberculosis* conditional knockdown (cKD) mutants (31). Attempts to silence different essential genes in *M. smegmatis* or *M. tuberculosis* with antisense RNAs of varying lengths in our own unpublished work have unfortunately all failed. The reasons for this failure are unclear to us and might be technical in nature. However, it is noteworthy that several research groups resorted to gene silencing approaches that are more complicated and time-consuming than antisense-mediated gene inactivation. Antisense-mediated gene silencing thus likely failed frequently, which suggests either that expression of only a few genes is susceptible to antisense inhibition or that some of the factors important for the functionality of an antisense RNA remain to be identified.

An alternative to expressing antisense RNAs is to exchange the targeted gene's promoter so that its transcription can be regulated directly. Promoter exchange can be achieved *in situ*, i.e., in the native chromosomal location, either by integrating a suicide plasmid immediately upstream of the targeted gene (15), by selecting for a double-crossover event that deletes the native promoter and replaces it with a regulated promoter (59), or by transposon insertion (67, 68). These strategies have been applied in many cases, and most cKD mutants of *M. tuberculosis* or *M. smegmatis* published to date employed direct transcriptional repression (Table 1). Obtaining phenotypically well-regulated cKD mutants can, however, be challenging, especially for genes that only need to be expressed at a low level to be functional.

In *M. tuberculosis*, *bioA* represents such a gene, whose mRNA is of low abundance during logarithmic growth (69, 70). It encodes the biotin biosynthetic enzyme 7,8-diaminopelargonic acid synthase, which is dispensable with extracellular biotin but essential for growth when biotin cannot be scavenged from the environment. The first *BioA* TetON mutant constructed with the Tn10-derived TetON system overexpressed BioA protein ~10-fold compared to wt *M. tuberculosis* (55). Removal of inducer decreased BioA expression by ~100-fold, yet only mildly reduced growth. In its original form the Tn10-derived TetON system contains a strong P_{tet} located upstream of a strong translation initiation site. Strength of the promoter and the translational initiation site were likely both responsible for overexpression of BioA. It was unclear if decreasing promoter strength would sufficiently reduce *bioA* transcription without inducer, but weaker translational initiation sites were expected to decrease both BioA overexpression with inducer and leaky expression without inducer. Accordingly, cKD mutants containing a weak translational initiation signal upstream of the *bioA* open reading frame reproduced the phenotype of a *bioA* deletion and only grew with inducer when growth depended on biotin synthesis (55). In our hands, this strategy of minimizing the phenotypic consequences of transcriptional leakiness with weak translation initiation signals has been successful for several other targets (unpublished data) and is generally useful to improve the efficiency of transcriptional gene silencing.

Another elegant use of direct transcriptional silencing is its combination with transposon mutagenesis. This depends on a transposon carrying a regulated promoter at one end in the outward-facing direction and allows the identification of well-regulated mutants based on their growth phenotypes (68).

cKD mutants that utilize transcriptional repression can be constructed by *in situ* promoter exchange, and similarly, cKD mutants that utilize controlled proteolysis can be generated by modifying a gene's 3′ end within its native location in the genome. This strategy has so far only been applied to the construction of cKD mutants in *M. smegmatis* but has shown good success in this species (22, 23, 71, 72). Nevertheless, for some targets depletion by controlled proteolysis was insufficient to produce the expected phenotypic consequences. For example, inactivation by controlled proteolysis of dihydrofolate reductase (DHFR) or alanine racemase (Alr), which are both essential for growth, depleted these enzymes by more than 97% but only modestly decreased growth of *M. smegmatis* (22).

Controlling Gene Expression during Infections

Mutations that attenuate *M. tuberculosis* can cause *growth in vivo* (*giv*), *severe giv* (*sgiv*), and *persistence* (*per*) phenotypes in mice (73). *giv* mutants replicate substantially less than wt, and *sgiv* mutants do not grow at all in mice, whereas *per* mutants replicate normally but fail to persist. Genes required for growth and persistence, i.e., genes whose inactivation causes *sgiv* and *per* phenotypes, can only be identified by conditional inactivation. The mycobacterial Tet systems helped demonstrate that the three *sgiv* genes *bioA*, *pckA* (encoding phosphenolpyruvate carboxy kinase), and *icl* (which encodes isocitrate lyase) are required by *M. tuberculosis* not only to grow in mice and establish an infection, but also to persist during the chronic phase of the infection (55, 57, 58). A cKD mutant of the *in vitro* essential CarD revealed that *M. tuberculosis* depends on this transcriptional regulator for replication and persistence in mice (60). Similarly, 4′-phosphopantetheinyl transferase PptT was shown to be required for the replication and survival of *M. tuberculosis* during the acute and chronic phases of infection in mice and helped validate these enzymes as a potential new drug target (56). The appearance of revertants, which are unresponsive to TetR-mediated transcriptional control, can complicate the analysis of essential genes *in vitro* and *in vivo* (reference 60 and our unpublished observations). A careful analysis of the bacterial population expressing the regulated gene under investigation is therefore necessary for conclusive data interpretation.

Target-Based Whole-Cell Screens

The application of regulated expression systems that can impact drug development most directly is their use in target-based whole-cell screens. Such screens employ mutants in which expression of the target protein has been decreased to the extent that it limits the growth rate, which increases sensitivity toward small molecule inhibitors of that protein. This principle was initially established with *Staphylococcus aureus* strains, which were engineered to express growth-limiting amounts of FabF and showed an increased susceptibility to FabF inhibitors but not to other antibiotics (74). Whole-cell screens against this FabF underexpressor identified platencin and platensimycin, the founding members of a new class of fatty acid biosynthesis inhibitors with broad-spectrum activity against Gram-positive bacteria (75–78). *M. tuberculosis* mutants expressing lower than wt levels of PanC, LysA, Icl1, or LepB have recently been constructed and also show target-specific changes in their susceptibility to different small molecule inhibitors (79, 80). Whole-cell screens with these strains promise to identify new inhibitors of pantothenate synthase, diaminopimelate decarboxylase, isocitrate lyase, and the type I signal peptidase, respectively.

CONCLUSIONS AND FUTURE PERSPECTIVES

When the first edition of this book was published, the only regulated expression system available was the acetamide system. Since then a dozen new regulatory systems have been developed that together utilize six transcription factors (TetR, revTetR, AraC, LacI, NitR, and Pip) and eight regulated promoters (P_{BAD}, P_{T7Lac}, $P_{trc}lacO$, P_{nitA}, P_{ptr}, and three different P_{tet} promoters). They were applied not only to facilitate purification of correctly folded proteins but also to study mycobacterial gene functions within their native hosts either by ectopic expression or conditional inactivation. By now several mycobacterial expression systems function so efficiently that their use in most applications is straightforward. However, the isolation of phenotypically well-regulated cKD mutants remains challenging, irrespective of the regulatory system one chooses for mutant construction. Reducing expression with antisense RNAs has been successful for some genes but failed to silence at least as many. This is unfortunate, because antisense-based gene silencing does not require manipulation of the host chromosome by homologous recombination. It would thus become the most straightforward approach to generate cKD mutants if its success rate could be improved.

Direct transcriptional silencing was often but not always successful. Due to the inherent leakiness of most regulated promoters, direct transcriptional silencing is most inefficient for genes whose products are only needed in small amounts. The opposite is likely true for controlled proteolysis because highly expressed proteins will burden the host's proteolytic machinery more than proteins expressed at a lower level. That transcriptional silencing and controlled proteolysis can both fail to produce phenotypically well-regulated cKD is essentially a consequence of their limited dynamic range, which spans only two orders of magnitude. In contrast, *M. tuberculosis* gene expression, as measured by RNA sequencing, spans at least four to five orders of magnitude (69, 70). One of the main remaining challenges in the development of regulated expression systems for mycobacteria is thus to expand their dynamic range. In ongoing work we observed that this can be achieved by combining transcriptional repression with controlled proteolysis. This strategy of combining existing regulatory systems that differ in their mechanism of regulation could be further extended. For example, it should be possible to combine the theophylline riboswitch with any of the transcriptional regulation systems to reduce their effective leakiness yet still allow high-level expression when necessary.

We are grateful for the support we have received from the National Institutes of Health, the Bill & Melinda Gates Foundation, the Wellcome Trust, the Heiser Program for Research in Leprosy and Tuberculosis, the Ellison Medical Foundation, and Cornell University.

Citation. Schnappinger D, Ehrt S. 2014. Regulated expression systems for mycobacteria and their applications. Microbiol Spectrum 2(1):MGM2-0018-2013.

References

1. **Draper P.** 1967. Aliphatic acylamide amidohydrolase of *Mycobacterium smegmatis*: its inducible nature and relation to acyl-transfer to hydroxylamine. *J Gen Microbiol* **46**:111–123.

2. **Mahenthiralingam E, Draper P, Davis EO, Colston MJ.** 1993. Cloning and sequencing of the gene which encodes the highly inducible acetamidase of *Mycobacterium smegmatis*. *J Gen Microbiol* **139**:575–583.

3. **Parish T, Mahenthiralingam E, Draper P, Davis EO, Colston MJ.** 1997. Regulation of the inducible acetamidase gene of *Mycobacterium smegmatis*. *Microbiology* **143**:2267–2276.

4. **Triccas JA, Parish T, Britton WJ, Gicquel B.** 1998. An inducible expression system permitting the efficient purification of a recombinant antigen from *Mycobacterium smegmatis*. *FEMS Microbiol Lett* **167**:151–156.

5. **Daugelat S, Kowall J, Mattow J, Bumann D, Winter R, Hurwitz R, Kaufmann SH.** 2003. The RD1 proteins of *Mycobacterium tuberculosis*: expression in *Mycobacterium smegmatis* and biochemical characterization. *Microbes Infect* **5**:1082–1095.

6. **Gomez JE, Bishai WR.** 2000. *whmD* is an essential mycobacterial gene required for proper septation and cell division. *Proc Natl Acad Sci USA* **97**:8554–8559.

7. **Greendyke R, Rajagopalan M, Parish T, Madiraju MV.** 2002. Conditional expression of *Mycobacterium smegmatis dnaA*: an essential DNA replication gene. *Microbiology* **148**:3887–3900.

8. **Brown AC, Parish T.** 2006. Instability of the acetamide-inducible expression vector pJAM2 in *Mycobacterium tuberculosis*. *Plasmid* **55**:81–86.

9. **Parish T, Turner J, Stoker NG.** 2001. *amiA* is a negative regulator of acetamidase expression in *Mycobacterium smegmatis*. *BMC Microbiol* **1**:19.

10. **Roberts G, Muttucumaru DG, Parish T.** 2003. Control of the acetamidase gene of *Mycobacterium smegmatis* by multiple regulators. *FEMS Microbiol Lett* **221**:131–136.

11. **Wang F, Jain P, Gulten G, Liu Z, Feng YC, Ganesula K, Motiwala AS, Ioerger TR, Alland D, Vilcheze C, Jacobs WR, Sacchettini JC.** 2010. *Mycobacterium tuberculosis* dihydrofolate reductase is not a target relevant to the antitubercular activity of isoniazid. *Antimicrob Agents Chemother* **54**:3776–3782.

12. **Hillen W, Berens C.** 1994. Mechanisms underlying expression of Tn10 encoded tetracycline resistance. *Annu Rev Microbiol* **48**:345–369.

13. **Lederer T, Kintrup M, Takahashi M, Sum PE, Ellestad GA, Hillen W.** 1996. Tetracycline analogs affecting binding to Tn10-encoded Tet repressor trigger the same mechanism of induction. *Biochemistry* **35**:7439–7446.

14. **Blokpoel MC, Murphy HN, O'Toole R, Wiles S, Runn ES, Stewart GR, Young DB, Robertson BD.** 2005. Tetracycline-inducible gene regulation in mycobacteria. *Nucleic Acids Res* **33**:e22.

15. **Ehrt S, Guo XV, Hickey CM, Ryou M, Monteleone M, Riley LW, Schnappinger D.** 2005. Controlling gene expression in mycobacteria with anhydrotetracycline and Tet repressor. *Nucleic Acids Res* **33**:e21.

16. **Carroll P, Muttucumaru DG, Parish T.** 2005. Use of a tetracycline-inducible system for conditional expression in *Mycobacterium tuberculosis* and *Mycobacterium smegmatis*. *Appl Environ Microbiol* **71**:3077–3084.

17. **Williams KJ, Joyce G, Robertson BD.** 2010. Improved mycobacterial tetracycline inducible vectors. *Plasmid* **64**:69–73.

18. **Wissmann A, Wray LV Jr, Somaggio U, Baumeister R, Geissendorfer M, Hillen W.** 1991. Selection for Tn10 tet repressor binding to tet operator in *Escherichia coli*: isolation of temperature-sensitive mutants and combinatorial mutagenesis in the DNA binding motif. *Genetics* **128**:225–232.

19. **Hecht B, Muller G, Hillen W.** 1993. Noninducible Tet repressor mutations map from the operator binding motif to the C terminus. *J Bacteriol* **175**:1206–1210.

20. **Guo XV, Monteleone M, Klotzsche M, Kamionka A, Hillen W, Braunstein M, Ehrt S, Schnappinger D.** 2007. Silencing *Mycobacterium smegmatis* by using tetracycline repressors. *J Bacteriol* **189**:4614–4623.

21. **Klotzsche M, Ehrt S, Schnappinger D.** 2009. Improved tetracycline repressors for gene silencing in mycobacteria. *Nucleic Acids Res* **37**:1778–1788.

22. **Wei JR, Krishnamoorthy V, Murphy K, Kim JH, Schnappinger D, Alber T, Sassetti CM, Rhee KY, Rubin EJ.** 2011. Depletion of antibiotic targets has widely varying effects on growth. *Proc Natl Acad Sci USA* **108**:4176–4181.

23. **Kim JH, Wei JR, Wallach JB, Robbins RS, Rubin EJ, Schnappinger D.** 2011. Protein inactivation in mycobacteria by controlled proteolysis and its application to deplete the beta subunit of RNA polymerase. *Nucleic Acids Res* **39**:2210–2220.

24. **Boldrin F, Casonato S, Dainese E, Sala C, Dhar N, Palu G, Riccardi G, Cole ST, Manganelli R.** 2010. Development of a repressible mycobacterial promoter system based on two transcriptional repressors. *Nucleic Acids Res* **38**:e134.

25. **Chang Y, Mead D, Dhodda V, Brumm P, Fox BG.** 2009. One-plasmid tunable coexpression for mycobacterial protein-protein interaction studies. *Protein Sci* **18**:2316–2325.

26. **Guzman LM, Belin D, Carson MJ, Beckwith J.** 1995. Tight regulation, modulation, and high-level expression by vectors containing the arabinose PBAD promoter. *J Bacteriol* **177**:4121–4130.

27. Schleif R. 2010. AraC protein, regulation of the l-arabinose operon in *Escherichia coli*, and the light switch mechanism of AraC action. *FEMS Microbiol Rev* **34**:779–796.

28. Carroll P, Brown AC, Hartridge AR, Parish T. 2007. Expression of *Mycobacterium tuberculosis* Rv1991c using an arabinose-inducible promoter demonstrates its role as a toxin. *FEMS Microbiol Lett* **274**:73–82.

29. Terpe K. 2006. Overview of bacterial expression systems for heterologous protein production: from molecular and biochemical fundamentals to commercial systems. *Applied Microbiol Biotechnol* **72**:211–222.

30. Lee BY, Clemens DL, Horwitz MA. 2008. The metabolic activity of *Mycobacterium tuberculosis*, assessed by use of a novel inducible GFP expression system, correlates with its capacity to inhibit phagosomal maturation and acidification in human macrophages. *Mol Microbiol* **68**:1047–1060.

31. Kaur P, Agarwal S, Datta S. 2009. Delineating bacteriostatic and bactericidal targets in mycobacteria using IPTG inducible antisense expression. *PLoS One* **4**:e5923.

32. Kobayashi M, Shimizu S. 1994. Versatile nitrilases: nitrile-hydrolyzing enzymes. *FEMS Microbiol Lett* **120**:217–223.

33. Nagasawa T, Kobayashi M, Yamada H. 1988. Optimum culture conditions for the production of benzonitrilase by *Rhodococcus rhodochrous* J1. *Arch Microbiol* **150**:89–94.

34. Komeda H, Hori Y, Kobayashi M, Shimizu S. 1996. Transcriptional regulation of the *Rhodococcus rhodochrous* J1 nitA gene encoding a nitrilase. *Proc Natl Acad Sci USA* **93**:10572–10577.

35. Herai S, Hashimoto Y, Higashibata H, Maseda H, Ikeda H, Omura S, Kobayashi M. 2004. Hyperinducible expression system for streptomycetes. *Proc Natl Acad Sci USA* **101**:14031–14035.

36. Pandey AK, Raman S, Proff R, Joshi S, Kang CM, Rubin EJ, Husson RN, Sassetti CM. 2009. Nitrile-inducible gene expression in mycobacteria. *Tuberculosis* **89**:12–16.

37. Mukhtar TA, Wright GD. 2005. Streptogramins, oxazolidinones, and other inhibitors of bacterial protein synthesis. *Chem Rev* **105**:529–542.

38. Blanc V, Salah-Bey K, Folcher M, Thompson CJ. 1995. Molecular characterization and transcriptional analysis of a multidrug resistance gene cloned from the pristinamycin-producing organism, *Streptomyces pristinaespiralis*. *Mol Microbiol* **17**:989–999.

39. Salah-Bey K, Blanc V, Thompson CJ. 1995. Stress-activated expression of a *Streptomyces pristinaespiralis* multidrug resistance gene (*ptr*) in various *Streptomyces* spp. and *Escherichia coli*. *Mol Microbiol* **17**:1001–1012.

40. Folcher M, Morris RP, Dale G, Salah-Bey-Hocini K, Viollier PH, Thompson CJ. 2001. A transcriptional regulator of a pristinamycin resistance gene in *Streptomyces coelicolor*. *J Biol Chem* **276**:1479–1485.

41. Forti F, Crosta A, Ghisotti D. 2009. Pristinamycin-inducible gene regulation in mycobacteria. *J Biotechnol* **140**:270–277.

42. Hernandez-Abanto SM, Woolwine SC, Jain SK, Bishai WR. 2006. Tetracycline-inducible gene expression in mycobacteria within an animal host using modified *Streptomyces* tcp830 regulatory elements. *Arch Microbiol* **186**:459–464.

43. Gur E, Biran D, Ron EZ. 2011. Regulated proteolysis in Gram-negative bacteria: how and when? *Nat Rev Microbiol* **9**:839–848.

44. Gottesman S. 2003. Proteolysis in bacterial regulatory circuits. *Annu Rev Cell Dev Biol* **19**:565–587.

45. Keiler KC. 2008. Biology of trans-translation. *Annu Rev Microbiol* **62**:133–151.

46. Flynn JM, Levchenko I, Seidel M, Wickner SH, Sauer RT, Baker TA. 2001. Overlapping recognition determinants within the *ssrA* degradation tag allow modulation of proteolysis. *Proc Natl Acad Sci USA* **98**:10584–10589.

47. Levchenko I, Seidel M, Sauer RT, Baker TA. 2000. A specificity-enhancing factor for the ClpXP degradation machine. *Science* **289**:2354–2356.

48. Lessner FH, Venters BJ, Keiler KC. 2007. Proteolytic adaptor for transfer-messenger RNA-tagged proteins from alpha-proteobacteria. *J Bacteriol* **189**:272–275.

49. McGinness KE, Baker TA, Sauer RT. 2006. Engineering controllable protein degradation. *Mol Cell* **22**:701–707.

50. Griffith KL, Grossman AD. 2008. Inducible protein degradation in *Bacillus subtilis* using heterologous peptide tags and adaptor proteins to target substrates to the protease ClpXP. *Mol Microbiol* **70**:1012–1025.

51. Nudler E, Mironov AS. 2004. The riboswitch control of bacterial metabolism. *Trends Biochem Sci* **29**:11–17.

52. Topp S, Reynoso CM, Seeliger JC, Goldlust IS, Desai SK, Murat D, Shen A, Puri AW, Komeili A, Bertozzi CR, Scott JR, Gallivan JP. 2010. Synthetic riboswitches that induce gene expression in diverse bacterial species. *Appl Environ Microbiol* **76**:7881–7884.

53. Seeliger JC, Topp S, Sogi KM, Previti ML, Gallivan JP, Bertozzi CR. 2012. A riboswitch-based inducible gene expression system for mycobacteria. *PLoS One* **7**:e29266.

54. Barnes PJ. 2003. Theophylline: new perspectives for an old drug. *Am J Respir Crit Care Med* **167**:813–818.

55. Woong Park S, Klotzsche M, Wilson DJ, Boshoff HI, Eoh H, Manjunatha U, Blumenthal A, Rhee K, Barry CE 3rd, Aldrich CC, Ehrt S, Schnappinger D. 2011. Evaluating the sensitivity of *Mycobacterium tuberculosis* to biotin deprivation using regulated gene expression. *PLoS Pathog* **7**:e1002264.

56. Leblanc C, Prudhomme T, Tabouret G, Ray A, Burbaud S, Cabantous S, Mourey L, Guilhot C, Chalut C. 2012. 4'-Phosphopantetheinyl transferase PptT, a new drug target required for *Mycobacterium tuberculosis* growth and persistence in vivo. *PLoS Pathog* **8**:e1003097.

57. Marrero J, Rhee KY, Schnappinger D, Pethe K, Ehrt S. 2010. Gluconeogenic carbon flow of tricarboxylic acid cycle intermediates is critical for *Mycobacterium tuberculosis* to establish and maintain infection. *Proc Natl Acad Sci USA* **107**:9819–9824.

58. Blumenthal A, Trujillo C, Ehrt S, Schnappinger D. 2010. Simultaneous analysis of multiple *Mycobacterium tuberculosis* knockdown mutants *in vitro* and *in vivo*. *PLoS One* **5**:e15667.

59. Gandotra S, Schnappinger D, Monteleone M, Hillen W, Ehrt S. 2007. *In vivo* gene silencing identifies the *Mycobacterium tuberculosis* proteasome as essential for the bacteria to persist in mice. *Nat Med* **13**:1515–1520.

60. Stallings CL, Stephanou NC, Chu L, Hochschild A, Nickels BE, Glickman MS. 2009. CarD is an essential regulator of rRNA transcription required for *Mycobacterium tuberculosis* persistence. *Cell* **138**:146–159.

61. Bashiri G, Squire CJ, Baker EN, Moreland NJ. 2007. Expression, purification and crystallization of native and selenomethionine labeled *Mycobacterium tuberculosis* FGD1 (Rv0407) using a *Mycobacterium smegmatis* expression system. *Protein Expr Purif* **54**:38–44.

62. Noens EE, Williams C, Anandhakrishnan M, Poulsen C, Ehebauer MT, Wilmanns M. 2011. Improved mycobacterial protein production using a *Mycobacterium smegmatis* groEL1DeltaC expression strain. *BMC Biotechnol* **11**:27.

63. Ramage HR, Connolly LE, Cox JS. 2009. Comprehensive functional analysis of *Mycobacterium tuberculosis* toxin-antitoxin systems: implications for pathogenesis, stress responses, and evolution. *PLoS Genet* **5**:e1000767.

64. Rodrigue S, Brodeur J, Jacques PE, Gervais AL, Brzezinski R, Gaudreau L. 2007. Identification of mycobacterial sigma factor binding sites by chromatin immunoprecipitation assays. *J Bacteriol* **189**:1505–1513.

65. Galagan JE, Minch K, Peterson M, Lyubetskaya A, Azizi E, Sweet L, Gomes A, Rustad T, Dolganov G, Glotova I, Abeel T, Mahwinney C, Kennedy AD, Allard R, Brabant W, Krueger A, Jaini S, Honda B, Yu WH, Hickey MJ, Zucker J, Garay C, Weiner B, Sisk P, Stolte C, Winkler JK, Van de Peer Y, Iazzetti P, Camacho D, Dreyfuss J, Liu Y, Dorhoi A, Mollenkopf HJ, Drogaris P, Lamontagne J, Zhou Y, Piquenot J, Park ST, Raman S, Kaufmann SH, Mohney RP, Chelsky D, Moody DB, Sherman DR, Schoolnik GK. 2013. The *Mycobacterium tuberculosis* regulatory network and hypoxia. *Nature* **499**:178–183.

66. Wilson T, de Lisle GW, Marcinkeviciene JA, Blanchard JS, Collins DM. 1998. Antisense RNA to *ahpC*, an oxidative stress defence gene involved in isoniazid resistance, indicates that AhpC of *Mycobacterium bovis* has virulence properties. *Microbiology* **144**(Pt 10):2687–2695.

67. Rubin EJ, Akerley BJ, Novik VN, Lampe DJ, Husson RN, Mekalanos JJ. 1999. In vivo transposition of *mariner*-based elements in enteric bacteria and mycobacteria. *Proc Natl Acad Sci USA* **96**:1645–1650.

68. Forti F, Mauri V, Deho G, Ghisotti D. 2011. Isolation of conditional expression mutants in *Mycobacterium tuberculosis* by transposon mutagenesis. *Tuberculosis* **91**:569–578.

69. Arnvig KB, Comas I, Thomson NR, Houghton J, Boshoff HI, Croucher NJ, Rose G, Perkins TT, Parkhill J, Dougan G, Young DB. 2011. Sequence-based analysis uncovers an abundance of non-coding RNA in the total transcriptome of *Mycobacterium tuberculosis*. *PLoS Pathog* **7**:e1002342.

70. Uplekar S, Rougemont J, Cole ST, Sala C. 2013. High-resolution transcriptome and genome-wide dynamics of RNA polymerase and NusA in *Mycobacterium tuberculosis*. *Nucleic Acids Res* **41**:961–977.

71. Raju RM, Unnikrishnan M, Rubin DH, Krishnamoorthy V, Kandror O, Akopian TN, Goldberg AL, Rubin EJ. 2012. *Mycobacterium tuberculosis* ClpP1 and ClpP2 function together in protein degradation and are required for viability in vitro and during infection. *PLoS Pathog* **8**:e1002511.

72. Gee CL, Papavinasasundaram KG, Blair SR, Baer CE, Falick AM, King DS, Griffin JE, Venghatakrishnan H, Zukauskas A, Wei JR, Dhiman RK, Crick DC, Rubin EJ, Sassetti CM, Alber T. 2012. A phosphorylated pseudokinase complex controls cell wall synthesis in mycobacteria. *Sci Signal* **5**:ra7.

73. Glickman MS, Jacobs WR Jr. 2001. Microbial pathogenesis of *Mycobacterium tuberculosis*: dawn of a discipline. *Cell* **104**:477–485.

74. Young K, Jayasuriya H, Ondeyka JG, Herath K, Zhang CW, Kodali S, Galgoci A, Painter R, Brown-Driver V, Yamamoto R, Silver LL, Zheng YC, Ventura JI, Sigmund J, Ha S, Basilio A, Vicente F, Tormo JR, Pelaez F, Youngman P, Cully D, Barrett JF, Schmatz D, Singh SB, Wang J. 2006. Discovery of FabH/FabF inhibitors from natural products. *Antimicrob Agents Chemother* **50**:519–526.

75. Jayasuriya H, Herath KB, Zhang C, Zink DL, Basilio A, Genilloud O, Diez MT, Vicente F, Gonzalez I, Salazar O, Pelaez F, Cummings R, Ha S, Wang J, Singh SB. 2007. Isolation and structure of platencin: a FabH and FabF dual inhibitor with potent broad-spectrum antibiotic activity. *Angew Chem Int Ed Engl* **46**:4684–4688.

76. Wang J, Kodali S, Lee SH, Galgoci A, Painter R, Dorso K, Racine F, Motyl M, Hernandez L, Tinney E, Colletti SL, Herath K, Cummings R, Salazar O, Gonzalez I, Basilio A, Vicente F, Genilloud O, Pelaez F, Jayasuriya H, Young K, Cully DF, Singh SB. 2007. Discovery of platencin, a dual FabF and FabH inhibitor with in vivo antibiotic properties. *Proc Natl Acad Sci USA* **104**:7612–7616.

77. Wang J, Soisson SM, Young K, Shoop W, Kodali S, Galgoci A, Painter R, Parthasarathy G, Tang YS, Cummings R, Ha S, Dorso K, Motyl M, Jayasuriya H, Ondeyka J, Herath K, Zhang C, Hernandez L, Allocco J, Basilio A, Tormo JR, Genilloud O, Vicente F, Pelaez F, Colwell L, Lee SH, Michael B, Felcetto T, Gill C, Silver LL, Hermes JD, Bartizal K, Barrett J, Schmatz D, Becker JW, Cully D, Singh SB. 2006. Platensimycin is a

selective FabF inhibitor with potent antibiotic properties. _Nature_ **441:**358–361.

78. **Fischbach MA, Walsh CT.** 2009. Antibiotics for emerging pathogens. _Science_ **325:**1089–1093.

79. **Abrahams GL, Kumar A, Savvi S, Hung AW, Wen S, Abell C, Barry CE 3rd, Sherman DR, Boshoff HI, Mizrahi V.** 2012. Pathway-selective sensitization of _Mycobacterium tuberculosis_ for target-based whole-cell screening. _Chem Biol_ **19:**844–854.

80. **Ollinger J, O'Malley T, Ahn J, Odingo J, Parish T.** 2012. Inhibition of the sole type I signal peptidase of _Mycobacterium tuberculosis_ is bactericidal under replicating and nonreplicating conditions. _J Bacteriol_ **194:** 2614–2619.

81. **Dziadek J, Rutherford SA, Madiraju MV, Atkinson MA, Rajagopalan M.** 2003. Conditional expression of _Mycobacterium smegmatisftsZ_, an essential cell division gene. _Microbiology_ **149:**1593–1603.

82. **Jani C, Eoh H, Lee JJ, Hamasha K, Sahana MB, Han JS, Nyayapathy S, Lee JY, Suh JW, Lee SH, Rehse SJ, Crick DC, Kang CM.** 2010. Regulation of polar peptidoglycan biosynthesis by Wag31 phosphorylation in mycobacteria. _BMC Microbiol_ **10:**327.

83. **Kang CM, Nyayapathy S, Lee JY, Suh JW, Husson RN.** 2008. Wag31, a homologue of the cell division protein DivIVA, regulates growth, morphology and polar cell wall synthesis in mycobacteria. _Microbiology_ **154:**725–735.

84. **Bhatt A, Kremer L, Dai AZ, Sacchettini JC, Jacobs WR Jr.** 2005. Conditional depletion of KasA, a key enzyme of mycolic acid biosynthesis, leads to mycobacterial cell lysis. _J Bacteriol_ **187:**7596–7606.

85. **Rao M, Liu H, Yang M, Zhao C, He ZG.** 2012. A copper-responsive global repressor regulates expression of diverse membrane-associated transporters and bacterial drug resistance in mycobacteria. _J Biol Chem_ **287:** 39721–39731.

86. **Frampton R, Aggio RB, Villas-Boas SG, Arcus VL, Cook GM.** 2012. Toxin-antitoxin systems of _Mycobacterium smegmatis_ are essential for cell survival. _J Biol Chem_ **287:**5340–5356.

87. **Huang F, He ZG.** 2010. Characterization of an interplay between a _Mycobacterium tuberculosis_ MazF homolog, Rv1495 and its sole DNA topoisomerase I. _Nucleic Acids Res_ **38:**8219–8230.

88. **Robson J, McKenzie JL, Cursons R, Cook GM, Arcus VL.** 2009. The vapBC operon from _Mycobacterium smegmatis_ is an autoregulated toxin-antitoxin module that controls growth via inhibition of translation. _J Mol Biol_ **390:**353–367.

89. **Yang M, Gao C, Wang Y, Zhang H, He ZG.** 2010. Characterization of the interaction and cross-regulation of three _Mycobacterium tuberculosis_ RelBE modules. _PLoS One_ **5:**e10672.

90. **Barik S, Sureka K, Mukherjee P, Basu J, Kundu M.** 2010. RseA, the SigE specific anti-sigma factor of _Mycobacterium tuberculosis_, is inactivated by phosphorylation-dependent ClpC1P2 proteolysis. _Mol Microbiol_ **75:**592–606.

91. **Nisa S, Blokpoel MC, Robertson BD, Tyndall JD, Lun S, Bishai WR, O'Toole R.** 2010. Targeting the chromosome partitioning protein ParA in tuberculosis drug discovery. _J Antimicrob Chemother_ **65:**2347–2358.

92. **Kang J, Xu L, Yang S, Yu W, Liu S, Xin Y, Ma Y.** 2013. Effect of phosphoglucosamine mutase on biofilm formation and antimicrobial susceptibilities in _M. smegmatis glmM_ gene knockdown strain. _PLoS One_ **8:**e61589.

93. **Sureka K, Dey S, Datta P, Singh AK, Dasgupta A, Rodrigue S, Basu J, Kundu M.** 2007. Polyphosphate kinase is involved in stress-induced _mprAB-sigE-rel_ signalling in mycobacteria. _Mol Microbiol_ **65:**261–276.

94. **Sureka K, Sanyal S, Basu J, Kundu M.** 2009. Polyphosphate kinase 2: a modulator of nucleoside diphosphate kinase activity in mycobacteria. _Mol Microbiol_ **74:** 1187–1197.

95. **Rao SP, Camacho L, Huat Tan B, Boon C, Russel DG, Dick T, Pethe K.** 2008. Recombinase-based reporter system and antisense technology to study gene expression and essentiality in hypoxic nonreplicating mycobacteria. _FEMS Microbiol Lett_ **284:**68–75.

96. **Goyal R, Das AK, Singh R, Singh PK, Korpole S, Sarkar D.** 2011. Phosphorylation of PhoP protein plays direct regulatory role in lipid biosynthesis of _Mycobacterium tuberculosis_. _J Biol Chem_ **286:**45197–45208.

97. **Minch K, Rustad T, Sherman DR.** 2012. _Mycobacterium tuberculosis_ growth following aerobic expression of the DosR regulon. _PLoS One_ **7:**e35935.

98. **Sinha KM, Stephanou NC, Gao F, Glickman MS, Shuman S.** 2007. Mycobacterial UvrD1 is a Ku-dependent DNA helicase that plays a role in multiple DNA repair events, including double-strand break repair. _J Biol Chem_ **282:** 15114–15125.

99. **Stephanou NC, Gao F, Bongiorno P, Ehrt S, Schnappinger D, Shuman S, Glickman MS.** 2007. Mycobacterial nonhomologous end joining mediates mutagenic repair of chromosomal double-strand DNA breaks. _J Bacteriol_ **189:**5237–5246.

100. **Baughn AD, Deng J, Vilcheze C, Riestra A, Welch JT, Jacobs WR Jr, Zimhony O.** 2010. Mutually exclusive genotypes for pyrazinamide and 5-chloropyrazinamide resistance reveal a potential resistance-proofing strategy. _Antimicrob Agents Chemother_ **54:**5323–5328.

101. **Duckworth BP, Geders TW, Tiwari D, Boshoff HI, Sibbald PA, Barry CE 3rd, Schnappinger D, Finzel BC, Aldrich CC.** 2011. Bisubstrate adenylation inhibitors of biotin protein ligase from _Mycobacterium tuberculosis_. _Chem Biol_ **18:**1432–1441.

102. **Hett EC, Chao MC, Rubin EJ.** 2010. Interaction and modulation of two antagonistic cell wall enzymes of mycobacteria. _PLoS Pathog_ **6:**e1001020.

103. **Raghavan S, Manzanillo P, Chan K, Dovey C, Cox JS.** 2008. Secreted transcription factor controls _Mycobacterium tuberculosis_ virulence. _Nature_ **454:**717–721.

104. **Korch SB, Contreras H, Clark-Curtiss JE.** 2009. Three _Mycobacterium tuberculosis_ Rel toxin-antitoxin modules inhibit mycobacterial growth and are expressed in infected human macrophages. _J Bacteriol_ **191:**1618–1630.

105. Ahidjo BA, Kuhnert D, McKenzie JL, Machowski EE, Gordhan BG, Arcus V, Abrahams GL, Mizrahi V. 2011. VapC toxins from *Mycobacterium tuberculosis* are ribonucleases that differentially inhibit growth and are neutralized by cognate VapB antitoxins. *PLoS One* 6:e21738.

106. Sharp JD, Cruz JW, Raman S, Inouye M, Husson RN, Woychik NA. 2012. Growth and translation inhibition through sequence-specific RNA binding by *Mycobacterium tuberculosis* VapC toxin. *J Biol Chem* 287:12835–12847.

107. Singh R, Barry CE 3rd, Boshoff HI. 2010. The three RelE homologs of *Mycobacterium tuberculosis* have individual, drug-specific effects on bacterial antibiotic tolerance. *J Bacteriol* 192:1279–1291.

108. Rigel NW, Gibbons HS, McCann JR, McDonough JA, Kurtz S, Braunstein M. 2009. The accessory SecA2 system of mycobacteria requires ATP binding and the canonical SecA1. *J Biol Chem* 284:9927–9936.

109. Chalut C, Botella L, de Sousa-D'Auria C, Houssin C, Guilhot C. 2006. The nonredundant roles of two 4′-phosphopantetheinyl transferases in vital processes of mycobacteria. *Proc Natl Acad Sci USA* 103:8511–8516.

110. Hett EC, Chao MC, Deng LL, Rubin EJ. 2008. A mycobacterial enzyme essential for cell division synergizes with resuscitation-promoting factor. *PLoS Pathog* 4:e1000001.

111. Rana AK, Singh A, Gurcha SS, Cox LR, Bhatt A, Besra GS. 2012. Ppm1-encoded polyprenyl monophospho_ mannose synthase activity is essential for lipoglycan synthesis and survival in mycobacteria. *PloS One* 7:e48211.

112. Trauner A, Lougheed KE, Bennett MH, Hingley-Wilson SM, Williams HD. 2012. The dormancy regulator DosR controls ribosome stability in hypoxic mycobacteria. *J Biol Chem* 287:24053–24063.

113. Garces A, Atmakuri K, Chase MR, Woodworth JS, Krastins B, Rothchild AC, Ramsdell TL, Lopez MF, Behar SM, Sarracino DA, Fortune SM. 2010. EspA acts as a critical mediator of ESX1-dependent virulence in *Mycobacterium tuberculosis* by affecting bacterial cell wall integrity. *PLoS Pathog* 6:e1000957.

114. de la Paz Santangelo M, Gest PM, Guerin ME, Coincon M, Pham H, Ryan G, Puckett SE, Spencer JS, Gonzalez-Juarrero M, Daher R, Lenaerts AJ, Schnappinger D, Therisod M, Ehrt S, Sygusch J, Jackson M. 2011. Glycolytic and non-glycolytic functions of *Mycobacterium tuberculosis* fructose-1,6-bisphosphate aldolase, an essential enzyme produced by replicating and non-replicating bacilli. *J Biol Chem* 286:40219–40231.

115. Siegrist MS, Unnikrishnan M, McConnell MJ, Borowsky M, Cheng TY, Siddiqi N, Fortune SM, Moody DB, Rubin EJ. 2009. Mycobacterial Esx-3 is required for mycobactin-mediated iron acquisition. *Proc Natl Acad Sci USA* 106:18792–18797.

116. Kumar A, Zhang M, Zhu L, Liao RP, Mutai C, Hafsat S, Sherman DR, Wang MW. 2012. High-throughput screening and sensitized bacteria identify an *M. tuberculosis* dihydrofolate reductase inhibitor with whole cell activity. *PLoS One* 7:e39961.

117. Carroll P, Faray-Kele MC, Parish T. 2011. Identifying vulnerable pathways in *Mycobacterium tuberculosis* by using a knockdown approach. *Appl Environ Microbiol* 77:5040–5043.

118. Cortes M, Singh AK, Reyrat JM, Gaillard JL, Nassif X, Herrmann JL. 2011. Conditional gene expression in *Mycobacterium abscessus*. *PLoS One* 6:e29306.

119. Di Luca M, Bottai D, Batoni G, Orgeur M, Aulicino A, Counoupas C, Campa M, Brosch R, Esin S. 2012. The ESX-5 associated *eccB-eccC* locus is essential for *Mycobacterium tuberculosis* viability. *PLoS One* 7:e52059.

120. Serafini A, Boldrin F, Palu G, Manganelli R. 2009. Characterization of a *Mycobacterium tuberculosis* ESX-3 conditional mutant: essentiality and rescue by iron and zinc. *J Bacteriol* 191:6340–6344.

The
Mycobacterial
Proteome

III

Molecular Genetics of Mycobacteria, 2nd Edition
Edited by Graham F. Hatfull and William R. Jacobs, Jr.
© 2014 American Society for Microbiology, Washington, DC
doi:10.1128/microbiolspec.MGM2-0020-2013

Martin Gengenbacher,[1] Jeppe Mouritsen,[2] Olga T. Schubert,[2] Ruedi Aebersold,[2,3] and Stefan H. E. Kaufmann[1]

Mycobacterium tuberculosis in the Proteomics Era

12

More than 60 years ago, Swedish biochemist Pehr Edman introduced the first technique for single peptide sequencing (1). The underlying principle involves a phenylisothiocyanate reaction with the free N-terminal amino group of a given peptide. The modified amino acid is then cleaved off and identified by chromatography or electrophoresis. Further cycles of the same process allow consecutive determination of up to 30 amino acids and thus the N-terminal amino acid sequence of a polypeptide. Major drawbacks of the so-called Edman degradation are that (i) N-terminal residues of a polypeptide must be freely accessible and unmodified and (ii) disulfide bonds cannot be directly identified. Nevertheless, Edman paved the way to modern protein identification. Proteomics has come a long way and is currently in transition from pure basic research to medical application. The reasons are obvious. The genome can be viewed as the blueprint of a cell; the transcriptome encompasses the first step, transcribing parts of the genome, which is active at a given time point. The proteome, however, describes the sum of the working parts of a cell. Thus, proteomics is the most direct platform for measuring cellular activity. Importantly, both during transcription from DNA to

RNA and during translation from RNA to protein, changes occur, which can multiply the different variants of the encoding gene. These include transcription errors, epigenetic changes, and other events, as well as translation errors, posttranslational modifications such as phosphorylation, and differential modes of protein folding. These changes increase complexity markedly, thereby allowing the most direct and most precise insight into a cell.

When the whole genome sequence of the human pathogen *Mycobacterium tuberculosis* became available in 1998, ~4,000 open reading frames were identified, and for 61% of them, explicit or putative functions could be assigned based on sequence homologies (2). The *M. tuberculosis* genome/proteome has been subdivided into 10 functional categories as depicted in Fig. 1A (TubercuList v2.6; http://tuberculist .epfl.ch/) (3). A recent study that inferred protein functions based on orthology and integrated genomic context analysis and literature mining reduced the number of hypothetical/unknown open reading frames to less than 12% (4). Soon after initial genome annotation, scientists used random whole-genome high-density mutagenesis to identify mycobacterial genes essential for

[1]Max Planck Institute for Infection Biology, Department of Immunology, Charitéplatz 1, 10117 Berlin, Germany; [2]Department of Biology, Institute of Molecular Systems Biology, ETH Zurich, Wolfgang-Pauli Strasse 16, 8093 Zurich, Switzerland; [3]Faculty of Science, University of Zurich, Winterthurerstrasse 190, 8057 Zurich, Switzerland.

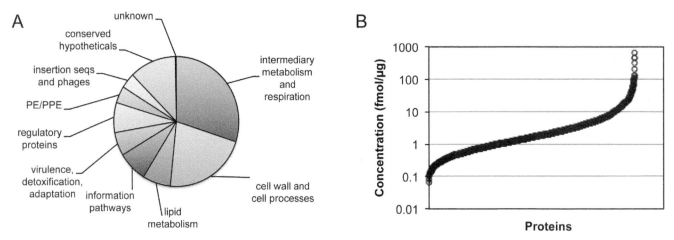

Figure 1 Functional annotation and abundance distribution of the *M. tuberculosis* proteome. (**A**) Distribution of functional classes of the *M. tuberculosis* proteome as annotated in TubercuList v2.6 release 27 (March 2013), updated with functional annotation for many of the "conserved hypotheticals" and "unknowns" (4). (**B**) Distribution of SRM-based absolute label-free abundance estimates for 2,195 proteins of *M. tuberculosis* H37Rv in a 1:1 mix of exponential and stationary cultures in rich medium (25). The concentration is given in femtomoles per microgram of extracted protein.
doi:10.1128/microbiolspec.MGM2-0020-2013.f1

survival *in vivo* (5, 6). However, so far only a minority of the characterized encoded proteins are known to be important for infection and virulence (7–12).

Proteomics aims to provide the most detailed insights into cellular processes by analyzing mature proteins, including modifications such as posttranslational processing or cleavage, which cannot be captured by genomics or transcriptomics. Furthermore, studies in numerous species and cell types have indicated that the cellular concentration of mRNA and protein encoded by the same locus do not strictly correlate and that this correlation is state specific (13, 14). First attempts toward the proteome of *M. tuberculosis* were made by analyzing bacterial fractions and culture supernatants by two-dimensional gel electrophoresis (2D-GE). While early approaches could resolve 50 to 170 proteins (15–22), improved methods including immobilized pH gradients in the first dimension of 2D-GE resulted in ~700 distinct protein spots in a single gel (19, 20), from which ~10% could be annotated (23) (Fig. 2).

Although 2D-GE maps proved to be useful for initial proteome annotation, it soon became clear that electrophoresis-based separation methods could not be further advanced due to physical resolution limits. By using liquid chromatography–tandem mass spectrometry (LC-MS/MS) for proteome analysis (Fig. 3), a much higher identification rate could be achieved. Two main LC-MS/MS strategies are applied in proteomics today: untargeted shotgun proteomics and the more recently

established targeted proteomics. Compared to 2D-GE, shotgun approaches enable superior characterization of proteins located in the mycobacterial cell wall or membrane, which are extremely difficult to resolve by 2D-GE due to their chemical properties. Shotgun proteomics with hybrid high-resolution MS of extensively fractionated mycobacterial lysates resulted in detection of over 3,000 proteins (24–26). Tremendous technological advances in the field of MS, notably the introduction of selected reaction monitoring (SRM), opened the field of targeted proteomics. Today, such techniques allow accurate detection and quantification of up to 80% of the *M. tuberculosis* proteome in unfractionated whole-cell lysates of liquid cultures without prior fractionation or separation (25) (Fig. 2 and Fig. 3B). Given that about 80% of the genome is expressed under such conditions, almost all proteins present are detectable, although their abundance range spans at least four orders of magnitude (25) (Fig. 1B). Further layers of complexity are added to proteomics by posttranslational modifications of proteins and their organization in complexes, both highly relevant for functionality, but impossible to deduce from genome or transcriptome data.

This chapter aims at providing a comprehensive overview of state-of-the-art knowledge in mycobacterial proteomics. First, studies describing the proteomic composition of subfractions of the tubercle bacillus, such as the cell wall and cytosol, as well as secreted

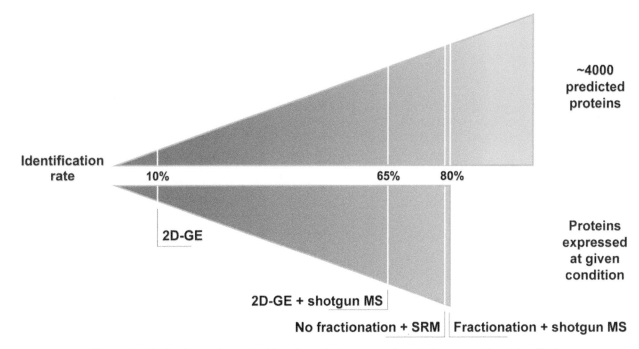

Figure 2 Technology-advanced *M. tuberculosis* proteomics. Early approaches identified only a small number of proteins. Combination of 2D-GE and untargeted shotgun mass spectrometry (MS), as well as extensive prefractionation of proteins or peptides, improved identification rates significantly. The latest targeted proteomics techniques, namely SRM, demonstrated identification of virtually all expressed proteins at given states in unfractionated *M. tuberculosis* cultures (25). doi:10.1128/microbiolspec.MGM2-0020-2013.f2

proteins found in culture supernatants, will be discussed. Then, various culture conditions, i.e., low pH, nutrient starvation, low oxygen tension, or nitric oxide exposure that likely reflect facets of the intracellular life of *M. tuberculosis*, move into the center of interest. A series of comparative studies provide first clues to the proteome relevant for phagosomal survival and host cell manipulation during infection. Further attempts have aimed at identifying the function of some of the hypothetical *M. tuberculosis* proteins and characterizing posttranslational modifications. The cutting-edge technique of SRM, combined with a public database of specific MS assays covering the entire proteome of *M. tuberculosis* now allows simple detection of mycobacterial proteins in different biological backgrounds. Using these techniques, scientists are about to move one step further by quantifying absolute amounts of single bacterial proteins in a complex mixture of proteins obtained from infected host cells without prior sample fractionation or enrichment. Finally, we discuss how proteomics can advance key application-related fields of mycobacterial research, such as vaccinology, drug discovery, and biomarker identification to improve tuberculosis (TB) prevention and intervention.

COMPARATIVE PROTEOMICS

M. tuberculosis Genome and Proteome Annotation

Most MS-based proteomic strategies rely on a high-quality protein sequence database to interpret the acquired mass spectra. However, comparison between two *M. tuberculosis* genome annotation exercises revealed that they are still incomplete and erroneous (27). High-throughput proteomics data can be harnessed to evaluate and refine genome annotation—a strategy called proteogenomics—by providing experimental evidence for missing genes, correcting translational start site annotations, and corroborating existing open reading frames (reviewed in reference 28). Typically, the first step of a proteogenomic analysis is the six-frame translation of the genome to capture all possible translated genomic regions. Extensive MS/MS data are then searched against this translated genome database to provide evidence for thus far unannotated open reading frames. To date, several such proteogenomic analyses have been carried out to improve *M. tuberculosis* genome annotations (24, 25, 29, 30). While conceptually simple, proteogenomic studies are

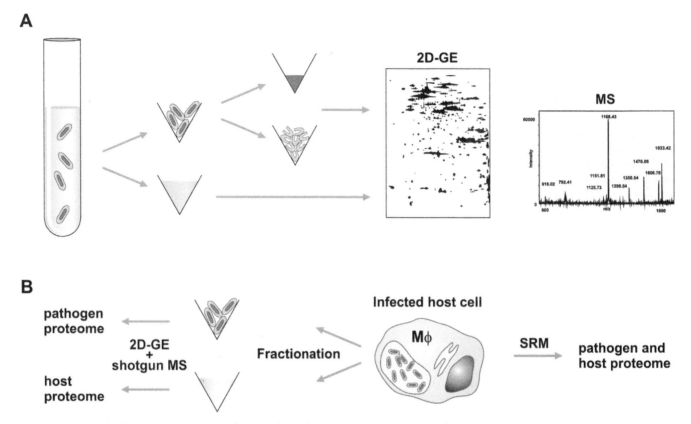

Figure 3 Uncovering the *M. tuberculosis* proteome. (**A**) Most of the current knowledge on *M. tuberculosis* proteomics has been generated by comparison of bacterial cultures, i.e., rich medium versus hypoxia or nutrient deprivation, conditions that *M. tuberculosis* might experience *in vivo*. Subfractions such as culture supernatant, cell wall debris, or the bacterial cytosol were separated by 2D-GE and subsequently analyzed by MS techniques. (**B**) Study of the *M. tuberculosis* proteome during infection remained difficult: due to the overwhelming protein abundance of the host as compared to the pathogen, enrichment for bacterial fractions was required prior to analysis by 2D-GE and shotgun MS. With the availability of the complete proteome libraries for *M. tuberculosis* (25) and the human host (U. Kusebauch et al., in preparation), SRM will allow simultaneous proteome analysis of the pathogen and the host in complex mixtures. doi:10.1128/microbiolspec.MGM2-0020-2013.f3

technically difficult, because the large search space generated by the six-frame translation of the entire genome challenges the error models of database search engines. Genes newly annotated by proteogenomics should therefore be treated with caution until orthogonal data, e.g., phylogenetic conservation, corroborates their existence. The online sequence annotation database TuberculList is updated several times annually with annotation corrections and new experimental evidence for existing genes and thus is a valuable reference resource for genome and proteome annotations (3, 31). The TB database (www.tbdb.org) is an equally helpful source for genome annotation and much other useful information as described elsewhere in this volume (10).

The *M. tuberculosis* Secretome

The driving forces in *M. tuberculosis* secretome research are the hunts for immunodominant antigens, biomarkers, and new drug targets (20, 30, 32, 33). Unlike most microbes, the intracellular pathogen *M. tuberculosis* has evolved several strategies to actively manipulate infected host cells and counteract their defense strategies (34). To discover proteins involved in these processes, scientists first analyzed mycobacterial culture supernatants. While very early 2D-GE studies identified only a few proteins by N-terminal sequencing (35) or immunodetection (19, 36), improved electrophoresis resolution has identified several hundred distinct protein spots in gels, of which roughly 38 could be identified by microsequencing and immunodetection

(37). The detection rate could be improved using MS-based techniques to detect protein spots obtained from 2D-GE (32). It must be kept in mind when analyzing secretome data that dying bacteria in a culture might "contaminate" supernatant fractions with intracellular proteins while they lyse.

A series of studies compared *M. tuberculosis* and the vaccine strain bacille Calmette-Guérin (BCG), the attenuated derivative of *Mycobacterium bovis*, the causative agent of bovine TB (38–40). Two secreted major virulence factors, ESAT-6 and CFP-10, were among the differential protein spots observed. The genes of both proteins are located in the region of difference 1 (RD1), which encodes a type VII secretion system limited to pathogenic mycobacteria and therefore absent in BCG (41). Further secretome differences include the ESAT-6-like proteins Rv1198 and Rv1793 as well as the acetyl-CoA-acetyltransferase Rv0243. Altogether, a set of potential antigens for vaccine development was identified (40). In a recent study a combination of 2D-GE and shotgun proteomics identified 1,176 proteins from culture filtrates of exponential phase and nutrient-starved cultures (42).

Proteomics of the Mycobacterial Cell Wall and Cytosol

The hallmark of *Mycobacterium* species is a thick and waxy cell wall (43). Its unique architecture renders the cell envelope very rigid and extremely impermeable, protecting the pathogen from dehydration and rendering it resistant to conventional antibiotics. Recent drug penetration studies have suggested that the cell wall contributes to phenotypic drug resistance observed in nonreplicating *M. tuberculosis* (44). In contrast to genetic drug resistance, which is based on mutations located in the genes encoding target proteins, phenotypic antibiotic resistance is related to the physiological state of dormancy and is thus reversed when growth resumes (34). Proteins of the mycobacterial cell wall are of particular interest for vaccine development and diagnostics. Enzymes of the antigen 85 (Ag85) complex, a family of mycolyl transferases vital for cell wall biogenesis, are among the most abundant immunodominant antigens of *M. tuberculosis* (32, 38) and thus were considered for improvement of the BCG vaccine by their overexpression and as central components of several subunit vaccines (45, 46). Although these transferases are actively directed to the cell wall compartment to fulfill their specific biological role, large quantities were found in culture supernatants (32, 38), which might have been partially extracted by

detergents commonly used in mycobacterial growth media. A comprehensive study identified 306 proteins (106 unique) in the cell envelope fraction of *M. tuberculosis* (47). More optimized protein extraction methods even led to the detection of 528 cell wall proteins, of which 87 were predicted to carry a signal peptide for secretion (48). A series of studies set out to describe the cytosolic proteome and/or membrane-associated proteins (39, 47, 49–54). In conclusion, definition of the proteome of subcellular compartments *per se* strongly depends on experimental conditions and remains difficult due to the unique properties of the mycobacterial cell wall. Nevertheless, many contributing researchers have laid the foundations of our current understanding of the *M. tuberculosis* proteome and paved the way for study of models of infection (see the next section).

Proteomics of Infection

Upon infection, *M. tuberculosis* experiences harsh environmental changes requiring adaptation and realignment of metabolic systems. To study the processes involved, cultures of *M. tuberculosis* have been exposed to adverse conditions such as acidic pH, nitric oxide, carbon starvation, and hypoxia that likely mimic facets of the pathogen's environment inside the host (Fig. 3A).

An attractive, widely studied model for understanding dormant *M. tuberculosis* uses gradual oxygen depletion of a culture incubated in sealed glass vials (55). The proteome of hypoxic, nonreplicating bacilli in this model is characterized by elevated levels of the heat shock protein HspX (Rv2031c), bacterioferritin (BfrB, Rv3841), L-alanine dehydrogenase (Ald, Rv2780), the chaperone GroEL2 (Rv0440), putative fructose-biphosphate aldolase (Fba, Rv0363c), and translation elongation factor EF-Tu (Tuf, Rv0685) (56–58). Another model of *in vitro* dormancy is based on carbon starvation of *M. tuberculosis* in physiological saline (59). Under these conditions, HspX and the hypothetical proteins Rv2557 and Rv2558 are strongly upregulated, while the secreted immune dominant antigens Mpt32 and Mpt64 and the membrane-associated trigger factor Tig are less abundant (60). Although Rv2557 and Rv2558 transcripts have been identified in human granulomas (61), both genes were dispensable for growth in the nutrient starvation model (62). HspX was also reported to be at high abundance in standing cultures and during long-term stationary phase, probably because both conditions reflect elements of both *in vitro* dormancy models (63, 64). More specifically, settled bacilli might experience low oxygen levels at the bottom of a culture flask, or organisms may face starvation inside large aggregates due to reduced oxygen/nutrient penetration. In line with these findings, HspX remained unchanged during *ex vivo* infection of host cells, which is a nonhypoxic system (65), while other results indicate elevated abundance of HspX under comparable conditions (66). The current data on the role of HspX warrants further investigation.

Employing isotope-coded affinity tagging techniques, the shift-down to dormancy by gradual oxygen depletion through two distinct phases of nonreplication was further dissected (67). Most differences observed were related to degradation and energy metabolism. A significant overlap of differentially expressed proteins exists in hypoxic nonreplicating *M. tuberculosis* and in the attenuated vaccine strain *M. bovis* BCG compared to metabolically active and replicating *M. tuberculosis* (38, 40, 66). It was concluded that HspX, BfrB, GroEL, Tuf, and Ald are among the proteins essential for pathogenesis.

To gain deeper insights into infection, the mycobacterial proteome during phagocytosis and its intracellular survival in the human monocytic cell line THP-1 were analyzed. In line with previous findings, elongation factor Tuf and the well-studied chaperones HspX, GroEL1, and GroEL2 were upregulated in *M. bovis* BCG upon phagocytosis (66). An approach involving metabolic labeling of *M. tuberculosis* prior to THP-1 infection revealed expression differences for 44 proteins (17). Another study identified enzymes of several metabolic pathways that are important during intracellular persistence of *M. tuberculosis* (65). Recent attempts aim at investigating the proteome of the tubercle bacillus *in vivo*. In a guinea pig infection model, *M. tuberculosis* showed continuous variation in protein abundance for some proteins, while others remained at high abundance (68). Although each animal model has its advantages, only the nonhuman primate model closely reproduces the complexity of human TB (34). Moreover, TB infection is a very dynamic process where bacilli reside in host tissues across a wide spectrum of physiological stages, due to their adaptation to the various microenvironments. Thus, data interpretation of whole-granuloma/organ approaches is challenging. One way to improve the problem of an "averaged proteome" could be collection of samples from different granuloma types or even different sections of a single granuloma, provided they are of sufficient size and bacterial load. As discussed in the sections below, "The Mtb Proteome Library" and "Quantification of *M. tuberculosis* Proteins in Complex Host Backgrounds," technological advances in MS have significantly improved detection rates and quantification accuracy for microbial proteins in complex host backgrounds, allowing us to refine our knowledge of the *M. tuberculosis* proteome *in vivo*.

FUNCTIONAL PROTEOMICS

Expression proteomics compares protein concentrations over different conditions to infer protein activity. In contrast, functional proteomics focuses on the regulation of proteins by posttranslational modifications and protein turnover, as well as the organization of proteins into multiprotein complexes, signaling pathways, and protein networks. Ultimately these factors all contribute to the regulation of cell function, adaptation to environmental stimuli, virulence, and pathogenicity.

Posttranslational Modifications

To date, using data from many species, more than 300 types of posttranslational modifications have been described, which potentially increase proteome complexity by orders of magnitude (69). They control numerous processes in a cell and thus cannot be neglected. MS/MS detection of posttranslational modifications is usually based on the specific mass increase of the modified amino acid residue or by the presence of a characteristic fragment resulting from the gas phase fractionation in MS. However, due to typical substoichiometric modifications, their analysis is challenging and often requires specific strategies to achieve sensitive detection, such as enrichment or purification of analytes prior to measurement. In this section we give an overview of the most important posttranslational modifications in mycobacteria and how they have been tackled by proteomics.

Phosphorylation

Protein phosphorylation is a ubiquitous mechanism for signal transduction in all three kingdoms of life. It is a reversible posttranslational modification that is catalyzed by protein kinases and removed by protein phosphatases (70). Different types of amino acids have been shown to be phosphorylated. These include the hydroxyl amino acids serine, threonine, and tyrosine, as well as histidine and aspartate. For technical reasons, most protein phosphorylation studies to date have focused on hydroxyl amino acids. The typical MS-based phosphoproteomics workflow includes a phospho-enrichment step, because the proportion of phosphorylated peptides in a whole proteome digest is usually very low. During LC-MS/MS analysis, the phosphate groups mostly remain attached to the peptide and thus can be identified and even assigned to a specific amino acid residue (71).

Two-component systems in most prokaryotes are the predominant form of microbial phospho-based signal transduction pathways (72). Principally, they consist of a histidine kinase as the membrane-bound sensor and a corresponding response regulator. Upon a specific stimulus the sensor kinase autophosphorylates at a histidine residue and then transfers the phosphoryl group to an aspartate residue on the response regulator. Typically, the activated response regulator directly binds DNA and acts as the transcription factor to regulate gene expression. In *M. tuberculosis*, 11 of these phospho-relay systems have been identified (2). However, because of the acid-labile nature of histidine and aspartate phosphorylations, they are currently out of reach of conventional MS-based proteomic workflows.

The *M. tuberculosis* genome encodes not only 11 two-component systems, but also an equal number of serine/threonine protein kinases (STPKs)—9 transmembrane and 2 soluble proteins—and 1 secreted tyrosine protein kinase; phosphorylation is reversed by one of the three identified protein phosphatases (73). Phosphorylation by STPKs has mostly been associated with eukaryotic organisms and has only recently been appreciated in prokaryotes. To date, the most comprehensive site-specific phosphoproteomic study of *M. tuberculosis* included more than 150 samples obtained from different growth and stress conditions (74). This analysis yielded 516 phosphorylation events in 301 phosphoproteins.

Pupylation

Pupylation is a recently discovered *M. tuberculosis* protein modification process (75). Specifically, covalent attachment of the small protein, prokaryotic ubiquitin-like protein (Pup), to a lysine residue of the target protein acts as a proteasomal degradation signal, similar to its eukaryotic counterpart ubiquitin (76). It remains to be determined whether pupylation has other, degradation-independent, regulatory functions. Three studies have employed an MS-based approach to search for potential target proteins of pupylation and identified partially overlapping sets of ~50 target proteins each (77–79). However, these studies have used expression of His-tagged Pup from an exogenous promoter to enrich pupylated proteins prior to MS analysis and thus do not precisely reflect true conditions. Functional pupylome studies where tagged Pup is solely expressed from its native promoter have yet to be performed.

Acetylation

N-Acetylation is a common posttranslational modification that occurs at lysine residues or at the N-terminus of proteins (80). Proteins with acetylated lysine residues may exhibit alterations in protein stability, interaction, localization, and function. Like many other posttranslational modifications, acetylation causes a shift in 2D-GE that can be used to distinguish modified proteins. Acetylation of ESAT-6 has, for instance, been shown to result in differential binding of CFP-10 (81). Acetylation can also be studied by LC-MS/MS, where acetylated peptides are enriched using specific antibodies and then identified by MS exploiting the characteristic mass shift of the modified amino acid (e.g., reference 82).

Lipidation

Covalent lipid modification of proteins allows the anchoring of proteins to the hydrophobic membrane. After attachment of a lipid residue and cleavage of the signal peptide, also called lipobox, the mature bacterial lipoprotein contains a diacylglycerol moiety at its N-terminal cysteine (83). The *M. tuberculosis* genome possesses 99 putative lipoproteins (84).

Glycosylation

The covalent attachment of carbohydrates to proteins is relatively rare in bacteria. However, during TB infection, human immune cells recognize not only mycobacterial lipoarabinomannans as antigens, but also glycosylated proteins, such as the secreted and cell surface protein Apa, highlighting the importance of mycobacterial glycoproteins (85–90). Proteomic analysis by 2D-GE and MS has led to the identification of 41 putative glycoproteins in *M. tuberculosis* by using ConA lectin affinity capture to enrich mannosylated proteins in *M. tuberculosis* culture filtrates (91). MS-based proteomics has also been applied to identify specific O-mannosylation sites, for instance, on an isolated culture filtrate protein (FasC) of *M. smegmatis* in a study on bacterial protein-O-mannosylating enzyme (92). Furthermore, glycosylation sites for an additional 13 *M. tuberculosis* culture filtrate proteins were recently reported (93).

Function and Interaction of Individual Proteins and Complexes

Understanding protein-protein interactions of *M. tuberculosis* is important (i) because complex formation is required for proteins to transmit cellular signals, (ii) because most biochemical functions are catalyzed by protein complexes, and (iii) because interactions with host proteins are involved in host adaptation and virulence.

Essential proteins in regulatory networks of *M. tuberculosis* and proteins directly interacting with the host are therefore potential drug targets (94, 95).

The knowledge base of *M. tuberculosis* protein-protein interactions is still incomplete due to the difficulty in engineering *M. tuberculosis* strains that express tagged bait proteins and limitations of existing technology for their analysis. Conventional yeast two-hybrid genetic screens have limited value for the study of mycobacterial interactions with their human host, due to the yeast-specific cellular machinery. Yet the system has been used as a rough screen upstream of more accurate bacterial pull-down experiments aiming to describe bacterial protein complexes (96–98). Co-immunoprecipitation (99), split protein sensor systems (100), and custom-designed shuttle vectors (101) are targeted and more accurate, but throughput is lower, and they cannot always deal with host-pathogen interactions, where computational prediction algorithms remain the preferred option today (102).

While studies of bacterial protein complexes and protein interactions with the host are challenging, recent findings in host-pathogen interaction experiments have provided significant new insight into the life cycle of *M. tuberculosis* in spite of the technological limitations. Upon first encounter with immune cells such as macrophages, *M. tuberculosis* secretes numerous soluble proteins including virulence factors which prime potential hosts (103). For instance, the abundantly secreted *M. tuberculosis* protein ESAT-6 directly binds to human TLR2. Here, it inhibits IRAK4-MyD88 and activates NFκB and Akt signaling (104). ESAT-6 is a known substrate of the ESX-1 secretion system, which is deleted from avirulent strains, such as *M. bovis* BCG (105). The proinflammatory action of ESAT-6 is further illustrated by its ability to cause endothelial secretion of matrix metalloprotease 9 (MMP9) in zebra fish. MMP9 has enzymatic activity required for the development of the hallmark inflammatory granulomatous tissue reaction in the host (106). Next, intracellular invasion is mediated by phagocytosis or macropinocytosis in an Mce protein family–dependent manner, involving direct molecular interaction with the host, although the targeted host proteins are still unknown (107, 108).

A selection of specific or redundant protein-protein interactions between *M. tuberculosis* and its host dictate the fate of the pathogen inside the phagosome (109, 110). To ensure survival in the macrophage, *M. tuberculosis* blocks fusion between the early phagosome and lysosome of the host macrophage. The underlying mechanism of action is complex, and at least three essential mycobacterial protein kinases—PknG,

SapM, and PtpA—are involved that directly interact with host proteins (12, 111, 112). The further development of infection is unclear, but it was recently demonstrated that *M. tuberculosis* escapes into the cytosol in an ESAT-6-dependent manner in later stages of macrophage infection, eventually resulting in host cell death. Necrosis leads to inflammation and allows *M. tuberculosis* to escape and infect a new host cell (113). The host protein interaction partner of ESAT-6 at this stage remains unknown.

High-throughput technologies for the study of protein-protein interactions between the pathogen and host are emerging (114, 115). These methods require preselection of interaction partners. One of these approaches involves bait proteins attached to one end of a tri-functional small molecule cross-linker. Intact human host cells are incubated with the bait-linker, which contains a reactive group that reacts with the N-linked sugar moieties of the target proteins on the host cell surface. In this way the covalently attached linker accurately reports specific host-bait protein interactions. The tri-functional cross-linker contains a biotin tag with which the target proteins can be purified for identification by LC-MS/MS (115). To date, only two outer membrane proteins, OmpA and MspA, that could be targets for this approach have been identified (116–118). Discovery of host-pathogen protein-protein interactions required for *M. tuberculosis* invasion of macrophages therefore remains limited to a computational prediction approach (119).

Protein Turnover

Protein turnover of a cell refers to the biochemical dynamics of protein synthesis and degradation. Mycobacteria regulate the abundance of many of their proteins at the transcriptional level in response to stress encountered by the bacillus (120). However, the overall correlation between transcript and protein abundances remains poor (13, 65, 121), and protein turnover is an important aspect to explain the discrepancy between transcriptional and protein levels.

To study protein turnover, bacterial cultures are typically transferred from standard growth media to a growth media supplemented with stable isotope-labeled carbon sources or amino acids, which are then incorporated into *de novo* synthesized proteins. Subsequently, LC-MS/MS can be used to determine the relative abundance of old and de novo synthesized proteins, reflecting the protein turnover rate (122). The first *in vitro* turnover study using conventional ^{15}N-labeled broth for the growth of *M. tuberculosis* revealed that *M. tuberculosis* regulates its protein turnover dramatically and in a stress-specific manner (123). To regulate protein degradation, mycobacteria have two known systems: the essential Clp proteases that actively degrade mistranslated proteins of *M. tuberculosis in vivo* (124) and the ubiquitin-like protein Pup that is covalently linked to proteins, marking them for degradation by the mycobacterial proteasome. The turnover dynamics and functional importance of pupylation remain unknown (75, 125).

A nonreplicating cell state is induced in response to human hypoxic stress, such as controlled, low oxygen tension in rich medium culture. Low oxygen tension is sensed by the two-component system DosR/DosS triggering transcriptional synthesis of the DosR regulon genes (126, 127) and synthesis of proteins such as HspX (64). While the transcriptional effect only lasts for 24 hours despite sustained low oxygen tension (126), in a standing culture hypoxia model, protein expression of the DosR regulon genes in *M. bovis* BCG is sustained over days (25), suggesting reduced turnover rather than continuous synthesis. Compared to continuous high synthesis and degradation, reduced turnover is a more energetically efficient way for the bacillus to protect itself via stress-induced proteins in the nonreplicating state.

Typical protein turnover experiments using stable isotope labeling of proteins are time-consuming and expensive. Therefore, more efficient analytical proteomics tools that consume little sample material, combined with high throughput, robust reproducibility, and sufficient analytical depth are needed. In the next section, we describe some promising examples.

TARGETED PROTEOMICS

The examples described thus far emphasize that robust identification and quantification of proteins and their modifications are critical to our understanding of the physiology of *M. tuberculosis* and the pathology of TB. 2D-GE coupled with MS analysis, as well as shotgun/ discovery proteomics, are currently the most widely used techniques for qualitative and quantitative measurements of the *M. tuberculosis* proteome. As described above, shotgun proteomics allows the detection of hundreds to several thousands of proteins in a single run. However this focus on high proteome coverage leads to some curtailments in reproducibility, quantitative accuracy, and sample throughput (128). The targeted MS technique SRM alleviates limitations by focusing the MS analysis on a defined set of proteins of interest (129, 130). In SRM mode, the instrument is instructed to monitor predefined combinations of peptide precursor and fragment ions, so-called transitions,

continuously over time. The optimal transitions, together with the chromatographic retention time of the peptides, have to be determined for each protein of interest prior to experimentation. An emerging novel approach is SWATH-MS, which provides targeted data extraction of MS/MS spectra generated by data-independent acquisition. It is comparable to SRM in terms of accuracy and consistency but unlimited in the number of proteins that can be analyzed per run (131, 132). Importantly, for SRM and SWATH-MS, the MS coordinates of the target proteins must be known prior to their identification and quantification. The Mtb Proteome Library is a publicly accessible research resource that provides these MS coordinates for all proteins of the *M. tuberculosis* proteome and thus supports unbiased protein-based research in *M. tuberculosis* by targeted proteomics (25).

The Mtb Proteome Library

The Mtb Proteome Library provides information at three different levels of a proteomic workflow (25): First, the library provides information regarding proteome mapping by untargeted shotgun/discovery proteomics of extensively fractionated *M. tuberculosis* and BCG lysates (Fig. 4A). Data sets of several research groups have been compiled in a dedicated *M. tuberculosis* "build" in the PeptideAtlas database (www.PeptideAtlas.org), which allows querying of MS spectra and peptide and protein identification, and visualizes protein coverage. Second, the library provides SRM and SWATH-MS assays for almost all annotated *M. tuberculosis* proteins. Here, synthetic peptides have been used to determine the most intensely fragmented ions, as well as the chromatographic retention times of each peptide (Fig. 4B). These assays can be downloaded from the *M. tuberculosis* build in the SRMAtlas database (www.SRMAtlas.org). Third, the library contains SRM data validating these assays for over 70% of the annotated proteins (Fig. 4C). All SRM signals can be browsed and inspected in the PASSEL database (www.PeptideAtlas.org/passel), permitting selection of the most specific and informative assays for proteins of interest, based on information on detectability.

Quantification of *M. tuberculosis* Proteins in Complex Host Backgrounds

Transcriptomics and proteomics have provided deep insights into mycobacterial responses to stress *in vitro*, but limited insights into the *in vivo* responses (133–137). Thus, the specific adaptation mechanisms for persistence in the infected host are still unknown.

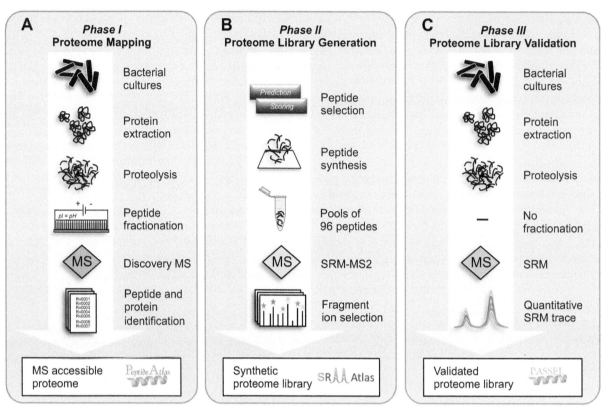

Figure 4 The Mtb Proteome Library. (**A**) Proteome mapping: After harvesting bacterial cultures, proteins were extracted and digested with the proteolytic enzyme trypsin. The resulting peptides were fractionated using off-gel isoelectric focusing to reduce sample complexity, and each fraction was analyzed by shotgun MS. The peptide and protein identifications, as well as the corresponding spectra, can be browsed interactively in the PeptideAtlas database (http://www.PeptideAtlas.org). (**B**) Proteome Library generation: From the collected data, the most MS-suited, unique peptides were selected for every annotated protein of *M. tuberculosis*. For proteins that had never been observed previously, representative peptides were predicted. The peptides were synthesized, pooled, and analyzed in SRM-triggered MS2 mode (SRM-MS2). From the resulting spectra the most intense fragment ions, as well as the chromatographic retention times, can be extracted. These MS coordinates, called SRM assays, constitute the synthetic Mtb Proteome Library and can be downloaded from the SRMAtlas database (http://www.SRMAtlas.org). (**C**) Proteome Library validation: The SRM assays in the synthetic Mtb Proteome Library were validated for the detection of proteins in unfractionated mycobacterial lysates by SRM. The resulting quantitative SRM traces and statistical scores can be viewed in the PASSEL database (http://www.PeptideAtlas.org/passel). Reprinted from reference 25 with permission from Elsevier. doi:10.1128/microbiolspec.MGM2-0020-2013.f4

The direct quantification of mycobacterial proteins in infected human cells is complicated by the complexity of the human background proteome that spans more than seven orders of magnitude in dynamic range and the usually low multiplicity of infection. *In vivo* expression of mycobacterial proteins in cells and tissues has been studied primarily with antibody-based assays and green fluorescent protein fusion constructs (138, 139). While these methods show high sensitivity, they suffer from low throughput and limited ability to multiplex. Furthermore, green fluorescent protein

assays cannot be applied to the study of human lesions, and few antibodies specific for *M. tuberculosis* proteins are known. To make bacteria in infected cells accessible to biochemical analysis, they were isolated from the phagosomal compartment in which they reside, and these isolates were subjected to 2D-GE-based separation for MS analysis. The success of such studies has been limited, mainly because of an excess of human proteins in the isolated phagosomes and the inherent variability of phagosomal populations (65, 66, 140).

Direct quantification of mycobacterial proteins in unfractionated human background has been challenging. *M. tuberculosis* proteins were quantified in infected macrophage cell lines (e.g., THP-1 cells) with the use of quantitative 2D-GE, but this only led to identification of the most abundant heat shock proteins in spite of high experimental multiplicity of infection (17). Classical shotgun-based protein quantification of full lysates of whole guinea pig lungs infected with *M. tuberculosis* at a high multiplicity of infection detected over 500 proteins (68). A recent study of cerebrospinal fluid from tuberculous meningitis patients using 2D-GE and spot quantification revealed differential regulation of a small set of *M. tuberculosis* proteins (141). Intriguingly, two independent studies using antibody-based techniques to detect mycobacterial proteins in TB patients indicated that secreted Ag85 protein was detectable in both serum and urine (142, 143). Furthermore, researchers studying urine from TB patients detected a number of cell membrane–associated metabolic *M. tuberculosis* proteins by shotgun MS (144). These results are encouraging because easily accessible human body fluids represent ideal specimens for the detection of *M. tuberculosis* biomarkers.

The above-mentioned studies using direct proteomic quantification of *M. tuberculosis* in human samples have provided new insight into the life and adaptation of *M. tuberculosis* in the host environment, but the robust quantification of mycobacterial proteins in the context of the host proteome remains a significant challenge. SRM is the most sensitive, selective, and reproducible LC-MS/MS-based technique currently available (128). Furthermore, targeted proteomics by SRM is easily multiplexed, and, in comparison with affinity reagent-based methods, assay development is fast and unambiguous (145). In fact, the Mtb Proteome Library resource described above contains specific SRM assays for 97% of the predicted open reading frames in the *M. tuberculosis* genome and thus supports, in principle, the quantification of essentially any *M. tuberculosis* protein in any background, including the human host. However, in reality, technical constraints, such as the dynamic range of the proteins of the human host matrix, the limited sample capacity of the chromatography columns used in LC-MS/MS measurements, and potential signal interferences and ion suppression, limit the scope of such analyses (146). It follows that any enrichment of the mycobacterial protein over the host protein background will improve bacterial protein detection. In a direct quantification workflow, the ratio of bacterial to host protein can be maximized by (i) an experimental design using a higher multiplicity of infection that

does not cause unwanted host cell pathology, (ii) optimizing the cell lysis and protein extraction protocol, and (iii) specific phagosome enrichment/purification methods. An optimal sample preparation protocol, combined with the use of optimized SRM assays, has the potential to make a significant fraction of the *M. tuberculosis* proteome directly detectable in the matrix of the human host cells. If these measurements are performed with the inclusion of stable isotope-labeled internal peptide standards, the detected proteins can also be accurately quantified by the same method (129, 147).

For optimal results, the specificity of the SRM assay should be asserted in the specific host background prior to making actual measurements. There are two levels of specificity for SRM-based quantification: the peptide and the fragment ion level. Very few SRM-suitable, tryptic *M. tuberculosis* peptides that are 7 to 21 amino acids in length have identical sequences in the two host species. These few peptides are predominantly derived from highly conserved central metabolic enzymes and should be excluded from SRM analyses. The fragment ion level specificity computed by the SRMCollider is depicted in Fig. 5 (148). The six most intense fragment ions of a peptide precursor of the respective protein are sufficient to select and develop a specific, scheduled SRM assay in human background for the vast majority of the proteome (25).

Direct quantification of mycobacterial proteins in infected cells and tissue inevitably captures all bacterial proteins in the sample, including proteins from nonreplicating bacteria. The interpretation of differential expression data in infected cells is influenced by inaccuracies in bacterial quantification when comparing two or more samples. Further complications arise in tissue analysis when extracellular proteins are secreted from live or shed from dead bacteria. The gold standard bacterial quantification method is plating of colony-forming units on solid, rich medium. However, only platable, replicating bacteria are detected. Additionally, the plating method has considerable technical variation, and other methods need to be considered for enumerating bacteria in complex samples (149). In infected cells the multiplicity of infection varies with infection efficiency and replication post-infection, and mycobacterial load varies considerably in human lesions (150). It follows that protein level sample normalization is a prerequisite for robust studies in human body fluids, infected cells, and tissues. Gene expression levels are predominantly normalized to a presumably stably expressed gene like *sigA* or 16S RNA (133, 136, 151). However, a single reference point is outlier-sensitive,

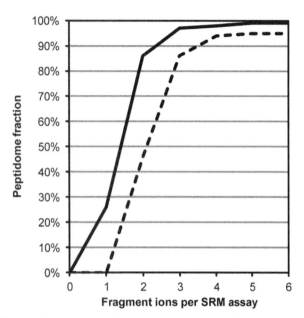

Figure 5 *M. tuberculosis* SRM assay specificity in myco-bacteria or host background. The theoretical specificity of SRM assays determined by the SRMCollider algorithm is shown as a cumulative plot of the number of peptides that can be uniquely identified with a given number of peptide-fragment ion pairs selected with decreasing intensity. Only background peptides with a chromatographic retention time close to that of the target peptide are considered as interfering background. The solid line indicates mycobacterial back-ground. The dashed line indicates human background. Reprinted from reference 25 with permission from Elsevier. doi:10.1128/microbiolspec.MGM2-0020-2013.f5

and more comprehensive strategies for protein quantifi-cation in complex host backgrounds are presumably required. These challenges notwithstanding, it can be expected that the increasingly powerful, targeted proteomics techniques will yield valuable information about the state of *M. tuberculosis* in the human host.

IMPACT OF PROTEOMICS ON TB INTERVENTION AND DIAGNOSTICS

Proteomics provides exciting new insights into the life cycle of *M. tuberculosis* under different conditions ranging from active metabolism and replication to dormancy. It also creates a sophisticated basis for the development of novel intervention measures against TB. This holds true for the three pillars of TB control: vaccines, drugs, and diagnostics.

TB Vaccine Strategies

The only TB vaccine currently in use is the live attenu-ated *M. bovis* strain, BCG. It protects against severe forms of childhood TB but not against pulmonary TB

at all stages of life. Thus, BCG has outlived its useful-ness, and new vaccines are needed for efficient TB con-trol (152). Current vaccine design against TB follows two strategies. First, improvement of BCG by genetic modification. These vaccine candidates are currently considered for BCG replacement. Second, subunit vac-cines composed of defined antigens, which are given as heterologous boosters on top of BCG prime. These vac-cines comprise defined antigens. Thus far, antigen selec-tion has been biased toward abundant antigens in broth cultures of *M. tuberculosis*. It has been proposed that the most abundant antigens of *M. tuberculosis* could benefit the pathogen rather than the host (153). As a corollary, novel strategies for vaccine-antigen discovery are warranted. Proteomics can provide clues to the discovery of better vaccine antigens. By means of proteomics, antigens expressed by *M. tuberculosis* un-der defined conditions with relevance for vaccine devel-opment can be identified. The following steps of experimental conditions, according to increasing com-plexity, could be pursued (Table 1):

- *M. tuberculosis* cultured under different conditions mimicking favorable or adverse conditions including stress, hypoxia, and nutrient deprivation.
- *M. tuberculosis*-infected host cells comprising (i) cells involved in *M. tuberculosis* control, i.e., profes-sional phagocytes and antigen-presenting cells, nota-bly, mononuclear phagocytes and dendritic cells, as well as (ii) cells that presumably serve as protective niches for *M. tuberculosis*, such as lung Clara cells, and alveolar epithelial cells of type I and type II.
- *M. tuberculosis* residing in cellular compartments, notably in the phagosome and the cytosol, could add further information.
- *M. tuberculosis* isolated from granulomatous lesions of different types, i.e., solid granulomas, necrotic granulomas, and caseating cavitary lesions, would best reflect the *in vivo* setting of different stages of infection and disease.

In addition to proteomics, identification of peptides from *M. tuberculosis* that are generated in professional antigen-presenting cells should be revitalized. These peptides are presented to T cells, the central mediators of protective immunity. Thus, following a strategy developed decades ago (154), modern proteomics technologies could help to define the immune-relevant peptidome of *M. tuberculosis* in its completeness. Peptides are generated in different subcellular compart-ments of antigen-presenting cells, notably, the phago-some and cytosol. Antigenic peptides generated in the phagosome are loaded onto gene products of the major

Table 1 Potential of proteomics for design of TB intervention measures

Experimental approach	Condition	Outcome
Proteome		
M. tuberculosis in broth culture	Different growth conditions: optimal, hypoxia, nutrient-deprivation, stress	Proteome of *M. tuberculosis* in cultures mimicking different infection stages
M. tuberculosis and drug in broth culture	Drug-induced perturbations	Proteome of *M. tuberculosis* under influence of novel drugs
M. tuberculosis in host cells	*M. tuberculosis* infection of professional phagocytes, antigen-presenting cells, cellular niches	Proteome of *M. tuberculosis* in different cellular habitats
M. tuberculosis in subcellular compartments	Phagosome, cytosol	Proteome of *M. tuberculosis* in subcellular compartments
M. tuberculosis in granulomatous lesions	Solid, necrotic, caseating lesions	Proteome of *M. tuberculosis* during different stages of infection (LTBI/active TB), severity of disease
Peptidome		
Elution of peptides from antigen-presenting molecules	MHC class II, MHC class I, unconventional antigen-presenting molecules	Identification of *M. tuberculosis* peptides relevant to T cell stimulation (CD4 T cells, CD8 T cells, unconventional T cells)

histocompatibility complex class II (MHC II) and presented to CD4 T cells. In contrast, antigenic peptides originating from the cytosol, or generated through cross-priming, are loaded on MHC class I peptides for CD8 T cell priming. These peptide epitopes are generally characterized by distinct anchor amino acids required for binding to the MHC cleft and are of ~8 or ~20 amino acids in length, for MHC class I or II, respectively (154).

After appropriate analysis and exclusion of human proteins/peptides, all *M. tuberculosis* protein antigens and peptide epitopes can be compiled and tested for their suitability as vaccine antigens.

TB Drug Discovery

Proteomics can guide rational drug discovery against TB (155). The increasing incidences of multidrug-resistant (MDR) TB, mounting to ~0.5 million cases annually, notification of extensively drug-resistant (XDR) TB in 84 countries globally, and the emergence of totally drug-resistant (TDR) TB urgently call for new TB drugs. Principally, the approach follows the one delineated for vaccine development. In addition, cultures of *M. tuberculosis* in broth under different conditions and in the presence of drug candidates will cause overall perturbations in the pathogen (Table 1). The available drugs against TB all target metabolically active and replicating *M. tuberculosis* organisms. Hence, dormant *M. tuberculosis* with a highly reduced metabolic and replicative activity is phenotypically resistant against these drugs. It is generally accepted that novel drugs should target not only active *M. tuberculosis*, but also dormant *M. tuberculosis*. Such drugs would not

only provide alternatives for treatment of drug-resistant *M. tuberculosis*, but also reduce treatment time. Current drug discovery is strongly biased toward a linear approach of one drug targeting one unique molecule in *M. tuberculosis*. Systems biologic analysis, however, has shown that even a single perturbation starting with a single target molecule causes complex overall changes in the proteome of *M. tuberculosis*, with profound consequences for the microbe. Proteomics of *M. tuberculosis* after drug perturbation will reveal this complex network and thus provide salient information for downstream drug development. The drug discovery platform would also benefit enormously from the proteome maps of *M. tuberculosis* cultured under different conditions, in different host cells, and within different types of lesions as described above. Comparisons of these signatures can facilitate identification of novel drug candidates targeting *M. tuberculosis* proteins differentially expressed under defined conditions such as dormant *M. tuberculosis* as it persists during latent *M. tuberculosis* infection (LTBI) and metabolically active and replicating *M. tuberculosis* as it ravages during active TB.

Obviously, proteomics need not be restricted to the pathogen but should also include the host response. Proteome analyses of host responses to *M. tuberculosis*, targeted by drugs, will lead to a better understanding of the consequences for host cells, and the macroorganism as a whole, and could therefore predict downstream events, either beneficial or adverse to the host. Finally, proteomics of host cells, under the pressure of *M. tuberculosis* infection, will promote identification of novel drug targets in the host for treatment of TB.

M. tuberculosis survives in host cells not only because of its robustness against aggressive defense mechanisms, but also because it actively subverts or modifies these mechanisms. Proteomics of *M. tuberculosis*–infected host cells will lead to the identification of host proteins that are differentially regulated during *M. tuberculosis* infection. By means of knockdown strategies, the relevance of these host molecules to control of intracellular *M. tuberculosis* can be revealed. It is expected that a number of signaling cascades involved (e.g., in the generation of reactive oxygen and nitrogen intermediates, in acidification of phagosomes, in phagosome-lysosome fusion, and in apoptosis/autophagy) are actively affected by *M. tuberculosis*. Such analyses can reveal targets for novel intervention strategies against TB aimed not directly at the pathogen, but indirectly affecting *M. tuberculosis* through modification of host effector mechanisms. This novel therapeutic approach combining chemotherapy and immunomodulation could open new avenues toward discovery of novel TB drug candidates against hitherto unexploited targets.

Biomarkers for TB

A third important area for proteomics is the design of biosignatures that allow discrimination between patients with active TB, individuals with LTBI, and uninfected healthy individuals. In the future, biosignatures that predict risk of active TB disease in LTBI subjects represent an ambitious approach, which would be of enormous value for TB control (156, 157). Proteome-based biomarker studies will likely focus on the host response rather than the pathogen. Probing the *M. tuberculosis* proteome with sera from individuals with LTBI and patients with active TB could allow identification of the antibody repertoire that is directed at all accessible *M. tuberculosis* proteins in the host and therefore lay the foundation for novel serodiagnostic tests. Although serodiagnosis of TB thus far has failed (158), novel tests comprising proteins that are reproducibly and specifically recognized by antibodies during distinct stages of infection (LTBI versus active TB disease) can be envisaged by exploiting unbiased analysis of the immunoproteome of *M. tuberculosis*. Following the experimental settings outlined above, namely, *M. tuberculosis* under different culture conditions, in different host cells, and derived from different types of lesions, proteomics of *M. tuberculosis* will create an in-depth overview of up- or downregulated proteins and thus provide guidelines for proteome-based *M. tuberculosis* diagnosis, be it in serum, sputum, or urine. High specificity of such a diagnostic assay would require

selection of proteins specific for *M. tuberculosis* and absent in the vaccine BCG and in nontuberculous mycobacteria. In addition, probing of these *M. tuberculosis* proteomes with serum antibodies from LTBI and TB patients generated under different conditions will provide helpful information that can lead to further refinement of antibody-based diagnostics for TB.

We thank Mary Louise Grossman and Diane Schad for outstanding editorial support and excellent graphic design, respectively. This work received financial support from the European 7th Framework Program SYSTEMTB (HEALTH-2009-2.1.2-1-241587).

Citation. Gengenbacher M, Mouritsen J, Schubert OT, Aebersold R, Kaufmann SH. 2014. *Mycobacterium tuberculosis* in the proteomics era. Microbiol Spectrum 2(2):MGM2-0020-2013.

References

1. Edman P. 1949. A method for the determination of amino acid sequence in peptides. *Arch Biochem* **22:**475.

2. Cole ST, Brosch R, Parkhill J, Garnier T, Churcher C, Harris D, Gordon SV, Eiglmeier K, Gas S, Barry CE III, Tekaia F, Badcock K, Basham D, Brown D, Chillingworth T, Connor R, Davies R, Devlin K, Feltwell T, Gentles S, Hamlin N, Holroyd S, Hornsby T, Jagels K, Krogh A, McLean J, Moule S, Murphy L, Oliver K, Osborne J, Quail MA, Rajandream MA, Rogers J, Rutter S, Seeger K, Skelton J, Squares R, Squares S, Sulston JE, Taylor K, Whitehead S, Barrell BG. 1998. Deciphering the biology of *Mycobacterium tuberculosis* from the complete genome sequence. *Nature* **393:**537–544.

3. Lew JM, Kapopoulou A, Jones LM, Cole ST. 2011. TubercuList: 10 years after. *Tuberculosis* (Edinb) **91:** 1–7.

4. Doerks T, van Noort V, Minguez P, Bork P. 2012. Annotation of the *M. tuberculosis* hypothetical orfeome: adding functional information to more than half of the uncharacterized proteins. *PLoS One* **7:**e34302.

5. Sassetti C, Rubin EJ. 2004. Genetic requirements for mycobacterial survival during infection. *Proc Natl Acad Sci USA* **100:**12989–12994.

6. Talaat AM, Lyons R, Howard ST, Johnston SA. 2004. The temporal expression profile of *Mycobacterium tuberculosis* infection in mice. *Proc Natl Acad Sci USA* **101:**4602–4607.

7. Hu Y, Movahedzadeh F, Stoker NG, Coates AR. 2006. Deletion of the *Mycobacterium tuberculosis* alpha-crystallin-like *hspX* gene causes increased bacterial growth in vivo. *Infect Immun* **74:**861–868.

8. Kumar A, Toledo JC, Patel RP, Lancaster JR Jr, Steyn AJ. 2007. *Mycobacterium tuberculosis* DosS is a redox sensor and DosT is a hypoxia sensor. *Proc Natl Acad Sci USA* **104:**11568–11573.

9. Noss EH, Pai RK, Sellati TJ, Radolf JD, Belisle J, Golenbock DT, Boom WH, Harding CV. 2001. Toll-like receptor 2-dependent inhibition of macrophage

class II MHC expression and antigen processing by 19-kDa lipoprotein of *Mycobacterium tuberculosis*. *J Immunol* **167**:910–918.

10. Reddy PV, Puri RV, Khera A, Tyagi AK. 2012. Iron storage proteins are essential for the survival and pathogenesis of *Mycobacterium tuberculosis* in THP-1 macrophages and the guinea pig model of infection. *J Bacteriol* **194**:567–575.

11. Vergne I, Chua J, Lee HH, Lucas M, Belisle J, Deretic V. 2005. Mechanism of phagolysosome biogenesis block by viable *Mycobacterium tuberculosis*. *Proc Natl Acad Sci USA* **102**:4033–4038.

12. Walburger A, Koul A, Ferrari G, Nguyen L, Prescianotto-Baschong C, Huygen K, Klebl B, Thompson C, Bacher G, Pieters J. 2004. Protein kinase G from pathogenic mycobacteria promotes survival within macrophages. *Science* **304**:1800–1804.

13. Gygi SP, Rochon Y, Franza BR, Aebersold R. 1999. Correlation between protein and mRNA abundance in yeast. *Mol Cell Biol* **19**:1720–1730.

14. Marguerat S, Schmidt A, Codlin S, Chen W, Aebersold R, Bahler J. 2012. Quantitative analysis of fission yeast transcriptomes and proteomes in proliferating and quiescent cells. *Cell* **151**:671–683.

15. Britton WJ, Hellqvist L, Ivanyi J, Basten A. 1987. Immunopurification of radiolabelled antigens of *Mycobacterium leprae* and *Mycobacterium bovis* (bacillus Calmette-Guérin) with monoclonal antibodies. *Scand J Immunol* **26**:149–159.

16. Daugelat S, Gulle H, Schoel B, Kaufmann SH. 1992. Secreted antigens of *Mycobacterium tuberculosis*: characterization with T lymphocytes from patients and contacts after two-dimensional separation. *J Infect Dis* **166**:186–190.

17. Lee BY, Horwitz MA. 1995. Identification of macrophage and stress-induced proteins of *Mycobacterium tuberculosis*. *J Clin Invest* **96**:245–249.

18. Mahairas GG, Sabo PJ, Hickey MJ, Singh DC, Stover CK. 1996. Molecular analysis of genetic differences between *Mycobacterium bovis* BCG and virulent *M. bovis*. *J Bacteriol* **178**:1274–1282.

19. Sonnenberg MG, Belisle JT. 1997. Definition of *Mycobacterium tuberculosis* culture filtrate proteins by two-dimensional polyacrylamide gel electrophoresis, N-terminal amino acid sequencing, and electrospray mass spectrometry. *Infect Immun* **65**:4515–4524.

20. Urquhart BL, Atsalos TE, Roach D, Basseal DJ, Bjellqvist B, Britton WL, Humphery-Smith I. 1997. "Proteomic contigs" of *Mycobacterium tuberculosis* and *Mycobacterium bovis* (BCG) using novel immobilised pH gradients. *Electrophoresis* **18**:1384–1392.

21. Wallis RS, Paranjape R, Phillips M. 1993. Identification by two-dimensional gel electrophoresis of a 58-kilodalton tumor necrosis factor-inducing protein of *Mycobacterium tuberculosis*. *Infect Immun* **61**:627–632.

22. Wong DK, Lee BY, Horwitz MA, Gibson BW. 1999. Identification of fur, aconitase, and other proteins expressed by *Mycobacterium tuberculosis* under conditions of low and high concentrations of iron by combined two-dimensional gel electrophoresis and mass spectrometry. *Infect Immun* **67**:327–336.

23. Mollenkopf HJ, Jungblut PR, Raupach B, Mattow J, Lamer S, Zimny-Arndt U, Schaible UE, Kaufmann SH. 1999. A dynamic two-dimensional polyacrylamide gel electrophoresis database: the mycobacterial proteome via Internet. *Electrophoresis* **20**:2172–2180.

24. Kelkar DS, Kumar D, Kumar P, Balakrishnan L, Muthusamy B, Yadav AK, Shrivastava P, Marimuthu A, Anand S, Sundaram H, Kingsbury R, Harsha HC, Nair B, Prasad TS, Chauhan DS, Katoch K, Katoch VM, Kumar P, Chaerkady R, Ramachandran S, Dash D, Pandey A. 2011. Proteogenomic analysis of *Mycobacterium tuberculosis* by high resolution mass spectrometry. *Mol Cell Proteomics* **10**:M111.

25. Schubert OT, Mouritsen JC, Ludwig C, Röst H, Rosenberger G, Arthur PK, Claassen M, Campbell DS, Sun Z, Farrah T, Gengenbacher M, Kaufmann SHE, Mortitz RL, Aebersold R. 2013. The Mtb Proteome Library: a resource of assays to quantify the complete proteome of *Mycobacterium tuberculosis*. *Cell Host Microbe* **13**:602–612.

26. Zheng J, Liu L, Wei C, Leng W, Yang J, Li W, Wang J, Jin Q. 2012. A comprehensive proteomic analysis of *Mycobacterium bovis* bacillus Calmette-Guérin using high resolution Fourier transform mass spectrometry. *J Proteomics* **77**:357–371.

27. de Souza GA, Malen H, Softeland T, Saelensminde G, Prasad S, Jonassen I, Wiker HG. 2008. High accuracy mass spectrometry analysis as a tool to verify and improve gene annotation using *Mycobacterium tuberculosis* as an example. *BMC Genomics* **9**:316.

28. Renuse S, Chaerkady R, Pandey A. 2011. Proteogenomics. *Proteomics* **11**:620–630.

29. de Souza GA, Arntzen MO, Fortuin S, Schurch AC, Malen H, McEvoy CR, van Soolingen D, Thiede B, Warren RM, Wiker HG. 2011. Proteogenomic analysis of polymorphisms and gene annotation divergences in prokaryotes using a clustered mass spectrometry-friendly database. *Mol Cell Proteomics* **10**:M110.

30. Jungblut PR, Muller EC, Mattow J, Kaufmann SH. 2001. Proteomics reveals open reading frames in *Mycobacterium tuberculosis* H37Rv not predicted by genomics. *Infect Immun* **69**:5905–5907.

31. Lew JM, Mao C, Shukla M, Warren A, Will R, Kuznetsov D, Xenarios I, Robertson BD, Gordon SV, Schnappinger D, Cole ST, Sobral B. 2013. Database resources for the tuberculosis community. *Tuberculosis* (Edinb) **93**:12–17.

32. Rosenkrands I, King A, Weldingh K, Moniatte M, Moertz E, Andersen P. 2000. Towards the proteome of *Mycobacterium tuberculosis*. *Electrophoresis* **21**:3740–3756.

33. Urquhart BL, Cordwell SJ, Humphery-Smith I. 1998. Comparison of predicted and observed properties of proteins encoded in the genome of *Mycobacterium tuberculosis* H37Rv. *Biochem Biophys Res Commun* **253**:70–79.

34. Gengenbacher M, Kaufmann SH. 2012. *Mycobacterium tuberculosis*: success through dormancy. *FEMS Microbiol Rev* **36**:514–532.

35. Nagai S, Wiker HG, Harboe M, Kinomoto M. 1991. Isolation and partial characterization of major protein antigens in the culture fluid of *Mycobacterium tuberculosis*. *Infect Immun* **59**:372–382.

36. Weldingh K, Rosenkrands I, Jacobsen S, Rasmussen PB, Elhay MJ, Andersen P. 1998. Two-dimensional electrophoresis for analysis of *Mycobacterium tuberculosis* culture filtrate and purification and characterization of six novel proteins. *Infect Immun* **66**:3492–3500.

37. Rosenkrands I, Weldingh K, Jacobsen S, Hansen CV, Florio W, Gianetri I, Andersen P. 2000. Mapping and identification of *Mycobacterium tuberculosis* proteins by two-dimensional gel electrophoresis, microsequencing and immunodetection. *Electrophoresis* **21**:935–948.

38. Jungblut PR, Schaible UE, Mollenkopf HJ, Zimny-Arndt U, Raupach B, Mattow J, Halada P, Lamer S, Hagens K, Kaufmann SH. 1999. Comparative proteome analysis of *Mycobacterium tuberculosis* and *Mycobacterium bovis* BCG strains: towards functional genomics of microbial pathogens. *Mol Microbiol* **33**:1103–1117.

39. Mattow J, Jungblut PR, Schaible UE, Mollenkopf HJ, Lamer S, Zimny-Arndt U, Hagens K, Muller EC, Kaufmann SH. 2001. Identification of proteins from *Mycobacterium tuberculosis* missing in attenuated *Mycobacterium bovis* BCG strains. *Electrophoresis* **22**:2936–2946.

40. Mattow J, Schaible UE, Schmidt F, Hagens K, Siejak F, Brestrich G, Haeselbarth G, Muller EC, Jungblut PR, Kaufmann SH. 2003. Comparative proteome analysis of culture supernatant proteins from virulent *Mycobacterium tuberculosis* H37Rv and attenuated *M. bovis* BCG Copenhagen. *Electrophoresis* **24**:3405–3420.

41. Simeone R, Bottai D, Brosch R. 2009. ESX/type VII secretion systems and their role in host-pathogen interaction. *Curr Opin Microbiol* **12**:4–10.

42. Albrethsen J, Agner J, Piersma SR, Hojrup P, Pham TV, Weldingh K, Jimenez CR, Andersen P, Rosenkrands I. 2013. Proteomic profiling of *Mycobacterium tuberculosis* identifies nutrient-starvation-responsive toxin-antitoxin systems. *Mol Cell Proteomics* **12**:1180–1191.

43. Brennan PJ. 2003. Structure, function, and biogenesis of the cell wall of *Mycobacterium tuberculosis*. *Tuberculosis* (Edinb) **83**:91–97.

44. Sarathy J, Dartois V, Dick T, Gengenbacher M. 2013. Reduced drug uptake in phenotypically resistant nutrient-starved nonreplicating *Mycobacterium tuberculosis*. *Antimicrob Agents Chemother* **57**:1648–1653.

45. Kaufmann SH, Gengenbacher M. 2012. Recombinant live vaccine candidates against tuberculosis. *Curr Opin Biotechnol* **23**:900–907.

46. Kaufmann SH. 2013. Tuberculosis vaccines: time to think about the next generation. *Semin Immunol* **25**(2):172–81.

47. Mawuenyega KG, Forst CV, Dobos KM, Belisle JT, Chen J, Bradbury EM, Bradbury AR, Chen X. 2005. *Mycobacterium tuberculosis* functional network analysis by global subcellular protein profiling. *Mol Biol Cell* **16**:396–404.

48. Wolfe LM, Mahaffey SB, Kruh NA, Dobos KM. 2010. Proteomic definition of the cell wall of *Mycobacterium tuberculosis*. *J Proteome Res* **9**:5816–5826.

49. Gu S, Chen J, Dobos KM, Bradbury EM, Belisle JT, Chen X. 2003. Comprehensive proteomic profiling of the membrane constituents of a *Mycobacterium tuberculosis* strain. *Mol Cell Proteomics* **2**:1284–1296.

50. Schmidt F, Donahoe S, Hagens K, Mattow J, Schaible UE, Kaufmann SH, Aebersold R, Jungblut PR. 2004. Complementary analysis of the *Mycobacterium tuberculosis* proteome by two-dimensional electrophoresis and isotope-coded affinity tag technology. *Mol Cell Proteomics* **3**:24–42.

51. Sinha S, Kosalai K, Arora S, Namane A, Sharma P, Gaikwad AN, Brodin P, Cole ST. 2005. Immunogenic membrane-associated proteins of *Mycobacterium tuberculosis* revealed by proteomics. *Microbiology* **151**:2411–2419.

52. Xiong Y, Chalmers MJ, Gao FP, Cross TA, Marshall AG. 2005. Identification of *Mycobacterium tuberculosis* H37Rv integral membrane proteins by one-dimensional gel electrophoresis and liquid chromatography electrospray ionization tandem mass spectrometry. *J Proteome Res* **4**:855–861.

53. Malen H, Pathak S, Softeland T, de Souza GA, Wiker HG. 2010. Definition of novel cell envelope associated proteins in Triton X-114 extracts of *Mycobacterium tuberculosis* H37Rv. *BMC Microbiol* **10**:132.

54. Bell C, Smith GT, Sweredoski MJ, Hess S. 2012. Characterization of the *Mycobacterium tuberculosis* proteome by liquid chromatography mass spectrometry-based proteomics techniques: a comprehensive resource for tuberculosis research. *J Proteome Res* **11**:119–130.

55. Wayne LG, Hayes LG. 1996. An in vitro model for sequential study of shiftdown of *Mycobacterium tuberculosis* through two stages of nonreplicating persistence. *Infect Immun* **64**:2062–2069.

56. Boon C, Li R, Qi R, Dick T. 2001. Proteins of *Mycobacterium bovis* BCG induced in the Wayne dormancy model. *J Bacteriol* **183**:2672–2676.

57. Rosenkrands I, Slayden RA, Crawford J, Aagaard C, Barry CE III, Andersen P. 2002. Hypoxic response of *Mycobacterium tuberculosis* studied by metabolic labeling and proteome analysis of cellular and extracellular proteins. *J Bacteriol* **184**:3485–3491.

58. Starck J, Kallenius G, Marklund BI, Andersson DI, Akerlund T. 2004. Comparative proteome analysis of *Mycobacterium tuberculosis* grown under aerobic and anaerobic conditions. *Microbiology* **150**:3821–3829.

59. Loebel RO, Shorr E, Richardson HB. 1933. The influence of adverse conditions upon the respiratory metabolism and growth of human tubercle bacilli. *J Bacteriol* **26**:167–200.

60. Betts JC, Lukey PT, Robb LC, McAdam RA, Duncan K. 2002. Evaluation of a nutrient starvation model of *Mycobacterium tuberculosis* persistence by gene and protein expression profiling. *Mol Microbiol* **43**: 717–731.

61. Fenhalls G, Stevens L, Moses L, Bezuidenhout J, Betts JC, Helden PP, Lukey PT, Duncan K. 2002. In situ detection of *Mycobacterium tuberculosis* transcripts in human lung granulomas reveals differential gene expression in necrotic lesions. *Infect Immun* **70**: 6330–6338.

62. Gordhan BG, Smith DA, Kana BD, Bancroft G, Mizrahi V. 2006. The carbon starvation-inducible genes Rv2557 and Rv2558 of *Mycobacterium tuberculosis* are not required for long-term survival under carbon starvation and for virulence in SCID mice. *Tuberculosis* (Edinb) **86**:430–437.

63. Florczyk MA, McCue LA, Stack RF, Hauer CR, McDonough KA. 2001. Identification and characterization of mycobacterial proteins differentially expressed under standing and shaking culture conditions, including Rv2623 from a novel class of putative ATP-binding proteins. *Infect Immun* **69**:5777–5785.

64. Yuan Y, Crane DD, Barry CE III. 1996. Stationary phase-associated protein expression in *Mycobacterium tuberculosis*: function of the mycobacterial alpha-crystallin homolog. *J Bacteriol* **178**:4484–4492.

65. Mattow J, Siejak F, Hagens K, Becher D, Albrecht D, Krah A, Schmidt F, Jungblut PR, Kaufmann SH, Schaible UE. 2006. Proteins unique to intraphagosomally grown *Mycobacterium tuberculosis*. *Proteomics* **6**:2485–2494.

66. Monahan IM, Betts J, Banerjee DK, Butcher PD. 2001. Differential expression of mycobacterial proteins following phagocytosis by macrophages. *Microbiology* **147**:459–471.

67. Cho SH, Goodlett D, Franzblau S. 2006. ICAT-based comparative proteomic analysis of non-replicating persistent *Mycobacterium tuberculosis*. *Tuberculosis* (Edinb) **86**:445–460.

68. Kruh NA, Troudt J, Izzo A, Prenni J, Dobos KM. 2010. Portrait of a pathogen: the *Mycobacterium tuberculosis* proteome in vivo. *PLoS One* **5**:e13938.

69. Witze ES, Old WM, Resing KA, Ahn NG. 2007. Mapping protein post-translational modifications with mass spectrometry. *Nat Methods* **4**:798–806.

70. Mijakovic I. 2010. Protein phosphorylation in bacteria. *Microbe* **5**:21–25.

71. Salih E. 2005. Phosphoproteomics by mass spectrometry and classical protein chemistry approaches. *Mass Spectrom Rev* **24**:828–846.

72. Casino P, Rubio V, Marina A. 2010. The mechanism of signal transduction by two-component systems. *Curr Opin Struct Biol* **20**:763–771.

73. Chao J, Wong D, Zheng X, Poirier V, Bach H, Hmama Z, Av-Gay Y. 2010. Protein kinase and phosphatase signaling in *Mycobacterium tuberculosis* physiology and pathogenesis. *Biochim Biophys Acta* **1804**: 620–627.

74. Prisic S, Dankwa S, Schwartz D, Chou MF, Locasale JW, Kang CM, Bemis G, Church GM, Steen H, Husson RN. 2010. Extensive phosphorylation with overlapping specificity by *Mycobacterium tuberculosis* serine/threonine protein kinases. *Proc Natl Acad Sci USA* **107**:7521–7526.

75. Pearce MJ, Mintseris J, Ferreyra J, Gygi SP, Darwin KH. 2008. Ubiquitin-like protein involved in the proteasome pathway of *Mycobacterium tuberculosis*. *Science* **322**:1104–1107.

76. Striebel F, Imkamp F, Ozcelik D, Weber-Ban E. 2013. Pupylation as a signal for proteasomal degradation in bacteria. *Biochim Biophys Acta* [Epub ahead of print.] doi:10.1016/j.bbamcr.2013.03.022.

77. Festa RA, McAllister F, Pearce MJ, Mintseris J, Burns KE, Gygi SP, Darwin KH. 2010. Prokaryotic ubiquitin-like protein (Pup) proteome of *Mycobacterium tuberculosis* [corrected]. *PLoS One* **5**:e8589.

78. Poulsen C, Akhter Y, Jeon AH, Schmitt-Ulms G, Meyer HE, Stefanski A, Stuhler K, Wilmanns M, Song YH. 2010. Proteome-wide identification of mycobacterial pupylation targets. *Mol Syst Biol* **6**:386.

79. Watrous J, Burns K, Liu WT, Patel A, Hook V, Bafna V, Barry CE III, Bark S, Dorrestein PC. 2010. Expansion of the mycobacterial "PUPylome." *Mol Biosyst* **6**: 376–385.

80. Hu LI, Lima BP, Wolfe AJ. 2010. Bacterial protein acetylation: the dawning of a new age. *Mol Microbiol* **77**: 15–21.

81. Okkels LM, Muller EC, Schmid M, Rosenkrands I, Kaufmann SH, Andersen P, Jungblut PR. 2004. CFP10 discriminates between nonacetylated and acetylated ESAT-6 of *Mycobacterium tuberculosis* by differential interaction. *Proteomics* **4**:2954–2960.

82. van Noort V, Seebacher J, Bader S, Mohammed S, Vonkova I, Betts MJ, Kuhner S, Kumar R, Maier T, O'Flaherty M, Rybin V, Schmeisky A, Yus E, Stulke J, Serrano L, Russell RB, Heck AJ, Bork P, Gavin AC. 2012. Cross-talk between phosphorylation and lysine acetylation in a genome-reduced bacterium. *Mol Syst Biol* **8**:571.

83. Kovacs-Simon A, Titball RW, Michell SL. 2011. Lipoproteins of bacterial pathogens. *Infect Immun* **79**: 548–561.

84. Sutcliffe IC, Harrington DJ. 2004. Lipoproteins of *Mycobacterium tuberculosis*: an abundant and functionally diverse class of cell envelope components. *FEMS Microbiol Rev* **28**:645–659.

85. Dobos KM, Swiderek K, Khoo KH, Brennan PJ, Belisle JT. 1995. Evidence for glycosylation sites on the 45-kilodalton glycoprotein of *Mycobacterium tuberculosis*. *Infect Immun* **63**:2846–2853.

86. Dobos KM, Khoo KH, Swiderek KM, Brennan PJ, Belisle JT. 1996. Definition of the full extent of glycosylation of the 45-kilodalton glycoprotein of *Mycobacterium tuberculosis*. *J Bacteriol* **178**:2498–2506.

87. Ragas A, Roussel L, Puzo G, Riviere M. 2007. The *Mycobacterium tuberculosis* cell-surface glycoprotein apa as a potential adhesin to colonize target cells via the

innate immune system pulmonary C-type lectin surfactant protein A. *J Biol Chem* 282:5133–5142.

88. Espitia C, Mancilla R. 1989. Identification, isolation and partial characterization of *Mycobacterium tuberculosis* glycoprotein antigens. *Clin Exp Immunol* 77: 378–383.

89. Horn C, Namane A, Pescher P, Riviere M, Romain F, Puzo G, Barzu O, Marchal G. 1999. Decreased capacity of recombinant 45/47-kDa molecules (Apa) of *Mycobacterium tuberculosis* to stimulate T lymphocyte responses related to changes in their mannosylation pattern. *J Biol Chem* 274:32023–32030.

90. Herrmann JL, O'Gaora P, Gallagher A, Thole JE, Young DB. 1996. Bacterial glycoproteins: a link between glycosylation and proteolytic cleavage of a 19 kDa antigen from *Mycobacterium tuberculosis*. *EMBO J* 15:3547–3554.

91. Gonzalez-Zamorano M, Mendoza-Hernandez G, Xolalpa W, Parada C, Vallecillo AJ, Bigi F, Espitia C. 2009. *Mycobacterium tuberculosis* glycoproteomics based on ConA-lectin affinity capture of mannosylated proteins. *J Proteome Res* 8:721–733.

92. Liu CF, Tonini L, Malaga W, Beau M, Stella A, Bouyssie D, Jackson MC, Nigou J, Puzo G, Guilhot C, Burlet-Schiltz O, Riviere M. 2013. Bacterial protein-O-mannosylating enzyme is crucial for virulence of *Mycobacterium tuberculosis*. *Proc Natl Acad Sci USA* 110:6560–6565.

93. Smith GT, Sweredoski MJ, Hess S. 2013. O-linked glycosylation sites profiling in *Mycobacterium tuberculosis* culture filtrate proteins. *J Proteomics* [Epub ahead of print.] doi:10.1016/j.jprot.2013.05.011.

94. Hase T, Tanaka H, Suzuki Y, Nakagawa S, Kitano H. 2009. Structure of protein interaction networks and their implications on drug design. *PLoS Comput Biol* 5: e1000550.

95. Loregian A, Palu G. 2005. Disruption of protein-protein interactions: towards new targets for chemotherapy. *J Cell Physiol* 204:750–762.

96. Steyn AJ, Joseph J, Bloom BR. 2003. Interaction of the sensor module of *Mycobacterium tuberculosis* H37Rv KdpD with members of the Lpr family. *Mol Microbiol* 47:1075–1089.

97. Tharad M, Samuchiwal SK, Bhalla K, Ghosh A, Kumar K, Kumar S, Ranganathan A. 2011. A three-hybrid system to probe in vivo protein-protein interactions: application to the essential proteins of the RD1 complex of *M. tuberculosis*. *PLoS One* 6: e27503.

98. Dyer MD, Neff C, Dufford M, Rivera CG, Shattuck D, Bassaganya-Riera J, Murali TM, Sobral BW. 2010. The human-bacterial pathogen protein interaction networks of *Bacillus anthracis*, *Francisella tularensis*, and *Yersinia pestis*. *PLoS One* 5:e12089.

99. Veyron-Churlet R, Guerrini O, Mourey L, Daffe M, Zerbib D. 2004. Protein-protein interactions within the fatty acid synthase-II system of *Mycobacterium tuberculosis* are essential for mycobacterial viability. *Mol Microbiol* 54:1161–1172.

100. O'Hare H, Juillerat A, Dianiskova P, Johnsson K. 2008. A split-protein sensor for studying protein-protein interaction in mycobacteria. *J Microbiol Methods* 73: 79–84.

101. Parikh A, Kumar D, Chawla Y, Kurthkoti K, Khan S, Varshney U, Nandicoori VK. 2013. Development of a new generation of vectors for gene expression, gene replacement, and protein-protein interaction studies in mycobacteria. *Appl Environ Microbiol* 79:1718–1729.

102. Davis FP, Barkan DT, Eswar N, McKerrow JH, Sali A. 2007. Host pathogen protein interactions predicted by comparative modeling. *Protein Sci* 16:2585–2596.

103. Malen H, Berven FS, Fladmark KE, Wiker HG. 2007. Comprehensive analysis of exported proteins from *Mycobacterium tuberculosis* H37Rv. *Proteomics* 7:1702–1718.

104. Pathak SK, Basu S, Basu KK, Banerjee A, Pathak S, Bhattacharyya A, Kaisho T, Kundu M, Basu J. 2007. Direct extracellular interaction between the early secreted antigen ESAT-6 of *Mycobacterium tuberculosis* and TLR2 inhibits TLR signaling in macrophages. *Nat Immunol* 8:610–618.

105. Fortune SM, Jaeger A, Sarracino DA, Chase MR, Sassetti CM, Sherman DR, Bloom BR, Rubin EJ. 2005. Mutually dependent secretion of proteins required for mycobacterial virulence. *Proc Natl Acad Sci USA* 102: 10676–10681.

106. Ramakrishnan L. 2012. Revisiting the role of the granuloma in tuberculosis. *Nat Rev Immunol* 12:352–366.

107. Zhang F, Xie JP. 2011. Mammalian cell entry gene family of *Mycobacterium tuberculosis*. *Mol Cell Biochem* 352:1–10.

108. Garcia-Perez BE, Mondragon-Flores R, Luna-Herrera J. 2003. Internalization of *Mycobacterium tuberculosis* by macropinocytosis in non-phagocytic cells. *Microb Pathog* 35:49–55.

109. Gaynor CD, McCormack FX, Voelker DR, McGowan SE, Schlesinger LS. 1995. Pulmonary surfactant protein A mediates enhanced phagocytosis of *Mycobacterium tuberculosis* by a direct interaction with human macrophages. *J Immunol* 155:5343–5351.

110. Bulut Y, Michelsen KS, Hayrapetian L, Naiki Y, Spallek R, Singh M, Arditi M. 2005. *Mycobacterium tuberculosis* heat shock proteins use diverse Toll-like receptor pathways to activate pro-inflammatory signals. *J Biol Chem* 280:20961–20967.

111. Bach H, Papavinasasundaram KG, Wong D, Hmama Z, Av-Gay Y. 2008. *Mycobacterium tuberculosis* virulence is mediated by PtpA dephosphorylation of human vacuolar protein sorting 33B. *Cell Host Microbe* 3:316–322.

112. Saleh MT, Belisle JT. 2000. Secretion of an acid phosphatase (SapM) by *Mycobacterium tuberculosis* that is similar to eukaryotic acid phosphatases. *J Bacteriol* 182:6850–6853.

113. Simeone R, Bobard A, Lippmann J, Bitter W, Majlessi L, Brosch R, Enninga J. 2012. Phagosomal rupture by *Mycobacterium tuberculosis* results in toxicity and host cell death. *PLoS Pathog* 8:e1002507.

114. Margarit I, Bonacci S, Pietrocola G, Rindi S, Ghezzo C, Bombaci M, Nardi-Dei V, Grifantini R, Speziale P, Grandi G. 2009. Capturing host-pathogen interactions by protein microarrays: identification of novel streptococcal proteins binding to human fibronectin, fibrinogen, and C4BP. *FASEB J* 23:3100–3112.

115. Frei AP, Jeon OY, Kilcher S, Moest H, Henning LM, Jost C, Pluckthun A, Mercer J, Aebersold R, Carreira EM, Wollscheid B. 2012. Direct identification of ligand-receptor interactions on living cells and tissues. *Nat Biotechnol* 30:997–1001.

116. Niederweis M, Ehrt S, Heinz C, Klocker U, Karosi S, Swiderek KM, Riley LW, Benz R. 1999. Cloning of the mspA gene encoding a porin from *Mycobacterium smegmatis*. *Mol Microbiol* 33:933–945.

117. Stahl C, Kubetzko S, Kaps I, Seeber S, Engelhardt H, Niederweis M. 2001. MspA provides the main hydrophilic pathway through the cell wall of *Mycobacterium smegmatis*. *Mol Microbiol* 40:451–464.

118. Senaratne RH, Mobasheri H, Papavinasasundaram KG, Jenner P, Lea EJ, Draper P. 1998. Expression of a gene for a porin-like protein of the OmpA family from *Mycobacterium tuberculosis* H37Rv. *J Bacteriol* 180:3541–3547.

119. Song H, Sandie R, Wang Y, Andrade-Navarro MA, Niederweis M. 2008. Identification of outer membrane proteins of *Mycobacterium tuberculosis*. *Tuberculosis* (Edinb) 88:526–544.

120. Smeulders MJ, Keer J, Speight RA, Williams HD. 1999. Adaptation of *Mycobacterium smegmatis* to stationary phase. *J Bacteriol* 181:270–283.

121. Schwanhausser B, Busse D, Li N, Dittmar G, Schuchhardt J, Wolf J, Chen W, Selbach M. 2011. Global quantification of mammalian gene expression control. *Nature* 473:337–342.

122. Rao PK, Roxas BA, Li Q. 2008. Determination of global protein turnover in stressed mycobacterium cells using hybrid-linear ion trap-Fourier transform mass spectrometry. *Anal Chem* 80:396–406.

123. Rao PK, Rodriguez GM, Smith I, Li Q. 2008. Protein dynamics in iron-starved *Mycobacterium tuberculosis* revealed by turnover and abundance measurement using hybrid-linear ion trap-Fourier transform mass spectrometry. *Anal Chem* 80:6860–6869.

124. Raju RM, Unnikrishnan M, Rubin DH, Krishnamoorthy V, Kandror O, Akopian TN, Goldberg AL, Rubin EJ. 2012. *Mycobacterium tuberculosis* ClpP1 and ClpP2 function together in protein degradation and are required for viability in vitro and during infection. *PLoS Pathog* 8:e1002511.

125. Burns KE, Liu WT, Boshoff HI, Dorrestein PC, Barry CE III. 2009. Proteasomal protein degradation in *Mycobacteria* is dependent upon a prokaryotic ubiquitin-like protein. *J Biol Chem* 284:3069–3075.

126. Park HD, Guinn KM, Harrell MI, Liao R, Voskuil MI, Tompa M, Schoolnik GK, Sherman DR. 2003. Rv3133c/dosR is a transcription factor that mediates the hypoxic response of *Mycobacterium tuberculosis*. *Mol Microbiol* 48:833–843.

127. Saini DK, Malhotra V, Dey D, Pant N, Das TK, Tyagi JS. 2004. DevR-DevS is a bona fide two-component system of *Mycobacterium tuberculosis* that is hypoxia-responsive in the absence of the DNA-binding domain of DevR. *Microbiology* 150:865–875.

128. Domon B, Aebersold R. 2010. Options and considerations when selecting a quantitative proteomics strategy. *Nat Biotechnol* 28:710–721.

129. Lange V, Picotti P, Domon B, Aebersold R. 2008. Selected reaction monitoring for quantitative proteomics: a tutorial. *Mol Syst Biol* 4:222.

130. Picotti P, Bodenmiller B, Mueller LN, Domon B, Aebersold R. 2009. Full dynamic range proteome analysis of *S. cerevisiae* by targeted proteomics. *Cell* 138:795–806.

131. Gillet LC, Navarro P, Tate S, Rost H, Selevsek N, Reiter L, Bonner R, Aebersold R. 2012. Targeted data extraction of the MS/MS spectra generated by data-independent acquisition: a new concept for consistent and accurate proteome analysis. *Mol Cell Proteomics* 11:O111.

132. Liu Y, Huttenhain R, Surinova S, Gillet LC, Mouritsen J, Brunner R, Navarro P, Aebersold R. 2013. Quantitative measurements of N-linked glycoproteins in human plasma by SWATH-MS. *Proteomics* 13:1247–1256.

133. Fontan P, Aris V, Ghanny S, Soteropoulos P, Smith I. 2008. Global transcriptional profile of *Mycobacterium tuberculosis* during THP-1 human macrophage infection. *Infect Immun* 76:717–725.

134. Rachman H, Strong M, Ulrichs T, Grode L, Schuchhardt J, Mollenkopf H, Kosmiadi GA, Eisenberg D, Kaufmann SH. 2006. Unique transcriptome signature of *Mycobacterium tuberculosis* in pulmonary tuberculosis. *Infect Immun* 74:1233–1242.

135. Rohde KH, Veiga DF, Caldwell S, Balazsi G, Russell DG. 2012. Linking the transcriptional profiles and the physiological states of *Mycobacterium tuberculosis* during an extended intracellular infection. *PLoS Pathog* 8:e1002769.

136. Schnappinger D, Ehrt S, Voskuil MI, Liu Y, Mangan JA, Monahan IM, Dolganov G, Efron B, Butcher PD, Nathan C, Schoolnik GK. 2003. Transcriptional adaptation of *Mycobacterium tuberculosis* within macrophages: insights into the phagosomal environment. *J Exp Med* 198:693–704.

137. Timm J, Post FA, Bekker LG, Walther GB, Wainwright HC, Manganelli R, Chan WT, Tsenova L, Gold B, Smith I, Kaplan G, McKinney JD. 2003. Differential expression of iron-, carbon-, and oxygen-responsive mycobacterial genes in the lungs of chronically infected mice and tuberculosis patients. *Proc Natl Acad Sci USA* 100:14321–14326.

138. Dhandayuthapani S, Via LE, Thomas CA, Horowitz PM, Deretic D, Deretic V. 1995. Green fluorescent protein as a marker for gene expression and cell biology of mycobacterial interactions with macrophages. *Mol Microbiol* 17:901–912.

139. Ramakrishnan L, Federspiel NA, Falkow S. 2000. Granuloma-specific expression of *Mycobacterium* virulence proteins from the glycine-rich PE-PGRS family. *Science* **288**:1436–1439.

140. Lee BY, Jethwaney D, Schilling B, Clemens DL, Gibson BW, Horwitz MA. 2010. The *Mycobacterium bovis* bacille Calmette-Guerin phagosome proteome. *Mol Cell Proteomics* **9**:32–53.

141. Kataria J, Rukmangadachar LA, Hariprasad G, O J, Tripathi M, Srinivasan A. 2011. Two dimensional difference gel electrophoresis analysis of cerebrospinal fluid in tuberculous meningitis patients. *J Proteomics* **74**:2194–2203.

142. Kashyap RS, Rajan AN, Ramteke SS, Agrawal VS, Kelkar SS, Purohit HJ, Taori GM, Daginawala HF. 2007. Diagnosis of tuberculosis in an Indian population by an indirect ELISA protocol based on detection of antigen 85 complex: a prospective cohort study. *BMC Infect Dis* **7**:74.

143. Bentley-Hibbert SI, Quan X, Newman T, Huygen K, Godfrey HP. 1999. Pathophysiology of antigen 85 in patients with active tuberculosis: antigen 85 circulates as complexes with fibronectin and immunoglobulin G. *Infect Immun* **67**:581–588.

144. Kashino SS, Pollock N, Napolitano DR, Rodrigues V Jr, Campos-Neto A. 2008. Identification and characterization of *Mycobacterium tuberculosis* antigens in urine of patients with active pulmonary tuberculosis: an innovative and alternative approach of antigen discovery of useful microbial molecules. *Clin Exp Immunol* **153**:56–62.

145. Picotti P, Rinner O, Stallmach R, Dautel F, Farrah T, Domon B, Wenschuh H, Aebersold R. 2010. High-throughput generation of selected reaction-monitoring assays for proteins and proteomes. *Nat Methods* **7**:43–46.

146. Taylor PJ. 2005. Matrix effects: the Achilles heel of quantitative high-performance liquid chromatography-electrospray-tandem mass spectrometry. *Clin Biochem* **38**:328–334.

147. Chang CY, Picotti P, Huttenhain R, Heinzelmann-Schwarz V, Jovanovic M, Aebersold R, Vitek O. 2012. Protein significance analysis in selected reaction monitoring (SRM) measurements. *Mol Cell Proteomics* **11**:M111.

148. Rost H, Malmstrom L, Aebersold R. 2012. A computational tool to detect and avoid redundancy in selected reaction monitoring. *Mol Cell Proteomics* **11**:540–549.

149. Pathak S, Awuh JA, Leversen NA, Flo TH, Asjo B. 2012. Counting mycobacteria in infected human cells and mouse tissue: a comparison between qPCR and CFU. *PLoS One* **7**:e34931.

150. Barry CE III, Boshoff HI, Dartois V, Dick T, Ehrt S, Flynn J, Schnappinger D, Wilkinson RJ, Young D. 2009. The spectrum of latent tuberculosis: rethinking the biology and intervention strategies. *Nat Rev Microbiol* **7**:845–855.

151. Dubnau E, Chan J, Mohan VP, Smith I. 2005. Responses of *Mycobacterium tuberculosis* to growth in the mouse lung. *Infect Immun* **73**:3754–3757.

152. Kaufmann SH, Hussey G, Lambert PH. 2010. New vaccines for tuberculosis. *Lancet* **375**:2110–2119.

153. Comas I, Chakravartti J, Small PM, Galagan J, Niemann S, Kremer K, Ernst JD, Gagneux S. 2010. Human T cell epitopes of *Mycobacterium tuberculosis* are evolutionarily hyperconserved. *Nat Genet* **42**:498–503.

154. Rammensee HG, Falk K, Rotzschke O. 1993. Peptides naturally presented by MHC class I molecules. *Annu Rev Immunol* **11**:213–244.

155. Ma Z, Lienhardt C, McIlleron H, Nunn AJ, Wang X. 2010. Global tuberculosis drug development pipeline: the need and the reality. *Lancet* **375**:2100–2109.

156. Weiner J, Maertzdorf J, Kaufmann SH. 2013. The dual role of biomarkers for understanding basic principles and devising novel intervention strategies in tuberculosis. *Ann NY Acad Sci* **1283**:22–29.

157. Wallis RS, Pai M, Menzies D, Doherty TM, Walzl G, Perkins MD, Zumla A. 2010. Biomarkers and diagnostics for tuberculosis: progress, needs, and translation into practice. *Lancet* **375**:1920–1937.

158. World Health Organization. 2011. *Tuberculosis serodiagnostic tests policy statement 2011*. WHO, Geneva, Switzerland.

Molecular Genetics of Mycobacteria, 2nd Edition
Edited by Graham F. Hatfull and William R. Jacobs, Jr.
© 2014 American Society for Microbiology, Washington, DC
doi:10.1128/microbiolspec.MGM2-0027-2013

Nagasuma Chandra[1]
Sankaran Sandhya[1]
Praveen Anand[1]

Structural Annotation of the *Mycobacterium tuberculosis* Proteome

13

Deciphering the complete genome sequence of *Mycobacterium tuberculosis* in 1998 (1) marked an important milestone in tuberculosis (TB) research and has triggered a whole array of downstream research in the area. A bewildering array of omics data of the organism and genome sequences of related species is now available (2–5). One of the major obstacles to ready exploitation of such huge volumes of genomics data is the lack of annotation of many of the gene products. With the architectural blueprint of the organism at hand, a logical next step is to comprehend the huge pool of data to identify and understand the function of the individual gene products.

From a protein function standpoint, the only practical way to convert vast quantities of raw sequence data into meaningful information is through transfer of annotation from known related proteins. Indeed, advances in bioinformatics approaches have become more reliable in transferring functional annotation and integrating sequence and protein family classifications (6, 7). However, a finer appreciation of the molecular

mechanisms within the cell is possible only through the elucidation of protein structure. It is clear that three-dimensional protein structures, where available, provide much better functional insights than one-dimensional sequences. Protein structures also pave the way for the use of much more sensitive approaches for detecting similarities among proteins (8, 9). Even for those sequences where the biochemical function is known, knowledge of the protein three-dimensional structure can provide further insight into the details of function and mechanism. The structure can reveal unique features of subunit and quaternary associations, substrate recognition, allosteric regulation, and the identification of putative active site residues. These aspects of structural information can be readily applied during the process of drug design (10).

Hence, in combination with sequence-based genome annotation efforts, structural genomics projects have launched large-scale studies to determine unique structures of proteins in an organism, primarily through X-ray crystallography (11). Toward the goal of obtaining

---

[1]Department of Biochemistry, Indian Institute of Science, Bangalore, Karnataka 560012, India.

structural information on a large number of proteins within the genome, a specific structural genomics consortium was formed for TB about a decade ago, the Tuberculosis Structural Genomics Consortium (TBSGC) (12). Efforts from such a consortium of several investigators across the globe, along with other TB structural projects, have led to the elucidation of structural information for about 10% of the proteome. The TBSGC has contributed 724 structures from 266 unique open reading frames to date, accounting for one-third of *M. tuberculosis* protein structures in the Protein Data Bank (PDB) (13). Although tremendous advances have been made in experimental methods to determine protein structure on a high-throughput scale, some bottlenecks still persist. These include experimental limitations of determining the right conditions for proteins that do not crystallize easily or occur in multidomain contexts or as large protein assemblies. Membrane proteins are also difficult to crystallize, and consequently, these efforts have been more successful for single-domain, cytoplasmic proteins (14).

On the other hand, the availability of a large number of structures of proteins from several organisms in the PDB (15), together with the advances in computational structure prediction algorithms, makes it feasible to obtain structural models for many more *M. tuberculosis* sequences with reasonable confidence (16). This approach has been utilized to obtain structural information for nearly 70% of the proteome (17). There are already a few excellent review articles on the structural biology of *M. tuberculosis* describing structural details of proteins involved in important biological processes (12, 13, 18, 19). This chapter focuses on a structural view of the entire proteome and the functional inferences obtained through this knowledge. Structure-based function annotation is discussed along with examples of insights into the mechanisms by which various biological events take place. The annotation includes identification of conserved residues in the family of proteins, identification of various sequence and substructural motifs, functional associations, compatible quaternary structures, and possible ligands and is expected to serve as a useful resource for TB researchers.

OBTAINING A STRUCTURAL PROTEOME

Sequence-based methods are used quite extensively for identifying homologous proteins (20). Homology-based comparative modeling has become a standard method to obtain the three-dimensional structural details of the protein (16). Broadly, homology models of the individual proteins are built based on rigorous, sometimes iterative steps of template identification for selecting the best template and alignment verification, followed by subsequent steps of structure validation. Methods in structure prediction have over the last two decades improved tremendously in applicability, speed, and confidence, as judged by the community-wide CASP experiments (21).

Where the extent of similarity is high, most methods usually detect homologs well. In cases of lower similarity, detection becomes a bit more complex and has to rely on sensitive approaches. Methods that use sequence-structure relationships can be broadly classified into two categories: profile-based and threading approaches. Profile-based approaches capture the residue environments of the proteins belonging to a particular superfamily and encode them in position-specific scoring matrices (22). Traditional sequence-based approaches are then adopted using position-specific scoring matrices to detect the homologs. In some cases where the homology is not readily detectable, HMMs (hidden Markov models) are used to find remote homologs (23). Threading, alternatively, fits the query sequence onto a database of known protein folds and evaluates the compatibility of the sequence with a structure using empirical scoring functions derived from a database of high-quality protein structures (24). Once the homology with a known structure is detected, the complete three-dimensional structure of the protein can be determined through homology modeling, a computational approach to determine the three-dimensional structure from the alignment of a query protein with a similar template protein whose structure has been solved. Whether homologous proteins are likely to perform similar functions can be further verified by examining conservation of residues that are involved in reaction chemistry.

Once the three-dimensional structure of the protein is predicted, rigorous evaluation must be carried out to check the quality of the models. The verification criteria generally used are (a) statistical potential scores (25) of the model, which reflect compatible residue environments of the query sequence and structural compactness; (b) statistical significance of alignments and extent of sequence similarity obtained from routinely employed sequence alignment tools such as BLAST (26) and PSI-BLAST (27); (c) deviation from ideality in geometric parameters (28), which considers bond lengths, bond angles, planarity, and dihedral angles; (d) deviation in the protein backbone dihedral angles ϕ, ψ from allowed regions in the Ramachandran plot; (e) statistics of nonbonded interactions between different atom types (29); and (f) conservation of residue contacts, solvent accessibility profiles, and secondary structure compatibility (30).

Using the above-mentioned computational protocols, it is now feasible to obtain structural information at a genome scale. Automatically generated models using the Modeller package are made available in a dedicated database called MODBASE (31). For around 70% of the *M. tuberculosis* genome, structural models with reasonable confidence have been generated (17) (Fig. 1). This information has enriched our understanding of the pathogen by providing insights in atomistic detail about various pathways and biological processes, as described below.

FUNCTIONAL INSIGHTS FROM PROTEIN STRUCTURES

Obtaining a protein's structure is the first step in the journey of obtaining mechanistic insights into the function of that protein. Experimental characterization of a protein function is available only for some proteins. However, the structural data resource is growing rapidly. The need to navigate and comprehend this large resource of both experimental and theoretical structural data has led to the emergence of a discipline called structural bioinformatics (11, 32, 33), which has truly come of age in the last decade (34). Structural bioinformatics is probably best thought of as an approach that rationalizes and classifies information contained in the three-dimensional structures of molecules, in terms of their functional capabilities, thus ultimately helping in understanding how they function in atomic detail. A main advantage of this approach over simpler sequence-based methods is that besides helping associate a molecule with a function, it also provides ultimate insights into the mechanisms by which various biological events take place. Thus, structural bioinformatics can serve as a bridge in transforming protein structures into biological insights.

An annotation exercise typically involves tasks of recognizing similarities and relationships among proteins, deriving structural patterns, correlating with function, and ultimately utilizing such patterns for prediction. This can be achieved at different levels. First, the protein structure itself can be classified and analyzed under different levels of structural organization, which are typically provided by classification schema as in the SCOP (Structural Classification of Proteins) (35) and CATH (36) databases. The SCOP database, a manually curated database, stores the structural information in four levels of hierarchy, namely, (a) class:

composition of secondary structures in proteins, (b) fold: a specific arrangement of secondary structural elements, (c) superfamily: domains of the fold that share similarities in function but poor sequence homology, and (d) family: domains in a superfamily that have recognizable sequence similarity and clear ancestry. Similarly, the CATH database classifies protein domains into class, architecture, topology, and homologous superfamily in a semiautomatic way. Once a new structural model is placed in one of the known structural classes and families, it can be explored if functional information from other well-characterized members in that family can be transferred to the new protein.

It must be noted that the term *function* itself can be defined at different levels of protein structure hierarchy, as mentioned above. The function of a protein could in order of ascending hierarchy be (a) the molecular function and (b) the biological process it is involved in. For example, the function of the RecA protein could be described as ATP-binding and DNA-binding at the first level and as a component of homologous recombination and DNA repair at the second level. "Fold to function" models, which have been the basis for the functional annotation of proteins in some cases, typically relate a structural fold to a predominant function known to be performed by proteins containing that fold. On the one hand, high structural similarity between two proteins along their entire polypeptide chains generally indicates high similarity in function at both the molecular function and biological process levels. On the other hand, when part similarity is exhibited by two proteins in their structures, their functions need not necessarily be the same. Detailed studies would be required to infer function in such cases.

Broadly, part similarity can mean that either medium-to-high similarity exists only in a portion of the polypeptide chain, indicating the presence of a common domain, or that low-to-medium level similarity exists in most of the polypeptide chain. In many cases, in the first category, inferring the molecular level function for that domain alone would be possible, but inferring the biological process level function would be much harder. For the second category, however, functional inference at either level would not be meaningful. This is because fold-level similarity does not necessarily indicate similarity in the functional regions of the molecule.

Gene ontology (37) can be very useful when describing the putative function of a protein through biologically relevant key terms. Various studies have been carried out to annotate proteins in terms of functional key words at different levels of structural hierarchy (38, 39). The functional terms are also obtained from

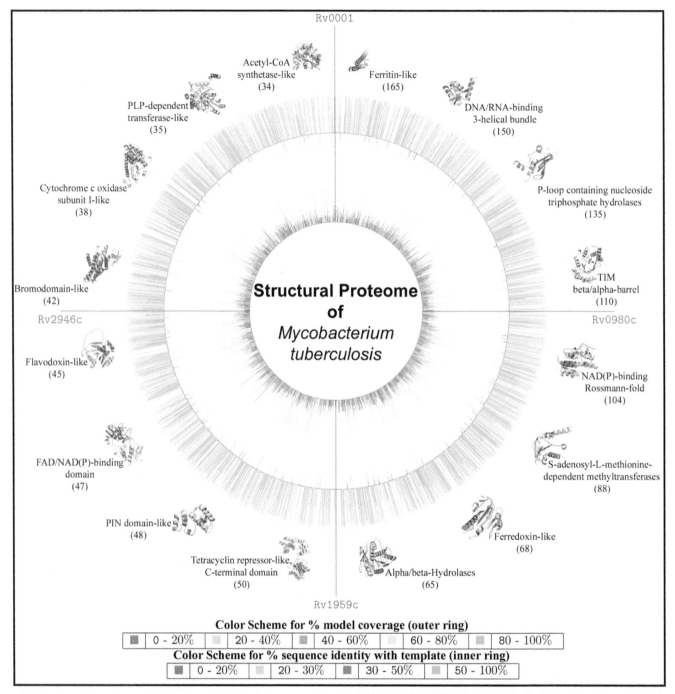

Figure 1 Schematic diagram of *M. tuberculosis* proteome structural annotation. The figure depicts the circular map of the *M. tuberculosis* H37Rv structural proteome, corresponding to the first genome view reported with its complete genome sequence (1). The outer circle represents the model coverage in terms of the percentage of the polypeptide chain, whereas the inner circle represents the percentage sequence identity shared by each model with its corresponding template. On both the outer and inner circles, radiating lines are drawn to indicate the parameters of the structural model for the corresponding protein in the genome view. The length of the lines in both cases is proportional to their values in percentages. The 100% mark is also shown for both the circles. In the outer circle, those models that had greater than 40% length coverage are drawn outside the circle, whereas those with coverage of less than that are drawn inside the circle (for clarity). Length coverage is divided into five classes and color coded as indicated, while the levels of sequence identity are divided into four classes and color coded as indicated. Predominantly occurring folds in the proteome are shown surrounding the outer circle, ordered clockwise by frequency of occurrence (indicated in parentheses). doi:10.1128/microbiolspec.MGM2-0027-2013.f1

other sequence databases such as InterPro, Pfam, and Swissprot. These associations help us analyze fold-based function annotation at a genome scale. In *M. tuberculosis* the majority of superfamilies (44%) are associated with metabolism, followed by superfamilies involved in general multiple functions (14%) such as interactions with proteins, ions, lipids, and small molecules; the third highest is superfamilies involved in intracellular processes (10%) such as cell motility, cell division, cell death, intracellular transport, and secretion.

Structural information has been useful in gaining a functional understanding of proteins in many cases (40, 41). The PE and PPE families—named after the identification of PE (proline-glutamate) and PPE (proline-proline-glutamate) motifs present at the N-terminal end of proteinsâ are associated with antigenic variation in mycobacteria. Based on a computational genomics prediction that some members of the PE and PPE families are functionally linked, experiments designed to test their interactions confirmed their interaction *in vitro* by coexpression and copurification, following which the crystal structures of the PE25/PPE41 complex were successfully determined. Furthermore, based on the structural features of the complex, the pair of proteins was suggested to represent a docking site for an as yet unidentified bacterial or host target and likely to be involved in signal transduction (42). Other important examples include (a) the identification of Rv3547 as a nitroreductase (43), (b) annotation of Rv2175c as a DNA-binding protein and a substrate of a protein kinase PknL, (c) understanding the subcellular localization of Rv2626c, a hypoxic response protein, and its involvement in modulating macrophage effector functions, (d) identification of Rv1155 as an enzyme in the pyridoxal-5′-phosphate biosynthesis pathway (44), (e) identification of a possible common mechanism of action for cyclopropane synthases (PcaA) (45) and methyl transferases (CmaA1 and CmaA2) (46), and (f) detection of Rv1347c as an enzyme in the mycobactin biosynthetic pathway, thus filling the missing link in the pathway (47). One of the most interesting insights concerns Rv3361c, MfpA, whose structure revealed a fold labeled as a right-handed quadrilateral beta-helix, which resembles standard B-DNA in size, shape, and electrostatics (48). Structural studies explained how the protein binds to DNA gyrase, mimicking DNA and affecting binding with fluoroquinolones, thereby conferring drug resistance (48).

Functional annotation from structures in a more automated manner can be quite complicated, since one-to-many and many-to-one relationships among fold types and functional categories are known to exist in a number of cases (49, 50). A more direct and insightful method is to extract functionally important regions in proteins and associate these regions with particular function(s) through substructure comparisons. The need to understand functional sites is not restricted to new sites. Even where protein structures are determined crystallographically as a complex with a ligand, a complete description of their binding sites is not always obtained, because they may not be cocrystallized with all the ligands required for the function of the molecule or because the bound ligands are often substitutes of the natural ligands. Identification of all relevant binding sites in protein molecules, therefore, forms a key step toward gaining functional insights from protein structures.

The detailsâ of functional inference that can be derived from protein structures through computational annotation are directly dependent upon the quality of the structure. Protein structures for about 1,064 proteins in the *M. tuberculosis* genome could be modeled based on a template with a sequence identity of ≥30%. For such structures, more detailed functional analysis can be performed to predict ligand-binding sites, active sites, conserved surfaces, ligand associations, and nucleic acid interactions. Since no single method can accurately predict the function of the protein from its given structure, a function annotation pipeline involving more than one algorithm is more useful. An example functional annotation for Rv1485 (*hemZ*) is depicted in Fig. 2.

COVERAGE OF *M. TUBERCULOSIS* STRUCTURAL PROTEOME AND ANNOTATION

The coverage of the structural information currently available for *M. tuberculosis* extends to about 70% of the proteome. Analysis of the distribution of structural information into various TubercuList (5) functional categories reveals that the topmost functional category with maximum structural information coverage is the information pathway (95.23%), followed by intermediary metabolism and respiration (94.76%) and lipid metabolism (91%). Interestingly, only 219 unique folds were associated with all the proteins involved in metabolism. Even in the category of conserved hypotheticals, models could be built for about 50% of them. The functional categories for which the structural information is limited are insertion elements and phages, due to lack of homology with any experimentally derived protein structures in the PDB. Around half of

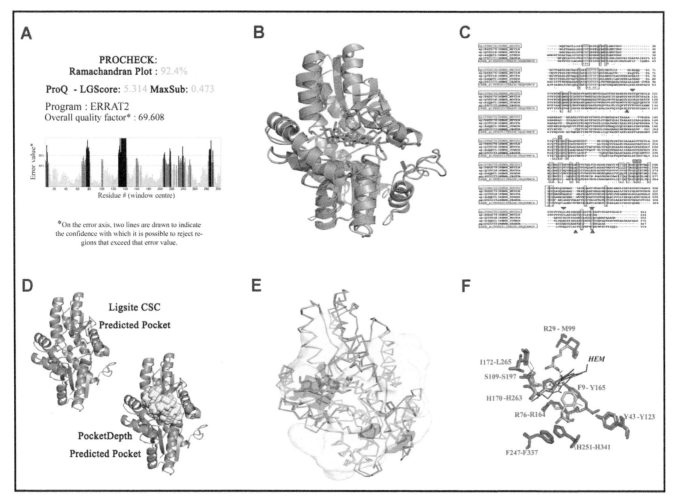

Figure 2 Aspects of annotation. (**A**) ERRAT output of Rv1485 showing an overall quality factor of ~70% by evaluating the nonbonded interactions between different atom types. (**B**) Superposition of Rv1485 with its template 1HRK:A. (**C**) Multiple sequence alignment with selected sequence neighbors, highlighting conserved catalytic site residues (in triangles). (**D**) Binding site prediction using LigsiteCSC and PocketDepth. (**E**) Predicted ligand-binding pockets in red surface. The expected ligand-binding site as determined by superposing the template is shown as sticks. (**F**) Association of the heme ligand to the predicted binding site (residues in red) based on high similarity to a known heme binding site by searching against PDB pockets (blue). doi:10.1128/microbiolspec.MGM2-0027-2013.f2

the genes involved in cell wall processes also do not have structural information because they would fall into the category of membrane proteins, which are inherently tough to crystallize and obtain structural information from.

Overall, there exists structural information for about 1,097 enzymes, and for 647 of them the active site residues could be determined through structure-based function annotation. Structure-based function annotation also revealed 1,728 ligand associations for the protein structures through binding site analysis and comparison. Around 740 possible DNA-interacting proteins were found to be present in the structural proteome.

FEATURES OF THE *M. TUBERCULOSIS* STRUCTURAL PROTEOME

The availability of the protein structures at a genome scale of *M. tuberculosis* presents a unique opportunity to understand the fold space covered by this proteome and to understand if there are particular biases or preferences. Figure 3 gives the distribution of various families of the proteins that have been modeled and the

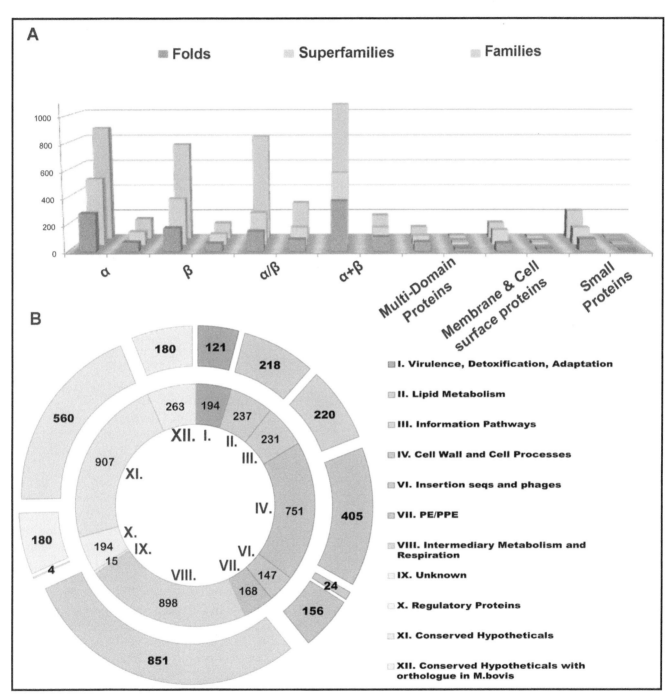

Figure 3 Coverage of a structural proteome. (A) Distribution of different structural classes as described through folds, superfamilies, and families occurring in SCOP and correspondingly in the *M. tuberculosis* proteome adjacent to it. (B) Distribution of structural information according to TubercuList functional categories. The inner circle represents the total number of genes in a particular functional category, and the outer circle represents the genes with structural information in the corresponding functional category.
doi:10.1128/microbiolspec.MGM2-0027-2013.f3

coverage of models depending upon the TubercuList classification.

Fold Distribution

The availability of the structural models of a particular organism, on a large scale, can give us the opportunity to analyze the distribution of structural classes and the most commonly occurring folds. A comparison with the currently existing SCOP database reveals that all the seven major structural classes of SCOP are observed among the available models of the *M. tuberculosis* proteome as well, although a marginal preponderance to alpha/beta class (36.45% after normalizing it to number of proteins when compared to 14.45% in SCOP) can be seen in *M. tuberculosis*. Out of the 1,195 known folds present in the SCOP database, 351 are seen to be present in the *M. tuberculosis* proteome, amounting to a coverage of known folds to about 30%. Similarly, at the family and superfamily levels, a significant coverage of 20% and 24.36% was observed in the *M. tuberculosis* proteome.

Around one-third (887 proteins) of the structural proteome is dominated by the most commonly occurring folds such as (Fig. 1) DNA/RNA-binding 3-helical bundle, P-loop containing nucleoside triphosphate hydrolases, TIM beta/alpha barrel, NAD(P)-binding Rossmann fold, SAM-dependent methyltransferases, ferredoxin-like, and alpha/beta-hydrolases. This is expected because it is known that distribution of copies of folds per genome in bacteria is known to follow power law (51), and these folds are commonly observed in most bacteria (52). But in the *M. tuberculosis* modeled proteome, ferritin fold, adopted by the N-termini of the PE and PPE proteins, dominates the list, which is not surprising given that nearly 180 of the proteins of this family exist in *M. tuberculosis*. Members of the PE and PPE families are known to be cell surface molecules involved in regulation of dendritic cell and macrophage immune-effector functions and also as antigens generating strong humoral responses that are possibly important for antigenic variability.

Fold Combination

The modeled structural proteome could also give us insights about combinations of protein folds that come together to form a polypeptide chain. Around 489 multidomain proteins could be identified in the modeled proteome. The fold combinations utilized can be represented in the form of a network with nodes representing a fold and edges representing the co-occurrence of folds within a polypeptide (Fig. 4A). The network revealed that the P-loop containing nucleoside triphos-

phate hydrolases is the fold with the highest number of co-occurrences. The most commonly occurring fold combination is observed between a tetracyline repressor-like fold and a DNA/RNA-binding 3-helical bundle. This combination is usually observed in the proteins that are known to play an important role in ribosomal protection and to help in the regulation of various efflux proteins (53, 54).

LEVELS OF ANNOTATION

The availability of protein structures for mycobacterial proteins provides an ideal starting point for arriving at functional insights at different levels. With improvements in bioinformatics approaches, many databases available online provide automated functional annotation of proteins. The availability of structures can complement such efforts to understand the molecular basis of biological function. Automated pipelines that impose strict quality filters to model the *M. tuberculosis* proteome, for instance, have resulted in structure assignment for 2,877 proteins. Using a case study, we describe how such pipelines can be applied to associate a structure with a protein and improve current understanding of its function.

Function Annotation through the Structural Annotation Pipeline

Rv1485 (P0A576: 344 residues, 4.99.1.1), annotated as a ferrochelatase, is known to catalyze ferrous insertion into Protophyrin IX to form Proto-heme (55). Based on sequence homology, Rv1485 was modeled using 1HRK (A chain, human ferrochelatase, 359 residues) as the template (Fig. 2B). The derived model, when superimposed on the template, showed less than 0.9 Å root mean square deviation, suggesting that the input alignment was of good quality. Parameters such as extent of residue conservation, compliance with the Ramachandran plot, and ERRAT scores could be determined to assess the quality of the generated models (Fig. 2A). For this example, these were all found to be satisfactory. Subsequently, one could perform a multiple-sequence alignment by including homologs from related bacteria, yeast, and humans, and such alignments showed a high conservation of residues in the protein core, indicating a well-conserved fold (Fig. 2C). Eukaryotic ferrochelatases typically possess three regions: an N-terminal organelle targeting region that is proteolytically cleaved, a second core region of 330 residues sharing homology with bacterial ferrochelatases, and a C-terminal region that contains the dimerization motif

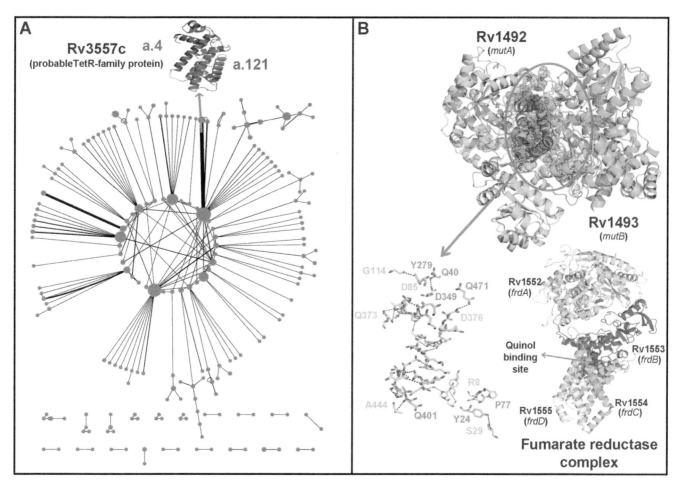

Figure 4 Fold combination and higher-order assemblies. (**A**) Network of fold combination observed in modeled multidomain proteins. Each node represents a fold, and an edge represents two folds occurring together within a polypeptide. The topmost occurring fold combination is the tetracycline-like repressor C-terminal (a.121) domain with DNA 3-helical bundle (a.4). An example protein, Rv3557c, is shown with both folds highlighted. (**B**) Examples of higher-order assemblies. The predicted assembly of methylmalonyl CoA mutase derived from the structural template 1REQ is shown. The assembly consists of Rv1492, MutA (cyan), and Rv1493, MutB (green). The conserved residues that could be involved in the interaction at the interface are shown in stick representation below. Similarly, the complex of fumarate reductase generated from 1KF6 is shown below, with residues involved in quinol binding that are conserved highlighted as spheres.
doi:10.1128/microbiolspec.MGM2-0027-2013.f4

as three of the four cysteine ligands of the 2Fe-2S cluster. These enzymes are known to function by involving the 2Fe-2S clusters. The availability of the alignments is then useful to compare the cysteine ligands of the 2Fe-2S clusters across the diverse species. In *S. pombe* ferrochelatase, for instance, cysteines analogous to the eukaryotic ferrochelatases are found. *M. tuberculosis* counterparts are seen to possess four cysteine-ligating residues, at C158, C332, C339, and C341.

Detailed structural analysis of the template has shown that it exists in an open substrate-free and closed substrate-bound form. In the two forms, it was shown that the active site pocket is closed around the porphyrin macrocycle. A number of active site residues are known to undergo reorientation of side chains due to a hydrogen bond network involving H263, H341, and E343. PocketDepth (56) and LigsiteCSC (57) predictions made on the modeled protein identified pockets that overlapped with the template pockets harboring the 2Fe-2S cluster and the cocrystallized ligand (cholic acid, 1HRK:A) (Fig. 2D). Comparison of the pockets in the modeled structure with that of the template also showed a similar network of residues in mycobacterial ferrochelatases. Methods such as ProFunc (58), when applied independently, also predicted these pockets. Finally, an overlap of the binding sites through methods such as PocketAlign (59) (Fig. 2F) could then be used to demonstrate the extent of structural overlap in the binding sites. In this example, for which function as a ferrochelatase was predicted, a combination of the results from multiple methods on the modeled structure could complement and enrich information pertaining to their binding pockets and re-affirm their role as potential ferrochelatases.

The level of detailed annotation possible for each protein, given a model, varies depending on the extent of homology to characterized proteins, homologous structures, and associated binding sites. In some cases, available information of the protein may be enriched further, while in others finer details of annotation such as identifying substrate specificities, etc., may be possible, especially where a template selected with high confidence and homology harbors a potential ligand. More importantly, the availability of models has permitted annotations of function for completely uncharacterized proteins in a few cases. In at least 39 cases, higher-order interactions and assemblies could also be inferred by examining gene neighborhoods of the target proteins. In the following section, different levels of function annotation are highlighted using representative examples.

Prediction of higher-order assemblies

Rv1552, a potential fumarate reductase, was modeled with our approach, using 1KF6:A as a template. The model generated for this protein covered 98.6% of the protein and showed 55% sequence identity with respect to 1KF6:A. Furthermore, 1KF6 was found to be one of the four subunits of the quinol-fumarate reductase respiratory complex of *Escherichia coli*, which is an integral-membrane complex. Interestingly, structural annotation of the gene neighbors of Rv1552, namely Rv1553, Rv1554, and Rv1555, identified 1KF6:B, 1KF6:C, and 1KF6:D as templates, suggesting that a similar four-subunit assembly is likely in *M. tuberculosis* as well.

It is known that in *E. coli*, 1KF6 catalyzes the final step of anaerobic respiration when fumarate is the terminal electron acceptor (60). During this type of anaerobic respiration, electrons are donated to quinol-fumarate reductase by menaquinol (MQH2) molecules in the membrane. The electrons are transferred to a covalently bound flavin adenine nucleotide at the active site through three distinct iron-sulfur clusters and ultimately are used to reduce fumarate to succinate. Two of these chains, the flavoprotein (FrdA) and the iron

protein (FrdB), comprise the soluble domain, which is involved in fumarate reduction, whereas the remaining two subunits (FrdC and FrdD) are membrane-spanning polypeptides involved in the electron transfer with quinones.

Results of the SCOP domain assignments of the *M. tuberculosis* proteins Rv1552, Rv1553, Rv1554, and Rv1555 show domain assignments and combinations that are observed in each of the template subunits with high statistical confidence (<10 to 20). Furthermore, alignments of the individual *M. tuberculosis* proteins with their closely related structural templates also show high similarity between these two proteins. Two quinol-binding sites in the membrane-spanning region of the template designated as Qp and Qd are known to lie proximal and distal to the 3Fe:4S cluster. Key residues involved in the quinol-binding site of the complex include C210, Q225, and K228 (B chain); F24, R28, and E29 (C chain); and R81, H84, H80, F17, E10, and W14 (D chain). These are also observed in the four-subunit assembly proposed by our pipeline (Fig. 4B), suggesting similar function and assembly in the *M. tuberculosis* proteins.

In another example, Rv1493 (P65487), currently annotated as a probable methylmalonyl-coenzyme A (CoA) mutase large subunit, was modeled using 1REQ:A as the template (71% identity and 98.27% model coverage). Methylmalonyl-CoA mutase catalyzes the isomerization of succinyl-CoA to methylmalonyl-CoA during synthesis of propionate using adenosylcobalamin as the cofactor (Fig. 4B). Alignments of the modeled protein with the template show conservation of active site residues such as Y89, H244, K604, D608, and H610 in the model as well. PocketDepth and SURFnet predictions of a highly conserved active site and a cobalamin-binding site in the model are seen in the template at equivalent positions. Gene neighborhood studies show that Rv1492, modeled with 1REQ:B as the template, lies upstream of Rv1493. With the availability of models from both these proteins, studies can be undertaken to predict the interacting residues lying at the interface of these two interacting chains. Detailed analysis of such interfaces may point to useful leads in the design of molecules that target such interfaces to disrupt protein function.

Improved annotation of conserved hypothetical protein

Rv3402c is a conserved hypothetical protein with an unknown function and structure. Our annotation pipeline identifies an aminotransferase, 1MDO, that catalyzes the transfer of an amino group from glutamic acid to a UDP-linked ketopyranose molecule as the template. This aminotransferase, a homodimer, is involved in a lipid A modification pathway that in some bacteria, such as *Salmonella,* confers resistance to cationic antimicrobial peptides. Crystal structures of ArnB show the presence of a large N-terminal cofactor-binding domain and a smaller three-stranded beta sheet. The template structure in the presence of cofactor, product, and inhibitor has revealed key residues that are implicated in the aminotransferase reaction. Comparison of the sequence of Rv3402c and the template shows high similarity, as does overall structural superposition (root mean square deviation of 0.49 Å). Residues that interact with the cofactor pyridoxal phosphate (PRP) in the template include Y136, H329, H163, K188, S183, H185, E194, T64, W89, D160, and F330. These residues are seen in structurally equivalent positions in Rv3402c. Likewise, residues that interact with the substrate L-glutamate in ArnB are also seen in Rv3402c, suggesting likely aspartate aminotransferase activity in *M. tuberculosis.* Figures 5A and B show an alignment of Rv3402c and 1MDO:A from the pipeline. Residues that are involved in cofactor binding in 1MDO:A and in the model are highlighted.

Enrichment of functional annotation

Rv0469, a possible mycolic acid synthase, has been modeled in Modbase with 1KPG:A, a cyclopropane-fatty-acyl-phospholipid synthase 1, as the template. Mycolic acids are major components of the cell wall of *M. tuberculosis,* and several efforts are under way to examine their potential as a drug target (61). Functional groups in the acyl chain of mycolic acids are understood to be important for bacterial pathogenesis and persistence. It is now known that there are at least three mycolic acid cyclopropane synthases (PcaA, CmaA1, and CmaA2) that produce these site-specific modifications. As in other methyltransferases, 1KPG:A adopts the seven-stranded alpha/beta fold and possesses a binding site for its cofactor, *S*-adenosyl-L-methionine. Multiple sequence alignments of Rv0469 with other known mycolic acid cyclopropane synthases such as CmaA2 (P0A5P0), PcaA (Q7D9R5), MmA2 (Q79FX6), MmA3 (P72027), MmA4 (Q79FX8), and CmaA1 (1KPG:A; P0C5C2) show a high conservation of residues involved in binding *S*-adenosylmethionine (cofactor), as do residues involved in ligand interactions. Ligand-binding-site predictions using PocketDepth and cleft predictions of the ProFunc server, when applied to the modeled Rv0469, identify potential substrate-binding and cofactor-binding sites with high confidence. Indeed, 86% of the cofactor-binding residues in 1KPG:A are seen at topologically equivalent positions in Rv0469. E124 that shows hydrogen bonds with N6 of *S*-adenosyl-L-homocysteine (SAH) is replaced with D123 in Rv0469. Assessments of the substrate-binding pocket show that 85% of the ligand-binding residues in 1KPG:A are also conserved in Rv0469 (Fig. 5C). Furthermore, cation-π interactions mediated by Y33 to stabilize the carbocation intermediate in the template are conserved (Y32) in Rv0469 as well. Since the ligand-binding pocket is 9 Å from the surface, it has been suggested that longer chain length mycolic acids can also be accommodated into the pocket. Such structures are thus ideal starting points for more detailed assessments involving the actual substrate (mycolic acids of varying chain length) within the predicted ligand-binding pockets of Rv0469.

Augmenting confidence of sequence-based annotation and confirming existing annotation

Rv2503c (P63650) is currently annotated as a probable succinyl-CoA:3-ketoacid-CoA transferase β subunit in TubercuList. This protein was modeled by considering 1OOY:A, with which it shows 58% identity, as the template. 1OOY also functions as a succinyl-CoA:3-ketoacid CoA transferase. Even though the template lacks a bound ligand, biochemical studies show that N281, Q99, G386, and A387 are conserved in CoA-transferases, and correspondingly, these residues were also conserved in the model of Rv2503c (62). Furthermore, binding site residues predicted by both PocketDepth (56) and SURFnet (63) show that these residues lie within the predicted binding pocket of Rv2503c (Fig. 5D).

STRUCTURAL PROTEOME AND ITS IMPACT ON PATHOGENESIS AND VIRULENCE

Interactions between proteins are important for a majority of biological functions. For instance, signals from the exterior of a cell are transmitted to the cell interior through interactions between proteins in processes such as signal transduction. Likewise, the process of transcription, its regulation, and a number of other processes in the cell are a consequence of protein-protein interactions that are either obligatory or transient in nature. Capturing such interactions is key to understanding which proteins interact with each other. A number of experimental procedures such as yeast two-hybrid assays and coaffinity purification have been developed to infer linkages between proteins on a genome-wide scale (64, 65). Computational methods that complement these techniques include the Rosetta stone (66), phylogenetic profile (67), conserved gene

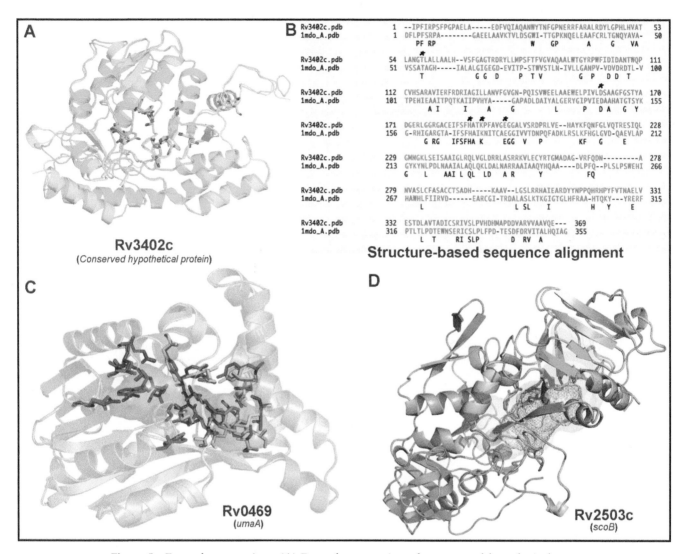

Figure 5 Example annotations. (**A**) Example annotation of a conserved hypothetical protein, Rv3402c. The superposition consists of a model of Rv3402c (shown in green), and the template 1MDO is shown as cyan. The conserved residues predicted to be involved in the interaction with PLP are represented as sticks. (**B**) The structure-based sequence alignment of Rv3402c with the template 1MDO. Functionally important residues are marked with (*). The amino acids are colored based on their chemical properties. (**C**) Superposition of Rv0469 (green) with the template 1KPG. The residues involved in cofactor recognition are shown in blue, and residues determining the substrate specificities are highlighted in red. The predicted pocket is shown in a surface representation. (**D**) Superposition of Rv2503c (green) with the template 2CTZ. The pocket predicted using PocketDepth and SURFnet enclosing the active site is shown in a mesh. doi:10.1128/microbiolspec.MGM2-0027-2013.f5

neighbors (68, 69), and operon or gene cluster methods. Protein linkages identified by these methods reveal proteins that participate in protein complexes, and therein further inferences about key interaction points in pathway(s) and related functions may be made. Therefore, such methods have been very useful in critically narrowing down the protein players involved in such interactions. Finer details on which specific parts

interact and what kinds of chemical bonds mediate that interaction can be gleaned through the availability of protein structures. Structures of such proteins and their complexes can assist in the design of molecules that target such interfaces and disrupt their formation, as has been shown in the case of class III adenylyl cyclases and other molecules involved in host-pathogen interactions such as NarL, Rv2413c, and LprG (70).

INSIGHTS INTO
TWO-COMPONENT SYSTEMS

A key feature of all living cells is the ability to sense environmental signals and implement adaptive changes. In pathogenic bacteria, this ability to adapt and respond to external stimuli, through simple phosphorylation events between membrane-localized kinases and response regulators that activate or repress transcription, cascades a series of signaling events in the pathogen referred to as two-component systems (71). These defense systems are believed to play a vital role in early intracellular survival of the pathogen as well as in various aspects of virulence. Mutation studies show that *in vivo* survival of the pathogen in the host macrophages is affected for mutants of these proteins. With the process of phosphotransfer from histidine to aspartate in such systems being absent in humans, such two-component systems are being increasingly viewed with interest in the area of structure-based drug design (72). Due to the nature of the interactions mediated by these proteins, such proteins are necessarily multidomain in nature, with long flexible linkers connecting the domain, and therefore, crystal structures of such proteins have always been challenging. Small-angle X-ray scattering approaches have been applied quite successfully to determine many of these structures. In mycobacteria, 13 structures are known to date in the PDB; these include MtrA (2GWR), NarL (3EUL), PdtaR (1SN8), PrrA (1YS6, 1YS7), and RegX3 (1ZKV).

The use of a number of bioinformatics approaches to predict the number of such components in mycobacteria has shown that they do not have a very high representation of such systems (71). Of the total 55 proteins in *M. tuberculosis* that are implicated in these systems in KEGG, 52 have been structurally modeled. Of the 12 complete two-component systems that are known to be encoded in the mycobacterial genome, 8 complete systems involving the sensor and the regulator have been modeled through such pipelines. This includes four of the five two-component systems, namely, SensX3-RegX3, PrrA-PrrB, MprA-MprB, MtrA-MtrB, and PdtaR-PdtaS, conserved in all mycobacterial species (19) sequenced to date. Furthermore, three of the five two-component systems, conserved in all except *Mycobacterium leprae*, namely PhoP-PhoR, KdpD-KdpE, TrcR-TrcS, DevR-DevS, and TrcX-TrcY, NarL have also been modeled through this approach. The availability of models for such complexes provides a useful starting point to determine the basis of such key molecular events and their biological roles in signal transduction systems and in adapting to host responses in pathogenesis.

Toxin-antitoxin (TA) gene pairs specify proteins that encode a stable toxin and an unstable antitoxin. Under normal physiological conditions both proteins are expressed and form a tight complex. However, under conditions of stress such as viral invasion, oxidative stress, or antibiotic attack, the unstable antitoxin is rapidly degraded by proteases, and thus the toxin is free to kill the host cell (13). The number of such TA systems in mycobacteria is high, and estimates range from 90 to over 120 genes. It is suggested that these protein pairs may play an important role in the persistence of *M. tuberculosis* and contribute to its survival within the phagosome subsequent to infection. Structural genomics efforts have already obtained the crystal structures for some TA pairs such as VapBC systems and RelBE systems, and in both these systems the toxins are RNases. Of the 122 TA genes extracted from the TADB database (73), 62 have been modeled (17), 38 of which were seen to adopt the PIN-domain-like fold, a characteristic of nuclease enzymes that cleave single-stranded RNA. Four others contain the lambda-repressor-like DNA-binding domain and YefM-like domain. The availability of structural information for such TA pairs should improve our understanding of how antitoxins inhibit their toxins and describe better the molecular basis of their mode of action.

APPLICATION IN DRUG DISCOVERY

The availability of the protein structure opens up avenues for structure-based drug discovery. The impact of structural-level understanding is felt at a number of steps in the discovery pipeline. To begin with, structures of the relevant protein molecules provide a much more detailed level of understanding of the underlying biological processes. Drug target identification benefits immensely from structural data. Properties desired in a target molecule, such as druggability, can be studied with the help of structures. *Druggability* in this context is a term that describes whether the given protein is amenable to modulation with a small molecule, which translates to asking whether a defined small molecule binding site can be identified in the protein and hence if the target is chemically tractable. Specificity to the pathogen can also be studied by combining structural information and informatics. The most direct and widespread application of structural knowledge, however, is in lead identification and lead refinement or lead optimization. Structural knowledge has also been used for searching through potential ligands among a database of approved drugs available for different diseases, setting the stage for possible drug repurposing as well. Each of these is described in the following section.

CURRENTLY EXPLORED TARGETS FOR STRUCTURE-BASED ANTI-TB DRUG DISCOVERY

The availability of a large number of structures in various functional classes from *M. tuberculosis* has provided the platform for the selection of targets from different pathways and evaluation of their druggability. These include protein targets involved in cell-wall biosynthesis such as alanine racemase (74); arabinogalactan and LAM biosynthesis pathways (61); and mycolic acid pathway players such as polyketide synthase Pks (75, 76), acyl-AMP ligase (77, 78), FadD32 (78, 79), the AccD4-containing acyl-CoA carboxylase (80), and others. Pathway comparison methods have highlighted a number of enzymes and pathways that are unique to mycobacteria and absent in the host. These include enzymes involved in essential amino acid synthesis such as enzymes from the shikimate pathway or pathways involved in vitamin B_2 and B_5 synthesis. Other targets include enzymes involved in regulatory processes, DNA metabolism–uracil DNA glycosylase (81), and carbon assimilatory processes such as the glyoxylate shunt pathways (18). TA pairs and other molecules that are thought to be involved in bacterial persistence are also of interest as drug targets (82). The availability of crystal structures of a number of such complexes such as VapBC and RelBE complexes (83) has not only shed light on the structural basis of such interactions, but have contributed to a better understanding of the competition exhibited by antitoxins for the RNA-binding substrates of the toxin. Likewise, comparison of toxins has shown that they differ extensively in substrate recognition sites, suggesting that different TA systems may be triggered by different regulatory controls and conditions.

NEW TARGET IDENTIFICATION: STRUCTURAL ASSESSMENT OF DRUGGABILITY

An important criterion to consider during drug target identification is that the target is chemically tractable and should be sufficiently different from that of any host protein (84). Chemical tractability, or druggability, can be translated into a problem of identifying whether a protein has a binding site capable of recognizing a small molecule ligand with reasonable affinity and whether the site and hence the binding is specific

enough, especially compared to host proteins. Considering specificity at an early step of target identification itself is likely to lead to significant reduction in failure rates due to cross-reactivity of the designed compound between host and pathogen. Similarity between proteins has been considered for some time now by comparison of their sequences, but where structural data for both proteins are available, similarity is better captured through structural comparisons. One of the ultimate requirements for determining the pharmacological profiles of drug molecules is which protein(s) in the given physiological environment can recognize the given drug molecule, a question that can be addressed with the help of structural comparisons of all possible binding sites of the pathogen's proteome with the counterpart set in the host. The goal here is to be able to weed out target candidates that share very high similarity with binding sites from the human "pocketome," since targeting these may lead to adverse drug reactions, due to inadvertent binding with human proteins. For this objective, it is not very intuitive to look at structural classes and overall properties such as the structural family or secondary structural types or a similar descriptor. Instead, the study of binding sites in proteins will provide information on which proteins could bind a given drug.

To render it meaningful, this type of analysis needs to be carried out at the proteome scale. About 229 and 3,515 experimentally derived structures are available in the PDB from *M. tuberculosis* and human, respectively. In addition, as described earlier, structural models for 70% of the *M. tuberculosis* proteome have been built. Similarly, homology-derived models are available in the ModBase database for 16,000 human proteins, making it feasible to carry out a proteome-scale structural assessment of targetability. A data set of about 3,500 binding pockets—comprising known ones from PDB structures combined with a large number of consensus binding sites predicted using three methods in putative drug targets in *M. tuberculosis*—were compared with about 70,000 similarly assimilated pockets from host proteins using algorithms (85) written specially for such a task (86). A mega task of about 245 million pairwise comparisons was performed on the IBM BlueGene supercomputer in a week. Of the 767 proteins tested in this manner, 145 were eliminated due to closely matching pockets in the human proteomes. It is possible that some of the eliminated proteins have sites that share part similarity only and hence could be targetable by using structural data to clearly identify common residues and different residues in each set, but this requires a close and more detailed analysis of all the pockets in the protein. As many as 622 proteins in *M. tuberculosis* that are flagged as essential for bacterial growth were found to contain proteins with binding pockets for small molecules that were unique to the pathogen. The screen thus identified unique druggable pockets in the pathogen and has yielded a shortlist of targets for designing antitubercular drugs.

STRUCTURE-BASED LEAD DESIGN

Knowledge of the structure of the target macromolecule and the functional regions in it provides a detailed definition of the binding site(s) that needs to be targeted with a drug molecule to achieve the desired manipulation. In the case of receptors involved in signaling cascades, the manipulation can be in the form of a small molecule that works as either an agonist or an antagonist, depending on both the disease and the receptor. For enzymes and other classes of targets, manipulation occurs predominantly through an inhibitor that prevents the natural substrate from binding to that protein. Lead design becomes a much more rational exercise with the knowledge of the target molecule. The approach has been successfully applied in the design of anti-influenza compounds (87, 88) and anti-HIV compounds targeting HIV protease (89, 90). The design can be performed through (a) virtual screening of large libraries of small molecules using docking methods or (b) fragment-based discovery combined with experimental structure determination and iteration of design or rational design of analogs of a known ligand with a specific aim to include or exclude interactions with particular residues in the binding site.

This has been shown in TB for the design of inhibitors against isocitrate lyase that play a key role in survival of the pathogen in the latent form during a chronic stage of infection (91, 92). The absence of the glyoxylate shunt pathway in mammals makes this an interesting target, and a number of interesting inhibitors—many of them mimics of metabolic intermediates in the pathway, such as succinate—have presented potential candidates showing inhibitory activity in the 0.10 μM range. Similarly, the structure of dehydroquinase from *M. tuberculosis* with bound ligand and inhibitors explained the molecular details of the reaction that could be used to design better inhibitors (93). Isoniazid, a frontline drug used in the treatment of TB, is known to be a prodrug, which is converted to the active species by catalase, which subsequently inhibits the function of InhA, a key enzyme in the mycolic acid biosynthetic pathway, by reacting nonenzymatically with NAD$^+$ and NADP$^+$ to form isonicotinoyl nucleotide

adducts. In addition to this, isoniazid was shown through crystallographic studies to bind to dihydrofolate reductase (DHFR) (94), and the structure of isoniazid-NADP-DHFR was employed in structure-based inhibitor design to maximize the interactions of substituted isonicotinoyl derivatives in the two subsites characteristic of the DHFR binding site.

Similarly, DNA ligases, which participate in replication and repair mechanisms, are broadly classified into NAD$^+$-dependent and ATP-dependent ligases. Due to their critical function in replication and repair, the crystal structure of the adenylation domain of the NAD$^+$-dependent ligase of *M. tuberculosis*, LigA, was employed to infer a different relative juxtaposition of the subdomains of the enzyme when compared to other bacterial ligases. Furthermore, the syn-conformation of AMP observed in this crystal structure appeared to mimic the conformation of AMP that is expected in the third step of the classical ligase reaction, because it is not covalently bound to the active site lysine. The availability of the crystal structure led to *in silico* screening of inhibitors and identification of a novel inhibitor class based on glycosylureides. These inhibitors are capable of distinguishing between the NAD$^+$- and ATP-dependent ligases, thus making them potential leads toward obtaining class-specific inhibitors of these enzymes (95). Similarly, ligand-based virtual screening efforts have also been shown with chorismate mutase (Rv1885c) to identify potential inhibitors (96).

Although a number of molecules can show inhibitory activity *in vitro*, transforming these activities to a nanomolar inhibitor range *in vivo* without impacting their ADME/Tox profile is often challenging. A number of interesting candidates have been explored to design more effective molecules against InhA (97). Several other examples are featured in the recent literature, such as the optimization of inhibitors of protein kinases such as protein kinase G by AX20017 (98). Fragment-based design has also been used to guide derivatization of a lead series of β-lactamase inhibitors against *M. tuberculosis* β-lactamase (BlaC) (99) and has been used successfully in the screening of fragment libraries against pantothenate synthetase to detect candidates that can be further modified to generate potent inhibitors (100).

Using a structural proteome-wide drug-target network of *M. tuberculosis* (TB-drugome), constructed by associating putative ligand-binding pockets in structural models of *M. tuberculosis* proteins with the known drug-binding sites, several new associations between approved drugs and proteins have been suggested (70), paving the way for possible drug repurposing. In this work, by combining molecular modeling, structural bioinformatics, and network approaches, a drug-target network was constructed, which the authors named the TB-drugome. Two proteins were connected if they exhibited similarity in their binding pockets. Drugs known for one protein were associated with another if the two proteins were connected with an edge in the network. Using such an approach, the anti-Parkinson's drugs entacapone and tolcapone were identified as possible inhibitors of InhA, a well-known target in *M. tuberculosis*, and were subsequently validated in an *in vitro* study as well (97).

CONCLUSIONS

Currently, there are structural models for nearly 70% of the proteome of *M. tuberculosis*. The availability of structural data has been invaluable in advancing our understanding of the biology of *M. tuberculosis*. Functional annotations through structures have been made to a large number of proteins, and new annotations have been obtained for a number of hypothetical proteins. The advantages of structure-based annotation are two-fold: (a) assigning a function based on the structural information, and that of the binding sites in it, makes the assignments more foolproof, and (b) they simultaneously provide a mechanistic basis for the assigned function by identification of the feasibility of binding, binding modes, and key residues involved in the process.

Getting a structural glimpse of a large portion of the proteome provides unique opportunities to explore the fold universe present in a single cell, fold frequencies, and preferences. For example, it is seen that the entire metabolism in the mycobacterial cell is achieved through a mere 219 folds. It is interesting to note that all seven major structural classes of proteins are represented in the *M. tuberculosis* proteome. A global perspective of folds that predominate in the genome, as well as the various fold combinations that make up multidomain proteins, is also obtained, throwing light on biologically significant protein assemblies.

Various stages of drug discovery benefit enormously from the availability of structural information of the *M. tuberculosis* proteome. Known or predicted targets can easily be characterized at a detailed level with the help of structural models. New target prediction is achieved not merely by considering the function of a molecule of interest but also by incorporating criteria of specificity and druggability, to pick targets that are unique to the pathogen and can be readily taken forward to the next steps in the pipeline. Through structural models, specificity is understood at a very

high level of resolution by considering uniqueness in the binding sites compared to thousands of proteins in the host proteome, setting a new trend of increasing likelihood of drug safety at a very early stage in the drug discovery pipeline. Identification of about 10% of the proteome as putative high-confidence drug targets, which are druggable, implies that there are still a large number of molecules whose potential as drug targets has not yet been tapped. Structure-based annotation and target prediction provide a wealth of information for TB drug discovery groups to explore.

Citation. Chandra N, Sandhya S, Anand P. 2014. Structural annotation of the *Mycobacterium tuberculosis* proteome. *Microbiol Spectrum* 2(2):MGM2-0027-2013.

References

1. Cole ST, Brosch R, Parkhill J, Garnier T, Churcher C, Harris D, Gordon SV, Eiglmeier K, Gas S, Barry CE 3rd Tekaia F, Badcock K, Basham D, Brown D, Chillingworth T, Connor R, Davies R, Devlin K, Feltwell T, Gentles S, Hamlin N, Holroyd S, Hornsby T, Jagels K, Krogh A, McLean J, Moule S, Murphy L, Oliver K, Osborne J, Quail MA, Rajandream MA, Rogers J, Rutter S, Seeger K, Skelton J, Squares R, Squares S, Sulston JE, Taylor K, Whitehead S, Barrell BG. 1998. Deciphering the biology of *Mycobacterium tuberculosis* from the complete genome sequence. *Nature* 393:537–544.

2. Reddy TB, Riley R, Wymore F, Montgomery P, DeCaprio D, Engels R, Gellesch M, Hubble J, Jen D, Jin H, Koehrsen M, Larson L, Mao M, Nitzberg M, Sisk P, Stolte C, Weiner B, White J, Zachariah ZK, Sherlock G, Galagan JE, Ball CA, Schoolnik GK. 2009. TB database: an integrated platform for tuberculosis research. *Nucleic Acids Res* 37:D499–D508.

3. Catanho M, Mascarenhas D, Degrave W, Miranda AB. 2006. GenoMycDB: a database for comparative analysis of mycobacterial genes and genomes. *Genet Mol Res* 5: 115–126.

4. Weniger T, Krawczyk J, Supply P, Niemann S, Harmsen D. 2010. MIRU-VNTRplus: a web tool for polyphasic genotyping of *Mycobacterium tuberculosis* complex bacteria. *Nucleic Acids Res* 38:W326–W331.

5. Lew JM, Kapopoulou A, Jones LM, Cole ST. 2011. TubercuList: 10 years after. *Tuberculosis* (Edinb) 91: 1–7.

6. Hunter S, Jones P, Mitchell A, Apweiler R, Attwood TK, Bateman A, Bernard T, Binns D, Bork P, Burge S, de Castro E, Coggill P, Corbett M, Das U, Daugherty L, Duquenne L, Finn RD, Fraser M, Gough J, Haft D, Hulo N, Kahn D, Kelly E, Letunic I, Lonsdale D, Lopez R, Madera M, Maslen J, McAnulla C, McDowall J, McMenamin C, Mi H, Mutowo-Muellenet P, Mulder N, Natale D, Orengo C, Pesseat S, Punta M, Quinn AF, Rivoire C, Sangrador-Vegas A, Selengut JD, Sigrist CJ, Scheremetjew M, Tate J, Thimmajanarthanan M, Thomas PD, Wu CH, Yeats C, Yong SY. 2012. InterPro

in 2011: new developments in the family and domain prediction database. *Nucleic Acids Res* 40:D306–D312.

7. Finn RD, Mistry J, Tate J, Coggill P, Heger A, Pollington JE, Gavin OL, Gunasekaran P, Ceric G, Forslund K, Holm L, Sonnhammer ELL, Eddy SR, Bateman A. 2010. The Pfam protein families database. *Nucleic Acids Res* 38:D211–D222.

8. Holm L, Sander C. 1998. Touring protein fold space with Dali/FSSP. *Nucleic Acids Res* 26:316–319.

9. Sujatha S, Balaji S, Srinivasan N. 2001. PALI: a database of alignments and phylogeny of homologous protein structures. *Bioinformatics* 17:375–376.

10. Blundell TL, Sibanda BL, Montalvao RW, Brewerton S, Chelliah V, Worth CL, Harmer NJ, Davies O, Burke D. 2006. Structural biology and bioinformatics in drug design: opportunities and challenges for target identification and lead discovery. *Philos Trans R Soc Lond B Biol Sci* 361:413–423.

11. Burley SK. 2000. An overview of structural genomics. *Nat Struct Biol* 7(Suppl):932–934.

12. Ioerger TR, Sacchettini JC. 2009. Structural genomics approach to drug discovery for *Mycobacterium tuberculosis*. *Curr Opin Microbiol* 12:318–325.

13. Chim N, Habel JE, Johnston JM, Krieger I, Miallau L, Sankaranarayanan R, Morse RP, Bruning J, Swanson S, Kim H, Kim CY, Li H, Bulloch EM, Payne RJ, Manos-Turvey A, Hung LW, Baker EN, Lott JS, James MN, Terwilliger TC, Eisenberg DS, Sacchettini JC, Goulding CW. 2011. The TB Structural Genomics Consortium: a decade of progress. *Tuberculosis* (Edinb) 91:155–172.

14. Canaves JM, Page R, Wilson IA, Stevens RC. 2004. Protein biophysical properties that correlate with crystallization success in *Thermotoga maritima*: maximum clustering strategy for structural genomics. *J Mol Biol* 344:977–991.

15. Berman HM, Westbrook J, Feng Z, Gilliland G, Bhat TN, Weissig H, Shindyalov IN, Bourne PE. 2000. The Protein Data Bank. *Nucleic Acids Res* 28:235–242.

16. Eswar N, Eramian D, Webb B, Shen MY, Sali A. 2008. Protein structure modeling with MODELLER. *Methods Mol Biol* 426:145–159.

17. Anand P, Sankaran S, Mukherjee S, Yeturu K, Laskowski R, Bhardwaj A, Bhagavat R, OSDD Consortium, Brahmachari SK, Chandra N. 2011. Structural annotation of *Mycobacterium tuberculosis* proteome. *PLoS One* 6:e27044.

18. Lou Z, Zhang X. 2010. Protein targets for structure-based anti-*Mycobacterium tuberculosis* drug discovery. *Protein Cell* 1:435–442.

19. Arora A, Chandra NR, Das A, Gopal B, Mande SC, Prakash B, Ramachandran R, Sankaranarayanan R, Sekar K, Suguna K, Tyagi AK, Vijayan M. 2011. Structural biology of *Mycobacterium tuberculosis* proteins: the Indian efforts. *Tuberculosis* (Edinb) 91:456–468.

20. Dunbrack RL Jr. 2006. Sequence comparison and protein structure prediction. *Curr Opin Struct Biol* 16:374–384.

21. MacCallum JL, Perez A, Schnieders MJ, Hua L, Jacobson MP, Dill KA. 2011. Assessment of protein structure refinement in CASP9. *Proteins* 79(Suppl 10):74–90.

22. Gribskov M, McLachlan AD, Eisenberg D. 1987. Profile analysis: detection of distantly related proteins. *Proc Natl Acad Sci USA* 84:4355–4358.

23. Eddy SR. 1998. Profile hidden Markov models. *Bioinformatics* 14:755–763.

24. Jones DT, Taylor WR, Thornton JM. 1992. A new approach to protein fold recognition. *Nature* 358:86–89.

25. Melo F, Sanchez R, Sali A. 2002. Statistical potentials for fold assessment. *Protein Sci* 11:430–448.

26. Altschul SF, Gish W, Miller W, Myers EW, Lipman DJ. 1990. Basic local alignment search tool. *J Mol Biol* 215:403–410.

27. Altschul SF, Madden TL, Schaffer AA, Zhang J, Zhang Z, Miller W, Lipman DJ. 1997. Gapped BLAST and PSI-BLAST: a new generation of protein database search programs. *Nucleic Acids Res* 25:3389–3402.

28. Laskowski RA, MacArthur MW, Moss DS, Thornton JM. 1993. {PROCHECK}: a program to check the stereochemical quality of protein structures. *J Appl Crystallogr* 26:283–291.

29. Colovos C, Yeates TO. 1993. Verification of protein structures: patterns of nonbonded atomic interactions. *Protein Sci* 2:1511–1519.

30. Mereghetti P, Ganadu ML, Papaleo E, Fantucci P, De Gioia L. 2008. Validation of protein models by a neural network approach. *BMC Bioinformatics* 9:66.

31. Pieper U, Webb BM, Barkan DT, Schneidman-Duhovny D, Schlessinger A, Braberg H, Yang Z, Meng EC, Pettersen EF, Huang CC, Ferrin TE, Sali A. 2011. ModBase, a database of annotated comparative protein structure models, and associated resources. *Nucleic Acids Res* 39:D465–D474.

32. Gutmanas A, Oldfield TJ, Patwardhan A, Sen S, Velankar S, Kleywegt GJ. 2013. The role of structural bioinformatics resources in the era of integrative structural biology. *Acta Crystallogr D Biol Crystallogr* 69:710–721.

33. Pal D, Eisenberg D. 2005. Inference of protein function from protein structure. *Structure* 13:121–130.

34. Chandra N, Anand P, Yeturu K. 2010. Structural bioinformatics: deriving biological insights from protein structures. *Interdiscip Sci* 2:347–366.

35. Lo Conte L, Ailey B, Hubbard TJ, Brenner SE, Murzin AG, Chothia C. 2000. SCOP: a structural classification of proteins database. *Nucleic Acids Res* 28:257–259.

36. Cuff AL, Sillitoe I, Lewis T, Clegg AB, Rentzsch R, Furnham N, Pellegrini-Calace M, Jones D, Thornton J, Orengo CA. 2011. Extending CATH: increasing coverage of the protein structure universe and linking structure with function. *Nucleic Acids Res* 39:D420–D426.

37. Ashburner M, Ball CA, Blake JA, Botstein D, Butler H, Cherry JM, Davis AP, Dolinski K, Dwight SS, Eppig JT, Harris MA, Hill DP, Issel-Tarver L, Kasarskis A, Lewis S, Matese JC, Richardson JE, Ringwald M, Rubin GM, Sherlock G. 2000. Gene ontology: tool for the unification of biology. The Gene Ontology Consortium. *Nat Genet* 25:25–29.

38. Vogel C, Chothia C. 2006. Protein family expansions and biological complexity. *PLoS Comput Biol* **2**:e48.

39. Vogel C, Berzuini C, Bashton M, Gough J, Teichmann SA. 2004. Supra-domains: evolutionary units larger than single protein domains. *J Mol Biol* **336**:809–823.

40. Baker EN. 2007. Structural genomics as an approach towards understanding the biology of tuberculosis. *J Struct Funct Genomics* **8**:57–65.

41. Hecker M. 2011. Microbial proteomics. *Proteomics* **11**: 2941–2942.

42. Strong M, Sawaya MR, Wang S, Phillips M, Cascio D, Eisenberg D. 2006. Toward the structural genomics of complexes: crystal structure of a PE/PPE protein complex from *Mycobacterium tuberculosis*. *Proc Natl Acad Sci USA* **103**:8060–8065.

43. Manjunatha UH, Boshoff H, Dowd CS, Zhang L, Albert TJ, Norton JE, Daniels L, Dick T, Pang SS, Barry CE 3rd. 2006. Identification of a nitroimidazo-oxazine-specific protein involved in PA-824 resistance in *Mycobacterium tuberculosis*. *Proc Natl Acad Sci USA* **103**:431–436.

44. Biswal BK, Cherney MM, Wang M, Garen C, James MN. 2005. Structures of *Mycobacterium tuberculosis* pyridoxine 5′-phosphate oxidase and its complexes with flavin mononucleotide and pyridoxal 5′-phosphate. *Acta Crystallogr D Biol Crystallogr* **61**: 1492–1499.

45. Glickman MS, Cox JS, Jacobs WR Jr. 2000. A novel mycolic acid cyclopropane synthetase is required for cording, persistence, and virulence of *Mycobacterium tuberculosis*. *Mol Cell* **5**:717–727.

46. Barkan D, Rao V, Sukenick GD, Glickman MS. 2010. Redundant function of cmaA2 and mmaA2 in *Mycobacterium tuberculosis* cis cyclopropanation of oxygenated mycolates. *J Bacteriol* **192**:3661–3668.

47. LaMarca BB, Zhu W, Arceneaux JE, Byers BR, Lundrigan MD. 2004. Participation of fad and mbt genes in synthesis of mycobactin in *Mycobacterium smegmatis*. *J Bacteriol* **186**:374–382.

48. Hegde SS, Vetting MW, Roderick SL, Mitchenall LA, Maxwell A, Takiff HE, Blanchard JS. 2005. A fluoroquinolone resistance protein from *Mycobacterium tuberculosis* that mimics DNA. *Science* **308**:1480–1483.

49. Nagano N, Orengo CA, Thornton JM. 2002. One fold with many functions: the evolutionary relationships between TIM barrel families based on their sequences, structures and functions. *J Mol Biol* **321**:741–765.

50. Dellus-Gur E, Toth-Petroczy A, Elias M, Tawfik DS. 2013. What makes a protein fold amenable to functional innovation? Fold polarity and stability trade-offs. *J Mol Biol* **425**:2609–2621.

51. Qian J, Luscombe NM, Gerstein M. 2001. Protein family and fold occurrence in genomes: power-law behaviour and evolutionary model. *J Mol Biol* **313**:673–681.

52. Wolf YI, Brenner SE, Bash PA, Koonin EV. 1999. Distribution of protein folds in the three superkingdoms of life. *Genome Res* **9**:17–26.

53. Hinrichs W, Kisker C, Duvel M, Muller A, Tovar K, Hillen W, Saenger W. 1994. Structure of the Tet repressor-tetracycline complex and regulation of antibiotic resistance. *Science* **264**:418–420.

54. Connell SR, Tracz DM, Nierhaus KH, Taylor DE. 2003. Ribosomal protection proteins and their mechanism of tetracycline resistance. *Antimicrob Agents Chemother* **47**:3675–3681.

55. Dailey TA, Dailey HA. 2002. Identification of [2Fe-2S] clusters in microbial ferrochelatases. *J Bacteriol* **184**: 2460–2464.

56. Kalidas Y, Chandra N. 2008. PocketDepth: a new depth based algorithm for identification of ligand binding sites in proteins. *J Struct Biol* **161**:31–42.

57. Huang B, Schroeder M. 2006. LIGSITEcsc: predicting ligand binding sites using the Connolly surface and degree of conservation. *BMC Struct Biol* **6**:19.

58. Laskowski RA, Watson JD, Thornton JM. 2005. ProFunc: a server for predicting protein function from 3D structure. *Nucleic Acids Res* **33**:W89–W93.

59. Yeturu K, Chandra N. 2011. PocketAlign a novel algorithm for aligning binding sites in protein structures. *J Chem Inf Model* **51**:1725–1736.

60. Iverson TM, Luna-Chavez C, Croal LR, Cecchini G, Rees DC. 2002. Crystallographic studies of the *Escherichia coli* quinol-fumarate reductase with inhibitors bound to the quinol-binding site. *J Biol Chem* **277**:16124–16130.

61. Chatterjee D. 1997. The mycobacterial cell wall: structure, biosynthesis and sites of drug action. *Curr Opin Chem Biol* **1**:579–588.

62. Coros AM, Swenson L, Wolodko WT, Fraser ME. 2004. Structure of the CoA transferase from pig heart to 1.7 A resolution. *Acta Crystallogr D Biol Crystallogr* **60**:1717–1725.

63. Laskowski RA. 1995. SURFNET: a program for visualizing molecular surfaces, cavities, and intermolecular interactions. *J Mol Graph* **13**:323–330, 307–308.

64. Uetz P, Giot L, Cagney G, Mansfield TA, Judson RS, Knight JR, Lockshon D, Narayan V, Srinivasan M, Pochart P, Qureshi-Emili A, Li Y, Godwin B, Conover D, Kalbfleisch T, Vijayadamodar G, Yang M, Johnston M, Fields S, Rothberg JM. 2000. A comprehensive analysis of protein-protein interactions in *Saccharomyces cerevisiae*. *Nature* **403**:623–627.

65. Gavin AC, Bosche M, Krause R, Grandi P, Marzioch M, Bauer A, Schultz J, Rick JM, Michon AM, Cruciat CM, Remor M, Höfert C, Schelder M, Brajenovic M, Ruffner H, Merino A, Klein K, Hudak M, Dickson D, Rudi T, Gnau V, Bauch A, Bastuck S, Huhse B, Leutwein C, Heurtier MA, Copley RR, Edelmann A, Querfurth E, Rybin V, Drewes G, Raida M, Bouwmeester T, Bork P, Seraphin B, Kuster B, Neubauer G, Superti-Furga G. 2002. Functional organization of the yeast proteome by systematic analysis of protein complexes. *Nature* **415**: 141–147.

66. Marcotte EM, Pellegrini M, Ng HL, Rice DW, Yeates TO, Eisenberg D. 1999. Detecting protein function and protein-protein interactions from genome sequences. *Science* **285**:751–753.

67. Pellegrini M, Marcotte EM, Thompson MJ, Eisenberg D, Yeates TO. 1999. Assigning protein functions by comparative genome analysis: protein phylogenetic profiles. *Proc Natl Acad Sci USA* **96**:4285–4288.

68. Overbeek R, Fonstein M, D'Souza M, Pusch GD, Maltsev N. 1999. The use of gene clusters to infer functional coupling. *Proc Natl Acad Sci USA* **96**:2896–2901.

69. Bowers PM, Pellegrini M, Thompson MJ, Fierro J, Yeates TO, Eisenberg D. 2004. Prolinks: a database of protein functional linkages derived from coevolution. *Genome Biol* **5**:R35.

70. Kinnings SL, Xie L, Fung KH, Jackson RM, Bourne PE. 2010. The *Mycobacterium tuberculosis* drugome and its polypharmacological implications. *PLoS Comput Biol* **6**:e1000976.

71. Tucker PA, Nowak E, Morth JP. 2007. Two-component systems of *Mycobacterium tuberculosis*: structure-based approaches. *Methods Enzymol* **423**:479–501.

72. Mizuno T. 2005. Two-component phosphorelay signal transduction systems in plants: from hormone responses to circadian rhythms. *Biosci Biotechnol Biochem* **69**:2263–2276.

73. Shao Y, Harrison EM, Bi D, Tai C, He X, Ou HY, Rajakumar K, Deng Z. 2011. TADB: a web-based resource for type 2 toxin-antitoxin loci in bacteria and archaea. *Nucleic Acids Res* **39**:D606–D611.

74. LeMagueres P, Im H, Ebalunode J, Strych U, Benedik MJ, Briggs JM, Kohn H, Krause KL. 2005. The 1.9 A crystal structure of alanine racemase from *Mycobacterium tuberculosis* contains a conserved entryway into the active site. *Biochemistry* **44**:1471–1481.

75. Raman K, Rajagopalan P, Chandra N. 2005. Flux balance analysis of mycolic acid pathway: targets for anti-tubercular drugs. *PLoS Comput Biol* **1**:e46.

76. Sankaranarayanan R, Saxena P, Marathe UB, Gokhale RS, Shanmugam VM, Rukmini R. 2004. A novel tunnel in mycobacterial type III polyketide synthase reveals the structural basis for generating diverse metabolites. *Nat Struct Mol Biol* **11**:894–900.

77. Trivedi OA, Arora P, Sridharan V, Tickoo R, Mohanty D, Gokhale RS. 2004. Enzymic activation and transfer of fatty acids as acyl-adenylates in mycobacteria. *Nature* **428**:441–445.

78. Portevin D, De Sousa-D'Auria C, Houssin C, Grimaldi C, Chami M, Daffe M, Guilhot C. 2004. A polyketide synthase catalyzes the last condensation step of mycolic acid biosynthesis in mycobacteria and related organisms. *Proc Natl Acad Sci USA* **101**:314–319.

79. Leger M, Gavalda S, Guillet V, van der Rest B, Slama N, Montrozier H, Mourey L, Quemard A, Daffe M, Marrakchi H. 2009. The dual function of the *Mycobacterium tuberculosis* FadD32 required for mycolic acid biosynthesis. *Chem Biol* **16**:510–519.

80. Portevin D, de Sousa-D'Auria C, Montrozier H, Houssin C, Stella A, Laneelle MA, Bardou F, Guilhot C, Daffe M. 2005. The acyl-AMP ligase FadD32 and AccD4-containing acyl-CoA carboxylase are required for the synthesis of mycolic acids and essential for mycobacterial growth: identification of the carboxylation product and determination of the acyl-CoA carboxylase components. *J Biol Chem* **280**:8862–8874.

81. Kaushal PS, Talawar RK, Krishna PD, Varshney U, Vijayan M. 2008. Unique features of the structure and interactions of mycobacterial uracil-DNA glycosylase: structure of a complex of the *Mycobacterium tuberculosis* enzyme in comparison with those from other sources. *Acta Crystallogr D Biol Crystallogr* **64**:551–560.

82. Lioy VS, Rey O, Balsa D, Pellicer T, Alonso JC. 2010. A toxin-antitoxin module as a target for antimicrobial development. *Plasmid* **63**:31–39.

83. Miallau L, Faller M, Chiang J, Arbing M, Guo F, Cascio D, Eisenberg D. 2009. Structure and proposed activity of a member of the VapBC family of toxin-antitoxin systems. VapBC-5 from *Mycobacterium tuberculosis*. *J Biol Chem* **284**:276–283.

84. Chandra N, Padiadpu J. 2013. Network approaches to drug discovery. *Expert Opin Drug Discov* **8**:7–20.

85. Yeturu K, Chandra N. 2008. PocketMatch: a new algorithm to compare binding sites in protein structures. *BMC Bioinformatics* **9**:543.

86. Raman K, Yeturu K, Chandra N. 2008. targetTB: a target identification pipeline for *Mycobacterium tuberculosis* through an interactome, reactome and genome-scale structural analysis. *BMC Syst Biol* **2**:109.

87. von Itzstein M. 2007. The war against influenza: discovery and development of sialidase inhibitors. *Nat Rev Drug Discov* **6**:967–974.

88. Wade RC. 1997. 'Flu' and structure-based drug design. *Structure* **5**:1139–1145.

89. Wlodawer A, Vondrasek J. 1998. Inhibitors of HIV-1 protease: a major success of structure-assisted drug design. *Annu Rev Biophys Biomol Struct* **27**:249–284.

90. Mastrolorenzo A, Rusconi S, Scozzafava A, Barbaro G, Supuran CT. 2007. Inhibitors of HIV-1 protease: current state of the art 10 years after their introduction. From antiretroviral drugs to antifungal, antibacterial and antitumor agents based on aspartic protease inhibitors. *Curr Med Chem* **14**:2734–2748.

91. Kratky M, Vinsova J. 2012. Advances in mycobacterial isocitrate lyase targeting and inhibitors. *Curr Med Chem* **19**:6126–6137.

92. Sharma V, Sharma S, Hoener zu Bentrup K, McKinney JD, Russell DG, Jacobs WR Jr, Sacchettini JC. 2000. Structure of isocitrate lyase, a persistence factor of *Mycobacterium tuberculosis*. *Nat Struct Biol* **7**:663–668.

93. Dias MV, Snee WC, Bromfield KM, Payne RJ, Palaninathan SK, Ciulli A, Howard NI, Abell C, Sacchettini JC, Blundell TL. 2011. Structural investigation of inhibitor designs targeting 3-dehydroquinate dehydratase from the shikimate pathway of *Mycobacterium tuberculosis*. *Biochem J* **436**:729–739.

94. Argyrou A, Vetting MW, Aladegbami B, Blanchard JS. 2006. *Mycobacterium tuberculosis* dihydrofolate reductase is a target for isoniazid. *Nat Struct Mol Biol* **13**:408–413.

95. Arora P, Goyal A, Natarajan VT, Rajakumara E, Verma P, Gupta R, Yousuf M, Trivedi OA, Mohanty D, Tyagi A, Sankaranarayanan R, Gokhale RS. 2009. Mechanistic and functional insights into fatty acid activation in *Mycobacterium tuberculosis*. *Nat Chem Biol* 5:166–173.

96. Agrawal H, Kumar A, Bal NC, Siddiqi MI, Arora A. 2007. Ligand based virtual screening and biological evaluation of inhibitors of chorismate mutase (Rv1885c) from *Mycobacterium tuberculosis* H37Rv. *Bioorg Med Chem Lett* 17:3053–3058.

97. Kinnings SL, Liu N, Buchmeier N, Tonge PJ, Xie L, Bourne PE. 2009. Drug discovery using chemical systems biology: repositioning the safe medicine Comtan to treat multi-drug and extensively drug resistant tuberculosis. *PLoS Comput Biol* 5:e1000423.

98. Scherr N, Honnappa S, Kunz G, Mueller P, Jayachandran R, Winkler F, Pieters J, Steinmetz MO. 2007. Structural basis for the specific inhibition of protein kinase G, a virulence factor of *Mycobacterium tuberculosis*. *Proc Natl Acad Sci USA* 104:12151–12156.

99. Eidam O, Romagnoli C, Dalmasso G, Barelier S, Caselli E, Bonnet R, Shoichet BK, Prati F. 2012. Fragment-guided design of subnanomolar beta-lactamase inhibitors active in vivo. *Proc Natl Acad Sci USA* 109:17448–17453.

100. Silvestre HL, Blundell TL, Abell C, Ciulli A. 2013. Integrated biophysical approach to fragment screening and validation for fragment-based lead discovery. *Proc Natl Acad Sci USA* 110:12984–12989.

Molecular Genetics of Mycobacteria, 2nd Edition
Edited by Graham F. Hatfull and William R. Jacobs, Jr.
© 2014 American Society for Microbiology, Washington, DC
doi:10.1128/microbiolspec.MGM2-0011-2013

Gwendowlyn S. Knapp[1]
Kathleen A. McDonough[1,2]

Cyclic AMP Signaling in Mycobacteria

14

CYCLIC AMP IS A UNIVERSAL SECOND MESSENGER USED BY PATHOGENS AND THEIR HOSTS

The ability to sense and respond to changing environments is essential for all organisms, and this process is mediated through signal transduction. The small molecules that relay signals from receptors to one or more effector proteins within the cell during signal transduction are called second messengers. Cyclic nucleotides, (p)ppGpp, Ca^{2+}, inositol triphosphate, and diacylglycerol function as second messengers in different types of cells. Cyclic $3',5'$-AMP (cAMP) is one of the most widely used second messengers, and its presence in bacteria, archaea, fungi, eukaryotic parasites, and mammals provides numerous opportunities for cAMP-mediated modulation of host-pathogen interactions (1–5).

cAMP signaling in mammals controls biological processes ranging from metabolism to memory formation and innate immunity, although it was first discovered for its role in hormone signal transduction (1, 6, 7). In bacteria, cAMP is best known for its role in mediating the "glucose response," or catabolite repression in *Escherichia coli* (2, 8). However, cAMP is also a critical

regulator of virulence for many bacterial and fungal pathogens (9). This chapter discusses the many roles of cAMP signaling in mycobacteria, including the regulation of gene expression and manipulation of host cell signaling during infection.

cAMP is generated from ATP by adenylyl cyclases (ACs) and hydrolytically degraded by phosphodiesterases (PDEs) (Fig. 1). ACs are distributed among six classes based on their primary amino acid sequences, with the well-studied bacterial AC from *E. coli* being a member of class I. The secreted AC toxins from *Pseudomonas aeruginosa*, *Bacillus anthracis*, and *Bordetella pertussis* (10–12) belong to class II, and classes IV to VI each contain a very small number of representatives from assorted bacteria (13–17). Class III comprises the largest and most diverse group of cyclases, including all known ACs from eukaryotes and many bacterial ACs (18).

Mycobacterium tuberculosis complex bacteria encode as many as 16 biochemically distinct class III ACs, which is highly unusual in the microbial world (Fig. 2) (19). This large number is especially striking when compared with the single AC of most bacteria and fungi, including *E. coli*, *Streptomyces griseus*,

[1]Wadsworth Center, New York State Department of Health, 120 New Scotland Avenue, PO Box 22002, Albany, NY 12222; [2]Department of Biomedical Sciences, University at Albany, Albany, NY 12222.

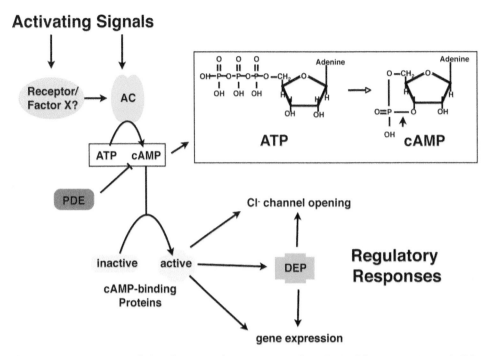

Figure 1 Environmental signals to regulatory outputs by cAMP. The conversion of ATP into cAMP and inorganic pyrophosphate and AMP is catalyzed by ACs. Degradation of cAMP is catalyzed by the phosphodiesterase. Activation of the AC can come from extracellular and intracellular signals that are relayed to the AC through membrane-bound or cytoplasmic receptors. The newly generated cAMP relays the activating signal to cAMP-binding proteins. doi:10.1128/microbiolspec.MGM2-0011-2013.f1

Corynebacterium glutamicum, *Candida albicans*, and *Cryptococcus neoformans* (20–23). In contrast to the abundance of ACs, only a single cAMP PDE, Rv0805, has been identified in mycobacteria, as discussed later in this article (24–27).

cAMP mediates its regulatory effects through its allosteric interactions with cAMP-binding proteins, which undergo conformational changes upon cAMP binding that alter their activation states. The best-studied outcome of cAMP signaling is regulation of gene expression. In the dominant bacterial paradigm, cAMP regulates transcription through cAMP-receptor protein (CRP) family transcription factors that are activated by direct binding of cAMP. Transcriptional regulation by cAMP in eukaryotes is less direct, because cAMP-mediated activation of eukaryotic transcription factors often occurs through a protein kinase A (PKA) complex intermediate. In this case, binding of cAMP to regulatory subunits in the PKA complex liberates catalytically active kinase subunits, which then activate downstream transcription factors by phosphorylation.

cAMP signaling in *M. tuberculosis* complex bacteria does not fit the classical catabolite repression paradigm that has been so well established in *E. coli*. This may

not be surprising, because metabolic profiling has recently shown that *M. tuberculosis*'s ability to cocatabolize different carbon sources reduces its need for catabolite repression (28). In *E. coli*, the *lac* operon codes for proteins that allow lactose to be used as a secondary carbon source when glucose is not available. Glucose depletion leads to an increase in cAMP levels, allowing induction of the *lac* operon in the presence of lactose through binding of the cAMP-CRP$_{Ec}$ complex to the *lac* operator. While cytoplasmic cAMP levels in *E. coli* drop 3- to 4-fold when ~0.2% glucose is substituted for glycerol as a carbon source (29, 30), cAMP levels in mycobacteria show little response to glucose. No significant change in the cytoplasmic cAMP levels of *Mycobacterium bovis* BCG occurred when cells were either provided 0.2% glucose (31) or starved for carbon (32). cAMP levels decrease in both fast- and slow-growing mycobacteria in response to very high levels of glucose (2%), but the biological significance of this result is not clear (33, 34). Rather, cAMP signaling in *M. tuberculosis* seems broadly important for metabolism, virulence, and host interactions, a trend that is being increasingly recognized in other bacterial pathogens as well.

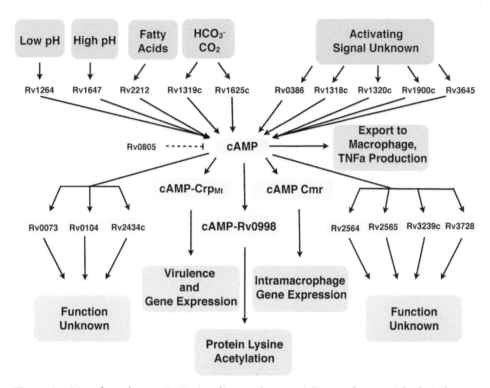

Figure 2 *M. tuberculosis* cAMP signaling pathways. ACs are shown with their known signals. Upon activation by the signal, the ACs generate cAMP and there are several fates of cAMP within *M. tuberculosis*. Most notable is cAMP binding to cNMP binding proteins to affect virulence, gene expression (including macrophage gene expression), and protein lysine acetylation. cAMP can be exported to the macrophage to affect TNFα production. Finally, Rv0805 can decrease cAMP levels by degrading the cAMP. Activating signals are shown in green, cAMP effector proteins are in yellow, and functional outcomes are designated in blue. doi:10.1128/microbiolspec.MGM2-0011-2013.f2

NUCLEOTIDE CYCLASES IN *M. TUBERCULOSIS*

Initial sequencing of the *M. tuberculosis* H37Rv genome predicted five ACs (35), and Bayesian analysis of sequenced genomes by McCue et al. expanded this number to 15 (19). Ten of these H37Rv ACs have been shown to be biochemically active: Rv0386, Rv1264, Rv1318, Rv1319, Rv1320, Rv3645, Rv1625c, Rv1647, Rv1900c, and Rv2212 (reviewed in reference 36). Sequencing of additional *Mycobacterium* genomes showed variable AC conservation across species and identified a pseudogene in *M. tuberculosis* H37Rv (Rv1120c), as well as an AC in *M. tuberculosis* CDC1551 (MT1360) that is not present in *M. tuberculosis* H37Rv (21, 37). MT1360 is thought to have arisen from expansion of the Rv1318c ortholog in CDC1551. Rv1120 is inactivated by a single base mutation that causes a frame shift in H37Rv, but its ortholog in *M. avium* MAP_2672 lacks this mutation and appears to be a functional gene. Table 1 lists the cyclases in *M. tuberculosis* H37Rv, as well as potential orthologs in *M. bovis* BCG, *M. avium*, and *Mycobacterium smegmatis*.

There is a great deal of diversity among *M. tuberculosis* cyclases with respect to their associated functional domains, and the group includes both soluble and membrane-associated proteins. Surprisingly, two *M. tuberculosis* ACs (Rv1625c and Rv2435) have mammalian-like catalytic sites (19). Rv1625c is particularly unusual in that its six transmembrane helices topologically resemble those of a mammalian-like integral membrane AC. Rv1625c activity has been demonstrated in both mammalian epithelial cells and *E. coli* (19, 38, 39), which is consistent with the sequence-based predictions. While four *M. tuberculosis* putative ACs (Rv0891c, Rv1359, Rv1647, and Rv2212) contain only a recognizable catalytic domain, the majority also have other functional motifs. Most of these associated domains are expected to add new effector function capabilities and/or regulate cyclase activity in response to

Table 1 ACs in *Mycobacterium*

Location	AC orthologs in *Mycobacterium* species[a]								Functionality		
	M. tuberculosis H37Rv	*M. tuberculosis* CDC1551	*M. bovis* AF2122/97	*M. bovis* BCG	*M. marinum* M	*M. avium* k10	*M. leprae* Tn	*M. smegmatis* mc²155	Activity[b]	Gene expression[c]	Signal[d]
Soluble	Rv0891c	MT0915	Mb0915c	BCG0943c	NID[e]	NID	NID	NID	ND[f]	NO/hypoxia (104)	ND
	Rv1120c	MT1152	Mb1151c	BCG1181c	NID	MAP2672	NID	NID	Pseudogene (24)	ND	Pseudogene
	Rv1264	MT1302	Mb1295	BCG1323	MMAR4173	MAP2507c	ML1111	MSMEG5018	Yes (40)	Hypoxia (46)	Low pH (41)
	Rv1359	MT1403	Mb1394	BCG1421	NID	NID	NID	MSMEG0545	ND	H2O2 (105)	ND
	Rv1647	MT1685	Mb1674	BCG1686	MMAR2454	MAP1357	ML1399	MSEMG3780	Yes (37)	Starvation (47)	High pH (37)
	Rv1900c	MT1951	Mb1935c	BCG1939c	MMAR0286	NID	ML2016	MSMEG4477	Yes (45)	ND	ND
	Rv2212	MT2268	Mb2235	BCG2228	MMAR3257	NID	NID	MSMEG4279	Yes (48)	Starvation (47)	Fatty acid (48)
Membrane-associated	Rv1318c	MT1359	Mb1352c	BCG1379c	MMAR4078	MAP2440	ML1154	MSMEG4924	Yes (42)	Starvation (47)	ND
	Rv1319c	MT1360	Mb1353c	BCG1380c	NID	NID	NID	MSMEG4924	Yes (42)	Starvation (47)	HCO3/CO2 (49)
		MT1361								Hypoxia (46)	
	Rv1320c	MT1362	Mb1354c	BCG1381c	MMAR4079	NID	NID	NID	Yes (42)	ND	ND
	Rv1625c	MT1661	Mb1651c	BCG1663c	MMAR2428	MAP1318c	ML1285	NID	Yes (38, 39)	ND	HCO3/CO2 (50)
	Rv2435c	MT2509	Mb2461c	BCG2454c	MMAR3757	MAP2250c	ML1460	MSMEG3578	ND	Hypoxia (46)	ND
	Rv3645	MT3748	Mb3669	BCG3703	MMAR5137	MAP0426c	ML0201	MSEMG6154	Yes (42)	ND	ND
Multi-domain	Rv0386	MT0399	Mb0393	BCG0424	MMAR2962	NID	NID	NID	Yes (44)	ND	ND
	Rv1358	MT1402	Mb1393	BCG1420	NID	NID	ML1753	NID	ND	ND	ND
	Rv2488c	MT2563	Mb2515c	BCG2507c	NID	NID	NID	NID	ND	Starvation (47)	ND

[a]Orthologs derived from *Mycobacterium tuberculosis* species as identified by http://tuberculist.epfl.ch/ and http://tbdb.org, accessed March 2013.
[b]Experimental evidence for AC activity.
[c]Conditions that affect expression of AC gene.
[d]Environmental conditions that affect AC functional activity.
[e]NID, not in database.
[f]ND, not determined.

environmental signals. For example, Rv2435c has a chemotaxis receptor-like extracellular domain (19), while Rv1264 contains an N-terminal auto-regulatory domain that inhibits cyclase activity above pH 6.0 (40, 41).

Five other multidomain ACs (Rv1318c, Rv1319c, Rv1320c, Rv2435c, and Rv3645c) are membrane-associated ACs that contain HAMP (histidine kinases, adenylyl cyclases, methyl binding proteins, and phosphatases) domains (42). HAMP domains are amphoteric alpha-helices that are often associated with two-component signal transduction pathways. These HAMP domains are thought to bridge the sensing of extracellular signals with the responding intracellular signaling domains (43). Three proteins (Rv0386, Rv1358, and Rv2488c) contain both ATPase and helix-turn-helix domains. While the role of its accessory domains is not clear, Rv0386 was identified as the *M. tuberculosis* AC responsible for elevating cAMP levels in infected macrophages, leading to an increase in tumor necrosis factor-α (TNF-α) production via the PKA and cAMP (44). Rv1900c contains an $\alpha\beta$-hydrolase domain commonly found in hydrolytic enzymes. Crystallographic studies showed that Rv1900c forms asymmetric homodimers that form a closed conformation upon binding substrate ATP (45). Rv1900c also showed some guanylyl cyclase activity (45).

REGULATION OF cAMP PRODUCTION

Expression of at least seven *M. tuberculosis* AC genes is likely modulated at the transcriptional level by environmental factors such as starvation and hypoxia (46, 47). However, cAMP production is often controlled largely at the posttranslational level through activation of the cyclases themselves (8, 18). Activity of *M. tuberculosis* ACs has been shown to respond to host-associated signals such as pH, fatty acids, ATP, and CO_2 levels (37, 41, 48–50) (Table 1), and it is likely that more signals will be identified. The activity of Rv1264 is directly responsive to pH, and the activity of Rv2212 is regulated *in vitro* by fatty acids, pH, and ATP concentration. Given the sheer number of ACs in *M. tuberculosis* H37Rv, it is not surprising that they respond to a wide variety of signals. Regulation by multiple signals provides an opportunity for ACs to coordinate *M. tuberculosis's* response to complex microenvironments within the host. Identifying the signals that affect cAMP signaling in *M. tuberculosis* during infection promises to be a rich area of future investigation.

DIVERSE ROLES FOR cAMP WITHIN MYCOBACTERIA

Three of ten predicted cyclic nucleotide monophosphate (cNMP) binding proteins have been functionally characterized to date (Fig. 2), and each of these proteins is discussed in greater detail later in this article. Rv0998 (and its ortholog in *M. smegmatis* [MSMEG_5458]) contains a domain with similarity to the GNAT (GCN5-related N-acetyltransferases) family and functions as a cAMP-responsive protein lysine acetylase in mycobacteria (51). Rv3676 (referred to as CRP$_{Mt}$, for cAMP responsive protein of *M. tuberculosis*) and Rv1675c (named Cmr, for cAMP and macrophage regulator) contain helix-turn-helix DNA binding domains and belong to the CRP-FNR family of transcription factors (19). Both CRP$_{Mt}$ and Cmr function as cAMP-responsive transcription factors, although only CRP$_{Mt}$ has been shown to directly bind cAMP (52–55).

Seven predicted cNMP binding proteins (Rv0073, Rv0104, Rv2434c, Rv2564, Rv2565, Rv3239c, and Rv3728) remain uncharacterized, although expression of Rv2565 is induced during human infection (56). Rv2565 contains a conserved esterase domain along with its putative cNMP-binding motif, but neither domain has been studied at the functional level. The remaining six predicted effector proteins also contain accessory domains with predicted transport and/or esterase activities (19, 36) that have not previously been associated with cAMP-mediated signal transduction in bacteria. Characterization of these proteins in mycobacteria is therefore likely to establish new roles for cAMP and/or cGMP signaling.

Coordinate regulation of some pathways by multiple cNMP binding proteins also seems likely. Recent studies from Stapleton et al. (57) showed that WhiB1 is a transcriptional repressor of *groEL2* and that expression of WhiB1 is regulated by CRP$_{Mt}$ in response to cAMP levels. Previous studies by Gazdik et al. (58) demonstrated Cmr binding to the upstream region of *groEL2*, as well as regulation within macrophages at 2 h postinfection. Stapleton et al. confirmed Cmr's binding and regulation of the *groEL2* promoter in *M. tuberculosis* (57). Thus, expression of *groEL2* is regulated directly by Cmr, while also being indirectly modulated by CRP$_{Mt}$ through WhiB1. Similarly, Pelly et al. suggested dual cAMP-mediated regulation by Cmr and CRP$_{Mt}$ of the small noncoding RNA ncRv11264c and/or Rv1265 (59), which was previously shown to be directly regulated by Cmr (58). A third possible example of multiple cNMP effector proteins affecting a regulatory pathway is illustrated in

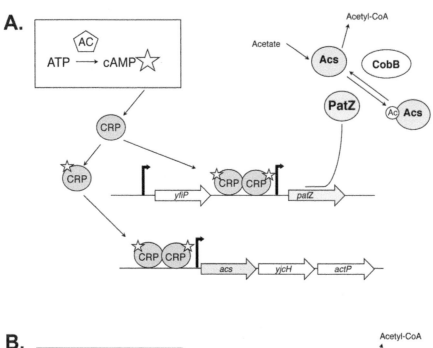

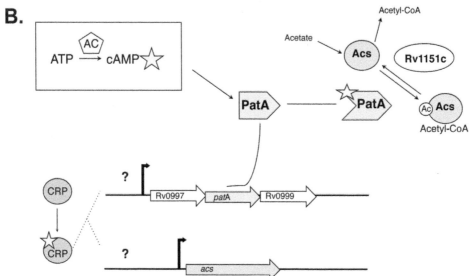

Figure 3 cAMP-dependent regulation of acetylation in *E. coli* (A) and *Mycobacterium* (B). In *E. coli*, cAMP's role in regulation is at the transcriptional level, whereas in *M. tuberculosis*, cAMP binds to the acetyltransferase, PatA, directly. cAMP-CRP complexes regulate *acs* and *patZ* at the transcriptional level in *E. coli*, while the role of CRP$_{Mt}$ in regulation of *acs* and *patA* is unknown at this time. CobB is a NAD$^+$-dependent sirtuin, as is Rv1151c. doi:10.1128/microbiolspec.MGM2-0011-2013.f3

Fig. 3. The upstream region of the acetyl CoA-synthase gene Rv3667 (*acs*) contains putative CRP$_{Mt}$ binding motifs (G. S. Knapp et al., unpublished), raising the possibility that expression of Rv3667 is also regulated by CRP$_{Mt}$. Such coordinate regulation by cAMP-dependent factors provides another exciting new area of investigation.

EMERGING ROLES FOR cAMP SIGNALING IN MYCOBACTERIAL GENE REGULATION AND METABOLISM

The role of cAMP in modulating gene expression in mycobacteria was first demonstrated by assessing changes in the *M. bovis* BCG proteome following the addition of dibutyryl cAMP (60). Dibutyryl cAMP

with *E. coli's* RNA polymerase (53, 67). Moreover, CRP$_{Mt}$ binds DNA strongly and specifically even in the absence of cAMP, shows only a modest increase in DNA binding affinity (2- to 10-fold) when bound with cAMP, and has relatively low affinity for cAMP compared with *E. coli* CRP (53, 55, 61). CRP$_{Mt}$ also appears to bind cAMP in a closed structure, while cAMP binding converts *E. coli* CRP to an open structure (53, 54, 61). Some of these features of CRP$_{Mt}$ are similar to those of the mutant cAMP-independent *E. coli* CRP* protein, although the structural bases of these differences are not clear (53, 78). Structural studies that directly compare apo and holo forms of CRP$_{Mt}$ with those of *E. coli* CRP would therefore be quite valuable in understanding the complexity and range of cAMP signaling mediated through this highly conserved transcription factor.

THE PROTEIN LYSINE ACETYLASE Rv0998

The recent finding that Rv0998 and its ortholog in *M. smegmatis*, MSMEG_5458, are cAMP-dependent lysine acetylases is an exciting new development in the cAMP signaling field (51). Protein lysine acetylation, a posttranslational modification, is evolutionarily conserved among eukaryotes and prokaryotes. In *E. coli* and *Salmonella*, acetylation of proteins is altered in response to carbon source availability, leading to differences in carbon and energy source usage and changing the metabolism of fatty acids and nucleotides (79–82). Additionally, acetylation plays a role in complex signal transduction systems that regulate cell division and flagellum synthesis (83).

Rv0998 and MSMEG_5458 share 56% identity at the amino acid level, and each contain a cyclic nucleotide binding domain fused to a GNAT family motif. Nambi et al. (51) identified a member of the universal stress protein (USP) family MSMEG_4207 as a target for acetylation in *M. smegmatis*. Purified MSMEG_5458 or Rv0998 acetylated the epsilon amino group of a lysine in this USP, and acetylation levels were higher in the presence of cAMP than in cGMP. They also demonstrated cAMP binding by both acetylases and showed that a point mutation, R95K, in the cNMP binding domain of MSMEG_5458 abolished cAMP binding (51).

The *M. smegmatis* USP MSMEG_4207 is not conserved in *M. tuberculosis*, but a later study identified acetyl-CoA synthetase (ACS) as another biologically relevant target of both acetylases that is also present in *M. tuberculosis* (84). ACS converts acetate into the high-energy compound acetyl-CoA, which is central

to many aspects of carbon metabolism. ACS is inactivated by acetylation at amino acid K167, and a reversible protein acetylation system was established when Rv1151c was shown to deactylate ACS. Crystallographic studies have also demonstrated acetyl-CoA binding to Rv0998 (85).

ACS is an interesting target for acetylation in the broad scope of metabolism. In *E. coli*, ACS fixes acetate into acetyl-CoA when environmental concentrations of acetate are low (Fig. 3) (86). In this case, PatZ is the acetyl transferase, while CobB is the deacetylase that converts ACS back into its active form (87). The overall activity of ACS in *E. coli* is also modulated by cAMP, because both PatZ and ACS are encoded by CRP-regulated genes, *patZ* and *acs*, respectively (Fig. 3) (88). This contrasts with cAMP's direct role in regulating the enzymatic activity of Rv0998 and MSMEG_5458 at the posttranslational level in mycobacteria. The role of CRP$_{Mt}$ in the regulation of Rv0998 or *acs* has not been investigated, but we have identified potential CRP$_{Mt}$ binding sites upstream of *acs* in *M. tuberculosis* (G. S. Knapp, unpublished).

Allosteric regulation of Rv0998 by cAMP involves conformational rearrangements that affect communication between the cNMP and GNAT domains, and Rv0998 activity is auto-inhibited in two ways in the absence of cAMP (85). In one case the C-terminal end of the peptide can insert as a helix within the cAMP-binding site, blocking cAMP binding. In the other, a large loop structure is formed that serves as a lid to block the catalytic active site. The major structural changes that occur in the presence of cAMP release the lid to expose the catalytic site while stabilizing the active form of the protein (85).

MORE MAKING THAN BREAKING OF cAMP IN MYCOBACTERIA?

Control of cytoplasmic cAMP levels can occur at the levels of production, degradation, or secretion; however, degradation is the primary way of balancing production in most organisms. Mycobacteria are unusual in this regard, because their cAMP secretion levels are high and reported degradation levels are weak. Despite the extraordinary number of ACs capable of generating cAMP in mycobacteria, Rv0805, a class III PDE that is structurally related to the metallophosphoesterase family (26, 36, 89), is the only cAMP PDE identified to date and is present only in slow-growing pathogenic mycobacteria. Rv0805 has significantly more activity (~150 times) against 2′,3′-cAMP than cAMP (26), and its overexpression increases the sensitivity of *M. smegmatis* to hydrophobic cytotoxic compounds, independent of its catalytic function (27). Thus, Rv0805's poorly understood role within mycobacteria represents an emerging story from two perspectives.

cAMP levels are reduced by ~30% in *M. smegmatis* (24), and ~50% in *M. tuberculosis* (44), compared with corresponding vector controls when Rv0805 is overexpressed. Seven amino acids contribute to Rv0805 dimerization and are required for coordination of active site metals (25). Asp21, His23, Asp63, and His209 mediate Fe^{3+} binding, while Asn97, His169, Asp63, and His207 coordinate Mn^{2+} binding (25), and a version of Rv0805 with an N97A mutation lacks cAMP PDE activity (24, 27). Rv0805's 2′,3′-cAMP PDE activity is greatly reduced by amino acid substitutions H98A or H98D (26), suggesting overlap of the functional sites for Rv0805's 2′,3′-cAMP and 3′,5′-cAMP PDE activities. While Rv0805 has not been successfully crystallized with cAMP, a structural model of Rv0805's active site accommodates cAMP docking (25, 27). However, Rv0805's ability to hydrolyze cAMP (V_{max} = 42 nmol cAMP hydrolyzed/min/mg protein; K_m = 153 M) (24) is only a fraction of *E. coli* CpdA's activity for cAMP (V_{max} = 2.0 μmole/min/mg; K_m = 0.5 mM) (90), making it difficult to understand how mycobacteria maintain cAMP homeostasis.

The biological significance of Rv0805's activity against 2′,3′-cNMPs is also unclear. It is possible that these nucleotides are used for signaling or nucleotide acquisition. However, RNA cleavage by endonucleases associated with some CRISPR and TA toxin mRNA interferase systems generates 2′,3′-cyclic ends (91–93), and it is also possible that such RNA molecules serve as natural substrates for Rv0805 in *M. tuberculosis*.

cAMP PRODUCTION, SECRETION, AND HOST CELL INTERACTIONS

The large ratio of active ACs to PDEs may present an opportunity for accumulation of high levels of cAMP relative to other organisms, and this remains an area of current investigation (36, 55, 61, 94, 95). However, cAMP levels can vary greatly with conditions, and the use of different normalization methods (e.g., wet versus dry weight [34, 95]) further complicates inter-study comparisons. cAMP levels reported for mycobacteria (4.5 pmol/mg wet weight and 18.7 to 1,000 pmol/mg dry weight) are surprisingly comparable to those of *E. coli* and *Salmonella typhimurium* (5.0 pmol/mg wet weight and 130 to 450 pmol/mg dry weight). However, mycobacterial cytoplasmic cAMP levels from recent

studies normalized to the number of bacteria rather than weight (0.5 to 400 pmol/10^8 bacteria) show that a tremendous range of cAMP accumulation is possible depending on the environmental conditions (31, 32, 44, 75).

Conditions that affect cAMP levels in mycobacteria include those associated with the host environment. For example, macrophage passage increases cAMP levels in TB-complex mycobacteria ~50-fold relative to bacteria in tissue culture medium alone (31). Low pH also increases cAMP levels in both *M. tuberculosis* and *M. bovis* BCG (58), while albumin is inhibitory for *M. bovis* BCG's cAMP production (31). Albumin's inhibitory effects can be overcome by the presence of oleic acid, but the cAMP produced in the presence of oleic acid is secreted rather than retained within the cells (31).

Mycobacteria appear to secrete far more cAMP than other bacteria (8, 95), and this secretion contributes to *M. tuberculosis* pathogenesis (44). However, the factors that regulate cAMP export and the outcomes of this secretion have not been fully defined (31, 75). Modulation of host cell cAMP levels is a common strategy of many bacterial pathogens, but *M. tuberculosis* is the first shown to secrete premade cAMP (9, 44). *B. anthracis*, *B. pertussis*, and *P. aeruginosa* export host-dependent ACs that serve as toxins (12, 96). An exported AC has also been reported for *Yersinia pestis*, although it is yet to be characterized (14). An alternative strategy is used by *Vibrio cholerae*, *E. coli*, and *B. pertussis*, which intoxicate their host cells by secreting enzymes that ADP-ribosylate the alpha subunits of heterotrimeric G proteins that regulate the activity of host ACs, leading to increased cAMP production in their host cells (7, 97, 98).

Live, but not heat-killed, *Mycobacterium microti*, *M. bovis* BCG, and *M. tuberculosis* directly export cAMP to increase the cAMP levels of their host macrophages (31, 44, 99, 100). The *M. tuberculosis* AC Rv0386 is responsible for this cAMP intoxication, which results in TNF-α production via the PKA and cAMP response-element-binding (CREB) protein pathway (44). Mutation of Rv0386 decreases survival and immunopathology in a mouse infection model, demonstrating the importance of cAMP intoxication for *M. tuberculosis* pathogenesis. Other reports indicate that CREB-mediated TNF-α production is also increased in macrophages infected with *M. smegmatis*, but not with *M. avium* (101, 102). These studies suggest that there are multiple ways in which mycobacteria can stimulate the CREB pathway in macrophages, because there is no Rv0386 ortholog in

M. smegmatis. M. tuberculosis may also use cAMP to modulate phagosome trafficking, because there is some evidence that elevated cAMP levels in infected macrophages can inhibit phagosome-lysosome fusion (100, 103). cAMP secretion and signaling is clearly a promising new frontier at the interface of mycobacteria-host interactions.

CONCLUSIONS AND FUTURE DIRECTIONS

cAMP signaling is a complex and exciting area of investigation in mycobacteria, and the pace of research has increased greatly in recent years. Nonetheless, many critical questions remain. It will be especially important to identify the factors that regulate levels of cAMP within mycobacteria and the environmental signals to which they respond, as well as the biological roles of Cmr and CRP_Mt genes. Similarly, the impact of cAMP-regulated lysine acetylation on mycobacterial metabolism has yet to be explored, and the functions of most putative cNMP-binding effector proteins await discovery. The unexpected biological roles of Rv0805 and its lack of robust cAMP phospodiesterase activity are particularly intriguing findings that warrant investigation. Elucidating the mechanism by which cAMP is secreted from mycobacterial cells, and the way in which this cAMP affects the host response to *M. tuberculosis* infection will be critical to our understanding of *M. tuberculosis* pathogenesis, as is the question of cAMP signaling specificity, given the large number of ACs in mycobacteria. There is clearly much to be done to unravel the multifactorial roles of cAMP signaling in mycobacteria at the molecular, genetic, and biological levels, and rewards of pursuing this journey hold great promise.

Citation. Knapp GS, McDonough KA. 2014. Cyclic AMP signaling in mycobacteria. Microbiol Spectrum 2(2):MGM2-0011-2013.

References

1. Altarejos JY, Montminy M. 2011. CREB and the CRTC co-activators: sensors for hormonal and metabolic signals. *Nat Rev Mol Cell Biol* 12:141–151.

2. Kolb A, Busby S, Buc H, Garges S, Adhya S. 1993. Transcriptional regulation by cAMP and its receptor protein. *Annu Rev Biochem* 62:749–795.

3. Kamenetsky M, Middelhaufe S, Bank EM, Levin LR, Buck J, Steegborn C. 2006. Molecular details of cAMP generation in mammalian cells: a tale of two systems. *J Mol Biol* 362:623–639.

4. Gorke B, Stulke J. 2008. Carbon catabolite repression in bacteria: many ways to make the most out of nutrients. *Nat Rev Microbiol* 6:613–624.

5. Botsford JL, Harman JG. 1992. Cyclic AMP in prokaryotes. *Microbiol Rev* 56:100–122.

6. Kresge N, Simoni RD, Hill RL. 2005. Earl W. Sutherland's discovery of cyclic adenine monophosphate and the second messenger system. *J Biol Chem* 280:e39–e40.

7. Serezani CH, Ballinger MN, Aronoff DM, Peters-Golden M. 2008. Cyclic AMP: master regulator of innate immune cell function. *Am J Respir Cell Mol Biol* 39:127–132.

8. Botsford JL. 1981. Cyclic nucleotides in procaryotes. *Microbiol Rev* 45:620–642.

9. McDonough KA, Rodriguez A. 2012. The myriad roles of cyclic AMP in microbial pathogens: from signal to sword. *Nat Rev Microbiol* 10:27–38.

10. Leppla SH. 1982. Anthrax toxin edema factor: a bacterial adenylate cyclase that increases cyclic AMP concentrations of eukaryotic cells. *Proc Natl Acad Sci USA* 79:3162–3166.

11. Weiss AA, Hewlett EL, Myers GA, Falkow S. 1984. Pertussis toxin and extracytoplasmic adenylate cyclase as virulence factors of *Bordetella pertussis*. *J Infect Dis* 150:219–222.

12. Yahr TL, Vallis AJ, Hancock MK, Barbieri JT, Frank DW. 1998. ExoY, an adenylate cyclase secreted by the *Pseudomonas aeruginosa* type III system. *Proc Natl Acad Sci USA* 95:13899–13904.

13. Gallagher DT, Kim SK, Robinson H, Reddy PT. 2011. Active-site structure of class IV adenylyl cyclase and transphyletic mechanism. *J Mol Biol* 405:787–803.

14. Smith N, Kim SK, Reddy PT, Gallagher DT. 2006. Crystallization of the class IV adenylyl cyclase from *Yersinia pestis*. *Acta Crystallogr Sect F Struct Biol Cryst Commun* 62:200–204.

15. Sismeiro O, Trotot P, Biville F, Vivares C, Danchin A. 1998. *Aeromonas hydrophila* adenylyl cyclase 2: a new class of adenylyl cyclases with thermophilic properties and sequence similarities to proteins from hyperthermophilic archaebacteria. *J Bacteriol* 180:3339–3344.

16. Cotta MA, Whitehead TR, Wheeler MB. 1998. Identification of a novel adenylate cyclase in the ruminal anaerobe, *Prevotella ruminicola* D31d. *FEMS Microbiol Lett* 164:257–260.

17. Tellez-Sosa J, Soberon N, Vega-Segura A, Torres-Marquez ME, Cevallos MA. 2002. The *Rhizobium etli cyaC* product: characterization of a novel adenylate cyclase class. *J Bacteriol* 184:3560–3568.

18. Tang WJ, Yan S, Drum CL. 1998. Class III adenylyl cyclases: regulation and underlying mechanisms. *Adv Second Messenger Phosphoprotein Res* 32:137–151.

19. McCue LA, McDonough KA, Lawrence CE. 2000. Functional classification of cNMP-binding proteins and nucleotide cyclases with implications for novel regulatory pathways in *Mycobacterium tuberculosis*. *Genome Res* 10:204–219.

20. Cha PH, Park SY, Moon MW, Subhadra B, Oh TK, Kim E, Kim JF, Lee JK. 2010. Characterization of an adenylate cyclase gene (*cyaB*) deletion mutant

of *Corynebacterium glutamicum* ATCC 13032. *Appl Microbiol Biotechnol* **85**:1061–1068.

21. **Shenoy AR, Sivakumar K, Krupa A, Srinivasan N, Visweswariah SS.** 2004. A survey of nucleotide cyclases in actinobacteria: unique domain organization and expansion of the class III cyclase family in *Mycobacterium tuberculosis*. *Comp Funct Genomics* **5**:17–38.

22. **Klengel T, Liang WJ, Chaloupka J, Ruoff C, Schroppel K, Naglik JR, Eckert SE, Mogensen EG, Haynes K, Tuite MF, Levin LR, Buck J, Muhlschlegel FA.** 2005. Fungal adenylyl cyclase integrates CO2 sensing with cAMP signaling and virulence. *Curr Biol* **15**:2021–2026.

23. **Mallet L, Renault G, Jacquet M.** 2000. Functional cloning of the adenylate cyclase gene of *Candida albicans* in *Saccharomyces cerevisiae* within a genomic fragment containing five other genes, including homologues of CHS6 and SAP185. *Yeast* **16**:959–966.

24. **Shenoy AR, Sreenath N, Podobnik M, Kovacevic M, Visweswariah SS.** 2005. The Rv0805 gene from *Mycobacterium tuberculosis* encodes a 3′,5′-cyclic nucleotide phosphodiesterase: biochemical and mutational analysis. *Biochemistry* **44**:15695–15704.

25. **Shenoy AR, Capuder M, Draskovic P, Lamba D, Visweswariah SS, Podobnik M.** 2007. Structural and biochemical analysis of the Rv0805 cyclic nucleotide phosphodiesterase from *Mycobacterium tuberculosis*. *J Mol Biol* **365**:211–225.

26. **Keppetipola N, Shuman S.** 2008. A phosphate-binding histidine of binuclear metallophosphodiesterase enzymes is a determinant of 2′,3′-cyclic nucleotide phosphodiesterase activity. *J Biol Chem* **283**:30942–30949.

27. **Podobnik M, Tyagi R, Matange N, Dermol U, Gupta AK, Mattoo R, Seshadri K, Visweswariah SS.** 2009. A mycobacterial cyclic AMP phosphodiesterase that moonlights as a modifier of cell wall permeability. *J Biol Chem* **284**:32846–32857.

28. **de Carvalho LP, Fischer SM, Marrero J, Nathan C, Ehrt S, Rhee KY.** 2010. Metabolomics of *Mycobacterium tuberculosis* reveals compartmentalized co-catabolism of carbon substrates. *Chem Biol* **17**:1122–1131.

29. **Buettner MJ, Spitz E, Rickenberg HV.** 1973. Cyclic adenosine 3′,5′-monophosphate in *Escherichia coli*. *J Bacteriol* **114**:1068–1073.

30. **Epstein W, Rothman-Denes LB, Hesse J.** 1975. Adenosine 3′:5′-cyclic monophosphate as mediator of catabolite repression in *Escherichia coli*. *Proc Natl Acad Sci USA* **72**:2300–2304.

31. **Bai G, Schaak DD, McDonough KA.** 2009. cAMP levels within *Mycobacterium tuberculosis* and *Mycobacterium bovis* BCG increase upon infection of macrophages. *FEMS Immunol Med Microbiol* **55**:68–73.

32. **Dass BK, Sharma R, Shenoy AR, Mattoo R, Visweswariah SS.** 2008. Cyclic AMP in mycobacteria: characterization and functional role of the Rv1647 ortholog in *Mycobacterium smegmatis*. *J Bacteriol* **190**:3824–3834.

33. **Padh H, Venkitasubramanian TA.** 1976. Adenosine 3′,5′-monophosphate in *Mycobacterium phlei* and *Mycobacterium tuberculosis* H37Ra. *Microbios* **16**:183–189.

34. **Lee CH.** 1979. Metabolism of cyclic AMP in non-pathogenic *Mycobacterium smegmatis*. *Arch Microbiol* **120**:35–37.

35. **Cole ST, Brosch R, Parkhill J, Garnier T, Churcher C, Harris D, Gordon SV, Eiglmeier K, Gas S, Barry CE 3rd, Tekaia F, Badcock K, Basham D, Brown D, Chillingworth T, Connor R, Davies R, Devlin K, Feltwell T, Gentles S, Hamlin N, Holroyd S, Hornsby T, Jagels K, Krogh A, McLean J, Moule S, Murphy L, Oliver K, Osborne J, Quail MA, Rajandream MA, Rogers J, Rutter S, Seeger K, Skelton J, Squares R, Squares S, Sulston JE, Taylor K, Whitehead S, Barrell BG.** 1998. Deciphering the biology of *Mycobacterium tuberculosis* from the complete genome sequence. *Nature* **393**:537–544.

36. **Shenoy AR, Visweswariah SS.** 2006. New messages from old messengers: cAMP and mycobacteria. *Trends Microbiol* **14**:543–550.

37. **Shenoy AR, Sreenath NP, Mahalingam M, Visweswariah SS.** 2005. Characterization of phylogenetically distant members of the adenylate cyclase family from mycobacteria: Rv1647 from *Mycobacterium tuberculosis* and its orthologue ML1399 from *M. leprae*. *Biochem J* **387**:541–551.

38. **Guo YL, Seebacher T, Kurz U, Linder JU, Schultz JE.** 2001. Adenylyl cyclase Rv1625c of *Mycobacterium tuberculosis*: a progenitor of mammalian adenylyl cyclases. *EMBO J* **20**:3667–3675.

39. **Reddy SK, Kamireddi M, Dhanireddy K, Young L, Davis A, Reddy PT.** 2001. Eukaryotic-like adenylyl cyclases in *Mycobacterium tuberculosis* H37Rv: cloning and characterization. *J Biol Chem* **276**:35141–35149.

40. **Linder JU, Schultz A, Schultz JE.** 2002. Adenylyl cyclase Rv1264 from *Mycobacterium tuberculosis* has an autoinhibitory N-terminal domain. *J Biol Chem* **277**:15271–15276.

41. **Tews I, Findeisen F, Sinning I, Schultz A, Schultz JE, Linder JU.** 2005. The structure of a pH-sensing mycobacterial adenylyl cyclase holoenzyme. *Science* **308**:1020–1023.

42. **Linder JU, Hammer A, Schultz JE.** 2004. The effect of HAMP domains on class IIIb adenylyl cyclases from *Mycobacterium tuberculosis*. *Eur J Biochem* **271**:2446–2451.

43. **Hulko M, Berndt F, Gruber M, Linder JU, Truffault V, Schultz A, Martin J, Schultz JE, Lupas AN, Coles M.** 2006. The HAMP domain structure implies helix rotation in transmembrane signaling. *Cell* **126**:929–940.

44. **Agarwal N, Lamichhane G, Gupta R, Nolan S, Bishai WR.** 2009. Cyclic AMP intoxication of macrophages by a *Mycobacterium tuberculosis* adenylate cyclase. *Nature* **460**:98–102.

45. **Sinha SC, Wetterer M, Sprang SR, Schultz JE, Linder JU.** 2005. Origin of asymmetry in adenylyl cyclases: structures of *Mycobacterium tuberculosis* Rv1900c. *EMBO J* **24**:663–673.

46. **Sherman DR, Voskuil M, Schnappinger D, Liao R, Harrell MI, Schoolnik GK.** 2001. Regulation of the *Mycobacterium tuberculosis* hypoxic response gene

encoding alpha-crystallin. *Proc Natl Acad Sci USA* **98:** 7534–7539.

47. Betts JC, Lukey PT, Robb LC, McAdam RA, Duncan K. 2002. Evaluation of a nutrient starvation model of *Mycobacterium tuberculosis* persistence by gene and protein expression profiling. *Mol Microbiol* **43:**717–731.

48. Abdel Motaal A, Tews I, Schultz JE, Linder JU. 2006. Fatty acid regulation of adenylyl cyclase Rv2212 from *Mycobacterium tuberculosis* H37Rv. *FEBS J* **273:**4219–4228.

49. Cann MJ, Hammer A, Zhou J, Kanacher T. 2003. A defined subset of adenylyl cyclases is regulated by bicarbonate ion. *J Biol Chem* **278:**35033–35038.

50. Townsend PD, Holliday PM, Fenyk S, Hess KC, Gray MA, Hodgson DR, Cann MJ. 2009. Stimulation of mammalian G-protein-responsive adenylyl cyclases by carbon dioxide. *J Biol Chem* **284:**784–791.

51. Nambi S, Basu N, Visweswariah SS. 2010. cAMP-regulated protein lysine acetylases in mycobacteria. *J Biol Chem* **285:**24313–24323.

52. Bai G, Gazdik MA, Schaak DD, McDonough KA. 2007. The *Mycobacterium bovis* BCG cyclic AMP receptor-like protein is a functional DNA binding protein in vitro and in vivo, but its activity differs from that of its *M. tuberculosis* ortholog, Rv3676. *Infect Immun* **75:**5509–5517.

53. Bai G, McCue LA, McDonough KA. 2005. Characterization of *Mycobacterium tuberculosis* Rv3676 (CRPMt), a cyclic AMP receptor protein-like DNA binding protein. *J Bacteriol* **187:**7795–7804.

54. Reddy MC, Palaninathan SK, Bruning JB, Thurman C, Smith D, Sacchettini JC. 2009. Structural insights into the mechanism of the allosteric transitions of *Mycobacterium tuberculosis* cAMP receptor protein. *J Biol Chem* **284:**36581–36591.

55. Stapleton M, Haq I, Hunt DM, Arnvig KB, Artymiuk PJ, Buxton RS, Green J. 2010. *Mycobacterium tuberculosis* cAMP receptor protein (Rv3676) differs from the *Escherichia coli* paradigm in its cAMP binding and DNA binding properties and transcription activation properties. *J Biol Chem* **285:**7016–7027.

56. Kumar M, Khan FG, Sharma S, Kumar R, Faujdar J, Sharma R, Chauhan DS, Singh R, Magotra SK, Khan IA. 2011. Identification of *Mycobacterium tuberculosis* genes preferentially expressed during human infection. *Microb Pathog* **50:**31–38.

57. Stapleton MR, Smith LJ, Hunt DM, Buxton RS, Green J. 2012. *Mycobacterium tuberculosis* WhiB1 represses transcription of the essential chaperonin GroEL2. *Tuberculosis (Edinb)* **92:**328–332.

58. Gazdik MA, Bai G, Wu Y, McDonough KA. 2009. Rv1675c (*cmr*) regulates intramacrophage and cyclic AMP-induced gene expression in *Mycobacterium tuberculosis*-complex mycobacteria. *Mol Microbiol* **71:**434–448.

59. Pelly S, Bishai WR, Lamichhane G. 2012. A screen for non-coding RNA in *Mycobacterium tuberculosis* reveals a cAMP-responsive RNA that is expressed during infection. *Gene* **500:**85–92.

60. Gazdik MA, McDonough KA. 2005. Identification of cyclic AMP-regulated genes in *Mycobacterium tuberculosis* complex bacteria under low-oxygen conditions. *J Bacteriol* **187:**2681–2692.

61. Rickman L, Scott C, Hunt DM, Hutchinson T, Menendez MC, Whalan R, Hinds J, Colston MJ, Green J, Buxton RS. 2005. A member of the cAMP receptor protein family of transcription regulators in *Mycobacterium tuberculosis* is required for virulence in mice and controls transcription of the rpfA gene coding for a resuscitation promoting factor. *Mol Microbiol* **56:**1274–1286.

62. Gottesman S, Storz G. 2011. Bacterial small RNA regulators: versatile roles and rapidly evolving variations. *Cold Spring Harb Perspect Biol* **3**(12).

63. DiChiara JM, Contreras-Martinez LM, Livny J, Smith D, McDonough KA, Belfort M. 2010. Multiple small RNAs identified in *Mycobacterium bovis* BCG are also expressed in *Mycobacterium tuberculosis* and *Mycobacterium smegmatis*. *Nucleic Acids Res* **38:**4067–4078.

64. Arnvig KB, Young DB. 2009. Identification of small RNAs in *Mycobacterium tuberculosis*. *Mol Microbiol* **73:**397–408.

65. Arnvig KB, Comas I, Thomson NR, Houghton J, Boshoff HI, Croucher NJ, Rose G, Perkins TT, Parkhill J, Dougan G, Young DB. 2011. Sequence-based analysis uncovers an abundance of non-coding RNA in the total transcriptome of *Mycobacterium tuberculosis*. *PLoS Pathog* **7:**e1002342.

66. Lamichhane G, Arnvig KB, McDonough KA. 2013. Definition and annotation of (myco)bacterial non-coding RNA. *Tuberculosis (Edinb)* **93:**26–29.

67. Spreadbury CL, Pallen MJ, Overton T, Behr MA, Mostowy S, Spiro S, Busby SJ, Cole JA. 2005. Point mutations in the DNA- and cNMP-binding domains of the homologue of the cAMP receptor protein (CRP) in *Mycobacterium bovis* BCG: implications for the inactivation of a global regulator and strain attenuation. *Microbiology* **151:**547–556.

68. Hunt DM, Saldanha JW, Brennan JF, Benjamin P, Strom M, Cole JA, Spreadbury CL, Buxton RS. 2008. Single nucleotide polymorphisms that cause structural changes in the cyclic AMP receptor protein transcriptional regulator of the tuberculosis vaccine strain *Mycobacterium bovis* BCG alter global gene expression without attenuating growth. *Infect Immun* **76:**2227–2234.

69. Clarke SJ, Low B, Konigsberg WH. 1973. Close linkage of the genes *serC* (for phosphohydroxy pyruvate transaminase) and *serS* (for seryl-transfer ribonucleic acid synthetase) in *Escherichia coli* K-12. *J Bacteriol* **113:** 1091–1095.

70. Bai G, Schaak DD, Smith EA, McDonough KA. 2011. Dysregulation of serine biosynthesis contributes to the growth defect of a *Mycobacterium tuberculosis crp* mutant. *Mol Microbiol* **82:**180–198.

71. Mukamolova GV, Turapov OA, Young DI, Kaprelyants AS, Kell DB, Young M. 2002. A family of autocrine growth factors in *Mycobacterium tuberculosis*. *Mol Microbiol* **46:**623–635.

72. Akhter Y, Yellaboina S, Farhana A, Ranjan A, Ahmed N, Hasnain SE. 2008. Genome scale portrait of cAMP-receptor protein (CRP) regulons in mycobacteria points to their role in pathogenesis. *Gene* **407**:148–158.

73. Krawczyk J, Kohl TA, Goesmann A, Kalinowski J, Baumbach J. 2009. From *Corynebacterium glutamicum* to *Mycobacterium tuberculosis*: towards transfers of gene regulatory networks and integrated data analyses with MycoRegNet. *Nucleic Acids Res* **37**:e97.

74. Agarwal N, Raghunand TR, Bishai WR. 2006. Regulation of the expression of *whiB1* in *Mycobacterium tuberculosis*: role of cAMP receptor protein. *Microbiology* **152**:2749–2756.

75. Bai G, Knapp GS, McDonough KA. 2011. Cyclic AMP signalling in mycobacteria: redirecting the conversation with a common currency. *Cell Microbiol* **13**:349–358.

76. Gallagher DT, Smith N, Kim SK, Robinson H, Reddy PT. 2009. Profound asymmetry in the structure of the cAMP-free cAMP receptor protein (CRP) from *Mycobacterium tuberculosis*. *J Biol Chem* **284**:8228–8232.

77. Kumar P, Joshi DC, Akif M, Akhter Y, Hasnain SE, Mande SC. 2010. Mapping conformational transitions in cyclic AMP receptor protein: crystal structure and normal-mode analysis of *Mycobacterium tuberculosis* apo-cAMP receptor protein. *Biophys J* **98**:305–314.

78. Aiba H, Nakamura T, Mitani H, Mori H. 1985. Mutations that alter the allosteric nature of cAMP receptor protein of *Escherichia coli*. *EMBO J* **4**:3329–3332.

79. Choudhary C, Kumar C, Gnad F, Nielsen ML, Rehman M, Walther TC, Olsen JV, Mann M. 2009. Lysine acetylation targets protein complexes and co-regulates major cellular functions. *Science* **325**:834–840.

80. Kim SC, Sprung R, Chen Y, Xu Y, Ball H, Pei J, Cheng T, Kho Y, Xiao H, Xiao L, Grishin NV, White M, Yang XJ, Zhao Y. 2006. Substrate and functional diversity of lysine acetylation revealed by a proteomics survey. *Mol Cell* **23**:607–618.

81. Zhao JY, Lu N, Yan Z, Wang N. 2010. SAHA and curcumin combinations co-enhance histone acetylation in human cancer cells but operate antagonistically in exerting cytotoxic effects. *J Asian Nat Prod Res* **12**:335–348.

82. Wang Q, Zhang Y, Yang C, Xiong H, Lin Y, Yao J, Li H, Xie L, Zhao W, Yao Y, Ning ZB, Zeng R, Xiong Y, Guan KL, Zhao S, Zhao GP. 2010. Acetylation of metabolic enzymes coordinates carbon source utilization and metabolic flux. **327**:1004–1007.

83. Thao S, Chen CS, Zhu H, Escalante-Semerena JC. 2010. Nepsilon-lysine acetylation of a bacterial transcription factor inhibits its DNA-binding activity. *PLoS One* **5**:e15123.

84. Xu H, Hegde SS, Blanchard JS. 2011. Reversible acetylation and inactivation of *Mycobacterium tuberculosis* acetyl-CoA synthetase is dependent on cAMP. *Biochemistry* **50**:5883–5892.

85. Lee HJ, Lang PT, Fortune SM, Sassetti CM, Alber T. 2012. Cyclic AMP regulation of protein lysine acetylation in *Mycobacterium tuberculosis*. *Nat Struct Mol Biol* **19**:811–818.

86. Starai VJ, Escalante-Semerena JC. 2004. Acetyl-coenzyme A synthetase (AMP forming). *Cell Mol Life Sci* **61**:2020–2030.

87. Zhao K, Chai X, Marmorstein R. 2004. Structure and substrate binding properties of cobB, a Sir2 homolog protein deacetylase from *Escherichia coli*. *J Mol Biol* **337**:731–741.

88. Castano-Cerezo S, Bernal V, Blanco-Catala J, Iborra JL, Canovas M. 2011. cAMP-CRP co-ordinates the expression of the protein acetylation pathway with central metabolism in *Escherichia coli*. *Mol Microbiol* **82**:1110–1128.

89. Richter W. 2002. 3′,5′ Cyclic nucleotide phosphodiesterases class III: members, structure, and catalytic mechanism. *Proteins* **46**:278–286.

90. Imamura R, Yamanaka K, Ogura T, Hiraga S, Fujita N, Ishihama A, Niki H. 1996. Identification of the *cpdA* gene encoding cyclic 3′,5′-adenosine monophosphate phosphodiesterase in *Escherichia coli*. *J Biol Chem* **271**:25423–25429.

91. Carte J, Wang R, Li H, Terns RM, Terns MP. 2008. Cas6 is an endoribonuclease that generates guide RNAs for invader defense in prokaryotes. *Genes Dev* **22**:3489–3496.

92. Zhang Y, Zhang J, Hara H, Kato I, Inouye M. 2005. Insights into the mRNA cleavage mechanism by MazF, an mRNA interferase. *J Biol Chem* **280**:3143–3150.

93. Zhang Y, Zhu L, Zhang J, Inouye M. 2005. Characterization of ChpBK, an mRNA interferase from *Escherichia coli*. *J Biol Chem* **280**:26080–26088.

94. Barba J, Alvarez AH, Flores-Valdez MA. 2010. Modulation of cAMP metabolism in *Mycobacterium tuberculosis* and its effect on host infection. *Tuberculosis (Edinb)* **90**:208–212.

95. Padh H, Venkitasubramanian TA. 1976. Cyclic adenosine 3′, 5′-monophosphate in mycobacteria. *Indian J Biochem Biophys* **13**:413–414.

96. Ahuja N, Kumar P, Bhatnagar R. 2004. The adenylate cyclase toxins. *Crit Rev Microbiol* **30**:187–196.

97. Krueger KM, Barbieri JT. 1995. The family of bacterial ADP-ribosylating exotoxins. *Clin Microbiol Rev* **8**:34–47.

98. Lory S, Wolfgang M, Lee V, Smith R. 2004. The multitalented bacterial adenylate cyclases. *Int J Med Microbiol* **293**:479–482.

99. Lowrie DB, Aber VR, Jackett PS. 1979. Phagosome-lysosome fusion and cyclic adenosine 3′:5′-monophosphate in macrophages infected with *Mycobacterium microti*, *Mycobacterium bovis* BCG or *Mycobacterium lepraemurium*. *J Gen Microbiol* **110**:431–441.

100. Lowrie DB, Jackett PS, Ratcliffe NA. 1975. Mycobacterium microti may protect itself from intracellular destruction by releasing cyclic AMP into phagosomes. *Nature* **254**:600–602.

101. Roach SK, Lee SB, Schorey JS. 2005. Differential activation of the transcription factor cyclic AMP response element binding protein (CREB) in macrophages following infection with pathogenic and

nonpathogenic mycobacteria and role for CREB in tumor necrosis factor alpha production. *Infect Immun* **73**: 514–522.

102. **Yadav M, Roach SK, Schorey JS.** 2004. Increased mitogen-activated protein kinase activity and TNF-alpha production associated with *Mycobacterium smegmatis*-but not *Mycobacterium avium*-infected macrophages requires prolonged stimulation of the calmodulin/calmodulin kinase and cyclic AMP/protein kinase A pathways. *J Immunol* **172**:5588–5597.

103. **Kalamidas SA, Kuehnel MP, Peyron P, Rybin V, Rauch S, Kotoulas OB, Houslay M, Hemmings BA, Gutierrez MG, Anes E, Griffiths G.** 2006. cAMP synthesis and degradation by phagosomes regulate actin assembly and fusion events: consequences for mycobacteria. *J Cell Sci* **119**:3686–3694.

104. **Ohno H, Zhu G, Mohan VP, Chu D, Kohno S, Jacobs WR Jr, Chan J.** 2003. The effects of reactive nitrogen intermediates on gene expression in *Mycobacterium tuberculosis*. *Cell Microbiol* **5**:637–648.

105. **Schnappinger D, Ehrt S, Voskuil MI, Liu Y, Mangan JA, Monahan IM, Dolganov G, Efron B, Butcher PD, Nathan C, Schoolnik GK.** 2003. Transcriptional adaptation of *Mycobacterium tuberculosis* within macrophages: insights into the phagosomal environment. *J Exp Med* **198**:693–704.

Metabolism

IV

Molecular Genetics of Mycobacteria, 2nd Edition
Edited by Graham F. Hatfull and William R. Jacobs, Jr.
© 2014 American Society for Microbiology, Washington, DC
doi:10.1128/microbiolspec.MGM2-0019-2013

Bridgette M. Cumming,[1] Dirk A. Lamprecht,[1] Ryan M. Wells,[1,2]
Vikram Saini,[2] James H. Mazorodze,[1] and Adrie J. C. Steyn[1,2]

The Physiology and Genetics of Oxidative Stress in Mycobacteria

15

Redox reactions are essential for life and play a role in both aerobic and anaerobic respiration. In aerobic microorganisms, the oxidants and reactive species are equalized by the antioxidants in order to maintain redox balance (1). *Mycobacterium tuberculosis* is an obligate aerobe, although it has been demonstrated that it can survive for more than a decade *in vitro* under anaerobic conditions. In the macrophage and the lung of the host, *M. tuberculosis* is exposed to a range of complex environments which can profoundly influence the physiology, including the redox homeostasis, of the mycobacterium. Thus, it is likely that the mechanisms to maintain redox homeostasis in *M. tuberculosis* are vital in determining disease outcome. As in other bacteria, *M. tuberculosis* has developed pathways that monitor and respond to gaseous signals, such as NO, CO, and O_2, and fluctuations in the intra- and extracellular redox status (2–4). In this article, we will explore the physiology and genetics of redox homeostasis in mycobacteria by considering the *in vivo* environments to which *M. tuberculosis* is exposed, the sensors whereby mycobacteria discern an imbalance in the redox balance both endogenously and in the extracellular environment, mechanisms utilized by mycobacteria to respond to redox stress in order to maintain the intracellular redox balance, and the means currently used to measure the redox state in mycobacteria.

REDOX HOMEOSTASIS

Oxidation is defined as a gain of oxygen, loss of electrons, or loss of hydrogen, whereas reduction is described as a loss of oxygen, gain of electrons, or gain of hydrogen. Redox homeostasis refers to the maintenance of the overall net balance of oxidative and reductive capacity within a biological system such as a single cell, organ, or organism. This equilibrium is crucial to life and provides the necessary redox environment to effectively utilize the reducing energy generated by the catabolism of cellular substrates for anabolism of DNA, lipids, and proteins.

One approach to maintain redox balance within living systems is by making use of redox couples, which function as redox intermediaries (or "redox buffers").

[1]KwaZulu-Natal Research Institute for Tuberculosis and HIV (K-RITH), Durban, South Africa; [2]Department of Microbiology, University of Alabama at Birmingham, AL 35294.

Reduction potential is a thermodynamic property of a redox couple and a measure of the tendency of the oxidized moiety to acquire electrons and therefore be reduced (1). The more positive the reduction potential of a redox pair, the greater the likelihood that the oxidized species will gain electrons and be reduced. Important redox couples for maintaining redox balance in living systems are NAD$^+$/NADH ($E^{O'}$ = −316 mV), NADP$^+$/NADPH ($E^{O'}$ = −315 mV), FAD/FADH$_2$ ($E^{O'}$ = −219 mV), ferredoxin (Fd$_{ox}$/Fd$_{red}$, $E^{O'}$ = −398 mV), and glutathione disulfide (GSSG)/2 glutathione (GSH) (E_{hc} = −250 mV [10 mM]) (1). The intracellular redox potential of a microorganism can theoretically be measured by the summation of the reduction potentials and reducing capacities of all the various linked redox couples (5). The reduction potential of these redox couples will be affected by the pH as well as the concentrations of the species that comprise the redox couple. The contribution of the redox couple (or reducing capacity) will be determined by how large the pool of the redox couple is for the redox buffering system of the cell.

However, Hansen et al. (6) demonstrated that protein thiols constitute a redox active pool in living cells that is just as essential as GSH, which is considered to be the major intracellular redox buffer in most bacteria (albeit not mycobacteria). Thus, determining the redox status of a bacterium would constitute measurement of all redox couples and protein thiols and their corresponding disulfides within the bacterium, which is probably technically impossible, because of the unknown redox couples. For this reason, quantification of a representative redox couple is used to infer changes in the redox environment. For example, the GSSG/2GSH redox couple is often used as an indicator of the status of the bacterial redox environment, except in mycobacteria (5). Methods for measuring the relative redox environment in mycobacteria will be discussed later.

Imbalances in the redox environment in both M. tuberculosis and the host environment may serve as stimuli in the bacillus and trigger mechanisms, which result in persistence, dormancy, or reactivation. Throughout the course of infection, M. tuberculosis is exposed to a range of microenvironments, each with its own unique set of redox conditions. The ability of M. tuberculosis to evade the immune system and cause disease may in large part be due to its capacity, by mostly unknown mechanisms, to sense environmental changes such as host-generated gases, available carbon sources, and pathological conditions such as the hypoxic granuloma, and respond by altering its metabolism and redox

balance. An understanding of redox homeostasis within M. tuberculosis is crucial because it may also help explain antimycobacterial drug efficacy, as isoniazid (INH) (7), ethionamide (ETA) (8), and PA-824 (9) are prodrugs which require bioreductive activation to exert their antimycobacterial effects. Therefore, the efficacy of the regimen may be compromised due to changes in cellular redox status. Also, reductive stress, as measured by an increased NADH/NAD$^+$ ratio within Mycobacterium smegmatis and Mycobacterium bovis, was shown to lead to INH resistance (10). Mutations in M. smegmatis type II NADH-menaquinone oxidoreductase (Ndh-2) increased NADH levels while reducing the frequency of INH-NAD (as well as ETA-NAD) adduct formation, which decreased its binding to InhA (10, 11). Furthermore, mutations in mycothiol (the main redox intermediary of M. tuberculosis) biosynthetic enzymes in M. smegmatis decreased INH and ETA efficacy (12). Thus far, few studies have examined the specific mechanisms of how M. tuberculosis maintains redox homeostasis in vivo and what role these mechanisms may play in drug efficacy. Such studies have the potential of providing insight into M. tuberculosis persistence as well as into the development of novel antimycobacterial intervention strategies.

OXIDATIVE STRESS

Oxidative stress can be defined as an imbalance in the redox environment in favor of prooxidants that is capable of causing damage to biological systems (1). From the perspective of M. tuberculosis inside the human host, endogenous sources of oxidative stress include natural byproducts of aerobic respiration and the electron transport chain (ETC) where O$_2$ is consumed as the terminal electron acceptor. Exogenous factors inducing oxidative stress in M. tuberculosis comprise the immune response of the host in addition to factors generating oxidative stress in the host, such as cigarette smoke and indoor air pollution.

By far the largest source of oxidative stress to M. tuberculosis inside the host is caused by the respiratory burst of phagocytic cells such as resident alveolar macrophages and recruited polymorphonuclear neutrophils (PMNs) that phagocytose M. tuberculosis upon inhalation into the lungs. NADPH oxidase (NOX2) of phagocytic cells was the first enzyme discovered with the primary function of generating reactive oxygen species (ROS) directly rather than as an unwanted byproduct (13). Once activated, NOX2 transports electrons from cytoplasmic NADPH in the phagocytic cells across the phagosomal membrane to oxygen to generate superoxide ion (equation 1):

$$2O_2 + NADPH \rightarrow O_2^{\bullet -} + NADP^+ + H^+ \quad (1)$$

Its activity is tightly controlled and dependent on the formation of a multiprotein complex composed of core and regulatory components. NOX2 is essentially a transmembrane electron bucket brigade that links the cytoplasmic electron donor, NADPH, with oxygen as the electron acceptor in the phagosomal lumen. Two superoxide ions can react in a dismutation reaction catalyzed by superoxide dismutase (SOD), resulting in production of hydrogen peroxide (H_2O_2).

The activity of NOX2 leads to the formation of ROS, which is a collective term that refers to O_2 radicals such as $O_2^{\bullet -}$, $HO_2^{\bullet}$, $HO^{\bullet}$, $RO_2^{\bullet}$, $RO^{\bullet \bullet}$, and $CO_2^{\bullet -}$ and nonradical derivatives such as H_2O_2, $ONOO^-$, $ONOOH$, $ONOOCO_2^-$, $HOCl$, $HOBr$, and O_3 (1). Reactive nitrogen species (RNS) are formed when $O_2^{\bullet -}$ reacts with nitric oxide (NO). Upon infection of macrophages with *M. tuberculosis*, inducible NO synthase (iNOS/NOS2) is expressed (14, 15), which leads to the generation of NO from L-arginine (equation 2).

$$L\text{-arginine} + NADPH + H^+ + O_2$$
$$\rightarrow L\text{-citrulline} + NADP^+ + H_2O + NO^{\bullet} \quad (2)$$

This NO then reacts with $O_2^{\bullet -}$ generated by NOX2 and results in the formation of free radicals such as $NO^{\bullet}$, $NO_2^{\bullet}$, and $NO_3^{\bullet}$ and nonradicals such as HNO_2, NO^+, NO^-, N_2O_4, N_2O_3, NO_2^+, $ROONO$, and RO_2ONO. The production of ROS/RNS by the host serves as the main microbicidal strategy for phagocytized bacteria; however, ROS/RNS can also be detrimental to the host by damaging DNA or peroxidation of lipids and proteins (1). In immunocompetent individuals the generation of ROS/RNS by innate immune cells is relatively efficient at containing and killing *M. tuberculosis* as exemplified by the fact that only 5 to 10% of those newly infected with *M. tuberculosis* will go on to develop tuberculosis (TB) in their lifetime (16).

REDOX ENVIRONMENT OF THE LUNG

The lung is the single organ of the human body that is solely dedicated to the uptake of atmospheric O_2 and the elimination of volatile metabolic waste products. Given this role and the lung's delicate architecture, which confers a vast surface area and blood supply, it is understandable that it is particularly susceptible to oxidative damage. Thus, redox homeostasis in the lung

is of pivotal importance. An average healthy adult breathing at a typical resting rate of 16 breaths/min inhales and exhales approximately 11,500 liters of air each day. Because of anatomical dead space within the lung, two-thirds of this volume (7,667 liters) takes part in gaseous exchange at the alveolar level (17). The lung is exposed to the highest O_2 concentrations in the body, but it is able to respond to oxidizing conditions that would threaten its structural and functional integrity and thereby maintain its redox state. It is now widely appreciated that oxidative stress plays an important role in the pathogenesis of various lung disorders including COPD, asthma, acute respiratory distress syndrome, pulmonary fibrosis, and even lung cancers (18, 19).

Antioxidants of the lung are crucial in maintaining redox balance and include GSH, ascorbate, β-carotene, albumin-SH, uric acid, superoxide dismutases, catalases, and peroxidases (20). GSH is the most abundant antioxidant in the lung, with concentrations in the extracellular lining fluid reaching over 400 μM and cellular cytoplasmic concentrations of 5 mM. The GSH:GSSG ratio in lungs of healthy adults is 30:1 (−210 to −240 mV), which represents a large antioxidant buffering capacity (21). Surprisingly, the most common risk factors for TB in decreasing order of population attributable risk—namely, malnutrition, indoor air pollution, smoking, heavy alcohol use, HIV infection, and diabetes—all induce decreased GSH levels in the lungs (18, 19, 21–25). These disturbances in the redox state of the host lung may act as triggers for reactivation of dormant *M. tuberculosis*.

In adults, the lungs have a total alveolar surface area of about 70 m^2, which is roughly the size of half of a tennis court. This large area is not homogenous and contains distinct geographical differences in ventilation and perfusion, resulting in regional blood oxygenation differences (17). Using resected lung tissue, it was shown that samples from lesions classified as "open" (oxygen rich) yielded actively growing *M. tuberculosis* that were predominantly drug resistant. Bacterial colonies that were "resurrected" from "closed" (oxygen-poor) lesions exhibited delayed growth yet were drug sensitive despite being refractory to antituberculosis treatment within the lung (26). These results indicated that *M. tuberculosis* drug resistance could be due in part to the variations in regional O_2 levels within the lung. The Wayne model of hypoxia gives further support for the role of variable O_2 levels in drug resistance by demonstrating that anaerobically exposed *M. tuberculosis* is a poor target for antimycobacterial drugs (27).

The ventilation perfusion ratio in the upper lung is greater than in the lower and ventral lung, resulting in higher O_2 tensions in the upper lung. It is widely believed that the lung apices are the preferred niche of *M. tuberculosis* replication as an obligate aerobe because of this relative excess of O_2 (28). However, in response to infection, the human immune system is able to wall off *M. tuberculosis* into granulomas that are most likely hypoxic and perhaps even anaerobic (29). Redox-active dyes that are reduced at pO_2 lower than 10 mmHg have shown that similar granulomas in monkeys and guinea pigs are indeed hypoxic (30). Clearly, more research is needed to elucidate the mechanisms employed by the "obligate aerobe" *M. tuberculosis* which enable its survival in such hypoxic conditions within the human host.

REDOX SENSORS OF *M. TUBERCULOSIS*

In order for *M. tuberculosis* to thrive as a pathogen it must be able to sense and adapt to the ever-changing redox stress that it experiences during the course of infection from its initial exposure to the lung alveoli, followed by the destructive effects of the phagocytes' respiratory burst, to the hypoxic environment of the granuloma. Furthermore, *M. tuberculosis* is exposed to anything a person inhales and thus possesses redox sensor systems that allow it to successfully navigate this dynamic redox landscape.

The DosR/S/T Dormancy Regulon

The DosR/S/T dormancy regulon is a "three-component" system that integrates two heme histidine kinase sensors, DosS and DosT, with a single response regulator, DosR. Briefly, DosS and DosT sense physiologically relevant gases by binding to O_2, NO, and CO, allowing *M. tuberculosis* to sense its extracellular redox environment and relay these signals through DosR to coordinate the expression of crucial genes in the Dos regulon involved in the metabolic shift of *M. tuberculosis* to the persistent state (31–36).

WhiB family

Another known redox sensor system of *M. tuberculosis* includes the intracellular WhiB Fe-S cluster family of proteins, particularly WhiB3 (37, 38). *M. tuberculosis* contains seven WhiB proteins (WhiB1 to 7). WhiB-like (Wbl) proteins are relatively small proteins (75 to 130 amino acids) found only in actinomycetes, and they contain a helix-turn-helix motif and four conserved cysteine residues that coordinate binding of the Fe-S cluster (39). Wbl proteins have also been shown to play

a role in pathogenesis and cell division in mycobacteria (38, 40), in oxidative stress in *Corynebacterium glutamicum* (41), and in antibiotic resistance in mycobacteria and *Streptomyces* (42). Experimental studies illustrating the mechanistic basis for how these Wbl proteins sense and respond to signals to exert their effects are lacking. The expression profiles of all seven *M. tuberculosis* *whiB* genes after exposure to antibiotic and *in vitro* stress conditions provided insights into the biological function of the *M. tuberculosis* WhiB family (43).

Using a two-hybrid screen, *M. tuberculosis* WhiB3 was shown to interact with RpoV (SigA, Rv2703) and play a role in virulence in mice and guinea pigs (40). Interestingly, in this study, mice infected with *M. tuberculosis* Δ*whiB3* showed a significant increase in survival compared to wild-type infected mice, even though *M. tuberculosis* Δ*whiB3* did not have an *in vivo* growth delay and animals (mice and guinea pigs) infected with *M. tuberculosis* Δ*whiB3* had identical bacterial organ burdens when compared with those infected with wild-type *M. tuberculosis*. Later, it was demonstrated that WhiB3 plays a role in maintaining intracellular redox homeostasis through its 4Fe-4S cluster by regulating catabolic metabolism and polyketide biosynthesis in *M. tuberculosis* (38, 44). The 4Fe-4S cluster of WhiB3 is crucial to its role as a redox sensor and has been shown to directly associate with NO and to be degraded by O_2 (38). It was further demonstrated that *M. tuberculosis* utilizes WhiB3 to link sensed changes in the intracellular redox state to a metabolic switchover to preferred *in vivo* carbon sources, namely fatty acids, which are likely to be in abundance inside host macrophages and are widely accepted as a major source of carbon and energy during chronic infection. This WhiB3-induced metabolic shift in *M. tuberculosis* differentially modulates the synthesis and incorporation of propionate into the complex virulence polyketides, polyacyltrehaloses, sulfolipids, phthiocerol dimycocerosates, and the storage lipid, triacylglycerol (TAG), under defined redox conditions (44). TAG production, which is under conditional WhiB3 control, is also induced on exposure to NO, CO, and hypoxia through the Dos dormancy system. Thus, TAG synthesis establishes a link between intracellular (WhiB3) and extracellular (Dos) redox-dependent signaling pathways (45–47).

WhiB4 is has been implicated in the pathophysiology of *M. tuberculosis*. The transcription of WhiB4 is induced by nutrient starvation (48), long-term hypoxia (49), oxidants such as cumene hydroperoxide and diamide, heat stress (30), and infection of macrophages

(50). However, *whiB4* expression is reduced after treatment with 2.5% ethanol, which denatures proteins and disrupts the membrane, and SDS, which resembles the surfactants that *M. tuberculosis* might encounter early in infection (30). WhiB4 was found to respond to O_2 and NO via its 4Fe-4S cluster, thereby functioning as a sensor of redox signals. Upon oxidation, the cysteine thiols in WhiB4 were oxidized to disulfide bonds that induced oligomerization, thereby enabling WhiB4 to bind to GC-rich sequences in DNA and repress transcription of WhiB4 while inducing the expression of antioxidant genes that were hyperinduced in *M. tuberculosis* Δ*whiB4*. Hence, *M. tuberculosis* Δ*whiB4* demonstrated increased resistance to oxidative stress and enhanced survival in macrophages in comparison to the wild type (51). Upon infection of guinea pigs, *M. tuberculosis* Δ*whiB4* showed hypervirulence in the lungs but was defective in disseminating to the spleen. Thus, WhiB4 appears to act as a redox sensor that activates a response to oxidative stress in *M. tuberculosis* by modulating growth, redox balance, and virulence in response to the enhanced antimycobacterial activity in macrophages (51).

WhiB7 contributes to intrinsic drug resistance in *M. tuberculosis* by activating its own expression and that of many drug-resistance genes in response to antibiotics (42). Transcription of WhiB7 is also activated by a diversity of stress conditions, such as nutrient starvation, entry into stationary phase, and heat shock (30). Furthermore, in clinical isolates of *M. tuberculosis*, WhiB7 is highly induced soon after the infection of resting and activated murine macrophages, which may be due to the reductive stress from the accumulation of NADH/NADPH generated by the catabolism of host fatty acids in macrophages (44, 52). It was found that combining reducing conditions with low concentrations of erythromycin synergistically increased WhiB7 transcription in *M. tuberculosis* (53). This provided further support that activation of WhiB7 is linked to cell metabolism and activated by a reducing environment.

Anti-σ factors

Another type of oxidative stress experienced by mycobacteria is termed disulfide stress (54) and results from the unwanted formation of disulfide bonds. The cytoplasm of mycobacteria is a highly reduced environment, which maintains protein cysteines in their reduced thiol forms. Disulfide stress sensors exist in *M. tuberculosis* in the form of anti-σ factors. SigH and SigE are extracytoplasmic σ factors that are important for *M. tuberculosis* survival under conditions of oxidative stress (55, 56). Although SigL does not appear to

be required for adaptation to oxidative stress, it has been shown to be regulated by redox conditions and required for full virulence of *M. tuberculosis* in mice (57, 58). All three of these σ factors are cotranscribed with genes that code for anti-σ factors, which specifically bind to and inhibit the σ factor–dependent transcription (58–60). These anti-σ factors contain an HXXXCXXC motif called the ZAS (zinc-associated anti-σ factors) motif. The thiols of the cysteines in the conserved ZAS motif bind Zn^{2+} under reducing conditions. Under conditions of oxidative stress, the thiols act as a redox switch by forming a disulfide bond, thereby releasing the zinc ion (61, 62). This leads to a conformational change in the anti-σ factors and decreases their affinity for their respective Sig proteins, resulting in transcriptional activation.

MosR

Recently, a redox-dependent transcriptional repressor of *M. tuberculosis* belonging to the MarR family, termed MosR (Rv1049, *M. tuberculosis* oxidation-sensing regulator), was described (63). In its reduced form, MosR binds DNA and represses transcription; however, upon oxidation a disulfide bond is formed between two cysteines in the N-terminus of the protein, causing a conformational change and its dissociation from DNA. MosR was previously shown to play a role in survival of *M. tuberculosis* in macrophages (64). The DNA binding sequence of the regulator was identified and found in the promoter regions of *narX*, *ndhA*, and *rv1050*, a secreted putative oxidoreductase (63). It's likely that other similar disulfide stress sensors remain to be discovered in *M. tuberculosis*.

M. TUBERCULOSIS DEFENSE STRATEGIES AGAINST ROS AND RNS

To survive the robust oxidative response inside macrophages, *M. tuberculosis* must withstand ROS/RNS. The thick cell envelope, consisting of large amounts of the oxygen radical scavengers lipoarabinomannan and phenolic glycolipid I, confers a degree of intrinsic resistance to ROS (65, 66). Similarly, sulfatides have been shown to interfere with the oxidative killing mechanisms of macrophages (67).

M. tuberculosis may also be capable of limiting its exposure to NO within the host and thus increasing its chance of survival. Activated macrophages containing live *M. tuberculosis* showed reduced iNOS recruitment to phagosomes compared to activated macrophages containing dead bacteria or latex beads (68, 69). *M. tuberculosis* is also able to induce host expression of arginase-1 (Arg-1), an enzyme that competes with iNOS for its substrate, L-arginine, leading to an overall decrease of NO production (70, 71). Virulence studies in mice with Arg-1-deficient macrophages produced higher NO levels and resulted in decreased lung bacterial loads of *M. tuberculosis* (70). It was also recently shown that Arg-1 is expressed in granuloma-associated macrophages and type II pneumocytes in human lung tissues from patients with TB (72). The Arg-1-induced expression in human lungs of patients with TB may be a defensive strategy of *M. tuberculosis* to avoid lethal NO of the immune response.

Superoxide Dismutases

SODs are metalloproteins that catalyze the dismutation reaction of $O_2^{\bullet-}$ into H_2O_2 and molecular O_2. *M. tuberculosis* contains two SODs, SodA and SodC, which belong to Fe-SOD and the CuZn-SOD families, respectively (73, 74). SodA is constitutively expressed in *M. tuberculosis*, indicating a role in detoxifying endogenous ROS, but its expression is increased under oxidative stress conditions (75) and nutrient starvation (48). SodA is efficiently secreted by the SecA2 secretion machinery into the culture filtrate (76), and it has been associated with virulence because *M. tuberculosis* secretes ~350-fold more enzyme than *M. smegmatis* (75). SodC, on the other hand, contains a putative signal peptide, and its homologues in other pathogenic bacteria have been found either in the periplasmic space or anchored to the membrane (77). In *M. tuberculosis*, using immunogold labeling, SodC has been shown to be located in the periphery of the bacilli (78). SodC also promotes survival of *M. tuberculosis* in activated macrophages (79).

Catalase-Peroxidase

Catalase-peroxidase (Kat) enzymes are hemoproteins that efficiently detoxify H_2O_2 by converting it into water and oxygen (catalase function) and catalyze the conversion of alkyl peroxides (ROOR′) into their respective alcohols (peroxidase function) (80). Both H_2O_2 and alkyl peroxides can be formed from the downstream effects of $O_2^{\bullet-}$. The *M. tuberculosis* genome encodes one catalase-peroxidase, KatG, which has catalase, peroxidase, and peroxynitritase activity (81, 82) and is a well-established virulence factor (83). INH is a prodrug that is bioreductively activated by KatG. High-level resistance to INH (~200× MIC) results from mutations in *katG*, often S315T, in approximately 50% of INH-resistant clinical isolates (84, 85). Furthermore, clinical isolates that were exposed to INH have higher levels of alkyl hydroperoxidases (AhpC) (86).

Alkyl Hydroperoxidase

Peroxides, if left unchecked, could react with cellular components such as lipids and lead to the generation of highly reactive alkyl hydroperoxides (ROOH). AhpC is a non-heme peroxiredoxin containing three active cysteines that converts these reactive organic peroxides into their respective alcohol derivatives (87). AhpC has also been demonstrated to confer protection against RNS by reducing peroxynitrite to nitrite (88). Peroxiredoxin family reductases typically use reduced thioredoxin (Trx) to restore the catalytic activity of the reductase; however, *M. tuberculosis* Trxs are incapable of reducing AhpC (89). Rather, *M. tuberculosis* AhpC forms a complex with dihydrolipoamide dehydrogenase (Ldp) and dihydrolipoamide succinyltransferase (SucB) via an adapter protein, AhpD, that reduces AhpC (90). AhpD has a thioredoxin-like active site that is sensitive to lipoamide. Thus, it has been proposed that Lpd, SucB, AhpC, and AhpD form a NADH-dependent peroxidase and peroxynitrate reductase complex that strategically links the peroxidase activity of AhpC to the metabolic state of *M. tuberculosis*. Four other peroxiredoxins exist in *M. tuberculosis*, namely, AhpE, TPx, Bcp, and BcpB, and they utilize the traditional Trx to regenerate their reduced forms (91, 92).

Thioredoxins

Thioredoxins (Trxs) are ubiquitous well-characterized low-molecular-weight (LMW) proteins that catalyze thiol-disulfide exchange reactions (93). The redox active cysteines of Trxs are found in the conserved catalytic motif, WCXXC, and allow the proteins to function as disulfide reductases by cycling between a reduced dithiol form and an oxidized disulfide bond form (94). The oxidized disulfide bond form of Trxs is reduced by the FAD-containing enzyme thioredoxin reductase (TrxR), which utilizes NADPH as an electron donor (95). Trxs are important for maintaining a reducing cytoplasmic environment as well as supplying the source of reducing power for deoxyribonucleotide synthesis, sulfur assimilation, detoxification of ROS/RNS, protein repair, and redox regulation (96).

M. tuberculosis contains three Trxs (TrxA, TrxB, and TrxC, with midpoint redox potentials of −248, −262, and −269 mV, respectively) that have all been shown to possess disulfide reductase activity using the insulin reduction assay, and one TrxR that was shown to reduce TrxB and TrxC only (97). The apparent redundancy in the *M. tuberculosis* Trxs suggests their importance in ensuring survival under oxidative conditions such as those that are encountered in the phagosomal environment (97). The expression of the genes that code for the Trxs and TrxR are under the control of SigH (59). As discussed previously, under oxidative stress a disulfide bond forms in the anti-σ factor, RshA, which then dissociates from SigH, allowing expression of the *trx* and *trxR* genes along with other components of the SigH regulon.

Truncated Hemoglobins

Truncated hemoglobins (trHbs) are small heme-binding globin proteins that are widely found among bacteria, protozoa, and plants (98). Based on the variability in the heme-binding pocket residues, trHbs have been divided into three groups, which share less than 30% sequence similarity with each other (98). Putative physiological functions of trHbs include detoxification of ROS/RNS. *M. tuberculosis* has two trHbs called trHbN and trHbO that belong to group I and group II trHbs, respectively. *M. tuberculosis* trHbO reacts with H_2O_2 and NO, suggesting a role in their detoxification (99), while trHbN has potent NO oxidizing activity (100). The group II trHbs from *Bacillus subtilis* and *Thermobifida fusca*, which are very similar to *M. tuberculosis* trHbO, form high-affinity complexes with H_2S (101). This may have physiological significance in thiol redox homeostasis in *M. tuberculosis*, which is known to produce H_2S during cysteine metabolism (102), but this remains to be investigated.

Methionine Sulfoxide Reductases

The atom within proteins that is most susceptible to oxidation is sulfur, in both cysteine and methionine residues (103). Exposure to ROS and RNS, such as *S*-nitrosothiols and nitrite, can lead to disulfide bonding (104) and methionine oxidation, probably through the formation of peroxynitrite (105). The oxidation of methionine, whether in proteins or as a free amino acid, leads to two stereomeric forms of methionine sulfoxide: methionine *S*-sulfoxide and methionine *R*-sulfoxide. Most prokaryotes and eukaryotes express a family of methionine sulfoxide reductases (Msr) that reduce methionine sulfoxide to methionine using thioredoxin, thioredoxin reductase, and NADPH as the reducing system (106). Many organisms express at least one MsrA that acts on both peptidyl and free methionine *S*-sulfoxide and one or more MsrB isoforms that reduce peptidyl (but not free) methionine *R*-sulfoxide. *M. tuberculosis* expresses both MsrA (Rv0137c) (105) and MsrB (Rv2674) (107), each one stereospecific for one of the epimers of methionine sulfoxide. In contrast to other organisms, single Msr mutants in *M. tuberculosis* were no more susceptible to ROI or RNI than the wild type. Only an *M. tuberculosis* mutant deficient in both MsrA and MsrB was susceptible to hypochlorite and nitrite (107). However, in contrast to the MsrA-deficient *M. smegmatis* (108), the dual-deficient mutant was not vulnerable to hydrogen peroxide and organic peroxides. This is possibly due to adaptations of the mycobacteria to the different oxidative environments in which the organisms reside. *M. tuberculosis* lives and persists in macrophage phagosomes, which have highly oxidative and nitrosative environments (109).

THIOLS AS REDOX BUFFERS AND CELLULAR DETOXIFIERS

Most organisms use an LMW thiol/thiol disulfide reductase/NAD(P)H redox pathway to keep the cytoplasm in a reduced state (Fig. 1A) (110–116). In this pathway, the LMW thiol (RSH) provides reducing equivalents (electrons) to cellular oxidants, reducing them and thereby rendering these oxidants unable to cause cellular damage. In doing this, the LMW thiol is oxidized to its corresponding disulfide (RSSR), which in turn, is cycled back to its reduced form by the NADPH-dependent flavoprotein disulfide reductase associated with the specific RSH/RSSR pair (Fig. 1B). Apart from the reduction of oxidizing agents, RSHs are also capable of detoxifying (mopping up) the cell from a wide range of oxidants and toxins, including antibiotics, alkylating agents, and nitrogen species. These detoxifying pathways can either proceed via a direct reaction between the RSH and oxidant/toxin or be enzyme mediated through an RSH-specific *S*-transferase (117–123). Although these RSHs are homologous in function, they are structurally diverse among the organisms in which they are produced and function in, as is evident from Fig. 1C.

Mycothiol

Mycothiol [1-D-*myo*-inosityl-2-(*N*-acetyl-L-cysteinyl)-amido-2-deoxy-α-D-glucopyranoside; MSH] is only produced in mycobacteria and other actinomycetes and serves as the major LMW thiol in most of these organisms (124). MSH was first detected as an unknown thiol during the HLPC analysis of GSH-deficient *Streptomyces* cell extracts (125). Its structure, which is unique when compared to the other LMW thiols, was elucidated shortly thereafter and consists of a central D-glucosamine moiety linked to *myo*-inositol through an α(1→1) glycoside bond to form a pseudo-disaccharide. *N*-Acetylcysteine is bound to the D-glucosamine moiety through an amide bond to provide the biologically active thiol to the MSH structure

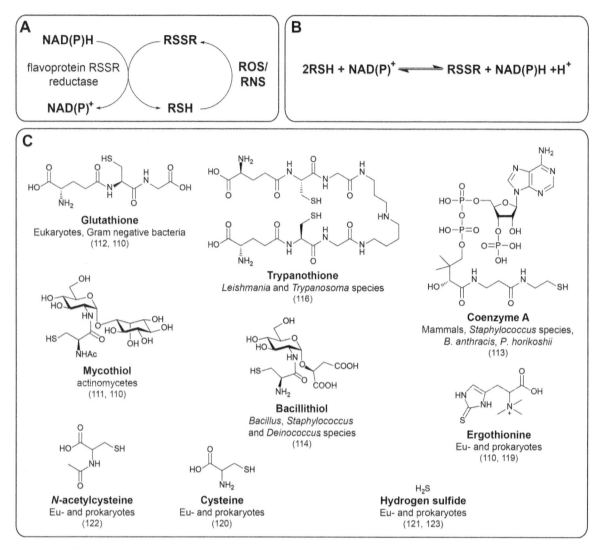

Figure 1   (**A**) The RSH/RSSR reductase/NAD(P)H redox pathway for the reduction of cellular oxidants. (**B**) The oxidation-reduction reaction of a typical flavoprotein disulfide reductase. (**C**) The commonly found intracellular low-molecular-weight thiols and the organisms in which they are found. doi:10.1128/microbiolspec.MGM2-0019-2013.f1

(126–128). For these reasons MSH, its related enzymes, and *in vivo* function have been the focus of numerous studies to determine, in most cases, if any of the MSH-related processes could provide suitable drug targets for new antitubercular drugs. In this regard, credit should be given to, among others, the Newton and Fahey, Steenkamp, Av-Gay, Jacobs, and Blanchard laboratories for their ground-breaking work in this area (124, 125, 127, 128, 131–134, 136–141, 146, 149–153, 158).

Biosynthesis of mycothiol

MSH is synthesized by a pathway involving five enzymes. The biosynthesis of MSH is illustrated in Fig. 2A. Inositol-1-phosphate synthase (tbINO) catalyzes the

formation of the MSH precursor, L-*myo*-inositol-1-phoshate (I-1-P), from glucose-6-phosphate (Glc-6-P). Although tbINO is not recognized as one of the MSH biosynthetic enzymes, it plays a vital role in the production of MSH and cell wall biosynthesis (129, 130).

In MSH biosynthesis, the glycosyltransferase MshA is responsible for the synthesis of 1-(D-*myo*-inosityl-3-phosphate)-2-acetamido-2-deoxy-α-D-glucopyranoside (GlcNAc-Ins-3P) by transferring N-acetylglucosamine (GlcNAc) from UDP-GlcNAc (131, 132). In the second step, MshA2, an as yet unidentified enzyme, catalyzes the dephosphorylation of GlcNAc-Ins-3-P to produce 1-D-*myo*-inosityl-2-acetamido-2-deoxy-α-D-glucopyranoside (GlcNAc-Ins) (132). GlcNAc-Ins is then deacetylized

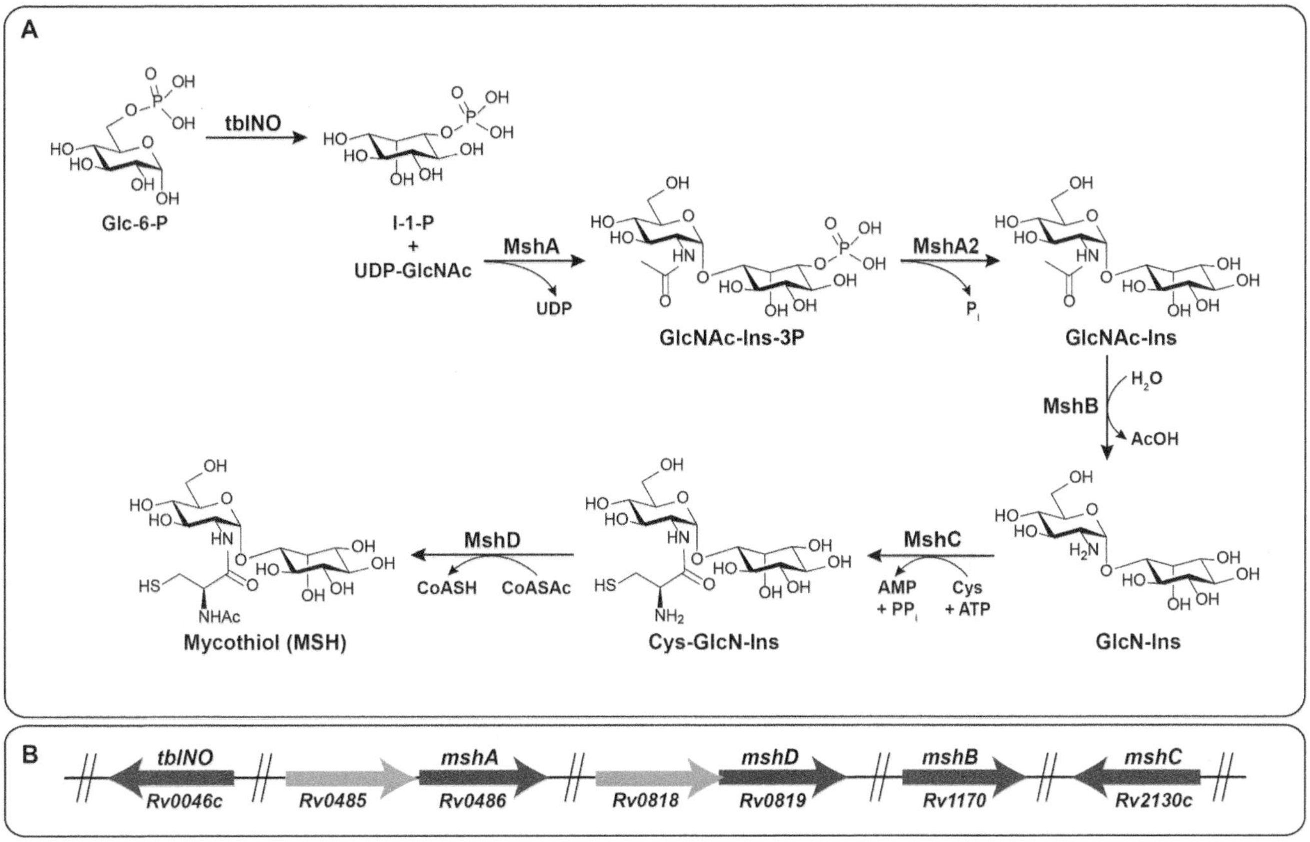

Figure 2 (A) The biosynthesis pathway of mycothiol (MSH), with the related biosynthesis of I-1-P by tbINO. The enzyme associated with MshA2 activity is yet to be identified. (B) The genomic organization of the MSH biosynthetic genes in *M. tuberculosis* H37Rv. See text for definition of abbreviations. doi:10.1128/microbiolspec.MGM2-0019-2013.f2

by the metal-dependent *N*-acetyl hydrolase, MshB, to produce 1-D-*myo*-inositol-2-amino-2-deoxy-α-D-glucopyranoside (GlcN-Ins) as the third intermediate (133–135). The ATP-dependent cysteine ligase, MshC, catalyzes the ligation of L-cysteine to GlcN-Ins through the formation of an amide bond to produce 1-D-*myo*-inositol-2-(L-cysteinyl)-amido-2-deoxy-α-D-glucopyranoside (Cys-GlcN-Ins) in the penultimate step (136, 137). In the final step, the acetyltransferase, MshD, transfers an acetyl group from the acetyl-CoA to Cys-GlcN-Ins, yielding MSH as the final product (138, 139).

Although different homologues of the four identified biosynthetic enzymes (MshA, MshB, MshC, and MshD) have been characterized (134, 137, 140, 141), MshA2 and its associated gene are still unidentified. The undiscovered phosphatase, MshA2, is crucial for the pathway to proceed to its final product (132). Numerous putative inositol monophosphatases have been investigated as possible candidates for MshA2 activity without any success (142–145). Thus, it has been pro-

posed that this enzyme activity might be shared by more than one enzyme (140).

Gene regulation and knockout studies of the MSH biosynthetic enzymes

Regulation of the expression of the MSH biosynthetic enzymes in mycobacteria is unknown (146). In *Streptomyces coelicolor*, the production of MSH is under the control of a σ factor, $σ^R$, which is regulated by the anti-σ factor RsrA (147). RsrA has a thiol/disulfide redox switch, which when oxidized, releases $σ^R$ for *mshA* and *mca* (encoding for mycothiol-*S*-conjugate amidase [Mca]) gene activation. MSH is able to modulate the system: A decrease in intracellular MSH initiates $σ^R$ activation through the oxidation of RsrA. When MSH levels are restored, the oxidized RsrA is reduced by MSH and $σ^R$ is inactivated through RsrA binding. However, in mycobacteria, the $σ^R$ homologue does not regulate the MSH biosynthetic genes. This is indicative of a different regulation system in mycobacteria than that present in *S. coelicolor*. The *mshA* and *mshD* genes in *M. tuberculosis* are both directly downstream of putative transcription factors, *Rv0485* and *Rv0818*, respectively (Fig. 2B), which may be involved in MSH biosynthesis regulation (148).

Numerous studies have investigated the effects of the knockouts of the MSH biosynthetic enzymes on the survival of different mycobacteria. Table 1 gives a brief overview of the major observations made during these studies. In *M. smegmatis*, chemically induced (149), transposon (131), and single deletion mutants (150) of *mshA* proved to be nonessential for growth under normal laboratory conditions, although the mutants were more susceptible to some antibiotics and oxidative stress. Less than 0.4% of the total MSH detected in the wild-type strains was detected in these mutants. Mutations in *M. smegmatis mshB* had the least effect on MSH production, with the *M. smegmatis mshB* mutant able to produce up to 25% of the total MSH produced in the parent strain (150, 151). A double *mshB/mca M. smegmatis* mutant showed a further decrease (<0.4% of the parent strain) in MSH levels, demonstrating that Mca can compensate for MshB activity in the *mshB* mutants (150). The *mshB* mutants were only slightly more susceptible to oxidative stress and the antibiotics tested than the wild type. The *M. smegmatis mshC* mutant was more susceptible to oxidizing agents and antibiotics than the *M. smegmatis*

mshD mutant. Both of these mutants had reduced MSH levels—less than 5% and less than 0.4% of the parent strain MSH levels for the *mshC* and *mshD* mutants, respectively (150, 152, 153). These studies also uncovered a link between MSH levels and *M. smegmatis* resistance to the prodrugs, isoniazid and ethionamide. Briefly, the lower the MSH levels in the mutant, the more resistant the mutant was to isoniazid and ethionamide relative to the parent strain. This suggests that MSH might be involved in the *in vivo* activation of these prodrugs (150, 154–156).

In comparison with the studies in *M. smegmatis* knockout mutants, studies with the *M. tuberculosis* mycothiol biosynthetic enzyme knockout mutants displayed surprisingly different results. An *M. tuberculosis* Erdman *mshA* knockout mutant proved to be lethal, indicating the essential cellular function of MSH in *M. tuberculosis* (157). An *mshA* mutant of *M. tuberculosis* H37Rv (158), also deficient in MSH, was able to grow *in vitro* in catalase-supplemented media and *in vivo* in immunocompromised and immunocompetent mice. This may be attributed to small but critical genetic differences between the two strains. The *M. tuberculosis* Erdman *mshB* mutant produced between 20 and 25% of MSH present in wild-type *M. tuberculosis* Erdman, and although it had increased sensitivity to rifampin, the mutant displayed otherwise normal growth under laboratory conditions (159). Disruption

Table 1   Mycothiol biosynthetic enzyme knockouts

Gene	Organism	MSH production (% of normal production)	Phenotype[a]	Reference(s)
mshA	*M. smegmatis* mc^{2}155	<0.1	RIF^S, ERM^S, ≈VA, INH^R, ETA^R, H$_2$O$_2$^{S}	131, 156
	M. smegmatis mc^{2}155	<0.4	INH^R, ETA^R	150
	M. bovis Ravenel	<0.9	INH^R, ETA^R	163
	M. tuberculosis Erdman	ND^b	Lethal	157
	M. tuberculosis H37Rv	≈0	Nonlethal, INH^R, ETA^R	158
mshB	*M. smegmatis* mc^{2}155	5 to 10	≈RIF, ERM^S, VA^S, ≈INH, ETA^R, H$_2$O$_2$^{S}	151, 156
	M. smegmatis mc^{2}155	<25	≈INH, ETA^R	150
	M. tuberculosis Erdman	<21	RIF^S, INH^R	159
mshC	*M. smegmatis* mc^{2}155	1 to 5	RIF^S, ERM^S, VA^S, INH^R, ETA^R, H$_2$O$_2$^{S}	152, 156
	M. smegmatis mc^{2}155	<0.4	INH^R, ETA^R	150
	M. bovis Ravenel	<7	INH^R, ETA^R	163
	M. tuberculosis Erdman	≈0	Lethal	160
mshD	*M. smegmatis* mc^{2}155	<1	≈RIF, ERM^S, VA^S, ≈INH, ≈ETA, H$_2$O$_2$^{S}	153, 156
	M. smegmatis mc^{2}155	<0.4	≈INH, ≈ETA	150
	M. tuberculosis Erdman	<2	H$_2$O$_2$^{S}	161
	M. tuberculosis H37Rv	ND	Limited survival in macrophages	162
mshB/mca	*M. smegmatis* mc^{2}155	<0.4	INH^R, ETA^R	150

[a]S, more susceptible; R, more resistant; ≈, similar to wild type; RIF, rifampin; ERM, erythromycin; VA, vancomycin; INH, isoniazid; ETA, ethionamide; H$_2$O$_2$, hydrogen peroxide.
[b]Not determined.

of *mshC* in *M. tuberculosis* Erdman was also lethal (160). The *M. tuberculosis* Erdman *mshD* mutant produced only 1% of the MSH present in the wild type and showed a heightened sensitivity to hydrogen peroxide (161). This mutant also produced the novel thiol *N*-formyl-Cys-GlcN-Ins, which behaved as a weak substitute for MSH but was unable to support normal growth of the mutant under stressful conditions. The *mshD* mutant also demonstrated poor survival inside macrophages (161, 162).

Mycothiol function and related enzymes

As stated previously, the major *in vivo* MSH function is to protect the cell against oxidative stress and toxins. When reacting with ROS, MSH is oxidized to mycothiol disulfide (mycothione, MSSM). The NADPH-dependent flavoprotein disulfide reductase (FDR), mycothiol disulfide reductase (Mtr), is responsible for recycling one equivalent of MSSM to two equivalents of MSH through the reduction of the MSSM disulfide bond (Fig. 3) (111, 163). Mtr reduces the MSSM disulfide bond by a catalytic cycle similar to those of other FDRs such as GSH and trypanothione reductase (164). When *M. bovis* BCG was treated with Mtr antisense oligonucleotides, there was a 66% reduction in growth of *M. bovis* BCG, demonstrating the essential nature of this enzyme (165).

MSH is also involved in the reduction of *S*-thiolated proteins through a pathway similar to the GSH/glutaredoxin (Grx) reduction pathway (Fig. 3). The oxidoreductase mycoredoxin-1 (Mrx1), a Grx-like protein, is responsible for the reduction of protein *S*-mycothiolated mixed disulfides through the transfer of electrons from MSH to the mixed disulfide (166). Van Laer and coworkers showed that oxidized Mrx1 (Mrx1$_{Ox}$) can only be reduced back to reduced Mrx1 (Mrx1$_{Red}$) by MSH with the formation of MSSM. An *M. smegmatis* Mrx1-deficient strain showed normal growth and a minor increase in hydrogen peroxide sensitivity when compared to the parent strain (166).

MSH is also essential for various cellular detoxification pathways. The detoxification of NO and formaldehyde both involve the direct reaction of MSH with the compounds, followed by the reaction of the resulting MSH-adduct with the bifunctional MscR (167). The reaction of MSH with NO produces *S*-nitrosomycothiol (MSNO), which is reduced by MscR to mycothiol-*N*-hydroxysulfenamide (MSNHOH) *in vitro* in an NADH-dependent reaction. The absence of MSNHOH *in vivo* points to its hydrolysis, most likely to MSSM and nitrate (NO−3), via intracellular

metabolism. Miller and coworker showed that MSH-producing *M. smegmatis* mc^{2}155 are more resistant to NO than MSH-deficient mutants and other GSH-producing bacteria (168). The hemithioacetal *S*-hydroxymethyl-mycothiol (MSCH$_2$OH) is formed from the direct reaction of MSH with formaldehyde. MSCH$_2$OH is then oxidized by MscR in an NAD$^+$-dependent reaction, at a rate 80 times slower than the MSNO reductase activity, to produce the MSH *S*-formyl thioester (MSCHO), which is subsequently hydrolyzed to produce MSH and formic acid (167).

Another MSH-dependent detoxification pathway involves the formation of MSH *S*-conjugates of cellular toxins (alkylating agents [X-R], free radicals, and xenobiotics), followed by the cleavage of the MSH *S*-conjugate glycosaminyl-amide bond and transport of the mercapturic acid–labeled toxin (AcCys-R) from the cell (Fig. 3). Mycothiol-*S*-transferase (MST) catalyzes the formation of the MSH *S*-conjugates (MS-R) (118), and Mca is responsible for the glycosaminyl-amide bond cleavage to produce the mercapturic acid–labeled toxin (AcCys-R) and GlcN-Ins, which is recycled back into the biosynthesis of MSH (169, 170). Mca, found in all MSH-producing bacteria, is a close functional homologue of MshB and has been credited with the explanation of why MSH is still produced in MshB-deficient mycobacteria (151, 159). Mca can also react with the MSH *S*-conjugates of various antibiotics, including cerulenin (170), rifamycin S (170, 171), streptomycin (171), and neocarzinostatin (172), to generate GlcN-Ins. The MSH *S*-conjugates of monobromobimane (a thiol-specific alkylating agent) and rifamycin S were shown to accumulate inside *M. smegmatis* mc^{2}155 deficient of Mca activity, which was reversed after complementation of Mca activity (171). The Mca-deficient mycobacteria were also more susceptible to rifamycin S and streptomycin, demonstrating that the Mca detoxification pathway plays a role in drug resistance of MSH-producing bacteria. This underscores the rationale of Mca being the focus of a number of inhibitor screening studies (173–176).

Intriguingly, a study also indicated that stationary-phase cultures of *M. smegmatis* mc^{2}155, deficient in either MshA or MshC and therefore producing no MSH, were able to actively transport MSH from the culture media to generate levels of MSH inside the cell comparable to those of the wild type (177). MSH itself also serves as a weak substrate for Mca (169, 178) and indicates the existence of a degradation pathway of MSH to produce *N*-acetylated cysteine (AcCys) and GlcN-Ins (170) (Fig. 3). MSH isotopomers were used to determine the fate of the MSH catabolites GlcN-Ins

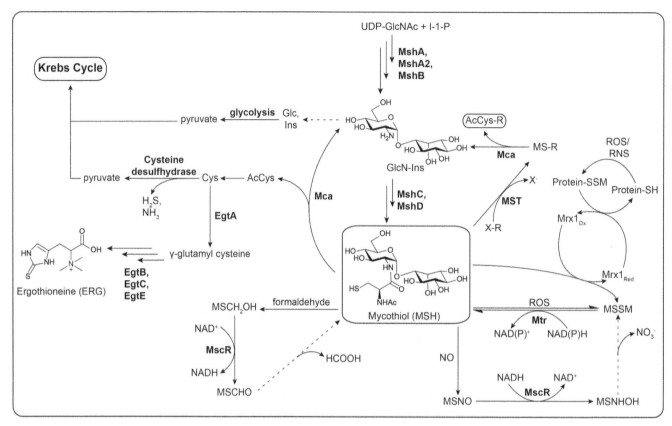

Figure 3 The intracellular functions of MSH: redox homeostasis, detoxification, source of carbon, and a cofactor in enzyme reactions. doi:10.1128/microbiolspec.MGM2-0019-2013.f3

and AcCys. It was found that GlcN-Ins either was ca-tabolized further and metabolized by glycolysis and the Krebs cycle, or it was cycled back into the MSH bio-synthetic pathway in the MshA-deficient mutants. AcCys was found to be deacetylated to cysteine and converted to pyruvate by a yet unknown cysteine desulfhydrase for metabolism by the Krebs cycle. Cys-teine can also be converted to γ-glutamyl cysteine by γ-glutamyl cysteine synthetase (EgtA) for ergothioneine (ERG) biosynthesis (179).

Ergothioneine
In 1909, Charles Tanret (180) isolated ERG (ergothio-neine, 2-mercaptohistidine trimethylbetaine) (Fig. 1) from the ergot fungus, *Claviceps purpurea*, and its structure was elucidated in 1911 (180, 181). ERG is synthesized not only in non-yeast fungi, but also in cyanobacteria (182), mycobacteria (183), and bacteria belonging to *Actinomycetales* (184). It is also detected in mammalian cells and higher plant tissues (185, 186), but there is no evidence to date for the direct biosyn-

thesis of ERG in animals and plants, and it is under-stood to be acquired through diet (1). ERG has drawn considerable interest due to its antioxidant properties, and many reports have shown that ERG may act as a cytoprotectant *in vitro*, but its role *in vivo* is uncertain.

Properties of ERG
ERG exists as a tautomer between its thiol and thione forms in solution, but it exists in its thione form at physiological pH (187). However, the thione is in equi-librium with two resonance forms, EGT-b and EGT-c, that are thiolate in character and contribute to the thiol-like activity of ERG (187). ERG does not undergo auto-oxidation as readily as other thiols considered to be antioxidants, for example, GSH, which generates free radicals in the process (188, 189). This stability of ERG accounts for its slow degeneration and its resis-tance to disulfide formation (185, 190). Disulfide for-mation in ERG has been demonstrated at very low pH in the presence of copper and H_2O_2 but not in alkaline or neutral solutions (191).

Biosynthesis of ERG

Seebeck (192) identified a putative five-gene cluster (Fig. 4B) for the biosynthesis of ERG in mycobacteria by identifying an S-adenosylmethionine (SAM)-dependent methyltransferase, an iron(II)-dependent oxidase, and a pyridoxal 5-phosphate–dependent β-lyase in *M. avium* that correspond to the homologous genes in organisms that produce ERG (e.g., *Neurospora crassa*) and are absent in nonproducers, such as *Escherichia coli* and *B. subtilis*. After cloning the corresponding genes for EgtB, EgtC, and EgtD in *M. smegmatis* and expressing them recombinantly in *E. coli*, in addition to using β-lyase from *Erwinia tasmaniensis, in vitro* reconstitution studies were used to verify that these enzymes did form part of the biosynthetic pathway of ERG in mycobacteria. Although awaiting genetic validation, the data suggested the following roles for the five Egt enzymes in the synthesis of ERG (Fig. 4A): (i) EgtA,

a γ-glutamyl cysteine synthetase, produces glutamyl cysteine, thought to be the sulfur donor; (ii) EgtB, a formylglycine-generating enzyme–like protein, is thought to catalyze the iron(II)-dependent oxidative sulfurization of hercynine; (iii) EgtC is a glutamine amidotransferase that catalyzes the hydrolysis of the γ-glutamyl amide bond in Intermediate 1; (iv) EgtD is a SAM-dependent methyltransferase that methylates histidine to form hercynine; and (v) EgtE is a pyridoxal 5-phosphate–dependent β-lyase that removes pyruvate and ammonia from Intermediate 2 to generate ERG. Homologues of EgtD (histidine-specific methyltransferase) and EgtB [iron(II)-dependent enzyme that catalyzes the sulfurization of hercynine] are often encoded by genes that are adjacent in prokaryotes, including many mycobacteria, cyanobacteria, bacteroidetes, β-proteobacteria, and, less frequently, in α-, γ-, and δ-proteobacteria and other actinobacteria (192).

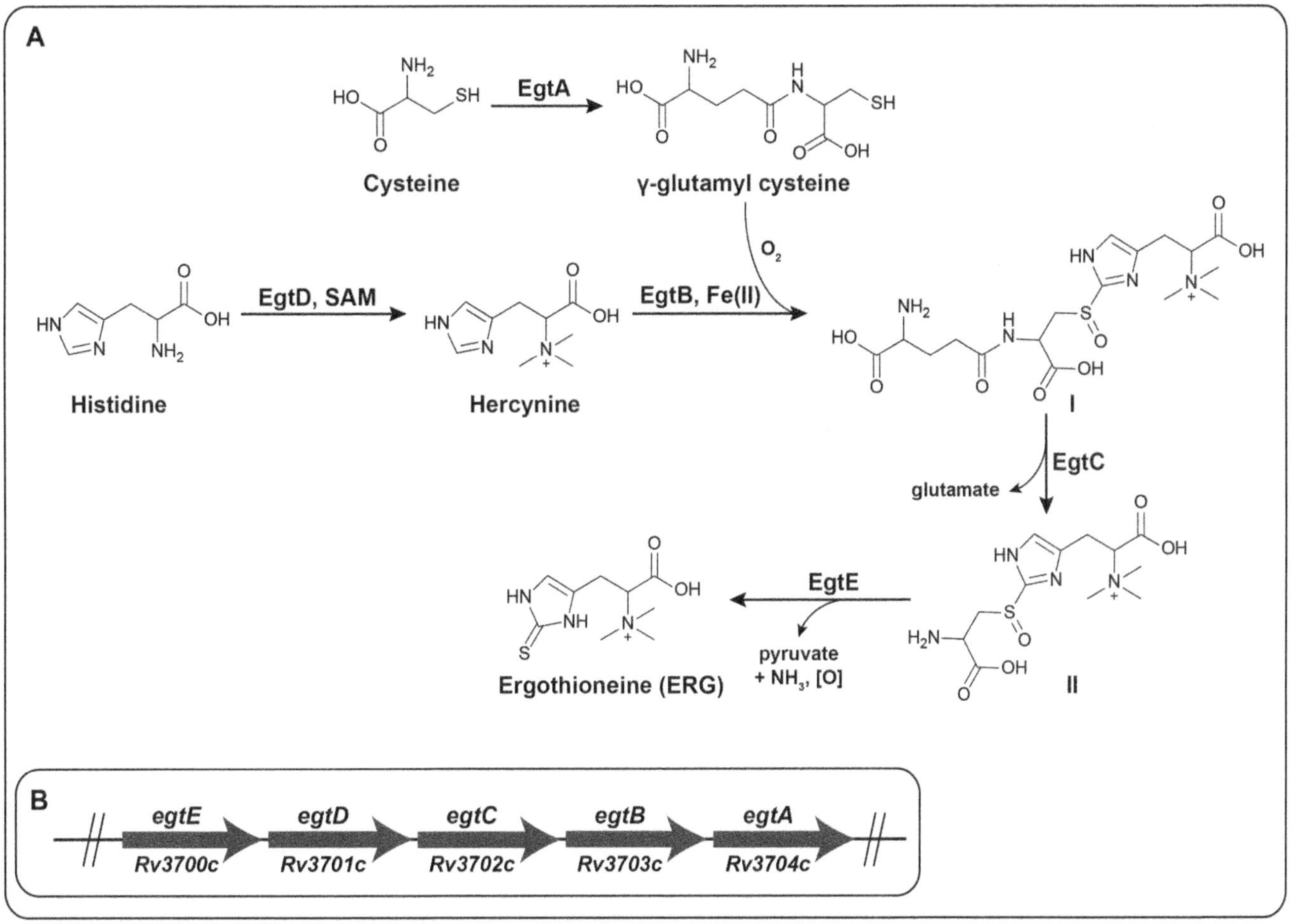

Figure 4   (A) Biosynthesis of ergothioneine as proposed by Seebeck (179). (B) Putative five-gene cluster in *M. tuberculosis* H37Rv encoding ERG biosynthetic enzymes.
doi:10.1128/microbiolspec.MGM2-0019-2013.f4

Functions of ERG

The role ERG plays in microbial cells is not well understood, but it does appear to have antioxidant and cytoprotectant properties in mammalian cells. In mammalian cells, the SLC22A4 gene encodes an integral membrane protein, the organic cation transporter (OCTN1) that facilitates Na^+ and pH-dependent transport of ERG across the membrane (193). To account for the ubiquitous occurrence of ERG and its accumulation in tissues, various functions have been proposed for ERG: a cation chelator (194–196), a factor in bioenergetics (197), a regulator of gene expression (198), an immune regulator (199), and the most cited role as an antioxidant and cytoprotectant (200–204). Although a large body of evidence demonstrates the antioxidant properties of ERG *in vitro*, Ey et al. (205) found that ERG only protected OCTN1-transfected HEK-293 against copper(II)-induced toxicity, but not against a range of other cellular stresses, although GSH at an equivalent concentration protected against all the cellular stresses tested. Yet, Kawano et al. (197) found that ERG accumulated in the mitochondria of hepatic cells of rats injected with radiolabeled ERG. In addition, Paul and Snyder (206) demonstrated increased mitochondrial DNA damage in OCTN1-silenced HeLa cells when exposed to H_2O_2.

In the case of microbes, an *NcΔEgt-1* mutant in *Neurospora crassa*—involving a knockout in the gene *NCU04343*, which encodes a protein with fused domains that are homologous to *M. avium* EgtB and EgtD—does not produce ERG, in contrast to the wild type. *NcΔEgt-1* was also found to be significantly more sensitive to *tert*-butyl hydroperoxide than the wild type. This suggested that ERG protects conidia from the toxicity of peroxide during the quiescent period between conidiogenesis and germination; however, similar protection was not observed with superoxide or Cu^{2+} (207). Seebeck (179) indicated that homologues of *M. smegmatis* EgtB and EgtD are encoded in the genomes of most cyanobacteria, suggesting that ERG is involved in the protection of DNA against visible and UV radiation damage in cyanobacteria. Soil bacteria, such as *M. smegmatis*, could also benefit from this protection when exposed on the surface. Furthermore, the *mshA* mutant of *M. smegmatis*, which is deficient in mycothiol, was found to overproduce ERG by 35-fold in the exponential growth phase, possibly as a compensatory mechanism (208). It is intriguing that mycobacteria have two redox intermediaries, although it is unknown if the production of either regulates the synthesis of the other redox buffer. However, they are unique in structure and have different redox properties,

suggesting that they may play different roles in maintaining redox balance.

Tuberculosis and ERG

In 1956, it was reported that higher levels of ERG were found in the blood of Native Americans and Eskimos suffering from pulmonary tuberculosis than in "normal subjects" (209). However, the "normal subjects" were sampled from two city hospitals where the Caucasian patients with tuberculosis were sampled. The Caucasian patients with pulmonary tuberculosis had mean ERG levels that were not significantly higher than those of the "normal subjects." Although the Native American diet was refuted as the cause of the unusually high ERG levels in the Native American and Eskimo patients, the "normal subjects" used as healthy controls were not from the same socioeconomic/racial group as the Native Americans and Eskimos with pulmonary tuberculosis, so that theoretically, the observations cannot be used to make an unbiased correlation with pulmonary tuberculosis.

In 2009, a genome-wide single nucleotide polymorphism (SNP)–based linkage analysis in 93 affected sib-pairs identified chromosome 5q23.2-31.3 as a region with suggestive evidence of linkage with tuberculosis susceptibility in the Asian population (210). Chromosome 5q31 contains the T helper 2 (Th2)–related cytokine gene cluster, which is potentially essential for the Th1/Th2 response. A family-based association analysis was used to fine-map a putative tuberculosis susceptibility locus in chromosome 5q23.2-31.3, and variants of a haplotype within the human gene SLC22A4, which codes for the ERG transporter OCTN1 (193), were shown to be associated with tuberculosis susceptibility among the Thai population (211).

MEASUREMENT OF THE REDOX STATE IN MYCOBACTERIA

All living organisms have protein thiols in the cytoplasm, which are kept in the reduced (-SH) state. Conventional methods used to measure the reduction potentials of $NAD(P)^+/NAD(P)H$, GSH, ascorbate, thioredoxin, or thiol status in a cell involve enzymatic assays, spectroscopic methods, HPLC, or gel mobility that usually require whole-cell extracts. Cell disruption has little buffering capacity, creates oxidation artifacts, prevents dynamic measurements, and implies an apparent redox potential, which has low accuracy. They are also not amenable to multiple sample analyses and have no time-dependent or spatial resolution (5, 6, 212, 213). A further drawback of these invasive approaches is that the subcellular compartments often have different redox environments (5, 214, 215). A noninvasive fluorescent-based method has been used to measure the intracellular thiol redox potential of *E. coli* involving a yellow fluorescent protein with a pair of redox-active cysteines introduced onto the surface of the protein (216). The noninvasive measurement of the redox potential, −259 mV, was fairly comparable to the redox potential inferred by the redox-couple-specific approach, which was estimated to be within −220 to −245 mV (217).

The intracellular redox environment of mycobacteria consists of known redox couples (218) and unknown redox couples, in addition to the combined intracellular protein thiol pool (6). Although various methods have been used to quantify the mycothiol redox couple (MSH:MSS ratio) and ERG levels in mycobacteria, the intracellular redox potential of mycobacteria has not yet been determined.

MSH derivatization with monobromobimane (mBBr), followed by HPLC analysis, is the most widely used method for MSH detection (118, 146, 158, 161, 177, 178, 219). However, this method failed to detect low quantities of ERG because the ERG moiety quenched the fluorescence of the bimane derivative, and the detection sensitivity was low. More recently, LC-MS/MS methods were used to quantify ERG in dietary sources (205), human plasma and erythrocytes (220), and MSH in the cell lysates of Mtr-deficient *M. smegmatis* mc²155 (221).

To date, the cellular redox ratio, reduced MSH:oxidized MSH (MSH:MSS), has been used to infer the redox status of the mycobacterial cell during different growth phases. In *M. smegmatis* mc²155, the redox ratio changes from 1,000:1 to 1,300:1 toward the end of lag phase to 200:1 in stationary phase (153, 161). In *M. tuberculosis* Erdman, the redox ratio decreases from 180:1 at the end of lag phase to 50:1 in stationary phase (161). The redox ratio of *M. bovis* BCG in stationary phase was also 50:1 (222). Ung and Av-Gay (222) also demonstrated a 2-fold decrease in the redox ratio of *M. bovis* BCG when treated with hydrogen peroxide, whereas the redox ratio of *M. smegmatis* mc²155 remained unchanged upon treatment with hydrogen peroxide, which may be explained by the higher reduced MSH levels present in *M. smegmatis* mc²155.

CONCLUSIONS AND FUTURE CHALLENGES

Control of redox pathways is crucial for bacteria, because redox homeostasis is central to metabolism and redistribution of cellular energy, ultimately controlling survival. What is the intrinsic redox and bioenergetic status of metabolically inactive *M. tuberculosis*, and how can this knowledge be exploited to devise new therapeutic intervention strategies? Unfortunately, to date, there are no clear answers to this question. *M. tuberculosis* is exposed to a wide spectrum of host redox stresses *in vivo*, including ROI, RNI, acidic pH, hypoxia, and nutrient deprivation, in addition to extraneous sources of redox imbalance such as cigarette smoke and indoor air pollution. Although there has been progress in the understanding of redox homeostasis in mycobacteria, the identity of all the main redox buffers and protein thiols, the behavior of these thiols and redox buffers under different environmental conditions, and all the intracellular mechanisms to maintain redox balance in *M. tuberculosis* are still unknown. Thus, it would be beneficial to define the *M. tuberculosis* "redoxome" using genome-wide technologies. We also lack knowledge of how oxidative, nitrosative, and carbonyl stress modulate the metabolism and hence pathogenesis of *M. tuberculosis*. Using noninvasive redox-sensitive green fluorescent proteins, it should be possible to generate numerical indicators of the intracellular redox state of *M. tuberculosis* and monitor how these indicators change when mimicking the complex milieu of the lung, in particular in the presence or absence of O_2, CO, and NO. Although mycothiol has been identified as a redox buffer in mycobacteria, the function of ERG in mycobacterial redox homeostasis and pathogenesis has yet to be elucidated. Furthermore, apart from known redox sensors in *M. tuberculosis*, such as the WhiB family members, DosS and DosT, σ factors, and thioredoxin-dependent redox-sensitive kinases, the redox sensors involved in reactivation of dormant bacilli have not yet been identified.

Redox homeostasis in mycobacteria also plays a role in drug efficacy, as isoniazid, ethionamide, and PA-824 require bioreductive activation for their antimycobacterial activity. However, our knowledge of the underlying mechanisms involved in this prodrug activation are largely unknown and would be of value in novel drug design. Likewise, treating latent *M. tuberculosis* infection requires a detailed knowledge of the physiology, and hence redox homeostasis, of the bacilli in response to their microenvironment, for instance, within the granuloma. Identification of terminal electron acceptors used in respiration of the mycobacteria and of mechanisms used to regenerate NAD(P)+ in this wide spectrum of microenvironments, together with the roles of CO, O_2, and NO in altering mycobacterial respiration, is fundamental in understanding *M. tuberculosis* pathogenesis and persistence.

When *in vitro* models are used to study redox homeostasis, the observed mycobacterial response is in relation to the redox stressors and conditions present in the experimental setting. However, this is usually an incomplete representation of the response of mycobacteria *in vivo*, within the diseased niche. As mentioned earlier, *M. tuberculosis* is exposed to complex environments both in the macrophage and the lung, and it is anticipated that redox homeostasis within the bacillus will vary considerably in these diverse environments. *In vivo* murine TB models have been used to investigate the virulence and pathophysiology of *M. tuberculosis* or its mutants to ascertain the role of mycobacterial proteins or host proteins in pathogenesis and TB disease progression. Yet it has recently been shown, using genomic responses, that inflammatory responses in mouse models show poor correlation with human responses (223). Thus, examining mycobacterial redox homeostasis within the varying environments of the human lung would ultimately give a more accurate, complete picture of how *M. tuberculosis* combats oxidative stress and subverts the host's immune response in order to establish persistence.

Comprehensive knowledge of the pathways involved in redox homeostasis in mycobacteria will give us a better understanding of the mechanisms whereby *M. tuberculosis* survives under diverse environmental extremes and induces pathogenesis, which will enable us to develop new therapeutic and infection control strategies.

Citation. Cumming BM, Lamprecht D, Wells RM, Saini V, Mazorodze JH, Steyn AJC. 2014. The physiology and genetics of oxidative stress in mycobacteria. Microbiol Spectrum 2(3):MGM2-0019-2013.

References

1. Halliwell B, Gutteridge JMC. 2007. *Free Radicals in Biology and Medicine*, 4th ed. Oxford University Press, Oxford/New York.

2. Sacchettini JC, Rubin EJ, Freundlich JS. 2008. Drugs versus bugs: in pursuit of the persistent predator *Mycobacterium tuberculosis*. *Nat Rev Microbiol* 6:41–52.

3. Kumar A, Toledo JC, Patel RP, Lancaster JR Jr, Steyn AJ. 2007. *Mycobacterium tuberculosis* DosS is a redox sensor and DosT is a hypoxia sensor. *Proc Natl Acad Sci USA* 104:11568–11573.

4. Singh A, Guidry L, Narasimhulu KV, Mai D, Trombley J, Redding KE, Giles GI, Lancaster JR Jr, Steyn AJ. 2007. *Mycobacterium tuberculosis* WhiB3 responds to O2 and nitric oxide via its [4Fe-4S] cluster and is essential for nutrient starvation survival. *Proc Natl Acad Sci USA* 104:11562–11567.

5. Schafer FQ, Buettner GR. 2001. Redox environment of the cell as viewed through the redox state of the glutathione disulfide/glutathione couple. *Free Radic Biol Med* 30:1191–1212.

6. Hansen RE, Roth D, Winther JR. 2009. Quantifying the global cellular thiol-disulfide status. *Proc Natl Acad Sci USA* 106:422–427.

7. Quemard A, Sacchettini JC, Dessen A, Vilcheze C, Bittman R, Jacobs WR Jr, Blanchard JS. 1995. Enzymatic characterization of the target for isoniazid in *Mycobacterium tuberculosis*. *Biochemistry* 34:8235–8241.

8. Baulard AR, Betts JC, Engohang-Ndong J, Quan S, McAdam RA, Brennan PJ, Locht C, Besra GS. 2000. Activation of the pro-drug ethionamide is regulated in mycobacteria. *J Biol Chem* 275:28326–28331.

9. Manjunatha UH, Boshoff H, Dowd CS, Zhang L, Albert TJ, Norton JE, Daniels L, Dick T, Pang SS, Barry CE 3rd. 2006. Identification of a nitroimidazooxazine-specific protein involved in PA-824 resistance in *Mycobacterium tuberculosis*. *Proc Natl Acad Sci USA* 103:431–436.

10. Vilcheze C, Weisbrod TR, Chen B, Kremer L, Hazbon MH, Wang F, Alland D, Sacchettini JC, Jacobs WR Jr. 2005. Altered NADH/NAD+ ratio mediates coresistance to isoniazid and ethionamide in mycobacteria. *Antimicrob Agents Chemother* 49:708–720.

11. Miesel L, Weisbrod TR, Marcinkeviciene JA, Bittman R, Jacobs WR Jr. 1998. NADH dehydrogenase defects confer isoniazid resistance and conditional lethality in *Mycobacterium smegmatis*. *J Bacteriol* 180:2459–2467.

12. Xu X, Vilcheze C, Av-Gay Y, Gomez-Velasco A, Jacobs WR Jr. 2011. Precise null deletion mutations of the mycothiol synthesis genes reveal their role in isoniazid and ethionamide resistance in *Mycobacterium smegmatis*. *Antimicrob Agents Chemother* 55:3133–3139.

13. Bedard K, Krause KH. 2007. The NOX family of ROS-generating NADPH oxidases: physiology and pathophysiology. *Physiol Rev* 87:245–313.

14. Chan J, Xing Y, Magliozzo RS, Bloom BR. 1992. Killing of virulent *Mycobacterium tuberculosis* by reactive nitrogen intermediates produced by activated murine macrophages. *J Exp Med* 175:1111–1122.

15. Chan J, Tanaka K, Carroll D, Flynn J, Bloom BR. 1995. Effects of nitric oxide synthase inhibitors on murine infection with *Mycobacterium tuberculosis*. *Infect Immun* 63:736–740.

16. Styblo K. 1980. Recent advances in epidemiological research in tuberculosis. *Adv Tuberc Res* 20:1–63.

17. DePalo VA, McCool FD. 2003. Pulmonary anatomy & physiology. *In* Hanley ME, Welsh CH (ed), *Current Diagnosis & Treatment in Pulmonary Medicine*. McGraw-Hill, New York, NY.

18. Park HS, Kim SR, Lee YC. 2009. Impact of oxidative stress on lung diseases. *Respirology* 14:2–38.

19. Ward PA. 2010. Oxidative stress: acute and progressive lung injury. *Ann NY Acad Sci* 1203:53–59.

20. Rahman I, Biswas SK, Kode A. 2006. Oxidant and antioxidant balance in the airways and airway diseases. *Eur J Pharmacol* 533:222–239.

21. Liang Y, Yeligar SM, Brown LA. 2012. Chronic-alcohol-abuse-induced oxidative stress in the development of acute respiratory distress syndrome. *Sci World J* **2012:** 740308.

22. Buhl R, Jaffe HA, Holroyd KJ, Wells FB, Mastrangeli A, Saltini C, Cantin AM, Crystal RG. 1989. Systemic glutathione deficiency in symptom-free HIV-seropositive individuals. *Lancet* **2:**1294–1298.

23. Narasimhan P, Wood J, Macintyre CR, Mathai D. 2013. Risk factors for tuberculosis. *Pulm Med* **2013:**828939.

24. Yeh MY, Burnham EL, Moss M, Brown LA. 2008. Non-invasive evaluation of pulmonary glutathione in the exhaled breath condensate of otherwise healthy alcoholics. *Respir Med* **102:**248–255.

25. Venketaraman V, Millman A, Salman M, Swaminathan S, Goetz M, Lardizabal A, David H, Connell ND. 2008. Glutathione levels and immune responses in tuberculosis patients. *Microb Pathog* **44:**255–261.

26. Vandiviere HM, Loring WE, Melvin I, Willis S. 1956. The treated pulmonary lesion and its tubercle bacillus. II. The death and resurrection. *Am J Med Sci* **232:**30–37; passim.

27. Wayne LG, Hayes LG. 1996. An in vitro model for sequential study of shiftdown of *Mycobacterium tuberculosis* through two stages of nonreplicating persistence. *Infect Immun* **64:**2062–2069.

28. Meylan PR, Richman DD, Kornbluth RS. 1992. Reduce intracellular growth of mycobacteria in human macrophages cultivated at physiologic oxygen pressure. *Am Rev Respir Dis* **145:**947–953.

29. Boshoff HI, Barry CE 3rd. 2005. Tuberculosis: metabolism and respiration in the absence of growth. *Nat Rev Microbiol* **3:**70–80.

30. Via LE, Lin PL, Ray SM, Carrillo J, Allen SS, Eum SY, Taylor K, Klein E, Manjunatha U, Gonzales J, Lee EG, Park SK, Raleigh JA, Cho SN, McMurray DN, Flynn JL, Barry CE 3rd. 2008. Tuberculous granulomas are hypoxic in guinea pigs, rabbits, and nonhuman primates. *Infect Immun* **76:**2333–2340.

31. Leistikow RL, Morton RA, Bartek IL, Frimpong I, Wagner K, Voskuil MI. 2010. The *Mycobacterium tuberculosis* DosR regulon assists in metabolic homeostasis and enables rapid recovery from nonrespiring dormancy. *J Bacteriol* **192:**1662–1670.

32. Dasgupta N, Kapur V, Singh KK, Das TK, Sachdeva S, Jyothisri K, Tyagi JS. 2000. Characterization of a two-component system, *devR-devS*, of *Mycobacterium tuberculosis*. *Tuber Lung Dis* **80:**141–159.

33. Kumar A, Deshane JS, Crossman DK, Bolisetty S, Yan BS, Kramnik I, Agarwal A, Steyn AJ. 2008. Heme oxygenase-1-derived carbon monoxide induces the *Mycobacterium tuberculosis* dormancy regulon. *J Biol Chem* **283:**18032–18039.

34. Shiloh MU, Manzanillo P, Cox JS. 2008. *Mycobacterium tuberculosis* senses host-derived carbon monoxide during macrophage infection. *Cell Host Microbe* **3:** 323–330.

35. Ioanoviciu A, Yukl ET, Moenne-Loccoz P, de Montellano PR. 2007. DevS, a heme-containing two-component oxygen sensor of *Mycobacterium tuberculosis*. *Biochemistry* **46:**4250–4260.

36. Sousa EH, Tuckerman JR, Gonzalez G, Gilles-Gonzalez MA. 2007. DosT and DevS are oxygen-switched kinases in *Mycobacterium tuberculosis*. *Protein Sci* **16:**1708–1719.

37. Kumar A, Toledo JC, Patel RP, Lancaster JR Jr, Steyn AJ. 2007. *Mycobacterium tuberculosis* DosS is a redox sensor and DosT is a hypoxia sensor. *Proc Natl Acad Sci USA* **104:**11568–11573.

38. Singh A, Guidry L, Narasimhulu KV, Mai D, Trombley J, Redding KE, Giles GI, Lancaster JR Jr, Steyn AJ. 2007. *Mycobacterium tuberculosis* WhiB3 responds to O2 and nitric oxide via its [4Fe-4S] cluster and is essential for nutrient starvation survival. *Proc Natl Acad Sci USA* **104:**11562–11567.

39. Soliveri JA, Gomez J, Bishai WR, Chater KF. 2000. Multiple paralogous genes related to the *Streptomyces coelicolor* developmental regulatory gene *whiB* are present in *Streptomyces* and other actinomycetes. *Microbiology* **146:**333–343.

40. Steyn AJ, Collins DM, Hondalus MK, Jacobs WR Jr, Kawakami RP, Bloom BR. 2002. *Mycobacterium tuberculosis* WhiB3 interacts with RpoV to affect host survival but is dispensable for in vivo growth. *Proc Natl Acad Sci USA* **99:**3147–3152.

41. Kim TH, Park JS, Kim HJ, Kim Y, Kim P, Lee HS. 2005. The *whcE* gene of *Corynebacterium glutamicum* is important for survival following heat and oxidative stress. *Biochem Biophys Res Commun* **337:**757–764.

42. Morris RP, Nguyen L, Gatfield J, Visconti K, Nguyen K, Schnappinger D, Ehrt S, Liu Y, Heifets L, Pieters J, Schoolnik G, Thompson CJ. 2005. Ancestral antibiotic resistance in *Mycobacterium tuberculosis*. *Proc Natl Acad Sci USA* **102:**12200–12205.

43. Larsson C, Luna B, Ammerman NC, Maiga M, Agarwal N, Bishai WR. 2012. Gene expression of *Mycobacterium tuberculosis* putative transcription factors whiB1-7 in redox environments. *PLoS One* **7:**e37516.

44. Singh A, Crossman DK, Mai D, Guidry L, Voskuil MI, Renfrow MB, Steyn AJ. 2009. *Mycobacterium tuberculosis* WhiB3 maintains redox homeostasis by regulating virulence lipid anabolism to modulate macrophage response. *PLoS Pathog* **5:**e1000545.

45. Sherman DR, Voskuil M, Schnappinger D, Liao R, Harrell MI, Schoolnik GK. 2001. Regulation of the *Mycobacterium tuberculosis* hypoxic response gene encoding alpha-crystallin. *Proc Natl Acad Sci USA* **98:** 7534–7539.

46. Ohno H, Zhu G, Mohan VP, Chu D, Kohno S, Jacobs WR Jr, Chan J. 2003. The effects of reactive nitrogen intermediates on gene expression in *Mycobacterium tuberculosis*. *Cell Microbiol* **5:**637–648.

47. Voskuil MI, Visconti KC, Schoolnik GK. 2004. *Mycobacterium tuberculosis* gene expression during adaptation to stationary phase and low-oxygen dormancy. *Tuberculosis (Edinb)* **84:**218–227.

48. Betts JC, Lukey PT, Robb LC, McAdam RA, Duncan K. 2002. Evaluation of a nutrient starvation model of

Mycobacterium tuberculosis persistence by gene and protein expression profiling. *Mol Microbiol* **43**:717–731.

49. Rustad TR, Harrell MI, Liao R, Sherman DR. 2008. The enduring hypoxic response of *Mycobacterium tuberculosis*. *PLoS One* **3**:e1502.

50. Rachman H, Strong M, Schaible U, Schuchhardt J, Hagens K, Mollenkopf H, Eisenberg D, Kaufmann SH. 2006. *Mycobacterium tuberculosis* gene expression profiling within the context of protein networks. *Microbes Infect* **8**:747–757.

51. Chawla M, Parikh P, Saxena A, Munshi M, Mehta M, Mai D, Srivastava AK, Narasimhulu KV, Redding KE, Vashi N, Kumar D, Steyn AJ, Singh A. 2012. *Mycobacterium tuberculosis* WhiB4 regulates oxidative stress response to modulate survival and dissemination in vivo. *Mol Microbiol* **85**:1148–1165.

52. Boshoff HI, Xu X, Tahlan K, Dowd CS, Pethe K, Camacho LR, Park TH, Yun CS, Schnappinger D, Ehrt S, Williams KJ, Barry CE 3rd. 2008. Biosynthesis and recycling of nicotinamide cofactors in *Mycobacterium tuberculosis*. An essential role for NAD in nonreplicating bacilli. *J Biol Chem* **283**:19329–19341.

53. Burian J, Ramon-Garcia S, Howes CG, Thompson CJ. 2012. WhiB7, a transcriptional activator that coordinates physiology with intrinsic drug resistance in *Mycobacterium tuberculosis*. *Expert Rev Anti Infect Ther* **10**:1037–1047.

54. Aslund F, Beckwith J. 1999. Bridge over troubled waters: sensing stress by disulfide bond formation. *Cell* **96**:751–753.

55. Voskuil MI, Bartek IL, Visconti K, Schoolnik GK. 2011. The response of *Mycobacterium tuberculosis* to reactive oxygen and nitrogen species. *Front Microbiol* **2**:105.

56. Humpel A, Gebhard S, Cook GM, Berney M. 2010. The SigF regulon in *Mycobacterium smegmatis* reveals roles in adaptation to stationary phase, heat, and oxidative stress. *J Bacteriol* **192**:2491–2502.

57. Hahn MY, Raman S, Anaya M, Husson RN. 2005. The *Mycobacterium tuberculosis* extracytoplasmic-function sigma factor SigL regulates polyketide synthases and secreted or membrane proteins and is required for virulence. *J Bacteriol* **187**:7062–7071.

58. Thakur KG, Praveena T, Gopal B. 2010. Structural and biochemical bases for the redox sensitivity of *Mycobacterium tuberculosis* RslA. *J Mol Biol* **397**:1199–1208.

59. Song T, Dove SL, Lee KH, Husson RN. 2003. RshA, an anti-sigma factor that regulates the activity of the mycobacterial stress response sigma factor SigH. *Mol Microbiol* **50**:949–959.

60. Barik S, Sureka K, Mukherjee P, Basu J, Kundu M. 2010. RseA, the SigE specific anti-sigma factor of *Mycobacterium tuberculosis*, is inactivated by phosphorylation-dependent ClpC1P2 proteolysis. *Mol Microbiol* **75**:592–606.

61. Li W, Bottrill AR, Bibb MJ, Buttner MJ, Paget MS, Kleanthous C. 2003. The role of zinc in the disulphide stress-regulated anti-sigma factor RsrA from *Streptomyces coelicolor*. *J Mol Biol* **333**:461–472.

62. Paget MS, Bae JB, Hahn MY, Li W, Kleanthous C, Roe JH, Buttner MJ. 2001. Mutational analysis of RsrA, a zinc-binding anti-sigma factor with a thiol-disulphide redox switch. *Mol Microbiol* **39**:1036–1047.

63. Brugarolas P, Movahedzadeh F, Wang Y, Zhang N, Bartek IL, Gao Y, Voskuil MI, Franzblau SG, He C. 2012. The oxidation-sensing regulator (MosR) is a new redox-dependent transcription factor in *Mycobacterium tuberculosis*. *J Biol Chem* **287**:37703–37712.

64. Rengarajan J, Bloom BR, Rubin EJ. 2005. Genome-wide requirements for *Mycobacterium tuberculosis* adaptation and survival in macrophages. *Proc Natl Acad Sci USA* **102**:8327–8332.

65. Chan J, Fujiwara T, Brennan P, McNeil M, Turco SJ, Sibille JC, Snapper M, Aisen P, Bloom BR. 1989. Microbial glycolipids: possible virulence factors that scavenge oxygen radicals. *Proc Natl Acad Sci USA* **86**:2453–2457.

66. Chan J, Fan XD, Hunter SW, Brennan PJ, Bloom BR. 1991. Lipoarabinomannan, a possible virulence factor involved in persistence of *Mycobacterium tuberculosis* within macrophages. *Infect Immun* **59**:1755–1761.

67. Pabst MJ, Gross JM, Brozna JP, Goren MB. 1988. Inhibition of macrophage priming by sulfatide from *Mycobacterium tuberculosis*. *J Immunol* **140**:634–640.

68. Miller BH, Fratti RA, Poschet JF, Timmins GS, Master SS, Burgos M, Marletta MA, Deretic V. 2004. Mycobacteria inhibit nitric oxide synthase recruitment to phagosomes during macrophage infection. *Infect Immun* **72**:2872–2878.

69. Davis AS, Vergne I, Master SS, Kyei GB, Chua J, Deretic V. 2007. Mechanism of inducible nitric oxide synthase exclusion from mycobacterial phagosomes. *PLoS Pathog* **3**:e186.

70. El Kasmi KC, Qualls JE, Pesce JT, Smith AM, Thompson RW, Henao-Tamayo M, Basaraba RJ, Konig T, Schleicher U, Koo MS, Kaplan G, Fitzgerald KA, Tuomanen EI, Orme IM, Kanneganti TD, Bogdan C, Wynn TA, Murray PJ. 2008. Toll-like receptor-induced arginase 1 in macrophages thwarts effective immunity against intracellular pathogens. *Nat Immunol* **9**:1399–1406.

71. Qualls JE, Neale G, Smith AM, Koo MS, DeFreitas AA, Zhang H, Kaplan G, Watowich SS, Murray PJ. 2010. Arginine usage in mycobacteria-infected macrophages depends on autocrine-paracrine cytokine signaling. *Sci Signal* **3**:ra62.

72. Pessanha AP, Martins RA, Mattos-Guaraldi AL, Vianna A, Moreira LO. 2012. Arginase-1 expression in granulomas of tuberculosis patients. *FEMS Immunol Med Microbiol* **66**:265–268.

73. Andersen P, Askgaard D, Ljungqvist L, Bennedsen J, Heron I. 1991. Proteins released from *Mycobacterium tuberculosis* during growth. *Infect Immun* **59**:1905–1910.

74. Zhang Y, Lathigra R, Garbe T, Catty D, Young D. 1991. Genetic analysis of superoxide dismutase, the 23 kilodalton antigen of *Mycobacterium tuberculosis*. *Mol Microbiol* **5**:381–391.

75. Harth G, Horwitz MA. 1999. Export of recombinant *Mycobacterium tuberculosis* superoxide dismutase is dependent upon both information in the protein and mycobacterial export machinery. A model for studying export of leaderless proteins by pathogenic mycobacteria. *J Biol Chem* **274**:4281–4292.

76. Braunstein M, Espinosa BJ, Chan J, Belisle JT, Jacobs WR Jr. 2003. SecA2 functions in the secretion of superoxide dismutase A and in the virulence of *Mycobacterium tuberculosis*. *Mol Microbiol* **48**:453–464.

77. Battistoni A. 2003. Role of prokaryotic Cu,Zn superoxide dismutase in pathogenesis. *Biochem Soc Trans* **31**:1326–1329.

78. Wu CH, Tsai-Wu JJ, Huang YT, Lin CY, Lioua GG, Lee FJ. 1998. Identification and subcellular localization of a novel Cu,Zn superoxide dismutase of *Mycobacterium tuberculosis*. *FEBS Lett* **439**:192–196.

79. Piddington DL, Fang FC, Laessig T, Cooper AM, Orme IM, Buchmeier NA. 2001. Cu,Zn superoxide dismutase of *Mycobacterium tuberculosis* contributes to survival in activated macrophages that are generating an oxidative burst. *Infect Immun* **69**:4980–4987.

80. Claiborne A, Malinowski DP, Fridovich I. 1979. Purification and characterization of hydroperoxidase II of *Escherichia coli* B. *J Biol Chem* **254**:11664–11668.

81. Diaz GA, Wayne LG. 1974. Isolation and characterization of catalase produced by *Mycobacterium tuberculosis*. *Am Rev Respir Dis* **110**:312–319.

82. Wengenack NL, Jensen MP, Rusnak F, Stern MK. 1999. *Mycobacterium tuberculosis* KatG is a peroxynitritase. *Biochem Biophys Res Commun* **256**:485–487.

83. Li Z, Kelley C, Collins F, Rouse D, Morris S. 1998. Expression of katG in *Mycobacterium tuberculosis* is associated with its growth and persistence in mice and guinea pigs. *J Infect Dis* **177**:1030–1035.

84. Nachamkin I, Kang C, Weinstein MP. 1997. Detection of resistance to isoniazid, rifampin, and streptomycin in clinical isolates of *Mycobacterium tuberculosis* by molecular methods. *Clin Infect Dis* **24**:894–900.

85. van Soolingen D, de Haas PE, van Doorn HR, Kuijper E, Rinder H, Borgdorff MW. 2000. Mutations at amino acid position 315 of the *katG* gene are associated with high-level resistance to isoniazid, other drug resistance, and successful transmission of *Mycobacterium tuberculosis* in the Netherlands. *J Infect Dis* **182**:1788–1790.

86. Wilson M, DeRisi J, Kristensen HH, Imboden P, Rane S, Brown PO, Schoolnik GK. 1999. Exploring drug-induced alterations in gene expression in *Mycobacterium tuberculosis* by microarray hybridization. *Proc Natl Acad Sci USA* **96**:12833–12838.

87. Wilson TM, Collins DM. 1996. ahpC, a gene involved in isoniazid resistance of the *Mycobacterium tuberculosis* complex. *Mol Microbiol* **19**:1025–1034.

88. Master SS, Springer B, Sander P, Boettger EC, Deretic V, Timmins GS. 2002. Oxidative stress response genes in *Mycobacterium tuberculosis*: role of ahpC in resistance to peroxynitrite and stage-specific survival in macrophages. *Microbiology* **148**:3139–3144.

89. Zhang Z, Hillas PJ, Ortiz de Montellano PR. 1999. Reduction of peroxides and dinitrobenzenes by *Mycobacterium tuberculosis* thioredoxin and thioredoxin reductase. *Arch Biochem Biophys* **363**:19–26.

90. Bryk R, Lima CD, Erdjument-Bromage H, Tempst P, Nathan C. 2002. Metabolic enzymes of mycobacteria linked to antioxidant defense by a thioredoxin-like protein. *Science* **295**:1073–1077.

91. Cole ST, Brosch R, Parkhill J, Garnier T, Churcher C, Harris D, Gordon SV, Eiglmeier K, Gas S, Barry CE 3rd, Tekaia F, Badcock K, Basham D, Brown D, Chillingworth T, Connor R, Davies R, Devlin K, Feltwell T, Gentles S, Hamlin N, Holroyd S, Hornsby T, Jagels K, Barrell BG. 1998. Deciphering the biology of *Mycobacterium tuberculosis* from the complete genome sequence. *Nature* **393**:537–544.

92. Jaeger T. 2007. Peroxiredoxin systems in mycobacteria. *Subcell Biochem* **44**:207–217.

93. Holmgren A. 1985. Thioredoxin. *Annu Rev Biochem* **54**:237–271.

94. Martin JL. 1995. Thioredoxin: a fold for all reasons. *Structure* **3**:245–250.

95. Williams CH Jr. 1995. Mechanism and structure of thioredoxin reductase from *Escherichia coli*. *FASEB J* **9**:1267–1276.

96. Kadokura H, Katzen F, Beckwith J. 2003. Protein disulfide bond formation in prokaryotes. *Annu Rev Biochem* **72**:111–135.

97. Akif M, Khare G, Tyagi AK, Mande SC, Sardesai AA. 2008. Functional studies of multiple thioredoxins from *Mycobacterium tuberculosis*. *J Bacteriol* **190**:7087–7095.

98. Wittenberg JB, Bolognesi M, Wittenberg BA, Guertin M. 2002. Truncated hemoglobins: a new family of hemoglobins widely distributed in bacteria, unicellular eukaryotes, and plants. *J Biol Chem* **277**:871–874.

99. Ouellet H, Ranguelova K, Labarre M, Wittenberg JB, Wittenberg BA, Magliozzo RS, Guertin M. 2007. Reaction of *Mycobacterium tuberculosis* truncated hemoglobin O with hydrogen peroxide: evidence for peroxidatic activity and formation of protein-based radicals. *J Biol Chem* **282**:7491–7503.

100. Ascenzi P, Visca P. 2008. Scavenging of reactive nitrogen species by mycobacterial truncated hemoglobins. *Methods Enzymol* **436**:317–337.

101. Nicoletti FP, Comandini A, Bonamore A, Boechi L, Boubeta FM, Feis A, Smulevich G, Boffi A. 2010. Sulfide binding properties of truncated hemoglobins. *Biochemistry* **49**:2269–2278.

102. Wheeler PR, Coldham NG, Keating L, Gordon SV, Wooff EE, Parish T, Hewinson RG. 2005. Functional demonstration of reverse transsulfuration in the *Mycobacterium tuberculosis* complex reveals that methionine is the preferred sulfur source for pathogenic mycobacteria. *J Biol Chem* **280**:8069–8078.

103. Weissbach H, Etienne F, Hoshi T, Heinemann SH, Lowther WT, Matthews B, St John G, Nathan C, Brot N. 2002. Peptide methionine sulfoxide reductase: structure, mechanism of action, and biological function. *Arch Biochem Biophys* **397**:172–178.

104. Rhee KY, Erdjument-Bromage H, Tempst P, Nathan CF. 2005. S-nitroso proteome of *Mycobacterium tuberculosis:* enzymes of intermediary metabolism and antioxidant defense. *Proc Natl Acad Sci USA* **102**:467–472.

105. St John G, Brot N, Ruan J, Erdjument-Bromage H, Tempst P, Weissbach H, Nathan C. 2001. Peptide methionine sulfoxide reductase from *Escherichia coli* and *Mycobacterium tuberculosis* protects bacteria against oxidative damage from reactive nitrogen intermediates. *Proc Natl Acad Sci USA* **98**:9901–9906.

106. Boschi-Muller S, Olry A, Antoine M, Branlant G. 2005. The enzymology and biochemistry of methionine sulfoxide reductases. *Biochim Biophys Acta* **1703**:231–238.

107. Lee WL, Gold B, Darby C, Brot N, Jiang X, de Carvalho LP, Wellner D, St John G, Jacobs WR Jr, Nathan C. 2009. *Mycobacterium tuberculosis* expresses methionine sulphoxide reductases A and B that protect from killing by nitrite and hypochlorite. *Mol Microbiol* **71**:583–593.

108. Douglas T, Daniel DS, Parida BK, Jagannath C, Dhandayuthapani S. 2004. Methionine sulfoxide reductase A (MsrA) deficiency affects the survival of *Mycobacterium smegmatis* within macrophages. *J Bacteriol* **186**:3590–3598.

109. Schnappinger D, Ehrt S, Voskuil MI, Liu Y, Mangan JA, Monahan IM, Dolganov G, Efron B, Butcher PD, Nathan C, Schoolnik GK. 2003. Transcriptional adaptation of *Mycobacterium tuberculosis* within macrophages: insights into the phagosomal environment. *J Exp Med* **198**:693–704.

110. Fahey RC. 2001. Novel thiols of prokaryotes. *Annu Rev Microbiol* **55**:333–356.

111. Patel MP, Blanchard JS. 1998. Synthesis of Des-myo-Inositol mycothiol and demonstration of a mycobacterial specific reductase activity. *J Am Chem Soc* **120**:11538–11539.

112. Wong KK, Vanoni MA, Blanchard JS. 1988. Glutathione reductase: solvent equilibrium and kinetic isotope effects. *Biochemistry* **27**:7091–7096.

113. del Cardayre SB, Stock KP, Newton GL, Fahey RC, Davies JE. 1998. Coenzyme A disulfide reductase, the primary low molecular weight disulfide reductase from *Staphylococcus aureus*. Purification and characterization of the native enzyme. *J Biol Chem* **273**:5744–5751.

114. Gaballa A, Newton GL, Antelmann H, Parsonage D, Upton H, Rawat M, Claiborne A, Fahey RC, Helmann JD. 2010. Biosynthesis and functions of bacillithiol, a major low-molecular-weight thiol in bacilli. *Proc Natl Acad Sci USA* **107**:6482–6486.

115. Antelmann H, Hamilton CJ. 2012. Bacterial mechanisms of reversible protein S-thiolation: structural and mechanistic insights into mycoredoxins. *Mol Microbiol* **86**:759–764.

116. Krauth-Siegel RL, Comini MA. 2008. Redox control in trypanosomatids, parasitic protozoa with trypanothione-based thiol metabolism. *Biochim Biophys Acta* **1780**:1236–1248.

117. Gutierrez-Lugo M-T, Newton GL, Fahey RC, Bewley CA. 2006. Cloning, expression and rapid purification of active recombinant mycothiol ligase as B1 immunoglobulin binding domain of streptococcal protein G, glutathione-S-transferase and maltose binding protein fusion proteins in *Mycobacterium smegmatis*. *Protein Expr Purif* **50**:128–136.

118. Newton GL, Leung SS, Wakabayashi JI, Rawat M, Fahey RC. 2011. The DinB superfamily includes novel mycothiol, bacillithiol, and glutathione S-transferases. *Biochemistry* **50**:10751–10760.

119. Cheah IK, Halliwell B. 2012. Ergothioneine: antioxidant potential, physiological function and role in disease. *Biochim Biophys Acta* **1822**:784–793.

120. Banerjee R. 2012. Redox outside the box: linking extracellular redox remodeling with intracellular redox metabolism. *J Biol Chem* **287**:4397–4402.

121. Benetti LR, Campos D, Gurgueira SA, Vercesi AE, Guedes CEV, Santos KL, Wallace JL, Teixeira SA, Florenzano J, Costa SKP, Muscara MN, Ferreira HHA. 2013. Hydrogen sulfide inhibits oxidative stress in lungs from allergic mice in vivo. *Eur J Pharmacol* **698**:463–469.

122. Parasassi T, Brunelli R, Costa G, De Spirito M, Krasnowska E, Lundeberg T, Pittaluga E, Ursini F. 2010. Thiol redox transitions in cell signaling: a lesson from N-acetylcysteine. *Sci World J* **10**:1192–1202.

123. Garcia-Mata C, Lamattina L. 2013. Gasotransmitters are emerging as new guard cell signaling molecules and regulators of leaf gas exchange. *Plant Sci* **201-202**:66–73.

124. Newton GL, Arnold K, Price MS, Sherrill C, Delcardayre SB, Aharonowitz Y, Cohen G, Davies J, Fahey RC, Davis C. 1996. Distribution of thiols in microorganisms: mycothiol is a major thiol in most actinomycetes. *J Bacteriol* **178**:1990–1995.

125. Newton GL, Fahey RC, Cohen G, Aharonowitz Y. 1993. Low-molecular-weight thiols in streptomycetes and their potential role as antioxidants. *J Bacteriol* **175**:2734–2742.

126. Sakuda S, Zhou Z-Y, Yamada Y. 1994. Structure of a novel disulfide of 2-(N-acetylcysteinyl)amido-2-deoxy-alpha-D-glucopyran-osyl-myo-inositol produced by *Streptomyces* sp. *Biosci Biotechnol Biochem* **58**:1347–1348.

127. Spies HSC, Steenkamp DJ. 1994. Thiols of intracellular pathogens. Identification of ovothiol A in *Leishmania donovani* and structural analysis of a novel thiol from *Mycobacterium bovis*. *Eur J Biochem* **224**:203–213.

128. Newton GL, Bewley CA, Dwyer TJ, Horn R, Aharonowitz Y, Cohen G, Davies J, Faulkner DJ, Fahey RC. 1995. The structure of U17 isolated from *Streptomyces clavuligerus* and its properties as an antioxidant thiol. *Eur J Biochem* **230**:821–825.

129. Norman RA, McAlister MSB, Murray-Rust J, Movahedzadeh F, Stoker NG, McDonald NQ. 2002. Crystal structure of inositol 1-phosphate synthase from *Mycobacterium tuberculosis*, a key enzyme in phosphatidylinositol synthesis. *Structure* **10**:393–402.

130. Morita YS, Fukuda T, Sena CBC, Yamaryo-Botte Y, McConville MJ, Kinoshita T. 2011. Inositol lipid metabolism in mycobacteria: biosynthesis and regulatory mechanisms. *Biochim Biophys Acta* **1810:**630–641.

131. Newton GL, Koledin T, Gorovitz B, Rawat M, Fahey RC, Av-Gay Y. 2003. The glycosyltransferase gene encoding the enzyme catalyzing the first step of mycothiol biosynthesis (mshA). *J Bacteriol* **185:**3476–3479.

132. Newton GL, Ta P, Bzymek KP, Fahey RC. 2006. Biochemistry of the initial steps of mycothiol biosynthesis. *J Biol Chem* **281:**33910–33920.

133. Newton GL, Av-Gay Y, Fahey RC. 2000. N-Acetyl-1-D-myo-inosityl-2-amino-2-deoxy-α-D-glucopyranoside deacetylase (MshB) is a key enzyme in mycothiol biosynthesis. *J Bacteriol* **182:**6958–6963.

134. Newton GL, Ko M, Ta P, Av-Gay Y, Fahey RC. 2006. Purification and characterization of *Mycobacterium tuberculosis* 1D-myo-inosityl-2-acetamido-2-deoxy-a-D-glucopyranoside deacetylase, MshB, a mycothiol biosynthetic enzyme. *Protein Expr Purif* **47:**542–550.

135. Huang X, Kocabas E, Hernick M. 2011. The activity and cofactor preferences of N-acetyl-1-D-myo-inosityl-2-amino-2-deoxy-α-D-glucopyranoside deacetylase (MshB) change depending on environmental conditions. *J Biol Chem* **286(23):**20275–20282.

136. Fan F, Blanchard JS. 2009. Toward the catalytic mechanism of a cysteine ligase (MshC) from *Mycobacterium smegmatis*: an enzyme involved in the biosynthetic pathway of mycothiol. *Biochemistry* **48:**7150–7159.

137. Fan F, Luxenburger A, Painter GF, Blanchard JS. 2007. Steady-state and pre-steady-state kinetic analysis of *Mycobacterium smegmatis* cysteine ligase (MshC). *Biochemistry* **46:**11421–11429.

138. Vetting MW, Roderick SL, Yu M, Blanchard JS. 2003. Crystal structure of mycothiol synthase (Rv0819) from *Mycobacterium tuberculosis* shows structural homology to the GNAT family of N-acetyltransferases. *Protein Sci* **12:**1954–1959.

139. Koledin T, Newton GL, Fahey RC. 2002. Identification of the mycothiol synthase gene (*mshD*) encoding the acetyltransferase producing mycothiol in actinomycetes. *Arch Microbiol* **178:**331–337.

140. Vetting MW, Frantom PA, Blanchard JS. 2008. Structural and enzymatic analysis of MshA from *Corynebacterium glutamicum*. *J Biol Chem* **283:**15834–15844.

141. Vetting MW, Yu M, Rendle PM, Blanchard JS. 2006. The substrate-induced conformational change of *Mycobacterium tuberculosis* mycothiol synthase. *J Biol Chem* **281:**2795–2802.

142. Bone R, Frank L, Springer JP, Pollack SJ, Osborne S, Atack JR, Knowles MR, McAllister G, Ragan CI. 1994. Structural analysis of inositol monophosphatase complexes with substrates. *Biochemistry* **33:**9460–9467.

143. Gu X, Chen M, Shen H, Jiang X, Huang Y, Wang H. 2006. Rv2131c gene product: an unconventional enzyme that is both inositol monophosphatase and fructose-1,6-bisphosphatase. *Biochem Biophys Res Commun* **339:**897–904.

144. Hatzios SK, Iavarone AT, Bertozzi CR. 2008. Rv2131c from *Mycobacterium tuberculosis* is a CysQ 3′-phosphoadenosine-5′-phosphatase. *Biochemistry* **47:**5823–5831.

145. Morgan AJ, Wang YK, Roberts MF, Miller SJ. 2004. Chemistry and biology of deoxy-myo-inositol phosphates: stereospecificity of substrate interactions within an archaeal and a bacterial IMPase. *J Am Chem Soc* **126:**15370–15371.

146. Newton GL, Fahey RC. 2008. Regulation of mycothiol metabolism by σR and the thiol redox sensor anti-sigma factor RsrA. *Mol Microbiol* **68:**805–809.

147. Park J-H, Roe J-H. 2008. Mycothiol regulates and is regulated by a thiol-specific antisigma factor RsrA and σR in *Streptomyces coelicolor*. *Mol Microbiol* **68:**861–870.

148. Trivedi A, Singh N, Bhat SA, Gupta P, Kumar A. 2012. Redox biology of tuberculosis pathogenesis. *Adv Microb Physiol* **60:**263–324.

149. Newton GL, Unson MD, Anderberg SJ, Aguilera JA, Oh NN, delCardayre SB, Av-Gay Y, Fahey RC. 1999. Characterization of *Mycobacterium smegmatis* mutants defective in 1-d-myo-Inosityl-2-amino-2-deoxy-α-d-glucopyranoside and mycothiol biosynthesis. *Biochem Biophys Res Commun* **255:**239–244.

150. Xu X, Vilcheze C, Av-Gay Y, Gomez-Velasco A, Jacobs WR Jr. 2011. Precise null deletion mutations of the mycothiol synthesis genes reveal their role in isoniazid and ethionamide resistance in *Mycobacterium smegmatis*. *Antimicrob Agents Chemother* **55:**3133–3139.

151. Rawat M, Kovacevic S, Billman-Jacobe H, Av-Gay Y. 2003. Inactivation of mshB, a key gene in the mycothiol biosynthesis pathway in *Mycobacterium smegmatis*. *Microbiology* **149:**1341–1349.

152. Rawat M, Newton GL, Ko M, Martinez GJ, Fahey RC, Av-Gay Y. 2002. Mycothiol-deficient *Mycobacterium smegmatis* mutants are hypersensitive to alkylating agents, free radicals, and antibiotics. *Antimicrob Agents Chemother* **46:**3348–3355.

153. Newton GL, Ta P, Fahey RC. 2005. A mycothiol synthase mutant of *Mycobacterium smegmatis* produces novel thiols and has an altered thiol redox status. *J Bacteriol* **187:**7309–7316.

154. Almeida Da Silva PE, Palomino JC. 2011. Molecular basis and mechanisms of drug resistance in *Mycobacterium tuberculosis*: classical and new drugs. *J Antimicrob Chemother* **66:**1417–1430.

155. Newton GL, Buchmeier N, Fahey RC. 2008. Biosynthesis and functions of mycothiol, the unique protective thiol of actinobacteria. *Microbiol Mol Biol Rev* **72:**471–494.

156. Rawat M, Johnson C, Cadiz V, Av-Gay Y. 2007. Comparative analysis of mutants in the mycothiol biosynthesis pathway in *Mycobacterium smegmatis*. *Biochem Biophys Res Commun* **363:**71–76.

157. Buchmeier N, Fahey RC. 2006. The mshA gene encoding the glycosyltransferase of mycothiol biosynthesis is essential in *Mycobacterium tuberculosis* Erdman. *FEMS Microbiol Lett* **264:**74–79.

158. Vilchèze C, Av-Gay Y, Attarian R, Zhen L, Hazbón MH, Colangeli R, Chen B, Weijun L, Alland D, Sacchettini JC, Jacobs WR Jr. 2008. Mycothiol biosynthesis is essential for ethionamide susceptibility in *Mycobacterium tuberculosis*. *Mol Microbiol* **69**:1316–1329.

159. Buchmeier NA, Newton GL, Koledin T, Fahey RC. 2003. Association of mycothiol with protection of *Mycobacterium tuberculosis* from toxic oxidants and antibiotics. *Mol Microbiol* **47**:1723–1732.

160. Sareen D, Newton GL, Fahey RC, Buchmeier NA. 2003. Mycothiol is essential for growth of *Mycobacterium tuberculosis* Erdman. *J Bacteriol* **185**:6736–6740.

161. Buchmeier NA, Newton GL, Fahey RC. 2006. A mycothiol synthase mutant of *Mycobacterium tuberculosis* has an altered thiol-disulfide content and limited tolerance to stress. *J Bacteriol* **188**:6245–6252.

162. Rengarajan J, Bloom BR, Rubin EJ. 2005. Genome-wide requirements for *Mycobacterium tuberculosis* adaptation and survival in macrophages. *Proc Natl Acad Sci USA* **102**:8327–8332.

163. Patel MP, Blanchard JS. 1999. Expression, purification, and characterization of *Mycobacterium tuberculosis* mycothione reductase. *Biochemistry* **38**:11827–11833.

164. Patel MP, Blanchard JS. 2001. *Mycobacterium tuberculosis* mycothione reductase: pH dependence of the kinetic parameters and kinetic isotope effects. *Biochemistry* **40**:5119–5126.

165. Hayward D, Wiid I, van Helden P. 2004. Differential expression of mycothiol pathway genes: are they affected by antituberculosis drugs? *IUBMB Life* **56**:131–138.

166. Van Laer K, Buts L, Foloppe N, Vertommen D, Van Belle K, Wahni K, Roos G, Nilsson L, Mateos LM, Rawat M, van Nuland NAJ, Messens J. 2012. Mycoredoxin-1 is one of the missing links in the oxidative stress defence mechanism of mycobacteria. *Mol Microbiol* **86**:787–804.

167. Vogt RN, Steenkamp DJ, Zheng RJ, Blanchard JS. 2003. The metabolism of nitrosothiols in the mycobacteria: identification and characterization of S-nitrosomycothiol reductase. *Biochem J* **375**:657–665.

168. Miller CC, Rawat M, Johnson T, Av-Gay Y. 2007. Innate protection of *Mycobacterium smegmatis* against the antimicrobial activity of nitric oxide is provided by mycothiol. *Antimicrob Agents Chemother* **51**:3364–3366.

169. Newton GL, Av-Gay Y, Fahey RC. 2000. A novel mycothiol-dependent detoxification pathway in mycobacteria involving mycothiol S-conjugate amidase. *Biochemistry* **39**:10739–10746.

170. Steffek M, Newton GL, Av-Gay Y, Fahey RC. 2003. Characterization of *Mycobacterium tuberculosis* mycothiol S-conjugate amidase. *Biochemistry* **42**:12067–12076.

171. Rawat M, Uppal M, Newton G, Steffek M, Fahey RC, Av-Gay Y. 2004. Targeted mutagenesis of the *Mycobacterium smegmatis* mca gene, encoding a mycothiol-dependent detoxification protein. *J Bacteriol* **186**:6050–6058.

172. Chi H-W, Huang C-C, Chin D-H. 2012. Thiols screened by the neocarzinostatin protein for preserving or detoxifying its bound enediyne antibiotic. *Chemistry* **18**:6238–6249.

173. Nicholas GM, Eckman LL, Newton GL, Fahey RC, Ray S, Bewley CA. 2003. Inhibition and kinetics of *Mycobacterium tuberculosis* and *Mycobacterium smegmatis* mycothiol-S-conjugate amidase by natural product inhibitors. *Bioorg Med Chem* **11**:601–608.

174. Nicholas GM, Eckman LL, Ray S, Hughes RO, Pfefferkorn JA, Barluenga S, Nicolaou KC, Bewley CA. 2002. Bromotyrosine-derived natural and synthetic products as inhibitors of mycothiol-S-conjugate amidase. *Bioorg Med Chem Lett* **12**:2487–2490.

175. Fetterolf B, Bewley CA. 2004. Synthesis of a bromotyrosine-derived natural product inhibitor of mycothiol-S-conjugate amidase. *Bioorg Med Chem Lett* **14**:3785–3788.

176. Metaferia BB, Ray S, Smith JA, Bewley CA. 2007. Design and synthesis of substrate-mimic inhibitors of mycothiol-S-conjugate amidase from *Mycobacterium tuberculosis*. *Bioorg Med Chem Lett* **17**:444–447.

177. Bzymek KP, Newton GL, Ta P, Fahey RC. 2007. Mycothiol import by *Mycobacterium smegmatis* and function as a resource for metabolic precursors and energy production. *J Bacteriol* **189**:6796–6805.

178. Anderberg SJ, Newton GL, Fahey RC. 1998. Mycothiol biosynthesis and metabolism. Cellular levels of potential intermediates in the biosynthesis and degradation of mycothiol in *Mycobacterium smegmatis*. *J Biol Chem* **273**:30391–30397.

179. Seebeck FP. 2010. In vitro reconstitution of mycobacterial ergothioneine biosynthesis. *J Am Chem Soc* **132**:6632–6633.

180. Tanret C. 1909. Sur une base novelle retirée du seigle ergote, l'ergothioneine. *Compt Rend* **149**:222–224.

181. Barger G, Ewins AJ. 1911. The constitution of ergothioneine: a betaine related to histidine. *J Chem Soc Trans* **99**:2336–2341.

182. Pfeiffer C, Bauer T, Surek B, Schomig E, Grundemann D. 2011. Cyanobacteria produce high levels of ergothioneine. *Food Chem* **129**:1766–1769.

183. Genghof DS, Vandamme O. 1964. Biosynthesis of ergothioneine and hercynine by mycobacteria. *J Bacteriol* **87**:852–862.

184. Genghof DS, Inamine E, Kovalenko V, Melville DB. 1956. Ergothioneine in microorganisms. *J Biol Chem* **223**:9–17.

185. Melville DB. 1959. Ergothioneine, p. 155–204. *In* Harris RH, Marrian GF, Thimann KV (ed), *Vitamins & Hormones*, vol. **17**. Academic Press, New York.

186. Melville DB, Eich S. 1956. The occurrence of ergothioneine in plant material. *J Biol Chem* **218**:647–651.

187. Fahey RC. 2013. Glutathione analogs in prokaryotes. *Biochim Biophys Acta* **1830**:3182–3198.

188. Misra HP. 1974. Generation of superoxide free radical during the autoxidation of thiols. *J Biol Chem* **249**:2151–2155.

189. Hua Long L, Halliwell B. 2001. Oxidation and generation of hydrogen peroxide by thiol compounds in commonly used cell culture media. *Biochem Biophys Res Commun* **286**:991–994.

190. Mayumi T, Kawano H, Sakamoto Y, Suehisa E, Kawai Y, Hama T. 1978. Studies on ergothioneine. V. Determination by high performance liquid chromatography and application to metabolic research. *Chem Pharm Bull* **26**:3772–3778.

191. Heath H, Toennies G. 1958. The preparation and properties of ergothioneine disulphide. *Biochem J* **68**:204–210.

192. Seebeck FP. 2010. In vitro reconstitution of mycobacterial ergothioneine biosynthesis. *J Am Chem Soc* **132**:6632–6633.

193. Grundemann D, Harlfinger S, Golz S, Geerts A, Lazar A, Berkels R, Jung N, Rubbert A, Schomig E. 2005. Discovery of the ergothioneine transporter. *Proc Natl Acad Sci USA* **102**:5256–5261.

194. Hanlon DP. 1971. Interaction of ergothioneine with metal ions and metalloenzymes. *J Med Chem* **14**:1084–1087.

195. Motohashi N, Mori I, Sugiura Y. 1976. Complexing of copper ion by ergothioneine. *Chemical Pharm Bull* **24**:2364–2368.

196. Zhu BZ, Mao L, Fan RM, Zhu JG, Zhang YN, Wang J, Kalyanaraman B, Frei B. 2011. Ergothioneine prevents copper-induced oxidative damage to DNA and protein by forming a redox-inactive ergothioneine-copper complex. *Chem Res Toxicol* **24**:30–34.

197. Kawano H, Higuchi F, Mayumi T, Hama T. 1982. Studies on ergothioneine. VII. Some effects on ergothioneine on glycolytic metabolism in red blood cells from rats. *Chem Pharm Bull* **30**:2611–2613.

198. Laurenza I, Colognato R, Migliore L, Del Prato S, Benzi L. 2008. Modulation of palmitic acid-induced cell death by ergothioneine: evidence of an anti-inflammatory action. *Biofactors* **33**:237–247.

199. Rahman I, Gilmour PS, Jimenez LA, Biswas SK, Antonicelli F, Aruoma OI. 2003. Ergothioneine inhibits oxidative stress- and TNF-alpha-induced NF-kappa B activation and interleukin-8 release in alveolar epithelial cells. *Biochem Biophys Res Commun* **302**:860–864.

200. Akanmu D, Cecchini R, Aruoma OI, Halliwell B. 1991. The antioxidant action of ergothioneine. *Arch Biochem Biophys* **288**:10–16.

201. Aruoma OI, Spencer JP, Mahmood N. 1999. Protection against oxidative damage and cell death by the natural antioxidant ergothioneine. *Food Chem Toxicol* **37**:1043–1053.

202. Mitsuyama H, May JM. 1999. Uptake and antioxidant effects of ergothioneine in human erythrocytes. *Clin Sci (Lond)* **97**:407–411.

203. Reglinski J, Smith WE, Sturrock RD. 1988. Spin-echo 1H NMR detected response of ergothioneine to oxidative stress in the intact human erythrocyte. *Magn Reson Med* **6**:217–223.

204. Hartman PE. 1990. Ergothioneine as antioxidant. *Methods Enzymol* **186**:310–318.

205. Ey J, Schomig E, Taubert D. 2007. Dietary sources and antioxidant effects of ergothioneine. *J Agric Food Chem* **55**:6466–6474.

206. Paul BD, Snyder SH. 2010. The unusual amino acid L-ergothioneine is a physiologic cytoprotectant. *Cell Death Differ* **17**:1134–1140.

207. Bello MH, Barrera-Perez V, Morin D, Epstein L. 2012. The *Neurospora crassa* mutant NcDeltaEgt-1 identifies an ergothioneine biosynthetic gene and demonstrates that ergothioneine enhances conidial survival and protects against peroxide toxicity during conidial germination. *Fungal Genet Biol* **49**:160–172.

208. Ta P, Buchmeier N, Newton GL, Rawat M, Fahey RC. 2011. Organic hydroperoxide resistance protein and ergothioneine compensate for loss of mycothiol in *Mycobacterium smegmatis* mutants. *J Bacteriol* **193**:1981–1990.

209. Fraser RS. 1951. Blood ergothioneine levels in disease. *J Lab Clin Med* **37**:199–206.

210. Mahasirimongkol S, Yanai H, Nishida N, Ridruechai C, Matsushita I, Ohashi J, Summanapan S, Yamada N, Moolphate S, Chuchotaworn C, Chaiprasert A, Manosuthi W, Kantipong P, Kanitwittaya S, Sura T, Khusmith S, Tokunaga K, Sawanpanyalert P, Keicho N. 2009. Genome-wide SNP-based linkage analysis of tuberculosis in Thais. *Genes Immun* **10**:77–83.

211. Ridruechai C, Mahasirimongkol S, Phromjai J, Yanai H, Nishida N, Matsushita I, Ohashi J, Yamada N, Moolphate S, Summanapan S, Chuchottaworn C, Manosuthi W, Kantipong P, Kanitvittaya S, Sawanpanyalert P, Keicho N, Khusmith S, Tokunaga K. 2010. Association analysis of susceptibility candidate region on chromosome 5q31 for tuberculosis. *Genes Immun* **11**:416–422.

212. Chen F, Xia Q, Ju LK. 2003. Aerobic denitrification of *Pseudomonas aeruginosa* monitored by online NAD(P)H fluorescence. *Appl Environ Microbiol* **69**:6715–6722.

213. Newton GL, Fahey RC. 2002. Mycothiol biochemistry. *Arch Microbiol* **178**:388–394.

214. Hwang C, Sinskey AJ, Lodish HF. 1992. Oxidized redox state of glutathione in the endoplasmic reticulum. *Science* **257**:1496–1502.

215. Austin CD, Wen X, Gazzard L, Nelson C, Scheller RH, Scales SJ. 2005. Oxidizing potential of endosomes and lysosomes limits intracellular cleavage of disulfide-based antibody-drug conjugates. *Proc Natl Acad Sci USA* **102**:17987–17992.

216. Ostergaard H, Henriksen A, Hansen FG, Winther JR. 2001. Shedding light on disulfide bond formation: engineering a redox switch in green fluorescent protein. *EMBO J* **20**:5853–5862.

217. Tuggle CK, Fuchs JA. 1985. Glutathione reductase is not required for maintenance of reduced glutathione in *Escherichia coli* K-12. *J Bacteriol* **162**:448–450.

218. Kumar A, Farhana A, Guidry L, Saini V, Hondalus M, Steyn AJ. 2011. Redox homeostasis in mycobacteria: the key to tuberculosis control? *Expert Rev Mol Med* **13**:e39.

219. Bornemann C, Jardine MA, Spies HSC, Steenkamp DJ. 1997. Biosynthesis of mycothiol: elucidation of the sequence of steps in *Mycobacterium smegmatis*. *Biochem J* 325:623–629.

220. Wang LZ, Thuya WL, Toh DS, Lie MG, Lau JY, Kong LR, Wan SC, Chua KN, Lee EJ, Goh BC. 2013. Quantification of L-ergothioneine in human plasma and erythrocytes by liquid chromatography-tandem mass spectrometry. *J Mass Spectrom* 48:406–412.

221. Holsclaw CM, Muse WB 3rd, Carroll KS, Leary JA. 2011. Mass spectrometric analysis of mycothiol levels in wild-type and mycothiol disulfide reductase mutant *Mycobacterium smegmatis*. *Int J Mass Spectrom* 305:151–156.

222. Ung KSE, Av-Gay Y. 2006. Mycothiol-dependent mycobacterial response to oxidative stress. *FEBS Lett* 580:2712–2716.

223. Seok J, Warren HS, Cuenca AG, Mindrinos MN, Baker HV, Xu W, Richards DR, McDonald-Smith GP, Gao H, Hennessy L, Finnerty CC, Lopez CM, Honari S, Moore EE, Minei JP, Cuschieri J, Bankey PE, Johnson JL, Sperry J, Nathens AB, Billiar TR, West MA, Jeschke MG, Klein MB, Gamelli RL, Gibran NS, Brownstein BH, Miller-Graziano C, Calvano SE, Mason PH, Cobb JP, Rahme LG, Lowry SF, Maier RV, Moldawer LL, Herndon DN, Davis RW, Xiao W, Tompkins RG. 2013. Genomic responses in mouse models poorly mimic human inflammatory diseases. *Proc Natl Acad Sci USA* 110:3507–3512.

Molecular Genetics of Mycobacteria, 2nd Edition
Edited by Graham F. Hatfull and William R. Jacobs, Jr.
© 2014 American Society for Microbiology, Washington, DC
doi:10.1128/microbiolspec.MGM2-0026-2013

Anthony D. Baughn[1]
Kyu Y. Rhee[2]

Metabolomics of Central Carbon Metabolism in *Mycobacterium tuberculosis*

16

Central carbon metabolism (CCM)—defined as the enzymatic transformation of carbon through glycolysis, the pentose phosphate pathway (PPP), the citric acid cycle, the glyoxylate shunt, the methylcitrate cycle, and gluconeogenesis—is a core feature of all cells that is used to provide energy, in the form of reducing equivalents and ATP, and essential biosynthetic precursors (Fig. 1). Remarkably, the same metabolic enzymes found in bacteria are also present in mammals, suggesting that the pathways of CCM have been fundamentally conserved. Intensive study has thus focused on the metabolic network of *Escherichia coli* as a model system with which to understand its basic principles. However, cells vary in their specific metabolic needs according to the selective pressures they encounter. Accordingly, growing evidence has established that metabolic enzymes are often capable of operating in a diverse array of configurations (1).

For organisms residing within polymicrobial ecosystems, it has generally been assumed that the objective goal of CCM is to meet the stoichiometric requirements needed to sustain maximal rates of replication with maximal efficiency. However, *Mycobacterium tuberculosis* is both microbiologically and ecologically unique. *M. tuberculosis* is a chronic intracellular pathogen that resides in humans as its only known host and reservoir, and within humans, the macrophage phagosome as its chief niche (2). *M. tuberculosis* has thus evolved, apart from other microbes, within an ultra-narrow ecologic niche corresponding to the cell type and compartment most committed to its eradication. Within its host niche, *M. tuberculosis* encounters diverse host-imposed stringencies, many of which are capable of inducing exit from the cell cycle. *M. tuberculosis* thus occupies the majority of its natural life cycle in a state of non- or slowly replicating physiology.

Relieved of the requirement to double biomass, non- or slowly replicating *M. tuberculosis* has generally been perceived to have minimal metabolic activity. However, even nonreplicating bacilli face the challenge of preserving the integrity of essential cellular components and maintaining the ability to reenter the cell cycle to ensure transmission to a new host. Yet its physiology has been most extensively studied when replicating at

[1]Department of Microbiology, University of Minnesota, Minneapolis, MN 55455; [2]Department of Medicine, Weill Cornell Medical College, New York, NY 10065.

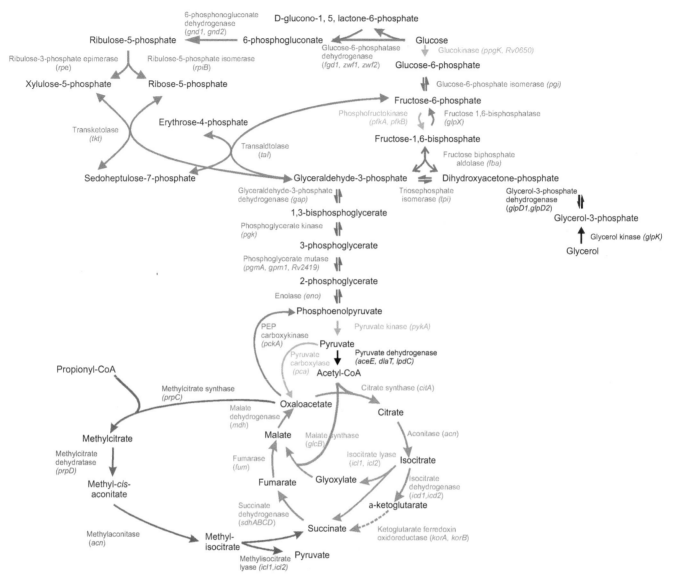

Figure 1   Schematic representation of the CCM network of *M. tuberculosis* based on bioinformatic reconstruction and published literature. Reversible steps of glycolysis/gluconeogenesis are shown in purple, dedicated steps of glycolysis are shown in light blue, dedicated steps of gluconeogenesis are shown in pink, the PPP is shown in red, the TCA cycle is shown in green, the glyoxylate shunt is shown in orange, and the methylcitrate cycle is shown in dark blue. The dotted green line indicates a standing question of connectivity in the TCA cycle via ketoglutarate ferredoxin oxidoreductase. Common enzyme names are shown next to the reactions they catalyze. Gene symbols are shown in parentheses. doi:10.1128/microbiolspec.MGM2-0026-2013.f1

maximal growth rates. Knowledge of the specific metabolic requirements and pathways used by both replicating and nonreplicating *M. tuberculosis* subpopulations thus represents major areas of unmet scientific need.

Like most bacteria, *M. tuberculosis* is capable of utilizing an array of organic substrates, including select carbohydrates, lipids, amino acids, and simple organic

acids, to populate its CCM. Decades of studies in the optimization of batch culture conditions for *in vitro* growth of *M. tuberculosis* revealed that glycerol supports maximal growth rates and yields (3). In contrast, studies with *ex vivo* bacilli recovered from the lungs of infected mice showed that lipids appeared to function as *M. tuberculosis*'s preferred respiratory substrate (4).

Multiple lines of evidence from *in vivo* studies, including gene expression analyses, as well as infection studies using defined mutant strains have further reinforced essential roles for lipid metabolism in both *M. tuberculosis* pathogenesis and persistence (1). Based on such studies, lipids such as fatty acids and cholesterol have come to be viewed as the predominant carbon and energy source for *M. tuberculosis* throughout infection, discounting a role for carbohydrates. Yet growing evidence has implicated an essential role for carbohydrate uptake and metabolism in the chronic phase of infection (5, 6).

In this article we review existing knowledge of *M. tuberculosis*'s CCM as reported by studies of its basic genetic and biochemical composition, regulation, and organization. This work is meant to build off a previous synthesis of the literature reported by Wheeler and Blanchard (7) and two recent minireviews by Rhee et al. (1) and Ehrt and Rhee (8) and thus focuses on complementing, rather than duplicating, information contained therein.

CARBOHYDRATE METABOLISM

Early work reported that *M. tuberculosis* was capable of utilizing both carbohydrate- and alcohol-based substrates during *in vitro* culture. However, in a systematic study of 24 substrates, Youmans and Youmans found that only glycerol, glucose, trehalose, and maltose (which may have contained trace amounts of trehalose, which is commonly used as a cryoprotectant) were capable of supporting its growth when provided as the primary carbon source in a chemically defined, basal medium (9). Gamble and Herrick further showed that the extent of glucose utilization by *M. tuberculosis* roughly paralleled its growth, while Loebel reported that metabolism of only glucose or glycerol was accompanied by increases in respiratory activity (10, 11). Youmans also reported the ability of gluconate, pyruvate, and lactate to support *in vitro* growth of *M. tuberculosis*, while Weinzirl and Knapton found that metabolism of only glycerol, but not glucose, lactose, or mannitol, led to acidification of the medium (12, 13). Together, these findings provided the first evidence for both an intact glycolytic pathway in *M. tuberculosis* and its coupling to an oxidative respiratory chain.

Work by Goldman, Murthy, and others subsequently adduced biochemical evidence for the presence of enzymes of the Embden-Meyerhof-Parnas (EMP) pathway and PPP in lysates of *M. tuberculosis* H37Rv (14, 15). Radiorespirometric studies using *M. tuberculosis* incubated in phosphate-buffered saline containing ^{14}C-labeled glucose labeled at the 1, 2, 3-4, and 6 positions further showed the preferential recovery of respired $^{14}CO_2$ from glucose labeled at the 3-4 positions, associated with an estimated ~94% metabolism through the EMP pathway that was accompanied by the detection of hexokinase, phosphofructokinase, fructose bisphosphate aldolase, and pyruvate kinase activities (16). The remaining 6% was attributed to metabolism through the PPP, owing to detection of glucose-6-phosphate dehydrogenase and 6-phosphogluconate dehydrogenase and absence of 6-phosphogluconate dehydratase and 2-keto-3-deoxy-6-phosphogluconate (KDPG) aldolase activities associated with the Entner-Doudoroff pathway. Of interest, Youmans reported that while gluconate could support *M. tuberculosis* growth in a minimal Proskauer-Beck medium, pentose sugars could not, raising the possibility of a functional, though as yet unsubstantiated, Entner-Doudoroff pathway (9, 13). Analogous studies revealed the presence of robust 6-phosphogluconate dehydratase and KDPG aldolase activities in the lysates of the saprophytic species *Mycobacterium smegmatis*, albeit only when grown on glucose (17).

Completion of the first *M. tuberculosis* genome sequence, coupled to homology-based sequence analysis, later provided an *in silico* inventory of *M. tuberculosis* CCM enzymes (18). This approach predicted the presence of candidates for all enzymes of both the EMP pathway and PPP, but also revealed unexpected apparent redundancies for some enzymes (such as its hexokinases, phosphofructokinases, etc.) while orphaning other previously detected activities (such as an alternative type I fructose bisphosphate aldolase).

Below, we provide a synthesis of the published literature concerning the specific enzymes of the *M. tuberculosis* glycolytic, gluconeogenic, and pentose phosphate pathways.

EMP PATHWAY AND GLUCONEOGENIC COUNTERPARTS

Glucokinase (E.C. 2.7.1.2) catalyzes the first committed step in glucose metabolism, the production of glucose-6-phosphate, an intermediate shared by both the EMP pathway and the PPP. The *M. tuberculosis* genome encodes two functional glucokinases: PPGK (*Rv2702*) (18) and GLKA (*Rv0650*) (6). Biochemical studies showed that PPGK catalyzes the direct phosphorylation of glucose using a broad range of phosphoryl donors with a preference for polyphosphate (PolyP) and, unlike most glucokinases, has a high affinity for its substrate, glucose (K_m = 0.28 mM) (19, 20). *glkA*

(*Rv0650*) encodes an annotated sugar kinase with only 22% amino acid identity to PPGK but 71% identity to a homolog from *M. smegmatis* which was shown to phosphorylate glucose (K_m = 9 mM) using ATP as the phosphoryl donor (21). Direct biochemical characterization of the *M. tuberculosis* enzyme, however, remains lacking. Genetic studies showed that targeted deletion of both PPGK and GLKA selectively abolished *M. tuberculosis*'s ability to grow and metabolize glucose, but not acetate, in a manner that could be restored by expression of either ortholog, albeit at different levels of expression. PPGK/GLKA-deficient *M. tuberculosis* strains were further found to exhibit a 1 to 2 $\log_{10}$ defect in bacterial burden during the chronic phase of infection in a mouse model of tuberculosis (TB) that could be attributed to neither a defect in the metabolism of trehalose nor enhanced susceptibility to nutrient starvation, glucose intoxication, or oxidative stress (6). Together, these data indicate that *M. tuberculosis* expresses two functional glucokinases whose activities are jointly required for *in vitro* growth and metabolism in glucose-containing media and *in vivo* persistence in a mouse model of TB, but whose specific physiologic roles in glycolysis remain to be resolved.

Glucose-6-phosphate isomerase (PGI; E.C. 5.3.1.9) catalyzes the reversible conversion of glucose-6-phosphate into fructose-6-phosphate, the second step of the EMP, which is required for stabilization of the carbanion required for the action of fructose bisphosphate aldolase (discussed below) and is encoded by *pgi* (*Rv0946c*). *In vitro* studies of purified recombinant *M. tuberculosis* PGI have confirmed its activity as a metal-independent PGI with kinetic parameters similar to those reported for other bacterial and eukaryotic PGIs (K_m = 0.318 mM for fructose-6-phosphate; K_i = 0.8 mM for 6-phosphogluconate) (22). Insertional transposon mutagenesis studies have provided genetic evidence that PGI is essential for optimal growth *in vitro* (23, 24).

In addition to PGI, *M. tuberculosis* also encodes a phosphoglucomutase (PGM; E.C. 5.4.2.2; *pgmA* [*Rv3068c*]), which can catalyze the reversible transfer of a phosphoryl group from the 1′ to the 6′ position on a glucose monomer (25). Glucose-1-phosphate is an intermediate of glycogen metabolism produced by the action of glycogen phosphorylase. PGM can thus generate glucose-6-phosphate as either a substrate or product for glycolytic or gluconeogenic metabolism, respectively. Insertional transposon mutagenesis data indicate that PGM is dispensable for optimal *in vitro* growth (23, 24).

Phosphofructokinase (PFK; E.C. 2.7.1.11) catalyzes the ATP-dependent phosphorylation of fructose-6-phosphate to produce fructose-1,6-bisphosphate, the first committed step of the EMP pathway, and is predicted to be encoded by two paralogous genes, *pfkA* (*Rv3010c*) and *pfkB* (*Rv2029c*), neither of which is predicted to be essential for *in vitro* and *in vivo* growth (5, 23, 24). Although both PFKA and PFKB are annotated as phosphofructokinases, each bears little sequence similarity to the other and belongs to a separate protein family, corresponding to the canonical PFK protein family and ribokinase superfamily, respectively. *In vitro* studies using hexahistidine-tagged recombinant forms of *M. tuberculosis* PFKA and PFKB confirmed that both were capable of phosphorylating fructose-6-phosphate, with PFKA exhibiting nearly 15 times as much activity as PFKB. Unlike the case for *E. coli* (where PFKB accounts for ~10% of total PFK activity), however, a genetically engineered *M. tuberculosis* PFKA deletion mutant was found to be unable to grow on glucose *in vitro* and to lack detectable PFK activity, while a PFKB deletion mutant exhibited no apparent defect in growth on glucose with wild-type levels of PFK activity in lysates. Moreover, neither the growth nor enzymatic defects observed in the PFKA mutant could be rescued by confirmed expression of *pfkB* under the control of a constitutive strong promoter. A PFK-deficient *E. coli* mutant could conversely be complemented only when expressing *M. tuberculosis* PFKA and not PFKB.

Together, these findings establish that *M. tuberculosis* PFKA is both necessary and sufficient to support *in vitro* growth and metabolism of glucose through the EMP pathway. PFKA-deficient *M. tuberculosis* was also found to exhibit no defect in its ability to establish or maintain a chronic infection in mice, suggesting that either *M. tuberculosis* encodes a cryptic or compensatory PFK activity triggered by factors specifically encountered in the host, or that glycolysis is dispensable *in vivo* (26). Of note, *M. tuberculosis* PFKB shares 40% identity with *E. coli* PFKB, including the conserved catalytic motif GXGD that is indicative of ribokinase superfamily members that have been reported to phosphorylate a variety of substrates besides fructose-6-phosphate, such as fructose-1-phosphate in *E. coli* (27, 28) and tagatose-6-phosphate in both *S. aureus* (29) and *E. coli*, albeit with a lower efficacy than fructose-6-phosphate (30). It is thus plausible that *M. tuberculosis* PFKB may also phosphorylate sugar-based substrates other than fructose-6-phosphate. More extensive characterization of the kinetic properties and substrate specificities of *M. tuberculosis*'s PFKs and PFKA/PFKB-deficient *M. tuberculosis* thus await further investigation.

Fructose-1,6-bisphosphatase (E.C. 3.1.3.11) is a key enzyme of gluconeogenesis that catalyzes the dephosphorylation of fructose-1,6-bisphosphate (FBP) to fructose-6-phosphate. The *M. tuberculosis* genome is annotated to encode a single type II, metal-ion-dependent FBPase, GlpX (*Rv1099c*), whose structure and activity have been characterized in purified recombinant form, with kinetic parameters similar to those of other type II FBPases (31, 32). Though not essential for *in vitro* growth on glycolytic carbon sources, insertional transposon mutagenesis studies have implicated *glpX* as being essential for *in vivo* growth in a mouse model of tuberculosis (5, 23, 24).

FBP aldolase (E.C. 4.1.2.13) catalyzes the reversible aldol cleavage of FBP and condensation of dihydroxyacetone phosphate (DHAP) and glyceraldehyde-3-phosphate (G3P) in glycolysis and gluconeogenesis, respectively. Though associated with structurally a conserved $(\beta/\alpha)_8$ barrel fold, FBP aldolases are divided into two groups according to reaction mechanism (33). Class I FBP aldolases, found in both higher eukaryotes and bacteria, form a Schiff-base intermediate between the carbonyl substrate (FBP or DHAP) and an active site lysine residue (34). Class II FBP aldolases, by contrast, are absent from mammals and require a divalent metal ion to polarize the carbonyl group of the substrate (FBP or DHAP) and to stabilize the carbanion intermediate during catalysis (35–37). Early biochemical studies by Venkitasubramanian and colleagues reported that *M. tuberculosis* encoded both class I and class II FBP aldolases, as distinguished by their sensitivity to borohydride inactivation and EDTA, respectively. The former activity was detected only in well-oxygenated, fermentor-grown cultures, while the latter was primarily associated with hypoxic conditions (38–40). Unfortunately, homology-based sequence analyses identified only a single paralog, corresponding to a class I FBP aldolase encoded by *fba* (*Rv0363c*). Both transposon mutagenesis data and unsuccessful gene deletion efforts in the attenuated *M. tuberculosis* strain H37Ra have suggested that *fba* is essential for *in vitro* growth. Biochemical and structural studies have confirmed its Zn cofactor and tetrameric (dimer of dimers) structure with an apparent turnover number ($21\,s^{-1}$ at 28°C) higher than that reported for other class II FBP aldolases, such as from *E. coli* ($10.5\,s^{-1}$ at 30°C) (37), and apparent K_m ($20\,\mu M$) approximately 10-fold lower than most FBP aldolases (33, 37, 41), resulting in one of the highest catalytic efficiencies (k_{cat}/K_m) for FBP cleavage. Efforts to develop cell-permeable, high-affinity inhibitors as potential novel TB drugs are thus under way.

Triose phosphate isomerase (TPI; E.C. 5.3.1.1) catalyzes the essential interconversion of DHAP and glyceraldehyde-3-phosphate and is one of the best-studied and catalytically most proficient enzymes ever characterized (42). In glycolysis, TPI channels these two products for the formation of pyruvate, while in gluconeogenesis, TPI ensures that both substrates are supplied in equimolar amounts to FBP aldolase. TPI is encoded by *tpiA* (*Rv1438*) and has been both structurally and kinetically characterized with features similar to those of other prototypic TPIs (43, 44).

In addition to its canonical roles in glycolysis and gluconeogenesis, TPI also serves an essential role in glycerol assimilation and/or glycerophospholipid metabolism. Early studies by Goldman demonstrated that cell-free extracts of *M. tuberculosis* H37Ra contained both glycerol dehydrogenase (E.C. 1.1.1.6) and dihydroxyacetone kinase (E.C. 2.7.1.29) activities and thus enable *M. tuberculosis* to assimilate glycerol into glycolysis and/or gluconeogenesis through its conversion to DHAP (45). Unfortunately, the genes encoding these orphan activities remain to be identified. Subsequent biochemical and genetic work provided additional evidence for the presence of a parallel pathway, consisting of glycerol kinase (E.C. 2.7.1.30; encoded by *glpK* [*Rv3696c*]) and both NAD(P)$^+$ and quinol-dependent glycerol-3-phosphate dehydrogenase (E.C. 1.1.1.94, 1.1.5.3; putatively encoded by *gpdA1* [*Rv0564c*] and *gpdA2* [*Rv2982c*] and by *glpD1* [*Rv2249c*] and *glpD2* [*Rv3302c*], respectively) orthologs, also capable of assimilating glycerol into glycolysis and/or gluconeogenesis through the production of DHAP. Not surprisingly, transposon mutagenesis data indicate that none of these genes is essential in isolation, though biochemical studies of these enzymes remain lacking (23, 24).

Glyceraldehyde-3-phosphate dehydrogenase (GAPDH; E.C. 1.2.1.12) catalyzes the conversion of glyceraldehyde-3-phosphate to D-glycerate-1,3-bisphosphate. GAPDH is encoded by a close sequence ortholog to other GAPDHs, *gap* (*Rv1436*), which is predicted to be essential by insertional transposon mutagenesis studies and genomically clustered with its TPI and phosphoglycerate kinase.

Unfortunately, specific biochemical data relating to *M. tuberculosis*'s GAPDH and lower glycolytic enzymes —including phosphoglycerokinase (E.C. 2.7.2.3; *pgk* [*Rv1437*]), phosphoglyceromutase (E.C. 5.4.2.1; *gpm1* [*Rv0489*, *Rv3837c*, *Rv2419c*]), enolase (E.C. 4.2.1.11; *eno* [*Rv1023*]), pyruvate kinase (E.C. 2.7.1.40; *pykA* [*Rv1617*]), and pyruvate phosphate dikinase (E.C. 2.7.9.1; *ppdK* [*Rv1127c*])—remain nearly absent. Notwithstanding, *M. tuberculosis* appears to encode an intact EMP pathway.

The EMP pathway is among the most conserved and extensively studied metabolic pathways in all of biology. Its main functions are to serve as a source of key biosynthetic precursors and, in the absence of respiratory chain activity, furnish cellular energy in the form of ATP. However, fundamental questions concerning its physiologic role and regulation remain. For example, unlike other bacteria, *M. tuberculosis* is not subject to classical catabolite repression and is instead capable of cocatabolizing multiple carbon sources (46). Moreover, isotopic labeling studies specifically showed that *M. tuberculosis* is capable of simultaneously metabolizing glucose and acetate through the segregated operation of its glycolytic and gluconeogenic pathways, a property potentially well adapted to the nutrient-poor conditions associated with the macrophage phagosome (1). However, a key question that arises is just what specific functions and precursors *M. tuberculosis*'s EMP has evolved to optimize production of? For organisms such as *M. tuberculosis* that occupy nutrient- and energy-poor niches and face little competition for resources, it is possible that their EMP pathway may have evolved to generate ATP at a higher yield at the expense of a lower metabolic rate using biochemical, rather than genetic, regulatory mechanisms.

THE PPP

Genomic analysis of all sequenced *M. tuberculosis* strains has revealed a fundamental conservation of orthologs for all genes of both the oxidative and nonoxidative branches. The canonical PPP metabolizes glucose to produce reducing equivalents (in the form of NADPH), nucleotide precursors (in the form of pentoses), and triose intermediates to fuel lower glycolysis, depending on the needs of the cell. The pathway itself consists of two different phases. One is an irreversible oxidative phase in which glucose-6-phosphate is converted to ribulose-5-phosphate by oxidative decarboxylation to generate NADPH and/or intermediates of the Entner-Doudoroff pathway discussed above. The other is a reversible nonoxidative phase in which phosphorylated sugars are interconverted to generate xylulose-5-phosphate, ribulose-5-phosphate, and ribose-5-phosphate, from which intermediates of lower glycolysis or phosphoribosyl pyrophosphate (PRPP), a key cofactor required for synthesis of histidine and purine/pyrimidine nucleotides, are formed.

In contrast to the EMP pathway, considerably less is known about the biochemical activities and metabolic roles of *M. tuberculosis*'s PPP enzymes. Apart from insertional transposon mutagenesis data (which suggest that only its transketolase and transaldolase are essential), biochemical knowledge of its enzymes is specifically limited to its glucose-6-phosphate dehydrogenase, ribose-5-phosphate isomerase, and transketolase, which are reviewed below (23, 24).

Glucose-6-phosphate dehydrogenase (G6PD; E.C. 1.1.1.49) converts glucose-6-phosphate into 6-phosphonoglucono-δ-lactone and is the rate-limiting enzyme of the PPP. Unlike most bacteria, including those containing the unusual flavin F420 cofactor, all mycobacterial species, including *M. tuberculosis*, encode two paralogous G6PD activities: a canonical NADP-dependent glucose-6-phosphate dehydrogenase (47, 48), which is found in most organisms ranging from bacteria to yeasts to humans (and in *M. tuberculosis* is encoded by two orthologs, *zwf1* [*Rv1121*]) and *zwf2* [*Rv1447c*]), and an evolutionarily distinct, F420-dependent G6PD (which in *M. tuberculosis* also consists of two orthologs encoded by *fgd1* [*Rv0407*] and *fgd2* [*Rv0132c*]) (48). Enzymatic assays using cell-free lysates suggested that the NADP and F420-dependent activities were similar, leaving their specific roles and apparent redundancy unresolved (49).

Ribose-5-phosphate isomerase (E.C. 5.3.1.6) catalyzes the reversible isomerization of ribulose-5-phosphate to D-ribose-5-phosphate (50) and is encoded by two nonhomologous genes, *rpiA* and *rpiB*. The *rpiA* type is most common and is found in all three domains of life. *rpiB*s, in contrast, have so far been restricted to the genomes of some bacteria and protozoa. *M. tuberculosis* encodes a single RpiB (*Rv2465*) that was crystallized and shown to catalyze the isomerization of ribulose-5-phosphate and D-ribose-5-phosphate with an efficiency very similar to that of the *E. coli* RpiB (50). Somewhat surprisingly, however, *M. tuberculosis*'s RpiB was not predicted to be essential for *in vitro* growth (23, 24).

Transketolase (TKT; E.C. 2.2.1.1) is encoded by *tkt* (*Rv1449*) and catalyzes the conversion of sedoheptulose-7-phosphate to D-ribose-5-phosphate. *In vitro* studies of purified recombinant *M. tuberculosis* TKT showed that, despite significant sequence differences from the TKTs of yeast, *E. coli*, maize, and spinach, *M. tuberculosis* TKT exhibits very similar kinetic constants and structurally consists of the same overall fold and domain topology as other TKTs (51). Compared to human TKT, *M. tuberculosis* TBTKT was also found to exhibit a broad substrate specificity for a range of phosphorylated sugars (52). Given that the apparent nonessentiality of its ribose-5-phosphate isomerase (24), TKT may play a key role in linking the nonoxidative part of the PPP to biosynthesis of pentose sugars.

Like its EMP, the *M. tuberculosis* PPP remains almost wholly undescribed. Key questions pertain to both the operation and regulation of its oxidative and nonoxidative branches as well as the differential biological roles of NADP and F420-dependent G6PD.

METABOLISM OF SHORT-CHAIN CARBOXYLIC ACIDS

Short-chain carboxylic acid metabolism, including pyruvate metabolism, the tricarboxylic acid (TCA) cycle, and its variants, serves as a hub for CCM, because it integrates glycolysis, β-oxidation, gluconeogenesis, energy metabolism, and many biosynthetic pathways. Early studies of carbon source utilization in *M. tuberculosis* indicated that cultured bacilli are capable of driving the near-complete respiratory oxidation of glucose and glycerol to CO_2 (53). Evidence of a conserved CCM network followed with several reports of glycolytic (14, 45), pyruvate dehydrogenase (PDH) (54), and TCA cycle (55–60) activities from crude and partially purified cellular extracts. Studies by Edson et al. regarding central carbon flux in *Mycobacterium butyricum* demonstrated that ^{14}C-labeled glycerol could be converted to α-ketoglutarate through pyruvate as an intermediate (61), solidifying a model for canonical oxidative connectivity between glycolysis and the TCA cycle in mycobacteria.

PDH COMPLEX

PDH, a ternary complex consisting of a thiamine pyrophosphate-dependent acetyl-transferring PDH (E1 subunit, E.C. 1.2.4.1), a dihydrolipoyllysine-residue acetyltransferase (E2 subunit, E.C. 2.3.1.12), and a flavin adenine dinucleotide-dependent dihydrolipoyl dehydrogenase (E3 subunit, E.C. 1.8.1.4), catalyzes the unidirectional oxidative decarboxylation of pyruvate to acetyl coenzyme A (CoA) with the reduction of NAD^+ to NADH. Preliminary annotation of the *M. tuberculosis* H37Rv genome sequence predicted that the E1 component of PDH was redundantly encoded by *Rv2241* and *Rv2496c/Rv2497c*, the E2 component by *Rv2495c*, and the E3 component by *Rv0462*, *Rv0794c*, and/or *Rv3303c* (18).

Biochemical assessment of purified recombinant PDH E3 candidate Rv0462 (Lpd) revealed that this protein is a multifunctional dihydrolipoamide:$NADP^+$ oxidoreductase for the *M. tuberculosis* PDH complex, as well as for branched-chain keto acid dehydrogenase (BCKAD) and peroxynitrite reductase/peroxidase (PNR/P) complexes (62–65). This unique multifunctionality links CCM with oxidative stress defense in *M. tuberculosis*. Consistent with its dual role in pyruvate and branched-chain keto acid metabolism, deletion of *Rv0462* was found to preclude growth of *M. tuberculosis* on carbohydrates and branched-chain amino acids (65). Moreover, owing to the essential role of Rv0462 in PNR/P activity, this mutant strain demonstrated hypersensitivity to oxidative stressors such as acidified nitrite and H_2O_2 (65). This compound phenotype also resulted in dramatic attenuation in a murine model of infection (65). Unlike that observed for Rv0462, enzymatic studies of purified recombinant Rv0794c and Rv3303c revealed that these E3 paralogs were nonfunctional in lipoamide reduction (62). However, as an indication of their dispensability in CCM, high-throughput mutational analysis of *M. tuberculosis* H37Rv has indicated that *Rv0794c* and *Rv3303c* are nonessential for normal growth in laboratory culture (24).

Assessment of the biochemical activity of purified recombinant PDH E2 candidate Rv2495c and α-ketoglutarate dehydrogenase (KDH) E2 candidate Rv2215 (DlaT) demonstrated that the latter is indeed the E2 component of both PDH (64) and of PNR/P (63), whereas the former is the E2 component of BCKAD (65). Like *Rv0462*, *Rv2215* has been shown to be critical for optimal growth of *M. tuberculosis* *in vitro* (24) and for growth and survival in murine bone marrow–derived macrophages, mice, and guinea pigs (23, 65–67). Based on its role in CCM and oxidative stress defense, Rv2215 has been the target of small molecule screening efforts that have revealed the selective activity of a set of drug-like compounds, rhodanines, that synergize with host immune effector functions to kill nonreplicating *M. tuberculosis* (67).

Enzymatic studies using a partially purified protein from *M. tuberculosis* H37Rv, and purified recombinant protein, have established Rv2241 (AceE) as the sole E1 component of PDH, consistent with its initial annotation (64). In contrast, the Rv2496c/Rv2497c complex has been shown to function as a component of BCKAD, rather than PDH (65). Like the other subunits of PDH, Rv2241 is essential for normal growth in standard laboratory medium (24), but, unlike the E2 and E3 components, is dispensable for growth and survival of *M. tuberculosis* in a murine model of infection (5). These data demonstrate that formation of acetyl-CoA from pyruvate may not be critical for *M. tuberculosis* pathogenesis, consistent with the observation that virulent *Mycobacterium bovis* lacks connectivity between glycolysis and pyruvate metabolism due to an inactive pyruvate kinase (PykA) (68, 69).

TCA CYCLE

In many aerobic organisms, a large proportion of acetyl-CoA generated from PDH enters the TCA cycle and is further oxidized to generate reducing equivalents, typically in the form of NAD(P)H and FADH$_2$, with the concurrent production of CO_2. These reducing equivalents can then participate in various oxidoreductive processes throughout central metabolism. In the presence of a suitable terminal electron acceptor, respiratory reoxidation of these reducing equivalents can be used in the generation of an electrochemical gradient across the cytoplasmic membrane. As described in reference 112, the potential energy of this gradient can be tapped into for the synthesis of ATP and for other processes that are driven by chemiosmosis. Beyond this role in providing an abundant source of reducing power, the TCA cycle is also critical for providing essential substrates for biosynthesis of many amino acids, cofactors, and nucleotides. This biosynthetic role for the TCA cycle mandates input of additional four-carbon units for sustained flux, a role that is often satisfied by pyruvate carboxylase, which catalyzes the ATP-dependent carboxylation of pyruvate to oxaloacetate.

While orthologs for TCA cycle enzymes are represented throughout the three domains of life, it is becoming increasingly apparent that the utility, structure, and enzymatic components of this central metabolic pathway can vary considerably from organism to organism. While the identities of enzymes of the TCA cycle and linked pathways, such as the glyoxylate shunt, have been inferred from the complete genome sequence of *M. tuberculosis* (18), functional roles for many of these paralogs have yet to be established. Indeed, despite evidence for oxidative decarboxylating enzymes of the TCA cycle, it is currently unclear whether there is significant carbon flow from isocitrate to succinyl-CoA via this pathway under physiological conditions. Further, until recently, it was unclear whether the glyoxylate shunt played an ancillary role to the TCA cycle or whether flux through the glyoxylate shunt represented the majority of traffic through the cycle. Given that the true enzymatic function, mechanisms of regulation, and directionality of metabolic flux cannot be reliably determined by homology-based modeling, the physiologic roles of most steps in the *M. tuberculosis* TCA cycle and related pathways await confirmation by functional analysis.

Citrate Synthase

Citrate synthase (E.C. 2.3.3.1) serves as the major entry point for carbon into the TCA cycle by catalyzing the sequential aldol condensation of oxaloacetate with acetyl-CoA to form citryl-CoA and hydrolysis to form citrate and free CoA. Ochoa et al. were the first to describe the presence of citrate synthase activity in cellular extracts of *M. tuberculosis* (55). Homology searches have indicated that *M. tuberculosis* encodes two paralogs of citrate synthase: *citA* and *gltA2* (*Rv0889c* and *Rv0896*). Based on its apparent essentiality for growth *in vitro*, *gltA2* is thought to encode the major citrate synthase of *M. tuberculosis* central metabolism (23). However, to date, biochemical characterization of the GltA2 protein and assessment of its specific role in *M. tuberculosis* physiology has yet to be reported. In contrast to *gltA2*, *citA* appears to be nonessential (23, 24); thus, it is unclear whether this gene encodes a functional citrate synthase and whether the corresponding gene product plays a role in CCM.

In addition to citrate synthase, *M. tuberculosis* also encodes a paralog of the β-subunit of citrate lyase (E.C. 4.1.3.6, encoded by *citE* or *Rv2498c*) that is predicted to catalyze the reverse reaction of citryl-CoA to oxaloacetate and acetyl-CoA. While CitE appears to be important for growth of *M. tuberculosis* within macrophages (23), its functional role in CCM has yet to be determined. Based on the proximity and positive correlation of gene expression of *citE* with genes for a putative acyl-CoA transferase (*Rv2503c* and *Rv2504c*), acyl-CoA carboxylase (Rv2501c and Rv2502c), and acyl-CoA dehydrogenase (*Rv2500c*), it is highly anticipated that the function of *M. tuberculosis* CitE is linked to fatty acid synthesis through the formation of some species of acyl-CoA (70). Because *M. tuberculosis* does not appear to utilize exogenous citrate (71, 72), a source for citrate to support such a pathway remains unclear. In some anaerobic microbes, such as *Chlorobium tepidum*, citrate lyase is associated with operation of a reductive TCA cycle that enables CO_2 assimilation and balancing of the cellular reduction/oxidation potential (73). While evidence for a full reductive TCA cycle in *M. tuberculosis* has not been reported, it is intriguing that the *citE* gene cluster is induced under anaerobic conditions (74–77).

Aconitase

Following citrate synthesis, the next step in the TCA cycle is the sequential dehydration of citrate to *cis*-aconitate and rehydration to isocitrate, which is catalyzed by the iron-sulfur cluster protein aconitase (E.C. 4.2.1.3). High-throughput gene essentiality screens indicate that *acn* (*Rv1475c*) is essential for growth of *M. tuberculosis* in culture (23, 24). Sequence analysis and biochemical characterization of purified recombinant *M. tuberculosis* aconitase indicates that this enzyme has the typical features of a dual function bacterial aconitase A/iron-responsive protein (AcnA/IRP) family member (78). Like other bacterial AcnA/IRPs, the *M. tuberculosis* protein contains a labile 4Fe-4S cluster that is essential for catalyzing interconversion of citrate and isocitrate. When this cluster is impaired due to oxidative damage or iron deficiency, the protein adopts the ability to bind RNA hairpins containing a sequence-specific loop of C-A-G-C/U-G. Gel shift assays have demonstrated that target RNAs for the *M. tuberculosis* apo-aconitase include messages for human ferritin, *M. tuberculosis* thioredoxin, and the iron regulator encoded by *ideR*. Based on this IRP activity, it is likely that *M. tuberculosis* apo-aconitase plays a role in the response to iron limitation and/or oxidative stress, yet assessment of this role awaits further analysis.

Isocitrate Dehydrogenase

Once formed, isocitrate can continue through the TCA cycle via oxidative decarboxylation to α-ketoglutarate catalyzed by isocitrate dehydrogenase (ICD; E.C. 1.1.1.42). *M. tuberculosis* encodes two orthologs of isocitrate dehydrogenase, ICD-2 (encoded by *Rv0066c*) and ICD-1 (encoded by *Rv3339c*). Recombinant proteins of each ortholog have been expressed, purified, and biochemically characterized as exclusively $NADP^+$-dependent enzymes (79–81). The precise reaction sequence has been elaborated for ICD-1, in which $NADP^+$ binds, followed by binding, oxidation, and decarboxylation of isocitrate and sequential release of CO_2, α-ketoglutarate, and NADPH. While deletions of the gene for each ICD ortholog have not been reported, transposon mutagenesis has demonstrated that each allele is dispensable for growth (23, 24, 82). Because *M. tuberculosis* can bypass the oxidative decarboxylation steps of the TCA cycle by diverting isocitrate through the glyoxylate shunt, it is unclear whether the nonessentiality of each ICD is due to redundancy or general nonessentiality for this node in CCM.

α-Ketoglutarate Decarboxylase

Depending on the metabolic demands of a cell, carbon flow can continue through the TCA cycle from isocitrate to α-ketoglutarate via a second oxidative decarboxylation. Most often, the unidirectional oxidative decarboxylation of α-ketoglutarate to succinyl-CoA is coupled to the reduction of NAD^+ and is catalyzed by KDH, a ternary complex that is closely related to PDH. This complex consists of a succinyl-transferring KDH (E.C. 1.2.4.2, E1 subunit), a dihydrolipoyllysine-residue succinyltransferase (E.C. 2.3.1.61, E2 subunit), and a dihydrolipoyl dehydrogenase (E.C. 1.8.1.4, E3 subunit). From the complete genome sequence of *M. tuberculosis* H37Rv, it was predicted that the E1

component of KDH was encoded by *Rv1248c*, E2 by *Rv2215* (*dlaT*), and E3 by *Rv0462* (*lpdC*), *Rv0794c*, and/or *Rv3303c* (18). However, consistent with previous reports of CoA-independent α-ketoglutarate decarboxylase (KGD) activity in *M. tuberculosis* cellular extracts (60), the predicted E1 component Rv1248c was demonstrated to have a nonoxidative KGD activity that yielded succinic semialdehyde rather than succinyl-CoA (64, 83). This activity was independent of the various *M. tuberculosis* E2 and E3 subunits and relied on ferric cyanide for reoxidation in cell free assays (83).

Subsequent work found that this KGD-mediated formation of succinic semialdehyde was the product of a slow side reaction whose physiologic role remains unknown (84). Using a novel approach termed activity-based metabolomic profiling, it was later demonstrated that Rv1248c was most proficient in catalyzing the synthesis of 2-hydroxy-3-oxoadipate from α-ketoglutarate and glyoxylate in reaction mixtures as well as in whole cells (84, 85). Interestingly, Rv1248c was also shown to be capable of functioning as a succinyl-transferring KDH when supplied with DlaT and Lpd as E2 and E2 components, respectively (86). Moreover, this KDH activity was shown to be allosterically activated by acetyl-CoA and inhibited by the forkhead-associated domain protein GarA (85, 86), the latter of which regulates many enzymes linked to glutamate metabolism in mycobacteria (87). In addition, *M. tuberculosis* strains deleted for *Rv1248c* show conditional growth defects in medium lacking fatty acids (88). These growth defects could be partially alleviated by inclusion of succinate in the growth medium. While these observations are consistent with a role for Rv1248c in succinate production, it is important to note that the metabolic basis for these phenotypes and their relevance for *in vivo* growth remains unresolved.

Many microaerophilic and strictly anaerobic organisms utilize an alternative enzyme for interconversion of α-ketoglutarate and succinyl-CoA, α-ketoglutarate:ferredoxin oxidoreductase (KOR; E.C. 1.2.7.3). KOR and other α-ketoic acid:ferredoxin oxidoreductase family members are composed of a CoA-coordinating α/γ subunit (KorA) and a TPP and iron-sulfur cluster containing β-subunit (KorB). In contrast to KGD, KOR is often utilized for reductive carboxylation of succinyl-CoA to α-ketoglutarate in a CO_2-assimilating reverse TCA cycle. Yet the presence of paralogs for this oxidoreductase in many aerobic organisms suggests that KOR might play a greater role in oxidative metabolism than previously thought. Corresponding activity for KOR was recently described for *M. tuberculosis* and was found to be dependent upon the *Rv2455c* (*korB*)

Rv2454c (*korA*) gene cluster (88). A mutant strain deleted for *korAB* was defective for growth in the absence of exogenously supplied CO_2, suggesting that KOR can function in the oxidative direction during aerobic growth *in vitro*. Interestingly, a mutant strain deleted for both *korAB* and *Rv1248c* showed a slow growth phenotype relative to the wild-type strain. Thus, if KOR and Rv1248c are involved in conversion of α-ketoglutarate to succinyl-CoA, this function is nonessential for growth of *M. tuberculosis* in laboratory culture. Further investigation is required to determine the extent of KOR's contribution to flux through the TCA cycle and to determine whether this function plays a role in *M. tuberculosis* pathogenesis.

Succinyl-CoA Synthetase

Succinyl-CoA synthetase (SCS; E.C. 6.2.1.5) is a heterodimeric enzyme responsible for the interconversion of succinyl-CoA and succinate. When functioning as a component of the oxidative TCA cycle, SCS-dependent conversion of succinyl-CoA to succinate yields ATP via substrate-level phosphorylation of ADP. Conversely, in organisms that drive a reductive TCA cycle, SCS can function in synthesis of succinyl-CoA at the expense of ATP. Tian et al. recently reported the detection of succinyl-CoA hydrolytic activity in crude cellular extracts of *M. tuberculosis* H37Rv using the colorimetric readout of adduct formation between thionitrobenzoate and free CoA (83). However, it is not clear whether this activity was distinguished from SCS-independent hydrolysis of succinyl-CoA. Thus, it remains to be determined whether *M. tuberculosis* expresses abundant SCS activity and, if so, whether this enzyme functions in the TCA cycle or perhaps serves as a source of succinyl-CoA from succinate derived from other pathways such as the glyoxylate shunt. Despite this uncertainty, genes for the putative SCS subunits (SucCD encoded by *Rv0951/Rv0952*) have been described as essential for growth of *M. tuberculosis* H37Rv in laboratory culture, consistent with a critical role for SCS in CCM (23, 24).

Succinate:Menaquinone Oxidoreductase

Succinate:quinone oxidoreductase (E.C. 1.3.5.1) couples the reversible oxidation of succinate to fumarate with the reduction of membrane-soluble quinones to quinols (menaquinone and menaquinol in the mycobacteria [89, 90]). When functioning in oxidative metabolism, succinate:quinone oxidoreductase is commonly referred to as succinate dehydrogenase (SDH) and serves as a direct link between the TCA cycle and electron transport. SDH complexes generally consist of an enzymatic flavin

adenine dinucleotide subunit SdhA, an iron-sulfur cluster subunit SdhB, an integral membrane *b*-type cytochrome subunit SdhC, and often an integral membrane anchor subunit SdhD.

SDH activity has been confirmed in *M. tuberculosis* H37Rv by following the succinate-dependent reduction of potassium ferricyanide by crude cellular extracts (60), as well as by following the succinate-dependent reduction of iodonitrotetrazolium by the membrane fraction of lysed bacilli (91). While the genes primarily responsible for encoding this activity have yet to be substantiated by direct evidence, SDH of *M. tuberculosis* is thought to be redundantly encoded by operons *Rv0247c-Rv0249c* and *Rv3316-Rv3319*. In these operons, *Rv0247c* (*sdhB2*) and *Rv3319* (*sdhB*) encode the putative iron-sulfur cluster subunits, *Rv0248c* (*sdhA2*) and *Rv3318* (*sdhA*) encode the flavin adenine dinucleotide-containing catalytic subunits, *Rv0249c* (*sdhC2*) and *Rv3316* (*sdhC*) encode the integral membrane cytochrome B subunits, and *Rv3317* (*sdhD*) encodes an integral membrane anchor. While the *Rv3316-Rv3319* operon has been predicted to be nonessential in both H37Rv and CDC1551 in high-throughput mutagenesis screens (23, 24, 82), essentiality data for the *Rv0247c-Rv0249c* operon have proven to be more variable, with indications of *in vitro* essentiality of all components in one report (23) and essentiality for growth in the spleen of C57BL/6J mice in another (5).

In addition to SDH, *M. tuberculosis* also expresses fumarate reductase (FRD) activity that is likely encoded by the *Rv1552-Rv1555* operon. In this operon, *Rv1552* encodes the enzymatic subunit FrdA, *Rv1553* encodes the iron-sulfur cluster subunit FrdB, *Rv1554* encodes the cytochrome subunit FrdC, and *Rv1555* encodes the anchor subunit FrdD. While these genes are nonessential for growth of *M. tuberculosis in vitro* and *in vivo* (23, 24, 82, 91), their expression is induced by more than 200-fold under hypoxic nonreplicating conditions (91). DNA binding studies have revealed that this regulation likely occurs through a cyclic AMP-mediated response with the binding of Crp (Rv3676) to the promoter region of *Rv1552* (92). Taken together with the observation that genes for the oxidative arm of the TCA cycle are downregulated under hypoxia, it has been suggested that FRD functions in a reductive half cycle that is designed to compensate for the lack of a sufficient exogenous terminal electron acceptor (91). By using stable isotope tracking methods, it has been demonstrated that *M. tuberculosis* can drive the reductive conversion of pyruvate to succinate, yet FrdA is nonessential for this pathway, likely due to the ability of SDH to compensate in the absence of FrdA (91).

With this metabolic redundancy, it is difficult to probe the role for FRD in supporting persistence of *M. tuberculosis in vivo*, and thus we await further examination.

Fumarase

Fumarase (FUM; E.C. 4.2.1.2) catalyzes the reversible hydration of fumarate to malate and comes in two general varieties, the class I iron-sulfur cluster-dependent type and the class II iron-sulfur cluster-independent type. Sequence analysis indicates that *M. tuberculosis* encodes a class II FUM (encoded by *Rv1098c*) and unlike many bacteria lacks a class I enzyme. Robust FUM activity from *M. tuberculosis* H37Rv cellular extracts has been reported (60, 83). The activity and structure of purified recombinant *M. tuberculosis* FUM have recently been characterized (93). Similar to other class II orthologs, this enzyme assembles into a homotetramer and shows absolute conservation of active site residues including the essential catalytic serine at position 318 (93). While *Rv1098c* deletion mutant strains have yet to be described, high-throughput assessments of gene essentiality indicate that this gene is required for growth of *M. tuberculosis* in culture (23, 24).

Malate Dehydrogenase

Malate dehydrogenase catalyzes the final step in the TCA cycle with the oxidation of malate to oxaloacetate. Malate dehydrogenase comes in two forms, one that is soluble and transfers reducing equivalents to NAD$^+$ (MDH; E.C. 1.1.1.37) and another that is a flavin-dependent membrane-associated enzyme that ultimately transfers reducing equivalents to quinones (MQO; E.C. 1.1.5.4). Malate dehydrogenase activity has been detected in *M. tuberculosis* H37Rv cell-free extracts (60, 83). Yet because *M. tuberculosis* is predicted to encode both MDH (*Rv1240*) and MQO (*Rv2852c*), the principal enzyme for this activity has yet to be determined. Of note, overexpression of MQO in *M. smegmatis* influences susceptibility to isoniazid through modulation of the NADH/NAD$^+$ ratio (94). Consistent with the possibility that these enzymes serve a semi-redundant function in CCM, *Rv1240* and *Rv2852c* are independently nonessential for growth in laboratory culture (23, 24, 82).

GLYOXYLATE SHUNT AND THE METHYLCITRATE CYCLE

In addition to serving as a source of reducing equivalents and ATP, the TCA cycle also serves as a key source of essential biosynthetic precursors. However, classical operation of the TCA cycle under conditions of aerobic respiration results in the oxidative decarboxylation of every acetyl-CoA unit consumed, without a net assimilation of additional carbon units. The use of TCA cycle intermediates for biosynthetic purposes thus requires the activity of anaplerotic reactions that replenish consumed intermediates and sustain their respiratory and/or bioenergetic functions.

During growth on glycolytic carbon sources, this anaplerotic function can be metabolically served by the catalytic activities of phosphoenolpyruvate (PEP) carboxylase/carboxykinase (PCKA) (E.C. 4.1.1.32), pyruvate carboxylase (PCA) (E.C.6.4.1.1), and/or malic enzyme (MEZ) (E.C.1.1.1.38), encoded by *pckA* (*Rv0211*), *pca* (*Rv2967c*), and *mez* (*Rv2332*), respectively. Activity of cellular and purified PckA and PCA has been characterized for *M. smegmatis* (95, 96) but not *M. tuberculosis*, while biochemical studies of MEZ remain lacking. Of these three enzymes, only *pca* is predicted to be essential for *in vitro* growth of strains H37Rv and CDC1551 (24, 82). Looking ahead, key unanswered questions concerning these enzymes will pertain to resolving their *in vitro* and *in vivo* regulation and directionalities (24, 82).

During growth on fatty acids, metabolite replenishment is accomplished by the linked enzymatic inactivation of isocitrate dehydrogenase and genetic upregulation of the glyoxylate shunt, which generates one molecule of malate from two molecules of acetyl-CoA via the successive actions of isocitrate lyase (ICL) (E.C. 4.1.3.1) and malate synthase (E.C. 2.3.3.9). The *M. tuberculosis* genome encodes an intact glyoxylate shunt that consists of two paralogous ICLs (encoded by *icl* [*Rv0467*] and *aceAa/aceAb* [*Rv1915-Rv1916*]) and a single malate synthase (encoded by *glcB* [*Rv1837c*]). Despite exhibiting only 27% sequence identity to one another, both ICLs share a highly conserved active site and catalytic activity that function to generate glyoxylate and succinate from cleavage of a single molecule of isocitrate, albeit with distinct catalytic efficiencies (K_m = 0.19 mM ICL1; 1.14 mM ICL2; k_{cat} = 5.24 s^{-1} ICL1; 1.38 s^{-1} ICL2) (97). Consistent with its canonical role in anaplerosis, *M. tuberculosis* mutants lacking both ICLs were shown to be incapable of growing on fatty acids *in vitro* and either establishing or maintaining a chronic infection in mice (98, 99). Moreover, ICL-deficient *M. tuberculosis* strains appear to be the most profoundly attenuated strains reported to date (99).

The *glcB*-encoded malate synthase G of *M. tuberculosis* catalyzes the Claisen-like condensation of acetyl-CoA and glyoxylate to enable the net assimilation of two acetyl-CoA units into the TCA cycle via the production of malate. The GlcB enzyme is an 80-kDa monomeric protein homologous (with ~60% identity) to the malate synthase (AceB) of *Corynebacterium glutamicum* (100) and *E. coli* (101) but distinct from a second group of malate synthases called malate synthase A found in *E. coli* K-12 (102), *Yersinia pestis* (103), *Vibrio cholerae* (104), yeast, and higher plants. Biochemical studies of the *M. tuberculosis* GlcB further reported evidence of sequential binding of glyoxylate followed by acetyl-CoA and ordered release of CoA followed by malate with the following rate constants (K_{gly} = 30 µM, $K_{acetyl\text{-}CoA}$ = 10 µM; k_{cat} = 23 s^{-1}). Consistent with its genomic nonredundancy, insertional transposon mutagenesis studies have predicted the *glcB* of *M. tuberculosis* to be essential for optimal growth *in vitro* (5).

In addition to their role in anaplerotic metabolism of even-numbered short-chain fatty acids, *M. tuberculosis* ICLs were also found to be essential for *in vitro* growth on odd-numbered short-chain fatty acids. This finding led to both structural and enzymatic evidence of ICL1 (but not ICL2) as a bifunctional ICL and methylisocitrate lyase (MCL). MCL is an enzyme of the methylcitrate cycle, the dominant pathway for metabolism of propionyl-CoA in eubacteria, for which the *M. tuberculosis* genome encodes orthologs of all enzymes except MCL (18, 97, 99). The ICL1 of *M. tuberculosis* is thus a dual substrate enzyme that participates in two separate pathways of fatty acid metabolism using a single active site chemistry.

Studies in *E. coli* showed that ICL also participated in the recently discovered PEP-glyoxylate cycle, which supports complete oxidation of carbohydrates to CO_2 during growth on limiting concentrations of glucose and balanced conversion of acetyl-CoA into CO_2 and pyruvate/PEP during growth on lipids (105). Evidence for such a cycle in *M. tuberculosis*, however, remains lacking. Nonetheless, a recent study by Beste et al. provided evidence of an analogous pathway for pyruvate dissimilation involving *M. tuberculosis* ICLs and carbon dioxide (106).

From a historical perspective, early studies by Wayne showed that *M. tuberculosis*, modeled to enter into a state of nonreplicative quiescence, exhibited a hypoxia- rather than fatty acid–induced increase in ICL activity that was linked to an increase in glycine dehydrogenase activity and thought to facilitate recycling of accumulated NADH reducing equivalents (107). More recent work, however, showed that this increase was tied to the biochemical ability of ICL to produce succinate as a multifunctional metabolic end product, rather than an intermediate, used to more broadly sustain membrane potential, respiratory (or ATP generating) activity, and

carbon flow during entry into, residence in, and exit from hypoxic quiescence (108). Such studies thus highlight the potential multiplicity of biochemical functions and physiologic roles encoded by canonical metabolic enzymes not foreseen by genetic or bioinformatic approaches.

CONCLUDING REMARKS

Despite the established connectivity between glycolysis and the TCA cycle in *M. tuberculosis*, Segal and Bloch made the striking observation that while glycerol, glucose, and fatty acids could stimulate respiration *in vitro*, only lipids could stimulate measurable respiration in *ex vivo*–derived bacilli (109). This result was the first indication that *M. tuberculosis* might preferentially use β-oxidation, rather than glycolysis, as a substrate entry pathway for CCM *in vivo*. The enzymatic basis for this ability to utilize β-oxidation units, in the form of acetyl-CoA, was later revealed with the detection of the glyoxylate shunt enzymes, ICL and malate synthase, in partially purified extracts of cultured *M. tuberculosis* (60, 110).

Through the use of site-directed mutagenesis, the importance of lipid utilization and its essential connectivity to gluconeogenesis for *in vivo* growth of *M. tuberculosis* has recently been established with the observation that independent mutant strains lacking the genes for ICL and phosphoenolpyruvate carboxykinase were defective for growth and persistence in a murine model of infection (98, 99, 105, 111). These observations, coupled with the observation that virulent *M. bovis* lacks integration between glycolysis and the TCA cycle due to a deficiency in pyruvate kinase (encoded by *pykA* or *Rv1617*) activity (68), have called into question whether carbohydrates, such as glucose, are relevant carbon sources for growth and survival of *M. tuberculosis in vivo*. Interestingly, it was recently revealed that while hexokinase, required for glucose utilization, is dispensable for the acute phase of *M. tuberculosis* infection, this enzyme plays an essential role in persistence during the chronic phase of infection, thereby indicating that acquisition and utilization of glucose are important for the fitness of *M. tuberculosis in vivo* (6). A major area of interest in the coming years will be to assess the specific roles of various CCM pathways and their connectivity in different microenvironments *in vivo*. Carbon transformations aside, a broader question to be addressed is how *M. tuberculosis* coordinates the need for biosynthetic intermediates with the accompanying alterations in respiratory equivalents and the extent to which *M. tuberculosis* can dissociate one from the other.

Citation. Baughn AD, Rhee KY. 2014. Metabolomics of central carbon metabolism in *Mycobacterium tuberculosis*. Microbiol Spectrum 2(3):MGM2-0026-2013.

References

1. Rhee KY, Carvalho LP, Bryk R, Ehrt S, Marrero J, Park SW, Schnappinger D, Venugopal A, Nathan C. 2011. Central carbon metabolism in *Mycobacterium tuberculosis*: an unexpected frontier. *Trends Microbiol* **19**:307–314.

2. Russell DG, Barry CE 3rd, Flynn JL. 2010. Tuberculosis: what we don't know can, and does, hurt us. *Science* **328**:852–856.

3. Long ER. 1954. *Chemistry and Chemotherapy of Tuberculosis*, 3rd ed. Williams & Wilkins, Baltimore, MD.

4. Bloch H, Segal W. 1956. Biochemical differentiation of *Mycobacterium tuberculosis* grown in vivo and in vitro. *J Bacteriol* **72**:132–141.

5. Sassetti CM, Rubin EJ. 2003. Genetic requirements for mycobacterial survival during infection. *Proc Natl Acad Sci USA* **100**:12989–12994.

6. Marrero J, Trujillo C, Rhee KY, Ehrt S. 2013. Glucose phosphorylation is required for *Mycobacterium tuberculosis* persistence in mice. *PLoS Pathog* **9**:e1003116.

7. Wheeler PR, Blanchard JS. 2005. General metabolism and biochemical pathways of tubercle bacilli, p 309–339. *In* Cole ST, Eisenach KD, McMurray DN, Jacobs WR Jr (ed), *Tuberculosis and the Tubercle Bacillus*. ASM Press, Washington, DC.

8. Ehrt S, Rhee K. 2013. *Mycobacterium tuberculosis* metabolism and host interaction: mysteries and paradoxes. *Curr Top Microbiol Immunol* **374**:163–188.

9. Youmans GP, Youmans AS. 1953. Studies on the metabolism of *Mycobacterium tuberculosis*. I. The effect of carbohydrates and alcohols on the growth of *Mycobacterium tuberculosis* var. *hominis. J Bacteriol* **65**:92–95.

10. Gamble CJ, Herrick MC. 1922. The utilization of dextrose by the tubercle bacillus. *Am Rev Tuberc* **6**:44–50.

11. Loebel RO, Shorr E, Richardson H. 1933. The influence of foodstuffs upon the respiratory metabolism and growth of human tubercle bacilli. *J Bacteriol* **26**:139–166.

12. Weinzirl J, Knapton F. 1927. The biology of the tubercle bacillus. I. Hydrogen-ion concentration produced by some members of the genus *Mycobacterium*. *Am Rev Tuberc* **15**:380–388.

13. Youmans AS, Youmans GP. 1953. Studies on the metabolism of *Mycobacterium tuberculosis*. III. The growth of *Mycobacterium tuberculosis* var. *hominis* in the presence of various intermediates of the dissimilation of glucose to pyruvic acid. *J Bacteriol* **65**:100–102.

14. Bastarrachea F, Anderson DG, Goldman DS. 1961. Enzyme systems in the mycobacteria. XI. Evidence for a functional glycolytic system. *J Bacteriol* **82**:94–100.

15. Ramakrishnan T, Murthy PS, Gopinathan KP. 1972. Intermediary metabolism of mycobacteria. *Bacteriol Rev* **36**:65–108.

16. Jayanthi Bai N, Ramachandra Pai M, Suryanarayana Murthy P, Venkitasubramanian TA. 1975. Pathways of carbohydrate metabolism in *Mycobacterium tuberculosis* H37Rv1. *Can J Microbiol* **21**:1688–1691.

17. Bai NJ, Pai MR, Murthy PS, Venkitasubramanian TA. 1976. Pathways of glucose catabolism in *Mycobacterium smegmatis. Can J Microbiol* **22**:1374–1380.

18. Cole ST, Brosch R, Parkhill J, Garnier T, Churcher C, Harris D, Gordon SV, Eiglmeier K, Gas S, Barry CE 3rd, Tekaia F, Badcock K, Basham D, Brown D, Chillingworth T, Connor R, Davies R, Devlin K, Feltwell T, Gentles S, Hamlin N, Holroyd S, Hornsby T, Jagels K, Krogh A, McLean J, Moule S, Murphy L, Oliver K, Osborne J, Quail MA, Rajandream MA, Rogers J, Rutter S, Seeger K, Skelton J, Squares R, Squares S, Sulston JE, Taylor K, Whitehead S, Barrell BG. 1998. Deciphering the biology of *Mycobacterium tuberculosis* from the complete genome sequence. *Nature* **393**:537–544.

19. Hsieh PC, Shenoy BC, Samols D, Phillips NF. 1996. Cloning, expression, and characterization of polyphosphate glucokinase from *Mycobacterium tuberculosis. J Biol Chem* **271**:4909–4915.

20. Hsieh PC, Kowalczyk TH, Phillips NF. 1996. Kinetic mechanisms of polyphosphate glucokinase from *Mycobacterium tuberculosis. Biochemistry* **35**:9772–9781.

21. Pimentel-Schmitt EF, Thomae AW, Amon J, Klieber MA, Roth H-M, Muller YA, Jahreis K, Burkovski A, Titgemeyer F. 2007. A glucose kinase from *Mycobacterium smegmatis. J Mol Microbiol Biotechnol* **12**:75–81.

22. Mathur D, Ahsan Z, Tiwari M, Garg LC. 2005. Biochemical characterization of recombinant phosphoglucose isomerase of *Mycobacterium tuberculosis. Biochem Biophys Res Commun* **337**:626–632.

23. Griffin JE, Gawronski JD, DeJesus MA, Ioerger TR, Akerley BJ, Sassetti CM. 2011. High-resolution phenotypic profiling defines genes essential for mycobacterial growth and cholesterol catabolism. *PLoS Pathog* **7**:e1002251.

24. Sassetti CM, Boyd DH, Rubin EJ. 2003. Genes required for mycobacterial growth defined by high density mutagenesis. *Mol Microbiol* **48**:77–84.

25. Chhabra G, Mathur D, Dixit A, Garg LC. 2012. Heterologous expression and biochemical characterization of recombinant alpha phosphoglucomutase from *Mycobacterium tuberculosis* H37Rv. *Protein Expr Purif* **85**:117–124.

26. Phong WY, Lin W, Rao SP, Dick T, Alonso S, Pethe K. 2013. Characterization of phosphofructokinase activity in *Mycobacterium tuberculosis* reveals that a functional glycolytic carbon flow is necessary to limit the accumulation of toxic metabolic intermediates under hypoxia. *PLoS One* **8**:e56037.

27. Orchard LMD, Kornberg HL. 1990. Sequence similarities between the gene specifying 1-phosphofructokinase (fruK), genes specifying other kinases in *Escherichia coli* K12, and lacC of *Staphylococcus aureus. Proc R Soc Lond B* **242**:87–90.

28. Buschmeier B, Hengstenberg W, Deutscher J. 1985. Purification and properties of 1-phosphofructokinase from *Escherichia coli. FEMS Microbiol Lett* **29**:231–235.

29. Miallau L, Hunter WN, McSweeney SM, Leonard GA. 2007. Structures of *Staphylococcus aureus* D-tagatose-6-phosphate kinase implicate domain motions in specificity and mechanism. *J Biol Chem* **282**:19948–19957.

30. Babul J. 1978. Phosphofructokinases from *Escherichia coli*. Purification and characterization of the nonallosteric isozyme. *J Biol Chem* **253**:4350–4355.

31. Gutka HJ, Franzblau SG, Movahedzadeh F, Abad-Zapatero C. 2011. Crystallization and preliminary X-ray characterization of the glpX-encoded class II fructose-1,6-bisphosphatase from *Mycobacterium tuberculosis*. *Acta Crystallogr Sect F Struct Biol Cryst Commun* **67**:710–713.

32. Gutka HJ, Rukseree K, Wheeler PR, Franzblau SG, Movahedzadeh F. 2011. glpX gene of *Mycobacterium tuberculosis*: heterologous expression, purification, and enzymatic characterization of the encoded fructose 1,6-bisphosphatase II. *Appl Biochem Biotechnol* **164**:1376–1389.

33. Rutter WJ. 1964. Evolution of aldolase. *Fed Proc* **23**:1248–1257.

34. Siebers B, Brinkmann H, Dörr C, Tjaden B, Lilie H, Van der Oost J, Verhees CH. 2001. Archaeal fructose-1,6-bisphosphate aldolases constitute a new family of archaeal type class I aldolase. *J Biol Chem* **276**:28710–28718.

35. Hall DR, Leonard GA, Reed CD, Watt CI, Berry A, Hunter WN. 1999. The crystal structure of *Escherichia coli* class II fructose-1,6-bisphosphate aldolase in complex with phosphoglycolohydroxamate reveals details of mechanism and specificity. *J Mol Biol* **287**:383–394.

36. Zgiby S, Plater AR, Bates MA, Thomson GJ, Berry A. 2002. A functional role for a flexible loop containing Glu182 in the class II fructose-1,6-bisphosphate aldolase from *Escherichia coli*. *J Mol Biol* **315**:131–140.

37. Zgiby SM, Thomson GJ, Qamar S, Berry A. 2000. Exploring substrate binding and discrimination in fructose 1,6-bisphosphate and tagatose 1,6-bisphosphate aldolases. *Eur J Biochem* **267**:1858–1868.

38. Rosenkrands I, Slayden RA, Crawford J, Aagaard C, Barry CE 3rd, Andersen P. 2002. Hypoxic response of *Mycobacterium tuberculosis* studied by metabolic labeling and proteome analysis of cellular and extracellular proteins. *J Bacteriol* **184**:3485–3491.

39. Bai NJ, Pai MR, Murthy PS, Venkitasubramanian TA. 1975. Fructose-1,6-diphosphate aldolase of *Mycobacterium tuberculosis* H37Rv. *Indian J Biochem Biophys* **12**:181–183.

40. Bai NJ, Pai MR, Murthy PS, Venkitasubramanian TA. 1974. Effect of oxygen tension on the aldolases of *Mycobacterium tuberculosis* H37Rv. *FEBS Lett* **45**:68–70.

41. Pelzer-Reith B, Wiegand S, Schnarrenberger C. 1994. Plastid class I and cytosol class II aldolase of *Euglena gracilis*. Purification and characterization. *Plant Physiol* **106**:1137–1144.

42. Alber T, Banner DW, Bloomer AC, Petsko GA, Phillips D, Rivers PS, Wilson IA. 1981. On the three-dimensional structure and catalytic mechanism of triose phosphate isomerase. *Philos Trans R Soc Lond B Biol Sci* **293**:159–171.

43. Connor SE, Capodagli GC, Deaton MK, Pegan SD. 2011. Structural and functional characterization of *Mycobacterium tuberculosis* triosephosphate isomerase. *Acta Crystallogr D Biol Crystallogr* **67**:1017–1022.

44. Mathur D, Malik G, Garg LC. 2006. Biochemical and functional characterization of triosephosphate isomerase from *Mycobacterium tuberculosis* H37Rv. *FEMS Microbiol Lett* **263**:229–235.

45. Goldman DS. 1963. Enzyme systems in the mycobacteria. XV. Initial steps in the metabolism of glycerol. *J Bacteriol* **86**:30–37.

46. de Carvalho LP, Fischer SM, Marrero J, Nathan C, Ehrt S, Rhee KY. 2010. Metabolomics of *Mycobacterium tuberculosis* reveals compartmentalized co-catabolism of carbon substrates. *Chem Biol* **17**:1122–1131.

47. Purwantini E, Daniels L. 1996. Purification of a novel coenzyme F420-dependent glucose-6-phosphate dehydrogenase from *Mycobacterium smegmatis*. *J Bacteriol* **178**:2861–2866.

48. Purwantini E, Gillis TP, Daniels L. 1997. Presence of F420-dependent glucose-6-phosphate dehydrogenase in *Mycobacterium* and *Nocardia* species, but absence from *Streptomyces* and *Corynebacterium* species and methanogenic *Archaea*. *FEMS Microbiol Lett* **146**:129–134.

49. Bashiri G, Squire CJ, Moreland NJ, Baker EN. 2008. Crystal structures of F420-dependent glucose-6-phosphate dehydrogenase FGD1 involved in the activation of the anti-tuberculosis drug candidate PA-824 reveal the basis of coenzyme and substrate binding. *J Biol Chem* **283**:17531–17541.

50. Roos AK, Andersson CE, Bergfors T, Jacobsson M, Karlen A, Unge T, Jones TA, Mowbray SL. 2004. *Mycobacterium tuberculosis* ribose-5-phosphate isomerase has a known fold, but a novel active site. *J Mol Biol* **335**:799–809.

51. Fullam E, Pojer F, Bergfors T, Jones TA, Cole ST. 2012. Structure and function of the transketolase from *Mycobacterium tuberculosis* and comparison with the human enzyme. *Open Biol* **2**:110026.

52. Mitschke L, Parthier C, Schröder-Tittmann K, Coy J, Lüdtke S, Tittmann K. 2010. The crystal structure of human transketolase and new insights into its mode of action. *J Biol Chem* **285**:31559–31570.

53. Novy FG, Soule M. 1925. Microbic respiration. II. Respiration of the tubercle bacillus. *J Infect Dis* **36**:168–232.

54. Goldman DS. 1959. Enzyme systems in the mycobacteria. VI. Further studies on the pyruvic dehydrogenase system. *Biochim Biophys Acta* **32**:80–95.

55. Ochoa S, Stern JR, Schneider MC. 1951. Enzymatic synthesis of citric acid. II. Crystalline condensing enzyme. *J Biol Chem* **193**:691–702.

56. Millman I, Youmans GP. 1954. Studies on the metabolism of *Mycobacterium tuberculosis*. VII. Terminal respiratory activity of an avirulent strain of *Mycobacterium tuberculosis*. *J Bacteriol* **68**:411–418.

57. Millman I, Youmans GP. 1955. The characterization of the terminal respiratory enzymes of the H37Ra strain of

Mycobacterium tuberculosis var. *hominis*. *J Bacteriol* **69**:320–325.

58. **Goldman DS.** 1956. Enzyme systems in the mycobacteria II. The malic dehydrogenase. *J Bacteriol* **72**:401–405.

59. **Goldman DS.** 1956. Enzyme systems in the mycobacteria. I. The isocitric dehydrogenase. *J Bacteriol* **71**: 732–736.

60. **Murthy PS, Sirsi M, Ramakrishnan T.** 1962. Tricarboxylic acid cycle and related enzymes in cell-free extracts of *Mycobacterium tuberculosis* H37Rv. *Biochem J* **84**: 263–269.

61. **Edson N, Hunter G, Kulka R, Wright D.** 1959. The metabolism of glycerol in *Mycobacterium butyricum*. *Biochem J* **72**:249.

62. **Argyrou A, Blanchard JS.** 2001. *Mycobacterium tuberculosis* lipoamide dehydrogenase is encoded by Rv0462 and not by the lpdA or lpdB genes. *Biochemistry* **40**: 11353–11363.

63. **Bryk R, Lima CD, Erdjument-Bromage H, Tempst P, Nathan C.** 2002. Metabolic enzymes of mycobacteria linked to antioxidant defense by a thioredoxin-like protein. *Science* **295**:1073–1077.

64. **Tian J, Bryk R, Shi S, Erdjument-Bromage H, Tempst P, Nathan C.** 2005. *Mycobacterium tuberculosis* appears to lack alpha-ketoglutarate dehydrogenase and encodes pyruvate dehydrogenase in widely separated genes. *Mol Microbiol* **57**:859–868.

65. **Venugopal A, Bryk R, Shi S, Rhee K, Rath P, Schnappinger D, Ehrt S, Nathan C.** 2011. Virulence of *Mycobacterium tuberculosis* depends on lipoamide dehydrogenase, a member of three multienzyme complexes. *Cell Host Microbe* **9**:21–31.

66. **Shi S, Ehrt S.** 2006. Dihydrolipoamide acyltransferase is critical for *Mycobacterium tuberculosis* pathogenesis. *Infect Immun* **74**:56–63.

67. **Bryk R, Gold B, Venugopal A, Singh J, Samy R, Pupek K, Cao H, Popescu C, Gurney M, Hotha S, Cherian J, Rhee K, Ly L, Converse PJ, Ehrt S, Vandal O, Jiang X, Schneider J, Lin G, Nathan C.** 2008. Selective killing of nonreplicating mycobacteria. *Cell Host Microbe* **3**: 137–145.

68. **Chavadi S, Wooff E, Coldham NG, Sritharan M, Hewinson RG, Gordon SV, Wheeler PR.** 2009. Global effects of inactivation of the pyruvate kinase gene in the *Mycobacterium tuberculosis* complex. *J Bacteriol* **191**: 7545–7553.

69. **Keating LA, Wheeler PR, Mansoor H, Inwald JK, Dale J, Hewinson RG, Gordon SV.** 2005. The pyruvate requirement of some members of the *Mycobacterium tuberculosis* complex is due to an inactive pyruvate kinase: implications for in vivo growth. *Mol Microbiol* **56**:163–174.

70. **Goulding CW, Bowers PM, Segelke B, Lekin T, Kim C-Y, Terwilliger TC, Eisenberg D.** 2007. The structure and computational analysis of *Mycobacterium tuberculosis* protein CitE suggest a novel enzymatic function. *J Mol Biol* **365**:275–283.

71. **Merrill MH.** 1931. Studies on carbon metabolism of organisms of the genus *Mycobacterium*. II. Utilization of

organic compounds in a synthetic medium. *J Bacteriol* **21**:361–374.

72. **Edson N, Hunter G.** 1943. Respiration and nutritional requirements of certain members of the genus *Mycobacterium*. *Biochem J* **37**:563.

73. **Wahlund TM, Tabita FR.** 1997. The reductive tricarboxylic acid cycle of carbon dioxide assimilation: initial studies and purification of ATP-citrate lyase from the green sulfur bacterium *Chlorobium tepidum*. *J Bacteriol* **179**:4859–4867.

74. **Boshoff HIM, Myers TG, Copp BR, McNeil MR, Wilson MA, Barry CE.** 2004. The transcriptional responses of *Mycobacterium tuberculosis* to inhibitors of metabolism: novel insights into drug mechanisms of action. *J Biol Chem* **279**:40174–40184.

75. **Honaker RW, Leistikow RL, Bartek IL, Voskuil MI.** 2009. Unique roles of DosT and DosS in DosR regulon induction and *Mycobacterium tuberculosis* dormancy. *Infect Immun* **77**:3258–3263.

76. **Rustad TR, Harrell MI, Liao R, Sherman DR.** 2008. The enduring hypoxic response of *Mycobacterium tuberculosis*. *PLoS One* **3**:e1502.

77. **Voskuil MI, Visconti KC, Schoolnik GK.** 2004. *Mycobacterium tuberculosis* gene expression during adaptation to stationary phase and low-oxygen dormancy. *Tuberculosis* **84**:218.

78. **Banerjee S, Nandyala AK, Raviprasad P, Ahmed N, Hasnain SE.** 2007. Iron-dependent RNA-binding activity of *Mycobacterium tuberculosis* aconitase. *J Bacteriol* **189**:4046–4052.

79. **Banerjee S, Nandyala A, Podili R, Katoch VM, Hasnain SE.** 2005. Comparison of *Mycobacterium tuberculosis* isocitrate dehydrogenases (ICD-1 and ICD-2) reveals differences in coenzyme affinity, oligomeric state, pH tolerance and phylogenetic affiliation. *BMC Biochem* **6**:20.

80. **Hatzopoulos GN, Kefala G, Mueller-Dieckmann J.** 2008. Cloning, expression, purification, crystallization and preliminary X-ray crystallographic analysis of isocitrate dehydrogenase 2 (Rv0066c) from *Mycobacterium tuberculosis*. *Acta Crystallogr Sect F Struct Biol Cryst Commun* **64**:1139–1142.

81. **Quartararo CE, Hazra S, Hadi T, Blanchard JS.** 2013. Structural, kinetic and chemical mechanism of isocitrate dehydrogenase-1 from *Mycobacterium tuberculosis*. *Biochemistry*. [Epub ahead of print.]

82. **Lamichhane G, Zignol M, Blades NJ, Geiman DE, Dougherty A, Grosset J, Broman KW, Bishai WR.** 2003. A postgenomic method for predicting essential genes at subsaturation levels of mutagenesis: application to *Mycobacterium tuberculosis*. *Proc Natl Acad Sci USA* **100**:7213–7218.

83. **Tian J, Bryk R, Itoh M, Suematsu M, Nathan C.** 2005. Variant tricarboxylic acid cycle in *Mycobacterium tuberculosis*: identification of alpha-ketoglutarate decarboxylase. *Proc Natl Acad Sci USA* **102**:10670–10675.

84. **de Carvalho LP, Zhao H, Dickinson CE, Arango NM, Lima CD, Fischer SM, Ouerfelli O, Nathan C, Rhee**

KY. 2010. Activity-based metabolomic profiling of enzymatic function: identification of Rv1248c as a mycobacterial 2-hydroxy-3-oxoadipate synthase. *Chem Biol* **17**:323–332.

85. Balakrishnan A, Jordan F, Nathan CF. 2013. Influence of allosteric regulators on individual steps in the reaction catalyzed by *Mycobacterium tuberculosis* 2-hydroxy-3-oxoadipate synthase. *J Biol Chem* **288**: 21688–21702.

86. Wagner T, Bellinzoni M, Wehenkel A, O'Hare HM, Alzari PM. 2011. Functional plasticity and allosteric regulation of alpha-ketoglutarate decarboxylase in central mycobacterial metabolism. *Chem Biol* **18**:1011–1020.

87. O'Hare HM, Durán R, Cerveñansky C, Bellinzoni M, Wehenkel AM, Pritsch O, Obal G, Baumgartner J, Vialaret J, Johnsson K, Alzari PM. 2008. Regulation of glutamate metabolism by protein kinases in mycobacteria. *Mol Microbiol* **70**:1408–1423.

88. Baughn AD, Garforth SJ, Vilcheze C, Jacobs WR Jr. 2009. An anaerobic-type alpha-ketoglutarate ferredoxin oxidoreductase completes the oxidative tricarboxylic acid cycle of *Mycobacterium tuberculosis*. *PLoS Pathog* **5**:e1000662.

89. Noll H. 1958. The chemistry of the native constituents of the acetone-soluble fat of *Mycobacterium tuberculosis* (Brevannes). II. Isolation and properties of a new crystalline naphthoquinone derivative related to vitamin K2. *J Biol Chem* **232**:919–930.

90. Lester RL, Crane FL. 1959. The natural occurrence of coenzyme Q and related compounds. *J Biol Chem* **234**: 2169–2175.

91. Watanabe S, Zimmermann M, Goodwin MB, Sauer U, Barry CE 3rd, Boshoff HI. 2011. Fumarate reductase activity maintains an energized membrane in anaerobic *Mycobacterium tuberculosis*. *PLoS Pathog* **7**: e1002287.

92. Bai G, Gazdik MA, Schaak DD, McDonough KA. 2007. The *Mycobacterium bovis* BCG cyclic AMP receptor-like protein is a functional DNA binding protein in vitro and in vivo, but its activity differs from that of its *M. tuberculosis* ortholog, Rv3676. *Infect Immun* **75**:5509–5517.

93. Mechaly AE, Haouz A, Miras I, Barilone N, Weber P, Shepard W, Alzari PM, Bellinzoni M. 2012. Conformational changes upon ligand binding in the essential class II fumarase Rv1098c from *Mycobacterium tuberculosis*. *FEBS Lett* **586**:1606–1611.

94. Miesel L, Weisbrod TR, Marcinkeviciene JA, Bittman R, Jacobs WR Jr. 1998. NADH dehydrogenase defects confer isoniazid resistance and conditional lethality in *Mycobacterium smegmatis*. *J Bacteriol* **180**:2459–2467.

95. Mukhopadhyay B, Concar EM, Wolfe RS. 2001. A GTP-dependent vertebrate-type phosphoenolpyruvate carboxykinase from *Mycobacterium smegmatis*. *J Biol Chem* **276**:16137–16145.

96. Mukhopadhyay B, Purwantini E. 2000. Pyruvate carboxylase from *Mycobacterium smegmatis*: stabilization,

rapid purification, molecular and biochemical characterization and regulation of the cellular level. *Biochim Biophys Acta* **1475**:191–206.

97. Gould TA, van de Langemheen H, Munoz-Elias EJ, McKinney JD, Sacchettini JC. 2006. Dual role of isocitrate lyase 1 in the glyoxylate and methylcitrate cycles in *Mycobacterium tuberculosis*. *Mol Microbiol* **61**:940–947.

98. McKinney JD, zu Bentrup KH, Munoz-Elias EJ, Miczak A, Chen B, Chan W-T, Swenson D, Sacchettini JC, Jacobs WR, Russell DG. 2000. Persistence of *Mycobacterium tuberculosis* in macrophages and mice requires the glyoxylate shunt enzyme isocitrate lyase. *Nature* **406**:735.

99. Munoz-Elias EJ, McKinney JD. 2005. *Mycobacterium tuberculosis* isocitrate lyases 1 and 2 are jointly required for in vivo growth and virulence. *Nat Med* **11**:638.

100. Reinscheid DJ, Eikmanns BJ, Sahm H. 1994. Malate synthase from *Corynebacterium glutamicum*: sequence analysis of the gene and biochemical characterization of the enzyme. *Microbiology* **140**:3099–3108.

101. Molina I, Pellicer M-T, Badia J, Aguilar J, Baldoma L. 1994. Molecular characterization of *Escherichia coli* malate synthase G. *Eur J Biochem* **224**:541–548.

102. Blattner FR, Plunkett G, Bloch CA, Perna NT, Burland V, Riley M, Collado-Vides J, Glasner JD, Rode CK, Mayhew GF, Gregor J, Davis NW, Kirkpatrick HA, Goeden MA, Rose DJ, Mau B, Shao Y. 1997. The complete genome sequence of *Escherichia coli* K-12. *Science* **277**:1453–1462.

103. Parkhill J, Wren B, Thomson N, Titball R, Holden M, Prentice M, Sebaihia M, James K, Churcher C, Mungall K. 2001. Genome sequence of *Yersinia pestis*, the causative agent of plague. *Nature* **413**:523–527.

104. Heidelberg JF, Eisen JA, Nelson WC, Clayton RA, Gwinn ML, Dodson RJ, Haft DH, Hickey EK, Peterson JD, Umayam L. 2000. DNA sequence of both chromosomes of the cholera pathogen *Vibrio cholerae*. *Nature* **406**:477–483.

105. Liu K, Yu J, Russell DG. 2003. pckA-deficient *Mycobacterium bovis* BCG shows attenuated virulence in mice and in macrophages. *Microbiology* **149**:1829–1835.

106. Beste DJ, Bonde B, Hawkins N, Ward JL, Beale MH, Noack S, Noh K, Kruger NJ, Ratcliffe RG, McFadden J. 2011. (1)(3)C metabolic flux analysis identifies an unusual route for pyruvate dissimilation in mycobacteria which requires isocitrate lyase and carbon dioxide fixation. *PLoS Pathog* **7**:e1002091.

107. Wayne LG, Lin K-Y. 1982. Glyoxylate metabolism and adaptation of *Mycobacterium tuberculosis* to survival under anaerobic conditions. *Infect Immun* **37**:1042–1049.

108. Eoh H, Rhee KY. 2013. Multifunctional essentiality of succinate metabolism in adaptation to hypoxia in *Mycobacterium tuberculosis*. *Proc Natl Acad Sci USA* **110**: 6554–6559.

109. Segal W, Bloch H. 1956. Biochemical differentiation of *Mycobacterium tuberculosis* grown *in vivo* and *in vitro*. *J Bacteriol* **72**:132–141.

110. Goldman DS, Wagner MJ. 1962. Enzyme systems in the mycobacteria. XIII. Glycine dehydrogenase and the glyoxylic acid cycle. *Biochim Biophys Acta* **65**:297–306.

111. Marrero J, Rhee KY, Schnappinger D, Pethe K, Ehrt S. 2010. Gluconeogenic carbon flow of tri-carboxylic acid cycle intermediates is critical for *Mycobacterium tuberculosis* to establish and maintain infection. *Proc Natl Acad Sci USA* **107**:9819–9824.

112. Cook GM, Hards K, Vilchèze C, Hartman T, Berney M. 2014. Energetics of respiration and oxidative phosphorylation in mycobacteria. *Microbiol Spectrum* **2**(3):MGM2-0015-2013.

Molecular Genetics of Mycobacteria, 2nd Edition
Edited by Graham F. Hatfull and William R. Jacobs, Jr.
© 2014 American Society for Microbiology, Washington, DC
doi:10.1128/microbiolspec.MGM2-0033-2013

Emilie Layre,[1] Reem Al-Mubarak,[2]
John T. Belisle,[2] and D. Branch Moody[1]

Mycobacterial Lipidomics

17

METABOLOMICS AS A DISCIPLINE

Genes to Proteins to Metabolites

Genomics, transcriptomics, proteomics, and metabolomics are disciplines that broadly measure an organism's molecular repertoire to provide a portrait of that organism at an instant in time or a series of pictures to describe organism response. Working downstream from genes and enzymes, metabolites comprise the majority of the nonwater biomass of the cell. Transcriptomics and proteomics study linear polymers, so the key information is the sequence of building blocks comprised of nucleotides or amino acids. Accordingly, proteins, DNA, or RNA can be studied by applying one detection method to one general type of molecule to determine its sequence.

In contrast, metabolites are small molecules that show extreme diversity in their atomic composition. Metabolites are composed of aliphatic hydrocarbons, peptides, sugars, purines, pyrimidines, and other constituents in which individual molecules vary in length, oxygen functions, and other defined chemical elements. A conceptual challenge to inventing metabolomics platforms is that any chemical organizing principle of metabolites is not obvious or self-defining. Therefore, the reporting of organism-wide metabolite profiles requires that all investigators in the field agree on organizational

systems for grouping metabolites together. Cellular metabolites can be organized based on the similarity of their chemical structures, as seen in the LIPID MAPS initiative, or by laying out the sequential relationships of substrates, enzymes, and products into biosynthetic pathways (1, 2). A further practical challenge in building effective metabolomics platforms relates to the atomic diversity of metabolites, which require different solvents and separation techniques for individual subclasses of metabolites, such as nucleotides, peptides, and lipids.

In the past decade, these basic problems in organizing and detecting metabolites in high throughput have been largely solved, giving rise to functioning experimental platforms for metabolomics, which represent the youngest of the major systems biology disciplines. Metabolomics platforms take advantage of nuclear magnetic resonance (NMR) spectroscopy and mass spectrometry (MS) as nearly universal detection methods. Based on the implementation of increasingly comprehensive metabolite databases, it is now possible to rapidly profile thousands of metabolites in one experiment to generate an organism's or cell's metabolome. Here we highlight the conceptual and practical basis of metabolomics as a discipline, focusing on the mass spectrometric detection of lipids in the pathogen *Mycobacterium tuberculosis* to solve basic questions about the virulence of this lipid-laden organism.

[1]Division of Rheumatology, Immunology and Allergy, Brigham and Women's Hospital, Harvard Medical School, Boston, MA 02115;
[2]Department of Microbiology, Immunology, and Pathology, Colorado State University, Fort Collins, CO 80523.

Lipidomics as a Subspecialty of Metabolomics

Metabolomics assumes a practical definition as the high-throughput study of nonprotein, non-DNA, non-RNA cellular intermediates, which emphasizes small molecules with atomic masses lower than 3,000. Lipidomics emerged as a distinct subspecialty of metabolomics, which recognizes that conventional metabolites and lipids differ in one essential property: solubility in aqueous solutions. Conventional metabolomics focuses on molecules that are soluble in the cytosol, whereas lipidomics is mainly concerned with water-insoluble molecules that organize into membranes. Practical issues related to differences in methods needed to recover cytosolic metabolites and lipids increasingly lead to splitting of metabolomics and lipidomics into meaningfully different subfields with different biological foci. As contrasted with cytosolic metabolites, lipids tend to have more aliphatic hydrocarbons, larger size, lower polarity, and slower turnover (Fig. 1). These differences translate into a need for hydrophobic solvents, distinct ion-pairing reagents, and chromatographic and MS ionization methods that are matched to the chemical properties of lipids. Lipids can be described in a general way as molecules composed of a polar head group of variable size linked to aliphatic hydrocarbon chains (Fig. 2). Each lipid subclass is often comprised of different acylforms or alkylforms that possess the same core structure but differ in the length, saturation state, or substitution of the aliphatic chains. Thus, a lipidome of a single organism might contain up to 100,000 individual molecular lipid species (3–5).

ORGANISM-WIDE LIPIDOMIC DETECTION

Early Methods

Although lipidomics is a young specialty of systems biology, development of new databases and methods now creates technically mature platforms, which can broadly detect the named and unnamed components of a cell or bacterium within hours. Several efficient separation and detection methodologies are commonly used for the profiling of lipid repertoires including gas chromatography–MS (GC-MS) (6), liquid chromatography–MS (LC-MS) (7), and NMR (8). Early methods for lipidomics were thin-layer chromatography (9) and high-performance liquid chromatography (HPLC) interfaced with optical detection systems (10). Such approaches are time-consuming and lack high sensitivity. Furthermore, these methodologies are mainly suitable for the characterization of global changes among lipid classes, rather than detection of detailed remodeling of individual molecular species in one lipid class. Detection of lipid fine structure is important because the biological activities of a given lipid class often depend on the structure and distribution of molecular species that differ in chain length, saturation, oxygenation, and other aspects of their fine molecular heterogeneity (11, 12).

NMR

Modern lipidomic approaches are based on MS or NMR. NMR is a nondestructive and quantitative methodology, which provides stereo- and regiochemical information (13). However, the disadvantage of NMR spectroscopy is its relatively low sensitivity, for which optimal analysis is limited to lipids analyzed in the micromolar range. Furthermore, NMR of cellular extracts emphasizes signals that can be unambiguously assigned to individual molecules in a mixture. Thus, NMR is typically used as a fingerprint analysis or to describe a defined set of molecules of a complex lipid extract rather than as a comprehensive description of all metabolites in a cell (14, 15).

MS

Based on its high sensitivity, broad detection of naturally occurring lipids, and high dynamic range, MS has emerged as the most widely used lipidomic methodology. MS has pushed the field forward so that thousands of distinct lipid species can be separately detected in one experiment as unnamed accurate-mass retention time (AMRT) values (16). Ionization of lipids can be performed on many types of platforms including electrospray ionization (ESI), atmospheric pressure chemical ionization (APCI), atmospheric pressure photoionization, and matrix-assisted laser desorption ionization (MALDI). These techniques differ in their efficiencies of ionization and ability to maintain intact molecular structures during ionization, and thus can be

	Conventional Metabolomics	Lipidomics
Solvent	Water	Organic
Usual Mass (amu)	< 500	200-3000
Turnover	Seconds	Minutes to hours
Key Function	Energy Generation	Barrier Function

Figure 1 Lipidomics versus metabolomics. doi:10.1128/microbiolspec.MGM2-0033-2013.f1

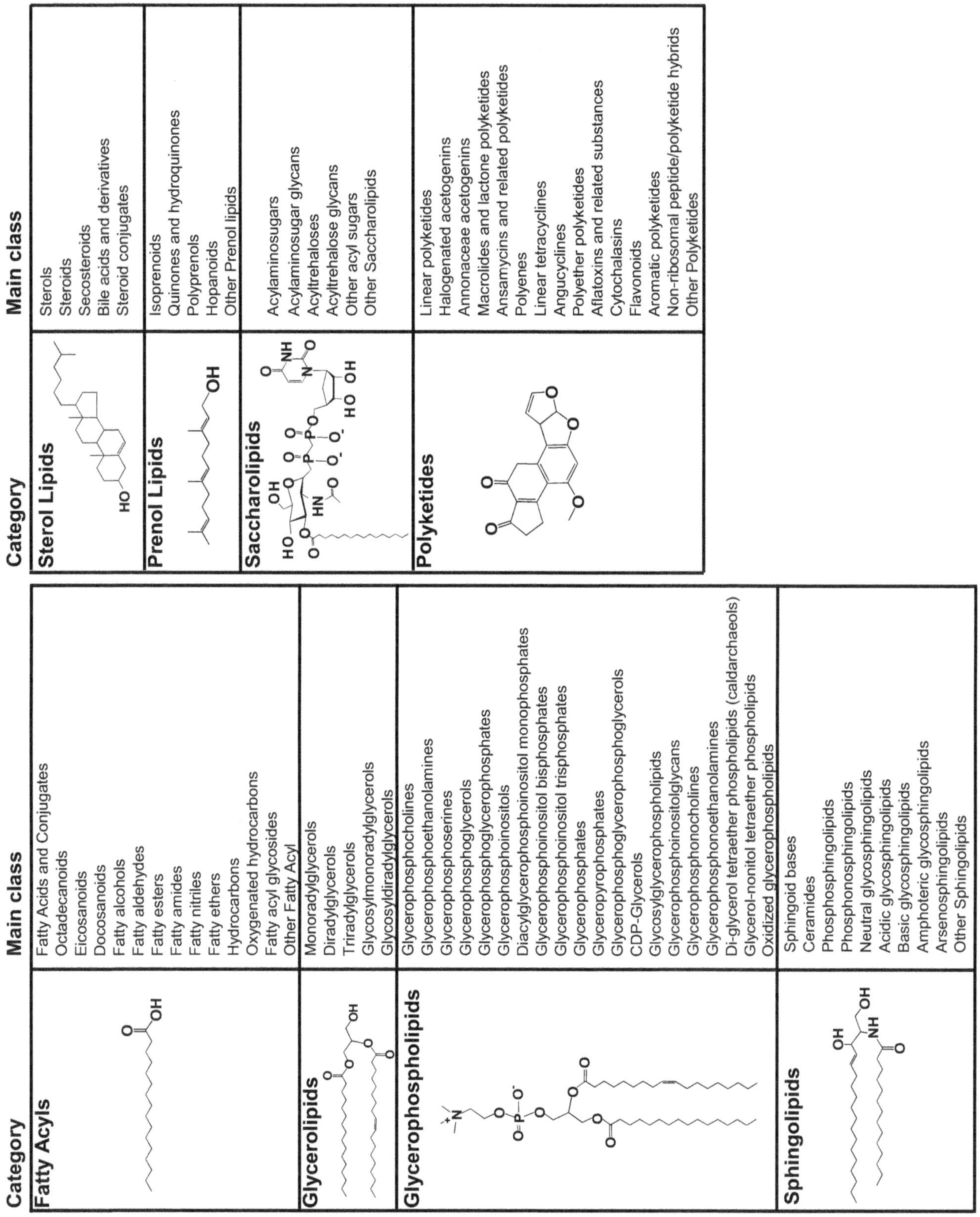

Category	Main class
Fatty Acyls	Fatty Acids and Conjugates
	Octadecanoids
	Eicosanoids
	Docosanoids
	Fatty alcohols
	Fatty aldehydes
	Fatty esters
	Fatty amides
	Fatty nitriles
	Fatty ethers
	Hydrocarbons
	Oxygenated hydrocarbons
	Fatty acyl glycosides
	Other Fatty Acyl
Glycerolipids	Monoradylglycerols
	Diradylglycerols
	Triradylglycerols
	Glycosylmonoradylglycerols
	Glycosyldiradylglycerols
Glycerophospholipids	Glycerophosphocholines
	Glycerophosphoethanolamines
	Glycerophosphoserines
	Glycerophosphoglycerols
	Glycerophosphoglycerophosphates
	Glycerophosphoinositols
	Glycerophosphoinositol monophosphates
	Glycerophosphoinositol bisphosphates
	Glycerophosphoinositol trisphosphates
	Glycerophosphates
	Glyceropyrophosphates
	Glycerophosphoglycerophosphoglycerols
	CDP-Glycerols
	Glycosylglycerophospholipids
	Glycerophosphoinositolglycans
	Glycerophosphonocholines
	Glycerophosphonoethanolamines
	Di-glycerol tetraether phospholipids (caldarchaeols)
	Glycerol-nonitol tetraether phospholipids
	Oxidized glycerophospholipids
Sphingolipids	Sphingoid bases
	Ceramides
	Phosphosphingolipids
	Phosphonosphingolipids
	Neutral glycosphingolipids
	Acidic glycosphingolipids
	Basic glycosphingolipids
	Amphoteric glycosphingolipids
	Arsenosphingolipids
	Other Sphingolipids

Category	Main class
Sterol Lipids	Sterols
	Steroids
	Secosteroids
	Bile acids and derivatives
	Steroid conjugates
Prenol Lipids	Isoprenoids
	Quinones and hydroquinones
	Polyprenols
	Hopanoids
	Other Prenol lipids
Saccharolipids	Acylaminosugars
	Acylaminosugar glycans
	Acyltrehaloses
	Acyltrehalose glycans
	Other acyl sugars
	Other Saccharolipids
Polyketides	Linear polyketides
	Halogenated acetogenins
	Annonaceae acetogenins
	Macrolides and lactone polyketides
	Ansamycins and related polyketides
	Polyenes
	Linear tetracyclines
	Angucyclines
	Polyether polyketides
	Aflatoxins and related substances
	Cytochalasins
	Flavonoids
	Aromatic polyketides
	Non-ribosomal peptide/polyketide hybrids
	Other Polyketides

Figure 2 LIPID MAPS classification system. doi:10.1128/microbiolspec.MGM2-0033-2013.f2

matched to the lipidomic application, depending on the size and the structure of the molecules of biological interest.

Currently, the most commonly employed configurations use ESI, a relatively gentle and sensitive ionization method that can be combined with many types of mass analyzers such as triple quadrupole, ion trap, Fourier transform (FT), and time of flight (TOF) devices, working in full scan mode or in collision-induced dissociation mode for tandem MS (MS/MS). ESI detects intrinsically charged molecules. Some lipids, such as polyketides and alcohols, are neutral in charge and thus are detected based on the formation of positive or negative adducts with ammonium, lithium, sodium, or other salts added to solvents. Ionization in the positive mode allows detection of the diverse lipids that possess an inherent positive charge or neutral species that adduct cations. The negative mode offers preferential detection of anionic free fatty acids, phospholipids, or sulfolipids. Accordingly, lipidomics of mammalian cells, which are predominantly composed of anionic membrane lipids, emphasizes negative mode detection. In contrast, the abundance of neutral polyketides and glycolipids in the mycobacterial cell envelope, especially its mycolate outer membrane (MOM), results in a greater analytical coverage of lipids by measuring ions in the positive mode (17, 18).

TARGETED AND GLOBAL LIPIDOMICS

Targeted Lipidomics

Targeted lipidomics involves conventional analytical chemistry experiments that emphasize analysis of the many molecular variants that belong to one lipid class. After extracting all classes of lipids from cells, each type of lipid is separated by specialized, class-specific chromatographic techniques. The goal of NMR or MS is to identify alterations in length, oxygen, or other chemical substitutions within one or a few related lipid classes. For example, the LIPID MAPS consortium uses a targeted approach for the study of all major membrane lipids in which lipid classes are separately analyzed using distinct LC-MS and MS/MS protocols (http://www.lipidmaps.org/protocols/index.html) (19). As an illustration, atmospheric pressure chemical ionization (APCI) favors the ionization of the most hydrophobic lipids such as triglycerides, whereas ESI with negative ion mode scanning is favored for anionic membrane phospholipids. Therefore, these two classes of lipids can be separated and then analyzed by chromatographic and ionization methods that are optimized for each class. Table 1 summarizes published examples of targeted lipidomics in which class-specific analyses are carried out to analyze micro-heterogeneity of lipid structures or enzyme products with limited diversity.

Table 1 Examples of lipidomic applications for targeted studies[a]

Lipid group	Enzyme/protein target	Enzyme function	Lipidomics techniques used[b]	References
Mycolic acids	MmaA1 and 4 (methyl mycolic acid synthase)	Methyltransferases (mycolic acid modifications)	ESI-MS, MALDI-TOF-MS, and NMR	97–99
	FadD32	Fatty acyl-AMP ligase	Applied ESI-MS to characterize enzymatic end products	99–100
SL	PapA2 and PapA1 (polyketide synthase associated protein)	Acyltransferases	Targeted analysis of SL by FT-ICR-MS	101
	Chp1 (cutinase-like hydrolase protein and Sap (sulfolipid 1–addressing protein).	Chp1-acyltransferase, Sap participates in SL transport	Targeted analysis of SL by FT-ICR-MS and MALDI-TOF-MS of cell-free enzymatic reaction	102
PDIM	Pks15/1 (polyketide synthase)	Synthesis of phenolphthiocerol	Targeted analysis by TLC, NMR, GC-MS, and MALDI-TOF-MS	103
	FadD26, FadD28, and FadD29	Fatty acyl-AMP ligases	Targeted analysis by TLC and MALDI-TOF-MS	104
Many of these PIMs	PimA (Rv2610c) (alpha-mannosyltransferase)	Mannosyltransferase	Targeted analysis by MALDI-TOF-MS	105
	Rv2611c	Acyltransferase	Targeted analyses by MALDI-TOF-MS and ESI-MS	106
GPL	Rv2962cc, Rv2958c, and Rv2957	Glycosyltrasferases	Targeted analyses by TLC, MALDI-TOF-MS, and GC-MS	107

[a]Many of these studies couple modern targeted lipidomic approaches with classical metabolic labeling of lipids and separation by thin-layer chromatography.
[b]Abbreviations: ESI/MS, electrospray ionization mass spectrometry; MALDI-TOF, matrix-assisted laser desorption/ionization-time of flight; NMR, nuclear magnetic resonance; SL, sulfolipid; FT-ICR-MS, Fourier transform ion cyclotron resonance-mass spectrometry; TLC, thin-layer chromatography; GC-MS, gas chromatography–mass spectrometry.

In targeted lipidomics, specific lipids can be detected and measured by searching for defined fragmentation patterns in parent ion scanning, daughter ion scanning, or neutral loss scanning (5). For LC-MS strategies performed on quadrupole or Q-TOF analyzers, specific measurements of targeted lipids eluting from a chromatographic column at any time are achieved by repeatedly scanning for selected reactions, a method called multiple reaction monitoring (20). A sensitive and specific scanning mode is called selected reaction monitoring, whereby both specified parent and daughter ions are monitored. These multidimensional MS analyses are emerging as a dominant tool in targeted lipidomic analyses of glycerolipids, phospholipids, and sphingolipids (21–24). These targeted methods are notable for their quantitative accuracy and high sensitivity. Such targeted methods are useful when an *a priori* goal assumes that only one or a few types of lipids are relevant to a biological problem.

Global Lipidomics

However, for biological questions in which the organism is perturbed by a drug, or growth condition or genetic deletion, the types of lipids that might change in response to the biological variable cannot typically be predicted from the outset of the experiment. Carrying out dozens of targeted analyses for each lipid class is impractical, so untargeted, or "global," methods that nearly simultaneously measure most or all lipid classes are needed. Global methods emphasize nearly comprehensive methods of lipid extraction, separation, and detection. Although truly universal metabolomic analysis in one detection system has not yet been realized, existing HPLC-MS methods provide broad detection of thousands of diverse metabolites in dozens of classes in a single experiment. Development of such untargeted methods of lipidomics has been optimized against a broad spectrum of lipids derived from one type of cell (17, 18). Untargeted lipidomics platforms typically use shotgun methods that avoid pre-separation of lipids prior to mass detection (25) or use normal phase chromatography with steep solvent gradients so that highly dissimilar molecules are rapidly analyzed using one chromatographic profile (17).

Shotgun and Separation Methods

Shotgun methods for untargeted lipidomics can use tandem quadrupole instruments for high throughput introduction of unfractionated lipid extract into the mass spectrometer (5), resulting in nearly simultaneous ionization of all analytes. Although rapid and simple, direct ionization of chemically diverse compounds present in an unfractionated preparation leads to cross-suppression effects and provides less quantifiable data (26, 27). To partially overcome this limitation, chromatographic separation of compounds prior to their ionization and MS detection provides advantages over shotgun methods for the analysis of complex lipid mixtures (17, 28). Chromatographic separation decreases the complexity of lipids being ionized at any moment in time, thus reducing the potential for ion suppression and increasing sensitive and quantitative detection of minor lipids or poorly ionizing lipid species within complex mixtures (29). Mycobacteria, however, create special challenges in performing untargeted lipidomic studies, as their lipid repertoire ranges from highly apolar polyketides, such as phthiocerol dimycocerosates (PDIM), to highly polar phosphatidylinositol mannosides (PIM) (Fig. 3). One normal phase chromatography method has been reported which uses one HPLC-MS method to separate 30 classes of mycobacterial lipids among more than 5,000 detected lipid species present in replicate HPLC-MS runs (17).

Comparative Lipidomics: Aligning Data Sets

Comparative lipidomics typically involves solving a lipidome in replicate before and after introducing a biological variable, such as gene deletion, nutrient deprivation, or drug treatment (Fig. 4). The resulting lipidomic data sets typically contain thousands of molecular features, also called molecular events. A molecular feature is a three-dimensional data point of connected m/z, retention time, and intensity data that describes one reproducibly detected ion, or when isotopic data are pooled, describes one compound (Fig. 5). Data sets collected before and after the biological intervention are compared so that any two events in different data sets with equivalent mass and retention time are considered to be the same molecule and thus are aligned. Next, the ion intensity for each aligned data point is compared to calculate a fold-change ratio. Typically, intensity values are measured in replicate so that statistics can be applied on each comparison to generate P values for each molecular feature. This process is repeated hundreds or thousands of times to provide fold-change values to all features in the data set. Plotting fold-change versus P value provides a description of all features (ions or molecules) in the data set that are changed or not changed after applying the biological variable (Fig. 4, middle panel). This approach can be applied to the role of lipids in diseases, environmental perturbations, response to drugs, biomarkers, and genetics (27). The sensitivity (femtomole) of modern mass spectrometers results in large data sets containing thousands of

Phthiocerol Dimycocerosates (PDIM)

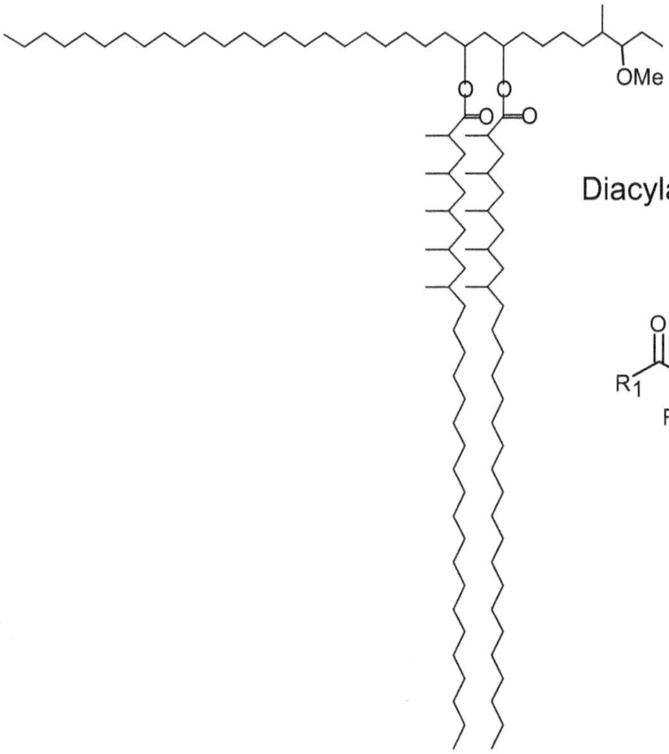

Diacylated-diacylglycerophosphoinositolmannosides
(Ac$_2$PIM$_2$)

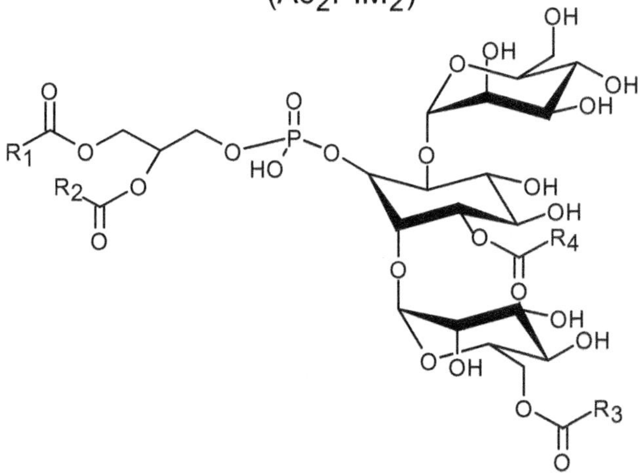

Phenolic glycolipid (PGL)

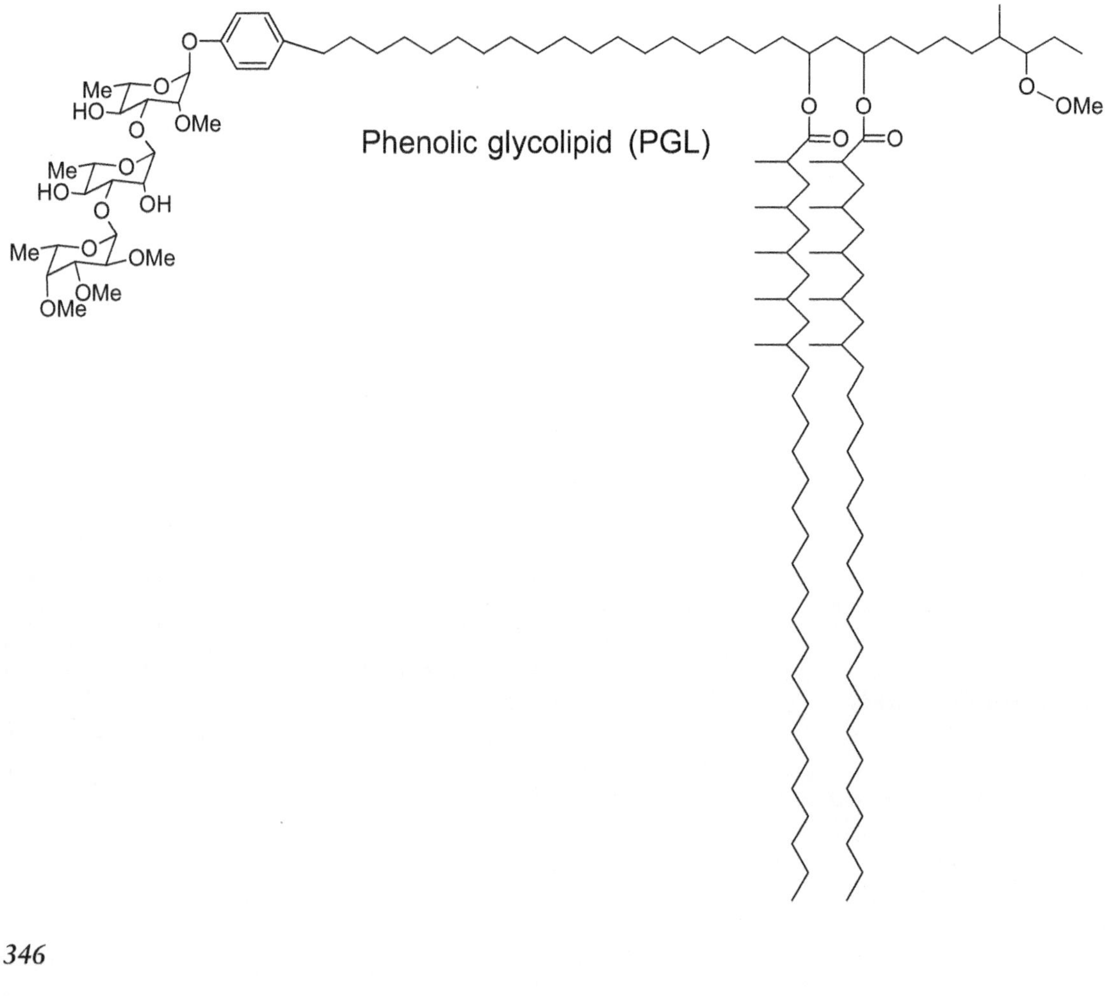

346

detected ions, which are analyzed in replicate and compared across multiple conditions such that the number of comparisons in a single experiment can exceed one million. Such analysis of large data sets to extract biological information is challenging and relies on bioinformatic tools previously developed in transcriptomics and other high-throughput biological platforms.

Quantitation

For targeted lipidomics, quantification is generally achieved by metabolic stable isotope labeling (30) or by addition of standards before extraction. Standards are often deuterated lipid species similar to the targeted lipid class, or a lipid species with odd-numbered carbon acyl chains that are naturally absent in the studied organism (31). Using calibration curves and internal standards, programs perform automatic conversion of intensity values of targeted lipids to absolute mass to determine lipid concentrations in the sample (32, 33).

For untargeted comparative lipidomics, XCMS and other software programs perform semiquantitative analyses by interrogating ion intensities or the volume of intensive-time curves measured from ion chromatograms (34). When analyzing hundreds or thousands of molecules in untargeted experiments, the use of internal standards is not feasible, so values are reported as ion intensity values of MS signals rather than absolute concentrations. The approach of reporting intensity values as a surrogate for measured concentrations relies on the high dynamic range of ESI-MS methods. Validation studies show that ion intensity values typically correlate with the concentration of the lipid in the sample across wide concentration ranges for most, but not all, lipids studied (17). Such untargeted lipid screens are biologically powerful based on their breadth of coverage and speed in identifying many compounds, whose detected intensity changes after introducing biological perturbations. Compounds reported as changed in such global lipidomic screens are typically subjected to detailed and quantitative analysis in a second round of confirmatory experiments.

Data Analysis and Display

Most approaches employ replicate biological analyses and automated ion finding algorithms that extract mass specific signals from noise, resulting in identification of discrete molecular features (Fig. 4 and 5). Molecular events can be reliably tracked in experiments, whether or not the mass and retention time values are connected to chemical names. Once all molecular features are assigned, mean intensity values with calculated variance, the data sets are systemically aligned to identify those molecular features that change from one biological condition to another. This is accomplished through software packages that use multivariate statistics and pattern recognition methods, which take advantage of statistical procedures and visualization methods previously developed for transcriptomics (reviewed in reference 35). Multiple testing corrections limit the false discovery rate (36). For example, MetaXCMS (R) allows comparison of multiple data sets from different biological conditions for the identification of metabolites that are specifically changed above threshold values after applying a biological variable (37).

The entire inventory of lipids can be displayed according to fold-change values versus P values, which appear in "volcano plots" (Fig. 4, middle panel). This common display visualizes the percent of all lipids in the inventory that meet threshold criteria for change and variance, and each feature encodes mass and retention time data that allow unnamed molecules of known mass to be named using mass databases or collisional MS. For example, comparing two mycobacterial strains before and after oxygen deprivation, gene deletion, or differing clinical origin allows detection of thousands of lipids from which only a few or many thousands can be shown to change (17, 18, 25, 38, 39). Comparative metabolomic experiments of this type show whether the biological variable results in narrow or very broad cellular remodeling, and it provides the information from which changed compounds can be identified. Unnamed compounds of known mass with high change values can be named using databases that match observed accurate mass values to databases of known compounds. For previously unknown lipids, MS/MS can be performed to assist in identifying any lipid of interest (Fig. 4, lower panel).

LIPID IDENTIFICATION

Databases of Lipid Masses

Over the years, targeted analyses of lipids have provided accurate mass, fragmentation patterns, and polarity information for thousands of lipid species appearing in the literature. This experimental information has been compiled into databases for automatic identification of

Figure 3 Atypical lipids produced by *M. tuberculosis*.
doi:10.1128/microbiolspec.MGM2-0033-2013.f3

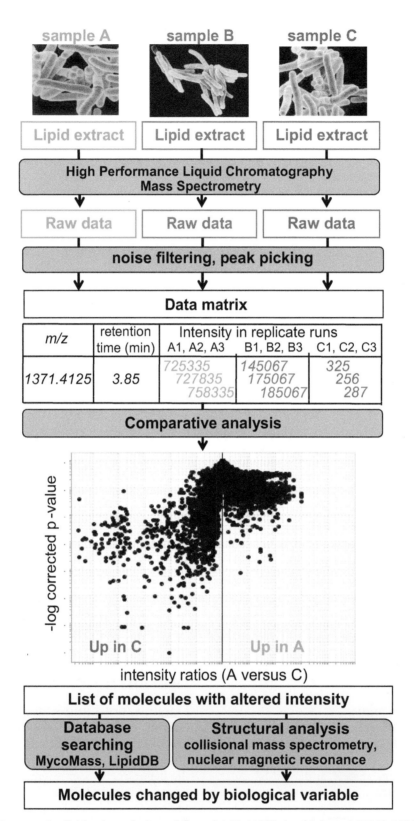

Figure 4 Comparative lipidomic analysis workflow. doi:10.1128/microbiolspec.MGM2-0033-2013.f4

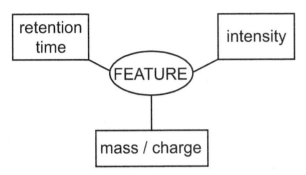

Figure 5 Molecular features. Comparative lipidomics relies on data sets in which thousands of molecular features from each biological condition are aligned based on mass and retention time. For aligned features, intensity ratios are reported to describe changes in abundance for each compound. Molecular features are also known as molecular events or linked accurate mass retention time intensity values. doi:10.1128/microbiolspec.MGM2-0033-2013.f5

previously described or "known" lipids in newly acquired data sets (32, 40, 41). Mass spectrometers with high resolution (10^{-5}) and mass accuracy (~2 parts per million) enable elemental composition determination based on detected m/z. However, two molecules of the same nominal masses (isobars) within a mixture of lipids cannot be differentiated based on the m/z information and require MS/MS experiments to be identified. Identification of lipids based on interrogation of MS/MS data is lower throughput than automated interrogation of databases with MS data but can result in the discovery of new types of lipids. Software has been optimized for automatic lipid identification based on database queries, including Lipid Profiler/LipidView (20), Lipid Inspector (42), Fatty Acid Analysis Tool (30), and LipidQA (32). LipidXplorer (43) carries out cross-platform interpretation of shotgun lipidomics data sets and identifies lipids via user-defined queries.

Mycobacterial Databases
Systematic organization of mycobacterial lipid classes shows that more than 90% of the defined lipid subclasses present in *M. tuberculosis* are found only in *Actinobacteria*. Mammalian cells, model eubacteria, and mycobacteria share a similar spectrum of phosphodiacylglycerols. However, the diverse mycolyl lipids, trehalose lipids, polyketides, and most other lipids found in mycobacteria are absent in non-mycobacterial model organisms (44). Therefore, widely used databases comprised of mass values of lipids, including LIPID MAPS (Fig. 2), provided little coverage for mycobacterial lipids. Therefore, rapid, automated naming of mycobacterial lipids was unavailable until recently.

In 2011, three databases that focus specifically on genus-specific mycobacterial lipids were published: Mtb LipidDB, MycoMass, and MycoMap (17, 18, 39). These tools have been integrated with LIPID MAPS and allow raw mass values to be compared to more than 100 classes of mycobacterial lipids (Mtb LipidDB, MycoMass) or detailed comparison to published mass, retention time, and collision spectra (MycoMap), facilitating rapid naming of mycobacterial lipids in global lipidomic profiling experiments.

MYCOBACTERIAL LIPIDS

Plasma Membrane
Mycobacteria are characterized by a complex cell wall structure that is unusually rich in lipids that constitute up to 60% of the dry weight of the organism. These lipids organize into two fully formed permeability barriers, the cytosolic membrane and the MOM, that are separated by the cell wall core (Fig. 6). The plasma membrane is composed mainly of anionic phospholipids in a bilayer arrangement with proteins. The most abundant fatty acids in the plasma membrane are palmitic (C16:0), octadecenoic (C18:1), and 10-methyloctadecanoic (tuberculoestearic), which are largely carried on glycerol backbones, as mycobacteria do not produce sphingolipids. The most common phospholipids are phosphatidylglycerol, diphosphatidylglycerol, and phosphatidylethanolamine, because mycobacteria do not produce phosphatidylcholine.

Mycobacterium species also produce phosphatidylinositol and an abundance of phosphatidylinositol mannosides (PIMs) that show different levels of mannosylation and acylation (Fig. 7) (45, 46). PIM serves as the anchor for the biologically important lipoglycans lipoarabinomannan and lipomannan (47–49). Despite their biological significance, lipoarabinomannan and lipomannan will not be discussed further due to their hydrophilic nature and high mass, which typically lead to their exclusion from apolar solvents and lipidomic detection methods that focus on lower-mass molecules.

The plasma membrane is also home to the menaquinones of the electron transport chain, including a unique sulfated menaquinone identified in *M. tuberculosis* (50). As with other prokaryotes, the plasma membrane of *M. tuberculosis* plays a central role in the biosynthesis of cell envelope structures and thus possesses a multitude of lipidic biosynthetic intermediates such as Lipid II, which carries subunits of peptidoglycan and is based on decaprenyl phosphate (51, 52). This polyprenyl phosphate also carries subunits of arabinose, galactose, and mannose for the synthesis of the arabinogalactan, lipoarabinomannan, and glycoproteins (53).

Periplasm

The cell wall skeleton of all mycobacteria is composed of peptidoglycan, a polymer of N-acetylglucosamine and N-muramic acid units (54, 55). The peptidoglycan is covalently attached to arabinogalactan, which is in turn esterified to mycolic acids at the nonreducing end of the arabinan. Thus, the inner segment of the MOM is more like a polymer than a conventional membrane. Due to the high mass and a tendency not to extract in chloroform-based solvents, most global lipidomics methods do not extract and analyze peptidoglycan, arabinogalactan, or covalently linked mycolates.

Mycolate outer membrane

In addition to the abundant cell wall esterified mycolic acids that comprise the inner section of the MOM, noncovalently linked lipids form the outer segment of the MOM (Fig. 6). These extractable glycolipids and highly apolar structures on the outer surface are thought to form a pseudobilayer, with the cell wall intrinsic mycolic acids in the inner segment. Mycolic acids are composed of α- and the β-hydroxy meromycolic chains that vary in length, oxygen modifications, and branching within and among *Mycobacterium* species (56). In many cases the length, unsaturations, and substitutions present on mycolic acids are sufficient to distinguish among mycobacterial species and thus represent useful chemotaxonomic information (57). *M. tuberculosis* has three major classes of mycolic acid: α-mycolic acid (di-*cis* cyclopropanated), keto-MA (α-methyl branched ketone and *cis*, *trans* cyclopropanated), and methoxy-MA (methyl ether and *cis*, *trans* cyclopropanated) (58).

One of the largest groups of extractable surface mycobacterial glycolipids are those based on trehalose (Fig. 8). These include trehalose dimycolate (TDM) and trehalose monomycolate (TMM). The former was originally described as a cord factor and was proposed

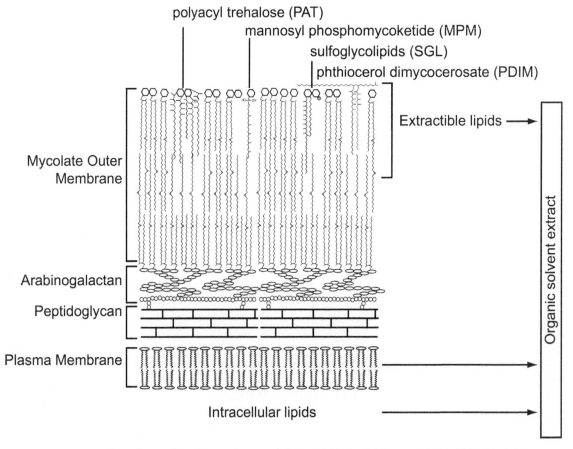

Figure 6   *M. tuberculosis* cell wall organization. doi:10.1128/microbiolspec.MGM2-0033-2013.f6

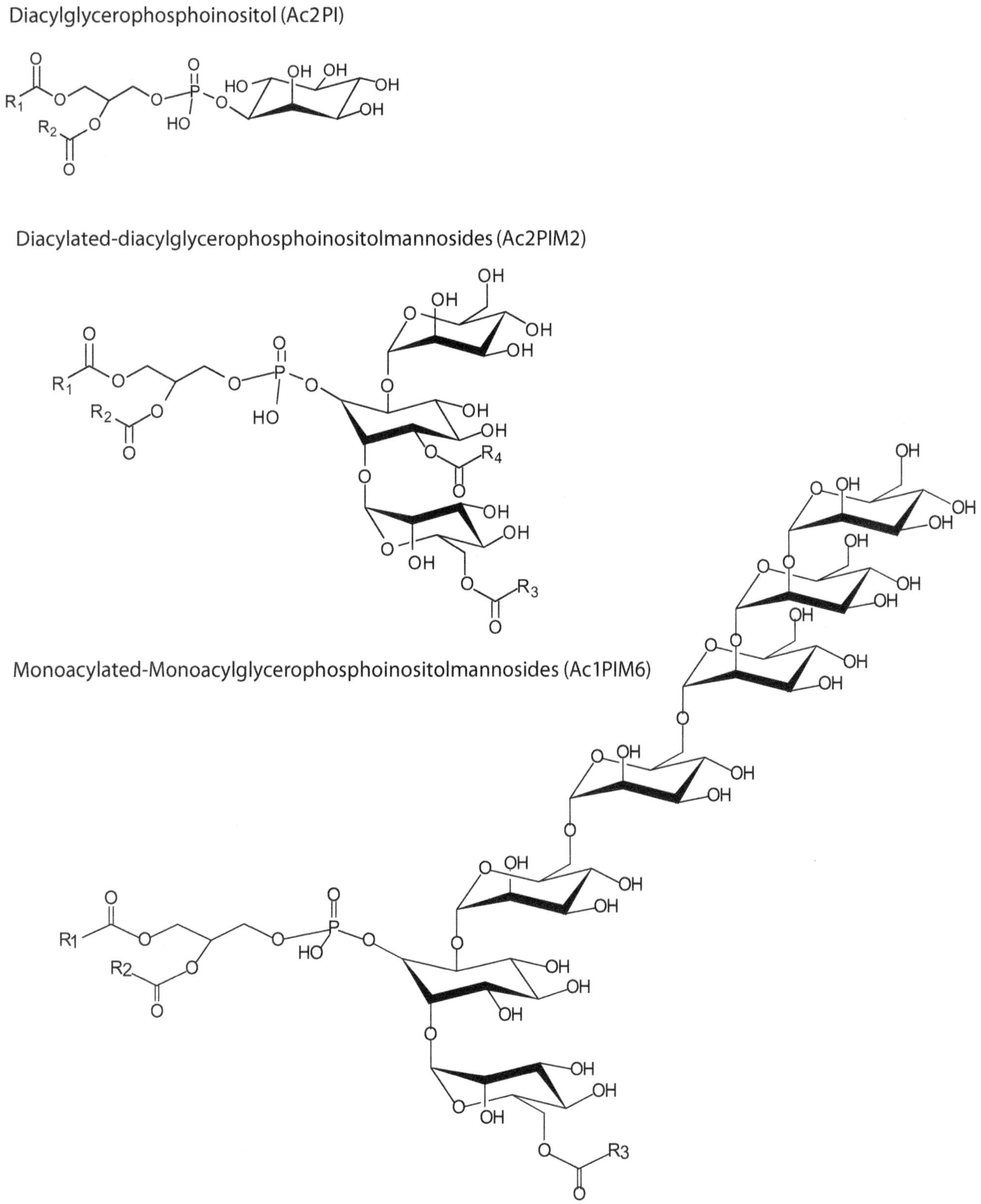

Diacylglycerophosphoinositol (Ac2PI)

Diacylated-diacylglycerophosphoinositolmannosides (Ac2PIM2)

Monoacylated-Monoacylglycerophosphoinositolmannosides (Ac1PIM6)

Figure 7 Structural diversity of glycerophosphoinositolmannosides. doi:10.1128/microbiolspec.MGM2-0033-2013.f7

Diacyltrehalose (DAT)　　　Triacyltrehalose (TAT)　　　Polyacyltrehalose (PAT)

Trehalose Monomycolate (TMM)　　　Trehalose Dimycolate (TDM)　　　Tetra-acylated sulfoglycolipids (SL-1)

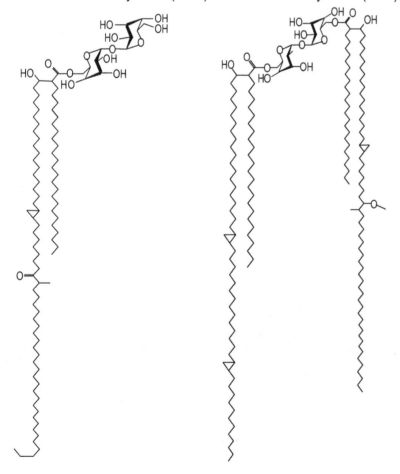
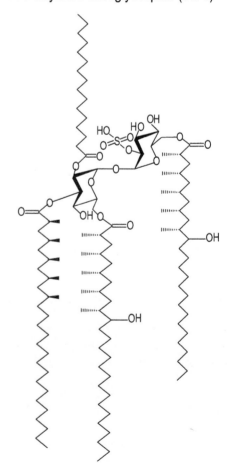

as one of the first virulence factors of *M. tuberculosis* (59–62). Trehalose dimycolate is a potent immunostimulatory molecule that acts through the lectin Mincle (63). A second group of acylated trehalose lipids includes diacyl-, triacyl-, and polyacyltrehalose (Fig. 8) (64–66). These structures are characterized by the presence of di- and tri-methyl branched fatty acids. The polyacyltrehalose has only been described in the *M. tuberculosis* complex (67). The sulfolipids are the third group of acylated trehalose molecules (Fig. 8). These sulfatides are defined by a 2′-sulfated trehalose that can be di-, tri-, or tetraacylated. The tetraacylated sulfolipid 1 (SL-1) is the most abundant form, and it is esterified with one palmitate or stearate, two hydroxyphthioceranate, and one phthioceranate group. Although this group of lipids was first recognized in 1959 (68) and was described as an *M. tuberculosis* complex–specific factor (69), a detailed understanding of the complexity and heterogeneity within this group was only recently achieved (70–73). The surface of *M. tuberculosis* is populated with other amphipathic lipids esterified with mycolic acids; these include glucose monomycolate (74) and glycerol monomycolate (75). These two lipids have not been well studied with regard to their physiological role in *M. tuberculosis*, but glucose monomycolate is produced *in vivo* from host glucose, and both of these lipids are well recognized as glycolipid antigens presented by CD1 proteins (75–77).

A final group of abundant cell surface lipids are the acylated polyketide diols that are based on a phthiocerol or the keto-modified phthiodiolone backbone (Fig. 3). These physiologically important lipids include the highly hydrophobic phthiocerol dimycocerosates (PDIM A) and phthiodiolone dimycocerosates (PDIM B). The phthiocerol dimycocerosates are structurally related to the phenolic glycolipids (PGL) (78, 79). *Mycobacterium bovis* produces a monoglycosylated PGL termed Mycoside B, while *M. tuberculosis* produces multiglycosylated PGL (80).

MYCOBACTERIAL LIPIDOMIC METHODS

Mycobacterial Lipids Differ from Those in Other Cellular Organisms

It is notable that the majority of *M. tuberculosis* lipids are limited in production to *Actinomycetales*, and many of these genus-specific lipids are large uncharged neutral lipids that accumulate in the MOM (Fig. 3, 6–8). Such mycolyl lipids, polyketides, and other neutral lipids require specialized extraction solvents, ionization conditions, and detection methods that are quite different than those used for anionic phospholipids that are the most abundant lipids in other organisms. For example, whereas the anionic phospholipids that dominate the plasma membranes of most cells are analyzed in the negative ion mode, neutral lipids that dominate the surface of mycobacteria form proton, sodium, and ammonium adducts that are detected in the positive mode. The hydrophobic nature and relatively high mass of mycolyl lipids and long chain polyketides generally require chloroform-based extraction solvents, addition of cations to run solvents, and ESI ionization conditions characterized by high voltage and low countercurrent gas (17, 18).

Bioinformatic Methods

To support rapid methods of compound naming, the observed and expected *m/z* of potential ionization adducts of nearly all mycobacterial lipid structures described in the literature has been assembled into databases: Mtb LipidDB (18) and MycoMass (17). They have cataloged more than 5,000 theoretical lipid species classified according to the LipidMaps standardized nomenclature based on eight lipid classes: fatty acyl lipids, glycerolipids, glycerophospholipids, sphingolipids, sterol lipids, prenol lipids, saccharolipids, and polyketides (40, 81) (Fig. 2). These databases allow matching of detected lipids to the *m/z* values for thousands of known and predicted lipid structures. Analysis of *M. tuberculosis* total lipid extracts by LC-(ESI)MS typically detects between 2,000 and 10,000 molecular features, of which 75 to 90% do not match mass values in these mycobacterial lipid databases (17, 18, 38, 82).

Annotating the *M. tuberculosis* Lipidome

This consistent finding indicates that, unlike mammalian cells and model Gram-negative bacteria, the annotation of the *M. tuberculosis* lipidome is far from complete. Sartain et al. noted that the lipids that could be identified using databases account for up to 75% of the total ion volume of a typical LC-MS data set (18). The detection of thousands of ions not matching database entries suggests that automated annotation methods generally fail to match database compounds to ions

Figure 8 *M. tuberculosis* produces diverse trehalose esters.
doi:10.1128/microbiolspec.MGM2-0033-2013.f8

of low intensity. In part, the high number of unannotated features in the LC-MS data files can be explained by the fact that existing databases focus on the terminal products of each lipid pathway and do not include all biosynthetic intermediates. For example, mycobactins are abundant siderophores annotated in all databases, but the stepwise generation of intermediates resulting from stepwise additions of fatty acids, peptide, or polyketide components are not yet identified and connected to mass values in data sets (82).

Minor lipid variants may also be missed since information used to populate the databases is based on earlier structural studies that likely describe only the most abundant products of a lipid group or use degradative methods such as hydrolysis to help define structures and that may remove highly labile and minor modifications such as acetylation of sugars. In addition, the products of certain lipid synthetic pathways remain unannotated, and little has been done to understand the turnover or products from the degradation of mycobacterial lipids. Unlike other widely studied model bacteria in which all key lipid biosynthetic genes are known, the functions of certain mycobacterial biosynthetic enzymes remain unknown. Thus, despite study of mycobacterial lipids dating back to the time of Robert Koch, there are likely additional mycobacterial lipids and biosynthetic pathways that remain to be discovered.

Needed Bioinformatic Tools

The identification of unannotated lipids can be facilitated by evaluating the products' retention time with respect to the annotated lipids. However, most laboratories use different chromatographic methods, so retention times are valid only within an individual method or research group. Thus, a long-term goal would be to perform annotation not only based on detected m/z and retention time, but based also on MS/MS in which parent ions and fragments provide a fingerprint of each molecule that would be useful in any laboratory. The MycoMap database reports collisional spectra for 30 mycobacterial lipid families, representing a start on this mapping project (17). Additionally, interrogation and extraction of collisional spectra from the literature, or the ability of individual investigators to contribute collisional spectra to an organized database, would greatly facilitate efforts to catalog all lipid fingerprints. Other bioinformatic tools that are still needed include a facile graphical user interface to further enhance the ability to perform automatic lipid identification with MS data sets. One graphical user interface, termed MS-LAMP, that includes data from

the existing mycobacterial lipid databases has been reported (83).

BIOLOGICAL STUDIES USING LIPIDOMICS

Lipid Fine Structure

An advantage of MS-based lipidomics is the ability to assess changes in the fine structure of lipid families in response to environmental stress. For example, Jain et al. reported that changes in the availability of methyl malonyl coenzyme A products among various *in vitro* and *in vivo* settings results in the production of sulfolipids and PDIM species with differing overall chain length (25). This study provided insight into the mechanism of polyketide lengthening and the role of substrate availability in the final structures of lipids. By separately tagging each chain length and phthiocerol/phthiodolone core variant to a distinct mass value and tracking all mass values in parallel through an experiment, it is possible to generate "fine molecular maps" that track changes in length and composition of every mass-based molecular variant in the family (Fig. 9) (17, 18, 38).

Another example of molecular diversity within one class of lipids is the PIM family found in *M. tuberculosis* (Fig. 3). This glycolipid family possesses species that can contain from 1 to 6 mannose units and 2 to 4 acyl chains. Thus, the Mtb LipidDB contains approximately 4,100 entries for PIM that take into account all possible combinations of acylation and glycosylation as well as ionization adducts. To provide a baseline for future analyses, Hsu and colleagues (72, 73) provided a comprehensive MS profile of the PIM species of *M. bovis* BCG including collisional MS spectra that can be used to develop selected reaction monitoring or MRM methodologies. Similarly, the many lipid modifications of sulfolipids create hundreds of natural variants of this molecule (70, 84).

Enzymology

Elucidation of biosynthetic pathways based on a stepwise understanding of enzyme and substrate function represents an area that is ideal for the application of lipidomics-based approaches. In the past, determining enzyme products in cells typically employed metabolic labeling and targeted *in vitro* assays with defined substrates to determine the products of enzymes. As noted in Table 1, conventional targeted lipid biochemistry approaches are being supplemented with more sensitive lipidomic methods that provide more detailed structural data with regard to end products or intermediates. A

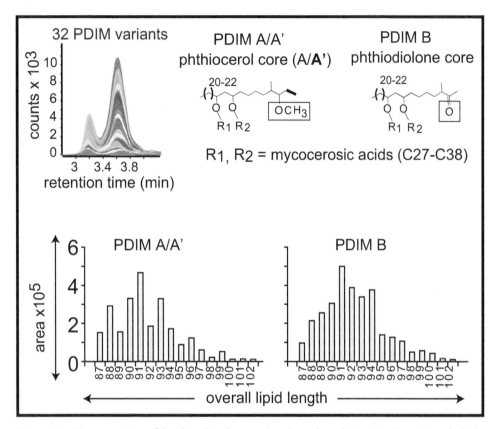

Figure 9 Fine mapping of lipid molecular species gives the relative ion intensity of chain length variant within each lipid class. doi:10.1128/microbiolspec.MGM2-0033-2013.f9

broader approach, whereby the deletion of biosynthetic enzyme genes in bacteria is followed by lipidomic profiling to detect both proximal products produced by the enzyme and distal products influenced indirectly, will represent an important step in defining the biosynthetic pathways for mycobacterial lipids. In particular, where multiple enzymes have similar or overlapping functions or perform multiple functions, gene deletion followed by lipidomic analysis can identify the nonredundant functions of any enzyme. Also, by tracking molecules of known masses but unknown chemical names or structures, it is possible to discover previously unknown molecules controlled by an enzyme. For example, previously unknown antigenic forms of deoxymycobactins were discovered by lipidomic profiling of *M. tuberculosis* lacking mycobactin synthases (39).

Drug Action
Similarly, lipidomic analysis of drug-treated bacteria can provide broader insights into the mechanisms of action. The action of isoniazid (INH) on the dehydratase step of fatty acid synthase II to prevent the formation of meromycolate precursors was confirmed by LC-MS and GC-MS of saponified fatty acids from treated *M. tuberculosis* cells. Specifically, these methodologies revealed a buildup of 3-hydroxy fatty acids and decreased meromycolate precursors (85, 86). A combination of metabolic labeling and LC-MS of lipid extracts from nonradiolabeled *M. tuberculosis* demonstrated that adamantyl urea inhibited the translocation of TMM through the inner membrane via the MmpL3 transporter. The function of MmpL3 was confirmed through LC-MS analyses of the lipidomes of the conditional mutants grown under nonpermissive conditions (85, 86).

In another example related to INH, early work assessed the incorporation of [14C]acetate or [14C]glycerol into the mycobacterial cell wall lipids and fatty acids by conventional methods (87, 88). These studies showed the direct effect of INH on mycolic acid synthesis. Subsequently, it was discovered that INH inhibited the formation of long fatty acids (C27-C40 and C39-C56) that were proposed as precursors in the synthesis of mycolic acids (89, 90). These early lipidomics studies ultimately allowed for the identification of InhA as the primary target of INH (91).

Clinical Biomarkers

Mycobacterial lipids are attractive candidates as biomarkers of tuberculosis infection because of their limited structural overlap with human lipids (Fig. 2). Also, compared to peptides, lipids are likely more refractory to rapid metabolism in the host. For example, mycolic acids are abundant, foreign, and chemically inert, making them a candidate as a diagnostic marker for tuberculosis (57, 92). Mycolic acid and mycocerosic acid components of PDIM (93) have been detected in the sputum of human patients as derivatized products by HPLC (94). More recently, MS was applied to underivatized mycolic acids isolated from human sputum, and MRM techniques were implemented to enhance the sensitivity and specificity of detection mycolic acid derivatives for tuberculosis infection (95, 96). These targeted approaches are based on what is understood with regard to the abundant products produced by *M. tuberculosis* in culture. A more global analysis of diagnostic samples such as sputum, serum, and urine may reveal additional mycobacterial lipids that can serve as diagnostic markers.

SUMMARY AND CONCLUSION

As lipidomics methods have become technically mature, a series of larger goals are now in sight. A substantially complete *M. tuberculosis* lipidome, based on the LIPID MAPS organizational scheme (Fig. 2) will likely emerge in coming years. To achieve this goal, the field needs to determine the functions of presumed lipid synthases that are currently of unknown function and to identify the stepwise intermediates that lead to more than 100 major lipid end products. This core goal will likely be followed by solving lipidomes of medically or evolutionarily important mycobacteria other than *M. tuberculosis*. It should be possible to achieve systematic understanding of lipid variation among clinical strains of mycobacteria from patients and to detect mycobacterial lipids within patient tissue or blood samples. With a complete lipid map, it will be possible to chart comprehensively the specific relationships among the genes and enzymes that regulate lipids. The prospect of integration of genomic, transcriptomic, proteomic, and metabolomic platforms is being realized. A recent study of *M. tuberculosis* entering and exiting an *in vitro* model of hypoxemia dormancy revealed connections among fatty acid and branched chain catabolism that broadly influence the metabolic state of mycobacteria (38). One lipidomic experiment of this type can generate more than one million data points, so development of better bioinformatics methods will be a key to success.

Citation. Layre E, Al-Mubarak R, Belisle JT, Moody DB. 2014. Mycobacterial lipidomics. Microbiol Spectrum 2(3): MGM2-0033-2013.

References

1. Beste DJ, Bonde B, Hawkins N, Ward JL, Beale MH, Noack S, Noh K, Kruger NJ, Ratcliffe RG, McFadden J. 2011. ^{13}C metabolic flux analysis identifies an unusual route for pyruvate dissimilation in mycobacteria which requires isocitrate lyase and carbon dioxide fixation. *PLoS Pathog* 7:e1002091.

2. Bonde BK, Beste DJ, Laing E, Kierzek AM, McFadden J. 2011. Differential producibility analysis (DPA) of transcriptomic data with metabolic networks: deconstructing the metabolic response of *M. tuberculosis*. *PLoS Comput Biol* 7:e1002060.

3. Murphy RC, Fiedler J, Hevko J. 2001. Analysis of nonvolatile lipids by mass spectrometry. *Chem Rev* 101: 479–526.

4. van Meer G. 2005. Cellular lipidomics. *EMBO J* 24: 3159–3165.

5. Han X, Gross RW. 2005. Shotgun lipidomics: electrospray ionization mass spectrometric analysis and quantitation of cellular lipidomes directly from crude extracts of biological samples. *Mass Spectrom Rev* 24: 367–412.

6. Quehenberger O, Armando AM, Dennis EA. 2011. High sensitivity quantitative lipidomics analysis of fatty acids in biological samples by gas chromatography-mass spectrometry. *Biochim Biophys Acta* 1811:648–656.

7. Del Boccio P, Pieragostino D, Di Ioia M, Petrucci F, Lugaresi A, De Luca G, Gambi D, Onofrj M, Di Ilio C, Sacchetta P, Urbani A. 2011. Lipidomic investigations for the characterization of circulating serum lipids in multiple sclerosis. *J Proteomics* 74:2826–2836.

8. Fernando H, Bhopale KK, Boor PJ, Ansari GA, Kaphalia BS. 2012. Hepatic lipid profiling of deer mice fed ethanol using ^{1}H and ^{31}P NMR spectroscopy: a dose-dependent subchronic study. *Toxicol Appl Pharmacol* 264:361–369.

9. Fontell K, Holman RT, Lambertsen G. 1960. Some new methods for separation and analysis of fatty acids and other lipids. *J Lipid Res* 1:391–404.

10. Bonanno LM, Denizot BA, Tchoreloff PC, Puisieux F, Cardot PJ. 1992. Determination of phospholipids from pulmonary surfactant using an on-line coupled silica/reversed-phase high-performance liquid chromatography system. *Anal Chem* 64:371–379.

11. Guiard J, Collmann A, Garcia-Alles LF, Mourey L, Brando T, Mori L, Gilleron M, Prandi J, De Libero G, Puzo G. 2009. Fatty acyl structures of *Mycobacterium tuberculosis* sulfoglycolipid govern T cell response. *J Immunol* 182:7030–7037.

12. Shinzawa-Itoh K, Aoyama H, Muramoto K, Terada H, Kurauchi T, Tadehara Y, Yamasaki A, Sugimura T, Kurono S, Tsujimoto K, Mizushima T, Yamashita E, Tsukihara T, Yoshikawa S. 2007. Structures and physiological roles of 13 integral lipids of bovine heart cytochrome c oxidase. *EMBO J* 26:1713–1725.

13. Reo NV. 2002. NMR-based metabolomics. *Drug Chem Toxicol* 25:375–382.

14. Mahrous EA, Lee RB, Lee RE. 2008. A rapid approach to lipid profiling of mycobacteria using 2D HSQC NMR maps. *J Lipid Res* 49:455–463.

15. Fernando H, Bhopale KK, Kondraganti S, Kaphalia BS, Shakeel Ansari GA. 2011. Lipidomic changes in rat liver after long-term exposure to ethanol. *Toxicol Appl Pharmacol* 255:127–137.

16. McLuckey SA, Wells JM. 2001. Mass analysis at the advent of the 21st century. *Chem Rev* 101:571–606.

17. Layre E, Sweet L, Hong S, Madigan CA, Desjardins D, Young DC, Cheng TY, Annand JW, Kim K, Shamputa IC, McConnell MJ, Debono CA, Behar SM, Minnaard AJ, Murray M, Barry CE 3rd, Matsunaga I, Moody DB. 2011. A comparative lipidomics platform for chemotaxonomic analysis of *Mycobacterium tuberculosis*. *Chem Biol* 18:1537–1549.

18. Sartain MJ, Dick DL, Rithner CD, Crick DC, Belisle JT. 2011. Lipidomic analyses of *Mycobacterium tuberculosis* based on accurate mass measurements and the novel "Mtb LipidDB." *J Lipid Res* 52:861–872.

19. Andreyev AY, Fahy E, Guan Z, Kelly S, Li X, McDonald JG, Milne S, Myers D, Park H, Ryan A, Thompson BM, Wang E, Zhao Y, Brown HA, Merrill AH, Raetz CR, Russell DW, Subramaniam S, Dennis EA. 2010. Subcellular organelle lipidomics in TLR-4-activated macrophages. *J Lipid Res* 51:2785–2797.

20. Ejsing CS, Duchoslav E, Sampaio J, Simons K, Bonner R, Thiele C, Ekroos K, Shevchenko A. 2006. Automated identification and quantification of glycerophospholipid molecular species by multiple precursor ion scanning. *Anal Chem* 78:6202–6214.

21. Yang K, Zhao Z, Gross RW, Han X. 2009. Systematic analysis of choline-containing phospholipids using multi-dimensional mass spectrometry-based shotgun lipidomics. *J Chromatogr B Analyt Technol Biomed Life Sci* 877:2924–2936.

22. Gross RW, Han X. 2009. Shotgun lipidomics of neutral lipids as an enabling technology for elucidation of lipid-related diseases. *Am J Physiol Endocrinol Metab* 297:E297–E303.

23. Han X, Gross RW. 2005. Shotgun lipidomics: multidimensional MS analysis of cellular lipidomes. *Exp Rev Proteomics* 2:253–264.

24. Han X, Yang K, Cheng H, Fikes KN, Gross RW. 2005. Shotgun lipidomics of phosphoethanolamine-containing lipids in biological samples after one-step in situ derivatization. *J Lipid Res* 46:1548–1560.

25. Jain M, Petzold CJ, Schelle MW, Leavell MD, Mougous JD, Bertozzi CR, Leary JA, Cox JS. 2007. Lipidomics reveals control of *Mycobacterium tuberculosis* virulence lipids via metabolic coupling. *Proc Natl Acad Sci USA* 104:5133–5138.

26. Han X, Yang K, Gross RW. 2012. Multi-dimensional mass spectrometry-based shotgun lipidomics and novel strategies for lipidomic analyses. *Mass Spectrom Rev* 31:134–178.

27. Harkewicz R, Dennis EA. 2011. Applications of mass spectrometry to lipids and membranes. *Annu Rev Biochem* 80:301–325.

28. Sandra K, Pereira Ados S, Vanhoenacker G, David F, Sandra P. 2010. Comprehensive blood plasma lipidomics by liquid chromatography/quadrupole time-of-flight mass spectrometry. *J Chromatogr A* 1217:4087–4099.

29. Griffiths WJ, Wang Y. 2009. Analysis of neurosterols by GC-MS and LC-MS/MS. *J Chromatogr B Analyt Technol Biomed Life Sci* 877:2778–2805.

30. Leavell MD, Leary JA. 2006. Fatty acid analysis tool (FAAT): an FT-ICR MS lipid analysis algorithm. *Anal Chem* 78:5497–5503.

31. Moore JD, Caufield WV, Shaw WA. 2007. Quantitation and standardization of lipid internal standards for mass spectroscopy. *Methods Enzymol* 432:351–367.

32. Song H, Hsu FF, Ladenson J, Turk J. 2007. Algorithm for processing raw mass spectrometric data to identify and quantitate complex lipid molecular species in mixtures by data-dependent scanning and fragment ion database searching. *J Am Soc Mass Spectrom* 18:1848–1858.

33. Song H, Ladenson J, Turk J. 2009. Algorithms for automatic processing of data from mass spectrometric analyses of lipids. *J Chromatogr B Analyt Technol Biomed Life Sci* 877:2847–2854.

34. Ivanova PT, Milne SB, Forrester JS, Brown HA. 2004. LIPID arrays: new tools in the understanding of membrane dynamics and lipid signaling. *Mol Interv* 4:86–96.

35. Niemela PS, Castillo S, Sysi-Aho M, Oresic M. 2009. Bioinformatics and computational methods for lipidomics. *J Chromatogr B Analyt Technol Biomed Life Sci* 877:2855–2862.

36. Benjamini Y, Drai D, Elmer G, Kafkafi N, Golani I. 2001. Controlling the false discovery rate in behavior genetics research. *Behav Brain Res* 125:279–284.

37. Tautenhahn R, Patti GJ, Kalisiak E, Miyamoto T, Schmidt M, Lo FY, McBee J, Baliga NS, Siuzdak G. 2011. metaXCMS: second-order analysis of untargeted metabolomics data. *Anal Chem* 83:696–700.

38. Galagan JE, Minch K, Peterson M, Lyubetskaya A, Azizi E, Sweet L, Gomes A, Rustad T, Dolganov G, Glotova I, Abeel T, Mahwinney C, Kennedy AD, Allard R, Brabant W, Krueger A, Jaini S, Honda B, Yu WH, Hickey MJ, Zucker J, Garay C, Weiner B, Sisk P, Stolte C, Winkler JK, Van de Peer Y, Iazzetti P, Camacho D, Dreyfuss J, Liu Y, Dorhoi A, Mollenkopf HJ, Drogaris P, Lamontagne J, Zhou Y, Piquenot J, Park ST, Raman S, Kaufmann SH, Mohney RP, Chelsky D, Moody DB, Sherman DR, Schoolnik GK. 2013. The *Mycobacterium tuberculosis* regulatory network and hypoxia. *Nature* 499:178–183.

39. Madigan CA, Cheng TY, Layre E, Young DC, McConnell MJ, Debono CA, Murry JP, Wei JR, Barry CE 3rd, Rodriguez GM, Matsunaga I, Rubin EJ, Moody DB. 2012. Lipidomic discovery of deoxysiderophores reveals a revised mycobactin biosynthesis pathway in *Mycobacterium tuberculosis*. *Proc Natl Acad Sci USA* 109:1257–1262.

40. Fahy E, Subramaniam S, Murphy RC, Nishijima M, Raetz CR, Shimizu T, Spener F, van Meer G, Wakelam MJ, Dennis EA. 2009. Update of the LIPID MAPS comprehensive classification system for lipids. *J Lipid Res* 50(Suppl):S9–S14.

41. Sud M, Fahy E, Cotter D, Brown A, Dennis EA, Glass CK, Merrill AH Jr, Murphy RC, Raetz CR, Russell DW, Subramaniam S. 2007. LMSD: LIPID MAPS structure database. *Nucleic Acids Res* 35:D527–D532.

42. Schwudke D, Oegema J, Burton L, Entchev E, Hannich JT, Ejsing CS, Kurzchalia T, Shevchenko A. 2006. Lipid profiling by multiple precursor and neutral loss scanning driven by the data-dependent acquisition. *Anal Chem* 78:585–595.

43. Herzog R, Schwudke D, Schuhmann K, Sampaio JL, Bornstein SR, Schroeder M, Shevchenko A. 2011. A novel informatics concept for high-throughput shotgun lipidomics based on the molecular fragmentation query language. *Genome Biol* 12:R8.

44. Layre E, Moody DB. 2013. Lipidomic profiling of model organisms and the world's major pathogens. *Biochimie* 95:109–115.

45. Hsu FF, Turk J, Owens RM, Rhoades ER, Russell DG. 2007. Structural characterization of phosphatidyl-myo-inositol mannosides from *Mycobacterium bovis* Bacillus Calmette Guerin by multiple-stage quadrupole ion-trap mass spectrometry with electrospray ionization. I. PIMs and lyso-PIMs. *J Am Soc Mass Spectrom* 18:466–478.

46. Hsu FF, Turk J, Owens RM, Rhoades ER, Russell DG. 2007. Structural characterization of phosphatidyl-myo-inositol mannosides from *Mycobacterium bovis* Bacillus Calmette Guerin by multiple-stage quadrupole ion-trap mass spectrometry with electrospray ionization. II. Monoacyl- and diacyl-PIMs. *J Am Soc Mass Spectrom* 18:479–492.

47. Pitarque S, Larrouy-Maumus G, Payre B, Jackson M, Puzo G, Nigou J. 2008. The immunomodulatory lipoglycans, lipoarabinomannan and lipomannan, are exposed at the mycobacterial cell surface. *Tuberculosis (Edinb)* 88:560–565.

48. Schlesinger LS, Hull SR, Kaufman TM. 1994. Binding of the terminal mannosyl units of lipoarabinomannan from a virulent strain of *Mycobacterium tuberculosis* to human macrophages. *J Immunol* 152:4070–4079.

49. Knutson KL, Hmama Z, Herrera-Velit P, Rochford R, Reiner NE. 1998. Lipoarabinomannan of *Mycobacterium tuberculosis* promotes protein tyrosine dephosphorylation and inhibition of mitogen-activated protein kinase in human mononuclear phagocytes. Role of the Src homology 2 containing tyrosine phosphatase 1. *J Biol Chem* 273:645–652.

50. Holsclaw CM, Sogi KM, Gilmore SA, Schelle MW, Leavell MD, Bertozzi CR, Leary JA. 2008. Structural characterization of a novel sulfated menaquinone produced by stf3 from *Mycobacterium tuberculosis*. ACS *Chem Biol* 3:619–624.

51. Mahapatra S, Yagi T, Belisle JT, Espinosa BJ, Hill PJ, McNeil MR, Brennan PJ, Crick DC. 2005. Mycobacterial lipid II is composed of a complex mixture of modi-

fied muramyl and peptide moieties linked to decaprenyl phosphate. *J Bacteriol* 187:2747–2757.

52. Kaur D, Brennan PJ, Crick DC. 2004. Decaprenyl diphosphate synthesis in *Mycobacterium tuberculosis*. *J Bacteriol* 186:7564–7570.

53. Berg S, Kaur D, Jackson M, Brennan PJ. 2007. The glycosyltransferases of *Mycobacterium tuberculosis*: roles in the synthesis of arabinogalactan, lipoarabinomannan, and other glycoconjugates. *Glycobiology* 17:35–56R.

54. Brennan PJ, Nikaido H. 1995. The envelope of mycobacteria. *Annu Rev Biochem* 64:29–63.

55. Bhamidi S, Shi L, Chatterjee D, Belisle JT, Crick DC, McNeil MR. 2012. A bioanalytical method to determine the cell wall composition of *Mycobacterium tuberculosis* grown in vivo. *Anal Biochem* 421:240–249.

56. Guenin-Mace L, Simeone R, Demangel C. 2009. Lipids of pathogenic mycobacteria: contributions to virulence and host immune suppression. *Transbound Emerg Dis* 56:255–268.

57. Butler WR, Guthertz LS. 2001. Mycolic acid analysis by high-performance liquid chromatography for identification of *Mycobacterium* species. *Clin Microbiol Rev* 14:704–726.

58. Barry CE 3rd, Lee RE, Mdluli K, Sampson AE, Schroeder BG, Slayden RA, Yuan Y. 1998. Mycolic acids: structure, biosynthesis and physiological functions. *Prog Lipid Res* 37:143–179.

59. Middlebrook G, Dubos RJ, Pierce C. 1947. Differential characteristics of virulent and avirulent variants of mammalian tubercle bacilli. *J Bacteriol* 54:66.

60. Indrigo J, Hunter RL Jr, Actor JK. 2003. Cord factor trehalose 6,6′-dimycolate (TDM) mediates trafficking events during mycobacterial infection of murine macrophages. *Microbiology* 149:2049–2059.

61. Indrigo J, Hunter RL Jr, Actor JK. 2002. Influence of trehalose 6,6′-dimycolate (TDM) during mycobacterial infection of bone marrow macrophages. *Microbiology* 148:1991–1998.

62. Hunter RL, Olsen MR, Jagannath C, Actor JK. 2006. Multiple roles of cord factor in the pathogenesis of primary, secondary, and cavitary tuberculosis, including a revised description of the pathology of secondary disease. *Ann Clin Lab Sci* 36:371–386.

63. Ishikawa E, Ishikawa T, Morita YS, Toyonaga K, Yamada H, Takeuchi O, Kinoshita T, Akira S, Yoshikai Y, Yamasaki S. 2009. Direct recognition of the mycobacterial glycolipid, trehalose dimycolate, by C-type lectin Mincle. *J Exp Med* 206:2879–2888.

64. Munoz M, Laneelle MA, Luquin M, Torrelles J, Julian E, Ausina V, Daffe M. 1997. Occurrence of an antigenic triacyl trehalose in clinical isolates and reference strains of *Mycobacterium tuberculosis*. *FEMS Microbiol Lett* 157:251–259.

65. Besra GS, Bolton RC, McNeil MR, Ridell M, Simpson KE, Glushka J, van Halbeek H, Brennan PJ, Minnikin DE. 1992. Structural elucidation of a novel family of acyltrehaloses from *Mycobacterium tuberculosis*. *Biochemistry* 31:9832–9837.

66. Daffe M, Lacave C, Laneelle MA, Gillois M, Laneelle G. 1988. Polyphthienoyl trehalose, glycolipids specific for virulent strains of the tubercle bacillus. *Eur J Biochem* **172**:579–584.

67. Hatzios SK, Schelle MW, Holsclaw CM, Behrens CR, Botyanszki Z, Lin FL, Carlson BL, Kumar P, Leary JA, Bertozzi CR. 2009. PapA3 is an acyltransferase required for polyacyltrehalose biosynthesis in *Mycobacterium tuberculosis*. *J Biol Chem* **284**:12745–12751.

68. Middlebrook G, Coleman CM, Schaefer WB. 1959. Sulfolipid from virulent tubercle bacilli. *Proc Natl Acad Sci USA* **45**:1801–1804.

69. Goren MB, Brokl O, Schaefer WB. 1974. Lipids of putative relevance to virulence in *Mycobacterium tuberculosis*: phthiocerol dimycocerosate and the attenuation indicator lipid. *Infect Immun* **9**:150–158.

70. Layre E, Paepe DC, Larrouy-Maumus G, Vaubourgeix J, Mundayoor S, Lindner B, Puzo G, Gilleron M. 2011. Deciphering sulfoglycolipids of *Mycobacterium tuberculosis*. *J Lipid Res* **52**:1098–1110.

71. Gilleron M, Stenger S, Mazorra Z, Wittke F, Mariotti S, Bohmer G, Prandi J, Mori L, Puzo G, De Libero G. 2004. Diacylated sulfoglycolipids are novel mycobacterial antigens stimulating CD1-restricted T cells during infection with *Mycobacterium tuberculosis*. *J Exp Med* **199**:649–659.

72. Hsu FF, Turk J, Owens RM, Rhoades ER, Russell DG. 2007. Structural characterization of phosphatidyl-myo-inositol mannosides from *Mycobacterium bovis* Bacillus Calmette Guerin by multiple-stage quadrupole ion-trap mass spectrometry with electrospray ionization. I. PIMs and lyso-PIMs. *J Am Soc Mass Spectrom* **18**:466–478.

73. Hsu FF, Turk J, Owens RM, Rhoades ER, Russell DG. 2007. Structural characterization of phosphatidyl-myo-inositol mannosides from *Mycobacterium bovis* Bacillus Calmette Guerin by multiple-stage quadrupole ion-trap mass spectrometry with electrospray ionization. II. Monoacyl- and diacyl-PIMs. *J Am Soc Mass Spectrom* **18**:479–492.

74. Brennan PJ, Lehane DP, Thomas DW. 1970. Acylglucoses of the corynebacteria and mycobacteria. *Eur J Biochem* **13**:117–123.

75. Layre E, Collmann A, Bastian M, Mariotti S, Czaplicki J, Prandi J, Mori L, Stenger S, De Libero G, Puzo G, Gilleron M. 2009. Mycolic acids constitute a scaffold for mycobacterial lipid antigens stimulating CD1-restricted T cells. *Chem Biol* **16**:82–92.

76. Moody DB, Reinhold BB, Guy MR, Beckman EM, Frederique DE, Furlong ST, Ye S, Reinhold VN, Sieling PA, Modlin RL, Besra GS, Porcelli SA. 1997. Structural requirements for glycolipid antigen recognition by CD1b-restricted T cells. *Science* **278**:283–286.

77. Moody DB, Guy MR, Grant E, Cheng TY, Brenner MB, Besra GS, Porcelli SA. 2000. CD1b-mediated T cell recognition of a glycolipid antigen generated from mycobacterial lipid and host carbohydrate during infection. *J Exp Med* **192**:965–976.

78. Reed MB, Domenech P, Manca C, Su H, Barczak AK, Kreiswirth BN, Kaplan G, Barry CE 3rd. 2004. A glycolipid of hypervirulent tuberculosis strains that inhibits the innate immune response. *Nature* **431**:84–87.

79. Brennan PJ. 1983. The phthiocerol-containing surface lipids of *Mycobacterium leprae*: a perspective of past and present work. *Int J Lepr Other Mycobact Dis* **51**:387–396.

80. Daffe M, Lacave C, Laneelle MA, Laneelle G. 1987. Structure of the major triglycosyl phenol-phthiocerol of *Mycobacterium tuberculosis* (strain Canetti). *Eur J Biochem* **167**:155–160.

81. Griffiths WJ, Wang Y. 2009. Mass spectrometry: from proteomics to metabolomics and lipidomics. *Chem Soc Rev* **38**:1882–1896.

82. Madigan CA, Cheng TY, Layre E, Young DC, McConnell MJ, Debono CA, Murry JP, Wei JR, Barry CE 3rd, Rodriguez GM, Matsunaga I, Rubin EJ, Moody DB. 2012. Lipidomic discovery of deoxysiderophores reveals a revised mycobactin biosynthesis pathway in *Mycobacterium tuberculosis*. *Proc Natl Acad Sci USA* **109**:1257–1262.

83. Sabareesh V, Singh G. 2013. Mass spectrometry based lipid(ome) analyzer and molecular platform: a new software to interpret and analyze electrospray and/or matrix-assisted laser desorption/ionization mass spectrometric data of lipids: a case study from *Mycobacterium tuberculosis*. *J Mass Spectrom* **48**:465–477.

84. Rhoades ER, Streeter C, Turk J, Hsu FF. 2011. Characterization of sulfolipids of *Mycobacterium tuberculosis* H37Rv by multiple-stage linear ion-trap high-resolution mass spectrometry with electrospray ionization reveals that the family of sulfolipid II predominates. *Biochemistry* **50**:9135–9147.

85. Grzegorzewicz AE, Kordulakova J, Jones V, Born SE, Belardinelli JM, Vaquie A, Gundi VA, Madacki J, Slama N, Laval F, Vaubourgeix J, Crew RM, Gicquel B, Daffe M, Morbidoni HR, Brennan PJ, Quemard A, McNeil MR, Jackson M. 2012. A common mechanism of inhibition of the *Mycobacterium tuberculosis* mycolic acid biosynthetic pathway by isoxyl and thiacetazone. *J Biol Chem* **287**:38434–38441.

86. Grzegorzewicz AE, Pham H, Gundi VA, Scherman MS, North EJ, Hess T, Jones V, Gruppo V, Born SE, Kordulakova J, Chavadi SS, Morisseau C, Lenaerts AJ, Lee RE, McNeil MR, Jackson M. 2012. Inhibition of mycolic acid transport across the *Mycobacterium tuberculosis* plasma membrane. *Nat Chem Biol* **8**:334–341.

87. Winder FG, Collins PB. 1970. Inhibition by isoniazid of synthesis of mycolic acids in *Mycobacterium tuberculosis*. *J Gen Microbiol* **63**:41–48.

88. Winder FG, Brennan P, Ratledge C. 1964. Synthesis of fatty acids by extracts of mycobacteria and the absence of inhibition by isoniazid. *Biochem J* **93**:635–640.

89. Davidson LA, Takayama K. 1979. Isoniazid inhibition of the synthesis of monounsaturated long-chain fatty acids in *Mycobacterium tuberculosis* H37Ra. *Antimicrob Agents Chemother* **16**:104–105.

90. Takayama K, Schnoes HK, Armstrong EL, Boyle RW. 1975. Site of inhibitory action of isoniazid in the

synthesis of mycolic acids in *Mycobacterium tuberculosis*. *J Lipid Res* **16**:308–317.

91. Vilcheze C, Wang F, Arai M, Hazbon MH, Colangeli R, Kremer L, Weisbrod TR, Alland D, Sacchettini JC, Jacobs WR Jr. 2006. Transfer of a point mutation in *Mycobacterium tuberculosis* inhA resolves the target of isoniazid. *Nat Med* **12**:1027–1029.

92. Kaneda K, Naito S, Imaizumi S, Yano I, Mizuno S, Tomiyasu I, Baba T, Kusunose E, Kusunose M. 1986. Determination of molecular species composition of C80 or longer-chain alpha-mycolic acids in *Mycobacterium* spp. by gas chromatography-mass spectrometry and mass chromatography. *J Clin Microbiol* **24**:1060–1070.

93. O'sullivan DM, Nicoara SC, Mutetwa R, Mungofa S, Lee OY, Minnikin DE, Bardwell MW, Corbett EL, McNerney R, Morgan GH. 2012. Detection of *Mycobacterium tuberculosis* in sputum by gas chromatography-mass spectrometry of methyl mycocerosates released by thermochemolysis. *PLoS One* **7**:e32836.

94. Viader-Salvado JM, Molina-Torres CA, Guerrero-Olazaran M. 2007. Detection and identification of mycobacteria by mycolic acid analysis of sputum specimens and young cultures. *J Microbiol Methods* **70**:479–483.

95. Shui G, Bendt AK, Jappar IA, Lim HM, Laneelle M, Herve M, Via LE, Chua GH, Bratschi MW, Zainul Rahim SZ, Michelle AL, Hwang SH, Lee JS, Eum SY, Kwak HK, Daffe M, Dartois V, Michel G, Barry CE 3rd, Wenk MR. 2012. Mycolic acids as diagnostic markers for tuberculosis case detection in humans and drug efficacy in mice. *EMBO Mol Med* **4**:27–37.

96. Szewczyk R, Kowalski K, Janiszewska-Drobinska B, Druszczynska M. 2013. Rapid method for *Mycobacterium tuberculosis* identification using electrospray ionization tandem mass spectrometry analysis of mycolic acids. *Diagn Microbiol Infect Dis* **76**:298–305.

97. Yuan Y, Crane DC, Musser JM, Sreevatsan S, Barry CE. 1997. MMAS-1, the branch point between cis- and trans-cyclopropane-containing oxygenated mycolates in *Mycobacterium tuberculosis*. *J Biol Chem* **272**:10041–10049.

98. Dinadayala P, Laval F, Raynaud C, Lemassu A, Laneelle MA, Laneelle G, Daffe M. 2003. Tracking the putative biosynthetic precursors of oxygenated mycolates of *Mycobacterium tuberculosis*. Structural analysis of fatty acids of a mutant strain deviod of methoxy- and ketomycolates. *J Biol Chem* **278**:7310–7319.

99. Takayama K, Wang C, Besra GS. 2005. Pathway to synthesis and processing of mycolic acids in *Mycobacterium tuberculosis*. *Clin Microbiol Rev* **18**:81–101.

100. Trivedi OA, Arora P, Sridharan V, Tickoo R, Mohanty D, Gokhale RS. 2004. Enzymic activation and transfer of fatty acids as acyl-adenylates in mycobacteria. *Nature* **428**:963.

101. Kumar P, Schelle MW, Jain M, Lin FL, Petzold CJ, Leavell MD, Leary JA, Cox JS, Bertozzi CR. 2007. PapA1 and PapA2 are acyltransferases essential for the biosynthesis of the *Mycobacterium tuberculosis* virulence factor sulfolipid-1. *Proc Natl Acad Sci USA* **104**:11221–11226.

102. Seeliger JC, Holsclaw CM, Schelle MW, Botyanszki Z, Gilmore SA, Tully SE, Niederweis M, Cravatt BF, Leary JA, Bertozzi CR. 2012. Elucidation and chemical modulation of sulfolipid-1 biosynthesis in *Mycobacterium tuberculosis*. *J Biol Chem* **287**:7990–8000.

103. Constant P, Perez E, Malaga W, Laneelle MA, Saurel O, Daffe M, Guilhot C. 2002. Role of the pks15/1 gene in the biosynthesis of phenolglycolipids in the *Mycobacterium tuberculosis* complex: evidence that all strains synthesize glycosylated p-hydroxybenzoic methyl esters and that strains devoid of phenolglycolipids harbor a frameshift mutation in the pks15/1 gene. *J Biol Chem* **277**:38148–38158.

104. Simeone R, Leger M, Constant P, Malaga W, Marrakchi H, Daffe M, Guilhot C, Chalut C. 2010. Delineation of the roles of FadD22, FadD26 and FadD29 in the biosynthesis of phthiocerol dimycocerosates and related compounds in *Mycobacterium tuberculosis*. *FEBS J* **277**:2715–2725.

105. Kordulakova J, Gilleron M, Mikusova K, Puzo G, Brennan PJ, Gicquel B, Jackson M. 2002. Definition of the first mannosylation step in phosphatidylinositol mannoside synthesis: PimA is essential for growth of mycobacteria. *J Biol Chem* **277**:31335–31344.

106. Kordulakova J, Gilleron M, Puzo G, Brennan PJ, Gicquel B, Mikusova K, Jackson M. 2003. Identification of the required acyltransferase step in the biosynthesis of the phosphatidylinositol mannosides of mycobacterium species. *J Biol Chem* **278**:36285–36295.

107. Perez E, Constant P, Lemassu A, Laval F, Daffe M, Guilhot C. 2004. Characterization of three glycosyltransferases involved in the biosynthesis of the phenolic glycolipid antigens from the *Mycobacterium tuberculosis* complex. *J Biol Chem* **279**:42574–42583.

Molecular Genetics of Mycobacteria, 2nd Edition
Edited by Graham F. Hatfull and William R. Jacobs, Jr.
© 2014 American Society for Microbiology, Washington, DC
doi:10.1128/microbiolspec.MGM2-0002-2013

Rainer Kalscheuer[1]
Hendrik Koliwer-Brandl[1]

Genetics of Mycobacterial Trehalose Metabolism

18

TREHALOSE: A UBIQUITOUS AND UNIVERSAL SUGAR

Trehalose is a natural nonreducing glucose disaccharide comprising an α,α-1,1-glycosidic linkage [α-D-glucopyranosyl-(1→1)-α-D-glucopyranoside]. The discovery of trehalose is credited to H. A. L. Wiggers (1832), who isolated a nonreducing sugar-like substance from the ergot fungus (*Claviceps pupurea*) of rye. Later this sugar was chemically characterized by Mitscherlich (1857), who termed it "mycose." About the same time (1858), the French chemist M. Berthelot isolated and characterized a sugar substance from "trehala manna," the sweet-tasting cocoon from the lepidopterous beetle *Larinus nidificans*, and coined the term "trehalose." In 1876, M. A. Müntz found that trehalose and mycose were identical (see reference 1 for an overview of the historical perspective). Since then, it has become clear that trehalose is a truly ubiquitous molecule that is synthesized by many different organisms including bacteria, yeasts, fungi, plants, and invertebrates. In the early days, trehalose was believed to exclusively serve as a carbon storage compound. Trehalose is nontoxic and, thus, accumulation to high intracellular concentrations is well tolerated. From this depot, glucose can rapidly be mobilized by hydrolysis mediated by the enzyme trehalase.

Today, however, it is evident that trehalose fulfills a plethora of potential biological functions that vary among the producing organisms (2, 3). Most notably, it can serve as a stress bioprotectant. Many organisms synthesize and accumulate high levels of trehalose in response to stimuli such as osmotic, heat, cold, dehydration, and other stresses in order to prevent denaturation and to maintain intracellular integrity. These characteristics are attributed to its peculiar physicochemical properties such as high hydrophilicity, chemical stability, formation of solid state polymorphs, and the absence of internal hydrogen bond formation. Due to these features, trehalose is capable of stabilizing macromolecular structures such as lipid membranes and proteins, partially by replacing hydration water (4). In fact, among other applications, trehalose is technically utilized as a preservative for enzyme preparations. In plants, where trehalose is typically found only at very low levels, trehalose-6-phosphate (a direct precursor in trehalose biosynthesis) is emerging as an important signaling molecule involved in the regulation of growth and development (5). An interesting notion is

[1]Institute for Medical Microbiology and Hospital Hygiene, Heinrich-Heine-University Duesseldorf, Universitaetsstr. 1, 40225 Duesseldorf, Germany.

that in most insects the main "blood sugar" in the hemolymph is not glucose but trehalose, serving as the energy and carbon transport form but simultaneously also acting as an antifreeze cryoprotectant (6). Probably, trehalose has more than one biological function in most synthesizing organisms. This is particularly true for the many roles of trehalose in mycobacteria, as discussed below.

FUNCTIONS OF TREHALOSE IN MYCOBACTERIA

Trehalose in mycobacteria is most widely recognized for being a structural component of cell wall glycolipids with important structural and/or immunomodulatory functions such as trehalose dimycolate (the cord factor). Less well known, free (i.e., nonconjugated) trehalose is also highly abundant intracellularly (7). In fact, it is probably the most prominent low-molecular-weight sugar compound present in mycobacteria during growth under regular laboratory culture conditions. Trehalose is an essential metabolite in *Mycobacterium tuberculosis* and *Mycobacterium smegmatis* and probably also in all other mycobacteria (8, 9). Mycobacteria can synthesize trehalose *de novo* from all kinds of unrelated carbon sources (such as glucose, glycerol, acetate, fatty acids, etc.), and they can also import and utilize exogenous trehalose. Trehalose auxotrophic mutants of *M. smegmatis* rapidly lose viability when they are starved for exogenous trehalose (9). Traditionally, free trehalose in microorganisms has been implicated in two main biological functions: as an intracellular carbon storage molecule and as a stress protectant. Likely, trehalose fulfills both of these functions in mycobacteria also. All mycobacteria tested so far can utilize trehalose as their sole source of carbon and energy, so it can serve as an intracellular storage compound that is mobilized during carbon source starvation. Furthermore, trehalose is probably the sole biocompatible stress protectant that *M. tuberculosis* is able to synthesize *de novo*. In line with its role in stress protection, trehalose auxotrophic mutants of *M. smegmatis* are sensitive to elevated temperatures when supplemented with suboptimal exogenous trehalose concentrations (9). The presence of an additional protective stress metabolite, ectoine, has been reported for *M. smegmatis*, but the corresponding biosynthetic genes are lacking in the *M. tuberculosis* genome (10).

However, beyond the two mentioned classical functions, trehalose is involved in many other biological processes in mycobacteria. Trehalose serves as the "sugar scaffold" for biosynthesis of numerous glycolipids, many of which play pivotal roles as structural cell wall components, carrier molecules in mycolic acid processing, and immune modulators influencing the interaction with infected host cells. Finally, it has recently been recognized that trehalose is also the substrate for a new metabolic four-step pathway in mycobacteria that converts it into glycogen-like branched alpha-glucans, which might be subject to further processing or export to yield molecules with different cellular destinations such as polymethylated alpha-glucan derivatives or extracellular capsular glucans. Our current knowledge of the genetic basis of the above-mentioned aspects of trehalose metabolism will be summarized below.

TREHALOSE *DE NOVO* BIOSYNTHESIS

To date, five natural metabolic routes for trehalose *de novo* biosynthesis have been described in the literature. Some organisms rely on only one pathway, whereas multiple pathways are present in others (5, 11). The most widespread route in prokaryotes and eukaryotes is the OtsA-OtsB pathway. This is the only biosynthetic route present in plants and has recently also been described in archaea (12). The trehalose-6-phosphate synthase OtsA catalyzes the transfer of nucleoside diphosphate-activated glucose (UDP-glucose) to glucose-6-phosphate to yield trehalose-6-phosphate with release of UDP. Subsequently, trehalose-6-phosphate phosphatase OtsB dephosphorylates trehalose-6-phosphate to trehalose (Fig. 1). The TreP pathway, which has been reported for fungi (13–16) and the protist *Euglena* (17), catalyzes the formation of trehalose from glucose-1-phosphate and glucose *in vitro*. It is uncertain, however, whether this pathway operates *in vivo* in the direction of trehalose production or rather in a reverse reaction in trehalose degradation. The other three pathways have been found exclusively in prokaryotes so far. In the TreY-TreZ pathway, first described in the thermophilic archaeon *Sulfolobus acidocaldarius* (18), the maltooligosyltrehalose synthase TreY converts the terminal α-1,4-glycosidic linkage at the reducing end of a linear α-1,4-glucan into an α-1,1-bond yielding maltooligosyltrehalose. Maltooligosyltrehalose trehalohydrolase, TreZ, then hydrolytically liberates trehalose (Fig. 1). Because this pathway requires linear glucans, branched alpha-glucans first need to be processed by glycogen phosphorylase GlgP, which reduces the branch length by releasing glucose-1-phosphate, followed by the debranching enzyme TreX, which hydrolyzes the α-1,6-glycosidic branch linkages. In the TreS pathway, trehalose synthase, TreS (first cloned from *Pimelobacter* [19, 20]), is a maltose

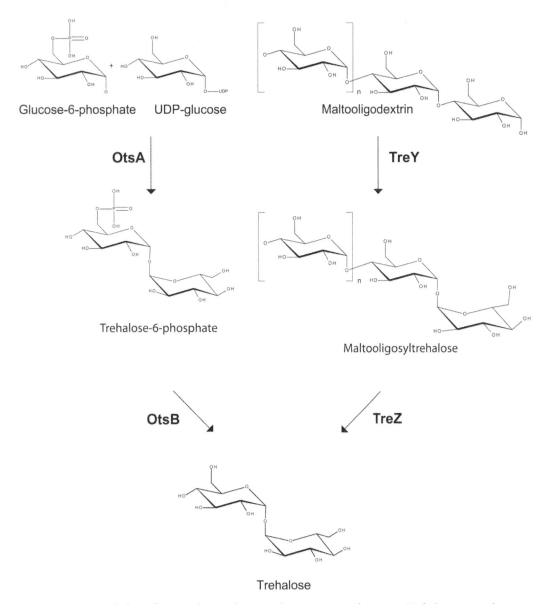

Glucose-6-phosphate UDP-glucose

Maltooligodextrin

OtsA

TreY

Trehalose-6-phosphate

Maltooligosyltrehalose

OtsB

TreZ

Trehalose

Figure 1 Trehalose *de novo* biosynthesis pathways in mycobacteria. Trehalose is synthesized either from glycolytic intermediates via the OtsA-OtsB pathway or from alpha-glucans via the TreY-TreZ pathway. UDP-glucose is formed from glucose-1-phosphate by the UTP-glucose-1-phosphate uridylyltransferase GalU (not depicted). This reaction, however, is not specific for trehalose biosynthesis. doi:10.1128/microbiolspec.MGM2-0002-2013.f1

α-D-glucosylmutase that reversibly interconverts maltose and trehalose by isomerizing the α-1,4- into an α-1,1-glycosidic linkage. Finally, in the TreT pathway, found in both bacteria and archaea and first described in the extremophilic archaea *Thermococcus litoralis* (21), trehalose glycosyltransferase, TreT, mediates the reversible formation of trehalose from ADP-glucose and glucose.

From bioinformatic analyses, it appeared that, of the five known trehalose biosynthesis pathways, three

operate in *M. tuberculosis*: the OtsA-OtsB, TreY-TreZ, and TreS pathways (22). Enzymatic *in vitro* characterizations confirmed that all three pathways are in principle functional in both fast-growing (*M. smegmatis*) and slow-growing (*Mycobacterium bovis* BCG) mycobacteria (22). Of the three alternative routes, the OtsA-OtsB pathway is the dominant one for trehalose formation in *M. tuberculosis*. A ΔotsA (Rv3490) gene deletion mutant was significantly attenuated for *in vitro* growth in trehalose-free medium (8).

Surprisingly, in contrast to the genetic dispensability of *otsA*, the *otsB2* gene (Rv3372) is obviously essential in *M. tuberculosis* and *M. bovis* BCG. The gene could not be inactivated even in the presence of exogenous trehalose to chemically complement for the biosynthetic defect (8). One explanation for this might be the potential toxic effect of the intermediate trehalose-6-phosphate that is expected to accumulate when OtsB is inactivated but OtsA is still active, similar to the toxicity of other phosphorylated sugar molecules, as will be discussed later for maltose-1-phosphate (23). The *M. tuberculosis* genome encodes a second gene with homology to bacterial trehalose-6-phosphate phosphatases, *otsB1* (Rv2006). The encoded OtsB1 protein is much larger than OtsB2 (1,327 amino acids [aa] vs. 391 aa). OtsB1 shares a conserved 260-aa core region with OtsB2 but contains additional N- and C-terminal domains. These domains are homologous to two other *M. tuberculosis* proteins, Rv3400 and Rv3401, which exhibit similarities to phosphoglucomutases and phosphorylases, respectively. Since *otsB1* cannot compensate for loss of *otsB2* and since a Δ*otsB1* mutant of *M. tuberculosis* showed no obvious phenotype, OtsB1 appears to play no role in trehalose biosynthesis (8). Its function, as well as those for Rv3400 and Rv3401, for which no mutants have been reported, is unclear. Deletion of *otsA* significantly attenuated growth of *M. tuberculosis* in a murine infection model (8). This would indicate that the OtsA-OtsB pathway constitutes the main trehalose biosynthetic pathway not only *in vitro* but also *in vivo* during infection in the mouse model. However, this study was not properly controlled by genetic complementation of the mutants. In fact, unpublished but publicly online available data from the same group (http://webhost.nts.jhu.edu/target/pdf/TARGET%20Report%20treS%20otsA.pdf) revealed that genetic complementation was not able to restore virulence of the *M. tuberculosis* Δ*otsA* mutant in C57BL/6 mice. This implies that, probably, inadvertent secondary mutations contribute to the observed *in vivo* phenotype, meaning that the definitive role of the trehalose biosynthetic pathways for virulence and viability of *M. tuberculosis* in mice is still elusive.

In contrast to the dominance of the OtsA-OtsB pathway in *M. tuberculosis*, genetic experiments in the fast-growing mycobacterium *M. smegmatis* appeared to show that all three pathways were individually sufficient to synthesize trehalose *de novo* (9). This result was puzzling regarding the role of TreS in trehalose formation. Contribution of TreS to trehalose *de novo* biosynthesis from defined mineral salts media would require a substantial intracellular source of free maltose

in *M. smegmatis*, which has never been reported in the literature so far. In order to unambiguously readdress the importance of TreS for trehalose biosynthesis in *M. smegmatis*, defined gene deletion mutants were generated with all possible combinatorial inactivations of the OtsA-OtsB, TreY-TreZ, and TreS pathways by targeting the genes *otsA*, *treY-Z*, and *treS*. These studies revealed that combined inactivation of the OtsA-OtsB and TreY-TreZ pathways was sufficient to result in trehalose auxotrophy in *M. smegmatis,* although in this genetic context the *treS* gene was still intact (24). This observation clearly defined that TreS plays no significant role in trehalose biosynthesis in *M. smegmatis* under the tested *in vitro* conditions. Furthermore, overexpression of TreS (Rv0126) in *M. tuberculosis* did not cause an elevation, but rather the opposite, a drastic reduction of the intracellular trehalose level, implying that TreS consumes rather than produces trehalose in mycobacteria (24). In fact, as described later in this article, TreS has recently been recognized as the first enzymatic step of a novel four-step pathway converting trehalose into branched alpha-glucans, which is widespread among prokaryotes (23, 25).

Although the equilibrium of purified TreS favors the formation of trehalose from maltose *in vitro*, flux through TreS *in vivo* is in the opposite direction (24), which is driven by the rapid and irreversible ATP-dependent phosphorylation of the formed maltose to maltose-1-phosphate by the maltose kinase Pep2 (Rv0127) (26, 27). The observed finding of the direction of flux through TreS for consumption of trehalose in *M. smegmatis* contradicts a previous study (9). The reason for this discrepancy is not obvious. However, in the previous study the authors did not employ a defined Δ*otsA*Δ*treY* double mutant to study the specific contribution of TreS in the *de novo* biosynthesis of trehalose, but rather, employed a surrogate strain (a Δ*otsA*Δ*treS*Δ*treY* triple mutant with a reconstituted *treS* gene constitutively expressed from an episomal multicopy plasmid) that likely exhibited a much higher *treS* expression level compared with the native gene. In this genetic context, TreS might suffice to form enough trehalose to support growth, given that a substantial source of maltose (e.g., from the medium) was present. The new findings in *M. smegmatis* are in agreement with the observed direction of flux in *M. tuberculosis* and with data from the closely related mycolic acid–producing actinomycete *Corynebacterium glutamicum*, in which the OtsA-OtsB and TreY-TreZ pathways, but not the TreS pathway, were demonstrated to be important for trehalose biosynthesis (28, 29). In conclusion, *M. tuberculosis, M. smegmatis,* and likely all other mycobacteria synthesize trehalose via two alternative routes, the OtsA-OtsB2 and the TreY-TreZ pathways (Fig. 1).

TREHALOSE DEGRADATION

All mycobacteria that have been tested for this metabolic capability so far (including *M. smegmatis*, *M. bovis*, *M. tuberculosis*, and *Mycobacterium marinum*) are able to grow on trehalose as the sole source of carbon and energy. Exogenous trehalose is imported by a highly specific and high-affinity ABC transporter (for ATP binding cassette), LpqY-SugABC (encoded by *Rv1235* to *Rv1238* in *M. tuberculosis*) (30), which will be discussed in detail below. It has been proposed that trehalose is hydrolyzed intracellularly by the enzyme trehalase (α,α-trehalose glucohydrolase; encoded by *Rv2402* in *M. tuberculosis*), which produces two glucose molecules that can then be further catabolized glycolytically (31). The enzyme has been biochemically characterized from *M. smegmatis*, but in an attempt to determine the role of trehalase for trehalose catabolism in this organism, the authors were unable to inactivate the respective gene, suggesting that it is essential (31). Our group, however, was able to delete the trehalase gene from *M. smegmatis* without noticing any negative effect on viability (32). According to its suggested role, the trehalase-deficient *M. smegmatis* mutant was unable to utilize trehalose as the sole carbon source, confirming its crucial role in intracellular trehalose breakdown (M. Alber and R. Kalscheuer, unpublished). However, the influence of trehalase on the regulation of the intracellular concentration of the endogenously *de novo* synthesized trehalose is currently not known.

BIOSYNTHESIS OF TREHALOSE-DERIVED GLYCOLIPIDS

The mycobacterial cell wall is characterized by an often high abundance of a rich variety of lipophilic molecules, including glycolipids with important structural functions and/or immunomodulatory properties. A major role of trehalose in mycobacteria is to serve as a "scaffold" for the buildup of many of these glycolipids. The biosynthesis of some of these glycoconjugates might start from free trehalose, whereas others might utilize the precursor trehalose-6-phosphate. Some of these trehalose-based glycolipids contain unusual an complex fatty acid derivatives synthesized by polyketide synthases. The metabolic pathways for production of the major trehalose-based glycolipids will be described in the following.

Trehalose Mycolates

The "cord factor" trehalose-6,6′-dimycolate (TDM) (Fig. 2) is probably the most well-known glycolipid of *M. tuberculosis*. The ability of *M. tuberculosis* to form rope-like structures in culture was first described by Robert Koch in 1884. Much later the substance responsible for this "cord" formation, the cord factor, was extracted with petroleum ether from *M. tuberculosis* cells and chemically identified as TDM (33, 34). TDM is now recognized as a critical structural component of the mycomembrane, forming the outer leaflet of this lipid bilayer along with other noncovalently linked (so-called extractable) lipids (35, 36). The biosynthesis of TDM is initiated in the cytoplasm, whereas the final steps take place extracellularly. The precursor of TDM, trehalose-6-monomycolate (TMM), not only is a

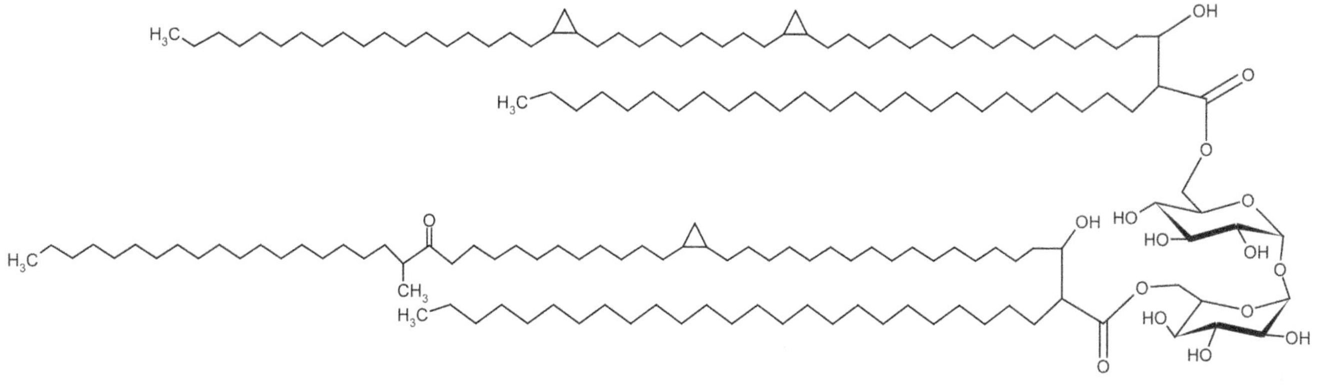

Figure 2 Structure of the cord factor trehalose-6,6′-dimycolate. In this example, *trans* cyclopropanated alpha- and ketomycolates are illustrated. doi:10.1128/microbiolspec.MGM2-0002-2013.f2

structural component of the cell wall, but it also fulfills a crucial role in the formation of the mycolic acid envelope layer.

Mature mycolates are first synthesized in the cytoplasm and remain coupled by a thioester linkage to a C-terminal domain of the polyketide synthase Pks13 (mycolyl-S-Pks13), as discussed in detail elsewhere in this book. The mycolyl group is then transferred from mycolyl-S-Pks13 to an isoprenoid carrier in the cytoplasm membrane (D-mannopyranosyl-1-phosphoheptaprenol) by an unknown cytoplasmic mycolyltransferase, yielding 6-O-mycolylmannopyranosyl phosphoheptaprenol (Myc-PL) (37). The mycolyl group is subsequently transferred from Myc-PL to trehalose-6-phosphate by a yet unidentified, probably membrane-associated mycolyltransferase, forming 6-mycolyl-trehalose-6′-phosphate, which is finally dephosphorylated to TMM by an unknown proposed membrane-associated phosphatase (38). TMM serves as the mycolate transport form which is exported outside the cell by the membrane transporter MmpL3, as has recently been discovered independently by several groups (39–41). Extracellularly, the periplasmic TMM is the substrate of the mycolyltransferases of the antigen 85 complex, which transfers the mycolate moiety to either the arabinogalactan layer to form cell-wall-bound mycolates or to another TMM molecule, resulting in formation of TDM (42, 43), whereas the concomitantly released trehalose is recycled and reimported into the cytoplasm, as will be discussed below (see Fig. 4). Trehalose thus functions as a carrier molecule to shuttle mycolates from the cytoplasm to the periplasm in the form of TMM. According to its transport function, pharmacological inhibition of MmpL3 in *M. tuberculosis* or conditional silencing of MmpL3 in *M. smegmatis* resulted in intracellular accumulation of TMM and abrogation of TDM formation (39–41).

The antigen 85 complex in *M. tuberculosis* is composed of Ag85A (FbpA; Rv3804c), Ag85B (FbpB; Rv1886c), and Ag85C (FbpC; Rv0129c) (42, 44). The mycolyltransferases of the antigen 85 complex are partially functionally redundant, implying overlapping substrate specificities. All three members are individually dispensable for viability of *M. tuberculosis*. An *M. tuberculosis* mutant lacking Ag85C exhibited a 40% reduction in the amount of cell-wall-linked mycolic acids (45). Inactivation of Ag85A or Ag85B had no significant effect on the cellular mycolic acid content, but their overexpression could compensate for the defect in mycolic acid content in the Δ*fbpC M. tuberculosis* mutant (43, 46).

Sulfolipids

The envelope of pathogenic mycobacteria contains complex trehalose-based glycolipids comprising up to five multiple methyl-branched long-chain fatty acids, including sulfolipids as well as di-, tri-, and polyacylated acyltrehaloses. Due to their high abundance in the cell wall, an important role of these abundant complex glycolipids for virulence of *M. tuberculosis* has been implicated. However, their definitive function has yet to be established. *M. tuberculosis* mutants lacking these cell envelope molecules typically exhibit no loss of virulence in the mouse infection model, but their role might be species specific and might reflect a specific adaptation to the human host. The most abundant sulfolipid in *M. tuberculosis* is SL-1, a tetraacyl-sulfotrehalose glycolipid. The biosynthesis of SL-1 is initiated by sulfation of trehalose at position 2 catalyzed by the sulfotransferase Stf0 (encoded by *Rv0295c* in *M. tuberculosis*), yielding trehalose-2-sulfate (Fig. 3A). Stf0 utilizes 3′-phosphoadenosine-5′-phosphosulfate (PAPS) as the sulfate donor. Deletion of *stf0* abrogates trehalose-2-sulfate biosynthesis and subsequent formation of SL-1 in *M. tuberculosis* (47). The acyltransferase PapA2 then catalyzes the esterification of trehalose-2-sulfate at the 2′-position to generate the monoacylated intermediate SL_{659}, which is further acylated by PapA1 at the 3′-position of SL_{659}, forming diacylated SL_{1278}. While PapA2 transfers a straight-chain fatty acid from acyl coenzyme A (CoA), PapA1 transfers a methyl-branched (hydroxy)phthioceranoyl chain likely directly from the acyl carrier protein domain of Pks2 (48). Pks2 synthesizes the required methyl-branched (hydroxy)phthioceranoyl chains using an activated fatty acid starter unit provided by the fatty acid AMP ligase FadD23 (Fig. 3A). Deletion of *pks2* in *M. tuberculosis* resulted in abrogation of SL_{1278} and SL-1 biosynthesis (49). Consistent with their sequential action in SL-1 synthesis, the *M. tuberculosis* Δ*papA1* mutant was deficient in SL_{1278} synthesis but still produced trehalose-2-sulfate and SL_{659}, whereas the Δ*papA2* mutant lacked SL_{659} and SL_{1278} while only retaining the ability to produce trehalose-2-sulfate (48). Additional acylations at the 6- and 6′-positions of SL_{1278} finally lead to synthesis of mature SL-1.

While these acylation steps are biochemically similar to the reaction catalyzed by PapA1, this activity is mediated by a different acyltransferase, Chp1, which has been identified recently. Chp1 (cutinase-like hydrolase protein; Rv3822) is a transacylase that can utilize SL_{1278} both as an acyl donor and acceptor and acylates SL_{1278} twice to reveal SL-1, while the deacylated side products are recycled (Fig. 3A) (50). Chp1 probably

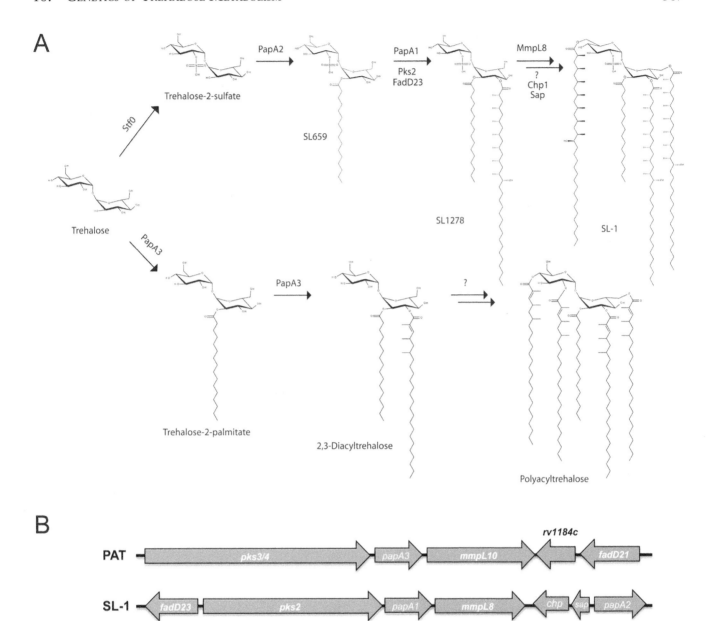

Figure 3 Biosynthesis of the major sulfolipid SL-1 and of polyacyltrehalose (PAT). (**A**) The synthesis of SL-1 and PAT is initiated by sulfation or acylation, respectively, at the 2-position of trehalose. The proposed reaction steps are depicted. (**B**) Molecular organization of the SL-1 and PAT biosynthesis gene clusters in *M. tuberculosis*.
doi:10.1128/microbiolspec.MGM2-0002-2013.f3

needs association with the large membrane protein MmpL8, which is part of the SL-1 biosynthesis gene cluster (Fig. 3B), for full enzymatic activity. MmpL8 is required both for proper SL-1 biosynthesis and for translocation across the cytoplasm membrane. Consistent with its transport function, inactivation of *mmpL8* in *M. tuberculosis* correlated with accumulation of the intermediate SL_{1278} retained in a deeper layer of the cell wall (51). A model has been proposed in which MmpL8 couples biosynthesis and transport by acting as a scaffold for the enzymatic machinery consisting of cytosolic PapA1, PapA2, Pks2, and membrane-associated Chp1 (50). By this macromolecular assembly, MmpL8 in a coordinated fashion might first facilitate biosynthesis and then translocate the final SL-1 synthesis product across the cytoplasm membrane. Another protein encoded within the SL-1 biosynthesis gene cluster, the integral membrane protein Sap (sulfolipid-1-addressing protein; Rv3821), probably associates with MmpL8 and modulates its transport function (50). While SL-1 deficiency *per se* does not impair virulence of *M. tuberculosis* in the mouse infection model as concluded from infection studies with the *M. tuberculosis* Δ*pks2* mutants (51, 52), deletion of *mmpL8* leads to attenuation (51, 53). This may imply that MmpL8, in addition to SL-1, transports other unidentified lipid molecules important for virulence of *M. tuberculosis*. Alternatively, the phenotype of the Δ*mmpL8* mutant might be explainable by the accumulation of a toxic intermediate that impairs *M. tuberculosis* viability *in vivo*.

Polyacylated Trehalose

Di-, tri-, and polyacylated acyltrehaloses are complex glycolipids containing methyl-branched fatty acids that are located in the outer layer of the mycobacterial cell envelope (54–58). They appear to be restricted to pathogenic mycobacteria. The major polyacyltrehalose (PAT) of *M. tuberculosis*, also referred to as pentaacyltrehalose, contains four mycolipenic acids and one saturated fatty acid (Fig. 3A). In contrast to the almost completely elucidated pathway for biosynthesis and export of SL-1, the available information on the molecular basis of biosynthesis of acytrehaloses is relatively scarce. However, the presumed PAT biosynthesis gene cluster of *M. tuberculosis* shows a molecular organization strongly resembling the SL-1 synthesis gene cluster (Fig. 3B). Hence, PAT formation likely follows a similar biochemical assembly series (Fig. 3A). PAT synthesis is initiated by the acyltransferase PapA3 (Rv1182), which transfers a straight-chain fatty acid moiety such as palmitate (C16), probably using acyl-CoA as the acyl donor to the 2-position of trehalose, yielding trehalose-2-palmitate (59). PapA3 then also catalyzes the subsequent acylation of the 3-position of trehalose-2-palmitate. Using palmitoyl-CoA as the donor substrate, purified PapA3 led to production of 2,3-dipalmitoyltrehalose *in vitro* (59). The mycolipenic acid moieties of PAT are produced by the polyketide synthase gene *pks3/4* (also referred to as *msl3*) (60). *In vivo*, PapA3 thus probably associates with Pks3/4 and might transfer a mycolipenoyl moiety directly from the acyl carrier protein domain of Pks3/4 to trehalose-2-palmitate to form 2,3-diacyltrehalose (Fig. 3A). Deletion of the *papA3* gene resulted in loss of PAT biosynthesis without affecting SL-1 formation in *M. tuberculosis* (59). Likewise, inactivation of *pks3/4* also led to loss of PAT production with no effect on SL-1 production (60). Interestingly, the laboratory strain *M. tuberculosis* H37Rv harbors a frameshift mutation in the *pks3/4* gene introducing a premature stop codon

leading to a truncated nonfunctional polyketide synthase that abolishes PAT biosynthesis. Other *M. tuberculosis* strains such as CDC1551 and Erdman, however, contain a single continuous open reading frame for *pks3/4* that encodes a functional enzyme (59, 61).

The sequential acylation steps to elaborate 2,3-diacyltrehalose with three further mycolipenoyl moieties to form PAT have not been identified yet. These may be mediated again by PapA3 and/or an unidentified acyltransferase. It is also not clear how PAT or its biosynthetic precursors are translocated across the cytoplasm membrane to reach their cell surface destination. However, by analogy to SL-1 biosynthesis, this transport likely involves the transmembrane protein MmpL11, which is encoded within the PAT gene cluster (Fig. 3B). Furthermore, the clustering of *fadD21* within the PAT biosynthesis gene cluster suggests a role for this acyl-CoA synthetase/acyl-AMP ligase. In analogy to SL-1 biosynthesis, FadD21 may provide an activated fatty acid starter unit that is then elongated by Pks3/4 to mycolipenic acids. However, so far there has been no experimental evidence for this. Recently, a novel locus was implicated in acyltrehalose biosynthesis. *M. tuberculosis* transposon mutants with disrupted *Rv1503c* and *Rv1506c* genes were found to be impaired in the synthesis of diacyltrehalose (62). The homologues of these genes in *M. marinum* belong to a locus involved in the synthesis of lipooligosaccharides (LOS) (63). However, as discussed below, *M. tuberculosis* strains do not synthesize LOS. The role of these genes and other genes of this cluster in acyltrehalose formation remains to be established.

Lipooligosaccharides

A fourth class of trehalose-containing mycobacterial glycolipids is LOS, which are structurally highly diverse. They contain a trehalose-containing tetraglucosyl "core" [β-D-glucose-(1→3)-β-D-glucose-(1→4)-α-D-glucose-(1→1)-α-D-glucose, with the latter two α,α-1,1-linked glucose residues representing the trehalose moiety], which is then further glycosylated (64). Most LOS-producing mycobacterial species synthesize a variety of LOS molecules that differ in the number and structure of their sugar residues. Furthermore, various acylation steps involving addition of methyl-branched fatty acids occur at the trehalose moiety (65). LOS-producing mycobacteria include *Mycobacterium kansasii* (66) and *M. marinum* (67), but among members of the *M. tuberculosis* complex, LOS have only been found in *Mycobacterium canetti* (68), a rather distant relative of *M. tuberculosis* (69). Due to their absence from *M. tuberculosis*, relatively little research has been published regarding the biosynthesis of LOS. Most genetic information regarding LOS biosynthesis is available for *M. marinum*. Mutants of *M. marinum* impaired in various glycosylation steps have been isolated, e.g., in MMAR_2333, which has been identified as a glycosyltransferase involved in addition of a caryophyllose moiety in LOS (70), or LosA, a glycosyltransferase involved in elongation of the tetraglucosyl core (67). Interestingly, *losA* is part of the supposed LOS biosynthesis gene cluster in *M. marinum* (63), which is partially conserved in *M. tuberculosis* and here encompasses the genes *Rv1496* to *Rv1505*. However, the corresponding gene cluster in *M. marinum* is much larger and contains nine additional genes. Thus, it can be speculated that the loss of one or more of these genes is responsible for the loss of LOS biosynthesis in *M. tuberculosis*. However, as mentioned above, genes of this gene cluster have also been found to be involved in diacyltrehalose formation, so that the precise function of the remnant LOS gene cluster in *M. tuberculosis* needs to be established (62).

TREHALOSE UPTAKE AND RECYCLING

As mentioned above, *M. tuberculosis* and other mycobacteria can utilize trehalose as their sole source of carbon and energy. Recently, an ABC transporter was identified in *M. tuberculosis* that mediates the active transport of trehalose across the cytoplasm membrane (30). This transporter, which is highly conserved among mycobacteria, is composed of the two transmembrane proteins SugA and SugB, the ATP-hydrolyzing protein SugC, and the periplasmic solute-binding lipoprotein LpqY, which is tethered by a lipid anchor to the cytoplasm membrane. The overall structure of this transporter is LpqY-SugA-SugB-(SugC)$_2$. The genes coding for the four components of this transporter are organized as an operon in the genomes of *M. tuberculosis* (*Rv1235-Rv1238*), *M. smegmatis,* as well as all other mycobacteria. Biochemical studies revealed that this ABC transporter is a high-affinity highly trehalose-specific importer. The substrate specificity is so pronounced that this transporter can discriminate between trehalose [α-D-glucopyranosyl-(1→1)-α-D-glucopyranoside] and maltose [α-D-glucopyranosyl-(1→4)-α-D-glucopyranoside] and even between the α,α-trehalose and α,β-trehalose stereoisomers. According to its transport function, gene deletion mutants with inactivated *lpqY* or *sugC* genes in *M. tuberculosis* and *M. smegmatis* were unable to grow *in vitro* with trehalose as the sole carbon source, whereas growth on

other substrates was not impaired (30). Since the mycomembrane is supposed to constitute a relatively selective permeability barrier, the presence of some sort of transport mechanism in this cell wall layer facilitating trehalose uptake must be postulated, e.g., maybe a pore-forming membrane channel such as a porin specific for low-molecular-weight hydrophilic molecules (Fig. 4). However, so far no candidate for such a permease has been identified.

The LpqY-SugABC transporter has previously been speculated to be involved in uptake and utilization of a carbohydrate compound from the host. However, trehalose is not present in mammals. Thus, it is unlikely to be involved in nutrient acquisition from the host. In contrast, it was demonstrated that this ABC transporter instead plays a role in recycling of trehalose that is extracellularly released from mycobacteria themselves as a by-product during cell wall biosynthesis (30). As outlined previously in this article, it has been established that trehalose functions as a carrier molecule in buildup of the mycolic acid cell wall layer (Fig. 4). Mycolic acids synthesized in the cytoplasm are first conjugated to trehalose to yield TMM. TMM subsequently serves as the mycolate transport form that is exported across the cytoplasm membrane by the membrane transporter MmpL3 (39–41). Extracellularly, the periplasmic TMM is the substrate of the mycolyltransferases of the antigen 85 complex, which transfers the mycolate moiety to either the arabinogalactan layer to yield cell-wall-bound mycolates or to another TMM molecule resulting in formation of TDM (42, 43). These enzymatic reactions lead to concomitant release of the trehalose moiety of TMM. The function of the LpqY-SugABC permease is the recycling of the released trehalose and its retrograde reimport into the cytoplasm, where it can be used as an acceptor for another mycolic acid moiety (Fig. 4). Inhibition of this recycling function in *M. tuberculosis* Δ*lpqY* or Δ*sugC* mutants led to secretion of substantial amounts of free trehalose into the culture medium *in vitro* (30). The importance of the LpqY-SugABC transporter for virulence of *M. tuberculosis* is highlighted by the strong attenuation of the Δ*lpqY* or Δ*sugC* mutants during the acute infection phase in the mouse model (30). The detailed molecular mechanisms causing attenuation during infection *in vivo* have not been elucidated yet. However, the primary niche for intracellular replication of *M. tuberculosis*, the arrested phagosome inside infected macrophages, is characterized as a nutrient-restricted microenvironment with very poor carbohydrate availability. In this regard, it is likely that continuous secretion of trehalose during intracellular growth of the

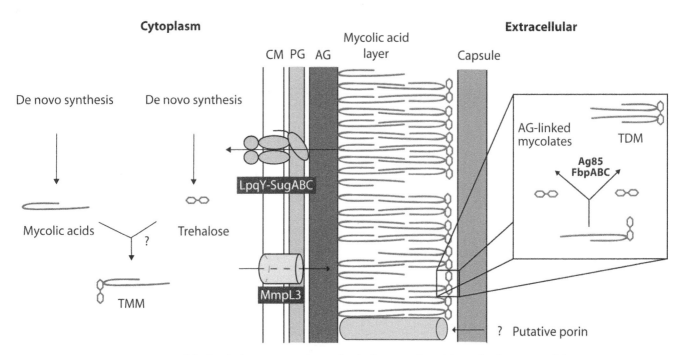

Figure 4 Model of trehalose as a carrier molecule for mycolic acids in the formation of the mycobacterial cell wall. Mycolic acids synthesized in the cytoplasm are first conjugated to trehalose and then exported as TMM via the transporter MmpL3. Extracellularly, TMM serves as the substrate of the antigen 85 complex comprising the mycolyltransferases FbpA-C, which transfer the mycolate moiety either to the arabinogalactan layer or to another TMM molecule, yielding cell-wall-bound mycolates or TDM, respectively. The released trehalose moiety of TMM is recycled by the ABC transporter LpqY-SugABC. A putative porin might facilitate transport across the mycomembrane for uptake of exogenous trehalose. doi:10.1128/microbiolspec.MGM2-0002-2013.f4

Δ*lpqY* or Δ*sugC* mutants would gradually cause carbon starvation. Therefore, LpqY-SugABC-mediated trehalose recycling might represent a specific adaptation of *M. tuberculosis* to its nutrient-limited intracellular lifestyle.

CONVERSION OF TREHALOSE TO ALPHA-GLUCANS

Previously, the trehalose synthase TreS was circumstantially implicated in glycogen formation in mycobacteria because it was observed that incubation of cells at high trehalose concentrations resulted in accumulation of large amounts of glycogen, and this effect was strictly TreS dependent. At this time, a direct conversion of trehalose to glycogen was speculated based on an alpha-amylase activity found to be associated with purified TreS (71). However, it has recently been established that TreS catalyzes the first step of a novel four-step biochemical pathway that converts trehalose into branched alpha-glucans (Fig. 5), which is now known as the GlgE pathway and which is widespread among

prokaryotes (23, 25). TreS interconverts trehalose and maltose with formation of the α-anomer of maltose (24), which is subsequently rapidly and quantitatively phosphorylated in an ATP-dependent reaction to maltose-1-phosphate by the maltose kinase Pep2 (Rv0127) (27). Maltose-1-phosphate then acts as the activated donor substrate for the key enzyme of this pathway, the maltosyltransferase GlgE (Rv1327c), which transfers the maltosyl moiety to the nonreducing 4′-hydroxyl group of an α-1,4-glucan acceptor molecule, yielding linear alpha-glucans. Finally, the branching enzyme GlgB (Rv1326c) introduces α-1,6-linked branches into the linear alpha-glucan, forming molecules structurally resembling glycogen (Fig. 5) (23). By using a phosphosugar as the activated substrate, GlgE [systematic name (1→4)-α-D-glucan:phosphate α-D-maltosyltransferase], which is a member of the glycoside hydrolase subfamily GH13_3, represents an unusual glycosyltransferase, while most known members of this enzyme class rely on nucleotide-bound sugars as donor substrates. In addition to the maltose-1-phosphate-dependent maltosyltransferase activity,

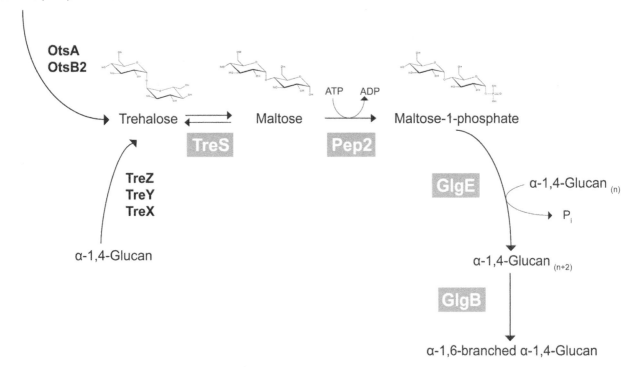

Figure 5 Conversion of trehalose to alpha-glucans. Trehalose is reversibly interconverted by the trehalose synthase TreS to α-maltose, which is subsequently phosphorylated to α-maltose-1-phosphate by the maltokinase Pep2. Maltose-1-phosphate serves as the activated donor substrate for the maltosyltransferase GlgE producing linear α-1,4-glucans by elongating the nonreducing end of an α-glucan acceptor molecule. Finally, the branching enzyme GlgB introduces α-1,6-linked branches into the linear molecule.
doi:10.1128/microbiolspec.MGM2-0002-2013.f5

GlgE also exhibits disproportionating transglucosidase activity; i.e., it can remove a maltosyl moiety from one alpha-glucan molecule and transfer it to another (e.g., α-1,4-glucan$_x$ + α-1,4-glucan$_y$ → α-1,4-glucan$_{x-2}$ + α-1,4-glucan$_{y+2}$, with x,y = number of glucose units). Therefore, not only can GlgE synthesize linear alpha-glucans, but it is also capable of shaping the length of the branches of branched glucans (23).

Although production of alpha-glucan molecules through the GlgE pathway is not required for viability of *M. tuberculosis in vitro* as indicated by the genetic dispensability of *treS* and *pep2*, both the *glgE* and *glgB* genes are strictly essential in *M. tuberculosis*. Inactivation of GlgE leads to an intracellular accumulation of the phosphorylated intermediate maltose-1-phosphate, which is toxic to the cells by eliciting pleiotropic stress responses. These maltose-1-phosphate-induced stress responses include DNA damage as indicated by global induction of the DNA damage-responsive SOS regulon, causing rapid cell death of *M. tuberculosis in vitro* as well as in lung and spleens of infected mice (23). Since

linear alpha-glucans rapidly become insoluble with increasing chain length, inactivation of GlgB indirectly retards GlgE activity by acceptor substrate depletion, eventually also causing maltose-1-phosphate accumulation. In agreement with the toxicity of maltose-1-phosphate as the basis of essentiality of GlgE and GlgB, prevention of maltose-1-phosphate formation through genetic or pharmacological inactivation of TreS allows deletion of both *glgE* and *glgB* in *M. tuberculosis* (23). Recently, it was demonstrated that GlgE activity can be negatively fine-tuned via phosphorylation by the serine/threonine protein kinase PknB (72), although it is unclear if this regulation is important for *M. tuberculosis* virulence. Given the high toxicity of maltose-1-phosphate, its synthesis rate also needs to be subject to one or more regulatory mechanisms in order to balance production and consumption. However, how this is mediated is currently unknown.

The final cellular destination(s) and function(s) of the branched alpha-glucans synthesized by the GlgE pathway are not fully clear. It is likely that the GlgE pathway products, like glycogen, are deposited in the cytosol and serve as an intracellular carbon storage compound that is hydrolytically remobilized during starvation. Furthermore, alpha-glucans are also found extracellularly in mycobacteria, where they constitute the major polysaccharide component of the capsule representing the outermost cell envelope layer, which has been implicated in *M. tuberculosis* persistence (73–75). Currently, however, a contribution of the GlgE pathway to production of capsular alpha-glucans has not been established. Finally, a synthetic lethal interaction of the GlgE pathway with an alternative pathway leading to production of methyl-branched alpha-glucan derivatives, methylglucose lipopolysaccharides (MGLPs), has been revealed (23). MGLPs are alpha-glucan oligomers consisting of 19 to 20 glucose residues, which are extensively methylated and acylated (76). MGLP biosynthesis involves the glucosyltransferase Rv3032, which can utilize UDP-glucose as well as ADP-glucose as activated donor substrates (76). The synthetic lethal interaction was demonstrated by the inability to delete the genes *treS* and *Rv3032* simultaneously in *M. tuberculosis*, while both genes individually are dispensable. This synthetic lethal interaction was further corroborated by demonstration that treatment of the *M. tuberculosis* ΔRv3032 mutant with the TreS inhibitor validamycin A resulted in rapid killing while having no effect on wild-type cells (23). Thus, it is likely that the branched alpha-glucan GlgE pathway products are at least partially subject to further chemical modifications, yielding derivatives that are structurally and/or functionally redundant to MGLP. However, these further modifications have yet to be identified. The combination of essentiality of GlgE within a synthetic lethal pathway makes GlgE an attractive drug target because it possibly enables two independent and synergistic killing mechanisms (77).

The question of the importance of alpha-glucans synthesized via the GlgE pathway for *M. tuberculosis* virulence has not conclusively been resolved yet. While the *M. tuberculosis* Δ*treS* mutant was shown to be not attenuated in BALB/c mice following infection via the high-dose intravenous route (23), studies from a different laboratory of C57BL/6 mice following aerosol challenge suggested a role of TreS in late-stage pathogenesis (8). However, since genetic complementation was not able to restore virulence of this *M. tuberculosis* Δ*treS* mutant, inadvertent secondary mutations likely contributed to the observed *in vivo* phenotype in C57BL/6 mice (unpublished data available at http://webhost.nts.jhu.edu/target/pdf/TARGET%20Report%20treS%20otsA.pdf).

CONCLUDING REMARKS

As outlined above, trehalose fulfills many roles in mycobacteria and has many metabolic fates. In this respect, the amount of the *de novo* synthesized trehalose is probably tightly regulated according to the intracellular demand. Likewise, the flux of trehalose into the many different metabolizing pathways probably requires tight regulation at multiple levels. However, while we now begin to understand the biochemical and genetic basis of trehalose-based metabolic pathways, the regulatory mechanisms underlying the balance of intracellular production and consumption of trehalose are completely unknown. Given the central importance of trehalose for *M. tuberculosis* pathogenesis, this aspect deserves increased attention in the future.

Research related to this chapter in the laboratory of R. Kalscheuer is supported by the Juergen Manchot Foundation, the Strategic Research Fund of the Heinrich-Heine-University Duesseldorf, and the German Research Foundation.

Citation. Kalscheuer R, Koliwer-Brandl H. 2014. Genetics of mycobacterial trehalose metabolism. Microbiol Spectrum 2(3):MGM2-0002-2013.

References

1. **Harding TS.** 1923. History of trehalose, its discovery and methods of preparation. *Sugar* **25:**476–478.

2. **Arguelles JC.** 2000. Physiological roles of trehalose in bacteria and yeasts: a comparative analysis. *Arch Microbiol* **174:**217–224.

3. **Elbein AD, Pan YT, Pastuszak I, Carroll D.** 2003. New insights on trehalose: a multifunctional molecule. *Glycobiology* **13:**17R–27R.

4. **Jain NK, Roy I.** 2009. Effect of trehalose on protein structure. *Protein Sci* **18:**24–36.

5. **Paul MJ, Primavesi LF, Jhurreea D, Zhang Y.** 2008. Trehalose metabolism and signaling. *Annu Rev Plant Biol* **59:**417–441.

6. **Vanin S, Bubacco L, Beltramini M.** 2008. Seasonal variation of trehalose and glycerol concentrations in winter snow-active insects. *Cryo Lett* **29:**485–491.

7. **Elbein AD, Mitchell M.** 1973. Levels of glycogen and trehalose in *Mycobacterium smegmatis* and the purification and properties of the glycogen synthetase. *J Bacteriol* **113:**863–873.

8. **Murphy HN, Stewart GR, Mischenko VV, Apt AS, Harris R, McAlister MS, Driscoll PC, Young DB, Robertson DB.** 2005. The OtsAB pathway is essential for trehalose biosynthesis in *Mycobacterium tuberculosis*. *J Biol Chem* **280:**14524–14529.

9. **Woodruff PJ, Carlson BL, Siridechadilok B, Pratt MR, Senaratne RH, Mougous JD, Riley LW, Williams SJ, Bertozzi CR.** 2004. Trehalose is required for growth of *Mycobacterium smegmatis*. *J Biol Chem* **279:**28835–28843.

10. **Ofer N, Wishkautzan M, Meijler M, Wang Y, Speer A, Niederweis M, Gur E.** 2012. Ectoine biosynthesis in

Mycobacterium smegmatis. Appl Environ Microbiol **78**: 7483–7486.

11. Avonce N, Mendoza-Vargas A, Morett E, Iturriaga G. 2006. Insights on the evolution of trehalose biosynthesis. *BMC Evol Biol* **6**:109.

12. Zaparty M, Tjaden B, Hensel R, Siebers B. 2008. The central carbohydrate metabolism of the hyperthermophilic crenarchaeote *Thermoproteus tenax:* pathways and insights into their regulation. *Arch Microbiol* **190**:231–245.

13. Eis C, Nidetzky B. 1999. Characterization of trehalose phosphorylase from *Schizophyllum commune. Biochem J* **341**(Pt 2):385–393.

14. Eis C, Watkins M, Prohaska T, Nidetzky B. 2001. Fungal trehalose phosphorylase: kinetic mechanism, pH-dependence of the reaction and some structural properties of the enzyme from *Schizophyllum commune. Biochem J* **356**:757–767.

15. Wannet WJ, Aben EM, van der Drift C, Van Griensven LJ, Vogels GD, Op den Camp HJ. 1999. Trehalose phosphorylase activity and carbohydrate levels during axenic fruiting in three *Agaricus bisporus* strains. *Curr Microbiol* **39**:205–210.

16. Wannet WJ, Op den Camp HJ, Wisselink HW, van der Drift C, Van Griensven LJ, Vogels GD. 1998. Purification and characterization of trehalose phosphorylase from the commercial mushroom *Agaricus bisporus. Biochim Biophys Acta* **1425**:177–188.

17. Belocopitow E, Marechal LR. 1970. Trehalose phosphorylase from *Euglena gracilis. Biochim Biophys Acta* **198**:151–154.

18. Maruta K, Mitsuzumi H, Nakada T, Kubota M, Chaen H, Fukuda S, Sugimoto T, Kurimoto M. 1996. Cloning and sequencing of a cluster of genes encoding novel enzymes of trehalose biosynthesis from thermophilic archaebacterium *Sulfolobus acidocaldarius. Biochim Biophys Acta* **1291**:177–181.

19. Nishimoto T, Nakano M, Nakada T, Chaen H, Fukuda S, Sugimoto T, Kurimoto M, Tsujisaka Y. 1996. Purification and properties of a novel enzyme, trehalose synthase, from *Pimelobacter* sp. R48. *Biosci Biotechnol Biochem* **60**:640–644.

20. Tsusaki K, Nishimoto T, Nakada T, Kubota M, Chaen H, Sugimoto T, Kurimoto M. 1996. Cloning and sequencing of trehalose synthase gene from *Pimelobacter* sp. R48. *Biochim Biophys Acta* **1290**:1–3.

21. Qu Q, Lee SJ, Boos W. 2004. TreT, a novel trehalose glycosyltransferring synthase of the hyperthermophilic archaeon *Thermococcus litoralis. J Biol Chem* **279**: 47890–47897.

22. De Smet KA, Weston A, Brown IN, Young DB, Robertson BD. 2000. Three pathways for trehalose biosynthesis in mycobacteria. *Microbiology* **146**(Pt 1):199–208.

23. Kalscheuer R, Syson K, Veeraraghavan U, Weinrick B, Biermann KE, Liu Z, Sacchettini JC, Besra G, Bornemann S, Jacobs WR Jr. 2010. Self-poisoning of *Mycobacterium tuberculosis* by targeting GlgE in an alpha-glucan pathway. *Nat Chem Biol* **6**:376–384.

24. Miah F, Koliwer-Brandl H, Rejzek M, Field RA, Kalscheuer R, Bornemann S. 2013. Flux through trehalose synthase flows from trehalose to the alpha anomer of maltose in mycobacteria. *Chem Biol* **20**:487–493.

25. Chandra G, Chater KF, Bornemann S. 2011. Unexpected and widespread connections between bacterial glycogen and trehalose metabolism. *Microbiology* **157**:1565–1572.

26. Syson K, Stevenson CE, Rejzek M, Fairhurst SA, Nair A, Bruton CJ, Field RA, Chater KF, Lawson DM, Bornemann S. 2011. Structure of *Streptomyces maltosyltransferase* GlgE, a homologue of a genetically validated anti-tuberculosis target. *J Biol Chem* **286**: 38298–38310.

27. Mendes V, Maranha A, Lamosa P, da Costa MS, Empadinhas N. 2010. Biochemical characterization of the maltokinase from *Mycobacterium bovis* BCG. *BMC Biochem* **11**:21.

28. Tzvetkov M, Klopprogge C, Zelder O, Liebl W. 2003. Genetic dissection of trehalose biosynthesis in *Corynebacterium glutamicum:* inactivation of trehalose production leads to impaired growth and an altered cell wall lipid composition. *Microbiology* **149**:1659–1673.

29. Wolf A, Kramer R, Morbach S. 2003. Three pathways for trehalose metabolism in *Corynebacterium glutamicum* ATCC13032 and their significance in response to osmotic stress. *Mol Microbiol* **49**:1119–1134.

30. Kalscheuer R, Weinrick B, Veeraraghavan V, Besra GS, Jacobs WR Jr. 2010. Trehalose-recycling ABC transporter LpqY-SugA-SugB-SugC is essential for virulence of *Mycobacterium tuberculosis. Proc Natl Acad Sci USA* **107**:21761–21766.

31. Carroll JD, Pastuszak I, Edavana VK, Pan YT, Elbein AD. 2007. A novel trehalase from *Mycobacterium smegmatis:* purification, properties, requirements. *FEBS J* **274**: 1701–1714.

32. Swarts BM, Holsclaw CM, Jewett JC, Alber M, Fox DM, Siegrist MS, Leary JA, Kalscheuer R, Bertozzi CR. 2012. Probing the mycobacterial trehalome with bioorthogonal chemistry. *J Am Chem Soc* **134**:16123–16126.

33. Middlebrook G, Dubos RJ, Pierce C. 1947. Virulence and morphological characteristics of mammalian tubercle bacilli. *J Exp Med* **86**:175–184.

34. Noll H, Bloch H, Asselineau J, Lederer E. 1956. The chemical structure of the cord factor of *Mycobacterium tuberculosis. Biochim Biophys Acta* **20**:299–309.

35. Hoffmann C, Leis A, Niederweis M, Plitzko JM, Engelhardt H. 2008. Disclosure of the mycobacterial outer membrane: cryo-electron tomography and vitreous sections reveal the lipid bilayer structure. *Proc Natl Acad Sci USA* **105**:3963–3967.

36. Zuber B, Chami M, Houssin C, Dubochet J, Griffiths G, Daffe M. 2008. Direct visualization of the outer membrane of mycobacteria and corynebacteria in their native state. *J Bacteriol* **190**:5672–5680.

37. Besra GS, Sievert T, Lee R, Slayden RA, Brennan PJ, Takayama K. 1994. Identification of the apparent carrier in mycolic acid synthesis. *Proc Natl Acad Sci USA* **91**: 12735–12739.

38. Takayama K, Wang C, Besra GS. 2005. Pathway to synthesis and processing of mycolic acids in *Mycobacterium tuberculosis*. *Clin Microbiol Rev* **18**:81–101.

39. Tahlan K, Wilson R, Kastrinsky DB, Arora K, Nair V, Fischer E, Barnes SW, Walker JR, Alland D, Barry CE 3rd, Boshoff HI. 2012. SQ109 targets MmpL3, a membrane transporter of trehalose monomycolate involved in mycolic acid donation to the cell wall core of *Mycobacterium tuberculosis*. *Antimicrob Agents Chemother* **56**:1797–1809.

40. Grzegorzewicz AE, Pham H, Gundi VA, Scherman MS, North EJ, Hess T, Jones V, Gruppo V, Born SE, Kordulakova J, Chavadi SS, Morisseau C, Lenaerts AJ, Lee RE, McNeil MR, Jackson M. 2012. Inhibition of mycolic acid transport across the *Mycobacterium tuberculosis* plasma membrane. *Nat Chem Biol* **8**:334–341.

41. Varela C, Rittmann D, Singh A, Krumbach K, Bhatt K, Eggeling L, Besra GS, Bhatt A. 2012. MmpL genes are associated with mycolic acid metabolism in mycobacteria and corynebacteria. *Chem Biol* **19**:498–506.

42. Belisle JT, Vissa VD, Sievert T, Takayama K, Brennan PJ, Besra GS. 1997. Role of the major antigen of *Mycobacterium tuberculosis* in cell wall biogenesis. *Science* **276**:1420–1422.

43. Puech V, Guilhot C, Perez E, Tropis M, Armitige LY, Gicquel B, Daffe M. 2002. Evidence for a partial redundancy of the fibronectin-binding proteins for the transfer of mycoloyl residues onto the cell wall arabinogalactan termini of *Mycobacterium tuberculosis*. *Mol Microbiol* **44**:1109–1122.

44. Wiker HG, Harboe M. 1992. The antigen 85 complex: a major secretion product of *Mycobacterium tuberculosis*. *Microbiol Rev* **56**:648–661.

45. Jackson M, Raynaud C, Laneelle MA, Guilhot C, Laurent-Winter C, Ensergueix D, Gicquel B, Daffe M. 1999. Inactivation of the antigen 85C gene profoundly affects the mycolate content and alters the permeability of the *Mycobacterium tuberculosis* cell envelope. *Mol Microbiol* **31**:1573–1587.

46. Armitige LY, Jagannath C, Wanger AR, Norris SJ. 2000. Disruption of the genes encoding antigen 85A and antigen 85B of *Mycobacterium tuberculosis* H37Rv: effect on growth in culture and in macrophages. *Infect Immun* **68**:767–778.

47. Mougous JD, Petzold CJ, Senaratne RH, Lee DH, Akey DL, Lin FL, Munchel SE, Pratt MR, Riley LW, Leary JA, Berger JM, Bertozzi CR. 2004. Identification, function and structure of the mycobacterial sulfotransferase that initiates sulfolipid-1 biosynthesis. *Nat Struct Mol Biol* **11**:721–729.

48. Kumar P, Schelle MW, Jain M, Lin FL, Petzold CJ, Leavell MD, Leary JA, Cox JS, Bertozzi CR. 2007. PapA1 and PapA2 are acyltransferases essential for the biosynthesis of the *Mycobacterium tuberculosis* virulence factor sulfolipid-1. *Proc Natl Acad Sci USA* **104**:11221–11226.

49. Sirakova TD, Thirumala AK, Dubey VS, Sprecher H, Kolattukudy PE. 2001. The *Mycobacterium tuberculosis* pks2 gene encodes the synthase for the hepta- and octamethyl-branched fatty acids required for sulfolipid synthesis. *J Biol Chem* **276**:16833–16839.

50. Seeliger JC, Holsclaw CM, Schelle MW, Botyanszki Z, Gilmore SA, Tully SE, Niederweis M, Cravatt BF, Leary JA, Bertozzi CR. 2012. Elucidation and chemical modulation of sulfolipid-1 biosynthesis in *Mycobacterium tuberculosis*. *J Biol Chem* **287**:7990–8000.

51. Converse SE, Mougous JD, Leavell MD, Leary JA, Bertozzi CR, Cox JS. 2003. MmpL8 is required for sulfolipid-1 biosynthesis and *Mycobacterium tuberculosis* virulence. *Proc Natl Acad Sci USA* **100**:6121–6126.

52. Rousseau C, Turner OC, Rush E, Bordat Y, Sirakova TD, Kolattukudy PE, Ritter S, Orme IM, Gicquel B, Jackson M. 2003. Sulfolipid deficiency does not affect the virulence of *Mycobacterium tuberculosis* H37Rv in mice and guinea pigs. *Infect Immun* **71**:4684–4690.

53. Domenech P, Reed MB, Dowd CS, Manca C, Kaplan G, Barry CE 3rd. 2004. The role of MmpL8 in sulfatide biogenesis and virulence of *Mycobacterium tuberculosis*. *J Biol Chem* **279**:21257–21265.

54. Lemassu A, Laneelle MS, Daffe M. 1991. Revised structure of a trehalose-containing immunoreactive glycolipid of *Mycobacterium tuberculosis*. *FEMS Microbiol Lett* **62**:171–175.

55. Besra GS, Bolton RC, McNeil MR, Ridell M, Simpson KE, Glushka J, van Halbeek H, Brennan PJ, Minnikin DE. 1992. Structural elucidation of a novel family of acyltrehaloses from *Mycobacterium tuberculosis*. *Biochemistry* **31**:9832–9837.

56. Munoz M, Laneelle MA, Luquin M, Torrelles J, Julian E, Ausina V, Daffe M. 1997. Occurrence of an antigenic triacyl trehalose in clinical isolates and reference strains of *Mycobacterium tuberculosis*. *FEMS Microbiol Lett* **157**:251–259.

57. Minnikin DE, Dobson G, Sesardic D, Ridell M. 1985. Mycolipenates and mycolipanolates of trehalose from *Mycobacterium tuberculosis*. *J Gen Microbiol* **131**:1369–1374.

58. Daffe M, Lacave C, Laneelle MA, Gillois M, Laneelle G. 1988. Polyphthienoyl trehalose, glycolipids specific for virulent strains of the tubercle bacillus. *Eur J Biochem* **172**:579–584.

59. Hatzios SK, Schelle MW, Holsclaw CM, Behrens CR, Botyanszki Z, Lin FL, Carlson BL, Kumar P, Leary JA, Bertozzi CR. 2009. PapA3 is an acyltransferase required for polyacyltrehalose biosynthesis in *Mycobacterium tuberculosis*. *J Biol Chem* **284**:12745–12751.

60. Dubey VS, Sirakova TS, Kolattukudy PE. 2002. Disruption of msl3 abolishes the synthesis of mycolipanoic and mycolipenic acids required for polyacyltrehalose synthesis in *Mycobacterium tuberculosis* H37Rv and causes cell aggregation. *Mol Microbiol* **45**:1451–1459.

61. Domenech P, Reed MB, Barry CE 3rd. 2005. Contribution of the *Mycobacterium tuberculosis* MmpL protein family to virulence and drug resistance. *Infect Immun* **73**:3492–3501.

62. Brodin P, Poquet Y, Levillain F, Peguillet I, Larrouy-Maumus G, Gilleron M, Ewann F, Christophe T, Fenistein D, Jang J, Jang MS, Park SJ, Rauzier J, Carralot JP, Shrimpton R, Genovesio A, Gonzalo-Asensio JA, Puzo G, Martin C, Brosch R, Stewart GR, Gicquel B,

Neyrolles O. 2010. High content phenotypic cell-based visual screen identifies *Mycobacterium tuberculosis* acyltrehalose-containing glycolipids involved in phagosome remodeling. *PLoS Pathog* 6:e1001100.

63. Ren H, Dover LG, Islam ST, Alexander DC, Chen JM, Besra GS, Liu J. 2007. Identification of the lipooligosaccharide biosynthetic gene cluster from *Mycobacterium marinum. Mol Microbiol* 63:1345–1359.

64. Hunter SW, Jardine I, Yanagihara DL, Brennan PJ. 1985. Trehalose-containing lipooligosaccharides from mycobacteria: structures of the oligosaccharide segments and recognition of a unique N-acylkanosamine-containing epitope. *Biochemistry* 24:2798–2805.

65. Gilleron M, Puzo G. 1995. Lipooligosaccharidic antigens from *Mycobacterium kansasii* and *Mycobacterium gastri. Glycoconj J* 12:298–308.

66. Hunter SW, Murphy RC, Clay K, Goren MB, Brennan PJ. 1983. Trehalose-containing lipooligosaccharides. A new class of species-specific antigens from *Mycobacterium. J Biol Chem* 258:10481–10487.

67. Burguiere A, Hitchen PG, Dover LG, Kremer L, Ridell M, Alexander DC, Liu J, Morris HR, Minnikin DE, Dell A, Besra GS. 2005. LosA, a key glycosyltransferase involved in the biosynthesis of a novel family of glycosylated acyltrehalose lipooligosaccharides from *Mycobacterium marinum. J Biol Chem* 280:42124–42133.

68. Daffe M, McNeil M, Brennan PJ. 1991. Novel type-specific lipooligosaccharides from *Mycobacterium tuberculosis. Biochemistry* 30:378–388.

69. van Soolingen D, Hoogenboezem T, de Haas PE, Hermans PW, Koedam MS, Teppema KS, Brennan PJ, Besra GS, Portaels F, Top J, Schouls LM, van Embden JD. 1997. A novel pathogenic taxon of the *Mycobacterium tuberculosis* complex, Canetti: characterization of an exceptional isolate from Africa. *Int J Syst Bacteriol* 47:1236–1245.

70. Sarkar D, Sidhu M, Singh A, Chen J, Lammas DA, van der Sar AM, Besra GS, Bhatt A. 2011. Identification of a glycosyltransferase from *Mycobacterium marinum* involved in addition of a caryophyllose moiety in lipo-oligosaccharides. *J Bacteriol* 193:2336–2340.

71. Pan YT, Carroll JD, Asano N, Pastuszak I, Edavana VK, Elbein AD. 2008. Trehalose synthase converts glycogen to trehalose. *FEBS J* 275:3408–3420.

72. Leiba J, Syson K, Baronian G, Zanella-Cleon I, Kalscheuer R, Kremer L, Bornemann S, Molle V. 2013. *Mycobacterium tuberculosis* maltosyltransferase GlgE, a genetically validated antituberculosis target, is negatively regulated by Ser/Thr phosphorylation. *J Biol Chem* 288: 16546–16556.

73. Dinadayala P, Sambou T, Daffe M, Lemassu A. 2008. Comparative structural analyses of the alpha-glucan and glycogen from *Mycobacterium bovis. Glycobiology* 18: 502–508.

74. Sambou T, Dinadayala P, Stadthagen G, Barilone N, Bordat Y, Constant P, Levillain F, Neyrolles O, Gicquel B, Lemassu A, Daffe M, Jackson M. 2008. Capsular glucan and intracellular glycogen of *Mycobacterium tuberculosis:* biosynthesis and impact on the persistence in mice. *Mol Microbiol* 70:762–774.

75. Sani M, Houben EN, Geurtsen J, Pierson J, de Punder K, van Zon M, Wever B, Piersma SR, Jimenez CR, Daffe M, Appelmelk BJ, Bitter W, van der Wel N, Peters PJ. 2010. Direct visualization by cryo-EM of the mycobacterial capsular layer: a labile structure containing ESX-1-secreted proteins. *PLoS Pathog* 6:e1000794.

76. Stadthagen G, Sambou T, Guerin M, Barilone N, Boudou F, Kordulakova J, Charles P, Alzari PM, Lemassu A, Daffe M, Puzo G, Gicquel B, Riviere M, Jackson M. 2007. Genetic basis for the biosynthesis of methylglucose lipopolysaccharides in *Mycobacterium tuberculosis. J Biol Chem* 282:27270–27276.

77. Kalscheuer R, Jacobs WR Jr. 2010. The significance of GlgE as a new target for tuberculosis. *Drug News Perspect* 23:619–624.

Molecular Genetics of Mycobacteria, 2nd Edition
Edited by Graham F. Hatfull and William R. Jacobs, Jr.
© 2014 American Society for Microbiology, Washington, DC
doi:10.1128/microbiolspec.MGM2-0012-2013

G. Marcela Rodriguez[1]
Olivier Neyrolles[2]

Metallobiology of Tuberculosis 19

THE BATTLE FOR IRON

Iron is absolutely required for the life of most organisms, including mycobacteria. Iron is incorporated into proteins, either as a mono- or binuclear species or as part of heme groups or iron-sulfur clusters. Iron undergoes reversible changes in its oxidation state, oscillating between the oxidized ferric (Fe^{3+}) and the reduced ferrous (Fe^{2+}) forms. In addition, depending on the local ligand environment, iron-containing compounds exhibit a wide range of oxidation-reduction potentials. These unique properties make this metal a very versatile prosthetic component as a biocatalyst and electro-carrier in essential cellular pathways including respiration, the trichloroacetic acid (TCA) cycle, oxygen transport, gene regulation, defense against oxidative stress, and DNA biosynthesis (1).

Iron is the fourth most plentiful element in the earth's crust. Before oxygenic photosynthesis it was found in its soluble ferrous form (solubility 0.1 M at pH 7.0); however, introduction of oxygen into the atmosphere caused a switch to the ferric form, which is insoluble as ferric hydroxide. In consequence, free iron became extremely scarce (solubility 10^{-18} M at pH 7.0). In host tissues, the concentration of this metal is lowered even further as Fe(III) is sequestered by iron binding proteins such as transferrin, lactoferritin, and ferritin (2, 3). In addition, the host produces proteins that either efflux iron from intracellular microbial compartments (NRAMP1) or bind heme and hemoglobin (e.g., hemopexin and haptoglobin) and reduce the availability of heme as an iron source (2, 4, 5). Thus, iron starvation in the host is a serious threat for infecting bacteria and has been recognized as such for decades (3). Pathogens are able to survive and multiply in the host in part because they have evolved numerous and often redundant high-affinity iron acquisition mechanisms, including (i) acquisition of iron directly from host iron binding proteins (e.g., transferrin and lactoferrin) by using receptor-mediated transport systems, (ii) uptake and utilization of heme, (iii) solubilization of ferric oxides by reduction of ferric iron and transport of soluble ferrous iron, and (iv) production of ferric iron chelators (siderophores) in conjunction with siderophore-based transport systems. *Mycobacterium tuberculosis* obtains iron by producing siderophores, and it also has the capacity to utilize heme as an iron source in a siderophore-independent manner.

Siderophore-Based Iron Acquisition in *M. tuberculosis*

M. tuberculosis produces mycobactin, a cell-associated, lipophilic siderophore, and a soluble amphiphilic variant of it named carboxymycobactin (6). *M. tuberculosis* does not produce or utilize exochelin, the peptidic

[1]Public Health Research Institute and New Jersey Medical School-Rutgers, the State University of New Jersey, Newark, NJ 07103; [2]Centre National de la Recherche Scientifique and Université de Toulouse, Université Paul Sabatier, Institut de Pharmacologie et de Biologie Structurale, Toulouse, France.

siderophore synthesized by nonpathogenic mycobacteria such as *Mycobacterium smegmatis*. Mycobactin and carboxymycobactin are composed of a hydroxyaromatic acid, an oxazoline moiety, a β-hydroxy acid, and two -*N*-hydroxylysines (Fig. 1). Genomic and biochemical analysis indicates that the *mbt-1* gene cluster (*mbt-IABCDEFG*) encodes the proteins necessary for the assembly of the siderophore core: MbtI for salicylate synthesis; the hybrid nonribosomal peptide synthase/polyketide synthase (MbtA-F); and the L-Lys hydroxylase (MbtG) (7–9). Mycobactin and carboxymycobactin differ mainly in an acyl group attached to the central L-Lys residue. Mycobactin has a long fatty acyl chain (10 to 21 carbons) that is capped by a methyl group, whereas carboxymycobactin has a shorter acyl chain (2 to 9 carbons) that is capped with either a carboxylate or methyl ester (10, 11). The presence of the long acyl chain makes mycobactin very hydrophobic and ensures retention within or in close proximity to the cytoplasmic membrane. Mycobactin may mediate uptake of iron donated by amphiphilic molecules that can penetrate the cell wall, for instance, carboxymycobactin and acinetoferrin (12, 13). Carboxymycobactin, being more polar than mycobactin, is water soluble and exported to the extracellular medium.

The process of carboxymycobactin export seems to be coupled—in an unknown way—to its synthesis and depends on two redundant systems composed of the MmpL4 and MmpL5 transporters and their associated proteins, MmpS4 and MmpS5 (14). These transport systems are postulated to mediate export of carboxymycobactin into the periplasm. However, the hypothetical outer membrane protein that mediates release of carboxymycobactin into the extracellular medium is unknown (Fig. 2). *M. tuberculosis* mutants unable to synthesize or export siderophores are drastically attenuated in a mouse model of tuberculosis infection, underscoring the importance of efficient iron acquisition for propagation of *M. tuberculosis* (14).

In the extracellular environment, carboxymycobactin avidly captures ferric iron. Ferric-carboxymycobactin can slowly transfer iron to mycobactin (12) or deliver this metal via the iron-regulated transporter, IrtAB. IrtAB is an ABC-type transporter synthesized in cells experiencing iron limitation and is necessary for Fe^{3+}-carboxymycobactin uptake. IrtAB mutants are iron deficient and fail to replicate normally in macrophages and in mice (15). Interestingly, the amino-terminal domain of the IrtA protein is located in the cytoplasm and has a functional flavin adenine dinucleotide (FAD) binding motif. Mutations that prevent FAD binding affect assimilation of iron imported by IrtAB. Since a common mechanism to dissociate iron-siderophore complexes is reduction of ferric iron to ferrous iron by cytoplasmic flavin reductases, it is possible that the amino-terminal domain of IrtA functions as a FAD-dependent ferric reductase that mediates the release of ferrous iron from imported Fe^{3+}-carboxymycobactin (16). Thus, IrtAB might couple iron transport and assimilation in *M. tuberculosis* (Fig. 2).

Considering the double membrane structure of the mycobacterial cell envelope, it is likely that Fe^{3+}-carboxymycobactin uptake is facilitated by siderophore binding proteins in the outer membrane and the periplasm. These proteins, however, are yet to be identified.

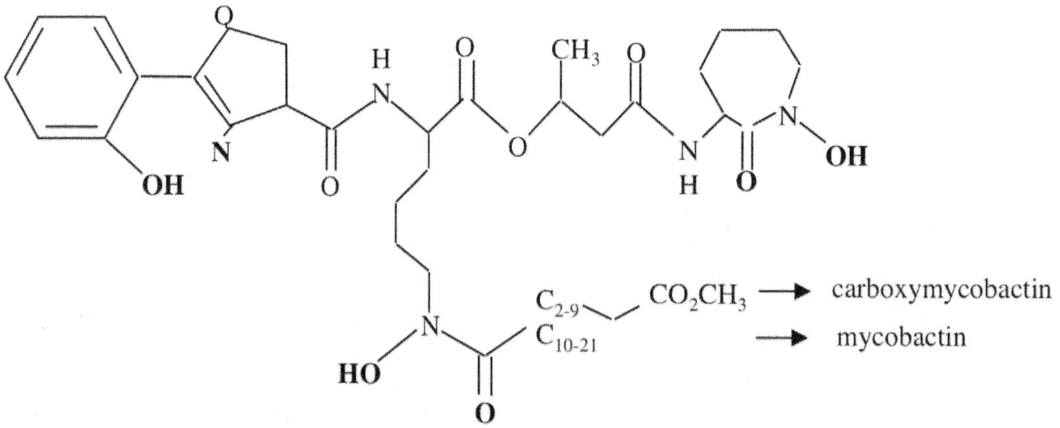

Figure 1 Carboxymycobactin and mycobactin share a common core structure but differ in the length of the alkyl substitution that determines their polarity and hence solubility. The groups involved in binding of Fe(III) are indicated in bold. doi:10.1128/microbiolspec.MGM2-0012-2013.f1

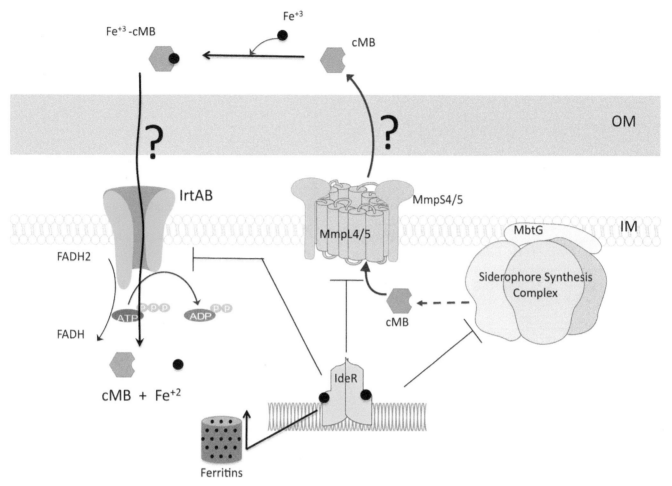

Figure 2 When experiencing iron limitation, *M. tuberculosis* produces carboxymycobactin (cMB) and mycobactin (MB). MB remains cell associated, although the precise location is not clear. cMB is secreted by a process dependent on the membrane proteins MmpL4 and MmpL5 and requiring the MmpS4 and MmpS5 membrane-associated proteins that function together with their cognate MmpL proteins. Proteins that mediate export of cMB across the outer membrane remain to be discovered. Once secreted, cMB chelates Fe^{3+} and possibly requires an outer membrane and periplasmic protein to reach the IrtAB importer in the inner membrane. In the cytosol, the FAD binding domain of IrtA may reduce ferric iron to ferrous iron and dissociate the iron-siderophore complex. Released ferrous iron can be utilized and stored in ferritins. Excess iron binds to the regulator IdeR and activates its DNA binding activity. Binding of IdeR to the promoters of siderophore synthesis, secretion, and transport represses the expression of those genes, turning off iron uptake. Meanwhile, IdeR-Fe^{2+} binding to the promoters of ferritins (ferritin and bacterioferritin) turns on iron storage, thereby preventing iron-mediated toxicity and maintaining iron homeostasis.
doi:10.1128/microbiolspec.MGM2-0012-2013.f2

The mycobacterial type VII secretion system, Esx-3, is induced under low iron conditions and has been shown to be necessary for growth of *M. tuberculosis* and *Mycobacterium bovis* BCG in iron-deficient medium and for utilization of exogenously added Fe^{3+}-carboxymycobactin (17, 18). These findings suggest that directly or indirectly, components of the Esx-3 system may contribute to Fe^{3+}-carboxymycobactin uptake.

Heme Utilization

Many pathogens have evolved strategies to obtain iron from heme, which is the most abundant source of iron in mammals. Bacteria can obtain heme using outer membrane receptors and periplasmic binding protein-dependent ABC transporters specific for heme, or they synthesize and secrete specialized proteins (hemophores) able to sequester heme and deliver it to a specific outer membrane receptor. *M. tuberculosis* is able to obtain iron from heme in the absence of siderophores (19). A genetic region encoding a secreted heme binding protein (hemophore) and two membrane transporters is necessary for normal iron acquisition from heme and hemoglobin (20). Once internalized, heme has to be degraded to release iron. This function is usually performed by heme oxygenases that degrade heme into iron, a tetrapyrrole product, and carbon monoxide (CO). In *M. tuberculosis* this role might be performed by the enzyme MhuD, a homolog of the *Staphylococcus aureus* heme oxygenases IsdG and IsdI. MhuD degrades heme in a unique way: it releases iron and a tetrapyrrole product named mycobilin but without CO generation (21). Presently, the role of heme uptake in the pathogenesis of *M. tuberculosis* is unknown. Studies in animal models will determine the relevance of heme utilization in *M. tuberculosis* virulence.

Iron Storage

The synthesis of iron storage proteins (ferritins) is central to iron homeostasis in most aerobic organisms and necessary for the virulence of many pathogens. Ferritin subunits form a hollow sphere where up to 4,500 atoms of iron can be sequestered as mineral, after being oxidized to Fe^{3+} at a ferroxidase center (22). Some bacteria and fungi synthesize ferritin-like proteins containing heme b known as bacterioferritins. *M. tuberculosis* has a bacterioferritin (BfrA) and a ferritin (BfrB). The crystal structure of these proteins shows the typical architecture of the ferritin superfamily of a cage-like hollow shell formed by 24 monomers with the characteristic fold of a four-helical bundle containing the ferroxidase catalytic center, and in bacterioferritin a heme group in each subunit-pair interface (23, 24). Analyses of single deletion mutants of *M. tuberculosis* showed that BfrA and BfrB are not functionally redundant. Deletion of *bfrB* drastically altered iron homeostasis, whereas no obvious defects were detected in a *bfrA*-deleted mutant (25). Iron stored by BfrB seems to be *M. tuberculosis*'s preferred reserve to overcome iron deficiency. In addition, *M. tuberculosis* lacking BfrB is highly sensitive to peroxide- and antibiotic-generated oxidative stress when cultured in iron-rich media. This indicates that BfrB is required to prevent excess free iron from catalyzing the generation of toxic reactive oxygen species (25). The significance of proper iron storage in the pathogenesis of *M. tuberculosis* has been demonstrated in animal models of infection. A mutant lacking *bfrB* is unable to persist in the lungs of mice and establish infection in the liver (25). Furthermore, a double *bfrA/bfrB* mutant is strongly attenuated in a guinea pig model of tuberculosis infection (26).

In addition to BfrA and BfrB, *M. tuberculosis* possesses a histone-like DNA binding protein (MDP1) that captures iron and also has ferroxidase activity. MDP1 may protect DNA by preventing the local generation of reactive oxygen radicals (27).

Regulation of Iron Metabolism

Iron can be very toxic because it catalyzes the generation of reactive oxygen species from normal products of aerobic respiration via the Harber-Weiss and Fenton reactions. Reactive oxygen species can damage most cellular components including DNA, lipids, and proteins. For this reason, aerobic organisms must tightly control intracellular iron levels. In bacteria, this control is generally achieved by regulating the uptake, utilization, and storage of this metal. Like other prokaryotes, *M. tuberculosis* regulates iron metabolism at the level of gene transcription. It induces the expression of iron uptake genes under iron deficiency and upregulates iron storage and oxidative stress defense genes when iron is readily available (28). *M. tuberculosis* achieves the delicate balance between the requirement for iron and its toxicity through the function of the iron-dependent regulator IdeR. IdeR is a metal and DNA binding protein, closely related to the *Corynebacterium diphtheriae* regulator of iron metabolism and toxin production DtxR (29). The structure of IdeR revealed two metal binding sites and three distinct functional domains: the amino-terminal containing a helix-turn-helix DNA binding motif, a dimerization domain that also bears most of the metal binding residues, and the carboxy-terminal domain characterized by adopting an SH3 (Src homology domain 3)–like folding, suggesting possible interactions with other proteins (30). Metal binding stabilizes dimer formation (31) and activates DNA binding. As two dimers, IdeR binds to both faces of the DNA at a unique 19-bp inverted repeat sequence, the "iron box" (TTAGGTTAGGCTAACCTAA), present in the promoter of iron-regulated genes, thereby modulating their transcription (32). Disruption of the *ideR* gene in *M. tuberculosis* is only possible in the presence of a second copy of the gene or when suppressor mutations arise. This indicates that IdeR is essential in *M. tuberculosis* (28). Approximately 150 genes respond to changes in iron availability in *M. tuberculosis*. IdeR controls the expression of about one-third of those genes including the siderophore synthesis and export genes, the siderophore transporter encoding genes *irtA* and *irtB*, genes in the *esx-3* cluster, and the iron storage genes *bfrA* and *bfrB* (28, 32). IdeR and Fe^{2+} turn off iron acquisition and turn on iron storage (Fig. 2).

These opposite effects of IdeR as a repressor of iron uptake and an activator of iron storage can be understood by considering the position of the iron box in repressed and activated promoters. Iron boxes on IdeR-repressed genes overlap the −10 region or the transcriptional start site; consequently, binding of IdeR to the iron box blocks access of the RNA polymerase and inhibits transcription of those genes. In the promoters of *bfrA* and *bfrB*, tandem iron boxes are located farther upstream (100 to 106 bp) from the transcriptional start site, suggesting a mechanism of activation by which IdeR-Fe^{2+} bound to these sites enhances access of the RNA polymerase to the promoter and initiation of transcription. In view of the strong attenuation of iron storage mutants in vivo (25), it is likely that IdeR-mediated activation of iron storage is essential for growth of *M. tuberculosis* during infection.

ZINC AND COPPER: NEVER TOO LITTLE OR TOO MUCH

Zinc and copper play vital functions in biological systems. The chemical properties of zinc, e.g., its Lewis acidity, coordination geometry, and rapid ligand exchange, allow it to form stable complexes with enzymes and proteins, where it functions in catalysis or as a structural factor. The majority of zinc-containing enzymes catalyze hydrolysis or related transfer reactions, some of which are essential for cell viability. The number of zinc-containing proteins identified in mycobacteria has increased significantly as more protein structures are resolved. Zinc is part of *M. tuberculosis* zinc-metallopeptidases (33, 34), carbonic anhydrase (35), fructose biphosphate aldolase Fba (36), the helicase RqlH (37), the cytidine deaminase Cda (38), the MshC ligase involved in mycothiol biosynthesis (39), the 2C-methyl-D-erythritol-2,4-cyclodiphosphate synthase IspF (40), the 2-isopropylmalate synthase LeuA involved in leucine biosynthesis (41), the superoxide dismutase (SOD) SodC (42, 43), the Esx-3 substrate EsxG-EsxH complex (44), the inositol-1-phosphate synthase (45), the RecA intein (46), and several more.

Because of its fast interconversion of Cu^+ and Cu^{2+}, copper is involved in several essential biochemical processes, such as oxygen-dependent electron transport reactions. In *M. tuberculosis*, at least two enzymes require copper as a cofactor, namely the SOD SodC (42, 43) and the cytochrome c oxidase subunits CtaC and CtaD.

Bioavailable levels of zinc are sufficiently low that most microbes have evolved high-affinity transport systems to capture this metal. Bacterial zinc transporters are generally ABC transporters consisting of a periplasmic binding protein, a membrane permease, and an ATPase. The periplasmic binding protein, which usually has a central His-, Asp-, and Glu-rich region, seems to allow specificity for zinc over manganese and other cations (47). Proteins involved in zinc import in mycobacteria have yet to be discovered. Regarding copper, as in most bacterial species, uptake systems for this metal have not been identified in *M. tuberculosis*.

The metallobiology of zinc and copper in *M. tuberculosis* recently provided insights into novel host defense mechanisms against bacterial infection involving intoxication by metal ions. To resist potential intoxication by metal ions, microbes express a range of metal efflux pumps and transporters belonging to three main families: heavy metal efflux members of the resistance–nodulation–cell division superfamily (HME-RND), the cation diffusion facilitator family, and the P-type ATPase family (48). A set of recent studies strikingly reported that some of these efflux systems are required for microbial virulence in various bacterial species, including *M. tuberculosis*, in order to resist newly described immune mechanisms relying on metal poisoning of microbes inside host cells.

In the *M. tuberculosis* genome, no putative heavy metal efflux system of the HME-RND family has been detected, while one putative cation diffusion facilitator transporter (Rv2025c) and 12 putative P-type ATPase members (CtpA-J, CtpV, and KdpB) are present (48, 49). The exact substrate specificity of these transporters is not known and is mostly inferred from indirect evidence such as similarity to known transporters and the presence of conserved metal binding motifs. For instance, the transcriptional regulator CmtR/Rv1994c, present in operons with the P-ATPase CtpG, responds to cadmium and lead (50) to alleviate *ctpG* transcriptional repression, suggesting that CtpG can efflux these two heavy metal cations; similarly, the ability of NmtR/Rv3744 to respond to nickel and cobalt and to bind the promoter region of the neighbor gene *ctpJ*/*nmtA*/Rv3743c to repress its expression in metal-free conditions (51) again suggests that CtpJ may efflux nickel and cobalt. Finally, the recent findings that *M. tuberculosis* mutants inactivated in *ctpV* and *ctpC* are highly sensitive to copper and zinc, respectively (52, 53), strongly suggest that these two P-ATPases may transport these metal ions, but again this is not a proof of their metal selectivity. Biochemical characterization of these transporters in recombinant biological systems and in reconstituted liposomal fractions will be required in order to understand their function.

In this context, a striking feature of three P-ATPase members in *M. tuberculosis*, namely CtpC, CtpG, and CtpV, is the presence of a putative metallochaperone-encoding gene, namely and respectively, Rv3269, Rv1993c, and Rv0968, upstream of the P-ATPase-encoding genes. The P-ATPase- and metallochaperone-encoding genes seem to be expressed in the operon. The function of these small proteins, predicted to be membrane bound and exposing putative metal binding motifs (e.g., DDGHDH in Rv0968) in their C-terminal cytoplasmic part, is not known; however, it is tempting to speculate that they may play a key part in metal selectivity and the transport mechanism of their cognate P-ATPase, as recently demonstrated for a similar transport system in *Streptococcus pneumoniae* (54).

A role for P-ATPase-mediated metal detoxification in *M. tuberculosis* has been recently suggested by several independent reports. In particular, *M. tuberculosis* mutants inactivated in the P-ATPase-encoding genes *ctpV* and *ctpC* were shown to be impaired in their ability to proliferate in model animals and/or host macrophages (52, 53). In a guinea pig model, Ward et al. reported that lung colonization by *M. tuberculo-sis* $\Delta ctpV$ was reduced by ≈ 1 $\log_{10}$ 3 weeks after inoculation, compared to the wild-type strain, and full virulence of the mutant was restored upon genetic complementation (53). The same authors reported a similar observation in mice, where the survival rate of animals infected with the mutant strain was increased by 16 weeks compared to those infected with the wild-type strain, although unlike in guinea pigs, no CFU difference was noticed in the mouse lungs. In both animal models, lung granulomatous inflammation was severely reduced in animals infected with the $\Delta ctpV$ mutant compared to the wild-type strain.

Although no direct demonstration has been provided yet regarding the metal selectivity of CtpV, it is most likely that this P-ATPase effluxes copper, because (i) the *ctpV* gene is induced by copper (55, 56), (ii) the CtpV protein contains typical motifs of the P_{1B1} family of copper-transporting P-ATPases (Table 1), (iii) the *ctpV* gene is encoded in operons with the copper-responsive transcriptional repressor CsoR (57), and most importantly, (iv) the $\Delta ctpV$ mutant is highly sensitive to copper *in vitro* (53). These results suggesting that *M. tuberculosis* faces copper intoxication *in vivo* during infection were further strengthened by a report that showed that the outer membrane channel protein

TABLE 1 P-ATPases in *M. tuberculosis*

Gene name	Group	Predicted substrate(s)	Comments	Reference(s)
ctpA/Rv0092	1B1	Cu$^+$/Ag$^+$	N-terminal CxxC; C-terminal MxxSS	
ctpB/Rv0103c	1B1	Cu$^+$/Ag$^+$	N-terminal CxxC; C-terminal MxxSS	
ctpC/Rv3270	1B?	Zn^{2+}; possibly others	Putative metallochaperone Rv3269; *M. tuberculosis* mutant sensitive to zinc	53
ctpD/Rv1469	1B4	Co^{2+}	C-terminal HEGxT; *M. smegmatis* homologue transports Co^{2+}	84
ctpE/	2A-like	Unknown	PEGL(P/V) motif found in calcium-transporting P-ATPases, such as SERCA	
ctpF/	2A	Ca^{2+}	PEGL(P/V) and Tm6-LWxNxxxd motifs found in calcium-transporting P-ATPases, such as SERCA	
ctpG/Rv1992c	1B?	Possibly Cd^{2+}/Pb^{2+} and others	Putative metallochaperone Rv1993c; repressed by Rv1994c/CmtR, unless Cd^{2+} or Pb^{2+} is present	81
ctpH/	2A-like	Unknown	Large N-terminal membrane-spanning domain; Tm6-PEGL(P/V) motif found in calcium-transporting P-ATPases, such as SERCA	
ctpI/	2A-like	Unknown	Large N-terminal membrane-spanning domain; Tm6-PEGL(P/V) motif found in calcium-transporting P-ATPases, such as SERCA	
ctpJ/Rv3743c	1B4	Co^{2+}	C-terminal HEGxT; repressed by Rv3744/NmtR, unless Ni^{2+} or Co^{2+} is present	51
ctpV/Rv0969	1B1	Cu$^+$	C-terminal MxxSS; *M. tuberculosis* mutant sensitive to copper; putative metallochaperone Rv0968; in operon with copper-responsive regulator *csoR/Rv0967*	53, 56
kdpB/Rv1030	1A	K$^+$	Homologous to many KdpB potassium transporters	

Rv1698/MctB is also required for both copper detoxification *in vitro* and for full virulence *in vivo* in guinea pigs (58). It was thus proposed that copper accumulation inside the mycobacterial phagosome may account for the phenotype of the Δ*ctpV* and Δ*mctB* mutants *in vivo* (59–61).

Phagosomal intoxication by copper has been suggested in other settings; in particular, an elegant study conducted in *Escherichia coli*–infected macrophages reported that copper enhances intracellular bacterial killing inside macrophages and that an *E. coli* mutant inactivated in the copper efflux P-ATPase CopA is killed faster in macrophages than its wild-type counterpart, unless the eukaryotic copper transporter ATP7A, which traffics to phago-lysosomes, is silenced through interference RNA (iRNA) (62). Although copper accumulation in the bacterial vacuole was not directly evidenced, this elegant study suggested for the first time that copper is an important mediator of microbial killing by immune cells and provided a mechanistic explanation for this phenomenon.

Regarding CtpC, we reported that genetic inactivation of this P-ATPase dramatically increases *M. tuberculosis* sensitivity to Zn^{2+}, which strongly suggested that CtpC might be involved in zinc efflux (52). However, a recent report suggested that CtpC may transport Mn^{2+} over Zn^{2+} and that the hypersensitivity of the *ctpC* mutant to zinc may be due to an increased sensitivity to oxidative stress following impaired Mn^{2+} loading of the SOD SodA and possibly other detoxification systems (63). Inside macrophages, we showed that zinc accumulates within *E. coli*– or *M. tuberculosis*–containing phagosomes and that bacterial strains impaired in resistance to zinc (a Δ*zntA* mutant in *E. coli* or a Δ*ctpC* mutant in *M. tuberculosis*) are impaired in intracellular survival. *In vivo* attenuation of the *M. tuberculosis* Δ*ctpC* mutant has yet to be clearly established (52, 63).

The requirement of P-ATPase-mediated copper resistance systems in bacterial virulence has been documented in several bacterial species, including *Listeria monocytogenes* (64), *Pseudomonas aeruginosa* (65), *S. pneumoniae* (66), and *Salmonella typhimurium* (67). Several mechanisms have been proposed to explain copper ion toxicity. These mechanisms include Fenton chemistry and generation of hydroxyl radicals (although this was challenged by data showing that there is no accumulation of hydroxyl radicals in copper-exposed *E. coli* [68]); degradation of iron-sulfur clusters in enzymes (69); and replacement of other metal ion cofactors, such as zinc ions, in proteins. The exact mechanism(s) of copper toxicity in *M. tuberculosis* remains to be identified.

The mechanism(s) of zinc ion toxicity may also include inactivation of iron-sulfur clusters (70) and inhibition of manganese uptake through transport competition in the bacterial periplasm (71). It was shown recently that P-ATPase-mediated copper export is required for the copper supply to periplasmic Cu,Zn-SOD and resistance to oxidative stress in *Salmonella enterica* (72). Whether copper and zinc export through CtpV, CtpC, and possibly other P-ATPases contributes to activation of the periplasmic Cu,Zn-SOD SodC in *M. tuberculosis* remains to be evaluated.

In summary, it is clear that *M. tuberculosis* uses the P-ATPases CtpC and CtpV to thrive inside macrophages and resist poisoning by Zn^{2+} and Cu^+. The function of the other *M. tuberculosis* P-ATPases, and their possible implication in mycobacterial virulence, remain to be understood. Equally important will be to understand the function of the putative metallochaperones associated with CtpC, CtpG, and CtpV.

Regulation of Metal Uptake

As stated above, although necessary, zinc and other metal ions can also be toxic if present at too high a concentration. For instance, zinc may interact with thiols or compete with other metals for protein binding, blocking essential reactions in the cells. Therefore, the quantity of zinc inside the cells is carefully regulated, usually by calibrating uptake and export. The genome of *M. tuberculosis* contains two genes encoding transcriptional regulators of the Fur family, FurA and FurB. Structural and functional characterization of FurB revealed it to be a Zn^{2+}-dependent repressor; hence, it has been renamed Zur (zinc uptake regulator) (73–75). Genes repressed by Zur-Zn^{2+} include the gene cluster encoding the Esx-3 secretion system, several ribosomal proteins, a protein similar to the *Bacillus subtilis* low-affinity zinc transporter YciC, and components of a putative ABC-type Zn^{2+}/Mn^{2+} transport system (74). Disruption of the *zur* gene did not affect the ability of *M. tuberculosis* to replicate in mice, suggesting that constitutive expression of Zur-regulated genes is not detrimental for *M. tuberculosis* in this model of infection (74).

The importance of sensing metal deficiency or excess is reflected in the multiple families of metalloregulatory proteins characterized in bacteria. These include Fur, DtxR, MerR, SmtB/ArsR, CsoR, CopY, and NikR. In general, these proteins are transcriptional regulators that sense specific metal ions via direct coordination. The DtxR, Fur, and NikR family proteins primarily regulate genes required for metal uptake, whereas members of the other families regulate mainly metal efflux. Fur was first described as an iron-responsive repressor of iron transport in *E. coli*. Since then, numerous studies have revealed functional specialization within the Fur family and a great diversity in metal selectivity and biological function. The Fur family includes sensors of iron (Fur), zinc (Zur), manganese (Mur), and nickel (Nur). Some members of the family use metal-catalyzed redox reactions to sense peroxide-mediated stress (Per) or heme (Irr).

M. tuberculosis has two Fur-like proteins, namely Zur (described above) and FurA. The FurA-encoding gene is located immediately upstream of *katG*, the gene encoding a catalase-peroxidase, a major virulence factor that also activates the prodrug isoniazid. *furA* and *katG* are cotranscribed from a common promoter upstream of *furA*. FurA auto-represses its expression and the expression of *katG* by binding to a unique sequence upstream of *furA* (76–78). FurA seems to have a very specialized biological role, as no other genes regulated by FurA have been identified to date.

Two regulators of the DtxR family are present in *M. tuberculosis*: the iron-dependent regulator IdeR (described above) and SirR (for staphylococcal iron-regulated repressor). In *Staphylococcus epidermidis*, SirR, in complex with Fe^{2+} or Mn^{2+}, binds to a unique sequence in the promoter of an operon encoding for a putative iron transporter (79). The biological role of the SirR homologue in *M. tuberculosis*, however, has not been determined.

Other metalloregulators characterized in *M. tuberculosis* include the Ni(II)/Co(II)-specific repressors NmtR (51) and KmtR (80), the copper sensors CsoR (57) and RicR (55), the Cd(II)/Pb(II) sensor CmtR (81), and the Zn(II)-responsive regulator encoded by the gene Rv2358 (82). In general, they regulate the transcription of membrane transporters that mediate cytoplasmic efflux of potentially toxic metals, as mentioned above.

FUTURE DIRECTIONS

Much remains to be understood regarding the mechanisms of transition metal uptake/efflux systems in *M. tuberculosis*, their regulation, and the biological impact of selective metal ion enrichment or depletion encountered in the mycobacterial phagosome inside host macrophages (59, 83). As mentioned above, the zinc and copper uptake systems still have to be identified in *M. tuberculosis*. Identification of the remaining components of the iron acquisition apparatus and a better understanding of the mechanisms that control iron sorting and assimilation in *M. tuberculosis* will reveal new possibilities of intervention. For instance,

ways to starve *M. tuberculosis* for iron or, alternatively, get it to intoxicate itself by corrupting its iron-sensing mechanisms may help develop novel treatments. Future work should also aim at deciphering the exact metal specificity and biological function of the many *M. tuberculosis* P-ATPases, and the use innate immune cells make of metal ion withdrawal or intoxication to contain mycobacterial infection.

The authors received no specific funding for this work. The Neyrolles laboratory is supported by the Centre National de la Recherche Scientifique (CNRS), the European Union (7th Framework Programme), the Agence Nationale de la Recherche (ANR), the Fondation Mérieux, and the Fondation pour la Recherche Médicale (FRM). Work from the Rodriguez laboratory discussed in this chapter was supported by NIH research grant AI44856 (GMR).

Citation. Rodriguez GM, Neyrolles O. 2014. Metallobiology of tuberculosis. Microbiol Spectrum **2**(3):MGM2-0012-2013.

References

1. **Andrews SC, Robinson AK, Rodriguez-Quinones F.** 2003. Bacterial iron homeostasis. *FEMS Microbiol Rev* **27**:215–237.

2. **Hood MI, Skaar EP.** 2012. Nutritional immunity: transition metals at the pathogen-host interface. *Nat Rev Microbiol* **10**:525–537.

3. **Weinberg ED.** 1974. Iron and susceptibility to infectious disease. *Science* **184**:952–956.

4. **Dobryszycka W.** 1997. Biological functions of haptoglobin: new pieces to an old puzzle. *Eur J Clin Chem Clin Biochem* **35**:647–654.

5. **Tolosano E, Altruda F.** 2002. Hemopexin: structure, function, and regulation. *DNA Cell Biol* **21**:297–306.

6. **Snow GA.** 1970. Mycobactins: iron-chelating growth factors from mycobacteria. *Bacteriol Rev* **34**:99–125.

7. **Madigan CA, Cheng TY, Layre E, Young DC, McConnell MJ, Debono CA, Murry JP, Wei JR, Barry CE 3rd, Rodriguez GM, Matsunaga I, Rubin EJ, Moody DB.** 2012. Lipidomic discovery of deoxysiderophores reveals a revised mycobactin biosynthesis pathway in *Mycobacterium tuberculosis*. *Proc Natl Acad Sci USA* **109**:1257–1262.

8. **McMahon MD, Rush JS, Thomas MG.** 2012. Analyses of MbtB, MbtE, and MbtF suggest revisions to the mycobactin biosynthesis pathway in *Mycobacterium tuberculosis*. *J Bacteriol* **194**:2809–2818.

9. **Quadri LE, Sello J, Keating TA, Weinreb PH, Walsh CT.** 1998. Identification of a *Mycobacterium tuberculosis* gene cluster encoding the biosynthetic enzymes for assembly of the virulence-conferring siderophore mycobactin. *Chem Biol* **5**:631–645.

10. **Gobin J, Moore CH, Reeve JR Jr, Wong DK, Gibson BW, Horwitz MA.** 1995. Iron acquisition by *Mycobacterium tuberculosis*: isolation and characterization of a family of iron-binding exochelins. *Proc Natl Acad Sci USA* **92**:5189–5193.

11. Ratledge C, Dover LG. 2000. Iron metabolism in pathogenic bacteria. *Annu Rev Microbiol* **54**:881–941.

12. Gobin J, Horwitz MA. 1996. Exochelins of *Mycobacterium tuberculosis* remove iron from human iron-binding proteins and donate iron to mycobactins in the *M. tuberculosis* cell wall. *J Exp Med* **183**:1527–1532.

13. Rodriguez GM, Gardner R, Kaur N, Phanstiel O 4th. 2008. Utilization of Fe3+-acinetoferrin analogs as an iron source by *Mycobacterium tuberculosis*. *Biometals* **21**:93–103.

14. Wells RM, Jones CM, Xi Z, Speer A, Danilchanka O, Doornbos KS, Sun P, Wu F, Tian C, Niederweis M. 2013. Discovery of a siderophore export system essential for virulence of *Mycobacterium tuberculosis*. *PLoS Pathog* **9**:e1003120.

15. Rodriguez GM, Smith I. 2006. Identification of an ABC transporter required for iron acquisition and virulence in *Mycobacterium tuberculosis*. *J Bacteriol* **188**:424–430.

16. Ryndak MB, Wang S, Smith I, Rodriguez GM. 2010. The *Mycobacterium tuberculosis* high-affinity iron importer, IrtA, contains an FAD-binding domain. *J Bacteriol* **192**:861–869.

17. Serafini A, Boldrin F, Palu G, Manganelli R. 2009. Characterization of a *Mycobacterium tuberculosis* ESX-3 conditional mutant: essentiality and rescue by iron and zinc. *J Bacteriol* **191**:6340–6344.

18. Siegrist MS, Unnikrishnan M, McConnell MJ, Borowsky M, Cheng TY, Siddiqi N, Fortune SM, Moody DB, Rubin EJ. 2009. Mycobacterial Esx-3 is required for mycobactin-mediated iron acquisition. *Proc Natl Acad Sci USA* **106**:18792–18797.

19. Jones CM, Niederweis M. 2011. *Mycobacterium tuberculosis* can utilize heme as an iron source. *J Bacteriol* **193**:1767–1770.

20. Tullius MV, Harmston CA, Owens CP, Chim N, Morse RP, McMath LM, Iniguez A, Kimmey JM, Sawaya MR, Whitelegge JP, Horwitz MA, Goulding CW. 2011. Discovery and characterization of a unique mycobacterial heme acquisition system. *Proc Natl Acad Sci USA* **108**:5051–5056.

21. Nambu S, Matsui T, Goulding CW, Takahashi S, Ikeda-Saito M. 2013. A new way to degrade heme: the *Mycobacterium tuberculosis* enzyme MhuD catalyzes heme degradation without generating CO. *J Biol Chem* **5**:10101–10109.

22. Chiancone E, Ceci P, Ilari A, Ribacchi F, Stefanini S. 2004. Iron and proteins for iron storage and detoxification. *Biometals* **17**:197–202.

23. Gupta V, Gupta RK, Khare G, Salunke DM, Tyagi AK. 2009. Crystal structure of Bfr A from *Mycobacterium tuberculosis*: incorporation of selenomethionine results in cleavage and demetallation of haem. *PLoS One* **4**:e8028.

24. Khare G, Gupta V, Nangpal P, Gupta RK, Sauter NK, Tyagi AK. 2011. Ferritin structure from *Mycobacterium tuberculosis*: comparative study with homologues identifies extended C-terminus involved in ferroxidase activity. *PLoS One* **6**:e18570.

25. Pandey R, Rodriguez GM. 2012. A ferritin mutant of *Mycobacterium tuberculosis* is highly susceptible to killing by antibiotics and is unable to establish a chronic infection in mice. *Infect Immun* **80**:3650–3659.

26. Reddy PV, Puri RV, Khera A, Tyagi AK. 2012. Iron storage proteins are essential for the survival and pathogenesis of *Mycobacterium tuberculosis* in THP-1 macrophages and the guinea pig model of infection. *J Bacteriol* **194**:567–575.

27. Takatsuka M, Osada-Oka M, Satoh EF, Kitadokoro K, Nishiuchi Y, Niki M, Inoue M, Iwai K, Arakawa T, Shimoji Y, Ogura H, Kobayashi K, Rambukkana A, Matsumoto S. 2011. A histone-like protein of mycobacteria possesses ferritin superfamily protein-like activity and protects against DNA damage by Fenton reaction. *PLoS One* **6**:e20985.

28. Rodriguez GM, Voskuil MI, Gold B, Schoolnik GK, Smith I. 2002. ideR, An essential gene in *Mycobacterium tuberculosis*: role of IdeR in iron-dependent gene expression, iron metabolism, and oxidative stress response. *Infect Immun* **70**:3371–3381.

29. Schmitt MP, Predich M, Doukhan L, Smith I, Holmes RK. 1995. Characterization of an iron-dependent regulatory protein (IdeR) of *Mycobacterium tuberculosis* as a functional homolog of the diphtheria toxin repressor (DtxR) from *Corynebacterium diphtheriae*. *Infect Immun* **63**:4284–4289.

30. Pohl E, Holmes RK, Hol WG. 1999. Crystal structure of the iron-dependent regulator (IdeR) from *Mycobacterium tuberculosis* shows both metal binding sites fully occupied. *J Mol Biol* **285**:1145–1156.

31. Semavina M, Beckett D, Logan TM. 2006. Metal-linked dimerization in the iron-dependent regulator from *Mycobacterium tuberculosis*. *Biochemistry* **45**:12480–12490.

32. Gold B, Rodriguez GM, Marras SA, Pentecost M, Smith I. 2001. The *Mycobacterium tuberculosis* IdeR is a dual functional regulator that controls transcription of genes involved in iron acquisition, iron storage and survival in macrophages. *Mol Microbiol* **42**:851–865.

33. Petrera A, Amstutz B, Gioia M, Hahnlein J, Baici A, Selchow P, Ferraris DM, Rizzi M, Sbardella D, Marini S, Coletta M, Sander P. 2012. Functional characterization of the *Mycobacterium tuberculosis* zinc metallopeptidase Zmp1 and identification of potential substrates. *Biol Chem* **393**:631–640.

34. Srinivasan R, Anilkumar G, Rajeswari H, Ajitkumar P. 2006. Functional characterization of AAA family FtsH protease of *Mycobacterium tuberculosis*. *FEMS Microbiol Lett* **259**:97–105.

35. Supuran CT. 2008. Carbonic anhydrases: an overview. *Curr Pharm Des* **14**:603–614.

36. Pegan SD, Rukseree K, Franzblau SG, Mesecar AD. 2009. Structural basis for catalysis of a tetrameric class IIa fructose 1,6-bisphosphate aldolase from *Mycobacterium tuberculosis*. *J Mol Biol* **386**:1038–1053.

37. Ordonez H, Unciuleac M, Shuman S. 2012. *Mycobacterium smegmatis* RqlH defines a novel clade of bacterial RecQ-like DNA helicases with ATP-dependent 3′-5′ translocase and duplex unwinding activities. *Nucleic Acids Res* **40**:4604–4614.

38. Sanchez-Quitian ZA, Schneider CZ, Ducati RG, de Azevedo WF Jr, Bloch C Jr, Basso LA, Santos DS. 2010. Structural and functional analyses of *Mycobacterium tuberculosis* Rv3315c-encoded metal-dependent homotetrameric cytidine deaminase. *J Struct Biol* **169**:413–423.

39. Tremblay LW, Fan F, Vetting MW, Blanchard JS. 2008. The 1.6 A crystal structure of *Mycobacterium smegmatis* MshC: the penultimate enzyme in the mycothiol biosynthetic pathway. *Biochemistry* **47**:13326–13335.

40. Buetow L, Brown AC, Parish T, Hunter WN. 2007. The structure of mycobacteria 2C-methyl-D-erythritol-2,4-cyclodiphosphate synthase, an essential enzyme, provides a platform for drug discovery. *BMC Struct Biol* **7**:68.

41. Koon N, Squire CJ, Baker EN. 2004. Crystal structure of LeuA from *Mycobacterium tuberculosis*, a key enzyme in leucine biosynthesis. *Proc Natl Acad Sci USA* **101**:8295–8300.

42. Piddington DL, Fang FC, Laessig T, Cooper AM, Orme IM, Buchmeier NA. 2001. Cu,Zn superoxide dismutase of *Mycobacterium tuberculosis* contributes to survival in activated macrophages that are generating an oxidative burst. *Infect Immun* **69**:4980–4987.

43. Wu CH, Tsai-Wu JJ, Huang YT, Lin CY, Lioua GG, Lee FJ. 1998. Identification and subcellular localization of a novel Cu,Zn superoxide dismutase of *Mycobacterium tuberculosis*. *FEBS Lett* **439**:192–196.

44. Ilghari D, Lightbody KL, Veverka V, Waters LC, Muskett FW, Renshaw PS, Carr MD. 2011. Solution structure of the *Mycobacterium tuberculosis* EsxG.EsxH complex: functional implications and comparisons with other *M. tuberculosis* Esx family complexes. *J Biol Chem* **286**:29993–30002.

45. Norman RA, McAlister MS, Murray-Rust J, Movahedzadeh F, Stoker NG, McDonald NQ. 2002. Crystal structure of inositol 1-phosphate synthase from *Mycobacterium tuberculosis*, a key enzyme in phosphatidylinositol synthesis. *Structure* **10**:393–402.

46. Zhang L, Xiao N, Pan Y, Zheng Y, Pan Z, Luo Z, Xu X, Liu Y. 2010. Binding and inhibition of copper ions to RecA inteins from *Mycobacterium tuberculosis*. *Chemistry* **16**:4297–4306.

47. Hantke K. 2005. Bacterial zinc uptake and regulators. *Curr Opin Microbiol* **8**:196–202.

48. Nies DH. 2003. Efflux-mediated heavy metal resistance in prokaryotes. *FEMS Microbiol Rev* **27**:313–339.

49. Cole ST, Brosch R, Parkhill J, Garnier T, Churcher C, Harris D, Gordon SV, Eiglmeier K, Gas S, Barry CE 3rd, Tekaia F, Badcock K, Basham D, Brown D, Chillingworth T, Connor R, Davies R, Devlin K, Feltwell T, Gentles S, Hamlin N, Holroyd S, Hornsby T, Jagels K, Krogh A, McLean J, Moule S, Murphy L, Oliver K, Osborne J, Quail MA, Rajandream MA, Rogers J, Rutter S, Seeger K, Skelton J, Squares R, Squares S, Sulston JE, Taylor K, Whitehead S, Barrell BG. 1998. Deciphering the biology of *Mycobacterium tuberculosis* from the complete genome sequence. *Nature* **393**:537–544.

50. Verkhovtseva NV, Filina N, Pukhov DE. 2001. Evolutionary role of iron in metabolism of prokaryotes and biogeochemical processes. *Zhurnal evoliutsionnoi biokhimii i fiziologii* **37**:338–343 (In Russian.)

51. Cavet JS, Meng W, Pennella MA, Appelhoff RJ, Giedroc DP, Robinson NJ. 2002. A nickel-cobalt-sensing ArsR-SmtB family repressor. Contributions of cytosol and effector binding sites to metal selectivity. *J Biol Chem* **277**:38441–38448.

52. Botella H, Peyron P, Levillain F, Poincloux R, Poquet Y, Brandli I, Wang C, Tailleux L, Tilleul S, Charriere GM, Waddell SJ, Foti M, Lugo-Villarino G, Gao Q, Maridonneau-Parini I, Butcher PD, Castagnoli PR, Gicquel B, de Chastellier C, Neyrolles O. 2011. Mycobacterial p(1)-type ATPases mediate resistance to zinc poisoning in human macrophages. *Cell Host Microbe* **10**:248–259.

53. Ward SK, Abomoelak B, Hoye EA, Steinberg H, Talaat AM. 2010. CtpV: a putative copper exporter required for full virulence of *Mycobacterium tuberculosis*. *Mol Microbiol* **77**:1096–1110.

54. Fu Y, Tsui HC, Bruce KE, Sham LT, Higgins KA, Lisher JP, Kazmierczak KM, Maroney MJ, Dann CE 3rd, Winkler ME, Giedroc DP. 2013. A new structural paradigm in copper resistance in *Streptococcus pneumoniae*. *Nat Chem Biol* **9**:177–183.

55. Festa RA, Jones MB, Butler-Wu S, Sinsimer D, Gerads R, Bishai WR, Peterson SN, Darwin KH. 2011. A novel copper-responsive regulon in *Mycobacterium tuberculosis*. *Mol Microbiol* **79**:133–148.

56. Ward SK, Hoye EA, Talaat AM. 2008. The global responses of *Mycobacterium tuberculosis* to physiological levels of copper. *J Bacteriol* **190**:2939–2946.

57. Liu T, Ramesh A, Ma Z, Ward SK, Zhang L, George GN, Talaat AM, Sacchettini JC, Giedroc DP. 2007. CsoR is a novel *Mycobacterium tuberculosis* copper-sensing transcriptional regulator. *Nat Chem Biol* **3**:60–68.

58. Wolschendorf F, Ackart D, Shrestha TB, Hascall-Dove L, Nolan S, Lamichhane G, Wang Y, Bossmann SH, Basaraba RJ, Niederweis M. 2011. Copper resistance is essential for virulence of *Mycobacterium tuberculosis*. *Proc Natl Acad Sci USA* **108**:1621–1626.

59. Botella H, Stadthagen G, Lugo-Villarino G, de Chastellier C, Neyrolles O. 2012. Metallobiology of host-pathogen interactions: an intoxicating new insight. *Trends Microbiol* **20**:106–112.

60. Rowland JL, Niederweis M. 2012. Resistance mechanisms of *Mycobacterium tuberculosis* against phagosomal copper overload. *Tuberculosis (Edinb)* **92**:202–210.

61. Samanovic MI, Ding C, Thiele DJ, Darwin KH. 2012. Copper in microbial pathogenesis: meddling with the metal. *Cell Host Microbe* **11**:106–115.

62. White C, Lee J, Kambe T, Fritsche K, Petris MJ. 2009. A role for the ATP7A copper-transporting ATPase in macrophage bactericidal activity. *J Biol Chem* **284**:33949–33956.

63. Padilla-Benavides T, Long JE, Raimunda D, Sassetti CM, Arguello JM. 2013. A novel P1B-type Mn2+ transporting ATPase is required for secreted protein metallation in mycobacteria. *J Biol Chem* **288**:11334–11347.

64. Francis MS, Thomas CJ. 1997. Mutants in the CtpA copper transporting P-type ATPase reduce virulence of *Listeria monocytogenes*. *Microbial Pathog* **22**:67–78.

65. Schwan WR, Warrener P, Keunz E, Stover CK, Folger KR. 2005. Mutations in the cueA gene encoding a copper homeostasis P-type ATPase reduce the pathogenicity of *Pseudomonas aeruginosa* in mice. *Int J Med Microbiol* **295**:237–242.

66. Shafeeq S, Yesilkaya H, Kloosterman TG, Narayanan G, Wandel M, Andrew PW, Kuipers OP, Morrissey JA. 2011. The cop operon is required for copper homeostasis and contributes to virulence in *Streptococcus pneumoniae*. *Mol Microbiol* **81**:1255–1270.

67. Osman D, Waldron KJ, Denton H, Taylor CM, Grant AJ, Mastroeni P, Robinson NJ, Cavet JS. 2010. Copper homeostasis in salmonella is atypical and copper-CueP is a major periplasmic metal complex. *J Biol Chem* **285**:25259–25268.

68. Macomber L, Rensing C, Imlay JA. 2007. Intracellular copper does not catalyze the formation of oxidative DNA damage in *Escherichia coli*. *J Bacteriol* **189**:1616–1626.

69. Chillappagari S, Seubert A, Trip H, Kuipers OP, Marahiel MA, Miethke M. 2010. Copper stress affects iron homeostasis by destabilizing iron-sulfur cluster formation in *Bacillus subtilis*. *J Bacteriol* **192**:2512–2524.

70. Xu FF, Imlay JA. 2012. Silver(I), mercury(II), cadmium (II), and zinc(II) target exposed enzymic iron-sulfur clusters when they toxify *Escherichia coli*. *Appl Environ Microbiol* **78**:3614–3621.

71. McDevitt CA, Ogunniyi AD, Valkov E, Lawrence MC, Kobe B, McEwan AG, Paton JC. 2011. A molecular mechanism for bacterial susceptibility to zinc. *PLoS Pathog* **7**:e1002357.

72. Osman D, Patterson CJ, Bailey K, Fisher K, Robinson NJ, Rigby SE, Cavet JS. 2013. The copper supply pathway to a salmonella Cu,Zn-superoxide dismutase (SodCII) involves P(1B)-type ATPase copper efflux and periplasmic CueP. *Mol Microbiol* **87**:466–477.

73. Lucarelli D, Russo S, Garman E, Milano A, Meyer-Klaucke W, Pohl E. 2007. Crystal structure and function of the zinc uptake regulator FurB from *Mycobacterium tuberculosis*. *J Biol Chem* **282**:9914–9922.

74. Maciag A, Dainese E, Rodriguez GM, Milano A, Provvedi R, Pasca MR, Smith I, Palu G, Riccardi G, Manganelli R. 2007. Global analysis of the *Mycobacterium tuberculosis* Zur (FurB) regulon. *J Bacteriol* **189**:730–740.

75. Milano A, Branzoni M, Canneva F, Profumo A, Riccardi G. 2004. The *Mycobacterium tuberculosis* Rv2358-furB operon is induced by zinc. *Res Microbiol* **155**:192–200.

76. Pym AS, Domenech P, Honore N, Song J, Deretic V, Cole ST. 2001. Regulation of catalase-peroxidase (KatG) expression, isoniazid sensitivity and virulence by furA of *Mycobacterium tuberculosis*. *Mol Microbiol* **40**:879–889.

77. Sala C, Forti F, Di Florio E, Canneva F, Milano A, Riccardi G, Ghisotti D. 2003. *Mycobacterium tuberculosis* FurA autoregulates its own expression. *J Bacteriol* **185**:5357–5362.

78. Zahrt TC, Song J, Siple J, Deretic V. 2001. Mycobacterial FurA is a negative regulator of catalase-peroxidase gene katG. *Mol Microbiol* **39**:1174–1185.

79. Massonet C, Pintens V, Merckx R, Anne J, Lammertyn E, Van Eldere J. 2006. Effect of iron on the expression of sirR and sitABC in biofilm-associated *Staphylococcus epidermidis*. *BMC Microbiol* **6**:103.

80. Campbell DR, Chapman KE, Waldron KJ, Tottey S, Kendall S, Cavallaro G, Andreini C, Hinds J, Stoker NG, Robinson NJ, Cavet JS. 2007. Mycobacterial cells have dual nickel-cobalt sensors: sequence relationships and metal sites of metal-responsive repressors are not congruent. *J Biol Chem* **282**:32298–32310.

81. Cavet JS, Graham AI, Meng W, Robinson NJ. 2003. A cadmium-lead-sensing ArsR-SmtB repressor with novel sensory sites. Complementary metal discrimination by NmtR and CmtR in a common cytosol. *J Biol Chem* **278**:44560–44566.

82. Canneva F, Branzoni M, Riccardi G, Provvedi R, Milano A. 2005. Rv2358 and FurB: two transcriptional regulators from *Mycobacterium tuberculosis* which respond to zinc. *J Bacteriol* **187**:5837–5840.

83. Wagner D, Maser J, Lai B, Cai Z, Barry CE 3rd, Honer Zu Bentrup K, Russell DG, Bermudez LE. 2005. Elemental analysis of *Mycobacterium avium*-, *Mycobacterium tuberculosis*-, and *Mycobacterium smegmatis*-containing phagosomes indicates pathogen-induced microenvironments within the host cell's endosomal system. *J Immunol* **174**:1491–1500.

84. Raimunda D, Long JE, Sassetti CM, Arguello JM. 2012. Role in metal homeostasis of CtpD, a Co(2)(+) transporting P(1B4)-ATPase of *Mycobacterium smegmatis*. *Mol Microbiol* **84**:1139–1149.

Molecular Genetics of Mycobacteria, 2nd Edition
Edited by Graham F. Hatfull and William R. Jacobs, Jr.
© 2014 American Society for Microbiology, Washington, DC
doi:10.1128/microbiolspec.MGM2-0015-2013

Gregory M. Cook,[1] Kiel Hards,[1] Catherine Vilchèze,[2]
Travis Hartman,[2] and Michael Berney[2]

Energetics of Respiration and Oxidative Phosphorylation in Mycobacteria

20

OVERVIEW OF MYCOBACTERIAL RESPIRATION AND OXIDATIVE PHOSPHORYLATION

The genus *Mycobacterium* comprises a group of obligately aerobic bacteria that have adapted to inhabit a wide range of intracellular and extracellular environments. A fundamental feature in this adaptation is the ability to respire and generate energy from variable sources or to sustain metabolism in the absence of growth. Early studies on respiration demonstrated that *Mycobacterium tuberculosis* H37Rv, grown in the lungs of infected mice, had high rates of endogenous respiration that were not stimulated by exogenous substrates (e.g., acetate, pyruvate, glucose, glycerol, lactate) (1). In contrast, cells grown *in vitro* respired these substrates at high rates. Fatty acids, however, stimulated the respiration of *in vivo*–grown *M. tuberculosis*, suggesting for the first time that *M. tuberculosis* switches to different energy sources in host tissues to fuel respiration. These early studies pointed to the fact that electron donor utilization and respiration are precisely controlled in *M. tuberculosis*, not only in response

to growth rate, but also in response to the carbon and energy sources used for growth. The pioneering work of Brodie and colleagues on *Mycobacterium phlei* established much of the primary information on the electron transport chain and oxidative phosphorylation system in mycobacteria (reviewed in reference 2).

Since these early studies, comparatively few biochemical studies have been performed on the electron transport components and the energetics of respiration in mycobacterial species. With the advent of microbial genome sequencing, sequence analyses have revealed that branched pathways exist in mycobacterial species for electron transfer from many low potential reductants via quinol, including H_2, to oxygen (Fig. 1). Unlike other bacteria, there appears to be little redundancy in the transfer of electrons to oxygen during growth with only two terminal respiratory oxidases present in mycobacteria, an aa_3-type cytochrome *c* oxidase (encoded by *ctaBCDE*) belonging to the heme-copper respiratory oxidase family and cytochrome *bd*-type menaquinol oxidase (*cydABCD*) (Fig. 1). Mycobacteria coordinate the expression of these oxidases

---

[1]University of Otago, Department of Microbiology and Immunology, Dunedin, New Zealand; [2]Department of Microbiology and Immunology, Albert Einstein College of Medicine, Bronx, NY 10461.

in response to oxygen supply to achieve maximal efficiency of oxygen utilization. The rationale for this coordinated expression is largely based on the assumption that the mycobacterial aa_3-type cytochrome c oxidase and cytochrome bd-type have markedly different affinities for oxygen in mycobacteria, but this remains to be experimentally shown. Furthermore, the molecular mechanisms governing the regulation of terminal oxidase expression in mycobacterial species are unknown. In the absence of oxygen, mycobacterial growth is inhibited, even if alternative electron acceptors are present (e.g., nitrate, fumarate). Despite growth being inhibited, mycobacteria are able to metabolize exogenous and endogenous energy sources under low oxygen for maintenance functions. Whether mycobacteria utilize energy spilling or overflow metabolism (3) under these conditions to balance the rate of catabolic versus anabolic reactions remains to be determined. The electron acceptors and mechanisms to recycle reducing equivalents under these conditions are poorly understood. Irrespective of the oxygen concentration or the proton motive force (PMF), ATP synthesis is obligatorily coupled to the electron transport chain and the F_1F_0-ATP synthase, but the reasons for this remain unexplained.

Research into mycobacterial respiration and oxidative phosphorylation has been energized by the discovery of a new drug (TMC207; Sirturo; Bedaquiline) that targets the ATP synthase of mycobacteria, suggesting that inhibitors of respiration and ATP synthesis will provide the next generation of front-line drugs to combat tuberculosis and nontuberculous mycobacterial disease. The aim of this article is to discuss recent advances in understanding the processes of respiration and oxidative phosphorylation in mycobacteria, with a particular focus on the obligate human pathogen *M. tuberculosis*.

ENERGETICS OF MYCOBACTERIA DURING GROWTH AND NONGROWTH CONDITIONS

Mycobacteria generally grow at neutral pH and under these conditions generate a PMF of approximately −180 mV (4). The PMF generated is used to drive proton-coupled energetic processes (e.g., ATP synthesis, solute transport, etc.) (4). The PMF-generating machinery in mycobacteria comprises three energy-conserving steps: complex I ($4H^+/2e^-$), complex III (menaquinol-cytochrome c oxidoreductase, $4H^+/2e^-$), and complex

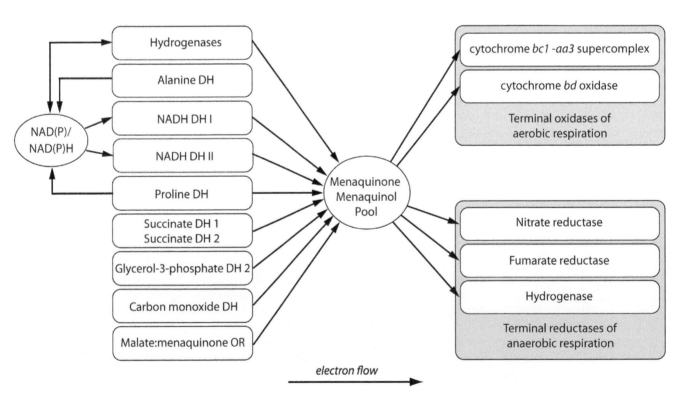

Figure 1 Organization and components of the electron transport chain in mycobacteria. doi:10.1128/microbiolspec.MGM2-0015-2013.f1

IV ($2H^+/2e^-$), suggesting that the overall proton translocation stoichiometry for the transfer of $2e^-$ from NADH to oxygen is $10H^+/2e^-$. A ratio of $3H^+$ utilized by the ATP synthase/ATP synthesized leads to a theoretical maximum P/O ratio (i.e., the number of moles of ADP phosphorylated to ATP per $2e^-$ passing to oxygen) of approximately 3.3 (theoretical maximum). If complex I is bypassed by the non-proton translocating type II NADH:menaquinone oxidoreductase, the P/O ratio would be approximately 2. Electron flow from the menaquinone pool to the cytochrome *bd* oxidase branch (bypassing complex III and IV) would produce a P/O ratio of 0.67. Measured experimental values for mycobacteria oxidizing NADH or succinate yield P/O ratios of 0.52 and 0.36, respectively (5). Variations in theoretical P/O ratios versus those determined experimentally are well accepted and can be explained by pathways that involve proton leakage (i.e., bypass ATP synthase), or the PMF is used to drive reverse electron transport. When succinate oxidation is coupled to menaquinone reduction, the energetics suggest a reverse electron flow from succinate (lower midpoint redox potential E_m = +30 mV) to menaquinone (E_m = −74 mV), resulting in lower P/O ratios.

Growth of mycobacteria is sensitive to compounds that dissipate the membrane potential (e.g., protonophores and valinomycin), and these compounds are bactericidal toward growing and nongrowing (aerobic or hypoxic) cells, further highlighting the importance of the membrane potential in mycobacterial viability (4, 6). Growth is also sensitive to the ionophore monensin, but the reasons for this are not clear. Interestingly, some solute transporters in mycobacteria are driven by a sodium-motive force (7), but the mechanism used for the generation of the primary sodium gradient in mycobacteria is not known. As the pH of the growth medium changes and becomes mildly acidic, mycobacteria are able to generate a considerable transmembrane pH gradient ($Z\Delta pH$) and maintain a constant PMF (4). While proton translocation via the respiratory chain generates the PMF during respiration with oxygen as the terminal electron acceptor, it is not clear how the PMF is established in the absence of oxygen under anaerobic growth conditions. Anaerobic bacteria are able to generate a significant PMF (−100 mV) using their membrane-bound F_1F_0-ATP synthase in the ATP hydrolysis direction (8). The ATPase activity (proton pumping) of the enzyme is fueled by ATP produced by substrate-level phosphorylation. This mechanism does not appear to operate in mycobacterial cells, where the F_1F_0-ATP synthase has been reported to have latent ATPase activity when measured in inverted membrane vesicles (9, 10). Whether the enzyme is also latent in actively growing cells is not known, and therefore the potential exists for this enzyme to function as a primary proton pump in the absence of oxygen and a functional respiratory chain to generate the PMF. Rao et al. (6) have reported that hypoxic, nonreplicating *M. tuberculosis* generates a total PMF of −113 mV, −73 mV of electrical potential ($\Delta\Psi$), and −41 mV of $Z\Delta pH$. The addition of thioridazine, a compound that targets NDH-2, results in dissipation of the $\Delta\Psi$ and significant cell death, suggesting that NADH is an important electron donor for the generation of the $\Delta\Psi$ under hypoxic conditions. The addition of TMC207, a specific inhibitor of the F_1F_0-ATP synthase, was bactericidal against hypoxic, nonreplicating *M. tuberculosis* but had no effect on the $\Delta\Psi$ (6).

ELECTRON DONORS FUELING RESPIRATION IN MYCOBACTERIA

Mycobacterial species use a variety of primary dehydrogenases to deliver electrons from central metabolism into the respiratory chain to generate energy (Fig. 1).

Proton-Pumping Type I NADH Dehydrogenase and Non-Proton-Pumping Type II NADH Dehydrogenase

The major entry point is the transfer of electrons from NADH oxidation to quinone reduction (e.g., ubiquinone or menaquinone). In bacteria, three different types of respiratory NADH dehydrogenases have been identified and characterized on the basis of reaction mechanism, subunit composition, and protein architecture (11). These include the proton-pumping type I NADH dehydrogenase (NDH-1, complex I), non-proton-pumping type II NADH dehydrogenase (NDH-2), and sodium-pumping NADH dehydrogenase (NQR). Weinstein et al. identified genes for two classes of NADH:menaquinone oxidoreductases in the genome of *M. tuberculosis* (12) (Table 1). NDH-1 is encoded by the *nuoABCDEFGHIJKLMN* operon and transfers electrons to menaquinone, conserving energy by translocating protons across the membrane to generate a PMF (Fig. 2). The second class is NDH-2, a non-proton-translocating type II NADH dehydrogenase that does not conserve energy and is present in two copies (Ndh and NdhA) in *M. tuberculosis* (12) (Table 1). Sodium-pumping NADH dehydrogenase has not been reported in mycobacterial genomes.

NDH-1 is composed of 14 subunits (*nuoA-N*), which represent a 15,704-bp operon in *M. tuberculosis*

Table 1 Electron transport chain components and energy-generating machinery of mycobacteria

Operon/gene[a]	Subunits	Rv #	Enzyme name	In vitro essentiality	
				Griffin et al. (26)	Zhang et al. (129)[b]
nuo	nuoABCDEFGHIJKLMN	Rv3145–Rv3158	Type I NADH:menaquinone oxidoreductase	nuoDFGHI = yes	nuoDFGL = yes
ndh		Rv1854c	Type II NADH:menaquinone oxidoreductase	Yes	No
ndhA		Rv0392c	Type II NADH:menaquinone oxidoreductase	No	No
sdh1	sdh1CD	Rv0249c	Succinate:menaquinone oxidoreductase I	Yes	Only sdh1A
	sdh1A	Rv0248c			
	sdh1B	Rv0247c			
sdh2	sdh2C	Rv3316	Succinate:menaquinone oxidoreductase II	No	No
	sdh2D	Rv3317			
	sdh2A	Rv3318			
	sdh2B	Rv3319			
mqo		Rv2852c	Malate:menaquinone oxidoreductase	No	No
pru	pruA	Rv1187	Proline dehydrogenase and pyrroline-5-carboxylate dehydrogenase	Yes	Yes
	pruB	Rv1188			
gpdA1		Rv0564c	Glycerol-3-phosphate dehydrogenase A1	No	No
gpdA2		Rv2982c	Glycerol-3-phosphate dehydrogenase A2	No	No
glpD1		Rv2249c (Rv2250c)	Glycerol-3-phosphate dehydrogenase D1	No	Yes
glpD2	glpD2	Rv3302c	Glycerol-3-phosphate dehydrogenase D2	Yes	Yes
	ldpA	Rv3303c	Dihydrolipoamide dehydrogenase		
cox	coxCMSLDEFG	Rv0376c–Rv0368c	Carbon monoxide dehydrogenase	coxLEF = Yes	coxCL = yes
ald		Rv2780	L-Alanine dehydrogenase	No	No
lldD1	?	Rv0692	L-Lactate dehydrogenase 1	No	lldD1 = yes
	pqqE	Rv0693			
	lldD1	Rv0694			
lldD2	lldD2	Rv1872c	L-Lactate dehydrogenase 2	No	No
		Rv1871c			
		Rv1870c			
Rv0843		(Rv0842)	Pyruvate dehydrogenase E1 component alpha subunit	No	No
		Rv0843			
pdb	pdbA	Rv2497c	Pyruvate dehydrogenase	pdbB = yes	pdbC = yes
	pdbB	Rv2496c			
	pdbC	Rv2495c			

Gene	Genes	Rv loci	Function		
ace		Rv2241	Pyruvate dehydrogenase E1 component	Yes	Yes
lpd		Rv0462	Dihydrolipoamide dehydrogenase	Yes	Yes
hyc	Rv0081 Rv0082 Rv0083 *hycDPQE* Rv0088	Rv0081–Rv0088	Formate hydrogenlyase OR energy-converting hydrogenase-related complex (Ehr) (a.k.a HydTB)	*hycE* and Rv0088 = yes	*hycQE* = yes
qcr	*qcrC* *qcrA* *qcrB*	Rv2194 Rv2195 Rv2196	Cytochrome *bc*$_1$	Yes	Yes
ctaB		Rv1451	Cytochrome *c* oxidase assembly factor	Yes	Yes
ctaC	*ctaC*	Rv2200c (*Rv2199c*)	Transmembrane cytochrome *c* subunit II	Yes	Rv2200c = Yes
ctaD		Rv3043c	Cytochrome *c* oxidase polypeptide I	Yes	Yes
ctaE		Rv2193	Cytochrome *c* oxidase subunit III	Yes	Yes
cyd	*cydA* *cydB* *cydD* *cydC*	Rv1623c Rv1622c Rv1621c Rv1620c	Cytochrome *bd* oxidase	Yes	*cydBC* = yes
nar	*narG* *narH* *narJ* *narI*	Rv1161 Rv1162 Rv1163 Rv1164	Menaquinone:nitrate reductase	No	No
sirA	*sirA* *cysH* *cbe1*	Rv2391 Rv2392 Rv2393	Sulfite reductase APS reductase Ferrochelatase	Yes	Rv2391 and Rv2392 = yes
nirBD	*nirB* *nirD*	Rv0252 Rv0253	Nitrite reductase [NAD(P)H]	No	No
frd	*frdA* *frdB* *frdC* *frdD*	Rv1552 Rv1553 Rv1554 Rv1555	Menaquinone:fumarate reductase	No	*frdA* = yes
atp	*atpIBEFHAGDC* and Rv1312	Rv1303–Rv1312	F$_1$F$_0$ ATP synthase operon	Yes	Yes except *atpI* and Rv1312

[a]In the absence of operons confirmed by previous publications, the TB Database operon prediction algorithms (www.tbdb.org) were used to identify potential operons. Rv loci enclosed in parentheses signify weak operon predictions.
[b]N.B. Genes identified as containing both essential and nonessential regions (a "D call") are referred to as essential here.

(Rv3145 to Rv3158) (Table 1). NuoB, C, D, E, F, and G are peripheral membrane proteins located in the cytoplasmic side, while NuoA, H, J, K, L, M, and N are in the membrane section of the complex with multiple predicted transmembrane domains (from 3 to 16). In contrast, the *nuo* operon has been lost from the genome of the intracellular parasite *Mycobacterium leprae* except for a single remaining *nuoN* pseudogene (13). NDH-1 uses flavin mononucleotide (FMN) and iron-sulfur clusters to transport electrons from NADH to the quinone pool (menaquinone). The release of the two electrons during the NADH oxidation produces

enough energy to pump four protons across the membrane to generate a PMF (Fig. 2). In *M. tuberculosis*, the *nuo* operon is essential for neither growth nor persistence in an *in vitro* Wayne model (6). *Mycobacterium smegmatis* also contains genes for a type I NADH: menaquinone oxidoreductase (*nuoA-N*). However, *M. smegmatis* NDH-I activity is very low, representing about 5% of the NDH-2 activity (14, 15). Rather, increased expression (15-fold) of the *nuo* operon in *M. smegmatis* was observed in a carbon-limited chemostat in response to a slowdown in growth rate (16). In the slow-growing *M. tuberculosis* and *Mycobacterium*

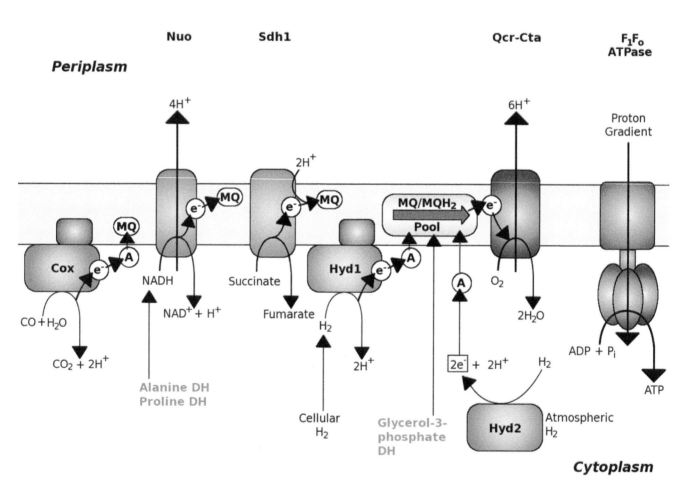

Figure 2 The core respiratory chain of mycobacteria and components upregulated under energy-limiting conditions. During *in vitro* exponential growth, mycobacteria use a classical respiratory chain composed of a type I NADH:menaquinone oxidoreductase (Nuo), succinate:menaquinone oxidoreductase 1 (SDH1), cytochrome *aa₃-bc* supercomplex (Qcr-Cta), and F₁F₀-ATPase. Menaquinone (MQ) is the only quinone present in mycobacterial membranes, and reverse electron transport driven by the PMF is proposed to facilitate the function of SDH1 and similar enzymes (see text). Components in light blue are upregulated in response to energy-limiting conditions (6). Catalysis and electron flow are indicated by arrows. Abbreviations: Cox, carbon monoxide dehydrogenase; Hyd, hydrogenase; DH, dehydrogenase; A, unidentified electron acceptor.
doi:10.1128/microbiolspec.MGM2-0015-2013.f2

bovis strains, NDH-1 activity is 28 to 50% lower than the NDH-2 activity when measured in membrane fractions of mycobacteria (C. Vilchèze and W. R. Jacobs, Jr., unpublished data), suggesting that the major NADH oxidizing activity is mediated by NDH-2. This is supported by the observation that NADH oxidation by mycobacterial membranes is relatively insensitive to complex I inhibitors (12). Notwithstanding this, *M. tuberculosis* mutants lacking only one of the subunits of NDH-1, *nuoG*, have an *in vivo* phenotype, where this gene was critical for host macrophage apoptosis inhibition and mouse virulence (17). This implies that *nuoG* and potentially other subunits of NDH-1 are anti-apoptosis factors and are attractive candidates for vaccine development.

Several studies have reported that *nuo* is downregulated in *M. tuberculosis* during mouse lung infection (18), during survival in macrophages (19), in both non-replicating persistence (NRP)-1 (1% oxygen saturation) and NRP-2 (0.06% oxygen saturation) relative to aerated mid-log growth (18), and upon starvation *in vitro* (20). The transcription of NDH-2 (*ndh*) is also downregulated in *M. tuberculosis* during mouse lung infection, but transcript levels for *ndh* peak during NRP-2 *in vitro*, demonstrating that the pattern of *ndh* regulation is different between *in vivo* and *in vitro* conditions (19). These data are in contrast to *Escherichia coli,* in which NDH-1 is usually associated with anaerobic respiratory pathways (e.g., fumarate) and noncoupling dehydrogenases such as NDH-2 are synthesized aerobically (21).

The non-proton-translocating NDH-2 is a small monotopic membrane protein (50 to 60 kDa) that catalyzes electron transfer from NADH via FAD (noncovalently bound redox prosthetic group) to quinone. No tertiary structural information exists for either the bacterial, plant, or protist NDH-2 enzymes, but the yeast NDH-2 structure was recently solved by two laboratories (22, 23). NDH-2 is widespread in bacteria and the mitochondria of fungi, plants, and some protists. In some cases more than one copy is present (24). The role(s) of multiple type II NADH dehydrogenases in prokaryotes, plants, and parasites is unclear. In some organisms with multiple type II NADH dehydrogenases, one copy appears more essential than the other, pointing to nonredundant functional differences (25, 26). *M. tuberculosis* harbors two copies of NDH-2 (*ndhRv1854c* and *ndhA Rv0392c*) (12) (Table 1), which are well conserved among slow-growing mycobacterial species. In *M. tuberculosis*, Ndh (1,392 bp) and NdhA (1,413 bp) share 65% identity; however, the FAD and NADH binding motifs, $G^{21}SGFGG^{26}$ and

$G^{177}AGPTG^{182}$, are highly conserved. Both proteins contain one transmembrane domain located at the amino acids 385 to 407 and 387 to 409, respectively. The Ndh and NdhA proteins of *M. tuberculosis* have been shown to be functional NADH dehydrogenases that transfer electrons to the quinone pool via a ping-pong reaction mechanism (27). NdhA is not present in *M. smegmatis*, yet the level of NADH oxidation by *M. smegmatis* NDH-2 is several-fold higher than in *M. tuberculosis* or *M. bovis* and represents 95% of the total NADH oxidation measured (15). This might correlate with higher NAD^+ concentrations in *M. smegmatis* compared to *M. bovis* (three times higher, as shown in reference 15). In addition, when *M. smegmatis* and *M. bovis* were transformed with a replicative plasmid containing *ndh,* the ratio NAD^+/NADH doubled (15), further confirming the involvement of *ndh* in the oxidation of NADH into NAD^+.

Several studies have suggested that *ndh* is essential for growth of *M. tuberculosis* (12, 26, 28, 29) (Table 1). Mutations in the *ndh* gene of *M. smegmatis* result in a pleiotropic effect: temperature sensitivity, amino acid auxotrophy, and resistance to the first-line anti-TB drug isoniazid (INH) and its analog and second-line anti-TB drug ethionamide (ETH) (14, 15). The *ndh* mutants had decreased NADH oxidase activity and increased NADH concentration (14, 15). Selection for *ndh* mutants grown on rich media (Mueller Hinton) led to the isolation of *ndh* mutants that were auxotrophic for serine and glycine; this auxotrophy was resolved with complementation by a wild-type copy of *ndh.* The correlation between *ndh* mutations and serine/glycine auxotrophy was attributed to the increase in NADH concentration, which might inhibit the first step in serine/glycine biosynthesis (14). The increase in NADH concentration was also responsible for the high resistance to INH and ETH by competitively inhibiting the binding of the INH-NAD or ETH-NAD adduct to the NADH-dependent enoyl-ACP reductase InhA (15). *ndh* mutants in both slow-growing (*M. bovis* BCG) and fast-growing (*M. smegmatis*) mycobacteria had no growth defect at permissive temperature, although they lost up to 90% of their ability to oxidize NADH. Interestingly, in *M. smegmatis* and *M. bovis,* the levels of NAD^+ cofactor stayed relatively constant despite overexpression of *ndh* or mutations in *ndh,* which highly reduced their NADH oxidation capability (15), suggesting that the NAD^+ pool is tightly regulated in mycobacteria to maintain essential biochemical functions.

NDH-2 has not been reported in mammalian mitochondria, leading to the proposal that these enzymes may represent a potential drug target for the treatment of human pathogens (6, 12, 27, 30, 31) and intracellular parasites (32). Despite the potential of NDH-2 as a drug target, no potent nanomolar inhibitors of NDH-2 are known. Several classes of compounds are proposed to target the enzyme at micromolar concentrations (e.g., phenothiazine analogues, platanetin, quinolinyl pyrimidines) (12, 27, 33), but the mechanism of inhibition remains unknown. Despite poor *in vitro* activity, drugs of the phenothiazine family (trifluoroperazine, chlorpromazine) have potent activity *in vitro* against drug-susceptible and drug-resistant *M. tuberculosis* strains (34, 35). A phenothiazine analog was also tested in a mouse model of acute *M. tuberculosis* infection and was found to reduce by 90% the *M. tuberculosis* bacterial load in the lungs after 11 days of treatment compared to a 3- to 4-log reduction in CFUs with the INH or rifampin control (12). From a library of microbial products, two new compounds, scopafungin and gramicidin S, were identified as inhibitors of *M. smegmatis* NDH-2, with IC_{50} values better than trifluoperazine (36). There is a crucial need for new drug targets to inhibit *M. tuberculosis,* and the electron transport chain is a very attractive avenue. However, reduction in NDH-2 activity has been linked to INH and ETH resistance in both slow- and fast-growing mycobacteria (15), and phenothiazines have been shown to be antagonistic with INH (30). Therefore, the development of NDH-2 inhibitors will have to ascertain that interference with current drug therapy does not occur.

Some interesting questions arise from these observations. Why do mycobacteria use type II NADH dehydrogenases to recycle NADH, when they could continue to use the energy-conserving and PMF-generating NDH-I? One potential explanation is that because type II NADH dehydrogenases are non-proton-translocating, they will not be impeded by a high PMF, which could ultimately slow down metabolic flux due to backpressure on the system. This mechanism is akin to a "relief valve" that would allow for a higher metabolic flux and ultimately higher rates of ATP synthesis, at the expense of low energetic efficiency of the respiratory chain. Second, why is *ndh* an essential gene when mycobacteria could also use *nuo*? The fact that *ndh* is essential implies that mycobacteria do not have another mechanism to recycle NADH during normal aerobic growth. Alternatively, this is the only NADH dehydrogenase that operates under these growth conditions, and the activity of this enzyme is essential for maintaining an energized membrane. Compounds that target NDH-2 are bactericidal toward hypoxic nonreplicating *M. tuberculosis,* suggesting that the respiratory chain is essential for the recycling of NADH under these conditions (6).

Multiple Succinate Dehydrogenases and Fumarate Reductase: An Essential Link Between Central Metabolism and Respiration in Mycobacteria

Succinate dehydrogenase forms complex II of the respiratory chain and couples oxidative phosphorylation to central carbon metabolism by being an integral part of the tricarboxylic acid (TCA) cycle (Fig. 2). This enzyme catalyzes the oxidation of succinate to fumarate wherein two electrons are transferred to quinol. The reverse reaction can be catalyzed by fumarate reductase (FRD), which is involved in anaerobic respiration (Fig. 3). FRD and succinate dehydrogenase are closely related enzymes, and the reaction catalyzed cannot be predicted based on the primary sequence alone. Most mycobacteria harbor two annotated succinate dehydrogenases, SDH1 and SDH2 (Table 1). SDH2 has high homology to SDH enzymes from other species and is encoded by four genes, *sdhC*, *sdhD*, *sdhA*, and *sdhB*. The genes *sdhA* and *sdhB* encode for the cytoplasmic part of the enzyme where the succinate to fumarate reaction takes place (SdhA), and electrons are shuttled via three iron-sulfur clusters (SdhB) to the membrane subunits SdhC and SdhD, which catalyze the electron transfer to menaquinone. The SdhA (encoded by *Rv0248c*) and SdhB (*Rv0247c*) subunits of SDH1 are similar to the common SDH enzyme. However, the other two genes (*Rv0250c* and *Rv0249c*) in the operon show no homology to membrane-bound components of SDH and FRD and are specific to the phylum *Actinobacteria*. Gene expression data show that all four genes of SDH1 in *M. smegmatis* are expressed in concert (16). As an exception among mycobacteria, *M. tuberculosis* encodes a third complex (*frdABCD*) that is annotated as an FRD (Table 1) and is absent in all other pathogenic strains such as *Mycobacterium avium paratuberculosis*, *Mycobacterium marinum*, *Mycobacterium ulcerans*, and *M. leprae*. Succinate dehydrogenase activity has been measured in *M. tuberculosis* as well as many other mycobacterial species (37, 38), but it is still unclear which of the three enzymes, or all, are responsible for this activity.

Mycobacteria utilize menaquinone/menaquinol (MQ/MQH₂) as a conduit between electron-donating and -accepting reactions (Fig. 1). Menaquinone has a lower midpoint redox potential ($E_m = -74$ mV) compared to ubiquinone ($E_m = +113$ mV) and is ideally poised to donate electrons to fumarate during anaerobic conditions

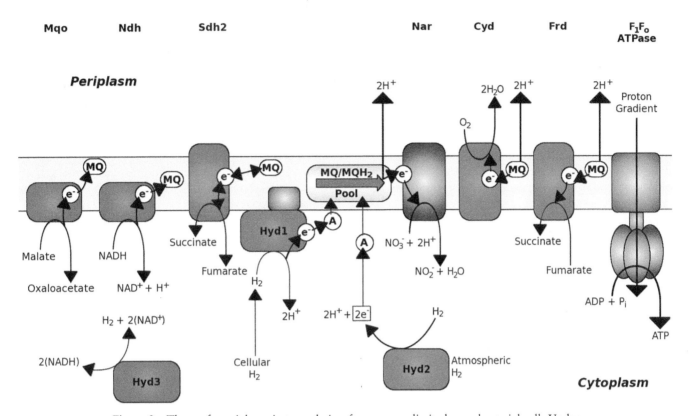

Figure 3 The preferential respiratory chain of an oxygen-limited mycobacterial cell. Under low-oxygen conditions, a diverse response utilizing alternate electron donors and acceptors, energy-conserving enzymes, and a high-affinity terminal oxidase permits survival under hypoxic conditions. Components in red are upregulated under microaerobic conditions (6). Catalysis and electron flow are indicated by arrows. The possible PMF-driven reverse electron flow of Sdh2 is not shown, for clarity. Abbreviations: Mqo, malate:menaquinone oxidoreductase; Ndh, type II NADH:menaquinone oxidoreductase; Sdh2, succinate:menaquinone oxidoreductase 2; Nar, nitrate reductase; Cyd, cytochrome *bd* oxidase; Frd, FRD; Hyd, hydrogenase; MQ, menaquinone; A, unidentified electron acceptor. doi:10.1128/microbiolspec.MGM2-0015-2013.f3

(39). This means that the SDH reaction in mycobacteria (succinate oxidation to fumarate) should have an unfavorable free energy due to reverse electron flow and proton uptake to drive this reaction. It is tempting to propose that the unusual subunits of SDH1 could be the result of structural specializations in the transmembrane region to facilitate reverse electron flow from succinate to menaquinone. In fact, SDH1 and SDH2 have been shown to be differentially expressed under energy- or oxygen-limiting conditions in *M. smegmatis* (16). Under energy-limiting conditions SDH1 was upregulated 4-fold, while SDH2 was downregulated 3-fold. Under oxygen-limiting conditions, SDH1 was downregulated 30-fold, while SDH2 was upregulated 2-fold. This indicates that SDH1 could be the dedicated succinate dehydrogenase, and SDH2 catalyzes FRD activity, which is important for survival under hypoxia. This hypothesis is supported by several reports in

the recent literature. A transposon-site hybridization (TraSH) screen in *M. tuberculosis* suggested that SDH1 but not SDH2 was essential for growth under aerobic conditions on standard medium (26). In contrast, in a TraSH screen selecting for mutants that continue to replicate under hypoxic conditions, Baek and coworkers (40) found SDH2 mutants overrepresented. These results suggest that SDH2 has a pivotal role in the transition of *M. tuberculosis* from aerobic to hypoxic conditions and supports the hypothesis of its being an FRD.

In fact, it has been proposed that fumarate may be an important endogenous electron acceptor for energy production and maintenance of redox balance (oxidation of NADH to NAD⁺) in hypoxic nonreplicating mycobacteria (6). Interestingly, the use of fumarate as an electron acceptor in *E. coli* requires complex I, and expression of the *nuo* operon is stimulated by the presence of fumarate (41). This stands in direct contrast to *M. tuberculosis*, where the *nuo* operon seems to be silent under anaerobic conditions (18). In a recent study it was shown by ^{13}C flux analysis that *M. tuberculosis* grown under hypoxia metabolizes glucose through a reverse TCA cycle to generate succinate as an excreted fermentation end product (42). However, the metabolic flux from fumarate to succinate was unchanged in an *M. tuberculosis* FRD deletion mutant, suggesting that one of the remaining putative succinate dehydrogenases (most likely SDH2) could catalyze this reaction. A more recent study suggests that the glyoxylate shunt, and not the reverse TCA, is used to metabolize both glycolytic and fatty acid carbon sources in response to oxygen limitation, and this route also produces succinate as its metabolic end product (43). The authors propose that during oxygen limitation large amounts of succinate are produced by this pathway that are used to sustain the membrane potential, ATP synthesis, and TCA cycle precursors akin to a "metabolic battery" (43). Moreover, because of the near neutral midpoint potential of the succinate/fumarate redox couple (+30 mV), succinate is able to bridge both oxidative and fermentative metabolic schemes depending on electron acceptor availability (43).

The role of the *frdABCD* operon in the *M. tuberculosis* complex is not clear. Increased expression of this operon has been shown during carbon starvation (20), oxygen depletion (44), and in macrophages (19), suggesting a role for this enzyme in persistence.

Alternative Electron Donor Utilization During Starvation and Hypoxia

During carbon starvation and slow growth, mycobacteria switch to alternative electron donors (16, 20, 45) (Fig. 2). Proline dehydrogenase is upregulated under both energy-limiting conditions and hypoxia (7, 16) and is increasingly being recognized as a critical amino acid in cellular bioenergetics and redox control (46). Proline can be utilized as an electron donor as well as a carbon and nitrogen source (Fig. 2). The degradation of proline occurs by means of two enzymes: proline dehydrogenase (PRODH) and pyrroline-5-carboxylate dehydrogenase (P5CDH). These two enzymes catalyze the oxidation of proline to glutamate with four electrons transferred to the respiratory chain (46). In the first step FAD is reduced to $FADH_2$, while in the second step NAD^+ is reduced to NADH. In some bacteria, PRODH and P5CDH are monofunctional enzymes, but in the majority of bacterial species they are fused into one protein called proline utilization A flavoenzyme PutA (47).

In mycobacteria, PRODH and P5CDH are predicted to be monofunctional enzymes (46). The genes encoding PRODH (*putB*, *Rv1188*) and P5CDH (*putA*, *rocA*, *Rv1187*) are expressed as part of an operon (7). It has been shown that *M. smegmatis* can grow on proline as the sole carbon and energy source and also that PRODH is an important electron donor, under both energy-limiting conditions and hypoxia (7, 16). The same authors showed that proline metabolism in mycobacteria is regulated by a unique membrane-associated transcriptional regulator called PruC (Rv1186c), encoded upstream of *pruA*. The *pruAB* operon with its upstream regulator *pruC* is highly conserved in mycobacteria, with the exception of *M. leprae*. Recent proteomics data on *M. avium paratuberculosis* show that protein levels of RocA (PutA) are elevated in the intestinal tissues of cows (48). The authors propose that *M. avium paratuberculosis* has adapted to utilize proline as a carbon and nitrogen source due to the high abundance of this amino acid in the plant material that is eaten by such ruminants. Mycobacteria also encode pyrroline-5-carboxylate reductase (encoded by *proC*) that catalyzes the reverse reaction of *putA*, converting pyrroline-5-carboxylate to proline. Interestingly, a *proC* mutant of *M. tuberculosis* was avirulent in immunocompetent mice, but this was not due to its inability to proliferate intracellulary because bacterial loads increased in the mouse lungs after 20 days postinfection (49).

Hydrogenases catalyze the reversible oxidation of molecular hydrogen: $2H^+ + 2e^- \rightarrow H_2$ and play a central role in energy metabolism of bacteria, archaea, and eukarya (50). Under physiological conditions, hydrogenases couple H_2 oxidation to respiration (Knallgas reaction) or reduce protons as a way to dispose of surplus reducing equivalents. Four different types of hydrogenases are found in mycobacteria and are annotated to be of the NiFe type. *M. smegmatis* harbors three (designated Hyd1, 2, and 3) of the four hydrogenase complexes (16). Hyd1 aligns closely with group 2a uptake hydrogenases of the cyanobacteria such as *Nostoc*, indicating that it oxidizes H_2 (51). In contrast, Hyd3 is closely related to the group 3 cytoplasmic bidirectional hydrogenases. Hyd2 appears to be a founding member of the group 5 high-affinity hydrogenases (52, 53). Hyd1, Hyd2, and Hyd3 are all soluble hydrogenases and are found in mycobacteria of the slow-growing and fast-growing type, as well as pathogenic and nonpathogenic mycobacteria (16). However, the fourth putative hydrogenase is only found in pathogenic mycobacteria (including *M. tuberculosis* complex) and seems to be restricted to slow growers (Table 1).

It shows homology to group 4 membrane-bound H_2 evolving hydrogenases. It has been shown that *M. smegmatis*, among other mycobacterial species, can oxidize molecular hydrogen in the presence of carbon monoxide (CO), implying that *M. smegmatis* expresses a functional hydrogenase (54). To date, no studies have reported on the ability of mycobacteria to produce hydrogen. Gene expression studies suggest that Hyd1 and Hyd2 are used during nutrient starvation as an alternative electron source (Fig. 2), while Hyd3 and the membrane-bound hydrogenase are more likely to have a function in disposing of electrons under anaerobic conditions (16) (Fig. 3). A knockout mutant of Hyd2 in *M. smegmatis* showed reduced biomass production when grown on complex medium under atmospheric conditions (16), and its homolog in *Streptomyces* sp. was shown to facilitate hydrogen oxidation (53). These data suggest that Hyd2 oxidizes hydrogen at very low concentrations, which fits with its purported role as a high-affinity hydrogenase (52).

Carbon monoxide dehydrogenase (CO-DH) is responsible for the oxidation of CO to carbon dioxide (CO_2) in carboxydobacteria, which grow on CO as a sole source of carbon and energy (55). Carboxydobacteria catalyze the oxidation of CO to CO_2 by the following reaction: $CO + H_2O \rightarrow CO_2 + 2H^+ + 2e^-$. Several pathogenic and nonpathogenic mycobacteria including *M. tuberculosis* are known to possess CO-DH genes. It has been shown that *M. tuberculosis* H37Ra, which possesses CO-DH activity, can grow on CO as a sole source of carbon and fuel for energy generation (56) (Fig. 2).

Glycerol-3-phosphate dehydrogenase catalyzes the oxidation of glycerol-3-phosphate to dihydroxy-acetone phosphate and reduces either quinone or NADP (57). In *E. coli*, glycerol-3-phosphate is used either as a precursor in the biosynthesis of phospholipids or as a carbon source for energy supply (58). *M. tuberculosis* possesses genes for four predicted glycerol-3-phosphate dehydrogenases, but their role and function in mycobacterial respiration remain unknown (Table 1).

The membrane-associated malate quinone oxidoreductase (MQO) oxidizes malate to oxaloacetate and transfers the reducing equivalents to menaquinone (59, 60) (Fig. 3). *M. tuberculosis* harbors a copy of MQO and a cytoplasmic NAD⁺-dependent malate dehydrogenase (MDH) (61, 62). The function and role of MQO in mycobacterial respiration is unknown. *M. smegmatis* lacks an MDH homolog, and it was shown that MQO responded to low oxygen concentration with a 4-fold increase in gene expression (16). Mutants of *Corynebacterium glutamicum* defective in NDH-2 activity are able to grow despite the loss of all membrane-bound NADH dehydrogenase activity (59). The authors propose a reaction scheme whereby electron transfer from NADH to menaquinone is catalyzed by the sequential action of MDH and MQO (membrane-bound). MDH can reduce oxaloacetate with NADH to malate, and then malate is reoxidized to oxaloacetate by MQO using menaquinone as an electron acceptor (59, 60). Support for this model comes from the observation that a Δ*mqo/ndh* double mutant failed to grow under conditions where the Δ*ndh* mutant grew. Furthermore, *M. smegmatis ndh* mutants could be complemented by *M. bovis* BCG *mdh*, suggesting that such a reaction scheme might operate in mycobacteria (15).

TERMINAL ELECTRON ACCEPTOR UTILIZATION

During aerobic respiration, energy is conserved by the generation of a PMF across a proton-impermeable membrane. The electron transport chain components are membrane bound and asymmetrically arranged across the membrane to achieve net consumption of protons from the cytoplasm and net release of protons on the outside of the cell. An important part of all electron transport chains is the terminal respiratory oxidases. In order for mycobacteria to utilize oxygen efficiently and obtain the maximum growth yield on a particular carbon and energy source, there must be coordinate regulation of terminal respiratory oxidase expression/activity. For example, in *E. coli*, cytochrome *bo* (K_m for oxygen in the micromolar range) and cytochrome *bd* (K_m for oxygen in the nanomolar range) (63, 64) are coordinately regulated by the ArcBA system and transcriptional regulator FNR (65). Cytochrome *bo* is synthesized at high oxygen tension (optimal from 15 to 100% air saturation) and repressed as the oxygen concentration decreases (66). This coincides with the induction of cytochrome *bd* at 7% air saturation, which is turned off (FNR-mediated repression) once the cells enter anaerobiosis (65, 66). Mycobacteria adopt regulation of oxidase expression to match oxygen supply. Under conditions of low oxygen tension (ca. 1% air saturation), cytochrome *bd* is induced in *M. smegmatis* as the transition to anaerobiosis is approached (67), a value that is 10-fold lower than that observed in *E. coli* (ca. 10% air saturation). In *M. tuberculosis*, cytochrome *bd* is upregulated in the early stages of NRP-1 (i.e., decreasing oxygen) (68), in NRP-2 (18), and in response to nitric oxide (NO) (18). Intriguingly, mRNA levels of *cydA* increased transiently about 7-fold after 30 days in mouse lungs (18).

Mycobacteria appear to adopt a strategy whereby downregulation or slowing of metabolism occurs as cells enter NRP-1 and NRP-2. Transcriptional analysis of *M. tuberculosis* in the macrophage (phagosomal environment) has revealed that NDH-1, menaquinol-cytochrome *c* oxidoreductase, and the ATP synthase are all downregulated when compared to cells growing exponentially, suggesting that there is a reduced need for energy generation during bacteriostasis, i.e., the growth state of intraphagosomal *M. tuberculosis* (19). Consistent with these observations is the repression of these operons during starvation. In contrast, FRD, nitrate reductase, and NDH-2 are all upregulated under these conditions (19). While these proteins do not appear to contribute to increased energy conservation, it has been suggested that they may play a pivotal role in the recycling of NAD$^+$ as a result of β-oxidation of fatty acids (19).

Complex III and IV of the *M. tuberculosis* Electron Transport Chain

M. tuberculosis harbors genes for a cytochrome *c* pathway consisting of a menaquinol-cytochrome *c* oxidoreductase termed the bc_1 complex (encoded by the *qcrCAB* operon, complex III) and an aa_3-type cytochrome *c* oxidase (encoded by the genes *ctaBCDE*, complex IV) belonging to the heme-copper respiratory oxidase family (69–71) (Table 1). The bc_1 complex (menaquinol-cytochrome *c* oxidoreductase) consists of redox groups comprising a 2Fe/2S center, located on a Rieske protein (QcrA); two *b*-type hemes (low and high potential) located on a single polypeptide (QcrB); and the heme of cytochrome c_1 (QcrC). For every two electrons passing from quinol to cytochrome *c*, complex III releases four protons into the periplasmic side of the membrane ($4H^+/2e^-$) (Fig. 2). The iron sulfur subunit QcrA has three transmembrane helices and sequence motifs characteristic of 2Fe-2S Rieske iron-sulfur proteins (e.g., CSHLGC and CPCH). In contrast to bovine heart cytochrome *b*, QcrB from *M. tuberculosis* has a 120-amino-acid extension at the C-terminus as noted for *C. glutamicum* and *Streptomyces* species (72). The QcrB subunit has been recently shown to be the target of novel imidazo[1,2-*a*]pyridine inhibitors that are active against *M. tuberculosis* at low micromolar concentrations (73).

As first reported by Niebisch and Bott (72), the QcrC subunit (cytochrome c_1) harbors two heme-binding signatures (CXXCH motifs) for *c*-type cytochromes (e.g., CVSCH and CASCH), suggesting a covalent di-heme(*bcc*) configuration. Mycobacterial

genomes do not appear to harbor genes for either a soluble or membrane-bound cytochrome *c*, suggesting that the *bcc* complex interacts directly with the aa_3-type cytochrome *c* oxidase to achieve electron transfer and function as a supercomplex. Niebisch and Bott (72, 74) suggest a supercomplex mechanism for *C. glutamicum* in which the heme group of QcrC takes over the function of a separate cytochrome *c* in electron transfer to cytochrome aa_3 oxidase, forging a close contact relationship between cytochrome c_1 and the Cu_A site in the subunit II of cytochrome aa_3 oxidase. Supercomplexes between bc_1 and aa_3 have been reported for *C. glutamicum* (72), *Paracoccus denitrificans* (75), and thermophilic strain PS3 (76). A functional association between *bcc* and aa_3 has been shown for the complex in the *M. smegmatis* family (71).

The aa_3-type cytochrome *c* oxidase in *M. tuberculosis* appears to be encoded by four genes, *ctaBCDE* (Table 1). The *ctaE* gene is located immediately upstream of the *qcrCAB* operon, but the other three genes are located in three different locations on the chromosome and in some cases appear to be in operons with other genes (Table 1). The four subunits of the *M. tuberculosis* aa_3-type cytochrome *c* oxidase include CtaB (a cytochrome *c* oxidase assembly factor), the catalytic subunits CtaD (cytochrome *c* oxidase subunit I containing heme *a*, a_3, and Cu_B) and CtaC (cytochrome *c* oxidase, subunit II location of copper A [Cu_A]), and CtaE (subunit III). For every two electrons passing through complex IV, four protons are taken up from the cytoplasm and two are released into the periplasm. The oxidase acts as a proton pump with a stoichiometry of $2H^+/2e^-$ (Fig. 2).

The bc_1-aa_3 pathway appears to be the major respiratory route in mycobacteria under standard aerobic culturing conditions (70). Matsoso et al. (70) have demonstrated that disruption of this pathway in *M. smegmatis* is accompanied by a constitutive upregulation of the cytochrome *bd*-type menaquinol oxidase. In *M. tuberculosis*, the bc_1-aa_3 pathway is essential for growth, suggesting an inability of this bacterium to adapt in a manner analogous to *M. smegmatis* (70). The aa_3 branch is proposed to contain three *ctaD* alleles in *M. smegmatis* versus the one in *M. tuberculosis*, suggesting the existence of alternative isoforms of cytochrome *c* oxidase in *M. smegmatis* (67).

Cytochrome *bd*-Type Oxidase

M. tuberculosis and other mycobacterial species harbor genes for the cytochrome *bd*-type menaquinol oxidase (*cydAB*) (67) (Table 1). The cytochrome *bd* branch is bioenergetically less efficient (non-proton-translocating) and is synthesized at low oxygen tensions in mycobacteria (67). In addition to cytochrome *bd*, the *M. smegmatis* respiratory chain has been proposed to contain a third possible respiratory branch terminating in the YthAB (*bd*-type) menaquinol oxidase (67). The existence of two cytochrome *bd*-type oxidases (I and II) is not unprecedented in bacteria (77), and recent work has reported that cytochrome *bd*-II in *E. coli* is able to generate a PMF with a H^+/e^- ratio of 1.0 (78). Cytochrome *bd* mutants of *E. coli* have been shown to have a pleiotropic phenotype: they are sensitive to H_2O_2, nitric oxide, temperature, and iron (III) chelators and are unable to exit from stationary phase and resume aerobic growth at 37°C (77). A *cydA* mutant of *M. smegmatis* has been generated, but the phenotypes are very subtle compared to *E. coli*. The *M. smegmatis* *cydA* mutant showed a reduced growth rate at 0.5 to 1% air saturation and also a 10-fold difference in CFU after 140 h of growth in the presence of cyanide at 21% air saturation when cocultured with wild-type *M. smegmatis* (67).

The *cydAB* genes appear to be in an operon with *cydDC* in mycobacterial species (Table 1), an arrangement similar to that found in *Bacillus*. In *E. coli*, *cydAB* and *cydDC* form two discrete operons, *cydDC* mutants are defective in cytochrome *bd* assembly, and the periplasmic space is more oxidized in the mutant versus wild type (79). In *E. coli*, the CydDC protein (ABC transporter) has been reported to pump glutathione and cysteine into the periplasm to maintain redox homeostasis (80). The role of the *cydDC* genes in mycobacteria is unknown, but evidence exists that CydDC plays a role during mycobacterial persistence *in vivo*. A *cydC* mutant of *M. tuberculosis* showed reduced ability to survive the transition from acute to chronic infection in mice (18), and Dhar and McKinney have reported that CydC plays a role in mycobacterial persistence in INH-treated mice (81).

ELECTRON ACCEPTOR UTILIZATION UNDER HYPOXIA

In the absence of oxygen, alternative electron acceptors (e.g., nitrate and fumarate) are available for mycobacterial metabolism, but none of these electron acceptors has been shown to support growth of mycobacterial species. These potential electron acceptors may play an important role in the disposal of reducing equivalents in the absence of oxygen.

NO is generated in large amounts within the macrophages and restricts the growth of *M. tuberculosis*. Nitrate can be produced by oxidation of NO and is an alternative source of nitrogen for bacteria within the human host. Early work in *E. coli* had suggested that *narK* was involved only in nitrite export (82), so the homologous *narK2* in *M. tuberculosis* was annotated as a "nitrite extrusion protein." More recent work with an *E. coli* *narK* *narU* double mutant indicated that the two proteins could transport nitrate into and nitrite out of the cell (83, 84). In *M. tuberculosis*, four genes, *narK1* through *narK3* and *narU*, are homologous to *narK* and *narU* (Table 1). Since *M. tuberculosis* is unable to reduce nitrite, which could accumulate to toxic levels, it must be exported out of the cell. The *M. tuberculosis* *narK2* was shown to complement this *E. coli* double mutant, supporting a role for *narK2* in nitrate reduction by coding for a transporter of nitrate into and nitrite out of the cell (85). Nitrate reduction by *M. tuberculosis* is regulated by control of nitrate transport into the cell by NarK2. It is proposed that NarK2 senses the redox state of the cell, possibly by monitoring the flow of electrons to cytochrome oxidase, and adjusts its activity so that nitrate is transported under reducing, but not under oxidizing, conditions (86). Inhibition of nitrate transport by oxygen has been documented in other bacteria (87). It is intriguing that *M. tuberculosis*, classified as an obligate aerobe, should have such intricate control of an anaerobic enzyme system. Transcription of *narK2* is controlled by DosR/DevR, which responds to O_2, NO, and CO (88–90). Both the transcription of the *narK2* gene and the activity of NarK2 are controlled by similar signals (86).

M. tuberculosis contains genes (*narGHJI*, Rv1161–1164) that code for a putative membrane-bound molybdenum-containing nitrate reductase complex similar to the corresponding *narGHJI* operon of *E. coli* (Table 1). Moreover, the *narGHJI* operon of *M. tuberculosis* is able to functionally complement a *nar* mutant of *E. coli* to grow on glycerol and reduce nitrate anaerobically (85). Importantly, however, the expression of the *narGHJI* operon in *M. tuberculosis* is not upregulated in response to either hypoxia or stationary phase (85). Sohaskey and Wayne demonstrated that overexpression of recombinant *M. tuberculosis* nitrate reductase in either *M. tuberculosis* or *M. smegmatis* (low nitrate reductase activity) does not confer the ability of these cells to grow anaerobically; i.e., no growth of either species is observed with nitrate anaerobically, even though the nitrate reductase activity of whole cells increases (85). The genome of *M. tuberculosis* also lacks orthologs of transcription regulator FNR which, in combination with NarL, are responsible for the transcriptional activation of the *narGHJI* operon by anaerobic conditions in *E. coli* (21). A putative NarL

(Rv0884c) has been identified in *M. tuberculosis*, but the promoter of the *narGHJI* lacks consensus-like binding sites for this regulatory protein. Based on these observations, it is apparent that this enzyme does not support anaerobic growth of mycobacteria, and therefore the role of this enzyme in the physiology of mycobacteria is unclear. Given the proposed membrane-bound location of the enzyme and the proton-translocating activity via a redox loop of the *E. coli* enzyme, perhaps the primary role of the mycobacterial enzyme is to generate a PMF when the concentration of oxygen is low, and hence its activity increases but not its expression.

An alternative role for nitrate reductase may be maintaining the redox balance of the cell during conditions of hypoxia. Sohaskey (91) has reported that exogenously supplied nitrate has no effect on long-term persistence during gradual oxygen depletion but played an important role during rapid adaptation to hypoxia (<18 h). This effect required a functional nitrate reductase, suggesting that nitrate reduction may play a role in protecting cells during sudden changes in oxygen concentration, leading to disruption of aerobic respiration. The same author (86) has proposed a role for NarK2 in sensing the redox state of the cell such that nitrate is transported into the cell under reducing, but not oxidizing, conditions. Lastly, Tan et al. (92) proposed a role for nitrate reductase in protecting *M. tuberculosis* from mild acid challenge under hypoxic conditions.

M. tuberculosis is unable to reduce nitrite, and the genome was proposed to harbor genes for a putative ferredoxin-dependent nitrite reductase (NirA) and a NAD(P)H nitrite reductase (NirBD). However, a subsequent study has reported that the *nirA* gene is misannotated and is in fact *sirA*, a sulfite reductase shown to be essential for growth on sulfate or sulfite as the sole sulfur source (93) (Table 1). The *nirBD* genes are operonic and nonessential for growth, and unlike the *narGHJI* genes, no role in resistance to acid stress under hypoxia was noted for NirBD (92). These data support the proposal that nitrite produced by nitrate reductase is not converted to ammonium, but instead is exported from the cell by NarK2.

The evolution of H_2 is a mechanism commonly employed by anaerobic bacteria to recycle reducing equivalents obtained from anaerobic degradation of organic substrates. However, even strictly aerobic bacteria have been shown to produce H_2 under anaerobic conditions (94). In the family of *Mycobacteriaceae*, two potential hydrogen-evolving hydrogenases are present. One belongs to group 2 cytoplasmic bidirectional hydrogenases and can be found in some slow-growing (e.g., *M. marinum* and *M. kansasii*) and fast-growing (e.g., *M. smegmatis*) mycobacteria. In *M. smegmatis*, this enzyme is termed Hyd3 and has been shown to respond strictly to oxygen-limiting conditions with up to a 50-fold increase in gene expression (16). The second type is a yet uncharacterized enzyme that shows homology to membrane-bound group 4 hydrogenases. This putative formate-hydrogen lyase is enocoded in *M. tuberculosis* (*viz.*, Rv0082 to Rv0087) and is upregulated during infection of human macrophage-like THP-1 cells (95). Moreover, the transcription of the early genes of the *hycP/hycQ*-containing operon was shown to be upregulated during anaerobic adaptation in several studies (44, 68, 90, 96, 97). This gene cluster shows homology to components of hydrogenase 4 and 3 complexes of *E. coli*, the latter of which has been shown to catalyze hydrogen evolution at acidic pH (98). Upon closer inspection, the predicted [NiFe]-hydrogenase of *M. tuberculosis* appears to form part of an Ehr (Ech hydrogenase-related) complex (99, 100). Ehr enzymes lack the cysteine residues needed to ligate a [NiFe] center, which is the catalytic center of hydrogenases. Additionally, no maturation factors to synthesize these complex centers are present in *M. tuberculosis*.

Recently, it was shown that the operon Rv0081-Rv0087 is positively regulated by the two two-component signal transduction systems DosRS-DosT and MprAB and negatively regulated by Rv0081, a member of the ArsR/SmtB family of metal-dependent transcriptional regulators (101). DosRS-DosT is a redox-sensing regulatory system that is important during the adaptation to hypoxic conditions. No function has yet been proposed for Ehr enzymes, but their homology to hydrogenases and to proton-pumping type I NADH dehydrogenases points to a role in energy metabolism.

ATP SYNTHESIS BY THE F_1F_0-ATP SYNTHASE

In *M. tuberculosis* and other mycobacterial species, ATP is synthesized via substrate level phosphorylation and oxidative phosphorylation using the membrane-bound F_1F_0-ATP synthases (encoded by the *atpBEFHAGDC* operon, Rv1304-1311) (Table 1). In *M. tuberculosis*, the *atp* operon is downregulated during growth in macrophages (19), in the mouse lung, and in cells exposed to NO or hypoxia (18). The *atp* operon of *M. bovis* BCG and *M. smegmatis* is downregulated in response to slow growth rate (16, 45). When slow-growing cells of *M. smegmatis* (70 h doubling time), with low levels of *atp* operon expression, are exposed to hypoxia (0.6% oxygen saturation), the *atp* operon is upregulated

3-fold, suggesting an important role for this enzyme during adaptation to hypoxia (16).

The F_1F_0-ATP synthase catalyzes ATP synthesis by utilizing the electrochemical gradient of protons to generate ATP from ADP and inorganic phosphate (P_i) and operates under conditions of a high PMF and low intracellular ATP. The enzyme is also capable of working as an ATPase under conditions of high intracellular ATP and an overall low PMF (102). As an ATPase, the enzyme hydrolyzes ATP while pumping protons from the cytoplasm to the outside of the cell. The ATP synthase of mycobacteria has been studied in detail at a biochemical level in *M. phlei* and shown to exhibit latent ATPase activity (10). ATPase activity could be activated by trypsin treatment and magnesium ions, but the mechanism of activation was not elucidated. Recent experiments with inverted membrane vesicles of *M. bovis* BCG and *M. smegmatis* demonstrate latent ATPase activity that could be activated by methanol and the PMF, suggesting regulation by the epsilon subunit and ADP inhibition (9). The reason for the extreme latency in ATP hydrolysis of the mycobacterial ATP synthase is unknown but may represent an adaptation to function at low PMF and under hypoxia. Hypoxic nonreplicating cells of *M. tuberculosis* generate a PMF in the order of −100 mV, and the ATP synthase inhibitor TMC207 is bactericidal toward these cells, demonstrating that the ATP synthase still continues to function at relatively low PMF (6).

The F_1F_0-ATP synthase in *M. tuberculosis* and *M. smegmatis* has been shown to be essential for optimal growth (29, 103). In other bacteria, the F_1F_0-ATP synthase is dispensable for growth on fermentable carbon sources (104, 105), where increased glycolytic flux can compensate for the loss of oxidative phosphorylation. This strategy does not appear to be exploited by *M. smegmatis*: the F_1F_0-ATP synthase is essential for growth even on fermentable substrates, suggesting that ATP production from substrate level phosphorylation alone, despite increased glycolytic flux, may be insufficient to sustain growth of these bacteria (103). This may be due to an extraordinarily high value for the amount of ATP required to synthesize a mycobacterial cell, a possibility that requires further investigation (106). Alternatively, in conjunction with a high ATP demand for growth, the ATP synthase may be an obligatory requirement for the oxidation of NADH by providing a sink for translocated protons during NADH oxidation coupled to oxygen reduction (103). Such strict coupling would imply that mycobacteria do not support uncoupled respiration; either they lack a conduit for proton reentry in the absence of the F_1F_0-ATP

synthase or they are unable to adjust the proton permeability of the cytoplasmic membrane to allow a futile cycle of protons to operate. In this context, the cytoplasmic membrane of *M. smegmatis* has been shown to be extremely impermeable to protons (107). The ATP synthase of the close mycobacterial phylogenetic relative *C. glutamicum* is nonessential for growth on fermentable carbon sources, and Δ*atp* mutants of this bacterium show enhanced rates of glucose uptake, oxygen consumption, and excretion of pyruvate into the growth medium, suggesting that substrate level phosphorylation alone can sustain growth of this bacterium (108).

Several new antitubercular compounds have been reported that target oxidative phosphorylation in mycobacteria (12, 109, 110). The most promising compounds clinically, the diarylquinolines, have been shown to target the F_1F_0-ATP synthase and inhibit ATP synthesis (109, 111, 112). Genome sequencing of both *M. tuberculosis* and *M. smegmatis* mutants that are resistant to diarylquinolines (i.e., TMC207) revealed that the target of these compounds is the oligomeric *c* ring (encoded by *atpE*) of the enzyme (109, 113, 114). The purified *c* ring from *M. smegmatis* binds TMC207 with a K_D of 500 nm, and modeling/docking and kinetic studies suggest that TMC207 blocks rotary movement of the *c* ring during catalysis by mimicking key residues in the proton transfer chain (115, 116). Further investigations with inverted membrane vesicles of *M. smegmatis* and TMC207 have revealed that TMC207 acts independently of the PMF and that electrostatic forces play an important role in the interaction of the drug with the ATP synthase (116).

When mycobacterial cells (growing or nongrowing) are treated with TMC207, time-dependent (not dose-dependent) killing is observed (109). The mechanism of killing is not clear, but it does not involve the dissipation of the membrane potential, which is lethal to all living cells. A dose-dependent decrease in intracellular ATP has been observed when *M. tuberculosis* cells are treated with TMC207 (111, 112), but these data do not explain cell death because mycobacterial cells can be depleted of ATP and yet remain viable (117). TMC207 is bactericidal toward most species of mycobacteria (109) but is only bacteriostatic against *M. avium* (118) and *M. smegmatis* (M. Berney and G. M. Cook, unpublished data). Even when *M. smegmatis* was grown at a doubling time of 70 h in glycerol-limited continuous culture, TMC207 was bacteriostatic (Berney and Cook, unpublished data). The identification of the mechanisms underlying this sensitivity will be important in understanding how TMC207 exerts its inhibitory effects on mycobacterial cells.

CONTROL OF ELECTRON TRANSPORT CHAIN EXPRESSION AND OXIDATIVE PHOSPHORYLATION

Mycobacteria are obligate aerobes and as such have to possess mechanisms to detect the ambient oxygen tension to enable them to adapt to changes in oxygen availability by adjusting their metabolism accordingly. Furthermore, gradual depletion of oxygen has been implicated in entry of *M. tuberculosis* into nonreplicating persistence and latency (119). The mechanisms by which mycobacteria sense oxygen have been extensively studied, and the two major sensory systems known to date are the DosT/DosS/DosR system and WhiB3.

The dormancy survival, or Dos, system consists of the three proteins DosR (or DevR), DosS (or DevS), and DosT. DosR has been shown to act as a transcriptional regulator, which is responsible for the induction of nearly all hypoxia-induced genes of *M. tuberculosis* (96). In addition to hypoxia, the Dos-regulon has also been shown to be activated by exposure of *M. tuberculosis* to NO or CO (88, 90, 120).

WhiB3 has been implicated in sensing of oxygen and redox state by mycobacteria. Singh and colleagues report that WhiB3 contains a 4Fe-4S cluster, which binds NO (121). Furthermore, in the presence of oxygen, the WhiB3 $[4Fe-4S]^{2+}$ cluster is degraded first to a $[3Fe-4S]^+$ cluster, then a $[2Fe-2S]^{2+}$ cluster and subsequently lost altogether, in a mechanism reminiscent of the one found in the *E. coli* oxygen sensor FNR (121, 122). Apo-WhiB3 was shown to have protein disulfide reductase activity, and it has been proposed that loss of the Fe-S cluster is required to gain this activity (123). WhiB3-mediated response to the presence of oxygen therefore may occur through direct control of the activity of metabolic proteins or through modification of transcriptional regulators (123). A role of WhiB3 in regulation of the transcriptional machinery in mycobacteria may be supported by the finding that WhiB3 interacts with the major sigma factor, SigA (RpoV) (124), but the effect of this interaction on SigA activity is not known, and further study is required to understand the precise role of WhiB3 in mediating any adaptation of mycobacteria to changes in oxygen tension.

A striking observation is that neither Dos nor WhiB3 controls any component of the electron transport machinery in *M. tuberculosis*. While the Dos system and WhiB3 are the most studied regarding the perception of oxygen tension by mycobacteria, they are likely not the only systems used by these bacteria. For example, DosR of *M. tuberculosis* H37Rv was shown not to be strictly required for survival of hypoxia *in vitro* (125). The authors further found that expression of Dos regulon genes in response to hypoxia was transient, with expression of about half of the 50 DosR-dependent genes returning to baseline levels within 24 h of hypoxia. In contrast, a set of 230 genes was significantly upregulated at 4 and 7 days of hypoxia, but not initially, and they were termed the enduring hypoxic response (EHR) (125). Induction of the EHR was independent of DosR, suggesting that other sensory and regulatory mechanisms must exist to signal prolonged exposure to hypoxia. Strikingly, the EHR genes contained an unusually high number of regulatory genes (FurA, FurB, PhoP, three WhiB family members, and two ECF sigma factors, SigH and SigE), but it is not yet known which of these, if any, are involved in entry into EHR (125). A study into the effect of addition of cyclic AMP (cAMP) to growing cultures of *M. bovis* BCG or *M. tuberculosis* H37Rv found that the number of genes affected by cAMP was larger under low oxygen, CO_2-enriched conditions than under ambient air (126). The authors proposed that cAMP may be used by mycobacteria as a signaling molecule in response to hypoxic conditions. However, the signal leading to cAMP synthesis, and which adenylyl cyclase might catalyze this reaction under hypoxia, remains unknown. The regulator of *cydABDC* expression in mycobacteria has not been identified. Several regulators have been implicated in controlling the expression of *cyd* expression, i.e., SenX3-RegX3 (127) and the cAMP receptor protein (CRP) (128), but the signal transduction pathway or signals sensed remain to be elucidated.

Research in the authors' laboratory is funded by the Health Research Council, Lottery Health, Marsden Fund, Royal Society New Zealand, the Maurice Wilkins Center, and National Institute of Health grant AI26170.

Citation. Cook GM, Hards K, Vilchèze C, Hartman T, Berney M. 2014. Energetics of respiration and oxidative phosphorylation in mycobacteria. Microbiol Spectrum 2(3): MGM2-0015-2013.

References

1. **Segal W, Bloch H.** 1956. Biochemical differentiation of *Mycobacterium tuberculosis* grown in vivo and in vitro. *J Bacteriol* **72:**132–141.

2. **Brodie AF, Gutnik DL.** 1972. Electron transport and oxidative phosphorylation in microbial systems, p 599–681. *In* King TE, Klingenberg M (ed), *Electron and Coupled Energy Transfer Systems*, vol. **1B**. Marcel Dekker, New York.

3. **Russell JB, Cook GM.** 1995. Energetics of bacterial growth: balance of anabolic and catabolic reactions. *Microbiol Rev* **59:**48–62.

4. **Rao M, Streur TL, Aldwell FE, Cook GM.** 2001. Intracellular pH regulation by *Mycobacterium smegmatis*

and *Mycobacterium bovis* BCG. *Microbiology* **147:** 1017–1024.

5. Ishaque M. 1992. Energy generation mechanisms in the in vitro-grown *Mycobacterium lepraemurium*. *Int J Lepr Other Mycobact Dis* **60:**61–70.

6. Rao SP, Alonso S, Rand L, Dick T, Pethe K. 2008. The protonmotive force is required for maintaining ATP homeostasis and viability of hypoxic, nonreplicating *Mycobacterium tuberculosis*. *Proc Natl Acad Sci USA* **105:**11945–11950.

7. Berney M, Weimar MR, Heikal A, Cook GM. 2012. Regulation of proline metabolism in mycobacteria and its role in carbon metabolism under hypoxia. *Mol Microbiol* **84:**664–681.

8. Dimroth P, Cook GM. 2004. Bacterial Na+ - or H+ -coupled ATP synthases operating at low electrochemical potential. *Adv Microb Physiol* **49:**175–218.

9. Haagsma AC, Driessen NN, Hahn MM, Lill H, Bald D. 2010. ATP synthase in slow- and fast-growing mycobacteria is active in ATP synthesis and blocked in ATP hydrolysis direction. *FEMS Microbiol Lett* **313:**68–74.

10. Higashi T, Kalra VK, Lee SH, Bogin E, Brodie AF. 1975. Energy-transducing membrane-bound coupling factor-ATPase from *Mycobacterium phlei*. I. Purification, homogeneity, and properties. *J Biol Chem* **250:** 6541–6548.

11. Kerscher S, Drose S, Zickermann V, Brandt U. 2008. The three families of respiratory NADH dehydrogenases. *Results Probl Cell Differ* **45:**185–222.

12. Weinstein EA, Yano T, Li LS, Avarbock D, Avarbock A, Helm D, McColm AA, Duncan K, Lonsdale JT, Rubin H. 2005. Inhibitors of type II NADH:menaquinone oxidoreductase represent a class of antitubercular drugs. *Proc Natl Acad Sci USA* **102:**4548–4553.

13. Cole ST, Eiglmeier K, Parkhill J, James KD, Thomson NR, Wheeler PR, Honore N, Garnier T, Churcher C, Harris D, Mungall K, Basham D, Brown D, Chillingworth T, Connor R, Davies RM, Devlin K, Duthoy S, Feltwell T, Fraser A, Hamlin N, Holroyd S, Hornsby T, Jagels K, Lacroix C, Maclean J, Moule S, Murphy L, Oliver K, Quail MA, Rajandream MA, Rutherford KM, Rutter S, Seeger K, Simon S, Simmonds M, Skelton J, Squares R, Squares S, Stevens K, Taylor K, Whitehead S, Woodward JR, Barrell BG. 2001. Massive gene decay in the leprosy bacillus. *Nature* **409:** 1007–1011.

14. Miesel L, Weisbrod TR, Marcinkeviciene JA, Bittman R, Jacobs WR Jr. 1998. NADH dehydrogenase defects confer isoniazid resistance and conditional lethality in *Mycobacterium smegmatis*. *J Bacteriol* **180:** 2459–2467.

15. Vilcheze C, Weisbrod TR, Chen B, Kremer L, Hazbon MH, Wang F, Alland D, Sacchettini JC, Jacobs WR Jr. 2005. Altered NADH/NAD+ ratio mediates coresistance to isoniazid and ethionamide in mycobacteria. *Antimicrob Agents Chemother* **49:**708–720.

16. Berney M, Cook GM. 2010. Unique flexibility in energy metabolism allows mycobacteria to combat starvation and hypoxia. *PLoS One* **5:**e8614.

17. Velmurugan K, Chen B, Miller JL, Azogue S, Gurses S, Hsu T, Glickman M, Jacobs WR Jr, Porcelli SA, Briken V. 2007. *Mycobacterium tuberculosis nuoG* is a virulence gene that inhibits apoptosis of infected host cells. *PLoS Pathog* **3:**e110.

18. Shi L, Sohaskey CD, Kana BD, Dawes S, North RJ, Mizrahi V, Gennaro ML. 2005. Changes in energy metabolism of *Mycobacterium tuberculosis* in mouse lung and under in vitro conditions affecting aerobic respiration. *Proc Natl Acad Sci USA* **102:**15629–15634.

19. Schnappinger D, Ehrt S, Voskuil MI, Liu Y, Mangan JA, Monahan IM, Dolganov G, Efron B, Butcher PD, Nathan C, Schoolnik GK. 2003. Transcriptional adaptation of *Mycobacterium tuberculosis* within macrophages: insights into the phagosomal environment. *J Exp Med* **198:**693–704.

20. Betts JC, Lukey PT, Robb LC, McAdam RA, Duncan K. 2002. Evaluation of a nutrient starvation model of *Mycobacterium tuberculosis* persistence by gene and protein expression profiling. *Mol Microbiol* **43:**717–731.

21. Unden G, Bongaerts J. 1997. Alternative respiratory pathways of *Escherichia coli*: energetics and transcriptional regulation in response to electron acceptors. *Biochim Biophys Acta* **1320:**217–234.

22. Iwata M, Lee Y, Yamashita T, Yagi T, Iwata S, Cameron AD, Maher MJ. 2012. The structure of the yeast NADH dehydrogenase (Ndi1) reveals overlapping binding sites for water- and lipid-soluble substrates. *Proc Natl Acad Sci USA* **109:**15247–15252.

23. Feng Y, Li WF, Li J, Wang JW, Ge JP, Xu D, Liu YJ, Wu KQ, Zeng QY, Wu JW, Tian CL, Zhou B, Yang MJ. 2012. Structural insight into the type-II mitochondrial NADH dehydrogenases. *Nature* **491:**478–482.

24. Melo AMP, Bandeiras TM, Teixeira M. 2004. New insights into type II NAD(P)H : quinone oxidoreductases. *Microbiol Mol Biol Rev* **68:**603–616.

25. Lin SS, Gross U, Bohne W. 2011. Two internal type II NADH dehydrogenases of *Toxoplasma gondii* are both required for optimal tachyzoite growth. *Mol Microbiol* **82:**209–221.

26. Griffin JE, Gawronski JD, Dejesus MA, Ioerger TR, Akerley BJ, Sassetti CM. 2011. High-resolution phenotypic profiling defines genes essential for mycobacterial growth and cholesterol catabolism. *PLoS Pathog* **7:** e1002251.

27. Yano T, Li LS, Weinstein E, Teh JS, Rubin H. 2006. Steady-state kinetics and inhibitory action of antitubercular phenothiazines on *Mycobacterium tuberculosis* type-II NADH-menaquinone oxidoreductase (NDH-2). *J Biol Chem* **281:**11456–11463.

28. McAdam RA, Quan S, Smith DA, Bardarov S, Betts JC, Cook FC, Hooker EU, Lewis AP, Woollard P, Everett MJ, Lukey PT, Bancroft GJ, Jacobs WR Jr, Duncan K. 2002. Characterization of a *Mycobacterium tuberculosis* H37Rv transposon library reveals insertions in 351 ORFs and mutants with altered virulence. *Microbiology* **148:**2975–2986.

29. Sassetti CM, Boyd DH, Rubin EJ. 2003. Genes required for mycobacterial growth defined by high density mutagenesis. *Mol Microbiol* **48**:77–84.

30. Warman AJ, Rito TS, Fisher NE, Moss DM, Berry NG, O'Neill PM, Ward SA, Biagini GA. 2013. Antitubercular pharmacodynamics of phenothiazines. *J Antimicrob Chemother* **68**:869–880.

31. Teh JS, Yano T, Rubin H. 2007. Type II NADH: menaquinone oxidoreductase of *Mycobacterium tuberculosis*. *Infectious Disorders Drug Targets* **7**:169–181.

32. Biagini GA, Viriyavejakul P, O'Neill PM, Bray PG, Ward SA. 2006. Functional characterization and target validation of alternative complex I of *Plasmodium falciparum* mitochondria. *Antimicrob Agents Chemother* **50**:1841–1851.

33. Shirude PS, Paul P, Choudhury NR, Kedari C, Bandodkar B, Ugarkar BG. 2012. Quinolinyl pyrimidines: potent inhibitors of NDH-2 as a novel class of anti-TB agents. *ACS Med Chem Lett* **3**:736–740.

34. Ordway D, Viveiros M, Leandro C, Bettencourt R, Almeida J, Martins M, Kristiansen JE, Molnar J, Amaral L. 2003. Clinical concentrations of thioridazine kill intracellular multidrug-resistant *Mycobacterium tuberculosis*. *Antimicrob Agents Chemother* **47**:917–922.

35. Amaral L, Kristiansen JE, Abebe LS, Millett W. 1996. Inhibition of the respiration of multi-drug resistant clinical isolates of *Mycobacterium tuberculosis* by thioridazine: potential use for initial therapy of freshly diagnosed tuberculosis. *J Antimicrob Chemother* **38**:1049–1053.

36. Mogi T, Matsushita K, Murase Y, Kawahara K, Miyoshi H, Ui H, Shiomi K, Omura S, Kita K. 2009. Identification of new inhibitors for alternative NADH dehydrogenase (NDH-II). *FEMS Microbiol Lett* **291**:157–161.

37. Tian J, Bryk R, Itoh M, Suematsu M, Nathan C. 2005. Variant tricarboxylic acid cycle in *Mycobacterium tuberculosis*: identification of alpha-ketoglutarate decarboxylase. *Proc Natl Acad Sci USA* **102**:10670–10675.

38. Youmans AS, Millman I, Youmans GP. 1956. The oxidation of compounds related to the tricarboxylic acid cycle by whole cells and enzyme preparations of *Mycobacterium tuberculosis* var. hominis. *J Bacteriol* **71**:565–570.

39. Cecchini G, Schroder I, Gunsalus RP, Maklashina E. 2002. Succinate dehydrogenase and fumarate reductase from *Escherichia coli*. *Biochimica Biophysica Acta* **1553**:140–157.

40. Baek SH, Li AH, Sassetti CM. 2011. Metabolic regulation of mycobacterial growth and antibiotic sensitivity. *PLoS Biol* **9**:e1001065.

41. Unden G, Schirawski J. 1997. The oxygen-responsive transcriptional regulator FNR of *Escherichia coli*: the search for signals and reactions. *Mol Microbiol* **25**:205–210.

42. Watanabe S, Zimmermann M, Goodwin MB, Sauer U, Barry CE 3rd, Boshoff HI. 2011. Fumarate reductase activity maintains an energized membrane in anaerobic *Mycobacterium tuberculosis*. *PLoS Pathog* **7**:e1002287.

43. Eoh H, Rhee KY. 2013. Multifunctional essentiality of succinate metabolism in adaptation to hypoxia in *Mycobacterium tuberculosis*. *Proc Natl Acad Sci USA* **110**:6554–6559.

44. Bacon J, James BW, Wernisch L, Williams A, Morley KA, Hatch GJ, Mangan JA, Hinds J, Stoker NG, Butcher PD, Marsh PD. 2004. The influence of reduced oxygen availability on pathogenicity and gene expression in *Mycobacterium tuberculosis*. *Tuberculosis (Edinb)* **84**:205–217.

45. Beste DJ, Laing E, Bonde B, Avignone-Rossa C, Bushell ME, McFadden JJ. 2007. Transcriptomic analysis identifies growth rate modulation as a component of the adaptation of mycobacteria to survival inside the macrophage. *J Bacteriol* **189**:3969–3976.

46. Tanner JJ. 2008. Structural biology of proline catabolism. *Amino Acids* **35**:719–730.

47. Menzel R, Roth J. 1981. Purification of the *putA* gene product. A bifunctional membrane-bound protein from *Salmonella typhimurium* responsible for the two-step oxidation of proline to glutamate. *J Biol Chem* **256**:9755–9761.

48. Weigoldt M, Meens J, Bange FC, Pich A, Gerlach GF, Goethe R. 2013. Metabolic adaptation of *Mycobacterium avium* subsp. *paratuberculosis* to the gut environment. *Microbiology* **159**:380–391.

49. Smith DA, Parish T, Stoker NG, Bancroft GJ. 2001. Characterization of auxotrophic mutants of *Mycobacterium tuberculosis* and their potential as vaccine candidates. *Infect Immun* **69**:1142–1150.

50. Vignais PM, Colbeau A. 2004. Molecular biology of microbial hydrogenases. *Curr Issues Mol Biol* **6**:159–188.

51. Tamagnini P, Leitao E, Oliveira P, Ferreira D, Pinto F, Harris DJ, Heidorn T, Lindblad P. 2007. Cyanobacterial hydrogenases: diversity, regulation and applications. *FEMS Microbiol Rev* **31**:692–720.

52. Constant P, Chowdhury SP, Hesse L, Pratscher J, Conrad R. 2011. Genome data mining and soil survey for the novel group 5 [NiFe]-hydrogenase to explore the diversity and ecological importance of presumptive high-affinity H(2)-oxidizing bacteria. *Appl Environ Microbiol* **77**:6027–6035.

53. Constant P, Chowdhury SP, Pratscher J, Conrad R. 2010. *Streptomycetes* contributing to atmospheric molecular hydrogen soil uptake are widespread and encode a putative high-affinity [NiFe]-hydrogenase. *Environ Microbiol* **12**:821–829.

54. King GM. 2003. Uptake of carbon monoxide and hydrogen at environmentally relevant concentrations by mycobacteria. *Appl Environ Microbiol* **69**:7266–7272.

55. Kim YM, Hegeman GD. 1983. Oxidation of carbon monoxide by bacteria. *Int Rev Cytol* **81**:1–32.

56. Park SW, Hwang EH, Park H, Kim JA, Heo J, Lee KH, Song T, Kim E, Ro YT, Kim SW, Kim YM. 2003. Growth of mycobacteria on carbon monoxide and methanol. *J Bacteriol* **185**:142–147.

57. Schryvers A, Lohmeier E, Weiner JH. 1978. Chemical and functional properties of the native and reconstituted forms of the membrane-bound, aerobic

glycerol-3-phosphate dehydrogenase of *Escherichia coli*. *J Biol Chem* **253**:783–788.

58. Boos W. 1998. Binding protein-dependent ABC transport system for glycerol 3-phosphate of *Escherichia coli*. *Methods Enzymol* **292**:40–51.

59. Molenaar D, van der Rest ME, Drysch A, Yucel R. 2000. Functions of the membrane-associated and cytoplasmic malate dehydrogenases in the citric acid cycle of *Corynebacterium glutamicum*. *J Bacteriol* **182**:6884–6891.

60. Molenaar D, van der Rest ME, Petrovic S. 1998. Biochemical and genetic characterization of the membrane-associated malate dehydrogenase (acceptor) from *Corynebacterium glutamicum*. *Eur J Biochem* **254**:395–403.

61. Prasada Reddy TL, Suryanarayana Murthy P, Venkitasubramanian TA. 1975. Respiratory chains of *Mycobacterium smegmatis*. *Indian J Biochem Biophys* **12**:255–259.

62. Prasada Reddy TL, Suryanarayana Murthy P, Venkitasubramanian TA. 1975. Variations in the pathways of malate oxidation and phosphorylation in different species of mycobacteria. *Biochim Biophys Acta* **376**:210–218.

63. D'Mello R, Hill S, Poole RK. 1995. The oxygen affinity of cytochrome bo' in *Escherichia coli* determined by the deoxygenation of oxyleghemoglobin and oxymyoglobin: Km values for oxygen are in the submicromolar range. *J Bacteriol* **177**:867–870.

64. D'Mello R, Hill S, Poole RK. 1996. The cytochrome bd quinol oxidase in *Escherichia coli* has an extremely high oxygen affinity and two oxygen-binding haems: implications for regulation of activity in vivo by oxygen inhibition. *Microbiology* **142**:755–763.

65. Cotter PA, Melville SB, Albrecht JA, Gunsalus RP. 1997. Aerobic regulation of cytochrome d oxidase (*cydAB*) operon expression in *Escherichia coli*: roles of Fnr and ArcA in repression and activation. *Mol Microbiol* **25**:605–615.

66. Tseng CP, Hansen AK, Cotter P, Gunsalus RP. 1994. Effect of cell growth rate on expression of the anaerobic respiratory pathway operons *frdABCD*, *dmsABC*, and *narGHJI* of *Escherichia coli*. *J Bacteriol* **176**:6599–6605.

67. Kana BD, Weinstein EA, Avarbock D, Dawes SS, Rubin H, Mizrahi V. 2001. Characterization of the *cydAB*-encoded cytochrome bd oxidase from *Mycobacterium smegmatis*. *J Bacteriol* **183**:7076–7086.

68. Voskuil MI, Visconti KC, Schoolnik GK. 2004. *Mycobacterium tuberculosis* gene expression during adaptation to stationary phase and low-oxygen dormancy. *Tuberculosis (Edinb)* **84**:218–227.

69. Boshoff HI, Barry CE 3rd. 2005. Tuberculosis: metabolism and respiration in the absence of growth. *Nat Rev Microbiol* **3**:70–80.

70. Matsoso LG, Kana BD, Crellin PK, Lea-Smith DJ, Pelosi A, Powell D, Dawes SS, Rubin H, Coppel RL, Mizrahi V. 2005. Function of the cytochrome bc1-aa3 branch of the respiratory network in mycobacteria and network adaptation occurring in response to its disruption. *J Bacteriol* **187**:6300–6308.

71. Megehee JA, Hosler JP, Lundrigan MD. 2006. Evidence for a cytochrome *bcc*-aa3 interaction in the respiratory chain of *Mycobacterium smegmatis*. *Microbiology* **152**:823–829.

72. Niebisch A, Bott M. 2001. Molecular analysis of the cytochrome bc1-aa3 branch of the *Corynebacterium glutamicum* respiratory chain containing an unusual diheme cytochrome c1. *Arch Microbiol* **175**:282–294.

73. Abrahams KA, Cox JA, Spivey VL, Loman NJ, Pallen MJ, Constantinidou C, Fernandez R, Alemparte C, Remuinan MJ, Barros D, Ballell L, Besra GS. 2012. Identification of novel imidazo[1,2-a]pyridine inhibitors targeting M. tuberculosis QcrB. *PLoS One* **7**:e52951.

74. Niebisch A, Bott M. 2003. Purification of a cytochrome bc-aa3 supercomplex with quinol oxidase activity from *Corynebacterium glutamicum*. Identification of a fourth subunity of cytochrome aa3 oxidase and mutational analysis of diheme cytochrome c1. *J Biol Chem* **278**:4339–4346.

75. Berry EA, Trumpower BL. 1985. Isolation of ubiquinol oxidase from *Paracoccus denitrificans* and resolution into cytochrome bc1 and cytochrome c-aa3 complexes. *J Biol Chem* **260**:2458–2467.

76. Sone N, Sekimachi M, Kutoh E. 1987. Identification and properties of a quinol oxidase super-complex composed of a bc1 complex and cytochrome oxidase in the thermophilic bacterium PS3. *J Biol Chem* **262**:15386–15391.

77. Poole RK, Cook GM. 2000. Redundancy of aerobic respiratory chains in bacteria? Routes, reasons and regulation. *Adv Microb Physiol* **43**:165–224.

78. Borisov VB, Murali R, Verkhovskaya ML, Bloch DA, Han H, Gennis RB, Verkhovsky MI. 2011. Aerobic respiratory chain of *Escherichia coli* is not allowed to work in fully uncoupled mode. *Proc Natl Acad Sci USA* **108**:17320–17324.

79. Goldman BS, Gabbert KK, Kranz RG. 1996. The temperature-sensitive growth and survival phenotypes of *Escherichia coli cydDC* and *cydAB* strains are due to deficiencies in cytochrome bd and are corrected by exogenous catalase and reducing agents. *J Bacteriol* **178**:6348–6351.

80. Pittman MS, Robinson HC, Poole RK. 2005. A bacterial glutathione transporter (*Escherichia coli* CydDC) exports reductant to the periplasm. *J Biol Chem* **280**:32254–32261.

81. Dhar N, McKinney JD. 2010. *Mycobacterium tuberculosis* persistence mutants identified by screening in isoniazid-treated mice. *Proc Natl Acad Sci USA* **107**:12275–12280.

82. Rowe JJ, Ubbink-Kok T, Molenaar D, Konings WN, Driessen AJ. 1994. NarK is a nitrite-extrusion system involved in anaerobic nitrate respiration by *Escherichia coli*. *Mol Microbiol* **12**:579–586.

83. Clegg S, Yu F, Griffiths L, Cole JA. 2002. The roles of the polytopic membrane proteins NarK, NarU and NirC in *Escherichia coli* K-12: two nitrate and three nitrite transporters. *Mol Microbiol* **44**:143–155.

84. Jia W, Cole JA. 2005. Nitrate and nitrite transport in *Escherichia coli*. *Biochem Soc Trans* **33**:159–161.

85. Sohaskey CD, Wayne LG. 2003. Role of *narK2X* and *narGHJI* in hypoxic upregulation of nitrate reduction by *Mycobacterium tuberculosis*. *J Bacteriol* **185**:7247–7256.

86. Sohaskey CD. 2005. Regulation of nitrate reductase activity in *Mycobacterium tuberculosis* by oxygen and nitric oxide. *Microbiology* **151**:3803–3810.

87. Moir JW, Wood NJ. 2001. Nitrate and nitrite transport in bacteria. *Cell Mol Life Sci* **58**:215–224.

88. Kumar A, Deshane JS, Crossman DK, Bolisetty S, Yan BS, Kramnik I, Agarwal A, Steyn AJ. 2008. Heme oxygenase-1-derived carbon monoxide induces the *Mycobacterium tuberculosis* dormancy regulon. *J Biol Chem* **283**:18032–18039.

89. Ohno H, Zhu G, Mohan VP, Chu D, Kohno S, Jacobs WR Jr, Chan J. 2003. The effects of reactive nitrogen intermediates on gene expression in *Mycobacterium tuberculosis*. *Cell Microbiol* **5**:637–648.

90. Voskuil MI, Schnappinger D, Visconti KC, Harrell MI, Dolganov GM, Sherman DR, Schoolnik GK. 2003. Inhibition of respiration by nitric oxide induces a *Mycobacterium tuberculosis* dormancy program. *J Exp Med* **198**:705–713.

91. Sohaskey CD. 2008. Nitrate enhances the survival of *Mycobacterium tuberculosis* during inhibition of respiration. *J Bacteriol* **190**:2981–2986.

92. Tan MP, Sequeira P, Lin WW, Phong WY, Cliff P, Ng SH, Lee BH, Camacho L, Schnappinger D, Ehrt S, Dick T, Pethe K, Alonso S. 2010. Nitrate respiration protects hypoxic *Mycobacterium tuberculosis* against acid- and reactive nitrogen species stresses. *PLoS One* **5**:e13356.

93. Pinto R, Harrison JS, Hsu T, Jacobs WR Jr, Leyh TS. 2007. Sulfite reduction in mycobacteria. *J Bacteriol* **189**:6714–6722.

94. Kuhn M, Steinbuchel A, Schlegel HG. 1984. Hydrogen evolution by strictly aerobic hydrogen bacteria under anaerobic conditions. *J Bacteriol* **159**:633–639.

95. Fontan P, Aris V, Ghanny S, Soteropoulos P, Smith I. 2008. Global transcriptional profile of *Mycobacterium tuberculosis* during THP-1 human macrophage infection. *Infect Immun* **76**:717–725.

96. Park HD, Guinn KM, Harrell MI, Liao R, Voskuil MI, Tompa M, Schoolnik GK, Sherman DR. 2003. Rv3133c/dosR is a transcription factor that mediates the hypoxic response of *Mycobacterium tuberculosis*. *Mol Microbiol* **48**:833–843.

97. Sherman DR, Voskuil M, Schnappinger D, Liao R, Harrell MI, Schoolnik GK. 2001. Regulation of the *Mycobacterium tuberculosis* hypoxic response gene encoding alpha-crystallin. *Proc Natl Acad Sci USA* **98**:7534–7539.

98. Mnatsakanyan N, Bagramyan K, Trchounian A. 2004. Hydrogenase 3 but not hydrogenase 4 is major in hydrogen gas production by *Escherichia coli* formate hydrogenlyase at acidic pH and in the presence of external formate. *Cell Biochem Biophys* **41**:357–366.

99. Coppi MV. 2005. The hydrogenases of *Geobacter sulfurreducens*: a comparative genomic perspective. *Microbiology* **151**:1239–1254.

100. Marreiros BC, Batista AP, Duarte AM, Pereira MM. 2013. A missing link between complex I and group 4 membrane-bound [NiFe] hydrogenases. *Biochimica Biophysica Acta* **1827**:198–209.

101. He H, Bretl DJ, Penoske RM, Anderson DM, Zahrt TC. 2011. Components of the Rv0081-Rv0088 locus, which encodes a predicted formate hydrogenlyase complex, are coregulated by Rv0081, MprA, and DosR in *Mycobacterium tuberculosis*. *J Bacteriol* **193**:5105–5118.

102. von Ballmoos C, Cook GM, Dimroth P. 2008. Unique rotary ATP synthase and its biological diversity. *Annu Rev Biophys* **37**:43–64.

103. Tran SL, Cook GM. 2005. The F_1F_0-ATP synthase of *Mycobacterium smegmatis* is essential for growth. *J Bacteriol* **187**:5023–5028.

104. Friedl P, Hoppe J, Gunsalus RP, Michelsen O, von Meyenburg K, Schairer HU. 1983. Membrane integration and function of the three F_0 subunits of the ATP synthase of *Escherichia coli* K12. *EMBO J* **2**:99–103.

105. Santana M, Ionescu MS, Vertes A, Longin R, Kunst F, Danchin A, Glaser P. 1994. *Bacillus subtilis* F_0F_1 ATPase: DNA sequence of the *atp* operon and characterization of *atp* mutants. *J Bacteriol* **176**:6802–6811.

106. Cox RA, Cook GM. 2007. Growth regulation in the mycobacterial cell. *Curr Mol Med* **7**:231–245.

107. Tran SL, Rao M, Simmers C, Gebhard S, Olsson K, Cook GM. 2005. Mutants of *Mycobacterium smegmatis* unable to grow at acidic pH in the presence of the protonophore carbonyl cyanide m-chlorophenylhydrazone. *Microbiology* **151**:665–672.

108. Koch-Koerfges A, Kabus A, Ochrombel I, Marin K, Bott M. 2012. Physiology and global gene expression of a *Corynebacterium glutamicum* DeltaF_1F_0-ATP synthase mutant devoid of oxidative phosphorylation. *Biochimica Biophysica Acta* **1817**:370–380.

109. Andries K, Verhasselt P, Guillemont J, Gohlmann HW, Neefs JM, Winkler H, Van Gestel J, Timmerman P, Zhu M, Lee E, Williams P, de Chaffoy D, Huitric E, Hoffner S, Cambau E, Truffot-Pernot C, Lounis N, Jarlier V. 2005. A diarylquinoline drug active on the ATP synthase of *Mycobacterium tuberculosis*. *Science* **307**:223–227.

110. Dhiman RK, Mahapatra S, Slayden RA, Boyne ME, Lenaerts A, Hinshaw JC, Angala SK, Chatterjee D, Biswas K, Narayanasamy P, Kurosu M, Crick DC. 2009. Menaquinone synthesis is critical for maintaining mycobacterial viability during exponential growth and recovery from non-replicating persistence. *Mol Microbiol* **72**:85–97.

111. Koul A, Dendouga N, Vergauwen K, Molenberghs B, Vranckx L, Willebrords R, Ristic Z, Lill H, Dorange I, Guillemont J, Bald D, Andries K. 2007. Diarylquinolines target subunit c of mycobacterial ATP synthase. *Nat Chem Biol* **3**:323–324.

112. Koul A, Vranckx L, Dendouga N, Balemans W, Van den Wyngaert I, Vergauwen K, Gohlmann HW, Willebrords R, Poncelet A, Guillemont J, Bald D, Andries K. 2008. Diarylquinolines are bactericidal for

dormant mycobacteria as a result of disturbed ATP homeostasis. *J Biol Chem* **283**:25273–25280.

113. Huitric E, Verhasselt P, Andries K, Hoffner SE. 2007. In vitro antimycobacterial spectrum of a diarylquinoline ATP synthase inhibitor. *Antimicrob Agents Chemother* **51**:4202–4204.

114. Huitric E, Verhasselt P, Koul A, Andries K, Hoffner S, Andersson DI. 2010. Rates and mechanisms of resistance development in *Mycobacterium tuberculosis* to a novel diarylquinoline ATP synthase inhibitor. *Antimicrob Agents Chemother* **54**:1022–1028.

115. de Jonge MR, Koymans LH, Guillemont JE, Koul A, Andries K. 2007. A computational model of the inhibition of *Mycobacterium tuberculosis* ATPase by a new drug candidate R207910. *Proteins* **67**:971–980.

116. Haagsma AC, Podasca I, Koul A, Andries K, Guillemont J, Lill H, Bald D. 2011. Probing the interaction of the diarylquinoline TMC207 with its target mycobacterial ATP synthase. *PLoS One* **6**:e23575.

117. Frampton R, Aggio RB, Villas-Boas SG, Arcus VL, Cook GM. 2012. Toxin-antitoxin systems of *Mycobacterium smegmatis* are essential for cell survival. *J Biol Chem* **287**:5340–5356.

118. Lounis N, Gevers T, Van den Berg J, Vranckx L, Andries K. 2009. ATP synthase inhibition of *Mycobacterium avium* is not bactericidal. *Antimicrob Agents Chemother* **53**:4927–4929.

119. Wayne LG, Sohaskey CD. 2001. Nonreplicating persistence of mycobacterium tuberculosis. *Annu Rev Microbiol* **55**:139–163.

120. Shiloh MU, Manzanillo P, Cox JS. 2008. *Mycobacterium tuberculosis* senses host-derived carbon monoxide during macrophage infection. *Cell Host Microbe* **3**: 323–330.

121. Singh A, Guidry L, Narasimhulu KV, Mai D, Trombley J, Redding KE, Giles GI, Lancaster JR Jr, Steyn AJ. 2007. *Mycobacterium tuberculosis* WhiB3 responds to O_2 and nitric oxide via its [4Fe-4S] cluster and is essential

for nutrient starvation survival. *Proc Natl Acad Sci USA* **104**:11562–11567.

122. Crack J, Green J, Thomson AJ. 2004. Mechanism of oxygen sensing by the bacterial transcription factor fumarate-nitrate reduction (FNR). *J Biol Chem* **279**: 9278–9286.

123. Suhail Alam M, Agrawal P. 2008. Matrix-assisted refolding and redox properties of WhiB3/Rv3416 of *Mycobacterium tuberculosis* H37Rv. *Protein Expr Purif* **61**:83–91.

124. Steyn AJ, Collins DM, Hondalus MK, Jacobs WR Jr, Kawakami RP, Bloom BR. 2002. *Mycobacterium tuberculosis* WhiB3 interacts with RpoV to affect host survival but is dispensable for *in vivo* growth. *Proc Natl Acad Sci USA* **99**:3147–3152.

125. Rustad TR, Harrell MI, Liao R, Sherman DR. 2008. The enduring hypoxic response of *Mycobacterium tuberculosis*. *PLoS One* **3**:e1502.

126. Gazdik MA, McDonough KA. 2005. Identification of cyclic AMP-regulated genes in *Mycobacterium tuberculosis* complex bacteria under low-oxygen conditions. *J Bacteriol* **187**:2681–2692.

127. Roberts G, Vadrevu IS, Madiraju MV, Parish T. 2011. Control of CydB and GltA1 expression by the SenX3 RegX3 two component regulatory system of *Mycobacterium tuberculosis*. *PLoS One* **6**:e21090.

128. Rickman L, Scott C, Hunt DM, Hutchinson T, Menendez MC, Whalan R, Hinds J, Colston MJ, Green J, Buxton RS. 2005. A member of the cAMP receptor protein family of transcription regulators in *Mycobacterium tuberculosis* is required for virulence in mice and controls transcription of the *rpfA* gene coding for a resuscitation promoting factor. *Mol Microbiol* **56**: 1274–1286.

129. Zhang YJ, Ioerger TR, Huttenhower C, Long JE, Sassetti CM, Sacchettini JC, Rubin EJ. 2012. Global assessment of genomic regions required for growth in *Mycobacterium tuberculosis*. *PLoS Pathog* **8**:e1002946.

Genetics of Drug Resistance

Molecular Genetics of Mycobacteria, 2nd Edition
Edited by Graham F. Hatfull and William R. Jacobs, Jr.
© 2014 American Society for Microbiology, Washington, DC
doi:10.1128/microbiolspec.MGM2-0036-2013

Keira A. Cohen[1,2]
William R. Bishai[3]
Alexander S. Pym[1]

Molecular Basis of Drug Resistance in *Mycobacterium tuberculosis*

21

In 2011, the World Health Organization (WHO) reported nearly 60,000 new cases of multidrug-resistant tuberculosis (MDR-TB) (1), and estimates of the annual global incidence are much higher. The emergence of drug-resistant strains has made the treatment of TB complex, costly, toxic, time-intensive, and less efficacious. Design of a treatment regimen for drug-resistant TB includes the administration of first-line drugs to which the strains remain susceptible together with second-line drugs. These second-line agents are more expensive, more difficult to administer (several require intravenous administration), and are often associated with severe toxicities, including hepatic and renal dysfunction. In comparison to the 6 months required to treat drug-susceptible TB, drug-resistant TB requires a prolonged treatment duration of 18 to 24 months. These logistics constitute considerable hardships for patients as well as for overburdened public health services. Too frequently, premature discontinuation of therapy occurs, leading to treatment failure and the emergence of *Mycobacterium tuberculosis* strains with additional drug resistance.

DEFINITIONS OF DRUG-RESISTANT TB

MDR-TB is defined as *M. tuberculosis* with *in vitro* resistance to two first-line medications: isoniazid (isonicotinic acid hydrazide, INH) and rifampin (RIF).

Extensively drug-resistant TB (XDR-TB) is defined as *M. tuberculosis* that is resistant to not only INH and RIF, but also to other medication classes that comprise the backbone of drug-resistant TB therapy, namely a quinolone and one of the second-line injectable drugs (kanamycin [KAN], amikacin [AMK], or capreomycin [CAP]) (2).

Totally drug-resistant TB (TDR-TB) refers to strains that are resistant to all available TB drugs, although the number and degree of resistance to each drug has not yet been precisely defined. The emergence of these strains further underscores the medical and public health urgency to control drug-resistant TB (3, 4).

[1]KwaZulu-Natal Research Institute for TB and HIV (K-RITH), Nelson R. Mandela School of Medicine, 719 Umbilo Road, Durban, 4001, South Africa; [2]Brigham and Women's Hospital, Harvard Medical School, 75 Francis Street, Boston, MA 02115; [3]Johns Hopkins School of Medicine, The Center for TB Research, 1550 Orleans St., CRBII, Room 103, Baltimore, MD 21287.

MECHANISMS OF DRUG RESISTANCE

The traditional mechanisms by which bacteria achieve antimicrobial resistance are (i) barrier mechanisms (decreased permeability/efflux), (ii) degrading/inactivating enzymes, (iii) modification of pathways involved in drug activation/metabolism, and (iv) drug target modification or target amplification. As will be discussed below, *M. tuberculosis* uses all of these mechanisms to achieve resistance. For the purposes of this chapter we define drug resistance as genetic changes that alter the phenotypic resistance levels. Antibiotic tolerance (not discussed herein) addresses mechanisms that do not require genetic changes in order to achieve an altered phenotypic resistance level and is usually reversible after removal of the drug or a change in the growth conditions.

Table 1 delineates the commonly used TB drugs with the genes associated with their respective resistance and major mechanism of resistance.

FIRST-LINE AGENTS

Rifamycins

Rifamycins are critical to sterilization and are the key drugs associated with the development of modern short-course therapy for human TB. Currently, three rifamcyins are used for the treatment of TB: RIF, rifapentine, and rifabutin (5). These agents differ primarily in their human pharmacology, but with respect to their antimycobacterial mechanisms of action and resistance they are similar.

Table 1 The commonly used TB drugs with the genes associated with their respective resistance and major mechanism of resistance

Drug or drug class	Resistance genes	Rv number	Gene function	Mechanism of drug resistance
Rifamycins	*rpoB*	Rv0667	RNA polymerase B	Target modification
Isoniazid	*katG*	Rv1908c	Catalase-peroxidase enzyme	Decreased drug activation
	inhA	Rv1484	NADH-dependent enoyl-acyl carrier protein	Target amplification or modification
Pyrazinamide	*pncA*	Rv2043c	Pyrazinamidase	Decreased drug activation
Ethambutol	*embCAB* operon	Rv3793-5	Arabinosyltransferases	Target modification
Streptomycin	*rpsL*	Rv0682	12S ribosomal protein	Target modification
	rrs	n/a	16S rRNA	Target modification
	gidB	Rv3919c	7-Methylguanosine methyltransferase	Target modification
Kanamycin/ amikacin	*rrs*	n/a	16S rRNA	Target modification
	eis	Rv2416c	Aminoglycoside acetyltransferase	Inactivating mutation
Capreomycin	*rrs*	n/a	16S rRNA	Target modification
	tylA	Rv1694	rRNA methyltransferase	Target modification
Quinolones	*gyrA*	Rv0006	DNA gyrase A	Target modification
	gyrB	Rv0005	DNA gyrase B	Target modification
Ethionamide	*ethA*	Rv3854c	Mono-oxygenase	Decreased drug activation
	ethR	Rv3855	Transcriptional regulatory repressor protein (TetR family)	Decreased drug activation
	inhA	Rv1484	NADH-dependent enoyl-acyl carrier protein	Target amplification and modification
Para-aminosalicylic acid	*thyA*	Rv2764c	Thymidylate synthase	
	ribD	Rv2671	Enzyme in riboflavin biosynthesis	
Cycloserine	*alr*	Rv3423c	Alanine racemase	
	ddl	Rv2981c	D-Alanine-D-alanine ligase	
	cycA	Rv1704c	Bacterial D-serine/L- and D-alanine/ glycine/D-cycloserine proton symporter	
Bedaquiline	*atpE*	Rv1305	ATP synthase	
Linezolid	*rrl*	n/a	23S rRNA	Target modification
	rplC	Rv0701	50S ribosomal protein L3	Target modification

The principal target of rifamycins is the β subunit of RNA polymerase (RNAP), which is encoded by the gene *rpoB* (Rv0667 in *M. tuberculosis* H37Rv or MT0695 in *M. tuberculosis* CDC1551). Binding of a rifamycin to the RpoB protein inhibits the activity of RNAP, thereby preventing transcription. Because a basal level of transcription occurs in stationary growth phases, rifamycins are active against mycobacterial quiescent states. It is this activity that is thought to underpin the sterilizing activity that has effectively shortened the duration of TB chemotherapy from 18 months to 9 months with the introduction of RIF.

Mechanism of action

RIF has long been known to have its impact after RNAP binds to DNA and initiates transcription. Studies have shown that rather than inhibiting RNAP initiation, RIF prevents the elongation of RNA when the transcript becomes 2 to 3 nucleotides in length. In 1999 Darst and colleagues completed the first crystal structure of an RNAP, namely, that for *Thermus aquaticus* (6), and in 2001 the structure of RIF-bound RNAP was reported (7). These structures showed that RIF binds deep within the DNA/RNA channel of RNAP but more than 12 Å away from the active site.

More recent studies using the crystal structure of *Escherichia coli* RNAP have revealed a yet more complex, multistep mechanism of RNAP action: (i) RNAP binds to promoter DNA, producing an RNAP-promoter closed complex where the DNA remains double-stranded and is not loaded into the RNAP active-center cleft; (ii) RNAP loads DNA into, and unwinds DNA in, the RNAP active-center cleft, yielding an RNAP-promoter open complex (RPoc); (iii) RNAP synthesizes the first ~10 nucleotides of RNA using a "scrunching" mechanism whereby RNAP remains stationary on promoter DNA, pulling downstream DNA into the cleft (called the RNAP promoter initial transcribing complex [RPitc]); and (iv) RNAP escapes the promoter and synthesizes the rest of the RNA using a "stepping" mechanism, in which RNAP translocates relative to DNA, as a transcription elongation complex. RIF has been shown to exert its action by preventing step iii, namely, the formation of the RNAP initial transcribing complex RPitc. These properties led to the proposal that rifamycins inhibit the formation of RPitc through a steric-occlusion mechanism, whereby RIF binds adjacent to the RNAP active center, along the path of the RNA product, and physically prevents synthesis or retention of RNA products >2 to 3 nucleotides in length (8, 9).

RIF bound within its β subunit pocket is 12 Å away from the active site of RNAP, where ribonucleotides complementary to the DNA template strand are added to the growing RNA chain. However, the RIF-binding pocket within the β subunit is situated in the main DNA/RNA channel of RNAP. The binding of RIF to its cognate pocket obstructs this channel, preventing the elongation of RNA molecules beyond 2 to 3 nucleotides in length.

Mechanism of resistance: target modification: *rpoB*

RNAP in *M. tuberculosis*, as in most bacterial species, is comprised of four polypeptides: α, β, β′, and σ, arranged in a five-subunit enzyme, α₂ββ′σ. Single genes (*rpoA*, *rpoB*, and *rpoC*) encode the α, β, and β′ subunits, respectively, while 13 sigma factors may serve as the promoter-recognition subunits (10).

The mechanism by which these mutations confer RIF resistance has been inferred from X-ray crystallography studies of RNAP from other bacterial species, in particular *T. aquaticus* and *E. coli*. Greater than 95% of RIF-resistant strains of *M. tuberculosis* harbor nonsynonymous mutations in the *rpoB* gene encoding the β subunit of RNAP (11). These mutations cluster in an 81-base-pair region of the *rpoB* gene known as the RIF resistance-determining region (RRDR) (11, 12). Across multiple prokaryotic species, numerous mutations in the RRDR have been identified (13), although mutations G498Q, G498E, D516V, D516Y, N518 deletion, H526R, H526D, H526P, H526Y, and S531L (corresponding to *E. coli* RNAP codons) are among the most common.

RIF binds between two structural domains of the RNAP β subunit. The mutations in the RRDR all occur in residues in or near this pocket. All of these mutations prevent effective binding of RIF within the β subunit pocket. Approximately half of the mutations accounting for RIF resistance occur in amino acids within this binding pocket that are in close enough proximity to directly interact with RIF when it is complexed with RNAP. Other residues known to account for RIF resistance occur one layer removed from the binding pocket and can lead to distortions of the binding pocket that also prevent RIF binding (7).

Fitness of RIF-resistant mutants and compensatory mutations

Mutations in the RRDR prevent binding of RIF to RNAP and at the same time permit effective function of RNAP. When *M. tuberculosis* is grown in the presence of RIF, mutants with classic mutations in the RRDR are easily selected. Such laboratory-derived *M. tuberculosis* *rpoB* mutant strains show a variable reduction in fitness as measured by their *in vitro* growth rate. In contrast, some clinical isolates that are RIF resistant with the same RRDR mutation displayed fewer fitness costs (using *in vitro* growth rates) when compared with their wild-type counterparts, raising the possibility of compensatory mechanisms to correct for loss of fitness accrued by the original RIF resistance-conferring mutation (14).

Recent studies have compared laboratory-derived *rpoB* RIF-resistant strains, passaged *in vitro* to select for compensatory mutations, with collections of MDR clinical isolates. These studies have shown that RIF-resistant strains with classic mutations in the RRDR also contain other mutations that are associated with a reduction in the fitness cost of the RIF resistance-conferring mutation. These putative fitness-compensatory mutations localized to the *rpoA* gene encoding the α subunit of RNAP and to the *rpoC* gene encoding the β′ subunit of RNAP (15). Although functional analysis of these mutations has not been completed in *M. tuberculosis*, studies in other RIF-resistant organisms have confirmed that secondary mutations in *rpoA* and *rpoC* do act in a compensatory manner. Some also confer low-level resistance to RIF (16). Hence, it appears that mutations occurring in other subunits of the multisubunit RNAP enzyme can compensate for the relative fitness loss of mutations in the *rpoB* RRDR that confer high-level RIF resistance.

INH

INH is a first-line drug for TB that has rapid bactericidal activity. During combination therapy of drug-susceptible TB infection, INH is the drug principally responsible for the 1 to 2 log decline of bacterial load in sputum seen within the first 14 days of treatment (17). INH is also widely used as a treatment for latent infection of *M. tuberculosis*. While its structure of a pyridine ring and hydrazide group is simple, the mechanism of action of INH is incredibly complex.

Mechanism of action

It has long been postulated that INH inhibits the biosynthesis of mycolic acids (18), resulting in accumulation of long-chain fatty acids (19) and cell death (20). However, the details of this complex process have only recently been elucidated.

INH is a prodrug that is activated by the mycobacterial catalase-peroxidase enzyme, KatG, encoded by *katG* (Rv1908c, MT1959) (21). Following activation by KatG, INH forms an adduct with NAD (22). This INH-NAD adduct then binds and inhibits a mycobacterial protein, InhA (23). *inhA* (Rv1484, MT1531) encodes InhA, an NADH-dependent enoyl-acyl carrier protein (ACP) reductase (24, 25), which is part of the fatty acid elongation system, fatty acid synthase type II (FASII), responsible for mycolic acid biosynthesis (26). Inhibition of InhA by the INH-NAD adduct results in intracellular accumulation of long-chain fatty acids, decreased mycolic acid biosynthesis, and eventual cell death.

Mechanism of resistance: loss of activation: *katG*

The two main molecular mechanisms of INH resistance are (i) loss of INH activation by *katG* (27) and (ii) increases in *inhA* expression or modification of the InhA target. Approximately 75 to 90% of INH resistance can be attributed to polymorphisms in the *katG* gene or the *inhA* promoter and gene (28).

The majority of clinical isolates that are resistant to INH harbor mutations in *katG* (29, 30). A broad range of polymorphisms has been reported worldwide including missense and nonsense mutations, insertions, deletions, truncation, and rarely, whole gene deletions (27, 31, 32). Mutations in *katG* result in a reduced ability to form the INH-NAD adduct (33) and a high level of INH resistance. The most common mutation occurring in clinical isolates in *katG* is S315T (30, 34–38).

Polymorphisms in *katG* may result in diminished or loss of catalase and peroxidase activity. Because these enzymatic activities are critical for *M. tuberculosis* defense against reactive oxygen species and virulence *in vivo* (39–41), mutations in *katG* can lead to a loss of fitness to the bacterium. However, unlike other polymorphisms, the S315T mutation renders *katG* unable to activate INH while still retaining some peroxidase and catalase activities, reducing the fitness cost to this particular mutation. It is likely that this retention of catalase-peroxidase activity and fitness with the S315T *katG* is responsible for the predominance of this mutation among drug-resistant isolates (41).

Mechanism of resistance: overexpression or target modification: *inhA*

While the majority of INH-resistant clinical strains contain *katG* mutations, a significant portion of INH-resistant clinical isolates have a wild-type *katG*, suggesting the existence of additional resistance mechanisms. Other such mechanisms involve changes in *inhA* expression and modification of the InhA target.

Association with INH resistance has been well documented for polymorphisms within the *inhA* promoter at the −15T and −8A loci. Mutations at this site result in overexpression of *inhA*, which confers low-level resistance to INH and cross-resistance to ethionamide (ETH) (42, 43). The prevalence of *inhA* promoter mutations varies geographically, but these mutations rarely account for greater than 20% of INH resistance (44, 45). While structural mutations in InhA are rare, the S94A mutation reduces affinity for the NADH cofactor and confers low-level resistance (25). Approximately 10% of INH resistance is unexplained by mutations in *katG* and *inhA* (28). Other genes including *kasA*, *ahpC*, *ndh*, and the *ahpC-oxyR* intragenic region have been associated with resistance to INH, but their impact on resistance among clinical isolates remains unclear (46–48).

Pyrazinamide

Pyrazinamide (PZA), a nicotinamide analogue, is a critically important first-line TB drug. The addition of PZA to the intensive phase of TB therapy has allowed treatment duration to be shortened from over 1 year to just 6 months of therapy (49). PZA has little bactericidal activity but has critical sterilizing activity during the later phases of treatment when relatively low numbers of persister or semi-dormant organisms remain (50).

PZA is the least understood drug in the current TB drug arsenal. It has no activity against *M. tuberculosis* grown under normal *in vitro* culture conditions, but killing can be demonstrated *in vitro* at acidic pH, typically pH 5.5 (51). Hence, phenotypic testing for PZA susceptibility is complicated and has not been fully standardized; many clinical laboratories do not routinely test for PZA susceptibility. Another curiosity of PZA is that in contrast to *M. tuberculosis*, *Mycobacterium bovis* and *M. bovis* BCG are naturally resistant to the drug. Recently, as described below, the molecular basis for this difference in susceptibility has been determined.

Mechanism of action

PZA is a prodrug that is converted by the cytoplasmic mycobacterial enzyme, pyrazinamidase, into pyrazinoic acid (POA) (52). As described below, loss of the bacterial-encoded activating enzyme confers resistance to PZA. However, the molecular targets of POA remain uncertain. Several possible mechanisms have been postulated including inhibition of trans-translation, inhibition of fatty acid synthase I, and modulation of host targets.

Mechanism of resistance: loss of activation: *pncA*

In 1996, Zhang and colleagues identified mutations in the *pncA* gene (Rv2043c, MT2103), which encodes the mycobacterial pyrazinamidase, as the source of PZA resistance. The gene product is an amidase that hydrolyzes PZA to POA and ammonia. Mutations in *pncA* are the major mechanism for PZA resistance in *M. tuberculosis* (52–55). As might be expected, a plethora of mutations has been found in the *pncA* genes of PZA-resistant clinical isolates since virtually any mutation that confers loss of enzyme function is sufficient to confer PZA resistance. PZA resistance-conferring alterations are widely dispersed throughout the 561-amino-acid open reading frame (ORF), while resistance-conferring changes in the 82-base-pair region of the putative promoter are rare (56). *M. bovis* BCG, which is naturally resistant to PZA, has an H57D amino acid substitution in PncA, which leads to a lack of activation. Interestingly, up to 15 to 30% of PZA-resistant clinical isolates harbor an intact *pncA* gene (57–59).

Other factors in PZA resistance

In addition to the importance of *pncA* in activating PZA, three possible mechanisms and bacterial targets have been implicated in the drug's action. The first is the weak acid activity of POA. Zhang and colleagues showed that following PZA entry into bacilli by passive diffusion and deamidation by cytoplasmic PncA, POA (pK_a 2.9) is trapped in the bacterial cytoplasm because it is a carboxylate anion with only a small amount of elimination by bacterial efflux pumps. Because the protonated acid form of POA may diffuse back out as a neutral molecule, accumulation of protonated POA external to the cell leads to collapse of the bacterial membrane potential, killing the bacillus (60). A second postulated mechanism is an inhibitory effect of POA on the microbial fatty acid synthase type 1 (FAS-1) system. Zimhony and colleagues used 5-Cl-PZA and showed a potent inhibitory effect on *Mycobacterium smegmatis* that could be reduced by overexpression of the FAS-1 gene. They also demonstrated 5-Cl-PZA inhibitory activity against *M. tuberculosis* FAS-1 and that under acidic conditions, PZA itself inhibits the mycobacterial FAS-1, findings that suggest that FAS-1 is the target of activated PZA (61). Later studies, however, confirmed the activity of 5-Cl-PZA against FAS-1 but could not reproduce that the effects are not observed with PZA or POA themselves (62).

Recently, Shi and colleagues studied PZA-resistant *M. tuberculosis* strains that lack *pncA* mutations (63). In three such strains they observed mutations in the *rpsA* gene, which encodes the ribosomal protein S1; these mutations were ΔA438, T5S, and D123A. Over-expression of *rpsA* increased *M. tuberculosis* resistance to PZA in wild-type bacteria but not in the *rpsA* ΔA438 mutant. RpsA in *E. coli* has been shown to bind to transfer-messenger RNA (tmRNA) in a process called trans-translation that rescues stalled ribosomes during periods of slow growth or nutrient depletion. *M. tuberculosis* RpsA was shown to have this tmRNA binding activity. This study suggests that PZA activity may interfere with *M. tuberculosis* trans-translation, which may account for the drug's unique activity against the persister forms.

Host-directed activity of PZA

In 2009, Mendez and colleagues (64) showed that PZA had an inhibitory effect on the parasitic organism *Leishmania major* both *in vitro* and in a murine model. In addition to controlling the proliferation of *L. major*, PZA treatment of macrophages in culture led to the upregulation of several cytokines including interleukin-12 (IL-12) in both the presence and absence of parasite infection. The authors concluded that PZA has a collat-eral host-directed effect that is beneficial in leishmania-sis. Recently, Manca and colleagues performed similar studies with *M. tuberculosis* (65). They showed signifi-cant reductions in the secretion of the proinflammatory cytokines tumor necrosis factor alpha, IL-6, IL-1β, and monocyte chemotactic protein-1 when *M. tuberculosis*-infected human monocytes were treated with PZA. The study also showed that on transcriptional analy-sis of the mouse lung transcriptome during *M. tuber-culosis* infection, treatment with PZA significantly upregulated the transcription of the adenylate cyclase and peroxisome-proliferator activated receptor (PPAR) genes that govern anti-inflammatory responses. The study suggests that PZA triggers an anti-inflammatory response in the host, which may play a role in the drug's action during chemotherapy.

To evaluate the hypothesis that PZA may have a host-targeted mechanism of action, Almeida and col-leagues tested PZA monotherapy against *M. bovis* (resistant to PZA due to a *pncA* mutation) in BALB/c mice and found no activity compared to untreated controls. They also evaluated the action of the drug in nude versus immunocompetent BALB/c mice (66). Over 8 weeks, BALB/c mice displayed a 4-log reduction on lung bacterial CFU counts with RIF-EMB-PZA and a 2-log reduction with RIF-EMB alone. Hence, PZA added 2 logs of killing in BALB/c mice. In contrast, nude mice showed only a 1.5-log reduction with both regimens, indicating that PZA did not add to the killing

potency of the other two drugs in nude mice. These results suggest that in addition to the need for PncA-mediated drug activation, an intact host immune sys-tem is needed for bacterial killing by PZA.

Ethambutol

Ethambutol (EMB) is the least effective of the first-line antitubercular drugs and is considered bacteriostatic against metabolically active bacteria. It is incorporated in first-line therapy mainly to protect against the emer-gence of drug resistance.

Mechanism of action

While the mechanism of action and genetic basis of re-sistance to EMB have not been fully elucidated, the arabinosyltransferases have been identified as molecu-lar targets for EMB (67, 68). These membrane-associat-ed enzymes are involved in the polymerization of arabinan, a component of arabinogalactan, the major polysaccharide of the mycobacterial cell wall (69–71). Interruption of arabinan biosynthesis leads to intra-cellular accumulation of mycolic acids and eventual cell death. *embCAB*, a three-gene operon encoding the arabinosyltransferases *embC*, *embA*, and *embB*, has been identified in *M. tuberculosis* (72). These enzymes share 65% homology with each other.

Mechanism of resistance: target modification: *embCAB* operon

EMB resistance is most frequently associated with mu-tations in the *embCAB* operon (Rv3793-5, MT3900-2). In particular, mutations in *embC* and *embB* have been associated with resistance, with codons 306, 406, and 497 in *embB* being the most common (72). In some samples, 50 to 70% of EMB-resistant clinical isolates contain polymorphisms in codon 306 of *embB*. While polymorphisms in codon 306 have also been identified among clinical isolates that are susceptible to EMB, this has previously been attributed to difficulties with phenotypic drug susceptibility testing for the low-level drug resistance conferred by *embB* mutations (73, 74). Further evidence of the importance of codon 306 to EMB resistance is that the variant amino acid motifs at this locus in *Mycobacterium leprae*, *Mycobacterium abscessus,* and *Mycobacterium chelonae* are thought to be responsible for the natural resistance of these orga-nisms to EMB (68).

Because a significant percentage of EMB-resistant isolates lack mutations in these genes, other yet un-identified resistance mechanisms must be important (75). One of these mechanisms may involve the gene *embR*, which has been noted to modulate the level of arabinosyltransferase activity *in vitro* (76), but its role in clinical resistance is unknown. More recently, genomic characterization of *in vitro*–selected EMB-resistant strains identified some additional loci (77). For example, mutations in Rv3806c, encoding *ubiA*, an enzyme that synthesizes a donor substrate for arabinosyltranferases, were identified in EMB-resistant clinical isolates as well as selected for *in vitro*. In combination with *embCAB* mutations, *ubiA* mutations are associated with higher-level resistance.

INJECTABLE AGENTS

Aminoglycosides

Aminoglycosides have been used to treat TB since the introduction of the first antituberculous agent, streptomycin (STR). Currently, STR remains a first-line agent, although its use is largely restricted to retreatment cases because like other aminoglycosides it must be given parenterally. AMK and KAN are the other major aminoglycosides used for TB. These may be given intravenously or intramuscularly and are second-line agents used to treat MDR- or XDR-TB. Aminoglycosides are considered bactericidal drugs and achieve rapid bacterial killing during the initiation phase of treatment. They have poor sterilizing activity and must be combined with agents such as rifamycins and PZA in order to achieve durable cure.

Mechanism of action

Aminoglycosides inhibit protein synthesis by binding to the 30S subunit of the mycobacterial ribosome. The majority of mutations that confer aminoglycoside resistance lead directly or indirectly to alterations in the aminoglycoside binding pockets of the ribosome, which prevent drug binding but preserve ribosome function as occurs with mutations in the *rpsL*, *rrs*, and *gidB* genes. A minor mechanism of aminoglycoside resistance is drug modification like that which occurs with the Eis protein, an aminoglycoside acetyltransferase.

STR is a streptidine aminoglycoside. On binding to the 30S ribosomal subunit, STR inhibits translational initiation and may also cause misreading of mRNA (78). AMK and KAN are deoxystreptamine aminoglycosides, which bind to a different locus on the 30S ribosome. Because of these differences in ribosome binding site between STR and AMK/KAN, different drug-conferring mutations are associated with these two drug groups. In contrast, owing to the structural and functional similarity between AMK and KAN, there is extensive cross-resistance between them.

Mechanism of resistance: target modification mutations

rpsL

The *rpsL* gene (Rv0682, or MT0710) encodes the 12S protein, which is a structural component of the ribosome. The RpsL protein serves to stabilize the pseudoknot that is formed by the 16S rRNA component of the 30S ribosome. While *rpsL* is an essential gene, nonsynonymous mutations are tolerated for ribosomal function but result in reduced aminoglycoside binding. Common mutations in aminoglycoside resistance that confer resistance to STR are RpsL K43R and K88R (79, 80).

rrs

The *rrs* gene encodes the 16S rRNA itself. The 16S rRNA contains key structures such as the 530 stem loop and the 915 turn; both of these rRNA structures form contacts for the binding of aminoglycosides. Like *rpsL*, *rrs* is an essential gene. Nonsynonymous SNPs in the *rrs* gene result in reduced aminoglycoside binding while allowing for preserved ribosome function. Many bacteria harbor multiple copies of the *rrs* gene, and consequently single-allele *rrs* mutations confer low-level aminoglycoside resistance. In contrast, *M. tuberculosis* has only one *rrs* gene, and mutations in the *M. tuberculosis rrs* gene are usually associated with high-level aminoglycoside resistance.

While more than 20 *rrs* mutations have been associated with aminoglycoside resistance, the common polymorphisms are A1401G, A514C, and C517T. Some *rrs* mutations also confer resistance to the cyclic peptide antibiotic CAP (see below). The most common mutation for second-line aminoglycoside resistance is A1401G, which was found in 78%, 56%, and 76% of AMK-, KAN-, and CAP-resistant strains, respectively, in one large systematic review (81).

gidB

The *gidB* gene (Rv3919c, MT4038) encodes a 7-methylguanosine methyltransferase that specifically modifies residues on the 16S rRNA (*rrs*). *gidB* is a nonessential gene, and loss-of-function mutations in *gidB* result in failure to methylate G527 within the 530 loop of the 16S rRNA molecule (82). Reduced ribosomal methylation confers low-level aminoglycoside resistance by reducing the affinity of the drugs for the 16S rRNA binding site. Many different *gidB* mutations including deletions are associated with aminoglycoside resistance, suggesting that loss of function confers resistance. Polymorphisms in *gidB* have also been identified in drug-susceptible strains, indicating that as for other resistance-conferring loci, the presence of a mutation in a gene is not necessarily indicative of resistance.

Mechanism of resistance: inactivating mutations: *eis*

The *eis* gene (Rv2416c, MT2489) encodes an aminoglycoside acetyltransferase, which has an affinity for KAN. KAN acetylation inactivates the drug by preventing it from binding to the 30S ribosome. Promoter upregulating mutations in the 5′ untranslated region of *eis* are associated with clinically relevant *M. tuberculosis* resistance to KAN. While the Eis protein is capable of acetylating AMK, its affinity for AMK is low, and studies of clinical isolates with *eis* promoter-up mutations reveal selective KAN resistance with relative preservation of AMK susceptibility (83). Nevertheless, clinical isolates with *eis* mutations that are KAN- and AMK-resistant have been described (79), so the specificity of *eis* mutations for KAN resistance remains uncertain.

Mechanism of resistance: upregulation of drug inactivator and drug efflux: *whiB7*

Recently, mutations in the promoter region of the transcriptional activator *whiB7* have been identified in clinical isolates. These promoter mutations result in upregulation of the *whiB7* regulon, leading to an increase in *eis* expression and resulting KAN resistance (84). The *whiB7* regulon also includes Rv1258c, which encodes an efflux pump, *tap*, that is upregulated in tandem with *eis*, resulting in STR resistance. *whiB7* promoter mutations therefore can result in both a target modification and efflux mechanism of resistance, leading to cross-resistance to two aminoglycoside drug groups (84).

STR resistance-conferring mutations: *rpsL, rrs, gidB*

High-level STR resistance is achieved predominantly by mutations in *rpsL* (~50% of STR-resistant strains) and *rrs* (~15% of STR-resistant strains). Important *rrs* mutations associated with STR resistance are *rrs* A514C and A908C. While the *rrs* A514C mutation is associated with high-level resistance to STR, the most common *rrs* mutation conferring AMK and KAN resistance (*rrs* A1401G) does not confer STR cross-resistance. *gidB* mutations are less frequent among STR-resistant strains (~20% of strains) and are associated with low-level STR resistance (79). Mechanisms of low-level resistance have not been well characterized but probably involve drug efflux.

AMK- and KAN-resistance-conferring mutations: *rrs, eis*

The *rrs* A1401G is the most frequent mutation conferring AMK and KAN resistance, occurring in ~85% or more of strains that are AMK or KAN resistant. The presence of A1401G appears to be 100% specific for coresistance to AMK and KAN. Of AMK- and KAN-resistant strains, 10 to 15% harbor *eis* mutations, indicating that *eis* is a minor determinant of AMK and KAN resistance (45, 79).

Capreomycin

CAP and viomycin (VIO) are cyclic peptide antibiotics with structural similarity. The drugs have uniform cross-resistance in *M. tuberculosis* and appear to have similar mechanisms of action. While VIO is rarely used due to high toxicity, CAP is an injectable drug commonly used as a second-line agent in the management of MDR- and XDR-TB that is resistant to aminoglycosides.

Like the structurally unrelated aminoglycosides, CAP and VIO are bactericidal drugs that inhibit protein synthesis. VIO has been shown to bind both the 30S and 50S ribosome subunits and to inhibit ribosomal translocation by interference with the peptidyl tRNA acceptor site (85, 86). Due to overlap in the binding region of CAP and the aminoglycosides, certain mutations confer cross-resistance to CAP and AMK/KAN. In contrast, cross-resistance between CAP and STR is rare. The major mechanisms of CAP resistance are mutations that result in ribosome modification, particularly *rrs* and *tlyA*. Interestingly, *tlyA* mutations uniquely affect CAP resistance and do not appear to play a role in resistance to aminoglycosides.

Mechanism of resistance: target modification mutations

rrs

The *rrs* A1401G mutation is found in ~85% of CAP-resistant XDR-TB strains. Other *rrs* mutations including C1402T and G1484T are also associated with CAP resistance (79).

tlyA

The *tlyA* gene (Rv1694, MT1733) is a nonessential gene found in many bacteria. Transposon mutants of *tlyA* as well as spontaneous point mutants display significant CAP and VIO resistance. Biochemical, genetic, and comparative genomics suggest that the *tlyA* gene is an rRNA methyltransferase and that loss of methyltransferase activity yields an unmethylated ribosome that is resistant to CAP inhibition (87). This mechanism is similar to that of the *eis* gene that confers resistance to KAN. Numerous *tlyA* mutations have been reported including L180R, S265T, S64W, frameshift at 218L, N236K, and L150P (79, 87, 88).

SECOND-LINE AGENTS

Quinolones

Fluoroquinolones are second-line antitubercular agents, and they currently form the backbone of MDR-TB therapy. In addition to use in drug-resistant TB, quinolones are indicated for the treatment of drug-susceptible TB among patients intolerant to components of first-line regimens.

Mechanism of action

Quinolones inhibit bacterial topoisomerases, enzymes that regulate the supercoiling of DNA and are thus essential for DNA replication, transcription, and recombination (89). While many bacterial species contain both topoisomerase II (also known as DNA gyrase) and topoisomerase IV, *M. tuberculosis* lacks an analogue of topoisomerase IV (90). Thus, DNA gyrase is the sole target for quinolone activity in *M. tuberculosis*. Clinically observed differences in efficacy among quinolones may be explained by specificity for different topoisomerases; ciprofloxacin, which is less effective in *M. tuberculosis*, primarily targets topoisomerase IV, which is lacking, whereas newer-generation fluoroquinolones that are also known to have improved efficacy, including moxifloxacin and levofloxacin, preferentially target DNA gyrase (91).

DNA gyrase is an ATP-dependent enzyme that cleaves and religates double-stranded DNA, which allows for the introduction of negative supercoils into DNA. DNA gyrase is a tetrameric protein, comprised of two α and two β subunits, encoded by the genes *gyrA* (Rv0006, MT0006) and *gyrB* (Rv0005, MT0005), respectively (92).

Mechanism of resistance:
target modification: *gyrA* and *gyrB*

A highly conserved area within *gyrA*, known as the quinolone-resistance-determining region (QRDR), contains mutations that have been shown to confer fluoroquinolone resistance in many bacterial species, including *M. tuberculosis* (92). In addition, *gyrB* mutations can also confer fluoroquinolone resistance. More than 90% of quinolone-resistant *M. tuberculosis* strains have mutations in *gyrA* or *gyrB* (93, 94). Classically, mutations in codons 90 and 94 of *gyrA* are most commonly associated with drug resistance, with A90V, D94G, and D94H frequently noted among clinical

isolates (45, 92, 94–97). Among quinolone-resistant isolates, double mutants in *gyrA* or additionally in *gyrB* have been noted to have higher MICs (98, 99).

Two additional potential mechanisms of quinolone resistance are efflux and DNA mimicry (100–103). However, to date these resistance mechanisms have been identified only in laboratory *M. tuberculosis* strains, so the clinical relevance and epidemiological impact of this resistance mechanism are unknown.

Ethionamide/Prothionamide

ETH is a commonly used oral, second-line agent used for the treatment of MDR- and XDR-TB. It is occasionally used in the treatment of drug-susceptible TB, primarily when patients cannot tolerate INH. Prothionamide is also in the thionamide drug class, and it contains the same active moiety as ETH. Because there is complete cross-resistance between ETH and prothionamide, these drugs are used interchangeably and can be considered functionally identical (104, 105).

Mechanism of action

ETH is a structural analog of INH, and like INH it inhibits mycolic acid biosynthesis. It has long been known that strains displaying low-level resistance to INH are also ETH-resistant but that strains with high-level INH resistance can remain ETH-susceptible (106). Like INH, ETH is a prodrug and requires activation, but in contrast to INH, which requires KatG activation, ETH is activated by a mono-oxygenase encoded by *ethA*. The EthA mono-oxygenase is unique for ETH and does not activate INH. Following EthA activation, ETH forms an adduct with NAD, and this adduct subsequently inhibits the NADH-dependent enoyl-ACP reductase, InhA, in the same manner as the INH-NAD adduct.

Hence, ETH and INH have a common mechanism of action but different bacterial activating enzymes. Thus, *katG* mutations confer exclusively high-level INH resistance, whereas mutants in the common target of INH and ETH (*inhA*) lead to low-level cross-resistance to INH and ETH. Resistance to ETH arises by two major mechanisms: (i) loss of activating enzyme activity (*ethA*) and (ii) target amplification (*inhA* promoter-up mutations).

Mechanism of resistance:
loss of activation: *ethA*

The *ethA* gene (Rv3854c, MT3969) encodes a nonessential mono-oxygenase. Its association with ETH resistance was identified in 2000 (107, 108). Expression of *ethA* is under the control of a TetR-like repressor known as *ethR* (Rv3855, MT3970), and overexpression of *ethR* leads to ETH resistance. Consistent with the fact that loss of function confers resistance, over 25 mutations in *ethA* have been associated with ETH resistance. In contrast to KatG, where the dominant INH-resistance-conferring mutation is S315T, a mutation that preserves catalase-peroxidase activity while failing to activate INH, EthA mutations are widely dispersed throughout the *ethA* gene. This suggests that there is little fitness cost for loss of EthA mono-oxygenase activity, possibly because there are over 30 other mono-oxygenases in the *M. tuberculosis* genome that may play compensatory roles (109).

In some surveys of clinical isolates, mutation in *ethA* occurs in 50% of strains with or without other ETH-resistance-associated mutations, while *inhA* mutations account for the remaining 50% (109). Other surveys have found that up to 50% of ETH resistance is unexplained by *inhA* or *ethA* mutations (48).

Mechanism of resistance: target modification: *inhA*

inhA encoding an NADH-dependent enoyl-ACP reductase, which is required for mycolic acid biosynthesis, is the target of both the ETH-NAD adduct and the INH-NAD adduct. Among strains showing ETH resistance without mutations in *ethA*, *inhA* promoter mutations account for about two-thirds of resistance, while promoters in combination with ORF mutations account for one-third. A small number of ETH-resistant strains have mutations only in the *inhA* ORF, usually with the S94A, S94W, or L11V alterations (48, 109).

mshA (Rv0486, MT0504), which encodes for a glycosyl-transferase involved in mycothiol biosynthesis, may also be associated with drug resistance to ETH in certain isolates (110).

OTHER DRUGS

Despite advances in defining the molecular targets of additional drugs and drug classes, reliable resistance mutations have not been fully characterized for the following agents.

Cycloserine

D-Cycloserine (CYS) is an analogue of the amino acid D-alanine, which competitively inhibits two essential enzymes required for peptidoglycan formation and thus cell wall biosynthesis. Terizidone (TER) combines two molecules of cycloserine and is thought to act similarly. Alanine racemase (*alr*, Rv3423c, MT3532) converts L-alanine to D-alanine, which is a substrate for the second enzyme, D-alanine-D-alanine ligase (*ddlA*, Rv2981c, MT3059), which incorporates D-alanine into the elongating pentapeptide necessary for peptidoglycan synthesis (111, 112). In *M. smegmatis*, overexpression of *alr* and *ddlA* results in resistance to CYS (113, 114).

CycA (Rv1704c, MT1744), a bacterial D-serine/L- and D-alanine/glycine/D-cycloserine proton symporter, has also been implicated in CYS susceptibility. A nonsynonymous SNP in *cycA*, G122S, has recently been identified in BCG, which has now been shown to be partially responsible for BCG's innate resistance to CYS (115).

Para-Aminosalicylic Acid

Para-aminosalicylic acid (PAS) is a second-line oral drug for TB. It was introduced in 1948 shortly after STR, and its early use without partner drugs was associated with the rapid emergence of resistance. PAS is an analogue of para-amino benzoic acid (PABA) and has long been known to interfere with folic acid biosynthesis. *thyA* (Rv2764c, MT2834) encoding thymidylate synthase was thought to be a resistance gene; however, large surveys have shown only a weak association between *thyA* polymorphisms and PAS resistance among clinical isolates (116). Recent studies have shown that the enzyme dihydropteroate synthase misincorporates PAS as if it were PABA at the dihydropteroate stage of the enzymatic pathway (117). This altered substrate serves to block the biosynthetic pathway downstream of dihydropteroate synthase including dihydrofolate reductase. RibD, a bifunctional enzyme involved in riboflavin biosynthesis (*ribD*, Rv2671, MT2745), was found to have dihydrofolate reductase activity, and clinical isolates that overexpress RibD due to mutation are PAS-resistant (118). To date, specific mutations in additional pathway enzymes that confer PAS resistance remain poorly defined in clinical isolates.

Oxazolidinones

Linezolid (LIN), an oxazolidinone, has shown promise among limited numbers of patients in treatment for drug-resistant TB (119, 120). LIN binds the bacterial 23S portion of the 50S subunit of rRNA, thereby inhibiting formation of the initiation complex. Mutations in the 23S rRNA gene have been documented with *in vitro* selection (121). More recently, mutations in the 50S ribosomal protein L3, *rplC*, at position T460C have been noted among *in vitro*–selected LIN-resistant isolates and also among several LIN-resistant clinical isolates (122).

β-Lactams

β-Lactam antibiotics bind transpeptidases that cross-link peptidoglycans and thus inhibit cell wall synthesis. *M. tuberculosis* possesses a highly active β-lactamase, BlaC, which hydrolyzes β-lactams, rendering them ineffective against *M. tuberculosis* except in the presence of a β-lactamase inhibitor (123). A β-lactam–β-lactamase inhibitor combination, meropenem-clavulanate, has been associated with encouraging clinical outcomes in a limited number of patients with XDR-TB (124). The genomic markers of β-lactam drug resistance in *M. tuberculosis* have yet to be elucidated.

Clofazimine

Clofazimine (CFZ) is a riminophenazine dye that has recently shown promise for use in MDR-TB and may allow for treatment regimen shortening (125). Though the precise mechanism of action of CFZ has remained elusive, the drug may target NDH-2, the primary NADH dehydrogenase involved in the respiratory chain, resulting in production of bacterial reactive oxygen species (126).

NOVEL AGENTS

Bedaquiline

Bedaquiline (BDQ, formerly known as TMC207) is a diarylquinoline drug approved for use against MDR-TB in 2012. It is the first new drug class to be introduced in the treatment of TB since the rifamycins in the 1960s. BDQ binds to and inhibits the *M. tuberculosis* ATP synthase gene encoded by the essential gene *atpE* (Rv1305, MT1345) (127). AtpE forms a part of the F_1F_0 proton ATP synthase, a transmembrane protein complex that generates ATP from proton translocation, and inhibition of ATP biosynthesis has been shown to kill both actively dividing and nondividing bacteria (128).

In vitro resistance to BDQ in mycobacteria has been shown to involve mutations in *atpE* (129). However, resistant mutants have been identified that do not have any mutations in *atpE* or in the other genes encoding components of ATP synthase, suggesting alternative mechanisms of resistance or perhaps other targets for BDQ (130).

Verapamil, an efflux pump inhibitor, was recently found to decrease profoundly the MIC of BDQ and CFZ to *M. tuberculosis* by 8- to 16-fold (131). Thus, efflux inhibition is an important sensitizer of BDQ and CFZ, and efflux may emerge as a resistance mechanism to these drugs.

DIAGNOSTIC TOOLS FOR DETECTING DRUG-RESISTANT *M. TUBERCULOSIS*

As discussed above, for each first-line anti-TB drug, there are several known *M. tuberculosis* genetic mutations commonly associated with bacterial drug resistance. Capitalizing on these genetic associations has allowed for the development of molecular diagnostics, including the GeneXpert and the Hain Line Probe Assay MTBDRplus, which are able to accelerate the identification of drug resistance to two of the most important TB drugs, RIF and INH.

GeneXpert is a PCR-based, point-of-care system that is able to quickly and accurately analyze a sample for the presence of the *M. tuberculosis* complex as well as assay for RIF resistance (132). Building upon the assumption that approximately 95% of RIF resistance is conferred by mutations in the *rpoB* gene, GeneXpert utilizes a molecular beacon to detect the presence of a mutated *rpoB* gene as an indicator of drug-resistant TB. Similarly, MTBDRplus uses *M. tuberculosis* complex–specific probes to assess for wild-type or mutant-specific probes to identify mutations conferring resistance in the RRDR of *rpoB*, as well as *katG* and *inhA* mutations that are known to confer INH drug resistance (133). A similar test that includes mutations to select second-line agents is also available.

However, the genetic mutations included in these assays do not account for all known drug resistance to their cognate anti-TB drugs. This genomic ambiguity translates into a relatively high false-negative rate of the above-mentioned genetics-based assays, which is a major limitation to their clinical utility. Indeed, MTBDRplus fails to identify 15 to 30% of INH resistance and 5% of RIF resistance. Improving the performance of these diagnostic tests will require identification of the additional molecular mechanisms of resistance and inclusion of additional polymorphisms that can be used to identify a minority of resistant phenotypes.

Genome sequencing shows promise as a reliable modality for the rapid identification of resistance. One recent study showed that by subjecting the primary growth of an isolate to genomic sequencing, XDR-TB could be predicted weeks before the culture-based diagnosis was made (134).

GENOMICS AND DRUG RESISTANCE

While significant progress has been made in identifying independent genes in which mutations confer resistance, some resistance cannot be accounted for by the current models. A possible explanation is that multiple incremental mutations combine to produce phenotypic resistance. Recently, there have been several genomic surveys seeking resistance "signatures" that may correlate with either individual resistance phenotypes or certain multidrug-resistant phenotypes. Several recent whole-genome sequencing projects involving resistant *M. tuberculosis* strains have led to the identification of over 100 genetic loci that are associated with resistance including genes involved in the synthesis or regulation of surface exposed lipids (135, 136). As sequencing of *M. tuberculosis* strains becomes more common, these signatures will have better definition and will have the potential to identify additional resistance mechanisms as well as form the basis of algorithmic prediction tools for diagnosing all drug-resistant TB.

Citation. Cohen KA, Bishai WR, Pym AS. 2014. Molecular basis of drug resistance in *Mycobacterium tuberculosis*. Microbiol Spectrum 2(3):MGM2-0036-2013.

References

1. **World Health Organization.** 2012. Global Tuberculosis Report. Geneva, Switzerland.

2. **Jassal M, Bishai WR.** 2009. Extensively drug-resistant tuberculosis. *Lancet Infect Dis* 9:19–30.

3. **Velayati AA, Masjedi MR, Farnia P, Tabarsi P, Ghanavi J, Ziazarifi AH, Hoffner SE.** 2009. Emergence of new forms of totally drug-resistant tuberculosis bacilli: super extensively drug-resistant tuberculosis or totally drug-resistant strains in Iran. *Chest* 136:420–425.

4. **Udwadia ZF, Amale RA, Ajbani KK, Rodrigues C.** 2012. Totally drug-resistant tuberculosis in India. *Clin Infect Dis* 54:579–581.

5. **Aristoff PA, Garcia GA, Kirchhoff PD, Hollis Showalter HD.** 2010. Rifamycins: obstacles and opportunities. *Tuberculosis (Edinb)* 90:94–118.

6. **Zhang G, Campbell EA, Minakhin L, Richter C, Severinov K, Darst SA.** 1999. Crystal structure of *Thermus aquaticus* core RNA polymerase at 3.3 A resolution. *Cell* 98:811–824.

7. **Campbell EA, Korzheva N, Mustaev A, Murakami K, Nair S, Goldfarb A, Darst SA.** 2001. Structural mechanism for rifampicin inhibition of bacterial RNA polymerase. *Cell* 104:901–912.

8. **Chakraborty A, Wang D, Ebright YW, Korlann Y, Kortkhonjia E, Kim T, Chowdhury S, Wigneshweraraj S, Irschik H, Jansen R, Nixon BT, Knight J, Weiss S, Ebright RH.** 2012. Opening and closing of the bacterial RNA polymerase clamp. *Science* 337:591–595.

9. **Zhang Y, Feng Y, Chatterjee S, Tuske S, Ho MX, Arnold E, Ebright RH.** 2012. Structural basis of transcription initiation. *Science* 338:1076–1080.

10. **Gomez JE, Chen JM, Bishai WR.** 1997. Sigma factors of *Mycobacterium tuberculosis*. *Tuber Lung Dis* 78:175–183.

11. **Telenti A, Imboden P, Marchesi F, Lowrie D, Cole S, Colston MJ, Matter L, Schopfer K, Bodmer T.** 1993.

Detection of rifampicin-resistance mutations in *Mycobacterium tuberculosis*. *Lancet* 341:647–650.

12. **Bodmer T, Zürcher G, Imboden P, Telenti A, Zurcher G.** 1995. Mutation position and type of substitution in the Beta-subunit of the RNA polymerase influence in-vitro activity of rifamycins in rifampicin-resistant *Mycobacterium tuberculosis*. *J Antimicrob Chemother* 35:345–348.

13. **Jin DJ, Gross CA.** 1988. Mapping and sequencing of mutations in the *Escherichia coli rpoB* gene that lead to rifampicin resistance. *J Mol Biol* 202:45–58.

14. **Gagneux S, DeRiemer K, Van T, Kato-Maeda M, de Jong BC, Narayanan S, Nicol M, Niemann S, Kremer K, Gutierrez MC, Hilty M, Hopewell PC, Small PM.** 2006. Variable host-pathogen compatibility in *Mycobacterium tuberculosis*. *Proc Natl Acad Sci USA* 103:2869–2873.

15. **Comas I, Borrell S, Roetzer A, Rose G, Malla B, Kato-Maeda M, Galagan J, Niemann S, Gagneux S.** 2012. Whole-genome sequencing of rifampicin-resistant *Mycobacterium tuberculosis* strains identifies compensatory mutations in RNA polymerase genes. *Nat Genet* 44:106–110.

16. **Brandis G, Wrande M, Liljas L, Hughes D.** 2012. Fitness-compensatory mutations in rifampicin-resistant RNA polymerase. *Mol Microbiol* 85:142–151.

17. **Jindani A, Aber VR, Edwards EA, Mitchison DA.** 1980. The early bactericidal activity of drugs in patients with pulmonary tuberculosis. *Am Rev Respir Dis* 121:939–949.

18. **Winder FG, Collins PB.** 1970. Inhibition by isoniazid of synthesis of mycolic acids in *Mycobacterium tuberculosis*. *J Gen Microbiol* 63:41–48.

19. **Takayama K, Schnoes HK, Armstrong EL, Boyle RW.** 1975. Site of inhibitory action of isoniazid in the synthesis of mycolic acids in *Mycobacterium tuberculosis*. *J Lipid Res* 16:308–317.

20. **Takayama K, Wang L, David HL.** 1972. Effect of isoniazid on the in vivo mycolic acid synthesis, cell growth, and viability of *Mycobacterium tuberculosis*. *Antimicrob Agents Chemother* 2:29–35.

21. **Johnsson K, Schultz PG.** 1994. Mechanistic studies of the oxidation of isoniazid by the catalase peroxidase from *Mycobacterium tuberculosis*. *J Am Chem Soc* 116:7425–7426.

22. **Rozwarski DA, Grant GA, Barton DH, Jacobs WR, Sacchettini JC.** 1998. Modification of the NADH of the isoniazid target (InhA) from *Mycobacterium tuberculosis*. *Science* 279:98–102.

23. **Vilchèze C, Weisbrod TR, Chen B, Kremer L, Hazbón MH, Wang F, Alland D, Sacchettini JC, Jacobs WR.** 2005. Altered NADH/NAD+ ratio mediates coresistance to isoniazid and ethionamide in mycobacteria. *Antimicrob Agents Chemother* 49:708–720.

24. **Dessen A, Quémard A, Blanchard JS, Jacobs WR, Sacchettini JC.** 1995. Crystal structure and function of the isoniazid target of *Mycobacterium tuberculosis*. *Science* 267:1638–1641.

25. **Quémard A, Sacchettini JC, Dessen A, Vilcheze C, Bittman R, Jacobs WR, Blanchard JS.** 1995. Enzymatic

characterization of the target for isoniazid in *Mycobacterium tuberculosis*. *Biochemistry* 34:8235–8241.

26. Marrakchi H, Lanéelle G, Quémard A. 2000. InhA, a target of the antituberculous drug isoniazid, is involved in a mycobacterial fatty acid elongation system, FAS-II. *Microbiology* 146(Pt 2):289–296.

27. Zhang Y, Heym B, Allen B, Young D, Cole S. 1992. The catalase-peroxidase gene and isoniazid resistance of *Mycobacterium tuberculosis*. *Nature* 358:591–593.

28. Heysell SK, Houpt ER. 2012. The future of molecular diagnostics for drug-resistant tuberculosis. *Expert Rev Mol Diagn* 12:395–405.

29. Gagneux S, Burgos MV, DeRiemer K, Encisco A, Muñoz S, Hopewell PC, Small PM, Pym AS. 2006. Impact of bacterial genetics on the transmission of isoniazid-resistant *Mycobacterium tuberculosis*. *PLoS Pathog* 2:e61.

30. Ramaswamy SV, Reich R, Dou S, Jasperse L, Pan X, Wanger A, Quitugua T, Graviss EA. 2003. Single nucleotide polymorphisms in genes associated with isoniazid resistance in *Mycobacterium tuberculosis*. *Antimicrob Agents Chemother* 47:1241–1250.

31. Heym B, Alzari PM, Honoré N, Cole ST. 1995. Missense mutations in the catalase-peroxidase gene, katG, are associated with isoniazid resistance in *Mycobacterium tuberculosis*. *Mol Microbiol* 15:235–245.

32. Heym B, Cole ST. 1992. Isolation and characterization of isoniazid-resistant mutants of *Mycobacterium smegmatis* and *M. aurum*. *Res Microbiol* 143:721–730.

33. Ghiladi RA, Cabelli DE, Ortiz de Montellano PR. 2004. Superoxide reactivity of KatG: insights into isoniazid resistance pathways in TB. *J Am Chem Soc* 126:4772–4773.

34. Brossier F, Veziris N, Truffot-Pernot C, Jarlier V, Sougakoff W. 2006. Performance of the genotype MTBDR line probe assay for detection of resistance to rifampin and isoniazid in strains of *Mycobacterium tuberculosis* with low- and high-level resistance. *J Clin Microbiol* 44:3659–3664.

35. Cardoso RF, Cooksey RC, Morlock GP, Barco P, Cecon L, Forestiero F, Leite CQF, Sato DN, Shikama M de L, Mamizuka EM, Hirata RDC, Hirata MH. 2004. Screening and characterization of mutations in isoniazid-resistant *Mycobacterium tuberculosis* isolates obtained in Brazil. *Antimicrob Agents Chemother* 48:3373–3381.

36. Guo H, Seet Q, Denkin S, Parsons L, Zhang Y. 2006. Molecular characterization of isoniazid-resistant clinical isolates of *Mycobacterium tuberculosis* from the USA. *J Med Microbiol* 55:1527–1531.

37. Zhang M, Yue J, Yang Y-P, Zhang H-M, Lei J-Q, Jin R-L, Zhang X-L, Wang H-H. 2005. Detection of mutations associated with isoniazid resistance in *Mycobacterium tuberculosis* isolates from China. *J Clin Microbiol* 43:5477–5482.

38. Hazbón MH, Brimacombe M, Bobadilla del Valle M, Cavatore M, Guerrero MI, Varma-Basil M, Billman-Jacobe H, Lavender C, Fyfe J, García-García L, León CI, Bose M, Chaves F, Murray M, Eisenach KD, Sifuentes-Osornio J, Cave MD, Ponce de León A, Alland D. 2006. Population genetics study of isoniazid resistance mutations and evolution of multidrug-resistant *Mycobacterium tuberculosis*. *Antimicrob Agents Chemother* 50:2640–2649.

39. Ng VH, Cox JS, Sousa AO, MacMicking JD, McKinney JD. 2004. Role of KatG catalase-peroxidase in mycobacterial pathogenesis: countering the phagocyte oxidative burst. *Mol Microbiol* 52:1291–1302.

40. Heym B, Stavropoulos E, Honoré N, Domenech P, Saint-Joanis B, Wilson TM, Collins DM, Colston MJ, Cole ST. 1997. Effects of overexpression of the alkyl hydroperoxide reductase AhpC on the virulence and isoniazid resistance of *Mycobacterium tuberculosis*. *Infect Immun* 65:1395–1401.

41. Pym AS, Saint-Joanis B, Cole ST. 2002. Effect of katG mutations on the virulence of *Mycobacterium tuberculosis* and the implication for transmission in humans. *Infect Immun* 70:4955–4960.

42. Banerjee A, Dubnau E, Quemard A, Balasubramanian V, Um KS, Wilson T, Collins D, de Lisle G, Jacobs WR. 1994. inhA, a gene encoding a target for isoniazid and ethionamide in *Mycobacterium tuberculosis*. *Science* 263:227–230.

43. Mdluli K, Sherman DR, Hickey MJ, Kreiswirth BN, Morris S, Stover CK, Barry CE. 1996. Biochemical and genetic data suggest that InhA is not the primary target for activated isoniazid in *Mycobacterium tuberculosis*. *J Infect Dis* 174:1085–1090.

44. Musser JM, Kapur V, Williams DL, Kreiswirth BN, van Soolingen D, van Embden JD. 1996. Characterization of the catalase-peroxidase gene (*katG*) and *inhA* locus in isoniazid-resistant and -susceptible strains of *Mycobacterium tuberculosis* by automated DNA sequencing: restricted array of mutations associated with drug resistance. *J Infect Dis* 173:196–202.

45. Campbell PJ, Morlock GP, Sikes RD, Dalton TL, Metchock B, Starks AM, Hooks DP, Cowan LS, Plikaytis BB, Posey JE. 2011. Molecular detection of mutations associated with first- and second-line drug resistance compared with conventional drug susceptibility testing of *Mycobacterium tuberculosis*. *Antimicrob Agents Chemother* 55:2032–2041.

46. Lee AS, Lim IH, Tang LL, Telenti A, Wong SY. 1999. Contribution of *kasA* analysis to detection of isoniazid-resistant *Mycobacterium tuberculosis* in Singapore. *Antimicrob Agents Chemother* 43:2087–2089.

47. Baker LV, Brown TJ, Maxwell O, Gibson AL, Fang Z, Yates MD, Drobniewski FA. 2005. Molecular analysis of isoniazid-resistant *Mycobacterium tuberculosis* isolates from England and Wales reveals the phylogenetic significance of the *ahpC* -46A polymorphism. *Antimicrob Agents Chemother* 49:1455–1464.

48. Boonaiam S, Chaiprasert A, Prammananan T, Leechawengwongs M. 2010. Genotypic analysis of genes associated with isoniazid and ethionamide resistance in MDR-TB isolates from Thailand. *Clin Microbiol Infect* 16:396–399.

49. Mitchison DA. 1985. The action of antituberculosis drugs in short-course chemotherapy. *Tubercle* **66**:219–225.

50. Heifets L, Lindholm-Levy P. 1992. Pyrazinamide sterilizing activity in vitro against semidormant *Mycobacterium tuberculosis* bacterial populations. *Am Rev Respir Dis* **145**:1223–1225.

51. Zhang Y, Mitchison D. 2003. The curious characteristics of pyrazinamide: a review. *Int J Tuberc Lung Dis* **7**:6–21.

52. Scorpio A, Zhang Y. 1996. Mutations in *pncA*, a gene encoding pyrazinamidase/nicotinamidase, cause resistance to the antituberculous drug pyrazinamide in tubercle bacillus. *Nat Med* **2**:662–667.

53. Scorpio A, Lindholm-Levy P, Heifets L, Gilman R, Siddiqi S, Cynamon M, Zhang Y. 1997. Characterization of *pncA* mutations in pyrazinamide-resistant *Mycobacterium tuberculosis*. *Antimicrob Agents Chemother* **41**:540–543.

54. Cheng SJ, Thibert L, Sanchez T, Heifets L, Zhang Y. 2000. *pncA* mutations as a major mechanism of pyrazinamide resistance in *Mycobacterium tuberculosis*: spread of a monoresistant strain in Quebec, Canada. *Antimicrob Agents Chemother* **44**:528–532.

55. Hirano K, Takahashi M, Kazumi Y, Fukasawa Y, Abe C. 1997. Mutation in *pncA* is a major mechanism of pyrazinamide resistance in *Mycobacterium tuberculosis*. *Tuber Lung Dis* **78**:117–122.

56. Stoffels K, Mathys V, Fauville-Dufaux M, Wintjens R, Bifani P. 2012. Systematic analysis of pyrazinamide-resistant spontaneous mutants and clinical isolates of *Mycobacterium tuberculosis*. *Antimicrob Agents Chemother* **56**:5186–5193.

57. Simons SO, van Ingen J, van der Laan T, Mulder A, Dekhuijzen PNR, Boeree MJ, van Soolingen D. 2012. Validation of *pncA* gene sequencing in combination with the mycobacterial growth indicator tube method to test susceptibility of *Mycobacterium tuberculosis* to pyrazinamide. *J Clin Microbiol* **50**:428–434.

58. Alexander DC, Ma JH, Guthrie JL, Blair J, Chedore P, Jamieson FB. 2012. Gene sequencing for routine verification of pyrazinamide resistance in *Mycobacterium tuberculosis*: a role for *pncA* but not *rpsA*. *J Clin Microbiol* **50**:3726–3728.

59. Sreevatsan S, Pan X, Zhang Y, Kreiswirth BN, Musser JM. 1997. Mutations associated with pyrazinamide resistance in *pncA* of *Mycobacterium tuberculosis* complex organisms. *Antimicrob Agents Chemother* **41**:636–640.

60. Zhang Y, Wade MM, Scorpio A, Zhang H, Sun Z. 2003. Mode of action of pyrazinamide: disruption of *Mycobacterium tuberculosis* membrane transport and energetics by pyrazinoic acid. *J Antimicrob Chemother* **52**:790–795.

61. Zimhony O, Cox JS, Welch JT, Vilchèze C, Jacobs WR. 2000. Pyrazinamide inhibits the eukaryotic-like fatty acid synthetase I (FASI) of *Mycobacterium tuberculosis*. *Nat Med* **6**:1043–1047.

62. Boshoff HI, Mizrahi V, Barry CE. 2002. Effects of pyrazinamide on fatty acid synthesis by whole mycobacterial cells and purified fatty acid synthase I. *J Bacteriol* **184**:2167–2172.

63. Shi W, Zhang X, Jiang X, Yuan H, Lee JS, Barry CE, Wang H, Zhang W, Zhang Y. 2011. Pyrazinamide inhibits trans-translation in *Mycobacterium tuberculosis*. *Science* **333**:1630–1632.

64. Mendez S, Traslavina R, Hinchman M, Huang L, Green P, Cynamon MH, Welch JT. 2009. The antituberculosis drug pyrazinamide affects the course of cutaneous leishmaniasis in vivo and increases activation of macrophages and dendritic cells. *Antimicrob Agents Chemother* **53**:5114–5121.

65. Manca C, Koo M-S, Peixoto B, Fallows D, Kaplan G, Subbian S. 2013. Host targeted activity of pyrazinamide in *Mycobacterium tuberculosis* infection. *PLoS One* **8**: e74082.

66. Almeida D, Tyagi S, Li S-YL, Wallengren K, Pym AS, Ammerman N, Bishai WR, Grosset JH. Revisiting anti-tuberculosis activity of pyrazinamide in mice. *Mycobacterial Dis*, in press.

67. Mikusová K, Slayden RA, Besra GS, Brennan PJ. 1995. Biogenesis of the mycobacterial cell wall and the site of action of ethambutol. *Antimicrob Agents Chemother* **39**:2484–2489.

68. Alcaide F, Pfyffer GE, Telenti A. 1997. Role of *embB* in natural and acquired resistance to ethambutol in mycobacteria. *Antimicrob Agents Chemother* **41**:2270–2273.

69. Wolucka BA, McNeil MR, de Hoffmann E, Chojnacki T, Brennan PJ. 1994. Recognition of the lipid intermediate for arabinogalactan/arabinomannan biosynthesis and its relation to the mode of action of ethambutol on mycobacteria. *J Biol Chem* **269**:23328–23335.

70. Takayama K, Kilburn JO. 1989. Inhibition of synthesis of arabinogalactan by ethambutol in *Mycobacterium smegmatis*. *Antimicrob Agents Chemother* **33**:1493–1499.

71. Brennan PJ, Nikaido H. 1995. The envelope of mycobacteria. *Annu Rev Biochem* **64**:29–63.

72. Telenti A, Philipp WJ, Sreevatsan S, Bernasconi C, Stockbauer KE, Wieles B, Musser JM, Jacobs WR. 1997. The *emb* operon, a gene cluster of *Mycobacterium tuberculosis* involved in resistance to ethambutol. *Nat Med* **3**:567–570.

73. Laszlo A, Rahman M, Espinal M, Raviglione M. 2002. Quality assurance programme for drug susceptibility testing of *Mycobacterium tuberculosis* in the WHO/IUATLD Supranational Reference Laboratory Network: five rounds of proficiency testing, 1994-1998. *Int J Tuberc Lung Dis* **6**:748–756.

74. Plinke C, Cox HS, Kalon S, Doshetov D, Rüsch-Gerdes S, Niemann S. 2009. Tuberculosis ethambutol resistance: concordance between phenotypic and genotypic test results. *Tuberculosis (Edinb)* **89**:448–452.

75. Ramaswamy S, Musser JM. 1998. Molecular genetic basis of antimicrobial agent resistance in *Mycobacterium tuberculosis*: 1998 update. *Tuber Lung Dis* **79**:3–29.

76. Belanger AE, Besra GS, Ford ME, Mikusová K, Belisle JT, Brennan PJ, Inamine JM. 1996. The *embAB* genes of *Mycobacterium avium* encode an arabinosyl

transferase involved in cell wall arabinan biosynthesis that is the target for the antimycobacterial drug ethambutol. *Proc Natl Acad Sci USA* 93:11919–11924.

77. Safi H, Lingaraju S, Amin A, Kim S, Jones M, Holmes M, McNeil M, Peterson SN, Chatterjee D, Fleischmann R, Alland D. 2013. Evolution of high-level ethambutol-resistant tuberculosis through interacting mutations in decaprenylphosphoryl-β-D-arabinose biosynthetic and utilization pathway genes. *Nat Genet* 1–10.

78. Noller HF. 1984. Structure of ribosomal RNA. *Annu Rev Biochem* 53:119–162.

79. Jnawali HN, Yoo H, Ryoo S, Lee K-J, Kim B-J, Koh W-J, Kim C-K, Kim H-J, Park YK. 2013. Molecular genetics of *Mycobacterium tuberculosis* resistant to aminoglycosides and cyclic peptide capreomycin antibiotics in Korea. *World J Microbiol Biotechnol* 29:975–982.

80. Katsukawa C, Tamaru A, Miyata Y, Abe C, Makino M, Suzuki Y. 1997. Characterization of the *rpsL* and *rrs* genes of streptomycin-resistant clinical isolates of *Mycobacterium tuberculosis* in Japan. *J Appl Microbiol* 83:634–640.

81. Georghiou SB, Magana M, Garfein RS, Catanzaro DG, Catanzaro A, Rodwell TC. 2012. Evaluation of genetic mutations associated with *Mycobacterium tuberculosis* resistance to amikacin, kanamycin and capreomycin: a systematic review. *PLoS One* 7:e33275.

82. Okamoto S, Tamaru A, Nakajima C, Nishimura K, Tanaka Y, Tokuyama S, Suzuki Y, Ochi K. 2007. Loss of a conserved 7-methylguanosine modification in 16S rRNA confers low-level streptomycin resistance in bacteria. *Mol Microbiol* 63:1096–1106.

83. Zaunbrecher MA, Sikes RD, Metchock B, Shinnick TM, Posey JE. 2009. Overexpression of the chromosomally encoded aminoglycoside acetyltransferase *eis* confers kanamycin resistance in *Mycobacterium tuberculosis. Proc Natl Acad Sci USA* 106:20004–20009.

84. Reeves AZ, Campbell PJ, Sultana R, Malik S, Murray M, Plikaytis BB, Shinnick TM, Posey JE. 2013. Aminoglycoside cross-resistance in *Mycobacterium tuberculosis* due to mutations in the 5′ untranslated region of *whiB7. Antimicrob Agents Chemother* 57:1857–1865.

85. Modolell J, Vázquez D. 1977. The inhibition of ribosomal translocation by viomycin. *Eur J Biochem* 81:491–497.

86. Yamada T, Mizugichi Y, Nierhaus KH, Wittmann HG. 1978. Resistance to viomycin conferred by RNA of either ribosomal subunit. *Nature* 275:460–461.

87. Maus CE, Plikaytis BB, Shinnick TM. 2005. Mutation of *tlyA* confers capreomycin resistance in *Mycobacterium tuberculosis. Antimicrob Agents Chemother* 49:571–577.

88. Maus CE, Plikaytis BB, Shinnick TM.. 2005. Molecular analysis of cross-resistance to capreomycin, kanamycin, amikacin, and viomycin in *Mycobacterium tuberculosis. Antimicrob Agents Chemother* 49:3192–3197.

89. Drlica K. 1999. Mechanism of fluoroquinolone action. *Curr Opin Microbiol* 2:504–508.

90. Cole ST, Brosch R, Parkhill J, Garnier T, Churcher C, Harris D, Gordon SV, Eiglmeier K, Gas S, Barry CE, Tekaia F, Badcock K, Basham D, Brown D, Chillingworth T, Connor R, Davies R, Devlin K, Feltwell T, Gentles S, Hamlin N, Holroyd S, Hornsby T, Jagels K, Krogh A, McLean J, Moule S, Murphy L, Oliver K, Osborne J, Quail MA, Rajandream MA, Rogers J, Rutter S, Seeger K, Skelton J, Squares R, Squares S, Sulston JE, Taylor K, Whitehead S, Barrell BG. 1998. Deciphering the biology of *Mycobacterium tuberculosis* from the complete genome sequence. *Nature* 393:537–544.

91. Ginsburg AS, Grosset JH, Bishai WR. 2003. Fluoroquinolones, tuberculosis, and resistance. *Lancet Infect Dis* 3:432–442.

92. Takiff HE, Salazar L, Guerrero C, Philipp W, Huang WM, Kreiswirth B, Cole ST, Jacobs WR, Telenti A. 1994. Cloning and nucleotide sequence of *Mycobacterium tuberculosis gyrA* and *gyrB* genes and detection of quinolone resistance mutations. *Antimicrob Agents Chemother* 38:773–780.

93. Maruri F, Sterling TR, Kaiga AW, Blackman A, van der Heijden YF, Mayer C, Cambau E, Aubry A. 2012. A systematic review of gyrase mutations associated with fluoroquinolone-resistant *Mycobacterium tuberculosis* and a proposed gyrase numbering system. *J Antimicrob Chemother* 67:819–831.

94. Aubry A, Veziris N, Cambau E, Truffot-Pernot C, Jarlier V, Fisher LM. 2006. Novel gyrase mutations in quinolone-resistant and -hypersusceptible clinical isolates of *Mycobacterium tuberculosis*: functional analysis of mutant enzymes. *Antimicrob Agents Chemother* 50:104–112.

95. Cambau E, Sougakoff W, Besson M, Truffot-Pernot C, Grosset J, Jarlier V. 1994. Selection of a *gyrA* mutant of *Mycobacterium tuberculosis* resistant to fluoroquinolones during treatment with ofloxacin. *J Infect Dis* 170:1351.

96. Huang T-S, Kunin CM, Shin-Jung Lee S, Chen Y-S, Tu H-Z, Liu Y-C. 2005. Trends in fluoroquinolone resistance of *Mycobacterium tuberculosis* complex in a Taiwanese medical centre: 1995–2003. *J Antimicrob Chemother* 56:1058–1062.

97. Siddiqi N, Shamim M, Hussain S, Choudhary RK, Ahmed N, Prachee, Banerjee S, Savithri GR, Alam M, Pathak N, Amin A, Hanief M, Katoch VM, Sharma SK, Hasnain SE. 2002. Molecular characterization of multidrug-resistant isolates of *Mycobacterium tuberculosis* from patients in North India. *Antimicrob Agents Chemother* 46:443–450.

98. Von Groll A, Martin A, Jureen P, Hoffner S, Vandamme P, Portaels F, Palomino JC, da Silva PA. 2009. Fluoroquinolone resistance in *Mycobacterium tuberculosis* and mutations in *gyrA* and *gyrB. Antimicrob Agents Chemother* 53:4498–4500.

99. Sun Z, Zhang J, Zhang X, Wang S, Zhang Y, Li C. 2008. Comparison of *gyrA* gene mutations between laboratory-selected ofloxacin-resistant *Mycobacterium tuberculosis* strains and clinical isolates. *Int J Antimicrob Agents* 31:115–121.

100. Hegde SS, Vetting MW, Roderick SL, Mitchenall LA, Maxwell A, Takiff HE, Blanchard JS. 2005. A fluoroquinolone resistance protein from *Mycobacterium tuberculosis* that mimics DNA. *Science* **308**:1480–1483.

101. Takiff HE, Cimino M, Musso MC, Weisbrod T, Martinez R, Delgado MB, Salazar L, Bloom BR, Jacobs WR. 1996. Efflux pump of the proton antiporter family confers low-level fluoroquinolone resistance in *Mycobacterium smegmatis*. *Proc Natl Acad Sci USA* **93**:362–366.

102. Pasca MR, Guglierame P, Arcesi F, Bellinzoni M, De Rossi E, Riccardi G. 2004. Rv2686c-Rv2687c-Rv2688c, an ABC fluoroquinolone efflux pump in *Mycobacterium tuberculosis*. *Antimicrob Agents Chemother* **48**:3175–3178.

103. Louw GE, Warren RM, Gey van Pittius NC, Leon R, Jimenez A, Hernandez-Pando R, McEvoy CRE, Grobbelaar M, Murray M, van Helden PD, Victor TC. 2011. Rifampicin reduces susceptibility to ofloxacin in rifampicin-resistant *Mycobacterium tuberculosis* through efflux. *Am J Respir Crit Care Med* **184**:269–276.

104. Di Perri G, Bonora S. 2004. Which agents should we use for the treatment of multidrug-resistant *Mycobacterium tuberculosis*? *J Antimicrob Chemother* **54**:593–602.

105. Wang F, Langley R, Gulten G, Dover LG, Besra GS, Jacobs WR, Sacchettini JC. 2007. Mechanism of thioamide drug action against tuberculosis and leprosy. *J Exp Med* **204**:73–78.

106. Canetti G. 1965. Present aspects of bacterial resistance in tuberculosis. *Am Rev Respir Dis* **92**:687–703.

107. DeBarber AE, Mdluli K, Bosman M, Bekker LG, Barry CE. 2000. Ethionamide activation and sensitivity in multidrug-resistant *Mycobacterium tuberculosis*. *Proc Natl Acad Sci USA* **97**:9677–9682.

108. Baulard AR, Betts JC, Engohang-Ndong J, Quan S, McAdam RA, Brennan PJ, Locht C, Besra GS. 2000. Activation of the pro-drug ethionamide is regulated in mycobacteria. *J Biol Chem* **275**:28326–28331.

109. Morlock GP, Metchock B, Sikes D, Crawford JT, Cooksey RC. 2003. *ethA*, *inhA*, and *katG* loci of ethionamide-resistant clinical *Mycobacterium tuberculosis* isolates. *Antimicrob Agents Chemother* **47**:3799–3805.

110. Vilchèze C, Av-Gay Y, Attarian R, Liu Z, Hazbón MH, Colangeli R, Chen B, Liu W, Alland D, Sacchettini JC, Jacobs WR. 2008. Mycothiol biosynthesis is essential for ethionamide susceptibility in *Mycobacterium tuberculosis*. *Mol Microbiol* **69**:1316–1329.

111. Bruning JB, Murillo AC, Chacon O, Barletta RG, Sacchettini JC. 2011. Structure of the *Mycobacterium tuberculosis* D-alanine:D-alanine ligase, a target of the antituberculosis drug D-cycloserine. *Antimicrob Agents Chemother* **55**:291–301.

112. Halouska S, Chacon O, Fenton RJ, Zinniel DK, Barletta RG, Powers R. 2007. Use of NMR metabolomics to analyze the targets of D-cycloserine in mycobacteria: role of D-alanine racemase. *J Proteome Res* **6**:4608–4614.

113. Cáceres NE, Harris NB, Wellehan JF, Feng Z, Kapur V, Barletta RG. 1997. Overexpression of the D-alanine racemase gene confers resistance to D-cycloserine in *Mycobacterium smegmatis*. *J Bacteriol* **179**:5046–5055.

114. Feng Z, Barletta RG. 2003. Roles of *Mycobacterium smegmatis* D-alanine:D-alanine ligase and D-alanine racemase in the mechanisms of action of and resistance to the peptidoglycan inhibitor D-cycloserine. *Antimicrob Agents Chemother* **47**:283–291.

115. Chen JM, Uplekar S, Gordon SV, Cole ST. 2012. A point mutation in *cycA* partially contributes to the D-cycloserine resistance trait of *Mycobacterium bovis* BCG vaccine strains. *PLoS One* **7**:e43467.

116. Mathys V, Wintjens R, Lefevre P, Singhal A, Kiass M, Kurepina N, Wang X-M, Mathema B, Baulard A, Barry N, Bifani P, Bertout J, Kreiswirth BN. 2009. Molecular genetics of para-aminosalicylic acid resistance in clinical isolates and spontaneous mutants of *Mycobacterium tuberculosis*. *Antimicrob Agents Chemother* **53**:2100–2109.

117. Chakraborty S, Gruber T, Barry CE, Boshoff HI, Rhee KY. 2013. Para-aminosalicylic acid acts as an alternative substrate of folate metabolism in *Mycobacterium tuberculosis*. *Science* **339**:88–91.

118. Zheng J, Rubin EJ, Bifani P, Mathys V, Lim V, Au M, Jang J, Nam J, Dick T, Walker JR, Pethe K, Camacho LR. 2013. *para*-Aminosalicylic acid is a prodrug targeting dihydrofolate reductase in *Mycobacterium tuberculosis*. *J Biol Chem* **288**:23447–23456.

119. Schecter GF, Scott C, True L, Raftery A, Flood J, Mase S. 2010. Linezolid in the treatment of multidrug-resistant tuberculosis. *Clin Infect Dis* **50**:49–55.

120. Migliori GB, Eker B, Richardson MD, Sotgiu G, Zellweger J-P, Skrahina A, Ortmann J, Girardi E, Hoffmann H, Besozzi G, Bevilacqua N, Kirsten D, Centis R, Lange C. 2009. A retrospective TBNET assessment of linezolid safety, tolerability and efficacy in multidrug-resistant tuberculosis. *Eur Respir J* **34**:387–393.

121. Hillemann D, Rüsch-Gerdes S, Richter E. 2008. In vitro-selected linezolid-resistant *Mycobacterium tuberculosis* mutants. *Antimicrob Agents Chemother* **52**:800–801.

122. Beckert P, Hillemann D, Kohl TA, Kalinowski J, Richter E, Niemann S, Feuerriegel S. 2012. *rplC* T460C identified as a dominant mutation in linezolid-resistant *Mycobacterium tuberculosis* strains. *Antimicrob Agents Chemother* **56**:2743–2745.

123. Wang F, Cassidy C, Sacchettini JC. 2006. Crystal structure and activity studies of the *Mycobacterium tuberculosis* beta-lactamase reveal its critical role in resistance to beta-lactam antibiotics. *Antimicrob Agents Chemother* **50**:2762–2771.

124. Payen MC, De Wit S, Martin C, Sergysels R, Muylle I, Van Laethem Y, Clumeck N. 2012. Clinical use of the meropenem-clavulanate combination for extensively drug-resistant tuberculosis. *Int J Tuberc Lung Dis* **16**:558–560.

125. Van Deun A, Maug AKJ, Salim MAH, Das PK, Sarker MR, Daru P, Rieder HL. 2010. Short, highly effective, and inexpensive standardized treatment of multidrug-resistant tuberculosis. *Am J Respir Crit Care Med* 182: 684–692.

126. Yano T, Kassovska-Bratinova S, Teh JS, Winkler J, Sullivan K, Isaacs A, Schechter NM, Rubin H. 2011. Reduction of clofazimine by mycobacterial type 2 NADH:quinone oxidoreductase: a pathway for the generation of bactericidal levels of reactive oxygen species. *J Biol Chem* 286:10276–10287.

127. Andries K, Verhasselt P, Guillemont J, Göhlmann HWH, Neefs J-M, Winkler H, Van Gestel J, Timmerman P, Zhu M, Lee E, Williams P, de Chaffoy D, Huitric E, Hoffner S, Cambau E, Truffot-Pernot C, Lounis N, Jarlier V. 2005. A diarylquinoline drug active on the ATP synthase of *Mycobacterium tuberculosis*. *Science* 307: 223–227.

128. Koul A, Vranckx L, Dendouga N, Balemans W, Van den Wyngaert I, Vergauwen K, Göhlmann HWH, Willebrords R, Poncelet A, Guillemont J, Bald D, Andries K. 2008. Diarylquinolines are bactericidal for dormant mycobacteria as a result of disturbed ATP homeostasis. *J Biol Chem* 283:25273–25280.

129. Biukovic G, Basak S, Manimekalai MSS, Rishikesan S, Roessle M, Dick T, Rao SPS, Hunke C, Grüber G. 2013. Variations of subunit ε of the *Mycobacterium tuberculosis* F1Fo ATP synthase and a novel model for mechanism of action of the tuberculosis drug TMC207. *Antimicrob Agents Chemother* 57:168–176.

130. Huitric E, Verhasselt P, Koul A, Andries K, Hoffner S, Andersson DI. 2010. Rates and mechanisms of resistance development in *Mycobacterium tuberculosis* to a novel diarylquinoline ATP synthase inhibitor. *Antimicrob Agents Chemother* 54:1022–1028.

131. Gupta S, Cohen KA, Winglee K, Maiga M, Diarra B, Bishai WR. 2013. Efflux inhibition with verapamil potentiates bedaquiline in *Mycobacterium tuberculosis*. *Antimicrob Agents Chemother* [Epub ahead of print.]

132. Helb D, Jones M, Story E, Boehme C, Wallace E, Ho K, Kop J, Owens MR, Rodgers R, Banada P, Safi H, Blakemore R, Lan NTN, Jones-López EC, Levi M, Burday M, Ayakaka I, Mugerwa RD, McMillan B, Winn-Deen E, Christel L, Dailey P, Perkins MD, Persing DH, Alland D. 2010. Rapid detection of *Mycobacterium tuberculosis* and rifampin resistance by use of on-demand, near-patient technology. *J Clin Microbiol* 48:229–237.

133. Miotto P, Piana F, Cirillo DM, Migliori GB. 2008. Genotype MTBDRplus: a further step toward rapid identification of drug-resistant *Mycobacterium tuberculosis*. *J Clin Microbiol* 46:393–394.

134. Köser CU, Bryant JM, Becq J, Török ME, Ellington MJ, Marti-Renom MA, Carmichael AJ, Parkhill J, Smith GP, Peacock SJ. 2013. Whole-genome sequencing for rapid susceptibility testing of *M. tuberculosis*. *N Engl J Med* 369:290–292.

135. Farhat MR, Shapiro BJ, Kieser KJ, Sultana R, Jacobson KR, Victor TC, Warren RM, Streicher EM, Calver A, Sloutsky A, Kaur D, Posey JE, Plikaytis B, Oggioni MR, Gardy JL, Johnston JC, Rodrigues M, Tang PKC, Kato-Maeda M, Borowsky ML, Muddukrishna B, Kreiswirth BN, Kurepina N, Galagan J, Gagneux S, Birren B, Rubin EJ, Lander ES, Sabeti PC, Murray M. 2013. Genomic analysis identifies targets of convergent positive selection in drug-resistant *Mycobacterium tuberculosis*. *Nat Genet* 45:1183–1189.

136. Zhang H, Li D, Zhao L, Fleming J, Lin N, Wang T, Liu Z, Li C, Galwey N, Deng J, Zhou Y, Zhu Y, Gao Y, Wang T, Wang S, Huang Y, Wang M, Zhong Q, Zhou L, Chen T, Zhou J, Yang R, Zhu G, Hang H, Zhang J, Li F, Wan K, Wang J, Zhang X-E, Bi L. 2013. Genome sequencing of 161 *Mycobacterium tuberculosis* isolates from China identifies genes and intergenic regions associated with drug resistance. *Nat Genet* 45:1255–1260.

Molecular Genetics of Mycobacteria, 2nd Edition
Edited by Graham F. Hatfull and William R. Jacobs, Jr.
© 2014 American Society for Microbiology, Washington, DC
doi:10.1128/microbiolspec.MGM2-0014-2013

Catherine Vilchèze[1]
William R. Jacobs Jr.[1]

Resistance to Isoniazid and Ethionamide in *Mycobacterium tuberculosis*: Genes, Mutations, and Causalities

22

THE EMERGENCE OF DRUG RESISTANCE

Tuberculosis (TB) chemotherapy started in the 1930s with the discovery by Domagk and colleagues of the anti-TB activity of sulfonamides. Since these compounds were very toxic and highly insoluble, analogs were synthesized, leading to the discovery of Tibione (thiacetazone, Fig. 1), a highly effective thiosemicarbazone against *Mycobacterium tuberculosis* (1, 2). In parallel, the natural product streptomycin (SM), discovered by Schatz and Waksman, showed activity against *M. tuberculosis* (3) and was used successfully to treat TB patients. Two new anti-TB drugs were discovered soon after: *para*-aminosalicylic acid (PAS) in 1946 (4) and isonicotinic acid hydrazine (isoniazid, INH) in 1952 (5, 6). Each drug had activity against *M. tuberculosis*; however, drug-resistant mutants emerged rapidly during clinical trials (7–9). To prevent drug resistance, in 1959, SM, PAS, and INH were combined to form the first successful multidrug, biphasic chemotherapy for TB (10). This combination

treatment was so impressive that Selman Waksman wrote, "the ancient foe of man, known as consumption, the great white plague, tuberculosis, or by whatever other name, is on the way to being reduced to a minor ailment of man. The future appears bright indeed, and the complete eradication of the disease is in sight" (185, p. 217). Nevertheless, the treatment was long and expensive, and patients often dropped out prior to completing chemotherapy. In 1984, a new short-course treatment was established that showed improved efficacy and patient compliance; the drug regimen consisted of two months on INH, rifampicin (RIF), pyrazinamide (PZA), and ethambutol, followed by four months on INH and RIF only.

Despite this successful chemotherapy, the rate of multidrug-resistant (MDR) *M. tuberculosis* strains, defined as strains resistant to INH and RIF, started to increase as early as 1985. Nowadays, the World Health Organization (WHO) estimates that 3.7% of new TB cases and 20% of previously treated TB cases are caused

[1]Howard Hughes Medical Institute, Department of Microbiology and Immunology, Albert Einstein College of Medicine, Bronx, NY 10461.

by MDR-TB (11). The highest incidence of MDR-TB (up to 76%) is found in Russia and the former Soviet republics. Extensively drug-resistant (XDR)-TB, defined as TB strains resistant to INH, RIF, fluoroquinolones, and one second-line injectable drug, is found in up to 9% of MDR-TB cases and has been reported in at least 84 countries so far. Furthermore, strains of *M. tuberculosis* that are resistant to up to 10 TB drugs, referred to as totally drug-resistant (TDR-TB), have been isolated in Europe, Africa, India, and Iran (12). One factor in the emergence and rapid spread of drug resistance is the paucity of rapid diagnostics. While new tools are available to quickly assess drug resistance, these methods are based on known drug resistance mechanisms. Understanding all these mechanisms is key to improving diagnosis and eradicating drug-resistant *M. tuberculosis*. In this article, we will discuss the mechanisms that *M. tuberculosis* developed to become resistant to the first-line anti-TB drug INH and its analog, the second-line anti-TB drug ethionamide (ETH).

MODE OF ACTION OF INH AND ETH

INH

The antimycobacterial activity of INH was published and commercialized simultaneously by three independent pharmaceutical companies: Bayer (Neoteben)

(13), Hoffman-La Roche (Rimifon) (6), and Squibb Institute for Medical Research (Nydrazid) (5). INH had been first synthesized 40 years earlier and reported in a doctoral thesis; therefore, none of the pharmaceutical companies could patent the discovery. Fox (6) described the discovery of INH as an attempt to combine the anti-TB activity of nicotinamide, which had been found to arrest *M. tuberculosis* growth in guinea pigs (14), and thiosemicarbazones. By replacing the benzene ring of Tibione by the pyridine ring found in nicotinamide (Fig. 1), *meta-* and *para*-pyridylaldehyde thiosemicarbazones were synthesized. All the intermediates in the synthesis of these pyridine thiosemicarbazones were tested for activity against TB, and among them, one intermediate, INH, had antimycobacterial activity far superior to any compound at the time (6). Notably, another analog synthesized, Marsilid (1-isonicotinoyl-2-isopropylhydrazine) (Fig. 1), also showed good anti-TB activity. The use of Marsilid as a TB drug was soon discontinued because it induced euphoria in TB patients. However, Marsilid did go on to become the first antidepressant. Interestingly, the earlier synthetic compounds active to some degree against *M. tuberculosis* share a similar chemical skeleton (Fig. 1), although their modes of action are different.

The mechanism of action of INH has been the subject of intensive research and controversies since its discovery in 1952. From 1953 to 1980, multiple modes of

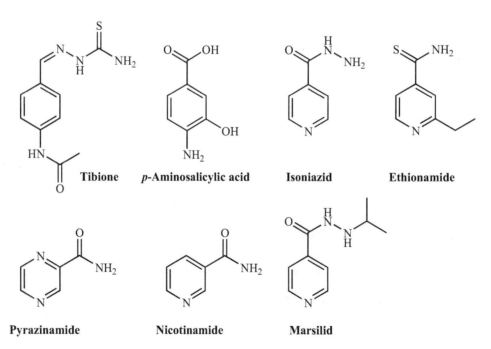

Figure 1 Early synthetic antituberculosis drugs. doi:10.1128/microbiolspec.MGM2-0014-2013.f1

action for INH were proposed: INH was thought to interfere with cell division (15), pyridoxal-dependent metabolic pathway(s) (16), lipid biosynthesis (17), fatty acid biosynthesis (18), nucleic acid biosynthesis (19), glycerol conversion to hexose phosphate (20), NAD^+ biosynthesis (21), and NADH dehydrogenase activity (22–23). A major breakthrough in understanding the mechanism of action of INH came with the study by Winder and Collins (186) in which they showed that INH inhibited synthesis of mycolic acids, long-chain α-alkyl β-hydroxy fatty acids that are a crucial component of the mycobacterial cell wall. The effect of INH on mycolic acids has been subsequently confirmed by numerous researchers. Among them, Takayama and coworkers were the first to demonstrate that inhibition of mycolic acid biosynthesis by INH correlated with cell death (24), accumulation of long-chain fatty acids (25), and inhibition of C24 and C26 monounsaturated fatty acid biosyntheses (26). Takayama and colleagues concluded that the mode of action of INH involved a desaturase required for the biosynthesis of these unsaturated fatty acids (26).

INH, like other TB drugs (ETH, PZA, isoxyl, thiacetazone) is a pro-drug. The catalase peroxidase KatG (encoded by *Rv1908c*) activates INH to form a hypothetical isonicotinoyl anion or radical (27–30). This entity reacts with NAD^+ to yield an INH-NAD adduct, which binds to the active site of the NADH-dependent enoyl-ACP reductase InhA (Rv1484) (Fig. 2) (31). This enzyme reduces monounsaturated acyl-ACP to acyl-ACP (32, 33) and is part of the fatty acid synthase type II (FASII) (34). FASII elongates fatty acids up to 56 carbons long where they are derivatized and coupled to a C24-C26 fatty acid generated by FASI to form mycolic acids. The INH-NAD adduct binds to and inhibits InhA (28, 35, 36), leading to disruption of mycolic acid biosynthesis and cell death (37, 38).

ETH

2-Ethylthioisonicotinamide, ETH (Fig. 1), is a structural analog of INH. ETH was first synthesized in 1956 by a French team trying to improve on the antimycobacterial properties of thioisonicotinamide (39). Grumbach and coworkers found that ETH was more

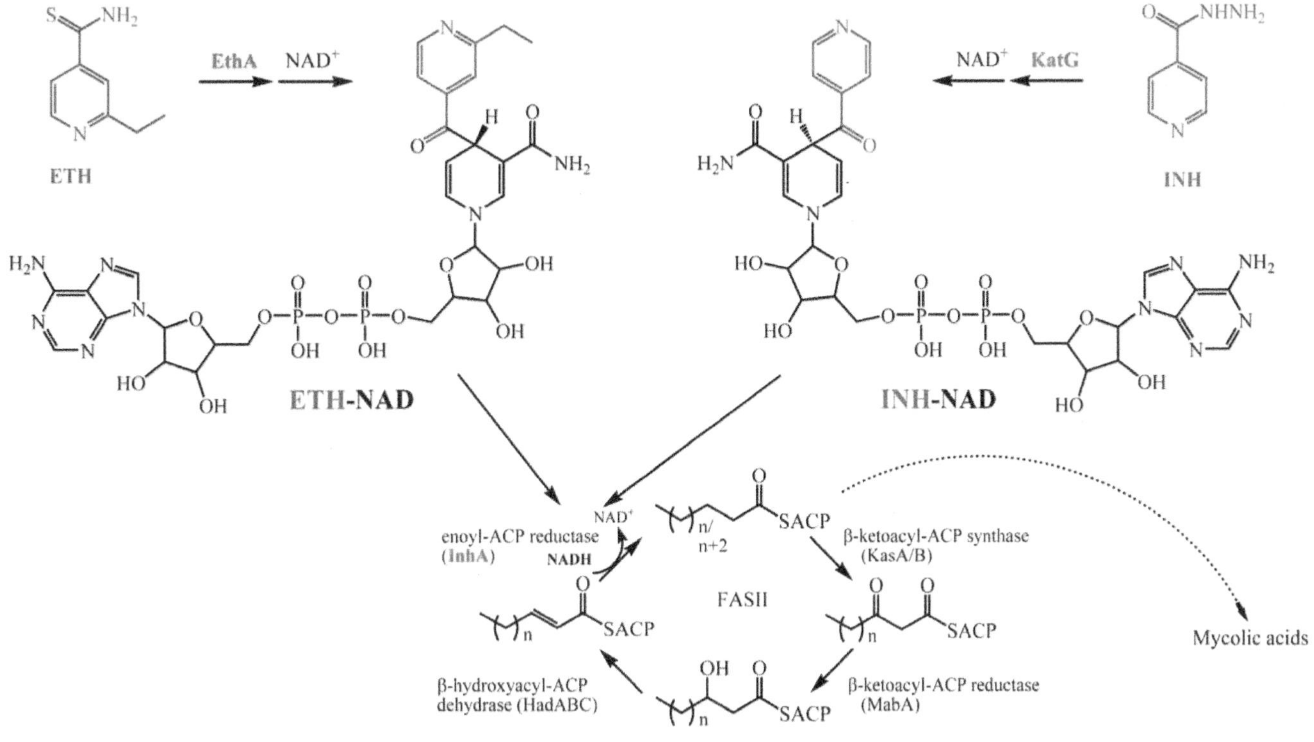

Figure 2 Mechanism of action of INH and ETH. INH and ETH are activated by the catalase peroxidase KatG and monooxygenase EthA, respectively, to form a reactive species that binds to NAD^+. The resulting adducts, INH-NAD or ETH-NAD, inhibit the enoyl-ACP reductase InhA of the FASII system, resulting in mycolic acid biosynthesis inhibition. doi:10.1128/microbiolspec.MGM2-0014-2013.f2

active than SM but less so than INH against *M. tuberculosis*. However, ETH was also active against INH-, PAS-, and SM-resistant *M. tuberculosis* strains. ETH was shown to be efficacious in combination with PZA and with or without INH in a clinical trial to treat TB patients infected with INH- and SM-resistant strains (40, 41). Nowadays, ETH is a second-line drug, mostly used to treat MDR-TB cases in South Africa.

Similar to INH (42, 43), *M. tuberculosis* treated with ETH loses its acid fastness (39) and its ability to synthesize mycolic acids (44). ETH is also a pro-drug, activated by the NADPH-specific flavin adenine dinucleotide–containing monooxygenase EthA (also called EtaA, encoded by *Rv3854c*) (45–48). Once activated, the mode of action of ETH is very similar to INH (Fig. 2). The active form of ETH reacts with NAD^+ to yield an ETH-NAD adduct (49), which inhibits InhA, leading to mycolic acid biosynthesis inhibition. Interestingly, while KatG only activates INH, EthA activates two other second-line anti-TB drugs: Tibione (thiacetazone) and isoxyl (46, 50).

MECHANISMS OF RESISTANCE TO INH AND ETH

Drug resistance in mycobacteria is due to the acquisition of mutations or efflux pump activation, not due to the acquisition of resistance plasmids or transposons, common resistance mechanisms in other bacterial species. The main mechanisms of resistance to INH and ETH can be divided into two categories. First, prevention of the activation of INH and ETH can be obtained by mutating the activators of the drugs *katG* and *ethA*, respectively, or by mutating regulators of their expression. For example, the *katG*(S315T) mutation is found in up to 94% of INH-resistant *M. tuberculosis* clinical isolates. Second, the inhibition of InhA by the INH-NAD or ETH-NAD adduct can be overcome by mutations in *inhA* or its promoter region. Other mechanisms of resistance exist such as drug inactivators, redox alteration, and efflux pumps. We will first describe the mechanisms of coresistance to INH and ETH and then the mechanisms of resistance specific to each drug.

Common Mechanisms of Resistance to INH and ETH

Clinical isolates coresistant to INH and ETH were isolated from TB patients who had received INH but had never been treated with ETH (51–53). This conundrum led to the hypothesis that INH and ETH shared a common mechanism of resistance, a hypothesis that could

not be tested until a plasmid transformation system was developed for mycobacteria (54, 55). It would take another three decades to discover that INH and ETH target the same enzyme in *M. tuberculosis*: the enoyl-ACP reductase InhA (56). Other mechanisms of coresistance to INH and ETH have been discovered. These mechanisms along with resistance mechanisms due to *inhA* mutations are listed below.

Alteration of InhA, the target of INH and ETH

The target of INH and ETH was discovered by isolating a *Mycobacterium smegmatis* mutant coresistant to INH and ETH (56). A genomic DNA cosmid library of this *M. smegmatis* mutant and of drug-susceptible *Mycobacterium avium*, *Mycobacterium bovis*, *M. smegmatis*, and *M. tuberculosis* strains was constructed and transformed into *M. smegmatis*. A single open reading frame (ORF) was found to be sufficient to confer coresistance to INH and ETH in *M. smegmatis* and was named *inhA*. These experiments demonstrated that a single amino acid mutation in *inhA*(S94A) or overexpression of *inhA* conferred coresistance to INH and ETH in mycobacteria. To prove that *inhA* inactivation was sufficient to lead to death in a manner similar to INH action, a temperature-sensitive mutant in *inhA* was isolated (37). Heat-inactivation of InhA(V238F) mimicked the effects of INH in *M. tuberculosis* described by Takayama and colleagues: inhibition of mycolic acid biosynthesis (24), alteration of the bacterium morphology (57), accumulation of long-chain fatty acids (25), and cell death (24), demonstrating that inhibition of InhA alone reproduced the biochemical characteristics of INH treatment of *M. tuberculosis*. Moreover, the S94A mutation identified in the INH- and ETH-resistant *M. smegmatis* mutant was transferred into wild-type *M. tuberculosis* by specialized linkage transduction, and the resulting strain was at least five times more resistant to INH and ETH than wild-type *M. tuberculosis* (38). This combined set of data confirms that InhA is the main target and the main mechanism of coresistance to INH and ETH.

The mechanism by which the S94A mutation leads to resistance to INH and ETH has been well studied. InhA is an NADH-dependent enoyl-ACP reductase, and the binding of the enoyl substrate to InhA is not disturbed by the S94A mutation; however, the mutation results in a 5-fold increase in the K_M for the InhA cofactor NADH (32). On the other hand, the ability of the INH-NAD adduct to inhibit InhA(S94A) is markedly reduced, because the IC_{50} and K_i are 17 and 30 times higher, respectively, for the mutated protein. Comparison of the crystal structures of InhA(S94A) to wild-type InhA revealed the loss of a water molecule and disruption of a hydrogen bonding network in the mutated protein, which was enough to reduce the binding of the INH-NAD adduct (38). Others disputed this conclusion and hypothesized that InhA interacts with FASII enzymes and that this interaction is perturbed by the S94A mutation, resulting in INH resistance (58). Overexpression of *inhA* is also a common factor of INH and ETH resistance in clinical isolates. The c-15t base pair change in the *inhA* regulatory region increases *inhA* mRNA levels by 20-fold, resulting in the overexpression of InhA. This leads to a titration of INH or ETH and consequently an eight-fold increase in INH and ETH MICs in *M. tuberculosis* (38).

Numerous point mutations in *inhA* and its promoter region have been identified in INH- and ETH-resistant *M. tuberculosis* clinical isolates (Tables 1 and 2); however, no base pair insertions or deletions have been observed. Mutations in the *inhA* promoter and ORF regions are associated with low-level resistance to INH, even when strains contained mutations in both regions (MIC <1 mg/liter) (59). The c-15t mutation in the promoter region of *inhA* is found in up to 35% of INH-resistant and 55% of ETH-resistant clinical isolates, but never in INH- or ETH-sensitive strains. This mutation was also overly represented in XDR-TB cases in South Africa. In a study on clinical isolates from the Western Cape Province, South Africa, the c-15t mutation was present in 30% of strains mono-resistant to INH, 48% of MDR-TB, and 85% of XDR-TB (60), suggesting that this mutation could be a marker for XDR-TB. Interestingly, in a survey of the Eastern Cape Province clinical isolates, the c-17t *inhA* promoter mutation was the predominant genetic modification present in 83% of XDR-TB cases. The combined *inhA* promoter mutations (at positions -8, -15, and -17) were found in 92% of the XDR-TB cases versus 62% of the MDR-TB cases in the Eastern Cape Province (60).

inhA is an essential gene; therefore, mutations in the coding region of *inhA* are rare. About 15 mutations in *inhA* have been identified in INH-resistant clinical isolates, although two of them were also found in INH-sensitive strains (Table 1). I21T, S94A, and I95P are the only amino acid changes found in both INH- and ETH-resistant clinical isolates. The first mutation identified in *inhA* in *M. smegmatis,* resulting in the S94A variant, has since been found in INH-resistant *M. tuberculosis* clinical isolates with no other mutation present in *katG* or the *fabG-inhA* intergenic region (59, 61). This confirms that the S94A mutation is sufficient to confer INH and ETH resistance in *M. tuberculosis* clinical isolates.

While mutations in the *inhA* promoter region can represent up to 35% of the INH-resistant cases, and mutations in the *inhA* gene are rare in INH-resistant clinical isolates, this is the predominant region where mutations are found in ETH-resistant clinical isolates. In one study (62), 62% of the ETH-resistant clinical isolates had mutations in *inhA* (gene and/or promoter region), while 47% had mutations in *ethA*. The *inhA* promoter mutation c-15t was therefore proposed as a marker for ETH resistance (63).

Alteration of redox potential

M. smegmatis and *M. bovis* BCG mutants coresistant to INH and ETH were isolated in *in vitro* experiments from nonmutagenized independent cultures. The mutants had mutations in *ndh* (*Rv1854c*), a gene encoding a type II NADH dehydrogenase, which oxidizes NADH into NAD$^+$. In *M. smegmatis*, the *ndh* mutants contained single base pair changes resulting in amino acid changes and a pleitropic phenotype: INH resistance, ETH resistance, temperature sensitivity, and for some mutants, Ser/Gly auxotrophy (64, 65). In *M. bovis* BCG, the mutants had either single base pair changes or insertions (65). The *ndh* mutants lost up to 95% of their Ndh activity compared to wild type and had higher levels of NADH, while their NAD$^+$ levels were similar to wild type. An increase in NADH concentration was shown to prevent the binding of the INH-NAD adduct to InhA by acting as a competitive inhibitor, leading to INH resistance (65).

In *M. tuberculosis* clinical isolates, *ndh* mutations have been found in both INH-sensitive and INH-resistant strains at a very low rate (66) (Table 1). However, two studies from Singapore and Brazil identified *ndh* mutations in 8 to 10% of INH-resistant *M. tuberculosis* clinical isolates and found no *ndh* mutations in their INH-sensitive clinical isolate strains (67, 68). The R13C mutation was found in a strain containing the *katG* (S315T) mutation, so the contribution to the INH resistance of this *ndh* mutation is uncertain (67). The *ndh*(T110A) mutation was only associated with an *ahpC* mutation resulting in the (T5I) variant (68). However, in that study, *katG* was only partially sequenced, and *inhA* was not (68, 69), so the INH resistance in that isolate may or may not be due to the *ndh*(T110A) mutation. The R268H mutation is the only mutation so far to have been identified in two independent studies and only in INH-resistant strains, but it was associated with *katG*, *inhA*, or *ahpC* mutations (66).

M. tuberculosis has an additional NADH dehydrogenase named *ndhA* (*Rv0392c*). Therefore, mutations in one NADH dehydrogenase such as *ndh* might not

Table 1   Identified mutations in genes other than *katG* in INH-resistant *M. tuberculosis* strains[a]

oxyR´-ahpC intergenic region
t-89a (130)
g-88a (130)
c-81t (131)
g-74a (132)
c-72t (131)
g-67a (66)
g-66a (133)
atgt-54 ins (92)
c-54t (132)
c-52t (136)
g-51a (131)
t-49g (137)
g-48a (135, 139)
g-46a* (102, 140)
g-46 del* (66)
c-45t (84)
t-44a (99)
t-40c (144)
c-39t (66, 135, 139, 140, 144)
t-34c,a (102, 140)
t34 del (102)
g-32a (140)
c-30t (66, 144)
c-20t* (102)
c-15t (66, 135, 139, 144)
c-12t (66, 135, 139, 140)
c-10a,g,t (66, 135, 139, 140)
g-9a* (66, 135, 139, 140, 144)
g-6a (66, 140, 144)
a-4g (140)

ahpC ORF
P2S (69)
L3K (140)
L4R (130)
T5I (66)
F10I (144)
D33N (84)
D73H* (66)
E76K (153)
L191K (66)

oxyR´
G12a (130)
g18a (69)
g27t (69)
c28a (69)
c37t (99)
bp67 ins ggcg (99)
g325t (99)
a331c (99)

furA-katG intergenic region
c-1 ins (102)
g-7a (96)
a-10c (96)
g-12a (96)

fabG-inhA regulatory region
g-147t (88)
a-113c (134)
g-102a (134)
a-92t (69)
g-67c (137)
g-47c,a (134, 138)
c-34t (138)
t-24g (141)
g-22c (142)
g-17t* (66, 135, 139, 143)
a-16g (84)
c-15t (66, 88, 135, 139, 140, 143)
t-12a (145)
a-11t (66)
t-8a*,g,c (66, 88, 140–142)
t-5a (146)
c-4a (140)
A5P (137)
V14L (137)
T21A (66)

inhA ORF
M1L (147)
K8N (148)
I16T (149)
I21T,V (66, 88, 135, 139, 149)
I25T (136)
I47T* (66, 149)
V78A (149)
S94A (59, 66, 144)
I95P (151)
A190S (152)
I194T* (66, 88)
R202G (133)
E217D (133)
T241M (142)
T253A* (152)
D256N (152)
I258T,V (135, 152)
Y259H (152)

kasA
D66N* (66, 109)
M77I* (88, 135)
R121K (69)
L245R (135)

Ndh
R13C (67)
V18A* (66–67)
T110A (68)
R268H (66, 68)
G313R* (116)

iniB
t198ins (135)
a211del (135)
222 12bp del (88)
N88S* (116)
G192* (135)
H481Q* (116)

iniA
P3A (88)
nt282-286 del (88)
H481Q* (88, 135)
R537H (88, 135)

iniC
t79ins (135)
a98ins (135)
W83G (88)

Rv0340
T143* (135)
G149* (135)
V163I (88, 135)

nat
G67R* (88)
Y177H (150)
G207R* (88, 103)

Rv1592c
P42L (88)
V430A (88)

fadE24
-64 2 bp ins (88)
a-23c* (88)

Rv1772
T4A (88)

efpA
T15R* (116)
I73T* (135)
Q513R (116)
E520V (135)

(Continued)

Table 1 *(Continued)*

furA	G269S* (66, 69, 88, 109, 135, 139)	*fabD*
S5P (88, 135)	G312S* (66, 69, 109, 135)	S275N* (88)
c34 del (96)	S341* (135)	A199T* (88)
A14V (96)	G387D (69)	
A46V* (116)	F413L (109)	*accD6*
L68F (94)		D229G* (88)
C97Y (94)	*srmR* homolog	
	D3G (88)	*fbpC*
	M323T* (88)	G158S* (88)

*a*Asterisk (*), found in INH-resistant and/or INH-sensitive strains.

alter the redox balance in *M. tuberculosis* and might not lead to INH and/or ETH resistance as long as the second NADH dehydrogenase is functional. *ndh* and *ndhA* are also present in *M. bovis* BCG, but *M. bovis* *ndhA* codes for a single amino acid change (V241A) relative to *M. tuberculosis*, which might explain why INH- and ETH-resistant *ndh* mutants were isolated in that strain. For that reason, *ndh* mutations might not be a mechanism of resistance to INH and ETH in *M. tuberculosis* unless both NADH dehydrogenase genes are mutated.

Alteration in mycothiol biosynthesis

Mutations in *ndh* in INH-resistant *M. bovis* BCG mutants were isolated *in vitro* by plating nonmutagenized, independent *M. bovis* BCG cultures on plates containing both INH and ETH to avoid mutants carrying mutations in the activator of INH or ETH. When the same experiment was repeated in *M. tuberculosis* H37Rv or the virulent *M. bovis* Ravenel strain, *ndh* mutants were not obtained. Instead, all of the INH- and ETH-resistant *M. tuberculosis* mutants had mutations in *mshA* (*Rv0486*) (70), a gene encoding a glycosyl transferase involved in the biosynthesis of mycothiol (*N*-acetylcysteine glucosamine inositol, MSH), while the *M. bovis* mutants coresistant to INH and ETH carried mutations either in *mshA* or in *mshC* (*Rv2130c*) encoding the cysteine ligase of the mycothiol biosynthesis (71). Five enzymes are required to synthesize mycothiol: MshA, MshA2, MshB (Rv1170), MshC, and MshD (Rv0819) (72). Mycothiol is the major thiol and the main reducing and detoxifying agent in mycobacteria (72), yet the role of mycothiol during infection is ambiguous since mycothiol-deficient *M. tuberculosis* strains do not have a growth defect *in vivo* (70). Eight *in vitro* *M. tuberculosis* *mshA* mutants were isolated containing a single base pair modification in *mshA*, resulting in amino acid changes, stop codons, and frameshifts, all of which caused a drastic decrease

in mycothiol levels (from 83 to 99.9%). The mutants had different levels of resistance to INH (2-fold to >10-fold) and ETH (4- to 8-fold increase). Complementation with wild-type *M. tuberculosis* *mshA* restored ETH susceptibility but not INH susceptibility in all the mutants. Interestingly, deletion of *mshA* in *M. tuberculosis* led to a strain that did not produce mycothiol and was highly resistant to ETH but fully sensitive to INH (70). This suggests that mycothiol is mostly involved in ETH resistance and might play a role in ETH activation. The observed INH resistance in *M. tuberculosis* *mshA* point mutants and INH susceptibility in *M. tuberculosis* Δ*mshA* strains might also indicate that *mshA* is required for INH resistance in mycothiol-deficient strains.

In *M. smegmatis*, a 4- to 8-fold increase in INH and ETH resistance was obtained when *mshA* or *mshC* was deleted, while deletion of *mshB* resulted in a strain resistant only to ETH, and deletion of *mshD* had no effect on INH or ETH resistance (73). *M. smegmatis* mycothiol mutants obtained from chemical mutagenesis or transposon insertion had slightly different INH and ETH resistance patterns (74). The role of mycothiol deficiency in INH resistance might be species-dependent. So far, we can only conclude that mutations in *mshA* will result in high-level ETH resistance and at most low-level INH resistance in *M. tuberculosis*.

In a highly INH- and ETH-resistant clinical isolate, a double mutation in *mshA* (V171G, A187V) was found (62). That strain had the *katG*(S315T) mutation to account for the INH resistance but no other mutation to explain its resistance to ETH. However, the *mshA*(A187V) mutation is present in wild-type *M. tuberculosis* Beijing strain. Other *mshA* mutations have also been found in drug-sensitive mycobacterial strains. The N111S mutation is present in *M. tuberculosis* Erdman and Haarlem strains (70, 75), while the *M. bovis* Ravenel and *M. bovis* ATCC19210 strains carry a g316a base pair change in *mshA*, resulting in a

Table 2 Identified mutations in ETH-resistant *M. tuberculosis* strains[a]

ethA	*ethA*	*ethA*
M1R (62)	A234D (154)	C403G (62)
I9T (154)	t703 del (61)	R404L (154)
G11A (62)	Q246STOP (61)	G413D (61)
g32 del (154)	A248D (154)	c1254 del (154)
A20 ins (62)	Y250STOP (154)	g1268 del (154)
a65 del (46)	cg754 ins (62)	c1290 del (61, 154)
H22P (62)	Q254P (154)	gc1322,1323 del (61)
Y32D (154)	Q254STOP (154)	T453I (154)
a110 del (61, 154)	g768 del (61)*	Y461H (62)
G43S (61), C (46)	S266R (62)	R463D (61)
T44N (154)	Q269STOP (62)	a1391 ins (62)
D49A (155)	Q271STOP (154)	
Y50C (154)	L272P (62)	*fabG-inhA* regulatory region
P51L (46)	P288R (154)	g-17t (61)
D55A (61)	Q291STOP (154)	c-15t (38, 61, 156)
D58A (46)	R292STOP (154)	
T61M (62)	C294F (154)	*inhA* ORF
Y84D (46)	F302L (154)	I21T(61)
1bp271 del (46)	T324 ins (154)	S94A (61)
cg282-283 del (154)	L328M (155)	I95P (151)
g337 del (154)	S329L (62)	
a338 ins (61)	L333R (155)	*mshA*
a342 del (154)	I338S (61)*	N111S* (62, 70)
G124D (62)	T342K (46)	Q128STOP[a](70)
G139S (154)	d1029 del (154)	V171G(62)
Y140 STOP (154)	N345K (155) (154)	A187V(62)
Q165P (62)	A352P (154)	R273C[b] (70)
W167STOP (154)	g1054 del (154)	G299C[b] (70)
T186K (46)	P378L (154)	Q331STOP[b] (70)
g593 del (154)	A381P (46)	G356D[b] (70)
c613 del (155)	t1152 del (154)	E361A[b] (70)
gt638-639 del (155)	G385D (61)	
D219G (154)	Y386C (62)	*ethR*
E223K (61)	S390F (154)	A95T (62)
g673 ins (154)	W391STOP (154)	F110L (62)
tc675 ins (62)	T392A (61)	
T232A (155)	L397R (155)	

[a]Asterisk (*), also found in ETH-sensitive strains.
[b]Mutations found in *in vitro* cultures.

G106R amino acid change (71, 75). Mycothiol genes such as *mshA* and *mshC* should be added to the list of candidates responsible for ETH resistance in clinical isolates, although correlation between *mshA* mutations and ETH resistance should be carefully analyzed since mutations in *mshA* might not lead to drug resistance in *M. tuberculosis*.

Degradation of the INH-NAD or ETH-NAD adduct

In a recent study, Wang and colleagues suggested that the NADH pyrophosphatase NudC (Rv3199c), an enzyme from the NAD+ recycling pathway, could hydrolyze the INH-NAD or ETH-NAD adduct, leading to INH and ETH resistance (76). *M. smegmatis* and *M. bovis* BCG NudC are functional enzymes, while *M. tuberculosis* H37Rv NudC has a point mutation (Q237P) that renders the enzyme inactive. The authors demonstrated that NudC from *M. smegmatis* and *M. bovis* BCG could release the adenosine monophosphate group from the INH-NAD and ETH-NAD adducts and that overexpression of *M. smegmatis* or *M. bovis* BCG *nudC* resulted in at least a 10-fold increased resistance to INH and ETH, while deletion of *nudC* rendered the strains more sensitive to the drugs. A small portion of *M. tuberculosis* clinical isolates (2%) were found to have the glutamine residue at position 237, suggesting that in these *M. tuberculosis* strains, NudC might be capable of hydrolyzing the adducts. However, no transfer of mutation was performed to prove that this mutation is sufficient to confer INH and ETH resistance in *M. tuberculosis*. The role of *nudC* in INH and/or ETH resistance in *M. tuberculosis* clinical isolates remains to be determined.

Mechanisms of Resistance Specific to INH

Alterations in KatG, the activator of INH

The first mutants isolated in *in vitro* cultures that were highly resistant to INH had the characteristic of being catalase-negative and avirulent in guinea pigs (77, 78). Winder hypothesized that the loss of catalase activity might imply that INH was activated by a catalase to yield some highly reactive species (79). The relationship between INH resistance and the catalase-negative phenotype was elucidated many years later when a highly INH-resistant strain, BHI, a mutant of *M. smegmatis* mc²155 (55), was complemented with an *M. tuberculosis* library (30). INH susceptibility was restored in BHI by the introduction of a single gene, *katG* (*Rv1908c*), encoding a catalase-peroxidase. Furthermore, the authors also found *katG* deletion or mutations in INH-resistant *M. tuberculosis* clinical isolates and demonstrated that transformation of these INH-resistant isolates with a wild-type copy of *katG* restored INH susceptibility (30, 80).

In INH-resistant *M. tuberculosis* clinical isolates, more than 300 mutations in *katG* have been identified throughout the ORF (Table 3). Complete deletion of the gene has been found in clinical isolates, including the first INH-resistant *M. tuberculosis* mutant identified in the Zhang study (30); it has subsequently been identified in other studies (Table 3). Point mutations causing a single amino acid substitution or premature termination, frameshift mutations after addition or deletion of bases, and partial or complete deletion of the gene have been identified. The incidence of *katG* mutations differs between geographical regions but represents at least 30% and up to 95% of INH-resistant clinical isolates.

The most frequent *katG* mutation is at codon S315, where each base (AGC) can be mutated to produce a Thr, Asn, Arg, Ile, Gly, or Leu residue. The S315T mutation can be found in up to 94% of the INH-resistant clinical isolates (81). Two independent biochemical analyses reported that KatG(S315T) has catalase-peroxidase activities, yet its ability to oxidize INH was significantly reduced (82, 83). Biochemical analyses of other KatG mutants showed a wide range of catalase-peroxidase and INH oxidase activities (Table 4). Nevertheless, there is a link between INH oxidase activity and INH resistance. *M. tuberculosis* strains with KatG proteins deficient in INH oxidase activity were highly resistant to INH, while *katG* mutants with INH oxidase activity close to wild-type levels had only 2- to 4-fold increases in the MIC for INH. However, the level of INH resistance might not be defined by location of a mutation: S315T, W341G, G494D, and R595STOP variants are highly resistant to INH (84), while L141F, E553K, and F658V variants are associated with low-level INH resistance (84). A very common polymorphism, R463L, is often found associated with other *katG* mutations and is more likely to be present in INH-sensitive strains than in INH-resistant strains (85).

The high-level resistance to INH associated with *katG*(S315T) (MIC >1 mg/liter) was reported to be specific to the Ser→Thr amino acid change (59, 86). Brossier and colleagues found *katG*(S315N) only in low-level INH-resistant clinical isolates (62). Curiously, KatG(S315N) had been shown to prevent the formation of the INH-NAD adduct, suggesting that INH cannot be activated by KatG(S315N), which conflicts with the above mentioned results (87). In a different study, a clinical isolate carrying *katG*(S315N) and *inhA* (c-15t), which is associated with low-level resistance to INH (see above), had an MIC for INH of >256 mg/liter (88). This was the same level of resistance found in clinical isolates where *katG* was missing or the mutation resulted in early termination of the protein (Q434STOP) (88). Furthermore, the INH-resistant clinical isolate containing only the *inhA* promoter mutation had an MIC of 0.19 mg/liter, confirming the low-level INH resistance associated with this mutation. This suggests that *katG*(S315N) might be associated with a high level of INH resistance in clinical isolates.

One of the first studies on the isolation of INH-resistant *M. tuberculosis* mutants *in vitro* found that the mutants could be classified into catalase-negative and highly INH-resistant or catalase-positive and weakly INH-resistant (MIC <10 mg/liter) (89). Catalase-negative mutants were unable to grow in guinea pigs and rabbits, while catalase-positive mutants grew relatively well *in vivo* (90). The relationship between catalase activity and fitness of an INH-resistant strain *in vivo* has been investigated. Pym and colleagues demonstrated that the INH-resistant *M. tuberculosis* KatG(T275P) variant had no detectable catalase peroxidase activity and was highly attenuated in a mouse model of infection, while *M. tuberculosis* carrying the KatG(S315T)

Table 3 Identified *katG* mutations in INH-resistant *M. tuberculosis* strains[a]

Complete deletion (30, 88, 134, 141, 157)	E217G,del (84, 158)	W397Y (88) STOP (66)
Partial deletion (138, 159–161)	N218K (153)	A409D,R,T,V (142, 152, 153, 162)
V1A (84) L (92)	Q224E (163)	Y413H (133)
P2S (163)	Y229F (164)	K414N (165)
a17 ins (139)	V230A (153)	R418Q (166)
c30 del (160)	P232R,S (139, 167)	D419A,Y,E, H (59, 139)
T11A (160)	G234E (102) R (153)	M420T (167)
T12P (160)	N236T (153)	A424E V (163)
S17N (163)	A243S (163)	G428R (168)
G19D (163)	A245V* (146)	P429S (163)
a98ins (160)	R249C (159)	c1297 ins, c1305 del (167)
N35D (160)	T251M (92)	Q434stop (88)
g109del (160)	F252L (84)	t1311 ins (139)
W38stop (66)	R254L (139)	W438R (88)
L48Q (146)	M257I,T (143, 146)	a1329 ins (139)
a149 del (102)	N258S (163)	c1339 del (139)
A61T (160)	E261Q (142)	L449F (169)
c185 ins (160)	T262R (84, 141)	S457I (165)
D63E (88, 141)	A264T,V* (152, 163)	K459STOP (168)
A65T (163)	H270Q (170)	R463L* (85, 141, 162, 171)
A65 or cccc ins (84, 157)	T271P (172)	W477stop (84)
A66P (163)	G273C,S (139, 173)	R484S (102)
I71N (84)	T275A,P, (84, 159, 162, 171)	G485V (84)
D72G (153) K (66)	G279D, A* (59, 139)	K488N (157)
D73N (160)	P280H,L*,P (92, 93, 152)	R489S (167)
D74Y, G (153)	A281V (163)	G490S (102, 139)
M84I (153)	G285C,D,V (116, 145, 152)	G491C (135, 143)
T85P (92)	E289D (153)	N493H (142)
Q88R (157)	A291P,V (152, 174)	G494D (59)
W90R (139) STOP (88, 135, 163)	L293V (139)	R496L (168)
W91R (88)	Q295K,P,STOP (130, 146, 157)	P501A (157)
D94A (84)	M296V (145)	W505S (175), R (163)
G96C (153)	Q297V (146)	D511del (84)
H97R (139)	G299S (139) C (153) A (133)	D513del (84)
G99E (84)	W300D,G,C,STOP (69, 84, 138, 157)	R515C (66, 84, 135)
R104L (141), Q (92)	gc900 ins (138)	L521del (84)
M105I (93)	S302R (139, 163)	Q525P (84)
A106V (92)	Y304S (165)	N529D (88)
W107R (92)	T308P (176)	D542H (167)
W107STOP (102)	G309C,S,V (177–179)	L546P (172)
H108E,Q (84, 141)	D311E (180) G (153) Y (179)	A550D (163)
A109V (139)	A312G,V (152)	E553K (59)
A110V (59, 84)	S315T,N,I,R,G,L (59, 66, 84, 88, 135, 139, 141–144, 157)	c1667del (66)
D117A (143)	G316D, S* (66)	F567S (84)
G120del (84)	I317L (153)	D573N (92)
G121C (92) V (157)	E318G,V (152, 180)	A574V (84) E (66)
A122del (84)	W321G (66) R (153)	L587I,M,P* (66, 84)
G123E (167)	T322A,I (136, 179)	P589T (163)
g371 del (167)	T324A,P (146, 179)	G593D (84)
G125C (153)	T326D (181)	R595stop (59)
H125 ins (171)	K327I (178)	E607K (92)
M126I (163)	W328L,C,F,S,G (84, 155, 172, 182, 183)	M609I (163)

(Continued)

Table 3 *(Continued)*

Q127E,P (153, 167)	D329A,C (138, 145)	L617del (84)
R128P (66) E (174)	S331C (180)	ac1849 ins (157)
F129S (116)	I334T (84)	L619P (84)
N133T (167)	I335T (84, 141) V (184)	G629S (88, 141)
N138S,H*,T (61, 66, 84, 88, 178)	L336R,P (88)	R632C (167)
A139P (84)	Y337F,C (163)	L634F (84)
S140A,N (84)	W341S,G (59, 139)	A636E (92)
L141F (59)	T344A,P (152, 175)	L653P (92)
D142A (84)	K345T (92)	F658V (59)
K143T (92)	P347L* (152)	L662R (139)
A144V (139)	A350S,T (84, 141)	G685R (135)
R146W (93)	Q352STOP (155)	D695A,G (162, 172)
L148A,P (84, 170)	A361D (133)	G699Q (102)
Y155C (92) S (157)	T363A (152)	S700P (84, 157)
S160L (84)	P365S* (152)	V708P (146)
cc478-479 del (167)	F368L* (152)	V710A (84)
A162T (92)	G372 ins (88)	c139 del (139)
G169A (163)	S374P* (152)	A713P (66)
A172T (66, 84) V (153)	L378P (133)	A714P (84)
M176I,T (153, 167)	A379T (142) V (133)	A716P (141)
T180C (84) K (66)	T380I (153)	Q717P (66)
G186V,H (153)	D381G (141)	V725A (146)
W191R, STOP (102, 172)	S383P (167)	A726T (139)
g572 del (153)	L384R (92)	A727D (159)
WE191-192 del (167)	D387H (167)	W728C (92)
E195K (61)	P388L,S (152)	D735 del (92)
W198stop (84)	T390I (153)	D735A (66, 84), N (92)
K200E (139), STOP (84)	L390 ins (146)	
W204R	I393N (84)	

[a]Asterisk (*), found in INH-resistant and/or INH-sensitive strains.

variant was found to be fully virulent in mice, to have no bacterial fitness cost, and to be fully transmissible (91). Consequently, the S315T variant is more often found in MDR-TB patients than in INH mono-resistant clinical isolates (86) and might be related to the higher transmission capabilities of this particular strain (92). A recent study of TB patients showed that those infected with non-*katG* INH-resistant strains (such as *inhA*[c-15t]) were more likely to exhibit sputum conversion after 1 month than those infected with *katG* mutant strains (93). Transmissibility, virulence, and response to chemotherapy seem to be affected by *katG* mutations.

Alterations of *katG* expression

katG is cotranscribed with its negative regulator *furA* (*Rv1909c*), a gene encoding a ferric uptake regulation protein (94, 95). Deletion of *furA* in *M. tuberculosis* results in overexpression of *katG* and hypersusceptibility to INH (94). Mutations have been identified in *furA* as well as in the 38-bp region between *furA* and *katG*

(Table 1). To assess the role of these mutations, Ando and colleagues constructed isogenic strains containing the mutations found in the intergenic region (g-12a, a-10c, g-7a) or in the *furA* coding sequence (A14V). *katG* expression in the g-12a strain (intergenic region) was only slightly lower than wild-type levels, and the A14V variant had no effect on KatG levels (96); however, the strains with the a-10c or g-7a mutation exhibited an 80% reduction in *katG* expression. This reduction was associated with a decrease in INH oxidase activity and a 2- to 4-fold increase in INH resistance, suggesting that the *furA-katG* intergenic region should be examined in low-level INH-resistant *M. tuberculosis* clinical strains.

Transcriptional analyses indicate that *katG* is also regulated by the sigma factor *sigI* (*Rv1189*) (97). *M. tuberculosis* Δ*sigI* had decreased catalase capabilities and was more resistant to INH than wild-type *M. tuberculosis in vitro* and *in vivo*. Furthermore, overexpression of *sigI* increased the susceptibility of *M. tuberculosis* to INH by 2-fold. Mutations in *sigI*

Table 4 Biochemical activity of KatG variants (87, 146, 167)

Activity	Variant		
	Similar to wt	Partial	None
Catalase peroxidase	A110V, A139P, A245V, S315N, S315T, R463L, L587M, L619P, L634F, D735A	L48Q	L141F, T275P, Q295P, G297V, T324P, L587P
INH oxidase	L48Q, A110V, A245V, R463L	A139P, Q127E, N133T, L141F, P232S, Q295P, G297V, T324P, S383P, D387H, D419H, M420T, R489S, L634F, D735A	M176T, S315(N,R,T), D542H, L619P, R632C

could therefore be another factor that modulates *katG* expression and induces low-level INH resistance in *M. tuberculosis* clinical isolates.

Compensatory mutations

The activity of INH against mycobacterial species is very specific. Other bacteria such as *Escherichia coli* and *Salmonella typhimurium* are not inhibited by a high dose of INH (500 mg/liter or higher; the MIC for *M. tuberculosis* is 0.05 mg/liter). Yet when *E. coli* and *S. typhimurium* have a deficient oxidative stress response regulator, encoded by *oxyR*, the strains become more sensitive to INH (MIC <50 mg/liter) (98). *M. tuberculosis oxyR´* is nonfunctional, because the coding region contains multiple frameshifts and deletions. Downstream of *oxyR´* is *ahpC* (*Rv2428*), which encodes an alkyl hydroperoxide reductase. Up to 29% of INH-resistant clinical isolates contain mutations in the *oxyR´-ahpC* region (99). Mutations in the *oxyR´-ahpC* intergenic region such as g-9a and c-15t have been shown to increase the expression of *ahpC* by 9- and 18-fold, respectively (100). This increase in *ahpC* expression is thought to compensate for the loss of KatG activity occurring in INH-resistant strains, which would render the strains more susceptible to hydrogen and organic peroxides (101) and to prevent further oxidative damage. However, most variant KatG enzymes, and in particular, KatG (S315T), are competent catalase-peroxidases, meaning that the organism is not deficient in its ability to detoxify peroxides or other compounds. With the present knowledge, the role of *ahpC* in INH resistance is a matter of debate. Baker et al. reported that mutations in the *oxyR´-ahpC* region

did not contribute to INH resistance since these mutations could be found in 20% of their INH-resistant clinical isolates but also in 8% of their INH-sensitive isolates (102).

Detoxification of INH

In humans, INH is acetylated by the arylamine *N*-acetyltransferase NAT2. The rate of this detoxification reaction varies between individuals, leading to the classification of TB patients as rapid or slow inactivators depending on their NAT activity. An enzyme similar to NAT is expressed in *M. tuberculosis* (103). When *M. tuberculosis nat* was overexpressed in *M. smegmatis*, the resulting strain was more resistant to INH, with an MIC 3-fold higher than wild-type *M. smegmatis*, suggesting that the mycobacterial arylamine *N*-acetyltransferase can acetylate and inactivate INH (104). Mutations in *nat* (*nhoA/Rv3566c*) have been identified in INH-resistant clinical isolates (Table 1); G67R and G207R variants were found in INH-resistant and INH-sensitive clinical isolates. Biochemical analyses revealed that the K_M for INH N-acetylation o Nat (G207R) was 10 times that of wild-type Nat (103), indicating that the variant Nat protein is mostly unable to acetylate and therefore inactivate INH. Thus, there is no obvious association between *nat* mutations and INH resistance. Nevertheless, deletion of *nat* in *M. bovis* BCG affected the biosynthesis of mycolic acids, glycolipids, and complex lipids as well as survival in mouse macrophages, indicating that *nat* might modulate other factors involved in INH resistance (105).

Genes induced upon INH treatment

When *M. tuberculosis* comes in contact with INH, numerous genes are upregulated, as first evidenced using a method that employed differential expression using customized amplification libraries (DECAL) (106). The availability of the *M. tuberculosis* genome sequence (107) led to the development of microarrays that have been used to explore the response of *M. tuberculosis* to INH and other drug treatments (108). Transcriptional analysis of INH-treated *M. tuberculosis* revealed that *M. tuberculosis* upregulated a set of genes encoding proteins involved in fatty acid biosynthesis (*fabD, acpM, kasA, kasB, accD6; Rv2243-2247*), trehalose dimycolyl transfer (*fbpC, Rv0129c*), fatty acid degradation (*fadE23, fadE24; Rv3139-3140*), peroxidase activity (*ahpC*), transport (*iniB, iniA; Rv0341-0342*), efflux pump (*efpA, Rv2846c*), and unknown functions (*Rv1592c, Rv1772*) (106, 108). Since genes that respond to drug treatment could be implicated in the mechanisms of resistance to the drug, Ramaswany and colleagues sequenced all the genes induced by INH except *kasB*, *fadE23*, and *acpM* in 38 INH-resistant and 86 INH-sensitive clinical isolates (88). The *kasA* operon is composed of *fabD*, *acpM*, *kasA*, *kasB*, and *accD6*. *Rv2242*, an *srmR* homolog, is located just upstream of the *kasA* operon and was added to this study. Mutations in *kasA*, *fadE24*, *Rv1592c*, and *Rv2242* were found in INH-resistant strains carrying *katG* mutations or in INH-sensitive strains. No mutations in *fbpC*, *fabD*, *accD6*, or *efpA* were found in INH-resistant clinical isolates. One T4A mutation was found in *Rv1772* in a low-level INH-resistant clinical isolate, with no mutation in the other 19 genes sequenced. This mutation was not found in the INH-sensitive isolates. Another study found a mutation in *efpA* resulting in a variant (E520V) found in INH-resistant strains only but associated with *katG* and *oxyR'-ahpC* mutations. In conclusion, most of the mutations identified in these genes (except for *Rv1772*) in INH-resistant clinical isolates were found either in INH-sensitive strains or in combination with other mutations known to confer INH resistance. Their roles in INH resistance cannot be assessed at this point.

KasA is a beta-ketoacyl-ACP synthase that condenses an elongating fatty acyl-ACP with malonyl-ACP and is the first enzyme in the FASII system. KasA was once considered a more likely target of INH than InhA based on the fact that INH treatment of *M. tuberculosis* resulted in the inhibition of mycolic acid biosynthesis and accumulation of the long-chain fatty acid hexacosanoic acid. The accumulation of fatty acids correlated better with the inhibition of a beta-ketoacyl-ACP synthase (KasA) than with the inhibition of an enoyl-ACP reductase (InhA), which should result in the accumulation of enoyl products (unsaturated fatty acids) (109). This biochemical red herring, however, did not consider the work of Takayama and coworkers, who demonstrated that a short exposure (5 min) of *M. tuberculosis* to INH resulted in the accumulation of unsaturated fatty acids, while longer INH exposures led to accumulation of hexacosanoic acid due to the total shutdown of the FASII system (26). Mdluli et al. demonstrated that in INH-treated *M. tuberculosis*, KasA formed a complex with INH and the acyl carrier protein AcpM and identified four variants in KasA (D66N, G269S, G312S, F413L) in INH-resistant clinical isolates (109). The authors thus concluded that KasA was the main target of INH. This conclusion was disproved by the demonstration that the complex between KasA, INH, and AcpM was formed only when InhA was inhibited (110) and by the discovery of

the KasA variants D66N, G269S, and G312S in INH-sensitive strains (66). Other *kasA* mutations have been identified in INH-resistant *M. tuberculosis* clinical isolates; however, these mutations are usually associated with *katG* mutations.

The INH-inducible gene *iniA* was shown to confer tolerance to INH when overexpressed in *M. bovis* BCG but not in *M. tuberculosis*, while deletion of *iniA* rendered *M. tuberculosis* more susceptible to INH (111). *iniA* is in an operon with the INH-inducible genes *iniB* and *iniC*. Immediately upstream of this operon is *Rv0340*, which is transcribed in the same direction. This operon is induced specifically by drugs inhibiting mycobacterial cell wall biosynthesis such as INH and ethambutol (112). The *iniBAC* operon encodes a membrane transporter, but it was shown not to transport INH (111). All the mutations in this cluster (Table 1) were only present in INH-resistant clinical isolates; however, they were always associated with mutations in *katG* and/or *inhA*. Hence, mutations in the *iniBAC* operon may only have a minor role in INH resistance.

Interestingly, the target of INH and ETH, *inhA*, is not upregulated upon INH or ETH treatment of *M. tuberculosis*, which may be an important determinant of successful drug targets.

Overexpression of NAD⁺/NADP⁺-binding enzymes

The INH-sensitive *E. coli oxyR* mutant described above was also used to identify other molecular determinants of INH resistance in *M. tuberculosis*. The *E. coli* mutant was transformed with an *M. tuberculosis* genomic plasmid library and screened for clones that became INH-resistant. Three genes, *glf* (*ceoA*), *ceoB*, and *ceoC*, restored INH resistance in *E. coli*. Each gene contains an NAD⁺ binding motif. Only overexpression of *M. tuberculosis glf* (*Rv3809c*), an NAD⁺- and flavin adenine dinucleotide–dependent UDP galactopyranose mutase, led to low-level INH resistance in BCG. However, a binding experiment with radioactive INH showed that Glf did not bind to INH. The authors concluded that upregulation of NAD⁺-binding proteins might play a role in INH resistance by either reducing the levels of NAD⁺ available for the formation of the INH-NAD adduct or by binding an unknown derivative of INH (113).

The finding that the NADPH-dependent β-ketoacyl-ACP reductase FabG (Rv1483), part of the FASII system, was inhibited *in vitro* by an INH-NADP adduct similarly to InhA (114) led others to investigate whether additional enzymes could be inhibited by this INH-NADP adduct. The enzymatic activity of the *M. tuberculosis* dihydrofolate reductase DHFR (Rv2763c), the target of the broad-spectrum antibiotic trimethoprim, was shown to be inhibited by the *4R*-INH-NADP adduct with subnanomolar affinity (115). Overexpression of *M. tuberculosis dfrA* in *M. smegmatis* increased INH resistance by 2-fold at 30°C (115). However, overexpression of *M. tuberculosis dfrA* in *M. tuberculosis* did not increase resistance to INH (116), and the sequences of 127 INH-resistant clinical isolates revealed no mutation in *dfrA*, suggesting that DHFR is not a marker for INH resistance in *M. tuberculosis* (117). Affinity chromatography identified 16 other proteins that could bind to the INH-NAD or INH-NADP adduct (118), but overexpression of the *M. tuberculosis* genes encoding these proteins did not confer resistance to INH or ETH in *M. smegmatis* (116).

Overexpression of efflux pumps

Active export of drugs from cells by efflux pumps was first described in *E. coli* for tetracycline (119). In mycobacteria, low-level resistance to tetracycline and other aminoglycosides has been attributed to efflux pumps (120, 121). *iniA*, a gene induced by INH treatment (106), was shown to be a component of an MDR-like efflux pump. Overexpression of *iniA* conferred tolerance to INH, while deletion of *iniA* increased the susceptibility of mycobacteria to INH (111). However, IniA does not pump INH out of the cells.

A microarray study of *M. tuberculosis* clinical isolates resistant to INH, RIF, SM, ethambutol, and ofloxacin revealed that upon INH treatment, expression of several predicted efflux pump genes was upregulated: *Rv1819c*, *Rv2459*, *Rv2846*, *Rv3065*, *and Rv3728* (122). Furthermore, administration of an efflux pump inhibitor decreased the MIC for INH in these strains by at least 4-fold, suggesting that these efflux pumps may play a role in drug resistance in *M. tuberculosis*.

Rv1217c-1218c, an operon encoding an ATP binding cassette transporter, was shown by RT-qPCR to be overexpressed in MDR- and XDR-TB strains; the overexpression of *Rv1218C* was associated with increased INH resistance (123). However, deletion of *Rv1218c* did not affect the MIC of INH in *M. tuberculosis* (124), nor did deletion of other efflux pumps such as *Rv1877, mmr,* and *mmpL7* (125, 126), although overexpression of *mmpL7* conferred resistance to INH (127). Conversely, deletion of *lfrA* led to a 2-fold decrease in the INH MIC (126). Unexpectedly, deletion of *efpA*, which encodes an efflux pump that is specifically induced upon INH treatment of *M. tuberculosis*, resulted in increased resistance to INH and RIF in *M. smegmatis* (126).

In light of these studies, the role of efflux pumps in INH resistance in *M. tuberculosis* needs to be further evaluated in order to assess their importance in the resistance mechanisms.

Mechanisms of Resistance to ETH Only

Alterations of EthA, the activator of ETH

The mechanisms of resistance to ETH are mutations in genes encoding its activator (*ethA*), its target (*inhA*), or the *ethA* regulator (*ethR*). So far, 85 *ethA* mutations have been identified, although some were also found in drug-susceptible or partially ETH-resistant *M. tuberculosis* strains (Table 2). Mutations have been identified throughout the length of the coding region. Approximately two-thirds of the nucleotide changes are missense mutations that result in amino acid changes, while the remaining mutations are insertions, deletions, or nonsense mutations. Unlike the *katG*(S315T) variant, which can be present in up to 94% of the INH-resistant clinical isolates, no dominant *ethA* mutation occurs in ETH-resistant clinical isolates. Morlock and colleagues (61) hypothesized that the lack of cluster or dominant *ethA* mutations in ETH-resistant clinical isolates could be attributed to the presence of numerous monooxygenase homologs in *M. tuberculosis* that could protect the cells against a loss of EthA activity.

In a study by Brossier and colleagues (62), 47 ETH-resistant clinical isolates were analyzed for mutations in *ethA*, *ethR*, *inhA*, *ndh*, or *mshA*. Of these, 22 clinical isolates (47%) had mutations in *ethA*, while 29 strains (62%) had a mutation in *inhA* (promoter region/gene). On average, the proportion of *inhA* mutations in ETH-resistant clinical isolates is 68%, suggesting that this is the main mechanism of ETH resistance in *M. tuberculosis*.

Mutations in *ethR*, the regulator of *ethA*

ethA is negatively regulated by EthR (Rv3855), a transcriptional repressor belonging to the TetR family. The two genes are oriented opposite to each other, separated by a 73-bp intergenic region that contains the *ethA* promoter and to which EthR binds. A strain with a transposon insertion in *ethR* was highly sensitive to ETH, while *ethR* overexpression increased the resistance to ETH (45). Two mutations in *ethR* have been identified (Table 2) in highly ETH-resistant clinical strains, which represent 4% of ETH-resistant clinical strains screened; however, these two strains also contained the c-15t mutation in the promoter region of *inhA,* and one of the two had a mutation in *ethA* as well (62). Since the majority of ETH-resistant clinical isolates have mutations in *ethA* and/or *inhA, ethR* might only play a minor role in ETH resistance in *M. tuberculosis* clinical isolates.

CONCLUSIONS AND THE FUTURE

The mechanisms of resistance to INH and ETH in *M. tuberculosis* are both simple and complex (Fig. 3). The main mechanisms of resistance are mutations in *katG* and *ethA,* the activators of INH and ETH, respectively, preventing the formation of the INH-NAD or ETH-NAD adduct, and mutations in *inhA,* the target of INH and ETH, leading to titration of the drug or reduced binding of the INH-NAD or ETH-NAD adduct to InhA. Numerous mutations in multiple genes have been identified in INH- or ETH-resistant *M. tuberculosis* clinical isolates (Tables 1 to 3). Ultimately,

the validation of a mutation responsible for INH or ETH resistance requires transferring the point mutation into wild-type *M. tuberculosis* and measuring the level of resistance. This is far beyond the scope of most studies that describe mutations associated with drug resistance in *M. tuberculosis.*

It is often reported that a certain percentage (up to 30%) of INH-resistant *M. tuberculosis* clinical isolates have no mutation in any of the genes studied, leading some to conclude that there is still much more to discover about INH and ETH mode of action and resistance. Yet most studies sequenced only a small fraction of the genes known to confer INH resistance such as *katG* and *inhA,* looking only at the region around codon 315 of *katG* or the regulatory region of *inhA.* However, more than 300 mutations have been identified in *katG* from amino acids 1 to 735 (Table 3), and mutations outside of the S315 region have also been shown to be highly defective in INH activation.

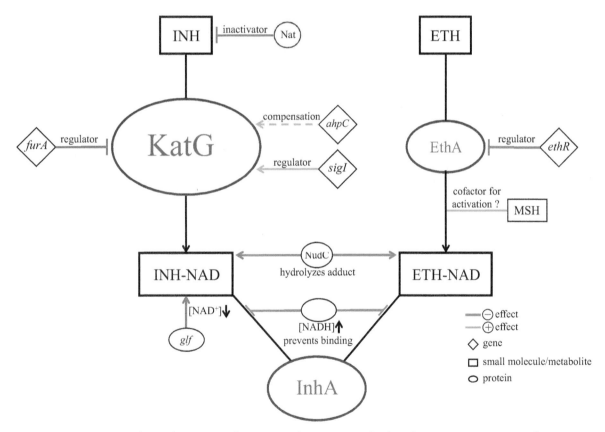

Figure 3 Relationship among the genes and proteins involved in the resistance to INH and ETH in *M. tuberculosis.* Connections in red indicate a negative relationship (degradation of an active molecule, negative regulator of an enzyme) that would lead to resistance to INH and/or ETH; in green are positive actions that would increase a strain fitness or susceptibility to the drugs. The dashed line points to an interaction that does not result directly in INH resistance or susceptibility. doi:10.1128/microbiolspec.MGM2-0014-2013.f3

Sequencing only a fraction of the *katG* gene around the S315 region might leave mutations responsible for the drug resistance phenotype undiscovered. Nevertheless, in studies where the *katG* and *inhA* genes and their regulatory regions are entirely sequenced, the mechanism of resistance is not identified in up to 5% of the INH-resistant clinical isolates.

Additional mechanisms of resistance likely remain to be identified, and a worthwhile endeavor may be to revisit the early studies on INH mechanisms of action to determine other factors of INH and ETH resistance in *M. tuberculosis*. Knowledge of resistance genes and mutations has provided the means to develop Geno-Type MTBDR*plus*, a rapid nucleic acid–based test for assessing INH and RIF resistance (128). Moreover, it may provide strategies to develop new drugs that bypass the known mechanisms of drug resistance. For example, the natural product pyridomycin inhibits InhA without requiring KatG activation and is therefore active against highly INH-resistant *M. tuberculosis* clinical isolates carrying a *katG* mutation (129). The identification of additional genes contributing to INH and ETH resistance will also expand our understanding of the mechanisms of drug action. Finally, in addition to specific mutations that confer resistance to every cell in a population, new studies that reveal the way in which a cell can become transiently phenotypically resistant to INH or ETH will be important in developing better ways to kill *M. tuberculosis* with INH and ETH and shorten chemotherapy.

Citation. Vilchèze C, Jacobs WR Jr. 2014. Resistance to isoniazid and ethionamide in *Mycobacterium tuberculosis*: genes, mutations, and causalities. Microbiol Spectrum 2(4):MGM2-0014-2013.

References

1. **Behnisch R, Mietzsch F, Schmidt H.** 1950. Chemical studies on thiosemicarbazones with particular reference to antituberculous activity. *Am Rev Tuberc* **61:** 1–7.

2. **Domagk G.** 1950. Investigations on the antituberculous activity of the thiosemicarbazones in vitro and in vivo. *Am Rev Tuberc* **61:**8–19.

3. **Schatz A, Waksman SA.** 1944. Effect of streptomycin upon *Mycobacterium tuberculosis* and related organisms. *Proc Soc Exp Biol Med* **57:**244–248.

4. **Lehmann J.** 1946. *para*-Aminosalicylic acid in the treatment of tuberculosis. *Lancet* **247:**15.

5. **Bernstein JW, Lott A, Steinberg BA, Yale HL.** 1952. Chemotherapy of experimental tuberculosis. *Am Rev Tuberc* **65:**357–374.

6. **Fox HH.** 1952. The chemical approach to the control of tuberculosis. *Science* **116:**129–134.

7. **Medical Research Council Investigation.** 1950. Treatment of pulmonary tuberculosis with streptomycin and *para*-amino-salicylic acid. *Br Med J* **2:**1073–1085.

8. **Medical Research Council Investigation.** 1952. The treatment of pulmonary tuberculosis with isoniazid. *Br Med J* **2:**735–746.

9. **Crofton J, Mitchison DA.** 1948. Streptomycin resistance in pulmonary tuberculosis. *Br Med J* **2:**1009–1015.

10. **Crofton J.** 1959. Chemotherapy of pulmonary tuberculosis. *Br Med J* **1:**1610–1614.

11. **WHO.** 2012. *Global Tuberculosis Report 2012.* Geneva, Switzerland.

12. **Cegielski P, Nunn P, Kurbatova EV, Weyer K, Dalton TL, Wares DF, Iademarco MF, Castro KG, Raviglione M.** 2012. Challenges and controversies in defining totally drug-resistant tuberculosis. *Emerg Infect Dis* **18:**e2.

13. **Domagk G, Offe HA, Siefken W.** 1952. [Additional investigations in experimental chemotherapy of tuberculosis (neoteban)]. *Dtsch Med Wochenschr* **77:**573–578.

14. **Chorine V.** 1945. Action of nicotinamide on bacilli of the species *Mycobacterium*. *Compt Ren* **220:**150–151.

15. **Barclay WR, Ebert RH, Kochweser D.** 1953. Mode of action of isoniazid. *Am Rev Tuberc* **67:**490–496.

16. **Pope H.** 1953. Antagonism of isoniazid by certain metabolites. *Am Rev Tuberc* **68:**938–939.

17. **Russe HP, Barclay WR.** 1955. The effect of isoniazid on lipids of the tubercle bacillus. *Am Rev Tuberc* **72:** 713–717.

18. **Ebina T, Motomiya M, Munakata K, Kobuya G.** 1961. Effect of isoniazid on fatty acids in *Mycobacterium*. *CR Seances Soc Biol Fil* **155:**1176–1178.

19. **Gangadharam PRJ, Harold FM, Schaefer W.** 1963. Selective inhibition of nucleic acid synthesis in *Mycobacterium tuberculosis* by isoniazid. *Nature* **198:**712–714.

20. **Winder FG, Brennan P.** 1965. Effect of isoniazid on lipid metabolism in *Mycobacterium tuberculosis*. *Biochem J* **96:**77P.

21. **Bekierkunst A.** 1966. Nicotinamide-adenine dinucleotide in tubercle bacilli exposed to isoniazid. *Science* **152:** 525–526.

22. **Davis WE, Weber MM.** 1977. Specificity of isoniazid on growth inhibition and competition for an oxidized nicotinamide adenine dinucleotide regulatory site on the electron transport pathway in *Mycobacterium phlei*. *Antimicrob Agents Chemother* **12:**213–218.

23. **Herman RP, Weber MM.** 1980. Site of action of isoniazid on the electron transport chain and its relationship to nicotinamide adenine dinucleotide regulation in *Mycobacterium phlei*. *Antimicrob Agents Chemother* **17:** 450–454.

24. **Takayama K, Wang L, David HL.** 1972. Effect of isoniazid on the in vivo mycolic acid synthesis, cell growth, and viability of *Mycobacterium tuberculosis*. *Antimicrob Agents Chemother* **2:**29–35.

25. **Takayama K, Schnoes HK, Armstrong EL, Boyle RW.** 1975. Site of inhibitory action of isoniazid in the synthesis of mycolic acids in *Mycobacterium tuberculosis*. *J Lipid Res* **16:**308–317.

26. Davidson LA, Takayama K. 1979. Isoniazid inhibition of the synthesis of monounsaturated long-chain fatty acids in *Mycobacterium tuberculosis* H37Ra. *Antimicrob Agents Chemother* **16**:104–105.

27. Johnsson K, Schultz PG. 1994. Mechanistic studies of the oxidation of isoniazid by the catalase peroxidase from *Mycobacterium tuberculosis*. *J Am Chem Soc* **116**:7425–7426.

28. Lei B, Wei CJ, Tu SC. 2000. Action mechanism of antitubercular isoniazid. Activation by *Mycobacterium tuberculosis* KatG, isolation, and characterization of *inhA* inhibitor. *J Biol Chem* **275**:2520–2526.

29. Wilming M, Johnsson K. 1999. Spontaneous formation of the bioactive form of the tuberculosis drug isoniazid. *Angew Chem Int Ed Engl* **38**:2588–2590.

30. Zhang Y, Heym B, Allen B, Young D, Cole S. 1992. The catalase-peroxidase gene and isoniazid resistance of *Mycobacterium tuberculosis*. *Nature* **358**:591–593.

31. Rozwarski DA, Grant GA, Barton DH, Jacobs WR Jr, Sacchettini JC. 1998. Modification of the NADH of the isoniazid target (InhA) from *Mycobacterium tuberculosis*. *Science* **279**:98–102.

32. Dessen A, Quemard A, Blanchard JS, Jacobs WR Jr, Sacchettini JC. 1995. Crystal structure and function of the isoniazid target of *Mycobacterium tuberculosis*. *Science* **267**:1638–1641.

33. Quemard A, Sacchettini JC, Dessen A, Vilcheze C, Bittman R, Jacobs WR Jr, Blanchard JS. 1995. Enzymatic characterization of the target for isoniazid in *Mycobacterium tuberculosis*. *Biochemistry* **34**:8235–8241.

34. Marrakchi H, Laneelle G, Quemard A. 2000. InhA, a target of the antituberculous drug isoniazid, is involved in a mycobacterial fatty acid elongation system, FAS-II. *Microbiology* **146**(Pt 2):289–296.

35. Nguyen M, Quemard A, Broussy S, Bernadou J, Meunier B. 2002. Mn(III) pyrophosphate as an efficient tool for studying the mode of action of isoniazid on the InhA protein of *Mycobacterium tuberculosis*. *Antimicrob Agents Chemother* **46**:2137–2144.

36. Rawat R, Whitty A, Tonge PJ. 2003. The isoniazid-NAD adduct is a slow, tight-binding inhibitor of InhA, the *Mycobacterium tuberculosis* enoyl reductase: adduct affinity and drug resistance. *Proc Natl Acad Sci USA* **100**:13881–13886.

37. Vilcheze C, Morbidoni HR, Weisbrod TR, Iwamoto H, Kuo M, Sacchettini JC, Jacobs WR Jr. 2000. Inactivation of the *inhA*-encoded fatty acid synthase II (FASII) enoyl-acyl carrier protein reductase induces accumulation of the FASI end products and cell lysis of *Mycobacterium smegmatis*. *J Bacteriol* **182**:4059–4067.

38. Vilcheze C, Wang F, Arai M, Hazbon MH, Colangeli R, Kremer L, Weisbrod TR, Alland D, Sacchettini JC, Jacobs WR Jr. 2006. Transfer of a point mutation in *Mycobacterium tuberculosis inhA* resolves the target of isoniazid. *Nat Med* **12**:1027–1029.

39. Grumbach F, Rist N, Libermann D, Moyeux M, Cals S, Clavel S. 1956. Experimental antituberculous activity of certain isonicotinic thiamides substituted on the nucleus. *CR Hebd Seances Acad Sci* **242**:2187–2189 (In French.)

40. Brouet G, Marche J, Rist N, Chevallier J, Lemeur G. 1959. Observations on the antituberculous effectiveness of alpha-ethyl-thioisonicotinamide in tuberculosis in humans. *Am Rev Tuberc* **79**:6–18.

41. Petty TL, Mitchell RS. 1962. Successful treatment of advanced isonizid- and streptomycin-resistant pulmonary tuberculosis with ethionamide, pyrazinamide, and isoniazid. *Am Rev Respir Dis* **86**:503–512.

42. Middlebrook G. 1952. Sterilization of tubercle bacilli by isonicotinic acid hydrazide and the incidence of variants resistant to the drug in vitro. *Am Rev Tuberc* **65**:765–767.

43. Schaefer WB. 1954. The effect of isoniazid on growing and resting tubercle bacilli. *Am Rev Tuberc* **69**:125–127.

44. Winder FG, Collins PB, Whelan D. 1971. Effects of ethionamide and isoxyl on mycolic acid synthesis in *Mycobacterium tuberculosis* BCG. *J Gen Microbiol* **66**:379–380.

45. Baulard AR, Betts JC, Engohang-Ndong J, Quan S, McAdam RA, Brennan PJ, Locht C, Besra GS. 2000. Activation of the pro-drug ethionamide is regulated in mycobacteria. *J Biol Chem* **275**:28326–28331.

46. DeBarber AE, Mdluli K, Bosman M, Bekker LG, Barry CE 3rd. 2000. Ethionamide activation and sensitivity in multidrug-resistant *Mycobacterium tuberculosis*. *Proc Natl Acad Sci USA* **97**:9677–9682.

47. Fraaije MW, Kamerbeek NM, Heidekamp AJ, Fortin R, Janssen DB. 2004. The prodrug activator EtaA from *Mycobacterium tuberculosis* is a Baeyer-Villiger monooxygenase. *J Biol Chem* **279**:3354–3360.

48. Vannelli TA, Dykman A, Ortiz de Montellano PR. 2002. The antituberculosis drug ethionamide is activated by a flavoprotein monooxygenase. *J Biol Chem* **277**:12824–12829.

49. Wang F, Langley R, Gulten G, Dover LG, Besra GS, Jacobs WR Jr, Sacchettini JC. 2007. Mechanism of thioamide drug action against tuberculosis and leprosy. *J Exp Med* **204**:73–78.

50. Dover LG, Alahari A, Gratraud P, Gomes JM, Bhowruth V, Reynolds RC, Besra GS, Kremer L. 2007. EthA, a common activator of thiocarbamide-containing drugs acting on different mycobacterial targets. *Antimicrob Agents Chemother* **51**:1055–1063.

51. Hok TT. 1964. A comparative study of the susceptibility to ethionamide, thiosemicarbazone, and isoniazid of tubercle bacilli from patients never treated with ethionamide or thiosemicarbazone. *Am Rev Respir Dis* **90**:468–469.

52. Stewart SM, Hall E, Riddell RW, Somner AR. 1962. Bacteriological aspects of the use of ethionamide, pyrazinamide and cycloserine in the treatment of chronic pulmonary tuberculosis. *Tubercle* **43**:417–431.

53. Lefford MJ. 1966. The ethionamide sensitivity of British pre-treatment strains of *Mycobacterium tuberculosis*. *Tubercle* **47**:198–206.

54. Snapper SB, Lugosi L, Jekkel A, Melton RE, Kieser T, Bloom BR, Jacobs WR Jr. 1988. Lysogeny and transformation in mycobacteria: stable expression of foreign genes. *Proc Natl Acad Sci USA* 85:6987–6991.

55. Snapper SB, Melton RE, Mustafa S, Kieser T, Jacobs WR Jr. 1990. Isolation and characterization of efficient plasmid transformation mutants of *Mycobacterium smegmatis*. *Mol Microbiol* 4:1911–1919.

56. Banerjee A, Dubnau E, Quemard A, Balasubramanian V, Um KS, Wilson T, Collins D, de Lisle G, Jacobs WR Jr. 1994. *inhA*, a gene encoding a target for isoniazid and ethionamide in *Mycobacterium tuberculosis*. *Science* 263:227–230.

57. Takayama K, Wang L, Merkal RS. 1973. Scanning electron microscopy of the H37Ra strain of *Mycobacterium tuberculosis* exposed to isoniazid. *Antimicrob Agents Chemother* 4:62–65.

58. Dias MV, Vasconcelos IB, Prado AM, Fadel V, Basso LA, de Azevedo WF Jr, Santos DS. 2007. Crystallographic studies on the binding of isonicotinyl-NAD adduct to wild-type and isoniazid resistant 2-trans-enoyl-ACP (CoA) reductase from *Mycobacterium tuberculosis*. *J Struct Biol* 159:369–380.

59. Brossier F, Veziris N, Truffot-Pernot C, Jarlier V, Sougakoff W. 2006. Performance of the genotype MTBDR line probe assay for detection of resistance to rifampin and isoniazid in strains of *Mycobacterium tuberculosis* with low- and high-level resistance. *J Clin Microbiol* 44:3659–3664.

60. Muller B, Streicher EM, Hoek KG, Tait M, Trollip A, Bosman ME, Coetzee GJ, Chabula-Nxiweni EM, Hoosain E, Gey van Pittius NC, Victor TC, van Helden PD, Warren RM. 2011. *inhA* promoter mutations: a gateway to extensively drug-resistant tuberculosis in South Africa? *Int J Tuberc Lung Dis* 15:344–351.

61. Morlock GP, Metchock B, Sikes D, Crawford JT, Cooksey RC. 2003. *ethA*, *inhA*, and *katG* loci of ethionamide-resistant clinical *Mycobacterium tuberculosis* isolates. *Antimicrob Agents Chemother* 47:3799–3805.

62. Brossier F, Veziris N, Truffot-Pernot C, Jarlier V, Sougakoff W. 2011. Molecular investigation of resistance to the antituberculous drug ethionamide in multidrug-resistant clinical isolates of *Mycobacterium tuberculosis*. *Antimicrob Agents Chemother* 55:355–360.

63. Vadwai V, Ajbani K, Jose M, Vineeth VP, Nikam C, Deshmukh M, Shetty A, Soman R, Rodrigues C. 2012. Can *inhA* mutation predict ethionamide resistance? *Int J Tuberc Lung Dis* 17:129–130.

64. Miesel L, Weisbrod TR, Marcinkeviciene JA, Bittman R, Jacobs WR Jr. 1998. NADH dehydrogenase defects confer isoniazid resistance and conditional lethality in *Mycobacterium smegmatis*. *J Bacteriol* 180:2459–2467.

65. Vilcheze C, Weisbrod TR, Chen B, Kremer L, Hazbon MH, Wang F, Alland D, Sacchettini JC, Jacobs WR Jr. 2005. Altered NADH/NAD$^+$ ratio mediates coresistance to isoniazid and ethionamide in mycobacteria. *Antimicrob Agents Chemother* 49:708–720.

66. Hazbon MH, Brimacombe M, Bobadilla del Valle M, Cavatore M, Guerrero MI, Varma-Basil M, Billman-Jacobe H, Lavender C, Fyfe J, Garcia-Garcia L, Leon CI, Bose M, Chaves F, Murray M, Eisenach KD, Sifuentes-Osornio J, Cave MD, Ponce de Leon A, Alland D. 2006. Population genetics study of isoniazid resistance mutations and evolution of multidrug-resistant *Mycobacterium tuberculosis*. *Antimicrob Agents Chemother* 50:2640–2649.

67. Cardoso RF, Cardoso MA, Leite CQ, Sato DN, Mamizuka EM, Hirata RD, de Mello FF, Hirata MH. 2007. Characterization of *ndh* gene of isoniazid resistant and susceptible *Mycobacterium tuberculosis* isolates from Brazil. *Mem Inst Oswaldo Cruz* 102:59–61.

68. Lee AS, Teo AS, Wong SY. 2001. Novel mutations in *ndh* in isoniazid-resistant *Mycobacterium tuberculosis* isolates. *Antimicrob Agents Chemother* 45:2157–2159.

69. Lee AS, Lim IH, Tang LL, Telenti A, Wong SY. 1999. Contribution of *kasA* analysis to detection of isoniazid-resistant *Mycobacterium tuberculosis* in Singapore. *Antimicrob Agents Chemother* 43:2087–2089.

70. Vilcheze C, Av-Gay Y, Attarian R, Liu Z, Hazbon MH, Colangeli R, Chen B, Liu W, Alland D, Sacchettini JC, Jacobs WR Jr. 2008. Mycothiol biosynthesis is essential for ethionamide susceptibility in *Mycobacterium tuberculosis*. *Mol Microbiol* 69:1316–1329.

71. Vilcheze C, Av-Gay Y, Barnes SW, Larsen MH, Walker JR, Glynne RJ, Jacobs WR Jr. 2011. Coresistance to isoniazid and ethionamide maps to mycothiol biosynthetic genes in *Mycobacterium bovis*. *Antimicrob Agents Chemother* 55:4422–4423.

72. Newton GL, Buchmeier N, Fahey RC. 2008. Biosynthesis and functions of mycothiol, the unique protective thiol of Actinobacteria. *Microbiol Mol Biol Rev* 72:471–494.

73. Xu X, Vilcheze C, Av-Gay Y, Gomez-Velasco A, Jacobs WR Jr. 2011. Precise null deletion mutations of the mycothiol synthesis genes reveal their role in isoniazid and ethionamide resistance in *Mycobacterium smegmatis*. *Antimicrob Agents Chemother* 55:3133–3139.

74. Rawat M, Johnson C, Cadiz V, Av-Gay Y. 2007. Comparative analysis of mutants in the mycothiol biosynthesis pathway in *Mycobacterium smegmatis*. *Biochem Biophys Res Commun* 363:71–76.

75. Projahn M, Koser CU, Homolka S, Summers DK, Archer JA, Niemann S. 2011. Polymorphisms in isoniazid and prothionamide resistance genes of the *Mycobacterium tuberculosis* complex. *Antimicrob Agents Chemother* 55:4408–4411.

76. Wang XD, Gu J, Wang T, Bi LJ, Zhang ZP, Cui ZQ, Wei HP, Deng JY, Zhang XE. 2011. Comparative analysis of mycobacterial NADH pyrophosphatase isoforms reveals a novel mechanism for isoniazid and ethionamide inactivation. *Mol Microbiol* 82:1375–1391.

77. Middlebrook G. 1954. Isoniazid resistance and catalase activity of tubercle bacilli. *Am Rev Tuberc* 69:471–472.

78. Middlebrook G, Cohn ML. 1953. Some observations on the pathogenicity of isoniazid-resistant variants of tubercle bacilli. *Science* 118:297–299.

79. Winder F. 1960. Catalase and peroxidase in myco-bacteria. Possible relationship to the mode of action of isoniazid. *Am Rev Respir Dis* **81**:68–78.

80. Zhang Y, Garbe T, Young D. 1993. Transformation with *katG* restores isoniazid-sensitivity in *Mycobacterium tuberculosis* isolates resistant to a range of drug concentrations. *Mol Microbiol* **8**:521–524.

81. Mokrousov I, Narvskaya O, Otten T, Limeschenko E, Steklova L, Vyshnevskiy B. 2002. High prevalence of KatG Ser315Thr substitution among isoniazid-resistant *Mycobacterium tuberculosis* clinical isolates from northwestern Russia, 1996 to 2001. *Antimicrob Agents Chemother* **46**:1417–1424.

82. Wengenack NL, Uhl JR, St Amand AL, Tomlinson AJ, Benson LM, Naylor S, Kline BC, Cockerill FR 3rd, Rusnak F. 1997. Recombinant *Mycobacterium tuberculosis* KatG(S315T) is a competent catalase-peroxidase with reduced activity toward isoniazid. *J Infect Dis* **176**: 722–727.

83. Saint-Joanis B, Souchon H, Wilming M, Johnsson K, Alzari PM, Cole ST. 1999. Use of site-directed muta-genesis to probe the structure, function and isonia-zid activation of the catalase/peroxidase, KatG, from *Mycobacterium tuberculosis*. *Biochem J* **338**(Pt 3):753–760.

84. Ramaswamy S, Musser JM. 1998. Molecular genetic ba-sis of antimicrobial agent resistance in *Mycobacterium tuberculosis*: 1998 update. *Tuber Lung Dis* **79**:3–29.

85. van Doorn HR, Kuijper EJ, van der Ende A, Welten AG, van Soolingen D, de Haas PE, Dankert J. 2001. The susceptibility of *Mycobacterium tuberculosis* to isoniazid and the Arg→Leu mutation at codon 463 of *katG* are not associated. *J Clin Microbiol* **39**:1591–1594.

86. van Soolingen D, de Haas PE, van Doorn HR, Kuijper E, Rinder H, Borgdorff MW. 2000. Mutations at amino acid position 315 of the *katG* gene are associated with high-level resistance to isoniazid, other drug resistance, and successful transmission of *Mycobacterium tubercu-losis* in the Netherlands. *J Infect Dis* **182**:1788–1790.

87. Wei CJ, Lei B, Musser JM, Tu SC. 2003. Isoniazid acti-vation defects in recombinant *Mycobacterium tubercu-losis* catalase-peroxidase (KatG) mutants evident in InhA inhibitor production. *Antimicrob Agents Chemother* **47**: 670–675.

88. Ramaswamy SV, Reich R, Dou SJ, Jasperse L, Pan X, Wanger A, Quitugua T, Graviss EA. 2003. Single nucle-otide polymorphisms in genes associated with isoniazid resistance in *Mycobacterium tuberculosis*. *Antimicrob Agents Chemother* **47**:1241–1250.

89. Cohn ML, Oda U, Kovitz C, Middlebrook G. 1954. Studies on isoniazid and tubercle bacilli. I. The isolation of isoniazid-resistant mutants in vitro. *Am Rev Tuberc* **70**:465–475.

90. Cohn ML, Kovitz C, Oda U, Middlebrook G. 1954. Studies on isoniazid and tubercle bacilli. II. The growth requirements, catalase activities, and pathogenic proper-ties of isoniazid-resistant mutants. *Am Rev Tuberc* **70**: 641–664.

91. Pym AS, Saint-Joanis B, Cole ST. 2002. Effect of *katG* mutations on the virulence of *Mycobacterium tubercu-losis* and the implication for transmission in humans. *Infect Immun* **70**:4955–4960.

92. Gagneux S, Burgos MV, DeRiemer K, Encisco A, Munoz S, Hopewell PC, Small PM, Pym AS. 2006. Impact of bacterial genetics on the transmission of isoniazid-resistant *Mycobacterium tuberculosis*. *PLoS Pathog* **2**:e61.

93. Escalante P, McKean-Cowdin R, Ramaswamy SV, Williams-Bouyer N, Teeter LD, Jones B, Graviss EA. 2013. Can mycobacterial katG genetic changes in isoniazid-resistant tuberculosis influence human disease features? *Int J Tuberc Lung Dis* **17**:641–51.

94. Pym AS, Domenech P, Honore N, Song J, Deretic V, Cole ST. 2001. Regulation of catalase-peroxidase (KatG) expression, isoniazid sensitivity and virulence by *furA* of *Mycobacterium tuberculosis*. *Mol Microbiol* **40**: 879–889.

95. Zahrt TC, Song J, Siple J, Deretic V. 2001. Mycobacte-rial FurA is a negative regulator of catalase-peroxidase gene *katG*. *Mol Microbiol* **39**:1174–1185.

96. Ando H, Kitao T, Miyoshi-Akiyama T, Kato S, Mori T, Kirikae T. 2011. Downregulation of *katG* expression is associated with isoniazid resistance in *Mycobacterium tuberculosis*. *Mol Microbiol* **79**:1615–1628.

97. Lee JH, Ammerman NC, Nolan S, Geiman DE, Lun S, Guo H, Bishai WR. 2012. Isoniazid resistance without a loss of fitness in *Mycobacterium tuberculosis*. *Nat Commun* **3**:753.

98. Rosner JL. 1993. Susceptibilities of *oxyR* regulon mutants of *Escherichia coli* and *Salmonella typhimu-rium* to isoniazid. *Antimicrob Agents Chemother* **37**: 2251–2253.

99. Sreevatsan S, Pan X, Zhang Y, Deretic V, Musser JM. 1997. Analysis of the *oxyR-ahpC* region in isoniazid-resistant and -susceptible *Mycobacterium tuberculosis* complex organisms recovered from diseased humans and animals in diverse localities. *Antimicrob Agents Chemother* **41**:600–606.

100. Sherman DR, Mdluli K, Hickey MJ, Arain TM, Morris SL, Barry CE 3rd, Stover CK. 1996. Compensatory *ahpC* gene expression in isoniazid-resistant *Mycobacte-rium tuberculosis*. *Science* **272**:1641–1643.

101. Ng VH, Cox JS, Sousa AO, MacMicking JD, McKinney JD. 2004. Role of KatG catalase-peroxidase in myco-bacterial pathogenesis: countering the phagocyte oxida-tive burst. *Mol Microbiol* **52**:1291–1302.

102. Baker LV, Brown TJ, Maxwell O, Gibson AL, Fang Z, Yates MD, Drobniewski FA. 2005. Molecular analy-sis of isoniazid-resistant *Mycobacterium tuberculosis* isolates from England and Wales reveals the phyloge-netic significance of the ahpC -46A polymorphism. *Antimicrob Agents Chemother* **49**:1455–1464.

103. Upton AM, Mushtaq A, Victor TC, Sampson SL, Sandy J, Smith DM, van Helden PV, Sim E. 2001. Arylamine N-acetyltransferase of *Mycobacterium tuberculosis* is a polymorphic enzyme and a site of isoniazid metabolism. *Mol Microbiol* **42**:309–317.

104. Payton M, Auty R, Delgoda R, Everett M, Sim E. 1999. Cloning and characterization of arylamine N-acetyltransferase genes from *Mycobacterium smegmatis* and *Mycobacterium tuberculosis*: increased expression results in isoniazid resistance. *J Bacteriol* 181:1343–1347.

105. Bhakta S, Besra GS, Upton AM, Parish T, Sholto-Douglas-Vernon C, Gibson KJ, Knutton S, Gordon S, DaSilva RP, Anderton MC, Sim E. 2004. Arylamine N-acetyltransferase is required for synthesis of mycolic acids and complex lipids in *Mycobacterium bovis* BCG and represents a novel drug target. *J Exp Med* 199:1191–1199.

106. Alland D, Kramnik I, Weisbrod TR, Otsubo L, Cerny R, Miller LP, Jacobs WR Jr, Bloom BR. 1998. Identification of differentially expressed mRNA in prokaryotic organisms by customized amplification libraries (DECAL): the effect of isoniazid on gene expression in *Mycobacterium tuberculosis*. *Proc Natl Acad Sci USA* 95:13227–13232.

107. Cole ST, Brosch R, Parkhill J, Garnier T, Churcher C, Harris D, Gordon SV, Eiglmeier K, Gas S, Barry CE 3rd, Tekaia F, Badcock K, Basham D, Brown D, Chillingworth T, Connor R, Davies R, Devlin K, Feltwell T, Gentles S, Hamlin N, Holroyd S, Hornsby T, Jagels K, Krogh A, McLean J, Moule S, Murphy L, Oliver K, Osborne J, Quail MA, Rajandream MA, Rogers J, Rutter S, Seeger K, Skelton J, Squares R, Squares S, Sulston JE, Taylor K, Whitehead S, Barrell BG. 1998. Deciphering the biology of *Mycobacterium tuberculosis* from the complete genome sequence. *Nature* 393:537–544.

108. Wilson M, DeRisi J, Kristensen HH, Imboden P, Rane S, Brown PO, Schoolnik GK. 1999. Exploring drug-induced alterations in gene expression in *Mycobacterium tuberculosis* by microarray hybridization. *Proc Natl Acad Sci USA* 96:12833–12838.

109. Mdluli K, Slayden RA, Zhu Y, Ramaswamy S, Pan X, Mead D, Crane DD, Musser JM, Barry CE 3rd. 1998. Inhibition of a *Mycobacterium tuberculosis* beta-ketoacyl ACP synthase by isoniazid. *Science* 280:1607–1610.

110. Kremer L, Dover LG, Morbidoni HR, Vilcheze C, Maughan WN, Baulard A, Tu S, Honore N, Deretic V, Sacchettini JC, Locht C, Jacobs WRJ, Besra G. 2003. Inhibition of InhA activity, but not KasA activity, induces formation of a KasA-containing complex in mycobacteria. *J Biol Chem* 278:20547–20554.

111. Colangeli R, Helb D, Sridharan S, Sun J, Varma-Basil M, Hazbon MH, Harbacheuski R, Megjugorac NJ, Jacobs WR Jr, Holzenburg A, Sacchettini JC, Alland D. 2005. The *Mycobacterium tuberculosis iniA* gene is essential for activity of an efflux pump that confers drug tolerance to both isoniazid and ethambutol. *Mol Microbiol* 55:1829–1840.

112. Alland D, Steyn AJ, Weisbrod T, Aldrich K, Jacobs WR Jr. 2000. Characterization of the *Mycobacterium tuberculosis iniBAC* promoter, a promoter that responds to cell wall biosynthesis inhibition. *J Bacteriol* 182:1802–1811.

113. Chen P, Bishai WR. 1998. Novel selection for isoniazid (INH) resistance genes supports a role for NAD$^+$-binding proteins in mycobacterial INH resistance. *Infect Immun* 66:5099–5106.

114. Ducasse-Cabanot S, Cohen-Gonsaud M, Marrakchi H, Nguyen M, Zerbib D, Bernadou J, Daffe M, Labesse G, Quemard A. 2004. In vitro inhibition of the *Mycobacterium tuberculosis* beta-ketoacyl-acyl carrier protein reductase MabA by isoniazid. *Antimicrob Agents Chemother* 48:242–249.

115. Argyrou A, Vetting MW, Aladegbami B, Blanchard JS. 2006. *Mycobacterium tuberculosis* dihydrofolate reductase is a target for isoniazid. *Nat Struct Mol Biol* 13:408–413.

116. Wang F, Jain P, Gulten G, Liu Z, Feng Y, Ganesula K, Motiwala AS, Ioerger TR, Alland D, Vilcheze C, Jacobs WR Jr, Sacchettini JC. 2010. *Mycobacterium tuberculosis* dihydrofolate reductase is not a target relevant to the antitubercular activity of isoniazid. *Antimicrob Agents Chemother* 54:3776–3782.

117. Ho YM, Sun YJ, Wong SY, Lee AS. 2009. Contribution of *dfrA* and *inhA* mutations to the detection of isoniazid-resistant *Mycobacterium tuberculosis* isolates. *Antimicrob Agents Chemother* 53:4010–4012.

118. Argyrou A, Jin L, Siconilfi-Baez L, Angeletti RH, Blanchard JS. 2006. Proteome-wide profiling of isoniazid targets in *Mycobacterium tuberculosis*. *Biochemistry* 45:13947–13953.

119. McMurry L, Petrucci RE Jr, Levy SB. 1980. Active efflux of tetracycline encoded by four genetically different tetracycline resistance determinants in *Escherichia coli*. *Proc Natl Acad Sci USA* 77:3974–3977.

120. Ainsa JA, Blokpoel MC, Otal I, Young DB, De Smet KA, Martin C. 1998. Molecular cloning and characterization of Tap, a putative multidrug efflux pump present in *Mycobacterium fortuitum* and *Mycobacterium tuberculosis*. *J Bacteriol* 180:5836–5843.

121. Silva PE, Bigi F, Santangelo MP, Romano MI, Martin C, Cataldi A, Ainsa JA. 2001. Characterization of P55, a multidrug efflux pump in *Mycobacterium bovis* and *Mycobacterium tuberculosis*. *Antimicrob Agents Chemother* 45:800–804.

122. Gupta AK, Katoch VM, Chauhan DS, Sharma R, Singh M, Venkatesan K, Sharma VD. 2010. Microarray analysis of efflux pump genes in multidrug-resistant *Mycobacterium tuberculosis* during stress induced by common anti-tuberculous drugs. *Microb Drug Resist* 16:21–28.

123. Wang K, Pei H, Huang B, Zhu X, Zhang J, Zhou B, Zhu L, Zhang Y, Zhou FF. 2013. The expression of ABC efflux pump, Rv1217c-Rv1218c, and its association with multidrug resistance of *Mycobacterium tuberculosis* in China. *Curr Microbiol* 66:222–226.

124. Balganesh M, Kuruppath S, Marcel N, Sharma S, Nair A, Sharma U. 2010. Rv1218c, an ABC transporter of *Mycobacterium tuberculosis* with implications in drug discovery. *Antimicrob Agents Chemother* 54:5167–5172.

125. Domenech P, Reed MB, Barry CE 3rd. 2005. Contribution of the *Mycobacterium tuberculosis* MmpL protein

family to virulence and drug resistance. *Infect Immun* 73:3492–3501.

126. Li XZ, Zhang L, Nikaido H. 2004. Efflux pump-mediated intrinsic drug resistance in *Mycobacterium smegmatis. Antimicrob Agents Chemother* 48:2415–2423.

127. Pasca MR, Guglierame P, De Rossi E, Zara F, Riccardi G. 2005. *mmpL7* gene of *Mycobacterium tuberculosis* is responsible for isoniazid efflux in *Mycobacterium smegmatis. Antimicrob Agents Chemother* 49:4775–4777.

128. Hillemann D, Rusch-Gerdes S, Richter E. 2007. Evaluation of the GenoType MTBDRplus assay for rifampin and isoniazid susceptibility testing of *Mycobacterium tuberculosis* strains and clinical specimens. *J Clin Microbiol* 45:2635–2640.

129. Hartkoorn RC, Sala C, Neres J, Pojer F, Magnet S, Mukherjee R, Uplekar S, Boy-Rottger S, Altmann KH, Cole ST. 2012. Towards a new tuberculosis drug: pyridomycin: nature's isoniazid. *EMBO Mol Med* 4:1032–1042.

130. Valvatne H, Syre H, Kross M, Stavrum R, Ti T, Phyu S, Grewal HM. 2009. Isoniazid and rifampicin resistance-associated mutations in *Mycobacterium tuberculosis* isolates from Yangon, Myanmar: implications for rapid molecular testing. *J Antimicrob Chemother* 64:694–701.

131. Caws M, Duy PM, Tho DQ, Lan NT, Hoa DV, Farrar J. 2006. Mutations prevalent among rifampin- and isoniazid-resistant *Mycobacterium tuberculosis* isolates from a hospital in Vietnam. *J Clin Microbiol* 44:2333–2337.

132. Rindi L, Bianchi L, Tortoli E, Lari N, Bonanni D, Garzelli C. 2005. Mutations responsible for *Mycobacterium tuberculosis* isoniazid resistance in Italy. *Int J Tuberc Lung Dis* 9:94–97.

133. Huang WL, Chen HY, Kuo YM, Jou R. 2009. Performance assessment of the GenoType MTBDRplus test and DNA sequencing in detection of multidrug-resistant *Mycobacterium tuberculosis. J Clin Microbiol* 47:2520–2524.

134. Fenner L, Egger M, Bodmer T, Altpeter E, Zwahlen M, Jaton K, Pfyffer GE, Borrell S, Dubuis O, Bruderer T, Siegrist HH, Furrer H, Calmy A, Fehr J, Stalder JM, Ninet B, Bottger EC, Gagneux S. 2012. Effect of mutation and genetic background on drug resistance in *Mycobacterium tuberculosis. Antimicrob Agents Chemother* 56:3047–3053.

135. Zhang M, Yue J, Yang YP, Zhang HM, Lei JQ, Jin RL, Zhang XL, Wang HH. 2005. Detection of mutations associated with isoniazid resistance in *Mycobacterium tuberculosis* isolates from China. *J Clin Microbiol* 43:5477–5482.

136. Luo T, Zhao M, Li X, Xu P, Gui X, Pickerill S, DeRiemer K, Mei J, Gao Q. 2010. Selection of mutations to detect multidrug-resistant *Mycobacterium tuberculosis* strains in Shanghai, China. *Antimicrob Agents Chemother* 54:1075–1081.

137. Doustdar F, Khosravi AD, Farnia P, Masjedi MR, Velayati AA. 2008. Molecular analysis of isoniazid resistance in different genotypes of *Mycobacterium tuberculosis* isolates from Iran. *Microb Drug Resist* 14:273–279.

138. Rahim Z, Nakajima C, Raqib R, Zaman K, Endtz HP, van der Zanden AG, Suzuki Y. 2012. Molecular mechanism of rifampicin and isoniazid resistance in *Mycobacterium tuberculosis* from Bangladesh. *Tuberculosis (Edinb)* 92:529–534.

139. Cardoso RF, Cooksey RC, Morlock GP, Barco P, Cecon L, Forestiero F, Leite CQ, Sato DN, Shikama Mde L, Mamizuka EM, Hirata RD, Hirata MH. 2004. Screening and characterization of mutations in isoniazid-resistant *Mycobacterium tuberculosis* isolates obtained in Brazil. *Antimicrob Agents Chemother* 48:3373–3381.

140. Kiepiela P, Bishop KS, Smith AN, Roux L, York DF. 2000. Genomic mutations in the *katG*, *inhA* and *aphC* genes are useful for the prediction of isoniazid resistance in *Mycobacterium tuberculosis* isolates from Kwazulu Natal, South Africa. *Tuber Lung Dis* 80:47–56.

141. Rouse DA, Li Z, Bai GH, Morris SL. 1995. Characterization of the *katG* and *inhA* genes of isoniazid-resistant clinical isolates of *Mycobacterium tuberculosis. Antimicrob Agents Chemother* 39:2472–2477.

142. Guo H, Seet Q, Denkin S, Parsons L, Zhang Y. 2006. Molecular characterization of isoniazid-resistant clinical isolates of *Mycobacterium tuberculosis* from the USA. *J Med Microbiol* 55:1527–1531.

143. Lavender C, Globan M, Sievers A, Billman-Jacobe H, Fyfe J. 2005. Molecular characterization of isoniazid-resistant *Mycobacterium tuberculosis* isolates collected in Australia. *Antimicrob Agents Chemother* 49:4068–4074.

144. Silva MS, Senna SG, Ribeiro MO, Valim AR, Telles MA, Kritski A, Morlock GP, Cooksey RC, Zaha A, Rossetti ML. 2003. Mutations in *katG*, *inhA*, and *ahpC* genes of Brazilian isoniazid-resistant isolates of *Mycobacterium tuberculosis. J Clin Microbiol* 41:4471–4474.

145. Poudel A, Nakajima C, Fukushima Y, Suzuki H, Pandey BD, Maharjan B, Suzuki Y. 2012. Molecular characterization of multidrug-resistant *Mycobacterium tuberculosis* isolated in Nepal. *Antimicrob Agents Chemother* 56:2831–2836.

146. Sekiguchi J, Miyoshi-Akiyama T, Augustynowicz-Kopec E, Zwolska Z, Kirikae F, Toyota E, Kobayashi I, Morita K, Kudo K, Kato S, Kuratsuji T, Mori T, Kirikae T. 2007. Detection of multidrug resistance in *Mycobacterium tuberculosis. J Clin Microbiol* 45:179–192.

147. N'Guessan KR, Dosso M, Ekaza E, Kouakou J, Jarlier V. 2008. Molecular characterisation of isoniazid-resistant *Mycobacterium tuberculosis* isolated from new cases in Lagunes region (Cote d'Ivoire). *Int J Antimicrob Agents* 31:498–500.

148. Lin HL, Kim H, Yun Y, Park CG, Kim B, Park Y, Kook Y. 2007. Mutations of katG and inhA in MDR *M. tuberculosis. Tuberc Respir Dis* 63:128–138.

149. Basso LA, Zheng R, Musser JM, Jacobs WR Jr, Blanchard JS. 1998. Mechanisms of isoniazid resistance in *Mycobacterium tuberculosis*: enzymatic characterization of enoyl reductase mutants identified in isoniazid-resistant clinical isolates. *J Infect Dis* 178:769–775.

150. Sholto-Douglas-Vernon C, Sandy J, Victor TC, Sim E, Helden PD. 2005. Mutational and expression analysis of *tbnat* and its response to isoniazid. *J Med Microbiol* 54:1189–1197.

151. Ristow M, Mohlig M, Rifai M, Schatz H, Feldmann K, Pfeiffer A. 1995. New isoniazid/ethionamide resistance gene mutation and screening for multidrug-resistant *Mycobacterium tuberculosis* strains. *Lancet* 346:502–503.

152. Yoon JH, Nam JS, Kim KJ, Choi Y, Lee H, Cho SN, Ro YT. 2012. Molecular characterization of drug-resistant and -susceptible *Mycobacterium tuberculosis* isolated from patients with tuberculosis in Korea. *Diagn Microbiol Infect Dis* 72:52–61.

153. Chan RC, Hui M, Chan EW, Au TK, Chin ML, Yip CK, AuYeang CK, Yeung CY, Kam KM, Yip PC, Cheng AF. 2007. Genetic and phenotypic characterization of drug-resistant *Mycobacterium tuberculosis* isolates in Hong Kong. *J Antimicrob Chemother* 59:866–873.

154. Leung KL, Yip CW, Yeung YL, Wong KL, Chan WY, Chan MY, Kam KM. 2010. Usefulness of resistant gene markers for predicting treatment outcome on second-line anti-tuberculosis drugs. *J Appl Microbiol* 109:2087–2094.

155. Boonaiam S, Chaiprasert A, Prammananan T, Leechawengwongs M. 2010. Genotypic analysis of genes associated with isoniazid and ethionamide resistance in MDR-TB isolates from Thailand. *Clin Microbiol Infect* 16:396–399.

156. Lee H, Cho SN, Bang HE, Lee JH, Bai GH, Kim SJ, Kim JD. 2000. Exclusive mutations related to isoniazid and ethionamide resistance among *Mycobacterium tuberculosis* isolates from Korea. *Int J Tuberc Lung Dis* 4:441–447.

157. Marttila HJ, Soini H, Huovinen P, Viljanen MK. 1996. *katG* mutations in isoniazid-resistant *Mycobacterium tuberculosis* isolates recovered from Finnish patients. *Antimicrob Agents Chemother* 40:2187–2189.

158. Khanna A, Raj VS, Tarai B, Sood R, Pareek PK, Upadhyay DJ, Sharma P, Rattan A, Saini KS, Singh H. 2010. Emergence and molecular characterization of extensively drug-resistant *Mycobacterium tuberculosis* clinical isolates from the Delhi Region in India. *Antimicrob Agents Chemother* 54:4789–4793.

159. Ramaswamy SV, Dou SJ, Rendon A, Yang Z, Cave MD, Graviss EA. 2004. Genotypic analysis of multidrug-resistant *Mycobacterium tuberculosis* isolates from Monterrey, Mexico. *J Med Microbiol* 53:107–113.

160. Siddiqi N, Shamim M, Hussain S, Choudhary RK, Ahmed N, Prachee, Banerjee S, Savithri GR, Alam M, Pathak N, Amin A, Hanief M, Katoch VM, Sharma SK, Hasnain SE. 2002. Molecular characterization of multidrug-resistant isolates of *Mycobacterium tuberculosis* from patients in North India. *Antimicrob Agents Chemother* 46:443–450.

161. Torres MJ, Criado A, Gonzalez N, Palomares JC, Aznar J. 2002. Rifampin and isoniazid resistance associated mutations in *Mycobacterium tuberculosis* clinical isolates in Seville, Spain. *Int J Tuberc Lung Dis* 6:160–163.

162. Pretorius GS, van Helden PD, Sirgel F, Eisenach KD, Victor TC. 1995. Mutations in *katG* gene sequences in isoniazid-resistant clinical isolates of *Mycobacterium tuberculosis* are rare. *Antimicrob Agents Chemother* 39:2276–2281.

163. Cockerill FR 3rd, Uhl JR, Temesgen Z, Zhang Y, Stockman L, Roberts GD, Williams DL, Kline BC. 1995. Rapid identification of a point mutation of the *Mycobacterium tuberculosis* catalase-peroxidase (*katG*) gene associated with isoniazid resistance. *J Infect Dis* 171:240–245.

164. Ghiladi RA, Medzihradszky KF, Rusnak FM, Ortiz de Montellano PR. 2005. Correlation between isoniazid resistance and superoxide reactivity in *Mycobacterium tuberculosis* KatG. *J Am Chem Soc* 127:13428–13442.

165. Bolotin S, Alexander DC, Chedore P, Drews SJ, Jamieson F. 2009. Molecular characterization of drug-resistant *Mycobacterium tuberculosis* isolates from Ontario, Canada. *J Antimicrob Chemother* 64:263–266.

166. Mo L, Zhang W, Wang J, Weng XH, Chen S, Shao LY, Pang MY, Chen ZW. 2004. Three-dimensional model and molecular mechanism of *Mycobacterium tuberculosis* catalase-peroxidase (KatG) and isoniazid-resistant KatG mutants. *Microb Drug Resist* 10:269–279.

167. Ando H, Kondo Y, Suetake T, Toyota E, Kato S, Mori T, Kirikae T. 2010. Identification of *katG* mutations associated with high-level isoniazid resistance in *Mycobacterium tuberculosis*. *Antimicrob Agents Chemother* 54:1793–1799.

168. Jnawali HN, Hwang SC, Park YK, Kim H, Lee YS, Chung GT, Choe KH, Ryoo S. 2013. Characterization of mutations in multi- and extensive drug resistance among strains of *Mycobacterium tuberculosis* clinical isolates in Republic of Korea. *Diagn Microbiol Infect Dis* 76:187–196.

169. Doustdar F, Khosravi AD, Farnia P, Masjedi MR, Velayati AA. 2008. Molecular analysis of isoniazid resistance in different genotypes of *Mycobacterium tuberculosis* isolates from Iran. *Microb Drug Resist* 14:273–279.

170. Rouse DA, DeVito JA, Li Z, Byer H, Morris SL. 1996. Site-directed mutagenesis of the *katG* gene of *Mycobacterium tuberculosis*: effects on catalase-peroxidase activities and isoniazid resistance. *Mol Microbiol* 22:583–592.

171. Heym B, Alzari PM, Honore N, Cole ST. 1995. Missense mutations in the catalase-peroxidase gene, *katG*, are associated with isoniazid resistance in *Mycobacterium tuberculosis*. *Mol Microbiol* 15:235–245.

172. Yuan X, Zhang T, Kawakami K, Zhu J, Li H, Lei J, Tu S. 2012. Molecular characterization of multidrug- and extensively drug-resistant *Mycobacterium tuberculosis* strains in Jiangxi, China. *J Clin Microbiol* 50:2404–2413.

173. Ramirez MV, Cowart KC, Campbell PJ, Morlock GP, Sikes D, Winchell JM, Posey JE. 2010. Rapid detection of multidrug-resistant *Mycobacterium tuberculosis* by use of real-time PCR and high-resolution melt analysis. *J Clin Microbiol* **48**:4003–4009.

174. Fang Z, Doig C, Rayner A, Kenna DT, Watt B, Forbes KJ. 1999. Molecular evidence for heterogeneity of the multiple-drug-resistant *Mycobacterium tuberculosis* population in Scotland (1990 to 1997). *J Clin Microbiol* **37**:998–1003.

175. Abe C, Kobayashi I, Mitarai S, Wada M, Kawabe Y, Takashima T, Suzuki K, Sng LH, Wang S, Htay HH, Ogata H. 2008. Biological and molecular characteristics of *Mycobacterium tuberculosis* clinical isolates with low-level resistance to isoniazid in Japan. *J Clin Microbiol* **46**:2263–2268.

176. Khadka DK, Eampokalap B, Panitchakorn J, Ramasoota P, Khusmith S. 2007. Multiple mutations in *katG* and *inhA* identified in Thai isoniazid-resistant *Mycobacterium tuberculosis* isolates. *Southeast Asian J Trop Med Public Health* **38**:376–382.

177. Kim SY, Park YJ, Kim WI, Lee SH, Ludgerus Chang C, Kang SJ, Kang CS. 2003. Molecular analysis of isoniazid resistance in *Mycobacterium tuberculosis* isolates recovered from South Korea. *Diagn Microbiol Infect Dis* **47**:497–502.

178. Minh NN, Van Bac N, Son NT, Lien VT, Ha CH, Cuong NH, Mai CT, Le TH. 2012. Molecular characteristics of rifampin- and isoniazid-resistant *Mycobacterium tuberculosis* strains isolated in Vietnam. *J Clin Microbiol* **50**:598–601.

179. Zakerbostanabad S, Titov LP, Bahrmand AR. 2008. Frequency and molecular characterization of isoniazid resistance in *katG* region of MDR isolates from tuberculosis patients in southern endemic border of Iran. *Infect Genet Evol* **8**:15–19.

180. Zenteno-Cuevas R, Zenteno JC, Cuellar A, Cuevas B, Sampieri CL, Riviera JE, Parissi A. 2009. Mutations in *rpoB* and *katG* genes in *Mycobacterium isolates* from the Southeast of Mexico. *Mem Inst Oswaldo Cruz* **104**:468–472.

181. Soudani A, Hadjfredj S, Zribi M, Messaadi F, Messaoud T, Masmoudi A, Fendri C. 2011. Genotypic and phenotypic characteristics of Tunisian isoniazid-resistant *Mycobacterium tuberculosis* strains. *J Microbiol* **49**:413–417.

182. Haas WH, Schilke K, Brand J, Amthor B, Weyer K, Fourie PB, Bretzel G, Sticht-Groh V, Bremer HJ. 1997. Molecular analysis of *katG* gene mutations in strains of *Mycobacterium tuberculosis* complex from Africa. *Antimicrob Agents Chemother* **41**:1601–1603.

183. Musser JM, Kapur V, Williams DL, Kreiswirth BN, van Soolingen D, van Embden JD. 1996. Characterization of the catalase-peroxidase gene (*katG*) and *inhA* locus in isoniazid-resistant and -susceptible strains of *Mycobacterium tuberculosis* by automated DNA sequencing: restricted array of mutations associated with drug resistance. *J Infect Dis* **173**:196–202.

184. Lipin MY, Stepanshina VN, Shemyakin IG, Shinnick TM. 2007. Association of specific mutations in *katG*, *rpoB*, *rpsL* and *rrs* genes with spoligotypes of multidrug-resistant *Mycobacterium tuberculosis* isolates in Russia. *Clin Microbiol Infect* **13**:620–626.

185. Waksman SA. 1964. *The Conquest of Tuberculosis*. University of California Press, Berkeley.

186. Winder FG, Collins PB. 1970. Inhibition by isoniazid of synthesis of mycolic acids in *Mycobacterium tuberculosis*. *J Gen Microbiol* **63**:41–48.

Molecular Genetics of Mycobacteria, 2nd Edition
Edited by Graham F. Hatfull and William R. Jacobs, Jr.
© 2014 American Society for Microbiology, Washington, DC
doi:10.1128/microbiolspec.MGM2-0009-2013

Claudine Mayer[1,2,3]
Howard Takiff[4]

The Molecular Genetics of Fluoroquinolone Resistance in *Mycobacterium tuberculosis*

23

The fluoroquinolones (FQs) are among the most widely prescribed antibiotics globally and until very recently were the only new antibiotics accepted for use against tuberculosis (TB) in the past 40 years. They are an important component of drug regimens for curing *Mycobacterium tuberculosis* strains resistant to other antibiotics, especially multidrug-resistant TB (MDR-TB)—strains resistant to at least isoniazid and rifampin (1, 2). They are also being proposed as first-line agents to reduce the duration of treatment for pan-susceptible strains, particularly when given in new combinations with other recently developed drugs (3–5). A major problem with the FQs is the development of resistant strains. MDR strains that have developed resistance to the FQs and also to any of the injectable drugs are termed extensively drug-resistant (XDR-TB)—strains that are extremely difficult to treat (6), with studies showing from 65% (7) to less than 50% having favorable outcomes (8). For the best chance of curing XDR-TB, resistance to FQs should be identified promptly so that alternative or additional antibiotics can be prescribed. So while there is intellectual interest in investigating the mechanisms and mutations through which *M. tuberculosis* develops FQ resistance, and the hope that this information will eventually lead to the design of more effective FQs or quinolone derivatives that are less susceptible to these resistance mechanisms, there is also an urgent need to define the mutations conferring resistance so they can be incorporated into rapid molecular tests to identify FQ-resistant strains.

This review will describe the current use of FQs to treat TB, their interaction with the drug target, their mechanism of action, and the mutations known to confer resistance. It will also describe limitations to the rapid detection of resistance mutations and additional potential mechanisms of resistance.

FQs AND THEIR INTRODUCTION INTO ANTI-TB THERAPY

The first quinolone, nalidixic acid, was discovered in 1962 (9) and introduced into clinical use in 1967 for the treatment of Gram-negative urinary tract infections (10). The quinolones are synthetic molecules that

[1]Unité de Microbiologie Structurale, Institut Pasteur; [2]UMR 3528 du CNRS; [3]Université Paris Diderot, Sorbonne Paris Cité, Cellule Pasteur, 75015, Paris, France; [4]Laboratorio de Genética Molecular, CMBC, IVIC, Caracas, Venezuela.

contain a 4-oxo-1,4-dihydroquinoline ring system with a carboxylic acid attached at position 3 (Fig. 1), and are amenable to multiple modifications. After it was found that the addition of a fluorine atom at the 6-position of the quinoline ring greatly improved their antibacterial potency and broadened their activity, the other side groups were modified, and thousands of different FQs were synthesized. They are classically divided into two subfamilies: the "older" agents, such as ciprofloxacin, ofloxacin, norfloxacin, and pefloxacin, and the "newer" FQs that were developed after 1990, such as levofloxacin, sparfloxacin, gatifloxacin, moxifloxacin, and gemifloxacin. Garenoxacin, one of the last quinolone generation, is a des-F(6)-quinolone. Among the side groups the three positions R1, R7, and R8 are the most variable and have been exploited to design FQs with increased efficacy.

When the first "blockbuster" FQ, ciprofloxacin, was introduced, its heralded broad-spectrum activity raised hopes that it might even be capable of replacing injected drugs for the treatment of serious infections such as *Staphylococcus aureus* osteomyelitis. Unfortunately, it soon became apparent that the usefulness of ciprofloxacin would be limited by the development of resistant strains. Within just a few years after it was introduced, a high percentage of nosocomial isolates (11), especially Gram-positive bacteria, were found to be resistant (12). Ciprofloxacin remained effective against some enteric Gram-negative bacteria, such as *Escherichia coli*, for much longer (13), probably because their innate MICs for ciprofloxacin were much lower than those of the Gram-positive bacteria. From *in vitro* studies, it appears that as the FQ concentration increases, the frequency of FQ-resistant colonies decreases, eventually reaching a "mutant prevention concentration," generally >8 times the MIC, at which point resistant colonies are rare (<1 in 10^{9-10}) (14–16).

When MDR-TB appeared in the early 1990s, the options for new antibiotics were very limited. Clinicians first tried using ciprofloxacin (17), which had been shown to have some activity against *M. tuberculosis in vitro* (18), but resistant strains appeared rapidly. Ciprofloxacin was known to have poor activity against Gram-positive bacteria (19), and the development of resistance could have been predicted from *in vitro* studies. The maximum ciprofloxacin serum

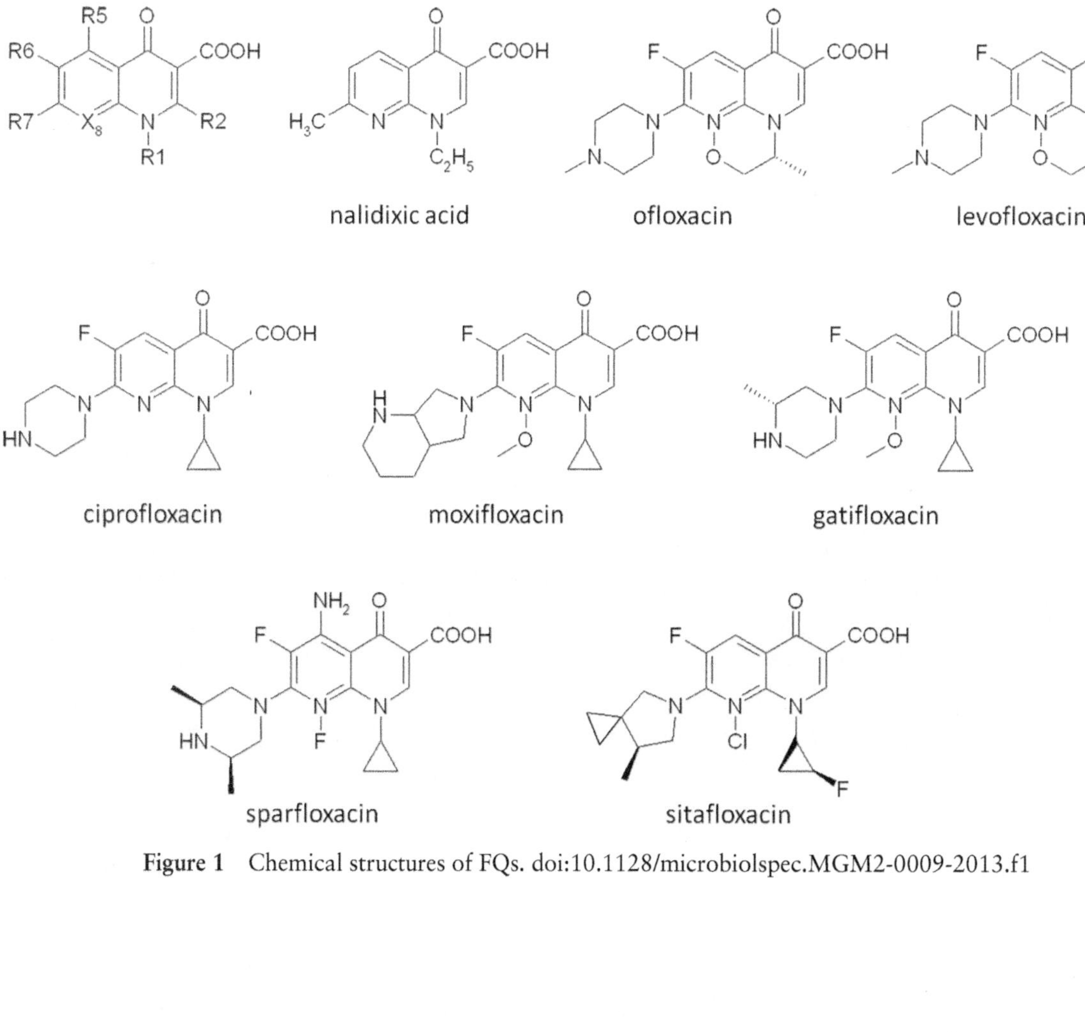

Figure 1   Chemical structures of FQs. doi:10.1128/microbiolspec.MGM2-0009-2013.f1

concentration is only about 1.5 µg/ml, only 2 to 3 times the MIC of most *M. tuberculosis* isolates; at this concentration mutant colonies can be isolated *in vitro* at relatively high frequencies (10^{6-7}) (20–22), and many of these mutants are also resistant to other FQs. Therefore, ciprofloxacin should be avoided in treating TB because not only is it ineffective, but it selects for mutations that confer resistance to other more effective FQs. Clinicians soon switched to ofloxacin and subsequently to its L-isomer, levofloxacin, the active member of the racemate, and several studies have shown that drug treatment regimens that include one of these FQs are significantly more likely to cure MDR-TB than regimens without them (1, 2, 23, 24). The MICs for ofloxacin are not much higher than for ciprofloxacin, but the advantage of ofloxacin and levofloxacin appears to include other factors, such as higher mutant prevention concentrations, more favorable pharmacokinetics, and better intramacrophage penetration (22).

After studies on the structure-activity relationships suggested that FQs with a methoxy group at the C_8 position would have more activity against Gram-positives and mycobacteria (25), newer-generation FQs appeared that were considerably more effective against *M. tuberculosis*; however, several had problems. Sitafloxacin and sparfloxacin have quite low MICs for *M. tuberculosis*, but both are phototoxic (26), and the promising gatifloxacin was associated with both hypoglycemia and hyperglycemia (27), especially in older patients. Therefore, moxifloxacin is currently the FQ most recommended for drug regimens to treat MDR-TB (27).

THE FQ DRUG TARGET

The targets of the FQs are type II topoisomerases. Topoisomerases are ubiquitous nucleic acid–dependent nanomachines essential to cell life that solve the topological problems of DNA during replication, transcription, or recombination (28). They have been divided into two classes, type I and type II, according to their basic mechanism of action (28, 29). Type I DNA topoisomerases introduce transient single-stranded breaks to force the passage of one DNA strand through the other, whereas type II DNA topoisomerases introduce transient double-stranded breaks to force the passage of one segment of double-stranded DNA through another. All organisms contain at least one type I and one type II topoisomerase. Most bacterial genomes encode two type II topoisomerases, DNA gyrase and topoisomerase IV, which together help manage chromosome integrity and topology (28). DNA gyrase tends to be the primary FQ target in Gram-negative bacteria,

whereas topoisomerase IV is preferentially targeted by most quinolones in Gram-positive bacteria. DNA gyrase is unique in introducing negative supercoils into DNA, an activity mediated by the carboxy-terminal domain of its DNA-binding GyrA subunit, and is therefore responsible for the DNA unwinding at replication forks. Although very similar to DNA gyrase, topoisomerase IV has a specialized function in mediating the decatenation of interlocked daughter chromosomes that result from replication of the closed circular DNA (30) and also relaxes positive supercoils. A few bacteria, such as *Treponema pallidum*, *Helicobacter pylori*, and mycobacteria, possess only one type II topoisomerase, the DNA gyrase, which is therefore the unique target of FQs in these organisms, as has been demonstrated for *M. tuberculosis* (31).

Based on evolutionary considerations, type II topoisomerases have been subclassified into two families, type IIA and type IIB (28, 29). Bacterial DNA gyrase and topoisomerase IV, together with eukaryal and viral topoisomerases, belong to type IIA, whereas type IIB includes archaeal topoisomerase VI and homologues in plants, a few protists, and a few bacteria (29, 32). Bacterial type IIA topoisomerases, DNA gyrase, and topoisomerase IV consist of two subunits, GyrA and GyrB, and ParC and ParE, respectively, which form the catalytic active heterotetrameric (A_2B_2 or C_2E_2) complex (Fig. 2A) of nearly 350 kDa. Subunit A consists of two domains, the N-terminal breakage-reunion domain (BRD) and the carboxy-terminal domain (represented in blue and green, respectively, in Fig. 2). Subunit B consists of the ATPase domain followed by the Toprim domain, a conserved catalytic domain in type IA and II topoisomerases, DnaG-type primases, OLD family nucleases, and RecR proteins (33) (represented in yellow and red, respectively, in Fig. 2).

Biochemical *in vitro* work combined with structural studies of isolated domains over the past 20 years has allowed several authors to propose a global quaternary structural model and a catalytic mechanism of the holoenzyme (34) (Fig. 2). The BRD binds a DNA segment termed the "gate" or G-segment at the DNA-gate created by the dimeric interface formed by the so-called catalytic core, constituted by the Toprim domain of GyrB and the BRD of GyrA. The GyrB N-terminal ATPase domains dimerize upon ATP binding, capturing the DNA duplex to be transported, termed the T-segment. A double-stranded break is then introduced into the G-segment by the BRDs, and each end of the cleaved DNA is covalently bound to a critical tyrosine residue of the GyrA subunits. The T-segment is then passed through the break in the G-segment that is

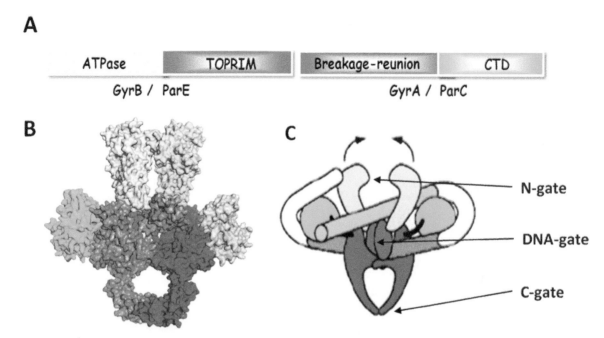

Figure 2 (**A**) Schematic representation of the sequence and domain organization of type IIA topoisomerases formed by the association of two subunits, A and B. Bacterial type IIA topoisomerases are A₂B₂ heterotetramers. The names of the four conserved domains are indicated. (**B, C**) Proposed atomic and schematic model of the type IIA topoisomerase architecture. The three gates are indicated. doi:10.1128/microbiolspec.MGM2-0009-2013.f2

opened, the DNA is resealed, and the T-segment is released through a protein gate at the other end of the complex, the C-gate, prior to resetting of the enzyme to the open clamp form to begin another cycle (Fig. 2C).

FQs POISON THE TYPE II TOPOISOMERASES

FQs exert their powerful antibacterial activity by interfering with the enzymatic reaction cycle and are classified as bacterial type II topoisomerase poisons. Specifically, they target the catalytic core, cooperatively formed by the Toprim domain of the GyrB subunit and the BRD of the GyrA subunit (Fig. 3A). Although the exact binding mechanism and amino acid specificities for the FQs are not yet known, available crystal structures show that two quinolones bind to a binary complex consisting of the gyrase and DNA, thereby stabilizing the covalent enzyme tyrosyl-DNA phosphate ester bond (Fig. 3). The resulting ternary complex, formed by the quinolone, gyrase protein, and DNA molecule, blocks DNA replication and leads to cell death (35). In addition, hydrolysis of the tyrosyl-DNA linkage without religation leads to the accumulation of double-stranded breaks in the chromosome, which is likely the principal cause of bacterial death.

The three-dimensional (3D) model of the *M. tuberculosis* gyrase catalytic core in complex with a 35-base-pair DNA oligonucleotide and moxifloxacin is a good tool for understanding in detail the mode of action of quinolones and their inhibition mechanism (36). Two FQ molecules, one for each protomer, bind at the ternary interface formed by the DNA, the BRD, and the Toprim domain (Fig. 3). FQs inhibit the religation of the broken DNA by intercalating in the bound DNA between the dinucleotides that form the covalent bond between the DNA and the enzyme—a phosphodiester bond between the oxygen of the tyrosine and the phosphor atom of a phosphate group of the DNA (Fig. 3D). Because the catalytic core is dimeric, two FQ molecules are bound and separated by 4 base pairs (Fig. 3E). The consequence of this intercalation is that the two ends of the DNA are moved farther away from each other— more than 9 Å—making it structurally impossible for the O3′-P religation to occur (Fig. 3).

INTERACTIONS WITH DIFFERENT FQs AND RELATION OF STRUCTURE TO RESISTANT DEVELOPMENT

The mutations causing FQ resistance have been investigated in many different bacteria. Most bacteria have gyrase and Topo IV, and resistance-conferring mutations can occur in both, with the stepwise accumulation

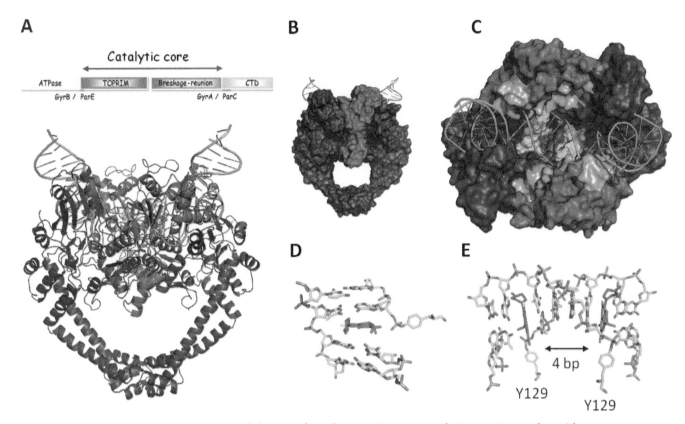

Figure 3 (A) Structure of the *M. tuberculosis* DNA gyrase catalytic core in complex with DNA and moxifloxacin in ribbon representation. (B) Side view and (C) top view of the molecular surface of the catalytic core. The Toprim domain is represented in red, the BRD in blue, the 35-base-pair DNA oligonucleotide in orange, and the moxifloxacin in green. Localization of the QRDR is indicated in pink and light blue (residues 500 to 538 for QRDR-B and 74 to 108 for QRDR-A). (D) Close view of the structure of the intercalated moxifloxacin (magenta) in the broken DNA double helix (green). The catalytic tyrosine (Y129 in the *M. tuberculosis* DNA gyrase sequence) is shown in green outside the DNA helix. (E) Close view of both moxifloxacin molecules in the broken DNA showing the 4 base pairs in between the two bound fluoroquinolones. Both catalytic tyrosines of each monomer are shown in green in the DNA major groove.
doi:10.1128/microbiolspec.MGM2-0009-2013.f3

of several mutations leading to progressively higher-level resistance. In Gram-negative bacteria, the mutations tend to occur first in the A subunit of the gyrase, within the quinolone-resistance-determining region (QRDR) of *gyrA* (QRDR-A), followed by mutations in the equivalent region of the gene for the A subunit of topoisomerase IV, *parC*. In Gram-positives, such as *Streptococcus pneumoniae*, the first mutations tend to be in the QRDR region of ParC, but this depends upon the specific FQ (37). Subsequent mutations can occur in the other A subunit (GyrA or ParC) and also in the QRDR-B of the B subunits of both enzymes (GyrB and ParE). Because mycobacteria do not have topoisomerase IV, the FQ resistance mutations occur in the genes encoding the gyrase, principally in the QRDR of *gyrA*

(20), but also less frequently in the QRDR of *gyrB*. Moxifloxacin appears to preferentially target the gyrase, which may contribute to its effectiveness against *M. tuberculosis* (38).

The substitutions conferring FQ resistance in the QRDR-A of *M. tuberculosis* occur most commonly in amino acid 94, but also in amino acids 89, 90, and 91 (39) and, less frequently, in amino acids 88 (40) and 74 (41) (Fig. 4). The only commercial test for resistance to second-line agents, the line probe assay MTBDR*sl* (Hain Lifescience GmbH, Germany) (42), is designed to detect the substitutions A90V, S91P, D94A, D94N/Y, D94G, and D94H. GyrB substitutions were initially described for an *in vitro* selected FQ-resistant strain (43), but for many years they were not reported in clinical

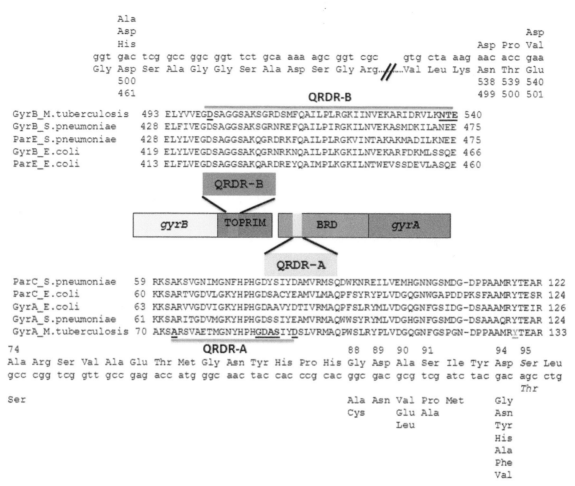

Figure 4 The schematic diagram in the center shows the arrangement of the *gyrA* and *gyrB* genes, encoding the GyrA and GyrB subunits of the *M. tuberculosis* DNA gyrase. Also shown are the locations of the TOPRIM region of GyrB, the BRD of GyrA, and the sites of the QRDRs of both subunits, QRDR-B or QRDR-A. Above the diagram is an alignment of the region of GyrB containing QRDR-B, illustrating that this region is highly conserved in the B subunits of the gyrases and the B subunits of the topoisomerase IV enzymes (ParE), as illustrated by the Gram-positive *S. pneumoniae* and the Gram-negative *E. coli*. Below the diagram is the alignment of segments including the QRDR-A for the A subunits of the gyrase and topoisomerase IV from the same bacteria. The underlined letters in bold indicate amino acids where mutations confer FQ resistance. The blue Y in the GyrA alignment indicates the tyrosine that is covalently bound to the cleaved G segment DNA (see text). On the top and bottom of the figure are the nucleotide and amino acid sequences of the QRDR-A and QRDR-B regions, with the amino acid substitutions shown to confer FQ resistance. Amino acid 95 of the QRDR-A is polymorphic and can be either serine or threonine depending upon the phylogeny of the strain, but has not been implicated in FQ resistance (20, 155). Below the sequence of the QRDR-B are the amino acid numbers in both the old and new numbering systems. doi:10.1128/microbiolspec.MGM2-0009-2013.f4

isolates of *M. tuberculosis*, and their relation to FQ resistance was questioned, although most studies did not look for *gyrB* mutations in FQ-resistant strains. With the increased use of FQs for TB, and more studies sequencing the QRDR-B, GyrB mutations have been reported more frequently (44), most consistently in amino acids 500, 538, 539, and 540 (45–47) (amino

acids 461, 499, 500, and 501 in the new proposed numbering system [48]).

Several studies have sequenced the entire *gyrA* and *gyrB* genes, some 4,500 nucleotides, from FQ-resistant strains and found a variety of mutations, mostly outside of the QRDR regions, which were possibly related to FQ resistance. Mutations in *gyrB* outside of the

QRDR-B appear to be more common than *gyrA* mutations outside of the QRDR-A. Part of the difficulty in clearly showing the effect of these mutations on FQ susceptibility, especially those in *gyrB*, was that they are often found in association with *gyrA* mutations known to cause resistance, and most of the confirmed *gyrB* mutations cause lower-level resistance than the validated QRDR-A mutations. The causative role of the mutations at the sites mentioned above, and the lack of association of several other mutations, were proven in four ways: (i) determining whether the mutations are also found in FQ-sensitive strains (48); (ii) careful studies of the FQ susceptibility of strains that contain only the mutations in question and no other mutations (45, 46); (iii) *in vitro* studies with overexpressed gyrase proteins containing the mutations in question to determine if their enzyme activity is inhibited by the FQs (46, 49); and (iv) allelic replacement to insert the mutations individually into two FQ-sensitive strain backgrounds (H37Rv and Erdmann) and then test the FQ sensitivity of the genetically engineered strains (45).

These studies have confirmed that FQ resistance is caused by the GyrA mutations in the QRDR-A and also by GyrB substitutions at amino acids 500, 538, 539, and 540 (45, 46) (Fig. 4). Although there were some differences in the levels of resistance to different FQs conferred by particular amino acid substitutions in the genetically engineered strains compared to the *in vitro* enzyme assay, there was good agreement on the amino acids involved in resistance. Both studies found that the N538D and E540V substitutions conferred high-level resistance to all FQs. The *in vitro* enzyme study also found resistance with the D500H, D500N, E540D, and T539P amino acid substitutions and low-level resistance with the D500A change (46), while the sensitivity of the engineered strains varied with the FQ tested and the strain background, and often the level of resistance did not meet the clinical definition of resistance with some or all of the FQs tested (2 μg/ml) (45). Reports that have tested many clinical isolates with different QRDR mutations against different FQs have shown that while the A90V substitution routinely confers resistance to ciprofloxacin and ofloxacin, with MICs >3 μg/ml, it only modestly increases resistance to moxifloxacin, with most strains showing moxifloxacin MICs of <0.5 to 1.0 μg/ml (50–52). A similar tendency has been observed in strains with the D94A substitution (50).

The model of the *M. tuberculosis* DNA gyrase catalytic core also establishes that the QRDR residues of both subunits are spatially close and form the quinolone-binding pocket (QBP) (Fig. 3). In this pocket, the FQ is maintained on one side by three residues of the QRDR-A (G88, D89, and A90) that are in close contact with the two conserved groups of the quinolone, the carboxylic and carbonyl functions (R3 and R4 group [Fig. 3]). On the other side, five residues of the QRDR-B (D500, R521, N538, T539, and E540 [D461, R482, N499, T500, and E501]) are in close contact with the most variable groups, R1, R7, and R8 (Fig. 3). Modification of any of these residues from either QRDR has been shown to affect the level of resistance to FQs (47, 51, 53–57). The three residues from the QRDR-A belong to the DNA-recognition helix from the CAP-like DNA binding domain (α4) and are localized in the bottom of the QBP, whereas the residues of the QRDR-B lie at the top entry of the QBP (Fig. 3). The 3D model shows that the geometry of the QBP is crucial for the recognition and the stability of the FQ molecule in the pocket. This could explain, for example, why nalidixic acid, a very small quinolone, has a very high MIC for *M. tuberculosis* and other bacteria, whereas several of the bulkiest quinolones (sparfloxacin, sitafloxacin, gatifloxacin, and moxifloxacin) are highly efficient gyrase inhibitors because they are perfectly adapted to fit snugly into the QBP. This also accounts for the observed difference in MIC between moxifloxacin and ciprofloxacin for most bacteria, because the R7 and R8 groups are very different for these two molecules (Fig. 1).

For a given FQ, moxifloxacin for example, any modification of the amino acids belonging to the QBP leads to either a direct or indirect change of the geometry of the pocket that could alter its affinity for the gyrase (Fig. 5). When the side chain of the modified amino acid is smaller, the pocket becomes too large to stabilize the FQ. In contrast, when the side chain of the modified amino acid is bulkier, the pocket becomes too small to fix the FQ. The most common substitutions change either D94 or A90 in GyrA. The substitution of a serine for the alanine at amino acid 90 generates a more adapted pocket for moxifloxacin, increasing the susceptibility to this inhibitor. In contrast, substitution to a valine, frequently observed in resistant strains (48), creates a smaller pocket with steric hindrance between the side chain and moxifloxacin (Fig. 5). However, there is no structural explanation yet of why strains carrying the A90V substitution are more resistant to ciprofloxacin than to moxifloxacin (48) as this residue interacts with the conserved part of the FQ (R3 and R4).

An indirect and more complex effect on the geometry of the pocket is observed when amino acid substitutions are localized in the DNA recognition α-helix (α4) that interacts with the major groove of DNA. As DNA

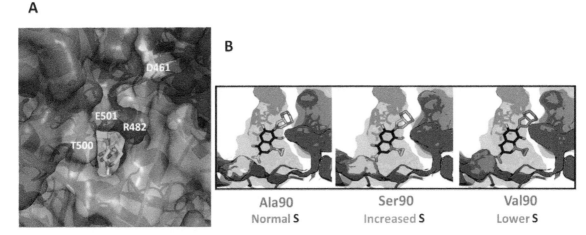

Figure 5 (**A**) Close view of the quinolone-binding pocket. The DNA-protein complex is represented in transparent molecular surface and moxifloxacin, in sticks (color code is the same as in Fig. 3). The residues of the QRDR-B (Toprim) belonging to the QBP are indicated in pink, purple, and yellow. Residue A90 of the QRDR-B is represented in light green in the background of the pocket. (**B**) Effect of the substitution of A90 (QRDR-A) on the geometry of the quinolone-binding pocket. (Left) Quinolone-binding pocket of the wild-type *M. tuberculosis* DNA gyrase. The A90 is colored in yellow. (Middle) Substitution of A90 to serine (S90 is represented in green). (Right) Substitution of A90 to valine (V90 is represented in magenta). doi:10.1128/microbiolspec.MGM2-0009-2013.f5

is complexed with the gyrase to form the QBP (Fig. 5), modification of the structure of the DNA also modifies the pocket geometry and can lead to destabilization of the FQs in the pocket. This mechanism could explain why substitutions of the amino acid D94 (the position most commonly mutated in FQ-resistant strains of *M. tuberculosis*) have paradoxical effects; e.g., the substitution of the aspartate by an amino acid with a smaller side chain, such as alanine or glycine, increases the resistance to the same level as substitutions by amino acids with bulky side chains, such as histidine (45).

There are a couple of interesting interactions for which there are as yet no good structural explanations. While the A74S substitution produces only a very modest increase in resistance to all FQs, when it occurs together with the D94G mutation there appears to be a synergistic effect, making the doubly mutant strain more resistant than strains carrying either mutation alone. Other reports have also suggested that strains with more than one mutation in the QRDR-A, or mutations in both QRDR-A and QRDR-B, have higher levels of resistance than strains with the corresponding single mutations (43, 50). Another unexplained observation is that while the T80A mutation by itself has no effect on FQ susceptibility, when it occurs together with the A90G mutation the strain becomes slightly more susceptible to ofloxacin than strains without any GyrA substitutions (45, 49).

The resistance of the *M. tuberculosis* DNA gyrase to FQs results from two mechanisms. The effect of mutations in the QRDR that confer "acquired" FQ resistance is superimposed upon the baseline "intrinsic" resistance due to the low natural affinity of the *M. tuberculosis* DNA gyrase catalytic core for the FQs. This is principally attributable to the amino acids at positions 81 and 90 in the QRDR-A and 500 (482 in the new numbering) in the QRDR-B (55). The amino acids at two of these positions in the QBP play a crucial role in both intrinsic and acquired resistance: position 90 in the QRDR-A, which is an alanine in the *M. tuberculosis* sequence, and position 500 (482), an arginine in the *M. tuberculosis* QRDR-B. Both amino acids are essential for the shape of the QBP and, as a result, for the binding of the quinolone. In the QRDR-B, the side chain of residue R521 (R482), which is located in a loop of the DNA binding Toprim domain, points toward the DNA minor groove and forms a "flap" that blocks the FQ in the pocket through contact with the R7 group (Fig. 5). The increase in the susceptibility to FQs by the substitution of a lysine for the arginine at position 521 (482) can be explained by the lower energy cost of moving an arginine rather than a lysine in the DNA minor groove (58). This means that the flap created by this amino acid would find it easier to open up and destabilize the FQ when the residue is an arginine rather than a lysine.

FQ RESISTANCE WITHOUT GYRASE MUTATIONS

While mutations in the QRDRs are found in most FQ-resistant isolates, almost all studies report that at least a few FQ-resistant clinical isolates of *M. tuberculosis* lacked mutations in both of these regions, and the percentage of these strains without QRDR mutations has varied from 0 to 50% (5). Some studies that sequenced the entire *gyrA* and *gyrB* genes found additional mutations outside of the QRDR regions, but most of these were shown not to have a role in FQ resistance (see above).

How can these strains be resistant if no mutations are found in either QRDR-A or QRDR-B? There are several possible explanations. A recent study found that strains with high-level resistance (ofloxacin 10 µg/ml) are more likely to have QRDR mutations than those with low-level resistance (ofloxacin 2 µg/ml), although a majority even of those with this lower level of resistance also had QRDR mutations (59). Strains with even lower levels of resistance, below the 2 µg/ml ofloxacin cutoff for the clinical definition of resistance, are more likely to lack QRDR-A mutations, but some QRDR-B mutations, such as T539P and E540D, conferred resistance to only 1 µg/ml ofloxacin (45). The MICs for FQs in "susceptible" strains of *M. tuberculosis* never exposed to FQs vary over an 8-fold range from 0.125 to 1 µg/ml ofloxacin (21), so perhaps some of the low-level "resistance" below 2 µg/ml merely represents strains at the upper limits of this range. The reasons for this variation of intrinsic MICs are unknown, as are the possible relationships between the innate MIC and the propensity to develop higher-level resistance, but the level of resistance conferred by a specific QRDR mutation can vary depending upon the strain background into which it is inserted (45). It is possible that FQ resistance in *M. tuberculosis* can be caused by mutations activating alternative resistance mechanisms, as described in other bacteria and discussed below, and perhaps some of these additional mechanisms may be more inducible in particular strains, accounting for the variation in baseline MICs. At this time, however, neither the putative nongyrase mutations, which would likely confer only low-level resistance, nor genes causing inducible resistance or tolerance, have been identified.

The success at finding gyrase mutations in FQ-resistant isolates can also depend upon the technique used and the definition of resistance. Clinical resistance is best defined by phenotypic resistance, tested by growing the bacteria in culture media in the presence of an FQ. The World Health Organization (WHO) recently revised its cutoff concentrations upward from 2 µg/ml to 4 µg/ml for ofloxacin and to 1 µg/ml for levofloxacin in 7H10 and 1.5 µg/ml with the MGIT 960 method (60). Two cutoffs were recommended for moxifloxacin, 0.5 and 2.0 µg/ml. The most accepted technique, the proportion method, takes 4 to 6 weeks to obtain results, but there are quicker methods on thin-layer agar, or using the nitrate reductase method, or in liquid media with the BACTEC-MGIT system, the Microscopic Observation Drug Susceptibility assay (MODS), or the Microplate Alamar Blue method (MAB), that require as few as 5 to 7 days (61, 62). Several of these methods require primary cultures and cannot be used directly on clinical material, although the green fluorescent protein–phage system shows promise for identifying resistant strains from sputum specimens in less than 48 hours (63).

Heteroresistance

All phenotypic drug sensitivity methods currently take more than a day to perform and therefore require a second visit by the patient to adjust therapy if resistance is found, which can result in critical delays before the patient receives an appropriate drug regimen. The goal for optimal treatment of resistant strains of MDR- or XDR-TB would be a point-of-care method that could both diagnose the presence of *M. tuberculosis* and also identify FQ resistance (in addition to resistance to other first- and second-line anti-TB drugs) within a couple of hours, something that the Xpert system has effectively done for rifampin resistance (64). To achieve this, several molecular techniques, such as the recently WHO-endorsed MTBDR*sl* line probe assay (42, 60), have been developed to rapidly detect the mutations most commonly associated with FQ resistance (65). An inherent problem with most molecular methods is that they only detect the resistance-conferring mutations if the mutations are present in the majority of the bacteria probed. Current methods do not have sufficient sensitivity to detect the 1% of resistant bacteria that defines resistance by the proportion method (66–70), although a couple of recent techniques can detect mutations in 10% or less of the bacteria probed (71, 72).

Primary resistance to the FQs appears to develop when FQ therapy selects for a resistant bacterium that then expands into a subpopulation that eventually becomes the dominant strain within a patient. During the course of this process, bacilli isolated from the patient may show "heteroresistance"—mixed populations of drug-resistant and drug-sensitive bacteria that are otherwise genetically homogenous (67, 73–77). Alternatively, patients may be infected with two genotypically different strains, only one carrying the resistance mutation (68), but this has been described less frequently. Heteroresistance has been found in 20% or more of FQ-resistant isolates (68, 78, 79). In such cases, isolates with FQ heteroresistance are designated as resistant by the proportion method (which determines a strain to be resistant when ≥1% of the bacterial population grows in the presence of the antibiotic) but may be characterized as sensitive by rapid molecular techniques and thereby produce confusing or "indeterminate" results. However, this depends upon the percentage of mutant bacteria present in the population. As a comparison, for the Xpert MTB/RIF assay to detect a mutation conferring rifampin resistance with a high degree of certainty, the mutation must be present in at least 65% of the bacteria tested (80).

A recent study that methodically searched for heteroresistance (78) by sequencing the QRDR of 171 FQ strains found that 21% had several sequencing peaks of either the wild-type and mutant nucleotides or more than one mutation present in the same codon (78) (Fig. 6). Spoligotyping (81) confirmed that only one genotype was present in all patients except for one who appeared to be infected with two genotypically different strains. In follow-up isolates taken during treatment with an FQ from the patients showing heteroresistance, one resistant genotype became dominant and some developed additional resistance mutations. In one patient whose FQ treatment was discontinued due to the detection of resistance, the pattern returned to wild-type, suggesting the reemergence of an underlying FQ-susceptible population. In 5% of the resistant isolates only wild-type sequence was detected in the first isolate, but when this isolate was cultured in media containing 2 μg/ml of ofloxacin, the colonies that appeared were found to contain QRDR mutations. This suggests that resistant bacteria were present in the original isolate but their proportion was too low for the mutation to be detected by sequencing. In two resistant isolates, no QRDR mutation was ever found. The results suggest that most of the phenotypically FQ-resistant strains without gyrase mutations are really heteroresistant and that although the resistant subpopulation is greater than 1% and can be detected by the proportion method, its percentage of the population is not enough to be detected by sequencing or rapid molecular methods.

It is thought that FQ resistance develops due to poor adherence when the patient does not take or does not absorb the antibiotics at the appropriate frequency or dosage, or when it is given inadvertently as monotherapy,

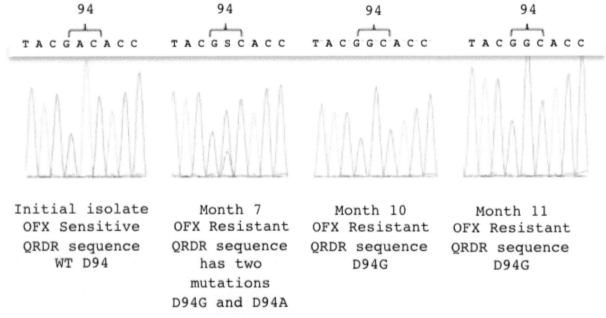

Figure 6 The sequence of the QRDR-A region from the initial patient isolate shows a wild-type GAC encoding aspartic acid at codon 94. In an isolate taken after 7 months of FQ therapy, the QRDR-A sequence shows that two mutant bacilli populations were present, one with a GCC encoding an alanine at codon 94 and one with GGC encoding glycine at codon 94. By month 10 the bacteria containing the D94G substitution predominated, and the population with the D94A substitution was no longer detected by sequencing. Figure modified from reference 78. doi:10.1128/microbiolspec.MGM2-0009-2013.f6

such as when an FQ is added to a failing drug regimen to which the strain is resistant (82, 83). However, there are reports of resistance developing despite excellent adherence (84, 85). This might be explained if the patient initially harbored a small subpopulation of resistant bacteria with a QRDR-A mutation prior to exposure to the antibiotic, which then had a growth advantage when the FQ was administered (86). One possibility for how resistance can develop was suggested by a recent study that found that serum concentrations of ofloxacin patients taking the prescribed dose of 800 mg/day did not appear to be sustained, even at the minimal therapeutic level of 1 µg/ml, for sufficient time to kill even some strains considered FQ "susceptible" (87).

Another possibility is that the mutation may have occurred if the patient received FQ therapy prior to the diagnosis of TB. FQs are used to treat infections at many sites (urinary and gastrointestinal tracts, paranasal sinuses, wounds, and sexually transmitted diseases [88]), and in some countries ciprofloxacin is freely available without a prescription. A patient in the early phases of incipient TB could have taken an FQ, which would have constituted FQ monotherapy for their TB. Resistance has been reported to develop after taking FQs for only 7 days (89). Also, FQs are commonly prescribed for community-acquired pneumonia, so when a patient initially presents with nonspecific respiratory symptoms that are due to undiagnosed TB, they may be given a course of an FQ. This may kill off some of the bacilli and lead to a transient improvement, but when the patient returns with a recrudescence of the symptoms there may be fewer bacilli, which make the diagnosis more difficult. Several studies have shown that taking an FQ for >5 days in the months before TB is diagnosed results in delays of 2 to 5 weeks in initiating anti-TB therapy and has been associated with increased mortality (5, 89).

Given the lack of sensitivity of current molecular tests to detect mutations in FQ heteroresistant isolates, the current WHO recommendation is that these tests, especially MTBDR*sl*, which has high specificity, should only be used to "rule in" a mutation: if a mutation is detected, the result should be believed (60). However, when no mutation is detected, the strain should be tested for FQ resistance with a phenotypic assay, such as the proportion method.

In the study cited above (78) that sequenced the QRDRs from several isolates of FQ-resistant strains and documented heteroresistance in a large percentage, there were two FQ-resistant strains in which no mutations were ever found. This suggests that there must be other sites for mutations that confer FQ resistance,

although they may be responsible for only a small percentage of resistance.

The rest of this article describes possible nongyrase mechanisms of FQ resistance that have been identified in other bacteria and the evidence that these might be related to resistance in *M. tuberculosis*.

Efflux Pumps

Most bacteria have a variety of efflux pumps belonging to different transporter families (90), and many antibiotics are substrates for one or more pumps in a given bacteria (91). However, not all pumps are always expressed at high levels, because some require their expression to be induced, and even if a pump is constitutively expressed and may partially determine the innate baseline MIC for a drug, it does not necessarily mean that the pump is involved in the development of acquired resistance to that drug.

For a pump to be involved as a mechanism of genetic drug resistance there should be a mutation that increases the expression of the pump and thereby reduces the drug concentration within the bacteria. This can occur through simple up-mutations in the promoter, such as is seen with mutations in the promoter region of *inhA* that increase the strength of the promoter and the expression of the isoniazid drug target (92) and isoniazid resistance. The expression of many efflux pumps is regulated, and the expression can be increased, or "induced," either by the presence of some of their substrates or as part of a global stress response. The classic example of a regulated efflux pump is TetA, which presumably originated as a self-protection mechanism in a *Streptomyces* species that produced tetracycline. Upstream of the *tetA* gene and divergently transcribed is the gene encoding its transcriptional repressor TetR, which binds to the DNA in the promoter region of *tetA* and blocks its transcription. When tetracycline binds to the TetR repressor, it induces a conformational change such that the repressor no longer binds and blocks the promoter region of *tetA*, leading to the expression of TetA and consequent extrusion of tetracycline out of the bacterium (93). In contrast, MarA is a transcriptional activator regulated by MarR in *E. coli* and an example of a global transcriptional regulator that induces the expression of different efflux pumps and confers resistance to several antibiotics (94). In *E. coli*, FQ efflux pumps are often part of the MarA regulon (82).

Mutations in the regulatory mechanisms involved with induction can occur in the repressor proteins or in their operators—the DNA sites to which they bind upstream of the genes they repress—and cause increased or unregulated, constitutive expression of the efflux pump (95). The increased number of pump proteins confers resistance by transporting more of the antibiotic out of the bacterium, thereby reducing the antibiotic concentration within the cytoplasm to levels below what is required to completely inhibit its target. Although efflux-mediated resistance generally has a smaller effect on MICs than mutations in the drug target, mutations increasing efflux can also occur subsequent to gyrase mutations and cause higher levels of resistance than the gyrase mutations alone (94).

In Gram-positive bacteria the most common efflux pumps associated with resistance to ciprofloxacin belong to the major facilitator family of proton antiporters (96) and are similar to the NorA efflux pump of *S. aureus* (82). Although they often have other substrates, such as ethidium bromide, acriflavine, or quartenary ammonium compounds, the pumps that confer FQ resistance often do not confer resistance to other antibiotics. When they are expressed, through either induction, a regulatory mutation in the promoter region, an inactivating mutation in a transcriptional regulator, or a mutation in the DNA where the regulator binds, these pumps generally confer low-level resistance to ciprofloxacin and ofloxacin, but not to the later-generation, more hydrophobic FQs such as moxifloxacin and gatifloxacin (97). Moxifloxacin and gatifloxacin (98) (and a few others that have been withdrawn because of toxicity) have better activity against Gram-positive bacteria, especially respiratory pathogens such as *S. pneumoniae* and mycobacteria, and have been called the "respiratory FQs" (10). The specificity of pumps such as the *S. aureus* pump NorA for different FQs may be more complex than simple hydrophobicity and may include elements of FQ structure and side groups (99). There are other similar pumps, such as the *S. aureus* NorB, which recognize moxifloxacin as a substrate and, when induced under stress conditions such as low oxygen, can increase the MIC for moxifloxacin and sparfloxacin as part of a generalized response involving several genes (100, 101).

Increased expression of ABC transporters, such as PatA and PatB in *S. pneumoniae*, has also been found in FQ-resistant strains (102). Their expression is induced by all FQs as well as DNA-damaging agents such as mitomycin, presumably as part of a global stress response inducing the competence system (102). These ABC transporters, like most of the major facilitator transporters, have a preference for hydrophilic FQs as substrates, so while all FQs may induce their expression, their effect on MICs varies with the FQ. They confer the most resistance to norfloxacin and

ciprofloxacin but have less activity on ofloxacin and levofloxacin, much less on sparfloxacin, and almost no effect on moxifloxacin (102), which may contribute to making moxifloxacin a very effective drug (5, 103).

There is a growing notion that the first step in the development of resistance to antibiotics may be the reversible induction of an efflux pump that makes the bacterium relatively tolerant to the antibiotic, which then increases the frequency for high-level resistance mutations in the drug target (104–106). In *S. pneumoniae* it has been shown that ciprofloxacin is an excellent inducer of an efflux pump that confers low-level resistance to ciprofloxacin and increases the frequency of *parC* mutations. Surprisingly, although induction of this pump only slightly increases the MIC for levofloxacin, from 0.6 µg/ml to 0.8 µg/ml, its overexpression dramatically increases the frequency at which *parC* mutations appear (107) in the presence of levofloxacin. Perhaps the efflux pump permits increased survival in the presence of the FQ, which allows chance mutations to appear. Additionally, it has been suggested that the FQs increase the mutation frequency, although this has only been demonstrated in *Mycobacterium fortuitum* (108).

The antiporter (90) LfrA of *M. smegmatis* was the first FQ-transporting mycobacterial efflux pump identified (109, 110). Like NorA, it confers low-level resistance to the hydrophilic FQs such as ciprofloxacin and levofloxacin, increasing the MIC by >4-fold, but has little effect on sensitivity to sparfloxacin or moxifloxacin. Upstream of LfrA is LfrR, a repressor of the TetR family, which normally represses the expression of LfrA to a low-baseline level (111). The repression is relieved and LfrA expression increased when the repressor LfrR binds to some of its substrates such as ethidium bromide or acriflavine but not the FQ (112, 113). FQ-resistant mutant strains of *M. smegmatis* selected on ciprofloxacin have increased constitutive expression of LfrA, presumably due to mutations that impede the repression by LfrR. In strains with a disrupted *lfrA* gene, the MICs increase significantly for ethidium bromide and acriflavine, but only 2-fold for ciprofloxacin, probably due to its low baseline expression (114). Resistant strains of *M. smegmatis* can also be selected that show low uptake of sparfloxacin, presumably due to a second, as yet unidentified, efflux pump (R. Rodriguez and H. Takiff, unpublished results). LfrA is a good example of an FQ resistance pump: its expression can be induced, although not by the FQs, and while it does not seem to significantly affect intrinsic MICs for ciprofloxacin, mutations altering its regulation cause low-level resistance to the hydrophilic FQs. However, it is not broadly conserved in mycobacteria and is not found in *M. tuberculosis* (115).

In contrast, the TAP efflux pump, originally found in *M. fortuitum*, is broadly conserved in mycobacteria (115) and has a homologue in *M. tuberculosis*: Rv1258c. When the *M. fortuitum* gene was overexpressed in *M. smegmatis*, it conferred resistance to tetracycline and some aminoglycosides. Its expression was induced in the presence of rifampin and isoniazid in one study, and of rifampin and ofloxacin but not isoniazid in another study (116, 117). It has also been suggested that the TAP efflux pump may play a role in the increase in the MICs for ofloxacin when *M. tuberculosis* is exposed to rifampin (106) and also when *Mycobacterium marinum* is taken up by macrophages (105). However, neither the overexpression of Rv1258c in *M. smegmatis* nor elimination of the gene from *Mycobacterium bovis* BCG had any effect on FQ MICs, although the MICs for tetracycline and aminoglycosides were altered (118). So while Rv1258c seems clearly to be inducible in *M. tuberculosis*, there is no evidence that the FQs are its substrates.

Much of the work on efflux pumps in mycobacteria has been done in rapid growers, especially *M. smegmatis*, and many of the putative pumps of *M. tuberculosis* have been expressed and studied in this genetically tractable avirulent bacterium (119, 120). Four *M. tuberculosis* efflux pumps have been found to confer at least low-level resistance to FQs when expressed in *M. smegmatis*: one from the major facilitator family and three ATPases. Rv1634 belongs to the major facilitator family and, when expressed in *M. smegmatis*, increases the MICs 2- to 4-fold for ciprofloxacin, norfloxacin, and ofloxacin (121). When the ABC efflux pump complex Rv2686c-Rv2627c-Rv2688c was cloned into *M. smegmatis*, the MICs increased 8-fold for ciprofloxacin and 2-fold for norfloxacin, sparfloxacin, and moxifloxacin but had no effect on the MICs for levofloxacin or ofloxacin (122). Expression of the *M. tuberculosis* DrrAB ATPase in *M. smegmatis* increased the MIC for norfloxacin by 4-fold, which was returned to normal by verapamil, an ATPase inhibitor (123). Finally, the phosphate-specific transporter *pst* is an ABC efflux pump that has been reported both as involved in ciprofloxacin resistance and as a determinant of baseline FQ sensitivity in *M. smegmatis* (124–126). A similar system exists in *M. tuberculosis* but has not been related to FQ resistance or baseline FQ MICs.

Are any *M. tuberculosis* efflux pumps involved in FQ resistance? Although no mutations affecting efflux pumps have been shown to play a role in FQ resistance

in *M. tuberculosis*, when *M. tuberculosis* was exposed to the efflux pump inhibitors reserpine and MC 207.110, there was a 2- to 6-fold decrease in the MICs for ciprofloxacin, levofloxacin, ofloxacin, gatifloxacin, and moxifloxacin in both FQ-sensitive and FQ-resistant strains, but the responsible efflux pumps were not identified (127). This suggests that *M. tuberculosis* expresses efflux pumps that extrude FQs and play a role in determining the baseline FQ MICs, but there is still no evidence that mutations involving efflux pumps are involved in the development of acquired FQ resistance. It is also probable that the expression of FQ pumping transporters is induced under certain conditions (105, 106) and that this can lead to a relative tolerance to FQs that may increase the frequency of a gyrase mutation, but an inducible FQ efflux pump has yet to be identified in *M. tuberculosis*.

A study looking at the effect of pump inhibitors on nonreplicating bacteria found that decreased drug permeability contributes to the phenotypic drug resistance of dormant *M. tuberculosis*, but the differences were independent of efflux processes (128). Studies in an *M. smegmatis* strain lacking porins, MspA, and MspC showed 2- to 4-fold increases in the MIC for norfloxacin, moxifloxacin, sitafloxacin, levofloxacin, and ofloxacin. While the porins may be important for entry of hydrophilic FQs into the bacteria, it is not clear if these are the primary entry pathway for FQs such as moxifloxacin, whose hydrophobicity may permit them to directly diffuse across the cell membrane (129). The Msp porins are not conserved in *M. tuberculosis*, and although channel-forming outer membrane proteins have been reported (130), true porin proteins have not yet been described. It seems probable that hydrophilic FQs such as ciprofloxacin require an entry vehicle to pass the hydrophobic, impermeable *M. tuberculosis* membrane and cell envelope (131), and the slight increases in MICs for moxifloxacin and sparfloxacin in *M. smegmatis* strains lacking the porins suggest that channels could also be an important entry vehicle for these agents.

Pentapeptide Repeat Proteins

MfpA was discovered by screening a plasmid genomic library for genes that would increase resistance to the FQ (132). Originally found in *M. smegmatis*, it appears to be present in all mycobacteria as well as in some other actinobacteria (133) but is likely a pseudogene in *M. leprae*. When MfpA was expressed on a plasmid in *M. smegmatis* or *M. bovis* BCG, the MICs for all the FQs increased by 2- to 8-fold, and when it was eliminated from the chromosome of *M. smegmatis*, the

MICs decreased by 2- to 4-fold, suggesting a role in determining the innate bacterial MICs. MfpA belongs to a protein family termed pentapeptide repeat proteins (PRPs) (134) that are composed of 5-amino-acid units, where every fifth amino acid is either a leucine or a phenylalanine and can be described as A(D/N)(L/F)XX (where X can be any amino acid) (135). The repeats form a regular right-handed quadrilateral β-helix structure known as the Rfr-fold (136) (see below).

Shortly after MfpA was described, it was realized that the Qnr proteins, which were found on transmissible plasmids that confer low-level FQ resistance, are also pentapeptides. Members of the Qnr family of pentapeptide proteins have been found in several pathogenic Gram-negative bacteria on a variety of different transferable plasmids (137), often within integrons or other mobile elements. Many of the Qnr proteins seem to originate from the chromosomes of environmental aquatic bacteria and were perhaps transferred onto plasmids and into pathogenic bacteria due to the selective pressure from extensive use of FQs in commercial fish farming. PRPs are also present in the chromosomes of several Gram-positive bacteria, including pathogenic species such as *Enterococcus faecalis* (138), *Enterococcus faecium*, *Listeria monocytogenes*, *Clostridium perfringens*, and *Clostridium difficile* (137), but these have not been found on plasmids, and their function is unknown. While the Qnr proteins do not confer high-level resistance, they may increase the mutant prevention concentration up to 10-fold, making it more likely that mutations conferring higher-level resistance mutations in the gyrase or topoisomerase IV are selected in the course of FQ therapy (139).

Today the PRP family includes more than 1,000 members (140). While the function of these repeats in their original chromosomal locations is uncertain for most proteins, the two bacterial PRPs, Qnr and MfpA, have been reported to interact with DNA gyrase or topoisomerase IV (141, 142), conferring a new mechanism of quinolone resistance by protecting DNA gyrase and topoisomerase IV from the inhibitory effect of quinolones. MfpA is dimeric in solution and in the crystal and folds as a right-handed quadrilateral β-helix with a size, shape, and negative charge that mimic 30 base pairs of B-form DNA (141). Each monomer has eight complete coils of square quadrilateral shape, with the conserved L/F residue in the middle of each side, pointing inward. The C-terminal α-helices interact to form the dimer. A model of interaction was proposed in which the MfpA dimer was docked into the gyrase groove formed by the dimer of the breakage-reunion domains (141) (Fig. 7). In this model, the MfpA protein represents a form of DNA mimicry and competes with DNA for binding to the breakage-reunion domain of DNA gyrase. Exactly what role MfpA plays in gyrase function and how it decreases the susceptibility to FQ in mycobacteria is still a puzzle. One idea is that its interaction with the gyrase could regulate the effective concentration of free gyrase available for binding to DNA, and thus also to the FQ, which associates with a DNA-gyrase complex. Another possibility is that MfpA may destabilize the FQ-DNA-gyrase complex and thus relieve FQ inhibition (143). Whether all bacterial PRPs have the same kind of interactions with DNA gyrase is still unknown. However, structural differences observed between other Qnr structures and MfpA could explain their possible differences in function, notably a loop that is essential for the protective effect of Qnr (4, 32, 143).

Although it has never been described, it is conceivable that any mutation that increases the expression of the chromosomal MfpA in *M. tuberculosis* could confer low-level FQ resistance, as is seen when *mfpA* is present on a multicopy plasmid (132, 144). The exact mechanism of the regulation of the *mfpA* expression and the signals that modulate its expression are unknown, but it seems likely that its expression is regulated, because in all mycobacteria it is found as the fifth gene in an operon, preceded by four genes that occur together in many actinomycetes as a unit termed a "conservon" (156). Although there is only one conservon in mycobacteria, this four-gene complex is present 13 times in the chromosome of *Streptomyces coelicolor* A3(2), 10 times in *Streptomyces avermitilis*, 3 times in different *Frankia* species, 5 times in *Nocardia farcinica*, 5 times in *Thermobifida fusca*, and 2 times in *Kinococcus radiotolerans*. In *Streptomyces venezuelensis* it appears to regulate the expression of the genes encoding the proteins that synthesize the antibiotic jadomycin B (145). The conservon units precede different proteins in the other bacteria, and only in mycobacteria, where they are named MfpB to MfpE, are they followed by a gene encoding a pentapeptide protein.

The first gene in the conservon, *Rv3365c*, encodes MfpE, a protein similar to other bacterial histidine kinases, which is likely membrane-bound and may serve as a sensor, although it is distinct from the histidine kinases in most two-component systems. It has the histidine residue that is phosphorylated in the classic histidine kinases, but the surrounding amino acids in the "H-box" are different. The next gene, *Rv3364c*, encodes MfpD, a protein of 130 amino acids belonging to the Roadblock/LC7 family. A recent study found

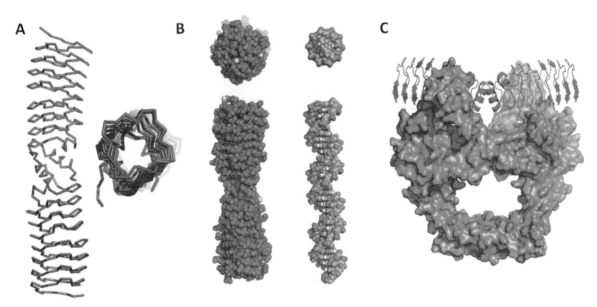

Figure 7 (A) Side and top views of the *M. tuberculosis* MfpA dimer shown in Cα trace (PDB code 2BM5). (B) Top and side views showing how MfpA mimics a 30-base pair B-form DNA. (C) Model of the interaction between the DNA gyrase catalytic core (represented in blue molecular surface) and MfpA (represented in magenta cartoon). doi:10.1128/microbiolspec.MGM2-0009-2013.f7

that a protein belonging to this family, MglB, served as a GAP (GTPase activating protein) for an adjacent small GTP binding protein, MglA (146). *Rv3363c* encodes MfpC, a 122-amino-acid peptide that belongs to the DUF 742 family (pfam05331.4) of proteins, whose function is unknown.

The fourth gene, *Rv3362c*, encodes MfpB, a small GTP binding protein that has been found to interact with MfpA, encoded by *Rv3361c*, the fifth and last gene in the operon (144). Overexpression of both MfpA and MfpB confers higher levels of resistance to ciprofloxacin than MfpA alone (132), and knock-out strains lacking either gene have reduced MICs for ciprofloxacin. Putting this together with what is known about the other proteins in the conservon, it is possible to imagine that the histidine kinase senses some unknown signal, which then is transmitted to the MfpA effector through the GAP protein (MfpD) that accelerates the GTPase activity of MfpB. This would then lead to an alteration of MfpA activity, perhaps changing its affinity for the gyrase and thereby regulating its inhibition of gyrase function, or toggling its function between supercoiling and decatenation, usually carried out by the topoisomerase IV that is lacking in mycobacteria. The protection of the gyrase from FQs and other poisons by MfpA (141) may be merely a side effect, or alternatively, the PRPs could have evolved as protection from environmental gyrase

poisons produced by other bacteria, as the pentapeptides McbG and AlbG serve as protection for bacteria producing the gyrase toxins B17 (147) and albicidin (148, 149), respectively. Whatever the true function of MfpA in mycobacteria may be, any mutation that increases its expression or alters its interaction with the gyrase might confer low-level FQ resistance or increase the level of resistance in strains with QRDR mutations. However, this has never been reported, neither *in vitro* nor in clinical isolates, although it is likely that the relevant mutations have not been sought in resistant isolates.

In Gram-negative bacteria, there are other plasmid-based mechanisms of resistance (137)—such as variants of aminoglycoside acetyltransferase AAC(6″)-Ib-*cr* that can acetylate ciprofloxacin, or efflux pumps of the major facilitator family such as QepA, or the RND family such as OqxAB—but neither transmissible FQ-resistance determinants nor FQ-inactivating enzymes have been described in mycobacteria.

CONCLUSIONS

The FQs are an important component in the treatment of MDR-TB, of which there are over 400,000 cases a year (150), and clinical trials are currently in progress to see if the use of FQs as first-line therapy can reduce the duration of treatment of drug-sensitive TB (5), so

FQs are likely to be used extensively in the treatment of TB for at least the next several years. Their use, however, will likely be accompanied by the development of resistance, which must be detected early so that effective alternative drugs can be prescribed. Although there are some newer, more rapid phenotypic tests, the proportion method is still used because of its reliability and high sensitivity for identifying resistant strains, even if the resistant bacteria represent only 1% of the total bacilli population. However, even if performed directly on clinical specimens, results take 4 to 6 weeks (61). Rapid molecular tests to detect resistance-conferring mutations can be performed in a few hours, but their sensitivity has been limited because in a variable percentage of resistant strains (5 to 50%) no *gyrA* mutations are found, or results are equivocal. Finding out why these rapid molecular tests have less-than-optimal sensitivity should help improve their sensitivity. Expanding the tests to include mutations in the QRDR of *gyrB* might increase the sensitivity, but only by a few percent. The intriguing, and still unanswered, question is whether there are mutations in additional sites, other than the genes encoding the gyrase, that can confer resistance. Mechanisms present in other bacteria include increased expression of efflux pumps that decreases the cytoplasmic FQ concentration, the presence of a plasmid with a pentapeptide that protects the gyrase or an enzyme that inactivates the FQ, and other as yet uncharacterized mechanisms. If these alternative non-gyrase mutations are causing resistance, improving the sensitivity of the rapid tests will require identification of these other sites for resistance-conferring mutations. Alternatively, if the explanation is heteroresistance—where resistance is caused by gyrase mutations that are initially present only in a minority of bacilli too small to be detected by the molecular tests in the initial isolates—improvements in the sensitivity of molecular tests will require techniques that can detect mutations present in only 1% or less of the bacterial population.

Because of the many FQ efflux pumps and Qnr proteins identified in other bacteria, it was thought that these non-gyrase mechanisms would eventually be found, but there are still no reports of non-gyrase mutations causing FQ resistance in *M. tuberculosis*, either in clinical isolates or in *in vitro*–selected resistant strains. Heteroresistance seems the more likely explanation, as has been well documented in several careful studies where it was shown to explain all, or almost all, instances where gyrase mutations were not detected in the initial isolates of strains that were determined to be resistant by the proportion method (78).

Alternative resistance mechanisms might be involved in FQ resistance, particularly if selected with ciprofloxacin, but are likely to cause only low-level resistance (151). They may play a role in the variation in baseline FQ MICs for different strains (21, 52) or the different levels of resistance in strains carrying the same gyrase mutations. These other mechanisms, and perhaps also the reversible induction of an FQ efflux pump yet to be identified, could cause a relative FQ tolerance that could increase the frequency at which higher-level gyrase mutations appear (104–106). Reports that efflux pump inhibitors reduce FQ MICs in both FQ-sensitive and -resistant strains suggest that an FQ efflux pump exists and could play a role in determining the baseline FQ MICs (105, 106, 127).

It is thought that FQ resistance develops in a step-wise fashion, at least in other bacteria, such that the level of resistance increases with each additional mutation, in the gyrase and the topoisomerase IV, or together with increased efflux pump expression and plasmid-based Qnr or inactivating genes. However, if the first-level mutation does not raise the level of resistance above the MIC for the FQ being used, it may still be effective. This may explain why ciprofloxacin remained effective against *E. coli*, because even with a first-level mutation, the MICs were perhaps still below the maximal serum level for ciprofloxacin (109, 151). This may also explain why ciprofloxacin is a useless drug for *M. tuberculosis*, and moxifloxacin is a very good drug.

With the usual doses, the maximum serum level for ciprofloxacin is only about 1.5 µg/ml, for ofloxacin 4.0, for levofloxacin 6, and for moxifloxacin 4. The half-life of ciprofloxacin is 4 hours, ofloxacin 4 to 5 hours, levofloxacin 6 to 8 hours, and moxifloxacin 10 to 13 hours (16). At the maximum ciprofloxacin serum concentration of ~1.5 µg/ml, resistant colonies can be selected *in vitro* at a frequency of ~10^{-6-7} (20, 151), and a tuberculous lung cavity contains about 10^8 bacteria (16). Early studies using ciprofloxacin to treat TB found that the high percentage of treatment failures or relapses was generally in patients carrying strains that were ciprofloxacin resistant (16). Although many of the mutations selected *in vitro* on this concentration of ciprofloxacin may not have substitutions in GyrA but only in GyrB or in neither gyrase subunit (151), even most of these with low-level resistance will not be effectively killed at the 1.5 µg/ml maximum ciprofloxacin serum concentration.

The baseline MICs for ofloxacin in wild-type strains range from <0.5 to 2.0 µg/ml (52), and strains with QRDR-A mutations can be selected at frequencies of 10^{-7} to 10^{-8} on ofloxacin concentrations of 1 to 2 µg/ml, just 2 to 4 times above the MIC. This drug level is probably close to the serum concentration at the end of a dosing interval, but it could be even lower if the patient misses or delays a dose. Almost all QRDR-A substitutions have MICs for ofloxacin and levofloxacin of at least 4 µg/ml, so these strains will not be effectively killed by ofloxacin (52, 151). Levofloxacin is probably better because of its advantageous pharmacokinetics—a longer half-life and a higher maximum serum concentration (16)—which will lower the frequency of resistance mutations and may be adequate to kill most strains with QRDR-B mutations and even some with QRDR-A mutations.

The moxifloxacin MIC for most strains of wild-type *M. tuberculosis* is ≤0.125 µg/ml, while the maximum moxifloxacin serum concentration is 4 µg/ml (16). At even half the maximum serum concentration, 2 µg/ml, the *in vitro* frequency for selecting resistant strains is less than 10^{-9} (151). While almost all the resistant strains selected on moxifloxacin have GyrA substitutions (151), more than 90% have MICs of ≤2 µg/ml (152). There is a variation in the level of resistance in strains carrying the same mutation, and substitutions in amino acids 94 and 88 seem prone to have higher MICs (52). It has been previously observed that strains with both QRDR-A and QRDR-B substitutions or two QRDR-A substitutions have higher MICs than strains with a single mutation (43). Therefore, while moxifloxacin might be effective against most strains with a single mutation, it could be expected that during moxifloxacin therapy some of these strains would develop additional mutations that would raise their MICs above the moxifloxacin serum concentration. Surprisingly, though, it was found that two clinical isolates with both the A90V and D94G mutations had MICs of 1 and 2 µg/ml for moxifloxacin, suggesting that even these would be effectively killed.

Despite the efforts to find and detect FQ-resistance-conferring mutations, the thinking about the clinical management of these patients is changing. Moxifloxacin is an excellent drug because it has good pharmacokinetics (16), a long half-life, and high affinity for the gyrase (38); is a poor substrate for most efflux pumps (97, 102); and has a low frequency of toxic side effects (tendon rupture and worsening of symptoms in patients with myasthenia gravis), so patient adherence is good. Moxifloxacin is such an effective drug that it may contribute to curing many patients whose *M. tuberculosis* strains contain gyrase mutations. Studies in mice suggested that moxifloxacin can be effective against strains with gyrase mutations if the MIC is <2 µg/ml: it was effective against a strain with

GyrB D500N (MIC 0.5 µg/ml), reduced mortality in strains with GyrA A09V (MIC 2 µg/ml), but had no effect on infections with a strain carrying the GyrA D94G substitution (MIC 4 µg/ml) (103). A meta-analysis of XDR-TB treatment outcomes found that despite the supposed FQ resistance, treatment with a later-generation FQ (levofloxacin, sparfloxacin, or moxifloxacin) was significantly associated with a better outcome (8). A WHO panel of experts recently recommended that resistance to moxifloxacin be tested at two cutoff levels: 0.5 µg/ml and 2.0 µg/ml (60). Given the very limited choices for treating XDR-TB, it may be advisable to give moxifloxacin to patients with XDR strains with moxifloxacin MICs less than 2 µg/ml.

Is it worthwhile to use molecular tests to find "FQ-resistance" mutations? Strains with GyrB substitutions, present in a small minority of "resistant" isolates, probably have MICs that are treatable with moxifloxacin. Although most GyrA substitutions might be susceptible to serum levels of moxifloxacin, rapid molecular tests could detect substitutions in amino acids 94 or 88 that, at least in some strains, have MICs above 2 µg/ml. However, not all strains with mutations in this codon have the higher moxifloxacin MICs (52). In a study of 51 clinical isolates with QRDR-A mutations and resistance to at least 2 µg/ml ofloxacin, only two had moxifloxacin MICs >2 µg/ml; the only G88C substitution had an MIC for moxifloxacin of 4 µg/ml, as did only one of six isolates with D94N. So even if these specific mutations are detected, withholding moxifloxacin might significantly reduce the patient's chance to be cured (8).

Price, however, is an important consideration in the low- and medium-resource countries where most MDR and XDR-TB is found. Moxifloxacin is expensive, while levofloxacin is more economical. So if levofloxacin is to be used, a rapid test for resistance seems more useful. If a gyrase mutation is detected, the patient might be given moxifloxacin, but if the result is equivocal or no mutation is detected, phenotypic drug sensitivity testing is indicated. If there is a single mutation detected that results in an amino acid substitution at codons 89, 90, or 91, moxifloxacin will probably be effective. If there is a mutation affecting codons 88 or 94, the effectiveness of moxifloxacin is less certain, and phenotypic FQ sensitivity testing is probably indicated, but maybe moxifloxacin should be given while awaiting these results. The Bayer patents for moxifloxacin expired in March 2014, which should make less costly generic versions available (153). This is an evolving therapeutic topic, and the purpose here is not to make clinical recommendations, but merely to review the literature related to resistance mechanisms and their significance. Definitive recommendations may await further follow-up outcome studies of patients with characterized "resistance" mutations who receive moxifloxacin.

New FQs or other modified quinolones continue to appear (154), and a super-agent may eventually be discovered, but for now moxifloxacin still seems to be the most effective and least toxic. Levofloxacin is fairly effective against *M. tuberculosis*, but resistance seems to develop more readily than resistance to moxifloxacin, at least *in vitro*. Clinical trials are currently in progress to determine whether and how moxifloxacin can best be used in combination with recently developed drugs to optimize and reduce the duration of therapy for both drug-sensitive and drug-resistant TB (4). The quinolones currently used for TB treatment were designed before structural data on the *M. tuberculosis* DNA gyrase catalytic core were available. Extensive 3D-QSAR analyses of existing structural data and new crystallographic data on the protein-DNA-FQ ternary complex will allow a more complete understanding of the relationships between the sequence, the structure, and the resistance phenotypes of this enzyme and should help to design more effective FQs whose affinity for the gyrase would be less susceptible to mutations and therefore less likely to generate resistance.

This work was supported by FONACIT proyecto G-2005-000393, LOCTI project, Nuevos Fármacos contra la Tuberculosis, the visiting professor program of the Université de Montpellier 2, France, and an Ecos Nord project between the Unité de Microbiologie Structurale, Institut Pasteur, Paris, and the Laboratorio de Genética Molecular, IVIC, Caracas, Venezuela. We thank Laurent Kremer, Mathew Vetting, and especially Robin Warren for critically reading the manuscript.

Citation. Mayer C, Takiff H. 2014. The molecular genetics of fluoroquinolone resistance in *Mycobacterium tuberculosis*. Microbiol Spectrum 2(4):MGM2-0009-2013.

References

1. Chan ED, Laurel V, Strand MJ, Chan JF, Huynh ML, Goble M, Iseman MD. 2004. Treatment and outcome analysis of 205 patients with multidrug-resistant tuberculosis. *Am J Respir Crit Care Med* **169:**1103–1109.

2. Tahaoglu K, Torun T, Sevim T, Atac G, Kir A, Karasulu L, Ozmen I, Kapakli N. 2001. The treatment of multidrug-resistant tuberculosis in Turkey. *N Engl J Med* **345:**170–174.

3. Nuermberger E, Tyagi S, Tasneen R, Williams KN, Almeida D, Rosenthal I, Grosset JH. 2008. Powerful bactericidal and sterilizing activity of a regimen containing PA-824, moxifloxacin, and pyrazinamide in a murine model of tuberculosis. *Antimicrob Agents Chemother* **52:**1522–1524.

4. Veziris N, Ibrahim M, Lounis N, Andries K, Jarlier V. 2011. Sterilizing activity of second-line regimens containing TMC207 in a murine model of tuberculosis. *PLoS One* **6:**e17556.

5. Takiff H, Guerrero E. 2011. Current prospects for the fluoroquinolones as first-line TB therapy. *Antimicrob Agents Chemother* **55:**5412–5419.

6. Gandhi NR, Nunn P, Dheda K, Schaaf HS, Zignol M, van Soolingen D, Jensen P, Bayona J. 2010. Multidrug-resistant and extensively drug-resistant tuberculosis: a threat to global control of tuberculosis. *Lancet* **375:**1830–1843.

7. Sotgiu G, Ferrara G, Matteelli A, Richardson MD, Centis R, Ruesch-Gerdes S, Toungoussova O, Zellweger JP, Spanevello A, Cirillo D, Lange C, Migliori GB. 2009. Epidemiology and clinical management of XDR-TB: a systematic review by TBNET. *Eur Respir J* **33:**871–881.

8. Jacobson KR, Tierney DB, Jeon CY, Mitnick CD, Murray MB. 2010. Treatment outcomes among patients with extensively drug-resistant tuberculosis: systematic review and meta-analysis. *Clin Infect Dis* **51:**6–14.

9. Lesher GY, Froelich EJ, Gruett MD, Bailey JH, Brundage RP. 1962. 1,8-Naphthyridine derivatives. A new class of chemotherapeutic agents. *J Med Pharm Chem* **91:**1063–1065.

10. Emmerson AM, Jones AM. 2003. The quinolones: decades of development and use. *J Antimicrob Chemother* **51(Suppl 1):**13–20.

11. Bouza E, Garcia-Garrote F, Cercenado E, Marin M, Diaz MS. 1999. *Pseudomonas aeruginosa*: a survey of resistance in 136 hospitals in Spain. The Spanish *Pseudomonas aeruginosa* Study Group. *Antimicrob Agents Chemother* **43:**981–982.

12. Blumberg HM, Rimland D, Carroll DJ, Terry P, Wachsmuth IK. 1991. Rapid development of ciprofloxacin resistance in methicillin-susceptible and -resistant *Staphylococcus aureus*. *J Infect Dis* **163:**1279–1285.

13. Livermore DM, Nichols T, Lamagni TL, Potz N, Reynolds R, Duckworth G. 2003. Ciprofloxacin-resistant *Escherichia coli* from bacteraemias in England; increasingly prevalent and mostly from men. *J Antimicrob Chemother* **52:**1040–1042.

14. Drlica K, Zhao X. 2007. Mutant selection window hypothesis updated. *Clin Infect Dis* **44:**681–688.

15. Almeida D, Nuermberger E, Tyagi S, Bishai WR, Grosset J. 2007. In vivo validation of the mutant selection window hypothesis with moxifloxacin in a murine model of tuberculosis. *Antimicrob Agents Chemother* **51:**4261–4266.

16. Ginsburg AS, Grosset JH, Bishai WR. 2003. Fluoroquinolones, tuberculosis, and resistance. *Lancet Infect Dis* **3:**432–442.

17. Sullivan EA, Kreiswirth BN, Palumbo L, Kapur V, Musser JM, Ebrahimzadeh A, Frieden TR. 1995. Emergence of fluoroquinolone-resistant tuberculosis in New York City. *Lancet* **345:**1148–1150.

18. Fenlon CH, Cynamon MH. 1986. Comparative in vitro activities of ciprofloxacin and other 4-quinolones

against *Mycobacterium tuberculosis* and *Mycobacterium intracellulare*. *Antimicrob Agents Chemother* **29**: 386–388.

19. Neu HC, Fang W, Gu JW, Chin NX. 1992. In vitro activity of OPC-17116. *Antimicrob Agents Chemother* **36**:1310–1315.

20. Takiff HE, Salazar L, Guerrero C, Philipp W, Huang WM, Kreiswirth B, Cole ST, Jacobs WR Jr, Telenti A. 1994. Cloning and nucleotide sequence of *Mycobacterium tuberculosis* gyrA and gyrB genes and detection of quinolone resistance mutations. *Antimicrob Agents Chemother* **38**:773–780.

21. Angeby KA, Jureen P, Giske CG, Chryssanthou E, Sturegard E, Nordvall M, Johansson AG, Werngren J, Kahlmeter G, Hoffner SE, Schon T. 2010. Wild-type MIC distributions of four fluoroquinolones active against *Mycobacterium tuberculosis* in relation to current critical concentrations and available pharmacokinetic and pharmacodynamic data. *J Antimicrob Chemother* **65**:946–952.

22. Shandil RK, Jayaram R, Kaur P, Gaonkar S, Suresh BL, Mahesh BN, Jayashree R, Nandi V, Bharath S, Balasubramanian V. 2007. Moxifloxacin, ofloxacin, sparfloxacin, and ciprofloxacin against *Mycobacterium tuberculosis*: evaluation of in vitro and pharmacodynamic indices that best predict in vivo efficacy. *Antimicrob Agents Chemother* **51**:576–582.

23. Park SK, Lee WC, Lee DH, Mitnick CD, Han L, Seung KJ. 2004. Self-administered, standardized regimens for multidrug-resistant tuberculosis in South Korea. *Int J Tuberc Lung Dis* **8**:361–368.

24. Yew WW, Chan CK, Leung CC, Chau CH, Tam CM, Wong PC, Lee J. 2003. Comparative roles of levofloxacin and ofloxacin in the treatment of multidrug-resistant tuberculosis: preliminary results of a retrospective study from Hong Kong. *Chest* **124**:1476–1481.

25. Renau TE, Gage JW, Dever JA, Roland GE, Joannides ET, Shapiro MA, Sanchez JP, Gracheck SJ, Domagala JM, Jacobs MR, Reynolds RC. 1996. Structure-activity relationships of quinolone agents against mycobacteria: effect of structural modifications at the 8 position. *Antimicrob Agents Chemother* **40**:2363–2368.

26. Dawe RS, Ibbotson SH, Sanderson JB, Thomson EM, Ferguson J. 2003. A randomized controlled trial (volunteer study) of sitafloxacin, enoxacin, levofloxacin and sparfloxacin phototoxicity. *Br J Dermatol* **149**:1232–1241.

27. Yadav V, Deopujari K. 2006. Gatifloxacin and dysglycemia in older adults. *N Engl J Med* **354**:2725–2726.

28. Champoux JJ. 2001. DNA topoisomerases: structure, function, and mechanism. *Annu Rev Biochem* **70**:369–413.

29. Forterre P, Gribaldo S, Gadelle D, Serre MC. 2007. Origin and evolution of DNA topoisomerases. *Biochimie* **89**:427–446.

30. Levine C, Hiasa H, Marians KJ. 1998. DNA gyrase and topoisomerase IV: biochemical activities, physiological roles during chromosome replication, and drug sensitivities. *Biochim Biophys Acta* **1400**:29–43.

31. Mdluli K, Ma Z. 2007. *Mycobacterium tuberculosis* DNA gyrase as a target for drug discovery. *Infect Disord Drug Targets* **7**:159–168.

32. Forterre P, Gadelle D. 2009. Phylogenomics of DNA topoisomerases: their origin and putative roles in the emergence of modern organisms. *Nucleic Acids Res* **37**: 679–692.

33. Aravind L, Leipe DD, Koonin EV. 1998. Toprim: a conserved catalytic domain in type IA and II topoisomerases, DnaG-type primases, OLD family nucleases and RecR proteins. *Nucleic Acids Res* **26**:4205–4213.

34. Schoeffler AJ, Berger JM. 2008. DNA topoisomerases: harnessing and constraining energy to govern chromosome topology. *Q Rev Biophys* **41**:41–101.

35. Hooper DC. 2002. Fluoroquinolone resistance among Gram-positive cocci. *Lancet Infect Dis* **2**:530–538.

36. Piton J, Petrella S, Delarue M, Andre-Leroux G, Jarlier V, Aubry A, Mayer C. 2010. Structural insights into the quinolone resistance mechanism of *Mycobacterium tuberculosis* DNA gyrase. *PLoS One* **5**:e12245.

37. Pan XS, Yague G, Fisher LM. 2001. Quinolone resistance mutations in *Streptococcus pneumoniae* GyrA and ParC proteins: mechanistic insights into quinolone action from enzymatic analysis, intracellular levels, and phenotypes of wild-type and mutant proteins. *Antimicrob Agents Chemother* **45**:3140–3147.

38. Pestova E, Millichap JJ, Noskin GA, Peterson LR. 2000. Intracellular targets of moxifloxacin: a comparison with other fluoroquinolones. *J Antimicrob Chemother* **45**: 583–590.

39. Yin X, Yu Z. 2010. Mutation characterization of gyrA and gyrB genes in levofloxacin-resistant *Mycobacterium tuberculosis* clinical isolates from Guangdong Province in China. *J Infect* **61**:150–154.

40. Ginsburg AS, Woolwine SC, Hooper N, Benjamin WH Jr, Bishai WR, Dorman SE, Sterling TR. 2003. The rapid development of fluoroquinolone resistance in M. tuberculosis. *N Engl J Med* **349**:1977–1978.

41. Lau RW, Ho PL, Kao RY, Yew WW, Lau TC, Cheng VC, Yuen KY, Tsui SK, Chen X, Yam WC. 2011. Molecular characterization of fluoroquinolone resistance in *Mycobacterium tuberculosis*: functional analysis of gyrA mutation at position 74. *Antimicrob Agents Chemother* **55**:608–614.

42. Feng Y, Liu S, Wang Q, Wang L, Tang S, Wang J, Lu W. 2013. Rapid diagnosis of drug resistance to fluoroquinolones, amikacin, capreomycin, kanamycin and ethambutol using genotype MTBDRsl assay: a meta-analysis. *PLoS One* **8**:e55292.

43. Kocagoz T, Hackbarth CJ, Unsal I, Rosenberg EY, Nikaido H, Chambers HF. 1996. Gyrase mutations in laboratory-selected, fluoroquinolone-resistant mutants of *Mycobacterium tuberculosis* H37Ra. *Antimicrob Agents Chemother* **40**:1768–1774.

44. Cui Z, Wang J, Lu J, Huang X, Hu Z. 2011. Association of mutation patterns in gyrA/B genes and ofloxacin resistance levels in *Mycobacterium tuberculosis* isolates from East China in 2009. *BMC Infect Dis* **11**:78.

45. Malik S, Willby M, Sikes S, Tsodikov OV, Posey JE. 2012. New insights into fluoroquinolone resistance in *Mycobacterium tuberculosis:* functional genetic analysis of gyrA and gyrB mutations. *PLoS One* 7:e39754.

46. Pantel A, Petrella S, Veziris N, Brossier F, Bastian S, Jarlier V, Mayer C, Aubry A. 2012. Extending the definition of the GyrB quinolone resistance-determining region in *Mycobacterium tuberculosis* DNA gyrase for assessing fluoroquinolone resistance in *M. tuberculosis*. *Antimicrob Agents Chemother* 56:1990–1996.

47. Kim H, Nakajima C, Yokoyama K, Rahim Z, Kim YU, Oguri H, Suzuki Y. 2011. Impact of the E540V amino acid substitution in GyrB of *Mycobacterium tuberculosis* on quinolone resistance. *Antimicrob Agents Chemother* 55:3661–3667.

48. Maruri F, Sterling TR, Kaiga AW, Blackman A, van der Heijden YF, Mayer C, Cambau E, Aubry A. 2012. A systematic review of gyrase mutations associated with fluoroquinolone-resistant *Mycobacterium tuberculosis* and a proposed gyrase numbering system. *J Antimicrob Chemother* 67:819–831.

49. Aubry A, Veziris N, Cambau E, Truffot-Pernot C, Jarlier V, Fisher LM. 2006. Novel gyrase mutations in quinolone-resistant and -hypersusceptible clinical isolates of *Mycobacterium tuberculosis:* functional analysis of mutant enzymes. *Antimicrob Agents Chemother* 50: 104–112.

50. Suzuki Y, Nakajima C, Tamaru A, Kim H, Matsuba T, Saito H. 2012. Sensitivities of ciprofloxacin-resistant *Mycobacterium tuberculosis* clinical isolates to fluoroquinolones: role of mutant DNA gyrase subunits in drug resistance. *Int J Antimicrob Agents* 39:435–439.

51. Von Groll A, Martin A, Jureen P, Hoffner S, Vandamme P, Portaels F, Palomino JC, da Silva PA. 2009. Fluoroquinolone resistance in *Mycobacterium tuberculosis* and mutations in gyrA and gyrB. *Antimicrob Agents Chemother* 53:4498–4500.

52. Sirgel FA, Warren RM, Streicher EM, Victor TC, van Helden PD, Bottger EC. 2012. gyrA mutations and phenotypic susceptibility levels to ofloxacin and moxifloxacin in clinical isolates of *Mycobacterium tuberculosis*. *J Antimicrob Chemother* 67:1088–1093.

53. Aubry A, Fisher LM, Jarlier V, Cambau C. 2006. First functional characterization of a singly expressed bacterial type II topoisomerase: the enzyme from *Mycobacterium tuberculosis*. *Biochem Biophys Res Commun* 348:158–165.

54. Aubry A, Pan XS, Fisher LM, Jarlier V, Cambau E. 2004. *Mycobacterium tuberculosis* DNA gyrase: interaction with quinolones and correlation with antimycobacterial drug activity. *Antimicrob Agents Chemother* 48:1281–1288.

55. Matrat S, Aubry A, Mayer C, Jarlier V, Cambau E. 2008. Mutagenesis in the alpha3alpha4 GyrA helix and in the Toprim domain of GyrB refines the contribution of *Mycobacterium tuberculosis* DNA gyrase to intrinsic resistance to quinolones. *Antimicrob Agents Chemother* 52:2909–2914.

56. Matrat S, Veziris N, Mayer C, Jarlier V, Truffot-Pernot C, Camuset J, Bouvet E, Cambau E, Aubry A. 2006.

Functional analysis of DNA gyrase mutant enzymes carrying mutations at position 88 in the A subunit found in clinical strains of *Mycobacterium tuberculosis* resistant to fluoroquinolones. *Antimicrob Agents Chemother* 50: 4170–4173.

57. Mokrousov I, Otten T, Manicheva O, Potapova Y, Vishnevsky B, Narvskaya O, Rastogi N. 2008. Molecular characterization of ofloxacin-resistant *Mycobacterium tuberculosis* strains from Russia. *Antimicrob Agents Chemother* 52:2937–2939.

58. Rohs R, West SM, Sosinsky A, Liu P, Mann RS, Honig B. 2009. The role of DNA shape in protein-DNA recognition. *Nature* 461:1248–1253.

59. Chernyaeva E, Fedorova E, Zhemkova G, Korneev Y, Kozlov A. 2013. Characterization of multiple and extensively drug resistant *Mycobacterium tuberculosis* isolates with different ofloxacin-resistance levels. *Tuberculosis (Edinb)* 93:291–295.

60. Gilpin C. 2012. *Summary of outcomes from the WHO Expert Group Meeting on Drug Susceptibility Testing.* World Health Organization, Geneva, Switzerland. http://www.finddiagnostics.org/export/sites/default/resource-centre/presentations/find_fifth_symposium_iuatld2012/04_ChristopherGilpin_WHO-ExpertGroupMeeting-DST.pdf.

61. World Health Organization Stop TB Dept. 2008. *Policy guidance on drug-susceptibility testing (DST) of second-line antituberculosis drugs.* World Health Organization, Geneva, Switzerland. http://whqlibdoc.who.int/hq/2008/WHO_HTM_TB_2008.392_eng.pdf.

62. Van Deun A, Martin A, Palomino JC. 2010. Diagnosis of drug-resistant tuberculosis: reliability and rapidity of detection. *Int J Tuberc Lung Dis* 14:131–140.

63. Jain P, Hartman TE, Eisenberg N, O'Donnell MR, Kriakov J, Govender K, Makume M, Thaler DS, Hatfull GF, Sturm AW, Larsen MH, Moodley P, Jacobs WR Jr. 2012. phi(2)GFP10, a high-intensity fluorophage, enables detection and rapid drug susceptibility testing of *Mycobacterium tuberculosis* directly from sputum samples. *J Clin Microbiol* 50:1362–1369.

64. Boehme CC, Nabeta P, Hillemann D, Nicol MP, Shenai S, Krapp F, Allen J, Tahirli R, Blakemore R, Rustomjee R, Milovic A, Jones M, O'Brien SM, Persing DH, Ruesch-Gerdes S, Gotuzzo E, Rodrigues C, Alland D, Perkins MD. 2010. Rapid molecular detection of tuberculosis and rifampin resistance. *N Engl J Med* 363:1005–1015.

65. Chang KC, Yew WW, Chan RC. 2010. Rapid assays for fluoroquinolone resistance in *Mycobacterium tuberculosis:* a systematic review and meta-analysis. *J Antimicrob Chemother* 65:1551–1561.

66. Folkvardsen DB, Svensson E, Thomsen VO, Rasmussen EM, Bang D, Werngren J, Hoffner S, Hillemann D, Rigouts L. 2013. Can molecular methods detect 1% isoniazid resistance in *Mycobacterium tuberculosis*? *J Clin Microbiol* 51:4220–4222.

67. Heep M, Brandstatter B, Rieger U, Lehn N, Richter E, Rusch-Gerdes S, Niemann S. 2001. Frequency of rpoB mutations inside and outside the cluster I region in rifampin-resistant clinical *Mycobacterium tuberculosis* isolates. *J Clin Microbiol* 39:107–110.

68. Hofmann-Thiel S, van Ingen J, Feldmann K, Turaev L, Uzakova GT, Murmusaeva G, van Soolingen D, Hoffmann H. 2009. Mechanisms of heteroresistance to isoniazid and rifampin of *Mycobacterium tuberculosis* in Tashkent, Uzbekistan. *Eur Respir J* 33:368–374.

69. de Oliveira MM, da Silva Rocha A, Cardoso Oelemann M, Gomes HM, Fonseca L, Werneck-Barreto AM, Valim AM, Rossetti ML, Rossau R, Mijs W, Vanderborght B, Suffys P. 2003. Rapid detection of resistance against rifampicin in isolates of *Mycobacterium tuberculosis* from Brazilian patients using a reverse-phase hybridization assay. *J Microbiol Methods* 53:335–342.

70. Cooksey RC, Morlock GP, Holloway BP, Mazurek GH, Abaddi S, Jackson LK, Buzard GS, Crawford JT. 1998. Comparison of two nonradioactive, single-strand conformation polymorphism electrophoretic methods for identification of rpoB mutations in rifampin-resistant isolates of *Mycobacterium tuberculosis*. *Mol Diagn* 3:73–79.

71. Chakravorty S, Aladegbami B, Thoms K, Lee JS, Lee EG, Rajan V, Cho EJ, Kim H, Kwak H, Kurepina N, Cho SN, Kreiswirth B, Via LE, Barry CE 3rd, Alland D. 2011. Rapid detection of fluoroquinolone-resistant and heteroresistant *Mycobacterium tuberculosis* by use of sloppy molecular beacons and dual melting-temperature codes in a real-time PCR assay. *J Clin Microbiol* 49:932–940.

72. Pholwat S, Stroup S, Foongladda S, Houpt E. 2013. Digital PCR to detect and quantify heteroresistance in drug resistant *Mycobacterium tuberculosis*. *PLoS One* 8:e57238.

73. Blaas SH, Mutterlein R, Weig J, Neher A, Salzberger B, Lehn N, Naumann L. 2008. Extensively drug resistant tuberculosis in a high income country: a report of four unrelated cases. *BMC Infect Dis* 8:60.

74. Tolani MP, D'souza DTB, Mistry NF. 2012. Drug resistance mutations and heteroresistance detected using the GenoType MTBDRPlus assay and their implication for treatment outcomes in patients from Mumbai, India. *BMC Infect Dis* 12:9.

75. Adjers-Koskela K, Katila ML. 2003. Susceptibility testing with the manual mycobacteria growth indicator tube (MGIT) and the MGIT 960 system provides rapid and reliable verification of multidrug-resistant tuberculosis. *J Clin Microbiol* 41:1235–1239.

76. Rinder H, Mieskes KT, Loscher T. 2001. Heteroresistance in *Mycobacterium tuberculosis*. *Int J Tuberc Lung Dis* 5:339–345.

77. Nikolayevskyy V, Balabanova Y, Timak T, Malomanova N, Fedorin I, Drobniewski F. 2009. Performance of the Genotype MTBDRPlus assay in the diagnosis of tuberculosis and drug resistance in Samara, Russian Federation. *BMC Clin Pathol* 9:2.

78. Streicher EM, Bergval I, Dheda K, Bottger EC, Gey van Pittius NC, Bosman M, Coetzee G, Anthony RM, van Helden PD, Victor TC, Warren RM. 2012. *Mycobacterium tuberculosis* population structure determines the outcome of genetics-based second-line drug resistance testing. *Antimicrob Agents Chemother* 56:2420–2427.

79. Zhang X, Zhao B, Liu L, Zhu Y, Zhao Y, Jin Q. 2012. Subpopulation analysis of heteroresistance to fluoroquinolone in *Mycobacterium tuberculosis* isolates from Beijing, China. *J Clin Microbiol* 50:1471–1474.

80. Blakemore R, Story E, Helb D, Kop J, Banada P, Owens MR, Chakravorty S, Jones M, Alland D. 2010. Evaluation of the analytical performance of the Xpert MTB/RIF assay. *J Clin Microbiol* 48:2495–2501.

81. Gori A, Bandera A, Marchetti G, Degli Esposti A, Catozzi L, Nardi GP, Gazzola L, Ferrario G, van Embden JD, van Soolingen D, Moroni M, Franzetti F. 2005. Spoligotyping and *Mycobacterium tuberculosis*. *Emerg Infect Dis* 11:1242–1248.

82. Woodford N, Ellington MJ. 2007. The emergence of antibiotic resistance by mutation. *Clin Microbiol Infect* 13:5–18.

83. Chen TC, Lu PL, Lin CY, Lin WR, Chen YH. 2011. Fluoroquinolones are associated with delayed treatment and resistance in tuberculosis: a systematic review and meta-analysis. *Int J Infect Dis* 15:e211–e216.

84. Cox HS, Sibilia K, Feuerriegel S, Kalon S, Polonsky J, Khamraev AK, Rusch-Gerdes S, Mills C, Niemann S. 2008. Emergence of extensive drug resistance during treatment for multidrug-resistant tuberculosis. *N Engl J Med* 359:2398–2400.

85. Calver AD, Falmer AA, Murray M, Strauss OJ, Streicher EM, Hanekom M, Liversage T, Masibi M, van Helden PD, Warren RM, Victor TC. 2010. Emergence of increased resistance and extensively drug-resistant tuberculosis despite treatment adherence, South Africa. *Emerg Infect Dis* 16:264–271.

86. Cullen MM, Sam NE, Kanduma EG, McHugh TD, Gillespie SH. 2006. Direct detection of heteroresistance in *Mycobacterium tuberculosis* using molecular techniques. *J Med Microbiol* 55:1157–1158.

87. Chigutsa E, Meredith S, Wiesner L, Padayatchi N, Harding J, Moodley P, Mac Kenzie WR, Weiner M, McIlleron H, Kirkpatrick CM. 2012. Population pharmacokinetics and pharmacodynamics of ofloxacin in South African patients with multidrug-resistant tuberculosis. *Antimicrob Agents Chemother* 56:3857–3863.

88. Ginsburg AS, Hooper N, Parrish N, Dooley KE, Dorman SE, Booth J, Diener-West M, Merz WG, Bishai WR, Sterling TR. 2003. Fluoroquinolone resistance in patients with newly diagnosed tuberculosis. *Clin Infect Dis* 37:1448–1452.

89. Wang JY, Hsueh PR, Jan IS, Lee LN, Liaw YS, Yang PC, Luh KT. 2006. Empirical treatment with a fluoroquinolone delays the treatment for tuberculosis and is associated with a poor prognosis in endemic areas. *Thorax* 61:903–908.

90. Piddock LJ. 2006. Clinically relevant chromosomally encoded multidrug resistance efflux pumps in bacteria. *Clin Microbiol Rev* 19:382–402.

91. Poole K. 2005. Efflux-mediated antimicrobial resistance. *J Antimicrob Chemother* 56:20–51.

92. Vilcheze C, Wang F, Arai M, Hazbon MH, Colangeli R, Kremer L, Weisbrod TR, Alland D, Sacchettini JC, Jacobs WR Jr. 2006. Transfer of a point mutation in

Mycobacterium tuberculosis inhA resolves the target of isoniazid. *Nat Med* **12**:1027–1029.

93. Meier I, Wray LV, Hillen W. 1988. Differential regulation of the Tn10-encoded tetracycline resistance genes tetA and tetR by the tandem tet operators O1 and O2. *Embo J* **7**:567–572.

94. Kern WV, Oethinger M, Jellen-Ritter AS, Levy SB. 2000. Non-target gene mutations in the development of fluoroquinolone resistance in *Escherichia coli. Antimicrob Agents Chemother* **44**:814–820.

95. Abouzeed YM, Baucheron B, Cloeckaert A. 2008. ramR mutations involved in efflux-mediated multidrug resistance in *Salmonella enterica* serovar *Typhimurium. Antimicrob Agents Chemother* **52**:2428–2434.

96. Fluman N, Bibi E. 2009. Bacterial multidrug transport through the lens of the major facilitator superfamily. *Biochim Biophys Acta* **1794**:738–747.

97. Godreuil S, Galimand M, Gerbaud G, Jacquet C, Courvalin P. 2003. Efflux pump Lde is associated with fluoroquinolone resistance in *Listeria monocytogenes. Antimicrob Agents Chemother* **47**:704–708.

98. Van Bambeke F, Michot JM, Van Eldere J, Tulkens PM. 2005. Quinolones in 2005: an update. *Clin Microbiol Infect* **11**:256–280.

99. Piddock LJ, Jin YF, Griggs DJ. 2001. Effect of hydrophobicity and molecular mass on the accumulation of fluoroquinolones by *Staphylococcus aureus. J Antimicrob Chemother* **47**:261–270.

100. Truong-Bolduc QC, Dunman PM, Strahilevitz J, Projan SJ, Hooper DC. 2005. MgrA is a multiple regulator of two new efflux pumps in *Staphylococcus aureus. J Bacteriol* **187**:2395–2405.

101. Truong-Bolduc QC, Hsing LC, Villet R, Bolduc GR, Estabrooks Z, Taguezem GF, Hooper DC. 2012. Reduced aeration affects the expression of the NorB efflux pump of *Staphylococcus aureus* by posttranslational modification of MgrA. *J Bacteriol* **194**:1823–1834.

102. El Garch F, Lismond A, Piddock LJ, Courvalin P, Tulkens PM, Van Bambeke F. 2010. Fluoroquinolones induce the expression of patA and patB, which encode ABC efflux pumps in *Streptococcus pneumoniae. J Antimicrob Chemother* **65**:2076–2082.

103. Poissy J, Aubry A, Fernandez C, Lott MC, Chauffour A, Jarlier V, Farinotti R, Veziris N. 2010. Should moxifloxacin be used for the treatment of extensively drug-resistant tuberculosis? An answer from a murine model. *Antimicrob Agents Chemother* **54**:4765–4771.

104. Schmalstieg AM, Srivastava S, Belkaya S, Deshpande D, Meek C, Leff R, van Oers NS, Gumbo T. 2012. The antibiotic resistance arrow of time: efflux pump induction is a general first step in the evolution of mycobacterial drug resistance. *Antimicrob Agents Chemother* **56**:4806–4815.

105. Adams KN, Takaki K, Connolly LE, Wiedenhoft H, Winglee K, Humbert O, Edelstein PH, Cosma CL, Ramakrishnan L. 2011. Drug tolerance in replicating mycobacteria mediated by a macrophage-induced efflux mechanism. *Cell* **145**:39–53.

106. Louw GE, Warren RM, Gey van Pittius NC, Leon R, Jimenez A, Pando RH, McEvoy CR, Grobbelaar M, Murray M, van Helden PD, Victor TC. 2011. Rifampicin reduces susceptibility to ofloxacin in rifampicin resistant *Mycobacterium tuberculosis* through efflux. *Am J Respir Crit Care Med* **184**:269–276.

107. Jumbe NL, Louie A, Miller MH, Liu W, Deziel MR, Tam VH, Bachhawat R, Drusano GL. 2006. Quinolone efflux pumps play a central role in emergence of fluoroquinolone resistance in *Streptococcus pneumoniae. Antimicrob Agents Chemother* **50**:310–317.

108. Gillespie SH, Basu S, Dickens AL, O'sullivan DM, McHugh TD. 2005. Effect of subinhibitory concentrations of ciprofloxacin on *Mycobacterium fortuitum* mutation rates. *J Antimicrob Chemother* **56**:344–348.

109. Takiff HE, Cimino M, Musso MC, Weisbrod T, Martinez R, Delgado MB, Salazar L, Bloom BR, Jacobs WR Jr. 1996. Efflux pump of the proton antiporter family confers low-level fluoroquinolone resistance in *Mycobacterium smegmatis. Proc Natl Acad Sci USA* **93**:362–366.

110. Liu J, Takiff HE, Nikaido H. 1996. Active efflux of fluoroquinolones in *Mycobacterium smegmatis* mediated by LfrA, a multidrug efflux pump. *J Bacteriol* **178**:3791–3795.

111. Li XZ, Zhang L, Nikaido H. 2004. Efflux pump-mediated intrinsic drug resistance in *Mycobacterium smegmatis. Antimicrob Agents Chemother* **48**:2415–2423.

112. Buroni S, Manina G, Guglierame P, Pasca MR, Riccardi G, De Rossi E. 2006. LfrR is a repressor that regulates expression of the efflux pump LfrA in *Mycobacterium smegmatis. Antimicrob Agents Chemother* **50**:4044–4052.

113. Bellinzoni M, Buroni S, Schaeffer F, Riccardi G, De Rossi E, Alzari PM. 2009. Structural plasticity and distinct drug-binding modes of LfrR, a mycobacterial efflux pump regulator. *J Bacteriol* **191**:7531–7537.

114. Sander P, De Rossi E, Boddinghaus B, Cantoni R, Branzoni M, Bottger EC, Takiff H, Rodriquez R, Lopez G, Riccardi G. 2000. Contribution of the multidrug efflux pump LfrA to innate mycobacterial drug resistance. *FEMS Microbiol Lett* **193**:19–23.

115. Esteban J, Martin-de-Hijas NZ, Ortiz A, Kinnari TJ, Bodas Sanchez A, Gadea I, Fernandez-Roblas R. 2009. Detection of lfrA and tap efflux pump genes among clinical isolates of non-pigmented rapidly growing mycobacteria. *Int J Antimicrob Agents* **34**:454–456.

116. Jiang X, Zhang W, Zhang Y, Gao F, Lu C, Zhang X, Wang H. 2008. Assessment of efflux pump gene expression in a clinical isolate *Mycobacterium tuberculosis* by real-time reverse transcription PCR. *Microb Drug Resist* **14**:7–11.

117. Siddiqi N, Das R, Pathak N, Banerjee S, Ahmed N, Katoch VM, Hasnain SE. 2004. *Mycobacterium tuberculosis* isolate with a distinct genomic identity over-expresses a tap-like efflux pump. *Infection* **32**:109–111.

118. Ramon-Garcia S, Mick V, Dainese E, Martin C, Thompson CJ, De Rossi E, Manganelli R, Ainsa JA. 2012. Functional and genetic characterization of the tap efflux pump in *Mycobacterium bovis* BCG. *Antimicrob Agents Chemother* **56**:2074–2083.

119. da Silva PE, Von Groll A, Martin A, Palomino JC. 2011. Efflux as a mechanism for drug resistance in *Mycobacterium tuberculosis*. *FEMS Immunol Med Microbiol* 63:1–9.

120. De Rossi E, Ainsa JA, Riccardi G. 2006. Role of mycobacterial efflux transporters in drug resistance: an unresolved question. *FEMS Microbiol Rev* 30:36–52.

121. De Rossi E, Arrigo P, Bellinzoni M, Silva PA, Martin C, Ainsa JA, Guglierame P, Riccardi G. 2002. The multidrug transporters belonging to major facilitator superfamily in *Mycobacterium tuberculosis*. *Mol Med* 8:714–724.

122. Pasca MR, Guglierame P, Arcesi F, Bellinzoni M, De Rossi E, Riccardi G. 2004. Rv2686c-Rv2687c-Rv2688c, an ABC fluoroquinolone efflux pump in *Mycobacterium tuberculosis*. *Antimicrob Agents Chemother* 48:3175–3178.

123. Choudhuri BS, Bhakta S, Barik R, Basu J, Kundu M, Chakrabarti P. 2002. Overexpression and functional characterization of an ABC (ATP-binding cassette) transporter encoded by the genes drrA and drrB of *Mycobacterium tuberculosis*. *Biochem J* 367:279–285.

124. Chakraborti PK, Bhatt K, Banerjee SK, Misra P. 1999. Role of an ABC importer in mycobacterial drug resistance. *Biosci Rep* 19:293–300.

125. Banerjee SK, Bhatt K, Misra P, Chakraborti PK. 2000. Involvement of a natural transport system in the process of efflux-mediated drug resistance in *Mycobacterium smegmatis*. *Mol Gen Genet* 262:949–956.

126. Bhatt K, Banerjee SK, Chakraborti PK. 2000. Evidence that phosphate specific transporter is amplified in a fluoroquinolone resistant *Mycobacterium smegmatis*. *Eur J Biochem* 267:4028–4032.

127. Escribano I, Rodriguez JC, Llorca B, Garcia-Pachon E, Ruiz M, Royo G. 2007. Importance of the efflux pump systems in the resistance of *Mycobacterium tuberculosis* to fluoroquinolones and linezolid. *Chemotherapy* 53:397–401.

128. Sarathy J, Dartois V, Dick T, Gengenbacher M. 2013. Reduced drug uptake in phenotypically resistant nutrient-starved nonreplicating *Mycobacterium tuberculosis*. *Antimicrob Agents Chemother* 57:1648–1653.

129. Danilchanka O, Pavlenok M, Niederweis M. 2008. Role of porins for uptake of antibiotics by *Mycobacterium smegmatis*. *Antimicrob Agents Chemother* 52:3127–3134.

130. Siroy A, Mailaender C, Harder D, Koerber S, Wolschendorf F, Danilchanka O, Wang Y, Heinz C, Niederweis M. 2008. Rv1698 of *Mycobacterium tuberculosis* represents a new class of channel-forming outer membrane proteins. *J Biol Chem* 283:17827–17837.

131. Brennan PJ, Nikaido H. 1995. The envelope of mycobacteria. *Annu Rev Biochem* 64:29–63.

132. Montero C, Mateu G, Rodriguez R, Takiff H. 2001. Intrinsic resistance of *Mycobacterium smegmatis* to fluoroquinolones may be influenced by new pentapeptide protein MfpA. *Antimicrob Agents Chemother* 45:3387–3392.

133. Jacoby GA, Hooper DC. 2013. Phylogenetic analysis of chromosomally determined qnr and related proteins. *Antimicrob Agents Chemother* 57:1930–1934.

134. Bateman A, Murzin AG, Teichmann SA. 1998. Structure and distribution of pentapeptide repeats in bacteria. *Protein Sci* 7:1477–1480.

135. Vetting MW, Hegde SS, Fajardo JE, Fiser A, Roderick SL, Takiff HE, Blanchard JS. 2006. Pentapeptide repeat proteins. *Biochemistry* 45:1–10.

136. Buchko GW, Ni S, Robinson H, Welsh EA, Pakrasi HB, Kennedy MA. 2006. Characterization of two potentially universal turn motifs that shape the repeated five-residues fold–crystal structure of a lumenal pentapeptide repeat protein from Cyanothece 51142. *Protein Sci* 15:2579–2595.

137. Poirel L, Cattoir V, Nordmann P. 2012. Plasmid-mediated quinolone resistance; interactions between human, animal, and environmental ecologies. *Front Microbiol* 3:24.

138. Vetting MW, Hegde SS, Blanchard JS. 2009. Crystallization of a pentapeptide-repeat protein by reductive cyclic pentylation of free amines with glutaraldehyde. *Acta Crystallogr D Biol Crystallogr* 65:462–469.

139. Rodriguez-Martinez JM, Velasco C, Garcia I, Cano ME, Martinez-Martinez L, Pascual A. 2007. Mutant prevention concentrations of fluoroquinolones for *Enterobacteriaceae* expressing the plasmid-carried quinolone resistance determinant qnrA1. *Antimicrob Agents Chemother* 51:2236–2239.

140. Buchko GW, Robinson H, Pakrasi HB, Kennedy MA. 2008. Insights into the structural variation between pentapeptide repeat proteins–crystal structure of Rfr23 from Cyanothece 51142. *J Struct Biol* 162:184–192.

141. Hegde SS, Vetting MW, Roderick SL, Mitchenall LA, Maxwell A, Takiff HE, Blanchard JS. 2005. A fluoroquinolone resistance protein from *Mycobacterium tuberculosis* that mimics DNA. *Science* 308:1480–1483.

142. Tran JH, Jacoby GA, Hooper DC. 2005. Interaction of the plasmid-encoded quinolone resistance protein Qnr with *Escherichia coli* DNA gyrase. *Antimicrob Agents Chemother* 49:118–125.

143. Vetting MW, Hegde SS, Wang M, Jacoby GA, Hooper DC, Blanchard JS. 2011. Structure of QnrB1, a plasmid-mediated fluoroquinolone resistance factor. *J Biol Chem* 286:25265–25273.

144. Tao J, Han J, Wu H, Hu X, Deng J, Fleming J, Maxwell A, Bi L, Mi K. 2013. *Mycobacterium fluoroquinolone* resistance protein B, a novel small GTPase, is involved in the regulation of DNA gyrase and drug resistance. *Nucleic Acids Res* 41:2370–2381.

145. Yang K, Han L, He J, Wang L, Vining LC. 2001. A repressor-response regulator gene pair controlling jadomycin B production in *Streptomyces venezuelae* ISP5230. *Gene* 279:165–173.

146. Miertzschke M, Koerner C, Vetter IR, Keilberg D, Hot E, Leonardy S, Sogaard-Andersen L, Wittinghofer A. 2011. Structural analysis of the Ras-like G protein MglA and its cognate GAP MglB and implications for bacterial polarity. *Embo J* 30:4185–4197.

147. Garrido MC, Herrero M, Kolter R, Moreno F. 1988. The export of the DNA replication inhibitor Microcin B17 provides immunity for the host cell. *Embo J* 7: 1853–1862.

148. Vetting MW, Hegde SS, Zhang Y, Blanchard JS. 2011. Pentapeptide-repeat proteins that act as topoisomerase poison resistance factors have a common dimer interface. *Acta Crystallogr Sect F Struct Biol Cryst Commun* 67:296–302.

149. Hashimi SM, Wall MK, Smith AB, Maxwell A, Birch RG. 2007. The phytotoxin albicidin is a novel inhibitor of DNA gyrase. *Antimicrob Agents Chemother* 51: 181–187.

150. World Health Organization. 2011. *Global tuberculosis control: WHO report 2011*. World Health Organization, Geneva.

151. Zhou J, Dong Y, Zhao X, Lee S, Amin A, Ramaswamy S, Domagala J, Musser JM, Drlica K. 2000. Selection of antibiotic-resistant bacterial mutants: allelic diversity among fluoroquinolone-resistant mutations. *J Infect Dis* 182:517–525.

152. Kam KM, Yip CW, Cheung TL, Tang HS, Leung OC, Chan MY. 2006. Stepwise decrease in moxifloxacin susceptibility amongst clinical isolates of multidrug-resistant *Mycobacterium tuberculosis:* correlation with ofloxacin susceptibility. *Microb Drug Resist* 12:7–11.

153. Wikipedia. *Moxifloxacin.* http://en.wikipedia.org/wiki /Moxifloxacin.

154. Gomez C, Ponien P, Serradji N, Lamouri A, Pantel A, Capton E, Jarlier V, Anquetin G, Aubry A. 2013. Synthesis of gatifloxacin derivatives and their biological activities against *Mycobacterium leprae* and *Mycobacterium tuberculosis. Bioorg Med Chem* 21: 948–956.

155. Sreevatsan S, Pan X, Stockbauer KE, Connell ND, Kreiswirth BN, Whittam TS, Musser JM. 1997. Restricted structural gene polymorphism in the *Mycobacterium tuberculosis* complex indicates evolutionarily recent global dissemination. *Proc Natl Acad Sci USA* 94:9869–9874.

156. Komatsu M, Takano H, Hiratsuka T, Ishigaki Y, Shimada K, Beppu T, Ueda K. 2006. Proteins encoded by the conservon of *Streptomyces coelicolor* A3(2) comprise a membrane-associated heterocomplex that resembles eukaryotic G protein-coupled regulatory system. *Mol Microbiol* 62:1534–1546.

Molecular Genetics of Mycobacteria, 2nd Edition
Edited by Graham F. Hatfull and William R. Jacobs, Jr.
© 2014 American Society for Microbiology, Washington, DC
doi:10.1128/microbiolspec.MGM2-0023-2013

Ying Zhang,[1,2] Wanliang Shi,[1]
Wenhong Zhang,[2] and Denis Mitchison[3]

Mechanisms of Pyrazinamide Action and Resistance

24

THE HISTORY: THE UNUSUAL DISCOVERY AND THE ROLLER COASTER OF PZA

Pyrazinamide (PZA), a nicotinamide analogue (Fig. 1), was first chemically synthesized in 1936 (1), but its antituberculosis (anti-TB) potential was not recognized until 1952 (2). Its discovery as a TB drug was based on a serendipitous observation that nicotinamide had certain activity against mycobacteria in animal models (3). Subsequent synthesis of nicotinamide analogues and direct testing in the mouse model of TB infection without *in vitro* testing led to the identification of PZA as an active agent (4, 5). Before the 1970s, PZA was mainly used as a second-line TB drug for the treatment of drug-resistant TB or in treatment of relapsed TB because of the hepatic toxicity caused by a higher PZA dosage (3.0 g daily) and longer treatment used in earlier clinical studies. However, largely encouraged by the impressive mouse studies by McDermott and colleagues that demonstrated high sterilizing activity of PZA in combination with isoniazid (INH) (6), the British Medical Research Council conducted clinical trials in East Africa with lower PZA doses (1.5 to 2.0 g daily), which are not significantly hepatotoxic. PZA

was found to be almost as effective as rifampin (RIF) as a sterilizing drug, as judged by more frequent sputum conversion at 2 months and by the relapse rates. Subsequent clinical studies showed that the effects of RIF and PZA were synergistic. These studies showed that treatment could be shortened from 12 months or more to 9 months if either RIF or PZA was added to the regimen, and to 6 months if both were included (7). PZA has since been used as a first-line agent for treatment of drug-susceptible TB with RIF, INH, and ethambutol, which is currently the best TB therapy. PZA is also an integral component of treatment regimens for multidrug-resistant (MDR) TB (8) and of any new regimens in conjunction with new TB drug candidates in clinical trials (9).

IMPORTANCE OF PZA IN SHORTENING TB THERAPY

PZA is a critical frontline TB drug that plays a unique role in shortening the treatment period from 9 to 12 months to 6 months (7, 10, 11). The inclusion of PZA with INH and RIF forms the basis for our current

[1]Department of Molecular Microbiology and Immunology, Bloomberg School of Public Health, Johns Hopkins University, Baltimore, MD 21205; [2]Department of Infectious Diseases, Huashan Hospital, Fudan University, Shanghai, China; [3]Centre for Infection, St. George's, University of London, Cranmer Terrace, London SW17 0RE, United Kingdom.

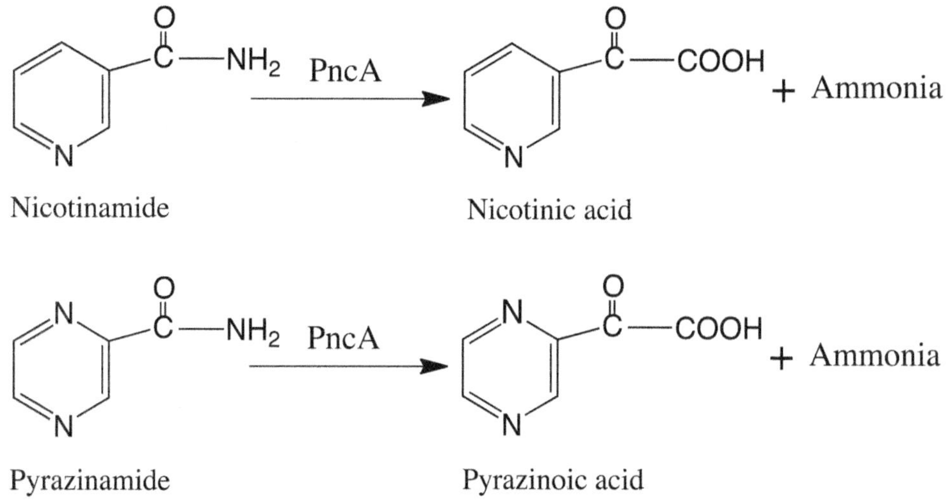

Figure 1 Conversion of nicotinamide and PZA to nicotinic acid and POA, respectively, by the enzyme PZase/nicotinamidase (PncA) encoded by the *pncA* gene. doi:10.1128/microbiolspec.MGM2-0023-2013.f1

short-course chemotherapy based on the work by McDermott and colleagues in a mouse model of TB infection (6, 12). This powerful sterilizing activity is due to PZA's ability to kill a population of *Mycobacterium tuberculosis* persisters that are not killed by other drugs (13). PZA is used during the first 2-month intensive phase of the 6-month therapy because giving PZA for longer than 2 months does not appear to add additional benefit (7). This is presumably because inflammation leading to an acid environment in the lesions decreases after 2 months. More recent efforts to find optimal drug combinations with new drug candidates for shortening TB treatment in the mouse model suggest that PZA is the only drug that cannot be replaced without compromising treatment efficacy (14–16). In view of its unique and indispensable sterilizing activity among all TB drugs, including new drug candidates in clinical trials, there is recent unprecedented interest in PZA as seen by three workshops on PZA in about a year (http://www.cdc.gov/tb/publications/newsletters/notes/TBN_4_12/labbranch_update.htm).

MECHANISMS OF PZA ACTION

PZA is a mysterious, unconventional, and paradoxical drug. The mode of action of PZA is unusual and has puzzled investigators ever since its clinical use began in 1952. The main reason for this is that PZA is very different from common antibiotics, which are primarily active against growing bacteria and have no or little activity against nongrowing persisters. However, PZA is exactly the opposite of common antibiotics because it

has no or little activity against growing tubercle bacilli and is primarily active against nongrowing persisters (17, 18).

Despite its powerful *in vivo* sterilizing activity, demonstrated both in the animal model (6, 12) and in humans in shortening TB chemotherapy (13), PZA has no activity *in vitro* under normal culture conditions at neutral pH (19) but is active only at an acid pH (e.g., pH 5.5) *in vitro* (20). Furthermore, unlike other TB drugs, the activity of PZA increases with decreasing metabolic activity. PZA only kills *M. tuberculosis* slowly *in vitro* at acid pH (21). *In vivo*, PZA has high sterilizing activity against persisters in an acidic environment that is present during inflammation (22, 23), which is responsible for its ability to shorten TB therapy. Despite its use for the past 70 years, and despite its importance as an irreplaceable frontline drug in shortening TB therapy, the mode of action of PZA is the least understood of all TB drugs (24). However, new progress has been made in our understanding of PZA in recent years. Much of the historical and clinical aspects of PZA were covered in a previous review article published in 2003 (24) and will not be discussed here. We will mainly focus on new developments in PZA since then while including the basic important information.

PZA is a prodrug that is converted to the active form pyrazinoic acid (POA) by pyrazinamidase (PZase)/nicotinamidase, encoded by the *pncA* gene in *M. tuberculosis* (25) (Fig. 1). The purified recombinant *M. tuberculosis* PncA is a Mn^{2+}- and Fe^{2+}-containing enzyme that is a monomer (26). The mechanism of PZA conversion to POA may be similar to that of the nitrilase superfamily, in which nucleophilic attack by active site cysteine generates a tetrahedral intermediate that collapses with the loss of ammonia and subsequent hydrolysis of the thioester bond by water (27, 28).

Based on various studies (17, 25, 29–31), the following model for the mode of action of PZA was proposed (Fig. 2) (24, 30). PZA enters bacilli through passive diffusion and is converted into POA (a moderately strong acid with pK_a of 2.9) by the cytoplasmic PZase encoded by *pncA*. POA then exits the cell via passive diffusion and a deficient efflux mechanism in *M. tuberculosis* (29). Once POA is outside the cell, if the extracellular pH is acidic (e.g., pH 5.5), a small proportion of POA will become uncharged protonated acid HPOA, which readily permeates through the membrane. The acid-facilitated POA influx can overcome the weak deficient POA efflux, which causes accumulation of POA in *M. tuberculosis* cells at acid pH over time (29). The HPOA brings protons into the cell, and this can eventually cause cytoplasmic acidification such that vital enzymes can be inhibited. In addition, POA can de-energize the membrane by collapsing proton motive force and affect membrane transport, inhibiting protein and RNA synthesis (30).

At neutral or alkaline pH, little POA is found in *M. tuberculosis* (29) because over 99.9% of POA is in charged anion form (17), which does not get into cells easily and remains outside the cells (29). This observation explains why PZA is active at acidic pH but not at neutral pH (20) and also explains the correlation between the MIC of PZA and acidic pH values, which can be expressed by the Henderson-Hasselbalch equation (17). It is worth noting that acid pH not only allows POA to re-enter and accumulate in the bacilli (29), but also decreases the membrane potential and inhibits the growth and metabolism required for drug action. The unique activity of PZA against *M. tuberculosis* appears to be due to a deficient POA efflux

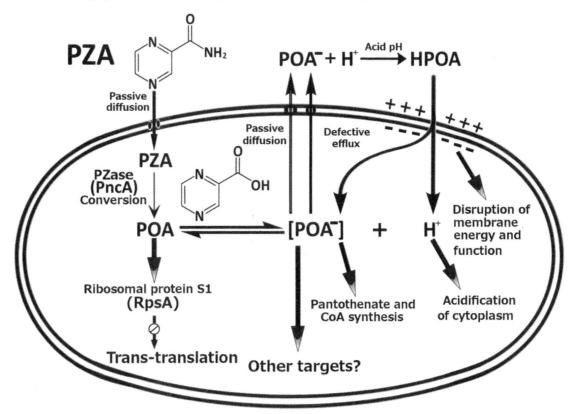

Figure 2 Mode of action of PZA. PZA enters tubercle bacilli by passive diffusion, where it is converted to POA by PZase/nicotinamidase encoded by the *pncA* gene. POA then reaches the cell surface through passive diffusion and a weak (deficient) efflux mechanism. At acid pH, the protonated POA (HPOA) enters the cell in a pH-dependent manner by passive diffusion and then accumulates to high levels intracellularly and kills by multiple mechanisms including disruption of membrane energy production, inhibition of trans-translation, possibly inhibition of pantothenate and CoA biosynthesis, and other as yet unidentified mechanisms. doi:10.1128/microbiolspec.MGM2-0023-2013.f2

mechanism (29) that is unable to counteract the acid-facilitated POA influx, which can cause increased accumulation of POA and eventual acidification of the cytoplasm, de-energized membrane (30), inhibition of various targets (see below on mechanisms of action), and cell death, especially in nongrowing persisters with low metabolism at acid pH.

Various conclusions can be derived from these hypotheses concerning the activity of PZA. Activity of PZA is strongly related to the pH of the microenvironment of the bacilli, with activity increasing with acidity (17, 20). Because diffusion of POA into the cell occurs passively but its removal requires energy to run the efflux pump (29, 30), bactericidal activity is greatest when bacterial energy sources are at their lowest. This conclusion was first demonstrated by the finding that old cultures were more susceptible to PZA than young, actively growing cultures (17). In addition, the above model predicted that energy inhibitors would enhance PZA activity, which was subsequently shown to be the case (30) (see below). The specificity of this PZA action on persisters was then demonstrated using the Hu/Coates models of persisters (18) and with work that showed increased bactericidal activity when low metabolic activity was produced by energy inhibitors (30, 32), by anaerobic conditions (33), and when incubation temperatures were reduced from 37°C to 15–25°C (34).

PZA Activity in Persister Models

The Hu/Coates models explore the action of drugs on persisters and therefore substantiate the hypothesis that PZA has unusual activity against persister subpopulations. The results obtained with these modes are therefore crucial to the overall thesis concerning the bacillary populations against which PZA is most effective. In model 1, cultures of M. tuberculosis incubated without shaking for up to 100 days were sampled, PZA was added, and bactericidal activity was measured. As the duration of incubation and starvation increased from 4 days (log phase) to 30 days and then 100 days, the bactericidal activity of PZA increased. In model 2, the cultures were sampled immediately after selection of the RIF-tolerant population when regrowth was occurring upon subculture into fresh RIF-free medium and when bacterial metabolism was high; PZA had little bactericidal activity against these actively growing bacilli. In model 3, the cultures were sampled at 3 days after inoculation in RIF-containing liquid medium, when growth and metabolism of the subpopulation would be expected to be minimal, and PZA was highly bactericidal against this population. The action of PZA

in these three models demonstrates clearly that PZA is most bactericidal when cultures of M. tuberculosis are the most static.

We found that starvation decreased membrane potential in old bacilli and enhanced PZA activity (35). In addition, it is worth noting that starved M. tuberculosis had increased expression of pncA (36), which could increase the PncA enzyme levels needed for enhanced conversion of PZA to the active form POA, and may thus contribute to increased killing of tubercle bacilli by PZA under starvation conditions.

Energy Inhibitors Enhance PZA Activity

Since PZA depletes membrane potential in M. tuberculosis (30), we reasoned that energy inhibitors could enhance PZA activity. Indeed, we found that the activity of PZA was significantly enhanced by energy inhibitors such as DCCD (F_1F_0-ATP synthase inhibitor), rotenone (NADH dehydrogenase I-complex I inhibitor), and azide (cytochrome c oxidase inhibitor) (30), and also by carbonyl cyanide m-chlorophenylhydrazone, dinitrophenol (DNP), valinomycin, and cyanide (32). The subsequent observation in 2005 that bedaquiline (or TMC207, a diarylquinoline), an inhibitor of F_1F_0 ATP synthase, could synergize with PZA activity (37)—a finding just like DCCD synergy with PZA shown earlier in 2003 (30)—provides further support for the model of PZA. These energy inhibitors deplete membrane energy, which renders tubercle bacilli more susceptible to the energy-depleting action of POA. This effect of energy inhibitors on enhancing PZA activity is specific to PZA, as they did not enhance the activity of other control drugs such as INH or RIF (32).

Mutations in Energy Production and NAD Pathways and Ion Homeostasis Cause Increased PZA Susceptibility

Based on the model of PZA (30), we predicted that energy production defects due to either chemical energy inhibitors or genetic mutations will lead to increased susceptibility to PZA. Indeed, this has proven to be true in the case of the bedaquiline and PZA synergy reported subsequently (37, 38). Furthermore, we found that M. tuberculosis mutants defective in energy production (made available through the Tuberculosis Animal Research and Gene Evaluation Taskforce [TARGET] at Johns Hopkins University) had higher PZA susceptibility (MIC = 10 µg/ml) than the parent strain (MIC = 50 µg/ml) (Table 1) (Y. Zhang, unpublished data). These mutants include mutations in NADH dehydrogenase subunits H and N (nuoH,

Table 1 Increased PZA susceptibility of *M. tuberculosis* mutants with a defect in energy metabolism

Strain	Gene/function	MIC (μg/ml) (pH 5.5)
CDC1551	Wild type	25–50
MT3240 mutant	*nuoH*, NADH dehydrogenase	10
MT3246 mutant	*nuoN*, NADH dehydrogenase	10
MT2968 mutant	*fdhF*, formate dehydrogenase α-subunit	10
MT1199 mutant	*narH*, nitrate reductase β-subunit	10
MT1372 mutant	*pncB1*, nicotinatephospho-ribosyltransferase	10
MT2345 mutant	*yjcE*, Na/H exchanger	10
MT1058 mutant	*kdpA*, potassium transporting ATPase	10

nuoN), nitrate reductase *narH*, formate dehydrogenase *fdhF*, *kdpA* (potassium transport ATPase) and *yjcE* (Na^+/H^+ exchanger) involved in potassium and sodium ion transport, and *pncB1* (Rv1330c) involved in NAD recycling. NADH dehydrogenase *nuoH* and *nuoN* mutants, nitrate reductase *narH* mutant, and formate dehydrogenase *fdhF* mutant, involved in energy production under anaerobic conditions, were also highly susceptible to PZA, with a 5-fold reduction in MIC from 50 μg/ml in the parent strain to 10 μg/ml in the mutants. It is of interest to note that mutation in *pncB1*, involved in NAD recycling and energy metabolism, is also more susceptible to PZA. The observation that mutations in *kdpA* and *yjcE*, involved in potassium and sodium ion transport, respectively, caused higher susceptibility to PZA suggests that potassium and sodium ion homeostasis involved in pH regulation may also be important for PZA action. However, mutations in transcription regulator *marR*, MT3006 (ATP binding protein), and MT3981 (putative ATPase) did not have a significant effect on PZA susceptibility.

The above findings are consistent with our current model of PZA (24, 30) and confirm that energy production pathways and pH homeostasis are important for PZA action. It is unlikely that all the above diverse energy production enzymes whose mutations cause increased PZA susceptibility represent targets of POA. It is more likely, as we predicted in the model of PZA, that since POA disrupts membrane energy (30), any defect in energy production pathways or ion homeostasis could potentiate PZA activity as shown previously (30, 32). These findings may have implications for developing new drugs that synergize with PZA for improved treatment of TB.

Anaerobic and Hypoxic Conditions Potentiate PZA Activity

PZA activity is significantly enhanced under hypoxic or anaerobic conditions compared with atmospheric conditions with ambient oxygen (33). Under microaerophilic or anaerobic conditions, bacteria produce less energy (ATP and membrane potential) due to less efficient nitrate or fumarate electron acceptor usage, compared with respiration with oxygen in aerobic conditions, which produces more energy using oxygen as an electron acceptor. The preferential activity of PZA against tubercle bacilli under hypoxic or anaerobic conditions is presumably due to low energy production under such conditions so that the bacilli are more prone to the energy-depleting effect of PZA. By contrast, supplementation of alternative electron donor nitrate to supply energy under anaerobic conditions antagonized PZA activity (33). While the energy inhibitors rotenone and azide enhanced PZA activity under anaerobic conditions (33), DCCD, which enhanced PZA activity under aerobic ambient oxygen conditions (30), failed to do so under anaerobic conditions (33). It is worth noting that PZA acts beyond a general weak acid effect, since although energy inhibitors such as DCCD, rotenone, and azide could increase the activity of both PZA and other weak acids, weak acids have no activity under anaerobic conditions, whereas PZA had increased activity for *M. tuberculosis* (33, 39).

PZA Activity at Different Incubation Temperatures

When the incubation temperature is below 28°C, cultures of *M. tuberculosis* survive but do not multiply (34). They would be expected to have low energy requirements, decreasing as the temperature drops toward 8°C when active metabolism is minimal. When incubation temperatures were reduced from 37°C to 25°C or 22°C, there was a considerable increase in the bactericidal activity of PZA (34).

Certain Weak Acids Enhance PZA Activity

Also predicted from our model of PZA, weak acids, which disrupt membrane energy, can enhance the activity of PZA. We have shown that weak acids including benzoic acid (2 mM), propyl hydroxybenzoic acid (1 mM), and sorbic acid (1 mM) can indeed enhance PZA activity *in vitro* (35, 39). However, the weak acid enhancement of PZA activity is mainly seen for old bacilli and is not as effective as energy inhibitors, which indicates an additive effect between weak acids and PZA rather than a synergistic effect between energy inhibitors and PZA. However, some weak acids such as lactic acid and fatty acid C10 had no effect on enhancing PZA activity (35). The common clinically used weak acids aspirin and ibuprofen enhanced the activity of PZA in the mouse model of TB infection (40).

Effects of Iron, Oxidative Stress, and DNA Damage on PZA Activity

The separation of PZA activity *in vitro* and *in vivo* prompted us to examine the effect of iron, which could potentially be elevated in local inflammatory lesions, on PZA activity *in vitro*. We found that iron enhanced PZA and POA activity against *M. tuberculosis* (41). Other metal ions such as Mg^{2+}, Ca^{2+}, and Zn^{2+} did not enhance the activity of PZA or POA (41). Iron is known to enhance oxidative stress by producing reactive oxygen, which can damage DNA and membranes, causing inhibition of cell division, a condition that allows PZA to act more effectively. Alternatively, iron has been shown to enhance the enzyme activity of PncA (26), which in turn results in increased PZA conversion to the active form POA, causing more effective killing. Sodium nitroprusside (1 mM), a reactive nitrogen nitric oxide (NO) producer, enhanced PZA activity (100 μg/ml) against old *M. tuberculosis* cultures but not young cultures (32). However, other oxidative stress agents such as menadione and hydrogen peroxide did not significantly enhance PZA activity.

UV, which is known to damage DNA and membrane, causing inhibition of cell division and depleting membrane energy, respectively, has been shown to enhance the activity of PZA against *M. tuberculosis* (32, 39). Since iron, UV, and acid stress all cause oxidative damage to DNA, it is quite likely that oxidative damage may be related to PZA action. It is possible that DNA damage leads to growth inhibition or cell stasis, which could then potentiate PZA activity. But the detailed mechanisms remain to be determined.

Overall, PZA is a peculiar TB drug whose activity not only is influenced by concentration, but is also increased by local acid pH (20, 29), by a low metabolic state of the bacilli, and by other factors such as hypoxia and iron concentration. Of particular importance, factors that reduce the metabolic activity of bacilli tend to enhance the activity of PZA; i.e., PZA activity is increased when bacterial metabolic activity is decreased.

TARGETS OF PZA

The target of PZA was suggested to be fatty acid synthase-I (Fas-I) in a study using *Mycobacterium smegmatis* and 5-chloro-PZA (5-Cl-PZA) (42). However, no mutations in Fas-I have been found in PZA-resistant *M. tuberculosis* strains. A subsequent study showed that Fas-I is the target of 5-Cl-PZA but not the target of PZA (43). In fact, recent studies have shown that 5-Cl-PZA and PZA act very differently because 5-Cl-PZA is converted by PzaA (a second PZase enzyme not related to PncA and not present in *M. tuberculosis*) to less active 5-Cl-POA (44) and is not converted by PncA, the enzyme involved in PZA conversion to POA (45). Despite some *in vitro* activity of 5-Cl-PZA, it has no activity against *M. tuberculosis* or *Mycobacterium bovis* in the mouse model (46). Overexpression of Fas-I (target of 5-Cl-PZA) and PzaA (involved in inactivating 5-Cl-PZA) caused 5-Cl-PZA resistance in *M. smegmatis*. However, overexpression of Fas-I in *M. tuberculosis* was toxic. The studies on Fas-I as a possible target of PZA in cell-free assays or in whole cells are questionable because extremely high concentrations of PZA or POA above physiological concentrations were used (47). Attempts to isolate POA-resistant mutants have failed (48), and clinical isolates of *M. tuberculosis* resistant to PZA with *pncA* mutations are still susceptible to POA (48). Our current model (Fig. 2) can best explain the various unusual features of PZA, including the requirement of acid pH for drug activity (29); the relationship between pH and PZA MIC (17); preferential activity against old nongrowing bacilli over growing bacilli (17); the unique susceptibility of PZA against *M. tuberculosis* (29); higher activity of PZA at hypoxic and anaerobic conditions than at normoxia (33); enhancement of PZA activity by energy inhibitors (30, 32), UV (32), and iron (41); and weak acid enhancement of PZA activity (32).

A new target of PZA, RpsA (ribosomal protein S1), which is involved in the process of trans-translation, was recently identified (31). Overexpression of RpsA caused resistance to PZA in *M. tuberculosis*, as seen by a 5-fold increase in the MIC of PZA (MIC = 500 μg/ml) compared with the vector control (MIC = 100 μg/ml, at pH 5.5). In addition, we found that a low-level PZA-resistant clinical strain DHM444 without *pncA* mutations (48) contained a deletion of amino acid alanine at the 438th residue (ΔA438) due to loss of a 3-nucleotide GCC near the C-terminus of the RpsA (31). Importantly, POA was found to bind to the wild-type RpsA but not the mutant RspAΔA438 from the PZA-resistant strain DHM444, or only weakly with the

RpsA from naturally PZA-resistant *M. smegmatis* or *Escherichia coli*. POA specifically inhibited the trans-translation of *M. tuberculosis* but not the canonical translation of *M. tuberculosis* or the *trans*-translation of *M. smegmatis* or *E. coli*.

Trans-translation is a process that removes toxic protein products formed under stress conditions by adding a transfer-messenger RNA (tmRNA) tag that is the protease recognition sequence, followed by toxic protein product degradation by proteases (49). Trans-translation is dispensable during active growth but becomes important for bacteria in managing stalled ribosomes or damaged mRNA and proteins under stress conditions (50, 51). It is required for stress survival and pathogenesis in some bacteria (49). More recently, we identified a new gene, *panD*, encoding aspartate decarboxylase, that is involved in PZA resistance (52) (see below). PanD is involved in synthesis of β-alanine, which is a precursor for pantothenate and CoA biosynthesis. It is likely that PanD is a target of PZA and that POA binding to PanD could inhibit synthesis of pantothenate and CoA, which may be critical for persister TB bacteria. The findings that POA inhibits the trans-translation and possibly pantothenate and CoA synthesis in *M. tuberculosis* help to explain why diverse stress conditions such as starvation, acid pH, hypoxia, and energy inhibitors and other drugs could all potentiate PZA activity (17, 30). Based on our current and previous studies, we propose a revised model of mechanisms of action of PZA that can better explain the peculiar features of this unique and paradoxical drug (Fig. 2).

MECHANISMS OF PZA RESISTANCE

Although PZA resistance in *M. tuberculosis* was shown by McDermott's group in 1967 to be related to loss of nicotinamidase and PZase (53), the mechanism of PZA resistance was not known until 1996 when a mutation in the *pncA* gene encoding nicotinamidase and PZase was demonstrated to cause PZA resistance (25).

Mutations in *pncA*

In vitro studies suggest that mutations leading to PZA resistance seem to occur frequently at a frequency of 10^{-5} (54). PZA-resistant *M. tuberculosis* strains typically lose PZase/nicotinamidase activity (53). PZase/nicotinamidase contains manganese and ferrous iron at a 1:1 ratio (26). There is a good correlation between loss of PZase activity and PZA resistance in *M. tuberculosis* (53, 55, 56). Mutations in the *pncA* gene encoding PZase/nicotinamidase are the major mechanism

of PZA resistance (25, 48). The identified *pncA* mutations are largely missense mutations causing amino acid substitutions, and in some cases nucleotide insertions or deletions, and nonsense mutations in the *pncA* structural gene or in the putative promoter region of *pncA* (e.g., at the −11 position) (48, 57). The *pncA* mutations are highly diverse and scattered along the gene (48) (Fig. 3). The diverse nature of *pncA* mutations is unique to PZA resistance and is not understood. The role of various *pncA* mutations in affecting PncA enzyme activity and contributing to PZA resistance was evaluated recently by expressing the mutant PncA enzymes and assessing their enzyme activity (58). It was found that mutations causing varying PncA enzyme activities in general correlated with the level of PZA resistance, but were not sufficient to explain a high variability of PZA resistance levels (58). The authors suggested that complementary mechanisms for PZA resistance with mutations in *pncA* might play a role (58). However, inaccurate PZA susceptibility testing may underlie some of the discrepancies between PZase activity and levels of PZA resistance. It would be of interest to validate the proposal of complementary mechanisms of PZA resistance with *pncA* mutations by transformation studies with the mutant *pncA* encoding varying enzyme activity in the background of a null-PZase *M. tuberculosis* strain.

Despite the highly diverse and scattered distribution of *pncA* mutations, there is some degree of clustering at three regions of PncA: 3–17, 61–85, and 132–142 (48) (Fig. 3). These regions happen to contain catalytic sites and metal-binding sites of the PZase enzyme (27, 59). The crystal structure of *Pyrococcus horikoshii* PncA (37% identity with *M. tuberculosis* PncA) has provided some insight into how *pncA* mutations in *M. tuberculosis* might cause PZA resistance (59). The three regions where *pncA* mutations appear to cluster correspond to three of the four loops that contribute to the scaffold of the active site. Mutations at C138, D8, K96, D49, H51, and H71 modify the active site triad and metal binding site. Residues F13, L19, H57 (position of the characteristic mutation of H57D in *M. bovis*), W68, G97, Y103, I113, A134, and H137 line up the active site, and mutations at these positions are predicted to cause loss of enzyme activity. Mutations at Q10, D12, S104, and T142 are predicted to disrupt hydrogen binding interactions between the side chain and main chain atoms. Loss of PZase activity due to mutations at other sites can be attributed to potential perturbation of the active site or disruption of the protein core. Recently, the crystal structure of *M. tuberculosis* PncA was solved and provided some insight into how very

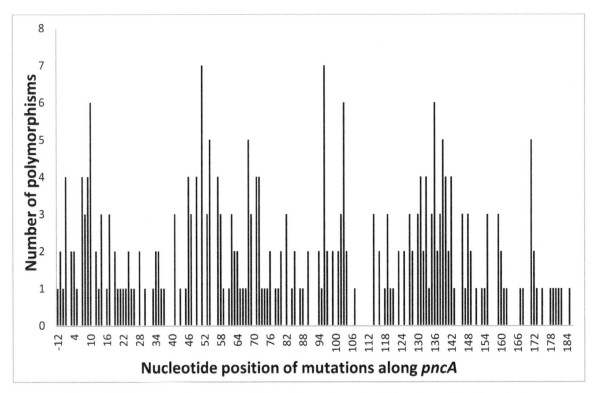

Figure 3 Diverse mutations scattered along the *pncA* gene in PZA-resistant *M. tuberculosis*. The *pncA* mutations were obtained from the Drug Resistance Mutation Database (http://www.tbdreamdb.com). doi:10.1128/microbiolspec.MGM2-0023-2013.f3

diverse mutations can contribute to PZA resistance (60). The wild-type *M. tuberculosis* PncA structure was found to be a monomer that contains manganous and ferrous ions, confirming earlier studies (26). It would be of interest to determine mutant PncA structures and see how mutant PncAs affect PZA activation and cause PZA resistance.

Most PZA-resistant *M. tuberculosis* strains (up to 99.9%) have mutations in *pncA* (24, 61, 62). However, some PZA-resistant strains without *pncA* mutations have been reported (48, 63–65). These could be due to false resistance as well as a small number of genuine PZA-resistant strains without *pncA* mutations. The average of PZA-resistant strains with *pncA* mutations from all published studies, including those that reported a low percentage of PZA-resistant strains without *pncA* mutations, is about 85%. The real percentage of PZA-resistant strains with *pncA* mutations could be higher after excluding false resistance. A few genuine PZA-resistant strains that do not have *pncA* mutations have the following phenotypes. One is PZase-negative with high levels of resistance (63, 66, 67), indicating that mutations in an undefined *pncA* regulatory gene may be involved in PZA resistance.

This is very rare. Another type of such strains has low-level PZA resistance and positive PZase activity, which are presumably due to alternative mechanisms of resistance such as mutations in the drug target *rpsA* gene (31) or other unknown genes. The clinical significance of the rare low-level PZA-resistant PZase-positive strains is unclear because they may still respond to PZA treatment *in vivo*. The virulence and fitness of PZA-resistant strains with *pncA* mutations seem to be unaltered, and such strains not only appear to be capable of causing active transmission of disease (63), but also seem to be more virulent, as shown in a more recent study (68).

Mutations in *rpsA*

Recently, it was shown that some PZA-resistant clinical isolates such as DHM444 without *pncA* mutations (48) and *Mycobacterium canettii* had mutations in the drug target RpsA (31, 69). Although it was initially thought that the C-terminus of RpsA, which harbors the alanine deletion in strain DHM444, might be the drug binding site, more recent studies suggest that mutations in the middle or near the N-terminal part of the RpsA (69) may also be involved in drug binding, because such strains have been found. For example, *M. canettii*, an *M. tuberculosis* complex organism that is naturally resistant to PZA without meaningful *pncA* mutations (62), has multiple *rpsA* mutations including Thr5Ala, Pro9Pro, Thr210Ala, and Glu457Glu (69) and R474L, R474W, and E433D (84). It is worth noting that RpsA target mutations are usually associated with a low level of PZA resistance (MIC = 200 to 300 μg/ml PZA). Future studies are needed to assess if the mutations identified in *rpsA* are responsible for low-level PZA resistance.

Mutations in *panD*

Although mutations in *pncA* and *rpsA* account for most PZA-resistant strains, some other PZA-resistant strains lack mutations in either *pncA* or *rpsA*. To identify potential new mechanisms of PZA resistance, we recently analyzed a large panel of 174 PZA-resistant mutants generated *in vitro* and found that 5 of them harbored mutations in a new gene, *panD*, encoding aspartate decarboxylase that is involved in PZA resistance (52). *panD* mutations were identified in naturally PZA-resistant *M. canettii* strains and a PZA-resistant MDR-TB clinical isolate. PanD is involved in synthesis of β-alanine, a precursor for pantothenate and CoA biosynthesis, which is known to be important for survival and pathogenesis *in vivo* (70). It is likely that PanD is a target of PZA and that POA binding to PanD could inhibit pantothenate and CoA biosynthesis, which is critical for the central metabolism required for energy production and fatty acid metabolism in *M. tuberculosis*. Studies are underway to address the role of *panD* mutations in PZA resistance and confirm PanD as a new target of PZA. However, there may still be a small number of PZA-resistant strains that do not have mutations in *pncA*, *rpsA*, or *panD*. The mechanism of PZA resistance in such strains remains to be determined.

The observation that most PZA-resistant *M. tuberculosis* strains have mutations in the *pncA* gene has implications for rapid detection of PZA resistance. The current phenotype-based PZA susceptibility testing is not routinely performed because it is not reliable due to false resistance (24, 71). Sequencing of the *pncA* gene represents a more rapid, cost-effective, and perhaps more reliable molecular test for PZA susceptibility testing and avoids the problems of phenotype-based susceptibility testing, which is slow and subject to false resistance. Rapid detection of PZA susceptibility by *pncA* sequencing should be used for guiding improved treatment of MDR- and XDR (extensively drug-resistant)-TB (72).

IMPLICATIONS OF PZA FOR DEVELOPING A NEW GENERATION OF PERSISTER DRUGS

PZA is an important frontline drug that plays a unique role in our fight against TB. The increasing emergence and outbreaks of MDR/XDR-TB call for urgent development of new drugs (73, 74). It is increasingly recognized that the new drugs should not only be active for drug-resistant TB but also, more importantly, shorten the current 6-month therapy (75–78). Developing new drugs that have activity for persister TB bacteria is critical for further shortening the current TB therapy. Although there are currently several drug candidates in clinical development (78, 79), none can replace PZA, and all the drug candidates, including the highly potent bedaquiline and PA-824 or Delamanid (OPC-67683), will have to be used together with PZA since any drug combination without PZA is invariably inferior in animal studies (14, 15, 37, 80). Since the demonstration in 2003 of PZA depleting membrane energy maintenance (30) and in 2005 of the synergy between PZA and bedaquiline, which itself inhibits ATP synthesis (37), there has been significant interest in developing new drugs targeting energy production pathways in *M. tuberculosis* for improved treatment of TB (81). It is interesting to note that PZA has recently been shown to enhance autophagy of host cells to facilitate clearance or killing of intracellular *M. tuberculosis* (82), a finding that may partly explain its high sterilizing activity *in vivo*. This finding has implications for the design of enhancers of autophagy as a novel approach to increasing PZA activity *in vivo* for improved treatment of TB.

PZA validates the principle that drugs active against nonreplicating persisters are important for improved treatment of persistent infections. PZA is a prototype model persister drug that plays an indispensable role in any new drug combination. Indeed, in a recent study on the sterilizing activities of new drug regimens in the mouse model, all regimens contained PZA (83). Improved understanding of how PZA works is important for the design of new drugs that further shorten the therapy. From the prototype persister drug PZA, one sees the future of antibiotic and cancer drug development. Drugs like PZA should be developed to target persisters and cancer stem cells (equivalent to persisters in cancer) for improved treatment of not only TB but also other persistent bacterial infections and cancers.

CONCLUSIONS

PZA is a unique anti-TB drug that plays a key role in shortening TB therapy. PZA is particularly effective in killing nonreplicating persisters that other TB drugs fail to kill, making it an essential drug for inclusion in any current or new drug combinations for treating both drug-susceptible and drug-resistant TB such as MDR-TB. PZA acts quite differently from common antibiotics by inhibiting multiple targets such as energy production, trans-translation, and perhaps pantothenate/CoA, which is required for persister survival. Resistance to PZA is mostly caused by mutations in the *pncA* gene encoding PZase, which is involved in conversion of the prodrug PZA to the active form POA. Mutations in the drug target RpsA are also found in some PZA-resistant strains. The recent finding that *panD* mutations are found in some PZA-resistant strains without *pncA* or *rpsA* mutations may suggest a third PZA resistance gene and a potential new target of PZA. Current phenotype-based PZA susceptibility testing is not reliable due to false resistance, and sequencing of the *pncA* gene represents a more rapid, cost-effective, and reliable molecular test for PZA susceptibility testing; avoids the problem of phenotype-based susceptibility testing; and should be used for guiding improved treatment of MDR/XDR-TB. Finally, the story of PZA has important implications for not only TB therapy but also chemotherapy in general (85): that is, persister drugs, by killing the nongrowing bacterial persisters, are critical for shortening therapy and reducing relapse. PZA serves as a model prototype persister drug and hopefully a tipping point that inspires new efforts at developing a new type of antibiotic or drug that targets nonreplicating persisters, i.e., bacterial persisters or cancer stem cells (the equivalent of cancer persisters), for improved treatment of not only TB but also other persistent bacterial infections such as persistent Lyme disease and urinary tract infections, biofilm infections, and even cancer.

This work was supported by NIH grants AI099512 and AI108535.

Citation. Zhang Y, Shi W, Zhang W, Mitchison D. 2014. Mechanisms of pyrazinamide action and resistance. Microbiol Spectrum 2(4):MGM2-0023-2013.

References

1. Dalmer O, Walter E, Firma E. 1936. *Merck in Darmstadt. Verfahren zur Herstellung von Abkömmlingen der Pyrazinmonocarbonsäure.* Patentiert im Deutschen Reiche vom 8. Juli 1934 ab. Germany patent 632 257 Klasse 12 p Gruppe 6 M 127990 IV a/12 p.

2. Yeager R, Munroe W, Dessau F. 1952. Pyrazinamide (Aldinamide) in the treatment of pulmonary tuberculosis. *Am Rev Tuberc* 65:523–534.

3. Chorine V. 1945. Action de l'amide nicotinique sur les bacilles du genre *Mycobacterium*. *CR Acad Sci (Paris)* 220:150–151.

4. Malone L, Schurr A, Lindh H, McKenzie D, Kiser JS, Williams JH. 1952. The effect of pyrazinamide (Aldinamide) on experimental tuberculosis in mice. *Am Rev Tuberc* 65:511–518.

5. Solotorovsky M, Gregory FJ, Ironson EJ, Bugie EJ, Oneill RC, Pfister K. 1952. Pyrazinoic acid amide: an agent active against experimental murine tuberculosis. *Proc Soc Exp Biol Med* 79:563–565.

6. McCune RM Jr, Tompsett R. 1956. Fate of *Mycobacterium tuberculosis* in mouse tissues as determined by the microbial enumeration technique. I. The persistence of drug-susceptible tubercle bacilli in the tissues despite prolonged antimicrobial therapy. *J Exp Med* 104:737–762.

7. Fox W, Ellard GA, Mitchison DA. 1999. Studies on the treatment of tuberculosis undertaken by the British Medical Research Council tuberculosis units, 1946-1986, with relevant subsequent publications. *Int J Tuberc Lung Dis* 3:S231–S279.

8. World Health Organization. 2011. *Guidelines for the Programmatic Management of Drug-Resistant Tuberculosis.* World Health Organization, Geneva, Switzerland.

9. Tasneen R, Li SY, Peloquin CA, Taylor D, Williams KN, Andries K, Mdluli KE, Nuermberger EL. 2011. Sterilizing activity of novel TMC207- and PA-824-containing regimens in a murine model of tuberculosis. *Antimicrobial Agents Chemother* 55:5485–5492.

10. British Thoracic and Tuberculosis Association. 1976. Short-course chemotherapy in pulmonary tuberclosis. *Lancet* ii:1102–1104.

11. British Thoracic Association. 1982. A controlled trial of six months chemotherapy in pulmonary tuberculosis. Second report: results during the 24 months after the end of chemotherapy. *Am Rev Respir Dis* 126:460–462.

12. McCune RM Jr, McDermott W, Tompsett R. 1956. The fate of *Mycobacterium tuberculosis* in mouse tissues as determined by the microbial enumeration technique. II. The conversion of tuberculous infection to the latent state by the administration of pyrazinamide and a companion drug. *J Exp Med* 104:763–802.

13. Mitchison DA. 1985. The action of antituberculosis drugs in short course chemotherapy. *Tubercle* 66:219–225.

14. Nuermberger E, Tyagi S, Tasneen R, Williams KN, Almeida D, Rosenthal I, Grosset JH. 2008. Powerful bactericidal and sterilizing activity of a regimen containing PA-824, moxifloxacin, and pyrazinamide in a murine model of tuberculosis. *Antimicrob Agents Chemother* 52:1522–1524.

15. Rosenthal IM, Zhang M, Williams KN, Peloquin CA, Tyagi S, Vernon AA, Bishai WR, Chaisson RE, Grosset JH, Nuermberger EL. 2007. Daily dosing of rifapentine cures tuberculosis in three months or less in the murine model. *PLoS Med* 4:e344.

16. Veziris N, Ibrahim M, Lounis N, Chauffour A, Truffot-Pernot C, Andries K, Jarlier V. 2009. A once-weekly R207910-containing regimen exceeds activity of the standard daily regimen in murine tuberculosis. *Am J Respir Crit Care Med* 179:75–79.

17. Zhang Y, Permar S, Sun Z. 2002. Conditions that may affect the results of susceptibility testing of *Mycobacterium tuberculosis* to pyrazinamide. *J Med Microbiol* **51:** 42–49.

18. Hu Y, Coates AR, Mitchison DA. 2006. Sterilising action of pyrazinamide in models of dormant and rifampicin-tolerant *Mycobacterium tuberculosis*. *Int J Tuberc Lung Dis* **10:**317–322.

19. Tarshis MS, Weed WA Jr. 1953. Lack of significant in vitro sensitivity of *Mycobacterium tuberculosis* to pyrazinamide on three different solid media. *Am Rev Tuberc* **67:**391–395.

20. McDermott W, Tompsett R. 1954. Activation of pyrazinamide and nicotinamide in acidic environment in vitro. *Am Rev Tuberc* **70:**748–754.

21. Heifets LB, Lindholm-Levy PJ. 1990. Is pyrazinamide bactericidal against *Mycobacterium tuberculosis*? *Am Rev Respir Dis* **141:**250–252.

22. McCune RM, Feldmann FM, Lambert HP, McDermott W. 1966. Microbial persistence. I. The capacity of tubercle bacilli to survive sterilization in mouse tissues. *J Exp Med* **123:**445–468.

23. McCune RM, Feldmann FM, McDermott W. 1966. Microbial persistence. II. Characteristics of the sterile state of tubercle bacilli. *J Exp Med* **123:**469–486.

24. Zhang Y, Mitchison D. 2003. The curious characteristics of pyrazinamide: a review. *Int J Tuberc Lung Dis* **7:**6–21.

25. Scorpio A, Zhang Y. 1996. Mutations in *pncA*, a gene encoding pyrazinamidase/nicotinamidase, cause resistance to the antituberculous drug pyrazinamide in tubercle bacillus. *Nat Med* **2:**662–667.

26. Zhang H, Deng JY, Bi LJ, Zhou YF, Zhang ZP, Zhang CG, Zhang Y, Zhang XE. 2008. Characterization of *Mycobacterium tuberculosis* nicotinamidase/pyrazinamidase. *FEBS J* **275:**753–762.

27. Fyfe PK, Rao VA, Zemla A, Cameron S, Hunter WN. 2009. Specificity and mechanism of *Acinetobacter baumanii* nicotinamidase: implications for activation of the front-line tuberculosis drug pyrazinamide. *Angew Chem Int Ed Engl* **48:**9176–9179.

28. Seiner DR, Hegde SS, Blanchard JS. 2010. Kinetics and inhibition of nicotinamidase from *Mycobacterium tuberculosis*. *Biochemistry* **49:**9613–9619.

29. Zhang Y, Scorpio A, Nikaido H, Sun Z. 1999. Role of acid pH and deficient efflux of pyrazinoic acid in unique susceptibility of *Mycobacterium tuberculosis* to pyrazinamide. *J Bacteriol* **181:**2044–2049.

30. Zhang Y, Wade MM, Scorpio A, Zhang H, Sun Z. 2003. Mode of action of pyrazinamide: disruption of *Mycobacterium tuberculosis* membrane transport and energetics by pyrazinoic acid. *J Antimicrob Chemother* **52:**790–795.

31. Shi W, Zhang X, Jiang X, Yuan H, Lee JS, Barry CE 3rd, Wang H, Zhang W, Zhang Y. 2011. Pyrazinamide inhibits trans-translation in *Mycobacterium tuberculosis*. *Science* **333:**1630–1632.

32. Wade MM, Zhang Y. 2006. Effects of weak acids, UV and proton motive force inhibitors on pyrazinamide activity against *Mycobacterium tuberculosis* in vitro. *J Antimicrob Chemother* **58:**936–941.

33. Wade MM, Zhang Y. 2004. Anaerobic incubation conditions enhance pyrazinamide activity against *Mycobacterium tuberculosis*. *J Med Microbiol* **53:**769–773.

34. Coleman D, Waddell SJ, Mitchison DA. 2011. Effects of low incubation temperatures on the bactericidal activity of anti-tuberculosis drugs. *J Antimicrob Chemother* **66:** 146–150.

35. Huang Q, Chen ZF, Li YY, Zhang Y, Ren Y, Fu Z, Xu SQ. 2007. Nutrient-starved incubation conditions enhance pyrazinamide activity against *Mycobacterium tuberculosis*. *Chemotherapy* **53:**338–343.

36. Betts J, Lukey P, Robb L, McAdam R, Duncan K. 2002. Evaluation of a nutrient starvation model of *Mycobacterium tuberculosis* persistence by gene and protein expression profiling. *Mol Microbiol* **43:**717–731.

37. Andries K, Verhasselt P, Guillemont J, Gohlmann H, Neefs J, Winkler H, Van Gestel J, Timmerman P, Zhu M, Lee E, Williams P, de Chaffoy D, Huitric E, Hoffner S, Cambau E, Truffot-Pernot C, Lounis N, Jarlier V. 2005. A diarylquinoline drug active on the ATP synthase of *Mycobacterium tuberculosis*. *Science* **307:**223–227.

38. Ibrahim M, Andries K, Lounis N, Chauffour A, Truffot-Pernot C, Jarlier V, Veziris N. 2007. Synergistic activity of R207910 combined with pyrazinamide against murine tuberculosis. *Antimicrob Agents Chemother* **51:**1011–1015.

39. Gu P, Constantino L, Zhang Y. 2008. Enhancement of the antituberculosis activity of weak acids by inhibitors of energy metabolism but not by anaerobiosis suggests that weak acids act differently from the front-line tuberculosis drug pyrazinamide. *J Med Microbiol* **57:**1129–1134.

40. Byrne ST, Denkin SM, Zhang Y. 2007. Aspirin and ibuprofen enhance pyrazinamide treatment of murine tuberculosis. *J Antimicrob Chemother* **59:**313–316.

41. Somoskovi A, Wade MM, Sun Z, Zhang Y. 2004. Iron enhances the antituberculous activity of pyrazinamide. *J Antimicrob Chemother* **53:**192–196.

42. Zimhony O, Cox JS, Welch JT, Vilcheze C, Jacobs WR Jr. 2000. Pyrazinamide inhibits the eukaryotic-like fatty acid synthetase I (FASI) of *Mycobacterium tuberculosis*. *Nat Med* **6:**1043–1047.

43. Boshoff HI, Mizrahi V, Barry CE 3rd. 2002. Effects of pyrazinamide on fatty acid synthesis by whole mycobacterial cells and purified fatty acid synthase I. *J Bacteriol* **184:**2167–2172.

44. Cynamon MH, Speirs RJ, Welch JT. 1998. In vitro antimycobacterial activity of 5-chloropyrazinamide. *Antimicrob Agents Chemother* **42:**462–463.

45. Baughn AD, Deng J, Vilcheze C, Riestra A, Welch JT, Jacobs WR Jr, Zimhony O. 2010. Mutually exclusive genotypes for pyrazinamide and 5-chloropyrazinamide resistance reveal a potential resistance proofing strategy. *Antimicrob Agents Chemother* **54:**5323–53238.

46. Ahmad Z, Tyagi S, Minkowsk A, Almeida D, Nuermberger EL, Peck KM, Welch JT, Baughn AS, Jacobs WR Jr, Grosset JH. 2012. Activity of 5-chloropyrazinamide in mice infected with *Mycobacterium tuberculosis* or *Mycobacterium bovis*. *Indian J Med Res* **136:**808–814.

47. Ngo SC, Zimhony O, Chung WJ, Sayahi H, Jacobs WR Jr, Welch JT. 2007. Inhibition of isolated *Mycobacterium tuberculosis* fatty acid synthase I by pyrazinamide analogs. *Antimicrob Agents Chemother* **51:**2430–2435.

48. Scorpio A, Lindholm-Levy P, Heifets L, Gilman R, Siddiqi S, Cynamon M, Zhang Y. 1997. Characterization of pncA mutations in pyrazinamide-resistant *Mycobacterium tuberculosis. Antimicrob Agents Chemother* **41:**540–543.

49. Keiler KC. 2008. Biology of trans-translation. *Annu Rev Microbiol* **62:**133–151.

50. Thibonnier M, Thiberge JM, De Reuse H. 2008. Trans-translation in *Helicobacter pylori:* essentiality of ribosome rescue and requirement of protein tagging for stress resistance and competence. *PLoS One* **3:**e3810.

51. Muto A, Fujihara A, Ito KI, Matsuno J, Ushida C, Himeno H. 2000. Requirement of transfer-messenger RNA for the growth of *Bacillus subtilis* under stresses. *Genes Cells* **5:**627–635.

52. Zhang S, Chen J, Shi W, Liu W, Zhang WH, Zhang Y. 2013. Mutations in *panD* encoding aspartate decarboxylase are associated with pyrazinamide resistance in *Mycobacterium tuberculosis. Emerg Microbes Infect* (Nature Publishing Group) **2:**e34. doi:10.1038/emi.2013.1038.

53. Konno K, Feldmann FM, McDermott W. 1967. Pyrazinamide susceptibility and amidase activity of tubercle bacilli. *Am Rev Respir Dis* **95:**461–469.

54. Stoffels K, Mathys V, Fauville-Dufaux M, Wintjens R, Bifani P. 2012. Systematic analysis of pyrazinamide-resistant spontaneous mutants and clinical isolates of *Mycobacterium tuberculosis. Antimicrobial Agents Chemother* **56:**5186–5193.

55. McClatchy JK, Tsang AY, Cernich MS. 1981. Use of pyrazinamidase activity on *Mycobacterium tuberculosis* as a rapid method for determination of pyrazinamide susceptibility. *Antimicrob Agents Chemother* **20:**556–557.

56. Trivedi SS, Desai SG. 1987. Pyrazinamidase activity of *Mycobacterium tuberculosis:* a test of sensitivity to pyrazinamide. *Tubercle* **68:**221–224.

57. Zhang Y, Telenti A. 2000. Genetics of drug resistance in *Mycobacterium tuberculosis,* p 235–254. *In* Hatfull G, Jacobs WR (eds), *Molecular Genetics of Mycobacteria.* ASM Press, Washington, DC.

58. Sheen P, Ferrer P, Gilman RH, Lopez-Llano J, Fuentes P, Valencia E, Zimic MJ. 2009. Effect of pyrazinamidase activity on pyrazinamide resistance in *Mycobacterium tuberculosis. Tuberculosis (Edinb)* **89:**109–113.

59. Du X, Wang W, Kim R, Yakota H, Nguyen H, Kim SH. 2001. Crystal structure and mechanism of catalysis of a pyrazinamidase from *Pyrococcus horikoshii. Biochemistry* **40:**14166–14172.

60. Petrella S, Gelus-Ziental N, Maudry A, Laurans C, Boudjelloul R, Sougakoff W. 2011. Crystal structure of the pyrazinamidase of *Mycobacterium tuberculosis:* insights into natural and acquired resistance to pyrazinamide. *PLoS One* **6:**e15785.

61. Shenai S, Rodrigues C, Sadani M, Sukhadia N, Mehta A. 2009. Comparison of phenotypic and genotypic methods for pyrazinamide susceptibility testing. *Indian J Tuberc* **56:**82–90.

62. Somoskovi A, Dormandy J, Parsons LM, Kaswa M, Goh KS, Rastogi N, Salfinger M. 2007. Sequencing of the pncA gene in members of the *Mycobacterium tuberculosis* complex has important diagnostic applications: identification of a species-specific *pncA* mutation in *"Mycobacterium canettii"* and the reliable and rapid predictor of pyrazinamide resistance. *J Clin Microbiol* **45:** 595–599.

63. Cheng SJ, Thibert L, Sanchez T, Heifets L, Zhang Y. 2000. *pncA* mutations as a major mechanism of pyrazinamide resistance in *Mycobacterium tuberculosis:* spread of a monoresistant strain in Quebec, Canada. *Antimicrob Agents Chemother* **44:**528–532.

64. Sreevatsan S, Pan X, Zhang Y, Kreiswirth BN, Musser JM. 1997. Mutations associated with pyrazinamide resistance in *pncA* of *Mycobacterium tuberculosis* complex organisms. *Antimicrob Agents Chemother* **41:** 636–640.

65. Mestdagh M, Fonteyne PA, Realini L, Rossau R, Jannes G, Mijs W, De Smet KA, Portaels F, Van den Eeckhout E. 1999. Relationship between pyrazinamide resistance, loss of pyrazinamidase activity, and mutations in the *pncA* locus in multidrug-resistant clinical isolates of *Mycobacterium tuberculosis. Antimicrob Agents Chemother* **43:**2317–2319.

66. Lemaitre N, Sougakoff W, Truffot-Pernot C, Jarlier V. 1999. Characterization of new mutations in pyrazinamide-resistant strains of *Mycobacterium tuberculosis* and identification of conserved regions important for the catalytic activity of the pyrazinamidase PncA. *Antimicrob Agents Chemother* **43:**1761–1763.

67. Marttila HJ, Marjamaki M, Vyshnevskaya E, Vyshnevskiy BI, Otten TF, Vasilyef AV, Viljanen MK. 1999. *pncA* mutations in pyrazinamide-resistant *Mycobacterium tuberculosis* isolates from northwestern Russia. *Antimicrob Agents Chemother* **43:**1764–1766.

68. Yee DP, Menzies D, Brassard P. 2012. Clinical outcomes of pyrazinamide-monoresistant *Mycobacterium tuberculosis* in Quebec. *Int J Tuberc Lung Dis* **16:**604–609.

69. Feuerriegel S, Koser CU, Richter E, Niemann S. 2013. *Mycobacterium canettii* is intrinsically resistant to both pyrazinamide and pyrazinoic acid. *J Antimicrob Chemother* **68:**1439–1440.

70. Sambandamurthy VK, Wang X, Chen B, Russell RG, Derrick S, Collins FM, Morris SL, Jacobs WR Jr. 2002. A pantothenate auxotroph of *Mycobacterium tuberculosis* is highly attenuated and protects mice against tuberculosis. *Nat Med* **8:**1171–1174.

71. Simons SO, van Ingen J, van der Laan T, Mulder A, Dekhuijzen PN, Boeree MJ, van Soolingen D. 2012. Validation of *pncA* gene sequencing in combination with the MGIT method to test susceptibility of *Mycobacterium tuberculosis* to pyrazinamide. *J Clin Microbiol* **50:**428–434.

72. Zhang Y, Chang K, Leung C, Yew W, Gicquel G, Fallows D, Kaplan G, Chaisson R, Zhang W. 2012. "ZS-MDR-TB" versus "ZR-MDR-TB": improving treatment of MDR-TB by identifying pyrazinamide susceptibility. *Emerg Microbes Infect* (Nature Publishing Group) **1:**e5. doi:10.1038/emi.2012.1018.

73. **World Health Organization.** 2008. *Anti-Tuberculosis Drug Resistance in the World*, Report No. 4. http://www.who.int/tb/publications/2008/drs_report4_26feb08.pdf. World Health Organization, Geneva, Switzerland.

74. **World Health Organization.** 2006. *XDR-TB, Extensively Drug-Resistant Tuberculosis*: recommendations for prevention and control. *Weekly Epidemiol Record* **81:**430–432.

75. **GATB.** 2001. Tuberculosis. Scientific blueprint for tuberculosis drug development. *Tuberculosis (Edinb)* **81**(Suppl 1): 1–52.

76. **Mitchison D.** 2004. The search for new sterilizing anti-tuberculosis drugs. *Front Biosci* **9:**1059–1072.

77. **Zhang Y.** 2005. The magic bullets and tuberculosis drug targets. *Annu Rev Pharmacol Toxicol* **45:**529–564.

78. **Yew WW, Cynamon M, Zhang Y.** 2011. Emerging drugs for the treatment of tuberculosis. *Expert Opin Emerg Drugs* **16:**1–21.

79. **Nuermberger EL, Spigelman MK, Yew WW.** 2010. Current development and future prospects in chemotherapy of tuberculosis. *Respirology* **15:**764–778.

80. **Tasneen R, Tyagi S, Williams K, Grosset J, Nuermberger E.** 2008. Enhanced bactericidal activity of rifampin and/or pyrazinamide when combined with PA-824 in a murine model of tuberculosis. *Antimicrob Agents Chemother* **52:**3664–3668.

81. **Rao SP, Alonso S, Rand L, Dick T, Pethe K.** 2008. The protonmotive force is required for maintaining ATP homeostasis and viability of hypoxic, nonreplicating *Mycobacterium tuberculosis*. *Proc Natl Acad Sci USA* **105:** 11945–11950.

82. **Kim JJ, Lee HM, Shin DM, Kim W, Yuk JM, Jin HS, Lee SH, Cha GH, Kim JM, Lee ZW, Shin SJ, Yoo H, Park YK, Park JB, Chung J, Yoshimori T, Jo EK.** 2012. Host cell autophagy activated by antibiotics is required for their effective antimycobacterial drug action. *Cell Host Microbe* **11:**457–468.

83. **Andries K, Gevers T, Lounis N.** 2010. Bactericidal potencies of new regimens are not predictive of their sterilizing potencies in a murine model of tuberculosis. *Antimicrob Agents Chemother* **54:**4540–4544.

84. **Tan Y, Hu Z, Zhang T, Cai X, Kuang H, Liu Y, Chen J, Yang F, Zhang K, Tan S, Zhao Y.** 2014. Role of *pncA* and *rpsA* gene sequencing in detection of pyrazinamide resistance in *Mycobacterium tuberculosis* isolates from southern China. *J Clin Microbiol* **52:**291–297.

85. **Zhang Y.** 2014. Persisters, persistent infections and the yin-yang model. *Emerg Microbes Infect* (Nature Publishing Group) **3:**e3. doi:10.1038/emi.2014.3.

Molecular Genetics of Mycobacteria, 2nd Edition
Edited by Graham F. Hatfull and William R. Jacobs, Jr.
© 2014 American Society for Microbiology, Washington, DC
doi:10.1128/microbiolspec.MGM2-0030-2013

Andrej Trauner[1]
Christopher M. Sassetti[2]
Eric J. Rubin[1]

Genetic Strategies for Identifying New Drug Targets

25

Bedaquiline was the first novel antitubercular to be approved by the FDA in over 4 decades (1); delamanid is likely to follow soon (2), and there are a number of others in the pipeline (SQ109, linezolid analogues [3–5]). In fact, tuberculosis drug development is one of the few dynamic branches of an otherwise stagnating field. Although there is some basic research into antibiotics in general, little progress is being made in bringing suitable leads into the clinic (6). Only two systemic antibiotics were approved by the FDA between 2008 and 2012, compared to six from 1998 to 2002 and 13 from 1988 to 1992, a number that must increase if we are to retain the upper hand over infectious diseases (7). While this is, in part, due to stringent FDA regulations and the withdrawal of large pharmaceutical companies from antibiotic research (8, 9), it is also due to the high attrition rate in antibiotic discovery, which is further exacerbated by the so-called discovery void (10).

While rational drug design has led to the successful identification of many potent antivirals (maraviroc, raltegravir, zanamivir), antitumor drugs (imatinib, gefitinib), and anti-inflammatory drugs (celecoxib, rofecoxib), target-based approaches played no role in the development of any of the antibacterials that are currently in trials. Our understanding of medicinal chemistry, structural biology, and genomics has delivered unprecedented insights into structure-activity relationships (SARs), detailed knowledge of the three-dimensional structure of proteins, and extensive catalogues of genetic information, but the success of target-based drug development remains elusive. The extent of the problem is illustrated by the failure of a large and intensive target-based campaign to develop novel antibiotics. Seventy high-throughput screens designed to identify inhibitors of essential enzymes were able to identify leads against only five targets: phenylanalyl-, methioninyl-tRNA synthetase (PheRS, MetRS); peptide deformylase (Pdf); enoyl-acyl carrier protein reductase (FabI); and 3-ketoacyl-acyl carrier protein III (FabH) (11). As a result, rational drug design is still confined to highly specialized applications such as optimization and development of next-generation versions of existing drugs (12, 13).

In this article we will focus on recent advances in the field that point to ways in which genetics could be used to improve the success and efficiency of drug development. It is important that we recognize lessons from past

---

[1]Harvard School of Public Health, 677 Huntington Avenue, Boston, MA 02115; [2]University of Massachusetts Medical School, 55 Lake Avenue North, Worcester, MA 01655.

gene-oriented efforts, particularly the need for making sure that all approaches be based on a whole-cell system (to incorporate *in vivo* activity and cell permeability from the start), and that we focus on *in vivo*–validated targets. By doing so there are significant advantages to be had when combining genetic approaches and more traditional screening methods. Key among these are:

1. *Targeting biology that is relevant to infection.* We will describe how unbiased genetic approaches can be used to identify key mediators of virulence and identify novel targets that are necessary for survival in the host.
2. *Strategies to identify drug synergy from the start.* Synergy, while difficult to attain, is highly desirable, because it allows treatment to be more efficient at lower drug concentrations, therefore improving the safety profile of any antibiotic. We propose an approach that may lead to the identification of drug synergy from first principles, which could result in shorter drug treatments and possibly antibiotic combinations with longer "shelf-lives" in the clinic.
3. *Compound-driven selection of targets.* We will discuss advances in methodologies that would allow researchers to identify compound mechanisms of action (MOAs) concurrently with performing screens for inhibitors. This should lead to the identification of novel validated targets in conjunction with new drug scaffolds.
4. *Exploring chemical space populated by weak inhibitors.* Screening hits need to undergo several rounds of chemical iteration to become optimized leads; we will show how a biology-driven identification and selection of hits can allow us to identify compounds with novel MOAs that would be missed by standard screening strategies.

While the current pace of antibiotic development has been disappointing, utilizing these strategies could expand the number of potential compound candidates and reveal new mechanisms that could be exploited to enhance drug efficacy.

CURRENT TARGET-BASED DRUG DISCOVERY IN *MYCOBACTERIUM TUBERCULOSIS*

The tuberculosis field has long evaluated basic research with an eye on the identification of novel antibiotic targets. Genes are regularly assessed for their contribution to virulence in the mouse model of infection, and multiple systematic assessments of gene essentiality both *in vitro* and *in vivo* have already been used to gain

a global perspective (14–19). As a result, there is already a growing body of knowledge about potential targets and metabolic pathways whose disruption should have a profound effect on *Mycobacterium tuberculosis* viability *in vivo*. The glyoxylate shunt enzymes isocitrate lyase (20, 21) and malate synthase (22) are good examples and are currently being explored as drug targets. Similarly, Beste et al. recently showed that anaplerotic reactions are likely to be important during infection (23). This is corroborated by the fact that inhibitors of carbonic anhydrase were shown to have an antibiotic effect on *M. tuberculosis* (24). Peptide deformylase, mentioned in the introduction as one of the few chemically validated drug targets (25), seems to be tractable in *M. tuberculosis* as well (26). Similarly, methionine aminopeptidase is showing promise as a target for antimycobacterial intervention (27).

In addition to these novel enzymes, there are also a number of known targets whose inhibition constitutes the backbone of current chemotherapeutic interventions, e.g., RNA polymerase and enoyl-acyl carrier protein reductase (InhA). By analogy to what is common practice in other parts of the antibiotic discovery field, targeting these with novel chemical entities is another possible strategy to overcome drug resistance. An example was recently reported in which Hartkoorn et al. showed that pyridomycin—a compound of known antitubercular activity—inhibits InhA (28). Unfortunately, pre-existing resistance to isoniazid interfered with the compound's efficacy, particularly in strains where InhA overexpression was the mechanism of resistance. Encouragingly, though, mutations in InhA leading to pyridomycin resistance did not influence isoniazid activity, and perhaps more importantly, catalase (KatG), which is required for the activation of isoniazid and whose mutation is the most common mechanism of resistance to isoniazid, did not affect pyridomycin MIC. These findings therefore provide an encouraging example of targeting known targets in the future.

Augmenting Existing Drugs through Target-Based Approaches

Due to extensive research into antibiotic resistance in *M. tuberculosis,* we currently understand many of the molecular mechanisms that underpin it. Loss of prodrug activation is a common theme among *M. tuberculosis* mechanisms of resistance, because many antituberculars are prodrugs (isoniazid, ethionamide, pyrazinamide, delamanid). As a result, ensuring prodrug activation is an interesting avenue to pursue to augment existing therapies. An example of where this was done successfully in *M. tuberculosis* is the EthR inhibitor—BDM31343 (29). EthR is a negative regulator of EthA, whose activity is necessary for the activation of ethionamide. The authors took advantage of the knowledge of the structure of EthR and the extensive understanding of the biology of transcriptional regulators belonging to the TetR-family of proteins. Starting from 131 different inhibitor candidates identified based on crystallographic data, the authors were able to optimize the molecules and finally identify a BD31343, which had good pharmacological properties and was able to synergize strongly with ethionamide in the mouse model of infection. BD31343 is still in preclinical development, but it is likely to be the first "designer" antimycobacterial compound to enter clinical trials (30).

TARGETING BIOLOGY THAT IS RELEVANT FOR INFECTION

The key determinant of the success of any target-based approach is the "quality" of the selected targets. Ideally, we would like to have a long list of protein targets whose inhibition is specific (not affecting the host), physiologically relevant, and chemically tractable so we can readily bypass resistance if and when it occurs. However, this is not an easy task. As mentioned at the beginning of this article, gene essentiality and conservation across microbial species are not sufficient attributes for the successful identification of suitable antibiotic targets. Additional parameters that should be considered are protein "druggability" (31, 32) and vulnerability (33). Can a small molecule interfere with the process we want to target? Does inhibiting that pathway sufficiently perturb the normal activity of a cell to result in stasis or preferably death? Each drug development program defines its own criteria through which it selects suitable targets (34), but the general theme is the same: while we can use structural biology to try to determine whether we can develop an inhibitor to an enzyme (35), we cannot predict whether inhibition will have the desired biological effect *in vivo*. The only way to identify optimal targets is to systematically perturb relevant pathways experimentally. One way to do this is to use random mutagenesis with transposons as described below.

Essentiality Screens as Tools for Drug Development

An essential gene can be broadly defined as a gene whose product is required for survival in a specific environment. This is an operational definition; in fact, different mutations produce different extents of growth

attenuation. Nevertheless, "essential" is a good shorthand term for those genes that encode cellular processes that are virtually indispensable for cell growth under any condition. Many of these are known sites of antibiotic action, e.g., translation, transcription, and cell wall biosynthesis. In addition to these genes that produce attenuation under the most basic conditions, there are also conditionally essential genes, whose function is dispensable under some conditions but essential in others; virulence determinants fall into this category. Genetic tests of essentiality rely on the inability to disrupt a given gene using either targeted or random approaches. Phage-mediated transposon mutagenesis has become the gold standard for random mutagenesis and has been successfully applied to all domains of life (36–38) including mycobacteria (14, 16, 19, 39). The distinct benefit of using transposon mutants rather than other approaches is the ability to interrogate the genomes of a highly complex library in a single experiment, particularly when using next-generation sequencing technologies to acquire data. At the simplest level, the frequency of gene disruption correlates with its essentiality: nonessential genes tolerate many insertions, while none are found in essential genes (see Fig. 1).

However, as pointed out earlier, gene essentiality is not a sufficiently powerful discriminator for target selection. One way to populate the list of genes of interest in a meaningful way is to focus on conditionally essential genes whose function is interwoven with pathogenesis. For example, transposon mutagenesis techniques were used to study the animal models of infection (15–18) and for the identification of genes required for survival in the macrophage (40). The results of these studies afford an eagle-eye view of the biological processes that are likely to be important during infection and provide important clues on the aspects of metabolism that should be investigated further as potential targets for intervention. Unlike screens focusing on core processes, where gene essentiality is statistically defined through the analysis of negative data (absence of specific transposon mutants in the population), conditional essentiality screens provide a direct measurement of the mutant fitness during infection, which allows target prioritization based on the degree of the mutant's growth or survival defect. While no specific gene identified in these studies has been selected for a concerted drug development effort to date, we already have a wealth of information to help choose possible targets and enhance the pace of drug development.

This point is illustrated very clearly by cholesterol metabolism. The Mce4 gene cluster is homologous to genes first identified during studies of mycobacterial

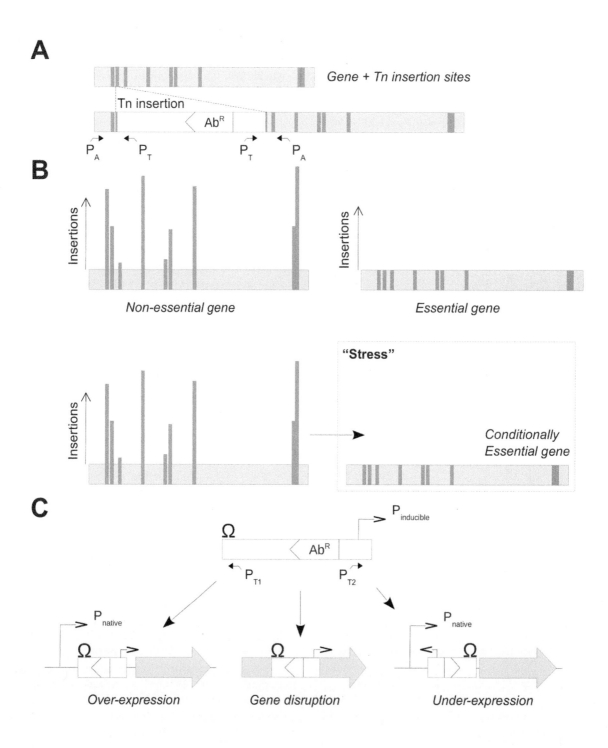

cell entry (41) and is essential for *M. tuberculosis* survival in mice (15), probably due to its role in cholesterol uptake by the bacterium (42). It was later shown to be required for survival of *M. tuberculosis* in the macaque model of tuberculosis (18). Cholesterol is therefore an important source of carbon for *M. tuberculosis* during infection (43), but it is also a source of a number of intermediates, such as catechols and propionyl-coenzyme A, which are toxic if not cleared by the cell (43–46). Griffin et al. (45) successfully used transposon mutagenesis to single out genes involved in cholesterol uptake, catabolism, and central metabolism adaptations required to sustain growth on this unusual carbon source. Pathways such as this, which perform an essential function and generate toxic intermediates, could be targeted in two ways. One could simply inhibit the essential function. Alternatively, knowledge of the pathway gained from genetic studies can be exploited to inhibit steps that create toxic metabolites, leading to more dramatic drug effects.

The validity of combining transposon mutagenesis with growth conditions that mimic infection for the identification of new drug targets is further illustrated by investigations into key mediators of pH homeostasis in *M. tuberculosis*. Vandal et al. used this approach to single out Rv3671c, a transmembrane serine protease, as crucial for the maintenance of intracellular pH (47). Furthermore, they were also able to confirm that loss of Rv3671c compromises the ability of *M. tuberculosis* to establish chronic infection in mice. They proposed that Rv3671c would be a good target, despite not being essential for growth *in vivo* (48). Protease inhibition is one of the best-understood mechanisms of drug action, and it is therefore likely that Rv3671c may well prove to be druggable.

Engineering Drug Synergy and Limiting Drug Resistance

Another valuable use of transposon mutagenesis is to screen for synthetic lethality. Drug perturbations of a cell do not always have a unique effect on the system (49, 50). Therefore, using transposon mutagenesis in the context of drug treatment may give rise to the identification of companion targets. While not the same as synthetic lethality in the classical sense of the word, such targets could lead to potent synergistic compounds. At the simplest level, transposon mutagenesis in this context can identify prodrug activators, e.g., KatG for isoniazid and EthA for ethionamide (14). An example of a companion drug that interferes with prodrug activation is BDM31343 (29), discussed above. A systematic drug-induced conditional essentiality screen should result in the identification of multiple mediators of drug toxicity (50, 51) as well as shed light on common death pathways in the cell, provided, of course, that these enzymes do not form part of the core essential genes. Genes identified in this way would be unlikely to represent valid drug targets in their own right because they are, by definition, not essential under the screening conditions. Nonetheless they should provide compelling candidates for the identification of synergistic

Figure 1 High-frequency transposon mutagenesis. (**A**) Any gene can have multiple potential transposon insertion sites (marked with dark gray bars). Transposon insertion is usually selected for by using antibiotic markers encoded within it and on a basic level results in the disruption of gene function. Identifying the site of insertion relies on the same principle as genome sequencing. Genomic DNA is sheared and an adaptor of known sequence is ligated to the fragments. In the case of transposon insertion site scoring, the resulting pool is amplified using primers specific for the transposon (in the simplest case these are the same for both flanks of the transposon; P_T) and a primer specific for the adapter (P_A). The number of reads mapping to each genomic locus is proportional to the abundance of the strain carrying this insertion. (**B**) The frequency of transposon insertion reflects gene essentiality. Genes that can tolerate insertions in multiple sites throughout their coding sequence are deemed nonessential. An organism cannot tolerate the disruption of an essential gene; therefore, no insertions can be detected. Genes that are not essential in a wild-type background under "normal" conditions but become essential once the system is suitably perturbed (e.g., low pH, presence of another mutation, drug treatment) provide a special case and are considered conditionally essential. Statistical methods should be used to determine whether a gene has a significantly low number of insertions. (**C**) More elaborate transposon architectures may include the presence of transcriptional terminators (Ω) or outward-facing inducible promoters (adapted from reference 69). Using such systems provides greater information because transposon insertion gains additional modalities that go beyond simple gene disruption. Since the directionality of insertion carries information, it is important to be able to use different primers for each transposon flank (P_{T1}, P_{T2}) during insertion scoring. doi:10.1128/microbiolspec.MGM2-0030-2013.f1

interactions, which in turn would have obvious benefits when devising next-generation combinations for treatment of tuberculosis.

In a similar vein, and perhaps more ambitiously, it is possible to perform transposon mutagenesis screens for true synthetic lethality in a mutant background. In this case one would first identify a novel target candidate and then perform a screen for genes whose activity becomes indispensable under conditions of diminished activity (or absence) of the primary target. Such approaches have been used successfully in the study of nutrient uptake (52) and propionyl-coenzyme A toxicity (53, 54). A systematic search may identify further interactions of a similar kind. In the first instance, one could focus on gene pairs (or networks) that can be inhibited using a single inhibitor for multiple targets—a feature that appears to be singly paramount for stemming antibiotic resistance (10, 55). Even if a single molecular entity would be impossible to generate, genetic screens of this nature could prove invaluable for devising future combination treatments. An important advantage of such a strategy is the fact that drug synergy would be built in at the developmental stages, much like β-lactamase inhibitors augmenting β-lactam antibiotic action, and not through trial and error, as was the case with trimethoprim and sulfamethoxazole. A powerful set of examples illustrating the appeal of this approach is provided by research aiming to resensitize methicillin-resistant *Staphylococcus aureus* (MRSA) to β-lactam antibiotics (reviewed in reference 56). Of particular interest is the identification (57) and further elaboration (58–60) of the synergistic action of β-lactams and teichoic acid inhibitors. Furthermore, in a comprehensive screen based on antisense-mediated gene silencing (61), Roemer and coworkers identified a number of auxiliary factors mediating β-lactam resistance (62) including SpsB (a signal peptidase) and the Z-ring proteins. Crucially, both of these were recently shown to be druggable, and the compounds that interfere with their function were strongly synergistic with β-lactams (63, 64).

So far we have focused entirely on the identification of targets based on genetic manipulation of cells. An alternative approach would be to use the well-established approach of whole-cell chemical screens to identify active compounds and to use genetic approaches to systematically identify their targets, ideally, concurrently with the screen for inhibitors. The key benefit of this compound-centric approach is that we are able to generate targets whose role in viability has already been validated, and we already have a chemical entity on which to build our search for inhibitors.

USING DRUG SCREENS TO PROSPECT FOR NEW TARGETS

One of the major advantages of whole-cell screening programs over target-based approaches is the immediate validation inherent in the experimental design. Compounds are selected based on the very clear endpoint readout of "Does it inhibit bacterial growth?" In a recent screen for antimycobacterial compounds (65), testing in excess of 100,000 compounds led to the identification of 1,549 hits. The prioritization of these for further development is currently based almost entirely on chemoinformatic approaches (66, 67). However, knowledge of a compound's specific target accelerates the medicinal chemistry required to improve lead compounds. Experience from the pharmaceutical industry shows that approximately 5% of screening hits inhibit a specific protein (13), suggesting that a screen like the one mentioned above could identify as many as 77 targets. Identifying compounds with a specific MOA remains a resource-intensive undertaking, but a number of approaches have been developed in a multitude of systems that strive to enrich the quality and depth of information we get from a drug screen by adding a biological component beyond the simple growth inhibition endpoint. The importance of knowing the biology of a compound and its relevance for bacterial survival during infection as soon as possible cannot be overstated and is discussed later in this article.

In order to identify targets through the characterization of their inhibitors, it is essential that a methodology is available that allows the systematic, rapid, and scalable interrogation of the entire genome. The simplest approach is the established identification of spontaneous resistance mutants. With the advent of powerful (and relatively cheap) next-generation sequencing platforms, such as Illumina, it has become feasible to rapidly identify resistance-mediating point mutations through whole-genome sequencing. An early example of the power of such an approach is the discovery of ATP synthase as the target for bedaquiline (68). When resistant mutants can be isolated, this approach has proven generally useful. However, not all compounds have a single target, and isolation of resistance mutants in cases where a single molecule affects multiple targets has very limited success, even if the constellation of mutations present within the tested bacterial population is enhanced by exposing it to ethyl methanesulfonate (EMS) prior to selection. More importantly, it is unfeasible to adapt this approach to high-throughput target identification.

More suitable approaches to fulfill the aforementioned criteria are those based on specialized libraries where either each gene in the genome is perturbed or its expression is controlled. Broadly speaking, such methodologies rely either on gene disruption through transposon mutagenesis (51, 69) and knockout libraries (13, 50, 70), or modulating expression using knockdown (71–73) and overexpression libraries (74, 75). Different approaches lead to different types of information. For example, gene disruption libraries are not compatible with the identification of essential genes (which, by definition, cannot be disrupted) and are therefore unlikely to lead to the identification of the target as such, but they have been used to describe the phenotypic landscape of a cell whose integrity has been perturbed by a drug (50) and can indirectly identify the inhibited pathway (76). As mentioned earlier, such approaches would be very well suited to identify conditionally essential targets, e.g., targets that are required specifically during infection. Similarly, they could also lead to the identification of companion drugs and consequently companion drug targets. An exception to this rule is the specific application of transposon mutagenesis to the identification of *S. aureus* growth inhibitors described by Wang et al. (69) and the identification of essential genes through promoter replacement in *Vibrio cholerae* (77). Normally, the transposon-based approaches used in such screens usually only carry an antibiotic resistance marker. Wang and colleagues modified their transposon to carry an outward-facing promoter as well as a transcriptional terminator. By doing so they can simultaneously screen for resistance mechanisms that arise through gene disruption and early termination (polar effects), as well as gene overexpression. Given that their transposon has a clearly defined directionality, they can correlate the site of transposon insertion as well as the orientation of the transposon to deduce which mechanism is driving resistance when the library is exposed to a specific drug, thus generating a multilayered biological output. However, modifying transposon sequences is not trivial, and the adaptation of their technique to mycobacteria can provide a significant challenge.

Target overexpression leads to an increase in the effective MIC for a drug, while target knockdown leads to the reverse effect. Measuring the relative enrichment or depletion of a particular strain within a highly complex population can therefore lead to the identification of the actual target for a compound—provided a single target exists. A key difference between the two is that overexpression is easily achieved by introducing an inducible copy of the gene on a plasmid that is transformed *en masse* into the target population (70). The generation of a knockdown library, on the other hand, requires a significantly larger resource investment because it relies on the generation of promoter replacement strains and has therefore been done in a systematic manner only in a handful of microorganisms (13, 70, 73, 78). Similarly to gene disruption libraries, expression modulation libraries could also be used to identify the broader impact of a compound on cell physiology, possibly paving the way for the identification of novel synergistic compounds. Importantly, modulating the expression of transcription and sigma factors may lead to more pleiotropic effects and could potentially give rise to the identification of multiple targets simultaneously. Another important technical distinction between overexpression and knockdown libraries is the differential ability to identify protein complexes as targets. Overexpressing a single component of a multiprotein complex is unlikely to lead to a significant change in MIC. Depleting a component, however, should lead to a shift in MIC consistent with the depletion of the target. It is important, therefore, to think of these approaches as complementary, and the choice of system should be influenced largely by the availability of the tools.

This approach has not been extensively developed in mycobacterial species. There is a wealth of proof-of-concept evidence in the literature showing that all of the above approaches could be implemented in mycobacteria (28, 33, 79–81), but there has been no real effort to date to perform such screens in a systematic manner. One of the few attempts at generating a high-throughput system to use for gene overexpression is random inducible controlled expression (RICE), but it has not been applied to the identification of potential antibiotic targets (82).

A different modality of this approach based on using substrate analogues, rather than growth inhibitors, could be used to identify families of targets based on the presence of a specific enzymatic activity. The benefit of performing such a screen would be to pinpoint groups of targets that could be inhibited using a single molecule, therefore potentially leading to the identification of compounds to which resistance would occur more slowly. Activity-based protein profiling approaches (83) are based on the ability to derivatize molecules such that they form a covalent bond with their target when the latter exhibits enzymatic activity against the bait. In this manner it may be possible to identify *M. tuberculosis*-specific homologues of targets that have been validated in other species. An example of such a study has been reported recently for the identification of ATPases in *M. tuberculosis* (84).

BIOLOGICALLY RELEVANT
TARGET VALIDATION

Biological validation increases the confidence in a target and can decrease the risk associated with investing in further chemical development of compounds. The exact approach to validation varies but should take into consideration multiple parameters such as gene essentiality *in vivo*, target druggability, and selectivity as well as vulnerability. A cautionary tale comes from a screen that led to the identification and extensive optimization of a very promising antimycobacterial lead by Novartis—a specific glycerol kinase inhibitor. The company committed a full team to the project, achieving excellent *in vitro* and pharmacological results, although the compound was found to be ineffective when tested in the mouse model of infection. Its activity was shown to be specific to growth on glycerol—a carbon source regularly used for culture but not encountered by the bacterium *in vivo* (85). While our knowledge of biology at the time would probably not have prevented the company from pursuing the lead, one could argue that knowing the target immediately and validating it *in vivo* either by creating a knockout or a conditional knockout strain would have allowed the researchers to terminate the program earlier and incur fewer losses.

Another important determinant of target viability is vulnerability. This concept can be defined as the proportion of an enzymatic activity that needs to be inhibited in order to achieve growth arrest and death. In principle, more vulnerable targets, i.e., those requiring less inhibition, are more attractive targets. This cannot be known *a priori*, and not all essential targets are equally as vulnerable. For example, mycobacteria are exquisitely sensitive to RNA polymerase and InhA depletion, but they can withstand great levels of dihydrofolate reductase (DHFR) and alanine racemase (Alr) depletion. Alr and dihydrofolate reductase are present in great excess within cells, and mycobacteria can easily withstand almost complete depletion of the protein in the cell without significant growth attenuation (33). The targets of isoniazid and rifampin are very vulnerable, which could be one reason why these compounds form the backbone of current tuberculosis chemotherapy. Cycloserine, which targets Alr, is a less effective antibiotic (though an antibiotic nonetheless, suggesting that vulnerability is only one component in prioritizing targets).

Loss-of-function mutants are not suitable tools to assess the vulnerability of potential targets. Thus, we must use dose-dependent target depletion to quantify the extent to which a gene product is required for normal cellular function. Strains in which the levels of a gene product are controlled, through either direct promoter replacement or inducible protein degradation, are key for validating drug targets both *in vitro* and *in vivo*, and they have been used successfully in the case of, among others, isocitrate lyase, Rv3671c (81), Clp protease (86–88), and DNA gyrase B (89). In addition, glutamine synthease (GlnA1) was shown to be essential *in vitro* and *in vivo* (90), although depletion of GlnA1 was not sufficient to arrest bacterial growth (91), therefore failing to validate GlnA1 as a target. Currently, the main obstacle to overcome for a systematic study and validation of knockdown mutants is the lengthy and technically difficult strategy to generate promoter replacement and inducible degradation strains (see Fig. 2).

Identifying potential antibiotic targets is a useful but adjunctive approach to developing new antibiotics. Ultimately, however, we need compounds that inhibit these targets. How do we proceed to getting new drugs?

FINDING MORE AND
BETTER COMPOUNDS

As stated earlier, genome-driven large-scale screening campaigns for potent bacterial growth inhibitors to be developed into new antibiotics have yielded a disappointing number of hits and only a handful of antibiotic targets. One important drawback is that until recently, existing chemical libraries were derived from existing drug development programs. Unfortunately, because pharmaceutical development is strongly oriented toward chronic diseases, antibiotics have not been a big part of chemical synthesis campaigns. Thus, many of the compounds available in current libraries are minor modifications of compounds effective in cancer or diabetes. And the rules that govern most synthetic efforts (92) do not apply to many of the most effective antibiotics. The vast majority of potential mycobacterial targets bear little resemblance to those that are of current pharmaceutical interest, and it is likely that there will be only a limited number of inhibitors among them. Moreover, compounds must both enter the bacterial cell and avoid inactivation and efflux, properties that seem restricted to a rather small set of chemicals. Thus, increasing the chemical space among potential inhibitors could have a considerable effect on the success rate of screens in general and perhaps even target-based strategies in particular.

Significant steps have been taken to improve the quality of compound libraries used for screening (65, 93, 94). While there have been a number of hits identified during these campaigns, the problem of compound prioritization remains, delaying lead optimization in resource-limited settings. The logical progression of a screening campaign is toward a more focused approach based on specific chemical scaffolds. Picking diverse, previously underexplored chemistries is producing good results, and many of the clinically most advanced drug candidates have been initially identified starting from a select set of chemical scaffolds such as diaryl-quinolines (68), nitroimidazoles (2, 95), and ethylene-diamine analogues (96). In addition, others, including benzothiazinones (97) and diphenyl-pyrroles (98), are currently going through preclinical development with very promising results.

It is important to keep in mind that focusing solely on inhibitors of exponential growth may not produce the most effective tuberculosis therapies. The clinical significance of bacterial populations recalcitrant to treatment during active tuberculosis has been appreciated for a long time, and historically, treatments were developed and evaluated with the aim to limit the rate of relapse thought to occur due to failure to clear nongrowing bacteria (99, 100). Two drugs were shown to be particularly successful in reducing the rate of relapse following isoniazid treatment: rifampin and pyrazinamide (101). The latter was also shown to greatly improve the early bactericidal activity of multiple drug combinations (102), leading to a higher proportion of culture-negative patients 2 months into treatment when compared to combination therapies without it (103). In fact, pyrazinamide is one of the four drugs currently administered during the directly observed treatment for tuberculosis endorsed by the World Health Organization (WHO) (104). It is also likely to remain relevant for future combination therapies because it greatly improves the efficacy of both bedaquiline and PA-824 (105). However, pyrazinamide itself is virtually inactive against actively growing *M. tuberculosis* under routine culture (and screening) conditions and would therefore not be detected under current screening approaches. Unfortunately, the mechanism of pyrazinamide action remains elusive (106, 107), limiting its use as a tool for identifying new drug targets. Perhaps the most important aspect of pyrazinamide as a drug is its ability to kill *M. tuberculosis* that is not actively growing (108). This is an area of tuberculosis drug discovery that has received increasing attention (e.g., bedaquiline and PA-824 are both active against nongrowing bacilli), and exploring it further is essential to guarantee the success of future treatment regimens that rapidly eradicate reservoirs of persistent bacteria.

Expanding Screening Conditions

As discussed above, the *in vitro* conditions that produce maximal bacterial growth rates do not necessarily reflect those found during infection (109). It is clear that nutrient sources, pH, and oxygen availability might all differ. Moreover, treatment occurs in the presence of host immunity, which might itself alter responsiveness to antibiotics; e.g., vitamin D synergizes with pyrazinamide to improve macrophage killing of *M. tuberculosis* (110).

Several groups have tried to replicate facets of the *in vivo* growth milieu *in vitro* (111). For example, nutrient limitation can be modeled with starvation, incubating cells in the absence of any carbon source (112), gradually depleting nutrients (113), or combining carbon limitation with other stresses such as low oxygen tension and low pH (114). Bacteria can be grown with alternative carbon sources, such as lipids (115), at low oxygen tension (116–118), under acidic conditions (119), or in the presence of reactive oxygen or nitrogen species (120, 121). Some of these conditions have already been adapted for inhibitor screens (122) and have led to the identification of new chemical entities whose activity can synergize with host responses to kill nonreplicating bacteria (121) or that can target both actively growing and nongrowing *M. tuberculosis* (89, 123).

Conditions mimicking the *in vivo* environment produce different physiologic responses, but one common characteristic they share is slowing or stopping bacterial replication. An alternative to many different experimental conditions would be to prevent or limit replication using genetic rather than physiologic manipulation, as has been done using a streptomycin-dependent strain of *M. tuberculosis* (124, 125). Bacteria in this model show a modified drug susceptibility profile, suggesting that they could provide an important tool in identifying the next generation of antibiotics whose activity would be exerted through the disruption of steady-state homeostasis rather than short-circuiting growth processes.

How will exploring these conditions affect drug development? Thus far, it remains unclear. Meta analysis of different models has led to the suggestion of specific targets (126), and chemoinformatic analysis of performed screens may point the way toward future chemical scaffolds to be explored (127). However, we do not know if any of these environments accurately replicate those found in the host. And, despite pyrazinamide, it is not clear if there would be a lot of enthusiasm for investing heavily in compounds with limited activity against replicating organisms.

Forward Genetic Approaches (e.g. Tn, EMS, γ-radiation)

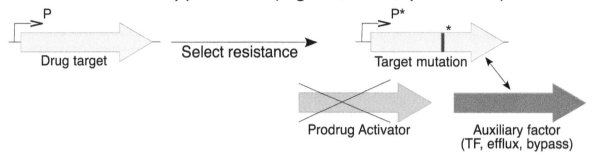

Advantages:
Minimal background knowledge required, robust and versatile, can lead to MOA and mapping of the whole drug landscape.

Limitations:
Hard to miniaturize, MOA requires significant amount of compound, limited information on essential genes, no information on target vulnerability.

Reverse Genetic Approaches (e.g. expression modulation)

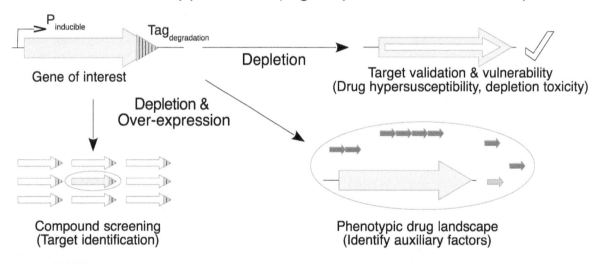

Advantages:
Complete target validation (including essential genes), compatible with high-throughput screening, provides gene-specific as well as whole genome information.

Limitations:
Very resource- and knowledge-intensive to set up, limited by existing knowledge of genome organisation.

Nevertheless, the concept that compounds that are effective under a broader array of conditions might lead to better efficacy against disease is attractive. In that respect, testing bacterial mutants that model drug activity under multiple growth conditions and infection models could be a valuable part of a target prioritization scheme.

Leveraging Biology

One of the primary problems in chemical screening is an inability to prioritize compounds for further development. The criteria that are currently used, including potency and chemistry, do not take any biology into account. Strategies that incorporate biological insights into the initial stages of screening might have a big impact on how decisions are made.

One such approach utilizes hypersensitive strains. The basic premise for the approach was described in the previous section and can be summarized as follows: decreasing the abundance of a target should result in hypersusceptibility of the strain to inhibitors of the target. Therefore, using attenuation of gene expression would achieve two goals. In the first instance, compounds that specifically inhibit strains under-expressing a gene of interest, but not a nonrelated control, would already demonstrate a potential for target specificity. Second, by using hypersusceptible strains, we could identify inhibitors whose initial activity is too weak to be identified in screens where wild-type cells are used, therefore allowing the screeners to explore the part of chemical space populated by weak but specific inhibitors. At the end of such a campaign we should have a number of hits that appear to have a specific activity against biologically validated targets—thus combining the advantages of target-based approaches, namely target specificity with whole-cell activity inherent in empirical approaches.

The viability of such approaches to identify novel inhibitors has already been extensively demonstrated in *S. aureus* (71, 72, 128) and *Candida albicans* (73). In these settings researchers were able to screen entire libraries of under-expression strains against whole chemical libraries and successfully identify inhibitors of growth and their MOA in a single screen, greatly reducing both the time and resources necessary for selecting interesting hits. The first steps providing proof-of-concept studies toward applying this approach to mycobacteria have been made already: Abrahams and coworkers used promoter replacement to decrease the expression of *icl1*, *lysA*, and *panC*, and they were able to validate these genes as important for *M. tuberculosis* survival *in vitro* (129). They then used the *panC* knockdown strain in a target-based whole-cell screen (TB-WCS) of a small compound library (600 compounds) to identify 37 hits. The authors also used specific culture conditions to maximize the susceptibility of the *panC* mutant strain. Kumar et al. took a similar approach focused on *dfrA* (130). They first screened a library of 32,000 compounds in an *in vitro* system and then used the hits to screen against a DfrA-depleted strain. They were able to identify 52 primary hits and only 3 confirmed hits (showing suitable dose-response kinetics), with a single confirmed lead with specific DfrA activity in *M. tuberculosis*. This hit rate is consistent with those previously reported for strategies using biochemical screens as starting points (11). The choice of target may have been unfortunate since DfrA is not a very vulnerable target in mycobacteria (see above).

CONCLUDING REMARKS

Every successful drug development program has a specific target product profile that serves as a roadmap to guide the developmental effort toward the ultimate goal. A similar list of desirable properties could be drawn up for the ideal drug target. A target should be vulnerable, as defined in this article, and druggable, with the potential for good structure-activity relationship analysis to facilitate the development of next-generation drugs in the same vein as β-lactams. We should aim to identify groups of interconnected drug targets whose inhibition has a synergistic effect. The goal would be to generate tailored combinations of drugs that would limit the evolution of drug resistance as well as decrease the necessity for high doses of drug to achieve inhibition. The former would improve the shelf life of new treatments, while the latter would contribute significantly toward their safety profile. It may even be possible to assemble multiple such combinations. We should aim to find targets whose mutation

Figure 2 Summary of genetic approaches to studying drug targets. Both forward- and reverse-genetic approaches can be used to provide overlapping and complementary tools for the investigation of antibiotic targets. The limitations and advantages are summarized in the boxes. The asterisk refers to a mutation within the promoter or coding region resulting in drug resistance. P, promoter; Tn, transposon; EMS, ethyl methanesulfonate; TF, transcription factor; MOA, mechanism of action. doi:10.1128/microbiolspec.MGM2-0030-2013.f2

carries a great fitness cost in the presence of its companion drug, either by penalizing the resistance mechanism itself, by blocking metabolic bypass routes, or by any other mechanism.

The starting point could be a validated drug target or a compound with a highly desirable pharmacological profile. By harnessing the combination of genetic approaches described here and using whole-cell-based screens mimicking the native environment found during infection, we should be able to identify companion drugs and drug targets that fit the above criteria.

It could be argued that detailed biological knowledge is not essential for the development of novel antibiotics. A whole suite of antimycobacterials was developed in the absence of such knowledge. While this is true for current drugs, the incorporation of biological goals can provide the right conceptual framework to facilitate the progression of future drug development programs. Furthermore, if biological considerations form the backbone of a project from the earliest stages of drug development, it may be possible to identify a constellation of targets whose vulnerability could be renewed in the face of drug resistance.

We thank Dr. Justin Pritchard for helpful discussions and critical reading of the evolving drafts.

Citation. Trauner A, Sassetti CM, and Rubin EJ. 2014. Genetic strategies for identifying new drug targets. Microbiol Spectrum 2(4):MGM2-0030-2013.

References

1. Osborne R. 2013. First novel anti-tuberculosis drug in 40 years. *Nat Biotechnol* **31**:89–91.

2. Gler MTM, Skripconoka VV, Sanchez-Garavito EE, Xiao HH, Cabrera-Rivero JLJ, Vargas-Vasquez DED, Gao MM, Awad MM, Park S-KS, Shim TST, Suh GYG, Danilovits MM, Ogata HH, Kurve AA, Chang JJ, Suzuki KK, Tupasi TT, Koh W-JW, Seaworth BB, Geiter LJL, Wells CDC. 2012. Delamanid for multidrug-resistant pulmonary tuberculosis. *New Engl J Med* **366**: 2151–2160.

3. Lamichhane G. 2011. Novel targets in *M. tuberculosis*: search for new drugs. *Trends Mol Med* **17**:25–33.

4. Villemagne B, Crauste C, Flipo M, Baulard AR, Déprez B, Willand N. 2012. Tuberculosis: the drug development pipeline at a glance. *Eur J Med Chem* **51**:1–16.

5. Cole ST, Riccardi G. 2011. New tuberculosis drugs on the horizon. *Curr Opin Microbiol* **14**:570–576.

6. Livermore DM, on behalf of the British Society for Antimicrobial Chemotherapy Working Party on The Urgent Need: Regenerating Antibacterial Drug Discovery and Development, Blaser M, Carrs O, Cassell G, Fishman N, Guidos R, Levy S, Powers J, Norrby R, Tillotson G, Davies R, Projan S, Dawson M, Monnet D, Keogh-Brown M, Hand K, Garner S, Findlay D, Morel C, Wise R, Bax R, Burke F, Chopra I, Czaplewski L, Finch R, Livermore D, Piddock LJV, White T. 2011. Discovery research: the scientific challenge of finding new antibiotics. *J Antimicrobial Chemother* **66**:1941–1944.

7. ISDA, PhRMA, Pew Research Institute. 2012. Reviving the pipeline of life-saving antibiotics: exploring solutions to spur innovation. Health Intiatives, The Pew Charitable Trusts. http://www.pewhealth.org/reports-analysis/issue-briefs/reviving-the-pipeline-of-life-saving-antibiotics-85899381282

8. Projan SJ. 2003. Why is big Pharma getting out of antibacterial drug discovery? *Curr Opin Microbiol* **6**:427–430.

9. Ledford H. 2012. FDA under pressure to relax drug rules. *Nature* **492**:19.

10. Silver LL. 2011. Challenges of antibacterial discovery. *Clin Microbiol Rev* **24**:71–109.

11. Payne DJ, Gwynn MN, Holmes DJ, Pompliano DL. 2006. Drugs for bad bugs: confronting the challenges of antibacterial discovery. *Nat Rev Drug Discov* **6**:29–40.

12. Swinney DC, Anthony J. 2011. How were new medicines discovered? *Nat Rev Drug Discov* **10**:507–519.

13. Roemer T, Boone C. 2013. Systems-level antimicrobial drug and drug synergy discovery. *Nat Chem Biol* **9**: 222–231.

14. Sassetti CM, Boyd DH, Rubin EJ. 2001. Comprehensive identification of conditionally essential genes in mycobacteria. *Proc Natl Acad Sci USA* **98**:12712–12717.

15. Sassetti C, Rubin E. 2003. Genetic requirements for mycobacterial survival during infection. *Proc Natl Acad Sci USA* **100**:12989–12994.

16. Lamichhane G, Tyagi S, Bishai WR. 2005. Designer arrays for defined mutant analysis to detect genes essential for survival of *Mycobacterium tuberculosis* in mouse lungs. *Infect Immun* **73**:2533–2540.

17. Jain S, Hernandez-Abanto S, Cheng Q-J, Singh P, Ly L, Klinkenberg L, Morrison N, Converse P, Nuermberger E, Grosset J, McMurray D, Karakousis P, Lamichhane G, Bishai W. 2007. Accelerated detection of *Mycobacterium tuberculosis* genes essential for bacterial survival in guinea pigs, compared with mice. *J Infect Dis* **195**: 1634–1642.

18. Dutta NK, Mehra S, Didier PJ, Roy CJ, Doyle LA, Alvarez X, Ratterree M, Be NA, Lamichhane G, Jain SK, Lacey MR, Lackner AA, Kaushal D. 2010. Genetic requirements for the survival of tubercle bacilli in primates. *J Infect Dis* **201**:1743–1752.

19. Griffin JE, Gawronski JD, DeJesus MA, Ioerger TR, Akerley BJ, Sassetti CM. 2011. High-resolution phenotypic profiling defines genes essential for mycobacterial growth and cholesterol catabolism. *PLoS Pathog* **7**:e1002251.

20. Mckinney JD, zu Bentrup KH, Muñoz-Elías EJ, Miczak A, Chen B, Chan W-T, Swenson D, Sacchettini JC, Jacobs WR, Russell DG. 2000. Persistence of *Mycobacterium tuberculosis* in macrophages and mice requires the glyoxylate shunt enzyme isocitrate lyase. *Nature* **406**:735–738.

21. Liu M, Vinšová J, Novotná E, Mandíková J, Wsól V, Trejtnar F, Ulmann V, Stolaříková J, Fernandes S, Bhat S, Liu JO. 2012. Salicylanilide derivatives block *Mycobacterium tuberculosis* through inhibition of isocitrate lyase and methionine aminopeptidase. *Tuberculosis (Edinb)* **92:**434–439.

22. Krieger IV, Freundlich JS, Gawandi VB, Roberts JP, Gawandi VB, Sun Q, Owen JL, Fraile MT, Huss SI, Lavandera J-L, Ioerger TR, Sacchettini JC. 2012. Structure-guided discovery of phenyl-diketo acids as potent inhibitors of *M. tuberculosis* malate synthase. *Chem Biol* **19:**1556–1567.

23. Beste DJV, Bonde B, Hawkins N, Ward JL, Beale MH, Noack S, Nöh K, Kruger NJ, Ratcliffe RG, McFadden J. 2011. ^{13}C Metabolic flux analysis identifies an unusual route for pyruvate dissimilation in mycobacteria which requires isocitrate lyase and carbon dioxide fixation. *PLoS Pathog* **7:**e1002091.

24. Buchieri MV, Riafrecha LE, Rodríguez OM, Vullo D, Morbidoni HR, Supuran CT, Colinas PA. 2013. Inhibition of the β-carbonic anhydrases from *Mycobacterium tuberculosis* with C-cinnamoyl glycosides: identification of the first inhibitor with anti-mycobacterial activity. *Bioorg Med Chem Lett* **23:**740–743.

25. Apfel CM, Locher H, Evers S, Takács B, Hubschwerlen C, Pirson W, Page MG, Keck W. 2001. Peptide deformylase as an antibacterial drug target: target validation and resistance development. *Antimicrob Agents Chemother* **45:**1058–1064.

26. Teo JWP, Thayalan P, Beer D, Yap ASL, Nanjundappa M, Ngew X, Duraiswamy J, Liung S, Dartois V, Schreiber M, Hasan S, Cynamon M, Ryder NS, Yang X, Weidmann B, Bracken K, Dick T, Mukherjee K. 2006. Peptide deformylase inhibitors as potent antimycobacterial agents. *Antimicrob Agents Chemother* **50:**3665–3673.

27. Olaleye O, Raghunand TR, Bhat S, He J, Tyagi S, Lamichhane G, Gu P, Zhou J, Zhang Y, Grosset J, Bishai WR, Liu JO. 2010. Methionine aminopeptidases from *Mycobacterium tuberculosis* as novel antimycobacterial targets. *Chem Biol* **17:**86–97.

28. Hartkoorn RC, Sala C, Neres J, Pojer F, Magnet S, Mukherjee R, Uplekar S, Boy-Röttger S, Altmann K-H, Cole ST. 2012. Towards a new tuberculosis drug: pyridomycin - nature's isoniazid. *EMBO Mol Med* **4:**1032–1042.

29. Willand N, Dirié B, Carette X, Bifani P, Singhal A, Desroses M, Leroux F, Willery E, Mathys V, Déprez-Poulain R, Delcroix G, Frénois F, Aumercier M, Locht C, Villeret V, Déprez B, Baulard AR. 2009. Synthetic EthR inhibitors boost antituberculous activity of ethionamide. *Nat Med* **15:**537–544.

30. Engohang-Ndong J. 2012. Antimycobacterial drugs currently in phase II clinical trials and preclinical phase for tuberculosis treatment. *Expert Opin Investig Drugs* **21:**1789–1800.

31. Hopkins AL, Groom CR. 2002. The druggable genome. *Nat Rev Drug Discov* **1:**727–730.

32. Cheng AC, Coleman RG, Smyth KT, Cao Q, Soulard P, Caffrey DR, Salzberg AC, Huang ES. 2007. Structure-based maximal affinity model predicts small-molecule druggability. *Nat Biotechnol* **25:**71–75.

33. Wei J-R, Krishnamoorthy V, Murphy K, Kim J-H, Schnappinger D, Alber T, Sassetti CM, Rhee KY, Rubin EJ. 2011. Depletion of antibiotic targets has widely varying effects on growth. *Proc Natl Acad Sci USA* **108:**4176–4181.

34. Wyatt PG, Gilbert IH, Read KD, Fairlamb AH. 2011. Target validation: linking target and chemical properties to desired product profile. *Curr Top Med Chem* **11:**1275.

35. Simmons KJ, Chopra I, Fishwick CWG. 2010. Structure-based discovery of antibacterial drugs. *Nat Rev Microbiol* **8:**501–510.

36. Akerley BJ, Rubin EJ, Camilli A, Lampe DJ, Robertson HM, Mekalanos JJ. 1998. Systematic identification of essential genes by in vitro mariner mutagenesis. *Proc Natl Acad Sci USA* **95:**8927–8932.

37. Brune W, Ménard C, Hobom U, Odenbreit S, Messerle M, Koszinowski UH. 1999. Rapid identification of essential and nonessential herpesvirus genes by direct transposon mutagenesis. *Nat Biotechnol* **17:**360–364.

38. Peter A, Schöttler P, Werner M, Beinert N, Dowe G, Burkert P, Mourkioti F, Dentzer L, He Y, Deak P, Benos PV, Gatt MK, Murphy L, Harris D, Barrell B, Ferraz C, Vidal S, Brun C, Demaille J, Cadieu E, Dreano S, Gloux S, Lelaure V, Mottier S, Galibert F, Borkova D, Miñana B, Kafatos FC, Bolshakov S, Sidén-Kiamos I, Papagiannakis G, Spanos L, Louis C, Madueño E, de Pablos B, Modolell J, Bucheton A, Callister D, Campbell L, Henderson NS, McMillan PJ, Salles C, Tait E, Valenti P, Saunders RDC, Billaud A, Pachter L, Klapper R, Janning W, Glover DM, Ashburner M, Bellen HJ, Jäckle H, Schäfer U. 2002. Mapping and identification of essential gene functions on the X chromosome of *Drosophila*. *EMBO Rep* **3:**34–38.

39. Zhang YJ, Ioerger TR, Huttenhower C, Long JE, Sassetti CM, Sacchettini JC, Rubin EJ. 2012. Global assessment of genomic regions required for growth in *Mycobacterium tuberculosis*. *PLoS Pathog* **8:**e1002946.

40. Rengarajan J, Bloom BR, Rubin EJ. 2005. Genome-wide requirements for *Mycobacterium tuberculosis* adaptation and survival in macrophages. *Proc Natl Acad Sci USA* **102:**8327–8332.

41. Chitale S, Ehrt S, Kawamura I, Fujimura T, Shimono N, Anand N, Lu S, Cohen-Gould L, Riley LW. 2001. Recombinant *Mycobacterium tuberculosis* protein associated with mammalian cell entry. *Cell Microbiol* **3:**247–254.

42. Pandey AK, Sassetti CM. 2008. Mycobacterial persistence requires the utilization of host cholesterol. *Proc Natl Acad Sci USA* **105:**4376–4380.

43. Chang JC, Miner MD, Pandey AK, Gill WP, Harik NS, Sassetti CM, Sherman DR. 2009. *igr* Genes and *Mycobacterium tuberculosis* cholesterol metabolism. *J Bacteriol* **191:**5232–5239.

44. Muñoz-Elías EJE, McKinney JDJ. 2005. *Mycobacterium tuberculosis* isocitrate lyases 1 and 2 are jointly required for in vivo growth and virulence. *Nat Med* **11:**638–644.

45. Griffin JE, Pandey AK, Gilmore SA, Mizrahi V, Mckinney JD, Bertozzi CR, Sassetti CM. 2012. Cholesterol catabolism by *Mycobacterium tuberculosis* requires transcriptional and metabolic adaptations. *Chem Biol* **19**:218–227.

46. Yam KC, D'Angelo I, Kalscheuer R, Zhu H, Wang J-X, Snieckus V, Ly LH, Converse PJ, Jacobs WR, Strynadka N, Eltis LD. 2009. Studies of a ring-cleaving dioxygenase illuminate the role of cholesterol metabolism in the pathogenesis of *Mycobacterium tuberculosis*. *PLoS Pathog* **5**:e1000344.

47. Vandal OH, Pierini LM, Schnappinger D, Nathan CF, Ehrt S. 2008. A membrane protein preserves intrabacterial pH in intraphagosomal *Mycobacterium tuberculosis*. *Nat Med* **14**:849–854.

48. Nathan C, Gold B, Lin G, Stegman M, de Carvalho LPS, Vandal O, Venugopal A, Bryk R. 2008. A philosophy of anti-infectives as a guide in the search for new drugs for tuberculosis. *Tuberculosis (Edinb)* **88**:S25–S33.

49. Kohanski M, DePristo M, Collins J. 2010. Sublethal antibiotic treatment leads to multidrug resistance via radical-induced mutagenesis. *Mol Cell* **37**:311–320.

50. Nichols RJ, Sen S, Choo YJ, Beltrao P, Zietek M, Chaba R, Lee S, Kazmierczak KM, Lee KJ, Wong A, Shales M, Lovett S, Winkler ME, Krogan NJ, Typas A, Gross CA. 2011. Phenotypic landscape of a bacterial cell. *Cell* **144**:143–156.

51. Girgis HS, Hottes AK, Tavazoie S, Herman C. 2009. Genetic architecture of intrinsic antibiotic susceptibility. *PLoS One* **4**:e5629.

52. Joshi SM, Pandey AK, Capite N, Fortune SM, Rubin EJ, Sassetti CM. 2006. Characterization of mycobacterial virulence genes through genetic interaction mapping. *Proc Natl Acad Sci USA* **103**:11760–11765.

53. Lee W, VanderVen BC, Fahey RJ, Russell DG. 2013. Intracellular *Mycobacterium tuberculosis* exploits host-derived fatty acids to limit metabolic stress. *J Biol Chem* **288**:6788–6800.

54. Gopinath K, Venclovas C, Ioerger TR, Sacchettini JC, McKinney JD, Mizrahi V, Warner DF. 2013. A vitamin B12 transporter in *Mycobacterium tuberculosis*. *Open Biol* **3**:120175.

55. Silver LL. 2007. Multi-targeting by monotherapeutic antibacterials. *Nat Rev Drug Discov* **6**:41–55.

56. Roemer T, Schneider T, Pinho MG. 2013. Auxiliary factors: a chink in the armor of MRSA resistance to β-lactam antibiotics. *Curr Opin Microbiol* **16**:1–11.

57. Komatsuzawa H, Suzuki J, Sugai M, Miyake Y, Suginaka H. 1994. Effect of combination of oxacillin and non-β-lactam antibiotics on methicillin-resistant *Staphylococcus aureus*. *J Antimicrob Chemother* **33**:1155–1163.

58. Swoboda JG, Meredith TC, Campbell J, Brown S, Suzuki T, Bollenbach T, Malhowski AJ, Kishony R, Gilmore MS, Walker S. 2009. Discovery of a small molecule that blocks wall teichoic acid biosynthesis in *Staphylococcus aureus*. *ACS Chem Biol* **4**:875–883.

59. Campbell J, Singh AK, Santa Maria JP, Kim Y, Brown S, Swoboda JG, Mylonakis E, Wilkinson BJ, Walker S.

2010. Synthetic lethal compound combinations reveal a fundamental connection between wall teichoic acid and peptidoglycan biosyntheses in *Staphylococcus aureus*. *ACS Chem Biol* **6**:106–116.

60. Wang H, Gill CJ, Lee SH, Mann P, Zuck P, Meredith TC, Murgolo N, She X, Kales S, Liang L, Liu J, Wu J, Maria JS, Su J, Pan J, Hailey J, Mcguinness D, Tan CM, Flattery A, Walker S, Black T, Roemer T. 2013. Discovery of wall teichoic acid inhibitors as potential anti-MRSA β-lactam combination agents. *Chem Biol* **20**:272–284.

61. Donald RGK, Skwish S, Forsyth RA, Anderson JW, Zhong T, Burns C, Lee S, Meng X, LoCastro L, Jarantow LW, Martin J, Lee SH, Taylor I, Robbins D, Malone C, Wang L, Zamudio CS, Youngman PJ, Phillips JW. 2009. A *Staphylococcus aureus* fitness test platform for mechanism-based profiling of antibacterial compounds. *Chem Biol* **16**:826–836.

62. Lee SH, Jarantow LW, Wang H, Sillaots S, Cheng H, Meredith TC, Thompson J, Roemer T. 2011. Antagonism of chemical genetic interaction networks resensitize MRSA to β-lactam antibiotics. *Chem Biol* **18**:1379–1389.

63. Tan CM, Therien AG, Lu J, Lee SH, Caron A, Gill CJ, Lebeau-Jacob C, Benton-Perdomo L, Monteiro JM, Pereira PM, Elsen NL, Wu J, Deschaps K, Petcu M, Wong S, Daigneault E, Kramer S, Liang L, Maxwell E, Claveau D, Vaillancourt JP, Skorey K, Tam J, Wang H, Meredith TC, Sillaots S, Wang-Jarantow L, Ramtohul Y, Langlois E, Landry F, Reid JC, Parthasarathy G, Sharma S, Baryshnikova A, Lumb KJ, Pinho MG, Soisson SM, Roemer T. 2012. Restoring methicillin-resistant *Staphylococcus aureus* susceptibility to β-lactam antibiotics. *Sci Trans Med* **4**:1–12.

64. Therien AG, Huber JL, Wilson KE, Beaulieu P, Caron A, Claveau D, Deschaps K, Donald RGK, Galgoci AM, Gallant M, Gu X, Kevin NJ, Lafleur J, Leavitt PS, Lebeau-Jacob C, Lee SS, Lin MM, Michels AA, Ogawa AM, Painter RE, Parish CA, Park Y-W, Benton-Perdomo L, Petcu M, Phillips JW, Powles MA, Skorey KI, Tam J, Tan CM, Young K, Wong S, Waddell ST, Miesel L. 2012. Broadening the spectrum of β-lactam antibiotics through inhibition of signal peptidase type I. *Antimicrob Agents Chemother* **56**:4662–4670.

65. Ananthan S, Faaleolea ER, Goldman RC, Hobrath JV, Kwong CD, Laughon BE, Maddry JA, Mehta A, Rasmussen L, Reynolds RC, Secrist JA III, Shindo N, Showe DN, Sosa MI, Suling WJ, White EL. 2009. High-throughput screening for inhibitors of *Mycobacterium tuberculosis* H37Rv. *Tuberculosis (Edinb)* **89**:334–353.

66. Ekins S, Kaneko T, Lipinski CA, Bradford J, Dole K, Spektor A, Gregory K, Blondeau D, Ernst S, Yang J, Goncharoff N, Hohman MM, Bunin BA. 2010. Analysis and hit filtering of a very large library of compounds screened against *Mycobacterium tuberculosis*. *Mol BioSyst* **6**:2316–2324.

67. Ekins S, Freundlich JS, Choi I, Sarker M, Talcott C. 2011. Computational databases, pathway and cheminformatics tools for tuberculosis drug discovery. *Trends Microbiol* **19**:65–74.

68. Andries K. 2005. A diarylquinoline drug active on the ATP synthase of *Mycobacterium tuberculosis*. *Science* 307:223–227.

69. Wang H, Claveau D, Vaillancourt JP, Roemer T, Meredith TC. 2011. High-frequency transposition for determining antibacterial mode of action. *Nat Chem Biol* 7:720–729.

70. Smith AM, Ammar R, Nislow C, Giaever G. 2010. A survey of yeast genomic assays for drug and target discovery. *Pharmacol Ther* 127:156–164.

71. DeVito JA, Mills JA, Liu VG, Agarwal A, Sizemore CF, Yao Z, Stoughton DM, Cappiello MG, Barbosa MDFS, Foster LA, Pompliano DL. 2002. An array of target-specific screening strains for antibacterial discovery. *Nat Biotechnol* 20:478–483.

72. Xu HH, Trawick JD, Haselbeck RJ, Forsyth RA, Yamamoto RT, Archer R, Patterson J, Allen M, Froelich JM, Taylor I, Nakaji D, Maile R, Kedar GC, Pilcher M, Brown-Driver V, McCarthy M, Files A, Robbins D, King P, Sillaots S, Malone C, Zamudio CS, Roemer T, Wang L, Youngman PJ, Wall D. 2010. *Staphylococcus aureus* TargetArray: comprehensive differential essential gene expression as a mechanistic tool to profile antibacterials. *Antimicrob Agents Chemother* 54:3659–3670.

73. Roemer T, Xu D, Singh SB, Parish CA, Harris G, Wang H, Davies JE, Bills GF. 2011. Confronting the challenges of natural product-based antifungal discovery. *Chem Biol* 18:148–164.

74. Lum PY, Armour CD, Stepaniants SB, Cavet G, Wolf MK, Butler JS, Hinshaw JC, Garnier P, Prestwich GD, Leonardson A, Garrett-Engele P, Rush CM, Bard M, Schimmack G, Phillips JW, Roberts CJ, Shoemaker DD. 2004. Discovering modes of action for therapeutic compounds using a genome-wide screen of yeast heterozygotes. *Cell* 116:121–137.

75. Luesch H, Wu TYH, Ren P, Gray NS, Schultz PG, Supek F. 2005. A genome-wide overexpression screen in yeast for small-molecule target identification. *Chem Biol* 12:55–63.

76. Rengarajan J, Sassetti CM, Naroditskaya V, Sloutsky A, Bloom BR, Rubin EJ. 2004. The folate pathway is a target for resistance to the drug *para*-aminosalicylic acid (PAS) in mycobacteria. *Mol Microbiol* 53:275–282.

77. Judson N, Mekalanos JJ. 2000. TnAraOut, a transposon-based approach to identify and characterize essential bacterial genes. *Nat Biotechnol* 18:740–745.

78. Warner JR, Reeder PJ, Karimpour-Fard A, Woodruff LBA, Gill RT. 2010. Rapid profiling of a microbial genome using mixtures of barcoded oligonucleotides. *Nat Biotechnol* 28:856–862.

79. Woong Park S, Klotzsche M, Wilson DJ, Boshoff HI, Eoh H, Manjunatha U, Blumenthal A, Rhee K, Barry CE, Aldrich CC, Ehrt S, Schnappinger D. 2011. Evaluating the sensitivity of *Mycobacterium tuberculosis* to biotin deprivation using regulated gene expression. *PLoS Pathog* 7:e1002264.

80. Kim JH, Wei JR, Wallach JB, Robbins RS, Rubin EJ, Schnappinger D. 2011. Protein inactivation in myco-bacteria by controlled proteolysis and its application to deplete the beta subunit of RNA polymerase. *Nucleic Acids Res* 39:2210–2220.

81. Blumenthal A, Trujillo C, Ehrt S, Schnappinger D. 2010. Simultaneous analysis of multiple *Mycobacterium tuberculosis* knockdown mutants in vitro and in vivo. *PLoS One* 5:e15667.

82. Janagama HK, Hassounah HA, Cirillo SLG, Cirillo JD. 2011. Random inducible controlled expression (RICE) for identification of mycobacterial virulence genes. *Tuberculosis (Edinb)* 91:S66–S68.

83. Speers AE, Adam GC, Cravatt BF. 2003. Activity-based protein profiling in vivo using a copper(I)-catalyzed azide-alkyne [3 + 2] cycloaddition. *J Am Chem Soc* 125:4686–4687.

84. Ansong C, Ortega C, Payne SH, Haft DH, Chauvignè-Hines LM, Lewis MP, Ollodart AR, Purvine SO, Shukla AK, Fortuin S, Smith RD, Adkins JN, Grundner C, Wright AT. 2013. Identification of widespread adenosine nucleotide binding in *Mycobacterium tuberculosis*. *Chem Biol* 20:123–133.

85. Pethe K, Sequeira PC, Agarwalla S, Rhee K, Kuhen K, Phong WY, Patel V, Beer D, Walker JR, Duraiswamy J, Jiricek J, Keller TH, Chatterjee A, Tan MP, Ujjini M, Rao SPS, Camacho L, Bifani P, Mak PA, Ma I, Barnes SW, Chen Z, Plouffe D, Thayalan P, Ng SH, Au M, Lee BH, Tan BH, Ravindran S, Nanjundappa M, Lin X, Goh A, Lakshminarayana SB, Shoen C, Cynamon M, Kreiswirth B, Dartois V, Peters EC, Glynne R, Brenner S, Dick T. 2010. A chemical genetic screen in *Mycobacterium tuberculosis* identifies carbon-source-dependent growth inhibitors devoid of in vivo efficacy. *Nat Commun* 1:1–8.

86. Gandotra S, Schnappinger D, Monteleone M, Hillen W, Ehrt S. 2007. In vivo gene silencing identifies the *Mycobacterium tuberculosis* proteasome as essential for the bacteria to persist in mice. *Nat Med* 13:1515–1520.

87. Raju RM, Goldberg AL, Rubin EJ. 2012. Bacterial proteolytic complexes as therapeutic targets. *Nat Rev Drug Discov* 11:777–789.

88. Ollinger J, O'Malley T, Kesicki EA, Odingo J, Parish T. 2012. Validation of the essential ClpP protease in *Mycobacterium tuberculosis* as a novel drug target. *J Bacteriol* 194:663–668.

89. Chopra S, Matsuyama K, Tran T, Malerich JP, Wan B, Franzblau SG, Lun S, Guo H, Maiga MC, Bishai WR, Madrid PB. 2012. Evaluation of gyrase B as a drug target in *Mycobacterium tuberculosis*. *J Antimicrob Chemother* 67:415–421.

90. Tullius MV, Harth G, Horwitz MA. 2003. Glutamine synthetase GlnA1 is essential for growth of *Mycobacterium tuberculosis* in human THP-1 macrophages and guinea pigs. *Infect Immun* 71:3927–3936.

91. Carroll P, Faray-Kele M-C, Parish T. 2011. Identifying vulnerable pathways in *Mycobacterium tuberculosis* by using a knockdown approach. *Appl Environ Microbiol* 77:5040–5043.

92. Lipinski CA, Lombardo F, Dominy BW, Feeney PJ. 1997. Experimental and computational approaches to

estimate solubility and permeability in drug discovery and development settings. *Adv Drug Deliv Rev* **23**: 3–25.

93. Brown JR, North EJ, Hurdle JG, Morisseau C, Scarborough JS, Sun D, Korduláková J, Scherman MS, Jones V, Grzegorzewicz A, Crew RM, Jackson M, McNeil MR, Lee RE. 2011. The structure-activity relationship of urea derivatives as anti-tuberculosis agents. *Bioorg Med Chem* **19**:5585–5595.

94. Stanley SA, Grant SS, Kawate T, Iwase N, Shimizu M, Wivagg C, Silvis M, Kazyanskaya E, Aquadro J, Golas A, Fitzgerald M, Dai H, Zhang L, Hung DT. 2012. Identification of novel inhibitors of *M. tuberculosis* growth using whole cell based high-throughput screening. *ACS Chem Biol* **7**:1377–1384.

95. Stover CK, Warrener P, VanDevanter DR, Sherman DR, Arain TM, Langhorne MH, Anderson SW, Towell JA, Yuan Y, McMurray DN, Kreiswirth BN, Barry CE, Baker WR. 2000. A small-molecule nitroimidazopyran drug candidate for the treatment of tuberculosis. *Nature* **405**:962–966.

96. Tahlan K, Wilson R, Kastrinsky DB, Arora K, Nair V, Fischer E, Barnes SW, Walker JR, Alland D, Barry CE, Boshoff HI. 2012. SQ109 targets MmpL3, a membrane transporter of trehalose monomycolate involved in mycolic acid donation to the cell wall core of *Mycobacterium tuberculosis*. *Antimicrob Agents Chemother* **56**: 1797–1809.

97. Makarov V, Manina G, Mikusova K, Mollmann U, Ryabova O, Saint-Joanis B, Dhar N, Pasca MR, Buroni S, Lucarelli AP, Milano A, De Rossi E, Belanova M, Bobovska A, Dianiskova P, Kordulakova J, Sala C, Fullam E, Schneider P, McKinney JD, Brodin P, Christophe T, Waddell S, Butcher P, Albrethsen J, Rosenkrands I, Brosch R, Nandi V, Bharath S, Gaonkar S, Shandil RK, Balasubramanian V, Balganesh T, Tyagi S, Grosset J, Riccardi G, Cole ST. 2009. Benzothiazinones kill *Mycobacterium tuberculosis* by blocking arabinan synthesis. *Science* **324**:801–804.

98. Poce G, Bates RH, Alfonso S, Cocozza M, Porretta GC, Ballell L, Rullas J, Ortega F, De Logu A, Agus E, La Rosa V, Pasca MR, De Rossi E, Wae B, Franzblau SG, Manetti F, Botta M, Biava M. 2013. Improved BM212 MmpL3 inhibitor analogue shows efficacy in acute murine model of tuberculosis infection. *PLoS One* **8**: e56980. doi:10.1371/journal.pone.0056980.

99. Mitchison DA. 1979. Basic mechanisms of chemotherapy. *Chest* **76**:771–781.

100. Baek S-H, Li AH, Sassetti CM. 2011. Metabolic regulation of mycobacterial growth and antibiotic sensitivity. *PLoS Biol* **9**:e1001065.

101. Mitchison DA. 2000. Role of individual drugs in the chemotherapy of tuberculosis. *Int J Tuberc Lung Dis* **4**: 796–806.

102. Jindani A, Doré CJ, Mitchison DA. 2003. Bactericidal and sterilizing activities of antituberculosis drugs during the first 14 days. *Am J Respir Crit Care Med* **167**: 1348–1354.

103. Fox W. 1979. The chemotherapy of pulmonary tuberculosis: a review. *Chest* **76**:785–796.

104. World Health Organization. *Global Tuberculosis Report 2012*. WHO Press, Geneva.

105. Diacon AH, Dawson R, von Groote-Bidlingmaier F, Symons G, Venter A, Donald PR, van Niekerk C, Everitt D, Winter H, Becker P, Mendel CM, Spigelman MK. 2012. 14-day bactericidal activity of PA-824, bedaquiline, pyrazinamide, and moxifloxacin combinations: a randomised trial. *Lancet* **380**:986–993.

106. Zhang Y, Mitchison D. 2003. The curious characteristics of pyrazinamide: a review. *Int J Tuberc Lung Dis* **7**:6–21.

107. Shi W, Zhang X, Jiang X, Yuan H, Lee JS, Barry CE, Wang H, Zhang W, Zhang Y. 2011. Pyrazinamide inhibits trans-translation in *Mycobacterium tuberculosis*. *Science* **333**:1630–1632.

108. Heifets L, Lindholm-Levy P. 1992. Pyrazinamide sterilizing activity *in vitro* against semidormant *Mycobacterium tuberculosis* bacterial populations. *Am Rev Respir Dis* **145**:1223–1225.

109. Warner DF, Mizrahi V. 2006. Tuberculosis chemotherapy: the influence of bacillary stress and damage response pathways on drug efficacy. *Clin Microbiol Rev* **19**:558–570.

110. Crowle AJ, Salfinger M, May MH. 1989. $^{1,25}(OH)_2$-vitamin D3 synergizes with pyrazinamide to kill tubercle bacilli in cultured human macrophages. *Am Rev Respir Dis* **139**:549–552.

111. Schnappinger D, Ehrt S, Voskuil M, Liu Y, Mangan J, Monahan I, Dolganov G, Efron B, Butcher P, Nathan C, Schoolnik G. 2003. Transcriptional adaptation of *Mycobacterium tuberculosis* within macrophages: insights into the phagosomal environment. *J Exp Med* **198**:693–704.

112. Betts JC, Lukey PT, Robb LC, McAdam RA, Duncan K. 2002. Evaluation of a nutrient starvation model of *Mycobacterium tuberculosis* persistence by gene and protein expression profiling. *Mol Microbiol* **43**:717–731.

113. Hampshire T, Soneji S, Bacon J, James B, Hinds J, Laing K, Stabler R, Marsh P, Butcher P. 2004. Stationary phase gene expression of *Mycobacterium tuberculosis* following a progressive nutrient depletion: a model for persistent organisms? *Tuberculosis (Edinb)* **84**:228–238.

114. Deb C, Lee C-M, Dubey VS, Daniel J, Bassam A, Pawar S, Rogers L, Kolattukudy PE. 2009. A novel in vitro multiple-stress dormancy model for *Mycobacterium tuberculosis* generates a lipid-loaded, drug-tolerant, dormant pathogen. *PLoS One* **4**:e6077.

115. Archuleta R, Yvonne Hoppes P, Primm T. 2005. *Mycobacterium avium* enters a state of metabolic dormancy in response to starvation. *Tuberculosis (Edinb)* **85**:147–158.

116. Wayne L, Sohaskey C. 2001. Nonreplicating persistence of *Mycobacterium tuberculosis*. *Annu Rev Microbiol* **55**:139–163.

117. Rosenkrands I, Slayden RA, Crawford J, Aagaard C, Barry CE, Andersen P. 2002. Hypoxic response of *Mycobacterium tuberculosis* studied by metabolic labeling

and proteome analysis of cellular and extracellular proteins. *J Bacteriol* **184**:3485–3491.

118. Park H, Guinn K, Harrell M, Liao R, Voskuil M, Tompa M, Schoolnik G, Sherman D. 2003. Rv3133c/dosR is a transcription factor that mediates the hypoxic response of *Mycobacterium tuberculosis*. *Mol Microbiol* **48**:833–843.

119. Fisher MA, Plikaytis BB, Shinnick TM. 2002. Microarray analysis of the *Mycobacterium tuberculosis* transcriptional response to the acidic conditions found in phagosomes. *J Bacteriol* **184**:4025–4032.

120. Voskuil M, Visconti K, Schoolnik G. 2004. *Mycobacterium tuberculosis* gene expression during adaptation to stationary phase and low-oxygen dormancy. *Tuberculosis (Edinb)* **84**:218–227.

121. Bryk R, Gold B, Venugopal A, Singh J, Samy R, Pupek K, Cao H, Popescu C, Gurney M, Hotha S, Cherian J, Rhee K, Ly L, Converse PJ, Ehrt S, Vandal O, Jiang X, Schneider J, Lin G, Nathan C. 2008. Selective killing of nonreplicating mycobacteria. *Cell Host Microbe* **3**:137–145.

122. Cho SH, Warit S, Wan B, Hwang CH, Pauli GF, Franzblau SG. 2007. Low-oxygen-recovery assay for high-throughput screening of compounds against nonreplicating *Mycobacterium tuberculosis*. *Antimicrob Agents Chemother* **51**:1380–1385.

123. Grant SS, Kawate T, Nag PP, Silvis MR, Gordon K, Stanley SA, Kazyanskaya E, Nietupski R, Golas A, Fitzgerald M, Cho S, Franzblau SG, Hung DT. 2013. Identification of novel inhibitors of non-replicating *M. tuberculosis* using a carbon starvation model. *ACS Chem Biol* **8**:2224–2234.

124. Sala C, Dhar N, Hartkoorn RC, Zhang M, Ha YH, Schneider P, Cole ST. 2010. Simple model for testing drugs against nonreplicating *Mycobacterium tuberculosis*. *Antimicrob Agents Chemother* **54**:4150–4158.

125. Zhang M, Sala C, Hartkoorn RC, Dhar N, Mendoza-Losana A, Cole ST. 2012. Streptomycin-starved *Mycobacterium tuberculosis* 18b, a drug discovery tool for latent tuberculosis. *Antimicrob Agents Chemother* **56**:5782–5789.

126. Murphy DJ, Brown JR. 2007. Identification of gene targets against dormant phase *Mycobacterium tuberculosis* infections. *BMC Infect Dis* **7**:84.

127. Singla D, Tewari R, Kumar A, Raghava GP, Open Source Drug Discovery Consortium. 2013. Designing of inhibitors against drug tolerant *Mycobacterium tuberculosis* (H37Rv). *Chem Central J* **7**:49.

128. Wang J, Soisson SM, Young K, Shoop W, Kodali S, Galgoci A, Painter R, Parthasarathy G, Tang YS, Cummings R, Ha S, Dorso K, Motyl M, Jayasuriya H, Ondeyka J, Herath K, Zhang C, Hernandez L, Allocco J, Basilio A, Tormo JR, Genilloud O, Vicente F, Pelaez F, Colwell L, Lee SH, Michael B, Felcetto T, Gill C, Silver LL, Hermes JD, Bartizal K, Barrett J, Schmatz D, Becker JW, Cully D, Singh SB. 2006. Platensimycin is a selective FabF inhibitor with potent antibiotic properties. *Nature* **441**:358–361.

129. Abrahams GL, Kumar A, Savvi S, Hung AW, Wen S, Abell C, Barry CE III, Sherman DR, Boshoff HIM, Mizrahi V. 2012. Pathway-selective sensitization of *Mycobacterium tuberculosis* for target-based whole-cell screening. *Chem Biol* **19**:844–854.

130. Kumar A, Zhang M, Zhu L, Liao RP, Mutai C, Hafsat S, Sherman DR, Wang M-W. 2012. High-throughput screening and sensitized bacteria identify an *M. tuberculosis* dihydrofolate reductase inhibitor with whole cell activity. *PLoS One* **7**:e39961.

Genetics of Membrane and Cell Wall Biosynthesis

VI

Molecular Genetics of Mycobacteria, 2nd Edition
Edited by Graham F. Hatfull and William R. Jacobs, Jr.
© 2014 American Society for Microbiology, Washington, DC
doi:10.1128/microbiolspec.MGM2-0034-2013

Martin S. Pavelka Jr.[1]
Sebabrata Mahapatra[2]
Dean C. Crick[2]

Genetics of Peptidoglycan Biosynthesis

26

The central core of the mycobacterial cell envelope consists of the mycolyl-arabinogalactan-peptidoglycan complex, also known as the MAPc. Anchoring the entire MAPc is the peptidoglycan (PG), which is composed of a glycan chain with alternating N-acylated glucosamine (GlcNAc) and muramic acid (MurNAcyl) residues bearing peptide chains, which may be cross-linked (1, 2). The PG of mycobacteria belongs to the A1γ chemotype along with that of *Escherichia coli* and a number of other organisms, but it has modifications to the monomeric units and other structural aspects that are likely related to a role for the PG in stabilizing the mycobacterial MAPc (2, 3). In this article, we will review the genetics of several aspects of PG biosynthesis in mycobacteria, including the production of monomeric precursors in the cytoplasm, assembly of the monomers into the mature wall, wall turnover, and cell division. Finally, we will touch upon the resistance of mycobacteria to β-lactam antibiotics, an important class of drugs that, until recently, have not been extensively investigated as potential antimycobacterial agents.

The glycan chain of the mycobacterial PG consists of alternating GlcNAc and MurNAcyl residues in a β-(1,4) linkage as seen in other bacteria, except that the MurNAcyl residues can be either N-glycolylated (MurNGlyc) or N-acetylated (MurNAc) (Fig. 1) (4, 5). The peptide chain attached to the lactyl moiety of MurNAcyl is similar to that of *E. coli* and many rod-shaped Gram-positive bacteria, typically consisting of L-alanyl-D-glutamine-*meso*-diaminopimelyl-D-alanyl-D-alanine (Fig. 1) (3). Variations in the peptide include the replacement of L-Ala with Gly in *Mycobacterium leprae*, amidation of the free carboxyl group of D-Glu (generating D-*iso*-Glu), and the amidation of the free carboxyl of the side chain of *meso*-diaminopimelate (DAP) (Fig. 2A and 2B) (2, 6, 7). The N-glycolylation of MurNAcyl is uncommon, only present in mycobacteria (an exception is *M. leprae*) and other closely related actinobacteria (4); this modification contributes to the lysozyme-resistant phenotype of mycobacteria (8) and potentiates the recognition of PG by the innate immunity protein Nod2 (9). The significance of the other modifications to mycobacterial PG biology is unknown.

The mature PG architecture is also marked by a high degree of direct peptide cross-links. Overall, 70 to 80%

[1]University of Rochester Medical Center, Department of Microbiology and Immunology, Rochester, NY 14642; [2]Mycobacteria Research Laboratories, Department of Microbiology, Immunology and Pathology, Colorado State University, Fort Collins, CO 80523.

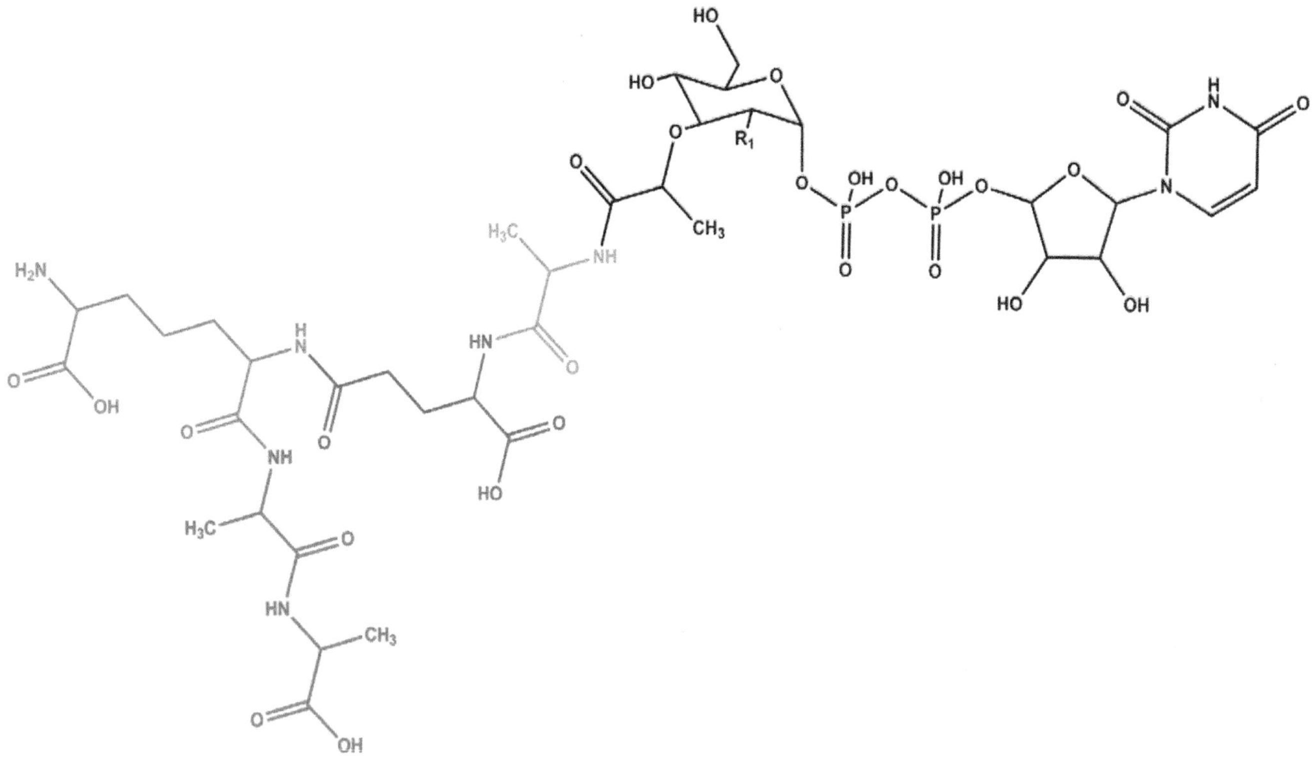

Figure 1 PG nucleotide precursor (Park's nucleotide). Basic structure of the PG monomer precursor with the muropeptide L-alanyl-D-glutaminyl-*meso*-DAP-D-alanyl-D-alanine. R_1 denotes the presence of either an *N*-acetyl or *N*-glycolyl modification of the muramic acid moiety. L-Ala, D-Glu, *meso*-DAP, and D-Ala are depicted in gold, blue, green, and red, respectively. doi:10.1128/microbiolspec.MGM2-0034-2013.f1

of the peptides are cross-linked, in two kinds of linkages. One type is between the D-Ala at position 4 of one peptide and the *meso*-DAP at position 3 of an adjacent peptide (Fig. 2A). These "4-3" linkages are catalyzed by typical D,D-transpeptidases, also known as penicillin-binding proteins (PBPs), which can be inhibited by various classes of β-lactam antibiotics (10). The other type of linkage is between two DAP residues, also known as a "3-3" linkage (Fig. 2B), that are catalyzed by the concerted activity of D,D-carboxypeptidases and L,D-transpeptidases, the latter of which have been found to be resistant to most β-lactam antibiotics, except for the carbapenem class (11–15). The role of 3-3 linkages is chiefly unknown (16), but they were first discovered in mycobacteria and they exist in a high proportion compared to most other bacteria, suggesting a correlation between the degree and type of cross-linking with the complexity of the mycobacterial cell envelope. It has subsequently been shown that 3-3 linkages are present in many other species of bacteria and that the relative percentage of linkage type can shift depending upon growth conditions or growth

phase (17, 18). Early studies examining mycobacterial PG showed that about one-third of the cross-links are 3-3 (19), but more recent work suggests that the percentage is much higher, in the 60 to 80% range. The percentage of 3-3 to 4-3 linkages appears to be constant, regardless of growth stage, in *M. tuberculosis* (20) and *Mycobacterium abscessus* (21).

Mature mycobacterial PG is covalently connected to the galactan portion of the arabinogalactan of the MAPc via a disaccharide linker unit consisting of a rhamnose residue and *N*-acetylglucosamine 1-phosphate attached to carbon 6 of some of the MurNAcyl moieties (Fig. 2A and 2B) (22).

GENETICS OF PG PRECURSOR SYNTHESIS

Most of the cytoplasmic steps for the synthesis of PG precursors in mycobacteria are shared with those of other bacteria (see Fig. 3 for the full cytoplasmic precursor pathway). The mycobacterial genes involved in this pathway have been found primarily by homology to known genes in other bacteria (23, 24). By virtue of the critical nature of the PG, the genes encoding the enzymes for precursor synthesis are essential for mycobacterial survival since there are no redundancies. In a few instances, the function of a mycobacterial gene was confirmed by complementation of *E. coli* mutants, either auxotrophs (*asd*, *dapB*) (25, 26) or a temperature-sensitive mutant (*murG*) (27). In addition, there have been some direct biochemical and genetic experiments in mycobacteria to examine the enzymes involved in this stage of PG biosynthesis, as described below.

Cytosolic precursor (UDP-N-acetylmuramyl-L-alanyl-D-glutamyl-*meso*-diaminopimelyl-D-alanyl-D-alanine, or Park's nucleotide; Fig.1) synthesis begins with the conversion of UDP-GlcNAc to UDP-MurNAc via the products of the *murA* and *murB* genes (28). Subsequent steps in this part of the pathway add each amino acid to the growing peptide chain attached to the MurNAcyl. This requires the generation of D-amino acids at several steps to produce D-Glu, *meso*-DAP, and D-Ala (Fig. 3) (28). Mutation of genes involved with the biosynthesis of these amino acids is generally permitted in bacteria, given that the appropriate isomers of the amino acids are supplied to the mutants. The MurI racemase that produces D-Glu has a moonlighting function in many bacteria (including mycobacteria) in that it can bind to DNA gyrase and prevent its inactivation by ciprofloxacin (29, 30). Consistent with this view, overexpression of *murI* in *M. tuberculosis* and *M. smegmatis* can confer resistance to ciprofloxacin (30).

DAP is produced from L-aspartate in a series of reactions beginning with aspartokinase, encoded by the *ask* gene (Fig. 3) (25, 31). Deletion of this gene leads to DAP auxotophy in *M. smegmatis*, and cell cultures of an *ask* mutant that are deprived of DAP will lyse within one generation after starvation ("DAP-less death") (32). Surprisingly, spontaneous mutation of the ribosome-binding site of the *cbs* gene, encoding cystathionine β-synthase, can suppress DAP-less death in an *M. smegmatis ask* mutant (33, 34). This occurs by subsequent overexpression of cystathionine β-synthase activity, resulting in the biosynthesis of lanthionine, an analog of DAP that can be inserted into PG precursors in place of DAP (34, 35). However, such suppressor strains with a lanthionine-containing PG are hypersusceptible to β-lactam antibiotics.

In spite of the ability to construct a DAP auxotroph by deletion of the *ask* gene, it has not been possible to delete genes further down the pathway. We have made several attempts to delete the *dapB* gene in *M. smegmatis* as well as *M. bovis* BCG to no avail. However, a *dapB* mutant of *M. smegmatis* was isolated from a transposon library screened for mutants that are

hypersusceptible to β-lactam antibiotics, but the mutant has an uncharacterized mutation, not in *cbs*, that suppresses death from DAP starvation (36). The reason why other *dapB* deletions have not been successful is unknown, but perhaps it may be due to suppression by an unrecognized *dapB* paralog residing in the genomes of these bacteria. Another potential barrier to the construction of DAP auxotrophs in other mycobacterial species is that DAP is the precursor to lysine, and mycobacteria exhibit differences in their ability to both utilize exogenous lysine and balance the DAP pool to support both lysine and PG requirements (32, 37). These factors may impact the viability of certain DAP auxotrophic strains.

The production of D-Ala in many bacteria is catalyzed by either alanine racemases or D-amino transaminases (38). Some bacteria have both types of enzymes, but mycobacterial genomes appear to encode only an alanine racemase (*alr*). Earlier work suggested that there is a redundant system for D-Ala production in *M. smegmatis*, since an *alr* insertion mutant did not exhibit a requirement for D-Ala (39). However, subsequent work showed that deletion of *alr* in *M. smegmatis* or *M. tuberculosis* resulted in mutants that required D-Ala supplementation for growth and, in the latter case, attenuated growth of the mutant in macrophages and mice (40, 41). The discrepancies between these studies likely resulted from partial activity of a truncated Alr protein being produced from the insertion mutation in the earlier work with *M. smegmatis*. Understanding the metabolism of D-alanine is important, since alanine racemase is the target of D-cycloserine, a potent second-line antitubercular drug. Overexpression of *alr* and the D-alanine ligase encoded by *ddl* confers increased resistance to D-cycloserine in both *M. smegmatis* and *M. tuberculosis* (42).

BIOSYNTHESIS OF PG CYTOSOLIC INTERMEDIATES

As noted above, the PG biosynthetic machinery bears a resemblance to the basic *E. coli* PG biosynthesis pathway. The PG biosynthesis in *E. coli* and likewise in mycobacteria occurs in three subcellular locations (16, 43, 44). The final cytosolic intermediate of PG biosynthesis is UDP-N-acetylmuramyl-L-alanyl-D-isoglutamyl-*meso*-diaminopimelyl-D-alanyl-D-alanine (UDP-MurNAc pentapeptide) precursor, or Park's nucleotide (Fig. 1). Synthesis of UDP-MurNAc pentapeptide initiates with the formation of UDP-MurNAc from UDP-GlcNAc in a two-step reaction catalyzed by MurA and MurB (45). MurA transfers the enolpyruvyl moiety of phosphoenol

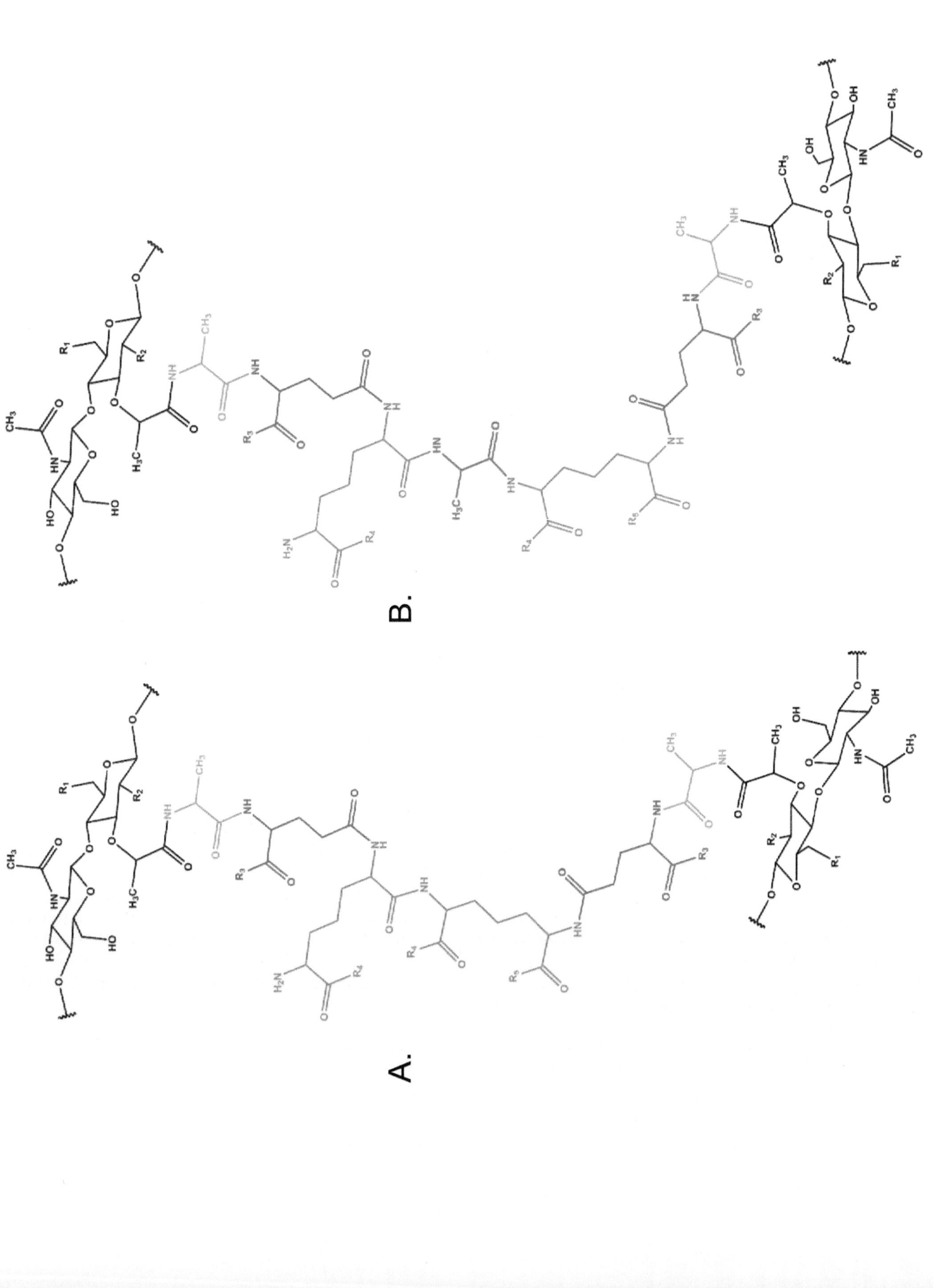
A.
B.

pyruvate to the 3′-hydroxyl group of UDP-GlcNAc, forming enolpyruvyl-UDP-GlcNAc, which is the first committed step of bacterial cell wall biosynthesis. The *murA* gene of *M. tuberculosis* has a point mutation that changes a canonical cysteine in the active site to an aspartate that confers resistance to the antibiotic fosfomycin (46). Eschenburg et al. proposed that the substitution of Cys115 with another amino acid has no effect on the catalytic activity of MurA (47). In the second step of the reaction, MurB (Rv0482) reduces the enolpyruvate residue of enolpyruvyl-UDP-GlcNAc to D-lactate, forming UDP-MurNAc (45). This reaction is NADH dependent. No direct biochemical study on *M. tuberculosis* MurA has been performed, but *in silico* studies to develop a three-dimensional structure and identify specific inhibitors have been performed (48). In mycobacteria the N-acetyl groups of the MurNAc residues can be oxidized to an N-glycolyl function by a UDP-MurNAc hydroxylase (NamH) encoded by Rv3818 (8). This gene is nonessential in *M. tuberculosis* H37Rv (49), and deletion of *namH* in *M. smegmatis* resulted in loss of N-glycolylated PG precursors (8). The *namH* gene in *M. leprae* is a pseudogene, which explains the absence of MurNGlyc from the PG of *M. leprae* (6). However, the formation of MurNGlyc has not been extensively studied, and UDP-N-acetyl-muramyl-tripeptide and lipid-linked intermediates are potential substrates of NamH (2, 5).

The biosynthesis of UDP-MurNAcyl-pentapeptide occurs via stepwise additions of L-Ala, D-Glu, *meso*-DAP, and D-Ala-D-Ala to the UDP-MurNAcyl. These reactions are catalyzed by the Mur family of ligases (45), which are nonribosomal peptide synthases requiring ATP and a divalent cation such as Mg^{2+} or Mn^{2+} for activity. Biosynthesis of the peptide side chain initiates with the addition of an L-Ala to the D-lactoyl group of MurNAcyl residue by MurC (50). MurC enzymes from *M. tuberculosis* (Rv2152c) and *M. leprae* (ML0915) have been overexpressed and partially characterized (51), showing similar substrate specificities and kinetics. Both enzymes demonstrated the ability to use L-Ala and Gly as substrates (51, 52). Despite the high degree of similarity between *M. tuberculosis* and *M. leprae* MurC enzymes in both sequence and specificity, the first amino acid in the *M. leprae* PG stem

peptide is exclusively Gly; the reason for this substitution is not clear and may result from the intracellular growth of *M. leprae* (53). In a subsequent reaction, catalyzed by MurD, a D-Glu residue is added to the carboxy terminus of the L-Ala residue of the UDP-MurNAcyl-L-Ala. The MurD enzyme (Rv2155c) from *M. tuberculosis* has been overexpressed, and the crystal structure has been evaluated and compared with the structure of MurD ligases of other organisms (52, 54). MurE catalyzes the synthesis of UDP-MurNAcyl-L-Ala-D-Glu-*m*-DAP from UDP-MurNAcyl-L-Ala and *meso*-DAP. The *M. tuberculosis* MurE (Rv2158c) is highly specific for *meso*-DAP, like its *E. coli* counterpart (55). The last two amino acids are added by MurF as a dipeptide (D-Ala-D-Ala) to the UDP-MurNAcyl-L-Ala-D-Glu-*m*-DAP, yielding Park's nucleotide. MurF (Rv2157c) from *M. tuberculosis* has been overexpressed and partially characterized; this enzyme has much lower K_M values for UDP-MurNAcyl-L-Ala-D-Glu-*m*-DAP and ATP than *E. coli* and *Staphylococcus aureus* MurF enzymes (52).

FORMATION OF LIPID LINKED INTERMEDIATES

Once Park's nucleotide is formed, the sugar peptide is transferred to decaprenyl phosphate (Dec-P), forming Lipid I by the gene product of *mraY* (also known as *murX*), a polytopic membrane transferase (Fig. 3) (7, 28). The use of Dec-P (synthesis described in reference 155) as a carrier of activated sugars is an unusual feature of mycobacterial PG synthesis, because other bacteria utilize undecaprenyl phosphate, and the significance of this difference is unknown. The sugar-peptide moiety of Lipid I lacks modification, but after it is coupled to GlcNAc by the glycosyltransferase encoded by *murG* (forming Lipid II; see Fig. 3 and 4), it is modified by amidation, methylation, and glycylation of the peptides, as well as N-glycolylation and deacylation of the MurNAcyl residues (7). This results in a pool of differentially modified Lipid II molecules, but it is not clear if all these modifications are found in mature PG. With the exception of the *namH* gene, the genes responsible for these modifications have not been experimentally identified in mycobacteria (8). However, genes

Figure 2 PG cross-links. (**A**) The direct 4-3 cross-link between D-Ala and *meso*-DAP. (**B**) The direct 3-3 cross-link between two *meso*-DAP residues. Also shown are various modifications of the PG: R_1 = H or disaccharide linker connecting the PG to the arabinan of the arabinogalactan; R_2 = N-acetyl or N-glycolyl on the muramic acid residue; R_3 = OH, NH_2 or glycine; R_4 = OH or NH_2. L-Ala, D-Glu, *meso*-DAP, and D-Ala are depicted in gold, blue, green, and red, respectively. doi:10.1128/microbiolspec.MGM2-0034-2013.f2

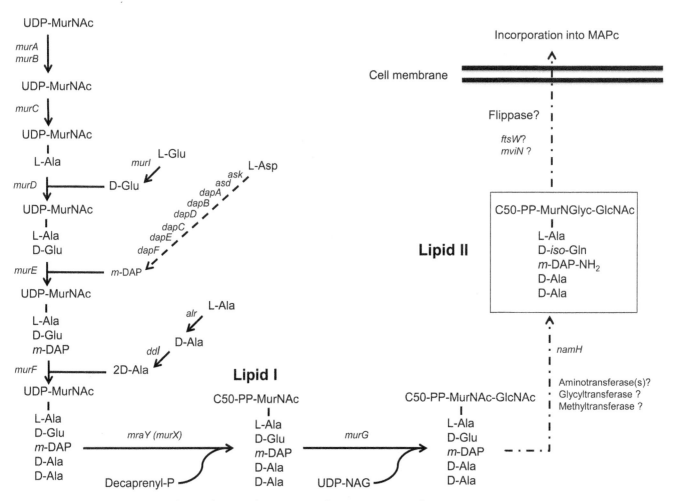

Figure 3 Pathways for cytoplasmic steps of PG precursor synthesis.
doi:10.1128/microbiolspec.MGM2-0034-2013.f3

homologous to those encoding enzymes involved in the amidation of DAP in *Lactobacillus plantarum*, *asnB1* (56), and the amidation of D-Glu in *S. aureus* (*murT*, *gatD*) (57, 58) are present in mycobacteria, represented by Rv2201, Rv3712, and Rv3713, respectively.

TRANSLOCATION OF LIPID-LINKED INTERMEDIATES

The formation, and likely the modification, of Lipid II occurs at the cytoplasmic face of the plasma membrane, and thus, the sugar peptide moiety of the mature Lipid II must be transported to the outer leaflet of the plasma membrane. The transfer of Lipid II across the membrane is not well understood but requires a protein transporter, or "flippase." The protein-facilitated transmembrane diffusion of lipid-linked biosynthetic intermediates and the lipids themselves is an essential and

ubiquitous feature in the synthesis and assembly of a wide variety of extracellular glycoconjugates in both prokaryotes and eukaryotes (59–61). The highly hydrophilic disaccharide-pentapeptide (PG monomer) moiety of Lipid II prevents spontaneous flipping across the membrane (62), and a dedicated flippase is essential for this process (63). In addition, the relatively low amount of decaprenyl diphosphate (Dec-PP) that is released by the transglycosylation reaction must be recycled back to the cytoplasmic face of the plasma membrane (64) in order to generate more Lipid II. Inhibition of the recycling reaction by the antibiotic bacitracin is lethal in mycobacteria (65). Like Lipid II flipping, the recycling of phosphorylated carrier-lipids (Dec-P or Dec-PP) is likely to require a flippase-mediated transport of this amphipathic molecule across the bilayer. Thus, the sustained export of PG precursors requires at least two distinct membrane translocation events: the

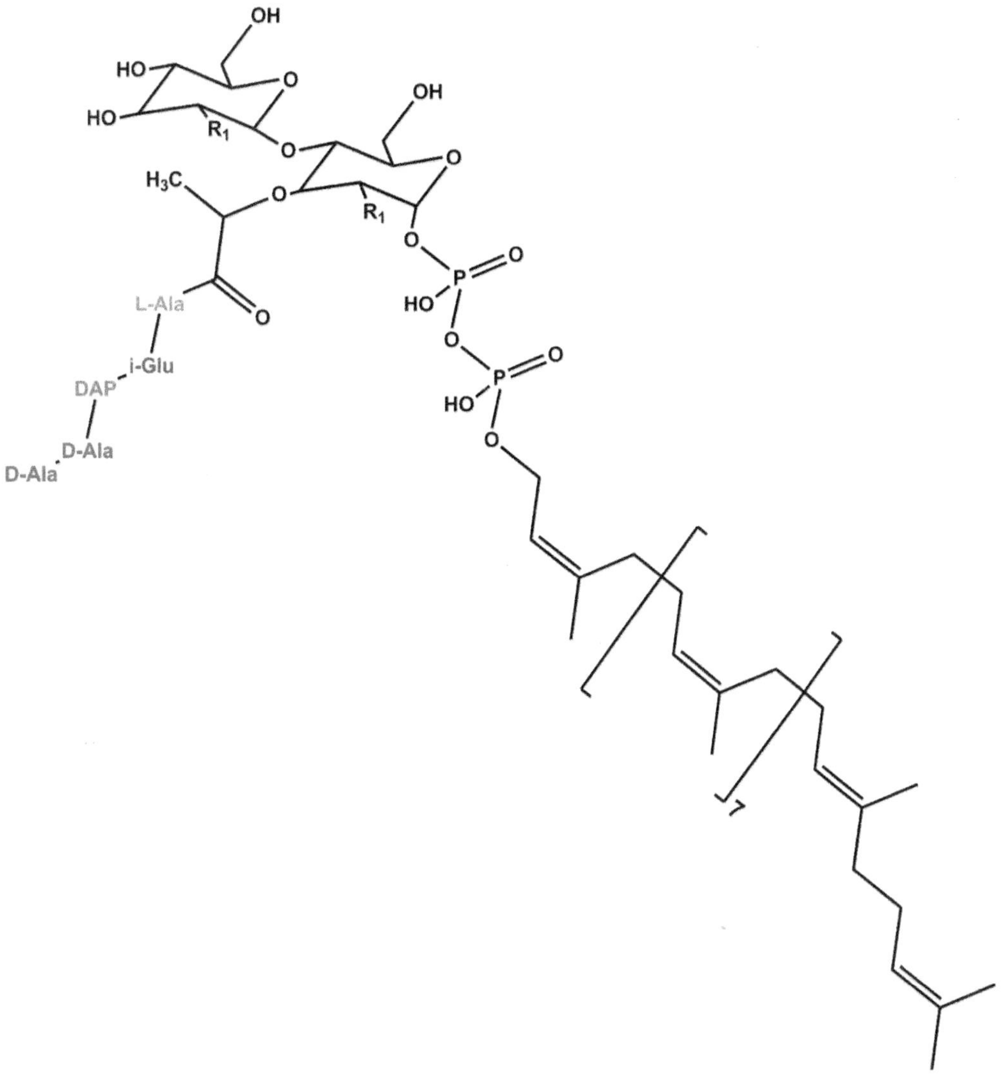

Figure 4 Structure of the mycobacterial Lipid II PG precursor. R_1 = N-acetyl or N-glycolyl on the muramic acid residue. doi:10.1128/microbiolspec.MGM2-0034-2013.f4

eversion of Lipid II to the outside of the plasma membrane and the inversion of the carrier-lipid to allow re-utilization. The structure of the recycled lipid carrier molecule has not been elucidated, leading to several possible scenarios for converting periplasmically localized Dec-PP into cytoplasmically oriented Dec-P. Each of these scenarios would require distinct accessory kinases and/or phosphatases.

The proteins that mediate the transbilayer movement of Lipid II and the reutilization of Dec-PP remain unidentified. However, both genetic and bioinformatic evidence suggests that the FtsW and MviN proteins (66–68) could be flippases involved in PG synthesis in several bacteria; in *M. tuberculosis*, the Rv3910 gene encodes a protein that has an N-terminal MviN-like

domain, which is essential, and a C-terminal pseudo-kinase domain (69). Although not universally essential, homologs of the large MviN integral membrane protein are found in the genomes of PG-producing organisms and in several cases are located in operons that also encode the Lipid I-generating MraY enzyme (70). While direct sequence identity between flippases is generally very weak, MviN proteins are members of the multidrug/oligosaccharidyl-lipid/polysaccharide (MOP) exporter superfamily (68, 71), which includes RfbX, a protein proposed to flip prenylphosphate-linked O-antigen units (72), and WxzE, which has been shown to flip the Lipid III intermediate of the enterobacterial common antigen pathway (73). MviN orthologs are required for PG synthesis in *E. coli* (66, 68) and mycobacteria (69), and the inhibition of MviN causes an accumulation of the mature lipid- and nucleotide-linked PG precursors that would be expected from a block in export. The pseudokinase domain of Rv3910 can be phosphorylated by the essential cell division kinase, PknB, which results in the recruitment of the regulatory protein FhaA to MviN and subsequent inhibition of PG biosynthesis (69). Although it has been shown that MviN is essential for mycobacterial PG synthesis and growth, its specific role in PG synthesis remains elusive.

PG ASSEMBLY

Incorporation of the Lipid II precursors into the mature PG in the MAPc has essentially three steps: polymerization of the glycan chains, peptide cross-linking, and coupling of the PG to the galactan portion of the arabinogalactan via the N-acetylglucosamine-rhamnose linker. The last step is not clearly understood, but the genes involved with synthesis of the linker are essential (74). The enzyme involved with the actual coupling of the PG to the linker has not been identified, but it is known that the linkage of the PG to the arabinogalactan requires concomitant cross-linking of the PG, implying that the incorporation of newly synthesized precursors is coordinated with the assembly of the overall MAPc structure (75, 76).

The polymerization of the sugar backbone and cross-linking of the peptides of the PG are carried out by a variety of transglycosylases, transpeptidases, and carboxypeptidases that are schematically depicted in Fig. 5. *M. tuberculosis* has eight PBPs likely involved with transglycosylation, D,D-transpeptidation, and carboxypeptidation reactions, as well as five L,D-transpeptidases, but the number of these enzymes can vary between species (see the Fig. 5 legend) (10, 23, 24). The mycobacterial PBPs involved with transglycosylation and D,D-transpeptidation fall into different classes (10, 77). The PonA1 and PonA2 proteins are large, class A PBPs that have both transglycosylation and transpeptidation domains (10, 78, 79). An *M. smegmatis ponA1* transposon mutant was isolated in a screen for mutants with an altered dye binding phenotype, and this mutant had reduced growth rate, altered permeability, and increased susceptibility to β-lactam antibiotics (80). Subsequent work has shown that PonA1 is involved with cell division and has been described as essential, since depletion of the operon mRNA containing *ponA1* stops the growth of *M. smegmatis* (81). However, the gene is not essential to *M. tuberculosis* (49), and the isolation of a *ponA1*

transposon mutant of *M. smegmatis* suggests that it is not essential to the latter organism, but this discrepancy may be due to the presence of a suppressor mutation in the *M. smegmatis* transposon mutant or differences between *M. smegmatis* and *M. tuberculosis*. PonA1 has also been shown to interact with RipA, an endopeptidase that is important for cell division (81). Interestingly, the intracellular N-terminus of PonA1 is phosphorylated in *M. tuberculosis*, but the role of this modification is unknown (82).

The other class A PBP, PonA2, is similar to PonA1 but seems to play no role in cell division. Instead, it has an important role during times of stress. A *ponA2* transposon mutant of *M. smegmatis* was isolated in a screen for mutants defective for long-term starvation (83), while *ponA2* transposon mutants of *M. tuberculosis* were isolated from independent screens for β-lactam hypersusceptibility (36) or sensitivity to low pH (84). In the latter case, the mutant was also found to be hypersensitive to reactive nitrogen and oxygen species and attenuated in the mouse model (85). An unmarked, in-frame *ponA2* deletion mutant of *M. smegmatis* exhibits decreased susceptibility to β-lactam antibiotics and increased susceptibility to rifampin and has morphological and survival defects under conditions of nonreplication (86). These data suggest that PonA2 has a regulatory role for maintaining cell wall integrity. Some soil mycobacteria contain an additional copy of PonA2, called PonA3, which is missing from pathogenic mycobacteria. An *M. smegmatis* mutant lacking *ponA3* has no phenotype, but when expressed from a constitutive promoter on a multicopy plasmid, the *ponA3* gene can partially complement a Δ*ponA2* mutant, suggesting some overlap of function (86). All soil mycobacteria that bear the *ponA3* gene were isolated from environments contaminated with polycyclic aromatic hydrocarbons, and it is possible that the PonA3 protein may have a specific role in responding to cell envelope stresses in this particular niche.

Class B PBPs are generally involved with cell division/morphology and typically have only a transpeptidase domain. Mycobacteria have three class B proteins (see Fig. 5); two of them, PbpA and PbpB (FtsI), will be described in more detail in the cell division section below (10, 24). Note that PbpB is essential (49) and belongs to the B3 subgroup, owing to its additional transglycosylase domain (10). The remaining class B PBP is encoded by the Rv3627c gene in *M. tuberculosis* or the MSMEG_2584 gene in *M. smegmatis* (10). It is annotated as a potential lipoprotein that is not essential to *M. tuberculosis* (24, 49), but little else is known about this particular PBP.

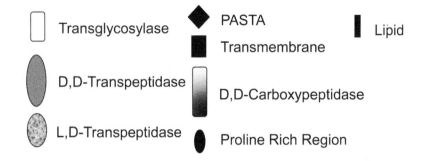

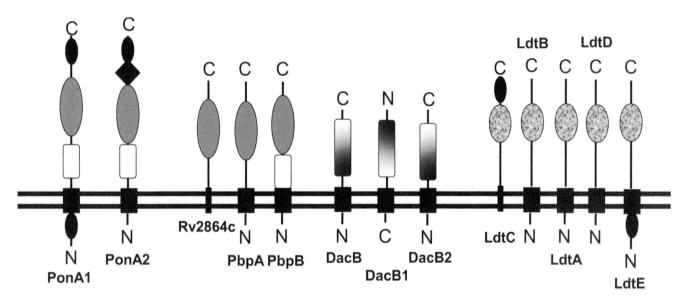

Figure 5 PG assembly proteins. The various PBPs and L,D-transpeptidases are shown as they exist in the genome of *M. tuberculosis*. Gene designations from the H37Rv genome: *ponA1* (*Rv0051*), *ponA2* (*Rv3682*), *pbpA* (*Rv0016c*), *pbpB* (PBP3, *ftsI*, *Rv2163c*), *dacB* (PBP4, *dacB*, *Rv3627c*), *dacB1* (*Rv3330*), *dacB2* (*Rv2911*), *ldtC* (*lprQ*, *Rv0483*), *ldtB* (*Rv2518c*), *ldtA* (*Rv0116c*), *ldtD* (*Rv1433*), *ldtE* (*Rv0192*). These genes are also present in *M. leprae*, with the exception of *dacB2* and *ldtE*, the latter of which is a pseudogene. *M. smegmatis* and other soil organisms have the novel *ponA3* gene, an extra variant of *ldtB*, and an additional copy of the *dacB2* gene as described in the text. The various domains in each protein are also indicated. Note that PonA2 is unique because it bears a single PASTA domain, which likely binds unlinked PG precursors, and that PonA1, PonA2, LdtC, and LdtE bear extensive proline-rich regions. The class B PBP encoded by Rv2864c and the L,D-transpeptidase encoded by *ldtC* are putative lipoproteins.
doi:10.1128/microbiolspec.MGM2-0034-2013.f5

The remaining PBPs are the carboxypeptidases DacB (PBP4), DacB1, and DacB2 (10, 24). *M. smegmatis* has an additional copy of DacB2, encoded by MSMEG_2433 (*dacB2b*), which is immediately downstream of the DacB2 (aka the *dacB2a* gene, MSMEG_2432). The role of these Dac proteins in mycobacterial PG biosynthesis is not entirely clear. An attenuated *dacB1* transposon mutant of *M. tuberculosis* was identified in a nonhuman primate signature-tagged mutagenesis (STM) screen (87). A *dacB2* mutant

of *M. tuberculosis* has reduced growth in Sauton's medium under acidic and microaerobic conditions but increased survival in host cells (88). Overexpression of the *M. tuberculosis dacB2* gene in *M. smegmatis* results in reduced growth and altered colony morphology with concomitant changes in biofilm formation and sliding motility (88).

The last set of PG assembly enzymes are the L,D-transpeptidase (Ldt) enzymes. These proteins belong to the YkuD superfamily and break L-D PG peptide bonds in order to form 3-3 cross-links in PG or link the PG peptides to other proteins (15, 17, 89). The latter reaction has only been seen in *E. coli*, in which a subset of L,D-transpeptidases couple Braun's lipoprotein to the PG (90). Overall, these proteins are distinct from classical PBPs and are evolutionarily related to sortase enzymes that are found in Gram-positive bacteria (91). As mentioned earlier, the presence of 3-3 linkages was first discovered in mycobacteria, but since then these linkages have been found in many other bacteria. A specific role for these linkages has not been discovered, but they may function to stabilize the cell envelope (16).

The Ldt enzymes are encoded by several genes in mycobacteria, but the number varies according to species (24, 92). *M. tuberculosis* has five noncontiguous *ldt* genes (Fig. 5), while *M. smegmatis* has an additional variant (MSMEG_1322, *ldtF*) of the *ldtB* gene (MSMEG_47457), for a total of six. *M. leprae* has the same genes as *M. tuberculosis*, but the *ldtE* gene is a pseudogene. The LdtA, LdtB, and LdtD proteins are very similar to each other, while the LdtC protein appears to be a lipoprotein and has a C-terminal proline-rich region like that of the class A PBPs. The LdtE protein is notable because it has an N-terminal proline-rich region. The function of these proline-rich regions is unknown, but they may be involved with protein-protein interactions. All five Ldts of *M. tuberculosis* have been expressed in *E. coli*, but only the LdtA and LdtB proteins exhibit L,D-transpeptidase activity (12, 92). An *M. tuberculosis ldtB* transposon mutant was isolated in a screen for mutants with altered colony morphology (92); this mutant exhibited increased susceptibility to amoxicillin and was attenuated in the mouse model, while an *M. smegmatis ldtB* transposon mutant was isolated in a screen for resistance to ubiquinated peptides (93); this latter mutant had a reduced envelope permeability phenotype. The *ldt* genes are not individually essential, since transposon mutants of *M. tuberculosis* have been isolated with insertions in each of the genes, although only the *ldtB* mutant appears to have a phenotype (92). There may be issues of redundancy that prevent the identification of additional phenotypes, so further analysis of these genes is warranted.

The formation of 3-3 linkages requires the generation of tetrapeptide PG precursors from the pentapeptide precursors via the action of D,D-carboxypeptidases (11, 15). The tetrapeptides are the substrate for the L,D-transpeptidases. The role of the DacB, DacB1, and DacB2 carboxypeptidase enzymes described above has not been thoroughly investigated in the formation

of 3-3 linkages, although it was recently shown that recombinant *M. tuberculosis* DacB2, purified from *E. coli*, has D,D-carboxypeptidase activity (20).

PG TURNOVER

The disruption of preexisting bonds is required for the insertion of newly synthesized PG into the maturing cell wall. These two opposing activities of breakdown and assembly must be balanced in order to maintain cellular integrity. Furthermore, because PG synthesis is energetically costly, bacteria have evolved mechanisms to recycle cell wall material turned over as a result of growth and cell division. Gram-negative bacteria such as *E. coli* have a complex series of enzymes that comprise a sophisticated PG recycling pathway capable of recovering and reusing the sugars and peptides of the PG and, in some cases, regulating the expression of transcriptionally repressed β-lactamases in response to PG damage (94). Not as much is known about the PG turnover systems in Gram-positive bacteria, and even less is known about such systems in mycobacteria. However, the mist is beginning to clear with some recent work that has examined various PG glycan hydrolases and endopeptidases (Table 1) that have been identified in mycobacteria.

The *cwlM* gene is essential to *M. tuberculosis* and encodes an amidase that cleaves between the muramyl acid residues and the L-alanine at the first position in the PG peptide (95). This is a typical autolysin activity

TABLE 1   Genes involved with PG turnover

H37Rv#	Gene	Description of gene product
Rv3915	*cwlM*	N-Acetylmuramoyl-L-alanine amidase
Rv0867c	*rpfA*	Resuscitation-promoting factor
Rv1009	*rpfB*	Resuscitation-promoting factor
Rv1884c	*rpfC*	Resuscitation-promoting factor
Rv2389c	*rpfD*	Resuscitation-promoting factor
Rv2450c	*rpfE*	Resuscitation-promoting factor
Rv1477	*ripA*	L,D-Peptidase (NLP/P60 family member)
Rv1478	*ripB*	L,D-Peptidase (NLP/P60 family member)
Rv0024		Homologous to RipA/B
Rv2190c		Homologous to RipA/B
Rv1566c		Homologous to RipA/B, but lacks active site residue
Rv1728c		NLP/P60 family member; *M. smegmatis* protein has PG hydrolytic activity
Rv0320		NLP/P60 family member; *M. smegmatis* protein has PG hydrolytic activity

that is often also found encoded in bacteriophage genomes. A similar enzyme, encoded by the *lysA* gene of mycobacteriophage Ms6, has found utility as a reagent in the analysis of mycobacterial PG structure (96). A proteomic analysis of phosphoproteins in *M. tuberculosis* showed that CwlM is phosphorylated at several positions, but the significance of this is unknown (82).

A group of enzymes that also act on the glycan chain, known as "resuscitation-promoting factors," encoded by the *rpf* genes (Table 1), play an enigmatic role in mycobacterial PG biology (97–100). The first Rpf protein was discovered in *Micrococcus luteus* and shown to resuscitate dormant cultures (97, 101). Rpf proteins were then found in a variety of Gram-positive bacteria and shown to have a lysozyme-like structure that could cleave the glycan chain of the PG (99, 102). The Rpf protein of *M. luteus* is essential, does cleave the PG, and has been shown to resuscitate dormant, nonculturable mycobacterial cultures (102–105). There are two hypotheses about the function of the Rpf proteins: they are responsible for the release of PG fragments that act as signaling messengers on other bacterial cells, or the cleavage of the PG is required to initiate replication after a period of dormancy (106). These functions may not be mutually exclusive in bacteria with multiple *rpf* genes. In either case, the action of the Rpf proteins is thought to be on PG, resulting in activation from dormancy. There are five *rpf* genes in *M. tuberculosis*, but only three of the genes found in *M. tuberculosis* (*rpfA*, *rpfB*, and *rpfC*) are present in *M. leprae* (24, 107, 108). *M. tuberculosis* mutants with single or multiple *rpf* mutations are viable (106, 108), and although the genes seem to have a certain degree of redundancy, some mutants do have altered phenotypes. Global transcriptional analysis of single unmarked deletion mutants of H37Rv showed that an *rpfC* mutant had the largest number of genes with altered expression, with a large transcriptional overlap with the *rpfB*, *rpfD*, and *rpfE* mutants and less of an overlap with an *rpfA* mutant (109).

Comparisons between different *rpf* studies are complicated by variations in mouse strains, route of infection, and background *M. tuberculosis* strains. Some studies used *M. tuberculosis* Erdman (108, 110), while others used an H37Rv derivative that is deficient in phthiocerol dimycocerosate (PDIM) production (111, 112). These studies have significantly increased our understanding of the influence of these proteins on mycobacterial physiology and pathogenesis, although the mechanism(s) by which these proteins function has remained elusive.

Single *rpf* marked deletion mutants of strain Erdman have no *in vitro* phenotype, except for a small colony phenotype for a Δ*rpfB* mutant that is the result of a polar effect on the downstream *ksgA* gene, encoding a 16S rRNA methyltransferase (108). The single mutants had no phenotype in an aerosol challenge using C57Bl/6 mice. Subsequently, the same group showed that the Δ*rpfB* mutant was defective for reactivation in the mouse model following immune suppression in the chronic stage of infection using aminoguanidine (110). Further work (113) using double deletion mutants showed that a Δ*rpfA* Δ*rpfB* mutant had a colony morphology change *in vitro* and was more deficient for resuscitation *in vivo* than the Δ*rpfB* mutant. A similar, if less pronounced, *in vivo* phenotype was seen for a Δ*rpfA* Δ*rpfD* mutant. The double Δ*rpfA* Δ*rpfB* mutant grew less well in murine bone marrow-derived macrophages and was more proinflammatory than wild type. With the exception of the macrophage survival defect, all phenotypes could be rescued by complementation with both the *rpfA* and *rpfB* genes.

Another study used mutants of H37Rv with multiple, unmarked deletion alleles of the *rpf* genes and the mouse B6 background using an intravenous administration route (111, 112). In this system, a Δ*rpfA* Δ*rpfD* Δ*rpfE* triple mutant had no phenotype, while a Δ*rpfA* Δ*rpfC* Δ*rpfB* and a Δ*rpfA* Δ*rpfC* Δ*rpfD* triple mutant were both attenuated in the mouse and were defective for spontaneous resuscitation *in vitro*. Another study, by Biketov et al. (111), using the same H37Rv strain background and mutants but using C57Bl/6 mice and an intraperitoneal route of entry, showed that a double mutant (Δ*rpfA* Δ*rpfC*) was less able to establish an infection in the lungs compared to wild type or a (Δ*rpfA* Δ*rpfB*) double mutant. Both mutants persisted in the lungs at lower amounts than wild type and were defective in reactivation via aminoguanine treatment. These results for the Δ*rpfA* Δ*rpfB* mutant are in agreement with those of Russell-Goldman et al. (113), but the latter group did not see a phenotype with a Δ*rpfA* Δ*rpfC* mutant; this may be due to differences in bacterial strain background and route of entry. Biketov et al. (111) also showed that the Δ*rpfA* Δ*rpfC* Δ*rpfB* and Δ*rpfA* Δ*rpfC* Δ*rpfD* strains that were previously shown to have an *in vitro* resuscitation defect also have difficulty establishing a lung infection after intraperitoneal injection, with the Δ*rpfA* Δ*rpfC* Δ*rpfD* mutant exhibiting a greater defect. Furthermore, both strains were unable to reactivate in the mouse model after treatment with either aminoguanine or anti-TNFα antibody.

Further work with this strain set and the construction of quadruple Δ*rpfA* Δ*rpfC* Δ*rpfB* Δ*rpfE* and quintuple Δ*rpfA* Δ*rpfC* Δ*rpfB* Δ*rpfE* Δ*rpfD* mutants showed that these mutants had delayed growth on solid media, with a greater delay seen on Middlebrook 7H11 than 7H10 (106). These mutants were also hypersusceptible to detergent, and it was shown that this was due to loss of both *rpfB* and *rpfE*. Multiple deletion mutants Δ*rpfA* Δ*rpfC* Δ*rpfB* Δ*rpfD* and Δ*rpfA* Δ*rpfC* Δ*rpfB* Δ*rpfE* were defective for *in vitro* resuscitation as well as persistence in B6 mice following aerosol challenge, with the latter mutant exhibiting a more pronounced defect. These mutants differed little in their ability to grow in human peripheral blood mononuclear cells, in contrast to the intracellular growth defect observed for a Δ*rpfA* Δ*rpfB* mutant in murine bone marrow-derived macrophages as described above. Additional studies showed that the Δ*rpfA* Δ*rpfC* Δ*rpfB* Δ*rpfD* and Δ*rpfA* Δ*rpfC* Δ*rpfD* Δ*rpfE* mutants are attenuated after aerosol infection of mice and that they cannot reactivate after aminoguanidine treatment. Interestingly, the mutants can persist in mice after subcutaneous injection and can be as protective as BCG in vaccination experiments (114).

It is clear from these studies that the Rpf proteins have important roles in the physiology of dormancy and in pathogenesis. There is some overlap in function, but it seems that RpfB and RpfE have the most significant roles, with RpfD and RpfC in smaller roles that have redundancies with RpfA. The lack of *rpfD* and *rpfE* genes in *M. leprae* and the lack of *rpfD* in *M. smegmatis* may suggest adaptation to different niches, or variations in PG physiology for these bacteria, compared to *M. tuberculosis*.

Several mycobacterial genes encode enzymes belonging to the NPL/P60 protein family (115). This family, an early member of which was identified in *Listeria monocytogenes* as an invasin protein, commonly share a PG degradation domain that includes a conserved catalytic cysteine residue (115, 116). There are seven such genes in the *M. tuberculosis* genome (Table 1); all but two (*Rv0024*, *Rv1728c*) are present in *M. leprae* (24, 107). The best studied are the *Rv1477* and *Rv1478* genes, also known as *ripA* and *ripB*, or alternatively, *iipA* and *iipB*. A transposon mutant with an insertion in this operon of genes was first identified in *Mycobacterium marinum*, and the genes were named *iip* because the mutant was defective for entry into, and survival in, host cells (117). A constructed *iipA::kan* mutant had an invasion and intracellular survival defect; was attenuated in zebrafish; was hypersusceptible to rifampin, ciprofloxacin, erythromycin, and lysozyme;

and had a septation defect. These phenotypes could be complemented by the *M. marinum* or *M. tuberculosis* genes, but some phenotypes could only be partially complemented by *iipB* (*ripB*) and some required both genes. This suggests that both genes are not entirely redundant. Both RipA and RipB from *M. tuberculosis* have been crystallized and their structures determined (118). While they share a common core, there are substantial changes in their N-termini. Both enzymes are L-D endopeptidases, cleaving PG fragments between the D-*iso*-glutamine and *meso*-DAP, with RipA being capable of degrading intact PG while RipB cannot (118).

RipA has been shown to interact with RpfE and RpfB, and RipA/RpfB localize to the septa where RipA is responsible for digesting the PG in the process of septation (119). Furthermore, RipA depletion results in a severe septation defect, which is *ripA* specific because *ripB* cannot complement this defect. In addition, the degradative activity of RipA on the PG appears to synergize with that of RpfB (120). RipA also interacts with the class A PBP PonA1, which regulates the activity of the protein during vegetative growth (81). More recent work has shown that RipA is produced as a zymogen and that a complex series of protein-protein interactions serves to keep the protein tightly regulated (121). Consistent with this view, the dysregulation of this control circuit results in morphological changes and defects in cell division (121).

The remaining five endopeptidases are not as well studied as IipA/RipA and IipB/RipB. Three of them, Rv0024, Rv2190c, and Rv1566c, are homologous to RipA/B, but the Rv1566c protein lacks the active site cysteine and may have a different function. Rv2109c has a role in cell envelope physiology and pathogenesis. An *M. tuberculosis* Rv2109c transposon mutant has a growth and colony morphology defect, has altered levels of phthiocerol dimycocerosate, is slightly more susceptible to lysozyme, and is attenuated in the mouse model (122). The endopeptidases Rv1728c and Rv0320 have not been studied in *M. tuberculosis*, but recombinant versions of the *M. smegmatis* proteins have been shown by zymographic analysis to degrade PG (123). Additional work is needed to define the substrate specificities of these enzymes and to determine what role they play in PG metabolism.

PG GENES AND CELL DIVISION
Mycobacteria have most of the cell division genes commonly seen in other rod-shaped bacteria, with the notable lack of the Mre proteins for coordinating lateral cell wall growth. Most of the cell division genes have been

identified by homology; the complete list, thus far, is shown in Table 2 (24). The *fts* genes, involved with development of the septal FtsZ ring and DNA partitioning, are present, along with the essential class B3 PbpB (aka FstI or PBP3) and the class B PbpA. PbpB is essential (49), as it is in other bacteria, and it has been shown to interact with FtsW and form a trimolecular complex of PbpB, FtsW, and FtsZ to regulate cell septation (124). Interestingly, PbpB can be cleaved by a membrane protease (Rv2869c) in oxidatively stressed cells, and PbpB can be protected from this cleavage by the Wag31 protein (Rv2145c), a DivIVA protein that is important for regulating cell division (125). Wag31 is an essential protein that controls growth and polar wall synthesis, and itself is regulated by phosphorylation, as disruption of the phosphorylation of Wag31 leads to alterations in cell morphology and PG synthesis (126–128). The other class B PBP, PbpA, is not essential, but its loss leads to a division defect in *M. smegmatis* (129) and hypersensitivity to low pH in *M. tuberculosis* (84). PbpA can be phosphorylated by the essential cell division serine/threonine kinase PknB, and this phosphorylation is required for proper localization and function of the protein (129). The crystal structure of PbpA suggests that the active site is not in the correct conformation to accept substrate and requires

movement to allow peptide entry into the binding pocket (130, 131).

One of the challenges in studying cell division genes is the essential nature of many of the genes. However, some information can be gained by overexpression studies. Such work has shown that FstH appears to respond to oxidative stress and that it may regulate FtsZ levels and inhibit cell division (132), while the Rv2719c protein, which has some PG degradative activity, seems to regulate cell division by an indirect effect of altering the midcell location of the FtsZ ring (133).

Several other cell division genes unique to mycobacteria have been identified. One of them, *crgA* (*Rv0011c*), which encodes a nonessential homolog of a protein involved with hyphal wall regulation in *Streptomyces*, was shown to interact with FtsZ and FtsQ, as well as PbpA and PbpB (134, 135). CrgA appears to have a role in septum production, as a deletion mutant of *M. smegmatis* grew as elongated cells with polar bulges, and overexpression of CrgA seems to stabilize the localization of PbpB at the septum. CrgA also interacts with another nonessential protein called CwsA (Rv0008c), which is involved with cell shape and cell wall biosynthesis (135). A *cwsA* deletion mutant of *M. smegmatis* exhibits rounding, bulges, and a decrease in PG synthesis. Loss of both *crgA* and *cwsA* in a double mutant leads to a synergistic cell wall defect with a loss of viability and increased autolysis with detergent. Another protein, ChiZ (Rv2719c), is an unspecified cell wall hydrolase that has a role in regulating division indirectly by influencing the stability of FtsZ ring formation, possibly by its interactions with FtsQ and PbpB (136). Another protein that influences cell division is the Ssd, or septum site-determining protein, encoded by the *Rv3660c* gene (137). Overexpression of this gene in *M. smegmatis* and *M. tuberculosis* results in longer, smoother cells, while an *M. tuberculosis ssd* tranposon insertion mutant exhibits shorter cells (137). Interestingly, expression of *ssd* was linked to a transcriptional program related to dormancy and stress, suggesting a connection with cell division regulation and an adaptive environmental response.

TABLE 2 Genes involved with cell division

H37Rv#	Gene	Description of gene product
Rv0008c	*cwsA*	Cell wall synthesis and cell shape protein
Rv0011c	*crgA*	Facilitates cell septation
Rv0017c	*rodA* (*ftsW*)	Facilitates cell septation
Rv0016c	*pbpA*	Class B PBP
Rv2163c	*pbpB* (*ftsI*)	Class B PBP
Rv2150c	*ftsZ*	Cell septation ring initiator/scaffold
Rv2151c	*ftsQ*	Involved with coordinating septation/ PG synthesis
Rv2748c	*ftsK*	DNA motor involved with chromosome partitioning
Rv2921c	*ftsY*	Signal recognition particle receptor
Rv3101c	*ftsX*	ABC transporter component
Rv3102c	*ftsE*	ABC transporter component
Rv3610c	*ftsH*	Quality control membrane protease
Rv2154c	*ftsW*-like	Facilitates cell septation?
Rv2145c	*wag31*	DivIVA family member, mediates cell division
Rv3660c	*ssd*	Septum site determining protein
Rv2719c	*chiZ*	Hydrolase involved with regulation of cell division

OTHER GENES INVOLVED WITH ANTIBIOTIC RESISTANCE AND CELL WALL METABOLISM

Mycobacteria, and *M. tuberculosis* in particular, are inherently resistant to β-lactam antibiotics such as penicillins, cephalosporins, and carbapenems, and thus these drugs have not been used to treat infections. A notable exception is the use of cefoxitin or imipenem in conjunction with clarithromycin or amikacin in the treatment of *M. abscessus* infections (138). In general, resistance to the β-lactams has been thought to result from (i) reduced permeability of the cell envelope to these drugs, (ii) differences in the susceptibility of PG assembly enzymes, and (iii) enzymatic degradation of the antibiotics by β-lactamases. Efflux pumps may also play a smaller role in resistance (139). Several groups have shown that the use of a β-lactamase inhibitor such as clavulanic acid in conjunction with a β-lactam antibiotic is lethal to mycobacteria, suggesting that the primary mechanism of resistance is enzymatic degradation (140–145). In *M. tuberculosis*, the BlaC enzyme, encoded by the *blaC* gene, or *RV2068c* (146), has been shown to have a broad spectrum of activity (147), and a *blaC* deletion mutant of *M. tuberculosis* is hypersusceptible to penicillins (148). However, the mutant has little or no change in susceptibility to cephalosporins. A similar situation was observed in an *M. smegmatis* mutant lacking the major β-lactamase, BlaS. *M. smegmatis* has another β-lactamase, encoded by *blaE*, but this enzyme makes a very minor contribution to the resistance profile of this organism (148).

The expression of β-lactamase activity in *M. tuberculosis* is constitutive, but the *blaC* gene can be induced in response to β-lactam antibiotics via a regulatory system encoded by the *blaR-bliA* operon (*Rv1845c-Rv1846c*), which is similar to the *mecI-mecR* system of *S. aureus* (149). The system is remarkable in that it also regulates the ATP synthase operon, suggesting a control circuit between synthesis of the cell wall and ATP generation. The *blaC* gene is absent in *M. leprae*, but the regulatory *blaR-blaI* genes appear to be intact, suggesting that the β-lactam response may be more important for linking cell wall damage to ATP synthesis control rather than direct protection of the cell from β-lactam damage by the β-lactamase (149). This raises the general question, Why does *M. tuberculosis* still have a β-lactamase? The presence of β-lactamase genes in environmental mycobacteria can easily be explained by their ecological niches being replete with β-lactam-producing organisms, but for a pathogen such as *M. tuberculosis* with no environmental niche, one can only speculate. Perhaps the *blaC* gene hasn't yet had sufficient time to degenerate as it has in *M. leprae*, or perhaps the protein has some other function in PG metabolism such that there is pressure for its continued existence in the *M. tuberculosis* genome.

Searching the *M. tuberculosis* H37Rv genome for the term "lactamase" results in several genes that are annotated as β-lactamases or esterases or carboxypeptidases (see Table 3: *Rv0339c, Rv0406c, Rv3677c,*

Rv0907, Rv1703c, Rv1923, Rv1497, and *Rv3762*), but very little is known about the function of the corresponding proteins (24). The *Rv0406c* and *Rv3677c* genes have been expressed in *E. coli*, purified, and shown to be able to degrade antibiotic (150). β-lactamases and carboxypeptidases are related to each other, and thus, some carboxypeptidases that are involved with PG metabolism may have weak β-lactamase activity, so such results should be interpreted with caution (151). A mutant of *M. smegmatis* with a deletion of the gene corresponding to *Rv0907* is less susceptible to vancomycin, although the significance of this is not known (123). The Rv1923 and Rv1497 proteins are annotated as putative lipases and thus may not have any role in PG metabolism. Further investigation into these genes is warranted to achieve a better understanding of the interaction between β-lactam antibiotics and mycobacteria, particularly since the idea of using β-lactamase inhibitors with these antibiotics is gaining traction in the treatment of tuberculosis.

Several other genes have been identified in a transposon screen of β-lactamase-deficient mutants of *M. tuberculosis* and *M. smegmatis* looking for mutants with increased susceptibility to cephalosporins (36). Several

TABLE 3 Genes involved with β-lactam antibiotic resistance

H37Rv#	Gene	Description of gene product
Rv2068c	*blaC*	Class A β-lactamase
Rv0399c	*lpqK*	Homologous to β-lactamases/esterases/carboxypeptidases
Rv0406c		Homologous to β-lactamases/esterases/carboxypeptidases
Rv3677c		Homologous to β-lactamases/esterases/carboxypeptidases
Rv0907		Homologous to β-lactamases/esterases/carboxypeptidases
Rv1703c		Homologous to β-lactamases/esterases/carboxypeptidases
Rv1923	*lipD*	Homologous to β-lactamases/esterases/carboxypeptidases (may be a lipase)
Rv1497	*lipL*	Homologous to β-lactamases/esterases/carboxypeptidases (may be a lipase)
Rv3762		Homologous to β-lactamases/esterases/carboxypeptidases
Rv1024	*cdpA*	DivIC family member, septum formation initiator
Rv2927c	*cdpC*	DivIVA family member, cell division initiator protein
Rv2198c	*mmpS3*	Mycobacterial membrane protein, small
Rv2224c-Rv2223c		Secreted proteases

mutants were found, all with increased susceptibility to penicillins and cephalosporins, as well as lysozyme. Some of the affected genes were already known to be involved with PG biosynthesis (*ponA2*, *namH*, *dapB*), but a few other genes are worthy of note. Two mutants of *M. smegmatis* had cell division defects, with insertions that affected genes encoding proteins with homology to proteins involved with cell division. One gene, *MSMEG_5414* (*cdpA*), homologous to *Rv1024*, is similar to DivIVC, which is involved with the initiation of septum formation in Gram-positive bacteria, while the other gene, *MSMEG_2416* (*cdpC*), homologous to *Rv2927c*, is similar to the DivIVA protein, which is important for the initiation of division in Gram-positive bacteria. Another *M. smegmatis* mutant with an insertion in the first gene in a two-gene operon (*MSMEG_4296-MSMEG_4295*), homologous to *Rv2224c-Rv2223c*, has swollen poles and is highly susceptible to lysozyme. The proteins encoded by these genes appear to be secreted proteases that may have a role in regulating the levels of proteins involved with PG metabolism. Transposon mutants of *M. tuberculosis* with an insertion in *Rv2224c* were isolated in two independent screens for hypersensitivity to low pH (84) and defective intracellular survival (152); a defined Rv2224c mutant was attenuated in the mouse, promoted a stronger innate immune response, and was hypersusceptible to lysozyme (153). The Rv2224c protein was also shown to process the GroEl2 chaperon (153). The link between protease activity, PG metabolism, and pathogenesis is not entirely clear, but this protein (and Rv2225c) should be further investigated. Lastly, an *M. tuberculosis* mutant with an insertion in the putative promoter region of the *mmpS3* gene had normal cell morphology with increased β-lactam susceptibility (36). Little is known about the MmpS3 protein, but it belongs to a family of small membrane proteins that are involved with transport functions (154). This protein may play a role in assembly or some other aspect of the cell envelope that may allow better penetration of the β-lactam, but the mutant was not hypersusceptible to other cell envelope-specific antibiotics such as isoniazid or ethambutol.

FUTURE RESEARCH

While much is known about the basic pathways involved with mycobacterial PG monomer synthesis, there are several areas in which additional work is needed. What are the enzymes involved with PG modifications such as amidation, and what role do these modifications play in mycobacterial PG biology? How

does the cell move PG monomers across the cell membrane? Why are there so many L,D-transpeptidases in mycobacteria? How are the 3-3 linkages synthesized, and what is their function? How is assembly of the PG coupled to that of the arabinogalactan? How is the division machinery linked to PG wall metabolism and the pathways involved with assembly of the MAPc? With regard to turnover of the PG: What is the role of Rpf proteins in PG metabolism, and how does this influence the ability of *M. tuberculosis* to persist and reactivate in an infection? How do the cell wall hydrolases interact with PBPs and Ldts to balance PG biosynthesis with degradation? Many proteins involved with PG metabolism appear to be phosphorylated, and this is likely to be a global mechanism of control that may be important in coordinating all these aspects of cell wall biosynthesis. Investigating the interactions with mycobacteria and β-lactam antibiotics, the role accessory proteins may play in this, and the significance of β-lactamases in mycobacteria will not only provide a clearer picture of PG metabolism, but also help in the rational development of potential new drug therapies to treat intractable mycobacterial infections.

We thank Fabio L. Fontes for the preparation of Fig. 1, 2, and 4.

Citation. Pavelka MS Jr, Mahapatra S, Crick DC. 2014. Genetics of peptidoglycan biosynthesis. Microbiol Spectrum 2(4):MGM2-0034-2013.

References

1. Lederer E. 1971. The mycobacterial cell wall. *Pure Appl Chem* **25**:135–165.

2. Mahapatra S, Basu J, Brennan PJ, Crick DC. 2005. Structure, biosynthesis, and genetics of the mycolic acid-arabinogalactan-peptidoglycan complex, p 275–277. *In* Cole ST, Eisenbach KD, McMurray DN, Jacobs WR Jr (eds), *Tuberculosis and the Tubercle Bacillus.* ASM Press, Washington, DC.

3. Schleifer KH, Kandler O. 1972. Peptidoglycan types of bacterial cell walls and their taxonomic implications. *Bacteriol Rev* **36**:407–477.

4. Azuma I, Thomas DW, Adam A, Ghuysen JM, Bonaly R, Petit JF, Lederer E. 1970. Occurrence of N-glycolylmuramic acid in bacterial cell walls. A preliminary survey. *Biochim Biophys Acta* **208**:444–451.

5. Mahapatra S, Scherman H, Brennan PJ, Crick DC. 2005. N glycolylation of the nucleotide precursors of peptidoglycan biosynthesis of *Mycobacterium* spp. is altered by drug treatment. *J Bacteriol* **187**:2341–2347.

6. Mahapatra S, Crick DC, McNeil MR, Brennan PJ. 2008. Unique structural features of the peptidoglycan of *Mycobacterium leprae. J Bacteriol* **190**:655–661.

7. Mahapatra S, Yagi T, Belisle JT, Espinosa BJ, Hill PJ, McNeil MR, Brennan PJ, Crick DC. 2005. Mycobacterial

lipid II is composed of a complex mixture of modified muramyl and peptide moieties linked to decaprenyl phosphate. *J Bacteriol* **187**:2747–2757.

8. Raymond JB, Mahapatra S, Crick DC, Pavelka MS Jr. 2005. Identification of the namH gene, encoding the hydroxylase responsible for the *N*-glycolylation of the mycobacterial peptidoglycan. *J Biol Chem* **280**:326–333.

9. Coulombe F, Divangahi M, Veyrier F, de Leseleuc L, Gleason JL, Yang Y, Kelliher MA, Pandey AK, Sassetti CM, Reed MB, Behr MA. 2009. Increased NOD2-mediated recognition of *N*-glycolyl muramyl dipeptide. *J Exp Med* **206**:1709–1716.

10. Goffin C, Ghuysen JM. 2002. Biochemistry and comparative genomics of SxxK superfamily acyltransferases offer a clue to the mycobacterial paradox: presence of penicillin-susceptible target proteins versus lack of efficiency of penicillin as therapeutic agent. *Microbiol Mol Biol Rev* **66**:702–738.

11. Lavollay M, Arthur M, Fourgeaud M, Dubost L, Marie A, Riegel P, Gutmann L, Mainardi JL. 2009. The beta-lactam-sensitive D,D-carboxypeptidase activity of Pbp4 controls the L,D and D,D transpeptidation pathways in *Corynebacterium jeikeium*. *Mol Microbiol* **74**:650–661.

12. Lavollay M, Arthur M, Fourgeaud M, Dubost L, Marie A, Veziris N, Blanot D, Gutmann L, Mainardi JL. 2008. The peptidoglycan of stationary phase *Mycobacterium tuberculosis* predominantly contains cross-links generated by L,D-transpeptidation. *J Bacteriol* **190**:4360–4366.

13. Mainardi JL, Fourgeaud M, Hugonnet JE, Dubost L, Brouard JP, Ouazzani J, Rice LB, Gutmann L, Arthur M. 2005. A novel peptidoglycan cross-linking enzyme for a beta-lactam-resistant transpeptidation pathway. *J Biol Chem* **280**:38146–38152.

14. Mainardi JL, Hugonnet JE, Rusconi F, Fourgeaud M, Dubost L, Moumi AN, Delfosse V, Mayer C, Gutmann L, Rice LB, Arthur M. 2007. Unexpected inhibition of peptidoglycan L,D-transpeptidase from *Enterococcus faecium* by the beta-lactam imipenem. *J Biol Chem* **282**:30414–30422.

15. Mainardi JL, Legrand R, Arthur M, Schoot B, van Heijenoort J, Gutmann L. 2000. Novel mechanism of beta-lactam resistance due to bypass of D,D-transpeptidation in *Enterococcus faecium*. *J Biol Chem* **275**:16490–16496.

16. Sanders AN, Pavelka MS. 2013. Phenotypic analysis of *Eschericia coli* mutants lacking L,D-transpeptidases. *Microbiology* **159**:1842–1852.

17. Magnet S, Arbeloa A, Mainardi JL, Hugonnet JE, Fourgeaud M, Dubost L, Marie A, Delfosse V, Mayer C, Rice LB, Arthur M. 2007. Specificity of L,D-transpeptidases from Gram-positive bacteria producing different peptidoglycan chemotypes. *J Biol Chem* **282**:13151–13159.

18. Quintela JC, Caparros M, de Pedro MA. 1995. Variability of peptidoglycan structural parameters in gram-negative bacteria. *FEMS Microbiol Lett* **125**:95–100.

19. Wietzerbin J, Das BC, Petit J-F, Lederer E, Leyh-Bouille M, Ghuysen J-M. 1974. Occurence of D-alanyl-(D)-*meso*-diaminopimelic acid and *meso*-diaminopimelyl-*meso*-diaminopimelic acid interpeptide linkages in the peptidoglycan of mycobacteria. *Biochemistry* **13**:3471–3476.

20. Kumar P, Arora K, Lloyd JR, Lee IY, Nair V, Fischer E, Boshoff HI, Barry CE 3rd. 2012. Meropenem inhibits D,D-carboxypeptidase activity in *Mycobacterium tuberculosis*. *Mol Microbiol* **86**:367–381.

21. Lavollay M, Fourgeaud M, Herrmann JL, Dubost L, Marie A, Gutmann L, Arthur M, Mainardi JL. 2011. The peptidoglycan of *Mycobacterium abscessus* is predominantly cross-linked by L,D-transpeptidases. *J Bacteriol* **193**:778–782.

22. McNeil M, Daffe M, Brennan PJ. 1990. Evidence for the nature of the link between the arabinogalactan and peptidoglycan of mycobacterial cell walls. *J Biol Chem* **265**:18200–18206.

23. Cole ST, Brosch R, Parkhill J, Garnier T, Churcher C, Harris D, Gordon SV, Eiglmeier K, Gas S, Barry CE 3rd, Tekaia F, Badcock K, Basham D, Brown D, Chillingworth T, Connor R, Davies R, Devlin K, Feltwell T, Gentles S, Hamlin N, Holroyd S, Hornsby T, Jagels K, Krogh A, McLean J, Moule S, Murphy L, Oliver K, Osborne J, Quail MA, Rajandream MA, Rogers J, Rutter S, Seeger K, Skelton J, Squares R, Squares S, Sulston JE, Taylor K, Whitehead S, Barrell BG. 1998. Deciphering the biology of *Mycobacterium tuberculosis* from the complete genome sequence. *Nature* **393**:537–544.

24. Slayden RA, Jackson M, Zucker J, Ramirez MV, Dawson CC, Crew R, Sampson NS, Thomas ST, Jamshidi N, Sisk P, Caspi R, Crick DC, McNeil MR, Pavelka MS, Niederweis M, Siroy A, Dona V, McFadden J, Boshoff H, Lew JM. 2013. Updating and curating metabolic pathways of TB. *Tuberculosis* **93**:47–59.

25. Cirillo JD, Weisbrod TR, Pascopella L, Jacobs WR Jr. 1994. Isolation and characterization of the aspartate semialdehyde dehydrogenase and aspartokinase genes from mycobacteria. *Mol Microbiol* **11**:629–639.

26. Pavelka MS Jr, Weisbrod TR, Jacobs WR Jr. 1997. Cloning of the *dapB* gene, encoding dihydrodipicolinate reductase, from *Mycobacterium tuberculosis*. *J Bacteriol* **179**:2777–2782.

27. Jha RK, Katagihallimath N, Hota SK, Das KS, de Sousa SM. 2012. An assay for exogenous sources of purified MurG, enabled by the complementation of *Escherichia coli murG*(Ts) by the *Mycobacterium tuberculosis* homologue. *FEMS Microbiol Lett* **326**:161–167.

28. van Heijenoort J. 1998. Assembly of the monomer unit of bacterial peptidoglycan. *Cell Mol Life Sci* **54**:300–304.

29. Ashiuchi M, Kuwana E, Komatsu K, Soda K, Misono H. 2003. Differences in effects on DNA gyrase activity between two glutamate racemases of *Bacillus subtilis*, the poly-gamma-glutamate synthesis-linking Glr enzyme and the YrpC (MurI) isozyme. *FEMS Microbiol Lett* **223**:221–225.

30. Sengupta S, Ghosh S, Nagaraja V. 2008. Moonlighting function of glutamate racemase from *Mycobacterium tuberculosis*: racemization and DNA gyrase inhibition are two independent activities of the enzyme. *Microbiology* **154**:2796–2803.

31. Umbarger HE. 1978. Amino acid biosynthesis and its regulation. *Annu Rev Biochem* **47**:533–606.

32. Pavelka MS Jr, Jacobs WR Jr. 1996. Biosynthesis of diaminopimelate (DAP), the precursor of lysine and a component of the peptidoglycan, is an essential function of *Mycobacterium smegmatis*. *J Bacteriol* **178**:6496–6507.

33. Consaul SA, Jacobs WR Jr, Pavelka MS Jr. 2003. Extragenic suppression of the requirement for diaminopimelate in diaminopimelate auxotrophs of *Mycobacterium smegmatis*. *FEMS Microbiol Lett* **225**:131–135.

34. Consaul SA, Wright LF, Mahapatra S, Crick DC, Pavelka MS Jr. 2005. An unusual mutation results in the replacement of diaminopimelate with lanthionine in the peptidoglycan of a mutant strain of *Mycobacterium smegmatis*. *J Bacteriol* **187**:1612–1620.

35. Mengin-Lecreulx D, Blanot D, van Heijenoort J. 1994. Replacement of diaminopimelic acid by cystathionine or lanthionine in the peptidoglycan of *Escherichia coli*. *J Bacteriol* **176**:4321–4327.

36. Flores AR, Parsons LM, Pavelka MS Jr. 2005. Characterization of novel *Mycobacterium tuberculosis* and *Mycobacterium smegmatis* mutants hypersusceptible to beta-lactam antibiotics. *J Bacteriol* **187**:1892–1900.

37. Pavelka MS Jr, Jacobs WR Jr. 1999. Comparison of the construction of unmarked deletion mutations in *Mycobacterium smegmatis*, *Mycobacterium bovis* bacillus Calmette-Guerin, and *Mycobacterium tuberculosis* H37Rv by allelic exchange. *J Bacteriol* **181**:4780–4789.

38. Liu L, Yoshimura T, Endo K, Kishimoto K, Fuchikami Y, Manning JM, Esaki N, Soda K. 1998. Compensation for D-glutamate auxotrophy of *Escherichia coli* WM335 by D-amino acid aminotransferase gene and regulation of murI expression. *Biosci Biotechnol Biochem* **62**:193–195.

39. Chacon O, Feng Z, Harris NB, Caceres NE, Adams LG, Barletta RG. 2002. *Mycobacterium smegmatis* D-alanine racemase mutants are not dependent on D-alanine for growth. *Antimicrobial Agents Chemother* **46**:47–54.

40. Awasthy D, Bharath S, Subbulakshmi V, Sharma U. 2012. Alanine racemase mutants of *Mycobacterium tuberculosis* require D-alanine for growth and are defective for survival in macrophages and mice. *Microbiology* **158**:319–327.

41. Milligan DL, Tran SL, Strych U, Cook GM, Krause KL. 2007. The alanine racemase of *Mycobacterium smegmatis* is essential for growth in the absence of D-alanine. *J Bacteriol* **189**:8381–8386.

42. Feng Z, Barletta RG. 2003. Roles of *Mycobacterium smegmatis* D-alanine:D-alanine ligase and D-alanine racemase in the mechanisms of action of and resistance to the peptidoglycan inhibitor D-cycloserine. *Antimicrobial Agents Chemother* **47**:283–291.

43. van Heijenoort J. 2001. Recent advances in the formation of the bacterial peptidoglycan monomer unit. *Nat Prod Rep* **18**:503–519.

44. van Heijenoort J. 2001. Formation of the glycan chains in the synthesis of bacterial peptidoglycan. *Glycobiology* **11**:25R–36R.

45. Mengin-Lecreulx D, Flouret B, van Heijenoort J. 1982. Cytoplasmic steps of peptidoglycan synthesis in *Escherichia coli*. *J Bacteriol* **151**:1109–1117.

46. De Smet KA, Kempsell KE, Gallagher A, Duncan K, Young DB. 1999. Alteration of a single amino acid residue reverses fosfomycin resistance of recombinant MurA from *Mycobacterium tuberculosis*. *Microbiology* **145**:3177–3184.

47. Eschenburg S, Priestman M, Schonbrunn E. 2005. Evidence that the fosfomycin target Cys115 in UDP-N-acetylglucosamine enolpyruvyl transferase (MurA) is essential for product release. *J Biol Chem* **280**:3757–3763.

48. Kumar V, Saravanan P, Arvind A, Mohan CG. 2011. Identification of hotspot regions of MurB oxidoreductase enzyme using homology modeling, molecular dynamics and molecular docking techniques. *J Mol Model* **17**:939–953.

49. Sassetti CM, Boyd DH, Rubin EJ. 2001. Comprehensive identification of conditionally essential genes in mycobacteria. *Proc Natl Acad Sci USA* **98**:12712–12717.

50. Mengin-Lecreulx D, van Heijenoort J, Park JT. 1996. Identification of the mpl gene encoding UDP-N-acetylmuramate: L-alanyl-gamma-d-glutamyl-meso-diaminopimelate ligase in *Escherichia coli* and its role in recycling of cell wall peptidoglycan. *J Bacteriol* **178**:5347–5352.

51. Mahapatra S, Crick DC, Brennan PJ. 2000. Comparison of the UDP-N-acetylmuramate:L-alanine ligase enzymes from *Mycobacterium tuberculosis* and *Mycobacterium leprae*. *J Bacteriol* **182**:6827–6830.

52. Munshi T, Gupta A, Evangelopoulos D, Guzman JD, Gibbons S, Keep NH, Bhakta S. 2013. Characterisation of ATP-dependent Mur ligases involved in the biogenesis of cell wall peptidoglycan in *Mycobacterium tuberculosis*. *PLoS One* **8**:e60143.

53. Draper P. 1976. Cell walls of *Mycobacterium leprae*. *Int J Lepr Other Mycobact Dis* **44**:95–98.

54. Barreteau H, Sosic I, Turk S, Humljan J, Tomasic T, Zidar N, Herve M, Boniface A, Peterlin-Masic L, Kikelj D, Mengin-Lecreulx D, Gobec S, Blanot D. 2012. MurD enzymes from different bacteria: evaluation of inhibitors. *Biochem Pharmacol* **84**:625–632.

55. Basavannacharya C, Robertson G, Munshi T, Keep NH, Bhakta S. 2010. ATP-dependent MurE ligase in *Mycobacterium tuberculosis*: biochemical and structural characterisation. *Tuberculosis* **90**:16–24.

56. Bernard E, Rolain T, Courtin P, Hols P, Chapot-Chartier MP. 2011. Identification of the amidotransferase AsnB1 as being responsible for meso-diaminopimelic acid amidation in *Lactobacillus plantarum* peptidoglycan. *J Bacteriol* **193**:6323–6330.

57. Figueiredo TA, Sobral RG, Ludovice AM, Almeida JM, Bui NK, Vollmer W, de Lencastre H, Tomasz A. 2012. Identification of genetic determinants and enzymes involved with the amidation of glutamic acid residues in the peptidoglycan of *Staphylococcus aureus*. *PLoS Pathog* **8**:e1002508.

58. Munch D, Roemer T, Lee SH, Engeser M, Sahl HG, Schneider T. 2012. Identification and in vitro analysis of the GatD/MurT enzyme-complex catalyzing lipid II amidation in *Staphylococcus aureus*. *PLoS Pathog* **8**: e1002509.

59. Menon AK. 1995. Flippases. *Trends Cell Biol* **5**: 355–360.

60. Pomorski T, Holthuis JC, Herrmann A, van Meer G. 2004. Tracking down lipid flippases and their biological functions. *J Cell Sci* **117**:805–813.

61. Raetz CR, Whitfield C. 2002. Lipopolysaccharide endotoxins. *Annu Rev Biochem* **71**:635–700.

62. Weppner WA, Neuhaus FC. 1978. Biosynthesis of peptidoglycan. Definition of the microenvironment of undecaprenyl diphosphate-N-acetylmuramyl-(5-dimethyl-aminonaphthalene-1-sulfonyl) pentapeptide by fluorescence spectroscopy. *J Biol Chem* **253**:472–478.

63. van Dam V, Sijbrandi R, Kol M, Swiezewska E, de Kruijff B, Breukink E. 2007. Transmembrane transport of peptidoglycan precursors across model and bacterial membranes. *Mol Microbiol* **64**:1105–1114.

64. Siewert G, Strominger JL. 1967. Bacitracin: an inhibitor of the dephosphorylation of lipid pyrophosphate, an intermediate in the biosynthesis of the peptidoglycan of bacterial cell walls. *Proc Natl Acad Sci USA* **57**:767–773.

65. Bosne-David S, Barros V, Verde SC, Portugal C, David HL. 2000. Intrinsic resistance of *Mycobacterium tuberculosis* to clarithromycin is effectively reversed by subinhibitory concentrations of cell wall inhibitors. *J Antimicrob Chemother* **46**:391–395.

66. Inoue A, Murata Y, Takahashi H, Tsuji N, Fujisaki S, Kato J. 2008. Involvement of an essential gene, mviN, in murein synthesis in *Escherichia coli*. *J Bacteriol* **190**: 7298–7301.

67. Mohammadi T, van Dam V, Sijbrandi R, Vernet T, Zapun A, Bouhss A, Diepeveen-de Bruin M, Nguyen-Disteche M, de Kruijff B, Breukink E. 2011. Identification of FtsW as a transporter of lipid-linked cell wall precursors across the membrane. *EMBO J* **30**:1425–1432.

68. Ruiz N. 2008. Bioinformatics identification of MurJ (MviN) as the peptidoglycan lipid II flippase in *Escherichia coli*. *Proc Natl Acad Sci USA* **105**:15553–15557.

69. Gee CL, Papavinasasundaram KG, Blair SR, Baer CE, Falick AM, King DS, Griffin JE, Venghatakrishnan H, Zukauskas A, Wei JR, Dhiman RK, Crick DC, Rubin EJ, Sassetti CM, Alber T. 2012. A phosphorylated pseudokinase complex controls cell wall synthesis in mycobacteria. *Sci Signal* **5**:ra7.

70. Szklarczyk D, Franceschini A, Kuhn M, Simonovic M, Roth A, Minguez P, Doerks T, Stark M, Muller J, Bork P, Jensen LJ, von Mering C. 2011. The STRING database in 2011: functional interaction networks of proteins, globally integrated and scored. *Nucleic Acids Res* **39**:D561–D568.

71. Hvorup RN, Winnen B, Chang AB, Jiang Y, Zhou XF, Saier MH Jr. 2003. The multidrug/oligosaccharidyl-lipid/polysaccharide (MOP) exporter superfamily. *Eur J Biochem* **270**:799–813.

72. Liu D, Cole RA, Reeves PR. 1996. An O-antigen processing function for Wzx (RfbX): a promising candidate for O-unit flippase. *J Bacteriol* **178**:2102–2107.

73. Rick PD, Barr K, Sankaran K, Kajimura J, Rush JS, Waechter CJ. 2003. Evidence that the wzxE gene of *Escherichia coli* K-12 encodes a protein involved in the transbilayer movement of a trisaccharide-lipid intermediate in the assembly of enterobacterial common antigen. *J Biol Chem* **278**:16534–16542.

74. Ma Y, Stern RJ, Scherman MS, Vissa VD, Yan W, Jones VC, Zhang F, Franzblau SG, Lewis WH, McNeil MR. 2001. Drug targeting *Mycobacterium tuberculosis* cell wall synthesis: genetics of dTDP-rhamnose synthetic enzymes and development of a microtiter plate-based screen for inhibitors of conversion of dTDP-glucose to dTDP-rhamnose. *Antimicrob Agents Chemother* **45**: 1407–1416.

75. Hancock IC, Carman S, Besra GS, Brennan PJ, Waite E. 2002. Ligation of arabinogalactan to peptidoglycan in the cell wall of *Mycobacterium smegmatis* requires concomitant synthesis of the two wall polymers. *Microbiology* **148**:3059–3067.

76. Yagi T, Mahapatra S, Mikusova K, Crick DC, Brennan PJ. 2003. Polymerization of mycobacterial arabinogalactan and ligation to peptidoglycan. *J Biol Chem* **278**:26497–26504.

77. Goffin C, Ghuysen JM. 1998. Multimodular penicillin-binding proteins: an enigmatic family of orthologs and paralogs. *Microbiol Mol Biol Rev* **62**:1079–1093.

78. Basu J, Mahapatra S, Kundu M, Mukhopadhyay S, Nguyen-Disteche M, Dubois P, Joris B, Van Beeumen J, Cole ST, Chakrabarti P, Ghuysen JM. 1996. Identification and overexpression in *Escherichia coli* of a *Mycobacterium leprae* gene, pon1, encoding a high-molecular-mass class A penicillin-binding protein, PBP1. *J Bacteriol* **178**:1707–1711.

79. Mahapatra S, Bhakta S, Ahamed J, Basu J. 2000. Characterization of derivatives of the high-molecular-mass penicillin-binding protein (PBP) 1 of *Mycobacterium leprae*. *Biochem J* **350**(Pt 1):75–80.

80. Billman-Jacobe H, Haites RE, Coppel RL. 1999. Characterization of a *Mycobacterium smegmatis* mutant lacking penicillin binding protein 1. *Antimicrob Agents Chemother* **43**:3011–3013.

81. Hett EC, Chao MC, Rubin EJ. 2010. Interaction and modulation of two antagonistic cell wall enzymes of mycobacteria. *PLoS Pathog* **6**:e1001020.

82. Prisic S, Dankwa S, Schwartz D, Chou MF, Locasale JW, Kang CM, Bemis G, Church GM, Steen H, Husson RN. 2010. Extensive phosphorylation with overlapping specificity by *Mycobacterium tuberculosis* serine/threonine protein kinases. *Proc Nat Acad Sci USA* **107**:7521–7526.

83. Keer J, Smeulders MJ, Gray KM, Williams HD. 2000. Mutants of *Mycobacterium smegmatis* impaired in stationary-phase survival. *Microbiology* **146**(Pt 9): 2209–2217.

84. Vandal OH, Pierini LM, Schnappinger D, Nathan CF, Ehrt S. 2008. A membrane protein preserves intrabacterial pH in intraphagosomal *Mycobacterium tuberculosis*. *Nat Med* 14:849–854.

85. Vandal OH, Roberts JA, Odaira T, Schnappinger D, Nathan CF, Ehrt S. 2009. Acid-susceptible mutants of *Mycobacterium tuberculosis* share hypersusceptibility to cell wall and oxidative stress and to the host environment. *J Bacteriol* 191:625–631.

86. Patru MM, Pavelka MS Jr. 2010. A role for the class A penicillin-binding protein PonA2 in the survival of *Mycobacterium smegmatis* under conditions of nonreplication. *J Bacteriol* 192:3043–3054.

87. Dutta NK, Mehra S, Didier PJ, Roy CJ, Doyle LA, Alvarez X, Ratterree M, Be NA, Lamichhane G, Jain SK, Lacey MR, Lackner AA, Kaushal D. 2010. Genetic requirements for the survival of tubercle bacilli in primates. *J Infect Dis* 201:1743–1752.

88. Bourai N, Jacobs WR Jr, Narayanan S. 2012. Deletion and overexpression studies on DacB2, a putative low molecular mass penicillin binding protein from *Mycobacterium tuberculosis* H(37)Rv. *Microb Pathog* 52:109–116.

89. Magnet S, Dubost L, Marie A, Arthur M, Gutmann L. 2008. Identification of the L,D-transpeptidases for peptidoglycan cross-linking in *Escherichia coli*. *J Bacteriol* 190:4782–4785.

90. Magnet S, Bellais S, Dubost L, Fourgeaud M, Mainardi JL, Petit-Frere S, Marie A, Mengin-Lecreulx D, Arthur M, Gutmann L. 2007. Identification of the L,D-transpeptidases responsible for attachment of the Braun lipoprotein to *Escherichia coli* peptidoglycan. *J Bacteriol* 189:3927–3931.

91. Dramsi S, Magnet S, Davison S, Arthur M. 2008. Covalent attachment of proteins to peptidoglycan. *FEMS Microbiol Rev* 32:307–320.

92. Gupta R, Lavollay M, Mainardi JL, Arthur M, Bishai WR, Lamichhane G. 2010. The *Mycobacterium tuberculosis* protein LdtMt2 is a nonclassical transpeptidase required for virulence and resistance to amoxicillin. *Nat Med* 16:466–469.

93. Purdy GE, Niederweis M, Russell DG. 2009. Decreased outer membrane permeability protects mycobacteria from killing by ubiquitin-derived peptides. *Mol Microbiol* 73:844–857.

94. Park JT, Uehara T. 2008. How bacteria consume their own exoskeletons (turnover and recycling of cell wall peptidoglycan). *Microbiol Mol Biol Rev* 72:211–227.

95. Deng LL, Humphries DE, Arbeit RD, Carlton LE, Smole SC, Carroll JD. 2005. Identification of a novel peptidoglycan hydrolase CwlM in *Mycobacterium tuberculosis*. *Biochim Biophys Acta* 1747:57–66.

96. Mahapatra S, Piechota C, Gil F, Ma Y, Huang H, Scherman MS, Jones V, Pavelka MS Jr, Moniz-Pereira J, Pimentel M, McNeil MR, Crick DC. 2013. Mycobacteriophage Ms6 LysA: a peptidoglycan amidase and a useful analytical tool. *Appl Environ Microbiol* 79:768–773.

97. Mukamolova GV, Kaprelyants AS, Young DI, Young M, Kell DB. 1998. A bacterial cytokine. *Proc Natl Acad Sci USA* 95:8916–8921.

98. Mukamolova GV, Turapov OA, Young DI, Kaprelyants AS, Kell DB, Young M. 2002. A family of autocrine growth factors in *Mycobacterium tuberculosis*. *Mol Microbiol* 46:623–635.

99. Kana BD, Mizrahi V. 2010. Resuscitation-promoting factors as lytic enzymes for bacterial growth and signaling. *FEMS Immunol Med Microbiol* 58:39–50.

100. Zhu W, Plikaytis BB, Shinnick TM. 2003. Resuscitation factors from mycobacteria: homologs of *Micrococcus luteus* proteins. *Tuberculosis (Edinb)* 83:261–269.

101. Mukamolova GV, Kormer SS, Kell DB, Kaprelyants AS. 1999. Stimulation of the multiplication of *Micrococcus luteus* by an autocrine growth factor. *Arch Microbiol* 172:9–14.

102. Mukamolova GV, Murzin AG, Salina EG, Demina GR, Kell DB, Kaprelyants AS, Young M. 2006. Muralytic activity of *Micrococcus luteus* Rpf and its relationship to physiological activity in promoting bacterial growth and resuscitation. *Mol Microbiol* 59:84–98.

103. Mukamolova GV, Turapov OA, Kazarian K, Telkov M, Kaprelyants AS, Kell DB, Young M. 2002. The *rpf* gene of *Micrococcus luteus* encodes an essential secreted growth factor. *Mol Microbiol* 46:611–621.

104. Shleeva M, Mukamolova GV, Young M, Williams HD, Kaprelyants AS. 2004. Formation of "non-culturable" cells of *Mycobacterium smegmatis* in stationary phase in response to growth under suboptimal conditions and their Rpf-mediated resuscitation. *Microbiology* 150:1687–1697.

105. Shleeva MO, Bagramyan K, Telkov MV, Mukamolova GV, Young M, Kell DB, Kaprelyants AS. 2002. Formation and resuscitation of "non-culturable" cells of *Rhodococcus rhodochrous* and *Mycobacterium tuberculosis* in prolonged stationary phase. *Microbiology* 148:1581–1591.

106. Kana BD, Gordhan BG, Downing KJ, Sung N, Vostroktunova G, Machowski EE, Tsenova L, Young M, Kaprelyants A, Kaplan G, Mizrahi V. 2008. The resuscitation-promoting factors of *Mycobacterium tuberculosis* are required for virulence and resuscitation from dormancy but are collectively dispensable for growth in vitro. *Mol Microbiol* 67:672–684.

107. Cole ST, Eiglmeier K, Parkhill J, James KD, Thomson NR, Wheeler PR, Honore N, Garnier T, Churcher C, Harris D, Mungall K, Basham D, Brown D, Chillingworth T, Connor R, Davies RM, Devlin K, Duthoy S, Feltwell T, Fraser A, Hamlin N, Holroyd S, Hornsby T, Jagels K, Lacroix C, Maclean J, Moule S, Murphy L, Oliver K, Quail MA, Rajandream MA, Rutherford KM, Rutter S, Seeger K, Simon S, Simmonds M, Skelton J, Squares R, Squares S, Stevens K, Taylor K, Whitehead S, Woodward JR, Barrell BG. 2001. Massive gene decay in the leprosy bacillus. *Nature* 409:1007–1011.

108. Tufariello JM, Jacobs WR Jr, Chan J. 2004. Individual *Mycobacterium tuberculosis* resuscitation-promoting

factor homologues are dispensable for growth in vitro and in vivo. *Infect Immun* 72:515–526.

109. Downing KJ, Betts JC, Young DI, McAdam RA, Kelly F, Young M, Mizrahi V. 2004. Global expression profiling of strains harbouring null mutations reveals that the five *rpf*-like genes of *Mycobacterium tuberculosis* show functional redundancy. *Tuberculosis (Edinb)* 84:167–179.

110. Tufariello JM, Mi K, Xu J, Manabe YC, Kesavan AK, Drumm J, Tanaka K, Jacobs WR Jr, Chan J. 2006. Deletion of the *Mycobacterium tuberculosis* resuscitation-promoting factor Rv1009 gene results in delayed reactivation from chronic tuberculosis. *Infect Immun* 74:2985–2995.

111. Biketov S, Potapov V, Ganina E, Downing K, Kana BD, Kaprelyants A. 2007. The role of resuscitation promoting factors in pathogenesis and reactivation of *Mycobacterium tuberculosis* during intra-peritoneal infection in mice. *BMC Infect Dis* 7:146.

112. Downing KJ, Mischenko VV, Shleeva MO, Young DI, Young M, Kaprelyants AS, Apt AS, Mizrahi V. 2005. Mutants of *Mycobacterium tuberculosis* lacking three of the five *rpf*-like genes are defective for growth in vivo and for resuscitation in vitro. *Infect Immun* 73:3038–3043.

113. Russell-Goldman E, Xu J, Wang X, Chan J, Tufariello JM. 2008. A *Mycobacterium tuberculosis* Rpf double-knockout strain exhibits profound defects in reactivation from chronic tuberculosis and innate immunity phenotypes. *Infect Immun* 76:4269–4281.

114. Kondratieva T, Rubakova E, Kana BD, Biketov S, Potapov V, Kaprelyants A, Apt A. 2011. *Mycobacterium tuberculosis* attenuated by multiple deletions of *rpf* genes effectively protects mice against TB infection. *Tuberculosis* 91:219–223.

115. Anantharaman V, Aravind L. 2003. Evolutionary history, structural features and biochemical diversity of the NlpC/P60 superfamily of enzymes. *Genome Biol* 4:R11.

116. Wuenscher MD, Kohler S, Bubert A, Gerike U, Goebel W. 1993. The *iap* gene of *Listeria monocytogenes* is essential for cell viability, and its gene product, p60, has bacteriolytic activity. *J Bacteriol* 175:3491–3501.

117. Gao LY, Pak M, Kish R, Kajihara K, Brown EJ. 2006. A mycobacterial operon essential for virulence in vivo and invasion and intracellular persistence in macrophages. *Infection Immun* 74:1757–1767.

118. Both D, Schneider G, Schnell R. 2011. Peptidoglycan remodeling in *Mycobacterium tuberculosis*: comparison of structures and catalytic activities of RipA and RipB. *J Mol Biol* 413:247–260.

119. Hett EC, Chao MC, Steyn AJ, Fortune SM, Deng LL, Rubin EJ. 2007. A partner for the resuscitation-promoting factors of *Mycobacterium tuberculosis*. *Mol Microbiol* 66:658–668.

120. Hett EC, Chao MC, Deng LL, Rubin EJ. 2008. A mycobacterial enzyme essential for cell division synergizes with resuscitation-promoting factor. *PLoS Pathog* 4:e1000001.

121. Chao MC, Kieser KJ, Minami S, Mavrici D, Aldridge BB, Fortune SM, Alber T, Rubin EJ. 2013. Protein complexes and proteolytic activation of the cell wall hydrolase RipA regulate septal resolution in mycobacteria. *PLoS Pathog* 9:e1003197.

122. Parthasarathy G, Lun S, Guo H, Ammerman NC, Geiman DE, Bishai WR. 2012. Rv2190c, an NlpC/P60 family protein, is required for full virulence of *Mycobacterium tuberculosis*. *PLoS One* 7:e43429.

123. Piuri M, Hatfull GF. 2006. A peptidoglycan hydrolase motif within the mycobacteriophage TM4 tape measure protein promotes efficient infection of stationary phase cells. *Mol Microbiol* 62:1569–1585.

124. Datta P, Dasgupta A, Singh AK, Mukherjee P, Kundu M, Basu J. 2006. Interaction between FtsW and penicillin-binding protein 3 (PBP3) directs PBP3 to mid-cell, controls cell septation and mediates the formation of a trimeric complex involving FtsZ, FtsW and PBP3 in mycobacteria. *Mol Microbiol* 62:1655–1673.

125. Mukherjee P, Sureka K, Datta P, Hossain T, Barik S, Das KP, Kundu M, Basu J. 2009. Novel role of Wag31 in protection of mycobacteria under oxidative stress. *Mol Microbiol* 73:103–119.

126. Hamasha K, Sahana MB, Jani C, Nyayapathy S, Kang CM, Rehse SJ. 2010. The effect of Wag31 phosphorylation on the cells and the cell envelope fraction of wild-type and conditional mutants of *Mycobacterium smegmatis* studied by visible-wavelength Raman spectroscopy. *Biochem Biophys Res Commun* 391:664–668.

127. Jani C, Eoh H, Lee JJ, Hamasha K, Sahana MB, Han JS, Nyayapathy S, Lee JY, Suh JW, Lee SH, Rehse SJ, Crick DC, Kang CM. 2010. Regulation of polar peptidoglycan biosynthesis by Wag31 phosphorylation in mycobacteria. *BMC Microbiol* 10:327.

128. Kang CM, Nyayapathy S, Lee JY, Suh JW, Husson RN. 2008. Wag31, a homologue of the cell division protein DivIVA, regulates growth, morphology and polar cell wall synthesis in mycobacteria. *Microbiology* 154:725–735.

129. Dasgupta A, Datta P, Kundu M, Basu J. 2006. The serine/threonine kinase PknB of *Mycobacterium tuberculosis* phosphorylates PBPA, a penicillin-binding protein required for cell division. *Microbiology* 152:493–504.

130. Fedarovich A, Nicholas RA, Davies C. 2010. Unusual conformation of the SxN motif in the crystal structure of penicillin-binding protein A from *Mycobacterium tuberculosis*. *J Mol Biol* 398:54–65.

131. Fedarovich A, Nicholas RA, Davies C. 2012. The role of the beta5-alpha11 loop in the active-site dynamics of acylated penicillin-binding protein A from *Mycobacterium tuberculosis*. *J Mol Biol* 418:316–330.

132. Kiran M, Chauhan A, Dziedzic R, Maloney E, Mukherji SK, Madiraju M, Rajagopalan M. 2009. *Mycobacterium tuberculosis ftsH* expression in response to stress and viability. *Tuberculosis* 89(Suppl 1):S70–S73.

133. Chauhan A, Lofton H, Maloney E, Moore J, Fol M, Madiraju MV, Rajagopalan M. 2006. Interference of *Mycobacterium tuberculosis* cell division by Rv2719c, a cell wall hydrolase. *Mol Microbiol* 62:132–147.

134. Plocinski P, Ziolkiewicz M, Kiran M, Vadrevu SI, Nguyen HB, Hugonnet J, Veckerle C, Arthur M, Dziadek J, Cross TA, Madiraju M, Rajagopalan M. 2011. Characterization of CrgA, a new partner of the *Mycobacterium tuberculosis* peptidoglycan polymerization complexes. *J Bacteriol* **193**:3246–3256.

135. Plocinski P, Arora N, Sarva K, Blaszczyk E, Qin H, Das N, Plocinska R, Ziolkiewicz M, Dziadek J, Kiran M, Gorla P, Cross TA, Madiraju M, Rajagopalan M. 2012. *Mycobacterium tuberculosis* CwsA interacts with CrgA and Wag31, and the CrgA-CwsA complex is involved in peptidoglycan synthesis and cell shape determination. *J Bacteriol* **194**:6398–6409.

136. Vadrevu IS, Lofton H, Sarva K, Blaszczyk E, Plocinska R, Chinnaswamy J, Madiraju M, Rajagopalan M. 2011. ChiZ levels modulate cell division process in mycobacteria. *Tuberculosis* **91**(Suppl 1):S128–S135.

137. England K, Crew R, Slayden RA. 2011. *Mycobacterium tuberculosis* septum site determining protein, Ssd encoded by *rv3660c*, promotes filamentation and elicits an alternative metabolic and dormancy stress response. *BMC Microbiol* **11**:79.

138. Griffith DE, Aksamit TR. 2012. Therapy of refractory nontuberculous mycobacterial lung disease. *Curr Opin Infect Dis* **25**:218–227.

139. Li XZ, Zhang L, Nikaido H. 2004. Efflux pump-mediated intrinsic drug resistance in *Mycobacterium smegmatis. Antimicrob Agents Chemother* **48**:2415–2423.

140. Chambers HF, Moreau D, Yajko D, Miick C, Wagner C, Hackbarth C, Kocagoz S, Rosenberg E, Hadley WK, Nikaido H. 1995. Can penicillins and other beta-lactam antibiotics be used to treat tuberculosis? *Antimicrob Agents Chemother* **39**:2620–2624.

141. Cynamon MH, Palmer GS. 1983. In vitro activity of amoxicillin in combination with clavulanic acid against *Mycobacterium tuberculosis. Antimicrob Agents Chemother* **24**:429–431.

142. Hugonnet JE, Tremblay LW, Boshoff HI, Barry CE 3rd, Blanchard JS. 2009. Meropenem-clavulanate is effective against extensively drug-resistant *Mycobacterium tuberculosis. Science* **323**:1215–1218.

143. Segura C, Salvado M, Collado I, Chaves J, Coira A. 1998. Contribution of beta-lactamases to beta-lactam susceptibilities of susceptible and multidrug-resistant *Mycobacterium tuberculosis* clinical isolates. *Antimicrob Agents Chemother* **42**:1524–1526.

144. Sorg TB, Cynamon MH. 1987. Comparison of four beta-lactamase inhibitors in combination with ampicillin against *Mycobacterium tuberculosis. J Antimicrob Chemother* **19**:59–64.

145. Wong CS, Palmer GS, Cynamon MH. 1988. In-vitro susceptibility of *Mycobacterium tuberculosis, Mycobacterium bovis* and *Mycobacterium kansasii* to amoxycillin and ticarcillin in combination with clavulanic acid. *J Antimicrob Chemother* **22**:863–866.

146. Voladri RK, Lakey DL, Hennigan SH, Menzies BE, Edwards KM, Kernodle DS. 1998. Recombinant expression and characterization of the major beta-lactamase of *Mycobacterium tuberculosis. Antimicrob Agents Chemother* **42**:1375–1381.

147. Hugonnet JE, Blanchard JS. 2007. Irreversible inhibition of the *Mycobacterium tuberculosis* beta-lactamase by clavulanate. *Biochemistry* **46**:11998–12004.

148. Flores AR, Parsons LM, Pavelka MS Jr. 2005. Genetic analysis of the beta-lactamases of *Mycobacterium tuberculosis* and *Mycobacterium smegmatis* and susceptibility to beta-lactam antibiotics. *Microbiology* **151**:521–532.

149. Sala C, Haouz A, Saul FA, Miras I, Rosenkrands I, Alzari PM, Cole ST. 2009. Genome-wide regulon and crystal structure of BlaI (Rv1846c) from *Mycobacterium tuberculosis. Mol Microbiol* **71**:1102–1116.

150. Nampoothiri KM, Rubex R, Patel AK, Narayanan SS, Krishna S, Das SM, Pandey A. 2008. Molecular cloning, overexpression and biochemical characterization of hypothetical beta-lactamases of *Mycobacterium tuberculosis* H37Rv. *J Appl Microbiol* **105**:59–67.

151. Galleni M, Raquet X, Lamotte-Brasseur J, Fonze E, Amicosante G, Frere JM. 1995. DD-peptidases and beta-lactamases: catalytic mechanisms and specificities. *J Chemother* **7**:3–7.

152. Rengarajan J, Bloom BR, Rubin EJ. 2005. Genome-wide requirements for *Mycobacterium tuberculosis* adaptation and survival in macrophages. *Proc Natl Acad Sci USA* **102**:8327–8332.

153. Rengarajan J, Murphy E, Park A, Krone CL, Hett EC, Bloom BR, Glimcher LH, Rubin EJ. 2008. *Mycobacterium tuberculosis* Rv2224c modulates innate immune responses. *Proc Natl Acad Sci USA* **105**:264–269.

154. Deshayes C, Bach H, Euphrasie D, Attarian R, Coureuil M, Sougakoff W, Laval F, Av-Gay Y, Daffe M, Etienne G, Reyrat JM. 2010. MmpS4 promotes glycopeptidolipids biosynthesis and export in *Mycobacterium smegmatis. Mol Microbiol* **78**:989–1003.

155. Daffé M, Crick DC, Jackson M. 2014. Genetics of capsular polysaccharides and cell envelope (glyco)lipids. *Microbiol Spectrum* **2**(4):MGM2-0021-2013.

Molecular Genetics of Mycobacteria, 2nd Edition
Edited by Graham F. Hatfull and William R. Jacobs, Jr.
© 2014 American Society for Microbiology, Washington, DC
doi:10.1128/microbiolspec.MGM2-0013-2013

Monika Jankute,[1],[†] Shipra Grover,[1],[†]
Helen L. Birch,[1] and Gurdyal S. Besra[1]

Genetics of Mycobacterial Arabinogalactan and Lipoarabinomannan Assembly

27

The cell wall of *Mycobacterium tuberculosis* is unique in that it differs significantly from both Gram-negative and Gram-positive bacteria. The thick, carbohydrate- and lipid-rich cell wall with distinct lipoglycans enables mycobacteria to survive under hostile conditions such as shortage of nutrients and antimicrobial exposure. The key features of this highly complex cell wall are the mycolyl-arabinogalactan-peptidoglycan (mAGP)–based and phosphatidyl-*myo*-inositol–based macromolecular structures, with the latter possessing potent immuno-modulatory properties. These structures are crucial for the growth, viability, and virulence of *M. tuberculosis* and therefore are often the targets of effective chemotherapeutic agents against tuberculosis (TB). Over the past decade, sophisticated genomic and molecular tools have advanced our understanding of the primary structure and biosynthesis of these macromolecules (1, 2). The availability of the full-genome sequences of various mycobacterial species, including *M. tuberculosis* (3), *Mycobacterium marinum* (4), and *Mycobacterium bovis* BCG (5), have greatly facilitated the identification of large numbers of drug targets and antigens specific to TB. Techniques to manipulate mycobacteria have also improved extensively; the conditional expression-specialized transduction essentiality test (CESTET) is currently used to determine the essentiality of individual genes (6). Finally, various biosynthetic assays using either purified proteins or synthetic cell wall acceptors have been developed to study enzyme function. This article focuses on the recent advances in determining the structural details and biosynthesis of arabinogalactan (AG), lipoarabinomannan (LAM), and related glycoconjugates.

ARABINOGALACTAN

Structural Features of AG

AG is a major cell wall heteropolysaccharide of mycobacteria. This highly branched structure is covalently attached to peptidoglycan (PG) via a phosphodiester bond to approximately 10 to 12% of the muramic acid residues (7). Collectively, PG and AG form a covalently linked network positioned between the plasma

[1]School of Biosciences, University of Birmingham, Edgbaston, Birmingham, B15 2TT, United Kingdom. [†]Authors contributed equally to this work.

membrane and the mycolic acid layer, resulting in an exceptionally robust cell wall. AG is composed predominantly of arabinose and galactose residues, both in their furanose ring form, that are extremely rare in nature (8). Unlike most bacterial polysaccharides, AG lacks repeating units and is composed of a few distinct structural motifs (9–12). The whole mycolyl-AG structure is attached to PG via a specific linkage unit, and thus it is believed to be the most vulnerable part of the complex. Detailed characterization of per-O-alkylated oligosaccharide alditols together with gas chromatography–mass spectrometry, fast atom bombardment–mass spectrometry, and nuclear magnetic resonance (NMR) analysis established the detailed structure of AG (Fig. 1).

The galactan domain is composed of approximately 30 alternating β(1→5) and β(1→6) galactofuranosyl (Galf) residues connected in a linear fashion. At the reducing end of AG, the galactan chain is linked to the C-6 position of selected N-glycolylmuramic acid residues of PG via an α-L-Rhap-(1→3)-α-D-GlcNAc-1-phosphate linkage unit (11). Three similar D-arabinan chains comprising roughly 30 arabinofuranosyl (Araf) residues each are attached to the C-5 of specific β(1→6) linked-Galf residues (9). Since the AG structure is essential to M. tuberculosis, many gene deletion studies investigating AG have been performed in the Corynebacterium genus, in which aspects of AG biosynthesis are nonessential. Knockout mutants in Corynebacterium glutamicum together with mass spectrometry determined that the arabinan chains of AG are attached distinctively to the 8th, 10th, and 12th residues of the linear galactan chain (13) (Fig. 1). Previous work demonstrated that the arabinan domain is present as a highly branched network built on a backbone of α(1→5) linked sugars with branching introduced by the presence of 3,5-α-D-Araf residues. Further α(1→5) linked Araf sugars are attached subsequent to this branch point, with the nonreducing ends terminating with β(1→2) Araf residues. The final structural motif is a distinct hexa-arabinofuranoside (14), present as [β-D-Araf-(1→2)-α-D-Araf]2-3,5-α-D-Araf-(1→5)-α-D-Araf. Analysis of per-O-methylated mAGP and per-O-alkylated oligoglycosyl alditols determined that position 5 of both the terminal β-D-Araf and the penultimate 2-α-D-Araf are the attachment sites for the mycolic acids (12). Follow-up studies determined that the mycolyl residues are located in clusters of four on the terminal hexa-arabinofuranoside motifs, with only two-thirds of these being mycolated.

An endogenous arabinase, which can cleave the arabinan, has been partially purified from Mycobacterium smegmatis (15). The use of this enzyme together with matrix-assisted laser desorption/ionization–time of flight–mass spectrometry and NMR allowed the sequencing of very large fragments of arabinan chains released from the mycobacterial cell wall. Significantly, galactosamine (GalN) residues, previously detected as a minor covalently bound sugar residue of the cell envelope of slow-growing mycobacteria such as M. tuberculosis and Mycobacterium avium (16), were shown to be located on the C-2 position of some of the internal 3,5-α-D-branched Araf residues (17), and the stereochemistry of the GalN moiety was confirmed to be an α-anomer (18) (Fig. 1). In addition, the succinyl groups were found on the interior branched arabinosyl residues (19). Approximately one of the three arabinan chains linked to the linear galactan contains a GalN group, and one of three is also succinylated (19). In addition, the succinyl residues were also shown to be present only on the nonmycolated chains. It is speculated that the GalN residue of AG may serve a specific function during host infection (20).

Precursor Formation

The biosynthesis of the linkage unit employs two high-energy substrates, UDP-GlcNAc and dTDP-Rha (Fig. 2). UDP-GlcNAc, a sugar donor for both the AG linkage unit and the biosynthesis of PG, is formed via a four-step reaction. Three enzymes (glutamine fructose-6-phosphate transferase, GlmS; phosphoglucosamine mutase, GlmM; and glucosamine-1-phosphate acetyl transferase/N-acetylglucosamine-1-phosphate urididyl transferase, GlmU) catalyze the conversion of fructose-6-phosphate to UDP-GlcNAc in Escherichia coli (21–24). Analysis of the genome sequence of M. tuberculosis determined that the proteins encoded by Rv3436c, Rv3441c, and Rv1018c are homologous to the E. coli GlmS, GlmM, and GlmU enzymes, respectively (25). GlmS is responsible for conversion of fructose-6-phosphate to glucosamine-6-phosphate, which is then converted to glucosamine-1-phosphate by GlmM. Recent gene deletion studies demonstrated that MSMEG_1556, an M. smegmatis gene encoding the homologue of glmM from E. coli, is essential for survival (26). Furthermore, it was shown that M. tuberculosis Rv3441c possesses phosphoglucosamine mutase activity and was able to compensate for the loss of MSMEG_1556 in the conditional mutant (26), thus demonstrating that they share the same function. Mycobacterial GlmU is a bifunctional enzyme involved in the last two sequential steps of UDP-GlcNAc synthesis (Fig. 2). Disruption of glmU in M. smegmatis resulted in gross morphological changes and loss of viability

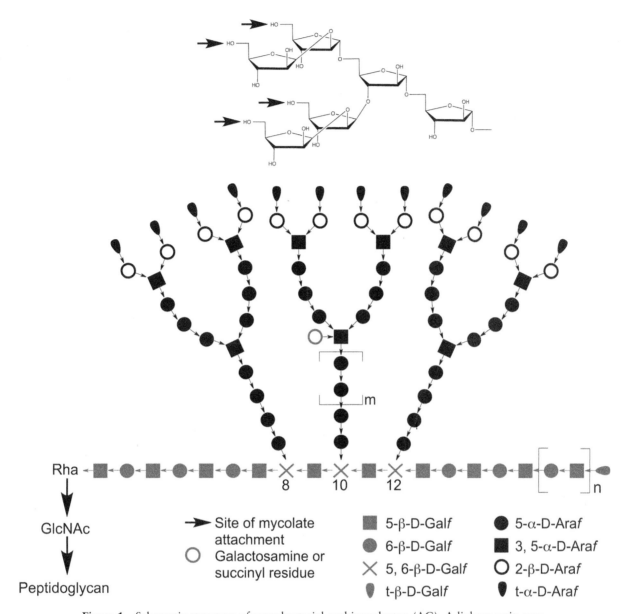

Figure 1 Schematic structure of mycobacterial arabinogalactan (AG). A linkage unit, composed of rhamnose and N-acetyl-glucosamine residues, anchors the whole AG structure to peptidoglycan. The galactan domain is composed of alternating β(1→5) and β(1→6) galactofuranose residues with three chains of arabinan attached to each linear galactan chain at positions 8, 10, and 12. The highly branched nonreducing end of AG terminates with a hexa-arabinofuranoside motif, two-thirds of which is substituted with mycolic acids. doi:10.1128/microbiolspec.MGM2-0013-2013.f1

(25). Biochemical characterization as well as the structure of *M. tuberculosis* GlmU has recently been established (27–29).

The second nucleotide donor utilized in the biosynthesis of the linkage unit is dTDP-Rha (Fig. 2). The presence of L-rhamnose, a sugar absent in humans, makes the biosynthetic machinery of the mycobacterial linkage unit an attractive drug target. As a result, the rhamnosyl biosynthetic pathway has come under close scrutiny, and hence a number of inhibitors targeting this pathway have been described (30–32). Synthesis of dTDP-Rha occurs via a linear four-step reaction. Recognition of the genes involved in this pathway was revealed by comparison to known polysaccharide biosynthetic enzymes found in other bacteria, namely *E. coli* (33). RmlA (*Rv0334*) sets in motion a sequence

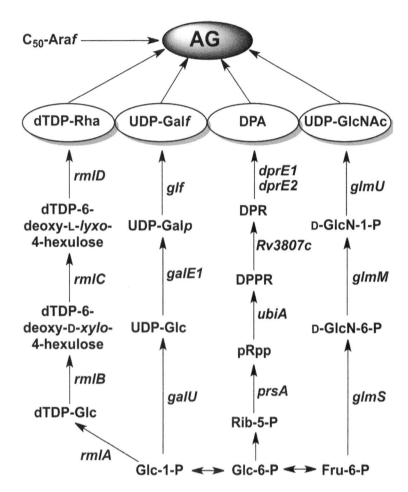

Figure 2 The biosynthesis of sugar donors required for mycobacterial AG biosynthesis. Both UDP-GlcNAc and dTDP-Rha are utilized in the formation of the linkage unit. UDP-Gal*f* is the sugar donor of the galactofuranosyl residues used in the galactan chain formation. Decaprenylphosphoryl-D-arabinofuranose (DPA) is the only known high-energy nucleotide providing arabinofuranosyl residues to the arabinan domain of AG.
doi:10.1128/microbiolspec.MGM2-0013-2013.f2

of reactions, converting dTTP and α-D-glucose-1-P into dTDP-glucose. A strain of *E. coli* lacking four rhamnose biosynthetic genes was complemented with *rmlA* from *M. tuberculosis*. Analysis of cellular extracts revealed an abundance of α-D-Glc-P thymidylyltransferase activity, thus confirming its proposed function (33). The product of RmlA activity is then shuttled through three sequential reactions catalyzed by dTDP-D-glucose-4,6-dehydratase RmlB (*Rv3464*), dTDP-4-oxo-6-deoxyglucose-3,5-epimerase RmlC (*Rv3465*), and dTDP-6-deoxy-L-*lyxo*-4-hexulose reductase RmlD (*Rv3266*) to finally form the nucleotide donor dTDP-Rha. Gene deletion studies in the presence of a rescue plasmid with a temperature-sensitive origin of replication determined that the *rmlA* (34), *rmlB* (35), *rmlC* (35), and *rmlD* (36) genes were all essential for *M. smegmatis*. Hence, dTDP-Rha is an essential sugar donor for mycobacterial growth, and enzymes involved in its synthesis are potential chemotherapeutic targets. Finally, enzyme assays employing RmlA-D from *M. tuberculosis* to screen inhibitors for developing novel anti-TB therapeutics have been established (30, 37).

The high-energy nucleotide substrate UDP-Gal*f* is formed via a three-step reaction (Fig. 2). GalU (*Rv0993*), a glucose-1-phosphate uridylyltransferase, catalyzes the formation of UDP-Glc*p* from UTP and glucose-1-P (38). Recently, *galU* from *M. tuberculosis* was successfully expressed, purified, and biochemically characterized (38). The second enzyme, GalE, is responsible for the epimerization reaction, which forms UDP-Gal*p* from UDP-Glc*p* in *E. coli* (39). Studies in *M. smegmatis* examined the reverse reaction using radiolabeled UDP-Gal*p* and observed UDP-glucose-4-epimerase activity. Sequentially, the *M. smegmatis* protein was purified, and its N-terminal sequence was shown to be similar to that of the *M. tuberculosis* product GalE1 (*Rv3634c*) (40). The further conversion of UDP-Gal*p* to the furanose form occurs via a ring contraction catalyzed by the enzyme UDP-galactopyranose mutase Glf, which was identified initially in *E. coli* (41) and subsequently in *M. smegmatis* (*MSMEG_6404*) and *M. tuberculosis* (*Rv3809c*) (40). Allelic exchange experiments highlighted the essentiality of *glf* to *M. smegmatis* (42). In addition, the crystal structures of Glf mutases from *M. tuberculosis*, *Escherichia coli*, and *Klebsiella pneumoniae* have been solved (43, 44).

Arabinan biosynthesis utilizes β-D-arabinofuranosyl-1-monophosphodecaprenol (DPA), the only known donor of Ara*f* residues in mycobacteria and corynebacteria (45). Recently, its membrane-linked synthesis was investigated in detail (13, 46) (Fig. 2). The initial reaction involves activation of ribose-5-phosphate by a phosphoribosyl-1-pyrophosphate synthetase, PrsA (*Rv1017c*), to yield 5-phosphoribosyl-1-pyrophosphate (pRpp) (46). UbiA (*Rv3806c*) then transfers pRpp to a decaprenylmonophosphate, producing decaprenyl-phosphoryl-5-phosphoribose (DPPR) (46). Disruption of *ubiA* (*NCgl2781*) in *C. glutamicum* resulted in a complete loss of cell wall arabinan, demonstrating that DPA is the only Ara*f* sugar donor used in AG biosynthesis (13). Remarkably, the mutant still generated a modified LAM version, which was arabinosylated even in the absence of DPA. An alternative source and mechanism by which these Ara*f* residues are added to this glycolipid is yet to be resolved (47). DPPR is then dephosphorylated to decaprenyl-5-phosphoribose (DPR) by the putative phospholipid phosphatase encoded by *Rv3807c*. Its homologue in *M. smegmatis* (*MSMEG_6402*) was shown to be a nonessential gene (48). The DprE1 (*Rv3790*) and DprE2 (*Rv3791*) heterodimer catalyzes the epimerization of DPR to DPA, which occurs via an oxidation-reduction mechanism. DPR is initially oxidized at the C_2-OH group to form the keto-sugar intermediate decaprenol-1-monophosphoryl-2-keto-β-erythro-pentofuranose (DPX), which is subsequently reduced to DPA (49). Deletion studies in *C. glutamicum* showed that *dprE1* (*NCgl0187*) is essential to bacterial growth, whereas *dprE2* (*NCgl0186*) is not (50). In the absence of *dprE2*, a different enzyme encoded by *NCgl1429* was proposed to carry out the function of DprE2 since *NCgl1429* showed a similar function *in vivo* and appeared to be essential in the *NCgl0186*-inactivated mutant (50). Further investigation demonstrated that *dprE1* (*MSMEG_6382*) is also an essential gene in *M. smegmatis* (51). These results highlighted DprE1 as a novel drug target. Indeed, recent studies led to the discovery of two classes of potent compounds with specific activities against mycobacteria: dinitrobenzamide derivatives (DNBs) and nitro-compounds related to DNBs—nitrobenzothiazinones (BTZs)—both of which were revealed to target the decaprenylphosphoryl-β-D-ribose 2′ epimerase encoded by *dprE1* (52, 53). The structural complex of DrpE1-BTZ has been determined, revealing the mode of inhibitor binding (52, 54).

Biosynthesis of AG

The biosynthesis of AG begins with the formation of the linkage unit synthesized on a decaprenyl phosphate (C_{50}-P) lipid carrier. WecA (*Rv1302*) catalyzes the first reaction by transferring GlcNAc-1-P from the sugar donor UDP-GlcNAc to the lipid carrier (55, 56). Lipopolysaccharide analysis of a *wecA*-defective strain of *E. coli* complemented with either *M. tuberculosis* (*Rv1302*) or *M. smegmatis* (*MSMEG_4947*) homologue showed restoration of lipopolysaccharide biosynthesis, thus providing evidence that it has the same function as the WecA protein from *E. coli* (55). In addition, inactivation of *wecA* from *M. smegmatis* using a homologous recombination strategy resulted in drastic morphological changes and loss of viability (55). Rhamnosyltransferase WbbL (*Rv3265c*) is responsible for the transfer of the rhamnose residue from the dTDP-Rha substrate to the 3-position of the GlcNAc of C_{50}-P-P-GlcNAc, thus yielding the full linkage unit C_{50}-P-P-GlcNAc-Rha of AG. The key to the discovery of mycobacterial WbbL was the successful complementation of an *E. coli* mutant lacking WbbL activity with the *Rv3265c* gene from *M. tuberculosis* (57). *M. tuberculosis wbbL* was expressed in *E. coli* and was used together with bioinformatics analysis to establish its preliminary structure and characteristics (58). Moreover, it was demonstrated that *wbbL* (*MSMEG_1826*) is crucial to the growth and viability of *M. smegmatis* (57).

The previously synthesized linkage unit serves as an acceptor for the addition of Gal*f* residues from the sugar donor UDP-Gal*f*. GlfT1 (*Rv3782*) recognizes the linkage unit and transfers the initial two Gal*f* residues to C_{50}-P-P-GlcNAc-Rha, resulting in C_{50}-P-P-GlcNAc-Rha-Gal*f*$_2$ (59–61) (Fig. 3). Further galactan polymerization is carried out by the second transferase GlfT2 (*Rv3808c*), identified through the use of a neoglycolipid acceptor assay together with UDP-Gal*f* and isolated *E. coli* membranes expressing *glfT2*. It was demonstrated that the enzyme had dual functionality, acting as both a UDP-Gal*f*:β-D-(1→5) galactofuranosyltransferase (Gal*f*T) and the UDP-Gal*f*:β-D-(1→6) Gal*f*T (62–64). Structural data together with site-directed mutagenesis and kinetic studies provided evidence for a mechanism that explains the unique ability of GlfT2 to generate β(1→5) and β(1→6) linkages using a single active site (64).

An arabinofuranosyltransferase (Ara*f*T) from the *emb* locus, AftA (*Rv3792*), is responsible for addition of the first key Ara*f* residue to the 8th, 10th, and 12th Gal*f* residues, thus "priming" the galactan chain for further attachment of α(1→5)-linked Ara*f* units

(13, 65) (Fig. 3). A homologue of *aftA* in *M. smegmatis*, *MSMEG_6386*, was shown to be essential for survival of mycobacteria. However, deletion of *aftA* in *C. glutamicum* resulted in a slow-growing but viable mutant. Cell wall analysis revealed the complete loss of arabinose leading to a truncated cell wall structure containing only a galactan chain and greatly diminished cell wall–bound mycolic acids (65). EmbA (*Rv3794*) and EmbB (*Rv3795*) catalyze further polymerization of arabinan. Emb proteins were first discovered as a target for ethambutol (EMB), a first-line TB drug. Individual inactivation of *embA* and *embB* in *M. smegmatis* resulted in diminished incorporation of arabinose into AG, specifically the terminal disaccharide β-D-Ara*f*-(1→2)-α-D-Ara*f*, normally situated on the 3-OH of the 3,5-linked Ara*f* residue (66). Efforts to generate a viable *embA*/*embB* mutant in *M. tuberculosis* and

an *embAB* double mutant in *M. smegmatis* have so far proven unfruitful, highlighting their essentiality to mycobacteria. However, a singular *emb* gene (*NCgl0184*) was successfully disrupted in *C. glutamicum* (13). *Corynebacterium* is deemed the archetype of *Corynebacterineae* since it maintains a low frequency of gene duplications and modifications, and thus it is reasonable that *C. glutamicum* possesses only one *emb* gene. Surprisingly, *NCgl0184* exhibited higher identity to *embC*, which encodes Ara*f*T, involved exclusively in LAM biosynthesis, even though *C. glutamicum* lacks an elaborately arabinosylated lipomannan (LM) product (47, 67).

Chemical analyses of the tolerable *emb* deletion mutant from *C. glutamicum* revealed an almost total loss of cell wall arabinan, except for terminal t-Ara*f* residues decorating the galactan backbone (13). Moreover,

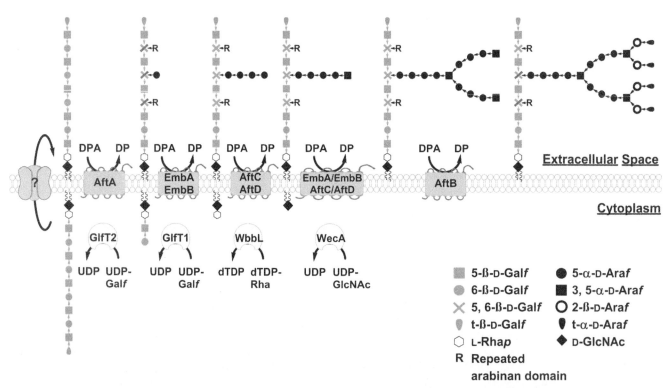

Figure 3 Schematic representation of mycobacterial arabinogalactan biosynthesis. WecA catalyzes the transfer of GlcNAc to decaprenyl phosphate, which is then used as an acceptor for addition of rhamnosyl residue by WbbL, thereby forming the full linkage unit. The first two galactofuranosyl (Gal*f*) residues are added to the linkage unit via GlfT1. The bifunctional GlfT2 adds the remaining Gal*f* residues forming a linear galactan chain. Before the polymerization with arabinofuranosyl (Ara*f*) residues, the galactan domain is thought to be translocated across the plasma membrane by the unknown flippase. AftA initiates the transfer of Ara*f* residues from the sugar donor DPA to the 8th, 10th and 12th β(1→6)-linked Gal*f* residues of the galactan chain. EmbA and EmbB proteins act as α-1,5-arabinosyltransferases utilizing the same nucleotide donor DPA. The 3,5-linked Ara*f* branching is introduced by AftC and AftD enzymes. Finally, the terminal Ara*f* residues are added to the arabinan domain by a "capping" enzyme AftB. doi:10.1128/microbiolspec.MGM2-0013-2013.f3

EMB treatment of wild-type *C. glutamicum* produced a profile identical to that of the mutant, illustrating that *emb* is indeed the target of EMB. Recent deletion studies in *M. smegmatis* identified a branching enzyme, AftC (*Rv2673*), that is responsible for the transfer of Araf residues from DPA to the arabinan domain to form α(1→3)-linked Araf residues of the internal arabinan domain at the nonreducing end of AG and LAM (68, 69). Yet another functional ArafT encoded by *aftD* (*Rv0236c*) has been shown to have α-1,3-branching activity on linear α-1,5-linked synthetic acceptors *in vitro*. Inactivation of the *aftD* homolog in *M. smegmatis*, MSMEG_0359, was shown to be lethal to mycobacteria, while overexpression of *aftD* in *M. smegmatis* resulted in an overall increase of Araf residues (70). Finally, AftB (*Rv3805c*) catalyzes the transfer of Araf residues to the arabinan domain to form the terminal β(1→2)-linked Araf residues (Fig. 3). Disruption of *aftB* in *C. glutamicum* resulted in a viable mutant with complete absence of terminal β(1→2)-linked arabinofuranosyl residues and decreased abundance of cell-wall-bound mycolic acids, consistent with a partial loss of mycolylation sites (71).

Decoration of AG and Its Attachment to PG and Mycolic Acids

Two decorating structures, succinyl and GalN residues, have been identified in the interior AG-arabinan domain of *M. tuberculosis*, thus concluding a model of the complete primary structure of mycobacterial AG. Enzymes involved in succinylation of arabinan chains toward the nonreducing end are yet to be determined (19), but the key components of the biosynthetic pathway of GalN have recently been elucidated (20). Glycosyltransferase PpgS (*Rv3631*) catalyzes the transfer of GalNAc from UDP-GalNAc to polyprenyl-P, yielding a sugar donor polyprenyl-P-GalNAc. This high-energy substrate is then presumably deacylated by an as yet unknown deacetylase before or after being translocated to the extracellular space where the membrane-associated enzyme Rv3779 transfers the GalN*p* (or GalNAc) residue to the C2 position of a portion of the internal 3,5-branched D-Araf residues of AG. The synthesis of GalN was demolished in both *ppgS* and *Rv3779* deletion mutants in *M. tuberculosis*. It is worth noting that the GalN residue is only found in slow-growing mycobacteria. Hence, expression of *ppgS* in the fast-growing *M. smegmatis* species, otherwise devoid of the *ppgS* orthologue and any detectable polyprenyl-P-GalNAc synthase activity, allowed mycobacteria to synthesize polyprenyl-P-GalNAc *in vivo*. The physio-

logical role and pathogenesis of both succinyl and GalN residues as well as the biosynthetic origin of succinylation remain to be elucidated.

Very little evidence has been obtained that shows how AG is ligated to PG to generate the complete cell wall core. *In vitro* assay in *M. smegmatis* utilizing cell-free extracts and radiolabeled substrates demonstrated the formation of simpler polyprenyl-P-P-GlcNAc-Rha-(Galf)$_n$ intermediates, followed by addition of AG and, finally, ligation to PG (72). However, there are no enzymes reported to show the attachment of AG to PG. It is not fully understood when the AG structure is mycolylated: before or after the attachment to PG. *In vitro* enzymatic assays have identified members of the antigen 85 complex, FbpA, FbpB, and FbpC (*Rv3804c*, *Rv1886c* and *Rv0129c*, respectively), that are responsible for the transfer of mycolic acids onto trehalose that leads to the formation of trehalose monomycolate (TMM) and trehalose dimycolate (TDM) (73). Inactivation of antigen 85 by transposon mutagenesis resulted in a mutant with reduced capacity to transfer mycolic acids to the mycobacterial cell wall (74). Similar mycolyltransferases from *C. glutamicum*, encoded by *cmytA* and *cmytB* genes, were deleted, leading to a viable double mutant with significantly impaired ability to transfer corynomycolates to AG (75).

PHOSPHATIDYL-*MYO*-INOSITOL BASED GLYCOLIPIDS

The plasma membrane of mycobacteria is rich in free lipids that play a crucial role in its pathogenesis. The phospholipids—phosphatidyl ethanolamine, phosphatidyl-*myo*-inositol (PI), phosphatidyl inositol mannosides (PIMs), LM, LAM, and cardiolipin (CL)—are among the major structural components of the mycobacterial plasma membrane, with PI/PIMs alone representing 28% of the total phospholipids in *M. smegmatis* (76). PIM, LM, and LAM are glycophospholipids built on a PI backbone by a series of modifications including glycosylation and acylation. It is still unclear whether these PI-based glycophospholipids are embedded in the plasma membrane or found in the outer membrane of mycobacteria. However, using surface labeling experiments it has recently been shown that the lipoglycans are exposed at the surface of mycobacteria, indicating their presence in the outer leaflet of the outer membrane (77). The surface exposed lipoglycans, PIMs, LM, and LAM possess significant immunomodulatory effects in macrophages, including cytokine production, inhibition of phagosome maturation, and apoptosis and also account for cross-protective immunity of mycobacteria (2).

Recent biochemical and genetic studies have indicated that the modifications of the PI anchor follow the order PI → PIM → LM → LAM (2), wherein mannosylation of the PI anchor with up to six mannose residues produces PIMs that are further modified with mannose and arabinose residues to generate LM and LAM. The enzymes for biogenesis of early PIM species are encoded by a gene cluster consisting of an operon of six open reading frames (ORFs), conserved in the *Corynebacterineae* family (3). These include *Rv2614c*, which encodes a probable threonyl-tRNA synthase. The second gene, *Rv2613c*, encodes a protein of unknown function that is predicted to be involved in nucleotide biosynthesis. The third gene of the gene cluster is *Rv2612c*, which encodes PgsA, a protein involved in PI synthesis, and the fourth gene, *Rv2611c*, encodes an acyltransferase (78) that acylates both PIM_1 and PIM_2. The fifth gene, *Rv2610c*, codes for PimA, an α-mannopyranosyltransferase of the GT-B superfamily and the first enzyme of the PIM biosynthetic pathway responsible for PIM_1 production (79). *Rv2609c* encodes a putative guanosine diphosphate (GDP)-Man*p* hydrolase with a conserved MutT domain (80). The genes encoding proteins involved in biogenesis of higher PIMs, LM, and LAM are all found scattered in the genome. Interestingly, Ara*f*Ts such as EmbA, EmbB, and EmbC are found in the *embCAB* operon (81, 82) in mycobacteria; of these, only EmbC is employed in biosynthesis of LAM (83), while EmbA and EmbB are specific for AG biosynthesis (66). The core framework of the mycobacterial cell wall provides a template for the insertion of mannosylated molecules such as LAM and its structurally related glycolipids, LM, and PIMs (84). These highly complex immunomodulatory lipoglycans are found ubiquitously in the envelopes of all mycobacterial species, noncovalently associated to the plasma membrane and/or the mycolic acid layer via a conserved mannosyl-phosphate inositol (MPI) anchor (85), which extends to the exterior of the cell wall (84, 86).

Structural Features of PIMs

The PI unit is based on *sn*-glycero-3-phosphate-(1-D-*myo*-inositol), whereby the glycerol phosphate component is attached to the L-1-position of *myo*-inositol (87, 88). The PI unit is sequentially substituted with mannose residues at positions C-2 and C-6 of the inositol ring, thus forming the MPI anchor (89, 90). This anchor is highly heterogenous with respect to the number, location, and nature of the fatty acids attached. There

are a total of four potential acylation sites on the MPI anchor, of which two are present on the glycerol unit, one is on the Man*p* unit linked to C-2 of *myo*-inositol, and the fourth is at the C-3 position of *myo*-inositol (91, 92). Of the initial PIMs, the PI in PIM_1 is glycosylated with α-D-mannopyranosyl (Man*p*) at position O-2, whereas PIM_2 is glycosylated with Man*p* at both O-2 and O-6. The terminal PIM species PIM_6 consists of Man*p*-α(1→2)-Man*p*-α(1→2)-Man*p*-α(1→6)-Man*p*-α(1→6)-Man*p*-α(1→) attached to the MPI anchor (49) (Fig. 4). Acylated forms of the PIMs, such as Ac_1PIM_2 and Ac_1PIM_6, are the major PIM species, with acylation occurring at the Man*p* residue attached to C-2 of *myo*-inositol. The diacyl forms of PIMs, such as Ac_2PIM_2 and Ac_2PIM_6, also exist, but Ac_1PIM_2 is believed to be the preferred precursor for building higher PIMs, such as PIM_6, LM, and LAM (93).

Structural Features of LM and LAM

The mannan of LM and LAM is an extension of PIMs, containing on average 20 to 30 residues, emanating from the C-6 position of the inositol ring (92, 94) (Fig. 4). The α(1→6)-mannose residues are linked linearly to form a linear backbone with branching introduced by 5 to 10 units of single α(1→2) Man*p* residues at C-2 of the occasional α(1→6)-linked mannose in all mycobacterial species. However, in the case of *Mycobacterium chelonae*, the branching is introduced by single α(1→2) Man*p* residues at the C-3 position (93, 95). The arabinan domain of LAM consists of 55 to 70 Ara*f* residues (92) arranged as a linear α(1→5)-linked arabinosylfuranosyl backbone with branching introduced at the third position on some residues (69, 96) (Fig. 4). The branches in the arabinan domain of LAM consist of two distinct motifs: a linear tetra-arabinoside comprising β-D-Ara*f* (1→2)-α-D-Ara*f* (1→5)-α-D-Ara*f* (1→5)-α-D-Ara*f* and a hexa-arabinoside comprising [β-D-Ara*f* (1→2)-α-D-Ara*f*]₂-3,5-α-D-Ara*f* (1→5)-α-D-Ara*f* (94, 97). Both tetra- and hexa-arabinoside moieties are decorated with a disaccharide unit, Ara*f*-β(1→2)-Ara*f*-α(1→) at the nonreducing end (14, 94, 97).

Mycobacterial species differ in the nature and extent of the capping motifs modifying the nonreducing termini of the arabinan chains, specifically the β(1→2)-linked terminal Ara*f* units. To date, three structural families have been recognized: mannose-capped LAM (Man-LAM), PI-capped LAM (PI-LAM), and noncapped LAM (Ara-LAM), of which "Man-caps" are an important feature of pathogenic species of slow-growing mycobacteria such as *M. tuberculosis*, *M. bovis*, *M. bovis* BCG, *M. leprae*, *M. avium*, *M. xenopi*, *M. marinum*, and *M. kansasii* (97). The caps may be

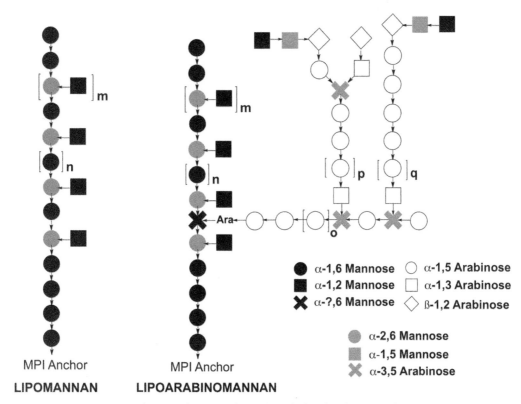

Figure 4 Structure of LM and LAM. The PI-based glycolipids LM and LAM consist of 20 to 30 mannose residues that form the mannose core. This mannose core is futher glycosylated with 55 to 70 arabinose residues arranged in hexa or tetra-motifs that form the arabinan domain of LAM. The terminal arabinose residues in LAM serve as the sites for attachment of mannose residues or phosphatidyl inositol, thus forming the Man-LAM and PI-LAM. These arabinose residues, if uncapped, form the Ara-LAM.
doi:10.1128/microbiolspec.MGM2-0013-2013.f4

present as single Man*p* capping residues—di- or tri-α(1→2)-mannosides, of which dimannosides predominate (93, 97). Fast-growing mycobacteria such as *M. smegmatis* and *M. fortuitum* possess PI caps (98), and *M. chelonae* is the only known example of Ara-LAM (95). In addition to the caps, NMR studies have demonstrated the presence of succinyl residues on the C-2 of the 3,5-di-α-D-Ara*f* units in Man-LAM isolated from *M. bovis* BCG (Pasteur, Glaxo, Copenhagen, and Japanese strains). The average number of succinyl residues varies from one to four per LAM molecule (99).

Biosynthesis of PIMs, LM, and LAM

Precursor formation

In mycobacteria, mannose is the key component of glycolipids, lipoglycans, and a number of glycosylated proteins, which makes it essential for growth and viability. The polymethylated polysaccharides are the components that contain mannose in mycobacteria (100).

Mycobacteria obtains mannose from two sources: first, the extracellular environment, from which mannose is obtained via sugar transporters and is phosphorylated with the aid of hexokinase (*Rv2702*) (101) to yield mannose-1-phosphate; and second, via the glycolytic pathway, where fructose-6-phosphate is converted to mannose-6-phosphate by a phosphomannose isomerase, ManA (*Rv3255c*) (102) (Fig. 5). Subsequently, mannose-1-phosphate is converted to mannose-6-phosphate in a reaction catalyzed by a phosphomannomutase, ManB (*Rv3257c*) (103). This mannose-1-phosphate generated from mannose obtained from both sources is loaded onto GDP to generate the mannose donor, GDP-Man*p*, by ManC (30, 104). GDP-Man*p* serves as an intracellular nucleotide-derived mannose donor for the synthesis of several glycolipids and mannosylated proteins by the GT-A/B superfamily of glycosyltransferases (105).

The GT-C superfamily of glycosyltransferases uses polyprenyl-phosphate-based mannose donors for periplasmic synthesis of lipoglycans, such as higher PIMs, LM, and LAM. The presence of a C_{50}-polyprenol-based mannolipid, C_{50}-decaprenol-phospho-mannose (C_{50}-P-Man*p*, polyprenol monophosphomannose [PPM]), in *M. tuberculosis* was first reported by Takayama and Goldman in 1970 (106). However, in *M. smegmatis*, a C_{35}-octahydroheptaprenyl-phospho-mannose, C_{35}-P-Man*p*, was later identified by Wolucka and de Hoffmann in 1998 (107).

In mycobacteria, PPM is synthesized using GDP-Man and C_{35}/C_{50}-P in a reaction catalyzed by a PPM synthase, Ppm1 (*Rv2051c*) (108) (Fig. 5). However, more recently, a transmembrane glycosyltransferase, *Rv3779*, was identified and suggested to be involved in the synthesis of $C_{35/50}$-P-Man*p* as a second PPM synthase (109). However, in a study conducted by Skovierova et al. (20), *Rv3779* was found to have glycosyltransferase activity, and it was suggested to be involved in transferring GalN from a polyprenyl-phospho-N-acetylgalactosamine to AG in *M. tuberculosis*. The roles of *Rv2051c* and *Rv3779* were recently reinvestigated by Rana et al. (110) using genetic and biochemical methods in *M. smegmatis*. This study revealed that a conditional mutant of *ppm1* generated using CESTET in *M. smegmatis* could only be rescued in the presence of plasmid encoding *Rv2051c*, while the second PPM synthase (*Rv3779*) failed to substitute for the loss of gene function *in vivo*. Therefore, *ppm1* (*Rv2051*) is the sole gene responsible for generating the mannose donor PPM in *M. tuberculosis*.

Biosynthesis of PI

The PI moiety plays a dual role, because it forms the backbone for synthesis of PIMs, LM, and LAM and also anchors the high-molecular-weight lipoglycans LM and LAM to the inner and outer membranes in the cell envelope. Synthesis of PI is a three-step process that begins with cyclization of glucose-6-phosphate by inositol phosphate synthase (InO1), encoded by *Rv0046c*, to generate *myo*-inositol-1-phosphate, which is subsequently dephosphorylated by inositol monophosphatase (IMP) to yield *myo*-inositol (111, 112) (Fig. 5). In the last step, *myo*-inositol is esterified to diacylglycerol (DAG), which is transferred from cytidine diphosphate-diacylglycerol (CDP-DAG) by

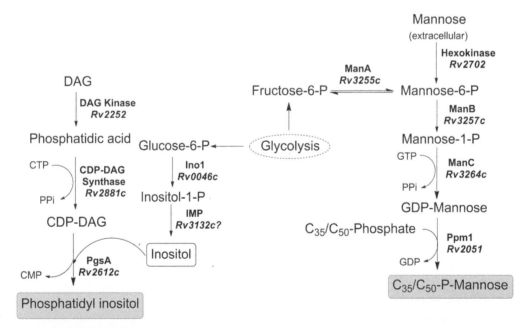

Figure 5 Biosynthesis of phosphatidyl inositol anchor and sugar donors involved in synthesis of PIMs, LM, and LAM. The synthesis of precursor molecules and sugar donors of the LAM biosynthesis pathway uses the products of the glycolytic pathway. However, the mannose utilized in the synthesis of GDP-Man and PPM can also be exogenously obtained. The pathway for PPM and GDP-Man biosynthesis is interlinked because the prenyl-based sugar donor PPM is synthesized by direct transfer of mannose from GDP-Man to the prenyl phosphate mediated by Ppm1 (*Rv2051*). The PI anchor on which the PIMs, LM, and LAM are based is synthesized by transfer of inositol to CDP-DAG, a reaction catalyzed by PgsA (*Rv2612c*). doi:10.1128/microbiolspec.MGM2-0013-2013.f5

a PI synthase, PgsA (*Rv2612c*). Recently, gene mutational studies in *M. tuberculosis* and *M. smegmatis* have revealed that the inositol phosphate synthase (Ino1) and the PI synthetase (PgsA) are essential for the viability of mycobacteria (111, 113). Interestingly, ImpC (*Rv3137*), an inositol mono-phosphatase, has been demonstrated to be essential for the growth of *M. tuberculosis* and *M. smegmatis* (114).

Biosynthesis of PIMs

PIM biogenesis follows a linear pathway that consists of a series of mannosylations of the PI anchor in the order PI → PIM_2 → PIM_4 → PIM_6 (Fig. 6). The higher PIM species have prominent structural and physiological roles; PIM_4 forms the structural basis for LM and LAM, while PIM_6 is required to maintain the integrity of the plasma membrane. The early steps of PIM biogenesis occur on the cytoplasmic side of the plasma membrane and utilize GDP-Man*p* as the sugar donor. The pathway for PIM biogenesis is initiated by transfer of Man*p* from GDP-Man to position O-2 of the inositol ring of PI to yield PIM_1. This reaction is catalyzed by PimA (*Rv2601*), a GDP-Man-dependent α-mannosyltransferase of the GT-B superfamily of glycosyltransferases (78–80). Recent biochemical and genetic studies conducted in *M. smegmatis* have demonstrated PimA to be an essential enzyme required for growth of mycobacteria.

The next step in the pathway is generation of the metabolic intermediate PIM_2, which serves as a scaffold on which higher PIMs, LM, and LAM are synthesized (4, 29). PIM_2 can exist in mono- and diacylated forms in *M. tuberculosis*, while only the monoacylated form is observed to accumulate in corynebacteria (78). Two separate hypotheses have been suggested for the formation of $AcPIM_2$, which is a two-step reaction catalyzed by a mannosyltransferase and an acyltransferase. According to the first hypothesis, PIM_1 is first acylated by *Rv2611* at the sixth position of the Man*p* residue to yield Ac_1/Ac_2PIM_1, which is then subsequently mannosylated by PimB' at position O-6 of the inositol ring to form Ac_1PIM_2 (78, 80). This model has been supported by recent studies conducted in *C. glutamicum*, where the gene deletion mutant of *pimB'* (*Rv2188c*), α-D-mannose-α-(1→6) phosphatidyl *myo*-inositol-mannopyranosyltransferase, which adds Man*p* to inositol to yield Ac/Ac_2PIM_2, was found to accumulate $AcPIM_1$, implying precedence over the acylation of PIM_1 and the second mannosylation step (78, 115, 116). The second hypothesis suggests

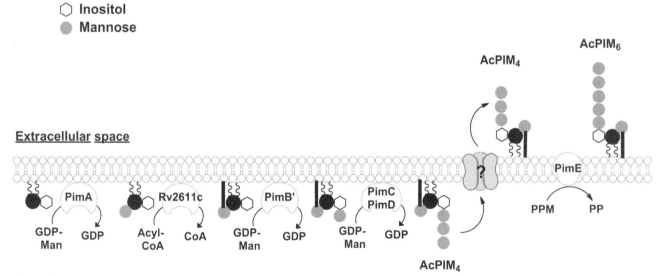

Figure 6 Biosynthesis of PIMs. The PI anchor synthesized by PgsA undergoes multiple mannosylations to produce PIMs. Of the PIM species synthesized, $AcPIM_2$ and $AcPIM_4$ are the most abundant. The production of $AcPIM_4$ serves as the branch point in the PIM biosynthesis, with one branch leading to formation of higher PIM species such as $AcPIM_6$ and the other leading to LM and LAM production. The $AcPIM_4$ is synthesized by the mannosyltransferases PimC and/or PimD, both of which remain unidentified. The flippase required for translocating the $AcPIM_4$ from the cytosolic to extracellular side is also unknown. doi:10.1128/microbiolspec.MGM2-0013-2013.f6

that acylation of PIM_1 and PIM_2 can occur together. However, the cell-free experiments conducted in *M. smegmatis* favor generation of $AcPIM_2$ after PIM_2 has been formed by the action of PimB′ (117). The acyltransferase involved in the transfer of the fourth acyl group to the third position of *myo*-inositol is yet to be identified.

The gene *Rv2611c* is required for growth of mycobacteria, because its disruption causes severe growth defects and the mutant accumulates nonacylated PIM_1 and PIM_2 (78). The enzyme PimB′, encoded by *Rv2188c*, is also an essential enzyme in *M. smegmatis* (118). The function of *Rv2188c* was initially assigned to *Rv0557*, earlier designated as PimB. However, genetic experiments in *M. tuberculosis* identified no changes to PIM biosynthesis due to disruption of the gene. The results obtained suggested that the role of PimB could either be substituted by complementary genes or that the function of PimB is different from that of PimB′ (119). In addition, recent investigations in *C. glutamicum* have identified *Rv2188c* as the mannopyranosyl transferase involved in the second mannosylation step for generation of $AcPIM_2$ (115, 116, 120) and caused *Rv0557* to be renamed as MgtA, due to the α-mannosyl-glucopyranosylglucuronic acid transferase activity responsible for production of Cg-LM-B, a glucuronic acid–based LM variant, and $ManGlcAGroAc_2$, a glucuronic acid diacylglycerol–based glycolipid found in *C. glutamicum* (120).

The next major step in the pathway is synthesis of $AcPIM_4$; however, intermediary PIM forms such as Ac_1PIM_3 and Ac_2PIM_3 can also exist in mycobacteria. PimC (RvD2-ORF1) is a mannosyltransferase identified in *M. tuberculosis* strain CDC1551 and is responsible for synthesis of trimannosylated PIMs, but no strong homologues of *pimC* have been found in *M. tuberculosis* H37Rv or *M. smegmatis*. Additionally, *M. bovis* BCG with a deleted chromosomal copy of *pimC* showed normal PIM, LM, and LAM levels, suggesting redundancy of the gene or the presence of compensatory pathways (121). The nonreducing end of Ac_1/Ac_2PIM_3 is mannosylated by an α(1→6) mannosyltransferase to form Ac_1/Ac_2PIM_4. This reaction is catalyzed either by PimC or by the unidentified putative "PimD" protein (122). Formation of Ac_1/Ac_2PIM_4 marks the "junction" point as the pathway now diverges into two branches, with one branch leading to the formation of polar PIM species, such as Ac_1/Ac_2PIM_5 and Ac_1/Ac_2PIM_6, by two consecutive additions of α(1→2) Man*p*, probably by PimE (*Rv1159*), and the second branch leading to the formation of LM and LAM (123, 124). The intermediate $Ac_1/$

Ac_2PIM_4 is translocated across the membrane to the extracellular space using an unknown flippase for subsequent glycosylation. The glycosyltransferases involved in further modifications of Ac_1/Ac_2PIM_4 employ polyprenylphosphate-based sugar donors, PPM and DPA, and belong to the GT-C superfamily (105). PimE, an α(1→2)-mannopyranosyltransferase, is an example of a GT-C superfamily glycosyltransferase, which utilizes PPM for addition of one or more mannose residues to Ac_1/Ac_2PIM_4 (124). Although the steps for biosynthesis of the final product Ac_1/Ac_2PIM_6 have yet to be deduced, involvement of PimE cannot be ruled out.

Biosynthesis of LM and LAM

The lipoglycans LM and LAM are generated by extensive glycosylation of the PIMs, specifically, Ac_1/Ac_2PIM_4 (Fig. 7). Recent studies conducted in *M. smegmatis* indicated the involvement of LpqW (*Rv1166*) in LM and LAM biogenesis (125, 126). The protein LpqW has a regulatory function where it promotes the channeling of Ac_1/Ac_2PIM_4 into LM and LAM biosynthesis by interacting with MptB, a mannosyltransferase involved in synthesis of the mannan core of LM and LAM (125, 127). The mannan of LM/LAM is composed of approximately 25 to 30 mannose residues associated linearly to form the α(1→6) mannan backbone that is punctuated occasionally at C-2 with single α(1→2)-linked mannose units in *M. tuberculosis*, *M. leprae*, and *M. smegmatis*. However, in the case of *M. chelonae* the mannan core is substituted at C-3 with single Man*p* units. Addition of mannose residues to build the mannan core in LM and LAM is mediated by mannosyltransferases of the GT-C superfamily that utilize PPM as the mannose donor. The α(1→6) mannosyltransferases MptA (*Rv2174*) and MptB (*Rv1459c*) are the key enzymes, which synthesize the characteristic mannan core of LM and LAM. Recent studies in *C. glutamicum* suggested the involvement of MptB in the synthesis of the proximal end through the addition of 12 to 15 Man*p* residues to the backbone (116) and designated MptA to be responsible for the synthesis of the distal end of the core (118, 128). However, *M. tuberculosis* and *M. smegmatis* MptB proteins were unable to complement the *C. glutamicum* Δ*mptB* mutant, indicating that either differences in substrate specificity or redundancy of the gene function was responsible for the contradictory observations.

The α(1→6)-mannan core is further branched by the addition of α(1→2)-Man*p* residues, which are catalyzed by MptC (*Rv2181*) (2). Kaur et al. (129)

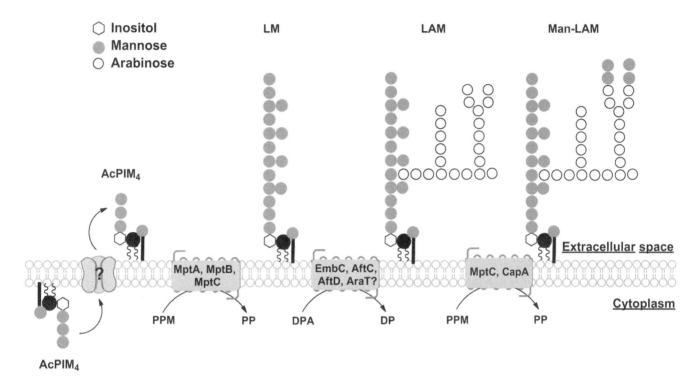

Figure 7 Biosynthesis of LM and LAM. Hyperglycosylation of AcPIM$_4$ produces LM and LAM. The mannosyltransferases MptB, MptA, and MptC are involved in the synthesis of the mannan core, while the arabinosyltransferases EmbC, AftC, and AftD and an unknown transferase are responsible for the synthesis of the arabinan domain. The arabinan in LAM is capped with mannose residues in *M. tuberculosis* at the nonreducing termini referred to as Man-LAM. The enzymes, MptC and CapA, mediate this reaction.
doi:10.1128/microbiolspec.MGM2-0013-2013.f7

recently identified the putative integral membrane proteins, MSMEG_4250 in *M. smegmatis* and Rv2181 in *M. tuberculosis*, as potential polyprenol-dependent glycosyltransferases based on shared characteristics with previously identified enzymes. A knockout of *MSMEG_4250* in *M. smegmatis* possessed a truncated version of LAM with a decrease in the number of α(1→2)-Man*p* branching residues and altered growth with an inability to synthesize LM. Complementation of the mutant with the corresponding orthologue of *M. tuberculosis* (*Rv2181*) restored normal LM/LAM synthesis. However, regulation of LM and LAM biosynthesis in *M. smegmatis* appears to differ somewhat with *M. tuberculosis*, as *M. tuberculosis* Δ*Rv2181* produced truncated versions of LM and Man-LAM (96).

LAM is generated via the further elaboration of LM with 55 to 70 Ara*f* units forming an arabinan domain akin to that found in AG (84). Addition of arabinofuranosyl residues to LM is mediated by the Ara*f*Ts: EmbC (*Rv3793*), AftC (*Rv2673*), AftD (*Rv0236c*), and an uncharacterized Ara*f*T that is considered to prime LM for LAM synthesis (65). The transition of LM to

LAM occurs by the addition of 12 to 16 α(1→5)-Ara*f* residues to the primed LM, a reaction exclusively catalyzed by EmbC, an α(1→5)Ara*f* transferase (69, 130, 131). Recently, the crystal structure of the truncated C-terminal domain of EmbC was solved (130). The linear Ara polymer of LAM is branched similarly to AG by addition of α(1→3) Ara*f* residues using AftC, resulting in 3,5 Ara*f* branch points (68, 69). Recently, AftD was designated as the second branching enzyme with α(1→3) Ara*f* transferase activity that extends the Ara branches (70). *In vitro* assays using artificial chemical acceptors and cell-free extracts from *M. smegmatis* and *C. glutamicum* showed that the enzyme was able to add α(1→3) Ara*f* residues to the linear α-1,5-linked neoglycolipid acceptor, resulting in branching of the linear arabinan core. Therefore, its function is considered similar to AftC. The branched Ara motif is further extended to the tetra-arabinoside and hexa-arabinoside motifs. The terminal β(1→2) Ara*f* residues present in both tetra and hexa motifs are likely to be added by AftB (*Rv3805c*) and considered to have a dual role similar to AftC in AG and LAM biosynthesis (69).

The arabinose domains, though similar in AG and LAM, differ in several respects. The arabinan domain in AG is more complicated and terminates in hexa-arabinoside motifs (10), with the terminal and penultimate Araf residues serving as attachment points for mycolic acids (12). In contrast, LAM consists of both tetra-arabinoside and hexa-arabinoside motifs, and the terminal Araf residue is modified with α-mannose residues in all pathogenic strains of mycobacteria, thus generating Man-LAM. Mannose capping of LAM consists of one, two, or three Manp residues, with di-mannose caps being the most abundant (93, 97). The addition of mannose caps is predicted to be a two-step process that requires two mannosyltransferases, one for identification of the arabinan domain and addition of the primary Araf residue following recognition and a second for elongation of the cap with a second α(1→2) mannose. Subtractive genomics of the *M. tuberculosis* genome, with those species of the genus that do not contain mannose-capped LAM, such as *M. smegmatis*, highlighted the GT-C enzyme CapA (*Rv1635c*) (132). The *MT1671* (*Rv1635c*) *M. tuberculosis* CDC1551 transposon mutant produced LAM devoid of Manp capping. Further studies supported this finding, with *Rv1635c* mutants in *M. marinum* and *M. bovis* BCG showing that *Rv1635c* encoded a mannopyranosyltransferase involved in the addition of the first Manp residue on the nonreducing arabinan termini of LAM (133). More recently, Kaur et al. (96) have shown that MptC (*Rv2181c*) possesses varied substrate specificity, capable of adding α(1→2)-Manp residues onto the mannan backbone, as well as the nonreducing end of LAM in combination with *Rv1635c*, thus generating Man-LAM.

DRUGS TARGETING AG AND LAM

EMB and Its Mechanism of Action

EMB [(S,S)-2,2′(ethylenediimino)di-1-butanol] is a bacteriostatic agent widely used as a frontline drug for the treatment of TB. It is a synthetic compound that was first recognized as an antimycobacterial agent in 1961 (134). Early work by Kilburn and Greenberg observed an unanticipated increase in viable cells during the initial 4 hours after addition of EMB to *M. smegmatis* cultures. It was postulated that large bacillary clusters disaggregated due to a possible reduction in lipid content, which would lead to the apparent increase in CFUs (135). This theory was supported by a series of studies of the effects of EMB on *M. smegmatis*, demonstrating that inhibition of mycolic acid transfer into the cell wall and the simultaneous accumulation of TMM, TDM, and free mycolic acids occurred within 15 minutes of drug administration (136, 137). This suggested that the target might be a mycolyltransferase responsible for the transfer of mycolic acids onto the arabinan polymer. However, it was later discovered that the earliest point of drug inhibition occurred during arabinan synthesis, demonstrated by the immediate inhibition of incorporated label from [^{14}C]glucose into the cell wall D-arabinose (138), while synthesis of the galactan of AG remained unaffected (139). Furthermore, the arabinans of both AG and LAM were inhibited, although reduction of label into the latter was less pronounced and at a later stage of its biosynthesis (139, 140).

The generation of EMB-resistant *M. smegmatis* mutants greatly aided the discovery of the primary EMB target, implicating it as an arabinan-specific inhibitor. This was illustrated by a study in which an EMB-resistant strain subjected to subinhibitory levels of EMB possessed a normal cell wall AG structure but a truncated version of LAM due to arabinan inhibition (139). Extending this observation using higher concentrations of EMB, it was shown that the degree of truncation in LAM was dose dependent, and at higher concentrations, the arabinan of AG was also impaired (92). Collectively, these studies indicate that the effects of EMB on the synthesis of both arabinan moieties are uncoupled, and the time difference of inhibition implies that synthesis occurs via distinct pathways, involving multiple ArafT targets with varying EMB sensitivities. A number of concurrent EMB studies provided evidence of an accumulation of DPA, the source of Araf residues in arabinan biosynthesis, confirming that EMB effects were due not to inhibition of Araf donor synthesis but rather to its utilization (141, 142). Taken together, all the evidence points to arabinan polymerization, specifically the arabinan of AG, as the primary target of EMB.

A major breakthrough in the discovery of the precise EMB cellular target arose through exploitation of a moderately resistant strain from the related *M. avium* species. The genomic library from the aforementioned strain was screened and overexpressed in an otherwise susceptible *M. smegmatis* host, leading to the identification of a resistance-conferring region encompassing three complete ORFs, *embR*, *embA*, and *embB* (81). Interestingly, neither *embA* nor *embB* alone was sufficient to confer multicopy resistance, thus supporting the supposition that they are translationally coupled, forming a multienzyme complex (81). EMB resistance was also used to identify the *embCAB* gene cluster from *M. smegmatis*, which was subsequently characterized in *M. tuberculosis* and *M. leprae*, all of which possess the same syntenic organization (82, 143) and encode homologues of *embA* and *embB* genes from *M. avium*. Individual genetic knockouts of *embC*, *embA*, and *embB* were generated in *M. smegmatis*, all of which were viable, with the most profound affects observed in the *embB* mutant (66). Chemical analysis revealed that there was a decrease in the arabinose content of AG in both *embA* and *embB* mutants, specifically, the terminal disaccharide β-D-Ara*f* (1→2)-α-D-Ara*f*. Thus, a substantial amount of the otherwise hexa-arabinofuranoside motifs were present as terminal linear tetra-arabinofuranoside structures akin to the terminal motif of LAM (66), also leading to a loss of cell wall-bound mycolates. Based on the above observations, it appears that the Emb proteins are involved in hexa-arabinofuranoside biosynthesis. Indeed, cell-free assays using wild-type *M. smegmatis* membranes were used to investigate the putative activity of EmbA/EmbB and demonstrated the formation of the nonreducing terminal disaccharide, which was absent in both *embA* and *embB* mutants (144). Moreover, the transferase activity was re-established upon mixing the membrane preparations from the disrupted strains with wild-type membranes (144).

Identification of the *emb* gene cluster has provided the opportunity to analyze the molecular basis of resistance of mycobacteria to EMB. Telenti et al. (82) demonstrated that high-level resistance to EMB in *M. smegmatis* could be related to either overproduction of the Emb protein(s), a structural mutation in a conserved region of EmbB, or both. A number of reports presented additional genetic evidence for a key role of the EmbB protein in cell wall biosynthesis, highlighting the fact it is the most EMB-sensitive protein in the gene cluster (145). Mutations in EmbB have been recorded in up to 65% of EMB-resistant clinical isolates of *M. tuberculosis*, with the majority of mutations present at codon 306 or in the immediate surrounding area (143, 146). This region is highly conserved among mycobacteria, and topological analysis of the Emb proteins (82) positioned this EMB resistance-determining region (ERDR) in the second intracellular loop of EmbB (147). Five distinct mutations have been recognized at codon 306, resulting in a substitution of the wild-type methionine with isoleucine, leucine, or valine (147). Other mutations have been identified in the second intracellular loop region and the large C-terminal globular region of EmbB (146). It should be noted that there are a number of EMB-resistant strains that do not possess ERDR mutations, so other genes may be involved in EMB resistance.

SQ109: Potent EMB Analog

SQ109 {N′-(2-adamantyl)-N-[(2E)-3,7-dimethylocta-2,6-dienyl]ethane-1,2-diamine} was identified through a high-throughput screen from a chemical library of EMB derivatives sharing the same ethylenediamine core (148). SQ109 has improved *in vivo* efficacy against *M. tuberculosis* as well as low cytotoxicity (148). More importantly, it is active against multidrug-resistant and extensively drug-resistant TB clinical strains (148). Substitution of EMB with SQ109 in drug combination regimens demonstrated additive effects with EMB and strong synergistic interactions when combined with isoniazid, rifampin, and the newly discovered TMC207 compound (149, 150). SQ109 has been reported to be a safe and well-tolerated agent that is currently in phase IIa clinical trials to evaluate its efficacy and safety in patients with pulmonary TB (151).

Recently, the molecular target of SQ109 was determined to be an essential *mmpL3* (*Rv0206c*) gene that encodes a conserved transmembrane transporter protein. Treatment of *M. tuberculosis* with SQ109 resulted in a rapid decrease in the attachment of mycolic acids to both the cell wall AG and TDM. The pool of mycolates remained unaffected, but the levels of TMM, the precursor of TDM, and mycolic acids accumulated in SQ109-treated cells (152), pointing to TMM transport as the SQ109 target. Efforts to generate a spontaneous SQ109-resistant mutant in *M. tuberculosis* were unsuccessful. However, similar ethylenediamine compounds related to SQ109 were then used to spontaneously generate resistant mutants that were shown to have cross-resistance to SQ109 (152). The whole-genome sequencing of these mutants identified mutations in the essential *mmpL3* gene from *M. tuberculosis*, suggesting that the target of SQ109 is likely to be MmpL3 (152).

BTZs and DNBs

BTZ and DNB are novel and highly potent antimycobacterials with high bactericidal activity against mycobacteria (53). Both DNBs and BTZs are nitro compounds that target decaprenylphosphoryl-β-D-ribose 2′ epimerase, which is encoded by *dprE1* (*Rv3790*) and *dprE2* (*Rv3791*) and is responsible for epimerization of DPR to DPA (52, 53). The lipid-based sugar donor DPA is a precursor molecule supplying Ara*f* residues to the AG and LAM of mycobacteria (45, 49). The nitro group in BTZ and DNB is proposed to be reduced to a nitroso group by the action of DprE1. This nitroso group forms a semi-mercaptal linkage to the conserved cysteine residue (Cys387) in the active site of the enzyme, thus rendering it nonfunctional (153, 154).

Therefore, BTZs are also classified as suicide substrates of the enzyme DprE1 (153). Recent studies have demonstrated that mis-sense mutations in Cys387 residue can confer resistance to both drugs (52, 53). BTZs and DNBs can also be inactivated by the nitroreductase NfnB in *M. smegmatis* (155). While orthologue of *nfnB* is absent in *M. tuberculosis*, its presence in eukaryotes raises questions about the efficacy of BTZ and DNB as antitubercular drugs (155). The lead compound of the BTZ series, BTZ043, is in the late stages of preclinical trials and has successfully demonstrated high bactericidal activity against the multidrug-resistant strains, thus paving the way for the clinical trials (156).

CONCLUDING REMARKS

Overall, the sequencing of several mycobacterial genomes, progress toward genetic tools to manipulate mycobacteria, and the use of surrogate systems such as *C. glutamicum* and *M. smegmatis* have contributed significantly to the identification of enzymes involved in the biosynthesis of AG and LAM. However, there are still some missing pieces to be found in these complex biosynthetic pathways. For example, the biosynthetic origin and function of succinyl residues found in both AG and LAM are not clearly understood and have merely been hypothesized. Galactosamine residue of AG, found only in pathogenic slow-growing mycobacterial species, is speculated to play a role during host infection, but further *in vivo* studies have to be undertaken to confirm this theory. Furthermore, very little is known about the transporters required for the translocation of lipid-linked sugar donors and oligosaccharide intermediates from the cytoplasm to the extracellular space of mycobacteria. Only one study so far proposed a small multidrug-resistance-like gene, *Rv3789*, to encode a transporter, which reorients sugar donor DPA to the extracellular space, thus allowing further polymerization of arabinan (157). Finally, our understanding of the regulatory mechanisms in cell wall biosynthesis is lacking and requires comprehensive research.

We have attempted in this article to describe the latest progress of biochemical and genetic studies of mycobacterial cell wall assembly. Deciphering the biosynthetic pathways of AG and LAM has increased our knowledge of the biology of pathogenic *M. tuberculosis* and set the stage for the next research step—high-throughput screening assays for identifying potent inhibitors against essential enzymes of mycobacteria, protein-protein complexes, and the structural elucidation of the transmembrane glycosyltransferases involved in AG and LAM biosynthesis.

Citation. Jankute M, Grover S, Birch HL, Besra GS. 2014. Genetics of mycobacterial arabinogalactan and lipoarabinomannan assembly. Microbiol Spectrum 2(4):MGM2-0013-2013.

References

1. Jankute M, Grover S, Rana AK, Besra GS. 2012. Arabinogalactan and lipoarabinomannan biosynthesis: structure, biogenesis and their potential as drug targets. *Future Microbiol* 7:129–147.

2. Mishra AK, Driessen NN, Appelmelk BJ, Besra GS. 2011. Lipoarabinomannan and related glycoconjugates: structure, biogenesis and role in *Mycobacterium tuberculosis* physiology and host-pathogen interaction. *FEMS Microbiol Rev* 35:1126–1157.

3. Cole ST, Brosch R, Parkhill J, Garnier T, Churcher C, Harris D, Gordon SV, Eiglmeier K, Gas S, Barry CE 3rd, Tekaia F, Badcock K, Basham D, Brown D, Chillingworth T, Connor R, Davies R, Devlin K, Feltwell T, Gentles S, Hamlin N, Holroyd S, Hornsby T, Jagels K, Krogh A, McLean J, Moule S, Murphy L, Oliver K, Osborne J, Quail MA, Rajandream MA, Rogers J, Rutter S, Seeger K, Skelton J, Squares R, Squares S, Sulston JE, Taylor K, Whitehead S, Barrell BG. 1998. Deciphering the biology of *Mycobacterium tuberculosis* from the complete genome sequence. *Nature* 393:537–544.

4. Stinear TP, Seemann T, Harrison PF, Jenkin GA, Davies JK, Johnson PD, Abdellah Z, Arrowsmith C, Chillingworth T, Churcher C, Clarke K, Cronin A, Davis P, Goodhead I, Holroyd N, Jagels K, Lord A, Moule S, Mungall K, Norbertczak H, Quail MA, Rabbinowitsch E, Walker D, White B, Whitehead S, Small PL, Brosch R, Ramakrishnan L, Fischbach MA, Parkhill J, Cole ST. 2008. Insights from the complete genome sequence of *Mycobacterium marinum* on the evolution of *Mycobacterium tuberculosis*. *Genome Res* 18:729–741.

5. Garnier T, Eiglmeier K, Camus JC, Medina N, Mansoor H, Pryor M, Duthoy S, Grondin S, Lacroix C, Monsempe C, Simon S, Harris B, Atkin R, Doggett J, Mayes R, Keating L, Wheeler PR, Parkhill J, Barrell BG, Cole ST, Gordon SV, Hewinson RG. 2003. The complete genome sequence of *Mycobacterium bovis*. *Proc Natl Acad Sci USA* 100:7877–7882.

6. Bhatt A, Jacobs WR Jr. 2009. Gene essentiality testing in *Mycobacterium smegmatis* using specialized transduction. *Methods Mol Biol* 465:325–336.

7. Amar C, Vilkas E. 1973. Isolation of arabinose phosphate from the walls of *Mycobacterium tuberculosis* H37Ra. *CR Acad Sci Hebd Seances Acad Sci D* 277:1949–1951.

8. McNeil M, Wallner SJ, Hunter SW, Brennan PJ. 1987. Demonstration that the galactosyl and arabinosyl residues in the cell-wall arabinogalactan of *Mycobacterium leprae* and *Mycobacterium tuberculosis* are furanoid. *Carbohydr Res* 166:299–308.

9. Besra GS, Khoo KH, McNeil MR, Dell A, Morris HR, Brennan PJ. 1995. A new interpretation of the

structure of the mycolyl-arabinogalactan complex of *Mycobacterium tuberculosis* as revealed through characterization of oligoglycosylalditol fragments by fast-atom bombardment mass spectrometry and ^{1}H nuclear magnetic resonance spectroscopy. *Biochemistry* **34**:4257–4266.

10. Daffe M, Brennan PJ, McNeil M. 1990. Predominant structural features of the cell wall arabinogalactan of *Mycobacterium tuberculosis* as revealed through characterization of oligoglycosyl alditol fragments by gas chromatography/mass spectrometry and by ^{1}H and ^{13}C NMR analyses. *J Biol Chem.* **265**:6734–6743.

11. McNeil M, Daffe M, Brennan PJ. 1990. Evidence for the nature of the link between the arabinogalactan and peptidoglycan of mycobacterial cell walls. *J Biol Chem* **265**:18200–18206.

12. McNeil M, Daffe M, Brennan PJ. 1991. Location of the mycolyl ester substituents in the cell walls of mycobacteria. *J Biol Chem* **266**:13217–13223.

13. Alderwick LJ, Radmacher E, Seidel M, Gande R, Hitchen PG, Morris HR, Dell A, Sahm H, Eggeling L, Besra GS. 2005. Deletion of *Cg-emb* in *Corynebacterianeae* leads to a novel truncated cell wall arabinogalactan, whereas inactivation of *Cg-ubiA* results in an arabinan-deficient mutant with a cell wall galactan core. *J Biol Chem* **280**:32362–32371.

14. McNeil MR, Robuck KG, Harter M, Brennan PJ. 1994. Enzymatic evidence for the presence of a critical terminal hexa-arabinoside in the cell walls of *Mycobacterium tuberculosis*. *Glycobiology* **4**:165–173.

15. Dong X, Bhamidi S, Scherman M, Xin Y, McNeil MR. 2006. Development of a quantitative assay for mycobacterial endogenous arabinase and ensuing studies of arabinase levels and arabinan metabolism in *Mycobacterium smegmatis*. *Appl Environ Microbiol* **72**:2601–2605.

16. Draper P, Khoo KH, Chatterjee D, Dell A, Morris HR. 1997. Galactosamine in walls of slow-growing mycobacteria. *Biochem J.* **327**:519–525.

17. Lee A, Wu SW, Scherman MS, Torrelles JB, Chatterjee D, McNeil MR, Khoo KH. 2006. Sequencing of oligoarabinosyl units released from mycobacterial arabinogalactan by endogenous arabinanase: identification of distinctive and novel structural motifs. *Biochemistry* **45**:15817–15828.

18. Peng W, Zou L, Bhamidi S, McNeil MR, Lowary TL. 2012. The galactosamine residue in mycobacterial arabinogalactan is α-linked. *J Org Chem* **77**:9826–9832.

19. Bhamidi S, Scherman MS, Rithner CD, Prenni JE, Chatterjee D, Khoo KH, McNeil MR. 2008. The identification and location of succinyl residues and the characterization of the interior arabinan region allow for a model of the complete primary structure of *Mycobacterium tuberculosis* mycolyl arabinogalactan. *J Biol Chem* **283**:12992–13000.

20. Skovierova H, Larrouy-Maumus G, Pham H, Belanova M, Barilone N, Dasgupta A, Mikusova K, Gicquel B, Gilleron M, Brennan PJ, Puzo G, Nigou J, Jackson M. 2010. Biosynthetic origin of the galactosamine substituent of arabinogalactan in *Mycobacterium tuberculosis*. *J Biol Chem* **285**:41348–41355 .

21. Mengin-Lecreulx D, van Heijenoort J. 1996. Characterization of the essential gene *glmM* encoding phosphoglucosamine mutase in *Escherichia coli*. *J Biol Chem* **271**:32–39.

22. Mengin-Lecreulx D, van Heijenoort J. 1994. Copurification of glucosamine-1-phosphate acetyltransferase and N-acetylglucosamine-1-phosphate uridyltransferase activities of *Escherichia coli*: characterization of the *glmU* gene product as a bifunctional enzyme catalyzing two subsequent steps in the pathway for UDP-N-acetylglucosamine synthesis. *J Bacteriol.* **176**:5788–5795.

23. Mengin-Lecreulx D, van Heijenoort J. 1993. Identification of the *glmU* gene encoding N-acetylglucosamine-1-phosphate uridyltransferase in *Escherichia coli*. *J Bacteriol* **175**:6150–6157.

24. Klein DJ, Ferre-D'Amare AR. 2006. Structural basis of *glmS* ribozyme activation by glucosamine-6-phosphate. *Science* **313**:1752–1756.

25. Zhang W, Jones VC, Scherman MS, Mahapatra S, Crick D, Bhamidi S, Xin Y, McNeil MR, Ma Y. 2008. Expression, essentiality, and a microtiter plate assay for mycobacterial GlmU, the bifunctional glucosamine-1-phosphate acetyltransferase and N-acetylglucosamine-1-phosphate uridyltransferase. *Int J Biochem Cell Biol* **40**:2560–2571.

26. Li S, Kang J, Yu W, Zhou Y, Zhang W, Xin Y, Ma Y. 2012. Identification of *M. tuberculosis* Rv3441c and *M. smegmatis* MSMEG_1556 and essentiality of *M. smegmatis* MSMEG_1556. *PLoS One* **7**:e42769.

27. Zhang Z, Bulloch EM, Bunker RD, Baker EN, Squire CJ. 2009. Structure and function of GlmU from *Mycobacterium tuberculosis*. *Acta Crystallogr D Biol Crystallogr* **65**:275–283.

28. Zhou Y, Xin Y, Sha S, Ma Y. 2011. Kinetic properties of *Mycobacterium tuberculosis* bifunctional GlmU. *Arch Microbiol* **193**:751–757.

29. Zhou Y, Yu W, Zheng Q, Xin Y, Ma Y. 2012. Identification of amino acids involved in catalytic process of *M. tuberculosis* GlmU acetyltransferase. *Glycoconj J* **29**:297–303.

30. Ma Y, Stern RJ, Scherman MS, Vissa VD, Yan W, Jones VC, Zhang F, Franzblau SG, Lewis WH, McNeil MR. 2001. Drug targeting *Mycobacterium tuberculosis* cell wall synthesis: genetics of dTDP-rhamnose synthetic enzymes and development of a microtiter plate-based screen for inhibitors of conversion of dTDP-glucose to dTDP-rhamnose. *Antimicrob Agents Chemother* **45**:1407–1416.

31. Kantardjieff KA, Kim CY, Naranjo C, Waldo GS, Lekin T, Segelke BW, Zemla A, Park MS, Terwilliger TC, Rupp B. 2004. *Mycobacterium tuberculosis* RmlC epimerase (*Rv3465*): a promising drug-target structure in the rhamnose pathway. *Acta Crystallogr D Biol Crystallogr* **60**:895–902.

32. Babaoglu K, Page MA, Jones VC, McNeil MR, Dong C, Naismith JH, Lee RH. 2003. Novel inhibitors of an emerging target in *Mycobacterium tuberculosis*; substituted thiazolidinones as inhibitors of dTDP-rhamnose synthesis. *Bioorg Med Chem Lett* **13**:3227–3230.

33. Ma Y, Mills JA, Belisle JT, Vissa V, Howell M, Bowlin K, Scherman MS, McNeil M. 1997. Determination of the pathway for rhamnose biosynthesis in mycobacteria: cloning, sequencing and expression of the *Mycobacterium tuberculosis* gene encoding α-D-glucose-1-phosphate thymidylyltransferase. *Microbiology* 143:937–945.

34. Qu H, Xin Y, Dong X, Ma Y. 2007. An *rmlA* gene encoding D-glucose-1-phosphate thymidylyltransferase is essential for mycobacterial growth. *FEMS Microbiol Lett* 275:237–243.

35. Li W, Xin Y, McNeil MR, Ma Y. 2006. *rmlB* and *rmlC* genes are essential for growth of mycobacteria. *Biochem Biophys Res Commun* 342:170–178.

36. Ma Y, Pan F, McNeil M. 2002. Formation of dTDP-rhamnose is essential for growth of mycobacteria. *J Bacteriol* 184:3392–3395.

37. Sha S, Zhou Y, Xin Y, Ma Y. 2012. Development of a colorimetric assay and kinetic analysis for *Mycobacterium tuberculosis* D-glucose-1-phosphate thymidylyltransferase. *J Biomol Screen* 17:252–257.

38. Lai X, Wu J, Chen S, Zhang X, Wang H. 2008. Expression, purification, and characterization of a functionally active *Mycobacterium tuberculosis* UDP-glucose pyrophosphorylase. *Protein Expr Purif* 61:50–56.

39. Lemaire HG, Muller-Hill B. 1986. Nucleotide sequences of the *galE* gene and the *galT* gene of *E. coli*. *Nucleic Acids Res* 14:7705–7711.

40. Weston A, Stern RJ, Lee RE, Nassau PM, Monsey D, Martin SL, Scherman MS, Besra GS, Duncan K, McNeil MR. 1997. Biosynthetic origin of mycobacterial cell wall galactofuranosyl residues. *Tuber Lung Dis* 78:123–131.

41. Nassau PM, Martin SL, Brown RE, Weston A, Monsey D, McNeil MR, Duncan K. 1996. Galactofuranose biosynthesis in *Escherichia coli* K-12: identification and cloning of UDP-galactopyranose mutase. *J Bacteriol* 178:1047–1052.

42. Pan F, Jackson M, Ma Y, McNeil M. 2001. Cell wall core galactofuran synthesis is essential for growth of mycobacteria. *J Bacteriol* 183:3991–3998.

43. Sanders DA, Staines AG, McMahon SA, McNeil MR, Whitfield C, Naismith JH. 2001. UDP-galactopyranose mutase has a novel structure and mechanism. *Nat Struct Biol* 8:858–863.

44. Beis K, Srikannathasan V, Liu H, Fullerton SW, Bamford VA, Sanders DA, Whitfield C, McNeil MR, Naismith JH. 2005. Crystal structures of *Mycobacterium tuberculosis* and *Klebsiella pneumoniae* UDP-galactopyranose mutase in the oxidised state and *Klebsiella pneumoniae* UDP-galactopyranose mutase in the (active) reduced state. *J Mol Biol* 348:971–982.

45. Wolucka BA. 2008. Biosynthesis of D-arabinose in mycobacteria: a novel bacterial pathway with implications for antimycobacterial therapy. *FEBS J* 275:2691–2711.

46. Alderwick LJ, Lloyd GS, Lloyd AJ, Lovering AL, Eggeling L, Besra GS. 2011. Biochemical characterization of the *Mycobacterium tuberculosis* phosphoribosyl-1-pyrophosphate synthetase. *Glycobiology* 21:410–425.

47. Tatituri RV, Alderwick LJ, Mishra AK, Nigou J, Gilleron M, Krumbach K, Hitchen P, Giordano A, Morris HR, Dell D, Eggeling L, Besra GS. 2007. Structural characterization of a partially arabinosylated lipoarabinomannan variant isolated from a *Corynebacterium glutamicum ubiA* mutant. *Microbiology* 153:2621–2629.

48. Jiang T, He L, Zhan Y, Zang S, Ma Y, Zhao X, Zhang C, Xin Y. 2011. The effect of *MSMEG_6402* gene disruption on the cell wall structure of *Mycobacterium smegmatis*. *Microb Pathog* 51:156–160.

49. Mikusova K, Huang H, Yagi T, Holsters M, Vereecke D, D'Haeze W, Scherman MS, Brennan PJ, McNeil MR, Crick DC. 2005. Decaprenylphosphoryl arabinofuranose, the donor of the D-arabinofuranosyl residues of mycobacterial arabinan, is formed via a two-step epimerization of decaprenylphosphoryl ribose. *J Bacteriol* 187:8020–8025.

50. Meniche X, de Sousa-d'Auria C, Van-der-Rest B, Bhamidi S, Huc E, Huang H, De Paepe D, Tropis M, McNeil M, Daffe M, Houssin C. 2008. Partial redundancy in the synthesis of the D-arabinose incorporated in the cell wall arabinan of *Corynebacterineae*. *Microbiology* 154:2315–2326.

51. Crellin PK, Brammananth R, Coppel RL. 2011. Decaprenylphosphoryl-β-D-ribose 2′-epimerase, the target of benzothiazinones and dinitrobenzamides, is an essential enzyme in *Mycobacterium smegmatis*. *PLoS One* 6:e16869.

52. Christophe T, Jackson M, Jeon HK, Fenistein D, Contreras-Dominguez M, Kim J, Genovesio A, Carralot JP, Ewann F, Kim EH, Lee SY, Kang A, Seo MJ, Park EJ, Skovierova H, Pham H, Riccardi G, Nam JY, Marsollier L, Kempf M, Joly-Guillou ML, Oh T, Shin WK, No Z, Nehrbass U, Brosch R, Cole ST, Brodin P. 2009. High content screening identifies decaprenylphosphoribose 2′ epimerase as a target for intracellular antimycobacterial inhibitors. *PLoS Pathog* 5:e1000645.

53. Makarov V, Manina G, Mikusova K, Mollmann U, Ryabova O, Saint-Joanis B, Dhar N, Pasca MR, Buroni S, Lucarelli AP, Milano A, De Rossi E, Belanova M, Bobovska A, Dianiskova P, Kordulakova J, Sala C, Fullam E, Schneider P, McKinney JD, Brodin P, Christophe T, Waddell S, Butcher P, Albrethsen J, Rosenkrands I, Brosch R, Nandi V, Bharath S, Gaonkar S, Shandil RK, Balasubramanian V, Balganesh T, Tyagi S, Grosset S, Riccardi G, Cole ST. 2009. Benzothiazinones kill *Mycobacterium tuberculosis* by blocking arabinan synthesis. *Science* 324:801–804.

54. Batt SM, Jabeen T, Bhowruth V, Quill L, Lund PA, Eggeling L, Alderwick LJ, Futterer K, Besra GS. 2012. Structural basis of inhibition of *Mycobacterium tuberculosis* DprE1 by benzothiazinone inhibitors. *Proc Natl Acad Sci USA* 109:11354–11359.

55. Jin Y, Xin Y, Zhang W, Ma Y. 2010. *Mycobacterium tuberculosis* Rv1302 and *Mycobacterium smegmatis* MSMEG_4947 have WecA function and MSMEG_4947 is required for the growth of *M. smegmatis*. *FEMS Microbiol Lett* 310:54–61.

56. Mikusova K, Mikus M, Besra GS, Hancock I, Brennan PJ. 1996. Biosynthesis of the linkage region of the mycobacterial cell wall. *J Biol Chem* 271:7820–7828.

57. Mills JA, Motichka K, Jucker M, Wu HP, Uhlik BC, Stern RJ, Scherman MS, Vissa VD, Pan F, Kundu M, Ma YF, McNeil M. 2004. Inactivation of the mycobacterial rhamnosyltransferase, which is needed for the formation of the arabinogalactan-peptidoglycan linker, leads to irreversible loss of viability. *J Biol Chem* **279:** 43540–43546.

58. Wu Q, Zhou P, Qian S, Qin X, Fan Z, Fu Q, Zhan Z, Pei H. 2011. Cloning, expression, identification and bioinformatics analysis of *Rv3265c* gene from *Mycobacterium tuberculosis* in *Escherichia coli*. *Asian Pac J Trop Med* **4:**266–270.

59. Alderwick LJ, Dover LG, Veerapen N, Gurcha SS, Kremer L, Roper DL, Pathak AK, Reynolds RC, Besra GS. 2008. Expression, purification and characterisation of soluble GlfT and the identification of a novel galactofuranosyltransferase *Rv3782* involved in priming GlfT-mediated galactan polymerisation in *Mycobacterium tuberculosis*. *Protein Expr Purif* **58:**332–341.

60. Mikusova K, Belanova M, Kordulakova J, Honda K, McNeil MR, Mahapatra S, Crick DC, Brennan PJ. 2006. Identification of a novel galactosyl transferase involved in biosynthesis of the mycobacterial cell wall. *J Bacteriol* **188:**6592–6598.

61. Belanova M, Dianiskova P, Brennan PJ, Completo GC, Rose NL, Lowary TL, Mikusova K. 2008. Galactosyl transferases in mycobacterial cell wall synthesis. *J Bacteriol* **190:**1141–1145.

62. Kremer L, Dover LG, Morehouse C, Hitchin P, Everett M, Morris HR, Dell A, Brennan PJ, McNeil MR, Flaherty C, Duncan K, Besra GS. 2001. Galactan biosynthesis in *Mycobacterium tuberculosis*. Identification of a bifunctional UDP-galactofuranosyltransferase. *J Biol Chem* **276:**26430–26440.

63. Rose NL, Completo GC, Lin SJ, McNeil M, Palcic MM, Lowary TL. 2006. Expression, purification, and characterization of a galactofuranosyltransferase involved in *Mycobacterium tuberculosis* arabinogalactan biosynthesis. *J Am Chem Soc* **128:**6721–6729.

64. Wheatley RW, Zheng RB, Richards MR, Lowary TL, Ng KK. 2012. Tetrameric structure of the GlfT2 galactofuranosyltransferase reveals a scaffold for the assembly of mycobacterial arabinogalactan. *J Biol Chem* **287:**28132–28143.

65. Alderwick LJ, Seidel M, Sahm H, Besra GS, Eggeling L. 2006. Identification of a novel arabinofuranosyltransferase (AftA) involved in cell wall arabinan biosynthesis in *Mycobacterium tuberculosis*. *J Biol Chem* **281:** 15653–15661.

66. Escuyer VE, Lety MA, Torrelles JB, Khoo KH, Tang JB, Rithner CD, Frehel C, McNeil MR, Brennan PJ, Chatterjee D. 2001. The role of the *embA* and *embB* gene products in the biosynthesis of the terminal hexaarabinofuranosyl motif of *Mycobacterium smegmatis* arabinogalactan. *J Biol Chem* **276:**48854–48862.

67. Dover LG, Cerdeno-Tarraga AM, Pallen MJ, Parkhill J, Besra GS. 2004. Comparative cell wall core biosynthesis in the mycolated pathogens, *Mycobacterium tuberculosis* and *Corynebacterium diphtheriae*. *FEMS Microbiol Rev* **28:**225–250.

68. Birch HL, Alderwick LJ, Bhatt A, Rittmann D, Krumbach K, Singh A, Bai Y, Lowary TL, Eggeling L, Besra GS. 2008. Biosynthesis of mycobacterial arabinogalactan: identification of a novel α(1→3) arabinofuranosyltransferase. *Mol Microbiol* **69:**1191–1206.

69. Birch HL, Alderwick LJ, Appelmelk BJ, Maaskant J, Bhatt A, Singh A, Nigou J, Eggeling L, Geurtsen J, Besra GS. 2010. A truncated lipoglycan from mycobacteria with altered immunological properties. *Proc Natl Acad Sci USA* **107:**2634–2639.

70. Skovierova H, Larrouy-Maumus G, Zhang J, Kaur D, Barilone N, Kordulakova J, Gilleron M, Guadagnini S, Belanova M, Prevost MC, Gicquel B, Puzo G, Chatterjee D, Brennan PJ, Nigou J, Jackson M. 2009. AftD, a novel essential arabinofuranosyltransferase from mycobacteria. *Glycobiology* **19:**1235–1247.

71. Seidel M, Alderwick LJ, Birch HL, Sahm H, Eggeling L, Besra GS. 2007. Identification of a novel arabinofuranosyltransferase AftB involved in a terminal step of cell wall arabinan biosynthesis in *Corynebacterianeae*, such as *Corynebacterium glutamicum* and *Mycobacterium tuberculosis*. *J Biol Chem* **282:**14729–14740.

72. Yagi T, Mahapatra S, Mikusova K, Crick DC, Brennan PJ. 2003. Polymerization of mycobacterial arabinogalactan and ligation to peptidoglycan. *J Biol Chem* **278:**26497–26504.

73. Belisle JT, Vissa VD, Sievert T, Takayama K, Brennan PJ, Besra GS. 1997. Role of the major antigen of *Mycobacterium tuberculosis* in cell wall biogenesis. *Science* **276:**1420–1422.

74. Jackson M, Raynaud C, Laneelle MA, Guilhot C, Laurent-Winter C, Ensergueix D, Gicquel B, Daffe M. 1999. Inactivation of the antigen 85C gene profoundly affects the mycolate content and alters the permeability of the *Mycobacterium tuberculosis* cell envelope. *Mol Microbiol* **31:**1573–1587.

75. Kacem R, De Sousa-D'Auria C, Tropis M, Chami M, Gounon P, Leblon G, Houssin C, Daffe M. 2004. Importance of mycoloyltransferases on the physiology of *Corynebacterium glutamicum*. *Microbiology* **150:** 73–84.

76. Zhang J, Angala SK, Pramanik PK, Li K, Crick DC, Liav A, Jozwiak A, Swiezewska E, Jackson M, Chatterjee D. 2011. Reconstitution of functional mycobacterial arabinosyltransferase AftC proteoliposome and assessment of decaprenylphosphorylarabinose analogues as arabinofuranosyl donors. *ACS Chem Biol* **6:**819–828.

77. Pitarque S, Larrouy-Maumus G, Payre B, Jackson M, Puzo G, Nigou J. 2008. The immunomodulatory lipoglycans, lipoarabinomannan and lipomannan, are exposed at the mycobacterial cell surface. *Tuberculosis (Edinb)* **88:**560–565.

78. Kordulakova J, Gilleron M, Puzo G, Brennan PJ, Gicquel B, Mikusova K, Jackson M. 2003. Identification of the required acyltransferase step in the biosynthesis of the phosphatidylinositol mannosides of mycobacterium species. *J Biol Chem* **278:**36285–36295.

79. Guerin ME, Kordulakova J, Schaeffer F, Svetlikova Z, Buschiazzo A, Giganti D, Gicquel B, Mikusova K, Jackson M, Alzari PM. 2007. Molecular recognition

and interfacial catalysis by the essential phosphatidyl-inositol mannosyltransferase PimA from mycobacteria. *J Biol Chem* 282:20705–20714.

80. Kordulakova J, Gilleron M, Mikusova K, Puzo G, Brennan PJ, Gicquel B, Jackson M. 2002. Definition of the first mannosylation step in phosphatidylinositol mannoside synthesis. PimA is essential for growth of mycobacteria. *J Biol Chem* 277:31335–31344.

81. Belanger AE, Besra GS, Ford ME, Mikusova K, Belisle JT, Brennan PJ, Inamine JM. 1996. The *embAB* genes of *Mycobacterium avium* encode an arabinosyl transferase involved in cell wall arabinan biosynthesis that is the target for the antimycobacterial drug ethambutol. *Proc Natl Acad Sci USA* 93:11919–11924.

82. Telenti A, Philipp WJ, Sreevatsan S, Bernasconi C, Stockbauer KE, Wieles B, Musser JM, Jacobs WR Jr. 1997. The *emb* operon, a gene cluster of *Mycobacterium tuberculosis* involved in resistance to ethambutol. *Nat Med* 3:567–570.

83. Zhang N, Torrelles JB, McNeil MR, Escuyer VE, Khoo KH, Brennan PJ, Chatterjee D. 2003. The Emb proteins of mycobacteria direct arabinosylation of lipoarabino-mannan and arabinogalactan via an N-terminal recognition region and a C-terminal synthetic region. *Mol Microbiol* 50:69–76.

84. Besra GS, Brennan PJ. 1997. The mycobacterial cell wall: biosynthesis of arabinogalactan and lipoarab-inomannan. *Biochem Soc Trans* 25:845–850.

85. Hunter SW, Brennan PJ. 1990. Evidence for the presence of a phosphatidylinositol anchor on the lipo-arabinomannan and lipomannan of *Mycobacterium tuberculosis*. *J Biol Chem* 265:9272–9279.

86. Nigou J, Gilleron M, Puzo G. 2003. Lipoarabino-mannans: from structure to biosynthesis. *Biochimie* 85: 153–166.

87. Ballou CE, Vilkas E, Lederer E. 1963. Structural studies on the myo-inositol phospholipids of *Mycobacterium tuberculosis* (var. *bovis*, strain BCG). *J Biol Chem* 238: 69–76.

88. Ballou CE, Lee YC. 1964. The structure of a myoinosi-tol mannoside from *Mycobacterium tuberculosis* glyco-lipid. *Biochemistry* 3:682–685.

89. Nigou J, Gilleron M, Brando T, Puzo G. 2004. Struc-tural analysis of mycobacterial lipoglycans. *Appl Biochem Biotechnol* 118:253–267.

90. Severn WB, Furneaux RH, Falshaw R, Atkinson PH. 1998. Chemical and spectroscopic characterisation of the phosphatidylinositol manno-oligosaccharides from *Mycobacterium bovis* AN5 and WAg201 and *Myco-bacterium smegmatis* MC² 155. *Carbohydr Res* 308: 397–408.

91. Brennan P, Ballou CE. 1968. Biosynthesis of manno-phosphoinositides by *Mycobacterium phlei*. Enzymatic acylation of the dimannophosphoinositides. *J Biol Chem* 243:2975–2984.

92. Khoo KH, Douglas E, Azadi P, Inamine JM, Besra GS, Mikusova K, Brennan PJ, Chatterjee D. 1996. Trun-cated structural variants of lipoarabinomannan in ethambutol drug-resistant strains of *Mycobacterium*

smegmatis. Inhibition of arabinan biosynthesis by eth-ambutol. *J Biol Chem* 271:28682–28690.

93. Chatterjee D, Hunter SW, McNeil M, Brennan PJ. 1992. Lipoarabinomannan. Multiglycosylated form of the mycobacterial mannosylphosphatidylinositols. *J Biol Chem* 267:6228–6233.

94. Chatterjee D, Bozic CM, McNeil M, Brennan PJ. 1991. Structural features of the arabinan component of the lipoarabinomannan of *Mycobacterium tuberculosis*. *J Biol Chem* 266:9652–9660.

95. Guerardel Y, Maes E, Elass E, Leroy Y, Timmerman P, Besra GS, Locht C, Strecker G, Kremer L. 2002. Struc-tural study of lipomannan and lipoarabinomannan from *Mycobacterium chelonae*. Presence of unusual compo-nents with α1,3-mannopyranose side chains. *J Biol Chem* 277:30635–30648.

96. Kaur D, Obregon-Henao A, Pham H, Chatterjee D, Brennan PJ, Jackson M. 2008. Lipoarabinomannan of *Mycobacterium*: mannose capping by a multifunctional terminal mannosyltransferase. *Proc Natl Acad Sci USA* 105:17973–17977.

97. Chatterjee D, Khoo KH, McNeil MR, Dell A, Morris HR, Brennan PJ. 1993. Structural definition of the non-reducing termini of mannose-capped LAM from *Mycobacterium tuberculosis* through selective enzymatic degradation and fast atom bombardment-mass spectro-metry. *Glycobiology* 3:497–506.

98. Khoo KH, Dell A, Morris HR, Brennan PJ, Chatterjee D. 1995. Structural definition of acylated phosphatidyl-inositol mannosides from *Mycobacterium tuberculosis*: definition of a common anchor for lipomannan and lipoarabinomannan. *Glycobiology* 5:117–127.

99. Delmas C, Gilleron M, Brando T, Vercellone A, Gheorghui M, Riviere M, Puzo G. 1997. Comparative structural study of the mannosylated-lipoarabinomannans from *Mycobacterium bovis* BCG vaccine strains: charac-terization and localization of succinates. *Glycobiology* 7:811–817.

100. Jackson M, Brennan PJ. 2009. Polymethylated polysac-charides from *Mycobacterium* species revisited. *J Biol Chem* 284:1949–1953.

101. Kowalska H, Pastuszak I, Szymona M. 1980. A manno-glucokinese of *Mycobacterium tuberculosis* H37Ra. *Acta Microbiol Pol* 29:249–257.

102. Patterson JH, Waller RF, Jeevarajah D, Billman-Jacobe H, McConville MJ. 2003. Mannose metabolism is required for mycobacterial growth. *Biochem J* 372:77–86.

103. McCarthy TR, Torrelles JB, MacFarlane AS, Katawczik M, Kutzbach B, Desjardin LE, Clegg S, Goldberg JB, Schlesinger LS. 2005. Overexpression of *Mycobacte-rium tuberculosis manB*, a phosphomannomutase that increases phosphatidylinositol mannoside biosynthesis in *Mycobacterium smegmatis* and mycobacterial associ-ation with human macrophages. *Mol Microbiol* 58: 774–790.

104. Ning B, Elbein AD. 1999. Purification and properties of mycobacterial GDP-mannose pyrophosphorylase. *Arch Biochem Biophys* 362:339–345.

105. Liu J, Mushegian A. 2003. Three monophyletic super-families account for the majority of the known glycosyl-transferases. *Protein Sci* 12:1418–1431.

106. Takayama K, Goldman DS. 1970. Enzymatic synthesis of mannosyl-1-phosphoryl-decaprenol by a cell-free system of *Mycobacterium tuberculosis*. *J Biol Chem* 245:6251–6257.

107. Wolucka BA, de Hoffmann E. 1998. Isolation and char-acterization of the major form of polyprenyl-phospho-mannose from *Mycobacterium smegmatis*. *Glycobiology* 8:955–962.

108. Gurcha SS, Baulard AR, Kremer L, Locht C, Moody DB, Muhlecker W, Costello CE, Crick DC, Brennan PJ, Besra GS. 2002. Ppm1, a novel polyprenol monophos-phomannose synthase from *Mycobacterium tuberculo-sis*. *Biochem J* 365:441–450.

109. Scherman H, Kaur D, Pham H, Skovierova H, Jackson M, Brennan PJ. 2009. Identification of a polyprenyl-phosphomannosyl synthase involved in the synthesis of mycobacterial mannosides. *J Bacteriol* 191:6769–6772.

110. Rana AK, Singh A, Gurcha SS, Cox LR, Bhatt A, Besra GS. 2012. Ppm1-encoded polyprenyl monophospho-mannose synthase activity is essential for lipoglycan synthesis and survival in mycobacteria. *PLoS One* 7: e48211.

111. Movahedzadeh F, Smith DA, Norman RA, Dinadayala P, Murray-Rust J, Russell DG, Kendall SL, Rison SC, McAlister MS, Bancroft GJ, McDonald NQ, Daffe M, Av-Gay Y, Stoker NG. 2004. The *Mycobacterium tu-berculosis ino1* gene is essential for growth and viru-lence. *Mol Microbiol* 51:1003–1014.

112. Bachhawat N, Mande SC. 1999. Identification of the INO1 gene of *Mycobacterium tuberculosis* H37Rv reveals a novel class of inositol-1-phosphate synthase enzyme. *J Mol Biol* 291:531–536.

113. Jackson M, Crick DC, Brennan PJ. 2000. Phosphatidyl-inositol is an essential phospholipid of mycobacteria. *J Biol Chem* 275:30092–30099.

114. Movahedzadeh F, Wheeler PR, Dinadayala P, Av-Gay Y, Parish T, Daffe M, Stoker NG. 2010. Inositol monophosphate phosphatase genes of *Mycobacterium tuberculosis*. *BMC Microbiol* 10:50.

115. Lea-Smith DJ, Martin KL, Pyke JS, Tull D, McConville MJ, Coppel RL, Crellin PK. 2008. Analysis of a new mannosyltransferase required for the synthesis of phos-phatidylinositol mannosides and lipoarbinomannan re-veals two lipomannan pools in corynebacterineae. *J Biol Chem* 283:6773–6782.

116. Mishra AK, Alderwick LJ, Rittmann D, Wang C, Bhatt A, Jacobs WR Jr, Takayama K, Eggeling L, Besra GS. 2008. Identification of a novel α(1→6) mannopyrano-syltransferase MptB from *Corynebacterium glutamicum* by deletion of a conserved gene, *NCgl1505*, affords a lipomannan- and lipoarabinomannan-deficient mutant. *Mol Microbiol* 68:1595–1613.

117. Guerin ME, Kaur D, Somashekar BS, Gibbs S, Gest P, Chatterjee D, Brennan PJ, Jackson M. 2009. New insights into the early steps of phosphatidylinositol mannoside biosynthesis in mycobacteria: PimB′ is an essential enzyme of *Mycobacterium smegmatis*. *J Biol Chem* 284:25687–25696.

118. Kaur D, McNeil MR, Khoo KH, Chatterjee D, Crick DC, Jackson M, Brennan PJ. 2007. New insights into the biosynthesis of mycobacterial lipomannan arising from deletion of a conserved gene. *J Biol Chem* 282: 27133–27140.

119. Torrelles JB, DesJardin LE, MacNeil J, Kaufman TM, Kutzbach B, Knaup R, McCarthy TR, Gurcha SS, Besra GS, Clegg S, Schlesinger LS. 2009. Inactivation of *My-cobacterium tuberculosis* mannosyltransferase *pimB* reduces the cell wall lipoarabinomannan and lipo-mannan content and increases the rate of bacterial-induced human macrophage cell death. *Glycobiology* 19:743–755.

120. Mishra AK, Batt S, Krumbach K, Eggeling L, Besra GS. 2009. Characterization of the *Corynebacterium gluta-micum* ΔpimB′ ΔmgtA double deletion mutant and the role of *Mycobacterium tuberculosis* orthologues *Rv2188c* and *Rv0557* in glycolipid biosynthesis. *J Bacteriol* 191:4465–4472.

121. Kremer L, Gurcha SS, Bifani P, Hitchen PG, Baulard A, Morris HR, Dell A, Brennan PJ, Besra GS. 2002. Characterization of a putative α-mannosyltransferase involved in phosphatidylinositol trimannoside biosyn-thesis in *Mycobacterium tuberculosis*. *Biochem J* 363: 437–447.

122. Guerin ME, J Kordulakova, PM Alzari, PJ Brennan, M Jackson. 2010. Molecular basis of phosphatidyl-myo-inositol mannoside biosynthesis and regulation in mycobacteria. *J Biol Chem* 285:33577–33583.

123. Morita YS, Patterson JH, Billman-Jacobe H, McConville MJ. 2004. Biosynthesis of mycobacterial phosphatidyl-inositol mannosides. *Biochem J* 378:589–597.

124. Morita YS, Sena CB, Waller RF, Kurokawa K, Sernee MF, Nakatani M, Haites RE, Billman-Jacobe H, McConville MJ, Maeda Y, Kinoshita T. 2006. PimE is a polyprenol-phosphate-mannose-dependent mannosyl-transferase that transfers the fifth mannose of phos-phatidylinositol mannoside in mycobacteria. *J Biol Chem* 281:25143–25155.

125. Crellin PK, Kovacevic S, Martin KL, Brammananth R, Morita YS, Billman-Jacobe H, McConville MJ, Coppel RL. 2008. Mutations in *pimE* restore lipoarabino-mannan synthesis and growth in a *Mycobacterium smegmatis lpqW* mutant. *J Bacteriol* 190:3690–3699.

126. Kovacevic S, Anderson D, Morita YS, Patterson J, Haites R, McMillan BN, Coppel R, McConville MJ, Billman-Jacobe H. 2006. Identification of a novel pro-tein with a role in lipoarabinomannan biosynthesis in mycobacteria. *J Biol Chem* 281:9011–9017.

127. Rainczuk AK, Yamaryo-Botte Y, Brammananth R, Stinear TP, Seemann T, Coppel RL, McConville MJ, Crellin PK. 2012. The lipoprotein LpqW is essential for the mannosylation of periplasmic glycolipids in *Corynebacteria*. *J Biol Chem* 287:42726–42738.

128. Mishra AK, Alderwick LJ, Rittmann D, Tatituri RV, Nigou J, Gilleron M, Eggeling L, Besra GS. 2007. Identification of an α(1→6) mannopyranosyltransferase (MptA), involved in *Corynebacterium glutamicum*

lipomanann biosynthesis, and identification of its orthologue in *Mycobacterium tuberculosis*. *Mol Microbiol* **65**:1503–1517.

129. Kaur D, Berg S, Dinadayala P, Gicquel B, Chatterjee D, McNeil MR, Vissa VD, Crick DC, Jackson M, Brennan PJ. 2006. Biosynthesis of mycobacterial lipoarabinomannan: role of a branching mannosyltransferase. *Proc Natl Acad Sci USA* **103**:13664–13669.

130. Alderwick LJ, Lloyd GS, Ghadbane H, May JW, Bhatt A, Eggeling L, Futterer K, Besra GS. 2011. The C-terminal domain of the arabinosyltransferase *Mycobacterium tuberculosis* EmbC is a lectin-like carbohydrate binding module. *PLoS Pathog* **7**:e1001299.

131. Shi L, Berg S, Lee A, Spencer JS, Zhang J, Vissa V, McNeil MR, Khoo KH, Chatterjee D. 2006. The carboxy terminus of EmbC from *Mycobacterium smegmatis* mediates chain length extension of the arabinan in lipoarabinomannan. *J Biol Chem* **281**:19512–19526.

132. Dinadayala P, Kaur D, Berg S, Amin AG, Vissa VD, Chatterjee D, Brennan PJ, Crick DC. 2006. Genetic basis for the synthesis of the immunomodulatory mannose caps of lipoarabinomannan in *Mycobacterium tuberculosis*. *J Biol Chem* **281**:20027–20035.

133. Appelmelk BJ, den Dunnen J, Driessen NN, Ummels R, Pak M, Nigou J, Larrouy-Maumus G, Gurcha SS, Movahedzadeh F, Geurtsen J, Brown EJ, Eysink Smeets MM, Besra GS, Willemsen PT, Lowary TL, van Kooyk Y, Maaskant JJ, Stoker NG, van der Ley P, Puzo G, Vandenbroucke-Grauls CM, Wieland CW, van der Poll T, Geijtenbeek TB, van der Sar AM, Bitter W. 2008. The mannose cap of mycobacterial lipoarabinomannan does not dominate the mycobacterium-host interaction. *Cell Microbiol* **10**:930–944.

134. Thomas JP, Baughn CO, Wilkinson RG, Shepherd RG. 1961. A new synthetic compound with antituberculous activity in mice: ethambutol (dextro-2,2′-(ethylenediimino)-di-l-butanol). *Am Rev Respir Dis* **83**:891–893.

135. Kilburn JO, Greenberg J. 1977. Effect of ethambutol on the viable cell count in *Mycobacterium smegmatis*. *Antimicrob Agents Chemother* **11**:534–540.

136. Kilburn JO, Takayama K, Armstrong EL, Greenberg J. 1981. Effects of ethambutol on phospholipid metabolism in *Mycobacterium smegmatis*. *Antimicrob Agents Chemother* **19**:346–348.

137. Takayama K, Armstrong EL, Kunugi KA, Kilburn JO. 1979. Inhibition by ethambutol of mycolic acid transfer into the cell wall of *Mycobacterium smegmatis*. *Antimicrob Agents Chemother* **16**:240–242.

138. Takayama K, Kilburn JO. 1989. Inhibition of synthesis of arabinogalactan by ethambutol in *Mycobacterium smegmatis*. *Antimicrob Agents Chemother* **33**:1493–1499.

139. Mikusova K, Slayden RA, Besra GS, Brennan PJ. 1995. Biogenesis of the mycobacterial cell wall and the site of action of ethambutol. *Antimicrob Agents Chemother* **39**:2484–2489.

140. Deng L, Mikusova K, Robuck KG, Scherman M, Brennan PJ, McNeil MR. 1995. Recognition of multiple

effects of ethambutol on metabolism of mycobacterial cell envelope. *Antimicrob Agents Chemother* **39**:694–701.

141. Lee M, Mikusova K, Brennan PJ, Besra GS. 1995. Synthesis of the arabinose donor β-D-arabinofuranosyl-1-monophosphoryldecaprenol, development of a basic arabinosyl-transferase assay, and identification of ethambutol as an arabinosyl transferase inhibitor. *J Am Chem Soc* **117**:11829–11832.

142. Wolucka BA, McNeil MR, de Hoffmann E, Chojnacki T, Brennan PJ. 1994. Recognition of the lipid intermediate for arabinogalactan/arabinomannan biosynthesis and its relation to the mode of action of ethambutol on mycobacteria. *J Biol Chem* **269**:23328–23335.

143. Lety MA, Nair S, Berche P, Escuyer V. 1997. A single point mutation in the *embB* gene is responsible for resistance to ethambutol in *Mycobacterium smegmatis*. *Antimicrob Agents Chemother* **41**:2629–2633.

144. Khasnobis S, Zhang J, Angala SK, Amin AG, McNeil MR, Crick DC, Chatterjee D. 2006. Characterization of a specific arabinosyltransferase activity involved in mycobacterial arabinan biosynthesis. *Chem Biol* **13**:787–795.

145. Alcaide F, Pfyffer GE, Telenti A. 1997. Role of *embB* in natural and acquired resistance to ethambutol in mycobacteria. *Antimicrob Agents Chemother* **41**:2270–2273.

146. Ramaswamy SV, Amin AG, Goksel S, Stager CE, Dou SJ, El Sahly H, Moghazeh SL, Kreiswirth BN, Musser JM. 2000. Molecular genetic analysis of nucleotide polymorphisms associated with ethambutol resistance in human isolates of *Mycobacterium tuberculosis*. *Antimicrob Agents Chemother* **44**:326–336.

147. Sreevatsan S, Stockbauer KE, Pan X, Kreiswirth BN, Moghazeh SL, Jacobs WR Jr, Telenti A, Musser JM. 1997. Ethambutol resistance in *Mycobacterium tuberculosis*: critical role of *embB* mutations. *Antimicrob Agents Chemother* **41**:1677–1681.

148. Protopopova M, Hanrahan C, Nikonenko B, Samala R, Chen P, Gearhart J, Einck L, Nacy CA. 2005. Identification of a new antitubercular drug candidate, SQ109, from a combinatorial library of 1,2-ethylenediamines. *J Antimicrob Chemother* **56**:968–974.

149. Nikonenko BV, Protopopova M, Samala R, Einck L, Nacy CA. 2007. Drug therapy of experimental tuberculosis (TB): improved outcome by combining SQ109, a new diamine antibiotic, with existing TB drugs. *Antimicrob Agents Chemother* **51**:1563–1565.

150. Reddy VM, Einck L, Andries K, Nacy CA. 2010. In vitro interactions between new antitubercular drug candidates SQ109 and TMC207. *Antimicrob Agents Chemother* **54**:2840–2846.

151. Engohang-Ndong J. 2012. Antimycobacterial drugs currently in phase II clinical trials and preclinical phase for tuberculosis treatment. *Expert Opin Investig Drugs* **21**:1789–1800.

152. Tahlan K, Wilson R, Kastrinsky DB, Arora K, Nair V, Fischer E, Barnes SW, Walker JR, Alland D, Barry CE 3rd, Boshoff HI. 2012. SQ109 targets MmpL3, a

membrane transporter of trehalose monomycolate involved in mycolic acid donation to the cell wall core of *Mycobacterium tuberculosis*. *Antimicrob Agents Chemother* **56**:1797–1809.

153. Trefzer C, Skovierova H, Buroni S, Bobovska A, Nenci S, Molteni E, Pojer F, Pasca MR, Makarov V, Cole ST, Riccardi G, Mikusova K, Johnsson K. 2012. Benzothiazinones are suicide inhibitors of mycobacterial decaprenylphosphoryl-β-D-ribofuranose 2′-oxidase DprE1. *J Am Chem Soc* **134**:912–915.

154. Trefzer C, Rengifo-Gonzalez M, Hinner MJ, Schneider P, Makarov V, Cole ST, Johnsson K. 2010. Benzothiazinones: prodrugs that covalently modify the decaprenylphosphoryl-β-D-ribose 2′-epimerase DprE1 of *Mycobacterium tuberculosis*. *J Am Chem Soc* **132**:13663–13665.

155. Manina G, Bellinzoni M, Pasca MR, Neres J, Milano A, Ribeiro AL, Buroni S, Skovierova H, Dianiskova P, Mikusova K, Marak J, Makarov V, Giganti D, Haouz A, Lucarelli AP, Degiacomi G, Piazza A, Chiarelli LR, De Rossi E, Salina E, Cole ST, Alzari PM, Riccardi G. 2010. Biological and structural characterization of the *Mycobacterium smegmatis* nitroreductase NfnB, and its role in benzothiazinone resistance. *Mol Microbiol* **77**:1172–1185.

156. Pasca MR, Degiacomi G, Ribeiro AL, Zara F, De Mori P, Heym B, Mirrione M, Brerra R, Pagani L, Pucillo L, Troupioti P, Makarov V, Cole ST, Riccardi G. 2010. Clinical isolates of *Mycobacterium tuberculosis* in four European hospitals are uniformly susceptible to benzothiazinones. *Antimicrob Agents Chemother* **54**:1616–1618.

157. Larrouy-Maumus G, Skovierova H, Dhouib R, Angala SK, Zuberogoitia S, Pham H, Villela AD, Mikusova K, Noguera A, Gilleron M, Valentinova L, Kordulakova J, Brennan PJ, Puzo G, Nigou J, Jackson M. 2012. A small multidrug resistance-like transporter involved in the arabinosylation of arabinogalactan and lipoarabinomannan in mycobacteria. *J Biol Chem* **287**:39933–39941.

Molecular Genetics of Mycobacteria, 2nd Edition
Edited by Graham F. Hatfull and William R. Jacobs, Jr.
© 2014 American Society for Microbiology, Washington, DC
doi:10.1128/microbiolspec.MGM2-0021-2013

Mamadou Daffé[1]
Dean C. Crick[2]
Mary Jackson[2]

Genetics of Capsular Polysaccharides and Cell Envelope (Glyco)lipids

28

GLOBAL STRUCTURE AND COMPOSITION OF THE MYCOBACTERIAL CELL ENVELOPE

The compositional and architectural complexity of the mycobacterial cell envelope is probably the most distinctive feature of the *Mycobacterium* genus. It is the basis of many of the physiological and pathogenic features of these bacteria and the site of susceptibility and resistance to many antimycobacterial drugs (1, 2). In the context of the increasing incidence of multidrug-resistant strains of *Mycobacterium tuberculosis*, elucidating the complex pathways allowing mycobacteria to synthesize and assemble this complex structure represents a crucial area of research.

The mycobacterial cell envelope is made up of three major segments: the plasma membrane, the cell wall core, and the outermost layer. The cell wall core consists of peptidoglycan (PG) in covalent attachment via phosphoryl-N-acetylglucosaminosyl-rhamnosyl linkage units with the heteropolysaccharide arabinogalactan, which is in turn esterified at its nonreducing ends to α-alkyl, β-hydroxy long-chain (C_{70} to C_{90}) mycolic acids. The latter form the bulk of the inner leaflet of the outer membrane, with the outer layer consisting of a variety of noncovalently attached (glyco)lipids, polysaccharides, lipoglycans, and proteins (1, 3, 4) (Fig. 1). Only recently have developments in cryo-electron microscopy techniques allowed the different layers of the mycobacterial cell envelope to be visualized in their native state (3–5). These studies provided direct evidence of the existence of an outer bilayer and periplasmic space in *M. tuberculosis*, *Mycobacterium bovis* BCG, *Mycobacterium marinum*, *Mycobacterium smegmatis*, and the closely related *Corynebacterium glutamicum* (Fig. 1) (3–5). Together with classical subfractionation and biochemical approaches, they also provided significant insights into the compositional diversity of the outermost layers of the cell envelopes of mycobacteria (5–9). All *Mycobacterium* species studied to date elaborate more or less abundant "capsule"-like structures both *in vitro* and during host infection that primarily

[1]CNRS, Institut de Pharmacologie et de Biologie Structurale, Département Mécanismes Moléculaires des Infections Mycobactériennes, and the Université de Toulouse Paul Sabatier, F-31077 Toulouse, France; [2]Mycobacteria Research Laboratories, Department of Microbiology, Immunology and Pathology, Colorado State University, Fort Collins CO 80523-1682.

consist of polysaccharides and proteins with generally minor amounts of lipids (7, 9). In some cases, however (e.g., *Mycobacterium lepraemurium*, *Mycobacterium leprae*, *Mycobacterium avium*), abundant quantities of species-specific glycolipids may be found (glyco-peptidolipids [GPLs] and phenolic glycolipids [PGLs] in particular). Many of the proteins and lipids typically found in the capsules of mycobacteria also occur in the outer membrane and periplasm, and their relative distribution between these three compartments seems to be species-dependent (8, 10). This diversity in terms of surface composition most likely reflects differences in the cell envelope organization of mycobacteria and is likely to significantly impact the way that *Mycobacterium* species interact with the host (11, 12).

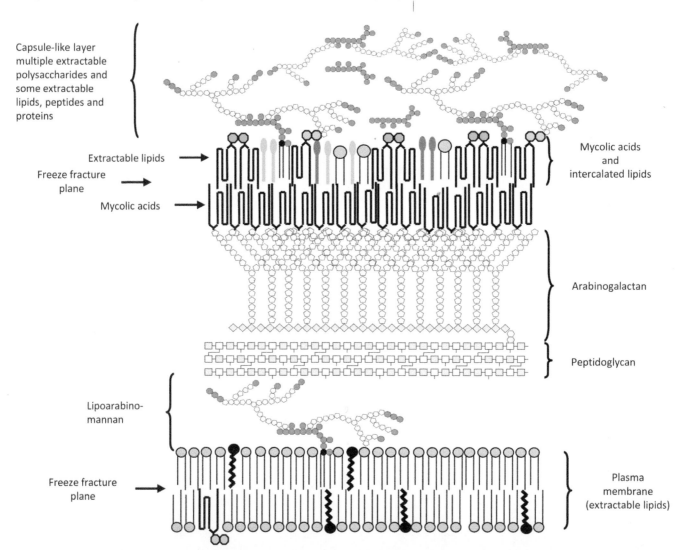

Figure 1 Schematic representation of the *M. tuberculosis* cell envelope. Many of the classes of lipids and glycolipids discussed in the text are represented schematically and are shown in probable locations in the cell envelope. The structures with light and dark green hexagons represent trehalose mono- and dimycolates, respectively; the red lollipops represent phthiocerol dimycocerosates, and the gold ones represent sulfolipids, diacyltrehaloses, and polyacyltrehaloses. Gray circles represent phospholipid headgroups; black circles, isoprenoids; light blue squares, GlcNAc; white squares, MurNAc; white pentagons, arabinofuranose; yellow diamonds, galactofuranose; and blue hexagons, mannose. The overall schematic and individual structures are not drawn to scale, and the numbers of carbohydrate residues shown are not representative of the actual molecules. Proteins and peptides are not shown for the sake of clarity. doi:10.1128/microbiolspec.MGM2-0021-2013.f1

Developments in the genetic manipulation of myco-bacteria in the 1990s and the publication of the com-plete genome sequence of *M. tuberculosis* in 1998, followed later by that of several other fast-growing and slow-growing mycobacteria, have provided a major im-petus to the study of cell envelope biosynthesis in vari-ous *Mycobacterium* species, with the result that many of the enzymes involved in their synthesis have now been identified. The molecular genetics of the cell wall core proper (PG, arabinogalactan, mycolic acids) is reviewed in other articles. This article focuses on what is known of the biosynthesis and translocation of the major noncovalently bound (extractable) lipid and glycoconjugate constituents populating the inner and outer membranes and capsule-like structures of mycobacteria. For those constituents ubiquitously distributed in mycobacteria, the gene nomen-clature used is that of *M. tuberculosis* H37Rv.

PHOSPHOLIPIDS, PHOSPHATIDYLINOSITOL MANNOSIDES, AND TRIGLYCERIDES

Phospholipids and Triacylglycerols of Mycobacteria

The mycobacterial phospholipids include phosphatidyl-glycerol, diphosphatidylglycerol (i.e, cardiolipin; CL),

phosphatidylethanolamine (PE), phosphatidyl-*myo*-ino-sitol (PI), and mannosylated forms of PI known col-lectively as the phosphatidyl-*myo*-inositol mannosides (PIMs) (Fig. 2). Phosphatidylserine also occurs in limited amounts (Fig. 2), but phosphatidylcholine is apparently not produced by mycobacteria (13). Phos-pholipids represent the main structural amphipathic polar lipids of the mycobacterial inner membrane and also populate the outer membrane (Fig. 1). PE and PIMs, in particular, were identified in the surface-exposed lipids of all *Mycobacterium* species investigat-ed (*M. tuberculosis*, *M. avium*, *M. kansasii*, *M. gastri*, *M. smegmatis*, and *M. aurum*) (8). Palmitic ($C_{16:0}$), oleic ($C_{18:1}$), and tuberculostearic (C_{19}) acids appear to be the major fatty acid substituents in the phos-pholipids of mycobacteria, with the unsaturated or branched $C_{18:1}$ and C_{19} fatty acids principally ester-ifying position 1 of glycerol, and $C_{16:0}$ preferentially occupying position 2.

Triacylglycerols (TAGs), triglycerides, have similarly been isolated from all mycobacterial species examined and represent the main apolar lipids when glycerol is the major carbon source in the culture medium (14). Mycobacteria grown *in vitro* or recovered from human samples essentially accumulate TAG in the form of intracellular lipid droplets, but TAGs have also been

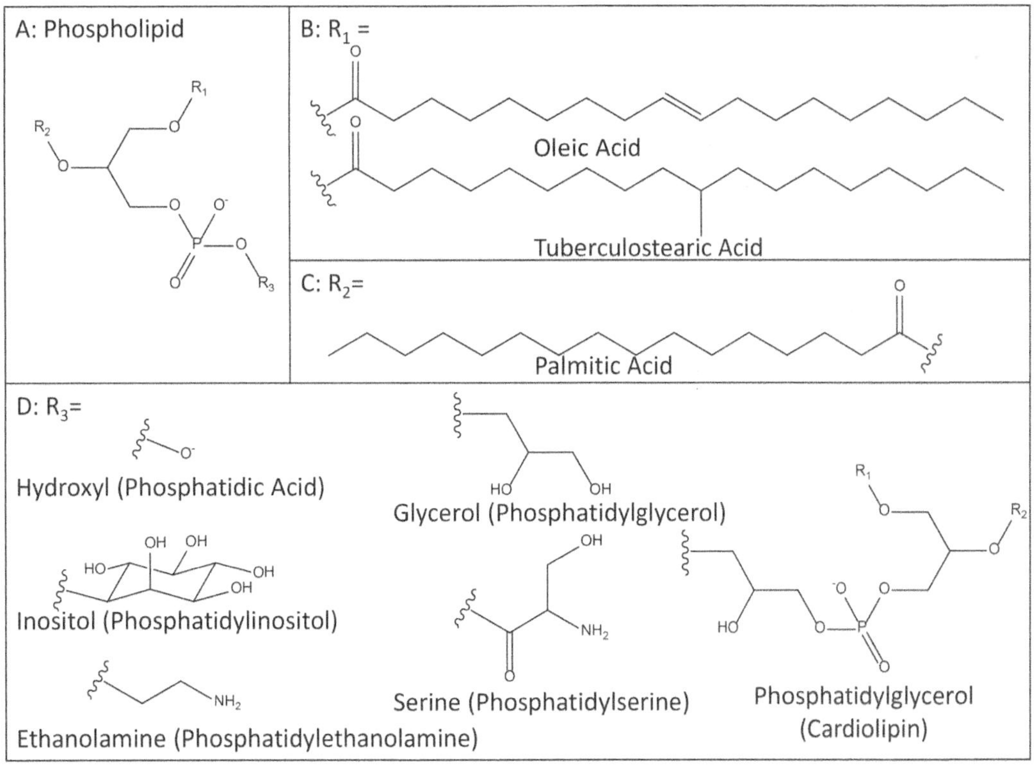

Figure 2 Structures of mycobacterial phospholipids. doi:10.1128/microbiolspec.MGM2-0021-2013.f2

identified in the surface-exposed lipids of *M. smegmatis* and *M. avium* (8). They are proposed to act as a source of energy for actively replicating bacteria as well as a means by which free fatty acids are detoxified. TAGs are also proposed to serve as an energy reserve for the long-term survival of *M. tuberculosis* during the persistence phase of infection (14, 15). In *M. bovis* BCG and *M. smegmatis*, position 1 of TAG is occupied principally by stearic ($C_{18:0}$), $C_{18:1}$, and C_{19} fatty acids; position 2 is mostly esterified with C_{16} fatty acids, whereas the third position predominantly bears fatty acids with more than 20 carbons (C_{20} to C_{33}) (16). The fatty acids acylating phospholipids and triglycerides in axenically grown bacteria are thought to be synthesized by fatty acid synthase I (FAS-I) (Rv2524c) (17, 18).

Phosphatidic Acid Synthesis

Phosphatidic acid (Fig. 2) is a common intermediate in the biosynthesis of both TAG and phospholipids. The pathway begins with glycerol-3-phosphate, which is formed by reduction of dihydroxyacetone phosphate by the glycerol-3-phosphate synthase GpsA. Two gene candidates were annotated for this function in the genome of *M. tuberculosis* H37Rv, *gpdA1* (*Rv0564c*) and *gpdA2* (*Rv2982c*), but neither of them has been confirmed biochemically. Glycerol-3-phosphate is first acylated by acyl-coenzyme A (CoA), acyl-ACP, or acyl-phosphate to form lysophosphatidate and is then acylated again by acyl-CoA or acyl-ACP to yield phosphatidate (19). Again, based on sequence similarities, two putative glycerol-3-phosphate acyltransferase genes, *plsB1* (*Rv1551*) and *plsB2* (*Rv2482c*), and one putative lysophosphatidate acyltransferase gene, *plsC* (*Rv2483c*), have been proposed to be involved in those acyl transfer reactions, but they have not yet been biochemically validated (Table 1).

TAG Synthesis

In the synthesis of TAG, phosphatidate is hydrolyzed by a specific phosphatase to yield diacylglycerol (DAG). This intermediate is then acylated to TAG in a reaction catalyzed by diglyceride acyltransferases (or triglyceride synthases). Although no phosphatidic acid phosphatases have yet been identified in mycobacteria, two proteins displaying this activity were recently characterized in *Streptomyces coelicolor* (20), one of which (SCO1102) displays sequence similarity with Rv0308 of *M. tuberculosis* H37Rv (H. Gramajo, personal communication). In the genome of *M. tuberculosis* H37Rv 15 genes were identified whose protein products display

triglyceride synthase activity *in vitro*, generating triolein from diolein and oleyl-CoA (15, 21). Interestingly, Ag85A (FbpA, Rv3804c) is also endowed with a similar acyltransferase activity, transferring long-chain acyl-CoA onto DAG (22) (Table 1).

Phospholipid Biosynthesis

CDP-DAG appears to be the common precursor for the biosynthesis of phospholipids in mycobacteria and is synthesized from phosphatidic acid and CTP by the CDP-DAG synthase (CTP:phosphatidate cytidylyltransferase). Such enzymatic activity was detected in *M. smegmatis* and found to be membrane-associated (23). The structural gene for CDP-DAG synthase in the genome of *M. tuberculosis* H37Rv is predicted to be *cdsA* (*Rv2881c*). Phosphatidyl-*myo*-inositol (PI) is made *de novo* from CDP-DAG and *myo*-inositol (24) in a reaction catalyzed by the PI synthase, PgsA1 (Rv2612c) (25). However, an alternative pathway for PI synthesis has been suggested wherein *myo*-inositol is first phosphorylated to form *myo*-inositol 3-phosphate, which then reacts with CDP-DAG to form PI 3-phosphate (PI3P). It was proposed that *pgsA1* encodes a PI3P synthase rather than a PI synthase and that PI3P is subsequently dephosphorylated (by an as yet unknown enzyme) to yield PI (26). Evidence based on sequence homology or changes in the phospholipid composition of *M. smegmatis* upon gene overexpression strongly suggests that the *pgsA3* (*Rv2746c*) and *pssA* (*Rv0436c*) genes of *M. tuberculosis* encode the phosphatidylglycerophosphate synthase and phosphatidylserine synthase involved in the formation of, respectively, phosphatidylglycerol and phosphatidylserine (25). As in other bacteria, PE is likely to arise from the decarboxylation of phosphatidylserine in a reaction catalyzed by the product of *psd* (*Rv0437c*). Cardiolipin may be formed from the condensation of two phosphatidylglycerol molecules by a cardiolipin synthase as in most prokaryotes or through the transfer of a phosphatidyl group from CDP-DAG onto phosphatidylglycerol as in yeast and as recently shown in *S. coelicolor* (27). *M. tuberculosis* H37Rv possesses a eukaryotic-type cardiolipin synthase bearing sequence similarity to the *Streptomyces* enzyme (PgsA2; Rv1822), whereas proteins displaying the characteristic phospholipase D-type features of classical prokaryotic cardiolipin synthases are missing, suggesting that the second pathway may be the one used by mycobacteria (25, 27). However, whether PgsA2 is endowed with such enzymatic activity remains to be established (Table 1).

PIMs

The PI dimannosides (PIM$_2$) are considered both metabolic end products and intermediates in the biosynthesis of polar PIM (PIM$_5$, PIM$_6$), lipomannan (LM), and lipoarabinomannan (LAM) (for more details about these molecules and their biosynthetic pathways, see reference 363). We will only briefly describe the initial steps of PIM synthesis leading to the formation of PIM$_2$ and PIM$_6$, the two most abundant forms of PIM found in mycobacteria. The first step in PIM synthesis involves the transfer of a mannose residue from GDP-Man$p$ to the 2-position of the *myo*-inositol ring of PI to form PI monomannoside, PIM$_1$. We have identified PimA (Rv2610c) as the α-mannosyltransferase responsible for this catalytic step and found it to be an essential enzyme (28–30). The second step involves the action of another essential α-mannosyltransferase, PimB′ (Rv2188c), which transfers a Man$p$ residue from GDP-Man$p$ to the 6-position of the *myo*-inositol ring of PIM$_1$ (31). Both PIM$_1$ and PIM$_2$ can be acylated with palmitate at position 6 of the Man$p$ residue transferred by PimA by the acyltransferase Rv2611c to form Ac$_1$PIM$_1$ and Ac$_1$PIM$_2$, respectively (32). The acyltransferase responsible for the transfer of a fourth acyl group to position 3 of the *myo*-inositol ring has not yet been identified. Likewise, the identity of the enzymes involved in the mannosylation of the dimannosylated forms of PIM to form PIM$_3$ and PIM$_4$ is at present unclear (33). PimE (Rv1159) has been identified as the α-1,2-mannosyltransferase involved in the biosynthesis of PIM$_5$ from PIM$_4$ (34). PimE belongs to the GT-C superfamily of glycosyltransferases, which comprises integral membrane proteins that use polyprenyl-linked sugars as donors (33, 35). Whether PimE also catalyzes the transfer of the second α-1,2-linked Man residue onto PIM$_5$ to yield PIM$_6$ or whether the formation of PIM$_6$ results from the action of an independent mannosyltransferase is at present not known.

Translocation of Phospholipids, PIM, and TAG to the Outer Membrane and Cell Surface

Phospholipids and TAG are synthesized in the cytoplasm or at the periphery of the inner leaflet of the plasma membrane. Likewise, the early steps of PIM biosynthesis take place on the cytosolic face of the plasma membrane until PIM intermediates, believed to be PIM$_2$ or PIM$_3$, are translocated across the plasma membrane by an as yet unknown flippase to serve as substrates for further mannosylation reactions catalyzed by PimE and other

GT-C polyprenyl-phosphate mannose-dependent glycosyltransferases (33, 35, 36). Beyond their translocation across the plasma membrane, the further export of phospholipids, TAG, and PIM to the outer membrane and cell surface most likely requires dedicated translocation machineries. Thus far, none of the flippases and transporters involved have been formally identified. Evidence based on physical interactions and cocrystallography suggests that the lipoprotein LprG (Rv1411c), which shares structural resemblance to LppX, a lipoprotein thought to carry phthiocerol dimycocerosates (PDIM) across the periplasm (37), may participate in the transport of PIM, LM, and LAM to the cell surface (38). This exciting hypothesis awaits further genetic and biochemical validation.

ISOPRENOIDS AND RELATED LIPIDS

Biosynthesis of Isoprenoid Precursors

A number of isoprenoids have been observed and characterized in *Mycobacterium* species including polyprenyl diphosphates, polyprenyl phosphates, lipid I and lipid II, carotenoids, menaquinones, sulfomenaquinones, and cyclic isoprenoids. These molecules have diverse and in some cases multiple functions. For example, polyisoprenyl phosphate (Pol-P) is involved in the biosynthesis of the arabinan portion of arabinogalactan, arabinomannan, and LAM (39), and in that of the PG precursors lipid I and lipid II (40, 41), as a lipid carrier of the activated saccharide subunits. Pol-P is also involved in the biosynthesis of the "linker unit" between two essential cell wall components, arabinogalactan and PG (42).

All isoprenoids are derived from the repetitive condensation of isopentenyl diphosphate (IPP) and allylic diphosphates (43) catalyzed by enzymes known as prenyldiphosphate synthases or prenyltransferases. To date, two distinct pathways for the biosynthesis of the IPP and dimethylallyl diphosphate (DMAPP; the smallest allylic diphosphate) have been identified: the mevalonate (MVA) pathway and the 2C-methyl-D-erythritol 4-phosphate (MEP) pathway. In mycobacteria, IPP and DMAPP (Fig. 3) are biosynthesized exclusively via the MEP pathway.

The MEP pathway of *M. tuberculosis*

The initial enzyme in the MEP pathway, 1-deoxy-D-xylulose-5-phosphate synthase (DXS), catalyzes the condensation of glyceraldehyde-3-phosphate and pyruvate, forming 1-deoxy-D-xylulose-5-phosphate (DXP)

Table 1   *M. tuberculosis* H37Rv genes involved or thought to be involved in the biogenesis of phospholipids, triglycerides, isoprenoids, and related lipids

Rv number	Gene name	Function	Evidence[a]	Reference
Rv0221		Putative acyl-CoA:diacylglycerol acyltransferase	E	15
Rv0308		Putative phosphatidic acid phosphatase	H	20
Rv0436c	*pssA*	Putative phosphatidylserine synthase	P, H	25
Rv0437c	*psd*	Putative phosphatidylserine decarboxylase	H	
Rv0534c	*menA*	Demethylmenaquinone synthase	E	114
Rv0542c	*menE*	o-Succinylbenzoyl-CoA synthase	E	111–113
Rv0548c	*menB*	1,4-Dihydroxy2-naphthoic acid synthase	E	108–110
Rv0562		ω,E,E,E-Geranylgeranyldiphosphate synthase	E	87
Rv0564c	*gpdA1*	Putative glycerol-3-phosphate synthase	H	
Rv0654		Carotenoid oxygenase	E	132
Rv0895		Putative acyl-CoA:diacylglycerol acyltransferase	E	15
Rv0989c		Geranyldiphosphate synthase	E	85
Rv1011	*ispE*	4-Diphosphocytidyl-2C-methyl-D-erythritol kinase	E	61, 62
Rv1086		ω,E,Z-Farnesyldiphosphate synthase	E	79, 80
Rv1159	*pimE*	Polyprenol phosphomannose-dependent α-1,2-mannosyltransferase	E, P	34
Rv1411c	*lprG*	Lipoprotein; putative PIM, LM, and LAM transporter	P	38
Rv1425		Putative acyl-CoA:diacylglycerol acyltransferase	E	15
Rv1551	*plsB1*	Putative glycerol-3-phosphate acyltransferase	H	
Rv1760		Putative acyl-CoA:diacylglycerol acyltransferase	E	15
Rv1822	*pgsA2*	Putative cardiolipin synthase	H	
Rv2188c	*pimB'*	GDP-Man-dependent α-1,6-phosphatidylinositol mannosyltransferase	E, P	31
Rv2267c	*stf3*	Putative sulfotransferase	P, H	117
Rv2285		Putative acyl-CoA:diacylglycerol acyltransferase	E	15
Rv2361c		ω,E,poly-Z-Decaprenyldiphosphate synthase	E	83
Rv2482c	*plsB2*	Putative glycerol-3-phosphate acyltransferase	H	
Rv2483c	*plsC*	Putative lysophosphatidate acyltransferase	H	
Rv2484c		Putative acyl-CoA:diacylglycerol acyltransferase	E	15
Rv2524c	*fas*	Fatty acid synthetase type I	E, P	17, 18
Rv2610c	*pimA*	GDP-Man-dependent α-1,2-phosphatidylinositol mannosyltransferase	E, P	28, 30
Rv2611c		Acyltransferase involved in the 6-O-acylation of the Man*p* residue linked to the 2-position of *myo*-inositol in PIM$_1$ and PIM$_2$	E, P	32
Rv2612c	*pgsA1*	Phosphatidyl-*myo*-inositol synthase and/or phosphatidyl-*myo*-inositol phosphate synthase	E, P	25, 26
Rv2682c	*dxs*	1-Deoxy-D-xylulose-5-phosphate synthase	E	49
Rv2746c	*pgsA3*	Phosphatidylglycerophosphate synthase	P, H	25
Rv2868c	*ispG*	1-Hydroxy-2-methyl-2(E)-butenyl 4-diphosphate synthase	H	50
Rv2870c	*ispC*	1-Deoxy-D-xylulose 5-phosphate reductoisomerase	E	54, 55
Rv2881c	*cdsA*	Putative CDP-diacylglycerol synthase	H	23
Rv2982c	*gpdA2*	Putative glycerol-3-phosphate synthase	H	
Rv3087		Putative acyl-CoA:diacylglycerol acyltransferase	E	15
Rv3088	*tgs4*	Putative acyl-CoA:diacylglycerol acyltransferase	E	15
Rv3130c	*tgs1*	Acyl-CoA:diacylglycerol acyltransferase	E, P	15, 21
Rv3233c		Putative acyl-CoA:diacylglycerol acyltransferase	E	15

(Continued)

Table 1 *(Continued)*

Rv number	Gene name	Function	Evidence[a]	Reference
Rv3234c	*tgs3*	Putative acyl-CoA:diacylglycerol acyltransferase	E	15
Rv3371		Putative acyl-CoA:diacylglycerol acyltransferase	E	15
Rv3377c		Tuberculosinyldiphosphate synthase	E	138
Rv3378c		Isotuberculosinol synthase	E	135, 139, 140
Rv3383c		ω,*E,E*-Geranylgeranyldiphosphate synthase	E	87
Rv3398c		ω,*E,E*-Farnesyldiphosphate synthase	E	86
Rv3480c		Putative acyl-CoA:diacylglycerol acyltransferase	E	15
Rv3581c	*ispF*	2C-Methyl-D-erythritol 2,4-cyclodiphosphate	E	63
Rv3582c	*ispD*	4-Diphosphocytidyl-2C-methyl-D-erythritol	E	58
Rv3734c	*tgs2*	Putative acyl-CoA:diacylglycerol acyltransferase	E	15
Rv3740c		Putative acyl-CoA:diacylglycerol acyltransferase	E	15
Rv3804c	*fbpA*	Acyl-CoA:diacylglycerol acyltransferase	E, P	22

[a]The experimental evidence for the annotation of a gene may either be "enzymatic" (E) (i.e., an enzymatic activity was associated with the gene's product *in vitro*) or "phenotypic" (P) (i.e., the annotation results from the biochemical analysis of mycobacterial recombinant strains—e.g., knockout/knockdown mutants, complemented mutant strains, overexpressors—or from the functional complementation of defined *E. coli* mutants). In some cases, the function of a gene is exclusively based on its homology to other known (myco)bacterial genes (H).

(44). The product of DXS is used not only as a biosynthetic intermediate of IPP but also as the precursor of thiamin (vitamin B_1) and of pyridoxol (vitamin B_6) in *Escherichia coli* (45–47); thus, DXS is not a committed step in the MEP pathway.

The *dxs* gene was first identified in *E. coli* (45, 46). Sequence alignment with *E. coli* DXS demonstrated that Rv2682c has approximately 38% identity with a conserved DRAG motif and a key amino acid (His49) required for catalytic activity (48) (Table 1). The function of Rv2682c was demonstrated empirically, because the purified recombinant enzyme is capable of producing DXS by condensation of pyruvate and glyceraldehyde-3-phosphate in the presence of thiamine pyrophosphate (49). Interestingly, *M. tuberculosis*

Isopentenyldiphosphate (IPP)

Dimethylallyldiphosphate (DMAPP)

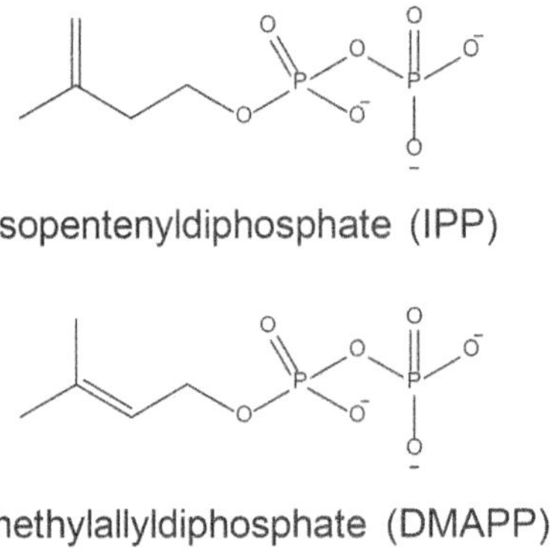

Figure 3 Structures of IPP and DMAPP. These molecules are precursors of all isoprenoid compounds.
doi:10.1128/microbiolspec.MGM2-0021-2013.f3

contains a second ortholog of *E. coli* DXS, Rv3379c. However, an alignment with *E. coli* DXS indicated that Rv3379c, despite a relatively high level of identity (38%), was truncated due to the positioning of an insertion element (IS6110) and, more importantly, the His49 residue is missing and the recombinant protein showed no DXS activity (49). This and the fact that *Rv2682c* is essential for bacterial survival (50) suggest that it encodes the only functional *M. tuberculosis* DXS.

1-Deoxy-D-xylulose-5-phosphate reductoisomerase (IspC), the second enzyme in the MEP biosynthetic pathway, catalyzes the rearrangement and reduction of DXP in the presence of NADPH to generate MEP (51). As mentioned above, DXP is a precursor not only of IPP and DMAPP but also of thiamine and pyridoxol; therefore, IspC catalyzes the first committed step for biosynthesis of IPP and DMAPP (52).

Alignments with *E. coli* IspC indicated that the primary structure of Rv2870c of *M. tuberculosis* is 25% identical to that of the *E. coli* IspC with conserved amino acid residues (53, 54). Recombinant Rv2870c efficiently catalyzes the conversion of DXP to MEP in the presence of NADPH and the reverse reaction in the presence of $NADP^+$ (54–56).

Incubation of MEP with crude, cell-free extracts of *E. coli* in the presence of CTP produces 4-diphosphocytidyl-2C-methyl-D-erythritol (CDP-ME), and the gene encoding the activity was identified as *ygbP* (57), which was later renamed *ispD*. The *Rv3582c* gene product has approximately 31% identity with *E. coli* IspD, and the recombinant Rv3582c protein was shown to be a functional IspD in *M. tuberculosis* (58).

The fourth step in the MEP pathway involves the conversion of CDP-ME to 4-diphosphocytidyl-2C-methyl-D-erythritol 2-phosphate (CDP-ME2P) in the presence of ATP catalyzed by IspE, which was initially identified in *E. coli* and tomatoes (59, 60). Alignment of *E. coli* IspE with genes of the *M. tuberculosis* genome indicated that *Rv1011* encodes a protein that harbors around 22% identity with conserved amino acids involved in forming the CDP-ME and ATP binding and crucial active sites and catalyzes CDP-ME phosphorylation in an ATP-dependent manner (61, 62).

The fifth step of the MEP pathway involves the formation of a metabolite containing a cyclodiphosphate moiety. The product of IspE, CDP-ME2P, is converted into 2C-methyl-D-erythritol 2,4-cyclodiphosphate (MECDP) with corresponding release of CMP by the *ispF* gene product (59). *Rv3581c* encodes *M. tuberculosis* IspF (63) and is essential for bacterial survival (64). The crystal structure of *M. smegmatis* IspF, harboring around 73% amino acid sequence identity with *M. tuberculosis* IspF, has been solved (64).

Recombinant *E. coli ispC*, *ispD*, *ispE*, *ispF*, and *ispG* were shown to catalyze the conversion of 1-deoxy-D-xylulose (DX) into 1-hydroxy-2-methyl-2(E)-butenyl-4-diphosphate (HMBPP) (65), and the *ispH* gene product is responsible for the conversion of HMBPP into IPP and DMAPP (66, 67). Recombinant IspG catalyzes the reduction of MECDP, resulting in opening of the cyclodiphosphate ring structure using a photoreduced deazaflavin derivative as an artificial electron donor (68, 69). Interestingly, *in vivo* experiments using an *E. coli* strain overexpressing *ispH* resulted in the formation of IPP and DMAPP from HMBPP in a molar ratio of 5:1 (70). Blast searches of *E. coli* IspG or IspH with the *M. tuberculosis* genome indicate that *Rv2868c*, an essential gene (50), is the likely *M. tuberculosis* IspG, and either *Rv1110* or *Rv3382c* is the candidate gene encoding *M. tuberculosis* IspH.

IPP isomerase
Upon biosynthesis of IPP and DMAPP by IspH, IPP isomerase (Idi) catalyzes the interconversion of the two isoforms (71), but the equilibrium favors the forward reaction, from IPP to DMAPP (72). In organisms capable of synthesizing isoprenoid units by the MVA pathway, Idi is reported to be essential (73), as pyrophosphomevalonate decarboxylase in the MVA pathway produces only IPP and both DMAPP and IPP (Fig. 3) are required for further biosyntheses of

isoprenoids. Orthologs of *idi* are also found in many organisms that utilize the MEP pathway, most of which are reported to encode nonessential enzymes (74), presumably because IspH of the MEP pathway produces both IPP and DMAPP. Two forms of bacterial Idi have been discovered to date: type I, which includes Idi from *E. coli*, and type II, which was identified in *Streptomyces* species strain CL190 (75); *M. tuberculosis* has an ortholog of a type I Idi, while *M. smegmatis* has an ortholog of the type II Idi.

Prenyldiphosphate synthases
As mentioned above, the universal precursors of all isoprenoid compounds are synthesized from IPP, DMAPP, or linear IPPs that are synthesized by sequential 1′-4 condensations of IPP with DMAPP. The enzymes catalyzing this sequential process are known as prenyltransferases or prenyldiphosphate synthases. These enzymes can be divided into two families depending on the stereochemistry of the double bonds formed during product formation and the chain length of the final product. Thus, prenyldiphosphate synthases can be categorized as E-prenyldiphosphate synthases or Z-prenyldiphosphate synthases, and there is no similarity between the two in terms of amino acid sequence. The E-prenyldiphosphate synthases can be further characterized as short-chain, with a product containing 10 to 25 carbons, medium-chain, 30 to 35 carbons, and long-chain, 40 to 50 carbons (76). Similarly, the Z-prenyldiphosphate synthases can be characterized as short-chain, medium-chain, and long-chain (77). Both the E- and Z-prenyldiphosphate synthase families generate products with the correct chain lengths via a molecular ruler mechanism, where one or two bulky amino acids occupy the bottom of each of the enzyme active sites to block extra chain elongation of the products, thereby determining the ultimate chain lengths (78). Both E- and Z-prenyldiphosphate synthases have been identified and characterized in *Mycobacterium* species.

Mycobacterium species are unusual in that they harbor two or three Z-prenyldiphosphate synthases, whereas most bacteria have only one of these enzymes. In *M. tuberculosis*, *Rv1086* encodes a short-chain Z-prenyldiphosphate synthase that generates ω,E,Z-farnesyldiphosphate (Fig. 4, Table 1). This gene has been cloned and expressed and the enzyme activity characterized (79, 80), and it was the first representative of this class of enzyme described. The crystal structure and mechanism of chain length determination have been solved (81, 82). Rv2361c has been identified as a long-chain Z-prenyldiphosphate that synthesizes

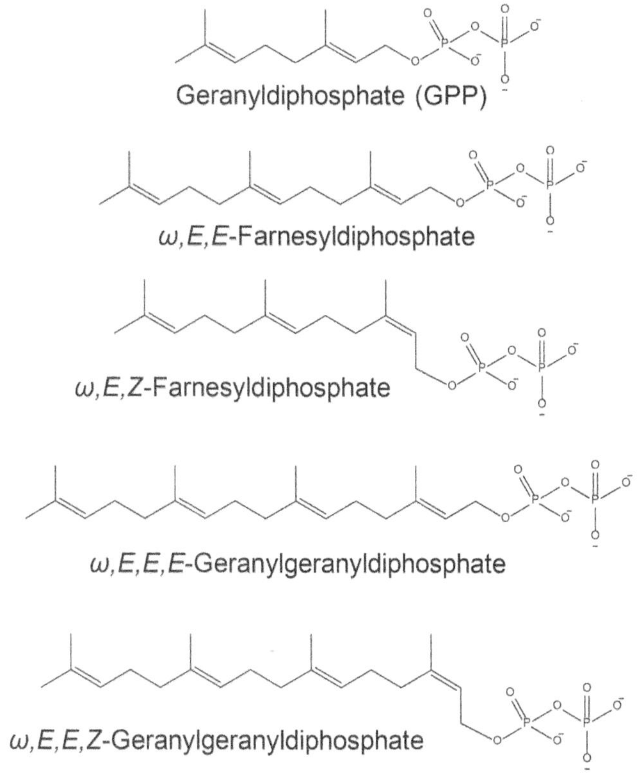

Geranyldiphosphate (GPP)

ω,E,E-Farnesyldiphosphate

ω,E,Z-Farnesyldiphosphate

ω,E,E,E-Geranylgeranyldiphosphate

ω,E,E,Z-Geranylgeranyldiphosphate

Figure 4 Structures of representative short-chain IPPs synthesized by mycobacteria. The sterochemical conformation is shown. doi:10.1128/microbiolspec.MGM2-0021-2013.f4

ω,E,poly-Z-decaprenyldiphosphate (79, 83) (Fig. 5). In *Mycobacterium vanbaalenii*, three Z-prenyldiphosphate synthases were identified and characterized (84). Mvan_4662 accepts only geranyldiphosphate as the allylic primer, producing only ω,E,Z-farnesyldiphosphate, indicating a function similar to Rv1086. Mvan_1705 accepts only ω,E,E-farnesyldiphosphate, synthesizing ω,Z,E,E-geranylgeranyldiphosphate whereas Mvan_3822 is a bifunctional Z-prenyldiphosphate synthase that preferentially synthesizes C$_{35}$ or C$_{50}$ products, depending on the allylic reaction primer.

A number of E-prenyldiphosphate synthases have also been identified in mycobacteria, which synthesize E-prenyldiphosphates of various chain lengths (Fig. 4, Table 1). These include *Rv0989c*, which is reported to synthesize geranyldiphosphate (85); *Rv3398c*, which encodes an ω,E,E-farnesyldiphosphate synthase (86); and *Rv0562* and *Rv3383c*, both of which are reported to encode ω,E,E,E-geranylgeranyldiphosphate synthases (87). It should be noted that stereochemistry of the products of the E-prenyldiphosphate synthases is assumed based on the amino acid sequence of the enzyme, not on empirical observation.

Polyprenyl Phosphate
Structures of mycobacterial polyprenyl phosphates

The most common structures of polyisoprenol (and therefore Pol-P) found in nature tend to be confined to four main groups: (i) ω,E-polyisoprenol, (ii) ω,di-E, poly-Z-polyisoprenol, (iii) ω,tri-E,poly-Z-polyisoprenol, and (iv) ω,Z-polyisoprenol (88). Most bacteria utilize undecaprenylphosphate (or bactoprenylphosphate), a ω,di-E,octa-Z-prenylphosphate, as a carrier of activated sugars primarily for synthesis of oligo- and polysaccharides on the outside of the plasma membrane as is seen in PG synthesis. However, mycobacteria synthesize and utilize at least two and perhaps three forms of Pol-P. In *M. smegmatis* two forms of Pol-P have been reported (Fig. 4): (i) decaprenyl phosphate (Dec-P) containing one ω, one E-, and eight Z-isoprene units (ω,E,poly-Z) (39) and (ii) a heptaprenyl phosphate (89) containing four saturated isoprene units on the omega end of the molecule and two E- and one Z-isoprene units (90) or four saturated and three Z-isoprene units (91). *M. tuberculosis*, however, appears to utilize a single predominant Pol-P (Dec-P). To date, the stereochemistry of the individual isoprene units of Dec-P from *M. tuberculosis* have not been determined (92), but it is likely that they are the same as those of the *M. smegmatis* Dec-P. In all three cases described

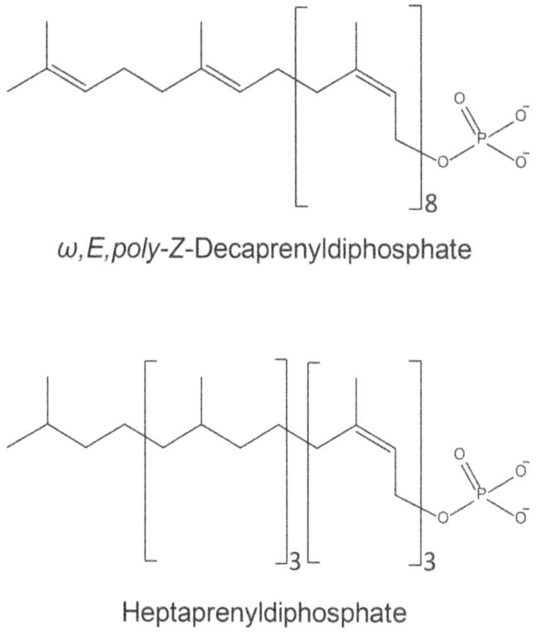

ω,E,poly-Z-Decaprenyldiphosphate

Heptaprenyldiphosphate

Figure 5 Structures of isoprenylphosphates reported from *M. smegmatis*.
doi:10.1128/microbiolspec.MGM2-0021-2013.f5

above, the mycobacterial Pol-P molecules are structurally unusual.

Pol-P biosynthesis

In general, all Pol-P molecules are synthesized via sequential condensation of IPP with allylic diphosphates catalyzed by the prenyldiphosphate synthases described above, forming polyisoprenyldiphosphates (Pol-PP) that are subsequently dephosphorylated. In mycobacteria, Rv1086 and Rv2361c (Table 1) can catalyze the addition of IPP to ω,E-GPP; however, kinetic analyses (80, 83) suggest that Rv1086 and Rv2361c act sequentially in the synthesis of ω,E,poly-Z-decaprenyl diphosphate (Dec-PP), the precursor of the ω,E,poly-Z-Dec-P found in mycobacteria (39, 89, 91–93), with Rv2361c adding seven isoprene units to the ω,E,Z-FPP synthesized by Rv1086. Thus, it seems likely that Rv0989c, Rv1086, and Rv2361c act in concert to generate decaprenyldiphosphate (Dec-PP), with isoprene units of the required stereochemistry. Once the Dec-PP has been synthesized, it must be dephosphorylated to form Dec-P (Fig. 5). Currently, there is little information regarding this biosynthetic transformation in mycobacteria; however, an ortholog of BacA, a phosphatase reported to be involved in dephosphorylation of Pol-PP in *E. coli* (94), may be involved.

Menaquinones

Structure of mycobacterial lipoquinones

The lipoquinones involved in the respiratory chains of bacteria consist of menaquinones and ubiquinones (95), while mammals have only ubiquinone. Menaquinones (2-methyl-3-polyprenyl-1,4-naphthoquinones) are the predominant isoprenoid lipoquinones of mycobacteria and many Gram-positive bacteria, whereas Gram-negative bacteria typically utilize both menaquinone and ubiquinone or only ubiquinone (which has a benzoquinone ring rather than a napthoquinone ring) (96–100).

Menaquinones are identified by the variable portions of the molecules. Generally, the only variation seen in the naphthyl ring structure is whether or not the C-2 position is methylated (Fig. 6). The most variant portion of the molecule is the polyisoprenyl side chain found at the C-3 position. Menaquinones (and ubiquinones) are identified by the length and chemical structure of this side chain. For example, a menaquinone with a side chain of eight isoprene units as seen in *E. coli* is identified as MK-8. The predominant form of menaquinone in mycobacteria has nine isoprene units, with the β position being saturated (96) (Fig. 6). Hence, this menaquinone is identified as MK-9 (II-H$_2$).

Functional significance of the menaquinone structure

Historically, respiratory quinones have been utilized for taxonomic purposes because the length and degree of saturation of the isoprenoid chain often reflect the phylogenetic affiliation of bacteria (101). The taxonomic distribution of structural features suggests that they are both functional and evolutionarily conserved. A great

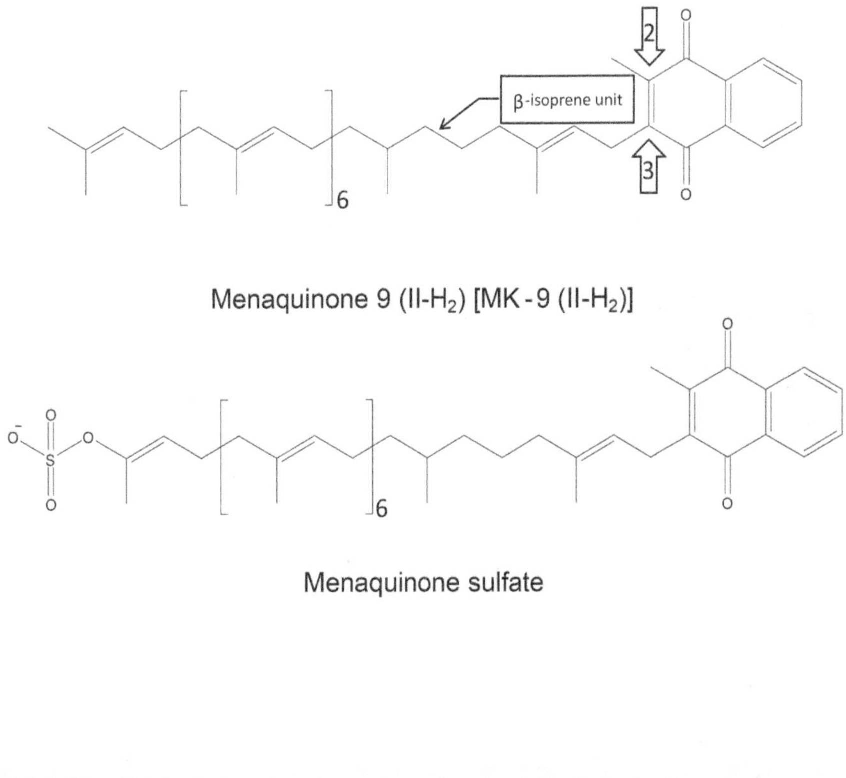

Figure 6 Structures of the predominant menaquinone and menaquinone sulfate reported from *M. tuberculosis*. Carbon positions 2 and 3 and the β-isoprene unit are indicated by the arrows and call-out. doi:10.1128/microbiolspec.MGM2-0021-2013.f6

Menaquinone 9 (II-H$_2$) [MK-9 (II-H$_2$)]

Menaquinone sulfate

deal of effort was put into defining the significance of the various structural variations in the 1960s, but this area of research has been largely ignored since then. In 1970, Brodie et al. summarized the state of the knowledge (102). Thus, it is known that the substitution at C-2 of the naphthyl ring is required for both oxidation and phosphorylation and must be a methyl group because conversion to a hydroxyl permits oxidation but not phosphorylation. The C-3 position must be substituted with an isoprenoid chain to function as a membrane-bound electron transporter. The double bond in the α-isoprene unit must be in the E-configuration; the Z-isomer does not allow phosphorylation. Thus, it appears that menaquinone in the electron transport is more than a simple electron transporter, because structural modifications allow uncoupling of oxidation and phosphorylation, suggesting that the menaquinone structure may regulate ATP synthesis. The single bond in the β-isoprene unit is conserved in many Gram-positive bacteria. However, the function of this modification is unknown. Recently, a novel sulfated menaquinone was isolated from *M. tuberculosis*, which appears to regulate virulence in mouse infection studies (103), but the precise function of this molecule is also unknown.

Menaquinone biosynthesis

The biosynthesis of menaquinone takes place via the intersection of two separate pathways. 1,4-Dihydroxy-2-naphthoate is synthesized via the shikimate pathway. The naphthoate ring is then prenylated with a prenyldiphosphate, derived from a series of prenyl transferase reactions, to form demethylmenaquinone and, subsequently, the C-2 position of the ring structure is methylated. The details of the biosynthesis of menaquinone studied in species other than *Mycobacterium* species have been reviewed (96, 104–106). In mycobacteria, the β-isoprene unit of the prenyl group is reduced to form MK-9 (II-H$_2$) after the formation of menaquinone (107).

In *E. coli*, the synthesis of menaquinone is accomplished by seven enzymes (*menA* to *menG*). These enzymes are encoded by two clusters of genes: the *men* cluster consisting of the *menB,C,D,E,F* and a separate cluster containing *menA* and *menG*. Menaquinone synthesis in Gram-positive bacteria in general has largely been ignored; however, the general pathway in *M. tuberculosis* appears to be similar. In *M. tuberculosis*, the *menA-E* genes appear to be found in a single cluster, whereas, the gene with the most homology to *menF* in *E. coli* is *Rv3215*, annotated as *entC* (isochorismate synthase).

Although menaquinone synthesis has been relatively extensively studied in *E. coli* (due in part to the availability of the *men* mutants, which can easily be generated in this organism, because it can utilize ubiquinone as an electron carrier in aerobic conditions), the synthesis of this compound in other organisms has received relatively little attention; however, MenB (1,4-dihydroxy2-naphthoic acid synthase, Rv0548c) (108–110) (Table 1), MenE (o-succinylbenzoyl-CoA synthase, Rv0542c) (111–113), and MenA (Rv0534c) (114, 115) from mycobacteria have been studied as potential drug targets.

The isoprenoid tail of the menaquinone must be generated by a prenyldiphosphate synthase as described above, and together with 1,4-dihydroxy-2-naphthoic acid is the substrate for MenA (Rv0534c). However, the specific prenyldiphosphate synthase generating this prenyldiphosphate has yet to be identified. As noted above, other functions have been assigned to the potential candidates, suggesting that additional study is required.

In addition, the saturation of the second isoprene unit from the head group of menaquinone in mycobacteria (Fig. 6) is not seen in *E. coli* or *Bacillus subtilis*. However, this modification is seen in many Gram-positive bacteria (100, 116). Based on the chemical mechanism of the prenyl diphosphate synthases, it is likely that this modification is introduced after the mature prenyldiphosphate is synthesized and potentially after the formation of demethylmenaquinone or menaquinone. There is a single report that cell free extracts of *Mycobacterium phlei* are capable of reducing MK-9 to MK-9 (II-H$_2$) (107). The reduction required either NADH or NADPH, but nothing further has been reported regarding the nature of this enzyme, and it is, as yet, unknown whether this modification is required for function in mycobacteria.

Sulfated menaquinone

Sulfated menaquinone, where the sulfate is found on the ω-end of the isoprenoid tail (Fig. 6), has been isolated from *M. tuberculosis* (103). The function of this unique lipid is, as yet, unknown. However, it has been reported that sulfated menaquinone, previously known as S881, negatively regulates the virulence of the organism in mouse infection models (117). It has been postulated that this molecule is synthesized from MK-9 (II-H$_2$) in at least two steps: (i) oxidation of the terminal position of the isoprenoid tail and (ii) sulfation of the resulting hydroxyl residue. It has been shown that the putative sulfotransferase encoded by *stf3* (Rv2267c) (Table 1), is required for the production of S881 (117) and has been hypothesized that Cyp128, encoded by

Rv2268c, hydroxylates the MK-9 (II-H$_2$). However, this remains to be definitively demonstrated, and Cyp124, encoded by *Rv2266*, has been shown to have appropriate ω-hydroxylase activity and a marked preference for lipids containing methyl branching such as isoprenoid compounds (118).

Carotenoids

The carotenoids of mycobacteria

Carotenoids are a diverse family of isoprenoids that typically have six to eight isoprene units. These molecules are structurally diverse but are similar in general structure with a long chain of conjugated double bonds. More than 700 carotenoids have been identified and are widespread among bacteria, including mycobacteria (Fig. 7). These often pigmented compounds play significant roles in protecting the organisms from oxidative damage and modify membrane fluidity (119, 120). The carotenoids can be divided into two classes based on the presence or absence of oxygen atoms. Carotenoids without oxygen atoms in the molecule are known as carotenes, whereas those with oxygen atoms in their structure are known as xanthophylls.

Many *Mycobacterium* species produce yellow, orange, or pink pigments in the dark (scotochromogens) or in the light (photochromogens), although these pigments may not be visible in culture. Very early on, mycobacteria were shown to contain carotenoid pigments (see reference 121 for a review). Chargaff reported the presence of carotenoid pigments in *M. phlei* in 1930, and subsequent analysis showed that the major carotenoid in *M. phei* was leprotene (or isoneriatene), a carotene that was first isolated from an organism mistakenly identified as *M. leprae* (96). In addition, many bacteria, including mycobacteria, produce carotenoid glycosides, which act as surfactants, stabilize membranes, and possibly contribute to regulating the permeability of membranes to oxygen (122–125). The first complete structure of glycosylated carotenoids, phleixanthophyll and 4-keto-phleixanthophyll isolated from *M. phlei*, was determined in 1967 (126).

Carotenoid biosynthesis

Carotenoid synthesis is well understood in many microorganisms (reviewed in reference 127) but has received limited attention in mycobacteria; however, the generally accepted pathway for carotenoid synthesis in mycobacteria, reviewed by Minnikin (96), appears to be similar to that of most nonphotosynthetic microbes (127). That is, the pathway consists of a geranylgeranyldiphosphate synthase, phytoene synthase, phytoene

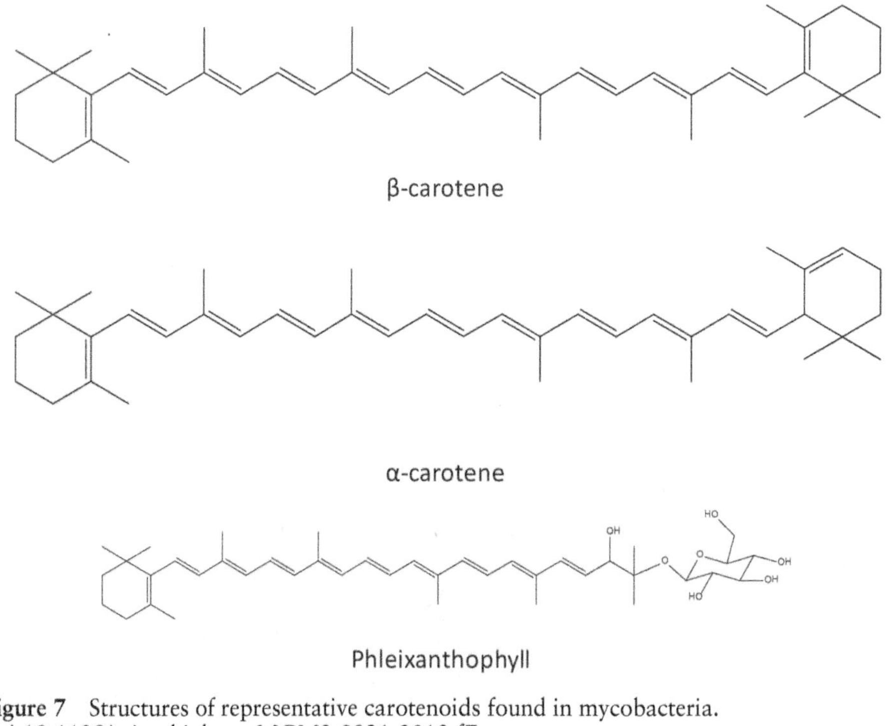

β-carotene

α-carotene

Phleixanthophyll

Figure 7 Structures of representative carotenoids found in mycobacteria.
doi:10.1128/microbiolspec.MGM2-0021-2013.f7

dehydrogenase, and lycopene cyclase. In the carotenoid literature these enzymes are designated CrtE, CrtB, CrtI, and CrtY, respectively. It should be noted that in nonphotosynthetic bacteria, CrtI catalyzes multiple dehydrogenations (usually two to four) that generate the conjugated double bond system and that there are multiple CrtY type cyclases with multiple designations (127). Once lycopene has been generated in mycobacteria, the pathway splits to form α- and β-carotene (96), one of which is presumably the precursor of leprotene.

As described above, orthologs of *Rv0562* and *Rv3383c*, both of which are reported to encode *E,E,E*-geranylgeranyldiphosphate synthases (87), have the potential to provide the CrtE functionality in mycobacteria. Studies, aimed primarily at the development of genetic tools for manipulating mycobacteria, have provided information about other genes and enzymes involved in carotenoid synthesis in mycobacteria as well. Thus, orthologs of CrtB, CrtI, and CrtY have been identified in *M. marinum* (128, 129) and *M. aurum* (130, 131). In addition, an ortholog of CrtU, a β-carotene desaturase, has been reported in *M. aurum* (131), and a carotenoid oxygenase, Rv0654, has been identified in *M. tuberculosis* (132). In terms of regulation of carotenoid synthesis in mycobacteria, orthologs of *crtR* and *crtP* encode a putative repressor and a positive regulator, respectively, in *M. marinum* and *M. tuberculosis* (128), and SigF controls carotenoid production in *M. smegmatis* (133). Details regarding

carotenoid synthesis in *M. tuberculosis* are not clear. The *M. tuberculosis* H37Rv genome encodes an ortholog of CrtB (PhyA), which may be nonfunctional (129).

Noncarotenoid Cyclic Isoprenoids

A novel class of cyclic C_{35} terpenes isolated from nonpathogenic *Mycobacterium aichiense, Mycobacterium chlorophenolicum, Mycobacterium parafortuitum, M. smegmatis, Mycobacterium thermoresistible*, and *Mycobacterium vanbaalenii* has been described (84, 134). These compounds, designated heptaprenylcyclines (Fig. 8), are synthesized via the cyclization of ω,*E,polyZ*-heptaprenyldiphosphate or ω,*E,E,polyZ*-heptaprenyldiphosphate; thus, the prenyldiphosphate synthases described in these species are likely involved in the production of these molecules, but little else is currently known about their synthesis or function.

A labdane-related diterpenoid compound, isotuberculosinol (Fig. 8), is produced by *M. tuberculosis*. This molecule appears to be immunomodulatory because it has been shown to block phagosome maturation in macrophages (135, 136). This role was first suggested when genes encoding enzymes involved in isotuberculosinol synthesis, *Rv3377c* and *Rv3378c*, were identified in a screen for mutants defective in arresting phagosome maturation (137). Rv3377c (Table 1) was demonstrated to be a class II diterpene cyclase, catalyzing bicyclization and rearrangement of geranylgeranyldiphosphate to form halimadienyl/

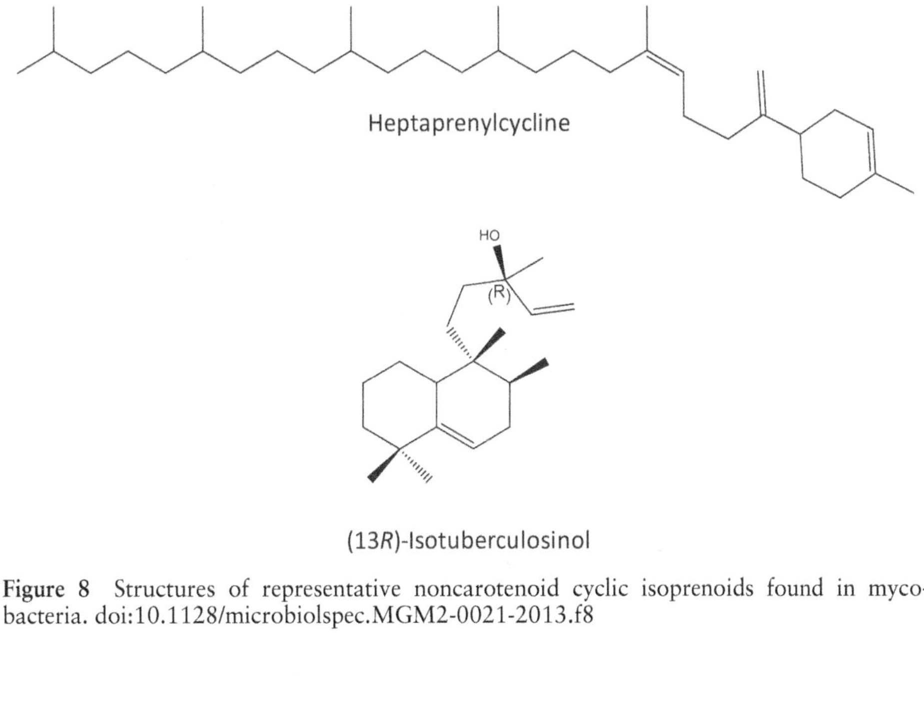

Heptaprenylcycline

(13*R*)-Isotuberculosinol

Figure 8 Structures of representative noncarotenoid cyclic isoprenoids found in mycobacteria. doi:10.1128/microbiolspec.MGM2-0021-2013.f8

tuberculosinyldiphosphate (138). It was then shown that halimadienyl/tuberculosinyldiphosphate was hydrolyzed to tuberculosinol and isotuberculosinol by Rv3378c (135, 136, 139, 140).

ACYLTREHALOSES

The outer membrane of mycobacteria contains a number of trehalose esters. Among them, trehalose monomycolates (TMMs) and trehalose dimycolates (TDMs; cord factor) are ubiquitously found across the *Mycobacterium* genus. Species-specific trehalose esters include di-, tri-, and poly-acyltrehaloses (DATs, TATs, and PATs); sulfolipids (SLs); and lipooligosaccharides (LOSs). Species-specific trehalose esters are found in the outermost capsule in addition to the outer membrane (8). TMM and TDM, in contrast, were identified in the surface-exposed capsular materials of *M. avium* and *M. smegmatis* but not in those of *M. tuberculosis*, *M. kansasii*, and *M. gastri*, indicating that they may be more deeply buried in the cell envelope of some *Mycobacterium* species (8). Interest in trehalose esters stems from their demonstrated or postulated roles in host-pathogen interactions and from their potential as diagnostic tools (for reviews see references 1, 141). The presence and abundance of species-specific acyltrehaloses (SL, DAT, TAT, and PAT) and phthiocerol dimycocerosates (PDIMs; see "PDIMs, PGLs, and Related Compounds," below) in the cell envelope of *M. tuberculosis* impact the ability of the bacilli to stain with the cationic dye neutral red (142, 143), a property known since Dubos and Middlebrook's early studies in the 1940s to correlate with virulence (144).

Here, we focus on steps in the formation of acyltrehaloses, including the biosynthesis of the fatty acyl substituents, their transfer onto trehalose, and what is known of the translocation of biosynthetic intermediates and end products across the cell envelope.

TMMs and TDM (Cord Factor)

In TMM and TDM, trehalose is esterified with long-chain α-branched β-hydroxy fatty acids known as the mycolic acids. The structure and biosynthesis of mycolic acids is reviewed in reference 364. Any structural type of mycolic acid may esterify positions 6 and 6′ of TDM and position 6 of TMM (Fig. 9). The biosynthesis of mycolic acids occurs in the cytoplasm, and so does that of trehalose. We recently identified MmpL3 (Rv0206c) (Table 2) as an inner membrane transporter required for the translocation of TMM to the periplasm, where TMM can then serve as a mycolic acid donor for the mycolylation of arabinogalactan and

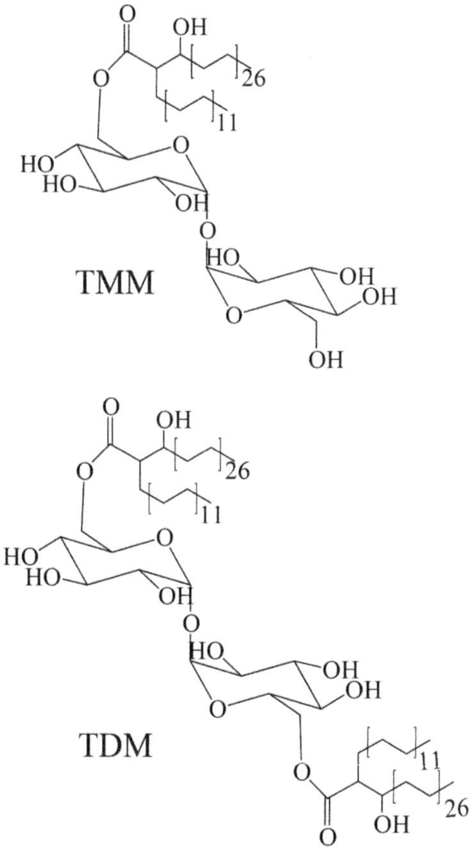

Figure 9   Structures of TMM and TDM.
doi:10.1128/microbiolspec.MGM2-0021-2013.f9

the formation of TDM (Fig. 1) (145, 146). This finding indicates that TMM is most likely the form under which mycolic acids are exported to the cell wall and outer membrane and, therefore, that TMM is probably made on the cytosolic side of the plasma membrane. The catalytic process underlying the cytoplasmic formation of TMM from fully elongated and functionalized mycolic acid chains and trehalose has not yet been elucidated. The subsequent synthesis of TDM from two TMM molecules and the transfer of mycolates to the nonreducing ends of arabinogalactan have been shown to involve antigens 85A (Rv3804c; FbpA), 85B (Rv1886c; FbpB), and 85C (Rv0129c; FbpC) (Table 2) (147–149). *In vitro*, these three mycolyltransferases display apparent redundant catalytic activities (147). Consistent with this finding, none of the *fbpA*, *B*, or *C* genes are individually required for the growth of *M. tuberculosis*. Their combined inactivation or chemical inhibition, however, leads to cell death (147, 150) (our unpublished data). Although the phenotypic characterization of *fbpA*, *B*, or *C* null mutants of *M. tuberculosis* and *M. smegmatis* indicates

Table 2 *M. tuberculosis* H37Rv genes involved in the biogenesis of trehalose mono- and dimycolates, sulfolipids, di- and poly-acyltrehaloses, and mannosyl-β-1-phosphomycoketides

Gene number	Gene name	Function	Evidence[a]	Reference
Rv0129c	*fbpC*	Mycolyltransferase (antigen 85C)	E, P	147–149
Rv0206c	*mmpL3*	Inner membrane transporter of the RND superfamily involved in the translocation of TMM	P	145, 146
Rv0295c	*Sft0*	Sulfotransferase responsible for the formation of the trehalose-2-sulfate moiety of sulfolipids	E, P	164
Rv0757-Rv0758	*phoP-phoR*	Two-component transcriptional regulator involved in the regulation of SL, DAT, and PAT	P	143, 176
Rv1180-Rv1181	*pks3-pks4*	Polyketide synthase responsible for the elongation of the methyl-branched mycosanoic and mycolipenic acids found in DAT, TAT, and PAT	P	196, 197
Rv1182	*papA3*	Acyltransferase catalyzing the sequential transfer of the first straight-chain saturated fatty acyl chain followed by the first mycolipenoyl group onto the 2- and 3-positions of trehalose, respectively, in the biosynthesis of DAT and PAT	E, P	199
Rv1183	*mmpL10*	Inner membrane transporter of the RND superfamily thought to be involved in the translocation of DAT and PAT	H	
Rv1184c	*chp2*	Acyltransferase thought to catalyze the last three acylations leading to the formation of PAT from DAT	H	
Rv1185c	*fadD21*	Putative fatty acid AMP ligase providing Pks3/4 with activated long-chain fatty acid starter units	H	
Rv1662-Rv1663	*pks8-pks17*	Polyketide synthase responsible for the elongation of the monomethyl-branched unsaturated C_{16} to C_{20} fatty acids found in DAT and PAT	P	162
Rv1886c	*fbpB*	Mycolyltransferase (antigen 85B)	E	147, 149
Rv2048c	*pks12*	Polyketide synthase involved in the elongation of the alkyl backbone of mycoketides	E, P	218, 220
Rv3416	*whiB3*	Regulator of SL, DAT, and PAT synthesis	P	174
Rv3804c	*fbpA*	Mycolyltransferase (antigen 85A)	E, P	147, 149, 152
Rv3820c	*papA2*	Acyltransferase catalyzing the transfer of the first straight-chain saturated fatty acyl chain onto trehalose-2-sulfate in the biosynthesis of sulfolipids	E, P	165
Rv3821	*sap*	Integral membrane protein thought to facilitate the translocation of SL-1 to the cell surface	P	168
Rv3822	*chp1*	Acyltransferase catalyzing the acylation at the 6- and 6′-positions of sulfolipids	E, P	168
Rv3823c	*mmpL8*	Inner membrane transporter of the RND superfamily involved in the translocation of sulfolipids	P	168, 170, 171
Rv3824c	*papA1*	Acyltransferase catalyzing the transfer of the first (hydroxy)phthioceranoyl group at the 3′-position of the product of PapA2	E, P	165
Rv3825c	*pks2*	Polyketide synthase responsible for the elongation of the methyl-branched phthioceranic and hydroxyphthioceranic acids found in sulfolipids	P	166
Rv3826	*fadD23*	Putative fatty acid AMP ligase providing Pks2 with activated long-chain fatty acid starter units	H	

[a]E, P, H: see Table 1 footnote.

that the function of these genes may in fact only partially overlap in whole cells, to date, the precise contribution of each of the three paralogs to the transfer of mycolic acids to their cell wall and outer membrane glycolipid acceptors remains unclear. FbpC appears to be essentially involved in the transfer of mycolic acids to arabinogalactan, and FbpA, in the formation of TDM (148, 149, 151–153).

Numerous biological activities have been associated with the TDM from tuberculous and nontuberculous mycobacteria both *in vitro* and *in vivo* (for reviews see references 1, 154–157). In fact, TDM seems to be a major contributor to the inflammation seen in mycobacterial infections. TDM contributes to protecting *M. tuberculosis* from killing by macrophages, is a potent modulator of the activation of macrophages, stimulates the formation of lung granulomas, and enhances the resistance of mycobacteria to antibiotics (152, 154, 156, 158, 159). The binding of TDM from *M. tuberculosis* to the C-type lectin Mincle is required for activation of macrophages and granuloma formation (158, 160). It is important to note that the biological activities of TDM are very dependent on the fine structure of their mycolyl substituents (156, 161).

Sulfolipids

Sulfolipids (SLs), also known as sulfatides and sulfoglycolipids, are sulfated trehalose esters that are acylated with three or four acyl groups consisting of one middle-chain saturated fatty acid (palmitic or stearic acid) at the 2-position and different combinations of the hepta- and octa-methyl-branched phthioceranic and hydroxyphthioceranic acids (C_{31} to C_{46}) at the 3-, 6-, and 6′-positions. Monomethyl-branched unsaturated C_{16} to C_{20} fatty acids have also been found as minor constituents of SL (162). Sulfolipid-1 (SL-1), whose structure is shown in Fig. 10, is the most abundant form of sulfolipid produced by *M. tuberculosis* (163). This family of lipids is exclusively found in the human pathogen *M. tuberculosis*.

The genes involved in the biogenesis of SL-1 have for the most part been identified and, with the exception of the sulfotransferase Sft0, found to cluster on the chromosome of *M. tuberculosis* (Table 2). The sulfotransferase Sft0 (Rv0295c) catalyzes the first committed step in the pathway by sulfating trehalose to form trehalose-2-sulfate (164). The acyltransferase PapA2 (Rv3820c) then catalyzes the esterification of trehalose-2-sulfate with a straight-chain saturated fatty acid (e.g., palmitic acid) at the 2′-position to generate a monoacyl intermediate, SL_{659} (165). The polyketide synthase Pks2 (Rv3825c) synthesizes the methyl-branched

phthioceranic and hydroxyphthioceranic acids (166), most likely using an activated long-chain fatty acid starter unit (an acyl-adenylate) provided by the fatty acid AMP ligase FadD23 (Rv3826) (167). The polyketide-associated protein-1 (PapA1; Rv3824c) catalyzes the transfer of the first (hydroxy)phthioceranoyl group at the 3′-position of the product of PapA2, yielding a diacylated form of SL known as SL_{1278} (165). The additional two acylations at the 6- and 6′-positions of SL_{1278} are catalyzed by the acyltransferase Chp1 (Rv3822) (168). PapA1 and PapA2 are related to the acyltransferase PapA5, which esterifies phthiocerol with mycocerosic acids in the biosynthesis of PDIM (see "PDIMs, PGLs, and Related Compounds," below). Chp1 (cutinase-like hydrolase protein-1), in contrast, more closely resembles cutinase-like proteins (168). All three acyltransferases are essential for the synthesis of SL-1 as demonstrated by the absence of fully elaborated SL-1 from the corresponding knockout mutants (165, 168, 169).

Evidence of the involvement of MmpL8 (Rv3823c), an inner membrane transporter of the RND (resistance, nodulation, and division) superfamily, in the translocation of SL-1 to the cell surface was provided in 2003–2004 by two independent research groups (170, 171). *M. tuberculosis* mmpL8 knockout mutants failed to produce SL-1 and instead accumulated the diacylated SL_{1278} intracellularly. A possible interpretation of this finding was that the first two acylation steps catalyzed by PapA2 and PapA1 occurred on the cytoplasmic side of the plasma membrane, whereas the two subsequent acylations catalyzed by Chp1 and yielding SL-1 required the prior MmpL8-mediated translocation of the diacylated SL_{1278} precursor across the plasma membrane. This model was, however, recently revised in light of the finding that the catalytic domain of the membrane-associated acyltransferase Chp1 is cytosolic and that its activity is potentiated by MmpL8 (168). These observations are consistent with a model similar to that proposed for PDIM (see "PDIMs, PGLs, and Related Compounds," below) wherein the biosynthesis and transport of SL-1 is coupled and MmpL8 acts as a scaffold for a cytoplasmically oriented macromolecular complex consisting of the SL biosynthetic machinery. Further support for this assumption was recently obtained by Zheng et al. (172) in identifying MmpL8 among the components of a membrane-associated protein complex containing Pks2, PapA1, and FadD23 in *M. bovis* BCG. Sap (sulfolipid-1-addressing protein) (Rv3821) is an integral membrane protein that appears to facilitate the translocation of SL-1 to the cell surface. Its disruption in *M. tuberculosis* causes the intracellular

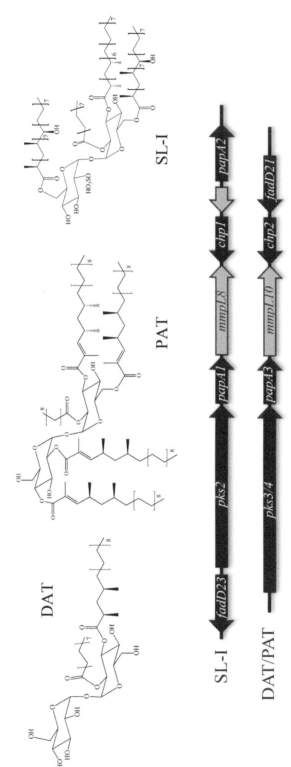

Figure 10 Structures of SLs, DATs, and PATs and biosynthetic gene clusters. The major sulfolipid, SL-I (2,3,6,6'-tetraacyl α-α'-trehalose-2'-sulfate), is represented. In SL-I, trehalose is sulfated at the 2' position and esterified with palmitic acid and the multimethyl-branched phthioceranic and hydroxyphthioceranic acids. In DAT (2,3-di-O-acyltrehalose), trehalose is esterified with palmitic acid and the multimethyl-branched mycosanoic acid. In PAT, trehalose is esterified with palmitic acid and the multimethyl-branched mycolipenic acids. doi:10.1128/microbiolspec.MGM2-0021-2013.f10

build-up of SL_{1278} similar to that observed in *mmpL8* knockouts, although the *sap* mutant retains the ability to synthesize small amounts of SL-1 (168). Beyond MmpL8 and Sap, it is likely that the translocation of SL-1 to the cell surface requires additional periplasmic and/or outer membrane transporters, but their identity is at present not known.

SL production appears to be regulated in *M. tuberculosis*, but the environmental factors governing the synthesis of these glycolipids are still poorly understood. Supporting a role for SL during host infection, the expression of the *pks2* gene was found to be strongly upregulated upon phagocytosis of *M. tuberculosis* by human primary macrophages (173). It appears that one of the roles of methyl-branched fatty acid–containing lipids such as PDIM, SL, DAT, and PAT during infection is to alleviate the propionate-mediated stress undergone by *M. tuberculosis* when the bacterium switches to host cholesterol as a major carbon source (174, 175). The propionyl-CoA generated upon β-oxidation of cholesterol is converted to methylmalonyl-CoA by the propionyl-CoA carboxylase, which is then used by dedicated polyketide synthases such as Pks2, Mas, and Pks3/4 (see below) in the elongation of the methyl-branched fatty acids found in PDIM, SL, DAT, and PAT. The regulator facilitating this metabolic switching to fatty acids was identified as WhiB3 (Rv3416), which binds the promoter region of *pks2* (174). Another important regulator of SL production is the two-component transcriptional regulator PhoP-PhoR (*Rv0757-Rv0758*). PhoP-PhoR positively regulates the synthesis of SL, and *M. tuberculosis* mutants deficient in the expression of this regulator are totally deficient in SL-1 production (143, 176). It was shown that a mutation in the *phoP* gene of *M. tuberculosis* H37Ra accounts for the inability of this avirulent strain to produce SL-1 (177). PhoP binds the promoter region of *pks2 in vitro* (178, 179).

The restriction of SL-1 to the human pathogen *M. tuberculosis* together with the observation some 50 years ago of a positive correlation between the levels of SL-1 produced by *M. tuberculosis* clinical isolates and their virulence in animal models has prompted extensive research aimed at elucidating the biological functions of sulfolipids during host infection (for reviews see references 141, 163, 180, 181). Numerous and sometimes controversial activities were associated with purified SL-1 molecules. Among these, the ability of SL-1 to potentiate the toxicity of TDM in mice, to inhibit mitochondrial oxidative phosphorylation, to prevent phagosome-lysosome fusion in cultured macrophages, and to modulate the oxidative and cytokine

responses of human monocytes and neutrophils has probably received the most attention. In more recent years, the diacylated SL biosynthetic precursor SL_{1278} was shown to stimulate CD1b-restricted T cells through mechanisms dependent on the number of C-methyl substituents on the fatty acyl chains, the configuration of the chiral centers, and the length and respective localization of the two acyl chains on the sugar moiety (182, 183). In the last decade, the elucidation of the biosynthetic pathway of SL finally allowed the generation of isogenic mutants of *M. tuberculosis* specifically deficient in their synthesis and an evaluation of the roles of these glycolipids during infection when carried by whole bacilli.

Unexpectedly, *pks2*, *papA1*, and *papA2* knockout mutants, which all lack fully elaborated SL-1 while in some cases retaining the ability to synthesize sulfated trehalose and mono- and/or di-acylated forms of SL, were found to be undistinguishable from their wild-type parents in their ability to replicate and persist in mice or guinea pigs (165, 184). In contrast, three independent studies indicated that *mmpL8* knockout mutants that accumulate diacylated SL_{1278} at the periphery of the plasma membrane display some level of attenuation in mice, although the attenuation phenotypes considerably differed between studies, possibly as a result of the different *M. tuberculosis* strains and models of infection that were used (170, 185, 186). Recently, Gilmore et al. (187) provided evidence that a *sftO* null mutant of *M. tuberculosis* survives better than its wild-type parent in human but not in murine macrophages, possibly as a result of the increased resistance of this strain to human antimicrobial peptides. These results suggest that SL may only have a detectable impact on infection in the human host.

DATs and PATs

The 2,3-di-*O*-acyltrehaloses (DATs) consist of trehalose acylated at the 2-position with one middle-chain saturated fatty acid (C_{16} to C_{19}) and at the 3-position with the di-methyl-branched mycosanoic acids (C_{21} to C_{25}) (Fig. 10). In other less common forms of DAT, the tri-methyl-branched C_{25} to C_{27} mycolipenic (phthienoic) or mono-hydroxylated tri-methyl-branched C_{24} to C_{28} mycolipanolic acids replace the mycosanoic acids (188–190). 2,3,6-Triacyltrehaloses (TATs) harboring stearic, palmitic, and mycolipenic acyl substituents have also been reported in *M. tuberculosis* (191). Polyacyltrehaloses (PATs) are trehalose esters acylated with five acyl groups consisting of one middle-chain saturated fatty acid (C_{16} to C_{19}) at the 2-position and different combinations of the tri-methyl-branched

C_{27}-mycolipenic and C_{27}-mycolipanolic acids at the 2′, 3′, 4, and 6′-positions (Fig. 10) (188, 192). Monomethyl-branched unsaturated C_{16} to C_{20} fatty acids have also been found as minor constituents esterifying PAT and DAT (162). So far, the mycolipenic acyl substituents found in DAT, TAT, and PAT have only been isolated from virulent isolates of the *M. tuberculosis* complex species *M. tuberculosis*, *M. bovis*, and *M. africanum* but were not found in the avirulent laboratory strain *M. tuberculosis* H37Ra or in the vaccine strain *M. bovis* BCG. While 2,3-diacyltrehaloses and 2,3,4- and 2,3,6-triacyltrehaloses may be found in nonpathogenic species of mycobacteria such as *M. fortuitum*, the fatty acyl substituents identified in this species consist of straight-chain (C_{14} to C_{18}) and mono-methyl-branched unsaturated C_{16} to C_{20} fatty acids (193, 194).

As their relative distribution to pathogenic species of the *M. tuberculosis* complex may suggest, DAT, TAT, and PAT are biologically active molecules capable of modulating a number of host immune responses *in vitro* (141, 195). Their precise role during host infection remains, however, poorly understood. Phenotypic observations made on a mutant of *M. tuberculosis* deficient in the biosynthesis of DAT and PAT indicated a role for these lipids in the retention of the capsular material at the cell surface (196, 197). The modification of the surface properties of the mutant affected its binding and uptake by phagocytic and nonphagocytic cells, but preliminary infection studies indicated that the mutant did not significantly differ from its wild-type parent in its ability to replicate and persist in cultured macrophages and in mice (197). Interestingly, increased binding to phagocytic cells was also reported in the case of an SL-deficient mutant of *M. tuberculosis* (198). It is thus likely that the different families of acyltrehaloses produced by *M. tuberculosis* have partially redundant activities in whole cells hampering the clear delineation of their individual contribution to virulence and other physiological functions. Independent from their binding or immunomodulatory properties and as noted above, methyl-branched fatty acid–containing lipids such as PDIM, SL, DAT, and PAT appear to play an important role in alleviating the propionate-mediated stress undergone by *M. tuberculosis* when the bacterium utilizes host cholesterol as a major carbon source during infection (174, 175). Consistently, WhiB3 acts as a positive transcriptional regulator of *pks3/4* in addition to *pks2* (174).

Gene knockout studies indicated that the polyketide synthase encoded by *pks3/4* (*Rv1180/Rv1181*) is responsible for the elongation of mycosanoic and mycolipenic acids, while *pks8* and *pks17* (*Rv1662* and *Rv1663*) together encode the polyketide synthase producing monomethyl-branched unsaturated C_{16} to C_{20} fatty acids (162, 196, 197) (Table 2). An *M. tuberculosis* mutant deficient in the expression of *pks3/4* failed to produce PAT and DAT (196, 197). In some *M. tuberculosis* strains, an intervening stop codon in *pks3/4* results in two separate open reading frames (ORFs) (annotated as *pks3* and *pks4*). Strains containing this mutation do not synthesize PAT (186). Striking resemblance in the genetic organization of the regions encompassing the polyketide synthase gene *pks3/4* and that involved in SL (Fig. 10) and, to a lesser extent, PDIM biosynthesis (see "PDIMs, PGLs, and Related Compounds," below) are suggestive of the involvement of *fadD21* (*Rv1185c*), *mmpL10* (*Rv1183*), and *Rv1184c* (*chp2*) in the assembly and export of DAT and PAT (Table 2). To date, however, only *papA3* (*Rv1182*) has been characterized (199). It encodes the acyltransferase responsible for the sequential transfer of a palmitoyl group at the 2-position of DAT/PAT followed by a mycolipenoyl group at the 3-position (Fig. 10). As is the case for SL, the two-component transcriptional regulator PhoP-PhoR (*Rv0757-Rv0758*) positively regulates the synthesis of DAT and PAT, and *M. tuberculosis* mutants deficient in the expression of this regulator are totally deficient in DAT and PAT production (143, 176). The same mutation in the *phoP* gene of *M. tuberculosis* H37Ra that accounts for the inability of this strain to produce SL also accounts for the absence of DAT and PAT from this avirulent *M. tuberculosis* isolate (177). PhoP was shown to bind the promoter regions of *pks3/4* and *fadD21* (178, 179).

Lipooligosaccharides

Lipooligosaccharides (LOSs) are surface-exposed glycolipids (8) produced by a number of *Mycobacterium* species (200). They were first found in *M. kansasii* (201) and *M. smegmatis* (202), then in nine other mycobacterial species (200), including "*M. canettii*" and related strains of the *M. tuberculosis* complex (203). LOSs are otherwise virtually absent from *M. tuberculosis* strains *sensu stricto* such as H37Rv (203).

LOSs (Fig. 11A and 11B) share a poly-O-acylated trehalose core further glycosylated by a mono- or, more frequently, an oligosaccharidyl unit (200). Similar to the situation in other trehalose-based mycobacterial glycolipids such as sulfolipids and di- or tri-acyltrehaloses, the trehalose moiety of LOS is invariably acylated by polymethyl-branched fatty acids that can be either saturated, e.g., in "*M. canettii*," or unsaturated, e.g., in *M. smegmatis*.

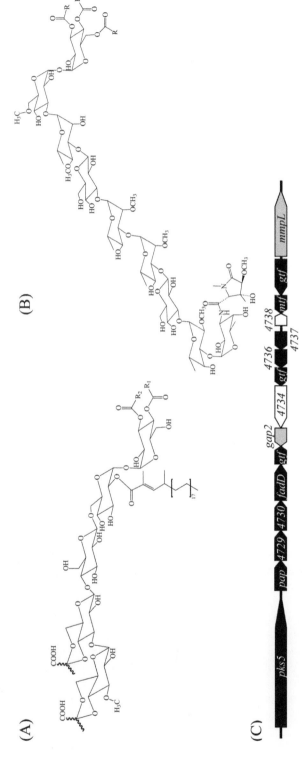

Figure 11 Structures of (A) major LOS (LOS-A) of *M. smegmatis* ATCC 356 (R₁ and/or R₂ : octanoic acid and tetra- or hexa-decanoic acid) and (B) *M. tuberculosis* "canettii"; R = Ac. (C) LOS biosynthetic gene cluster of *M. smegmatis* mc²155. Shown is the 25.15-kb region spanning MSMEG_4727 (*pks5*) to MSMEG_4741 (*mmpL*). ORFs are depicted as arrows. Black arrows indicate genes encoding biosynthetic enzymes; gray arrows show indicate putative transporter genes; white arrows show hypothetical genes of unknown function. Abbreviations: Pks5, Mas-like polyketide synthase; Pap, putative acyltransferase; MSMEG_4729 and MSMEG_4730, putative acyltransferases; FadD, putative acyl-CoA synthase; Gtf (MSMEG_4732), putative glycosyltransferase; Gap2, putative transmembrane protein involved in glycolipid translocation; MSMEG_4734, hypothetical PE/PPE-like protein; Gtf (MSMEG_4735), putative glycosyltransferase; MSMEG_4736 and MSMEG_4737, putative pyrruvylyl transferases; MSMEG_4738, hypothetical protein; Mtf, possible O-methyltransferase; Gtf (MSMEG_4740), putative glycosyltransferase; MmpL, putative inner membrane transporter. doi:10.1128/microbiolspec.MGM2-0021-2013.f11

The biosynthesis of LOS molecules is still poorly understood, with only a few genes experimentally demonstrated to be involved in their elongation and assembly (204–206). The synthesis of polymethyl-branched fatty acids invariably requires a polyketide synthase (Pks) that uses methylmalonyl-CoA instead of malonyl-CoA as the elongation unit, resulting in the formation of a polymethyl branched aliphatic chain. The *MSMEG_4727* (*pks5*) gene, whose sequence is 65.6% identical to that of the *M. tuberculosis* Mas-like gene *Rv1527c*, was involved in the biosynthesis of LOS in *M. smegmatis* (207). The genomic surroundings of *pks5* from *M. smegmatis* resemble those described earlier for other acyltrehaloses (see subsections covering SLs, DATs, and PATs in "Acyltrehaloses," above, and Fig. 9) in that *pap*- and *fadD*-like genes likely to be required for the activation and transfer of the acyl groups of LOS (206), and an *mmpL* gene putatively involved in the translocation of these lipids, are found (Fig. 11C). In addition, genes whose products were tentatively annotated as polysaccharide pyruvyltransferases are found in the biosynthetic cluster of pyruvylated LOS-producing species such as *M. smegmatis* (202, 208). Consistent with the finding of various methylated glucosyl residues in LOS (Fig. 11A), genes encoding putative glycosyltranferases and O-methyltransferases also map in the vicinity of *pks5* (Fig. 11C). In *M. marinum*, several of these have been characterized (204, 205, 209). It is noteworthy that orthologs of five of the *M. smegmatis* LOS-related genes (*pks*, *pap*, *fadD*, *mmpL*, and *gap*) are conserved in the corresponding biosynthetic gene clusters of *M. marinum* and *M. tuberculosis* (205). Interestingly, homologous genes are also found in the glycopeptidolipid (GPL) biosynthetic gene clusters of *M. smegmatis* (210), *M. abscessus* (211), *M. chelonae* (211), and *M. avium* (212) (see "Glycopeptidolipids," below). This conserved set of genes may delineate the minimum biosynthetic machinery required for the synthesis and export of GPL- and LOS-type glycolipids

in mycobacteria. The remaining ORFs identified in the confirmed or putative mycobacterial LOS biosynthetic clusters are less conserved, an observation consistent with the fact that LOSs differ from other mycobacterial glycolipids in terms of the number and nature of their sugar constituents (200). Recently, the regulatory protein WhiB4 from *M. marinum* was associated with LOS biosynthesis, but its precise function is not known (213).

LOSs are highly antigenic molecules (203). Recent observations suggest that they play an important role in retaining proteins at the cell surface of some *Mycobacterium* species such as *M. marinum* (213). Their precise role in the colony morphology of mycobacteria is still a matter of debate and seems to be species-specific (205, 214, 215). In *M. marinum*, for instance, LOSs have clearly been associated with colony morphology, sliding motility, biofilm formation, and the ability of this *Mycobacterium* species to enter macrophages (205). The *M. marinum* LOSs are also endowed with immunomodulatory activities (216) and modulate virulence in the zebrafish embryo model of infection (213).

MANNOSYL-β-1-PHOSPHOMYCOKETIDES

Mannosyl-β-1-phosphomycoketides consist of a mannosyl-β-1-phosphate moiety reminiscent of polyprenol phosphomannose (the lipid-linked mannose donor) and an alkyl chain of varying length (C_{30} to C_{34}) made of a fully saturated 4,8,12,16,20-pentamethylpentacosyl unit (Fig. 12). Mycoketides were first isolated from *M. avium* based on their ability to activate human CD1c-restricted T-cells (217). This family of lipids was later identified in the slow-growing pathogenic species *M. tuberculosis* and *M. bovis* BCG but not in the rapidly growing saprophytes, *M. phlei*, *Mycobacterium fallax*, and *M. smegmatis* (218). Under standard liquid culture conditions, mycoketides are produced in minute amounts and are found both inside the cells and released in the culture medium (219). Their restricted

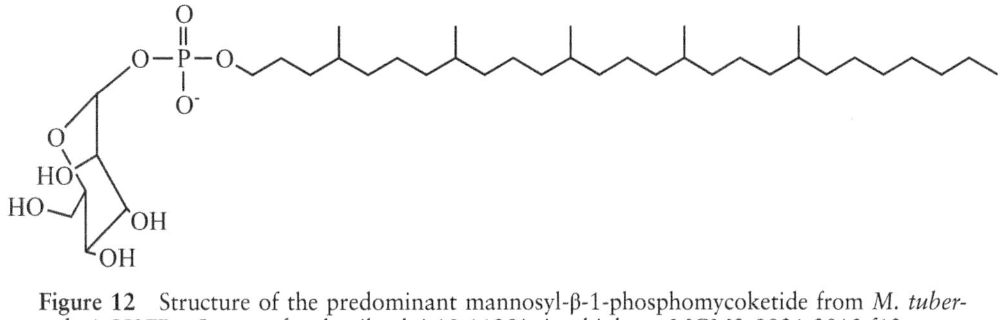

Figure 12 Structure of the predominant mannosyl-β-1-phosphomycoketide from *M. tuberculosis* H37Rv. See text for details. doi:10.1128/microbiolspec.MGM2-0021-2013.f12

distribution to pathogenic slow-growing *Mycobacterium* species are suggestive of an involvement in pathogenicity, and several studies aimed at comparing the virulence of mycoketide-deficient mutants of *M. tuberculosis*, *M. avium*, and *M. marinum* to that of their wild-type parents in animal models of infection have provided support for this assumption (219). In addition to their potential role in modulating the host immune response, mycoketides were proposed to be mycobacterial secondary metabolites acting as signaling factors to regulate cell division and virulence and to contribute to the suppression of phagosomal acidification (219).

The alkyl backbone of mycoketides is elongated by the polyketide synthase Pks12 (Rv2048c) (Table 2) (218, 220). Pks12 is the largest predicted protein of *M. tuberculosis* (430 KDa) and consists of two complete sets of fatty acid synthase–like catalytic domains capable together of using alternating C_2 (malonyl-CoA) and C_3 (methylmalonyl-CoA) units to elongate the alkyl backbone of mycoketides. After five cycles of C_3 and C_2 chain elongation, it is believed that the alkyl chain is released from the polyketide synthase upon hydrolysis, yielding mycoketidic acid, which is further reduced to the corresponding long-chain alcohol, mycoketide, and finally phosphorylated and mannosylated to generate mannosyl-β-1-phosphomycoketides (219). The enzymes catalyzing the hydrolysis, reduction, phosphorylation, and mannosylation steps have not yet been identified. The finding of orthologs of *pks12* in *M. marinum*, *M. ulcerans*, *M. avium paratuberculosis*, and several species of the *M. tuberculosis* complex suggests that the production mannosyl-β-1-phosphomycoketides may be a common feature of slow-growing mycobacterial pathogens.

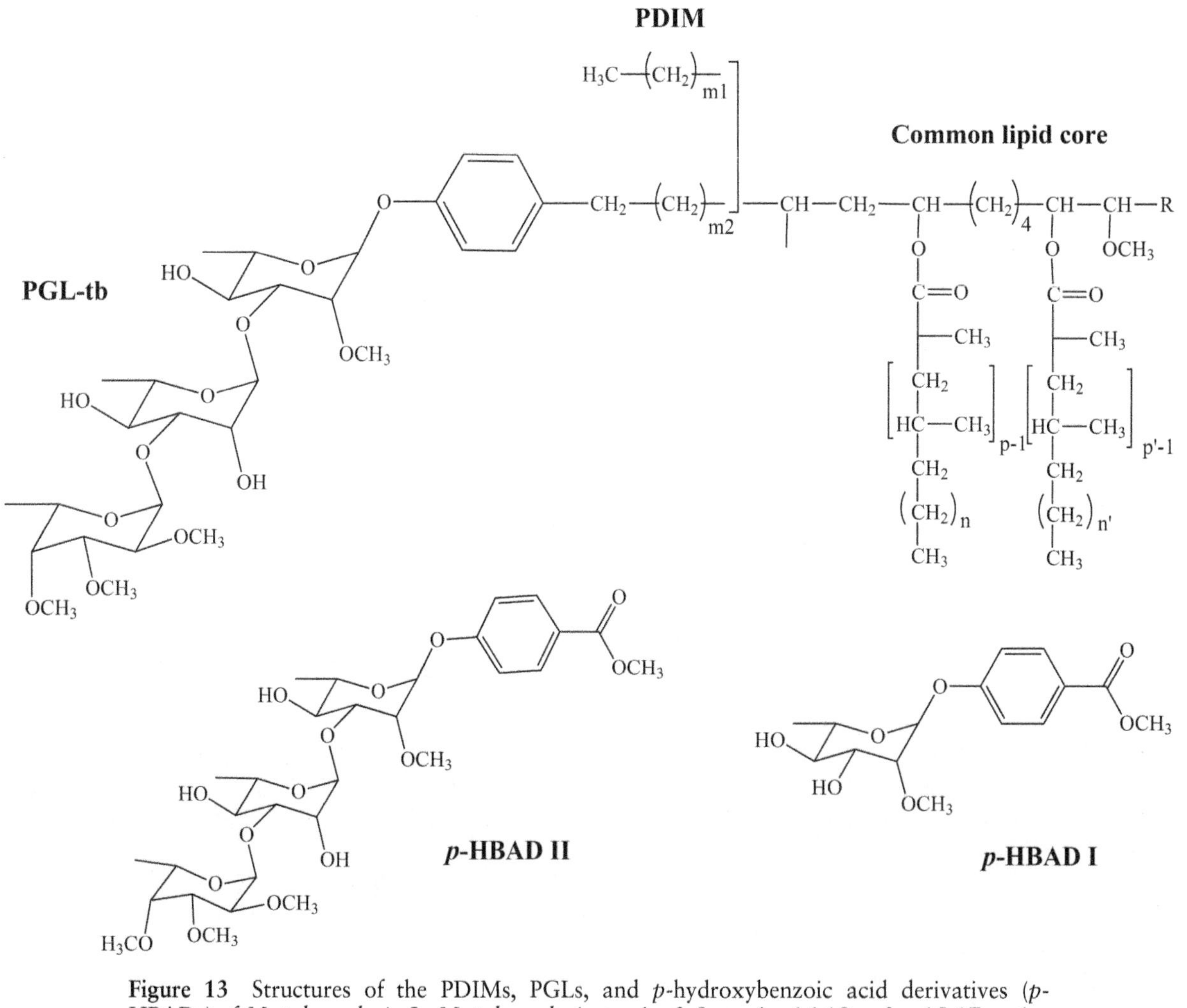

Figure 13 Structures of the PDIMs, PGLs, and *p*-hydroxybenzoic acid derivatives (*p*-HBADs) of *M. tuberculosis*. In *M. tuberculosis*, p, p′ = 3-5; n, n′ = 16-18; m2 = 15-17; m1 = 20-22; R = CH₂-CH₃ or CH₃. doi:10.1128/microbiolspec.MGM2-0021-2013.f13

PDIMs, PGLs, AND RELATED COMPOUNDS

Phthiocerol Diesters and Related Compounds: Structures, Distribution, and Cell Localization

Phthiocerol dimycocerosates (PDIMs) and diphthioceranates are part of a family of long-chain C_{33}-C_{41} β-diols (phthiocerols) esterified by two moles of polymethyl-branched (C_{27}-C_{34}) fatty acids. When the configuration of the asymmetric centers bearing the methyl branches belong to the D series, the fatty acids are called mycocerosic acids, whereas those of the L series are known as phthioceranic acids (14) (Fig. 13). The major β-diols (phthiocerol A) are usually accompanied by structural variants of these alcohols containing either a keto group in place of the methoxy group (phthiodiolone A) or a methyl group rather than an ethyl group at the terminus of the molecules and near the methoxyl group (phthiocerol B) (Fig. 13). To date, PDIMs have been found in *M. tuberculosis*, *M. bovis*, *M. leprae*, *M. microti*, *M. kansasii*, *M. gastri*, and *M. haemophilum*, whereas diphthioceranates have been found in *M. ulcerans* and *M. marinum* (221).

Glycosylated phenolic derivatives of PDIM and DIP, called phenolic glycolipids (PGLs), are found in the same species, although they may not be present in all strains; for instance, the PGL from *M. tuberculosis* (PGL-tb) has only been identified in the "canettii" strain (222, 223) and in some East-Asian/Beijing isolates (224, 225). In PGLs, the β-diols (phenolphthiocerols) are esterified by two moles of polymethyl-branched (C_{27}-C_{34}) fatty acids, except in Beijing strains, where a palmitic acid is found esterifying the additional hydroxyl group occurring in the aliphatic core of phenolphthiotriol (225). The glycosyl moiety of PGL is composed of one to four sugar residues depending on the species, most of which are O-methylated deoxysugars (14, 200). Identical PGL structures may be found in phylogenetically related mycobacterial species, for instance, in species of the *M. tuberculosis* complex (*M. bovis*, *M. microti*, *M. pinnipeddii*, and *M. africanum*), *M. kansasii* and *M. gastri*, and *M. marinum* and *M. ulcerans* (200, 226). The glycosyl moiety of PGL was also found attached to *p*-hydroxybenzoic acid, i.e., as methyl esters, to form *p*-hydroxybenzoic acid derivatives (*p*-HBADs) in *M. tuberculosis* and *M. bovis* BCG (Fig. 13) (223). In an attempt to correlate the lipid content with the virulence of *M. tuberculosis* isolates, Goren and collaborators characterized a methoxylated phenolphthiocerol, the so-called attenuation indicator lipid (227). The correlation between the occurrence of this lipid and reduced virulence remains, however, unclear. This lipid and its unmethylated form were detected in East-Asian/Beijing strains and accumulated in all of the Indo-Oceanic strains of *M. tuberculosis* examined (228).

PDIMs and PGLs are found in the capsules of *M. tuberculosis* and other pathogenic mycobacteria (8). PDIMs are otherwise abundant components of the outer membrane of *M. tuberculosis*, where they contribute to its well-known impermeability (229). *p*-HBADs, in contrast, are released in culture filtrates and tend not to remain associated with the cell envelope (223).

Biosynthesis of PDIM and PGL

Biosynthesis of phthiocerol and related compounds

Common enzymes participate in the biosynthesis of the lipid core of PDIM and PGL (Fig. 13), where *n*-C_{22}-C_{24} fatty-acyl chains and *p*-hydroxyphenylalkanoates, respectively, are elongated to form the long-chain β-diols, phthiocerol or phenolphthiocerol. Coupled genetic and biochemical strategies have allowed much of the biosynthetic pathways of PDIM and PGL to be elucidated. On the basis of mutant phenotypes, genes such as *pks11* (230), *pks12* (231), *pks10* (232), *mb0100* (233), and *pks7* (234) have been associated with the biosynthesis of PDIM; however, in the absence of genetic complementation, definitive proof of their involvement in the pathway is lacking, and no clear biosynthetic roles have yet been assigned to these genes. The genes unambiguously demonstrated to participate in PDIM and PGL biosynthesis are shown in Fig. 14, and their specific function in the pathway is detailed in Fig. 15 and Table 3. They are clustered on a 73-kb fragment of the *M. tuberculosis* chromosome (Fig. 14), and the organization of this locus is apparently highly conserved in all PDIM/PGL-producing mycobacteria, with the exception of *M. leprae*, in which this locus is split into two loci.

The *M. tuberculosis* genome encodes 36 FadD proteins with homology to acyl-CoA synthases. As noted in "Acyltrehaloses," above, some of them map in the vicinity of *pks* genes. FadD26 is required for the synthesis of PDIM but not that of PGL in *M. tuberculosis* and *M. bovis* (229, 235, 236). The role of the FadD proteins in activating Pks substrates was demonstrated by Trivedi et al. (167), who established that FadD26 and FadD28 belong to a large family of fatty-acyl AMP ligases responsible for the activation of long-chain *n*-C_{22}-C_{24} fatty acids as acyl-adenylates. They showed that FadD26 loads the activated substrates directly onto PpsA. These substrates are then elongated with

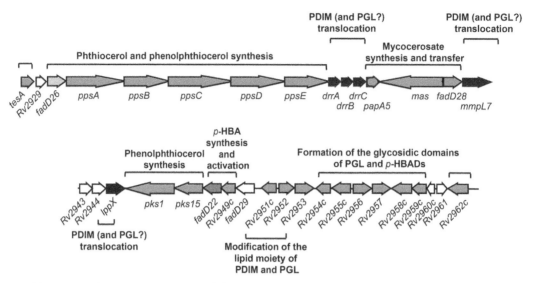

Figure 14 Genetic organization of the PDIM and PGL locus of *M. tuberculosis* H37Rv. ORFs are depicted as arrows. Black arrows indicate genes encoding biosynthetic enzymes; gray arrows indicate putative transporter genes; white arrows indicate hypothetical genes of unknown function. More details about the function of each gene are provided in Table 3 and Fig. 15. Adapted from reference 260. doi:10.1128/microbiolspec.MGM2-0021-2013.f14

malonyl-CoA and methylmalonyl-CoA by PpsA-PpsE to yield phthiocerol (Table 3; Figs. 14, 15).

The enzyme encoded by *Rv2949c* catalyzes the formation of *p*-hydroxybenzoic acid from chorismate (237). *p*-Hydroxybenzoic acid is activated by FadD22, which displays *p*-hydroxybenzoyl-AMP ligase activity (236) and is subsequently elongated by the type 1 polyketide synthase Pks15/1 to form *p*-hydroxyphenyl-alkanoates; the reaction may involve eight or nine elongation cycles using malonyl-CoA as the extender unit. A frameshift mutation within the *pks15/1* gene accounts for the lack of production of PGL by the *M. tuberculosis* reference strains H37Rv, Erdman, and CDC1551 (223). The fatty acyl-AMP ligase FadD29 activates *p*-hydroxyphenylalkanoates that are then transferred onto PpsA and finally elongated with malonyl-CoA and methylmalonyl-CoA by PpsA-PpsE to yield phenolphthiocerol (236) (Table 3). The type II thioesterase TesA is thought to be involved in the release of the phthiocerol and phenolphthiocerol moieties of PDIM and PGL, respectively, from the polyketide synthase PpsE (230, 238). The demonstrated interaction of TesA with the C-terminal half of PpsE tends to support this assumption (239).

Biosynthesis and transfer of mycocerosates

The mycocerosic acids that esterify the β-diols of phthiocerol and phenolphthiocerol are elongated from C_{16} and C_{20} fatty acids with three or four propionate

units by a dedicated type I polyketide synthase known as Mas (for mycocerosic acid synthase) (240–244) (Fig. 15; Table 3). Mas preferentially uses methyl-malonyl-CoA instead of malonyl-CoA for fatty acid elongation, thereby introducing methyl branches into the mycocerosic acid chain. In addition, the keto-reductase and enoylreductase activities of this enzyme require NADPH as a cofactor (241). FadD28 is the fatty-acyl AMP ligase responsible for the activation of the C_{16} and C_{20} fatty acid starter units of Mas as acyl-AMP and their transfer onto the polyketide synthase (167, 229, 235, 245, 246) (Fig. 15; Table 3). The synthesized mycocerosates are not released from Mas by a conventional thioesterase but rather are directly transferred by PapA5 onto their phthiocerol or phenolphthiocerol acceptors through an interaction with Mas to catalyze the final esterification step (244) (Figs. 14, 15; Table 3).

Synthesis of the saccharide moiety of PGL-tb and *p*-HBADs

Consistent with their conserved structures (Fig. 13), the biosynthesis of the glycosyl moiety of PGL-tb and *p*-HBADs involves the same set of enzymes (Table 3). In the case of *M. tuberculosis* "canetti," four genes (*Rv2962c*, *Rv2957*, *Rv2958c*, and *Rv2959c*) encoding three glycosyltransferases and one methyltransferase are involved in the formation of this structure (247, 248) (Fig. 14; Table 3). The glycosyltransferase

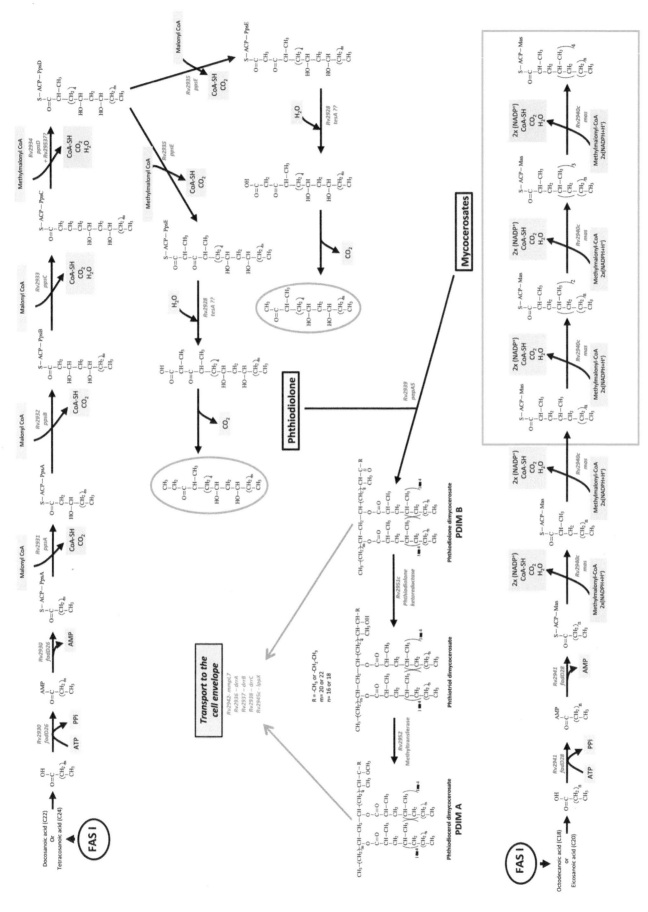

Figure 15 The PDIM biosynthetic pathway. See text for details. doi:10.1128/microbiolspec.MGM2-0021-2013.f15

Table 3 *M. tuberculosis* H37Rv genes involved in the biogenesis of phthiocerol dimycocerosates, phenolic glycolipids, and *p*-hydroxybenzoic acids

Gene number	Gene name	Function	Evidence[a]	Reference
Rv2928	*tesA*	Type II thioesterase thought to be involved in the release of phthiocerol and phenolphthiocerol from PpsE in the biosynthesis of PDIM and PGL	P	230, 238, 239
Rv2930	*fadD26*	Long-chain fatty acyl-AMP ligase providing PpsA with activated long-chain fatty acid starter units for the generation of phthiocerol in the biosynthesis of PDIM	P	167, 229, 236, 244
Rv2931	*ppsA*	Type 1 polyketide synthase responsible with PpsB-C-D-E for the elongation of C_{22} to C_{24} fatty acids and *p*-hydroxyphenylalkanoates with malonyl-CoA and methylmalonyl-CoA to yield phthiocerol and phenolphthiocerol derivatives, respectively	E, P	235, 244
Rv2932	*ppsB*	Type 1 polyketide synthase responsible with PpsA-C-D-E for the elongation of C_{22} to C_{24} fatty acids and *p*-hydroxyphenylalkanoates with malonyl-CoA and methylmalonyl-CoA to yield phthiocerol and phenolphthiocerol derivatives, respectively	E, P	235, 244, 349
Rv2933	*ppsC*	Type 1 polyketide synthase responsible with PpsA-B-D-E for the elongation of C_{22} to C_{24} fatty acids and *p*-hydroxyphenylalkanoates with malonyl-CoA and methylmalonyl-CoA to yield phthiocerol and phenolphthiocerol derivatives, respectively	P	235, 349
Rv2934	*ppsD*	Type 1 polyketide synthase responsible with PpsA-B-C-E for the elongation of C_{22} to C_{24} fatty acids and *p*-hydroxyphenylalkanoates with malonyl-CoA and methylmalonyl-CoA to yield phthiocerol and phenolphthiocerol derivatives, respectively	P	235
Rv2935	*ppsE*	Type 1 polyketide synthase responsible with PpsA-B-C-D for the elongation of C_{22} to C_{24} fatty acids and *p*-hydroxyphenylalkanoates with malonyl-CoA and methylmalonyl-CoA to yield phthiocerol and phenolphthiocerol derivatives, respectively	E, P	235, 244
Rv2936	*drrA*	Component of the DrrABC ABC-transporter involved in the translocation of PDIM (and PGL?) across the plasma membrane; ATP-binding protein	P	350
Rv2937	*drrB*	Component of the DrrABC ABC-transporter involved in the translocation of PDIM (and PGL?) across the plasma membrane; integral membrane protein	P	230, 350
Rv2938	*drrC*	Component of the DrrABC ABC-transporter involved in the translocation of PDIM (and PGL?) across the plasma membrane; integral membrane protein	P	229
Rv2939	*papA5*	Acyltransferase responsible for the transfer of mycocerosic acids to phthiocerol to form PDIM	E, P	244, 351, 352
Rv2940c	*mas*	Mycocerosic acid synthase (type I polyketide synthase) involved in the biosynthesis of PDIM and PGL	E, P	240–244
Rv2941	*fadD28*	Long-chain fatty acyl-AMP ligase responsible for providing the long-chain fatty acid starter unit to Mas for the generation of mycocerosic acids	E, P	167, 229, 235, 245, 246
Rv2942	*mmpL7*	RND superfamily inner membrane transporter involved in PDIM translocation to the periplasm (and PGL?)	P	229, 235, 252
Rv2945	*lppX*	Lipoprotein involved in the transport of PDIM (and PGL?) to the cell surface	P	253
Rv2946c	*pks1*	Together with Pks15, type I polyketide synthase involved in the elongation of *p*-hydroxybenzoic acid derivatives with malonyl-CoA to form *p*-hydroxyphenylalkanoates (precursors of PGL)	P	223
Rv2947c	*pks15*	Together with Pks1, type I polyketide synthase involved in the elongation of *p*-hydroxybenzoic acid derivatives with malonyl-CoA to form *p*-hydroxyphenylalkanoates (precursors of PGL)	P	223
Rv2948c	*fadD22*	*p*-Hydroxybenzoyl-AMP ligase involved in the biosynthesis of PGL; catalyzes the activation of *p*-hydroxybenzoic acid and its subsequent transfer onto Pks15/1 for the production of *p*-hydroxyphenylalkanoates	E, P	236, 353, 354

(Continued)

Table 3 (Continued)

Gene number	Gene name	Function	Evidence[a]	Reference
Rv2949c		p-Hydroxybenzoic acid synthase	E, P	237
Rv2950c	fadD29	Fatty acyl-AMP ligase involved in the biosynthesis of PGL; catalyzes the activation of hydroxyphenylalkanoates that are then transferred onto PpsA to yield phenolphthiocerol	E, P	236
Rv2951c		Ketoreductase catalyzing the reduction of (phenol)phthiodiolone to yield (phenol)phthiotriol in the biosynthesis of PDIM and PGL	P	355, 356
Rv2952		SAM-dependent O-methyltransferase involved in the formation of (phenol)phthiocerol dimycocerosates from (phenol)phthiodiolone dimycocerosates. Catalyzes the transfer of a methyl group to the third hydroxyl group of (phenol)phthiotriol in PDIM and glycosylated phenolphthiotriol dimycocerosates	P	247, 356
Rv2953		Enoyl-reductase acting in concert with PpsD in the biosynthesis of the (phenol)phthiocerol moiety of PDIM and PGL	P	357
Rv2954c		SAM-dependent methyltransferase responsible for the O-methylation of the hydroxyl group at position 3 of the fucosyl residue of PGL (and possibly p-HBAD)	P	249
Rv2955c		SAM-dependent methyltransferase responsible for the O-methylation of the hydroxyl group at position 4 of the fucosyl residue of PGL (and possibly p-HBAD)	P	249
Rv2956		SAM-dependent methyltransferase responsible for the O-methylation of the hydroxyl group at position 2 of the fucosyl residue of PGL and (and possibly p-HBAD)	P	249
Rv2957		Fucosyltransferase responsible for the transfer of the third glycosyl residue of the triglycosyl appendage of PGL and p-HBAD	P	248
Rv2958c		Rhamnosyltransferase responsible for the transfer of the second rhamnosyl residue of the triglycosyl appendage of PGL and p-HBAD	P	248
Rv2959c		SAM-dependent methyltransferase responsible for the O-methylation of the hydroxyl group at position 2 of the rhamnosyl residue linked to the phenolic group of PGL and p-HBAD	P	247
Rv2962c		Rhamnosyltransferase responsible for the transfer of a rhamnosyl residue onto p-hydroxybenzoic ethyl ester and/or phenolphthiocerol dimycocerosates	P	248

[a]E, P, H: see Table 1 footnote.

encoded by Rv2962c is involved in the transfer of the first rhamnosyl residue onto p-hydroxy-phenolphthiocerol dimycocerosates and p-hydroxy-phenolmethylester. A single nucleotide polymorphism (SNP) at position 880 of Rv2962c in the Indo-Oceanic isolates of M. tuberculosis results in a truncated ORF accounting for the accumulation of phenolphthiocerol dimycocerosates and the related "attenuation lipid" observed in this lineage (228). Rv2958c encodes the rhamnosyltransferase responsible for the transfer of the second rhamnosyl residue onto the mono-rhamnosylated PGL or p-hydroxy-phenolmethylester (248), and a frameshift mutation within this gene explains the lack of production of triglycosylated PGL by M. bovis, M. microti, M. pinnipeddii, and M. africanum (226). Rv2957 encodes the fucosyltransferase responsible for

the transfer of the third glycosyl residue of the triglycosyl appendage of PGL and p-HBAD-II (Fig. 13) (248). Rv2959c encodes a methyltransferase involved in the methylation of position 2 of the first rhamnosyl residue of PGL-tb and p-HBADs (247). Rv2954c, Rv2955c, and Rv2956 encode the methyltransferases that catalyze the O-methylation of the hydroxyl groups located, respectively, at positions 3, 4, and 2 of the terminal fucosyl residue of PGL-tb in a sequential process, starting with methylation at position 2, followed by positions 4 and 3 (249). The genes involved in the production of the glycosyl moiety of the PGL of M. leprae were identified through genetic complementation of M. bovis BCG, leading to the synthesis of M. leprae–specific PGL-1 by the vaccine strain (250).

Translocation of PDIM and Related Molecules

PDIM and PGL-tb are found in the outermost layers of the *M. tuberculosis* cell envelope (8), and *p*-HBADs are secreted in the culture medium (223). Since at least some of the enzymes involved in the biosynthesis of PDIM and PGL are cytosolic (e.g., polyketide synthases and FadD enzymes), the presence of these two lipids at the surface implies the existence of a translocation machinery. All the published work on this topic to date has focused on PDIM because the *M. tuberculosis* strains used to generate knockout mutants were naturally deficient in the production of PGL-tb.

The *mmpL7* or *drrC* genes, both in the PDIM and PGL locus (Fig. 14), have been involved in the translocation of PDIM. *drrC* and *mmpL7* null mutants synthesized PDIMs structurally identical to those of the wild-type strain but failed to translocate these compounds to the cell surface (229, 235). PDIMs in these mutants were apparently retained in deeper layers of the cell envelope. DrrC is an integral membrane protein belonging to an ABC transporter involving two other subunits encoded by *drrA* and *drrB* (Fig. 14; Table 3). The MmpL7 protein belongs to the RND superfamily of transporters (251). Like other members of this family, MmpL7 is predicted to consist of 12 transmembrane domains and two large soluble periplasmic loops. Using a yeast two-hybrid system, Jain and Cox (252) showed that the loop between the seventh and eighth transmembrane domains interacts with the polyketide synthase PpsE involved in PDIM and PGL synthesis. Based on this finding, a model was proposed wherein the synthesis and transport of PDIM are coupled (252). Another gene, *lppX*, which encodes a lipoprotein, has been found to be required for PDIM to reach the cell surface (253). Interestingly, LppX shares a similar fold with the periplasmic chaperone LolA and the outer membrane lipoprotein LolB, which in Gram-negative bacteria are involved in the localization of lipoproteins to the outer membrane. The crystal structure of LppX revealed a large hydrophobic cavity suitable to accommodate a single PDIM molecule (253). It is possible that LppX acts downstream from MmpL7 and DrrABC, carrying PDIM across the periplasm to the outer membrane once the two membrane transporters have translocated the fully synthesized lipid products across the plasma membrane. The exact role of each of these transporters in the translocation process remains, however, to be determined. Given the nature of the enzymes involved in the biosynthesis of PGL and *p*-HBADs, it is likely that most if not all of their biosynthesis takes place in the cytoplasm or at the periphery of the plasma membrane. The same transporters as those involved in the export of PDIM may be involved in their translocation to the cell surface.

Roles of PDIM, *p*-HBADs, and PGL-tb in the Organization of the Cell Envelope and Virulence

As glycosylated capsular or secreted components (8, 223), *p*-HBADs and PGLs are serologically active. Accordingly, several studies have explored the potential of PGLs as serodiagnostic tools for the detection of tuberculosis and leprosy (14, 254–258). Because most clinical isolates of *M. tuberculosis* do not produce PGL-tb (221–223), it is likely that the antibodies detected in patients were in fact directed against *p*-HBADs (223).

PDIMs have been found in all *M. tuberculosis* clinical isolates tested (221, 227). Their nonamphipathic character and abundance in the cell envelope have long suggested that they play a structural role, providing a hydrophobic barrier around *M. tuberculosis* cells and possibly a platform for anchoring other components of the cell envelope (259). The roles of PDIM in the permeability barrier, intracellular survival, and virulence of *M. tuberculosis* have been extensively discussed in previous reviews and will therefore not be detailed here (141, 260). Likewise, the reader is referred to earlier reviews for details on the roles of PGL-tb and *p*-HBADs in the modulation of the host immune response and pathogenicity of *M. tuberculosis* (260).

Numerous biological activities have also been associated with the PGLs of nontuberculous mycobacteria in general. The PGLs from *M. leprae* and *M. kansasii*, like those of *M. bovis* BCG, seem to nonspecifically inhibit lymphoproliferative responses to various stimuli, including several antigens and mitogens (261). Other biological activities are dependent on the nature of the carbohydrate moiety. For instance, specific suppression of cell-mediated immunity is a feature of lepromatous leprosy, and the PGL-1 from *M. leprae* has been involved in many aspects of this process. In contrast to the PGLs from *M. microti* and *M. kansasii*, PGL-1 from *M. leprae* has the ability to suppress the "oxidative response" of human macrophages (262, 263), probably explaining why this response is abnormally low in leprosy patients. PGL-1, but not the PGLs from *M. bovis* and *M. kansasii*, also has been reported to be active in an indirect test of specific immunosuppression, inhibiting the concanavalin-A stimulation of lymphocytes from patients with lepromatous leprosy (264). PGL-1 also can neutralize hydroxyl and superoxide radicals *in vitro*, a property shared by deacylated

PGL-1 and to some extent by the carbohydrate moiety of the molecule (265).

GLYCOPEPTIDOLIPIDS

Structure and Subcellular Location

Mycobacteria synthesize type- or species-specific GPLs that differ from one another by their sensitivity to alkali. Alkali-stable GPLs, also known as C-mycosides (266), are produced by a number of both fast- and slow-growing mycobacterial species, including *M. avium*, *M. abscessus*, and *M. smegmatis* (for detailed reviews, see references 200, 267, 268). Alkali-labile GPLs have thus far only been described in *Mycobacterium xenopi* (269, 270). Their structures greatly differ from those of the C-type GPLs. The C-type GPLs share a common lipopeptidyl core that consists of a mixture of 3-hydroxylated and 3-methoxylated long-chain (C_{26} to C_{34}) fatty acids (271), amidated by a tripeptide (D-Phe-D-*allo*Thr-D-Ala) and terminated by an aminoalcohol (L-alaninol) (Fig. 16A). The position of hydroxyl/methoxyl group A on the fatty acyl chain has been recently questioned, and position 5 has been proposed (272). C-type GPLs differ by the number and the nature of the glycosyl residues that substitute the lipopeptidyl core. In the most abundant molecular species, the apolar C-type GPLs (Fig. 16A), also called nonspecific GPLs (nsGPLs), the alaninol is glycosylated by a mono- or di-O-methylated rhamnosyl residue, while a di-O-acetylated 6-deoxytalosyl unit is attached to the *allo*Thr residue. In the polar C-type GPLs, also known as serospecific GPLs (ssGPLs), additional sugar units are attached to the 6-deoxytalosyl residue; at least 14 out of the 28 described ssGPLs from the *M. avium-intracellulare* complex have been structurally characterized (200, 268). The serological variance among the members of the *M. avium-intracellulare* complex is due to subtle differences in the structure of the oligosaccharide chain that substitutes the communal C-type nsGPL core. The oligosaccharide haptens from several polar C-type GPLs contain unusual sugars: glucuronic acid and variants, acetamido-dideoxy-hexosyl residues, and other branched sugars.

The correlation between the presence of C-type GPLs, smooth colony morphotype, and staining of the outer membrane with Ruthenium Red (273–275), and the fact that polar GPLs correspond to "Schaefer typing antigens" used in the identification of isolates of the *M. avium-intracellulare* complex (276) suggested that GPLs were present at the cell surface. Consistently, the capsular materials of *M. lepraemurium* (277) and

(A)

(B)

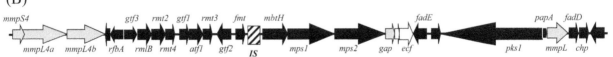

Figure 16 (A) Structure of the nonspecific glycopeptidolipids of *M. smegmatis*. R_1 = –H or –CH_3; R_2 = –H or –Ac; R_3, –CH_3, -succinyl, -rhamnosyl or -2-O-succinylrhamnosyl; m = 12-14; n, 6-10. (B) GPL biosynthetic gene cluster of *M. smegmatis* mc^{2}155. Shown is the 64.97-kb region spanning *MSMEG_0380* (*mmpS4*) to *MSMEG_0413*. ORFs are depicted as arrows. Black arrows indicate genes encoding biosynthetic enzymes; gray arrows indicate putative transporter genes; white arrows indicate putative regulatory genes. Chp, putative acyltransferase; FadE, putative acyl-CoA dehydrogenase; PapA, putative acyltransferase. Other genes are described in the text. doi:10.1128/microbiolspec.MGM2-0021-2013.f16

M. avium (278) have been shown to consist of C-mycosides.

Biological Properties of GPLs

The antigenic properties of ssGPLs and the relationships between GPL production and colonial morphotype or drug resistance have been abundantly reviewed (268, 279). Other properties associated with GPL production are as follows. Although mycobacteria are nonflagellated, *M. smegmatis* and the slow-growing *M. avium* can spread on the surface of solid media by a sliding mechanism (280). Rough strains lacking GPLs appear to be devoid of such sliding motility (210, 275). Consistently, all of the nonsliding mutants isolated and analyzed by Recht et al. (281) had a rough morphotype and showed no detectable levels of GPLs. These nonsliding mutants were also defective for attachment and biofilm formation on PVC plastic (281). This observation emphasizes the importance of GPLs in determining the cell surface properties of *M. smegmatis*. Suggestive of the important role played by GPLs in the permeability of the cell envelope, the absence of nsGPLs from the cell envelope of a defined knockout mutant of *M. smegmatis* was shown to have a profound effect on the uptake of chenodeoxycholate (275), a hydrophobic molecule that diffuses through lipid domains of the mycobacterial cell envelope.

The species *M. avium-intracellulare* has received attention as a major opportunistic pathogen in AIDS patients. Although the specific mechanisms that define its pathogenicity have not been entirely clarified, it is becoming apparent that GPL antigens have a variety of biological activities that could influence host responses. During infection, *M. avium* synthesizes GPLs that accumulate in macrophages (278, 282–284). Early studies have shown that *M. avium* ssGPLs suppress the mitogen-induced proliferative responses of murine splenic cells (285–287) but not those of human peripheral blood mononuclear cells (288). More recently, GPLs from *M. avium* serovar 4 have been proposed to participate in the ability of *M. avium* to invade human macrophages and escape bactericidal responses (289). GPLs are also thought to impact adaptive immunity. Pretreatment of human blood mononuclear cells with serovar-specific GPLs, for instance, suppresses the production of Th1 cytokines including interleukin 2 (IL-2) and interferon γ (IFNγ) (290). In contrast, ssGPLs induce the production of two important immunomodulatory substances, tumor necrosis factor α (TNF-α) and prostaglandin E2 (PGE$_2$) (288, 291). A group of polar GPLs from *M. chelonae* (pGPL-Mc) has also been reported to increase the resistance of mice to disseminated

candidiasis (292) and to enhance the immune response to influenza vaccination (293). Moreover, pGPL-Mc molecules exhibit the properties of haematopoietic growth factor (294–296).

Biosynthesis of GPLs

nsGPL biosynthesis in *M. smegmatis*

The GPL biosynthetic gene cluster of *M. smegmatis* (Fig. 16B) is currently thought to encompass 24 genes. Twelve of them have been experimentally characterized using a combination of genetic approaches (210, 274, 281, 297–301). The nonribosomal peptide synthase genes, *mps1* and *msp2*, encode the enzymes responsible for the synthesis of the peptidic moiety of D-Phe-D-*allo*-Thr-D-Ala-L-alaninol (210, 274). Mps1 and Mps2 are each composed of two modules catalyzing the incorporation of an amino acid. The last module of Mps2, however, is devoid of a racemase domain (210), suggesting that it is responsible for the synthesis of the distal alaninol-containing moiety of the pseudo-tetrapeptide. The putative nonribosomal peptide synthases of *M. avium* share the same genetic organization consisting of two ORFs (302). The *mbtH* gene has been shown to be required for GPL production (303), but no exact function has yet been attributed to its protein product.

The main acyl residue of the *M. smegmatis* GPLs is a monounsaturated hydroxylated C$_{30}$ fatty acid (271, 304) whose synthesis is thought to involve the polyketide synthase product of *pks1* (210) (Fig. 16B). Three glycosyltransferase genes are also found in the GPL cluster, a number in agreement with the structure of the *M. smegmatis* GPLs (Fig. 16A and 16B) (299, 301, 305). Two other genes of the cluster, namely, *rmlA* (aka *rfbA*), which encodes a putative glucose-1-phosphate thymidylyltransferase, and *rmlB*, which encodes a putative dTDP-glucose-4,6-dehydrogenase, are likely to be involved in the synthesis of the deoxyhexoses, rhamnose and 6-deoxytalose, which would subsequently be incorporated into nsGPLs (305).

The *mtf1* (aka *rmt3*) gene (300) encodes an S-adenosylmethionine-dependent rhamnosyl-3-O-methyltransferase (297). This enzyme is required for the O-methylation of position 3 of the rhamnosyl unit that glycosylates the alaninol. Disruption of *rmt3* virtually abolishes the further methylation of the rhamnosyl unit, suggesting that this enzme is the first methyltransferase to act on the GPL precursors (297). Three other methyltransferase genes are found in the GPL gene cluster (305). The *fmt* gene encodes a fatty acid O-methyltransferase that modifies the hydroxyl group of the GPL fatty acid (299), whereas *rmt4* encodes a rhamnosyl-4-O-methyltransferase, and *rmt2* encodes a rhamnosyl-2-O-methyltransferase (300); all of these methyltransferases have orthologs in *M. avium*. The gene *atf1* is predicted to encode a 6-deoxytalose acetyltransferase (306). The methylation of the rhamnosyl residue occurs independently of the acetylation of the 6-deoxytalose residue, since the GPLs from an *atf1* knockout mutant are normally methylated.

ssGPL biosynthesis in *M. avium*

Limited information is currently available about GPL biosynthesis in *M. avium*, except for the serotype 2 (ser2) ssGPLs recently reviewed by Billman-Jacobe (305) and Chatterjee and Khoo (268). The rhamnosyltransferase gene *rtfA* is the first gene whose function was determined experimentally (307, 308). When produced in *M. smegmatis*, RtfA catalyzed the addition of a rhamnosyl unit to the 6-deoxytalosyl residue of the GPL core (308), thus showing that the simpler nsGPLs can serve as biosynthetic precursors in the synthesis of ssGPLs. This result was confirmed by showing that the targeted disruption of *rtfA* in *M. avium* led to the loss of ser2-specific GPL (307). Complementing *M. smegmatis* methyltransferase mutants with *M. avium* genes from the ser2 gene cluster, Jeevarajah et al. provided evidence of the 4-O-methyltransferase activity of the *M. avium* MtfC and MtfB proteins (300). In addition, they showed that MtfD displays 3-O-methyltransferase activity on the rhamnosyl residue of the *M. smegmatis* GPLs (300). The specificity of this methyltransferase was recently confirmed *via* the construction of a *mtfD* knock-out mutant of *M. avium* (309, 310). Interestingly, the virulence of this mutant was attenuated in mice (309).

On the basis of an altered colony morphotype, Laurent et al. identified other genes likely to be involved in the biosynthesis of GPLs in *M. avium* (302). Two nonribosomal peptide synthase genes (*pstA* and *pstB*) and a probable polyketide synthase gene located downstream of the ser2 cluster are orthologous to the *mps* and *pks* genes found in the GPL biosynthetic gene cluster of *M. smegmatis* mc²155, but a direct involvement of these genes in the mutant phenotypes remains to be established. The ser2 gene clusters of two ser2 strains of *M. avium* were also sequenced and compared with the homologous regions of *M. avium* ser1 strain 104, *M. avium* subspecies *paratuberculosis*, and *M. avium* subspecies *silvaticum* (311). Fifteen ORFs were identified and their putative functions in GPL biosynthesis determined: five encode glycosyltransferases (including RtfA), six encode O-methyltransferases (including MtfB,C,D), one encodes an O-acetyltransferase, and three encode hexose synthetases (D-glucose dehydrogenase, mannose dehydratase and 6-deoxy-4-keto-D-mannose reductase/epimerase). A biosynthetic model in which ser2-specific GPLs are synthesized from a serovar-1-specific GPL intermediate, itself derived from a nonspecific GPL precursor, was proposed (311).

Regulation and Transport of GPLs

The C-type GPL biosynthetic gene cluster begins with a triplet of transmembrane protein encoding genes possibly forming an operon: *tmtpA*, *B*, and *C* (now named *mmpS4*, *mmpL4a*, and *mmpL4b*). TmtpABC belongs to the MmpL and MmpS families of mycobacterial proteins (210) (Fig. 16B). Both MmpL4a and MmpL4b have 12 putative transmembrane domains, whereas the smaller MmpS4 protein displays only one. *mmpL4a* and *mmpL4b* transposon mutants have been reported to have a rough colony morphology, to lack sliding motility, and to be devoid of GPLs (210, 281); the precise role of the MmpL4 proteins in this phenotype, however, has yet to be determined. Interestingly, the biochemical characterization of an *mmpS4* mutant of *M. smegmatis* established that this protein is required for the production and export of large amounts of GPLs but is dispensable for biosynthesis *per se*. Cross-complementation experiments demonstrated that the MmpS4 proteins from *M. smegmatis*, *M. avium*, *M. tuberculosis*, and *M. abscessus* are exchangeable and thus not specific for a particular GPL species (312). MmpS4 requires the formation of a protein complex at the pole of the bacillus to function. It was suggested that MmpS proteins facilitate lipid biosynthesis by acting as a scaffold for a coupled biosynthetic and transport machinery (312). A similar mechanism has also been proposed for the transport of PDIM and SL in *M. tuberculosis* (see "Acyltrehaloses" and "PDIMs, PGLs, and Related Compounds, above") and is thus emerging as a common trait in the biogenesis of mycobacterial complex lipids.

While screening an *M. smegmatis* transposon mutant library for mutants with changes in cell surface properties, strains that failed to stain with Ruthenium Red were isolated (210). All of these mutants harbored a transposon insertion in the *gap* gene (Fig. 16B) and produced GPLs chemically identical to those of the wild-type strain. *gap* mutants, however, had many fewer GPLs at their surface, suggestive of a role for Gap in the export of these lipids. The precise role of Gap in the biogenesis of GPLs—particularly in relation to the MmpL4 and MmpS4 proteins—remains to be determined. Gap may be required for the transport of GPLs across the periplasmic space upon their translocation across the plasma membrane in a process involving the MmpL4-MmpS4 proteins.

Nutrient starvation was reported to induce the production of triglycosylated C-type GPLs in *M. smegmatis* mc^{2}155 (313). The accumulation of polar GPLs in *M. smegmatis* mc^{2}155 seems to be dependent on SigB, because the overexpression of *sigB* induces the production of these lipids, while the disruption of this gene leads, in contrast, to the abolition of their production (314). The expression of *gtf3*, the glycosyltransferase responsible for the addition of the last sugar moiety of triglycosylated GPLs (301), is directly or indirectly controlled by *sigB*, at least during certain stages of growth (314). Another gene potentially coding for an extracytoplasmic sigma factor, *ecf*, is present in the GPL biosynthetic cluster (Fig. 16B), but evidence of the involvement of this gene in the regulation of GPL biosynthesis is lacking. *ecf* is located upstream of a gene encoding a putative sigma factor–associated protein. *M. smegmatis* displays a low frequency of spontaneous morphological variation that correlates with the production of larger amounts of GPLs (315). The transposition of insertion elements into two GPL loci accounts for these morphological changes. One locus is the promoter region of the *mps* operon. The other locus is the *lsr2* gene, which encodes a small basic protein that likely plays a regulatory role.

CAPSULAR POLYSACCHARIDES

As mentioned above, and with a few exceptions, the "capsule"-like structures produced by *Mycobacterium* species primarily consist of polysaccharides and proteins with generally minor amounts of lipids. The ratio of protein to polysaccharide varies according to the species. While in *M. tuberculosis*, *M. kansasii*, and *M. gastri*, the major surface capsular constituents consist of polysaccharides, they mainly are proteins in *M. phlei* and *M. smegmatis* (7, 9). Capsular polysaccharides, like other capsular components, are not covalently bound to the rest of the cell envelope. The three types of capsular polysaccharides identified in the capsular material of tuberculous and nontuberculous mycobacteria are a high-molecular-weight (>1,000,000 Da) α-D-glucan composed of a →4-α-D-Glc-1→ core branched every five or six residues by oligoglucosides; a D-arabino-D-mannan (AM) similar in structure to LAM; and a D-mannan composed of a →6-α-D-Man-1→ core substituted at some of the 2 positions with an α-D-Man residue (6, 7, 316, 317). All are neutral compounds, devoid of acyl substituents.

The structure of AM appears to be identical to that of LAM except for the loss of the phosphatidyl-*myo*-inositol anchor, suggesting that it may be formed from LAM by a specific hydrolytic enzyme. Likewise, the structure of D-mannan appears to be identical to that of the mannan domain of LM and LAM. It is therefore reasonable to assume that the same enzymes participate in the biosynthesis of LM/LAM and in that of the two extracellular polysaccharides. The reader is referred to reference 363 for details about this biosynthetic pathway. D-Mannan and AM are expected to share with LM and LAM common properties in their interactions with the host.

α-D-Glucan is structurally very similar to the intracellular glycogen of *M. tuberculosis* and *M. bovis* BCG, although its three-dimensional structure appears to be more compact and its molecular mass slightly higher (13 × 10^6 versus 7.5 × 10^6 Da). Capsular α-D-glucan was shown to be a ligand of the C-type lectin DC-SIGN of dendritic cells, to modulate the effector functions of monocyte-derived dendritic cells and to mediate the nonopsonic binding of *M. tuberculosis* to complement receptor 3 (318–320). It was also postulated to contribute to the antiphagocytic properties of the capsule of *M. tuberculosis*, thereby possibly controlling the interactions of *M. tuberculosis* with macrophages and promoting uptake *via* complement receptor 3 (321). Altogether, these biological properties may contribute to the survival of *M. tuberculosis* in the host. The structural similarity between α-D-glucan and glycogen has allowed some of the genes involved in the biosynthesis of the capsular polysaccharide to be identified, among them the α-1,4-glucosyltransferases Rv3032 and GlgA (Rv1212c), the ADP-glucose pyrophosphorylase GlgC (Rv1213), and the branching enzyme GlgB (Rv1326c), which is responsible for introducing α-1,6-linked branches into linear α-1,4-glucans (322) (Table 4, Fig. 17).

The phenotypic analysis of *M. tuberculosis* recombinant strains affected by or totally deficient in the expression of these genes confirmed their involvement in the elongation and branching of the capsular α-D-glucan and a partial redundancy between the two α-1,4-glucosyltransferases Rv3032 and GlgA. These analyses further revealed the participation of GlgC, GlgB, and Rv3032 in the biosynthesis of other intracellular *M. tuberculosis* α-1,4-glucans, namely, glycogen and the methylglucose lipopolysaccharides (MGLPs) (322, 323). Attempts to knock out both *glgA* and *Rv3032* in *M. tuberculosis* mutants were unsuccessful, indicating that a functional copy of at least one of the two α-1,4-glucosyltransferases is required for growth. The apparent

Table 4 *M. tuberculosis* H37Rv genes involved in the biogenesis of capsular α-D-glucan

Rv number	Gene name	Function	Evidence[a]	Reference
Rv0126	*treS*	Trehalose synthase; displays maltose <-> trehalose interconverting activity and glycogen amylase activity	E, P	324, 358–360
Rv0127	*pep2*	Maltokinase	E, P	324, 361
Rv1212c	*glgA*	α-1,4-Glucosyltransferase	P	322
Rv1213	*glgC*	ADP-glucose pyrophosphorylase	P	322
Rv1326c	*glgB*	α-1,4-Glucan branching enzyme	E, P	322, 362
Rv1327c	*glgE*	Maltose-1-phosphate maltosyltransferase	E, P	324, 325
Rv1328	*glgP*	Putative glycogen phosphorylase	H	
Rv3032	-	α-1,4-Glucosyltransferase	E, P	322, 323

[a]E, P, H: see Table 1 footnote.

essentiality of *glgB* (322), in contrast, is believed to be related to the toxic accumulation of maltose-1-phosphate that follows the inactivation of this gene (324) (Fig. 17). It is important to note that mycobacterial α-1,4-glucans can also be synthesized from trehalose by a four-step pathway comprising the trehalose synthase TreS, the maltokinase Pep2, the maltose-1-phosphate maltosyltransferase GlgE, and GlgB (Fig. 17; Table 4) (324, 325). Disruption of *glgE*, like that of *glgB*, is lethal because of the toxic accumulation of maltose-1-phosphate that ensues.

As evidenced by their Rv numbers, the genes involved in the metabolism of glycogen, capsular α-D-glucan, and MGLP are clustered in four major locations on the chromosome of *M. tuberculosis* H37Rv (Table 4). The first cluster (*Rv3030-Rv3037c*) encompasses the

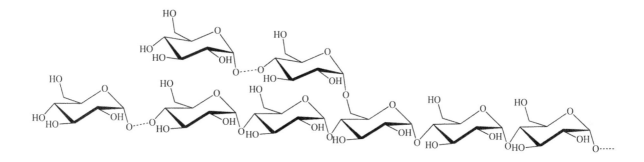

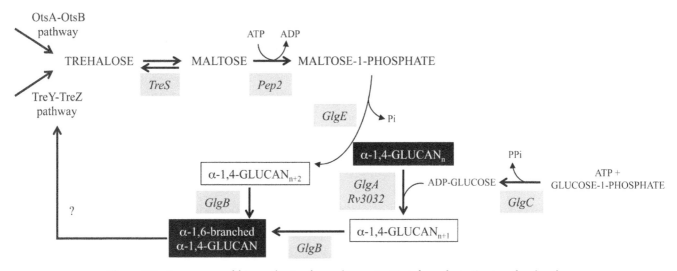

Figure 17 Structure and biosynthesis of α-D-glucans in *M. tuberculosis*. See text for details.
doi:10.1128/microbiolspec.MGM2-0021-2013.f17

α-1,4-glucosyltransferase gene *Rv3032* and other genes likely to be involved in the modifications of MGLPs (326). The second cluster (*Rv1208-Rv1213*) carries *glgA*, *glgC*, and the glucosyl-3-phosphoglycerate synthase gene, *Rv1208*, required for the initiation of MGLPs (33). The third cluster (*Rv1326c-Rv1328*) carries *glgB*, *glgE*, and the probable glycogen phosphorylase gene *glgP* (35). The fourth region (*Rv0126-Rv0127*) harbors *pep2* and *treS*.

Consistent with the intracellular localization of glycogen and MGLPs, GlgA and Rv3032 are nucleotide sugar-utilizing glucosyltransferases predicted to catalyze the elongation of polysaccharides on the cytosolic face of the plasma membrane. GlgB, GlgC, GlgE, Pep2, and TreS are also predicted to be cytosolic enzymes. It is therefore reasonable to assume that capsular α-D-glucan, like glycogen and MGLPs, is synthesized in the cytoplasm. Nothing is known of the translocation machinery responsible for its export to the cell surface.

OTHER LIPOPHILIC COMPOUNDS

Other lipophilic compounds found in the cell envelope of some mycobacterial species include siderophores known as mycobactins and the *M. ulcerans* toxin, mycolactone. Their structure is shown in Figures 18 and 19. The reader is referred to recent reviews for complete details of their biosynthetic pathways (327, 328).

Mycobactins

Pathogenic and nonpathogenic mycobacteria rely on siderophores with high affinity for the ferric ion as the primary mechanism for iron acquisition (for a review, see reference 327). Two classes of siderophores are produced by mycobacteria: the exochelins and the (carboxy)mycobactins. The exochelins are water-soluble peptidic molecules and are secreted into the medium. Mycobactins and carboxymycobactins are both salicyl-capped peptide polyketide-based molecules

but vary in the length of an alkyl substitution and hence in polarity and solubility (Fig. 18). The lipid-soluble mycobactins have long-chain acyl chains on the first lysine residue, whereas the water-soluble carboxymycobactins have a shorter side chain that terminates with a carboxylic acid or methyl ester. Mycobactins tend to remain cell-associated, while carboxymycobactins are secreted in the culture medium. Mycobacteria fall into four groups based upon the production of these molecules. *M. tuberculosis* produces only (carboxy)mycobactins, which are essential for its virulence (327, 329); *M. vaccae* produces only the exochelin type; *M. smegmatis* produces both types, and *M. leprae* produces none. The biosynthesis of mycobacterial siderophores has been reviewed recently (327) and will therefore not be detailed here. Briefly, a cluster of 10 genes (annotated *mbtA* through *mbtJ*; *Rv2377c-Rv2386c*—the *mbt* locus) encompassing approximately 24 kb, another cluster of 6 genes referred to as *mbt-2*, the phosphopantetheinyl gene *pptT* (*Rv2794c*), the *esx3* cluster (*Rv0282-Rv0292*), and the transport genes *mmpS4/mmpL4* (*Rv0450c-Rv0451c*) and *mmpS5/mmpL5* (*Rv0676c-Rv0677c*) encode the proteins required for the synthesis, export, utilization, and uptake of (carboxy)mycobactins (327, 329). The regulator IdeR (Rv2711) represses the expression of the *mbtA-N* genes.

Mycolactones

Mycolactones are a family of lipophilic macrocyclic polyketide molecules that is the primary virulence factor produced by *M. ulcerans*, the etiologic agent of Buruli ulcer in humans, and some closely related aquatic mycobacteria (Fig. 19). Mycolactones display cytotoxic, analgesic, and immunosuppressive activities (328). They are found in abundant quantities in the extracellular matrix surrounding *M. ulcerans* under certain *in vitro* growth conditions and during host infection (330). A 174-kb megaplasmid named pMUM001 in the *M. ulcerans* strain Agy99 carries all of the genes

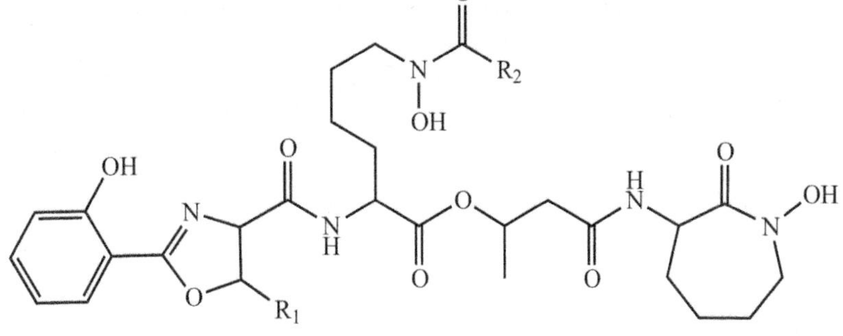

Figure 18 Representative structures of mycobactins and carboxymycobactins from *M. tuberculosis*. See text for details. Mycobactins: R_1 = H; R_2 = $(CH_2)_nCH_3$, n = 16-19; $(CH_2)_xCH = CH(CH_2)_yCH_3$, x+y = 14-17. Carboxymycobactins: R_1 = H, CH_3; R_2 = $(CH_2)_nCOOCH_3/COOH$, n = 1-7; $(CH_2)_xCH = CH(CH_2)_yCOOCH_3/COOH$, x+y = 1-5.
doi:10.1128/microbiolspec.MGM2-0021-2013.f18

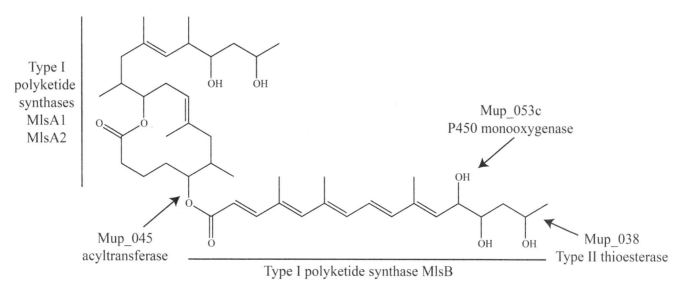

Figure 19 Representative structure of a mycolactone from *M. ulcerans*. The genes involved in the biosynthesis of the various constituents of mycolactone are indicated on the structure. doi:10.1128/microbiolspec.MGM2-0021-2013.f19

required for mycolactone synthesis (331). These include two very large type I polyketide synthase genes, *mlsA1* (51 kb) and *mlsA2* (7 kb), responsible for the synthesis of the upper side chain and macrolactone core; another giant type I polyketide synthase gene, *mlsB* (42 kb), involved in the elongation of the acyl side chain; *mup_045*, thought to encode the transferase linking the acyl side chain and the macrolactone core; *mup_053c*, a P450 monooxygenase gene most likely responsible for hydroxylation of the mycolactone acyl side chain at C12′; and *mup_038*, a type II thioesterase gene predicted to play a role in maintaining the fidelity of the polyketide synthases by removing acyl chains from modules where synthesis has stalled (328).

CONCLUSIONS AND FUTURE PROSPECTS

As illustrated in this article, knowledge of cell envelope biosynthesis in *M. tuberculosis* has greatly benefited from the publication of the complete genome sequence of this bacterium and developments in the genetic manipulation of mycobacteria in the late 1990s. As more genomes from slow- and fast-growing mycobacteria are sequenced, this impetus is progressively extending to other tuberculous and nontuberculous mycobacterial species, with the result that the processes leading to the biosynthesis of more and more species-specific cell envelope constituents are now being elucidated. Beyond the opportunities offered by some of these pathways for drug development, interest in the biosynthesis of species-specific cell envelope

constituents stems from their antigenicity and potential for serodiagnosis and as biomarkers. GPL, DAT, LOS, PGL, pHBAD, and TDM in particular are potent B-cell antigens and have been the object of extensive studies aimed at assessing their potential for the diagnosis of tuberculosis, leprosy, and other mycobacterial diseases (14, 223, 256, 257, 332–334). Key to their widespread application for therapeutic or diagnostic purposes, however, will be a precise understanding of their role in the physiology and virulence of the bacterium and regulatory processes governing their synthesis during the various stages of the lifecycle of the producing mycobacterium. While it is now well established that mycobacteria adjust the composition of their cell envelope in response to the nutrients available in the environment (e.g., carbon and nitrogen source, iron concentration, etc.), physical conditions to which they are exposed (pH, oxygen tension), and age of the culture, knowledge of the regulatory processes involved is more limited.

Yet these changes affect all major cell envelope constituents including phospholipids, PIMs, triglycerides, capsular polysaccharides, lipoglycans, PG, arabinogalactan, and mycolic acids (14, 15, 141, 180, 335–341). Regulation appears to occur both at the transcriptional and the posttranslational levels. The two-component transcriptional regulator PhoP-PhoR, for instance, stands out as a major regulator of polymethyl-branched acyltrehalose production in *M. tuberculosis* (SL, DAT, and PAT) (143, 177). Other important regulators controlling cell division and cell envelope biogenesis are found among the serine/threonine kinase family (for a review, see reference 342). These enzymes regulate through phosphorylation the activity of multiple enzymes and transporters involved in the biosynthesis of mycolic acids, PG, arabinogalactan, LAM, and PDIM.

Despite the considerable advances made in deciphering the metabolic pathways of mycobacterial cell envelope constituents, most of them are not yet complete. The clustered genetic organization of many of these pathways (e.g., lipoglycans, acyltrehaloses, GPLs, LOSs, PDIMs, etc.) raises the hope that some of the missing genes will be found on the basis of their colocalization with known clusters. In the case of a few other minor or species-specific cell envelope constituents, however, biosynthetic pathways have hardly started to be explored. This is, for instance, the case for the glycosyl DAGs described by Hunter et al. (343) and the mycobacterial carotenoids, whose structural definition extends back to the work of E. Chargaff in 1930 (259). Likewise, although there is at present no chemical evidence of LOS in *M. tuberculosis*, preliminary data indicate that some of the unannotated glycosyltransferases of the GT-A, -B, or -C classes may participate in the synthesis of chemically undetectable amounts of these products (204, 207).

Beyond the identification of the missing biosynthetic and regulatory proteins will be the identification of the transporters required for the translocation of biosynthetic precursors or end products from their site of production, for the most part cytoplasmic, to their final periplasmic, outer membrane or capsular locations. More than 148 transport-associated proteins belonging to 33 major transporter families were identified in the genome of *M. tuberculosis* H37Rv (http://www.membranetransport.org/). More transporters are typically found in environmental *Mycobacterium* species. The latest *M. tuberculosis* genome annotation was updated with 134 bioinformatically predicted outer membrane proteins (344). The transporters required for the building of the cell envelope are thus most likely to be found in this long and diverse list of candidate genes with or without homologs in other prokaryotes. Indeed, searches for mycobacterial transporters sharing sequence similarity with known (lipo)polysaccharide or glycolipid transporters from Gram-positive or Gram-negative bacteria typically yield limited if any meaningful candidates. Yet it is becoming increasingly evident that, similar to the biogenesis of other prokaryotic (lipo)polysaccharides, the biosynthesis and translocation of many mycobacterial cell envelope constituents (e.g., mycolic acids, PDIMs, acyltrehaloses, mycobactins, lipoglycans, and arabinogalactan) are temporally and spatially coupled by multiprotein complexes that possibly span the cell envelope. Reasons for their elusive nature may be found in the unusual structure and composition of the mycobacterial cell envelope, which may have driven mycobacteria to evolve specialized transport systems mechanistically related to but structurally divergent from those of other prokaryotes. Examples of these include the type VII secretion system ESX-3 for the uptake of mycobactins (345), the Mce4 proteins for the import of cholesterol (346), the RND-like inner membrane transporters of the MmpL family involved in the export of complex lipids and mycobactins (145, 170, 171, 210, 229, 235, 252, 281, 329, 347), the periplasmic Lol-like lipoprotein carriers LppX and LprG for the export of PDIM and PIM/lipoglycans (38, 253), and the SMR-like (small multidrug resistance) lipid-linked arabinose translocase Rv3789 (348).

Finally, another area where much remains to be done is in understanding the genetic bases underlying the cell envelope's constant remodeling (including degradation and recycling), which accompanies cell division or any changes in the metabolism of the bacterium following host infection or exposure to various environmental stresses.

The authors' research on cell envelope biogenesis is supported through NIH/NIAID grants AI064798, AI063054, AI049151, and AI097550.

We thank Hugo Gramajo (Universidad Nacional de Rosario, Rosario, Argentina) for sharing unpublished results, and Victoria Jones (CSU), Michael McNeil (CSU), Gilles Etienne, Anne Lemassu, and Lucie Spina (IPBS-CNRS, Toulouse) for critical reading of the manuscript and their help with the preparation of the figures.

Citation. Daffé M, Crick DC, Jackson M. 2014. Genetics of capsular polysaccharides and cell envelope (glyco)lipids. Microbiol Spectrum 2(4):MGM2-0021-2013.

References

1. Daffé M, Draper P. 1998. The envelope layers of mycobacteria with reference to their pathogenicity. *Adv Microb Physiol* **39**:131–203.

2. Jackson M, McNeil MR, Brennan PJ. 2013. Progress in targeting cell envelope biogenesis in *Mycobacterium tuberculosis*. *Future Microbiol* **8**:855–875.

3. Hoffmann C, Leis A, Niederweis M, Plitzko JM, Engelhardt H. 2008. Disclosure of the mycobacterial outer membrane: cryo-electron tomography and vitreous sections reveal the lipid bilayer structure. *Proc Natl Acad Sci USA* **105**:3963–3967.

4. Zuber B, Chami M, Houssin C, Dubochet J, Griffiths G, Daffe M. 2008. Direct visualization of the outer membrane of mycobacteria and corynebacteria in their native state. *J Bacteriol* **190**:5672–5680.

5. Sani M, Houben ENG, Geurtsen J, Pierson J, de Punder K, van Zon M, Wever B, Piersma SR, Jimenez CR, Daffe M, Appelmelk BJ, Bitter W, van der Wel N, Peters PJ. 2010. Direct visualization by cryo-EM of the mycobacterial capsular layer: a labile structure containing ESX-1-secreted proteins. *PLoS Pathog* 6:e1000794.

6. Lemassu A, Daffé M. 1994. Structural features of the exocellular polysaccharides of *Mycobacterium tuberculosis*. *Biochem J* 297:351–357.

7. Ortalo-Magné A, Dupont MA, Lemassu A, Andersen AB, Gounon P, Daffé M. 1995. Molecular composition of the outermost capsular material of the tubercle bacillus. *Microbiology* 141:1609–1620.

8. Ortalo-Magné A, Lemassu A, Lanéelle MA, Bardou F, Silve G, Gounon P, Marchal G, Daffé M. 1996. Identification of the surface-exposed lipids on the cell envelopes of *Mycobacterium tuberculosis* and other mycobacterial species. *J Bacteriol* 178:456–461.

9. Lemassu A, Ortalo-Magne A, Bardou F, Silve G, Laneelle M-A, Daffe M. 1996. Extracellular and surface-exposed polysaccharides of non-tuberculous mycobacteria. *Microbiology* 142:1513–1520.

10. Raynaud C, Etienne G, Peyron P, Laneelle MA, Daffe M. 1998. Extracellular enzyme activities potentially involved in the pathogenicity of *Mycobacterium tuberculosis*. *Microbiology* 144:577–587.

11. Ehlers MRW, Daffé M. 1998. Interactions between *Mycobacterium tuberculosis* and host cells: are mycobacterial sugars the key? *Trends Microbiol* 6:328–335.

12. Daffé M, Etienne G. 1999. The capsule of *Mycobacterium tuberculosis* and its implications for pathogenicity. *Tuber Lung Dis* 79:153–169.

13. Goren MB. 1984. Biosynthesis and structures of phospholipids and sulfatides, p 379–415. *In* Kubica GP, Wayne LG (ed), *The Mycobacteria. A Sourcebook*, vol. **1**. Marcel Dekker, New York/Basel.

14. Brennan PJ. 1988. *Mycobacterium* and other actinomycetes, p 203–298. *In* Ratledge C, Wilkinson SG (ed), *Microbial Lipids*, vol. **1**. Academic Press, London.

15. Daniel J, Deb C, Dubey VS, Sirakova TD, Abomoelak B, Morbidoni HR, Kolattukudy PE. 2004. Induction of a novel class of diacylglycerol acyltransferases and triacylglycerol accumulation in *Mycobacterium tuberculosis* as it goes into a dormancy-like state in culture. *J Bacteriol* 186:5017–5030.

16. Walker RW, Barakat H, Hung JGC. 1970. The positional distribution of fatty acids in the phospholipids and triglycerides of *Mycobacterium smegmatis* and *M. bovis* BCG. *Lipids* 5:684–691.

17. Fernandes ND, Kolattukudy PE. 1996. Cloning, sequencing and characterization of a fatty acid synthase-encoding gene from *Mycobacterium tuberculosis* var. *bovis* BCG. *Gene* 170:95–99.

18. Zimhony O, Vilcheze C, Jacobs WR Jr. 2004. Characterization of *Mycobacterium smegmatis* expressing the *Mycobacterium tuberculosis* fatty acid synthase I (*fas1*) gene. *J Bacteriol* 186:4051–4055.

19. Yao J, Rock CO. 2013. Phosphatidic acid synthesis in bacteria. *Biochim Biophys Acta* 1831:495–502.

20. Comba S, Menendez-Bravo S, Arabolaza A, Gramajo H. 2013. Identification and physiological characterization of phosphatidic acid phosphatase enzymes involved in triacylglycerol biosynthesis in *Streptomyces coelicolor*. *Microbial Cell Factories* 12:9.

21. Sirakova TD, Dubey VS, Deb C, Daniel J, Korotkova TA, Abomoelak B, Kolattukudy PE. 2006. Identification of a diacylglycerol acyltransferase gene involved in accumulation of triacylglycerol in *Mycobacterium tuberculosis* under stress. *Microbiology* 152:2717–2725.

22. Elamin AA, Stehr M, Spallek R, Rohde M, Singh M. 2011. The *Mycobacterium tuberculosis* Ag85A is a novel diacylglycerol acyltransferase involved in lipid body formation. *Mol Microbiol* 81:1577–1592.

23. Nigou J, Besra GS. 2002. Cytidine diphosphate-diacylglycerol synthesis in *Mycobacterium smegmatis*. *Biochem J* 367:157–162.

24. Salman M, Lonsdale JT, Besra GS, Brennan PJ. 1999. Phosphatidylinositol synthesis in mycobacteria. *Biochim Biophys Acta* 1436:437–450.

25. Jackson M, Crick DC, Brennan PJ. 2000. Phosphatidylinositol is an essential phospholipid of mycobacteria. *J Biol Chem* 275:30092–30099.

26. Morii H, Ogawa M, Fukuda K, Taniguchi H, Koga Y. 2010. A revised biosynthetic pathway for phosphatidylinositol in mycobacteria. *J Biochem* 148:593–602.

27. Sandoval-Calderon M, Geiger O, Guan Z, Barona-Gomez F, Sohlenkamp C. 2009. A eukaryote-like cardiolipin synthase is present in *Streptomyces coelicolor* and in most actinobacteria. *J Biol Chem* 284:17383–17390.

28. Korduláková J, Gilleron M, Mikusova K, Puzo G, Brennan PJ, Gicquel B, Jackson M. 2002. Definition of the first mannosylation step in phosphatidylinositol synthesis: PimA is essential for growth of mycobacteria. *J Biol Chem* 277:31335–31344.

29. Guerin ME, Kordulakova J, Schaeffer F, Svetlikova Z, Buschiazzo A, Giganti D, Gicquel B, Mikusova K, Jackson M, Alzari PM. 2007. Molecular recognition and interfacial catalysis by the essential phosphatidylinositol mannosyltransferase PimA from mycobacteria. *J Biol Chem* 282:20705–20714.

30. Guerin ME, Schaeffer F, Chaffotte A, Gest P, Giganti D, Kordulakova J, van der Woerd M, Jackson M, Alzari PM. 2009. Substrate-induced conformational changes in the essential peripheral membrane-associated mannosyltransferase PimA from mycobacteria: implications for catalysis. *J Biol Chem* 284:21613–21625.

31. Guerin ME, Kaur D, Somashekar BS, Gibbs S, Gest P, Chatterjee D, Brennan PJ, Jackson M. 2009. New insights into the early steps of phosphatidylinositol mannosides biosynthesis in mycobacteria. PimB′ is an essential enzyme of *Mycobacterium smegmatis*. *J Biol Chem* 284:25687–25696.

32. Korduláková J, Gilleron M, Puzo G, Brennan PJ, Gicquel B, Mikusova K, Jackson M. 2003. Identification of the required acyltransferase step in the biosynthesis of the phosphatidylinositol mannosides of *Mycobacterium* species. *J Biol Chem* 278:36285–36295.

33. Kaur D, Guerin ME, Skovierova H, Brennan PJ, Jackson M. 2009. Chapter 2: biogenesis of the cell wall and other glycoconjugates of *Mycobacterium tuberculosis*. *Adv Appl Microbiol* **69**:23–78.

34. Morita YS, Sena CCB, Waller RF, Kurokawa K, Sernee MF, Nakatani F, Haites RE, Billman-Jacobe H, McConville MJ, Maeda Y, Kinoshita T. 2006. PimE is a polyprenol-phosphate-mannose-dependent mannosyl-transferase that transfers the fifth mannose of phosphatidylinositol mannoside in mycobacteria. *J Biol Chem* **281**:25143–25155.

35. Berg S, Kaur D, Jackson M, Brennan PJ. 2007. The glycosyltransferases of *Mycobacterium tuberculosis*: roles in the synthesis of arabinogalactan, lipoarabinomannan, and other glycoconjugates. *Glycobiology* **17**:35R–56R.

36. Morita YS, Velasquez R, Taig E, Waller RF, Patterson JH, Tull D, Williams SJ, Billman-Jacobe H, McConville MJ. 2005. Compartmentalization of lipid biosynthesis in mycobacteria. *J Biol Chem* **280**:21645–21652.

37. Sulzenbacher G, Canaan S, Bordat Y, Neyrolles O, Stadthagen G, Roig-Zamboni V, Rauzier J, Maurin D, Laval F, Daffe M, Cambillau C, Gicquel B, Bourne Y, Jackson M. 2006. LppX is a lipoprotein required for the translocation of phthiocerol dimycocerosates to the surface of *Mycobacterium tuberculosis*. *EMBO J* **25**: 1436–1444.

38. Drage MG, Tsai HC, Pecora ND, Cheng TY, Arida AR, Shukla S, Rojas RE, Seshadri C, Moody DB, Boom WH, Sacchettini JC, Harding CV. 2010. *Mycobacterium tuberculosis* lipoprotein LprG (Rv1411c) binds triacylated glycolipid agonists of Toll-like receptor 2. *Nat Struct Mol Biol* **17**:1088–1095.

39. Wolucka BA, McNeil MR, de Hoffmann E, Chojnacki T, Brennan PJ. 1994. Recognition of the lipid intermediate for arabinogalactan/arabinomannan biosynthesis and its relation to the mode of action of ethambutol on mycobacteria. *J Biol Chem* **269**:23328–23335.

40. Mahapatra S, Yagi T, Belisle JT, Espinosa BJ, Hill PJ, McNeil MR, Brennan PJ, Crick DC. 2005. Mycobacterial lipid II is composed of a complex mixture of modified muramyl and peptide moieties linked to decaprenyl phosphate. *J Bacteriol* **187**:2747–2757.

41. Anderson RG, Hussey H, Baddiley J. 1972. The mechanism of wall synthesis in bacteria. The organization of enzymes and isoprenoid phosphates in the membrane. *Biochem J* **127**:11–25.

42. Mikusova K, Mikus M, Besra GS, Hancock I, Brennan PJ. 1996. Biosynthesis of the linkage region of the mycobacterial cell wall. *J Biol Chem* **271**:7820–7828.

43. Sacchettini JC, Poulter CD. 1997. Creating isoprenoid diversity. *Science* **277**:1788–1789.

44. Schwender J, Seemann M, Lichtenthaler HK, Rohmer M. 1996. Biosynthesis of isoprenoids (carotenoids, sterols, prenyl side-chains of chlorophylls and plastoquinone) via a novel pyruvate/glyceraldehyde 3-phosphate non-mevalonate pathway in the green alga *Scenedesmus obliquus*. *Biochem J* **316**:73–80.

45. Sprenger GA, Schorken U, Wiegert T, Grolle S, De Graaf AA, Taylor SV, Begley TP, Bringer-Meyer S, Sahm H. 1997. Identification of a thiamin-dependent synthase in *Escherichia coli* required for the formation of the 1-deoxy-D-xylulose 5-phosphate precursor to isoprenoids, thiamin, and pyridoxol. *Proc Natl Acad Sci USA* **94**:12857–12862.

46. Lois LM, Campos N, Putra SR, Danielsen K, Rohmer M, Boronat A. 1998. Cloning and characterization of a gene from *Escherichia coli* encoding a transketolase-like enzyme that catalyzes the synthesis of D-1-deoxyxylulose 5-phosphate, a common precursor for isoprenoid, thiamin, and pyridoxol biosynthesis. *Proc Natl Acad Sci USA* **95**:2105–2110.

47. Hill RE, Himmeldirk K, Kennedy IA, Pauloski RM, Sayer BG, Wolf E, Spenser ID. 1996. The biogenetic anatomy of vitamin B-6. A C-13NMR investigation of the biosynthesis of pyridoxol in *Escherichia coli*. *J Biol Chem* **271**:30426–30435.

48. Querol J, Rodriguez-Concepcion M, Boronat A, Imperial S. 2001. Essential role of residue H49 for activity of *Escherichia coli* 1-deoxy-D-xylulose 5-phosphate synthase, the enzyme catalyzing the first step of the 2-C-methyl-D-erythritol 4-phosphate pathway for isoprenoid synthesis. *Biochem Biophys Res Commun* **289**: 155–160.

49. Bailey AM, Mahapatra S, Brennan PJ, Crick DC. 2002. Identification, cloning, purification, and enzymatic characterization of *Mycobacterium tuberculosis* 1-deoxy-D-xylulose 5-phosphate synthase. *Glycobiology* **12**:813–820.

50. Brown AC, Eberl M, Crick DC, Jomaa H, Parish T. 2010. The nonmevalonate pathway of isoprenoid biosynthesis in *Mycobacterium tuberculosis* is essential and transcriptionally regulated by Dxs. *J Bacteriol* **192**: 2424–2433.

51. Takahashi S, Kuzuyama T, Watanabe H, Seto H. 1998. A 1-deoxy-D-xylulose 5-phosphate reductoisomerase catalyzing the formation of 2-C-methyl-D-erythritol 4-phosphate in an alternative nonmevalonate pathway for terpenoid biosynthesis. *Proc Natl Acad Sci USA* **95**: 9879–9884.

52. Arigoni D, Sagner S, Latzel C, Eisenreich W, Bacher A, Zenk MH. 1997. Terpenoid biosynthesis from 1-deoxy-D-xylulose in higher plants by intramolecular skeletal rearrangement. *Proc Natl Acad Sci USA* **94**:10600–10605.

53. Reuter K, Sanderbrand S, Jomaa H, Wiesner J, Steinbrecher I, Beck E, Hintz M, Klebe G, Stubbs MT. 2002. Crystal structure of 1-deoxy-D-xylulose-5-phosphate reductoisomerase, a crucial enzyme in the non-mevalonate pathway of isoprenoid biosynthesis. *J Biol Chem* **277**:5378–5384.

54. Argyrou A, Blanchard JS. 2004. Kinetic and chemical mechanism of *Mycobacterium tuberculosis* 1-deoxy-D-xylulose-5-phosphate isomeroreductase. *Biochemistry* **43**:4375–4384.

55. Dhiman RK, Schaeffer ML, Bailey AM, Testa CA, Scherman H, Crick DC. 2005. 1-Deoxy-D-xylulose 5-phosphate reductoisomerase (IspC) from *Mycobacterium tuberculosis*: towards understanding mycobacterial resistance to fosmidomycin. *J Bacteriol* **187**:8395–8402.

56. Henriksson LM, Unge T, Carlsson J, Aqvist J, Mowbray SL, Jones TA. 2007. Structures of *Mycobacterium tuberculosis* 1-deoxy-D-xylulose-5-phosphate reductoisomerase provide new insights into catalysis. *J Biol Chem* **282**:19905–19916.

57. Rohdich F, Wungsintaweekul J, Fellermeier M, Sagner S, Herz S, Kis K, Eisenreich W, Bacher A, Zenk MH. 1999. Cytidine 5′-triphosphate-dependent biosynthesis of isoprenoids: YgbP protein of *Escherichia coli* catalyzes the formation of 4-diphosphocytidyl-2-C-methylerythritol. *Proc Natl Acad Sci USA* **96**:11758–11763.

58. Eoh H, Brown AC, Buetow L, Hunter WN, Parish T, Kaur D, Brennan PJ, Crick DC. 2007. Characterization of the *Mycobacterium tuberculosis* 4-diphosphocytidyl-2-C-methyl-D-erythritol synthase: potential for drug development. *J Bacteriol* **189**:8922–8927.

59. Herz S, Wungsintaweekul J, Schuhr CA, Hecht S, Luttgen H, Sagner S, Fellermeier M, Eisenreich W, Zenk MH, Bacher A, Rohdich F. 2000. Biosynthesis of terpenoids: YgbB protein converts 4-diphosphocytidyl-2C- methyl-D-erythritol 2-phosphate to 2C-methyl-D-erythritol 2,4- cyclodiphosphate. *Proc Natl Acad Sci USA* **97**:2486–2490.

60. Rohdich F, Wungsintaweekul J, Luttgen H, Fischer M, Eisenreich W, Schuhr CA, Fellermeier M, Schramek N, Zenk MH, Bacher A. 2000. Biosynthesis of terpenoids: 4-diphosphocytidyl-2-C-methyl-D-erythritol kinase from tomato. *Proc Natl Acad Sci USA* **97**:8251–8256.

61. Eoh H, Narayanasamy P, Brown AC, Parish T, Brennan PJ, Crick DC. 2009. Expression and characterization of soluble 4-diphosphocytidyl-2-C-methyl-D-erythritol kinase from bacterial pathogens. *Chem Biol* **16**:1230–1239.

62. Narayanasamy P, Eoh H, Crick DC. 2008. Chemoenzymatic synthesis of 4-diphosphocytidyl-2-C-methyl-D-erythritol: a substrate for IspE. *Tetrahedron Lett* **49**: 4461–4463.

63. Narayanasamy P, Eoh H, Brennan PJ, Crick DC. 2010. Synthesis of 4-diphosphocytidyl-2-C-methyl-D-erythritol-2-phosphate and kinetic studies of *Mycobacterium tuberculosis* IspF. *Chem Biol* **17**:117–122.

64. Buetow L, Brown AC, Parish T, Hunter WN. 2007. The structure of mycobacteria 2C-methyl-D-erythritol-2,4-cyclodiphosphatesynthase, an essential enzyme, provides a platform for drug discovery. *BMC Struct Biol* **7**:68.

65. Hecht S, Eisenreich W, Adam P, Amslinger S, Kis K, Bacher A, Arigoni D, Rohdich F. 2001. Studies on the nonmevalonate pathway to terpenes: the role of the GcpE (IspG) protein. *Proc Natl Acad Sci USA* **98**: 14837–14842.

66. Altincicek B, Duin EC, Reichenberg A, Hedderich R, Kollas AK, Hintz M, Wagner S, Wiesner J, Beck E, Jomaa H. 2002. LytB protein catalyzes the terminal step of the 2-C-methyl-D-erythritol-4-phosphate pathway of isoprenoid biosynthesis. *FEBS Lett* **532**:437–440.

67. Altincicek B, Kollas A, Eberl M, Wiesner J, Sanderbrand S, Hintz M, Beck E, Jomaa H. 2001. LytB, a novel gene of the 2-C-methyl-D-erythritol 4-phosphate pathway of isoprenoid biosynthesis in *Escherichia coli*. *FEBS Lett* **499**:37–40.

68. Seemann M, Bui BTS, Wolff M, Tritsch D, Campos N, Boronat A, Marquet A, Rohmer M. 2002. Isoprenoid biosynthesis through the methylerythritol phosphate pathway: the (E)-4-hydroxy-3-methylbut-2-enyl diphosphate synthase (GcpE) is a [4Fe-4S] protein. *Angew Chem Int Edu Engl* **41**:4337–4339.

69. Rohdich F, Zepeck F, Adam P, Hecht S, Kaiser J, Laupitz R, Grawert T, Amslinger S, Eisenreich W, Bacher A, Arigoni D. 2003. The deoxyxylulose phosphate pathway of isoprenoid biosynthesis: studies on the mechanisms of the reactions catalyzed by IspG and IspH protein. *Proc Natl Acad Sci USA* **100**:1586–1591.

70. Rohdich F, Hecht S, Gartner K, Adam P, Krieger C, Amslinger S, Arigoni D, Bacher A, Eisenreich W. 2002. Studies on the nonmevalonate terpene biosynthetic pathway: metabolic role of IspH (LytB) protein. *Proc Natl Acad Sci USA* **99**:1158–1163.

71. Agranoff BW, Eggerer H, Henning U, Lynen F. 2013. Biosynthesis of terpenes. VII. Isopentenyl pyrophosphate isomerase. *J Biol Chem* **235**:326–332.

72. Phillips MA, D'Auria JC, Gershenzon J, Pichersky E. 2008. The *Arabidopsis thaliana* type I isopentenyl diphosphate isomerases are targeted to multiple subcellular compartments and have overlapping functions in isoprenoid biosynthesis. *Plant Cell* **20**:677–696.

73. Kuzuyama T, Seto H. 2003. Diversity of the biosynthesis of the isoprene units. *Nat Prod Rep* **20**:171–183.

74. Hahn FM, Hurlburt AP, Poulter CD. 1999. *Escherichia coli* open reading frame 696 is idi, a nonessential gene encoding isopentenyl diphosphate isomerase. *J Bacteriol* **181**:4499–4504.

75. Kaneda K, Kuzuyama T, Takagi M, Hayakawa Y, Seto H. 2001. An unusual isopentenyl diphosphate isomerase found in the mevalonate pathway gene cluster from *Streptomyces* sp strain CL190. *Proc Natl Acad Sci USA* **98**:932–937.

76. Vandermoten S, Haubruge E, Cusson M. 2009. New insights into short-chain prenyltransferases: structural features, evolutionary history and potential for selective inhibition. *Cell Mol Life Sci* **66**:3685–3695.

77. Ambo T, Noike M, Kurokawa H, Koyama T. 2008. Cloning and functional analysis of novel short-chain cis-prenyltransferases. *Biochem Biophys Res Commun* **375**:536–540.

78. Liang PH. 2009. Reaction kinetics, catalytic mechanisms, conformational changes, and inhibitor design for prenyltransferases. *Biochemistry* **48**:6562–6570.

79. Schulbach MC, Brennan PJ, Crick DC. 2000. Identification of a short (C_{15}) chain Z-isoprenyl diphosphate synthase and a homologous long (C_{50}) chain isoprenyl diphosphate synthase in *Mycobacterium tuberculosis*. *J Biol Chem* **275**:22876–22881.

80. Schulbach MC, Mahapatra S, Macchia M, Barontini S, Papi C, Minutolo F, Bertini S, Brennan PJ, Crick DC. 2001. Purification, enzymatic characterization, and inhibition of the Z-farnesyl diphosphate synthase from

Mycobacterium tuberculosis. J Biol Chem **276**:11624–11630.

81. Noike M, Ambo T, Kikuchi S, Suzuki T, Yamashita S, Takahashi S, Kurokawa H, Mahapatra S, Crick DC, Koyama T. 2008. Product chain-length determination mechanism of Z,E-farnesyl diphosphate synthase. *Biochem Biophys Res Commun* **377**:17–22.

82. Wang W, Dong C, McNeil M, Kaur D, Mahapatra S, Crick DC, Naismith JH. 2008. The structural basis of chain length control in Rv1086. *J Mol Biol* **381**:129–140.

83. Kaur D, Brennan PJ, Crick DC. 2004. Decaprenyl diphosphate synthesis in *Mycobacterium tuberculosis. J Bacteriol* **186**:7564–7570.

84. Sato T, Takizawa K, Orito Y, Kudo H, Hoshino T. 2010. Insight into C$_{35}$ terpene biosyntheses by nonpathogenic *Mycobacterium* species: functional analyses of three Z-prenyltransferases and identification of dehydroheptaprenylcyclines. *Chembiochem* **11**:1874–1881.

85. Mann FM, Thomas JA, Peters RJ. 2011. *Rv0989c* encodes a novel (E)-geranyl diphosphate synthase facilitating decaprenyl diphosphate biosynthesis in *Mycobacterium tuberculosis. FEBS Lett* **585**:549–554.

86. Dhiman RK, Schulbach MC, Mahapatra S, Baulard AR, Vissa V, Brennan PJ, Crick DC. 2004. Identification of a novel class of {omega}, E,E-farnesyl diphosphate synthase from *Mycobacterium tuberculosis. J Lipid Res* **45**:1140–1147.

87. Mann FM, Xu M, Davenport EK, Peters RJ. 2012. Functional characterization and evolution of the isotuberculosinol operon in *Mycobacterium tuberculosis* and related mycobacteria. *Front Microbiol* **3**:368.

88. IUPAC-IUB Joint Commission on Biochemical Nomenclature (JCBN). 1987. Prenol nomenclature. Recommendations 1986. *Eur J Biochem* **167**:181–184.

89. Takayama K, Schnoes HK, Semmler EJ. 1973. Characterization of the alkali-stable mannophospholipids of *Mycobacterium smegmatis. Biochim Biophys Acta* **316**:212–221.

90. Besra GS, Sievert T, Lee RE, Slayden RA, Brennan PJ, Takayama K. 1994. Identification of the apparent carrier in mycolic acid synthesis. *Proc Natl Acad Sci USA* **91**:12735–12739.

91. Wolucka BA, de Hoffmann E. 1998. Isolation and characterization of the major form of polyprenyl-phospho-mannose from *Mycobacterium smegmatis. Glycobiology* **8**:955–962.

92. Takayama K, Goldman DS. 1970. Enzymatic synthesis of mannosyl-1-phosphoryl-decaprenol by a cell-free system of *Mycobacterium tuberculosis. J Biol Chem* **245**:6251–6257.

93. Wolucka BA, de Hoffmann E. 1995. The presence of beta-D-ribosyl-1-monophosphodecaprenol in mycobacteria. *J Biol Chem* **270**:20151–20155.

94. El Ghachi M, Bouhss A, Blanot D, Mengin-Lecreulx D. 2004. The *bacA* gene of *Escherichia coli* encodes a undecaprenyl pyrophosphate phosphatase activity. *J Biol Chem* **279**:30106–30113.

95. Sherman MM, Petersen LA, Poulter CD. 1989. Isolation and characterization of isoprene mutants of *Escherichia coli. J Bacteriol* **171**:3619–3628.

96. Minnikin DE. 1982. Lipids: complex lipids, their chemistry, biosynthesis and roles, p 95–184. *In* Ratledge C, Stanford J (ed), *The Biology of Mycobacteria*. Academic Press, London.

97. Embley TM, Stackebrandt E. 1994. The molecular phylogeny and systematics of the *Actinomycetes. Annu Rev Microbiol* **48**:257–289.

98. Pandya KP, King HK. 1966. Ubiquinone and menaquinone in bacteria: a comparative study of some bacterial respiratory systems. *Arch Biochem Biophys* **114**:154–157.

99. Meganathan R. 2001. Biosynthesis of menaquinone (vitamin K-2) and ubiquinone (coenzyme Q): a perspective on enzymatic mechanisms. *Vitam Horm* **61**:173–218.

100. Collins MD, Jones D. 1981. Distribution of isoprenoid quinone structural types in bacteria and their taxonomic implications. *Microbiol Rev* **45**:316–354.

101. da Costa MS, Albuquerque L, Nobre MF, Wait R. 2011. The extraction and identification of respiratory lipoquinones of prokaryotes and their use in taxonomy. *Methods Microbiol.* **38**:197–206.

102. Brodie AF, Revsin B, Kalra V, Phillips P, Bogin E, Higashi T, Murti CR, Cavari BZ, Marquez E. 1970. Biological function of terpenoid quinones. *Biochem Soc Symp* **29**:119–143.

103. Holsclaw CM, Sogi KM, Gilmore SA, Schelle MW, Leavell MD, Bertozzi CR, Leary JA. 2008. Structural characterization of a novel sulfated menaquinone produced by stf3 from *Mycobacterium tuberculosis. ACS Chem Biol* **3**:619–624.

104. Bentley R. 1975. Biosynthesis of vitamin-K and other natural naphthoquinones. *Pure Appl Chem* **41**:47–68.

105. Bentley R, Meganathan R. 1982. Biosynthesis of vitamin K (menaquinone) in bacteria. *Microbiol Rev* **46**:241–280.

106. Meganathan R. 1996. Biosynthesis of vitamin K (menaquinone) and ubiquinone (coenzyme Q), p 642–656. *In* Neihardt FC (ed), *Escherichia coli and Salmonella*. ASM Press, Washington, DC.

107. Azerad R, Bleiler-Hill R, Lederer E. 1965. Biosynthesis of a vitamin K2 by cell-free extracts of *Mycobacterium phlei. Biochem Biophys Res Commun* **19**:194–197.

108. Li HJ, Li X, Liu N, Zhang H, Truglio JJ, Mishra S, Kisker C, Garcia-Diaz M, Tonge PJ. 2011. Mechanism of the intramolecular Claisen condensation reaction catalyzed by MenB, a crotonase superfamily member. *Biochemistry* **50**:9532–9544.

109. Li X, Liu N, Zhang H, Knudson SE, Li HJ, Lai CT, Simmerling C, Slayden RA, Tonge PJ. 2011. CoA adducts of 4-oxo-4-phenylbut-2-enoates: inhibitors of MenB from the *M. tuberculosis* menaquinone biosynthesis pathway. *ACS Med Chem Lett* **2**:818–823.

110. Li X, Liu N, Zhang H, Knudson SE, Slayden RA, Tonge PJ. 2010. Synthesis and SAR studies of 1,4-benzoxazine MenB inhibitors: novel antibacterial agents against *Mycobacterium tuberculosis. Bioorg Med Chem Lett* **20**:6306–6309.

111. Lu X, Zhou R, Sharma I, Li X, Kumar G, Swaminathan S, Tonge PJ, Tan DS. 2012. Stable analogues of OSB-AMP: potent inhibitors of MenE, the o-succinylbenzoate-CoA synthetase from bacterial menaquinone biosynthesis. *Chembiochem* **13**:129–136.

112. Lu X, Zhang H, Tonge PJ, Tan DS. 2008. Mechanism-based inhibitors of MenE, an acyl-CoA synthetase involved in bacterial menaquinone biosynthesis. *Bioorg Med Chem Lett* **18**:5963–5966.

113. Truglio JJ, Theis K, Feng Y, Gajda R, Machutta C, Tonge PJ, Kisker C. 2003. Crystal structure of *Mycobacterium tuberculosis* MenB, a key enzyme in vitamin K2 biosynthesis. *J Biol Chem* **278**:42352–42360.

114. Dhiman RK, Mahapatra S, Slayden RA, Boyne ME, Lenaerts A, Hinshaw JC, Angala SK, Chatterjee D, Biswas K, Narayanasamy P, Kurosu M, Crick DC. 2009. Menaquinone synthesis is critical for maintaining mycobacterial viability during exponential growth and recovery from non-replicating persistence. *Mol Microbiol* **72**:85–97.

115. Debnath J, Siricilla S, Wan B, Crick DC, Lenaerts AJ, Franzblau SG, Kurosu M. 2012. Discovery of selective menaquinone biosynthesis inhibitors against *Mycobacterium tuberculosis*. *J Med Chem* **55**:3739–3755.

116. Collins MD, Goodfellow M, Minnikin DE, Alderson G. 1985. Menaquinone composition of mycolic acid-containing actinomycetes and some sporoactinomycetes. *J Appl Bacteriol* **58**:77–86.

117. Mougous JD, Senaratne RH, Petzold CJ, Jain M, Lee DH, Schelle MW, Leavell MD, Cox JS, Leary JA, Riley LW, Bertozzi CR. 2006. A sulfated metabolite produced by stf3 negatively regulates the virulence of *Mycobacterium tuberculosis*. *Proc Natl Acad Sci USA* **103**:4258–4263.

118. Johnston JB, Kells PM, Podust LM, Ortiz de Montellano PR. 2009. Biochemical and structural characterization of CYP124: a methyl-branched lipid omega-hydroxylase from *Mycobacterium tuberculosis*. *Proc Natl Acad Sci USA* **106**:20687–20692.

119. Vershinin A. 1999. Biological functions of carotenoids: diversity and evolution. *Biofactors* **10**:99–104.

120. Mathews MM, Krinsky NI. 1965. The relationship between carotenoid pigments and resistance to radiation in non-photosynthetic bacteria. *Photochem Photobiol* **4**:813–817.

121. Goodwin TW. 1972. Carotenoids in fungi and non-photosynthetic bacteria. *Prog Ind Microbiol* **11**:29–88.

122. Dembitsky VM. 2005. Astonishing diversity of natural surfactants. 3. Carotenoid glycosides and isoprenoid glycolipids. *Lipids* **40**:535–557.

123. Subczynsk WK, Markowska E, Sieiewiesiuk J. 1991. Effect of polar carotenoids on the oxygen diffusion-concentration product in lipid bilayers. An EPR spin label study. *Biochim Biophys Acta* **1068**:68–72.

124. Woodall AA, Britton G, Jackson MJ. 1997. Carotenoids and protection of phospholipids in solution or in liposomes against oxidation by peroxyl radicals: relationship between carotenoid structure and protective ability. *Biochim Biophys Acta* **1336**:575–586.

125. Woodall AA, Britton G, Jackson MJ. 1995. Antioxidant activity of carotenoids in phosphatidylcholine vesicles: chemical and structural considerations. *Biochem Soc Trans* **23**:133S.

126. Hertzberg S, Liaaen JS. 1967. Bacterial carotenoids. XX. The carotenoids of *Mycobacterium phlei* strain Vera. 2. The structures of the phlei-xanthophylls: two novel tertiary glucosides. *Acta Chem Scand* **21**:15–41.

127. Sieiro C, Poza M, de MT, Villa TG. 2003. Genetic basis of microbial carotenogenesis. *Int Microbiol* **6**:11–16.

128. Gao LY, Groger R, Cox JS, Beverley SM, Lawson EH, Brown EJ. 2003. Transposon mutagenesis of *Mycobacterium marinum* identifies a locus linking pigmentation and intracellular survival. *Infect Immun.* **71**:922–929.

129. Ramakrishnan L, Tran HT, Federspiel NA, Falkow S. 1997. A *crtB* homolog essential for photochromogenicity in *Mycobacterium marinum*: isolation, characterization, and gene disruption via homologous recombination. *J Bacteriol* **179**:5862–5868.

130. Houssaini-Iraqui M, Lazraq MH, Clavel-Seres S, Rastogi N, David HL. 1992. Cloning and expression of *Mycobacterium aurum* carotenogenesis genes in *Mycobacterium smegmatis*. *FEMS Microbiol Lett* **69**:239–244.

131. Viveiros M, Krubasik P, Sandmann G, Houssaini-Iraqui M. 2000. Structural and functional analysis of the gene cluster encoding carotenoid biosynthesis in *Mycobacterium aurum* A+. *FEMS Microbiol Lett* **187**:95–101.

132. Scherzinger D, Scheffer E, Bar C, Ernst H, Al-Babili S. 2010. The *Mycobacterium tuberculosis* ORF Rv0654 encodes a carotenoid oxygenase mediating central and excentric cleavage of conventional and aromatic carotenoids. *FEBS J* **277**:4662–4673.

133. Provvedi R, Kocincova D, Dona V, Euphrasie D, Daffe M, Etienne G, Manganelli R, Reyrat JM. 2008. SigF controls carotenoid pigment production and affects transformation efficiency and hydrogen peroxide sensitivity in *Mycobacterium smegmatis*. *J Bacteriol* **190**:7859–7863.

134. Sato T, Kigawa A, Takagi R, Adachi T, Hoshino T. 2008. Biosynthesis of a novel cyclic C$_{35}$-terpene via the cyclisation of a Z-type C$_{35}$-polyprenyl diphosphate obtained from a nonpathogenic *Mycobacterium* species. *Org Biomol Chem* **6**:3788–3794.

135. Mann FM, Xu M, Chen X, Fulton DB, Russell DG, Peters RJ. 2009. Edaxadiene: a new bioactive diterpene from *Mycobacterium tuberculosis*. *J Am Chem Soc* **131**:17526–17527.

136. Hoshino T, Nakano C, Ootsuka T, Shinohara Y, Hara T. 2011. Substrate specificity of Rv3378c, an enzyme from *Mycobacterium tuberculosis*, and the inhibitory activity of the bicyclic diterpenoids against macrophage phagocytosis. *Org Biomol Chem* **9**:2156–2165.

137. Pethe K, Swenson DL, Alonso S, Anderson J, Wang C, Russell DG. 2004. Isolation of *Mycobacterium tuberculosis* mutants defective in the arrest of phagosome maturation. *Proc Natl Acad Sci USA* **101**:13642–13647.

138. Nakano C, Okamura T, Sato T, Dairi T, Hoshino T. 2005. *Mycobacterium tuberculosis* H37Rv3377c encodes the

diterpene cyclase for producing the halimane skeleton. *Chem Commun (Camb)* **8**:1016–1018.

139. Maugel N, Mann FM, Hillwig ML, Peters RJ, Snider BB. 2010. Synthesis of (+/-)-nosyberkol (iso-tuberculosinol, revised structure of edaxadiene) and (+/-)-tuberculosinol. *Org Lett* **12**:2626–2629.

140. Spangler JE, Carson CA, Sorensen EJ. 2010. Synthesis enables a structural revision of the *Mycobacterium tuberculosis*-produced diterpene, edaxadiene. *Chem Sci* **1**:202–205.

141. Jackson M, Stadthagen G, Gicquel B. 2007. Long-chain multiple methyl-branched fatty acid-containing lipids of *Mycobacterium tuberculosis*: biosynthesis, transport, regulation and biological activities. *Tuberculosis* **87**: 78–86.

142. Cardona P-J, Soto CY, Martin C, Gicquel B, Agusti G, Guirado E, Sirakova TD, Kolattukudy PE, Julian E, Luquin M. 2006. Neutral red reaction is related to virulence and cell wall methyl-branched lipids in *Mycobacterium tuberculosis*. *Microbes Infect* **8**:183–190.

143. Gonzalo Asensio J, Maia C, Ferrer NL, Barilone N, Laval F, Soto CY, Winter N, Daffe M, Gicquel B, Martin C, Jackson M. 2006. The virulence-associated two-component PhoP-PhoR system controls the biosynthesis of polyketide-derived lipids in *Mycobacterium tuberculosis*. *J Biol Chem* **281**:1313–1316.

144. Dubos RJ, Middlebrook G. 1948. Cytochemical reaction of virulent tubercle bacilli. *Am Rev Tuberc* **58**: 698–699.

145. Grzegorzewicz AE, Pham H, Gundi VA, Scherman MS, North EJ, Hess T, Jones V, Gruppo V, Born SE, Korduláková J, Chavadi SS, Morisseau C, Lenaerts AJ, Lee RE, McNeil MR, Jackson M. 2012. Inhibition of mycolic acid transport across the *Mycobacterium tuberculosis* plasma membrane. *Nat Chem Biol* **8**:334–341.

146. Tahlan K, Wilson R, Kastrinsky DB, Arora K, Nair V, Fischer E, Barnes SW, Walker JR, Alland D, Barry CE 3rd, Boshoff HI. 2012. SQ109 targets MmpL3, a membrane transporter of trehalose monomycolate involved in mycolic acid donation to the cell wall core of *Mycobacterium tuberculosis*. *Antimicrobial Agents Chemother* **56**:1797–1809.

147. Belisle JT, Vissa VD, Sievert T, Takayama K, Brennan PJ, Besra GS. 1997. Role of the major antigen of *Mycobacterium tuberculosis* in the cell wall biogenesis. *Science* **276**:1420–1422.

148. Jackson M, Raynaud C, Lanéelle MA, Guilhot C, Laurent-Winter C, Ensergueix D, Gicquel B, Daffé M. 1999. Inactivation of the antigen 85C gene profoundly affects the mycolate content and alters the permeability of the *Mycobacterium tuberculosis* cell envelope. *Mol Microbiol* **31**:1573–1587.

149. Puech V, Guilhot C, Perez E, Tropis M, Armitige LY, Gicquel B, Daffe M. 2002. Evidence for a partial redundancy of the fibronectin-binding proteins for the transfer of mycoloyl residues onto the cell wall arabinogalactan termini of *Mycobacterium tuberculosis*. *Mol Microbiol* **44**:1109–1122.

150. Harth G, Horwitz MA, Tabatadze D, Zamecnik PC. 2002. Targeting the *Mycobacterium tuberculosis* 30/32-kDa mycolyl transferase complex as a therapeutic strategy against tuberculosis: proof of principle by using antisense technology. *Proc Natl Acad Sci USA* **99**: 15614–15619.

151. Armitige LY, Jagannath C, Wanger AR, Norris SJ. 2000. Disruption of the genes encoding antigen 85A and antigen 85B of *Mycobacterium tuberculosis* H37Rv: effect on growth in culture and in macrophages. *Infect Immun* **68**:767–778.

152. Nguyen L, Chinnapapagari S, Thompson CJ. 2005. FbpA-dependent biosynthesis of trehalose dimycolate is required for the intrinsic multidrug resistance, cell wall structure, and colonial morphology of *Mycobacterium smegmatis*. *J Bacteriol* **187**:6603–6611.

153. Katti MK, Dai G, Armitige LY, Rivera Marrero C, Daniel S, Singh CR, Lindsey DR, Dhandayuthapani S, Hunter RL, Jagannath C. 2008. The Delta fbpA mutant derived from *Mycobacterium tuberculosis* H37Rv has an enhanced susceptibility to intracellular antimicrobial oxidative mechanisms, undergoes limited phagosome maturation and activates macrophages and dendritic cells. *Cell Microbiol* **10**:1286–1303.

154. Hunter RL, Armitige L, Jagannath C, Actor JK. 2009. TB research at UT-Houston: a review of cord factor: new approaches to drugs, vaccines and the pathogenesis of tuberculosis. *Tuberculosis (Edinb)* **89**(Suppl 1):S18–S25.

155. Li C, Du Q, Deng W, Xie J. 2012. The biology of *Mycobacterium* cord factor and roles in pathogen-host interaction. *Crit Rev Eukaryotic Gene Expr* **22**:289–297.

156. Linares C, Bernabeu A, Luquin M, Valero-Guillen PL. 2012. Cord factors from atypical mycobacteria (*Mycobacterium alvei*, *Mycobacterium brumae*) stimulate the secretion of some pro-inflammatory cytokines of relevance in tuberculosis. *Microbiology* **158**:2878–2885.

157. Glickman MS. 2008. Cording, cord factors and trehalose dimycolate, p 63–73. *In* Daffé M, Reyrat J-M (ed), *The Mycobacterial Cell Envelope*. ASM Press, Washington, DC.

158. Ishikawa E, Ishikawa T, Morita YS, Toyonaga K, Yamada H, Takeuchi O, Kinoshita T, Akira S, Yoshikai Y, Yamasaki S. 2009. Direct recognition of the mycobacterial glycolipid, trehalose dimycolate, by C-type lectin Mincle. *J Exp Med* **206**:2879–2888.

159. Sakamoto K, Kim MJ, Rhoades ER, Allavena RE, Ehrt S, Wainwright HC, Russell DG, Rohde KH. 2013. Mycobacterial trehalose dimycolate reprograms macrophage global gene expression and activates matrix metalloproteinases. *Infect Immun* **81**:764–776.

160. Lang R. 2013. Recognition of the mycobacterial cord factor by Mincle: relevance for granuloma formation and resistance to tuberculosis. *Front Immunol* **4**:5.

161. Rao V, Fujiwara N, Porcelli SA, Glickman MS. 2005. *Mycobacterium tuberculosis* controls host innate immune activation through cyclopropane modification of a glycolipid effector molecule. *J Exp Med* **201**:535–543.

162. Dubey VS, Sirakova TD, Cynamon MH, Kolattukudy PE. 2003. Biochemical function of *msl5* (*pks8* plus *pks17*) in *Mycobacterium tuberculosis* H37Rv: biosynthesis of monomethyl branched unsaturated fatty acids. *J Bacteriol* 185:4620–4625.

163. Goren MB. 1990. Mycobacterial fatty acid esters of sugars and sulfosugars, p 363–461. *In* Kates M (ed), *Handbook of Lipid Research. Glycolipids, Phosphoglycolipids and Sulfoglycolipids*, vol. 6. Plenum Press, New York/London.

164. Mougous JD, Petzold CJ, Seraratne RH, Lee DH, Akey DL, Lin FL, Munchel SE, Pratt MR, Riley LW, Leary JA, Berger JM, Bertozzi CR. 2004. Identification, function and structure of the mycobacterial sulfotransferase that initiates sulfolipid-1 biosynthesis. *Nat Struct Mol Biol* 11:721–729.

165. Kumar P, Schelle MW, Jain M, Lin FL, Petzold CJ, Leavell MD, Leary JA, Cox JS, Bertozzi CR. 2007. PapA1 and PapA2 are acyltransferases essential for the biosynthesis of the *Mycobacterium tuberculosis* virulence factor sulfolipid-1. *Proc Natl Acad Sci USA* 104:11221–11226.

166. Sirakova TD, Thirumala AK, Dubey VS, Sprecher H, Kolattukudy PE. 2001. The *Mycobacterium tuberculosis pks2* gene encodes the synthase for the hepta- and octamethyl branched fatty acids required for sulfolipid synthesis. *J Biol Chem* 276:16833–16839.

167. Trivedi OA, Arora P, Sridharan V, Tickoo R, Mohanty D, Gokhale RS. 2004. Enzymic activation and transfer of fatty acids as acyl-adenylates in mycobacteria. *Nature* 428:441–445.

168. Seeliger JC, Holsclaw CM, Schelle MW, Botyanszki Z, Gilmore SA, Tully SE, Niederweis M, Cravatt BF, Leary JA, Bertozzi CR. 2012. Elucidation and chemical modulation of sulfolipid-1 biosynthesis in *Mycobacterium tuberculosis*. *J Biol Chem* 287:7990–8000.

169. Bhatt K, Gurcha SS, Bhatt A, Besra GS, Jacobs WR Jr. 2007. Two polyketide-synthase-associated acyltransferases are required for sulfolipid biosynthesis in *Mycobacterium tuberculosis*. *Microbiology* 153:513–520.

170. Converse SE, Mougous JD, Leavell MD, Leary JA, Bertozzi CR, Cox JS. 2003. MmpL8 is required for sulfolipid-1 biosynthesis and *Mycobacterium tuberculosis* virulence. *Proc Natl Acad Sci USA* 100:6121–6126.

171. Domenech P, Reed MB, Dowd CS, Manca C, Kaplan G, Barry CE III. 2004. The role of MmpL8 in sulfatide biogenesis and virulence of *Mycobacterium tuberculosis*. *J Biol Chem* 279:21257–21265.

172. Zheng J, Wei C, Zhao L, Liu L, Leng W, Li W, Jin Q. 2011. Combining blue native polyacrylamide gel electrophoresis with liquid chromatography tandem mass spectrometry as an effective strategy for analyzing potential membrane protein complexes of *Mycobacterium bovis* bacillus Calmette-Guerin. *BMC Genomics* 12:40.

173. Graham JE, Clark-Curtiss JE. 1999. Identification of *Mycobacterium tuberculosis* RNAs synthesized in response to phagocytosis by human macrophages by selective capture of transcribed sequences (SCOTS). *Proc Natl Acad Sci USA* 96:11554–11559.

174. Singh A, Crossman DK, Mai D, Guidry L, Voskuil MI, Renfrow MB, Steyn AJ. 2009. *Mycobacterium tuberculosis* WhiB3 maintains redox homeostasis by regulating virulence lipid anabolism to modulate macrophage response. *PLoS Pathog* 5:e1000545.

175. Lee W, Vanderven BC, Fahey RJ, Russell DG. 2013. Intracellular *Mycobacterium tuberculosis* exploits host-derived fatty acids to limit metabolic stress. *J Biol Chem* 288:6788–6800.

176. Walters SB, Dubnau E, Kolesnikova I, Laval F, Daffé M, Smith I. 2006. The *Mycobacterium tuberculosis* PhoPR two-component system regulates genes essential for virulence and complex lipid biosynthesis. *Mol Microbiol* 60:312–330.

177. Chesne-Seck M-L, Barilone N, Boudou F, Gonzalo Asensio J, Kolattukudy PE, Martin C, Cole ST, Gicquel B, Gopaul DN, Jackson M. 2008. A point mutation in the two-component regulator PhoP-PhoR accounts for the absence of polyketide-derived acyltrehaloses but not that of phthiocerol dimycocerosates in *Mycobacterium tuberculosis* H37Ra. *J Bact* 190:1329–1334.

178. Goyal R, Das AK, Singh R, Singh PK, Korpole S, Sarkar D. 2011. Phosphorylation of PhoP protein plays direct regulatory role in lipid biosynthesis of *Mycobacterium tuberculosis*. *J Biol Chem* 286:45197–45208.

179. Cimino M, Thomas C, Namouchi A, Dubrac S, Gicquel B, Gopaul DN. 2012. Identification of DNA binding motifs of the *Mycobacterium tuberculosis* PhoP/PhoR two-component signal transduction system. *PloS One* 7:e42876.

180. Goren MB, Brennan PJ. 1979. Mycobacterial lipids: chemistry and biologic activities, p 63–193. *In* Youmans GP (ed), *Tuberculosis*. W. B. Saunders, Philadelphia, PA.

181. Bertozzi CR, Schelle MW. 2008. Sulfated metabolites from *Mycobacterium tuberculosis*: sulfolipid-1 and beyond, p 291–304. *In* Daffé M, Reyrat J-M (ed), *The Mycobacterial Cell Envelope*. ASM Press, Washington, DC.

182. Gilleron M, Stenger S, Mazorra Z, Wittke F, Mariotti S, Böhmer G, Prandi J, Mori L, Puzo G, De Libero G. 2004. Diacylated sulfoglycolipids are novel mycobacterial antigens stimulating CD1-restricted T cells during infection with *Mycobacterium tuberculosis*. *J Exp Med* 199:649–659.

183. Guiard J, Collmann A, Garcia-Alles LF, Mourey L, Brando T, Mori L, Gilleron M, Prandi J, De Libero G, Puzo G. 2009. Fatty acyl structures of *Mycobacterium tuberculosis* sulfoglycolipid govern T cell response. *J Immunol* 182:7030–7037.

184. Rousseau C, Turner OC, Rush E, Bordat Y, Sirakova TD, Kolattukudy PE, Ritter S, Orme IM, Gicquel B, Jackson M. 2003. Sulfolipid deficiency does not affect the virulence of *Mycobacterium tuberculosis* H37Rv in mice and guinea pigs. *Infect Immun* 71:4684–4690.

185. Lamichhane G, Tyagi S, Bishai WR. 2005. Designer arrays for defined mutant analysis to detect genes essential for survival of *Mycobacterium tuberculosis* in mouse lungs. *Infect Immun* 73:2533–2540.

186. Domenech P, Reed MB, Barry CE III. 2005. Contribution of the *Mycobacterium tuberculosis* MmpL protein family to virulence and drug resistance. *Infect Immun* 73:3492–3501.

187. Gilmore SA, Schelle MW, Holsclaw CM, Leigh CD, Jain M, Cox JS, Leary JA, Bertozzi CR. 2012. Sulfolipid-1 biosynthesis restricts *Mycobacterium tuberculosis* growth in human macrophages. *ACS Chem Biol* 7:863–870.

188. Minnikin DE, Dobson G, Sesardic D, Ridell M. 1985. Mycolipenates and mycolipanolates of trehalose from *Mycobacterium tuberculosis*. *J Gen Microbiol* 131:1369–1374.

189. Lemassu A, Lanéelle M-A, Daffé M. 1991. Revised structure of a trehalose-containing immunoreactive glycolipid of *Mycobacterium tuberculosis*. *FEMS Microbiol Lett* 78:171–176.

190. Besra GS, Bolton R, McNeil MR, Ridell M, Simpson KE, Glushka J, van Halbeek H, Brennan PJ, Minnikin DE. 1992. Structure elucidation and antigenicity of a novel family of glycolipid antigens from *Mycobacterium tuberculosis* H37Rv. *Biochemistry* 31:9832–9837.

191. Munoz M, Lanéelle M-A, Luquin M, Torrelles J, Julian E, Ausina V, Daffé M. 1997. Occurence of an antigenic triacyl trehalose in clinical isolates and reference strains of *Mycobacterium tuberculosis*. *FEMS Microbiol Lett* 157:251–259.

192. Daffé M, Lacave C, Lanéelle M-A, Gillois M, Lanéelle G. 1988. Polyphthienoyl trehalose, glycolipids specific for virulent strains of the tubercle bacillus. *Eur J Biochem* 172:579–584.

193. Gautier N, Lopez Marin LM, Lanéelle M-A, Daffé M. 1992. Structure of mycoside F, a family of trehalose-containing glycolipids of *Mycobacterium fortuitum*. *FEMS Microbiol Lett* 98:81–88.

194. Ariza MA, Martin-Luengo F, Valero-Guillen PL. 1994. A family of diacyltrehaloses isolated from *Mycobacterium fortuitum*. *Microbiology* 140:1989–1994.

195. Lee KS, Dubey VS, Kolattukudy PE, Song CH, Shin AR, Jung SB, Yang CS, Kim SY, Jo EK, Park JK, Kim HJ. 2007. Diacyltrehalose of *Mycobacterium tuberculosis* inhibits lipopolysaccharide- and mycobacteria-induced proinflammatory cytokine production in human monocytic cells. *FEMS Microbiol Lett* 267:121–128.

196. Dubey VS, Sirakova TD, Kolattukudy PE. 2002. Disruption of *msl3* abolishes the synthesis of mycolipanoic and mycolipenic acids required for polyacyltrehalose synthesis in *Mycobacterium tuberculosis* H37Rv and causes cell aggregation. *Mol Microbiol* 45:1451–1459.

197. Rousseau C, Neyrolles O, Bordat Y, Giroux S, Sirakova TD, Prevost M-C, Kolattukudy PE, Gicquel B, Jackson M. 2003. Deficiency in mycolipenate- and mycosanoate-derived acyltrehaloses enhances early interactions of *Mycobacterium tuberculosis* with host cells. *Cell Microbiol* 5:405–415.

198. Lynett J, Stokes RW. 2007. Selection of transposon mutants of *Mycobacterium tuberculosis* with increased macrophage infectivity identifies fadD23 to be involved in sulfolipid production and association with macrophages. *Microbiology* 153:3133–3140.

199. Hatzios SK, Schelle MW, Holsclaw CM, Behrens CR, Botyanszki Z, Lin FL, Carlson BL, Kumar P, Leary JA, Bertozzi CR. 2009. PapA3 is an acyltransferase required for polyacyltrehalose biosynthesis in *Mycobacterium tuberculosis*. *J Biol Chem* 284:12745–12751.

200. Daffé M, Lemassu A. 2000. Glycobiology of the mycobacterial surface. Structures and biological activities of the cell envelope glycoconjugates. *In* Doyle RJ (ed), *Glycomicrobiology*. Kluwer Academic/Plenum, New York.

201. Hunter SW, Murphy RC, Clay K, Goren MB, Brennan PJ. 1983. Trehalose-containing lipooligosaccharides. A new class of species-specific antigens from *Mycobacterium*. *J Biol Chem* 258:10481–10487.

202. Saadat S, Ballou CE. 1983. Pyruvylated glycolipids from *Mycobacterium smegmatis*. Structures of two oligosaccharide components. *J Biol Chem* 258:1813–1818.

203. Daffé M, McNeil MR, Brennan PJ. 1991. Novel type-specific lipooligosaccharides from *Mycobacterium tuberculosis*. *Biochemistry* 30:378–388.

204. Burguiere A, Hitchen P, Dover LG, Kremer L, Ridell M, Alexander DC, Liu J, Morris HR, Minnikin DE, Dell A, Besra GS. 2005. LosA, a key glycosyltransferase involved in the biosynthesis of a novel family of glycosylated acyltrehalose lipooligosaccharides from *Mycobacterium marinum*. *J Biol Chem* 280:42124–42133.

205. Ren H, Dover LG, Islam ST, Alexander DC, Chen JM, Besra GS, Liu J. 2007. Identification of the lipooligosaccharide biosynthetic gene cluster from *Mycobacterium marinum*. *Mol Microbiol* 63:1345–1359.

206. Rombouts Y, Alibaud L, Carrere-Kremer S, Maes E, Tokarski C, Elass E, Kremer L, Guerardel Y. 2011. Fatty acyl chains of *Mycobacterium marinum* lipooligosaccharides: structure, localization and acylation by PapA4 (MMAR_2343) protein. *J Biol Chem* 286:33678–33688.

207. Etienne G, Malaga W, Laval F, Lemassu A, Guilhot C, Daffe M. 2009. Identification of the polyketide synthase involved in the biosynthesis of the surface-exposed lipooligosaccharides in mycobacteria. *J Bacteriol* 191:2613–2621.

208. Besra GS, Khoo KH, Belisle JT, McNeil MR, Morris HR, Dell A, Brennan PJ. 1994. New pyruvylated, glycosylated acyltrehaloses from *Mycobacterium smegmatis* strains, and their implications for phage resistance in mycobacteria. *Carbohydr Res* 251:99–114.

209. Sarkar D, Sidhu M, Singh A, Chen J, Lammas DA, van der Sar AM, Besra GS, Bhatt A. 2011. Identification of a glycosyltransferase from *Mycobacterium marinum* involved in addition of a caryophyllose moiety in lipooligosaccharides. *J Bacteriol* 193:2336–2340.

210. Sonden B, Kocincova D, Deshayes C, Euphrasie D, Rhayat L, Laval F, Frehel C, Daffe M, Etienne G, Reyrat J-M. 2005. Gap, a mycobacterial specific integral membrane protein, is required for glycolipid transport to the cell surface. *Mol Microbiol* 58:426–440.

211. Ripoll F, Deshayes C, Pasek S, Laval F, Beretti JL, Biet F, Risler JL, Daffe M, Etienne G, Gaillard JL, Reyrat JM. 2007. Genomics of glycopeptidolipid biosynthesis in *Mycobacterium abscessus* and *M. chelonae*. *BMC Genomics* 8:114.

212. Deshayes C, Kocinkova D, Etienne G, Reyrat J-M. 2008. Glycopeptidolipids: a complex pathway for small pleitotropic molecules. *In* Daffé M, Reyrat J-M (ed), *The Mycobacterial Cell Envelope*. ASM Press, Washington, DC.

213. van der Woude AD, Sarkar D, Bhatt A, Sparrius M, Raadsen SA, Boon L, Geurtsen J, van der Sar AM, Luirink J, Houben EN, Besra GS, Bitter W. 2012. Unexpected link between lipooligosaccharide biosynthesis and surface protein release in *Mycobacterium marinum*. *J Biol Chem* 287:20417–20429.

214. Belisle JT, Brennan PJ. 1989. Chemical basis of rough and smooth variation in mycobacteria. *J Bacteriol* 171:3465–3470.

215. Lemassu A, Levy-Frebault VV, Laneelle MA, Daffe M. 1992. Lack of correlation between colony morphology and lipooligosaccharide content in the *Mycobacterium tuberculosis* complex. *J Gen Microbiol* 138:1535–1541.

216. Rombouts Y, Burguiere A, Maes E, Coddeville B, Elass E, Guerardel Y, Kremer L. 2009. *Mycobacterium marinum* lipooligosaccharides are unique caryophyllose-containing cell wall glycolipids that inhibit tumor necrosis factor-alpha secretion in macrophages. *J Biol Chem* 284:20975–20988.

217. Moody DB, Ulrichs T, Muhlecker W, Young DC, Gurcha SS, Grant E, Rosat JP, Brenner MB, Costello CE, Besra GS, Porcelli SA. 2000. CD1c-mediated T-cell recognition of isoprenoid glycolipids in *Mycobacterium tuberculosis* infection. *Nature* 404:884–888.

218. Matsunaga I, Bhatt A, Young DC, Cheng T-Y, Eyles SJ, Besra GS, Briken V, Porcelli SA, Costello CE, Jacobs WR Jr, Moody DB. 2004. *Mycobacterium tuberculosis pks12* produces a novel polyketide presented by CD1c to T cells. *J Exp Med* 200:1559–1569.

219. Matsunaga I, Sugita M. 2012. Mycoketide: a CD1c-presented antigen with important implications in mycobacterial infection. *Clin Dev Immunol* 2012:981821.

220. Chopra T, Banerjee S, Gupta S, Yadav G, Anand S, Surolia A, Roy RP, Mohanty D, Gokhale RS. 2008. Novel intermolecular iterative mechanism for biosynthesis of mycoketide catalyzed by a bimodular polyketide synthase. *PLoS Biol* 6:1584–1598.

221. Daffé M, Lanéelle M-A. 1988. Distribution of phthiocerol diester, phenolic mycosides and related compounds in mycobacteria. *J Gen Microbiol* 134:2049–2055.

222. Daffé M, Lacave C, Laneelle MA, Laneelle G. 1987. Structure of the major triglycosyl phenol-phthiocerol of *Mycobacterium tuberculosis* (strain Canetti). *Eur J Biochem* 167:155–160.

223. Constant P, Pérez E, Malaga W, Lanéelle M-A, Saurel O, Daffé M, Guilhot C. 2002. Role of the *pks15/1* gene in the biosynthesis of phenolglycolipids in the *Mycobacterium tuberculosis* complex. *J Biol Chem* 277:38148–38158.

224. Reed MB, Domenech P, Manca C, Su H, Barczak AK, Kreiswirth BN, Kaplan G, Barry CE III. 2004. A glycolipid of hypervirulent tuberculosis strains that inhibits the innate immune response. *Nature* 431:84–87.

225. Huet G, Constant P, Malaga W, Laneelle MA, Kremer K, van Soolingen D, Daffe M, Guilhot C. 2009. A lipid profile typifies the Beijing strains of *Mycobacterium tuberculosis*: identification of a mutation responsible for a modification of the structures of phthiocerol dimycocerosates and phenolic glycolipids. *J Biol Chem* 284:27101–27113.

226. Malaga W, Constant P, Euphrasie D, Cataldi A, Daffé M, Reyrat J-M, Guilhot C. 2008. Deciphering the genetic bases of the structural diversity of phenolic glycolipids in strains of the *Mycobacterium tuberculosis* complex. *J Biol Chem* 283:15177–15184.

227. Goren MB, Brokl O, Schaefer WB. 1974. Lipids of putative relevance to virulence in *Mycobacterium tuberculosis*: phthiocerol dimycocerosate and the attenuation indicator lipid. *Infect Immun* 9:150–158.

228. Krishnan N, Malaga W, Constant P, Caws M, Tran TH, Salmons J, Nguyen TN, Nguyen DB, Daffe M, Young DB, Robertson BD, Guilhot C, Thwaites GE. 2011. *Mycobacterium tuberculosis* lineage influences innate immune response and virulence and is associated with distinct cell envelope lipid profiles. *PLoS One* 6:e23870.

229. Camacho LR, Constant P, Raynaud C, Lanéelle M-A, Triccas JA, Gicquel B, Daffé M, Guilhot C. 2001. Analysis of the phthiocerol dimycocerosate locus of *Mycobacterium tuberculosis*. Evidence that this lipid is involved in the cell wall permeability barrier. *J Biol Chem* 276:19845–19854.

230. Waddell SJ, Chung GA, Gibson KJ, Everett MJ, Minnikin DE, Besra GS, Butcher PD. 2005. Inactivation of polyketide synthase and related genes results in the loss of complex lipids in *Mycobacterium tuberculosis* H37Rv. *Lett Appl Microbiol* 40:201–206.

231. Sirakova TD, Dubey VS, Kim H-J, Cynamon MH, Kolattukudy PE. 2003. The largest open reading frame (*pks12*) in the *Mycobacterium tuberculosis* genome is involved in pathogenesis and dimycocerosyl phthiocerol synthesis. *Infect Immun* 71:3794–3801.

232. Sirakova TD, Dubey VS, Cynamon MH, Kolattukudy PE. 2003. Attenuation of *Mycobacterium tuberculosis* by disruption of a *mas*-like gene or a chalcone synthase-like gene, which causes deficiency in dimycocerosyl phthiocerol synthesis. *J Bacteriol* 185:2999–3008.

233. Hotter GS, Wards BJ, Mouat P, Besra GS, Gomes J, Singh M, Bassett S, Kawakami P, Wheeler PR, de Lisle GW, Collins DM. 2005. Transposon mutagenesis of Mb0100 at the *ppe1-nrp* locus in *Mycobacterium bovis* disrupts phthiocerol dimycocerosate (PDIM) and glycosylphenol-PDIM biosynthesis, producing an avirulent strain with vaccine properties at least equal to those of *M. bovis* BCG. *J Bacteriol* 187:2267–2277.

234. Rousseau C, Sirakova TD, Dubey VS, Bordat Y, Kolattukudy PE, Gicquel B, Jackson M. 2003. Virulence attenuation of two Mas-like polyketide synthase mutants of *Mycobacterium tuberculosis*. *Microbiology* 149:1837–1847.

235. Cox JS, Chen B, McNeil M, Jacobs WR Jr. 1999. Complex lipid determines tissue-specific replication of *Mycobacterium tuberculosis* in mice. *Nature* **402**: 79–83.

236. Simeone R, Leger M, Constant P, Malaga W, Marrakchi H, Daffe M, Guilhot C, Chalut C. 2010. Delineation of the roles of FadD22, FadD26 and FadD29 in the biosynthesis of phthiocerol dimycocerosates and related compounds in *Mycobacterium tuberculosis*. *FEBS J* **277**:2715–2725.

237. Stadthagen G, Kordulakova J, Griffin R, Constant P, Bottova I, Barilone N, Gicquel B, Daffe M, Jackson M. 2005. p-Hydroxybenzoic acid synthesis in *Mycobacterium tuberculosis*. *J Biol Chem* **280**:40699–40706.

238. Chavadi SS, Edupuganti UR, Vergnolle O, Fatima I, Singh SM, Soll CE, Quadri LE. 2011. Inactivation of tesA reduces cell wall lipid production and increases drug susceptibility in mycobacteria. *J Biol Chem* **286**: 24616–24625.

239. Rao A, Ranganathan A. 2004. Interaction studies on proteins encoded by the phthiocerol dimycocerosate locus of *Mycobacterium tuberculosis*. *Mol Genet Genomics* **272**:571–579.

240. Rainwater DL, Kolattukudy PE. 1983. Synthesis of mycocerosic acids from methylmalonyl coenzyme A by cell-free extracts of *Mycobacterium tuberculosis* var. *bovis* BCG. *J Biol Chem* **258**:2979–2985.

241. Rainwater DL, Kolattukudy PE. 1985. Fatty acid biosynthesis in *Mycobacterium tuberculosis* var. *bovis* Bacillus Calmette-Guerin. Purification and characterization of a novel fatty acid synthase, mycocerosic acid synthase, which elongates n-fatty acyl-CoA with methylmalonyl-CoA. *J Biol Chem* **260**:616–623.

242. Mathur M, Kolattukudy PE. 1992. Molecular cloning and sequencing of the gene for mycocerosic acid synthase, a novel fatty acid elongating multifunctional enzyme, from *Mycobacterium tuberculosis* var. *bovis* Bacillus Calmette-Guerin. *J Biol Chem* **267**:19388–19395.

243. Azad AK, Sirakova TD, Rogers LM, Kolattukudy PE. 1996. Targeted replacement of the mycocerosic acid synthase gene in *Mycobacterium bovis* BCG produces a mutant that lacks mycosides. *Proc Natl Acad Sci USA* **93**:4787–4792.

244. Trivedi OA, Arora P, Vats A, Ansari MZ, Tickoo R, Sridharan V, Mohanty D, Gokhale RS. 2005. Dissecting the mechanism and assembly of a complex virulence mycobacterial lipid. *Mol Cell* **17**:631–643.

245. Fitzmaurice AM, Kolattukudy PE. 1997. Open reading frame 3, which is adjacent to the mycocerosic acid synthase gene, is expressed as an acyl coenzyme A synthase in *Mycobacterium bovis* BCG. *J Bacteriol* **179**:2608–2615.

246. Fitzmaurice AM, Kolattukudy PE. 1998. An acyl-CoA synthase (acoas) gene adjacent to the mycocerosic acid synthase (mas) locus is necessary for mycocerosyl lipid synthesis in *Mycobacterium tuberculosis* var. *bovis* BCG. *J Biol Chem* **273**:8033–8039.

247. Perez E, Constant P, Laval F, Lemassu A, Laneelle MA, Daffe M, Guilhot C. 2004. Molecular dissection of the role of two methyltransferases in the biosynthesis of phenolglycolipids and phthiocerol dimycoserosate in the *Mycobacterium tuberculosis* complex. *J Biol Chem* **279**: 42584–42592.

248. Perez E, Constant P, Lemassu A, Laval F, Daffe M, Guilhot C. 2004. Characterization of three glycosyltransferases involved in the biosynthesis of the phenolic glycolipid antigens from the *Mycobacterium tuberculosis* complex. *J Biol Chem* **279**:42574–42583.

249. Simeone R, Huet G, Constant P, Malaga W, Lemassu A, Laval F, Daffe M, Guilhot C, Chalut C. 2013. Functional characterisation of three O-methyltransferases involved in the biosynthesis of phenolglycolipids in *Mycobacterium tuberculosis*. *PloS One* **8**: e58954.

250. Tabouret G, Astarie-Dequeker C, Demangel C, Malaga W, Constant P, Ray A, Honore N, Bello NF, Perez E, Daffe M, Guilhot C. 2010. *Mycobacterium leprae* phenolglycolipid-1 expressed by engineered *M. bovis* BCG modulates early interaction with human phagocytes. *PLoS Pathog* **6**:e1001159.

251. Jain M, Chow ED, Cox JS. 2008. The MmpL protein family, p 201–210. *In* Daffe M, Reyrat J-M (ed), *The Mycobacterial Cell Envelope*. ASM Press, Washington, DC.

252. Jain M, Cox JS. 2005. Interaction between polyketide synthase and transporter suggests coupled synthesis and export of virulence lipid in *M. tuberculosis*. *PLoS Pathog* **1**:12–19.

253. Sulzenbacher G, Canaan S, Bordat Y, Neyrolles O, Stadthagen G, Roig-Zamboni V, Rauzier J, Maurin D, Laval F, Daffé M, Cambillau C, Gicquel B, Bourne Y, Jackson M. 2006. LppX is a lipoprotein required for the translocation of phthiocerol dimycocerosates to the surface of *Mycobacterium tuberculosis*. *EMBO J* **25**: 1436–1444.

254. Daffé M, Cho SN, Chatterjee D, Brennan PJ. 1991. Chemical synthesis and seroreactivity of a neoantigen containing the oligosaccharide hapten of the *Mycobacterium tuberculosis*-specific phenolic glycolipid. *J Infect Dis* **163**:161–168.

255. Cho SN, Shin JS, Daffe M, Chong Y, Kim SK, Kim JD. 1992. Production of monoclonal antibody to a phenolic glycolipid of *Mycobacterium tuberculosis* and its use in detection of the antigen in clinical isolates. *J Clin Microbiol* **30**:3065–3069.

256. Simonney N, Molina JM, Molimard M, Oksenhendler E, Lagrange PH. 1996. Comparison of A60 and three glycolipid antigens in an ELISA test for tuberculosis. *Clin Microbiol Infect* **2**:214–222.

257. Simonney N, Molina JM, Molimard M, Oksenhendler E, Perronne C, Lagrange PH. 1995. Analysis of the immunological humoral response to *Mycobacterium tuberculosis* glycolipid antigens (DAT, PGLTb1) for diagnosis of tuberculosis in HIV-seropositive and -seronegative patients. *Eur J Clin Microbiol Infect Dis* **14**: 883–891.

258. Puzo G. 1990. The carbohydrate- and lipid- containing cell wall of mycobacteria, phenolic glycolipids: structure and immunological properties. *Crit Rev Microbiol* **17:**305–327.

259. Minnikin DE. 1982. Lipids: complex lipids, their chemistry, biosynthesis and roles, p 95–184. *In* Ratledge C, Stanford J (ed), *The Biology of Mycobacteria*, vol. **1.** Academic Press, London.

260. Guilhot C, Chalut C, Daffé M. 2008. Biosynthesis and roles of phenolic glycolipids and related molecules in *Mycobacterium tuberculosis*, p 273–289. *In* Daffé M, Reyrat J-M (ed), *The Mycobacterial Cell Envelope.* ASM Press, Washington, DC.

261. Fournié JJ, Adams E, Mullins RJ, Basten A. 1989. Inhibition of human lymphoproliferative responses by mycobacterial phenolic glycolipids. *Infect Immun* **57:** 3653–3659.

262. Vachula M, Holzer TJ, Andersen BR. 1989. Suppression of monocyte oxidative response by phenolic glycolipid I of *Mycobacterium leprae*. *J Immunol* **60:**203–206.

263. Vachula M, Holzer TJ, Kizlaitis L, Andersen BR. 1990. Effect of *Mycobacterium leprae*'s phenolic glycolipid-I on interferon-gamma augmentation of monocyte oxidative responses. *Int J Lepr Other Mycobact Dis* **58:** 342–346.

264. Mehra V, Brennan PJ, Rada E, Convit J, Bloom BR. 1984. Lymphocyte suppression in leprosy induced by unique *M. leprae* glycolipid. *Nature* **308:**194–196.

265. Chan J, Fujiwara T, Brennan PJ, McNeil M, Turco SJ, Sibille J-C, Snapper M, Aisen P, Bloom BR. 1989. Microbial glycolipids: possible virulence factors that scavenge oxygen radicals. *Proc Natl Acad Sci USA* **86:** 2453–2457.

266. Smith DW, Randall HM, Maclennan AP, Lederer E. 1960. Mycosides: a new class of type-specific glycolipids of mycobacteria. *Nature* **186:**887–888.

267. Aspinall GO, Chatterjee D, Brennan PJ. 1995. The variable surface glycolipids of mycobacteria: structures, synthesis of epitopes, and biological properties. *Adv Carbohydr Chem Biochem* **51:**169–242.

268. Chatterjee D, Khoo KH. 2001. The surface glycopeptidolipids of mycobacteria: structures and biological properties. *Cell Mol Life Sci* **58:**2018–2042.

269. Riviere M, Puzo G. 1991. A new type of serine-containing glycopeptidolipid from *Mycobacterium xenopi*. *J Biol Chem* **266:**9057–9063.

270. Besra GS, McNeil MR, Rivoire B, Khoo KH, Morris HR, Dell A, Brennan PJ. 1993. Further structural definition of a new family of glycopeptidolipids from *Mycobacterium xenopi*. *Biochemistry* **32:**347–355.

271. Daffé M, Laneelle MA, Puzo G. 1983. Structural elucidation by field desorption and electron-impact mass spectrometry of the C-mycosides isolated from *Mycobacterium smegmatis*. *Biochim Biophys Acta* **751:** 439–443.

272. Vats A, Singh AK, Mukherjee R, Chopra T, Ravindran MS, Mohanty D, Chatterji D, Reyrat JM, Gokhale RS.

273. Barrow WW, Brennan PJ. 1982. Isolation in high frequency of rough variants of *Mycobacterium intracellulare* lacking C-mycoside glycopeptidolipid antigens. *J Bacteriol* **150:**381–384.

274. Billman-Jacobe H, McConville MJ, Haites RE, Kovacevic S, Coppel RL. 1999. Identification of a peptide synthetase involved in the biosynthesis of glycopeptidolipids of *Mycobacterium smegmatis*. *Mol Microbiol* **33:**1244–1253.

275. Etienne G, Villeneuve C, Billman-Jacobe H, Astarie-Dequeker C, Dupont M-A, Daffé M. 2002. The impact of the absence of glycopeptidolipids on the ultrastructure, cell surface and cell wall properties, and phagocytosis of *Mycobacterium smegmatis*. *Microbiology* **148:** 3089–3100.

276. Schaefer WB. 1965. Serologic identification and classification of the atypical mycobacteria by their agglutination. *Am Rev Respir Dis* **92:**85–93.

277. Draper P, Rees RJ. 1973. The nature of the electron-transparent zone that surrounds *Mycobacterium lepraemurium* inside host cells. *J Gen Microbiol* **77:** 79–87.

278. Draper P. 1974. The mycoside capsule of *Mycobacterium avium* 357. *J Gen Microbiol* **83:**431–433.

279. Vergne I, Daffe M. 1998. Interaction of mycobacterial glycolipids with host cells. *Front Biosci* **3:**d865–d876.

280. Martinez A, Torello S, Kolter R. 1999. Sliding motility in mycobacteria. *J Bacteriol* **181:**7331–7338.

281. Recht J, Martinez A, Torello S, Kolter R. 2000. Genetic analysis of sliding motility in *Mycobacterium smegmatis*. *J Bacteriol* **182:**4348–4351.

282. Draper P, Rees RJ. 1970. Electron-transparent zone of mycobacteria may be a defence mechanism. *Nature* **228:**860–861.

283. Tereletsky MJ, Barrow WW. 1983. Postphagocytic detection of glycopeptidolipids associated with the superficial L1 layer of *Mycobacterium intracellulare*. *Infect Immun* **41:**1312–1321.

284. Rulong S, Aguas AP, da Silva PP, Silva MT. 1991. Intramacrophagic *Mycobacterium avium* bacilli are coated by a multiple lamellar structure: freeze fracture analysis of infected mouse liver. *Infect Immun* **59:** 3895–3902.

285. Brownback PE, Barrow WW. 1988. Modified lymphocyte response to mitogens after intraperitoneal injection of glycopeptidolipid antigens from *Mycobacterium avium* complex. *Infect Immun* **56:**1044–1050.

286. Hooper LC, Barrow WW. 1988. Decreased mitogenic response of murine spleen cells following intraperitoneal injection of serovar-specific glycopeptidolipid antigens from the *Mycobacterium avium* complex. *Adv Exp Med Biol* **239:**309–325.

287. Pourshafie M, Ayub Q, Barrow WW. 1993. Comparative effects of *Mycobacterium avium* glycopeptidolipid

2012. Retrobiosynthetic approach delineates the biosynthetic pathway and the structure of the acyl chain of mycobacterial glycopeptidolipids. *J Biol Chem* **287:** 30677–30687.

and lipopeptide fragment on the function and ultra-structure of mononuclear cells. *Clin Exp Immunol* **93**: 72–79.

288. Barrow WW, de Sousa JP, Davis TL, Wright EL, Bachelet M, Rastogi N. 1993. Immunomodulation of human peripheral blood mononuclear cell functions by defined lipid fractions of *Mycobacterium avium*. *Infect Immun* **61**:5286–5293.

289. Kano H, Doi T, Fujita Y, Takimoto H, Yano I, Kumazawa Y. 2005. Serotype-specific modulation of human monocyte functions by glycopeptidolipid (GPL) isolated from *Mycobacterium avium* complex. *Biol Pharm Bull* **28**:335–339.

290. Horgen L, Barrow EL, Barrow WW, Rastogi N. 2000. Exposure of human peripheral blood mononuclear cells to total lipids and serovar-specific glycopeptidolipids from *Mycobacterium avium* serovars 4 and 8 results in inhibition of TH1-type responses. *Microbial Pathog* **29**: 9–16.

291. Barrow WW, Davis TL, Wright EL, Labrousse V, Bachelet M, Rastogi N. 1995. Immunomodulatory spectrum of lipids associated with *Mycobacterium avium* serovar 8. *Infect Immun* **63**:126–133.

292. Lagrange PH, Fourgeaud M, Neway T, Pilet C. 1994. Enhanced resistance against lethal disseminated *Candida albicans* infection in mice treated with polar glycopeptidolipids from *Mycobacterium chelonae* (pGPL-Mc). *CR Acad Sci III* **317**:1107–1113.

293. Gjata B, Hannoun C, Boulouis HJ, Neway T, Pilet C. 1994. Adjuvant activity of polar glycopeptidolipids of *Mycobacterium chelonae* (pGPL-Mc) on the immunogenic and protective effects of an inactivated influenza vaccine. *CR Acad Sci III* **317**:257–263.

294. Vincent-Naulleau S, Neway T, Thibault D, Barrat F, Boulouis HJ, Pilet C. 1995. Effects of polar glycopeptidolipids of *Mycobacterium chelonae* (pGPL-Mc) on granulomacrophage progenitors. *Res Immunol* **146**: 363–371.

295. Vincent-Naulleau S, Thibault D, Neway T, Pilet C. 1997. Stimulatory effects of polar glycopeptidolipids of *Mycobacterium chelonae* on murine haematopoietic stem cells and megakaryocyte progenitors. *Res Immunol* **148**:127–136.

296. de Souza Matos DC, Marcovistz R, Neway T, Vieira da Silva AM, Alves EN, Pilet C. 2000. Immunostimulatory effects of polar glycopeptidolipids of *Mycobacterium chelonae* for inactivated rabies vaccine. *Vaccine* **18**: 2125–2131.

297. Patterson JH, McConville MJ, Haites RE, Coppel RL, Billman-Jacobe H. 2000. Identification of a methyltransferase from *Mycobacterium smegmatis* involved in glycopeptidolipid synthesis. *J Biol Chem* **275**:24900–24906.

298. Recht J, Kolter R. 2001. Glycopeptidolipid acetylation affects sliding motility and biofilm formation in *Mycobacterium smegmatis*. *J Bacteriol* **183**:5718–5724.

299. Jeevarajah D, Patterson JH, McConville MJ, Billman-Jacobe H. 2002. Modification of glycopeptidolipids by an O-methyltransferase of *Mycobacterium smegmatis*. *Microbiology* **148**:3079–3087.

300. Jeevarajah D, Patterson JH, Taig E, Sargeant T, McConville MJ, Billman-Jacobe H. 2004. Methylation of GPLs in *Mycobacterium smegmatis* and *Mycobacterium avium*. *J Bacteriol* **186**:6792–6799.

301. Deshayes C, Laval F, Montrozier H, Daffe M, Etienne G, Reyrat JM. 2005. A glycosyltransferase involved in biosynthesis of triglycosylated glycopeptidolipids in *Mycobacterium smegmatis*: impact on surface properties. *J Bacteriol* **187**:7283–7291.

302. Laurent JP, Hauge K, Burnside K, Cangelosi G. 2003. Mutational analysis of cell wall biosynthesis in *Mycobacterium avium*. *J Bacteriol* **185**:5003–5006.

303. Tatham E, Sundaram Chavadi S, Mohandas P, Edupuganti UR, Angala SK, Chatterjee D, Quadri LE. 2012. Production of mycobacterial cell wall glycopeptidolipids requires a member of the MbtH-like protein family. *BMC Microbiol* **12**:118.

304. Villeneuve C, Etienne G, Abadie V, Montrozier H, Bordier C, Laval F, Daffe M, Maridonneau-Parini I, Astarie-Dequeker C. 2003. Surface-exposed glycopeptidolipids of *Mycobacterium smegmatis* specifically inhibit the phagocytosis of mycobacteria by human macrophages. Identification of a novel family of glycopeptidolipids. *J Biol Chem* **278**:51291–51300.

305. Billman-Jacobe H. 2004. Glycopeptidolipid synthesis in mycobacteria. *Curr Sci* **86**:111–114.

306. Bull TJ, Sheridan JM, Martin H, Sumar N, Tizard M, Hermon-Taylor J. 2000. Further studies on the GS element. A novel mycobacterial insertion sequence (IS1612), inserted into an acetylase gene (*mpa*) in *Mycobacterium avium* subsp. *silvaticum* but not in *Mycobacterium avium* subsp. *paratuberculosis*. *Vet Microbiol* **77**:453–463.

307. Maslow JN, Irani VR, Lee SH, Eckstein TM, Inamine JM, Belisle JT. 2003. Biosynthetic specificity of the rhamnosyltransferase gene of *Mycobacterium avium* serovar 2 as determined by allelic exchange mutagenesis. *Microbiology* **149**:3193–3202.

308. Eckstein TM, Silbaq FS, Chatterjee D, Kelly NJ, Brennan PJ, Belisle JT. 1998. Identification and recombinant expression of a *Mycobacterium avium* rhamnosyltransferase gene (*rtfA*) involved in glycopeptidolipid biosynthesis. *J Bacteriol* **180**:5567–5573.

309. Krzywinska E, Bhatnagar S, Sweet L, Chatterjee D, Schorey JS. 2005. *Mycobacterium avium* 104 deleted of the methyltransferase D gene by allelic replacement lacks serotype-specific glycopeptidolipids and shows attenuated virulence in mice. *Mol Microbiol* **56**:1262–1273.

310. Irani VR, Lee SH, Eckstein TM, Inamine JM, Belisle JT, Maslow JN. 2004. Utilization of a *ts-sacB* selection system for the generation of a *Mycobacterium avium* serovar-8 specific glycopeptidolipid allelic exchange mutant. *Ann Clin Microbiol Antimicrob* **3**:18.

311. Eckstein TM, Belisle JT, Inamine JM. 2003. Proposed pathway for the biosynthesis of serovar-specific glyco-

peptidolipids in *Mycobacterium avium* serovar 2. *Microbiology* 149:2797–2807.

312. Deshayes C, Bach H, Euphrasie D, Attarian R, Coureuil M, Sougakoff W, Laval F, Av-Gay Y, Daffe M, Etienne G, Reyrat JM. 2010. MmpS4 promotes glycopeptidolipids biosynthesis and export in *Mycobacterium smegmatis*. *Mol Microbiol* 78:989–1003.

313. Ojha AK, Varma S, Chatterji D. 2002. Synthesis of an unusual polar glycopeptidolipid in glucose-limited culture of *Mycobacterium smegmatis*. *Microbiology* 148:3039–3048.

314. Mukherjee R, Chatterji D. 2005. Evaluation of the role of sigma B in *Mycobacterium smegmatis*. *Biochem Biophys Res Commun* 338:964–972.

315. Kocincova D, Singh AK, Beretti JL, Ren H, Euphrasie D, Liu J, Daffe M, Etienne G, Reyrat JM. 2008. Spontaneous transposition of IS1096 or ISMsm3 leads to glycopeptidolipid overproduction and affects surface properties in *Mycobacterium smegmatis*. *Tuberculosis (Edinb)* 88:390–398.

316. Dinadayala P, Lemassu A, Granovski P, Cérantola S, Winter N, Daffé M. 2004. Revisiting the structure of the anti-neoplastic glucans of *Mycobacterium bovis* bacille Calmette-Guérin. *J Biol Chem* 279:12369–12378.

317. Dinadayala P, Sambou T, Daffé M, Lemassu A. 2008. Comparative structural analyses of the alpha-glucan and glycogen from *Mycobacterium bovis*. *Glycobiology* 18:502–508.

318. Cywes C, Hoppe HC, Daffé M, Ehlers MRW. 1997. Nonopsonic binding of *Mycobacterium tuberculosis* to human complement receptor type 3 is mediated by capsular polysaccharides and is strain dependent. *Infect Immun* 65:4258–4266.

319. Gagliardi MC, Lemassu A, Teloni R, Mariotti S, Sargentini V, Pardini M, Daffe M, Nisini R. 2007. Cell wall-associated alpha-glucan is instrumental for *Mycobacterium tuberculosis* to block CD1 molecule expression and disable the function of dendritic cells derived from infected monocyte. *Cell Microbiol* 9:2081–2092.

320. Geurtsen J, Chedammi S, Mesters J, Cot M, Driessen NN, Sambou T, Kakutani R, Ummels R, Maaskant J, Takata H, Baba O, Terashima T, Bovin N, Vandenbroucke-Grauls CMJE, Nigou J, Puzo G, Lemassu A, Daffe M, Appelmelk BJ. 2009. Identification of mycobacterial α-glucan as a novel ligand for DC-SIGN: involvement of mycobacterial capsular polysaccharides in host immune modulation. *J Immunol* 183:5221–5231.

321. Stokes RW, Norris-Jones R, Brooks DE, Beveridge TJ, Doxsee D, Thorson LM. 2004. The glycan-rich outer layer of the cell wall of *Mycobacterium tuberculosis* acts as an antiphagocytic capsule limiting the association of the bacterium with macrophages. *Infect Immun* 72:5676–5686.

322. Sambou T, Dinadayala P, Stadthagen G, Barilone N, Bordat Y, Constant P, Levillain F, Neyrolles O, Gicquel B, Lemassu A, Daffé M, Jackson M. 2008. Capsular glucan and intracellular glycogen of *Mycobacterium*

tuberculosis: biosynthesis and impact on the persistence in mice. *Mol Microbiol* 70:762–774.

323. Stadthagen G, Sambou T, Guerin M, Barilone N, Boudou F, Kordulakova J, Charles P, Alzari PM, Lemassu A, Daffé M, Puzo G, Gicquel B, Riviere M, Jackson M. 2007. Genetic basis for the biosynthesis of methylglucose lipopolysaccharides in *Mycobacterium tuberculosis*. *J Biol Chem* 282:27270–27276.

324. Kalscheuer R, Syson K, Veeraraghavan U, Weinrick B, Biermann KE, Liu Z, Sacchettini JC, Besra GS, Bornemann S, Jacobs WR Jr. 2010. Self-poisoning of *Mycobacterium tuberculosis* by targeting GlgE in an alpha-glucan pathway. *Nat Chem Biol* 6:376–384.

325. Elbein AD, Pastuszak I, Tackett AJ, Wilson T, Pan YT. 2010. The last step in the conversion of trehalose to glycogen: a mycobacterial enzyme that transfers maltose from maltose-1-phosphate to glycogen. *J Biol Chem* 285:9803–9812.

326. Jackson M, Brennan PJ. 2009. Polymethylated polysaccharides from *Mycobacterium* species revisited. *J Biol Chem* 284:1949–1953.

327. Quadri LEN. 2008. Iron uptake in mycobacteria, p 167–184. *In* Daffé M, Reyrat J-M (ed), *The Mycobacterial Cell Envelope*. ASM Press, Washington, DC.

328. Stinear TP, Small P. 2008. The mycolactones: biologically active polyketides produced by *Mycobacterium ulcerans* and related aquatic mycobacteria, p 367–377. *In* Daffé M, Reyrat J-M (ed), *The Mycobacterial Cell Envelope*. ASM Press, Washington, DC.

329. Wells RM, Jones CM, Xi Z, Speer A, Danilchanka O, Doornbos KS, Sun P, Wu F, Tian C, Niederweis M. 2013. Discovery of a siderophore export system essential for virulence of *Mycobacterium tuberculosis*. *PLoS Pathog* 9:e1003120.

330. Marsollier L, Brodin P, Jackson M, Korduláková J, Tafelmeyer P, Carbonnelle E, Aubry J, Milon G, Legras P, Andre JP, Leroy C, Cottin J, Guillou ML, Reysset G, Cole ST. 2007. Impact of *Mycobacterium ulcerans* biofilm on transmissibility to ecological niches and Buruli ulcer pathogenesis. *PLoS Pathog* 3:e62.

331. Stinear TP, Mve-Obiang A, Small PLC, Frigui W, Pryor MJ, Brosch R, Jenkin GA, Johnson PDR, Davies JK, Lee RE, Adusumilli S, Granier T, Haydock SF, Leadlay PF, Cole ST. 2004. Giant plasmid-encoded polyketide synthases produce the macrolide toxin of *Mycobacterium ulcerans*. *Proc Natl Acad Sci USA* 101:1345–1349.

332. He H, Oka S, Han YK, Yamamura Y, Kusunose E, Kusunose M, Yano I. 1991. Rapid serodiagnosis of human mycobacteriosis by ELISA using cord factor (trehalose-6,6′-dimycolate) purified from *Mycobacterium tuberculosis* as antigen. *FEMS Microbiol Immunol* 3:201–204.

333. Vera-Cabrera L, Handzel V, Laszlo A. 1994. Development of an enzyme-linked immunosorbent assay (ELISA) combined with a streptavidin-biotin and enzyme amplification method to detect anti-2,3-di-O-acyltrehalose (DAT) antibodies in patients with tuberculosis. *J Immunol Methods* 177:69–77.

334. Julian E, Matas L, Perez A, Alcaide J, Laneelle MA, Luquin M. 2002. Serodiagnosis of tuberculosis: comparison of immunoglobulin A (IgA) response to sulfolipid I with IgG and IgM responses to 2,3-diacyltrehalose, 2,3,6-triacyltrehalose, and cord factor antigens. *J Clin Microbiol* 40:3782–3788.

335. Cunningham AF, Spreadbury CL. 1998. Mycobacterial stationary phase induced by low oxygen tension: cell wall thickening and localization of the 16-kilodalton α-crystallin homolog. *J Bacteriol* 180:801–808.

336. de Chastellier C, Thilo L. 1997. Phagosome maturation and fusion with lysosomes in relation to surface property and size of the phagocytic particle. *Eur J Cell Biol* 74:49–62.

337. Guerin ME, Kordulakova J, Alzari PM, Brennan PJ, Jackson M. 2010. Molecular basis of phosphatidyl-myo-inositol mannoside biosynthesis and regulation in mycobacteria. *J Biol Chem* 285:33577–33583.

338. Gupta R, Lavollay M, Mainardi JL, Arthur M, Bishai WR, Lamichhane G. 2010. The *Mycobacterium tuberculosis* protein LdtMt2 is a nonclassical transpeptidase required for virulence and resistance to amoxicillin. *Nat Med* 16:466–469.

339. Ryan GJ, Hoff DR, Driver ER, Voskuil ML, Gonzalez-Juarrero M, Basaraba RJ, Crick DC, Spencer JS, Lenaerts AJ. 2010. Multiple *M. tuberculosis* phenotypes in mouse and guinea pig lung tissue revealed by a dual-staining approach. *PloS One* 5:e11108.

340. Dhiman RK, Dinadayala P, Ryan GJ, Lenaerts AJ, Schenkel AR, Crick DC. 2011. Lipoarabinomannan localization and abundance during growth of *Mycobacterium smegmatis*. *J Bacteriol* 193:5802–5809.

341. Bhamidi S, Shi L, Chatterjee D, Belisle JT, Crick DC, McNeil MR. 2012. A bioanalytical method to determine the cell wall composition of *Mycobacterium tuberculosis* grown *in vivo*. *Anal Biochem* 421:240–249.

342. Molle V, Kremer L. 2010. Division and cell envelope regulation by Ser/Thr phosphorylation: *Mycobacterium* shows the way. *Mol Microbiol* 75:1064–1077.

343. Hunter SW, Neil MRM, Brennan PJ. 1986. Diglycosyl diacylglycerol of *Mycobacterium tuberculosis*. *J Bacteriol* 168:917–922.

344. Slayden RA, Jackson M, Zucker J, Ramirez MV, Dawson CC, Crew R, Sampson NS, Thomas ST, Jamshidi N, Sisk P, Caspi R, Crick DC, McNeil MR, Pavelka MS, Niederweis M, Siroy A, Dona V, McFadden J, Boshoff H, Lew JM. 2013. Updating and curating metabolic pathways of TB. *Tuberculosis (Edinb)* 93:47–59.

345. Siegrist MS, Unnikrishnan M, McConnell MJ, Borowsky M, Cheng TY, Siddiqi N, Fortune SM, Moody DB, Rubin EJ. 2009. Mycobacterial Esx-3 is required for mycobactin-mediated iron acquisition. *Proc Natl Acad Sci USA* 106:18792–18797.

346. Pandey AK, Sassetti CM. 2008. Mycobacterial persistence requires the utilization of host cholesterol. *Proc Natl Acad Sci USA* 105:4376–4380.

347. Camacho LR, Ensergueix D, Pérez E, Gicquel B, Guilhot C. 1999. Identification of a virulence gene cluster of *Mycobacterium tuberculosis* by signature-tagged transposon mutagenesis. *Mol Microbiol* 34:257–267.

348. Larrouy-Maumus G, Škovierová H, Dhouib R, Angala SK, Zuberogoitia S, Pham H, Drumond Villela A, Mikušová K, Noguera A, Gilleron M, Valentinova L, Korduláková J, Brennan PJ, Puzo G, Nigou J, Jackson M. 2012. A small multidrug resistance-like transporter involved in the arabinosylation of arabinogalactan and lipoarabinomannan in mycobacteria. *J Biol Chem* 287:39933–39941.

349. Azad AK, Sirakova TD, Fernandes ND, Kolattukudy PE. 1997. Gene knockout reveals a novel gene cluster for the synthesis of a class of cell wall lipids unique to pathogenic mycobacteria. *J Biol Chem* 272:16741–16745.

350. Choudhuri BS, Bhakta S, Barik R, Basu J, Kundu M, Chakrabarti P. 2002. Overexpression and functional characterization of an ABC (ATP-binding cassette) transporter encoded by the genes drrA and drrB of *Mycobacterium tuberculosis*. *Biochem J* 367:279–285.

351. Onwueme KC, Ferreras JA, Buglino J, Lima CD, Quadri LE. 2004. Mycobacterial polyketide-associated proteins are acyltransferases: proof of principle with *Mycobacterium tuberculosis* PapA5. *Proc Natl Acad Sci USA* 101:4608–4613.

352. Buglino J, Onwueme KC, Ferreras JA, Quadri LE, Lima CD. 2004. Crystal structure of PapA5, a phthiocerol dimycocerosyl transferase from *Mycobacterium tuberculosis*. *J Biol Chem* 279:30634–30642.

353. Ferreras JA, Stirrett KL, Lu X, Ryu JS, Soll CE, Tan DS, Quadri LE. 2008. Mycobacterial phenolic glycolipid virulence factor biosynthesis: mechanism and small-molecule inhibition of polyketide chain initiation. *Chem Biol* 15:51–61.

354. He W, Soll CE, Chavadi SS, Zhang G, Warren JD, Quadri LE. 2009. Cooperation between a coenzyme A-independent stand-alone initiation module and an iterative type I polyketide synthase during synthesis of mycobacterial phenolic glycolipids. *J Am Chem Soc* 131:16744–16750.

355. Onwueme KC, Vos CJ, Zurita J, Soll CE, Quadri LE. 2005. Identification of phthiodiolone ketoreductase, an enzyme required for production of mycobacterial diacyl phthiocerol virulence factors. *J Bacteriol* 187:4760–4766.

356. Simeone R, Constant P, Malaga W, Guilhot C, Daffe M, Chalut C. 2007. Molecular dissection of the biosynthetic relationship between phthiocerol and phthiodiolone dimycocerosates and their critical role in the virulence and permeability of *Mycobacterium tuberculosis*. *FEBS J* 274:1957–1969.

357. Simeone R, Constant P, Guilhot C, Daffe M, Chalut C. 2007b. Identification of the missing trans-acting enoyl reductase required for phthiocerol dimycocerosate and phenolglycolipid biosynthesis in *Mycobacterium tuberculosis*. *J Bacteriol* 189:4597–4602.

358. De Smet KAL, Weston A, Brown IN, Young DB, Robertson BD. 2000. Three pathways for trehalose biosynthesis in mycobacteria. *Microbiology* 146:199–208.

359. Pan Y-T, Carroll JD, Asano N, Pastuszak I, Edavana VK, Elbein AD. 2008. Trehalose synthase converts glycogen to trehalose. *FEBS J* **275:**3408–3420.

360. Pan YT, Koroth Edavana V, Jourdian WJ, Edmondson R, Carroll JD, Pastuszak I, Elbein AD. 2004. Trehalose synthase of *Mycobacterium smegmatis*: purification, cloning, expression, and properties of the enzyme. *Eur J Biochem* **271:**4259–4269.

361. Mendes V, Maranha A, Lamosa P, da Costa MS, Empadinhas N. 2010. Biochemical characterization of the maltokinase from *Mycobacterium bovis* BCG. *BMC Biochem* **11:**21.

362. Garg SK, Alam MS, Kishan KVR, Agrawal P. 2007. Expression and characterization of α-(1,4)-glucan branching enzyme Rv1326c of *Mycobacterium tuberculosis* H37Rv. *Protein Expr Purif* **51:**198–208.

363. Jankute M, Grover S, Birch HL, Besra GS. 2014. Genetics of mycobacterial arabinogalactan and lipoarabinomannan assembly. *Microbiol Spectrum* **2**(4): MGM2-0013-2013.

364. Pawelczyk J, Kremer L. 2014. The molecular genetics of mycolic acid biosynthesis. *Microbiol Spectrum* **2**(4): MGM2-0003-2013.

Molecular Genetics of Mycobacteria, 2nd Edition
Edited by Graham F. Hatfull and William R. Jacobs, Jr.
© 2014 American Society for Microbiology, Washington, DC
doi:10.1128/microbiolspec.MGM2-0003-2013

Jakub Pawełczyk[1]
Laurent Kremer[2,3]

The Molecular Genetics of Mycolic Acid Biosynthesis

29

The mycobacterial cell wall is essential for *Mycobacterium tuberculosis* growth and survival. It is lipid-rich and highly impermeable and thereby provides protection from many antibiotics, and also allows the pathogen to proliferate within macrophages and to persist for extended periods of time in the infected host (1). Mycolic acids, which are long-chain α-alkyl β-hydroxy fatty acids, constitute up to 60% of the cell wall and are principally responsible for the low permeability of the waxy cell envelope (2). They are found primarily as esters of the nonreducing arabinan terminus of arabinogalactan (AG) but are also present as extractable "free" lipids within the cell wall, mainly associated with trehalose to form trehalose dimycolate (TDM), also known as cord factor (3). Recent studies have also demonstrated the presence of free mycolates associated with *M. tuberculosis* biofilms (4). The crucial importance of the cell envelope integrity for the viability of *M. tuberculosis* has raised interest in understanding the enzymatic pathway for mycolic acid biosynthesis (5). Our knowledge of the biosynthesis of mycolic acid is important for finding new therapeutic targets to combat tuberculosis as well as for unraveling the mode of action of several existing antitubercular drugs (6–8). Indeed, the inhibition

of mycolic acid biosynthesis is the primary effect of the frontline drug isoniazid (INH) (9). This unique metabolic pathway represents an important and attractive reservoir of targets for future chemotherapy, whose development is particularly urgent in the context of multidrug-resistant (MDR) tuberculosis and the nearly untreatable (10) extensively drug-resistant (XDR) strains of *M. tuberculosis*.

Through a combination of genetic and biochemical studies over the past 15 years, impressive progress has been made in understanding the metabolism, structure, and regulation of many cell envelope components including mycolic acids, whose biosynthesis remains an active field of research. This article focuses on the principal metabolic steps and enzymatic components in the biogenesis of mycolic acids, highlighting the key recent advances in this very dynamic area of research.

STRUCTURE AND DIVERSITY OF MYCOLIC ACIDS

Mycolic Acid Structures
It has been 75 years since R. J. Anderson demonstrated that the prolonged saponification of the wax fraction

[1]Institute for Medical Biology, Polish Academy of Sciences, Lodz, Poland; [2]Laboratoire de Dynamique des Interactions Membranaires Normales et Pathologiques, Université de Montpellier 2 et 1, CNRS, UMR 5235, case 107; [3]INSERM, DIMNP, Place Eugène Bataillon, 34095 Montpellier Cedex 05, France.

from the human tubercle bacillus yielded large amounts of very high molecular weight hydroxy fatty acids that were named "mycolic acids" (11). The first structural information came from heating the mycolic acid extract under reduced pressure to 250 to 300°C (Fig. 1), which generated hexacosanoic acid and an unidentified long-chain component that together showed an empirical formula of $C_{88}H_{172}O_4$. Intensive development of chromatography in the late 1950s and 1960s made it possible to define the general structure of mycolic acids as very long-chain, α-alkyl β-hydroxy fatty acids (Fig. 1). This definition explained the nature of the two products of mycolic acid pyrolytic cleavage: a short alkyl chain called the α branch (placed in the α position according to the carboxylic acid group) and a long-chain meroaldehyde called the meromycolic acid chain

(the part of the molecule from the terminal methyl to the carbon atom bearing the hydroxyl group) (12). The two stereocenters in the α and β positions relative to the carboxylic group have both been found to be in the R configuration for all the mycolic acids examined, irrespective of the other functional groups (13).

The use of reversed-phase high-pressure liquid chromatography (HPLC), nuclear magnetic resonance spectroscopy, and mass spectrometry permitted the identification of the variety of mycolic acids with respect to their length and chemical modifications. In contrast to the nonfunctionalized, fully saturated α branch, the meromycolate chain is usually composed of three polymethylenic parts (Fig. 1), linked by two (sometimes three) carbon atoms with either a double bond, cyclopropane rings, or various oxygen functional groups (2).

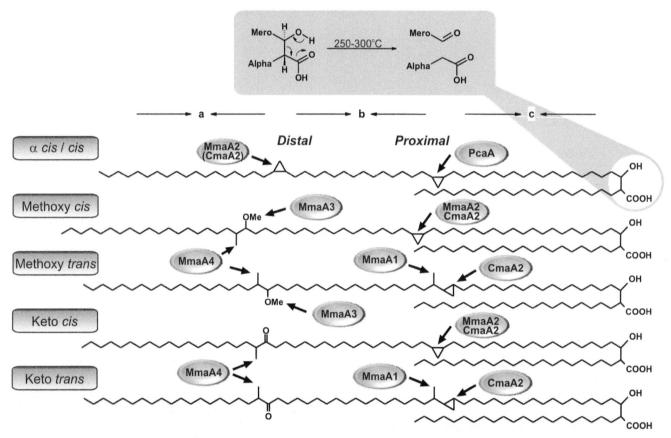

Figure 1 Structures of representative mycolic acids in *M. tuberculosis*. General scheme of mycolic acid pyrolytic cleavage to form the α branch and meroaldehyde (top panel). Representative mycolic acid structures are named (left). Polymethylenic parts of the meromycolate are marked with thin arrows and letters: a, b, c. The chain modifications are shown and annotated with the methyltransferase responsible for their synthesis. Two enzymes are listed without parentheses when the modification is lost only in a double mutant of the genes encoding the listed enzymes and not in either of the single mutants. The parenthetical enzyme plays a secondary role that is evident only when the gene encoding the primary enzyme is deleted. doi:10.1128/microbiolspec.MGM2-0003-2013.f1

Structural modifications occur in two positions of the meromycolate chain: proximal and distal, with respect to the β-hydroxy group (2). Polar modifications are usually restricted to the distal position and include keto-, epoxy-, and methoxy-functional groups. Nonpolar modifications occur at both proximal and distal positions and include double bonds or cyclopropane rings in *cis* or *trans* conformation. Nonpolar modifications in *trans* conformation are always accompanied by an adjacent methyl group (2). Each mycobacterial species is characterized by a specific mycolic acid profile. *M. tuberculosis* possesses three major classes of mycolic acids: α-mycolic, keto-mycolic, and methoxy-mycolic acids (Fig. 1). The α-mycolic acid is a *cis*, *cis*-dicyclopropyl fatty acid that can occur in two variants, depending on the strain. These variations occur in the length of the terminal alkyl group and the number of methylene groups between the cyclopropane rings and the carboxyl group. Methoxy- and keto-mycolic acids can also contain either *cis*- or *trans*-cyclopropane rings at the proximal position.

In addition to mycobacteria, mycolic acids are also found in other genera named "mycolata" of the *Corynebacterineae* suborder: *Corynebacterium*, *Dietzia*, *Rhodococcus*, *Nocardia*, *Gordonia*, *Tsukamurella*, and *Segniliparus*. The common feature of the nonmycobacterial mycolic acids is the presence of a double bond as the only modification within the meromycolate chain. The clear distinctions in chain size between genera can be utilized in their differentiation (14). In *Corynebacterium* the mycolic acids are the smallest (C_{22} to C_{36}) (15), and a recent finding dethroned *M. tuberculosis* as the bacteria producing the longest mycolate chains. The rapidly growing, acid-fast staining bacteria isolated from the human respiratory tract, assigned to the *Segniliparus* genera—*Segniliparus rotundus*—produces mycolates (termed segnilomycolates) that range from C_{58} to C_{100} (16). The overall chain length, the chain length diversity, and the degree of *cis* unsaturation of segnilomycolates are larger than previously described for other mycolic acid–producing organisms (16).

Originally, α-mycolic acid was the first fraction eluted from a column of adsorption chromatography during separation of structural classes of mycolic acids. This subspecies represents the most widespread class of mycolic acids, containing cyclopropane rings or double bonds in *cis* or *trans* with respect to the adjacent methyl group conformation. It is the major mycolic acid in most mycobacterial species and commonly possesses 76 to 86 carbons (17). Subsequently, the shorter α-mycolates with about 60 carbon atoms,

designated α′-mycolic acids, were identified in mycobacterial species such as *Mycobacterium smegmatis* (C_{62} to C_{64}), *Mycobacterium chelonae* (C_{64}), *Mycobacterium fortuitum* (C_{68}), and *Mycobacterium vaccae* (C_{58} to C_{60}), but not in *M. tuberculosis*. Mycolic acids containing keto-, epoxy-, and methoxy-functional groups are eluted from a column after the α-mycolates. Like the α-mycolates, the keto-mycolates are also a widely distributed class of mycolic acids, being present in pathogenic and saprophytic strains regardless of their growth rate. Epoxy-mycolic acids contain a *trans*-epoxy ring in the meromycolate chain (18). Their presence is restricted to several slow- and fast-growing species such as *Mycobacterium farcinogenes*, *Mycobacterium senegalense*, *Mycobacterium chitae*, *Mycobacterium aurum*, and *M. smegmatis*. However, neither the growth rate nor the pathogenic status correlates with epoxy-mycolic acid distribution (2).

With only a few exceptions, the presence of methoxy-mycolates is restricted to pathogenic, slow-growing species. More than 20 years after the initial description of methoxylated mycolates (19), Luquin et al. identified a novel mycolic acid containing a methoxy group at the ω-1 position and two double bonds in the long mero-aldehyde chain (20). The last fraction of mycolates eluted from a chromatographic column contains highly adsorbed carboxy-mycolic acids and their oxygenated precursors—wax-mycolic acids. Carboxy-mycolic acids have a second carboxyl group and a lower molecular weight (C_{60} to C_{68}) (21).

The variability of mycolic acids is not limited to the structural class. Heterogeneity is also observed within each class and can be described on three levels: the length of the α branch, the length of the meromycolate chain, and the internal position of the functional groups (2). Mycolic acid species- and strain-specific patterns of *Mycobacterium* and other *Corynebacterineae* provide specific fingerprints that have been valuable in chemotaxonomic assignments. Because mycolates are bacteria-derived (not synthesized by humans), chemically inert, and directly involved in the host-pathogen interaction, they are attractive diagnostic markers for tuberculosis. Early methods for profiling the different classes of mycolates used one- and two-dimensional thin-layer chromatography (22), but as HPLC gained widespread application in biochemistry, this was proposed as an aid in mycobacterial classification and offered as a standard test for the identification of mycobacteria in clinical specimens. Although HPLC-based methods are constantly being improved (23), recent work suggests that the future of clinically useful methods of *Mycobacterium* speciation belongs to mass spectrometry (24). These authors used electrospray ionization/mass spectrometry to develop a selective and sensitive diagnostic strategy that involves targeted quantification of mycolates in poorly defined heterogeneous biological material such as sputum. The approach is safe and fast, and because no culturing or chemical derivatization is needed, sample preparation and handling are simplified.

Although useful in taxonomy and diagnostics, the complexity of natural mixtures of mycolic acids is problematic when attempting to characterize a single enantiomer molecule and elucidate its physical and chemical properties or function(s). For these reasons, extensive work was done recently in the field of chemical synthesis of mycolic acids. Alpha-mycolates were the first synthetic mycolate reported (25), but the synthesis of methoxy-, keto-, and epoxy-mycolic acids as well as single enantiomers of cis- and trans-alkenes-containing mycolates have also been reported and recently reviewed (26).

Mycolic Acid Biological Functions

Mycolic acids exist in mycobacteria in three forms: (i) covalently bound to the AG, a peptidoglycan-linked polysaccharide in the inner leaflet of cell wall lipid bilayer; (ii) esterified to a variety of carbohydrate-containing molecules (mostly as trehalose mono- and di-mycolate) in the extractable lipid fraction of the cell wall outer leaflet; and (iii) secreted as free mycolic acids. Depending upon the organization in the cell envelope, mycolic acids give rise to important characteristics including resistance to chemical injury and dehydration, low permeability to hydrophobic antibiotics, the ability to form biofilms (4, 27, 28), and the capacity to persist within the host (28–30). In the mycobacterial cell wall, mycolic acid hydrocarbon chains of the inner leaflet are tightly packed in a parallel fashion, perpendicular to the cell surface (1, 3). They esterify hexaarabinose motifs of AG at the C-5 position, thereby limiting their lateral mobility. Simultaneously, 1,5-linked arabinofuranose units allow for exceptional conformational flexibility of the hexaarabinose "head groups" that likely facilitates closer packing of the mycolate chains (31). Functionalities localized in the proximal part of the meromycolate chain are generally responsible for maintaining the viscosity of the cell wall at an appropriate level, while the trans- proximal unsaturations or cyclopropane rings make this part of the meromycolate chain more rigid. Conversely, the distal part of the mycolate layer is the region that interacts with extractable lipids; thus, cis unsaturations and methyl groups at the distal position disrupt tight local packing of mycolates and allow the extractable lipids to intercalate their acyl chains into the inner leaflet, where they associate with AG-linked mycolates. As shown by Yuan et al., distal cis-cyclopropanation renders Mycobacterium more resistant to killing by hydrogen peroxide (32). Oxygenated mycolates (mainly keto-mycolates) are very active as hydrogen bond acceptors, which promotes the association of peripheral cell surface molecules (2).

Mycolic acids of the extractable lipid fraction exist predominantly as mono- and di-mycolyl trehalose (TMM, TDM). TDM represents the most abundant, granulomagenic, and significantly toxic lipid extractable from the cell surface of virulent M. tuberculosis (33). It has immunostimulating (34) and adjuvant properties (35) and potent antitumor activity (36) and actively participates in blocking mycobacterial phagosome maturation (37). It was recently shown that macrophage-inducible C-type lectin (Mincle) (38) on the macrophage surface recognizes M. tuberculosis TDM, and by working together with the Fcγ receptor transmembrane segment, it induces inflammation (39). In addition, mycolic acids can also be found in dimycolyl diarabinoglycerol (DMAG) (40), an extractable, cell wall–associated glycolipid capable of inducing the expression of proinflammatory cytokines and promoting the expression of the ICAM-1 and CD40 cell surface antigens through a TLR2-dependent mechanism (41).

There is a growing body of evidence of the significance of secreted free mycolates as potential players in the host-pathogen interaction during M. tuberculosis infection. The presence of free mycolic acids in the extracellular matrix of M. tuberculosis biofilms was confirmed in vitro (4), and Ojha et al. provided evidence that newly synthesized TDM is one of the precursors of free mycolates. TDM is processed by a TDM-specific esterase releasing mycolic acids that are subsequently secreted to form the biofilm matrix (42). The role of mycolic acids in immune regulation was reported by Beckman et al., who showed that presentation of mycolates on CD1b of human dendritic cells stimulated CD4/CD8 double negative T cell lines (43). Administration of purified mycolic acids into mice airways elicited an acute neutrophilic airway inflammation that was accompanied by a moderate and chronic IL-12 production (44). Working on modified synthetic mycolic acids, Vander Beken et al. confirmed that the type of distal group oxygenation in the meromycolate chain is the main determinant of pulmonary inflammatory potency: oxygenated methoxy- and keto-mycolic acids with cyclopropane rings in the cis conformation exhibited strong and mild inflammatory responses, respectively, whereas α-mycolates did not cause any inflammation (45).

MYCOLIC ACID BIOSYNTHESIS

Synthesis of Malonyl-CoA

Malonyl-coenzyme A (malonyl-CoA) is the universal, two-carbon substrate for the synthesis of mycolic and other fatty acids in mycobacteria. It is generated by the carboxylation of acetyl-CoA in a biotin-dependent, two-step reaction catalyzed by acetyl-CoA carboxylase (ACC) (46) and incorporated into the growing acyl chain during the repetitive cycle of the fatty acid synthase I and II (FAS I/FAS II) reactions. Each half-reaction is catalyzed by a specific ACC subunit: the first step by biotin carboxylase and the second step by carboxyltransferase, each catalytic subunit being encoded by a separate gene (47). Three genes thought to encode the α subunit or biotin carboxylase (*accA1* to *accA3*) and six genes believed to encode the β subunit or carboxyltransferase (*accD1* to *accD6*) have been identified in the *M. tuberculosis* genome (48). Since the β subunits confer the substrate specificity of ACC, the large number of *accD* genes in mycobacterial genomes reflects the ability of mycobacteria to carboxylate not only acetyl-CoA but also several other distinct substrates. Among the six carboxyltransferase genes in *M. tuberculosis*, *accD4*, *accD5*, and *accD6* are

essential for cell survival (49, 50) and are expressed at high levels during intensive mycolate biosynthesis (51). For several years, AccD4 and AccD5 were the only carboxyltransferase subunits purified from mycobacterial cell extracts, so they were initially considered to be major constituents of ACC complexes in tubercle bacilli, but *in vitro* analysis showed that neither of them can be considered the subunit dedicated exclusively to acetyl-CoA carboxylation (52–56).

Recent studies have focused attention on the third essential carboxyltransferase gene, *accD6* (Rv2247), which surprisingly remained the least characterized carboxyltransferase gene, despite being the only ACC candidate of the FAS II locus (Fig. 2). An *M. tuberculosis* mutant exhibiting 10-fold reduction of *accD6* expression presented restricted growth, inhibition of proper fatty and mycolic acid biosynthesis, and an altered cell morphology (50). Inhibition of fatty acid synthesis occurred at a very early stage, likely reflecting the impaired activity of acetyl-CoA carboxylase. These results provided the first *in vivo* evidence that AccD6 is a key player in mycolate biosynthesis and implicated AccD6 as the critical component of *M. tuberculosis* ACC, in agreement with *in vitro* findings demonstrating that the AccD6 (Rv2247) protein (β subunit), together with AccA3 (Rv3285) (α subunit), reconstitutes an enzyme that preferentially carboxylates acetyl-CoA (51). Strikingly, parallel studies demonstrated that although essential for *M. tuberculosis*, the *M. smegmatis accD6* homologue—*MSMEG_4329*—can be deleted without

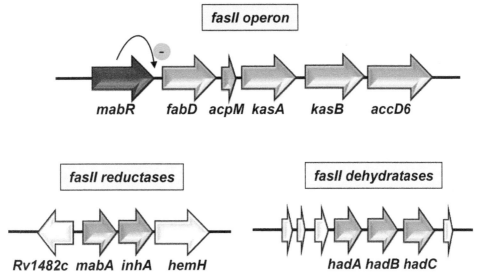

Figure 2 Genomic organization of *fas II* operons in *M. tuberculosis*. Genes encoding the FAS II components are shown in gray, whereas the transcriptional repressor MabR is represented in black. doi:10.1128/microbiolspec.MGM2-0003-2013.f2

affecting the cell envelope integrity and permeability (50). This finding, contradicting that of Kurth et al. (57), suggests that another AccD subunit fulfills the function of AccD6 in *M. smegmatis*. In addition, although in both species *accD6* is a member of the FAS II transcriptional unit and its expression is controlled by the P_{fasII} promoter, it was found that *accD6* of *M. tuberculosis*, but not *M. smegmatis*, possesses its own, additional promoter (P_{acc}). This implies that in the pathogenic strain, *accD6* expression may be controlled by two regulatory sequences, P_{fasII} and P_{acc}. Although the additional promoter seems not to participate in supporting the physiological expression level of *accD6*$_{Mtb}$ under standard growth conditions, it is able to sustain the expression of this gene on a level allowing for cell survival in the absence of P_{fasII} (50). The cause of the difference in *accD6* essentiality between pathogenic and nonpathogenic species as well as the role of possible differences in regulation of its expression remain to be investigated, but insights into the enzymatic mechanism of the *M. tuberculosis* AccD6 should be forthcoming thanks to the recently published crystallographic analysis (58).

FAS I and Short-Chain Fatty Acid Biosynthesis

The *fasI* gene (Rv2524c) encodes the multifunctional type I fatty acid synthase (FAS I) that carries all the functional domains required for *de novo* fatty acid synthesis (59). These domains are organized in the following order: acyltransferase, enoyl reductase, dehydratase, malonyl/palmitoyl transferase, acyl carrier protein, ketoacyl reductase, and ketoacyl synthase. All intermediates that are generated during the process of elongation remain enzyme-bound and undergo transacylation to other catalytic sites within the enzyme. FAS I catalyzes the *de novo* synthesis of $C_{16:0}$ and $C_{18:0}$ acyl-CoAs from acetyl-CoA using malonyl-CoA as the extender unit (Fig. 3). These acyl-CoAs either can be used for the synthesis of membrane phospholipids or can be further elongated by FAS I to produce $C_{24:0}$-CoA in the fast-growing organism *M. smegmatis* (60) and $C_{26:0}$-CoA products in the slow-growing *Mycobacterium bovis* and *M. tuberculosis* (61).

This bimodal product distribution characterizing FAS I has been further investigated by generating a recombinant *M. smegmatis* strain in which the native *fasI* gene was deleted and replaced with the *M. tuberculosis fasI* gene (62). Surprisingly, the *in vivo* elongation of $C_{16:0}$ did not follow the *M. tuberculosis* profile, but

this recombinant strain contained both $C_{24:0}$ and $C_{26:0}$ fatty acids, thus challenging the simple concept that the $C_{16:0}$ elongation to $C_{24:0}$ in *M. smegmatis* or $C_{26:0}$ in *M. tuberculosis* is exclusively dependent on FAS I; this is likely to result from a more complex interaction between the FAS I enzyme and the FAS II cycle. The bimodal product distribution may also rely on intrinsic determinants of the FAS polypeptides and/or external factors affecting the elongation process. For example, studies on the product-regulation mechanisms for fatty acid biosynthesis catalyzed by *M. smegmatis* FAS I demonstrated that the formation of palmitic acid in the presence of synthetic O-alkyl polysaccharide is regulated by the removal of the end product via a tight stoichiometric complex between the $C_{16:0}$ fatty acid and the O-alkyl polysaccharide (63). In *M. tuberculosis* the $C_{26:0}$ fatty acids synthesized by FAS I, known as the α branch, will become the substrate of a dedicated acyl-CoA carboxylase to generate the α-carboxy $C_{26:0}$ fatty acid used as one of the substrates by the polyketide synthase Pks13. The recent cryo-electron microscopy reconstruction of the 2-MDa *M. smegmatis* FAS I at 7.5 Å revealed that there is a high degree of conservation between FAS I multifunctional enzymes from different kingdoms not only in the amino acid sequence but also on the overall architectural level (64).

FabH: the Pivotal Link Between FAS I and FAS II

The FAS I and FAS II systems are interconnected by the β-ketoacyl-ACP synthase III or FabH (Rv0533c) (Fig. 3). This enzyme catalyzes a decarboxylative condensation of malonyl-AcpM with the acyl-CoA ($C_{16:0}$ to $C_{18:0}$) products of the type I FAS. The resulting 3-ketoacyl-AcpM derivatives are then reduced to an acyl-AcpM, extended by two carbons, and shuffled into the FAS II cycle. FabH represents the pivotal link between the two FAS systems and thus initiates meromycolic acid biosynthesis (65). This proposed function of FabH is supported by the presence of a conserved catalytic triad (Cys112, His244, and Asn274) characteristic of β-ketoacyl-ACP synthase III enzymes, by its ability to produce β-ketoacyl-AcpM derivatives (65), and by its marked preference for acyl-CoA as a substrate, rather than acyl-ACP primers. Compared with the general *Escherichia coli* FabB/FabH type of β-ketoacyl synthases, the *M. tuberculosis* enzyme exhibits high substrate specificity, consistent with structural studies (66). It contains a CoA/malonyl-ACP binding channel that runs from the surface of the protein to the cysteine residue within the active site. FabH also contains a second

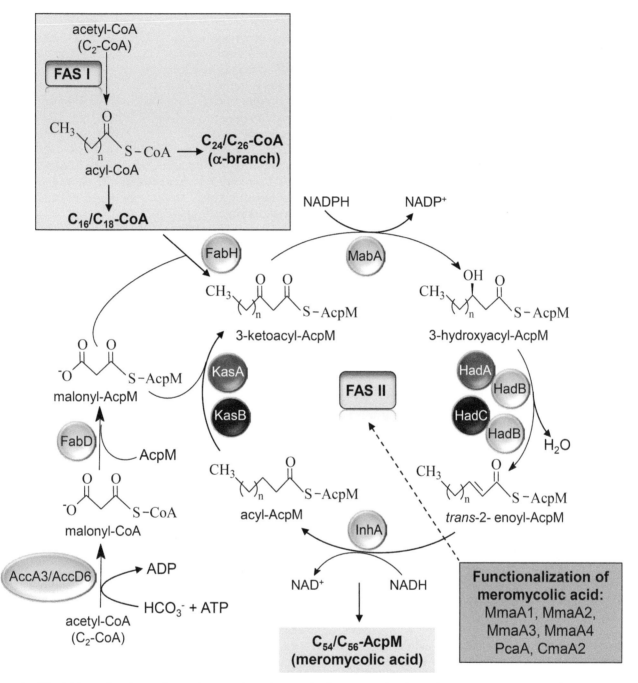

Figure 3 The FAS I and FAS II pathways in *M. tuberculosis*. In both systems, the chain elongation steps consist of an iterative series of reactions built on successive addition of a two-carbon unit to a nascent acyl group, and reaction intermediates are covalently linked to the acyl carrier protein AcpM. FAS I is capable of *de novo* synthesis from acetyl-CoA producing acyl-CoA either used to synthesize the α-branch or C_{16}/C_{18}-CoA that are directly shuttled into FAS II for the production of the meromycolic acid. FAS II is primed by the CoA-dependent β-ketoacyl-AcpM synthase FabH, which condenses the acyl-CoA with malonyl-AcpM to generate a β-ketoacyl-AcpM, subsequently converted into a saturated enoyl-AcpM by the sequential actions of a β-ketoacyl-AcpM reductase (MabA), a β-hydroxyacyl-AcpM dehydratase complex (HadABC), and a *trans*-2-enoyl-AcpM reductase (InhA). Subsequent rounds of elongation are initiated by either the KasA or KasB β-ketoacyl AcpM synthases. KasA is thought to be responsible for the early rounds of elongations, whereas KasB is involved in later stages. Also shown are the acetyl-CoA carboxylase (AccA3/AccD6) that produces malonyl-CoA, the malonyl-CoA:AcpM transacylase (FabD) responsible for the synthesis of malonyl-AcpM, as well as the set of SAM-dependent methyltransferases involved in functionalization of the meromycolic acid. doi:10.1128/microbiolspec.MGM2-0003-2013.f3

hydrophobic pocket leading from the active site, which is blocked by Thr87. The difference in these amino acid residues is postulated to account for the difference in substrate specificities between the two enzymes; Phe87 constrains specificity to acetyl-CoA in *E. coli*, while Thr87 allows binding of long-chain acyl-CoAs in FabH. This second hydrophobic pocket in FabH is capped by an α helix, which restricts the bound acyl chain length to 16 carbons, thus excluding longer-chain acyl-CoA products (C_{24} to C_{26}) from chain elongation. In addition, several residues influencing catalysis and substrate specificity of FabH have been assigned through the combination of structural studies and site-directed mutagenesis (67). These observations are consistent with the proposed role of FabH as the initiator of mycolic acid elongation and clearly distinguish it from the chain extension steps catalyzed by KasA/KasB.

FAS II and Long-Chain Fatty Acid Biosynthesis

The genes encoding the FAS II enzymes lie in three independent transcription units (Fig. 2). The first is formed by *fabD-acpM-kasA-kasB-accD6*, the second by *mabA-inhA,* and the third consists of three genes: *hadA-hadB-hadC*. The first open reading frame of the five-gene operon, *fadD*, encodes a malonyl-CoA:AcpM transacylase (MCAT) that catalyzes the formation of malonyl-AcpM, the building block used by the condensing enzymes for the extension of the fatty acid chain (Fig. 3). FabD catalyzes *in vitro* the transacylation of malonate from malonyl-CoA to phosphopantothenylated *holo*-AcpM (68). The second open reading frame of this operon encodes the acyl carrier protein AcpM, a C-terminally extended homologue of bacterial ACPs that shuttles the growing acyl chain between the discrete monofunctional enzymes that catalyze the individual steps (68, 69). The elongation process that follows the FabH-dependent condensation reaction of the FAS II system is carried out by the β-ketoacyl-AcpM enzymes encoded by *kasA* (Rv2245) and *kasB* (Rv2246) that are also in the five-gene operon (Fig. 2 and Fig. 3).

Both KasA and KasB catalyze the condensation of acyl-AcpM and malonyl-AcpM, hence elongating the growing meromycolate chain by a further two carbon units (8, 70, 71). KasA and KasB share many similarities, including the specificity for long chain acyl-AcpM primers. The Kas enzymes of *E. coli*, ecFabB/ecFabF, and KasA/KasB of *M. tuberculosis* share substantial sequence similarity (70): KasA and KasB are 67%

identical and are 92% and 91% identical with their respective homologues in *Mycobacterium leprae*. These enzymes first transfer the acyl chain to the active-site cysteine, resulting in an acylated KasA intermediate. Subsequently, the acyl chain is elongated by two more carbon atoms derived from the second substrate malonyl-AcpM in a condensation reaction with the KasA intermediate (72). Participation of Cys171, His311, Lys340, and His345 in KasA catalysis was subsequently confirmed by replacing these residues with alanine, which abolishes the overall elongation activity of KasA (72). Both the KasA and KasB crystal structures have been solved (73, 74), and by using polyethylene glycol to mimic a 40-carbon fatty acid substrate, it was possible to characterize the hydrophobic acyl-binding cavity of KasA, which is lined with numerous hydrophobic amino acids and perfectly accommodates the growing fatty acid chain (73). Although appearing highly redundant, KasA might catalyze the initial elongation reactions, while KasB extends the elongation to full-length mycolates. Cell lysates of *M. smegmatis* overproducing KasA generate C_{40}-monounsaturated fatty acids, whereas cell lysates of *M. smegmatis* overproducing both KasA and KasB generate C_{54}-multiunsaturated fatty acids (75). In addition, disruption of *kasB* in *M. tuberculosis* resulted in meromycolate chains that were four to six carbons shorter than wild-type mycolates and also in the loss of the *trans*-cyclopropanated oxygenated mycolates, which are usually produced by the *trans*-cyclopropane synthase CmaA2 (29).

These results indicate that KasA and KasB function independently on separate sets of substrates and that KasB is required for the full extension of the mycolic acids. In contrast to KasB, KasA is indispensable (76), representing the only essential β-ketoacyl-ACP synthase II. Overall, these findings suggest that KasA is very likely the condensase involved in initial extension of the mycolate chains, whereas KasB is primarily involved in the full-length extension of these molecules. Interestingly, the *M. tuberculosis kasB* mutant lost its acid-fast staining and was unable to persist in immunocompetent infected mice (29). Together, these findings suggest that both KasA and KasB could be potent drug targets, which seems feasible because both enzymes have been shown to be inhibited by thiolactomycin and related analogues (70, 77).

KasA has recently been shown to link together the FAS II and the phthiocerol dimycocerosate (PDIM) biosynthesis pathways through a specific interaction between PpsB and PpsD. This raises the possibility that lipids could be transferred between the two pathways as a means of increasing mycobacterial lipid diversity (78).

After the condensation step, the β-ketoacyl-AcpM product undergoes a cycle of keto-reduction, dehydration, and enoyl-reduction catalyzed by the β-ketoacyl-AcpM reductase (MabA, FabG1) (79), the β-hydroxyacyl-AcpM dehydratase complex HadABC (80, 81), and the enoyl-AcpM reductase InhA (82), respectively. The genes encoding the functionally and structurally related FAS II reductases, *mabA* (Rv1483) and *inhA* (Rv1484), are found in a cluster in the *M. tuberculosis* genome (Fig. 2). The *inhA* gene was identified as the putative target for both INH and ethionamide in *M. tuberculosis* (83), and the InhA protein was demonstrated to catalyze the 2-*trans*-enoyl-ACP reduction, with a preference for long-chain substrates (82) (Fig. 3). MabA encodes a β-ketoacyl-AcpM reductase with preference for long-chain substrates and was shown to be involved in the elongation activity of FAS II (79) (Fig. 3). In comparison with other bacterial reductases, MabA has unique functional and structural properties, such as a large hydrophobic substrate-binding pocket, which is consistent with its preference for long-chain substrates and its role in mycolic acid biosynthesis (79, 84). Despite intensive investigation over several years, there have been no reports of the successful generation of *mabA* or *inhA* gene knockout mutants, strongly suggesting that they are essential for mycobacterial viability.

Dehydration of the β-hydroxyacyl-ACP intermediate into *trans*-2-enoyl-ACP is catalyzed by the β-hydroxyacyl-AcpM dehydratases encoded in the gene cluster *hadA-hadB-hadC* (Rv0635, Rv0636, Rv0637, respectively) (80) (Fig. 2 and Fig. 3). Heterologous expression of these proteins in *E. coli* led to the formation of two heterodimers, HadAB and HadBC, both showing the enzymatic properties expected for the mycobacterial FAS II dehydratases, with a marked specificity for both long-chain and AcpM-bound substrates. Interestingly, several recent studies indicated that mutations within the HadA or HadC components were associated with resistance to the antitubercular drugs thiacetazone (85–87) and isoxyl (87, 88). Future work defining the three-dimensional structures of HadAB and HadBC heterocomplexes should help in understanding how these mutations generate drug resistance and should also delineate new structural scaffolds for drug development.

Functionalization of the Meromycolic Acid

As mentioned above, the different mycolic acid species are determined by the chemical modifications of the unsaturations within the meromycolyl chain and the nature of chemical groups introduced. *M. tuberculosis* carries three types of mycolic acids (Fig. 1) containing cyclopropane rings, methoxy and keto functions, or methyl branches (3, 89). The chemical modifications are performed by a series of eight *S*-adenosyl methionine (SAM)-dependent methyltransferases, encoded at four genetic loci (89) (Table 1). These enzymes have extensive sequence and structural similarities (90–92), but genetic studies have defined the distinct functional roles of each in the biosynthesis of mycolic acids, which were not revealed when the enzymes were overexpressed in *M. smegmatis* (32, 93, 94).

Deletion of *pcaA* greatly reduces synthesis of the proximal cyclopropane ring of the α-mycolates (28), whereas deletion of *mmaA2* reduces the distal cyclopropane of the same lipid and also causes a mild impairment of methoxy-mycolate, but not keto-mycolate, *cis*-cyclopropanation (95). Similar genetic approaches established *cmaA2* as the only *trans*-cyclopropane synthase of oxygenated mycolates, because inactivation of this gene was associated with double bonds in place of the proximal *trans*-cyclopropane rings (96). To examine potential redundancy between mycolic acid methyltransferases, *M. tuberculosis* mutants were generated lacking *mmaA2* and *cmaA2*, *mmaA2* and *cmaA1*, or *mmaA1* alone (97). The strain lacking both *cmaA2* and *mmaA2* failed to *cis*-cyclopropanate methoxy-mycolates or keto-mycolates, phenotypes not shared by the *mmaA2* and *cmaA2* single mutants. In contrast to the loss of *mmaA2* alone, disruption of both *cmaA1* and *mmaA2* had no effect on mycolic acid modification. Deletion of *mmaA1* from *M. tuberculosis* abolishes *trans*-cyclopropanation without accumulation of *trans*-unsaturated oxygenated mycolates, placing MmaA1 in the biosynthetic pathway for *trans*-cyclopropanated oxygenated mycolates before CmaA2. These results suggest a substantial redundancy of function for MmaA2 and CmaA2, the latter of which can function as both a *cis*- and *trans*-cyclopropane synthase for the oxygenated mycolates (97). No mycolic acid modification function has been elucidated for the *umaA1* or *cmaA1* gene products in *M. tuberculosis* (95, 97, 98), despite the cyclopropanating activity of the latter when expressed in *M. smegmatis* (32).

Deletion of *mmaA4* abolishes synthesis of both methoxy- and keto-mycolates, while still able to synthesize α-mycolates (27, 99, 100). MmaA4 is capable of methylation of the *cis* double bond and, in the presence of a water molecule, catalyzes hydroxylation. The hydroxy-mycolate intermediates are at the branch point between keto- and methoxy-mycolates, the latter resulting from the action of the methoxylase MmaA3 (27, 94, 101–103). Many *M. bovis* BCG strains have point mutations within the *mmaA3* gene that inactivate its activity and thus abolish the production of methoxy-mycolates (102, 104).

Several studies have implicated individual cyclopropane modifications as important factors in the *M. tuberculosis* host-pathogen interactions. Inactivation of *pcaA* was associated with a loss of bacterial cording, defective persistence, less-severe granulomatous pathology, and attenuation in a mouse tuberculosis model (28, 105). Recent work has also highlighted the unexpected role of PcaA in intracellular survival and the inhibition of phagosome maturation in *M. bovis* BCG–infected human monocyte-derived macrophages (106). In contrast to *pcaA*, deletion of *cmaA2* had no effect on bacterial loads during mouse infection but appeared to produce hypervirulence while stimulating a more severe granulomatous pathology (107). Inactivation of *mmaA4*, which leads to an absence of methoxy- and

keto-mycolates, causes a severe growth defect during the first 3 weeks of infection (27, 99). Overall, these studies suggest that the fine structure of mycolic acids plays a role in the pathogenesis of *M. tuberculosis*. One mechanism by which cyclopropanation mediates virulence is through altered inflammatory activity of TDM. The cyclopropane content of TDM has been shown to be a major determinant of its inflammatory activity and *M. tuberculosis* virulence (99, 105, 107). From studies on mutant strains constructed to contain multiple gene deletions, it was found that *M. tuberculosis* is viable either without cyclopropanation or without both cyclopropanation and oxygenated mycolates (108). A complete lack of cyclopropanation confers severe attenuation during the first week after aerosol infection of the mouse, whereas complete loss of all methyltransferases confers attenuation in the second week of

Table 1 Genes involved in *M. tuberculosis* fatty/mycolic acid biosynthesis

Activity	Gene	Designation	Product
De novo synthesis of fatty acids	*fasI*	Rv2524c	Type I fatty acid synthase
Fatty acid elongation	*fabD*	Rv2243	Malonyl-CoA:AcpM transacylase
	acpM	Rv2244	Acyl carrier protein
	kasA	Rv2245	β-Ketoacyl-AcpM synthase I
	kasB	Rv2246	β-Ketoacyl-AcpM synthase II
	mabA	Rv1483	β-Ketoacyl-AcpM reductase
	inhA	Rv1484	2-*trans*-Enoyl-AcpM reductase
	hadA	Rv0635	β-Hydroxyacyl-AcpM dehydratase
	hadB	Rv0636	β-Hydroxyacyl-AcpM dehydratase
	hadC	Rv0637	β-Hydroxyacyl-AcpM dehydratase
	FabH	Rv0533	β-Ketoacyl-AcpM synthase III
Synthesis of fatty/mycolic acid precursors	*accD6*	Rv2247	Carboxyltransferase subunit of acetyl-CoA carboxylase
	accD5	Rv3280	Carboxyltransferase subunit of propionyl-CoA carboxylase
	accA3	Rv3285	Biotin carboxylase subunit of acetyl-, propionyl-, and fatty acyl-CoA carboxylase
Meromycolic acid functionalization[a]	*mmaA1*	Rv0645c	Methyl branch in *trans*-cyclopropanated, oxygenated mycolates
	mmaA2	Rv0644c	*cis*-Cyclopropane synthase of α and oxygenated mycolates
	mmaA3	Rv0643c	Methoxy mycolic acid synthase
	mmaA4	Rv0642c	Hydroxy mycolic acid synthase
	cmaA1	Rv3392c	Methyltransferase of unknown specificity
	cmaA2	Rv0503c	*trans*-Cyclopropane synthase of oxygenated mycolates
			cis-Cyclopropane synthase of oxygenated mycolates (redundant with mmaA2)
			Distal cyclopropane synthase of α-mycolates
	pcaA	Rv0470c	*cis*-Cyclopropane synthase of α-mycolates
	umaA1	Rv0469	Methyl transferase of unknown specificity
Mycolic acid condensation	*accD4*	Rv3799c	Carboxyltransferase subunit of fatty acyl-CoA carboxylase
	Pks13	Rv3800c	Polyketide synthase
	fadD32	Rv3801c	Fatty acyl-AMP ligase/fatty acyl-ACP synthetase
	cmrA	Rv2509	β-Ketoacyl reductase

[a]Roles of the various methyltransferases are depicted in Fig. 1.

infection. Characterization of immune responses to the cyclopropane- and methyltransferase-deficient strains indicated that the net effect of mycolate cyclopropanation is to dampen host immunity.

These findings establish the immunomodulatory function of the mycolic acid modifications and their role in *M. tuberculosis* pathogenesis, which makes this enzyme family an attractive target for antitubercular drug development. Indeed, compounds that inhibit this class of enzyme have been shown to kill mycobacteria, thus emphasizing their importance for mycobacterial viability (92, 109). In addition, the activity of thiacetazone, a second-line antitubercular drug, was shown to inhibit mycolic acid cyclopropanation (110), and this activity was dependent on an intact *mmaA4* gene. Several thiacetazone-resistant strains of *M. bovis* BCG or *M. tuberculosis* harboring mutations within *mmaA4* were found to lack keto-mycolates (85, 86, 100). Overall, these results suggest that the combined inhibition of the whole family of methyltransferases involved in mycolic acid functionalization is highly detrimental to *M. tuberculosis*, thus supporting this enzyme family as a valid target for future antimycobacterial drug development.

Although all mycobacteria synthesize mycolic acids, it had been thought that only pathogenic mycobacteria produce significant quantities of mycolic acids with cyclopropane rings. Recently, however, the presence of cyclopropanated cell wall mycolates was demonstrated in the nonpathogenic species *M. smegmatis* (98, 111). *MSMEG_1351* was identified as a gene encoding a *pcaA* homologue, and disruption of *MSMEG_1351* produced a marked deficiency in cyclopropanation of α-mycolates. Unexpectedly, α-mycolic acid cyclopropanation in *M. smegmatis* was induced at low growing temperatures (111). The functional equivalence of PcaA and MSMEG_1351 was established by cross-complementation of the MSMEG_1351 knockout mutant with the *M. tuberculosis* gene. In addition, complementation of an *M. bovis* BCG *pcaA* mutant strain with MSMEG_1351 restored the wild-type mycolic acid profile and the cording phenotype. Although the biological significance of mycolic acid cyclopropanation in nonpathogenic mycobacteria remains obscure, it may represent a mechanism of adaptation of the cell wall structure and composition to cope with environmental stresses.

Mycolic Acid Condensation

The ultimate step in the synthesis of full-length, mature mycolic acid relies on the condensation of its two constituents—a C_{26}-CoA acyl chain (α branch) released from FAS I and C_{52}-meromycolyl-AcpM, the FAS II end-product. It has been shown that this process requires at least three enzymes encoded within the putative *fadD32-pks13-accD4* operon (Table 1) that is only present in mycolic acid–producing bacterial species (53, 112, 113). Prior to condensation by the polyketide synthase Pks13, the two acyl chains must be activated by AccD4 and FadD32, respectively (53, 114, 115). AccD4 is a putative carboxyltransferase that associates with the AccA3 subunit to form a complex that activates the α branch (C_{26}-CoA) through carboxylation, yielding 2-carboxyl-C_{26}-CoA (53). The *M. tuberculosis* AccD4 substrate specificity for long-chain C_{24} to C_{26} acyl-CoAs was confirmed *in vitro* (52). It had been proposed that the AccD5 carboxyltransferase might also play a role in the activation of condensation substrates, but solving its crystal structure refuted this notion by showing that the substrate binding pocket of AccD5 is too small to accommodate even the 16-carbon palmitoyl-CoA (116).

FadD32, a fatty acyl-AMP ligase (FAAL) converts the second condensation substrate—meromycolyl-AcpM—to meromycolyl-AMP (53, 115). Following activation, both acyl substrates are loaded onto Pks13, a type I polyketide synthase (PKS) composed of several domains that allow completion of the mature mycolate synthesis (117) (Fig. 4). In addition to its fatty acyl-AMP ligase activity, FadD32 functions as a fatty acyl-ACP synthetase (FAAS) with the N-terminal ACP domain of Pks13 being the natural and most efficient acceptor (118). After meromycolyl-AMP/ACP transacylation, the meromycolyl chain is transferred onto the Pks13 ketoacyl synthase (KS) domain. Simultaneously, 2-carboxyl-C_{26}-CoA is loaded onto the acyl transferase (AT) domain, which catalyzes its covalent attachment to the enzyme active site and then transfers it onto the Pks13 C-terminal ACP domain (117, 119). The ketoacyl synthase domain catalyzes the Claisen-type condensation between the meromycolyl and the carboxyacyl chains to produce α-alkyl, β-keto thioester, which remains bound to the C-terminal ACP domain. Subsequently, the thioesterase (TE) domain catalyzes the release of the α-alkyl, β-keto acyl chain from Pks13. A recent report identified a new class of thiophene compounds that efficiently kill *M. tuberculosis* by specifically inhibiting fatty acyl-AMP loading onto Pks13 (120), thus confirming the essential requirement of Pks13 in mycolic acid biosynthesis and validating Pks13 as a druggable target.

The final step of mycolic acid synthesis involves reduction of the β-ketoacyl product to yield the mature mycolic acid (Fig. 4). There are a large number of putative reductases in mycobacterial genomes, but comparative studies between *C. glutamicum* and *M. tuberculosis* led to the identification of the putative

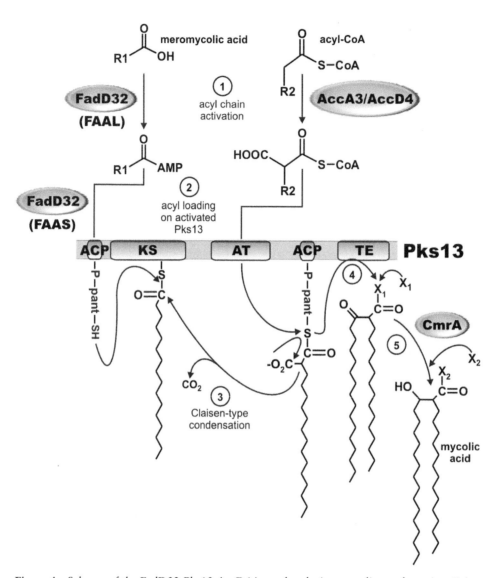

Figure 4 Scheme of the FadD32-Pks13-AccD4 interplay during mycolic condensation. Prior to condensation the two acyl chains have to be activated (step 1). FadD32, a fatty acyl-AMP ligase (FAAL) converts the meromycolyl-AcpM to meromycolyl-AMP. AccD4 associating with the AccA3 carboxylates acyl-CoA yielding carboxyacyl-CoA. Both substrates are then loaded onto Pks13 (step 2). The meromycolyl-AMP is transacylated onto the Pks13 N-terminal ACP domain by FadD32 fatty acyl-ACP synthetase (FAAS) activity and subsequently transferred onto the ketoacyl synthase (KS) domain. The carboxyacyl-CoA initially binds to the Pks13 acyl transferase (AT) domain, which catalyzes its transfer onto the Pks13 C-terminal ACP. The KS domain catalyzes the Claisen-type condensation between the meromycolyl and the carboxyacyl chains to produce an α-alkyl β-keto thioester, which remains bound to the C-terminal ACP domain (step 3). The thioesterase (TE) catalyzes the release of the α-alkyl β-ketoacyl chain and its transfer onto an unknown acceptor (X_1) (step 4). Finally, the reduction of the β-ketoacyl product by the CmrA reductase leads to the mature mycolic acid, which is transferred onto another unknown acceptor (X_2) (step 5). doi:10.1128/microbiolspec.MGM2-0003-2013.f4

reductase encoding gene, Rv2509, which is highly homologous to *cmrA* of *C. glutamicum* (121). Deletion of this gene in *Corynebacterium* resulted in a slow-growing strain that was deficient in AG-linked mycolates and synthesized abnormal forms of the trehalose dicorynomycolate and trehalose monocorynomycolate. Analysis of the aberrant glycolipids by matrix-assisted laser desorption ionization–time of flight (MALDI-TOF) mass spectrometry indicated that they contained an un-reduced β-keto ester. Thus, these data confirmed the role of CmrA and its mycobacterial homologue in the final step of (coryno)mycolic acid maturation and demonstrated the importance of the β-keto ester reduction in the subsequent attachment of mycolic acids to AG in *Corynebacterineae*.

Because the genes belonging to the *pks13* cluster are essential in mycobacteria, they represent attractive drug targets, and a recent screening identified novel classes of FadD32 inhibitors (122). Similarly, an assay was recently described to screen for inhibitors of PptT, the phosphopantetheinyl transferase that activates Pks13 and was shown to be essential for *in vitro* growth and persistence of *M. tuberculosis* (123). Following synthesis, MmpL3 transports the mycolic acids outside of the cell in the form of TMM (124–127), which is then used as a substrate by the mycolyltransferase enzymes of the Ag85 complex (128, 129) to transfer the mycolate residue either to the AG complex or to another TMM to form TDM.

Protein-Protein Interactions within the "Mycolome"

The EccA1 component of the ESX-1 protein secretion system has recently been shown to be required for optimal synthesis of mycolic acids (130). Increased mycolic acid synthesis defects were observed in an *eccA1* mutant of *M. marinum* and correlated with decreased *in vivo* virulence and intracellular growth. This phenotype was proposed to be linked by the interaction between EccA1 with Pks13, KasA, KasB, and MmaA4. This prompted speculation that a normal function of EccA1 is to make mycolic acid biosynthesis more efficient, perhaps by shuttling relevant enzymes to sites of mycolate production (130). These findings indicate that two mycobacterial virulence hallmarks, ESX-1-dependent protein secretion and mycolic acid synthesis, are critically linked via EccA1. The concept of a "mycolome" has recently emerged from elegant work demonstrating that the FAS II system of *M. tuberculosis* is organized in specialized interconnected complexes composed of the condensing enzymes, dehydratase heterodimers, and methyltransferases (131–133). This led the authors to propose that because the heterotypic interactions among these enzymes are crucial and their disruption detrimental for *M. tuberculosis* survival, the protein interactions could represent attractive drug targets. In this model of interactome, three types of FAS II specialized complexes are interconnected together: (i) the "initiation FAS II" (I-FAS II) is formed by a core

consisting of the reductases, FabD, and FabH, thus linking together FAS I and FAS II; (ii) two "elongation FAS II" (E-FAS II) complexes consisting of a core and either KasA (E1-FAS II) or KasB (E2-FAS II), which are capable of elongating acyl-AcpM to produce full-length meromycolyl-AcpM; (iii) the "termination FAS II" (T-FAS II) involving Pks13, which interacts with KasB and condenses the α branch with the meromycolic branch. The specialized and interconnected complexes of the "dissociated" FAS II system of *M. tuberculosis* may adopt a composition and architecture similar to a multifunctional FAS I protein.

REGULATION OF MYCOLIC ACID BIOSYNTHESIS

Although a large amount of information has been generated on the genetics, enzymology, and biochemical characterization of the FAS II components, little is known about the regulation of lipid homeostasis in mycobacteria. Since fatty/mycolic acids are not only essential but also energetically expensive for mycobacteria, it is very likely that these microorganisms have developed mechanisms that tightly regulate lipid concentrations. In this context, important advances have been made recently, allowing better understanding of how mycobacteria regulate their mycolic acid metabolism and composition. At least two levels of regulation have been proposed, at a transcriptional and at a posttranslational level, indicating the need for a high degree of control over the synthesis of mycolic acids, to efficiently adapt its composition to the various environmental conditions encountered during the complex lifestyle within the host and/or during the infection process.

Transcriptional Control by MabR

MabR (mycolic acid biosynthesis regulator) was originally identified as a putative transcriptional regulator encoded by *Rv2242* and located immediately upstream of the main *fasII* operon (Fig. 2) (134). *In vitro* characterization indicated that MabR functions as a transcriptional repressor of the *fasII* genes by binding directly to the region upstream of *fabD*. Moreover, overexpression of MabR in *M. smegmatis* represses transcription of *fabD*, *acpM*, *kasA*, and *kasB* and was accompanied by reduced levels of mycolic acids and changes in the colony morphotype. Knockdown of *mabR* expression using antisense RNA increased transcription of both *fasII* and *fasI* genes. MabR does not directly regulate *fasI* expression, but the reduced *fasI* transcripts in a MabR-overexpressing strain illustrate the coordination of phospholipid, triglyceride, and mycolic acid synthesis in mycobacteria (134). These observations, along with the fact that a *mabR* knockout mutant could only be generated in a merodiploid strain of *M. smegmatis*, confirm the predicted essential role of this repression in controlling mycobacterial mycolic acid metabolism. The discovery of MabR raises several important questions concerning how this regulator is integrated into mycobacterial physiology, and a key issue will be to identify the ligand(s) that modulate the DNA binding properties of this transcription factor.

Posttranslational Regulation by Ser/Thr Phosphorylation

Reversible protein phosphorylation is a key mechanism by which extracellular signals can be translated into cellular responses through the modulation of protein expression or activity. Signaling through Ser/Thr phosphorylation has emerged as a critical regulatory mechanism in various microorganisms and particularly in mycobacteria. *M. tuberculosis* possesses 11 Ser/Thr protein kinases (STPKs) (48, 135) that have been shown to regulate various biological functions ranging from central metabolism to environmental adaptive responses and pathogenicity. Recent studies have emphasized the importance of Ser/Thr phosphorylation in regulating mycobacterial cell shape/division and also the synthesis of major cell envelope components, including mycolic acids (136).

It was originally demonstrated that KasA and KasB were phosphorylated *in vitro* by multiple STPKs and that phosphorylation modulates their condensing activity (137). This suggests that environmental conditions might directly influence the phosphorylation profile of the condensing enzymes and thus modulate mycolic acid biosynthesis in order to adapt to various stresses. This view was further supported by the fact that, in addition to KasA and KasB, the other condensase, FabH, is also a substrate for mycobacterial STPKs and is efficiently phosphorylated at Thr45, by PknF and PknA (138). Interestingly, Thr45 is located at the entrance of the substrate channel on the crystal structure of FabH, suggesting that the phosphate group may affect FabH enzymatic activity by altering substrate accessibility. Importantly, a T45D variant of FabH, designed to mimic constitutive phosphorylation, exhibited markedly decreased transacylation, malonyl-AcpM decarboxylation, and condensing activities compared with the wild-type protein (138). MabA was also reported to be efficiently phosphorylated *in vitro* and *in vivo*, with Thr191 being the primary phosphor-acceptor (139). A MabA T191D mutant, designed to mimic constitutive

phosphorylation, exhibited markedly decreased keto-acyl reductase activity compared to the wild-type protein, as well as impaired binding of the NADPH cofactor. The negative effect of phosphorylation on MabA enzymatic activity and the consequent effect on mycolic acid biosynthesis was also shown when constitutive overexpression of the *mabA (T191D)* allele strongly impaired mycobacterial growth, and conditional expression of the phosphomimetic MabA T191D led to a significant inhibition of *de novo* biosynthesis of mycolic acids. Studies on the reductases culminated with the recent demonstration that InhA, the primary target of INH, is also controlled via phosphorylation on Thr266 (140). The physiological relevance of Thr266 phosphorylation was demonstrated using an *inhA* phosphomimetic (T266D) mutant. The *in vitro*, enoyl reductase activity was severely impaired in the mimetic mutant, and the introduction of the *inhA (T266D)* allele failed to complement an *inhA*-thermosensitive *M. smegmatis* strain, demonstrating that the *inhA* phosphorylation inhibited mycolic acid in a manner similar to that seen with INH and growth inhibition (140). Furthermore, the activity of the HadAB and HadBC dehydratases also decreased when these enzymes where phosphorylated (141).

Altogether, these studies strongly suggest that all essential enzymes forming the central core of type II fatty acid synthase are regulated by STPK phosphorylation, and *M. tuberculosis* may subtly control its FAS II system by regulating each step of the elongation cycle. Because phosphorylation of HadAB and HadBC enzymes was found to be increased during stationary growth phase (141), it is tempting to speculate that mycobacteria shut down meromycolic acid chain production under nonreplicating conditions, a view that is supported by the fact that the mycolic acid biosynthesis is growth phase–dependent and abrogated during the stationary phase (142). However, whether the control of the meromycolic acid chain length by phosphorylation of FAS II components contributes to *M. tuberculosis* virulence and/or persistence remains to be investigated.

There is also some evidence that STPK-dependent phosphorylation may additionally regulate the activity of enzymes involved in functional modification of mycolic acids, such as cyclopropane synthases. The phosphorylation of cyclopropane synthase PcaA on Thr168 and Thr183 by mycobacterial Ser/Thr kinases was reported both *in vitro* and *in vivo* (106), although phosphorylation of MmaA2 was not found. Phosphorylation of PcaA was associated with a significant decrease in the methyltransferase activity, and its physiological relevance was further assessed by generating the corresponding PcaA phosphoablative (T168A/T183A) or phosphomimetic (T168D/T183D) *M. bovis* BCG strains. In contrast to the wild-type and phosphoablative *pcaA* alleles, introduction of the phosphomimetic *pcaA* allele in a Δ*pcaA* mutant failed to restore the parental mycolic acid profile and cording morphotype. Importantly, the PcaA phosphomimetic mutant strain exhibited reduced survival in human macrophages and was unable to prevent phagosome maturation (106), thus providing a first link between a Ser/Thr kinase-dependent mechanism for modulating mycolic acid composition and intra-macrophage survival, a hallmark of mycobacterial virulence.

CONCLUDING REMARKS AND FUTURE PROSPECTS

Considerable progress has been made over the past two decades in determining the role of mycolic acids in mycobacterial physiology and virulence, as well as in identifying the genes that participate in mycolic acid biosynthesis. Genetics has clearly established that several genes in this pathway are essential for mycobacterial growth, while others participate in the interaction with the host immune system and/or are required for persistence, thus playing a critical role in the physiopathology of the disease. Because these genes are absent from humans, they represent valuable targets for future drug development. Several independent laboratories are currently developing large panels of chemical analogues of INH, thiolactomycin, thiocarlide or thiacetazone to specifically target the enzymes of the mycolic acid synthetic pathway. Combined with genetics to identify the mutations conferring drug resistance and structural studies of the target enzymes, these studies should foster the design of new lead compounds with potent antitubercular activity that could be particularly valuable against multidrug-resistant or extensively drug-resistant strains of *M. tuberculosis*. Recent work has also identified proteins that participate in the postbiosynthetic processing and transport of mycolic acids, which represent an additional source of potential drug targets, as exemplified by the discovery of inhibitors that target the mycolic acid transporter MmpL3 (124–127, 143).

The important breakthroughs that have begun to unravel the molecular mechanisms involved in the regulation of mycolic acid biosynthesis should also help in understanding how *M. tuberculosis* modulates its cell wall lipid composition to respond to the changes of environment encountered during infection. Among the key features that remain to be resolved are the signaling events that activate the different Ser/Thr protein kinases and how their phosphorylation of the different components modulates the activity of the FAS II enzymes and the transport of mycolates to the cell surface during the active and chronic phases of the disease. A distinguishing feature of *M. tuberculosis* is its acid-fast staining, and this property in Ziehl-Neelsen staining remains the cornerstone for diagnosing tuberculosis, particularly in poor countries where the infection is highly prevalent (144). However, dormant bacilli have distinct structural alterations in the cell wall and become Ziehl-Neelsen negative (145). This loss of acid-fastness during dormancy may be linked, at least partially, to changes within the mycolic acid profile, but this remains to be conclusively addressed experimentally. Therefore, investigating the role and contribution of regulatory mechanisms that tightly control mycolic acid biosynthesis under nonreplicating conditions where there is a loss of acid-fast staining could potentially affect our thinking on how to diagnose latent tuberculosis.

The authors wish to thank Howard Takiff for critically reading the manuscript.

JP is supported by a grant from the National Science Centre (contract 2012/05/N/NZ2/00622) and European Regional Development Fund under the Operational Programme Innovative Economy, grant POIG.01.01.02-10-107/09.

Citation. Pawełczyk J, Kremer L. 2014. The molecular genetics of mycolic acid biosynthesis. Microbiol Spectrum 2(4): MGM2-0003-2013.

References

1. Daffe M, Draper P. 1998. The envelope layers of mycobacteria with reference to their pathogenicity. *Adv Microb Physiol* **39:**131–203.

2. Barry CE 3rd, Lee RE, Mdluli K, Sampson AE, Schroeder BG, Slayden RA, Yuan Y. 1998. Mycolic acids: structure, biosynthesis and physiological functions. *Prog Lipid Res* **37:**143–179.

3. Brennan PJ, Nikaido H. 1995. The envelope of mycobacteria. *Annu Rev Biochem* **64:**29–63.

4. Ojha AK, Baughn AD, Sambandan D, Hsu T, Trivelli X, Guerardel Y, Alahari A, Kremer L, Jacobs WR Jr, Hatfull GF. 2008. Growth of *Mycobacterium tuberculosis* biofilms containing free mycolic acids and harbouring drug-tolerant bacteria. *Mol Microbiol* **69:** 164–174.

5. Takayama K, Wang C, Besra GS. 2005. Pathway to synthesis and processing of mycolic acids in *Mycobacterium tuberculosis*. *Clin Microbiol Rev* **18:**81–101.

6. Singh V, Mani I, Chaudhary DK, Somvanshi P. 2011. The beta-ketoacyl-ACP synthase from *Mycobacterium tuberculosis* as potential drug targets. *Curr Med Chem* **18:**1318–1324.

7. Barry CE, Crick DC, McNeil MR. 2007. Targeting the formation of the cell wall core of *M. tuberculosis*. *Infect Disord Drug Targets* **7**:182–202.

8. Bhatt A, Molle V, Besra GS, Jacobs WR Jr, Kremer L. 2007. The *Mycobacterium tuberculosis* FAS-II condensing enzymes: their role in mycolic acid biosynthesis, acid-fastness, pathogenesis and in future drug development. *Mol Microbiol* **64**:1442–1454.

9. Takayama K, Wang L, David HL. 1972. Effect of isoniazid on the in vivo mycolic acid synthesis, cell growth, and viability of *Mycobacterium tuberculosis*. *Antimicrob Agents Chemother* **2**:29–35.

10. Jain A, Mondal R. 2008. Extensively drug-resistant tuberculosis: current challenges and threats. *FEMS Immunol Med Microbiol* **53**:145–150.

11. Stodola FH, Lesuk A, Anderson RJ. 1938. The chemistry of the lipids of tubercle bacilli. The isolation and properties of mycolic acid. *J Biol Chem* **126**:505–513.

12. Asselineau J, Lederer E. 1950. Structure of the mycolic acids of mycobacteria. *Nature* **166**:782–783.

13. Etemadi AH, Lederer E. 1965. On the structure of the alpha-mycolic acids of the human test strain of *Mycobacterium tuberculosis*. *Bull Soc Chim Fr* **9**:2640–2645.

14. Goodfellow M, Weaver CR, Minnikin DE. 1982. Numerical classification of some *Rhodococci*, *Corynebacteria* and related organisms. *J Gen Microbiol* **128**:731–745.

15. Collins MD, Goodfellow M, Minnikin DE. 1982. A survey of the structures of mycolic acids in *Corynebacterium* and related taxa. *J Gen Microbiol* **128**:129–149.

16. Hong S, Cheng TY, Layre E, Sweet L, Young DC, Posey JE, Butler WR, Moody DB. 2012. Ultralong C100 mycolic acids support the assignment of *Segniliparus* as a new bacterial genus. *PLoS One* **7**:e39017.

17. Kaneda K, Naito S, Imaizumi S, Yano I, Mizuno S, Tomiyasu I, Baba T, Kusunose E, Kusunose M. 1986. Determination of molecular species composition of C80 or longer-chain alpha-mycolic acids in *Mycobacterium* spp. by gas chromatography-mass spectrometry and mass chromatography. *J Clin Microbiol* **24**:1060–1070.

18. Daffe M, Lanéelle MA, Puzo G, Asselineau C. 1981. Acide mycolique époxydique, un nouveau type d'acide mycolique. *Tetrahed Lett* **22**:4515–4516.

19. Minnikin DE, Polgar N. 1967. The methoxymycolic and ketomycolic acids from human tubercle bacilli. *Chem Commun* **22**:1172–1174.

20. Luquin M, Roussel J, Lopez-Calahorra F, Laneelle G, Ausina V, Laneelle MA. 1990. A novel mycolic acid in a *Mycobacterium* sp. from the environment. *Eur J Biochem* **192**:753–759.

21. Markovits J, Pinte F, Etemadi AH. 1966. Sur la structure des acides mycoliques dicarboxyliques insaturés isolés de *Mycobacterium phlei*. *CR Acad Sci (Paris)* **263**:960–962.

22. Minnikin DE, Minnikin SM, Parlett JH, Goodfellow M, Magnusson M. 1984. Mycolic acid patterns of some species of *Mycobacterium*. *Arch Microbiol* **139**:225–231.

23. Viader-Salvado JM, Molina-Torres CA, Guerrero-Olazaran M. 2007. Detection and identification of mycobacteria by mycolic acid analysis of sputum specimens and young cultures. *J Microbiol Methods* **70**:479–483.

24. Shui G, Bendt AK, Jappar IA, Lim HM, Laneelle M, Herve M, Via LE, Chua GH, Bratschi MW, Zainul Rahim SZ, Michelle AL, Hwang SH, Lee JS, Eum SY, Kwak HK, Daffe M, Dartois V, Michel G, Barry CE 3rd, Wenk MR. 2012. Mycolic acids as diagnostic markers for tuberculosis case detection in humans and drug efficacy in mice. *EMBO Mol Med* **4**:27–37.

25. Al Dulayymi JR, Baird MS, Roberts E. 2003. The synthesis of a single enantiomer of a major alpha-mycolic acid of *Mycobacterium tuberculosis*. *Chem Commun (Camb)* Jan. **21**:228–229.

26. Verschoor JA, Baird MS, Grooten J. 2012. Towards understanding the functional diversity of cell wall mycolic acids of *Mycobacterium tuberculosis*. *Prog Lipid Res* **51**:325–339.

27. Dubnau E, Chan J, Raynaud C, Mohan VP, Laneelle MA, Yu K, Quemard A, Smith I, Daffe M. 2000. Oxygenated mycolic acids are necessary for virulence of *Mycobacterium tuberculosis* in mice. *Mol Microbiol* **36**:630–637.

28. Glickman MS, Cox JS, Jacobs WR Jr. 2000. A novel mycolic acid cyclopropane synthetase is required for cording, persistence, and virulence of *Mycobacterium tuberculosis*. *Mol Cell* **5**:717–727.

29. Bhatt A, Fujiwara N, Bhatt K, Gurcha SS, Kremer L, Chen B, Chan J, Porcelli SA, Kobayashi K, Besra GS, Jacobs WR Jr. 2007. Deletion of kasB in *Mycobacterium tuberculosis* causes loss of acid-fastness and subclinical latent tuberculosis in immunocompetent mice. *Proc Natl Acad Sci USA* **104**:5157–5262.

30. Yuan Y, Zhu Y, Crane D, Barry CE 3rd. 1998. The effect of oxygenated mycolic acid composition on cell wall function and macrophage growth in *Mycobacterium tuberculosis*. *Mol Microbiol* **29**:1449–1458.

31. Hong X, Hopfinger AJ. 2004. Construction, molecular modeling, and simulation of *Mycobacterium tuberculosis* cell walls. *Biomacromolecules* **5**:1052–1065.

32. Yuan Y, Lee RE, Besra GS, Belisle JT, Barry CE 3rd. 1995. Identification of a gene involved in the biosynthesis of cyclopropanated mycolic acids in *Mycobacterium tuberculosis*. *Proc Natl Acad Sci USA* **92**:6630–6634.

33. Hunter RL, Olsen M, Jagannath C, Actor JK. 2006. Trehalose 6,6'-dimycolate and lipid in the pathogenesis of caseating granulomas of tuberculosis in mice. *Am J Pathol* **168**:1249–1261.

34. Bekierkunst A. 1968. Acute granulomatous response produced in mice by trehalose-6,6-dimycolate. *J Bacteriol* **96**:958–961.

35. Bekierkunst A, Levij IS, Yarkoni E, Vilkas E, Lederer E. 1971. Cellular reaction in the footpad and draining lymph nodes of mice induced by mycobacterial fractions and BCG bacilli. *Infect Immun* **4**:245–255.

36. Bekierkunst A, Levij IS, Yarkoni E. 1971. Suppression of urethan-induced lung adenomas in mice treated with

trehalose-6,6-dimycolate (cord factor) and living bacillus Calmette Guerin. *Science* 174:1240–1242.

37. Indrigo J, Hunter RL Jr, Actor JK. 2003. Cord factor trehalose 6,6′-dimycolate (TDM) mediates trafficking events during mycobacterial infection of murine macrophages. *Microbiology* 149:2049–2059.

38. Yamasaki S, Ishikawa E, Sakuma M, Hara H, Ogata K, Saito T. 2008. Mincle is an ITAM-coupled activating receptor that senses damaged cells. *Nat Immunol* 9: 1179–1188.

39. Ishikawa E, Ishikawa T, Morita YS, Toyonaga K, Yamada H, Takeuchi O, Kinoshita T, Akira S, Yoshikai Y, Yamasaki S. 2009. Direct recognition of the mycobacterial glycolipid, trehalose dimycolate, by C-type lectin Mincle. *J Exp Med* 206:2879–2888.

40. Rombouts Y, Brust B, Ojha AK, Maes E, Coddeville B, Elass-Rochard E, Kremer L, Guerardel Y. 2012. Exposure of mycobacteria to cell wall-inhibitory drugs decreases production of arabinoglycerolipid related to mycolyl-arabinogalactan-peptidoglycan metabolism. *J Biol Chem* 287:11060–11069.

41. Elass-Rochard E, Rombouts Y, Coddeville B, Maes E, Blervaque R, Hot D, Kremer L, Guerardel Y. 2012. Structural determination and Toll-like receptor 2-dependent proinflammatory activity of dimycolyl-diarabino-glycerol from *Mycobacterium marinum*. *J Biol Chem* 287:34432–34444.

42. Ojha AK, Trivelli X, Guerardel Y, Kremer L, Hatfull GF. 2010. Enzymatic hydrolysis of trehalose dimycolate releases free mycolic acids during mycobacterial growth in biofilms. *J Biol Chem* 285:17380–17389.

43. Beckman EM, Porcelli SA, Morita CT, Behar SM, Furlong ST, Brenner MB. 1994. Recognition of a lipid antigen by CD1-restricted alpha beta+ T cells. *Nature* 372:691–694.

44. Korf J, Stoltz A, Verschoor J, De Baetselier P, Grooten J. 2005. The *Mycobacterium tuberculosis* cell wall component mycolic acid elicits pathogen-associated host innate immune responses. *Eur J Immunol* 35:890–900.

45. Vander Beken S, Al Dulayymi JR, Naessens T, Koza G, Maza-Iglesias M, Rowles R, Theunissen C, De Medts J, Lanckacker E, Baird MS, Grooten J. Molecular structure of the *Mycobacterium tuberculosis* virulence factor, mycolic acid, determines the elicited inflammatory pattern. *Eur J Immunol* 41:450–460.

46. Wakil SJ, Stoops JK, Joshi VC. 1983. Fatty acid synthesis and its regulation. *Annu Rev Biochem* 52:537–579.

47. Cronan JE Jr, Waldrop GL. 2002. Multi-subunit acetyl-CoA carboxylases. *Prog Lipid Res* 41:407–435.

48. Cole ST, Brosch R, Parkhill J, Garnier T, Churcher C, Harris D, Gordon SV, Eiglmeier K, Gas S, Barry CE 3rd, Tekaia F, Badcock K, Basham D, Brown D, Chillingworth T, Connor R, Davies R, Devlin K, Feltwell T, Gentles S, Hamlin N, Holroyd S, Hornsby T, Jagels K, Krogh A, McLean J, Moule S, Murphy L, Oliver K, Osborne J, Quail MA, Rajandream MA, Rogers J, Rutter S, Seeger K, Skelton J, Squares R, Squares S, Sulston JE, Taylor K, Whitehead S, Barrell BG. 1998. Deciphering the biology of *Mycobacterium*

tuberculosis from the complete genome sequence. *Nature* 393:537–544.

49. Sassetti CM, Boyd DH, Rubin EJ. 2003. Genes required for mycobacterial growth defined by high density mutagenesis. *Mol Microbiol* 48:77–84.

50. Pawelczyk J, Brzostek A, Kremer L, Dziadek B, Rumijowska-Galewicz A, Fiolka M, Dziadek J. 2011. AccD6, a key carboxyltransferase essential for mycolic acid synthesis in *Mycobacterium tuberculosis*, is dispensable in a nonpathogenic strain. *J Bacteriol* 193:6960–6972.

51. Daniel J, Oh TJ, Lee CM, Kolattukudy PE. 2007. AccD6, a member of the Fas II locus, is a functional carboxyltransferase subunit of the acyl-coenzyme A carboxylase in *Mycobacterium tuberculosis*. *J Bacteriol* 189:911–917.

52. Oh TJ, Daniel J, Kim HJ, Sirakova TD, Kolattukudy PE. 2006. Identification and characterization of Rv3281 as a novel subunit of a biotin-dependent Acyl-CoA carboxylase in *Mycobacterium tuberculosis* H37Rv. *J Biol Chem* 281:3899–3908.

53. Portevin D, de Sousa-D'Auria C, Montrozier H, Houssin C, Stella A, Laneelle MA, Bardou F, Guilhot C, Daffe M. 2005. The acyl-AMP ligase FadD32 and AccD4-containing acyl-CoA carboxylase are required for the synthesis of mycolic acids and essential for mycobacterial growth: identification of the carboxylation product and determination of the acyl-CoA carboxylase components. *J Biol Chem* 280:8862–8874.

54. Gago G, Kurth D, Diacovich L, Tsai SC, Gramajo H. 2006. Biochemical and structural characterization of an essential acyl coenzyme A carboxylase from *Mycobacterium tuberculosis*. *J Bacteriol* 188:477–486.

55. Holton SJ, King-Scott S, Nasser Eddine A, Kaufmann SH, Wilmanns M. 2006. Structural diversity in the six-fold redundant set of acyl-CoA carboxyltransferases in *Mycobacterium tuberculosis*. *FEBS Lett* 580:6898–6902.

56. Lin TW, Melgar MM, Kurth D, Swamidass SJ, Purdon J, Tseng T, Gago G, Baldi P, Gramajo H, Tsai SC. 2006. Structure-based inhibitor design of AccD5, an essential acyl-CoA carboxylase carboxyltransferase domain of *Mycobacterium tuberculosis*. *Proc Natl Acad Sci USA* 103:3072–3077.

57. Kurth DG, Gago GM, de la Iglesia A, Bazet Lyonnet B, Lin TW, Morbidoni HR, Tsai SC, Gramajo H. 2009. ACCase 6 is the essential acetyl-CoA carboxylase involved in fatty acid and mycolic acid biosynthesis in mycobacteria. *Microbiology* 155:2664–2675.

58. Niu C, Yin J, Cherney MM, James MN. 2011. Expression, purification and preliminary crystallographic analysis of Rv2247, the beta subunit of acyl-CoA carboxylase (ACCD6) from *Mycobacterium tuberculosis*. *Acta Crystallogr Sect F Struct Biol Cryst Commun* 67:1637–1640.

59. Smith S, Witkowski A, Joshi AK. 2003. Structural and functional organization of the animal fatty acid synthase. *Prog Lipid Res* 42:289–317.

60. Bloch K, Vance D. 1977. Control mechanisms in the synthesis of saturated fatty acids. *Annu Rev Biochem* 46:263–298.

61. Kikuchi S, Rainwater DL, Kolattukudy PE. 1992. Purification and characterization of an unusually large fatty acid synthase from *Mycobacterium tuberculosis* var. *bovis* BCG. *Arch Biochem Biophys* **295**:318–326.

62. Zimhony O, Vilcheze C, Jacobs WR Jr. 2004. Characterization of *Mycobacterium smegmatis* expressing the *Mycobacterium tuberculosis* fatty acid synthase I (fas1) gene. *J Bacteriol* **186**:4051–4055.

63. Papaioannou N, Cheon HS, Lian Y, Kishi Y. 2007. Product-regulation mechanisms for fatty acid biosynthesis catalyzed by *Mycobacterium smegmatis* FAS I. *Chembiochem* **8**:1775–1780.

64. Boehringer D, Ban N, Leibundgut M. 2013. 7.5-Å Cryo-EM structure of the mycobacterial fatty acid synthase. *J Mol Biol* **425**:841–849.

65. Choi KH, Kremer L, Besra GS, Rock CO. 2000. Identification and substrate specificity of beta-ketoacyl (acyl carrier protein) synthase III (mtFabH) from *Mycobacterium tuberculosis*. *J Biol Chem* **275**:28201–28207.

66. Scarsdale JN, Kazanina G, He X, Reynolds KA, Wright HT. 2001. Crystal structure of the *Mycobacterium tuberculosis* beta-ketoacyl-acyl carrier protein synthase III. *J Biol Chem* **276**:20516–20522.

67. Brown AK, Sridharan S, Kremer L, Lindenberg S, Dover LG, Sacchettini JC, Besra GS. 2005. Probing the mechanism of the *Mycobacterium tuberculosis* beta-ketoacyl-acyl carrier protein synthase III mtFabH: factors influencing catalysis and substrate specificity. *J Biol Chem* **280**:32539–32547.

68. Kremer L, Nampoothiri KM, Lesjean S, Dover LG, Graham S, Betts J, Brennan PJ, Minnikin DE, Locht C, Besra GS. 2001. Biochemical characterization of acyl carrier protein (AcpM) and malonyl-CoA:AcpM transacylase (mtFabD), two major components of *Mycobacterium tuberculosis* fatty acid synthase II. *J Biol Chem* **276**:27967–27974.

69. Wong HC, Liu G, Zhang YM, Rock CO, Zheng J. 2002. The solution structure of acyl carrier protein from *Mycobacterium tuberculosis*. *J Biol Chem* **277**:15874–15880.

70. Kremer L, Douglas JD, Baulard AR, Morehouse C, Guy MR, Alland D, Dover LG, Lakey JH, Jacobs WR Jr, Brennan PJ, Minnikin DE, Besra GS. 2000. Thiolactomycin and related analogues as novel antimycobacterial agents targeting KasA and KasB condensing enzymes in *Mycobacterium tuberculosis*. *J Biol Chem* **275**:16857–16864.

71. Schaeffer ML, Agnihotri G, Volker C, Kallender H, Brennan PJ, Lonsdale JT. 2001. Purification and biochemical characterization of the *Mycobacterium tuberculosis* beta-ketoacyl-acyl carrier protein synthases KasA and KasB. *J Biol Chem* **276**:47029–47037.

72. Kremer L, Dover LG, Carrere S, Nampoothiri KM, Lesjean S, Brown AK, Brennan PJ, Minnikin DE, Locht C, Besra GS. 2002. Mycolic acid biosynthesis and enzymic characterization of the beta-ketoacyl-ACP synthase A-condensing enzyme from *Mycobacterium tuberculosis*. *Biochem J* **364**:423–430.

73. Luckner SR, Machutta CA, Tonge PJ, Kisker C. 2009. Crystal structures of *Mycobacterium tuberculosis* KasA show mode of action within cell wall biosynthesis and its inhibition by thiolactomycin. *Structure* **17**:1004–1013.

74. Sridharan S, Wang L, Brown AK, Dover LG, Kremer L, Besra GS, Sacchettini JC. 2007. X-ray crystal structure of *Mycobacterium tuberculosis* beta-ketoacyl acyl carrier protein synthase II (mtKasB). *J Mol Biol* **366**:469–480.

75. Slayden RA, Barry CE 3rd. 2002. The role of KasA and KasB in the biosynthesis of meromycolic acids and isoniazid resistance in *Mycobacterium tuberculosis*. *Tuberculosis (Edinb)* **82**:149–160.

76. Bhatt A, Kremer L, Dai AZ, Sacchettini JC, Jacobs WR Jr. 2005. Conditional depletion of KasA, a key enzyme of mycolic acid biosynthesis, leads to mycobacterial cell lysis. *J Bacteriol* **187**:7596–7606.

77. Kapilashrami K, Bommineni GR, Machutta CA, Kim P, Lai CT, Simmerling C, Picart F, Tonge PJ. 2013. Thiolactomycin-based beta-ketoacyl-AcpM synthase A (KasA) inhibitors: fragment-based inhibitor discovery using transient one-dimensional nuclear overhauser effect NMR spectroscopy. *J Biol Chem* **288**:6045–6052.

78. Kruh NA, Borgaro JG, Ruzsicska BP, Xu H, Tonge PJ. 2008. A novel interaction linking the FAS-II and phthiocerol dimycocerosate (PDIM) biosynthetic pathways. *J Biol Chem* **283**:31719–31725.

79. Marrakchi H, Ducasse S, Labesse G, Montrozier H, Margeat E, Emorine L, Charpentier X, Daffe M, Quemard A. 2002. MabA (FabG1), a *Mycobacterium tuberculosis* protein involved in the long-chain fatty acid elongation system FAS-II. *Microbiology* **148**:951–960.

80. Sacco E, Covarrubias AS, O'Hare HM, Carroll P, Eynard N, Jones TA, Parish T, Daffe M, Backbro K, Quemard A. 2007. The missing piece of the type II fatty acid synthase system from *Mycobacterium tuberculosis*. *Proc Natl Acad Sci USA* **104**:14628–14633.

81. Brown AK, Bhatt A, Singh S, Saparia E, Evans AF, Besra GS. 2007. Identification of the dehydratase component of the mycobacterial mycolic acid-synthesizing fatty acid synthase-II complex. *Microbiology* **153**:4166–4173.

82. Quemard A, Sacchettini JC, Dessen A, Vilcheze C, Bittman R, Jacobs WR Jr, Blanchard JS. 1995. Enzymatic characterization of the target for isoniazid in *Mycobacterium tuberculosis*. *Biochemistry* **34**:8235–8241.

83. Banerjee A, Dubnau E, Quemard A, Balasubramanian V, Um KS, Wilson T, Collins D, de Lisle G, Jacobs WR Jr. 1994. inhA, a gene encoding a target for isoniazid and ethionamide in *Mycobacterium tuberculosis*. *Science* **263**:227–230.

84. Cohen-Gonsaud M, Ducasse S, Hoh F, Zerbib D, Labesse G, Quemard A. 2002. Crystal structure of MabA from *Mycobacterium tuberculosis*, a reductase involved in long-chain fatty acid biosynthesis. *J Mol Biol* **320**:249–261.

85. Belardinelli JM, Morbidoni HR. 2012. Mutations in the essential FAS II beta-hydroxyacyl ACP dehydratase complex confer resistance to thiacetazone in *Mycobacterium tuberculosis* and *Mycobacterium kansasii*. *Mol Microbiol* **86**:568–579.

86. Coxon GD, Craig D, Corrales RM, Vialla E, Gannoun-Zaki L, Kremer L. 2013. Synthesis, antitubercular activity and mechanism of resistance of highly effective thiacetazone analogues. *PLoS One* 8:e53162.

87. Grzegorzewicz AE, Kordulakova J, Jones V, Born SE, Belardinelli JM, Vaquie A, Gundi VA, Madacki J, Slama N, Laval F, Vaubourgeix J, Crew RM, Gicquel B, Daffe M, Morbidoni HR, Brennan PJ, Quemard A, McNeil MR, Jackson M. 2012. A common mechanism of inhibition of the *Mycobacterium tuberculosis* mycolic acid biosynthetic pathway by isoxyl and thiacetazone. *J Biol Chem* 287:38434–38441.

88. Gannoun-Zaki L, Alibaud L, Kremer L. 2013. Point mutations within the fatty acid synthase type II dehydratase components HadA or HadC contribute to isoxyl resistance in *Mycobacterium tuberculosis*. *Antimicrob Agents Chemother* 57:629–632.

89. Kremer L, Baulard AR, Besra GS. 2000. Genetics of mycolic acid biosynthesis, p 173–190. *In* Hatfull GF, Jacobs WR Jr (ed), *Molecular Genetics of Mycobacteria*. ASM Press, Washington, DC.

90. Huang CC, Smith CV, Glickman MS, Jacobs WR Jr, Sacchettini JC. 2002. Crystal structures of mycolic acid cyclopropane synthases from *Mycobacterium tuberculosis*. *J Biol Chem* 277:11559–11569.

91. Boissier F, Bardou F, Guillet V, Uttenweiler-Joseph S, Daffe M, Quemard A, Mourey L. 2006. Further insight into S-adenosylmethionine-dependent methyltransferases: structural characterization of Hma, an enzyme essential for the biosynthesis of oxygenated mycolic acids in *Mycobacterium tuberculosis*. *J Biol Chem* 281:4434–4445.

92. Barkan D, Liu Z, Sacchettini JC, Glickman MS. 2009. Mycolic acid cyclopropanation is essential for viability, drug resistance, and cell wall integrity of *Mycobacterium tuberculosis*. *Chem Biol* 16:499–509.

93. George KM, Yuan Y, Sherman DR, Barry CE 3rd. 1995. The biosynthesis of cyclopropanated mycolic acids in *Mycobacterium tuberculosis*. Identification and functional analysis of CMAS-2. *J Biol Chem* 270:27292–27298.

94. Yuan Y, Barry CE 3rd. 1996. A common mechanism for the biosynthesis of methoxy and cyclopropyl mycolic acids in *Mycobacterium tuberculosis*. *Proc Natl Acad Sci USA* 93:12828–12833.

95. Glickman MS. 2003. The mmaA2 gene of *Mycobacterium tuberculosis* encodes the distal cyclopropane synthase of the alpha-mycolic acid. *J Biol Chem* 278:7844–7849.

96. Glickman MS, Cahill SM, Jacobs WR Jr. 2001. The *Mycobacterium tuberculosis* cmaA2 gene encodes a mycolic acid trans-cyclopropane synthetase. *J Biol Chem* 276:2228–2233.

97. Barkan D, Rao V, Sukenick GD, Glickman MS. 2010. Redundant function of cmaA2 and mmaA2 in *Mycobacterium tuberculosis* cis cyclopropanation of oxygenated mycolates. *J Bacteriol* 192:3661–3668.

98. Laval F, Haites R, Movahedzadeh F, Lemassu A, Wong CY, Stoker N, Billman-Jacobe H, Daffe M. 2008. Investigating the function of the putative mycolic acid methyltransferase UmaA: divergence between the *Mycobacterium smegmatis* and *Mycobacterium tuberculosis* proteins. *J Biol Chem* 283:1419–1427.

99. Dao DN, Sweeney K, Hsu T, Gurcha SS, Nascimento IP, Roshevsky D, Besra GS, Chan J, Porcelli SA, Jacobs WR. 2008. Mycolic acid modification by the mmaA4 gene of *M. tuberculosis* modulates IL-12 production. *PLoS Pathog* 4:e1000081.

100. Alahari A, Alibaud L, Trivelli X, Gupta R, Lamichhane G, Reynolds RC, Bishai WR, Guerardel Y, Kremer L. 2009. Mycolic acid methyltransferase, MmaA4, is necessary for thiacetazone susceptibility in *Mycobacterium tuberculosis*. *Mol Microbiol* 71:1263–1277.

101. Dubnau E, Laneelle MA, Soares S, Benichou A, Vaz T, Prome D, Prome JC, Daffe M, Quemard A. 1997. *Mycobacterium bovis* BCG genes involved in the biosynthesis of cyclopropyl keto- and hydroxy-mycolic acids. *Mol Microbiol* 23:313–322.

102. Dubnau E, Marrakchi H, Smith I, Daffe M, Quemard A. 1998. Mutations in the cmaB gene are responsible for the absence of methoxymycolic acid in *Mycobacterium bovis* BCG Pasteur. *Mol Microbiol* 29:1526–1528.

103. Dinadayala P, Laval F, Raynaud C, Lemassu A, Laneelle MA, Laneelle G, Daffe M. 2003. Tracking the putative biosynthetic precursors of oxygenated mycolates of *Mycobacterium tuberculosis*. Structural analysis of fatty acids of a mutant strain devoid of methoxy- and ketomycolates. *J Biol Chem* 278:7310–7319.

104. Behr MA, Schroeder BG, Brinkman JN, Slayden RA, Barry CE 3rd. 2000. A point mutation in the mma3 gene is responsible for impaired methoxymycolic acid production in *Mycobacterium bovis* BCG strains obtained after 1927. *J Bacteriol* 182:3394–3399.

105. Rao V, Fujiwara N, Porcelli SA, Glickman MS. 2005. *Mycobacterium tuberculosis* controls host innate immune activation through cyclopropane modification of a glycolipid effector molecule. *J Exp Med* 201:535–543.

106. Corrales RM, Molle V, Leiba J, Mourey L, de Chastellier C, Kremer L. 2012. Phosphorylation of mycobacterial PcaA inhibits mycolic acid cyclopropanation: consequences for intracellular survival and for phagosome maturation block. *J Biol Chem* 287:26187–26199.

107. Rao V, Gao F, Chen B, Jacobs WR Jr, Glickman MS. 2006. Trans-cyclopropanation of mycolic acids on trehalose dimycolate suppresses *Mycobacterium tuberculosis*-induced inflammation and virulence. *J Clin Invest* 116:1660–1667.

108. Barkan D, Hedhli D, Yan HG, Huygen K, Glickman MS. 2012. *Mycobacterium tuberculosis* lacking all mycolic acid cyclopropanation is viable but highly attenuated and hyperinflammatory in mice. *Infect Immun* 80:1958–1968.

109. Vaubourgeix J, Bardou F, Boissier F, Julien S, Constant P, Ploux O, Daffe M, Quemard A, Mourey L. 2009. S-adenosyl-N-decyl-aminoethyl, a potent bisubstrate inhibitor of *Mycobacterium tuberculosis* mycolic acid methyltransferases. *J Biol Chem* 284:19321–19330.

110. Alahari A, Trivelli X, Guerardel Y, Dover LG, Besra GS, Sacchettini JC, Reynolds RC, Coxon GD, Kremer L. 2007. Thiacetazone, an antitubercular drug that inhibits cyclopropanation of cell wall mycolic acids in mycobacteria. *PLoS One* 2:e1343.

111. Alibaud L, Alahari A, Trivelli X, Ojha AK, Hatfull GF, Guerardel Y, Kremer L. 2010. Temperature-dependent regulation of mycolic acid cyclopropanation in saprophytic mycobacteria: role of the *Mycobacterium smegmatis* 1351 gene (MSMEG_1351) in cis-cyclopropanation of alpha-mycolates. *J Biol Chem* 285:21698–21707.

112. Portevin D, De Sousa-D'Auria C, Houssin C, Grimaldi C, Chami M, Daffe M, Guilhot C. 2004. A polyketide synthase catalyzes the last condensation step of mycolic acid biosynthesis in mycobacteria and related organisms. *Proc Natl Acad Sci USA* 101:314–319.

113. Gande R, Gibson KJ, Brown AK, Krumbach K, Dover LG, Sahm H, Shioyama S, Oikawa T, Besra GS, Eggeling L. 2004. Acyl-CoA carboxylases (accD2 and accD3), together with a unique polyketide synthase (Cg-pks), are key to mycolic acid biosynthesis in *Corynebacterianeae* such as *Corynebacterium glutamicum* and *Mycobacterium tuberculosis*. *J Biol Chem* 279: 44847–44857.

114. Gande R, Dover LG, Krumbach K, Besra GS, Sahm H, Oikawa T, Eggeling L. 2007. The two carboxylases of *Corynebacterium glutamicum* essential for fatty acid and mycolic acid synthesis. *J Bacteriol* 189:5257–5264.

115. Trivedi OA, Arora P, Sridharan V, Tickoo R, Mohanty D, Gokhale RS. 2004. Enzymic activation and transfer of fatty acids as acyl-adenylates in mycobacteria. *Nature* 428:441–445.

116. Lin TW, Melgar MM, Kurth D, Swamidass SJ, Purdon J, Tseng T, Gago G, Baldi P, Gramajo H, Tsai SC. 2006. Structure-based inhibitor design of AccD5, an essential acyl-CoA carboxylase carboxyltransferase domain of *Mycobacterium tuberculosis*. *Proc Natl Acad Sci USA* 103:3072–3077.

117. Gavalda S, Leger M, van der Rest B, Stella A, Bardou F, Montrozier H, Chalut C, Burlet-Schiltz O, Marrakchi H, Daffe M, Quemard A. 2009. The Pks13/FadD32 crosstalk for the biosynthesis of mycolic acids in *Mycobacterium tuberculosis*. *J Biol Chem* 284:19255–19264.

118. Leger M, Gavalda S, Guillet V, van der Rest B, Slama N, Montrozier H, Mourey L, Quemard A, Daffe M, Marrakchi H. 2009. The dual function of the *Mycobacterium tuberculosis* FadD32 required for mycolic acid biosynthesis. *Chem Biol* 16:510–519.

119. Bergeret F, Gavalda S, Chalut C, Malaga W, Quemard A, Pedelacq JD, Daffe M, Guilhot C, Mourey L, Bon C. 2012. Biochemical and structural study of the atypical acyltransferase domain from the mycobacterial polyketide synthase Pks13. *J Biol Chem* 287:33675–33690.

120. Wilson R, Kumar P, Parashar V, Vilcheze C, Veyron-Churlet R, Freundlich JS, Barnes SW, Walker JR, Szymonifka MJ, Marchiano E, Shenai S, Colangeli R, Jacobs WR Jr, Neiditch MB, Kremer L, Alland D. 2013. Antituberculosis thiophenes define a requirement for Pks13 in mycolic acid biosynthesis. *Nat Chem Biol* 9:499–506.

121. Lea-Smith DJ, Pyke JS, Tull D, McConville MJ, Coppel RL, Crellin PK. 2007. The reductase that catalyzes mycolic motif synthesis is required for efficient attachment of mycolic acids to arabinogalactan. *J Biol Chem* 282:11000–11008.

122. Galandrin S, Guillet V, Rane RS, Leger M, Radha N, Eynard N, Das K, Balganesh TS, Mourey L, Daffe M, Marrakchi H. 2013. Assay development for identifying inhibitors of the mycobacterial FadD32 activity. *J Biomol Screen* 18:576–587.

123. Leblanc C, Prudhomme T, Tabouret G, Ray A, Burbaud S, Cabantous S, Mourey L, Guilhot C, Chalut C. 2012. 4′-Phosphopantetheinyl transferase PptT, a new drug target required for *Mycobacterium tuberculosis* growth and persistence in vivo. *PLoS Pathog* 8:e1003097.

124. Varela C, Rittmann D, Singh A, Krumbach K, Bhatt K, Eggeling L, Besra GS, Bhatt A. 2012. MmpL genes are associated with mycolic acid metabolism in mycobacteria and corynebacteria. *Chem Biol* 19:498–506.

125. La Rosa V, Poce G, Canseco JO, Buroni S, Pasca MR, Biava M, Raju RM, Porretta GC, Alfonso S, Battilocchio C, Javid B, Sorrentino F, Ioerger TR, Sacchettini JC, Manetti F, Botta M, De Logu A, Rubin EJ, De Rossi E. 2012. MmpL3 is the cellular target of the antitubercular pyrrole derivative BM212. *Antimicrob Agents Chemother* 56:324–331.

126. Tahlan K, Wilson R, Kastrinsky DB, Arora K, Nair V, Fischer E, Barnes SW, Walker JR, Alland D, Barry CE 3rd, Boshoff HI. 2012. SQ109 targets MmpL3, a membrane transporter of trehalose monomycolate involved in mycolic acid donation to the cell wall core of *Mycobacterium tuberculosis*. *Antimicrob Agents Chemother* 56:1797–1809.

127. Grzegorzewicz AE, Pham H, Gundi VA, Scherman MS, North EJ, Hess T, Jones V, Gruppo V, Born SE, Kordulakova J, Chavadi SS, Morisseau C, Lenaerts AJ, Lee RE, McNeil MR, Jackson M. 2012. Inhibition of mycolic acid transport across the *Mycobacterium tuberculosis* plasma membrane. *Nat Chem Biol* 8:334–341.

128. Belisle JT, Vissa VD, Sievert T, Takayama K, Brennan PJ, Besra GS. 1997. Role of the major antigen of *Mycobacterium tuberculosis* in cell wall biogenesis. *Science* 276:1420–1422.

129. Puech V, Guilhot C, Perez E, Tropis M, Armitige LY, Gicquel B, Daffe M. 2002. Evidence for a partial redundancy of the fibronectin-binding proteins for the transfer of mycoloyl residues onto the cell wall arabinogalactan termini of *Mycobacterium tuberculosis*. *Mol Microbiol* 44:1109–1122.

130. Joshi SA, Ball DA, Sun MG, Carlsson F, Watkins BY, Aggarwal N, McCracken JM, Huynh KK, Brown EJ. 2012. EccA1, a component of the *Mycobacterium marinum* ESX-1 protein virulence factor secretion pathway, regulates mycolic acid lipid synthesis. *Chem Biol* 19:372–380.

131. Veyron-Churlet R, Guerrini O, Mourey L, Daffe M, Zerbib D. 2004. Protein-protein interactions within the fatty acid synthase-II system of *Mycobacterium tuberculosis* are essential for mycobacterial viability. *Mol Microbiol* 54:1161–1172.

132. Veyron-Churlet R, Bigot S, Guerrini O, Verdoux S, Malaga W, Daffe D, Zerbib D. 2005. The biosynthesis of mycolic acids in *Mycobacterium tuberculosis* relies on multiple specialized elongation complexes interconnected by specific protein-protein interactions. *J Mol Biol* **353**:847–858.

133. Cantaloube S, Veyron-Churlet R, Haddache N, Daffe M, Zerbib D. 2011. The *Mycobacterium tuberculosis* FAS-II dehydratases and methyltranserases define the specificity of the mycolic acid elongation complexes. *PLoS One* **6**:e29564.

134. Salzman V, Mondino S, Sala C, Cole ST, Gago G, Gramajo H. 2010. Transcriptional regulation of lipid homeostasis in mycobacteria. *Mol Microbiol* **78**:64–77.

135. Av-Gay Y, Everett M. 2000. The eukaryotic-like Ser/Thr protein kinases of *Mycobacterium tuberculosis*. *Trends Microbiol* **8**:238–244.

136. Molle V, Kremer L. 2010. Division and cell envelope regulation by Ser/Thr phosphorylation: mycobacterium shows the way. *Mol Microbiol* **75**:1064–1077.

137. Molle V, Brown AK, Besra GS, Cozzone AJ, Kremer L. 2006. The condensing activities of the *Mycobacterium tuberculosis* type II fatty acid synthase are differentially regulated by phosphorylation. *J Biol Chem* **281**:30094–30103.

138. Veyron-Churlet R, Molle V, Taylor RC, Brown AK, Besra GS, Zanella-Cleon I, Futterer K, Kremer L. 2009. The *Mycobacterium tuberculosis* beta-ketoacyl-acyl carrier protein synthase III activity is inhibited by phosphorylation on a single threonine residue. *J Biol Chem* **284**:6414–6424.

139. Veyron-Churlet R, Zanella-Cleon I, Cohen-Gonsaud M, Molle V, Kremer L. 2010. Phosphorylation of the *Mycobacterium tuberculosis* beta-ketoacyl-acyl carrier protein reductase MabA regulates mycolic acid biosynthesis. *J Biol Chem* **285**:12714–12725.

140. Molle V, Gulten G, Vilcheze C, Veyron-Churlet R, Zanella-Cleon I, Sacchettini JC, Jacobs WR Jr, Kremer L. 2010. Phosphorylation of InhA inhibits mycolic acid biosynthesis and growth of *Mycobacterium tuberculosis*. *Mol Microbiol* **78**:1591–1605.

141. Slama N, Leiba J, Eynard N, Daffe M, Kremer L, Quemard A, Molle V. 2011. Negative regulation by Ser/Thr phosphorylation of HadAB and HadBC dehydratases from *Mycobacterium tuberculosis* type II fatty acid synthase system. *Biochem Biophys Res Commun* **412**:401–406.

142. Lacave C, Laneelle MA, Daffe M, Montrozier H, Laneelle G. 1989. Mycolic acid metabolic filiation and location in *Mycobacterium aurum* and *Mycobacterium phlei*. *Eur J Biochem* **181**:459–466.

143. Stanley SA, Grant SS, Kawate T, Iwase N, Shimizu M, Wivagg C, Silvis M, Kazyanskaya E, Aquadro J, Golas A, Fitzgerald M, Dai H, Zhang L, Hung DT. 2012. Identification of novel inhibitors of *M. tuberculosis* growth using whole cell based high-throughput screening. *ACS Chem Biol* **7**:1377–1384.

144. Trebucq A. 2004. Revisiting sputum smear microscopy. *Int J Tuberc Lung Dis* **8**:805.

145. Seiler P, Ulrichs T, Bandermann S, Pradl L, Jorg S, Krenn V, Morawietz L, Kaufmann SH, Aichele P. 2003. Cell-wall alterations as an attribute of *Mycobacterium tuberculosis* in latent infection. *J Infect Dis* **188**:1326–1331.

Genetics of Macromolecular Biosynthesis

VII

Molecular Genetics of Mycobacteria, 2nd Edition
Edited by Graham F. Hatfull and William R. Jacobs, Jr.
© 2014 American Society for Microbiology, Washington, DC
doi:10.1128/microbiolspec.MGM2-0001-2013

Digby F. Warner[1]
Joanna C. Evans[1]
Valerie Mizrahi[1]

Nucleotide Metabolism and DNA Replication

30

The first edition of *Molecular Genetics of Mycobacteria* (1) was published very shortly after the release of the complete genome sequence of *Mycobacterium tuberculosis* H37Rv (2). Armed with that resource, we searched for genes that might be involved in mycobacterial nucleotide metabolism and DNA replication (3). Our analysis at the time relied entirely on the homology-based identification of genes that had been discovered and characterized in other organisms; however, by confirming the presence, or suggesting the absence, in *M. tuberculosis* of homologs of genes of known function, it provided a useful framework for subsequent studies of the reactions and pathways underlying nucleotide metabolism and DNA replication in this major human pathogen. At that stage, the field of mycobacterial genetics was in its infancy, and little was known about the function of individual mycobacterial genes and their encoded proteins in these or other metabolic pathways. However, over the past 13 years, spectacular technical advances have been made that have had a massive impact on the broader field of general bacteriology and, more importantly in the context of this book, have driven the postgenomic revolution in our understanding of fundamental mycobacterial physiology and metabolism. The development, in particular, of a powerful toolkit for random, targeted, and conditional mutagenesis of mycobacterial genomes has allowed gene function to be probed under a variety of conditions. In turn, this has enabled the compilation of catalogs of genes (conditionally) essential for mycobacterial growth and/or survival, while providing new insights into the biology of mycobacteria.

More recently, genetic advances have been matched by parallel developments in the use of "omics" approaches to elucidate the structure, function, and regulation of mycobacterial proteins. Therefore, in this article, we provide an update of the genetics of nucleotide metabolism and DNA replication in mycobacteria, highlighting key findings from the past decade or so. As far as possible, section headings from the first edition have been retained in order to enable direct assessment of progress in each area—and to highlight those aspects that have received little attention in the intervening period. As with the previous version, this article is

[1]MRC/NHLS/UCT Molecular Mycobacteriology Research Unit, DST/NRF Center of Excellence for Biomedical TB Research, Institute of Infectious Disease and Molecular Medicine, and Division of Medical Microbiology, University of Cape Town, P/Bag X3 Rondebosch 7700, South Africa.

focused primarily on observations in *M. tuberculosis*; however, comparisons with, and insights from, other mycobacterial species as well as better characterized bacterial models such as *Escherichia coli* are included where appropriate. Finally, a common theme underlying almost all investigations of mycobacterial metabolic function is the potential to identify, or validate, functions or pathways that can be exploited for tuberculosis (TB) drug development; for this reason, we have attempted to highlight processes in mycobacterial DNA replication that might satisfy this criterion.

BIOSYNTHESIS OF dNTPs

Purine Ribonucleotide Synthesis

The first half of the *de novo* purine biosynthetic pathway in mycobacteria involves the generation of an aminoimidazole moiety attached to a ribose. In the second half of the pathway, the 4-C and 5-C atoms of the imidazole are modified with the appropriate substituents, leading to the final cyclization reaction that generates inosine 5′-monophosphate (IMP) (4). The first step in the synthesis of purines involves the transfer of the β,γ-diphosphoryl moiety of ATP to the C1-hydroxyl group of α-D-ribose-5-phosphate by a class I ribose-phosphate pyrophosphokinase (PRPPase) (*prsA*; Rv1017c) to yield 5-phospho-α-D-ribose 1-diphosphate (PRPP) (5, 6). In addition to functioning as a central metabolite in both *de novo* and salvage nucleoside metabolism, PRPP is also required for the biosynthesis of NAD, NADP, histidine, and tryptophan (7, 8). Furthermore, members of the *Corynebacteriaceae* family, which includes mycobacteria, have evolved a mechanism for utilizing PRPP in the biosynthesis of cell wall arabinogalactan (9). As the sole PRPPase in *M. tuberculosis*, PrsA is responsible for the intracellular provision of PRPP and displays a significantly higher specific activity than other bacterial PRPPases (6). While most class I PRPPases are allosterically inhibited by purine nucleosides, the *M. tuberculosis* enzyme is only inhibited by ADP, with GDP displaying weak inhibition (10). It is likely that the less stringent repression of PPRPase activity in *M. tuberculosis* reflects the crucial role that this enzyme plays in the provision of PRPP for use in multiple metabolic pathways.

Following the generation of PRPP, the amidophosphoribosyltransferase, PurF (*purF*; Rv0808), catalyzes the first committed step in purine biosynthesis by replacing the pyrophosphate group of PRPP with the amide nitrogen of L-glutamine, yielding 5-phospho-β-D-ribosyl-amine and L-glutamate (11). Interestingly,

loss of *purF* function results in a transient loss in culturability of *Mycobacterium smegmatis* in stationary phase—a phenotype that has yet to be fully explained (11). In an ATP-dependent reaction, phosphoribosylamine-glycine ligase (*purD*; Rv0772) then ligates glycine to the amino group of 5-phospho-β-D-ribosyl-amine, producing 5-phosphoribosylglycinamide (GAR), and thereby providing atoms C-4, C-5, and N-7 of the purine base. The formyl group from N^{10}-formyltetrahydrofolate (fTHF) is transferred to GAR by 5-phosphoribosylglycinamide formyltransferase (*purN*; Rv0956) to produce *N*-formyl glycinamide ribonucleotide (fGAR) and tetrahydrofolate, thereby providing the C-8 atom (12). The structure of the *M. tuberculosis* PurN enzyme revealed distinct features that could potentially be exploited for drug discovery in other organisms (12). In *M. tuberculosis*, this step can also be performed by a second ATP-dependent phosphoribosylglycinamide formyltransferase II (*purT*; Rv0389), which uses formate and Mg^{2+} to form a formyl phosphate intermediate before transferring the formyl group to GAR. In another ATP-driven reaction, phosphoribosylformylglycinamide synthase II (*purL*; Rv0803) catalyzes the addition of an amide group derived from glycine to the C-5 ring position to form fGAR and glutamate. The presence of a phosphoribosylformylglycinamide synthase I (*purQ*; Rv0788) homolog in *M. tuberculosis* suggests that, as in other Gram-positive bacteria, including *M. leprae* (13), PurL probably requires two additional products, PurQ and PurS, for activity (14). Under such circumstances, the PurL domain possesses the phosphoribosylformylglycinamide synthase activity and PurQ acts as the glutaminase (15, 16). The *M. tuberculosis* PurS homolog is yet to be identified but may be the conserved hypothetical protein encoded by *Rv0708A*, located upstream of *purQ* in an operon (2).

The fifth step in the pathway involves the cyclization of fGAR to form the five-membered ring of the purine base, and it is carried out by the ATP-dependent 5′-phosphoribosyl-5-aminoimidazole synthase (*purM*; Rv0809). The addition of C-6 from bicarbonate requires the coordinated activity of two enzymes: first, the phosphoribosylaminoimidazole carboxylase ATPase subunit (*purK*; Rv3276c) catalyzes the ATP-dependent ligation of bicarbonate and the N-5 amino group of aminoimidazole ribonucleotide, yielding N^5-carboxyaminoimidazole ribonucleotide (CAIR); CAIR is then converted to carboxyaminoimidazole ribonucleotide by phosphoribosylaminoimidazole carboxylase (*purE*; Rv3275c), with the carboxylate carbon becoming C-6 of the purine ring and one of the carboxylate oxygens becoming atom O-6. Phosphoribosylaminoimidazole-succinocarboxamide synthase (*purC*; Rv0780) catalyzes the ATP-dependent conversion of 5′-phosphoribosyl-5-aminoimidazole-carboxylic acid to 5′-phosphoribosyl-4-(*N*-succino-carboxamide)-5-aminoimidazole by ligation of the carboxylate group of carboxyaminoimidazole ribonucleotide to the amino group of aspartate, thereby providing the N-1 atom of the final purine base. Fumarate is then released from N-1 through the activity of adenylosuccinate lyase (*purB*; Rv0777). The bifunctional enzyme, 5-aminoimidazole-4-carboxamide ribonucleotide transformylase/inosine monophosphate cyclohydrolase (*purH*; Rv0957) catalyzes the final steps in the pathway to form IMP, which is ultimately converted to AMP or GMP (4). Structural studies of *M. tuberculosis* PurH revealed a bound nucleotide, 4-carboxy-5-aminoimidazole ribonucleotide, in the cyclohydrolase active site, suggesting that it may be a cyclohydrolase inhibitor (4). Interestingly, neither *M. tuberculosis* nor *M. leprae* contains a homolog of the purine repressor, PurR (2, 13). The absence of PUR box-like sequences in the region upstream of *purB* in *M. tuberculosis* suggests that PurB may be subject only to allosteric regulation by purine pools.

IMP can be converted into AMP by adenylosuccinate synthase (*purA*; Rv0357c). Fumarate is then released from adenylosuccinate through β-elimination by adenylosuccinate lyase (*purB*; Rv0777) (17). Adenylate kinase (*adk*; Rv0733) catalyzes the reversible, Mg^{2+}-dependent transfer of a terminal phosphate from ATP/GTP to AMP to form ADP (18), which is then converted to ATP by nucleoside diphosphate kinase (*ndkA*; Rv2445c) (19). *M. tuberculosis* NdkA has both nucleoside mono- and diphosphate kinase activity, suggesting that this enzyme may potentially be involved in RNA and DNA metabolism in addition to nucleotide metabolism (18). The conversion of IMP to GMP begins with the NAD$^+$-dependent oxidation of IMP to xanthosine 5′-monophosphate by IMP dehydrogenase (IMPDH, GuaB). Although *M. tuberculosis* possesses three *guaB* homologs, only *guaB2* (Rv3411c) was found to encode a functional IMPDH enzyme when expressed in *E. coli* (20). This enzyme was susceptible to inhibition by diphenyl urea-based derivatives. These compounds displayed potent antimycobacterial activity that was diminished by overexpression of all three *guaB* genes, suggesting that IMPDH is the target (20). GMP synthase (*guaA*; Rv3396c) then catalyzes the addition of an amide nitrogen from L-glutamine to yield GMP, which is converted to GDP by guanylate kinase (*gmk*; Rv1389) (21) and to GTP by NdkA (Rv2445c) (19). The domain structure of *M. tuberculosis* Gmk revealed

differences in domain dynamics and GMP binding of relevance to the design of specific inhibitors (21). In addition to its role in purine nucleoside biosynthesis, *M. tuberculosis* NdkA has been shown to localize within the nucleus of mammalian cells and cause superoxide radical-mediated DNA cleavage (22, 23). This protein also appears to play a role in inhibiting phagosome biogenesis processes (24), possibly due to its ability to sequester the extracellular ATP that accumulates at the site of inflammation as part of the host cell defense mechanism (25).

The stringent response

The stringent response is a broadly conserved mechanism that operates in bacteria in response to nutritional stress (26), and its role in *M. tuberculosis* pathogenesis has been the subject of intense investigation (27, 28). It is mediated by (p)ppGpp, which is produced by the stringent response regulator, designated Rel$_{MTb}$ in *M. tuberculosis*. Rel$_{MTb}$ (Rv2583c) (29, 30) is a bifunctional enzyme possessing both RelA-like ATP:GTP/GDP/ITP 3′-pyrophosphoryltransferase and SpoT-like Mn^{2+}-dependent (p)ppGpp 3′-pyrophosphorylhydrolase activities on a single polypeptide (31, 32). Rel$_{MTb}$ is critical for the establishment and maintenance of chronic *M. tuberculosis* infection in both mice (28) and guinea pigs (33), and functional inactivation of this protein is associated with a significant alteration in the *M. tuberculosis* transcriptome, with affected genes including those associated with cell wall synthesis and immunogenicity (28, 29). In *M. tuberculosis*, expression of *rel*$_{MTb}$ is regulated by SigE, which, in turn, is regulated by the PPK1-regulated MprAB two-component system (34). Populations of actively growing *M. smegmatis* cells have been found to display a bimodal distribution of low and high *rel* expression levels, with high expression proposed to confer an enhanced ability to respond to stress (35). In *E. coli*, increased concentrations of (p)ppGpp mediate large-scale transcriptional changes by directly altering the stability of the RNA polymerase (RNAP) complex at regulated promoters. Together with the general transcription factor DksA, which interacts directly with RNAP in *E. coli*, (p)ppGpp redirects the RNAP to exert its effects on the expression of certain genes (36). While there are no homologs of DksA in mycobacteria, the transcriptional regulator CarD (Rv3583c) may perform an analogous function, because (p)ppGpp is ineffective at inducing stringent control in the absence of CarD, which forms a complex with the RNAP at rRNA and ribosomal protein loci (37).

Recently, a novel, (p)ppGpp synthetase was identified in *M. smegmatis* on a bifunctional protein (MSMEG_5849) that also possesses the ability to hydrolyze DNA:RNA duplexes through the activity of an RNase HII catalytic domain (38). As discussed below ("Elongation and Termination"), this represents another example of the fusion of an RNase H domain to a separate catalytic function in mycobacteria (*M. tuberculosis* contains the *Rv2228c*-encoded RNase HI-CobC protein; see Table 2, below) and, in the case of *M. smegmatis*, implies a role for (p)ppGpp in coordinating DNA replication with the stringent response via inhibition of the major replisome component, DnaG primase.

Pyrimidine Ribonucleotide Synthesis

The *de novo* synthesis of pyrimidine nucleotides is carried out in six enzymatic steps that culminate in the formation of UMP. In *M. tuberculosis*, the genes encoding five of the enzymes involved in pyrimidine synthesis, as well as the regulatory protein PyrR (*pyrR*; Rv1379), are located on the *pyr* operon. Some of these genes are essential for growth of *M. tuberculosis in vitro* (39). However, much of what is known about pyrimidine biosynthesis in *M. tuberculosis* is inferred from studies in other bacteria.

Pyrimidine biosynthesis begins with the formation of carbamoyl phosphate from glutamine, bicarbonate, and two molecules of MgATP, by carbamoyl-phosphate synthase, an enzyme comprising a small subunit (*carA*; Rv1383) and a large subunit (*carB*; Rv1384). Aspartate carbamoyltransferase (ATCase) (*pyrB*; Rv1380) catalyzes the condensation of aspartate and carbamoyl phosphate, yielding *N*-carbamoylaspartate, after which dihydroorotase (*pyrC*; Rv1381) catalyzes the closure of the pyrimidine ring structure to form dihydroorotate. It is likely that the mechanism of allosteric regulation of mycobacterial ATCase is similar to that of *E. coli*, in which ATP activates and CTP inhibits the activity of ATCase, thereby regulating the intracellular ratios of purine and pyrimidine nucleotides. The oxidation of dihydroorotate to orotate is carried out by dihydroorotate dehydrogenase (*pyrD*; Rv2139), resulting in the formation of the 5,6-double bond of the pyrimidine base. The fifth step in the pathway is then catalyzed by orotate phosphoribosyltransfrase (OPRT) (*pyrE*; Rv0382c), a type I PRTase that converts orotate to orotidine 5′-monophosphate (OMP) (40). This is the first committed step in *de novo* pyrimidine biosynthesis because it is the last step for which there is no chemical intermediate that can be derived from the pyrimidine salvage pathway. Submicromolar pyrimidin-2(1H)-one-

based inhibitors of the *M. tuberculosis* orotate phosphoribosyltransferase enzyme have been identified but, as yet, have not been tested for antimycobacterial activity (40). OMP is further decarboxylated to UMP by OMP decarboxylase (*pyrF*; Rv1385). Subsequent reversible phosphorylation of UMP by uridylate kinase (UMPK) (*pyrH*; Rv2883c) yields UDP, which is then used in the synthesis of all other pyrimidine nucleotides (41). *M. tuberculosis* UMPK differs from homohexameric UMPKs in other bacteria in that it is tetrameric, although its specificity for UMP as the phosphoryl group acceptor is similar to that of other UMPKs (42). UDP is then converted to UTP by NdkA, and CTP is formed from UTP via the activity of CTP synthase (*pyrG*; Rv1699).

In *M. smegmatis*, regulation of the *pyr* operon occurs via translational repression and involves the nucleotide-regulated binding of PyrR to occlude the first ribosome binding site in the *pyr* mRNA (43). This is consistent with evidence from *M. tuberculosis*, in which PyrR has been shown to bind the mRNA-binding loop in a UMP/UTP-dependent manner, and probably as a dimer (44). This regulatory mechanism differs from that in Gram-positive bacteria such as *Bacillus subtilis* (45) owing to the absence of transcription antiterminator and attenuation terminator sequences in mycobacterial 5′-*pyr* regions (43). *M. tuberculosis pyrR* expression is upregulated by hypoxia, suggesting that an attenuation of pyrimidine biosynthesis accompanies adaptation of *M. tuberculosis* to the hypoxic conditions prevailing in granulomas (46). In addition to its regulatory function, the PyrR from *M. tuberculosis* may also display weak UPRTase activity (44), as observed in other organisms (45).

Salvage and Interconversion Pathways

In contrast to most intracellular pathogens, which utilize either one pathway or the other, pathogenic mycobacteria express enzymes of *de novo* biosynthesis as well as salvage pathways for purines and pyrimidines (2, 47, 48). Since the *de novo* synthesis of nucleosides is energy-intensive, it is tempting to speculate that organisms such as *M. tuberculosis* preferentially engage salvage pathways during chronic, persistent infection when energy stores are likely to be low; however, experimental data in support of this notion are lacking. The majority of genes involved in the salvage pathways of *M. tuberculosis* are individually dispensable for growth *in vitro* (39, 49), in macrophages (50) and in animal tissue (51). There are, however, notable exceptions, as outlined below.

Purine salvage

The *de novo* synthesis of AMP and GMP from IMP is irreversible, but the purine salvage pathway facilitates the reconversion of preformed purine bases and nucleosides from the external environment to their corresponding purine nucleotides (52). The *M. tuberculosis* genome encodes a number of enzymes capable of interconverting purine bases, nucleosides, and nucleotides (Table 1). Adenosine deaminase (*add*; Rv3133c) catalyzes the irreversible hydrolytic deamination of (deoxy)adenosine to (deoxy)inosine. The reversible phosphorolysis of the *N*-glycosidic bond of β-purine (deoxy)ribonucleotides, generating α-(deoxy)ribose 1-phosphate and the corresponding purine base, is carried out by purine nucleoside phosphorylase (PNP) (*deoD*; Rv3307) (53). In contrast to the human PNP, the *M. tuberculosis* enzyme has greater specificity for guanosine N7 methyl analogs than natural nucleosides

(53). However, in keeping with the trimeric structure of both the human and *M. tuberculosis* enzymes, as opposed to the hexameric structure of most prokaryotic PNPs, the *M. tuberculosis* PNP is unable to phosphorylate adenosine (53). The identification of unique hydrogen-bonding patterns upon inhibition of *M. tuberculosis* PNP by immucillin-H suggests that the rational design of mycobacterial PNP-specific inhibitors may be feasible (54).

Adenosine kinase (AK) (*adoK*; Rv2202c) catalyzes the Mg^{2+}-dependent phosphorylation of adenosine to AMP. Although AKs are found in most eukaryotes, fungi, plants, and parasites, they are seldom found in bacteria (47). The *M. tuberculosis* enzyme was the first bacterial AK to be characterized (47) and functions as a homodimer, unlike most AKs, which are monomeric. Moreover, the activity of *M. tuberculosis* AK is stimulated by monovalent metal ions such as K^+, but not by

Table 1 Genes involved in purine and pyrimidine salvage pathways in *M. tuberculosis*

Gene	Rv no.	Protein	Reaction catalyzed	Essentiality[a]
Purine salvage				
add	Rv3133c	Adenosine deaminase	(d)Ado + H_2O ⟺ (d)Ino + NH_3	Nonessential[b]
deoD	Rv3307	Purine nucleoside phosphorylase	Purine nucleoside + P_i ⟺ purine base + α-ribose-1-phosphate	Nonessential
adoK	Rv2202c	Adenosine kinase	Ado + ATP ⟺ AMP + ADP	Nonessential
hpt	Rv3624c	Hypoxanthine-guanine phosphoribosyl transferase	Hypoxanthine (guanine) + PRPP ⟺ IMP (GMP) + PP_i	Nonessential
apt	Rv2584c	Adenine PRTase	Adenine + PRPP ⟺ AMP + PP_i	Nonessential
iunH	Rv3393	Nucleoside hydrolase	Inosine (uridine) + H_2O ⟺ D-ribose + hypoxanthine (uracil)	Nonessential
Pyrimidine salvage				
dut	Rv2697c	dUTPase	dUTP + H_2O ⟺ dUMP + PP_i + H^+	Essential *in vitro*
dcd	Rv0321	dCTP deaminase:dUTPase	dCTP + H_2O ⟺ dUTP + NH_3 and dUTP + H_2O ⟺ dUMP + PP_i + H^+	Nonessential
thyA	Rv2764c	Thymidylate synthase	CH_2H_4folate + dUMP ⟺ H_2folate + dTMP	Nonessential
thyX	Rv2754c	Flavin-dependent thymidylate synthase	CH_2H_4folate + dUMP + (FADH2) + NADPH ⟺ H3folate + dTMP + $NADP^+$	Essential *in vitro*
tmk	Rv3247c	Thymidylate kinase	dTMP + ATP ⟺ ADP + dTDP	Nonessential
ndkA	Rv2445c	Nucleoside diphosphate kinase	Ribonucleoside diphosphate + ATP ⟺ ribonucleoside triphosphate + ADP	Nonessential
cdd	Rv3315c	Cytidine deaminase	(d)Cytidine + H_2O ⟺ (d)uridine + NH_3	Nonessential
deoA	Rv3314c	Thymidine phosphorylase	Thymidine + P_i ⟺ 2-deoxyribose-1-phosphate + thymine	Nonessential
upp	Rv3309c	Uracil PRTase	Uracil + PRPP ⟺ UMP + PP_i	Nonessential
cmk	Rv1712	Cytidylate kinase	(d)CMP + ATP ⟺ (d)CDP + ADP	Essential *in vitro*

[a]*In vitro* essentiality, as determined by TraSH (39, 49)
[b]All of the genes listed as nonessential for growth *in vitro* are also dispensable for adaptation and survival in macrophages (50) or for survival during infection in mice (51).

inorganic phosphate, as for other AK enzymes (47, 55, 56). Interestingly, despite the presence of adenosine deaminase and PNP enzymes, mycobacteria preferentially use AK to dephosphorylate adenosine (55). Since the proinflammatory response is likely to result in adenosine accumulation (elevated levels of host adenosine deaminase are a diagnostic marker of tuberculous pleurisy [57]), it is possible that *M. tuberculosis* expresses an AK that preferentially utilizes adenosine over deoxyadenosine in order to exploit this purine reservoir (58). The observation that AK catalyzes the first step in the conversion of 2-methyladenosine into an analog that is active against *M. tuberculosis* (58) has provided a proof-of-concept for targeting the purine salvage pathway for TB drug discovery.

The hypoxanthine-guanine and adenine PRTases are encoded by *hpt* (Rv3624c) and *apt* (Rv2584c), respectively. Both are type I enzymes that catalyze the Mg^{2+}-dependent, reversible transfer of the 5′-phosphoribosyl moiety from PRPP to the N-9 position of the purine. Guanine and hypoxanthine are substrates for Hpt, giving rise to GMP and IMP, respectively (52), whereas Apt utilizes adenine as a substrate to produce AMP (59). *M. tuberculosis* Hpt has a preference for hypoxanthine over guanine as a substrate and, unlike the Hpt from most other organisms, is unable to utilize xanthine as a substrate

(52). Finally, *M. tuberculosis* also possesses a homolog of nucleoside hydrolase (*iunH*; Rv3393), an enzyme that preferentially utilizes inosine and uridine as substrates but can likely catalyze irreversible hydrolysis of all of the common purine and pyrimidine nucleosides. These enzymes are of particular importance in parasitic protozoa, such as *Crithidia fasciculate*, which lack a *de novo* purine synthetic pathway and therefore rely solely on salvage mechanisms for purine supply (60).

Pyrimidine salvage

As in the case of purines, mycobacteria are able to reutilize pyrimidine bases and nucleosides derived from preformed nucleotides via the pyrimidine salvage pathway (61). The majority of enzymes involved in salvage pathways are dispensable *in vitro* and during growth in macrophages and in animal tissue (Table 1); in contrast, the mono-functional dUTPase (*dut*; Rv2697c) is distinguished by virtue of its essentiality in *M. tuberculosis* (39, 49) and *M. smegmatis* (62). This homotrimeric enzyme (63) converts dUTP to dUMP, thus providing the immediate precursor of thymidine nucleotide. Since *M. tuberculosis* lacks both a dCMP deaminase and a thymidine kinase (2), dUTPase activity is of particular importance because it provides the only source of d(U)TMP for DNA biosynthesis (64) (Fig. 1).

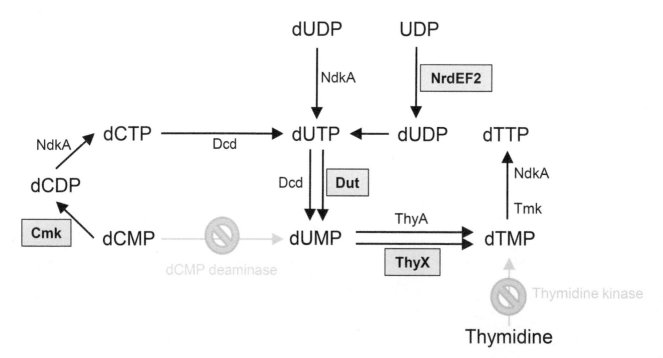

Figure 1 Pathways for the formation of dTTP in *M. tuberculosis*. The steps for which homologs are absent are shown in gray, and enzymes that are essential for growth *in vitro* are shown in shaded boxes. Adapted from references 64 and 62.
doi:10.1128/microbiolspec.MGM2-0001-2013.f1

Interestingly, although Dut is essential for growth of *M. tuberculosis in vitro*, it is not the only enzyme with dUTPase activity: dCTP deaminase (*dcd*, Rv0321), a homotrimeric enzyme that converts dCTP to dUTP and is dispensable *in vitro* (39, 49), also displays dUTPase activity—a feature only observed previously in archaea (65). Biochemical studies on recombinant Dcd from *M. tuberculosis* showed that, although its dUTPase activity is lower than that of Dut (Rv2697c), the bifunctionality of this enzyme allows the formation of dUMP from both dCTP and dUTP, with dUMP being formed at the same rate that dCTP is deaminated. In light of these findings, the essentiality of Dut will need to be established *in vivo* in order to validate this enzyme as a novel TB drug target.

Thymidylate synthase (TS) converts dUMP to dTMP. *M. tuberculosis* contains both the classical TS enzyme, ThyA (Rv2764c), and the evolutionarily unrelated flavin-dependent TS, ThyX (Rv2754c) (66). Both enzymes catalyze the reductive methylation of dUMP by N^5, N^{10}-methylenetetrahydrofolate to produce dTMP and dihydrofolate, but ThyX additionally requires $FADH_2$ and NAD(P)H as reductant (67). ThyA is dispensable for growth *in vitro* and *in vivo*, whereas ThyX is essential *in vitro* (68). Although the biological significance of distinct TS enzymes in a single organism remains obscure (69), the *in vitro* essentiality of *M. tuberculosis* ThyX in the presence of a functional ThyA enyzme suggests that ThyX may serve other cellular functions in addition to providing dTMP (68, 70). This uncertainty notwithstanding, ThyX has been the subject of intense investigation and has emerged as a popular target for structure-guided TB drug discovery efforts (67, 71, 72). However, as in the case of Dut, the essentiality of ThyX *in vivo* must still be established in order to validate this target. Following the conversion of dUMP to dTMP by TS, thymidylate kinase (*tmk*; Rv3247c) catalyzes the phosphorylation of dTMP to form dTDP, and nucleoside diphosphate kinase (*ndkA*; Rv2445c) phosphorylates dTDP to produce dTTP.

Cytidine deaminase (*cdd*; Rv3315c) catalyzes the Zn^{2+}-dependent hydrolytic deamination of cytidine or 2′-deoxycytidine to uridine or 2′-deoxyuridine, respectively (73, 74). Located downstream of *cdd*, on an operon, is the gene encoding thymidine phosphorylase (*deoA*; Rv3314c), another pyrimidine salvage enzyme, which catalyzes the reversible hydrolysis of thymidine to D-2-deoxyribose-1-phosphate and thymine. Mycobacteria also express a cytidylate kinase (*cmk*; Rv1712), which catalyzes the reversible γ-phosphoryl transfer from ATP to CMP or dCMP, generating CDP or dCDP, respectively (75, 76). The *M. tuberculosis* enzyme,

which is essential for growth *in vitro* (39, 49), preferentially phosphorylates dCMP and is thought to play a role in the recycling of nucleotides derived from RNA degradation (75).

As described previously, PyrR displays weak UPRTase activity (44). However, the major UPRTase in *M. tuberculosis* is probably provided by Upp (Rv3309c), which catalyzes the conversion of uracil and PRPP to UMP, the common precursor of all pyrimidine nucleotides. This enzyme may play a particularly important role in pyrimidine salvage in *M. tuberculosis* because the organism lacks both uridine nucleosidase and uridine phosphorylase enzymes, which catalyze conversion of uracil to uridine, as well as uridine kinase and uridine monophosphatase, which convert uridine to UMP (2).

Formation of Deoxyribonucleotides from Ribonucleotides

The reduction of ribonucleotides to their corresponding deoxyribonucleotides is carried out by ribonucleotide reductase (RNR). *M. tuberculosis* and other mycobacteria contain *nrdE* (Rv3051c) and *nrdF2* (Rv3048c) genes, which encode an essential class Ib RNR that is regulated by the regulator NrdR (Rv2718c) (77, 78). Class Ib RNRs are composed of two homodimeric proteins (the large R1 or α subunit, NrdE, and the small R2 or β subunit, NrdF) in an $\alpha_2\beta_2$ configuration (79). Association of the subunits is essential for catalytic activity, with both the substrate and allosteric effector binding sites located on R1, and the stable free radical on a tyrosine residue on R2 that is involved in the reduction of the ribonucleotide. In addition, the presence of an Fe(III)-Fe(III) radical diiron site on R2 is necessary for activity of the *M. tuberculosis* NrdEF2 enzyme (79). The C-terminus of the small subunit is required for holoenzyme association, which has facilitated the design of small peptides that competitively inhibit binding of R2 to R1, thereby abrogating enzymatic activity (80–82). Interestingly, *M. tuberculosis* contains another hypothetical class Ib small subunit encoded by *nrdF1* (Rv1981c). Although this protein contains the critical tyrosine residue, biochemical and genetic studies have confirmed that it is unable to substitute for NrdF2 to form a functional class Ib enzyme (77, 78). The essentiality of NrdEF2, together with its vulnerability to chemical inhibition, has validated this enzyme as a TB drug target (78). In contrast, the role—if any—of the vitamin B_{12}-dependent class II RNR, NrdZ, in deoxynucleoside triphosphate (dNTP) provision in *M. tuberculosis* remains obscure. As part of the DosR regulon (83, 84), *nrdZ* is induced in response to hypoxia; however, loss of *nrdZ* function was found to confer no discernible phenotype *in vitro* or *in vivo* (77).

NrdH, which is encoded by the first gene in the *nrdHIE* operon (Rv3053c), is a member of the thioredoxin family of proteins that are involved in the maintenance of redox homeostasis. NrdH redoxins are widespread among prokaryotes that are not capable of producing glutathione; in mycobacteria, these proteins function as hydrogen donors, providing the electrons that are required for reconstitution of the RNR active site radicals (85). NrdH sequences are conserved throughout mycobacteria, but the genomic arrangement of the *nrdHIE* operon varies between pathogenic and nonpathogenic mycobacteria. In nonpathogenic mycobacteria, NADPH-dependent FMN reductase is located upstream of the *nrdHIE* operon, while hypothetical proteins are found in this location in pathogenic mycobacteria. The significance of this is as yet unknown but may become clearer once the function of these hypothetical proteins is elucidated. The fourth gene in the *nrd* operon encodes NrdI (Rv3052c), a stimulatory protein that affects the activity of class Ib RNRs in *E. coli* by reduction of ribonucleotides (86). NrdI, which represents a novel class of flavodoxin (87), is essential for class Ib RNR activity in *Streptococcus pyogenes* (88) but is not required for growth of *M. tuberculosis in vitro*.

DNA REPLICATION

In contrast to eukaryotes—in which replication and segregation of DNA are separated temporally—the cell cycle in bacteria is not divided into discrete stages; instead, cell growth, DNA replication, chromosome segregation, and the initial assembly of the cell division machinery can overlap (89). As might be expected, this requires that the activity of the replication machinery is carefully regulated to ensure that the generation of a second chromosomal copy is coordinated with the pathways that function in chromosomal segregation and cell division. In the remaining sections of this article, we will provide some insight into the genes involved in the replication of the 4.4-Mb *M. tuberculosis* chromosome, with a special focus on the composition and operation of the mycobacterial replisome. In addition, we will highlight briefly the key pathways and interactions that coordinate the initiation of DNA replication with growth of the mycobacterial cell. As will become evident, in contrast to many other aspects of mycobacterial metabolism—including the biosynthesis and salvage of nucleotides described above—there is

limited biochemical and mechanistic insight into many aspects of mycobacterial DNA replication, with some notable exceptions. Therefore, wherever possible, the analysis will be informed by current knowledge from other bacterial systems, in particular the *E. coli* model.

The Mycobacterial Replisome

A large, multiprotein complex and numerous interacting partners function in bacterial DNA replication (90). Many components of the replication machinery are conserved and have been most thoroughly characterized in *E. coli* (91) and *B. subtilis* (92). Based on evidence from these organisms, the model bacterial "replisome" comprises two main subcomplexes: (i) the primosome, which itself includes the replicative helicase that unwinds the duplex DNA, and the primase, which is activated by DnaB to produce short RNA primers for discontinuous lagging-strand synthesis (93); and (ii) the DNA polymerase III holoenzyme (Pol III HE), which is made up of the Pol III core, the β_2-sliding clamp, and the clamp-loader complex (93).

Until very recently, it was thought that the Pol III HE consisted of two core polymerase subassemblies (each comprising an α polymerase subunit, $3' \rightarrow 5'$ proofreading exonuclease, ϵ, and the ϵ stabilizer, θ), which were required for concurrent leading and lagging strand synthesis at a single replication fork. However, new evidence has established that a third ($\alpha\epsilon\theta$) core is tethered to the clamp loader to form a tripolymerase replisome (91, 94): a single Pol III core is involved in leading strand synthesis, while the other two Pol III cores extend multiple Okazaki fragments on the lagging strand (95). Current models of the *E. coli* replisome are therefore based on a Pol III HE that contains three separate subcomplexes: a seven-subunit DnaX clamp loader complex ($\tau_3\delta\delta'\chi\psi$) binds three Pol III cores through the τ-α interaction, while the β_2 sliding-clamp processivity factor tethers the Pol III core to the DNA through a separate interaction with α. Of the remaining DnaX subunits, δ and δ' serve to load and unload the β_2 sliding-clamp, while χ and ψ stabilize the interactions between DnaX, δ, and δ' (93).

The genes known to be involved in DNA replication in *M. tuberculosis* are summarized in Table 2. On inspection, it is apparent that the mycobacterial replisome lacks several components that perform key functions in the *E. coli* system. However, comparative genomic analyses have established that this gene complement is typical of many bacteria and has contributed to the definition of a basic bacterial replication module that contains the replication initiator protein, DnaA, the DnaB helicase, DnaG primase, Pol IIIα, the β_2

Table 2 Genes known to be involved in DNA replication in *M. tuberculosis*

Gene	Rv Number	Function	Catalytic activity	Essential for growth *in vitro*[a]
Initiation complex				
dnaA	*Rv0001*	Replication initiator	ATPase	Essential
dnaB	*Rv0058*	Replicative helicase	ATPase	Essential
Primosome				
dnaG	*Rv2343c*	DNA primase	RNA primase	Essential
priA	*Rv1402*	Primosome replication factor; replication restart	ATPase	Essential
Elongation – the DNA Pol III core				
dnaE1	*Rv1547*	α Subunit, polymerase activity	DNA polymerase	Essential
dnaQ	*Rv3711c*	ε Subunit, proofreading activity	Exonuclease	Nonessential
dnaQ-uvrC	*Rv2191*	Chimeric protein; N-terminal 3′-5′ exonuclease domain, **C-terminal UvrC-like endonuclease domain	Exonuclease/**Endonuclease	Nonessential
Elongation – the DNA Pol III clamp loader complex				
dnaZX	*Rv3721c*	τ and γ subunits	ATPase	Essential
holA	*Rv2413c*	δ subunit		Nonessential
holB	*Rv3644c*	δ' subunit	ATPase	Essential
Elongation – the mycobacterial replisome				
dnaN	*Rv0002*	β sliding clamp		Essential
ssb	*Rv0054*	Single-stranded DNA binding protein		(Essential)[b]
polA	*Rv1629*	DNA Pol I	DNA polymerase	Essential
ligA	*Rv3014c*	DNA ligase, DNA replication	NAD-dependent DNA ligase	Essential
ligB	*Rv3062*	DNA ligase	ATP-dependent DNA ligase	Nonessential
ligC	*Rv3731*	DNA ligase	ATP-dependent DNA ligase	Nonessential
ligD	*Rv0938*	DNA ligase, DSB repair	ATP-dependent DNA ligase	Essential
gyrA	*Rv0006*	DNA gyrase, subunit A (DNA topoisomerase II)	Topoisomerase, ATPase	Essential
gyrB	*Rv0005*	DNA gyrase, subunit B (DNA topoisomerase II)	Topoisomerase, ATPase	Essential
topA	*Rv3646c*	DNA topoisomerase I	Topoisomerase, ATPase	Essential
rnhA-cobC	*Rv2228c*	Chimeric protein; N-terminal RNase HI domain, **C-terminal CobC-like α-ribazole phosphatase domain	RNase/**α-ribazole phosphatase	Essential
rnhB	*Rv2902c*	RNase HII	Rnase	Nonessential
Other DNA polymerases				
dnaE2	*Rv3370c*	α Subunit, TLS polymerase activity	DNA polymerase	Nonessential
dinB1	*Rv1537*	DNA Pol IV.I	DNA polymerase	Nonessential
dinB2	*Rv3056*	DNA Pol IV.II	DNA polymerase	Nonessential
polX	*Rv3856c*	DNA Pol X	Unknown[c]	Nonessential

[a]In vitro essentiality, as determined by TraSH (39, 49).
[b]Rv0054 did not satisfy the strict criterion for essentiality in the study by Griffin et al. (49) owing to a paucity of TA sites within the open reading frame; no transposon (Tn) insertions were identified in any of the five possible TA dinucleotides, suggesting that it is likely to be essential.
[c]A truncation in the polymerase domain of Rv3856c is predicted to preclude catalytic activity (98).

sliding clamp, the ε proofreading subunit, $τ_3$, δ and δ′, single-stranded DNA (ssDNA)-binding protein, DNA ligase, and Pol I (93, 96). In the sections below, we highlight some of the major advances that have been made since the previous edition of this article was published in identifying and understanding the components and functions of the conserved mycobacterial replisome, as well as the genes required for the initiation of DNA replication.

The replicative Pol IIIα subunit in *M. tuberculosis*

A feature of the *M. tuberculosis* genome is that it encodes two *dnaE*-type Pol IIIα subunits, DnaE1 and DnaE2. In a study that redefined notions of the strict role of C-family DNA polymerases in high-fidelity DNA replication, Boshoff et al. (97) established that the DnaE1 and DnaE2 subunits fulfill separate roles: DnaE1 is the essential replicative polymerase in *M. tuberculosis*, whereas DnaE2 functions in nonessential, error-prone translesion synthesis (TLS). This work identified DnaE2 as a central player in the DNA damage response in *M. tuberculosis* and, importantly, demonstrated a critical role for the TLS polymerase in virulence and the emergence of drug resistance *in vivo*. Subsequently, DnaE2 was shown to operate with two accessory factors as part of a split *imuA′-imuB/dnaE2* mutagenic cassette, with ImuB acting as a potential hub protein that interacts functionally with the other two cassette proteins, as well as the $β_2$ sliding clamp (98). The elucidation of a functional analog of *E. coli* Pol V suggested that the components of the tripartite mycobacterial mutasome might offer compelling targets for novel drugs designed to limit mutagenesis (99). In addition, it reinforced the idea that, together with potential differences in the architecture of the respective DnaE active sites (100), differential protein interactions might contribute to the specialist function of the *dnaE1*- and *dnaE2*-encoded subunits, which remains poorly understood (93).

All major DNA Pol IIIα structural features are readily identifiable in both DnaE1 and DnaE2, except for the C-terminal τ-interacting domain, which is absent in DnaE2 (98). Since the α-τ interaction enables simultaneous leading and lagging strand synthesis by Pol III HE in *E. coli*, the absence of this region might account for the inability of DnaE2 to function as a chromosomal replicase in *M. tuberculosis*. Another critical interaction in the *E. coli* replisome is that which occurs between Pol IIIα and the *dnaQ*-encoded 5′→3″ exonuclease proofreading subunit. The *M. tuberculosis* genome encodes two DnaQ-like proteins, Rv3711c and

Rv2191; although antigenic evidence suggests that Rv3711c is actively expressed during host infection (101), the contribution of either DnaQ homolog to replicative fidelity, or differential α subunit function, is unknown. Considering the inferred role of DnaE2 in TLS across DNA damage lesions, it seems very likely that structural determinants such as active site architecture contribute significantly to the polymerase function (100). Again, however, this possibility requires further investigation. Recent evidence suggests that the related DnaE3-type α subunit in *B. subtilis* is essential for the extension of RNA primers to enable PolC-mediated DNA synthesis (92); therefore, while the nonessentiality of *dnaE2* precludes a similar role for the second α subunit in *M. tuberculosis*, it remains to be determined whether DnaE2 has a function in mycobacterial DNA replication separate from its key role as a damage-responsive translesion polymerase.

A role for other DNA polymerases at the replication fork?

Although the α subunit functions as dominant replicase in bacterial DNA synthesis, there is increasing evidence to suggest that chromosomal replication is characterized by dynamic DNA polymerase exchange (102). For example, of the four accessory DNA polymerases in *E. coli*, Pol I (*polA*) and Pol II (*polB*) contribute directly to replication fidelity during normal chromosomal replication (102) through their respective roles in high-fidelity maturation of Okazaki fragments during lagging strand synthesis (Pol I), and as a back-up replicative polymerase during transient dissociation of the Pol III HE from either leading or lagging strand (Pol II). The remaining accessory polymerases, DNA Pol IV (*dinB*) and Pol V (*umuDC*), are both members of the Y family of specialist translesion polymerases that function primarily in the DNA damage response but can also access the lagging strand under conditions of elevated expression (102).

M. tuberculosis does not possess a Pol II enzyme but does encode two homologs of Pol IV, DinB1 and DinB2. However, unlike the *E. coli* model in which both Pol IV and Pol V are induced as part of the SOS regulon, the mycobacterial DNA damage response does not include either *dinB1* or *dinB2* and is instead limited to the two α subunits, DnaE1 and DnaE2 (98). Moreover, only DinB1 possesses a consensus β clamp binding motif, suggesting that DinB2 must interact with another protein(s) in order to access the replication fork. Therefore, while recent work has established that the *M. smegmatis* homolog of *M. tuberculosis* DinB2 is a functional DNA polymerase with a tendency to promote G:T and T:G mismatches (103), the activity and function of the DinB-type Y-family polymerases in *M. tuberculosis* remain poorly understood: deletion of *dinB1* and/or *dinB2* does not result in any discernible phenotype in *M. tuberculosis* in various assays *in vitro* or during mouse infection *in vivo* (104). Further research will, therefore, be required to elucidate the contribution, if any, of either mycobacterial Pol IV homolog to replication fidelity.

Initiation of Replication

The replication of the chromosome constitutes a major event in the bacterial cell cycle and must be carefully coordinated with cell growth and division. For this reason, bacteria have evolved rigorous control mechanisms to regulate the initiation of DNA replication and ensure that it does not occur at random sequences throughout the chromosome (105). Instead, replication is initiated at a single site—termed the origin of replication, *oriC*—and proceeds bidirectionally around the chromosome until the two replication forks meet in the replication terminus (*ter*), a region located approximately opposite *oriC*. Moreover, since it must occur only once during the cell cycle, a diverse array of regulatory mechanisms—many of which are centered on the activity of the initiator protein, DnaA—ensure that the assembly of the replication machinery is triggered at the appropriate stage (106).

The prereplication complex

Oligomerization of DnaA on *oriC* is critical for replication initiation since it results in the opening of the DNA duplex to allow loading of DnaB. In *E. coli*, DnaA proteins bind throughout *oriC* at specific 9-bp binding sites (DnaA boxes) that are closely spaced to facilitate the formation of a higher-order nucleoprotein scaffold comprising over 20 DnaA monomers (106). The formation of this "prereplication complex" (pre-RC) causes the melting of a short AT-rich region to create an open DNA-DnaA platform for loading the replicative helicase and the other replisome components. For this reason, mechanisms regulating replication initiation in *E. coli* are focused on the pre-RC complex, in particular the activity of DnaA, and include (i) the prevention of *de novo* DnaA synthesis via sequestration of the *dnaA* promoter, (ii) the depletion of available DnaA levels through the location of alternative DnaA binding sites elsewhere in the chromosome, (iii) the inactivation of DnaA by DnaN- and/or Hda-dependent stimulation of intrinsic DnaA ATPase activity, and (iv) SeqA-dependent sequestration of *oriC* (106).

The key features of the *E. coli* model are retained in *M. tuberculosis*: the DNA-ATP interaction is critical for replication initiation and, as in *E. coli*, the *dnaA* promoter remains active during replication to ensure progression through the cell cycle (107). *M. tuberculosis oriC* is located in the 527-bp intergenic region between *dnaA* and *dnaN* and contains multiple predicted and confirmed DnaA binding sites. Interestingly, this region also serves as a common locus for the insertion of *IS6110* transposable elements. To date, however, there is no evidence to suggest the insertions have any effect on the replication process, including the timing of initiation. Instead, these sites have proved practically useful as markers for RFLP fingerprinting of clinical *M. tuberculosis* isolates (108).

There are 13 closely spaced DnaA boxes in the *oriC* region of *M. tuberculosis*, of which 11 have been shown to interact with DnaA (109). Since *M. tuberculosis* DnaA appears to associate weakly with isolated DnaA box elements (110), it is thought that the close spacing of the DNA boxes favors formation of the nucleoprotein complex through cooperative binding of DnaA monomers (111). Moreover, in an important departure from the *E. coli* model (where binding of DnaA to ATP is sufficient to promote DnaA binding, oligomerization, and DNA strand separation), both ATP- and ADP-bound DnaA exhibit similar affinities for DnaA boxes and, in *M. tuberculosis*, the hydrolysis of ATP is required for rapid oligomerization of DnaA on *oriC* (112). This observation is consistent with the absence in *M. tuberculosis* of an Hda homolog as well as an equivalent of the *datA* locus, which in *E. coli*, act as principle negative regulators of ATP-DnaA (113). In combination, these observations further support the idea that the intrinsic ATPase activity of *M. tuberculosis* DnaA is critical in regulating replication, although a mycobacterial IciA (inhibitor of chromosomal initiation) protein has been shown to inhibit DnaA-mediated open complex formation by binding to a specific AT-rich region of *oriC* (114).

It has recently been shown that MtrA, the response regulator in the essential MtrAB two-component regulatory system in *M. tuberculosis*, binds a 7-bp sequence motif within the *dnaA* promoter region (115). This suggests a direct link between the initiation of replication and the processes involved in septation and cell division (116, 117) and is further supported by the observation that DnaA colocalizes with cardiolipin during cell cycle progression (118). In addition to MtrA, there is some evidence to suggest that multiple transcriptional regulators might control expression of *dnaA* in response to a variety of environmental conditions, including metal availability, oxygen limitation, and low pH (119). These analyses have also identified Mce1R as a putative *dnaA* regulator, thereby elucidating a possible functional association between cell growth and the activity of the mammalian cell entry module (120).

The mycobacterial primosome

In *E. coli*, DnaA recruits the hexameric DnaB replicative helicase to the origin in complex with the loader protein, DnaC, which suppresses the ATPase activity of DnaB and so remains bound until loading is complete. Following loading, the DnaG primase triggers dissociation of DnaC from DnaB, unlocking the helicase activity of DnaB for the initiation of strand separation. Recent work has confirmed the physical interaction of *M. tuberculosis* DnaA and DnaB and, further, has implicated DnaB in controlling DnaA complex formation and the interaction with *oriC* (121). However, like many other organisms, *M. tuberculosis* does not possess a homolog of the DnaC helicase loader (Table 2), suggesting that this activity is provided by a nonorthologous protein, or that a PriA/PriB/DnaT-type replication restart complex functions as a helicase loader at the origin (96). Alternatively, DnaA alone might be sufficient for DnaB loading, an idea that is supported by recent insights into the structure of DnaB from *H. pylori*—another organism lacking a DnaC helicase loader (122).

Elongation and Termination

Following strand separation and early unwinding of the parental chromosomal DNA by the DnaB replicative helicase, Pol III HE is loaded onto each of the two replication forks. As detailed above, the mycobacterial replisome consists of three core polymerases connected to the central clamp loader via a τ subunit, which also acts as a link to the DnaB helicase. The previous edition of this article (3) noted that *M. tuberculosis* *dnaZX* does not contain a programmed ribosomal frameshift signal for the production of both τ and the shorter γ subunit, a feature common to several other bacteria. When viewed in the light of recent evidence that three Pol III replicases are tethered to the τ subunit of the clamp loader complex to form a tripolymerase replisome (91, 94, 95), this observation appears consistent with the idea that all bacterial replisomes comprise trimeric Pol III replicases (95, 96).

Leading strand synthesis is highly processive and involves the continuous extension of DNA; in contrast, lagging-strand replication is rate-limiting and requires discontinuous replication via the extension and ligation of Okazaki fragments. Here, DNA primase fulfills an essential role in producing short RNA primers for extension by Pol III. *M. tuberculosis* DnaG was recently expressed and applied in a novel high-throughput enzyme assay to identify inhibitors of the priming reaction (123). Following completion of Okazaki fragment synthesis, a second switch occurs so that Pol III is replaced by Pol I, which catalyzes high-fidelity DNA synthesis across the resulting gap. Consistent with previous predictions, *M. tuberculosis* Rv2228c has been shown to be essential for growth of *M. tuberculosis in vitro* (49) and encodes a bifunctional protein that fuses RNase H and CobC-like α-ribazole phosphatase activities in a single polypeptide. The protein is unusual in that, in addition to functioning as a classic RNase HI in cleaving RNA/DNA hybrids, it also has double-stranded RNase activity (124).

DNA Unwinding: Helicases and Topoisomerases

In order to replicate the chromosome, the replication complex must gain access to separate leading and lagging strands for template-directed DNA synthesis. This requires the concerted action of helicases and topoisomerases: helicases are motor enzymes that unwind the DNA duplex into the separate single-stranded templates, whereas topoisomerases control supercoiling and resolve topological stress to enable translocation of the replisome (125).

In the *E. coli* replisome, the τ complex connects leading- and lagging-strand polymerases to the replicative helicase, DnaB, which acts on the lagging strand to unwind the DNA and, through its interaction with DnaG, stimulates primase activity. The *M. tuberculosis* genome encodes more than 15 putative helicase or helicase-related proteins, including an essential DnaB helicase that, like its *E. coli* counterpart, forms a hexameric helicase complex (126). Single-molecule analyses of replisome components have shown that *E. coli* DnaB is highly processive (127). This processivity is not maintained in the context of a full replisome, however, where lagging-strand synthesis is rate-limiting during replication of a double-stranded template (127). While similar analyses have not been performed for the corresponding *M. tuberculosis* proteins, accurate measurements of replisome processivity could provide critical insight into the inferred relationship between the mycobacterial replication rate and intrinsic fidelity in the absence of postreplicative mismatch repair, as discussed below.

The first edition of this article noted that, unlike *E. coli*, which encodes four topoisomerases, the

M. tuberculosis genome contains only a type I topo-isomerase, *topA*, and the type II topoisomerase, *gyrAB*, which introduce transient single-strand (type I) or double-strand (type II) breaks in DNA to relax super-coiling. Owing to their essentiality for the growth of *M. tuberculosis*, both enzymes represent attractive drug targets. However, whereas DNA gyrase has been heavi-ly exploited as a target for the various fluoroquinolone antibiotics, inhibitors of bacterial type I topoisomerases are rare (128).

Separate physical interactions have been elucidated between TopA and ribokinase (129) and between TopA and MazF (130) in *M. tuberculosis*. Ribokinase is a close homolog of the AK enzyme described above (purine salvage) and catalyzes the phosphorylation of D-ribose to yield ribose-5-phosphate, a key intermediate in multiple cellular processes including the synthesis of DNA, RNA, and ATP. Notably, the interaction between the *M. tuberculosis* TopA and ribokinase proteins was reported to inhibit the TopA-catalyzed relaxation of supercoiled DNA. In contrast, the same TopA-ribokinase interaction stimulated ribokinase activity. The significance of this interaction for mycobacterial physiology and cell function remains to be determined; however, it has been suggested that cross-regulatory mechanisms control D-ribose utilization in *M. tuberculosis*, particularly during growth under nutrient-limited conditions (129). The second reported interaction was with MazF, the stable toxin protein of the *mazEF*-encoded toxin-antitoxin pair (131). Recently, Huang and He (130) showed that the interaction between TopA and MazF inhibits MazF ribonuclease activity. More-over, as with ribokinase, this interaction in turn elimi-nated the ability of TopA to relax supercoiled DNA, suggesting MazF-mediated inhibition of TopA as a pos-sible mechanism to regulate mycobacterial growth.

Replication Fidelity and the *M. tuberculosis* Mutation Rate

In most bacteria, the fidelity of DNA replication depends on the selectivity of the replicative polymerase for the correct nucleotide during template-directed syn-thesis, the removal of any misincorporated nucleotides by the $3' \rightarrow 5'$ exonuclease activity of the replicative po-lymerase itself or its interacting proofreading subunit, and the operation of a postreplicative mismatch repair (MMR) system that detects polymerase errors that have escaped proofreading (102). In *E. coli*, these three highly conserved mechanisms are estimated to contrib-ute 10^{-5} (polymerase selectivity), 10^{-2} (proofreading), and 10^{-3} (MMR), respectively, to the observed overall

error rate of $\sim 10^{-10}$ per base pair per round of replica-tion (3). The absence of MMR would be expected, therefore, to lead to an elevated mutation rate, part-icularly in an organism such as *M. tuberculosis* whose genome is characterized by a high GC content. For this reason, the discovery that *M. tuberculosis* lacks an identifiable MMR system provoked considerable speculation and prompted several experimental in-vestigations since it was first made by Mizrahi and Andersen (132) 15 years ago through comparative ge-nomic analyses.

Despite the prediction that loss of MMR should compromise replication fidelity, *M. tuberculosis* is not a natural "mutator": consistent with previous estimates (97), the mycobacterial mutation rate was recently cal-culated at $\sim 2 \times 10^{-10}$ per base pair per round of repli-cation (133), a value that corresponds well with other bacterial systems including *E. coli*. Although there is a possibility that mycobacteria encode a nonorthologous alternative to MMR (132), it seems more likely that the activity of other repair components has enabled *M. tu-berculosis* to mitigate the lack of this function (134–136). For example, it has recently been shown that the key nucleotide excision repair helicase, UvrD1, fulfills a critical role in limiting recombination-associated mis-matches (137). In addition, intrinsic features of the ge-nome itself might limit the risk of replication errors: short sequences of nucleotide repeats pose a significant problem to replicative DNA polymerases and can result in frameshift mutations, but the GC-rich *M. tuberculo-sis* genome appears to have been under strong selec-tion to restrict the number of repeat regions through context-dependent codon choice (135).

As noted above, in addition to MMR, mechanisms ensuring the insertion of the correct nucleotide by the replicative polymerase represent another critical factor in replication fidelity. The first part of this article high-lighted advances in our understanding of the functions and regulation of the mycobacterial RNRs in supplying dNTPs for various DNA metabolic pathways. In this context, it is interesting to consider very recent evi-dence that a reduction of the intracellular dNTP pool as a consequence of ATP depletion can enhance replica-tion fidelity in *E. coli* (138): it seems possible, for ex-ample, that the closely regulated supply of dNTPs might limit the risk of insertion errors—an effect that would be further enhanced in combination with an in-herently slower replicative polymerase (139). Again, more research will be required to establish accurate estimates of the intracellular dNTP concentration in the different phases of the mycobacterial cell cycle, as well as to determine the speed and processivity of the mycobacterial replicase. Certainly, the current lack of structural data for most of the replisome components—with a few notable exceptions (126, 140)—implies that a major effort will be needed to produce sufficient soluble protein for the requisite assays.

The final fidelity mechanism relates to proofreading function. Here, it is tempting to speculate that a putative intrinsic $3' \rightarrow 5'$ exonuclease activity in the PHP domain of DnaE-type polymerases provides an additional fidelity measure (93). However, while the PHP domains of both *M. tuberculosis* DnaE1 and DnaE2 contain conserved amino acids predicted to be required for exonuclease activity, it is uncertain if this function is preserved in either of the mycobacterial *dnaE*-type subunits and, further, whether it contributes to relative α subunit fidelity.

Replication During States of Low Metabolic Activity

The replicative status of *M. tuberculosis* during the various stages of infection is of critical importance in understanding the immunobiology of the host-pathogen interaction and is a key determinant of the susceptibility of the bacillus to antibiotics and host immune effectors. Despite the absence of definitive evidence, slow growth is often cited as a fundamental determinant of mycobacterial virulence. In contrast, the demonstrated association of reduced growth and metabolic activity with a state(s) in which the organism shows reduced susceptibility to antimycobacterial agents is an established component of mycobacterial pathogenesis (141). This is an important concept invoked in attempts to elucidate the biological mechanisms subverting antibiotic efficacy as well as in understanding of the observed utility of isoniazid prevention therapy for the prevention of relapse: as a frontline anti-TB agent and inhibitor of cell wall biosynthesis, isoniazid requires active replication for maximal activity (142).

The well-defined Wayne model of nonreplicating persistence is based on the notion that hypoxia recapitulates a major characteristic of some TB lesions (143). When cultured under oxygen-limiting conditions, *M. tuberculosis* ceases replication through gradual and programmed shutdown of macromolecular synthesis, including DNA, which ceases as the oxygen tension falls below 1%. However, the dependence on continued ATP production for viability during this process suggests that cellular metabolism remains active (144, 145). Moreover, mycobacteria in hypoxia-induced dormancy appear to be diploid, and the resumption of growth on re-exposure to oxygen is

characterized by the synchronous initiation of RNA synthesis, followed by cell division and then finally DNA replication (146).

In the mouse model, several studies have identified a correlation between total and viable bacterial counts that is suggestive of the existence of nonreplicating (or very slowly replicating) bacilli in "static equilibrium" with the host immune system during chronic infection (147). More recently, however, the application of an unstable "clock" plasmid by Gill et al. (148) has instead revealed that infecting bacilli maintain a slow but constant replication, with total bacterial population numbers under the control of a continuously active immune response. This is a key observation and is supported by evidence from whole-genome sequencing of latent *M. tuberculosis* infection in the non-human primate model (133). It remains unclear, however, whether replication is limited to a specific subpopulation that is able to generate sufficient ATP (and biomass) to drive chromosomal replication and cell division.

Targeting the Replication Machinery for Drug Discovery

As illustrated in Table 2, most genes involved in DNA replication are essential for the growth of *M. tuberculosis*. However, while this is a general feature of all bacterial pathogens, the replication machinery remains relatively under-exploited as a source of potential targets for novel antibacterial agents, with inhibitors of DNA gyrase representing the major exception (96, 149). Recent reports indicate that this gap is being addressed. For example, Dallmann et al. (150) developed an assay that enabled identification of both species-specific and broad-spectrum inhibitors of bacterial replicases. In addition, compounds have been identified that target RctB, the protein essential for replication of the second chromosome, *chrII*, in *Vibrio cholerae* (151), as well as *E. coli* RecA (152). For *M. tuberculosis*, small-molecule inhibitors of DnaG have been identified from biochemical screens *in vitro* (123) but have not been tested in whole-cell assays. In contrast, napthoquinones derived from natural product extracts with demonstrated activity against *M. tuberculosis* have been shown to inhibit DNA gyrase B by a novel mechanism (153). However, it is not known whether the biochemical and whole-cell inhibitory activities are related.

There is also increasing evidence to suggest that disruption of essential protein-protein interactions might offer a viable alternative to targeting enzyme function. For example, multiple studies have described the identification of small molecules and peptides that target the β-clamp processivity factor, suggesting the potential to block access of essential DNA replication and repair components to the replication fork, including the Pol IIIα subunit (154–156). Similarly, others have explored the use of small-molecule inhibitors of single-stranded DNA-binding protein interactions in various organisms including *M. tuberculosis* (157). As noted below, however, the design of DNA replication-specific protein-protein interaction inhibitors for *M. tuberculosis* will require the construction of comprehensive maps describing the critical interactions involving replisome components (158, 159), as has been done for organisms such as *E. coli* and *B. subtilis* (160, 161).

DIRECTIONS FOR FUTURE RESEARCH

The past decade or so has witnessed exciting progress in our understanding of the pathways and functions involved in mycobacterial DNA replication. However, as with all areas of mycobacterial metabolism, the central question that continues to motivate fundamental research of this nature is whether key pathways and functions have adapted specifically to pathogenesis and, in turn, the extent to which these components might be targeted for novel interventions. The increasing availability of whole-genome sequencing data from clinical *M. tuberculosis* strains isolated within populations, and from individual *M. tuberculosis*-infected patients through the course of infection and/or relapse, raises the prospect of identifying specific genotype-phenotype correlations that have contributed to lineage divergence. Comparative genomics have already been used to infer the operation of strong diversifying selection in the adaptive evolution of specific DNA replication and repair components (162, 163); however, as noted previously for similar analyses of DNA repair genes (99), the demonstration that an observed DNA replication polymorphism results in a measurable phenotypic consequence remains elusive. Instead, a significant component of future research should involve the generation of a complete and multilayered picture of the different functional and regulatory interactions that govern DNA replication in *M. tuberculosis* and ensure that chromosomal replication is coordinated with mycobacterial cell division—*in vitro* as well as within relevant disease models and, ultimately, within the obligate human host.

This article has detailed the significant progress made toward understanding the operation and regulation of nucleotide biosynthetic and salvage pathways in maintaining intracellular pools of dNTPs and precursors for DNA synthesis and repair, as well as in identifying potential vulnerabilities for drug targeting. The number of genes involved, as well as the presence in some cases of alternative mechanisms for key steps, nevertheless suggests the potential capacity of *M. tuberculosis* to respond to dynamic and hostile environments. An important question therefore relates to the differential regulation of specific components, and the potential interaction between different pathways, as well as their roles in different disease stages. For example, do dNTP levels fluctuate as a consequence of growth phase in *M. tuberculosis*, as reported for *E. coli* (164)? Does the interplay between replisome function and dNTP provision determine overall replication fidelity? And is this modulated under specific conditions (e.g., hypoxia, stress) or in response to exogenous (host-derived) stimuli (e.g., nitric oxide, low pH)?

As highlighted above, advances in optical imaging have provided critical data to complement genetic analyses in refining models of DNA replication (91, 95). It is expected, therefore, that the application of similar technology to *M. tuberculosis* will elucidate the components and subcellular localization of the mycobacterial replisome and its interacting partners during cell division, including chromosomal replication and segregation. Questions here include the following: What is the composition of the mycobacterial replisome? What is the inherent processivity of the fully constituted replisome, and how does it compare with other known bacterial systems? How does the observed rate of cell division correlate with the replication rate, and is there a link between replication rate and error rate? Alternatively, are there other factors that mediate replication fidelity? Similarly, do the accessory polymerases function in chromosomal replication and, if so, what is their role in determining the overall fidelity of replication? This last question is of special relevance given the inferred dependence of *M. tuberculosis* on chromosomal rearrangements and point mutations for its microevolution within the host, including the development of drug resistance. Moreover, given the absence of predicted β clamp-binding motifs in some replication and repair proteins, what are the additional factors that might regulate access to the replication fork?

The recent application of single-cell time-lapse imaging has yielded conflicting evidence of the timing and nature of mycobacterial cell division (165, 166), which will need to be resolved. Moreover, the cryptic role of a diploid state in nonreplicating persistence requires further attention, since it remains unclear whether this is functionally significant and, if so, how it is regulated: which components are critical for the completion of

DNA synthesis during metabolic shutdown, and what are the factors involved in organizing the separate chromosomes? In turn, this highlights the need for further research into the factors that coordinate replication and cell division in *M. tuberculosis*. Where does DNA replication occur within the mycobacterial cell, particularly considering that the size of the replisome almost certainly precludes the possibility that replication occurs within the mycobacterial nucleoid (105)? Is the *M. tuberculosis ori* tethered to the cell pole? Also, given evidence of transcriptional repression of replication components during stationary phase (167), is there a direct link between the stringent response and replication shutdown?

The previous edition of this article noted the presence in the *M. tuberculosis* genome of putative components of DNA restriction modification systems. However, despite increasing evidence implicating differential DNA methylation status in bacterial physiology and virulence (168), the contribution of epigenetic modifications to *M. tuberculosis* pathogenesis remains under-explored. It is expected, therefore, that the application of novel sequencing technologies (168) to bacilli isolated from clinical and experimental TB infections will be a significant research area in the near future. Finally, there is emerging (and resurgent) interest in the potential utility of DNA replication as a source of new antibacterial drug targets (96, 149). Realizing this potential will require the development of novel screens for potential enzyme and protein-protein interaction inhibitors (123); in addition, it will necessitate the generation of structural data for the mycobacterial DNA replication proteins. It is impossible, therefore, to predict what the next version of this article will contain; however, we look forward to another exciting decade in the molecular genetics of DNA replication in *M. tuberculosis*.

This work was funded by grants from the South African Medical Research Council (to V. M.), the National Research Foundation (to V. M. and D. F. W), and the Howard Hughes Medical Institute (Senior International Research Scholar's grant to V. M.).

Citation. Warner DF, Evans JC, Mizrahi V. 2014. Nucleotide metabolism and DNA replication. Microbiol Spectrum 2(5):MGM2-0001-2013.

References

1. Hatfull GF, Jacobs WR Jr (ed). 2000. *Molecular Genetics of Mycobacteria.* ASM Press, Washington, DC.

2. Cole ST, Brosch R, Parkhill J, Garnier T, Churcher C, Harris D, Gordon SV, Eiglmeier K, Gas S, Barry CE 3rd, Tekaia F, Badcock K, Basham D, Brown D, Chillingworth T, Connor R, Davies R, Devlin K, Feltwell T, Gentles S, Hamlin N, Holroyd S, Hornsby T, Jagels K, Krogh A, McLean J, Moule S, Murphy L, Oliver K, Osborne J, Quail MA, Rajandream MA, Rogers J, Rutter S, Seeger K, Skelton J, Squares R, Squares S, Sulston JE, Taylor K, Whitehead S, Barrell BG. 1998. Deciphering the biology of *Mycobacterium tuberculosis* from the complete genome sequence. *Nature* 393:537–544.

3. Mizrahi V, Dawes SS, Rubin H. 2000. DNA replication, p 159–172. *In* Hatfull G, Jacobs WR Jr (ed), *Molecular Genetics of Mycobacteria.* ASM Press, Washington, DC.

4. Le Nours J, Bulloch EM, Zhang Z, Greenwood DR, Middleditch MJ, Dickson JM, Baker EN. 2011. Structural analyses of a purine biosynthetic enzyme from *Mycobacterium tuberculosis* reveal a novel bound nucleotide. *J Biol Chem* 286:40706–40716.

5. Breda A, Martinelli LK, Bizarro CV, Rosado LA, Borges CB, Santos DS, Basso LA. 2012. Wild-type phosphoribosylpyrophosphate synthase (PRS) from *Mycobacterium tuberculosis*: a bacterial class II PRS? *PLoS One* 7:e39245.

6. Alderwick LJ, Lloyd GS, Lloyd AJ, Lovering AL, Eggeling L, Besra GS. 2011. Biochemical characterization of the *Mycobacterium tuberculosis* phosphoribosyl-1-pyrophosphate synthetase. *Glycobiology* 21:410–425.

7. Hove-Jensen B. 1988. Mutation in the phosphoribosyl-pyrophosphate synthetase gene (prs) that results in simultaneous requirements for purine and pyrimidine nucleosides, nicotinamide nucleotide, histidine, and tryptophan in *Escherichia coli. J Bacteriol* 170:1148–1152.

8. Hove-Jensen B, Rosenkrantz TJ, Haldimann A, Wanner BL. 2003. *Escherichia coli* phnN, encoding ribose 1,5-bisphosphokinase activity (phosphoribosyl diphosphate forming): dual role in phosphonate degradation and NAD biosynthesis pathways. *J Bacteriol* 185:2793–2801.

9. Wolucka BA. 2008. Biosynthesis of D-arabinose in mycobacteria: a novel bacterial pathway with implications for antimycobacterial therapy. *FEBS J* 275:2691–2711.

10. Lucarelli AP, Buroni S, Pasca MR, Rizzi M, Cavagnino A, Valentini G, Riccardi G, Chiarelli LR. 2010. *Mycobacterium tuberculosis* phosphoribosylpyrophosphate synthetase: biochemical features of a crucial enzyme for mycobacterial cell wall biosynthesis. *PLoS One* 5:e15494.

11. Keer J, Smeulders MJ, Williams HD. 2001. A *purF* mutant of *Mycobacterium smegmatis* has impaired survival during oxygen-starved stationary phase. *Microbiology* 147:473–481.

12. Zhang Z, Caradoc-Davies TT, Dickson JM, Baker ENSquire CJ. 2009. Structures of glycinamide ribonucleotide transformylase (PurN) from *Mycobacterium tuberculosis* reveal a novel dimer with relevance to drug discovery. *J Mol Biol* 389:722–733.

13. Dawes SS, Mizrahi V. 2001. DNA replication in *Mycobacterium leprae. Lepr Rev* 72:408.

14. Hoskins AA, Anand R, Ealick SE, Stubbe J. 2004. The formylglycinamide ribonucleotide amidotransferase complex from *Bacillus subtilis*: metabolite-mediated complex formation. *Biochemistry* **43**:10314–10327.

15. Anand R, Hoskins AA, Bennett EM, Sintchak MD, Stubbe J, Ealick SE. 2004. A model for the *Bacillus subtilis* formylglycinamide ribonucleotide amidotransferase multiprotein complex. *Biochemistry* **43**:10343–10352.

16. Anand R, Hoskins AA, Stubbe J, Ealick SE. 2004. Domain organization of *Salmonella typhimurium* formylglycinamide ribonucleotide amidotransferase revealed by X-ray crystallography. *Biochemistry* **43**:10328–10342.

17. Toth EA, Yeates TO. 2000. The structure of adenylosuccinate lyase, an enzyme with dual activity in the de novo purine biosynthetic pathway. *Structure* **8**:163–174.

18. Meena LS, Chopra P, Bedwal RS, Singh Y. 2003. Nucleoside diphosphate kinase-like activity in adenylate kinase of *Mycobacterium tuberculosis*. *Biotechnol Appl Biochem* **38**:169–174.

19. Chopra P, Singh A, Koul A, Ramachandran S, Drlica K, Tyagi AK, Singh Y. 2003. Cytotoxic activity of nucleoside diphosphate kinase secreted from *Mycobacterium tuberculosis*. *Eur J Biochem* **270**:625–634.

20. Usha V, Gurcha S, Lovering AL, Lloyd AJ, Papaemmanouil A, Reynolds RC, Besra GS. 2011. Identification of novel diphenyl urea inhibitors of Mt-GuaB2 active against *Mycobacterium tuberculosis*. *Microbiology* **157**:290–299.

21. Hible G, Christova P, Renault L, Seclaman E, Thompson A, Girard E, Munier-Lehmann H, Cherfils J. 2006. Unique GMP-binding site in *Mycobacterium tuberculosis* guanosine monophosphate kinase. *Proteins* **62**:489–500.

22. Kumar P, Verma A, Saini AK, Chopra P, Chakraborti PK, Singh Y, Chowdhury S. 2005. Nucleoside diphosphate kinase from *Mycobacterium tuberculosis* cleaves single strand DNA within the human c-myc promoter in an enzyme-catalyzed reaction. *Nucleic Acids Res* **33**:2707–2714.

23. Saini AK, Maithal K, Chand P, Chowdhury S, Vohra R, Goyal A, Dubey GP, Chopra P, Chandra R, Tyagi AK, Singh Y, Tandon V. 2004. Nuclear localization and in situ DNA damage by *Mycobacterium tuberculosis* nucleoside-diphosphate kinase. *J Biol Chem* **279**:50142–50149.

24. Sun J, Wang X, Lau A, Liao TY, Bucci C, Hmama Z. 2010. Mycobacterial nucleoside diphosphate kinase blocks phagosome maturation in murine RAW 264.7 macrophages. *PLoS One* **5**:e8769.

25. Dar HH, Prasad D, Varshney GC, Chakraborti PK. 2011. Secretory nucleoside diphosphate kinases from both intra- and extracellular pathogenic bacteria are functionally indistinguishable. *Microbiology* **157**:3024–3035.

26. Potrykus K, Cashel M. 2008. (p)ppGpp: still magical? *Annu Rev Microbiol* **62**:35–51.

27. Rifat D, Bishai WR, Karakousis PC. 2009. Phosphate depletion: a novel trigger for *Mycobacterium tuberculosis* persistence. *J Infect Dis* **200**:1126–1135.

28. Dahl JL, Kraus CN, Boshoff HI, Doan B, Foley K, Avarbock D, Kaplan G, Mizrahi V, Rubin H, Barry CE 3rd. 2003. The role of RelMtb-mediated adaptation to stationary phase in long-term persistence of *Mycobacterium tuberculosis* in mice. *Proc Natl Acad Sci USA* **100**:10026–10031.

29. Dahl JL, Arora K, Boshoff HI, Whiteford DC, Pacheco SA, Walsh OJ, Lau-Bonilla D, Davis WB, Garza AG. 2005. The relA homolog of *Mycobacterium smegmatis* affects cell appearance, viability, and gene expression. *J Bacteriol* **187**:2439–2447.

30. Primm TP, Andersen SJ, Mizrahi V, Avarbock D, Rubin H, Barry CE 3rd. 2000. The stringent response of *Mycobacterium tuberculosis* is required for long-term survival. *J Bacteriol* **182**:4889–4898.

31. Avarbock D, Avarbock A, Rubin H. 2000. Differential regulation of opposing RelMtb activities by the aminoacylation state of a tRNA.ribosome.mRNA.RelMtb complex. *Biochemistry* **39**:11640–11648.

32. Avarbock A, Avarbock D, Teh JS, Buckstein M, Wang ZM, Rubin H. 2005. Functional regulation of the opposing (p)ppGpp synthetase/hydrolase activities of RelMtb from *Mycobacterium tuberculosis*. *Biochemistry* **44**:9913–9923.

33. Klinkenberg LG, Lee JH, Bishai WR, Karakousis PC. 2010. The stringent response is required for full virulence of *Mycobacterium tuberculosis* in guinea pigs. *J Infect Dis* **202**:1397–1404.

34. Sureka K, Dey S, Datta P, Singh AK, Dasgupta A, Rodrigue S, Basu J, Kundu M. 2007. Polyphosphate kinase is involved in stress-induced mprAB-sigE-rel signalling in mycobacteria. *Mol Microbiol* **65**:261–276.

35. Sureka K, Ghosh B, Dasgupta A, Basu J, Kundu M, Bose I. 2008. Positive feedback and noise activate the stringent response regulator rel in mycobacteria. *PLoS One* **3**:e1771.

36. Perederina A, Svetlov V, Vassylyeva MN, Tahirov TH, Yokoyama S, Artsimovitch I, Vassylyev DG. 2004. Regulation through the secondary channel—structural framework for ppGpp-DksA synergism during transcription. *Cell* **118**:297–309.

37. Stallings CL, Stephanou NC, Chu L, Hochschild A, Nickels BE, Glickman MS. 2009. CarD is an essential regulator of rRNA transcription required for *Mycobacterium tuberculosis* persistence. *Cell* **138**:146–159.

38. Murdeshwar MS, Chatterji D. 2012. MS_RHII-RSD, a dual-function RNase HII-(p)ppGpp synthetase from *Mycobacterium smegmatis*. *J Bacteriol* **194**:4003–4014.

39. Sassetti CM, Boyd DH, Rubin EJ. 2003. Genes required for mycobacterial growth defined by high density mutagenesis. *Mol Microbiol* **48**:77–84.

40. Breda A, Machado P, Rosado LA, Souto AA, Santos DS, Basso LA. 2012. Pyrimidin-2(1H)-ones based inhibitors of *Mycobacterium tuberculosis* orotate phosphoribosyltransferase. *Eur J Med Chem* **54**:113–122.

41. Labesse G, Benkali K, Salard-Arnaud I, Gilles AM, Munier-Lehmann H. 2011. Structural and functional

characterization of the *Mycobacterium tuberculosis* uridine monophosphate kinase: insights into the allosteric regulation. *Nucleic Acids Res* **39:**3458–3472.

42. Rostirolla DC, Breda A, Rosado LA, Palma MS, Basso LA, Santos DS. 2011. UMP kinase from *Mycobacterium tuberculosis*: mode of action and allosteric interactions, and their likely role in pyrimidine metabolism regulation. *Arch Biochem Biophys* **505:**202–212.

43. Fields CJ, Switzer RL. 2007. Regulation of *pyr* gene expression in *Mycobacterium smegmatis* by PyrR-dependent translational repression. *J Bacteriol* **189:**6236–6245.

44. Kantardjieff KA, Vasquez C, Castro P, Warfel NM, Rho BS, Lekin T, Kim CY, Segelke BW, Terwilliger TC, Rupp B. 2005. Structure of pyrR (Rv1379) from *Mycobacterium tuberculosis*: a persistence gene and protein drug target. *Acta Crystallogr D Biol Crystallogr* **61:**355–364.

45. Turnbough CL Jr, Switzer RL. 2008. Regulation of pyrimidine biosynthetic gene expression in bacteria: repression without repressors. *Microbiol Mol Biol Rev* **72:**266–300.

46. Via LE, Lin PL, Ray SM, Carrillo J, Allen SS, Eum SY, Taylor K, Klein E, Manjunatha U, Gonzales J, Lee EG, Park SK, Raleigh JA, Cho SN, McMurray DN, Flynn JL, Barry CE 3rd. 2008. Tuberculous granulomas are hypoxic in guinea pigs, rabbits, and nonhuman primates. *Infect Immun* **76:**2333–2340.

47. Long MC, Escuyer V, Parker WB. 2003. Identification and characterization of a unique adenosine kinase from *Mycobacterium tuberculosis*. *J Bacteriol* **185:**6548–6555.

48. Wheeler PR. 1987. Enzymes for purine synthesis and scavenging in pathogenic mycobacteria and their distribution in *Mycobacterium leprae*. *J Gen Microbiol* **133:**3013–3018.

49. Griffin JE, Gawronski JD, Dejesus MA, Ioerger TR, Akerley BJ, Sassetti CM. 2011. High-resolution phenotypic profiling defines genes essential for mycobacterial growth and cholesterol catabolism. *PLoS Pathog* **7:**e1002251.

50. Rengarajan J, Bloom BR, Rubin EJ. 2005. Genome-wide requirements for *Mycobacterium tuberculosis* adaptation and survival in macrophages. *Proc Natl Acad Sci USA* **102:**8327–8332.

51. Sassetti CM, Rubin EJ. 2003. Genetic requirements for mycobacterial survival during infection. *Proc Natl Acad Sci USA* **100:**12989–12994.

52. Biazus G, Schneider CZ, Palma MS, Basso LA, Santos DS. 2009. Hypoxanthine-guanine phosphoribosyltransferase from *Mycobacterium tuberculosis* H37Rv: cloning, expression, and biochemical characterization. *Protein Expr Purif* **66:**185–190.

53. Ducati RG, Santos DS, Basso LA. 2009. Substrate specificity and kinetic mechanism of purine nucleoside phosphorylase from *Mycobacterium tuberculosis*. *Arch Biochem Biophys* **486:**155–164.

54. Shi W, Basso LA, Santos DS, Tyler PC, Furneaux RH, Blanchard JS, Almo SC, Schramm VL. 2001. Structures

of purine nucleoside phosphorylase from *Mycobacterium tuberculosis* in complexes with immucillin-H and its pieces. *Biochemistry* **40:**8204–8215.

55. Reddy MC, Palaninathan SK, Shetty ND, Owen JL, Watson MD, Sacchettini JC. 2007. High resolution crystal structures of *Mycobacterium tuberculosis* adenosine kinase: insights into the mechanism and specificity of this novel prokaryotic enzyme. *J Biol Chem* **282:**27334–27342.

56. Wang Y, Long MC, Ranganathan S, Escuyer V, Parker WB, Li R. 2005. Overexpression, purification and crystallographic analysis of a unique adenosine kinase from *Mycobacterium tuberculosis*. *Acta Crystallogr Sect F Struct Biol Cryst Commun* **61:**553–557.

57. Krenke R, Korczynski P. 2010. Use of pleural fluid levels of adenosine deaminase and interferon gamma in the diagnosis of tuberculous pleuritis. *Curr Opin Pulm Med* **16:**367–375.

58. Long MC, Shaddix SC, Moukha-Chafiq O, Maddry JA, Nagy L, Parker WB. 2008. Structure-activity relationship for adenosine kinase from *Mycobacterium tuberculosis* II. Modifications to the ribofuranosyl moiety. *Biochem Pharmacol* **75:**1588–1600.

59. Bashor C, Denu JM, Brennan RG, Ullman B. 2002. Kinetic mechanism of adenine phosphoribosyltransferase from *Leishmania donovani*. *Biochemistry* **41:**4020–4031.

60. Degano M, Gopaul DN, Scapin G, Schramm VL, Sacchettini JC. 1996. Three-dimensional structure of the inosine-uridine nucleoside N-ribohydrolase from *Crithidia fasciculata*. *Biochemistry* **35:**5971–5981.

61. Villela AD, Sanchez-Quitian ZA, Ducati RG, Santos DS, Basso LA. 2011. Pyrimidine salvage pathway in *Mycobacterium tuberculosis*. *Curr Med Chem* **18:**1286–1298.

62. Pecsi I, Hirmondo R, Brown AC, Lopata A, Parish T, Vertessy BG, Toth J. 2012. The dUTPase enzyme is essential in *Mycobacterium smegmatis*. *PLoS One* **7:**e37461.

63. Chan S, Segelke B, Lekin T, Krupka H, Cho US, Kim MY, So M, Kim CY, Naranjo CN, Rogers YC, Park MS, Waldo GS, Pashkov I, Cascio D, Perry JL, Sawaya MR. 2004. Crystal structure of the *Mycobacterium tuberculosis* dUTPase: insights into the catalytic mechanism. *J Mol Biol* **341:**503–517.

64. Vertessy BG, Toth J. 2009. Keeping uracil out of DNA: physiological role, structure and catalytic mechanism of dUTPases. *Acc Chem Res* **42:**97–106.

65. Helt SS, Thymark M, Harris P, Aagaard C, Dietrich J, Larsen S, Willemoes M. 2008. Mechanism of dTTP inhibition of the bifunctional dCTP deaminase:dUTPase encoded by *Mycobacterium tuberculosis*. *J Mol Biol* **376:**554–569.

66. Myllykallio H, Lipowski G, Leduc D, Filee J, Forterre P, Liebl U. 2002. An alternative flavin-dependent mechanism for thymidylate synthesis. *Science* **297:**105–107.

67. Sampathkumar P, Turley S, Ulmer JE, Rhie HG, Sibley CH, Hol WG. 2005. Structure of the *Mycobacterium tuberculosis* flavin dependent thymidylate synthase

(MtbThyX) at 2.0A resolution. *J Mol Biol* **352**:1091–1104.

68. Fivian-Hughes AS, Houghton J, Davis EO. 2012. *Mycobacterium tuberculosis* thymidylate synthase gene thyX is essential and potentially bifunctional, while thyA deletion confers resistance to p-aminosalicylic acid. *Microbiology* **158**:308–318.

69. Escartin F, Skouloubris S, Liebl U, Myllykallio H. 2008. Flavin-dependent thymidylate synthase X limits chromosomal DNA replication. *Proc Natl Acad Sci USA* **105**:9948–9952.

70. Hunter JH, Gujjar R, Pang CK, Rathod PK. 2008. Kinetics and ligand-binding preferences of *Mycobacterium tuberculosis* thymidylate synthases, ThyA and ThyX. *PLoS One* **3**:e2237.

71. Basta T, Boum Y, Briffotaux J, Becker HF, Lamarre-Jouenne I, Lambry JC, Skouloubris S, Liebl U, Graille M, van Tilbeurgh H, Myllykallio H. 2012. Mechanistic and structural basis for inhibition of thymidylate synthase ThyX. *Open Biol* **2**:120120.

72. Kogler M, Vanderhoydonck B, De Jonghe S, Rozenski J, Van Belle K, Herman J, Louat T, Parchina A, Sibley C, Lescrinier E, Herdewijn P. 2011. Synthesis and evaluation of 5-substituted 2′-deoxyuridine monophosphate analogues as inhibitors of flavin-dependent thymidylate synthase in *Mycobacterium tuberculosis*. *J Med Chem* **54**:4847–4862.

73. Sanchez-Quitian ZA, Schneider CZ, Ducati RG, de Azevedo WF Jr, Bloch C Jr, Basso LA, Santos DS. 2010. Structural and functional analyses of *Mycobacterium tuberculosis* Rv3315c-encoded metal-dependent homotetrameric cytidine deaminase. *J Struct Biol* **169**:413–423.

74. Sanchez-Quitian ZA, Timmers LF, Caceres RA, Rehm JG, Thompson CE, Basso LA, de Azevedo WF Jr, Santos DS. 2011. Crystal structure determination and dynamic studies of *Mycobacterium tuberculosis* cytidine deaminase in complex with products. *Arch Biochem Biophys* **509**:108–115.

75. Thum C, Schneider CZ, Palma MS, Santos DS, Basso LA. 2009. The Rv1712 locus from *Mycobacterium tuberculosis* H37Rv codes for a functional CMP kinase that preferentially phosphorylates dCMP. *J Bacteriol* **191**:2884–2887.

76. Caceres RA, Macedo Timmers LF, Vivan AL, Schneider CZ, Basso LA, De Azevedo WF Jr, Santos DS. 2008. Molecular modeling and dynamics studies of cytidylate kinase from *Mycobacterium tuberculosis* H37Rv. *J Mol Model* **14**:427–434.

77. Dawes SS, Warner DF, Tsenova L, Timm J, McKinney JD, Kaplan G, Rubin H, Mizrahi V. 2003. Ribonucleotide reduction in *Mycobacterium tuberculosis*: function and expression of genes encoding class Ib and class II ribonucleotide reductases. *Infect Immun* **71**:6124–6131.

78. Mowa MB, Warner DF, Kaplan G, Kana BD, Mizrahi V. 2009. Function and regulation of class I ribonucleotide reductase-encoding genes in mycobacteria. *J Bacteriol* **191**:985–995.

79. Georgieva ER, Narvaez AJ, Hedin N, Graslund A. 2008. Secondary structure conversions of *Mycobacterium tuberculosis* ribonucleotide reductase protein R2 under varying pH and temperature conditions. *Biophys Chem* **137**:43–48.

80. Ericsson DJ, Nurbo J, Muthas D, Hertzberg K, Lindeberg G, Karlen A, Unge T. 2010. Identification of small peptides mimicking the R2 C-terminus of *Mycobacterium tuberculosis* ribonucleotide reductase. *J Pept Sci* **16**:159–164.

81. Nurbo J, Ericsson DJ, Rosenstrom U, Muthas D, Jansson AM, Lindeberg G, Unge T, Karlen A. 2013. Novel pseudopeptides incorporating a benzodiazepine-based turn mimetic-targeting *Mycobacterium tuberculosis* ribonucleotide reductase. *Bioorg Med Chem* **21**:1992–2000.

82. Nurbo J, Roos AK, Muthas D, Wahlstrom E, Ericsson DJ, Lundstedt T, Unge T, Karlen A. 2007. Design, synthesis and evaluation of peptide inhibitors of *Mycobacterium tuberculosis* ribonucleotide reductase. *J Pept Sci* **13**:822–832.

83. Schnappinger D, Ehrt S, Voskuil MI, Liu Y, Mangan JA, Monahan IM, Dolganov G, Efron B, Butcher PD, Nathan C, Schoolnik GK. 2003. Transcriptional adaptation of *Mycobacterium tuberculosis* within macrophages: insights into the phagosomal environment. *J Exp Med* **198**:693–704.

84. Voskuil MI, Schnappinger D, Visconti KC, Harrell MI, Dolganov GM, Sherman DR, Schoolnik GK. 2003. Inhibition of respiration by nitric oxide induces a *Mycobacterium tuberculosis* dormancy program. *J Exp Med* **198**:705–713.

85. Leiting WU, Jianping XI. 2010. Comparative genomics analysis of mycobacterium NrdH-redoxins. *Microb Pathog* **48**:97–102.

86. Jordan A, Aslund F, Pontis E, Reichard P, Holmgren A. 1997. Characterization of *Escherichia coli* NrdH. A glutaredoxin-like protein with a thioredoxin-like activity profile. *J Biol Chem* **272**:18044–18050.

87. Johansson R, Torrents E, Lundin D, Sprenger J, Sahlin M, Sjoberg BM, Logan DT. 2010. High-resolution crystal structures of the flavoprotein NrdI in oxidized and reduced states: an unusual flavodoxin. Structural biology. *FEBS J* **277**:4265–4277.

88. Roca I, Torrents E, Sahlin M, Gibert I, Sjoberg BM. 2008. NrdI essentiality for class Ib ribonucleotide reduction in *Streptococcus pyogenes*. *J Bacteriol* **190**:4849–4858.

89. Chien AC, Hill NS, Levin PA. 2012. Cell size control in bacteria. *Curr Biol* **22**:R340–R349.

90. Johnson A, O'Donnell M. 2005. Cellular DNA replicases: components and dynamics at the replication fork. *Annu Rev Biochem* **74**:283–315.

91. Reyes-Lamothe R, Sherratt DJ, Leake MC. 2010. Stoichiometry and architecture of active DNA replication machinery in *Escherichia coli*. *Science* **328**:498–501.

92. Sanders GM, Dallmann HG, McHenry CS. 2010. Reconstitution of the *B. subtilis* replisome with 13 proteins including two distinct replicases. *Mol Cell* **37**:273–281.

93. McHenry CS. 2011. DNA replicases from a bacterial perspective. *Annu Rev Biochem* 80:403–436.

94. McInerney P, Johnson A, Katz F, O'Donnell M. 2007. Characterization of a triple DNA polymerase replisome. *Mol Cell* 27:527–538.

95. Georgescu RE, Kurth I, O'Donnell ME. 2012. Single-molecule studies reveal the function of a third polymerase in the replisome. *Nat Struct Mol Biol* 19:113–116.

96. Robinson A, Causer RJ, Dixon NE. 2012. Architecture and conservation of the bacterial DNA replication machinery, an underexploited drug target. *Curr Drug Targets* 13:352–372.

97. Boshoff HI, Reed MB, Barry CE 3rd, Mizrahi V. 2003. DnaE2 polymerase contributes to in vivo survival and the emergence of drug resistance in *Mycobacterium tuberculosis*. *Cell* 113:183–193.

98. Warner DF, Ndwandwe DE, Abrahams GL, Kana BD, Machowski EE, Venclovas Č, Mizrahi V. 2010. Essential roles for *imuA'*- and *imuB*-encoded accessory factors in DnaE2-dependent mutagenesis in *Mycobacterium tuberculosis*. *Proc Natl Acad Sci USA* 107:13093–13098.

99. Warner DF. 2010. The role of DNA repair in *M. tuberculosis* pathogenesis. *Drug Discov Today Dis Mech* 7: e5–e11.

100. McHenry CS. 2011. Breaking the rules: bacteria that use several DNA polymerase IIIs. *EMBO Rep* 12:408–414.

101. Deb DK, Dahiya P, Srivastava KK, Srivastava R, Srivastava BS. 2002. Selective identification of new therapeutic targets of *Mycobacterium tuberculosis* by IVIAT approach. *Tuberculosis (Edinb)* 82:175–182.

102. Fijalkowska IJ, Schaaper RM, Jonczyk P. 2012. DNA replication fidelity in *Escherichia coli*: a multi-DNA polymerase affair. *FEMS Microbiol Rev* 36:1105–1121.

103. Sharma A, Nair DT. 2012. MsDpo4-a DinB homolog from *Mycobacterium smegmatis* is an error-prone DNA polymerase that can promote G:T and T:G mismatches. *J Nucleic Acids* 2012:285481.

104. Kana BD, Abrahams GL, Sung N, Warner DF, Gordhan BG, Machowski EE, Tsenova L, Sacchettini JC, Stoker NG, Kaplan G, Mizrahi V. 2010. Role of the DinB homologs Rv1537 and Rv3056 in *Mycobacterium tuberculosis*. *J Bacteriol* 192:2220–2227.

105. Reyes-Lamothe R, Nicolas E, Sherratt DJ. 2012. Chromosome replication and segregation in bacteria. *Annu Rev Genet* 46:121–143.

106. Leonard AC, Grimwade JE. 2011. Regulation of DnaA assembly and activity: taking directions from the genome. *Annu Rev Microbiol* 65:19–35.

107. Nair N, Dziedzic R, Greendyke R, Muniruzzaman S, Rajagopalan M, Madiraju MV. 2009. Synchronous replication initiation in novel *Mycobacterium tuberculosis dnaA* cold-sensitive mutants. *Mol Microbiol* 71:291–304.

108. Turcios L, Casart Y, Florez I, de Waard J, Salazar L. 2009. Characterization of IS6110 insertions in the *dnaA-dnaN* intergenic region of *Mycobacterium tuberculosis* clinical isolates. *Clin Microbiol Infect* 15:200–203.

109. Zawilak A, Kois A, Konopa G, Smulczyk-Krawczyszyn A, Zakrzewska-Czerwinska J. 2004. *Mycobacterium tuberculosis* DnaA initiator protein: purification and DNA-binding requirements. *Biochem J* 382:247–252.

110. Tsodikov OV, Biswas T. 2011. Structural and thermodynamic signatures of DNA recognition by *Mycobacterium tuberculosis* DnaA. *J Mol Biol* 410:461–476.

111. Mott ML, Berger JM. 2007. DNA replication initiation: mechanisms and regulation in bacteria. *Nat Rev Microbiol* 5:343–354.

112. Madiraju MV, Moomey M, Neuenschwander PF, Muniruzzaman S, Yamamoto K, Grimwade JE, Rajagopalan M. 2006. The intrinsic ATPase activity of *Mycobacterium tuberculosis* DnaA promotes rapid oligomerization of DnaA on *oriC*. *Mol Microbiol* 59: 1876–1890.

113. Camara JE, Breier AM, Brendler T, Austin S, Cozzarelli NR, Crooke E. 2005. Hda inactivation of DnaA is the predominant mechanism preventing hyperinitiation of *Escherichia coli* DNA replication. *EMBO Rep* 6:736–741.

114. Kumar S, Farhana A, Hasnain SE. 2009. *In-vitro* helix opening of *M. tuberculosis oriC* by DnaA occurs at precise location and is inhibited by IciA like protein. *PLoS One* 4:e4139.

115. Li Y, Zeng J, Zhang H, He ZG. 2010. The characterization of conserved binding motifs and potential target genes for *M. tuberculosis* MtrAB reveals a link between the two-component system and the drug resistance of *M. smegmatis*. *BMC Microbiol* 10:242.

116. Plocinska R, Purushotham G, Sarva K, Vadrevu IS, Pandeeti EV, Arora N, Plocinski P, Madiraju MV, Rajagopalan M. 2012. Septal localization of the *Mycobacterium tuberculosis* MtrB sensor kinase promotes MtrA regulon expression. *J Biol Chem* 287:23887–23899.

117. Nguyen HT, Wolff KA, Cartabuke RH, Ogwang S, Nguyen L. 2010. A lipoprotein modulates activity of the MtrAB two-component system to provide intrinsic multidrug resistance, cytokinetic control and cell wall homeostasis in *Mycobacterium*. *Mol Microbiol* 76: 348–364.

118. Maloney E, Madiraju SC, Rajagopalan M, Madiraju M. 2011. Localization of acidic phospholipid cardiolipin and DnaA in mycobacteria. *Tuberculosis (Edinb)* 91 (Suppl 1):S150–S155.

119. Zeng J, Li Y, Zhang S, He ZG. 2012. A novel high-throughput B1H-ChIP method for efficiently validating and screening specific regulator-target promoter interactions. *Appl Microbiol Biotechnol* 93: 1257–1269.

120. Zeng J, Cui T, He ZG. 2012. A genome-wide regulator-DNA interaction network in the human pathogen *Mycobacterium tuberculosis* H37Rv. *J Proteome Res* 11:4682–4692.

121. Xie Y, He ZG. 2009. Characterization of physical interaction between replication initiator protein DnaA and replicative helicase from *Mycobacterium tuberculosis* H37Rv. *Biochemistry (Mosc)* 74:1320–1327.

122. Stelter M, Gutsche I, Kapp U, Bazin A, Bajic G, Goret G, Jamin M, Timmins J, Terradot L. 2012. Architecture of a dodecameric bacterial replicative helicase. *Structure* 20:554–564.

123. Biswas T, Resto-Roldan E, Sawyer SK, Artsimovitch I, Tsodikov OV. 2013. A novel non-radioactive primase-pyrophosphatase activity assay and its application to the discovery of inhibitors of *Mycobacterium tuberculosis* primase DnaG. *Nucleic Acids Res* 41:e56.

124. Watkins HA, Baker EN. 2010. Structural and functional characterization of an RNase HI domain from the bifunctional protein Rv2228c from *Mycobacterium tuberculosis*. *J Bacteriol* 192:2878–2886.

125. Drlica K, Zhao X. 1997. DNA gyrase, topoisomerase IV, and the 4-quinolones. *Microbiol Mol Biol Rev* 61: 377–392.

126. Biswas T, Tsodikov OV. 2008. Hexameric ring structure of the N-terminal domain of *Mycobacterium tuberculosis* DnaB helicase. *FEBS J* 275:3064–3071.

127. Yao NY, Georgescu RE, Finkelstein J, O'Donnell ME. 2009. Single-molecule analysis reveals that the lagging strand increases replisome processivity but slows replication fork progression. *Proc Natl Acad Sci USA* 106: 13236–13241.

128. Godbole AA, Leelaram MN, Bhat AG, Jain P, Nagaraja V. 2012. Characterization of DNA topoisomerase I from *Mycobacterium tuberculosis*: DNA cleavage and religation properties and inhibition of its activity. *Arch Biochem Biophys* 528:197–203.

129. Yang Q, Liu Y, Huang F, He ZG. 2011. Physical and functional interaction between D-ribokinase and topoisomerase I has opposite effects on their respective activity in *Mycobacterium smegmatis* and *Mycobacterium tuberculosis*. *Arch Biochem Biophys* 512:135–142.

130. Huang F, He ZG. 2010. Characterization of an interplay between a *Mycobacterium tuberculosis* MazF homolog, Rv1495 and its sole DNA topoisomerase I. *Nucleic Acids Res* 38:8219–8230.

131. Zhu L, Zhang Y, Teh JS, Zhang J, Connell N, Rubin H, Inouye M. 2006. Characterization of mRNA interferases from *Mycobacterium tuberculosis*. *J Biol Chem* 281: 18638–18643.

132. Mizrahi V, Andersen SJ. 1998. DNA repair in *Mycobacterium tuberculosis*. What have we learnt from the genome sequence? *Mol Microbiol* 29:1331–1339.

133. Ford CB, Lin PL, Chase MR, Shah RR, Iartchouk O, Galagan J, Mohaideen N, Ioerger TR, Sacchettini JC, Lipsitch M, Flynn JL, Fortune SM. 2011. Use of whole genome sequencing to estimate the mutation rate of *Mycobacterium tuberculosis* during latent infection. *Nat Genet* 43:482–486.

134. Springer B, Sander P, Sedlacek L, Hardt WD, Mizrahi V, Schar P, Bottger EC. 2004. Lack of mismatch correction facilitates genome evolution in mycobacteria. *Mol Microbiol* 53:1601–1609.

135. Wanner RM, Guthlein C, Springer B, Bottger D, Ackermann M. 2008. Stabilization of the genome of the mismatch repair deficient *Mycobacterium tuberculosis* by context-dependent codon choice. *BMC Genomics* 9:249.

136. Machowski EE, Barichievy S, Springer B, Durbach SI, Mizrahi V. 2007. In vitro analysis of rates and spectra of mutations in a polymorphic region of the Rv0746 PE_PGRS gene of *Mycobacterium tuberculosis*. *J Bacteriol* 189:2190–2195.

137. Guthlein C, Wanner RM, Sander P, Davis EO, Bosshard M, Jiricny J, Bottger EC, Springer B. 2009. Characterization of the mycobacterial NER system reveals novel functions of the *uvrD1* helicase. *J Bacteriol* 191:555–562.

138. Laureti L, Selva M, Dairou J, Matic I. 2013. Reduction of dNTP levels enhances DNA replication fidelity *in vivo*. *DNA Repair (Amst)* 12:300–305.

139. Hiriyanna KT, Ramakrishnan T. 1986. Deoxyribonucleic acid replication time in *Mycobacterium tuberculosis* H37 Rv. *Arch Microbiol* 144:105–109.

140. Gui WJ, Lin SQ, Chen YY, Zhang XE, Bi LJ, Jiang T. 2011. Crystal structure of DNA polymerase III β sliding clamp from *Mycobacterium tuberculosis*. *Biochem Biophys Res Commun* 405:272–277.

141. Keren I, Minami S, Rubin E, Lewis K. 2011. Characterization and transcriptome analysis of *Mycobacterium tuberculosis* persisters. *MBio* 2:e00100–e00111.

142. Zhang M, Sala C, Hartkoorn RC, Dhar N, Mendoza-Losana A, Cole ST. 2012. Streptomycin-starved *Mycobacterium tuberculosis* 18b, a drug discovery tool for latent tuberculosis. *Antimicrob Agents Chemother* 56: 5782–5789.

143. Wayne LG, Hayes LG. 1996. An *in vitro* model for sequential study of shiftdown of *Mycobacterium tuberculosis* through two stages of nonreplicating persistence. *Infect Immun* 64:2062–2069.

144. Rao SP, Alonso S, Rand L, Dick T, Pethe K. 2008. The protonmotive force is required for maintaining ATP homeostasis and viability of hypoxic, nonreplicating *Mycobacterium tuberculosis*. *Proc Natl Acad Sci USA* 105: 11945–11950.

145. Eoh H, Rhee KY. 2013. Multifunctional essentiality of succinate metabolism in adaptation to hypoxia in *Mycobacterium tuberculosis*. *Proc Natl Acad Sci USA* 110: 6554–6559.

146. Wayne LG. 1977. Synchronized replication of *Mycobacterium tuberculosis*. *Infect Immun* 17:528–530.

147. Munoz-Elias EJ, Timm J, Botha T, Chan WT, Gomez JE, McKinney JD. 2005. Replication dynamics of *Mycobacterium tuberculosis* in chronically infected mice. *Infect Immun* 73:546–551.

148. Gill WP, Harik NS, Whiddon MR, Liao RP, Mittler JE, Sherman DR. 2009. A replication clock for *Mycobacterium tuberculosis*. *Nat Med* 15:211–214.

149. Sanyal G, Doig P. 2012. Bacterial DNA replication enzymes as targets for antibacterial drug discovery. *Expert Opin Drug Discov* 7:327–339.

150. Dallmann HG, Fackelmayer OJ, Tomer G, Chen J, Wiktor-Becker A, Ferrara T, Pope C, Oliveira MT, Burgers PM, Kaguni LS, McHenry CS. 2010. Parallel multiplicative target screening against divergent bacterial replicases: identification of specific inhibitors with broad spectrum potential. *Biochemistry* 49:2551–2562.

151. Yamaichi Y, Duigou S, Shakhnovich EA, Waldor MK. 2009. Targeting the replication initiator of the second *Vibrio* chromosome: towards generation of *Vibrionaceae*-specific antimicrobial agents. *PLoS Pathog* 5:e1000663.

152. Sexton JZ, Wigle TJ, He Q, Hughes MA, Smith GR, Singleton SF, Williams AL, Yeh LA. 2010. Novel inhibitors of *E. coli* RecA ATPase activity. *Curr Chem Genomics* 4:34–42.

153. Karkare S, Chung TT, Collin F, Mitchenall LA, McKay AR, Greive SJ, Meyer JJ, Lall N, Maxwell A. 2013. The naphthoquinone diospyrin is an inhibitor of DNA gyrase with a novel mechanism of action. *J Biol Chem* 288:5149–5156.

154. Georgescu RE, Yurieva O, Kim SS, Kuriyan J, Kong XP, O'Donnell M. 2008. Structure of a small-molecule inhibitor of a DNA polymerase sliding clamp. *Proc Natl Acad Sci USA* 105:11116–11121.

155. Wolff P, Olieric V, Briand JP, Chaloin O, Dejaegere A, Dumas P, Ennifar E, Guichard G, Wagner J, Burnouf DY. 2011. Structure-based design of short peptide ligands binding onto the *E. coli* processivity ring. *J Med Chem* 54:4627–4637.

156. Wijffels G, Johnson WM, Oakley AJ, Turner K, Epa VX, Briscoe SJ, Polley M, Liepa AJ, Hofmann A, Buchardt J, Christensen C, Prosselkov P, Dalrymple BP, Alewood PF, Jennings PA, Dixon NE, Winkler DA. 2011. Binding inhibitors of the bacterial sliding clamp by design. *J Med Chem* 54:4831–4838.

157. Marceau AH, Bernstein DA, Walsh BW, Shapiro W, Simmons LA, Keck JL. 2013. Protein interactions in genome maintenance as novel antibacterial targets. *PLoS One* 8:e58765.

158. Cui T, Zhang L, Wang X, He ZG. 2009. Uncovering new signaling proteins and potential drug targets through the interactome analysis of *Mycobacterium tuberculosis*. *BMC Genomics* 10:118.

159. Wang Y, Cui T, Zhang C, Yang M, Huang Y, Li W, Zhang L, Gao C, He Y, Li Y, Huang F, Zeng J, Huang C, Yang Q, Tian Y, Zhao C, Chen H, Zhang H, He ZG. 2010. Global protein-protein interaction network in the human pathogen *Mycobacterium tuberculosis* H37Rv. *J Proteome Res* 9:6665–6677.

160. Schaeffer PM, Headlam MJ, Dixon NE. 2005. Protein-protein interactions in the eubacterial replisome. *IUBMB Life* 57:5–12.

161. Noirot-Gros MF, Dervyn E, Wu LJ, Mervelet P, Errington J, Ehrlich SD, Noirot P. 2002. An expanded view of bacterial DNA replication. *Proc Natl Acad Sci USA* 99:8342–8347.

162. Osorio NS, Rodrigues F, Gagneux S, Pedrosa J, Pinto-Carbo M, Castro AG, Young D, Comas I, Saraiva M. 2013. Evidence for diversifying selection in a set of *Mycobacterium tuberculosis* genes in response to antibiotic- and nonantibiotic-related pressure. *Mol Biol Evol* 30:1326–1336.

163. Dos Vultos T, Mestre O, Rauzier J, Golec M, Rastogi N, Rasolofo V, Tonjum T, Sola C, Matic I, Gicquel B. 2008. Evolution and diversity of clonal bacteria: the paradigm of *Mycobacterium tuberculosis*. *PLoS One* 3:e1538.

164. Buckstein MH, He J, Rubin H. 2008. Characterization of nucleotide pools as a function of physiological state in *Escherichia coli*. *J Bacteriol* 190:718–726.

165. Aldridge BB, Fernandez-Suarez M, Heller D, Ambravaneswaran V, Irimia D, Toner M, Fortune SM. 2012. Asymmetry and aging of mycobacterial cells lead to variable growth and antibiotic susceptibility. *Science* 335:100–104.

166. Wakamoto Y, Dhar N, Chait R, Schneider K, Signorino-Gelo F, Leibler S, McKinney JD. 2013. Dynamic persistence of antibiotic-stressed mycobacteria. *Science* 339:91–95.

167. Hu Y, Coates AR. 2001. Increased levels of *sigJ* mRNA in late stationary phase cultures of *Mycobacterium tuberculosis* detected by DNA array hybridisation. *FEMS Microbiol Lett* 202:59–65.

168. Fang G, Munera D, Friedman DI, Mandlik A, Chao MC, Banerjee O, Feng Z, Losic B, Mahajan MC, Jabado OJ, Deikus G, Clark TA, Luong K, Murray IA, Davis BM, Keren-Paz A, Chess A, Roberts RJ, Korlach J, Turner SW, Kumar V, Waldor MK, Schadt EE. 2012. Genome-wide mapping of methylated adenine residues in pathogenic *Escherichia coli* using single-molecule real-time sequencing. *Nat Biotechnol* 30:1232–1239.

Molecular Genetics of Mycobacteria, 2nd Edition
Edited by Graham F. Hatfull and William R. Jacobs, Jr.
© 2014 American Society for Microbiology, Washington, DC
doi:10.1128/microbiolspec.MGM2-0024-2013

Michael S. Glickman[1]

Double-Strand DNA Break Repair in Mycobacteria

31

Repair of double-strand DNA breaks (DSBs) is critical to all living organisms. Scission of the phosphodiester backbone of both DNA strands is lethal if not repaired because such loss of linear chromosome integrity compromises chromosome replication and thereby prevents genome duplication. In contrast to some other types of DNA lesions which can be bypassed by damage-tolerant DNA polymerases, there is no known mechanism for the replication or transcription machinery to bypass a DSB, mandating their repair before replication or transcription can proceed. As such, multiple systems have evolved to repair DSBs, from bacterial to human cells (1–6). In the past decade, mycobacterial DNA repair systems in general, and mycobacterial DSB repair systems in particular, have received increasing attention. It has become clear that mycobacterial DSB repair differs substantially from the standard models of prokaryotic DSB repair derived from work in the *Escherichia coli* system. Most prominent among these differences is the existence of two additional DSB repair pathways that are not present in *E. coli* and were previously thought not to exist in bacteria: nonhomologous end joining (NHEJ) and single-strand annealing (SSA). Multiple other novel features of mycobacterial DSB repair have also been elucidated, making mycobacteria a new model system for the study of prokaryotic DSB repair. As now conceptualized, mycobacterial DSB repair actually most resembles DSB repair in budding yeast rather than other prokaryotes (Table 1). In addition to its emerging place as a model system, studies of mycobacterial DNA repair also are of great importance for understanding mechanisms of mutagenesis and genome diversification in *Mycobacterium tuberculosis*, the ultimate cause of antimicrobial resistance in *M. tuberculosis* (7). In addition to the information and references contained in this article, the reader is pointed to several excellent recently published reviews of mycobacterial DNA repair and mutagenesis (8–10).

Although DSBs are induced experimentally by ionizing radiation or clastogenic chemicals such as bleomycin, the doses of ionizing radiation used in laboratory experiments are not often encountered in nature. The major sources of physiologic DSBs that arise *in vivo* are likely to be (i) DNA replication across a single-strand nick, (ii) oxidative damage, (iii) desiccation, and potentially, (iv) ribonucleotides incorporated into DNA (11, 12). The frequency and number of DSBs that arise in a mycobacterial cell are difficult to estimate, and in particular, it is difficult to know the frequency and character of DNA lesions that *M. tuberculosis* experiences within its human host. We will review the

[1]Immunology Program, Sloan Kettering Institute and Infectious Diseases, Memorial Sloan Kettering Cancer Center, 1275 York Ave., New York, NY 10065.

Table 1 Comparison of DSB repair systems in bacteria and yeast[a]

Pathway	Function	E. coli	B. subtilis	M. smegmatis	Saccharomyces cerevisiae
HR	End resection	RecBCD	AddAB	AdnAB + ?	MRX/Sae2/Dna2/Exo1
	Single-strand protection	SSB	SSB1 (SSB2)	SSB1 (SSB2)	Rpa
	Mediator (RecA loading)	RecBCD (RecFOR)	RecFOR (?)	RecFOR (?)	Rad52
	Strand exchange	RecA	RecA	RecA	Rad51
NHEJ	End binding		Ku (ykoV)	Ku	Ku70/Ku80
	End sealing		LigD (ykoU)	LigD-Lig (LigC)	Lig4
	End remodeling: nucleotide addition		LigD-POL	LigD-POL (PolD1/PolD2)	Pol4
	End remodeling: phosphatase		? (No PE domain in BSuLigD)	LigD-PE	Tpp1
	End remodeling: nuclease		Unknown	Unknown	?
SSA	End resection	Pathway not described	Pathway not described	RecBCD	ExoI/MRX/SgsI/Dna2
	Strand annealing			RecO	Rad52

[a]Major factors for each pathway are listed with secondary factors listed in parentheses. Please see text for details.

three pathways of DSB repair in mycobacteria, with emphasis on recent genetic studies.

HOMOLOGOUS RECOMBINATION

Homologous recombination (HR) is a universally distributed mechanism of DSB repair throughout all domains of life. From bacterial to human cells, HR proceeds through a common set of biochemical transactions that share common themes. By definition, homology-directed repair of a DSB requires an unbroken homologous chromosome to direct repair and therefore must occur after chromosome replication has occurred. The DNA structure ultimately used for homology search and strand invasion into the homologous duplex is a 3′ single-stranded DNA coated with a strand invasion protein, RecA, in bacteria. Therefore, the initiating step in HR is recognition of the break by proteins that initiate resection to generate this 3′ single-stranded substrate. In E. coli, this resection is accomplished by the RecBCD enzyme complex (2, 13). RecBCD resects double-stranded DNA from the DSB until it encounters a Chi site, an 8-bp recognition sequence distributed throughout the E. coli chromosome. Upon encountering Chi, RecBCD converts to a single-stranded 5′-to-3′ exonuclease, thereby generating a 3′ single-stranded DNA that is coated by the single-strand binding protein SSB. Exchange of SSB for RecA to form the RecA nucleoprotein filament is facilitated by RecBCD (2).

END RESECTION IN MYCOBACTERIAL RECOMBINATION

Mycobacteria encode an ortholog of RecBCD, but the role of this enzyme complex in mycobacterial recombination differs substantially from that in E. coli. Mycobacterium smegmatis lacking RecBCD is not sensitive to a variety of clastogens (14, 15), which is inconsistent with a dominant role in HR because M. smegmatis lacking recA is highly sensitized to these same clastogens (14, 15). A natural hypothesis of redundancy between resection nucleases arose when mycobacteria were found to have a novel heterodimeric helicase that was named AdnAB (16), which is similar to the AddAB enzyme complex of Bacillus subtilis (6). This enzyme complex resects double-stranded DNA in vitro and has helicase activity (16), suggesting it may participate in DSB resection. In contrast to the RecBCD null strain, M. smegmatis lacking AdnAB is sensitive to ionizing radiation, although not to the level of a ΔrecA strain (14). A ΔrecBCD ΔadnAB strain was identical to the ΔadnAB strain in clastogen sensitivity, a finding inconsistent with redundancy between these two enzyme complexes (14). Although these studies provided some insight into the epistatic relationships between recBCD and adnAB, clastogen sensitivity assays invariably involve multiple types of DNA damage and therefore do not specifically interrogate the function of a particular factor in repair of specific DNA lesions.

Clarification of the roles of mycobacterial AdnAB and RecBCD in recombination came from studies using a chromosomal reporter system in which a single chromosomal break is induced by the homing endonuclease I-SceI. In this system, two 18-bp recognition sites for I-SceI were placed into a defective *lacZ* allele with a downstream LacZ donor segment. A key feature of this system is that one of the two I-SceI sites is inverted such that I-SceI cleavage at both sites creates a DSB with incompatible ends. This break cannot be directly ligated by NHEJ without end modification. This configuration seeks to minimize repeated cycles of I-SceI cutting and resealing that can occur with a single I-SceI site and which obscure conclusions about the frequency of each repair pathway. Chromosomal breakage is induced by transfection with an I-SceI encoding plasmid. Three repair outcomes are possible in this system, all of which can be distinguished by scoring the surviving colonies on X-Gal- and kanamycin-containing media (14): NHEJ, HR, and SSA (Fig. 1). When this assay is performed in strains

lacking putative repair factors, the role of each can be deduced by examining the repair outcomes in this genetic background.

When repair outcomes were assayed using the I-SceI system in strains lacking *recBCD*, *adnAB*, or both, surprising findings emerged. First, RecBCD was not required for HR in mycobacteria, because the percentage of HR outcomes in the I-SceI system was actually increased above wild type in Δ*recBCD M. smegmatis* (14). Deletion of *adnAB* resulted in an approximately 50% decrement in HR outcomes, supporting a role for *adnAB* in DSB resection *in vivo*. Notably, some residual HR events, which were all *recA* dependent, were preserved in the Δ*recBCD* Δ*adnAB* strain, indicating an additional pathway of DSB resection that is yet to be defined (14). Additional conclusions about the function of RecBCD in mycobacteria were gleaned from these experiments, which are discussed below in "SSA Pathway."

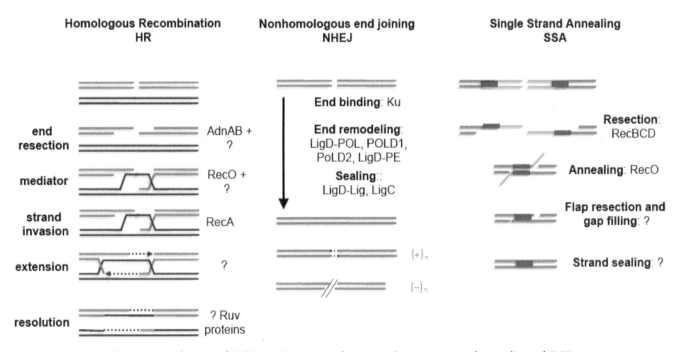

Figure 1 Pathways of DSB repair in mycobacteria. Our present understanding of DSB repair in mycobacteria. The three pathways shown are HR, NHEJ, and SSA. For each pathway, the major DNA processing events are depicted with the factors required for each step, when known. A question mark indicates that no specific experimental genetic data is available about that step, despite the presence of predicted proteins in mycobacterial chromosomes that may mediate these steps, or even biochemical activities consistent with a role in these pathways. In the NHEJ column the three outcomes below the arrow indicate faithful repair, nucleotide addition, and nucleotide trimming, respectively. In the SSA column, the blue rectangles indicate repeat sequences that flank the DSB. Please see text for further details and references. doi:10.1128/microbiolspec.MGM2-0024-2013.f1

MEDIATOR FUNCTION IN MYCOBACTERIAL RECOMBINATION

After resection of the DSB to create the 3′ single-stranded tail, the single strand is coated by SSB. SSB has a very high affinity to single-stranded DNA, so loading of RecA onto the 3′ single strand requires exchange of SSB for RecA to create the RecA nucleoprotein filament, which is the agent of homology search and strand invasion. This exchange reaction is facilitated by a mediator complex, a function that is conserved in all domains of life. In *E. coli*, mediator function is supplied by RecBCD, which has a RecA loading function. An additional mediator function is supplied by the RecFOR complex, which is evident in *E. coli* when RecBCD is inactivated but is a more prominent mediator in wild-type *B. subtilis* cells (17). In yeast, the major mediator is Rad52, which participates in Rad51 loading; in human cells, BRCA2 is the major mediator (Table 1).

Mediator function in mycobacteria has not been investigated in depth. The finding that RecBCD in mycobacteria does not participate in RecA-dependent HR would seem to exclude a mediator function for RecBCD. AdnAB, as a resection nuclease in mycobacterial HR, is a candidate to have RecA loading function, analogous to the mediator function of RecBCD in *E. coli*, but this function has not yet been demonstrated. A recent publication examined the function of the RecO protein in mycobacteria. RecO is a widely conserved bacterial protein that is the central component of the RecFOR complex (18). *M. smegmatis* lacking *recO* was found to be viable but highly sensitive to a variety of clastogens, including ionizing radiation. Importantly, the Δ*recO* strain was nearly as sensitive as the Δ*recA* strain, and more sensitive than the Δ*adnAB* strain, suggesting a major role for the RecFOR system in mycobacterial recombination (18).

The phenotype of the *M. smegmatis* Δ*recO* strain in the I-SceI recombination system confirmed a major role in HR. HR events in the Δ*recO* strain were significantly reduced compared to the wild-type strain, although some residual HR was present (18). Examination of the Δ*recO*Δ*adnAB* strain in this assay revealed a complete lack of HR events, indicating that RecO and AdnAB define two parallel pathways that lead to RecA-dependent recombination in mycobacteria. These results leave several questions unanswered. Although the results clearly support two parallel pathways of end resection and mediator function in RecA-dependent HR, multiple components remain to be identified.

In the RecO pathway, the identity of the resection nuclease is unknown. In the RecFOR system of *E. coli*, the nuclease that resects the DSB is RecJ, but no RecJ protein has been identified in mycobacteria. In addition, the functions of the RecF and RecR proteins in mycobacterial HR have not yet been examined. In the AdnAB pathway, the mediator for RecA loading after AdnAB-mediated resection is not known and will be the subject of future experimentation.

OTHER RECOMBINATION FACTORS

Unfortunately, the focus of this article on genetic studies of DSB repair does not permit a full discussion of extensive studies that have been performed on recombination proteins from mycobacterium and the regulation of their expression. These include the biochemical feature of RecA (19–22), the regulation of the DNA damage response in mycobacteria (23–27), and studies on Holliday junction resolvases (28–31). The reader is referred to these articles for in-depth discussion of these important topics.

NHEJ IN BACTERIA

Historical Perspective

The standard model of prokaryotic double-strand DNA break repair was derived from seminal studies in *E. coli* that defined, over many decades, the molecular requirements for HR. These studies established the primacy of RecA as the dominant strand exchange protein in bacterial recombination. However, the existence of alternative DSB repair pathways in *E. coli* was not documented, and such pathways were presumed not to exist. This situation is in stark contrast to the plethora of DSB repair pathways that have been intensely studied in eukaryotic cells and which include HR, NHEJ, and SSA. As genomic sequencing of broad bacterial phyla revealed new prokaryotic protein families, bioinformatic analysis of these genomes revealed clues to a greater diversity of DSB repair pathways than had been deduced from studies in *E. coli*. Foremost among these observations was the finding that some bacterial genomes, including those of *M. tuberculosis* and *M. smegmatis*, encoded apparent orthologs of the Ku protein (32). In eukaryal organisms, the Ku70/Ku80 heterodimer binds DNA ends and acts as an organizing hub for NHEJ. The existence of Ku in bacteria suggested the existence of an NHEJ pathway. Tantalizing additional bioinformatic observations identified ATP-dependent DNA ligases, sometimes in close genomic proximity to the gene encoding Ku, in these bacteria. The conserved bacterial DNA ligase, LigA (33), is an

NAD-dependent enzyme, whereas eukaryotic ligases, such as DNA ligase IV, which mediates NHEJ, are ATP dependent.

In Vivo Evidence for a Bacterial NHEJ Pathway

Early biochemical studies of the putative NHEJ proteins *M. tuberculosis* LigD and Ku demonstrated that these proteins can cooperate to seal DNA ends *in vitro* (34, 35). Definitive evidence of the existence of an NHEJ pathway *in vivo* initially came from studies using transfection of linear plasmids into *M. tuberculosis* and *M. smegmatis* and mutant strains lacking the genes encoding LigD and Ku. In contrast to *E. coli*, *M. tuberculosis* and *M. smegmatis* can recircularize linear plasmid DNA with either blunt or cohesive ends. For 5′ and blunt-end linear plasmids, this recircularization is reduced by approximately 500-fold in strains with null mutations in Ku or LigD (36–38), indicating an active NHEJ pathway that depends on Ku and LigD, which physically interact. The linear plasmid transformation assay is useful for interrogating NHEJ, but because the DNA sequence flanking the plasmid DSB does not have homology to the chromosome, HR-mediated repair is not possible. Therefore, the plasmid assay does not interrogate the relative frequency of NHEJ when alternative repair pathways are possible. Additional evidence for the existence and characteristics of the mycobacterial NHEJ pathway came from the I-SceI system. In wild-type cells carrying the I-SceI recombination substrate in which all repair outcomes are possible, NHEJ-mediated repair constitutes 20% of the repair outcomes (14). These NHEJ events are abolished in strains lacking Ku or LigD, and loss of NHEJ leads to a compensatory increase in HR (14).

As noted above, NHEJ in bacteria was first demonstrated using linear plasmid transfection and homing endonuclease-induced DSBs. In yeast and human cells, NHEJ is critical for DSB repair when a chromosomal homolog is not available to provide a template for strand invasion (3). By analogy, NHEJ in bacteria should be critical for clastogen resistance in stationary phase or other states in which chromosome replication has ceased and/or the cell is monochromosomal. Spores of *B. subtilis* are monochromosomal and non-replicating and thereby provide a physiologic state in which NHEJ should be important for genomic integrity. Accordingly, Ku-deficient *B. subtilis* spores are highly sensitive to dry heat or hydrogen peroxide (in an α/β small acid-soluble spore protein [SASP] mutant background; 39). Similarly Ku- or LigD-deficient

M. smegmatis in late stationary phase is sensitive to ionizing radiation, a sensitivity that is not observed in log-phase cells or in stationary-phase cells grown in rich media (15). Additional studies have also shown that NHEJ protects stationary-phase mycobacteria from desiccation (40).

In summary, *M. tuberculosis* and *M. smegmatis* encode a pathway of NHEJ that repairs DSBs in late stationary-phase cells and is dependent upon Ku and LigD. These experiments indicated that these two proteins, Ku and LigD, perform many (although not all; see below) of the repair functions of NHEJ. In contrast, eukaryotic NHEJ has multiple additional proteins that participate in end remodeling during NHEJ, including polymerases and nucleases that remodel the broken ends into ligatable termini (3). In the mycobacterial NHEJ system, some of these end remodeling functions are supplied by autonomous enzymatic functions encoded with the LigD protein.

Biochemical Functions of the LigD Protein and Their Role in NHEJ *In Vivo*

The LigD protein from *M. smegmatis* and *M. tuberculosis* has three identifiable enzymatic domains, polymerase (POL), phosphoesterase (PE), and ligase (LIG). Detailed biochemical, structural, and genetic studies have elucidated the NHEJ functions of each of these domains.

LigD-LIG

The LigD-LIG domain is an ATP-dependent DNA ligase with a classic three-step ligase mechanism (38, 41). As noted above, deletion of the *ligD* gene from *M. tuberculosis* or *M. smegmatis* reduces the efficiency of NHEJ for both plasmid and chromosomal substrates. Surprisingly, ablation of the LigD-LIG active site *in vivo* (through replacement of the active site lysine with alanine) surprisingly did not phenocopy the severe NHEJ defect observed in the Δ*ligD* strain (36, 41). NHEJ efficiency was substantially preserved, although these LigD-LIG-independent NHEJ events were low fidelity (36). This result indicated that another DNA ligase can supply strand sealing activity when LigD-LIG is inactivated, and further studies indicated that this activity is likely supplied by Ligase C (36).

LigD-POL

A remarkable feature of mycobacterial NHEJ is the frequent addition of template and nontemplated nucleotides to the repaired ends (36, 37). These end modifications suggest the participation of a polymerase in mycobacterial NHEJ. The POL domain of LigD is a primase-like polymerase that can add both templated and nontemplated nucleotides to DNA *in vitro* (37, 42, 43). LigD-POL is also proficient at adding ribonucleotides, a property it shares with polymerases that participate in eukaryotic NHEJ (44). Ablation of the LigD-POL domain activity by mutation of the diaspartate metal binding site abolishes mutagenic NHEJ at blunt ends during plasmid NHEJ but has relatively little effect on the templated fill-in events observed at resealed 5′ overhang plasmid termini (36, 37, 43). The necessity of the LigD protein for NHEJ, which is independent of its ligase activity, is supplied by the POL domain, which, in addition to catalyzing NHEJ additions, also appears to play a structural role in the NHEJ complex (36, 45).

LigD-PE

The LigD-PE domain is the third enzymatic module of the LigD protein that participates in end remodeling. The PE domain from *Pseudomonas* LigD has been most extensively characterized *in vitro*. LigD-PE is the founding member of a novel 3′ end healing enzyme family present in bacteria, archaea, and fungi. PE resects the ribonucleotide tract on an RNA-DNA hybrid until a single ribonucleotide with a 3′ OH group remains (46–49). This ribonucleotide-terminated strand is the preferred substrate for the LigD ligase, indicating that the PE and POL domains may cooperate to produce a ribonucleotide-terminated DNA strand that is the ultimate substrate for LigD ligase activity. The role of the PE domain in NHEJ *in vivo* is less well understood because it does not participate in the efficiency or fidelity of plasmid NHEJ (36).

Additional Potential NHEJ Factors

Several additional factors have been identified in mycobacteria that may participate in NHEJ. As noted above, the LigD-POL domain is a primase-like polymerase that is the direct catalyst of nontemplated mutagenic NHEJ *in vivo*. The persistence of templated fill-in events when LigD-POL activity is inactivated indicated the participation of additional polymerases in mycobacterial NHEJ. Prime candidates for this activity were the PolD1 and PolD2 enzymes. PolD1 and PolD2 are freestanding LigD-POL like enzymes without the Lig and PE domains contained in the LigD polypeptide. PolD1 and PolD2 have biochemical activities similar to LigD-POL, including a preference for ribonucleotide addition over deoxyribonucleotides and both template and nontemplated polymerase activity (50). Despite the suspicion that PolD1 and PolD2 participate in NHEJ based on their activities and genomic proximity (in the

case of PolD1) to the backup NHEJ ligase LigC, ablation of PolD1 and PolD2 in combination *in vivo* did not affect NHEJ fidelity of 5′ overhangs, even when LigD-POL was inactivated (50). These results indicate that there are as yet unidentified polymerases that participate in NHEJ, which may be redundant with LigD-POL, PolD1, and PolD2.

Experiments seeking mycobacterial proteins interacting with Ku identified UvrD1 as a Ku interacting protein (51). Ku stimulates UvrD1 helicase activity *in vitro*, and ablation of UvrD1 sensitizes *M. smegmatis* to ionizing radiation in all growth phases (51), but deletion of UvrD1 does not impair NHEJ-mediated resealing of linear plasmids. The exact function of the Ku-UvrD1 interaction in mycobacterial DSB repair remains to be elucidated. Additional efforts to identify Ku interacting proteins identified the protein encoded by MSmeg_5175, which has homology to NAD-dependent deacetylases of the Sir2 family. Deletion of MSmeg_5175 in *M. smegmatis* reduced NHEJ efficiency and conferred sensitivity to ionizing radiation, although not to the degree of the *M. smegmatis* Δ*ku* strain (52).

SSA PATHWAY

Single-strand annealing (SSA) is a mechanism of DSB repair that occurs when the DSB is flanked by repeats. Bidirectional single-strand resection from the break reveals complementary single strands that can anneal. Subsequent flap removal, fill-in synthesis, and ligation result in repair of the break with a deletion and single remaining repeat (Fig. 1). SSA has been described in yeast and is independent of the eukaryal stand exchange protein Rad51 but dependent on the Rad52 protein (53). However, SSA was not described as a DSB repair pathway in bacteria. Use of the I-SceI recombination system described above allowed detection of recombination events consistent with an SSA mechanism of repair. Specifically, approximately 15% of repair events at an I-SceI-induced DSB resulted in a deletion of the intervening DNA between two *lacZ* repeats. Importantly, these repair events were independent of RecA, consistent with an SSA mechanism of repair (14). Surprisingly, although deletion of RecBCD did not reduce HR events (see above), SSA events were abolished in a RecBCD null strain (14). In conjunction with the lack of any clastogen sensitivity of the RecBCD strain, this finding strongly indicated that RecBCD in mycobacteria has a function distinct from its well-established role as a resection nuclease and RecA mediator in the *E. coli* HR pathway.

Additional studies with an *M. smegmatis* strain lacking RecO indicated a substantial role for RecO in the mycobacterial SSA pathway. A deletion mutant of *M. smegmatis* lacking RecO was severely compromised in executing SSA compared to wild-type cells (18). Consistent with this involvement in the SSA pathway, purified RecO protein displayed a zinc-dependent DNA binding activity and accelerated the annealing of single-stranded DNA (18).

As a pathway that repairs DSBs that arise between repeats, the SSA pathway may have particular relevance to *M. tuberculosis* genome evolution. The *M. tuberculosis* chromosome contains repetitive DNA in the form of PE, PPE (Pro-Pro-Glu), and PGRS (polymorphic GC-rich sequence) elements, which account for a substantial fraction of the *M. tuberculosis* chromosome. DSBs that may arise within these regions are prime candidates to be repaired by the SSA pathway (or NHEJ for that matter) but are problematic for repair by HR because of difficulty in finding the appropriate homologous sequence for RecA-dependent strand invasion. Whether SSA-mediated repair would preferentially operate within repetitive regions of the mycobacterial chromosome remains to be tested, but it is potentially relevant that studies of the *M. tuberculosis* genome variation have found that PE, PPE, and PGRS genes do contain sequence polymorphisms that could be consistent with either SSA- or NHEJ-mediated repair (54–57), and studies of instability of direct and inverted repeats in mycobacteria have implicated these pathways (58).

ROLE OF DNA REPAIR IN *M. TUBERCULOSIS* PATHOGENESIS

It is clear that the products of the mammalian immune system can damage DNA through base damage, sugar damage, or phosphodiester strand breakage. It is also clear that several bacterial pathogens rely on DNA repair systems to resist the clastogenic environment of the host, including *Helicobacter pylori*, *Salmonella enterica*, and *Neisseria meningitidis* (see references in reference 8). However, despite the strong suspicion that *M. tuberculosis* uses DNA repair pathways to survive *in vivo*, demonstration of the importance of repair pathways in animal models of infection has been difficult. As has been suggested (8), the lack of dramatic pathogenesis phenotypes of some DNA repair mutant strains may reflect redundancy among repair mechanisms or reflect the relative gross nature of bacterial load assays to detect significant changes in bacterial chromosome structure that may occur independently of net bacterial survival. Nevertheless, some important findings

have implicated specific DNA repair pathways in *M. tuberculosis* pathogenesis and *in vivo* mutagenesis.

Several studies have examined the role of nucleotide excision repair (NER) in *M. tuberculosis* pathogenesis. After its identification as a factor conferring resistance to nitric oxide (59), an *M. tuberculosis* ΔuvrB strain was tested in the mouse model of infection and found to be mildly attenuated by CFU and time to death assays (60). The attenuation phenotype of the ΔuvrB strain was reversed in phox/iNOS$^{-/-}$ mice, indicating that the NER system does defend against host immune pressure *in vivo*. A recent study examined the role of the UvrD1 helicase in *M. tuberculosis* pathogenesis (61). UvrD is a component of the NER machinery and in this function removes the incised oligonucleotide after strand cleavage by UvrABC. However, UvrD also has additional functions in recombination, and in mycobacteria there are two UvrD-like proteins, UvrD1 and UvrD2. As noted above, UvrD1 interacts directly with Ku, whereas UvrD2 is essential (62, 63), although this essential function is not supplied by its helicase activity (63). The Davis group examined ΔuvrA, ΔuvrD1, and ΔuvrAΔuvrD1 strains in the mouse model of infection. The ΔuvrA strain was mildly attenuated, comparable to previously reported findings with the ΔuvrB strain. The ΔuvrD1 strain was attenuated by ~1.5 logs at 50 days postinfection and by ~3 logs at 150 days. Dramatically and somewhat surprisingly, the ΔuvrAΔuvrD1 strain was more attenuated than either single mutant and barely replicated in mice, a phenotype that was complemented by *uvrD1* (61). The attenuation phenotype of this double mutant is the most dramatic attenuation phenotype reported for a DNA repair–deficient strain of *M. tuberculosis* and clearly indicates the value of combining mutations from distinct repair pathways (in this case presumably reflecting inactivation of NER through loss of *uvrA* and the non-NER functions of UvrD1).

Another dramatic example of the effect of a DNA repair pathway on *M. tuberculosis* infection and *in vivo* mutagenesis came from the examination of the *dnaE2* mutant by the Mizrahi group. Inactivation of *dnaE2* conferred attenuation to *M. tuberculosis* in mice as measured by CFU and time to death analyses (64). In addition, the Δ*dnaE2* strain failed to evolve resistance to rifampin, providing a clear example that an error-prone repair pathway is essential to the evolution of drug resistance *in vivo* (64).

Despite the substantial advances in understanding the pathways of double-strand DNA break repair in mycobacteria, including the discovery of the NHEJ and SSA pathways as detailed above, the role of DSB repair in *M. tuberculosis* pathogenesis has been difficult to document. Early studies tested a BCG strain lacking *recA* and found no attenuation in either BALB/c mice or nude mice with intravenous infection (65). As noted above, mycobacteria elaborate three pathways of DSB repair: HR, NHEJ, and SSA. Two of these pathways, NHEJ and SSA, do not require a replicated chromosome to direct repair and therefore may be relevant to stages of *M. tuberculosis* infection during which replication has ceased. In addition, the presence of three repair pathways may suggest redundancy that would mask the phenotypes of *M. tuberculosis* strains lacking a single DSB repair pathway. Extensive experimentation in the Glickman lab has examined the pathogenesis phenotypes of *M. tuberculosis* Δ*ku*, Δ*recA*, and Δ*ku* Δ*recA* in multiple mouse models of infection including C57BL/6 and C3HeB/FeJ mice (66) and in guinea pigs. Surprisingly, these experiments did not demonstrate any change in bacterial load in any of these mutant strains, in any of the animal models, compared to wild-type *M. tuberculosis* (M. Glickman, in preparation).

These surprising findings fail to document a role for DSB repair in *M. tuberculosis* pathogenesis. The lack of phenotype of NHEJ-deficient strains in the mouse model could be attributed to the lack of true growth arrest/latency in these models (67), although the lack of phenotype in the *rec Aku* double mutant cannot be explained by ongoing replication. The possibility exists that the murine immune response to *M. tuberculosis* is not capable of inflicting DSB-inducing genotoxic stress on the *M. tuberculosis* chromosome. The lack of phenotype of these DSB repair–deficient strains fits with other studies that show a similar lack of phenotype for DNA repair–deficient strains (68) and, as has been suggested by others, may reflect the relative gross measure of bacterial load as an outcome (8). More granular data about the role of individual repair pathways in pathogenesis may come from examination of chromosomal mutagenesis and rearrangements by next generation sequencing of DNA repair pathway mutant strains passaged through animal models (69).

Citation. Glickman MS. 2014. Double-strand DNA break repair in mycobacteria. Microbiol Spectrum 2(5):MGM2-0024-2013.

References

1. Daley JM, Palmbos PL, Wu D, Wilson TE. 2005. Nonhomologous end joining in yeast. *Annu Rev Genet* **39:**431–451.

2. Dillingham MS, Kowalczykowski SC. 2008. RecBCD enzyme and the repair of double-stranded DNA breaks. *Microbiol Mol Biol Rev* **72:**642–671.

3. Lieber MR. 2010. The mechanism of double-strand DNA break repair by the nonhomologous DNA end-joining pathway. *Annu Rev Biochem* **79:**181–211.

4. Shuman S, Glickman MS. 2007. Bacterial DNA repair by nonhomologous end joining. *Nat Rev Microbiol* **5:**852–861.

5. Symington LS, Gautier J. 2011. Double-strand break end resection and repair pathway choice. *Annu Rev Genet* **45:**247–271.

6. Yeeles JT, Dillingham MS. 2010. The processing of double-stranded DNA breaks for recombinational repair by helicase-nuclease complexes. *DNA Repair* **9:**276–285.

7. Ford CB, Shah RR, Maeda MK, Gagneux S, Murray MB, Cohen T, Johnston JC, Gardy J, Lipsitch M, Fortune SM. 2013. *Mycobacterium tuberculosis* mutation rate estimates from different lineages predict substantial differences in the emergence of drug-resistant tuberculosis. *Nat Genet* **45:**784–790.

8. Warner DF, Tonjum T, Mizrahi V. 2013. DNA metabolism in mycobacterial pathogenesis. *Curr Top Microbiol Immunol* **374:**27–51.

9. Kurthkoti K, Varshney U. 2012. Distinct mechanisms of DNA repair in mycobacteria and their implications in attenuation of the pathogen growth. *Mech Ageing Dev* **133:**138–146.

10. Gorna AE, Bowater RP, Dziadek J. 2010. DNA repair systems and the pathogenesis of *Mycobacterium tuberculosis*: varying activities at different stages of infection. *Clin Sci* **119:**187–202.

11. Reijns MA, Rabe B, Rigby RE, Mill P, Astell KR, Lettice LA, Boyle S, Leitch A, Keighren M, Kilanowski F, Devenney PS, Sexton D, Grimes G, Holt IJ, Hill RE, Taylor MS, Lawson KA, Dorin JR, Jackson AP. 2012. Enzymatic removal of ribonucleotides from DNA is essential for mammalian genome integrity and development. *Cell* **149:**1008–1022.

12. Hiller B, Achleitner M, Glage S, Naumann R, Behrendt R, Roers A. 2012. Mammalian RNase H2 removes ribonucleotides from DNA to maintain genome integrity. *J Exp Med* **209:**1419–1426.

13. Wigley DB. 2013. Bacterial DNA repair: recent insights into the mechanism of RecBCD, AddAB and AdnAB. *Nat Rev Microbiol* **11:**9–13.

14. Gupta R, Barkan D, Redelman-Sidi G, Shuman S, Glickman MS. 2011. Mycobacteria exploit three genetically distinct DNA double-strand break repair pathways. *Mol Microbiol* **79:**316–330.

15. Stephanou NC, Gao F, Bongiorno P, Ehrt S, Schnappinger D, Shuman S, Glickman MS. 2007. Mycobacterial nonhomologous end joining mediates mutagenic repair of chromosomal double-strand DNA breaks. *J Bacteriol* **189:**5237–5246.

16. Sinha KM, Unciuleac MC, Glickman MS, Shuman S. 2009. AdnAB: a new DSB-resecting motor-nuclease from mycobacteria. *Genes Dev* **23:**1423–1437.

17. Fernandez S, Kobayashi Y, Ogasawara N, Alonso JC. 1999. Analysis of the *Bacillus subtilis* recO gene: RecO forms part of the RecFLOR function. *Mol Gen Genet* **261:**567–573.

18. Gupta R, Ryzhikov M, Koroleva O, Unciuleac M, Shuman S, Korolev S, Glickman MS. 2013. A dual role for mycobacterial RecO in RecA-dependent homologous recombination and RecA-independent single-strand annealing. *Nucleic Acids Res* **41**:2284–2295.

19. Patil KN, Singh P, Muniyappa K. 2011. DNA binding, coprotease, and strand exchange activities of mycobacterial RecA proteins: implications for functional diversity among RecA nucleoprotein filaments. *Biochemistry* **50**:300–311.

20. Ganesh N, Muniyappa K. 2003. Characterization of DNA strand transfer promoted by *Mycobacterium smegmatis* RecA reveals functional diversity with *Mycobacterium tuberculosis* RecA. *Biochemistry* **42**:7216–7225.

21. Datta S, Krishna R, Ganesh N, Chandra NR, Muniyappa K, Vijayan M. 2003. Crystal structures of *Mycobacterium smegmatis* RecA and its nucleotide complexes. *J Bacteriol* **185**:4280–4284.

22. Reddy MS, Guhan N, Muniyappa K. 2001. Characterization of single-stranded DNA-binding proteins from mycobacteria. The carboxyl-terminal of domain of SSB is essential for stable association with its cognate RecA protein. *J Biol Chem* **276**:45959–45968.

23. Dawson LF, Dillury J, Davis EO. 2010. RecA-independent DNA damage induction of *Mycobacterium tuberculosis* ruvC despite an appropriately located SOS box. *J Bacteriol* **192**:599–603.

24. Yang M, Gao C, Cui T, An J, He ZG. 2012. A TetR-like regulator broadly affects the expressions of diverse genes in *Mycobacterium smegmatis*. *Nucleic Acids Res* **40**:1009–1020.

25. Davis EO, Springer B, Gopaul KK, Papavinasasundaram KG, Sander P, Bottger EC. 2002. DNA damage induction of recA in *Mycobacterium tuberculosis* independently of RecA and LexA. *Mol Microbiol* **46**:791–800.

26. Davis EO, Dullaghan EM, Rand L. 2002. Definition of the mycobacterial SOS box and use to identify LexA-regulated genes in *Mycobacterium tuberculosis*. *J Bacteriol* **184**:3287–3295.

27. Brooks PC, Movahedzadeh F, Davis EO. 2001. Identification of some DNA damage-inducible genes of *Mycobacterium tuberculosis*: apparent lack of correlation with LexA binding. *J Bacteriol* **183**:4459–4467.

28. Thakur RS, Basavaraju S, Somyajit K, Jain A, Subramanya S, Muniyappa K, Nagaraju G. 2013. Evidence for the role of *Mycobacterium tuberculosis* RecG helicase in DNA repair and recombination. *FEBS J* **280**:1841–1860.

29. Prabu JR, Thamotharan S, Khanduja JS, Alipio EZ, Kim CY, Waldo GS, Terwilliger TC, Segelke B, Lekin T, Toppani D, Hung LW, Yu M, Bursey E, Muniyappa K, Chandra NR, Vijayan M. 2006. Structure of *Mycobacterium tuberculosis* RuvA, a protein involved in recombination. *Acta Crystallogr Sect F Struct Biol Cryst Commun* **62**:731–734.

30. Khanduja JS, Muniyappa K. 2012. Functional analysis of DNA replication fork reversal catalyzed by *Mycobacterium tuberculosis* RuvAB proteins. *J Biol Chem* **287**:1345–1360.

31. Khanduja JS, Tripathi P, Muniyappa K. 2009. *Mycobacterium tuberculosis* RuvA induces two distinct types of structural distortions between the homologous and heterologous Holliday junctions. *Biochemistry* **48**:27–40.

32. Aravind L, Koonin EV. 2001. Prokaryotic homologs of the eukaryotic DNA-end-binding protein Ku, novel domains in the Ku protein and prediction of a prokaryotic double-strand break repair system. *Genome Res* **11**:1365–1374.

33. Nandakumar J, Nair PA, Shuman S. 2007. Last stop on the road to repair: structure of *E. coli* DNA ligase bound to nicked DNA-adenylate. *Mol Cell* **26**:257–271.

34. Della M, Palmbos PL, Tseng HM, Tonkin LM, Daley JM, Topper LM, Pitcher RS, Tomkinson AE, Wilson TE, Doherty AJ. 2004. Mycobacterial Ku and ligase proteins constitute a two-component NHEJ repair machine. *Science* **306**:683–685.

35. Weller GR, Kysela B, Roy R, Tonkin LM, Scanlan E, Della M, Devine SK, Day JP, Wilkinson A, d'Adda di Fagagna F, Devine KM, Bowater RP, Jeggo PA, Jackson SP, Doherty AJ. 2002. Identification of a DNA nonhomologous end-joining complex in bacteria. *Science* **297**:1686–1689.

36. Aniukwu J, Glickman MS, Shuman S. 2008. The pathways and outcomes of mycobacterial NHEJ depend on the structure of the broken DNA ends. *Genes Dev* **22**:512–527.

37. Gong C, Bongiorno P, Martins A, Stephanou NC, Zhu H, Shuman S, Glickman MS. 2005. Mechanism of nonhomologous end-joining in mycobacteria: a low-fidelity repair system driven by Ku, ligase D and ligase C. *Nat Struct Mol Biol* **12**:304–312.

38. Gong C, Martins A, Bongiorno P, Glickman M, Shuman S. 2004. Biochemical and genetic analysis of the four DNA ligases of mycobacteria. *J Biol Chem* **279**:20594–20606.

39. Wang ST, Setlow B, Conlon EM, Lyon JL, Imamura D, Sato T, Setlow P, Losick R, Eichenberger P. 2006. The forespore line of gene expression in *Bacillus subtilis*. *J Mol Biol* **358**:16–37.

40. Pitcher RS, Green AJ, Brzostek A, Korycka-Machala M, Dziadek J, Doherty AJ. 2007. NHEJ protects mycobacteria in stationary phase against the harmful effects of desiccation. *DNA Repair* **6**:1271–1276.

41. Akey D, Martins A, Aniukwu J, Glickman MS, Shuman S, Berger JM. 2006. Crystal structure and nonhomologous end-joining function of the ligase component of mycobacterium DNA ligase D. *J Biol Chem* **281**:13412–13423.

42. Pitcher RS, Brissett NC, Picher AJ, Andrade P, Juarez R, Thompson D, Fox GC, Blanco L, Doherty AJ. 2007. Structure and function of a mycobacterial NHEJ DNA repair polymerase. *J Mol Biol* **366**:391–405.

43. Zhu H, Nandakumar J, Aniukwu J, Wang LK, Glickman MS, Lima CD, Shuman S. 2006. Atomic structure and nonhomologous end-joining function of the polymerase component of bacterial DNA ligase D. *Proc Natl Acad Sci USA* **103**:1711–1716.

44. Bebenek K, Garcia-Diaz M, Patishall SR, Kunkel TA. 2005. Biochemical properties of *Saccharomyces cerevisiae* DNA polymerase IV. *J Biol Chem* **280**:20051–20058.

45. Brissett NC, Pitcher RS, Juarez R, Picher AJ, Green AJ, Dafforn TR, Fox GC, Blanco L, Doherty AJ. 2007. Structure of a NHEJ polymerase-mediated DNA synaptic complex. *Science* **318**:456–459.

46. Nair PA, Smith P, Shuman S. 2010. Structure of bacterial LigD 3′-phosphoesterase unveils a DNA repair superfamily. *Proc Natl Acad Sci USA* **107**:12822–12827.

47. Zhu H, Shuman S. 2006. Substrate specificity and structure-function analysis of the 3′-phosphoesterase component of the bacterial NHEJ protein, DNA ligase D. *J Biol Chem* **281**:13873–13881.

48. Zhu H, Wang LK, Shuman S. 2005. Essential constituents of the 3′-phosphoesterase domain of bacterial DNA ligase D, a nonhomologous end-joining enzyme. *J Biol Chem* **280**:33707–33715.

49. Zhu H, Shuman S. 2005. Novel 3′-ribonuclease and 3′-phosphatase activities of the bacterial non-homologous end-joining protein, DNA ligase D. *J Biol Chem* **280**:25973–25981.

50. Zhu H, Bhattarai H, Yan HG, Shuman S, Glickman MS. 2012. Characterization of *Mycobacterium smegmatis* PolD2 and PolD1 as RNA/DNA polymerases homologous to the POL domain of bacterial DNA ligase D. *Biochemistry* **51**:10147–10158.

51. Sinha KM, Stephanou NC, Gao F, Glickman MS, Shuman S. 2007. Mycobacterial UvrD1 is a Ku-dependent DNA helicase that plays a role in multiple DNA repair events, including double-strand break repair. *J Biol Chem* **282**:15114–15125.

52. Li Z, Wen J, Lin Y, Wang S, Xue P, Zhang Z, Zhou Y, Wang X, Sui L, Bi LJ, Zhang XE. 2011. A Sir2-like protein participates in mycobacterial NHEJ. *PloS One* **6**:e20045.

53. Ivanov EL, Sugawara N, Fishman-Lobell J, Haber JE. 1996. Genetic requirements for the single-strand annealing pathway of double-strand break repair in *Saccharomyces cerevisiae*. *Genetics* **142**:693–704.

54. McEvoy CR, Cloete R, Muller B, Schurch AC, van Helden PD, Gagneux S, Warren RM, Gey van Pittius NC. 2012. Comparative analysis of *Mycobacterium tuberculosis* pe and ppe genes reveals high sequence variation and an apparent absence of selective constraints. *PloS One* **7**:e30593.

55. McEvoy CR, van Helden PD, Warren RM, Gey van Pittius NC. 2009. Evidence for a rapid rate of molecular evolution at the hypervariable and immunogenic *Mycobacterium tuberculosis* PPE38 gene region. *BMC Evol Biol* **9**:237.

56. Talarico S, Cave MD, Marrs CF, Foxman B, Zhang L, Yang Z. 2005. Variation of the *Mycobacterium tuberculosis* PE_PGRS 33 gene among clinical isolates. *J Clin Microbiol* **43**:4954–4960.

57. Talarico S, Zhang L, Marrs CF, Foxman B, Cave MD, Brennan MJ, Yang Z. 2008. *Mycobacterium tuberculosis* PE_PGRS16 and PE_PGRS26 genetic polymorphism among clinical isolates. *Tuberculosis* **88**:283–294.

58. Wojcik EA, Brzostek A, Bacolla A, Mackiewicz P, Vasquez KM, Korycka-Machala M, Jaworski A, Dziadek J. 2012. Direct and inverted repeats elicit genetic instability by both exploiting and eluding DNA double-strand break repair systems in mycobacteria. *PloS One* **7**:e51064.

59. Darwin KH, Ehrt S, Gutierrez-Ramos JC, Weich N, Nathan CF. 2003. The proteasome of *Mycobacterium tuberculosis* is required for resistance to nitric oxide. *Science* **302**:1963–1966.

60. Darwin KH, Nathan CF. 2005. Role for nucleotide excision repair in virulence of *Mycobacterium tuberculosis*. *Infect Immun* **73**:4581–4587.

61. Houghton J, Townsend C, Williams AR, Rodgers A, Rand L, Walker KB, Bottger EC, Springer B, Davis EO. 2012. Important role for *Mycobacterium tuberculosis* UvrD1 in pathogenesis and persistence apart from its function in nucleotide excision repair. *J Bacteriol* **194**:2916–2923.

62. Sinha KM, Stephanou NC, Unciuleac MC, Glickman MS, Shuman S. 2008. Domain requirements for DNA unwinding by mycobacterial UvrD2, an essential DNA helicase. *Biochemistry* **47**:9355–9364.

63. Williams A, Guthlein C, Beresford N, Bottger EC, Springer B, Davis EO. 2011. UvrD2 is essential in *Mycobacterium tuberculosis*, but its helicase activity is not required. *J Bacteriol* **193**:4487–4494.

64. Boshoff HI, Reed MB, Barry CE 3rd, Mizrahi V. 2003. DnaE2 polymerase contributes to in vivo survival and the emergence of drug resistance in *Mycobacterium tuberculosis*. *Cell* **113**:183–193.

65. Sander P, Papavinasasundaram KG, Dick T, Stavropoulos E, Ellrott K, Springer B, Colston MJ, Bottger EC. 2001. *Mycobacterium bovis* BCG recA deletion mutant shows increased susceptibility to DNA-damaging agents but wild-type survival in a mouse infection model. *Infect Immun* **69**:3562–3568.

66. Harper J, Skerry C, Davis SL, Tasneen R, Weir M, Kramnik I, Bishai WR, Pomper MG, Nuermberger EL, Jain SK. 2012. Mouse model of necrotic tuberculosis granulomas develops hypoxic lesions. *J Infect Dis* **205**:595–602.

67. Gill WP, Harik NS, Whiddon MR, Liao RP, Mittler JE, Sherman DR. 2009. A replication clock for *Mycobacterium tuberculosis*. *Nat Med* **15**:211–214.

68. Kana BD, Abrahams GL, Sung N, Warner DF, Gordhan BG, Machowski EE, Tsenova L, Sacchettini JC, Stoker NG, Kaplan G, Mizrahi V. 2010. Role of the DinB homologs Rv1537 and Rv3056 in *Mycobacterium tuberculosis*. *J Bacteriol* **192**:2220–2227.

69. Ford CB, Lin PL, Chase MR, Shah RR, Iartchouk O, Galagan J, Mohaideen N, Ioerger TR, Sacchettini JC, Lipsitch M, Flynn JL, Fortune SM. 2011. Use of whole genome sequencing to estimate the mutation rate of *Mycobacterium tuberculosis* during latent infection. *Nat Genet* **43**:482–486.

Molecular Genetics of Mycobacteria, 2nd Edition
Edited by Graham F. Hatfull and William R. Jacobs, Jr.
© 2014 American Society for Microbiology, Washington, DC
doi:10.1128/microbiolspec.MGM2-0008-2013

Nadine J. Bode[1]
K. Heran Darwin[1]

The Pup-Proteasome System of Mycobacteria

32

Murine models of tuberculosis have implicated the production of nitric oxide (NO) by activated macrophages as a pivotal part of the immune response, because mice lacking inducible nitric oxide synthase (iNOS$^{-/-}$) readily succumb to infection with *M. tuberculosis* (1). Formation of reactive nitrogen and oxygen intermediates (RNIs and ROIs) is toxic to a variety of microbes (reviewed in reference 2). The free radical NO is neutral and hydrophobic, allowing it to pass cellular and bacterial membranes. Reaction with superoxide generated by NADPH phagocyte oxidase results in the formation of the particularly destructive product peroxynitrite. The cytotoxic effects of RNI and ROI include DNA strand breakage, lipid peroxidation, and protein damage (reviewed in references 3, 4, and 5). Although the importance of NO has not yet been irrevocably demonstrated in humans, several *in vitro* studies suggest a role of host-derived RNI in control of tuberculosis (reviewed in references 2 and 6).

In an effort to deepen the understanding of the bacterial mechanisms in place to avoid elimination by the host immune system, Darwin et al. performed a genetic screen to discover genes that confer resistance to RNI *in vitro*. After screening 10,100 transposon mutants of H37Rv for hypersensitivity to acidified nitrite, the authors identified five mutants with independent insertions in Rv2097c and Rv2115c, two genes encoding putative accessory factors of a protease called the proteasome (7). Rv2115c was predicted to encode a AAA (ATPase associated with various cellular activities) ATPase related to those associated with the proteasome in eukaryotes. On the basis of its genomic location and a high degree of sequence similarity (81%) to ARC (AAA ATPase forming ring-shaped complexes) from *Rhodococcus erythropolis*, the authors named Rv2115c *mpa* (mycobacterial proteasome ATPase). ARC was the first characterized bacterial ATPase with suggested proteasomal function (8). In contrast, Rv2097c was annotated as a hypothetical protein (9) and shared no homology with any protein of known function. Mutations in Rv2097c and *mpa* caused similar phenotypes *in vivo* and *in vitro*, suggesting that both gene products participate in the same pathway. As a result of this observation, as well as its genomic association with the proteasome core protease-encoding *prcBA* genes, Rv2097c was named *pafA* for proteasome accessory factor A (7).

The identification of proteasomal components as potential antituberculosis targets that would sensitize *M. tuberculosis* to host immunity immediately

[1]New York University School of Medicine, Department of Microbiology, 550 First Avenue, MSB 236, New York, NY 10016.

attracted the interest of the scientific community. Previously, no function had been assigned to the bacterial proteasome, and attempts to reconstitute protein degradation by a bacterial proteasome *in vitro* had proven unsuccessful (8), likely due to the lack of additional required cofactors. Over the course of the last decade, seminal experiments demonstrated the existence of a bacterial posttranslational protein modifier called Pup (prokaryotic ubiquitin-like protein), which serves as a signal for degradation by the mycobacterial proteasome. Here we summarize what is currently known about the genetics and function of the *M. tuberculosis* Pup-proteasome system (PPS) and its role in pathogenesis.

PROKARYOTIC PROTEASOMES

Proteasomes are self-compartmentalizing, ATP-dependent multisubunit proteases that are responsible for the majority of nonlysosomal protein degradation in eukaryotes. The barrel-shaped complexes have a critical function in protein quality control, stress response, cell cycle control, transcription, metabolism, signal transduction, immune response, and many other biological processes (reviewed in references 10, 11). The "26S proteasome" is composed of two general substructures: the catalytically active 20S core particle (CP) and a 19S regulatory particle (RP) involved in substrate recognition, unfolding, and translocation into the inner core. The CP is assembled of 14 structurally related alpha (α1 to α7) and beta (β1 to β7) subunits, which associate as four stacked hetero-heptameric rings. The outer α-rings provide an entry gate to the inner proteolytic chamber composed of β-rings. Of the seven different β-subunits, three have been shown to have distinct catalytic activities as threonine proteases, cleaving peptides after hydrophobic ("chymotrypsin-like activity"), basic ("trypsin-like activity"), or acidic ("peptidylglutamyl-hydrolyzing activity") residues (reviewed in reference 12). Proteasomes degrade proteins in a highly processive fashion, releasing peptides with a median length of five residues (13). All cleavage sites are sequestered on the inside of the CP, and access is controlled by the RP attached to the α-rings on one or both ends (reviewed in reference 14). The RP is made up of 19 subunits, which form 2 subcomplexes referred to as lid and base. Six AAA+ (ATPase associated with various activities) ATPases that assemble into a hexameric ring at the base of the RP function as the driving force in substrate unfolding and translocation. Other non-ATPase subunits are involved in various aspects of substrate recognition and processing.

Prokaryotic proteasomes were first discovered in the thermoacidophilic archaeon *Thermoplasma acidophilus*, where electron microscopy revealed the existence of proteolytically active particles similar in shape and size to eukaryotic proteasomes (15). In comparison to the eukaryotic 20S CP, the subunit composition of the archaeal proteasome is less complex, which facilitated the determination of the first crystal structure of a proteasome from any of the three domains of life (16). The first bacterial proteasome was identified in the actinomycete *R. erythropolis* (17). Later, 20S proteasomes were biochemically or genetically characterized in other members of this order, namely *Mycobacterium smegmatis* (18), *Streptomyces coelicolor* (19), and *Frankia* (20). Additionally, sequence information suggested the presence of proteasomes in *M. tuberculosis* (9) and *Mycobacterium leprae* (reviewed in reference 21). The notion that bacterial proteasomes are limited to actinomycetes was challenged when the Banfield group discovered peptides encoded by proteasomal genes in biofilm samples of the uncultivated, Gram-negative bacterium *Leptospirillum* using a mass-spectrometry-based approach (22). Thus far, actinomycetes and *Leptospirillum* remain the only known bacterial lineages with a proteasome system. It is hypothesized that both have independently acquired the gene clusters encoding this complex protease via lateral gene transfer events (23, 24).

Overall, the architecture of proteasomes from all three domains of life is remarkably similar: cylinder-shaped particles 15 nm in height and 11 nm in diameter are assembled in the $\alpha_7\beta_7\beta_7\alpha_7$ fashion. While cognate ATPases have been identified and genetically linked to 20S CPs in both archaea and bacteria, robust physical interactions have not been detected, which is suggestive of other factors required for proteolysis (25). In contrast, the eukaryotic 26S holocomplex can be purified. Most prokaryotic CPs consist of only one α (PrcA) and one β (PrcB) subunit, both of which form homoheptameric rings. This arrangement generally limits prokaryotic proteasomes to chymotryptic activity. The *M. tuberculosis* proteasome, which shares modest sequence similarity with the *Thermoplasma* CP (32% for both α- and β-subunits) and a high degree of identity (65%) with the *Rhodococcus* proteasome, is compliant with these features. However, several observations distinguish the *M. tuberculosis* proteasome from other prokaryotic 20S CPs. All β-subunits are synthesized with N-terminal pro-peptides, which undergo autocatalytic processing to reveal the catalytic threonine residue. Unlike the *Rhodococcus* propeptide, which acts as an assembly-promoting factor, the *M. tuberculosis* pro-peptides represent a thermodynamic barrier to this process. Additionally, the *M. tuberculosis* proteasome shows broad substrate specificity despite the presence of a single type of catalytic β-chain. Topological features of the substrate-binding pocket that mimic that of corresponding surfaces in *Saccharomyces* cause this unusual characteristic (26, 27). It is notable that although proteasomes share certain properties with the four classes of energy-dependent proteases commonly found in bacteria, they are structurally and enzymatically distinct from ClpXP, HslUV, FtsH, and Lon (reviewed in reference 28).

PUPYLATION

Almost all proteins targeted for proteasomal degradation in eukaryotes are tagged with ubiquitin (Ub). Ub is a small protein of 76 amino acids that can be covalently linked to other proteins via its C-terminal di-glycine

motif. Produced as an inactive precursor, Ub is proteolytically processed by deubiquitylases (DUBs) to expose the site of substrate attachment. "Ubiquitylation" (also referred to as ubiquitination) then proceeds in a multistep cascade (Fig. 1). Initially, Ub is activated through adenylation at the C-terminus. A high-energy thioester bond to the active site cysteine of a Ub activating enzyme (E1) quickly replaces the mixed anhydride bond. Subsequently, Ub is transferred to the catalytic cysteine of a Ub conjugating enzyme (E2). Ub can then be transferred to its substrates by a Ub ligase (E3), resulting in the formation of an isopeptide bond between the C-terminal glycine of Ub and the ε-amino group of substrate lysines (reviewed in references 29, 30). Substrate specificity is largely determined by the presence of more than 600 E3 ligases (31) that are subdivided into the HECT (homologous to the E6-AP C-terminus) and RING (really interesting new gene) classes (reviewed in references 32, 33). Ub has seven lysine (K) residues,

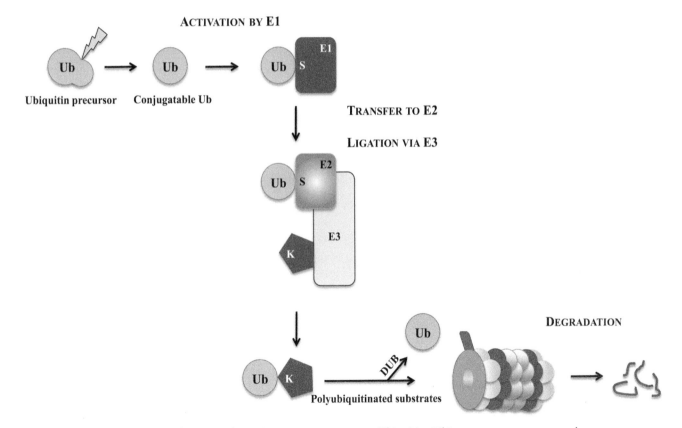

Figure 1 Eukaryotic ubiquitin-proteasome system. Ubiquitin (Ub) precursors are processed to expose a C-terminal di-glycine motif. The conjugation-competent Ub is adenylated and subsequently bound in a high-energy thioester bond by the E1-activating enzyme. Ub is then transferred to the catalytic cysteine of an E2-conjugating enzyme. Ub can be ligated to substrates with the help of E3-ligases. Typically, tetra-Ub chains linked at lysine 48 are recognized by the 26S proteasome. Deubiquitylases can remove Ub from substrates. doi:10.1128/microbiolspec.MGM2-0008-2013.f1

which allows for the generation of branched poly-ubiquitin chains. Classically, tetra-Ub chains linked at K48 are regarded as the minimal signal for degradation (34), although an increasing number of studies report monoubiquitylation to be sufficient for recognition (35). Polyubiquitylated substrates can be recognized by specific Ub receptors located at the proteasome. Before substrates are fed into the CP, proteasome-associated DUBs can cleave and recycle the poly-Ub chain (reviewed in reference 11).

Although there are striking structural and biochemical similarities between eukaryotic and prokaryotic CPs, it was unclear how proteins are targeted for proteasomal degradation in bacteria, which do not have Ub. Unsuccessful attempts to reconstitute the ATPase-dependent degradation of proteasomal substrates *in vitro* suggested the involvement of additional cofactors. An important first step was the identification of endogenous mycobacterial proteasome substrates (36). Two proteins, malonyl CoA-acyl carrier protein transacylase (FabD) and ketopantoate hydroxymethyltransferase (PanB), accumulate in an *mpa* null strain or in wild-type *M. tuberculosis* treated with the mammalian proteasome inhibitor epoxomicin. Transcriptional analysis showed no difference in *fabD* or *panB* transcript levels, suggesting that increased protein abundance is not caused by induced gene expression. To further test these results, epitope-tagged *fabD* and *panB* were placed under the control of a heterologous *Mycobacterium bovis hsp60* promoter and an *Escherichia coli* ribosome binding site to control for changes in transcription and translation initiation. Heterologously produced FabD and PanB accumulate in *mpa* and *pafA* mutant strains, as well as in wild-type *M. tuberculosis* treated with proteasome inhibitor (36), thus further strengthening the functional association between Mpa, PafA, and the proteasome. Interestingly, the same study also demonstrated that Mpa levels are regulated by the concerted efforts of Mpa, PafA, and the proteasome. Mpa accumulates in strains producing mutant Mpa lacking its ATPase activity (37), as well as in the *pafA* null strain or in wild-type *M. tuberculosis* after chemical inhibition of the proteasome (36). Thus, Mpa is autoregulated via proteasomal degradation.

In order to determine additional components of the *M. tuberculosis* proteasome system, Darwin and coworkers performed a genetic screen for interaction partners of Mpa, utilizing an *E. coli* two-hybrid system (38). Screening an *M. tuberculosis* genomic library of ~100,000 fragments for positive interactions with an Mpa bait fusion, the authors identified Rv2111c, which is the gene located immediately upstream of *prcBA* and predicted to form an operon with the CP genes. However, addition of Rv2111c to CP and Mpa did not stimulate degradation of FabD *in vitro*, suggesting the requirement of additional factors for proteolysis. The authors hypothesized that mycobacteria-specific cofactors were required to promote proteolysis. Consequently, they employed a mycobacterial two-hybrid system (39), which enables the study of protein-protein associations in *M. smegmatis*. Surprisingly, Rv2111c and a proteasome substrate (FabD) showed a strong interaction in this system, which was further tested by copurification of the recombinant proteins from *M. smegmatis*. Unexpectedly, these experiments yielded a stable covalent complex between the two proteins. Mass spectrometry revealed the formation of an isopeptide bond between the C-terminal residue of Rv2111c and a particular FabD lysine residue (K173). Unlike ubiquitin, Rv2111c does not encode a C-terminal di-glycine motif, but instead contains this motif at the penultimate position, followed by glutamine. Moreover, the C-terminal residue is deamidated to glutamate prior to conjugation. Mutation of the pupylated lysine of FabD (K173A) results in stabilization of the protein in *M. smegmatis*, indicating that modification of proteins with Rv2111c is a signal for degradation. Based on its functional, albeit not biochemical, analogy to ubiquitin, Rv2111c was named Pup. Interestingly, while numerous pupylated proteins can be observed by immunoblot in total lysates from wild-type and *mpa* *M. tuberculosis* strains, no anti-Pup reactive bands are detectable in a *pafA* null strain, indicating the involvement of PafA in substrate conjugation (40).

In an independent study, Burns et al. showed that a homologue of Rv2111c (MSMEG_3896) acts as a protein modifier in *M. smegmatis*. Production of epitope-tagged Pup in *M. smegmatis* was sufficient to modify numerous proteins, demonstrating that the signal for proteasomal degradation is conserved in both pathogenic and nonpathogenic mycobacteria (41). Taken together, these studies showed for the first time that some bacteria are able to utilize a posttranslational tagging system reminiscent of ubiquitylation in eukaryotes, thereby marking a turning point in the understanding of prokaryotic proteasome biology.

Significant progress has been made in elucidating the biochemistry of this posttranslational modifier (Fig. 2). Pup is an intrinsically disordered protein of 64 amino acids in length (42–44), which is in stark contrast to Ub and other related eukaryotic modifiers that have a conserved β-grasp fold (reviewed in reference 45). Intrinsically disordered proteins lack a compact structure under physiological conditions, although

DEAMIDATION **LIGATION** **DEPUPYLATION** **DEGRADATION**

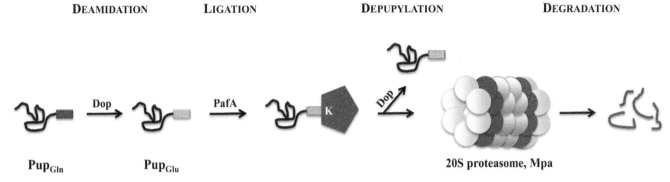

Pup_Gln **Pup_Glu** **20S proteasome, Mpa**

Figure 2 Mycobacterial pupylation pathway. Pup is deamidated at the C-terminal glutamine by Dop (deamidase of Pup). The Pup ligase PafA phosphorylates the C-terminal γ-carboxylate of Pup and then conjugates Pup to lysine residues of target proteins via an isopeptide bond. The mycobacterial proteasome and its cognate ATPase Mpa degrade pupylated proteins. Dop also acts as a depupylase, allowing for Pup to be recycled. doi:10.1128/microbiolspec.MGM2-0008-2013.f2

binding-induced folding of the C-terminal half of Pup aids its recognition by Mpa (see below).

At the time of its discovery, *pafA* did not share any homology with proteins of known function (7, 46). Later, sensitive sequence analysis classified PafA as a member of the carboxylate-amine/ammonia ligase superfamily. Proteins of this class catalyze a two-step ligase reaction that results in an amide linkage between an amino and a carboxylate group. Aravind and colleagues proposed a model in which PafA phosphorylates the γ-carboxylate of the C-terminal glutamate of Pup (Pup_Glu), priming it for nucleophilic attack by the ε-side-chain amino group of a target lysine, which results in formation of an isopeptide bond (47). *In vitro*, PafA is sufficient to conjugate Pup_Glu to mycobacterial proteasome substrates in an ATP-dependent fashion (48). Site-directed mutagenesis of any residues predicted to coordinate Mg^{2+} or ATP abrogated pupylation *in vivo*, supporting the predicted model (49). A report by Weber-Ban and coworkers tested the Aravind lab's hypothesis and found that PafA indeed catalyzes ligation of Pup to substrates via a two-step reaction mechanism. The authors were able to detect a phosphorylated Pup intermediate bound to PafA following ATP hydrolysis (50). Nuclear magnetic resonance spectroscopy revealed that coupling to substrates occurs via the side chain carboxylate of Pup, ruling out any possible involvement of the α-carboxylate of its C-terminus (51).

Purification of epitope-tagged Pup from *M. smegmatis* yields both Pup_Glu and Pup_Gln. The observation that the majority of Pup is Pup_Glu when purified from *M. smegmatis* but not from *E. coli* suggested that an enzymatic activity is responsible for this conversion

(40). Aravind and colleagues noticed that the *M. tuberculosis* genome encodes two paralogues of PafA and thus proposed that the ligase assembles into a heterodimer (47). This hypothesis was refuted in a series of experiments that demonstrated that the paralogue of PafA has a distinct function: Dop (deamidase of Pup) converts the C-terminal glutamine residue of Pup to glutamate. Unlike PafA, Dop appeared to utilize ATP only as a cofactor, because no hydrolysis and accompanying release of ADP or AMP was detected. Consolidating this finding, deamidation also proceeds in the presence of a nonhydrolyzable nucleotide analogue, such as ATPγS, albeit at a lower rate (48).

Sequential action of Dop and PafA constitutes a two-step pupylation pathway. However, *in vitro*, the requirement for Dop can be bypassed, because PafA is sufficient to conjugate Pup_Glu to target substrates. Actinomycetes encode *pup* either with a C-terminal glutamine residue, like mycobacteria, or with glutamate (e.g., *Corynebacterium* spp.), thus eliminating the need for a deamidation step (reviewed in reference 52). This observation hinted at another function for Dop.

DEPUPYLATION

In humans, over 70 DUBs are responsible for removing Ub chains from modified substrates. Their action prevents the turnover of Ub together with proteins targeted to the 26S proteasome, so that Ub can be recycled for subsequent conjugation reactions. Under certain circumstances, removal of Ub rescues a protein from degradation (reviewed in reference 53).

It was hypothesized that mycobacterial cells have the ability to resolve the bond between Pup and its substrate. A series of experiments reported in two independent studies identified Dop, the deamidase of Pup, as a bifunctional enzyme with depupylase (DPUP) activity. Darwin and coworkers noted that numerous pupylated substrates (the "pupylome") disappeared over time in the presence of ATP but absence of proteasomes. In addition, specific recombinant pupylated proteasome substrates (Pup~FabD and Pup~Ino1 [myo-inositol-1-phosphate synthase]) were "depupylated" in mycobacterial lysates. However, *M. tuberculosis* lysates of a *dop* mutant lack the ability to remove Pup from substrates. The authors reasoned that a putative depupylation reaction is chemically identical to the previously described deamidation reaction: both activities involve hydrolysis of an amide bond at the C-terminal residue of Pup. Indeed, purified Dop is able to depupylate different proteasomal substrates *in vitro*, releasing Pup_{Glu}. As for deamidation, ATP binding appears to be critical for functionality of the enzyme (54). Similarly, Weber-Ban and coworkers demonstrated that Dop functions as a depupylase in *M. smegmatis*. The proteasome substrate Pup~PanB is stable in lysates of an *M. smegmatis dop* mutant, while depupylation is restored in a complemented strain. Mass spectrometry analysis of the products released after incubation of Dop with Pup~PanB identifies Pup with a C-terminal glutamate residue, demonstrating that the enzyme breaks the isopeptide bond between Pup and a substrate (55). Dop is a strict isopeptidase: unlike certain DUBs, which can cleave linear Ub chains, Dop does not show any reactivity against linear Pup-substrate fusions (54–56). Going forward, it will be critical to determine if depupylation of substrates can rescue them from degradation by the proteasome or if this step facilitates their delivery into the catalytic core of the protease.

Assessment of Dop's DPUP activity *in vivo* was complicated by its requirement for deamidation of Pup. In *M. smegmatis*, ectopic expression of pup_{Glu} in a *dop* mutant is sufficient to restore pupylation to wild-type levels (57). Interestingly, the pupylation defect of an *M. tuberculosis dop* mutant cannot fully be restored by production of Pup_{Glu}. However, treatment of an *M. tuberculosis dop* strain expressing pup_{Glu} with the proteasome-inhibitor epoxomicin restores a robust pupylome. This indicates that the DPUP activity of Dop is essential to maintain steady-state levels of pupylation through the recycling of Pup (49).

Interestingly, Mpa contributes to depupylation. Overproduction of Pup results in the rapid degradation of Ino1 in *M. smegmatis*. In an *M. smegmatis mpa* mutant, Pup~Ino1 accumulates robustly but is virtually undetectable in wild-type or proteasome deletion strains (54, 58). This suggested Mpa helps facilitate depupylation. Along these lines, Weber-Ban and colleagues demonstrated that Mpa enhances depupylation *in vitro*. Substrate unfolding by Mpa appears to increase the accessibility of the isopeptide bond for subsequent cleavage (54, 55). This result implies that Dop interacts with pupylated proteins subsequent to or concurrent with Mpa.

The crystal structures of Dop and PafA homologues from *Acidothermus cellulolyticus* and *Corynebacterium glutamicum*, respectively, have been solved (59). Notwithstanding moderate conservation on a primary sequence level (38% sequence identity), the two proteins show high structural similarity. It appears that PafA and Dop arose from a gene duplication event that results in the production of enzymes with opposing activities. One might assume that the shared structural elements of PafA and Dop were conserved to accommodate Pup binding. As previously predicted, the large N-terminal domain of both PafA and Dop adopts a carboxylate-amine ligase fold. A smaller C-terminal domain is unique to these proteins. Even though the active site of both enzymes is located in a particular structural element (referred to as β-sheet cradle) present in the N-terminal domain, mutation analysis has demonstrated that the C-termini of PafA and Dop are critical for function (49, 59, 60).

While multiple differences in the molecular architecture of the enzymes have been identified, site-directed mutagenesis and structural analysis did not identify a singular determinant of Dop activity (59). In another study, Burns et al. utilized an electrophilic trap consisting of Pup modified with a C-terminal glutamine mimic to identify a nucleophilic residue in Dop. Labeling of aspartate 95, an atypical nucleophile for a protease, classified Dop as an unusual amidase whereby the deamidation or depupylation reaction proceeds via an anhydride intermediate (61). Although this model is supported by the crystal structure, further experimentation is necessary to definitively prove this unusual protease activity.

GENOMIC ORGANIZATION OF THE PUP-PROTEASOME LOCUS

All PPS-associated genes identified to date are localized in direct proximity to the core proteasome (Fig. 3). In mycobacteria, *pup* is encoded immediately upstream of *prcBA*, and the overlap of the stop and start codons of *pup* and *prcB*, as well as *prcB* and *prcA*, suggest translational coupling. Usually, *dop* precedes this operon,

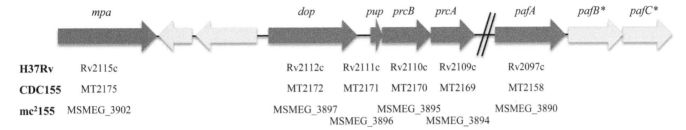

Figure 3 Genomic organization of PPS genes in mycobacteria. Gene data are from http:// tuberculist.epfl.ch/. *pafB and pafC are cotranscribed with *pafA* in *M. tuberculosis* H37Rv (46). However, no apparent contribution to pupylation has been demonstrated for PafB or PafC (40). doi:10.1128/microbiolspec.MGM2-0008-2013.f3

while *pafA* is encoded further downstream, separated by several presumably unrelated genes. The *mpa* gene is found several open reading frames (ORFs) upstream of *dop*.

It is notable that corynebacteria encode homologues of all genes associated with (de)pupylation, despite lacking the proteasomal core genes (reviewed in reference 62). The presence of *mpa* in these organisms is particularly intriguing and may suggest that pupylation can serve as a degradation tag in the absence of the proteasome. It is tempting to speculate that in these organisms, Mpa interacts with other degradation machines to catalyze turnover of pupylated substrates.

Perhaps somewhat surprisingly, little to nothing is known about the expression of proteasome-associated genes. Future studies will need to more closely examine if the PPS is differentially regulated under certain conditions.

SUBSTRATE RECOGNITION BY THE MYCOBACTERIAL PROTEASOME

The RP of the eukaryotic proteasome contains two subunits designated as intrinsic ubiquitin receptors (63). Although a similar arrangement is missing from the *M. tuberculosis* 20S proteasome, Mpa interacts strongly with Pup (40). Optimal degradation of substrates by the 26S proteasome requires both a proteasome binding signal (e.g., modification with Ub) and an unstructured region to initiate degradation (reviewed in reference 14). Nuclear magnetic resonance studies determined that the C-terminal half of Pup interacts with Mpa (42, 44). Cocrystallization experiments of Mpa and Pup further elucidated the mechanistic details of recognition: Pup binds the N-terminal coiled coil domains of an Mpa hexamer through hydrophobic interactions. This contact induces the formation of an α-helix in the central part of Pup (64). Burns et al. demonstrated that this region of Pup also facilitates

interaction with enzymes involved in pupylation, while the N-terminal half is required to initiate unfolding and degradation. Thus, Pup itself provides the two components of the degron necessary for proteasomal degradation (58).

LACK OF A PUPYLATION MOTIF

Several proteomics studies identified targets of pupylation in *M. tuberculosis* and *M. smegmatis*. Purification of proteins conjugated to Pup coupled with mass spectrometry identified 604 *M. tuberculosis* proteins, representing ~15% of the predicted proteome. A Pup attachment site was identified for 55 of these proteins (65). Two separate *M. smegmatis* pupylome analyses identified 52 (66) and 41 (67) proteins, respectively, with confirmed pupylation sites. In all cases Pup was attached to substrates via a lysine residue, corroborating initial observations. Furthermore, Pup does not appear to form polymeric chains as seen in ubiquitylation. Determining the functional assignment of proteins identified in the *M. tuberculosis* pupylome, three categories appeared to be overrepresented when compared to their presence in the genome: intermediary metabolism, lipid metabolism, and detoxification/virulence (65). This functional clustering of Pup targets is preserved in *M. smegmatis* (67). Interestingly, the authors of one study identified significant overlap between the *M. smegmatis* pupylome and the *M. tuberculosis* nitrosoproteome, suggesting that the PPS could have an important role in the turnover of damaged proteins in response to RNI challenge (66).

The pupylome studies did not reveal a consensus primary sequence recognition motif at the site of modification. Generally, the attachment sites for posttranslational protein modifiers are too diverse to define a consensus sequence. SUMO (small ubiquitin-related modifier) is the only notable exception, because many of its target proteins contain an acceptor lysine within a ψKxD/E motif, where ψ is a large, hydrophobic residue (reviewed in reference 68). Although Dorrestein and colleagues mapped the pupylation sites on available crystal structures to the periphery of the target subunits (66), it remains a mystery how a single ligase PafA can act on potentially all pupylation substrates and discriminate them from nontargets in the cell.

Darwin and coworkers reconstituted the pupylation pathway in *E. coli*, a surrogate host that does not natively encode the Pup machinery. Coexpression of *pafA* and pup_{Glu} is sufficient to pupylate mycobacterial substrates in *E. coli*. Surprisingly, anti-Pup immunoblots revealed that numerous *E. coli* proteins were also modified in this system, suggesting that no additional mycobacteria-specific cofactors are required in the process of pupylation. The *E. coli* pupylome consists of 51 proteins with identified Pup attachment sites. The *E. coli* substrate phosphoenolpyruvate protein phosphotransferase I (PtsI), which does not have a homologue in mycobacteria, could be pupylated in *M. smegmatis*, demonstrating that a completely foreign protein can be recognized by a native pupylation system. However, pupylation is not an arbitrary event, because not every protein can be pupylated. For example, dihydrolipoamide acyltransferase (DlaT) is pupylated neither in mycobacteria nor in the *E. coli* system, despite having 29 lysines. Furthermore, there is a distinct preference for the modified lysine residue in mycobacteria, which is particularly striking in FabD, where all eight lysines are surface-exposed but only one is targeted for pupylation in the native system. The authors speculate that the unmodified lysines are shielded from PafA by other components of the multienzyme fatty acid synthesis pathway in which FabD participates (60). It remains to be determined how protein-protein interactions shape pupylation in general, because the current knowledge cannot rule out involvement of additional factors in target selection or delivery of substrates to the proteasome.

CHARACTERIZATION OF PUP-PROTEASOME MUTANTS

M. tuberculosis mutants in *mpa* and *pafA* are highly attenuated in a mouse model of infection as measured by reduced bacterial load, gross pathology, and histologically alleviated pneumonitis after a low-dose aerosol challenge (~200 bacilli/mouse) (7). Time-to-death experiments showed that unlike mice infected with wild-type *M. tuberculosis*, which show a median survival of ~1.5 years after infection, mice infected with *mpa* or *pafA* remained alive and healthy while maintaining

robust body weight and exhibiting no symptoms beyond normal aging (36, 37). As explained earlier, Nathan and coworkers observed increased *in vitro* sensitivity to acidified nitrite for *mpa* and *pafA* mutants, raising the question if higher susceptibility to RNI *in vivo* was the reason for their attenuation in mouse infections. In mammals, NO is produced by three isoforms of nitric oxide synthase: endothelial (eNOS), neuronal (nNOS), and inducible/immune (iNOS) (reviewed in reference 69). Phagocytes are prolific generators of NO, contributing to the control of various infections, such as *Leishmania major* and Coxsackie B3 virus (reviewed in reference 2). Mice genetically inactivated for iNOS (iNOS$^{-/-}$) or treated with a chemical inhibitor of iNOS are extremely susceptible to *M. tuberculosis* infection compared to wild-type mice (1). Low-dose aerosol infection of iNOS$^{-/-}$ mice with wild-type *M. tuberculosis* results in death after two to three months (7, 36, 37). In contrast, iNOS$^{-/-}$ mice infected with *mpa* or *pafA* mutant strains live significantly longer than iNOS$^{-/-}$ mice infected with wild-type *M. tuberculosis*, but nonetheless eventually all die (36, 37). Hence, genetic inactivation of iNOS partially restores susceptibility to *mpa* and *pafA* mutants, indicating the importance of these genes to resisting NO.

In another study, Bishai and colleagues screened a library of transposon mutants in the CDC1551 background for mutants displaying morphological changes on agar media. Three independent mutants of *mpa* (MT2175) consistently formed smaller colonies than wild-type *M. tuberculosis* on solid media. To facilitate a more in-depth analysis of the consequences caused by absence of this gene, the authors generated a deletion disruption mutant, in which a hygromycin resistance cassette replaced more than 90% of the coding sequence of *mpa*. Growth of the Δ*mpa::hyg* mutant in rich medium (7H9) was impaired: the average *in vitro* doubling time was determined to be 22 h, compared to 18 h for wild-type CDC1551. High-dose aerosol infection of BALB/c mice shows an infection profile similar to that obtained with the H37Rv *mpa* mutant. The authors also observed significantly less pathology in mice infected with the Δ*mpa::hyg* mutant, characterized by reduced levels of granuloma-like immune infiltrates in the lungs and no changes in body weight. Furthermore, mice infected with the Δ*mpa::hyg* mutant did not succumb to infection until the experiment was terminated at 180 days, while the median survival time for wild-type *M. tuberculosis*–infected mice was 25 days (70).

Disruption of *dop* also sensitizes *M. tuberculosis* to RNI *in vitro* and severely attenuates survival and growth in mice. The degree of attenuation, as assessed by histopathology and bacterial load, is similar among *dop*, *mpa*, and *pafA* mutants, supporting the hypothesis that a common pathway functionally links all three genes (7, 49). Mice infected with any of these mutants do not show the severe tissue destruction and infiltration of lymphocytes characteristic of wild-type *M. tuberculosis*. Bishai and coworkers dubbed their Δ*mpa::hyg* strain as an "immunopathology mutant" due to its ability to persist in the lungs without causing significant immune-induced disease. Interferon gamma production is not induced in mice infected with a Δ*mpa::hyg* mutant compared to wild-type *M. tuberculosis* at 3 weeks postinfection. Interestingly, their Δ*mpa::hyg* strain offered some immunological protection against virulent *M. tuberculosis* challenge in mice (70).

The reduced pathology caused by infections with PPS mutants is likely a direct result of a 100- to 1,000-fold lower bacterial burden during the persistent stage of the infection; however, we cannot exclude the possibility that it is the dysregulation of a specific protein or subset of proteins in proteasomal degradation-deficient mutants that causes the changes in pathology.

The genetic screen for mutants hypersensitive to NO did not yield any mutants with insertions in the core proteasomal genes *prcBA*. Employing transposon *in situ* hybridization (TraSH) of a saturating mutant library, Rubin and coworkers identified *prcBA* but not *mpa* or *pafA* among a set of genes required for cellular growth *in vitro* (71), suggesting that *prcBA* are essential. A conditional gene silencing system using a tetracycline-responsive promoter facilitated studies to test the essentiality of *prcBA* (72). This study indeed showed that *prcBA* is required for normal growth *in vitro*. Later, a deletion-disruption strain in which *prcBA* was replaced with a hygromycin resistance cassette (Δ*prcBA::hyg*) was characterized (73). Although the mutant is viable, it grows very slowly. Not surprisingly, infection with either of the *prcBA* mutants results in fewer bacilli in the lungs of C57BL/6 mice compared to infection with wild-type *M. tuberculosis*. Importantly, the number of bacteria decreased dramatically in later weeks, suggesting that the proteasome core protease is essential for mycobacterial persistence in mice (72, 73).

A *pup* mutant has never been reported. Like *prcBA*, it is predicted to be essential based on TraSH (71). Because *pup* is predicted to be cotranscribed and translated with *prcBA*, disruption of *pup* would likely result in polar effects on *prcBA* expression. Pup is undetectable in either a *dop* or *pafA* mutant, presumably because unconjugated Pup is highly unstable (40, 49). Thus, one would presume that these mutants would be phenotypically similar if not identical to a hypothetical *pup* mutant.

It is striking that *prcBA* mutants display an *in vitro* growth defect absent from *mpa*, *pafA*, and *dop* null strains (7, 49). To discern if the mycobacterial proteasome has functionality beyond its proteolytic activity, Ehrt and colleagues mutated the catalytic threonine residue of PrcB to alanine (T1A). The active-site mutant assembles into a 20S complex with PrcA, but production of this recombinant proteasome in a Δ*prcBA* mutant does not rescue proteasomal activity. Surprisingly, production of this inactive proteasome complex is sufficient to complement the hypersusceptibility to NO. In contrast, proteasomal proteolysis is essential for persistence in mice (73). It is unclear how proteasomal activity is connected to certain phenotypes but not others. It is possible that an inactive 20S core structure could act as a "trap" to contain proteins that would normally be degraded. Trapping could result in (partial) inactivation of these proteins and may thus explain how an active site mutant of the *M. tuberculosis* 20S can still protect against certain stresses but not others.

It has been hypothesized that the mechanism of protection by the proteasome against NO damage involves the degradation of irreversibly oxidized or nitrated proteins to prevent toxicity to the bacterium. However, *mpa*-, *pafA*-, and *prcBA*-deficient strains are more resistant than wild-type *M. tuberculosis* to hydrogen peroxide (7, 72). Although this could be due to compensatory induction of additional antioxidant pathways, these data suggest that mere turnover of accumulating damaged proteins does not explain the hypersensitivity of PPS mutants to NO.

The role of the CP in virulent mycobacteria appears to be distinct from that in nonpathogenic *M. smegmatis*, in which disruption of the proteasome does not result in any discernible phenotype (18) (Table 1). Interestingly, unlike *M. tuberculosis*, *M. smegmatis* encodes a homologue of the compartmentalized serine protease Lon, which is the primary protease responsible for protein quality control in *E. coli* (reviewed in references 74, 75). Hence, it is possible that production of Lon compensates for the effects caused by deletion of the proteasome in *M. smegmatis*, but no current experimental evidence has tested this assumption.

ROLE OF THE PROTEASOME IN TRANSCRIPTIONAL REGULATION

A microarray study was conducted to investigate a potential role of the proteasome in transcriptional regulation. Self-compartmentalizing proteases in both prokaryotes and eukaryotes can directly or indirectly regulate the stability of transcription factors (reviewed

Table 1 Summary of phenotypes observed for mutants of the PPS[a]

Genomic background	Mutation	*In vitro* growth	NO resistance	H_2O_2 resistance	*In vivo* phenotype	Reference
M. tuberculosis H37Rv	mpa::ΦMycoMarT7	Comparable to wild type	Hyper-susceptible	Hyper-resistant	Reduced replication in acute phase; bacterial load in lungs declines during chronic phase; mild pathology	7, 36, 37
	pafA::ΦMycoMarT7	Comparable to wild type				
	P_{tetO}-prcBA	Impaired for growth on solid media, small growth defect in liquid media				72
	ΔprcBA::hyg^R			ND		73
	ΔprcBA::hyg^R + PrcBA-T1A	Comparable to wild type	Comparable to wild type	ND	Persistence defect	
M. tuberculosis CDC155	Δmpa::hyg^R	Smaller colony size, increased doubling time (22 h)	ND	ND	Reduced replication in acute phase; persistence defect; immunopathology mutant	70
	dop::ΦMycoMarT7	Comparable to wild type	Hyper-susceptible	ND	Reduced replication in acute phase; bacterial load in lungs declines during chronic phase; mild pathology	49
M. smegmatis mc²155	Δdop::kan^R	Clumping in liquid media	ND	ND	ND	57
	prcBA::kan^R	Comparable to wild type	ND	ND	ND	18

[a]ND, not determined.

in references 74, 76). Comparison of the early stationary phase transcriptomes from wild-type *M. tuberculosis* to *mpa* and *pafA* mutants revealed that a shared set of genes (less than 2% of all predicted ORFs) is differentially regulated in the strains deficient in proteasome-mediated degradation, suggesting that Mpa and PafA do not have independent roles affecting gene expression.

Among the genes upregulated in the *mpa* and *pafA* mutants are members of the Zur (zinc uptake regulator) regulon. Zur acts as a zinc-sensing transcriptional repressor that directly binds to a conserved palindromic regulatory element found in eight *M. tuberculosis* promoters. Under zinc-limiting conditions, Zur is released from its operators and gene expression is induced (77). Among the Zur-regulated genes identified in the microarray was the *esx-3* (esat-6, region 3) operon. The Esx-3 cluster, which is required for optimal growth of *M. tuberculosis in vitro*, encodes a type VII secretion system involved in iron and zinc uptake (78, 79). Under

normal growth conditions, the levels of free zinc in the cytosol are almost negligible, because zinc-coordinating enzymes and ribosomes largely sequester zinc. Several ribosomal genes are controlled by the Zur regulon in *M. tuberculosis*. These genes encode paralogues of zinc-binding proteins, which lack the metal-binding motif (77). In *Bacillus subtilis*, zinc-binding ribosomal proteins are replaced under zinc-limiting conditions, allowing for mobilization of stored zinc, as well as *de novo* synthesis of ribosomes (80, 81). Transcript levels of *zur* itself are unchanged in *mpa*- and *pafA*-deficient mycobacteria, and it is unclear how the absence of proteasomal degradation results in the deregulation of this regulon. Maintaining cellular metal homeostasis is critical to sustain key metabolic processes without exposing cells to the toxic effects exhibited by excess zinc or free iron (reviewed in reference 82). Hence, altered intracellular zinc levels throughout the course of infection could contribute to the attenuation of *mpa* and *pafA* mutants *in vivo*.

Microarray analysis further identified a novel copper-responsive regulon, expression of which is downregulated in the *mpa* and *pafA* mutants. A common repressor, RicR (regulated in copper repressor), binds to a palindromic sequence (5'-TACCC-N$_5$-G/AGGTA) located between the -10/-35 sites in the promoter of several physically unlinked genes. In the presence of copper, RicR dissociates from promoters, resulting in transcriptional derepression of *ricR* itself, *mymT* (a copper metallothionein), *lpqS* (a putative lipoprotein), Rv2963 (a putative permease), and an operon (*socAB* [small ORF induced by copper A and B]) of unknown function. A ΦMycoMarT7 transposon mutant of *ricR* in the CDC1551 background grows slower than wild-type *M. tuberculosis* and never reaches the same culture density under routine culture conditions. Perhaps not surprisingly, a *ricR* mutant is hyper-resistant to copper *in vitro*, most likely because one or more of the gene products of the regulon combat toxicity associated with excess levels of this transition metal. For example, MymT is robustly produced in a *ricR* null mutant (83) and binds up to six reduced copper ions to protect *M. tuberculosis* from copper-mediated toxicity (84).

It is noteworthy that several of the genes underlying regulation by RicR (*lpqS*, *mymT*, and *socAB*) are only found in pathogenic mycobacteria, suggesting that their function is critical to adapt to the specific niche inside the animal host. Interestingly, nonproteasomal ATP-dependent proteases have been implicated in the regulation of virulence gene expression in other bacteria (reviewed in reference 28). This study, as well as others that recently reported the identification of copper-responsive regulators in different pathogens, contribute to an emerging notion of copper as an important antimicrobial defense mechanism exhibited by macrophages (reviewed in reference 85).

The mechanism for deregulation of the RicR and Zur regulons in proteasome-mediated degradation-deficient mutants of *M. tuberculosis* is unknown at this point. However, it is interesting that the transcriptome of these strains appears to respond to low metal conditions, possibly as a result of accumulation of one or more metal-binding proteins in degradation-deficient strains. In response to inflammation, plasma nutrient availability is altered to restrict the growth of pathogenic invaders: as a result of the acute phase response, zinc levels are decreased, while the amount of copper in the plasma increases (reviewed in reference 86). Hence, adaptation to the particular environment inside the host to achieve metal homeostasis is critical during infection.

OUTLOOK AND REMAINING QUESTIONS

The discovery of the PPS in *M. tuberculosis* began a period of exciting findings in mycobacteria. Posttranslational protein modifiers, previously thought limited to eukaryotes, have since been shown to exist in a similar fashion in archaea, where conjugation of two ubiquitin-like proteins termed SAMP1/SAMP2 (small archaeal modifier proteins) presumably targets proteins for proteasomal degradation (87). While much progress can be denoted in elucidating the biochemistry of pupylation, some major questions remain and others have emerged.

For one, the regulation of pupylation is poorly understood. While a myriad of enzymes is devoted to target selection in the eukaryotic ubiquitylation pathway, one enzyme appears to fulfill this role in mycobacteria. Future studies will have to be directed at understanding whether or not additional cellular factors aid substrate recognition, because no structural or primary sequence motif unites all pupylation targets. Elucidating the principles that regulate modification with Pup might also shed light on whether pupylation has nonproteasomal functions. Several pupylated proteins do not accumulate in *M. tuberculosis* strains deficient in proteasome-mediated degradation (65), and a recent report suggests that pupylation can inactivate Mpa function *in vitro* (88). It remains unknown at this time if this finding has physiological relevance and how a degradation-independent regulatory function of pupylation could be distinguished in a cellular context.

The presence of a proteasome provides a fitness advantage to mycobacteria during infection. However, the link between proteasome function and pathogenesis has not yet been fully elucidated. Is the accumulation of an individual substrate toxic to the cell, resulting in the observed NO hypersensitivity? Or is the large overlap between nitrosoproteome and pupylome indicative of the proteasome's role to relieve multifactorial insults in response to stress? Although the mechanism of how the mycobacterial proteasome affects transcription is not understood, it is remarkable that it appears connected to metal homeostasis. Reaction of NO with iron-sulfur clusters has been shown to destabilize the transition metal center. Release of free iron can promote the production of hydroxyl radicals through Fenton chemistry (reviewed in references 3, 4). Additionally, NO can displace copper from *M. tuberculosis* proteins (84). Hence, it is tempting to speculate that a combination of proteolysis-dependent and independent actions of the proteasome influence the bacterial transcriptional response to oxidative and nitrosative stress.

Completing the characterization of the PPS will undoubtedly broaden our understanding of *M. tuberculosis* biology. The importance of this pathway for virulence makes the pupylation machinery an attractive drug target. One of the exciting new endeavors of the young field of bacterial proteasome biology will be the design of chemotherapeutics against one of the world's most successful pathogens.

Pup-proteasome research in the Darwin lab is supported by NIH grants AI088075 and HL92774 and by the Irma T. Hirschl Trust. K.H.D. holds an Investigators in the Pathogenesis of Infectious Disease Award from the Burroughs Wellcome Fund.

Citation. Bode NJ, Darwin KH. 2014. The Pup-proteasome system of mycobacteria. Microbiol Spectrum 2(5):MGM2-0008-2013.

References

1. **MacMicking JD, North RJ, LaCourse R, Mudgett JS, Shah SK, Nathan CF.** 1997. Identification of nitric oxide synthase as a protective locus against tuberculosis. *Proc Natl Acad Sci USA* **94:**5243–5248.

2. **Nathan C, Shiloh MU.** 2000. Reactive oxygen and nitrogen intermediates in the relationship between mammalian hosts and microbial pathogens. *Proc Natl Acad Sci USA* **97:**8841–8848.

3. **Alvarez B, Radi R.** 2003. Peroxynitrite reactivity with amino acids and proteins. *Amino Acids* **25:**295–311.

4. **Fang FC.** 2004. Antimicrobial reactive oxygen and nitrogen species: concepts and controversies. *Nat Rev Microbiol* **2:**820–832.

5. **Szabo C.** 2003. Multiple pathways of peroxynitrite cytotoxicity. *Toxicol Lett* **140–141:**105–112.

6. **Ernst JD.** 2012. The immunological life cycle of tuberculosis. *Nat Rev Immunol* **12:**581–591.

7. **Darwin KH, Ehrt S, Gutierrez-Ramos JC, Weich N, Nathan CF.** 2003. The proteasome of *Mycobacterium tuberculosis* is required for resistance to nitric oxide. *Science* **302:**1963–1966.

8. **Wolf S, Nagy I, Lupas A, Pfeifer G, Cejka Z, Muller SA, Engel A, De Mot R, Baumeister W.** 1998. Characterization of ARC, a divergent member of the AAA ATPase family from *Rhodococcus erythropolis*. *J Mol Biol* **277:**13–25.

9. **Cole ST, Brosch R, Parkhill J, Garnier T, Churcher C, Harris D, Gordon SV, Eiglmeier K, Gas S, Barry CE 3rd, Tekaia F, Badcock K, Basham D, Brown D, Chillingworth T, Connor R, Davies R, Devlin K, Feltwell T, Gentles S, Hamlin N, Holroyd S, Hornsby T, Jagels K, Krogh A, McLean J, Moule S, Murphy L, Oliver K, Osborne J, Quail MA, Rajandream MA, Rogers J, Rutter S, Seeger K, Skelton J, Squares R, Squares S, Sulston JE, Taylor K, Whitehead S, Barrell BG.** 1998. Deciphering the biology of *Mycobacterium tuberculosis* from the complete genome sequence. *Nature* **393:**537–544.

10. **Murata S, Yashiroda H, Tanaka K.** 2009. Molecular mechanisms of proteasome assembly. *Nat Rev Mol Cell Biol* **10:**104–115.

11. **Finley D.** 2009. Recognition and processing of ubiquitin-protein conjugates by the proteasome. *Annu Rev Biochem* **78:**477–513.

12. **Rubin DM, Finley D.** 1995. Proteolysis. The proteasome: a protein-degrading organelle? *Curr Biol* **5:**854–858.

13. **Kisselev AF, Akopian TN, Woo KM, Goldberg AL.** 1999. The sizes of peptides generated from protein by mammalian 26 and 20 S proteasomes. Implications for understanding the degradative mechanism and antigen presentation. *J Biol Chem* **274:**3363–3371.

14. **Schrader EK, Harstad KG, Matouschek A.** 2009. Targeting proteins for degradation. *Nat Chem Biol* **5:**815–822.

15. **Dahlmann B, Kopp F, Kuehn L, Niedel B, Pfeifer G, Hegerl R, Baumeister W.** 1989. The multicatalytic proteinase (prosome) is ubiquitous from eukaryotes to archaebacteria. *FEBS Lett* **251:**125–131.

16. **Lowe J, Stock D, Jap B, Zwickl P, Baumeister W, Huber R.** 1995. Crystal structure of the 20S proteasome from the archaeon *T. acidophilum* at 3.4 A resolution. *Science* **268:**533–539.

17. **Tamura T, Nagy I, Lupas A, Lottspeich F, Cejka Z, Schoofs G, Tanaka K, De Mot R, Baumeister W.** 1995. The first characterization of a eubacterial proteasome: the 20S complex of *Rhodococcus*. *Curr Biol* **5:**766–774.

18. **Knipfer N, Shrader TE.** 1997. Inactivation of the 20S proteasome in *Mycobacterium smegmatis*. *Mol Microbiol* **25:**375–383.

19. **Nagy I, Tamura T, Vanderleyden J, Baumeister W, De Mot R.** 1998. The 20S proteasome of *Streptomyces coelicolor*. *J Bacteriol* **180:**5448–5453.

20. **Pouch MN, Cournoyer B, Baumeister W.** 2000. Characterization of the 20S proteasome from the actinomycete *Frankia*. *Mol Microbiol* **35:**368–377.

21. **Lupas A, Zuhl F, Tamura T, Wolf S, Nagy N, De Mot R, Baumeister W.** 1997. Eubacterial proteasomes. *Mol Biol Rep* **24:**125–131.

22. **Ram RJ, Verberkmoes NC, Thelen MP, Tyson GW, Baker BJ, Blake RC 2nd, Shah M, Hettich RL, Banfield JF.** 2005. Community proteomics of a natural microbial biofilm. *Science* **308:**1915–1920.

23. **Gille C, Goede A, Schloetelburg C, Preissner R, Kloetzel PM, Gobel UB, Frommel C.** 2003. A comprehensive view on proteasomal sequences: implications for the evolution of the proteasome. *J Mol Biol* **326:**1437–1448.

24. **De Mot R.** 2007. Actinomycete-like proteasomes in a Gram-negative bacterium. *Trends Microbiol* **15:**335–338.

25. **Wang T, Li H, Lin G, Tang C, Li D, Nathan C, Darwin KH, Li H.** 2009. Structural insights on the *Mycobacterium tuberculosis* proteasomal ATPase Mpa. *Structure* **17:**1377–1385.

26. **Hu G, Lin G, Wang M, Dick L, Xu RM, Nathan C, Li H.** 2006. Structure of the *Mycobacterium tuberculosis* proteasome and mechanism of inhibition by a peptidyl boronate. *Mol Microbiol* **59:**1417–1428.

27. Lin G, Hu G, Tsu C, Kunes YZ, Li H, Dick L, Parsons T, Li P, Chen Z, Zwickl P, Weich N, Nathan C. 2006. *Mycobacterium tuberculosis* prcBA genes encode a gated proteasome with broad oligopeptide specificity. *Mol Microbiol* 59:1405–1416.

28. Butler SM, Festa RA, Pearce MJ, Darwin KH. 2006. Self-compartmentalized bacterial proteases and pathogenesis. *Mol Microbiol* 60:553–562.

29. Husnjak K, Dikic I. 2012. Ubiquitin-binding proteins: decoders of ubiquitin-mediated cellular functions. *Annu Rev Biochem* 81:291–322.

30. Kerscher O, Felberbaum R, Hochstrasser M. 2006. Modification of proteins by ubiquitin and ubiquitin-like proteins. *Annu Rev Cell Dev Biol* 22:159–180.

31. Li W, Bengtson MH, Ulbrich A, Matsuda A, Reddy VA, Orth A, Chanda SK, Batalov S, Joazeiro CA. 2008. Genome-wide and functional annotation of human E3 ubiquitin ligases identifies MULAN, a mitochondrial E3 that regulates the organelle's dynamics and signaling. *PLoS One* 3:e1487.

32. Deshaies RJ, Joazeiro CA. 2009. RING domain E3 ubiquitin ligases. *Annu Rev Biochem* 78:399–434.

33. Rotin D, Kumar S. 2009. Physiological functions of the HECT family of ubiquitin ligases. *Nat Rev Mol Cell Biol* 10:398–409.

34. Thrower JS, Hoffman L, Rechsteiner M, Pickart CM. 2000. Recognition of the polyubiquitin proteolytic signal. *EMBO J* 19:94–102.

35. Shabek N, Herman-Bachinsky Y, Buchsbaum S, Lewinson O, Haj-Yahya M, Hejjaoui M, Lashuel HA, Sommer T, Brik A, Ciechanover A. 2012. The size of the proteasomal substrate determines whether its degradation will be mediated by mono- or polyubiquitylation. *Mol Cell* 48:87–97.

36. Pearce MJ, Arora P, Festa RA, Butler-Wu SM, Gokhale RS, Darwin KH. 2006. Identification of substrates of the *Mycobacterium tuberculosis* proteasome. *EMBO J* 25:5423–5432.

37. Darwin KH, Lin G, Chen Z, Li H, Nathan CF. 2005. Characterization of a *Mycobacterium tuberculosis* proteasomal ATPase homologue. *Mol Microbiol* 55:561–571.

38. Karimova G, Pidoux J, Ullmann A, Ladant D. 1998. A bacterial two-hybrid system based on a reconstituted signal transduction pathway. *Proc Natl Acad Sci USA* 95:5752–5756.

39. Singh A, Mai D, Kumar A, Steyn AJ. 2006. Dissecting virulence pathways of *Mycobacterium tuberculosis* through protein-protein association. *Proc Natl Acad Sci USA* 103:11346–11351.

40. Pearce MJ, Mintseris J, Ferreyra J, Gygi SP, Darwin KH. 2008. Ubiquitin-like protein involved in the proteasome pathway of *Mycobacterium tuberculosis*. *Science* 322:1104–1107.

41. Burns KE, Liu WT, Boshoff HI, Dorrestein PC, Barry CE 3rd. 2009. Proteasomal protein degradation in mycobacteria is dependent upon a prokaryotic ubiquitin-like protein. *J Biol Chem* 284:3069–3075.

42. Chen X, Solomon WC, Kang Y, Cerda-Maira F, Darwin KH, Walters KJ. 2009. Prokaryotic ubiquitin-like protein pup is intrinsically disordered. *J Mol Biol* 392:208–217.

43. Liao S, Shang Q, Zhang X, Zhang J, Xu C, Tu X. 2009. Pup, a prokaryotic ubiquitin-like protein, is an intrinsically disordered protein. *Biochem J* 422:207–215.

44. Sutter M, Striebel F, Damberger FF, Allain FH, Weber-Ban E. 2009. A distinct structural region of the prokaryotic ubiquitin-like protein (Pup) is recognized by the N-terminal domain of the proteasomal ATPase Mpa. *FEBS Lett* 583:3151–3157.

45. Hochstrasser M. 2009. Origin and function of ubiquitin-like proteins. *Nature* 458:422–429.

46. Festa RA, Pearce MJ, Darwin KH. 2007. Characterization of the proteasome accessory factor (paf) operon in *Mycobacterium tuberculosis*. *J Bacteriol* 189:3044–3050.

47. Iyer LM, Burroughs AM, Aravind L. 2008. Unraveling the biochemistry and provenance of pupylation: a prokaryotic analog of ubiquitination. *Biol Direct* 3:45.

48. Striebel F, Imkamp F, Sutter M, Steiner M, Mamedov S, Weber-Ban E. 2009. Bacterial ubiquitin-like modifier Pup is deamidated and conjugated to substrates by distinct but homologous enzymes. *Nat Struct Mol Biol* 16:647–651.

49. Cerda-Maira FA, Pearce MJ, Fuortes M, Bishai WR, Hubbard SR, Darwin KH. 2010. Molecular analysis of the prokaryotic ubiquitin-like protein (Pup) conjugation pathway in *Mycobacterium tuberculosis*. *Mol Microbiol* 77:1123–1135.

50. Guth E, Thommen M, Weber-Ban E. 2011. Mycobacterial ubiquitin-like protein ligase PafA follows a two-step reaction pathway with a phosphorylated pup intermediate. *J Biol Chem* 286:4412–4419.

51. Sutter M, Damberger FF, Imkamp F, Allain FH, Weber-Ban E. 2010. Prokaryotic ubiquitin-like protein (Pup) is coupled to substrates via the side chain of its C-terminal glutamate. *J Am Chem Soc* 132:5610–5612.

52. Burns KE, Darwin KH. 2010. Pupylation: a signal for proteasomal degradation in *Mycobacterium tuberculosis*. *Subcell Biochem* 54:149–157.

53. Komander D, Clague MJ, Urbe S. 2009. Breaking the chains: structure and function of the deubiquitinases. *Nat Rev Mol Cell Biol* 10:550–563.

54. Burns KE, Cerda-Maira FA, Wang T, Li H, Bishai WR, Darwin KH. 2010. "Depupylation" of prokaryotic ubiquitin-like protein from mycobacterial proteasome substrates. *Mol Cell* 39:821–827.

55. Imkamp F, Striebel F, Sutter M, Ozcelik D, Zimmermann N, Sander P, Weber-Ban E. 2010. Dop functions as a depupylase in the prokaryotic ubiquitin-like modification pathway. *EMBO Rep* 11:791–797.

56. Komander D, Reyes-Turcu F, Licchesi JD, Odenwaelder P, Wilkinson KD, Barford D. 2009. Molecular discrimination of structurally equivalent Lys 63-linked and linear polyubiquitin chains. *EMBO Rep* 10:466–473.

57. Imkamp F, Rosenberger T, Striebel F, Keller PM, Amstutz B, Sander P, Weber-Ban E. 2010. Deletion of dop in *Mycobacterium smegmatis* abolishes pupylation of protein substrates in vivo. *Mol Microbiol* 75:744–754.

58. Burns KE, Pearce MJ, Darwin KH. 2010. Prokaryotic ubiquitin-like protein provides a two-part degron to *Mycobacterium* proteasome substrates. *J Bacteriol* 192: 2933–2935.

59. Ozcelik D, Barandun J, Schmitz N, Sutter M, Guth E, Damberger FF, Allain FH, Ban N, Weber-Ban E. 2012. Structures of Pup ligase PafA and depupylase Dop from the prokaryotic ubiquitin-like modification pathway. *Nat Commun* 3:1014.

60. Cerda-Maira FA, McAllister F, Bode NJ, Burns KE, Gygi SP, Darwin KH. 2011. Reconstitution of the *Mycobacterium tuberculosis* pupylation pathway in *Escherichia coli*. *EMBO Rep* 12:863–870.

61. Burns KE, McAllister FE, Schwerdtfeger C, Mintseris J, Cerda-Maira F, Noens EE, Wilmanns M, Hubbard SR, Melandri F, Ovaa H, Gygi SP, Darwin KH. 2012. *Mycobacterium tuberculosis* prokaryotic ubiquitin-like protein-deconjugating enzyme is an unusual aspartate amidase. *J Biol Chem* 287:37522–37529.

62. Barandun J, Delley CL, Weber-Ban E. 2012. The pupylation pathway and its role in mycobacteria. *BMC Biol* 10:95.

63. Lander GC, Estrin E, Matyskiela ME, Bashore C, Nogales E, Martin A. 2012. Complete subunit architecture of the proteasome regulatory particle. *Nature* 482: 186–191.

64. Wang T, Darwin KH, Li H. 2010. Binding-induced folding of prokaryotic ubiquitin-like protein on the *Mycobacterium* proteasomal ATPase targets substrates for degradation. *Nat Struct Mol Biol* 17:1352–1357.

65. Festa RA, McAllister F, Pearce MJ, Mintseris J, Burns KE, Gygi SP, Darwin KH. 2010. Prokaryotic ubiquitin-like protein (Pup) proteome of *Mycobacterium tuberculosis* [corrected]. *PLoS One* 5:e8589.

66. Watrous J, Burns K, Liu WT, Patel A, Hook V, Bafna V, Barry CE 3rd, Bark S, Dorrestein PC. 2010. Expansion of the mycobacterial "PUPylome." *Mol Biosyst* 6:376–385.

67. Poulsen C, Akhter Y, Jeon AH, Schmitt-Ulms G, Meyer HE, Stefanski A, Stuhler K, Wilmanns M, Song YH. 2010. Proteome-wide identification of mycobacterial pupylation targets. *Mol Syst Biol* 6:386.

68. Gareau JR, Lima CD. 2010. The SUMO pathway: emerging mechanisms that shape specificity, conjugation and recognition. *Nat Rev Mol Cell Biol* 11:861–871.

69. MacMicking J, Xie QW, Nathan C. 1997. Nitric oxide and macrophage function. *Annu Rev Immunol* 15:323–350.

70. Lamichhane G, Raghunand TR, Morrison NE, Woolwine SC, Tyagi S, Kandavelou K, Bishai WR. 2006. Deletion of a *Mycobacterium tuberculosis* proteasomal ATPase homologue gene produces a slow-growing strain that persists in host tissues. *J Infect Dis* 194:1233–1240.

71. Sassetti CM, Boyd DH, Rubin EJ. 2003. Genes required for mycobacterial growth defined by high density muta-genesis. *Mol Microbiol* 48:77–84.

72. Gandotra S, Schnappinger D, Monteleone M, Hillen W, Ehrt S. 2007. In vivo gene silencing identifies the *Mycobacterium tuberculosis* proteasome as essential for the bacteria to persist in mice. *Nat Med* 13:1515–1520.

73. Gandotra S, Lebron MB, Ehrt S. 2010. The *Mycobacterium tuberculosis* proteasome active site threonine is essential for persistence yet dispensable for replication and resistance to nitric oxide. *PLoS Pathog* 6:e1001040.

74. Gottesman S. 2003. Proteolysis in bacterial regulatory circuits. *Annu Rev Cell Dev Biol* 19:565–587.

75. Gur E, Biran D, Ron EZ. 2011. Regulated proteolysis in Gram-negative bacteria: how and when? *Nat Rev Microbiol* 9:839–848.

76. Collins GA, Tansey WP. 2006. The proteasome: a utility tool for transcription? *Curr Opin Genet Dev* 16:197–202.

77. Maciag A, Dainese E, Rodriguez GM, Milano A, Provvedi R, Pasca MR, Smith I, Palu G, Riccardi G, Manganelli R. 2007. Global analysis of the *Mycobacterium tuberculosis* Zur (FurB) regulon. *J Bacteriol* 189: 730–740.

78. Serafini A, Boldrin F, Palu G, Manganelli R. 2009. Characterization of a *Mycobacterium tuberculosis* ESX-3 conditional mutant: essentiality and rescue by iron and zinc. *J Bacteriol* 191:6340–6344.

79. Siegrist MS, Unnikrishnan M, McConnell MJ, Borowsky M, Cheng TY, Siddiqi N, Fortune SM, Moody DB, Rubin EJ. 2009. Mycobacterial Esx-3 is required for mycobactin-mediated iron acquisition. *Proc Natl Acad Sci USA* 106:18792–18797.

80. Nanamiya H, Akanuma G, Natori Y, Murayama R, Kosono S, Kudo T, Kobayashi K, Ogasawara N, Park SM, Ochi K, Kawamura F. 2004. Zinc is a key factor in controlling alternation of two types of L31 protein in the *Bacillus subtilis* ribosome. *Mol Microbiol* 52:273–283.

81. Natori Y, Nanamiya H, Akanuma G, Kosono S, Kudo T, Ochi K, Kawamura F. 2007. A fail-safe system for the ribosome under zinc-limiting conditions in *Bacillus subtilis*. *Mol Microbiol* 63:294–307.

82. Pruteanu M, Baker TA. 2009. Proteolysis in the SOS response and metal homeostasis in *Escherichia coli*. *Res Microbiol* 160:677–683.

83. Festa RA, Jones MB, Butler-Wu S, Sinsimer D, Gerads R, Bishai WR, Peterson SN, Darwin KH. 2011. A novel copper-responsive regulon in *Mycobacterium tuberculosis*. *Mol Microbiol* 79:133–148.

84. Gold B, Deng H, Bryk R, Vargas D, Eliezer D, Roberts J, Jiang X, Nathan C. 2008. Identification of a copper-binding metallothionein in pathogenic mycobacteria. *Nat Chem Biol* 4:609–616.

85. Samanovic MI, Ding C, Thiele DJ, Darwin KH. 2012. Copper in microbial pathogenesis: meddling with the metal. *Cell Host Microbe* 11:106–115.

86. Prentice AM, Ghattas H, Cox SE. 2007. Host-pathogen interactions: can micronutrients tip the balance? *J Nutr* 137:1334–1337.

87. Humbard MA, Miranda HV, Lim JM, Krause DJ, Pritz JR, Zhou G, Chen S, Wells L, Maupin-Furlow JA. 2010. Ubiquitin-like small archaeal modifier proteins (SAMPs) in *Haloferax volcanii*. *Nature* 463:54–60.

88. Delley CL, Striebel F, Heydenreich FM, Ozcelik D, Weber-Ban E. 2012. Activity of the mycobacterial proteasomal ATPase Mpa is reversibly regulated by pupylation. *J Biol Chem* 287:7907–7914.

Molecular Genetics of Mycobacteria, 2nd Edition
Edited by Graham F. Hatfull and William R. Jacobs, Jr.
© 2014 American Society for Microbiology, Washington, DC
doi:10.1128/microbiolspec.MGM2-0006-2013

Sladjana Prisic[1]
Robert N. Husson[1]

Mycobacterium tuberculosis Serine/Threonine Protein Kinases

33

Signal transduction is an essential activity of all living cells. Broadly defined, signal transduction is the sensing of a signal or input and its conversion into an output or response that alters cell physiology. The sensor is the molecule or domain of a molecule (typically a protein) that senses the signal. The transducer is the molecule or domain that converts the signal into a response. Most commonly, signal transduction refers to the sensing of an extracellular signal that is transduced across the cytoplasmic membrane and converted into an intracellular response. Thus, signal transduction is critical for cellular adaptation to changes in the extracellular environment. In the case of bacterial pathogens, including *Mycobacterium tuberculosis*, these adaptive responses allow growth and/or survival in the environments encountered by the pathogen during the course of infection in the human host.

The most widely distributed and intensively studied transmembrane signaling systems in bacteria are the two-component systems (1). In these systems the sensor and transducer (referred to as the response regulator) are separate proteins, in which the sensor protein spans the cytoplasmic membrane and the response regulator is a cytoplasmic protein, usually a transcription factor that is activated in response to this phosphorylation event. Two-component systems are discussed in depth in reference 138. So-called one-component systems are cytoplasmic proteins that contain both a sensor domain and an output domain (2). The sensor domain typically senses intracellular signals via binding of small molecules, leading to effects on transcription by the output domain.

Another group of transcription regulators, the extra-cytoplasmic function (ECF), or group IV, sigma factors, has been referred to as the "third pillar" of bacterial signal transduction (3). The ECF sigma factors were originally described as sensing and/or regulating ECFs (4). Members of the ECF subfamily are often negatively regulated by direct interaction with an anti-sigma factor protein that serves as the sensor. Anti-sigma factors may be transmembrane proteins, e.g., *M. tuberculosis* RslA (the anti-sigma factor of SigL), or cytoplasmic proteins, e.g., *M. tuberculosis* RshA (anti-sigma factor of SigH), and thus may transduce either extracytoplasmic or intracellular signals (5, 6). Sigma factors and their regulatory mechanisms are discussed in depth in reference 139.

---

[1]Division of Infectious Diseases, Boston Children's Hospital, Harvard Medical School, 300 Longwood Ave., Boston, MA 02115.

The other major mechanism of transmembrane signaling in *M. tuberculosis* is via the serine/threonine protein kinases (STPKs), the focus of this review. Unlike two-component systems, which are a major signaling mechanism in nearly all phyla of bacteria, STPKs are less widely distributed among different groups of bacteria. STPKs are most abundant among *Acidobacteria*, *Actinobacteria* (which includes mycobacteria), some cyanobacteria, and one order of the *Deltaproteobacteria* (the *Myxococcales*, the first bacteria in which STPKs were identified) (7, 8). In contrast to many of the widely studied bacterial pathogens and model organisms that have few or no STPKs but many two-component systems, the *M. tuberculosis* genome encodes 11 STPKs and a similar number of two-component systems, indicating that these two mechanisms both play important roles in signal transduction in this organism.

Of the 11 *M. tuberculosis* STPKs, all but 2 have a single transmembrane domain with an extracellular sensor domain and an intracellular kinase domain (KD)

(Fig. 1). These nine transmembrane proteins can thus be classified as receptor-type kinases, in which the extracellular sensor domain senses extracytoplasmic signals and transduces this information to the intracellular KD, leading to activation of the kinase and phosphorylation of Ser or Thr residues on substrate proteins. This phosphorylation may alter protein function directly or by affecting interactions between specific pairs of proteins or within multiprotein complexes. In contrast to two-component, one-component, and ECF sigma factor signal transduction, where the usual primary output is changes in transcription, the output of Ser/Thr phosphorylation is rarely direct regulation of transcription.

As will be discussed below, for several of the *M. tuberculosis* STPKs, at least some of the signals sensed and some of the proteins targeted are known. There remains a great deal to be learned, however, about the exact mechanisms and functions of the STPKs in regulating *M. tuberculosis* physiology. In this article we

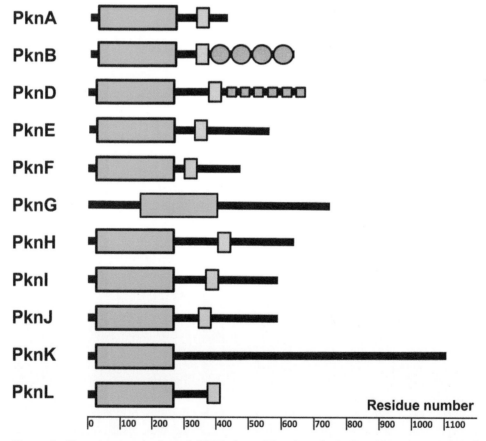

Figure 1 Domain organization of STPKs from *M. tuberculosis*. Domains were predicted using the SMART algorithm (136, 137). Kinase domains are shown as green boxes, transmembrane portions are in blue, and some known extracellular domains are in light red. doi:10.1128/microbiolspec.MGM2-0006-2013.f1

highlight some of the major findings and current state of knowledge regarding the role of Ser/Thr and Tyr phosphorylation-mediated signal transduction in *M. tuberculosis*. The rapid expansion of the literature in this field, however, makes it impossible to note every observation regarding protein phosphorylation and its functional effects in mycobacteria.

SEQUENCE CHARACTERISTICS AND COMPARATIVE GENOMICS OF MYCOBACTERIAL STPKs

The *M. tuberculosis* STPKs were first described as "eukaryotic-like" protein kinases based on their sequence similarity to eukaryotic STPKs (9). The protein sequence similarity among the KDs of eukaryotic kinases led Hanks and Hunter in 1995 to identify a "superfamily" of protein kinases containing 11 subdomains (10). These subdomains contain conserved residues and motifs present in members of the superfamily, with specific functions attributable to each subdomain. With the massive expansion in the number of eukaryotic protein kinase sequences in the genomic era, this subdomain organization has remained valid, and sequence alignments have indicated the presence of many subfamilies of functionally and/or structurally related kinases. Subdomains 1 through 4 and part of 5 comprise the N-terminal lobe of the KD (see below), responsible for ATP binding and alignment, while subdomains 5 through 11 are responsible for substrate binding and phosphate transfer. Comparison of the *M. tuberculosis* STPKs to eukaryotic protein kinases demonstrates that the *M. tuberculosis* proteins incorporate each of the 11 Hanks subdomains, despite relatively limited sequence identity (Fig. 2).

Sequence alignment of the *M. tuberculosis* STPK KDs shows that the 11 STPKs can be grouped into three clusters of three kinases and two KDs that are less similar to any of the nine clustered domains (Fig. 3). The unclustered outliers, PknG and PknK, are also the two kinases that lack a transmembrane domain. This observation suggests that the genes encoding the nine receptor-type kinases may be derived from a single common ancestral gene via gene duplication, whereas PknG and PknK may have been acquired separately. In contrast to the intracellular KDs, the extracellular domains of the nine transmembrane STPKs show no sequence similarity, indicating that they likely bind to and respond to distinct extracytoplasmic molecular signals. As discussed below for each kinase, motifs are present within the protein sequence of some of the extracellular domains and in the non-KD regions for

PknG and PknK. For a few of the extracellular domains, candidate ligands have been identified.

As noted above, STPKs are not evenly distributed among different bacterial phyla. Within the mycobacteria and closely related actinomycetes, their distribution is also uneven. Table 1, which compares the genomes of the pathogens *M. tuberculosis* and *Mycobacterium leprae*, the opportunistic pathogens *Mycobacterium avium* and *M. avium* subspecies *paratuberculosis*, the nonpathogen *Mycobacterium smegmatis*, and the more distantly related nonpathogenic actinomycete *Corynebacterium glutamicum*, shows these differences in the distribution of the STPKs. The presence of PknA, PknB, PknG, and PknL in all species suggests that these kinases play important roles in regulating key aspects of mycobacterial physiology, though only PknA and PknB are essential in *M. tuberculosis* (11). The other STPKs likely have more specialized regulatory roles corresponding to the niches occupied by these different species.

STRUCTURAL ANALYSIS OF STPK KINASE DOMAINS

Structures of three KDs from *M. tuberculosis* STPKs (PknB, PknE, and PknG) have been solved (Table 2). These structures have provided important insights into mechanisms by which the *M. tuberculosis* STPKs are activated and regulated. The PknB KD was the first bacterial kinase structure described and was published by two groups independently in 2003 (12, 13). Though they were cocrystallized with different ATP analogs, the two PknB KD structures are virtually identical. As in other members of the eukaryotic protein kinase superfamily, the PknB KD (as shown in the 1MRU crystal structure) has amino (N)- and carboxy (C)-terminal lobes with a nucleotide/Mg^{2+} binding site in the cleft between them (Fig. 4A) (13). The N-terminal lobe is mostly composed of β-sheets with a single long α-helix, designated helix C. In contrast, the C-terminal lobe contains only α-helices. Even with very low sequence conservation, this structure closely matches the overall structure of eukaryotic KDs. For example, human Clk1 KD has less than 20% sequence identity with PknB KD but has extensive overlap in structure (Fig. 4B). The KDs from PknE and the more divergent PknG also share this highly conserved kinase fold (Fig. 4C).

The PknB KD structures have all of the common elements found in eukaryotic protein kinases, as was predicted from the PknB amino acid sequence (Fig. 2 and 4). Both PknB KD structures (1MRU and 1O6Y) appear to represent the active or "closed" conformation

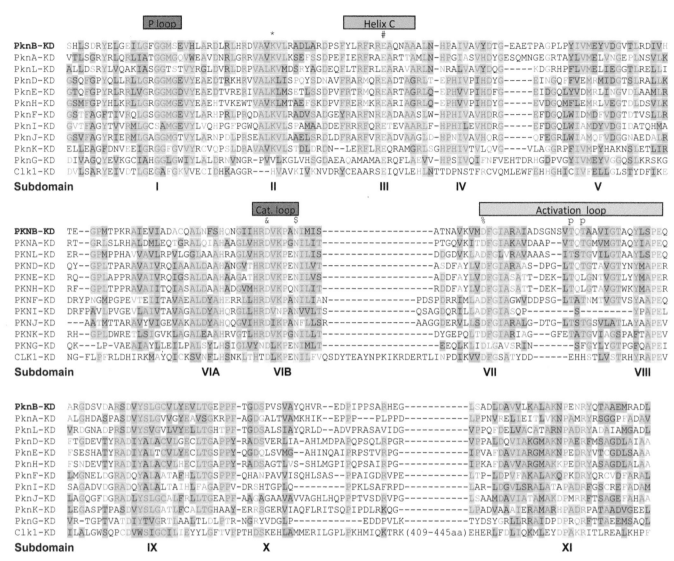

Figure 2 Sequence alignment of STPKs from *M. tuberculosis*. Kinase domains predicted by the SMART algorithm (see Fig. 1) were aligned and grouped using AlignX software (Life Technologies). Human Clk1 kinase is also included for comparison. Major features are noted. Selected conserved residues are labeled with the following symbols (residue numbers from PknB): *, Lys40; #, Glu59; &, Asp138; $, Asn143; %, Asp156; p, major phosphorylation sites in the activation loop. doi:10.1128/microbiolspec.MGM2-0006-2013.f2

of the kinase (12, 13). In most protein kinases, the conserved C helix is farther from the nucleotide binding cleft in the "open" position, while it is shifted toward this site in the "closed" position. The close proximity of the C loop to the nucleotide binding cleft of PknB in the closed position is shown in Fig. 5. The apo-PknE KD, was crystallized in the open or inactive form (Fig. 5A) (14). Positioning of the C helix away from the active site in the open conformation causes unfavorable placement of the conserved Glu64 (Glu59 in PknB) (Fig. 5A) (14). In contrast, in the PknB KD active conformation

Glu59 is close enough to the invariant Lys-40 of the nucleotide binding domain to allow proper positioning toward the α and β phosphates of ATP (Fig. 5B).

Additional conserved features in the PknB structure are the Gly-rich P-loop present in the N-terminal lobe, which interacts with ATP (and ATP analogs) and the highly conserved Asp-Phe-Gly (DFG) residues at the amino-terminal boundary of the activation loop. In particular, Asp156 of the DFG triplet in PknB contributes to positioning of Mg^{2+} (Fig. 5B). In addition, two important residues in the catalytic loop, Asp138 (catalytic

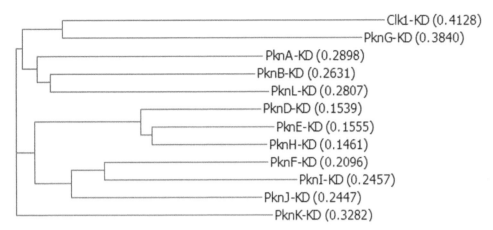

Figure 3 Dendrogram of KDs of *M. tuberculosis* STPKs. KDs identified by the SMART algorithm were aligned and grouped using the AlignX software (Life Technologies). Human Clk1 kinase is also included for comparison. Distance scores as given by AlignX are shown in parentheses. doi:10.1128/microbiolspec.MGM2-0006-2013.f3

base) and Asn143, are properly paired in this active conformation structure for attack on the substrate (Thr or Ser) hydroxyl group and transfer of γ phosphate from ATP (Fig. 5B). The hydrophobic pocket that binds adenine is also structurally similar to those observed in eukaryotic kinases (12, 13).

In addition to these conserved features, the PknB structure has interesting distinct characteristics. The activation loop sequence between the conserved DFG and APE (SPE in PknB) motifs (Fig. 2) is, surprisingly, disordered in both PknB structures (1MRU and 1O6Y), although it is usually visible in structures of protein kinases in the active form. The activation loop is also missing in the crystal structure of the inactive conformation of apo-PknE KD, indicating that it is also disordered, as is typical in open KD structures (14). There are no reported crystal structures of any *M. tuberculosis* STPKs in complex with a substrate, but a working model for substrate binding by PknB involves a large portion of the activation loop that likely provides kinase specificity (15). This loop was shown to be phosphorylated on at least four residues in PknB, two of which are important for kinase activation (T171 and T173), while two others may make minor contributions (Ser166 and Ser169) (13, 16). Activation loops of many eukaryotic kinases are phosphorylated *in vivo*, and most *M. tuberculosis* kinases have been shown to be autophosphorylated or trans-phosphorylated on activation loop residues *in vitro* (15, 17–20).

Although phosphorylation of the activation loop is required for PknB to be fully activated, "back-to-back" interaction of the unphosphorylated kinase is thought to be the first step in activation (21–24) (Fig. 6A). Although inferred previously from the PknB KD structural data

(1MRU and 2FUM), this mechanism of *M. tuberculosis* kinase activation was first demonstrated in PknD KD fusion constructs that could be brought into proximity by rapamycin binding tags (23). Although PknD KD structures were not obtained, this study clearly showed that bringing KDs together stimulated auto- and trans-phosphorylation and that this activation depended on residues in a predicted dimer interface homologous to the interface observed in PknB KD crystal structures (13, 23, 25). Subsequent studies of PknB KD back-to-back dimerization provided additional insights into this activation mechanism (21, 22). Conserved residues at the interface of the back of the two N-terminal lobes adjacent to the C terminus of the C helix, such as the conserved Leu33 and the Arg10/Asp76 salt bridge, were demonstrated to be essential for this allosteric activation. When those contact sites were mutated, monomeric structures were obtained in multiple conformational states showing greater flexibility in the N-terminal lobe and misplacement of the C helix and its essential Glu59, resulting in lower enzymatic activity (Fig. 6B,C) (21, 22). Though not typical of many eukaryotic kinases, examples of back-to-back dimerization-induced activation are well described, e.g., for the human kinase PKR (26). The PknE KD was also shown to form a dimer with a similar, although not identical, interface (14).

It is important to note that the dimers observed in the PknB crystal structures do not appear to be formed with high affinity in solution (22). Also, the PknB and PknE extracellular domains do not dimerize on their own, so ligand binding is likely required, directly or indirectly, to facilitate dimerization of the KDs (22, 27, 28).

Table 1 Closest orthologs of *M. tuberculosis* STPKs[a]

Species	PknA	PknB	PknD	PknE	PknF	PknG	PknH	PknI	PknJ	PknK	PknL
Mtb	Rv0015c	Rv0014c	Rv0931c	Rv1743	Rv1746	Rv0410c	Rv1266c	Rv2914c	Rv2088	Rv3080c	Rv2176
Mmar	MMAR_0017	MMAR_0016	MMAR_4577	MMAR_2581	MMAR_2606 MMAR_4174	MMAR_0713	MMAR_1982 MMAR_2444 MMAR_4156 MMAR_4171	MMAR_1794	MMAR_1423 MMAR_2408 MMAR_2941	MMAR_2576	MMAR_3211
Mav	MAV_0019	MAV_0017	MAV_4238		MAV_3145	MAV_4751	MAV_1417 MAV_2158				MAV_2318
Map	MAP0018c	MAP0016c	MAP3387c		MAP1332	MAP3893c	MAP2026 MAP2031c MAP2504				MAP1914
Msmeg	MSMEG_0030	MSMEG_0028			MSMEG_0886 MSMEG_3677	MSMEG_0786	MSMEG_4366		MSMEG_5513	MSMEG_0529	MSMEG_4243
Mle	ML0017	ML0016				ML0304					ML0897
Cgl	cg0059	cg0057				cg3046					cg2388

[a]Selected bacteria are shown: Mtb, *M. tuberculosis* H37Rv; Mmar, *M. marinum*; Mav, *M. avium* 104; Map, *M. avium paratuberculosis* k10; Msmeg, *M. smegmatis* MC^2-155; Mle, *M. leprae* TN; Cgl, *C. glutamicum* ATCC 13032.

To examine subsequent steps of PknB activation, structural analysis was performed on a form of the PknB KD with substitutions at two Met residues that allow the KD to accommodate the inhibitor Kt5720, plus a single mutation that prevents back-to-back dimerization. This structure revealed an asymmetric "front-to-front" dimer (3f69) (Fig. 7A) (21). The interface between the two PknB KD monomers in this structure was mostly comprised of conserved G helix contacts. The activation loop of one KD monomer was ordered and in contact with the second monomer, while this second monomer had only a partially ordered activation loop with a small portion entering the active site of the first monomer, demonstrating a mechanism for activation by transphosphorylation (Fig. 7A). Mutational analysis of G helix residues that provide the contact surface in the front-to-front dimer showed their requirement for activation loop phosphorylation and demonstrated that they act synergistically with back-to-back contacts to allow full activation (21). Importantly, once activated by the allosteric conformational change to allow autophosphorylation, the primed KD with a fully phosphorylated activation loop can then remain active as a monomer, allowing continuing phosphorylation of substrate proteins even after the initial signal causing dimerization is gone (21–23).

In addition to the KD, the intracellular portion of PknB includes a short juxtamembrane linker that connects the KD to the membrane-spanning segment (Fig. 1). Crystallization of the complete intracellular region of PknB including this region has been unsuccessful, indicating that the linker is disordered. The linker has been found to be phosphorylated in several *M. tuberculosis* STPKs, however, suggesting that it may have a regulatory role (12, 13, 15, 16, 18, 20).

The cytoplasmic protein kinase PknG is the only *M. tuberculosis* kinase that has been successfully crystallized as a nearly full-length protein (69 residues were removed from the N-terminus to allow crystallization) (Fig. 7B) (29). This construct includes both kinase and sensor domains: amino-terminal to the PknG KD is an iron binding rubredoxin domain, and carboxy-terminal to the KD is a tetratricopeptide repeat (TPR)–containing domain. TPRs are short repeats that are usually present in multiple copies and are involved in protein-protein interactions (Fig. 7B). The PknG dimer is formed through extensive contacts between the TPR domains of each monomer and not through N-lobe interactions, as was seen in the PknB KD and PknE KD dimer structures. Another difference compared to these other two kinases is that the PknG activation loop is ordered and stabilized, although not phosphorylated,

Table 2 Structures of *M. tuberculosis* STPKs available in the Research Collaboratory for Structural Bioinformatics (RCSB) Protein Database (PDB)

Kinase	PDB ID	Notes	Reference
PknB	1MRU	KD + short linker (1-307) in complex with ATPγS, "back-to-back" dimer	13
	1O6Y	KD (1-279) in complex with AMP-PCP, monomer	12
	2FUM	KD (1-279) in complex with mitoxantrone, "back-to-back" dimer	25
	3F69	L33D/M145L/M155V mutant (PknD surrogate) with Kt5720 "front-to-front" dimer	21
	3F61	L33D / V222D double mutant	
	2KUI	Extracellular sensor (PASTA)	28
	2KUE	Extracellular sensor (PASTA)-NMR structures	
	2KUF 2KUD		
	3ORI	L33D mutant, various conformations	22
	3ORK 3ORL		
	3ORO	L33D mutant, no metal ion bound	
	3ORP 3ORT		
	3ORM	D76A mutant	
PknD	1RWI 1RWL	Extracellular sensor domain	53
PknE	2H34	Apo-KD	14
PknG	2PZI	Rubredoxin + KD + TPR domain (69 aa truncated from N-terminus) in complex with Ax20017, dimer	29
PknH	4ESQ	Extracellular sensor domain, dimer	70

suggesting a distinct mechanism of activation. Further, PknG is the only *M. tuberculosis* STPK that lacks Arg in front of the invariant Asp in the catalytic loop, and it was suggested that this difference results in the unique conformation of the activation loop in the absence of phosphorylation (Fig. 2) (29). PknG was found to be phosphorylated at the N-terminus in front of the rubredoxin domain, both *in vivo* and *in vitro* (15, 19). Although this modification is not required for PknG kinase activity, it was suggested that it might aid in substrate recognition (19). The rubredoxin domain, which contains two conserved Cys-X-X-Cys-Gly iron binding motifs, was shown to interact with both N- and C-terminal lobes of the PknG KD. Importantly, mutational analysis of the Cys residues showed that they are required for PknG kinase activity, suggesting that rubredoxin might sense the redox status of the environment to regulate PknG function (29–31).

These structural data for PknB, PknE, and PknG have provided key insights into mechanisms of STPK activation and regulation. Structures of the extracellular regions of PknB, PknD, and PknH have also been determined and are discussed below in the sections for each of these individual kinases. Going forward, structures of these domains bound to ligands and structures of KDs interacting with substrates may provide further insights into kinase regulation and substrate specificity.

THE *M. TUBERCULOSIS* PHOSPHOPROTEOME

In eukaryotes, STPKs and Tyr protein kinases are highly abundant. The human genome encodes over 500 of these protein kinases, and it is currently estimated that over two-thirds of human proteins are modified by this posttranslational modification, often at multiple sites (32–34). In contrast, *M. tuberculosis* encodes 11 STPKs and no typical Tyr kinases, though a protein with Tyr kinase activity has been identified (35). The number of proteins that are phosphorylated within the mycobacterial cell is unknown. In 2010, Prisic et al. published a large-scale analysis of the *M. tuberculosis* phosphoproteome, identifying over 500 phosphorylation sites in over 300 *M. tuberculosis* proteins using state-of-the-art tandem mass-spectrometry methods (15). These data represent the minimal *M. tuberculosis* phosphoproteome, and many examples have been published of *in vivo* phosphorylation of additional *M. tuberculosis* proteins that were not identified in this phosphoproteomic study (Table 3), suggesting that the total *M. tuberculosis* phosphoproteome, when examined under multiple environmental conditions, is likely to include at least several hundred proteins.

Several interesting findings emerged from this work. First, in contrast to human Ser/Thr phosphorylation, where phosphorylation on Ser and Thr account for 90% and 10% of the identified phosphorylation sites on human proteins, respectively, in *M. tuberculosis* Thr

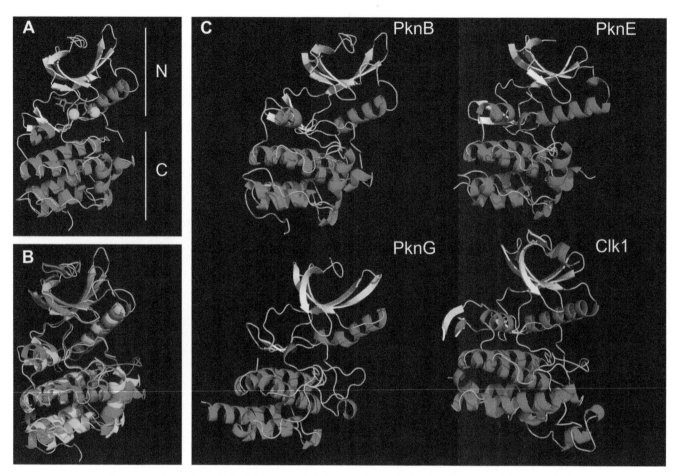

Figure 4 Overview of the *M. tuberculosis* STPK's KD. (**A**) Major features of the PknB KD (1MRU_B): N-terminal (upper) and C-terminal (lower) lobes are labeled. The ATP analog is in blue, and two Mg^{2+} ions are in green. (**B**) Overlap of PknB (green) and Clk1 (magenta). Clk1 was a top hit when the PknB structure was used to search similar three-dimensional structures using the NCBI VAST program. For clarity, residues 298 to 319 and 395 to 443 in Clk1 that are absent in *M. tuberculosis* STPKs (see Fig. 2) are truncated in Clk1. (**C**) PknB (1MRU_B), PknE (2H34_B), PknG (2PZI_A), and Clk1 (1Z57). α-Helix is in red, β-sheet is in yellow. Figures were made using PyMOL (Schrödinger) and POV-Ray (povray.org).
doi:10.1128/microbiolspec.MGM2-0006-2013.f4

phosphorylation is predominant, with a 60:40 ratio of phosphorylation on Thr versus Ser. Second, phosphorylation was identified on proteins involved in all aspects of *M. tuberculosis* physiology (Fig. 8). Third, based on *in vitro* phosphorylation of peptides corresponding to *in vivo* phosphorylation sites, a conserved Thr-centered phosphorylation motif was identified in which acidic residues are prominent amino-terminal to the phosphoacceptor, particularly at the −2 and −3 positions, and hydrophobic residues are dominant at the +3 and to a lesser extent at the +5 positions. The core components of this motif are shared by six of the *M. tuberculosis* STPKs (PknA, PknB, PknD, PknE, PknF, and PknH), with less prominent differences

among the motifs of each STPK that may provide additional substrate specificity.

While these data have provided an important resource for further investigation, for most of these phosphoproteins the kinase(s) responsible for their phosphorylation and the effects of this posttranslational modification on protein function are not known. Identifying cognate kinase-substrate pairs and determining the functional effects of protein phosphorylation of individual protein substrates are essential for understanding the regulatory role of specific *M. tuberculosis* kinases. For *M. tuberculosis* STPKs, as for eukaryotic kinases, where well-characterized kinase-substrate pairs have been characterized and functional effects of phosphorylation have been elucidated for

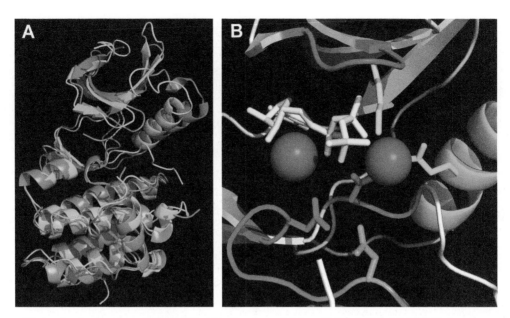

Figure 5 Active site of PknB KD. (**A**) Overlap of "closed" PknB KD (1MRU_B) in green and "open" apo-PknE-KD (2H34) in blue, with the PknE C helix labeled in red. (**B**) PknB active site (1MRU_B) P loop (GFGGMS), magenta; Mg^{2+}, red balls; ATPγS, yellow; C-helix, green (Glu59-green); Lys40, aqua; catalytic loop, red (Asp138-orange, Asn143-red); DFG motif, purple (Asp156-purple). Figures were made using PyMOL (Schrödinger) and POV-Ray (povray.org). doi:10.1128/microbiolspec.MGM2-0006-2013.f5

only a tiny fraction of phosphoproteins, these goals are experimentally challenging. In the following sections, we will summarize the current state of knowledge for each of the *M. tuberculosis* STPKs, including candidate substrates and physiologic pathways potentially regulated by each kinase.

PknA AND PknB

These two STPKs are encoded by adjacent genes in an operon that includes genes for the cell wall synthesis enzyme PBPA and the cell-shape-determining protein RodA, as well as the protein phosphatase PspA. Both *pknA* and *pknB* are essential for growth based on transposon mutagenesis experiments (11). Based on their linkage to *rodA* and *pbpA*, PknA and PknB were predicted to regulate cell shape and cell wall synthesis. An early study by Kang et al. confirmed this prediction, showing that overexpression of these kinases had marked effects on cell shape, including branching, elongation, and incomplete septation (17). Further support for the role of these kinases in cell wall synthesis and cell morphology were provided by the characterization of proteins involved in cell wall synthesis and its regulation that are likely substrates of one or both of these STPKs. Examples of these phosphoproteins include PBPA, a bifunctional penicillin binding protein that is a

possible substrate of PknB; the DivIVA homologue Wag31, a substrate of PknA that has been shown to be required for peptidoglycan (PGN) synthesis in mycobacteria at the growing cell pole; and MviN, an essential protein required for late stages of PGN synthesis that is a likely substrate of PknB (17, 36, 37). In addition to these examples, several other proteins involved in PGN synthesis have been identified as phosphoproteins, in some cases with functional effects of phosphorylation shown (Table 3) (15). Further supporting a role for PknB in regulating cell division and cell wall synthesis, Mir et al. demonstrated that PknB is localized to the cell poles and the mid-cell, the sites of PGN turnover and assembly of the divisome (27). Consistent with PknB regulating cell wall synthesis during *M. tuberculosis* growth and with data showing 10-fold decreased expression of the *pknA/pknB* operon in stationary phase compared to log phase, a model was recently proposed in which PknB activity is decreased in hypoxia-induced stasis and PknB activity is required for oxygen-induced re-growth (17, 38).

The sequence of the extracellular region of PknB also suggested a role for this kinase in regulating PGN turnover. This region comprises four penicillin binding protein and serine/threonine kinase associated (PASTA) domains. These domains were first identified bioinformatically and predicted to bind PGN fragments

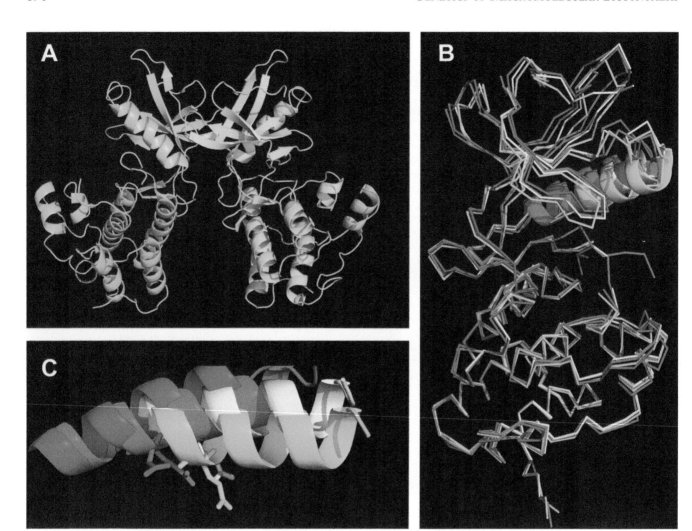

Figure 6 Back-to-back dimerization of KDs. (**A**) PknB-KD dimer showing "back-to-back" interaction. (**B**) Overlap of PknB-KD in active form (1MRU_B-blue) and conformations of the PknB-KD L33D mutant that perturbs the dimer interface (3ORK, yellow; 3ORI_A, red; 3ORL, green). The C helix is shown in ribbon, while the rest of the structure is shown in wire. (**C**) C helix from the PknB structures in panel B magnified to highlight differences in the position of Glu59. Figures were made using PyMOL (Schrödinger) and POV-Ray (povray.org). doi:10.1128/microbiolspec.MGM2-0006-2013.f6

(39). The incorporation of PASTA domains in some PBPs and in PknB-like STPKs led to the prediction that PknB and its homologues, which are widely distributed in Gram-positive bacteria, would regulate PGN synthesis. Evidence that the PASTA domains bind PGN fragments was first obtained in *Bacillus subtilis*, where spore germination was potently stimulated by muropeptides, PGN fragments derived from cell wall hydrolysis that contain a three- to five-residue stem peptide linked to the *N*-acetylglucosamine-*N*-acetymuramic acid disaccharide (40).

In *M. tuberculosis*, *in vitro* binding assays to the extracellular domain of PknB, using a comprehensive library of synthetic muropeptides, demonstrated that specific residues at the second and third positions of the stem peptide were required for binding (27). These residues, D-isoglutamine (versus D-isoglutamate) at position 2 and diaminopimelic acid (versus lysine) at position 3, are predominant in *M. tuberculosis* PGN, indicating that the PknB extracellular domain is adapted to recognize autologous muropeptides. A muropeptide that bound the *M. tuberculosis* PASTA domain *in vitro* was able to resuscitate dormant *M. tuberculosis* cells, albeit with less potency than spent medium, suggesting a link between PASTA binding, PknB activation, and cell growth.

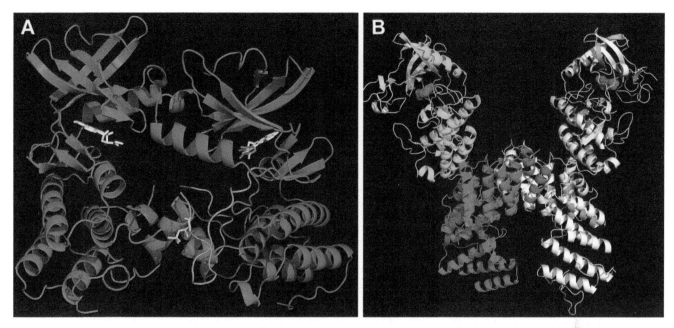

Figure 7 Distinct modes of monomer interaction in dimers of PknB versus PknG (**A**) "Front-to-front" dimer of mutant PknB KD (3F69) in complex with Kt5720 inhibitor (yellow). The "substrate" subunit (magenta) has most of its activation loop disordered (red), while the "enzyme" subunit (blue) has a well-defined activation loop (orange) with visible phosphorylated Thr171 (green). (**B**) Structure of PknG (2PZI) in complex with inhibitor Ax20017 (magenta). Three domains: rubredoxin (yellow), KD (green), and TPR domain (red) are shown only in one subunit. The second subunit is depicted in gray. Figures were made in PyMOL (Schrödinger) and POV-Ray (povray.org).
doi:10.1128/microbiolspec.MGM2-0006-2013.f7

Structural analysis of the PknB extracellular domain using nuclear magnetic resonance showed it to be an elongated structure, with rigid links between each of the four PASTA domains (28). This structure suggests that the PASTA domains protrude from the external surface of the cytoplasmic membrane into the PGN layer, where they may encounter muropeptides produced by cell wall hydrolases. This finding led the authors to propose a model of how ligand binding might activate PknB, in which muropeptide binding serves to cross-link the extracellular domains of two PknB proteins, leading directly to PknB dimerization and activation. In an alternative model, binding of muropeptides to extracellular domains of individual PknB molecules localizes PknB to sites of cell wall turnover, i.e., the septum and cell poles, resulting in high local PknB concentrations allowing dimerization of the intracellular KDs and their activation (27). Data from *B. subtilis* PrkC show that the PASTA-containing extracellular region does not dimerize *in vitro*, whether or not muropeptides are present (41). This result, together with the demonstrated ability of the KDs alone to dimerize and activate, favors the latter model (21), though it is

possible that additional proteins might play a role in dimerization of the extracellular domain in a manner that would support the model of extracellular domain dimerization.

Though phosphorylation by PknA and PknB of several proteins involved in cell shape and cell wall synthesis has been shown or suggested (Table 3), here we will highlight two that play key roles in PGN synthesis and localization: Wag31 and MviN. Wag31, the *M. tuberculosis* homologue of the cell division protein DivIVA, was shown to be phosphorylated in *M. tuberculosis* cells and, through a combination of *in vivo* and *in vitro* experiments, to be phosphorylated by PknA (17). Subsequent analysis showed that Wag31 localizes to the cell poles, with preference for the old cell pole, the site of new PGN synthesis in mycobacteria and other actinomycetes (42, 43). Depletion of Wag31 led to delocalized PGN synthesis, asymmetric bulging, and ultimately lysis of the cells, indicating a critical role for this protein in proper localization of PGN synthesis. Through the use of phosphomimetic (T73E) or phosphoablative (T73A) substitutions in Wag31, phosphorylation of Wag31 appears to positively affect growth

Table 3 Phosphorylated proteins in *M. tuberculosis* in addition to those identified in the phosphoproteomic study of Prisic et al. 15)ᵃ

Name	Function	Phosphorylation, candidate cognate kinase, and effect	References
DacB1	Cell wall synthesis	*In vitro* (PknH)	(116)
DosR	Dormancy	*In vitro* (PknH), enhances DNA binding	(77)
EF-Tu	Protein synthesis	*In vitro* (PknB), *in vivo*; reduces interaction with GTP	(15, 117)
EmbR	Cell wall synthesis (arabinan)	*In vitro* (multiple kinases), *in vivo* (*M. smegmatis*); FHA domain required for phosphorylation; phosphorylation activates ATPase activity and enhances binding to *embCAB* promoter	(71, 74, 79)
EmbR2	Cell wall synthesis?	*In vitro* (PknE, PknF), inhibits PknH	(118)
FabD	Cell wall synthesis (mycolic acid)	*In vitro* (multiple kinases), *in vivo* (BCG)	(49, 119)
FabH	Cell wall synthesis (mycolic acid)	*In vitro* (multiple kinases), decreases activity	(51)
FhaA	Cell wall synthesis	*In vitro* (multiple kinases), *in vivo*, FHA interacts with phosphorylated juxtamembrane region of PknB	(15, 120, 121)
FipA	Cell division	*In vitro* (PknA, PknB), *in vivo*, depends on FHA, required for activity under oxidative stress	(122)
FtsZ	Cell division	*In vitro* (PknA) (and in *E. coli* when coexpressed); impairs GTP hydrolysis and polymerization	(123)
GarA	Central metabolism	*In vitro* (PknB, PknG), *in vivo*; N-terminal tail binds to FHA, regulates interaction with TCA enzymes	(15, 19, 69, 124)
GlgE	Cell wall synthesis	*In vitro* (PknB), *in vivo* (BCG), decreases maltosyltransferase activity	(125)
GlmU	Cell wall synthesis	*In vitro* (PknB), decreases acetyltransferase activity	(126)
GroEL1	Heat shock protein	*In vitro* (multiple kinases)	(127)
InhA	Cell wall synthesis (mycolic acid)	*In vitro* (multiple kinases) (also when coexpressed in *E. coli*), *in vivo* (*M. smegmatis*, BCG), decreases activity	(47, 48)
KasA	Cell wall synthesis (mycolic acid)	*In vitro* (multiple kinases), *in vivo* (BCG), decreases activity	(49)
KasB	Cell wall synthesis (mycolic acid)	*In vitro* (multiple kinases), *in vivo* (BCG); increases activity	(49)
MabA	Cell wall synthesis (mycolic acid)	*In vitro* (multiple kinases), *in vivo* (BCG); decreases activity	(50)
MmA4	Cell wall synthesis (mycolic acid)	*In vitro* (PknJ)	(79)
MmpL7	Transporter? Virulence factor	Possibly PknD substrate, *in vivo*	(128)
MurD	Cell wall synthesis	*In vitro* (PknA) (and in *E. coli* when coexpressed)	(129)
MviN	Cell wall synthesis	*In vitro* (PknB), *in vivo*, enhances binding to FhaA	(15, 37)
PapA5	Cell wall synthesis	*In vitro* (PknB)	(130)
PbpA	Cell wall synthesis	*In vitro* (PknB), possibly regulates localization	(36)
PepE	Dipeptidase	*In vitro* (PknJ)	(79)
PcaA	Cell wall synthesis (mycolic acid)	*In vitro* (multiple kinases), decreases activity	(131)
PstP	Dephosphorylation	*In vitro* (PknA, PknB) (and in *E. coli* when coexpressed), activates	(132)
PykA	Glycolysis	*In vitro* (PknJ)	(133)
RshA	Stress response	*In vitro* (PknB), *in vivo*, prevents interaction with SigH	(134)
Rv0516c	Anti-anti-sigma factor?	*In vitro* (PknD), *in vivo*, inhibits protein interactions	(55)

(Continued)

Table 3 *(Continued)*

Name	Function	Phosphorylation, candidate cognate kinase, and effect	References
Rv0681	Transcriptional regulator	*In vitro* (PknH)	(116)
Rv1422	Cell wall synthesis?	*In vitro* (PknA and PknB), *in vivo*	(17)
Rv1747	ABC-transporter? Virulence factor	*In vitro* (multiple kinases), *in vivo*, FHA domain interacts with kinases, activates	(15, 60, 61, 63, 120)
Rv2175c	DNA binding?	*In vitro* (PknL), *in vivo* (*M. smegmatis*), inhibits DNA binding	(85, 135)
SigH	Stress response	*In vitro* (PknB), *in vivo*	(134)
VirS	Cell wall synthesis	*In vitro* (PknK) (and by coexpression in *E. coli*), increases binding to *mym* promoter	(81)
Wag31	Cell division	*In vitro* (PknA), *in vivo* (PknA, PknB), enhances activity	(17, 42)

*a*Whether phosphorylation was shown *in vitro* or in *M. tuberculosis* (*in vivo*) is indicated, as are the kinases shown to phosphorylate the protein. Where known, the effects of phosphorylation on protein function are listed.

rate, though no clear morphologic differences were observed with expression of the different alleles (42).

MviN, a widely conserved multipass membrane protein, has been shown in *Escherichia coli* to be the protein that flips the lipid-linked PGN precursor lipid II from the cytoplasm to the periplasmic space, where the muropeptide can be incorporated into PGN (44). The

M. tuberculosis MviN homologue, Rv3910, has an essential amino-terminal domain similar to MviN proteins in other bacteria but also has a nonessential carboxy-terminal region with sequence similarities to protein kinases (37). In the *M. tuberculosis* phosphoproteome study, this protein was found to be phosphorylated on as many as five distinct residues, including

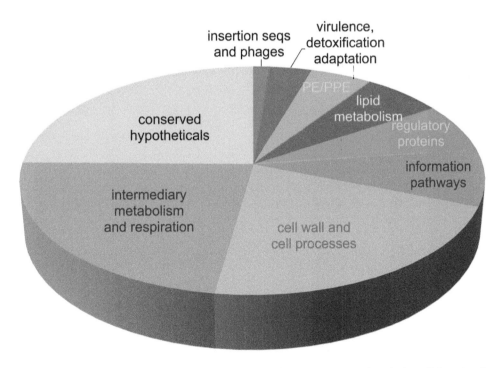

Figure 8 *M. tuberculosis* phosphoproteome. Phosphoproteins were identified in all functional categories of *M. tuberculosis* proteins (15). doi:10.1128/microbiolspec.MGM2-0006-2013.f8

one well-localized site at Thr947 in the kinase homology domain (15). Gee et al. demonstrated that although this domain folds and dimerizes in a manner similar to protein kinases such as PknB, it lacks critical residues and motifs required for enzyme activity and is a pseudokinase that lacks both ATP binding and catalytic activities (37). The Thr947 residue was found to be phosphorylated *in vitro* by PknB and to a lesser extent by PknA, PknD, PknE, and PknH. Phosphorylated MviN was found to bind FhaA tightly; though the direct effect of this binding on MviN function was not shown, analysis of MviN and FhaA depletion strains suggested a model in which PknB phosphorylation of the kinase homology domain of MviN recruits FhaA, leading to inhibition of the terminal steps of PGN synthesis. This negative regulation of PGN synthesis by phosphorylation of MviN, in contrast to positive regulation of cell growth by PknA-mediated phosphorylation of Wag31, suggests a complex regulatory role of phosphorylation, likely by multiple kinases, in controlling *M. tuberculosis* growth and PGN synthesis.

In addition to cell wall synthesis, PknA and PknB have been linked, directly or indirectly, to regulation of several additional cellular processes, including lipid synthesis, cell division, and transcription regulation, among others (Table 3). The regulation of multiple processes by these kinases suggests that they may function to broadly coordinate cell physiology with cell growth. Linking a specific kinase to a validated *in vivo* phosphorylation event is challenging, however, and in many cases more than one STPK may target the same protein substrate. For many of the phosphoproteins involved in the processes noted above, whether PknA and/or PknB is the primary cognate kinase is uncertain. In other cases, rather than starting from well-defined *in vivo* phosphorylation in mycobacteria, investigators have started with a protein (enzyme) of interest, demonstrated *in vitro* phosphorylation by one or more *M. tuberculosis* STPKs *in vitro* or in *E. coli*, and then characterized effects of phosphorylation on enzyme function *in vitro*. In some cases the role of phosphorylation of the protein of interest *in vivo* remains unknown, while in other cases, additional work was undertaken to demonstrate *in vivo* phosphorylation and effects of phosphorylation on bacterial phenotypes.

An interesting example of a phosphoprotein first characterized *in vitro* and then shown to be phosphorylated *in vivo* with important functional effects of phosphorylation on lipid synthesis is the enoyl-acyl carrier protein reductase InhA, an essential component of the FASII fatty acid synthesis pathway in mycobacteria and the primary target of the first-line antitubercular,

isoniazid (45, 46). Two groups independently examined *in vitro* phosphorylation of InhA and then pursued functional effects of phosphorylation *in vitro* and *in vivo* (47, 48). InhA was found to be phosphorylated by multiple kinases, including PknA and PknB. Elucidation of the sites of phosphorylation by mutagenesis and mass spectrometry yielded similar results, with one group observing phosphorylation at the carboxy-terminus of the protein at Thr253, Thr254, and Thr266, while the other group identified Thr266 as a unique phosphorylation site. The Thr254 and Thr266 sites have features of the preferred phosphorylation site motifs of PknA and PknB (15, 17). *In vivo* phosphorylation of InhA was investigated in *M. smegmatis* wild type and an InhA overexpression strain (47) and in *Mycobacterium bovis* BCG overexpressing InhA (48), with both studies confirming Thr266 as the primary *in vivo* site of phosphorylation. Using phosphorylated InhA and phosphomimetic (T266D or T266E) or phosphoablative (T266A) substituted proteins, enzyme activity of InhA was shown to be markedly decreased by phosphorylation. Structural analysis showed that these effects were not the result of gross disruption of InhA folding, and in combination with binding and kinetic analyses suggested a mechanism by which the effects on enzyme activity are the result of decreased affinity of the phosphomimetic-substituted InhA for NADH (48).

In vivo, it was observed that expression of the phosphoablative form of InhA was well tolerated in *M. smegmatis* but that the phosphomimetic form led to severe growth inhibition. Similarly, *inhA* conditional expression strains were complemented by *inhA* expressing the native or phosphoablative proteins but not the phosphomimetic form. These data provide strong evidence for regulation of mycolic acid synthesis by Ser/Thr phosphorylation, though it remains uncertain which specific STPK(s) target InhA. Consistent with this observation, other FASII enzymes, including KasA, KasB, and MabA, have been shown to be phosphorylated *in vitro* and in mycobacteria with direct effects of phosphorylation on enzyme activity (49, 50); FabD and FabH have also been shown to be phosphorylated *in vitro*, though not yet in *M. tuberculosis* (49, 51).

PknD

PknD is encoded by a nonessential gene at a chromosomal locus encoding multiple genes involved in phosphate transport (11, 52). The KD of PknD is most similar to the KDs of PknH and PknE and appears to be activated by allosteric effects of dimerization of the KDs as described above (23). The PknD extracellular domain has been shown to form a highly symmetrical six-bladed β-propeller structure, with variation in the blades concentrated in the membrane-distal "cup" region that is the likely site of ligand binding (53).

Though the ligands of PknD and its function are not known, the linkage of *pknD* to phosphate transport genes suggests a role in phosphate uptake, and survival of a *pknD* deletion mutant has been shown to be compromised in a phosphate-deficient growth medium (54). PknD has also been shown to phosphorylate the amino-terminal extension of Rv0516c, a putative regulator of the sigma factor SigF (55), and recent data have linked Rv0516c, SigF, and PknD in an osmosensory signaling pathway (56).

Interestingly, *pknD* has been linked to central nervous system tuberculosis. A screen for genes required for central nervous system infection by *M. tuberculosis* identified a *pknD* deletion strain as defective for central nervous system invasion (57). PknD was shown to be required for invasion of brain endothelial cells *in vitro*, and the extracytoplasmic domain of the molecule was sufficient to stimulate invasion of these cells. The mechanism by which PknD may contribute to central nervous system invasion and the protein substrates of PknD that may contribute to this phenotype are not known.

PknE

Relatively little is known regarding the substrates or function of PknE. It appears to be a receptor-type kinase, with an extracellular domain, transmembrane domain, and intracellular KD. The extracytoplasmic domain of PknE has not been characterized and does not contain known protein motifs based on amino acid sequence. The gene encoding this STPK is adjacent to genes of unknown function and does not appear to be part of an operon. Interestingly, *pknE* is separated from *pknF* by just two genes, suggesting a possible functional link between these two STPKs. A possible link to regulation of lipid synthesis by the type II fatty acid synthase system has been suggested, based on functional effects of phosphorylation or substitutions of phosphoacceptors and *in vitro* phosphorylation of some of these enzymes by PknE, as well as other kinases (49). PknE has also been suggested to regulate apoptosis in *M. tuberculosis*–infected macrophages (58).

PknF

As noted above, *pknF* is separated from *pknE* by less than 2 kb in the *M. tuberculosis* chromosome. Immediately 3′ of *pknF* is Rv1747, which encodes an ABC transporter that contains two forkhead associated (FHA) domains in addition to the transmembrane transporter and ATP binding domains (52, 59). Several studies point to functional and physical interactions between PknF and Rv1747, in which autophosphorylated PknF recruits Rv1747 via its amino-terminal FHA domain, leading to phosphorylation of Rv1747 by PknF (60, 61). Though the transport role of Rv1747 and the effects of phosphorylation on transport were not determined, an Rv1747 deletion mutant was shown to be attenuated for growth in macrophages and in lungs and spleens of mice (61). In a study using *pknF* overexpression and antisense expressing strains, marked effects on growth, morphology and septum placement were observed, suggesting a role for *pknF* in regulating growth and cell shape (62).

More recently, it was shown that *pknF* and Rv1747 are cotranscribed in an operon and that PknF phosphorylates Rv1747 at two sites, Thr150 and Thr208, located between the two FHA domains of this protein (63). In addition to confirming that the Rv1747 deletion strain is attenuated in mice and macrophages, this study demonstrated that phosphoablative mutants of Rv1747 did not complement the virulence defect, indicating a role for phosphorylation in Rv1747 function. Supporting this interpretation, a construct expressing Rv1747 with a mutation in the first FHA domain that is required for phosphothreonine binding failed to complement the growth defect of the Rv1747 deletion strain in macrophages. Interestingly, however, a *pknF* deletion strain was not clearly defective for growth in macrophages. Given the requirement for phosphorylation in complementing the intracellular growth defects, this result suggests that other kinases may also phosphorylate Rv1747, consistent with the strong similarity between the phosphorylation motifs of PknF and several other STPKs (15, 63). Despite substantial molecular and pathogenic insights into the interactions of PknF and Rv1747, the transport activity of Rv1747 and the effect of phosphorylation on this activity remain unknown. In addition, the extracellular domain of PknF has not been characterized to date.

PknG

PknG is one of the two *M. tuberculosis* STPKs, PknK being the other, that lack a transmembrane region. Thus, PknG does not have the structure of a typical receptor type kinase that functions in transmembrane signal transduction (Fig. 7B). In addition to its KD, which does not cluster with any of the other *M. tuberculosis* STPKs based on amino acid sequence (Fig. 2

and 3), PknG has amino and carboxy-terminal sequences that are important for its function. The amino-terminal region contains a rubredoxin redox-sensing domain, and the carboxy-terminus contains a tetratricopeptide (TPR) repeat domain, which may function in protein-protein interactions and regulation of kinase activity (30, 31, 64). As described in detail above, the PknG KD has the characteristic two-lobed structure of Ser/Thr protein kinases, though with differences from the PknB structure including absence of phosphorylation of the activation loop, absence of the highly conserved Arg preceding the catalytic Asp in the catalytic loop, and unusual residues in the ATP-binding pocket (29). Mutagenesis and activity studies based on this structure indicate that the amino-terminal rubredoxin motif likely regulates PknG activity. Deletion of the rubredoxin domain resulted in decreased kinase activity and substitution of Cys residues in the iron binding motif caused reduction or complete elimination of kinase activity, depending on the nature of the substitution (30, 31). PknG is encoded by a gene that is linked in an operon to *glnH*, a glutamine-binding lipoprotein, and a conserved membrane protein of unknown function (52).

An early publication, using *M. bovis* BCG and *M. smegmatis*, suggested that PknG was a virulence factor that was secreted into the macrophage phagosome and functioned by modulating host signaling to prevent phagosome-lysosome fusion (65). In the first publication that analyzed a *pknG* deletion mutant, Cowley et al. found that the Δ*pknG* strain grew poorly in liquid and on solid media, with decreased growth compared to wild type that was most notable after mid-log phase (66). These investigators also demonstrated that the Δ*pknG* strain was attenuated in SCID mice and had elevated intracellular levels of glutamate plus glutamine, associated with decreased *de novo* glutamine synthase.

Further insight regarding the functions of PknG emerged with the identification of its primary substrate, the FHA domain-containing protein GarA (Rv1827), and the role of GarA in regulating central metabolic pathways (67). O'Hare and colleagues first demonstrated three phosphorylation sites in the amino-terminal region of PknG (19). In contrast to other *M. tuberculosis* STPKs, there was no evidence of phosphorylation of the activation or catalytic loops of the KD, and autophosphorylation was not required for kinase activity of PknG. Building on results obtained in *C. glutamicum* (68), a related actinomycete, these authors identified GarA as a substrate of PknG that is phosphorylated at Thr22, adjacent to the Thr21 residue of this protein that is phosphorylated by PknB.

In this and subsequent work by Nott et al., an elegant intramolecular phospho-switch mechanism was elucidated as the mechanism by which the phosphorylation state of GarA regulates central carbon metabolism (19, 69). In this research, it was shown that unphosphorylated GarA binds to three enzymes: α-ketoglutarate dehydrogenase (KGD, Rv1248c), NAD-dependent glutamate dehydrogenase (GDH, Rv2476c), and the α-subunit of the glutamate synthase complex (GltB, Rv3859c). In each case, binding regulates their enzymatic activity. Phosphorylation of GarA by either PknG (shown for KGD and GDH) (19) or PknB (shown for KGD, GDH, and GltB) (69) abolished the binding of GarA to these enzymes. Though the amino-terminal region containing the Thr21 or Thr22 phosphoacceptors was not required for binding of GarA to the enzyme ligands, these residues were essential for the phosphorylation-mediated abrogation of binding. Structural and mutagenesis analysis demonstrated that the FHA domain of GarA was required for binding of each of the enzyme ligands. Further, phosphorylation of GarA on Thr21 resulted in binding of the amino-terminal region of GarA to the FHA domain, blocking interaction of the FHA domain with the enzymes and thereby increasing their activity in the cell. Unlike most other *M. tuberculosis* STPKs, PknG has limited *in vitro* kinase activity toward a wide range of peptide and protein substrates, suggesting that this regulation of metabolic activity by phosphorylation of GarA may be its primary role and/or that additional regulatory inputs are required for its activity.

PknH

PknH is encoded 3′ of the gene encoding the transcriptional regulator EmbR. Sequence analysis indicates that PknH is a typical receptor-type kinase with an intracellular KD and an extracellular receptor domain (52). The extracytoplasmic domain contains no previously described motifs, and its ligand(s) are not known. A recent structural analysis, however, identified unusual features of the structure of this region. The crystal structure revealed the presence of two disulfide bonds that contribute rigidity to the structure and the presence of a deep cleft, the likely site of ligand binding, lined by a mix of hydrophobic and polar residues (70). The presumed ligand binding cleft is highly conserved among PknH orthologues and bears some similarity to binding sites of lipoproteins LprG and LppX. However, the greater polarity of the PknH cleft was thought to make binding of hydrophobic glycolipids unlikely.

The physical proximity between the genes encoding PknH and the transcriptional regulator EmbR suggests a functional interaction between these proteins. The presence of an FHA domain in EmbR, in addition to winged helix-turn-helix and bacterial activation domains that are characteristic of this family of transcription factors led to investigation of the role of phosphorylation, and PknH specifically, in regulating EmbR activity. In an initial study, Molle and colleagues demonstrated Thr phosphorylation of EmbR by PknH *in vitro* (71). Using substitutions of conserved residues in the EmbR FHA domain, these authors also demonstrated that this domain is essential for EmbR phosphorylation. Singh and colleagues subsequently identified the *embCAB* operon, which encodes proteins required for arabinosylation of lipoarabinomannan (*embC*) and arabinogalactan (*embA* and *embB*) (72, 73), as a target of EmbR, with apparent binding sites 5′ of *embC*, *embA*, and *embB* (74). EmbR phosphorylation by PknH enhanced its binding to each of these regions. Expression of *M. tuberculosis pknH* in *M. smegmatis* resulted in increased phosphorylation of *M. smegmatis* EmbR. In a strain expressing native PknH, but not a kinase inactive form, semiquantitative reverse transcription PCR (RT-PCR) showed increased transcription of *embC*, *embA*, and *embB*. Consistent with increased expression of these enzymes, the ratio of lipoarabinomannan to lipomannan was increased in the *pknH* overexpression strain. Though expression of *M. tuberculosis* PknH in *M. smegmatis*, and the absence of evidence that the same sites on EmbR are phosphorylated in *M. tuberculosis* and *M. smegmatis*, complicates interpretation of these data, taken together they do suggest that phosphorylation of EmbR by PknH positively regulates expression of the *embA*, *embB*, and *embC* genes. Decreased expression of embC and embB in a *pknH* deletion mutant (Δ*pknH*) further supports this regulatory mechanism (75).

In addition to its regulation of cell envelope glycolipids, proteomic and lipid analysis of a Δ*pknH* strain compared to wild type suggests a possible role for PknH in regulating lipids in the *M. tuberculosis* cell envelope (76). Specific phthiocerol dimycoserosates (PDIMs) were found to be decreased in the Δ*pknH* strain. Surprisingly, enzymes involved in PDIM synthesis were expressed at similar or higher levels in the mutant strain. In contrast to *M. smegmatis*, where overexpression of PknH led to higher lipoarabinomannan:lipomannan ratios, in this study higher lipoarabinomannan:lipomannan ratios were observed in the *M. tuberculosis pknH* deletion strain. The mechanism by which PDIM production is affected by *pknH*

deletion, and how lower levels of a virulence-associated surface lipid would lead to hypervirulence, remain unknown.

The roles of *pknH* in response to stress and during infection are complex. The Δ*pknH* strain was found to be more susceptible to peroxide and superoxide (paraquat) stresses but more resistant to nitric oxide (acidified nitrite) stresses (75). The mutant showed decreased survival in THP-1 macrophages but, surprisingly, was hypervirulent in mice, with greater replication in the Δ*pknH* strain starting at 4 weeks post-infection, reaching 1 to 2 log greater bacterial burden of the deletion strain in both lungs and spleen of Balb/C mice after 4 months of infection.

Further insight into PknH signaling emerged from a proteomic study comparing protein levels in wild type and the Δ*pknH* strain, untreated or after 48 hours of acidified nitrite treatment (77). The striking finding of this study was the identification of several DosR regulon proteins that were increased in response to acidified nitrite treatment in the *pknH* mutant. Following this lead, Chao et al. (77) then demonstrated that PknH can phosphorylate DosR *in vitro* on two Thr residues and that this phosphorylation enhances the binding of DosR to DNA containing a DosR recognition sequence. Subsequent analysis of transcription of DosR regulon genes in response to hypoxia and nitric oxide stress indicated decreased induction of these genes in the Δ*pknH* strain, though this effect was modest. These data suggest a role for PknH phosphorylation as a mechanism for modulating the *M. tuberculosis* dormancy response, though *in vivo* confirmation of DosR phosphorylation and the mechanism by which this affects DosR activity remain to be shown.

PknI

Relatively little is known regarding the substrates or function of PknI. This STPK has the domain organization of a receptor-type kinase, with an extracellular domain, transmembrane domain, and intracellular KD. The extracytoplasmic domain of PknI has not been characterized and is not similar to known protein domains based on amino acid sequence. Immediately 3′ of the gene encoding PknI (Rv2914c) is a gene that encodes a predicted D-amino acid aminohydrolase (Rv2913c) followed by a transcriptional regulator (Rv2912c) and dacB2 (Rv2911), predicted to encode a D-alanyl-D-alanine hydrolase, raising the possibility of a role for PknI in regulating cell wall turnover. Immediately 5′ of *pknI* is a gene of unknown function

(Rv2915c) and a gene for a predicted signal recognition particle (SRP) protein (52). No well-documented substrates of PknI have been identified, but a *pknI* deletion strain was found to be hypervirulent for replication in macrophages and for morbidity in SCID mice (78).

PknJ

Like PknI, relatively little is known regarding the substrates or function of PknJ. Jang et al. performed an initial characterization of this protein (79). They demonstrated that the KD can dimerize and that it autophosphorylates on three Thr residues in the activation loop (79), consistent with the mode of activation of the other *M. tuberculosis* STPKs that have been studied. PknJ is a receptor-type kinase and was shown to have a single transmembrane region, an intracellular KD, and an extracellular domain (79). The extracellular domain has not been characterized and has no recognizable motifs. Several *M. tuberculosis* proteins were shown to be phosphorylated by PknJ *in vitro*, but *in vivo* phosphorylation was not demonstrated (79). The several genes 5′ of *pknJ* encode conserved hypothetical proteins; a gene encoding a predicted dipeptidase is located immediately 3′ of *pknJ* in the opposite orientation. Thus, the chromosomal locus provides few clues as to a likely function of PknJ. A *pknJ* transposon insertion mutant did not have a growth phenotype in macrophages or in BALB/c mice (79).

PknK

PknK is one of two *M. tuberculosis* STPKs that lack a transmembrane segment, the other being PknG, and is therefore predicted to be a cytoplasmic protein. PknK is a large (119 kDa) protein with the KD in the amino-terminal region and a long carboxy-terminal region that shows similarity to the MalT family of ATP-dependent transcription regulators. Within this region are a P-loop ATP binding motif characteristic of AAA+ ATPases, a PDZ domain, and a single tetratricopeptide (TPR) repeat sequence (64). PDZ domains and TPR domains are often involved in protein-protein interactions. TPR sequences typically occur in clusters of three or more repeats and mediate protein-protein interactions (80). The function and interaction partners of the PDZ domain and the TPR motif in PknK are not known.

Kumar et al. demonstrated that, like most other *M. tuberculosis* STPKs, PknK autophosphorylates on two Thr residues in a TXT motif in the activation loop, and this autophosphorylation is required for kinase activity (81). In addition, they provided evidence for phosphorylation of the carboxy-terminal region of the protein and suggest a positive regulatory role for this region. Despite its lack of a predicted transmembrane domain, probing of subcellular fractions of *M. tuberculosis* lysates showed PknK to be present in the cell wall/membrane fraction rather than in the cytosol. These investigators demonstrated the ability of PknK to phosphorylate *in vitro* VirS, a transcription factor encoded by a gene that is separated from *pknK* by a single gene in the *M. tuberculosis* chromosome. PknK also phosphorylated several proteins encoded in the *mym* operon that is regulated by VirS and that is adjacent to *virS*. Phosphorylation of VirS increased binding to the intergenic region that likely contains the virS and *mym* promoters, and increased transcriptional activity from a *mym* promoter-luciferase transcriptional fusion construct when active PknK and VirS were coexpressed in *M. smegmatis*. Though these data suggest a role for PknK-mediated phosphorylation in regulating VirS activity, whether VirS is phosphorylated in *M. tuberculosis* and whether PknK might be a cognate kinase for VirS were not shown.

Malhotra et al. have investigated the expression of *pknK* and phenotypes of a *pknK* deletion strain (Δ*pknK*) (82). Expression of *pknK* was found to be greater in virulent *M. tuberculosis* H37Rv compared to avirulent H37Ra, and PknK protein was shown to be much more abundant in stationary versus log phase cells grown in broth culture *in vitro*. The Δ*pknK* strain was found to grow to higher levels during stationary phase, and the cells of the mutant strain were shorter than wild type. The Δ*pknK* strain was slightly more resistant to acidic, oxidative, and hypoxic stress, and in a low-dose aerosol infection of C57BL/6 the mutant strain showed a transient early growth defect. In a follow-up report the same group showed striking effects on the expression of several tRNAs in the Δ*pknK* strain compared to wild type, with a large subset of tRNAs expressed at lower levels in the mutant in logarithmic growth and at higher levels during stationary phase (83). These phenotypes and expression data suggest a role for PknK in regulating translation in relation to growth and environmental conditions. These investigators also demonstrated dimerization of the PknK KD and, based on indirect evidence of effects on kinase activity, proposed a model in which the carboxy terminal domain negatively regulates the activity of the amino-terminal KD through steric hindrance.

PknL

PknL is encoded by a gene adjacent to a transposase and a putative DNA binding protein (52). Additional genes in the region encode several conserved hypothetical proteins and biosynthetic enzymes for aromatic amino acids, mannan, and lipid biosynthesis. Located further 5′ is a division and cell wall cluster (dcw) that includes several genes that encode enzymes for PGN precursor synthesis, a penicillin binding protein, and several cell division proteins, suggesting the possibility that PknL regulates cell wall synthesis. Though the *M. tuberculosis* PknL lacks an extracellular domain, the PknL orthologue in *C. glutamicum* has an extracellular domain comprised of three PASTA domains, further supporting a role for this STPK in regulating the mycobacterial cell wall (84).

There has been relatively little investigation of PknL in mycobacteria. The intracellular domain has been shown to autophosphorylate on two Thr residues in the activation loop, similar to the phosphorylation required for activation of PknB and several other *M. tuberculosis* STPKs (16, 85). Rv2175c, a putative transcription factor encoded by the gene adjacent to pknL in *M. tuberculosis*, was shown to be phosphorylated *in vitro* by recombinant PknL. Overexpression of Rv2175c in *M. smegmatis* allowed detection of three isoforms of the protein with different pIs, consistent with one and two phosphorylations, but direct evidence of Rv2175c phosphorylation in *M. tuberculosis*, and whether PknL is the cognate kinase, has not been shown.

TYROSINE PHOSPHORYLATION IN *M. TUBERCULOSIS*

No tyrosine kinases were annotated in the *M. tuberculosis* H37Rv genome sequence or in other sequenced strains. Speculating that a Tyr kinase might be linked to a Tyr phosphatase, Bach et al. examined Rv2232, a protein of unknown function encoded by the gene adjacent to the gene encoding the Tyr phosphatase PtpA, for kinase activity. These investigators determined that recombinant Rv2232, which they renamed PtkA, could autophosphorylate and that the site of phosphorylation was a Tyr residue (35). They further demonstrated that PtkA interacted with PtpA and phosphorylated PtpA *in vitro* on two Tyr residues. Surprisingly, PtpA did not have phosphatase activity toward PtkA. PtkA has no characteristic Tyr kinase motifs and thus would be an atypical Tyr kinase, whose role in *M. tuberculosis* remains to be defined.

M. TUBERCULOSIS PROTEIN PHOSPHATASES

Phosphorylation of Ser, Thr, or Tyr is a relatively stable modification, so that reversal of phosphorylation requires the enzymatic activity of a phosphatase that can bind to and act on specific phosphorylated residues on proteins. In most organisms, the number of protein kinases substantially exceeds the number of protein phosphatases. In *M. tuberculosis* there is one annotated phospho-Ser/phospho-Thr phosphatase, PstP, encoded by the first gene in the operon that includes *pknA* and *pknB*. There are two annotated Tyr phosphatases, the low-molecular-weight phosphatase PtpA and the high-molecular-weight phosphatase PtpB.

PstP is a member of the PPM family of Ser/Thr phosphatases, which utilize Mg^{2+} or Mn^{2+} to catalyze dephosphorylation (86). PstP was initially shown to dephosphorylate Thr residues that had been phosphorylated by PknB (16) and has since been shown to have broad activity in dephosphorylating Ser and Thr residues targeted by several *M. tuberculosis* kinases. The X-ray crystal structure of PstP was determined and found to be highly similar to the structure of PP2Cα, the prototype of the PPM phosphatases, with additional features of interest (87). In addition to the two Mn^{2+} ions present in PP2Cα, a third Mn^{2+} was found to be present in the PstP active site, with positioning of a large flap domain to bring it in contact with this metal ion. Given the broad reactivity of PstP, how dephosphorylation by this enzyme may contribute to regulating specific Ser/Thr signaling pathways is not clear and remains to be elucidated.

The two Tyr phosphatases have attracted a great deal of interest based on the hypothesis that one or both may target host phosphotyrosine signaling pathways. In an early paper, Koul et al. demonstrated that these enzymes specifically target phospho-Tyr and not phospho-Ser or phospho-Thr (88). Examination of subcellular fractions by this group indicated that these proteins were present in culture supernatants, suggesting that they are secreted proteins. These proteins lack SecI or TAT secretion signals, however, and the mechanism by which they are secreted is not known.

Both *ptpA* and *ptpB* are widely distributed among pathogenic and nonpathogenic mycobacteria. Using a combination of bioinformatic and experimental approaches, Beresford et al. identified features of the conserved active site motif of PtpB, suggesting that it may be a dual-specificity phosphatase (89). While confirming greater activity against phospho-Tyr-containing substrates, these investigators demonstrated phosphatase activity against phospho-Ser and phospho-Thr residues as well. Importantly, they also demonstrated potent phosphatase activity against phosphoinositides, demonstrating that PtpB may be a triple specificity phosphatase that could target multiple host signaling pathways to alter macrophage activity against *M. tuberculosis.*

Crystal and solution structures of PtpA have been determined and demonstrate that PtpA has a structure and active site characteristic of eukaryotic low-molecular-protein phosphatases despite limited sequence similarity (90, 91). Specific differences in residues of the substrate binding site suggested differences in specificity. The d-loop in the apo form of the protein was found to be in a more open orientation compared to ligand bound PtpA, and phosphorylation of two adjacent Tyr residues that are likely to play regulatory roles was demonstrated.

In the case of PtpB, crystal structures demonstrated preservation of the typical phosphatase fold plus two unique features: a disordered loop that may play a role in substrate binding or regulation and a mobile lid over the active site that may serve to protect this site from oxidative damage encountered in the phagosome of macrophages or neutrophils that have ingested *M. tuberculosis* (92, 93).

Supporting the hypothesis that these Tyr phosphatases act on host Tyr signaling pathways, PtpA has been implicated in modulating phagosome acidification. In an initial report Bach et al. demonstrated that a *ptpA* deletion mutant was defective for survival in macrophages (94). Using a substrate trap form of PtpA, vacuolar protein sorting 33B (VPS33B) was identified as a candidate substrate of PtpA. This protein was found to autophosphorylate on Tyr residues and to be bound and dephosphorylated by PtpA. PtpA and VPS33B were found to colocalize in *M. tuberculosis* macrophages. In a subsequent study from the same group, Wong et al. demonstrated that PtpA binds to subunit H of the host vacuolar ATPase (V-ATPase) in the phagosome membrane and that the macrophage class C sorting complex associates with the V-ATPase during phagosome maturation (95). These interactions are required for the dephosphorylation of VPS33B and result in exclusion of V-ATPase from the *M. tuberculosis*–containing phagosomes, thereby preventing acidification.

PtpB has also been implicated in modulating host responses, apparently by targeting specific Tyr signaling pathways to alter host immune function (96). Zhou et al. demonstrated decreased interferon-gamma-stimulated production of IL-6 from a macrophage cell line stably transfected with *ptpB* compared to non-transfected cells and demonstrated decreased phospho-

rylation of Erk2 and p38 in the MptB-expressing cells, but not in cells expressing an inactive form of PtpB. These investigators also noted decreased apoptosis in PtpB-expressing cells and an associated decrease in Caspase 3 activity. Though these data are indirect, i.e., experiments were not performed using cells infected with *M. tuberculosis*, they suggest a mechanism that may explain the decreased viability of *ptpB* deletion strains in activated macrophages and guinea pigs (97).

A lipid phosphatase of *M. tuberculosis*, SapM, has also been identified and shown to be a secreted protein (98). In macrophages this protein appears to target phosphatidyl inositol-3-phosphate in a manner that prevents phagosome maturation (99).

M. TUBERCULOSIS STPKS AND PHOSPHATASES AS POTENTIAL DRUG TARGETS

Inhibition of protein kinases and phosphatases has been one the most active areas of current pharmaceutical research and drug development to target human diseases. The critical role of protein phosphorylation in controlling key physiologies that are dysregulated in diseases such as cancer and autoimmune diseases, the extensive detailed structural information on proteins and complexes involved in signal transduction, and the ability to develop potent inhibitors of these enzymes has led to the development of several approved drugs targeting protein phosphorylation, with many more in development (100). The importance of protein phosphorylation in regulating key processes for *M. tuberculosis* viability and virulence, together with substantial structural information on *M. tuberculosis* kinases and phosphatases and the extensive expertise in design and development of small molecule inhibitors of these enzymes, make them highly attractive targets for the development of new antituberculars.

The wealth of experience in targeting human Ser/Thr and Tyr kinases provides a huge advantage in developing inhibitors of *M. tuberculosis* kinases and phosphatases. The similarity of the mycobacterial proteins to their eukaryotic counterparts, however, has led to concern that inhibitors of *M. tuberculosis* enzymes will not be sufficiently selective relative to human kinases and phosphatases to avoid important toxicities. While this is a valid concern, the success in developing specific inhibitors of individual human kinases suggests that this challenge can be overcome, even with ATP-competitive inhibitors that target the conserved ATP-binding site of kinases. Both type I inhibitors that bind to the active conformation through interactions with the purine binding site and adjacent hydrophobic pockets and type 2 inhibitors that bind to the inactive conformation, in which an additional binding site is exposed, have been developed as FDA-approved selective inhibitors of human kinases (100). Other classes of kinase inhibitors, e.g., allosteric inhibitors that bind at sites separate from the ATP binding, are of great interest in that they may target unique features of a kinase and offer the potential for very high selectivity.

While the conservation of the KDs of *M. tuberculosis* kinases offers advantages and disadvantages for the inhibitor identification and candidate drug development described above, other regions of the *M. tuberculosis* kinases may also serve as targets for inhibitor development. The extracellular domains of the receptor-type *M. tuberculosis* kinases, and the regulatory and protein interaction domains of PknG and PknK, may provide targets that are unrelated to human proteins and thus allow development of highly selective inhibitors of *M. tuberculosis* kinases.

A number of groups have undertaken screens to identify inhibitors of *M. tuberculosis* kinases, with most efforts focusing on the essential kinases PknA and PknB and on PknG, which is required for virulence and regulates central metabolism (29, 65, 101–103). A number of *M. tuberculosis* kinase inhibitors have advanced through medicinal chemistry efforts to optimize their activity. In some cases enzyme inhibitory activity in the low nanomolar range has been achieved. Some compounds also show substantial antibacterial activity, with MICs in the low micromolar range (102–104), but none has achieved potency sufficient to lead to evaluation in clinical trials.

The protein phosphatases also present important opportunities for antitubercular drug development (105). Phosphatases have advantages and challenges as targets for drug development that are similar to those of the *M. tuberculosis* kinases; i.e., the presence of similar enzymes in humans provides a wealth of knowledge but raises concerns regarding selectivity and potential for toxicity (106). The smaller number of protein phosphatases relative to kinases suggests that these phosphatases will be less specific than protein kinases, and lack of selectivity has been a challenge in developing phosphatase inhibitors to treat human diseases. In addition, the natural substrate includes the charged phosphate residue; active site inhibitors that share this charge are unlikely to enter the host cell, a problem that has slowed phosphatase inhibitor development for human diseases and will be a challenge in targeting the secreted protein phosphatases of *M. tuberculosis*. Potentially offsetting these challenges, the fact that

PtpA and PtpB are both secreted proteins means that a major impediment to developing antituberculars that target proteins in the bacterial cytoplasm, passage across the relatively impermeable mycobacterial cell envelope, does not have to be overcome. The PtpA structure is highly similar to that of human protein phosphatases, so that development of highly selective inhibitors may be difficult. In contrast, though PtpB retains the phosphatase active site signature sequence, it is otherwise highly divergent from known human phosphatases and thus may be especially attractive as a target for inhibitors. Further, as described above, both PtpA and PtpB have unique structural features that may allow more specific targeting of these proteins.

Several groups have identified compounds that inhibit PtpA. Manger et al. identified two classes of natural compounds with IC50s in the low micromolar range and also screened a fragment-based library to identify synthetic molecules that inhibit PtpA activity, with the most potent having K_is of 1 to 2 μM (107). The compounds that were tested, however, also had significant activity against one or more human phosphatases. Chiradia et al. identified synthetic chalcones as inhibitors of PtpA, and in follow-up work elucidated structure-activity relationships and demonstrated inhibition by one compound of *M. bovis* BCG in human macrophages but not in axenic culture (108–110). A group that had previously identified PtpB inhibitors using a fragment-based approach applied this method to identify inhibitors of PtpA (111). The most potent of these (K_i = 1.4 μM) was also selective relative to human Tyr and dual-specificity kinases.

In the case of PtpB, a number of groups have identified potent inhibitory compounds. Grundner et al., using scaffolds from which inhibitors of eukaryotic Tyr phosphatases had been developed, identified a competitive inhibitor of PtpB with IC$_{50}$ of 0.44 μM, with good selectivity versus several human tyrosine phosphatases (112). Crystal structure analysis of this compound in complex with PtpB demonstrated conformational changes and key residues that may guide further inhibitor development. Soellner et al. subsequently used a fragment-based approach to identify a more potent inhibitor that maintained good selectivity (113). Beresford et al. identified inhibitors with IC$_{50}$s in the low micromolar range with good selectivity (114). One of these compounds was shown to inhibit growth of *M. bovis* BCG in resting murine macrophages, but not in axenic culture, consistent with an effect on macrophage function rather than a direct antibacterial effect.

SUMMARY AND FUTURE RESEARCH

Since the identification of 11 STPKs in the *M. tuberculosis* genome sequence in 1998 (115), substantial progress has been made in gaining insight into the role of Ser/Thr phosphorylation in regulating *M. tuberculosis* physiology. The effect of phosphorylation on protein function has been elucidated for several *M. tuberculosis* proteins, a number of which have been highlighted in this article. In particular, important insights have been gained into the roles of protein phosphorylation in regulating the synthesis of several components of the mycobacterial cell envelope, including PGN, glycans, and lipids. To a lesser extent, knowledge has been acquired regarding the cognate STPKs that target specific enzymes to control these and other cell functions. Structural analyses have also provided key insights into STPK function, first demonstrating the structural and mechanistic similarity between the mycobacterial and eukaryotic protein kinases, and then providing important and novel insights regarding activation and regulation of *M. tuberculosis* kinases and phosphatases, as well as potential ways to target these molecules with inhibitors.

Despite these advances, in many ways we have barely scratched the surface of the regulatory roles of protein phosphorylation in *M. tuberculosis*. Going forward, we expect that many new phosphoproteins will be identified, and the effects of phosphorylation on protein function will be discerned for many of these. Identification of cognate kinases for individual phosphoproteins, though challenging, is critical for deepening our understanding of the regulatory roles of Ser/Thr phosphorylation in *M. tuberculosis*. Identifying ligands for the receptor-type STPKs and how ligand binding affects kinase function will provide key new information on how these signaling molecules sense the external environment. Ultimately, integration of these data to construct functional signal transduction networks will provide deep insights into the mechanisms by which *M. tuberculosis* adapts during the course of infection of the mammalian host. Finally, we hope to see these insights translated into novel and potent antituberculars to improve the treatment of drug-susceptible and drug-resistant tuberculosis.

The authors thank Benoit Smagghe for help with the protein structure figures. The authors' work on the M. tuberculosis *kinases has been supported by grants from the National Institute of Allergy and Infectious Diseases and from Vertex Pharmaceuticals Incorporated.*

Citation. Prisic S, Husson RN. 2014. *Mycobacterium tuberculosis* serine/threonine protein kinases. Microbiol Spectrum 2(5):MGM2-0006-2013.

References

1. **West AH, Stock AM.** 2001. Histidine kinases and response regulator proteins in two-component signaling systems. *Trends Biochem Sci* **26:**369–376.
2. **Ulrich LE, Koonin EV, Zhulin IB.** 2005. One-component systems dominate signal transduction in prokaryotes. *Trends Microbiol* **13:**52–56.
3. **Staroń A, Sofia H, Dietrich S, Ulrich L, Liesegang H, Mascher T.** 2009. The third pillar of bacterial signal transduction: classification of the extracytoplasmic function (ECF) sigma factor protein family. *Mol Microbiol* **74:**557–581.
4. **Lonetto M, Brown K, Rudd K, Buttner M.** 1994. Analysis of the *Streptomyces coelicolor sigE* gene reveals the existence of a subfamily of eubacterial RNA polymerase σ factors involved in the regulation of extracytoplasmic functions. *Proc Natl Acad Sci USA* **91:**7573–7577.
5. **Song T, Dove SL, Lee KH, Husson RN.** 2003. RshA, an anti-sigma factor that regulates the activity of the mycobacterial stress response sigma factor SigH. *Mol Microbiol* **50:**949–959.
6. **Hahn MY, Raman S, Anaya M, Husson RN.** 2005. The *Mycobacterium tuberculosis* ECF sigma factor SigL regulates polyketide synthases and membrane/secreted proteins, and is required for virulence. *J Bacteriol* **187:** 7062–7071.
7. **Perez J, Castaneda-Garcia A, Jenke-Kodama H, Muller R, Munoz-Dorado J.** 2008. Eukaryotic-like protein kinases in the prokaryotes and the myxobacterial kinome. *Proc Natl Acad Sci USA* **105:**15950–15955.
8. **Munoz-Dorado J, Inouye S, Inouye M.** 1991. A gene encoding a protein serine/threonine kinase is required for normal development of *M. xanthus*, a gram-negative bacterium. *Cell* **67:**995–1006.
9. **Av-Gay Y, Everett M.** 2000. The eukaryotic-like Ser/Thr protein kinases of *Mycobacterium tuberculosis*. *Trends Microbiol* **8:**238–244.
10. **Hanks S, Hunter T.** 1995. The eukaryotic protein kinase superfamily: kinase (catalytic) domain structure and classification. *FASEB J* **9:**576–596.
11. **Sassetti CM, Boyd DH, Rubin EJ.** 2003. Genes required for mycobacterial growth defined by high density mutagenesis. *Mol Microbiol* **48:**77–84.
12. **Ortiz-Lombardia M, Pompeo F, Boitel B, Alzari PM.** 2003. Crystal structure of the catalytic domain of the PknB serine/threonine kinase from *Mycobacterium tuberculosis*. *J Biol Chem* **278:**13094–13100.
13. **Young TA, Delagoutte B, Endrizzi JA, Falick AM, Alber T.** 2003. Structure of *Mycobacterium tuberculosis* PknB supports a universal activation mechanism for Ser/Thr protein kinases. *Nat Struct Biol* **10:**168–174.
14. **Gay LM, Ng HL, Alber T.** 2006. A conserved dimer and global conformational changes in the structure of apo-PknE Ser/Thr protein kinase from *Mycobacterium tuberculosis*. *J Mol Biol* **360:**409–420.
15. **Prisic S, Dankwa S, Schwartz D, Chou MF, Locasale JW, Kang CM, Bemis G, Church GM, Steen H, Husson RN.** 2010. Extensive phosphorylation with overlapping specificity by *Mycobacterium tuberculosis*

serine/threonine protein kinases. *Proc Natl Acad Sci USA* 107:7521–7526.

16. Boitel B, Ortiz-Lombardia M, Duran R, Pompeo F, Cole ST, Cervenansky C, Alzari PM. 2003. PknB kinase activity is regulated by phosphorylation in two Thr residues and dephosphorylation by PstP, the cognate phospho-Ser/Thr phosphatase, in *Mycobacterium tuberculosis*. *Mol Microbiol* 49:1493–1508.

17. Kang CM, Abbott DW, Park ST, Dascher CC, Cantley LC, Husson RN. 2005. The *Mycobacterium tuberculosis* serine/threonine kinases PknA and PknB: substrate identification and regulation of cell shape. *Genes Dev* 19:1692–1704.

18. Molle V, Zanella-Cleon I, Robin JP, Mallejac S, Cozzone AJ, Becchi M. 2006. Characterization of the phosphorylation sites of *Mycobacterium tuberculosis* serine/threonine protein kinases, PknA, PknD, PknE, and PknH by mass spectrometry. *Proteomics* 6:3754–3766.

19. O'Hare HM, Duran R, Cervenansky C, Bellinzoni M, Wehenkel AM, Pritsch O, Obal G, Baumgartner J, Vialaret J, Johnsson K, Alzari PM. 2008. Regulation of glutamate metabolism by protein kinases in mycobacteria. *Mol Microbiol* 70:1408–1423.

20. Duran R, Villarino A, Bellinzoni M, Wehenkel A, Fernandez P, Boitel B, Cole ST, Alzari PM, Cervenansky C. 2005. Conserved autophosphorylation pattern in activation loops and juxtamembrane regions of *Mycobacterium tuberculosis* Ser/Thr protein kinases. *Biochem Biophys Res Comm* 333:858–867.

21. Mieczkowski C, Iavarone AT, Alber T. 2008. Autoactivation mechanism of the *Mycobacterium tuberculosis* PknB receptor Ser/Thr kinase. *Embo J* 27:3186–3197.

22. Lombana TN, Echols N, Good MC, Thomsen ND, Ng H-L, Greenstein AE, Falick AM, King DS, Alber T. 2010. Allosteric activation mechanism of the *Mycobacterium tuberculosis* receptor Ser/Thr protein kinase, PknB. *Structure* 18:1667–1677.

23. Greenstein AE, Echols N, Lombana TN, King DS, Alber T. 2007. Allosteric activation by dimerization of the PknD receptor Ser/Thr protein kinase from *Mycobacterium tuberculosis*. *J Biol Chem* 282:11427–11435.

24. Wehenkel A, Bellinzoni M, Graña M, Duran R. 2008. Mycobacterial Ser/Thr protein kinases and phosphatases: physiological roles and therapeutic potential. *Biochim Biophys Acta* 1784:193–202.

25. Wehenkel A, Fernandez P, Bellinzoni M, Catherinot V, Barilone N, Labesse G, Jackson M, Alzari PM. 2006. The structure of PknB in complex with mitoxantrone, an ATP-competitive inhibitor, suggests a mode of protein kinase regulation in mycobacteria. *FEBS Lett* 580:3018–3022.

26. Dey M, Cao C, Dar AC, Tamura T, Ozato K, Sicheri F, Dever TE. 2005. Mechanistic link between PKR dimerization, autophosphorylation, and eIF2alpha substrate recognition. *Cell* 122:901–913.

27. Mir M, Asong J, Li X, Cardot J, Boons GJ, Husson RN. 2011. The extracytoplasmic domain of the *Mycobacterium tuberculosis* Ser/Thr kinase PknB binds specific muropeptides and is required for PknB localization. *PLoS Pathog* 7:e1002182.

28. Barthe P, Mukamolova GV, Roumestand C, Cohen-Gonsaud M. 2010. The structure of PknB extracellular PASTA domain from *Mycobacterium tuberculosis* suggests a ligand-dependent kinase activation. *Structure* 18:606–615.

29. Scherr N, Honnappa S, Kunz G, Mueller P, Jayachandran R, Winkler F, Pieters J, Steinmetz MO. 2007. Structural basis for the specific inhibition of protein kinase G, a virulence factor of *Mycobacterium tuberculosis*. *Proc Natl Acad Sci USA* 104:12151–12156.

30. Tiwari D, Singh RK, Goswami K, Verma SK, Prakash B, Nandicoori VK. 2009. Key residues in *Mycobacterium tuberculosis* protein kinase G play a role in regulating kinase activity and survival in the host. *Journal of Biological Chemistry* 284:27467–27479.

31. Gil M, Graña M, Schopfer FJ, Wagner T, Denicola A, Freeman BA, Alzari PM, Batthyány C, Duran R. 2013. Inhibition of *Mycobacterium tuberculosis* PknG by non-catalytic rubredoxin domain specific modification: reaction of an electrophilic nitro-fatty acid with the Fe-S center. *Free Radic Biol Med* 65:150–161.

32. Braconi Quintaje S, Orchard S. 2008. The annotation of both human and mouse kinomes in UniProtKB/Swiss-Prot: one small step in manual annotation, one giant leap for full comprehension of genomes. *Mol Cell Proteomics* 7:1409–1419.

33. Manning G, Whyte DB, Martinez R, Hunter T, Sudarsanam S. 2002. The protein kinase complement of the human genome. *Science* 298:1912–1934.

34. Kinexus. March 1, 2013, posting date. *Phosphonet: Human Phospho-Site KnowledgeBase*. [Online.]

35. Bach H, Wong D, Av-Gay Y. 2009. *Mycobacterium tuberculosis* PtkA is a novel protein tyrosine kinase whose substrate is PtpA. *Biochem J* 420:155–160.

36. Dasgupta A, Datta P, Kundu M, Basu J. 2006. The serine/threonine kinase PknB of *Mycobacterium tuberculosis* phosphorylates PBPA, a penicillin-binding protein required for cell division. *Microbiology* 152:493–504.

37. Gee CL, Papavinasasundaram KG, Blair SR, Baer CE, Falick AM, King DS, Griffin JE, Venghatakrishnan H, Zukauskas A, Wei JR, Dhiman RK, Crick DC, Rubin EJ, Sassetti CM, Alber T. 2012. A phosphorylated pseudokinase complex controls cell wall synthesis in mycobacteria. *Sci Signal* 5:ra7.

38. Ortega C, Liao R, Anderson LN, Rustad T, Ollodart AR, Wright AT, Sherman DR, Grundner C. 2014. *Mycobacterium tuberculosis* Ser/Thr protein kinase B mediates an oxygen-dependent replication switch. *PLoS Biol* 12:e1001746.

39. Yeats C, Finn RD, Bateman A. 2002. The PASTA domain: a beta-lactam-binding domain. *Trends Biochem Sci* 27:438–440.

40. Shah IM, Laaberki MH, Popham DL, Dworkin J. 2008. A eukaryotic-like Ser/Thr kinase signals bacteria to exit dormancy in response to peptidoglycan fragments. *Cell* 135:486–496.

41. Squeglia F, Marchetti R, Ruggiero A, Lanzetta R, Marasco D, Dworkin J, Petoukhov M, Molinaro A, Berisio R, Silipo A. 2011. Chemical basis of peptidoglycan discrimination by PrkC, a key kinase involved in bacterial resuscitation from dormancy. *J Am Chem Soc* **133**:20676–20679.

42. Kang CM, Nyayapathy S, Lee JY, Suh JW, Husson RN. 2008. Wag31, a homologue of the cell division protein DivIVA, regulates growth, morphology and polar cell wall synthesis in mycobacteria. *Microbiology* **154**:725–735.

43. Daniel RA, Errington J. 2003. Control of cell morphogenesis in bacteria: two distinct ways to make a rod-shaped cell. *Cell* **113**:767–776.

44. Ruiz N. 2008. Bioinformatics identification of MurJ (MviN) as the peptidoglycan lipid II flippase in *Escherichia coli. Proc Natl Acad Sci USA* **105**:15553–15557.

45. Larsen MH, Vilchèze C, Kremer L, Besra GS, Parsons L, Salfinger M, Heifets L, Hazbon MH, Alland D, Sacchettini JC, Jacobs WR. 2002. Overexpression of *inhA*, but not *kasA*, confers resistance to isoniazid and ethionamide in *Mycobacterium smegmatis, M. bovis* BCG and *M. tuberculosis. Mol Microbiol* **46**:453–466.

46. Vilchèze C, Wang F, Arai M, Hazbón MH, Colangeli R, Kremer L, Weisbrod TR, Alland D, Sacchettini JC, Jacobs WR. 2006. Transfer of a point mutation in *Mycobacterium tuberculosis inhA* resolves the target of isoniazid. *Nat Med* **12**:1027–1029.

47. Khan S, Nagarajan SN, Parikh A, Samantaray S, Singh A, Kumar D, Roy RP, Bhatt A, Nandicoori VK. 2010. Phosphorylation of enoyl-acyl carrier protein reductase InhA impacts mycobacterial growth and survival. *J Biol Chem* **285**:37860–37871.

48. Molle V, Gulten G, Vilchèze C, Veyron-Churlet R, Zanella-Cleon I, Sacchettini JC, Jacobs WR, Kremer L. 2010. Phosphorylation of InhA inhibits mycolic acid biosynthesis and growth of *Mycobacterium tuberculosis. Mol Microbiol* **78**:1591–1605.

49. Molle V, Brown AK, Besra GS, Cozzone AJ, Kremer L. 2006. The condensing activities of the *Mycobacterium tuberculosis* type II fatty acid synthase are differentially regulated by phosphorylation. *J Biol Chem* **281**:30094–30103.

50. Veyron-Churlet R, Zanella-Cleon I, Cohen-Gonsaud M, Molle V, Kremer L. 2010. Phosphorylation of the *Mycobacterium tuberculosis* beta-ketoacyl-acyl carrier protein reductase MabA regulates mycolic acid biosynthesis. *J Biol Chem* **285**:12714–12725.

51. Veyron-Churlet R, Molle V, Taylor RC, Brown AK, Besra GS, Zanella-Cleon I, Fütterer K, Kremer L. 2009. The *Mycobacterium tuberculosis* beta-ketoacyl-acyl carrier protein synthase III activity is inhibited by phosphorylation on a single threonine residue. *J Biol Chem* **284**:6414–6424.

52. Camus JC, Pryor MJ, Medigue C, Cole ST. 2002. Re-annotation of the genome sequence of *Mycobacterium tuberculosis* H37Rv. *Microbiology* **148**:2967–2973.

53. Good MC, Greenstein AE, Young TA, Ng HL, Alber T. 2004. Sensor domain of the *Mycobacterium tuberculosis* receptor Ser/Thr protein kinase, PknD, forms a highly symmetric beta propeller. *J Mol Biol* **339**:459–469.

54. Vanzembergh F, Peirs P, Lefèvre P, Celio N, Mathys V, Content J, Kalai M. 2010. Effect of PstS sub-units or PknD deficiency on the survival of *Mycobacterium tuberculosis. Tuberculosis (Edinb)* **90**:338–345.

55. Greenstein AE, Macgurn JA, Baer CE, Falick AM, Cox JS, Alber T. 2007. *M. tuberculosis* Ser/Thr protein kinase D phosphorylates an anti-anti-sigma factor homolog. *PLoS Pathog* **3**:e49.

56. Hatzios SK, Baer CE, Rustad TR, Siegrist MS, Pang JM, Ortega C, Alber T, Grundner C, Sherman DR, Bertozzi CR. 2013. Osmosensory signaling in *Mycobacterium tuberculosis* mediated by a eukaryotic-like Ser/Thr protein kinase. *Proceedings of the National Academy of Sciences* **110**:E5069–E5077.

57. Be NA, Bishai WR, Jain SK. 2012. Role of *Mycobacterium tuberculosis pknD* in the pathogenesis of central nervous system tuberculosis. *BMC Microbiol* **12**:7.

58. Kumar D, Narayanan S. 2012. PknE, a serine/threonine kinase of *Mycobacterium tuberculosis*, modulates multiple apoptotic paradigms. *Infect Genet Evol* **12**:737–747.

59. Hofmann K, Bucher P. 1995. The FHA domain: a putative nuclear signalling domain found in protein kinases and transcription factors. *Trends Biochem Sci* **20**:347–349.

60. Molle V, Soulat D, Jault JM, Grangeasse C, Cozzone AJ, Prost JF. 2004. Two FHA domains on an ABC transporter, Rv1747, mediate its phosphorylation by PknF, a Ser/Thr protein kinase from *Mycobacterium tuberculosis. FEMS Microbiol Lett* **234**:215–223.

61. Curry JM, Whalan R, Hunt DM, Gohil K, Strom M, Rickman L, Colston MJ, Smerdon SJ, Buxton RS. 2005. An ABC transporter containing a forkhead-associated domain interacts with a serine-threonine protein kinase and is required for growth of *Mycobacterium tuberculosis* in mice. *Infect Immun* **73**:4471–4477.

62. Deol P, Vohra R, Saini AK, Singh A, Chandra H, Chopra P, Das TK, Tyagi AK, Singh Y. 2005. Role of *Mycobacterium tuberculosis* Ser/Thr kinase PknF: implications in glucose transport and cell division. *J Bacteriol* **187**:3415–3420.

63. Spivey VL, Molle V, Whalan RH, Rodgers A, Leiba J, Stach L, Walker KB, Smerdon SJ, Buxton RS. 2011. Forkhead-associated (FHA) domain containing ABC transporter Rv1747 is positively regulated by Ser/Thr phosphorylation in *Mycobacterium tuberculosis. J Biol Chem* **286**:26198–26209.

64. Marchler-Bauer A, Lu S, Anderson JB, Chitsaz F, Derbyshire MK, DeWeese-Scott C, Fong JH, Geer LY, Geer RC, Gonzales NR, Gwadz M, Hurwitz DI, Jackson JD, Ke Z, Lanczycki CJ, Lu F, Marchler GH, Mullokandov M, Omelchenko MV, Robertson CL, Song JS, Thanki N, Yamashita RA, Zhang D, Zhang N, Zheng C, Bryant SH. 2011. CDD: a Conserved Domain Database for the functional annotation of proteins. *Nucleic Acids Res* **39**:D225–D229.

65. Walburger A, Koul A, Ferrari G, Nguyen L, Prescianotto-Baschong C, Huygen K, Klebl B, Thompson C, Bacher

G, Pieters J. 2004. Protein kinase G from pathogenic mycobacteria promotes survival within macrophages. *Science* 304:1800–1804.

66. Cowley S, Ko M, Pick N, Chow R, Downing KJ, Gordhan BG, Betts JC, Mizrahi V, Smith DA, Stokes RW, Av-Gay Y. 2004. The *Mycobacterium tuberculosis* protein serine/threonine kinase PknG is linked to cellular glutamate/glutamine levels and is important for growth in vivo. *Mol Microbiol* 52:1691–1702.

67. Ventura M, Rieck B, Boldrin F, Degiacomi G, Bellinzoni M, Barilone N, Alzaidi F, Alzari PM, Manganelli R, O'Hare HM. 2013. GarA is an essential regulator of metabolism in *Mycobacterium tuberculosis*. *Mol Microbiol* 90:356–366.

68. Niebisch A, Kabus A, Schultz C, Weil B, Bott M. 2006. Corynebacterial protein kinase G controls 2-oxoglutarate dehydrogenase activity via the phosphorylation status of the OdhI protein. *J Biol Chem* 281:12300–12307.

69. Nott TJ, Kelly G, Stach L, Li J, Westcott S, Patel D, Hunt DM, Howell S, Buxton RS, O'Hare HM, Smerdon SJ. 2009. An intramolecular switch regulates phosphoindependent FHA domain interactions in *Mycobacterium tuberculosis*. *Sci Signaling* 2:1–9.

70. Cavazos A, Prigozhin DM, Alber T. 2012. Structure of the sensor domain of *Mycobacterium tuberculosis* PknH receptor kinase reveals a conserved binding cleft. *J Mol Biol* 422:488–494.

71. Molle V, Kremer L, Girard-Blanc C, Besra GS, Cozzone AJ, Prost JF. 2003. An FHA phosphoprotein recognition domain mediates protein EmbR phosphorylation by PknH, a Ser/Thr protein kinase from *Mycobacterium tuberculosis*. *Biochemistry* 42:15300–15309.

72. Zhang N, Torrelles JB, McNeil MR, Escuyer VE, Khoo K-H, Brennan PJ, Chatterjee D. 2003. The Emb proteins of mycobacteria direct arabinosylation of lipoarabinomannan and arabinogalactan via an N-terminal recognition region and a C-terminal synthetic region. *Mol Microbiol* 50:69–76.

73. Escuyer VE, Lety MA, Torrelles JB, Khoo KH, Tang JB, Rithner CD, Frehel C, Mcneil MR, Brennan PJ, Chatterjee D. 2001. The role of the *embA* and *embB* gene products in the biosynthesis of the terminal hexaarabinofuranosyl motif of *Mycobacterium smegmatis* arabinogalactan. *J Biol Chem* 276:48854–48862.

74. Sharma K, Gupta M, Pathak M, Gupta N, Koul A, Sarangi S, Baweja R, Singh Y. 2006. Transcriptional control of the mycobacterial *embCAB* operon by PknH through a regulatory protein, EmbR, in vivo. *J Bacteriol* 188:2936–2944.

75. Papavinasasundaram KG, Chan B, Chung JH, Colston MJ, Davis EO, Av-Gay Y. 2005. Deletion of the *Mycobacterium tuberculosis pknH* gene confers a higher bacillary load during the chronic phase of infection in BALB/c mice. *J Bacteriol* 187:5751–5760.

76. Gomez-Velasco A, Bach H, Rana AK, Cox LR, Bhatt A, Besra GS, Av-Gay Y. 2013. Disruption of the serine/threonine protein kinase H affects phthiocerol dimycocerosates synthesis in *Mycobacterium tuberculosis*. *Microbiology* 159:726–736.

77. Chao JD, Papavinasasundaram KG, Zheng X, Chávez-Steenbock A, Wang X, Lee GQ, Av-Gay Y. 2010. Convergence of Ser/Thr and two-component signaling to coordinate expression of the dormancy regulon in *Mycobacterium tuberculosis*. *J Biol Chem* 285:29239–29246.

78. Gopalaswamy R, Narayanan S, Chen B, Jacobs WR, Av-Gay Y. 2009. The serine/threonine protein kinase PknI controls the growth of *Mycobacterium tuberculosis* upon infection. *FEMS Microbiol Lett* 295:23–29.

79. Jang J, Stella A, Boudou F, Levillain F, Darthuy E, Vaubourgeix J, Wang C, Bardou F, Puzo G, Gilleron M, Burlet-Schiltz O, Monsarrat B, Brodin P, Gicquel B, Neyrolles O. 2010. Functional characterization of the *Mycobacterium tuberculosis* serine/threonine kinase PknJ. *Microbiology* 156:1619–1631.

80. Blatch GL, Lässle M. 1999. The tetratricopeptide repeat: a structural motif mediating protein-protein interactions. *Bioessays* 21:932–939.

81. Kumar P, Kumar D, Parikh A, Rananaware D, Gupta M, Singh Y, Nandicoori VK. 2009. The *Mycobacterium tuberculosis* protein kinase K modulates activation of transcription from the promoter of mycobacterial monooxygenase operon through phosphorylation of the transcriptional regulator VirS. *J Biol Chem* 284:11090–11099.

82. Malhotra V, Arteaga-Cortés LT, Clay G, Clark-Curtiss JE. 2010. *Mycobacterium tuberculosis* protein kinase K confers survival advantage during early infection in mice and regulates growth in culture and during persistent infection: implications for immune modulation. *Microbiology* 156:2829–2841.

83. Malhotra V, Okon BP, Clark-Curtiss JE. 2012. *Mycobacterium tuberculosis* protein kinase K enables growth adaptation through translation control. *J Bacteriol* 194:4184–4196.

84. Narayan A, Sachdeva P, Sharma K, Saini AK, Tyagi AK, Singh Y. 2007. Serine threonine protein kinases of mycobacterial genus: phylogeny to function. *Physiol Genomics* 29:66–75.

85. Canova MJ, Veyron-Churlet R, Zanella-Cleon I, Cohen-Gonsaud M, Cozzone AJ, Becchi M, Kremer L, Molle V. 2008. The *Mycobacterium tuberculosis* serine/threonine kinase PknL phosphorylates Rv2175c: mass spectrometric profiling of the activation loop phosphorylation sites and their role in the recruitment of Rv2175c. *Proteomics* 8:521–533.

86. Barford D, Das AK, Egloff MP. 1998. The structure and mechanism of protein phosphatases: insights into catalysis and regulation. *Annu Rev Biophys Biomol Struct* 27:133–164.

87. Pullen KE, Ng H-L, Sung P-Y, Good MC, Smith SM, Alber T. 2004. An alternate conformation and a third metal in PstP/Ppp, the *M. tuberculosis* PP2C-family Ser/Thr protein phosphatase. *Structure* 12:1947–1954.

88. Koul A, Choidas A, Treder M, Tyagi AK, Drlica K, Singh Y, Ullrich A. 2000. Cloning and characterization of secretory tyrosine phosphatases of *Mycobacterium tuberculosis*. *J Bacteriol* 182:5425–5432.

89. Beresford N, Patel S, Armstrong J, Szöor B, Fordham-Skelton AP, Tabernero L. 2007. MptpB, a virulence factor from *Mycobacterium tuberculosis,* exhibits triple-specificity phosphatase activity. *Biochem J* 406:13–18.

90. Stehle T, Sreeramulu S, Löhr F, Richter C, Saxena K, Jonker HRA, Schwalbe H. 2012. The apo-structure of the low molecular weight protein-tyrosine phosphatase A (MptpA) from *Mycobacterium tuberculosis* allows for better target-specific drug development. *J Biol Chem* 287:34569–34582.

91. Madhurantakam C, Rajakumara E, Mazumdar PA, Saha B, Mitra D, Wiker HG, Sankaranarayanan R, Das AK. 2005. Crystal structure of low-molecular-weight protein tyrosine phosphatase from *Mycobacterium tuberculosis* at 1.9-A resolution. *J Bacteriol* 187:2175–2181.

92. Flynn EM, Hanson JA, Alber T, Yang H. 2010. Dynamic active-site protection by the *M. tuberculosis* protein tyrosine phosphatase PtpB lid domain. *J Am Chem Soc* 132:4772–4780.

93. Grundner C, Ng H-L, Alber T. 2005. *Mycobacterium tuberculosis* protein tyrosine phosphatase PtpB structure reveals a diverged fold and a buried active site. *Structure* 13:1625–1634.

94. Bach H, Papavinasasundaram KG, Wong D, Hmama Z, Av-Gay Y. 2008. *Mycobacterium tuberculosis* virulence is mediated by PtpA dephosphorylation of human vacuolar protein sorting 33B. *Cell Host Microbe* 3:316–322.

95. Wong D, Bach H, Sun J, Hmama Z, Av-Gay Y. 2011. *Mycobacterium tuberculosis* protein tyrosine phosphatase (PtpA) excludes host vacuolar-H+-ATPase to inhibit phagosome acidification. *Proc Natl Acad Sci USA* 108:19371–19376.

96. Zhou B, He Y, Zhang X, Xu J, Luo Y, Wang Y, Franzblau SG, Yang Z, Chan RJ, Liu Y, Zheng J, Zhang Z-Y. 2010. Targeting mycobacterium protein tyrosine phosphatase B for antituberculosis agents. *Proc Natl Acad Sci USA* 107:4573–4578.

97. Singh R, Rao V, Shakila H, Gupta R, Khera A, Dhar N, Singh A, Koul A, Singh Y, Naseema M, Narayanan PR, Paramasivan CN, Ramanathan VD, Tyagi AK. 2003. Disruption of *mptpB* impairs the ability of *Mycobacterium tuberculosis* to survive in guinea pigs. *Mol Microbiol* 50:751–762.

98. Saleh MT, Belisle JT. 2000. Secretion of an acid phosphatase (SapM) by *Mycobacterium tuberculosis* that is similar to eukaryotic acid phosphatases. *J Bacteriol* 182:6850–6853.

99. Vergne I, Chua J, Lee H-H, Lucas M, Belisle J, Deretic V. 2005. Mechanism of phagolysosome biogenesis block by viable *Mycobacterium tuberculosis*. *Proc Natl Acad Sci USA* 102:4033–4038.

100. Zhang J, Yang PL, Gray NS. 2009. Targeting cancer with small molecule kinase inhibitors. *Nat Rev Cancer* 9:28–39.

101. Chapman TM, Bouloc N, Buxton RS, Chugh J, Lougheed KEA, Osborne SA, Saxty B, Smerdon SJ, Taylor DL, Whalley D. 2012. Substituted aminopyrimidine protein kinase B (PknB) inhibitors show activity against *Myco-*

bacterium tuberculosis. *Bioorg Med Chem Lett* 22:3349–3353.

102. Lougheed KEA, Osborne SA, Saxty B, Whalley D, Chapman T, Bouloc N, Chugh J, Nott TJ, Patel D, Spivey VL, Kettleborough CA, Bryans JS, Taylor DL, Smerdon SJ, Buxton RS. 2011. Effective inhibitors of the essential kinase PknB and their potential as antimycobacterial agents. *Tuberculosis (Edinb)* 91:277–286.

103. Hanzelka BL, Wang T, Bemis G, Zuccola H, Doyle T, Stuver-Moody C, Huang YN, Sears C, Fleming M, Erwin AL, Locher C, Muh U. 2009. *Evidence that inhibition of the serine-threonine protein kinases of* Mycobacterium tuberculosis *holds promise for developing novel anti-tuberculosis drugs*. Gordon Research Conference on Tuberculosis Drug Development, Oxford University, Oxford, U.K.

104. Székely R, Wáczek F, Szabadkai I, Németh G, Hegymegi-Barakonyi B, Eros D, Szokol B, Pató J, Hafenbradl D, Satchell J, Saint-Joanis B, Cole ST, Orfi L, Klebl BM, Kéri G. 2008. A novel drug discovery concept for tuberculosis: inhibition of bacterial and host cell signalling. *Immunol Lett* 116:225–231.

105. Wong D, Chao JD, Av-Gay Y. 2013. *Mycobacterium tuberculosis*-secreted phosphatases: from pathogenesis to targets for TB drug development. *Trends Microbiol* 21:100–109.

106. Vintonyak VV, Antonchick AP, Rauh D, Waldmann H. 2009. The therapeutic potential of phosphatase inhibitors. *Curr Opin Chem Biol* 13:272–283.

107. Manger M, Scheck M, Prinz H, von Kries JP, Langer T, Saxena K, Schwalbe H, Fürstner A, Rademann J, Waldmann H. 2005. Discovery of *Mycobacterium tuberculosis* protein tyrosine phosphatase A (MptpA) inhibitors based on natural products and a fragment-based approach. *Chembiochem* 6:1749–1753.

108. Chiaradia LD, Martins PGA, Cordeiro MNS, Guido RVC, Ecco G, Andricopulo AD, Yunes RA, Vernal J, Nunes RJ, Terenzi H. 2012. Synthesis, biological evaluation, and molecular modeling of chalcone derivatives as potent inhibitors of *Mycobacterium tuberculosis* protein tyrosine phosphatases (PtpA and PtpB). *J Med Chem* 55:390–402.

109. Chiaradia LD, Mascarello A, Purificação M, Vernal J, Cordeiro MNS, Zenteno ME, Villarino A, Nunes RJ, Yunes RA, Terenzi H. 2008. Synthetic chalcones as efficient inhibitors of *Mycobacterium tuberculosis* protein tyrosine phosphatase PtpA. *Bioorg Med Chem Lett* 18:6227–6230.

110. Mascarello A, Chiaradia LD, Vernal J, Villarino A, Guido RVC, Perizzolo P, Poirier V, Wong D, Martins PGA, Nunes RJ, Yunes RA, Andricopulo AD, Av-Gay Y, Terenzi H. 2010. Inhibition of *Mycobacterium tuberculosis* tyrosine phosphatase PtpA by synthetic chalcones: kinetics, molecular modeling, toxicity and effect on growth. *Bioorg Med Chem* 18:3783–3789.

111. Rawls KA, Lang PT, Takeuchi J, Imamura S, Baguley TD, Grundner C, Alber T, Ellman JA. 2009. Fragment-based discovery of selective inhibitors of the *Mycobacterium tuberculosis* protein tyrosine phosphatase PtpA. *Bioorg Med Chem Lett* 19:6851–6854.

112. Grundner C, Perrin D, Hooft van Huijsduijnen R, Swinnen D, Gonzalez J, Gee CL, Wells TN, Alber T. 2007. Structural basis for selective inhibition of *Mycobacterium tuberculosis* protein tyrosine phosphatase PtpB. *Structure* 15:499–509.

113. Soellner MB, Rawls KA, Grundner C, Alber T, Ellman JA. 2007. Fragment-based substrate activity screening method for the identification of potent inhibitors of the *Mycobacterium tuberculosis* phosphatase PtpB. *J Am Chem Soc* 129:9613–9615.

114. Beresford NJ, Mulhearn D, Szczepankiewicz B, Liu G, Johnson ME, Fordham-Skelton A, Abad-Zapatero C, Cavet JS, Tabernero L. 2009. Inhibition of MptpB phosphatase from *Mycobacterium tuberculosis* impairs mycobacterial survival in macrophages. *J Antimicrob Chemother* 63:928–936.

115. Cole ST, Brosch R, Parkhill J, Garnier T, Churcher C, Harris D, Gordon SV, Eiglmeier K, Gas S, Barry CE 3rd, Tekaia F, Badcock K, Basham D, Brown D, Chillingworth T, Connor R, Davies R, Devlin K, Feltwell T, Gentles S, Hamlin N, Holroyd S, Hornsby T, Jagels K, Barrell BG. 1998. Deciphering the biology of *Mycobacterium tuberculosis* from the complete genome sequence. *Nature* 393:537–544.

116. Zheng X, Papavinasasundaram KG, Av-Gay Y. 2007. Novel substrates of *Mycobacterium tuberculosis* PknH Ser/Thr kinase. *Biochem Biophys Res Comm* 355:162–168.

117. Sajid A, Arora G, Gupta M, Singhal A, Chakraborty K, Nandicoori VK, Singh Y. 2011. Interaction of *Mycobacterium tuberculosis* elongation factor Tu with GTP is regulated by phosphorylation. *J Bacteriol* 193:5347–5358.

118. Molle V, Reynolds RC, Alderwick LJ, Besra GS, Cozzone AJ, Fütterer K, Kremer L. 2008. EmbR2, a structural homologue of EmbR, inhibits the *Mycobacterium tuberculosis* kinase/substrate pair PknH/EmbR. *Biochem J* 410:309–317.

119. Sinha I, Boon C, Dick T. 2003. Apparent growth phase-dependent phosphorylation of malonyl coenzyme A:acyl carrier protein transacylase (MCAT), a major fatty acid synthase II component in *Mycobacterium bovis* BCG. *FEMS Microbiol Lett* 227:141–147.

120. Grundner C, Gay LM, Alber T. 2005. *Mycobacterium tuberculosis* serine/threonine kinases PknB, PknD, PknE, and PknF phosphorylate multiple FHA domains. *Protein Sci* 14:1918–1921.

121. Roumestand C, Leiba J, Galophe N, Margeat E, Padilla A, Bessin Y, Barthe P, Molle V, Cohen-Gonsaud M. 2011. Structural insight into the *Mycobacterium tuberculosis* Rv0020c protein and its interaction with the PknB kinase. *Structure* 19:1525–1534.

122. Sureka K, Hossain T, Mukherjee P, Chatterjee P, Datta P, Kundu M, Basu J. 2010. Novel role of phosphorylation-dependent interaction between FtsZ and FipA in mycobacterial cell division. *PLoS One* 5:e8590.

123. Thakur M, Chakraborti PK. 2006. GTPase activity of mycobacterial FtsZ is impaired due to its transphosphorylation by the eukaryotic-type Ser/Thr kinase, PknA. *J Biol Chem* 281:40107–40113.

124. Villarino A, Duran R, Wehenkel A, Fernandez P, England P, Brodin P, Cole ST, Zimny-Arndt U, Jungblut PR, Cervenansky C, Alzari PM. 2005. Proteomic identification of *M. tuberculosis* protein kinase substrates: PknB recruits GarA, a FHA domain-containing protein, through activation loop-mediated interactions. *J Mol Biol* 350:953–963.

125. Leiba J, Syson K, Baronian G, Zanella-Cleon I, Kalscheuer R, Kremer L, Bornemann S, Molle V. 2013. *Mycobacterium tuberculosis* maltosyltransferase GlgE, a genetically validated antituberculosis target, is negatively regulated by Ser/Thr phosphorylation. *Journal of Biological Chemistry* 288:16546–16556.

126. Parikh A, Verma SK, Khan S, Prakash B, Nandicoori VK. 2009. PknB-mediated phosphorylation of a novel substrate, N-acetylglucosamine-1-phosphate uridyltransferase, modulates its acetyltransferase activity. *J Mol Biol* 386:451–464.

127. Canova MJ, Kremer L, Molle V. 2009. The *Mycobacterium tuberculosis* GroEL1 chaperone is a substrate of Ser/Thr protein kinases. *J Bacteriol* 191:2876–2883.

128. Perez J, Garcia R, Bach H, de Waard JH, Jacobs WR Jr, Av-Gay Y, Bubis J, Takiff HE. 2006. *Mycobacterium tuberculosis* transporter MmpL7 is a potential substrate for kinase PknD. *Biochem Biophys Res Commun* 348:6–12.

129. Thakur M, Chakraborti PK. 2008. Ability of PknA, a mycobacterial eukaryotic-type serine/threonine kinase, to transphosphorylate MurD, a ligase involved in the process of peptidoglycan biosynthesis. *Biochem J* 415:27–33.

130. Gupta M, Sajid A, Arora G, Tandon V, Singh Y. 2009. Forkhead-associated domain-containing protein Rv0019c and polyketide-associated protein PapA5, from substrates of serine/threonine protein kinase PknB to interacting proteins of *Mycobacterium tuberculosis*. *J Biol Chem* 284:34723–34734.

131. Corrales RM, Molle V, Leiba J, Mourey L, de Chastellier C, Kremer L. 2012. Phosphorylation of mycobacterial PcaA inhibits mycolic acid cyclopropanation: consequences for intracellular survival and for phagosome maturation block. *J Biol Chem* 287:26187–26199.

132. Sajid A, Arora G, Gupta M, Upadhyay S, Nandicoori VK, Singh Y. 2011. Phosphorylation of *Mycobacterium tuberculosis* Ser/Thr phosphatase by PknA and PknB. *PLoS One* 6:e17871.

133. Arora G, Sajid A, Gupta M, Bhaduri A, Kumar P, Basu-Modak S, Singh Y. 2010. Understanding the role of PknJ in *Mycobacterium tuberculosis*: biochemical characterization and identification of novel substrate pyruvate kinase A. *PLoS One* 5:e10772.

134. Park ST, Kang CM, Husson RN. 2008. Regulation of the SigH stress response regulon by an essential protein kinase in *Mycobacterium tuberculosis*. *Proc Natl Acad Sci USA* 105:13105–13110.

135. Cohen-Gonsaud M, Barthe P, Canova MJ, Stagier-Simon C, Kremer L, Roumestand C, Molle V. 2009. The *Mycobacterium tuberculosis* Ser/Thr kinase substrate Rv2175c is a DNA-binding protein regulated by phosphorylation. *J Biol Chem* 284:19290–19300.

136. Schultz J, Milpetz F, Bork P, Ponting CP. 1998. SMART, a simple modular architecture research tool: identification of signaling domains. *Proc Natl Acad Sci USA* **95:** 5857–5864.

137. Letunic I, Doerks T, Bork P. 2012. SMART 7: recent updates to the protein domain annotation resource. *Nucleic Acids Res* **40:**D302–D305.

138. Parish T. 2014. Two-component regulatory systems of mycobacteria. *Microbiol Spectrum* **2**(1):MGM2-0010-2013. doi:10.1128/microbiolspec.MGM2-0010-2013.

139. Manganelli R. 2014. Sigma factors: key molecules in *Mycobacterium tuberculosis* physiology and virulence. *Microbiol Spectrum* **2**(1):MGM2-0007-2013. doi: 10.1128/microbiolspec.MGM2-0007-2013.

The Mycobacterial Lifestyle, Persistence, and Macrophage Survival

VIII

Molecular Genetics of Mycobacteria, 2nd Edition
Edited by Graham F. Hatfull and William R. Jacobs, Jr.
© 2014 American Society for Microbiology, Washington, DC
doi:10.1128/microbiolspec.MGM2-0031-2013

Bree B. Aldridge[1]
Iris Keren[2]
Sarah M. Fortune[3]

The Spectrum of Drug Susceptibility in Mycobacteria

34

One of the major clinical challenges facing the tuberculosis (TB) field is the fact that long durations of treatment are required to fully clear infection. Patients with drug-susceptible TB are treated with 6 to 9 months of a multidrug regimen, while treatment of drug-resistant TB, which necessitates the use of less effective drugs, can take years. The lengths of the standard antibiotic regimens for TB have been empirically determined and have evolved as new TB drugs have been developed. Thus, initial regimens (with streptomycin [SM] or SM and 4-aminosalicylic acid [PAS]) were well over a year in duration; the long duration of therapy was deemed necessary based on patients' symptoms and long times until sputum conversion. The development of rifampin allowed the treatment course of uncomplicated, pulmonary TB to be shortened to 6 months. However, this is still substantially longer than the length of treatment required for other chronic infections; bacterial endocarditis, for example, which is considered difficult to clear, requires only 4 to 6 weeks of antibiotics.

What makes TB so difficult to treat? Several theories have been put forth, which we will broadly group into two categories. First, mycobacteria may be less easily cleared by antibiotics because of unique aspects of their physiology such as their slow growth rate and impermeable cell wall. We consider these to be generalized factors that describe the entire population of bacteria and, thus, can be studied at a population level. Second, an increasingly compelling explanation for the long duration of treatment is that within a mycobacterial population there are specialized subpopulations of cells that are extremely tolerant of antibiotics. These subpopulations of cells may be small, and their biology may be obscured in population-level analyses. Thus, identifying and studying the molecular biology of these subpopulations requires distinct experimental approaches. Importantly, however, there are conceptual similarities in the mechanisms underlying drug tolerance that act across entire populations or only within a subpopulation of cells. That is, drug tolerance typically reflects a few basic aspects of cell physiology—growth rate, metabolic state, cell wall composition, and permeability—that may change either globally or in subsets of cells.

[1]Department of Molecular Biology and Microbiology, Tufts University, Boston, MA 02111, and Department of Biomedical Engineering, Tufts University, Medford, MA 02155; [2]Antimicrobial Discovery Center and Department of Biology, Northeastern University, Boston, MA 02115; [3]Department of Immunology and Infectious Diseases, Harvard School of Public Health, Boston, MA 02115.

DEFINITIONS

In a clinical diagnostic laboratory, *Mycobacterium tuberculosis* is tested for drug susceptibility, which is reported in binary terms: *drug sensitive* or *drug resistant*. These terms seem deceptively straightforward. Drug resistance is commonly understood to mean genetically encoded changes in the bacterium—either in the drug target or the drug activation pathway—that significantly reduce the activity of the given drug in a manner that cannot be overcome by simply increasing drug concentration. In mycobacteria, drug resistance is achieved by changes in the drug target or activating enzyme such that the drug can no longer act on its target or is not converted to its active form.

Drug susceptibility also seems self-evident. However, it is clear that drug susceptibility is more complicated than the binary definition from the clinical microbiology lab implies. First, there is a large body of literature demonstrating that for a given antibiotic, the efficacy of even a highly drug-susceptible strain of *M. tuberculosis* varies significantly according to bacterial growth conditions. This has been well demonstrated for environmental stressors such as hypoxia, nutrient limitation, and low pH, all of which are relevant conditions faced by *M. tuberculosis* in the infected host (1–4).

In addition, population-level measures of drug susceptibility obscure the fact that within a population of genetically identical cells, there are subpopulations of cells that can vary significantly in their drug susceptibility. The best known example of this is *persister cells*, which were identified as nongrowing and able to withstand drug concentrations many times higher than the MIC (5–8). However, more recent work in both model organisms and mycobacteria suggests that persister cells might be more accurately viewed as the end of a continuum of physiologic states, and thus drug susceptibilities, within a seemingly homogeneous population, arise through a variety of mechanisms.

Often, the convention in the field is to refer to these states of altered drug susceptibility as reflecting *phenotypic drug tolerance*. The term *phenotypic* is used to signify changes in drug activity that are temporary, dependent on a specific intrinsic or environmentally programmed physiologic state. *Phenotypic* is used in distinction to genetic effects, that is, effects that are encoded for on the chromosome.

We found it more difficult to rigorously define the term *drug tolerance* to distinguish it from *drug resistance*. The term *drug tolerance* can be used to denote instances where increased concentrations of drug are required to achieve the same effect as under the reference condition (typically logarithmic-phase growth). However, this definition covers many instances of what would be considered true drug resistance. For example, *M. tuberculosis* carrying genetic mutations in *rpoB* that confer what is considered to be genetic resistance to rifampin are killed by extremely high levels of rifampin exposure. Moreover, many studies do not elucidate the extent to which a change in mycobacterial susceptibility to drug can be overcome at high drug concentrations.

We therefore adopted an alternative definition of *drug tolerance*, which is the bacterial state that allows the bacterium to survive but not grow in the presence of drug, where changes in the bacterium that allow growth in the presence of drug with an increase in MIC would be considered drug resistance. Under this definition, drug tolerance becomes nearly synonymous with growth-arrested conditions. The temporary states that alter drug susceptibility but still allow growth (induction of efflux pump expression or variable KatG expression in some cells) will be considered *phenotypic drug resistance*. Importantly, the changes in drug susceptibility required to affect clinical outcome are poorly defined and likely to be different for each drug.

Here we explore mechanisms by which cells with phenotypic changes in drug susceptibility arise within a mycobacterial population (Fig. 1). We include mechanisms identified through population-level studies, specifically growth arrest, efflux pump induction, and biofilm formation. We then consider the different mechanisms of altered drug susceptibility identified in single-cell studies. We anticipate that population-level studies will elucidate paths to phenotypic changes in drug susceptibility that will also be important to the durable survival of small subpopulations of mycobacterial cells under conditions where the bulk of the population is eliminated by drug. We expect that the importance of biologically distinct subpopulations will be better elucidated in future studies.

POPULATION-LEVEL DRUG TOLERANCE THROUGH GROWTH ARREST AND METABOLIC REMODELING

One of the pathognomonic features of TB is the chronic nature of the infection, as evidenced both by the slow progression of infection even in cases of active disease and, more clearly, by clinically latent infection. These clinical aspects of TB reflect significant metabolic changes in the organism in the host environment, the most obvious of which is growth slowing.

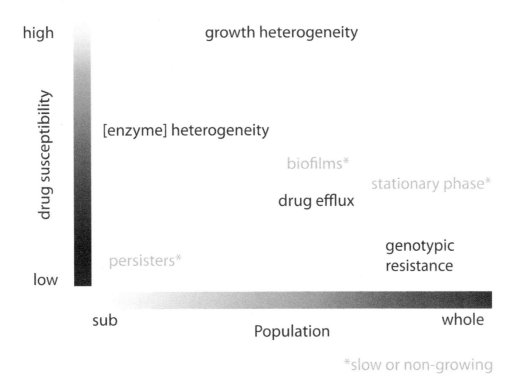

Figure 1 Landscape of drug susceptibility. doi:10.1128/microbiolspec.MGM2-0031-2013.f1

The doubling time of *M. tuberculosis* (which is somewhere between 18 and 24 hours depending on the strain, under ideal growth conditions) slows dramatically *in vivo*, though there is some debate as to whether it fully stops (9, 10).

The purpose of this article is not to comprehensively review the biology of dormancy in *M. tuberculosis*, as this has been reviewed in depth elsewhere. It is important to recognize, however, that growth slowing and the metabolic remodeling associated with growth arrest profoundly affect the susceptibility of *M. tuberculosis* to antibiotics. This has been shown most clearly in *in vitro* models of growth arrest that are thought to mimic specific aspects of the host environment including stationary phase, starvation, hypoxia, and low pH (5–8).

Nongrowing bacteria are no longer susceptible to the classes of antibiotics that specifically target growth machinery. Thus, cell wall biosynthesis inhibitors such as isoniazid are not active against nonreplicating organisms (1). Likewise, *M. tuberculosis* that is growth arrested because of nutrient limitation is not susceptible to ciprofloxacin, an effect that cannot be overcome with increasing concentrations of drug (4, 6).

However, even antibiotics that target processes essential for survival under all conditions, such as rifam-

pin, display reduced efficacy against bacteria that are replicating only slowly or are in a low-activity metabolic state, though the relative change in drug efficacy is less profound than with the cell-wall-acting drugs. For example, in a study by de Steenwinkel et al., examining the time-kill kinetics of antimycobacterial drugs as a function of metabolic state, isoniazid showed rapid, concentration-dependent killing activity against replicating mycobacteria, while rifampin exhibited slow killing but was able to fully sterilize the mycobacterial culture (11). In contrast, to achieve equivalent activity against a culture of mycobacteria characterized by a low metabolic state required a 64-fold increase in isoniazid concentration but only a 4-fold increase in rifampin concentration (11).

The changes in growth rate and metabolic state do not, of course, imply reduced activity of every drug. As the examples of rifampin and isoniazid make clear, the effects are compound- and target-specific. Indeed, there are some antibiotics that have increased efficacy against dormant mycobacteria. The most obvious examples of this are the diarylquinolones, targeting ATP synthesis, which have increased activity against dormant mycobacteria (generated through hypoxia) as compared to actively growing cells (12), and drugs that target the altered metabolic state specifically, such as metronidazole, which is active only under hypoxic conditions (13, 14).

Importantly, the path by which mycobacteria arrive in a growth-arrested state is likely to significantly impact the degree of drug tolerance. For example, subjecting mycobacteria to multiple stresses simultaneously (low oxygen, high carbon dioxide, low nutrient, low pH) produced cells that were more tolerant of rifampin than the cells reported by Wayne and Hayes (1, 15) to arise under hypoxic conditions alone.

While changes in growth rate and metabolic state are clearly important determinants of drug sensitivity in mycobacteria, recent work has demonstrated that other factors are also likely to play important roles in determining antibiotic efficacy. Advances in microscopy and single-cell biology have allowed investigators to ask questions about drug tolerance from new perspectives. Based on pioneering studies in other organisms, researchers have sought to identify drug-tolerant subpopulations of cells within an otherwise drug-susceptible population. These studies were initially driven by the identification of persister cells in model organisms. However, rather than simply identifying one mechanism by which drug-tolerant cells are generated, these studies have found multiple mechanisms by which mycobacterial populations generate cells of different drug susceptibilities. The characteristics of these subpopulations of cells differ in the degree of drug tolerance conferred and the frequency with which drug-tolerant cells arise. The data therefore suggest a spectrum of physiologic states within what is typically considered to be a homogeneous population. Below, we consider some major mechanisms by which these drug-tolerant states arise, beginning with high-frequency mechanisms and concluding with a discussion of persister cells, as the term has been classically used in the literature. Importantly, to date, many of these studies have been done in the model mycobacterium *Mycobacterium smegmatis* rather than *M. tuberculosis*, though as live cell imaging becomes more accessible in the biosafety level 3 environment, we expect these approaches to be applied to *M. tuberculosis* as well (16–19).

DRUG RESISTANCE THROUGH EFFLUX PUMP INDUCTION

The induction of multidrug-resistant efflux pumps is a well-defined mechanism by which bacterial cells lower their exposure to toxins and achieve phenotypic drug tolerance (20). In Gram-negative bacteria, efflux pumps consist of multiple components spanning the periplasm to transport molecules across both the inner and outer membrane. *Pseudomonas aeruginosa*, a Gram-negative opportunistic pathogen, is a well-studied example of efflux pump–mediated drug resistance. *P. aeruginosa* is "intrinsically" resistant to many antibiotics including fluoroquinolones, chloramphenicol, tetracycline, and beta-lactams (21–23). The cell wall of *P. aeruginosa* is relatively impermeable compared to other Gram-negative bacteria, and it was thought that this impermeability caused the relative drug tolerance. However, using a combination of theoretical and experimental evidence, Livermore and Davy (24) and Nikaido (25) showed that the timescale of drug permeation is on the order of seconds, casting doubt on this hypothesis. Instead, Li et al. (22) postulated that efflux pumps caused this apparent intrinsic resistance to a wide variety of molecules. They found that *P. aeruginosa* isolates actively efflux tetracycline, chloramphenical, norfloxacin, and beta-lactams, and highly drug-resistant isolates transport more drug. The MexAB-OprA efflux pump system had recently been described by Poole et al. using knockouts of the multicomponent pump to generate drug-sensitive strains (26). Li et al. found upregulation of these components in highly drug-resistant bacterial isolates (22). Since then, multidrug efflux pumps have been implicated in drug resistance of many other Gram-negative pathogens including *Escherichia coli*, *Salmonella enterica*, *Acinetobacter baumannii*, *Campylobacter jejuni*, and *Neisseria gonorrhoeae* (reviewed in references 21 and 27).

Gram-positive bacteria express three classes of single-component multidrug-resistant efflux pumps. Multidrug-resistant efflux pumps have been implicated in drug resistance in pathogenic Gram-positives including *Staphylococcus aureus* and *Streptococcus pneumoniae* (21, 28). For example, overexpression of NorA, a member of the major facilitation superfamily (MFS) of efflux pumps, has been identified in some clinical isolates of methicillin-resistant *S. aureus* (MRSA) (21, 29–31).

Both slow- and fast-growing mycobacterial species express several single-component drug efflux pumps (membranetransport.org) belonging to both the MFS and the ATP-binding cassette (ABC) superfamily. Reminiscent of the intrinsic drug tolerance of *P. aeruginosa*, it was first hypothesized that drug resistance in mycobacteria is caused by cell wall impermeability from the thick mycolic acid layer of mycobacteria (32). However, again like *P. aeruginosa*, drug access occurs within minutes, suggesting that mycobacteria achieve drug tolerance through a combination of impermeability and active efflux (33, 34).

The first transporter characterized in mycobacteria was LfrA, an MFS pump (35). Takiff et al. found that *M. smegmatis* expressing LfrA-containing plasmids exhibited increased tolerance to fluoroquinolones and was more able to acquire genetic fluoroquinolone resistance (35). Importantly, a single transporter does not target just one drug, but commonly is capable of effluxing several molecules. Therefore, expression of one type of pump can create resistance to multiple drugs in parallel. Srivastava et al. recently found that shortly after treatment of *M. tuberculosis* with ethambutol, resistance to both isoniazid and ethambutol emerges (36). Cotreatment with reserpine, an efflux pump inhibitor, decreased the rate of isoniazid and ethambutol resistance.

Since this identification, increased expression of efflux pumps has been identified in drug-tolerant clinical isolates, suggesting that efflux pumps contribute to drug tolerance in a clinically meaningful way (36–45). In addition, efflux pump expression has recently been shown to be induced *in vivo* (during macrophage infection), where it can contribute to rifampin and isoniazid resistance. Adams and colleagues showed efflux pump induction in *Mycobacterium marinum* and *M. tuberculosis* during macrophage infection (40). The authors showed that almost 50% of *M. marinum* cells were tolerant of rifampin treatment 4 days post-infection. Sub-MIC levels of the efflux pump inhibitors verapamil and reserpine decreased the number of rifampin-resistant cells, providing a potential clinically applicable solution to treatment failure in TB.

Thus, host- or drug-induced stress may induce expression of efflux pumps that act to quickly lower the concentration of drug in cells, creating temporary states of drug resistance. Increasing attention is being drawn to the role of efflux pumps in virulence and survival during hostile host conditions in many pathogens including *P. aeruginosa*, *C. jejuni*, *N. gonorrhoeae*, and *E. coli* (reviewed in reference 21). Induction of multidrug resistance efflux pumps in pathogenic mycobacteria upon infection suggests that the virulence capacity and drug efficacy in individual cells may be connected by efflux pumps. Understanding the dynamics of drug efflux, molecular cross-talk and specificity among efflux pumps, and transporter expression and induction may be a critical step toward designing improved therapeutic regimens that include the use of efflux pump inhibitors. Inhibitors that target host cell efflux pumps may also increase the concentration of drug to which mycobacteria residing in protected niches are exposed, such as those recently found to reside in mesenchymal bone marrow stem cells (46).

POPULATION-LEVEL CHANGES IN DRUG EFFICACY THROUGH BIOFILM FORMATION

Many species of bacteria form biofilms, creating an organized conglomerate of cells that are physically linked, usually by an extracellular matrix. Several bacterial biofilms have been studied extensively including those formed by *P. aeruginosa*, *E. coli*, *Bacillus subtilis*, and *S. aureus*. Biofilm provides distinct environments for different subpopulations, stratifying and physically organizing access to signaling molecules, metabolites, and stressors (28, 47–50). The biofilm lifestyle also induces changes to cell morphology, gene expression, and, of importance when considering drug activity, changes in growth rate, efflux pump expression, and metabolic state (28, 49, 51–54). Moreover, as discussed below, specialized subpopulations of cells including persister cells are generated at high rates under biofilm conditions. Consequently, bacteria in biofilms can be profoundly insensitive to drug—likely encompassing both phenotypically drug-tolerant and phenotypically drug-resistant populations (28, 47–50).

Mycobacteria are well known for their waxy cell walls, which make them clump and be difficult to grow in liquid media without detergents. Without detergents and surfactants, mycobacteria form biofilms in liquid culture (pellicles) and on solid media (55–60). *In vivo*, extracellular mycobacteria form cords, which can be viewed as a type of biofilm, in the caseum of granulomas, and convex biofilms in extracellular regions inside of granulomas (61, 62).

Unlike many biofilm-forming bacteria, mycobacteria do not excrete exopolysaccharides as an extracellular matrix to develop and maintain biofilms (63). In an elegant series of experiments, Ojha and colleagues found that short-chain fatty acids in the mycobacterial mycolic acids are required for biofilm formation and that this process is regulated by GroEL1 (55). In essence, changing the property of their waxy cell wall determines the biofilm structure. Biofilm formation is inhibited by disruption of mycolic acid synthesis by mutation of *inhA* or *kasB*, loss of GroEL1 function by infection with Bxb1 mycobacteriophage, or depletion of iron (55, 59, 64). *In vitro*, Ojha et al. have observed enrichment of isoniazid and rifampin-tolerant *M. tuberculosis* in biofilms compared to biofilm-free culture conditions or biofilm-deficient mutant strains (64).

The spatial localization and different cellular properties of drug-tolerant and drug-resistant mycobacteria in biofilms have yet to be characterized. Properties such as growth rate, permeability (correlating with mycolic acid composition), and drug efflux may be important determinants of drug efficacy that are supported by these three-dimensional structures. Live cell imaging in bioreactors and microfluidic devices may help us understand how biofilms are generated and how they contribute to the generation of protection of a significant subpopulation from antibiotic stress in these environments (53, 65). Biofilm formation, structure, and relationship to immune stress tolerance *in vivo* also remain to be studied.

RAPID GENERATION OF DRUG-TOLERANT CELLS THROUGH ASYMMETRIC GROWTH

Slow growth is commonly cited as a determinant of drug tolerance (66–68), with nongrowing persisters at the extreme end of the spectrum. Aldridge et al. (16) have recently demonstrated that cell-to-cell variation in growth properties contributes to mycobacteria's variable response to drug treatment. They found that mycobacteria exhibit an asymmetric growth pattern, whereby cells elongate predominantly from their older pole (Fig. 2A). Because cells are polarized, each sister cell inherits a different pole. The daughter cell that inherits the older, faster-growing pole is born larger and elongates faster than its sister and is called an "accelerator." The sister inheriting the newer, slower-growing pole from the mother cell must convert this slow-growing pole into a fast-growing pole and is therefore called an "alternator." As growth poles mature, the cell size and growth rate also increase. Thus, the asymmetric growth of mycobacteria innately instills diversity in the growth rate and size of cells (Fig. 2B and 2C). This diversity arises in even very small populations, because each microcolony (seeded from a single cell) contains cells with growth poles of different ages.

The variation in growth rate between alternator and accelerator cells is modest (~10%), yet the physiological differences between these cell types are enough to correlate with differences in drug effect. The faster-growing accelerator cells exhibit reduced survival in the face of treatment with cell wall-acting antibiotics (cycloserine, meropenem, and isoniazid) at minimal inhibitory concentrations, whereas alternator cells are more susceptible to rifampin, an inhibitor of transcription (Fig. 2D). Together, these findings suggest that subtle variations in growth characteristics translate into differences in drug tolerance at a single-cell level.

However, it is important to note that the spectrum of growth phenotypes is not sufficient to fully describe variation in drug response. Aldridge et al. (16) observed microcolony-to-microcolony variation in drug effect, notably in response to isoniazid and rifampin treatment. This variability suggests that there are additional semi-heritable factors of drug tolerance.

HERITABLE DRUG TOLERANCE THROUGH VARIATION IN ENZYME CONCENTRATION

M. smegmatis cells capable of surviving sustained, high doses of isoniazid treatment have recently been characterized. By tracking *M. smegmatis* in a microfluidic device, Wakamoto and colleagues recently observed turnover of isoniazid-resistant mycobacteria (17). These cells grow and divide, and a small proportion of their progeny continue to resist drug treatment. While the number of resistant cells remains relatively constant, there is a dynamic growth and death process underlying this apparent steady state. These single-cell observations are consistent with bulk measurements of replication (9) and mutation (69) of *M. tuberculosis* in latency *in vivo*.

Under high-dose treatment conditions, Wakamoto et al. (17) observed that sister cell pairs exhibited a correlated response to isoniazid treatment, suggesting that an epigenetic factor may play a key role in isoniazid resistance. Interestingly, the possible contribution of semi-heritable factors was also suggested by the microcolony-to-microcolony variation observed following MIC-level treatment with isoniazid (16).

Isoniazid is delivered as a prodrug and relies on conversion to the active form by catalase-peroxidase, expressed by *katG* in mycobacteria. Because small (~2-fold) increases in *katG* induction produced significant (~1,000-fold) increases in bacterial death, Wakamoto et al. (17) explored how variation in *katG* expression at a single-cell level contributed to isoniazid resistance. Using a KatG-fluorescent fusion, they were able to simultaneously measure the dynamics of *katG* expression, lineage, and response to isoniazid treatment. They observed a surprising pulsing of KatG (on the scale of ~5 hours) and a correlation between *katG* expression and isoniazid susceptibility: that is, cells scored as *katG* positive were less capable of surviving isoniazid, scored by the ability to divide after the drug was removed.

PROLONGED DRUG TOLERANCE THROUGH THE FORMATION OF A PERSISTER CELL POPULATION

The idea that some cells are phenotypically drug tolerant was first raised by Joseph Bigger in 1944 (5). Bigger was not able to sterilize a culture of *Staphylococcus* using penicillin, explaining that "The only hypothesis

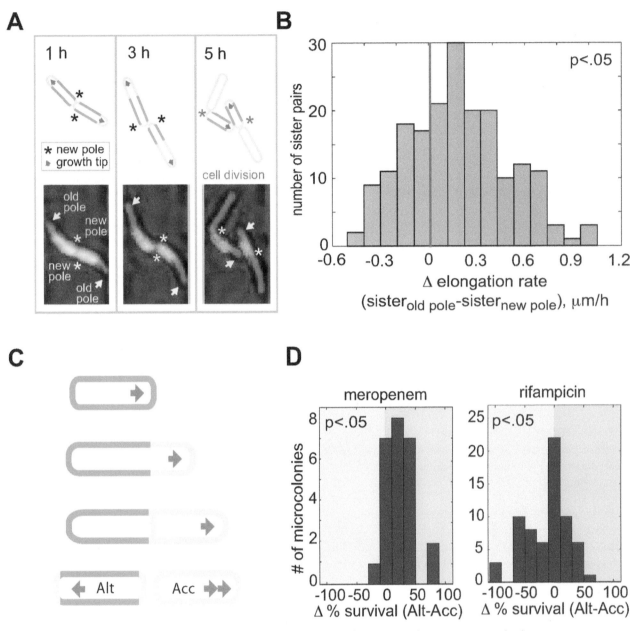

Figure 2 Mycobacterial growth is asymmetric and generates diversity in growth characteristics and drug susceptibility. (**A**) Time-lapse imaging of pulse-labeled *M. smegmatis*. An amine-active dye (green) stains old cell wall, revealing new growth at old cell poles. (**B**) Distribution of the difference in absolute elongation rate (averaged over the course of a full division cycle) in paired sister *M. smegmatis* cells. In most pairs, the sister inheriting the old pole elongates faster. (**C**) Schematic model of mycobacterial growth. Most of the new growth (light green) occurs from the old growing pole (red arrow). Following division, the two sister cells exhibit different growth properties. The cell inheriting the growth pole (called an "accelerator") elongates faster and is born larger than its sister (called an "alternator"). (**D**) Distribution of differential drug susceptibility in accelerator and alternator cells in individual microcolonies. Using microfluidics, *M. smegmatis* microcolonies were challenged with MIC levels of antibiotics (meropenem and rifampicin shown here) and scored for survivors by their ability to elongate in recovery media. All figures were adapted from Aldridge et al. (16). doi:10.1128/microbiolspec.MGM2-0031-2013.f2

which seems to explain the occurrence of a small number of survivors out of the millions of cocci originally present is that these differ from the majority of their fellows in that they are capable of surviving a concentration of penicillin which, in the time of action allowed, kills the others." Bigger named these surviving cells "persisters," meaning a small phenotypic variant in all bacterial populations that do not die in the presence of high doses of bactericidal drugs for long periods of time that are capable of killing the bulk of the population (Fig. 3A). When the antibiotic is removed and the bacterial population is reestablished, the new population has a similar level of persisters, indicating that persisters are not genetic mutants but, rather, phenotypic variants (5, 70).

Persisters have typically been observed by adding bactericidal drugs to a bacterial population and enumerating the number of surviving cells; a clear biphasic curve emerges, with the bulk of the population dying rapidly, while persisters die slowly or do not die at all (Fig. 3A). The majority of research has concentrated on *E. coli* persisters, but persisters were also identified in many other bacterial species including *P. aeruginosa*, *M. tuberculosis*, *S. enterica* serovar Typhimurium, *S. aureus*, and *Streptococcus mutans*, and even in the fungal pathogen *Candida albicans* (6, 71–76).

High-Persistence (hip) Mutants

The importance of persisters was not immediately appreciated, and little research was done until the early 1980s, when Harris Moyed and coworkers isolated high-persistence (hip) mutants (77, 78). They identified the first gene, *hipA*, to be known to contribute to persisters' formation; this gene was later identified as the toxin of a toxin-antitoxin (TA) module (79). The *hipA7* allele of the gene allowed mutated cells to form 1,000 times more persisters to ampicillin and increased persisters to other antibiotic classes such as fluoroquinolone and aminoglycosides (7, 80). This was the first clue that persisters are multidrug tolerant. Importantly, the *hipA7* mutant has the same MIC as the parent strain, indicating that the increased level of persisters is not due to acquisition of resistance.

Persisters as Dormant Cells

The *hipA7* mutant was used for single-cell time-lapse microscopy which showed that persisters are slow- or nongrowing cells, stochastically formed and preexisting in the population (81). This work could not be done with wild-type *E. coli* strains, because the number of persisters is very low. In a less direct observation, a wild-type *E. coli* culture was diluted several times and challenged with either ampicillin or ofloxacin. A gradual decline and eventual elimination of persisters was observed, despite keeping the initial population size constant (70). In another investigation, the authors used an unstable green fluorescent protein variant expressed by a ribosomal promoter to distinguish between growing and nongrowing cells (82). They demonstrated that nongrowing cells exist in an untreated *E. coli* population and that those cells are more tolerant to ofloxacin, a fluoroquinolone antibiotic capable of killing nongrowing cells. These results suggest that there is a population of persisters that are pre-existing in the population and are not formed in a response to the antibiotic treatment. However, later work demonstrated that persisters can also be formed as a response to an environmental stress such as DNA damage or oxidative stress (83, 84).

The first transcriptome of persisters was generated by taking advantage of the fact that β-lactam antibiotics such as ampicillin are bacteriolytic, lysing the dying cells (7). An *E. coli* culture was treated with a high concentration of ampicillin until only persisters survived. A simple centrifugation step was used to collect the persisters, and expression profiles were determined using a microarray. Several TA modules were among the genes overexpressed in persisters. Ectopic overexpression of RelE, the toxin of the *relBA* TA module, generated multidrug-tolerant cells, similarly to that observed with the *hipA7* mutant. Based on these observations, it was suggested that stochastic variation in the expression of toxin genes generates dormant persister cells that are not killed by the antibiotic. Bactericidal antibiotics kill by binding to their target (such as ribosomes or DNA gyrase) and causing the production of corrupted toxic products (85–90). In dormant persisters the targets are inactive, so the cells survive (91).

Genetic Determinants of Persisters

TA genes were first identified as addiction modules playing a role in plasmid maintenance (92–95). The toxin and antitoxin genes usually form an operon, with the antitoxin situated upstream of the toxin (96). The antitoxin is less stable and is degraded by cellular proteases; if a daughter cell loses the plasmid, the antitoxin is rapidly degraded and the toxin stops the cells from growing. It was surprising to find homologues of TA genes on bacterial chromosomes, because their function there is not immediately clear (97). A possible role in bacterial persistence was suggested when several of the known TAs in *E. coli* were found to

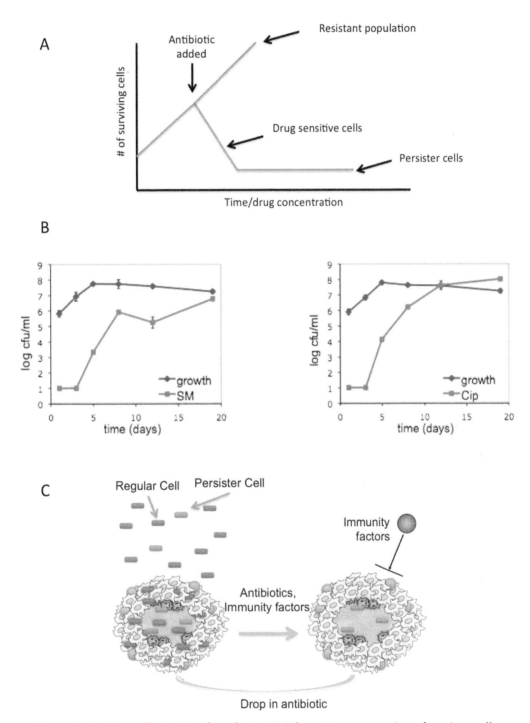

Figure 3 Persister cells in *M. tuberculosis*. (**A**) Schematic representation of persister cells. (**B**) Persister cell levels in a growing population of *M. tuberculosis*. Stationary-phase culture of *M. tuberculosis* was diluted 1:100, and growth and persister levels were followed over time. Persister levels were determined after 1 week of challenge with streptomycin (SM) (40 µg/ml) or ciprofloxacin (Cip) (5 g/ml). (**C**) Model explaining the need for lengthy antibiotic treatment of *M. tuberculosis* infection. Panels B and C were adapted from Keren et al. (6). doi:10.1128/microbiolspec.MGM2-0031-2013.f3

be upregulated in the persisters' fraction (7, 82). The best-known persister gene, *hipA* of the *hipBA* operon, was suggested to be a TA module (98); more recent work has identified a novel mechanism of action. HipA is a kinase that phosphorylates elongation factor TU and halts protein synthesis (79). The HipB antitoxin can neutralize HipA by binding to it and is regulated by the cellular protease Lon (99–102).

Although there were several clues suggesting a connection between TAs and persisters, a direct cause and effect was not established until recently. Single deletion of several of the TA modules did not result in reduction of persisters' levels. Gerdes and coworkers serially deleted up to 10 of the recognized type II TA modules (Δ10) in *E. coli* (103). They found that at least five modules needed to be deleted before a significant reduction in persister levels was observed. When more TAs were deleted, a progressive reduction in persister levels was observed, up to about 2 logs in the Δ10 strain. These results clearly demonstrate a cause-and-effect relationship between TA modules and persisters in *E. coli*.

Several genome-wide approaches have been used to identify genetic determinants of persisters. A selection for the hip phenotype in an overexpression library identified *glpD* and *plsB* (104). Both genes are involved in glycerol metabolism. Recent work has suggested a mechanism by which the *glpD* mutant has increased levels of methylglyoxal, a bacteriostatic metabolite (105). Other screens have identified a number of regulators in both *E. coli* (106, 107) and *P. aeruginosa* (71, 108, 109). No mutant has been identified in which no persisters are present. This suggests redundancy of the different mechanisms of persisters' formation. This view is consistent with recent work with *S. aureus*, where the authors suggested that persisters could be generated by many mechanisms, all of which result in a transient growth arrest (110).

Persisters in *M. tuberculosis*

The current "short" course of antibiotic treatment for M. tuberculosis infections is 6 months long, despite the fact that the majority of the cells die in the first 2 weeks. Biphasic killing typical of persisters is easy to identify in the *M. tuberculosis* literature. They can be seen both *in vitro* and *in vivo* (6, 111–113). Currently, very little is known about persisters in *M. tuberculosis*, and no persister genes have been identified to date. Recently, *M. tuberculosis* persisters were studied *in vitro*. The dynamics of persister cell formation were found to be similar to what was described for *E. coli*, *P. aeruginosa*, and *S. aureus* (6); the levels of persisters are low in lag and early log phase and then shoot up sharply in mid-log phase, reaching a high-level plateau in stationary state (Fig. 3B). This indicates that a persister cell formation is stochastic early in the growth phase but has some deterministic component as the culture enters stationary phase (6, 114).

The method used for isolating *E. coli* persisters was modified, and *M. tuberculosis* persisters were isolated by treating the bacterial population with D-cycloserine, a bacteriolytic antibiotic. A transcriptome of the isolated persisters was generated and analyzed. The most striking finding was the massive shutdown of expression: 282 genes were upregulated, while 1,408 were downregulated more than 2-fold. Many of the energy metabolism pathways were downregulated, indicating some level of dormancy (6).

The *M. tuberculosis* genome has a large number of TA loci (97, 115, 116). Their physiological roles are unknown, but some have been shown to be upregulated under stressful conditions and are suspected to be a part of the bacterial stress response (116). As discussed above, in *E. coli*, TA modules were identified as being involved in persister formation (103). In the *M. tuberculosis* persisters' transcriptome, there were 10 TA modules that were upregulated (6). Three *M. tuberculosis* RelE toxins were overexpressed, and the level of persisters was determined using four antibiotics (117). Unlike *E. coli*, there was not a strong multidrug-tolerance phenotype. Each toxin increased tolerance for one drug specifically, with a very modest increase to some of the other drugs. These results could indicate that TAs in *M. tuberculosis* cannot generate multidrug tolerance or that only some unidentified TAs are capable of generating such tolerance. It is also possible that *M. tuberculosis* uses its large arsenal of TAs to generate many different populations, each with cells that overexpress one or more TAs and are specialized for survival under specific conditions.

As discussed earlier, the biofilm generated by many bacterial species is thought to contribute to the difficulty to cure, and the recalcitrant nature of, many infections. During *M. tuberculosis* infection, the immune response generates lesions called granulomas, where many of the bacilli reside (118). Similar to a biofilm, the granulomas create a physical barrier between the residing bacteria and the immune system. Antibiotics kill sensitive cells in the granuloma, but the persisters survive and can repopulate when favorable conditions resume (Fig. 3C). This scenario can explain the need for long treatment durations and the relapses common in *M. tuberculosis* infections.

INTERSECTION OF POPULATION-LEVEL AND SINGLE-CELL STUDIES

As described above, biofilm formation during infection appears to be a major cause of drug tolerance in many infections (119). While this is often conceived of as population-level drug tolerance, it was also demonstrated that biofilms of *P. aeruginosa* harbor persister cells (72). Lewis suggested a model explaining the recalcitrant nature of biofilm infection (120). When a biofilm infection is treated with antibiotics, the majority of the bacterial cells die and only persisters survive. In the biofilm, persisters are protected from the immune system, and upon removal of the antibiotic pressure they can repopulate the biofilm culture. A similar role was recently suggested for the granuloma in *M. tuberculosis* infection: that it acts as a physical barrier protecting persisters from the immune system (6).

We anticipate that in the future, there will be further convergence of population-level and single-cell studies. It is likely that the unique biology of specialized subpopulations of cells generated under specific conditions will contribute to biologically relevant outcomes in these settings. However, the mechanisms of drug tolerance will continue to reflect the importance of drug activation, growth rate, metabolic state, and permeability in dictating drug activity.

Citation. Aldridge BB, Keren I, Fortune SM. 2014. The spectrum of drug susceptibility in mycobacteria. Microbiol Spectrum 2(5):MGM2-0031-2013.

References

1. Wayne LG, Hayes LG. 1996. An in vitro model for sequential study of shiftdown of *Mycobacterium tuberculosis* through two stages of nonreplicating persistence. *Infect Immun* 64:2062–2069.

2. Rustad TR, Harrell MI, Liao R, Sherman DR. 2008. The enduring hypoxic response of *Mycobacterium tuberculosis*. *PLoS One* 3:e1502.

3. Betts JC, Lukey PT, Robb LC, McAdam RA, Duncan K. 2002. Evaluation of a nutrient starvation model of *Mycobacterium tuberculosis* persistence by gene and protein expression profiling. *Mol Microbiol* 43:717–731.

4. Voskuil MI, Schnappinger D, Rutherford R, Liu Y, Schoolnik GK. 2004. Regulation of the *Mycobacterium tuberculosis* PE/PPE genes. *Tuberculosis* 84:256–262.

5. Bigger JW. 1944. Treatment of staphylococcal infections with penicillin. *Lancet* ii:497–500.

6. Keren I, Minami S, Rubin E, Lewis K. 2011. Characterization and transcriptome analysis of *Mycobacterium tuberculosis* persisters. *MBio* 2:e00100–11.

7. Keren I, Shah D, Spoering A, Kaldalu N, Lewis K. 2004. Specialized persister cells and the mechanism of multidrug tolerance in *Escherichia coli*. *J Bacteriol* 186:8172–8180.

8. Moyed HS, Bertrand KP. 1983. *hipA*, a newly recognized gene of *Escherichia coli* K-12 that affects frequency of persistence after inhibition of murein synthesis. *J Bacteriol* 155:768–775.

9. Gill WP, Harik NS, Whiddon MR, Liao RP, Mittler JE, Sherman DR. 2009. A replication clock for *Mycobacterium tuberculosis*. *Nat Med* 15:211–214.

10. Muñoz-Elías EJ, Timm J, Botha T, Chan WT, Gomez JE, McKinney JD. 2004. Replication dynamics of *Mycobacterium tuberculosis* in chronically infected mice. *Infect Immun* 73:546–551.

11. de Steenwinkel JEM, Kate ten MT, de Knegt GJ, Kremer K, Aarnoutse RE, Boeree MJ, Verbrugh HA, van Soolingen D, Bakker-Woudenberg IAJM. 2012. Drug susceptibility of *Mycobacterium tuberculosis* Beijing genotype and association with MDR TB. *Emerg Infect Dis* 18:660–663.

12. Koul A, Vranckx L, Dendouga N, Balemans W, Van den Wyngaert I, Vergauwen K, Göhlmann HWH, Willebrords R, Poncelet A, Guillemont J, Bald D, Andries K. 2008. Diarylquinolines are bactericidal for dormant mycobacteria as a result of disturbed ATP homeostasis. *J Biol Chem* 283:25273–25280.

13. Wayne LG, Sramek HA. 1994. Metronidazole is bactericidal to dormant cells of *Mycobacterium tuberculosis*. *Antimicrob Agents Chemother* 38:2054–2058.

14. Brooks JV, Furney SK, Orme IM. 1999. Metronidazole therapy in mice infected with tuberculosis. *Antimicrob Agents Chemother* 43:1285–1288.

15. Deb C, Lee C-M, Dubey VS, Daniel J, Abomoelak B, Sirakova TD, Pawar S, Rogers L, Kolattukudy PE. 2009. A novel in vitro multiple-stress dormancy model for *Mycobacterium tuberculosis* generates a lipid-loaded, drug-tolerant, dormant pathogen. *PLoS One* 4:e6077.

16. Aldridge BB, Fernandez-Suarez M, Heller D, Ambravaneswaran V, Irimia D, Toner M, Fortune SM. 2012. Asymmetry and aging of mycobacterial cells lead to variable growth and antibiotic susceptibility. *Science* 335:100–104.

17. Wakamoto Y, Dhar N, Chait R, Schneider K, Signorino-Gelo F, Leibler S, McKinney JD. 2013. Dynamic persistence of antibiotic-stressed mycobacteria. *Science* 339:91–95.

18. Joyce G, Williams KJ, Robb M, Noens E, Tizzano B, Shahrezaei V, Robertson BD. 2012. Cell division site placement and asymmetric growth in mycobacteria. *PLoS One* 7:e44582.

19. Golchin SA, Stratford J, Curry RJ, McFadden J. 2012. A microfluidic system for long-term time-lapse microscopy studies of mycobacteria. *Tuberculosis (Edinb)* 92:489–496.

20. Li X-Z, Nikaido H. 2009. Efflux-mediated drug resistance in bacteria. *Drugs* 69:1555–1623.

21. Piddock LJV. 2006. Multidrug-resistance efflux pumps —not just for resistance. *Nat Rev Microbiol* 4:629–636.

22. Li XZ, Livermore DM, Nikaido H. 1994. Role of efflux pump(s) in intrinsic resistance of *Pseudomonas aeruginosa*: resistance to tetracycline, chloramphenicol, and

norfloxacin. *Antimicrob Agents Chemother* **38**:1732–1741.

23. Li XZ, Ma D, Livermore DM, Nikaido H. 1994. Role of efflux pump(s) in intrinsic resistance of *Pseudomonas aeruginosa*: active efflux as a contributing factor to beta-lactam resistance. *Antimicrob Agents Chemother* **38**:1742–1752.

24. Livermore DM, Davy KW. 1991. Invalidity for *Pseudomonas aeruginosa* of an accepted model of bacterial permeability to beta-lactam antibiotics. *Antimicrob Agents Chemother* **35**:916–921.

25. Nikaido H. 1989. Outer membrane barrier as a mechanism of antimicrobial resistance. *Antimicrob Agents Chemother* **33**:1831–1836.

26. Poole K, Krebes K, McNally C, Neshat S. 1993. Multiple antibiotic resistance in *Pseudomonas aeruginosa*: evidence for involvement of an efflux operon. *J Bacteriol* **175**:7363–7372.

27. Piddock LJV. 2006. Clinically relevant chromosomally encoded multidrug resistance efflux pumps in bacteria. *Clin Microbiol Rev* **19**:382–402.

28. Soto SM. 2013. Role of efflux pumps in the antibiotic resistance of bacteria embedded in a biofilm. *Virulence* **4**:223–229.

29. Noguchi N, Okada H, Narui K, Sasatsu M. 2004. Comparison of the nucleotide sequence and expression of *norA* genes and microbial susceptibility in 21 strains of *Staphylococcus aureus*. *Microb Drug Resist* **10**:197–203.

30. Kaatz GW, Seo SM, Ruble CA. 1993. Efflux-mediated fluoroquinolone resistance in *Staphylococcus aureus*. *Antimicrob Agents Chemother* **37**:1086–1094.

31. Jones ME, Boenink NM, Verhoef J, Kohrer K, Schmitz FJ. 2000. Multiple mutations conferring ciprofloxacin resistance in *Staphylococcus aureus* demonstrate long-term stability in an antibiotic-free environment. *J Antimicrob Chemother* **45**:353–356.

32. Hett EC, Rubin EJ. 2008. Bacterial growth and cell division: a mycobacterial perspective. *Microbiol Mol Biol Rev* **72**:126–156.

33. Nikaido H. 2001. Preventing drug access to targets: cell surface permeability barriers and active efflux in bacteria. *Semin Cell Dev Biol* **12**:215–223.

34. Jarlier V, Nikaido H. 1990. Permeability barrier to hydrophilic solutes in *Mycobacterium chelonei*. *J Bacteriol* **172**:1418–1423.

35. Takiff HEH, Cimino MM, Musso MCM, Weisbrod TT, Martinez RR, Delgado MBM, Salazar LL, Bloom BRB, Jacobs WRW. 1996. Efflux pump of the proton antiporter family confers low-level fluoroquinolone resistance in *Mycobacterium smegmatis*. *Proc Natl Acad Sci USA* **93**:362–366.

36. Srivastava S, Musuka S, Sherman C, Meek C, Leff R, Gumbo T. 2010. Efflux-pump-derived multiple drug resistance to ethambutol monotherapy in *Mycobacterium tuberculosis* and the pharmacokinetics and pharmacodynamics of ethambutol. *J Infect Dis* **201**:1225–1231.

37. Li X-ZX, Zhang LL, Nikaido HH. 2004. Efflux pump-mediated intrinsic drug resistance in *Mycobacterium smegmatis*. *Antimicrob Agents Chemother* **48**:2415–2423.

38. De Rossi EE, Aínsa JAJ, Riccardi GG. 2006. Role of mycobacterial efflux transporters in drug resistance: an unresolved question. *FEMS Microbiol Rev* **30**:36–52.

39. Danilchanka OO, Mailaender CC, Niederweis MM. 2008. Identification of a novel multidrug efflux pump of *Mycobacterium tuberculosis*. *Antimicrob Agents Chemother* **52**:2503–2511.

40. Adams KN, Takaki K, Connolly LE, Wiedenhoft H, Winglee K, Humbert O, Edelstein PH, Cosma CL, Ramakrishnan L. 2011. Drug tolerance in replicating mycobacteria mediated by a macrophage-induced efflux mechanism. *Cell* **145**:39–53.

41. Gupta AK, Katoch VM, Chauhan DS, Sharma R, Singh M, Venkatesan K, Sharma VD. 2010. Microarray analysis of efflux pump genes in multidrug-resistant *Mycobacterium tuberculosis* during stress induced by common anti-tuberculous drugs. *Microb Drug Resist* **16**:21–28.

42. Amaral L, Martins M, Viveiros M. 2007. Enhanced killing of intracellular multidrug-resistant *Mycobacterium tuberculosis* by compounds that affect the activity of efflux pumps. *J Antimicrob Chemother* **59**:1237–1246.

43. Rodrigues L, Ramos J, Couto I, Amaral L, Viveiros M. 2011. Ethidium bromide transport across *Mycobacterium smegmatis* cell-wall: correlation with antibiotic resistance. *BMC Microbiol* **11**:35.

44. Jiang X, Zhang W, Zhang Y, Gao F, Lu C, Zhang X, Wang H. 2008. Assessment of efflux pump gene expression in a clinical isolate *Mycobacterium tuberculosis* by real-time reverse transcription PCR. *Microb Drug Resist* **14**:7–11.

45. Spies FS, Almeida da Silva PE, Ribeiro MO, Rossetti ML, Zaha A. 2008. Identification of mutations related to streptomycin resistance in clinical isolates of *Mycobacterium tuberculosis* and possible involvement of efflux mechanism. *Antimicrob Agents Chemother* **52**:2947–2949.

46. Das B, Kashino SS, Pulu I, Kalita D, Swami V, Yeger H, Felsher DW, Campos-Neto A. 2013. CD271(+) bone marrow mesenchymal stem cells may provide a niche for dormant *Mycobacterium tuberculosis*. *Sci Transl Med* **5**:170ra13.

47. Lewis K. 2005. Persister cells and the riddle of biofilm survival. *Biochemistry (Moscow)* **70**:267–274.

48. Liao J, Schurr MJ, Sauer K. 2013. The MerR-like regulator BrlR confers biofilm tolerance by activating multidrug efflux pumps in *Pseudomonas aeruginosa* biofilms. *J Bacteriol* **195**:3352–3363.

49. Lopez D, Vlamakis H, Kolter R. 2010. Biofilms. *Cold Spring Harbor Perspect Biol* **2**:a000398.

50. Levin BR, Rozen DE. 2006. Non-inherited antibiotic resistance. *Nat Rev Microbiol* **4**:556–562.

51. Whiteley M, Bangera MG, Bumgarner RE, Parsek MR, Teitzel GM, Lory S, Greenberg EP. 2001. Gene expression in *Pseudomonas aeruginosa*: biofilms. *Nature* **413**:860–864.

52. Veening JW, Kuipers OP, Brul S, Hellingwerf KJ, Kort R. 2006. Effects of phosphorelay perturbations on architecture, sporulation, and spore resistance in biofilms of *Bacillus subtilis*. *J Bacteriol* 188:3099–3109.

53. Stewart PS, Franklin MJ. 2008. Physiological heterogeneity in biofilms. *Nat Rev Microbiol* 6:199–210.

54. Mah T-F, Pitts B, Pellock B, Walker GC, Stewart PS, O'Toole GA. 2003. A genetic basis for *Pseudomonas aeruginosa* biofilm antibiotic resistance. *Nature* 426:306–310.

55. Ojha A, Anand M, Bhatt A, Kremer L, Jacobs WR Jr, Hatfull GF. 2005. GroEL1: a dedicated chaperone involved in mycolic acid biosynthesis during biofilm formation in mycobacteria. *Cell* 123:861–873.

56. Nguyen KT, Piastro K, Gray TA. 2010. Mycobacterial biofilms facilitate horizontal DNA transfer between strains of *Mycobacterium smegmatis*. *J Bacteriol* 192:5134–5142.

57. Hall-Stoodley L, Stoodley P. 2005. Biofilm formation and dispersal and the transmission of human pathogens. *Trends Microbiol* 13:7–10.

58. Bardouniotis EE, Huddleston WW, Ceri HH, Olson MEM. 2001. Characterization of biofilm growth and biocide susceptibility testing of *Mycobacterium phlei* using the MBEC assay system. *FEMS Microbiol Lett* 203:263–267.

59. Ojha A, Hatfull GF. 2007. The role of iron in *Mycobacterium smegmatis* biofilm formation: the exochelin siderophore is essential in limiting iron conditions for biofilm formation but not for planktonic growth. *Mol Microbiol* 66:468–483.

60. Schulze-Röbbecke R, Fischeder R. 1989. Mycobacteria in biofilms. *Zentralbl Hyg Umweltmed* 188:385–390.

61. Tobin DM, Vary JC, Ray JP, Walsh GS, Dunstan SJ, Bang ND, Hagge DA, Khadge S, King M-C, Hawn TR, Moens CB, Ramakrishnan L. 2010. The lta4h locus modulates susceptibility to mycobacterial infection in zebrafish and humans. *Cell* 140:717–730.

62. Lenaerts AJ, Hoff D, Aly S, Ehlers S, Andries K, Cantarero L, Orme IM, Basaraba RJ. 2007. Location of persisting mycobacteria in a guinea pig model of tuberculosis revealed by r207910. *Antimicrob Agents Chemother* 51:3338–3345.

63. Zambrano MM, Kolter R. 2005. Mycobacterial biofilms: a greasy way to hold it together. *Cell* 123:762–764.

64. Ojha AK, Baughn AD, Sambandan D, Hsu T, Trivelli X, Guerardel Y, Alahari A, Kremer L, Jacobs WR Jr, Hatfull GF. 2008. Growth of *Mycobacterium tuberculosis* biofilms containing free mycolic acids and harbouring drug-tolerant bacteria. *Mol Microbiol* 69:164–174.

65. Asally M, Kittisopikul M, Rué P, Du Y, Hu Z, Çağatay T, Robinson AB, Lu H, Garcia-Ojalvo J, Süel GM. 2012. Localized cell death focuses mechanical forces during 3D patterning in a biofilm. *Proc Natl Acad Sci USA* 109:18891–18896.

66. Connolly LE, Edelstein PH, Ramakrishnan L. 2007. Why is long-term therapy required to cure tuberculosis? *PLoS Med* 4:e120.

67. Rocco A, Kierzek AM, McFadden J. 2013. Slow protein fluctuations explain the emergence of growth phenotypes and persistence in clonal bacterial populations. *PLoS One* 8:e54272.

68. Xie Z, Siddiqi N, Rubin EJ. 2005. Differential antibiotic susceptibilities of starved *Mycobacterium tuberculosis* isolates. *Antimicrob Agents Chemother* 49:4778–4780.

69. Ford CB, Lin PL, Chase MR, Shah RR, Iartchouk O, Galagan J, Mohaideen N, Ioerger TR, Sacchettini JC, Lipsitch M, Flynn JL, Fortune SM. 2011. Use of whole genome sequencing to estimate the mutation rate of *Mycobacterium tuberculosis* during latent infection. *Nat Genet* 43:482–486.

70. Keren I, Kaldalu N, Spoering A, Wang Y, Lewis K. 2004. Persister cells and tolerance to antimicrobials. *FEMS Microbiol Lett* 230:13–18.

71. Moker N, Dean CR, Tao J. *Pseudomonas aeruginosa* increases formation of multidrug-tolerant persister cells in response to quorum-sensing signaling molecules. *J Bacteriol* 192:1946–1955.

72. Spoering AL, Lewis K. 2001. Biofilms and planktonic cells of *Pseudomonas aeruginosa* have similar resistance to killing by antimicrobials. *J Bacteriol* 183:6746–6751.

73. Slattery A, Victorsen AH, Brown A, Hillman K, Phillips GJ. 2013. Isolation of highly persistent mutants of *Salmonella enterica* serovar *typhimurium* reveals a new toxin-antitoxin module. *J Bacteriol* 195:647–657.

74. Lechner S, Lewis K, Bertram R. 2012. *Staphylococcus aureus* persisters tolerant to bactericidal antibiotics. *J Mol Microbiol Biotechnol* 22:235–244.

75. Leung V, Levesque CM. 2012. A stress-inducible quorum-sensing peptide mediates the formation of persister cells with noninherited multidrug tolerance. *J Bacteriol* 194:2265–2274.

76. LaFleur MD, Kumamoto CA, Lewis K. 2006. *Candida albicans* biofilms produce antifungal-tolerant persister cells. *Antimicrob Agents Chemother* 50:3839–3846.

77. Moyed HS, Bertrand KP. 1983. hipA, a newly recognized gene of *Escherichia coli* K-12 that affects frequency of persistence after inhibition of murein synthesis. *J Bacteriol* 155:768–775.

78. Moyed HS, Broderick SH. 1986. Molecular cloning and expression of *hipA*, a gene of *Escherichia coli* K-12 that affects frequency of persistence after inhibition of murein synthesis. *J Bacteriol* 166:399–403.

79. Correia FF, D'Onofrio A, Rejtar T, Li L, Karger BL, Makarova K, Koonin EV, Lewis K. 2006. Kinase activity of overexpressed HipA is required for growth arrest and multidrug tolerance in *Escherichia coli*. *J Bacteriol* 188:8360–8367.

80. Falla TJ, Chopra I. 1998. Joint tolerance to beta-lactam and fluoroquinolone antibiotics in *Escherichia coli* results from overexpression of *hipA*. *Antimicrob Agents Chemother* 42:3282–3284.

81. Balaban NQ. 2004. Bacterial persistence as a phenotypic switch. *Science* 305:1622–1625.

82. Shah D, Zhang Z, Khodursky A, Kaldalu N, Kurg K, Lewis K. 2006. Persisters: a distinct physiological state of *E. coli*. *BMC Microbiol* 6:53.

83. Wu Y, Vulic M, Keren I, Lewis K. 2012. Role of oxidative stress in persister tolerance. *Antimicrob Agents Chemother* **56**:4922–4926.

84. Dörr T, Lewis K, Vulic M. 2009. SOS response induces persistence to fluoroquinolones in *Escherichia coli*. *PLoS Genet* **5**:e1000760.

85. Keren I, Wu Y, Innocencio J, Mulcahy L, Lewis K. 2013. Killing by antibiotics does not depend on reactive oxygen species. *Science* **339**:1213–1216.

86. Hooper DC. 2001. Mechanisms of action of antimicrobials: focus on fluoroquinolones. *Clin Infect Dis* **32** (Suppl 1):S9–S15.

87. Davis BD, Chen LL, Tai PC. 1986. Misread protein creates membrane channels: an essential step in the bactericidal action of aminoglycosides. *Proc Natl Acad Sci USA* **83**:6164–6168.

88. Kohanski MA, Dwyer DJ, Wierzbowski J, Cottarel G, Collins JJ. 2008. Mistranslation of membrane proteins and two-component system activation trigger antibiotic-mediated cell death. *Cell* **135**:679–690.

89. Bayles KW. 2007. The biological role of death and lysis in biofilm development. *Nat Rev Microbiol* **5**:721–726.

90. Uehara T, Dinh T, Bernhardt TG. 2009. LytM-domain factors are required for daughter cell separation and rapid ampicillin-induced lysis in *Escherichia coli*. *J Bacteriol* **191**:5094–5107.

91. Lewis K. 2010. Persister cells. *Annu Rev Microbiol* **64**:357–372.

92. Ogura T, Hiraga S. 1983. Mini-F plasmid genes that couple host cell division to plasmid proliferation. *Proc Natl Acad Sci USA* **80**:4784–4788.

93. Gerdes K, Bech FW, Jorgensen ST, Løbner-Olesen A, Rasmussen PB, Atlung T, Boe L, Karlstrom O, Molin S, von Meyenburg K. 1986. Mechanism of postsegregational killing by the *hok* gene product of the *parB* system of plasmid R1 and its homology with the *relF* gene product of the *E. colirelB* operon. *EMBO J* **5**:2023–2029.

94. Gerdes K, Rasmussen PB, Molin S. 1986. Unique type of plasmid maintenance function: postsegregational killing of plasmid-free cells. *Proc Natl Acad Sci USA*, **83**:3116–3120.

95. Greenfield TJ, Ehli E, Kirshenmann T, Franch T, Gerdes K, Weaver KE. 2000. The antisense RNA of the *par* locus of pAD1 regulates the expression of a 33-amino-acid toxic peptide by an unusual mechanism. *Mol Microbiol* **37**:652–660.

96. Gerdes K, Christensen SK, Løbner-Olesen A. 2005. Prokaryotic toxin-antitoxin stress response loci. *Nat Rev Microbiol* **3**:371–382.

97. Pandey DP, Gerdes K. 2005. Toxin-antitoxin loci are highly abundant in free-living but lost from host-associated prokaryotes. *Nucleic Acids Res* **33**:966–976.

98. Lewis K. 2000. Programmed death in bacteria. *Microbiol Mol Biol Rev* **64**:503–514.

99. Hansen S, Vulic M, Min J, Yen TJ, Schumacher MA, Brennan RG, Lewis K. 2012. Regulation of the *Escherichia coli* HipBA toxin-antitoxin system by proteolysis. *PLoS One* **7**:e39185.

100. Black DS, Irwin B, Moyed HS. 1994. Autoregulation of *hip*, an operon that affects lethality due to inhibition of peptidoglycan or DNA synthesis. *J Bacteriol* **176**:4081–4091.

101. Schumacher MA, Piro KM, Xu W, Hansen S, Lewis K, Brennan RG. 2009. Molecular mechanisms of HipA-mediated multidrug tolerance and its neutralization by HipB. *Science* **323**:396–401.

102. Black DS, Kelly AJ, Mardis MJ, Moyed HS. 1991. Structure and organization of *hip*, an operon that affects lethality due to inhibition of peptidoglycan or DNA synthesis. *J Bacteriol* **173**:5732–5739.

103. Maisonneuve E, Shakespeare LJ, Jorgensen MG, Gerdes K. 2011. Bacterial persistence by RNA endonucleases. *Proc Natl Acad Sci USA* **108**:13206–13211.

104. Spoering A, Vulic M, Lewis K. 2006. GlpD and PlsB participate in persister cell formation in *Escherichia coli*. *J Bacteriol* **188**:5136–5144.

105. Girgis HS, Harris K, Tavazoie S. 2012. Large mutational target size for rapid emergence of bacterial persistence. *Proc Natl Acad Sci USA* **109**:12740–12745.

106. Li Y, Zhang Y. 2007. PhoU is a persistence switch involved in persister formation and tolerance to multiple antibiotics and stresses in *Escherichia coli*. *Antimicrob Agents Chemother* **51**:2092–2099.

107. Hansen S, Lewis K, Vulic M. 2008. Role of global regulators and nucleotide metabolism in antibiotic tolerance in *Escherichia coli*. *Antimicrob Agents Chemother* **52**:2718–2726.

108. Murakami K, Ono T, Viducic D, Kayama S, Mori M, Hirota K, Nemoto K, Miyake Y. 2005. Role for rpoS gene of *Pseudomonas aeruginosa* in antibiotic tolerance. *FEMS Microbiol Lett* **242**:161–167.

109. Viducic D, Ono T, Murakami K, Susilowati H, Kayama S, Hirota K, Miyake Y. 2006. Functional analysis of *spoT*, *relA* and *dksA* genes on quinolone tolerance in *Pseudomonas aeruginosa* under nongrowing condition. *Microbiol Immunol* **50**:349–357.

110. Johnson PJ, Levin BR. 2013. Pharmacodynamics, population dynamics, and the evolution of persistence in *Staphylococcus aureus*. *PLoS Genet* **9**:e1003123.

111. McCune RM, McDermott W, Tompsett R. 1956. The fate of *Mycobacterium tuberculosis* in mouse tissues as determined by the microbial enumeration technique. II. The conversion of tuberculous infection to the latent state by the administration of pyrazinamide and a companion drug. *J Exp Med* **104**:763–802.

112. Ahmad Z, Klinkenberg LG, Pinn ML. 2009. Biphasic kill curve of isoniazid reveals the presence of drug-tolerant, not drug-resistant, *Mycobacterium tuberculosis* in the guinea pig. *J Infect Dis* **200**:1136–1143.

113. Jindani A, Doré CJ, Mitchison DA. 2003. Bactericidal and sterilizing activities of antituberculosis drugs during the first 14 days. *Am J Respir Crit Care Med* **167**:1348–1354.

114. Lewis K. 2010. Persister cells. *Annu Rev Microbiol* **64**:357–372.

115. **Ramage HR, Connolly LE, Cox JS.** 2009. Comprehensive functional analysis of *Mycobacterium tuberculosis* toxin-antitoxin systems: implications for pathogenesis, stress responses, and evolution. *PLoS Genet* 5: e1000767.

116. **Sala A, Calderon V, Bordes P, Genevaux P.** 2013. TAC from *Mycobacterium tuberculosis*: a paradigm for stress-responsive toxin-antitoxin systems controlled by SecB-like chaperones. *Cell Stress Chaperones* 18:129–135.

117. **Singh R, Barry CE 3rd, Boshoff HI.** 2010. The three RelE homologs of *Mycobacterium tuberculosis* have individual, drug-specific effects on bacterial antibiotic tolerance. *J Bacteriol* 192:1279–1291.

118. **Barry CE 3rd, Boshoff HI, Dartois V, Dick T, Ehrt S, Flynn J, Schnappinger D, Wilkinson RJ, Young D.** 2009. The spectrum of latent tuberculosis: rethinking the biology and intervention strategies. *Nat Rev Microbiol* 7:845–855.

119. **Lewis K.** 2001. Riddle of biofilm resistance. *Antimicrob Agents Chemother* 45:999–1007.

120. **Lewis K.** 2005. Persister cells and the riddle of biofilm survival. *Biochemistry(Mosc)* 70:267–274.

Molecular Genetics of Mycobacteria, 2nd Edition
Edited by Graham F. Hatfull and William R. Jacobs, Jr.
© 2014 American Society for Microbiology, Washington, DC
doi:10.1128/microbiolspec.MGM2-0016-2013

David G. Russell,[1] Wonsik Lee,[1] Shumin Tan,[1] Neelima Sukumar,[1]
Maria Podinovskaia,[1] Ruth J. Fahey,[1] and Brian C. VanderVen[1]

35

The Sculpting of the *Mycobacterium tuberculosis* Genome by Host Cell–Derived Pressures

INTRODUCING THE HOST CELL

The macrophage plays host to *Mycobacterium tuberculosis* for much of the duration of its infection cycle (1, 2). And while the bacterium is not an obligate intracellular pathogen, the macrophage has undoubtedly played a dominant role in shaping the genome of *M. tuberculosis*.

The macrophage is an extremely plastic cell that fulfills multiple roles in the body. Macrophages are strongly influenced by their environment and the cytokines and chemokines released by cells in their vicinity. Tissue-resident macrophages are noninflammatory and maintain tissue homeostasis through the phagocytosis and degradation of dead cells. These resting or non-activated cells have extremely high proteolytic capabilities to facilitate breakdown of phagocytic cargo. In the face of infection, the macrophage becomes activated through both innate and adaptive immune pathways. Innate immune activation is initiated by ligation of either transmembrane sensors, such as the toll-like

receptors (TLRs), or cytosolic sensors, which include the inflammasome or NALP3 sensors (3–6). Innate immune activation leads to enhancement of the microbicidal capacities of the macrophage, most notably the increased intensity of the superoxide burst. The production of reactive oxygen intermediates by the membrane-associated NADPH oxidase complex is a relatively short-lived event that is induced upon ligation of certain phagocytic receptors (7). Adaptive immune activation is predominantly due to the release of interferon-γ (IFN-γ) by CD4[+] or CD8[+] T-cells. Stimulation by IFN-γ leads to an extensive reprogramming of the macrophage through the induction of "IFN-γ response genes" that impacts phagosome maturation, autophagy and the delivery of antimicrobial peptides, the superoxide burst, and the expression of the inducible nitric oxide synthase (NOS2) (8–11). In mice, NOS2 is clearly the dominant IFN-γ-mediated microbial response that restricts the growth of *M. tuberculosis*. The upregulation of these microbicidal activities is

[1]Microbiology and Immunology, College of Veterinary Medicine, Cornell University, Ithaca, NY 14853.

accompanied by a downregulation in the degradative capabilities of the phagocyte (12). It is hypothesized that this downregulation of proteolysis enhances the antigen-presentation capabilities of the cell by extending the epitope half-life.

The macrophage's role in host defense means that it is the cell type most likely to encounter microbes, placing considerable evolutionary pressure on pathogens such as *M. tuberculosis* to evolve a range of strategies to subvert killing by its potential host cell. The enduring nature of the infection, and the fact that most infections remain in a latent disease state, suggests a dynamic equilibrium that may move in favor of the pathogen should the host be perturbed by an immune-compromising situation such as age, malnutrition, or HIV. This article discusses the pressures within the host cell that have shaped the genome of *M. tuberculosis*.

TRANSCRIPTIONAL PROFILING AND TRANSPOSON INSERTION-SITE MAPPING

Two methods of global genome analysis have had a major impact on our understanding of the genetic basis of *M. tuberculosis*'s survival in the macrophage. Transcriptional profiling enables simultaneous quantitation of all transcripts in the bacterium at any given time. While the data generated are correlative, the approach provides invaluable insights, particularly into the metabolic pathways that are upregulated in the intracellular environment. In contrast, transposon insertion-site mapping, or TraSH as it was named in the first study (13, 14), facilitates the functional interrogation of all conditionally essential genes that are required for survival in a given context, such as in the macrophage or mouse versus growth in rich medium.

In the original, full-genome transcriptional analysis study published by Schnappinger and colleagues (15), the value of the approach to identify physiological themes was immediately apparent. The study was performed on resting and activated murine bone marrow–derived macrophages infected with *M. tuberculosis* for up to 48 h. The authors reported that *M. tuberculosis* perceives the phagosome as an environment that is DNA- and cell-wall-damaging and is replete in fatty acids rather than carbohydrates as a possible carbon source. The authors compared their data with similar profiling performed on *Salmonella*, where this bacterium appeared to favor gluconate as its primary carbon source. They hypothesized that *M. tuberculosis* infection may affect macrophage physiology in a manner that impacts nutrient availability in the endosomal

system—a hypothesis that has turned out to be particularly insightful and is discussed in greater depth in the section on metabolism of intracellular *M. tuberculosis*.

The initial transposon-based screens performed on *M. tuberculosis* utilized signature-tagged mutagenesis (16–18), an approach developed first in *Salmonella* (19). It is difficult to assess the saturation of these screens, but both groups performing the screens identified loci involved in the synthesis and export of the cell wall lipid phthiocerol dimycocerosate (PDIM) as being critical to the maintenance of infection in the mouse. The first transposon insertion-site mapping screen on an infection model was performed by Sassetti and Rubin on a murine infection (13, 14, 20). This approach allowed analysis of a much larger and complex input mutant pool and therefore had much greater genome coverage. The screen led to the generation of a list of 194 "functionally essential" genes required for establishment and maintenance of infection in the mouse that were identified through their depletion from the output mutant pool, isolated postinfection. When the genes were clustered according to known, or annotated, function, the largest gene cluster was associated with lipid metabolism, which is consistent with the transcriptional profiling analysis of Schnappinger and colleagues. Along a related vein, Lamichhane and colleagues used a subsaturation mutagenesis approach to generate a defined library of transposon-insertion mutants to model the conditional essentiality of specific metabolic pathways (21, 22) and demonstrated the significance of polyketide, mycolic acid, and fatty acid synthesis to *M. tuberculosis* in culture. This mutant collection was then interrogated in a mouse model and recapitulated the identification of genes involved in PDIM synthesis as being important to *M. tuberculosis* survival in a murine model of infection. The same collection of mutants has more recently been screened in an experimental primate infection model, which appears to exert a stronger selection pressure than the murine infection. Mutants defective in the transport of virulence factors, the synthesis and transport of cell wall intermediates, as well as DNA repair and sterol metabolism appeared particularly impaired for survival in macaques (23).

THE CORE TRANSCRIPTOME OF INTRACELLULAR *M. TUBERCULOSIS*

Since the development of microarray and, more recently, deep sequencing technology, several transcriptional profiles have been generated on *M. tuberculosis* under a variety of conditions ranging from drug pressure to host cell models, including both human and murine macrophages (24–27). In broad terms there is a strong degree of overlap across studies on intracellular bacteria that establishes lipid metabolism as a recurrent theme. However, variation in experimental conditions and the diversity of bacterial strains limit the power of comparison. In an attempt to reduce strain-specific limitations and to allow identification of a "core" intracellular transcriptional profile shared by all *M. tuberculosis* strains, Homolka and colleagues looked at the transcriptional profiles of two laboratory strains (CDC1551 and H37Rv) and 15 clinical isolates that represented the genetic diversity of the *M. tuberculosis* complex across the planet (28). The authors chose a single time point at 24 h infection of both resting and IFN-γ-activated macrophages, because this appears to be one of the more stressful time points in the infection process and induces maximal transcriptional response in the bacterium. Condition tree analysis of the transcriptional profiles was encouraging because the profiles segregated according to the genotype of the 17 strains (Fig. 1). Analysis of shared transcriptional responses identified a core transcriptome of 280 genes, of which 168 were universally induced and 112 were universally repressed. The induction of expression of genes with roles in hypoxia, oxidative and nitrosative stress, cell wall remodeling, and fatty acid metabolism across the panel of clinical strains confirmed previous studies using single strains. One-way analysis of variance facilitated the identification of 719 genes whose expression was significantly modulated by macrophage activation. Responses to known intraphagosomal stresses such as nitric oxide production (*dosR* regulon) and iron limitation (mycobactin synthesis operon) were evident among those upregulated genes. Among the genes more highly expressed in resting macrophages were many genes linked to protein synthesis and growth and genes indicative of an iron-replete environment, such as *bfrA* and *bfrB*. It was particularly interesting to note that markers of general stress response such as the heat shock proteins GrpE, GroEL, GroEL2, and GroES were all upregulated in resting macrophages, suggesting that these chaperones were required to sustain growth.

The intracellular growth characteristics of the different clinical isolates in murine macrophages showed considerable heterogeneity, with the West African strains consistently less "fit" and the most robust growth exhibited by the Beijing and Haarlem strains. There were also marked genotype- and strain-specific transcriptional signatures, demonstrating that the genetic diversity did lead to distinct transcriptional responses when the panel of strains was interrogated under defined, common infection conditions.

TRANSITIONAL PHYSIOLOGICAL STATES IN A PROLONGED EXPERIMENTAL INFECTION

More recently, Rohde and colleagues employed the same intensive profiling approach to delineate the different physiological states experienced by *M. tuberculosis* during an extended period of infection (14 days) in macrophages (29). Analysis of *M. tuberculosis* CFUs following experimental infection of macrophages revealed a reproducible trend. Immediately following infection, the initial inoculum clearly experiences extreme stress, and many of the infecting bacilli are killed (Fig. 2). The CFU count then stabilizes as the bacteria appear to undergo a period of adaptation to the intracellular environment, before the CFU number increases as the bacteria enter into a period of intracellular replication. These experiments were performed with *M. tuberculosis* strain CDC1551, which shows little propensity to escape from the vacuole or to induce cell death in its host cell. To look at the relative replication rates of the bacteria during the infection, Rohde and colleagues exploited a "clock" plasmid developed by David Sherman's laboratory (30). The plasmid is lost from bacteria at a fixed rate determined by the rate of bacterial replication. Intriguingly, when the rate of plasmid loss during the course of the infection was analyzed, it was determined that the bacteria were actually replicating fastest during initial infection when they were being killed, and the death rate diminished as bacterial replication slowed (Fig. 2). These data indicate that rapid replication is inconsistent with intracellular survival.

Transcriptional profiling of changes in bacterial gene expression through these transitions was followed by isolation and microarray analysis of bacterial mRNA at 2-day intervals from day 0 to day 14. Analysis of the global transcriptional profile demonstrated that the largest change in expression, up and down, was observed at day 2 postinfection and that subsequently, the transcriptional activity trended toward baseline (1:1). However, within the changing profiles one can observe distinct patterns of gene clusters or regulons linked to known environmental cues, metabolic pathways, and stresses (Fig. 2). The *dosR* regulon is of particular interest (31–34) because of its linkage to hypoxia and nitric oxide. These genes were upregulated markedly in the first 2 days postinfection. The elevated level of expression was sustained until day 8, at which time expression

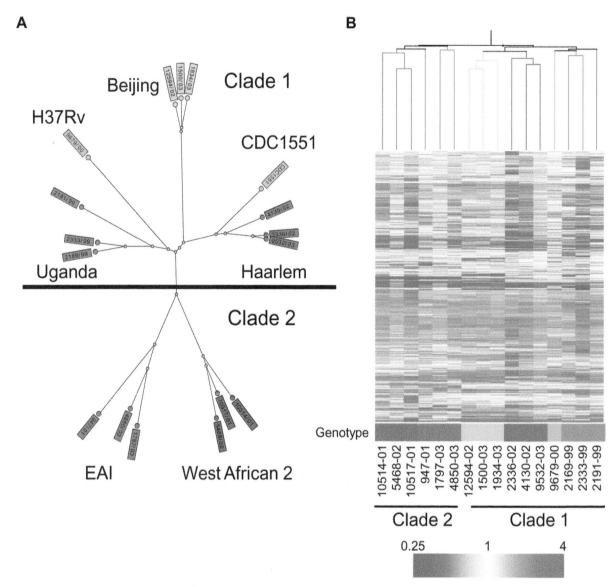

Figure 1 Genetic and transcriptome diversity of *M. tuberculosis* complex (MTC) clinical isolates. (**A**) Radial neighbor-joining tree based on 24 loci MIRU-VNTR and 43 spacer spoligotyping showing the phylogenetic relationship of strains in this study. Three strains each from the five distinct lineages pathogenic to humans plus two sequenced reference strains (H37Rv and CDC1551) were chosen to represent the global diversity of MTC. (**B**) Condition tree of MTC clinical isolate transcription profiles *in vitro* during log phase growth in 7H9 medium relative to the CDC1551 reference strain (three biological replicates). Expression profiles for genes passing quality filters (flagged as present in 42 of 48 samples) with differential expression in at least one strain (up or down >1.5-fold in at least 1 of 16) were clustered using the Spearman correlation. Each column represents the global transcription profile of a single strain. Genes were clustered vertically based on the distance measure. Red and blue indicate higher or lower gene expression than the CDC1551 control, respectively. This figure is reproduced from Homolka et al. (28). doi:10.1128/microbiolspec.MGM2-0016-2013.f1

was downregulated to close to baseline. The period of upregulation of transcription coincided with stress and adaptation, while the period of downregulation correlated with the restoration of growth.

Clearly, the bacterium responds to macrophage-derived pressures, and these pressures are dynamic and change from the time of uptake by the phagocyte to the establishment of a productive infection.

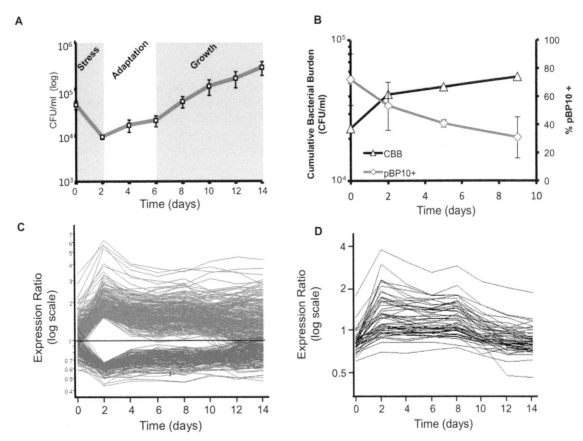

Figure 2 Life and death dynamics during long-term intracellular survival of *M. tuberculosis*. (**A**) Survival assays. Resting murine bone-marrow-derived macrophages were infected at low multiplicity of infection (~1:1) with *M. tuberculosis* CDC1551. Viable CFUs were quantified at day 0 and at 2-day intervals postinfection over a 14-day time course by lysis of monolayers, serial dilution, and plating on 7H10 medium. Error bars indicate standard error of the mean from two independent biological replicates, each consisting of three technical replicates per strain (total of six wells/strain). (**B**) Replication clock plasmid. The percentage of bacteria containing the pBP10 plasmid during growth in resting macrophages was determined by comparing CFUs (mean ± standard deviation) recovered on kanamycin vs. nonselective media (red). The cumulative bacterial burden (CBB) (black) was determined by mathematical modeling based on total viable CFUs and plasmid frequency data. Data shown represent two independent experiments, with each sample performed in quadruplicate (eight total wells/time point). (**C**) The "bottleneck" response. Temporal expression profiles of genes differentially regulated at day 2 postinfection, including genes from (**A**) that were up- (red) or downregulated (blue) >1.5-fold (shown as ratio of signal intensity relative to control). Note the maximal change in transcript levels at day 2 postinfection followed by the majority trending back toward control levels. (**D**) "Guilt by association" analysis. Genes regulated in synch with known virulence regulons (i.e., the DosR regulon) were identified by using a highly regulated member of this regulon, *hspX*, in place of synthetic profiles. This figure is reproduced from Rohde et al. (29). doi:10.1128/microbiolspec.MGM2-0016-2013.f2

MYCOBACTERIUM SENSES AND RESPONDS TO ENVIRONMENTAL CUES

Phagosome maturation represents a continuum, a series of environments that, for the most part, become increasingly hostile as the phagosome acquires more of the characteristics of the lysosome. This is the niche that *M. tuberculosis* has evolved to exploit, and to do so it must sense and respond appropriately to environmental cues. Evolution appears to have generated responses that fall into two distinct categories: those that are physiological imperatives that are linked directly to the environmental cue or stress and "off-target" responses that confer an advantage to *M. tuberculosis* for reasons unrelated to the nature of the cue itself. The latter response appears to be linked predominantly to alterations in bacterial metabolism critical for intracellular survival.

There is convincing evidence that some strains of *M. tuberculosis* escape from their vacuole into the host cell cytosol with varying degrees of efficiency (35). *M. tuberculosis* mutants defective in the secretory system ESX-1 or in the secreted effector ESAT-6 are unable to access the cytosol and are attenuated in mice (36). Recent studies indicate that vacuolar egress leads to death of the host cell (37, 38). These data imply that existence in the cytosol of the host cell is likely to be a transient state that may be very important for the spread of the infection, but because the bulk of the intracellular, replicative phase appears to be intravacuolar, this article focuses predominantly on *M. tuberculosis*'s strategies to survive within this membrane-bound compartment.

pH

Obviously, it would be a great advantage for *M. tuberculosis* to be able to sense its intracellular environment and to modify its transcriptional profile to enhance its survival. The pH of the intracellular compartment in which the bacterium resides has long been known to constitute an important environmental cue to *M. tuberculosis* (39–41). The notable significance of pH is that it is an inevitable consequence of phagosome maturation (40, 42, 43). The more mature the phagosome, the lower the pH. The pH of the *M. tuberculosis*–containing phagosome is around 6.4 (39, 42), which is lower than the neutral pH outside the cell but higher than the lysosomal pH (4.5 to 5.0). For *M. tuberculosis*, this means that the ability to quantify pH would act as an intracellular locator signal. *M. tuberculosis* possesses a PhoPR two-component sensor/effector kinase homolog, and mutants in this locus exhibit a marked drop in virulence to the extent that they have been exploited as experimental vaccine constructs (44).

In screening for *M. tuberculosis* mutants that were defective in resistance to pH stress, Vandal and colleagues identified Rv3671c, a membrane-bound protease required to maintain pH homostasis of the bacterial cytosol (45). These data are consistent with several other screens that have identified mutations conferring hypersensitivity to acid pH in that the majority of these genes encode proteins associated with cell wall integrity.

Using transcriptional profiling, Rohde and colleagues determined *M. tuberculosis*'s response to pH change in the macrophage by infecting murine macrophages with *M. tuberculosis* in the presence and absence of the host proton ATPase inhibitor concanamycin A (27). Over the first 2 h postinvasion of the macrophage, *M. tuberculosis* upregulated expression of 68 genes. In the absence of the pH drop, the increased expression of 30 of these 68 genes was no longer detectable. Among these genes was a locus, *aprABC*, that is unique to the tuberculosis complex and was probably acquired by horizontal gene transfer at the time the tuberculosis complex diverged from less pathogenic mycobacteria (46). The first gene in the operon, *aprA*, behaves like a transcriptional regulator, and mutants deficient in this gene show extensive alterations in transcriptional profile. Expression of *aprA* is PhoPR dependent, and *aprA* regulates expression of genes linked to triacylglycerol (TAG) and PDIM biosynthesis. When wild-type *M. tuberculosis* is exposed to low pH, it shows increased synthesis of PDIMs, most notably phthiodiolone and phthiocerol A, but this response is lost in the *aprA*-deficient mutant. In contrast, mutants lacking *phoP* are extreme overproducers of PDIMs at low pH, indicating that the two regulators operate in the same pathway, but with *phoP* functionally upstream of *aprA*. These data appear consistent with the link between pH and cell wall integrity, but recent data indicate that an increase in the production of methyl-branched cell wall lipids, such as PDIMs, may also provide relief from other intracellular stresses associated with the major carbon sources exploited by intracellular *M. tuberculosis*. These data are discussed in depth later in this article.

To understand the link between environmental cues, such as pH, and the immune status of the host cell during the course of infection, Tan and colleagues generated reporter bacterial strains that exhibit conditional expression of green fluorescent protein (GFP) (47). The promoter upstream of the gene *Rv2390c* responds to both pH and chloride concentration as synergistic cues. This is significant because, as a phagosome matures, the chloride concentration increases with the drop in pH. Chloride is one of the counterions that balance the activity of the proton ATPase responsible for phagosomal acidification (48). The induction of expression of GFP in *M. tuberculosis* was more marked in IFN-γ-activated compared to resting macrophages, which is consistent with increased phagosome maturation (40, 41). This reporter strain was then interrogated in a murine infection model in wild-type and IFN-γ-deficient mice, and the level of induction of GFP expression was found to be much lower in the immune-compromised animals. A second reporter strain was generated that encoded GFP driven by the *hspX* promoter. Expression of *hspX* is controlled through *dosRS*, which is strongly induced under conditions of low oxygen tension or in the presence of nitric oxide. The *hspX* reporter strain was shown to respond to both conditions in culture and exhibited strong induction upon infection of wild-type versus IFN-γ-deficient mice. Interestingly, the kinetics of expression of GFP differed in the two reporter strains. The chloride/pH responders showed high induction early, at 14 days, and reduced expression at later time points. In contrast, the *dosRS*-dependent reporter showed low induction early and strong induction at 28 days. The late induction of the *dosRS*-dependent reporter may correlate with the development of an acquired immune response and the expression of the inducible nitric oxide synthase, NOS2. These reporter strains have provided novel insights into the dynamic nature of the intracellular environment during the course of experimental infection in mice.

Reactive Oxygen and Nitrogen Intermediates

As mentioned earlier, macrophages have the capacity to generate reactive oxygen and nitrogen intermediates (ROIs and RNIs). Superoxide production by the NADPH oxidase complex is increased in intensity in macrophages activated by IFN-γ(7) However, *M. tuberculosis* does not appear to be particularly sensitive to reactive oxygen intermediates or their downstream products. The course of infection in wild-type versus $p47^{phox}$-deficient mice was impacted minimally by the defect (49, 50). The absence of a phenotype indicates that *M. tuberculosis* is particularly adept at dealing with reactive oxygen intermediates, and there are several different resistance mechanisms that have been demonstrated. *katG* encodes a catalase-peroxidase-peroxynitritase, which processes H_2O_2, and a mutant lacking this gene was hypersusceptible to H_2O_2 *in vitro* (51). This mutant was also seriously attenuated in wild-type mice and mice deficient in NOS2 but exhibited a phenotype comparable to wild-type *M. tuberculosis* in $gp91^{phox}$-deficient mice, indicating that the basis of the attenuation was dependent on the host $gp91^{phox}$ NADPH oxidase. *M. tuberculosis* also possess two genes encoding superoxide dismutases, *sodA* and *sodC*, capable of converting superoxide to H_2O_2 (52). Mutants defective in each superoxide dismutase exhibited reduced virulence under different conditions, but the iron-dependent SodA-deficient strains exhibited the strongest phenotype in a murine infection model (53). Finally, low-molecular-weight thiols are important antioxidants in many biological systems, and *M. tuberculosis* synthesizes mycothiol instead of glutathione (54). Deletion of the gene *mshA*, which encodes a glycosyltransferase that is the first step in mycothiol synthesis,

renders the bacterium hypersensitive to oxidative stress (55), but these mutants exhibit minimal phenotype in murine infection models. These data all indicate that *M. tuberculosis* has evolved multiple, overlapping pathways of avoiding or mitigating the consequences of oxidative stress due to the action of the host NADPH oxidase complex.

This is in marked contrast to the impact of NOS2 on controlling *M. tuberculosis* infection in mice. A NOS2-deficient mouse is almost as impaired in control of tuberculosis as an IFN-γ-deficient mouse (10), which, acknowledging the probable existence of compensatory microbicidal mechanisms, argues that nitric oxide and its products are the dominant means of immune-mediated control of *M. tuberculosis* infection in mice. The demonstration of nitric oxide production in humans under various conditions of infection argues comparable significance in human tuberculosis (56).

If nitric oxide and its products are the primary mechanism of immune-mediated control of *M. tuberculosis*, it would be logical for *M. tuberculosis* to have evolved strategies to try to minimize the impact of these radicals on its survival. *M. tuberculosis* possesses an NADH-dependent peroxidase/peroxynitrite reductase that consists of four subunits: an alkylhydroperoxide reductase (AhpC), a thioredoxin-related oxidoreductase (AhpD), a dihydrolipoamide transferase (DlaT), and a lipoamide dehydrogenase (Lpd) (57). DlaT and Lpd are dual-function proteins that also serve as the E2 and E3 subunits of pyruvate dehydrogenase (58), which generates acetyl-coenzyme A (acetyl-CoA). This enzyme complex serves to detoxify both reactive nitrogen and oxygen intermediates. *M. tuberculosis* deficient in DlaT exhibits a carbon source–dependent phenotype, growing well in fatty acids and poorly in glucose or glycerol, revealing the function of DlaT in the pyruvate dehydrogenase. This mutant was also hypersusceptible to RNI *in vitro*, but the mutant was avirulent in both wild-type and NOS2-deficient mice, indicating that the phenotype was not related purely to detoxification of RNI (59).

In addition to the machinery for detoxification of RNI and ROI, *M. tuberculosis* also possesses responses linked to the repair or degradation of proteins and DNA damaged by oxidative mechanisms. The sulfurs of cysteine and methionine are known to be highly susceptible to oxidation, which leads to impairment of protein function. *M. tuberculosis* possesses two methionine sulfoxide reductases, MsrA and MsrB, capable of restoring oxidized methionine to its reduced state (60). *M. tuberculosis* deficient in *msrA* and *msrB* is hypersusceptible to killing by acidified nitrite and hypochlorite but not by H_2O_2. Furthermore, the double mutant showed minimal phenotype in a murine infection model, indicating that this pathway was not central to survival within the mouse.

Darwin and colleagues performed a transposon screen to identify mutants defective in resistance to RNI stress (61, 62). Among the mutants that were most susceptible were those featuring insertions in genes that encode components of the bacterial proteasome, including the proteasome ATPase *mpa* and the proteasome accessory factor *pafA*. Further analysis employing chemical inhibitors of proteasome function, and the characterization of mutants lacking the core proteasome genes *prcBA*, revealed that these bacteria were now highly sensitized to RNI-mediated damage. The proteasome-deficient mutants were also attenuated in mice and the loss of virulence was restored in NOS2-deficient mice, demonstrating that the phenotype correlated with the ability to produce RNI *in vivo*.

Metal Ions

The presence of metal ions in the endosomal-lysosomal continuum has been known for a while through both direct demonstration and the identification and characterization of different metal ion transporters. However, the role(s) that manipulation of metal ion concentration plays in controlling the growth of intracellular microbes is an emerging field. Cells can sequester trace metal ions from pathogens to deny them essential enzyme function, or deliver excess metal ions into phagosomes to combine with ROIs to generate highly toxic intermediates.

Wagner and coworkers utilized hard X-ray probe analysis to determine the concentration of 10 individual elements inside the phagosomes containing *M. tuberculosis* (63). They observed an early decrease in the intraphagosomal concentrations of calcium and potassium that was coincident with an increase in sulfur and chloride concentrations. These data on chloride concentrations match the recent reporter data of Tan et al. (47), and the further elevation of chloride concentration upon macrophage activation is consistent with increased maturation of the *M. tuberculosis*–containing phagosome.

Wagner et al. also documented extremely high concentrations of iron in the *M. tuberculosis* phagosome that reached maximal levels 24 h postinfection (63). This demonstration of the *M. tuberculosis*–containing vacuole as an iron-replete environment is consistent with more recent transcriptional profiling that reports the upregulation of expression of *bfrB*, which encodes an iron-chelating protein that protects and detoxifies

the metal inside the bacterium. Homolka and colleagues, who generated the "universal" core intracellular transcriptome, also reported upregulated expression of *bfrB* in infection of resting macrophages (28). In contrast, genes involved in the synthesis of the iron-acquisition molecule mycobactin (*mbtA-J*), which is produced in conditions of iron scarcity, were upregulated in activated macrophages.

The natural-resistance-associated membrane protein (NRAMP-1/SLC11A1) has long been implicated in regulating *M. tuberculosis* infection through the manipulation of the concentration of Fe^{2+} or Mn^{2+} in the phagosome of the infected macrophage (64). Although there is still considerable controversy as to its mode of action, the transport of manganese and iron from the phagosome into the cytosol and out of the cell appears to provide the most plausible mechanism for NRAMP-1-mediated restriction of *M. tuberculosis* growth (65). This is consistent with the transcriptional response of *M. tuberculosis* and the known concentrations of iron in the phagosomes of resting versus activated macrophages.

Wagner et al. also reported that macrophage activation leads to increases in the concentration of both copper and zinc in the *M. tuberculosis*–containing phagosomes (63). Several recent studies have demonstrated that these metal ions play important roles in controlling *M. tuberculosis* infection in its host cell. Excess copper is toxic to *Escherichia coli* through destabilization of the Fe-S clusters in some bacterial enzymes and the displacement of iron or zinc in others. *M. tuberculosis* senses and responds transcriptionally to copper through the upregulation of a range of genes encoding copper detoxification mechanisms, many of which are under the control of the transcriptional regulator *csoR* (66, 67). Among the upregulated genes is *ctpV*, encoding a putative cation-transporting P_1-type ATPase thought to be active in copper efflux (68–70). Mutants defective in *ctpV* show increased sensitivity to copper intoxication, and together with the copper transporter, MctB, are required for full virulence in mice. The macrophage controls the levels of copper in the phagosome through the activity of the transporters ATP7A and ATP7B. Their activity is increased in inflammatory conditions through the increased expression of Ctr1, a cytosolic copper-binding protein that acts as a chaperone and delivers the metal ion to the transporters (71), facilitating enhanced transport into the phagosome.

Zinc is delivered to the phagosome through the activity of several metal ion transporters including the SLC39/ZIP family and the ZnT proteins of the SLC/

CDF family (72). These transporters liaise with the metal-chelating protein metallothionein to balance the cytosolic and vacuolar concentration of zinc. Expression of metallothionein is also upregulated in inflammatory conditions. In a recent study by Botella and colleagues it was demonstrated that, much like copper, exposure of *M. tuberculosis* to zinc increases the expression of another P_1-type ATPase, CtpC (73, 74). Mutants lacking the *ctpC* gene retain excess zinc and are growth-impaired. Transcriptional profiling of *M. tuberculosis* in macrophages revealed time-dependent upregulation of expression of genes associated with zinc detoxication (*ctpC* and *Rv3269*), along with other P1-type ATPases associated with metal ion transport (*ctpG*, *ctpV*, and *ctpF*). The intraphagosomal and intrabacterial accumulation of zinc was visualized at the electron microscopy level by a method developed by Stoltenberg and coworkers (75) and was shown to correlate with the immune status of the host cell (74), illustrated in Fig. 3.

While the macrophage has evolved to exploit these metal ions to control intracellular infections, *M. tuberculosis* has evolved concomitantly to attempt to minimize their impact, predominantly from the expression of metal efflux pumps.

Antimicrobial Peptides

The lysosomal system of macrophages contains antimicrobial peptides capable of killing a range of different bacterial pathogens. While it was known that the induction of autophagy leads to death of *M. tuberculosis* (11, 76), the mechanism(s) that leads to the actual demise of the bacteria was unknown. Alonso and colleagues demonstrated that the most active microbial peptide that could be isolated from the lysosomes of murine macrophage was a ubiquitin-derived peptide generated through the degradation of ubiquitinated cargo delivered to the lysosome through the autophagous pathway (77). The induction of autophagy enhanced both the production of the peptide and the delivery of live *M. tuberculosis* to the peptide-containing compartments (Fig. 4). This observation was confirmed by Ponpuak and colleagues, who also identified peptides generated from the degradation of the autophagic chaperone Fau as contributing to the microbicidal activity of the autophagosome (78).

Purdy and coworkers have since studied a synthetic peptide, Ub2, based on a 12-amino-acid, positively charged region of ubiquitin that has anti–*M. tuberculosis* activity (79–81). Using an assay based on the maintenance of cytosolic pH homeostasis in the bacterium, they demonstrated that the peptide compromises

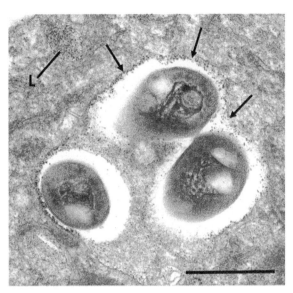

Figure 3 Distribution of zinc within intraphagosomal *Mycobacterium avium*. Bone-marrow-derived mouse macrophages were infected with *M. avium* and treated for capturing of free zinc ions by an auto-metallography (AMG) procedure. Cells were then processed for electron microscopy observation. The zinc crystals formed during this procedure appeared as small dense deposits along the inner face of the phagosome membrane (arrows). The mycobacterial cytoplasm was devoid of zinc. Lysosomes also displayed zinc crystals. Comparable images were observed in *M. tuberculosis*–infected macrophages. Scale bar, 0.5 m. This picture was provided by Dr. Chantal de Chastellier and was prepared as detailed in reference 74. doi:10.1128/microbiolspec.MGM2-0016-2013.f3

membrane integrity. In a genetic screen conducted on *Mycobacterium smegmatis*, the porin MspA was identified as a major susceptibility factor to killing by Ub2. Expression of the *M. smegmatis mspA* gene in *M. tuberculosis* rendered the bacteria exquisitely sensitive to Ub2 peptide-mediated killing and markedly reduced their ability to survive in macrophages. More recent studies indicate that the peptide adopts a β-sheet conformation in artificial membranes that leads to loss of membrane integrity (79). Interestingly, *M. tuberculosis* has lost genes encoding MspA-like porins from its genome, which may represent an evolutionary step toward survival in an environment rich in antimicrobial peptides.

Recently, the mechanism underlying the mode of action of vitamin D (1α,25-dihydroxycholecalciferol) in the enhancement of killing of *M. tuberculosis* at the level of the infected macrophage has been resolved and shown also to rely on the upregulated delivery of microbicidal peptides to *M. tuberculosis*–containing vacuoles through the process of autophagy (82, 83).

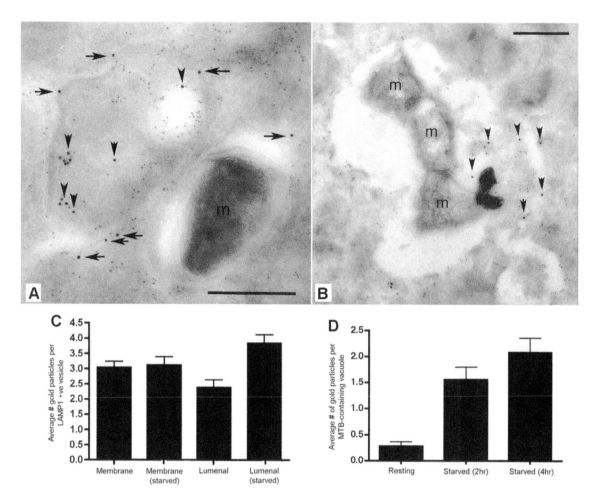

Figure 4 Immunoelectronmicroscopy of *M. tuberculosis*–infected macrophages. All sections were probed with mouse antiubiquitin (12 nm gold) and rat anti-LAMP1 (6 nm gold). (**A**) Untreated, infected macrophage demonstrating that the bacteria-containing vacuoles lack ubiquitin signal and that the ubiquitin signal is predominantly associated with either the membranes (horizontal arrows) or the lumen (vertical arrowheads) of LAMP1-positive, dense lysosomes. (**B**) In cells treated by starvation for 2 h, the *M. tuberculosis*–containing vacuole has acquired a dense, LAMP1-positive, flocculent matrix that is also positive for ubiquitin (arrowed), demonstrating delivery of lysosomal ubiquitin to the bacteria-containing compartment. The relative distribution of ubiquitin-associated label in *M. tuberculosis*–containing vacuoles in untreated and starved bone marrow-derived macrophages is detailed in Fig. 4D, with standard error bars. This figure is reproduced from Alonso et al. (77). doi:10.1128/microbiolspec.MGM2-0016-2013.f4

Campbell and Spector showed that treatment of human macrophages with physiological levels of vitamin D leads to an upregulation in expression of the antimicrobial peptide cathelicidin and the induction of autophagy, which restricted the growth of both *M. tuberculosis* and HIV (84, 85). HIV infection leads to downregulation of expression of Beclin-1 and microtubule-associated protein 1 light chain 3B (LC3B), which are required for autophagy, and vitamin D reverses this process. Treatment of the cells with RNAi targeting the mRNA encoding cathelicidin markedly reduced expression of the antimicrobial peptide and inhibited vitamin D–mediated killing of *M. tuberculosis*.

These studies demonstrate the importance of antimicrobial peptides in controlling *M. tuberculosis* infection and also demonstrate that enhanced maturation of the *M. tuberculosis*–containing vacuole and delivery of the bacterium to the lysosome through the induction of autophagy is important in promoting this control.

Hypoxia, Host Tissue Metabolism, and the Granuloma

The macrophage does not exist in isolation, and in an *M. tuberculosis* infection it is part of a complex tissue-remodeling program that leads to the generation of a granuloma (86, 87). In an immune-competent individual, whether the infection will progress to active disease and transmission or remain under immune control is determined at the level of the granuloma. If the granuloma becomes fibrotic and necrotic, it also becomes increasingly hypoxic, and this drop in oxygen tension is part of a major realignment of host tissue metabolism that impacts on bacterial metabolism.

Transcriptional profiling of human tuberculosis granulomas isolated from infected lung tissue by laser-capture microdissection revealed a marked dysregulation in lipid metabolism with an upregulation in genes involved in both lipid processing and lipid sequestration (88). The lipid sequestration phenotype was confirmed by immune histology showing abundant expression of peripilin 2 or adipophilin (ADFP; a protein associated with lipid droplet formation in foamy macrophages) in the phagocytes subtending the caseous center of the granuloma (Fig. 5). Kim and colleagues isolated the caseum, performed lipid analysis on its contents, and demonstrated that the major lipid species were cholesterol ester, cholesterol, and triacylglycerol (88). Because cholesterol from low-density lipoproteins (LDL) becomes esterified when it is incorporated into lipid droplets in macrophages, this strongly suggests that the accumulation of the caseum in tuberculosis granulomas most likely depends on the generation and subsequent death of foamy, lipid-laden macrophages. The foamy macrophage phenotype can be induced by hypoxia (89), by *M. tuberculosis* infection (90, 91), or by the exposure of macrophages to plastic beads coated in the bacterial lipid trehalose dimycolate (TDM) (88), which has been shown to be the most bioactive of the cell wall lipids released by intracellular *M. tuberculosis* (91–93). This latter response is likely due to chronic stimulation of proinflammatory pathways similar to those activated in atherosclerosis. Murine granulomas induced by TDM-coated beads exhibit many of the characteristics observed in human tuberculosis granulomas including neovascularization, foamy macrophage formation, fibrosis and extracellular matrix remodeling, and necrosis (92, 94, 95). Transcriptional analysis of the murine TDM granuloma reveals that 74% of the genes upregulated in this model are also upregulated in human tuberculosis granulomas, suggesting that the response to TDM is a significant component of human tuberculosis (95). Comparable modulation of lipid metabolism is observed in infected macrophages in culture (90, 91). *M. tuberculosis*–infected cells show downregulation of lipid hydrolysis and increased lipid retention.

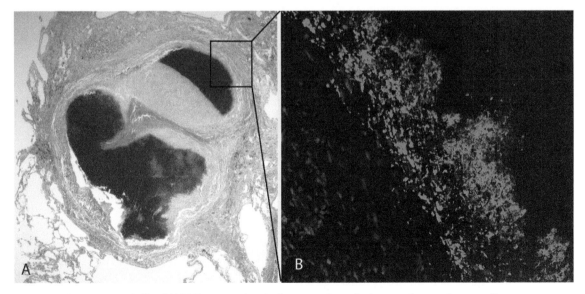

Figure 5 Peripilin 2/ADFP expression in human tuberculosis granulomas. Immunofluorescence signals were obtained for each granuloma (**B**), and the corresponding region from a hematoxylin and eosin stained slide (**A**) is shown. Nuclei are shown in blue and antigens in red. The macrophages subtending the caseum of this fibrocaseous granuloma label strongly for peripilin2/ADFP expression. This figure is reproduced from Kim et al. (88).
doi:10.1128/microbiolspec.MGM2-0016-2013.f5

TDM is recognized by the signaling receptor Mincle (96) and TLR2 (97, 98), though recognition by TLR2 requires the activity of the scavenger receptor MARCO to fulfill a tethering capacity. MARCO is expressed to low levels on resting macrophages but is highly upregulated upon innate or adaptive immune activation.

The realignment of the host macrophage and the granuloma tissue has a major impact on the metabolism of *M. tuberculosis* and its success as a pathogen. Many studies, from transcriptional profiling of intracellular bacteria to genetic screens of transposon libraries, have implicated the metabolism of host lipids in bacterial survival in the host. Pandey and Sassetti identified a locus, *mce4*, that functions as a bacterial cholesterol transporter (14, 99). Mutants defective in this gene show a phenotype similar to that reported for *icl1*, the inducible isocitrate lyase (100), in their inability to sustain a chronic infection in mice. *M. tuberculosis* appears to rely on cholesterol as a significant carbon source inside its host cell, and this dependence is enhanced in immune-competent mice. The degradation of cholesterol generates a considerable amount of propionyl-CoA, which is potentially toxic to *M. tuberculosis* (Fig. 6). Assimilation of propionyl-CoA is achieved through three possible routes (101): (i) the methyl citrate cycle, which is dependent on isocitrate lyase *icl1* functioning as a methylisocitrate lyase to generate succinate to feed into the tricarboxylic acid (TCA) cycle, (ii) the methyl malonyl pathway (MMP) that produces succinyl-CoA, which can also feed into the TCA cycle, or (iii) methyl malonyl-CoA intermediates, which can be incorporated directly into methyl-branched polyketide lipids such as PDIMs, sulfolipids, and lipooligosaccharides. Recent data indicate that *M. tuberculosis* inside its host macrophages utilizes the synthesis of these cell wall lipids to relieve stress from propionyl-CoA intermediates (102). An *icl1* mutant, which lacks a functional methyl citrate cycle, was attenuated inside murine macrophages in culture but could be rescued either by the addition of vitamin B_{12}, which is a cofactor required for the MMP, or through the induction of a foamy macrophage phenotype by addition of oleate to the macrophages prior to infection (102) (Fig. 7). Oleate likely acts through expansion of the acetyl-CoA pool following its degradation by β-oxidation, or directly as an acyl-CoA primer, to facilitate additional synthesis of methyl-branched lipids. The ability of *M. tuberculosis* to access and utilize the lipid stores of the host cell may also be visualized through the use of fluorescent fatty acid analogs (89, 90).

There is a satisfying sense of symmetry to these metabolic adaptations. The metabolism of the bacterium appears geared to exploit the altered metabolism of the host cell and tissue. The altered metabolism of the host can be induced by the *M. tuberculosis* cell wall lipids, and those cell wall lipids contain methyl branched lipids and act as a sink for potentially toxic cholesterol intermediates.

When bacterial metabolism is fueled primarily by fatty acids, it is necessary for *M. tuberculosis* to synthesize sugars from intermediates generated in the TCA cycle. This process, gluconeogenesis, is initiated by phosphoenolpyruvate carboxykinase (PckA). Early experiments on *Mycobacterium bovis* bacillus Calmette-Guérin demonstrated that PckA expression was upregulated on growth on acetate and palmitate but not on glucose (103). Furthermore, *pckA*-deficient mutants showed impaired survival in macrophages and reduced virulence in mice. More recently, Marrero and colleagues used a Tet^r inducible-expression construct encoding *pckA* to demonstrate that *M. tuberculosis* could only grow on fatty acids under conditions of induction of *pckA* expression (104). They went on to show that mutants of *M. tuberculosis* lacking *pckA* were attenuated in macrophages and mice.

More recently, the same investigators analyzed the phenotype of *M. tuberculosis* that is impaired in its ability to utilize glucose directly (105). The first step in glycolysis is the phosphorylation of glucose. *M. tuberculosis* encodes two functional glucokinase genes, one of which, *ppgK*, is required for normal growth on glucose, while the other, *glkA*, appears dispensable. Metabolomic analysis revealed that both genes encode enzymes that facilitate entry of glucose into *M. tuberculosis*'s central carbon metabolism, and a double mutant was unable to utilize glucose as a substrate for growth or metabolism. The double mutant established an infection in murine lungs to levels comparable to wild type but was mildly attenuated when the infection transitioned into a chronic state. These data indicate that bacterial metabolism has to adapt during the course of an infection and that while mutants unable to utilize fatty acids are severely attenuated, those that are defective in glycolysis are impaired in survival in later stages of infection in the murine model.

At the heart of much of *M. tuberculosis*'s metabolic flexibility is the enzyme isocitrate lyase, Icl1. In early studies, Icl1 activity was shown to be required for growth on acetate, and mutants of *M. tuberculosis* showed impaired growth on various fatty acid sources (100, 106). Mutants deficient in *icl1* also showed an attenuated phenotype in infection models that was most marked in activated macrophages, or in immune-competent mice once the infection was controlled. In

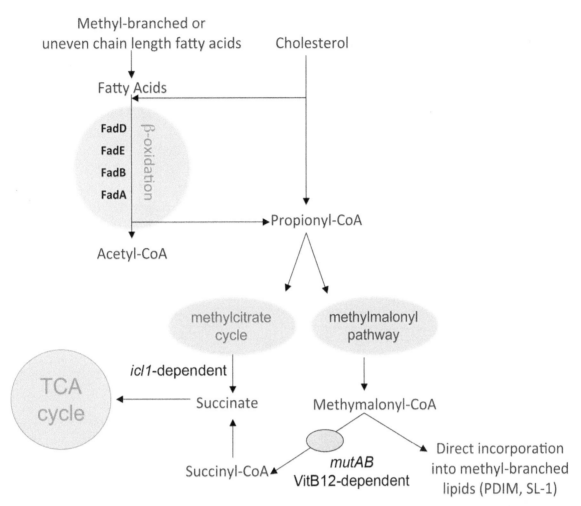

Figure 6 Outline of the pathways relevant to C$_3$ metabolism. The metabolism of cholesterol, methyl-branched fatty acids, and odd-chain length lipids raises the intracellular levels of the C$_3$ compounds propionate or propionyl-CoA, which *M. tuberculosis* finds highly toxic. The bacterium has developed three strategies to detoxify propionyl-CoA. Isocitrate lyase (*icl1*) activity has been suborned to fulfill the function of methylcitrate lyase in the last step of the methyl citrate cycle to generate the TCA cycle intermediate succinate. The methylmalonyl pathway has also been mobilized to metabolize propionyl-CoA to produce succinyl-CoA via the vitamin B$_{12}$-dependent activity of methylmalonyl-CoA mutase (*mutAB*). Finally, intermediates from the methylmalonyl pathway can be incorporated directly into the abundant, methyl-branched lipids of the bacterial cell wall, such as PDIM and SL-1. doi:10.1128/microbiolspec.MGM2-0016-2013.f6

keeping with our knowledge at the time, it was thought that Icl1 functioned as the gating enzyme into the glyoxylate shunt that would enable *M. tuberculosis* to retain carbon while growing on fatty acids. Subsequently, McKinney and collaborators showed that it is the ability of Icl1 to sustain the methyl citrate cycle for detoxification of propionyl-CoA that is critical to the success of *M. tuberculosis* in a host environment (101, 107, 108). Most recently, Eoh and Rhee used a metabolomics approach to demonstrate that Icl1 activity is also necessary under hypoxic conditions where the en-

zyme is critical to the sustained production of the flexible metabolic intermediate succinate, which is required to maintain membrane potential, for ATP synthesis, and for anaplerosis (109). The requirement for Icl1 activity is independent of fatty acid catabolism. The authors propose that this activity, coupled with the ability of *M. tuberculosis* to operate a reductive TCA cycle (110), enables the bacteria to generate ATP under low oxygen tension. The ability to adopt such a reduced yet sustainable metabolic state is likely critical to *M. tuberculosis*'s capacity to maintain a subclinical infection for decades.

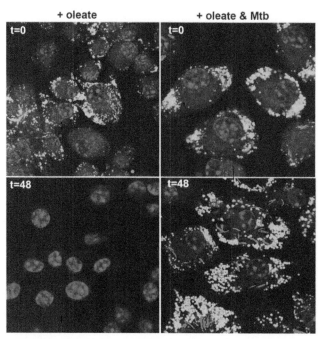

Macrophages : stained w/ BODIPY 493/503 after 2 days induction with oleate

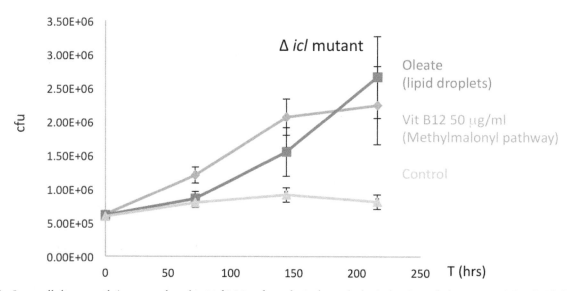

Figure 7   Intracellular growth is restored to the Δ*icl1 M. tuberculosis* through the induction of oleate-containing lipid droplets in the infected cell. The survival of the Δ*icl1* mutant is impaired in macrophages, but growth could be restored by preloading the host cells with exogenous oleate or by adding vitamin B_{12} to the cell culture medium to facilitate operation of the methyl malonyl pathway. (**Upper**) Induction of macrophage lipid droplets, detected with BODIPY 493/503 (green), *M. tuberculosis* expressing pVV16-mCherry (*smyc´::mCherry*) (red), and nuclei (blue) in oleate-loaded macrophages versus untreated macrophages. In the absence of *M. tuberculosis* infection, the lipid droplets are lost from the macrophages in 48 h. However, in the presence of *M. tuberculosis*, the lipid droplets persist. (**Lower**) Bacterial CFUs were determined at 2-day intervals across an 8-day period for the Δ*icl1 M. tuberculosis* in untreated, control macrophages (triangle), in lipid droplet–containing, oleate-loaded macrophages (square), and in macrophages supplemented by the addition of vitamin B_{12} to the medium (diamond). Error bars indicate the standard error of the mean from three representative replicates. This figure is reproduced from Podinovskaia et al. (90) and Lee et al. (102). doi:10.1128/microbiolspec.MGM2-0016-2013.f7

CONCLUDING REMARKS

M. tuberculosis is an incredibly successful pathogen with an extraordinary penetrance of its target host population. The ability to infect many yet cause disease in few is undoubtedly central to this success (111). This ability relies on sensing and responding to the changing environments encountered during the course of disease in the human host. This article discussed these environmental cues and stresses and explored how the genome of *M. tuberculosis* has evolved under the purifying selections that they exert. In analyzing the response of *M. tuberculosis* to a broad range of intracellular pressures, it is clear that, despite genome downsizing, *M. tuberculosis* has retained an extraordinary flexibility in central carbon metabolism. We believe that it is this metabolic plasticity, more than any of the virulence factors, that is the foundation for *M. tuberculosis*'s qualities of endurance.

This research was supported by the U.S. National Institutes of Health grants AI 067027, AI095519, and HL055939 to DGR and AI099569 to BCV. We would like to thank Chantal de Chastellier, Center d'Immunologie de Marseille-Luminy, France, for generously providing the electron micrograph used in Fig. 3.

Citation. Russell DG, Lee W, Tan S, Sukumar N, Podinovskaia M, Fahey RJ, VanderVen BC. 2014. The sculpting of the *Mycobacterium tuberculosis* genome by host cell-derived pressures. Microbiol Spectrum 2(5):MGM2-0016-2013.

References

1. **Russell DG.** 2001. *Mycobacterium tuberculosis*: here today, and here tomorrow. *Nat Rev Mol Cell Biol* **2:** 569–577.

2. **Russell DG.** 2011. *Mycobacterium tuberculosis* and the intimate discourse of a chronic infection. *Immunol Rev* **240:**252–268.

3. **Manzanillo PS, Shiloh MU, Portnoy DA, Cox JS.** 2012. *Mycobacterium tuberculosis* activates the DNA-dependent cytosolic surveillance pathway within macrophages. *Cell Host Microbe* **11:**469–480.

4. **Reiling N, Holscher C, Fehrenbach A, Kroger S, Kirschning CJ, Goyert S, Ehlers S.** 2002. Cutting edge: Toll-like receptor (TLR)2- and TLR4-mediated pathogen recognition in resistance to airborne infection with *Mycobacterium tuberculosis*. *J Immunol* **169:**3480–3484.

5. **Heitmann L, Schoenen H, Ehlers S, Lang R, Holscher C.** 2013. Mincle is not essential for controlling *Mycobacterium tuberculosis* infection. *Immunobiology* **218:**506–516.

6. **McElvania Tekippe E, Allen IC, Hulseberg PD, Sullivan JT, McCann JR, Sandor M, Braunstein M, Ting JP.** 2010. Granuloma formation and host defense in chronic *Mycobacterium tuberculosis* infection requires PYCARD/ASC but not NLRP3 or caspase-1. *PloS One* **5:**e12320.

7. **VanderVen BC, Yates RM, Russell DG.** 2009. Intraphagosomal measurement of the magnitude and duration of the oxidative burst. *Traffic* **10:**372–378.

8. **Kim BH, Shenoy AR, Kumar P, Das R, Tiwari S, MacMicking JD.** 2011. A family of IFN-gamma-inducible 65-kD GTPases protects against bacterial infection. *Science* **332:**717–721.

9. **MacMicking JD.** 2005. Immune control of phagosomal bacteria by p47 GTPases. *Curr Opin Microbiol* **8:**74–82.

10. **MacMicking JD, North RJ, LaCourse R, Mudgett JS, Shah SK, Nathan CF.** 1997. Identification of nitric oxide synthase as a protective locus against tuberculosis. *Proc Natl Acad Sci USA* **94:**5243–5248.

11. **Gutierrez MG, Master SS, Singh SB, Taylor GA, Colombo MI, Deretic V.** 2004. Autophagy is a defense mechanism inhibiting BCG and *Mycobacterium tuberculosis* survival in infected macrophages. *Cell* **119:**753–766.

12. **Yates RM, Hermetter A, Taylor GA, Russell DG.** 2007. Macrophage activation downregulates the degradative capacity of the phagosome. *Traffic* **8:**241–250.

13. **Sassetti CM, Boyd DH, Rubin EJ.** 2001. Comprehensive identification of conditionally essential genes in mycobacteria. *Proc Natl Acad Sci USA* **98:**12712–12717.

14. **Sassetti CM, Rubin EJ.** 2003. Genetic requirements for mycobacterial survival during infection. *Proc Natl Acad Sci USA* **100:**12989–12994.

15. **Schnappinger D, Ehrt S, Voskuil MI, Liu Y, Mangan JA, Monahan IM, Dolganov G, Efron B, Butcher PD, Nathan C, Schoolnik GK.** 2003. Transcriptional adaptation of *Mycobacterium tuberculosis* within macrophages: insights into the phagosomal environment. *J Exp Med* **198:**693–704.

16. **Camacho LR, Constant P, Raynaud C, Laneelle MA, Triccas JA, Gicquel B, Daffe M, Guilhot C.** 2001. Analysis of the phthiocerol dimycocerosate locus of *Mycobacterium tuberculosis*. Evidence that this lipid is involved in the cell wall permeability barrier. *J Biol Chem* **276:**19845–19854.

17. **Camacho LR, Ensergueix D, Perez E, Gicquel B, Guilhot C.** 1999. Identification of a virulence gene cluster of *Mycobacterium tuberculosis* by signature-tagged transposon mutagenesis. *Mol Microbiol* **34:**257–267.

18. **Cox JS, Chen B, McNeil M, Jacobs WR Jr.** 1999. Complex lipid determines tissue-specific replication of *Mycobacterium tuberculosis* in mice. *Nature* **402:**79–83.

19. **Hensel M, Shea JE, Gleeson C, Jones MD, Dalton E, Holden DW.** 1995. Simultaneous identification of bacterial virulence genes by negative selection. *Science* **269:** 400–403.

20. **Sassetti CM, Boyd DH, Rubin EJ.** 2003. Genes required for mycobacterial growth defined by high density mutagenesis. *Mol Microbiol* **48:**77–84.

21. **Lamichhane G, Bishai W.** 2007. Defining the "survivasome" of *Mycobacterium tuberculosis*. *Nat Med* **13:** 280–282.

22. **Lamichhane G, Zignol M, Blades NJ, Geiman DE, Dougherty A, Grosset J, Broman KW, Bishai WR.**

2003. A postgenomic method for predicting essential genes at subsaturation levels of mutagenesis: application to *Mycobacterium tuberculosis*. *Proc Natl Acad Sci USA* **100**:7213–7218.

23. Dutta NK, Mehra S, Didier PJ, Roy CJ, Doyle LA, Alvarez X, Ratterree M, Be NA, Lamichhane G, Jain SK, Lacey MR, Lackner AA, Kaushal D. 2010. Genetic requirements for the survival of tubercle bacilli in primates. *J Infect Dis* **201**:1743–1752.

24. Boshoff HI, Myers TG, Copp BR, McNeil MR, Wilson MA, Barry CE 3rd. 2004. The transcriptional responses of *Mycobacterium tuberculosis* to inhibitors of metabolism: novel insights into drug mechanisms of action. *J Biol Chem* **279**:40174–40184.

25. Talaat AM, Lyons R, Howard ST, Johnston SA. 2004. The temporal expression profile of *Mycobacterium tuberculosis* infection in mice. *Proc Natl Acad Sci USA* **101**:4602–4607.

26. Waddell SJ, Stabler RA, Laing K, Kremer L, Reynolds RC, Besra GS. 2004. The use of microarray analysis to determine the gene expression profiles of *Mycobacterium tuberculosis* in response to anti-bacterial compounds. *Tuberculosis (Edinb)* **84**:263–274.

27. Rohde KH, Abramovitch RB, Russell DG. 2007. *Mycobacterium tuberculosis* invasion of macrophages: linking bacterial gene expression to environmental cues. *Cell Host Microbe* **2**:352–364.

28. Homolka S, Niemann S, Russell DG, Rohde KH. 2010. Functional genetic diversity among *Mycobacterium tuberculosis* complex clinical isolates: delineation of conserved core and lineage-specific transcriptomes during intracellular survival. *PLoS Pathog* **6**:e1000988.

29. Rohde KH, Veiga DF, Caldwell S, Balazsi G, Russell DG. 2012. Linking the transcriptional profiles and the physiological states of *Mycobacterium tuberculosis* during an extended intracellular infection. *PLoS Pathog* **8**:e1002769.

30. Gill WP, Harik NS, Whiddon MR, Liao RP, Mittler JE, Sherman DR. 2009. A replication clock for *Mycobacterium tuberculosis*. *Nat Med* **15**:211–214.

31. Park HD, Guinn KM, Harrell MI, Liao R, Voskuil MI, Tompa M, Schoolnik GK, Sherman DR. 2003. Rv3133c/dosR is a transcription factor that mediates the hypoxic response of *Mycobacterium tuberculosis*. *Mol Microbiol* **48**:833–843.

32. Roberts DM, Liao RP, Wisedchaisri G, Hol WG, Sherman DR. 2004. Two sensor kinases contribute to the hypoxic response of *Mycobacterium tuberculosis*. *J Biol Chem* **279**:23082–23087.

33. Rustad TR, Harrell MI, Liao R, Sherman DR. 2008. The enduring hypoxic response of *Mycobacterium tuberculosis*. *PLoS One* **3**:e1502.

34. Leistikow RL, Morton RA, Bartek IL, Frimpong I, Wagner K, Voskuil MI. 2010. The *Mycobacterium tuberculosis* DosR regulon assists in metabolic homeostasis and enables rapid recovery from nonrespiring dormancy. *J Bacteriol* **192**:1662–1670.

35. van der Wel N, Hava D, Houben D, Fluitsma D, van Zon M, Pierson J, Brenner M, Peters PJ. 2007.

M. tuberculosis and *M. leprae* translocate from the phagolysosome to the cytosol in myeloid cells. *Cell* **129**:1287–1298.

36. Houben D, Demangel C, van Ingen J, Perez J, Baldeon L, Abdallah AM, Caleechurn L, Bottai D, van Zon M, de Punder K, van der Laan T, Kant A, Bossers-de Vries R, Willemsen P, Bitter W, van Soolingen D, Brosch R, van der Wel N, Peters PJ. 2012. ESX-1-mediated translocation to the cytosol controls virulence of mycobacteria. *Cell Microbiol* **14**:1287–1298.

37. Abdallah AM, Bestebroer J, Savage ND, de Punder K, van Zon M, Wilson L, Korbee CJ, van der Sar AM, Ottenhoff TH, van der Wel NN, Bitter W, Peters PJ. 2011. Mycobacterial secretion systems ESX-1 and ESX-5 play distinct roles in host cell death and inflammasome activation. *J Immunol* **187**:4744–4753.

38. Simeone R, Bobard A, Lippmann J, Bitter W, Majlessi L, Brosch R, Enninga J. 2012. Phagosomal rupture by *Mycobacterium tuberculosis* results in toxicity and host cell death. *PLoS Pathog* **8**:e1002507.

39. Pethe K, Swenson DL, Alonso S, Anderson J, Wang C, Russell DG. 2004. Isolation of *Mycobacterium tuberculosis* mutants defective in the arrest of phagosome maturation. *Proc Natl Acad Sci USA* **101**:13642–13647.

40. Schaible UE, Sturgill-Koszycki S, Schlesinger PH, Russell DG. 1998. Cytokine activation leads to acidification and increases maturation of *Mycobacterium avium*-containing phagosomes in murine macrophages. *J Immunol* **160**:1290–1296.

41. Via LE, Fratti RA, McFalone M, Pagan-Ramos E, Deretic D, Deretic V. 1998. Effects of cytokines on mycobacterial phagosome maturation. *J Cell Sci* **111**(Pt 7): 897–905.

42. Sturgill-Koszycki S, Schlesinger PH, Chakraborty P, Haddix PL, Collins HL, Fok AK, Allen RD, Gluck SL, Heuser J, Russell DG. 1994. Lack of acidification in *Mycobacterium* phagosomes produced by exclusion of the vesicular proton-ATPase. *Science* **263**:678–681.

43. Yates RM, Hermetter A, Russell DG. 2005. The kinetics of phagosome maturation as a function of phagosome/lysosome fusion and acquisition of hydrolytic activity. *Traffic* **6**:413–420.

44. Martin C, Williams A, Hernandez-Pando R, Cardona PJ, Gormley E, Bordat Y, Soto CY, Clark SO, Hatch GJ, Aguilar D, Ausina V, Gicquel B. 2006. The live *Mycobacterium tuberculosis* phoP mutant strain is more attenuated than BCG and confers protective immunity against tuberculosis in mice and guinea pigs. *Vaccine* **24**:3408–3419.

45. Vandal OH, Pierini LM, Schnappinger D, Nathan CF, Ehrt S. 2008. A membrane protein preserves intrabacterial pH in intraphagosomal *Mycobacterium tuberculosis*. *Nat Med* **14**:849–854.

46. Abramovitch RB, Rohde KH, Hsu FF, Russell DG. 2011. aprABC: a *Mycobacterium tuberculosis* complex-specific locus that modulates pH-driven adaptation to the macrophage phagosome. *Mol Microbiol* **80**:678–694.

47. Tan S, Sukumar N, Abramovitch RB, Parish T, Russell DG. 2013. *Mycobacterium tuberculosis* responds to chloride and pH as synergistic cues to the immune status of its host cell. *PLoS Pathog* 9:e1003282.

48. Sonawane ND, Thiagarajah JR, Verkman AS. 2002. Chloride concentration in endosomes measured using a ratioable fluorescent Cl- indicator: evidence for chloride accumulation during acidification. *J Biol Chem* 277:5506–5513.

49. Adams LB, Dinauer MC, Morgenstern DE, Krahenbuhl JL. 1997. Comparison of the roles of reactive oxygen and nitrogen intermediates in the host response to *Mycobacterium tuberculosis* using transgenic mice. *Tuber Lung Dis* 78:237–246.

50. Cooper AM, Segal BH, Frank AA, Holland SM, Orme IM. 2000. Transient loss of resistance to pulmonary tuberculosis in p47(phox-/-) mice. *Infect Immun* 68:1231–1234.

51. Ng VH, Cox JS, Sousa AO, MacMicking JD, McKinney JD. 2004. Role of KatG catalase-peroxidase in mycobacterial pathogenesis: countering the phagocyte oxidative burst. *Mol Microbiol* 52:1291–1302.

52. Wu CH, Tsai-Wu JJ, Huang YT, Lin CY, Lioua GG, Lee FJ. 1998. Identification and subcellular localization of a novel Cu,Zn superoxide dismutase of *Mycobacterium tuberculosis*. *FEBS Lett* 439:192–196.

53. Edwards KM, Cynamon MH, Voladri RK, Hager CC, DeStefano MS, Tham KT, Lakey DL, Bochan MR, Kernodle DS. 2001. Iron-cofactored superoxide dismutase inhibits host responses to *Mycobacterium tuberculosis*. *Am J Respir Crit Care Med* 164:2213–2219.

54. Buchmeier N, Fahey RC. 2006. The *mshA* gene encoding the glycosyltransferase of mycothiol biosynthesis is essential in *Mycobacterium tuberculosis* Erdman. *FEMS Microbiol Lett* 264:74–79.

55. Vilcheze C, Av-Gay Y, Attarian R, Liu Z, Hazbon MH, Colangeli R, Chen B, Liu W, Alland D, Sacchettini JC, Jacobs WR Jr. 2008. Mycothiol biosynthesis is essential for ethionamide susceptibility in *Mycobacterium tuberculosis*. *Mol Microbiol* 69:1316–1329.

56. Nicholson S, Bonecini-Almeida Mda G, Lapa e Silva JR, Nathan C, Xie QW, Mumford R, Weidner JR, Calaycay J, Geng J, Boechat N, Linhares C, Rom W, Ho JL. 1996. Inducible nitric oxide synthase in pulmonary alveolar macrophages from patients with tuberculosis. *J Exp Med* 183:2293–2302.

57. Bryk R, Lima CD, Erdjument-Bromage H, Tempst P, Nathan C. 2002. Metabolic enzymes of mycobacteria linked to antioxidant defense by a thioredoxin-like protein. *Science* 295:1073–1077.

58. Tian J, Bryk R, Shi S, Erdjument-Bromage H, Tempst P, Nathan C. 2005. *Mycobacterium tuberculosis* appears to lack alpha-ketoglutarate dehydrogenase and encodes pyruvate dehydrogenase in widely separated genes. *Mol Microbiol* 57:859–868.

59. Shi S, Ehrt S. 2006. Dihydrolipoamide acyltransferase is critical for *Mycobacterium tuberculosis* pathogenesis. *Infect Immun* 74:56–63.

60. Lee WL, Gold B, Darby C, Brot N, Jiang X, de Carvalho LP, Wellner D, St John G, Jacobs WR Jr, Nathan C. 2009. *Mycobacterium tuberculosis* expresses methionine sulphoxide reductases A and B that protect from killing by nitrite and hypochlorite. *Mol Microbiol* 71:583–593.

61. Darwin KH, Lin G, Chen Z, Li H, Nathan CF. 2005. Characterization of a *Mycobacterium tuberculosis* proteasomal ATPase homologue. *Mol Microbiol* 55:561–571.

62. Darwin KH, Ehrt S, Gutierrez-Ramos JC, Weich N, Nathan CF. 2003. The proteasome of *Mycobacterium tuberculosis* is required for resistance to nitric oxide. *Science* 302:1963–1966.

63. Wagner D, Maser J, Lai B, Cai Z, Barry CE 3rd, Honer Zu Bentrup K, Russell DG, Bermudez LE. 2005. Elemental analysis of *Mycobacterium avium*-, *Mycobacterium tuberculosis*-, and *Mycobacterium smegmatis*-containing phagosomes indicates pathogen-induced microenvironments within the host cell's endosomal system. *J Immunol* 174:1491–1500.

64. Vidal SM, Malo D, Vogan K, Skamene E, Gros P. 1993. Natural resistance to infection with intracellular parasites: isolation of a candidate for Bcg. *Cell* 73:469–485.

65. Forbes JR, Gros P. 2003. Iron, manganese, and cobalt transport by Nramp1 (Slc11a1) and Nramp2 (Slc11a2) expressed at the plasma membrane. *Blood* 102:1884–1892.

66. Ward SK, Hoye EA, Talaat AM. 2008. The global responses of *Mycobacterium tuberculosis* to physiological levels of copper. *J Bacteriol* 190:2939–2946.

67. Liu T, Ramesh A, Ma Z, Ward SK, Zhang L, George GN, Talaat AM, Sacchettini JC, Giedroc DP. 2007. CsoR is a novel *Mycobacterium tuberculosis* copper-sensing transcriptional regulator. *Nat Chem Biol* 3:60–68.

68. Rowland JL, Niederweis M. 2012. Resistance mechanisms of *Mycobacterium tuberculosis* against phagosomal copper overload. *Tuberculosis (Edinb)* 92:202–210.

69. Ward SK, Abomoelak B, Hoye EA, Steinberg H, Talaat AM. 2010. CtpV: a putative copper exporter required for full virulence of *Mycobacterium tuberculosis*. *Mol Microbiol* 77:1096–1110.

70. Wolschendorf F, Ackart D, Shrestha TB, Hascall-Dove L, Nolan S, Lamichhane G, Wang Y, Bossmann SH, Basaraba RJ, Niederweis M. 2011. Copper resistance is essential for virulence of *Mycobacterium tuberculosis*. *Proc Natl Acad Sci USA* 108:1621–1626.

71. Soldati T, Neyrolles O. 2012. Mycobacteria and the intraphagosomal environment: take it with a pinch of salt(s)! *Traffic* 13:1042–1052.

72. Trost M, English L, Lemieux S, Courcelles M, Desjardins M, Thibault P. 2009. The phagosomal proteome in interferon-gamma-activated macrophages. *Immunity* 30:143–154.

73. Botella H, Stadthagen G, Lugo-Villarino G, de Chastellier C, Neyrolles O. 2012. Metallobiology of host-pathogen interactions: an intoxicating new insight. *Trends Microbiol* 20:106–112.

74. Botella H, Peyron P, Levillain F, Poincloux R, Poquet Y, Brandli I, Wang C, Tailleux L, Tilleul S, Charriere GM, Waddell SJ, Foti M, Lugo-Villarino G, Gao Q, Maridonneau-Parini I, Butcher PD, Castagnoli PR, Gicquel B, de Chastellier C, Neyrolles O. 2011. Mycobacterial p(1)-type ATPases mediate resistance to zinc poisoning in human macrophages. *Cell Host Microbe* 10:248–259.

75. Stoltenberg M, Bruhn M, Sondergaard C, Doering P, West MJ, Larsen A, Troncoso JC, Danscher G. 2005. Immersion autometallographic tracing of zinc ions in Alzheimer beta-amyloid plaques. *Histochem Cell Biol* 123:605–611.

76. Deretic V, Delgado M, Vergne I, Master S, De Haro S, Ponpuak M, Singh S. 2009. Autophagy in immunity against *Mycobacterium tuberculosis*: a model system to dissect immunological roles of autophagy. *Curr Top Microbiol Immunol* 335:169–188.

77. Alonso S, Pethe K, Russell DG, Purdy GE. 2007. Lysosomal killing of mycobacterium mediated by ubiquitin-derived peptides is enhanced by autophagy. *Proc Natl Acad Sci USA* 104:6031–6036.

78. Ponpuak M, Delgado MA, Elmaoued RA, Deretic V. 2009. Monitoring autophagy during *Mycobacterium tuberculosis* infection. *Methods Enzymol* 452:345–361.

79. Foss MH, Powers KM, Purdy GE. 2012. Structural and functional characterization of mycobactericidal ubiquitin-derived peptides in model and bacterial membranes. *Biochemistry* 51:9922–9929.

80. Purdy GE. 2011. Taking out TB-lysosomal trafficking and mycobactericidal ubiquitin-derived peptides. *Front Microbiol* 2:7.

81. Purdy GE, Niederweis M, Russell DG. 2009. Decreased outer membrane permeability protects mycobacteria from killing by ubiquitin-derived peptides. *Mol Microbiol* 73:844–857.

82. Jo EK. 2010. Innate immunity to mycobacteria: vitamin D and autophagy. *Cell Microbiol* 12:1026–1035.

83. Yuk JM, Shin DM, Lee HM, Yang CS, Jin HS, Kim KK, Lee ZW, Lee SH, Kim JM, Jo EK. 2009. Vitamin D3 induces autophagy in human monocytes/ macrophages via cathelicidin. *Cell Host Microbe* 6:231–243.

84. Campbell GR, Spector SA. 2012. Autophagy induction by vitamin D inhibits both *Mycobacterium tuberculosis* and human immunodeficiency virus type 1. *Autophagy* 8:1523–1525.

85. Campbell GR, Spector SA. 2012. Vitamin D inhibits human immunodeficiency virus type 1 and *Mycobacterium tuberculosis* infection in macrophages through the induction of autophagy. *PLoS Pathog* 8:e1002689.

86. Russell DG. 2007. Who puts the tubercle in tuberculosis? *Nat Rev* 5:39–47.

87. Russell DG, Cardona PJ, Kim MJ, Allain S, Altare F. 2009. Foamy macrophages and the progression of the human tuberculosis granuloma. *Nat Immunol* 10:943–948.

88. Kim MJ, Wainwright HC, Locketz M, Bekker LG, Walther GB, Dittrich C, Visser A, Wang W, Hsu FF, Wiehart U, Tsenova L, Kaplan G, Russell DG. 2010. Caseation of human tuberculosis granulomas correlates with elevated host lipid metabolism. *EMBO Mol Med* 2:258–274.

89. Daniel J, Maamar H, Deb C, Sirakova TD, Kolattukudy PE. 2011. *Mycobacterium tuberculosis* uses host triacylglycerol to accumulate lipid droplets and acquires a dormancy-like phenotype in lipid-loaded macrophages. *PLoS Pathog* 7:e1002093.

90. Podinovskaia M, Lee W, Caldwell S, Russell DG. 2013. Infection of macrophages with *Mycobacterium tuberculosis* induces global modifications to phagosomal function. *Cell Microbiol* 15:843–859.

91. Peyron P, Vaubourgeix J, Poquet Y, Levillain F, Botanch C, Bardou F, Daffe M, Emile JF, Marchou B, Cardona PJ, de Chastellier C, Altare F. 2008. Foamy macrophages from tuberculous patients' granulomas constitute a nutrient-rich reservoir for M. *tuberculosis* persistence. *PLoS Pathog* 4:e1000204.

92. Geisel RE, Sakamoto K, Russell DG, Rhoades ER. 2005. In vivo activity of released cell wall lipids of *Mycobacterium bovis* bacillus Calmette-Guerin is due principally to trehalose mycolates. *J Immunol* 174:5007–5015.

93. Rhoades E, Hsu F, Torrelles JB, Turk J, Chatterjee D, Russell DG. 2003. Identification and macrophage-activating activity of glycolipids released from intracellular *Mycobacterium bovis* BCG. *Mol Microbiol* 48:875–888.

94. Rhoades ER, Geisel RE, Butcher BA, McDonough S, Russell DG. 2005. Cell wall lipids from *Mycobacterium bovis* BCG are inflammatory when inoculated within a gel matrix: characterization of a new model of the granulomatous response to mycobacterial components. *Tuberculosis (Edinb)* 85:159–176.

95. Sakamoto K, Kim MJ, Rhoades ER, Allavena RE, Ehrt S, Wainwright HC, Russell DG, Rohde KH. 2013. Mycobacterial trehalose dimycolate reprograms macrophage global gene expression and activates matrix metalloproteinases. *Infect Immun* 81:764–776.

96. Ishikawa E, Ishikawa T, Morita YS, Toyonaga K, Yamada H, Takeuchi O, Kinoshita T, Akira S, Yoshikai Y, Yamasaki S. 2009. Direct recognition of the mycobacterial glycolipid, trehalose dimycolate, by C-type lectin Mincle. *J Exp Med* 206:2879–2888.

97. Bowdish DM, Sakamoto K, Kim MJ, Kroos M, Mukhopadhyay S, Leifer CA, Tryggvason K, Gordon S, Russell DG. 2009. MARCO, TLR2, and CD14 are required for macrophage cytokine responses to mycobacterial trehalose dimycolate and *Mycobacterium tuberculosis*. *PLoS Pathog* 5:e1000474.

98. Desel C, Werninghaus K, Ritter M, Jozefowski K, Wenzel J, Russkamp N, Schleicher U, Christensen D, Wirtz S, Kirschning C, Agger EM, da Costa CP, Lang R. 2013. The Mincle-activating adjuvant TDB induces MyD88-dependent Th1 and Th17 responses through IL-1R signaling. *PLoS One* 8:e53531.

99. Pandey AK, Sassetti CM. 2008. Mycobacterial persistence requires the utilization of host cholesterol. *Proc Natl Acad Sci USA* 105:4376–4380.

100. McKinney JD, Honer zu Bentrup K, Munoz-Elias EJ, Miczak A, Chen B, Chan WT, Swenson D, Sacchettini JC, Jacobs WR Jr, Russell DG. 2000. Persistence of *Mycobacterium tuberculosis* in macrophages and mice requires the glyoxylate shunt enzyme isocitrate lyase. *Nature* 406:735–738.

101. Savvi S, Warner DF, Kana BD, McKinney JD, Mizrahi V, Dawes SS. 2008. Functional characterization of a vitamin B12-dependent methylmalonyl pathway in *Mycobacterium tuberculosis*: implications for propionate metabolism during growth on fatty acids. *J Bacteriol* 190:3886–3895.

102. Lee W, VanderVen BC, Fahey RJ, Russell DG. 2013. Intracellular *Mycobacterium tuberculosis* exploits host-derived fatty acids to limit metabolic stress. *J Biol Chem* 288:6788–6800.

103. Liu K, Yu J, Russell DG. 2003. pckA-deficient *Mycobacterium bovis* BCG shows attenuated virulence in mice and in macrophages. *Microbiology* 149:1829–1835.

104. Marrero J, Rhee KY, Schnappinger D, Pethe K, Ehrt S. 2010. Gluconeogenic carbon flow of tricarboxylic acid cycle intermediates is critical for *Mycobacterium tuberculosis* to establish and maintain infection. *Proc Natl Acad Sci USA* 107:9819–9824.

105. Marrero J, Trujillo C, Rhee KY, Ehrt S. 2013. Glucose phosphorylation is required for *Mycobacterium tuberculosis* persistence in mice. *PLoS Pathog* 9:e1003116.

106. Honer Zu Bentrup K, Miczak A, Swenson DL, Russell DG. 1999. Characterization of activity and expression of isocitrate lyase in *Mycobacterium avium* and *Mycobacterium tuberculosis*. *J Bacteriol* 181:7161–7167.

107. Gould TA, van de Langemheen H, Munoz-Elias EJ, McKinney JD, Sacchettini JC. 2006. Dual role of isocitrate lyase 1 in the glyoxylate and methylcitrate cycles in *Mycobacterium tuberculosis*. *Mol Microbiol* 61:940–947.

108. Munoz-Elias EJ, McKinney JD. 2005. *Mycobacterium tuberculosis* isocitrate lyases 1 and 2 are jointly required for in vivo growth and virulence. *Nat Med* 11:638–644.

109. Eoh H, Rhee KY. 2013. Multifunctional essentiality of succinate metabolism in adaptation to hypoxia in *Mycobacterium tuberculosis*. *Proc Natl Acad Sci USA* 110:6554–6559.

110. Watanabe S, Zimmermann M, Goodwin MB, Sauer U, Barry CE 3rd, Boshoff HI. 2011. Fumarate reductase activity maintains an energized membrane in anaerobic *Mycobacterium tuberculosis*. *PLoS Pathog* 7:e1002287.

111. Russell DG. 2013. The evolutionary pressures that have molded *Mycobacterium tuberculosis* into an infectious adjuvant. *Curr Opin Microbiol* 16:78–84.

Molecular Genetics of Mycobacteria, 2nd Edition
Edited by Graham F. Hatfull and William R. Jacobs, Jr.
© 2014 American Society for Microbiology, Washington, DC
doi:10.1128/microbiolspec.MGM2-0005-2013

Michael F. Goldberg[1]
Neeraj K. Saini[1]
Steven A. Porcelli[1,2]

Evasion of Innate and Adaptive Immunity by *Mycobacterium tuberculosis*

36

Mycobacterium tuberculosis is an extremely successful pathogen that appears to have coevolved with humans as its specific host species for thousands of years (1). This persistent relationship has uniquely shaped the mycobacterial genome to encode mechanisms that enable the bacilli to resist attack and elimination by the human immune system. Although both innate and adaptive immunity clearly modify the course of *M. tuberculosis* infection, this organism can persist and cause disease even in fully immunocompetent hosts. In addition, the currently available vaccine for prevention of tuberculosis, the attenuated *Mycobacterium bovis* strain known as bacille Calmette-Guérin (BCG), has proven largely ineffective despite widespread use. This resistance to host immunity most likely reflects a highly evolved and multifactorial ability of pathogenic mycobacteria to prevent or evade effective host responses. A more complete understanding of how this occurs will likely be crucial to the design and production of better vaccines for prevention of tuberculosis. In this review we summarize current knowledge

and recent advances in the study of immune evasion by *M. tuberculosis*.

EVASION OF INNATE IMMUNITY

The cellular arm of the innate immune system relies primarily on an array of germline-encoded receptors for pathogen recognition (2). These pattern recognition receptors are composed of members of the Toll-like receptor (TLR), nuclear oligomerization domain (NOD) or NOD-like receptors (NLR), C-type lectin receptor, complement receptor, and mannose receptor families (3). Each of these have been implicated in the initial recognition and uptake of *M. tuberculosis* by phagocytic cells of the innate immune system and contribute to the initiation of different programmed cell death pathways including apoptosis and pyroptosis (4, 5). Additionally, *M. tuberculosis* has developed intricate cellular mechanisms to survive and persist in host macrophages. These include the active blockade of mycobacteria-resident phagosome fusion with lysosomes

[1]Department of Microbiology and Immunology; [2]Department of Medicine, Albert Einstein College of Medicine, Bronx, NY 10461.

(6) and have recently been expanded to include the activation of intracellular autophagic pathways (7) (Fig. 1). Inhibition of various pathways of innate immune recognition and pathogen clearance not only ensures the survival of *M. tuberculosis* in the host environment, but also limits the development of potentially more potent adaptive immune responses (8).

EVASION OF TLR SIGNALING

M. tuberculosis has many diverse lipoprotein and lipoglycan moieties that can be recognized by TLR2 heterodimeric complexes, including products encoded by *lpqH* (19-kDa lipoprotein) and the *lpr* gene family (9–12). These are expressed largely on the bacterial cell wall and have been shown to be a significant

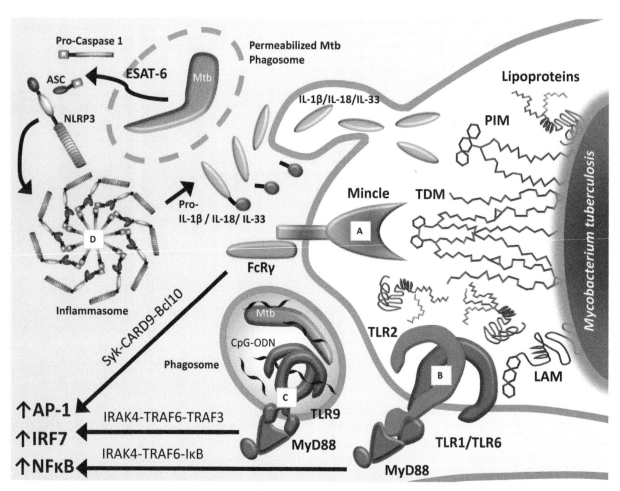

Figure 1 Dominant pattern recognition receptor pathways for sensing *M. tuberculosis*. Cell wall lipids and lipoproteins, which are associated with the external surface of the bacteria, are probably the initial stimuli for pattern recognition receptors of innate phagocytic cells. The proximal interaction between the macrophage engulfing an *M. tuberculosis* bacillus most likely begins with recognition of trehalose-6,6-dimycolate (TDM) by the C-type lectin mincle (**A**), which leads to a signaling cascade that initiates inflammatory cytokine gene transcription. Heterodimers of TLR2 with TLR1 or TLR6 at the plasma membrane recognize di- and tri-acylated lipoproteins, lipomannan, and lipoarabinomannan (LAM) from *M. tuberculosis* (**B**), resulting in the activation of NFκB and cytokine expression. Fragments of hypomethylated DNA from *M. tuberculosis* lead to dimerization of TLR9 within endosomes (**C**), which promotes type I IFN production through the activation of IRF7. Permeabilization of the phagosomal membrane, which is driven by the secretion of the ESAT-6 protein by *M. tuberculosis*, activates NLRP3 and recruits ASC and pro-caspase-1 to form the inflammasome (**D**), which activates caspase-1 and generates active forms of IL-1β, IL-18, and IL-33 that are subsequently secreted.
doi:10.1128/microbiolspec.MGM2-0005-2013.f1

component of secreted membrane vesicles implicated in TLR2-dependent macrophage activation, cytokine production, and granuloma formation (13). Roles, possibly redundant, for TLR4 and TLR9 in the innate response to *M. tuberculosis* have been described in mouse models (14–16). These receptors share a common signal transduction pathway, which depends on the adaptor molecule myeloid differentiation factor-88 (MyD88). Mice lacking MyD88 are highly susceptible to aerogenic *M. tuberculosis* infection, with a mean time to death of approximately 42 days (17), but this appeared to be due mainly to the role of MyD88 in interleukin-1R (IL-1R) signaling, and not in the TLR2, TLR4, or TLR9 activation pathways (18, 19). The importance of this MyD88-dependent pathway to *M. tuberculosis* immunity, while clearly established in mouse models, remains uncertain in humans because rare individuals with an inborn deficiency in MyD88 or IRAK-4 (a necessary MyD88 signaling intermediate) have not yet been found to display enhanced susceptibility or morbidity to mycobacterial infections (20).

Some reports have suggested that TLR9 signaling, which promotes T_H1 immunity, is actively antagonized by the TLR2-dependent depletion of intracellular IL-1 receptor associated kinase-1 (IRAK1) (21, 22). TLR2 signaling may also constrain optimal immune activation during *M. tuberculosis* infection and eventually downmodulate antigen-specific CD4 T cell priming through the production of IL-10 (23, 24). The TLR2 agonist lipoarabinomannan (LAM) from nonpathogenic mycobacterial cell walls is highly immunogenic, but mannose-capped LAM (ManLAM) from virulent mycobacterial species is not. This modification confounds the capacity of this abundant cell wall glycolipid to initiate interferon-γ (IFN-γ) receptor signaling in macrophages, IL-12p70 production by dendritic cells, phagosomal maturation, and apoptosis of infected cells (25). In accordance with these observations, some studies of TLR2 polymorphisms in humans have found associations with enhanced susceptibility to pulmonary tuberculosis, although this has not been observed in all cases (26–34). Together, these data suggest that *M. tuberculosis* has taken advantage of the TLR2 pathway to modify the host environment to be more amenable for latent disease and persistence.

EVASION OF NLR SIGNALING

For many years it was generally believed that *M. tuberculosis* resided exclusively within a modified endosomal compartment after entering host phagocytic cells and did not gain access to the cytosol at any stage

during its intracellular life cycle. However, as early as 1993 data began to appear that suggested an ability of virulent *M. tuberculosis* to escape from phagosomes into the cytosol of infected macrophages in a coordinated and controlled fashion (35, 36). This has been strongly confirmed in more recent studies of infected human monocyte-derived dendritic cells, which used elegant electron-microscopy imaging to show convincingly that virulent mycobacteria periodically escape from phagolysosomes via a pathway that is dependent on the Early Secretory Antigenic Target-6 (ESAT-6) secretion system-1 (ESX-1) to replicate in the cytoplasm (37, 38). Given these findings, it is not surprising that an array of cytosolic pattern recognition receptors is involved in the innate sensing of *M. tuberculosis* and can shape both the innate and adaptive immune responses (39). Specifically, NOD2 senses muramyl dipeptide (MDP) fragments of bacterial peptidoglycan (PGN) and can oligomerize with multiple signaling intermediates to trigger cellular responses ranging from autophagy to inflammatory cytokine production (40). It is notable that *M. tuberculosis* has developed an unusual variant of MDP, which is N-glycolyl modified as opposed to the N-acetylated MDP produced by most other types of bacteria. This structural difference imparts the unique capacity of mycobacterial MDP to trigger type I IFN production (i.e., IFN-α/β) in macrophages through an alternative receptor-interacting protein kinase-2 (RIP2) and interferon regulatory factor-5 (IRF5) pathway. Importantly, this recognition is dependent on ESX-1-mediated phagosomal membrane perturbation and escape of MDP fragments into the cytosol (41). It is likely that *M. tuberculosis* promotes type I IFN signaling, which antagonizes host-protective IL-1β and IFN-γ pathways. Furthermore, studies of type I IFN receptor-deficient mice have shown that IFN-α/β has no effect on *M. tuberculosis* growth in the lungs and even promotes bacterial replication at extrapulmonary sites (42–44).

Recently, considerable attention has been paid to the role of NLRs in microbial infections. Upon activation, these intracellular sensors are recruited to a large macromolecular complex termed the "inflammasome" via their pyrin (PYD)-containing domains by the PYD and caspase activation and recruitment (CARD) domain containing adaptor protein ASC, resulting in the processing of pro-caspase-1 into its enzymatically active form that leads to processing and secretion of IL-1β, IL-18, and IL-33. In many cases this form of activation leads to a type of programmed cell death termed pyroptosis (45, 46). There have been a number of studies examining the role of NLRs, particularly NLRP3, and terminal effectors of the inflammasome during

M. tuberculosis infection *in vitro* and *in vivo*. While *M. tuberculosis* has been shown to potently activate the inflammasome in an ESAT-6-dependent fashion in both mouse and human macrophages, leading to IL-1β release and pyroptosis (47, 48), similar studies using mouse dendritic cells have revealed NLRP3- and ASC-independent IL-1β secretion without pyroptotic cell death (49). Mice that are genetically deficient in production or response to IL-1β have increased susceptibility to acute disease following aerogenic infection with virulent *M. tuberculosis*, and this phenotype appears to be independent of TLRs, NLRP3, caspase-1, and ASC (50–52). On the other hand, these sensors may cooperate in maintaining granuloma architecture and stabilizing *M. tuberculosis* virulence during the chronic stage of the disease in mice. If so, then it is possible that inflammasome activation may be deliberately triggered in some situations by virulent mycobacteria as part of a program for promoting persistent infection rather than runaway virulence.

EVASION OF C-TYPE LECTIN RECEPTOR SIGNALING

The immunogenic properties of mycobacterial cord factor, trehalose-6,6′-dimycolate (TDM), have been known since TDM was purified from pathogenic mycobacteria in the early 1950s by Bloch and colleagues (53). This led to the development of a cell-free model of granuloma development by which TDMs emulsified in oil droplets were injected into experimental animals, resulting in granuloma-like structures with cellular composition quite similar to natural *M. tuberculosis* granulomas (54). Recently, a number of specific genes responsible for the inflammatory effects of TDMs from pathogenic mycobacteria were identified. These included the cyclopropyl synthetase genes *pcaA* and *cmaA2*, which upon deletion led to different modifications of mycolate structure and altered host immune responses (55, 56). TDM from the *pcaA* mutant of *M. tuberculosis*, containing α-mycolates that lack cyclopropyl groups along with an overabundance of ketomycolates, had a significantly reduced capacity to elicit pro-inflammatory cytokine production from macrophages and form nascent granulomas in wild-type mice (57). In contrast, the *cmaA2* mutant of *M. tuberculosis*, which selectively lacks mycolate cyclopropyl groups in the *trans*-configuration, is hypervirulent in mice, and TDMs isolated from this strain were 5-fold more potent in stimulating tumor necrosis factor-α (TNF-α) production from macrophages (58).

Another gene involved in mycolate modification was discovered to have a significant impact on the regulation of IL-12p40 in murine macrophages. Identified during a genome-wide transposon mutagenesis screen, the enzyme encoded by the *mmaA4* gene catalyzes the introduction of a hydroxyl group at the distal carbon of the meromycolate chain during mycolic acid synthesis to produce a hydroxymycolate intermediate. TDMs from this mutant are unusually potent in stimulating IL-12p40 production from macrophages, which is inhibited by wild-type TDM. In addition, *M. tuberculosis* with a deletion of the *mmaA4* gene is highly attenuated for growth and virulence in mice, and this phenotype is directly linked to the enhanced production of IL-12p40 (59).

In 2009 the mammalian receptor for TDMs was identified as the macrophage-inducible C-type lectin receptor mincle (60). This receptor initiates cellular activation and cytokine production through a Syk-CARD9-Bcl10 intracellular pathway (61). It is important to note that CARD9 knockout mice, while able to mount an efficient adaptive immune response, are acutely susceptible to aerogenic *M. tuberculosis* infection with uncontrolled bacterial growth and aggressive pulmonary neutrophilia (62). Mincle activation by mycolates leads to the production of cytokines important for the development of T_H1 and T_H17 CD4 T cells, in part by amplifying IL-1R signaling (63, 64). While mincle appears to be critical for the adjuvant effect of TDMs both *in vitro* and *in vivo* and has been implicated in TDM-induced neutrophilia (65), an extensive survey of aerogenic infection in mincle-deficient mice revealed no significant differences compared to wild-type littermates in organ pathology or bacillary burden (66).

These data appear to demonstrate that *M. tuberculosis* has fine-tuned mycolate biosynthesis to achieve a level of inflammation sufficient to promote granuloma development without triggering excessive IL-12p40 production. The modulation of the IL-12p40 response by *M. tuberculosis* may represent a strategy for achieving an ideal balance that allows the bacterium to persist or proliferate slowly without bringing about rapid death of the host organism.

INHIBITION OF MACROPHAGE EFFECTOR FUNCTION AND PHAGOSOME MATURATION

Macrophages are the primary host cell type in which *M. tuberculosis* resides during infection, and the bacilli have evolved a number of sophisticated mechanisms to persist and replicate in this environment (67). One of the earliest virulence traits described for *M. tuberculosis* is its ability to block the fusion of phagosomes with lysosomes in infected macrophages (68–71). This mechanism is generally believed to be important for allowing the organism to avoid exposure to lysosomal hydrolases, low pH, and other components of lysosomes with bactericidal properties. The mechanisms by which *M. tuberculosis* inhibits phagosome maturation and phagosome-lysosome fusion has been an area of active investigation but remains incompletely understood. Initial molecular and cell biological studies showed that mycobacterial phagosomes failed to incorporate vacuolar ATPases, which could account for their inability to acidify (72). In addition, studies showed that mycobacterial phagosomes retained a protein known as coronin-1 or TACO on the cytoplasmic face of their limiting membranes, whereas phagosomes that undergo fusion with lysosomes tended to rapidly lose the association with this protein (73). Such studies provided confirmation and molecular details of the inhibition of the normal process of phagosome-lysosome fusion in *M. tuberculosis*–infected macrophages but did not offer direct insight into the mechanism by which the pathogen controls this process.

More recently, studies have identified a serine/threonine kinase encoded by the *M. tuberculosis pknG* gene as a potential effector of the inhibition of phagosome-lysosome fusion. Expression of the product of this gene (*M. tuberculosis* protein kinase G) in the nonpathogenic species *Mycobacterium smegmatis* was sufficient to cause this inhibition, and deletion of *pknG* from *M. tuberculosis* caused the bacteria to localize to lysosome-like structures (74). Subsequent studies have identified a small secreted protein tyrosine phosphatase encoded by the *ptpA* gene to inhibit vacuolar acidification by binding to the H-subunit of the macrophage vacuolar-H$^+$-ATPase (75). In addition, an acid- and phagosome-regulated (*aprABC*) locus was discovered in virulent mycobacteria that is uniquely involved in responding to acidic stress within the phagosome. These genes require acid sensing by the *phoPR* operon, which leads to *aprABC* expression and the modulation of cell wall lipid synthesis and sequestration, enabling survival at low pH (76). *M. tuberculosis* also uses a specialized adenylate cyclase encoded by the gene *Rv0386* that causes a transient cyclic AMP burst in the macrophage cytosol that promotes protein kinase A (PKA)–dependent phosphorylation of the transcriptional element CREB (cyclic AMP response element binding) and confounds inflammatory signaling (77).

PREVENTION OF AUTOPHAGY BY *M. TUBERCULOSIS*

Autophagy is a homeostatic and inducible process by which the cells can eliminate damaged organelles or other unwanted structures within the cell. The process

is initiated by the formation of a double membrane-enclosed structure that forms an autophagic vacuole that engulfs organelles or other inclusions in the cytosol. The material engulfed by the vacuole can then be digested following fusion with lysosomes (78). Moreover, it has been shown that many pathogens, including intracellular bacteria such as *M. tuberculosis*, can become localized to autophagic vacuoles within host cells (79).

In the case of *M. tuberculosis*–infected cells, it has been shown that autophagy can contribute to the killing and clearance of the bacteria (79–81). Nutrient starvation or IFN-γ treatment can induce autophagy in *M. tuberculosis*–infected cells, and phagosomes containing the bacilli acquire lysosomal markers and become acidified during this process. In addition, *M. tuberculosis* localization to autophagosomes in infected macrophages is markedly increased by treatment of the cells with lipopolysaccharide, defining a TLR4-mediated pathway for induction of autophagy (82). Several reports also indicate that through ESX-1-dependent phagosomal permeabilization, *M. tuberculosis* initiates activation of macrophage autophagy by intracellular DNA sensing (STING), vitamin D, and IL-1R-dependent pathways (83–85). Furthermore, these effects have been shown to decrease the survival of *M. tuberculosis* in cells induced to undergo autophagy, and animals that have conditionally knocked out autophagic functionality in myeloid cells (Atg5$^{\text{flox/flox}}$ LysM^{Cre}) are acutely susceptible to aerogenic infection with *M. tuberculosis* and have unrestrained bacterial replication, elevated inflammatory cytokine production, and large focal pulmonary abscesses (85, 86). In light of these findings, it is reasonable to speculate that *M. tuberculosis* may target this pathway to increase its intracellular survival, and recently data have begun to appear that support this possibility (87). Additionally, several reports have indicated that the enhanced intracellular survival (*eis*) gene of *M. tuberculosis* encodes a temperature-stable hexameric protein that inhibits phagosome maturation, reactive oxygen species (ROS) production, and autophagy through the direct acetylation of a specific c-Jun N-terminal kinase (JNK)–specific phosphatase (88–90).

PE AND PE_PGRS PROTEINS AS POTENTIAL MEDIATORS OF IMMUNE EVASION

Sequencing of the genome of *M. tuberculosis* H37Rv revealed the presence of a large family of related genes encoding what have come to be known as PE proteins, with 99 member genes accounting for nearly 4% of the organism's whole genome (91, 92). This enigmatic gene family encodes proteins that are primarily characterized by the presence of a characteristic proline and glutamic acid (PE) motif. A large subfamily of these proteins is made up of the 61 PE_PGRS proteins, which also contain C-terminal domains characterized by polymorphic guanosine-cytosine-rich sequences (PGRSs), which vary in size, sequence, and repeat copy numbers (93). There is much interest in this gene family and its role in mycobacterial virulence since genomic analyses show that the marked expansion of PE and PE_PGRS proteins is seen only in *M. tuberculosis* and other pathogenic species of mycobacteria (94). Therefore, it is reasonable to speculate that they play an important role in the pathogenesis and persistence of *M. tuberculosis*.

Many of these proteins are located at the mycobacterial cell wall and membrane (95–97), and this localization at the *M. tuberculosis*–phagosomal interface may indicate that these proteins play an important role in host-pathogen interaction. This localization also suggests a requirement for secretion, and while these proteins do not contain the canonical N-terminal signal peptide, specialized secretion systems may play a role in the export of PE_PGRS proteins from the bacterial cell. Consistent with this, a recent study identified ESX-5, one of the type VII secretion systems of *M. tuberculosis*, as the most likely candidate for the translocation of some PE_PGRS proteins across the bacterial cell membrane (98). Early work by Ramakrishnan and colleagues implicated PE_PGRS proteins from *Mycobacterium marinum* as virulence factors in a *Xenopus* model of infection. They demonstrated that two PE_PGRS genes were upregulated within granuloma-like structures of infected frogs and that mutant strains replicate poorly in macrophages (99). Initial studies on *M. tuberculosis* PE_PGRS genes have focused on PE_PGRS33 as the prototypical family member (100). Constitutive expression of PE_PGRS33 in nonpathogenic *M. smegmatis* was shown to enhance intramacrophage survival, necrosis of infected macrophages, and TNF-α expression and to reduce nitric oxide (NO) production (101). Ectopic expression of PE_PGRS33 in mammalian Jurkat T cells induces apoptosis and can be blocked by Bcl-2 (102). Subsequently, it was shown that induction of apoptosis by PE_PGRS33 is mediated by TLR-2-dependent release of TNF-α (103).

A limited comparative study has shown that PE_PGRS33, -16, and -26 are differentially regulated during *M. tuberculosis* infection of macrophages and in the lungs of mice (104). These have also been shown to differentially modulate the host cytokine response depending on the specific PE_PGRS protein evaluated. While PE_PGRS33 expressing *M. smegmatis* enhances production of TNF-α and IL-10, it reduces IL-12p40 and NO production (101). On the other hand, expression of *M. tuberculosis* PE_PGRS16 in *M. smegmatis* has been shown to enhance IL-12p40 and NO while suppressing the level of IL-10, in contrast to PE_PGRS26, which has the opposite effect (105). Subsequent examinations of another family member, *M. tuberculosis* PE_PGRS62, demonstrated an ability to suppress the expression of IL-1β, IL-6, and the inducible nitric oxide synthase (iNOS) and arrest phagosomal maturation (106, 107). Collectively, these data demonstrate specific functions for individual PE_PGRS family members in modulating macrophage effector functions.

It is well known that proximal interactions of *M. tuberculosis* within macrophages result in a number of Ca^{2+} signaling events critical for phagosome maturation (6, 108, 109). It has been suggested from *in silico* analysis of the PE_PGRS family that the PGRS domain forms a Ca^{2+}-binding motif known as a parallel β-roll or parallel β-helix typical of many calcium-binding proteins. Interestingly, of the entire PE_PGRS family, only five members (PE_PGRS8, -12, -17, -60, and -62) lack this putative domain (110). These observations suggest that PE_PGRS proteins could be involved in controlling Ca^{2+}-dependent events within host cells or tissues, although direct studies of this possibility are not yet available. Overall, it appears that the PE and PE_PGRS proteins are likely to represent major components of immune evasion by *M. tuberculosis*, and a great deal remains to be learned about the precise functions of these proteins during different stages of mycobacterial infection.

EVASION OF ADAPTIVE IMMUNITY

Many of the innate immune evasion mechanisms described above also have major downstream effects on the development of adaptive immunity. In addition, it is well recognized that successful intracellular pathogens possess multiple, highly evolved mechanisms to limit T cell–mediated immunity by disrupting antigen processing and presentation (111, 112). This is well established to be the case for *M. tuberculosis*, which likely uses multiple sophisticated and precisely targeted mechanisms for inhibiting or manipulating all of the known pathways for antigen presentation to T cells (113, 114).

INFECTION OF ANTIGEN PRESENTING CELLS BY *M. TUBERCULOSIS*

The fact that *M. tuberculosis* directly and efficiently infects professional antigen-presenting cells is likely to have a major impact on how this pathogen subverts and evades antigen-specific T cell responses. Macrophages are the major cell type infected by *M. tuberculosis in vivo* and provide the most important site for intracellular replication. Macrophages also have the capability of presenting peptides to effector T cells on both major histocompatibility complex (MHC) class I and class II molecules, and they depend on activation by T cells to exert their microbicidal functions against mycobacteria such as nitric oxide production (115). While antigen presentation by macrophages is likely to be critical to expression of the antimycobacterial functions of previously primed effector or memory T cells, it is unlikely that macrophages are effective at priming the initial T cell responses of naïve T cells against mycobacterial antigens. This function is provided by various subsets of dendritic cells, which are highly specialized to process and present all types of antigens to naïve CD4$^+$ and CD8$^+$ T cells. Recent studies have shown that dendritic cells are also infected at high levels during *M. tuberculosis* infection in mice, suggesting that direct manipulation of dendritic cell function by intracellular mycobacteria is likely to be a prominent feature of infection (116).

The use of *M. tuberculosis* engineered to express green fluorescent protein (GFP) has allowed the accurate localization of the bacteria *in vivo* after infection. This technique has been used to show that lung myeloid dendritic cells (CD11c^{high}CD11b^{high}) are among the major cell types that become infected after introduction of aerosolized *M. tuberculosis* into the lungs, along with alveolar macrophages (CD11c^{high}CD11b^{low}), recruited interstitial macrophages (CD11c^{low}CD11b^{low}), monocytes, and neutrophils (116). Dendritic cells infected in the lung are then able to transport mycobacterial antigens to the mediastinal lymph nodes to initiate the adaptive immune response. However, macrophages and dendritic cells that have not been sufficiently activated do not seem capable of killing the intracellular bacteria and serve as a reservoir of *M. tuberculosis* that could be responsible for the dissemination and delay of the adaptive immune response. These infected phagocytes may serve the role of "Trojan horse" by carrying the bacteria away from the primary focus of infection to widely seed many other tissues. Because of the marked migratory potential of tissue dendritic cells, these cells may be particularly likely to play a significant role in bacterial dissemination (117, 118).

DISRUPTION OF DENDRITIC CELL MATURATION BY *M. TUBERCULOSIS*

A number of studies have shown that *M. tuberculosis* infection of dendritic cells can alter the normal process of maturation of these cells, which is crucial to their ability to efficiently prime antigen-specific T cells. At least two studies using cultured human monocyte-derived dendritic cells have found that these cells fail to show normal maturation following infection with *M. tuberculosis*, as indicated by a lack of rapid mobilization of MHC class II molecules to the cell surface that is normally associated with the maturation process (119, 120). Other data indicate that immune evasion may occur not through a blockade of dendritic cell maturation, but rather by stimulating a poorly coordinated maturation that causes effective antigen processing to cease before *M. tuberculosis* antigen production begins (121). According to this model, rapid maturation of dendritic cells infected by *M. tuberculosis* results in the movement of the great majority of MHC class II molecules to the cell surface, coupled with the cessation of new MHC class II molecule synthesis. This suggests that by the time peptide antigens secreted by *M. tuberculosis* become available for processing and loading onto MHC class II molecules in endocytic compartments, the majority of MHC class II molecules may already have moved to the cell surface to limit the pool of these molecules available for peptide loading. This "kinetic model" of immune evasion may explain how *M. tuberculosis* can severely restrict the ability of directly infected dendritic cells to present bacterial antigens. However, such a mechanism would not clearly apply to cross-presentation of secreted antigens by uninfected dendritic cells, which may offer a potential explanation for the apparent bias toward recognition of secreted protein antigens by T cells in *M. tuberculosis*–infected animals.

MODULATION OF CD4 T-HELPER CELL DIFFERENTIATION

CD4 T cells have a high level of developmental plasticity, which is likely due in part to their central involvement in immune responses to a litany of diverse pathogens. To counter the differential nature of these diverse threats, helper T cells make use of an array of differentiation programs to either limit or compromise pathogen survival (122) (Fig. 2). It was recognized fairly early on that CD4 T cells are critical for the control of *M. tuberculosis* infection in mouse models and in humans (123–126). It was also quickly recognized that activation of macrophage effector functions by IFN-γ-producing T$_H$1 cells is central to this protective effect (127–130). Subsequent studies demonstrated significant roles for FoxP3$^+$ regulatory T cells (T$_{reg}$) and IL-17-producing T$_H$17 cells in the development of protective immunity to *M. tuberculosis* (131, 132).

It is likely that the array of pattern recognition receptors expressed by *M. tuberculosis* triggers the development of a priming milieu that initially favors the expansion of T_{reg} cells, and later T_H1 and to a lesser extent T_H17 cells. Interestingly, T_{reg} cells have been shown to expand in the lungs of *M. tuberculosis*–infected mice and appear to constrain the full effectiveness of T_H1 cells by delaying their homing to the lung

from the mediastinal lymph node. This may be due in part to the ability of *M. tuberculosis* to recruit CD103+ dendritic cells to infected tissues (132–134). A role for IL-17-producing T_H17 cells in pulmonary inflammation following aerogenic *M. tuberculosis* infection in mice has been demonstrated, and these cells appear to play a positive role in conferring vaccine-mediated protection (131, 135). Such T_H17 cells are differentiated in

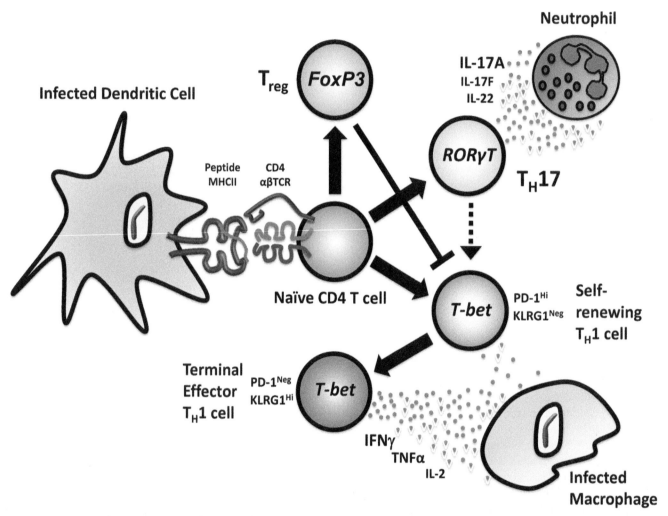

Figure 2   CD4 T helper cell development in the context of *M. tuberculosis* infection. Dendritic cells displaying *M. tuberculosis* antigens through the expression of MHC class II activate *M. tuberculosis*–specific naïve CD4 T cells. This can lead to a number of different outcomes, which are largely determined by which cytokines predominate in the T cell–priming environment. Following activation, CD4 T cells can differentiate into FoxP3+ regulatory T cells (T_{regs}) that strongly inhibit many immune responses. A related but distinct pathway of T cell differentiation upregulates the transcription factor RORγt, leading to T_H17 cells that produce IL-17A, IL-17F, and IL-22, which promote neutrophil chemotaxis. Another possible outcome is the development into TH1 cells that express the transcription factor Tbet. These may be either self-renewing Tbet+ T_H1 cells that are PD-1+ and produce low amounts of IFN-γ and TNF-α, or terminal effector T_H1 cells that are KLRG1+ and produce high levels of IFN-γ and TNF-α that stimulate macrophage effector function during *M. tuberculosis* infection. doi:10.1128/microbiolspec.MGM2-0005-2013.f2

the presence of IL-6 and IL-1β and maintained by IL-23, which are abundantly produced cytokines during active *M. tuberculosis* infection and following the administration of killed organisms (24, 136, 137). Surprisingly, IL-23, a critical factor for the maintenance of T_H17 cells, appears to mediate control of *M. tuberculosis* infection at least in part by stimulating the CXCL13-dependent organization of B-cell follicles in the infected lung (138). IL-17 indirectly serves as a chemoattractant for neutrophils, and recent reports have shown that apoptosis of infected neutrophils is a potent vehicle for the delivery of mycobacterial antigens to dendritic cells, facilitating the priming of naïve CD4 T cells (139, 140). In this respect, it is intriguing to speculate that *M. tuberculosis*, in addition to actively inhibiting neutrophil apoptosis, may inhibit the priming of T_H17 cells to reduce neutrophil infiltration of lung tissue during disease.

The evaluation of T_H1 immunity during mycobacterial infection has been a particularly active area of investigation for more than 2 decades. Recently, considerable interest has focused on the ability of individual T_H1 cells to produce multiple cytokines (IFN-γ, TNF-α, and IL-2) upon recognition of their cognate antigens (141, 142). The role of these so-called multifunctional T cells remains controversial during *M. tuberculosis* infection and also in the context of vaccine-mediated protection (143–146). However, to date, a large pool of multifunctional antigen-specific CD4 T cells in the tissues of *M. tuberculosis*–infected mice may represent the strongest correlate of mycobactericidal immunity yet identified, at least in mouse models (147).

The ability to analyze antigen-specific CD4 T cells using tetramer technology has led to several interesting observations about the *M. tuberculosis*–specific effector memory CD4 T cell population. One such study segregated effector CD4 T cells into three populations based on surface expression of programmed death receptor-1 (PD-1) and the killer cell lectin-like receptor G1 (KLRG1) (148). These markers delineate both self-renewing and terminal effector subsets, as KLRG1^{neg} CD4 T cells that are either low or high for PD-1 expression are both weak cytokine producers that are capable of self-renewal, whereas CD4 T cells expressing high KLRG1 levels and lacking PD-1 subset produce high amounts of IFN-γ and TNF-α but do not proliferate. Interestingly, mice deficient for KLRG1 mount more efficient T cell responses against *M. tuberculosis* and have lower bacterial burdens compared to wild-type mice 90 days postinfection and beyond (149). Other reports have shown that PD-1-deficient mice succumb rapidly to aerogenic *M. tuberculosis* infection,

with large disorganized lesions composed predominantly of infiltrating neutrophils. CD4 T cells from these mice had poor multifunctionality in terms of cytokine production, and the early mortality following *M. tuberculosis* infection was abrogated by CD4 T cell depletion (150, 151). While no studies to date have linked these two observations, it is reasonable to assume that KLRG1 and PD-1, rather than just serving as phenotypic markers, might have a specific role in the self-renewing to terminal effector development of *M. tuberculosis*–specific CD4 T cells. These data suggest that control of *M. tuberculosis* infection requires a delicate balance of CD4 differentiation phenotypes, and direct dysregulation of this process is likely to be an important component of immune evasion by *M. tuberculosis*.

SECRETED PROTEINS AS IMMUNOLOGICAL DECOYS

The highly coordinated expression of one or a small set of antigens that dominates the immune response may represent the use of immunological "decoys" by *M. tuberculosis* to distract the immune response and prevent it from responding against target antigens that are more relevant to host protection. In theory, these decoy proteins would contain peptides with high affinity for MHC class I and II molecules that out-compete other subdominant but potentially more important epitopes for the formation of protective effector and memory T cells. In some cases, such as with the immunodominant secreted Ag85B antigen, the protein is not essential for bacterial survival, and its expression is shut off after infection is established (152). In this way, a subversion of the T cell response can occur in which strong responses are primed that subsequently fail to recognize the pathogen. Effector T cells primed against such decoy antigens would thus survey the body for antigenic targets that are no longer present, and would be unable to recognize infected macrophages containing bacteria that do not express the decoy epitope but instead express other antigens to which the immune system has not been primed. Once the initial targets of the primary immune response have been established, it is possible that *M. tuberculosis* may actively inhibit the priming of new responses against subdominant epitopes or those that are expressed in a delayed manner, thus preventing truly protective responses from developing (153). Regulation of antigen expression in this way has been suggested to result in part from feedback loops that allow the bacterium to regulate its protein secretion systems through a process related to quorum sensing (154).

Some support for the concept of immunological decoys in *M. tuberculosis* infection has recently been provided from comparative genomic studies of mycobacteria. By performing deep sequencing of the genomes of 21 strains from the *M. tuberculosis* complex, Gagneux and colleagues made the surprising discovery that sequences encoding human T cell epitopes were hyperconserved across strains (155). In the same study, the sequences of 16 human T cell antigens were compared over 99 *M. tuberculosis* complex strains and again showed an unexpectedly high level of sequence conservation. Thus these results suggest that, rather than undergoing mutational variation and selection to escape from immunological pressure, *M. tuberculosis* may actually benefit from the recognition of its immunodominant epitopes by T cells.

Nonsecreted or cytoplasmic proteins of mycobacteria have generally not been found to be major antigenic targets of T cell responses in *M. tuberculosis*–infected animals, although humoral responses have sometimes been found against such antigens (156–158). However, the fact that nonsecreted mycobacterial proteins do not strongly stimulate T cell responses during infection with live mycobacteria does not necessarily mean that these antigens cannot be the targets for protective T cell responses. In fact, a few studies have shown that vaccination of animals to specifically stimulate immunity against nonsecreted protein antigens can lead to a significant level of protection against subsequent mycobacterial challenge (159, 160). This suggests that the failure of nonsecreted proteins to effectively prime MHC class I or class II restricted T cell responses during infection with *M. tuberculosis* could be the result of active immune evasion mechanisms. On the other hand, the observation that killed mycobacterial vaccines are less efficient than live attenuated vaccines for eliciting protective immunity argues that protein secretion may be needed to generate effective immunity (161). Another recent report has assessed CD4 T cell recognition of all *in silico*-predicted HLA-DR, -DB, and -DQ binding epitopes in the *M. tuberculosis* genome by a large cohort of humans with latent *M. tuberculosis* infection. This revealed strong preferential responses to three "antigenic islands" corresponding in part to ESX secreted proteins, but also to associated components of the ESX secretion apparatus that are not themselves believed to be secreted (162). These findings raise important issues for ongoing efforts to produce more effective vaccines, particularly since most of the target antigens that are currently being used in vaccine development are immunodominant secreted antigens that could potentially represent relatively

nonprotective decoy antigens (163, 164). More analysis of the impact of immunizing against secreted as opposed to nonsecreted target antigens will be needed to resolve this important point.

DELAYED PRIMING OF ADAPTIVE IMMUNITY IN *M. TUBERCULOSIS* INFECTION

A key feature of the adaptive immune response to *M. tuberculosis* in the lung is that it is remarkably delayed and slow to develop, which in mouse models allows the bacteria to grow uncontrolled for approximately 21 days (165). The recent development of transgenic mice expressing TCRs specific for MHC class II–presented peptides of *M. tuberculosis* Ag85B and ESAT-6 has significantly enhanced the study of the early events of the adaptive immune response in the lungs after low-dose aerosol infection (166–168). Ag85B and ESAT6 are both secreted antigens that are highly expressed during the initial stages of infection and then decline after approximately 3 weeks. Studies using adoptive transfer of T cells from these TCR transgenic mice have emphasized the slow pace of the evolution of the T cell response even against strongly expressed secreted antigens during the initial phase of pulmonary tuberculosis (166–168). These studies also show that T cell responses in mice following aerosol infection with *M. tuberculosis* are initiated in the draining mediastinal lymph nodes and that this requires the transport of mycobacteria into the lymph nodes by migrating lung dendritic cells. The delay in T cell priming can be accounted for by the finding that migration of infected dendritic cells from the lung to lymph nodes is also delayed, requiring 11 to 12 days to first become detectable (116, 168).

These findings suggest that one way in which *M. tuberculosis* avoids elimination by the host immune system involves delaying the onset of adaptive responses until the bacteria have established a protected niche in which they can persist permanently. Thus, while *M. tuberculosis* produces many secreted antigens that are highly immunogenic during the early phase of infection, the host immune system is not able to mount a rapid protective response against these to control bacterial growth. It is noteworthy that T cell responses following an aerosol *M. tuberculosis* infection seem to be significantly more delayed than those that develop after intravenous infection (153), suggesting that the delay in T cell priming could be largely due to inhibitory effects of the pathogen that are exerted specifically on antigen-presenting cells of the lungs. This may reflect a specialized adaptation of this bacterium to its natural route of infection. The slow development of adaptive immunity following *M. tuberculosis* infection is in marked contrast to the kinetics typically observed for acute viral infections of the lung or with other intracellular bacteria such as *Listeria monocytogenes* (169). In the sections that follow, we review known or postulated mechanisms that subvert, evade, or alter the kinetics of the antimycobacterial immune response by direct inhibition of key steps in antigen presentation.

MANIPULATION OF MHC CLASS II ANTIGEN PRESENTATION BY *M. TUBERCULOSIS*

MHC class II restricted CD4[+] T cells are well known to be critical for control of *M. tuberculosis* infection in experimental animals and humans (165), and it is therefore extremely probable that the MHC class II pathway for antigen presentation is targeted by the bacterium to enhance its intracellular survival and persistence. In fact, multiple mechanisms have been described by which *M. tuberculosis* can potentially affect antigen presentation by MHC class II molecules (Fig. 3). As previously discussed, *M. tuberculosis* can efficiently block phagosome maturation and fusion with lysosomes, resulting in the maintenance of an intracellular compartment that is favorable for bacterial survival. This process also undoubtedly plays a role in restricting access of mycobacterial protein antigens to the processing machinery that generates MHC class II complexes bearing *M. tuberculosis* epitopes, and thus blunts the CD4 T cell response to the pathogen. In addition, a number of recent studies have provided strong connections between the process of autophagy and MHC class II presentation of peptides from intracellular proteins (80, 170–173). Since *M. tuberculosis* appears to actively inhibit autophagy by infected cells, this likely represents an additional mechanism for limiting the generation of MHC class II presented epitopes. Consistent with such a proposed mechanism of immune evasion, a recent study showed that autophagy enhances the efficacy of BCG vaccination by increasing mycobacterial peptide presentation in mouse dendritic cells (174).

Studies focusing on the 19-kDa lipoprotein antigen (LpqH), a major triacylated exported cell wall protein of *M. tuberculosis*, demonstrated the ability of this ligand for mammalian TLR2 to globally reduce the surface levels of MHC class II molecules expressed by *M. tuberculosis*–infected macrophages (12, 175, 176). This effect appears to be linked to excessive or prolonged TLR2 signaling, which reduces activity of the MHC class II transcriptional transactivator (CIITA) and also interferes with IFN-γ signaling. Detailed *in vitro* studies of this phenomenon revealed that isoforms of the transcription factor C/EBP that suppress transcription of CIITA are induced following TLR2-dependent stimulation of macrophages by 19-kDa lipoprotein. These C/EBP isoforms are further modified by a mitogen-activated protein kinase (MAPK) that is also activated by TLR2 signaling, and this increases their binding affinity to promoter regions I and IV of the CIITA gene to inhibit its expression (177–179).

The reduction of MHC class II expression induced by 19-kDa lipoprotein has been shown clearly in experiments *in vitro* with cultured cells, but the *in vivo* relevance of this mechanism during infection remains unclear. Experiments *in vivo* showed indistinguishable levels of MHC class II on cells either expressing or lacking TLR2 in mixed bone marrow chimeras that were reconstituted with a mixture of wild-type and TLR2-deficient bone marrow (180). This result indicates that TLR2 expression is not required on macrophages for inhibition of their MHC class II synthesis and that other mechanisms that may be independent of TLR2 could account for this effect during *M. tuberculosis* infection *in vivo* (181, 182). Other recent *in vivo* data show that the decrease in MHC class II expression on lung antigen-presenting cells during infection is restricted mainly or exclusively to cells that are directly infected with mycobacteria (183). This raises the question of how the 19-kDa lipoprotein–induced downregulation of MHC class II is restricted to infected but not uninfected cells, given that this lipoprotein is secreted by the bacterium. It is possible that the lipoprotein is unstable and short-lived, so that only secretion in an autocrine manner (i.e., by cells that are directly infected) is sufficient to give TLR2 activation, or that secretion directly within the phagosomal lumen is required.

In addition, experiments have not yet fully clarified what the impact of the 19-kDa lipoprotein-induced MHC class II downmodulation is on the presentation of mycobacterial protein antigens. Although inhibition of the presentation of exogenously added ovalbumin has been shown for various antigen-presenting cells purified from the lungs of *M. tuberculosis*–infected mice (183), it remains to be determined whether this effect is directly related to activities of the 19-kDa lipoprotein. Likewise, data are not yet available for the effect of 19-kDa lipoprotein on presentation of actual bacterial proteins produced in the infected cell. It should be noted that 19-kDa lipoprotein is only one of multiple

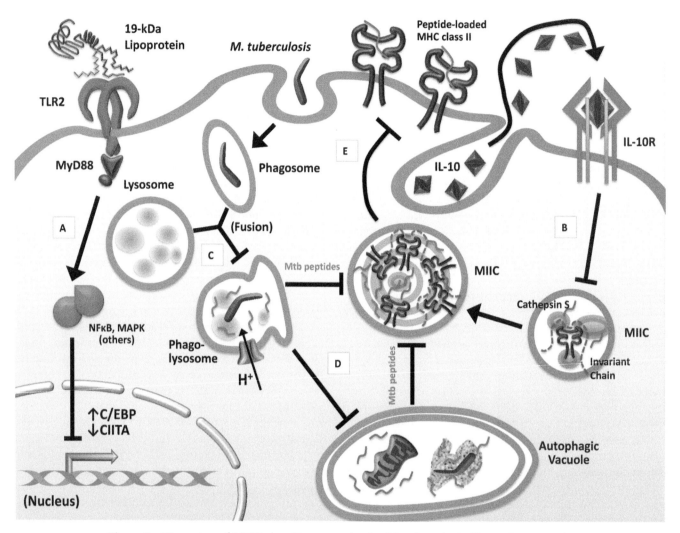

Figure 3 Disruption of MHC class II presentation by *M. tuberculosis*. The various steps in the MHC class II processing and presentation pathway that are known or postulated to be influenced by *M. tuberculosis* infection are illustrated. (**A**) New synthesis of MHC class II molecules is blocked by TLR2 signaling due to mycobacterial products such as the 19-kDa lipoprotein. (**B**) Intracellular trafficking of MHC class II is disrupted by the suppression of cathepsin S, which is due to induction of IL-10 by mycobacterial infection. (**C**) Generation of peptide antigens for loading onto MHC class II in relevant endocytic compartments (MIIC) is also inhibited by several effects of mycobacterial infection, including inhibition of phagosome-lysosome fusion, by neutralization of phagosomal pH by bacterial urease, and by blockade of recruitment of the vacuolar proton ATPase. (**D**) Proposed inhibition of autophagy and autophagic vacuole formation also eliminates a potential source of antigenic peptides that can load MHC class II molecules. (**E**) The reduction of peptide antigen availability and incomplete cleavage of MHC class II associated invariant chain (Ii) resulting from cathepsin S suppression result in a reduced transport of stable peptide-loaded MHC class II molecules to the APC surface. doi:10.1128/microbiolspec.MGM2-0005-2013.f3

known TLR2 ligands in *M. tuberculosis*, others of which include the lipoproteins LprA, LprG, and 38-kDa lipoprotein (also known as PstS-1, phoS, or phoS1), as well as the small secreted protein ESAT-6 and complex glycolipids in the LAM family (184–187).

All of these *M. tuberculosis*–derived ligands for TLR2 could potentially contribute to the modulation of MHC class II levels, and may synergize with numerous other microbial products to influence the outcome of the adaptive immune response to *M. tuberculosis*.

M. tuberculosis can affect the intracellular trafficking of MHC class II molecules by interfering with cathepsins and by alkalinization of intracellular compartments (188). Cathepsins are cysteine proteases that are essential for processing of the MHC class II–associated invariant chain (Ii), which is a necessary step in peptide loading and subsequent cell surface expression of MHC class II molecules. Studies of cathepsin S (CatS) have shown this particular member of the cathepsin family to be the most important or possibly the only protease that is capable of mediating the late steps of Ii cleavage needed to generate MHC class II molecules that can be efficiently loaded with peptide antigens (189, 190). *In vitro* experiments showed that mycobacterial infection of human macrophages caused reduction of CatS activity and gene expression through the induction of the inhibitory cytokine IL-10 (191). This was associated with an intracellular retention of MHC class II molecules and a reduced expression of peptide-loaded MHC class II complexes at the cell surface. This effect was reversed by addition of anti-IL-10 antibodies, which restored expression of active CatS and export of mature MHC class II molecules to the surface of infected cells. Other experiments showed that a recombinant BCG strain that was engineered to secrete CatS could also restore cell surface levels of MHC class II molecules (192).

In related studies, it was reported that intraphagosomal production of urease encoded by the mycobacteria *ureC* gene, which hydrolyzes urea into carbon dioxide and ammonia, is also involved in disruption of MHC class II trafficking leading to intracellular retention. This is believed to be due to the effect of urease on preventing the acidification of MHC class II processing and loading compartments, which may also prevent the activity of proteases such as CatS (193). Other reports have also implicated inhibition of expression or activity of other cathepsin proteases, such as cathepsins L and D, in the impaired processing and presentation of antigens by MHC class II in mycobacteria-infected cells (68, 194).

EFFECTS OF *M. TUBERCULOSIS* ON PRESENTATION BY MHC CLASS I MOLECULES

Peptides presented by MHC class I molecules are the principal target structures of CD8+ cytolytic T cells, which are important for immunity against viruses and many other intracellular pathogens. Earlier dogmas held that peptides presented by MHC class I molecules were derived only from protein antigens that were produced in the cytosol of target cells or that somehow gained access to the cytosolic compartment. This seemed to rule out a role for MHC class I presentation of antigens from *M. tuberculosis*, since that organism and its antigens were believed to be confined either to endosomal compartments or to the extracellular space. This idea is now known to be incorrect for several reasons. First, it is now becoming accepted that *M. tuberculosis* does gain access directly to the cytosol of infected cells, as shown originally in a mouse macrophage cell line and human monocyte-derived macrophages (195) and more recently confirmed in elegant studies of infected human monocyte-derived dendritic cells (196). Second, mechanisms by which endocytosed antigens can be processed and presented by the MHC class I system, a process generally referred to as cross-presentation, are now well documented and are strongly believed to be relevant to *M. tuberculosis* infection (Fig. 4) (197, 198).

One potential pathway for cross-presentation by MHC class I molecules of mycobacterial proteins originating in the lumen of a phagosome or phagolysosome involves their translocation to the cytosol (199–203), which is proposed to involve a transporter in the phagosomal membrane referred to as the dislocon. Mycobacterial proteins transported into the cytosol in this way would be expected to undergo cytosolic processing, which involves polyubiquitination, proteolysis by the proteosome, and ultimately transport into the lumen of the endoplasmic reticulum (ER) by the transporter associated with antigen processing (TAP), where binding to nascent MHC class I molecules occurs with high efficiency (204). The identification of the dislocon remains controversial, although it is postulated to be related or possibly identical to protein-transporting complexes found on the ER membrane, such as Sec61, which translocates proteins into the ER lumen, or the AAA-ATPase p97, which dislocates proteins out of the ER as part of the ER-associated degradation (ERAD) pathway (205). In support of this, proteomic analysis of isolated phagosomes has detected the presence of Sec61 and the AAA-ATPase p97 (206), although these findings are of uncertain significance due to possible contamination of phagosome preparations with ER membranes (207).

In the case of *M. tuberculosis*–infected cells, it is possible that the bacterium itself may cause leakiness of the phagosomal membrane, and this could allow bacterial proteins to enter the cytosolic processing pathway. For example, the secreted *M. tuberculosis* protein ESAT-6 and its chaperone partner CFP-10 have been found to have membrane-disrupting properties (208),

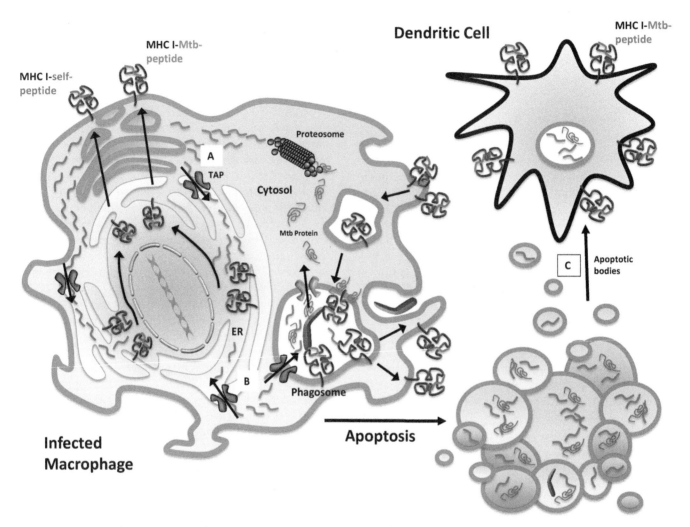

Figure 4 MHC class I presentation pathways in *M. tuberculosis* infection. The large cell on the left of the figure represents a macrophage infected with *M. tuberculosis*. Newly synthesized MHC class I molecules in the endoplasmic reticulum (ER) are loaded with peptides that are produced by the cytosolic proteosome complex and transported into the ER lumen by TAP (transporter associated with antigen presentation). Additional trimming of the cytosol-derived peptides can occur as a result of aminopeptidase activity in the ER lumen. Escape of mycobacterial proteins from the phagosome into the cytosol can lead to peptide presentation by this classical MHC class I pathway (**A**). Mechanisms for loading of peptides onto MHC class I molecules in the lumen of the phagosome are also likely to exist. This vacuolar pathway for cross presentation (**B**) may involve transfer of ER membrane components (e.g., newly synthesized MHC class I complexes and TAP) to the phagosome membrane, enabling the loading of peptides generated in the cytosol. Alternatively, peptides may be generated by proteases in the phagosome lumen, and these may be loaded by a process of peptide exchange onto MHC class I molecules recycling from the plasma membrane. The so-called detour pathway (**C**) is a third way that peptides from a vacuolar intracellular pathogen such as *M. tuberculosis* can be cross-presented by MHC class I. In this case, an infected cell must first die by apoptosis, and the released apoptotic vesicles carry the mycobacterial antigens into uninfected dendritic cells. Current evidence suggests that all of these pathways are likely to be actively inhibited or effectively bypassed during *M. tuberculosis* infection (see text for details). doi:10.1128/microbiolspec.MGM2-0005-2013.f4

suggesting that production of these proteins in the phagosome lumen could be responsible for leakage into the cytosol. In support of such a mechanism, it was shown that *M. tuberculosis* infection of cells facilitated the MHC class I presentation of endosomally delivered ovalbumin through a process that required the presence of live bacteria (200). In addition, the previously documented escape of *M. tuberculosis* bacilli into the cytoplasm of human cultured dendritic cells was shown to require an intact RD1 locus, which encodes the ESAT-6 and CFP-10 proteins and their secretion apparatus (196). However, whether dendritic cells remain competent to process and present antigens after *M. tuberculosis* has undergone escape into the cytosol is unclear, since these cells are likely to rapidly undergo necrotic cell death that is triggered by the bacteria (209, 210). Presumably, this pathway would be less effective for nonsecreted bacterial protein antigens, which would remain within the phagosome and would thus be predicted to favor the presentation of secreted bacterial proteins.

An alternative mechanism that does not depend on phagosomal escape of antigens into the cytosol is the vacuolar model for cross-presentation. This proposes that peptides derived from bacterial proteins are generated within the phagosomal lumen and loaded directly on MHC class I molecules that recycle through the endocytic system (Fig. 4) (198, 211). This process requires the presence of proteases within the phagosome that can cleave the mycobacterial proteins to generate the peptides required for loading MHC class I molecules, and could potentially be relevant for both secreted and nonsecreted bacterial proteins if sufficient degradation of the bacterial cells occurs within the phagosome lumen. Although it is unclear to what extent this vacuolar pathway may account for cross-presentation of antigens by MHC class I in *M. tuberculosis*–infected cells, there is some evidence that peptides from Ag85B can be loaded and presented in this way (68, 212). Whether *M. tuberculosis* actively interferes with MHC class I presentation mediated by either the cytosolic or vacuolar cross-presentation pathways is unknown. However, one study has reported that the *M. tuberculosis* 19-kDa lipoprotein can reduce processing of protein antigens by the vacuolar pathway for MHC class I presentation (213). In addition, it would be anticipated that the previously discussed inhibition of phagosome-lysosome fusion, prevention of phagosome acidification, and downmodulation of cathepsins associated with *M. tuberculosis* infection would all interfere with a vacuolar MHC class I cross-presentation pathway.

Another pathway for cross-presentation of antigens by MHC class I molecules exists that has been more strongly implicated in CD8+ T cell responses to *M. tuberculosis*. This is known as the "detour pathway," and it depends upon the uptake by uninfected antigen-presenting cells of bacterial antigens originating from infected cells that have undergone death by apoptosis (Fig. 4) (214–216). Although the mechanisms involved in this antigen transfer are not completely resolved, it has been shown that *M. tuberculosis*–infected macrophages that are induced to undergo apoptosis by serum starvation shed subcellular vesicles that are thought to be derived from membrane blebs of the dying cells. These contain bacterial proteins and lipids and can be efficiently taken up by healthy, uninfected dendritic cells (216). Antigens taken up this way are delivered to a TAP-dependent pathway for MHC class I presentation and can therefore prime CD8+ T cell responses. Interestingly, the detour pathway also appears to be effective for uptake of lipid antigens for CD1 presentation but does not seem to enhance MHC class II restricted antigen presentation (214, 216).

Arguing in favor of the relevance of the detour pathway to immune responses against *M. tuberculosis* is the well-documented observation that pathogenic mycobacteria actively block the ability of infected macrophages to undergo apoptosis (217–220) and most likely escape eventually from infected cells by causing a form of necrotic cell death (210, 221, 222). *M. tuberculosis* and other pathogenic mycobacteria appear to have evolved powerful mechanisms for controlling the process of host cell death, and this is likely to be a key factor in the survival and persistence of these bacteria in immunocompetent hosts. Recently, two genes of *M. tuberculosis* that are directly involved in blocking apoptosis of infected macrophages have been identified. One of these is *secA2 (Rv1821)*, which encodes a component of a virulence-related secretion system and is known to be involved in the transport of the enzyme superoxide dismutase (SodA) out of the bacterial cell. Given that SodA has also been implicated in the inhibition of apoptosis (223), it seems likely that secretion of this enzyme by intracellular mycobacteria is important for controlling the levels of reactive oxygen intermediates that may act as a trigger for apoptosis if present at sufficiently high levels (224). The other identified *M. tuberculosis* antiapoptosis gene is *nuoG (Rv3151)*, which encodes a subunit of the large, proton-extruding type I NADH dehydrogenase complex in the mycobacterial membrane (225). In support of the relevance of anti-apoptosis mechanisms to the ability of *M. tuberculosis* to evade presentation by the detour pathway, it has been shown that a mutant strain of *M. tuberculosis* with a deletion of the *secA2* gene not only fails to block macrophage apoptosis, but also concurrently stimulates markedly enhanced CD8[+] T cell priming (226). Along similar lines, it has been reported that a recombinant strain of BCG that induces increased levels of apoptosis of infected macrophages produces greater protective immunity against *M. tuberculosis* challenge when used as a vaccine in mice (227). Thus, the blockade of apoptosis and induction of necrosis may be one of the main strategies by which *M. tuberculosis* evades or delays antigen presentation by MHC class I molecules.

Several observations also suggest a role for the PE_PGRS family of proteins in the suppression of MHC class I-mediated antigen presentation to CD8[+] T cells in responses to *M. tuberculosis* infection. It has been observed that these transfections of mammalian cells with PE_PGRS16 and -62 inhibit ubiquitin-dependent proteosomal degradation, potentially reducing the generation of MHC class I-presented peptides (228). Additionally, it has been reported that both in-frame and frameshift mutations are frequently observed in a number of genes encoding various PE_PGRS proteins in clinical isolates of *M. tuberculosis*, with the majority occurring within the PGRS repeat domain (229, 230). It has been speculated that this high level of mutation could reflect a mechanism for generating rapid antigenic variation to facilitate escape from CD8[+] T cells or other adaptive responses (91).

INFLUENCE OF *M. TUBERCULOSIS* ON LIPID ANTIGEN PRESENTATION BY CD1

M. tuberculosis contains abundant and diverse lipids in its cell wall, and the presentation of some of these by CD1 molecules may account for a significant component of T cell–mediated immunity in humans (231). Consistent with a proposed role in host resistance to mycobacteria, there have been suggestive data that *M. tuberculosis* may interfere with lipid antigen presentation by CD1 molecules. Since macrophages in general do not express the group 1 CD1 molecules (CD1a, CD1b, and CD1c), the transfer of lipids from infected cells to uninfected dendritic cells may be critical for mycobacterial lipid presentation by CD1. Interestingly, one study has shown that this lipid transfer may occur by the detour pathway described above for MHC class I cross-presentation, which depends on the ability of an infected macrophage to undergo death by apoptosis (214). Thus, the ability of *M. tuberculosis* to block apoptosis of infected cells may also have important

implications for evasion of lipid antigen presentation by CD1. It has also been suggested that apolipoprotein E produced by macrophages could function as a carrier to shuttle lipids from infected cells to dendritic cells (232), although whether this occurs in *M. tuberculosis* infection or is in any way blocked by the bacterium remains to be determined.

Direct evidence for immune evasion mechanisms targeting CD1 presentation was provided by one provocative study showing a dramatic decrease of group 1 CD1 surface expression following infection of human monocyte-derived dendritic cells *in vitro* with virulent *M. tuberculosis*. In this case, the reduced CD1a, CD1b, and CD1c expression was determined to be associated with reduced mRNA transcripts, suggesting a mechanism by which the bacterium represses the transcription of CD1 genes (233). In addition, a few studies have shown that *M. tuberculosis* infection of monocytes may impair their upregulation of CD1 expression during subsequent differentiation into dendritic cells (234, 235). Although some of these findings appear quite dramatic and biologically significant, the mechanisms responsible have not been explored, and the potential significance has not yet been extended to *in vivo* models. Given that CD1 molecules are generally loaded with their antigens within endocytic compartments (236), the ability of *M. tuberculosis* to arrest phagosome maturation and prevent acidification of endosomal compartments in which it resides could potentially interfere with CD1-dependent antigen presentation.

Natural killer T cells (NKT cells) recognize lipid antigens presented by the CD1d molecule (also known as group 2 CD1). These cells can produce large amounts of IFN-γ when activated and have been proposed to serve as part of the first line of defense against many pathogens. NKT cells appear to be activated during mycobacterial infection in mice, and when activated early in *M. tuberculosis* infection by the glycolipid ligand α-galactosylceramide, they can slow the progression of disease (237). It has also been shown that the transfer of additional NKT cells into wild-type mice can increase their resistance to *M. tuberculosis* (238). Surprisingly, mice that are NKT cell deficient due to deletion of genes for CD1d (CD1d[-/-] mice) or for the Jα segment of the NKT cell TCR (Jα18[-/-] mice) have generally not been found to have altered susceptibility to *M. tuberculosis* infection (239–241). This raises the possibility that CD1d presentation or other factors involved in the activation of NKT cells may be downmodulated by *M. tuberculosis* infection, although this has not yet been directly demonstrated.

SUMMARY AND CONCLUSIONS

It is clear from a wide array of experimental evidence and clinical observations that the remarkable success of *M. tuberculosis* as a human pathogen requires a sophisticated program for control of host immune responses by the bacilli. Given that the genome of *M. tuberculosis* has most likely evolved over millennia to meet the challenges of an aggressive mammalian immune system, it is not surprising that immune evasion programs expressed by this organism interfere with many of the known pathways of mammalian innate and adaptive immunity. While the examples of immune evasion mechanisms discussed in this chapter are quite numerous, they most likely represent only a fraction of the processes by which this remarkable pathogen achieves its extraordinary ability to persist within the human host. The integration of high-throughput platforms for the study of host and microbial responses, along with recent advances in computational analyses, will likely aid in the development of a comprehensive map of the immune evasion networks at play during *M. tuberculosis* infection and facilitate our understanding of this extraordinarily complex host-pathogen relationship. As *M. tuberculosis* infection remains a global public health problem of the greatest magnitude, the development of a comprehensive understanding of immune evasion by this organism remains a central goal in the quest for more effective vaccines and treatments for tuberculosis.

Citation. Goldberg MF, Saini NK, Porcelli SA. 2014. Evasion of innate and adaptive immunity by *Mycobacterium tuberculosis*. Microbiol Spectrum 2(5):MGM2-0005-2013.

References

1. Gagneux S. 2012. Host-pathogen coevolution in human tuberculosis. *Philos Trans R Soc Lond Ser B Biol Sci* 367:850–859.

2. Kawai T, Akira S. 2010. The role of pattern-recognition receptors in innate immunity: update on Toll-like receptors. *Nat Immunol* 11:373–384.

3. Coll RC, O'Neill LA. 2010. New insights into the regulation of signalling by toll-like receptors and nod-like receptors. *J Innate Immun* 2:406–421.

4. Kleinnijenhuis J, Oosting M, Joosten LA, Netea MG, Van Crevel R. 2011. Innate immune recognition of *Mycobacterium tuberculosis*. *Clin Dev Immunol* 2011:405310.

5. Saiga H, Shimada Y, Takeda K. 2011. Innate immune effectors in mycobacterial infection. *Clin Dev Immunol* 2011:347594.

6. Vergne I, Chua J, Singh SB, Deretic V. 2004. Cell biology of *Mycobacterium tuberculosis* phagosome. *Annu Rev Cell Dev Biol* 20:367–394.

7. Deretic V. 2012. Autophagy as an innate immunity paradigm: expanding the scope and repertoire of pattern recognition receptors. *Curr Opin Immunol* 24:21–31.

8. Tsolaki AG. 2009. Innate immune recognition in tuberculosis infection. *Adv Exp Med Biol* 653:185–197.

9. Drage MG, Tsai HC, Pecora ND, Cheng TY, Arida AR, Shukla S, Rojas RE, Seshadri C, Moody DB, Boom WH, Sacchettini JC, Harding CV. 2010. *Mycobacterium tuberculosis* lipoprotein LprG (Rv1411c) binds triacylated glycolipid agonists of Toll-like receptor 2. *Nat Struct Mol Biol* 17:1088–1095.

10. Gehring AJ, Dobos KM, Belisle JT, Harding CV, Boom WH. 2004. *Mycobacterium tuberculosis* LprG (Rv1411c): a novel TLR-2 ligand that inhibits human macrophage class II MHC antigen processing. *J Immunol* 173:2660–2668.

11. Pecora ND, Gehring AJ, Canaday DH, Boom WH, Harding CV. 2006. *Mycobacterium tuberculosis* LprA is a lipoprotein agonist of TLR2 that regulates innate immunity and APC function. *J Immunol* 177:422–429.

12. Noss EH, Pai RK, Sellati TJ, Radolf JD, Belisle J, Golenbock DT, Boom WH, Harding CV. 2001. Toll-like receptor 2-dependent inhibition of macrophage class II MHC expression and antigen processing by 19-kDa lipoprotein of *Mycobacterium tuberculosis*. *J Immunol* 167:910–918.

13. Prados-Rosales R, Baena A, Martinez LR, Luque-Garcia J, Kalscheuer R, Veeraraghavan U, Camara C, Nosanchuk JD, Besra GS, Chen B, Jimenez J, Glatman-Freedman A, Jacobs WR Jr, Porcelli SA, Casadevall A. 2011. Mycobacteria release active membrane vesicles that modulate immune responses in a TLR2-dependent manner in mice. *J Clin Invest* 121:1471–1483.

14. Abel B, Thieblemont N, Quesniaux VJ, Brown N, Mpagi J, Miyake K, Bihl F, Ryffel B. 2002. Toll-like receptor 4 expression is required to control chronic *Mycobacterium tuberculosis* infection in mice. *J Immunol* 169:3155–3162.

15. Reiling N, Holscher C, Fehrenbach A, Kroger S, Kirschning CJ, Goyert S, Ehlers S. 2002. Cutting edge: Toll-like receptor (TLR)2- and TLR4-mediated pathogen recognition in resistance to airborne infection with *Mycobacterium tuberculosis*. *J Immunol* 169:3480–3484.

16. Bafica A, Scanga CA, Feng CG, Leifer C, Cheever A, Sher A. 2005. TLR9 regulates Th1 responses and cooperates with TLR2 in mediating optimal resistance to *Mycobacterium tuberculosis*. *J Exp Med* 202:1715–1724.

17. Scanga CA, Bafica A, Feng CG, Cheever AW, Hieny S, Sher A. 2004. MyD88-deficient mice display a profound loss in resistance to *Mycobacterium tuberculosis* associated with partially impaired Th1 cytokine and nitric oxide synthase 2 expression. *Infect Immun* 72:2400–2404.

18. Fremond CM, Togbe D, Doz E, Rose S, Vasseur V, Maillet I, Jacobs M, Ryffel B, Quesniaux VF. 2007. IL-1 receptor-mediated signal is an essential component of MyD88-dependent innate response to *Mycobacterium tuberculosis* infection. *J Immunol* 179:1178–1189.

19. Holscher C, Reiling N, Schaible UE, Holscher A, Bathmann C, Korbel D, Lenz I, Sonntag T, Kroger S, Akira S, Mossmann H, Kirschning CJ, Wagner H,

Freudenberg M, Ehlers S. 2008. Containment of aerogenic *Mycobacterium tuberculosis* infection in mice does not require MyD88 adaptor function for TLR2, -4 and -9. *Eur J Immunol* 38:680–694.

20. von Bernuth H, Picard C, Puel A, Casanova JL. 2012. Experimental and natural infections in MyD88- and IRAK-4-deficient mice and humans. *Eur J Immunol* 42:3126–3135.

21. Simmons DP, Canaday DH, Liu Y, Li Q, Huang A, Boom WH, Harding CV. 2010. *Mycobacterium tuberculosis* and TLR2 agonists inhibit induction of type I IFN and class I MHC antigen cross processing by TLR9. *J Immunol* 185:2405–2415.

22. Liu YC, Simmons DP, Li X, Abbott DW, Boom WH, Harding CV. 2012. TLR2 signaling depletes IRAK1 and inhibits induction of type I IFN by TLR7/9. *J Immunol* 188:1019–1026.

23. Nair S, Ramaswamy PA, Ghosh S, Joshi DC, Pathak N, Siddiqui I, Sharma P, Hasnain SE, Mande SC, Mukhopadhyay S. 2009. The PPE18 of *Mycobacterium tuberculosis* interacts with TLR2 and activates IL-10 induction in macrophage. *J Immunol* 183:6269–6281.

24. Jang S, Uematsu S, Akira S, Salgame P. 2004. IL-6 and IL-10 induction from dendritic cells in response to *Mycobacterium tuberculosis* is predominantly dependent on TLR2-mediated recognition. *J Immunol* 173:3392–3397.

25. Briken V, Porcelli SA, Besra GS, Kremer L. 2004. Mycobacterial lipoarabinomannan and related lipoglycans: from biogenesis to modulation of the immune response. *Mol Microbiol* 53:391–403.

26. Shah JA, Vary JC, Chau TT, Bang ND, Yen NT, Farrar JJ, Dunstan SJ, Hawn TR. 2012. Human TOLLIP regulates TLR2 and TLR4 signaling and its polymorphisms are associated with susceptibility to tuberculosis. *J Immunol* 189:1737–1746.

27. Biswas D, Gupta SK, Sindhwani G, Patras A. 2009. TLR2 polymorphisms, Arg753Gln and Arg677Trp, are not associated with increased burden of tuberculosis in Indian patients. *BMC Res Notes* 2:162.

28. Velez DR, Hulme WF, Myers JL, Stryjewski ME, Abbate E, Estevan R, Patillo SG, Gilbert JR, Hamilton CD, Scott WK. 2009. Association of SLC11A1 with tuberculosis and interactions with NOS2A and TLR2 in African-Americans and Caucasians. *Int J Tuberc Lung Dis* 13:1068–1076.

29. Ma MJ, Xie LP, Wu SC, Tang F, Li H, Zhang ZS, Yang H, Chen SL, Liu N, Liu W, Cao WC. 2010. Toll-like receptors, tumor necrosis factor-alpha, and interleukin-10 gene polymorphisms in risk of pulmonary tuberculosis and disease severity. *Hum Immunol* 71:1005–1010.

30. Moller M, Hoal EG. 2010. Current findings, challenges and novel approaches in human genetic susceptibility to tuberculosis. *Tuberculosis (Edinb)* 90:71–83.

31. Motsinger-Reif AA, Antas PR, Oki NO, Levy S, Holland SM, Sterling TR. 2010. Polymorphisms in IL-1beta, vitamin D receptor Fok1, and Toll-like receptor 2 are associated with extrapulmonary tuberculosis. *BMC Med Genet* 11:37.

32. Xue Y, Jin L, Li AZ, Wang HJ, Li M, Zhang YX, Wang Y, Li JC. 2010. Microsatellite polymorphisms in intron 2 of the Toll-like receptor 2 gene and their association with susceptibility to pulmonary tuberculosis in Han Chinese. *Clin Chem Lab Med* 48:785–789.

33. Dalgic N, Tekin D, Kayaalti Z, Soylemezoglu T, Cakir E, Kilic B, Kutlubay B, Sancar M, Odabasi M. 2011. Arg753Gln polymorphism of the human Toll-like receptor 2 gene from infection to disease in pediatric tuberculosis. *Hum Immunol* 72:440–445.

34. Sanchez D, Lefebvre C, Rioux J, Garcia LF, Barrera LF. 2012. Evaluation of Toll-like receptor and adaptor molecule polymorphisms for susceptibility to tuberculosis in a Colombian population. *Int J Immunogenet* 39:216–223.

35. McDonough KA, Kress Y, Bloom BR. 1993. The interaction of *Mycobacterium tuberculosis* with macrophages: a study of phagolysosome fusion. *Infect Agents Dis* 2:232–235.

36. McDonough KA, Kress Y, Bloom BR. 1993. Pathogenesis of tuberculosis: interaction of *Mycobacterium tuberculosis* with macrophages. *Infect Immun* 61:2763–2773.

37. van der Wel N, Hava D, Houben D, Fluitsma D, van Zon M, Pierson J, Brenner M, Peters PJ. 2007. *M. tuberculosis* and *M. leprae* translocate from the phagolysosome to the cytosol in myeloid cells. *Cell* 129:1287–1298.

38. Houben D, Demangel C, van Ingen J, Perez J, Baldeon L, Abdallah AM, Caleechurn L, Bottai D, van Zon M, de Punder K, van der Laan T, Kant A, Bossers-de Vries R, Willemsen P, Bitter W, van Soolingen D, Brosch R, van der Wel N, Peters PJ. 2012. ESX-1-mediated translocation to the cytosol controls virulence of mycobacteria. *Cell Microbiol* 14:1287–1298.

39. Koizumi Y, Toma C, Higa N, Nohara T, Nakasone N, Suzuki T. 2012. Inflammasome activation via intracellular NLRs triggered by bacterial infection. *Cell Microbiol* 14:149–154.

40. Juarez E, Carranza C, Hernandez-Sanchez F, Leon-Contreras JC, Hernandez-Pando R, Escobedo D, Torres M, Sada E. 2012. NOD2 enhances the innate response of alveolar macrophages to *Mycobacterium tuberculosis* in humans. *Eur J Immunol* 42:880–889.

41. Pandey AK, Yang Y, Jiang Z, Fortune SM, Coulombe F, Behr MA, Fitzgerald KA, Sassetti CM, Kelliher MA. 2009. NOD2, RIP2 and IRF5 play a critical role in the type I interferon response to *Mycobacterium tuberculosis*. *PLoS Pathog* 5:e1000500.

42. Stanley SA, Johndrow JE, Manzanillo P, Cox JS. 2007. The type I IFN response to infection with *Mycobacterium tuberculosis* requires ESX-1-mediated secretion and contributes to pathogenesis. *J Immunol* 178:3143–3152.

43. Rayamajhi M, Humann J, Kearney S, Hill KK, Lenz LL. 2010. Antagonistic crosstalk between type I and II interferons and increased host susceptibility to bacterial infections. *Virulence* 1:418–422.

44. Novikov A, Cardone M, Thompson R, Shenderov K, Kirschman KD, Mayer-Barber KD, Myers TG, Rabin RL, Trinchieri G, Sher A, Feng CG. 2011. *Mycobacterium tuberculosis* triggers host type I IFN signaling to

regulate IL-1beta production in human macrophages. *J Immunol* 187:2540–2547.

45. Keller M, Ruegg A, Werner S, Beer HD. 2008. Active caspase-1 is a regulator of unconventional protein secretion. *Cell* 132:818–831.

46. Fietta P, Delsante G. 2009. The inflammasomes: the key regulators of inflammation. *Riv Biol* 102:365–384.

47. Mishra BB, Moura-Alves P, Sonawane A, Hacohen N, Griffiths G, Moita LF, Anes E. 2010. *Mycobacterium tuberculosis* protein ESAT-6 is a potent activator of the NLRP3/ASC inflammasome. *Cell Microbiol* 12:1046–1063.

48. Wong KW, Jacobs WR Jr. 2011. Critical role for NLRP3 in necrotic death triggered by *Mycobacterium tuberculosis*. *Cell Microbiol* 13:1371–1384.

49. Abdalla H, Srinivasan L, Shah S, Mayer-Barber KD, Sher A, Sutterwala FS, Briken V. 2012. *Mycobacterium tuberculosis* infection of dendritic cells leads to partially caspase-1/11-independent IL-1beta and IL-18 secretion but not to pyroptosis. *PLoS One* 7:e40722.

50. Mayer-Barber KD, Barber DL, Shenderov K, White SD, Wilson MS, Cheever A, Kugler D, Hieny S, Caspar P, Nunez G, Schlueter D, Flavell RA, Sutterwala FS, Sher A. 2010. Caspase-1 independent IL-1beta production is critical for host resistance to *Mycobacterium tuberculosis* and does not require TLR signaling in vivo. *J Immunol* 184:3326–3330.

51. McElvania Tekippe E, Allen IC, Hulseberg PD, Sullivan JT, McCann JR, Sandor M, Braunstein M, Ting JP. 2010. Granuloma formation and host defense in chronic *Mycobacterium tuberculosis* infection requires PYCARD/ASC but not NLRP3 or caspase-1. *PLoS One* 5:e12320.

52. Dorhoi A, Nouailles G, Jorg S, Hagens K, Heinemann E, Pradl L, Oberbeck-Muller D, Duque-Correa MA, Reece ST, Ruland J, Brosch R, Tschopp J, Gross O, Kaufmann SH. 2012. Activation of the NLRP3 inflammasome by *Mycobacterium tuberculosis* is uncoupled from susceptibility to active tuberculosis. *Eur J Immunol* 42:374–384.

53. Bloch H, Sorkin E, Erlenmeyer H. 1953. A toxic lipid component of the tubercle bacillus (cord factor). I. Isolation from petroleum ether extracts of young bacterial cultures. *Am Rev Tuberc* 67:629–643.

54. Hunter RL, Olsen MR, Jagannath C, Actor JK. 2006. Multiple roles of cord factor in the pathogenesis of primary, secondary, and cavitary tuberculosis, including a revised description of the pathology of secondary disease. *Ann Clin Lab Sci* 36:371–386.

55. Glickman MS, Cahill SM, Jacobs WR Jr. 2001. The *Mycobacterium tuberculosis* cmaA2 gene encodes a mycolic acid trans-cyclopropane synthetase. *J Biol Chem* 276:2228–2233.

56. Glickman MS, Cox JS, Jacobs WR Jr. 2000. A novel mycolic acid cyclopropane synthetase is required for cording, persistence, and virulence of *Mycobacterium tuberculosis*. *Mol Cell* 5:717–727.

57. Rao V, Fujiwara N, Porcelli SA, Glickman MS. 2005. *Mycobacterium tuberculosis* controls host innate immune activation through cyclopropane modification of a glycolipid effector molecule. *J Exp Med* 201:535–543.

58. Rao V, Gao F, Chen B, Jacobs WR Jr, Glickman MS. 2006. Trans-cyclopropanation of mycolic acids on trehalose dimycolate suppresses *Mycobacterium tuberculosis*-induced inflammation and virulence. *J Clin Invest* 116:1660–1667.

59. Dao DN, Sweeney K, Hsu T, Gurcha SS, Nascimento IP, Roshevsky D, Besra GS, Chan J, Porcelli SA, Jacobs WR. 2008. Mycolic acid modification by the mmaA4 gene of *M. tuberculosis* modulates IL-12 production. *PLoS Pathog* 4:e1000081.

60. Ishikawa E, Ishikawa T, Morita YS, Toyonaga K, Yamada H, Takeuchi O, Kinoshita T, Akira S, Yoshikai Y, Yamasaki S. 2009. Direct recognition of the mycobacterial glycolipid, trehalose dimycolate, by C-type lectin Mincle. *J Exp Med* 206:2879–2888.

61. Marakalala MJ, Graham LM, Brown GD. 2010. The role of Syk/CARD9-coupled C-type lectin receptors in immunity to *Mycobacterium tuberculosis* infections. *Clin Dev Immunol* 2010:567571.

62. Dorhoi A, Desel C, Yeremeev V, Pradl L, Brinkmann V, Mollenkopf HJ, Hanke K, Gross O, Ruland J, Kaufmann SH. 2010. The adaptor molecule CARD9 is essential for tuberculosis control. *J Exp Med* 207:777–792.

63. Schoenen H, Bodendorfer B, Hitchens K, Manzanero S, Werninghaus K, Nimmerjahn F, Agger EM, Stenger S, Andersen P, Ruland J, Brown GD, Wells C, Lang R. 2010. Cutting edge: Mincle is essential for recognition and adjuvanticity of the mycobacterial cord factor and its synthetic analog trehalose-dibehenate. *J Immunol* 184:2756–2760.

64. Desel C, Werninghaus K, Ritter M, Jozefowski K, Wenzel J, Russkamp N, Schleicher U, Christensen D, Wirtz S, Kirschning C, Agger EM, da Costa CP, Lang R. 2013. The Mincle-activating adjuvant TDB induces MyD88-dependent Th1 and Th17 responses through IL-1R signaling. *PLoS One* 8:e53531.

65. Lee WB, Kang JS, Yan JJ, Lee MS, Jeon BY, Cho SN, Kim YJ. 2012. Neutrophils promote mycobacterial trehalose dimycolate-induced lung inflammation via the Mincle pathway. *PLoS Pathog* 8:e1002614.

66. Heitmann L, Schoenen H, Ehlers S, Lang R, Holscher C. 2013. Mincle is not essential for controlling *Mycobacterium tuberculosis* infection. *Immunobiology* 218:506–516.

67. Soldati T, Neyrolles O. 2012. Mycobacteria and the intraphagosomal environment: take it with a pinch of salt(s)! *Traffic* 13:1042–1052.

68. Singh CR, Moulton RA, Armitige LY, Bidani A, Snuggs M, Dhandayuthapani S, Hunter RL, Jagannath C. 2006. Processing and presentation of a mycobacterial antigen 85B epitope by murine macrophages is dependent on the phagosomal acquisition of vacuolar proton ATPase and in situ activation of cathepsin D. *J Immunol* 177:3250–3259.

69. Rohde K, Yates RM, Purdy GE, Russell DG. 2007. *Mycobacterium tuberculosis* and the environment within the phagosome. *Immunol Rev* 219:37–54.

70. Kyei GB, Vergne I, Chua J, Roberts E, Harris J, Junutula JR, Deretic V. 2006. Rab14 is critical for maintenance of *Mycobacterium tuberculosis* phagosome maturation arrest. *EMBO J* **25**:5250–5259.

71. Clemens DL, Horwitz MA. 1996. The *Mycobacterium tuberculosis* phagosome interacts with early endosomes and is accessible to exogenously administered transferrin. *J Exp Med* **184**:1349–1355.

72. Sturgill-Koszycki S, Schlesinger PH, Chakraborty P, Haddix PL, Collins HL, Fok AK, Allen RD, Gluck SL, Heuser J, Russell DG. 1994. Lack of acidification in mycobacterium phagosomes produced by exclusion of the vesicular proton-ATPase. *Science* **263**:678–681.

73. Ferrari G, Langen H, Naito M, Pieters J. 1999. A coat protein on phagosomes involved in the intracellular survival of mycobacteria. *Cell* **97**:435–447.

74. Walburger A, Koul A, Ferrari G, Nguyen L, Prescianotto-Baschong C, Huygen K, Klebl B, Thompson C, Bacher G, Pieters J. 2004. Protein kinase G from pathogenic mycobacteria promotes survival within macrophages. *Science* **304**:1800–1804.

75. Wong D, Bach H, Sun J, Hmama Z, Av-Gay Y. 2011. *Mycobacterium tuberculosis* protein tyrosine phosphatase (PtpA) excludes host vacuolar-H+-ATPase to inhibit phagosome acidification. *Proc Natl Acad Sci USA* **108**:19371–19376.

76. Abramovitch RB, Rohde KH, Hsu FF, Russell DG. 2011. aprABC: a *Mycobacterium tuberculosis* complex-specific locus that modulates pH-driven adaptation to the macrophage phagosome. *Mol Microbiol* **80**:678–694.

77. Agarwal N, Lamichhane G, Gupta R, Nolan S, Bishai WR. 2009. Cyclic AMP intoxication of macrophages by a *Mycobacterium tuberculosis* adenylate cyclase. *Nature* **460**:98–102.

78. Mizushima N, Levine B, Cuervo AM, Klionsky DJ. 2008. Autophagy fights disease through cellular self-digestion. *Nature* **451**:1069–1075.

79. Gorvel JP, de Chastellier C. 2005. Bacteria spurned by self-absorbed cells. *Nature Med* **11**:18–19.

80. Singh SB, Davis AS, Taylor GA, Deretic V. 2006. Human IRGM induces autophagy to eliminate intracellular mycobacteria. *Science* **313**:1438–1441.

81. Gutierrez MG, Master SS, Singh SB, Taylor GA, Colombo MI, Deretic V. 2004. Autophagy is a defense mechanism inhibiting BCG and *Mycobacterium tuberculosis* survival in infected macrophages. *Cell* **119**:753–766.

82. Xu Y, Jagannath C, Liu X-D, Sharafkhaneh A, Kolodziejska KE, Eissa NT. 2007. Toll-like receptor 4 is a sensor for autophagy associated with innate immunity. *Immunity* **27**:135–144.

83. Campbell GR, Spector SA. 2012. Vitamin D inhibits human immunodeficiency virus type 1 and *Mycobacterium tuberculosis* infection in macrophages through the induction of autophagy. *PLoS Pathog* **8**:e1002689.

84. Castillo EF, Dekonenko A, Arko-Mensah J, Mandell MA, Dupont N, Jiang S, Delgado-Vargas M, Timmins GS, Bhattacharya D, Yang H, Hutt J, Lyons CR, Dobos KM, Deretic V. 2012. Autophagy protects against active tuberculosis by suppressing bacterial burden and inflammation. *Proc Natl Acad Sci USA* **109**:E3168–E3176.

85. Watson RO, Manzanillo PS, Cox JS. 2012. Extracellular *M. tuberculosis* DNA targets bacteria for autophagy by activating the host DNA-sensing pathway. *Cell* **150**:803–815.

86. Pilli M, Arko-Mensah J, Ponpuak M, Roberts E, Master S, Mandell MA, Dupont N, Ornatowski W, Jiang S, Bradfute SB, Bruun JA, Hansen TE, Johansen T, Deretic V. 2012. TBK-1 promotes autophagy-mediated antimicrobial defense by controlling autophagosome maturation. *Immunity* **37**:223–234.

87. Romagnoli A, Etna MP, Giacomini E, Pardini M, Remoli ME, Corazzari M, Falasca L, Goletti D, Gafa V, Simeone R, Delogu G, Piacentini M, Brosch R, Fimia GM, Coccia EM. 2012. ESX-1 dependent impairment of autophagic flux by *Mycobacterium tuberculosis* in human dendritic cells. *Autophagy* **8**:1357–1370.

88. Ganaie AA, Lella RK, Solanki R, Sharma C. 2011. Thermostable hexameric form of Eis (Rv2416c) protein of *M. tuberculosis* plays an important role for enhanced intracellular survival within macrophages. *PLoS One* **6**:e27590.

89. Kim KH, An DR, Song J, Yoon JY, Kim HS, Yoon HJ, Im HN, Kim J, Kim do J, Lee SJ, Kim KH, Lee HM, Kim HJ, Jo EK, Lee JY, Suh SW. 2012. *Mycobacterium tuberculosis* Eis protein initiates suppression of host immune responses by acetylation of DUSP16/MKP-7. *Proc Natl Acad Sci USA* **109**:7729–7734.

90. Shin DM, Jeon BY, Lee HM, Jin HS, Yuk JM, Song CH, Lee SH, Lee ZW, Cho SN, Kim JM, Friedman RL, Jo EK. 2010. *Mycobacterium tuberculosis eis* regulates autophagy, inflammation, and cell death through redox-dependent signaling. *PLoS Pathog* **6**:e1001230.

91. Cole ST, Brosch R, Parkhill J, Garnier T, Churcher C, Harris D, Gordon SV, Eiglmeier K, Gas S, Barry CE 3rd, Tekaia F, Badcock K, Basham D, Brown D, Chillingworth T, Connor R, Davies R, Devlin K, Feltwell T, Gentles S, Hamlin N, Holroyd S, Hornsby T, Jagels K, Krogh A, McLean J, Moule S, Murphy L, Oliver K, Osborne J, Quail MA, Rajandream MA, Rogers J, Rutter S, Seeger K, Skelton J, Squares R, Squares S, Sulston JE, Taylor K, Whitehead S, Barrell BG. 1998. Deciphering the biology of *Mycobacterium tuberculosis* from the complete genome sequence. *Nature* **393**:537–544.

92. Brennan MJ, Delogu G. 2002. The PE multigene family: a 'molecular mantra' for mycobacteria. *Trends Microbiol* **10**:246–249.

93. Bottai D, Brosch R. 2009. Mycobacterial PE, PPE and ESX clusters: novel insights into the secretion of these most unusual protein families. *Mol Microbiol* **73**:325–328.

94. Brennan MJ, Espitia C, Gey van Pittus N. 2004. The PE and PPE multigene families of *Mycobacterium tuberculosis*, p 513–525. *In* Cole ST, McMurray DN, Eisenach K, Gicquel B, Jacobs WR (ed), *Tuberculosis*, 2nd ed. ASM Press, Washington, DC.

95. Banu S, Honore N, Saint-Joanis B, Philpott D, Prevost MC, Cole ST. 2002. Are the PE-PGRS proteins of *Mycobacterium tuberculosis* variable surface antigens? *Mol Microbiol* 44:9–19.

96. Brennan MJ, Delogu G, Chen Y, Bardarov S, Kriakov J, Alavi M, Jacobs WR Jr. 2001. Evidence that mycobacterial PE_PGRS proteins are cell surface constituents that influence interactions with other cells. *Infect Immun* 69:7326–7333.

97. Espitia C, Laclette JP, Mondragon-Palomino M, Amador A, Campuzano J, Martens A, Singh M, Cicero R, Zhang Y, Moreno C. 1999. The PE-PGRS glycine-rich proteins of *Mycobacterium tuberculosis*: a new family of fibronectin-binding proteins? *Microbiology* 145(Pt 12): 3487–3495.

98. Abdallah AM, Verboom T, Weerdenburg EM, Gey van Pittius NC, Mahasha PW, Jimenez C, Parra M, Cadieux N, Brennan MJ, Appelmelk BJ, Bitter W. 2009. PPE and PE_PGRS proteins of *Mycobacterium marinum* are transported via the type VII secretion system ESX-5. *Mol Microbiol* 73:329–340.

99. Ramakrishnan L, Federspiel NA, Falkow S. 2000. Granuloma-specific expression of *Mycobacterium* virulence proteins from the glycine-rich PE-PGRS family. *Science* 288:1436–1439.

100. Delogu G, Brennan MJ. 2001. Comparative immune response to PE and PE_PGRS antigens of *Mycobacterium tuberculosis*. *Infect Immun* 69:5606–5611.

101. Dheenadhayalan V, Delogu G, Brennan MJ. 2006. Expression of the PE_PGRS 33 protein in *Mycobacterium smegmatis* triggers necrosis in macrophages and enhanced mycobacterial survival. *Microbes Infect* 8:262–272.

102. Balaji KN, Goyal G, Narayana Y, Srinivas M, Chaturvedi R, Mohammad S. 2007. Apoptosis triggered by Rv1818c, a PE family gene from *Mycobacterium tuberculosis* is regulated by mitochondrial intermediates in T cells. *Microbes Infect* 9:271–281.

103. Basu S, Pathak SK, Banerjee A, Pathak S, Bhattacharyya A, Yang Z, Talarico S, Kundu M, Basu J. 2007. Execution of macrophage apoptosis by PE_PGRS33 of *Mycobacterium tuberculosis* is mediated by Toll-like receptor 2-dependent release of tumor necrosis factor-alpha. *J Biol Chem* 282:1039–1050.

104. Dheenadhayalan V, Delogu G, Sanguinetti M, Fadda G, Brennan MJ. 2006. Variable expression patterns of *Mycobacterium tuberculosis* PE_PGRS genes: evidence that PE_PGRS16 and PE_PGRS26 are inversely regulated in vivo. *J Bacteriol* 188:3721–3725.

105. Singh PP, Parra M, Cadieux N, Brennan MJ. 2008. A comparative study of host response to three *Mycobacterium tuberculosis* PE_PGRS proteins. *Microbiology* 154:3469–3479.

106. Huang Y, Zhou X, Bai Y, Yang L, Yin X, Wang Z, Zhao D. 2012. Phagolysosome maturation of macrophages was reduced by PE_PGRS 62 protein expressing in *Mycobacterium smegmatis* and induced in IFN-gamma priming. *Vet Microbiol* 160:117–125.

107. Thi EP, Hong CJ, Sanghera G, Reiner NE. 2012. Identification of the *Mycobacterium tuberculosis* protein PE-PGRS62 as a novel effector that functions to block phagosome maturation and inhibit iNOS expression. *Cell Microbiol* 15:795–808.

108. Kusner DJ. 2005. Mechanisms of mycobacterial persistence in tuberculosis. *Clin Immunol* 114:239–247.

109. Malik ZA, Denning GM, Kusner DJ. 2000. Inhibition of Ca(2+) signaling by *Mycobacterium tuberculosis* is associated with reduced phagosome-lysosome fusion and increased survival within human macrophages. *J Exp Med* 191:287–302.

110. Bachhawat N, Singh B. 2007. Mycobacterial PE_PGRS proteins contain calcium-binding motifs with parallel beta-roll folds. *Genomics Proteomics Bioinformatics* 5: 236–241.

111. Brodsky FM, Lem L, Solache A, Bennett EM. 1999. Human pathogen subversion of antigen presentation. *Immunol Rev* 168:199–215.

112. Harding CV, Ramachandra L, Wick MJ. 2003. Interaction of bacteria with antigen presenting cells: influences on antigen presentation and antibacterial immunity. *Curr Opin Immunol* 15:112–119.

113. Russell DG. 1995. *Mycobacterium* and *Leishmania*: stowaways in the endosomal network. *Trends Cell Biol* 5:125–128.

114. Flannagan RS, Cosio G, Grinstein S. 2009. Antimicrobial mechanisms of phagocytes and bacterial evasion strategies. *Nat Rev Microbiol* 7:355–366.

115. Orme I. 2004. Adaptive immunity to mycobacteria. *Curr Opin Microbiol* 7:58–61.

116. Wolf AJ, Linas B, Trevejo-Nunez GJ, Kincaid E, Tamura T, Takatsu K, Ernst JD. 2007. *Mycobacterium tuberculosis* infects dendritic cells with high frequency and impairs their function in vivo. *J Immunol* 179: 2509–2519.

117. Flynn JL, Chan J. 2003. Immune evasion by *Mycobacterium tuberculosis*: living with the enemy. *Curr Opin Immunol* 15:450–455.

118. Humphreys IR, Stewart GR, Turner DJ, Patel J, Karamanou D, Snelgrove RJ, Young DB. 2006. A role for dendritic cells in the dissemination of mycobacterial infection. *Microbes Infect* 8:1339–1346.

119. Henderson RA, Watkins SC, Flynn JL. 1997. Activation of human dendritic cells following infection with *Mycobacterium tuberculosis*. *J Immunol* 159:635–643.

120. Hanekom WA, Mendillo M, Manca C, Haslett PAJ, Siddiqui MR, Barry C, Kaplan G. 2003. *Mycobacterium tuberculosis* inhibits maturation of human monocyte-derived dendritic cells in vitro. *J Infect Dis* 188:257–266.

121. Hava DL, van der Wel N, Cohen N, Dascher CC, Houben D, Leon L, Agarwal S, Sugita M, van Zon M, Kent SC, Shams H, Peters PJ, Brenner MB. 2008. Evasion of peptide, but not lipid antigen presentation, through pathogen-induced dendritic cell maturation. *Proc Natl Acad Sci USA* 105:11281–11286.

122. Yamane H, Paul WE. 2012. Memory CD4+ T cells: fate determination, positive feedback and plasticity. *Cell Mol Life Sci* 69:1577–1583.

123. Crowe SM, Carlin JB, Stewart KI, Lucas CR, Hoy JF. 1991. Predictive value of CD4 lymphocyte numbers for

the development of opportunistic infections and malignancies in HIV-infected persons. *J Acquir Immune Defic Syndr* **4**:770–776.

124. Flory CM, Hubbard RD, Collins FM. 1992. Effects of in vivo T lymphocyte subset depletion on mycobacterial infections in mice. *J Leukoc Biol* **51**:225–229.

125. Kaufmann SH, Flesch IE. 1988. The role of T cell-macrophage interactions in tuberculosis. *Springer Semin Immunopathol* **10**:337–358.

126. Mogues T, Goodrich ME, Ryan L, LaCourse R, North RJ. 2001. The relative importance of T cell subsets in immunity and immunopathology of airborne *Mycobacterium tuberculosis* infection in mice. *J Exp Med* **193**:271–280.

127. Huygen K, Abramowicz D, Vandenbussche P, Jacobs F, De Bruyn J, Kentos A, Drowart A, Van Vooren JP, Goldman M. 1992. Spleen cell cytokine secretion in *Mycobacterium bovis* BCG-infected mice. *Infect Immun* **60**:2880–2886.

128. Jung YJ, LaCourse R, Ryan L, North RJ. 2002. Evidence inconsistent with a negative influence of T helper 2 cells on protection afforded by a dominant T helper 1 response against *Mycobacterium tuberculosis* lung infection in mice. *Infect Immun* **70**:6436–6443.

129. Sullivan BM, Jobe O, Lazarevic V, Vasquez K, Bronson R, Glimcher LH, Kramnik I. 2005. Increased susceptibility of mice lacking T-bet to infection with *Mycobacterium tuberculosis* correlates with increased IL-10 and decreased IFN-gamma production. *J Immunol* **175**:4593–4602.

130. Surcel HM, Troye-Blomberg M, Paulie S, Andersson G, Moreno C, Pasvol G, Ivanyi J. 1994. Th1/Th2 profiles in tuberculosis, based on the proliferation and cytokine response of blood lymphocytes to mycobacterial antigens. *Immunology* **81**:171–176.

131. Khader SA, Bell GK, Pearl JE, Fountain JJ, Rangel-Moreno J, Cilley GE, Shen F, Eaton SM, Gaffen SL, Swain SL, Locksley RM, Haynes L, Randall TD, Cooper AM. 2007. IL-23 and IL-17 in the establishment of protective pulmonary CD4+ T cell responses after vaccination and during *Mycobacterium tuberculosis* challenge. *Nat Immunol* **8**:369–377.

132. Larson RP, Shafiani S, Urdahl KB. 2013. Foxp3(+) regulatory T cells in tuberculosis. *Adv Exp Med Biol* **783**:165–180.

133. Kursar M, Koch M, Mittrucker HW, Nouailles G, Bonhagen K, Kamradt T, Kaufmann SH. 2007. Cutting edge: regulatory T cells prevent efficient clearance of *Mycobacterium tuberculosis*. *J Immunol* **178**:2661–2665.

134. Leepiyasakulchai C, Ignatowicz L, Pawlowski A, Kallenius G, Skold M. 2012. Failure to recruit anti-inflammatory CD103+ dendritic cells and a diminished CD4+ Foxp3+ regulatory T cell pool in mice that display excessive lung inflammation and increased susceptibility to *Mycobacterium tuberculosis*. *Infect Immun* **80**:1128–1139.

135. Khader SA, Pearl JE, Sakamoto K, Gilmartin L, Bell GK, Jelley-Gibbs DM, Ghilardi N, deSauvage F,

Cooper AM. 2005. IL-23 compensates for the absence of IL-12p70 and is essential for the IL-17 response during tuberculosis but is dispensable for protection and antigen-specific IFN-gamma responses if IL-12p70 is available. *J Immunol* **175**:788–795.

136. Sutton CE, Lalor SJ, Sweeney CM, Brereton CF, Lavelle EC, Mills KH. 2009. Interleukin-1 and IL-23 induce innate IL-17 production from gammadelta T cells, amplifying Th17 responses and autoimmunity. *Immunity* **31**:331–341.

137. Lalor SJ, Dungan LS, Sutton CE, Basdeo SA, Fletcher JM, Mills KH. 2011. Caspase-1-processed cytokines IL-1beta and IL-18 promote IL-17 production by gammadelta and CD4 T cells that mediate autoimmunity. *J Immunol* **186**:5738–5748.

138. Khader SA, Guglani L, Rangel-Moreno J, Gopal R, Junecko BA, Fountain JJ, Martino C, Pearl JE, Tighe M, Lin YY, Slight S, Kolls JK, Reinhart TA, Randall TD, Cooper AM. 2011. IL-23 is required for long-term control of *Mycobacterium tuberculosis* and B cell follicle formation in the infected lung. *J Immunol* **187**:5402–5407.

139. Bold TD, Banaei N, Wolf AJ, Ernst JD. 2011. Suboptimal activation of antigen-specific CD4+ effector cells enables persistence of *M. tuberculosis* in vivo. *PLoS Pathog* **7**:e1002063.

140. Blomgran R, Desvignes L, Briken V, Ernst JD. 2012. *Mycobacterium tuberculosis* inhibits neutrophil apoptosis, leading to delayed activation of naive CD4 T cells. *Cell Host Microbe* **11**:81–90.

141. Darrah PA, Patel DT, De Luca PM, Lindsay RW, Davey DF, Flynn BJ, Hoff ST, Andersen P, Reed SG, Morris SL, Roederer M, Seder RA. 2007. Multifunctional TH1 cells define a correlate of vaccine-mediated protection against *Leishmania major*. *Nat Med* **13**:843–850.

142. Seder RA, Darrah PA, Roederer M. 2008. T-cell quality in memory and protection: implications for vaccine design. *Nat Rev Immunol* **8**:247–258.

143. Forbes EK, Sander C, Ronan EO, McShane H, Hill AV, Beverley PC, Tchilian EZ. 2008. Multifunctional, high-level cytokine-producing Th1 cells in the lung, but not spleen, correlate with protection against *Mycobacterium tuberculosis* aerosol challenge in mice. *J Immunol* **181**:4955–4964.

144. Lindenstrom T, Agger EM, Korsholm KS, Darrah PA, Aagaard C, Seder RA, Rosenkrands I, Andersen P. 2009. Tuberculosis subunit vaccination provides long-term protective immunity characterized by multifunctional CD4 memory T cells. *J Immunol* **182**:8047–8055.

145. Derrick SC, Yabe IM, Yang A, Morris SL. 2011. Vaccine-induced anti-tuberculosis protective immunity in mice correlates with the magnitude and quality of multifunctional CD4 T cells. *Vaccine* **29**:2902–2909.

146. Kaveh DA, Bachy VS, Hewinson RG, Hogarth PJ. 2011. Systemic BCG immunization induces persistent lung mucosal multifunctional CD4 T(EM) cells which expand following virulent mycobacterial challenge. *PloS One* **6**:e21566.

147. Sweeney KA, Dao DN, Goldberg MF, Hsu T, Venkataswamy MM, Henao-Tamayo M, Ordway D, Sellers RS, Jain P, Chen B, Chen M, Kim J, Lukose R, Chan J, Orme IM, Porcelli SA, Jacobs WR Jr. 2011. A recombinant *Mycobacterium smegmatis* induces potent bactericidal immunity against *Mycobacterium tuberculosis*. *Nat Med* **17**:1261–1268.

148. Reiley WW, Shafiani S, Wittmer ST, Tucker-Heard G, Moon JJ, Jenkins MK, Urdahl KB, Winslow GM, Woodland DL. 2010. Distinct functions of antigen-specific CD4 T cells during murine *Mycobacterium tuberculosis* infection. *Proc Natl Acad Sci USA* **107**: 19408–19413.

149. Cyktor JC, Carruthers B, Stromberg P, Flano E, Pircher H, Turner J. 2013. Killer cell lectin-like receptor g1 deficiency significantly enhances survival after *Mycobacterium tuberculosis* infection. *Infect Immun* **81**: 1090–1099.

150. Lazar-Molnar E, Chen B, Sweeney KA, Wang EJ, Liu W, Lin J, Porcelli SA, Almo SC, Nathenson SG, Jacobs WR Jr. 2010. Programmed death-1 (PD-1)-deficient mice are extraordinarily sensitive to tuberculosis. *Proc Natl Acad Sci USA* **107**:13402–13407.

151. Barber DL, Mayer-Barber KD, Feng CG, Sharpe AH, Sher A. 2011. CD4 T cells promote rather than control tuberculosis in the absence of PD-1-mediated inhibition. *J Immunol* **186**:1598–1607.

152. Rogerson BJ, Jung Y-J, LaCourse R, Ryan L, Enright N, North RJ. 2006. Expression levels of *Mycobacterium tuberculosis* antigen-encoding genes versus production levels of antigen-specific T cells during stationary level lung infection in mice. *Immunology* **118**:195–201.

153. Winslow GM, Cooper A, Reiley W, Chatterjee M, Woodland DL. 2008. Early T-cell responses in tuberculosis immunity. *Immunol Rev* **225**:284–299.

154. Raghavan S, Manzanillo P, Chan K, Dovey C, Cox JS. 2008. Secreted transcription factor controls *Mycobacterium tuberculosis* virulence. *Nature* **454**:717–721.

155. Comas I, Chakravartti J, Small PM, Galagan J, Niemann S, Kremer K, Ernst JD, Gagneux S. 2010. Human T cell epitopes of *Mycobacterium tuberculosis* are evolutionarily hyperconserved. *Nat Genet* **42**:498–503.

156. Achkar JM, Dong Y, Holzman RS, Belisle J, Kourbeti IS, Sherpa T, Condos R, Rom WN, Laal S. 2006. *Mycobacterium tuberculosis* malate synthase- and MPT51-based serodiagnostic assay as an adjunct to rapid identification of pulmonary tuberculosis. *Clin Vaccine Immunol* **13**:1291–1293.

157. Verbon A, Kuijper S, Jansen HM, Speelman P, Kolk AH. 1992. Antibodies against secreted and non-secreted antigens in mice after infection with live *Mycobacterium tuberculosis*. *Scand J Immunol* **36**:371–384.

158. Deshpande RG, Khan MB, Bhat DA, Navalkar RG. 1996. Isolation of a 33-kDa protein antigen from delipidified *Mycobacterium tuberculosis* H37Rv. *Med Microbiol Immunol* **185**:153–155.

159. Reed SG, Coler RN, Dalemans W, Dalemans W, Tan EV, DeLa Cruz EC, Basaraba RJ, Orme IM, Skeiky YAW, Alderson MR, Cowgill KD, Prieels J-P, Abalos RM, Dubois M-C, Cohen J, Mettens P, Lobet Y. 2009. Defined tuberculosis vaccine, Mtb72F/AS02A, evidence of protection in cynomolgus monkeys. *Proc Natl Acad Sci USA* **106**:2301–2306.

160. Bonato VL, Lima VM, Tascon RE, Lowrie DB, Silva CL. 1998. Identification and characterization of protective T cells in hsp65 DNA-vaccinated and *Mycobacterium tuberculosis*-infected mice. *Infect Immun* **66**:169–175.

161. Orme IM. 1988. Induction of nonspecific acquired resistance and delayed-type hypersensitivity, but not specific acquired resistance in mice inoculated with killed mycobacterial vaccines. *Infect Immun* **56**:3310–3312.

162. Lindestam Arlehamn CS, Gerasimova A, Mele F, Henderson R, Swann J, Greenbaum JA, Kim Y, Sidney J, James EA, Taplitz R, McKinney DM, Kwok WW, Grey H, Sallusto F, Peters B, Sette A. 2013. Memory T cells in latent *Mycobacterium tuberculosis* infection are directed against three antigenic islands and largely contained in a CXCR3+CCR6+ Th1 subset. *PLoS Pathog* **9**:e1003130.

163. Horwitz MA, Harth G, Dillon BJ, Maslesa-Galic S. 2000. Recombinant bacillus Calmette-Guerin (BCG) vaccines expressing the *Mycobacterium tuberculosis* 30-kDa major secretory protein induce greater protective immunity against tuberculosis than conventional BCG vaccines in a highly susceptible animal model. *Proc Natl Acad Sci USA* **97**:13853–13858.

164. Agger EM, Rosenkrands I, Olsen AW, Hatch G, Williams A, Kritsch C, Lingnau K, von Gabain A, Andersen CS, Korsholm KS, Andersen P. 2006. Protective immunity to tuberculosis with Ag85B-ESAT-6 in a synthetic cationic adjuvant system IC31. *Vaccine* **24**: 5452–5460.

165. Cooper AM. 2009. Cell-mediated immune responses in tuberculosis. *Annu Rev Immunol* **27**:393–422.

166. Gallegos AM, Pamer EG, Glickman MS. 2008. Delayed protection by ESAT-6-specific effector CD4+ T cells after airborne *M. tuberculosis* infection. *J Exp Med* **205**: 2359–2368.

167. Reiley WW, Calayag MD, Wittmer ST, Huntington JL, Pearl JE, Fountain JJ, Martino CA, Roberts AD, Cooper AM, Winslow GM, Woodland DL. 2008. ESAT-6-specific CD4 T cell responses to aerosol *Mycobacterium tuberculosis* infection are initiated in the mediastinal lymph nodes. *Proc Natl Acad Sci USA* **105**: 10961–10966.

168. Wolf AJ, Desvignes L, Linas B, Banaiee N, Tamura T, Takatsu K, Ernst JD. 2008. Initiation of the adaptive immune response to *Mycobacterium tuberculosis* depends on antigen production in the local lymph node, not the lungs. *J Exp Med* **205**:105–115.

169. Lara-Tejero M, Pamer EG. 2004. T cell responses to *Listeria monocytogenes*. *Curr Opin Microbiol* **7**:45–50.

170. Dengjel J, Schoor O, Fischer R, Reich M, Kraus M, Muller M, Kreymborg K, Altenberend F, Brandenburg J, Kalbacher H, Brock R, Driessen C, Rammensee H-G, Stevanovic S. 2005. Autophagy promotes MHC class II presentation of peptides from intracellular source proteins. *Proc Natl Acad Sci USA* **102**:7922–7927.

171. Paludan C, Schmid D, Landthaler M, Vockerodt M, Kube D, Tuschl T, Munz C. 2005. Endogenous MHC class II processing of a viral nuclear antigen after autophagy. *Science* **307:**593–596.

172. Schmid D, Pypaert M, Munz C. 2007. Antigen-loading compartments for major histocompatibility complex class II molecules continuously receive input from autophagosomes. *Immunity* **26:**79–92.

173. Zhou D, Li P, Lin Y, Lott JM, Hislop AD, Canaday DH, Brutkiewicz RR, Blum JS. 2005. Lamp-2a facilitates MHC class II presentation of cytoplasmic antigens. *Immunity* **22:**571–581.

174. Jagannath C, Lindsey DR, Dhandayuthapani S, Xu Y, Hunter RL Jr, Eissa NT. 2009. Autophagy enhances the efficacy of BCG vaccine by increasing peptide presentation in mouse dendritic cells. *Nat Med* **15:**267–276.

175. Fulton SA, Reba SM, Pai RK, Pennini M, Torres M, Harding CV, Boom WH. 2004. Inhibition of major histocompatibility complex II expression and antigen processing in murine alveolar macrophages by *Mycobacterium bovis* BCG and the 19-kilodalton mycobacterial lipoprotein. *Infect Immun* **72:**2101–2110.

176. Torres M, Ramachandra L, Rojas RE, Bobadilla K, Thomas J, Canaday DH, Harding CV, Boom WH. 2006. Role of phagosomes and major histocompatibility complex class II (MHC-II) compartment in MHC-II antigen processing of *Mycobacterium tuberculosis* in human macrophages. *Infect Immun* **74:**1621–1630.

177. Pai RK, Convery M, Hamilton TA, Boom WH, Harding CV. 2003. Inhibition of IFN-gamma-induced class II transactivator expression by a 19-kDa lipoprotein from *Mycobacterium tuberculosis*: a potential mechanism for immune evasion. *J Immunol* **171:**175–184.

178. Pennini ME, Liu Y, Yang J, Croniger CM, Boom WH, Harding CV. 2007. CCAAT/enhancer-binding protein beta and delta binding to CIITA promoters is associated with the inhibition of CIITA expression in response to *Mycobacterium tuberculosis* 19-kDa lipoprotein. *J Immunol* **179:**6910–6918.

179. Pennini ME, Pai RK, Schultz DC, Boom WH, Harding CV. 2006. *Mycobacterium tuberculosis* 19-kDa lipoprotein inhibits IFN-gamma-induced chromatin remodeling of MHC2TA by TLR2 and MAPK signaling. *J Immunol* **176:**4323–4330.

180. Kincaid EZ, Wolf AJ, Desvignes L, Mahapatra S, Crick DC, Brennan PJ, Pavelka MS, Ernst JD. 2007. Codominance of TLR2-dependent and TLR2-independent modulation of MHC class II in *Mycobacterium tuberculosis* infection in vivo. *J Immunol* **179:**3187–3195.

181. Banaiee N, Kincaid EZ, Buchwald U, Jacobs WR, Ernst JD. 2006. Potent inhibition of macrophage responses to IFN-gamma by live virulent *Mycobacterium tuberculosis* is independent of mature mycobacterial lipoproteins but dependent on TLR2. *J Immunol* **176:**3019–3027.

182. Fortune SM, Solache A, Jaeger A, Hill PJ, Belisle JT, Bloom BR, Rubin EJ, Ernst JD. 2004. *Mycobacterium tuberculosis* inhibits macrophage responses to IFN-gamma through myeloid differentiation factor 88-dependent and -independent mechanisms. *J Immunol* **172:**6272–6280.

183. Pecora ND, Fulton SA, Reba SM, Drage MG, Simmons DP, Urankar-Nagy NJ, Boom WH, Harding CV. 2009. *Mycobacterium bovis* BCG decreases MHC-II expression *in vivo* on murine lung macrophages and dendritic cells during aerosol infection. *Cell Immunol* **254:**94–104.

184. Drage MG, Pecora ND, Hise AG, Febbraio M, Silverstein RL, Golenbock DT, Boom WH, Harding CV. 2009. TLR2 and its co-receptors determine responses of macrophages and dendritic cells to lipoproteins of *Mycobacterium tuberculosis*. *Cell Immunol* **258:**29–37.

185. Gehring AJ, Dobos KM, Belisle JT, Harding CV, Boom WH. 2004. *Mycobacterium tuberculosis* LprG (Rv1411c): a novel TLR-2 ligand that inhibits human macrophage class II MHC antigen processing. *J Immunol* **173:**2660–2668.

186. Pathak SK, Basu S, Basu KK, Banerjee A, Pathak S, Bhattacharyya A, Kaisho T, Kundu M, Basu J. 2007. Direct extracellular interaction between the early secreted antigen ESAT-6 of *Mycobacterium tuberculosis* and TLR2 inhibits TLR signaling in macrophages. *Nat Immunol* **8:**610–618.

187. Tapping RI, Tobias PS. 2003. Mycobacterial lipoarabinomannan mediates physical interactions between TLR1 and TLR2 to induce signaling. *J Endotoxin Res* **9:**264–268.

188. Hestvik AL, Hmama Z, Av-Gay Y. 2005. Mycobacterial manipulation of the host cell. *FEMS Microbiol Rev* **29:**1041–1050.

189. Driessen C, Bryant RA, Lennon-Dumenil AM, Villadangos JA, Bryant PW, Shi GP, Chapman HA, Ploegh HL. 1999. Cathepsin S controls the trafficking and maturation of MHC class II molecules in dendritic cells. *J Cell Biol* **147:**775–790.

190. Chow AY, Mellman I. 2005. Old lysosomes, new tricks: MHC II dynamics in DCs. *Trends Immunol* **26:**72–78.

191. Sendide K, Deghmane A-E, Pechkovsky D, Av-Gay Y, Talal A, Hmama Z. 2005. *Mycobacterium bovis* BCG attenuates surface expression of mature class II molecules through IL-10-dependent inhibition of cathepsin S. *J Immunol* **175:**5324–5332.

192. Soualhine H, Deghmane AE, Sun J, Mak K, Talal A, Av-Gay Y, Hmama Z. 2007. *Mycobacterium bovis* bacillus Calmette-Guerin secreting active cathepsin S stimulates expression of mature MHC class II molecules and antigen presentation in human macrophages. *J Immunol* **179:**5137–5145.

193. Sendide K, Deghmane A-E, Reyrat J-M, Talal A, Hmama Z. 2004. *Mycobacterium bovis* BCG urease attenuates major histocompatibility complex class II trafficking to the macrophage cell surface. *Infect Immun* **72:**4200–4209.

194. Nepal RM, Mampe S, Shaffer B, Erickson AH, Bryant P. 2006. Cathepsin L maturation and activity is impaired in macrophages harboring *M. avium* and *M. tuberculosis*. *Int Immunol* **18:**931–939.

195. McDonough KA, Kress Y, Bloom BR. 1993. The interaction of *Mycobacterium tuberculosis* with macrophages: a study of phagolysosome fusion. *Infect Agents Dis* **2:**232–235.

196. van der Wel N, Hava D, Houben D, Fluitsma D, van Zon M, Pierson J, Brenner M, Peters PJ. 2007. *M. tuberculosis* and *M. leprae* translocate from the phagolysosome to the cytosol in myeloid cells. *Cell* 129:1287–1298.

197. Jensen PE. 2007. Recent advances in antigen processing and presentation. *Nat Immunol* 8:1041–1048.

198. Rock KL, Shen L. 2005. Cross-presentation: underlying mechanisms and role in immune surveillance. *Immunol Rev* 207:166–183.

199. Lewinsohn DM, Grotzke JE, Heinzel AS, Zhu L, Ovendale PJ, Johnson M, Alderson MR. 2006. Secreted proteins from *Mycobacterium tuberculosis* gain access to the cytosolic MHC class-I antigen-processing pathway. *J Immunol* 177:437–442.

200. Mazzaccaro RJ, Gedde M, Jensen ER, van Santen HM, Ploegh HL, Rock KL, Bloom BR. 1996. Major histocompatibility class I presentation of soluble antigen facilitated by *Mycobacterium tuberculosis* infection. *Proc Natl Acad Sci USA* 93:11786–11791.

201. Tobian AA, Canaday DH, Boom WH, Harding CV. 2004. Bacterial heat shock proteins promote CD91-dependent class I MHC cross-presentation of chaperoned peptide to CD8+ T cells by cytosolic mechanisms in dendritic cells versus vacuolar mechanisms in macrophages. *J Immunol* 172:5277–5286.

202. Tobian AA, Harding CV, Canaday DH. 2005. *Mycobacterium tuberculosis* heat shock fusion protein enhances class I MHC cross-processing and -presentation by B lymphocytes. *J Immunol* 174:5209–5214.

203. Vyas JM, Van der Veen AG, Ploegh HL. 2008. The known unknowns of antigen processing and presentation. *Nat Rev. Immunol* 8:607–618.

204. Johnstone C, Del Val M. 2007. Traffic of proteins and peptides across membranes for immunosurveillance by CD8(+) T lymphocytes: a topological challenge. *Traffic* 8:1486–1494.

205. Schnell DJ, Hebert DN. 2003. Protein translocons: multifunctional mediators of protein translocation across membranes. *Cell* 112:491–505.

206. Houde M, Bertholet S, Gagnon E, Brunet S, Goyette G, Laplante A, Princiotta MF, Thibault P, Sacks D, Desjardins M. 2003. Phagosomes are competent organelles for antigen cross-presentation. *Nature* 425:402–406.

207. Touret N, Paroutis P, Terebiznik M, Harrison RE, Trombetta S, Pypaert M, Chow A, Jiang A, Shaw J, Yip C, Moore H-P, van der Wel N, Houben D, Peters PJ, de Chastellier C, Mellman I, Grinstein S. 2005. Quantitative and dynamic assessment of the contribution of the ER to phagosome formation. *Cell* 123:157–170.

208. Hsu T, Hingley-Wilson SM, Chen B, Chen M, Dai AZ, Morin PM, Marks CB, Padiyar J, Goulding C, Gingery M, Eisenberg D, Russell RG, Derrick SC, Collins FM, Morris SL, King CH, Jacobs WR. 2003. The primary mechanism of attenuation of bacillus Calmette-Guerin is a loss of secreted lytic function required for invasion of lung interstitial tissue. *Proc Natl Acad Sci USA* 100:12420–12425.

209. Duan L, Gan H, Golan DE, Remold HG. 2002. Critical role of mitochondrial damage in determining outcome of macrophage infection with *Mycobacterium tuberculosis*. *J Immunol* 169:5181–5187.

210. Gan H, Lee J, Ren F, Chen M, Kornfeld H, Remold HG. 2008. *Mycobacterium tuberculosis* blocks crosslinking of annexin-1 and apoptotic envelope formation on infected macrophages to maintain virulence. *Nat Immunol* 9:1189–1197.

211. Rock KL. 2006. Exiting the outside world for cross-presentation. *Immunity* 25:523–525.

212. Ramachandra L, Noss E, Boom WH, Harding CV. 2001. Processing of *Mycobacterium tuberculosis* antigen 85B involves intraphagosomal formation of peptide-major histocompatibility complex II complexes and is inhibited by live bacilli that decrease phagosome maturation. *J Exp Med* 194:1421–1432.

213. Tobian AAR, Potter NS, Ramachandra L, Pai RK, Convery M, Boom WH, Harding CV. 2003. Alternate class I MHC antigen processing is inhibited by Toll-like receptor signaling pathogen-associated molecular patterns: *Mycobacterium tuberculosis* 19-kDa lipoprotein, CpG DNA, and lipopolysaccharide. *J Immunol* 171:1413–1422.

214. Schaible UE, Winau F, Sieling PA, Fischer K, Collins HL, Hagens K, Modlin RL, Brinkmann V, Kaufmann SHE. 2003. Apoptosis facilitates antigen presentation to T lymphocytes through MHC-I and CD1 in tuberculosis. *Nat Med* 9:1039–1046.

215. Winau F, Hegasy G, Kaufmann SHE, Schaible UE. 2005. No life without death: apoptosis as prerequisite for T cell activation. *Apoptosis* 10:707–715.

216. Winau F, Weber S, Sad S, de Diego J, Hoops SL, Breiden B, Sandhoff K, Brinkmann V, Kaufmann SHE, Schaible UE. 2006. Apoptotic vesicles crossprime CD8 T cells and protect against tuberculosis. *Immunity* 24:105–117.

217. Danelishvili L, McGarvey J, Li Y-J, Bermudez LE. 2003. *Mycobacterium tuberculosis* infection causes different levels of apoptosis and necrosis in human macrophages and alveolar epithelial cells. *Cell Microbiol* 5:649–660.

218. Keane J, Remold HG, Kornfeld H. 2000. Virulent *Mycobacterium tuberculosis* strains evade apoptosis of infected alveolar macrophages. *J Immunol* 164:2016–2020.

219. Riendeau CJ, Kornfeld H. 2003. THP-1 cell apoptosis in response to mycobacterial infection. *Infect Immun* 71:254–259.

220. Sly LM, Hingley-Wilson SM, Reiner NE, McMaster WR. 2003. Survival of *Mycobacterium tuberculosis* in host macrophages involves resistance to apoptosis dependent upon induction of antiapoptotic Bcl-2 family member Mcl-1. *J Immunol* 170:430–437.

221. Chen M, Divangahi M, Gan H, Shin DS, Hong S, Lee DM, Serhan CN, Behar SM, Remold HG. 2008. Lipid mediators in innate immunity against tuberculosis: opposing roles of PGE2 and LXA4 in the induction of macrophage death. *J Exp Med* 205:2791–2801.

222. Porcelli SA, Jacobs WR Jr. 2008. Tuberculosis: unsealing the apoptotic envelope. *Nat Immunol* **9:**1101–1102.

223. Edwards KM, Cynamon MH, Voladri RK, Hager CC, DeStefano MS, Tham KT, Lakey DL, Bochan MR, Kernodle DS. 2001. Iron-cofactored superoxide dismutase inhibits host responses to *Mycobacterium tuberculosis. Am J Respir Crit Care Med* **164:**2213–2219.

224. Kahl R, Kampkotter A, Watjen W, Chovolou Y. 2004. Antioxidant enzymes and apoptosis. *Drug Metab Rev* **36:**747–762.

225. Velmurugan K, Chen B, Miller JL, Azogue S, Gurses S, Hsu T, Glickman M, Jacobs WR, Porcelli SA, Briken V. 2007. *Mycobacterium tuberculosis* nuoG is a virulence gene that inhibits apoptosis of infected host cells. *PLoS Pathog* **3:**e110.

226. Hinchey J, Lee S, Jeon BY, Basaraba RJ, Venkataswamy MM, Chen B, Chan J, Braunstein M, Orme IM, Derrick SC, Morris SL, Jacobs WR Jr, Porcelli SA. 2007. Enhanced priming of adaptive immunity by a proapoptotic mutant of *Mycobacterium tuberculosis. J Clin Invest* **117:**2279–2288.

227. Grode L, Seiler P, Baumann S, Hess J, Brinkmann V, Nasser Eddine A, Mann P, Goosmann C, Bandermann S, Smith D, Bancroft GJ, Reyrat J-M, van Soolingen D, Raupach B, Kaufmann SHE. 2005. Increased vaccine efficacy against tuberculosis of recombinant *Mycobacterium bovis* bacille Calmette-Guerin mutants that secrete listeriolysin. *J Clin Invest* **115:**2472–2479.

228. Koh KW, Lehming N, Seah GT. 2009. Degradation-resistant protein domains limit host cell processing and immune detection of mycobacteria. *Mol Immunol* **46:**1312–1318.

229. Talarico S, Cave MD, Marrs CF, Foxman B, Zhang L, Yang Z. 2005. Variation of the *Mycobacterium tuberculosis* PE_PGRS 33 gene among clinical isolates. *J Clin Microbiol* **43:**4954–4960.

230. Talarico S, Zhang L, Marrs CF, Foxman B, Cave MD, Brennan MJ, Yang Z. 2008. *Mycobacterium tuberculosis* PE_PGRS16 and PE_PGRS26 genetic polymorphism among clinical isolates. *Tuberculosis (Edinb)* **88:**283–294.

231. Dutronc Y, Porcelli SA. 2002. The CD1 family and T cell recognition of lipid antigens. *Tissue Antigens* **60:**337–353.

232. van den Elzen P, Garg S, Leon L, Brigl M, Leadbetter EA, Gumperz JE, Dascher CC, Cheng T-Y, Sacks FM, Illarionov PA, Besra GS, Kent SC, Moody DB, Brenner MB. 2005. Apolipoprotein-mediated pathways of lipid antigen presentation. *Nature* **437:**906–910.

233. Stenger S, Niazi KR, Modlin RL. 1998. Down-regulation of CD1 on antigen-presenting cells by infection with *Mycobacterium tuberculosis. J Immunol* **161:**3582–3588.

234. Gagliardi MC, Lemassu A, Teloni R, Mariotti S, Sargentini V, Pardini M, Daffe M, Nisini R. 2007. Cell wall-associated alpha-glucan is instrumental for *Mycobacterium tuberculosis* to block CD1 molecule expression and disable the function of dendritic cell derived from infected monocyte. *Cell Microbiol* **9:**2081–2092.

235. Mariotti S, Teloni R, Iona E, Fattorini L, Giannoni F, Romagnoli G, Orefici G, Nisini R. 2002. *Mycobacterium tuberculosis* subverts the differentiation of human monocytes into dendritic cells. *Eur J Immunol* **32:**3050–3058.

236. Barral DC, Brenner MB. 2007. CD1 antigen presentation: how it works. *Nat Rev Immunol* **7:**929–941.

237. Chackerian A, Alt J, Perera V, Behar SM. 2002. Activation of NKT cells protects mice from tuberculosis. *Infect Immun* **70:**6302–6309.

238. Sada-Ovalle I, Chiba A, Gonzales A, Brenner MB, Behar SM. 2008. Innate invariant NKT cells recognize *Mycobacterium tuberculosis*-infected macrophages, produce interferon-gamma, and kill intracellular bacteria. *PLoS Pathog* **4:**e1000239.

239. Behar SM, Dascher CC, Grusby MJ, Wang CR, Brenner MB. 1999. Susceptibility of mice deficient in CD1D or TAP1 to infection with *Mycobacterium tuberculosis. J Exp Med* **189:**1973–1980.

240. Kawakami K, Kinjo Y, Uezu K, Yara S, Miyagi K, Koguchi Y, Nakayama T, Taniguchi M, Saito A. 2002. Minimal contribution of Valpha14 natural killer T cells to Th1 response and host resistance against mycobacterial infection in mice. *Microbiol Immunol* **46:**207–210.

241. Sugawara I, Yamada H, Mizuno S, Li CY, Nakayama T, Taniguchi M. 2002. Mycobacterial infection in natural killer T cell knockout mice. *Tuberculosis (Edinb)* **82:**97–104.

Molecular Genetics of Mycobacteria, 2nd Edition
Edited by Graham F. Hatfull and William R. Jacobs, Jr.
© 2014 American Society for Microbiology, Washington, DC
doi:10.1128/microbiolspec.MGM2-0004-2013

Jacobs P. Richards[1]
Anil K. Ojha[1]

Mycobacterial Biofilms

37

In a review published in 1896 on early bacteriological studies of the tubercle bacilli, *Mycobacterium tuberculosis*, A. Coppen Jones (1) wrote:

> When old cultures are examined by means of sections it is found that the growth does not consist of separated rod-like forms, isolated from one another and lying at angles, but of strands of parallel filaments, frequently showing dichotomous branching. These facts indicate that the so-called "tubercle bacillus" is really a stage in the life history of some higher form of fungus with definite mycellial growth. From a systemic point of view, it cannot be regarded as coming within any definition of the genus Bacillus, and it is suggested that a more appropriate name would be Tuberculomyces.

Despite an outrageous suggestion of calling *M. tuberculosis* a fungus, Jones's thought of "separated rod-like forms" and "strands of parallel filament" as alternative lifestyles of the tubercle bacilli is remarkably consistent with contemporary views of most bacteria: switching their lifestyle from single-cell planktonic forms to sessile, multicellular communities, called biofilms (2–4). Moreover, biofilms are phenotypically unique from planktonic forms of their constituent cells in many ways, most notable of which is their extraordinary resistance to environmental challenges (5, 6). Formation of biofilms therefore is considered a universal persistent strategy for microbes in diverse growth conditions.

The long-term persistence and extraordinary drug tolerance of mycobacterial infections, particularly *M. tuberculosis*, are strikingly similar to the characteristics of biofilm infections associated with many bacterial pathogens. Furthermore, many mycobacterial species, including *M. tuberculosis*, form drug-tolerant biofilms under *in vitro* conditions through genetically controlled developmental processes. Together, these raise pertinent questions as to whether mycobacterial infections could also be associated with a multicellular lifestyle of mycobacteria. In this article, while reviewing the characteristics of *in vitro* biofilms of mycobacteria we will discuss the relevance of biofilm models to developing an understanding of the persistence characteristics of mycobacterial infections.

ORIGINS OF THE BIOFILM HYPOTHESIS

Microbes were perhaps first described as communities of individual cells by Antonie van Leeuwenhoek in 1684 when he reported the microscopic observation of human dental "scuff" (plaques) as "animalcules." However, bacteriology, blooming under the influence of Koch's postulates, had a strong purpose of obtaining clonally pure strains from uniformly dispersed cultures. Thus, widespread culturing in nutrient-rich medium on shakers perhaps led to the common notion that bacteria

---

[1]Department of Infectious Diseases and Microbiology, Graduate School of Public Health, University of Pittsburgh, Pittsburgh, PA 15261.

predominantly grow in unicellular planktonic forms, although occasional studies in the early 20th century raised caution on this view. In an attempt to grow algae on a submerged glass surface, Henrici observed the surface to be fully colonized with bacteria (7). Subsequently, Heukelekian and Heller, as well as Claude Zobell, confirmed Henrici's observation (8, 9). Planktonic cultures of bacteria, however, remained the mainstream culture technique, perhaps influenced by the urgencies of anti-infective discoveries. Not surprisingly, Tween-80 as a dispersing agent, first used by Dubos and colleagues (10), continues to be a key medium ingredient for dispersed broth cultures of *M. tuberculosis* and other mycobacteria, despite a direct influence of the detergent on physical and biological characteristics of the pathogen (11, 12).

Between the late 1970s and early 1980s Costerton and colleagues published a series of electron micrographs of a wide range of specimens, from rumen fluids to medical implants, showing bacterial populations as adherent microcolonies encapsulated by exopolysaccharide (EPS) (13–15). It was further observed that these sessile communities are distinct in terms of their recalcitrance to antimicrobial agents (16). These discoveries were followed by extensive microscopic observations of biotic and abiotic surfaces, and it subsequently became apparent that biofilms are the predominant lifestyle of microbes in both environmental and clinical settings and pose significant challenges in treatment of microbial infections.

BASIC CHARACTERISTICS OF MICROBIAL BIOFILMS

Dynamic Architecture Encapsulated within Extracellular Matrix

Using confocal scanning laser microscopy (CSLM), Caldwell and colleagues showed that the biofilms of *Pseudomonas fluorescence* were not random aggregates, but were uniformly raised in mushroom-shaped structures (17). Subsequently, Stoodley et al. demonstrated the presence of water channels in the interiors of *Pseudomonas aeruginosa* biofilms that were big enough to allow unrestricted flow of particles greater than 5 μm (18). These water channels are thought to deliver oxygen and nutrients to the microbial population residing in the inner structures of the biofilms (19).

One of the invariable structural components of biofilm architecture is the extracellular matrix (ECM), which physically holds the individual cells together. Although the molecular composition of the ECM differs

significantly across species, and even among strains within a species, EPS is one of the most predominant matrix components in both Gram-positive and Gram-negative species (20). Extracellular DNA (eDNA) is also a structural component of the matrix in biofilms of *P. aeruginosa* and *Staphylococcus aureus* (21, 22). In addition to EPS and eDNA, matrices of most species also contain adhesive proteins. For example, multiple biofilm-associated proteins (Bap) in *S. aureus* are sufficient to produce matured biofilms even without EPS (23). Similarly, TasA in *Bacillus subtilis* (24), type IV pili in *P. aeruginosa* (25), and type I fimbria as well as curli in *Escherichia coli* (26, 27) are some of the major adhesive proteins implicated in cell-cell and cell-surface attachment.

Genetically Programmed Distinct Developmental Stages

Biofilm formation is a developmental process that proceeds through distinct stages of (i) surface attachment, (ii) sessile growth, (iii) matrix synthesis, and (iv) dispersal (Fig. 1). Each stage appears to be associated with distinct sets of genetic factors, expressions of which are regulated through master regulators and signaling molecules. For example, mutation in Clp protease leads to defective attachment of *P. aeruginosa* to substratum (28), suggesting that protein processing could be critical during attachment. Surface attachment of *P. aeruginosa* also induces expression of *algC* and, therefore, alginate synthesis (29). In *E. coli*, surface attachment triggers activation of CsgD, followed by cyclic-di-GMP-dependent reprogramming of gene expression that suppresses motility while inducing sessile growth and matrix synthesis (30). Many other Gram-negative species, as well as *B. subtilis*, also utilize cyclic-di-GMP as a secondary messenger during biofilm development (31–33). Intercellular communication through a quorum-sensing phenomenon, involving autoinducers, is also widely implicated in biofilm formation of both Gram-negative and Gram-positive species (34–37). The intricacies of regulatory networks in biofilm development are further revealed by the implications of many transcriptional activators and suppressors in biofilm-specific traits, such as loss of motility and increased synthesis of EPS (30, 38–40).

Phenotypic Heterogeneity

One of the most fascinating features of biofilms is that they harbor phenotypically heterogeneous cells even in a genetically clonal population (2). Such phenotypic diversity presumably originates from cell-to-cell

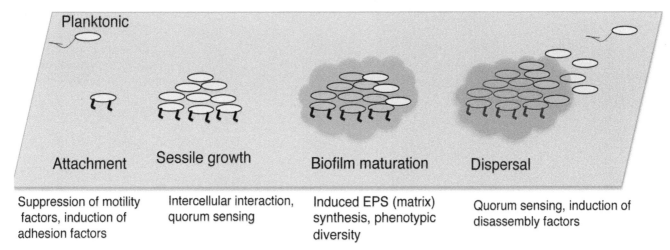

Figure 1 A schematic representation of distinct developmental stages of microbial biofilms. Transition from one stage has specific genetic requirements. This scheme was originally published by the authors in *Expert Review of Anti-Infective Therapy* (106) and is reproduced here in accordance with the publisher's policy.
doi:10.1128/microbiolspec.MGM2-0004-2013.f1

differences in gene expression, as a result of nonuniform microenvironments of the biofilms (41–43). Promoter fusions with fluorescent reporters have been useful in the analysis of phenotypic heterogeneity in biofilms. By fusing green fluorescent protein to the growth-rate-dependent promoter (*rrnBP1*) of *P. aeruginosa*, Stewart and colleagues demonstrated that the promoter activity and therefore growth was limited to the cells in the outer regions of biofilms (44). Similarly, Vlamakis et al. observed distinct spatiotemporal organization of a motile, sessile, and sporulating subpopulation of bacilli within a colony-biofilm of *B. subtilis* (43).

Genetically Controlled Dispersal

Both Gram-positive and Gram-negative species utilize genetically controlled mechanisms for programmed self-dispersal of biofilms. From an ecological perspective, dispersal can be thought to be an integral part of microbial dissemination (and recolonization) under immediate stress. Dispersal factors are of particular interest from clinical perspectives because these could lead to potential therapeutics for biofilm infections. In *P. aeruginosa*, dispersal is preceded by localized lysis of bacteria in the interior subpopulation, which then are believed to provide nutrients to immediate cells, which activate the expression of their motility genes to exit the community (45, 46). Matured biofilms of *B. subtilis* secrete D-amino acids and norspermidine as dispersal factors, which interestingly, are also effective against other Gram-positive and Gram-negative species

(47–49). Potential of dispersal factors in the treatment of biofilm infections, however, remains to be evaluated.

BIOFILM AS A PERSISTENCE STRATEGY OF MICROBES

Formation of biofilms is arguably a significant commitment from microbes not only because a sessile form restricts their spatial freedom, but also because architectural development requires energy expenditure. In return, the encapsulated microbes in biofilms enjoy greater protection from various kinds of environmental threats such as antibiotics, protozoan predation, and host immunity (4, 5, 50–53). For many bacterial pathogens, biofilms are implicated in their survival against both host defense mechanisms and antibiotics (13, 54–57).

Recalcitrance to stress is likely a combined effect of both physical protection by ECM and physiological adaptation to limiting growth conditions of inner microenvironments (58). Drug tolerance in *P. aeruginosa* biofilms is considered to be primarily due to a large proportion of stationary-phase-like metabolically inactive cells (44, 59). However, such analyses are inherently complicated by the complex relationship between the bacterial physiology and components of the biofilm microenvironment, including ECM.

The role of biofilms in subverting host immunity during chronic infection is evident in the pathogenesis of *P. aeruginosa* and *S. aureus*. In *P. aeruginosa*, two biofilm-associated extracellular materials, alginate and

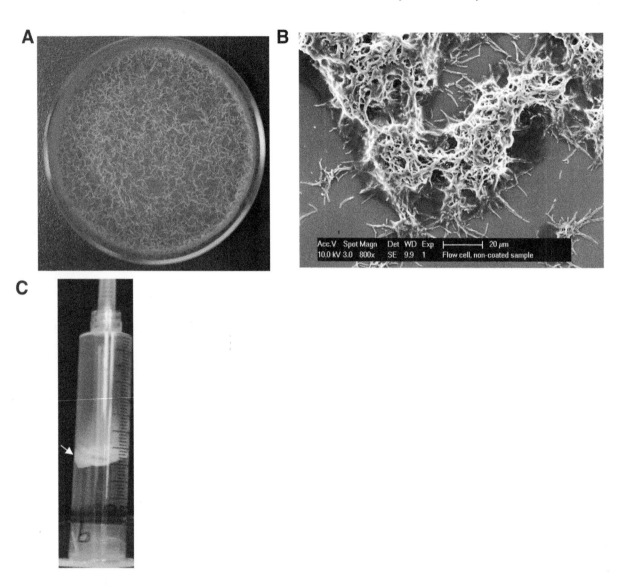

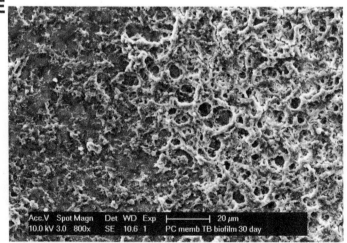

rhamnolipid, have been implicated in protection against host immunity. Whereas alginate protects the bacteria from the macrophages, rhamnolipid protects by inducing necrosis of polymorphonuclear leukocytes (PMNs) (51, 60). *S. aureus* biofilms appear to misguide the host response to facilitate greater persistence of microbes. In the later stages of infection, the biofilms induce Th2-dependent humoral immune responses that are less effective than the Th1-dependent stronger T-cell response induced during early stages of infection (5).

MYCOBACTERIA SPONTANEOUSLY FORM BIOFILMS

Unless detergent is added, most mycobacterial species in liquid culture display a strong propensity to attach to substrata, and to each other, to form a variety of microscopic as well as macroscopic multicellular structures (Fig. 2). Besides such homospecies biofilms *in vitro*, many nontuberculous mycobacteria (NTM), including opportunistic pathogens, have been ubiquitously found as heterospecies biofilms in environmental specimens. *Mycobacterium chelonae*, *Mycobacterium fortuitum*, *Mycobacterium gordonae*, and *Mycobacterium tarrae/nonchromogenicum* were detected in 90% of polymicrobial biofilms obtained from domestic water supplies and water treatment plants (61). Similarly, water distribution systems including showerheads and faucets are the predominant habitat for the polymicrobial biofilms containing *Mycobacterium avium*, *Mycobacterium intracellulare*, and *Mycobacterium xenopi* (62–66). Moreover, biofilms of NTM are highly tolerant to disinfectant chlorine and thus pose a significant public health challenge to effective control strategies for nosocomial infections (67, 68).

The propensity of mycobacteria to form biofilms raises many fundamental questions such as (i) how the multicellular architectures are developed, (ii) what unique phenotypes are associated with mycobacterial biofilms, (iii) how such phenotypes originate, and most importantly, (iv) whether there is any relationship between biofilms and virulence as well as characteristic persistence of mycobacterial infections. The most reasonable approach to these questions involves genetic analysis of mycobacterial biofilms. Not surprisingly, novel molecular insights have been gained in the last 10 years, which also witnessed major breakthroughs in the development of mycobacterial genetic tools.

GENETICS OF MYCOBACTERIAL BIOFILMS

Structural Development

In a genetic screen of transposon insertion mutants of *Mycobacterium smegmatis*, Recht and Kolter isolated attachment-defective mutants with disruptions in genes involved in acetylation of glycopeptidolipid (GPL) (69). GPL was subsequently found to be also required for the development of *M. avium* biofilms (70, 71). Because GPL is also a potent immunomodulator (72), the question arises whether biofilms directly influence the pathology and clinical symptoms of *M. avium* infections. A direct association between biofilms and virulence of *M. avium* is further consistent with the fact that biofilm-defective mutants of *M. avium* cannot colonize and translocate through bronchial epithelial cells (73). Similar linkage between virulence and biofilms exists for another NTM, *Mycobacterium ulcerans*. *M. ulcerans* colonizes an aquatic insect, *Naucoris cimicoides*, primarily as matrix-encapsulated multicellular structures containing large amounts of an extracellular toxin, mycolactone (74, 75). Because mycolactone is the major virulence factor of *M. ulcerans*, its biofilms can be thought to directly cause the pathological features of Buruli ulcers.

Evidence of a programmed development of biofilms proceeding through distinct stages emerges from the phenotype of a Δ*groEL1* mutant of *M. smegmatis* (76). While the mutant seems indistinguishable from the wild type during the first 3 days (early stage) of growth, it fails to form a mature pellicle, typically seen after 5

Figure 2 Various models of mycobacterial biofilms grown in our laboratory. (A) Pellicles biofilms of *M. smegmatis* on air-liquid interface in a petri dish. (B) Scanning electron micrograph of flow-cell biofilms of *M. smegmatis* on silicon surface, developed against the shear fluid flow of 1 ml/minute. (C) Pellicle biofilms of *M. smegmatis* in syringes (marked by arrow). This technique is amenable to screening mutants that remain exclusively in planktonic suspension beneath biofilms. (D) Pellicle biofilms of *M. tuberculosis* on liquid-air interface grown in a 12-well plate. (E) Scanning electron micrograph of *M. tuberculosis* biofilms grown on the surface of a polycarbonate membrane. Images in panels (B) and (E) were generated with help from Curtis Larimer and Ian Nettleship from the Swanson School of Engineering, University of Pittsburgh. doi:10.1128/microbiolspec.MGM2-0004-2013.f2

days (late stage) of wild-type growth (76). Further-more, GroEL-1 directly interacts with the fatty acid synthase complex II (FAS II) to modulate mycolic acid biosynthesis during maturation of pellicles (76). This regulated interaction appears to induce the synthesis of free mycolic acids (FM), an abundant extracellular lipid, in the later stages of biofilms (77). Given the abundance and extracellular location of FM, as well as its association with the matured architecture of biofilms, it can be reasonably considered one of the components of the matrix. Interestingly, FM is also an abundant extracellular lipid of *M. tuberculosis* pellicles (78), although unlike the alpha- and epoxy-FM produced by *M. smegmatis*, *M. tuberculosis* predominantly produces methoxy-FM (77, 78) (Fig. 3). The structural distinction between the FMs of *M. smeg-matis* and *M. tuberculosis* raises the possibility that the biofilms of the two species, despite being morphologi-cally similar, could have distinct properties. Nonethe-less, the abundance of FM in biofilms is of particular interest because the majority of the mycolic acids syn-thesized by planktonic bacteria are esterified to either mycolyl-arabinogalactan-peptidoglycan complex, tre-halose, or other sugars (79). Thus, a question arises as to whether FM is generated by cleavage of the mycolyl esters or is synthesized and secreted *de novo* through a dedicated FAS II complex. Discovery of a cutinase-like serine esterase in *M. smegmatis*, Msmeg_1529, with an ability to hydrolyze trehalose dimycolate into FM sup-ports the former mechanism (77). However, modula-tion of FAS II complex by GroEL1 during FM synthesis in *M. smegmatis* also implies regulation in the biosyn-thetic pathways of mycolic acids.

In addition to FM and GPL, mycolyl diacyl glycerol (MDAG) is also implicated in biofilm formation of *M. smegmatis* (80), indicating that the multicellular architecture of mycobacteria is predominantly waxy. Loss of biofilm development in *M. tuberculosis* by dis-ruption of putative polyketide synthases, *pks16* and *pks1/15*, or altered synthesis of mycolic acids through deletion in a nucleoid-associated protein, further sup-ports the critical contribution of surface lipids in self-assemblage of mycobacteria (78, 81, 82).

It is noteworthy that mycobacteria in detergent-free media often produce a polysaccharide-rich capsular material on the surface (83, 84). Although a clear role of capsular structures in biofilm architectures remains undetermined, lipooligosaccharides on the *M. marinum* surface appear to be required for its biofilm develop-ment (85). In summary, the structures of mycobacterial biofilms are likely assembled by various kinds of mole-cules, although a subset of these could be shared as core components by multiple species.

Physiological Adaptations

Genetic evidence for mycobacterial adaptation to the limiting environments of biofilms can be derived from exclusive induction of 82 genes in *M. smegmatis* bio-films (86). These can be classified into various func-tional categories such as lipid biosynthesis, nutrient transport, toxin efflux, iron sequestration, DNA dam-age repair, transcriptional regulation, etc. (86). The gene-expression pattern is further complemented by ge-netic analysis of mutants involved in the process. For example, the upregulation of iron sequestration ma-chinery in *M. smegmatis* biofilms is consistent with the impaired biofilms of a mutant that fails to sequester iron from the environment, despite its normal growth in planktonic culture (86). Increased iron sequestra-tion in biofilms could possibly be due to either limited supply of the metal in the inner regions of biofilms or specialized metabolic requirement or both. Similar arguments apply to the induced expression of trans-porters as well as efflux pumps in *M. smegmatis* bio-films (86). Although the mechanisms underlying the regulation of gene expression during mycobacterial growth in biofilms remain unclear, a mutation in *rpoZ* appears to affect the developmental process of *M. smeg-matis* biofilms (87).

The roles of environmental stimulants and inter-cellular interactions in development of mycobacterial biofilms remain largely unexplored, although the likeli-hood of such mechanisms is supported by multiple studies. Growth of *M. avium* biofilms is enhanced in re-sponse to the exogenous signaling molecule autoinducer-2, as well as subinhibitory concentrations of antibiotics (71, 88). Similarly, intercellular interaction is evident from the fact that biofilms of *M. smegmatis* are essen-tial for the conjugal transfer of DNA from a donor to a recipient strain (89). Interestingly, the recipient

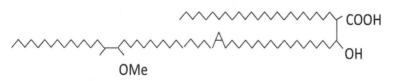

Figure 3 A representative structure of methoxy-free mycolic acids, predominantly accumulated in the biofilms of *M. tuberculosis*.
doi:10.1128/microbiolspec.MGM2-0004-2013.f3

strain is incapable of forming biofilms on its own and therefore is likely to be actively recruited in the multicellular structure by the donor strain for conjugation (89).

PHENOTYPIC RESISTANCE OF MYCOBACTERIAL BIOFILMS

Mycobacterial biofilms *in vitro* display greater phenotypic resistance to antibiotics than planktonic cultures (78, 90, 91). Although the mechanisms of drug tolerance remain unclear, both extrinsic factors such as ECM and intrinsic physiology of the bacteria are likely to contribute to the phenomenon. While the waxy ECM can restrict the exposure to antibiotics, limiting growth conditions in biofilm microenvironments could lead to a drug-refractory physiology of the constituent bacteria. This scenario is further supported by the fact that depleted nutrients and oxygen promote the development of nonreplicating but viable drug-tolerant persisters of mycobacteria (92, 93). Recently, phenotypic drug tolerance in mycobacteria has also been linked to heterogeneity in cell length, as a result of asymmetric cell division, and cell-to-cell stochastic fluctuation in gene expression (94, 95). Drug tolerance through these mechanisms can also be envisaged to occur in mycobacterial biofilms because of the asynchronous growth of the cells in nonuniform microenvironments.

Importantly, impaired biofilms of genetically unrelated mutants of *M. tuberculosis* as well as *M. smegmatis* are relatively more sensitive to antibiotics than their corresponding parent wild type, implying that maturation of structural assembly appears to be critical for the phenotypic tolerance (Fig. 4). This provides a strong incentive for discovery of chemical inhibitors of mycobacterial biofilms. Such inhibitors in conjunction with conventional drugs can potentially facilitate effective and shorter treatment of mycobacterial infections.

RELEVANCE OF BIOFILM FORMATION TO TUBERCULOSIS

In a 7-year-long histopathological study of lung lesions from hundreds of active tuberculosis (TB) cases, Georges Canetti documented numerous micrographs with biofilm-like multicellular aggregates of *M. tuberculosis* (96). Given that *M. tuberculosis* has a natural tendency to form drug-tolerant biofilms *in vitro*, the *in vivo* aggregates certainly open up important questions about the conditions in which these are formed and their possible involvement in long-term persistence of the pathogen against host defense mechanisms and antibiotics.

In vivo persistence of *M. tuberculosis* against antibiotics is manifested by a prolonged and complicated treatment regimen of TB involving multiple antibiotics administered daily for at least 6 to 9 months (97). It is clinically demonstrated that the prolonged regimen is necessary for clearing a small number (<5%) of phenotypically resistant bacilli, while the majority are cleared within days of treatment (98). Interestingly, similar biphasic clearance patterns are observed when biofilms are exposed to antibiotics (78). Although where and how these rare persisters survive in TB patients remains unknown, using the guinea pig model, Orme and colleagues found that the persisters were predominantly located in multicellular microcolonies located in the acellular rim of granulomas (99). Further extension of this observation with multiple biofilm-defective mutants of *M. tuberculosis*, such as *pks16*, *helY*, *pks1/15*, etc. (78, 81), is likely to strengthen the possible association between *in vivo* persistence and the multicellular growth of the pathogens.

Besides persistence against antibiotics, biofilms could also be envisioned as a key persistence strategy of *M. tuberculosis* against the host immune system in chronic infections, particularly those without any clinical symptoms. Asymptomatic infections are prevalent in more than 90% of an estimated two billion people infected with *M. tuberculosis* worldwide (97). Long-term asymptomatic persistence of *M. tuberculosis*, clinically defined as latent TB (LTB), has long been associated with a metabolically dormant pathogen "hidden" from the host immune system (100). However, this view is contradicted by recent studies in which LTB is associated with (i) active engagement of immune cells (101), (ii) the presence of drug-responsive lesions (102, 103), (iii) active replication of bacilli (104), and (iv) a pathology that overlaps with active disease in terms of heterogeneity in lesion distribution, size, and morphology (105). Together, these findings suggest that at least a subset of LTB might represent a dynamic host-pathogen interaction that maintains host inflammation below a symptomatic threshold despite active growth of *M. tuberculosis*. A possible path to such a scenario could involve immune subversion by actively replicating *M. tuberculosis* in biofilms, coated with lipids that fail to elicit any inflammatory response in the host. This idea gains support from the fact that unlike many other surface lipids of *M. tuberculosis*, purified FM fails to induce a pro-inflammatory response in macrophages (Y. Yang and A. K. Ojha, unpublished result).

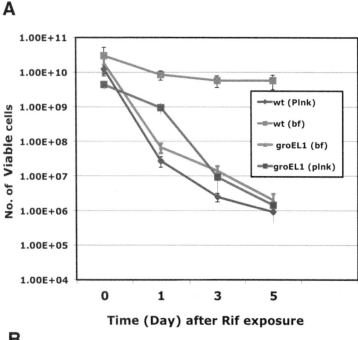

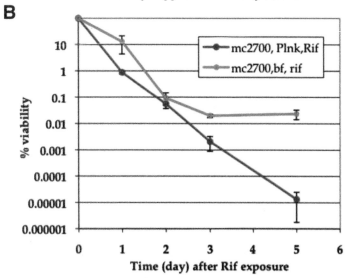

Figure 4 Biofilms of *M. smegmatis* (**A**) and *M. tuberculosis* (**B**) harbor higher numbers of rifampin-tolerant persisters than their planktonic counterparts. The frequency of such persisters is diminished in the impaired biofilms of Δ*groEL1 M. smegmatis*. Data in panel (**B**) were originally published in *Molecular Microbiology* (78) and reproduced here in accordance with the publisher's policy.
doi:10.1128/microbiolspec.MGM2-0004-2013.f4

Novel mutants with specific defects in biofilm-associated extracellular lipids including FM could be invaluable in testing this hypothesis.

CONCLUSION

An increasing body of evidence suggests that biofilms are not only the spontaneous manifestations of mycobacterial growth, but also promote their persistence against exogenous threats. However, a significant knowledge gap exists in the molecular mechanisms underlying structural as well as phenotypic developments of mycobacterial biofilms. With new high-throughput genetic and biochemical tools as well as advanced microscopic techniques, there is an enormous opportunity to explore the biological complexities and intercellular dynamics of multicellular structures and their likely involvement in the extraordinary recalcitrance of mycobacterial infections.

Citation. Richards JP, Ojha AK. 2014. Mycobacterial biofilms. Microbiol Spectrum 2(5):MGM2-0004-2013.

References

1. **Jones AC.** 1896. On the so-called tubercle bacilli. Report of the sixty-sixth meeting of the British Association for the Advancements of Science, p 1015–1016.

2. Kolter R, Losick R. 1998. One for all and all for one. *Science* 280:226–227.

3. Stoodley P, Sauer K, Davies DG, Costerton JW. 2002. Biofilms as complex differentiated communities. *Annu Rev Microbiol* 56:187–209.

4. Hoiby N, Bjarnsholt T, Givskov M, Molin S, Ciofu O. 2010. Antibiotic resistance of bacterial biofilms. *Int J Antimicrob Agents* 35:322–332.

5. Prabhakara R, Harro JM, Leid JG, Harris M, Shirtliff ME. 2011. Murine immune response to a chronic *Staphylococcus aureus* biofilm infection. *Infect Immun* 79:1789–1796.

6. Moser C, Kjaergaard S, Pressler T, Kharazmi A, Koch C, Hoiby N. 2000. The immune response to chronic *Pseudomonas aeruginosa* lung infection in cystic fibrosis patients is predominantly of the Th2 type. *APMIS* 108:329–335.

7. Henrici AT. 1933. Studies of freshwater bacteria. I. A direct microscopic technique. *J Bacteriol* 25:277–287.

8. Zobell CE. 1943. The effect of solid surfaces upon bacterial activity. *J Bacteriol* 46:39–56.

9. Heukelekian H, Heller A. 1940. Relation between food concentration and surface for bacterial growth. *J Bacteriol* 40:547–558.

10. Dubos RJ, Middlebrook G. 1948. The effect of wetting agents on the growth of tubercle bacilli. *J Exp Med* 88:81–88.

11. Bloch H, Noll H. 1953. Studies on the virulence of tubercle bacilli: variations in virulence effected by tween 80 and thiosemicarbazone. *J Exp Med* 97:1–16.

12. Van Boxtel RM, Lambrecht RS, Collins MT. 1990. Effects of colonial morphology and tween 80 on antimicrobial susceptibility of *Mycobacterium paratuberculosis*. *Antimicrob Agents Chemother* 34:2300–2303.

13. Lam J, Chan R, Lam K, Costerton JW. 1980. Production of mucoid microcolonies by *Pseudomonas aeruginosa* within infected lungs in cystic fibrosis. *Infect Immun* 28:546–556.

14. Geesey GG, Richardson WT, Yeomans HG, Irvin RT, Costerton JW. 1977. Microscopic examination of natural sessile bacterial populations from an alpine stream. *Can J Microbiol* 23:1733–1736.

15. Costerton JW, Irvin RT, Cheng KJ. 1981. The role of bacterial surface structures in pathogenesis. *Crit Rev Microbiol* 8:303–338.

16. Nickel JC, Ruseska I, Wright JB, Costerton JW. 1985. Tobramycin resistance of *Pseudomonas aeruginosa* cells growing as a biofilm on urinary catheter material. *Antimicrob Agents Chemother* 27:619–624.

17. Lawrence JR, Korber DR, Hoyle BD, Costerton JW, Caldwell DE. 1991. Optical sectioning of microbial biofilms. *J Bacteriol* 173:6558–6567.

18. Stoodley P, Debeer D, Lewandowski Z. 1994. Liquid flow in biofilm systems. *Appl Environ Microbiol* 60:2711–2716.

19. de Beer D, Stoodley P, Roe F, Lewandowski Z. 1994. Effects of biofilm structures on oxygen distribution and mass transport. *Biotechnol Bioeng* 43:1131–1138.

20. Branda SS, Vik S, Friedman L, Kolter R. 2005. Biofilms: the matrix revisited. *Trends Microbiol* 13:20–26.

21. Whitchurch CB, Tolker-Nielsen T, Ragas PC, Mattick JS. 2002. Extracellular DNA required for bacterial biofilm formation. *Science* 295:1487.

22. Rice KC, Mann EE, Endres JL, Weiss EC, Cassat JE, Smeltzer MS, Bayles KW. 2007. The cidA murein hydrolase regulator contributes to DNA release and biofilm development in *Staphylococcus aureus*. *Proc Natl Acad Sci USA* 104:8113–8118.

23. Lasa I, Penades JR. 2006. Bap: a family of surface proteins involved in biofilm formation. *Res Microbiol* 157:99–107.

24. Branda SS, Chu F, Kearns DB, Losick R, Kolter R. 2006. A major protein component of the *Bacillus subtilis* biofilm matrix. *Mol Microbiol* 59:1229–1238.

25. O'Toole GA, Kolter R. 1998. Flagellar and twitching motility are necessary for *Pseudomonas aeruginosa* biofilm development. *Mol Microbiol* 30:295–304.

26. Chapman MR, Robinson LS, Pinkner JS, Roth R, Heuser J, Hammar M, Normark S, Hultgren SJ. 2002. Role of *Escherichia coli* curli operons in directing amyloid fiber formation. *Science* 295:851–855.

27. Pratt LA, Kolter R. 1998. Genetic analysis of *Escherichia coli* biofilm formation: roles of flagella, motility, chemotaxis and type I pili. *Mol Microbiol* 30:285–293.

28. O'Toole GA, Kolter R. 1998. Initiation of biofilm formation in *Pseudomonas fluorescens* WCS365 proceeds via multiple, convergent signalling pathways: a genetic analysis. *Mol Microbiol* 28:449–461.

29. Davies DG, Chakrabarty AM, Geesey GG. 1993. Exopolysaccharide production in biofilms: substratum activation of alginate gene expression by *Pseudomonas aeruginosa*. *Appl Environ Microbiol* 59:1181–1186.

30. Ogasawara H, Yamamoto K, Ishihama A. 2011. Role of the biofilm master regulator CsgD in cross-regulation between biofilm formation and flagellar synthesis. *J Bacteriol* 193:2587–2597.

31. Cotter PA, Stibitz S. 2007. c-di-GMP-mediated regulation of virulence and biofilm formation. *Curr Opin Microbiol* 10:17–23.

32. Sondermann H, Shikuma NJ, Yildiz FH. 2012. You've come a long way: c-di-GMP signaling. *Curr Opin Microbiol* 15:140–146.

33. Chen Y, Chai Y, Guo JH, Losick R. 2012. Evidence for cyclic Di-GMP-mediated signaling in *Bacillus subtilis*. *J Bacteriol* 194:5080–5090.

34. Waters CM, Lu W, Rabinowitz JD, Bassler BL. 2008. Quorum sensing controls biofilm formation in *Vibrio cholerae* through modulation of cyclic di-GMP levels and repression of vpsT. *J Bacteriol* 190:2527–2536.

35. Lopez D, Vlamakis H, Kolter R. 2010. Biofilms. *Cold Spring Harbor Perspect Biol* 2:a000398.

36. Novick RP, Geisinger E. 2008. Quorum sensing in staphylococci. *Annu Rev Genet* 42:541–564.

37. Davies DG, Parsek MR, Pearson JP, Iglewski BH, Costerton JW, Greenberg EP. 1998. The involvement of cell-to-cell signals in the development of a bacterial biofilm. *Science* 280:295–298.

38. Kearns DB, Chu F, Branda SS, Kolter R, Losick R. 2005. A master regulator for biofilm formation by *Bacillus subtilis*. *Mol Microbiol* **55**:739–749.

39. Chu F, Kearns DB, Branda SS, Kolter R, Losick R. 2006. Targets of the master regulator of biofilm formation in *Bacillus subtilis*. *Mol Microbiol* **59**:1216–1228.

40. Chai Y, Chu F, Kolter R, Losick R. 2008. Bistability and biofilm formation in *Bacillus subtilis*. *Mol Microbiol* **67**:254–263.

41. Xu KD, Stewart PS, Xia F, Huang CT, McFeters GA. 1998. Spatial physiological heterogeneity in *Pseudomonas aeruginosa* biofilm is determined by oxygen availability. *Appl Environ Microbiol* **64**:4035–4039.

42. Rani SA, Pitts B, Beyenal H, Veluchamy RA, Lewandowski Z, Davison WM, Buckingham-Meyer K, Stewart PS. 2007. Spatial patterns of DNA replication, protein synthesis, and oxygen concentration within bacterial biofilms reveal diverse physiological states. *J Bacteriol* **189**:4223–4233.

43. Vlamakis H, Aguilar C, Losick R, Kolter R. 2008. Control of cell fate by the formation of an architecturally complex bacterial community. *Genes Dev* **22**:945–953.

44. Werner E, Roe F, Bugnicourt A, Franklin MJ, Heydorn A, Molin S, Pitts B, Stewart PS. 2004. Stratified growth in *Pseudomonas aeruginosa* biofilms. *Appl Environ Microbiol* **70**:6188–6196.

45. Purevdorj-Gage B, Costerton WJ, Stoodley P. 2005. Phenotypic differentiation and seeding dispersal in nonmucoid and mucoid *Pseudomonas aeruginosa* biofilms. *Microbiology* **151**:1569–1576.

46. McDougald D, Rice SA, Barraud N, Steinberg PD, Kjelleberg S. 2012. Should we stay or should we go: mechanisms and ecological consequences for biofilm dispersal. *Nat Rev Microbiol* **10**:39–50.

47. Kolodkin-Gal I, Romero D, Cao S, Clardy J, Kolter R, Losick R. 2010. D-amino acids trigger biofilm disassembly. *Science* **328**:627–629.

48. Hochbaum AI, Kolodkin-Gal I, Foulston L, Kolter R, Aizenberg J, Losick R. 2011. Inhibitory effects of D-amino acids on *Staphylococcus aureus* biofilm development. *J Bacteriol* **193**:5616–5622.

49. Kolodkin-Gal I, Cao S, Chai L, Bottcher T, Kolter R, Clardy J, Losick R. 2012. A self-produced trigger for biofilm disassembly that targets exopolysaccharide. *Cell* **149**:684–692.

50. Jensen PO, Givskov M, Bjarnsholt T, Moser C. 2010. The immune system vs. *Pseudomonas aeruginosa* biofilms. *FEMS Immunol Med Microbiol* **59**:292–305.

51. Leid JG, Willson CJ, Shirtliff ME, Hassett DJ, Parsek MR, Jeffers AK. 2005. The exopolysaccharide alginate protects *Pseudomonas aeruginosa* biofilm bacteria from IFN-gamma-mediated macrophage killing. *J Immunol* **175**:7512–7518.

52. Matz C, Kjelleberg S. 2005. Off the hook: how bacteria survive protozoan grazing. *Trends Microbiol* **13**:302–307.

53. Mah TF, O'Toole GA. 2001. Mechanisms of biofilm resistance to antimicrobial agents. *Trends Microbiol* **9**:34–39.

54. Post JC, Stoodley P, Hall-Stoodley L, Ehrlich GD. 2004. The role of biofilms in otolaryngologic infections. *Curr Opin Otolaryngol Head Neck Surg* **12**:185–190.

55. Marrie TJ, Nelligan J, Costerton JW. 1982. A scanning and transmission electron microscopic study of an infected endocardial pacemaker lead. *Circulation* **66**:1339–1341.

56. Davis LE, Cook G, Costerton JW. 2002. Biofilm on ventriculo-peritoneal shunt tubing as a cause of treatment failure in coccidioidal meningitis. *Emerg Infect Dis* **8**:376–379.

57. Anderson GG, Palermo JJ, Schilling JD, Roth R, Heuser J, Hultgren SJ. 2003. Intracellular bacterial biofilm-like pods in urinary tract infections. *Science* **301**:105–107.

58. Anderson GG, O'Toole GA. 2008. Innate and induced resistance mechanisms of bacterial biofilms. *Curr Top Microbiol Immunol* **322**:85–105.

59. Spoering AL, Lewis K. 2001. Biofilms and planktonic cells of *Pseudomonas aeruginosa* have similar resistance to killing by antimicrobials. *J Bacteriol* **183**:6746–6751.

60. Van Gennip M, Christensen LD, Alhede M, Phipps R, Jensen PO, Christophersen L, Pamp SJ, Moser C, Mikkelsen PJ, Koh AY, Tolker-Nielsen T, Pier GB, Hoiby N, Givskov M, Bjarnsholt T. 2009. Inactivation of the rhlA gene in *Pseudomonas aeruginosa* prevents rhamnolipid production, disabling the protection against polymorphonuclear leukocytes. *APMIS* **117**:537–546.

61. Schulze-Röbbecke R, Janning B, Fischeder R. 1992. Occurrence of mycobacteria in biofilm samples. *Tuber Lung Dis* **73**:141–144.

62. Hall-Stoodley L, Lappin-Scott H. 1998. Biofilm formation by the rapidly growing mycobacterial species *Mycobacterium fortuitum*. *FEMS Microbiol Lett* **168**:77–84.

63. Falkinham JO, Norton CD, LeChevallier MW. 2001. Factors influencing numbers of *Mycobacterium avium*, *Mycobacterium intracellulare*, and other mycobacteria in drinking water distribution systems. *Appl Environ Microbiol* **67**:1225–1231.

64. Angenent LT, Kelley ST, Amand AS, Pace NR, Hernandez MT. 2005. Molecular identification of potential pathogens in water and air of a hospital therapy pool. *Proc Natl Acad Sci USA* **102**:4860–4865.

65. Dailloux M, Albert M, Laurain C, Andolfatto S, Lozniewski A, Hartemann P, Mathieu L. 2003. *Mycobacterium xenopi* and drinking water biofilms. *Appl Environ Microbiol* **69**:6946–6948.

66. Feazel LM, Baumgartner LK, Peterson KL, Frank DN, Harris JK, Pace NR. 2009. Opportunistic pathogens enriched in showerhead biofilms. *Proc Natl Acad Sci USA* **106**:16393–16399.

67. Steed KA, Falkinham JO. 2006. Effect of growth in biofilms on chlorine susceptibility of *Mycobacterium avium* and *Mycobacterium intracellulare*. *Appl Environ Microbiol* **72**:4007–4011.

68. Vaerewijck MJ, Huys G, Palomino JC, Swings J, Portaels F. 2005. Mycobacteria in drinking water distribution systems: ecology and significance for human health. *FEMS Microbiol Rev* **29**:911–934.

69. Recht J, Kolter R. 2001. Glycopeptidolipid acetylation affects sliding motility and biofilm formation in *Mycobacterium smegmatis*. *J Bacteriol* 183:5718–5724.

70. Yamazaki Y, Danelishvili L, Wu M, Macnab M, Bermudez LE. 2006. *Mycobacterium avium* genes associated with the ability to form a biofilm. *Appl Environ Microbiol* 72:819–825.

71. Geier H, Mostowy S, Cangelosi GA, Behr MA, Ford TE. 2008. Autoinducer-2 triggers the oxidative stress response in *Mycobacterium avium*, leading to biofilm formation. *Appl Environ Microbiol* 74:1798–1804.

72. Schorey JS, Sweet L. 2008. The mycobacterial glycopeptidolipids: structure, function, and their role in pathogenesis. *Glycobiology* 18:832–841.

73. Carter G, Wu M, Drummond DC, Bermudez LE. 2003. Characterization of biofilm formation by clinical isolates of *Mycobacterium avium*. *J Med Microbiol* 52:747–752.

74. Marsollier L, Aubry J, Coutanceau E, André J-PS, Small PL, Milon G, Legras P, Guadagnini S, Carbonnelle B, Cole ST. 2005. Colonization of the salivary glands of *Naucoris cimicoides* by *Mycobacterium ulcerans* requires host plasmatocytes and a macrolide toxin, mycolactone. *Cell Microbiol* 7:935–943.

75. Marsollier L, Brodin P, Jackson M, Kordulakova J, Tafelmeyer P, Carbonnelle E, Aubry J, Milon G, Legras P, Andre JP, Leroy C, Cottin J, Guillou ML, Reysset G, Cole ST. 2007. Impact of *Mycobacterium ulcerans* biofilm on transmissibility to ecological niches and Buruli ulcer pathogenesis. *PLoS Pathog* 3:e62.

76. Ojha A, Anand M, Bhatt A, Kremer L, Jacobs WR Jr, Hatfull GF. 2005. GroEL1: a dedicated chaperone involved in mycolic acid biosynthesis during biofilm formation in mycobacteria. *Cell* 123:861–873.

77. Ojha AK, Trivelli X, Guerardel Y, Kremer L, Hatfull GF. 2010. Enzymatic hydrolysis of trehalose dimycolate releases free mycolic acids during mycobacterial growth in biofilms. *J Biol Chem* 285:17380–17389.

78. Ojha AK, Baughn AD, Sambandan D, Hsu T, Trivelli X, Guerardel Y, Alahari A, Kremer L, Jacobs WR Jr, Hatfull GF. 2008. Growth of *Mycobacterium tuberculosis* biofilms containing free mycolic acids and harbouring drug-tolerant bacteria. *Mol Microbiol* 69:164–174.

79. Takayama K, Wang C, Besra GS. 2005. Pathway to synthesis and processing of mycolic acids in *Mycobacterium tuberculosis*. *Clin Microbiol Rev* 18:81–101.

80. Chen JM, German GJ, Alexander DC, Ren H, Tan T, Liu J. 2006. Roles of Lsr2 in colony morphology and biofilm formation of *Mycobacterium smegmatis*. *J Bacteriol* 188:633–641.

81. Pang JM, Layre E, Sweet L, Sherrid A, Moody DB, Ojha A, Sherman DR. 2012. The polyketide Pks1 contributes to biofilm formation in *Mycobacterium tuberculosis*. *J Bacteriol* 194:715–721.

82. Ghosh S, Indi SS, Nagaraja V. 2013. Regulation of lipid biosynthesis, sliding motility and biofilm formation by a membrane-anchored nucleoid associated protein of *Mycobacterium tuberculosis*. *J Bacteriol* 195:1769–1778.

83. Ortalo-Magne A, Dupont MA, Lemassu A, Andersen AB, Gounon P, Daffe M. 1995. Molecular composition of the outermost capsular material of the tubercle bacillus. *Microbiology* 141(Pt 7):1609–1620.

84. Sani M, Houben EN, Geurtsen J, Pierson J, de Punder K, van Zon M, Wever B, Piersma SR, Jimenez CR, Daffe M, Appelmelk BJ, Bitter W, van der Wel N, Peters PJ. 2010. Direct visualization by cryo-EM of the mycobacterial capsular layer: a labile structure containing ESX-1-secreted proteins. *PLoS Pathog* 6:e1000794.

85. Ren H, Dover LG, Islam ST, Alexander DC, Chen JM, Besra GS, Liu J. 2007. Identification of the lipooligosaccharide biosynthetic gene cluster from *Mycobacterium marinum*. *Mol Microbiol* 63:1345–1359.

86. Ojha A, Hatfull GF. 2007. The role of iron in *Mycobacterium smegmatis* biofilm formation: the exochelin siderophore is essential in limiting iron conditions for biofilm formation but not for planktonic growth. *Mol Microbiol* 66:468–483.

87. Mathew R, Mukherjee R, Balachandar R, Chatterji D. 2006. Deletion of the rpoZ gene, encoding the omega subunit of RNA polymerase, results in pleiotropic surface-related phenotypes in *Mycobacterium smegmatis*. *Microbiology* 152:1741–1750.

88. McNabe M, Tennant R, Danelishvili L, Young L, Bermudez LE. 2011. *Mycobacterium avium* ssp. *hominissuis* biofilm is composed of distinct phenotypes and influenced by the presence of antimicrobials. *Clin Microbiol Infect* 17:697–703.

89. Nguyen KT, Piastro K, Gray TA, Derbyshire KM. 2010. Mycobacterial biofilms facilitate horizontal DNA transfer between strains of *Mycobacterium smegmatis*. *J Bacteriol* 192:5134–5142.

90. Teng R, Dick T. 2003. Isoniazid resistance of exponentially growing *Mycobacterium smegmatis* biofilm culture. *FEMS Microbiol Lett* 227:171–174.

91. McNabe M, Tennant R, Danelishvili L, Young L, Bermudez LE. 2011. *Mycobacterium avium* ssp. *hominissuis* biofilm is composed of distinct phenotypes and influenced by the presence of antimicrobials. *Clin Microbiol Infect* 17:697–703.

92. Gengenbacher M, Rao SP, Pethe K, Dick T. 2010. Nutrient-starved, non-replicating *Mycobacterium tuberculosis* requires respiration, ATP synthase and isocitrate lyase for maintenance of ATP homeostasis and viability. *Microbiology* 156:81–87.

93. Wayne LG, Hayes LG. 1996. An in vitro model for sequential study of shiftdown of *Mycobacterium tuberculosis* through two stages of nonreplicating persistence. *Infect Immun* 64:2062–2069.

94. Aldridge BB, Fernandez-Suarez M, Heller D, Ambravaneswaran V, Irimia D, Toner M, Fortune SM. 2012. Asymmetry and aging of mycobacterial cells lead to variable growth and antibiotic susceptibility. *Science* 335:100–104.

95. Wakamoto Y, Dhar N, Chait R, Schneider K, Signorino-Gelo F, Leibler S, McKinney JD. 2013. Dynamic persistence of antibiotic-stressed mycobacteria. *Science* 339:91–95.

96. Canetti G. 1955. *Tubercle Bacillus in the Pulmonary Lesion of Man: Histobacteriology and Its Bearing on the Therapy of Pulmonary Tuberculosis.* Springer Publishing Company, New York, NY.

97. WHO. 2012. The burden of diseases caused by TB. Global Tuberculosis Report:8–28.

98. Jindani A, Dore CJ, Mitchison DA. 2003. Bactericidal and sterilizing activities of antituberculosis drugs during the first 14 days. *Am J Respir Crit Care Med* **167**:1348–1354.

99. Lenaerts AJ, Hoff D, Aly S, Ehlers S, Andries K, Cantarero L, Orme IM, Basaraba RJ. 2007. Location of persisting mycobacteria in a guinea pig model of tuberculosis revealed by r207910. *Antimicrob Agents Chemother* **51**:3338–3345.

100. Parrish NM, Dick JD, Bishai WR. 1998. Mechanisms of latency in *Mycobacterium tuberculosis. Trends Microbiol* **6**:107–112.

101. Ulrichs T, Kosmiadi GA, Jorg S, Pradl L, Titukhina M, Mishenko V, Gushina N, Kaufmann SH. 2005. Differential organization of the local immune response in patients with active cavitary tuberculosis or with nonprogressive tuberculoma. *J Infect Dis* **192**:89–97.

102. Barry CE 3rd, Boshoff HI, Dartois V, Dick T, Ehrt S, Flynn J, Schnappinger D, Wilkinson RJ, Young D. 2009. The spectrum of latent tuberculosis: rethinking the biology and intervention strategies. *Nat Rev Microbiol* **7**:845–855.

103. Park IN, Ryu JS, Shim TS. 2008. Evaluation of therapeutic response of tuberculoma using F-18 FDG positron emission tomography. *Clin Nucl Med* **33**:1–3.

104. Ford CB, Lin PL, Chase MR, Shah RR, Iartchouk O, Galagan J, Mohaideen N, Ioerger TR, Sacchettini JC, Lipsitch M, Flynn JL, Fortune SM. 2011. Use of whole genome sequencing to estimate the mutation rate of *Mycobacterium tuberculosis* during latent infection. *Nat Genet* **43**:482–486.

105. Lin PL, Rodgers M, Smith L, Bigbee M, Myers A, Bigbee C, Chiosea I, Capuano SV, Fuhrman C, Klein E, Flynn JL. 2009. Quantitative comparison of active and latent tuberculosis in the cynomolgus macaque model. *Infect Immun* **77**:4631–4642.

106. Islam MS, Richards JP, Ojha AK. 2012. Targeting drug tolerance in mycobacteria: a perspective from mycobacterial biofilms. *Expert Rev Anti Infect Ther* **10**:1055–1066.

Index

A

ABC transporter, in trehalose transport, 365, 369
Abrahams, G. L., 503
Acadian phage, 128
acc genes, 615–616
Accelerator, in asymmetric growth, 716
Accurate-mass retention time values, 342
aceA gene, 146
Acetamidase system, switches, 225, 228
Acetylated PIMs, 545–546
Acetylation, 247, 289
Acetyl-CoA carboxylase, 615–616
N-Acetylglucosamine, in lysis, 122
N-Acetylglucosamine rhamnose linker, in peptidoglycan synthesis, 520
N-Acetylglycosylases, in lysis, 122
N-Acetylmuramidases, in lysis, 122
N-Acetyl-muramyl-L-alanine amidases, in lysis, 122, 124
Acetyltransferase, 307
Acid stress, 193
  genome changes due to, 732
  PZA activity and, 484
Acidobacteria, STPKs of, 682
Acidothermus cellulolyticus, Pup-proteasome system of, 669
Acinetobacter baumannii, drug susceptibility in, 714

Acinetoferrin, 378
Aconitase, 330–331
AcpM protein, 443
Acr2 protein, 213
Actinobacteria, 27
  energetics of, 396
  STPKs of, 682
  stress response of, 193
Actinomycetales, 310
Activation loops, in STPKs, 685–687
Active site pockets, 269–270
N-Acylated glucosamine (GlcNAc), in peptidoglycan, 513, 515–517
Acyltransferases, 563, 574
Acyltrehaloses, 572–579
Adams, K. N., 715
Adaptive immunity, evasion of, 756–762
Adaptive response, in signal transduction, 681
AddAB protein, 658
Adenosine deaminase, 639
Adenosine kinase, 639–640
S-Adenosylmethionine, 311
S-Adenosylmethionine-dependent methyltransferase, 619
Adenylate cyclases, in cyclic AMP signaling, 281, 283–285
Adenylosuccinate synthase, 637
Adephagia phage, 106

AdnAB protein, 658, 660
Ag85 antigens, 755–756
ahp genes and Ahp proteins, 151, 232, 304, 417, 436, 733
Airborne pathogens, biosafety requirements for, 4–5
Akhter, Y., 188
Alanine dehydrogenase, 245–246
Alanine ligase, 515
Alanine racemase, 232, 500, 515
L-Alanyl-D-glutamine-*meso*-diaminopimelyl-D-alahine, 513
Alber, M., 365
AlbG protein, 469
Albicidin, 469
Aldridge, B. B., 716
Alkyl hydroperoxidase, 304
Alkyl hydroperoxide reductase, 733
Allelic exchange reactions, 17–18
Alma phage, 97
Almeida, D., 418
Alonso, S., 735
Alpha/beta hydrolases, 268
Alpha-glucans, 362, 370–372
Alpha-mycolic acids, 613
alr gene, 414
Alternative sigma factor density, 138
Alternator, in asymmetric growth, 716
Alveolar surface area, 301